Cut here and keep for reference

Algebra

Arithmetic Operations

$$a(b + c) = ab + ac \qquad \frac{a}{b} + \frac{c}{d} = \frac{ad + bc}{bd}$$

$$\frac{a + c}{b} = \frac{a}{b} + \frac{c}{b} \qquad \frac{\frac{a}{b}}{\frac{c}{d}} = \frac{a}{b} \times \frac{d}{c} = \frac{ad}{bc}$$

Exponents and Radicals

$$x^m x^n = x^{m+n} \qquad \frac{x^m}{x^n} = x^{m-n}$$

$$(x^m)^n = x^{mn} \qquad x^{-n} = \frac{1}{x^n}$$

$$(xy)^n = x^n y^n \qquad \left(\frac{x}{y}\right)^n = \frac{x^n}{y^n}$$

$$x^{1/n} = \sqrt[n]{x} \qquad x^{m/n} = \sqrt[n]{x^m} = (\sqrt[n]{x})^m$$

$$\sqrt[n]{xy} = \sqrt[n]{x}\sqrt[n]{y} \qquad \sqrt[n]{\frac{x}{y}} = \frac{\sqrt[n]{x}}{\sqrt[n]{y}}$$

Factoring Special Polynomials

$$x^2 - y^2 = (x + y)(x - y)$$

$$x^3 + y^3 = (x + y)(x^2 - xy + y^2)$$

$$x^3 - y^3 = (x - y)(x^2 + xy + y^2)$$

Binomial Theorem

$$(x + y)^2 = x^2 + 2xy + y^2 \qquad (x - y)^2 = x^2 - 2xy + y^2$$

$$(x + y)^3 = x^3 + 3x^2y + 3xy^2 + y^3$$

$$(x - y)^3 = x^3 - 3x^2y + 3xy^2 - y^3$$

$$(x + y)^n = x^n + nx^{n-1}y + \frac{n(n-1)}{2}x^{n-2}y^2 + \cdots + \binom{n}{k}x^{n-k}y^k + \cdots + nxy^{n-1} + y^n$$

where $\binom{n}{k} = \frac{n(n-1)\cdots(n-k+1)}{1 \cdot 2 \cdot 3 \cdot \cdots \cdot k}$

Quadratic Formula

If $ax^2 + bx + c = 0$, then $x = \dfrac{-b \pm \sqrt{b^2 - 4ac}}{2a}$.

Inequalities and Absolute Value

If $a < b$ and $b < c$, then $a < c$.

If $a < b$, then $a + c < b + c$.

If $a < b$ and $c > 0$, then $ca < cb$.

If $a < b$ and $c < 0$, then $ca > cb$.

If $a > 0$, then

$|x| = a$ means $x = a$ or $x = -a$

$|x| < a$ means $-a < x < a$

$|x| > a$ means $x > a$ or $x < -a$

Geor

Geometric F

Formulas for

Triangle

$A = \frac{1}{2}bh$

$= \frac{1}{2}ab \sin\theta$

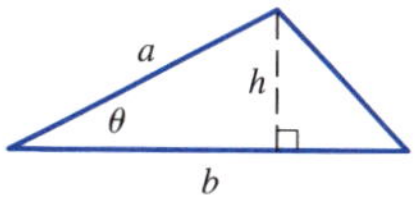

Circle

$A = \pi r^2$

$C = 2\pi r$

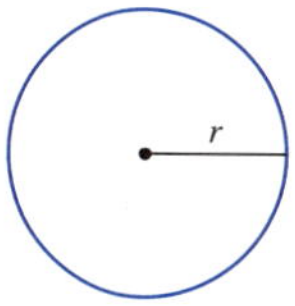

Sector of Circle

$A = \frac{1}{2}r^2\theta$

$s = r\theta$ (θ in radians)

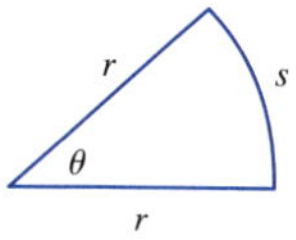

Sphere

$V = \frac{4}{3}\pi r^3$

$A = 4\pi r^2$

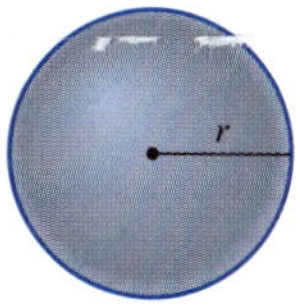

Cylinder

$V = \pi r^2 h$

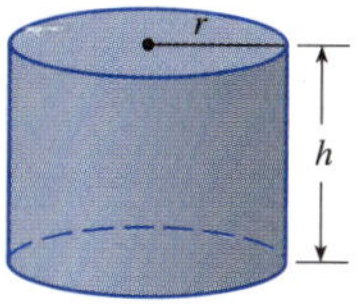

Cone

$V = \frac{1}{3}\pi r^2 h$

$A = \pi r\sqrt{r^2 + h^2}$

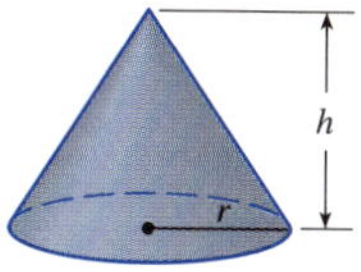

Distance and Midpoint Formulas

Distance between $P_1(x_1, y_1)$ and $P_2(x_2, y_2)$:

$$d = \sqrt{(x_2 - x_1)^2 + (y_2 - y_1)^2}$$

Midpoint of $\overline{P_1P_2}$: $\left(\dfrac{x_1 + x_2}{2}, \dfrac{y_1 + y_2}{2}\right)$

Lines

Slope of line through $P_1(x_1, y_1)$ and $P_2(x_2, y_2)$:

$$m = \frac{y_2 - y_1}{x_2 - x_1}$$

Point-slope equation of line through $P_1(x_1, y_1)$ with slope m:

$$y - y_1 = m(x - x_1)$$

Slope-intercept equation of line with slope m and y-intercept b:

$$y = mx + b$$

Circles

Equation of the circle with center (h, k) and radius r:

$$(x - h)^2 + (y - k)^2 = r^2$$

Trigonometry

Angle Measurement

π radians $= 180°$

$1° = \frac{\pi}{180}$ rad $\quad$ 1 rad $= \frac{180°}{\pi}$

$s = r\theta$

(θ in radians)

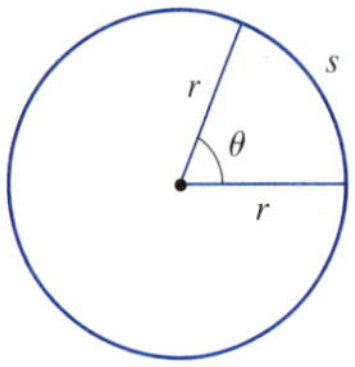

Right Angle Trigonometry

$\sin\theta = \frac{\text{opp}}{\text{hyp}}$ $\quad$ $\csc\theta = \frac{\text{hyp}}{\text{opp}}$

$\cos\theta = \frac{\text{adj}}{\text{hyp}}$ $\quad$ $\sec\theta = \frac{\text{hyp}}{\text{adj}}$

$\tan\theta = \frac{\text{opp}}{\text{adj}}$ $\quad$ $\cot\theta = \frac{\text{adj}}{\text{opp}}$

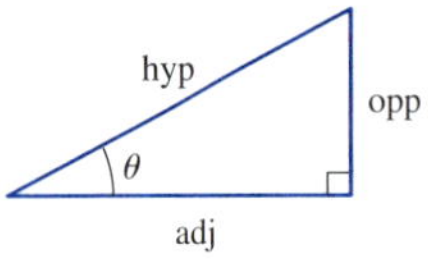

Trigonometric Functions

$\sin\theta = \frac{y}{r}$ $\quad$ $\csc\theta = \frac{r}{y}$

$\cos\theta = \frac{x}{r}$ $\quad$ $\sec\theta = \frac{r}{x}$

$\tan\theta = \frac{y}{x}$ $\quad$ $\cot\theta = \frac{x}{y}$

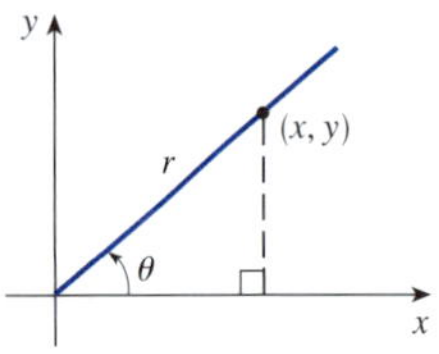

Graphs of Trigonometric Functions

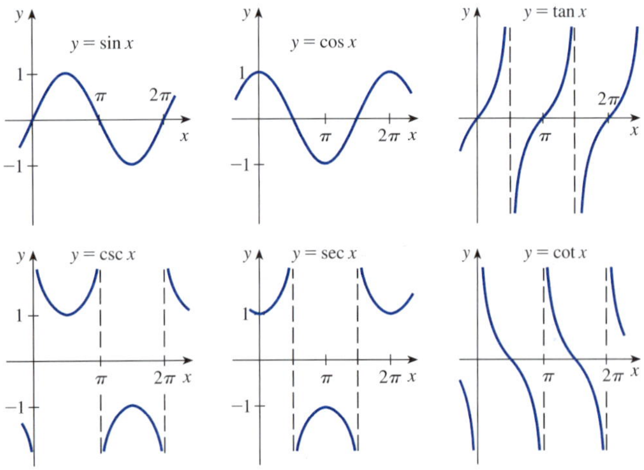

Trigonometric Functions of Important Angles

θ	radians	$\sin\theta$	$\cos\theta$	$\tan\theta$
0°	0	0	1	0
30°	$\pi/6$	$1/2$	$\sqrt{3}/2$	$\sqrt{3}/3$
45°	$\pi/4$	$\sqrt{2}/2$	$\sqrt{2}/2$	1
60°	$\pi/3$	$\sqrt{3}/2$	$1/2$	$\sqrt{3}$
90°	$\pi/2$	1	0	—

Fundamental Identities

$\csc\theta = \frac{1}{\sin\theta}$ $\quad$ $\sec\theta = \frac{1}{\cos\theta}$

$\tan\theta = \frac{\sin\theta}{\cos\theta}$ $\quad$ $\cot\theta = \frac{\cos\theta}{\sin\theta}$

$\cot\theta = \frac{1}{\tan\theta}$ $\quad$ $\sin^2\theta + \cos^2\theta = 1$

$1 + \tan^2\theta = \sec^2\theta$ $\quad$ $1 + \cot^2\theta = \csc^2\theta$

$\sin(-\theta) = -\sin\theta$ $\quad$ $\cos(-\theta) = \cos\theta$

$\tan(-\theta) = -\tan\theta$ $\quad$ $\sin\left(\frac{\pi}{2} - \theta\right) = \cos\theta$

$\cos\left(\frac{\pi}{2} - \theta\right) = \sin\theta$ $\quad$ $\tan\left(\frac{\pi}{2} - \theta\right) = \cot\theta$

The Law of Sines

$$\frac{\sin A}{a} = \frac{\sin B}{b} = \frac{\sin C}{c}$$

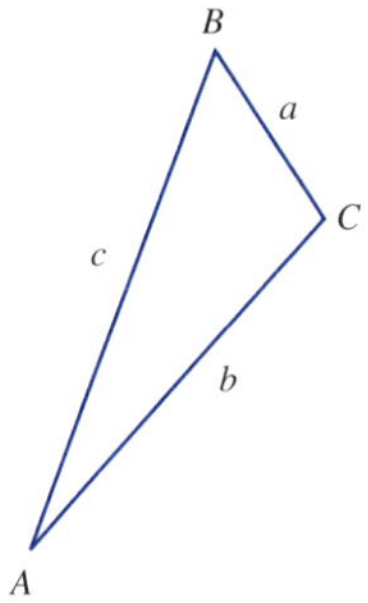

The Law of Cosines

$a^2 = b^2 + c^2 - 2bc\cos A$

$b^2 = a^2 + c^2 - 2ac\cos B$

$c^2 = a^2 + b^2 - 2ab\cos C$

Addition and Subtraction Formulas

$\sin(x + y) = \sin x\cos y + \cos x\sin y$

$\sin(x - y) = \sin x\cos y - \cos x\sin y$

$\cos(x + y) = \cos x\cos y - \sin x\sin y$

$\cos(x - y) = \cos x\cos y + \sin x\sin y$

$\tan(x + y) = \frac{\tan x + \tan y}{1 - \tan x\tan y}$

$\tan(x - y) = \frac{\tan x - \tan y}{1 + \tan x\tan y}$

Double-Angle Formulas

$\sin 2x = 2\sin x\cos x$

$\cos 2x = \cos^2 x - \sin^2 x = 2\cos^2 x - 1 = 1 - 2\sin^2 x$

$\tan 2x = \frac{2\tan x}{1 - \tan^2 x}$

Half-Angle Formulas

$\sin^2 x = \frac{1 - \cos 2x}{2}$ $\quad$ $\cos^2 x = \frac{1 + \cos 2x}{2}$

Calculus: Early Transcendentals, 5th Edition

Volume I

James Stewart
McMaster University

THOMSON
BROOKS/COLE

Australia • Canada • Mexico • Singapore • Spain • United Kingdom • United States

THOMSON
BROOKS/COLE

Calculus Early Transcendentals, 5th Edition
James Stewart

Executive Editors:
Michele Baird, Maureen Staudt & Michael Stranz

Project Development Manager:
Linda de Stefano

Marketing Coordinators:
Lindsay Annett and Sara Mercurio

Production/Manufacturing Supervisor:
Donna M. Brown

Pre-Media Services Supervisor:
Dan Plofchan

Rights and Permissions Specialists:
Kalina Hintz and Bahman Naraghi

Cover Image
Getty Images*

Printed in the United States of America
1 2 3 4 5 6 7 8 08 07 06 05

For more information, please contact Thomson Custom Solutions, 5191 Natorp Boulevard, Mason, OH 45040. Or you can visit our Internet site at www.thomsoncustom.com

The Adaptable Courseware Program consists of products and additions to existing Brooks/Cole products that are produced from camera-ready copy. Peer review, class testing, and accuracy are primarily the responsibility of the author(s).

Calculus Early Transcendentals, 5th Edition
/ James Stewart
ISBN 0-495-27755-X

International Divisions List

Asia (Including India):
Thomson Learning
(a division of Thomson Asia Pte Ltd)
5 Shenton Way #01-01
UIC Building
Singapore 068808
Tel: (65) 6410-1200
Fax: (65) 6410-1208

Australia/New Zealand:
Thomson Learning Australia
102 Dodds Street
Southbank, Victoria 3006
Australia

Latin America:
Thomson Learning
Seneca 53
Colonia Polano
11560 Mexico, D.F., Mexico
Tel (525) 281-2906
Fax (525) 281-2656

Canada:
Thomson Nelson
1120 Birchmount Road
Toronto, Ontario
Canada M1K 5G4
Tel (416) 752-9100
Fax (416) 752-8102

UK/Europe/Middle East/Africa:
Thomson Learning
High Holborn House
50-51 Bedford Row
London, WC1R 4LS
United Kingdom
Tel 44 (020) 7067-2500
Fax 44 (020) 7067-2600

Spain (Includes Portugal):
Thomson Paraninfo
Calle Magallanes 25
28015 Madrid
España
Tel 34 (0)91 446-3350
Fax 34 (0)91 445-6218

*Unless otherwise noted, all cover images used by Thomson Custom Solutions have been supplied courtesy of Getty Images with the exception of the *Earthview* cover image, which has been supplied by the National Aeronautics and Space Administration (NASA).

Custom Table of Contents

PREFACE

A great discovery solves a great problem but there is a grain of discovery in the solution of any problem. Your problem may be modest; but if it challenges your curiosity and brings into play your inventive faculties, and if you solve it by your own means, you may experience the tension and enjoy the triumph of discovery.

GEORGE POLYA

The art of teaching, Mark Van Doren said, is the art of assisting discovery. I have tried to write a book that assists students in discovering calculus—both for its practical power and its surprising beauty. In this edition, as in the first four editions, I aim to convey to the student a sense of the utility of calculus and develop technical competence, but I also strive to give some appreciation for the intrinsic beauty of the subject. Newton undoubtedly experienced a sense of triumph when he made his great discoveries. I want students to share some of that excitement.

The emphasis is on understanding concepts. I think that nearly everybody agrees that this should be the primary goal of calculus instruction. In fact, the impetus for the current calculus reform movement came from the Tulane Conference in 1986, which formulated as their first recommendation:

Focus on conceptual understanding.

I have tried to implement this goal through the *Rule of Three*: "Topics should be presented geometrically, numerically, and algebraically." Visualization, numerical and graphical experimentation, and other approaches have changed how we teach conceptual reasoning in fundamental ways. More recently, the Rule of Three has been expanded to become the *Rule of Four* by emphasizing the verbal, or descriptive, point of view as well.

In writing the fifth edition my premise has been that it is possible to achieve conceptual understanding and still retain the best traditions of traditional calculus. The book contains elements of reform, but within the context of a traditional curriculum. (Instructors who prefer a more streamlined curriculum should look at my book *Calculus: Concepts and Contexts, Second Edition.*)

What's New in the Fifth Edition

By way of preparing to write the fifth edition of this text, I spent a year teaching calculus from the fourth edition at the University of Toronto. I listened carefully to my students' questions and my colleagues' suggestions. And as I prepared each lecture I sometimes realized that an additional example was needed, or a sentence could be clarified, or a section could use a few more exercises of a certain type. In addition, I paid attention to the suggestions sent to me by many users and to the comments of the reviewers.

An unusual source of new problems was a phone call I received from a friend of mine, Richard Armstrong. Richard is a partner in an engineering consulting firm and advises clients who build hospitals and hotels. He told me that, in certain parts of the world, sprinkler systems for large buildings are supplied with water by tanks located on the roofs of the buildings. Of course he knew that the water pressure decreases as the water level decreases, but he needed to be able to quantify this effect so his clients could guarantee a certain pressure for a certain period of time. I told him how he could solve his problem by solving a separable differential equation, but it occurred to me that his problem could be developed into a rather nice project when combined with other ideas. (See the project on page 609).

The structure of *Calculus, Early Transcendentals, Fifth Edition,* remains largely unchanged, but there are hundreds of improvements, small and large:

- The review of inverse trigonometric functions has been moved from an appendix to Section 1.6.
- Two sections in Chapter 10 have been combined.
- I have rewritten Section 12.2 to give more prominence to the geometric description of vectors.
- New phrases and margin notes have been added to clarify the exposition.
- A number of pieces of art have been redrawn.
- The data in examples and exercises have been updated to be more timely.
- Examples have been added. For instance, I added the new Example 1 in Section 5.3 (page 394) because students have a tough time grasping the idea of a function defined by an integral with a variable limit of integration. I think it helps to look at Example 1 before considering the Fundamental Theorem of Calculus.
- Extra steps have been provided in some of the existing examples.
- More than 25% of the exercises in each chapter are new. Here are a few of my favorites:

Exercise	Page	Exercise	Page	Exercise	Page
2.8.34	164	3.9.55	256	4.4.74	315
5.4.52	412	7.7.36	529	9.1.11–12	592
10.3.47–48	678	11.9.40	760	11.12.35	784
13.3.32–34	869	14.3.5–6	920	14.5.15–16	938

  I've also added new problems to the Problems Plus sections. See, for instance, Problems 20 and 21 on page 277, Problems 9 and 10 on page 585, and Problems 20 and 22 on page 791.
- Five new projects have been added. The one on page 243 asks students to design a roller coaster so the track is smooth at transition points. The project on page 554, the idea for which I thank Larry Riddle, is actually a contest in which the winning curve has the smallest arc length (within a certain class of curves).
- A CD called *Tools for Enriching Calculus* (*TEC*) is included with every copy of the fifth edition. See the description on page xvii.
- Conscious of the need to control the size of the book, I've put additional topics (with exercises) on the revamped web site www.stewartcalculus.com (see the description on page xvii) rather than in the text itself. These include the new topics Fourier Series and Formulas for the Remainder Term in Taylor Series, as well as topics that appeared in previous editions: Review of Basic Algebra, Rotation of Axes, and Lies My Calculator and Computer Told Me.

Features

Conceptual Exercises The most important way to foster conceptual understanding is through the problems that we assign. To that end I have devised various types of new problems. Some exercise sets begin with requests to explain the meanings of the basic concepts of the section. (See, for instance, the first few exercises in Sections 2.2, 2.5, 2.7, 11.2, 14.2, and 14.3.) Similarly, all the review sections begin with a *Concept Check* and a *True-False Quiz*. Other exercises test conceptual understanding through graphs or tables (see Exercises 2.8.1–3, 2.9.35–38, 3.7.1–4, 9.1.11–12, 10.1.24–27, 11.10.2, 13.2.1–2, 13.3.29–33, 14.1.1–2, 14.1.30–36, 14.3.3–8, 14.6.1–2, 14.7.3–4, 15.1.5–10, 16.1.11–18, 16.2.17–18, and 16.3.1–2).

Another type of exercise uses verbal description to test conceptual understanding (see Exercises 2.5.8, 2.9.48, 4.3.59–60, and 7.8.67). I particularly value problems that combine and compare graphical, numerical, and algebraic approaches (see Exercises 2.6.35–36, 3.3.23, and 9.5.2).

Graded Exercise Sets Each exercise set is carefully graded, progressing from basic conceptual exercises and skill-development problems to more challenging problems involving applications and proofs.

Real-World Data My assistants and I spent a great deal of time looking in libraries, contacting companies and government agencies, and searching the Internet for interesting real-world data to introduce, motivate, and illustrate the concepts of calculus. As a result, many of the examples and exercises deal with functions defined by such numerical data or graphs. See, for instance, Figures 1, 11, and 12 in Section 1.1 (seismograms from the Northridge earthquake), Exercise 2.9.36 (percentage of the population under age 18), Exercise 5.1.14 (velocity of the space shuttle *Endeavour*), and Figure 4 in Section 5.4 (San Francisco power consumption). Functions of two variables are illustrated by a table of values of the wind-chill index as a function of air temperature and wind speed (Example 2 in Section 14.1). Partial derivatives are introduced in Section 14.3 by examining a column in a table of values of the heat index (perceived air temperature) as a function of the actual temperature and the relative humidity. This example is pursued further in connection with linear approximations (Example 3 in Section 14.4). Directional derivatives are introduced in Section 14.6 by using a temperature contour map to estimate the rate of change of temperature at Reno in the direction of Las Vegas. Double integrals are used to estimate the average snowfall in Colorado on December 24, 1982 (Example 4 in Section 15.1). Vector fields are introduced in Section 16.1 by depictions of actual velocity vector fields showing San Francisco Bay wind patterns.

Projects One way of involving students and making them active learners is to have them work (perhaps in groups) on extended projects that give a feeling of substantial accomplishment when completed. I have included four kinds of projects: *Applied Projects* involve applications that are designed to appeal to the imagination of students. The project after Section 9.3 asks whether a ball thrown upward takes longer to reach its maximum height or to fall back to its original height. (The answer might surprise you.) The project after Section 14.8 uses Lagrange multipliers to determine the masses of the three stages of a rocket so as to minimize the total mass while enabling the rocket to reach a desired velocity. *Laboratory Projects* involve technology; the one following Section 10.2 shows how to use Bézier curves to design shapes that represent letters for a laser printer. *Writing Projects* ask students to compare present-day methods with those of the founders of calculus—Fermat's method for finding tangents, for instance. Suggested references are supplied. *Discovery Projects* anticipate results to be discussed later or encourage discovery through pattern recognition (see the one following Section 7.6). Others explore aspects of geometry: tetrahedra (after Section 12.4), hyperspheres (after Section 15.7), and intersections of three cylinders (after Section 15.8). Additional projects can be found in the *Instructor's Guide* (see, for instance, Group Exercise 5.1: Position from Samples) and also in the *CalcLabs* supplements.

Problem Solving

Students usually have difficulties with problems for which there is no single well-defined procedure for obtaining the answer. I think nobody has improved very much on George Polya's four-stage problem-solving strategy and, accordingly, I have included a version of his problem-solving principles following Chapter 1. They are applied, both explicitly and implicitly, throughout the book. After the other chapters I have placed sections called *Problems Plus,* which feature examples of how to tackle challenging calculus problems. In selecting the varied problems for these sections I kept in mind the following advice from David Hilbert: "A mathematical problem should be difficult in order to entice us, yet not inaccessible lest it mock our efforts." When I put these challenging problems on assignments and tests I grade them in a different way. Here I reward a student significantly for ideas toward a solution and for recognizing which problem-solving principles are relevant.

Technology

The availability of technology makes it not less important but more important to clearly understand the concepts that underlie the images on the screen. But, when properly used, graphing calculators and computers are powerful tools for discovering and understanding those concepts. This textbook can be used either with or without technology and I use two special symbols to indicate clearly when a particular type of machine is required. The icon indicates an exercise that definitely requires the use of such technology, but that is not to say that it can't be used on the other exercises as well. The symbol CAS is reserved for problems in which the full resources of a computer algebra system (like Derive, Maple, Mathematica, or the TI-89/92) are required. But technology doesn't make pencil and paper obsolete. Hand calculation and sketches are often preferable to technology for illustrating and reinforcing some concepts. Both instructors and students need to develop the ability to decide where the hand or the machine is appropriate.

Tools for Enriching™ Calculus

The CD-ROM called *TEC* included with every copy of this book is a companion to the text and is intended to enrich and complement its contents. Developed by Harvey Keynes at the University of Minnesota and Dan Clegg at Palomar College, *TEC* uses a discovery and exploratory approach. In sections of the book where technology is particularly appropriate, marginal icons direct students to *TEC* modules that provide a laboratory environment in which they can explore the topic in different ways and at different levels. Instructors can choose to become involved at several different levels, ranging from simply encouraging students to use the modules for independent exploration, to assigning specific exercises from those included with each module, or to creating additional exercises, labs, and projects that make use of the modules.

TEC also includes *homework hints* for representative exercises (usually odd-numbered) in every section of the text, indicated by printing the exercise number in red. These hints are usually presented in the form of questions and try to imitate an effective teaching assistant by functioning as a silent tutor. They are constructed so as not to reveal any more of the actual solution than is minimally necessary to make further progress.

Web Site: www.stewartcalculus.com

This site has been renovated and now includes the following.

- Algebra Review, with tutorial
- Additional Topics (complete with exercise sets):

  Fourier Series, Formulas for the Remainder Term in Taylor Series,
  Rotation of Axes, Lies My Calculator and Computer Told Me

- Drill exercises that appeared in previous editions, together with their solutions
- Problems Plus from prior editions
- Links, for particular topics, to outside web resources
- History of Mathematics, with links to the better historical web sites
- Downloadable versions of *CalcLabs* for Derive and TI graphing calculators

Content

A Preview of Calculus The book begins with an overview of the subject and includes a list of questions to motivate the study of calculus.

1 Functions and Models From the beginning, multiple representations of functions are stressed: verbal, numerical, visual, and algebraic. A discussion of mathematical models leads to a review of the standard functions, including exponential and logarithmic functions, from these four points of view.

2 Limits and Derivatives The material on limits is motivated by a prior discussion of the tangent and velocity problems. Limits are treated from descriptive, graphical, numerical, and algebraic points of view. Section 2.4, on the precise ε-δ definition of a limit, is an optional section. Sections 2.8 and 2.9 deal with derivatives (especially with functions defined graphically and numerically) before the differentiation rules are covered in Chapter 3. Here the examples and exercises explore the meanings of derivatives in various contexts.

3 Differentiation Rules All the basic functions, including exponential, logarithmic, and inverse trigonometric functions, are differentiated here. When derivatives are computed in applied situations, students are asked to explain their meanings.

4 Applications of Differentiation The basic facts concerning extreme values and shapes of curves are deduced from the Mean Value Theorem. Graphing with technology emphasizes the interaction between calculus and calculators and the analysis of families of curves. Some substantial optimization problems are provided, including an explanation of why you need to raise your head 42° to see the top of a rainbow.

5 Integrals The area problem and the distance problem serve to motivate the definite integral, with sigma notation introduced as needed. (Full coverage of sigma notation is provided in Appendix E.) Emphasis is placed on explaining the meanings of integrals in various contexts and on estimating their values from graphs and tables.

6 Applications of Integration Here I present the applications of integration—area, volume, work, average value—that can reasonably be done without specialized techniques of integration. General methods are emphasized. The goal is for students to be able to divide a quantity into small pieces, estimate with Riemann sums, and recognize the limit as an integral.

7 Techniques of Integration All the standard methods are covered but, of course, the real challenge is to be able to recognize which technique is best used in a given situation. Accordingly, in Section 7.5, I present a strategy for integration. The use of computer algebra systems is discussed in Section 7.6.

8 Further Applications of Integration Here are the applications of integration—arc length and surface area—for which it is useful to have available all the techniques of integration, as well as applications to biology, economics, and physics (hydrostatic force and centers of mass). I have also included a section on probability. There are more applications here than can realistically be covered in a given course. Instructors should select applications suitable for their students and for which they themselves have enthusiasm.

9 ▫ Differential Equations

Modeling is the theme that unifies this introductory treatment of differential equations. Direction fields and Euler's method are studied before separable and linear equations are solved explicitly, so that qualitative, numerical, and analytic approaches are given equal consideration. These methods are applied to the exponential, logistic, and other models for population growth. The first five or six sections of this chapter serve as a good introduction to first-order differential equations. An optional final section uses predator-prey models to illustrate systems of differential equations.

10 ▫ Parametric Equations and Polar Coordinates

The sections on areas and tangents for parametric curves and arc length and surface area have been streamlined and combined as *Calculus with Parametric Curves.* Such curves are well suited to laboratory projects; the two presented here involve families of curves and Bézier curves. A brief treatment of conic sections in polar coordinates prepares the way for Kepler's Laws in Chapter 13.

11 ▫ Infinite Sequences and Series

The convergence tests have intuitive justifications (see page 723) as well as formal proofs. Numerical estimates of sums of series are based on which test was used to prove convergence. The emphasis is on Taylor series and polynomials and their applications to physics. Error estimates include those from graphing devices.

12 ▫ Vectors and the Geometry of Space

The material on three-dimensional analytic geometry and vectors is divided into two chapters. Chapter 12 deals with vectors, the dot and cross products, lines, planes, surfaces, and cylindrical and spherical coordinates.

13 ▫ Vector Functions

This chapter covers vector-valued functions, their derivatives and integrals, the length and curvature of space curves, and velocity and acceleration along space curves, culminating in Kepler's laws.

14 ▫ Partial Derivatives

Functions of two or more variables are studied from verbal, numerical, visual, and algebraic points of view. In particular, I introduce partial derivatives by looking at a specific column in a table of values of the heat index (perceived air temperature) as a function of the actual temperature and the relative humidity. Directional derivatives are estimated from contour maps of temperature, pressure, and snowfall.

15 ▫ Multiple Integrals

Contour maps and the Midpoint Rule are used to estimate the average snowfall and average temperature in given regions. Double and triple integrals are used to compute probabilities, surface areas, and (in projects) volumes of hyperspheres and volumes of intersections of three cylinders.

16 ▫ Vector Calculus

Vector fields are introduced through pictures of velocity fields showing San Francisco Bay wind patterns. The similarities among the Fundamental Theorem for line integrals, Green's Theorem, Stokes' Theorem, and the Divergence Theorem are emphasized.

17 ▫ Second-Order Differential Equations

Since first-order differential equations are covered in Chapter 9, this final chapter deals with second-order linear differential equations, their application to vibrating springs and electric circuits, and series solutions.

Ancillaries

Calculus, Early Transcendentals, Fifth Edition, is supported by a complete set of ancillaries developed under my direction. Each piece has been designed to enhance student understanding and to facilitate creative instruction. The tables on pages xxiv–xxv describe each of these ancillaries.

Acknowledgments

The preparation of this and previous editions has involved much time spent reading the reasoned (but sometimes contradictory) advice from a large number of astute reviewers. I greatly appreciate the time they spent to understand my motivation for the approach taken. I have learned something from each of them.

Fifth Edition Reviewers

Martina Bode, *Northwestern University*
Philip L. Bowers, *Florida State University*
Scott Chapman, *Trinity University*
Charles N. Curtis, *Missouri Southern State College*
John W. Davenport, *Georgia Southern University*
Elias Deeba, *University of Houston–Downtown*
Greg Dresden, *Washington and Lee University*
Martin Erickson, *Truman State University*
Laurene V. Fausett, *Georgia Southern University*
Norman Feldman, *Sonoma State University*
José D. Flores, *The University of South Dakota*
Kevin A. Grasse, *The University of Oklahoma*
Howard B. Hamilton, *California State University, Sacramento*
Gary W. Harrison, *College of Charleston*
Randall R. Holmes, *Auburn University*
James F. Hurley, *University of Connecticut*
Matthew A. Isom, *Arizona State University*
Zsuzsanna M. Kadas, *St. Michael's College*
Frederick W. Keene, *Pasadena City College*
Robert L. Kelley, *University of Miami*
John C. Lawlor, *University of Vermont*
Christopher C. Leary, *State University of New York at Geneseo*
William O. Martin, *North Dakota State University*
Gerald Y. Matsumoto, *American River College*
Gordon Melrose, *Old Dominion University*
Michael Montaño, *Riverside Community College*
Hussain S. Nur, *California State University, Fresno*
Mike Penna, *Indiana University–Purdue University Indianapolis*
John Ringland, *State University of New York at Buffalo*
E. Arthur Robinson, Jr., *The George Washington University*
Rob Root *Lafayette College*
Teresa Morgan Smith, *Blinn College*
Donald W. Solomon, *University of Wisconsin–Milwaukee*
Kristin Stoley, *Blinn College*
Paul Xavier Uhlig, *St. Mary's University, San Antonio*
Dennis H. Wortman, *University of Massachusetts, Boston*
Xian Wu, *University of South Carolina*

Previous Edition Reviewers

B. D. Aggarwala,
University of Calgary
John Alberghini,
Manchester Community College
Michael Albert,
Carnegie-Mellon University
Daniel Anderson,
University of Iowa
Donna J. Bailey,
Northeast Missouri State University
Wayne Barber,
Chemeketa Community College
Neil Berger,
University of Illinois, Chicago
David Berman,
University of New Orleans
Richard Biggs,
University of Western Ontario
Robert Blumenthal,
Oglethorpe University
Barbara Bohannon,
Hofstra University
Jay Bourland,
Colorado State University
Stephen W. Brady,
Wichita State University
Michael Breen,
Tennessee Technological University
Stephen Brown
Robert N. Bryan,
University of Western Ontario
David Buchthal,
University of Akron
Jorge Cassio,
Miami-Dade Community College
Jack Ceder,
University of California, Santa Barbara
James Choike,
Oklahoma State University
Barbara Cortzen,
DePaul University
Carl Cowen,
Purdue University
Philip S. Crooke,
Vanderbilt University
Daniel Cyphert,
Armstrong State College
Robert Dahlin
Gregory J. Davis,
University of Wisconsin–Green Bay
Daniel DiMaria,
Suffolk Community College
Seymour Ditor,
University of Western Ontario
Daniel Drucker,
Wayne State University
Kenn Dunn,
Dalhousie University
Dennis Dunninger,
Michigan State University
Bruce Edwards,
University of Florida
David Ellis,
San Francisco State University
John Ellison,
Grove City College
Garret Etgen,
University of Houston
Theodore G. Faticoni,
Fordham University
Newman Fisher,
San Francisco State University
William Francis,
Michigan Technological University
James T. Franklin,
Valencia Community College, East
Stanley Friedlander,
Bronx Community College
Patrick Gallagher,
Columbia University–New York
Paul Garrett,
University of Minnesota–Minneapolis
Frederick Gass,
Miami University of Ohio
Bruce Gilligan,
University of Regina
Matthias K. Gobbert,
University of Maryland, Baltimore County
Gerald Goff,
Oklahoma State University
Stuart Goldenberg,
California Polytechnic State University
John A. Graham,
Buckingham Browne & Nichols School
Richard Grassl,
University of New Mexico
Michael Gregory,
University of North Dakota
Charles Groetsch,
University of Cincinnati
Salim M. Haïdar,
Grand Valley State University
D. W. Hall,
Michigan State University
Robert L. Hall,
University of Wisconsin–Milwaukee
Darel Hardy,
Colorado State University
Melvin Hausner,
New York University/Courant Institute
Curtis Herink,
Mercer University
Russell Herman,
University of North Carolina at Wilmington
Allen Hesse,
Rochester Community College
Gerald Janusz,
University of Illinois at Urbana-Champaign
John H. Jenkins,
Embry-Riddle Aeronautical University, Prescott Campus
Clement Jeske,
University of Wisconsin, Platteville
Carl Jockusch,
University of Illinois at Urbana-Champaign
Jan E. H. Johansson,
University of Vermont
Jerry Johnson,
Oklahoma State University
Matt Kaufman
Matthias Kawski,
Arizona State University
Virgil Kowalik,
Texas A&I University

Kevin Kreider,
University of Akron
Leonard Krop,
DePaul University
Mark Krusemeyer,
Carleton College
David Leeming,
University of Victoria
Sam Lesseig,
Northeast Missouri State University
Phil Locke,
University of Maine
Joan McCarter,
Arizona State University
Phil McCartney,
Northern Kentucky University
Igor Malyshev,
San Jose State University
Larry Mansfield,
Queens College
Mary Martin,
Colgate University
Nathaniel F. G. Martin,
University of Virginia
Tom Metzger,
University of Pittsburgh
Teri Jo Murphy,
University of Oklahoma
Richard Nowakowski,
Dalhousie University
Wayne N. Palmer,
Utica College
Vincent Panico,
University of the Pacific
F. J. Papp,
University of Michigan–Dearborn
Mark Pinsky,
Northwestern University
Lothar Redlin,
The Pennsylvania State University
Tom Rishel,
Cornell University
Joel W. Robbin,
University of Wisconsin–Madison
Richard Rockwell,
Pacific Union College
Richard Ruedemann,
Arizona State University
David Ryeburn,
Simon Fraser University
Richard St. Andre,
Central Michigan University
Ricardo Salinas,
San Antonio College
Robert Schmidt,
South Dakota State University
Eric Schreiner,
Western Michigan University
Mihr J. Shah,
Kent State University–Trumbull
Theodore Shifrin,
University of Georgia
Wayne Skrapek,
University of Saskatchewan
Larry Small,
Los Angeles Pierce College
William Smith,
University of North Carolina
Edward Spitznagel,
Washington University
Joseph Stampfli,
Indiana University
M. B. Tavakoli,
Chaffey College
Stan Ver Nooy,
University of Oregon
Andrei Verona,
California State University–Los Angeles
Russell C. Walker,
Carnegie Mellon University
William L. Walton,
McCallie School
Jack Weiner,
University of Guelph
Alan Weinstein,
University of California, Berkeley
Theodore W. Wilcox,
Rochester Institute of Technology
Steven Willard,
University of Alberta
Robert Wilson,
University of Wisconsin–Madison
Jerome Wolbert,
University of Michigan–Ann Arbor
Mary Wright,
Southern Illinois University–Carbondale
Paul M. Wright,
Austin Community College

In addition, I would like to thank George Bergman, Stuart Goldenberg, Emile LeBlanc, Gerald Leibowitz, Charles Pugh, Marina Ratner, Peter Rosenthal, and Alan Weinstein for their suggestions; Dan Clegg for his research in libraries and on the Internet; Arnold Good for his treatment of optimization problems with implicit differentiation; Al Shenk and Dennis Zill for permission to use exercises from their calculus texts; COMAP for permission to use project material; George Bergman, David Bleecker, Dan Clegg, Victor Kaftal, Anthony Lam, Jamie Lawson, Ira Rosenholtz, Lowell Smylie, and Larry Wallen for ideas for exercises; Dan Drucker for the roller derby project; Tom Farmer, Fred Gass, John Ramsay, Larry Riddle, and Philip Straffin for ideas for projects; Dan Anderson and Dan Drucker for solving the new exercises; and Jeff Cole and Dan Clegg for their careful preparation and proofreading of the answer manuscript. I'm grateful to Jeff Cole for suggesting ways to improve the exercises. Dan Clegg acted as my assistant throughout; he proofread, made suggestions, and contributed many of the new exercises.

In addition, I thank those who have contributed to past editions: Ed Barbeau, Fred Brauer, Andy Bulman-Fleming, Bob Burton, Tom DiCiccio, Garret Etgen, Chris Fisher, Gene Hecht, Harvey Keynes, Kevin Kreider, E. L. Koh, Zdislav Kovarik, David Leep, Lothar Redlin, Carl Riehm, Doug Shaw, and Saleem Watson.

I also thank Stephanie Kuhns, Kathi Townes, and Brian Betsill of TECHarts for their production services and the following Brooks/Cole staff: Kirk Bomont, editorial production project manager; Karin Sandberg, Stephanie Taylor, and Bryan Vann, marketing team; Stacy Green, assistant editor, and Jessica Zimmerman, editorial assistant; Earl Perry, technology project manager, and Jessica Reed, print/media buyer. They have all done an outstanding job.

I have been very fortunate to have worked with some of the best mathematics editors in the business over the past two decades: Ron Munro, Harry Campbell, Craig Barth, Jeremy Hayhurst, Gary Ostedt, and now Bob Pirtle. Bob continues in that tradition of editors who, while offering sound advice and ample assistance, trust my instincts and allow me to write the books that I want to write.

JAMES STEWART

Ancillaries for Instructors

Instructor's Resource CD-ROM
ISBN 0-534-39463-7

Contains Electronic Instructor's Guide, Electronic Solutions, BCA Testing, Instructions for BCA Homework, and electronic transparencies (CalcLink).

▫ http://bca.brookscole.com

Tools for Enriching™ Calculus CD-ROM
by Harvey B. Keynes, James Stewart, and Dan Clegg
ISBN 0-534-39731-X

TEC *provides a laboratory environment in which students can explore selected topics.* TEC *also includes homework hints for representative exercises.*

Instructor's Guide
by Douglas Shaw, Harvey B. Keynes, and James Stewart
ISBN 0-534-39334-9

Each section of the main text is discussed from several viewpoints and contains suggested time to allot, points to stress, text discussion topics, core materials for lecture, workshop/discussion suggestions, group work exercises in a form suitable for handout, and suggested homework problems. An electronic version is available on the Instructor's Resource CD-ROM.

Instructor's Guide for AP® Calculus
by Douglas Shaw and Robert Gerver, contributing author
ISBN 0-534-39341-1

Taking the perspective of optimizing preparation for the AP exam, each section of the main text is discussed from several viewpoints and contains suggested time to allot, points to stress, daily quizzes, core materials for lecture, workshop/discussion suggestions, group work exercises in a form suitable for handout, tips for the AP exam, and suggested homework problems.

Complete Solutions Manual

Single Variable Early Transcendentals
by Daniel Anderson, Jeffery A. Cole, and Daniel Drucker
ISBN 0-534-39332-2

Multivariable Early Transcendentals
by Dan Clegg and Barbara Frank
ISBN 0-534-39359-4

Includes worked-out solutions to all exercises in the text.

Printed Test Items
ISBN 0-534-39336-5

Contains multiple-choice and short-answer test items that key directly to the text.

iLrn Assessment
ISBN 0-534-39335-7

Available online or on CD-ROM, browser-based iLrn is fully integrated text-specific testing, tutorial, and course management software. With no need for plug-ins or downloads, iLrn offers algorithmically generated problem values and machine-graded free response mathematics.

▫ http://www.iLrn.com

Text-Specific Videos
ISBN 0-534-39325-X

Text-specific videotape sets, available at no charge to adopters, consisting of one tape per text chapter. Each tape features a 10- to 20-minute problem-solving lesson for each section of the chapter. Covers both single- and multivariable calculus.

Transparencies by James Stewart

Single Variable ISBN 0-534-39337-3

Multivariable ISBN 0-534-39383-7

Full-color, large-scale sheets of reproductions of material from the text.

Ancillaries for Instructors and Students

Stewart Specialty Web Site
www.stewartcalculus.com

Contents: *Algebra Review with tutorial* ▫ *Additional Topics* ▫ *Drill exercises* ▫ *Problems Plus* ▫ *Web Links* ▫ *History of Mathematics* ▫ *Downloadable versions of* CalcLabs *for Derive and TI graphing calculators*

iLrn Homework
ISBN 0-534-39380-2

iLrn Homework allows instructors to assign machine-gradable homework problems that help students identify where they need additional help. That assistance is available through worked-out solutions that guide students through the steps of problem solving, or via live online tutoring at vMentor. *The tutors at this online service will skillfully guide students through a problem, using unique two-way audio and whiteboard features.*

▫ http://www.iLrn.com

The Brooks/Cole Mathematics Resource Center Web Site
http://mathematics.brookscole.com

When you adopt a Thomson–Brooks/Cole mathematics text, you and your students will have access to a variety of teaching and learning resources. This Web site features everything from

|||| Electronic items |||| Printed items

book-specific resources to newsgroups. It's a great way to make teaching and learning an interactive and intriguing experience.

WebTutor™ on WebCT

ISBN 0-534-39328-4

Lecture notes, discussion threads, and quizzes on WebCT.

Journey Through Calculus

by Bill Ralph in conjunction with James Stewart

Student Version ISBN 0-534-26220-1

Instructor's Version ISBN 0-534-36823-9

A Calculus CD-ROM that brings together activities, tutorials, testing, a computer algebra system, and calculus content into one unified environment.

Student Resources

Tools for Enriching™ Calculus CD-ROM

by Harvey B. Keynes, James Stewart, and Dan Clegg

ISBN 0-534-39731-X

TEC *provides a laboratory environment in which students can explore selected topics.* TEC *also includes homework hints for representative exercises.*

Interactive Video SkillBuilder CD-ROM

ISBN 0-534-39326-8

Think of it as portable office hours! The Interactive Video Skillbuilder CD-ROM *contains more than eight hours of video instruction. The problems worked during each video lesson are shown next to the viewing screen so that students can try working them before watching the solution. To help students evaluate their progress, each section contains a ten-question Web quiz (the results of which can be emailed to the instructor) and each chapter contains a chapter test, with answers to each problem. Also included is* MathCue Tutorial *software. This dual-platform software presents and scores problems and tutors students by displaying annotated, step-by-step solutions. Problem sets may be customized.*

Study Guide

Single Variable Early Transcendentals by Richard St. Andre

ISBN 0-534-39331-4

Multivariable by Richard St. Andre

ISBN 0-534-39358-6

Contains a short list of key concepts, a short list of skills to master, a brief introduction to the ideas of the section, an elaboration of the concepts and skills, including extra worked-out examples, and links in the margin to earlier and later material in the text and Study Guide.

Student Solutions Manual

Single Variable Early Transcendentals

by Daniel Anderson, Jeffery A. Cole, and Daniel Drucker

ISBN 0-534-39333-0

Multivariable

by Dan Clegg and Barbara Frank

ISBN 0-534-39360-8

Provides completely worked-out solutions to all odd-numbered exercises within the text, giving students a way to check their answers and ensure that they took the correct steps to arrive at an answer.

CalcLabs with Maple

Single Variable

by Philip Yasskin, Albert Boggess, David Barrow, Maurice Rahe, Jeffery Morgan, Michael Stecher, Art Belmonte, and Kirby Smith

ISBN 0-534-39370-5

Multivariable

by Philip Yasskin and Art Belmonte

ISBN 0-534-39361-6

CalcLabs with Mathematica

Single Variable by Selwyn Hollis

ISBN 0-534-39371-3

Multivariable by Selwyn Hollis

ISBN 0-534-39362-4

Each of these comprehensive lab manuals will help students learn to effectively use the technology tools available to them. Each lab contains clearly explained exercises and a variety of labs and projects to accompany the text.

A Companion to Calculus

by Dennis Ebersole, Doris Schattschneider, Alicia Sevilla, and Kay Somers

ISBN 0-534-26592-8

Written to improve algebra and problem-solving skills of students taking a calculus course, every chapter in this companion is keyed to a calculus topic, providing conceptual background and specific algebra techniques needed to understand and solve calculus problems related to that topic. It is designed for calculus courses that integrate the review of precalculus concepts or for individual use.

Linear Algebra for Calculus

by Konrad J. Heuvers, William P. Francis, John H. Kuisti, Deborah F. Lockhart, Daniel S. Moak, and Gene M. Ortner

ISBN 0-534-25248-6

This comprehensive book, designed to supplement the calculus course, provides an introduction to and review of the basic ideas of linear algebra.

|||| Electronic items |||| Printed items

To the Student

Reading a calculus textbook is different from reading a newspaper or a novel, or even a physics book. Don't be discouraged if you have to read a passage more than once in order to understand it. You should have pencil and paper and calculator at hand to sketch a diagram or make a calculation.

Some students start by trying their homework problems and read the text only if they get stuck on an exercise. I suggest that a far better plan is to read and understand a section of the text before attempting the exercises. In particular, you should look at the definitions to see the exact meanings of the terms. And before you read each example, I suggest that you cover up the solution and try solving the problem yourself. You'll get a lot more from looking at the solution if you do so.

Part of the aim of this course is to train you to think logically. Learn to write the solutions of the exercises in a connected, step-by-step fashion with explanatory sentences—not just a string of disconnected equations or formulas.

The answers to the odd-numbered exercises appear at the back of the book, in Appendix H. Some exercises ask for a verbal explanation or interpretation or description. In such cases there is no single correct way of expressing the answer, so don't worry that you haven't found the definitive answer. In addition, there are often several different forms in which to express a numerical or algebraic answer, so if your answer differs from mine, don't immediately assume you're wrong. For example, if the answer given in the back of the book is $\sqrt{2} - 1$ and you obtain $1/(1 + \sqrt{2})$, then you're right and rationalizing the denominator will show that the answers are equivalent.

The icon indicates an exercise that definitely requires the use of either a graphing calculator or a computer with graphing software. (Section 1.4 discusses the use of these graphing devices and some of the pitfalls that you may encounter.) But that doesn't mean that graphing devices can't be used to check your work on the other exercises as well. The symbol CAS is reserved for problems in which the full resources of a computer algebra system (like Derive, Maple, Mathematica, or the TI-89/92) are required.

You will also encounter the symbol ⊘, which warns you against committing an error. I have placed this symbol in the margin in situations where I have observed that a large proportion of my students tend to make the same mistake.

The icon JTC indicates a reference to the CD-ROM *Journey Through Calculus.* The symbols in the margin refer you to the location in *Journey* where a concept is introduced through an interactive exploration or animation.

The CD-ROM *Tools for Enriching Calculus,* which is included with this textbook, is referred to by means of the symbol TEC. It directs you to modules in which you can explore aspects of calculus for which the computer is particularly useful. *TEC* also provides *Homework Hints* for representative exercises that are indicated by printing the exercise number in red: **23.** These homework hints ask you questions that allow you to make progress toward a solution without actually giving you the answer. You need to pursue each hint in an active manner with pencil and paper to work out the details. If a particular hint doesn't enable you to solve the problem, you can click to reveal the next hint.

The other CD-ROM that comes with this book is *Interactive Video Skillbuilder,* which contains videos of instructors explaining two or three of the examples in every section of the text. Also on the CD is a video in which I offer advice on how to succeed in your calculus course.

I recommend that you keep this book for reference purposes after you finish the course. Because you will likely forget some of the specific details of calculus, the book will serve as a useful reminder when you need to use calculus in subsequent courses. And, because this book contains more material than can be covered in any one course, it can also serve as a valuable resource for a working scientist or engineer.

Calculus is an exciting subject, justly considered to be one of the greatest achievements of the human intellect. I hope you will discover that it is not only useful but also intrinsically beautiful.

JAMES STEWART

CALCULUS

early transcendentals

By the time you finish this course, you will be able to explain the formation and location of rainbows, compute the force exerted by water on a dam, analyze the population cycles of predators and prey, and calculate the escape velocity of a rocket.

A Preview of Calculus

Calculus is fundamentally different from the mathematics that you have studied previously. Calculus is less static and more dynamic. It is concerned with change and motion; it deals with quantities that approach other quantities. For that reason it may be useful to have an overview of the subject before beginning its intensive study. Here we give a glimpse of some of the main ideas of calculus by showing how the concept of a limit arises when we attempt to solve a variety of problems.

The Area Problem

The origins of calculus go back at least 2500 years to the ancient Greeks, who found areas using the "method of exhaustion." They knew how to find the area A of any polygon by dividing it into triangles as in Figure 1 and adding the areas of these triangles.

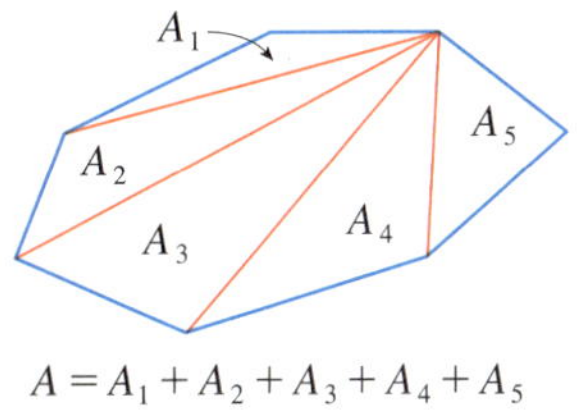

FIGURE 1

It is a much more difficult problem to find the area of a curved figure. The Greek method of exhaustion was to inscribe polygons in the figure and circumscribe polygons about the figure and then let the number of sides of the polygons increase. Figure 2 illustrates this process for the special case of a circle with inscribed regular polygons.

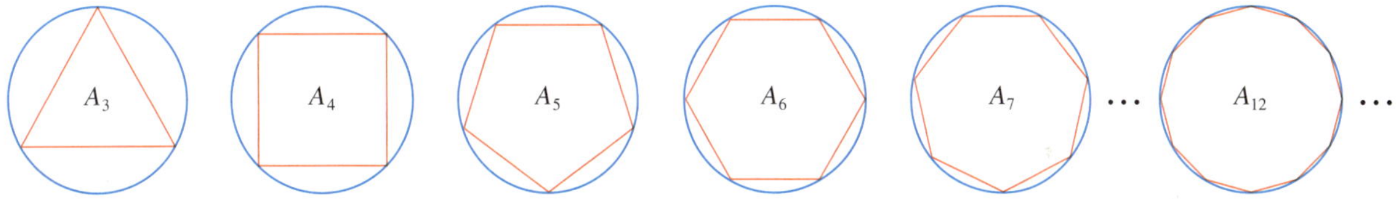

FIGURE 2

The Preview Module is a numerical and pictorial investigation of the approximation of the area of a circle by inscribed and circumscribed polygons.

Let A_n be the area of the inscribed polygon with n sides. As n increases, it appears that A_n becomes closer and closer to the area of the circle. We say that the area of the circle is the *limit* of the areas of the inscribed polygons, and we write

$$A = \lim_{n \to \infty} A_n$$

The Greeks themselves did not use limits explicitly. However, by indirect reasoning, Eudoxus (fifth century B.C.) used exhaustion to prove the familiar formula for the area of a circle: $A = \pi r^2$.

We will use a similar idea in Chapter 5 to find areas of regions of the type shown in Figure 3. We will approximate the desired area A by areas of rectangles (as in Figure 4), let the width of the rectangles decrease, and then calculate A as the limit of these sums of areas of rectangles.

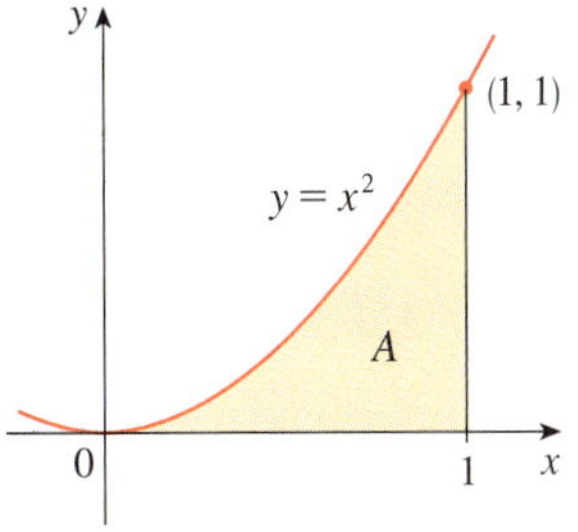

FIGURE 3

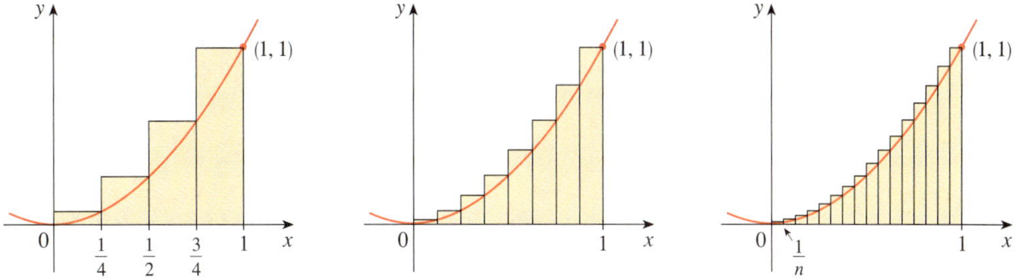

FIGURE 4

Is it possible to fill a circle with rectangles? Try it for yourself.

Resources / Module 1
/ Area
/ Rectangles in Circles

The area problem is the central problem in the branch of calculus called *integral calculus.* The techniques that we will develop in Chapter 5 for finding areas will also enable us to compute the volume of a solid, the length of a curve, the force of water against a dam, the mass and center of gravity of a rod, and the work done in pumping water out of a tank.

The Tangent Problem

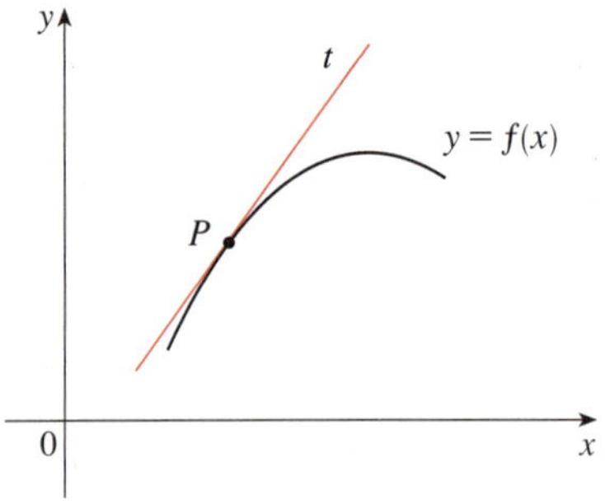

FIGURE 5
The tangent line at P

Consider the problem of trying to find an equation of the tangent line t to a curve with equation $y = f(x)$ at a given point P. (We will give a precise definition of a tangent line in Chapter 2. For now you can think of it as a line that touches the curve at P as in Figure 5.) Since we know that the point P lies on the tangent line, we can find the equation of t if we know its slope m. The problem is that we need two points to compute the slope and we know only one point, P, on t. To get around the problem we first find an approximation to m by taking a nearby point Q on the curve and computing the slope m_{PQ} of the secant line PQ. From Figure 6 we see that

1 $$m_{PQ} = \frac{f(x) - f(a)}{x - a}$$

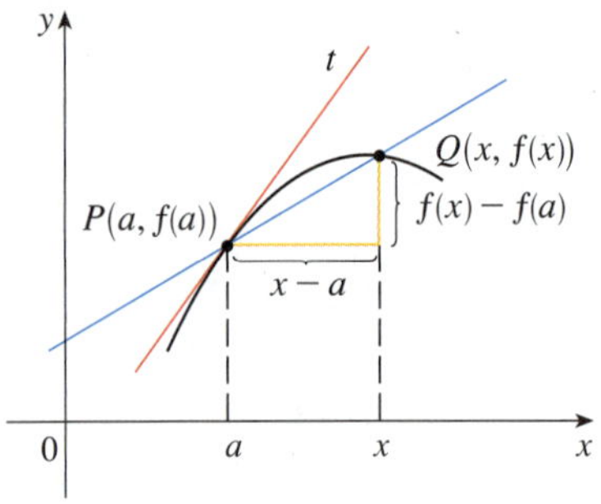

FIGURE 6
The secant line PQ

Now imagine that Q moves along the curve toward P as in Figure 7. You can see that the secant line rotates and approaches the tangent line as its limiting position. This means that the slope m_{PQ} of the secant line becomes closer and closer to the slope m of the tangent line. We write

$$m = \lim_{Q \to P} m_{PQ}$$

and we say that m is the limit of m_{PQ} as Q approaches P along the curve. Since x approaches a as Q approaches P, we could also use Equation 1 to write

2 $$m = \lim_{x \to a} \frac{f(x) - f(a)}{x - a}$$

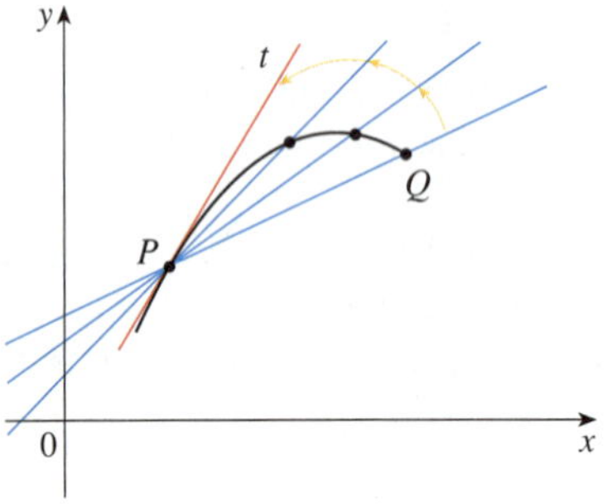

FIGURE 7
Secant lines approaching the tangent line

Specific examples of this procedure will be given in Chapter 2.

The tangent problem has given rise to the branch of calculus called *differential calculus,* which was not invented until more than 2000 years after integral calculus. The main ideas behind differential calculus are due to the French mathematician Pierre Fermat (1601–1665) and were developed by the English mathematicians John Wallis (1616–1703), Isaac Barrow (1630–1677), and Isaac Newton (1642–1727) and the German mathematician Gottfried Leibniz (1646–1716).

The two branches of calculus and their chief problems, the area problem and the tangent problem, appear to be very different, but it turns out that there is a very close connection between them. The tangent problem and the area problem are inverse problems in a sense that will be described in Chapter 5.

Velocity

When we look at the speedometer of a car and read that the car is traveling at 48 mi/h, what does that information indicate to us? We know that if the velocity remains constant, then after an hour we will have traveled 48 mi. But if the velocity of the car varies, what does it mean to say that the velocity at a given instant is 48 mi/h?

In order to analyze this question, let's examine the motion of a car that travels along a straight road and assume that we can measure the distance traveled by the car (in feet) at l-second intervals as in the following chart:

t = Time elapsed (s)	0	1	2	3	4	5
d = Distance (ft)	0	2	10	25	43	78

As a first step toward finding the velocity after 2 seconds have elapsed, we find the average velocity during the time interval $2 \leqslant t \leqslant 4$:

$$\begin{aligned} \text{average velocity} &= \frac{\text{distance traveled}}{\text{time elapsed}} \\ &= \frac{43 - 10}{4 - 2} \\ &= 16.5 \text{ ft/s} \end{aligned}$$

Similarly, the average velocity in the time interval $2 \leqslant t \leqslant 3$ is

$$\text{average velocity} = \frac{25 - 10}{3 - 2} = 15 \text{ ft/s}$$

We have the feeling that the velocity at the instant $t = 2$ can't be much different from the average velocity during a short time interval starting at $t = 2$. So let's imagine that the distance traveled has been measured at 0.l-second time intervals as in the following chart:

t	2.0	2.1	2.2	2.3	2.4	2.5
d	10.00	11.02	12.16	13.45	14.96	16.80

Then we can compute, for instance, the average velocity over the time interval [2, 2.5]:

$$\text{average velocity} = \frac{16.80 - 10.00}{2.5 - 2} = 13.6 \text{ ft/s}$$

The results of such calculations are shown in the following chart:

Time interval	[2, 3]	[2, 2.5]	[2, 2.4]	[2, 2.3]	[2, 2.2]	[2, 2.1]
Average velocity (ft/s)	15.0	13.6	12.4	11.5	10.8	10.2

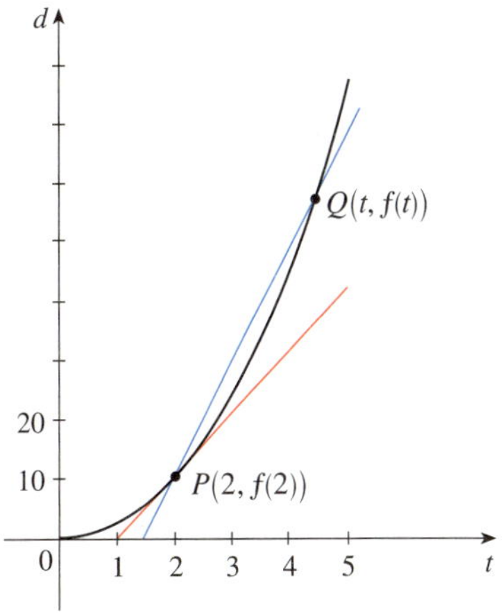

FIGURE 8

The average velocities over successively smaller intervals appear to be getting closer to a number near 10, and so we expect that the velocity at exactly $t = 2$ is about 10 ft/s. In Chapter 2 we will define the instantaneous velocity of a moving object as the limiting value of the average velocities over smaller and smaller time intervals.

In Figure 8 we show a graphical representation of the motion of the car by plotting the distance traveled as a function of time. If we write $d = f(t)$, then $f(t)$ is the number of feet traveled after t seconds. The average velocity in the time interval $[2, t]$ is

$$\text{average velocity} = \frac{\text{distance traveled}}{\text{time elapsed}} = \frac{f(t) - f(2)}{t - 2}$$

which is the same as the slope of the secant line PQ in Figure 8. The velocity v when $t = 2$ is the limiting value of this average velocity as t approaches 2; that is,

$$v = \lim_{t \to 2} \frac{f(t) - f(2)}{t - 2}$$

and we recognize from Equation 2 that this is the same as the slope of the tangent line to the curve at P.

Thus, when we solve the tangent problem in differential calculus, we are also solving problems concerning velocities. The same techniques also enable us to solve problems involving rates of change in all of the natural and social sciences.

The Limit of a Sequence

In the fifth century B.C. the Greek philosopher Zeno of Elea posed four problems, now known as *Zeno's paradoxes,* that were intended to challenge some of the ideas concerning space and time that were held in his day. Zeno's second paradox concerns a race between the Greek hero Achilles and a tortoise that has been given a head start. Zeno argued, as follows, that Achilles could never pass the tortoise: Suppose that Achilles starts at position a_1 and the tortoise starts at position t_1. (See Figure 9.) When Achilles reaches the point $a_2 = t_1$, the tortoise is farther ahead at position t_2. When Achilles reaches $a_3 = t_2$, the tortoise is at t_3. This process continues indefinitely and so it appears that the tortoise will always be ahead! But this defies common sense.

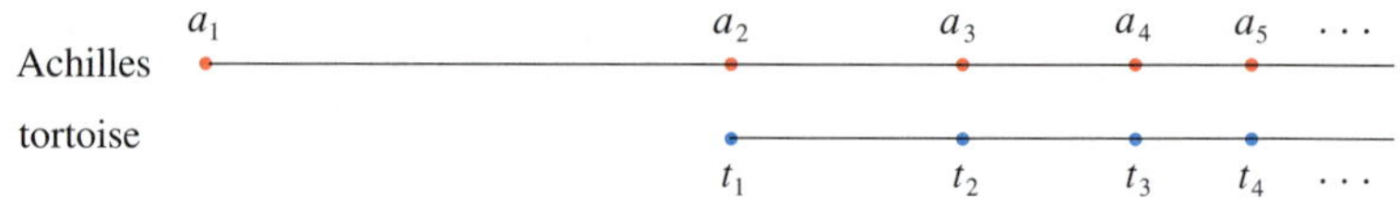

FIGURE 9

One way of explaining this paradox is with the idea of a *sequence.* The successive positions of Achilles $(a_1, a_2, a_3, \ldots)$ or the successive positions of the tortoise $(t_1, t_2, t_3, \ldots)$ form what is known as a sequence.

In general, a sequence $\{a_n\}$ is a set of numbers written in a definite order. For instance, the sequence

$$\left\{1, \tfrac{1}{2}, \tfrac{1}{3}, \tfrac{1}{4}, \tfrac{1}{5}, \ldots\right\}$$

can be described by giving the following formula for the nth term:

$$a_n = \frac{1}{n}$$

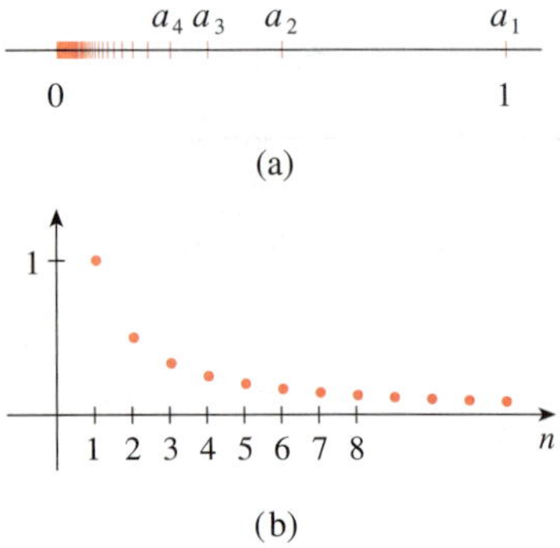

FIGURE 10

We can visualize this sequence by plotting its terms on a number line as in Figure 10(a) or by drawing its graph as in Figure 10(b). Observe from either picture that the terms of the sequence $a_n = 1/n$ are becoming closer and closer to 0 as n increases. In fact, we can find terms as small as we please by making n large enough. We say that the limit of the sequence is 0, and we indicate this by writing

$$\lim_{n \to \infty} \frac{1}{n} = 0$$

In general, the notation

$$\lim_{n \to \infty} a_n = L$$

is used if the terms a_n approach the number L as n becomes large. This means that the numbers a_n can be made as close as we like to the number L by taking n sufficiently large.

The concept of the limit of a sequence occurs whenever we use the decimal representation of a real number. For instance, if

$$a_1 = 3.1$$
$$a_2 = 3.14$$
$$a_3 = 3.141$$
$$a_4 = 3.1415$$
$$a_5 = 3.14159$$
$$a_6 = 3.141592$$
$$a_7 = 3.1415926$$
$$\vdots$$

then $$\lim_{n\to\infty} a_n = \pi$$

The terms in this sequence are rational approximations to π.

Let's return to Zeno's paradox. The successive positions of Achilles and the tortoise form sequences $\{a_n\}$ and $\{t_n\}$, where $a_n < t_n$ for all n. It can be shown that both sequences have the same limit:

$$\lim_{n\to\infty} a_n = p = \lim_{n\to\infty} t_n$$

It is precisely at this point p that Achilles overtakes the tortoise.

The Sum of a Series

Watch a movie of Zeno's attempt to reach the wall.
Resources / Module 1 / Introduction / Zeno's Paradox

Another of Zeno's paradoxes, as passed on to us by Aristotle, is the following: "A man standing in a room cannot walk to the wall. In order to do so, he would first have to go half the distance, then half the remaining distance, and then again half of what still remains. This process can always be continued and can never be ended." (See Figure 11.)

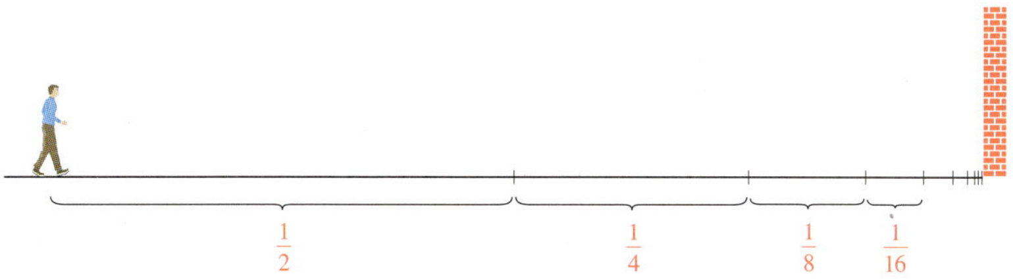

FIGURE 11

Of course, we know that the man can actually reach the wall, so this suggests that perhaps the total distance can be expressed as the sum of infinitely many smaller distances as follows:

3 $$1 = \frac{1}{2} + \frac{1}{4} + \frac{1}{8} + \frac{1}{16} + \cdots + \frac{1}{2^n} + \cdots$$

Zeno was arguing that it doesn't make sense to add infinitely many numbers together. But there are other situations in which we implicitly use infinite sums. For instance, in decimal notation, the symbol $0.\overline{3} = 0.3333\ldots$ means

$$\frac{3}{10} + \frac{3}{100} + \frac{3}{1000} + \frac{3}{10{,}000} + \cdots$$

and so, in some sense, it must be true that

$$\frac{3}{10} + \frac{3}{100} + \frac{3}{1000} + \frac{3}{10{,}000} + \cdots = \frac{1}{3}$$

More generally, if d_n denotes the nth digit in the decimal representation of a number, then

$$0.d_1d_2d_3d_4\ldots = \frac{d_1}{10} + \frac{d_2}{10^2} + \frac{d_3}{10^3} + \cdots + \frac{d_n}{10^n} + \cdots$$

Therefore, some infinite sums, or infinite series as they are called, have a meaning. But we must define carefully what the sum of an infinite series is.

Returning to the series in Equation 3, we denote by s_n the sum of the first n terms of the series. Thus

$$s_1 = \tfrac{1}{2} = 0.5$$

$$s_2 = \tfrac{1}{2} + \tfrac{1}{4} = 0.75$$

$$s_3 = \tfrac{1}{2} + \tfrac{1}{4} + \tfrac{1}{8} = 0.875$$

$$s_4 = \tfrac{1}{2} + \tfrac{1}{4} + \tfrac{1}{8} + \tfrac{1}{16} = 0.9375$$

$$s_5 = \tfrac{1}{2} + \tfrac{1}{4} + \tfrac{1}{8} + \tfrac{1}{16} + \tfrac{1}{32} = 0.96875$$

$$s_6 = \tfrac{1}{2} + \tfrac{1}{4} + \tfrac{1}{8} + \tfrac{1}{16} + \tfrac{1}{32} + \tfrac{1}{64} = 0.984375$$

$$s_7 = \tfrac{1}{2} + \tfrac{1}{4} + \tfrac{1}{8} + \tfrac{1}{16} + \tfrac{1}{32} + \tfrac{1}{64} + \tfrac{1}{128} = 0.9921875$$

$$\vdots$$

$$s_{10} = \tfrac{1}{2} + \tfrac{1}{4} + \cdots + \tfrac{1}{1024} \approx 0.99902344$$

$$\vdots$$

$$s_{16} = \frac{1}{2} + \frac{1}{4} + \cdots + \frac{1}{2^{16}} \approx 0.99998474$$

Observe that as we add more and more terms, the partial sums become closer and closer to 1. In fact, it can be shown that by taking n large enough (that is, by adding sufficiently many terms of the series), we can make the partial sum s_n as close as we please to the number 1. It therefore seems reasonable to say that the sum of the infinite series is 1 and to write

$$\frac{1}{2} + \frac{1}{4} + \frac{1}{8} + \cdots + \frac{1}{2^n} + \cdots = 1$$

In other words, the reason the sum of the series is 1 is that

$$\lim_{n \to \infty} s_n = 1$$

In Chapter 11 we will discuss these ideas further. We will then use Newton's idea of combining infinite series with differential and integral calculus.

Summary

We have seen that the concept of a limit arises in trying to find the area of a region, the slope of a tangent to a curve, the velocity of a car, or the sum of an infinite series. In each case the common theme is the calculation of a quantity as the limit of other, easily calculated quantities. It is this basic idea of a limit that sets calculus apart from other areas of mathematics. In fact, we could define calculus as the part of mathematics that deals with limits.

Sir Isaac Newton invented his version of calculus in order to explain the motion of the planets around the Sun. Today calculus is used in calculating the orbits of satellites and spacecraft, in predicting population sizes, in estimating how fast coffee prices rise, in forecasting weather, in measuring the cardiac output of the heart, in calculating life insurance premiums, and in a great variety of other areas. We will explore some of these uses of calculus in this book.

In order to convey a sense of the power of the subject, we end this preview with a list of some of the questions that you will be able to answer using calculus:

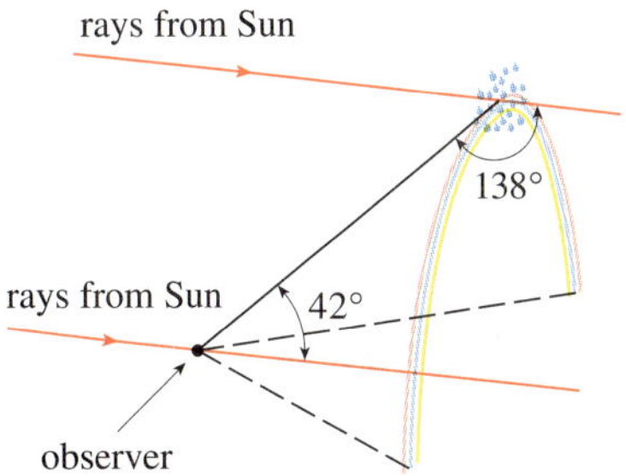

FIGURE 12

1. How can we explain the fact, illustrated in Figure 12, that the angle of elevation from an observer up to the highest point in a rainbow is 42°? (See page 288.)
2. How can we explain the shapes of cans on supermarket shelves? (See page 341.)
3. Where is the best place to sit in a movie theater? (See page 468.)
4. How far away from an airport should a pilot start descent? (See page 243.)
5. How can we fit curves together to design shapes to represent letters on a laser printer? (See page 669.)
6. Where should an infielder position himself to catch a baseball thrown by an outfielder and relay it to home plate? (See page 622.)
7. Does a ball thrown upward take longer to reach its maximum height or to fall back to its original height? (See page 610.)
8. How can we explain the fact that planets and satellites move in elliptical orbits? (See page 880.)
9. How can we distribute water flow among turbines at a hydroelectric station so as to maximize the total energy production? (See page 973.)
10. If a marble, a squash ball, a steel bar, and a lead pipe roll down a slope, which of them reaches the bottom first? (See page 1039.)

CHAPTER 1

A graphical representation of a function—here the number of hours of daylight as a function of the time of year at various latitudes— is often the most natural and convenient way to represent the function.

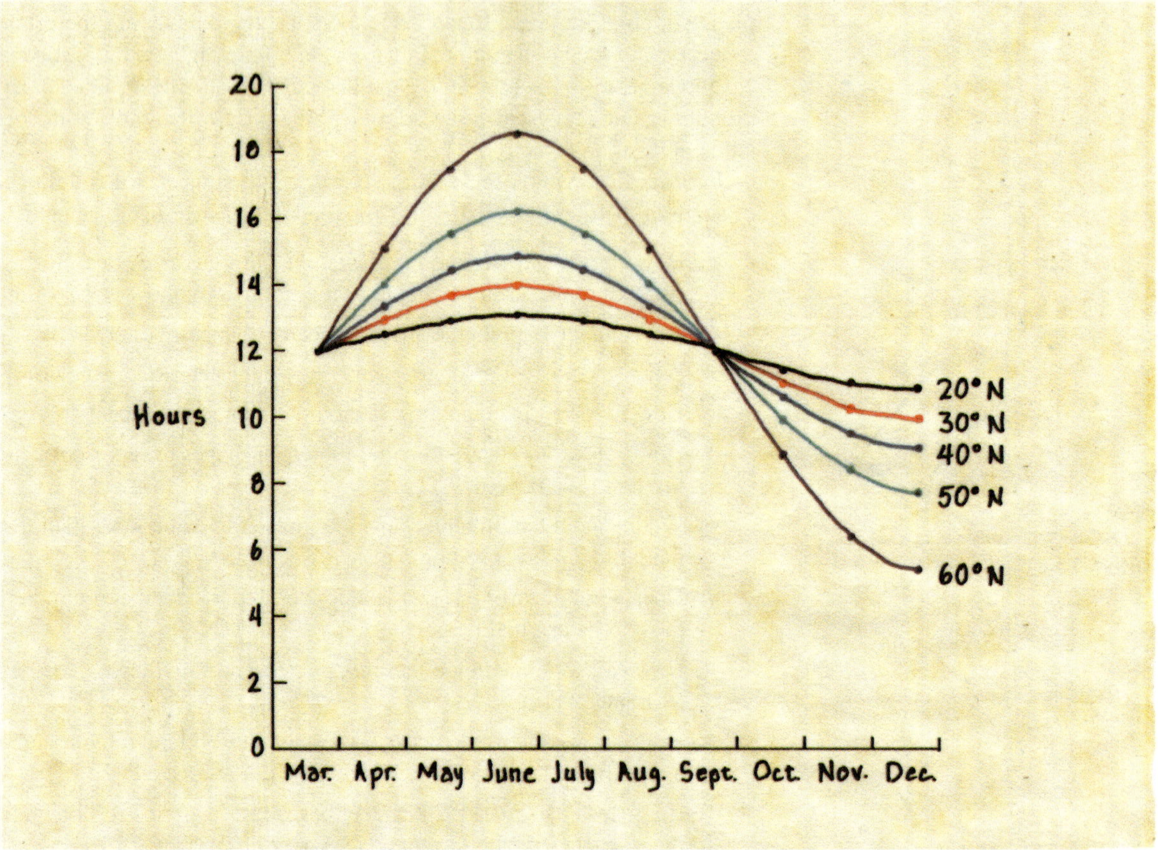

Functions and Models

The fundamental objects that we deal with in calculus are functions. This chapter prepares the way for calculus by discussing the basic ideas concerning functions, their graphs, and ways of transforming and combining them. We stress that a function can be represented in different ways: by an equation, in a table, by a graph, or in words. We look at the main types of functions that occur in calculus and describe the process of using these functions as mathematical models of real-world phenomena. We also discuss the use of graphing calculators and graphing software for computers.

1.1 Four Ways to Represent a Function

Functions arise whenever one quantity depends on another. Consider the following four situations.

A. The area A of a circle depends on the radius r of the circle. The rule that connects r and A is given by the equation $A = \pi r^2$. With each positive number r there is associated one value of A, and we say that A is a *function* of r.

B. The human population of the world P depends on the time t. The table gives estimates of the world population $P(t)$ at time t, for certain years. For instance,

$$P(1950) \approx 2{,}560{,}000{,}000$$

But for each value of the time t there is a corresponding value of P, and we say that P is a function of t.

Year	Population (millions)
1900	1650
1910	1750
1920	1860
1930	2070
1940	2300
1950	2560
1960	3040
1970	3710
1980	4450
1990	5280
2000	6080

C. The cost C of mailing a first-class letter depends on the weight w of the letter. Although there is no simple formula that connects w and C, the post office has a rule for determining C when w is known.

D. The vertical acceleration a of the ground as measured by a seismograph during an earthquake is a function of the elapsed time t. Figure 1 shows a graph generated by seismic activity during the Northridge earthquake that shook Los Angeles in 1994. For a given value of t, the graph provides a corresponding value of a.

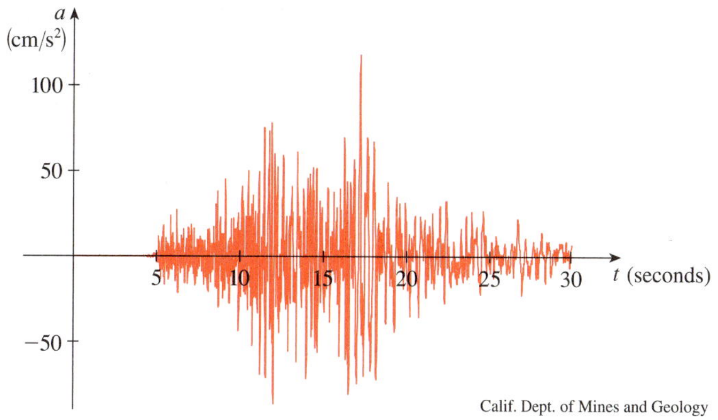

FIGURE 1
Vertical ground acceleration during the Northridge earthquake

Each of these examples describes a rule whereby, given a number (r, t, w, or t), another number (A, P, C, or a) is assigned. In each case we say that the second number is a function of the first number.

> A **function** f is a rule that assigns to each element x in a set A exactly one element, called $f(x)$, in a set B.

We usually consider functions for which the sets A and B are sets of real numbers. The set A is called the **domain** of the function. The number $f(x)$ is the **value of f at x** and is read "f of x." The **range** of f is the set of all possible values of $f(x)$ as x varies throughout the domain. A symbol that represents an arbitrary number in the *domain* of a function f is called an **independent variable**. A symbol that represents a number in the *range* of f is called a **dependent variable**. In Example A, for instance, r is the independent variable and A is the dependent variable.

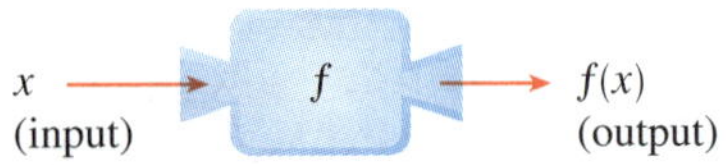

FIGURE 2
Machine diagram for a function f

It's helpful to think of a function as a **machine** (see Figure 2). If x is in the domain of the function f, then when x enters the machine, it's accepted as an input and the machine produces an output $f(x)$ according to the rule of the function. Thus, we can think of the domain as the set of all possible inputs and the range as the set of all possible outputs.

The preprogrammed functions in a calculator are good examples of a function as a machine. For example, the square root key on your calculator computes such a function. You press the key labeled $\sqrt{\ }$ (or $\sqrt{x}$) and enter the input x. If $x < 0$, then x is not in the domain of this function; that is, x is not an acceptable input, and the calculator will indicate an error. If $x \geqslant 0$, then an *approximation* to $\sqrt{x}$ will appear in the display. Thus, the $\sqrt{x}$ key on your calculator is not quite the same as the exact mathematical function f defined by $f(x) = \sqrt{x}$.

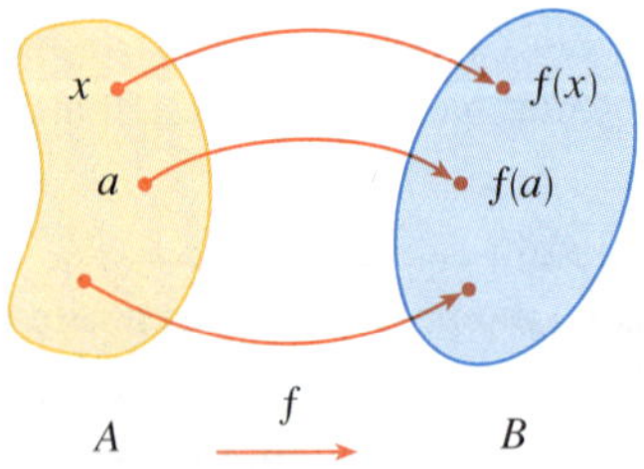

FIGURE 3
Arrow diagram for f

Another way to picture a function is by an **arrow diagram** as in Figure 3. Each arrow connects an element of A to an element of B. The arrow indicates that $f(x)$ is associated with x, $f(a)$ is associated with a, and so on.

The most common method for visualizing a function is its graph. If f is a function with domain A, then its **graph** is the set of ordered pairs

$$\{(x, f(x)) \mid x \in A\}$$

(Notice that these are input-output pairs.) In other words, the graph of f consists of all points (x, y) in the coordinate plane such that $y = f(x)$ and x is in the domain of f.

The graph of a function f gives us a useful picture of the behavior or "life history" of a function. Since the y-coordinate of any point (x, y) on the graph is $y = f(x)$, we can read the value of $f(x)$ from the graph as being the height of the graph above the point x (see Figure 4). The graph of f also allows us to picture the domain of f on the x-axis and its range on the y-axis as in Figure 5.

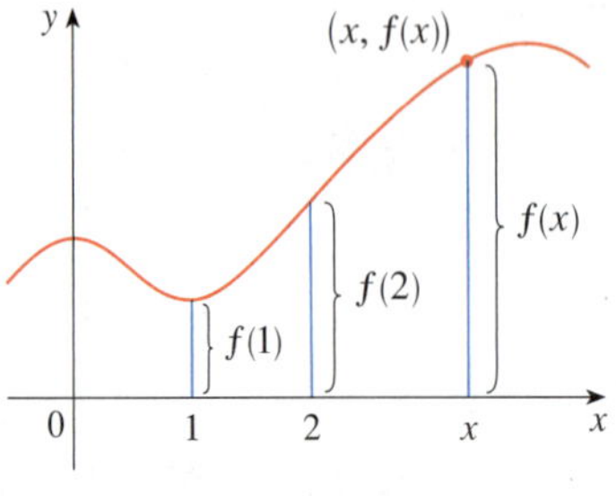

FIGURE 4

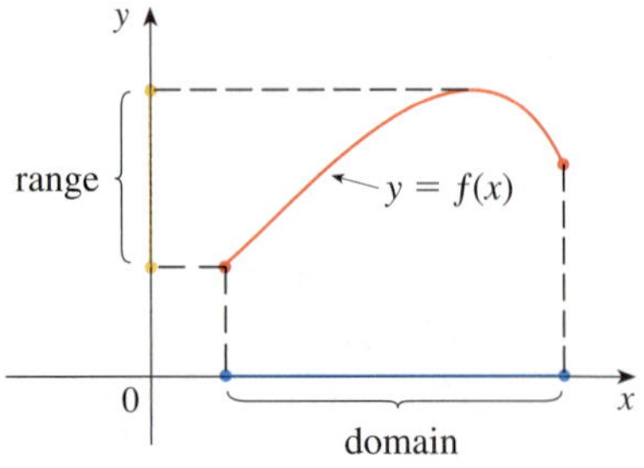

FIGURE 5

EXAMPLE 1 The graph of a function f is shown in Figure 6.
(a) Find the values of $f(1)$ and $f(5)$.
(b) What are the domain and range of f?

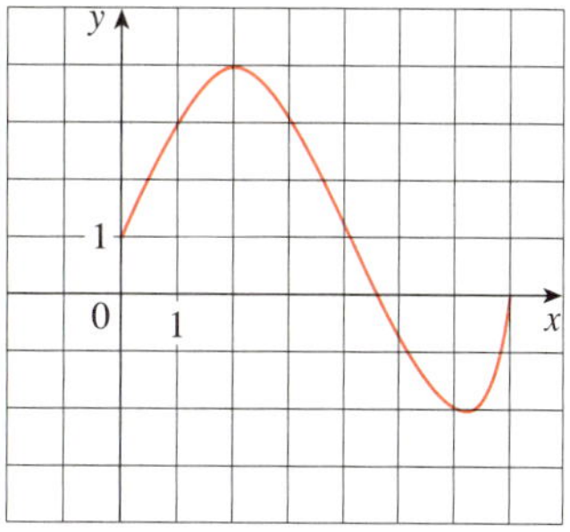

FIGURE 6

SOLUTION
(a) We see from Figure 6 that the point $(1, 3)$ lies on the graph of f, so the value of f at 1 is $f(1) = 3$. (In other words, the point on the graph that lies above $x = 1$ is 3 units above the x-axis.)

When $x = 5$, the graph lies about 0.7 unit below the x-axis, so we estimate that $f(5) \approx -0.7$.

|||| The notation for intervals is given in Appendix A.

(b) We see that $f(x)$ is defined when $0 \leqslant x \leqslant 7$, so the domain of f is the closed interval $[0, 7]$. Notice that f takes on all values from -2 to 4, so the range of f is

$$\{y \mid -2 \leqslant y \leqslant 4\} = [-2, 4]$$

EXAMPLE 2 Sketch the graph and find the domain and range of each function.
(a) $f(x) = 2x - 1$ (b) $g(x) = x^2$

SOLUTION
(a) The equation of the graph is $y = 2x - 1$, and we recognize this as being the equation of a line with slope 2 and y-intercept -1. (Recall the slope-intercept form of the equation of a line: $y = mx + b$. See Appendix B.) This enables us to sketch the graph of f in Figure 7. The expression $2x - 1$ is defined for all real numbers, so the domain of f is the set of all real numbers, which we denote by $\mathbb{R}$. The graph shows that the range is also $\mathbb{R}$.

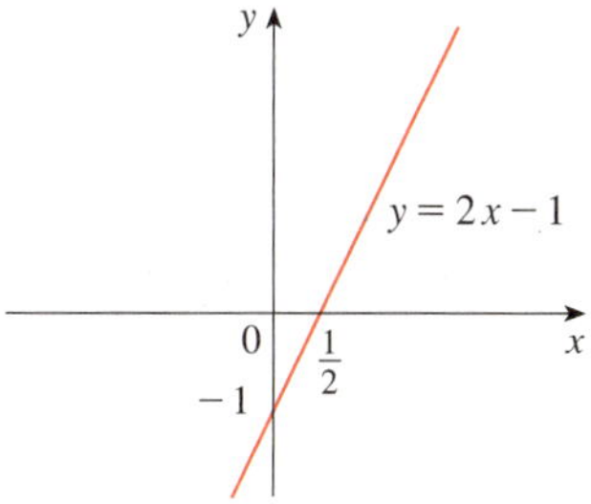

FIGURE 7

(b) Since $g(2) = 2^2 = 4$ and $g(-1) = (-1)^2 = 1$, we could plot the points $(2, 4)$ and $(-1, 1)$, together with a few other points on the graph, and join them to produce the graph (Figure 8). The equation of the graph is $y = x^2$, which represents a parabola (see Appendix C). The domain of g is $\mathbb{R}$. The range of g consists of all values of $g(x)$, that is, all numbers of the form x^2. But $x^2 \geqslant 0$ for all numbers x and any positive number y is a square. So the range of g is $\{y \mid y \geqslant 0\} = [0, \infty)$. This can also be seen from Figure 8.

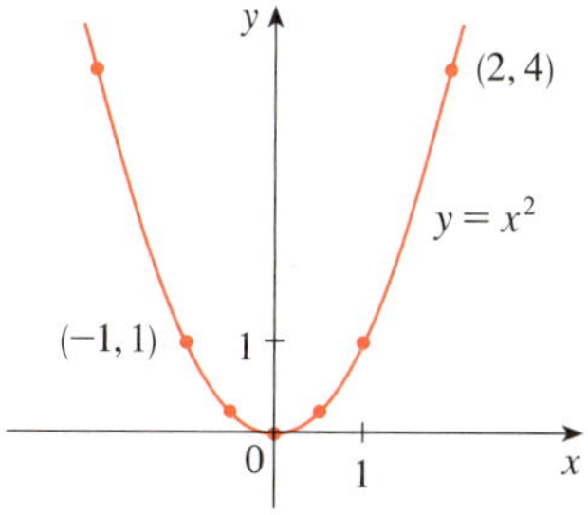

FIGURE 8

Representations of Functions

There are four possible ways to represent a function:

- verbally (by a description in words)
- numerically (by a table of values)
- visually (by a graph)
- algebraically (by an explicit formula)

If a single function can be represented in all four ways, it is often useful to go from one representation to another to gain additional insight into the function. (In Example 2, for instance, we started with algebraic formulas and then obtained the graphs.) But certain functions are described more naturally by one method than by another. With this in mind, let's reexamine the four situations that we considered at the beginning of this section.

A. The most useful representation of the area of a circle as a function of its radius is probably the algebraic formula $A(r) = \pi r^2$, though it is possible to compile a table of values or to sketch a graph (half a parabola). Because a circle has to have a positive radius, the domain is $\{r \mid r > 0\} = (0, \infty)$, and the range is also $(0, \infty)$.

B. We are given a description of the function in words: $P(t)$ is the human population of the world at time t. The table of values of world population on page 11 provides a convenient representation of this function. If we plot these values, we get the graph (called a *scatter plot*) in Figure 9. It too is a useful representation; the graph allows us to absorb all the data at once. What about a formula? Of course, it's impossible to devise an explicit formula that gives the exact human population $P(t)$ at any time t. But it is possible to find an expression for a function that *approximates* $P(t)$. In fact, using methods explained in Section 1.5, we obtain the approximation

$$P(t) \approx f(t) = (0.008079266) \cdot (1.013731)^t$$

and Figure 10 shows that it is a reasonably good "fit." The function f is called a *mathematical model* for population growth. In other words, it is a function with an explicit formula that approximates the behavior of our given function. We will see, however, that the ideas of calculus can be applied to a table of values; an explicit formula is not necessary.

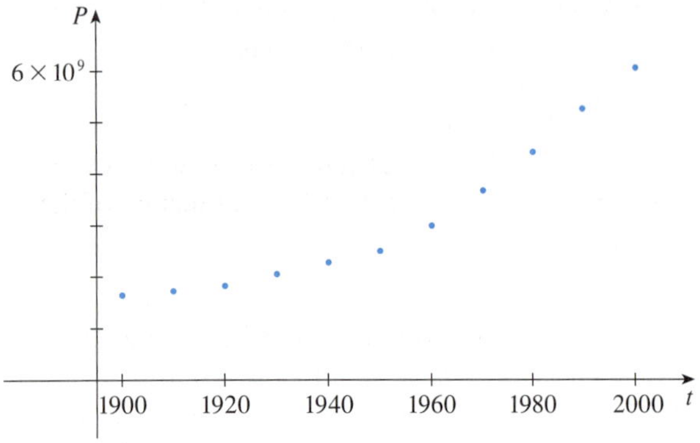

FIGURE 9

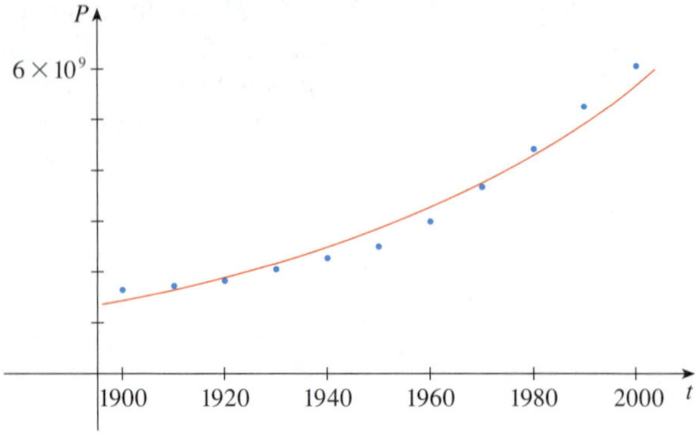

FIGURE 10

The function P is typical of the functions that arise whenever we attempt to apply calculus to the real world. We start with a verbal description of a function. Then we may be able to construct a table of values of the function, perhaps from instrument readings in a scientific experiment. Even though we don't have complete knowledge of the values of the function, we will see throughout the book that it is still possible to perform the operations of calculus on such a function.

|||| A function defined by a table of values is called a *tabular* function.

w (ounces)	$C(w)$ (dollars)
$0 < w \leqslant 1$	0.37
$1 < w \leqslant 2$	0.60
$2 < w \leqslant 3$	0.83
$3 < w \leqslant 4$	1.06
$4 < w \leqslant 5$	1.29
⋮	⋮

C. Again the function is described in words: $C(w)$ is the cost of mailing a first-class letter with weight w. The rule that the U.S. Postal Service used as of 2002 is as follows: The cost is 37 cents for up to one ounce, plus 23 cents for each successive ounce up to 11 ounces. The table of values shown in the margin is the most convenient representation for this function, though it is possible to sketch a graph (see Example 10).

D. The graph shown in Figure 1 is the most natural representation of the vertical acceleration function $a(t)$. It's true that a table of values could be compiled, and it is even possible to devise an approximate formula. But everything a geologist needs to know—amplitudes and patterns—can be seen easily from the graph. (The same is true for the patterns seen in electrocardiograms of heart patients and polygraphs for lie-detection.) Figures 11 and 12 show the graphs of the north-south and east-west accelerations for the Northridge earthquake; when used in conjunction with Figure 1, they provide a great deal of information about the earthquake.

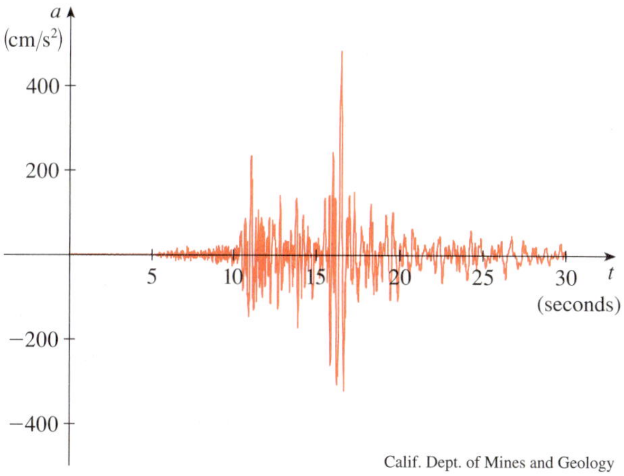

FIGURE 11 North-south acceleration for the Northridge earthquake

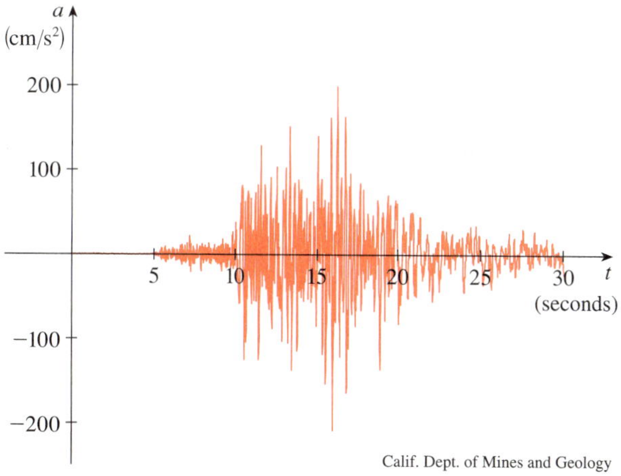

FIGURE 12 East-west acceleration for the Northridge earthquake

In the next example we sketch the graph of a function that is defined verbally.

EXAMPLE 3 When you turn on a hot-water faucet, the temperature T of the water depends on how long the water has been running. Draw a rough graph of T as a function of the time t that has elapsed since the faucet was turned on.

SOLUTION The initial temperature of the running water is close to room temperature because of the water that has been sitting in the pipes. When the water from the hot-water tank starts coming out, T increases quickly. In the next phase, T is constant at the temperature of the heated water in the tank. When the tank is drained, T decreases to the temperature of the water supply. This enables us to make the rough sketch of T as a function of t in Figure 13.

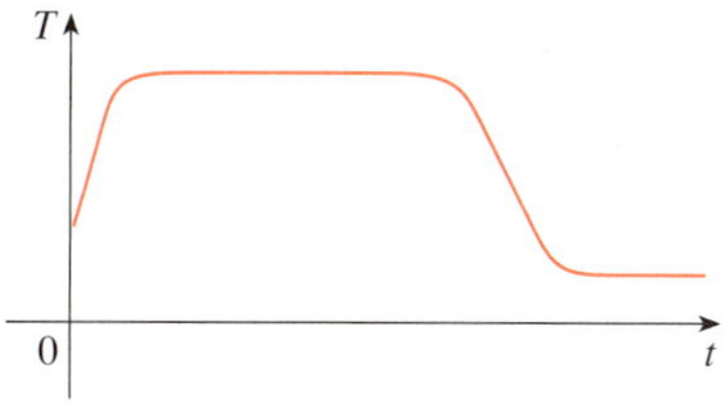

FIGURE 13

A more accurate graph of the function in Example 3 could be obtained by using a thermometer to measure the temperature of the water at 10-second intervals. In general, scientists collect experimental data and use them to sketch the graphs of functions, as the next example illustrates.

t	$C(t)$
0	0.0800
2	0.0570
4	0.0408
6	0.0295
8	0.0210

EXAMPLE 4 The data shown in the margin come from an experiment on the lactonization of hydroxyvaleric acid at 25°C. They give the concentration $C(t)$ of this acid (in moles per liter) after t minutes. Use these data to draw an approximation to the graph of the concentration function. Then use this graph to estimate the concentration after 5 minutes.

SOLUTION We plot the five points corresponding to the data from the table in Figure 14. The curve-fitting methods of Section 1.2 could be used to choose a model and graph it. But the data points in Figure 14 look quite well behaved, so we simply draw a smooth curve through them by hand as in Figure 15.

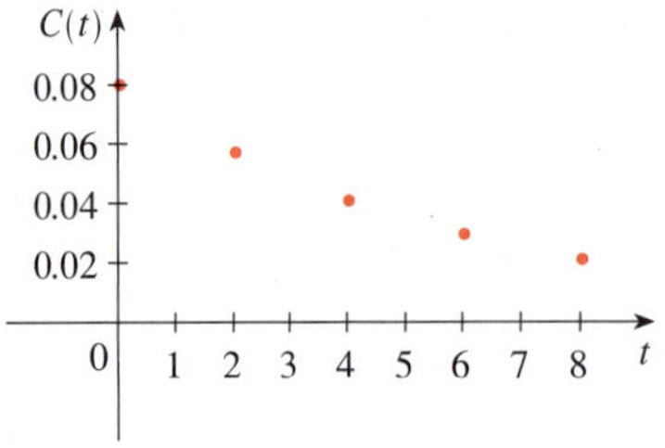

FIGURE 14

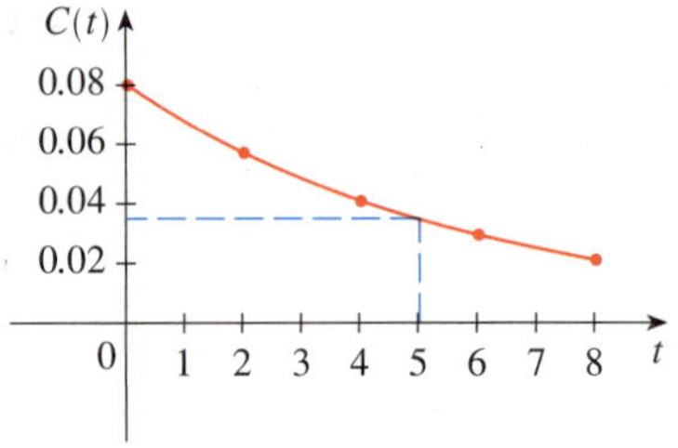

FIGURE 15

Then we use the graph to estimate that the concentration after 5 minutes is

$$C(5) \approx 0.035 \text{ mole/liter}$$

In the following example we start with a verbal description of a function in a physical situation and obtain an explicit algebraic formula. The ability to do this is a useful skill in solving calculus problems that ask for the maximum or minimum values of quantities.

EXAMPLE 5 A rectangular storage container with an open top has a volume of 10 m^3. The length of its base is twice its width. Material for the base costs \$10 per square meter; material for the sides costs \$6 per square meter. Express the cost of materials as a function of the width of the base.

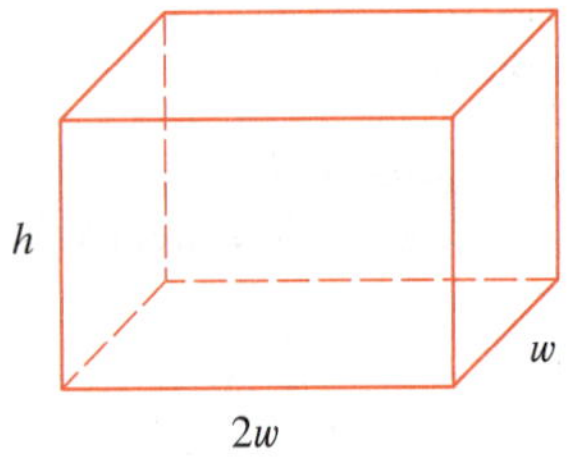

FIGURE 16

SOLUTION We draw a diagram as in Figure 16 and introduce notation by letting w and $2w$ be the width and length of the base, respectively, and h be the height.

The area of the base is $(2w)w = 2w^2$, so the cost, in dollars, of the material for the base is $10(2w^2)$. Two of the sides have area wh and the other two have area $2wh$, so the cost of the material for the sides is $6[2(wh) + 2(2wh)]$. The total cost is therefore

$$C = 10(2w^2) + 6[2(wh) + 2(2wh)] = 20w^2 + 36wh$$

To express C as a function of w alone, we need to eliminate h and we do so by using the fact that the volume is 10 m^3. Thus

$$w(2w)h = 10$$

which gives

$$h = \frac{10}{2w^2} = \frac{5}{w^2}$$

|||| In setting up applied functions as in Example 5, it may be useful to review the principles of problem solving as discussed on page 80, particularly *Step 1: Understand the Problem.*

Substituting this into the expression for C, we have

$$C = 20w^2 + 36w\left(\frac{5}{w^2}\right) = 20w^2 + \frac{180}{w}$$

Therefore, the equation

$$C(w) = 20w^2 + \frac{180}{w} \qquad w > 0$$

expresses C as a function of w.

EXAMPLE 6 Find the domain of each function.

(a) $f(x) = \sqrt{x + 2}$ (b) $g(x) = \dfrac{1}{x^2 - x}$

SOLUTION

|||| If a function is given by a formula and the domain is not stated explicitly, the convention is that the domain is the set of all numbers for which the formula makes sense and defines a real number.

(a) Because the square root of a negative number is not defined (as a real number), the domain of f consists of all values of x such that $x + 2 \geqslant 0$. This is equivalent to $x \geqslant -2$, so the domain is the interval $[-2, \infty)$.

(b) Since

$$g(x) = \frac{1}{x^2 - x} = \frac{1}{x(x - 1)}$$

and division by 0 is not allowed, we see that $g(x)$ is not defined when $x = 0$ or $x = 1$. Thus, the domain of g is

$$\{x \mid x \neq 0, x \neq 1\}$$

which could also be written in interval notation as

$$(-\infty, 0) \cup (0, 1) \cup (1, \infty)$$

The graph of a function is a curve in the xy-plane. But the question arises: Which curves in the xy-plane are graphs of functions? This is answered by the following test.

The Vertical Line Test A curve in the xy-plane is the graph of a function of x if and only if no vertical line intersects the curve more than once.

The reason for the truth of the Vertical Line Test can be seen in Figure 17. If each vertical line $x = a$ intersects a curve only once, at (a, b), then exactly one functional value is defined by $f(a) = b$. But if a line $x = a$ intersects the curve twice, at (a, b) and (a, c), then the curve can't represent a function because a function can't assign two different values to a.

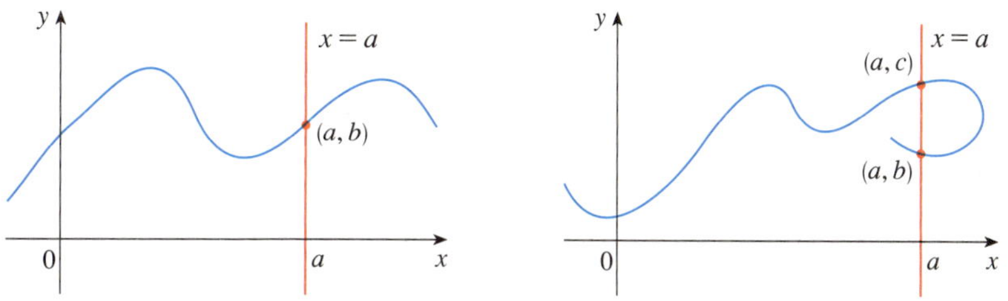

FIGURE 17

For example, the parabola $x = y^2 - 2$ shown in Figure 18(a) is not the graph of a function of x because, as you can see, there are vertical lines that intersect the parabola twice. The parabola, however, does contain the graphs of *two* functions of x. Notice that the equation $x = y^2 - 2$ implies $y^2 = x + 2$, so $y = \pm\sqrt{x+2}$. Thus, the upper and lower halves of the parabola are the graphs of the functions $f(x) = \sqrt{x+2}$ [from Example 6(a)] and $g(x) = -\sqrt{x+2}$. [See Figures 18(b) and (c).] We observe that if we reverse the roles of x and y, then the equation $x = h(y) = y^2 - 2$ does define x as a function of y (with y as the independent variable and x as the dependent variable) and the parabola now appears as the graph of the function h.

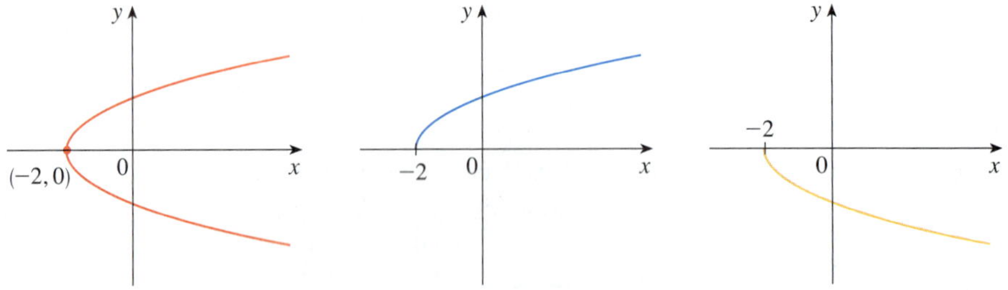

FIGURE 18 (a) $x = y^2 - 2$ (b) $y = \sqrt{x+2}$ (c) $y = -\sqrt{x+2}$

Piecewise Defined Functions

The functions in the following four examples are defined by different formulas in different parts of their domains.

EXAMPLE 7 A function f is defined by

$$f(x) = \begin{cases} 1 - x & \text{if } x \leqslant 1 \\ x^2 & \text{if } x > 1 \end{cases}$$

Evaluate $f(0)$, $f(1)$, and $f(2)$ and sketch the graph.

SOLUTION Remember that a function is a rule. For this particular function the rule is the following: First look at the value of the input x. If it happens that $x \leqslant 1$, then the value of $f(x)$ is $1 - x$. On the other hand, if $x > 1$, then the value of $f(x)$ is x^2.

Since $0 \leqslant 1$, we have $f(0) = 1 - 0 = 1$.

Since $1 \leqslant 1$, we have $f(1) = 1 - 1 = 0$.

Since $2 > 1$, we have $f(2) = 2^2 = 4$.

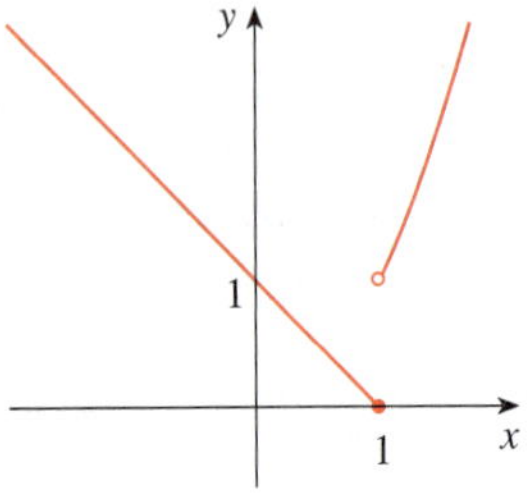

FIGURE 19

How do we draw the graph of f? We observe that if $x \leqslant 1$, then $f(x) = 1 - x$, so the part of the graph of f that lies to the left of the vertical line $x = 1$ must coincide with the line $y = 1 - x$, which has slope -1 and y-intercept 1. If $x > 1$, then $f(x) = x^2$, so the part of the graph of f that lies to the right of the line $x = 1$ must coincide with the graph of $y = x^2$, which is a parabola. This enables us to sketch the graph in Figure 19. The solid dot indicates that the point $(1, 0)$ is included on the graph; the open dot indicates that the point $(1, 1)$ is excluded from the graph.

(c) $$h(-x) = 2(-x) - (-x)^2 = -2x - x^2$$

Since $h(-x) \neq h(x)$ and $h(-x) \neq -h(x)$, we conclude that h is neither even nor odd.

The graphs of the functions in Example 11 are shown in Figure 25. Notice that the graph of h is symmetric neither about the y-axis nor about the origin.

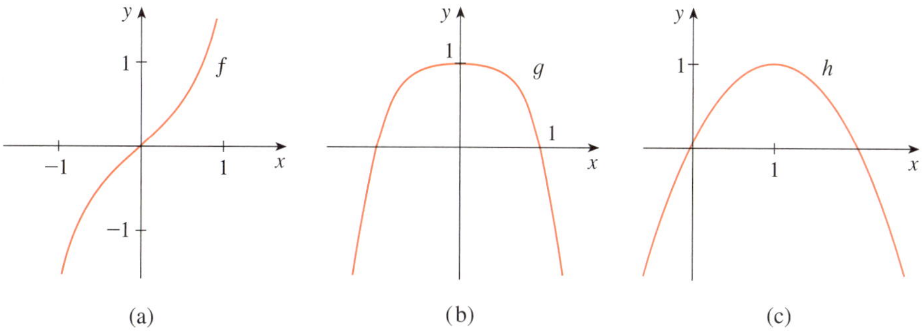

FIGURE 25

Increasing and Decreasing Functions

The graph shown in Figure 26 rises from A to B, falls from B to C, and rises again from C to D. The function f is said to be increasing on the interval $[a, b]$, decreasing on $[b, c]$, and increasing again on $[c, d]$. Notice that if x_1 and x_2 are any two numbers between a and b with $x_1 < x_2$, then $f(x_1) < f(x_2)$. We use this as the defining property of an increasing function.

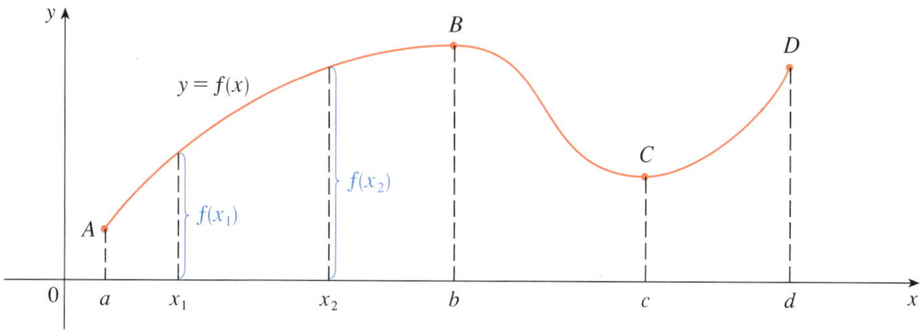

FIGURE 26

A function f is called **increasing** on an interval I if

$$f(x_1) < f(x_2) \qquad \text{whenever } x_1 < x_2 \text{ in } I$$

It is called **decreasing** on I if

$$f(x_1) > f(x_2) \qquad \text{whenever } x_1 < x_2 \text{ in } I$$

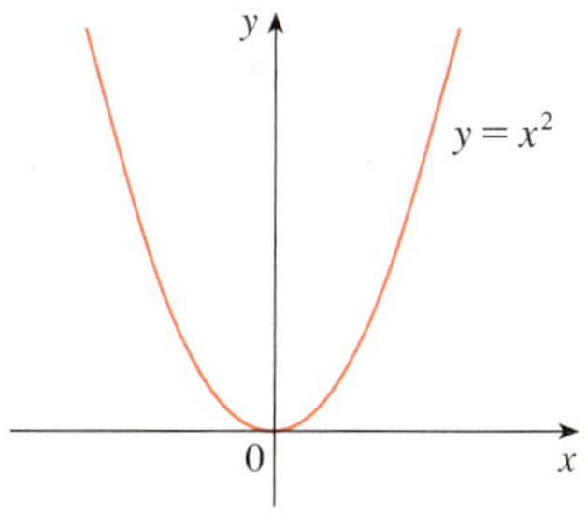

FIGURE 27

In the definition of an increasing function it is important to realize that the inequality $f(x_1) < f(x_2)$ must be satisfied for *every* pair of numbers x_1 and x_2 in I with $x_1 < x_2$.

You can see from Figure 27 that the function $f(x) = x^2$ is decreasing on the interval $(-\infty, 0]$ and increasing on the interval $[0, \infty)$.

1.1 Exercises

1. The graph of a function f is given.
(a) State the value of $f(-1)$.
(b) Estimate the value of $f(2)$.
(c) For what values of x is $f(x) = 2$?
(d) Estimate the values of x such that $f(x) = 0$.
(e) State the domain and range of f.
(f) On what interval is f increasing?

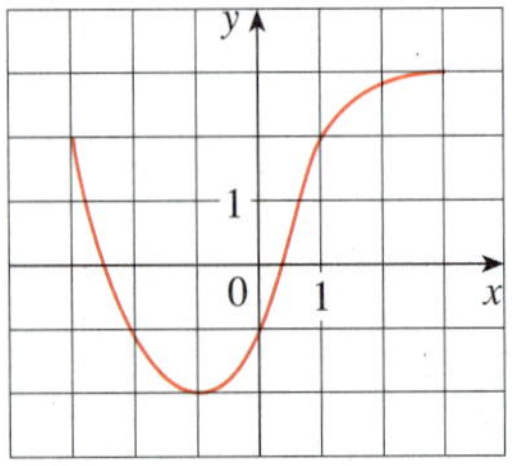

2. The graphs of f and g are given.
(a) State the values of $f(-4)$ and $g(3)$.
(b) For what values of x is $f(x) = g(x)$?
(c) Estimate the solution of the equation $f(x) = -1$.
(d) On what interval is f decreasing?
(e) State the domain and range of f.
(f) State the domain and range of g.

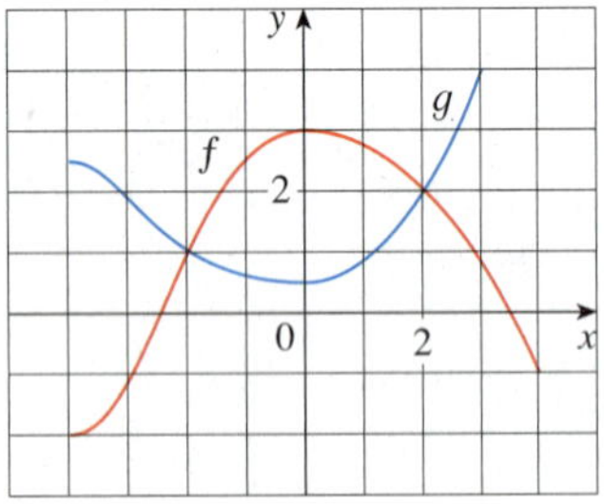

3. Figures 1, 11, and 12 were recorded by an instrument operated by the California Department of Mines and Geology at the University Hospital of the University of Southern California in Los Angeles. Use them to estimate the ranges of the vertical, north-south, and east-west ground acceleration functions at USC during the Northridge earthquake.

4. In this section we discussed examples of ordinary, everyday functions: Population is a function of time, postage cost is a function of weight, water temperature is a function of time. Give three other examples of functions from everyday life that are described verbally. What can you say about the domain and range of each of your functions? If possible, sketch a rough graph of each function.

5–8 Determine whether the curve is the graph of a function of x. If it is, state the domain and range of the function.

5.

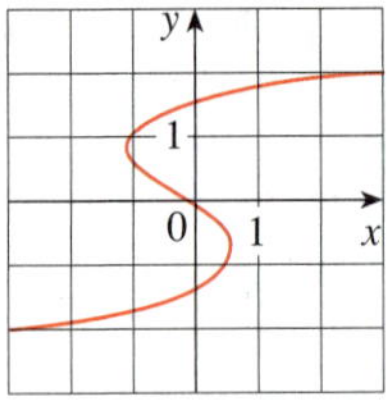

6.

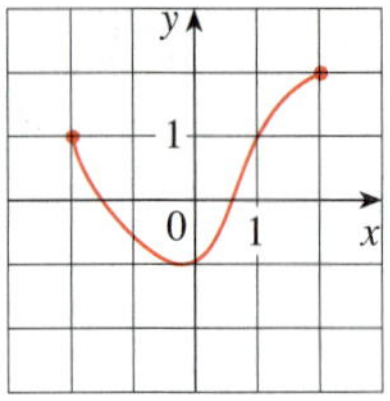

7.

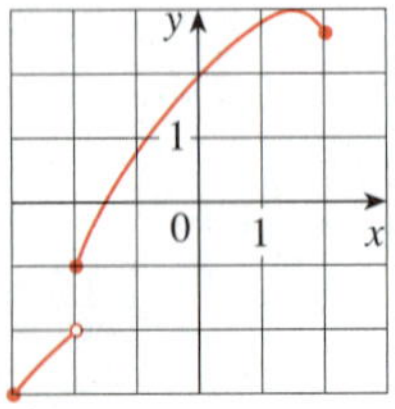

8.

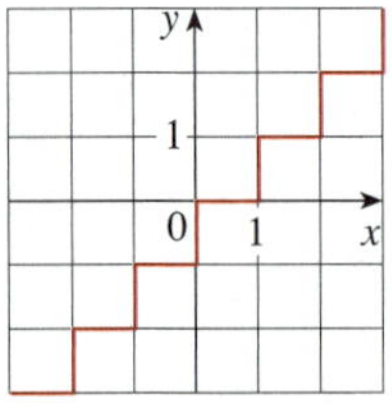

9. The graph shown gives the weight of a certain person as a function of age. Describe in words how this person's weight varies over time. What do you think happened when this person was 30 years old?

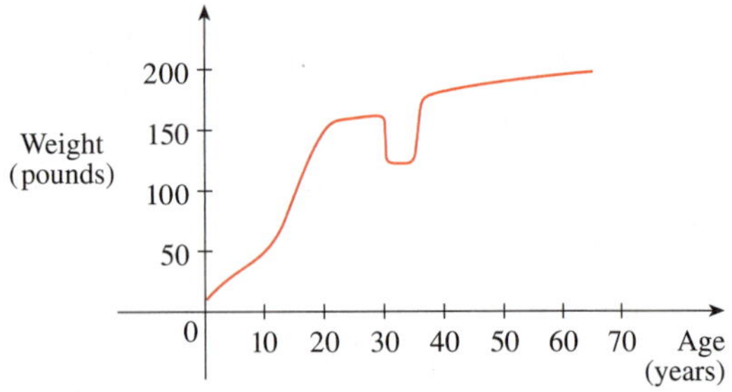

10. The graph shown gives a salesman's distance from his home as a function of time on a certain day. Describe in words what the graph indicates about his travels on this day.

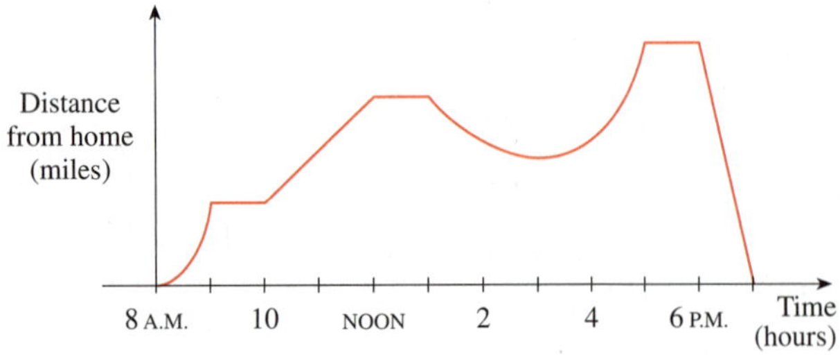

11. You put some ice cubes in a glass, fill the glass with cold water, and then let the glass sit on a table. Describe how the temperature of the water changes as time passes. Then sketch a rough graph of the temperature of the water as a function of the elapsed time.

12. Sketch a rough graph of the number of hours of daylight as a function of the time of year.

13. Sketch a rough graph of the outdoor temperature as a function of time during a typical spring day.

14. You place a frozen pie in an oven and bake it for an hour. Then you take it out and let it cool before eating it. Describe how the temperature of the pie changes as time passes. Then sketch a rough graph of the temperature of the pie as a function of time.

15. A homeowner mows the lawn every Wednesday afternoon. Sketch a rough graph of the height of the grass as a function of time over the course of a four-week period.

16. An airplane flies from an airport and lands an hour later at another airport, 400 miles away. If t represents the time in minutes since the plane has left the terminal building, let $x(t)$ be the horizontal distance traveled and $y(t)$ be the altitude of the plane.
(a) Sketch a possible graph of $x(t)$.
(b) Sketch a possible graph of $y(t)$.
(c) Sketch a possible graph of the ground speed.
(d) Sketch a possible graph of the vertical velocity.

17. The number N (in thousands) of cellular phone subscribers in Malaysia is shown in the table. (Midyear estimates are given.)

t	1991	1993	1995	1997
N	132	304	873	2461

(a) Use the data to sketch a rough graph of N as a function of t.
(b) Use your graph to estimate the number of cell-phone subscribers in Malaysia at midyear in 1994 and 1996.

18. Temperature readings T (in °F) were recorded every two hours from midnight to 2:00 P.M. in Dallas on June 2, 2001. The time t was measured in hours from midnight.

t	0	2	4	6	8	10	12	14
T	73	73	70	69	72	81	88	91

(a) Use the readings to sketch a rough graph of T as a function of t.
(b) Use your graph to estimate the temperature at 11:00 A.M.

19. If $f(x) = 3x^2 - x + 2$, find $f(2)$, $f(-2)$, $f(a)$, $f(-a)$, $f(a+1)$, $2f(a)$, $f(2a)$, $f(a^2)$, $[f(a)]^2$, and $f(a+h)$.

20. A spherical balloon with radius r inches has volume $V(r) = \frac{4}{3}\pi r^3$. Find a function that represents the amount of air required to inflate the balloon from a radius of r inches to a radius of $r + 1$ inches.

21–22 ||| Find $f(2+h)$, $f(x+h)$, and $\dfrac{f(x+h) - f(x)}{h}$, where $h \neq 0$.

21. $f(x) = x - x^2$

22. $f(x) = \dfrac{x}{x+1}$

23–27 ||| Find the domain of the function.

23. $f(x) = \dfrac{x}{3x - 1}$

24. $f(x) = \dfrac{5x + 4}{x^2 + 3x + 2}$

25. $f(t) = \sqrt{t} + \sqrt[3]{t}$

26. $g(u) = \sqrt{u} + \sqrt{4 - u}$

27. $h(x) = \dfrac{1}{\sqrt[4]{x^2 - 5x}}$

28. Find the domain and range and sketch the graph of the function $h(x) = \sqrt{4 - x^2}$.

29–40 ||| Find the domain and sketch the graph of the function.

29. $f(x) = 5$

30. $F(x) = \frac{1}{2}(x + 3)$

31. $f(t) = t^2 - 6t$

32. $H(t) = \dfrac{4 - t^2}{2 - t}$

33. $g(x) = \sqrt{x - 5}$

34. $F(x) = |2x + 1|$

35. $G(x) = \dfrac{3x + |x|}{x}$

36. $g(x) = \dfrac{|x|}{x^2}$

37. $f(x) = \begin{cases} x & \text{if } x \leqslant 0 \\ x + 1 & \text{if } x > 0 \end{cases}$

38. $f(x) = \begin{cases} 2x + 3 & \text{if } x < -1 \\ 3 - x & \text{if } x \geqslant -1 \end{cases}$

39. $f(x) = \begin{cases} x + 2 & \text{if } x \leqslant -1 \\ x^2 & \text{if } x > -1 \end{cases}$

40. $f(x) = \begin{cases} -1 & \text{if } x \leqslant -1 \\ 3x + 2 & \text{if } |x| < 1 \\ 7 - 2x & \text{if } x \geqslant 1 \end{cases}$

41–46 ||| Find an expression for the function whose graph is the given curve.

41. The line segment joining the points $(-2, 1)$ and $(4, -6)$

42. The line segment joining the points $(-3, -2)$ and $(6, 3)$

43. The bottom half of the parabola $x + (y - 1)^2 = 0$

44. The top half of the circle $(x - 1)^2 + y^2 = 1$

45.

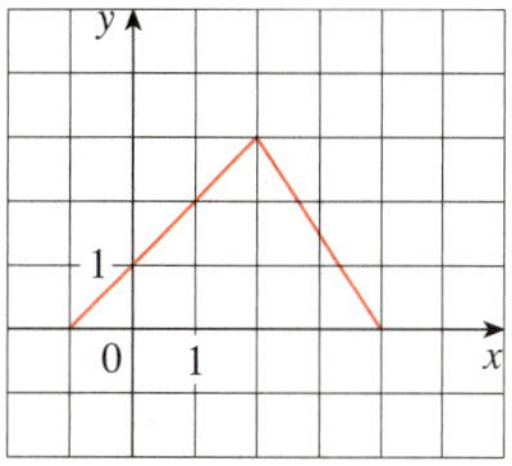

46.

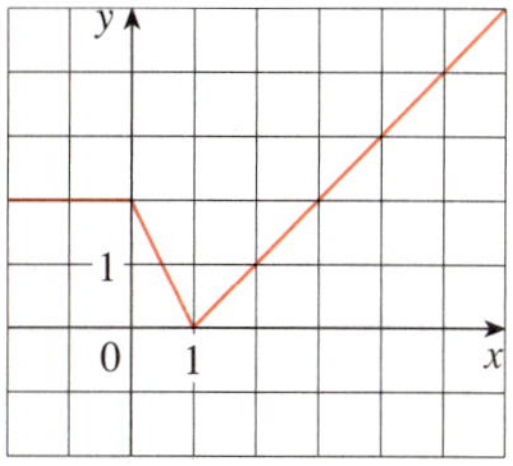

47–51 Find a formula for the described function and state its domain.

47. A rectangle has perimeter 20 m. Express the area of the rectangle as a function of the length of one of its sides.

48. A rectangle has area 16 m^2. Express the perimeter of the rectangle as a function of the length of one of its sides.

49. Express the area of an equilateral triangle as a function of the length of a side.

50. Express the surface area of a cube as a function of its volume.

51. An open rectangular box with volume 2 m^3 has a square base. Express the surface area of the box as a function of the length of a side of the base.

52. A Norman window has the shape of a rectangle surmounted by a semicircle. If the perimeter of the window is 30 ft, express the area A of the window as a function of the width x of the window.

53. A box with an open top is to be constructed from a rectangular piece of cardboard with dimensions 12 in. by 20 in. by cutting out equal squares of side x at each corner and then folding up the sides as in the figure. Express the volume V of the box as a function of x.

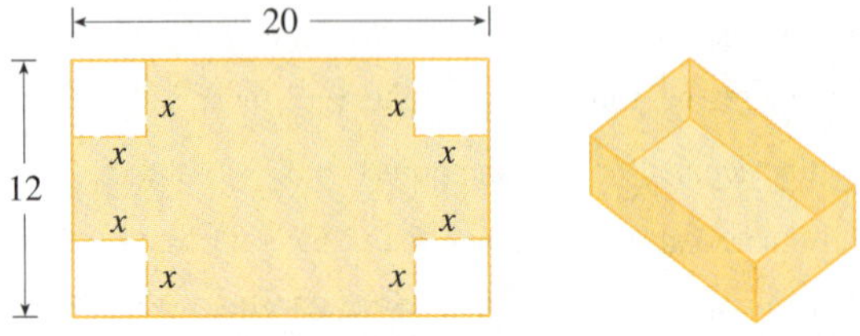

54. A taxi company charges two dollars for the first mile (or part of a mile) and 20 cents for each succeeding tenth of a mile (or part). Express the cost C (in dollars) of a ride as a function of the distance x traveled (in miles) for $0 < x < 2$, and sketch the graph of this function.

55. In a certain country, income tax is assessed as follows. There is no tax on income up to \$10,000. Any income over \$10,000 is taxed at a rate of 10%, up to an income of \$20,000. Any income over \$20,000 is taxed at 15%.

(a) Sketch the graph of the tax rate R as a function of the income I.

(b) How much tax is assessed on an income of \$14,000? On \$26,000?

(c) Sketch the graph of the total assessed tax T as a function of the income I.

56. The functions in Example 10 and Exercises 54 and 55(a) are called *step functions* because their graphs look like stairs. Give two other examples of step functions that arise in everyday life.

57–58 Graphs of f and g are shown. Decide whether each function is even, odd, or neither. Explain your reasoning.

57.

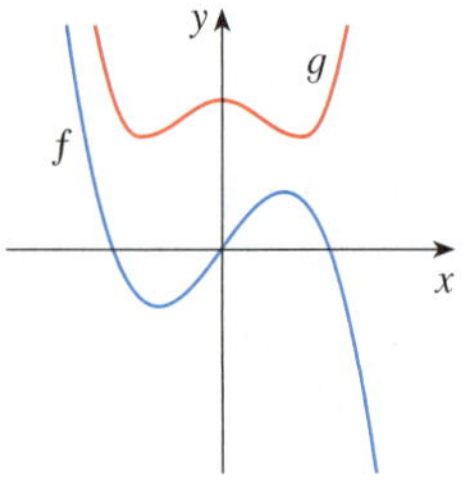

58.

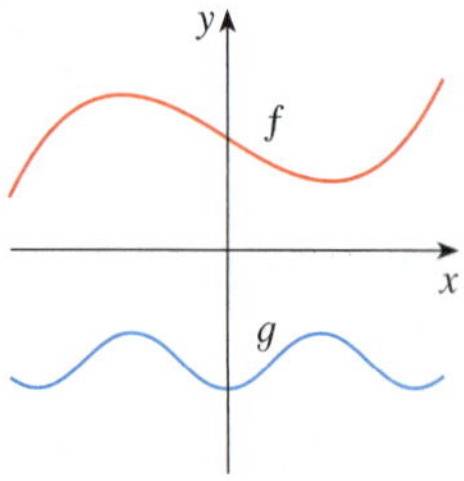

59. (a) If the point (5, 3) is on the graph of an even function, what other point must also be on the graph?

(b) If the point (5, 3) is on the graph of an odd function, what other point must also be on the graph?

60. A function f has domain $[-5, 5]$ and a portion of its graph is shown.

(a) Complete the graph of f if it is known that f is even.

(b) Complete the graph of f if it is known that f is odd.

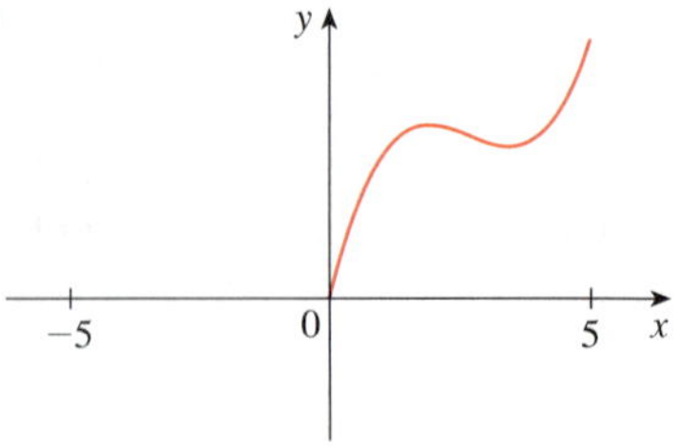

61–66 Determine whether f is even, odd, or neither. If f is even or odd, use symmetry to sketch its graph.

61. $f(x) = x^{-2}$

62. $f(x) = x^{-3}$

63. $f(x) = x^2 + x$

64. $f(x) = x^4 - 4x^2$

65. $f(x) = x^3 - x$

66. $f(x) = 3x^3 + 2x^2 + 1$

1.2 Mathematical Models: A Catalog of Essential Functions

A **mathematical model** is a mathematical description (often by means of a function or an equation) of a real-world phenomenon such as the size of a population, the demand for a product, the speed of a falling object, the concentration of a product in a chemical reaction, the life expectancy of a person at birth, or the cost of emission reductions. The purpose of the model is to understand the phenomenon and perhaps to make predictions about future behavior.

Figure 1 illustrates the process of mathematical modeling. Given a real-world problem, our first task is to formulate a mathematical model by identifying and naming the independent and dependent variables and making assumptions that simplify the phenomenon enough to make it mathematically tractable. We use our knowledge of the physical situation and our mathematical skills to obtain equations that relate the variables. In situations where there is no physical law to guide us, we may need to collect data (either from a library or the Internet or by conducting our own experiments) and examine the data in the form of a table in order to discern patterns. From this numerical representation of a function we may wish to obtain a graphical representation by plotting the data. The graph might even suggest a suitable algebraic formula in some cases.

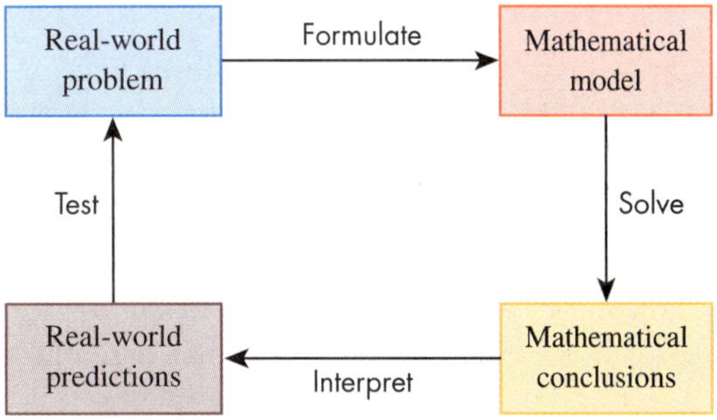

FIGURE 1
The modeling process

The second stage is to apply the mathematics that we know (such as the calculus that will be developed throughout this book) to the mathematical model that we have formulated in order to derive mathematical conclusions. Then, in the third stage, we take those mathematical conclusions and interpret them as information about the original real-world phenomenon by way of offering explanations or making predictions. The final step is to test our predictions by checking against new real data. If the predictions don't compare well with reality, we need to refine our model or to formulate a new model and start the cycle again.

A mathematical model is never a completely accurate representation of a physical situation—it is an *idealization.* A good model simplifies reality enough to permit mathematical calculations but is accurate enough to provide valuable conclusions. It is important to realize the limitations of the model. In the end, Mother Nature has the final say.

There are many different types of functions that can be used to model relationships observed in the real world. In what follows, we discuss the behavior and graphs of these functions and give examples of situations appropriately modeled by such functions.

Linear Models

|||| The coordinate geometry of lines is reviewed in Appendix B.

When we say that y is a **linear function** of x, we mean that the graph of the function is a line, so we can use the slope-intercept form of the equation of a line to write a formula for the function as

$$y = f(x) = mx + b$$

where m is the slope of the line and b is the y-intercept.

A characteristic feature of linear functions is that they grow at a constant rate. For instance, Figure 2 shows a graph of the linear function $f(x) = 3x - 2$ and a table of sample values. Notice that whenever x increases by 0.1, the value of $f(x)$ increases by 0.3. So $f(x)$ increases three times as fast as x. Thus, the slope of the graph $y = 3x - 2$, namely 3, can be interpreted as the rate of change of y with respect to x.

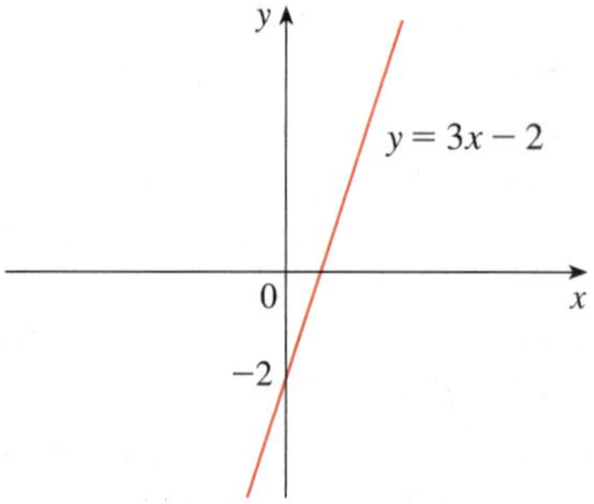

x	$f(x) = 3x - 2$
1.0	1.0
1.1	1.3
1.2	1.6
1.3	1.9
1.4	2.2
1.5	2.5

FIGURE 2

EXAMPLE 1

(a) As dry air moves upward, it expands and cools. If the ground temperature is 20°C and the temperature at a height of 1 km is 10°C, express the temperature T (in °C) as a function of the height h (in kilometers), assuming that a linear model is appropriate.
(b) Draw the graph of the function in part (a). What does the slope represent?
(c) What is the temperature at a height of 2.5 km?

SOLUTION

(a) Because we are assuming that T is a linear function of h, we can write

$$T = mh + b$$

We are given that $T = 20$ when $h = 0$, so

$$20 = m \cdot 0 + b = b$$

In other words, the y-intercept is $b = 20$.

We are also given that $T = 10$ when $h = 1$, so

$$10 = m \cdot 1 + 20$$

The slope of the line is therefore $m = 10 - 20 = -10$ and the required linear function is

$$T = -10h + 20$$

(b) The graph is sketched in Figure 3. The slope is $m = -10$°C/km, and this represents the rate of change of temperature with respect to height.

(c) At a height of $h = 2.5$ km, the temperature is

$$T = -10(2.5) + 20 = -5\text{°C}$$

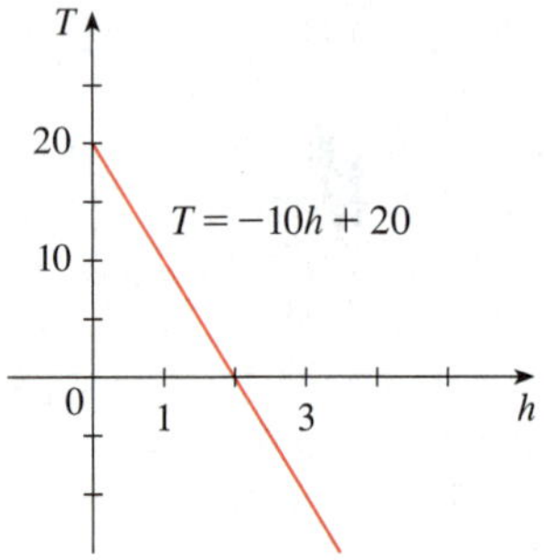

FIGURE 3

If there is no physical law or principle to help us formulate a model, we construct an **empirical model**, which is based entirely on collected data. We seek a curve that "fits" the data in the sense that it captures the basic trend of the data points.

TABLE 1

Year	CO_2 level (in ppm)
1980	338.7
1982	341.1
1984	344.4
1986	347.2
1988	351.5
1990	354.2
1992	356.4
1994	358.9
1996	362.6
1998	366.6
2000	369.4

EXAMPLE 2 Table 1 lists the average carbon dioxide level in the atmosphere, measured in parts per million at Mauna Loa Observatory from 1980 to 2000. Use the data in Table 1 to find a model for the carbon dioxide level.

SOLUTION We use the data in Table 1 to make the scatter plot in Figure 4, where t represents time (in years) and C represents the CO_2 level (in parts per million, ppm).

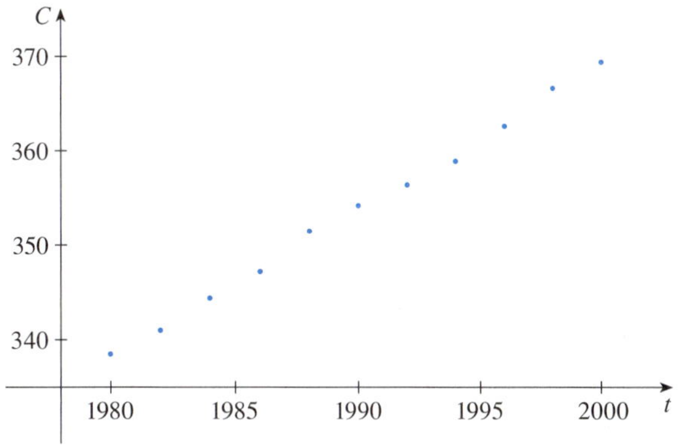

FIGURE 4
Scatter plot for the average CO_2 level

Notice that the data points appear to lie close to a straight line, so it's natural to choose a linear model in this case. But there are many possible lines that approximate these data points, so which one should we use? From the graph, it appears that one possibility is the line that passes through the first and last data points. The slope of this line is

$$\frac{369.4 - 338.7}{2000 - 1980} = \frac{30.7}{20} = 1.535$$

and its equation is

$$C - 338.7 = 1.535(t - 1980)$$

or

$$C = 1.535t - 2700.6 \quad \boxed{1}$$

Equation 1 gives one possible linear model for the carbon dioxide level; it is graphed in Figure 5.

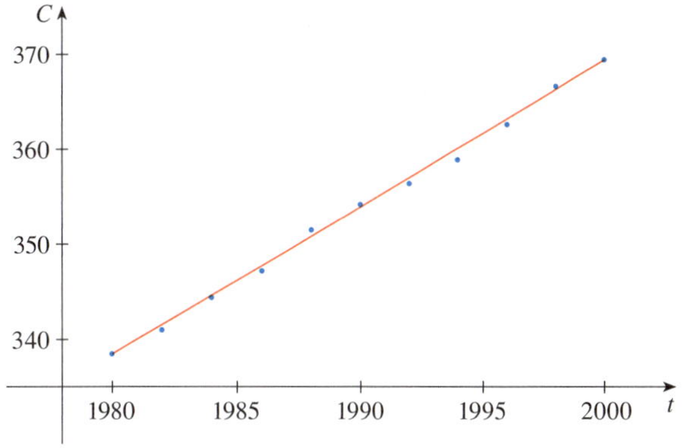

FIGURE 5
Linear model through first and last data points

Although our model fits the data reasonably well, it gives values higher than most of the actual CO_2 levels. A better linear model is obtained by a procedure from statistics

|||| A computer or graphing calculator finds the regression line by the method of **least squares**, which is to minimize the sum of the squares of the vertical distances between the data points and the line. The details are explained in Section 14.7.

called *linear regression.* If we use a graphing calculator, we enter the data from Table 1 into the data editor and choose the linear regression command. (With Maple we use the fit[leastsquare] command in the stats package; with Mathematica we use the Fit command.) The machine gives the slope and y-intercept of the regression line as

$$m = 1.53818 \qquad b = -2707.25$$

So our least squares model for the CO_2 level is

$$C = 1.53818t - 2707.25 \tag{2}$$

In Figure 6 we graph the regression line as well as the data points. Comparing with Figure 5, we see that it gives a better fit than our previous linear model.

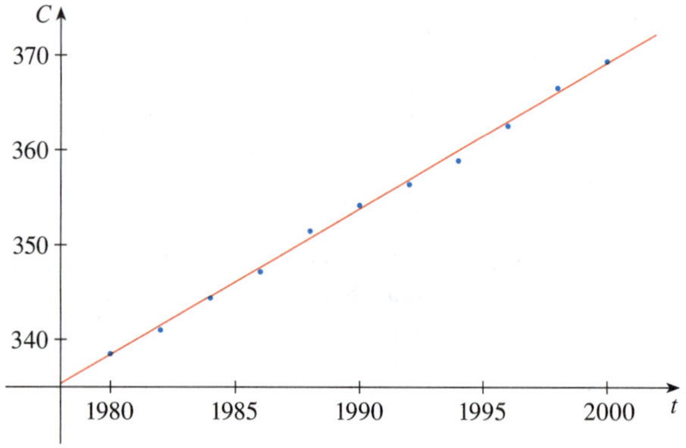

FIGURE 6
The regression line

EXAMPLE 3 Use the linear model given by Equation 2 to estimate the average CO_2 level for 1987 and to predict the level for the year 2010. According to this model, when will the CO_2 level exceed 400 parts per million?

SOLUTION Using Equation 2 with $t = 1987$, we estimate that the average CO_2 level in 1987 was

$$C(1987) = (1.53818)(1987) - 2707.25 \approx 349.11$$

This is an example of *interpolation* because we have estimated a value *between* observed values. (In fact, the Mauna Loa Observatory reported that the average CO_2 level in 1987 was 348.93 ppm, so our estimate is quite accurate.)

With $t = 2010$, we get

$$C(2010) = (1.53818)(2010) - 2707.25 \approx 384.49$$

So we predict that the average CO_2 level in the year 2010 will be 384.5 ppm. This is an example of *extrapolation* because we have predicted a value *outside* the region of observations. Consequently, we are far less certain about the accuracy of our prediction.

Using Equation 2, we see that the CO_2 level exceeds 400 ppm when

$$1.53818t - 2707.25 > 400$$

Solving this inequality, we get

$$t > \frac{3107.25}{1.53818} \approx 2020.08$$

We therefore predict that the CO_2 level will exceed 400 ppm by the year 2020. This prediction is somewhat risky because it involves a time quite remote from our observations.

Polynomials

A function P is called a **polynomial** if

$$P(x) = a_n x^n + a_{n-1} x^{n-1} + \cdots + a_2 x^2 + a_1 x + a_0$$

where n is a nonnegative integer and the numbers $a_0, a_1, a_2, \ldots, a_n$ are constants called the **coefficients** of the polynomial. The domain of any polynomial is $\mathbb{R} = (-\infty, \infty)$. If the leading coefficient $a_n \neq 0$, then the **degree** of the polynomial is n. For example, the function

$$P(x) = 2x^6 - x^4 + \tfrac{2}{5}x^3 + \sqrt{2}$$

is a polynomial of degree 6.

A polynomial of degree 1 is of the form $P(x) = mx + b$ and so it is a linear function. A polynomial of degree 2 is of the form $P(x) = ax^2 + bx + c$ and is called a **quadratic function**. Its graph is always a parabola obtained by shifting the parabola $y = ax^2$, as we will see in the next section. The parabola opens upward if $a > 0$ and downward if $a < 0$. (See Figure 7.)

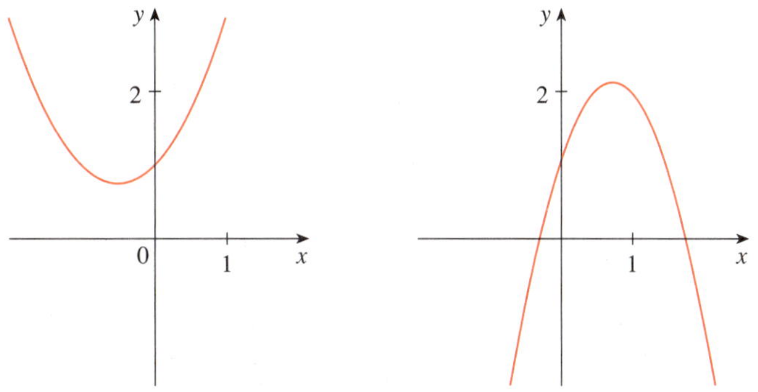

(a) $y = x^2 + x + 1$ (b) $y = -2x^2 + 3x + 1$

FIGURE 7 The graphs of quadratic functions are parabolas.

A polynomial of degree 3 is of the form

$$P(x) = ax^3 + bx^2 + cx + d$$

and is called a **cubic function**. Figure 8 shows the graph of a cubic function in part (a) and graphs of polynomials of degrees 4 and 5 in parts (b) and (c). We will see later why the graphs have these shapes.

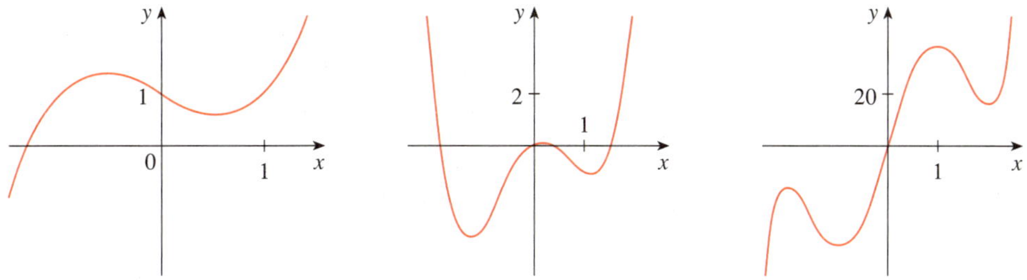

FIGURE 8 (a) $y = x^3 - x + 1$ (b) $y = x^4 - 3x^2 + x$ (c) $y = 3x^5 - 25x^3 + 60x$

Polynomials are commonly used to model various quantities that occur in the natural and social sciences. For instance, in Section 3.3 we will explain why economists often use a polynomial $P(x)$ to represent the cost of producing x units of a commodity. In the following example we use a quadratic function to model the fall of a ball.

TABLE 2

Time (seconds)	Height (meters)
0	450
1	445
2	431
3	408
4	375
5	332
6	279
7	216
8	143
9	61

EXAMPLE 4 A ball is dropped from the upper observation deck of the CN Tower, 450 m above the ground, and its height h above the ground is recorded at 1-second intervals in Table 2. Find a model to fit the data and use the model to predict the time at which the ball hits the ground.

SOLUTION We draw a scatter plot of the data in Figure 9 and observe that a linear model is inappropriate. But it looks as if the data points might lie on a parabola, so we try a quadratic model instead. Using a graphing calculator or computer algebra system (which uses the least squares method), we obtain the following quadratic model:

$$h = 449.36 + 0.96t - 4.90t^2 \tag{3}$$

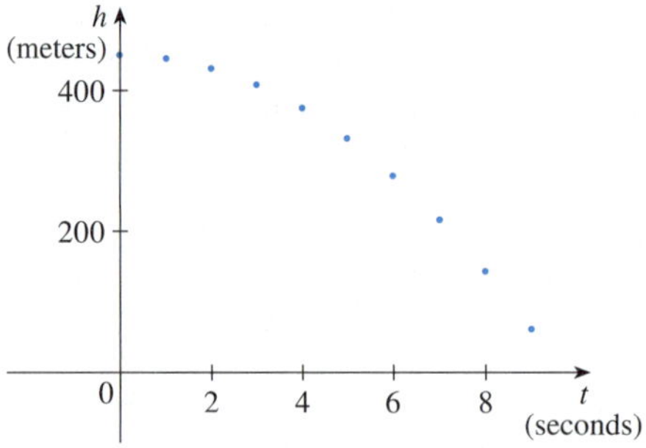

FIGURE 9
Scatter plot for a falling ball

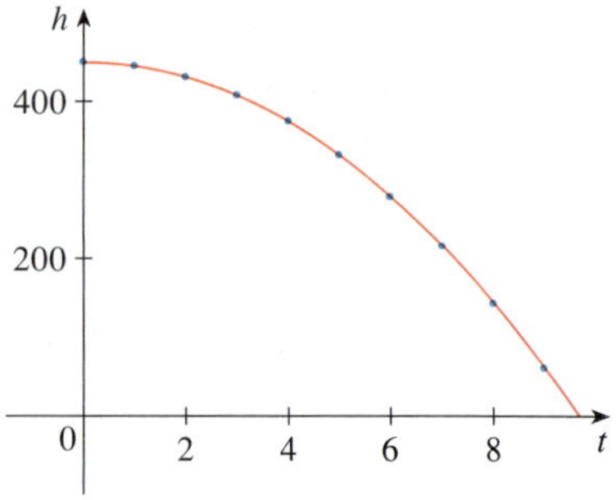

FIGURE 10
Quadratic model for a falling ball

In Figure 10 we plot the graph of Equation 3 together with the data points and see that the quadratic model gives a very good fit.

The ball hits the ground when $h = 0$, so we solve the quadratic equation

$$-4.90t^2 + 0.96t + 449.36 = 0$$

The quadratic formula gives

$$t = \frac{-0.96 \pm \sqrt{(0.96)^2 - 4(-4.90)(449.36)}}{2(-4.90)}$$

The positive root is $t \approx 9.67$, so we predict that the ball will hit the ground after about 9.7 seconds.

Power Functions

A function of the form $f(x) = x^a$, where a is a constant, is called a **power function**. We consider several cases.

(i) $a = n$, where n is a positive integer

The graphs of $f(x) = x^n$ for $n = 1, 2, 3, 4$, and 5 are shown in Figure 11. (These are polynomials with only one term.) We already know the shape of the graphs of $y = x$ (a line through the origin with slope 1) and $y = x^2$ [a parabola, see Example 2(b) in Section 1.1].

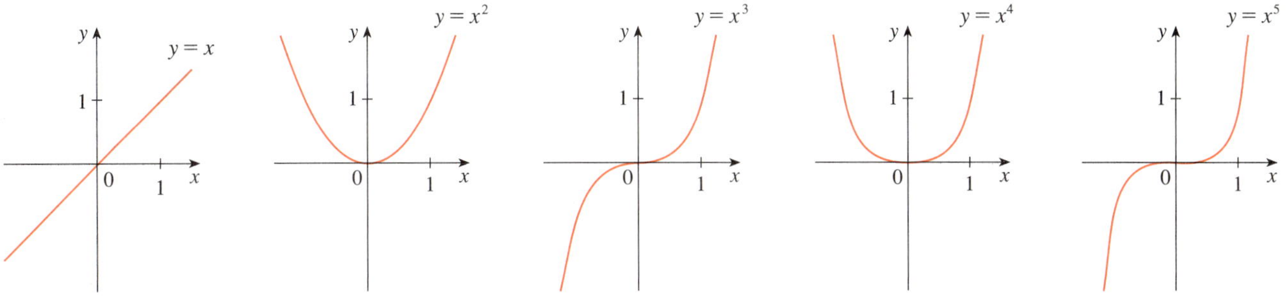

FIGURE 11 Graphs of $f(x) = x^n$ for $n = 1, 2, 3, 4, 5$

The general shape of the graph of $f(x) = x^n$ depends on whether n is even or odd. If n is even, then $f(x) = x^n$ is an even function and its graph is similar to the parabola $y = x^2$. If n is odd, then $f(x) = x^n$ is an odd function and its graph is similar to that of $y = x^3$. Notice from Figure 12, however, that as n increases, the graph of $y = x^n$ becomes flatter near 0 and steeper when $|x| \geqslant 1$. (If x is small, then x^2 is smaller, x^3 is even smaller, x^4 is smaller still, and so on.)

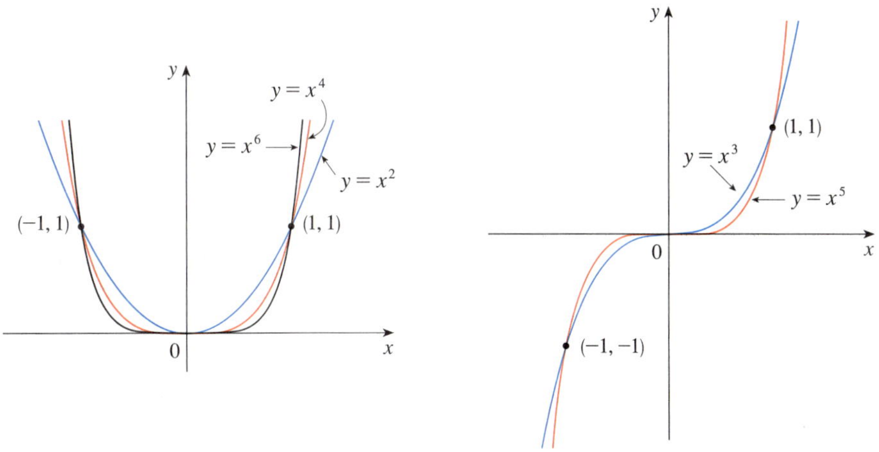

FIGURE 12
Families of power functions

(ii) $a = 1/n$, where n is a positive integer
The function $f(x) = x^{1/n} = \sqrt[n]{x}$ is a **root function**. For $n = 2$ it is the square root function $f(x) = \sqrt{x}$, whose domain is $[0, \infty)$ and whose graph is the upper half of the parabola $x = y^2$. [See Figure 13(a).] For other even values of n, the graph of $y = \sqrt[n]{x}$ is similar to that of $y = \sqrt{x}$. For $n = 3$ we have the cube root function $f(x) = \sqrt[3]{x}$ whose domain is $\mathbb{R}$ (recall that every real number has a cube root) and whose graph is shown in Figure 13(b). The graph of $y = \sqrt[n]{x}$ for n odd $(n > 3)$ is similar to that of $y = \sqrt[3]{x}$.

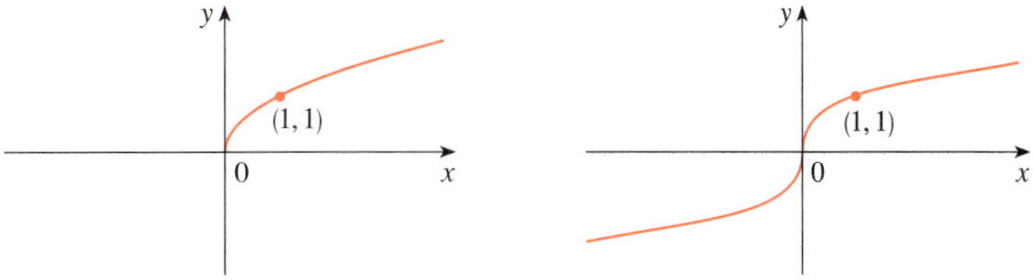

FIGURE 13
Graphs of root functions

(a) $f(x) = \sqrt{x}$ (b) $f(x) = \sqrt[3]{x}$

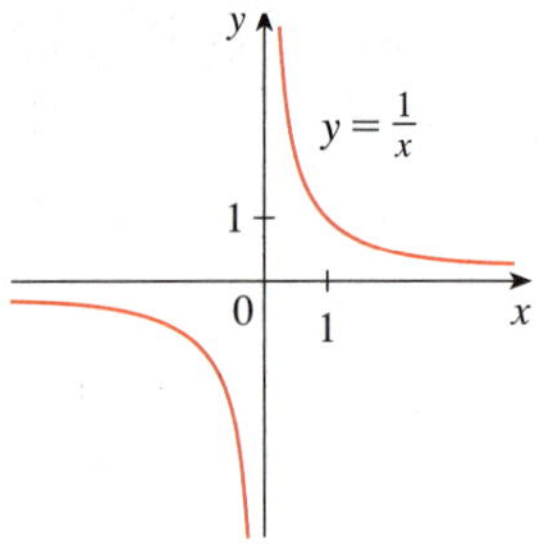

FIGURE 14
The reciprocal function

(iii) $a = -1$

The graph of the **reciprocal function** $f(x) = x^{-1} = 1/x$ is shown in Figure 14. Its graph has the equation $y = 1/x$, or $xy = 1$, and is a hyperbola with the coordinate axes as its asymptotes.

This function arises in physics and chemistry in connection with Boyle's Law, which says that, when the temperature is constant, the volume V of a gas is inversely proportional to the pressure P:

$$V = \frac{C}{P}$$

where C is a constant. Thus, the graph of V as a function of P (see Figure 15) has the same general shape as the right half of Figure 14.

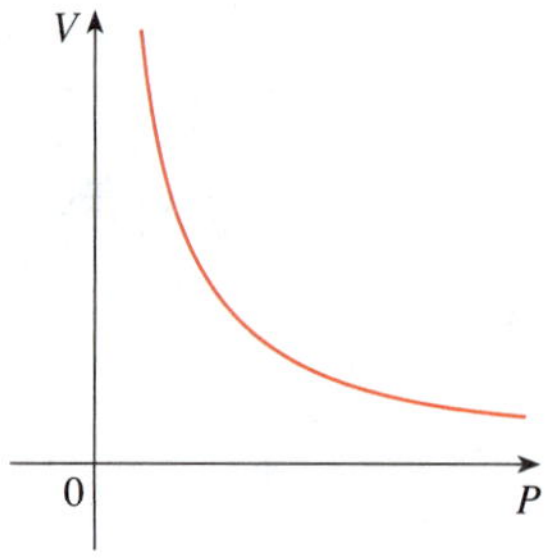

FIGURE 15
Volume as a function of pressure at constant temperature

Another instance in which a power function is used to model a physical phenomenon is discussed in Exercise 22.

Rational Functions

A **rational function** f is a ratio of two polynomials:

$$f(x) = \frac{P(x)}{Q(x)}$$

where P and Q are polynomials. The domain consists of all values of x such that $Q(x) \neq 0$. A simple example of a rational function is the function $f(x) = 1/x$, whose domain is $\{x \mid x \neq 0\}$; this is the reciprocal function graphed in Figure 14. The function

$$f(x) = \frac{2x^4 - x^2 + 1}{x^2 - 4}$$

is a rational function with domain $\{x \mid x \neq \pm 2\}$. Its graph is shown in Figure 16.

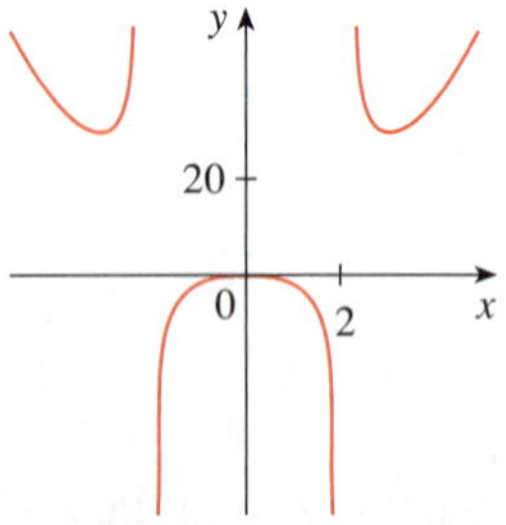

FIGURE 16
$f(x) = \dfrac{2x^4 - x^2 + 1}{x^2 - 4}$

Algebraic Functions

A function f is called an **algebraic function** if it can be constructed using algebraic operations (such as addition, subtraction, multiplication, division, and taking roots) starting with polynomials. Any rational function is automatically an algebraic function. Here are two more examples:

$$f(x) = \sqrt{x^2 + 1} \qquad g(x) = \frac{x^4 - 16x^2}{x + \sqrt{x}} + (x - 2)\sqrt[3]{x + 1}$$

When we sketch algebraic functions in Chapter 4, we will see that their graphs can assume a variety of shapes. Figure 17 illustrates some of the possibilities.

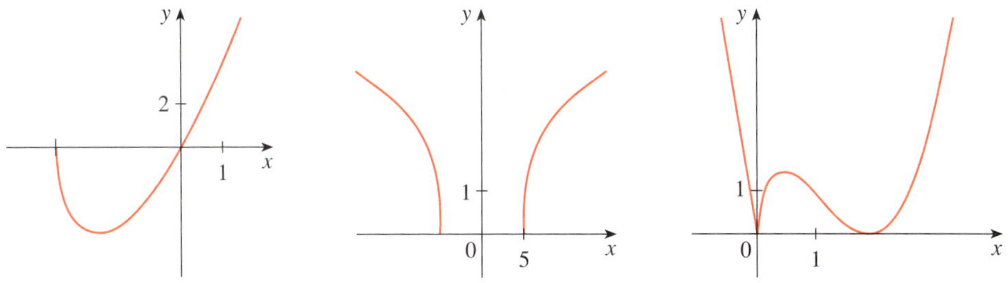

FIGURE 17 (a) $f(x) = x\sqrt{x+3}$ (b) $g(x) = \sqrt[4]{x^2 - 25}$ (c) $h(x) = x^{2/3}(x-2)^2$

An example of an algebraic function occurs in the theory of relativity. The mass of a particle with velocity v is

$$m = f(v) = \frac{m_0}{\sqrt{1 - v^2/c^2}}$$

where m_0 is the rest mass of the particle and $c = 3.0 \times 10^5$ km/s is the speed of light in a vacuum.

Trigonometric Functions

Trigonometry and the trigonometric functions are reviewed on Reference Page 2 and also in Appendix D. In calculus the convention is that radian measure is always used (except when otherwise indicated). For example, when we use the function $f(x) = \sin x$, it is understood that $\sin x$ means the sine of the angle whose radian measure is x. Thus, the graphs of the sine and cosine functions are as shown in Figure 18.

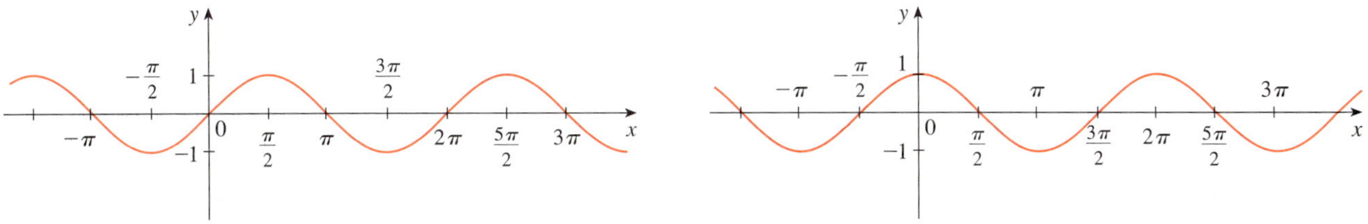

(a) $f(x) = \sin x$ (b) $g(x) = \cos x$

FIGURE 18

Notice that for both the sine and cosine functions the domain is $(-\infty, \infty)$ and the range is the closed interval $[-1, 1]$. Thus, for all values of x, we have

$$-1 \leqslant \sin x \leqslant 1 \qquad -1 \leqslant \cos x \leqslant 1$$

or, in terms of absolute values,

$$|\sin x| \leqslant 1 \qquad |\cos x| \leqslant 1$$

Also, the zeros of the sine function occur at the integer multiples of π; that is,

$$\sin x = 0 \qquad \text{when} \qquad x = n\pi \quad n \text{ an integer}$$

An important property of the sine and cosine functions is that they are periodic functions and have period 2π. This means that, for all values of x,

$$\sin(x + 2\pi) = \sin x \qquad \cos(x + 2\pi) = \cos x$$

The periodic nature of these functions makes them suitable for modeling repetitive phenomena such as tides, vibrating springs, and sound waves. For instance, in Example 4 in Section 1.3 we will see that a reasonable model for the number of hours of daylight in Philadelphia t days after January 1 is given by the function

$$L(t) = 12 + 2.8 \sin\left[\frac{2\pi}{365}(t - 80)\right]$$

The tangent function is related to the sine and cosine functions by the equation

$$\tan x = \frac{\sin x}{\cos x}$$

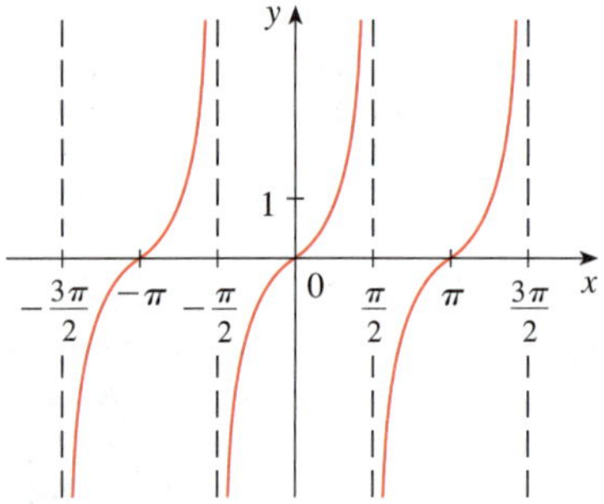

FIGURE 19
$y = \tan x$

and its graph is shown in Figure 19. It is undefined whenever $\cos x = 0$, that is, when $x = \pm\pi/2, \pm 3\pi/2, \ldots$. Its range is $(-\infty, \infty)$. Notice that the tangent function has period π:

$$\tan(x + \pi) = \tan x \qquad \text{for all } x$$

The remaining three trigonometric functions (cosecant, secant, and cotangent) are the reciprocals of the sine, cosine, and tangent functions. Their graphs are shown in Appendix D.

Exponential Functions

The **exponential functions** are the functions of the form $f(x) = a^x$, where the base a is a positive constant. The graphs of $y = 2^x$ and $y = (0.5)^x$ are shown in Figure 20. In both cases the domain is $(-\infty, \infty)$ and the range is $(0, \infty)$.

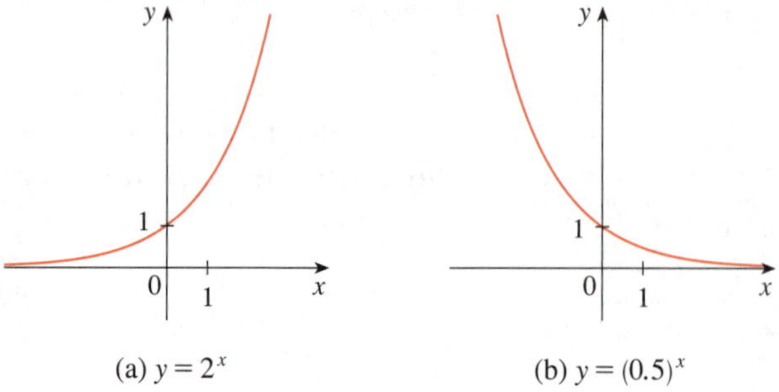

FIGURE 20

Exponential functions will be studied in detail in Section 1.5, and we will see that they are useful for modeling many natural phenomena, such as population growth (if $a > 1$) and radioactive decay (if $a < 1$).

Logarithmic Functions

The **logarithmic functions** $f(x) = \log_a x$, where the base a is a positive constant, are the inverse functions of the exponential functions. They will be studied in Section 1.6. Figure 21 shows the graphs of four logarithmic functions with various bases. In each case the domain is $(0, \infty)$, the range is $(-\infty, \infty)$, and the function increases slowly when $x > 1$.

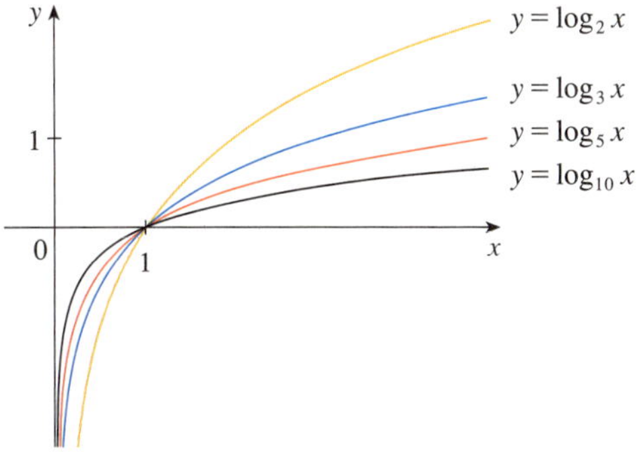

FIGURE 21

Transcendental Functions

These are functions that are not algebraic. The set of transcendental functions includes the trigonometric, inverse trigonometric, exponential, and logarithmic functions, but it also includes a vast number of other functions that have never been named. In Chapter 11 we will study transcendental functions that are defined as sums of infinite series.

EXAMPLE 5 Classify the following functions as one of the types of functions that we have discussed.

(a) $f(x) = 5^x$ (b) $g(x) = x^5$

(c) $h(x) = \dfrac{1 + x}{1 - \sqrt{x}}$ (d) $u(t) = 1 - t + 5t^4$

SOLUTION

(a) $f(x) = 5^x$ is an exponential function. (The x is the exponent.)

(b) $g(x) = x^5$ is a power function. (The x is the base.) We could also consider it to be a polynomial of degree 5.

(c) $h(x) = \dfrac{1 + x}{1 - \sqrt{x}}$ is an algebraic function.

(d) $u(t) = 1 - t + 5t^4$ is a polynomial of degree 4.

1.2 Exercises

1–2 Classify each function as a power function, root function, polynomial (state its degree), rational function, algebraic function, trigonometric function, exponential function, or logarithmic function.

1. (a) $f(x) = \sqrt[5]{x}$ (b) $g(x) = \sqrt{1 - x^2}$

(c) $h(x) = x^9 + x^4$ (d) $r(x) = \dfrac{x^2 + 1}{x^3 + x}$

(e) $s(x) = \tan 2x$ (f) $t(x) = \log_{10} x$

2. (a) $y = \dfrac{x - 6}{x + 6}$ (b) $y = x + \dfrac{x^2}{\sqrt{x - 1}}$

(c) $y = 10^x$ (d) $y = x^{10}$

(e) $y = 2t^6 + t^4 - \pi$ (f) $y = \cos\theta + \sin\theta$

3–4 Match each equation with its graph. Explain your choices. (Don't use a computer or graphing calculator.)

3. (a) $y = x^2$ (b) $y = x^5$ (c) $y = x^8$

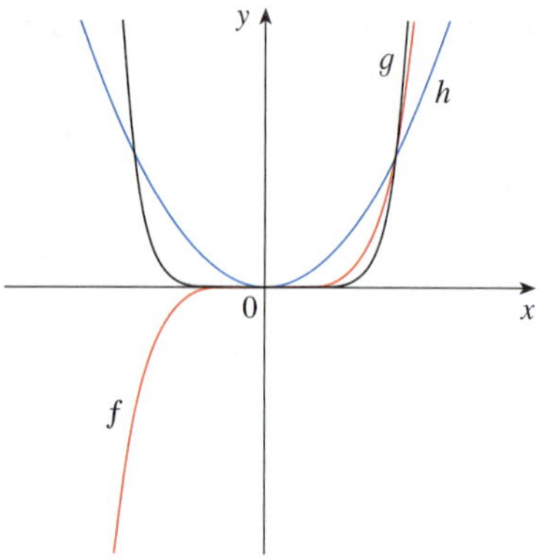

4. (a) $y = 3x$ (b) $y = 3^x$
(c) $y = x^3$ (d) $y = \sqrt[3]{x}$

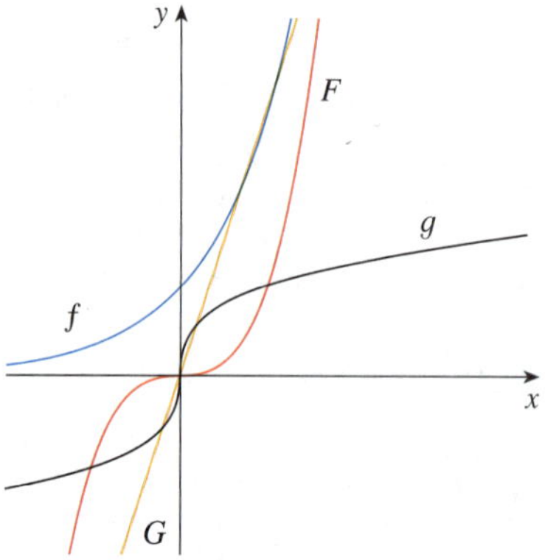

5. (a) Find an equation for the family of linear functions with slope 2 and sketch several members of the family.
(b) Find an equation for the family of linear functions such that $f(2) = 1$ and sketch several members of the family.
(c) Which function belongs to both families?

6. What do all members of the family of linear functions $f(x) = 1 + m(x + 3)$ have in common? Sketch several members of the family.

7. What do all members of the family of linear functions $f(x) = c - x$ have in common? Sketch several members of the family.

8. The manager of a weekend flea market knows from past experience that if he charges x dollars for a rental space at the flea market, then the number y of spaces he can rent is given by the equation $y = 200 - 4x$.
(a) Sketch a graph of this linear function. (Remember that the rental charge per space and the number of spaces rented can't be negative quantities.)
(b) What do the slope, the y-intercept, and the x-intercept of the graph represent?

9. The relationship between the Fahrenheit (F) and Celsius (C) temperature scales is given by the linear function $F = \frac{9}{5}C + 32$.
(a) Sketch a graph of this function.
(b) What is the slope of the graph and what does it represent? What is the F-intercept and what does it represent?

10. Jason leaves Detroit at 2:00 P.M. and drives at a constant speed west along I-96. He passes Ann Arbor, 40 mi from Detroit, at 2:50 P.M.
(a) Express the distance traveled in terms of the time elapsed.
(b) Draw the graph of the equation in part (a).
(c) What is the slope of this line? What does it represent?

11. Biologists have noticed that the chirping rate of crickets of a certain species is related to temperature, and the relationship appears to be very nearly linear. A cricket produces 113 chirps per minute at 70°F and 173 chirps per minute at 80°F.
(a) Find a linear equation that models the temperature T as a function of the number of chirps per minute N.
(b) What is the slope of the graph? What does it represent?
(c) If the crickets are chirping at 150 chirps per minute, estimate the temperature.

12. The manager of a furniture factory finds that it costs \$2200 to manufacture 100 chairs in one day and \$4800 to produce 300 chairs in one day.
(a) Express the cost as a function of the number of chairs produced, assuming that it is linear. Then sketch the graph.
(b) What is the slope of the graph and what does it represent?
(c) What is the y-intercept of the graph and what does it represent?

13. At the surface of the ocean, the water pressure is the same as the air pressure above the water, 15 lb/in^2. Below the surface, the water pressure increases by 4.34 lb/in^2 for every 10 ft of descent.
(a) Express the water pressure as a function of the depth below the ocean surface.
(b) At what depth is the pressure 100 lb/in^2?

14. The monthly cost of driving a car depends on the number of miles driven. Lynn found that in May it cost her \$380 to drive 480 mi and in June it cost her \$460 to drive 800 mi.
(a) Express the monthly cost C as a function of the distance driven d, assuming that a linear relationship gives a suitable model.
(b) Use part (a) to predict the cost of driving 1500 miles per month.
(c) Draw the graph of the linear function. What does the slope represent?
(d) What does the y-intercept represent?
(e) Why does a linear function give a suitable model in this situation?

15–16 For each scatter plot, decide what type of function you might choose as a model for the data. Explain your choices.

15. (a)

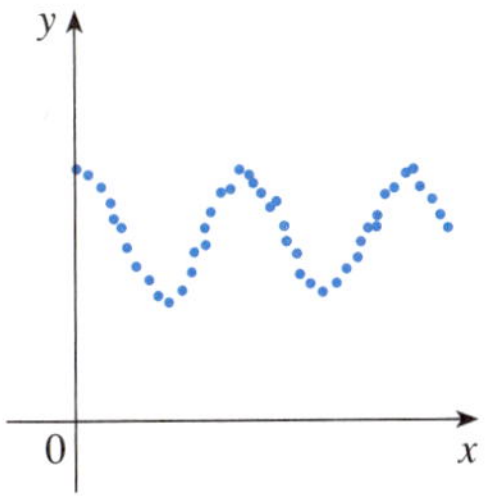

(b)

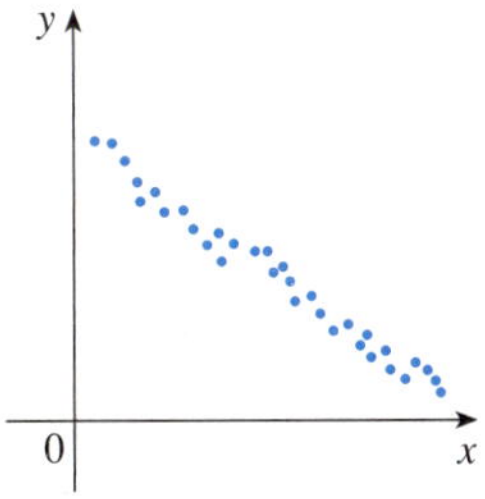

16. (a)

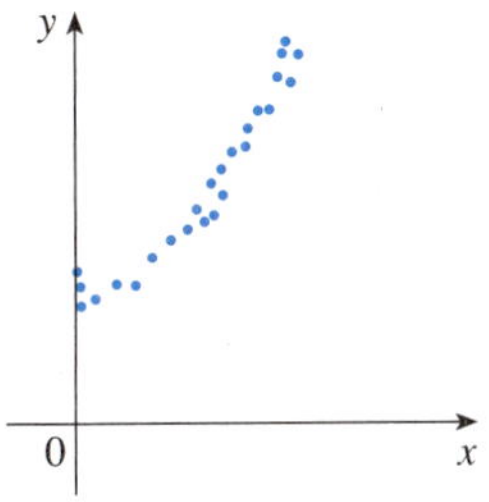

(b)

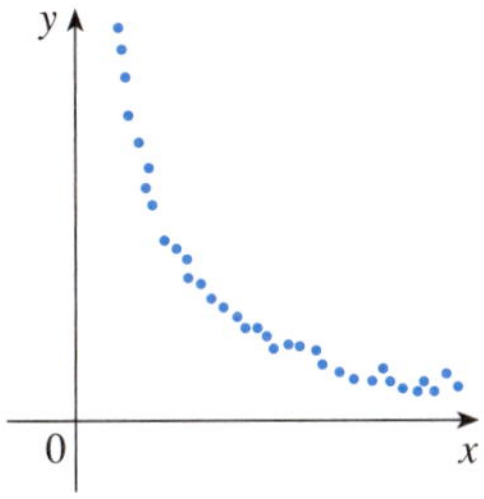

17. The table shows (lifetime) peptic ulcer rates (per 100 population) for various family incomes as reported by the 1989 National Health Interview Survey.

Income	Ulcer rate (per 100 population)
\$4,000	14.1
\$6,000	13.0
\$8,000	13.4
\$12,000	12.5
\$16,000	12.0
\$20,000	12.4
\$30,000	10.5
\$45,000	9.4
\$60,000	8.2

(a) Make a scatter plot of these data and decide whether a linear model is appropriate.
(b) Find and graph a linear model using the first and last data points.
(c) Find and graph the least squares regression line.
(d) Use the linear model in part (c) to estimate the ulcer rate for an income of \$25,000.
(e) According to the model, how likely is someone with an income of \$80,000 to suffer from peptic ulcers?
(f) Do you think it would be reasonable to apply the model to someone with an income of \$200,000?

18. Biologists have observed that the chirping rate of crickets of a certain species appears to be related to temperature. The table shows the chirping rates for various temperatures.

Temperature (°F)	Chirping rate (chirps/min)
50	20
55	46
60	79
65	91
70	113
75	140
80	173
85	198
90	211

(a) Make a scatter plot of the data.
(b) Find and graph the regression line.
(c) Use the linear model in part (b) to estimate the chirping rate at 100°F.

19. The table gives the winning heights for the Olympic pole vault competitions in the 20th century.

Year	Height (ft)	Year	Height (ft)
1900	10.83	1956	14.96
1904	11.48	1960	15.42
1908	12.17	1964	16.73
1912	12.96	1968	17.71
1920	13.42	1972	18.04
1924	12.96	1976	18.04
1928	13.77	1980	18.96
1932	14.15	1984	18.85
1936	14.27	1988	19.77
1948	14.10	1992	19.02
1952	14.92	1996	19.42

(a) Make a scatter plot and decide whether a linear model is appropriate.
(b) Find and graph the regression line.
(c) Use the linear model to predict the height of the winning pole vault at the 2000 Olympics and compare with the winning height of 19.36 feet.
(d) Is it reasonable to use the model to predict the winning height at the 2100 Olympics?

20. A study by the U. S. Office of Science and Technology in 1972 estimated the cost (in 1972 dollars) to reduce automobile emissions by certain percentages:

Reduction in emissions (%)	Cost per car (in \$)	Reduction in emissions (%)	Cost per car (in \$)
50	45	75	90
55	55	80	100
60	62	85	200
65	70	90	375
70	80	95	600

Find a model that captures the “diminishing returns” trend of these data.

21. Use the data in the table to model the population of the world in the 20th century by a cubic function. Then use your model to estimate the population in the year 1925.

Year	Population (millions)
1900	1650
1910	1750
1920	1860
1930	2070
1940	2300
1950	2560
1960	3040
1970	3710
1980	4450
1990	5280
2000	6080

22. The table shows the mean (average) distances d of the planets from the Sun (taking the unit of measurement to be the distance from Earth to the Sun) and their periods T (time of revolution in years).

Planet	d	T
Mercury	0.387	0.241
Venus	0.723	0.615
Earth	1.000	1.000
Mars	1.523	1.881
Jupiter	5.203	11.861
Saturn	9.541	29.457
Uranus	19.190	84.008
Neptune	30.086	164.784
Pluto	39.507	248.350

(a) Fit a power model to the data.

(b) Kepler's Third Law of Planetary Motion states that "The square of the period of revolution of a planet is proportional to the cube of its mean distance from the Sun." Does your model corroborate Kepler's Third Law?

1.3 New Functions from Old Functions

In this section we start with the basic functions we discussed in Section 1.2 and obtain new functions by shifting, stretching, and reflecting their graphs. We also show how to combine pairs of functions by the standard arithmetic operations and by composition.

Transformations of Functions

By applying certain transformations to the graph of a given function we can obtain the graphs of certain related functions. This will give us the ability to sketch the graphs of many functions quickly by hand. It will also enable us to write equations for given graphs. Let's first consider **translations**. If c is a positive number, then the graph of $y = f(x) + c$ is just the graph of $y = f(x)$ shifted upward a distance of c units (because each y-coordinate is increased by the same number c). Likewise, if $g(x) = f(x - c)$, where $c > 0$, then the value of g at x is the same as the value of f at $x - c$ (c units to the left of x). Therefore, the graph of $y = f(x - c)$ is just the graph of $y = f(x)$ shifted c units to the right (see Figure 1).

Vertical and Horizontal Shifts Suppose $c > 0$. To obtain the graph of

$y = f(x) + c$, shift the graph of $y = f(x)$ a distance c units upward

$y = f(x) - c$, shift the graph of $y = f(x)$ a distance c units downward

$y = f(x - c)$, shift the graph of $y = f(x)$ a distance c units to the right

$y = f(x + c)$, shift the graph of $y = f(x)$ a distance c units to the left

Now let's consider the **stretching** and **reflecting** transformations. If $c > 1$, then the graph of $y = cf(x)$ is the graph of $y = f(x)$ stretched by a factor of c in the vertical direction (because each y-coordinate is multiplied by the same number c). The graph of

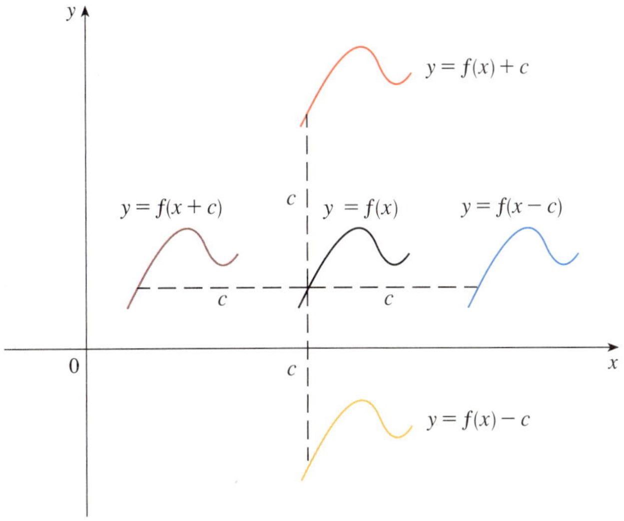

FIGURE 1
Translating the graph of f

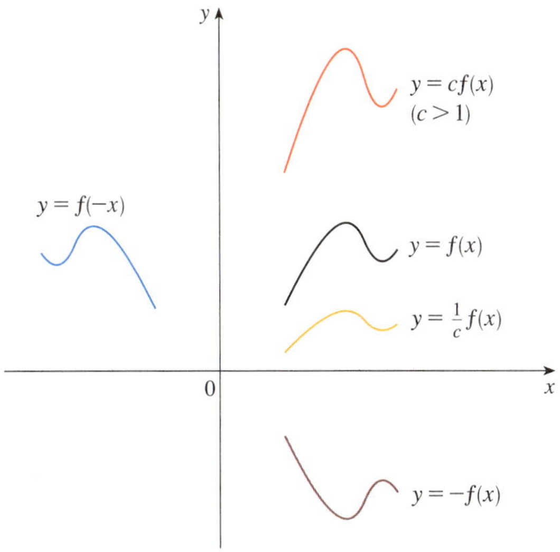

FIGURE 2
Stretching and reflecting the graph of f

$y = -f(x)$ is the graph of $y = f(x)$ reflected about the x-axis because the point (x, y) is replaced by the point $(x, -y)$. (See Figure 2 and the following chart, where the results of other stretching, compressing, and reflecting transformations are also given.)

In Module 1.3 you can see the effect of combining the transformations of this section.

Vertical and Horizontal Stretching and Reflecting Suppose $c > 1$. To obtain the graph of

- $y = cf(x)$, stretch the graph of $y = f(x)$ vertically by a factor of c
- $y = (1/c)f(x)$, compress the graph of $y = f(x)$ vertically by a factor of c
- $y = f(cx)$, compress the graph of $y = f(x)$ horizontally by a factor of c
- $y = f(x/c)$, stretch the graph of $y = f(x)$ horizontally by a factor of c
- $y = -f(x)$, reflect the graph of $y = f(x)$ about the x-axis
- $y = f(-x)$, reflect the graph of $y = f(x)$ about the y-axis

Figure 3 illustrates these stretching transformations when applied to the cosine function with $c = 2$. For instance, in order to get the graph of $y = 2\cos x$ we multiply the y-coordinate of each point on the graph of $y = \cos x$ by 2. This means that the graph of $y = \cos x$ gets stretched vertically by a factor of 2.

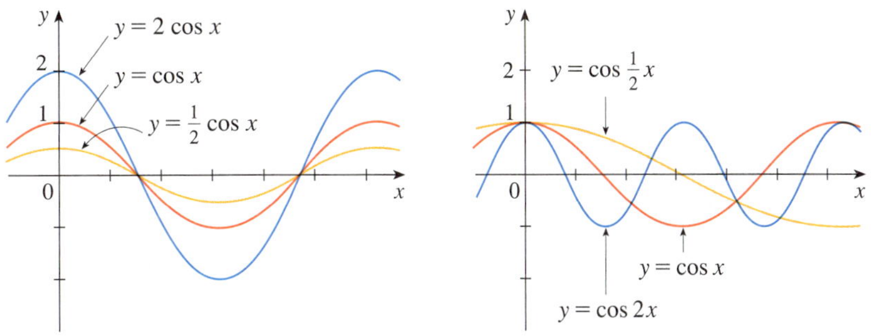

FIGURE 3

EXAMPLE 1 Given the graph of $y = \sqrt{x}$, use transformations to graph $y = \sqrt{x} - 2$, $y = \sqrt{x - 2}$, $y = -\sqrt{x}$, $y = 2\sqrt{x}$, and $y = \sqrt{-x}$.

SOLUTION The graph of the square root function $y = \sqrt{x}$, obtained from Figure 13 in Section 1.2, is shown in Figure 4(a). In the other parts of the figure we sketch $y = \sqrt{x} - 2$ by shifting 2 units downward, $y = \sqrt{x - 2}$ by shifting 2 units to the right, $y = -\sqrt{x}$ by reflecting about the x-axis, $y = 2\sqrt{x}$ by stretching vertically by a factor of 2, and $y = \sqrt{-x}$ by reflecting about the y-axis.

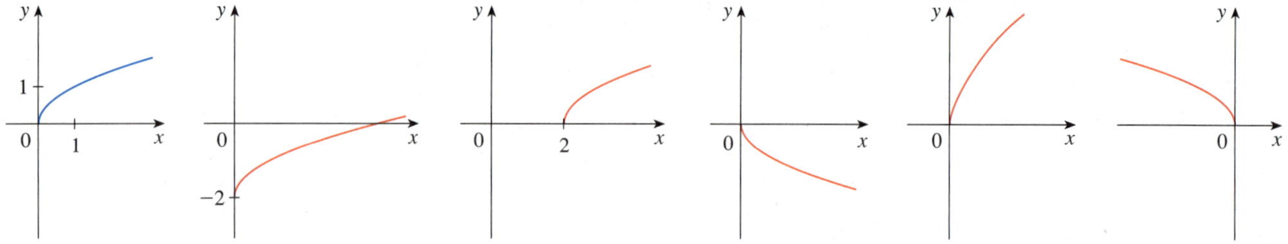

(a) $y = \sqrt{x}$ (b) $y = \sqrt{x} - 2$ (c) $y = \sqrt{x - 2}$ (d) $y = -\sqrt{x}$ (e) $y = 2\sqrt{x}$ (f) $y = \sqrt{-x}$

FIGURE 4

EXAMPLE 2 Sketch the graph of the function $f(x) = x^2 + 6x + 10$.

SOLUTION Completing the square, we write the equation of the graph as

$$y = x^2 + 6x + 10 = (x + 3)^2 + 1$$

This means we obtain the desired graph by starting with the parabola $y = x^2$ and shifting 3 units to the left and then 1 unit upward (see Figure 5).

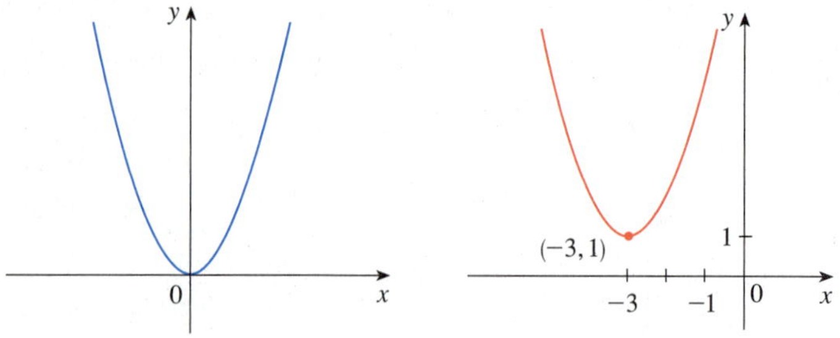

FIGURE 5 (a) $y = x^2$ (b) $y = (x + 3)^2 + 1$

EXAMPLE 3 Sketch the graphs of the following functions.

(a) $y = \sin 2x$ (b) $y = 1 - \sin x$

SOLUTION

(a) We obtain the graph of $y = \sin 2x$ from that of $y = \sin x$ by compressing horizontally by a factor of 2 (see Figures 6 and 7). Thus, whereas the period of $y = \sin x$ is 2π, the period of $y = \sin 2x$ is $2\pi/2 = \pi$.

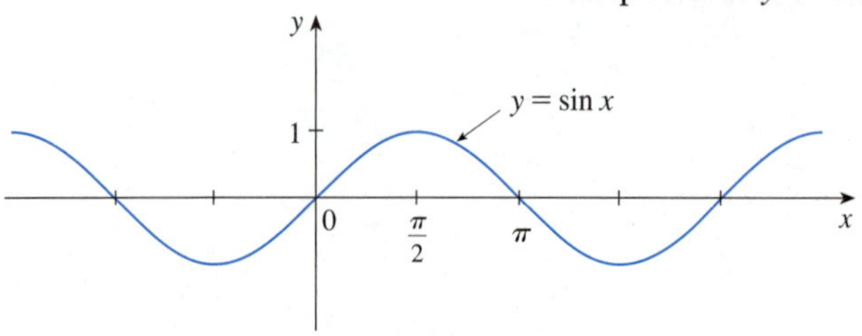

FIGURE 6

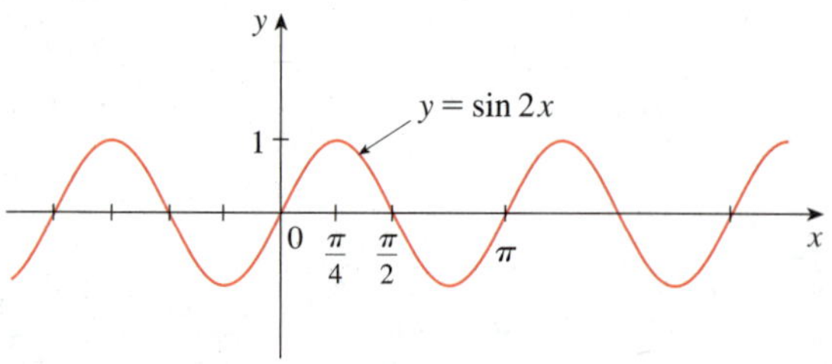

FIGURE 7

(b) To obtain the graph of $y = 1 - \sin x$, we again start with $y = \sin x$. We reflect about the x-axis to get the graph of $y = -\sin x$ and then we shift 1 unit upward to get $y = 1 - \sin x$. (See Figure 8.)

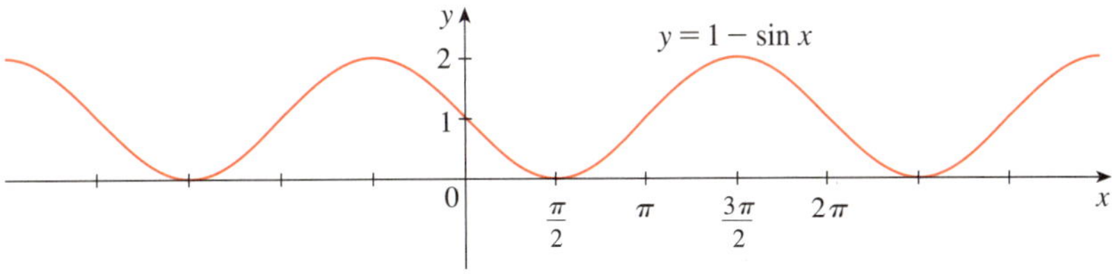

FIGURE 8

EXAMPLE 4 Figure 9 shows graphs of the number of hours of daylight as functions of the time of the year at several latitudes. Given that Philadelphia is located at approximately 40°N latitude, find a function that models the length of daylight at Philadelphia.

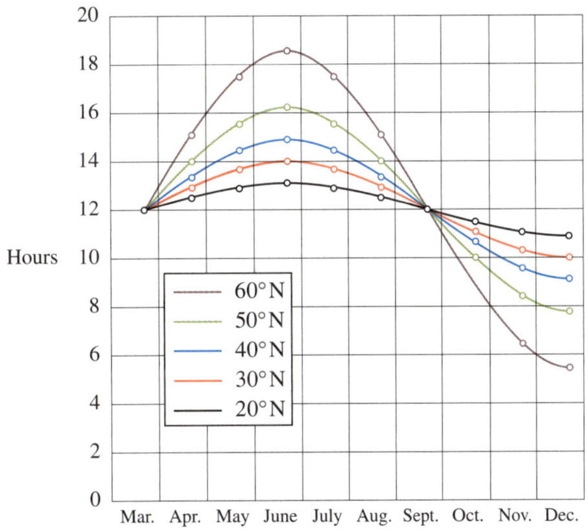

FIGURE 9

Graph of the length of daylight from March 21 through December 21 at various latitudes

Source: Lucia C. Harrison, *Daylight, Twilight, Darkness and Time* (New York: Silver, Burdett, 1935) page 40.

SOLUTION Notice that each curve resembles a shifted and stretched sine function. By looking at the blue curve we see that, at the latitude of Philadelphia, daylight lasts about 14.8 hours on June 21 and 9.2 hours on December 21, so the amplitude of the curve (the factor by which we have to stretch the sine curve vertically) is $\frac{1}{2}(14.8 - 9.2) = 2.8$.

By what factor do we need to stretch the sine curve horizontally if we measure the time t in days? Because there are about 365 days in a year, the period of our model should be 365. But the period of $y = \sin t$ is 2π, so the horizontal stretching factor is $c = 2\pi/365$.

We also notice that the curve begins its cycle on March 21, the 80th day of the year, so we have to shift the curve 80 units to the right. In addition, we shift it 12 units upward. Therefore, we model the length of daylight in Philadelphia on the tth day of the year by the function

$$L(t) = 12 + 2.8 \sin\left[\frac{2\pi}{365}(t - 80)\right]$$

Another transformation of some interest is taking the absolute value of a function. If $y = |f(x)|$, then according to the definition of absolute value, $y = f(x)$ when $f(x) \geqslant 0$ and $y = -f(x)$ when $f(x) < 0$. This tells us how to get the graph of $y = |f(x)|$ from the graph of $y = f(x)$: The part of the graph that lies above the x-axis remains the same; the part that lies below the x-axis is reflected about the x-axis.

EXAMPLE 5 Sketch the graph of the function $y = |x^2 - 1|$.

SOLUTION We first graph the parabola $y = x^2 - 1$ in Figure 10(a) by shifting the parabola $y = x^2$ downward 1 unit. We see that the graph lies below the x-axis when $-1 < x < 1$, so we reflect that part of the graph about the x-axis to obtain the graph of $y = |x^2 - 1|$ in Figure 10(b).

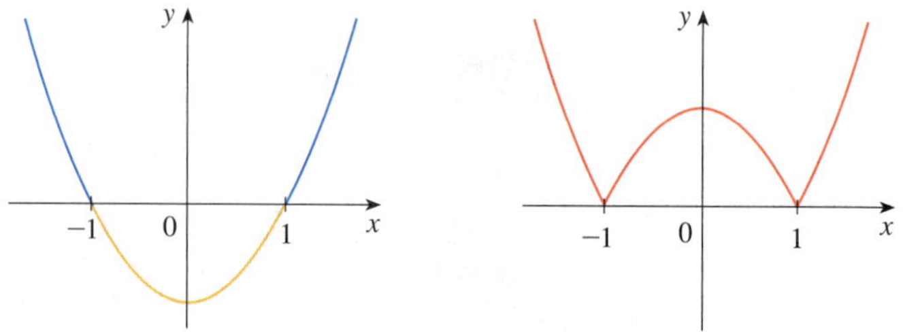

FIGURE 10 (a) $y = x^2 - 1$ (b) $y = |x^2 - 1|$

Combinations of Functions

Two functions f and g can be combined to form new functions $f + g$, $f - g$, fg, and f/g in a manner similar to the way we add, subtract, multiply, and divide real numbers.

If we define the sum $f + g$ by the equation

1 $$(f + g)(x) = f(x) + g(x)$$

then the right side of Equation 1 makes sense if both $f(x)$ and $g(x)$ are defined, that is, if x belongs to the domain of f and also to the domain of g. If the domain of f is A and the domain of g is B, then the domain of $f + g$ is the intersection of these domains, that is, $A \cap B$.

Notice that the $+$ sign on the left side of Equation 1 stands for the operation of addition of *functions,* but the $+$ sign on the right side of the equation stands for addition of the *numbers* $f(x)$ and $g(x)$.

Similarly, we can define the difference $f - g$ and the product fg, and their domains are also $A \cap B$. But in defining the quotient f/g we must remember not to divide by 0.

Algebra of Functions Let f and g be functions with domains A and B. Then the functions $f + g$, $f - g$, fg, and f/g are defined as follows:

$$(f + g)(x) = f(x) + g(x) \qquad \text{domain} = A \cap B$$

$$(f - g)(x) = f(x) - g(x) \qquad \text{domain} = A \cap B$$

$$(fg)(x) = f(x)g(x) \qquad \text{domain} = A \cap B$$

$$\left(\frac{f}{g}\right)(x) = \frac{f(x)}{g(x)} \qquad \text{domain} = \{x \in A \cap B \mid g(x) \neq 0\}$$

EXAMPLE 6 If $f(x) = \sqrt{x}$ and $g(x) = \sqrt{4 - x^2}$, find the functions $f + g$, $f - g$, fg, and f/g.

|||| Another way to solve $4 - x^2 \geqslant 0$:

$$(2 - x)(2 + x) \geqslant 0$$

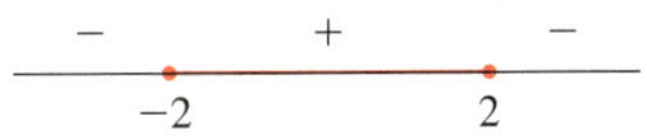

SOLUTION The domain of $f(x) = \sqrt{x}$ is $[0, \infty)$. The domain of $g(x) = \sqrt{4 - x^2}$ consists of all numbers x such that $4 - x^2 \geqslant 0$, that is, $x^2 \leqslant 4$. Taking square roots of both sides, we get $|x| \leqslant 2$, or $-2 \leqslant x \leqslant 2$, so the domain of g is the interval $[-2, 2]$. The intersection of the domains of f and g is

$$[0, \infty) \cap [-2, 2] = [0, 2]$$

Thus, according to the definitions, we have

$$(f + g)(x) = \sqrt{x} + \sqrt{4 - x^2} \qquad 0 \leqslant x \leqslant 2$$

$$(f - g)(x) = \sqrt{x} - \sqrt{4 - x^2} \qquad 0 \leqslant x \leqslant 2$$

$$(fg)(x) = \sqrt{x}\sqrt{4 - x^2} = \sqrt{4x - x^3} \qquad 0 \leqslant x \leqslant 2$$

$$\left(\frac{f}{g}\right)(x) = \frac{\sqrt{x}}{\sqrt{4 - x^2}} = \sqrt{\frac{x}{4 - x^2}} \qquad 0 \leqslant x < 2$$

Notice that the domain of f/g is the interval $[0, 2)$; we have to exclude $x = 2$ because $g(2) = 0$.

The graph of the function $f + g$ is obtained from the graphs of f and g by **graphical addition**. This means that we add corresponding y-coordinates as in Figure 11. Figure 12 shows the result of using this procedure to graph the function $f + g$ from Example 6.

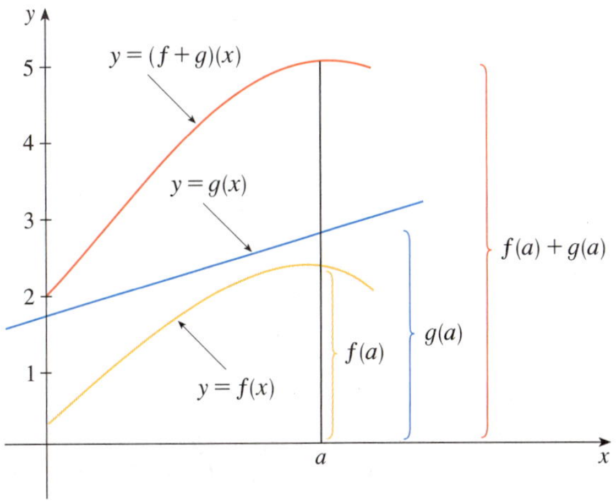

FIGURE 11

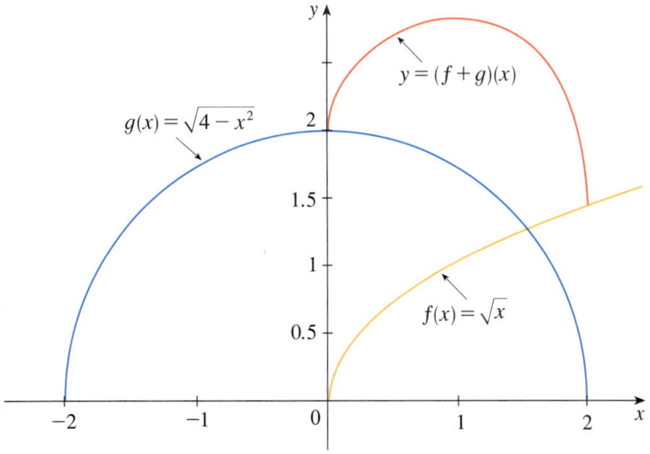

FIGURE 12

|||| Composition of Functions

There is another way of combining two functions to get a new function. For example, suppose that $y = f(u) = \sqrt{u}$ and $u = g(x) = x^2 + 1$. Since y is a function of u and u is, in turn, a function of x, it follows that y is ultimately a function of x. We compute this by substitution:

$$y = f(u) = f(g(x)) = f(x^2 + 1) = \sqrt{x^2 + 1}$$

The procedure is called *composition* because the new function is *composed* of the two given functions f and g.

In general, given any two functions f and g, we start with a number x in the domain of g and find its image $g(x)$. If this number $g(x)$ is in the domain of f, then we can calculate the value of $f(g(x))$. The result is a new function $h(x) = f(g(x))$ obtained by substituting g into f. It is called the *composition* (or *composite*) of f and g and is denoted by $f \circ g$ ("f circle g").

Definition Given two functions f and g, the **composite function** $f \circ g$ (also called the **composition** of f and g) is defined by

$$(f \circ g)(x) = f(g(x))$$

The domain of $f \circ g$ is the set of all x in the domain of g such that $g(x)$ is in the domain of f. In other words, $(f \circ g)(x)$ is defined whenever both $g(x)$ and $f(g(x))$ are defined. The best way to picture $f \circ g$ is by either a machine diagram (Figure 13) or an arrow diagram (Figure 14).

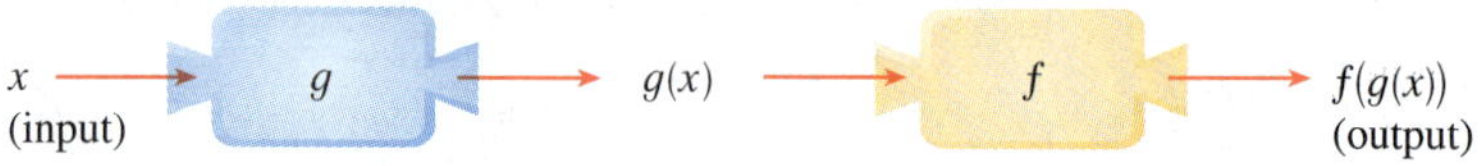

FIGURE 13
The $f \circ g$ machine is composed of the g machine (first) and then the f machine.

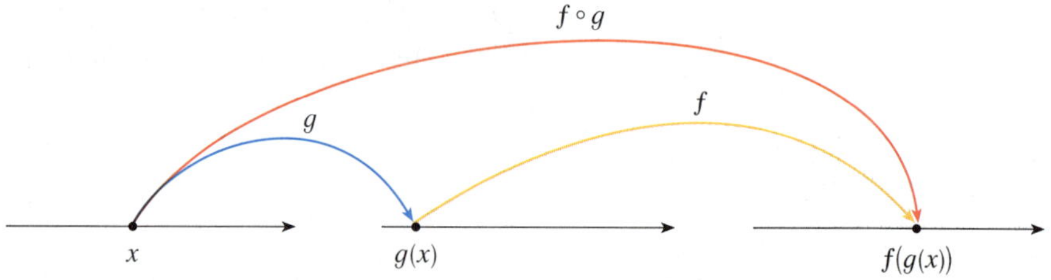

FIGURE 14
Arrow diagram for $f \circ g$

EXAMPLE 7 If $f(x) = x^2$ and $g(x) = x - 3$, find the composite functions $f \circ g$ and $g \circ f$.

SOLUTION We have

$$(f \circ g)(x) = f(g(x)) = f(x - 3) = (x - 3)^2$$

$$(g \circ f)(x) = g(f(x)) = g(x^2) = x^2 - 3$$

NOTE ▫ You can see from Example 7 that, in general, $f \circ g \neq g \circ f$. Remember, the notation $f \circ g$ means that the function g is applied first and then f is applied second. In Example 7, $f \circ g$ is the function that *first* subtracts 3 and *then* squares; $g \circ f$ is the function that *first* squares and *then* subtracts 3.

EXAMPLE 8 If $f(x) = \sqrt{x}$ and $g(x) = \sqrt{2 - x}$, find each function and its domain.
(a) $f \circ g$ (b) $g \circ f$ (c) $f \circ f$ (d) $g \circ g$

SOLUTION

(a) $$(f \circ g)(x) = f(g(x)) = f\left(\sqrt{2 - x}\right) = \sqrt{\sqrt{2 - x}} = \sqrt[4]{2 - x}$$

The domain of $f \circ g$ is $\{x \mid 2 - x \geq 0\} = \{x \mid x \leq 2\} = (-\infty, 2]$.

(b) $$(g \circ f)(x) = g(f(x)) = g(\sqrt{x}) = \sqrt{2 - \sqrt{x}}$$

If $0 \leqslant a \leqslant b$, then $a^2 \leqslant b^2$.

For $\sqrt{x}$ to be defined we must have $x \geqslant 0$. For $\sqrt{2 - \sqrt{x}}$ to be defined we must have $2 - \sqrt{x} \geqslant 0$, that is, $\sqrt{x} \leqslant 2$, or $x \leqslant 4$. Thus, we have $0 \leqslant x \leqslant 4$, so the domain of $g \circ f$ is the closed interval $[0, 4]$.

(c) $$(f \circ f)(x) = f(f(x)) = f(\sqrt{x}) = \sqrt{\sqrt{x}} = \sqrt[4]{x}$$

The domain of $f \circ f$ is $[0, \infty)$.

(d) $$(g \circ g)(x) = g(g(x)) = g(\sqrt{2 - x}) = \sqrt{2 - \sqrt{2 - x}}$$

This expression is defined when $2 - x \geqslant 0$, that is, $x \leqslant 2$, and $2 - \sqrt{2 - x} \geqslant 0$. This latter inequality is equivalent to $\sqrt{2 - x} \leqslant 2$, or $2 - x \leqslant 4$, that is, $x \geqslant -2$. Thus, $-2 \leqslant x \leqslant 2$, so the domain of $g \circ g$ is the closed interval $[-2, 2]$.

It is possible to take the composition of three or more functions. For instance, the composite function $f \circ g \circ h$ is found by first applying h, then g, and then f as follows:

$$(f \circ g \circ h)(x) = f(g(h(x)))$$

EXAMPLE 9 Find $f \circ g \circ h$ if $f(x) = x/(x + 1)$, $g(x) = x^{10}$, and $h(x) = x + 3$.

SOLUTION

$$(f \circ g \circ h)(x) = f(g(h(x))) = f(g(x + 3))$$
$$= f((x + 3)^{10}) = \frac{(x + 3)^{10}}{(x + 3)^{10} + 1}$$

So far we have used composition to build complicated functions from simpler ones. But in calculus it is often useful to be able to decompose a complicated function into simpler ones, as in the following example.

EXAMPLE 10 Given $F(x) = \cos^2(x + 9)$, find functions f, g, and h such that $F = f \circ g \circ h$.

SOLUTION Since $F(x) = [\cos(x + 9)]^2$, the formula for F says: First add 9, then take the cosine of the result, and finally square. So we let

$$h(x) = x + 9 \qquad g(x) = \cos x \qquad f(x) = x^2$$

Then

$$(f \circ g \circ h)(x) = f(g(h(x))) = f(g(x + 9)) = f(\cos(x + 9))$$
$$= [\cos(x + 9)]^2 = F(x)$$

1.3 Exercises

1. Suppose the graph of f is given. Write equations for the graphs that are obtained from the graph of f as follows.
   (a) Shift 3 units upward.
   (b) Shift 3 units downward.
   (c) Shift 3 units to the right.
   (d) Shift 3 units to the left.
   (e) Reflect about the x-axis.
   (f) Reflect about the y-axis.
   (g) Stretch vertically by a factor of 3.
   (h) Shrink vertically by a factor of 3.

2. Explain how the following graphs are obtained from the graph of $y = f(x)$.

(a) $y = 5f(x)$ (b) $y = f(x - 5)$
(c) $y = -f(x)$ (d) $y = -5f(x)$
(e) $y = f(5x)$ (f) $y = 5f(x) - 3$

3. The graph of $y = f(x)$ is given. Match each equation with its graph and give reasons for your choices.

(a) $y = f(x - 4)$ (b) $y = f(x) + 3$
(c) $y = \frac{1}{3}f(x)$ (d) $y = -f(x + 4)$
(e) $y = 2f(x + 6)$

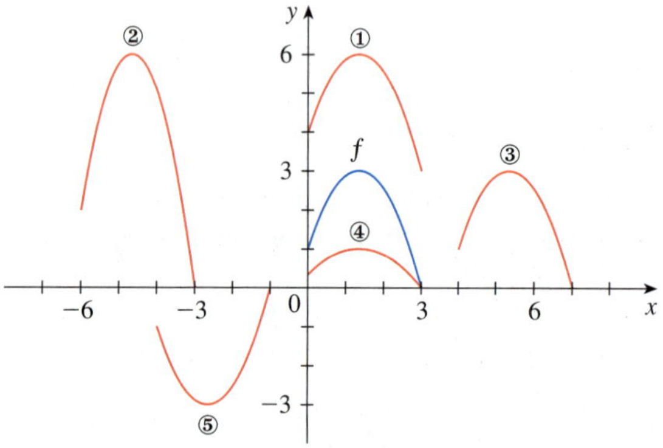

4. The graph of f is given. Draw the graphs of the following functions.

(a) $y = f(x + 4)$ (b) $y = f(x) + 4$
(c) $y = 2f(x)$ (d) $y = -\frac{1}{2}f(x) + 3$

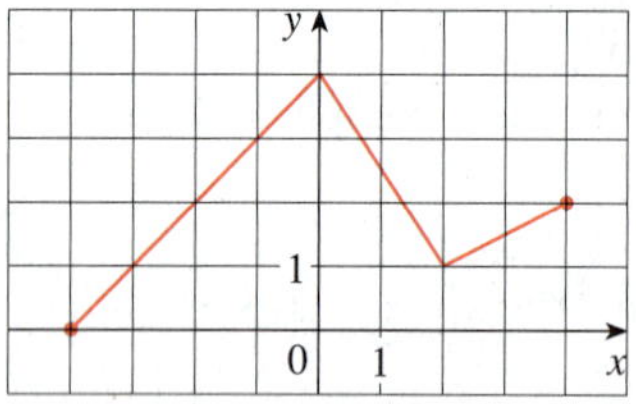

5. The graph of f is given. Use it to graph the following functions.

(a) $y = f(2x)$ (b) $y = f(\frac{1}{2}x)$
(c) $y = f(-x)$ (d) $y = -f(-x)$

6–7 |||| The graph of $y = \sqrt{3x - x^2}$ is given. Use transformations to create a function whose graph is as shown.

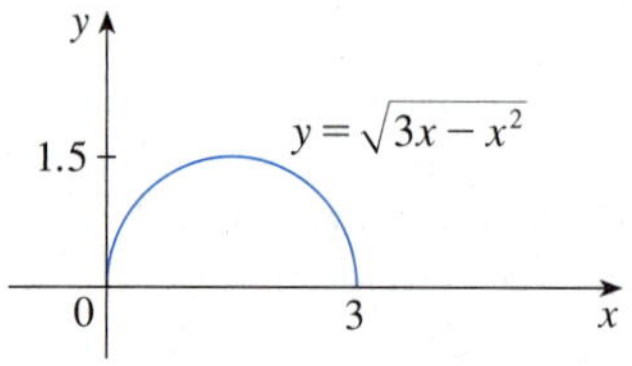

6.

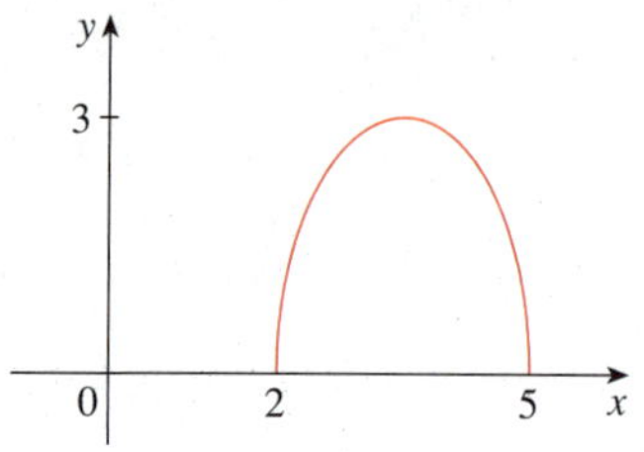

7.

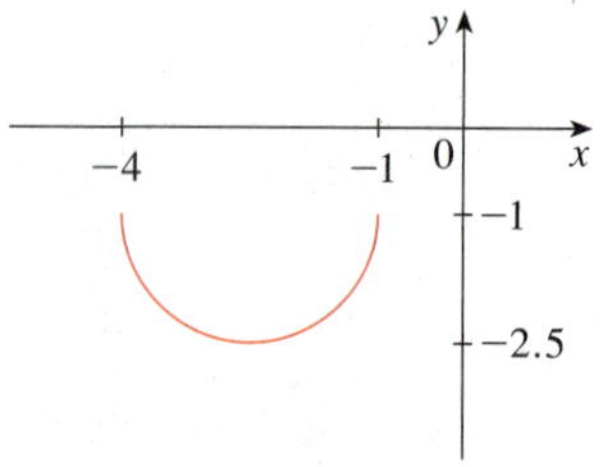

8. (a) How is the graph of $y = 2 \sin x$ related to the graph of $y = \sin x$? Use your answer and Figure 6 to sketch the graph of $y = 2 \sin x$.

(b) How is the graph of $y = 1 + \sqrt{x}$ related to the graph of $y = \sqrt{x}$? Use your answer and Figure 4(a) to sketch the graph of $y = 1 + \sqrt{x}$.

9–24 |||| Graph the function, not by plotting points, but by starting with the graph of one of the standard functions given in Section 1.2, and then applying the appropriate transformations.

9. $y = -x^3$
10. $y = 1 - x^2$
11. $y = (x + 1)^2$
12. $y = x^2 - 4x + 3$
13. $y = 1 + 2 \cos x$
14. $y = 4 \sin 3x$
15. $y = \sin(x/2)$
16. $y = \dfrac{1}{x - 4}$
17. $y = \sqrt{x + 3}$
18. $y = (x + 2)^4 + 3$
19. $y = \frac{1}{2}(x^2 + 8x)$
20. $y = 1 + \sqrt[3]{x - 1}$
21. $y = \dfrac{2}{x + 1}$
22. $y = \dfrac{1}{4} \tan\left(x - \dfrac{\pi}{4}\right)$
23. $y = |\sin x|$
24. $y = |x^2 - 2x|$

25. The city of New Orleans is located at latitude 30°N. Use Figure 9 to find a function that models the number of hours of daylight at New Orleans as a function of the time of year. Use the fact that on March 31 the Sun rises at 5:51 A.M. and sets at 6:18 P.M. in New Orleans to check the accuracy of your model.

26. A variable star is one whose brightness alternately increases and decreases. For the most visible variable star, Delta Cephei, the time between periods of maximum brightness is 5.4 days, the average brightness (or magnitude) of the star is 4.0, and its brightness varies by ±0.35 magnitude. Find a function that models the brightness of Delta Cephei as a function of time.

27. (a) How is the graph of $y = f(|x|)$ related to the graph of f?
(b) Sketch the graph of $y = \sin|x|$.
(c) Sketch the graph of $y = \sqrt{|x|}$.

28. Use the given graph of f to sketch the graph of $y = 1/f(x)$. Which features of f are the most important in sketching $y = 1/f(x)$? Explain how they are used.

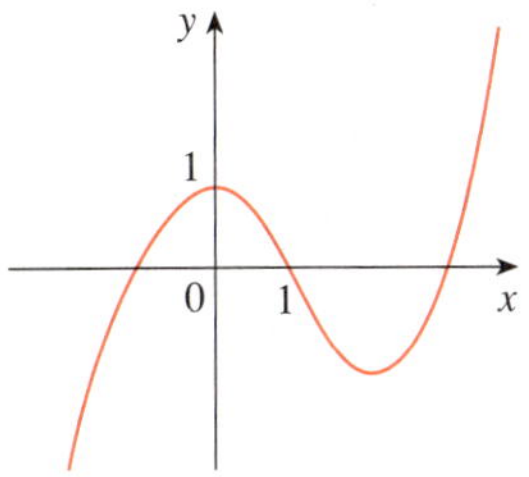

29–30 |||| Use graphical addition to sketch the graph of $f + g$.

29.

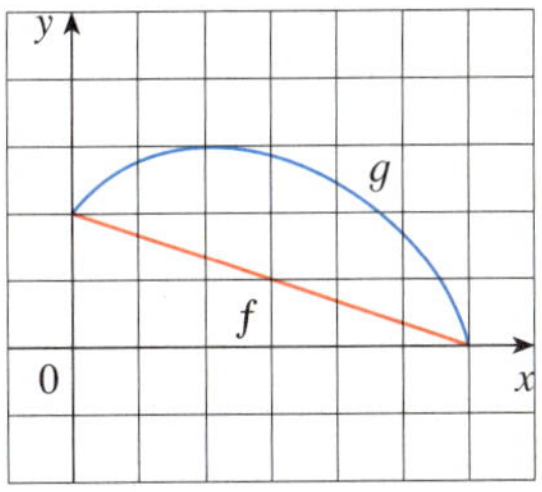

30.

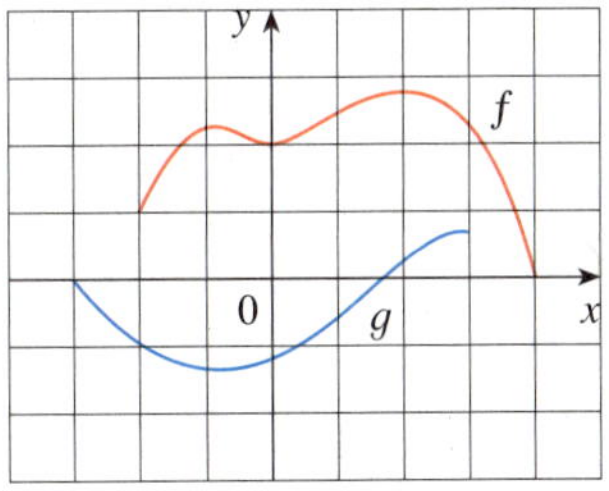

31–32 |||| Find $f + g$, $f - g$, fg, and f/g and state their domains.

31. $f(x) = x^3 + 2x^2, \quad g(x) = 3x^2 - 1$

32. $f(x) = \sqrt{1 + x}, \quad g(x) = \sqrt{1 - x}$

33–34 |||| Use the graphs of f and g and the method of graphical addition to sketch the graph of $f + g$.

33. $f(x) = x, \quad g(x) = 1/x$

34. $f(x) = x^3, \quad g(x) = -x^2$

35–40 |||| Find the functions $f \circ g$, $g \circ f$, $f \circ f$, and $g \circ g$ and their domains.

35. $f(x) = 2x^2 - x, \quad g(x) = 3x + 2$

36. $f(x) = 1 - x^3, \quad g(x) = 1/x$

37. $f(x) = \sin x, \quad g(x) = 1 - \sqrt{x}$

38. $f(x) = 1 - 3x, \quad g(x) = 5x^2 + 3x + 2$

39. $f(x) = x + \dfrac{1}{x}, \quad g(x) = \dfrac{x + 1}{x + 2}$

40. $f(x) = \sqrt{2x + 3}, \quad g(x) = x^2 + 1$

41–44 |||| Find $f \circ g \circ h$.

41. $f(x) = x + 1, \quad g(x) = 2x, \quad h(x) = x - 1$

42. $f(x) = 2x - 1, \quad g(x) = x^2, \quad h(x) = 1 - x$

43. $f(x) = \sqrt{x - 1}, \quad g(x) = x^2 + 2, \quad h(x) = x + 3$

44. $f(x) = \dfrac{2}{x + 1}, \quad g(x) = \cos x, \quad h(x) = \sqrt{x + 3}$

45–50 |||| Express the function in the form $f \circ g$.

45. $F(x) = (x^2 + 1)^{10}$

46. $F(x) = \sin(\sqrt{x})$

47. $G(x) = \dfrac{x^2}{x^2 + 4}$

48. $G(x) = \dfrac{1}{x + 3}$

49. $u(t) = \sqrt{\cos t}$

50. $u(t) = \dfrac{\tan t}{1 + \tan t}$

51–53 |||| Express the function in the form $f \circ g \circ h$.

51. $H(x) = 1 - 3^{x^2}$

52. $H(x) = \sqrt[3]{\sqrt{x} - 1}$

53. $H(x) = \sec^4(\sqrt{x})$

54. Use the table to evaluate each expression.
(a) $f(g(1))$ (b) $g(f(1))$ (c) $f(f(1))$
(d) $g(g(1))$ (e) $(g \circ f)(3)$ (f) $(f \circ g)(6)$

x	1	2	3	4	5	6
$f(x)$	3	1	4	2	2	5
$g(x)$	6	3	2	1	2	3

55. Use the given graphs of f and g to evaluate each expression, or explain why it is undefined.
(a) $f(g(2))$ (b) $g(f(0))$ (c) $(f \circ g)(0)$
(d) $(g \circ f)(6)$ (e) $(g \circ g)(-2)$ (f) $(f \circ f)(4)$

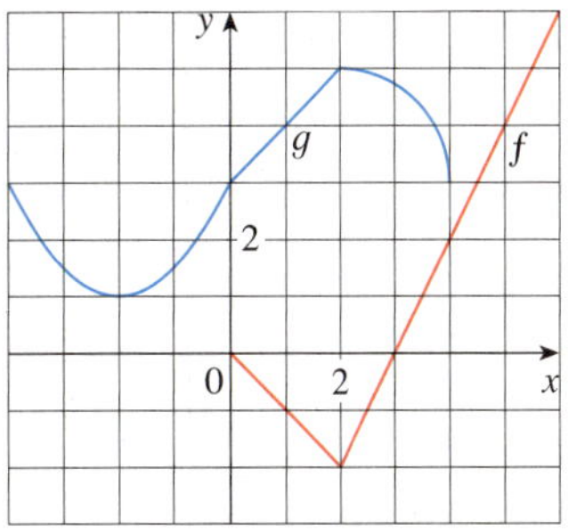

56. Use the given graphs of f and g to estimate the value of $f(g(x))$ for $x = -5, -4, -3, \ldots, 5$. Use these estimates to sketch a rough graph of $f \circ g$.

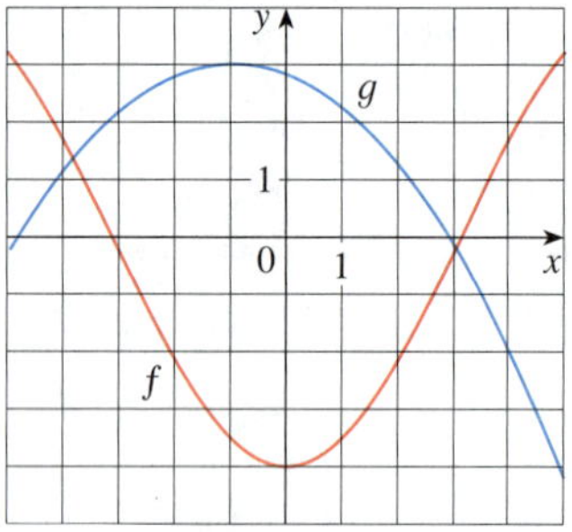

57. A stone is dropped into a lake, creating a circular ripple that travels outward at a speed of 60 cm/s.
(a) Express the radius r of this circle as a function of the time t (in seconds).
(b) If A is the area of this circle as a function of the radius, find $A \circ r$ and interpret it.

58. An airplane is flying at a speed of 350 mi/h at an altitude of one mile and passes directly over a radar station at time $t = 0$.
(a) Express the horizontal distance d (in miles) that the plane has flown as a function of t.
(b) Express the distance s between the plane and the radar station as a function of d.
(c) Use composition to express s as a function of t.

59. The **Heaviside function** H is defined by

$$H(t) = \begin{cases} 0 & \text{if } t < 0 \\ 1 & \text{if } t \geq 0 \end{cases}$$

It is used in the study of electric circuits to represent the sudden surge of electric current, or voltage, when a switch is instantaneously turned on.
(a) Sketch the graph of the Heaviside function.
(b) Sketch the graph of the voltage $V(t)$ in a circuit if the switch is turned on at time $t = 0$ and 120 volts are applied instantaneously to the circuit. Write a formula for $V(t)$ in terms of $H(t)$.
(c) Sketch the graph of the voltage $V(t)$ in a circuit if the switch is turned on at time $t = 5$ seconds and 240 volts are applied instantaneously to the circuit. Write a formula for $V(t)$ in terms of $H(t)$. (Note that starting at $t = 5$ corresponds to a translation.)

60. The Heaviside function defined in Exercise 59 can also be used to define the **ramp function** $y = ctH(t)$, which represents a gradual increase in voltage or current in a circuit.
(a) Sketch the graph of the ramp function $y = tH(t)$.
(b) Sketch the graph of the voltage $V(t)$ in a circuit if the switch is turned on at time $t = 0$ and the voltage is gradually increased to 120 volts over a 60-second time interval. Write a formula for $V(t)$ in terms of $H(t)$ for $t \leq 60$.
(c) Sketch the graph of the voltage $V(t)$ in a circuit if the switch is turned on at time $t = 7$ seconds and the voltage is gradually increased to 100 volts over a period of 25 seconds. Write a formula for $V(t)$ in terms of $H(t)$ for $t \leq 32$.

61. (a) If $g(x) = 2x + 1$ and $h(x) = 4x^2 + 4x + 7$, find a function f such that $f \circ g = h$. (Think about what operations you would have to perform on the formula for g to end up with the formula for h.)
(b) If $f(x) = 3x + 5$ and $h(x) = 3x^2 + 3x + 2$, find a function g such that $f \circ g = h$.

62. If $f(x) = x + 4$ and $h(x) = 4x - 1$, find a function g such that $g \circ f = h$.

63. Suppose g is an even function and let $h = f \circ g$. Is h always an even function?

64. Suppose g is an odd function and let $h = f \circ g$. Is h always an odd function? What if f is odd? What if f is even?

1.4 Graphing Calculators and Computers

In this section we assume that you have access to a graphing calculator or a computer with graphing software. We will see that the use of such a device enables us to graph more complicated functions and to solve more complex problems than would otherwise be possible. We also point out some of the pitfalls that can occur with these machines.

Graphing calculators and computers can give very accurate graphs of functions. But we will see in Chapter 4 that only through the use of calculus can we be sure that we have uncovered all the interesting aspects of a graph.

A graphing calculator or computer displays a rectangular portion of the graph of a function in a **display window** or **viewing screen**, which we refer to as a **viewing rectangle**. The default screen often gives an incomplete or misleading picture, so it is important to choose the viewing rectangle with care. If we choose the x-values to range from a minimum value of $Xmin = a$ to a maximum value of $Xmax = b$ and the y-values to range from

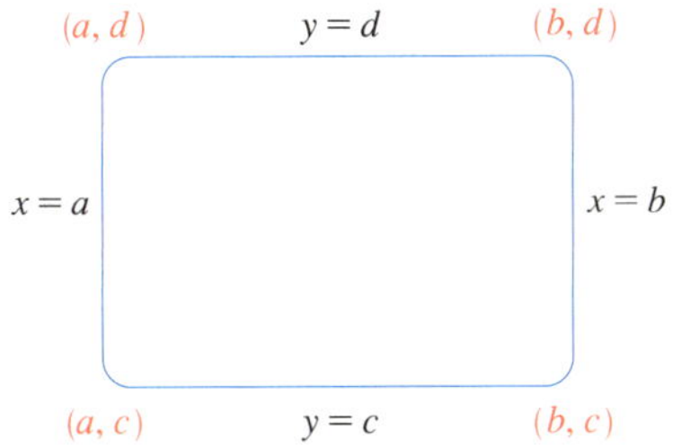

FIGURE 1
The viewing rectangle $[a, b]$ by $[c, d]$

a minimum of *Ymin* $= c$ to a maximum of *Ymax* $= d$, then the visible portion of the graph lies in the rectangle

$$[a, b] \times [c, d] = \{(x, y) \mid a \leqslant x \leqslant b, c \leqslant y \leqslant d\}$$

shown in Figure 1. We refer to this rectangle as the $[a, b]$ *by* $[c, d]$ *viewing rectangle.*

The machine draws the graph of a function f much as you would. It plots points of the form $(x, f(x))$ for a certain number of equally spaced values of x between a and b. If an x-value is not in the domain of f, or if $f(x)$ lies outside the viewing rectangle, it moves on to the next x-value. The machine connects each point to the preceding plotted point to form a representation of the graph of f.

EXAMPLE 1 Draw the graph of the function $f(x) = x^2 + 3$ in each of the following viewing rectangles.

(a) $[-2, 2]$ by $[-2, 2]$ (b) $[-4, 4]$ by $[-4, 4]$
(c) $[-10, 10]$ by $[-5, 30]$ (d) $[-50, 50]$ by $[-100, 1000]$

SOLUTION For part (a) we select the range by setting *Xmin* $= -2$, *Xmax* $= 2$, *Ymin* $= -2$, and *Ymax* $= 2$. The resulting graph is shown in Figure 2(a). The display window is blank! A moment's thought provides the explanation: Notice that $x^2 \geqslant 0$ for all x, so $x^2 + 3 \geqslant 3$ for all x. Thus, the range of the function $f(x) = x^2 + 3$ is $[3, \infty)$. This means that the graph of f lies entirely outside the viewing rectangle $[-2, 2]$ by $[-2, 2]$.

The graphs for the viewing rectangles in parts (b), (c), and (d) are also shown in Figure 2. Observe that we get a more complete picture in parts (c) and (d), but in part (d) it is not clear that the y-intercept is 3.

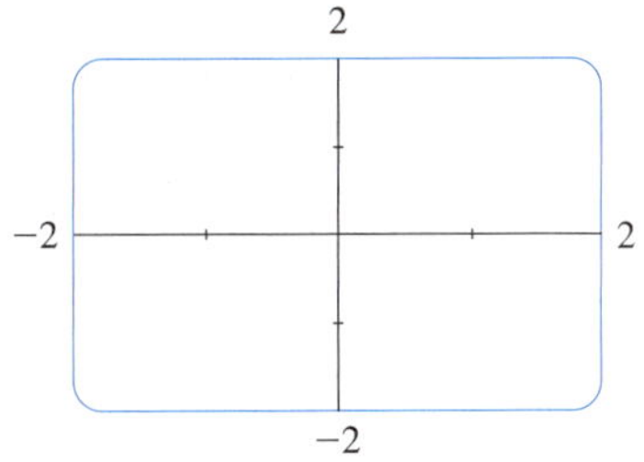

(a) $[-2, 2]$ by $[-2, 2]$

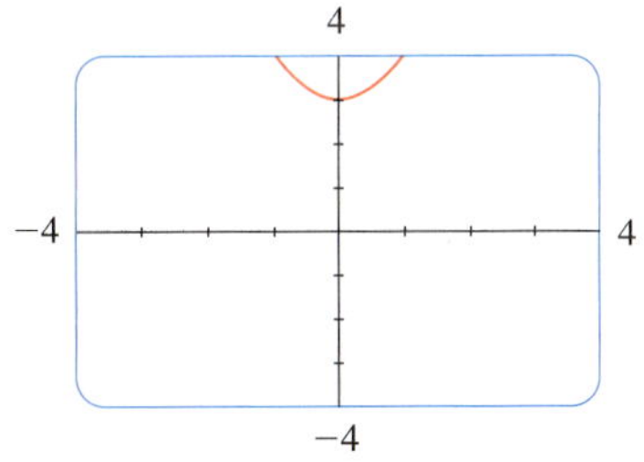

(b) $[-4, 4]$ by $[-4, 4]$

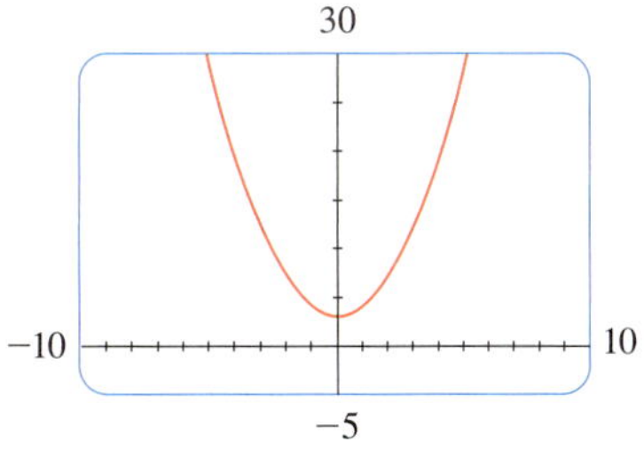

(c) $[-10, 10]$ by $[-5, 30]$

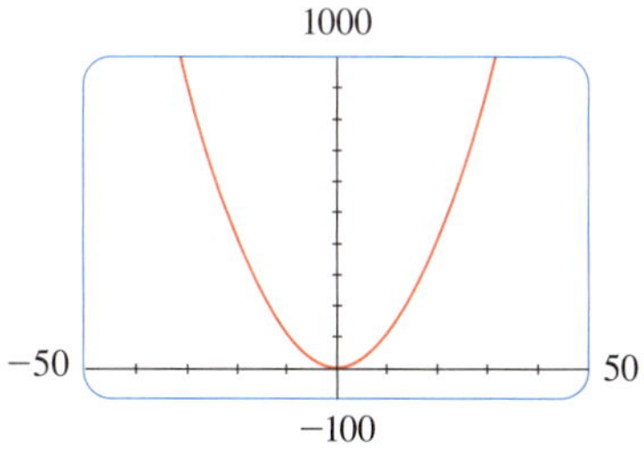

(d) $[-50, 50]$ by $[-100, 1000]$

FIGURE 2 Graphs of $f(x) = x^2 + 3$

We see from Example 1 that the choice of a viewing rectangle can make a big difference in the appearance of a graph. Sometimes it's necessary to change to a larger viewing rectangle to obtain a more complete picture, a more global view, of the graph. In the next example we see that knowledge of the domain and range of a function sometimes provides us with enough information to select a good viewing rectangle.

EXAMPLE 2 Determine an appropriate viewing rectangle for the function $f(x) = \sqrt{8 - 2x^2}$ and use it to graph f.

SOLUTION The expression for $f(x)$ is defined when

$$8 - 2x^2 \geqslant 0 \iff 2x^2 \leqslant 8 \iff x^2 \leqslant 4$$
$$\iff |x| \leqslant 2 \iff -2 \leqslant x \leqslant 2$$

Therefore, the domain of f is the interval $[-2, 2]$. Also,

$$0 \leqslant \sqrt{8 - 2x^2} \leqslant \sqrt{8} = 2\sqrt{2} \approx 2.83$$

so the range of f is the interval $[0, 2\sqrt{2}]$.

We choose the viewing rectangle so that the x-interval is somewhat larger than the domain and the y-interval is larger than the range. Taking the viewing rectangle to be $[-3, 3]$ by $[-1, 4]$, we get the graph shown in Figure 3.

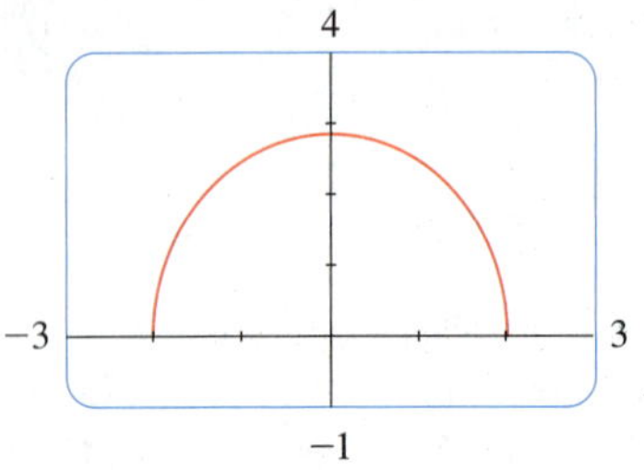

FIGURE 3

EXAMPLE 3 Graph the function $y = x^3 - 150x$.

SOLUTION Here the domain is $\mathbb{R}$, the set of all real numbers. That doesn't help us choose a viewing rectangle. Let's experiment. If we start with the viewing rectangle $[-5, 5]$ by $[-5, 5]$, we get the graph in Figure 4. It appears blank, but actually the graph is so nearly vertical that it blends in with the y-axis.

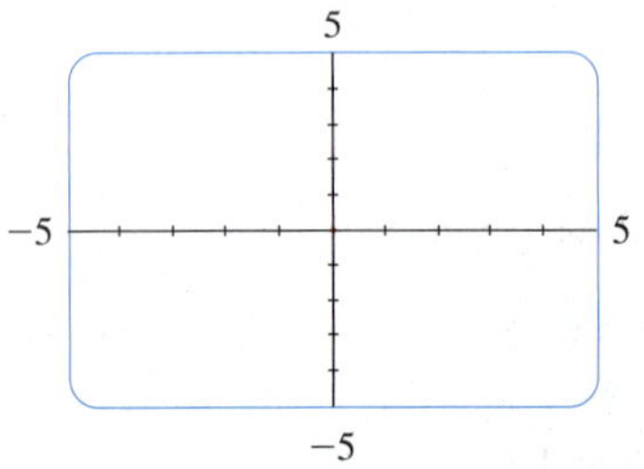

FIGURE 4

If we change the viewing rectangle to $[-20, 20]$ by $[-20, 20]$, we get the picture shown in Figure 5(a). The graph appears to consist of vertical lines, but we know that can't be correct. If we look carefully while the graph is being drawn, we see that the graph leaves the screen and reappears during the graphing process. This indicates that we need to see more in the vertical direction, so we change the viewing rectangle to $[-20, 20]$ by $[-500, 500]$. The resulting graph is shown in Figure 5(b). It still doesn't quite reveal all the main features of the function, so we try $[-20, 20]$ by $[-1000, 1000]$ in Figure 5(c). Now we are more confident that we have arrived at an appropriate viewing rectangle. In Chapter 4 we will be able to see that the graph shown in Figure 5(c) does indeed reveal all the main features of the function.

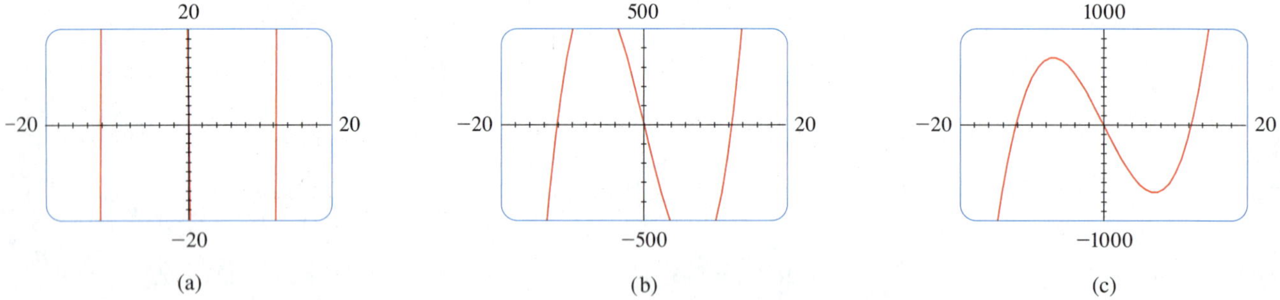

FIGURE 5 $f(x) = x^3 - 150x$

EXAMPLE 4 Graph the function $f(x) = \sin 50x$ in an appropriate viewing rectangle.

SOLUTION Figure 6(a) shows the graph of f produced by a graphing calculator using the viewing rectangle $[-12, 12]$ by $[-1.5, 1.5]$. At first glance the graph appears to be reasonable. But if we change the viewing rectangle to the ones shown in the following parts of Figure 6, the graphs look very different. Something strange is happening.

In order to explain the big differences in appearance of these graphs and to find an appropriate viewing rectangle, we need to find the period of the function $y = \sin 50x$. We know that the function $y = \sin x$ has period 2π and the graph of $y = \sin 50x$ is compressed horizontally by a factor of 50, so the period of $y = \sin 50x$ is

$$\frac{2\pi}{50} = \frac{\pi}{25} \approx 0.126$$

||| The appearance of the graphs in Figure 6 depends on the machine used. The graphs you get with your own graphing device might not look like these figures, but they will also be quite inaccurate.

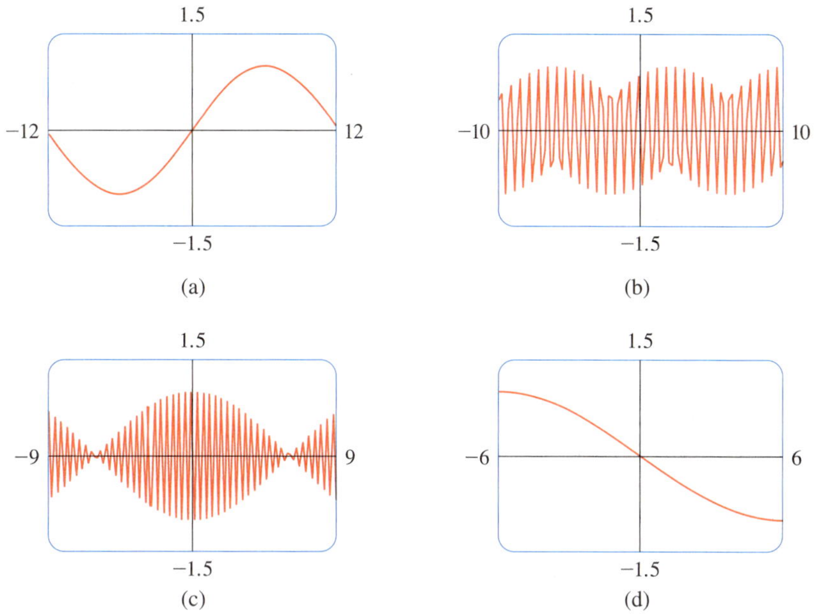

FIGURE 6
Graphs of $f(x) = \sin 50x$ in four viewing rectangles

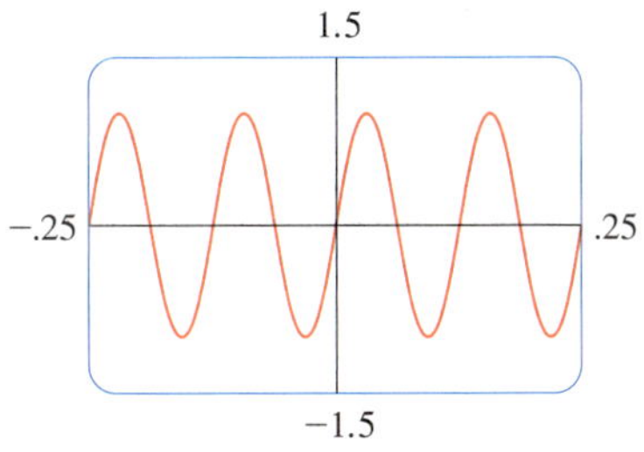

FIGURE 7
$f(x) = \sin 50x$

This suggests that we should deal only with small values of x in order to show just a few oscillations of the graph. If we choose the viewing rectangle $[-0.25, 0.25]$ by $[-1.5, 1.5]$, we get the graph shown in Figure 7.

Now we see what went wrong in Figure 6. The oscillations of $y = \sin 50x$ are so rapid that when the calculator plots points and joins them, it misses most of the maximum and minimum points and therefore gives a very misleading impression of the graph.

We have seen that the use of an inappropriate viewing rectangle can give a misleading impression of the graph of a function. In Examples 1 and 3 we solved the problem by changing to a larger viewing rectangle. In Example 4 we had to make the viewing rectangle smaller. In the next example we look at a function for which there is no single viewing rectangle that reveals the true shape of the graph.

EXAMPLE 5 Graph the function $f(x) = \sin x + \frac{1}{100} \cos 100x$.

SOLUTION Figure 8 shows the graph of f produced by a graphing calculator with viewing rectangle $[-6.5, 6.5]$ by $[-1.5, 1.5]$. It looks much like the graph of $y = \sin x$, but perhaps with some bumps attached. If we zoom in to the viewing rectangle $[-0.1, 0.1]$ by $[-0.1, 0.1]$, we can see much more clearly the shape of these bumps in Figure 9. The reason for this behavior is that the second term, $\frac{1}{100} \cos 100x$, is very small in comparison with the first term, $\sin x$. Thus, we really need two graphs to see the true nature of this function.

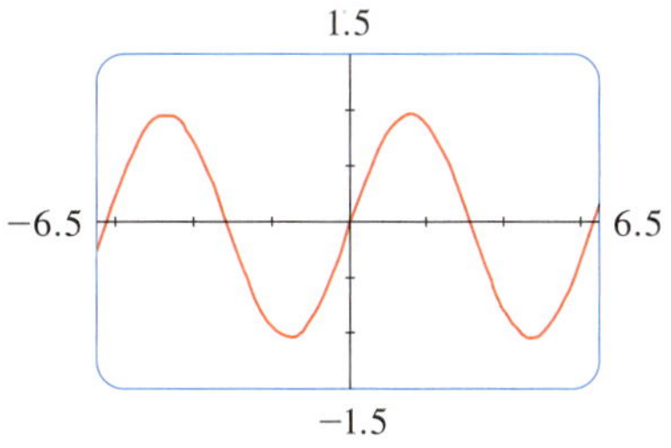

FIGURE 8

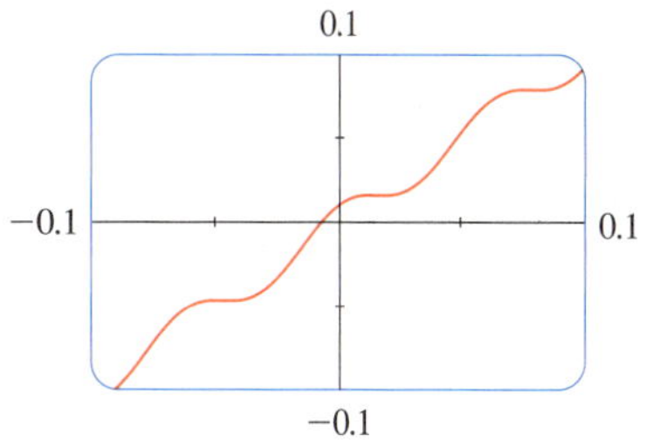

FIGURE 9

EXAMPLE 6 Draw the graph of the function $y = \dfrac{1}{1 - x}$.

SOLUTION Figure 10(a) shows the graph produced by a graphing calculator with viewing rectangle $[-9, 9]$ by $[-9, 9]$. In connecting successive points on the graph, the calculator produced a steep line segment from the top to the bottom of the screen. That line segment is not truly part of the graph. Notice that the domain of the function $y = 1/(1 - x)$ is $\{x \mid x \neq 1\}$. We can eliminate the extraneous near-vertical line by experimenting with a change of scale. When we change to the smaller viewing rectangle $[-4.7, 4.7]$ by $[-4.7, 4.7]$ on this particular calculator, we obtain the much better graph in Figure 10(b).

|||| Another way to avoid the extraneous line is to change the graphing mode on the calculator so that the dots are not connected. Alternatively, we could zoom in using the Zoom Decimal mode.

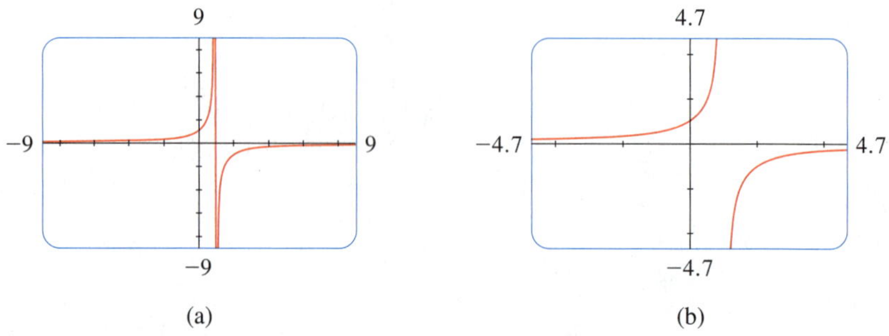

FIGURE 10

$y = \dfrac{1}{1 - x}$

EXAMPLE 7 Graph the function $y = \sqrt[3]{x}$.

SOLUTION Some graphing devices display the graph shown in Figure 11, whereas others produce a graph like that in Figure 12. We know from Section 1.2 (Figure 13) that the graph in Figure 12 is correct, so what happened in Figure 11? The explanation is that some machines compute the cube root of x using a logarithm, which is not defined if x is negative, so only the right half of the graph is produced.

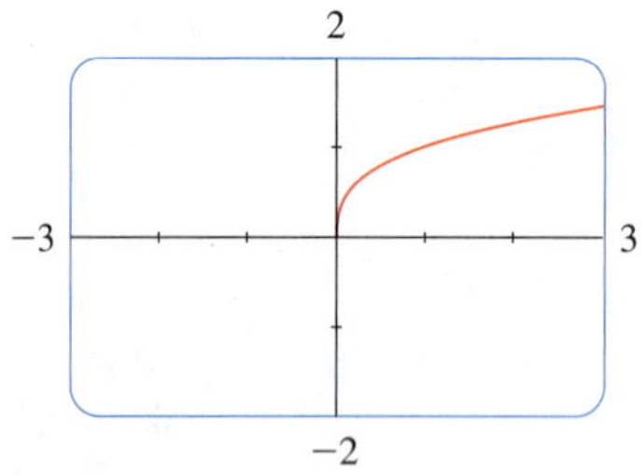

FIGURE 11

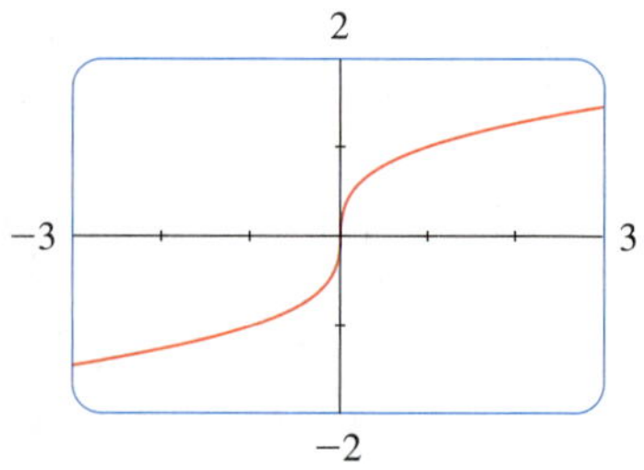

FIGURE 12

You should experiment with your own machine to see which of these two graphs is produced. If you get the graph in Figure 11, you can obtain the correct picture by graphing the function

$$f(x) = \frac{x}{|x|} \cdot |x|^{1/3}$$

Notice that this function is equal to $\sqrt[3]{x}$ (except when $x = 0$).

To understand how the expression for a function relates to its graph, it's helpful to graph a **family of functions**, that is, a collection of functions whose equations are related. In the next example we graph members of a family of cubic polynomials.

EXAMPLE 8 Graph the function $y = x^3 + cx$ for various values of the number c. How does the graph change when c is changed?

SOLUTION Figure 13 shows the graphs of $y = x^3 + cx$ for $c = 2, 1, 0, -1$, and -2. We see that, for positive values of c, the graph increases from left to right with no maximum or minimum points (peaks or valleys). When $c = 0$, the curve is flat at the origin. When c is negative, the curve has a maximum point and a minimum point. As c decreases, the maximum point becomes higher and the minimum point lower.

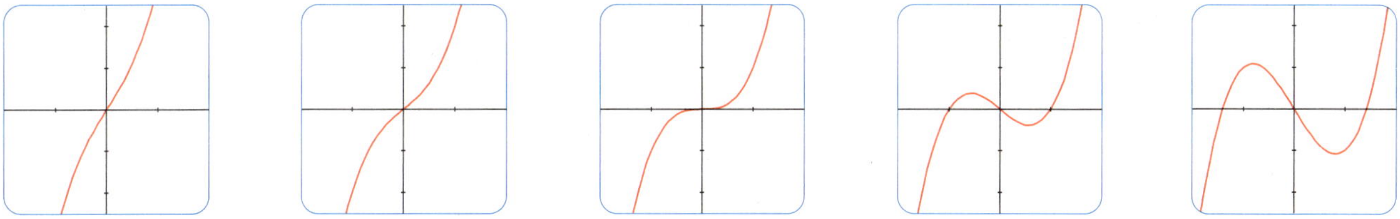

(a) $y = x^3 + 2x$ (b) $y = x^3 + x$ (c) $y = x^3$ (d) $y = x^3 - x$ (e) $y = x^3 - 2x$

FIGURE 13
Several members of the family of functions $y = x^3 + cx$, all graphed in the viewing rectangle $[-2, 2]$ by $[-2.5, 2.5]$

EXAMPLE 9 Find the solution of the equation $\cos x = x$ correct to two decimal places.

SOLUTION The solutions of the equation $\cos x = x$ are the x-coordinates of the points of intersection of the curves $y = \cos x$ and $y = x$. From Figure 14(a) we see that there is only one solution and it lies between 0 and 1. Zooming in to the viewing rectangle $[0, 1]$ by $[0, 1]$, we see from Figure 14(b) that the root lies between 0.7 and 0.8. So we zoom in further to the viewing rectangle $[0.7, 0.8]$ by $[0.7, 0.8]$ in Figure 14(c). By moving the cursor to the intersection point of the two curves, or by inspection and the fact that the x-scale is 0.01, we see that the root of the equation is about 0.74. (Many calculators have a built-in intersection feature.)

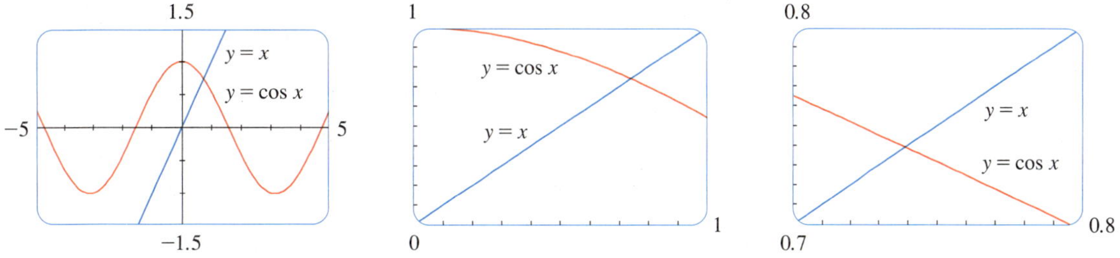

(a) $[-5, 5]$ by $[-1.5, 1.5]$
x-scale = 1

(b) $[0, 1]$ by $[0, 1]$
x-scale = 0.1

(c) $[0.7, 0.8]$ by $[0.7, 0.8]$
x-scale = 0.01

FIGURE 14
Locating the roots of $\cos x = x$

1.4 Exercises

1. Use a graphing calculator or computer to determine which of the given viewing rectangles produces the most appropriate graph of the function $f(x) = x^4 + 2$.
(a) $[-2, 2]$ by $[-2, 2]$
(b) $[0, 4]$ by $[0, 4]$
(c) $[-4, 4]$ by $[-4, 4]$
(d) $[-8, 8]$ by $[-4, 40]$
(e) $[-40, 40]$ by $[-80, 800]$

2. Use a graphing calculator or computer to determine which of the given viewing rectangles produces the most appropriate graph of the function $f(x) = x^2 + 7x + 6$.
(a) $[-5, 5]$ by $[-5, 5]$
(b) $[0, 10]$ by $[-20, 100]$
(c) $[-15, 8]$ by $[-20, 100]$
(d) $[-10, 3]$ by $[-100, 20]$

3. Use a graphing calculator or computer to determine which of the given viewing rectangles produces the most appropriate graph of the function $f(x) = 10 + 25x - x^3$.
(a) $[-4, 4]$ by $[-4, 4]$
(b) $[-10, 10]$ by $[-10, 10]$
(c) $[-20, 20]$ by $[-100, 100]$
(d) $[-100, 100]$ by $[-200, 200]$

4. Use a graphing calculator or computer to determine which of the given viewing rectangles produces the most appropriate graph of the function $f(x) = \sqrt{8x - x^2}$.
(a) $[-4, 4]$ by $[-4, 4]$
(b) $[-5, 5]$ by $[0, 100]$
(c) $[-10, 10]$ by $[-10, 40]$
(d) $[-2, 10]$ by $[-2, 6]$

5–18 Determine an appropriate viewing rectangle for the given function and use it to draw the graph.

5. $f(x) = 5 + 20x - x^2$

6. $f(x) = x^3 + 30x^2 + 200x$

7. $f(x) = 0.01x^3 - x^2 + 5$

8. $f(x) = x(x + 6)(x - 9)$

9. $f(x) = \sqrt[4]{81 - x^4}$

10. $f(x) = \sqrt{0.1x + 20}$

11. $f(x) = x^2 + \dfrac{100}{x}$

12. $f(x) = \dfrac{x}{x^2 + 100}$

13. $f(x) = \cos 100x$

14. $f(x) = 3 \sin 120x$

15. $f(x) = \sin(x/40)$

16. $y = \tan 25x$

17. $y = 3^{\cos(x^2)}$

18. $y = x^2 + 0.02 \sin 50x$

19. Graph the ellipse $4x^2 + 2y^2 = 1$ by graphing the functions whose graphs are the upper and lower halves of the ellipse.

20. Graph the hyperbola $y^2 - 9x^2 = 1$ by graphing the functions whose graphs are the upper and lower branches of the hyperbola.

21–23 Find all solutions of the equation correct to two decimal places.

21. $x^3 - 9x^2 - 4 = 0$

22. $x^3 = 4x - 1$

23. $x^2 = \sin x$

24. We saw in Example 9 that the equation $\cos x = x$ has exactly one solution.
(a) Use a graph to show that the equation $\cos x = 0.3x$ has three solutions and find their values correct to two decimal places.
(b) Find an approximate value of m such that the equation $\cos x = mx$ has exactly two solutions.

25. Use graphs to determine which of the functions $f(x) = 10x^2$ and $g(x) = x^3/10$ is eventually larger (that is, larger when x is very large).

26. Use graphs to determine which of the functions $f(x) = x^4 - 100x^3$ and $g(x) = x^3$ is eventually larger.

27. For what values of x is it true that $|\sin x - x| < 0.1$?

28. Graph the polynomials $P(x) = 3x^5 - 5x^3 + 2x$ and $Q(x) = 3x^5$ on the same screen, first using the viewing rectangle $[-2, 2]$ by $[-2, 2]$ and then changing to $[-10, 10]$ by $[-10{,}000, 10{,}000]$. What do you observe from these graphs?

29. In this exercise we consider the family of root functions $f(x) = \sqrt[n]{x}$, where n is a positive integer.
(a) Graph the functions $y = \sqrt{x}$, $y = \sqrt[4]{x}$, and $y = \sqrt[6]{x}$ on the same screen using the viewing rectangle $[-1, 4]$ by $[-1, 3]$.
(b) Graph the functions $y = x$, $y = \sqrt[3]{x}$, and $y = \sqrt[5]{x}$ on the same screen using the viewing rectangle $[-3, 3]$ by $[-2, 2]$. (See Example 7.)
(c) Graph the functions $y = \sqrt{x}$, $y = \sqrt[3]{x}$, $y = \sqrt[4]{x}$, and $y = \sqrt[5]{x}$ on the same screen using the viewing rectangle $[-1, 3]$ by $[-1, 2]$.
(d) What conclusions can you make from these graphs?

30. In this exercise we consider the family of functions $f(x) = 1/x^n$, where n is a positive integer.
(a) Graph the functions $y = 1/x$ and $y = 1/x^3$ on the same screen using the viewing rectangle $[-3, 3]$ by $[-3, 3]$.
(b) Graph the functions $y = 1/x^2$ and $y = 1/x^4$ on the same screen using the same viewing rectangle as in part (a).
(c) Graph all of the functions in parts (a) and (b) on the same screen using the viewing rectangle $[-1, 3]$ by $[-1, 3]$.
(d) What conclusions can you make from these graphs?

31. Graph the function $f(x) = x^4 + cx^2 + x$ for several values of c. How does the graph change when c changes?

32. Graph the function $f(x) = \sqrt{1 + cx^2}$ for various values of c. Describe how changing the value of c affects the graph.

33. Graph the function $y = x^n 2^{-x}$, $x \geqslant 0$, for $n = 1, 2, 3, 4, 5$, and 6. How does the graph change as n increases?

34. The curves with equations

$$y = \frac{|x|}{\sqrt{c - x^2}}$$

are called **bullet-nose curves**. Graph some of these curves to see why. What happens as c increases?

35. What happens to the graph of the equation $y^2 = cx^3 + x^2$ as c varies?

36. This exercise explores the effect of the inner function g on a composite function $y = f(g(x))$.
(a) Graph the function $y = \sin(\sqrt{x})$ using the viewing rectangle $[0, 400]$ by $[-1.5, 1.5]$. How does this graph differ from the graph of the sine function?
(b) Graph the function $y = \sin(x^2)$ using the viewing rectangle $[-5, 5]$ by $[-1.5, 1.5]$. How does this graph differ from the graph of the sine function?

37. The figure shows the graphs of $y = \sin 96x$ and $y = \sin 2x$ as displayed by a TI-83 graphing calculator.

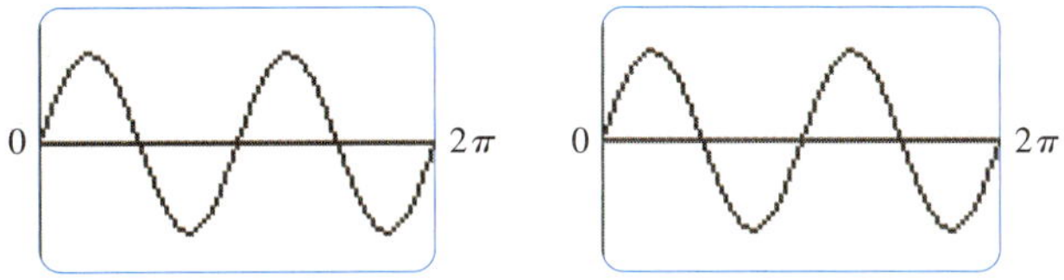

$y = \sin 96x$ $y = \sin 2x$

The first graph is inaccurate. Explain why the two graphs appear identical. [*Hint:* The TI-83's graphing window is 95 pixels wide. What specific points does the calculator plot?]

38. The first graph in the figure is that of $y = \sin 45x$ as displayed by a TI-83 graphing calculator. It is inaccurate and so, to help explain its appearance, we replot the curve in dot mode in the second graph.

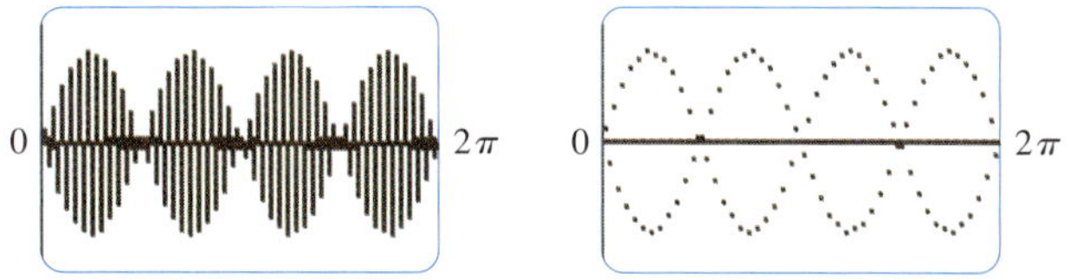

What two sine curves does the calculator appear to be plotting? Show that each point on the graph of $y = \sin 45x$ that the TI-83 chooses to plot is in fact on one of these two curves. (The TI-83's graphing window is 95 pixels wide.)

1.5 Exponential Functions

The function $f(x) = 2^x$ is called an *exponential function* because the variable, x, is the exponent. It should not be confused with the power function $g(x) = x^2$, in which the variable is the base.

In general, an **exponential function** is a function of the form

$$f(x) = a^x$$

where a is a positive constant. Let's recall what this means.

If $x = n$, a positive integer, then

$$a^n = \underbrace{a \cdot a \cdot \cdots \cdot a}_{n \text{ factors}}$$

If $x = 0$, then $a^0 = 1$, and if $x = -n$, where n is a positive integer, then

$$a^{-n} = \frac{1}{a^n}$$

If x is a rational number, $x = p/q$, where p and q are integers and $q > 0$, then

$$a^x = a^{p/q} = \sqrt[q]{a^p} = \left(\sqrt[q]{a}\right)^p$$

But what is the meaning of a^x if x is an irrational number? For instance, what is meant by $2^{\sqrt{3}}$ or 5^{π}?

To help us answer this question we first look at the graph of the function $y = 2^x$, where x is rational. A representation of this graph is shown in Figure 1. We want to enlarge the domain of $y = 2^x$ to include both rational and irrational numbers.

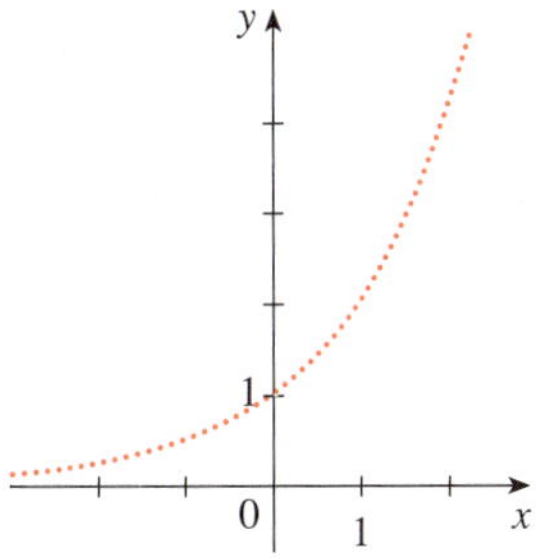

FIGURE 1
Representation of $y = 2^x$, x rational

There are holes in the graph in Figure 1 corresponding to irrational values of x. We want to fill in the holes by defining $f(x) = 2^x$, where $x \in \mathbb{R}$, so that f is an increasing function. In particular, since the irrational number $\sqrt{3}$ satisfies

$$1.7 < \sqrt{3} < 1.8$$

we must have

$$2^{1.7} < 2^{\sqrt{3}} < 2^{1.8}$$

and we know what $2^{1.7}$ and $2^{1.8}$ mean because 1.7 and 1.8 are rational numbers. Similarly, if we use better approximations for $\sqrt{3}$, we obtain better approximations for $2^{\sqrt{3}}$:

$$\begin{array}{rcl} 1.73 < \sqrt{3} < 1.74 & \Rightarrow & 2^{1.73} < 2^{\sqrt{3}} < 2^{1.74} \\ 1.732 < \sqrt{3} < 1.733 & \Rightarrow & 2^{1.732} < 2^{\sqrt{3}} < 2^{1.733} \\ 1.7320 < \sqrt{3} < 1.7321 & \Rightarrow & 2^{1.7320} < 2^{\sqrt{3}} < 2^{1.7321} \\ 1.73205 < \sqrt{3} < 1.73206 & \Rightarrow & 2^{1.73205} < 2^{\sqrt{3}} < 2^{1.73206} \\ \vdots \quad \vdots & & \vdots \quad \vdots \end{array}$$

|||| A proof of this fact is given in J. Marsden and A. Weinstein, *Calculus Unlimited* (Menlo Park, CA: Benjamin/Cummings, 1980).

It can be shown that there is exactly one number that is greater than all of the numbers

$$2^{1.7}, \quad 2^{1.73}, \quad 2^{1.732}, \quad 2^{1.7320}, \quad 2^{1.73205}, \quad \ldots$$

and less than all of the numbers

$$2^{1.8}, \quad 2^{1.74}, \quad 2^{1.733}, \quad 2^{1.7321}, \quad 2^{1.73206}, \quad \ldots$$

We define $2^{\sqrt{3}}$ to be this number. Using the preceding approximation process we can compute it correct to six decimal places:

$$2^{\sqrt{3}} \approx 3.321997$$

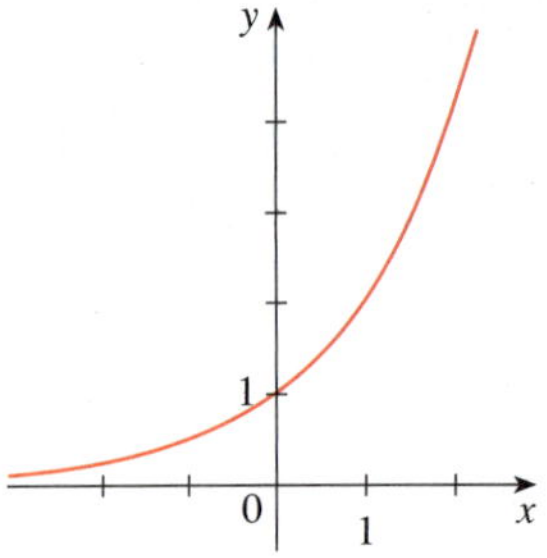

FIGURE 2
$y = 2^x$, x real

Similarly, we can define 2^x (or a^x, if $a > 0$) where x is any irrational number. Figure 2 shows how all the holes in Figure 1 have been filled to complete the graph of the function $f(x) = 2^x$, $x \in \mathbb{R}$.

The graphs of members of the family of functions $y = a^x$ are shown in Figure 3 for various values of the base a. Notice that all of these graphs pass through the same point (0, 1) because $a^0 = 1$ for $a \neq 0$. Notice also that as the base a gets larger, the exponential function grows more rapidly (for $x > 0$).

|||| If $0 < a < 1$, then a^x approaches 0 as x becomes large. If $a > 1$, then a^x approaches 0 as x decreases through negative values. In both cases the x-axis is a horizontal asymptote. These matters are discussed in Section 2.6.

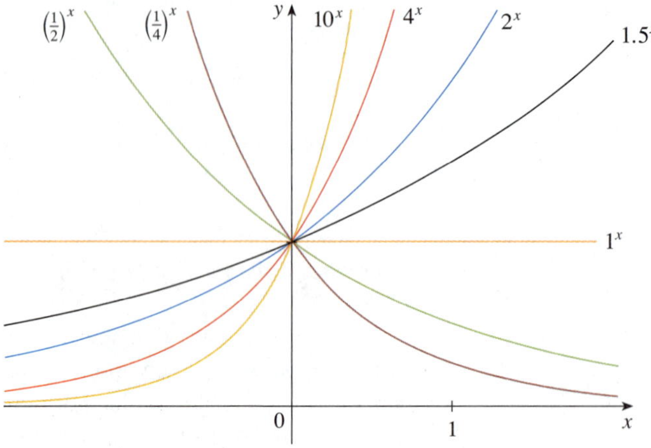

FIGURE 3

You can see from Figure 3 that there are basically three kinds of exponential functions $y = a^x$. If $0 < a < 1$, the exponential function decreases; if $a = 1$, it is a constant; and if $a > 1$, it increases. These three cases are illustrated in Figure 4. Observe that if $a \neq 1$, then the exponential function $y = a^x$ has domain $\mathbb{R}$ and range $(0, \infty)$. Notice also that, since $(1/a)^x = 1/a^x = a^{-x}$, the graph of $y = (1/a)^x$ is just the reflection of the graph of $y = a^x$ about the y-axis.

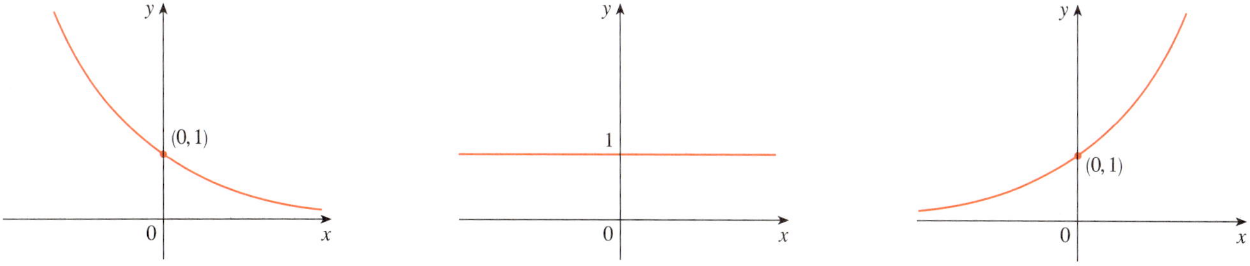

(a) $y = a^x$, $0 < a < 1$ (b) $y = 1^x$ (c) $y = a^x$, $a > 1$

FIGURE 4

One reason for the importance of the exponential function lies in the following properties. If x and y are rational numbers, then these laws are well known from elementary algebra. It can be proved that they remain true for arbitrary real numbers x and y.

|||| In Section 5.6 we will present a definition of the exponential function that will enable us to give an easy proof of the Laws of Exponents.

Laws of Exponents If a and b are positive numbers and x and y are any real numbers, then

1. $a^{x+y} = a^x a^y$ **2.** $a^{x-y} = \dfrac{a^x}{a^y}$ **3.** $(a^x)^y = a^{xy}$ **4.** $(ab)^x = a^x b^x$

EXAMPLE 1 Sketch the graph of the function $y = 3 - 2^x$ and determine its domain and range.

|||| For a review of reflecting and shifting graphs, see Section 1.3.

SOLUTION First we reflect the graph of $y = 2^x$ (shown in Figure 2) about the x-axis to get the graph of $y = -2^x$ in Figure 5(b). Then we shift the graph of $y = -2^x$ upward 3 units to obtain the graph of $y = 3 - 2^x$ in Figure 5(c). The domain is $\mathbb{R}$ and the range is $(-\infty, 3)$.

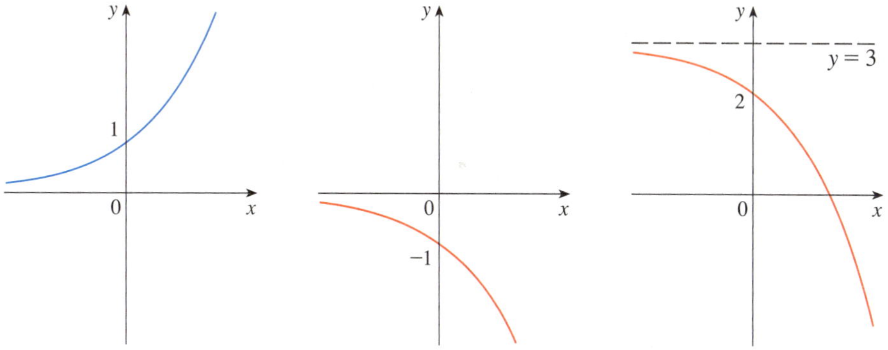

FIGURE 5   (a) $y = 2^x$ (b) $y = -2^x$ (c) $y = 3 - 2^x$

EXAMPLE 2 Use a graphing device to compare the exponential function $f(x) = 2^x$ and the power function $g(x) = x^2$. Which function grows more quickly when x is large?

SOLUTION Figure 6 shows both functions graphed in the viewing rectangle $[-2, 6]$ by $[0, 40]$. We see that the graphs intersect three times, but for $x > 4$ the graph of

Example 2 shows that $y = 2^x$ increases more quickly than $y = x^2$. To demonstrate just how quickly $f(x) = 2^x$ increases, let's perform the following thought experiment. Suppose we start with a piece of paper a thousandth of an inch thick and we fold it in half 50 times. Each time we fold the paper in half, the thickness of the paper doubles, so the thickness of the resulting paper would be $2^{50}/1000$ inches. How thick do you think that is? It works out to be more than 17 million miles!

$f(x) = 2^x$ stays above the graph of $g(x) = x^2$. Figure 7 gives a more global view and shows that for large values of x, the exponential function $y = 2^x$ grows far more rapidly than the power function $y = x^2$.

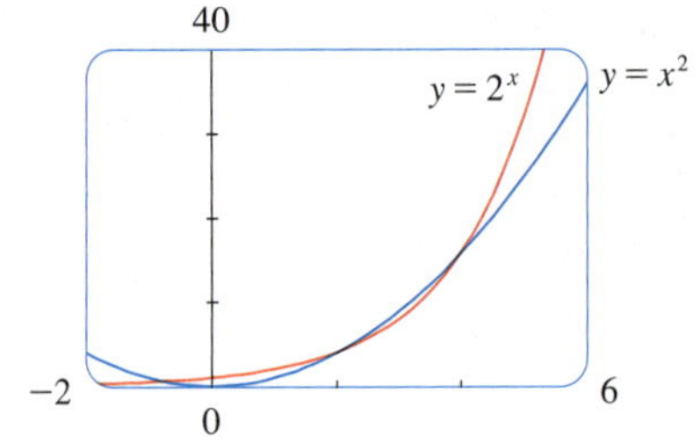

FIGURE 6

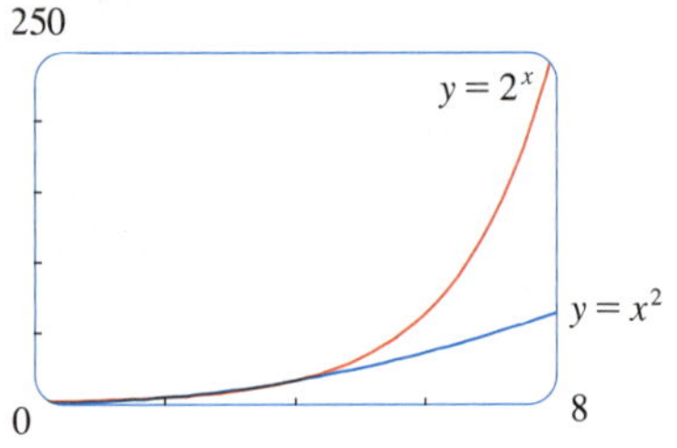

FIGURE 7

Applications of Exponential Functions

The exponential function occurs very frequently in mathematical models of nature and society. Here we indicate briefly how it arises in the description of population growth and radioactive decay. In later chapters we will pursue these and other applications in greater detail.

First we consider a population of bacteria in a homogeneous nutrient medium. Suppose that by sampling the population at certain intervals it is determined that the population doubles every hour. If the number of bacteria at time t is $p(t)$, where t is measured in hours, and the initial population is $p(0) = 1000$, then we have

$$p(1) = 2p(0) = 2 \times 1000$$

$$p(2) = 2p(1) = 2^2 \times 1000$$

$$p(3) = 2p(2) = 2^3 \times 1000$$

It seems from this pattern that, in general,

$$p(t) = 2^t \times 1000 = (1000)2^t$$

This population function is a constant multiple of the exponential function $y = 2^t$, so it exhibits the rapid growth that we observed in Figures 2 and 7. Under ideal conditions (unlimited space and nutrition and freedom from disease) this exponential growth is typical of what actually occurs in nature.

What about the human population? Table 1 shows data for the population of the world in the 20th century and Figure 8 shows the corresponding scatter plot.

TABLE 1

Year	Population (millions)
1900	1650
1910	1750
1920	1860
1930	2070
1940	2300
1950	2560
1960	3040
1970	3710
1980	4450
1990	5280
2000	6080

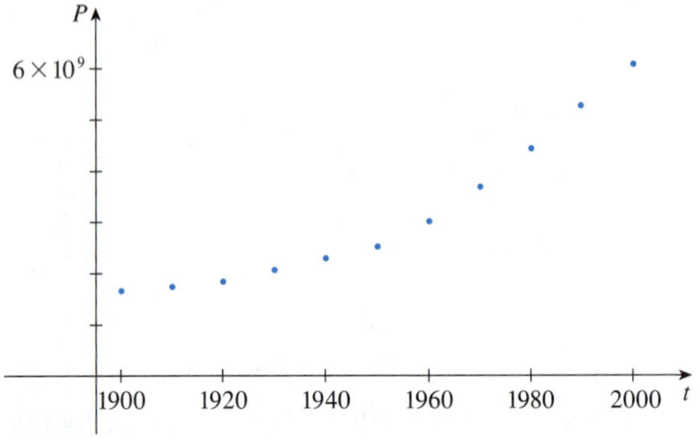

FIGURE 8 Scatter plot for world population growth

The pattern of the data points in Figure 8 suggests exponential growth, so we use a graphing calculator with exponential regression capability to apply the method of least squares and obtain the exponential model

$$P = (0.008079266) \cdot (1.013731)^t$$

Figure 9 shows the graph of this exponential function together with the original data points. We see that the exponential curve fits the data reasonably well. The period of relatively slow population growth is explained by the two world wars and the Great Depression of the 1930s.

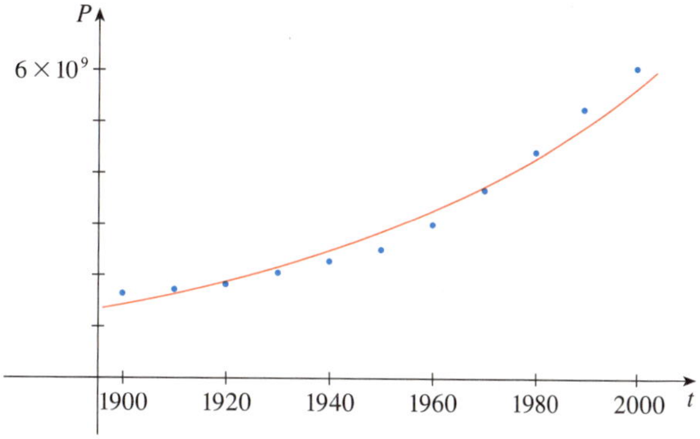

FIGURE 9
Exponential model for population growth

EXAMPLE 3 The *half-life* of strontium-90, ^{90}Sr, is 25 years. This means that half of any given quantity of ^{90}Sr will disintegrate in 25 years.

(a) If a sample of ^{90}Sr has a mass of 24 mg, find an expression for the mass $m(t)$ that remains after t years.

(b) Find the mass remaining after 40 years, correct to the nearest milligram.

(c) Use a graphing device to graph $m(t)$ and use the graph to estimate the time required for the mass to be reduced to 5 mg.

SOLUTION

(a) The mass is initially 24 mg and is halved during each 25-year period, so

$$m(0) = 24$$

$$m(25) = \frac{1}{2}(24)$$

$$m(50) = \frac{1}{2} \cdot \frac{1}{2}(24) = \frac{1}{2^2}(24)$$

$$m(75) = \frac{1}{2} \cdot \frac{1}{2^2}(24) = \frac{1}{2^3}(24)$$

$$m(100) = \frac{1}{2} \cdot \frac{1}{2^3}(24) = \frac{1}{2^4}(24)$$

From this pattern, it appears that the mass remaining after t years is

$$m(t) = \frac{1}{2^{t/25}}(24) = 24 \cdot 2^{-t/25}$$

This is an exponential function with base $a = 2^{-1/25} = 1/2^{1/25}$.

(b) The mass that remains after 40 years is

$$m(40) = 24 \cdot 2^{-40/25} \approx 7.9 \text{ mg}$$

(c) We use a graphing calculator or computer to graph the function $m(t) = 24 \cdot 2^{-t/25}$ in Figure 10. We also graph the line $m = 5$ and use the cursor to estimate that $m(t) = 5$ when $t \approx 57$. So the mass of the sample will be reduced to 5 mg after about 57 years.

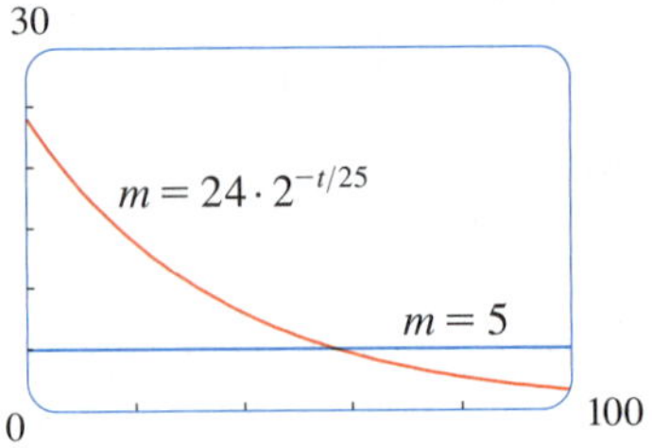

FIGURE 10

|||| The Number *e*

Of all possible bases for an exponential function, there is one that is most convenient for the purposes of calculus. The choice of a base a is influenced by the way the graph of $y = a^x$ crosses the y-axis. Figures 11 and 12 show the tangent lines to the graphs of $y = 2^x$ and $y = 3^x$ at the point (0, 1). (Tangent lines will be defined precisely in Section 2.7. For present purposes, you can think of the tangent line to an exponential graph at a point as the line that touches the graph only at that point.) If we measure the slopes of these tangent lines at (0, 1), we find that $m \approx 0.7$ for $y = 2^x$ and $m \approx 1.1$ for $y = 3^x$.

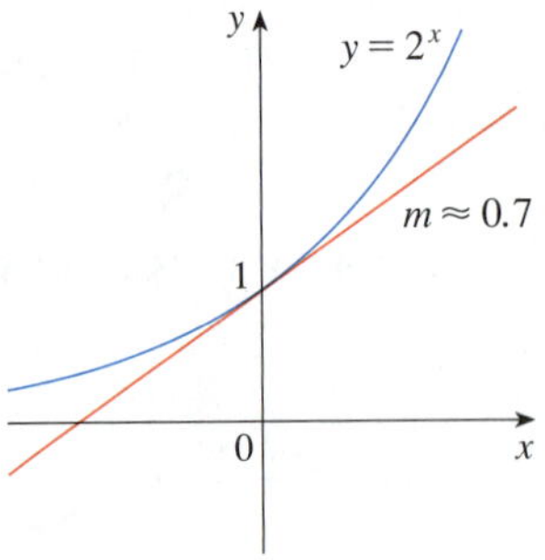

FIGURE 11

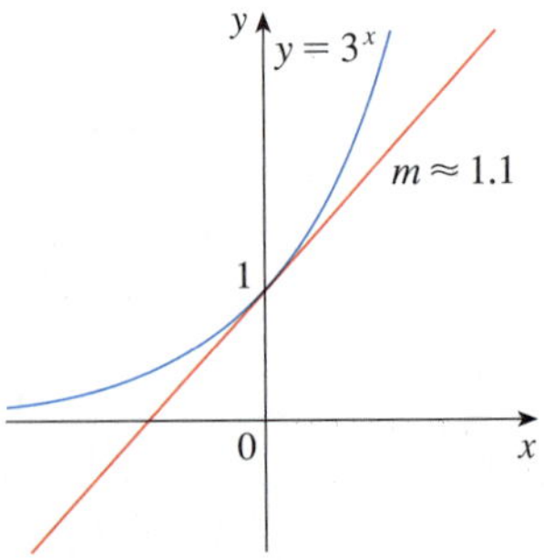

FIGURE 12

It turns out, as we will see in Chapter 3, that some of the formulas of calculus will be greatly simplified if we choose the base a so that the slope of the tangent line to $y = a^x$ at (0, 1) is *exactly* 1 (see Figure 13). In fact, there *is* such a number (as we will see in Section 5.6) and it is denoted by the letter e. (This notation was chosen by the Swiss mathematician Leonhard Euler in 1727, probably because it is the first letter of the word *exponential.*) In view of Figures 11 and 12, it comes as no surprise that the number e lies between 2 and 3 and the graph of $y = e^x$ lies between the graphs of $y = 2^x$ and $y = 3^x$. (See Figure 14.) In Chapter 3 we will see that the value of e, correct to five decimal places, is

$$e \approx 2.71828$$

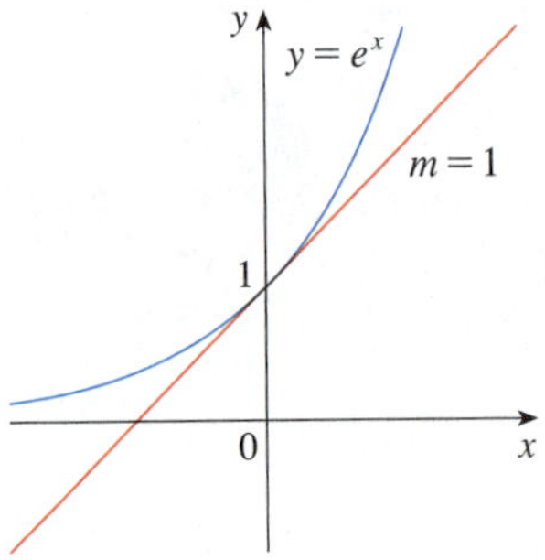

FIGURE 13
The natural exponential function crosses the y-axis with a slope of 1.

TEC Module 1.5 enables you to graph exponential functions with various bases and their tangent lines in order to estimate more closely the value of a for which the tangent has slope 1.

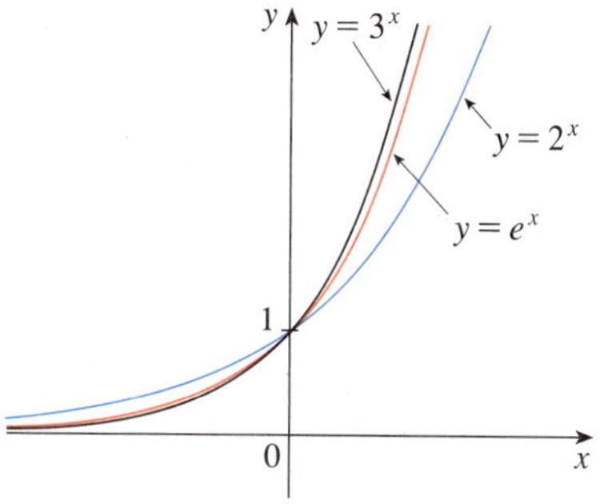

FIGURE 14

EXAMPLE 4 Graph the function $y = \frac{1}{2}e^{-x} - 1$ and state the domain and range.

SOLUTION We start with the graph of $y = e^x$ from Figures 13 and 15(a) and reflect about the y-axis to get the graph of $y = e^{-x}$ in Figure 15(b). (Notice that the graph crosses the y-axis with a slope of -1). Then we compress the graph vertically by a factor of 2 to obtain the graph of $y = \frac{1}{2}e^{-x}$ in Figure 15(c). Finally, we shift the graph downward one unit to get the desired graph in Figure 15(d). The domain is $\mathbb{R}$ and the range is $(-1, \infty)$.

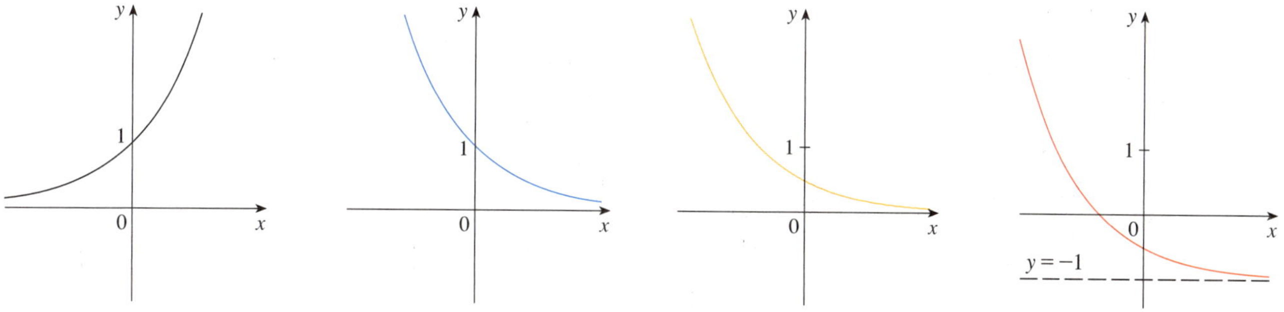

(a) $y = e^x$ (b) $y = e^{-x}$ (c) $y = \frac{1}{2}e^{-x}$ (d) $y = \frac{1}{2}e^{-x} - 1$

FIGURE 15

How far to the right do you think we would have to go for the height of the graph of $y = e^x$ to exceed a million? The next example demonstrates the rapid growth of this function by providing an answer that might surprise you.

EXAMPLE 5 Use a graphing device to find the values of x for which $e^x > 1{,}000{,}000$.

SOLUTION In Figure 16 we graph both the function $y = e^x$ and the horizontal line $y = 1{,}000{,}000$. We see that these curves intersect when $x \approx 13.8$. Thus, $e^x > 10^6$ when $x > 13.8$. It is perhaps surprising that the values of the exponential function have already surpassed a million when x is only 14.

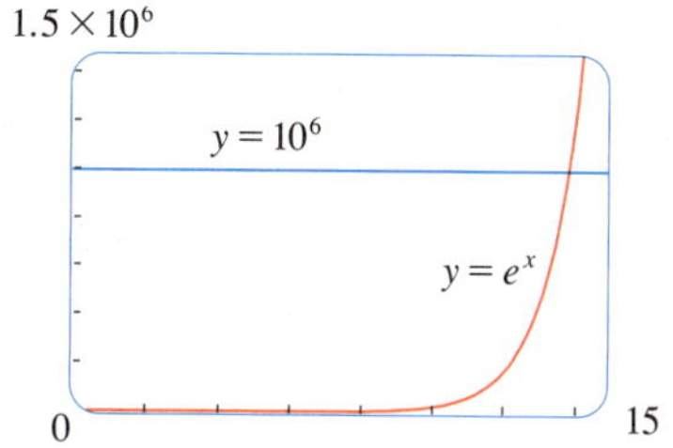

FIGURE 16

1.5 Exercises

1. (a) Write an equation that defines the exponential function with base $a > 0$.
(b) What is the domain of this function?
(c) If $a \neq 1$, what is the range of this function?
(d) Sketch the general shape of the graph of the exponential function for each of the following cases.
(i) $a > 1$ (ii) $a = 1$ (iii) $0 < a < 1$

2. (a) How is the number e defined?
(b) What is an approximate value for e?
(c) What is the natural exponential function?

3–6 Graph the given functions on a common screen. How are these graphs related?

3. $y = 2^x$, $y = e^x$, $y = 5^x$, $y = 20^x$

4. $y = e^x$, $y = e^{-x}$, $y = 8^x$, $y = 8^{-x}$

5. $y = 3^x$, $y = 10^x$, $y = \left(\frac{1}{3}\right)^x$, $y = \left(\frac{1}{10}\right)^x$

6. $y = 0.9^x$, $y = 0.6^x$, $y = 0.3^x$, $y = 0.1^x$

7–12 Make a rough sketch of the graph of the function. Do not use a calculator. Just use the graphs given in Figures 3 and 14 and, if necessary, the transformations of Section 1.3.

7. $y = 4^x - 3$

8. $y = 4^{x-3}$

9. $y = -2^{-x}$

10. $y = 1 + 2e^x$

11. $y = 3 - e^x$

12. $y = 2 + 5(1 - e^{-x})$

13. Starting with the graph of $y = e^x$, write the equation of the graph that results from
(a) shifting 2 units downward
(b) shifting 2 units to the right
(c) reflecting about the x-axis
(d) reflecting about the y-axis
(e) reflecting about the x-axis and then about the y-axis

14. Starting with the graph of $y = e^x$, find the equation of the graph that results from
(a) reflecting about the line $y = 4$
(b) reflecting about the line $x = 2$

15–16 Find the domain of each function.

15. (a) $f(x) = \dfrac{1}{1 + e^x}$ (b) $f(x) = \dfrac{1}{1 - e^x}$

16. (a) $g(t) = \sin(e^{-t})$ (b) $g(t) = \sqrt{1 - 2^t}$

17–18 Find the exponential function $f(x) = Ca^x$ whose graph is given.

17.

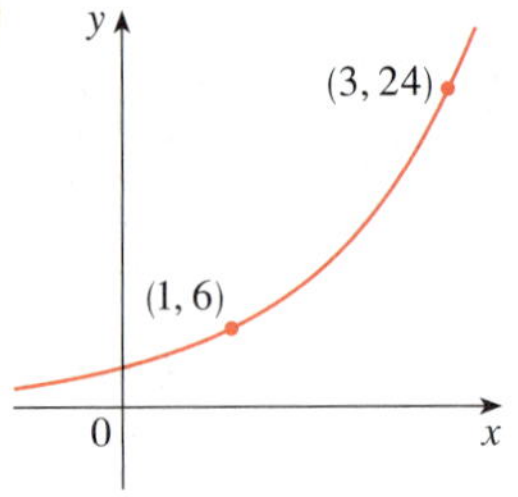

18.

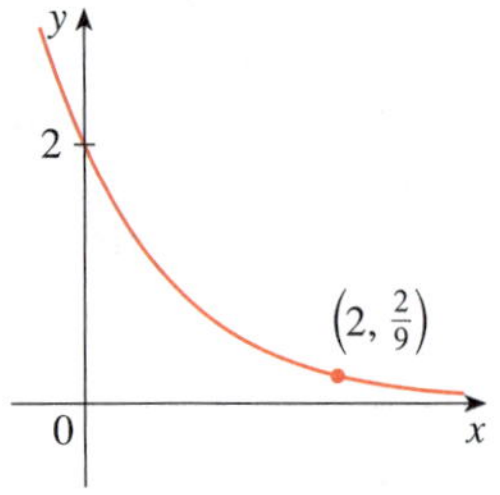

19. If $f(x) = 5^x$, show that

$$\frac{f(x + h) - f(x)}{h} = 5^x\left(\frac{5^h - 1}{h}\right)$$

20. Suppose you are offered a job that lasts one month. Which of the following methods of payment do you prefer?
I. One million dollars at the end of the month.
II. One cent on the first day of the month, two cents on the second day, four cents on the third day, and, in general, 2^{n-1} cents on the nth day.

21. Suppose the graphs of $f(x) = x^2$ and $g(x) = 2^x$ are drawn on a coordinate grid where the unit of measurement is 1 inch. Show that, at a distance 2 ft to the right of the origin, the height of the graph of f is 48 ft but the height of the graph of g is about 265 mi.

22. Compare the functions $f(x) = x^5$ and $g(x) = 5^x$ by graphing both functions in several viewing rectangles. Find all points of intersection of the graphs correct to one decimal place. Which function grows more rapidly when x is large?

23. Compare the functions $f(x) = x^{10}$ and $g(x) = e^x$ by graphing both f and g in several viewing rectangles. When does the graph of g finally surpass the graph of f?

24. Use a graph to estimate the values of x such that $e^x > 1{,}000{,}000{,}000$.

25. Under ideal conditions a certain bacteria population is known to double every three hours. Suppose that there are initially 100 bacteria.
(a) What is the size of the population after 15 hours?
(b) What is the size of the population after t hours?

(c) Estimate the size of the population after 20 hours.

(d) Graph the population function and estimate the time for the population to reach 50,000.

26. An isotope of sodium, ^{24}Na, has a half-life of 15 hours. A sample of this isotope has mass 2 g.

(a) Find the amount remaining after 60 hours.

(b) Find the amount remaining after t hours.

(c) Estimate the amount remaining after 4 days.

(d) Use a graph to estimate the time required for the mass to be reduced to 0.01 g.

27. Use a graphing calculator with exponential regression capability to model the population of the world with the data from 1950 to 2000 in Table 1 on page 58. Use the model to estimate the population in 1993 and to predict the population in the year 2010.

28. The table gives the population of the United States, in millions, for the years 1900–2000.

Year	Population	Year	Population
1900	76	1960	179
1910	92	1970	203
1920	106	1980	227
1930	123	1990	250
1940	131	2000	281
1950	150		

Use a graphing calculator with exponential regression capability to model the U.S. population since 1900. Use the model to estimate the population in 1925 and to predict the population in the years 2010 and 2020.

1.6 Inverse Functions and Logarithms

Table 1 gives data from an experiment in which a bacteria culture started with 100 bacteria in a limited nutrient medium; the size of the bacteria population was recorded at hourly intervals. The number of bacteria N is a function of the time t: $N = f(t)$.

Suppose, however, that the biologist changes her point of view and becomes interested in the time required for the population to reach various levels. In other words, she is thinking of t as a function of N. This function is called the *inverse function* of f, denoted by f^{-1}, and read "f inverse." Thus, $t = f^{-1}(N)$ is the time required for the population level to reach N. The values of f^{-1} can be found by reading Table 1 from right to left or by consulting Table 2. For instance, $f^{-1}(550) = 6$ because $f(6) = 550$.

TABLE 1 N as a function of t

t (hours)	$N = f(t)$ = population at time t
0	100
1	168
2	259
3	358
4	445
5	509
6	550
7	573
8	586

TABLE 2 t as a function of N

N	$t = f^{-1}(N)$ = time to reach N bacteria
100	0
168	1
259	2
358	3
445	4
509	5
550	6
573	7
586	8

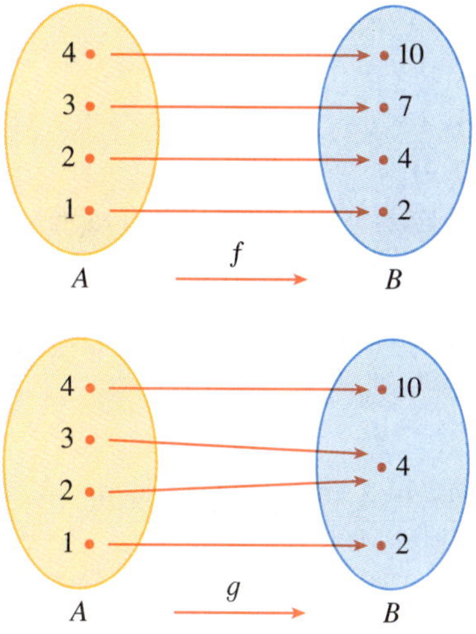

FIGURE 1

Not all functions possess inverses. Let's compare the functions f and g whose arrow diagrams are shown in Figure 1.

Note that f never takes on the same value twice (any two inputs in A have different outputs), whereas g does take on the same value twice (both 2 and 3 have the same output, 4).

In symbols,

$$g(2) = g(3)$$

but $$f(x_1) \neq f(x_2) \qquad \text{whenever } x_1 \neq x_2$$

Functions that have this property are called *one-to-one functions.*

|||| In the language of inputs and outputs, this definition says that f is one-to-one if each output corresponds to only one input.

1 Definition A function f is called a **one-to-one function** if it never takes on the same value twice; that is,

$$f(x_1) \neq f(x_2) \qquad \text{whenever } x_1 \neq x_2$$

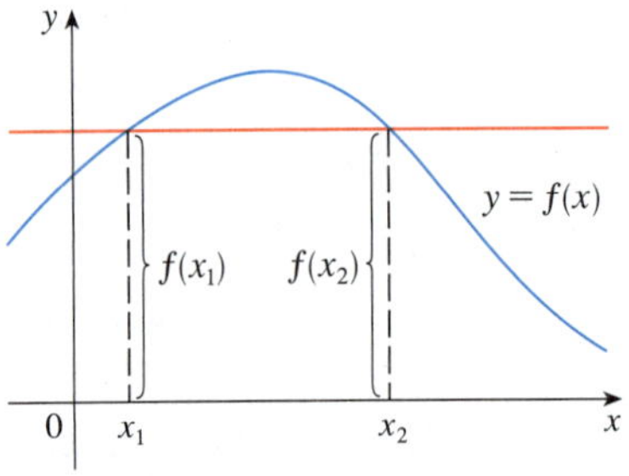

FIGURE 2
This function is not one-to-one because $f(x_1) = f(x_2)$.

If a horizontal line intersects the graph of f in more than one point, then we see from Figure 2 that there are numbers x_1 and x_2 such that $f(x_1) = f(x_2)$. This means that f is not one-to-one. Therefore, we have the following geometric method for determining whether a function is one-to-one.

Horizontal Line Test A function is one-to-one if and only if no horizontal line intersects its graph more than once.

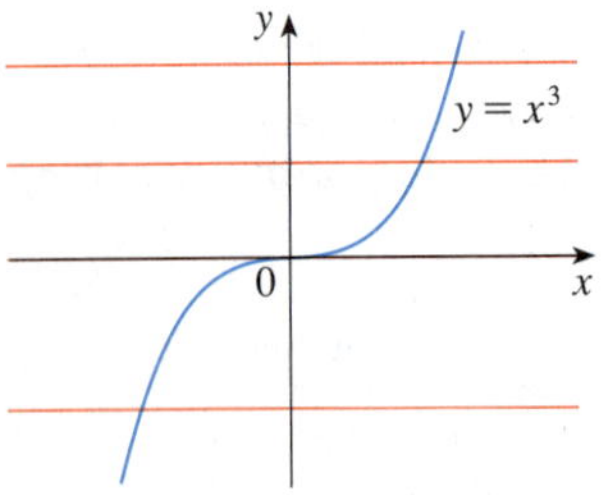

FIGURE 3
$f(x) = x^3$ is one-to-one.

EXAMPLE 1 Is the function $f(x) = x^3$ one-to-one?

SOLUTION 1 If $x_1 \neq x_2$, then $x_1^3 \neq x_2^3$ (two different numbers can't have the same cube). Therefore, by Definition 1, $f(x) = x^3$ is one-to-one.

SOLUTION 2 From Figure 3 we see that no horizontal line intersects the graph of $f(x) = x^3$ more than once. Therefore, by the Horizontal Line Test, f is one-to-one.

EXAMPLE 2 Is the function $g(x) = x^2$ one-to-one?

SOLUTION 1 This function is not one-to-one because, for instance,

$$g(1) = 1 = g(-1)$$

and so 1 and -1 have the same output.

SOLUTION 2 From Figure 4 we see that there are horizontal lines that intersect the graph of g more than once. Therefore, by the Horizontal Line Test, g is not one-to-one.

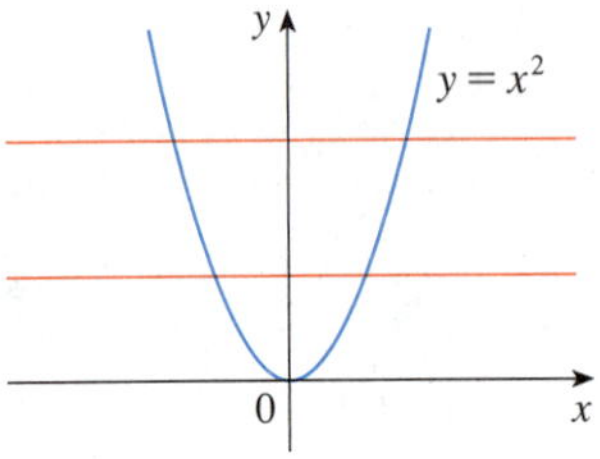

FIGURE 4
$g(x) = x^2$ is not one-to-one.

One-to-one functions are important because they are precisely the functions that possess inverse functions according to the following definition.

2 Definition Let f be a one-to-one function with domain A and range B. Then its **inverse function** f^{-1} has domain B and range A and is defined by

$$f^{-1}(y) = x \iff f(x) = y$$

for any y in B.

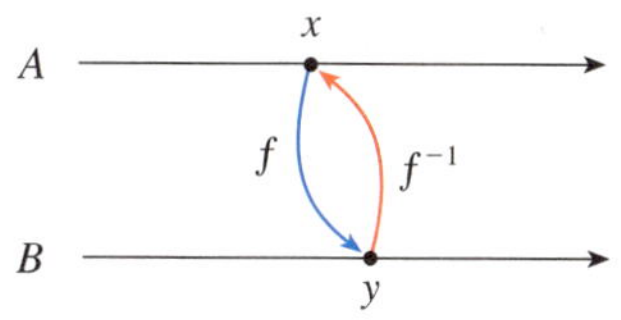

FIGURE 5

This definition says that if f maps x into y, then f^{-1} maps y back into x. (If f were not one-to-one, then f^{-1} would not be uniquely defined.) The arrow diagram in Figure 5 indicates that f^{-1} reverses the effect of f. Note that

$$\text{domain of } f^{-1} = \text{range of } f$$
$$\text{range of } f^{-1} = \text{domain of } f$$

For example, the inverse function of $f(x) = x^3$ is $f^{-1}(x) = x^{1/3}$ because if $y = x^3$, then

$$f^{-1}(y) = f^{-1}(x^3) = (x^3)^{1/3} = x$$

CAUTION ◦ Do not mistake the -1 in f^{-1} for an exponent. Thus

$$f^{-1}(x) \quad \text{does } \textit{not} \text{ mean} \quad \frac{1}{f(x)}$$

The reciprocal $1/f(x)$ could, however, be written as $[f(x)]^{-1}$.

EXAMPLE 3 If $f(1) = 5$, $f(3) = 7$, and $f(8) = -10$, find $f^{-1}(7)$, $f^{-1}(5)$, and $f^{-1}(-10)$.

SOLUTION From the definition of f^{-1} we have

$$f^{-1}(7) = 3 \quad \text{because} \quad f(3) = 7$$
$$f^{-1}(5) = 1 \quad \text{because} \quad f(1) = 5$$
$$f^{-1}(-10) = 8 \quad \text{because} \quad f(8) = -10$$

The diagram in Figure 6 makes it clear how f^{-1} reverses the effect of f in this case.

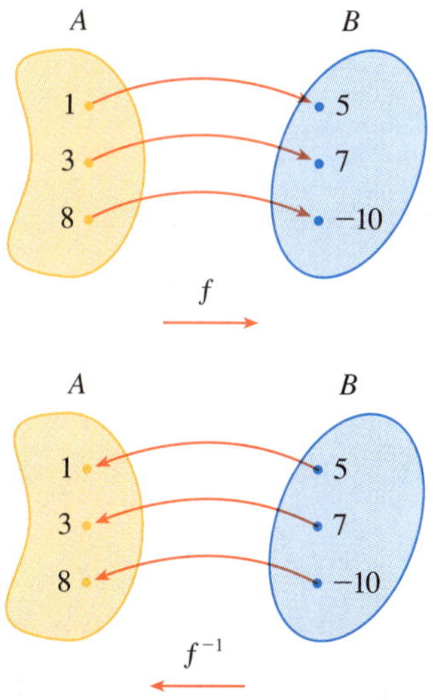

FIGURE 6
The inverse function reverses inputs and outputs.

The letter x is traditionally used as the independent variable, so when we concentrate on f^{-1} rather than on f, we usually reverse the roles of x and y in Definition 2 and write

3 $$f^{-1}(x) = y \iff f(y) = x$$

By substituting for y in Definition 2 and substituting for x in (3), we get the following **cancellation equations**:

4
$$f^{-1}(f(x)) = x \quad \text{for every } x \text{ in } A$$
$$f(f^{-1}(x)) = x \quad \text{for every } x \text{ in } B$$

The first cancellation equation says that if we start with x, apply f, and then apply f^{-1}, we arrive back at x, where we started (see the machine diagram in Figure 7). Thus, f^{-1} undoes what f does. The second equation says that f undoes what f^{-1} does.

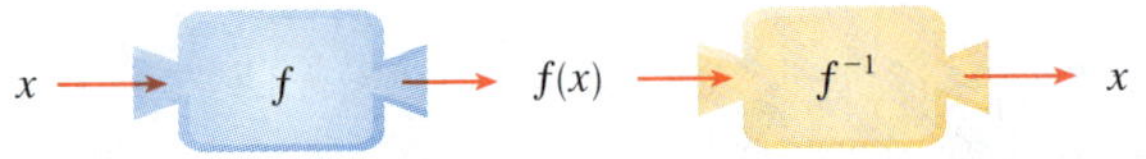

FIGURE 7

For example, if $f(x) = x^3$, then $f^{-1}(x) = x^{1/3}$ and so the cancellation equations become

$$f^{-1}(f(x)) = (x^3)^{1/3} = x$$
$$f(f^{-1}(x)) = (x^{1/3})^3 = x$$

These equations simply say that the cube function and the cube root function cancel each other when applied in succession.

Now let's see how to compute inverse functions. If we have a function $y = f(x)$ and are able to solve this equation for x in terms of y, then according to Definition 2 we must have $x = f^{-1}(y)$. If we want to call the independent variable x, we then interchange x and y and arrive at the equation $y = f^{-1}(x)$.

5 How to Find the Inverse Function of a One-to-One Function f

STEP 1 Write $y = f(x)$.

STEP 2 Solve this equation for x in terms of y (if possible).

STEP 3 To express f^{-1} as a function of x, interchange x and y. The resulting equation is $y = f^{-1}(x)$.

EXAMPLE 4 Find the inverse function of $f(x) = x^3 + 2$.

SOLUTION According to (5) we first write

$$y = x^3 + 2$$

Then we solve this equation for x:

$$x^3 = y - 2$$
$$x = \sqrt[3]{y - 2}$$

Finally, we interchange x and y:

$$y = \sqrt[3]{x - 2}$$

Therefore, the inverse function is $f^{-1}(x) = \sqrt[3]{x - 2}$.

|||| In Example 4, notice how f^{-1} reverses the effect of f. The function f is the rule "Cube, then add 2"; f^{-1} is the rule "Subtract 2, then take the cube root."

The principle of interchanging x and y to find the inverse function also gives us the method for obtaining the graph of f^{-1} from the graph of f. Since $f(a) = b$ if and only if $f^{-1}(b) = a$, the point (a, b) is on the graph of f if and only if the point (b, a) is on the graph of f^{-1}. But we get the point (b, a) from (a, b) by reflecting about the line $y = x$. (See Figure 8.)

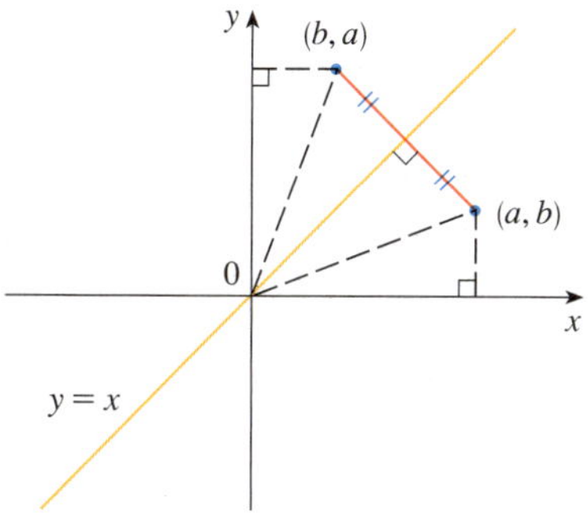

FIGURE 8

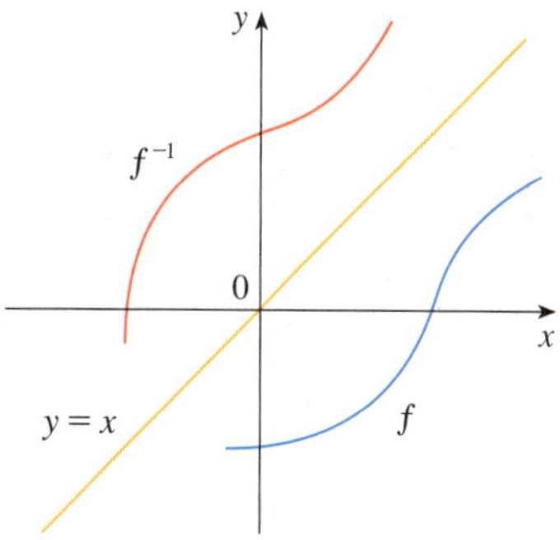

FIGURE 9

Therefore, as illustrated by Figure 9:

> The graph of f^{-1} is obtained by reflecting the graph of f about the line $y = x$.

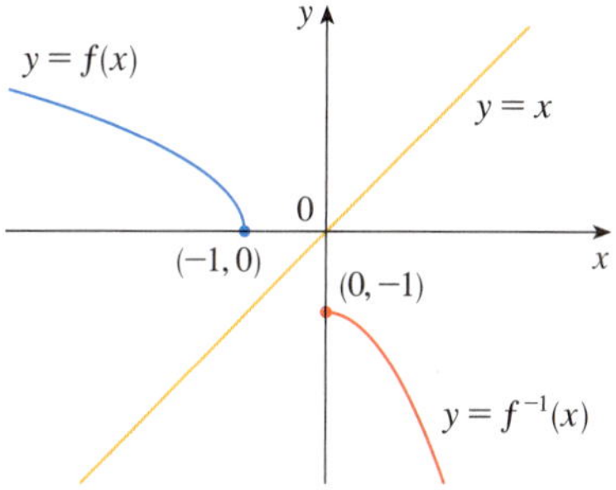

FIGURE 10

EXAMPLE 5 Sketch the graphs of $f(x) = \sqrt{-1 - x}$ and its inverse function using the same coordinate axes.

SOLUTION First we sketch the curve $y = \sqrt{-1 - x}$ (the top half of the parabola $y^2 = -1 - x$, or $x = -y^2 - 1$) and then we reflect about the line $y = x$ to get the graph of f^{-1}. (See Figure 10.) As a check on our graph, notice that the expression for f^{-1} is $f^{-1}(x) = -x^2 - 1$, $x \geqslant 0$. So the graph of f^{-1} is the right half of the parabola $y = -x^2 - 1$ and this seems reasonable from Figure 10.

Logarithmic Functions

If $a > 0$ and $a \neq 1$, the exponential function $f(x) = a^x$ is either increasing or decreasing and so it is one-to-one by the Horizontal Line Test. It therefore has an inverse function f^{-1}, which is called the **logarithmic function with base *a*** and is denoted by $\log_a$. If we use the formulation of an inverse function given by (3),

$$f^{-1}(x) = y \iff f(y) = x$$

then we have

6
$$\log_a x = y \iff a^y = x$$

Thus, if $x > 0$, then $\log_a x$ is the exponent to which the base a must be raised to give x. For example, $\log_{10} 0.001 = -3$ because $10^{-3} = 0.001$.

The cancellation equations (4), when applied to the functions $f(x) = a^x$ and $f^{-1}(x) = \log_a x$, become

7
$$\log_a(a^x) = x \quad \text{for every } x \in \mathbb{R}$$
$$a^{\log_a x} = x \quad \text{for every } x > 0$$

The logarithmic function $\log_a$ has domain $(0, \infty)$ and range $\mathbb{R}$. Its graph is the reflection of the graph of $y = a^x$ about the line $y = x$.

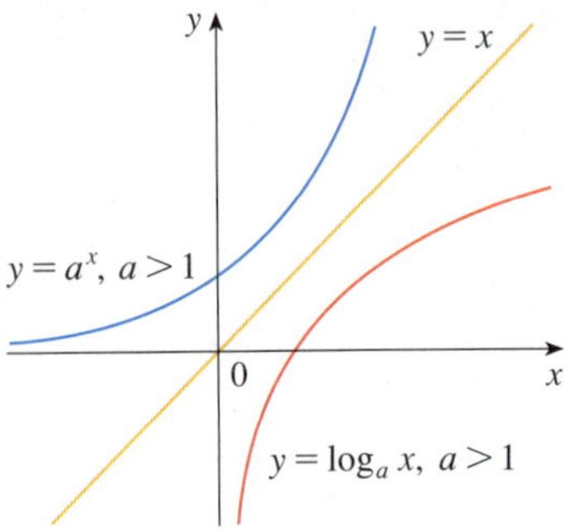

FIGURE 11

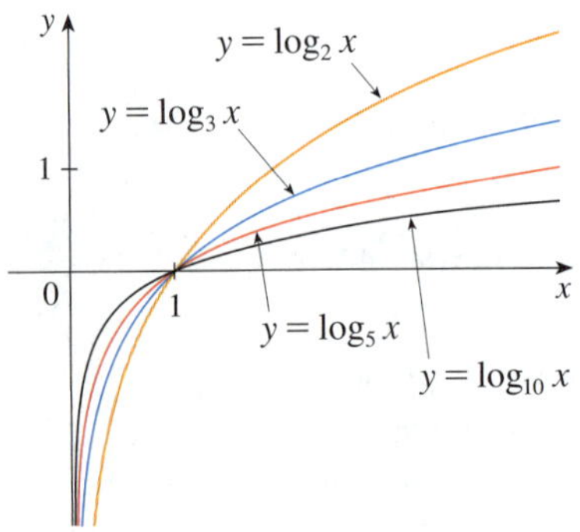

FIGURE 12

|||| NOTATION FOR LOGARITHMS

Most textbooks in calculus and the sciences, as well as calculators, use the notation $\ln x$ for the natural logarithm and $\log x$ for the "common logarithm," $\log_{10} x$. In the more advanced mathematical and scientific literature and in computer languages, however, the notation $\log x$ usually denotes the natural logarithm.

Figure 11 shows the case where $a > 1$. (The most important logarithmic functions have base $a > 1$.) The fact that $y = a^x$ is a very rapidly increasing function for $x > 0$ is reflected in the fact that $y = \log_a x$ is a very slowly increasing function for $x > 1$.

Figure 12 shows the graphs of $y = \log_a x$ with various values of the base a. Since $\log_a 1 = 0$, the graphs of all logarithmic functions pass through the point $(1, 0)$.

The following properties of logarithmic functions follow from the corresponding properties of exponential functions given in Section 1.5.

Laws of Logarithms If x and y are positive numbers, then

1. $\log_a(xy) = \log_a x + \log_a y$

2. $\log_a\left(\dfrac{x}{y}\right) = \log_a x - \log_a y$

3. $\log_a(x^r) = r\log_a x$ (where r is any real number)

EXAMPLE 6 Use the laws of logarithms to evaluate $\log_2 80 - \log_2 5$.

SOLUTION Using Law 2, we have

$$\log_2 80 - \log_2 5 = \log_2\left(\frac{80}{5}\right) = \log_2 16 = 4$$

because $2^4 = 16$.

|||| Natural Logarithms

Of all possible bases a for logarithms, we will see in Chapter 3 that the most convenient choice of a base is the number e, which was defined in Section 1.5. The logarithm with base e is called the **natural logarithm** and has a special notation:

$$\log_e x = \ln x$$

If we put $a = e$ and replace $\log_e$ with "ln" in (6) and (7), then the defining properties of the natural logarithm function become

8 $$\ln x = y \iff e^y = x$$

9 $$\ln(e^x) = x \qquad x \in \mathbb{R}$$
$$e^{\ln x} = x \qquad x > 0$$

In particular, if we set $x = 1$, we get

$$\ln e = 1$$

EXAMPLE 7 Find x if $\ln x = 5$.

SOLUTION 1 From (8) we see that

$$\ln x = 5 \qquad \text{means} \qquad e^5 = x$$

Therefore, $x = e^5$.

(If you have trouble working with the "ln" notation, just replace it by $\log_e$. Then the equation becomes $\log_e x = 5$; so, by the definition of logarithm, $e^5 = x$.)

SOLUTION 2 Start with the equation

$$\ln x = 5$$

and apply the exponential function to both sides of the equation:

$$e^{\ln x} = e^5$$

But the second cancellation equation in (9) says that $e^{\ln x} = x$. Therefore, $x = e^5$.

EXAMPLE 8 Solve the equation $e^{5-3x} = 10$.

SOLUTION We take natural logarithms of both sides of the equation and use (9):

$$\begin{aligned} \ln(e^{5-3x}) &= \ln 10 \\ 5 - 3x &= \ln 10 \\ 3x &= 5 - \ln 10 \\ x &= \tfrac{1}{3}(5 - \ln 10) \end{aligned}$$

Since the natural logarithm is found on scientific calculators, we can approximate the solution to four decimal places: $x \approx 0.8991$.

EXAMPLE 9 Express $\ln a + \frac{1}{2}\ln b$ as a single logarithm.

SOLUTION Using Laws 3 and 1 of logarithms, we have

$$\begin{aligned} \ln a + \tfrac{1}{2}\ln b &= \ln a + \ln b^{1/2} \\ &= \ln a + \ln\sqrt{b} \\ &= \ln\left(a\sqrt{b}\right) \end{aligned}$$

The following formula shows that logarithms with any base can be expressed in terms of the natural logarithm.

10 Change of Base Formula For any positive number a ($a \neq 1$), we have

$$\log_a x = \frac{\ln x}{\ln a}$$

Proof Let $y = \log_a x$. Then, from (6), we have $a^y = x$. Taking natural logarithms of both sides of this equation, we get $y \ln a = \ln x$. Therefore

$$y = \frac{\ln x}{\ln a}$$

Scientific calculators have a key for natural logarithms, so Formula 10 enables us to use a calculator to compute a logarithm with any base (as shown in the next example). Similarly, Formula 10 allows us to graph any logarithmic function on a graphing calculator or computer (see Exercises 43 and 44).

EXAMPLE 10 Evaluate $\log_8 5$ correct to six decimal places.

SOLUTION Formula 10 gives

$$\log_8 5 = \frac{\ln 5}{\ln 8} \approx 0.773976$$

EXAMPLE 11 In Example 3 in Section 1.5 we showed that the mass of ^{90}Sr that remains from a 24-mg sample after t years is $m = f(t) = 24 \cdot 2^{-t/25}$. Find the inverse of this function and interpret it.

SOLUTION We need to solve the equation $m = 24 \cdot 2^{-t/25}$ for t. We start by isolating the exponential and taking natural logarithms of both sides:

$$2^{-t/25} = \frac{m}{24}$$

$$\ln(2^{-t/25}) = \ln\left(\frac{m}{24}\right)$$

$$-\frac{t}{25}\ln 2 = \ln m - \ln 24$$

$$t = -\frac{25}{\ln 2}(\ln m - \ln 24) = \frac{25}{\ln 2}(\ln 24 - \ln m)$$

So the inverse function is

$$f^{-1}(m) = \frac{25}{\ln 2}(\ln 24 - \ln m)$$

This function gives the time required for the mass to decay to m milligrams. In particular, the time required for the mass to be reduced to 5 mg is

$$t = f^{-1}(5) = \frac{25}{\ln 2}(\ln 24 - \ln 5) \approx 56.58 \text{ years}$$

This answer agrees with the graphical estimate that we made in Example 3 in Section 1.5.

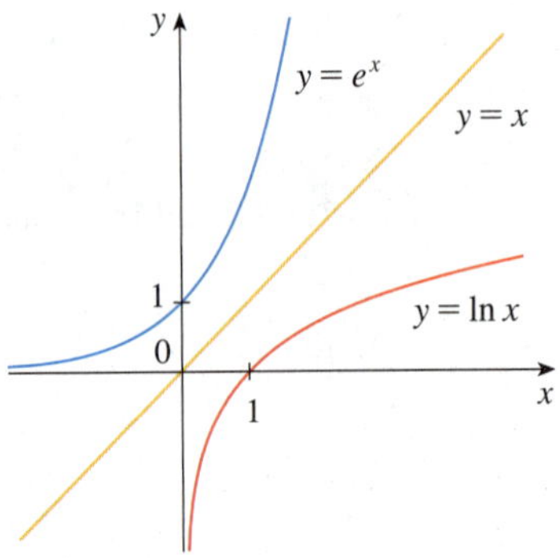

FIGURE 13

The graphs of the exponential function $y = e^x$ and its inverse function, the natural logarithm function, are shown in Figure 13. Because the curve $y = e^x$ crosses the y-axis with a slope of 1, it follows that the reflected curve $y = \ln x$ crosses the x-axis with a slope of 1.

In common with all other logarithmic functions with base greater than 1, the natural logarithm is an increasing function defined on $(0, \infty)$ and the y-axis is a vertical asymptote. (This means that the values of $\ln x$ become very large negative as x approaches 0.)

EXAMPLE 12 Sketch the graph of the function $y = \ln(x - 2) - 1$.

SOLUTION We start with the graph of $y = \ln x$ as given in Figure 13. Using the transformations of Section 1.3, we shift it 2 units to the right to get the graph of $y = \ln(x - 2)$ and then we shift it 1 unit downward to get the graph of $y = \ln(x - 2) - 1$. (See Figure 14.)

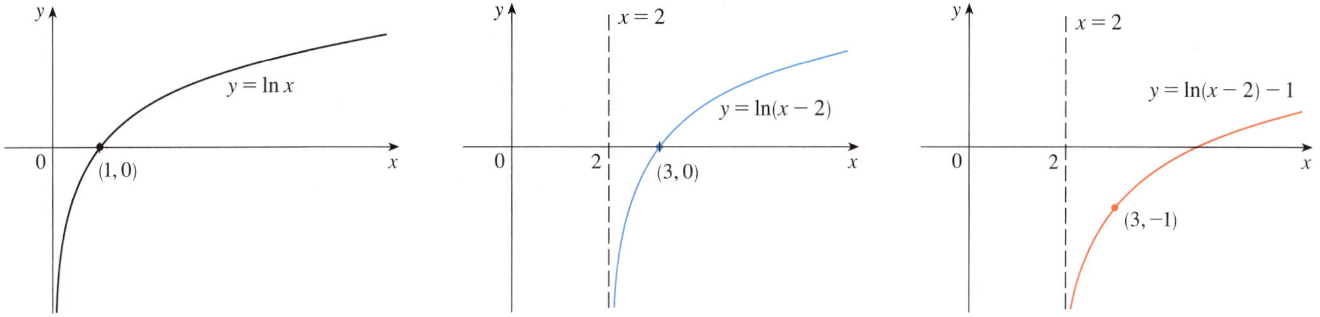

FIGURE 14

Although $\ln x$ is an increasing function, it grows *very* slowly when $x > 1$. In fact, $\ln x$ grows more slowly than any positive power of x. To illustrate this fact, we compare approximate values of the functions $y = \ln x$ and $y = x^{1/2} = \sqrt{x}$ in the following table and we graph them in Figures 15 and 16. You can see that initially the graphs of $y = \sqrt{x}$ and $y = \ln x$ grow at comparable rates, but eventually the root function far surpasses the logarithm.

x	1	2	5	10	50	100	500	1000	10,000	100,000
$\ln x$	0	0.69	1.61	2.30	3.91	4.6	6.2	6.9	9.2	11.5
$\sqrt{x}$	1	1.41	2.24	3.16	7.07	10.0	22.4	31.6	100	316
$\frac{\ln x}{\sqrt{x}}$	0	0.49	0.72	0.73	0.55	0.46	0.28	0.22	0.09	0.04

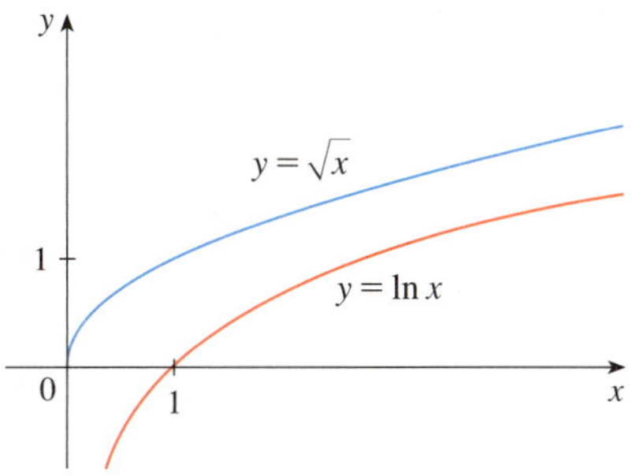

FIGURE 15

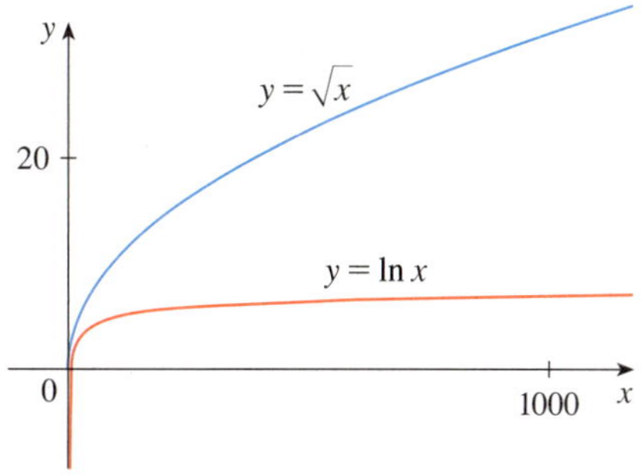

FIGURE 16

Inverse Trigonometric Functions

When we try to find the inverse trigonometric functions, we have a slight difficulty: Because the trigonometric functions are not one-to-one, they don't have inverse functions. The difficulty is overcome by restricting the domains of these functions so that they become one-to-one.

You can see from Figure 17 that the sine function $y = \sin x$ is not one-to-one (use the Horizontal Line Test). But the function $f(x) = \sin x$, $-\pi/2 \leqslant x \leqslant \pi/2$ (see Figure 18), *is* one-to-one. The inverse function of this restricted sine function f exists and is denoted by $\sin^{-1}$ or arcsin. It is called the **inverse sine function** or the **arcsine function**.

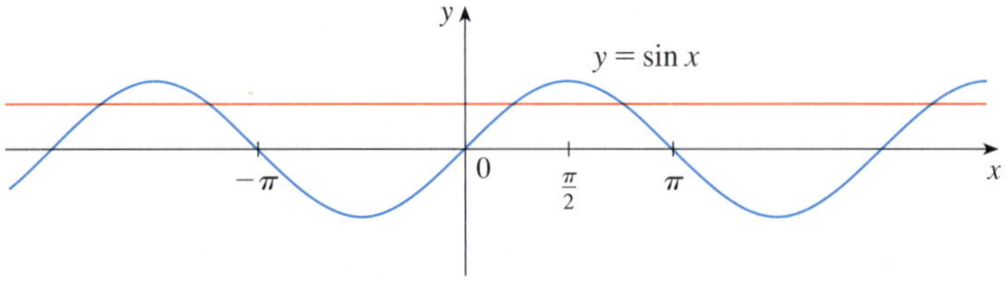

FIGURE 17

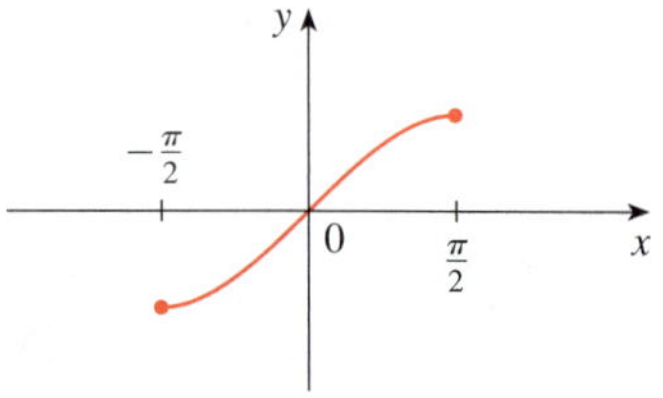

FIGURE 18 $y = \sin x, -\frac{\pi}{2} \leqslant x \leqslant \frac{\pi}{2}$

Since the definition of an inverse function says that

$$f^{-1}(x) = y \iff f(y) = x$$

we have

$$\sin^{-1}x = y \iff \sin y = x \quad \text{and} \quad -\frac{\pi}{2} \leqslant y \leqslant \frac{\pi}{2}$$

$\sin^{-1}x \neq \dfrac{1}{\sin x}$

Thus, if $-1 \leqslant x \leqslant 1$, $\sin^{-1}x$ is the number between $-\pi/2$ and $\pi/2$ whose sine is x.

EXAMPLE 13 Evaluate (a) $\sin^{-1}(\frac{1}{2})$ and (b) $\tan(\arcsin\frac{1}{3})$.

SOLUTION

(a) We have

$$\sin^{-1}(\tfrac{1}{2}) = \frac{\pi}{6}$$

because $\sin(\pi/6) = \frac{1}{2}$ and $\pi/6$ lies between $-\pi/2$ and $\pi/2$.

(b) Let $\theta = \arcsin\frac{1}{3}$, so $\sin\theta = \frac{1}{3}$. Then we can draw a right triangle with angle θ as in Figure 19 and deduce from the Pythagorean Theorem that the third side has length $\sqrt{9-1} = 2\sqrt{2}$. This enables us to read from the triangle that

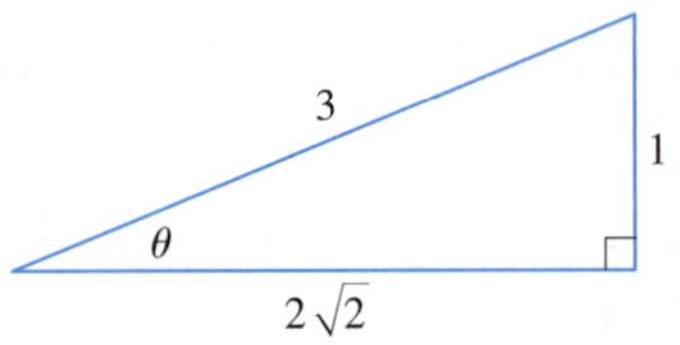

FIGURE 19

$$\tan(\arcsin\tfrac{1}{3}) = \tan\theta = \frac{1}{2\sqrt{2}}$$

The cancellation equations for inverse functions become, in this case,

$$\sin^{-1}(\sin x) = x \quad \text{for } -\frac{\pi}{2} \leqslant x \leqslant \frac{\pi}{2}$$

$$\sin(\sin^{-1}x) = x \quad \text{for } -1 \leqslant x \leqslant 1$$

The inverse sine function, $\sin^{-1}$, has domain $[-1, 1]$ and range $[-\pi/2, \pi/2]$, and its graph, shown in Figure 20, is obtained from that of the restricted sine function (Figure 18) by reflection about the line $y = x$.

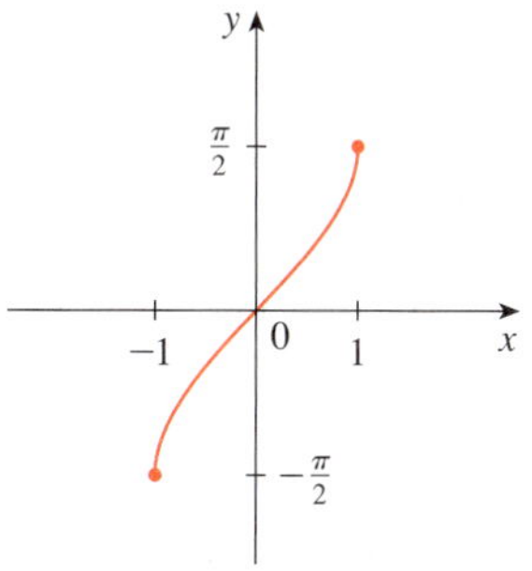

FIGURE 20
$y = \sin^{-1}x = \arcsin x$

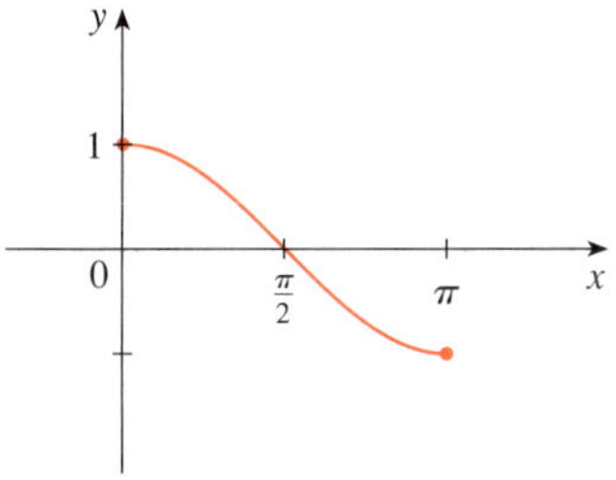

FIGURE 21
$y = \cos x,\ 0 \leqslant x \leqslant \pi$

The **inverse cosine function** is handled similarly. The restricted cosine function $f(x) = \cos x$, $0 \leqslant x \leqslant \pi$, is one-to-one (see Figure 21) and so it has an inverse function denoted by $\cos^{-1}$ or arccos.

$$\cos^{-1}x = y \iff \cos y = x \quad \text{and} \quad 0 \leqslant y \leqslant \pi$$

The cancellation equations are

$$\cos^{-1}(\cos x) = x \quad \text{for } 0 \leqslant x \leqslant \pi$$
$$\cos(\cos^{-1}x) = x \quad \text{for } -1 \leqslant x \leqslant 1$$

The inverse cosine function, $\cos^{-1}$, has domain $[-1, 1]$ and range $[0, \pi]$. Its graph is shown in Figure 22.

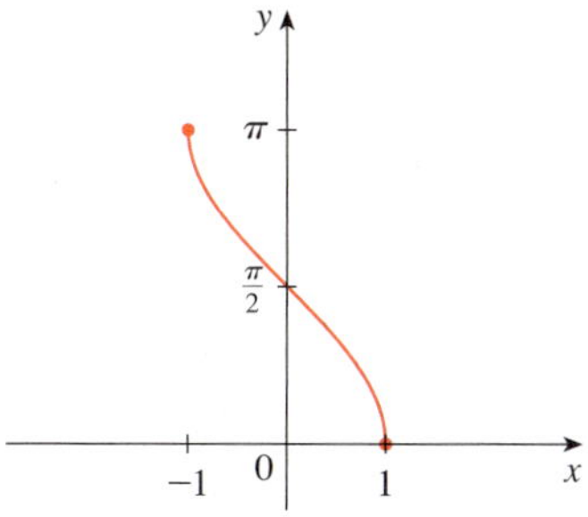

FIGURE 22
$y = \cos^{-1}x = \arccos x$

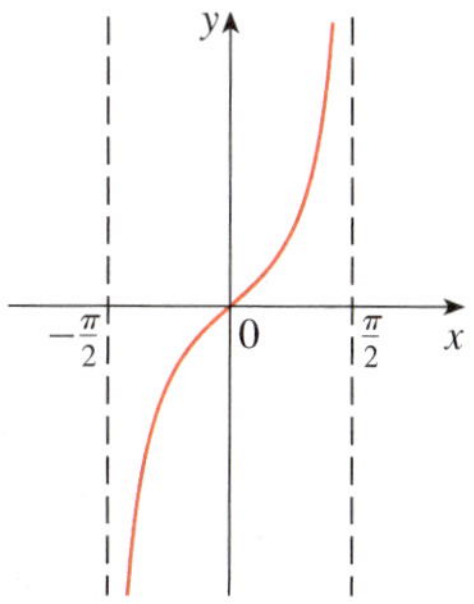

FIGURE 23
$y = \tan x,\ -\frac{\pi}{2} < x < \frac{\pi}{2}$

The tangent function can be made one-to-one by restricting it to the interval $(-\pi/2, \pi/2)$. Thus, the **inverse tangent function** is defined as the inverse of the function $f(x) = \tan x$, $-\pi/2 < x < \pi/2$. (See Figure 23.) It is denoted by $\tan^{-1}$ or arctan.

$$\tan^{-1}x = y \iff \tan y = x \quad \text{and} \quad -\frac{\pi}{2} < y < \frac{\pi}{2}$$

EXAMPLE 14 Simplify the expression $\cos(\tan^{-1}x)$.

SOLUTION 1 Let $y = \tan^{-1}x$. Then $\tan y = x$ and $-\pi/2 < y < \pi/2$. We want to find $\cos y$ but, since $\tan y$ is known, it is easier to find $\sec y$ first:

$$\sec^2 y = 1 + \tan^2 y = 1 + x^2$$

$$\sec y = \sqrt{1 + x^2} \qquad \text{(since } \sec y > 0 \text{ for } -\pi/2 < y < \pi/2\text{)}$$

Thus $$\cos(\tan^{-1}x) = \cos y = \frac{1}{\sec y} = \frac{1}{\sqrt{1 + x^2}}$$

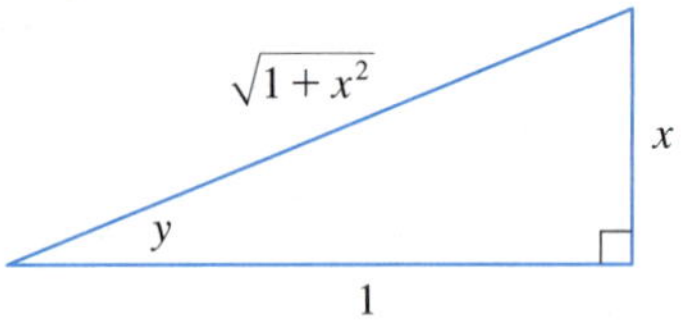

FIGURE 24

SOLUTION 2 Instead of using trigonometric identities as in Solution 1, it is perhaps easier to use a diagram. If $y = \tan^{-1}x$, then $\tan y = x$, and we can read from Figure 24 (which illustrates the case $y > 0$) that

$$\cos(\tan^{-1}x) = \cos y = \frac{1}{\sqrt{1 + x^2}}$$

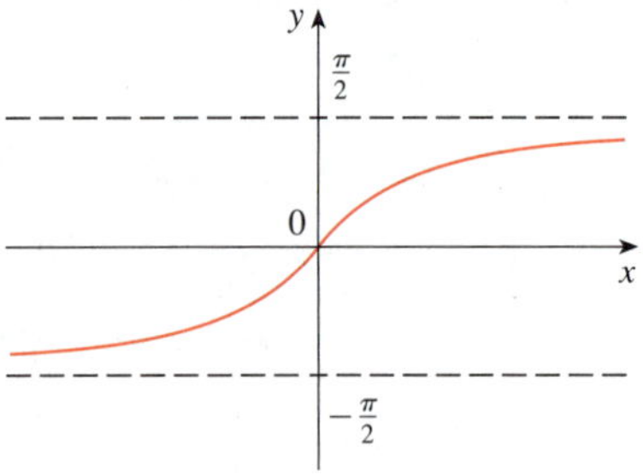

FIGURE 25
$y = \tan^{-1}x = \arctan x$

The inverse tangent function, $\tan^{-1} = \arctan$, has domain $\mathbb{R}$ and range $(-\pi/2, \pi/2)$. Its graph is shown in Figure 25.

We know that the lines $x = \pm\pi/2$ are vertical asymptotes of the graph of tan. Since the graph of $\tan^{-1}$ is obtained by reflecting the graph of the restricted tangent function about the line $y = x$, it follows that the lines $y = \pi/2$ and $y = -\pi/2$ are horizontal asymptotes of the graph of $\tan^{-1}$.

The remaining inverse trigonometric functions are not used as frequently and are summarized here.

11
$$y = \csc^{-1}x \; (|x| \geq 1) \iff \csc y = x \text{ and } y \in (0, \pi/2] \cup (\pi, 3\pi/2]$$

$$y = \sec^{-1}x \; (|x| \geq 1) \iff \sec y = x \text{ and } y \in [0, \pi/2) \cup [\pi, 3\pi/2)$$

$$y = \cot^{-1}x \; (x \in \mathbb{R}) \iff \cot y = x \text{ and } y \in (0, \pi)$$

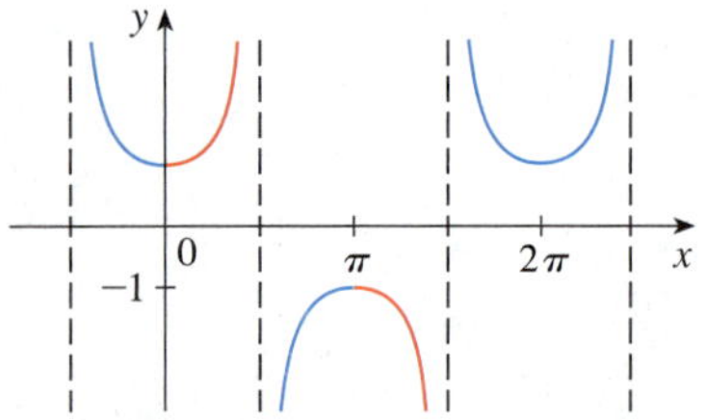

FIGURE 26
$y = \sec x$

The choice of intervals for y in the definitions of $\csc^{-1}$ and $\sec^{-1}$ is not universally agreed upon. For instance, some authors use $y \in [0, \pi/2) \cup (\pi/2, \pi]$ in the definition of $\sec^{-1}$. [You can see from the graph of the secant function in Figure 26 that both this choice and the one in (11) will work.]

1.6 Exercises

1. (a) What is a one-to-one function?
(b) How can you tell from the graph of a function whether it is one-to-one?
(c) If you are given the graph of f, how do you find the graph of f^{-1}?

2. (a) Suppose f is a one-to-one function with domain A and range B. How is the inverse function f^{-1} defined? What is the domain of f^{-1}? What is the range of f^{-1}?
(b) If you are given a formula for f, how do you find a formula for f^{-1}?

3–14 A function is given by a table of values, a graph, a formula, or a verbal description. Determine whether it is one-to-one.

3.

x	1	2	3	4	5	6
$f(x)$	1.5	2.0	3.6	5.3	2.8	2.0

4.

x	1	2	3	4	5	6
$f(x)$	1	2	4	8	16	32

5.

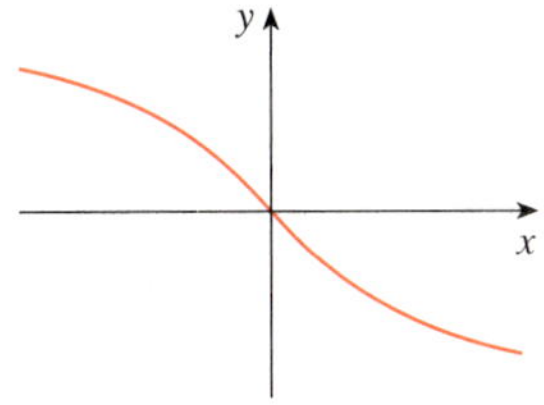

6.

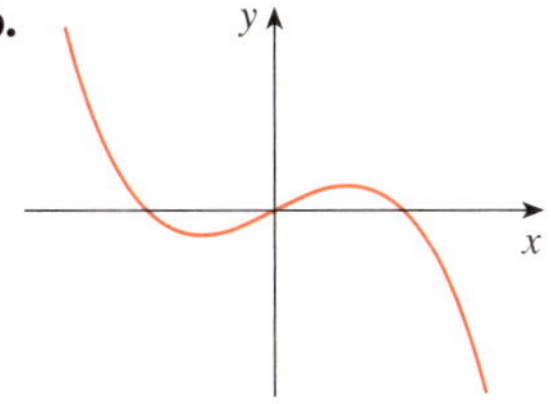

7.

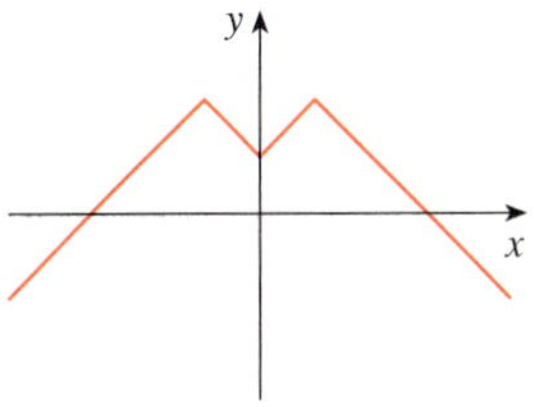

8.

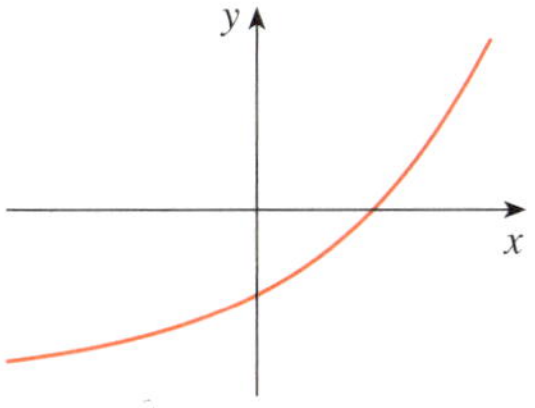

9. $f(x) = \frac{1}{2}(x + 5)$

10. $f(x) = 1 + 4x - x^2$

11. $g(x) = |x|$

12. $g(x) = \sqrt{x}$

13. $f(t)$ is the height of a football t seconds after kickoff.

14. $f(t)$ is your height at age t.

15–16 |||| Use a graph to decide whether f is one-to-one.

15. $f(x) = x^3 - x$

16. $f(x) = x^3 + x$

17. If f is a one-to-one function such that $f(2) = 9$, what is $f^{-1}(9)$?

18. Let $f(x) = 3 + x^2 + \tan(\pi x/2)$, where $-1 < x < 1$.
(a) Find $f^{-1}(3)$.
(b) Find $f(f^{-1}(5))$.

19. If $g(x) = 3 + x + e^x$, find $g^{-1}(4)$.

20. The graph of f is given.
(a) Why is f one-to-one?
(b) State the domain and range of f^{-1}.
(c) Estimate the value of $f^{-1}(1)$.

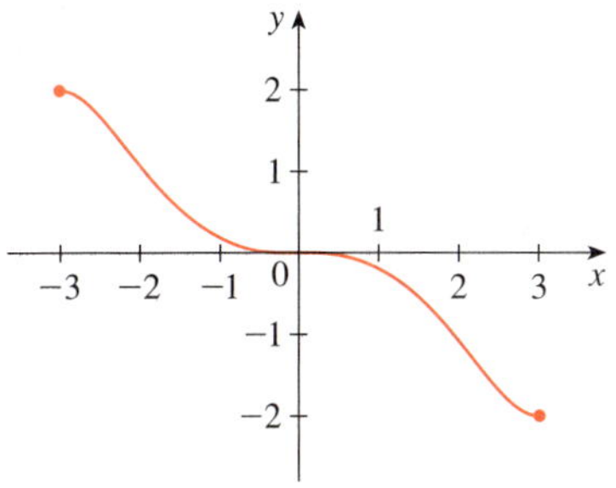

21. The formula $C = \frac{5}{9}(F - 32)$, where $F \geqslant -459.67$, expresses the Celsius temperature C as a function of the Fahrenheit temperature F. Find a formula for the inverse function and interpret it. What is the domain of the inverse function?

22. In the theory of relativity, the mass of a particle with speed v is

$$m = f(v) = \frac{m_0}{\sqrt{1 - v^2/c^2}}$$

where m_0 is the rest mass of the particle and c is the speed of light in a vacuum. Find the inverse function of f and explain its meaning.

23–28 |||| Find a formula for the inverse of the function.

23. $f(x) = \sqrt{10 - 3x}$

24. $f(x) = \dfrac{4x - 1}{2x + 3}$

25. $f(x) = e^{x^3}$

26. $y = 2x^3 + 3$

27. $y = \ln(x + 3)$

28. $y = \dfrac{1 + e^x}{1 - e^x}$

29–30 |||| Find an explicit formula for f^{-1} and use it to graph f^{-1}, f, and the line $y = x$ on the same screen. To check your work, see whether the graphs of f and f^{-1} are reflections about the line.

29. $f(x) = 1 - 2/x^2, \quad x > 0$

30. $f(x) = \sqrt{x^2 + 2x}, \quad x > 0$

31. Use the given graph of f to sketch the graph of f^{-1}.

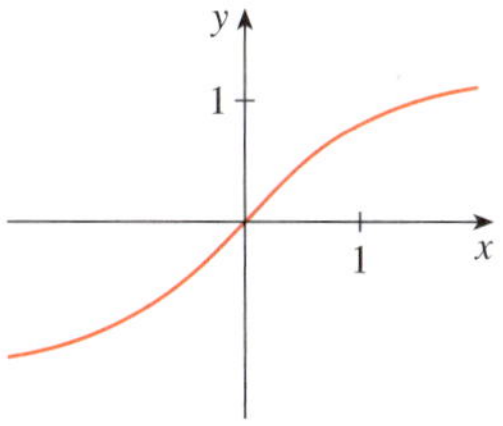

32. Use the given graph of f to sketch the graphs of f^{-1} and $1/f$.

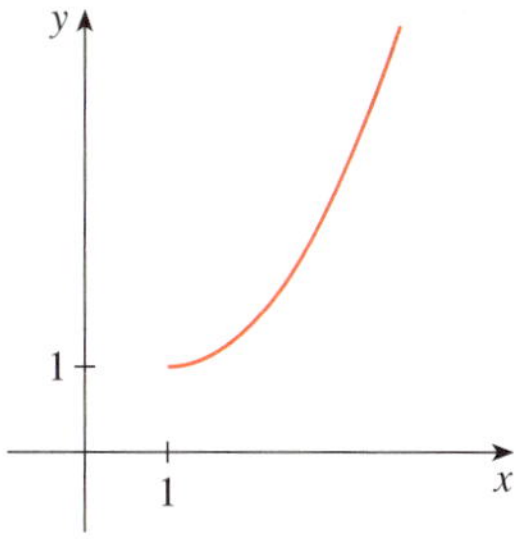

33. (a) How is the logarithmic function $y = \log_a x$ defined?
(b) What is the domain of this function?
(c) What is the range of this function?
(d) Sketch the general shape of the graph of the function $y = \log_a x$ if $a > 1$.

34. (a) What is the natural logarithm?
(b) What is the common logarithm?
(c) Sketch the graphs of the natural logarithm function and the natural exponential function with a common set of axes.

35–38 Find the exact value of each expression.

35. (a) $\log_2 64$ (b) $\log_6 \frac{1}{36}$

36. (a) $\log_8 2$ (b) $\ln e^{\sqrt{2}}$

37. (a) $\log_{10} 1.25 + \log_{10} 80$
(b) $\log_5 10 + \log_5 20 - 3 \log_5 2$

38. (a) $2^{(\log_2 3 + \log_2 5)}$ (b) $e^{3 \ln 2}$

39–41 Express the given quantity as a single logarithm.

39. $2 \ln 4 - \ln 2$

40. $\ln x + a \ln y - b \ln z$

41. $\ln(1 + x^2) + \frac{1}{2} \ln x - \ln \sin x$

42. Use Formula 10 to evaluate each logarithm correct to six decimal places.
(a) $\log_{12} 10$ (b) $\log_2 8.4$

43–44 Use Formula 10 to graph the given functions on a common screen. How are these graphs related?

43. $y = \log_{1.5} x$, $y = \ln x$, $y = \log_{10} x$, $y = \log_{50} x$

44. $y = \ln x$, $y = \log_{10} x$, $y = e^x$, $y = 10^x$

45. Suppose that the graph of $y = \log_2 x$ is drawn on a coordinate grid where the unit of measurement is an inch. How many miles to the right of the origin do we have to move before the height of the curve reaches 3 ft?

46. Compare the functions $f(x) = x^{0.1}$ and $g(x) = \ln x$ by graphing both f and g in several viewing rectangles. When does the graph of f finally surpass the graph of g?

47–48 Make a rough sketch of the graph of each function. Do not use a calculator. Just use the graphs given in Figures 12 and 13 and, if necessary, the transformations of Section 1.3.

47. (a) $y = \log_{10}(x + 5)$ (b) $y = -\ln x$

48. (a) $y = \ln(-x)$ (b) $y = \ln |x|$

49–52 Solve each equation for x.

49. (a) $2 \ln x = 1$ (b) $e^{-x} = 5$

50. (a) $e^{2x+3} - 7 = 0$ (b) $\ln(5 - 2x) = -3$

51. (a) $2^{x-5} = 3$ (b) $\ln x + \ln(x - 1) = 1$

52. (a) $\ln(\ln x) = 1$ (b) $e^{ax} = Ce^{bx}$, where $a \neq b$

53–54 Solve each inequality for x.

53. (a) $e^x < 10$ (b) $\ln x > -1$

54. (a) $2 < \ln x < 9$ (b) $e^{2-3x} > 4$

55–56 Find (a) the domain of f and (b) f^{-1} and its domain.

55. $f(x) = \sqrt{3 - e^{2x}}$

56. $f(x) = \ln(2 + \ln x)$

CAS **57.** Graph the function $f(x) = \sqrt{x^3 + x^2 + x + 1}$ and explain why it is one-to-one. Then use a computer algebra system to find an explicit expression for $f^{-1}(x)$. (Your CAS will produce three possible expressions. Explain why two of them are irrelevant in this context.)

CAS **58.** (a) If $g(x) = x^6 + x^4$, $x \geqslant 0$, use a computer algebra system to find an expression for $g^{-1}(x)$.
(b) Use the expression in part (a) to graph $y = g(x)$, $y = x$, and $y = g^{-1}(x)$ on the same screen.

59. If a bacteria population starts with 100 bacteria and doubles every three hours, then the number of bacteria after t hours is $n = f(t) = 100 \cdot 2^{t/3}$. (See Exercise 25 in Section 1.5.)
(a) Find the inverse of this function and explain its meaning.
(b) When will the population reach 50,000?

60. When a camera flash goes off, the batteries immediately begin to recharge the flash's capacitor, which stores electric charge given by

$$Q(t) = Q_0(1 - e^{-t/a})$$

(The maximum charge capacity is Q_0 and t is measured in seconds.)
(a) Find the inverse of this function and explain its meaning.
(b) How long does it take to recharge the capacitor to 90% of capacity if $a = 2$?

61. Starting with the graph of $y = \ln x$, find the equation of the graph that results from
(a) shifting 3 units upward
(b) shifting 3 units to the left
(c) reflecting about the x-axis
(d) reflecting about the y-axis
(e) reflecting about the line $y = x$
(f) reflecting about the x-axis and then about the line $y = x$
(g) reflecting about the y-axis and then about the line $y = x$
(h) shifting 3 units to the left and then reflecting about the line $y = x$

62. (a) If we shift a curve to the left, what happens to its reflection about the line $y = x$? In view of this geometric principle, find an expression for the inverse of $g(x) = f(x + c)$, where f is a one-to-one function.
(b) Find an expression for the inverse of $h(x) = f(cx)$, where $c \neq 0$.

63–68 Find the exact value of each expression.

63. (a) $\sin^{-1}(\sqrt{3}/2)$ (b) $\cos^{-1}(-1)$

64. (a) $\arctan(-1)$ (b) $\csc^{-1} 2$

65. (a) $\tan^{-1}\sqrt{3}$ (b) $\arcsin(-1/\sqrt{2})$

66. (a) $\sec^{-1}\sqrt{2}$ (b) $\arcsin 1$

67. (a) $\sin(\sin^{-1} 0.7)$ (b) $\tan^{-1}\left(\tan \frac{4\pi}{3}\right)$

68. (a) $\sec(\arctan 2)$ (b) $\cos\left(2 \sin^{-1}\left(\frac{5}{13}\right)\right)$

69. Prove that $\cos(\sin^{-1}x) = \sqrt{1 - x^2}$.

70–72 Simplify the expression.

70. $\tan(\sin^{-1}x)$

71. $\sin(\tan^{-1}x)$

72. $\sin(2 \cos^{-1}x)$

73–74 Graph the given functions on the same screen. How are these graphs related?

73. $y = \sin x, \ -\pi/2 \leqslant x \leqslant \pi/2; \quad y = \sin^{-1}x; \quad y = x$

74. $y = \tan x, \ -\pi/2 < x < \pi/2; \quad y = \tan^{-1}x; \quad y = x$

75. Find the domain and range of the function

$$g(x) = \sin^{-1}(3x + 1)$$

76. (a) Graph the function $f(x) = \sin(\sin^{-1}x)$ and explain the appearance of the graph.
(b) Graph the function $g(x) = \sin^{-1}(\sin x)$. How do you explain the appearance of this graph?

1 Review

CONCEPT CHECK

1. (a) What is a function? What are its domain and range?
(b) What is the graph of a function?
(c) How can you tell whether a given curve is the graph of a function?

2. Discuss four ways of representing a function. Illustrate your discussion with examples.

3. (a) What is an even function? How can you tell if a function is even by looking at its graph?
(b) What is an odd function? How can you tell if a function is odd by looking at its graph?

4. What is an increasing function?

5. What is a mathematical model?

6. Give an example of each type of function.
(a) Linear function (b) Power function
(c) Exponential function (d) Quadratic function
(e) Polynomial of degree 5 (f) Rational function

7. Sketch by hand, on the same axes, the graphs of the following functions.
(a) $f(x) = x$ (b) $g(x) = x^2$
(c) $h(x) = x^3$ (d) $j(x) = x^4$

8. Draw, by hand, a rough sketch of the graph of each function.
(a) $y = \sin x$ (b) $y = \tan x$
(c) $y = e^x$ (d) $y = \ln x$
(e) $y = 1/x$ (f) $y = |x|$
(g) $y = \sqrt{x}$ (h) $y = \tan^{-1}x$

9. Suppose that f has domain A and g has domain B.
(a) What is the domain of $f + g$?
(b) What is the domain of fg?
(c) What is the domain of f/g?

10. How is the composite function $f \circ g$ defined? What is its domain?

11. Suppose the graph of f is given. Write an equation for each of the graphs that are obtained from the graph of f as follows.
(a) Shift 2 units upward.
(b) Shift 2 units downward.
(c) Shift 2 units to the right.
(d) Shift 2 units to the left.
(e) Reflect about the x-axis.
(f) Reflect about the y-axis.
(g) Stretch vertically by a factor of 2.
(h) Shrink vertically by a factor of 2.
(i) Stretch horizontally by a factor of 2.
(j) Shrink horizontally by a factor of 2.

12. (a) What is a one-to-one function? How can you tell if a function is one-to-one by looking at its graph?
(b) If f is a one-to-one function, how is its inverse function f^{-1} defined? How do you obtain the graph of f^{-1} from the graph of f?

13. (a) How is the inverse sine function $f(x) = \sin^{-1}x$ defined? What are its domain and range?
(b) How is the inverse cosine function $f(x) = \cos^{-1}x$ defined? What are its domain and range?
(c) How is the inverse tangent function $f(x) = \tan^{-1}x$ defined? What are its domain and range?

TRUE-FALSE QUIZ

Determine whether the statement is true or false. If it is true, explain why. If it is false, explain why or give an example that disproves the statement.

1. If f is a function, then $f(s + t) = f(s) + f(t)$.

2. If $f(s) = f(t)$, then $s = t$.

3. If f is a function, then $f(3x) = 3f(x)$.

4. If $x_1 < x_2$ and f is a decreasing function, then $f(x_1) > f(x_2)$.

5. A vertical line intersects the graph of a function at most once.

6. If f and g are functions, then $f \circ g = g \circ f$.

7. If f is one-to-one, then $f^{-1}(x) = \dfrac{1}{f(x)}$.

8. You can always divide by e^x.

9. If $0 < a < b$, then $\ln a < \ln b$.

10. If $x > 0$, then $(\ln x)^6 = 6 \ln x$.

11. If $x > 0$ and $a > 1$, then $\dfrac{\ln x}{\ln a} = \ln \dfrac{x}{a}$.

EXERCISES

1. Let f be the function whose graph is given.
(a) Estimate the value of $f(2)$.
(b) Estimate the values of x such that $f(x) = 3$.
(c) State the domain of f.
(d) State the range of f.
(e) On what interval is f increasing?
(f) Is f one-to-one? Explain.
(g) Is f even, odd, or neither even nor odd? Explain.

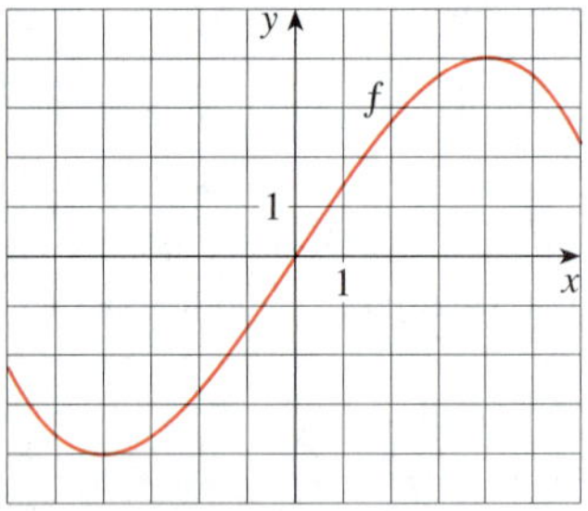

2. The graph of g is given.
(a) State the value of $g(2)$.
(b) Why is g one-to-one?
(c) Estimate the value of $g^{-1}(2)$.
(d) Estimate the domain of g^{-1}.
(e) Sketch the graph of g^{-1}.

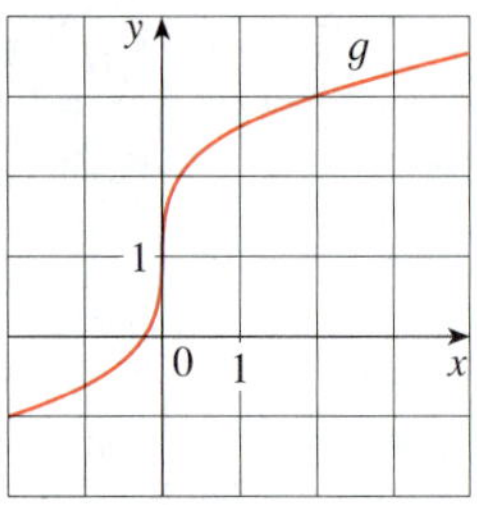

3. The distance traveled by a car is given by the values in the table.

t (seconds)	0	1	2	3	4	5
d (feet)	0	10	32	70	119	178

(a) Use the data to sketch the graph of d as a function of t.
(b) Use the graph to estimate the distance traveled after 4.5 seconds.

4. Sketch a rough graph of the yield of a crop as a function of the amount of fertilizer used.

5–8 |||| Find the domain and range of the function.

5. $f(x) = \sqrt{4 - 3x^2}$

6. $g(x) = 1/(x + 1)$

7. $y = 1 + \sin x$

8. $y = \ln \ln x$

9. Suppose that the graph of f is given. Describe how the graphs of the following functions can be obtained from the graph of f.
(a) $y = f(x) + 8$
(b) $y = f(x + 8)$
(c) $y = 1 + 2f(x)$
(d) $y = f(x - 2) - 2$
(e) $y = -f(x)$
(f) $y = f^{-1}(x)$

10. The graph of f is given. Draw the graphs of the following functions.
(a) $y = f(x - 8)$
(b) $y = -f(x)$
(c) $y = 2 - f(x)$
(d) $y = \frac{1}{2}f(x) - 1$
(e) $y = f^{-1}(x)$
(f) $y = f^{-1}(x + 3)$

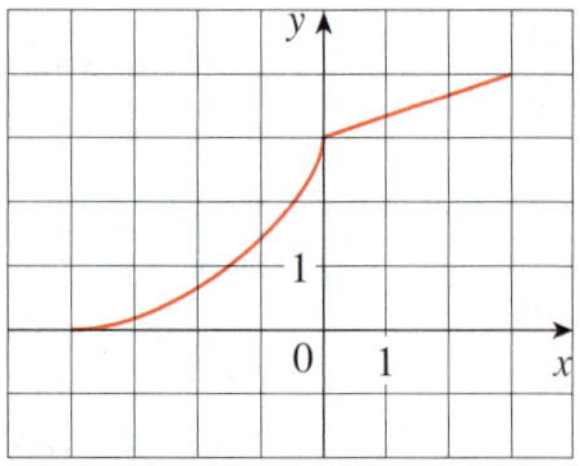

11–16 Use transformations to sketch the graph of the function.

11. $y = -\sin 2x$

12. $y = 3\ln(x - 2)$

13. $y = (1 + e^x)/2$

14. $y = 2 - \sqrt{x}$

15. $f(x) = \dfrac{1}{x + 2}$

16. $f(x) = \begin{cases} -x & \text{if } x < 0 \\ e^x - 1 & \text{if } x \geqslant 0 \end{cases}$

17. Determine whether f is even, odd, or neither even nor odd.
(a) $f(x) = 2x^5 - 3x^2 + 2$
(b) $f(x) = x^3 - x^7$
(c) $f(x) = e^{-x^2}$
(d) $f(x) = 1 + \sin x$

18. Find an expression for the function whose graph consists of the line segment from the point $(-2, 2)$ to the point $(-1, 0)$ together with the top half of the circle with center the origin and radius 1.

19. If $f(x) = \ln x$ and $g(x) = x^2 - 9$, find the functions $f \circ g$, $g \circ f$, $f \circ f$, $g \circ g$, and their domains.

20. Express the function $F(x) = 1/\sqrt{x + \sqrt{x}}$ as a composition of three functions.

21. Life expectancy improved dramatically in the 20th century. The table gives the life expectancy at birth (in years) of males born in the United States.

Birth year	Life expectancy
1900	48.3
1910	51.1
1920	55.2
1930	57.4
1940	62.5
1950	65.6
1960	66.6
1970	67.1
1980	70.0
1990	71.8
2000	73.0

Use a scatter plot to choose an appropriate type of model. Use your model to predict the life span of a male born in the year 2010.

22. A small-appliance manufacturer finds that it costs \$9000 to produce 1000 toaster ovens a week and \$12,000 to produce 1500 toaster ovens a week.
(a) Express the cost as a function of the number of toaster ovens produced, assuming that it is linear. Then sketch the graph.
(b) What is the slope of the graph and what does it represent?
(c) What is the y-intercept of the graph and what does it represent?

23. If $f(x) = 2x + \ln x$, find $f^{-1}(2)$.

24. Find the inverse function of $f(x) = \dfrac{x + 1}{2x + 1}$.

25. Find the exact value of each expression.
(a) $e^{2\ln 3}$
(b) $\log_{10} 25 + \log_{10} 4$
(c) $\tan\left(\arcsin \frac{1}{2}\right)$
(d) $\sin\left(\cos^{-1}\left(\frac{4}{5}\right)\right)$

26. Solve each equation for x.
(a) $e^x = 5$
(b) $\ln x = 2$
(c) $e^{e^x} = 2$
(d) $\tan^{-1} x = 1$

27. The half-life of palladium-100, ^{100}Pd, is four days. (So half of any given quantity of ^{100}Pd will disintegrate in four days.) The initial mass of a sample is one gram.
(a) Find the mass that remains after 16 days.
(b) Find the mass $m(t)$ that remains after t days.
(c) Find the inverse of this function and explain its meaning.
(d) When will the mass be reduced to 0.01 g?

28. The population of a certain species in a limited environment with initial population 100 and carrying capacity 1000 is

$$P(t) = \frac{100{,}000}{100 + 900e^{-t}}$$

where t is measured in years.
(a) Graph this function and estimate how long it takes for the population to reach 900.
(b) Find the inverse of this function and explain its meaning.
(c) Use the inverse function to find the time required for the population to reach 900. Compare with the result of part (a).

29. Graph members of the family of functions $f(x) = \ln(x^2 - c)$ for several values of c. How does the graph change when c changes?

30. Graph the three functions $y = x^a$, $y = a^x$, and $y = \log_a x$ on the same screen for two or three values of $a > 1$. For large values of x, which of these functions has the largest values and which has the smallest values?

PRINCIPLES OF PROBLEM SOLVING

There are no hard and fast rules that will ensure success in solving problems. However, it is possible to outline some general steps in the problem-solving process and to give some principles that may be useful in the solution of certain problems. These steps and principles are just common sense made explicit. They have been adapted from George Polya's book *How To Solve It*.

1 UNDERSTAND THE PROBLEM

The first step is to read the problem and make sure that you understand it clearly. Ask yourself the following questions:

What is the unknown?

What are the given quantities?

What are the given conditions?

For many problems it is useful to

draw a diagram

and identify the given and required quantities on the diagram.

Usually it is necessary to

introduce suitable notation

In choosing symbols for the unknown quantities we often use letters such as a, b, c, m, n, x, and y, but in some cases it helps to use initials as suggestive symbols; for instance, V for volume or t for time.

2 THINK OF A PLAN

Find a connection between the given information and the unknown that will enable you to calculate the unknown. It often helps to ask yourself explicitly: "How can I relate the given to the unknown?" If you don't see a connection immediately, the following ideas may be helpful in devising a plan.

Try to Recognize Something Familiar Relate the given situation to previous knowledge. Look at the unknown and try to recall a more familiar problem that has a similar unknown.

Try to Recognize Patterns Some problems are solved by recognizing that some kind of pattern is occurring. The pattern could be geometric, or numerical, or algebraic. If you can see regularity or repetition in a problem, you might be able to guess what the continuing pattern is and then prove it.

Use Analogy Try to think of an analogous problem, that is, a similar problem, a related problem, but one that is easier than the original problem. If you can solve the similar, simpler problem, then it might give you the clues you need to solve the original, more difficult problem. For instance, if a problem involves very large numbers, you could first try a similar problem with smaller numbers. Or if the problem involves three-dimensional geometry, you could look for a similar problem in two-dimensional geometry. Or if the problem you start with is a general one, you could first try a special case.

Introduce Something Extra It may sometimes be necessary to introduce something new, an auxiliary aid, to help make the connection between the given and the unknown. For instance, in a problem where a diagram is useful the auxiliary aid could be a new line drawn in a diagram. In a more algebraic problem it could be a new unknown that is related to the original unknown.

Take Cases We may sometimes have to split a problem into several cases and give a different argument for each of the cases. For instance, we often have to use this strategy in dealing with absolute value.

Work Backward Sometimes it is useful to imagine that your problem is solved and work backward, step by step, until you arrive at the given data. Then you may be able to reverse your steps and thereby construct a solution to the original problem. This procedure is commonly used in solving equations. For instance, in solving the equation $3x - 5 = 7$, we suppose that x is a number that satisfies $3x - 5 = 7$ and work backward. We add 5 to each side of the equation and then divide each side by 3 to get $x = 4$. Since each of these steps can be reversed, we have solved the problem.

Establish Subgoals In a complex problem it is often useful to set subgoals (in which the desired situation is only partially fulfilled). If we can first reach these subgoals, then we may be able to build on them to reach our final goal.

Indirect Reasoning Sometimes it is appropriate to attack a problem indirectly. In using proof by contradiction to prove that P implies Q, we assume that P is true and Q is false and try to see why this can't happen. Somehow we have to use this information and arrive at a contradiction to what we absolutely know is true.

Mathematical Induction In proving statements that involve a positive integer n, it is frequently helpful to use the following principle.

Principle of Mathematical Induction Let S_n be a statement about the positive integer n. Suppose that

1. S_1 is true.
2. S_{k+1} is true whenever S_k is true.

Then S_n is true for all positive integers n.

This is reasonable because, since S_1 is true, it follows from condition 2 (with $k = 1$) that S_2 is true. Then, using condition 2 with $k = 2$, we see that S_3 is true. Again using condition 2, this time with $k = 3$, we have that S_4 is true. This procedure can be followed indefinitely.

3 CARRY OUT THE PLAN

In Step 2 a plan was devised. In carrying out that plan we have to check each stage of the plan and write the details that prove that each stage is correct.

4 LOOK BACK

Having completed our solution, it is wise to look back over it, partly to see if we have made errors in the solution and partly to see if we can think of an easier way to solve the problem. Another reason for looking back is that it will familiarize us with the method of solution and this may be useful for solving a future problem. Descartes said, "Every problem that I solved became a rule which served afterwards to solve other problems."

These principles of problem solving are illustrated in the following examples. Before you look at the solutions, try to solve these problems yourself, referring to these Principles of Problem Solving if you get stuck. You may find it useful to refer to this section from time to time as you solve the exercises in the remaining chapters of this book.

EXAMPLE 1 Express the hypotenuse h of a right triangle with area 25 m^2 as a function of its perimeter P.

||| Understand the problem

SOLUTION Let's first sort out the information by identifying the unknown quantity and the data:

Unknown: hypotenuse h

Given quantities: perimeter P, area 25 m^2

||| Draw a diagram

It helps to draw a diagram and we do so in Figure 1.

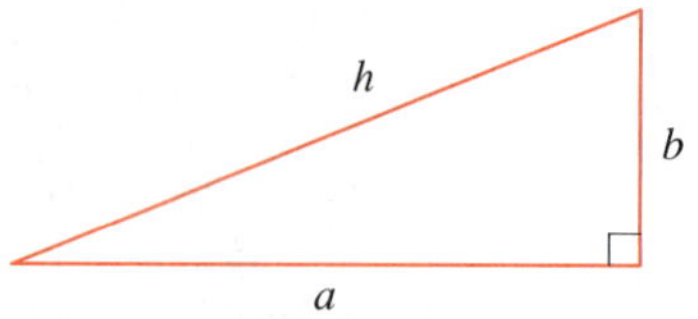

FIGURE 1

||| Connect the given with the unknown

||| Introduce something extra

In order to connect the given quantities to the unknown, we introduce two extra variables a and b, which are the lengths of the other two sides of the triangle. This enables us to express the given condition, which is that the triangle is right-angled, by the Pythagorean Theorem:

$$h^2 = a^2 + b^2$$

The other connections among the variables come by writing expressions for the area and perimeter:

$$25 = \tfrac{1}{2}ab \qquad P = a + b + h$$

Since P is given, notice that we now have three equations in the three unknowns a, b, and h:

$$\boxed{1} \qquad h^2 = a^2 + b^2$$

$$\boxed{2} \qquad 25 = \tfrac{1}{2}ab$$

$$\boxed{3} \qquad P = a + b + h$$

||| Relate to the familiar

Although we have the correct number of equations, they are not easy to solve in a straightforward fashion. But if we use the problem-solving strategy of trying to recognize something familiar, then we can solve these equations by an easier method. Look at the right sides of Equations 1, 2, and 3. Do these expressions remind you of anything familiar? Notice that they contain the ingredients of a familiar formula:

$$(a + b)^2 = a^2 + 2ab + b^2$$

Using this idea, we express $(a + b)^2$ in two ways. From Equations 1 and 2 we have

$$(a + b)^2 = (a^2 + b^2) + 2ab = h^2 + 4(25)$$

From Equation 3 we have

$$(a + b)^2 = (P - h)^2 = P^2 - 2Ph + h^2$$

Thus

$$h^2 + 100 = P^2 - 2Ph + h^2$$

$$2Ph = P^2 - 100$$

$$h = \frac{P^2 - 100}{2P}$$

This is the required expression for h as a function of P.

As the next example illustrates, it is often necessary to use the problem-solving principle of *taking cases* when dealing with absolute values.

EXAMPLE 2 Solve the inequality $|x - 3| + |x + 2| < 11$.

SOLUTION Recall the definition of absolute value:

$$|x| = \begin{cases} x & \text{if } x \geqslant 0 \\ -x & \text{if } x < 0 \end{cases}$$

It follows that

$$|x - 3| = \begin{cases} x - 3 & \text{if } x - 3 \geqslant 0 \\ -(x - 3) & \text{if } x - 3 < 0 \end{cases}$$

$$= \begin{cases} x - 3 & \text{if } x \geqslant 3 \\ -x + 3 & \text{if } x < 3 \end{cases}$$

Similarly

$$|x + 2| = \begin{cases} x + 2 & \text{if } x + 2 \geqslant 0 \\ -(x + 2) & \text{if } x + 2 < 0 \end{cases}$$

$$= \begin{cases} x + 2 & \text{if } x \geqslant -2 \\ -x - 2 & \text{if } x < -2 \end{cases}$$

|||| Take cases

These expressions show that we must consider three cases:

$$x < -2 \qquad -2 \leqslant x < 3 \qquad x \geqslant 3$$

CASE I ▫ If $x < -2$, we have

$$|x - 3| + |x + 2| < 11$$
$$-x + 3 - x - 2 < 11$$
$$-2x < 10$$
$$x > -5$$

CASE II ▫ If $-2 \leqslant x < 3$, the given inequality becomes

$$-x + 3 + x + 2 < 11$$
$$5 < 11 \qquad \text{(always true)}$$

CASE III ▫ If $x \geqslant 3$, the inequality becomes

$$x - 3 + x + 2 < 11$$
$$2x < 12$$
$$x < 6$$

Combining cases I, II, and III, we see that the inequality is satisfied when $-5 < x < 6$. So the solution is the interval $(-5, 6)$.

In the following example we first guess the answer by looking at special cases and recognizing a pattern. Then we prove it by mathematical induction.

In using the Principle of Mathematical Induction, we follow three steps:

STEP 1 Prove that S_n is true when $n = 1$.

STEP 2 Assume that S_n is true when $n = k$ and deduce that S_n is true when $n = k + 1$.

STEP 3 Conclude that S_n is true for all n by the Principle of Mathematical Induction.

EXAMPLE 3 If $f_0(x) = x/(x + 1)$ and $f_{n+1} = f_0 \circ f_n$ for $n = 0, 1, 2, \ldots$, find a formula for $f_n(x)$.

Analogy: Try a similar, simpler problem

SOLUTION We start by finding formulas for $f_n(x)$ for the special cases $n = 1, 2,$ and 3.

$$f_1(x) = (f_0 \circ f_0)(x) = f_0(f_0(x)) = f_0\left(\frac{x}{x+1}\right)$$

$$= \frac{\dfrac{x}{x+1}}{\dfrac{x}{x+1} + 1} = \frac{\dfrac{x}{x+1}}{\dfrac{2x+1}{x+1}} = \frac{x}{2x+1}$$

$$f_2(x) = (f_0 \circ f_1)(x) = f_0(f_1(x)) = f_0\left(\frac{x}{2x+1}\right)$$

$$= \frac{\dfrac{x}{2x+1}}{\dfrac{x}{2x+1} + 1} = \frac{\dfrac{x}{2x+1}}{\dfrac{3x+1}{2x+1}} = \frac{x}{3x+1}$$

Look for a pattern

$$f_3(x) = (f_0 \circ f_2)(x) = f_0(f_2(x)) = f_0\left(\frac{x}{3x+1}\right)$$

$$= \frac{\dfrac{x}{3x+1}}{\dfrac{x}{3x+1} + 1} = \frac{\dfrac{x}{3x+1}}{\dfrac{4x+1}{3x+1}} = \frac{x}{4x+1}$$

We notice a pattern: The coefficient of x in the denominator of $f_n(x)$ is $n + 1$ in the three cases we have computed. So we make the guess that, in general,

4
$$f_n(x) = \frac{x}{(n+1)x + 1}$$

To prove this, we use the Principle of Mathematical Induction. We have already verified that (4) is true for $n = 1$. Assume that it is true for $n = k$, that is,

$$f_k(x) = \frac{x}{(k+1)x + 1}$$

Then $$f_{k+1}(x) = (f_0 \circ f_k)(x) = f_0(f_k(x)) = f_0\left(\frac{x}{(k+1)x+1}\right)$$

$$= \frac{\dfrac{x}{(k+1)x+1}}{\dfrac{x}{(k+1)x+1} + 1} = \frac{\dfrac{x}{(k+1)x+1}}{\dfrac{(k+2)x+1}{(k+1)x+1}} = \frac{x}{(k+2)x+1}$$

This expression shows that (4) is true for $n = k + 1$. Therefore, by mathematical induction, it is true for all positive integers n.

PROBLEMS

1. One of the legs of a right triangle has length 4 cm. Express the length of the altitude perpendicular to the hypotenuse as a function of the length of the hypotenuse.

2. The altitude perpendicular to the hypotenuse of a right triangle is 12 cm. Express the length of the hypotenuse as a function of the perimeter.

3. Solve the equation $|2x - 1| - |x + 5| = 3$.

4. Solve the inequality $|x - 1| - |x - 3| \geqslant 5$.

5. Sketch the graph of the function $f(x) = |x^2 - 4|x| + 3|$.

6. Sketch the graph of the function $g(x) = |x^2 - 1| - |x^2 - 4|$.

7. Draw the graph of the equation $x + |x| = y + |y|$.

8. Draw the graph of the equation $x^4 - 4x^2 - x^2y^2 + 4y^2 = 0$.

9. Sketch the region in the plane consisting of all points (x, y) such that $|x| + |y| \leqslant 1$.

10. Sketch the region in the plane consisting of all points (x, y) such that

$$|x - y| + |x| - |y| \leqslant 2$$

11. Evaluate $(\log_2 3)(\log_3 4)(\log_4 5) \cdots (\log_{31} 32)$.

12. (a) Show that the function $f(x) = \ln\left(x + \sqrt{x^2 + 1}\right)$ is an odd function.
    (b) Find the inverse function of f.

13. Solve the inequality $\ln(x^2 - 2x - 2) \leqslant 0$.

14. Use indirect reasoning to prove that $\log_2 5$ is an irrational number.

15. A driver sets out on a journey. For the first half of the distance she drives at the leisurely pace of 30 mi/h; she drives the second half at 60 mi/h. What is her average speed on this trip?

16. Is it true that $f \circ (g + h) = f \circ g + f \circ h$?

17. Prove that if n is a positive integer, then $7^n - 1$ is divisible by 6.

18. Prove that $1 + 3 + 5 + \cdots + (2n - 1) = n^2$.

19. If $f_0(x) = x^2$ and $f_{n+1}(x) = f_0(f_n(x))$ for $n = 0, 1, 2, \ldots$, find a formula for $f_n(x)$.

20. (a) If $f_0(x) = \dfrac{1}{2 - x}$ and $f_{n+1} = f_0 \circ f_n$ for $n = 0, 1, 2, \ldots$, find an expression for $f_n(x)$ and use mathematical induction to prove it.
    (b) Graph f_0, f_1, f_2, f_3 on the same screen and describe the effects of repeated composition.

CHAPTER 2

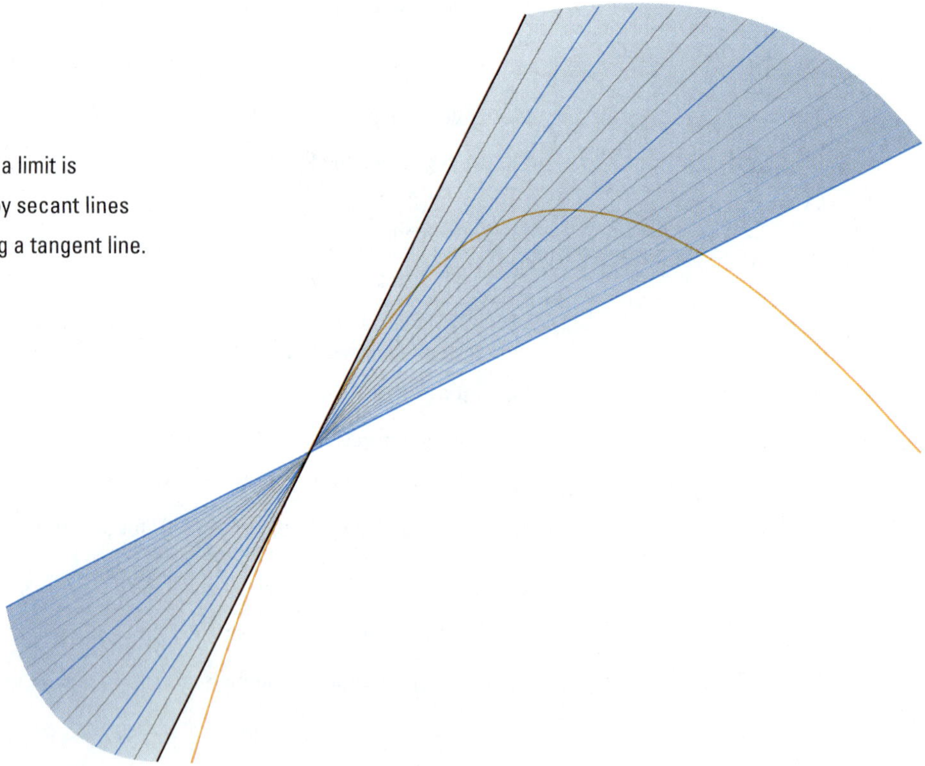

The idea of a limit is illustrated by secant lines approaching a tangent line.

Limits and Derivatives

In *A Preview of Calculus* (page 2) we saw how the idea of a limit underlies the various branches of calculus. It is therefore appropriate to begin our study of calculus by investigating limits and their properties. The special type of limit that is used to find tangents and velocities gives rise to the central idea in differential calculus, the derivative.

2.1 The Tangent and Velocity Problems

In this section we see how limits arise when we attempt to find the tangent to a curve or the velocity of an object.

The Tangent Problem

The word *tangent* is derived from the Latin word *tangens*, which means "touching." Thus, a tangent to a curve is a line that touches the curve. In other words, a tangent line should have the same direction as the curve at the point of contact. How can this idea be made precise?

Locate tangents interactively and explore them numerically.
Resources / Module 1
/ Tangents
/ What Is a Tangent?

For a circle we could simply follow Euclid and say that a tangent is a line that intersects the circle once and only once as in Figure 1(a). For more complicated curves this definition is inadequate. Figure 1(b) shows two lines l and t passing through a point P on a curve C. The line l intersects C only once, but it certainly does not look like what we think of as a tangent. The line t, on the other hand, looks like a tangent but it intersects C twice.

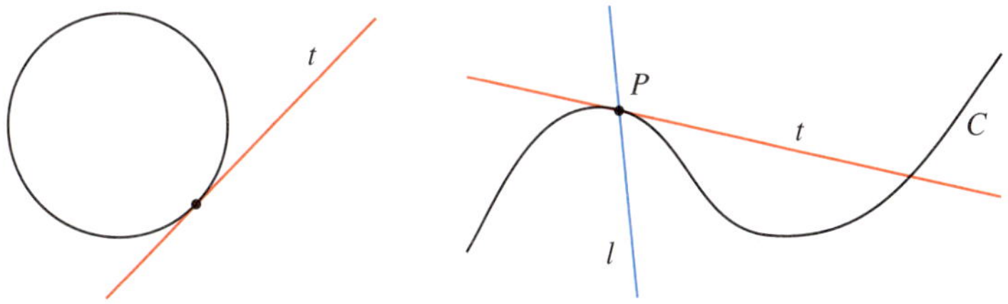

FIGURE 1 (a) (b)

To be specific, let's look at the problem of trying to find a tangent line t to the parabola $y = x^2$ in the following example.

EXAMPLE 1 Find an equation of the tangent line to the parabola $y = x^2$ at the point $P(1, 1)$.

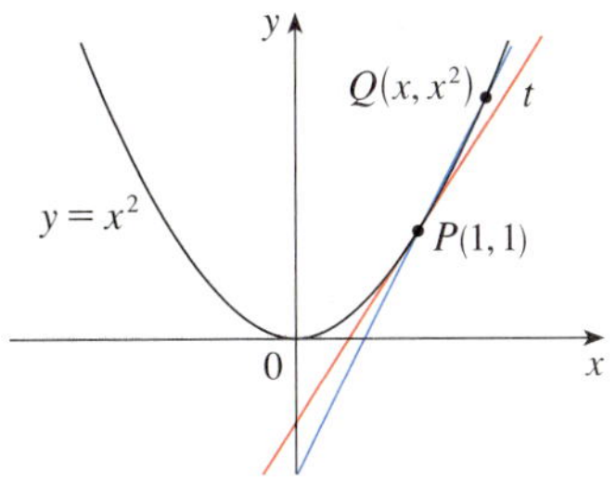

FIGURE 2

SOLUTION We will be able to find an equation of the tangent line t as soon as we know its slope m. The difficulty is that we know only one point, P, on t, whereas we need two points to compute the slope. But observe that we can compute an approximation to m by choosing a nearby point $Q(x, x^2)$ on the parabola (as in Figure 2) and computing the slope m_{PQ} of the secant line PQ.

We choose $x \neq 1$ so that $Q \neq P$. Then

$$m_{PQ} = \frac{x^2 - 1}{x - 1}$$

For instance, for the point $Q(1.5, 2.25)$ we have

$$m_{PQ} = \frac{2.25 - 1}{1.5 - 1} = \frac{1.25}{0.5} = 2.5$$

x	m_{PQ}
2	3
1.5	2.5
1.1	2.1
1.01	2.01
1.001	2.001

x	m_{PQ}
0	1
0.5	1.5
0.9	1.9
0.99	1.99
0.999	1.999

The tables in the margin show the values of m_{PQ} for several values of x close to 1. The closer Q is to P, the closer x is to 1 and, it appears from the tables, the closer m_{PQ} is to 2. This suggests that the slope of the tangent line t should be $m = 2$.

We say that the slope of the tangent line is the *limit* of the slopes of the secant lines, and we express this symbolically by writing

$$\lim_{Q \to P} m_{PQ} = m \qquad \text{and} \qquad \lim_{x \to 1} \frac{x^2 - 1}{x - 1} = 2$$

Assuming that the slope of the tangent line is indeed 2, we use the point-slope form of the equation of a line (see Appendix B) to write the equation of the tangent line through $(1, 1)$ as

$$y - 1 = 2(x - 1) \qquad \text{or} \qquad y = 2x - 1$$

Figure 3 illustrates the limiting process that occurs in this example. As Q approaches P along the parabola, the corresponding secant lines rotate about P and approach the tangent line t.

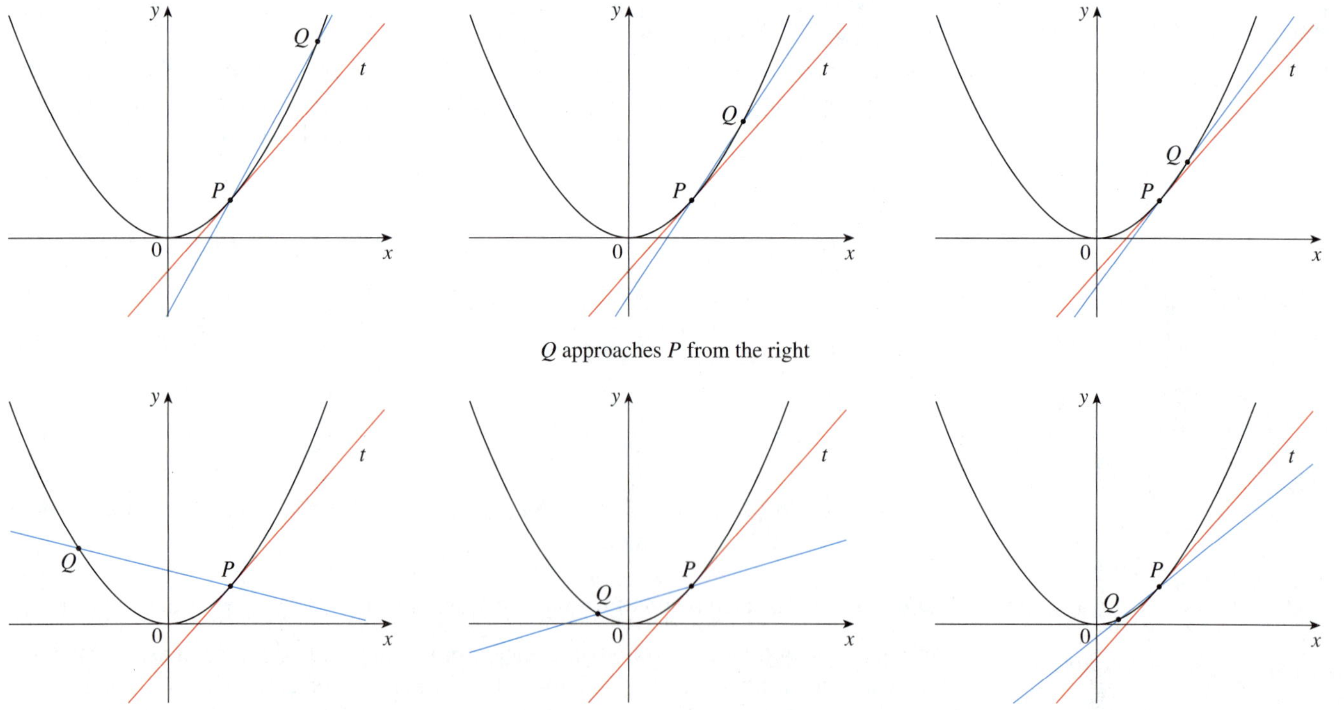

Q approaches P from the right

Q approaches P from the left

FIGURE 3

In Module 2.1 you can see how the process in Figure 3 works for five additional functions.

Many functions that occur in science are not described by explicit equations; they are defined by experimental data. The next example shows how to estimate the slope of the tangent line to the graph of such a function.

t	Q
0.00	100.00
0.02	81.87
0.04	67.03
0.06	54.88
0.08	44.93
0.10	36.76

EXAMPLE 2 The flash unit on a camera operates by storing charge on a capacitor and releasing it suddenly when the flash is set off. The data at the left describe the charge Q remaining on the capacitor (measured in microcoulombs) at time t (measured in seconds after the flash goes off). Use the data to draw the graph of this function and estimate the slope of the tangent line at the point where $t = 0.04$. [*Note:* The slope of the tangent line represents the electric current flowing from the capacitor to the flash bulb (measured in microamperes).]

SOLUTION In Figure 4 we plot the given data and use them to sketch a curve that approximates the graph of the function.

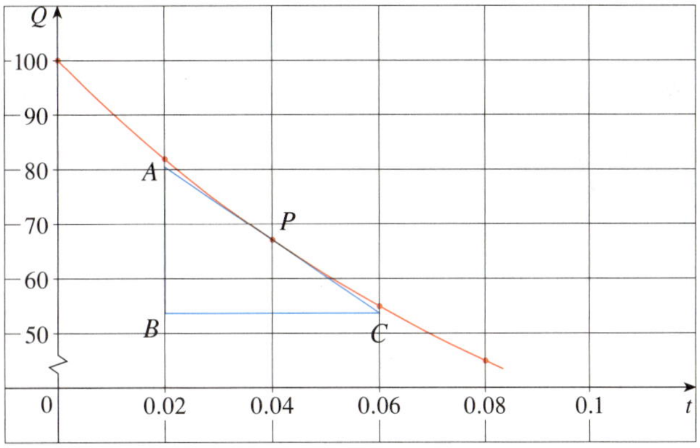

FIGURE 4

Given the points $P(0.04, 67.03)$ and $R(0.00, 100.00)$ on the graph, we find that the slope of the secant line PR is

$$m_{PR} = \frac{100.00 - 67.03}{0.00 - 0.04} = -824.25$$

R	m_{PR}
(0.00, 100.00)	−824.25
(0.02, 81.87)	−742.00
(0.06, 54.88)	−607.50
(0.08, 44.93)	−552.50
(0.10, 36.76)	−504.50

The table at the left shows the results of similar calculations for the slopes of other secant lines. From this table we would expect the slope of the tangent line at $t = 0.04$ to lie somewhere between -742 and -607.5. In fact, the average of the slopes of the two closest secant lines is

$$\tfrac{1}{2}(-742 - 607.5) = -674.75$$

So, by this method, we estimate the slope of the tangent line to be -675.

Another method is to draw an approximation to the tangent line at P and measure the sides of the triangle ABC, as in Figure 4. This gives an estimate of the slope of the tangent line as

$$-\frac{|AB|}{|BC|} \approx -\frac{80.4 - 53.6}{0.06 - 0.02} = -670$$

|||| The physical meaning of the answer in Example 2 is that the electric current flowing from the capacitor to the flash bulb after 0.04 second is about –670 microamperes.

|||| The Velocity Problem

If you watch the speedometer of a car as you travel in city traffic, you see that the needle doesn't stay still for very long; that is, the velocity of the car is not constant. We assume from watching the speedometer that the car has a definite velocity at each moment, but how is the "instantaneous" velocity defined? Let's investigate the example of a falling ball.

The CN Tower in Toronto is currently the tallest freestanding building in the world.

EXAMPLE 3 Suppose that a ball is dropped from the upper observation deck of the CN Tower in Toronto, 450 m above the ground. Find the velocity of the ball after 5 seconds.

SOLUTION Through experiments carried out four centuries ago, Galileo discovered that the distance fallen by any freely falling body is proportional to the square of the time it has been falling. (This model for free fall neglects air resistance.) If the distance fallen after t seconds is denoted by $s(t)$ and measured in meters, then Galileo's law is expressed by the equation

$$s(t) = 4.9t^2$$

The difficulty in finding the velocity after 5 s is that we are dealing with a single instant of time ($t = 5$), so no time interval is involved. However, we can approximate the desired quantity by computing the average velocity over the brief time interval of a tenth of a second from $t = 5$ to $t = 5.1$:

$$\text{average velocity} = \frac{\text{distance traveled}}{\text{time elapsed}}$$

$$= \frac{s(5.1) - s(5)}{0.1}$$

$$= \frac{4.9(5.1)^2 - 4.9(5)^2}{0.1} = 49.49 \text{ m/s}$$

The following table shows the results of similar calculations of the average velocity over successively smaller time periods.

Time interval	Average velocity (m/s)
$5 \leqslant t \leqslant 6$	53.9
$5 \leqslant t \leqslant 5.1$	49.49
$5 \leqslant t \leqslant 5.05$	49.245
$5 \leqslant t \leqslant 5.01$	49.049
$5 \leqslant t \leqslant 5.001$	49.0049

It appears that as we shorten the time period, the average velocity is becoming closer to 49 m/s. The **instantaneous velocity** when $t = 5$ is defined to be the limiting value of these average velocities over shorter and shorter time periods that start at $t = 5$. Thus, the (instantaneous) velocity after 5 s is

$$v = 49 \text{ m/s}$$

You may have the feeling that the calculations used in solving this problem are very similar to those used earlier in this section to find tangents. In fact, there is a close connection between the tangent problem and the problem of finding velocities. If we draw the graph of the distance function of the ball (as in Figure 5) and we consider the points $P(a, 4.9a^2)$ and $Q(a + h, 4.9(a + h)^2)$ on the graph, then the slope of the secant line PQ is

$$m_{PQ} = \frac{4.9(a + h)^2 - 4.9a^2}{(a + h) - a}$$

which is the same as the average velocity over the time interval $[a, a + h]$. Therefore, the velocity at time $t = a$ (the limit of these average velocities as h approaches 0) must be equal to the slope of the tangent line at P (the limit of the slopes of the secant lines).

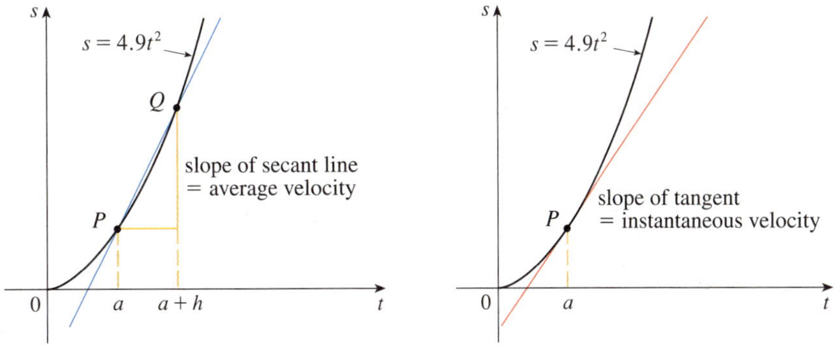

FIGURE 5

Examples 1 and 3 show that in order to solve tangent and velocity problems we must be able to find limits. After studying methods for computing limits in the next five sections, we will return to the problems of finding tangents and velocities in Section 2.7.

2.1 Exercises

1. A tank holds 1000 gallons of water, which drains from the bottom of the tank in half an hour. The values in the table show the volume V of water remaining in the tank (in gallons) after t minutes.

t (min)	5	10	15	20	25	30
V (gal)	694	444	250	111	28	0

(a) If P is the point (15, 250) on the graph of V, find the slopes of the secant lines PQ when Q is the point on the graph with $t = 5$, 10, 20, 25, and 30.

(b) Estimate the slope of the tangent line at P by averaging the slopes of two secant lines.

(c) Use a graph of the function to estimate the slope of the tangent line at P. (This slope represents the rate at which the water is flowing from the tank after 15 minutes.)

2. A cardiac monitor is used to measure the heart rate of a patient after surgery. It compiles the number of heartbeats after t minutes. When the data in the table are graphed, the slope of the tangent line represents the heart rate in beats per minute.

t (min)	36	38	40	42	44
Heartbeats	2530	2661	2806	2948	3080

The monitor estimates this value by calculating the slope of a secant line. Use the data to estimate the patient's heart rate after 42 minutes using the secant line between the points with the given values of t.

(a) $t = 36$ and $t = 42$ (b) $t = 38$ and $t = 42$

(c) $t = 40$ and $t = 42$ (d) $t = 42$ and $t = 44$

What are your conclusions?

3. The point $P(1, \frac{1}{2})$ lies on the curve $y = x/(1 + x)$.

(a) If Q is the point $(x, x/(1 + x))$, use your calculator to find the slope of the secant line PQ (correct to six decimal places) for the following values of x:

(i) 0.5 (ii) 0.9

(iii) 0.99 (iv) 0.999

(v) 1.5 (vi) 1.1

(vii) 1.01 (viii) 1.001

(b) Using the results of part (a), guess the value of the slope of the tangent line to the curve at $P(1, \frac{1}{2})$.

(c) Using the slope from part (b), find an equation of the tangent line to the curve at $P(1, \frac{1}{2})$.

4. The point $P(2, \ln 2)$ lies on the curve $y = \ln x$.

(a) If Q is the point $(x, \ln x)$, use your calculator to find the slope of the secant line PQ (correct to six decimal places) for the following values of x:

(i) 1.5 (ii) 1.9

(iii) 1.99 (iv) 1.999

(v) 2.5 (vi) 2.1

(vii) 2.01 (viii) 2.001

(b) Using the results of part (a), guess the value of the slope of the tangent line to the curve at $P(2, \ln 2)$.

(c) Using the slope from part (b), find an equation of the tangent line to the curve at $P(2, \ln 2)$.

(d) Sketch the curve, two of the secant lines, and the tangent line.

5. If a ball is thrown into the air with a velocity of 40 ft/s, its height in feet after t seconds is given by $y = 40t - 16t^2$.

(a) Find the average velocity for the time period beginning when $t = 2$ and lasting

(i) 0.5 second (ii) 0.1 second

(iii) 0.05 second (iv) 0.01 second

(b) Find the instantaneous velocity when $t = 2$.

6. If an arrow is shot upward on the moon with a velocity of 58 m/s, its height in meters after t seconds is given by $h = 58t - 0.83t^2$.

(a) Find the average velocity over the given time intervals:

(i) [1, 2] (ii) [1, 1.5] (iii) [1, 1.1]

(iv) [1, 1.01] (v) [1, 1.001]

(b) Find the instantaneous velocity after one second.

7. The displacement (in feet) of a certain particle moving in a straight line is given by $s = t^3/6$, where t is measured in seconds.

(a) Find the average velocity over the following time periods:

(i) [1, 3] (ii) [1, 2]

(iii) [1, 1.5] (iv) [1, 1.1]

(b) Find the instantaneous velocity when $t = 1$.

(c) Draw the graph of s as a function of t and draw the secant lines whose slopes are the average velocities found in part (a).

(d) Draw the tangent line whose slope is the instantaneous velocity from part (b).

8. The position of a car is given by the values in the table.

t (seconds)	0	1	2	3	4	5
s (feet)	0	10	32	70	119	178

(a) Find the average velocity for the time period beginning when $t = 2$ and lasting

(i) 3 seconds (ii) 2 seconds (iii) 1 second

(b) Use the graph of s as a function of t to estimate the instantaneous velocity when $t = 2$.

9. The point $P(1, 0)$ lies on the curve $y = \sin(10\pi/x)$.

(a) If Q is the point $(x, \sin(10\pi/x))$, find the slope of the secant line PQ (correct to four decimal places) for $x = 2$, 1.5, 1.4, 1.3, 1.2, 1.1, 0.5, 0.6, 0.7, 0.8, and 0.9. Do the slopes appear to be approaching a limit?

(b) Use a graph of the curve to explain why the slopes of the secant lines in part (a) are not close to the slope of the tangent line at P.

(c) By choosing appropriate secant lines, estimate the slope of the tangent line at P.

2.2 The Limit of a Function

Having seen in the preceding section how limits arise when we want to find the tangent to a curve or the velocity of an object, we now turn our attention to limits in general and numerical and graphical methods for computing them.

Let's investigate the behavior of the function f defined by $f(x) = x^2 - x + 2$ for values of x near 2. The following table gives values of $f(x)$ for values of x close to 2, but not equal to 2.

x	$f(x)$	x	$f(x)$
1.0	2.000000	3.0	8.000000
1.5	2.750000	2.5	5.750000
1.8	3.440000	2.2	4.640000
1.9	3.710000	2.1	4.310000
1.95	3.852500	2.05	4.152500
1.99	3.970100	2.01	4.030100
1.995	3.985025	2.005	4.015025
1.999	3.997001	2.001	4.003001

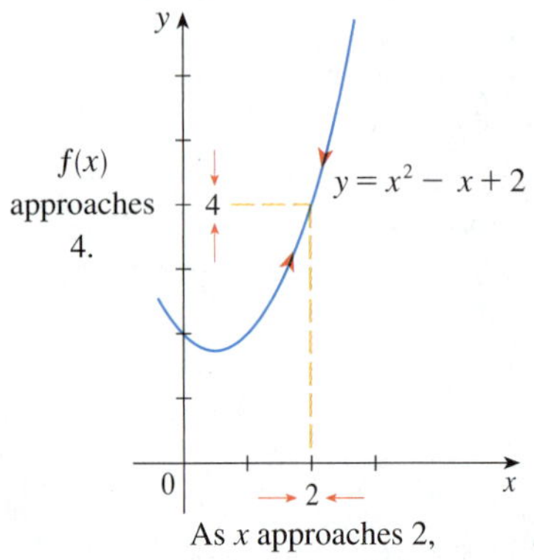

FIGURE 1

From the table and the graph of f (a parabola) shown in Figure 1 we see that when x is close to 2 (on either side of 2), $f(x)$ is close to 4. In fact, it appears that we can make the values of $f(x)$ as close as we like to 4 by taking x sufficiently close to 2. We express this by saying "the limit of the function $f(x) = x^2 - x + 2$ as x approaches 2 is equal to 4." The notation for this is

$$\lim_{x \to 2} (x^2 - x + 2) = 4$$

In general, we use the following notation.

1 Definition We write

$$\lim_{x \to a} f(x) = L$$

and say "the limit of $f(x)$, as x approaches a, equals L"

if we can make the values of $f(x)$ arbitrarily close to L (as close to L as we like) by taking x to be sufficiently close to a (on either side of a) but not equal to a.

Roughly speaking, this says that the values of $f(x)$ get closer and closer to the number L as x gets closer and closer to the number a (from either side of a) but $x \neq a$. A more precise definition will be given in Section 2.4.

An alternative notation for

$$\lim_{x \to a} f(x) = L$$

is

$$f(x) \to L \quad \text{as} \quad x \to a$$

which is usually read "$f(x)$ approaches L as x approaches a."

Notice the phrase "but $x \neq a$" in the definition of limit. This means that in finding the limit of $f(x)$ as x approaches a, we never consider $x = a$. In fact, $f(x)$ need not even be defined when $x = a$. The only thing that matters is how f is defined *near* a.

Figure 2 shows the graphs of three functions. Note that in part (c), $f(a)$ is not defined and in part (b), $f(a) \neq L$. But in each case, regardless of what happens at a, it is true that $\lim_{x \to a} f(x) = L$.

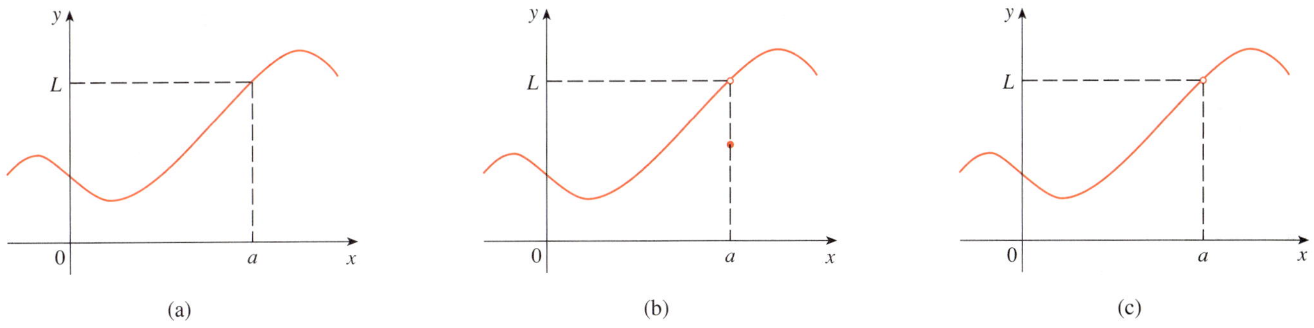

FIGURE 2 $\lim_{x \to a} f(x) = L$ in all three cases

EXAMPLE 1 Guess the value of $\displaystyle\lim_{x \to 1} \frac{x - 1}{x^2 - 1}$.

SOLUTION Notice that the function $f(x) = (x - 1)/(x^2 - 1)$ is not defined when $x = 1$, but that doesn't matter because the definition of $\lim_{x \to a} f(x)$ says that we consider values

$x < 1$	$f(x)$
0.5	0.666667
0.9	0.526316
0.99	0.502513
0.999	0.500250
0.9999	0.500025

$x > 1$	$f(x)$
1.5	0.400000
1.1	0.476190
1.01	0.497512
1.001	0.499750
1.0001	0.499975

of x that are close to a but not equal to a. The tables at the left give values of $f(x)$ (correct to six decimal places) for values of x that approach 1 (but are not equal to 1). On the basis of the values in the tables, we make the guess that

$$\lim_{x \to 1} \frac{x - 1}{x^2 - 1} = 0.5$$

Example 1 is illustrated by the graph of f in Figure 3. Now let's change f slightly by giving it the value 2 when $x = 1$ and calling the resulting function g:

$$g(x) = \begin{cases} \dfrac{x - 1}{x^2 - 1} & \text{if } x \neq 1 \\ 2 & \text{if } x = 1 \end{cases}$$

This new function g still has the same limit as x approaches 1 (see Figure 4).

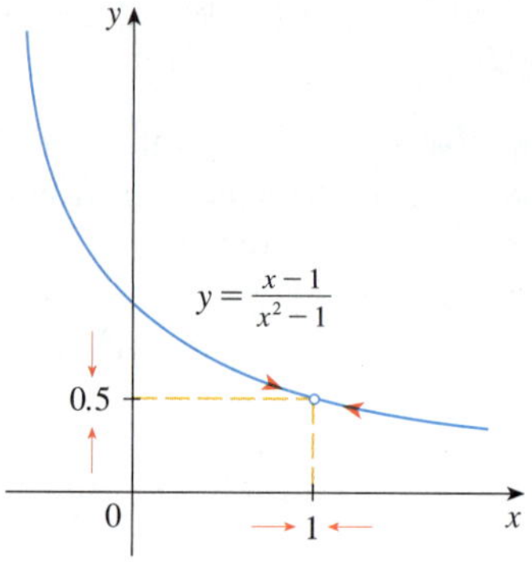

FIGURE 3

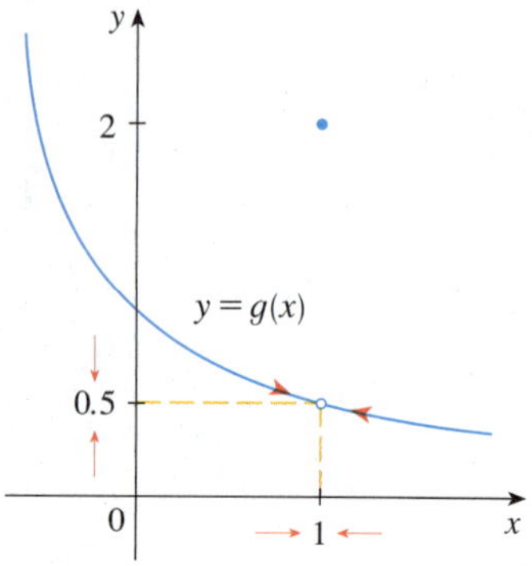

FIGURE 4

EXAMPLE 2 Estimate the value of $\displaystyle\lim_{t \to 0} \frac{\sqrt{t^2 + 9} - 3}{t^2}$.

SOLUTION The table lists values of the function for several values of t near 0.

t	$\dfrac{\sqrt{t^2 + 9} - 3}{t^2}$
± 1.0	0.16228
± 0.5	0.16553
± 0.1	0.16662
± 0.05	0.16666
± 0.01	0.16667

As t approaches 0, the values of the function seem to approach 0.1666666 . . . and so we guess that

$$\lim_{t \to 0} \frac{\sqrt{t^2 + 9} - 3}{t^2} = \frac{1}{6}$$

In Example 2 what would have happened if we had taken even smaller values of t? The table in the margin shows the results from one calculator; you can see that something strange seems to be happening.

t	$\dfrac{\sqrt{t^2 + 9} - 3}{t^2}$
± 0.0005	0.16800
± 0.0001	0.20000
± 0.00005	0.00000
± 0.00001	0.00000

If you try these calculations on your own calculator you might get different values, but eventually you will get the value 0 if you make t sufficiently small. Does this mean that the answer is really 0 instead of $\frac{1}{6}$? No, the value of the limit is $\frac{1}{6}$, as we will show in the next section. The problem is that the calculator gave false values because $\sqrt{t^2 + 9}$ is very close to 3 when t is small. (In fact, when t is sufficiently small, a calculator's value for $\sqrt{t^2 + 9}$ is 3.000 . . . to as many digits as the calculator is capable of carrying.)

|||| For a further explanation of why calculators sometimes give false values, see the web site

www.stewartcalculus.com

Click on *Additional Topics* and then on *Lies My Calculator and Computer Told Me*. In particular, see the section called The Perils of Subtraction.

Something similar happens when we try to graph the function

$$f(t) = \frac{\sqrt{t^2 + 9} - 3}{t^2}$$

of Example 2 on a graphing calculator or computer. Parts (a) and (b) of Figure 5 show quite accurate graphs of f, and when we use the trace mode (if available) we can estimate easily that the limit is about $\frac{1}{6}$. But if we zoom in too far, as in parts (c) and (d), then we get inaccurate graphs, again because of problems with subtraction.

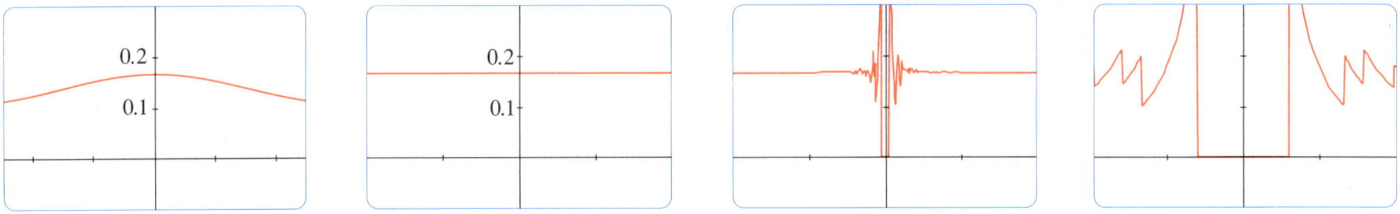

(a) $[-5, 5]$ by $[-0.1, 0.3]$ (b) $[-0.1, 0.1]$ by $[-0.1, 0.3]$ (c) $[-10^{-6}, 10^{-6}]$ by $[-0.1, 0.3]$ (d) $[-10^{-7}, 10^{-7}]$ by $[-0.1, 0.3]$

FIGURE 5

EXAMPLE 3 Guess the value of $\lim_{x \to 0} \frac{\sin x}{x}$.

SOLUTION The function $f(x) = (\sin x)/x$ is not defined when $x = 0$. Using a calculator (and remembering that, if $x \in \mathbb{R}$, $\sin x$ means the sine of the angle whose *radian* measure is x), we construct the following table of values correct to eight decimal places.

From the table and the graph in Figure 6 we guess that

$$\lim_{x \to 0} \frac{\sin x}{x} = 1$$

This guess is in fact correct, as will be proved in Chapter 3 using a geometric argument.

x	$\frac{\sin x}{x}$
± 1.0	0.84147098
± 0.5	0.95885108
± 0.4	0.97354586
± 0.3	0.98506736
± 0.2	0.99334665
± 0.1	0.99833417
± 0.05	0.99958339
± 0.01	0.99998333
± 0.005	0.99999583
± 0.001	0.99999983

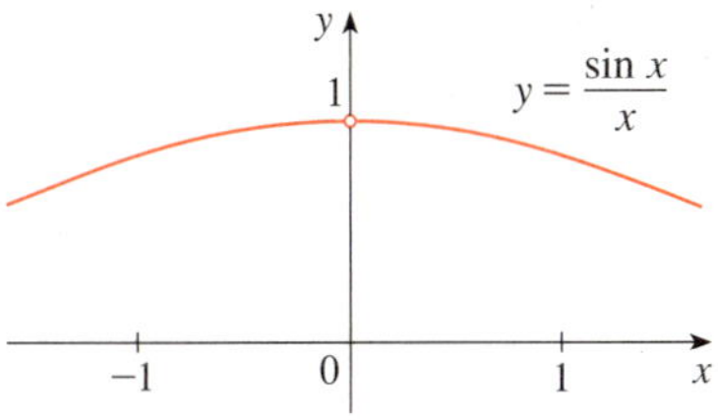

FIGURE 6

COMPUTER ALGEBRA SYSTEMS

Computer algebra systems (CAS) have commands that compute limits. In order to avoid the types of pitfalls demonstrated in Examples 2, 4, and 5, they don't find limits by numerical experimentation. Instead, they use more sophisticated techniques such as computing infinite series. If you have access to a CAS, use the limit command to compute the limits in the examples of this section and to check your answers in the exercises of this chapter.

EXAMPLE 4 Investigate $\lim_{x \to 0} \sin \dfrac{\pi}{x}$.

SOLUTION Again the function $f(x) = \sin(\pi/x)$ is undefined at 0. Evaluating the function for some small values of x, we get

$$f(1) = \sin \pi = 0 \qquad f\left(\tfrac{1}{2}\right) = \sin 2\pi = 0$$

$$f\left(\tfrac{1}{3}\right) = \sin 3\pi = 0 \qquad f\left(\tfrac{1}{4}\right) = \sin 4\pi = 0$$

$$f(0.1) = \sin 10\pi = 0 \qquad f(0.01) = \sin 100\pi = 0$$

Similarly, $f(0.001) = f(0.0001) = 0$. On the basis of this information we might be tempted to guess that

$$\lim_{x \to 0} \sin \frac{\pi}{x} = 0$$

but this time our guess is wrong. Note that although $f(1/n) = \sin n\pi = 0$ for any integer n, it is also true that $f(x) = 1$ for infinitely many values of x that approach 0. [In fact, $\sin(\pi/x) = 1$ when

$$\frac{\pi}{x} = \frac{\pi}{2} + 2n\pi$$

and, solving for x, we get $x = 2/(4n + 1)$.] The graph of f is given in Figure 7.

Listen to the sound of this function trying to approach a limit.

Resources / Module 2
/ Basics of Limits
/ Sound of a Limit that Does Not Exist

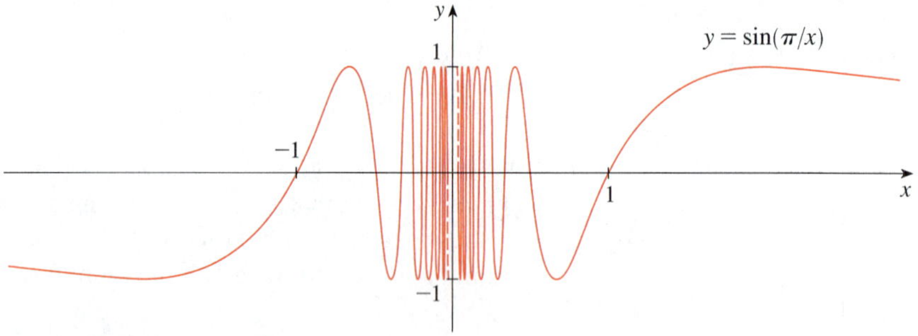

FIGURE 7

Module 2.2 helps you explore limits at points where graphs exhibit unusual behavior.

The dashed lines indicate that the values of $\sin(\pi/x)$ oscillate between 1 and -1 infinitely often as x approaches 0 (see Exercise 37). Since the values of $f(x)$ do not approach a fixed number as x approaches 0,

$$\lim_{x \to 0} \sin \frac{\pi}{x} \text{ does not exist}$$

x	$x^3 + \dfrac{\cos 5x}{10{,}000}$
1	1.000028
0.5	0.124920
0.1	0.001088
0.05	0.000222
0.01	0.000101

EXAMPLE 5 Find $\lim_{x \to 0} \left(x^3 + \dfrac{\cos 5x}{10{,}000} \right)$.

SOLUTION As before, we construct a table of values. From the table in the margin it appears that

$$\lim_{x \to 0} \left(x^3 + \frac{\cos 5x}{10{,}000} \right) = 0$$

x	$x^3 + \dfrac{\cos 5x}{10{,}000}$
0.005	0.00010009
0.001	0.00010000

But if we persevere with smaller values of x, the second table suggests that

$$\lim_{x \to 0} \left(x^3 + \frac{\cos 5x}{10{,}000} \right) = 0.000100 = \frac{1}{10{,}000}$$

Later we will see that $\lim_{x \to 0} \cos 5x = 1$; then it follows that the limit is 0.0001.

Examples 4 and 5 illustrate some of the pitfalls in guessing the value of a limit. It is easy to guess the wrong value if we use inappropriate values of x, but it is difficult to know when to stop calculating values. And, as the discussion after Example 2 shows, sometimes calculators and computers give the wrong values. In the next two sections, however, we will develop foolproof methods for calculating limits.

EXAMPLE 6 The Heaviside function H is defined by

$$H(t) = \begin{cases} 0 & \text{if } t < 0 \\ 1 & \text{if } t \geqslant 0 \end{cases}$$

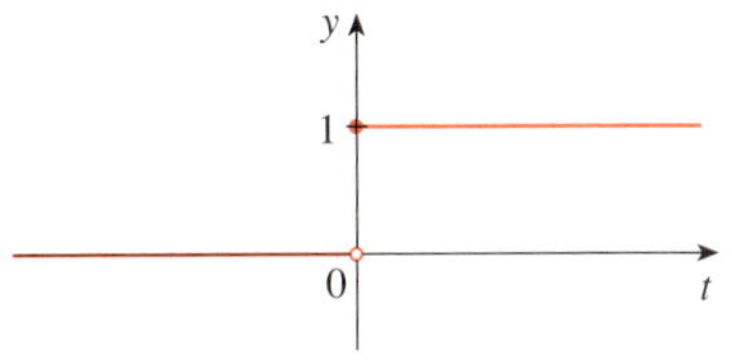

FIGURE 8

[This function is named after the electrical engineer Oliver Heaviside (1850–1925) and can be used to describe an electric current that is switched on at time $t = 0$.] Its graph is shown in Figure 8.

As t approaches 0 from the left, $H(t)$ approaches 0. As t approaches 0 from the right, $H(t)$ approaches 1. There is no single number that $H(t)$ approaches as t approaches 0. Therefore, $\lim_{t \to 0} H(t)$ does not exist.

One-Sided Limits

We noticed in Example 6 that $H(t)$ approaches 0 as t approaches 0 from the left and $H(t)$ approaches 1 as t approaches 0 from the right. We indicate this situation symbolically by writing

$$\lim_{t \to 0^-} H(t) = 0 \qquad \text{and} \qquad \lim_{t \to 0^+} H(t) = 1$$

The symbol "$t \to 0^-$" indicates that we consider only values of t that are less than 0. Likewise, "$t \to 0^+$" indicates that we consider only values of t that are greater than 0.

2 Definition We write

$$\lim_{x \to a^-} f(x) = L$$

and say the **left-hand limit of $f(x)$ as x approaches a** [or the **limit of $f(x)$ as x approaches a from the left**] is equal to L if we can make the values of $f(x)$ arbitrarily close to L by taking x to be sufficiently close to a and x less than a.

Notice that Definition 2 differs from Definition 1 only in that we require x to be less than a. Similarly, if we require that x be greater than a, we get "the **right-hand limit of $f(x)$ as x approaches a** is equal to L" and we write

$$\lim_{x \to a^+} f(x) = L$$

Thus, the symbol "$x \to a^+$" means that we consider only $x > a$. These definitions are illustrated in Figure 9.

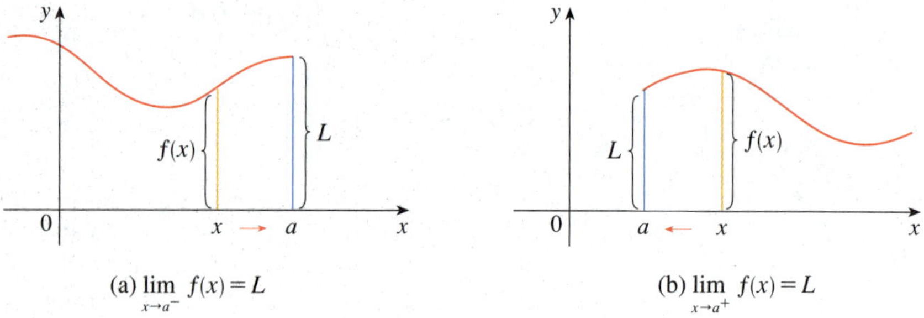

FIGURE 9 (a) $\lim_{x \to a^-} f(x) = L$ (b) $\lim_{x \to a^+} f(x) = L$

By comparing Definition 1 with the definitions of one-sided limits, we see that the following is true.

3 $\lim_{x \to a} f(x) = L$ if and only if $\lim_{x \to a^-} f(x) = L$ and $\lim_{x \to a^+} f(x) = L$

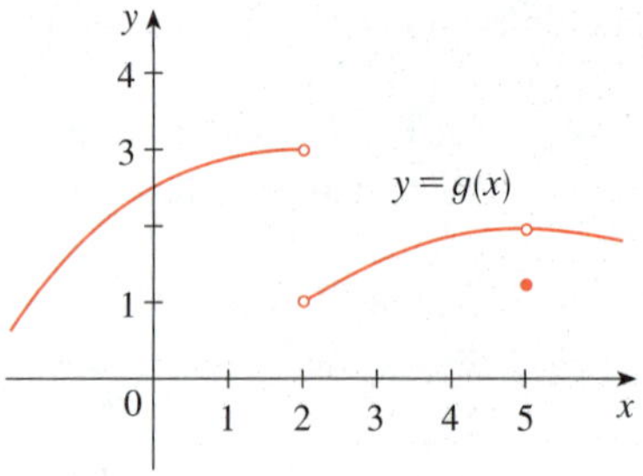

FIGURE 10

EXAMPLE 7 The graph of a function g is shown in Figure 10. Use it to state the values (if they exist) of the following:

(a) $\lim_{x \to 2^-} g(x)$ (b) $\lim_{x \to 2^+} g(x)$ (c) $\lim_{x \to 2} g(x)$

(d) $\lim_{x \to 5^-} g(x)$ (e) $\lim_{x \to 5^+} g(x)$ (f) $\lim_{x \to 5} g(x)$

SOLUTION From the graph we see that the values of $g(x)$ approach 3 as x approaches 2 from the left, but they approach 1 as x approaches 2 from the right. Therefore

$$\text{(a)} \lim_{x \to 2^-} g(x) = 3 \quad \text{and} \quad \text{(b)} \lim_{x \to 2^+} g(x) = 1$$

(c) Since the left and right limits are different, we conclude from (3) that $\lim_{x \to 2} g(x)$ does not exist.

The graph also shows that

$$\text{(d)} \lim_{x \to 5^-} g(x) = 2 \quad \text{and} \quad \text{(e)} \lim_{x \to 5^+} g(x) = 2$$

(f) This time the left and right limits are the same and so, by (3), we have

$$\lim_{x \to 5} g(x) = 2$$

Despite this fact, notice that $g(5) \neq 2$.

Infinite Limits

EXAMPLE 8 Find $\lim_{x \to 0} \frac{1}{x^2}$ if it exists.

SOLUTION As x becomes close to 0, x^2 also becomes close to 0, and $1/x^2$ becomes very large. (See the table on the next page.) In fact, it appears from the graph of the function $f(x) = 1/x^2$ shown in Figure 11 that the values of $f(x)$ can be made arbitrarily large

x	$\frac{1}{x^2}$
± 1	1
± 0.5	4
± 0.2	25
± 0.1	100
± 0.05	400
± 0.01	10,000
± 0.001	1,000,000

by taking x close enough to 0. Thus, the values of $f(x)$ do not approach a number, so $\lim_{x \to 0} (1/x^2)$ does not exist.

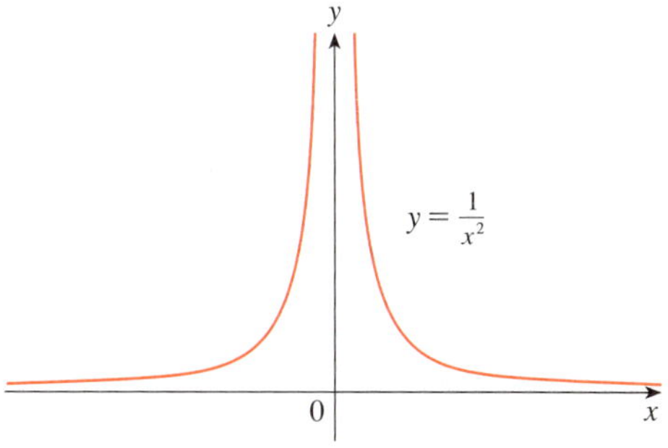

FIGURE 11

To indicate the kind of behavior exhibited in Example 8, we use the notation

$$\lim_{x \to 0} \frac{1}{x^2} = \infty$$

This does not mean that we are regarding ∞ as a number. Nor does it mean that the limit exists. It simply expresses the particular way in which the limit does not exist: $1/x^2$ can be made as large as we like by taking x close enough to 0.

In general, we write symbolically

$$\lim_{x \to a} f(x) = \infty$$

to indicate that the values of $f(x)$ become larger and larger (or "increase without bound") as x becomes closer and closer to a.

Explore infinite limits interactively.
Resources / Module 2
/ Limits that Are Infinite
/ Examples A and B

4 Definition Let f be a function defined on both sides of a, except possibly at a itself. Then

$$\lim_{x \to a} f(x) = \infty$$

means that the values of $f(x)$ can be made arbitrarily large (as large as we please) by taking x sufficiently close to a, but not equal to a.

Another notation for $\lim_{x \to a} f(x) = \infty$ is

$$f(x) \to \infty \qquad \text{as} \qquad x \to a$$

Again the symbol ∞ is not a number, but the expression $\lim_{x \to a} f(x) = \infty$ is often read as

"the limit of $f(x)$, as x approaches a, is infinity"

or "$f(x)$ becomes infinite as x approaches a"

or "$f(x)$ increases without bound as x approaches a"

This definition is illustrated graphically in Figure 12.

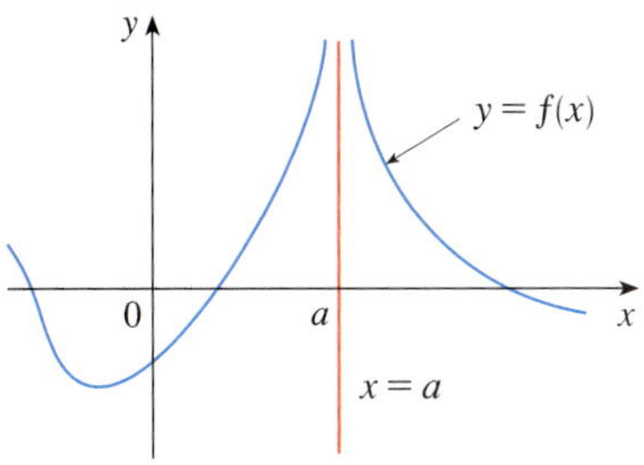

FIGURE 12
$\lim_{x \to a} f(x) = \infty$

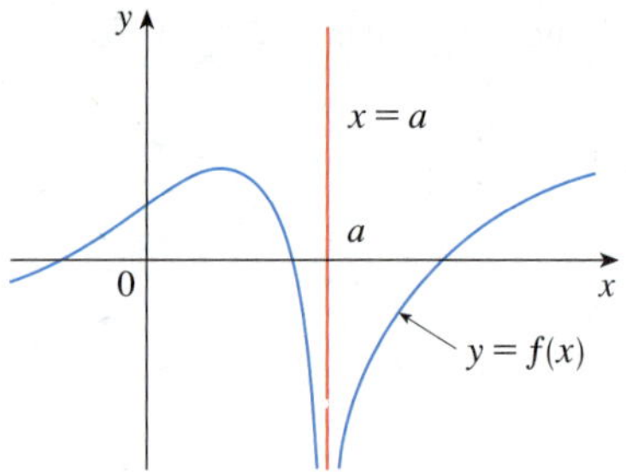

FIGURE 13

$\lim_{x \to a} f(x) = -\infty$

A similar sort of limit, for functions that become large negative as x gets close to a, is defined in Definition 5 and is illustrated in Figure 13.

5 Definition Let f be defined on both sides of a, except possibly at a itself. Then

$$\lim_{x \to a} f(x) = -\infty$$

means that the values of $f(x)$ can be made arbitrarily large negative by taking x sufficiently close to a, but not equal to a.

The symbol $\lim_{x \to a} f(x) = -\infty$ can be read as "the limit of $f(x)$, as x approaches a, is negative infinity" or "$f(x)$ decreases without bound as x approaches a." As an example we have

$$\lim_{x \to 0} \left(-\frac{1}{x^2} \right) = -\infty$$

Similar definitions can be given for the one-sided infinite limits

$$\lim_{x \to a^-} f(x) = \infty \qquad \lim_{x \to a^+} f(x) = \infty$$

$$\lim_{x \to a^-} f(x) = -\infty \qquad \lim_{x \to a^+} f(x) = -\infty$$

remembering that "$x \to a^-$" means that we consider only values of x that are less than a, and similarly "$x \to a^+$" means that we consider only $x > a$. Illustrations of these four cases are given in Figure 14.

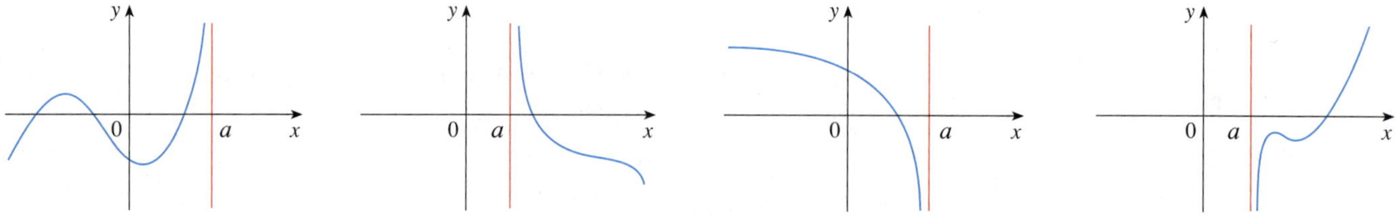

(a) $\lim_{x \to a^-} f(x) = \infty$ (b) $\lim_{x \to a^+} f(x) = \infty$ (c) $\lim_{x \to a^-} f(x) = -\infty$ (d) $\lim_{x \to a^+} f(x) = -\infty$

FIGURE 14

6 Definition The line $x = a$ is called a **vertical asymptote** of the curve $y = f(x)$ if at least one of the following statements is true:

$$\lim_{x \to a} f(x) = \infty \qquad \lim_{x \to a^-} f(x) = \infty \qquad \lim_{x \to a^+} f(x) = \infty$$

$$\lim_{x \to a} f(x) = -\infty \qquad \lim_{x \to a^-} f(x) = -\infty \qquad \lim_{x \to a^+} f(x) = -\infty$$

For instance, the y-axis is a vertical asymptote of the curve $y = 1/x^2$ because $\lim_{x \to 0} (1/x^2) = \infty$. In Figure 14 the line $x = a$ is a vertical asymptote in each of the four cases shown. In general, knowledge of vertical asymptotes is very useful in sketching graphs.

EXAMPLE 9 Find $\lim_{x \to 3^+} \dfrac{2x}{x - 3}$ and $\lim_{x \to 3^-} \dfrac{2x}{x - 3}$.

SOLUTION If x is close to 3 but larger than 3, then the denominator $x - 3$ is a small positive number and $2x$ is close to 6. So the quotient $2x/(x - 3)$ is a large *positive* number. Thus, intuitively we see that

$$\lim_{x \to 3^+} \frac{2x}{x - 3} = \infty$$

Likewise, if x is close to 3 but smaller than 3, then $x - 3$ is a small negative number but $2x$ is still a positive number (close to 6). So $2x/(x - 3)$ is a numerically large *negative* number. Thus

$$\lim_{x \to 3^-} \frac{2x}{x - 3} = -\infty$$

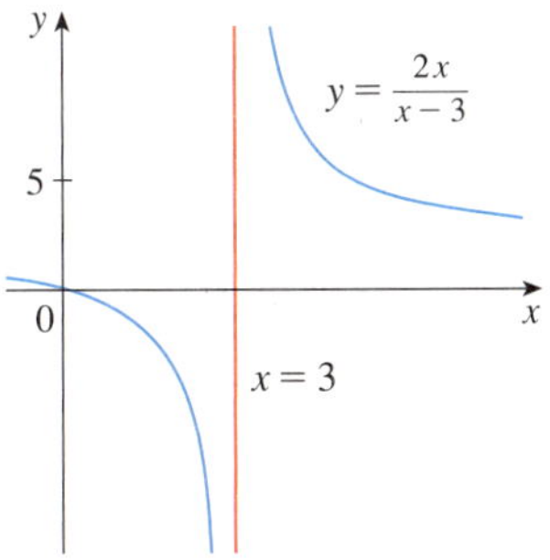

FIGURE 15

The graph of the curve $y = 2x/(x - 3)$ is given in Figure 15. The line $x = 3$ is a vertical asymptote.

EXAMPLE 10 Find the vertical asymptotes of $f(x) = \tan x$.

SOLUTION Because

$$\tan x = \frac{\sin x}{\cos x}$$

there are potential vertical asymptotes where $\cos x = 0$. In fact, since $\cos x \to 0^+$ as $x \to (\pi/2)^-$ and $\cos x \to 0^-$ as $x \to (\pi/2)^+$, whereas $\sin x$ is positive when x is near $\pi/2$, we have

$$\lim_{x \to (\pi/2)^-} \tan x = \infty \qquad \text{and} \qquad \lim_{x \to (\pi/2)^+} \tan x = -\infty$$

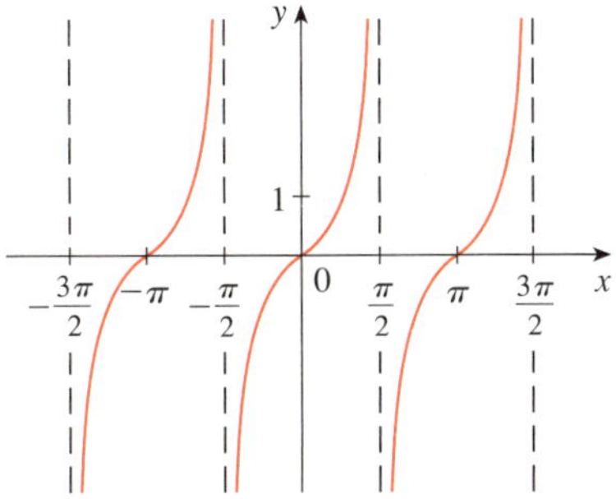

FIGURE 16
$y = \tan x$

This shows that the line $x = \pi/2$ is a vertical asymptote. Similar reasoning shows that the lines $x = (2n + 1)\pi/2$, where n is an integer, are all vertical asymptotes of $f(x) = \tan x$. The graph in Figure 16 confirms this.

Another example of a function whose graph has a vertical asymptote is the natural logarithmic function $y = \ln x$. From Figure 17 we see that

$$\lim_{x \to 0^+} \ln x = -\infty$$

and so the line $x = 0$ (the y-axis) is a vertical asymptote. In fact, the same is true for $y = \log_a x$ provided that $a > 1$. (See Figures 11 and 12 in Section 1.6.)

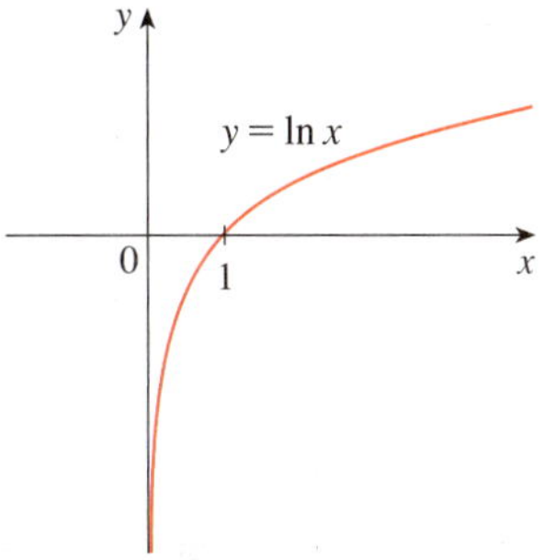

FIGURE 17
The y-axis is a vertical asymptote of the natural logarithmic function.

2.2 Exercises

1. Explain in your own words what is meant by the equation

$$\lim_{x \to 2} f(x) = 5$$

Is it possible for this statement to be true and yet $f(2) = 3$? Explain.

2. Explain what it means to say that

$$\lim_{x \to 1^-} f(x) = 3 \quad \text{and} \quad \lim_{x \to 1^+} f(x) = 7$$

In this situation is it possible that $\lim_{x \to 1} f(x)$ exists? Explain.

3. Explain the meaning of each of the following.

(a) $\lim_{x \to -3} f(x) = \infty$ (b) $\lim_{x \to 4^+} f(x) = -\infty$

4. For the function f whose graph is given, state the value of the given quantity, if it exists. If it does not exist, explain why.

(a) $\lim_{x \to 0} f(x)$ (b) $\lim_{x \to 3^-} f(x)$

(c) $\lim_{x \to 3^+} f(x)$ (d) $\lim_{x \to 3} f(x)$

(e) $f(3)$

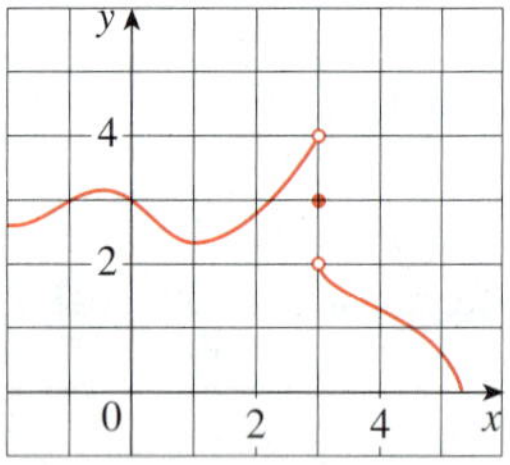

5. Use the given graph of f to state the value of each quantity, if it exists. If it does not exist, explain why.

(a) $\lim_{x \to 1^-} f(x)$ (b) $\lim_{x \to 1^+} f(x)$ (c) $\lim_{x \to 1} f(x)$

(d) $\lim_{x \to 5} f(x)$ (e) $f(5)$

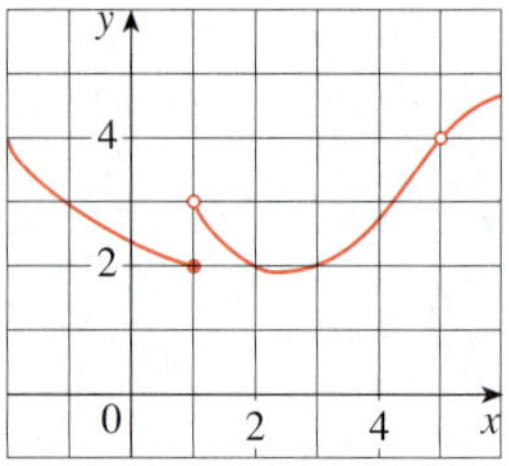

6. For the function g whose graph is given, state the value of each quantity, if it exists. If it does not exist, explain why.

(a) $\lim_{x \to -2^-} g(x)$ (b) $\lim_{x \to -2^+} g(x)$ (c) $\lim_{x \to -2} g(x)$

(d) $g(-2)$ (e) $\lim_{x \to 2^-} g(x)$ (f) $\lim_{x \to 2^+} g(x)$

(g) $\lim_{x \to 2} g(x)$ (h) $g(2)$ (i) $\lim_{x \to 4^+} g(x)$

(j) $\lim_{x \to 4^-} g(x)$ (k) $g(0)$ (l) $\lim_{x \to 0} g(x)$

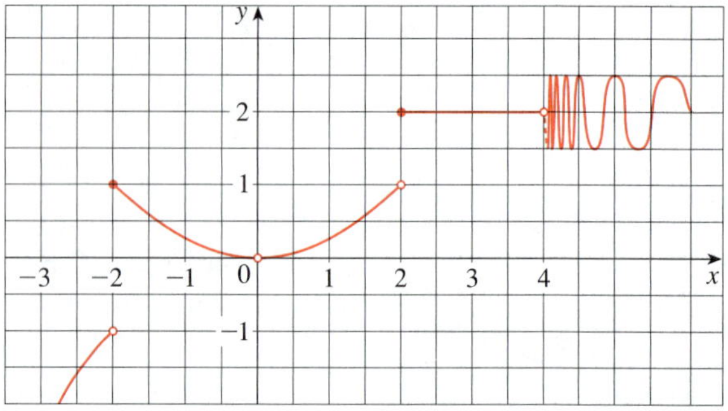

7. For the function g whose graph is given, state the value of each quantity, if it exists. If it does not exist, explain why.

(a) $\lim_{t \to 0^-} g(t)$ (b) $\lim_{t \to 0^+} g(t)$ (c) $\lim_{t \to 0} g(t)$

(d) $\lim_{t \to 2^-} g(t)$ (e) $\lim_{t \to 2^+} g(t)$ (f) $\lim_{t \to 2} g(t)$

(g) $g(2)$ (h) $\lim_{t \to 4} g(t)$

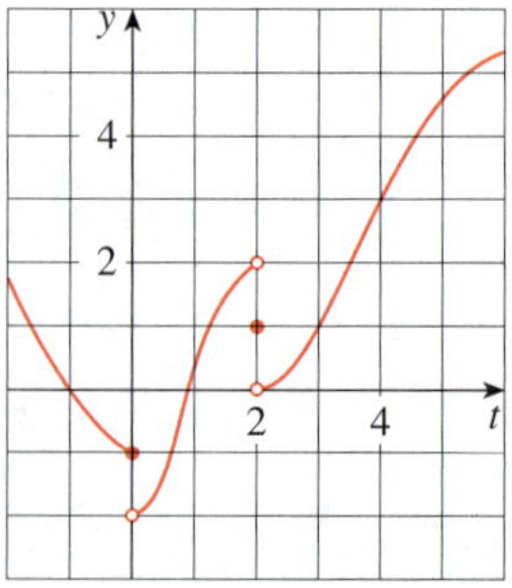

8. For the function R whose graph is shown, state the following.

(a) $\lim_{x \to 2} R(x)$ (b) $\lim_{x \to 5} R(x)$

(c) $\lim_{x \to -3^-} R(x)$ (d) $\lim_{x \to -3^+} R(x)$

(e) The equations of the vertical asymptotes.

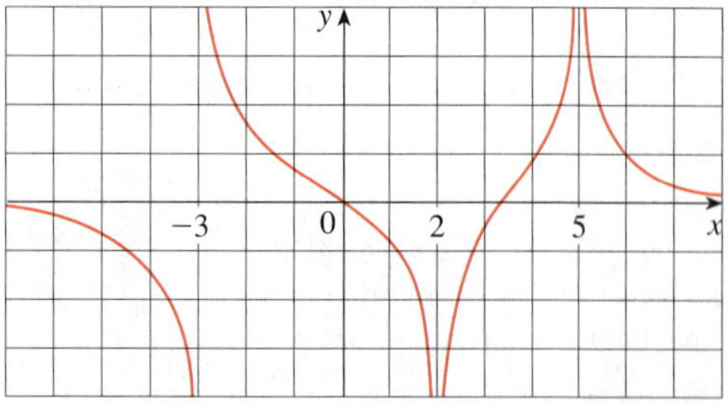

9. For the function f whose graph is shown, state the following.

(a) $\lim_{x\to-7} f(x)$ (b) $\lim_{x\to-3} f(x)$ (c) $\lim_{x\to0} f(x)$

(d) $\lim_{x\to6^-} f(x)$ (e) $\lim_{x\to6^+} f(x)$

(f) The equations of the vertical asymptotes.

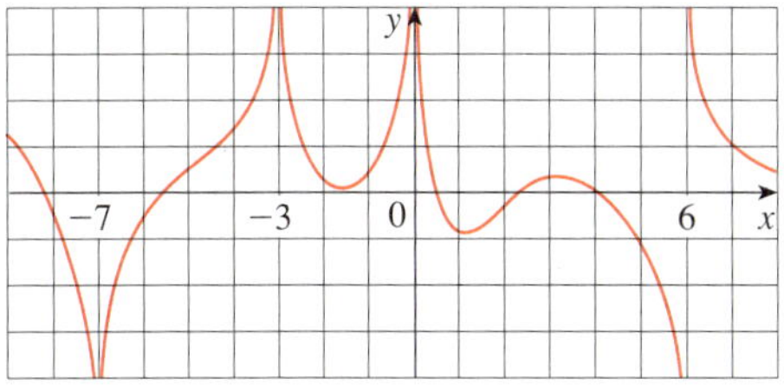

10. A patient receives a 150-mg injection of a drug every 4 hours. The graph shows the amount $f(t)$ of the drug in the bloodstream after t hours. (Later we will be able to compute the dosage and time interval to ensure that the concentration of the drug does not reach a harmful level.) Find

$$\lim_{t\to12^-} f(t) \quad \text{and} \quad \lim_{t\to12^+} f(t)$$

and explain the significance of these one-sided limits.

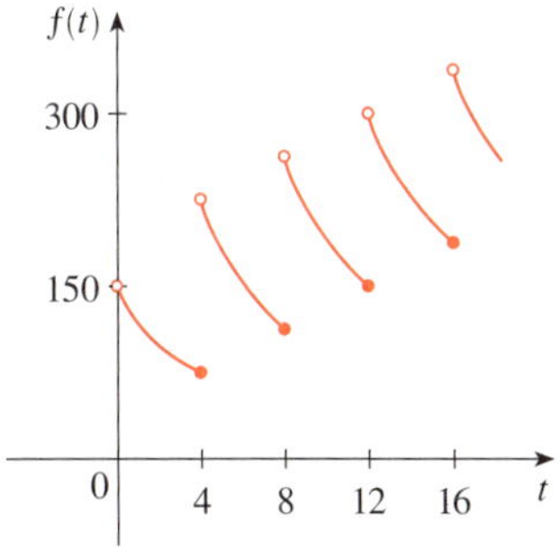

11. Use the graph of the function $f(x) = 1/(1 + e^{1/x})$ to state the value of each limit, if it exists. If it does not exist, explain why.

(a) $\lim_{x\to0^-} f(x)$ (b) $\lim_{x\to0^+} f(x)$ (c) $\lim_{x\to0} f(x)$

12. Sketch the graph of the following function and use it to determine the values of a for which $\lim_{x\to a} f(x)$ exists:

$$f(x) = \begin{cases} 2 - x & \text{if } x < -1 \\ x & \text{if } -1 \leqslant x < 1 \\ (x-1)^2 & \text{if } x \geqslant 1 \end{cases}$$

13–14 ||| Sketch the graph of an example of a function f that satisfies all of the given conditions.

13. $\lim_{x\to3^+} f(x) = 4$, $\lim_{x\to3^-} f(x) = 2$, $\lim_{x\to-2} f(x) = 2$,
$f(3) = 3$, $f(-2) = 1$

14. $\lim_{x\to0^-} f(x) = 1$, $\lim_{x\to0^+} f(x) = -1$, $\lim_{x\to2^-} f(x) = 0$
$\lim_{x\to2^+} f(x) = 1$, $f(2) = 1$, $f(0)$ is undefined

15–18 ||| Guess the value of the limit (if it exists) by evaluating the function at the given numbers (correct to six decimal places).

15. $\lim_{x\to2} \dfrac{x^2 - 2x}{x^2 - x - 2}$, $x = 2.5, 2.1, 2.05, 2.01, 2.005, 2.001,$
$1.9, 1.95, 1.99, 1.995, 1.999$

16. $\lim_{x\to-1} \dfrac{x^2 - 2x}{x^2 - x - 2}$, $x = 0, -0.5, -0.9, -0.95, -0.99,$
$-0.999, -2, -1.5, -1.1, -1.01, -1.001$

17. $\lim_{x\to0} \dfrac{e^x - 1 - x}{x^2}$, $x = \pm1, \pm0.5, \pm0.1, \pm0.05, \pm0.01$

18. $\lim_{x\to0^+} x \ln(x + x^2)$, $x = 1, 0.5, 0.1, 0.05, 0.01, 0.005, 0.001$

19–22 ||| Use a table of values to estimate the value of the limit. If you have a graphing device, use it to confirm your result graphically.

19. $\lim_{x\to0} \dfrac{\sqrt{x+4} - 2}{x}$

20. $\lim_{x\to0} \dfrac{\tan 3x}{\tan 5x}$

21. $\lim_{x\to1} \dfrac{x^6 - 1}{x^{10} - 1}$

22. $\lim_{x\to0} \dfrac{9^x - 5^x}{x}$

23–30 ||| Determine the infinite limit.

23. $\lim_{x\to5^+} \dfrac{6}{x - 5}$

24. $\lim_{x\to5^-} \dfrac{6}{x - 5}$

25. $\lim_{x\to1} \dfrac{2 - x}{(x - 1)^2}$

26. $\lim_{x\to0} \dfrac{x - 1}{x^2(x + 2)}$

27. $\lim_{x\to-2^+} \dfrac{x - 1}{x^2(x + 2)}$

28. $\lim_{x\to\pi^-} \csc x$

29. $\lim_{x\to(-\pi/2)^-} \sec x$

30. $\lim_{x\to5^+} \ln(x - 5)$

31. Determine $\lim_{x\to1^-} \dfrac{1}{x^3 - 1}$ and $\lim_{x\to1^+} \dfrac{1}{x^3 - 1}$

(a) by evaluating $f(x) = 1/(x^3 - 1)$ for values of x that approach 1 from the left and from the right,

(b) by reasoning as in Example 9, and

(c) from a graph of f.

32. (a) Find the vertical asymptotes of the function

$$y = \frac{x}{x^2 - x - 2}$$

(b) Confirm your answer to part (a) by graphing the function.

33. (a) Estimate the value of the limit $\lim_{x\to0} (1 + x)^{1/x}$ to five decimal places. Does this number look familiar?

(b) Illustrate part (a) by graphing the function $y = (1 + x)^{1/x}$.

34. The slope of the tangent line to the graph of the exponential function $y = 2^x$ at the point $(0, 1)$ is $\lim_{x\to0} (2^x - 1)/x$. Estimate the slope to three decimal places.

35. (a) Evaluate the function $f(x) = x^2 - (2^x/1000)$ for $x = 1$, 0.8, 0.6, 0.4, 0.2, 0.1, and 0.05, and guess the value of

$$\lim_{x \to 0} \left(x^2 - \frac{2^x}{1000} \right)$$

(b) Evaluate $f(x)$ for $x = 0.04$, 0.02, 0.01, 0.005, 0.003, and 0.001. Guess again.

36. (a) Evaluate $h(x) = (\tan x - x)/x^3$ for $x = 1$, 0.5, 0.1, 0.05, 0.01, and 0.005.

(b) Guess the value of $\lim_{x \to 0} \dfrac{\tan x - x}{x^3}$.

(c) Evaluate $h(x)$ for successively smaller values of x until you finally reach 0 values for $h(x)$. Are you still confident that your guess in part (b) is correct? Explain why you eventually obtained 0 values. (In Section 4.4 a method for evaluating the limit will be explained.)

(d) Graph the function h in the viewing rectangle $[-1, 1]$ by $[0, 1]$. Then zoom in toward the point where the graph crosses the y-axis to estimate the limit of $h(x)$ as x approaches 0. Continue to zoom in until you observe distortions in the graph of h. Compare with the results of part (c).

37. Graph the function $f(x) = \sin(\pi/x)$ of Example 4 in the viewing rectangle $[-1, 1]$ by $[-1, 1]$. Then zoom in toward the origin several times. Comment on the behavior of this function.

38. In the theory of relativity, the mass of a particle with velocity v is

$$m = \frac{m_0}{\sqrt{1 - v^2/c^2}}$$

where m_0 is the rest mass of the particle and c is the speed of light. What happens as $v \to c^-$?

39. Use a graph to estimate the equations of all the vertical asymptotes of the curve

$$y = \tan(2 \sin x) \qquad -\pi \leqslant x \leqslant \pi$$

Then find the exact equations of these asymptotes.

40. (a) Use numerical and graphical evidence to guess the value of the limit

$$\lim_{x \to 1} \frac{x^3 - 1}{\sqrt{x} - 1}$$

(b) How close to 1 does x have to be to ensure that the function in part (a) is within a distance 0.5 of its limit?

2.3 Calculating Limits Using the Limit Laws

In Section 2.2 we used calculators and graphs to guess the values of limits, but we saw that such methods don't always lead to the correct answer. In this section we use the following properties of limits, called the *Limit Laws,* to calculate limits.

Limit Laws Suppose that c is a constant and the limits

$$\lim_{x \to a} f(x) \qquad \text{and} \qquad \lim_{x \to a} g(x)$$

exist. Then

1. $\lim_{x \to a} [f(x) + g(x)] = \lim_{x \to a} f(x) + \lim_{x \to a} g(x)$

2. $\lim_{x \to a} [f(x) - g(x)] = \lim_{x \to a} f(x) - \lim_{x \to a} g(x)$

3. $\lim_{x \to a} [cf(x)] = c \lim_{x \to a} f(x)$

4. $\lim_{x \to a} [f(x)g(x)] = \lim_{x \to a} f(x) \cdot \lim_{x \to a} g(x)$

5. $\lim_{x \to a} \dfrac{f(x)}{g(x)} = \dfrac{\lim_{x \to a} f(x)}{\lim_{x \to a} g(x)} \quad \text{if } \lim_{x \to a} g(x) \neq 0$

These five laws can be stated verbally as follows:

Sum Law

1. The limit of a sum is the sum of the limits.

Difference Law

2. The limit of a difference is the difference of the limits.

Constant Multiple Law

3. The limit of a constant times a function is the constant times the limit of the function.

Product Law

4. The limit of a product is the product of the limits.

Quotient Law

5. The limit of a quotient is the quotient of the limits (provided that the limit of the denominator is not 0).

It is easy to believe that these properties are true. For instance, if $f(x)$ is close to L and $g(x)$ is close to M, it is reasonable to conclude that $f(x) + g(x)$ is close to $L + M$. This gives us an intuitive basis for believing that Law 1 is true. In Section 2.4 we give a precise definition of a limit and use it to prove this law. The proofs of the remaining laws are given in Appendix F.

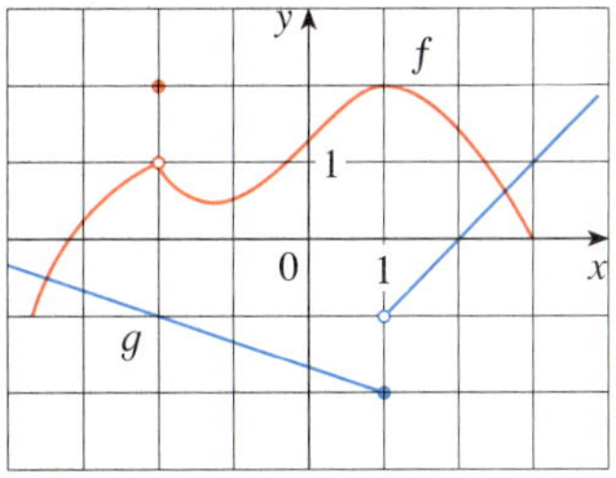

FIGURE 1

EXAMPLE 1 Use the Limit Laws and the graphs of f and g in Figure 1 to evaluate the following limits, if they exist.

(a) $\lim_{x \to -2} [f(x) + 5g(x)]$ (b) $\lim_{x \to 1} [f(x)g(x)]$ (c) $\lim_{x \to 2} \dfrac{f(x)}{g(x)}$

SOLUTION

(a) From the graphs of f and g we see that

$$\lim_{x \to -2} f(x) = 1 \qquad \text{and} \qquad \lim_{x \to -2} g(x) = -1$$

Therefore, we have

$$\begin{aligned} \lim_{x \to -2} [f(x) + 5g(x)] &= \lim_{x \to -2} f(x) + \lim_{x \to -2} [5g(x)] && \text{(by Law 1)} \\ &= \lim_{x \to -2} f(x) + 5 \lim_{x \to -2} g(x) && \text{(by Law 3)} \\ &= 1 + 5(-1) = -4 \end{aligned}$$

(b) We see that $\lim_{x \to 1} f(x) = 2$. But $\lim_{x \to 1} g(x)$ does not exist because the left and right limits are different:

$$\lim_{x \to 1^-} g(x) = -2 \qquad \lim_{x \to 1^+} g(x) = -1$$

So we can't use Law 4. The given limit does not exist, since the left limit is not equal to the right limit.

(c) The graphs show that

$$\lim_{x \to 2} f(x) \approx 1.4 \qquad \text{and} \qquad \lim_{x \to 2} g(x) = 0$$

Because the limit of the denominator is 0, we can't use Law 5. The given limit does not exist because the denominator approaches 0 while the numerator approaches a nonzero number.

If we use the Product Law repeatedly with $g(x) = f(x)$, we obtain the following law.

Power Law

6. $\lim_{x \to a} [f(x)]^n = \left[\lim_{x \to a} f(x)\right]^n$ where n is a positive integer

In applying these six limit laws, we need to use two special limits:

7. $\lim_{x \to a} c = c$ **8.** $\lim_{x \to a} x = a$

These limits are obvious from an intuitive point of view (state them in words or draw graphs of $y = c$ and $y = x$), but proofs based on the precise definition are requested in the exercises for Section 2.4.

If we now put $f(x) = x$ in Law 6 and use Law 8, we get another useful special limit.

9. $\lim_{x \to a} x^n = a^n$ where n is a positive integer

A similar limit holds for roots as follows. (For square roots the proof is outlined in Exercise 37 in Section 2.4.)

10. $\lim_{x \to a} \sqrt[n]{x} = \sqrt[n]{a}$ where n is a positive integer

(If n is even, we assume that $a > 0$.)

More generally, we have the following law, which is proved as a consequence of Law 10 in Section 2.5.

Root Law

11. $\lim_{x \to a} \sqrt[n]{f(x)} = \sqrt[n]{\lim_{x \to a} f(x)}$ where n is a positive integer

$\left[\text{If } n \text{ is even, we assume that } \lim_{x \to a} f(x) > 0.\right]$

Explore limits like these interactively.
Resources / Module 2
/ The Essential Examples
/ Examples D and E

EXAMPLE 2 Evaluate the following limits and justify each step.

(a) $\lim_{x \to 5} (2x^2 - 3x + 4)$ (b) $\lim_{x \to -2} \dfrac{x^3 + 2x^2 - 1}{5 - 3x}$

SOLUTION

(a)
$$\begin{aligned}
\lim_{x \to 5} (2x^2 - 3x + 4) &= \lim_{x \to 5} (2x^2) - \lim_{x \to 5} (3x) + \lim_{x \to 5} 4 && \text{(by Laws 2 and 1)} \\
&= 2 \lim_{x \to 5} x^2 - 3 \lim_{x \to 5} x + \lim_{x \to 5} 4 && \text{(by 3)} \\
&= 2(5^2) - 3(5) + 4 && \text{(by 9, 8, and 7)} \\
&= 39
\end{aligned}$$

(b) We start by using Law 5, but its use is fully justified only at the final stage when we see that the limits of the numerator and denominator exist and the limit of the denominator is not 0.

IIII NEWTON AND LIMITS

Isaac Newton was born on Christmas Day in 1642, the year of Galileo's death. When he entered Cambridge University in 1661 Newton didn't know much mathematics, but he learned quickly by reading Euclid and Descartes and by attending the lectures of Isaac Barrow. Cambridge was closed because of the plague in 1665 and 1666, and Newton returned home to reflect on what he had learned. Those two years were amazingly productive for at that time he made four of his major discoveries: (1) his representation of functions as sums of infinite series, including the binomial theorem; (2) his work on differential and integral calculus; (3) his laws of motion and law of universal gravitation; and (4) his prism experiments on the nature of light and color. Because of a fear of controversy and criticism, he was reluctant to publish his discoveries and it wasn't until 1687, at the urging of the astronomer Halley, that Newton published *Principia Mathematica*. In this work, the greatest scientific treatise ever written, Newton set forth his version of calculus and used it to investigate mechanics, fluid dynamics, and wave motion, and to explain the motion of planets and comets.

The beginnings of calculus are found in the calculations of areas and volumes by ancient Greek scholars such as Eudoxus and Archimedes. Although aspects of the idea of a limit are implicit in their "method of exhaustion," Eudoxus and Archimedes never explicitly formulated the concept of a limit. Likewise, mathematicians such as Cavalieri, Fermat, and Barrow, the immediate precursors of Newton in the development of calculus, did not actually use limits. It was Isaac Newton who was the first to talk explicitly about limits. He explained that the main idea behind limits is that quantities "approach nearer than by any given difference." Newton stated that the limit was the basic concept in calculus, but it was left to later mathematicians like Cauchy to clarify his ideas about limits.

$$\lim_{x\to-2}\frac{x^3+2x^2-1}{5-3x}=\frac{\lim_{x\to-2}(x^3+2x^2-1)}{\lim_{x\to-2}(5-3x)} \qquad \text{(by Law 5)}$$

$$=\frac{\lim_{x\to-2}x^3+2\lim_{x\to-2}x^2-\lim_{x\to-2}1}{\lim_{x\to-2}5-3\lim_{x\to-2}x} \qquad \text{(by 1, 2, and 3)}$$

$$=\frac{(-2)^3+2(-2)^2-1}{5-3(-2)} \qquad \text{(by 9, 8, and 7)}$$

$$=-\frac{1}{11}$$

NOTE ▫ If we let $f(x) = 2x^2 - 3x + 4$, then $f(5) = 39$. In other words, we would have gotten the correct answer in Example 2(a) by substituting 5 for x. Similarly, direct substitution provides the correct answer in part (b). The functions in Example 2 are a polynomial and a rational function, respectively, and similar use of the Limit Laws proves that direct substitution always works for such functions (see Exercises 53 and 54). We state this fact as follows.

Direct Substitution Property If f is a polynomial or a rational function and a is in the domain of f, then

$$\lim_{x\to a} f(x) = f(a)$$

Functions with the Direct Substitution Property are called *continuous at a* and will be studied in Section 2.5. However, not all limits can be evaluated by direct substitution, as the following examples show.

EXAMPLE 3 Find $\displaystyle\lim_{x\to1}\frac{x^2-1}{x-1}$.

SOLUTION Let $f(x) = (x^2 - 1)/(x - 1)$. We can't find the limit by substituting $x = 1$ because $f(1)$ isn't defined. Nor can we apply the Quotient Law because the limit of the denominator is 0. Instead, we need to do some preliminary algebra. We factor the numerator as a difference of squares:

$$\frac{x^2-1}{x-1}=\frac{(x-1)(x+1)}{x-1}$$

The numerator and denominator have a common factor of $x - 1$. When we take the limit as x approaches 1, we have $x \neq 1$ and so $x - 1 \neq 0$. Therefore, we can cancel the common factor and compute the limit as follows:

$$\lim_{x\to1}\frac{x^2-1}{x-1}=\lim_{x\to1}\frac{(x-1)(x+1)}{x-1}$$

$$=\lim_{x\to1}(x+1)$$

$$=1+1=2$$

The limit in this example arose in Section 2.1 when we were trying to find the tangent to the parabola $y = x^2$ at the point (1, 1).

NOTE ▫ In Example 3 we were able to compute the limit by replacing the given function $f(x) = (x^2 - 1)/(x - 1)$ by a simpler function, $g(x) = x + 1$, with the same limit. This is valid because $f(x) = g(x)$ except when $x = 1$, and in computing a limit as x approaches 1 we don't consider what happens when x is actually *equal* to 1. In general, if $f(x) = g(x)$ when $x \neq a$, then

$$\lim_{x \to a} f(x) = \lim_{x \to a} g(x)$$

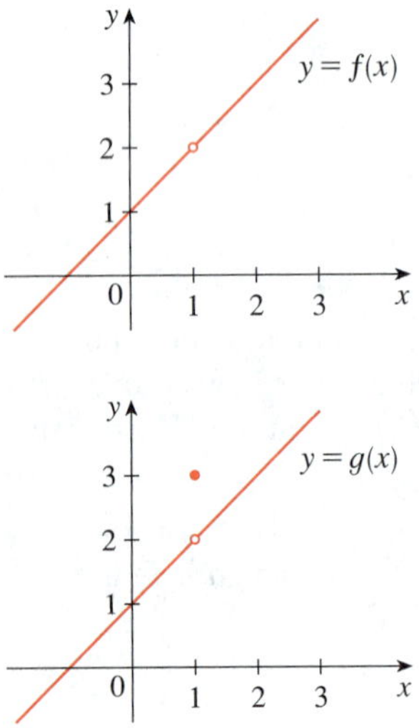

FIGURE 2
The graphs of the functions f (from Example 3) and g (from Example 4)

EXAMPLE 4 Find $\lim_{x \to 1} g(x)$ where

$$g(x) = \begin{cases} x + 1 & \text{if } x \neq 1 \\ \pi & \text{if } x = 1 \end{cases}$$

SOLUTION Here g is defined at $x = 1$ and $g(1) = \pi$, but the value of a limit as x approaches 1 does not depend on the value of the function at 1. Since $g(x) = x + 1$ for $x \neq 1$, we have

$$\lim_{x \to 1} g(x) = \lim_{x \to 1} (x + 1) = 2$$

Note that the values of the functions in Examples 3 and 4 are identical except when $x = 1$ (see Figure 2) and so they have the same limit as x approaches 1.

EXAMPLE 5 Evaluate $\displaystyle\lim_{h \to 0} \frac{(3 + h)^2 - 9}{h}$.

SOLUTION If we define

$$F(h) = \frac{(3 + h)^2 - 9}{h}$$

then, as in Example 3, we can't compute $\lim_{h \to 0} F(h)$ by letting $h = 0$ since $F(0)$ is undefined. But if we simplify $F(h)$ algebraically, we find that

$$F(h) = \frac{(9 + 6h + h^2) - 9}{h} = \frac{6h + h^2}{h} = 6 + h$$

(Recall that we consider only $h \neq 0$ when letting h approach 0.) Thus

$$\lim_{h \to 0} \frac{(3 + h)^2 - 9}{h} = \lim_{h \to 0} (6 + h) = 6$$

Explore a limit like this one interactively.
Resources / Module 2
/ The Essential Examples
/ Example C

EXAMPLE 6 Find $\displaystyle\lim_{t \to 0} \frac{\sqrt{t^2 + 9} - 3}{t^2}$.

SOLUTION We can't apply the Quotient Law immediately, since the limit of the denominator is 0. Here the preliminary algebra consists of rationalizing the numerator:

$$\begin{aligned} \lim_{t \to 0} \frac{\sqrt{t^2 + 9} - 3}{t^2} &= \lim_{t \to 0} \frac{\sqrt{t^2 + 9} - 3}{t^2} \cdot \frac{\sqrt{t^2 + 9} + 3}{\sqrt{t^2 + 9} + 3} \\ &= \lim_{t \to 0} \frac{(t^2 + 9) - 9}{t^2\left(\sqrt{t^2 + 9} + 3\right)} = \lim_{t \to 0} \frac{t^2}{t^2\left(\sqrt{t^2 + 9} + 3\right)} \\ &= \lim_{t \to 0} \frac{1}{\sqrt{t^2 + 9} + 3} = \frac{1}{\sqrt{\lim_{t \to 0} (t^2 + 9)} + 3} = \frac{1}{3 + 3} = \frac{1}{6} \end{aligned}$$

This calculation confirms the guess that we made in Example 2 in Section 2.2.

Some limits are best calculated by first finding the left- and right-hand limits. The following theorem is a reminder of what we discovered in Section 2.2. It says that a two-sided limit exists if and only if both of the one-sided limits exist and are equal.

1 Theorem $\lim_{x\to a} f(x) = L$ if and only if $\lim_{x\to a^-} f(x) = L = \lim_{x\to a^+} f(x)$

When computing one-sided limits, we use the fact that the Limit Laws also hold for one-sided limits.

EXAMPLE 7 Show that $\lim_{x\to 0} |x| = 0$.

SOLUTION Recall that

$$|x| = \begin{cases} x & \text{if } x \geqslant 0 \\ -x & \text{if } x < 0 \end{cases}$$

|||| The result of Example 7 looks plausible from Figure 3.

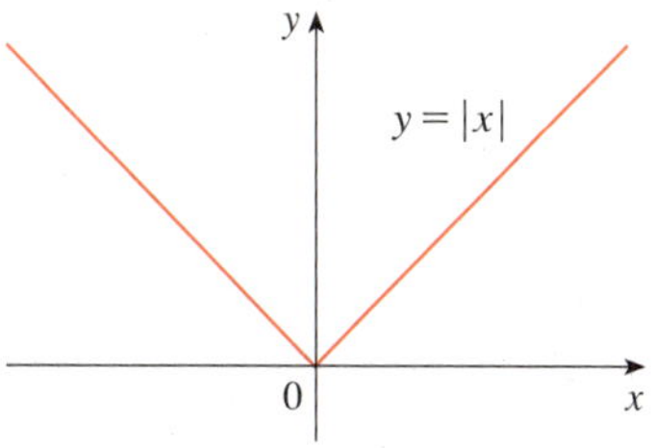

FIGURE 3

Since $|x| = x$ for $x > 0$, we have

$$\lim_{x\to 0^+} |x| = \lim_{x\to 0^+} x = 0$$

For $x < 0$ we have $|x| = -x$ and so

$$\lim_{x\to 0^-} |x| = \lim_{x\to 0^-} (-x) = 0$$

Therefore, by Theorem 1,

$$\lim_{x\to 0} |x| = 0$$

EXAMPLE 8 Prove that $\lim_{x\to 0} \dfrac{|x|}{x}$ does not exist.

SOLUTION

$$\lim_{x\to 0^+} \frac{|x|}{x} = \lim_{x\to 0^+} \frac{x}{x} = \lim_{x\to 0^+} 1 = 1$$

$$\lim_{x\to 0^-} \frac{|x|}{x} = \lim_{x\to 0^-} \frac{-x}{x} = \lim_{x\to 0^-} (-1) = -1$$

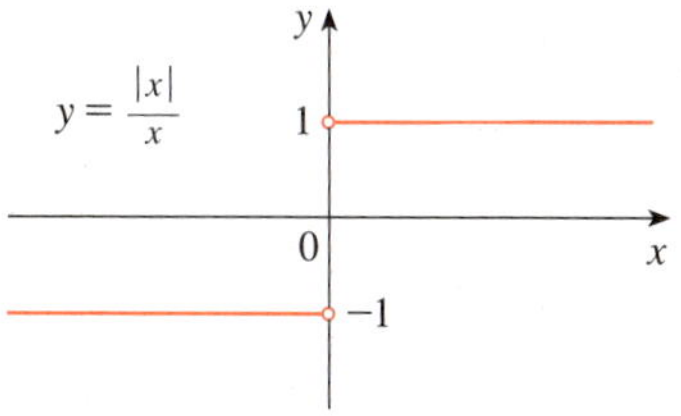

FIGURE 4

Since the right- and left-hand limits are different, it follows from Theorem 1 that $\lim_{x\to 0} |x|/x$ does not exist. The graph of the function $f(x) = |x|/x$ is shown in Figure 4 and supports the one-sided limits that we found.

EXAMPLE 9 If

$$f(x) = \begin{cases} \sqrt{x-4} & \text{if } x > 4 \\ 8 - 2x & \text{if } x < 4 \end{cases}$$

determine whether $\lim_{x\to 4} f(x)$ exists.

|||| It is shown in Example 3 in Section 2.4 that $\lim_{x\to 0^+} \sqrt{x} = 0$.

SOLUTION Since $f(x) = \sqrt{x-4}$ for $x > 4$, we have

$$\lim_{x\to 4^+} f(x) = \lim_{x\to 4^+} \sqrt{x-4} = \sqrt{4-4} = 0$$

Since $f(x) = 8 - 2x$ for $x < 4$, we have

$$\lim_{x \to 4^-} f(x) = \lim_{x \to 4^-} (8 - 2x) = 8 - 2 \cdot 4 = 0$$

The right- and left-hand limits are equal. Thus, the limit exists and

$$\lim_{x \to 4} f(x) = 0$$

The graph of f is shown in Figure 5.

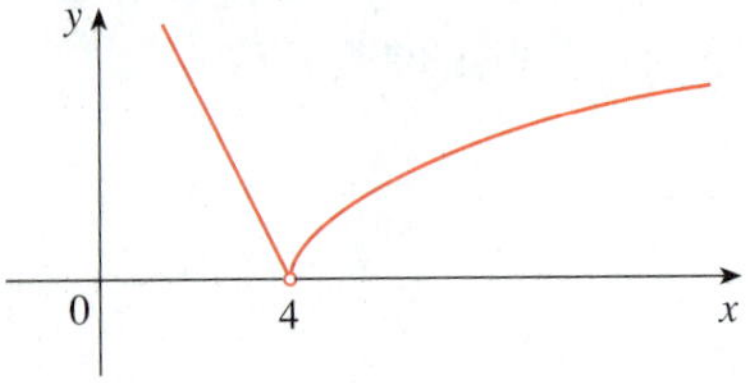

FIGURE 5

EXAMPLE 10 The **greatest integer function** is defined by $[\![x]\!]$ = the largest integer that is less than or equal to x. (For instance, $[\![4]\!] = 4$, $[\![4.8]\!] = 4$, $[\![\pi]\!] = 3$, $[\![\sqrt{2}]\!] = 1$, $[\![-\frac{1}{2}]\!] = -1$.) Show that $\lim_{x \to 3} [\![x]\!]$ does not exist.

|||| Other notations for $[\![x]\!]$ are $[x]$ and $\lfloor x \rfloor$.

SOLUTION The graph of the greatest integer function is shown in Figure 6. Since $[\![x]\!] = 3$ for $3 \leqslant x < 4$, we have

$$\lim_{x \to 3^+} [\![x]\!] = \lim_{x \to 3^+} 3 = 3$$

Since $[\![x]\!] = 2$ for $2 \leqslant x < 3$, we have

$$\lim_{x \to 3^-} [\![x]\!] = \lim_{x \to 3^-} 2 = 2$$

Because these one-sided limits are not equal, $\lim_{x \to 3} [\![x]\!]$ does not exist by Theorem 1.

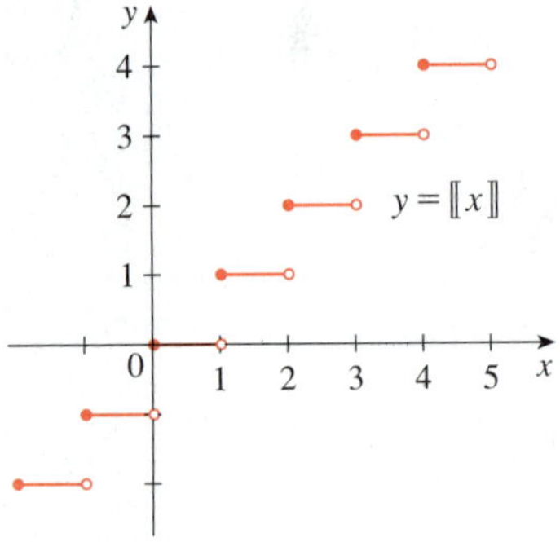

FIGURE 6
Greatest integer function

The next two theorems give two additional properties of limits. Their proofs can be found in Appendix F.

2 Theorem If $f(x) \leqslant g(x)$ when x is near a (except possibly at a) and the limits of f and g both exist as x approaches a, then

$$\lim_{x \to a} f(x) \leqslant \lim_{x \to a} g(x)$$

3 The Squeeze Theorem If $f(x) \leqslant g(x) \leqslant h(x)$ when x is near a (except possibly at a) and

$$\lim_{x \to a} f(x) = \lim_{x \to a} h(x) = L$$

then
$$\lim_{x \to a} g(x) = L$$

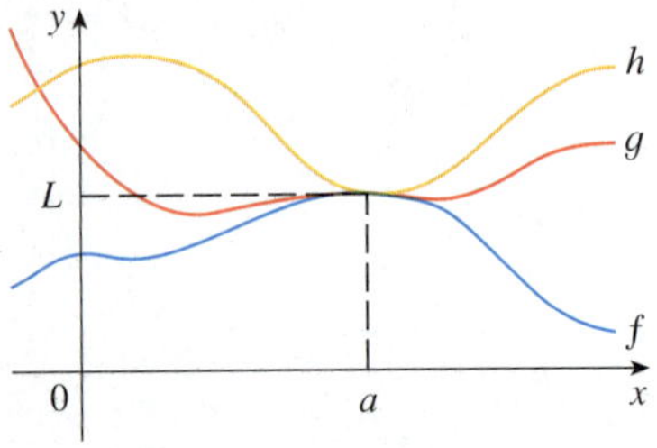

FIGURE 7

The Squeeze Theorem, which is sometimes called the Sandwich Theorem or the Pinching Theorem, is illustrated by Figure 7. It says that if $g(x)$ is squeezed between $f(x)$ and $h(x)$ near a, and if f and h have the same limit L at a, then g is forced to have the same limit L at a.

EXAMPLE 11 Show that $\lim_{x \to 0} x^2 \sin \dfrac{1}{x} = 0$.

SOLUTION First note that we *cannot* use

$$\lim_{x \to 0} x^2 \sin \frac{1}{x} = \lim_{x \to 0} x^2 \cdot \lim_{x \to 0} \sin \frac{1}{x}$$

because $\lim_{x \to 0} \sin(1/x)$ does not exist (see Example 4 in Section 2.2). However, since

$$-1 \leqslant \sin \frac{1}{x} \leqslant 1$$

we have, as illustrated by Figure 8,

$$-x^2 \leqslant x^2 \sin \frac{1}{x} \leqslant x^2$$

Watch an animation of a similar limit.
Resources / Module 2
/ Basics of Limits
/ Sound of a Limit that Exists

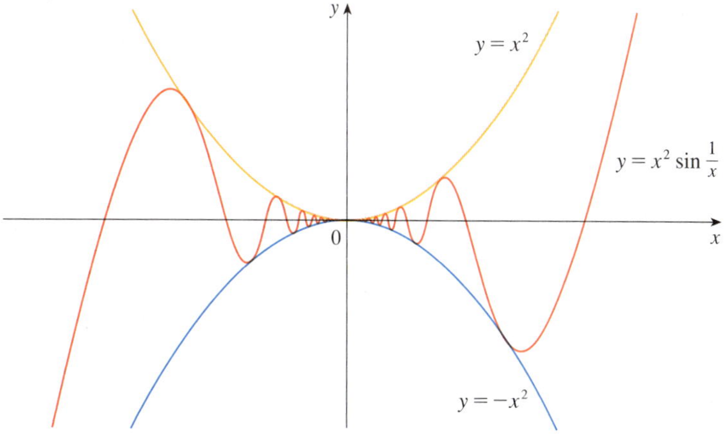

FIGURE 8

We know that

$$\lim_{x \to 0} x^2 = 0 \qquad \text{and} \qquad \lim_{x \to 0} (-x^2) = 0$$

Taking $f(x) = -x^2$, $g(x) = x^2 \sin(1/x)$, and $h(x) = x^2$ in the Squeeze Theorem, we obtain

$$\lim_{x \to 0} x^2 \sin \frac{1}{x} = 0$$

2.3 Exercises

1. Given that

$$\lim_{x \to a} f(x) = -3 \qquad \lim_{x \to a} g(x) = 0 \qquad \lim_{x \to a} h(x) = 8$$

find the limits that exist. If the limit does not exist, explain why.

(a) $\lim_{x \to a} [f(x) + h(x)]$

(b) $\lim_{x \to a} [f(x)]^2$

(c) $\lim_{x \to a} \sqrt[3]{h(x)}$

(d) $\lim_{x \to a} \dfrac{1}{f(x)}$

(e) $\lim_{x \to a} \dfrac{f(x)}{h(x)}$

(f) $\lim_{x \to a} \dfrac{g(x)}{f(x)}$

(g) $\lim_{x \to a} \dfrac{f(x)}{g(x)}$

(h) $\lim_{x \to a} \dfrac{2f(x)}{h(x) - f(x)}$

2. The graphs of f and g are given. Use them to evaluate each limit, if it exists. If the limit does not exist, explain why.

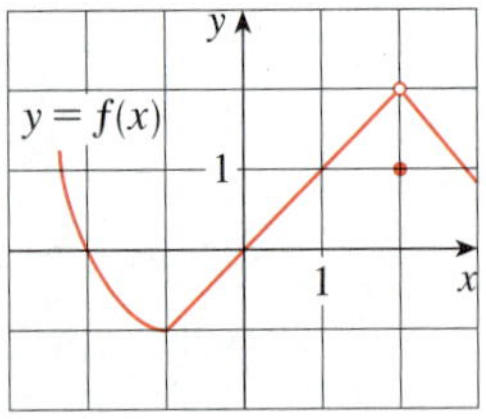

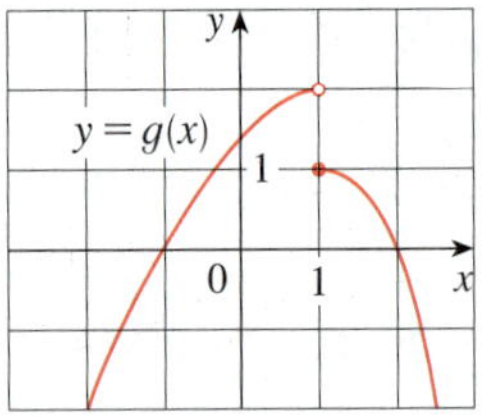

(a) $\lim_{x\to 2}[f(x)+g(x)]$ (b) $\lim_{x\to 1}[f(x)+g(x)]$

(c) $\lim_{x\to 0}[f(x)g(x)]$ (d) $\lim_{x\to -1}\dfrac{f(x)}{g(x)}$

(e) $\lim_{x\to 2}x^3 f(x)$ (f) $\lim_{x\to 1}\sqrt{3+f(x)}$

3–9 Evaluate the limit and justify each step by indicating the appropriate Limit Law(s).

3. $\lim_{x\to -2}(3x^4+2x^2-x+1)$

4. $\lim_{x\to 2}\dfrac{2x^2+1}{x^2+6x-4}$

5. $\lim_{x\to 3}(x^2-4)(x^3+5x-1)$

6. $\lim_{t\to -1}(t^2+1)^3(t+3)^5$

7. $\lim_{x\to 1}\left(\dfrac{1+3x}{1+4x^2+3x^4}\right)^3$

8. $\lim_{u\to -2}\sqrt{u^4+3u+6}$

9. $\lim_{x\to 4^-}\sqrt{16-x^2}$

10. (a) What is wrong with the following equation?

$$\frac{x^2+x-6}{x-2}=x+3$$

(b) In view of part (a), explain why the equation

$$\lim_{x\to 2}\frac{x^2+x-6}{x-2}=\lim_{x\to 2}(x+3)$$

is correct.

11–30 Evaluate the limit, if it exists.

11. $\lim_{x\to 2}\dfrac{x^2+x-6}{x-2}$

12. $\lim_{x\to -4}\dfrac{x^2+5x+4}{x^2+3x-4}$

13. $\lim_{x\to 2}\dfrac{x^2-x+6}{x-2}$

14. $\lim_{x\to 4}\dfrac{x^2-4x}{x^2-3x-4}$

15. $\lim_{t\to -3}\dfrac{t^2-9}{2t^2+7t+3}$

16. $\lim_{x\to -1}\dfrac{x^2-4x}{x^2-3x-4}$

17. $\lim_{h\to 0}\dfrac{(4+h)^2-16}{h}$

18. $\lim_{x\to 1}\dfrac{x^3-1}{x^2-1}$

19. $\lim_{h\to 0}\dfrac{(1+h)^4-1}{h}$

20. $\lim_{h\to 0}\dfrac{(2+h)^3-8}{h}$

21. $\lim_{t\to 9}\dfrac{9-t}{3-\sqrt{t}}$

22. $\lim_{h\to 0}\dfrac{\sqrt{1+h}-1}{h}$

23. $\lim_{x\to 7}\dfrac{\sqrt{x+2}-3}{x-7}$

24. $\lim_{x\to 2}\dfrac{x^4-16}{x-2}$

25. $\lim_{x\to -4}\dfrac{\frac{1}{4}+\frac{1}{x}}{4+x}$

26. $\lim_{t\to 0}\left(\dfrac{1}{t}-\dfrac{1}{t^2+t}\right)$

27. $\lim_{x\to 9}\dfrac{x^2-81}{\sqrt{x}-3}$

28. $\lim_{h\to 0}\dfrac{(3+h)^{-1}-3^{-1}}{h}$

29. $\lim_{t\to 0}\left(\dfrac{1}{t\sqrt{1+t}}-\dfrac{1}{t}\right)$

30. $\lim_{x\to 1}\dfrac{\sqrt{x}-x^2}{1-\sqrt{x}}$

31. (a) Estimate the value of

$$\lim_{x\to 0}\frac{x}{\sqrt{1+3x}-1}$$

by graphing the function $f(x)=x/(\sqrt{1+3x}-1)$.

(b) Make a table of values of $f(x)$ for x close to 0 and guess the value of the limit.

(c) Use the Limit Laws to prove that your guess is correct.

32. (a) Use a graph of

$$f(x)=\frac{\sqrt{3+x}-\sqrt{3}}{x}$$

to estimate the value of $\lim_{x\to 0}f(x)$ to two decimal places.

(b) Use a table of values of $f(x)$ to estimate the limit to four decimal places.

(c) Use the Limit Laws to find the exact value of the limit.

33. Use the Squeeze Theorem to show that $\lim_{x\to 0}x^2\cos 20\pi x=0$. Illustrate by graphing the functions $f(x)=-x^2$, $g(x)=x^2\cos 20\pi x$, and $h(x)=x^2$ on the same screen.

34. Use the Squeeze Theorem to show that

$$\lim_{x\to 0}\sqrt{x^3+x^2}\sin\frac{\pi}{x}=0$$

Illustrate by graphing the functions f, g, and h (in the notation of the Squeeze Theorem) on the same screen.

35. If $1\leqslant f(x)\leqslant x^2+2x+2$ for all x, find $\lim_{x\to -1}f(x)$.

36. If $3x\leqslant f(x)\leqslant x^3+2$ for $0\leqslant x\leqslant 2$, evaluate $\lim_{x\to 1}f(x)$.

37. Prove that $\lim_{x\to 0}x^4\cos\dfrac{2}{x}=0$.

38. Prove that $\lim_{x\to 0^+}\sqrt{x}\,e^{\sin(\pi/x)}=0$.

39–44 Find the limit, if it exists. If the limit does not exist, explain why.

39. $\lim_{x\to -4}|x+4|$

40. $\lim_{x\to -4^-}\dfrac{|x+4|}{x+4}$

41. $\lim_{x\to 2} \dfrac{|x-2|}{x-2}$

42. $\lim_{x\to 1.5} \dfrac{2x^2-3x}{|2x-3|}$

43. $\lim_{x\to 0^-} \left(\dfrac{1}{x} - \dfrac{1}{|x|}\right)$

44. $\lim_{x\to 0^+} \left(\dfrac{1}{x} - \dfrac{1}{|x|}\right)$

45. The *signum* (or sign) *function*, denoted by sgn, is defined by

$$\operatorname{sgn} x = \begin{cases} -1 & \text{if } x < 0 \\ 0 & \text{if } x = 0 \\ 1 & \text{if } x > 0 \end{cases}$$

(a) Sketch the graph of this function.
(b) Find each of the following limits or explain why it does not exist.
(i) $\lim_{x\to 0^+} \operatorname{sgn} x$ (ii) $\lim_{x\to 0^-} \operatorname{sgn} x$
(iii) $\lim_{x\to 0} \operatorname{sgn} x$ (iv) $\lim_{x\to 0} |\operatorname{sgn} x|$

46. Let

$$f(x) = \begin{cases} 4 - x^2 & \text{if } x \leqslant 2 \\ x - 1 & \text{if } x > 2 \end{cases}$$

(a) Find $\lim_{x\to 2^-} f(x)$ and $\lim_{x\to 2^+} f(x)$.
(b) Does $\lim_{x\to 2} f(x)$ exist?
(c) Sketch the graph of f.

47. Let $F(x) = \dfrac{x^2-1}{|x-1|}$.

(a) Find
(i) $\lim_{x\to 1^+} F(x)$ (ii) $\lim_{x\to 1^-} F(x)$
(b) Does $\lim_{x\to 1} F(x)$ exist?
(c) Sketch the graph of F.

48. Let

$$h(x) = \begin{cases} x & \text{if } x < 0 \\ x^2 & \text{if } 0 < x \leqslant 2 \\ 8 - x & \text{if } x > 2 \end{cases}$$

(a) Evaluate each of the following limits, if it exists.
(i) $\lim_{x\to 0^+} h(x)$ (ii) $\lim_{x\to 0} h(x)$ (iii) $\lim_{x\to 1} h(x)$
(iv) $\lim_{x\to 2^-} h(x)$ (v) $\lim_{x\to 2^+} h(x)$ (vi) $\lim_{x\to 2} h(x)$
(b) Sketch the graph of h.

49. (a) If the symbol $[\![\ ]\!]$ denotes the greatest integer function defined in Example 10, evaluate
(i) $\lim_{x\to -2^+} [\![x]\!]$ (ii) $\lim_{x\to -2} [\![x]\!]$ (iii) $\lim_{x\to -2.4} [\![x]\!]$
(b) If n is an integer, evaluate
(i) $\lim_{x\to n^-} [\![x]\!]$ (ii) $\lim_{x\to n^+} [\![x]\!]$
(c) For what values of a does $\lim_{x\to a} [\![x]\!]$ exist?

50. Let $f(x) = x - [\![x]\!]$.
(a) Sketch the graph of f.
(b) If n is an integer, evaluate
(i) $\lim_{x\to n^-} f(x)$ (ii) $\lim_{x\to n^+} f(x)$
(c) For what values of a does $\lim_{x\to a} f(x)$ exist?

51. If $f(x) = [\![x]\!] + [\![-x]\!]$, show that $\lim_{x\to 2} f(x)$ exists but is not equal to $f(2)$.

52. In the theory of relativity, the Lorentz contraction formula

$$L = L_0\sqrt{1 - v^2/c^2}$$

expresses the length L of an object as a function of its velocity v with respect to an observer, where L_0 is the length of the object at rest and c is the speed of light. Find $\lim_{v\to c^-} L$ and interpret the result. Why is a left-hand limit necessary?

53. If p is a polynomial, show that $\lim_{x\to a} p(x) = p(a)$.

54. If r is a rational function, use Exercise 53 to show that $\lim_{x\to a} r(x) = r(a)$ for every number a in the domain of r.

55. If

$$f(x) = \begin{cases} x^2 & \text{if } x \text{ is rational} \\ 0 & \text{if } x \text{ is irrational} \end{cases}$$

prove that $\lim_{x\to 0} f(x) = 0$.

56. Show by means of an example that $\lim_{x\to a} [f(x) + g(x)]$ may exist even though neither $\lim_{x\to a} f(x)$ nor $\lim_{x\to a} g(x)$ exists.

57. Show by means of an example that $\lim_{x\to a} [f(x)g(x)]$ may exist even though neither $\lim_{x\to a} f(x)$ nor $\lim_{x\to a} g(x)$ exists.

58. Evaluate $\lim_{x\to 2} \dfrac{\sqrt{6-x} - 2}{\sqrt{3-x} - 1}$.

59. Is there a number a such that

$$\lim_{x\to -2} \frac{3x^2 + ax + a + 3}{x^2 + x - 2}$$

exists? If so, find the value of a and the value of the limit.

60. The figure shows a fixed circle C_1 with equation $(x-1)^2 + y^2 = 1$ and a shrinking circle C_2 with radius r and center the origin. P is the point $(0, r)$, Q is the upper point of intersection of the two circles, and R is the point of intersection of the line PQ and the x-axis. What happens to R as C_2 shrinks, that is, as $r \to 0^+$?

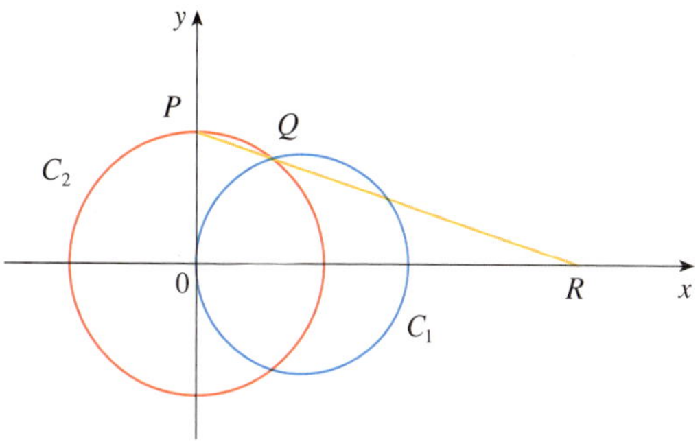

2.4 The Precise Definition of a Limit

The intuitive definition of a limit given in Section 2.2 is inadequate for some purposes because such phrases as "x is close to 2" and "$f(x)$ gets closer and closer to L" are vague. In order to be able to prove conclusively that

$$\lim_{x\to 0}\left(x^3 + \frac{\cos 5x}{10{,}000}\right) = 0.0001 \qquad \text{or} \qquad \lim_{x\to 0}\frac{\sin x}{x} = 1$$

we must make the definition of a limit precise.

To motivate the precise definition of a limit, let's consider the function

$$f(x) = \begin{cases} 2x - 1 & \text{if } x \neq 3 \\ 6 & \text{if } x = 3 \end{cases}$$

Intuitively, it is clear that when x is close to 3 but $x \neq 3$, then $f(x)$ is close to 5, and so $\lim_{x\to 3} f(x) = 5$.

To obtain more detailed information about how $f(x)$ varies when x is close to 3, we ask the following question:

How close to 3 does x have to be so that $f(x)$ differs from 5 by less than 0.1?

It is traditional to use the Greek letter δ (delta) in this situation.

The distance from x to 3 is $|x - 3|$ and the distance from $f(x)$ to 5 is $|f(x) - 5|$, so our problem is to find a number δ such that

$$|f(x) - 5| < 0.1 \quad \text{if} \quad |x - 3| < \delta \quad \text{but } x \neq 3$$

If $|x - 3| > 0$, then $x \neq 3$, so an equivalent formulation of our problem is to find a number δ such that

$$|f(x) - 5| < 0.1 \quad \text{if} \quad 0 < |x - 3| < \delta$$

Notice that if $0 < |x - 3| < (0.1)/2 = 0.05$, then

$$|f(x) - 5| = |(2x - 1) - 5| = |2x - 6| = 2|x - 3| < 0.1$$

that is, $$|f(x) - 5| < 0.1 \quad \text{if} \quad 0 < |x - 3| < 0.05$$

Thus, an answer to the problem is given by $\delta = 0.05$; that is, if x is within a distance of 0.05 from 3, then $f(x)$ will be within a distance of 0.1 from 5.

If we change the number 0.1 in our problem to the smaller number 0.01, then by using the same method we find that $f(x)$ will differ from 5 by less than 0.01 provided that x differs from 3 by less than $(0.01)/2 = 0.005$:

$$|f(x) - 5| < 0.01 \quad \text{if} \quad 0 < |x - 3| < 0.005$$

Similarly,

$$|f(x) - 5| < 0.001 \quad \text{if} \quad 0 < |x - 3| < 0.0005$$

The numbers 0.1, 0.01, and 0.001 that we have considered are *error tolerances* that we might allow. For 5 to be the precise limit of $f(x)$ as x approaches 3, we must not only be able to bring the difference between $f(x)$ and 5 below each of these three numbers; we

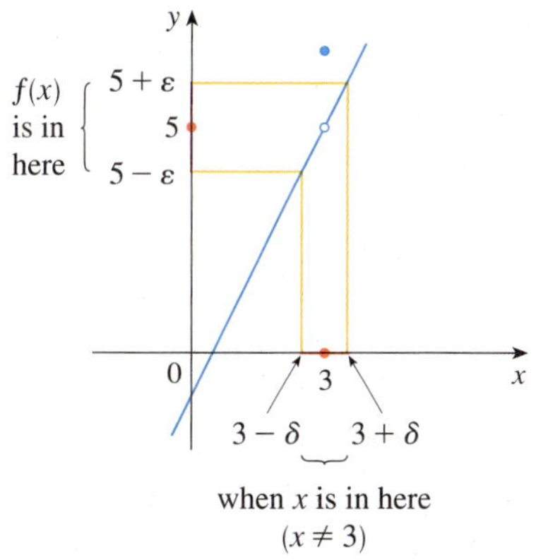

FIGURE 1

must be able to bring it below *any* positive number. And, by the same reasoning, we can! If we write ε (the Greek letter epsilon) for an arbitrary positive number, then we find as before that

$$\boxed{1} \qquad |f(x) - 5| < \varepsilon \quad \text{if} \quad 0 < |x - 3| < \delta = \frac{\varepsilon}{2}$$

This is a precise way of saying that $f(x)$ is close to 5 when x is close to 3 because (1) says that we can make the values of $f(x)$ within an arbitrary distance ε from 5 by taking the values of x within a distance $\varepsilon/2$ from 3 (but $x \neq 3$).

Note that (1) can be rewritten as

$$5 - \varepsilon < f(x) < 5 + \varepsilon \quad \text{whenever} \quad 3 - \delta < x < 3 + \delta \quad (x \neq 3)$$

and this is illustrated in Figure 1. By taking the values of x ($\neq 3$) to lie in the interval $(3 - \delta, 3 + \delta)$ we can make the values of $f(x)$ lie in the interval $(5 - \varepsilon, 5 + \varepsilon)$.

Using (1) as a model, we give a precise definition of a limit.

2 Definition Let f be a function defined on some open interval that contains the number a, except possibly at a itself. Then we say that the **limit of $f(x)$ as x approaches a is L**, and we write

$$\lim_{x \to a} f(x) = L$$

if for every number $\varepsilon > 0$ there is a number $\delta > 0$ such that

$$|f(x) - L| < \varepsilon \quad \text{whenever} \quad 0 < |x - a| < \delta$$

Another way of writing the last line of this definition is

$$\text{if} \quad 0 < |x - a| < \delta \quad \text{then} \quad |f(x) - L| < \varepsilon$$

Since $|x - a|$ is the distance from x to a and $|f(x) - L|$ is the distance from $f(x)$ to L, and since ε can be arbitrarily small, the definition of a limit can be expressed in words as follows:

> $\lim_{x \to a} f(x) = L$ means that the distance between $f(x)$ and L can be made arbitrarily small by taking the distance from x to a sufficiently small (but not 0).

Alternatively,

> $\lim_{x \to a} f(x) = L$ means that the values of $f(x)$ can be made as close as we please to L by taking x close enough to a (but not equal to a).

We can also reformulate Definition 2 in terms of intervals by observing that the inequality $|x - a| < \delta$ is equivalent to $-\delta < x - a < \delta$, which in turn can be written as $a - \delta < x < a + \delta$. Also $0 < |x - a|$ is true if and only if $x - a \neq 0$, that is, $x \neq a$. Similarly, the inequality $|f(x) - L| < \varepsilon$ is equivalent to the pair of inequalities $L - \varepsilon < f(x) < L + \varepsilon$. Therefore, in terms of intervals, Definition 2 can be stated as follows:

> $\lim_{x \to a} f(x) = L$ means that for every $\varepsilon > 0$ (no matter how small ε is) we can find $\delta > 0$ such that if x lies in the open interval $(a - \delta, a + \delta)$ and $x \neq a$, then $f(x)$ lies in the open interval $(L - \varepsilon, L + \varepsilon)$.

We interpret this statement geometrically by representing a function by an arrow diagram as in Figure 2, where f maps a subset of $\mathbb{R}$ onto another subset of $\mathbb{R}$.

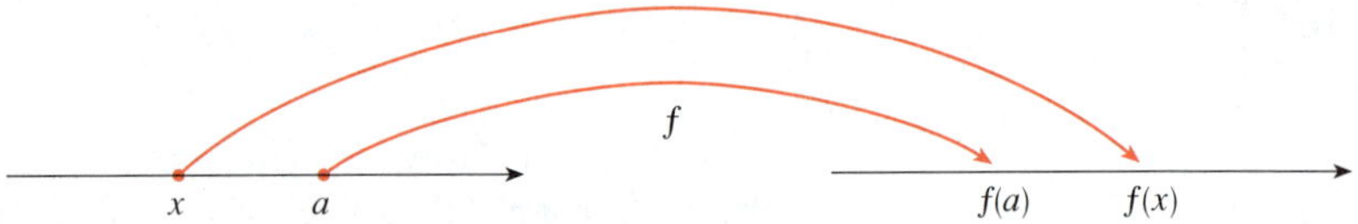

FIGURE 2

The definition of limit says that if any small interval $(L - \varepsilon, L + \varepsilon)$ is given around L, then we can find an interval $(a - \delta, a + \delta)$ around a such that f maps all the points in $(a - \delta, a + \delta)$ (except possibly a) into the interval $(L - \varepsilon, L + \varepsilon)$. (See Figure 3.)

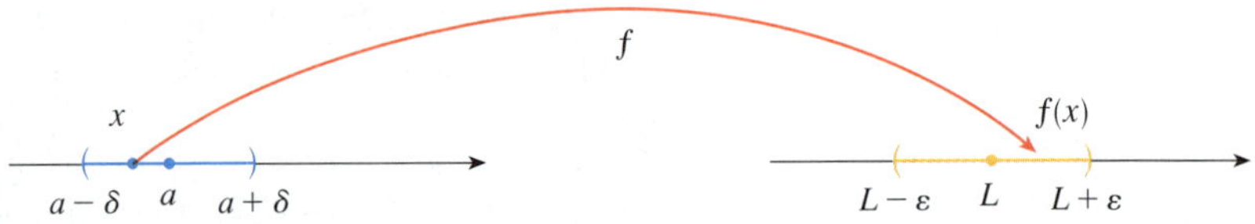

FIGURE 3

Another geometric interpretation of limits can be given in terms of the graph of a function. If $\varepsilon > 0$ is given, then we draw the horizontal lines $y = L + \varepsilon$ and $y = L - \varepsilon$ and the graph of f (see Figure 4). If $\lim_{x \to a} f(x) = L$, then we can find a number $\delta > 0$ such that if we restrict x to lie in the interval $(a - \delta, a + \delta)$ and take $x \neq a$, then the curve $y = f(x)$ lies between the lines $y = L - \varepsilon$ and $y = L + \varepsilon$. (See Figure 5.) You can see that if such a δ has been found, then any smaller δ will also work.

It is important to realize that the process illustrated in Figures 4 and 5 must work for *every* positive number ε no matter how small it is chosen. Figure 6 shows that if a smaller ε is chosen, then a smaller δ may be required.

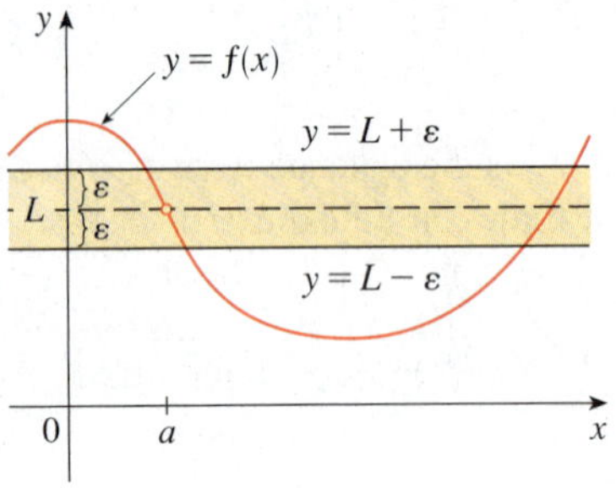

FIGURE 4

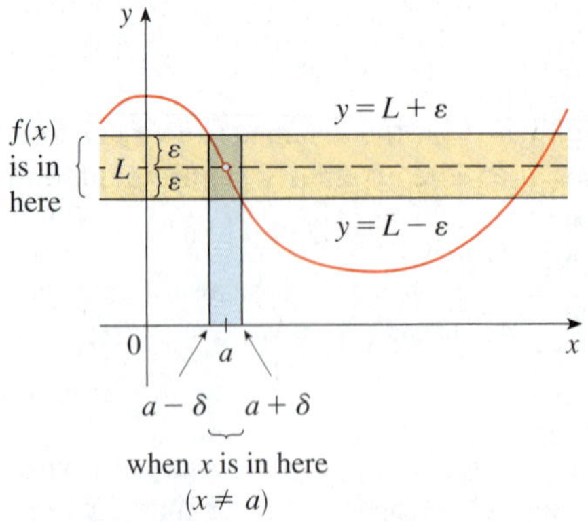

FIGURE 5

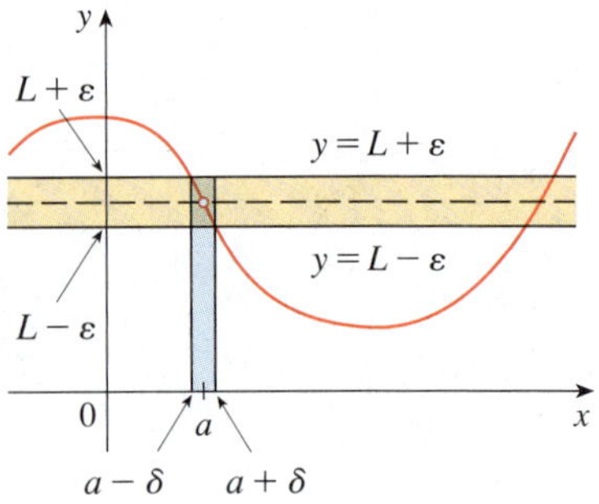

FIGURE 6

EXAMPLE 1 Use a graph to find a number δ such that

$$|(x^3 - 5x + 6) - 2| < 0.2 \quad \text{whenever} \quad |x - 1| < \delta$$

In other words, find a number δ that corresponds to $\varepsilon = 0.2$ in the definition of a limit for the function $f(x) = x^3 - 5x + 6$ with $a = 1$ and $L = 2$.

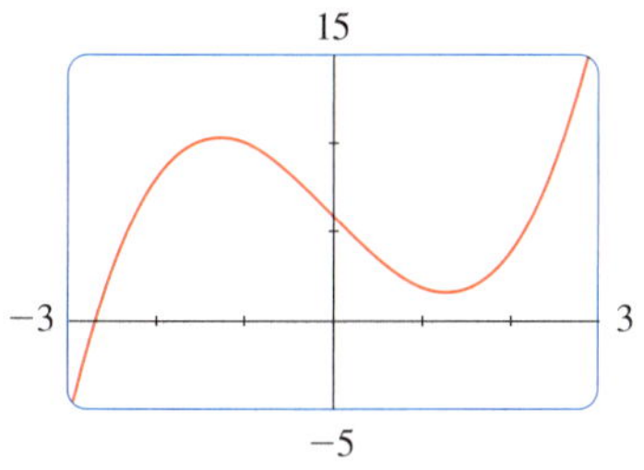

FIGURE 7

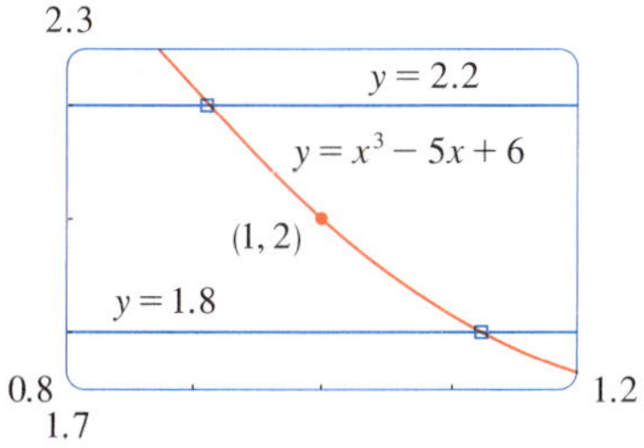

FIGURE 8

SOLUTION A graph of f is shown in Figure 7; we are interested in the region near the point (1, 2). Notice that we can rewrite the inequality

$$|(x^3 - 5x + 6) - 2| < 0.2$$

as

$$1.8 < x^3 - 5x + 6 < 2.2$$

So we need to determine the values of x for which the curve $y = x^3 - 5x + 6$ lies between the horizontal lines $y = 1.8$ and $y = 2.2$. Therefore, we graph the curves $y = x^3 - 5x + 6$, $y = 1.8$, and $y = 2.2$ near the point (1, 2) in Figure 8. Then we use the cursor to estimate that the x-coordinate of the point of intersection of the line $y = 2.2$ and the curve $y = x^3 - 5x + 6$ is about 0.911. Similarly, $y = x^3 - 5x + 6$ intersects the line $y = 1.8$ when $x \approx 1.124$. So, rounding to be safe, we can say that

$$1.8 < x^3 - 5x + 6 < 2.2 \qquad \text{whenever} \qquad 0.92 < x < 1.12$$

This interval (0.92, 1.12) is not symmetric about $x = 1$. The distance from $x = 1$ to the left endpoint is $1 - 0.92 = 0.08$ and the distance to the right endpoint is 0.12. We can choose δ to be the smaller of these numbers, that is, $\delta = 0.08$. Then we can rewrite our inequalities in terms of distances as follows:

$$|(x^3 - 5x + 6) - 2| < 0.2 \qquad \text{whenever} \qquad |x - 1| < 0.08$$

This just says that by keeping x within 0.08 of 1, we are able to keep $f(x)$ within 0.2 of 2.

Although we chose $\delta = 0.08$, any smaller positive value of δ would also have worked.

The graphical procedure in Example 1 gives an illustration of the definition for $\varepsilon = 0.2$, but it does not *prove* that the limit is equal to 2. A proof has to provide a δ for *every* ε.

In proving limit statements it may be helpful to think of the definition of limit as a challenge. First it challenges you with a number ε. Then you must be able to produce a suitable δ. You have to be able to do this for *every* $\varepsilon > 0$, not just a particular ε.

Imagine a contest between two people, A and B, and imagine yourself to be B. Person A stipulates that the fixed number L should be approximated by the values of $f(x)$ to within a degree of accuracy ε (say, 0.01). Person B then responds by finding a number δ such that $|f(x) - L| < \varepsilon$ whenever $0 < |x - a| < \delta$. Then A may become more exacting and challenge B with a smaller value of ε (say, 0.0001). Again B has to respond by finding a corresponding δ. Usually the smaller the value of ε, the smaller the corresponding value of δ must be. If B always wins, no matter how small A makes ε, then $\lim_{x \to a} f(x) = L$.

EXAMPLE 2 Prove that $\lim\limits_{x \to 3} (4x - 5) = 7$.

SOLUTION

1. *Preliminary analysis of the problem (guessing a value for δ).* Let ε be a given positive number. We want to find a number δ such that

$$|(4x - 5) - 7| < \varepsilon \qquad \text{whenever} \qquad 0 < |x - 3| < \delta$$

But $|(4x - 5) - 7| = |4x - 12| = |4(x - 3)| = 4|x - 3|$. Therefore, we want

$$4|x - 3| < \varepsilon \qquad \text{whenever} \qquad 0 < |x - 3| < \delta$$

that is, $$|x-3| < \frac{\varepsilon}{4} \quad \text{whenever} \quad 0 < |x-3| < \delta$$

This suggests that we should choose $\delta = \varepsilon/4$.

2. *Proof (showing that this δ works).* Given $\varepsilon > 0$, choose $\delta = \varepsilon/4$. If $0 < |x-3| < \delta$, then

$$|(4x-5)-7| = |4x-12| = 4|x-3| < 4\delta = 4\left(\frac{\varepsilon}{4}\right) = \varepsilon$$

Thus

$$|(4x-5)-7| < \varepsilon \quad \text{whenever} \quad 0 < |x-3| < \delta$$

Therefore, by the definition of a limit,

$$\lim_{x\to 3}(4x-5) = 7$$

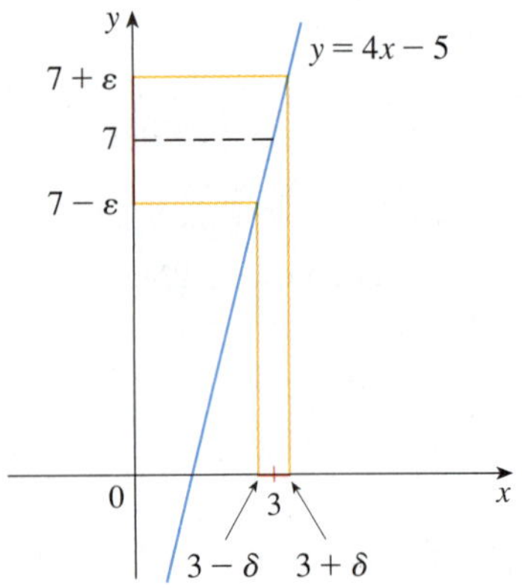

FIGURE 9

This example is illustrated by Figure 9.

Note that in the solution of Example 2 there were two stages—guessing and proving. We made a preliminary analysis that enabled us to guess a value for δ. But then in the second stage we had to go back and prove in a careful, logical fashion that we had made a correct guess. This procedure is typical of much of mathematics. Sometimes it is necessary to first make an intelligent guess about the answer to a problem and then prove that the guess is correct.

The intuitive definitions of one-sided limits that were given in Section 2.2 can be precisely reformulated as follows.

3 Definition of Left-Hand Limit

$$\lim_{x\to a^-} f(x) = L$$

if for every number $\varepsilon > 0$ there is a number $\delta > 0$ such that

$$|f(x)-L| < \varepsilon \quad \text{whenever} \quad a-\delta < x < a$$

4 Definition of Right-Hand Limit

$$\lim_{x\to a^+} f(x) = L$$

if for every number $\varepsilon > 0$ there is a number $\delta > 0$ such that

$$|f(x)-L| < \varepsilon \quad \text{whenever} \quad a < x < a+\delta$$

Notice that Definition 3 is the same as Definition 2 except that x is restricted to lie in the *left* half $(a-\delta, a)$ of the interval $(a-\delta, a+\delta)$. In Definition 4, x is restricted to lie in the *right* half $(a, a+\delta)$ of the interval $(a-\delta, a+\delta)$.

EXAMPLE 3 Use Definition 4 to prove that $\lim_{x\to 0^+}\sqrt{x} = 0$.

IIII CAUCHY AND LIMITS

After the invention of calculus in the 17th century, there followed a period of free development of the subject in the 18th century. Mathematicians like the Bernoulli brothers and Euler were eager to exploit the power of calculus and boldly explored the consequences of this new and wonderful mathematical theory without worrying too much about whether their proofs were completely correct.

The 19th century, by contrast, was the Age of Rigor in mathematics. There was a movement to go back to the foundations of the subject—to provide careful definitions and rigorous proofs. At the forefront of this movement was the French mathematician Augustin-Louis Cauchy (1789–1857), who started out as a military engineer before becoming a mathematics professor in Paris. Cauchy took Newton's idea of a limit, which was kept alive in the 18th century by the French mathematician Jean d'Alembert, and made it more precise. His definition of a limit reads as follows: "When the successive values attributed to a variable approach indefinitely a fixed value so as to end by differing from it by as little as one wishes, this last is called the *limit* of all the others." But when Cauchy used this definition in examples and proofs, he often employed delta-epsilon inequalities similar to the ones in this section. A typical Cauchy proof starts with: "Designate by δ and ε two very small numbers; . . ." He used ε because of the correspondence between epsilon and the French word *erreur*. Later, the German mathematician Karl Weierstrass (1815–1897) stated the definition of a limit exactly as in our Definition 2.

SOLUTION

1. *Guessing a value for δ.* Let ε be a given positive number. Here $a = 0$ and $L = 0$, so we want to find a number δ such that

$$|\sqrt{x} - 0| < \varepsilon \qquad \text{whenever} \qquad 0 < x < \delta$$

that is,

$$\sqrt{x} < \varepsilon \qquad \text{whenever} \qquad 0 < x < \delta$$

or, squaring both sides of the inequality $\sqrt{x} < \varepsilon$, we get

$$x < \varepsilon^2 \qquad \text{whenever} \qquad 0 < x < \delta$$

This suggests that we should choose $\delta = \varepsilon^2$.

2. *Showing that this δ works.* Given $\varepsilon > 0$, let $\delta = \varepsilon^2$. If $0 < x < \delta$, then

$$\sqrt{x} < \sqrt{\delta} = \sqrt{\varepsilon^2} = \varepsilon$$

so

$$|\sqrt{x} - 0| < \varepsilon$$

According to Definition 4, this shows that $\lim_{x\to 0^+} \sqrt{x} = 0$.

EXAMPLE 4 Prove that $\lim\limits_{x\to 3} x^2 = 9$.

SOLUTION

1. *Guessing a value for δ.* Let $\varepsilon > 0$ be given. We have to find a number $\delta > 0$ such that

$$|x^2 - 9| < \varepsilon \qquad \text{whenever} \qquad 0 < |x - 3| < \delta$$

To connect $|x^2 - 9|$ with $|x - 3|$ we write $|x^2 - 9| = |(x + 3)(x - 3)|$. Then we want

$$|x + 3||x - 3| < \varepsilon \qquad \text{whenever} \qquad 0 < |x - 3| < \delta$$

Notice that if we can find a positive constant C such that $|x + 3| < C$, then

$$|x + 3||x - 3| < C|x - 3|$$

and we can make $C|x - 3| < \varepsilon$ by taking $|x - 3| < \varepsilon/C = \delta$.

We can find such a number C if we restrict x to lie in some interval centered at 3. In fact, since we are interested only in values of x that are close to 3, it is reasonable to assume that x is within a distance 1 from 3, that is, $|x - 3| < 1$. Then $2 < x < 4$, so $5 < x + 3 < 7$. Thus, we have $|x + 3| < 7$, and so $C = 7$ is a suitable choice for the constant.

But now there are two restrictions on $|x - 3|$, namely

$$|x - 3| < 1 \qquad \text{and} \qquad |x - 3| < \frac{\varepsilon}{C} = \frac{\varepsilon}{7}$$

To make sure that both of these inequalities are satisfied, we take δ to be the smaller of the two numbers 1 and $\varepsilon/7$. The notation for this is $\delta = \min\{1, \varepsilon/7\}$.

2. *Showing that this δ works.* Given $\varepsilon > 0$, let $\delta = \min\{1, \varepsilon/7\}$. If $0 < |x - 3| < \delta$, then $|x - 3| < 1 \Rightarrow 2 < x < 4 \Rightarrow |x + 3| < 7$ (as in part l). We also have $|x - 3| < \varepsilon/7$, so

$$|x^2 - 9| = |x + 3||x - 3| < 7 \cdot \frac{\varepsilon}{7} = \varepsilon$$

This shows that $\lim_{x\to 3} x^2 = 9$.

As Example 4 shows, it is not always easy to prove that limit statements are true using the ε, δ definition. In fact, if we had been given a more complicated function such as $f(x) = (6x^2 - 8x + 9)/(2x^2 - 1)$, a proof would require a great deal of ingenuity. Fortunately this is unnecessary because the Limit Laws stated in Section 2.3 can be proved using Definition 2, and then the limits of complicated functions can be found rigorously from the Limit Laws without resorting to the definition directly.

For instance, we prove the Sum Law: If $\lim_{x \to a} f(x) = L$ and $\lim_{x \to a} g(x) = M$ both exist, then

$$\lim_{x \to a} [f(x) + g(x)] = L + M$$

The remaining laws are proved in the exercises and in Appendix F.

Proof of the Sum Law Let $\varepsilon > 0$ be given. We must find $\delta > 0$ such that

$$|f(x) + g(x) - (L + M)| < \varepsilon \qquad \text{whenever} \qquad 0 < |x - a| < \delta$$

|||| Triangle Inequality:

$$|a + b| \leqslant |a| + |b|$$

(See Appendix A.)

Using the Triangle Inequality we can write

5
$$\begin{aligned} |f(x) + g(x) - (L + M)| &= |(f(x) - L) + (g(x) - M)| \\ &\leqslant |f(x) - L| + |g(x) - M| \end{aligned}$$

We make $|f(x) + g(x) - (L + M)|$ less than ε by making each of the terms $|f(x) - L|$ and $|g(x) - M|$ less than $\varepsilon/2$.

Since $\varepsilon/2 > 0$ and $\lim_{x \to a} f(x) = L$, there exists a number $\delta_1 > 0$ such that

$$|f(x) - L| < \frac{\varepsilon}{2} \qquad \text{whenever} \qquad 0 < |x - a| < \delta_1$$

Similarly, since $\lim_{x \to a} g(x) = M$, there exists a number $\delta_2 > 0$ such that

$$|g(x) - M| < \frac{\varepsilon}{2} \qquad \text{whenever} \qquad 0 < |x - a| < \delta_2$$

Let $\delta = \min\{\delta_1, \delta_2\}$. Notice that

$$\text{if} \quad 0 < |x - a| < \delta \quad \text{then} \quad 0 < |x - a| < \delta_1 \quad \text{and} \quad 0 < |x - a| < \delta_2$$

and so
$$|f(x) - L| < \frac{\varepsilon}{2} \qquad \text{and} \qquad |g(x) - M| < \frac{\varepsilon}{2}$$

Therefore, by (5),

$$\begin{aligned} |f(x) + g(x) - (L + M)| &\leqslant |f(x) - L| + |g(x) - M| \\ &< \frac{\varepsilon}{2} + \frac{\varepsilon}{2} = \varepsilon \end{aligned}$$

To summarize,

$$|f(x) + g(x) - (L + M)| < \varepsilon \qquad \text{whenever} \qquad 0 < |x - a| < \delta$$

Thus, by the definition of a limit,

$$\lim_{x \to a} [f(x) + g(x)] = L + M$$

Infinite Limits

Infinite limits can also be defined in a precise way. The following is a precise version of Definition 4 in Section 2.2.

6 Definition Let f be a function defined on some open interval that contains the number a, except possibly at a itself. Then

$$\lim_{x \to a} f(x) = \infty$$

means that for every positive number M there is a positive number δ such that

$$f(x) > M \qquad \text{whenever} \qquad 0 < |x - a| < \delta$$

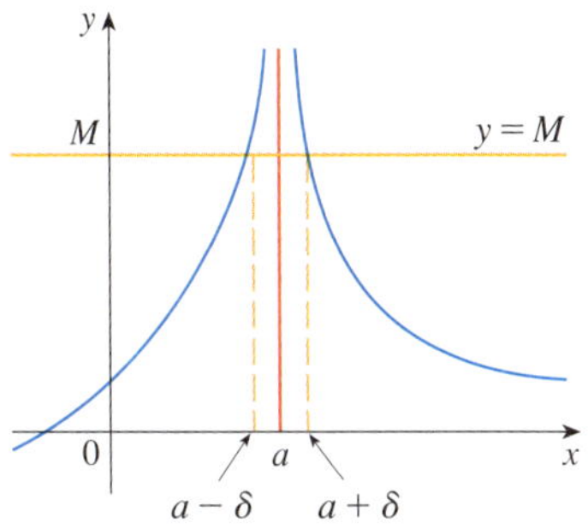

FIGURE 10

This says that the values of $f(x)$ can be made arbitrarily large (larger than any given number M) by taking x close enough to a (within a distance δ, where δ depends on M, but with $x \neq a$). A geometric illustration is shown in Figure 10.

Given any horizontal line $y = M$, we can find a number $\delta > 0$ such that if we restrict x to lie in the interval $(a - \delta, a + \delta)$ but $x \neq a$, then the curve $y = f(x)$ lies above the line $y = M$. You can see that if a larger M is chosen, then a smaller δ may be required.

EXAMPLE 5 Use Definition 6 to prove that $\lim_{x \to 0} \dfrac{1}{x^2} = \infty$.

SOLUTION

1. *Guessing a value for δ.* Given $M > 0$, we want to find $\delta > 0$ such that

$$\frac{1}{x^2} > M \qquad \text{whenever} \qquad 0 < |x - 0| < \delta$$

that is,

$$x^2 < \frac{1}{M} \qquad \text{whenever} \qquad 0 < |x| < \delta$$

or

$$|x| < \frac{1}{\sqrt{M}} \qquad \text{whenever} \qquad 0 < |x| < \delta$$

This suggests that we should take $\delta = 1/\sqrt{M}$.

2. *Showing that this δ works.* If $M > 0$ is given, let $\delta = 1/\sqrt{M}$. If $0 < |x - 0| < \delta$, then

$$|x| < \delta \;\Rightarrow\; x^2 < \delta^2$$

$$\Rightarrow\; \frac{1}{x^2} > \frac{1}{\delta^2} = M$$

Thus

$$\frac{1}{x^2} > M \qquad \text{whenever} \qquad 0 < |x - 0| < \delta$$

Therefore, by Definition 6,

$$\lim_{x \to 0} \frac{1}{x^2} = \infty$$

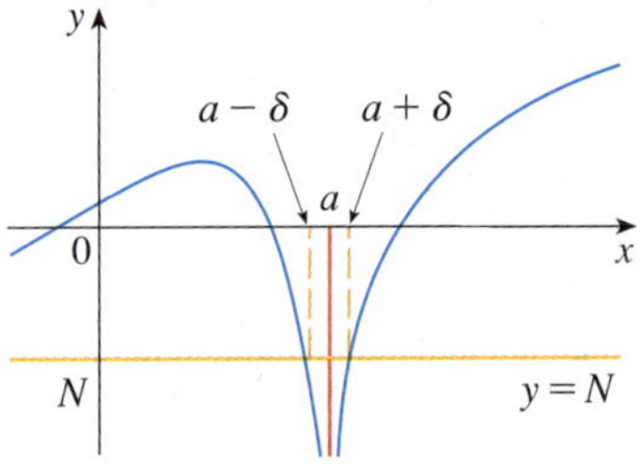

FIGURE 11

Similarly, the following is a precise version of Definition 5 in Section 2.2. It is illustrated by Figure 11.

7 Definition Let f be a function defined on some open interval that contains the number a, except possibly at a itself. Then

$$\lim_{x \to a} f(x) = -\infty$$

means that for every negative number N there is a positive number δ such that

$$f(x) < N \qquad \text{whenever} \qquad 0 < |x - a| < \delta$$

2.4 Exercises

1. How close to 2 do we have to take x so that $5x + 3$ is within a distance of (a) 0.1 and (b) 0.01 from 13?

2. How close to 5 do we have to take x so that $6x - 1$ is within a distance of (a) 0.01, (b) 0.001, and (c) 0.0001 from 29?

3. Use the given graph of $f(x) = 1/x$ to find a number δ such that

$$\left| \frac{1}{x} - 0.5 \right| < 0.2 \qquad \text{whenever} \qquad |x - 2| < \delta$$

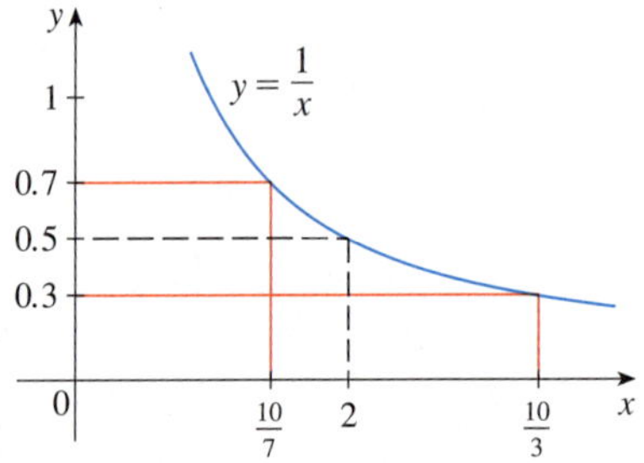

4. Use the given graph of f to find a number δ such that

$$|f(x) - 3| < 0.6 \qquad \text{whenever} \qquad 0 < |x - 5| < \delta$$

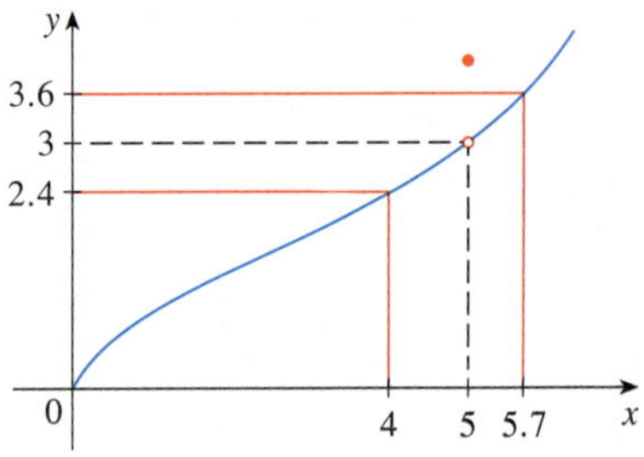

5. Use the given graph of $f(x) = \sqrt{x}$ to find a number δ such that

$$|\sqrt{x} - 2| < 0.4 \qquad \text{whenever} \qquad |x - 4| < \delta$$

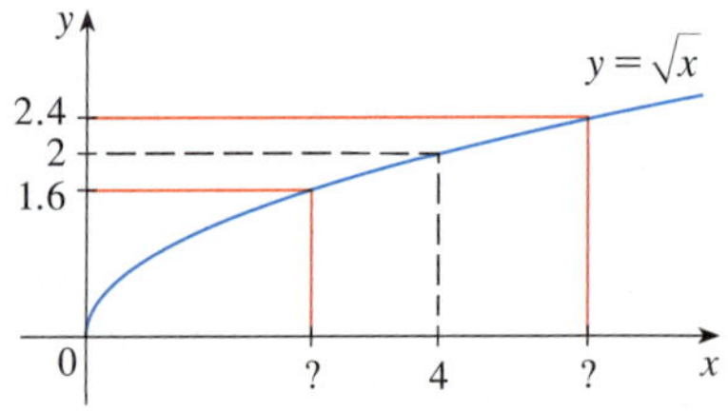

6. Use the given graph of $f(x) = x^2$ to find a number δ such that

$$|x^2 - 1| < \tfrac{1}{2} \qquad \text{whenever} \qquad |x - 1| < \delta$$

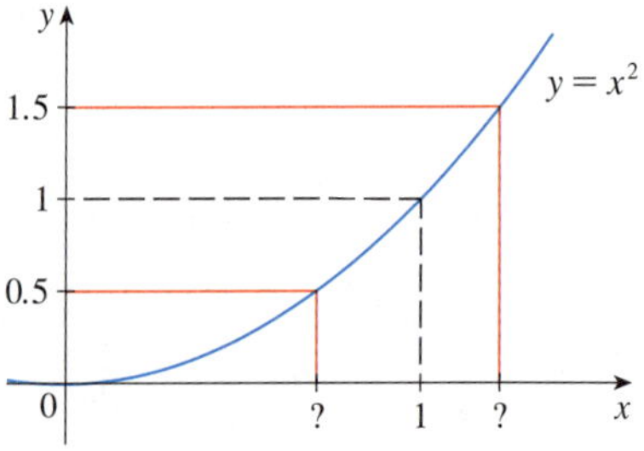

7. Use a graph to find a number δ such that

$$|\sqrt{4x + 1} - 3| < 0.5 \qquad \text{whenever} \qquad |x - 2| < \delta$$

8. Use a graph to find a number δ such that

$$\left| \sin x - \tfrac{1}{2} \right| < 0.1 \qquad \text{whenever} \qquad \left| x - \frac{\pi}{6} \right| < \delta$$

9. For the limit

$$\lim_{x \to 1} (4 + x - 3x^3) = 2$$

illustrate Definition 2 by finding values of δ that correspond to $\varepsilon = 1$ and $\varepsilon = 0.1$.

10. For the limit

$$\lim_{x \to 0} \frac{e^x - 1}{x} = 1$$

illustrate Definition 2 by finding values of δ that correspond to $\varepsilon = 0.5$ and $\varepsilon = 0.1$.

11. Use a graph to find a number δ such that

$$\frac{x}{(x^2+1)(x-1)^2} > 100 \qquad \text{whenever} \qquad 0 < |x - 1| < \delta$$

12. For the limit

$$\lim_{x \to 0} \cot^2 x = \infty$$

illustrate Definition 6 by finding values of δ that correspond to (a) $M = 100$ and (b) $M = 1000$.

13. A machinist is required to manufacture a circular metal disk with area $1000\ \text{cm}^2$.
(a) What radius produces such a disk?
(b) If the machinist is allowed an error tolerance of $\pm 5\ \text{cm}^2$ in the area of the disk, how close to the ideal radius in part (a) must the machinist control the radius?
(c) In terms of the ε, δ definition of $\lim_{x \to a} f(x) = L$, what is x? What is $f(x)$? What is a? What is L? What value of ε is given? What is the corresponding value of δ?

14. A crystal growth furnace is used in research to determine how best to manufacture crystals used in electronic components for the space shuttle. For proper growth of the crystal, the temperature must be controlled accurately by adjusting the input power. Suppose the relationship is given by

$$T(w) = 0.1w^2 + 2.155w + 20$$

where T is the temperature in degrees Celsius and w is the power input in watts.
(a) How much power is needed to maintain the temperature at 200°C?
(b) If the temperature is allowed to vary from 200°C by up to ±1°C, what range of wattage is allowed for the input power?
(c) In terms of the ε, δ definition of $\lim_{x \to a} f(x) = L$, what is x? What is $f(x)$? What is a? What is L? What value of ε is given? What is the corresponding value of δ?

15–18 Prove the statement using the ε, δ definition of limit and illustrate with a diagram like Figure 9.

15. $\lim_{x \to 1} (2x + 3) = 5$

16. $\lim_{x \to -2} \left(\frac{1}{2}x + 3\right) = 2$

17. $\lim_{x \to -3} (1 - 4x) = 13$

18. $\lim_{x \to 4} (7 - 3x) = -5$

19–32 Prove the statement using the ε, δ definition of limit.

19. $\lim_{x \to 3} \frac{x}{5} = \frac{3}{5}$

20. $\lim_{x \to 6} \left(\frac{x}{4} + 3\right) = \frac{9}{2}$

21. $\lim_{x \to -5} \left(4 - \frac{3x}{5}\right) = 7$

22. $\lim_{x \to 3} \frac{x^2 + x - 12}{x - 3} = 7$

23. $\lim_{x \to a} x = a$

24. $\lim_{x \to a} c = c$

25. $\lim_{x \to 0} x^2 = 0$

26. $\lim_{x \to 0} x^3 = 0$

27. $\lim_{x \to 0} |x| = 0$

28. $\lim_{x \to 9^-} \sqrt[4]{9 - x} = 0$

29. $\lim_{x \to 2} (x^2 - 4x + 5) = 1$

30. $\lim_{x \to 3} (x^2 + x - 4) = 8$

31. $\lim_{x \to -2} (x^2 - 1) = 3$

32. $\lim_{x \to 2} x^3 = 8$

33. Verify that another possible choice of δ for showing that $\lim_{x \to 3} x^2 = 9$ in Example 4 is $\delta = \min\{2, \varepsilon/8\}$.

34. Verify, by a geometric argument, that the largest possible choice of δ for showing that $\lim_{x \to 3} x^2 = 9$ is $\delta = \sqrt{9 + \varepsilon} - 3$.

CAS **35.** (a) For the limit $\lim_{x \to 1} (x^3 + x + 1) = 3$, use a graph to find a value of δ that corresponds to $\varepsilon = 0.4$.
(b) By using a computer algebra system to solve the cubic equation $x^3 + x + 1 = 3 + \varepsilon$, find the largest possible value of δ that works for any given $\varepsilon > 0$.
(c) Put $\varepsilon = 0.4$ in your answer to part (b) and compare with your answer to part (a).

36. Prove that $\lim_{x \to 2} \frac{1}{x} = \frac{1}{2}$.

37. Prove that $\lim_{x \to a} \sqrt{x} = \sqrt{a}$ if $a > 0$.

$$\left[\textit{Hint}: \text{Use } \left|\sqrt{x} - \sqrt{a}\right| = \frac{|x - a|}{\sqrt{x} + \sqrt{a}}.\right]$$

38. If H is the Heaviside function defined in Example 6 in Section 2.2, prove, using Definition 2, that $\lim_{t \to 0} H(t)$ does not exist. [*Hint*: Use an indirect proof as follows. Suppose that the limit is L. Take $\varepsilon = \frac{1}{2}$ in the definition of a limit and try to arrive at a contradiction.]

39. If the function f is defined by

$$f(x) = \begin{cases} 0 & \text{if } x \text{ is rational} \\ 1 & \text{if } x \text{ is irrational} \end{cases}$$

prove that $\lim_{x \to 0} f(x)$ does not exist.

40. By comparing Definitions 2, 3, and 4, prove Theorem 1 in Section 2.3.

41. How close to -3 do we have to take x so that

$$\frac{1}{(x + 3)^4} > 10{,}000$$

42. Prove, using Definition 6, that $\lim_{x \to -3} \frac{1}{(x + 3)^4} = \infty$.

43. Prove that $\lim_{x \to 0^+} \ln x = -\infty$.

44. Suppose that $\lim_{x \to a} f(x) = \infty$ and $\lim_{x \to a} g(x) = c$, where c is a real number. Prove each statement.

(a) $\lim_{x \to a} [f(x) + g(x)] = \infty$

(b) $\lim_{x \to a} [f(x)g(x)] = \infty$ if $c > 0$

(c) $\lim_{x \to a} [f(x)g(x)] = -\infty$ if $c < 0$

2.5 Continuity

Explore continuous functions interactively.
Resources / Module 2
/ Continuity
/ Start of Continuity

We noticed in Section 2.3 that the limit of a function as x approaches a can often be found simply by calculating the value of the function at a. Functions with this property are called *continuous at a*. We will see that the mathematical definition of continuity corresponds closely with the meaning of the word *continuity* in everyday language. (A continuous process is one that takes place gradually, without interruption or abrupt change.)

1 Definition A function f is **continuous at a number *a*** if

$$\lim_{x \to a} f(x) = f(a)$$

|||| As illustrated in Figure 1, if f is continuous, then the points $(x, f(x))$ on the graph of f approach the point $(a, f(a))$ on the graph. So there is no gap in the curve.

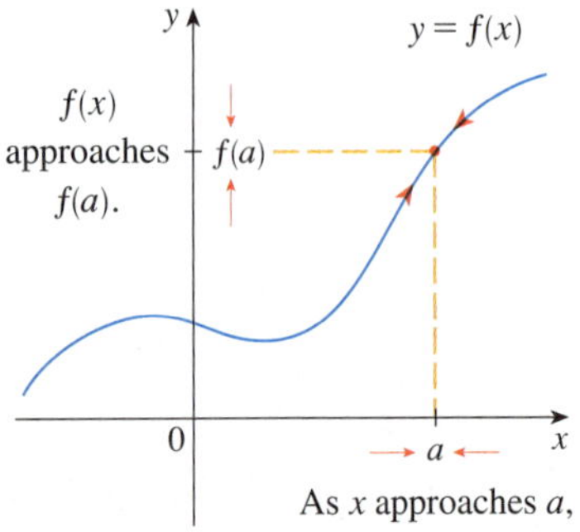

FIGURE 1

Notice that Definition 1 implicitly requires three things if f is continuous at a:

1. $f(a)$ is defined (that is, a is in the domain of f)

2. $\lim_{x \to a} f(x)$ exists

3. $\lim_{x \to a} f(x) = f(a)$

The definition says that f is continuous at a if $f(x)$ approaches $f(a)$ as x approaches a. Thus, a continuous function f has the property that a small change in x produces only a small change in $f(x)$. In fact, the change in $f(x)$ can be kept as small as we please by keeping the change in x sufficiently small.

If f is defined near a (in other words, f is defined on an open interval containing a, except perhaps at a), we say that f is **discontinuous at *a***, or f has a **discontinuity** at a, if f is not continuous at a.

Physical phenomena are usually continuous. For instance, the displacement or velocity of a vehicle varies continuously with time, as does a person's height. But discontinuities do occur in such situations as electric currents. [See Example 6 in Section 2.2, where the Heaviside function is discontinuous at 0 because $\lim_{t \to 0} H(t)$ does not exist.]

Geometrically, you can think of a function that is continuous at every number in an interval as a function whose graph has no break in it. The graph can be drawn without removing your pen from the paper.

EXAMPLE 1 Figure 2 shows the graph of a function f. At which numbers is f discontinuous? Why?

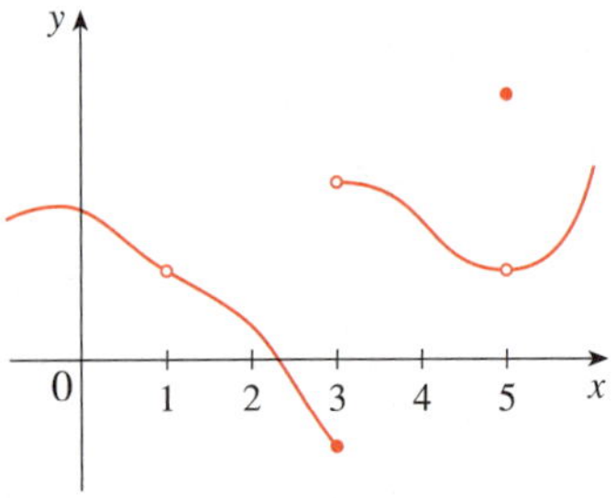

FIGURE 2

SOLUTION It looks as if there is a discontinuity when $a = 1$ because the graph has a break there. The official reason that f is discontinuous at 1 is that $f(1)$ is not defined.

The graph also has a break when $a = 3$, but the reason for the discontinuity is different. Here, $f(3)$ is defined, but $\lim_{x \to 3} f(x)$ does not exist (because the left and right limits are different). So f is discontinuous at 3.

What about $a = 5$? Here, $f(5)$ is defined and $\lim_{x \to 5} f(x)$ exists (because the left and right limits are the same). But

$$\lim_{x \to 5} f(x) \neq f(5)$$

So f is discontinuous at 5.

Now let's see how to detect discontinuities when a function is defined by a formula.

Resources / Module 2
/ Continuity
/ Problems and Tests

EXAMPLE 2 Where are each of the following functions discontinuous?

(a) $f(x) = \dfrac{x^2 - x - 2}{x - 2}$ (b) $f(x) = \begin{cases} \dfrac{1}{x^2} & \text{if } x \neq 0 \\ 1 & \text{if } x = 0 \end{cases}$

(c) $f(x) = \begin{cases} \dfrac{x^2 - x - 2}{x - 2} & \text{if } x \neq 2 \\ 1 & \text{if } x = 2 \end{cases}$ (d) $f(x) = [\![x]\!]$

SOLUTION

(a) Notice that $f(2)$ is not defined, so f is discontinuous at 2. Later we'll see why f is continuous at all other numbers.

(b) Here $f(0) = 1$ is defined but

$$\lim_{x \to 0} f(x) = \lim_{x \to 0} \frac{1}{x^2}$$

does not exist. (See Example 8 in Section 2.2.) So f is discontinuous at 0.

(c) Here $f(2) = 1$ is defined and

$$\lim_{x \to 2} f(x) = \lim_{x \to 2} \frac{x^2 - x - 2}{x - 2} = \lim_{x \to 2} \frac{(x - 2)(x + 1)}{x - 2} = \lim_{x \to 2} (x + 1) = 3$$

exists. But

$$\lim_{x \to 2} f(x) \neq f(2)$$

so f is not continuous at 2.

(d) The greatest integer function $f(x) = [\![x]\!]$ has discontinuities at all of the integers because $\lim_{x \to n} [\![x]\!]$ does not exist if n is an integer. (See Example 10 and Exercise 49 in Section 2.3.)

Figure 3 shows the graphs of the functions in Example 2. In each case the graph can't be drawn without lifting the pen from the paper because a hole or break or jump occurs in the graph. The kind of discontinuity illustrated in parts (a) and (c) is called **removable** because we could remove the discontinuity by redefining f at just the single number 2. [The function $g(x) = x + 1$ is continuous.] The discontinuity in part (b) is called an **infinite discontinuity**. The discontinuities in part (d) are called **jump discontinuities** because the function "jumps" from one value to another.

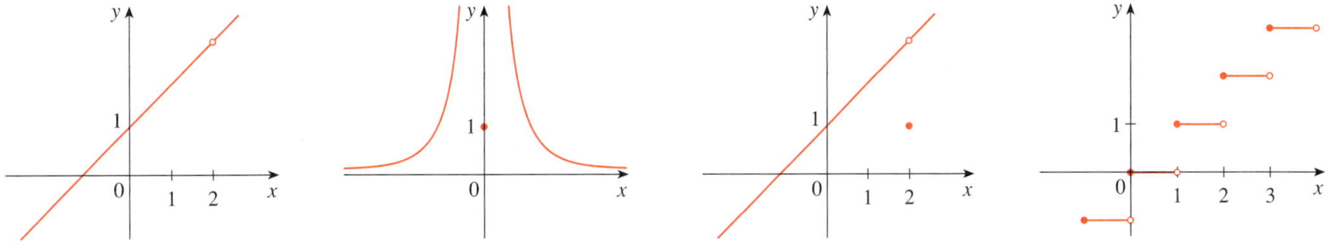

(a) $f(x) = \dfrac{x^2 - x - 2}{x - 2}$ (b) $f(x) = \begin{cases} 1/x^2 & \text{if } x \neq 0 \\ 1 & \text{if } x = 0 \end{cases}$ (c) $f(x) = \begin{cases} \dfrac{x^2 - x - 2}{x - 2} & \text{if } x \neq 2 \\ 1 & \text{if } x = 2 \end{cases}$ (d) $f(x) = [\![x]\!]$

FIGURE 3 Graphs of the functions in Example 2

2 Definition A function f is **continuous from the right at a number *a*** if

$$\lim_{x \to a^+} f(x) = f(a)$$

and f is **continuous from the left at *a*** if

$$\lim_{x \to a^-} f(x) = f(a)$$

EXAMPLE 3 At each integer n, the function $f(x) = [\![x]\!]$ [see Figure 3(d)] is continuous from the right but discontinuous from the left because

$$\lim_{x \to n^+} f(x) = \lim_{x \to n^+} [\![x]\!] = n = f(n)$$

but $$\lim_{x \to n^-} f(x) = \lim_{x \to n^-} [\![x]\!] = n - 1 \neq f(n)$$

3 Definition A function f is **continuous on an interval** if it is continuous at every number in the interval. (If f is defined only on one side of an endpoint of the interval, we understand *continuous* at the endpoint to mean *continuous from the right* or *continuous from the left.*)

EXAMPLE 4 Show that the function $f(x) = 1 - \sqrt{1 - x^2}$ is continuous on the interval $[-1, 1]$.

SOLUTION If $-1 < a < 1$, then using the Limit Laws, we have

$$\begin{aligned}
\lim_{x \to a} f(x) &= \lim_{x \to a} \left(1 - \sqrt{1 - x^2}\right) \\
&= 1 - \lim_{x \to a} \sqrt{1 - x^2} && \text{(by Laws 2 and 7)} \\
&= 1 - \sqrt{\lim_{x \to a} (1 - x^2)} && \text{(by 11)} \\
&= 1 - \sqrt{1 - a^2} && \text{(by 2, 7, and 9)} \\
&= f(a)
\end{aligned}$$

Thus, by Definition l, f is continuous at a if $-1 < a < 1$. Similar calculations show that

$$\lim_{x \to -1^+} f(x) = 1 = f(-1) \qquad \text{and} \qquad \lim_{x \to 1^-} f(x) = 1 = f(1)$$

so f is continuous from the right at -1 and continuous from the left at 1. Therefore, according to Definition 3, f is continuous on $[-1, 1]$.

The graph of f is sketched in Figure 4. It is the lower half of the circle

$$x^2 + (y - 1)^2 = 1$$

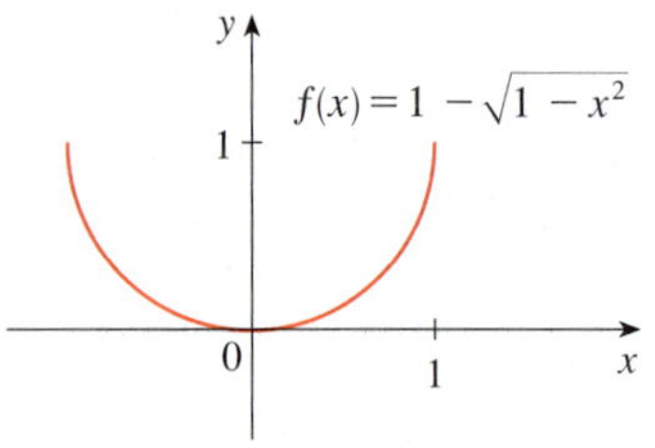

FIGURE 4

Instead of always using Definitions 1, 2, and 3 to verify the continuity of a function as we did in Example 4, it is often convenient to use the next theorem, which shows how to build up complicated continuous functions from simple ones.

4 Theorem If f and g are continuous at a and c is a constant, then the following functions are also continuous at a:

1. $f + g$ **2.** $f - g$ **3.** cf

4. fg **5.** $\dfrac{f}{g}$ if $g(a) \neq 0$

Proof Each of the five parts of this theorem follows from the corresponding Limit Law in Section 2.3. For instance, we give the proof of part 1. Since f and g are continuous at a, we have

$$\lim_{x \to a} f(x) = f(a) \qquad \text{and} \qquad \lim_{x \to a} g(x) = g(a)$$

Therefore

$$\begin{aligned} \lim_{x \to a} (f + g)(x) &= \lim_{x \to a} [f(x) + g(x)] \\ &= \lim_{x \to a} f(x) + \lim_{x \to a} g(x) \qquad \text{(by Law 1)} \\ &= f(a) + g(a) \\ &= (f + g)(a) \end{aligned}$$

This shows that $f + g$ is continuous at a. ■

It follows from Theorem 4 and Definition 3 that if f and g are continuous on an interval, then so are the functions $f + g$, $f - g$, cf, fg, and (if g is never 0) f/g. The following theorem was stated in Section 2.3 as the Direct Substitution Property.

5 Theorem
(a) Any polynomial is continuous everywhere; that is, it is continuous on $\mathbb{R} = (-\infty, \infty)$.
(b) Any rational function is continuous wherever it is defined; that is, it is continuous on its domain.

Proof
(a) A polynomial is a function of the form

$$P(x) = c_n x^n + c_{n-1} x^{n-1} + \cdots + c_1 x + c_0$$

where $c_0, c_1, \ldots, c_n$ are constants. We know that

$$\lim_{x \to a} c_0 = c_0 \qquad \text{(by Law 7)}$$

and

$$\lim_{x \to a} x^m = a^m \qquad m = 1, 2, \ldots, n \qquad \text{(by 9)}$$

This equation is precisely the statement that the function $f(x) = x^m$ is a continuous function. Thus, by part 3 of Theorem 4, the function $g(x) = cx^m$ is continuous. Since P is a sum of functions of this form and a constant function, it follows from part 1 of Theorem 4 that P is continuous.

(b) A rational function is a function of the form

$$f(x) = \frac{P(x)}{Q(x)}$$

where P and Q are polynomials. The domain of f is $D = \{x \in \mathbb{R} \mid Q(x) \neq 0\}$. We know from part (a) that P and Q are continuous everywhere. Thus, by part 5 of Theorem 4, f is continuous at every number in D.

As an illustration of Theorem 5, observe that the volume of a sphere varies continuously with its radius because the formula $V(r) = \frac{4}{3}\pi r^3$ shows that V is a polynomial function of r. Likewise, if a ball is thrown vertically into the air with a velocity of 50 ft/s, then the height of the ball in feet after t seconds is given by the formula $h = 50t - 16t^2$. Again this is a polynomial function, so the height is a continuous function of the elapsed time.

Knowledge of which functions are continuous enables us to evaluate some limits very quickly, as the following example shows. Compare it with Example 2(b) in Section 2.3.

EXAMPLE 5 Find $\lim_{x \to -2} \frac{x^3 + 2x^2 - 1}{5 - 3x}$.

SOLUTION The function

$$f(x) = \frac{x^3 + 2x^2 - 1}{5 - 3x}$$

is rational, so by Theorem 5 it is continuous on its domain, which is $\{x \mid x \neq \frac{5}{3}\}$. Therefore

$$\lim_{x \to -2} \frac{x^3 + 2x^2 - 1}{5 - 3x} = \lim_{x \to -2} f(x) = f(-2)$$

$$= \frac{(-2)^3 + 2(-2)^2 - 1}{5 - 3(-2)} = -\frac{1}{11}$$

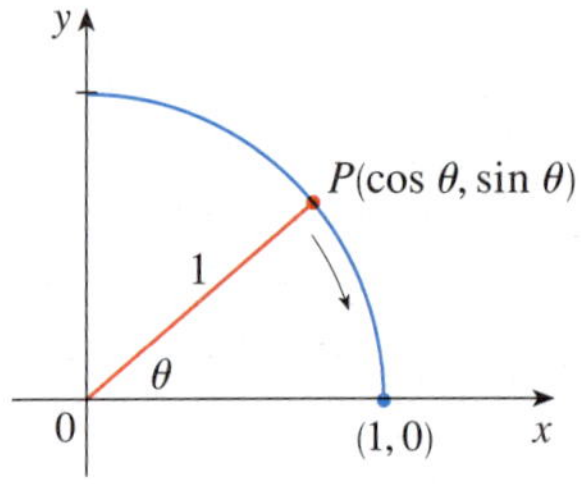

FIGURE 5

It turns out that most of the familiar functions are continuous at every number in their domains. For instance, Limit Law 10 (page 106) implies that root functions are continuous. [Example 3 in Section 2.4 shows that $f(x) = \sqrt{x}$ is continuous from the right at 0.]

From the appearance of the graphs of the sine and cosine functions (Figure 18 in Section 1.2), we would certainly guess that they are continuous. We know from the definitions of $\sin \theta$ and $\cos \theta$ that the coordinates of the point P in Figure 5 are $(\cos \theta, \sin \theta)$. As $\theta \to 0$, we see that P approaches the point $(1, 0)$ and so $\cos \theta \to 1$ and $\sin \theta \to 0$. Thus

6 $$\lim_{\theta \to 0} \cos \theta = 1 \qquad \lim_{\theta \to 0} \sin \theta = 0$$

|||| Another way to establish the limits in (6) is to use the Squeeze Theorem with the inequality $\sin \theta < \theta$ (for $\theta > 0$), which is proved in Section 3.4.

Since $\cos 0 = 1$ and $\sin 0 = 0$, the equations in (6) assert that the cosine and sine functions are continuous at 0. The addition formulas for cosine and sine can then be used to deduce that these functions are continuous everywhere (see Exercises 56 and 57).

It follows from part 5 of Theorem 4 that

$$\tan x = \frac{\sin x}{\cos x}$$

is continuous except where $\cos x = 0$. This happens when x is an odd integer multiple of $\pi/2$, so $y = \tan x$ has infinite discontinuities when $x = \pm\pi/2, \pm3\pi/2, \pm5\pi/2$, and so on (see Figure 6).

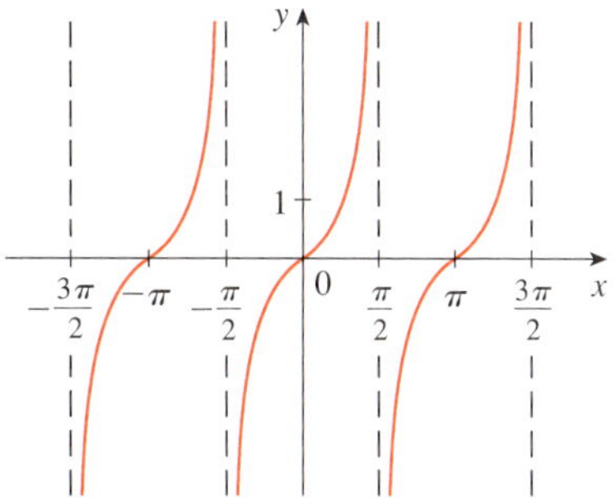

FIGURE 6
$y = \tan x$

|||| The inverse trigonometric functions are reviewed in Section 1.6.

The inverse function of any continuous function is also continuous. (The graph of f^{-1} is obtained by reflecting the graph of f about the line $y = x$. So if the graph of f has no break in it, neither does the graph of f^{-1}.) Thus, the inverse trigonometric functions are continuous.

In Section 1.5 we defined the exponential function $y = a^x$ so as to fill in the holes in the graph of $y = a^x$ where x is rational. In other words, the very definition of $y = a^x$ makes it a continuous function on $\mathbb{R}$. Therefore, its inverse function $y = \log_a x$ is continuous on $(0, \infty)$.

7 Theorem The following types of functions are continuous at every number in their domains:

polynomials	rational functions	root functions
trigonometric functions	inverse trigonometric functions	
exponential functions	logarithmic functions	

EXAMPLE 6 Where is the function $f(x) = \dfrac{\ln x + \tan^{-1}x}{x^2 - 1}$ continuous?

SOLUTION We know from Theorem 7 that the function $y = \ln x$ is continuous for $x > 0$ and $y = \tan^{-1}x$ is continuous on $\mathbb{R}$. Thus, by part 1 of Theorem 4, $y = \ln x + \tan^{-1}x$ is continuous on $(0, \infty)$. The denominator, $y = x^2 - 1$, is a polynomial, so it is continuous everywhere. Therefore, by part 5 of Theorem 4, f is continuous at all positive numbers x except where $x^2 - 1 = 0$. So f is continuous on the intervals $(0, 1)$ and $(1, \infty)$.

Another way of combining continuous functions f and g to get a new continuous function is to form the composite function $f \circ g$. This fact is a consequence of the following theorem.

|||| This theorem says that a limit symbol can be moved through a function symbol if the function is continuous and the limit exists. In other words, the order of these two symbols can be reversed.

8 Theorem If f is continuous at b and $\lim_{x\to a} g(x) = b$, then $\lim_{x\to a} f(g(x)) = f(b)$. In other words,

$$\lim_{x\to a} f(g(x)) = f\left(\lim_{x\to a} g(x)\right)$$

Intuitively, Theorem 8 is reasonable because if x is close to a, then $g(x)$ is close to b, and since f is continuous at b, if $g(x)$ is close to b, then $f(g(x))$ is close to $f(b)$. A proof of Theorem 8 is given in Appendix F.

EXAMPLE 7 Evaluate $\lim_{x\to 1} \arcsin\left(\dfrac{1-\sqrt{x}}{1-x}\right)$.

SOLUTION Because arcsin is a continuous function, we can apply Theorem 8:

$$\begin{aligned}\lim_{x\to 1} \arcsin\left(\frac{1-\sqrt{x}}{1-x}\right) &= \arcsin\left(\lim_{x\to 1}\frac{1-\sqrt{x}}{1-x}\right)\\ &= \arcsin\left(\lim_{x\to 1}\frac{1-\sqrt{x}}{(1-\sqrt{x})(1+\sqrt{x})}\right)\\ &= \arcsin\left(\lim_{x\to 1}\frac{1}{1+\sqrt{x}}\right)\\ &= \arcsin\frac{1}{2} = \frac{\pi}{6}\end{aligned}$$

Let's now apply Theorem 8 in the special case where $f(x) = \sqrt[n]{x}$, with n being a positive integer. Then

$$f(g(x)) = \sqrt[n]{g(x)}$$

and

$$f\left(\lim_{x\to a} g(x)\right) = \sqrt[n]{\lim_{x\to a} g(x)}$$

If we put these expressions into Theorem 8, we get

$$\lim_{x\to a} \sqrt[n]{g(x)} = \sqrt[n]{\lim_{x\to a} g(x)}$$

and so Limit Law 11 has now been proved. (We assume that the roots exist.)

9 Theorem If g is continuous at a and f is continuous at $g(a)$, then the composite function $f \circ g$ given by $(f \circ g)(x) = f(g(x))$ is continuous at a.

This theorem is often expressed informally by saying "a continuous function of a continuous function is a continuous function."

Proof Since g is continuous at a, we have

$$\lim_{x\to a} g(x) = g(a)$$

Since f is continuous at $b = g(a)$, we can apply Theorem 8 to obtain

$$\lim_{x\to a} f(g(x)) = f(g(a))$$

which is precisely the statement that the function $h(x) = f(g(x))$ is continuous at a; that is, $f \circ g$ is continuous at a.

EXAMPLE 8 Where are the following functions continuous?
(a) $h(x) = \sin(x^2)$ (b) $F(x) = \ln(1 + \cos x)$

SOLUTION
(a) We have $h(x) = f(g(x))$, where

$$g(x) = x^2 \qquad \text{and} \qquad f(x) = \sin x$$

Now g is continuous on $\mathbb{R}$ since it is a polynomial, and f is also continuous everywhere. Thus, $h = f \circ g$ is continuous on $\mathbb{R}$ by Theorem 9.

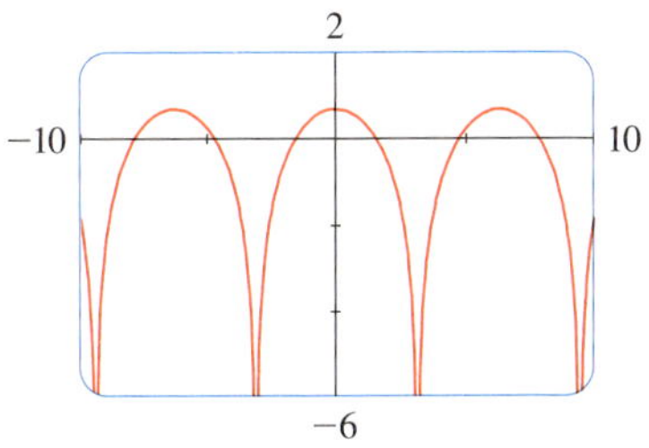

FIGURE 7
$y = \ln(1 + \cos x)$

(b) We know from Theorem 7 that $f(x) = \ln x$ is continuous and $g(x) = 1 + \cos x$ is continuous (because both $y = 1$ and $y = \cos x$ are continuous). Therefore, by Theorem 9, $F(x) = f(g(x))$ is continuous wherever it is defined. Now $\ln(1 + \cos x)$ is defined when $1 + \cos x > 0$. So it is undefined when $\cos x = -1$, and this happens when $x = \pm\pi, \pm 3\pi, \ldots$. Thus, F has discontinuities when x is an odd multiple of π and is continuous on the intervals between these values (see Figure 7).

An important property of continuous functions is expressed by the following theorem, whose proof is found in more advanced books on calculus.

10 The Intermediate Value Theorem Suppose that f is continuous on the closed interval $[a, b]$ and let N be any number between $f(a)$ and $f(b)$, where $f(a) \neq f(b)$. Then there exists a number c in (a, b) such that $f(c) = N$.

The Intermediate Value Theorem states that a continuous function takes on every intermediate value between the function values $f(a)$ and $f(b)$. It is illustrated by Figure 8. Note that the value N can be taken on once [as in part (a)] or more than once [as in part (b)].

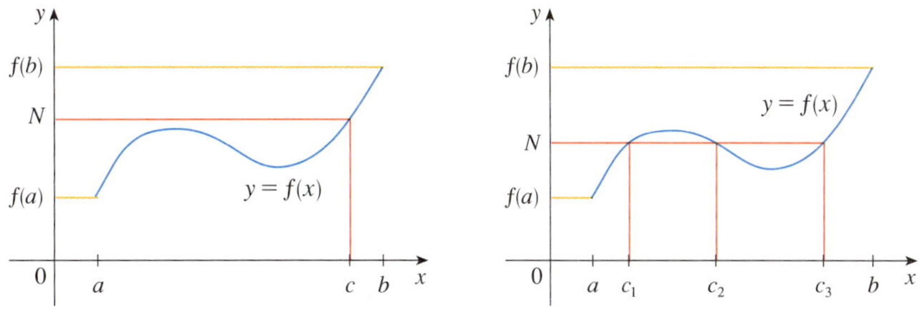

FIGURE 8
(a) (b)

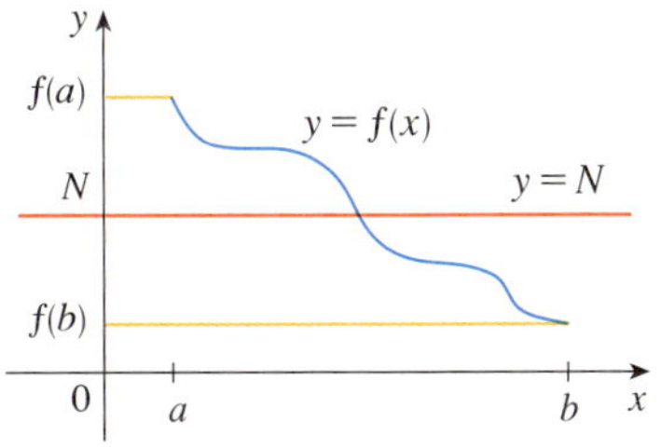

FIGURE 9

If we think of a continuous function as a function whose graph has no hole or break, then it is easy to believe that the Intermediate Value Theorem is true. In geometric terms it says that if any horizontal line $y = N$ is given between $y = f(a)$ and $y = f(b)$ as in Figure 9, then the graph of f can't jump over the line. It must intersect $y = N$ somewhere.

It is important that the function f in Theorem 10 be continuous. The Intermediate Value Theorem is not true in general for discontinuous functions (see Exercise 44).

One use of the Intermediate Value Theorem is in locating roots of equations as in the following example.

EXAMPLE 9 Show that there is a root of the equation

$$4x^3 - 6x^2 + 3x - 2 = 0$$

between 1 and 2.

SOLUTION Let $f(x) = 4x^3 - 6x^2 + 3x - 2$. We are looking for a solution of the given equation, that is, a number c between 1 and 2 such that $f(c) = 0$. Therefore, we take $a = 1$, $b = 2$, and $N = 0$ in Theorem 10. We have

$$f(1) = 4 - 6 + 3 - 2 = -1 < 0$$

and

$$f(2) = 32 - 24 + 6 - 2 = 12 > 0$$

Thus, $f(1) < 0 < f(2)$; that is, $N = 0$ is a number between $f(1)$ and $f(2)$. Now f is continuous since it is a polynomial, so the Intermediate Value Theorem says there is a number c between 1 and 2 such that $f(c) = 0$. In other words, the equation $4x^3 - 6x^2 + 3x - 2 = 0$ has at least one root c in the interval $(1, 2)$.

In fact, we can locate a root more precisely by using the Intermediate Value Theorem again. Since

$$f(1.2) = -0.128 < 0 \quad \text{and} \quad f(1.3) = 0.548 > 0$$

a root must lie between 1.2 and 1.3. A calculator gives, by trial and error,

$$f(1.22) = -0.007008 < 0 \quad \text{and} \quad f(1.23) = 0.056068 > 0$$

so a root lies in the interval $(1.22, 1.23)$.

We can use a graphing calculator or computer to illustrate the use of the Intermediate Value Theorem in Example 9. Figure 10 shows the graph of f in the viewing rectangle $[-1, 3]$ by $[-3, 3]$ and you can see that the graph crosses the x-axis between 1 and 2. Figure 11 shows the result of zooming in to the viewing rectangle $[1.2, 1.3]$ by $[-0.2, 0.2]$.

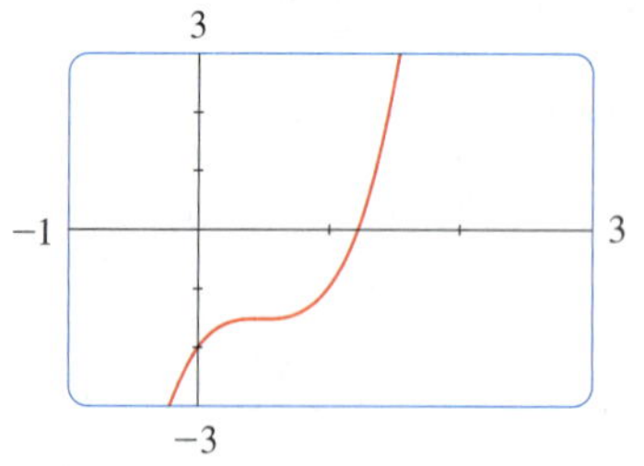

FIGURE 10

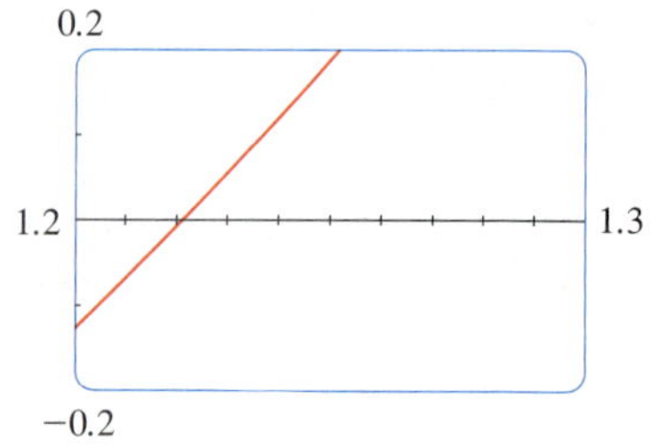

FIGURE 11

In fact, the Intermediate Value Theorem plays a role in the very way these graphing devices work. A computer calculates a finite number of points on the graph and turns on the pixels that contain these calculated points. It assumes that the function is continuous and takes on all the intermediate values between two consecutive points. The computer therefore connects the pixels by turning on the intermediate pixels.

2.5 Exercises

1. Write an equation that expresses the fact that a function f is continuous at the number 4.

2. If f is continuous on $(-\infty, \infty)$, what can you say about its graph?

3. (a) From the graph of f, state the numbers at which f is discontinuous and explain why.
(b) For each of the numbers stated in part (a), determine whether f is continuous from the right, or from the left, or neither.

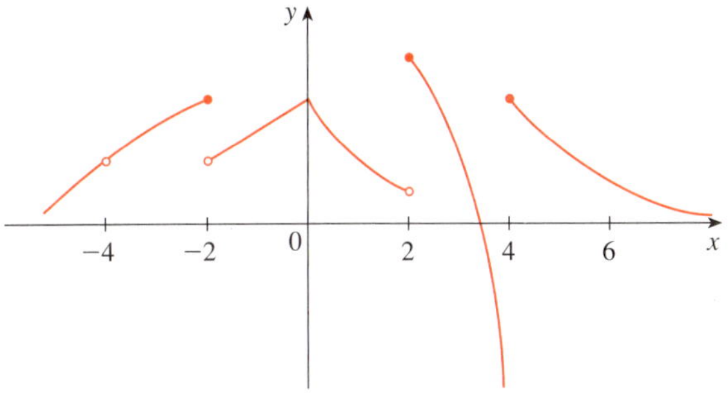

4. From the graph of g, state the intervals on which g is continuous.

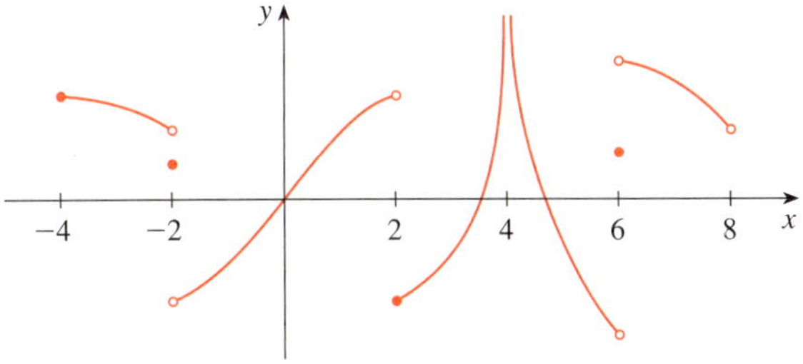

5. Sketch the graph of a function that is continuous everywhere except at $x = 3$ and is continuous from the left at 3.

6. Sketch the graph of a function that has a jump discontinuity at $x = 2$ and a removable discontinuity at $x = 4$, but is continuous elsewhere.

7. A parking lot charges \$3 for the first hour (or part of an hour) and \$2 for each succeeding hour (or part), up to a daily maximum of \$10.
(a) Sketch a graph of the cost of parking at this lot as a function of the time parked there.
(b) Discuss the discontinuities of this function and their significance to someone who parks in the lot.

8. Explain why each function is continuous or discontinuous.
(a) The temperature at a specific location as a function of time
(b) The temperature at a specific time as a function of the distance due west from New York City
(c) The altitude above sea level as a function of the distance due west from New York City
(d) The cost of a taxi ride as a function of the distance traveled
(e) The current in the circuit for the lights in a room as a function of time

9. If f and g are continuous functions with $f(3) = 5$ and $\lim_{x\to 3}[2f(x) - g(x)] = 4$, find $g(3)$.

10–12 Use the definition of continuity and the properties of limits to show that the function is continuous at the given number.

10. $f(x) = x^2 + \sqrt{7 - x}, \quad a = 4$

11. $f(x) = (x + 2x^3)^4, \quad a = -1$

12. $g(x) = \dfrac{x + 1}{2x^2 - 1}, \quad a = 4$

13–14 Use the definition of continuity and the properties of limits to show that the function is continuous on the given interval.

13. $f(x) = \dfrac{2x + 3}{x - 2}, \quad (2, \infty)$

14. $g(x) = 2\sqrt{3 - x}, \quad (-\infty, 3]$

15–20 Explain why the function is discontinuous at the given number a. Sketch the graph of the function.

15. $f(x) = \ln|x - 2| \qquad a = 2$

16. $f(x) = \begin{cases} \dfrac{1}{x - 1} & \text{if } x \neq 1 \\ 2 & \text{if } x = 1 \end{cases} \qquad a = 1$

17. $f(x) = \begin{cases} e^x & \text{if } x < 0 \\ x^2 & \text{if } x \geqslant 0 \end{cases} \qquad a = 0$

18. $f(x) = \begin{cases} \dfrac{x^2 - x}{x^2 - 1} & \text{if } x \neq 1 \\ 1 & \text{if } x = 1 \end{cases} \qquad a = 1$

19. $f(x) = \begin{cases} \dfrac{x^2 - x - 12}{x + 3} & \text{if } x \neq -3 \\ -5 & \text{if } x = -3 \end{cases} \qquad a = -3$

20. $f(x) = \begin{cases} 1 + x^2 & \text{if } x < 1 \\ 4 - x & \text{if } x \geqslant 1 \end{cases} \qquad a = 1$

21–28 Explain, using Theorems 4, 5, 7, and 9, why the function is continuous at every number in its domain. State the domain.

21. $F(x) = \dfrac{x}{x^2 + 5x + 6}$

22. $G(x) = \sqrt[3]{x}(1 + x^3)$

23. $R(x) = x^2 + \sqrt{2x - 1}$

24. $h(x) = \dfrac{\sin x}{x + 1}$

25. $f(x) = e^x \sin 5x$

26. $F(x) = \sin^{-1}(x^2 - 1)$

27. $G(t) = \ln(t^4 - 1)$

28. $H(x) = \cos(e^{\sqrt{x}})$

29–30 Locate the discontinuities of the function and illustrate by graphing.

29. $y = \dfrac{1}{1 + e^{1/x}}$

30. $y = \ln(\tan^2 x)$

31–34 Use continuity to evaluate the limit.

31. $\displaystyle\lim_{x \to 4} \frac{5 + \sqrt{x}}{\sqrt{5 + x}}$

32. $\displaystyle\lim_{x \to \pi} \sin(x + \sin x)$

33. $\displaystyle\lim_{x \to 1} e^{x^2 - x}$

34. $\displaystyle\lim_{x \to 2} \arctan\left(\frac{x^2 - 4}{3x^2 - 6x}\right)$

35–36 Show that f is continuous on $(-\infty, \infty)$.

35. $f(x) = \begin{cases} x^2 & \text{if } x < 1 \\ \sqrt{x} & \text{if } x \geqslant 1 \end{cases}$

36. $f(x) = \begin{cases} \sin x & \text{if } x < \pi/4 \\ \cos x & \text{if } x \geqslant \pi/4 \end{cases}$

37–39 Find the numbers at which f is discontinuous. At which of these numbers is f continuous from the right, from the left, or neither? Sketch the graph of f.

37. $f(x) = \begin{cases} 1 + x^2 & \text{if } x \leqslant 0 \\ 2 - x & \text{if } 0 < x \leqslant 2 \\ (x - 2)^2 & \text{if } x > 2 \end{cases}$

38. $f(x) = \begin{cases} x + 1 & \text{if } x \leqslant 1 \\ 1/x & \text{if } 1 < x < 3 \\ \sqrt{x - 3} & \text{if } x \geqslant 3 \end{cases}$

39. $f(x) = \begin{cases} x + 2 & \text{if } x < 0 \\ e^x & \text{if } 0 \leqslant x \leqslant 1 \\ 2 - x & \text{if } x > 1 \end{cases}$

40. The gravitational force exerted by Earth on a unit mass at a distance r from the center of the planet is

$$F(r) = \begin{cases} \dfrac{GMr}{R^3} & \text{if } r < R \\ \dfrac{GM}{r^2} & \text{if } r \geqslant R \end{cases}$$

where M is the mass of Earth, R is its radius, and G is the gravitational constant. Is F a continuous function of r?

41. For what value of the constant c is the function f continuous on $(-\infty, \infty)$?

$$f(x) = \begin{cases} cx + 1 & \text{if } x \leqslant 3 \\ cx^2 - 1 & \text{if } x > 3 \end{cases}$$

42. Find the constant c that makes g continuous on $(-\infty, \infty)$.

$$g(x) = \begin{cases} x^2 - c^2 & \text{if } x < 4 \\ cx + 20 & \text{if } x \geqslant 4 \end{cases}$$

43. Which of the following functions f has a removable discontinuity at a? If the discontinuity is removable, find a function g that agrees with f for $x \neq a$ and is continuous at a.

(a) $f(x) = \dfrac{x^2 - 2x - 8}{x + 2}, \quad a = -2$

(b) $f(x) = \dfrac{x - 7}{|x - 7|}, \quad a = 7$

(c) $f(x) = \dfrac{x^3 + 64}{x + 4}, \quad a = -4$

(d) $f(x) = \dfrac{3 - \sqrt{x}}{9 - x}, \quad a = 9$

44. Suppose that a function f is continuous on $[0, 1]$ except at 0.25 and that $f(0) = 1$ and $f(1) = 3$. Let $N = 2$. Sketch two possible graphs of f, one showing that f might not satisfy the conclusion of the Intermediate Value Theorem and one showing that f might still satisfy the conclusion of the Intermediate Value Theorem (even though it doesn't satisfy the hypothesis).

45. If $f(x) = x^3 - x^2 + x$, show that there is a number c such that $f(c) = 10$.

46. Use the Intermediate Value Theorem to prove that there is a positive number c such that $c^2 = 2$. (This proves the existence of the number $\sqrt{2}$.)

47–50 Use the Intermediate Value Theorem to show that there is a root of the given equation in the specified interval.

47. $x^4 + x - 3 = 0, \quad (1, 2)$

48. $\sqrt[3]{x} = 1 - x, \quad (0, 1)$

49. $\cos x = x, \quad (0, 1)$

50. $\ln x = e^{-x}, \quad (1, 2)$

51–52 (a) Prove that the equation has at least one real root. (b) Use your calculator to find an interval of length 0.01 that contains a root.

51. $e^x = 2 - x$

52. $x^5 - x^2 + 2x + 3 = 0$

53–54 (a) Prove that the equation has at least one real root. (b) Use your graphing device to find the root correct to three decimal places.

53. $x^5 - x^2 - 4 = 0$

54. $\sqrt{x - 5} = \dfrac{1}{x + 3}$

55. Prove that f is continuous at a if and only if

$$\lim_{h \to 0} f(a + h) = f(a)$$

56. To prove that sine is continuous, we need to show that $\lim_{x \to a} \sin x = \sin a$ for every real number a. By Exercise 55 an equivalent statement is that

$$\lim_{h \to 0} \sin(a + h) = \sin a$$

Use (6) to show that this is true.

57. Prove that cosine is a continuous function.

58. (a) Prove Theorem 4, part 3.
(b) Prove Theorem 4, part 5.

59. For what values of x is f continuous?

$$f(x) = \begin{cases} 0 & \text{if } x \text{ is rational} \\ 1 & \text{if } x \text{ is irrational} \end{cases}$$

60. For what values of x is g continuous?

$$g(x) = \begin{cases} 0 & \text{if } x \text{ is rational} \\ x & \text{if } x \text{ is irrational} \end{cases}$$

61. Is there a number that is exactly 1 more than its cube?

62. (a) Show that the absolute value function $F(x) = |x|$ is continuous everywhere.
(b) Prove that if f is a continuous function on an interval, then so is $|f|$.
(c) Is the converse of the statement in part (b) also true? In other words, if $|f|$ is continuous, does it follow that f is continuous? If so, prove it. If not, find a counterexample.

63. A Tibetan monk leaves the monastery at 7:00 A.M. and takes his usual path to the top of the mountain, arriving at 7:00 P.M. The following morning, he starts at 7:00 A.M. at the top and takes the same path back, arriving at the monastery at 7:00 P.M. Use the Intermediate Value Theorem to show that there is a point on the path that the monk will cross at exactly the same time of day on both days.

2.6 Limits at Infinity; Horizontal Asymptotes

In Sections 2.2 and 2.4 we investigated infinite limits and vertical asymptotes. There we let x approach a number and the result was that the values of y became arbitrarily large (positive or negative). In this section we let x become arbitrarily large (positive or negative) and see what happens to y.

Let's begin by investigating the behavior of the function f defined by

$$f(x) = \frac{x^2 - 1}{x^2 + 1}$$

as x becomes large. The table at the left gives values of this function correct to six decimal places, and the graph of f has been drawn by a computer in Figure 1.

x	$f(x)$
0	−1
±1	0
±2	0.600000
±3	0.800000
±4	0.882353
±5	0.923077
±10	0.980198
±50	0.999200
±100	0.999800
±1000	0.999998

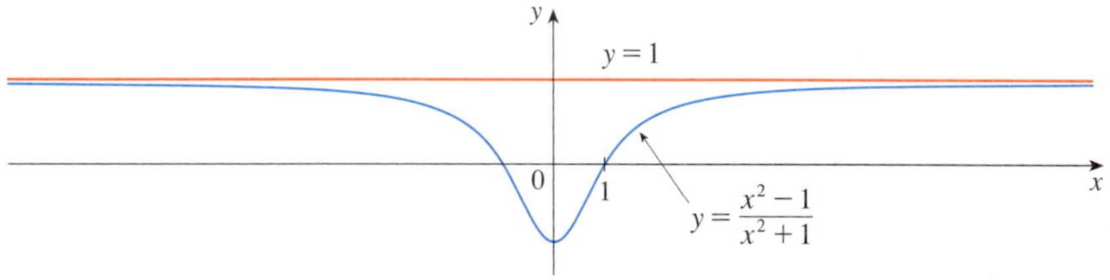

FIGURE 1

As x grows larger and larger you can see that the values of $f(x)$ get closer and closer to 1. In fact, it seems that we can make the values of $f(x)$ as close as we like to 1 by taking x sufficiently large. This situation is expressed symbolically by writing

$$\lim_{x \to \infty} \frac{x^2 - 1}{x^2 + 1} = 1$$

In general, we use the notation

$$\lim_{x \to \infty} f(x) = L$$

to indicate that the values of $f(x)$ become closer and closer to L as x becomes larger and larger.

1 Definition Let f be a function defined on some interval (a, ∞). Then

$$\lim_{x \to \infty} f(x) = L$$

means that the values of $f(x)$ can be made arbitrarily close to L by taking x sufficiently large.

Another notation for $\lim_{x \to \infty} f(x) = L$ is

$$f(x) \to L \quad \text{as} \quad x \to \infty$$

The symbol ∞ does not represent a number. Nonetheless, the expression $\lim_{x \to \infty} f(x) = L$ is often read as

"the limit of $f(x)$, as x approaches infinity, is L"

or "the limit of $f(x)$, as x becomes infinite, is L"

or "the limit of $f(x)$, as x increases without bound, is L"

The meaning of such phrases is given by Definition 1. A more precise definition, similar to the ε, δ definition of Section 2.4, is given at the end of this section.

Geometric illustrations of Definition 1 are shown in Figure 2. Notice that there are many ways for the graph of f to approach the line $y = L$ (which is called a *horizontal asymptote*) as we look to the far right of each graph.

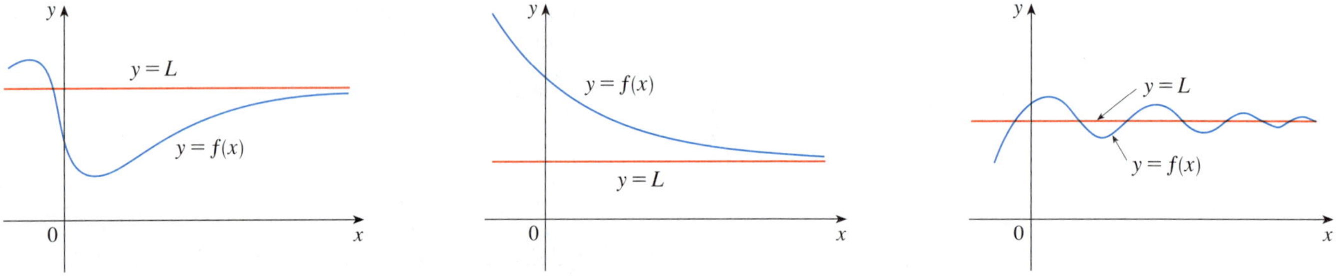

FIGURE 2
Examples illustrating $\lim_{x \to \infty} f(x) = L$

Referring back to Figure 1, we see that for numerically large negative values of x, the values of $f(x)$ are close to 1. By letting x decrease through negative values without bound, we can make $f(x)$ as close as we like to 1. This is expressed by writing

$$\lim_{x \to -\infty} \frac{x^2 - 1}{x^2 + 1} = 1$$

The general definition is as follows.

2 Definition Let f be a function defined on some interval $(-\infty, a)$. Then

$$\lim_{x \to -\infty} f(x) = L$$

means that the values of $f(x)$ can be made arbitrarily close to L by taking x sufficiently large negative.

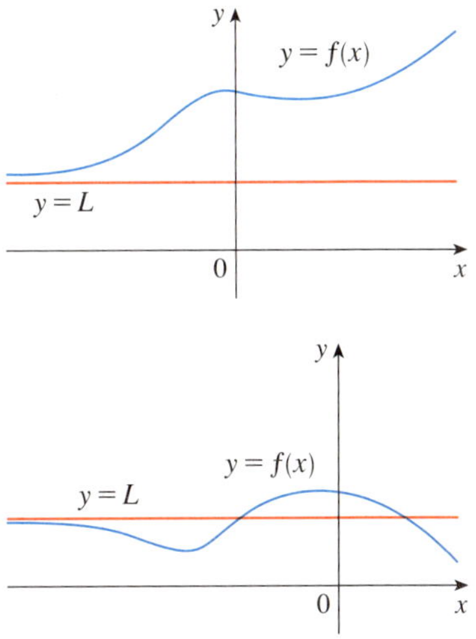

FIGURE 3
Examples illustrating $\lim_{x \to -\infty} f(x) = L$

Again, the symbol $-\infty$ does not represent a number, but the expression $\lim_{x \to -\infty} f(x) = L$ is often read as

"the limit of $f(x)$, as x approaches negative infinity, is L"

Definition 2 is illustrated in Figure 3. Notice that the graph approaches the line $y = L$ as we look to the far left of each graph.

3 Definition The line $y = L$ is called a **horizontal asymptote** of the curve $y = f(x)$ if either

$$\lim_{x \to \infty} f(x) = L \qquad \text{or} \qquad \lim_{x \to -\infty} f(x) = L$$

For instance, the curve illustrated in Figure 1 has the line $y = 1$ as a horizontal asymptote because

$$\lim_{x \to \infty} \frac{x^2 - 1}{x^2 + 1} = 1$$

An example of a curve with two horizontal asymptotes is $y = \tan^{-1}x$. (See Figure 4.) In fact,

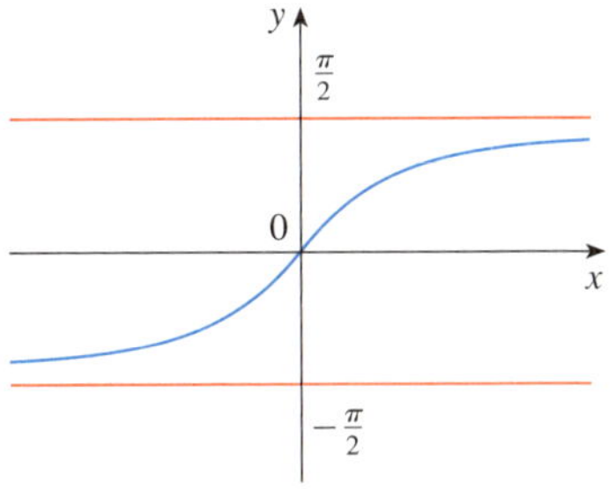

FIGURE 4
$y = \tan^{-1}x$

4
$$\lim_{x \to -\infty} \tan^{-1}x = -\frac{\pi}{2} \qquad \lim_{x \to \infty} \tan^{-1}x = \frac{\pi}{2}$$

so both of the lines $y = -\pi/2$ and $y = \pi/2$ are horizontal asymptotes. (This follows from the fact that the lines $x = \pm\pi/2$ are vertical asymptotes of the graph of tan.)

EXAMPLE 1 Find the infinite limits, limits at infinity, and asymptotes for the function f whose graph is shown in Figure 5.

SOLUTION We see that the values of $f(x)$ become large as $x \to -1$ from both sides, so

$$\lim_{x \to -1} f(x) = \infty$$

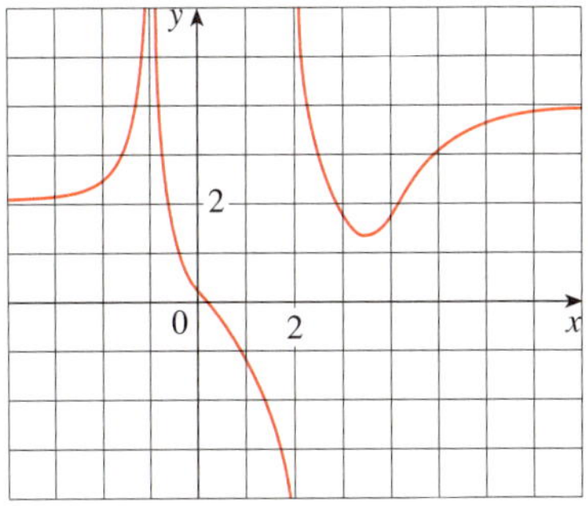

FIGURE 5

Notice that $f(x)$ becomes large negative as x approaches 2 from the left, but large positive as x approaches 2 from the right. So

$$\lim_{x \to 2^-} f(x) = -\infty \qquad \text{and} \qquad \lim_{x \to 2^+} f(x) = \infty$$

Thus, both of the lines $x = -1$ and $x = 2$ are vertical asymptotes.

As x becomes large, it appears that $f(x)$ approaches 4. But as x decreases through negative values, $f(x)$ approaches 2. So

$$\lim_{x \to \infty} f(x) = 4 \qquad \text{and} \qquad \lim_{x \to -\infty} f(x) = 2$$

This means that both $y = 4$ and $y = 2$ are horizontal asymptotes.

EXAMPLE 2 Find $\lim_{x \to \infty} \frac{1}{x}$ and $\lim_{x \to -\infty} \frac{1}{x}$.

SOLUTION Observe that when x is large, $1/x$ is small. For instance,

$$\frac{1}{100} = 0.01 \qquad \frac{1}{10{,}000} = 0.0001 \qquad \frac{1}{1{,}000{,}000} = 0.000001$$

In fact, by taking x large enough, we can make $1/x$ as close to 0 as we please. Therefore, according to Definition 1, we have

$$\lim_{x \to \infty} \frac{1}{x} = 0$$

Similar reasoning shows that when x is large negative, $1/x$ is small negative, so we also have

$$\lim_{x \to -\infty} \frac{1}{x} = 0$$

It follows that the line $y = 0$ (the x-axis) is a horizontal asymptote of the curve $y = 1/x$. (This is an equilateral hyperbola; see Figure 6.)

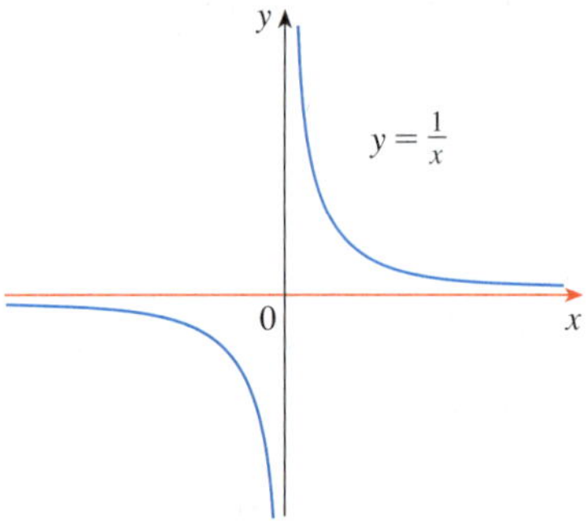

FIGURE 6

$\lim_{x \to \infty} \frac{1}{x} = 0, \ \lim_{x \to -\infty} \frac{1}{x} = 0$

Most of the Limit Laws that were given in Section 2.3 also hold for limits at infinity. It can be proved that the *Limit Laws listed in Section 2.3 (with the exception of Laws 9 and 10) are also valid if "$x \to a$" is replaced by "$x \to \infty$" or "$x \to -\infty$."* In particular, if we combine Laws 6 and 11 with the results of Example 2, we obtain the following important rule for calculating limits.

5 Theorem If $r > 0$ is a rational number, then

$$\lim_{x \to \infty} \frac{1}{x^r} = 0$$

If $r > 0$ is a rational number such that x^r is defined for all x, then

$$\lim_{x \to -\infty} \frac{1}{x^r} = 0$$

EXAMPLE 3 Evaluate

$$\lim_{x \to \infty} \frac{3x^2 - x - 2}{5x^2 + 4x + 1}$$

and indicate which properties of limits are used at each stage.

SOLUTION As x becomes large, both numerator and denominator become large, so it isn't obvious what happens to their ratio. We need to do some preliminary algebra.

To evaluate the limit at infinity of any rational function, we first divide both the numerator and denominator by the highest power of x that occurs in the denominator. (We may assume that $x \neq 0$, since we are interested only in large values of x.) In this case the highest power of x in the denominator is x^2, so we have

$$\lim_{x\to\infty}\frac{3x^2-x-2}{5x^2+4x+1}=\lim_{x\to\infty}\frac{\dfrac{3x^2-x-2}{x^2}}{\dfrac{5x^2+4x+1}{x^2}}=\lim_{x\to\infty}\frac{3-\dfrac{1}{x}-\dfrac{2}{x^2}}{5+\dfrac{4}{x}+\dfrac{1}{x^2}}$$

$$=\frac{\lim_{x\to\infty}\left(3-\dfrac{1}{x}-\dfrac{2}{x^2}\right)}{\lim_{x\to\infty}\left(5+\dfrac{4}{x}+\dfrac{1}{x^2}\right)} \qquad \text{(by Limit Law 5)}$$

$$=\frac{\lim_{x\to\infty}3-\lim_{x\to\infty}\dfrac{1}{x}-2\lim_{x\to\infty}\dfrac{1}{x^2}}{\lim_{x\to\infty}5+4\lim_{x\to\infty}\dfrac{1}{x}+\lim_{x\to\infty}\dfrac{1}{x^2}} \qquad \text{(by 1, 2, and 3)}$$

$$=\frac{3-0-0}{5+0+0} \qquad \text{(by 7 and Theorem 5)}$$

$$=\frac{3}{5}$$

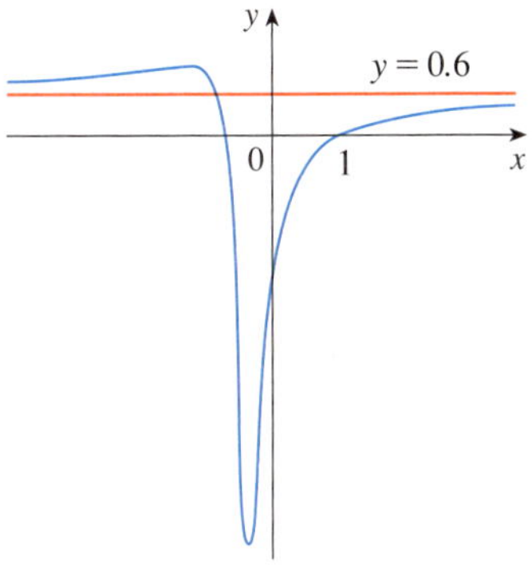

FIGURE 7
$y=\dfrac{3x^2-x-2}{5x^2+4x+1}$

A similar calculation shows that the limit as $x\to-\infty$ is also $\frac{3}{5}$. Figure 7 illustrates the results of these calculations by showing how the graph of the given rational function approaches the horizontal asymptote $y=\frac{3}{5}$.

EXAMPLE 4 Find the horizontal and vertical asymptotes of the graph of the function

$$f(x)=\frac{\sqrt{2x^2+1}}{3x-5}$$

SOLUTION Dividing both numerator and denominator by x and using the properties of limits, we have

$$\lim_{x\to\infty}\frac{\sqrt{2x^2+1}}{3x-5}=\lim_{x\to\infty}\frac{\sqrt{2+\dfrac{1}{x^2}}}{3-\dfrac{5}{x}} \qquad \left(\text{since } \sqrt{x^2}=x \text{ for } x>0\right)$$

$$=\frac{\lim_{x\to\infty}\sqrt{2+\dfrac{1}{x^2}}}{\lim_{x\to\infty}\left(3-\dfrac{5}{x}\right)}=\frac{\sqrt{\lim_{x\to\infty}2+\lim_{x\to\infty}\dfrac{1}{x^2}}}{\lim_{x\to\infty}3-5\lim_{x\to\infty}\dfrac{1}{x}}$$

$$=\frac{\sqrt{2+0}}{3-5\cdot 0}=\frac{\sqrt{2}}{3}$$

Therefore, the line $y=\sqrt{2}/3$ is a horizontal asymptote of the graph of f.

In computing the limit as $x\to-\infty$, we must remember that for $x<0$, we have $\sqrt{x^2}=|x|=-x$. So when we divide the numerator by x, for $x<0$ we get

$$\frac{1}{x}\sqrt{2x^2+1}=-\frac{1}{\sqrt{x^2}}\sqrt{2x^2+1}=-\sqrt{2+\frac{1}{x^2}}$$

Therefore
$$\lim_{x\to-\infty}\frac{\sqrt{2x^2+1}}{3x-5}=\lim_{x\to-\infty}\frac{-\sqrt{2+\dfrac{1}{x^2}}}{3-\dfrac{5}{x}}$$

$$=\frac{-\sqrt{2+\lim\limits_{x\to-\infty}\dfrac{1}{x^2}}}{3-5\lim\limits_{x\to-\infty}\dfrac{1}{x}}=-\frac{\sqrt{2}}{3}$$

Thus, the line $y=-\sqrt{2}/3$ is also a horizontal asymptote.

A vertical asymptote is likely to occur when the denominator, $3x-5$, is 0, that is, when $x=\frac{5}{3}$. If x is close to $\frac{5}{3}$ and $x>\frac{5}{3}$, then the denominator is close to 0 and $3x-5$ is positive. The numerator $\sqrt{2x^2+1}$ is always positive, so $f(x)$ is positive. Therefore

$$\lim_{x\to(5/3)^+}\frac{\sqrt{2x^2+1}}{3x-5}=\infty$$

If x is close to $\frac{5}{3}$ but $x<\frac{5}{3}$, then $3x-5<0$ and so $f(x)$ is large negative. Thus

$$\lim_{x\to(5/3)^-}\frac{\sqrt{2x^2+1}}{3x-5}=-\infty$$

The vertical asymptote is $x=\frac{5}{3}$. All three asymptotes are shown in Figure 8.

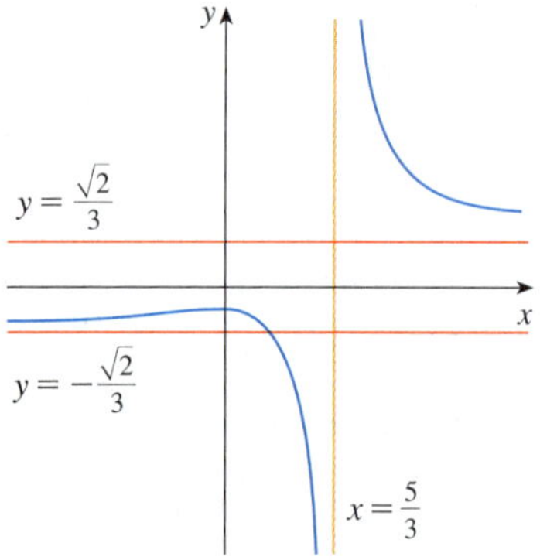

FIGURE 8
$y=\dfrac{\sqrt{2x^2+1}}{3x-5}$

EXAMPLE 5 Compute $\lim\limits_{x\to\infty}\left(\sqrt{x^2+1}-x\right)$.

SOLUTION Because both $\sqrt{x^2+1}$ and x are large when x is large, it's difficult to see what happens to their difference, so we use algebra to rewrite the function. We first multiply numerator and denominator by the conjugate radical:

|||| We can think of the given function as having a denominator of 1.

$$\lim_{x\to\infty}\left(\sqrt{x^2+1}-x\right)=\lim_{x\to\infty}\left(\sqrt{x^2+1}-x\right)\frac{\sqrt{x^2+1}+x}{\sqrt{x^2+1}+x}$$

$$=\lim_{x\to\infty}\frac{(x^2+1)-x^2}{\sqrt{x^2+1}+x}=\lim_{x\to\infty}\frac{1}{\sqrt{x^2+1}+x}$$

The Squeeze Theorem could be used to show that this limit is 0. But an easier method is to divide numerator and denominator by x. Doing this and using the Limit Laws, we obtain

$$\lim_{x\to\infty}\left(\sqrt{x^2+1}-x\right)=\lim_{x\to\infty}\frac{1}{\sqrt{x^2+1}+x}=\lim_{x\to\infty}\frac{\dfrac{1}{x}}{\dfrac{\sqrt{x^2+1}+x}{x}}$$

$$=\lim_{x\to\infty}\frac{\dfrac{1}{x}}{\sqrt{1+\dfrac{1}{x^2}}+1}=\frac{0}{\sqrt{1+0}+1}=0$$

Figure 9 illustrates this result.

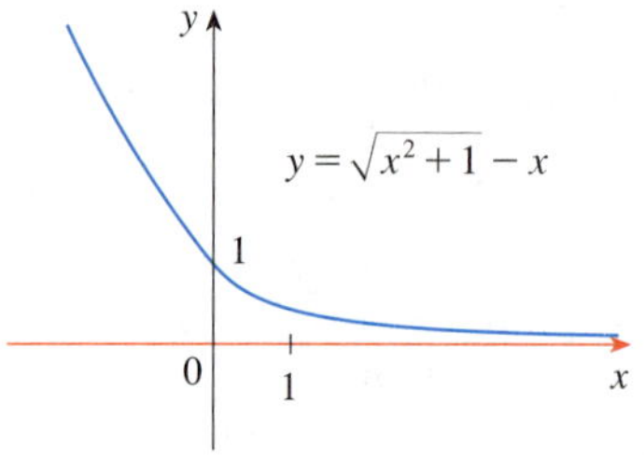

FIGURE 9

The graph of the natural exponential function $y=e^x$ has the line $y=0$ (the x-axis) as a horizontal asymptote. (The same is true of any exponential function with base $a>1$.) In fact, from the graph in Figure 10 and the corresponding table of values, we see that

6
$$\lim_{x \to -\infty} e^x = 0$$

Notice that the values of e^x approach 0 very rapidly.

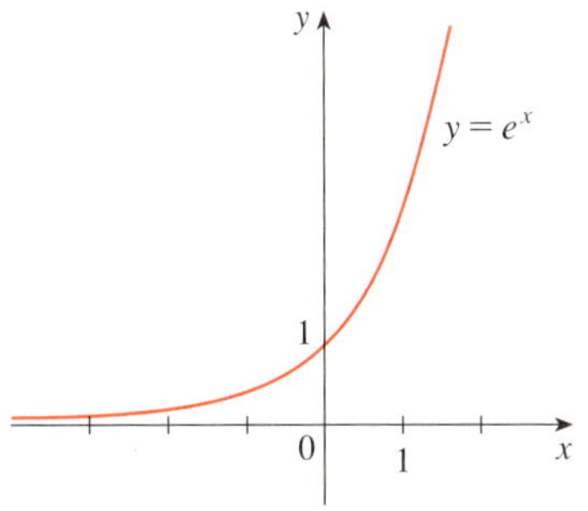

x	e^x
0	1.00000
−1	0.36788
−2	0.13534
−3	0.04979
−5	0.00674
−8	0.00034
−10	0.00005

FIGURE 10

EXAMPLE 6 Evaluate $\lim_{x \to 0^-} e^{1/x}$.

The problem-solving strategy for Example 6 is *introducing something extra* (see page 80). Here, the something extra, the auxiliary aid, is the new variable t.

SOLUTION If we let $t = 1/x$, we know that $t \to -\infty$ as $x \to 0^-$. Therefore, by (6),

$$\lim_{x \to 0^-} e^{1/x} = \lim_{t \to -\infty} e^t = 0$$

(See Exercise 67.)

EXAMPLE 7 Evaluate $\lim_{x \to \infty} \sin x$.

SOLUTION As x increases, the values of $\sin x$ oscillate between 1 and -1 infinitely often and so they don't approach any definite number. Thus, $\lim_{x \to \infty} \sin x$ does not exist.

Infinite Limits at Infinity

The notation

$$\lim_{x \to \infty} f(x) = \infty$$

is used to indicate that the values of $f(x)$ become large as x becomes large. Similar meanings are attached to the following symbols:

$$\lim_{x \to -\infty} f(x) = \infty \qquad \lim_{x \to \infty} f(x) = -\infty \qquad \lim_{x \to -\infty} f(x) = -\infty$$

EXAMPLE 8 Find $\lim_{x \to \infty} x^3$ and $\lim_{x \to -\infty} x^3$.

SOLUTION When x becomes large, x^3 also becomes large. For instance,

$$10^3 = 1000 \qquad 100^3 = 1{,}000{,}000 \qquad 1000^3 = 1{,}000{,}000{,}000$$

In fact, we can make x^3 as big as we like by taking x large enough. Therefore, we can write

$$\lim_{x \to \infty} x^3 = \infty$$

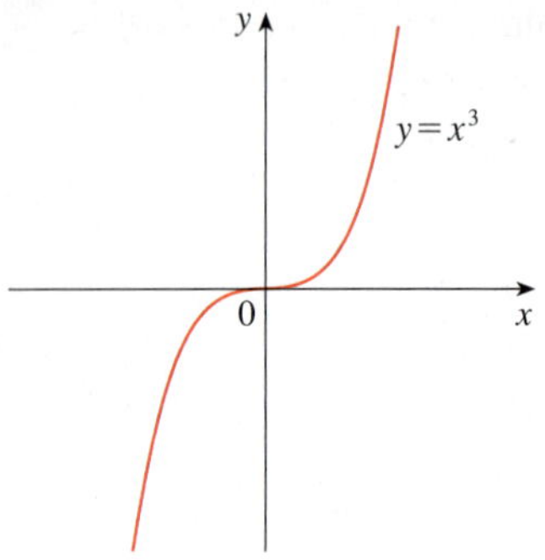

FIGURE 11
$\lim_{x\to\infty} x^3 = \infty$, $\lim_{x\to-\infty} x^3 = -\infty$

Similarly, when x is large negative, so is x^3. Thus

$$\lim_{x\to-\infty} x^3 = -\infty$$

These limit statements can also be seen from the graph of $y = x^3$ in Figure 11.

Looking at Figure 10 we see that

$$\lim_{x\to\infty} e^x = \infty$$

but, as Figure 12 demonstrates, $y = e^x$ becomes large as $x \to \infty$ at a much faster rate than $y = x^3$.

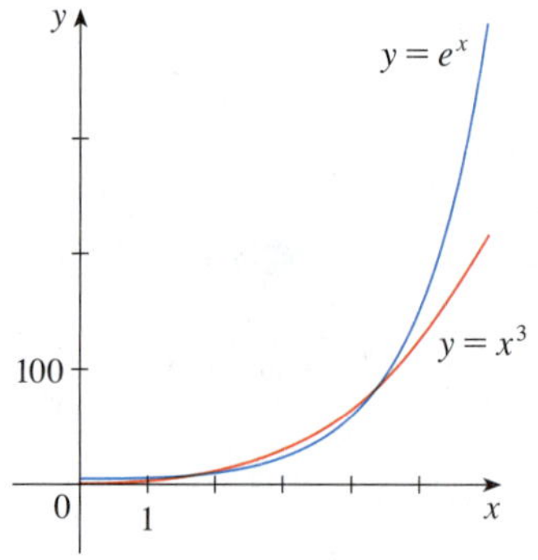

FIGURE 12
e^x is much larger than x^3 when x is large.

EXAMPLE 9 Find $\lim_{x\to\infty} (x^2 - x)$.

SOLUTION Note that we *cannot* write

$$\lim_{x\to\infty} (x^2 - x) = \lim_{x\to\infty} x^2 - \lim_{x\to\infty} x$$
$$= \infty - \infty$$

The Limit Laws can't be applied to infinite limits because ∞ is not a number ($\infty - \infty$ can't be defined). However, we can write

$$\lim_{x\to\infty} (x^2 - x) = \lim_{x\to\infty} x(x - 1) = \infty$$

because both x and $x - 1$ become arbitrarily large and so their product does too.

EXAMPLE 10 Find $\lim_{x\to\infty} \dfrac{x^2 + x}{3 - x}$.

SOLUTION As in Example 3, we divide the numerator and denominator by the highest power of x in the denominator, which is just x:

$$\lim_{x\to\infty} \frac{x^2 + x}{3 - x} = \lim_{x\to\infty} \frac{x + 1}{\dfrac{3}{x} - 1} = -\infty$$

because $x + 1 \to \infty$ and $3/x - 1 \to -1$ as $x \to \infty$.

The next example shows that by using infinite limits at infinity, together with intercepts, we can get a rough idea of the graph of a polynomial without having to plot a large number of points.

EXAMPLE 11 Sketch the graph of $y = (x - 2)^4(x + 1)^3(x - 1)$ by finding its intercepts and its limits as $x \to \infty$ and as $x \to -\infty$.

SOLUTION The y-intercept is $f(0) = (-2)^4(1)^3(-1) = -16$ and the x-intercepts are found by setting $y = 0$: $x = 2, -1, 1$. Notice that since $(x - 2)^4$ is positive, the function doesn't change sign at 2; thus, the graph doesn't cross the x-axis at 2. The graph crosses the axis at -1 and 1.

When x is large positive, all three factors are large, so

$$\lim_{x \to \infty} (x - 2)^4(x + 1)^3(x - 1) = \infty$$

When x is large negative, the first factor is large positive and the second and third factors are both large negative, so

$$\lim_{x \to -\infty} (x - 2)^4(x + 1)^3(x - 1) = \infty$$

Combining this information, we give a rough sketch of the graph in Figure 13.

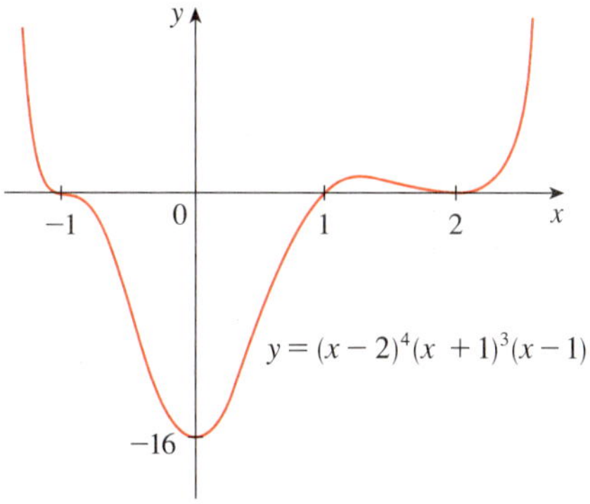

FIGURE 13

Precise Definitions

Definition 1 can be stated precisely as follows.

> **7 Definition** Let f be a function defined on some interval (a, ∞). Then
>
> $$\lim_{x \to \infty} f(x) = L$$
>
> means that for every $\varepsilon > 0$ there is a corresponding number N such that
>
> $$|f(x) - L| < \varepsilon \qquad \text{whenever} \qquad x > N$$

In words, this says that the values of $f(x)$ can be made arbitrarily close to L (within a distance ε, where ε is any positive number) by taking x sufficiently large (larger than N, where N depends on ε). Graphically it says that by choosing x large enough (larger than some number N) we can make the graph of f lie between the given horizontal lines $y = L - \varepsilon$ and $y = L + \varepsilon$ as in Figure 14. This must be true no matter how small we choose ε. Figure 15 shows that if a smaller value of ε is chosen, then a larger value of N may be required.

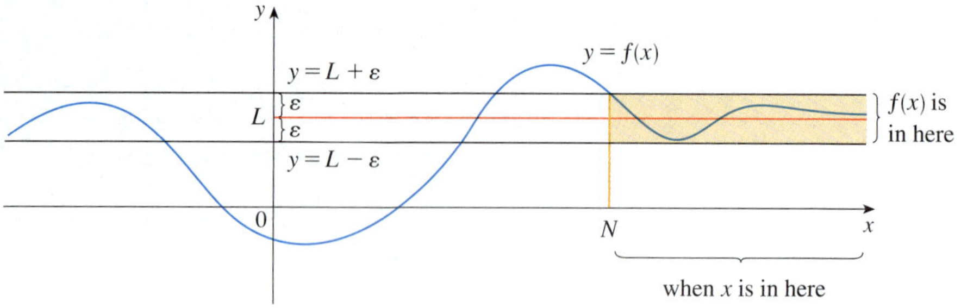

FIGURE 14
$\lim_{x \to \infty} f(x) = L$

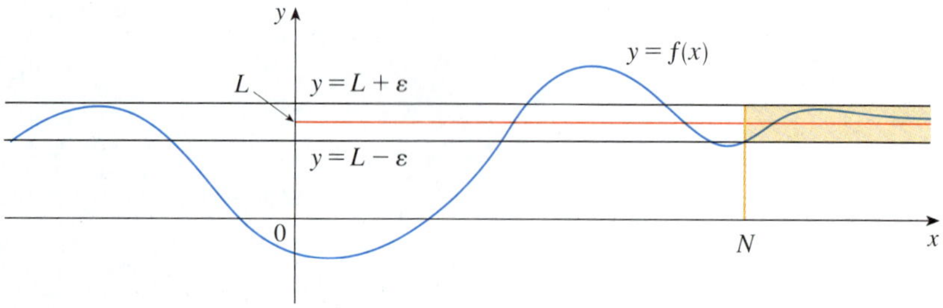

FIGURE 15
$\lim_{x \to \infty} f(x) = L$

Similarly, a precise version of Definition 2 is given by Definition 8, which is illustrated in Figure 16.

8 Definition Let f be a function defined on some interval $(-\infty, a)$. Then

$$\lim_{x \to -\infty} f(x) = L$$

means that for every $\varepsilon > 0$ there is a corresponding number N such that

$$|f(x) - L| < \varepsilon \qquad \text{whenever} \qquad x < N$$

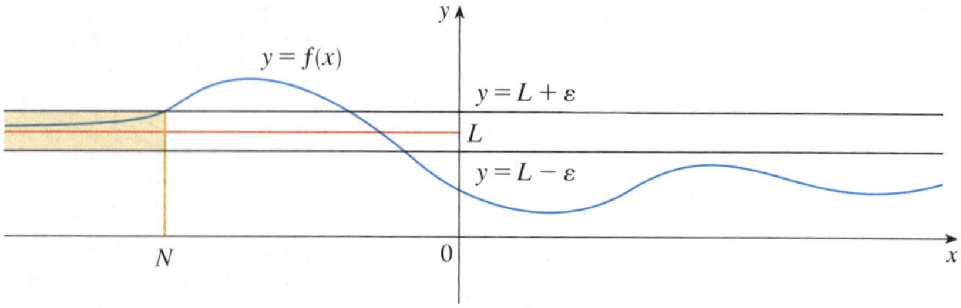

FIGURE 16
$\lim_{x \to -\infty} f(x) = L$

In Example 3 we calculated that

$$\lim_{x \to \infty} \frac{3x^2 - x - 2}{5x^2 + 4x + 1} = \frac{3}{5}$$

In the next example we use a graphing device to relate this statement to Definition 7 with $L = \frac{3}{5}$ and $\varepsilon = 0.1$.

EXAMPLE 12 Use a graph to find a number N such that

$$\left| \frac{3x^2 - x - 2}{5x^2 + 4x + 1} - 0.6 \right| < 0.1 \qquad \text{whenever} \qquad x > N$$

SOLUTION We rewrite the given inequality as

$$0.5 < \frac{3x^2 - x - 2}{5x^2 + 4x + 1} < 0.7$$

We need to determine the values of x for which the given curve lies between the horizontal lines $y = 0.5$ and $y = 0.7$. So we graph the curve and these lines in Figure 17. Then we use the cursor to estimate that the curve crosses the line $y = 0.5$ when $x \approx 6.7$. To the right of this number the curve stays between the lines $y = 0.5$ and $y = 0.7$. Rounding to be safe, we can say that

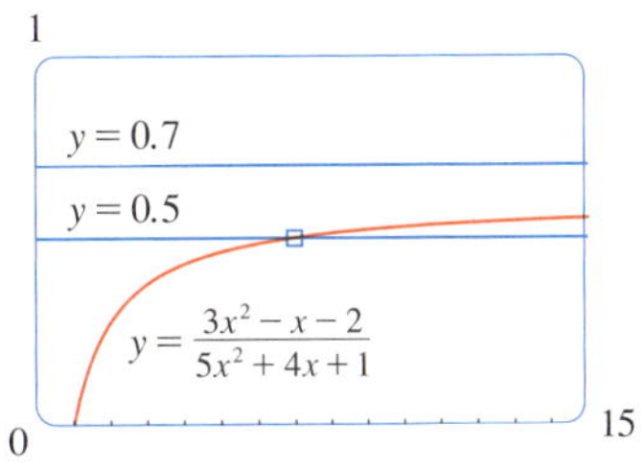

FIGURE 17

$$\left| \frac{3x^2 - x - 2}{5x^2 + 4x + 1} - 0.6 \right| < 0.1 \qquad \text{whenever} \qquad x > 7$$

In other words, for $\varepsilon = 0.1$ we can choose $N = 7$ (or any larger number) in Definition 7.

EXAMPLE 13 Use Definition 7 to prove that $\lim_{x \to \infty} \frac{1}{x} = 0$.

SOLUTION

1. *Preliminary analysis of the problem (guessing a value for N).* Given $\varepsilon > 0$, we want to find N such that

$$\left| \frac{1}{x} - 0 \right| < \varepsilon \qquad \text{whenever} \qquad x > N$$

In computing the limit we may assume $x > 0$, in which case

$$\left| \frac{1}{x} - 0 \right| = \left| \frac{1}{x} \right| = \frac{1}{x}$$

Therefore, we want

$$\frac{1}{x} < \varepsilon \qquad \text{whenever} \qquad x > N$$

that is,

$$x > \frac{1}{\varepsilon} \qquad \text{whenever} \qquad x > N$$

This suggests that we should take $N = 1/\varepsilon$.

2. *Proof (showing that this N works).* Given $\varepsilon > 0$, we choose $N = 1/\varepsilon$. Let $x > N$. Then

$$\left| \frac{1}{x} - 0 \right| = \frac{1}{|x|} = \frac{1}{x} < \frac{1}{N} = \varepsilon$$

Thus

$$\left| \frac{1}{x} - 0 \right| < \varepsilon \qquad \text{whenever} \qquad x > N$$

Therefore, by Definition 7,

$$\lim_{x \to \infty} \frac{1}{x} = 0$$

Figure 18 illustrates the proof by showing some values of ε and the corresponding values of N.

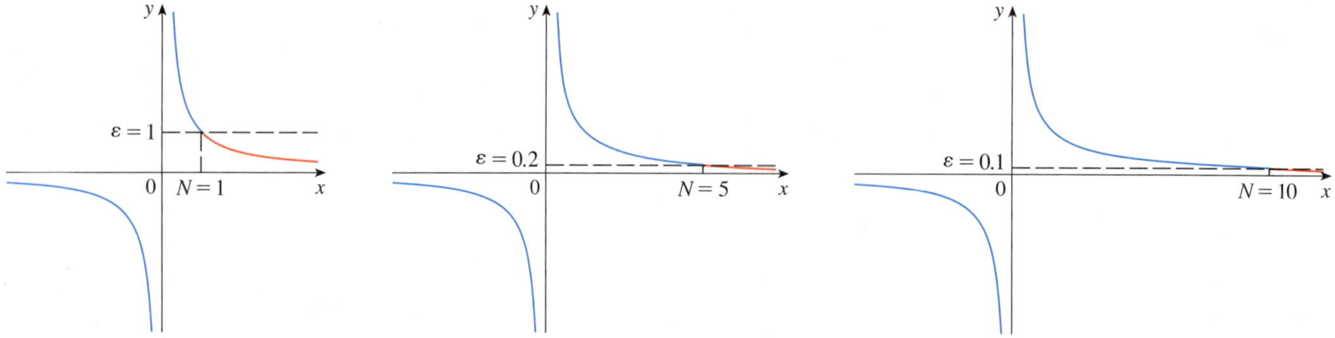

FIGURE 18

Finally we note that an infinite limit at infinity can be defined as follows. The geometric illustration is given in Figure 19.

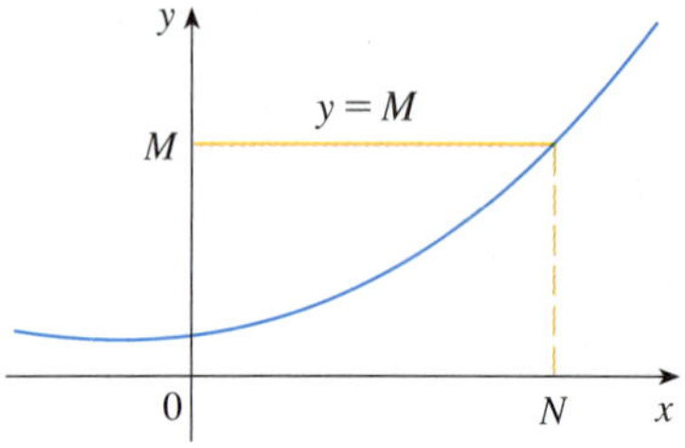

FIGURE 19
$\lim_{x \to \infty} f(x) = \infty$

> **9 Definition** Let f be a function defined on some interval (a, ∞). Then
>
> $$\lim_{x \to \infty} f(x) = \infty$$
>
> means that for every positive number M there is a corresponding positive number N such that
>
> $$f(x) > M \qquad \text{whenever} \qquad x > N$$

Similar definitions apply when the symbol ∞ is replaced by $-\infty$. (See Exercise 66.)

2.6 Exercises

1. Explain in your own words the meaning of each of the following.

(a) $\lim_{x \to \infty} f(x) = 5$ (b) $\lim_{x \to -\infty} f(x) = 3$

2. (a) Can the graph of $y = f(x)$ intersect a vertical asymptote? Can it intersect a horizontal asymptote? Illustrate by sketching graphs.

(b) How many horizontal asymptotes can the graph of $y = f(x)$ have? Sketch graphs to illustrate the possibilities.

3. For the function f whose graph is given, state the following.

(a) $\lim_{x \to 2} f(x)$ (b) $\lim_{x \to -1^-} f(x)$ (c) $\lim_{x \to -1^+} f(x)$

(d) $\lim_{x \to \infty} f(x)$ (e) $\lim_{x \to -\infty} f(x)$

(f) The equations of the asymptotes

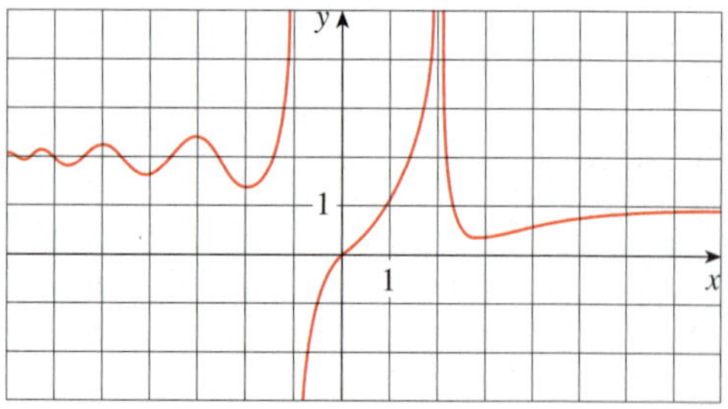

4. For the function g whose graph is given, state the following.

(a) $\lim_{x \to \infty} g(x)$ (b) $\lim_{x \to -\infty} g(x)$

(c) $\lim_{x \to 3} g(x)$ (d) $\lim_{x \to 0} g(x)$

(e) $\lim_{x \to -2^+} g(x)$ (f) The equations of the asymptotes

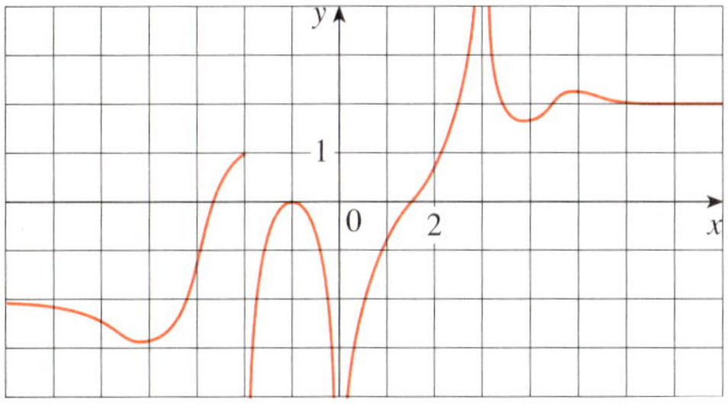

5–8 Sketch the graph of an example of a function f that satisfies all of the given conditions.

5. $f(0) = 0, \quad f(1) = 1, \quad \lim_{x \to \infty} f(x) = 0, \quad f$ is odd

6. $\lim_{x \to 0^+} f(x) = \infty, \quad \lim_{x \to 0^-} f(x) = -\infty, \quad \lim_{x \to \infty} f(x) = 1,$
$\lim_{x \to -\infty} f(x) = 1$

7. $\lim_{x \to 2} f(x) = -\infty, \quad \lim_{x \to \infty} f(x) = \infty, \quad \lim_{x \to -\infty} f(x) = 0,$
$\lim_{x \to 0^+} f(x) = \infty, \quad \lim_{x \to 0^-} f(x) = -\infty$

8. $\lim_{x \to -2} f(x) = \infty, \quad \lim_{x \to -\infty} f(x) = 3, \quad \lim_{x \to \infty} f(x) = -3$

9. Guess the value of the limit

$$\lim_{x \to \infty} \frac{x^2}{2^x}$$

by evaluating the function $f(x) = x^2/2^x$ for $x = 0, 1, 2, 3, 4, 5, 6, 7, 8, 9, 10, 20, 50$, and 100. Then use a graph of f to support your guess.

10. (a) Use a graph of

$$f(x) = \left(1 - \frac{2}{x}\right)^x$$

to estimate the value of $\lim_{x \to \infty} f(x)$ correct to two decimal places.
(b) Use a table of values of $f(x)$ to estimate the limit to four decimal places.

11–12 Evaluate the limit and justify each step by indicating the appropriate properties of limits.

11. $\lim_{x \to \infty} \dfrac{3x^2 - x + 4}{2x^2 + 5x - 8}$

12. $\lim_{x \to \infty} \sqrt{\dfrac{12x^3 - 5x + 2}{1 + 4x^2 + 3x^3}}$

13–34 Find the limit.

13. $\lim_{x \to \infty} \dfrac{1}{2x + 3}$

14. $\lim_{x \to \infty} \dfrac{3x + 5}{x - 4}$

15. $\lim_{x \to -\infty} \dfrac{1 - x - x^2}{2x^2 - 7}$

16. $\lim_{y \to \infty} \dfrac{2 - 3y^2}{5y^2 + 4y}$

17. $\lim_{x \to \infty} \dfrac{x^3 + 5x}{2x^3 - x^2 + 4}$

18. $\lim_{t \to -\infty} \dfrac{t^2 + 2}{t^3 + t^2 - 1}$

19. $\lim_{u \to \infty} \dfrac{4u^4 + 5}{(u^2 - 2)(2u^2 - 1)}$

20. $\lim_{x \to \infty} \dfrac{x + 2}{\sqrt{9x^2 + 1}}$

21. $\lim_{x \to \infty} \dfrac{\sqrt{9x^6 - x}}{x^3 + 1}$

22. $\lim_{x \to -\infty} \dfrac{\sqrt{9x^6 - x}}{x^3 + 1}$

23. $\lim_{x \to \infty} \left(\sqrt{9x^2 + x} - 3x\right)$

24. $\lim_{x \to -\infty} \left(x + \sqrt{x^2 + 2x}\right)$

25. $\lim_{x \to \infty} \left(\sqrt{x^2 + ax} - \sqrt{x^2 + bx}\right)$

26. $\lim_{x \to \infty} \cos x$

27. $\lim_{x \to \infty} \sqrt{x}$

28. $\lim_{x \to -\infty} \sqrt[3]{x}$

29. $\lim_{x \to \infty} \left(x - \sqrt{x}\right)$

30. $\lim_{x \to \infty} \dfrac{x^3 - 2x + 3}{5 - 2x^2}$

31. $\lim_{x \to -\infty} (x^4 + x^5)$

32. $\lim_{x \to \infty} \tan^{-1}(x^2 - x^4)$

33. $\lim_{x \to \infty} \dfrac{x + x^3 + x^5}{1 - x^2 + x^4}$

34. $\lim_{x \to (\pi/2)^+} e^{\tan x}$

35. (a) Estimate the value of

$$\lim_{x \to -\infty} \left(\sqrt{x^2 + x + 1} + x\right)$$

by graphing the function $f(x) = \sqrt{x^2 + x + 1} + x$.
(b) Use a table of values of $f(x)$ to guess the value of the limit.
(c) Prove that your guess is correct.

36. (a) Use a graph of

$$f(x) = \sqrt{3x^2 + 8x + 6} - \sqrt{3x^2 + 3x + 1}$$

to estimate the value of $\lim_{x \to \infty} f(x)$ to one decimal place.
(b) Use a table of values of $f(x)$ to estimate the limit to four decimal places.
(c) Find the exact value of the limit.

37–42 Find the horizontal and vertical asymptotes of each curve. Check your work by graphing the curve and estimating the asymptotes.

37. $y = \dfrac{x}{x + 4}$

38. $y = \dfrac{x^2 + 4}{x^2 - 1}$

39. $y = \dfrac{x^3}{x^2 + 3x - 10}$

40. $y = \dfrac{x^3 + 1}{x^3 + x}$

41. $h(x) = \dfrac{x}{\sqrt[4]{x^4 + 1}}$

42. $F(x) = \dfrac{x - 9}{\sqrt{4x^2 + 3x + 2}}$

43. Find a formula for a function f that satisfies the following conditions:

$$\lim_{x\to\pm\infty} f(x) = 0, \quad \lim_{x\to 0} f(x) = -\infty, \quad f(2) = 0,$$

$$\lim_{x\to 3^-} f(x) = \infty, \quad \lim_{x\to 3^+} f(x) = -\infty$$

44. Find a formula for a function that has vertical asymptotes $x = 1$ and $x = 3$ and horizontal asymptote $y = 1$.

45–48 Find the limits as $x \to \infty$ and as $x \to -\infty$. Use this information, together with intercepts, to give a rough sketch of the graph as in Example 11.

45. $y = x^2(x - 2)(1 - x)$

46. $y = (2 + x)^3(1 - x)(3 - x)$

47. $y = (x + 4)^5(x - 3)^4$

48. $y = (1 - x)(x - 3)^2(x - 5)^2$

49. (a) Use the Squeeze Theorem to evaluate $\lim_{x\to\infty} \dfrac{\sin x}{x}$.

(b) Graph $f(x) = (\sin x)/x$. How many times does the graph cross the asymptote?

50. By the *end behavior* of a function we mean the behavior of its values as $x \to \infty$ and as $x \to -\infty$.

(a) Describe and compare the end behavior of the functions

$$P(x) = 3x^5 - 5x^3 + 2x \qquad Q(x) = 3x^5$$

by graphing both functions in the viewing rectangles $[-2, 2]$ by $[-2, 2]$ and $[-10, 10]$ by $[-10{,}000, 10{,}000]$.

(b) Two functions are said to have the *same end behavior* if their ratio approaches 1 as $x \to \infty$. Show that P and Q have the same end behavior.

51. Let P and Q be polynomials. Find

$$\lim_{x\to\infty} \frac{P(x)}{Q(x)}$$

if the degree of P is (a) less than the degree of Q and (b) greater than the degree of Q.

52. Make a rough sketch of the curve $y = x^n$ (n an integer) for the following five cases:

(i) $n = 0$ (ii) $n > 0$, n odd

(iii) $n > 0$, n even (iv) $n < 0$, n odd

(v) $n < 0$, n even

Then use these sketches to find the following limits.

(a) $\lim_{x\to 0^+} x^n$ (b) $\lim_{x\to 0^-} x^n$

(c) $\lim_{x\to\infty} x^n$ (d) $\lim_{x\to -\infty} x^n$

53. Find $\lim_{x\to\infty} f(x)$ if

$$\frac{4x - 1}{x} < f(x) < \frac{4x^2 + 3x}{x^2}$$

for all $x > 5$.

54. (a) A tank contains 5000 L of pure water. Brine that contains 30 g of salt per liter of water is pumped into the tank at a rate of 25 L/min. Show that the concentration of salt after t minutes (in grams per liter) is

$$C(t) = \frac{30t}{200 + t}$$

(b) What happens to the concentration as $t \to \infty$?

55. In Chapter 9 we will be able to show, under certain assumptions, that the velocity $v(t)$ of a falling raindrop at time t is

$$v(t) = v^*(1 - e^{-gt/v^*})$$

where g is the acceleration due to gravity and v^* is the *terminal velocity* of the raindrop.

(a) Find $\lim_{t\to\infty} v(t)$.

(b) Graph $v(t)$ if $v^* = 1$ m/s and $g = 9.8$ m/s^2. How long does it take for the velocity of the raindrop to reach 99% of its terminal velocity?

56. (a) By graphing $y = e^{-x/10}$ and $y = 0.1$ on a common screen, discover how large you need to make x so that $e^{-x/10} < 0.1$.

(b) Can you solve part (a) without using a graphing device?

57. Use a graph to find a number N such that

$$\left| \frac{6x^2 + 5x - 3}{2x^2 - 1} - 3 \right| < 0.2 \qquad \text{whenever} \qquad x > N$$

58. For the limit

$$\lim_{x\to\infty} \frac{\sqrt{4x^2 + 1}}{x + 1} = 2$$

illustrate Definition 7 by finding values of N that correspond to $\varepsilon = 0.5$ and $\varepsilon = 0.1$.

59. For the limit

$$\lim_{x\to -\infty} \frac{\sqrt{4x^2 + 1}}{x + 1} = -2$$

illustrate Definition 8 by finding values of N that correspond to $\varepsilon = 0.5$ and $\varepsilon = 0.1$.

60. For the limit

$$\lim_{x\to\infty} \frac{2x + 1}{\sqrt{x + 1}} = \infty$$

illustrate Definition 9 by finding a value of N that corresponds to $M = 100$.

61. (a) How large do we have to take x so that $1/x^2 < 0.0001$?
(b) Taking $r = 2$ in Theorem 5, we have the statement

$$\lim_{x \to \infty} \frac{1}{x^2} = 0$$

Prove this directly using Definition 7.

62. (a) How large do we have to take x so that $1/\sqrt{x} < 0.0001$?
(b) Taking $r = \frac{1}{2}$ in Theorem 5, we have the statement

$$\lim_{x \to \infty} \frac{1}{\sqrt{x}} = 0$$

Prove this directly using Definition 7.

63. Use Definition 8 to prove that $\displaystyle\lim_{x \to -\infty} \frac{1}{x} = 0$.

64. Prove, using Definition 9, that $\displaystyle\lim_{x \to \infty} x^3 = \infty$.

65. Use Definition 9 to prove that

$$\lim_{x \to \infty} e^x = \infty$$

66. Formulate a precise definition of

$$\lim_{x \to -\infty} f(x) = -\infty$$

Then use your definition to prove that

$$\lim_{x \to -\infty} (1 + x^3) = -\infty$$

67. Prove that

$$\lim_{x \to \infty} f(x) = \lim_{t \to 0^+} f(1/t)$$

and $$\lim_{x \to -\infty} f(x) = \lim_{t \to 0^-} f(1/t)$$

if these limits exist.

2.7 Tangents, Velocities, and Other Rates of Change

In Section 2.1 we guessed the values of slopes of tangent lines and velocities on the basis of numerical evidence. Now that we have defined limits and have learned techniques for computing them, we return to the tangent and velocity problems with the ability to calculate slopes of tangents, velocities, and other rates of change.

Tangents

If a curve C has equation $y = f(x)$ and we want to find the tangent line to C at the point $P(a, f(a))$, then we consider a nearby point $Q(x, f(x))$, where $x \neq a$, and compute the slope of the secant line PQ:

$$m_{PQ} = \frac{f(x) - f(a)}{x - a}$$

Then we let Q approach P along the curve C by letting x approach a. If m_{PQ} approaches a number m, then we define the *tangent* t to be the line through P with slope m. (This amounts to saying that the tangent line is the limiting position of the secant line PQ as Q approaches P. See Figure 1.)

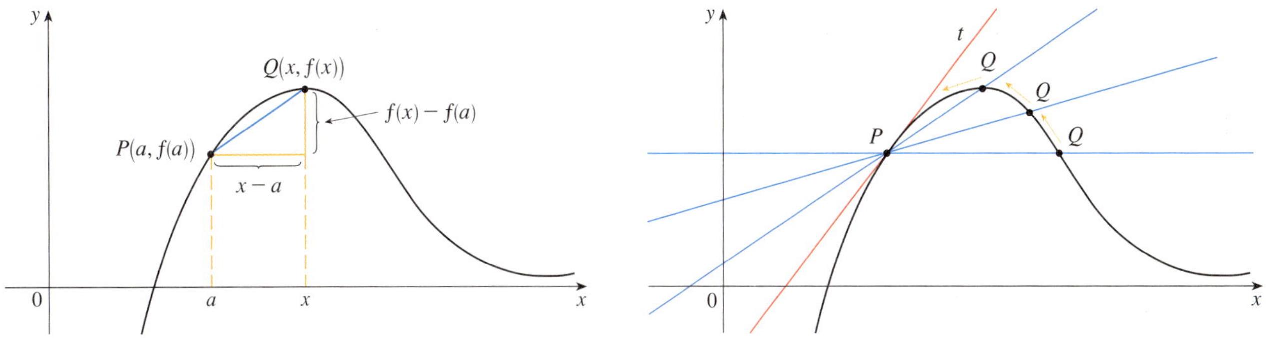

FIGURE 1

1 Definition The **tangent line** to the curve $y = f(x)$ at the point $P(a, f(a))$ is the line through P with slope

$$m = \lim_{x \to a} \frac{f(x) - f(a)}{x - a}$$

provided that this limit exists.

In our first example we confirm the guess we made in Example 1 in Section 2.1.

EXAMPLE 1 Find an equation of the tangent line to the parabola $y = x^2$ at the point $P(1, 1)$.

SOLUTION Here we have $a = 1$ and $f(x) = x^2$, so the slope is

$$\begin{aligned} m &= \lim_{x \to 1} \frac{f(x) - f(1)}{x - 1} = \lim_{x \to 1} \frac{x^2 - 1}{x - 1} \\ &= \lim_{x \to 1} \frac{(x - 1)(x + 1)}{x - 1} \\ &= \lim_{x \to 1} (x + 1) = 1 + 1 = 2 \end{aligned}$$

|||| Point-slope form for a line through the point (x_1, y_1) with slope m:

$$y - y_1 = m(x - x_1)$$

Using the point-slope form of the equation of a line, we find that an equation of the tangent line at $(1, 1)$ is

$$y - 1 = 2(x - 1) \qquad \text{or} \qquad y = 2x - 1$$

We sometimes refer to the slope of the tangent line to a curve at a point as the **slope of the curve** at the point. The idea is that if we zoom in far enough toward the point, the curve looks almost like a straight line. Figure 2 illustrates this procedure for the curve $y = x^2$ in Example 1. The more we zoom in, the more the parabola looks like a line. In other words, the curve becomes almost indistinguishable from its tangent line.

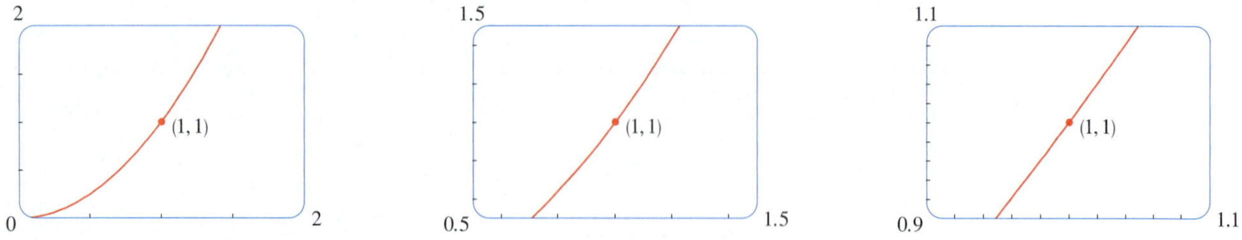

FIGURE 2 Zooming in toward the point (1, 1) on the parabola $y = x^2$

There is another expression for the slope of a tangent line that is sometimes easier to use. Let

$$h = x - a$$

Then

$$x = a + h$$

so the slope of the secant line PQ is

$$m_{PQ} = \frac{f(a+h) - f(a)}{h}$$

(See Figure 3 where the case $h > 0$ is illustrated and Q is to the right of P. If it happened that $h < 0$, however, Q would be to the left of P.)

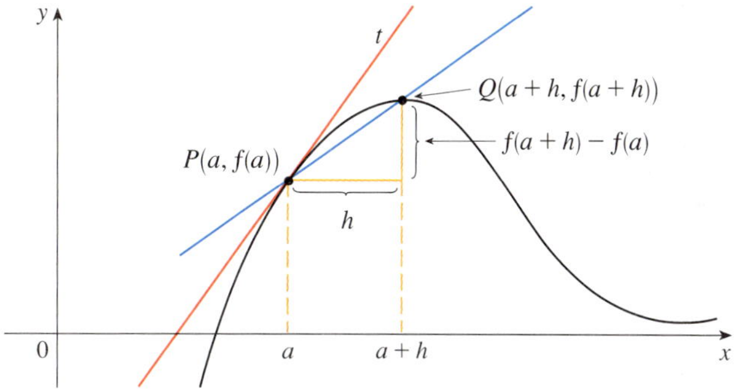

FIGURE 3

Notice that as x approaches a, h approaches 0 (because $h = x - a$) and so the expression for the slope of the tangent line in Definition 1 becomes

2
$$m = \lim_{h \to 0} \frac{f(a+h) - f(a)}{h}$$

EXAMPLE 2 Find an equation of the tangent line to the hyperbola $y = 3/x$ at the point (3, 1).

SOLUTION Let $f(x) = 3/x$. Then the slope of the tangent at (3, 1) is

$$m = \lim_{h \to 0} \frac{f(3+h) - f(3)}{h}$$

$$= \lim_{h \to 0} \frac{\dfrac{3}{3+h} - 1}{h} = \lim_{h \to 0} \frac{\dfrac{3 - (3+h)}{3+h}}{h}$$

$$= \lim_{h \to 0} \frac{-h}{h(3+h)} = \lim_{h \to 0} -\frac{1}{3+h}$$

$$= -\frac{1}{3}$$

Therefore, an equation of the tangent at the point (3, 1) is

$$y - 1 = -\tfrac{1}{3}(x - 3)$$

which simplifies to $\qquad x + 3y - 6 = 0$

The hyperbola and its tangent are shown in Figure 4.

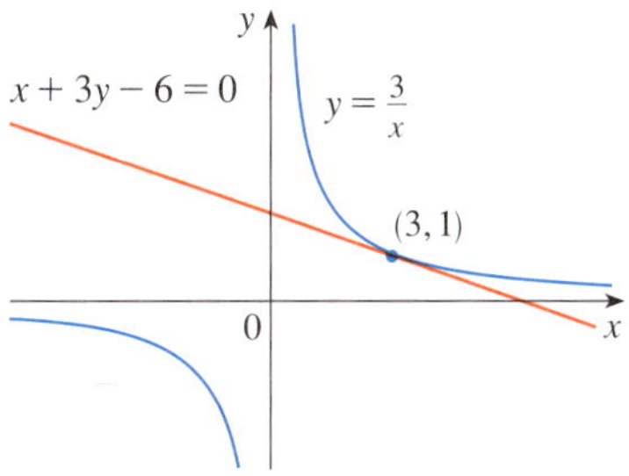

FIGURE 4

EXAMPLE 3 Find the slopes of the tangent lines to the graph of the function $f(x) = \sqrt{x}$ at the points (1, 1), (4, 2), and (9, 3).

SOLUTION Since three slopes are requested, it is efficient to start by finding the slope at the general point $(a, \sqrt{a})$:

$$m = \lim_{h \to 0} \frac{f(a+h) - f(a)}{h} = \lim_{h \to 0} \frac{\sqrt{a+h} - \sqrt{a}}{h}$$

Rationalize the numerator

$$= \lim_{h \to 0} \frac{\sqrt{a+h} - \sqrt{a}}{h} \cdot \frac{\sqrt{a+h} + \sqrt{a}}{\sqrt{a+h} + \sqrt{a}}$$

$$= \lim_{h \to 0} \frac{(a+h) - a}{h\left(\sqrt{a+h} + \sqrt{a}\right)} = \lim_{h \to 0} \frac{h}{h\left(\sqrt{a+h} + \sqrt{a}\right)}$$

Continuous function of h

$$= \lim_{h \to 0} \frac{1}{\sqrt{a+h} + \sqrt{a}} = \frac{1}{\sqrt{a} + \sqrt{a}} = \frac{1}{2\sqrt{a}}$$

At the point (1, 1), we have $a = 1$, so the slope of the tangent is $m = 1/\left(2\sqrt{1}\right) = \frac{1}{2}$. At (4, 2), we have $m = 1/\left(2\sqrt{4}\right) = \frac{1}{4}$; at (9, 3), $m = 1/\left(2\sqrt{9}\right) = \frac{1}{6}$.

Velocities

Learn about average and instantaneous velocity by comparing falling objects.
Resources / Module 3
/ Derivative at a Point
/ The Falling Robot

In Section 2.1 we investigated the motion of a ball dropped from the CN Tower and defined its velocity to be the limiting value of average velocities over shorter and shorter time periods.

In general, suppose an object moves along a straight line according to an equation of motion $s = f(t)$, where s is the displacement (directed distance) of the object from the origin at time t. The function f that describes the motion is called the **position function** of the object. In the time interval from $t = a$ to $t = a + h$ the change in position is $f(a + h) - f(a)$. (See Figure 5.) The average velocity over this time interval is

$$\text{average velocity} = \frac{\text{displacement}}{\text{time}} = \frac{f(a+h) - f(a)}{h}$$

which is the same as the slope of the secant line PQ in Figure 6.

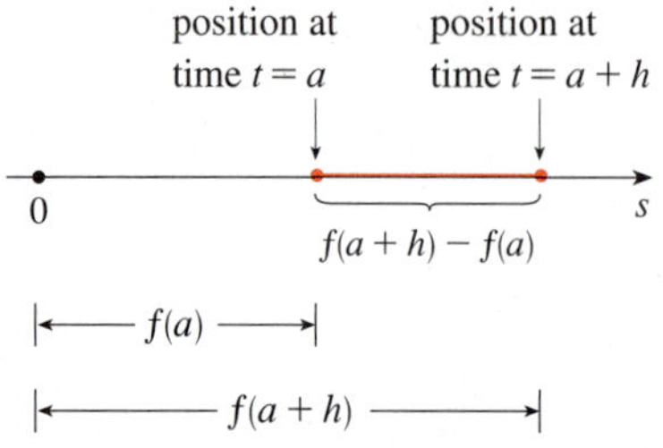

FIGURE 5

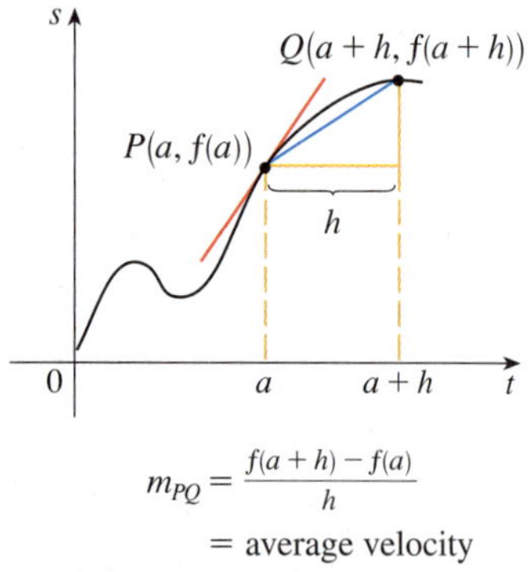

FIGURE 6

Now suppose we compute the average velocities over shorter and shorter time intervals $[a, a + h]$. In other words, we let h approach 0. As in the example of the falling ball, we define the **velocity** (or **instantaneous velocity**) $v(a)$ at time $t = a$ to be the limit of these average velocities:

3
$$v(a) = \lim_{h \to 0} \frac{f(a + h) - f(a)}{h}$$

This means that the velocity at time $t = a$ is equal to the slope of the tangent line at P (compare Equations 2 and 3).

Now that we know how to compute limits, let's reconsider the problem of the falling ball.

EXAMPLE 4 Suppose that a ball is dropped from the upper observation deck of the CN Tower, 450 m above the ground.
(a) What is the velocity of the ball after 5 seconds?
(b) How fast is the ball traveling when it hits the ground?

|||| Recall from Section 2.1: The distance (in meters) fallen after t seconds is $4.9t^2$.

SOLUTION We first use the equation of motion $s = f(t) = 4.9t^2$ to find the velocity $v(a)$ after a seconds:

$$v(a) = \lim_{h \to 0} \frac{f(a + h) - f(a)}{h} = \lim_{h \to 0} \frac{4.9(a + h)^2 - 4.9a^2}{h}$$

$$= \lim_{h \to 0} \frac{4.9(a^2 + 2ah + h^2 - a^2)}{h} = \lim_{h \to 0} \frac{4.9(2ah + h^2)}{h}$$

$$= \lim_{h \to 0} 4.9(2a + h) = 9.8a$$

(a) The velocity after 5 s is $v(5) = (9.8)(5) = 49$ m/s.

(b) Since the observation deck is 450 m above the ground, the ball will hit the ground at the time t_1 when $s(t_1) = 450$, that is,

$$4.9t_1^2 = 450$$

This gives

$$t_1^2 = \frac{450}{4.9} \quad \text{and} \quad t_1 = \sqrt{\frac{450}{4.9}} \approx 9.6 \text{ s}$$

The velocity of the ball as it hits the ground is therefore

$$v(t_1) = 9.8t_1 = 9.8\sqrt{\frac{450}{4.9}} \approx 94 \text{ m/s}$$

|||| Other Rates of Change

Suppose y is a quantity that depends on another quantity x. Thus, y is a function of x and we write $y = f(x)$. If x changes from x_1 to x_2, then the change in x (also called the **increment** of x) is

$$\Delta x = x_2 - x_1$$

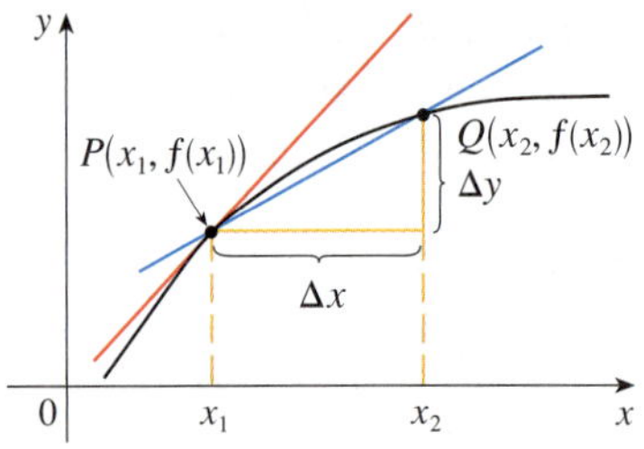

average rate of change $= m_{PQ}$

instantaneous rate of change = slope of tangent at P

FIGURE 7

and the corresponding change in y is

$$\Delta y = f(x_2) - f(x_1)$$

The difference quotient

$$\frac{\Delta y}{\Delta x} = \frac{f(x_2) - f(x_1)}{x_2 - x_1}$$

is called the **average rate of change of y with respect to x** over the interval $[x_1, x_2]$ and can be interpreted as the slope of the secant line PQ in Figure 7.

By analogy with velocity, we consider the average rate of change over smaller and smaller intervals by letting x_2 approach x_1 and therefore letting Δx approach 0. The limit of these average rates of change is called the **(instantaneous) rate of change of y with respect to x** at $x = x_1$, which is interpreted as the slope of the tangent to the curve $y = f(x)$ at $P(x_1, f(x_1))$:

4 $$\text{instantaneous rate of change} = \lim_{\Delta x \to 0} \frac{\Delta y}{\Delta x} = \lim_{x_2 \to x_1} \frac{f(x_2) - f(x_1)}{x_2 - x_1}$$

x (h)	T (°C)	x (h)	T (°C)
0	6.5	13	16.0
1	6.1	14	17.3
2	5.6	15	18.2
3	4.9	16	18.8
4	4.2	17	17.6
5	4.0	18	16.0
6	4.0	19	14.1
7	4.8	20	11.5
8	6.1	21	10.2
9	8.3	22	9.0
10	10.0	23	7.9
11	12.1	24	7.0
12	14.3		

EXAMPLE 5 Temperature readings T (in degrees Celsius) were recorded every hour starting at midnight on a day in April in Whitefish, Montana. The time x is measured in hours from midnight. The data are given in the table at the left.

(a) Find the average rate of change of temperature with respect to time

(i) from noon to 3 P.M. (ii) from noon to 2 P.M.

(iii) from noon to 1 P.M.

(b) Estimate the instantaneous rate of change at noon.

SOLUTION

(a) (i) From noon to 3 P.M. the temperature changes from 14.3°C to 18.2°C, so

$$\Delta T = T(15) - T(12) = 18.2 - 14.3 = 3.9°\text{C}$$

while the change in time is $\Delta x = 3$ h. Therefore, the average rate of change of temperature with respect to time is

$$\frac{\Delta T}{\Delta x} = \frac{3.9}{3} = 1.3°\text{C/h}$$

(ii) From noon to 2 P.M. the average rate of change is

$$\frac{\Delta T}{\Delta x} = \frac{T(14) - T(12)}{14 - 12} = \frac{17.3 - 14.3}{2} = 1.5°\text{C/h}$$

(iii) From noon to 1 P.M. the average rate of change is

$$\frac{\Delta T}{\Delta x} = \frac{T(13) - T(12)}{13 - 12} = \frac{16.0 - 14.3}{1} = 1.7°\text{C/h}$$

|||| A NOTE ON UNITS

The units for the average rate of change $\Delta T/\Delta x$ are the units for ΔT divided by the units for Δx, namely, degrees Celsius per hour. The instantaneous rate of change is the limit of the average rates of change, so it is measured in the same units: degrees Celsius per hour.

(b) We plot the given data in Figure 8 and use them to sketch a smooth curve that approximates the graph of the temperature function. Then we draw the tangent at the point P where $x = 12$ and, after measuring the sides of triangle ABC, we estimate that the slope of the tangent line is

|||| Another method is to average the slopes of two secant lines. See Example 2 in Section 2.1.

$$\frac{|BC|}{|AC|} = \frac{10.3}{5.5} \approx 1.9$$

Therefore, the instantaneous rate of change of temperature with respect to time at noon is about 1.9°C/h.

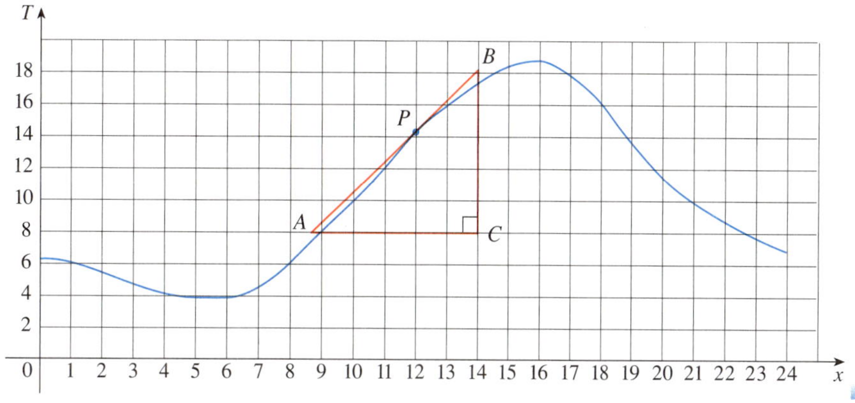

FIGURE 8

The velocity of a particle is the rate of change of displacement with respect to time. Physicists are interested in other rates of change as well—for instance, the rate of change of work with respect to time (which is called *power*). Chemists who study a chemical reaction are interested in the rate of change in the concentration of a reactant with respect to time (called the *rate of reaction*). A steel manufacturer is interested in the rate of change of the cost of producing x tons of steel per day with respect to x (called the *marginal cost*). A biologist is interested in the rate of change of the population of a colony of bacteria with respect to time. In fact, the computation of rates of change is important in all of the natural sciences, in engineering, and even in the social sciences. Further examples will be given in Section 3.3.

All these rates of change can be interpreted as slopes of tangents. This gives added significance to the solution of the tangent problem. Whenever we solve a problem involving tangent lines, we are not just solving a problem in geometry. We are also implicitly solving a great variety of problems involving rates of change in science and engineering.

2.7 Exercises

1. A curve has equation $y = f(x)$.
(a) Write an expression for the slope of the secant line through the points $P(3, f(3))$ and $Q(x, f(x))$.
(b) Write an expression for the slope of the tangent line at P.

2. Suppose an object moves with position function $s = f(t)$.
(a) Write an expression for the average velocity of the object in the time interval from $t = a$ to $t = a + h$.
(b) Write an expression for the instantaneous velocity at time $t = a$.

3. Consider the slope of the given curve at each of the five points shown. List these five slopes in decreasing order and explain your reasoning.

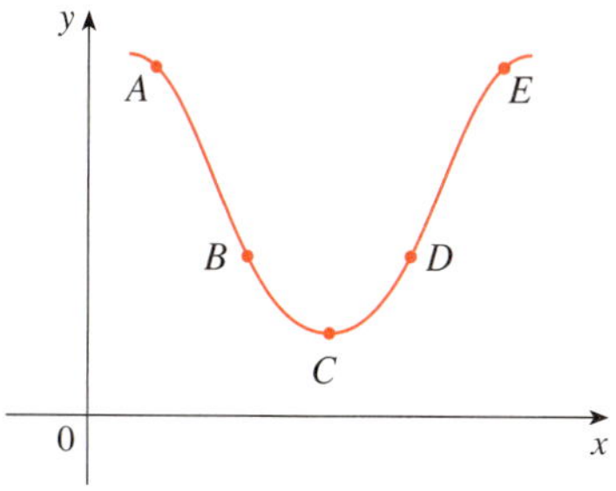

4. Graph the curve $y = e^x$ in the viewing rectangles $[-1, 1]$ by $[0, 2]$, $[-0.5, 0.5]$ by $[0.5, 1.5]$, and $[-0.1, 0.1]$ by $[0.9, 1.1]$. What do you notice about the curve as you zoom in toward the point $(0, 1)$?

5. (a) Find the slope of the tangent line to the parabola $y = x^2 + 2x$ at the point $(-3, 3)$
(i) using Definition 1
(ii) using Equation 2
(b) Find an equation of the tangent line in part (a).
(c) Graph the parabola and the tangent line. As a check on your work, zoom in toward the point $(-3, 3)$ until the parabola and the tangent line are indistinguishable.

6. (a) Find the slope of the tangent line to the curve $y = x^3$ at the point $(-1, -1)$
(i) using Definition 1
(ii) using Equation 2
(b) Find an equation of the tangent line in part (a).
(c) Graph the curve and the tangent line in successively smaller viewing rectangles centered at $(-1, -1)$ until the curve and the line appear to coincide.

7–10 Find an equation of the tangent line to the curve at the given point.

7. $y = 1 + 2x - x^3, \quad (1, 2)$

8. $y = \sqrt{2x + 1}, \quad (4, 3)$

9. $y = (x - 1)/(x - 2), \quad (3, 2)$

10. $y = 2x/(x + 1)^2, \quad (0, 0)$

11. (a) Find the slope of the tangent to the curve $y = 2/(x + 3)$ at the point where $x = a$.
(b) Find the slopes of the tangent lines at the points whose x-coordinates are (i) -1, (ii) 0, and (iii) 1.

12. (a) Find the slope of the tangent to the parabola $y = 1 + x + x^2$ at the point where $x = a$.
(b) Find the slopes of the tangent lines at the points whose x-coordinates are (i) -1, (ii) $-\frac{1}{2}$, and (iii) 1.
(c) Graph the curve and the three tangents on a common screen.

13. (a) Find the slope of the tangent to the curve $y = x^3 - 4x + 1$ at the point where $x = a$.
(b) Find equations of the tangent lines at the points $(1, -2)$ and $(2, 1)$.
(c) Graph the curve and both tangents on a common screen.

14. (a) Find the slope of the tangent to the curve $y = 1/\sqrt{x}$ at the point where $x = a$.
(b) Find equations of the tangent lines at the points $(1, 1)$ and $(4, \frac{1}{2})$.
(c) Graph the curve and both tangents on a common screen.

15. The graph shows the position function of a car. Use the shape of the graph to explain your answers to the following questions.
(a) What was the initial velocity of the car?
(b) Was the car going faster at B or at C?
(c) Was the car slowing down or speeding up at A, B, and C?
(d) What happened between D and E?

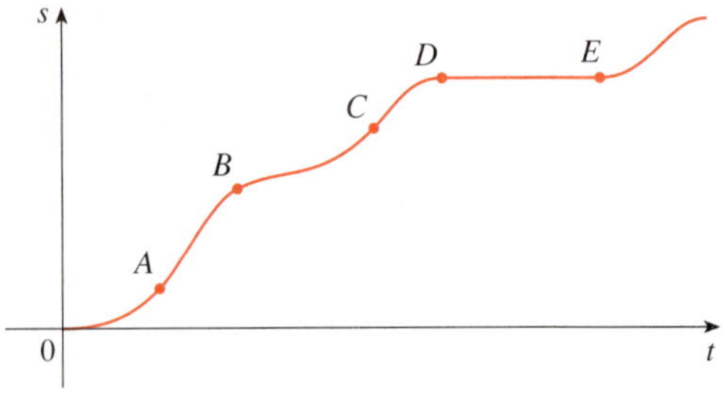

16. Valerie is driving along a highway. Sketch the graph of the position function of her car if she drives in the following manner: At time $t = 0$, the car is at mile marker 15 and is traveling at a constant speed of 55 mi/h. She travels at this speed for exactly an hour. Then the car slows gradually over a 2-minute period as Valerie comes to a stop for dinner. Dinner lasts 26 min; then she restarts the car, gradually speeding up to 65 mi/h over a 2-minute period. She drives at a constant 65 mi/h for two hours and then over a 3-minute period gradually slows to a complete stop.

17. If a ball is thrown into the air with a velocity of 40 ft/s, its height (in feet) after t seconds is given by $y = 40t - 16t^2$. Find the velocity when $t = 2$.

18. If an arrow is shot upward on the moon with a velocity of 58 m/s, its height (in meters) after t seconds is given by $H = 58t - 0.83t^2$.
(a) Find the velocity of the arrow after one second.
(b) Find the velocity of the arrow when $t = a$.
(c) When will the arrow hit the moon?
(d) With what velocity will the arrow hit the moon?

19. The displacement (in meters) of a particle moving in a straight line is given by the equation of motion $s = 4t^3 + 6t + 2$, where t is measured in seconds. Find the velocity of the particle at times $t = a$, $t = 1$, $t = 2$, and $t = 3$.

20. The displacement (in meters) of a particle moving in a straight line is given by $s = t^2 - 8t + 18$, where t is measured in seconds.
(a) Find the average velocity over each time interval:
(i) $[3, 4]$ (ii) $[3.5, 4]$
(iii) $[4, 5]$ (iv) $[4, 4.5]$
(b) Find the instantaneous velocity when $t = 4$.
(c) Draw the graph of s as a function of t and draw the secant lines whose slopes are the average velocities in part (a) and the tangent line whose slope is the instantaneous velocity in part (b).

21. A warm can of soda is placed in a cold refrigerator. Sketch the graph of the temperature of the soda as a function of time. Is the initial rate of change of temperature greater or less than the rate of change after an hour?

22. A roast turkey is taken from an oven when its temperature has reached 185°F and is placed on a table in a room where the temperature is 75°F. The graph shows how the temperature of the turkey decreases and eventually approaches room temperature. (In Section 9.4 we will be able to use Newton's Law of Cooling to find an equation for T as a function of time.) By measuring the slope of the tangent, estimate the rate of change of the temperature after an hour.

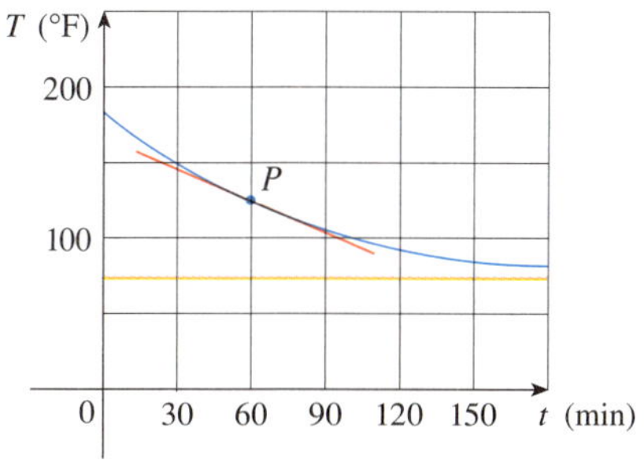

23. (a) Use the data in Example 5 to find the average rate of change of temperature with respect to time
(i) from 8 P.M. to 11 P.M.
(ii) from 8 P.M. to 10 P.M.
(iii) from 8 P.M. to 9 P.M.
(b) Estimate the instantaneous rate of change of T with respect to time at 8 P.M. by measuring the slope of a tangent.

24. The population P (in thousands) of Belgium from 1992 to 2000 is shown in the table. (Midyear estimates are given.)

Year	1992	1994	1996	1998	2000
P	10,036	10,109	10,152	10,175	10,186

(a) Find the average rate of growth
(i) from 1992 to 1996
(ii) from 1994 to 1996
(iii) from 1996 to 1998
In each case, include the units.
(b) Estimate the instantaneous rate of growth in 1996 by taking the average of two average rates of change. What are its units?
(c) Estimate the instantaneous rate of growth in 1996 by measuring the slope of a tangent.

25. The number N (in thousands) of cellular phone subscribers in Malaysia is shown in the table. (Midyear estimates are given.)

Year	1993	1994	1995	1996	1997
N	304	572	873	1513	2461

(a) Find the average rate of growth
(i) from 1995 to 1997
(ii) from 1995 to 1996
(iii) from 1994 to 1995
In each case, include the units.
(b) Estimate the instantaneous rate of growth in 1995 by taking the average of two average rates of change. What are its units?
(c) Estimate the instantaneous rate of growth in 1995 by measuring the slope of a tangent.

26. The number N of locations of a popular coffeehouse chain is given in the table. (The numbers of locations as of June 30 are given.)

Year	1996	1997	1998	1999	2000
N	1015	1412	1886	2135	3300

(a) Find the average rate of growth
(i) from 1996 to 1998
(ii) from 1997 to 1998
(iii) from 1998 to 1999
In each case, include the units.
(b) Estimate the instantaneous rate of growth in 1998 by taking the average of two average rates of change. What are its units?
(c) Estimate the instantaneous rate of growth in 1998 by measuring the slope of a tangent.

27. The cost (in dollars) of producing x units of a certain commodity is $C(x) = 5000 + 10x + 0.05x^2$.
(a) Find the average rate of change of C with respect to x when the production level is changed
(i) from $x = 100$ to $x = 105$
(ii) from $x = 100$ to $x = 101$
(b) Find the instantaneous rate of change of C with respect to x when $x = 100$. (This is called the *marginal cost*. Its significance will be explained in Section 3.3.)

28. If a cylindrical tank holds 100,000 gallons of water, which can be drained from the bottom of the tank in an hour, then Torricelli's Law gives the volume V of water remaining in the tank after t minutes as

$$V(t) = 100{,}000\left(1 - \frac{t}{60}\right)^2 \qquad 0 \leqslant t \leqslant 60$$

Find the rate at which the water is flowing out of the tank (the instantaneous rate of change of V with respect to t) as a function of t. What are its units? For times $t = 0, 10, 20, 30, 40, 50$, and 60 min, find the flow rate and the amount of water remaining in the tank. Summarize your findings in a sentence or two. At what time is the flow rate the greatest? The least?

2.8 Derivatives

In Section 2.7 we defined the slope of the tangent to a curve with equation $y = f(x)$ at the point where $x = a$ to be

1 $$m = \lim_{h \to 0} \frac{f(a+h) - f(a)}{h}$$

We also saw that the velocity of an object with position function $s = f(t)$ at time $t = a$ is

$$v(a) = \lim_{h \to 0} \frac{f(a+h) - f(a)}{h}$$

In fact, limits of the form

$$\lim_{h \to 0} \frac{f(a+h) - f(a)}{h}$$

arise whenever we calculate a rate of change in any of the sciences or engineering, such as a rate of reaction in chemistry or a marginal cost in economics. Since this type of limit occurs so widely, it is given a special name and notation.

2 Definition The **derivative of a function f at a number a**, denoted by $f'(a)$, is

$$f'(a) = \lim_{h \to 0} \frac{f(a+h) - f(a)}{h}$$

if this limit exists.

$f'(a)$ is read "f prime of a."

If we write $x = a + h$, then $h = x - a$ and h approaches 0 if and only if x approaches a. Therefore, an equivalent way of stating the definition of the derivative, as we saw in finding tangent lines, is

3 $$f'(a) = \lim_{x \to a} \frac{f(x) - f(a)}{x - a}$$

EXAMPLE 1 Find the derivative of the function $f(x) = x^2 - 8x + 9$ at the number a.

SOLUTION From Definition 2 we have

Try problems like this one.
Resources / Module 3
/ Derivative at a Point
/ Problem Wizard

$$\begin{aligned} f'(a) &= \lim_{h \to 0} \frac{f(a+h) - f(a)}{h} \\ &= \lim_{h \to 0} \frac{[(a+h)^2 - 8(a+h) + 9] - [a^2 - 8a + 9]}{h} \\ &= \lim_{h \to 0} \frac{a^2 + 2ah + h^2 - 8a - 8h + 9 - a^2 + 8a - 9}{h} \\ &= \lim_{h \to 0} \frac{2ah + h^2 - 8h}{h} = \lim_{h \to 0} (2a + h - 8) \\ &= 2a - 8 \end{aligned}$$

Interpretation of the Derivative as the Slope of a Tangent

In Section 2.7 we defined the tangent line to the curve $y = f(x)$ at the point $P(a, f(a))$ to be the line that passes through P and has slope m given by Equation 1. Since, by Definition 2, this is the same as the derivative $f'(a)$, we can now say the following.

> The tangent line to $y = f(x)$ at $(a, f(a))$ is the line through $(a, f(a))$ whose slope is equal to $f'(a)$, the derivative of f at a.

Thus, the geometric interpretation of a derivative [as defined by either (2) or (3)] is as shown in Figure 1.

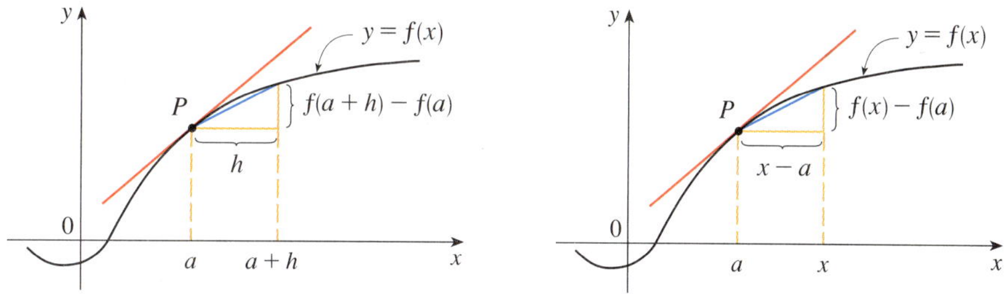

(a) $f'(a) = \lim_{h \to 0} \frac{f(a+h) - f(a)}{h}$
= slope of tangent at P
= slope of curve at P

(b) $f'(a) = \lim_{x \to a} \frac{f(x) - f(a)}{x - a}$
= slope of tangent at P
= slope of curve at P

FIGURE 1
Geometric interpretation of the derivative

If we use the point-slope form of the equation of a line, we can write an equation of the tangent line to the curve $y = f(x)$ at the point $(a, f(a))$:

$$y - f(a) = f'(a)(x - a)$$

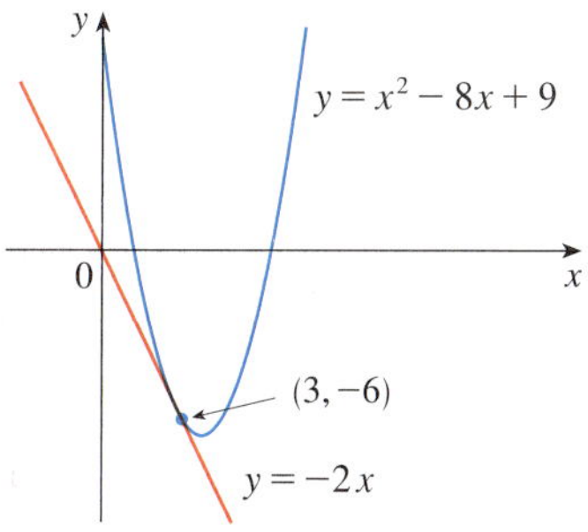

FIGURE 2

EXAMPLE 2 Find an equation of the tangent line to the parabola $y = x^2 - 8x + 9$ at the point $(3, -6)$.

SOLUTION From Example 1 we know that the derivative of $f(x) = x^2 - 8x + 9$ at the number a is $f'(a) = 2a - 8$. Therefore, the slope of the tangent line at $(3, -6)$ is $f'(3) = 2(3) - 8 = -2$. Thus, an equation of the tangent line, shown in Figure 2, is

$$y - (-6) = (-2)(x - 3) \qquad \text{or} \qquad y = -2x$$

EXAMPLE 3 Let $f(x) = 2^x$. Estimate the value of $f'(0)$ in two ways:
(a) By using Definition 2 and taking successively smaller values of h.
(b) By interpreting $f'(0)$ as the slope of a tangent and using a graphing calculator to zoom in on the graph of $y = 2^x$.

SOLUTION
(a) From Definition 2 we have

$$f'(0) = \lim_{h \to 0} \frac{f(h) - f(0)}{h} = \lim_{h \to 0} \frac{2^h - 1}{h}$$

h	$\dfrac{2^h - 1}{h}$
0.1	0.718
0.01	0.696
0.001	0.693
0.0001	0.693
−0.1	0.670
−0.01	0.691
−0.001	0.693
−0.0001	0.693

Since we are not yet able to evaluate this limit exactly, we use a calculator to approximate the values of $(2^h - 1)/h$. From the numerical evidence in the table at the left we see that as h approaches 0, these values appear to approach a number near 0.69. So our estimate is

$$f'(0) \approx 0.69$$

(b) In Figure 3 we graph the curve $y = 2^x$ and zoom in toward the point $(0, 1)$. We see that the closer we get to $(0, 1)$, the more the curve looks like a straight line. In fact, in Figure 3(c) the curve is practically indistinguishable from its tangent line at $(0, 1)$. Since the x-scale and the y-scale are both 0.01, we estimate that the slope of this line is

$$\frac{0.14}{0.20} = 0.7$$

So our estimate of the derivative is $f'(0) \approx 0.7$. In Section 3.5 we will show that, correct to six decimal places, $f'(0) \approx 0.693147$.

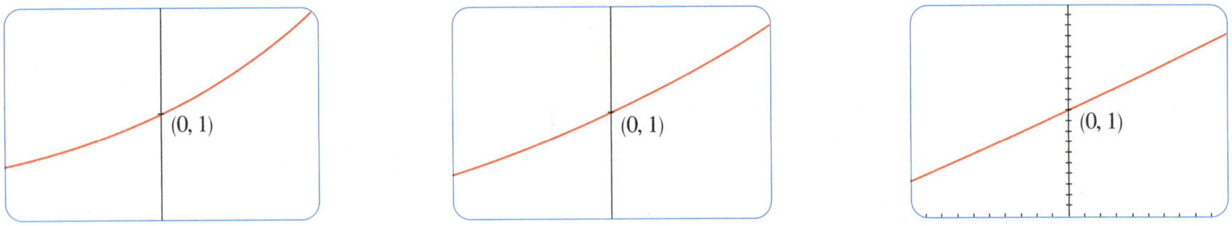

(a) $[-1, 1]$ by $[0, 2]$ (b) $[-0.5, 0.5]$ by $[0.5, 1.5]$ (c) $[-0.1, 0.1]$ by $[0.9, 1.1]$

FIGURE 3 Zooming in on the graph of $y = 2^x$ near $(0, 1)$

Interpretation of the Derivative as a Rate of Change

In Section 2.7 we defined the instantaneous rate of change of $y = f(x)$ with respect to x at $x = x_1$ as the limit of the average rates of change over smaller and smaller intervals. If the interval is $[x_1, x_2]$, then the change in x is $\Delta x = x_2 - x_1$, the corresponding change in y is

$$\Delta y = f(x_2) - f(x_1)$$

and

4 $$\text{instantaneous rate of change} = \lim_{\Delta x \to 0} \frac{\Delta y}{\Delta x} = \lim_{x_2 \to x_1} \frac{f(x_2) - f(x_1)}{x_2 - x_1}$$

From Equation 3 we recognize this limit as being the derivative of f at x_1, that is, $f'(x_1)$. This gives a second interpretation of the derivative:

> The derivative $f'(a)$ is the instantaneous rate of change of $y = f(x)$ with respect to x when $x = a$.

The connection with the first interpretation is that if we sketch the curve $y = f(x)$, then the instantaneous rate of change is the slope of the tangent to this curve at the point where

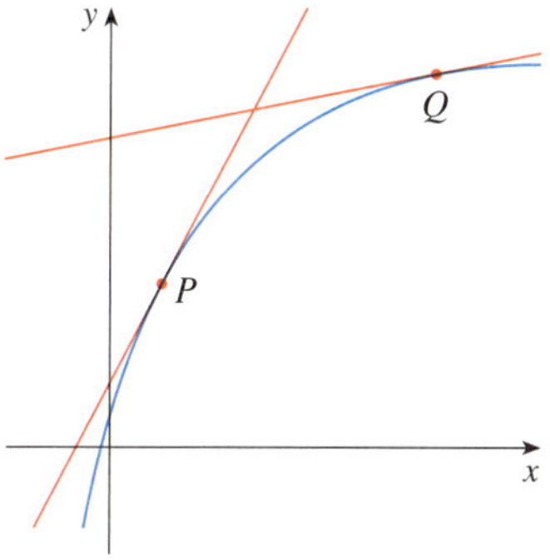

FIGURE 4
The y-values are changing rapidly at P and slowly at Q.

In Module 2.8 you are asked to compare and order the slopes of tangent and secant lines at several points on a curve.

$x = a$. This means that when the derivative is large (and therefore the curve is steep, as at the point P in Figure 4), the y-values change rapidly. When the derivative is small, the curve is relatively flat and the y-values change slowly.

In particular, if $s = f(t)$ is the position function of a particle that moves along a straight line, then $f'(a)$ is the rate of change of the displacement s with respect to the time t. In other words, *$f'(a)$ is the velocity of the particle at time $t = a$*. (See Section 2.7.) The *speed* of the particle is the absolute value of the velocity, that is, $|f'(a)|$.

EXAMPLE 4 The position of a particle is given by the equation of motion $s = f(t) = 1/(1 + t)$, where t is measured in seconds and s in meters. Find the velocity and the speed after 2 seconds.

SOLUTION The derivative of f when $t = 2$ is

$$f'(2) = \lim_{h \to 0} \frac{f(2 + h) - f(2)}{h} = \lim_{h \to 0} \frac{\dfrac{1}{1 + (2 + h)} - \dfrac{1}{1 + 2}}{h}$$

$$= \lim_{h \to 0} \frac{\dfrac{1}{3 + h} - \dfrac{1}{3}}{h} = \lim_{h \to 0} \frac{\dfrac{3 - (3 + h)}{3(3 + h)}}{h}$$

$$= \lim_{h \to 0} \frac{-h}{3(3 + h)h} = \lim_{h \to 0} \frac{-1}{3(3 + h)} = -\frac{1}{9}$$

Thus, the velocity after 2 seconds is $f'(2) = -\frac{1}{9}$ m/s, and the speed is $|f'(2)| = |-\frac{1}{9}| = \frac{1}{9}$ m/s.

EXAMPLE 5 A manufacturer produces bolts of a fabric with a fixed width. The cost of producing x yards of this fabric is $C = f(x)$ dollars.

(a) What is the meaning of the derivative $f'(x)$? What are its units?
(b) In practical terms, what does it mean to say that $f'(1000) = 9$?
(c) Which do you think is greater, $f'(50)$ or $f'(500)$? What about $f'(5000)$?

SOLUTION

(a) The derivative $f'(x)$ is the instantaneous rate of change of C with respect to x; that is, $f'(x)$ means the rate of change of the production cost with respect to the number of yards produced. (Economists call this rate of change the *marginal cost*. This idea is discussed in more detail in Sections 3.3 and 4.8.)

Because

$$f'(x) = \lim_{\Delta x \to 0} \frac{\Delta C}{\Delta x}$$

the units for $f'(x)$ are the same as the units for the difference quotient $\Delta C/\Delta x$. Since ΔC is measured in dollars and Δx in yards, it follows that the units for $f'(x)$ are dollars per yard.

(b) The statement that $f'(1000) = 9$ means that, after 1000 yards of fabric have been manufactured, the rate at which the production cost is increasing is \$9/yard. (When $x = 1000$, C is increasing 9 times as fast as x.)

Since $\Delta x = 1$ is small compared with $x = 1000$, we could use the approximation

|||| Here we are assuming that the cost function is well behaved; in other words, $C(x)$ doesn't oscillate rapidly near $x = 1000$.

$$f'(1000) \approx \frac{\Delta C}{\Delta x} = \frac{\Delta C}{1} = \Delta C$$

and say that the cost of manufacturing the 1000th yard (or the 1001st) is about \$9.

(c) The rate at which the production cost is increasing (per yard) is probably lower when $x = 500$ than when $x = 50$ (the cost of making the 500th yard is less than the cost of the 50th yard) because of economies of scale. (The manufacturer makes more efficient use of the fixed costs of production.) So

$$f'(50) > f'(500)$$

But, as production expands, the resulting large-scale operation might become inefficient and there might be overtime costs. Thus, it is possible that the rate of increase of costs will eventually start to rise. So it may happen that

$$f'(5000) > f'(500)$$

The following example shows how to estimate the derivative of a tabular function, that is, a function defined not by a formula but by a table of values.

t	$D(t)$
1980	930.2
1985	1945.9
1990	3233.3
1995	4974.0
2000	5674.2

EXAMPLE 6 Let $D(t)$ be the U.S. national debt at time t. The table in the margin gives approximate values of this function by providing end of year estimates, in billions of dollars, from 1980 to 2000. Interpret and estimate the value of $D'(1990)$.

SOLUTION The derivative $D'(1990)$ means the rate of change of D with respect to t when $t = 1990$, that is, the rate of increase of the national debt in 1990.

According to Equation 3,

$$D'(1990) = \lim_{t \to 1990} \frac{D(t) - D(1990)}{t - 1990}$$

So we compute and tabulate values of the difference quotient (the average rates of change) as follows.

t	$\frac{D(t) - D(1990)}{t - 1990}$
1980	230.31
1985	257.48
1995	348.14
2000	244.09

|||| Another method is to plot the debt function and estimate the slope of the tangent line when $t = 1990$. (See Example 5 in Section 2.7.)

From this table we see that $D'(1990)$ lies somewhere between 257.48 and 348.14 billion dollars per year. [Here we are making the reasonable assumption that the debt didn't fluctuate wildly between 1980 and 2000.] We estimate that the rate of increase of the national debt of the United States in 1990 was the average of these two numbers, namely

$$D'(1990) \approx 303 \text{ billion dollars per year}$$

2.8 Exercises

1. On the given graph of f, mark lengths that represent $f(2)$, $f(2+h)$, $f(2+h)-f(2)$, and h. (Choose $h>0$.) What line has slope $\dfrac{f(2+h)-f(2)}{h}$?

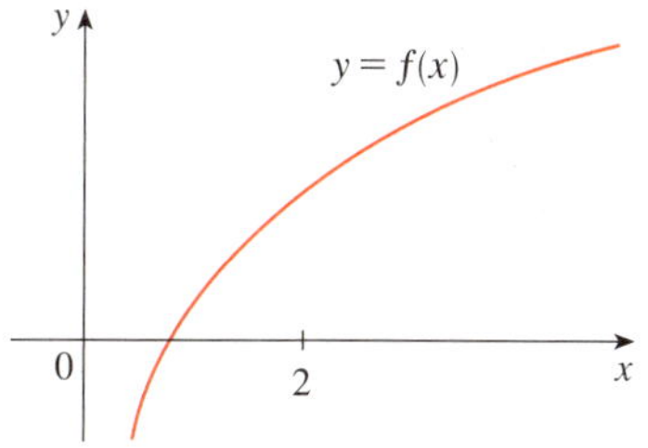

2. For the function f whose graph is shown in Exercise 1, arrange the following numbers in increasing order and explain your reasoning:

$$0 \qquad f'(2) \qquad f(3)-f(2) \qquad \tfrac{1}{2}[f(4)-f(2)]$$

3. For the function g whose graph is given, arrange the following numbers in increasing order and explain your reasoning:

$$0 \qquad g'(-2) \qquad g'(0) \qquad g'(2) \qquad g'(4)$$

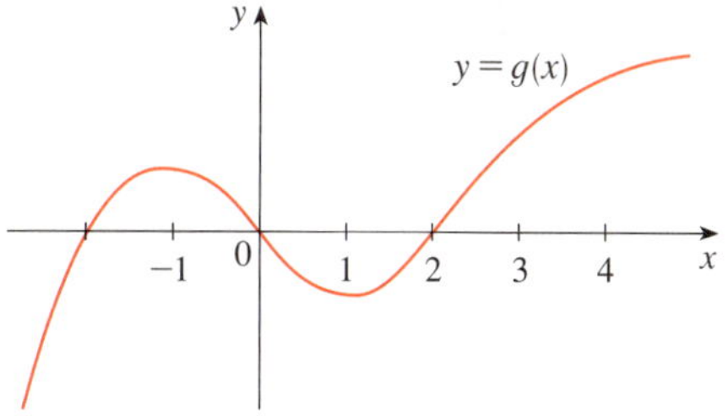

4. If the tangent line to $y=f(x)$ at $(4,3)$ passes through the point $(0,2)$, find $f(4)$ and $f'(4)$.

5. Sketch the graph of a function f for which $f(0)=0$, $f'(0)=3$, $f'(1)=0$, and $f'(2)=-1$.

6. Sketch the graph of a function g for which $g(0)=0$, $g'(0)=3$, $g'(1)=0$, and $g'(2)=1$.

7. If $f(x)=3x^2-5x$, find $f'(2)$ and use it to find an equation of the tangent line to the parabola $y=3x^2-5x$ at the point $(2,2)$.

8. If $g(x)=1-x^3$, find $g'(0)$ and use it to find an equation of the tangent line to the curve $y=1-x^3$ at the point $(0,1)$.

9. (a) If $F(x)=x^3-5x+1$, find $F'(1)$ and use it to find an equation of the tangent line to the curve $y=x^3-5x+1$ at the point $(1,-3)$.
(b) Illustrate part (a) by graphing the curve and the tangent line on the same screen.

10. (a) If $G(x)=x/(1+2x)$, find $G'(a)$ and use it to find an equation of the tangent line to the curve $y=x/(1+2x)$ at the point $\left(-\frac{1}{4},-\frac{1}{2}\right)$.
(b) Illustrate part (a) by graphing the curve and the tangent line on the same screen.

11. Let $f(x)=3^x$. Estimate the value of $f'(1)$ in two ways:
(a) By using Definition 2 and taking successively smaller values of h.
(b) By zooming in on the graph of $y=3^x$ and estimating the slope.

12. Let $g(x)=\tan x$. Estimate the value of $g'(\pi/4)$ in two ways:
(a) By using Definition 2 and taking successively smaller values of h.
(b) By zooming in on the graph of $y=\tan x$ and estimating the slope.

13–18 Find $f'(a)$.

13. $f(x)=3-2x+4x^2$

14. $f(t)=t^4-5t$

15. $f(t)=\dfrac{2t+1}{t+3}$

16. $f(x)=\dfrac{x^2+1}{x-2}$

17. $f(x)=\dfrac{1}{\sqrt{x+2}}$

18. $f(x)=\sqrt{3x+1}$

19–24 Each limit represents the derivative of some function f at some number a. State such an f and a in each case.

19. $\displaystyle\lim_{h\to 0}\frac{(1+h)^{10}-1}{h}$

20. $\displaystyle\lim_{h\to 0}\frac{\sqrt[4]{16+h}-2}{h}$

21. $\displaystyle\lim_{x\to 5}\frac{2^x-32}{x-5}$

22. $\displaystyle\lim_{x\to \pi/4}\frac{\tan x-1}{x-\pi/4}$

23. $\displaystyle\lim_{h\to 0}\frac{\cos(\pi+h)+1}{h}$

24. $\displaystyle\lim_{t\to 1}\frac{t^4+t-2}{t-1}$

25–26 A particle moves along a straight line with equation of motion $s=f(t)$, where s is measured in meters and t in seconds. Find the velocity when $t=2$.

25. $f(t)=t^2-6t-5$

26. $f(t)=2t^3-t+1$

27. The cost of producing x ounces of gold from a new gold mine is $C=f(x)$ dollars.
(a) What is the meaning of the derivative $f'(x)$? What are its units?
(b) What does the statement $f'(800)=17$ mean?
(c) Do you think the values of $f'(x)$ will increase or decrease in the short term? What about the long term? Explain.

28. The number of bacteria after t hours in a controlled laboratory experiment is $n = f(t)$.
(a) What is the meaning of the derivative $f'(5)$? What are its units?
(b) Suppose there is an unlimited amount of space and nutrients for the bacteria. Which do you think is larger, $f'(5)$ or $f'(10)$? If the supply of nutrients is limited, would that affect your conclusion? Explain.

29. The fuel consumption (measured in gallons per hour) of a car traveling at a speed of v miles per hour is $c = f(v)$.
(a) What is the meaning of the derivative $f'(v)$? What are its units?
(b) Write a sentence (in layman's terms) that explains the meaning of the equation $f'(20) = -0.05$.

30. The quantity (in pounds) of a gourmet ground coffee that is sold by a coffee company at a price of p dollars per pound is $Q = f(p)$.
(a) What is the meaning of the derivative $f'(8)$? What are its units?
(b) Is $f'(8)$ positive or negative? Explain.

31. Let $T(t)$ be the temperature (in °F) in Dallas t hours after midnight on June 2, 2001. The table shows values of this function recorded every two hours. What is the meaning of $T'(10)$? Estimate its value.

t	0	2	4	6	8	10	12	14
T	73	73	70	69	72	81	88	91

32. Life expectancy improved dramatically in the 20th century. The table gives values of $E(t)$, the life expectancy at birth (in years) of a male born in the year t in the United States. Interpret and estimate the values of $E'(1910)$ and $E'(1950)$.

t	$E(t)$	t	$E(t)$
1900	48.3	1960	66.6
1910	51.1	1970	67.1
1920	55.2	1980	70.0
1930	57.4	1990	71.8
1940	62.5	2000	74.1
1950	65.6		

33. The quantity of oxygen that can dissolve in water depends on the temperature of the water. (So thermal pollution influences the oxygen content of water.) The graph shows how oxygen solubility S varies as a function of the water temperature T.
(a) What is the meaning of the derivative $S'(T)$? What are its units?
(b) Estimate the value of $S'(16)$ and interpret it.

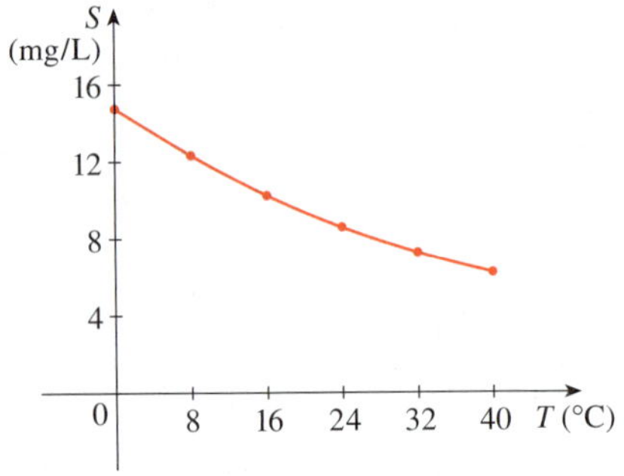

34. The graph shows the influence of the temperature T on the maximum sustainable swimming speed S of Coho salmon.
(a) What is the meaning of the derivative $S'(T)$? What are its units?
(b) Estimate the values of $S'(15)$ and $S'(25)$ and interpret them.

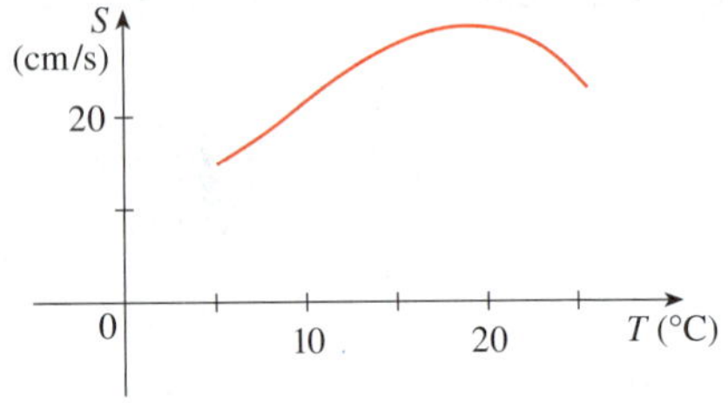

35–36 ||| Determine whether $f'(0)$ exists.

35. $f(x) = \begin{cases} x \sin \dfrac{1}{x} & \text{if } x \neq 0 \\ 0 & \text{if } x = 0 \end{cases}$

36. $f(x) = \begin{cases} x^2 \sin \dfrac{1}{x} & \text{if } x \neq 0 \\ 0 & \text{if } x = 0 \end{cases}$

WRITING PROJECT

Early Methods for Finding Tangents

The first person to formulate explicitly the ideas of limits and derivatives was Sir Isaac Newton in the 1660s. But Newton acknowledged that "If I have seen further than other men, it is because I have stood on the shoulders of giants." Two of those giants were Pierre Fermat (1601–1665) and Newton's teacher at Cambridge, Isaac Barrow (1630–1677). Newton was familiar with the methods that these men used to find tangent lines, and their methods played a role in Newton's eventual formulation of calculus.

The following references contain explanations of these methods. Read one or more of the references and write a report comparing the methods of either Fermat or Barrow to modern methods. In particular, use the method of Section 2.8 to find an equation of the tangent line to the curve $y = x^3 + 2x$ at the point (1, 3) and show how either Fermat or Barrow would have solved the same problem. Although you used derivatives and they did not, point out similarities between the methods.

1. Carl Boyer and Uta Merzbach, *A History of Mathematics* (New York: Wiley, 1989), pp. 389, 432.

2. C. H. Edwards, *The Historical Development of the Calculus* (New York: Springer-Verlag, 1979), pp. 124, 132.

3. Howard Eves, *An Introduction to the History of Mathematics*, 6th ed. (New York: Saunders, 1990), pp. 391, 395.

4. Morris Kline, *Mathematical Thought from Ancient to Modern Times* (New York: Oxford University Press, 1972), pp. 344, 346.

2.9 The Derivative as a Function

In the preceding section we considered the derivative of a function f at a fixed number a:

1
$$f'(a) = \lim_{h \to 0} \frac{f(a + h) - f(a)}{h}$$

Here we change our point of view and let the number a vary. If we replace a in Equation 1 by a variable x, we obtain

2
$$f'(x) = \lim_{h \to 0} \frac{f(x + h) - f(x)}{h}$$

Given any number x for which this limit exists, we assign to x the number $f'(x)$. So we can regard f' as a new function, called the **derivative of f** and defined by Equation 2. We know that the value of f' at x, $f'(x)$, can be interpreted geometrically as the slope of the tangent line to the graph of f at the point $(x, f(x))$.

The function f' is called the derivative of f because it has been "derived" from f by the limiting operation in Equation 2. The domain of f' is the set $\{x \mid f'(x) \text{ exists}\}$ and may be smaller than the domain of f.

EXAMPLE 1 The graph of a function f is given in Figure 1. Use it to sketch the graph of the derivative f'.

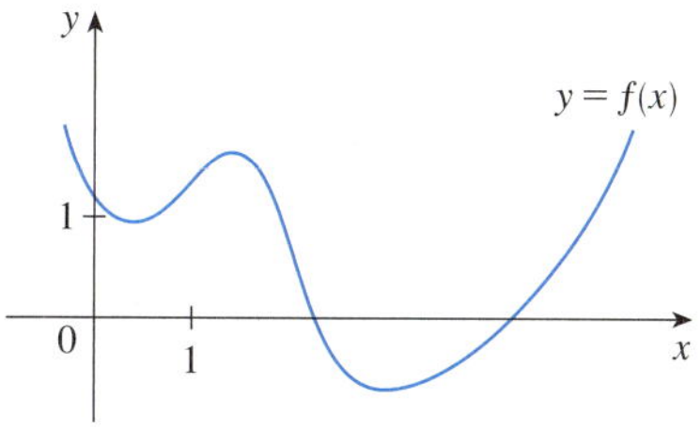

FIGURE 1

Watch an animation of the relation between a function and its derivative.
Resources / Module 3
/ Derivatives as Functions
/ Mars Rover
Resources / Module 3
/ Slope-a-Scope
/ Derivative of a Cubic

SOLUTION We can estimate the value of the derivative at any value of x by drawing the tangent at the point $(x, f(x))$ and estimating its slope. For instance, for $x = 5$ we draw the tangent at P in Figure 2(a) and estimate its slope to be about $\frac{3}{2}$, so $f'(5) \approx 1.5$. This allows us to plot the point $P'(5, 1.5)$ on the graph of f' directly beneath P. Repeating this procedure at several points, we get the graph shown in Figure 2(b). Notice that the tangents at A, B, and C are horizontal, so the derivative is 0 there and the graph of f' crosses the x-axis at the points A', B', and C', directly beneath A, B, and C. Between A and B the tangents have positive slope, so $f'(x)$ is positive there. But between B and C the tangents have negative slope, so $f'(x)$ is negative there.

|||| Notice that where the derivative is positive (to the right of C and between A and B), the function f is increasing. Where $f'(x)$ is negative (to the left of A and between B and C), f is decreasing. In Section 4.3 we will prove that this is true for all functions.

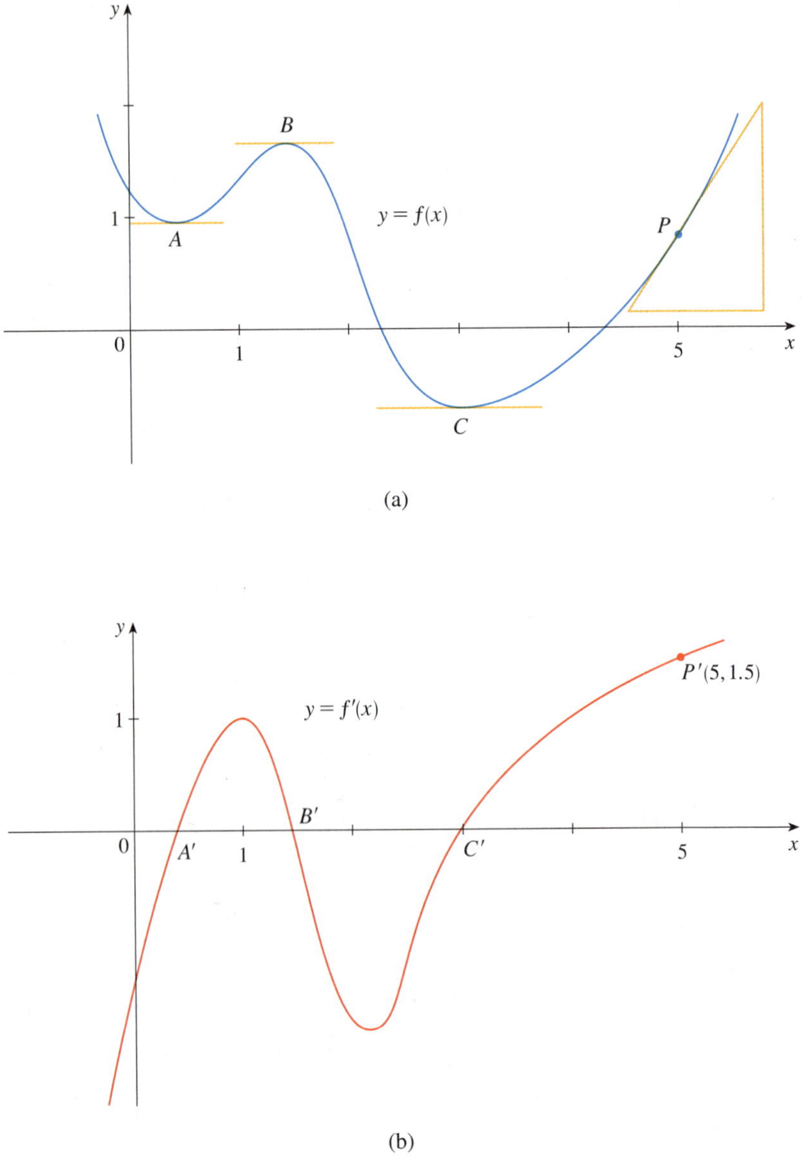

FIGURE 2

If a function is defined by a table of values, then we can construct a table of approximate values of its derivative, as in the next example.

t	$B(t)$
1980	9,847
1982	9,856
1984	9,855
1986	9,862
1988	9,884
1990	9,962
1992	10,036
1994	10,109
1996	10,152
1998	10,175
2000	10,186

EXAMPLE 2 Let $B(t)$ be the population of Belgium at time t. The table at the left gives midyear values of $B(t)$, in thousands, from 1980 to 2000. Construct a table of values for the derivative of this function.

SOLUTION We assume that there were no wild fluctuations in the population between the stated values. Let's start by approximating $B'(1988)$, the rate of increase of the population of Belgium in mid-1988. Since

$$B'(1988) = \lim_{h \to 0} \frac{B(1988 + h) - B(1988)}{h}$$

we have

$$B'(1988) \approx \frac{B(1988 + h) - B(1988)}{h}$$

for small values of h.

For $h = 2$, we get

$$B'(1988) \approx \frac{B(1990) - B(1988)}{2} = \frac{9962 - 9884}{2} = 39$$

(This is the average rate of increase between 1988 and 1990.) For $h = -2$, we have

$$B'(1988) \approx \frac{B(1986) - B(1988)}{-2} = \frac{9862 - 9884}{-2} = 11$$

which is the average rate of increase between 1986 and 1988. We get a more accurate approximation if we take the average of these rates of change:

$$B'(1988) \approx \tfrac{1}{2}(39 + 11) = 25$$

This means that in 1988 the population was increasing at a rate of about 25,000 people per year.

Making similar calculations for the other values (except at the endpoints), we get the table at the left, which shows the approximate values for the derivative.

t	$B'(t)$
1980	4.5
1982	2.0
1984	1.5
1986	7.3
1988	25.0
1990	38.0
1992	36.8
1994	29.0
1996	16.5
1998	8.5
2000	5.5

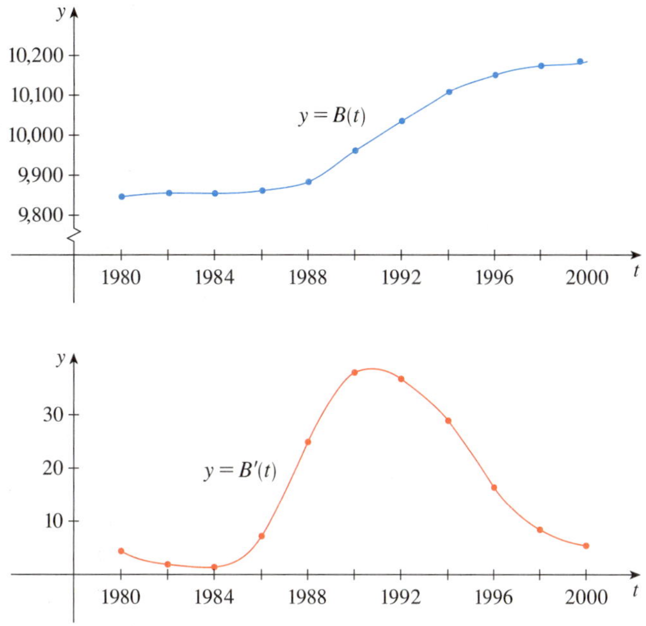

|||| Figure 3 illustrates Example 2 by showing graphs of the population function $B(t)$ and its derivative $B'(t)$. Notice how the rate of population growth increases to a maximum in 1990 and decreases thereafter.

FIGURE 3

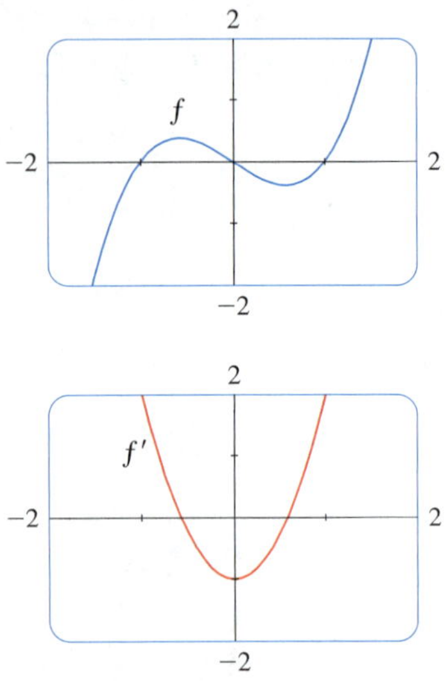

FIGURE 4

EXAMPLE 3

(a) If $f(x) = x^3 - x$, find a formula for $f'(x)$.

(b) Illustrate by comparing the graphs of f and f'.

SOLUTION

(a) When using Equation 2 to compute a derivative, we must remember that the variable is h and that x is temporarily regarded as a constant during the calculation of the limit.

$$\begin{aligned} f'(x) &= \lim_{h\to 0} \frac{f(x+h) - f(x)}{h} = \lim_{h\to 0} \frac{[(x+h)^3 - (x+h)] - [x^3 - x]}{h} \\ &= \lim_{h\to 0} \frac{x^3 + 3x^2h + 3xh^2 + h^3 - x - h - x^3 + x}{h} \\ &= \lim_{h\to 0} \frac{3x^2h + 3xh^2 + h^3 - h}{h} \\ &= \lim_{h\to 0} (3x^2 + 3xh + h^2 - 1) = 3x^2 - 1 \end{aligned}$$

(b) We use a graphing device to graph f and f' in Figure 4. Notice that $f'(x) = 0$ when f has horizontal tangents and $f'(x)$ is positive when the tangents have positive slope. So these graphs serve as a check on our work in part (a).

EXAMPLE 4 If $f(x) = \sqrt{x - 1}$, find the derivative of f. State the domain of f'.

See more problems like these.
Resources / Module 3
/ How to Calculate
/ The Essential Examples

SOLUTION

Here we rationalize the numerator.

$$\begin{aligned} f'(x) &= \lim_{h\to 0} \frac{f(x+h) - f(x)}{h} \\ &= \lim_{h\to 0} \frac{\sqrt{x+h-1} - \sqrt{x-1}}{h} \\ &= \lim_{h\to 0} \frac{\sqrt{x+h-1} - \sqrt{x-1}}{h} \cdot \frac{\sqrt{x+h-1} + \sqrt{x-1}}{\sqrt{x+h-1} + \sqrt{x-1}} \\ &= \lim_{h\to 0} \frac{(x+h-1) - (x-1)}{h\left(\sqrt{x+h-1} + \sqrt{x-1}\right)} \\ &= \lim_{h\to 0} \frac{1}{\sqrt{x+h-1} + \sqrt{x-1}} \\ &= \frac{1}{\sqrt{x-1} + \sqrt{x-1}} = \frac{1}{2\sqrt{x-1}} \end{aligned}$$

We see that $f'(x)$ exists if $x > 1$, so the domain of f' is $(1, \infty)$. This is smaller than the domain of f, which is $[1, \infty)$.

Let's check to see that the result of Example 4 is reasonable by looking at the graphs of f and f' in Figure 5. When x is close to 1, $\sqrt{x-1}$ is close to 0, so $f'(x) = 1/\left(2\sqrt{x-1}\right)$ is very large; this corresponds to the steep tangent lines near $(1, 0)$ in Figure 5(a) and the large values of $f'(x)$ just to the right of 1 in Figure 5(b). When x is large, $f'(x)$ is very small; this corresponds to the flatter tangent lines at the far right of the graph of f and the horizontal asymptote of the graph of f'.

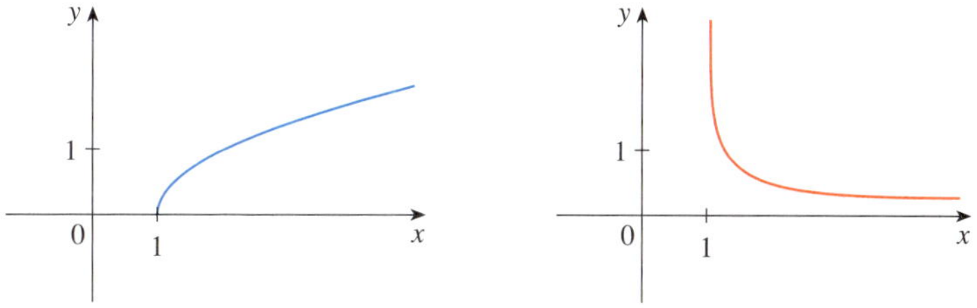

FIGURE 5

(a) $f(x) = \sqrt{x - 1}$

(b) $f'(x) = \dfrac{1}{2\sqrt{x-1}}$

EXAMPLE 5 Find f' if $f(x) = \dfrac{1 - x}{2 + x}$.

SOLUTION

$$\frac{\dfrac{a}{b} - \dfrac{c}{d}}{e} = \frac{ad - bc}{bd} \cdot \frac{1}{e}$$

$$\begin{aligned}
f'(x) &= \lim_{h \to 0} \frac{f(x + h) - f(x)}{h} \\
&= \lim_{h \to 0} \frac{\dfrac{1 - (x + h)}{2 + (x + h)} - \dfrac{1 - x}{2 + x}}{h} \\
&= \lim_{h \to 0} \frac{(1 - x - h)(2 + x) - (1 - x)(2 + x + h)}{h(2 + x + h)(2 + x)} \\
&= \lim_{h \to 0} \frac{(2 - x - 2h - x^2 - xh) - (2 - x + h - x^2 - xh)}{h(2 + x + h)(2 + x)} \\
&= \lim_{h \to 0} \frac{-3h}{h(2 + x + h)(2 + x)} \\
&= \lim_{h \to 0} \frac{-3}{(2 + x + h)(2 + x)} = -\frac{3}{(2 + x)^2}
\end{aligned}$$

Other Notations

If we use the traditional notation $y = f(x)$ to indicate that the independent variable is x and the dependent variable is y, then some common alternative notations for the derivative are as follows:

$$f'(x) = y' = \frac{dy}{dx} = \frac{df}{dx} = \frac{d}{dx}f(x) = Df(x) = D_x f(x)$$

The symbols D and d/dx are called **differentiation operators** because they indicate the operation of **differentiation**, which is the process of calculating a derivative.

The symbol dy/dx, which was introduced by Leibniz, should not be regarded as a ratio (for the time being); it is simply a synonym for $f'(x)$. Nonetheless, it is a very useful and suggestive notation, especially when used in conjunction with increment notation. Referring to Equation 2.8.4, we can rewrite the definition of derivative in Leibniz notation in the form

$$\frac{dy}{dx} = \lim_{\Delta x \to 0} \frac{\Delta y}{\Delta x}$$

|||| Gottfried Wilhelm Leibniz was born in Leipzig in 1646 and studied law, theology, philosophy, and mathematics at the university there, graduating with a bachelor's degree at age 17. After earning his doctorate in law at age 20, Leibniz entered the diplomatic service and spent most of his life traveling to the capitals of Europe on political missions. In particular, he worked to avert a French military threat against Germany and attempted to reconcile the Catholic and Protestant churches.

His serious study of mathematics did not begin until 1672 while he was on a diplomatic mission in Paris. There he built a calculating machine and met scientists, like Huygens, who directed his attention to the latest developments in mathematics and science. Leibniz sought to develop a symbolic logic and system of notation that would simplify logical reasoning. In particular, the version of calculus that he published in 1684 established the notation and the rules for finding derivatives that we use today.

Unfortunately, a dreadful priority dispute arose in the 1690s between the followers of Newton and those of Leibniz as to who had invented calculus first. Leibniz was even accused of plagiarism by members of the Royal Society in England. The truth is that each man invented calculus independently. Newton arrived at his version of calculus first but, because of his fear of controversy, did not publish it immediately. So Leibniz's 1684 account of calculus was the first to be published.

If we want to indicate the value of a derivative dy/dx in Leibniz notation at a specific number a, we use the notation

$$\left.\frac{dy}{dx}\right|_{x=a} \qquad \text{or} \qquad \left.\frac{dy}{dx}\right]_{x=a}$$

which is a synonym for $f'(a)$.

3 Definition A function f is **differentiable at a** if $f'(a)$ exists. It is **differentiable on an open interval** (a, b) [or (a, ∞) or $(-\infty, a)$ or $(-\infty, \infty)$] if it is differentiable at every number in the interval.

EXAMPLE 6 Where is the function $f(x) = |x|$ differentiable?

SOLUTION If $x > 0$, then $|x| = x$ and we can choose h small enough that $x + h > 0$ and hence $|x + h| = x + h$. Therefore, for $x > 0$ we have

$$f'(x) = \lim_{h \to 0} \frac{|x + h| - |x|}{h}$$

$$= \lim_{h \to 0} \frac{(x + h) - x}{h} = \lim_{h \to 0} \frac{h}{h} = \lim_{h \to 0} 1 = 1$$

and so f is differentiable for any $x > 0$.

Similarly, for $x < 0$ we have $|x| = -x$ and h can be chosen small enough that $x + h < 0$ and so $|x + h| = -(x + h)$. Therefore, for $x < 0$,

$$f'(x) = \lim_{h \to 0} \frac{|x + h| - |x|}{h}$$

$$= \lim_{h \to 0} \frac{-(x + h) - (-x)}{h} = \lim_{h \to 0} \frac{-h}{h} = \lim_{h \to 0} (-1) = -1$$

and so f is differentiable for any $x < 0$.

For $x = 0$ we have to investigate

$$f'(0) = \lim_{h \to 0} \frac{f(0 + h) - f(0)}{h}$$

$$= \lim_{h \to 0} \frac{|0 + h| - |0|}{h} \qquad \text{(if it exists)}$$

Let's compute the left and right limits separately:

$$\lim_{h \to 0^+} \frac{|0 + h| - |0|}{h} = \lim_{h \to 0^+} \frac{|h|}{h} = \lim_{h \to 0^+} \frac{h}{h} = \lim_{h \to 0^+} 1 = 1$$

and

$$\lim_{h \to 0^-} \frac{|0 + h| - |0|}{h} = \lim_{h \to 0^-} \frac{|h|}{h} = \lim_{h \to 0^-} \frac{-h}{h} = \lim_{h \to 0^-} (-1) = -1$$

Since these limits are different, $f'(0)$ does not exist. Thus, f is differentiable at all x except 0.

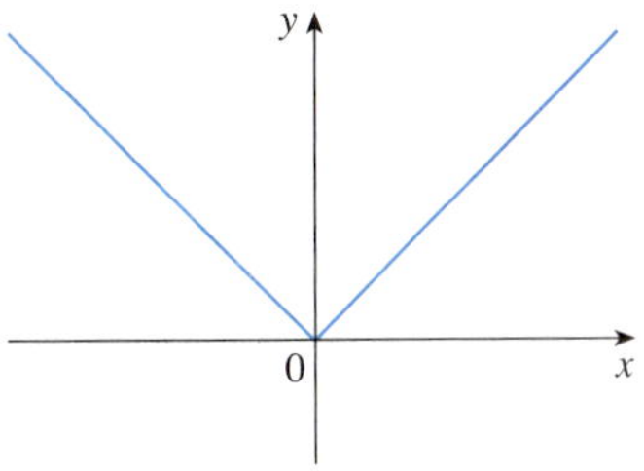

(a) $y = f(x) = |x|$

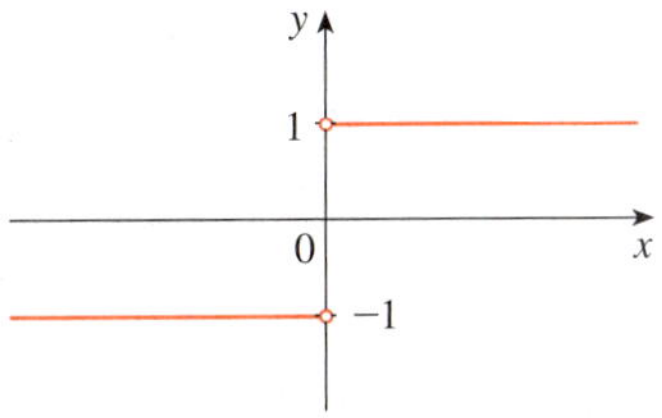

(b) $y = f'(x)$

FIGURE 6

A formula for f' is given by

$$f'(x) = \begin{cases} 1 & \text{if } x > 0 \\ -1 & \text{if } x < 0 \end{cases}$$

and its graph is shown in Figure 6(b). The fact that $f'(0)$ does not exist is reflected geometrically in the fact that the curve $y = |x|$ does not have a tangent line at (0, 0). [See Figure 6(a).]

Both continuity and differentiability are desirable properties for a function to have. The following theorem shows how these properties are related.

4 Theorem If f is differentiable at a, then f is continuous at a.

Proof To prove that f is continuous at a, we have to show that $\lim_{x \to a} f(x) = f(a)$. We do this by showing that the difference $f(x) - f(a)$ approaches 0.

The given information is that f is differentiable at a, that is,

$$f'(a) = \lim_{x \to a} \frac{f(x) - f(a)}{x - a}$$

exists (see Equation 2.8.3). To connect the given and the unknown, we divide and multiply $f(x) - f(a)$ by $x - a$ (which we can do when $x \neq a$):

$$f(x) - f(a) = \frac{f(x) - f(a)}{x - a}(x - a)$$

Thus, using the Product Law and (2.8.3), we can write

$$\begin{aligned} \lim_{x \to a} [f(x) - f(a)] &= \lim_{x \to a} \frac{f(x) - f(a)}{x - a}(x - a) \\ &= \lim_{x \to a} \frac{f(x) - f(a)}{x - a} \lim_{x \to a} (x - a) \\ &= f'(a) \cdot 0 = 0 \end{aligned}$$

To use what we have just proved, we start with $f(x)$ and add and subtract $f(a)$:

$$\begin{aligned} \lim_{x \to a} f(x) &= \lim_{x \to a} [f(a) + (f(x) - f(a))] \\ &= \lim_{x \to a} f(a) + \lim_{x \to a} [f(x) - f(a)] \\ &= f(a) + 0 = f(a) \end{aligned}$$

Therefore, f is continuous at a.

NOTE ▫ The converse of Theorem 4 is false; that is, there are functions that are continuous but not differentiable. For instance, the function $f(x) = |x|$ is continuous at 0 because

$$\lim_{x \to 0} f(x) = \lim_{x \to 0} |x| = 0 = f(0)$$

(See Example 7 in Section 2.3.) But in Example 6 we showed that f is not differentiable at 0.

How Can a Function Fail to Be Differentiable?

We saw that the function $y = |x|$ in Example 6 is not differentiable at 0 and Figure 6(a) shows that its graph changes direction abruptly when $x = 0$. In general, if the graph of a function f has a "corner" or "kink" in it, then the graph of f has no tangent at this point and f is not differentiable there. [In trying to compute $f'(a)$, we find that the left and right limits are different.]

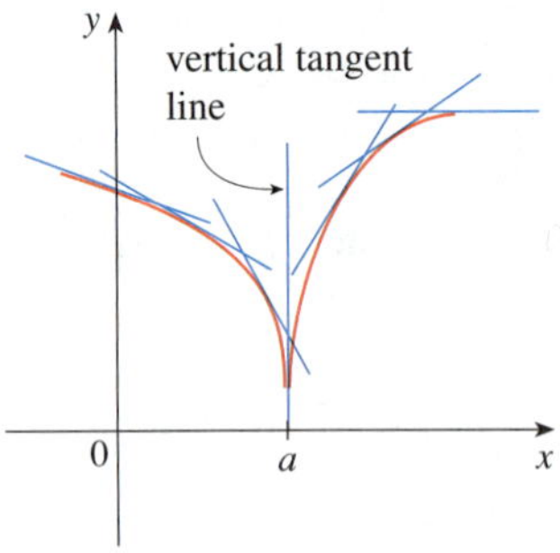

FIGURE 7

Theorem 4 gives another way for a function not to have a derivative. It says that if f is not continuous at a, then f is not differentiable at a. So at any discontinuity (for instance, a jump discontinuity) f fails to be differentiable.

A third possibility is that the curve has a **vertical tangent line** when $x = a$; that is, f is continuous at a and

$$\lim_{x \to a} |f'(x)| = \infty$$

This means that the tangent lines become steeper and steeper as $x \to a$. Figure 7 shows one way that this can happen; Figure 8(c) shows another. Figure 8 illustrates the three possibilities that we have discussed.

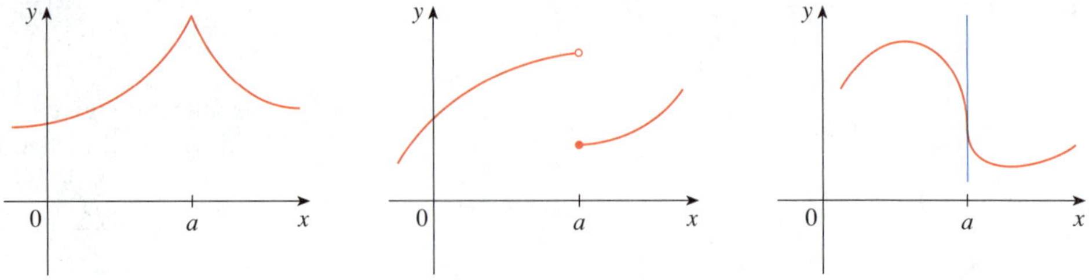

FIGURE 8
Three ways for f not to be differentiable at a

(a) A corner (b) A discontinuity (c) A vertical tangent

A graphing calculator or computer provides another way of looking at differentiability. If f is differentiable at a, then when we zoom in toward the point $(a, f(a))$ the graph straightens out and appears more and more like a line. (See Figure 9. We saw a specific example of this in Figure 3 in Section 2.8.) But no matter how much we zoom in toward a point like the ones in Figures 7 and 8(a), we can't eliminate the sharp point or corner (see Figure 10).

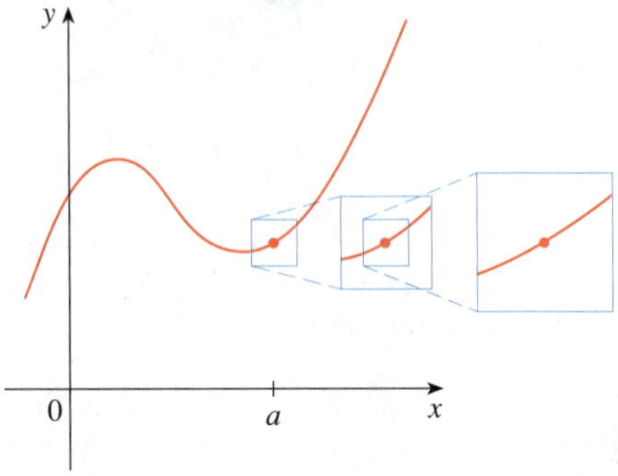

FIGURE 9
f is differentiable at a.

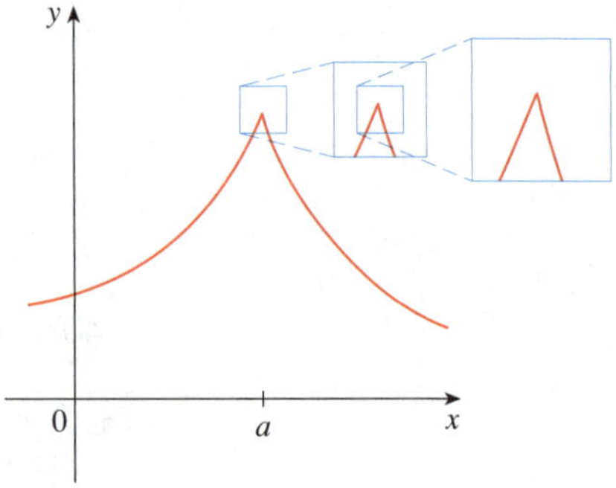

FIGURE 10
f is not differentiable at a.

2.9 Exercises

1–3 Use the given graph to estimate the value of each derivative. Then sketch the graph of f'.

1. (a) $f'(1)$
(b) $f'(2)$
(c) $f'(3)$
(d) $f'(4)$

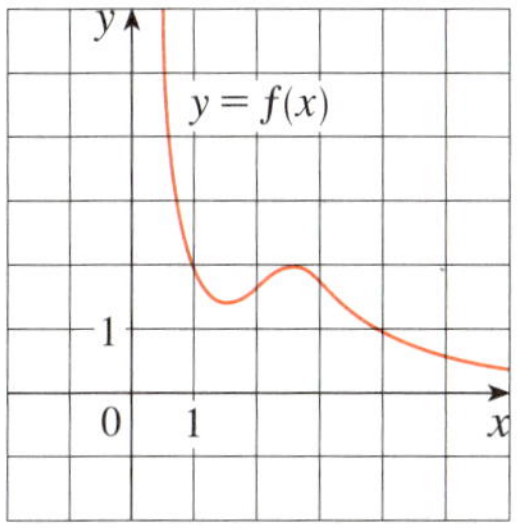

2. (a) $f'(0)$
(b) $f'(1)$
(c) $f'(2)$
(d) $f'(3)$
(e) $f'(4)$
(f) $f'(5)$

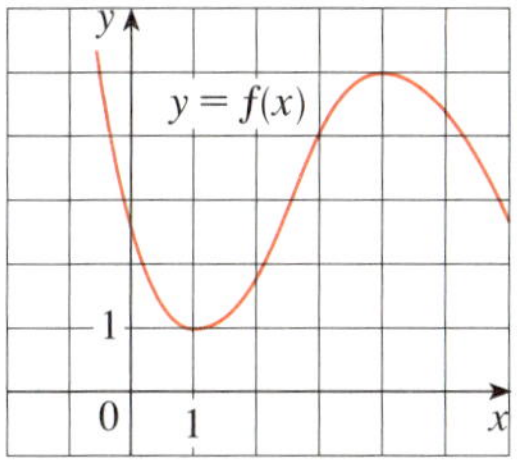

3. (a) $f'(-3)$
(b) $f'(-2)$
(c) $f'(-1)$
(d) $f'(0)$
(e) $f'(1)$
(f) $f'(2)$
(g) $f'(3)$

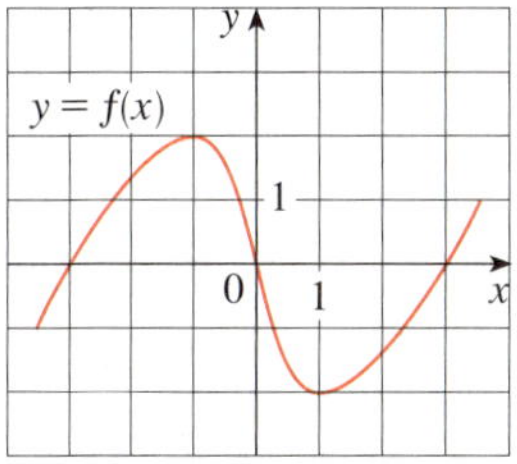

4. Match the graph of each function in (a)–(d) with the graph of its derivative in I–IV. Give reasons for your choices.

(a)
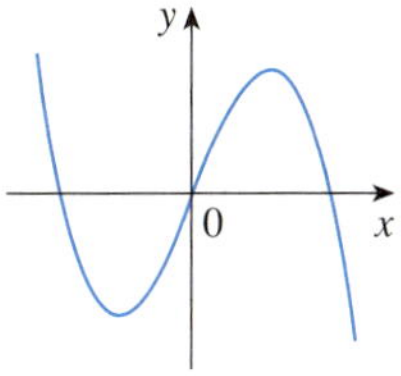

(b)
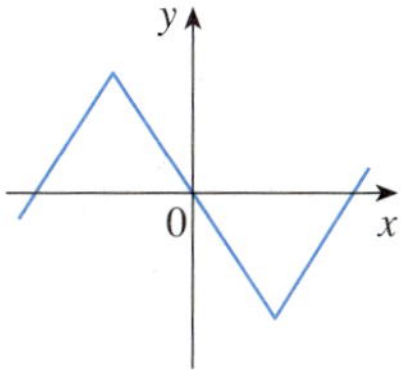

(c)
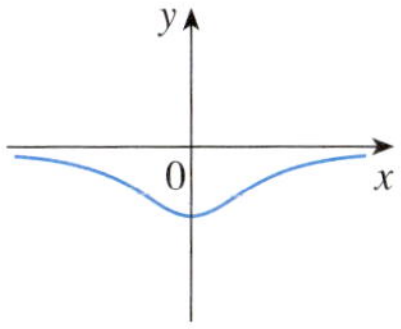

(d)
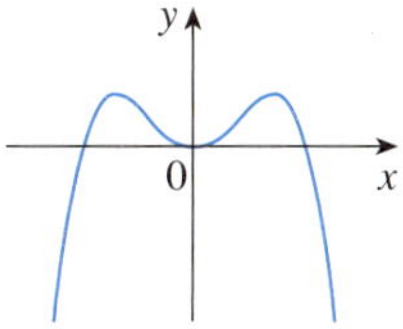

I
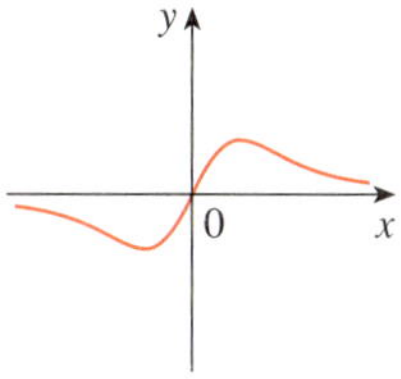

II
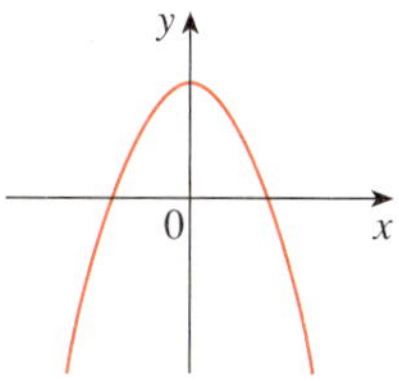

III
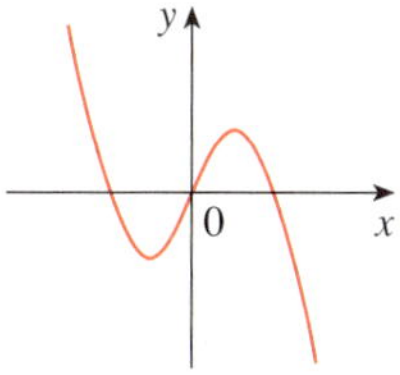

IV
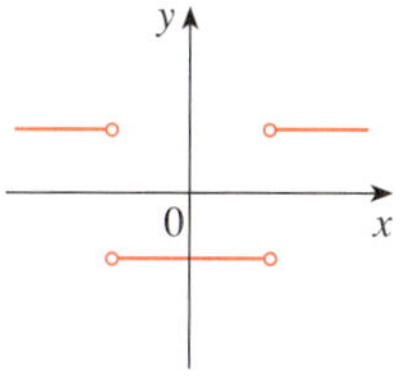

5–13 Trace or copy the graph of the given function f. (Assume that the axes have equal scales.) Then use the method of Example 1 to sketch the graph of f' below it.

5.
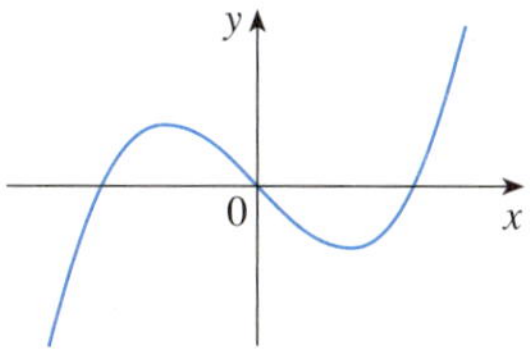

6.
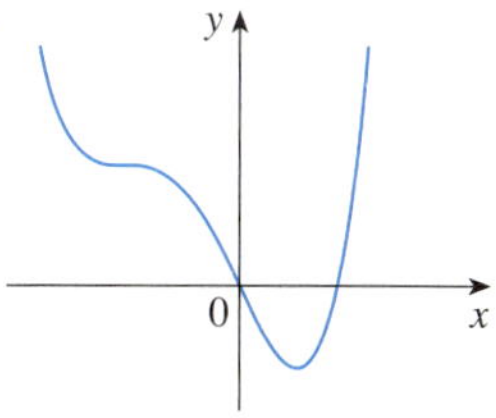

7.
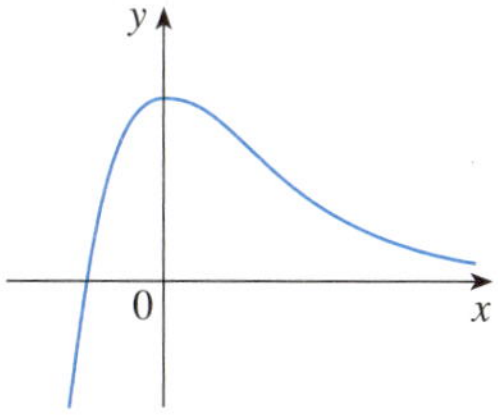

8.
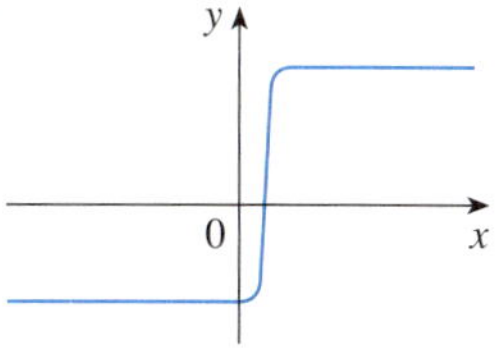

9.
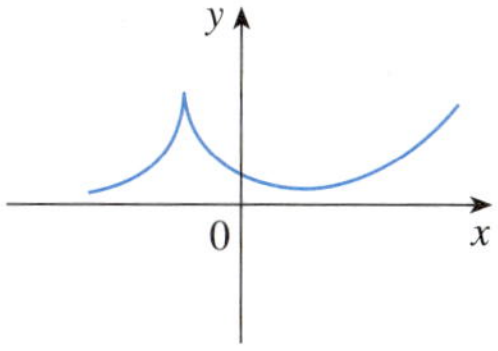

10.
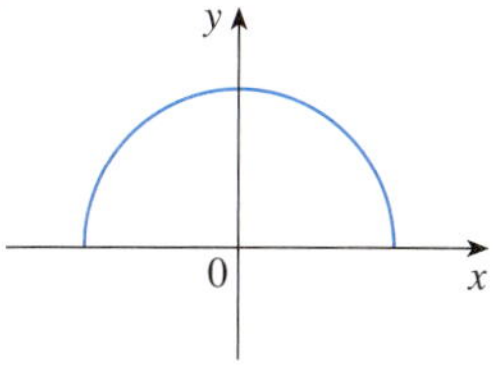

11.
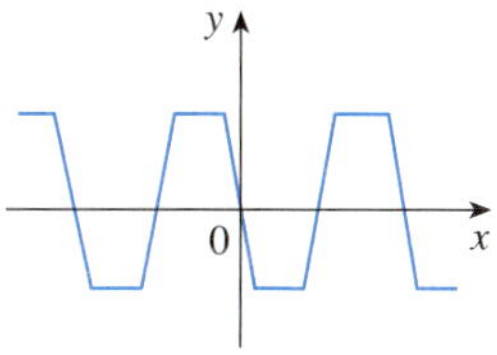

12.
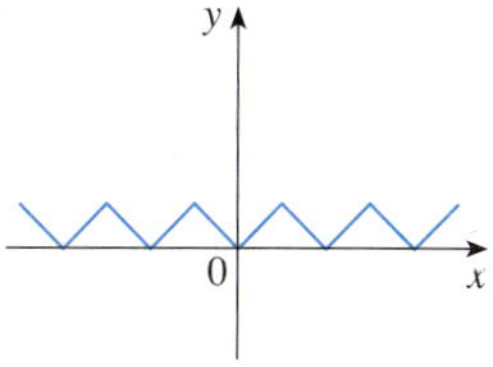

13.

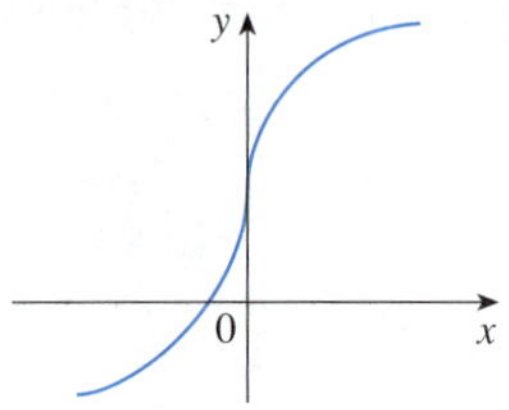

14. Shown is the graph of the population function $P(t)$ for yeast cells in a laboratory culture. Use the method of Example 1 to graph the derivative $P'(t)$. What does the graph of P' tell us about the yeast population?

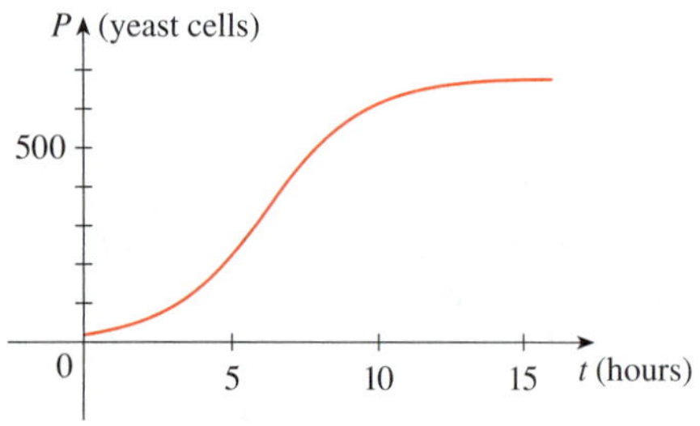

15. The graph shows how the average age of first marriage of Japanese men varied in the last half of the 20th century. Sketch the graph of the derivative function $M'(t)$. During which years was the derivative negative?

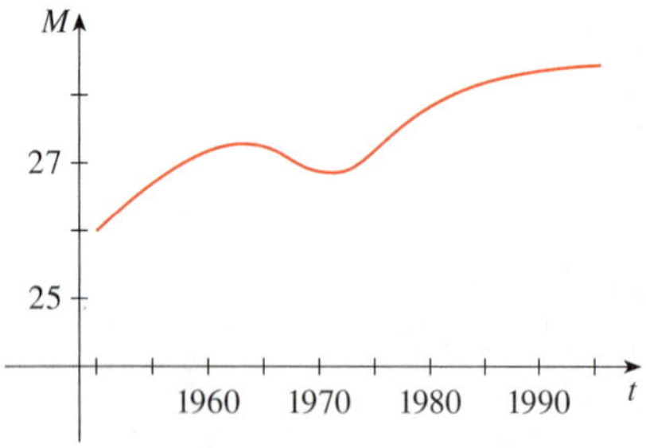

16–18 Make a careful sketch of the graph of f and below it sketch the graph of f' in the same manner as in Exercises 5–13. Can you guess a formula for $f'(x)$ from its graph?

16. $f(x) = \sin x$

17. $f(x) = e^x$

18. $f(x) = \ln x$

19. Let $f(x) = x^2$.

(a) Estimate the values of $f'(0)$, $f'(\frac{1}{2})$, $f'(1)$, and $f'(2)$ by using a graphing device to zoom in on the graph of f.

(b) Use symmetry to deduce the values of $f'(-\frac{1}{2})$, $f'(-1)$, and $f'(-2)$.

(c) Use the results from parts (a) and (b) to guess a formula for $f'(x)$.

(d) Use the definition of a derivative to prove that your guess in part (c) is correct.

20. Let $f(x) = x^3$.

(a) Estimate the values of $f'(0)$, $f'(\frac{1}{2})$, $f'(1)$, $f'(2)$, and $f'(3)$ by using a graphing device to zoom in on the graph of f.

(b) Use symmetry to deduce the values of $f'(-\frac{1}{2})$, $f'(-1)$, $f'(-2)$, and $f'(-3)$.

(c) Use the values from parts (a) and (b) to graph f'.

(d) Guess a formula for $f'(x)$.

(e) Use the definition of a derivative to prove that your guess in part (d) is correct.

21–31 Find the derivative of the function using the definition of derivative. State the domain of the function and the domain of its derivative.

21. $f(x) = 37$

22. $f(x) = 12 + 7x$

23. $f(x) = 1 - 3x^2$

24. $f(x) = 5x^2 + 3x - 2$

25. $f(x) = x^3 - 3x + 5$

26. $f(x) = x + \sqrt{x}$

27. $g(x) = \sqrt{1 + 2x}$

28. $f(x) = \dfrac{3 + x}{1 - 3x}$

29. $G(t) = \dfrac{4t}{t + 1}$

30. $g(x) = \dfrac{1}{x^2}$

31. $f(x) = x^4$

32. (a) Sketch the graph of $f(x) = \sqrt{6 - x}$ by starting with the graph of $y = \sqrt{x}$ and using the transformations of Section 1.3.

(b) Use the graph from part (a) to sketch the graph of f'.

(c) Use the definition of a derivative to find $f'(x)$. What are the domains of f and f'?

(d) Use a graphing device to graph f' and compare with your sketch in part (b).

33. (a) If $f(x) = x - (2/x)$, find $f'(x)$.

(b) Check to see that your answer to part (a) is reasonable by comparing the graphs of f and f'.

34. (a) If $f(t) = 6/(1 + t^2)$, find $f'(t)$.

(b) Check to see that your answer to part (a) is reasonable by comparing the graphs of f and f'.

35. The unemployment rate $U(t)$ varies with time. The table (from the Bureau of Labor Statistics) gives the percentage of unemployed in the U.S. labor force from 1991 to 2000.

t	$U(t)$	t	$U(t)$
1991	6.8	1996	5.4
1992	7.5	1997	4.9
1993	6.9	1998	4.5
1994	6.1	1999	4.2
1995	5.6	2000	4.0

(a) What is the meaning of $U'(t)$? What are its units?

(b) Construct a table of values for $U'(t)$.

36. Let $P(t)$ be the percentage of Americans under the age of 18 at time t. The table gives values of this function in census years from 1950 to 2000.

t	$P(t)$	t	$P(t)$
1950	31.1	1980	28.0
1960	35.7	1990	25.7
1970	34.0	2000	25.7

(a) What is the meaning of $P'(t)$? What are its units?
(b) Construct a table of values for $P'(t)$.
(c) Graph P and P'.
(d) How would it be possible to get more accurate values for $P'(t)$?

37. The graph of f is given. State, with reasons, the numbers at which f is not differentiable.

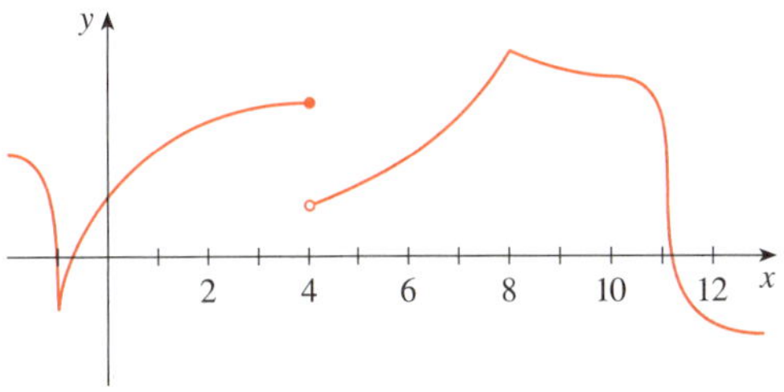

38. The graph of g is given.
(a) At what numbers is g discontinuous? Why?
(b) At what numbers is g not differentiable? Why?

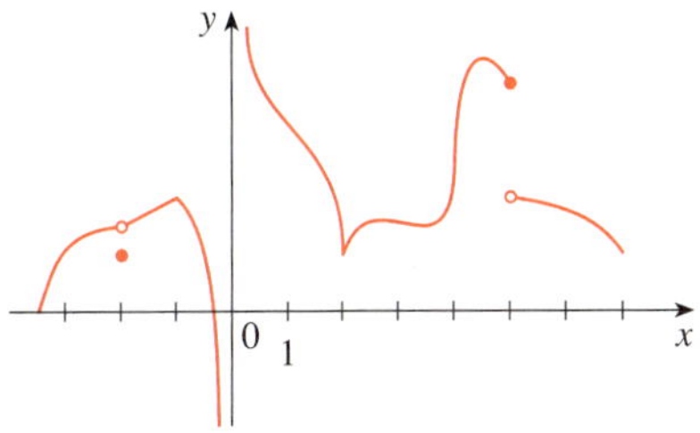

39. Graph the function $f(x) = x + \sqrt{|x|}$. Zoom in repeatedly, first toward the point $(-1, 0)$ and then toward the origin. What is different about the behavior of f in the vicinity of these two points? What do you conclude about the differentiability of f?

40. Zoom in toward the points $(1, 0)$, $(0, 1)$, and $(-1, 0)$ on the graph of the function $g(x) = (x^2 - 1)^{2/3}$. What do you notice? Account for what you see in terms of the differentiability of g.

41. Let $f(x) = \sqrt[3]{x}$.
(a) If $a \neq 0$, use Equation 2.8.3 to find $f'(a)$.
(b) Show that $f'(0)$ does not exist.
(c) Show that $y = \sqrt[3]{x}$ has a vertical tangent line at $(0, 0)$. (Recall the shape of the graph of f. See Figure 13 in Section 1.2.)

42. (a) If $g(x) = x^{2/3}$, show that $g'(0)$ does not exist.
(b) If $a \neq 0$, find $g'(a)$.
(c) Show that $y = x^{2/3}$ has a vertical tangent line at $(0, 0)$.
(d) Illustrate part (c) by graphing $y = x^{2/3}$.

43. Show that the function $f(x) = |x - 6|$ is not differentiable at 6. Find a formula for f' and sketch its graph.

44. Where is the greatest integer function $f(x) = [\![x]\!]$ not differentiable? Find a formula for f' and sketch its graph.

45. (a) Sketch the graph of the function $f(x) = x|x|$.
(b) For what values of x is f differentiable?
(c) Find a formula for f'.

46. The **left-hand** and **right-hand derivatives** of f at a are defined by

$$f'_-(a) = \lim_{h \to 0^-} \frac{f(a + h) - f(a)}{h}$$

and
$$f'_+(a) = \lim_{h \to 0^+} \frac{f(a + h) - f(a)}{h}$$

if these limits exist. Then $f'(a)$ exists if and only if these one-sided derivatives exist and are equal.
(a) Find $f'_-(4)$ and $f'_+(4)$ for the function

$$f(x) = \begin{cases} 0 & \text{if } x \leqslant 0 \\ 5 - x & \text{if } 0 < x < 4 \\ \dfrac{1}{5 - x} & \text{if } x \geqslant 4 \end{cases}$$

(b) Sketch the graph of f.
(c) Where is f discontinuous?
(d) Where is f not differentiable?

47. Recall that a function f is called *even* if $f(-x) = f(x)$ for all x in its domain and *odd* if $f(-x) = -f(x)$ for all such x. Prove each of the following.
(a) The derivative of an even function is an odd function.
(b) The derivative of an odd function is an even function.

48. When you turn on a hot-water faucet, the temperature T of the water depends on how long the water has been running.
(a) Sketch a possible graph of T as a function of the time t that has elapsed since the faucet was turned on.
(b) Describe how the rate of change of T with respect to t varies as t increases.
(c) Sketch a graph of the derivative of T.

49. Let ℓ be the tangent line to the parabola $y = x^2$ at the point $(1, 1)$. The *angle of inclination* of ℓ is the angle ϕ that ℓ makes with the positive direction of the x-axis. Calculate ϕ correct to the nearest degree.

2 Review

CONCEPT CHECK

1. Explain what each of the following means and illustrate with a sketch.
   (a) $\lim_{x \to a} f(x) = L$
   (b) $\lim_{x \to a^+} f(x) = L$
   (c) $\lim_{x \to a^-} f(x) = L$
   (d) $\lim_{x \to a} f(x) = \infty$
   (e) $\lim_{x \to \infty} f(x) = L$

2. Describe several ways in which a limit can fail to exist. Illustrate with sketches.

3. State the following Limit Laws.
   (a) Sum Law
   (b) Difference Law
   (c) Constant Multiple Law
   (d) Product Law
   (e) Quotient Law
   (f) Power Law
   (g) Root Law

4. What does the Squeeze Theorem say?

5. (a) What does it mean to say that the line $x = a$ is a vertical asymptote of the curve $y = f(x)$? Draw curves to illustrate the various possibilities.
   (b) What does it mean to say that the line $y = L$ is a horizontal asymptote of the curve $y = f(x)$? Draw curves to illustrate the various possibilities.

6. Which of the following curves have vertical asymptotes? Which have horizontal asymptotes?
   (a) $y = x^4$
   (b) $y = \sin x$
   (c) $y = \tan x$
   (d) $y = \tan^{-1}x$
   (e) $y = e^x$
   (f) $y = \ln x$
   (g) $y = 1/x$
   (h) $y = \sqrt{x}$

7. (a) What does it mean for f to be continuous at a?
   (b) What does it mean for f to be continuous on the interval $(-\infty, \infty)$? What can you say about the graph of such a function?

8. What does the Intermediate Value Theorem say?

9. Write an expression for the slope of the tangent line to the curve $y = f(x)$ at the point $(a, f(a))$.

10. Suppose an object moves along a straight line with position $f(t)$ at time t. Write an expression for the instantaneous velocity of the object at time $t = a$. How can you interpret this velocity in terms of the graph of f?

11. If $y = f(x)$ and x changes from x_1 to x_2, write expressions for the following.
    (a) The average rate of change of y with respect to x over the interval $[x_1, x_2]$.
    (b) The instantaneous rate of change of y with respect to x at $x = x_1$.

12. Define the derivative $f'(a)$. Discuss two ways of interpreting this number.

13. (a) What does it mean for f to be differentiable at a?
    (b) What is the relation between the differentiability and continuity of a function?
    (c) Sketch the graph of a function that is continuous but not differentiable at $a = 2$.

14. Describe several ways in which a function can fail to be differentiable. Illustrate with sketches.

TRUE-FALSE QUIZ

Determine whether the statement is true or false. If it is true, explain why. If it is false, explain why or give an example that disproves the statement.

1. $\lim_{x \to 4}\left(\dfrac{2x}{x-4} - \dfrac{8}{x-4}\right) = \lim_{x \to 4}\dfrac{2x}{x-4} - \lim_{x \to 4}\dfrac{8}{x-4}$

2. $\lim_{x \to 1}\dfrac{x^2 + 6x - 7}{x^2 + 5x - 6} = \dfrac{\lim_{x \to 1}(x^2 + 6x - 7)}{\lim_{x \to 1}(x^2 + 5x - 6)}$

3. $\lim_{x \to 1}\dfrac{x - 3}{x^2 + 2x - 4} = \dfrac{\lim_{x \to 1}(x - 3)}{\lim_{x \to 1}(x^2 + 2x - 4)}$

4. If $\lim_{x \to 5} f(x) = 2$ and $\lim_{x \to 5} g(x) = 0$, then $\lim_{x \to 5}[f(x)/g(x)]$ does not exist.

5. If $\lim_{x \to 5} f(x) = 0$ and $\lim_{x \to 5} g(x) = 0$, then $\lim_{x \to 5}[f(x)/g(x)]$ does not exist.

6. If $\lim_{x \to 6} f(x)g(x)$ exists, then the limit must be $f(6)g(6)$.

7. If p is a polynomial, then $\lim_{x \to b} p(x) = p(b)$.

8. If $\lim_{x \to 0} f(x) = \infty$ and $\lim_{x \to 0} g(x) = \infty$, then $\lim_{x \to 0}[f(x) - g(x)] = 0$.

9. A function can have two different horizontal asymptotes.

10. If f has domain $[0, \infty)$ and has no horizontal asymptote, then $\lim_{x \to \infty} f(x) = \infty$ or $\lim_{x \to \infty} f(x) = -\infty$.

11. If the line $x = 1$ is a vertical asymptote of $y = f(x)$, then f is not defined at 1.

12. If $f(1) > 0$ and $f(3) < 0$, then there exists a number c between 1 and 3 such that $f(c) = 0$.

13. If f is continuous at 5 and $f(5) = 2$ and $f(4) = 3$, then $\lim_{x \to 2} f(4x^2 - 11) = 2$.

14. If f is continuous on $[-1, 1]$ and $f(-1) = 4$ and $f(1) = 3$, then there exists a number r such that $|r| < 1$ and $f(r) = \pi$.

15. Let f be a function such that $\lim_{x\to 0} f(x) = 6$. Then there exists a number δ such that if $0 < |x| < \delta$, then $|f(x) - 6| < 1$.

16. If $f(x) > 1$ for all x and $\lim_{x\to 0} f(x)$ exists, then $\lim_{x\to 0} f(x) > 1$.

17. If f is continuous at a, then f is differentiable at a.

18. If $f'(r)$ exists, then $\lim_{x\to r} f(x) = f(r)$.

EXERCISES

1. The graph of f is given.

(a) Find each limit, or explain why it does not exist.

(i) $\lim_{x\to 2^+} f(x)$ (ii) $\lim_{x\to -3^+} f(x)$

(iii) $\lim_{x\to -3} f(x)$ (iv) $\lim_{x\to 4} f(x)$

(v) $\lim_{x\to 0} f(x)$ (vi) $\lim_{x\to 2^-} f(x)$

(vii) $\lim_{x\to \infty} f(x)$ (viii) $\lim_{x\to -\infty} f(x)$

(b) State the equations of the horizontal asymptotes.

(c) State the equations of the vertical asymptotes.

(d) At what numbers is f discontinuous? Explain.

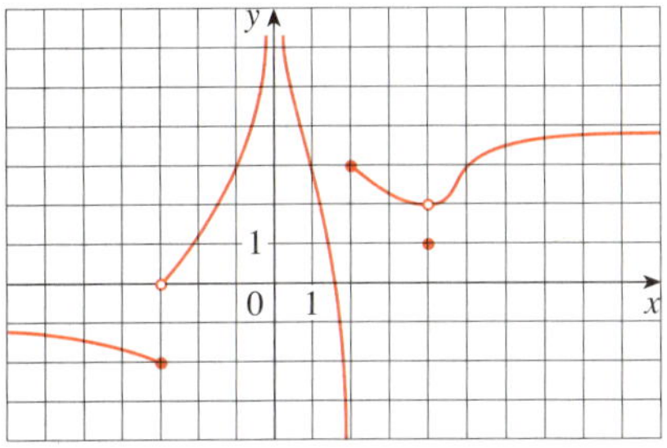

2. Sketch the graph of an example of a function f that satisfies all of the following conditions:

$$\lim_{x\to 0^+} f(x) = -2, \quad \lim_{x\to 0^-} f(x) = 1, \quad f(0) = -1,$$

$$\lim_{x\to 2^-} f(x) = \infty, \quad \lim_{x\to 2^+} f(x) = -\infty, \quad \lim_{x\to \infty} f(x) = 3,$$

$$\lim_{x\to -\infty} f(x) = 4$$

3–22 Find the limit.

3. $\lim_{x\to 1} e^{x^3 - x}$

4. $\lim_{x\to 3} \dfrac{x^2 - 9}{x^2 + 2x - 3}$

5. $\lim_{x\to -3} \dfrac{x^2 - 9}{x^2 + 2x - 3}$

6. $\lim_{x\to 1^+} \dfrac{x^2 - 9}{x^2 + 2x - 3}$

7. $\lim_{h\to 0} \dfrac{(h - 1)^3 + 1}{h}$

8. $\lim_{t\to 2} \dfrac{t^2 - 4}{t^3 - 8}$

9. $\lim_{r\to 9} \dfrac{\sqrt{r}}{(r - 9)^4}$

10. $\lim_{v\to 4^+} \dfrac{4 - v}{|4 - v|}$

11. $\lim_{s\to 16} \dfrac{4 - \sqrt{s}}{s - 16}$

12. $\lim_{v\to 2} \dfrac{v^2 + 2v - 8}{v^4 - 16}$

13. $\lim_{x\to 8^-} \dfrac{|x - 8|}{x - 8}$

14. $\lim_{x\to 9^+} (\sqrt{x - 9} + [\![x + 1]\!])$

15. $\lim_{x\to 0} \dfrac{1 - \sqrt{1 - x^2}}{x}$

16. $\lim_{x\to 2} \dfrac{\sqrt{x + 2} - \sqrt{2x}}{x^2 - 2x}$

17. $\lim_{x\to \infty} \dfrac{1 + 2x - x^2}{1 - x + 2x^2}$

18. $\lim_{x\to -\infty} \dfrac{5x^3 - x^2 + 2}{2x^3 + x - 3}$

19. $\lim_{x\to \infty} \dfrac{\sqrt{x^2 - 9}}{2x - 6}$

20. $\lim_{x\to 10^-} \ln(100 - x^2)$

21. $\lim_{x\to \infty} e^{-3x}$

22. $\lim_{x\to \infty} \arctan(x^3 - x)$

23–24 Use graphs to discover the asymptotes of the curve. Then prove what you have discovered.

23. $y = \dfrac{\cos^2 x}{x^2}$

24. $y = \sqrt{x^2 + x + 1} - \sqrt{x^2 - x}$

25. If $2x - 1 \leqslant f(x) \leqslant x^2$ for $0 < x < 3$, find $\lim_{x\to 1} f(x)$.

26. Prove that $\lim_{x\to 0} x^2 \cos(1/x^2) = 0$.

27–30 Prove the statement using the precise definition of a limit.

27. $\lim_{x\to 5} (7x - 27) = 8$

28. $\lim_{x\to 0} \sqrt[3]{x} = 0$

29. $\lim_{x\to 2} (x^2 - 3x) = -2$

30. $\lim_{x\to 4^+} \dfrac{2}{\sqrt{x - 4}} = \infty$

31. Let

$$f(x) = \begin{cases} \sqrt{-x} & \text{if } x < 0 \\ 3 - x & \text{if } 0 \leqslant x < 3 \\ (x - 3)^2 & \text{if } x > 3 \end{cases}$$

(a) Evaluate each limit, if it exists.

(i) $\lim_{x\to 0^+} f(x)$ (ii) $\lim_{x\to 0^-} f(x)$ (iii) $\lim_{x\to 0} f(x)$

(iv) $\lim_{x\to 3^-} f(x)$ (v) $\lim_{x\to 3^+} f(x)$ (vi) $\lim_{x\to 3} f(x)$

(b) Where is f discontinuous?

(c) Sketch the graph of f.

32. Let

$$g(x) = \begin{cases} 2x - x^2 & \text{if } 0 \leqslant x \leqslant 2 \\ 2 - x & \text{if } 2 < x \leqslant 3 \\ x - 4 & \text{if } 3 < x < 4 \\ \pi & \text{if } x \geqslant 4 \end{cases}$$

(a) For each of the numbers 2, 3, and 4, discover whether g is continuous from the left, continuous from the right, or continuous at the number.
(b) Sketch the graph of g.

33–34 |||| Show that the function is continuous on its domain. State the domain.

33. $h(x) = xe^{\sin x}$

34. $g(x) = \dfrac{\sqrt{x^2 - 9}}{x^2 - 2}$

35–36 |||| Use the Intermediate Value Theorem to show that there is a root of the equation in the given interval.

35. $2x^3 + x^2 + 2 = 0, \quad (-2, -1)$

36. $e^{-x^2} = x, \quad (0, 1)$

37. (a) Find the slope of the tangent line to the curve $y = 9 - 2x^2$ at the point $(2, 1)$.
(b) Find an equation of this tangent line.

38. Find equations of the tangent lines to the curve

$$y = \frac{2}{1 - 3x}$$

at the points with x-coordinates 0 and -1.

39. The displacement (in meters) of an object moving in a straight line is given by $s = 1 + 2t + t^2/4$, where t is measured in seconds.
(a) Find the average velocity over each time period.
(i) $[1, 3]$ (ii) $[1, 2]$
(iii) $[1, 1.5]$ (iv) $[1, 1.1]$
(b) Find the instantaneous velocity when $t = 1$.

40. According to Boyle's Law, if the temperature of a confined gas is held fixed, then the product of the pressure P and the volume V is a constant. Suppose that, for a certain gas, $PV = 800$, where P is measured in pounds per square inch and V is measured in cubic inches.
(a) Find the average rate of change of P as V increases from 200 in^3 to 250 in^3.
(b) Express V as a function of P and show that the instantaneous rate of change of V with respect to P is inversely proportional to the square of P.

41. (a) Use the definition of a derivative to find $f'(2)$, where $f(x) = x^3 - 2x$.
(b) Find an equation of the tangent line to the curve $y = x^3 - 2x$ at the point $(2, 4)$.
(c) Illustrate part (b) by graphing the curve and the tangent line on the same screen.

42. Find a function f and a number a such that

$$\lim_{h \to 0} \frac{(2 + h)^6 - 64}{h} = f'(a)$$

43. The total cost of repaying a student loan at an interest rate of $r\%$ per year is $C = f(r)$.
(a) What is the meaning of the derivative $f'(r)$? What are its units?
(b) What does the statement $f'(10) = 1200$ mean?
(c) Is $f'(r)$ always positive or does it change sign?

44–46 |||| Trace or copy the graph of the function. Then sketch a graph of its derivative directly beneath.

44.

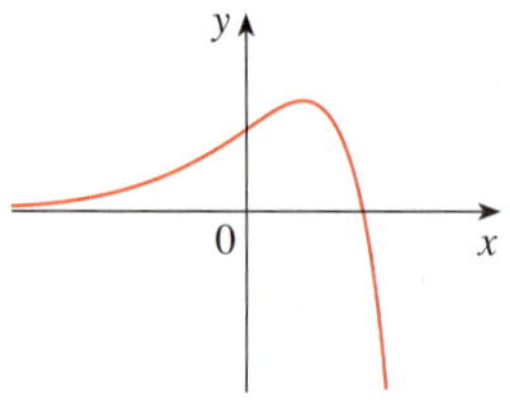

45.

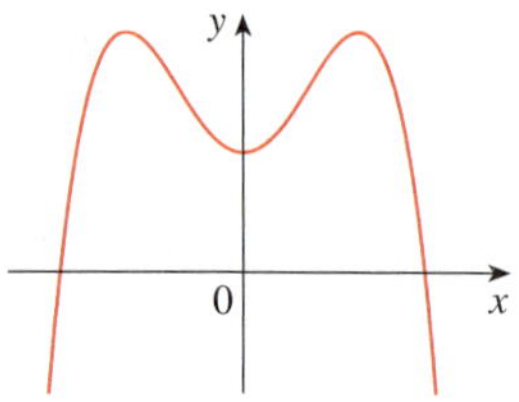

46.

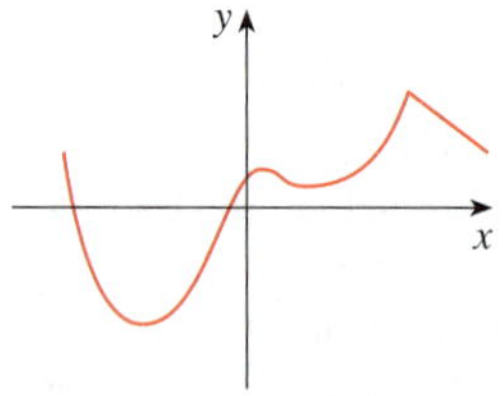

47. (a) If $f(x) = \sqrt{3 - 5x}$, use the definition of a derivative to find $f'(x)$.
(b) Find the domains of f and f'.
(c) Graph f and f' on a common screen. Compare the graphs to see whether your answer to part (a) is reasonable.

48. (a) Find the asymptotes of the graph of

$$f(x) = \frac{4 - x}{3 + x}$$

and use them to sketch the graph.
(b) Use your graph from part (a) to sketch the graph of f'.
(c) Use the definition of a derivative to find $f'(x)$.
(d) Use a graphing device to graph f' and compare with your sketch in part (b).

49. The graph of f is shown. State, with reasons, the numbers at which f is not differentiable.

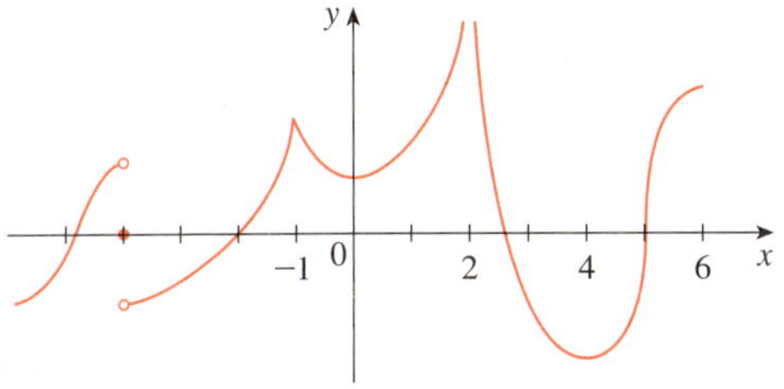

50. The *total fertility rate* at time t, denoted by $F(t)$, is an estimate of the average number of children born to each woman (assuming that current birth rates remain constant). The graph of the total fertility rate in the United States shows the fluctuations from 1940 to 1990.

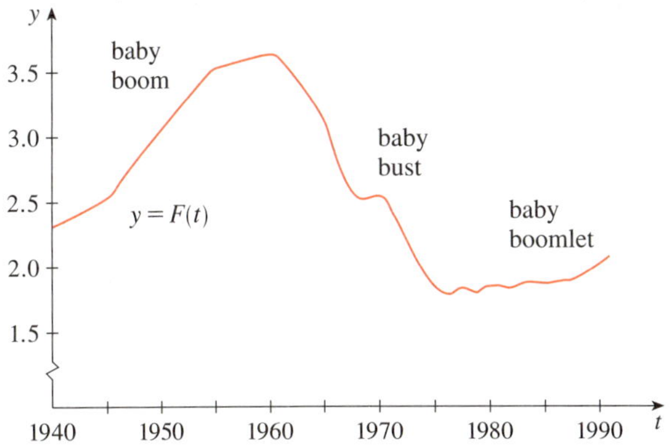

(a) Estimate the values of $F'(1950)$, $F'(1965)$, and $F'(1987)$.
(b) What are the meanings of these derivatives?
(c) Can you suggest reasons for the values of these derivatives?

51. Let $B(t)$ be the total value of U.S. banknotes in circulation at time t. The table gives values of this function from 1980 to 1998, at year end, in billions of dollars. Interpret and estimate the value of $B'(1990)$.

t	$B(t)$
1980	124.8
1985	182.0
1990	268.2
1995	401.5
1998	492.2

52. Graph the curve $y = (x + 1)/(x - 1)$ and the tangent lines to this curve at the points (2, 3) and (−1, 0).

53. Suppose that $|f(x)| \leqslant g(x)$ for all x, where $\lim_{x \to a} g(x) = 0$. Find $\lim_{x \to a} f(x)$.

54. Let $f(x) = [\![x]\!] + [\![-x]\!]$.
(a) For what values of a does $\lim_{x \to a} f(x)$ exist?
(b) At what numbers is f discontinuous?

PROBLEMS PLUS

In our discussion of the principles of problem solving we considered the problem-solving strategy of *introducing something extra* (see page 80). In the following example we show how this principle is sometimes useful when we evaluate limits. The idea is to change the variable—to introduce a new variable that is related to the original variable—in such a way as to make the problem simpler. Later, in Section 5.5, we will make more extensive use of this general idea.

EXAMPLE 1 Evaluate $\displaystyle\lim_{x \to 0} \frac{\sqrt[3]{1 + cx} - 1}{x}$, where c is a constant.

SOLUTION As it stands, this limit looks challenging. In Section 2.3 we evaluated several limits in which both numerator and denominator approached 0. There our strategy was to perform some sort of algebraic manipulation that led to a simplifying cancellation, but here it's not clear what kind of algebra is necessary.

So we introduce a new variable t by the equation

$$t = \sqrt[3]{1 + cx}$$

We also need to express x in terms of t, so we solve this equation:

$$t^3 = 1 + cx$$

$$x = \frac{t^3 - 1}{c}$$

Notice that $x \to 0$ is equivalent to $t \to 1$. This allows us to convert the given limit into one involving the variable t:

$$\lim_{x \to 0} \frac{\sqrt[3]{1 + cx} - 1}{x} = \lim_{t \to 1} \frac{t - 1}{(t^3 - 1)/c}$$

$$= \lim_{t \to 1} \frac{c(t - 1)}{t^3 - 1}$$

The change of variable allowed us to replace a relatively complicated limit by a simpler one of a type that we have seen before. Factoring the denominator as a difference of cubes, we get

$$\lim_{t \to 1} \frac{c(t - 1)}{t^3 - 1} = \lim_{t \to 1} \frac{c(t - 1)}{(t - 1)(t^2 + t + 1)}$$

$$= \lim_{t \to 1} \frac{c}{t^2 + t + 1} = \frac{c}{3}$$

The following problems are meant to test and challenge your problem-solving skills. Some of them require a considerable amount of time to think through, so don't be discouraged if you can't solve them right away. If you get stuck, you might find it helpful to refer to the discussion of the principles of problem solving on page 80.

PROBLEMS

1. Evaluate $\lim_{x \to 1} \dfrac{\sqrt[3]{x} - 1}{\sqrt{x} - 1}$.

2. Find numbers a and b such that $\lim_{x \to 0} \dfrac{\sqrt{ax + b} - 2}{x} = 1$.

3. Evaluate $\lim_{x \to 0} \dfrac{|2x - 1| - |2x + 1|}{x}$.

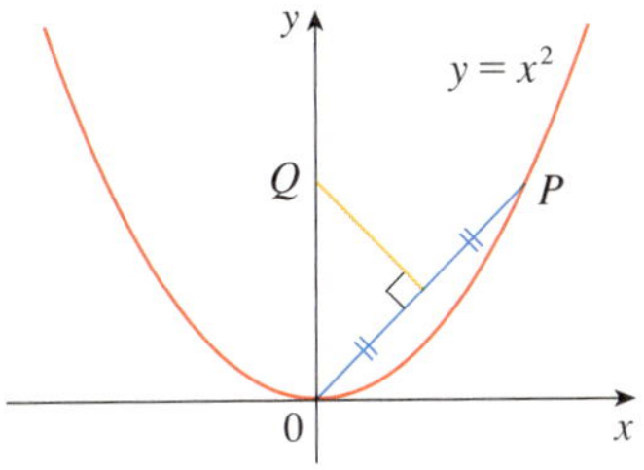

FIGURE FOR PROBLEM 4

4. The figure shows a point P on the parabola $y = x^2$ and the point Q where the perpendicular bisector of OP intersects the y-axis. As P approaches the origin along the parabola, what happens to Q? Does it have a limiting position? If so, find it.

5. If $[\![x]\!]$ denotes the greatest integer function, find $\lim_{x \to \infty} x/[\![x]\!]$.

6. Sketch the region in the plane defined by each of the following equations.
   (a) $[\![x]\!]^2 + [\![y]\!]^2 = 1$ (b) $[\![x]\!]^2 - [\![y]\!]^2 = 3$ (c) $[\![x + y]\!]^2 = 1$ (d) $[\![x]\!] + [\![y]\!] = 1$

7. Find all values of a such that f is continuous on $\mathbb{R}$:

$$f(x) = \begin{cases} x + 1 & \text{if } x \leq a \\ x^2 & \text{if } x > a \end{cases}$$

8. A **fixed point** of a function f is a number c in its domain such that $f(c) = c$. (The function doesn't move c; it stays fixed.)
   (a) Sketch the graph of a continuous function with domain $[0, 1]$ whose range also lies in $[0, 1]$. Locate a fixed point of f.
   (b) Try to draw the graph of a continuous function with domain $[0, 1]$ and range in $[0, 1]$ that does *not* have a fixed point. What is the obstacle?
   (c) Use the Intermediate Value Theorem to prove that any continuous function with domain $[0, 1]$ and range a subset of $[0, 1]$ must have a fixed point.

9. If $\lim_{x \to a} [f(x) + g(x)] = 2$ and $\lim_{x \to a} [f(x) - g(x)] = 1$, find $\lim_{x \to a} f(x)g(x)$.

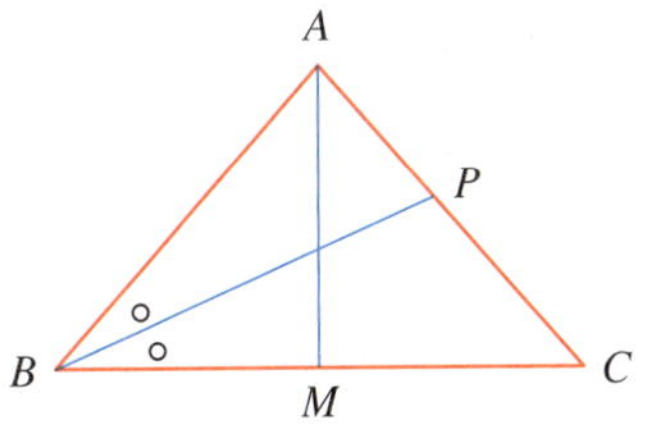

FIGURE FOR PROBLEM 10

10. (a) The figure shows an isosceles triangle ABC with $\angle B = \angle C$. The bisector of angle B intersects the side AC at the point P. Suppose that the base BC remains fixed but the altitude $|AM|$ of the triangle approaches 0, so A approaches the midpoint M of BC. What happens to P during this process? Does it have a limiting position? If so, find it.
    (b) Try to sketch the path traced out by P during this process. Then find the equation of this curve and use this equation to sketch the curve.

11. (a) If we start from 0° latitude and proceed in a westerly direction, we can let $T(x)$ denote the temperature at the point x at any given time. Assuming that T is a continuous function of x, show that at any fixed time there are at least two diametrically opposite points on the equator that have exactly the same temperature.
    (b) Does the result in part (a) hold for points lying on any circle on Earth's surface?
    (c) Does the result in part (a) hold for barometric pressure and for altitude above sea level?

12. If f is a differentiable function and $g(x) = xf(x)$, use the definition of a derivative to show that $g'(x) = xf'(x) + f(x)$.

13. Suppose f is a function that satisfies the equation $f(x + y) = f(x) + f(y) + x^2y + xy^2$ for all real numbers x and y. Suppose also that

$$\lim_{x \to 0} \frac{f(x)}{x} = 1$$

    (a) Find $f(0)$. (b) Find $f'(0)$. (c) Find $f'(x)$.

14. Suppose f is a function with the property that $|f(x)| \leq x^2$ for all x. Show that $f(0) = 0$. Then show that $f'(0) = 0$.

CHAPTER 3

By measuring slopes at points on the sine curve, we get strong visual evidence that the derivative of the sine function is the cosine function.

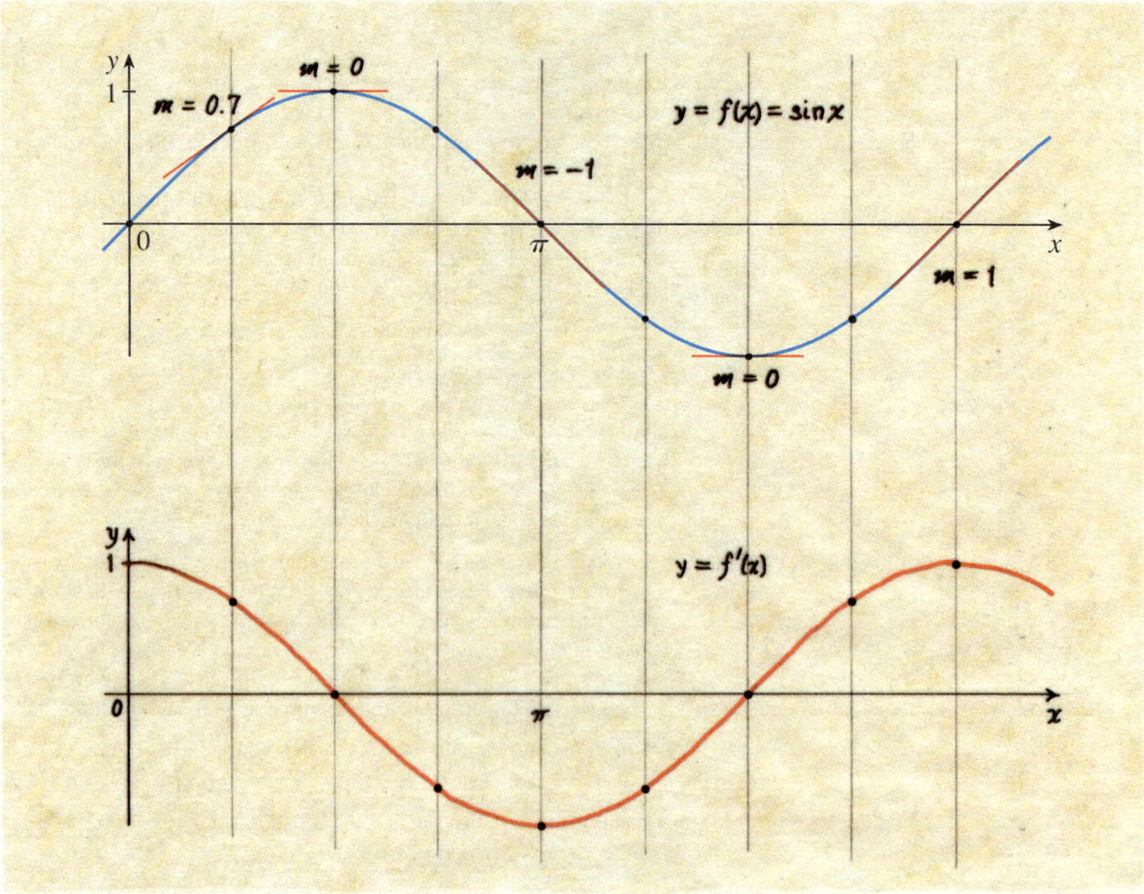

Differentiation Rules

We have seen how to interpret derivatives as slopes and rates of change. We have seen how to estimate derivatives of functions given by tables of values. We have learned how to graph derivatives of functions that are defined graphically. We have used the definition of a derivative to calculate the derivatives of functions defined by formulas. But it would be tedious if we always had to use the definition, so in this chapter we develop rules for finding derivatives without having to use the definition directly. These differentiation rules enable us to calculate with relative ease the derivatives of polynomials, rational functions, algebraic functions, exponential and logarithmic functions, and trigonometric and inverse trigonometric functions. We then use these rules to solve problems involving rates of change and the approximation of functions.

3.1 Derivatives of Polynomials and Exponential Functions

In this section we learn how to differentiate constant functions, power functions, polynomials, and exponential functions.

Let's start with the simplest of all functions, the constant function $f(x) = c$. The graph of this function is the horizontal line $y = c$, which has slope 0, so we must have $f'(x) = 0$. (See Figure 1.) A formal proof, from the definition of a derivative, is also easy:

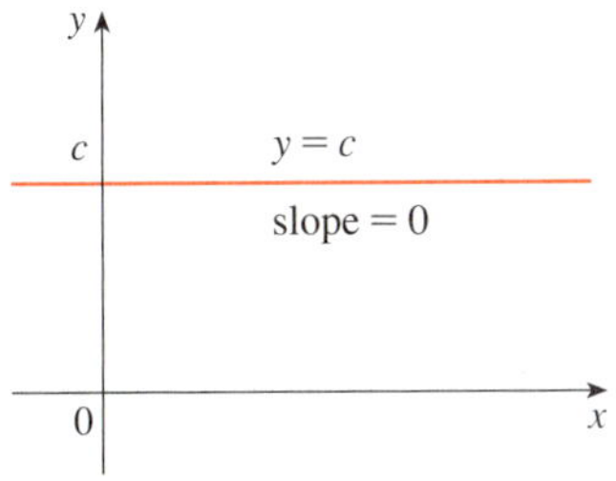

FIGURE 1
The graph of $f(x) = c$ is the line $y = c$, so $f'(x) = 0$.

$$f'(x) = \lim_{h \to 0} \frac{f(x + h) - f(x)}{h} = \lim_{h \to 0} \frac{c - c}{h}$$

$$= \lim_{h \to 0} 0 = 0$$

In Leibniz notation, we write this rule as follows.

Derivative of a Constant Function $\quad \dfrac{d}{dx}(c) = 0$

Power Functions

We next look at the functions $f(x) = x^n$, where n is a positive integer. If $n = 1$, the graph of $f(x) = x$ is the line $y = x$, which has slope 1 (see Figure 2). So

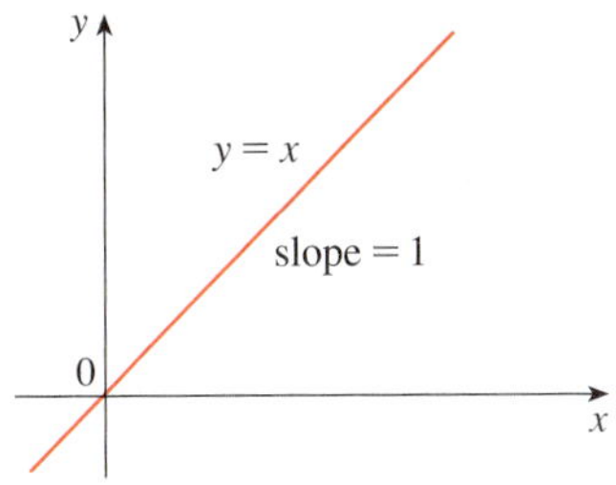

FIGURE 2
The graph of $f(x) = x$ is the line $y = x$, so $f'(x) = 1$.

1 $$\frac{d}{dx}(x) = 1$$

(You can also verify Equation 1 from the definition of a derivative.) We have already investigated the cases $n = 2$ and $n = 3$. In fact, in Section 2.9 (Exercises 19 and 20) we found that

2 $$\frac{d}{dx}(x^2) = 2x \qquad \frac{d}{dx}(x^3) = 3x^2$$

For $n = 4$ we find the derivative of $f(x) = x^4$ as follows:

$$f'(x) = \lim_{h \to 0} \frac{f(x+h) - f(x)}{h} = \lim_{h \to 0} \frac{(x+h)^4 - x^4}{h}$$

$$= \lim_{h \to 0} \frac{x^4 + 4x^3h + 6x^2h^2 + 4xh^3 + h^4 - x^4}{h}$$

$$= \lim_{h \to 0} \frac{4x^3h + 6x^2h^2 + 4xh^3 + h^4}{h}$$

$$= \lim_{h \to 0} (4x^3 + 6x^2h + 4xh^2 + h^3) = 4x^3$$

Thus

3 $$\frac{d}{dx}(x^4) = 4x^3$$

Comparing the equations in (1), (2), and (3), we see a pattern emerging. It seems to be a reasonable guess that, when n is a positive integer, $(d/dx)(x^n) = nx^{n-1}$. This turns out to be true. We prove it in two ways; the second proof uses the Binomial Theorem.

The Power Rule If n is a positive integer, then

$$\frac{d}{dx}(x^n) = nx^{n-1}$$

First Proof The formula

$$x^n - a^n = (x - a)(x^{n-1} + x^{n-2}a + \cdots + xa^{n-2} + a^{n-1})$$

can be verified simply by multiplying out the right-hand side (or by summing the second factor as a geometric series). If $f(x) = x^n$, we can use Equation 2.8.3 for $f'(a)$ and the equation above to write

$$f'(a) = \lim_{x \to a} \frac{f(x) - f(a)}{x - a} = \lim_{x \to a} \frac{x^n - a^n}{x - a}$$

$$= \lim_{x \to a} (x^{n-1} + x^{n-2}a + \cdots + xa^{n-2} + a^{n-1})$$

$$= a^{n-1} + a^{n-2}a + \cdots + aa^{n-2} + a^{n-1}$$

$$= na^{n-1}$$

Second Proof

$$f'(x) = \lim_{h \to 0} \frac{f(x+h) - f(x)}{h} = \lim_{h \to 0} \frac{(x+h)^n - x^n}{h}$$

|||| The Binomial Theorem is given on Reference Page 1.

In finding the derivative of x^4 we had to expand $(x + h)^4$. Here we need to expand $(x + h)^n$ and we use the Binomial Theorem to do so:

$$f'(x) = \lim_{h \to 0} \frac{\left[x^n + nx^{n-1}h + \dfrac{n(n-1)}{2}x^{n-2}h^2 + \cdots + nxh^{n-1} + h^n\right] - x^n}{h}$$

$$= \lim_{h \to 0} \frac{nx^{n-1}h + \dfrac{n(n-1)}{2}x^{n-2}h^2 + \cdots + nxh^{n-1} + h^n}{h}$$

$$= \lim_{h \to 0} \left[nx^{n-1} + \frac{n(n-1)}{2}x^{n-2}h + \cdots + nxh^{n-2} + h^{n-1} \right]$$

$$= nx^{n-1}$$

because every term except the first has h as a factor and therefore approaches 0.

We illustrate the Power Rule using various notations in Example 1.

EXAMPLE 1

(a) If $f(x) = x^6$, then $f'(x) = 6x^5$.

(b) If $y = x^{1000}$, then $y' = 1000x^{999}$.

(c) If $y = t^4$, then $\dfrac{dy}{dt} = 4t^3$.

(d) $\dfrac{d}{dr}(r^3) = 3r^2$

What about power functions with negative integer exponents? In Exercise 53 we ask you to verify from the definition of a derivative that

$$\frac{d}{dx}\left(\frac{1}{x}\right) = -\frac{1}{x^2}$$

We can rewrite this equation as

$$\frac{d}{dx}(x^{-1}) = (-1)x^{-2}$$

and so the Power Rule is true when $n = -1$. In fact, we will show in the next section [Exercise 44(c)] that it holds for all negative integers.

What if the exponent is a fraction? In Example 3 in Section 2.7 we found, in effect, that

$$\frac{d}{dx}\sqrt{x} = \frac{1}{2\sqrt{x}}$$

which can be written as

$$\frac{d}{dx}(x^{1/2}) = \tfrac{1}{2}x^{-1/2}$$

This shows that the Power Rule is true even when $n = \frac{1}{2}$. In fact, we will show in Section 3.8 that it is true for all real numbers n.

The Power Rule (General Version) If n is any real number, then

$$\frac{d}{dx}(x^n) = nx^{n-1}$$

EXAMPLE 2 Differentiate:

(a) $f(x) = \dfrac{1}{x^2}$ (b) $y = \sqrt[3]{x^2}$

Figure 3 shows the function y in Example 2(b) and its derivative y'. Notice that y is not differentiable at 0 (y' is not defined there). Observe that y' is positive when y increases and is negative when y decreases.

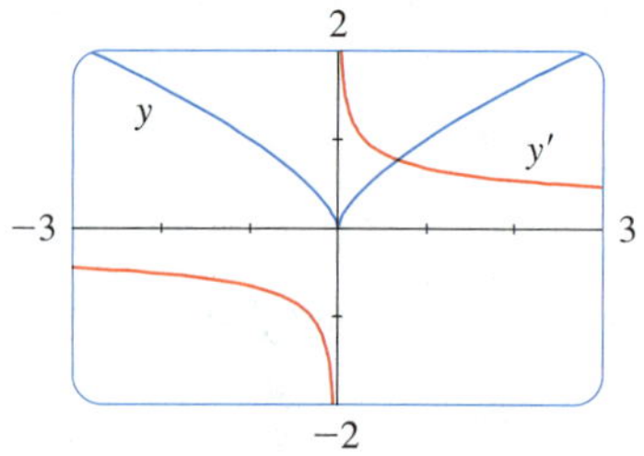

FIGURE 3
$y = \sqrt[3]{x^2}$

SOLUTION In each case we rewrite the function as a power of x.

(a) Since $f(x) = x^{-2}$, we use the Power Rule with $n = -2$:

$$f'(x) = \frac{d}{dx}(x^{-2}) = -2x^{-2-1} = -2x^{-3} = -\frac{2}{x^3}$$

(b) $$\frac{dy}{dx} = \frac{d}{dx}\left(\sqrt[3]{x^2}\right) = \frac{d}{dx}(x^{2/3}) = \tfrac{2}{3}x^{(2/3)-1} = \tfrac{2}{3}x^{-1/3}$$

EXAMPLE 3 Find an equation of the tangent line to the curve $y = x\sqrt{x}$ at the point (1, 1). Illustrate by graphing the curve and its tangent line.

SOLUTION The derivative of $f(x) = x\sqrt{x} = xx^{1/2} = x^{3/2}$ is

$$f'(x) = \tfrac{3}{2}x^{(3/2)-1} = \tfrac{3}{2}x^{1/2} = \tfrac{3}{2}\sqrt{x}$$

So the slope of the tangent line at (1, 1) is $f'(1) = \frac{3}{2}$. Therefore, an equation of the tangent line is

$$y - 1 = \tfrac{3}{2}(x - 1) \qquad \text{or} \qquad y = \tfrac{3}{2}x - \tfrac{1}{2}$$

We graph the curve and its tangent line in Figure 4.

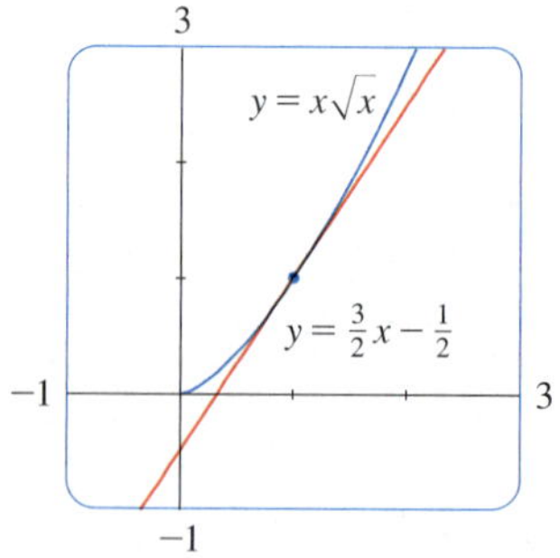

FIGURE 4

New Derivatives from Old

When new functions are formed from old functions by addition, subtraction, or multiplication by a constant, their derivatives can be calculated in terms of derivatives of the old functions. In particular, the following formula says that *the derivative of a constant times a function is the constant times the derivative of the function.*

The Constant Multiple Rule If c is a constant and f is a differentiable function, then

$$\frac{d}{dx}[cf(x)] = c\frac{d}{dx}f(x)$$

GEOMETRIC INTERPRETATION OF THE CONSTANT MULTIPLE RULE

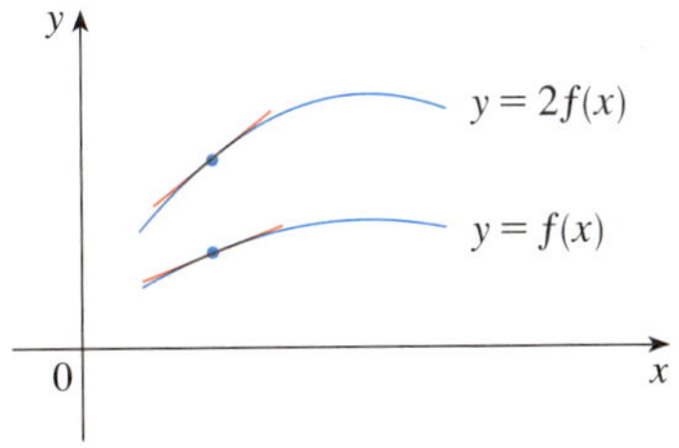

Multiplying by $c = 2$ stretches the graph vertically by a factor of 2. All the rises have been doubled but the runs stay the same. So the slopes are doubled, too.

Proof Let $g(x) = cf(x)$. Then

$$\begin{aligned} g'(x) &= \lim_{h \to 0} \frac{g(x+h) - g(x)}{h} = \lim_{h \to 0} \frac{cf(x+h) - cf(x)}{h} \\ &= \lim_{h \to 0} c \left[\frac{f(x+h) - f(x)}{h} \right] \\ &= c \lim_{h \to 0} \frac{f(x+h) - f(x)}{h} \qquad \text{(by Law 3 of limits)} \\ &= cf'(x) \end{aligned}$$

EXAMPLE 4

(a) $\dfrac{d}{dx}(3x^4) = 3\dfrac{d}{dx}(x^4) = 3(4x^3) = 12x^3$

(b) $\dfrac{d}{dx}(-x) = \dfrac{d}{dx}[(-1)x] = (-1)\dfrac{d}{dx}(x) = -1(1) = -1$

The next rule tells us that *the derivative of a sum of functions is the sum of the derivatives.*

Using prime notation, we can write the Sum Rule as

$$(f + g)' = f' + g'$$

The Sum Rule If f and g are both differentiable, then

$$\frac{d}{dx}[f(x) + g(x)] = \frac{d}{dx}f(x) + \frac{d}{dx}g(x)$$

Proof Let $F(x) = f(x) + g(x)$. Then

$$\begin{aligned} F'(x) &= \lim_{h \to 0} \frac{F(x+h) - F(x)}{h} \\ &= \lim_{h \to 0} \frac{[f(x+h) + g(x+h)] - [f(x) + g(x)]}{h} \\ &= \lim_{h \to 0} \left[\frac{f(x+h) - f(x)}{h} + \frac{g(x+h) - g(x)}{h} \right] \\ &= \lim_{h \to 0} \frac{f(x+h) - f(x)}{h} + \lim_{h \to 0} \frac{g(x+h) - g(x)}{h} \qquad \text{(by Law 1)} \\ &= f'(x) + g'(x) \end{aligned}$$

The Sum Rule can be extended to the sum of any number of functions. For instance, using this theorem twice, we get

$$(f + g + h)' = [(f + g) + h]' = (f + g)' + h' = f' + g' + h'$$

By writing $f - g$ as $f + (-1)g$ and applying the Sum Rule and the Constant Multiple Rule, we get the following formula.

The Difference Rule If f and g are both differentiable, then

$$\frac{d}{dx}[f(x) - g(x)] = \frac{d}{dx}f(x) - \frac{d}{dx}g(x)$$

The Constant Multiple Rule, the Sum Rule, and the Difference Rule can be combined with the Power Rule to differentiate any polynomial, as the following examples demonstrate.

EXAMPLE 5

Try more problems like this one.
Resources / Module 4
/ Polynomial Models
/ Basic Differentiation Rules and Quiz

$$\frac{d}{dx}(x^8 + 12x^5 - 4x^4 + 10x^3 - 6x + 5)$$

$$= \frac{d}{dx}(x^8) + 12\frac{d}{dx}(x^5) - 4\frac{d}{dx}(x^4) + 10\frac{d}{dx}(x^3) - 6\frac{d}{dx}(x) + \frac{d}{dx}(5)$$

$$= 8x^7 + 12(5x^4) - 4(4x^3) + 10(3x^2) - 6(1) + 0$$

$$= 8x^7 + 60x^4 - 16x^3 + 30x^2 - 6$$

EXAMPLE 6 Find the points on the curve $y = x^4 - 6x^2 + 4$ where the tangent line is horizontal.

SOLUTION Horizontal tangents occur where the derivative is zero. We have

$$\frac{dy}{dx} = \frac{d}{dx}(x^4) - 6\frac{d}{dx}(x^2) + \frac{d}{dx}(4)$$

$$= 4x^3 - 12x + 0 = 4x(x^2 - 3)$$

Thus, $dy/dx = 0$ if $x = 0$ or $x^2 - 3 = 0$, that is, $x = \pm\sqrt{3}$. So the given curve has horizontal tangents when $x = 0$, $\sqrt{3}$, and $-\sqrt{3}$. The corresponding points are $(0, 4)$, $(\sqrt{3}, -5)$, and $(-\sqrt{3}, -5)$. (See Figure 5.)

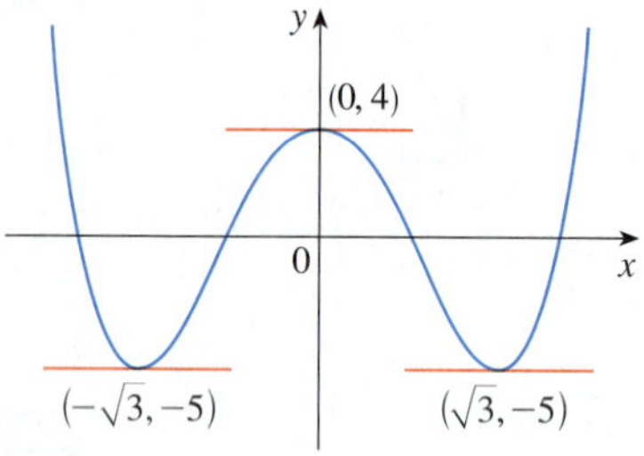

FIGURE 5
The curve $y = x^4 - 6x^2 + 4$ and its horizontal tangents

Exponential Functions

Let's try to compute the derivative of the exponential function $f(x) = a^x$ using the definition of a derivative:

$$f'(x) = \lim_{h \to 0} \frac{f(x+h) - f(x)}{h} = \lim_{h \to 0} \frac{a^{x+h} - a^x}{h}$$

$$= \lim_{h \to 0} \frac{a^x a^h - a^x}{h} = \lim_{h \to 0} \frac{a^x(a^h - 1)}{h}$$

The factor a^x doesn't depend on h, so we can take it in front of the limit:

$$f'(x) = a^x \lim_{h \to 0} \frac{a^h - 1}{h}$$

Notice that the limit is the value of the derivative of f at 0, that is,

$$\lim_{h \to 0} \frac{a^h - 1}{h} = f'(0)$$

Therefore, we have shown that if the exponential function $f(x) = a^x$ is differentiable at 0, then it is differentiable everywhere and

4 $$f'(x) = f'(0)a^x$$

This equation says that *the rate of change of any exponential function is proportional to the function itself.* (The slope is proportional to the height.)

Numerical evidence for the existence of $f'(0)$ is given in the table at the left for the cases $a = 2$ and $a = 3$. (Values are stated correct to four decimal places. For the case $a = 2$, see also Example 3 in Section 2.8.) It appears that the limits exist and

h	$\frac{2^h - 1}{h}$	$\frac{3^h - 1}{h}$
0.1	0.7177	1.1612
0.01	0.6956	1.1047
0.001	0.6934	1.0992
0.0001	0.6932	1.0987

$$\text{for } a = 2, \quad f'(0) = \lim_{h \to 0} \frac{2^h - 1}{h} \approx 0.69$$

$$\text{for } a = 3, \quad f'(0) = \lim_{h \to 0} \frac{3^h - 1}{h} \approx 1.10$$

In fact, we will show in Section 5.6 that these limits exist and, correct to six decimal places, the values are

$$\frac{d}{dx}(2^x)\bigg|_{x=0} \approx 0.693147 \qquad \frac{d}{dx}(3^x)\bigg|_{x=0} \approx 1.098612$$

Thus, from Equation 4 we have

5 $$\frac{d}{dx}(2^x) \approx (0.69)2^x \qquad \frac{d}{dx}(3^x) \approx (1.10)3^x$$

Of all possible choices for the base a in Equation 4, the simplest differentiation formula occurs when $f'(0) = 1$. In view of the estimates of $f'(0)$ for $a = 2$ and $a = 3$, it seems reasonable that there is a number a between 2 and 3 for which $f'(0) = 1$. It is traditional to denote this value by the letter e. (In fact, that is how we introduced e in Section 1.5.) Thus, we have the following definition.

||| In Exercise 1 we will see that e lies between 2.7 and 2.8. In Section 5.6 we will give a definition of e that will enable us to show that, correct to five decimal places,

$$e \approx 2.71828$$

Definition of the Number e

$$e \text{ is the number such that} \quad \lim_{h \to 0} \frac{e^h - 1}{h} = 1$$

Geometrically, this means that of all the possible exponential functions $y = a^x$, the function $f(x) = e^x$ is the one whose tangent line at (0, 1) has a slope $f'(0)$ that is exactly 1. (See Figures 6 and 7.)

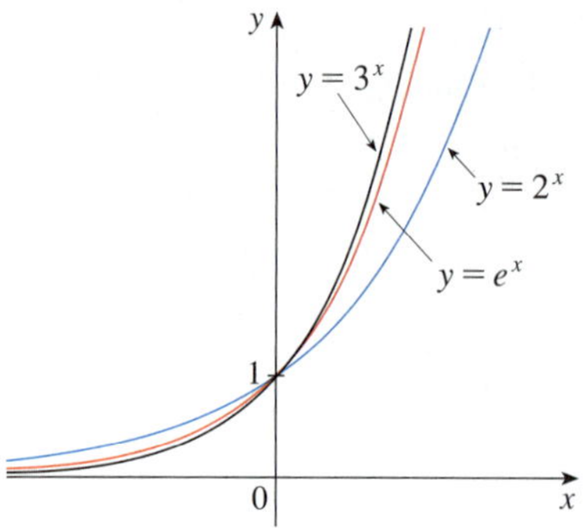

FIGURE 6

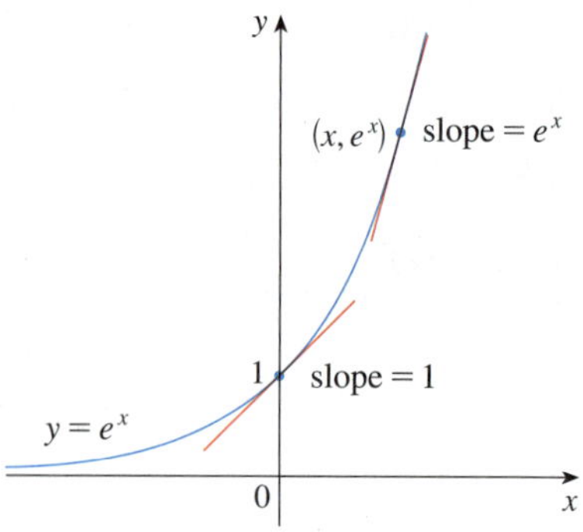

FIGURE 7

If we put $a = e$ and, therefore, $f'(0) = 1$ in Equation 4, it becomes the following important differentiation formula.

Derivative of the Natural Exponential Function

$$\frac{d}{dx}(e^x) = e^x$$

Thus, the exponential function $f(x) = e^x$ has the property that it is its own derivative. The geometrical significance of this fact is that the slope of a tangent line to the curve $y = e^x$ is equal to the y-coordinate of the point (see Figure 7).

EXAMPLE 7 If $f(x) = e^x - x$, find f'. Compare the graphs of f and f'.

SOLUTION Using the Difference Rule, we have

$$f'(x) = \frac{d}{dx}(e^x - x) = \frac{d}{dx}(e^x) - \frac{d}{dx}(x) = e^x - 1$$

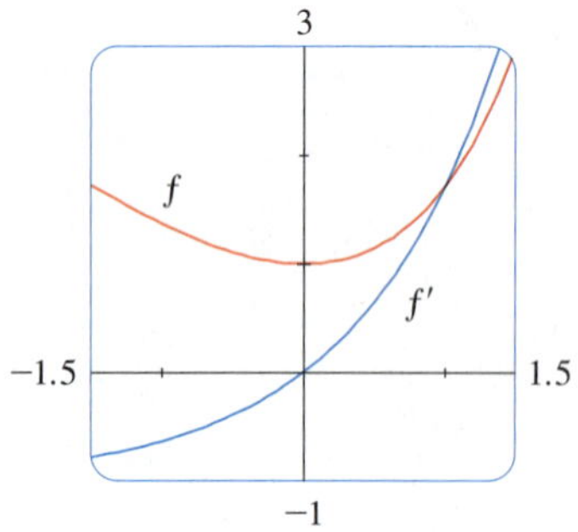

FIGURE 8

The function f and its derivative f' are graphed in Figure 8. Notice that f has a horizontal tangent when $x = 0$; this corresponds to the fact that $f'(0) = 0$. Notice also that, for $x > 0$, $f'(x)$ is positive and f is increasing. When $x < 0$, $f'(x)$ is negative and f is decreasing.

EXAMPLE 8 At what point on the curve $y = e^x$ is the tangent line parallel to the line $y = 2x$?

SOLUTION Since $y = e^x$, we have $y' = e^x$. Let the x-coordinate of the point in question be a. Then the slope of the tangent line at that point is e^a. This tangent line will be parallel to the line $y = 2x$ if it has the same slope, that is, 2. Equating slopes, we get

$$e^a = 2 \qquad a = \ln 2$$

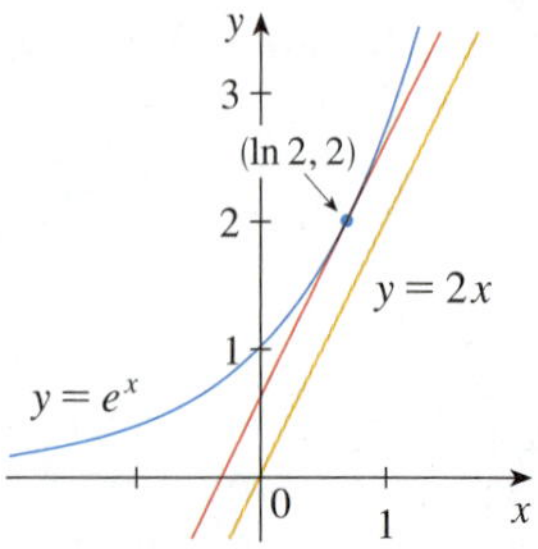

FIGURE 9

Therefore, the required point is $(a, e^a) = (\ln 2, 2)$. (See Figure 9.)

3.1 Exercises

1. (a) How is the number e defined?
(b) Use a calculator to estimate the values of the limits

$$\lim_{h\to 0}\frac{2.7^h - 1}{h} \quad \text{and} \quad \lim_{h\to 0}\frac{2.8^h - 1}{h}$$

correct to two decimal places. What can you conclude about the value of e?

2. (a) Sketch, by hand, the graph of the function $f(x) = e^x$, paying particular attention to how the graph crosses the y-axis. What fact allows you to do this?
(b) What types of functions are $f(x) = e^x$ and $g(x) = x^e$? Compare the differentiation formulas for f and g.
(c) Which of the two functions in part (b) grows more rapidly when x is large?

3–32 Differentiate the function.

3. $f(x) = 186.5$
4. $f(x) = \sqrt{30}$
5. $f(x) = 5x - 1$
6. $F(x) = -4x^{10}$
7. $f(x) = x^2 + 3x - 4$
8. $g(x) = 5x^8 - 2x^5 + 6$
9. $f(t) = \frac{1}{4}(t^4 + 8)$
10. $f(t) = \frac{1}{2}t^6 - 3t^4 + t$
11. $y = x^{-2/5}$
12. $y = 5e^x + 3$
13. $V(r) = \frac{4}{3}\pi r^3$
14. $R(t) = 5t^{-3/5}$
15. $Y(t) = 6t^{-9}$
16. $R(x) = \dfrac{\sqrt{10}}{x^7}$
17. $G(x) = \sqrt{x} - 2e^x$
18. $y = \sqrt[3]{x}$
19. $F(x) = (\frac{1}{2}x)^5$
20. $f(t) = \sqrt{t} - \dfrac{1}{\sqrt{t}}$
21. $g(x) = x^2 + \dfrac{1}{x^2}$
22. $y = \sqrt{x}\,(x - 1)$
23. $y = \dfrac{x^2 + 4x + 3}{\sqrt{x}}$
24. $y = \dfrac{x^2 - 2\sqrt{x}}{x}$
25. $y = 4\pi^2$
26. $g(u) = \sqrt{2}u + \sqrt{3u}$
27. $y = ax^2 + bx + c$
28. $y = ae^v + \dfrac{b}{v} + \dfrac{c}{v^2}$
29. $v = t^2 - \dfrac{1}{\sqrt[4]{t^3}}$
30. $u = \sqrt[3]{t^2} + 2\sqrt{t^3}$
31. $z = \dfrac{A}{y^{10}} + Be^y$
32. $y = e^{x+1} + 1$

33–36 Find $f'(x)$. Compare the graphs of f and f' and use them to explain why your answer is reasonable.

33. $f(x) = e^x - 5x$
34. $f(x) = 3x^5 - 20x^3 + 50x$
35. $f(x) = 3x^{15} - 5x^3 + 3$
36. $f(x) = x + \dfrac{1}{x}$

37–38 Estimate the value of $f'(a)$ by zooming in on the graph of f. Then differentiate f to find the exact value of $f'(a)$ and compare with your estimate.

37. $f(x) = 3x^2 - x^3$, $a = 1$
38. $f(x) = 1/\sqrt{x}$, $a = 4$

39–40 Find an equation of the tangent line to the curve at the given point.

39. $y = x^4 + 2e^x$, $(0, 2)$
40. $y = (1 + 2x)^2$, $(1, 9)$

41–42 Find an equation of the tangent line to the curve at the given point. Illustrate by graphing the curve and the tangent line on the same screen.

41. $y = 3x^2 - x^3$, $(1, 2)$
42. $y = x\sqrt{x}$, $(4, 8)$

43. (a) Use a graphing calculator or computer to graph the function $f(x) = x^4 - 3x^3 - 6x^2 + 7x + 30$ in the viewing rectangle $[-3, 5]$ by $[-10, 50]$.
(b) Using the graph in part (a) to estimate slopes, make a rough sketch, by hand, of the graph of f'. (See Example 1 in Section 2.9.)
(c) Calculate $f'(x)$ and use this expression, with a graphing device, to graph f'. Compare with your sketch in part (b).

44. (a) Use a graphing calculator or computer to graph the function $g(x) = e^x - 3x^2$ in the viewing rectangle $[-1, 4]$ by $[-8, 8]$.
(b) Using the graph in part (a) to estimate slopes, make a rough sketch, by hand, of the graph of g'. (See Example 1 in Section 2.9.)
(c) Calculate $g'(x)$ and use this expression, with a graphing device, to graph g'. Compare with your sketch in part (b).

45. Find the points on the curve $y = 2x^3 + 3x^2 - 12x + 1$ where the tangent is horizontal.

46. For what values of x does the graph of $f(x) = x^3 + 3x^2 + x + 3$ have a horizontal tangent?

47. Show that the curve $y = 6x^3 + 5x - 3$ has no tangent line with slope 4.

48. At what point on the curve $y = 1 + 2e^x - 3x$ is the tangent line parallel to the line $3x - y = 5$? Illustrate by graphing the curve and both lines.

49. Draw a diagram to show that there are two tangent lines to the parabola $y = x^2$ that pass through the point $(0, -4)$. Find the coordinates of the points where these tangent lines intersect the parabola.

50. Find equations of both lines through the point $(2, -3)$ that are tangent to the parabola $y = x^2 + x$.

51. The **normal line** to a curve C at a point P is, by definition, the line that passes through P and is perpendicular to the tangent line to C at P. Find an equation of the normal line to the parabola $y = 1 - x^2$ at the point $(2, -3)$. Sketch the parabola and its normal line.

52. Where does the normal line to the parabola $y = x - x^2$ at the point $(1, 0)$ intersect the parabola a second time? Illustrate with a sketch.

53. Use the definition of a derivative to show that if $f(x) = 1/x$, then $f'(x) = -1/x^2$. (This proves the Power Rule for the case $n = -1$.)

54. Find the parabola with equation $y = ax^2 + bx$ whose tangent line at $(1, 1)$ has equation $y = 3x - 2$.

55. Let

$$f(x) = \begin{cases} 2 - x & \text{if } x \leq 1 \\ x^2 - 2x + 2 & \text{if } x > 1 \end{cases}$$

Is f differentiable at 1? Sketch the graphs of f and f'.

56. At what numbers is the following function g differentiable?

$$g(x) = \begin{cases} -1 - 2x & \text{if } x < -1 \\ x^2 & \text{if } -1 \leq x \leq 1 \\ x & \text{if } x > 1 \end{cases}$$

Give a formula for g' and sketch the graphs of g and g'.

57. (a) For what values of x is the function $f(x) = |x^2 - 9|$ differentiable? Find a formula for f'.
(b) Sketch the graphs of f and f'.

58. Where is the function $h(x) = |x - 1| + |x + 2|$ differentiable? Give a formula for h' and sketch the graphs of h and h'.

59. For what values of a and b is the line $2x + y = b$ tangent to the parabola $y = ax^2$ when $x = 2$?

60. Let

$$f(x) = \begin{cases} x^2 & \text{if } x \leq 2 \\ mx + b & \text{if } x > 2 \end{cases}$$

Find the values of m and b that make f differentiable everywhere.

61. Find a cubic function

$$y = ax^3 + bx^2 + cx + d$$

whose graph has horizontal tangents at the points $(-2, 6)$ and $(2, 0)$.

62. A tangent line is drawn to the hyperbola $xy = c$ at a point P.
(a) Show that the midpoint of the line segment cut from this tangent line by the coordinate axes is P.
(b) Show that the triangle formed by the tangent line and the coordinate axes always has the same area, no matter where P is located on the hyperbola.

63. Evaluate $\lim_{x \to 1} \dfrac{x^{1000} - 1}{x - 1}$.

64. Draw a diagram showing two perpendicular lines that intersect on the y-axis and are both tangent to the parabola $y = x^2$. Where do these lines intersect?

3.2 The Product and Quotient Rules

The formulas of this section enable us to differentiate new functions formed from old functions by multiplication or division.

The Product Rule

By analogy with the Sum and Difference Rules, one might be tempted to guess, as Leibniz did three centuries ago, that the derivative of a product is the product of the derivatives. We can see, however, that this guess is wrong by looking at a particular example. Let $f(x) = x$ and $g(x) = x^2$. Then the Power Rule gives $f'(x) = 1$ and $g'(x) = 2x$. But $(fg)(x) = x^3$, so $(fg)'(x) = 3x^2$. Thus, $(fg)' \neq f'g'$. The correct formula was discovered by Leibniz (soon after his false start) and is called the Product Rule.

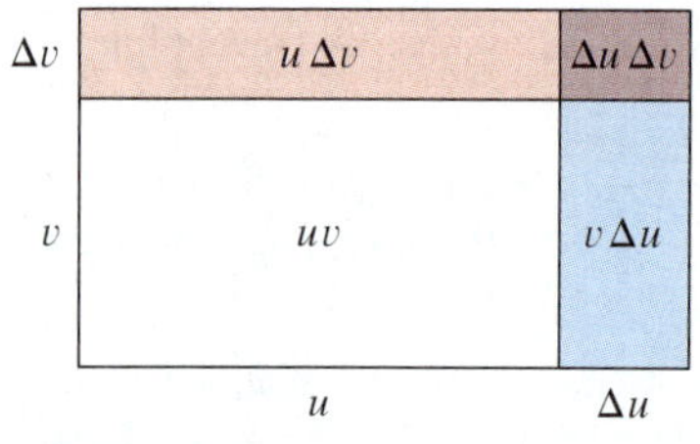

FIGURE 1
The geometry of the Product Rule

Before stating the Product Rule, let's see how we might discover it. We start by assuming that $u = f(x)$ and $v = g(x)$ are both positive differentiable functions. Then we can interpret the product uv as an area of a rectangle (see Figure 1). If x changes by an amount Δx, then the corresponding changes in u and v are

$$\Delta u = f(x + \Delta x) - f(x) \qquad \Delta v = g(x + \Delta x) - g(x)$$

and the new value of the product, $(u + \Delta u)(v + \Delta v)$, can be interpreted as the area of the large rectangle in Figure 1 (provided that Δu and Δv happen to be positive).

The change in the area of the rectangle is

1
$$\Delta(uv) = (u + \Delta u)(v + \Delta v) - uv = u\,\Delta v + v\,\Delta u + \Delta u\,\Delta v$$
$$= \text{the sum of the three shaded areas}$$

If we divide by Δx, we get

$$\frac{\Delta(uv)}{\Delta x} = u\frac{\Delta v}{\Delta x} + v\frac{\Delta u}{\Delta x} + \Delta u\frac{\Delta v}{\Delta x}$$

If we now let $\Delta x \to 0$, we get the derivative of uv:

||| Recall that in Leibniz notation the definition of a derivative can be written as

$$\frac{dy}{dx} = \lim_{\Delta x \to 0} \frac{\Delta y}{\Delta x}$$

$$\frac{d}{dx}(uv) = \lim_{\Delta x \to 0} \frac{\Delta(uv)}{\Delta x} = \lim_{\Delta x \to 0} \left(u\frac{\Delta v}{\Delta x} + v\frac{\Delta u}{\Delta x} + \Delta u\frac{\Delta v}{\Delta x}\right)$$
$$= u\lim_{\Delta x \to 0}\frac{\Delta v}{\Delta x} + v\lim_{\Delta x \to 0}\frac{\Delta u}{\Delta x} + \left(\lim_{\Delta x \to 0}\Delta u\right)\left(\lim_{\Delta x \to 0}\frac{\Delta v}{\Delta x}\right)$$
$$= u\frac{dv}{dx} + v\frac{du}{dx} + 0 \cdot \frac{dv}{dx}$$

2
$$\frac{d}{dx}(uv) = u\frac{dv}{dx} + v\frac{du}{dx}$$

(Notice that $\Delta u \to 0$ as $\Delta x \to 0$ since f is differentiable and therefore continuous.)

Although we started by assuming (for the geometric interpretation) that all the quantities are positive, we notice that Equation 1 is always true. (The algebra is valid whether u, v, Δu, and Δv are positive or negative.) So we have proved Equation 2, known as the Product Rule, for all differentiable functions u and v.

||| In prime notation:

$$(fg)' = fg' + gf'$$

The Product Rule If f and g are both differentiable, then

$$\frac{d}{dx}[f(x)g(x)] = f(x)\frac{d}{dx}[g(x)] + g(x)\frac{d}{dx}[f(x)]$$

||| Figure 2 shows the graphs of the function f of Example 1 and its derivative f'. Notice that $f'(x)$ is positive when f is increasing and negative when f is decreasing.

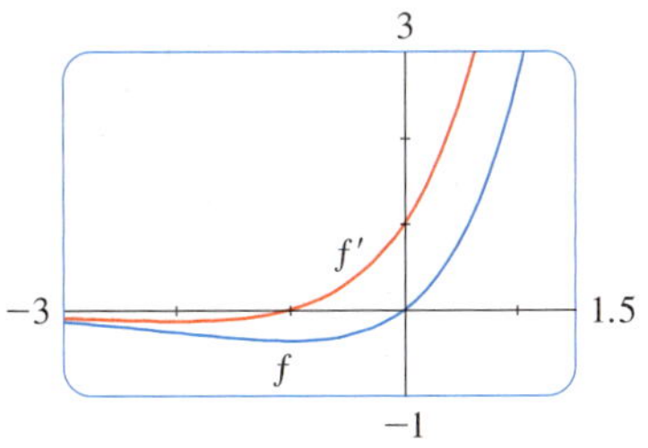

FIGURE 2

In words, the Product Rule says that *the derivative of a product of two functions is the first function times the derivative of the second function plus the second function times the derivative of the first function.*

EXAMPLE 1 If $f(x) = xe^x$, find $f'(x)$.

SOLUTION By the Product Rule, we have

$$f'(x) = \frac{d}{dx}(xe^x) = x\frac{d}{dx}(e^x) + e^x\frac{d}{dx}(x)$$
$$= xe^x + e^x \cdot 1 = (x + 1)e^x$$

EXAMPLE 2 Differentiate the function $f(t) = \sqrt{t}\,(1 - t)$.

SOLUTION 1 Using the Product Rule, we have

$$f'(t) = \sqrt{t}\,\frac{d}{dt}(1 - t) + (1 - t)\frac{d}{dt}\sqrt{t}$$
$$= \sqrt{t}\,(-1) + (1 - t)\cdot\tfrac{1}{2}t^{-1/2}$$
$$= -\sqrt{t} + \frac{1 - t}{2\sqrt{t}} = \frac{1 - 3t}{2\sqrt{t}}$$

SOLUTION 2 If we first use the laws of exponents to rewrite $f(t)$, then we can proceed directly without using the Product Rule.

$$f(t) = \sqrt{t} - t\sqrt{t} = t^{1/2} - t^{3/2}$$
$$f'(t) = \tfrac{1}{2}t^{-1/2} - \tfrac{3}{2}t^{1/2}$$

which is equivalent to the answer given in Solution 1.

Example 2 shows that it is sometimes easier to simplify a product of functions than to use the Product Rule. In Example 1, however, the Product Rule is the only possible method.

EXAMPLE 3 If $f(x) = \sqrt{x}\,g(x)$, where $g(4) = 2$ and $g'(4) = 3$, find $f'(4)$.

SOLUTION Applying the Product Rule, we get

$$f'(x) = \frac{d}{dx}\left[\sqrt{x}\,g(x)\right] = \sqrt{x}\,\frac{d}{dx}[g(x)] + g(x)\frac{d}{dx}\left[\sqrt{x}\right]$$
$$= \sqrt{x}\,g'(x) + g(x)\cdot\tfrac{1}{2}x^{-1/2}$$
$$= \sqrt{x}\,g'(x) + \frac{g(x)}{2\sqrt{x}}$$

So $$f'(4) = \sqrt{4}\,g'(4) + \frac{g(4)}{2\sqrt{4}} = 2\cdot 3 + \frac{2}{2\cdot 2} = 6.5$$

EXAMPLE 4 A telephone company wants to estimate the number of new residential phone lines that it will need to install during the upcoming month. At the beginning of January the company had 100,000 subscribers, each of whom had 1.2 phone lines, on average. The company estimated that its subscribership was increasing at the rate of 1000 monthly. By polling its existing subscribers, the company found that each intended to install an average of 0.01 new phone lines by the end of January. Estimate the number of new lines the company will have to install in January by calculating the rate of increase of lines at the beginning of the month.

SOLUTION Let $s(t)$ be the number of subscribers and let $n(t)$ be the number of phone lines per subscriber at time t, where t is measured in months and $t = 0$ corresponds to the beginning of January. Then the total number of lines is given by

$$L(t) = s(t)n(t)$$

and we want to find $L'(0)$. According to the Product Rule, we have

$$L'(t) = \frac{d}{dt}[s(t)n(t)] = s(t)\frac{d}{dt}n(t) + n(t)\frac{d}{dt}s(t)$$

We are given that $s(0) = 100{,}000$ and $n(0) = 1.2$. The company's estimates concerning rates of increase are that $s'(0) \approx 1000$ and $n'(0) \approx 0.01$. Therefore,

$$\begin{aligned} L'(0) &= s(0)n'(0) + n(0)s'(0) \\ &\approx 100{,}000 \cdot 0.01 + 1.2 \cdot 1000 = 2200 \end{aligned}$$

The company will need to install approximately 2200 new phone lines in January.

Notice that the two terms arising from the Product Rule come from different sources—old subscribers and new subscribers. One contribution to L' is the number of existing subscribers (100,000) times the rate at which they order new lines (about 0.01 per subscriber monthly). A second contribution is the average number of lines per subscriber (1.2 at the beginning of the month) times the rate of increase of subscribers (1000 monthly).

The Quotient Rule

We find a rule for differentiating the quotient of two differentiable functions $u = f(x)$ and $v = g(x)$ in much the same way that we found the Product Rule. If x, u, and v change by amounts Δx, Δu, and Δv, then the corresponding change in the quotient u/v is

$$\begin{aligned} \Delta\left(\frac{u}{v}\right) &= \frac{u + \Delta u}{v + \Delta v} - \frac{u}{v} = \frac{(u + \Delta u)v - u(v + \Delta v)}{v(v + \Delta v)} \\ &= \frac{v\Delta u - u\Delta v}{v(v + \Delta v)} \end{aligned}$$

so

$$\frac{d}{dx}\left(\frac{u}{v}\right) = \lim_{\Delta x \to 0} \frac{\Delta(u/v)}{\Delta x} = \lim_{\Delta x \to 0} \frac{v\dfrac{\Delta u}{\Delta x} - u\dfrac{\Delta v}{\Delta x}}{v(v + \Delta v)}$$

As $\Delta x \to 0$, $\Delta v \to 0$ also, because g is differentiable and therefore continuous. Thus, using the Limit Laws, we get

$$\frac{d}{dx}\left(\frac{u}{v}\right) = \frac{v \displaystyle\lim_{\Delta x \to 0} \frac{\Delta u}{\Delta x} - u \lim_{\Delta x \to 0} \frac{\Delta v}{\Delta x}}{v \displaystyle\lim_{\Delta x \to 0} (v + \Delta v)} = \frac{v\dfrac{du}{dx} - u\dfrac{dv}{dx}}{v^2}$$

The Quotient Rule If f and g are differentiable, then

$$\frac{d}{dx}\left[\frac{f(x)}{g(x)}\right] = \frac{g(x)\dfrac{d}{dx}[f(x)] - f(x)\dfrac{d}{dx}[g(x)]}{[g(x)]^2}$$

In words, the Quotient Rule says that the *derivative of a quotient is the denominator times the derivative of the numerator minus the numerator times the derivative of the denominator, all divided by the square of the denominator.*

The Quotient Rule and the other differentiation formulas enable us to compute the derivative of any rational function, as the next example illustrates.

|||| We can use a graphing device to check that the answer to Example 5 is plausible. Figure 3 shows the graphs of the function of Example 5 and its derivative. Notice that when y grows rapidly (near -2), y' is large. And when y grows slowly, y' is near 0.

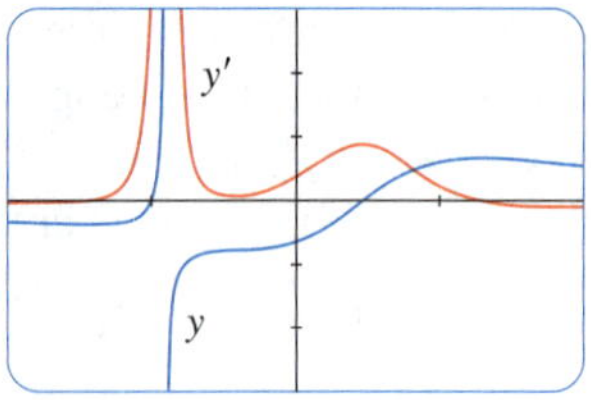

FIGURE 3

EXAMPLE 5 Let $y = \dfrac{x^2 + x - 2}{x^3 + 6}$.

Then

$$y' = \frac{(x^3 + 6)\dfrac{d}{dx}(x^2 + x - 2) - (x^2 + x - 2)\dfrac{d}{dx}(x^3 + 6)}{(x^3 + 6)^2}$$

$$= \frac{(x^3 + 6)(2x + 1) - (x^2 + x - 2)(3x^2)}{(x^3 + 6)^2}$$

$$= \frac{(2x^4 + x^3 + 12x + 6) - (3x^4 + 3x^3 - 6x^2)}{(x^3 + 6)^2}$$

$$= \frac{-x^4 - 2x^3 + 6x^2 + 12x + 6}{(x^3 + 6)^2}$$

EXAMPLE 6 Find an equation of the tangent line to the curve $y = e^x/(1 + x^2)$ at the point $(1, e/2)$.

SOLUTION According to the Quotient Rule, we have

$$\frac{dy}{dx} = \frac{(1 + x^2)\dfrac{d}{dx}(e^x) - e^x\dfrac{d}{dx}(1 + x^2)}{(1 + x^2)^2}$$

$$= \frac{(1 + x^2)e^x - e^x(2x)}{(1 + x^2)^2} = \frac{e^x(1 - x)^2}{(1 + x^2)^2}$$

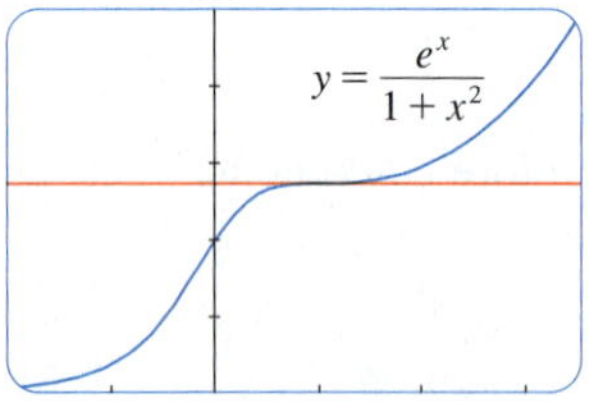

FIGURE 4

So the slope of the tangent line at $(1, e/2)$ is

$$\left.\frac{dy}{dx}\right|_{x=1} = 0$$

This means that the tangent line at $(1, e/2)$ is horizontal and its equation is $y = e/2$. [See Figure 4. Notice that the function is increasing and crosses its tangent line at $(1, e/2)$.]

NOTE ▫ Don't use the Quotient Rule *every* time you see a quotient. Sometimes it's easier to rewrite a quotient first to put it in a form that is simpler for the purpose of differentiation. For instance, although it is possible to differentiate the function

$$F(x) = \frac{3x^2 + 2\sqrt{x}}{x}$$

using the Quotient Rule, it is much easier to perform the division first and write the function as

$$F(x) = 3x + 2x^{-1/2}$$

before differentiating.

We summarize the differentiation formulas we have learned so far as follows.

Table of Differentiation Formulas

$\dfrac{d}{dx}(c) = 0$	$\dfrac{d}{dx}(x^n) = nx^{n-1}$	$\dfrac{d}{dx}(e^x) = e^x$
$(cf)' = cf'$	$(f + g)' = f' + g'$	$(f - g)' = f' - g'$
$(fg)' = fg' + gf'$	$\left(\dfrac{f}{g}\right)' = \dfrac{gf' - fg'}{g^2}$	

3.2 Exercises

1. Find the derivative of $y = (x^2 + 1)(x^3 + 1)$ in two ways: by using the Product Rule and by performing the multiplication first. Do your answers agree?

2. Find the derivative of the function

$$F(x) = \frac{x - 3x\sqrt{x}}{\sqrt{x}}$$

in two ways: by using the Quotient Rule and by simplifying first. Show that your answers are equivalent. Which method do you prefer?

3–22 Differentiate.

3. $f(x) = x^2e^x$

4. $g(x) = \sqrt{x}\, e^x$

5. $y = \dfrac{e^x}{x^2}$

6. $y = \dfrac{e^x}{1 + x}$

7. $g(x) = \dfrac{3x - 1}{2x + 1}$

8. $f(t) = \dfrac{2t}{4 + t^2}$

9. $V(x) = (2x^3 + 3)(x^4 - 2x)$

10. $Y(u) = (u^{-2} + u^{-3})(u^5 - 2u^2)$

11. $F(y) = \left(\dfrac{1}{y^2} - \dfrac{3}{y^4}\right)(y + 5y^3)$

12. $R(t) = (t + e^t)(3 - \sqrt{t})$

13. $y = \dfrac{t^2}{3t^2 - 2t + 1}$

14. $y = \dfrac{t^3 + t}{t^4 - 2}$

15. $y = (r^2 - 2r)e^r$

16. $y = \dfrac{1}{s + ke^s}$

17. $y = \dfrac{v^3 - 2v\sqrt{v}}{v}$

18. $z = w^{3/2}(w + ce^w)$

19. $y = \dfrac{1}{x^4 + x^2 + 1}$

20. $y = \dfrac{\sqrt{x} - 1}{\sqrt{x} + 1}$

21. $f(x) = \dfrac{x}{x + \dfrac{c}{x}}$

22. $f(x) = \dfrac{ax + b}{cx + d}$

23–26 Find an equation of the tangent line to the given curve at the specified point.

23. $y = \dfrac{2x}{x + 1}$, $(1, 1)$

24. $y = \dfrac{\sqrt{x}}{x + 1}$, $(4, 0.4)$

25. $y = 2xe^x$, $(0, 0)$

26. $y = \dfrac{e^x}{x}$, $(1, e)$

27. (a) The curve $y = 1/(1 + x^2)$ is called a **witch of Maria Agnesi**. Find an equation of the tangent line to this curve at the point $\left(-1, \frac{1}{2}\right)$.
(b) Illustrate part (a) by graphing the curve and the tangent line on the same screen.

28. (a) The curve $y = x/(1 + x^2)$ is called a **serpentine**. Find an equation of the tangent line to this curve at the point $(3, 0.3)$.
(b) Illustrate part (a) by graphing the curve and the tangent line on the same screen.

29. (a) If $f(x) = e^x/x^3$, find $f'(x)$.
(b) Check to see that your answer to part (a) is reasonable by comparing the graphs of f and f'.

30. (a) If $f(x) = x/(x^2 - 1)$, find $f'(x)$.
(b) Check to see that your answer to part (a) is reasonable by comparing the graphs of f and f'.

31. Suppose that $f(5) = 1$, $f'(5) = 6$, $g(5) = -3$, and $g'(5) = 2$. Find the following values.
(a) $(fg)'(5)$
(b) $(f/g)'(5)$
(c) $(g/f)'(5)$

32. If $f(3) = 4$, $g(3) = 2$, $f'(3) = -6$, and $g'(3) = 5$, find the following numbers.
(a) $(f + g)'(3)$
(b) $(fg)'(3)$
(c) $\left(\dfrac{f}{g}\right)'(3)$
(d) $\left(\dfrac{f}{f - g}\right)'(3)$

33. If $f(x) = e^xg(x)$, where $g(0) = 2$ and $g'(0) = 5$, find $f'(0)$.

34. If $h(2) = 4$ and $h'(2) = -3$, find

$$\left.\frac{d}{dx}\left(\frac{h(x)}{x}\right)\right|_{x=2}$$

35. If f and g are the functions whose graphs are shown, let $u(x) = f(x)g(x)$ and $v(x) = f(x)/g(x)$.
(a) Find $u'(1)$. (b) Find $v'(5)$.

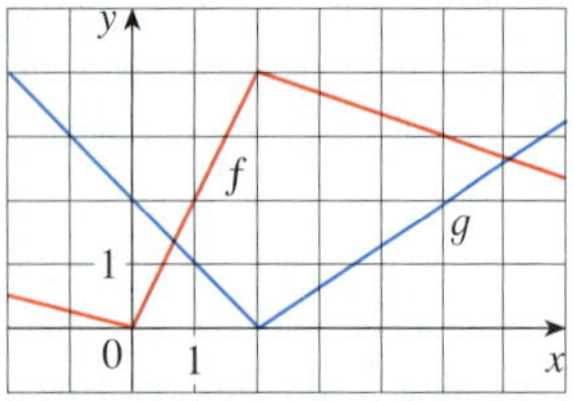

36. Let $P(x) = F(x)G(x)$ and $Q(x) = F(x)/G(x)$, where F and G are the functions whose graphs are shown.
(a) Find $P'(2)$. (b) Find $Q'(7)$.

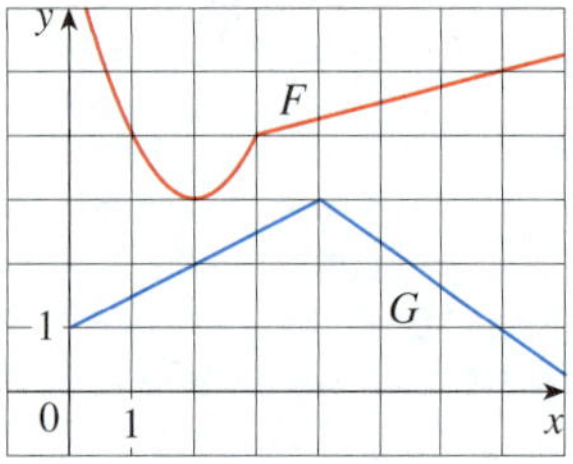

37. If g is a differentiable function, find an expression for the derivative of each of the following functions.

(a) $y = xg(x)$ (b) $y = \dfrac{x}{g(x)}$ (c) $y = \dfrac{g(x)}{x}$

38. If f is a differentiable function, find an expression for the derivative of each of the following functions.

(a) $y = x^2 f(x)$ (b) $y = \dfrac{f(x)}{x^2}$

(c) $y = \dfrac{x^2}{f(x)}$ (d) $y = \dfrac{1 + xf(x)}{\sqrt{x}}$

39. In this exercise we estimate the rate at which the total personal income is rising in the Richmond-Petersburg, Virginia, metropolitan area. In 1999, the population of this area was 961,400, and the population was increasing at roughly 9200 people per year. The average annual income was \$30,593 per capita, and this average was increasing at about \$1400 per year (a little above the national average of about \$1225 yearly). Use the Product Rule and these figures to estimate the rate at which total personal income was rising in the Richmond-Petersburg area in 1999. Explain the meaning of each term in the Product Rule.

40. A manufacturer produces bolts of a fabric with a fixed width. The quantity q of this fabric (measured in yards) that is sold is a function of the selling price p (in dollars per yard), so we can write $q = f(p)$. Then the total revenue earned with selling price p is $R(p) = pf(p)$.
(a) What does it mean to say that $f(20) = 10{,}000$ and $f'(20) = -350$?
(b) Assuming the values in part (a), find $R'(20)$ and interpret your answer.

41. How many tangent lines to the curve $y = x/(x + 1)$ pass through the point $(1, 2)$? At which points do these tangent lines touch the curve?

42. Find equations of the tangent lines to the curve

$$y = \frac{x - 1}{x + 1}$$

that are parallel to the line $x - 2y = 2$.

43. (a) Use the Product Rule twice to prove that if f, g, and h are differentiable, then

$$(fgh)' = f'gh + fg'h + fgh'$$

(b) Taking $f = g = h$ in part (a), show that

$$\frac{d}{dx}[f(x)]^3 = 3[f(x)]^2 f'(x)$$

(c) Use part (b) to differentiate $y = e^{3x}$.

44. (a) If g is differentiable, the **Reciprocal Rule** says that

$$\frac{d}{dx}\left[\frac{1}{g(x)}\right] = -\frac{g'(x)}{[g(x)]^2}$$

Use the Quotient Rule to prove the Reciprocal Rule.
(b) Use the Reciprocal Rule to differentiate the function in Exercise 19.
(c) Use the Reciprocal Rule to verify that the Power Rule is valid for negative integers, that is,

$$\frac{d}{dx}(x^{-n}) = -nx^{-n-1}$$

for all positive integers n.

3.3 Rates of Change in the Natural and Social Sciences

Recall from Section 2.8 that if $y = f(x)$, then the derivative dy/dx can be interpreted as the rate of change of y with respect to x. In this section we examine some of the applications of this idea to physics, chemistry, biology, economics, and other sciences.

Let's recall from Section 2.7 the basic idea behind rates of change. If x changes from x_1 to x_2, then the change in x is

$$\Delta x = x_2 - x_1$$

and the corresponding change in y is

$$\Delta y = f(x_2) - f(x_1)$$

The difference quotient

$$\frac{\Delta y}{\Delta x} = \frac{f(x_2) - f(x_1)}{x_2 - x_1}$$

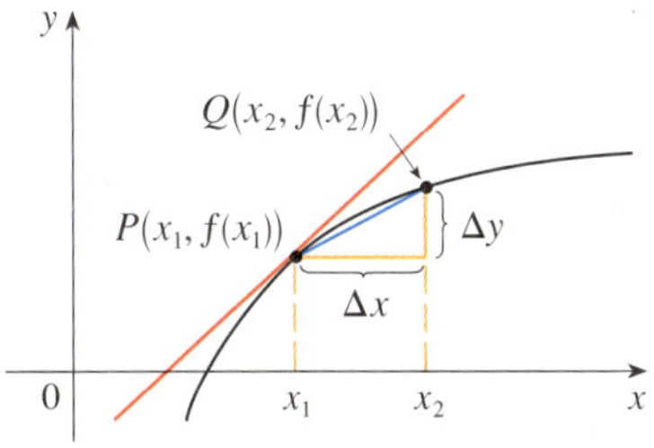

m_{PQ} = average rate of change

$m = f'(x_1)$ = instantaneous rate of change

FIGURE 1

is the **average rate of change of y with respect to x** over the interval $[x_1, x_2]$ and can be interpreted as the slope of the secant line PQ in Figure 1. Its limit as $\Delta x \to 0$ is the derivative $f'(x_1)$, which can therefore be interpreted as the instantaneous rate of change of y with respect to x or the slope of the tangent line at $P(x_1, f(x_1))$. Using Leibniz notation, we write the process in the form

$$\frac{dy}{dx} = \lim_{\Delta x \to 0} \frac{\Delta y}{\Delta x}$$

Whenever the function $y = f(x)$ has a specific interpretation in one of the sciences, its derivative will have a specific interpretation as a rate of change. (As we discussed in Section 2.7, the units for dy/dx are the units for y divided by the units for x.) We now look at some of these interpretations in the natural and social sciences.

Physics

Resources / Module 4 / Polynomial Models / Start of Polynomial Models

If $s = f(t)$ is the position function of a particle that is moving in a straight line, then $\Delta s/\Delta t$ represents the average velocity over a time period Δt, and $v = ds/dt$ represents the instantaneous **velocity** (the rate of change of displacement with respect to time). This was discussed in Sections 2.7 and 2.8, but now that we know the differentiation formulas, we are able to solve velocity problems more easily.

EXAMPLE 1 The position of a particle is given by the equation

$$s = f(t) = t^3 - 6t^2 + 9t$$

where t is measured in seconds and s in meters.

(a) Find the velocity at time t.

(b) What is the velocity after 2 s? After 4 s?

(c) When is the particle at rest?

(d) When is the particle moving forward (that is, in the positive direction)?
(e) Draw a diagram to represent the motion of the particle.
(f) Find the total distance traveled by the particle during the first five seconds.

SOLUTION
(a) The velocity function is the derivative of the position function.

$$s = f(t) = t^3 - 6t^2 + 9t$$

$$v(t) = \frac{ds}{dt} = 3t^2 - 12t + 9$$

(b) The velocity after 2 s means the instantaneous velocity when $t = 2$, that is,

$$v(2) = \left.\frac{ds}{dt}\right|_{t=2} = 3(2)^2 - 12(2) + 9 = -3 \text{ m/s}$$

The velocity after 4 s is

$$v(4) = 3(4)^2 - 12(4) + 9 = 9 \text{ m/s}$$

(c) The particle is at rest when $v(t) = 0$, that is,

$$3t^2 - 12t + 9 = 3(t^2 - 4t + 3) = 3(t - 1)(t - 3) = 0$$

and this is true when $t = 1$ or $t = 3$. Thus, the particle is at rest after 1 s and after 3 s.
(d) The particle moves in the positive direction when $v(t) > 0$, that is,

$$3t^2 - 12t + 9 = 3(t - 1)(t - 3) > 0$$

This inequality is true when both factors are positive ($t > 3$) or when both factors are negative ($t < 1$). Thus, the particle moves in the positive direction in the time intervals $t < 1$ and $t > 3$. It moves backward (in the negative direction) when $1 < t < 3$.

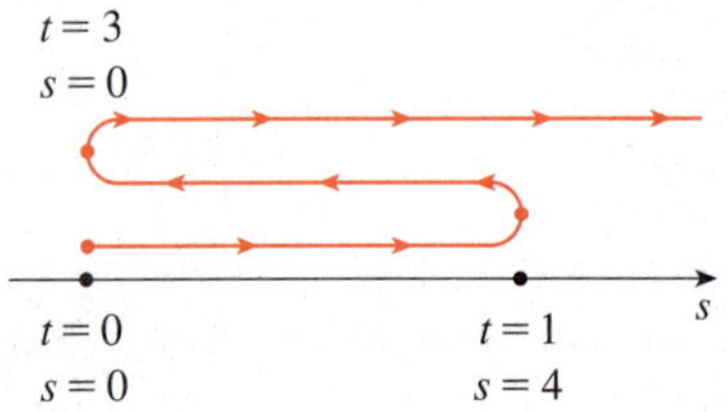

FIGURE 2

(e) Using the information from part (d), we make a schematic sketch as shown in Figure 2 of the motion of the particle back and forth along a line (the s-axis).

(f) Because of what we learned in parts (d) and (e), we need to calculate the distances traveled during the time intervals [0, 1], [1, 3], and [3, 5] separately.
The distance traveled in the first second is

$$|f(1) - f(0)| = |4 - 0| = 4 \text{ m}$$

From $t = 1$ to $t = 3$ the distance traveled is

$$|f(3) - f(1)| = |0 - 4| = 4 \text{ m}$$

From $t = 3$ to $t = 5$ the distance traveled is

$$|f(5) - f(3)| = |20 - 0| = 20 \text{ m}$$

The total distance is $4 + 4 + 20 = 28$ m.

EXAMPLE 2 If a rod or piece of wire is homogeneous, then its *linear density* is uniform and is defined as the mass per unit length ($\rho = m/l$) and measured in kilograms per meter. Suppose, however, that the rod is not homogeneous but that its mass measured from its left end to a point x is $m = f(x)$, as shown in Figure 3.

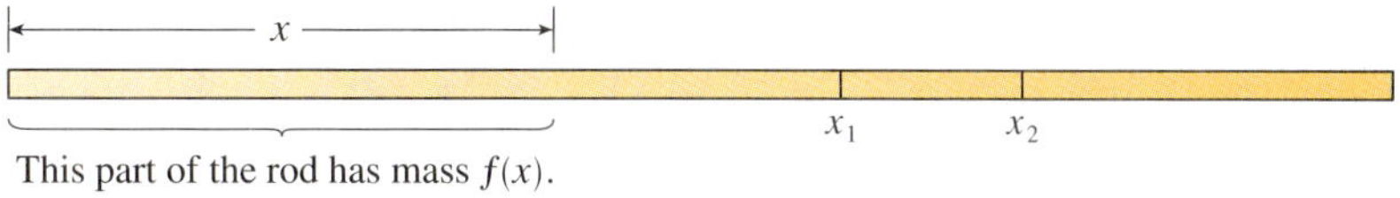

FIGURE 3

The mass of the part of the rod that lies between $x = x_1$ and $x = x_2$ is given by $\Delta m = f(x_2) - f(x_1)$, so the average density of that part of the rod is

$$\text{average density} = \frac{\Delta m}{\Delta x} = \frac{f(x_2) - f(x_1)}{x_2 - x_1}$$

If we now let $\Delta x \to 0$ (that is, $x_2 \to x_1$), we are computing the average density over smaller and smaller intervals. The **linear density** ρ at x_1 is the limit of these average densities as $\Delta x \to 0$; that is, the linear density is the rate of change of mass with respect to length. Symbolically,

$$\rho = \lim_{\Delta x \to 0} \frac{\Delta m}{\Delta x} = \frac{dm}{dx}$$

Thus, the linear density of the rod is the derivative of mass with respect to length.

For instance, if $m = f(x) = \sqrt{x}$, where x is measured in meters and m in kilograms, then the average density of the part of the rod given by $1 \leqslant x \leqslant 1.2$ is

$$\frac{\Delta m}{\Delta x} = \frac{f(1.2) - f(1)}{1.2 - 1} = \frac{\sqrt{1.2} - 1}{0.2} \approx 0.48 \text{ kg/m}$$

while the density right at $x = 1$ is

$$\rho = \left.\frac{dm}{dx}\right|_{x=1} = \left.\frac{1}{2\sqrt{x}}\right|_{x=1} = 0.50 \text{ kg/m}$$

FIGURE 4

EXAMPLE 3 A current exists whenever electric charges move. Figure 4 shows part of a wire and electrons moving through a shaded plane surface. If ΔQ is the net charge that passes through this surface during a time period Δt, then the average current during this time interval is defined as

$$\text{average current} = \frac{\Delta Q}{\Delta t} = \frac{Q_2 - Q_1}{t_2 - t_1}$$

If we take the limit of this average current over smaller and smaller time intervals, we get what is called the **current** I at a given time t_1:

$$I = \lim_{\Delta t \to 0} \frac{\Delta Q}{\Delta t} = \frac{dQ}{dt}$$

Thus, the current is the rate at which charge flows through a surface. It is measured in units of charge per unit time (often coulombs per second, called amperes).

Velocity, density, and current are not the only rates of change that are important in physics. Others include power (the rate at which work is done), the rate of heat flow, temperature gradient (the rate of change of temperature with respect to position), and the rate of decay of a radioactive substance in nuclear physics.

Chemistry

EXAMPLE 4 A chemical reaction results in the formation of one or more substances (called *products*) from one or more starting materials (called *reactants*). For instance, the "equation"

$$2H_2 + O_2 \rightarrow 2H_2O$$

indicates that two molecules of hydrogen and one molecule of oxygen form two molecules of water. Let's consider the reaction

$$A + B \rightarrow C$$

where A and B are the reactants and C is the product. The **concentration** of a reactant A is the number of moles (1 mole $= 6.022 \times 10^{23}$ molecules) per liter and is denoted by [A]. The concentration varies during a reaction, so [A], [B], and [C] are all functions of time (t). The average rate of reaction of the product C over a time interval $t_1 \leqslant t \leqslant t_2$ is

$$\frac{\Delta[C]}{\Delta t} = \frac{[C](t_2) - [C](t_1)}{t_2 - t_1}$$

But chemists are more interested in the **instantaneous rate of reaction**, which is obtained by taking the limit of the average rate of reaction as the time interval Δt approaches 0:

$$\text{rate of reaction} = \lim_{\Delta t \to 0} \frac{\Delta[C]}{\Delta t} = \frac{d[C]}{dt}$$

Since the concentration of the product increases as the reaction proceeds, the derivative $d[C]/dt$ will be positive. (You can see intuitively that the slope of the tangent to the graph of an increasing function is positive.) Thus, the rate of reaction of C is positive. The concentrations of the reactants, however, decrease during the reaction, so, to make the rates of reaction of A and B positive numbers, we put minus signs in front of the derivatives $d[A]/dt$ and $d[B]/dt$. Since [A] and [B] each decrease at the same rate that [C] increases, we have

$$\text{rate of reaction} = \frac{d[C]}{dt} = -\frac{d[A]}{dt} = -\frac{d[B]}{dt}$$

More generally, it turns out that for a reaction of the form

$$aA + bB \rightarrow cC + dD$$

we have

$$-\frac{1}{a}\frac{d[A]}{dt} = -\frac{1}{b}\frac{d[B]}{dt} = \frac{1}{c}\frac{d[C]}{dt} = \frac{1}{d}\frac{d[D]}{dt}$$

The rate of reaction can be determined by graphical methods (see Exercise 22). In some cases we can use the rate of reaction to find explicit formulas for the concentrations as functions of time (see Exercises 9.3).

EXAMPLE 5 One of the quantities of interest in thermodynamics is compressibility. If a given substance is kept at a constant temperature, then its volume V depends on its pressure P. We can consider the rate of change of volume with respect to pressure—namely, the derivative dV/dP. As P increases, V decreases, so $dV/dP < 0$. The **compressibility** is defined by introducing a minus sign and dividing this derivative by the volume V:

$$\text{isothermal compressibility} = \beta = -\frac{1}{V}\frac{dV}{dP}$$

Thus, β measures how fast, per unit volume, the volume of a substance decreases as the pressure on it increases at constant temperature.

For instance, the volume V (in cubic meters) of a sample of air at 25°C was found to be related to the pressure P (in kilopascals) by the equation

$$V = \frac{5.3}{P}$$

The rate of change of V with respect to P when $P = 50$ kPa is

$$\left.\frac{dV}{dP}\right|_{P=50} = \left.-\frac{5.3}{P^2}\right|_{P=50}$$

$$= -\frac{5.3}{2500} = -0.00212 \text{ m}^3/\text{kPa}$$

The compressibility at that pressure is

$$\beta = \left.-\frac{1}{V}\frac{dV}{dP}\right|_{P=50} = \frac{0.00212}{\dfrac{5.3}{50}} = 0.02 \text{ (m}^3/\text{kPa)/m}^3$$

Biology

EXAMPLE 6 Let $n = f(t)$ be the number of individuals in an animal or plant population at time t. The change in the population size between the times $t = t_1$ and $t = t_2$ is $\Delta n = f(t_2) - f(t_1)$, and so the average rate of growth during the time period $t_1 \leqslant t \leqslant t_2$ is

$$\text{average rate of growth} = \frac{\Delta n}{\Delta t} = \frac{f(t_2) - f(t_1)}{t_2 - t_1}$$

The **instantaneous rate of growth** is obtained from this average rate of growth by letting the time period Δt approach 0:

$$\text{growth rate} = \lim_{\Delta t \to 0} \frac{\Delta n}{\Delta t} = \frac{dn}{dt}$$

Strictly speaking, this is not quite accurate because the actual graph of a population function $n = f(t)$ would be a step function that is discontinuous whenever a birth or death occurs and, therefore, not differentiable. However, for a large animal or plant population, we can replace the graph by a smooth approximating curve as in Figure 5.

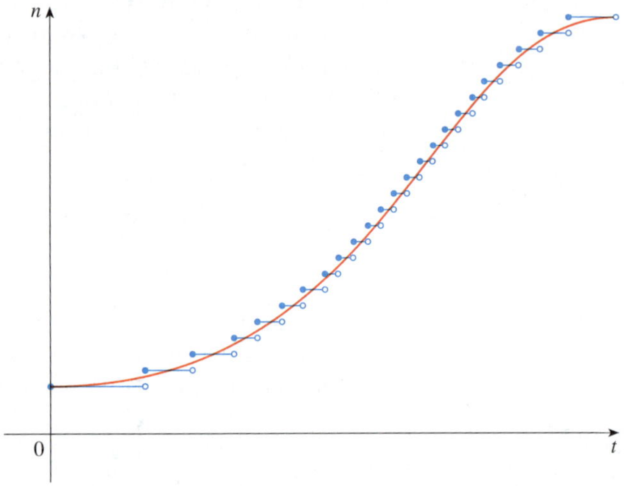

FIGURE 5
A smooth curve approximating a growth function

To be more specific, consider a population of bacteria in a homogeneous nutrient medium. Suppose that by sampling the population at certain intervals it is determined that the population doubles every hour. If the initial population is n_0 and the time t is measured in hours, then

$$f(1) = 2f(0) = 2n_0$$

$$f(2) = 2f(1) = 2^2n_0$$

$$f(3) = 2f(2) = 2^3n_0$$

and, in general,

$$f(t) = 2^t n_0$$

The population function is $n = n_0 2^t$.

In Section 3.1 we discussed derivatives of exponential functions and found that

$$\frac{d}{dx}(2^x) \approx (0.69)2^x$$

So the rate of growth of the bacteria population at time t is

$$\frac{dn}{dt} = \frac{d}{dt}(n_0 2^t) \approx n_0(0.69)2^t$$

For example, suppose that we start with an initial population of $n_0 = 100$ bacteria. Then the rate of growth after 4 hours is

$$\left.\frac{dn}{dt}\right|_{t=4} \approx 100(0.69)2^4 = 1104$$

This means that, after 4 hours, the bacteria population is growing at a rate of about 1100 bacteria per hour.

EXAMPLE 7 When we consider the flow of blood through a blood vessel, such as a vein or artery, we can take the shape of the blood vessel to be a cylindrical tube with radius R and length l as illustrated in Figure 6.

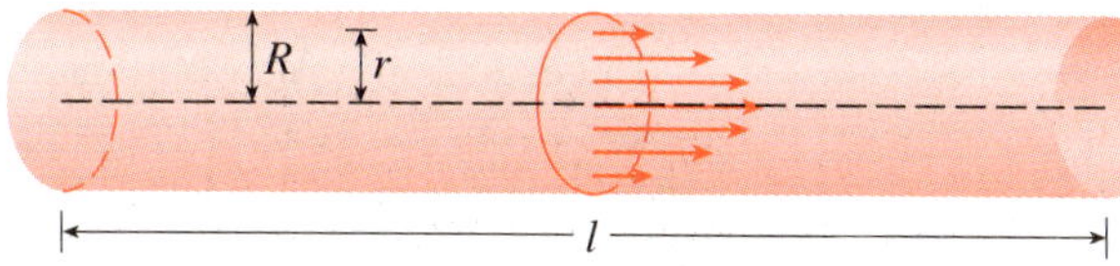

FIGURE 6
Blood flow in an artery

Because of friction at the walls of the tube, the velocity v of the blood is greatest along the central axis of the tube and decreases as the distance r from the axis increases until v becomes 0 at the wall. The relationship between v and r is given by the **law of laminar flow** discovered by the French physician Jean-Louis-Marie Poiseuille in 1840. This states that

$$\boxed{1} \qquad v = \frac{P}{4\eta l}(R^2 - r^2)$$

where η is the viscosity of the blood and P is the pressure difference between the ends of the tube. If P and l are constant, then v is a function of r with domain $[0, R]$. [For more detailed information, see W. Nichols and M. O'Rourke (eds.), *McDonald's Blood Flow in Arteries: Theoretic, Experimental, and Clinical Principles,* 4th ed. (New York: Oxford University Press, 1998).]

The average rate of change of the velocity as we move from $r = r_1$ outward to $r = r_2$ is given by

$$\frac{\Delta v}{\Delta r} = \frac{v(r_2) - v(r_1)}{r_2 - r_1}$$

and if we let $\Delta r \to 0$, we obtain the **velocity gradient**, that is, the instantaneous rate of change of velocity with respect to r:

$$\text{velocity gradient} = \lim_{\Delta r \to 0} \frac{\Delta v}{\Delta r} = \frac{dv}{dr}$$

Using Equation 1, we obtain

$$\frac{dv}{dr} = \frac{P}{4\eta l}(0 - 2r) = -\frac{Pr}{2\eta l}$$

For one of the smaller human arteries we can take $\eta = 0.027$, $R = 0.008$ cm, $l = 2$ cm, and $P = 4000$ dynes/cm^2, which gives

$$\begin{aligned} v &= \frac{4000}{4(0.027)2}(0.000064 - r^2) \\ &\approx 1.85 \times 10^4(6.4 \times 10^{-5} - r^2) \end{aligned}$$

At $r = 0.002$ cm the blood is flowing at a speed of

$$\begin{aligned} v(0.002) &\approx 1.85 \times 10^4(64 \times 10^{-6} - 4 \times 10^{-6}) \\ &= 1.11 \text{ cm/s} \end{aligned}$$

and the velocity gradient at that point is

$$\left.\frac{dv}{dr}\right|_{r=0.002} = -\frac{4000(0.002)}{2(0.027)2} \approx -74 \text{ (cm/s)/cm}$$

To get a feeling for what this statement means, let's change our units from centimeters to micrometers (1 cm = 10,000 μm). Then the radius of the artery is 80 μm. The velocity at the central axis is 11,850 μm/s, which decreases to 11,110 μm/s at a distance of $r = 20$ μm. The fact that $dv/dr = -74$ (μm/s)/μm means that, when $r = 20$ μm, the velocity is decreasing at a rate of about 74 μm/s for each micrometer that we proceed away from the center.

Economics

EXAMPLE 8 Suppose $C(x)$ is the total cost that a company incurs in producing x units of a certain commodity. The function C is called a **cost function**. If the number of items produced is increased from x_1 to x_2, the additional cost is $\Delta C = C(x_2) - C(x_1)$, and the average rate of change of the cost is

$$\frac{\Delta C}{\Delta x} = \frac{C(x_2) - C(x_1)}{x_2 - x_1} = \frac{C(x_1 + \Delta x) - C(x_1)}{\Delta x}$$

The limit of this quantity as $\Delta x \to 0$, that is, the instantaneous rate of change of cost with respect to the number of items produced, is called the **marginal cost** by economists:

$$\text{marginal cost} = \lim_{\Delta x \to 0} \frac{\Delta C}{\Delta x} = \frac{dC}{dx}$$

[Since x often takes on only integer values, it may not make literal sense to let Δx approach 0, but we can always replace $C(x)$ by a smooth approximating function as in Example 6.]

Taking $\Delta x = 1$ and n large (so that Δx is small compared to n), we have

$$C'(n) \approx C(n + 1) - C(n)$$

Thus, the marginal cost of producing n units is approximately equal to the cost of producing one more unit [the $(n + 1)$st unit].

It is often appropriate to represent a total cost function by a polynomial

$$C(x) = a + bx + cx^2 + dx^3$$

where a represents the overhead cost (rent, heat, maintenance) and the other terms represent the cost of raw materials, labor, and so on. (The cost of raw materials may be proportional to x, but labor costs might depend partly on higher powers of x because of overtime costs and inefficiencies involved in large-scale operations.)

For instance, suppose a company has estimated that the cost (in dollars) of producing x items is

$$C(x) = 10{,}000 + 5x + 0.01x^2$$

Then the marginal cost function is

$$C'(x) = 5 + 0.02x$$

The marginal cost at the production level of 500 items is

$$C'(500) = 5 + 0.02(500) = \$15/\text{item}$$

This gives the rate at which costs are increasing with respect to the production level when $x = 500$ and predicts the cost of the 501st item.

The actual cost of producing the 501st item is

$$\begin{aligned} C(501) - C(500) &= [10{,}000 + 5(501) + 0.01(501)^2] \\ &\qquad - [10{,}000 + 5(500) + 0.01(500)^2] \\ &= \$15.01 \end{aligned}$$

Notice that $C'(500) \approx C(501) - C(500)$.

Economists also study marginal demand, marginal revenue, and marginal profit, which are the derivatives of the demand, revenue, and profit functions. These will be considered in Chapter 4 after we have developed techniques for finding the maximum and minimum values of functions.

Other Sciences

Rates of change occur in all the sciences. A geologist is interested in knowing the rate at which an intruded body of molten rock cools by conduction of heat into surrounding rocks. An engineer wants to know the rate at which water flows into or out of a reservoir. An urban geographer is interested in the rate of change of the population density in a city as the distance from the city center increases. A meteorologist is concerned with the rate of change of atmospheric pressure with respect to height (see Exercise 17 in Section 9.4).

In psychology, those interested in learning theory study the so-called learning curve, which graphs the performance $P(t)$ of someone learning a skill as a function of the training time t. Of particular interest is the rate at which performance improves as time passes, that is, dP/dt.

In sociology, differential calculus is used in analyzing the spread of rumors (or innovations or fads or fashions). If $p(t)$ denotes the proportion of a population that knows a rumor by time t, then the derivative dp/dt represents the rate of spread of the rumor (see Exercise 70 in Section 3.5).

Summary

Velocity, density, current, power, and temperature gradient in physics, rate of reaction and compressibility in chemistry, rate of growth and blood velocity gradient in biology, marginal cost and marginal profit in economics, rate of heat flow in geology, rate of improvement of performance in psychology, rate of spread of a rumor in sociology—these are all special cases of a single mathematical concept, the derivative.

This is an illustration of the fact that part of the power of mathematics lies in its abstractness. A single abstract mathematical concept (such as the derivative) can have different interpretations in each of the sciences. When we develop the properties of the mathematical concept once and for all, we can then turn around and apply these results to all of the sciences. This is much more efficient than developing properties of special concepts in each separate science. The French mathematician Joseph Fourier (1768–1830) put it succinctly: "Mathematics compares the most diverse phenomena and discovers the secret analogies that unite them."

3.3 Exercises

1–6 A particle moves according to a law of motion $s = f(t)$, $t \geqslant 0$, where t is measured in seconds and s in feet.

(a) Find the velocity at time t.
(b) What is the velocity after 3 s?
(c) When is the particle at rest?
(d) When is the particle moving in the positive direction?
(e) Find the total distance traveled during the first 8 s.
(f) Draw a diagram like Figure 2 to illustrate the motion of the particle.

1. $f(t) = t^2 - 10t + 12$

2. $f(t) = t^3 - 9t^2 + 15t + 10$

3. $f(t) = t^3 - 12t^2 + 36t$

4. $f(t) = t^4 - 4t + 1$

5. $s = \dfrac{t}{t^2 + 1}$

6. $s = \sqrt{t}(3t^2 - 35t + 90)$

7. The position function of a particle is given by

$$s = t^3 - 4.5t^2 - 7t \qquad t \geqslant 0$$

When does the particle reach a velocity of 5 m/s?

8. If a ball is given a push so that it has an initial velocity of 5 m/s down a certain inclined plane, then the distance it has rolled after t seconds is $s = 5t + 3t^2$.
(a) Find the velocity after 2 s.
(b) How long does it take for the velocity to reach 35 m/s?

9. If a stone is thrown vertically upward from the surface of the moon with a velocity of 10 m/s, its height (in meters) after t seconds is $h = 10t - 0.83t^2$.
(a) What is the velocity of the stone after 3 s?
(b) What is the velocity of the stone after it has risen 25 m?

10. If a ball is thrown vertically upward with a velocity of 80 ft/s, then its height after t seconds is $s = 80t - 16t^2$.
(a) What is the maximum height reached by the ball?
(b) What is the velocity of the ball when it is 96 ft above the ground on its way up? On its way down?

11. (a) A company makes computer chips from square wafers of silicon. It wants to keep the side length of a wafer very close to 15 mm and it wants to know how the area $A(x)$ of a wafer changes when the side length x changes. Find $A'(15)$ and explain its meaning in this situation.
(b) Show that the rate of change of the area of a square with respect to its side length is half its perimeter. Try to explain geometrically why this is true by drawing a square whose side length x is increased by an amount Δx. How can you approximate the resulting change in area ΔA if Δx is small?

12. (a) Sodium chlorate crystals are easy to grow in the shape of cubes by allowing a solution of water and sodium chlorate to evaporate slowly. If V is the volume of such a cube with side length x, calculate dV/dx when $x = 3$ mm and explain its meaning.
(b) Show that the rate of change of the volume of a cube with respect to its edge length is equal to half the surface area of the cube. Explain geometrically why this result is true by arguing by analogy with Exercise 11(b).

13. (a) Find the average rate of change of the area of a circle with respect to its radius r as r changes from
(i) 2 to 3 (ii) 2 to 2.5 (iii) 2 to 2.1
(b) Find the instantaneous rate of change when $r = 2$.
(c) Show that the rate of change of the area of a circle with respect to its radius (at any r) is equal to the circumference of the circle. Try to explain geometrically why this is true by drawing a circle whose radius is increased by an amount Δr. How can you approximate the resulting change in area ΔA if Δr is small?

14. A stone is dropped into a lake, creating a circular ripple that travels outward at a speed of 60 cm/s. Find the rate at which the area within the circle is increasing after (a) 1 s, (b) 3 s, and (c) 5 s. What can you conclude?

15. A spherical balloon is being inflated. Find the rate of increase of the surface area ($S = 4\pi r^2$) with respect to the radius r when r is (a) 1 ft, (b) 2 ft, and (c) 3 ft. What conclusion can you make?

16. (a) The volume of a growing spherical cell is $V = \frac{4}{3}\pi r^3$, where the radius r is measured in micrometers (1 μm $= 10^{-6}$ m). Find the average rate of change of V with respect to r when r changes from
(i) 5 to 8 μm (ii) 5 to 6 μm (iii) 5 to 5.1 μm
(b) Find the instantaneous rate of change of V with respect to r when $r = 5$ μm.
(c) Show that the rate of change of the volume of a sphere with respect to its radius is equal to its surface area. Explain geometrically why this result is true. Argue by analogy with Exercise 13(c).

17. The mass of the part of a metal rod that lies between its left end and a point x meters to the right is $3x^2$ kg. Find the linear density (see Example 2) when x is (a) 1 m, (b) 2 m, and (c) 3 m. Where is the density the highest? The lowest?

18. If a tank holds 5000 gallons of water, which drains from the bottom of the tank in 40 minutes, then Torricelli's Law gives the volume V of water remaining in the tank after t minutes as

$$V = 5000\left(1 - \frac{t}{40}\right)^2 \qquad 0 \leqslant t \leqslant 40$$

Find the rate at which water is draining from the tank after (a) 5 min, (b) 10 min, (c) 20 min, and (d) 40 min. At what time is the water flowing out the fastest? The slowest? Summarize your findings.

19. The quantity of charge Q in coulombs (C) that has passed through a point in a wire up to time t (measured in seconds) is given by $Q(t) = t^3 - 2t^2 + 6t + 2$. Find the current when (a) $t = 0.5$ s and (b) $t = 1$ s. [See Example 3. The unit of current is an ampere (1 A = 1 C/s).] At what time is the current lowest?

20. Newton's Law of Gravitation says that the magnitude F of the force exerted by a body of mass m on a body of mass M is

$$F = \frac{GmM}{r^2}$$

where G is the gravitational constant and r is the distance between the bodies.

(a) Find dF/dr and explain its meaning. What does the minus sign indicate?

(b) Suppose it is known that Earth attracts an object with a force that decreases at the rate of 2 N/km when $r = 20{,}000$ km. How fast does this force change when $r = 10{,}000$ km?

21. Boyle's Law states that when a sample of gas is compressed at a constant temperature, the product of the pressure and the volume remains constant: $PV = C$.

(a) Find the rate of change of volume with respect to pressure.

(b) A sample of gas is in a container at low pressure and is steadily compressed at constant temperature for 10 minutes. Is the volume decreasing more rapidly at the beginning or the end of the 10 minutes? Explain.

(c) Prove that the isothermal compressibility (see Example 5) is given by $\beta = 1/P$.

22. The data in the table concern the lactonization of hydroxyvaleric acid at 25°C. They give the concentration $C(t)$ of this acid in moles per liter after t minutes.

t	0	2	4	6	8
$C(t)$	0.0800	0.0570	0.0408	0.0295	0.0210

(a) Find the average rate of reaction for the following time intervals:
(i) $2 \leq t \leq 6$ (ii) $2 \leq t \leq 4$ (iii) $0 \leq t \leq 2$

(b) Plot the points from the table and draw a smooth curve through them as an approximation to the graph of the concentration function. Then draw the tangent at $t = 2$ and use it to estimate the instantaneous rate of reaction when $t = 2$.

23. The table gives the population of the world in the 20th century.

Year	Population (in millions)	Year	Population (in millions)
1900	1650	1960	3040
1910	1750	1970	3710
1920	1860	1980	4450
1930	2070	1990	5280
1940	2300	2000	6080
1950	2560		

(a) Estimate the rate of population growth in 1920 and in 1980 by averaging the slopes of two secant lines.

(b) Use a graphing calculator or computer to find a cubic function (a third-degree polynomial) that models the data. (See Section 1.2.)

(c) Use your model in part (b) to find a model for the rate of population growth in the 20th century.

(d) Use part (c) to estimate the rates of growth in 1920 and 1980. Compare with your estimates in part (a).

(e) Estimate the rate of growth in 1985.

24. The table shows how the average age of first marriage of Japanese women varied in the last half of the 20th century.

t	$A(t)$	t	$A(t)$
1950	23.0	1975	24.7
1955	23.8	1980	25.2
1960	24.4	1985	25.5
1965	24.5	1990	25.9
1970	24.2	1995	26.3

(a) Use a graphing calculator or computer to model these data with a fourth-degree polynomial.

(b) Use part (a) to find a model for $A'(t)$.

(c) Estimate the rate of change of marriage age for women in 1990.

(d) Graph the data points and the models for A and A'.

25. If, in Example 4, one molecule of the product C is formed from one molecule of the reactant A and one molecule of the reactant B, and the initial concentrations of A and B have a common value $[A] = [B] = a$ moles/L, then

$$[C] = a^2kt/(akt + 1)$$

where k is a constant.

(a) Find the rate of reaction at time t.

(b) Show that if $x = [C]$, then

$$\frac{dx}{dt} = k(a - x)^2$$

(c) What happens to the concentration as $t \to \infty$?

(d) What happens to the rate of reaction as $t \to \infty$?

(e) What do the results of parts (c) and (d) mean in practical terms?

26. Suppose that a bacteria population starts with 500 bacteria and triples every hour.

(a) What is the population after 3 hours? After 4 hours? After t hours?

(b) Use (5) in Section 3.1 to estimate the rate of increase of the bacteria population after 6 hours.

27. Refer to the law of laminar flow given in Example 7. Consider a blood vessel with radius 0.01 cm, length 3 cm, pressure difference 3000 dynes/cm^2, and viscosity $\eta = 0.027$.

(a) Find the velocity of the blood along the centerline $r = 0$, at radius $r = 0.005$ cm, and at the wall $r = R = 0.01$ cm.

(b) Find the velocity gradient at $r = 0$, $r = 0.005$, and $r = 0.01$.

(c) Where is the velocity the greatest? Where is the velocity changing most?

28. The frequency of vibrations of a vibrating violin string is given by

$$f = \frac{1}{2L}\sqrt{\frac{T}{\rho}}$$

where L is the length of the string, T is its tension, and ρ is its linear density. [See Chapter 11 in Donald E. Hall, *Musical Acoustics,* 3d ed. (Pacific Grove, CA: Brooks/Cole, 2002).]

(a) Find the rate of change of the frequency with respect to
  (i) the length (when T and ρ are constant),
  (ii) the tension (when L and ρ are constant), and
  (iii) the linear density (when L and T are constant).

(b) The pitch of a note (how high or low the note sounds) is determined by the frequency f. (The higher the frequency, the higher the pitch.) Use the signs of the derivatives in part (a) to determine what happens to the pitch of a note
  (i) when the effective length of a string is decreased by placing a finger on the string so a shorter portion of the string vibrates,
  (ii) when the tension is increased by turning a tuning peg,
  (iii) when the linear density is increased by switching to another string.

29. Suppose that the cost (in dollars) for a company to produce x pairs of a new line of jeans is

$$C(x) = 2000 + 3x + 0.01x^2 + 0.0002x^3$$

(a) Find the marginal cost function.

(b) Find $C'(100)$ and explain its meaning. What does it predict?

(c) Compare $C'(100)$ with the cost of manufacturing the 101st pair of jeans.

30. The cost function for a certain commodity is

$$C(x) = 84 + 0.16x - 0.0006x^2 + 0.000003x^3$$

(a) Find and interpret $C'(100)$.

(b) Compare $C'(100)$ with the cost of producing the 101st item.

31. If $p(x)$ is the total value of the production when there are x workers in a plant, then the *average productivity* of the workforce at the plant is

$$A(x) = \frac{p(x)}{x}$$

(a) Find $A'(x)$. Why does the company want to hire more workers if $A'(x) > 0$?

(b) Show that $A'(x) > 0$ if $p'(x)$ is greater than the average productivity.

32. If R denotes the reaction of the body to some stimulus of strength x, the *sensitivity* S is defined to be the rate of change of the reaction with respect to x. A particular example is that when the brightness x of a light source is increased, the eye reacts by decreasing the area R of the pupil. The experimental formula

$$R = \frac{40 + 24x^{0.4}}{1 + 4x^{0.4}}$$

has been used to model the dependence of R on x when R is measured in square millimeters and x is measured in appropriate units of brightness.

(a) Find the sensitivity.

(b) Illustrate part (a) by graphing both R and S as functions of x. Comment on the values of R and S at low levels of brightness. Is this what you would expect?

33. The gas law for an ideal gas at absolute temperature T (in kelvins), pressure P (in atmospheres), and volume V (in liters) is $PV = nRT$, where n is the number of moles of the gas and $R = 0.0821$ is the gas constant. Suppose that, at a certain instant, $P = 8.0$ atm and is increasing at a rate of 0.10 atm/min and $V = 10$ L and is decreasing at a rate of 0.15 L/min. Find the rate of change of T with respect to time at that instant if $n = 10$ mol.

34. In a fish farm, a population of fish is introduced into a pond and harvested regularly. A model for the rate of change of the fish population is given by the equation

$$\frac{dP}{dt} = r_0\left(1 - \frac{P(t)}{P_c}\right)P(t) - \beta P(t)$$

where r_0 is the birth rate of the fish, P_c is the maximum population that the pond can sustain (called the *carrying capacity*), and β is the percentage of the population that is harvested.

(a) What value of dP/dt corresponds to a stable population?

(b) If the pond can sustain 10,000 fish, the birth rate is 5%, and the harvesting rate is 4%, find the stable population level.

(c) What happens if β is raised to 5%?

35. In the study of ecosystems, *predator-prey models* are often used to study the interaction between species. Consider populations of tundra wolves, given by $W(t)$, and caribou, given by $C(t)$, in northern Canada. The interaction has been modeled by the equations

$$\frac{dC}{dt} = aC - bCW \qquad \frac{dW}{dt} = -cW + dCW$$

(a) What values of dC/dt and dW/dt correspond to stable populations?

(b) How would the statement "The caribou go extinct" be represented mathematically?

(c) Suppose that $a = 0.05$, $b = 0.001$, $c = 0.05$, and $d = 0.0001$. Find all population pairs (C, W) that lead to stable populations. According to this model, is it possible for the two species to live in balance or will one or both species become extinct?

3.4 Derivatives of Trigonometric Functions

|||| A review of the trigonometric functions is given in Appendix D.

Before starting this section, you might need to review the trigonometric functions. In particular, it is important to remember that when we talk about the function f defined for all real numbers x by

$$f(x) = \sin x$$

it is understood that $\sin x$ means the sine of the angle whose *radian* measure is x. A similar convention holds for the other trigonometric functions cos, tan, csc, sec, and cot. Recall from Section 2.5 that all of the trigonometric functions are continuous at every number in their domains.

If we sketch the graph of the function $f(x) = \sin x$ and use the interpretation of $f'(x)$ as the slope of the tangent to the sine curve in order to sketch the graph of f' (see Exercise 16 in Section 2.9), then it looks as if the graph of f' may be the same as the cosine curve (see Figure 1 and also page 182).

See an animation of Figure 1.
Resources / Module 4
/ Trigonometric Models
/ Slope-A-Scope for Sine

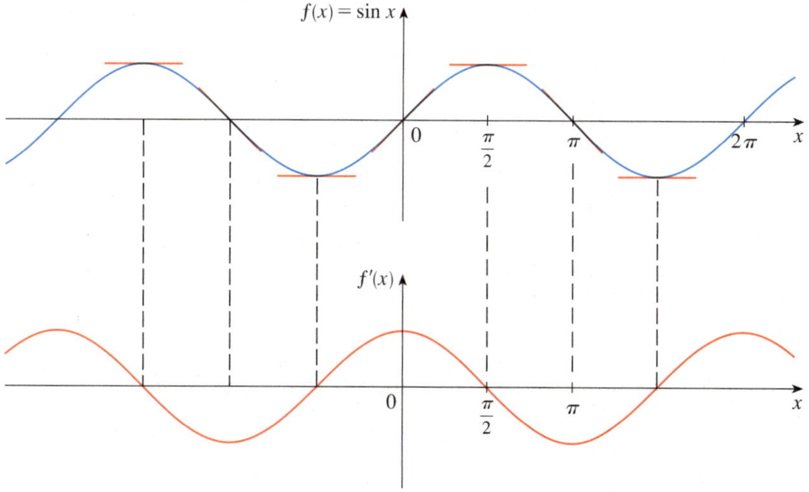

FIGURE 1

Let's try to confirm our guess that if $f(x) = \sin x$, then $f'(x) = \cos x$. From the definition of a derivative, we have

|||| We have used the addition formula for sine. See Appendix D.

$$\begin{aligned}
f'(x) &= \lim_{h \to 0} \frac{f(x+h) - f(x)}{h} \\
&= \lim_{h \to 0} \frac{\sin(x+h) - \sin x}{h} \\
&= \lim_{h \to 0} \frac{\sin x \cos h + \cos x \sin h - \sin x}{h} \\
&= \lim_{h \to 0} \left[\frac{\sin x \cos h - \sin x}{h} + \frac{\cos x \sin h}{h} \right] \\
&= \lim_{h \to 0} \left[\sin x \left(\frac{\cos h - 1}{h} \right) + \cos x \left(\frac{\sin h}{h} \right) \right]
\end{aligned}$$

1 $$= \lim_{h \to 0} \sin x \cdot \lim_{h \to 0} \frac{\cos h - 1}{h} + \lim_{h \to 0} \cos x \cdot \lim_{h \to 0} \frac{\sin h}{h}$$

Two of these four limits are easy to evaluate. Since we regard x as a constant when computing a limit as $h \to 0$, we have

$$\lim_{h \to 0} \sin x = \sin x \qquad \text{and} \qquad \lim_{h \to 0} \cos x = \cos x$$

The limit of $(\sin h)/h$ is not so obvious. In Example 3 in Section 2.2 we made the guess, on the basis of numerical and graphical evidence, that

$$\boxed{2} \qquad \lim_{\theta \to 0} \frac{\sin \theta}{\theta} = 1$$

We now use a geometric argument to prove Equation 2. Assume first that θ lies between 0 and $\pi/2$. Figure 2(a) shows a sector of a circle with center O, central angle θ, and radius 1. BC is drawn perpendicular to OA. By the definition of radian measure, we have arc $AB = \theta$. Also, $|BC| = |OB| \sin \theta = \sin \theta$. From the diagram we see that

$$|BC| < |AB| < \text{arc } AB$$

Therefore $$\sin \theta < \theta \qquad \text{so} \qquad \frac{\sin \theta}{\theta} < 1$$

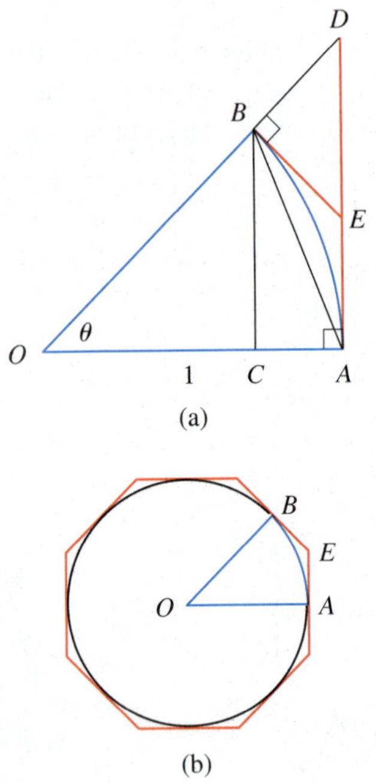

FIGURE 2

Let the tangent lines at A and B intersect at E. You can see from Figure 2(b) that the circumference of a circle is smaller than the length of a circumscribed polygon, and so arc $AB < |AE| + |EB|$. Thus

$$\begin{aligned} \theta = \text{arc } AB &< |AE| + |EB| \\ &< |AE| + |ED| \\ &= |AD| = |OA| \tan \theta \\ &= \tan \theta \end{aligned}$$

(In Appendix F the inequality $\theta \leqslant \tan \theta$ is proved directly from the definition of the length of an arc without resorting to geometric intuition as we did here.) Therefore, we have

$$\theta < \frac{\sin \theta}{\cos \theta}$$

so $$\cos \theta < \frac{\sin \theta}{\theta} < 1$$

We know that $\lim_{\theta \to 0} 1 = 1$ and $\lim_{\theta \to 0} \cos \theta = 1$, so by the Squeeze Theorem, we have

$$\lim_{\theta \to 0^+} \frac{\sin \theta}{\theta} = 1$$

But the function $(\sin \theta)/\theta$ is an even function, so its right and left limits must be equal. Hence, we have

$$\lim_{\theta \to 0} \frac{\sin \theta}{\theta} = 1$$

so we have proved Equation 2.

We can deduce the value of the remaining limit in (1) as follows:

|||| We multiply numerator and denominator by $\cos\theta + 1$ in order to put the function in a form in which we can use the limits we know.

$$\lim_{\theta\to 0}\frac{\cos\theta - 1}{\theta} = \lim_{\theta\to 0}\left(\frac{\cos\theta - 1}{\theta}\cdot\frac{\cos\theta + 1}{\cos\theta + 1}\right) = \lim_{\theta\to 0}\frac{\cos^2\theta - 1}{\theta(\cos\theta + 1)}$$

$$= \lim_{\theta\to 0}\frac{-\sin^2\theta}{\theta(\cos\theta + 1)} = -\lim_{\theta\to 0}\frac{\sin\theta}{\theta}\cdot\frac{\sin\theta}{\cos\theta + 1}$$

$$= -\lim_{\theta\to 0}\frac{\sin\theta}{\theta}\cdot\lim_{\theta\to 0}\frac{\sin\theta}{\cos\theta + 1}$$

$$= -1\cdot\left(\frac{0}{1+1}\right) = 0 \qquad \text{(by Equation 2)}$$

3
$$\lim_{\theta\to 0}\frac{\cos\theta - 1}{\theta} = 0$$

If we now put the limits (2) and (3) in (1), we get

$$f'(x) = \lim_{h\to 0}\sin x\cdot\lim_{h\to 0}\frac{\cos h - 1}{h} + \lim_{h\to 0}\cos x\cdot\lim_{h\to 0}\frac{\sin h}{h}$$

$$= (\sin x)\cdot 0 + (\cos x)\cdot 1 = \cos x$$

So we have proved the formula for the derivative of the sine function:

4
$$\frac{d}{dx}(\sin x) = \cos x$$

|||| Figure 3 shows the graphs of the function of Example 1 and its derivative. Notice that $y' = 0$ whenever y has a horizontal tangent.

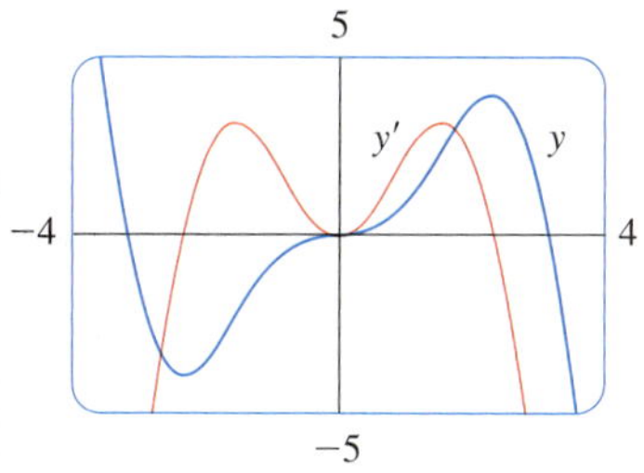

FIGURE 3

EXAMPLE 1 Differentiate $y = x^2 \sin x$.

SOLUTION Using the Product Rule and Formula 4, we have

$$\frac{dy}{dx} = x^2\frac{d}{dx}(\sin x) + \sin x\frac{d}{dx}(x^2)$$

$$= x^2\cos x + 2x\sin x$$

Using the same methods as in the proof of Formula 4, one can prove (see Exercise 20) that

5
$$\frac{d}{dx}(\cos x) = -\sin x$$

The tangent function can also be differentiated by using the definition of a derivative, but it is easier to use the Quotient Rule together with Formulas 4 and 5:

$$\begin{aligned}
\frac{d}{dx}(\tan x) &= \frac{d}{dx}\left(\frac{\sin x}{\cos x}\right) \\
&= \frac{\cos x \dfrac{d}{dx}(\sin x) - \sin x \dfrac{d}{dx}(\cos x)}{\cos^2 x} \\
&= \frac{\cos x \cdot \cos x - \sin x(-\sin x)}{\cos^2 x} \\
&= \frac{\cos^2 x + \sin^2 x}{\cos^2 x} \\
&= \frac{1}{\cos^2 x} = \sec^2 x
\end{aligned}$$

6

$$\frac{d}{dx}(\tan x) = \sec^2 x$$

The derivatives of the remaining trigonometric functions, csc, sec, and cot, can also be found easily using the Quotient Rule (see Exercises 17–19). We collect all the differentiation formulas for trigonometric functions in the following table. Remember that they are valid only when x is measured in radians.

Derivatives of Trigonometric Functions

$\frac{d}{dx}(\sin x) = \cos x$	$\frac{d}{dx}(\csc x) = -\csc x \cot x$
$\frac{d}{dx}(\cos x) = -\sin x$	$\frac{d}{dx}(\sec x) = \sec x \tan x$
$\frac{d}{dx}(\tan x) = \sec^2 x$	$\frac{d}{dx}(\cot x) = -\csc^2 x$

|||| When you memorize this table, it is helpful to notice that the minus signs go with the derivatives of the "cofunctions," that is, cosine, cosecant, and cotangent.

EXAMPLE 2 Differentiate $f(x) = \dfrac{\sec x}{1 + \tan x}$. For what values of x does the graph of f have a horizontal tangent?

SOLUTION The Quotient Rule gives

$$\begin{aligned}
f'(x) &= \frac{(1 + \tan x)\dfrac{d}{dx}(\sec x) - \sec x \dfrac{d}{dx}(1 + \tan x)}{(1 + \tan x)^2} \\
&= \frac{(1 + \tan x)\sec x \tan x - \sec x \cdot \sec^2 x}{(1 + \tan x)^2} \\
&= \frac{\sec x(\tan x + \tan^2 x - \sec^2 x)}{(1 + \tan x)^2} \\
&= \frac{\sec x(\tan x - 1)}{(1 + \tan x)^2}
\end{aligned}$$

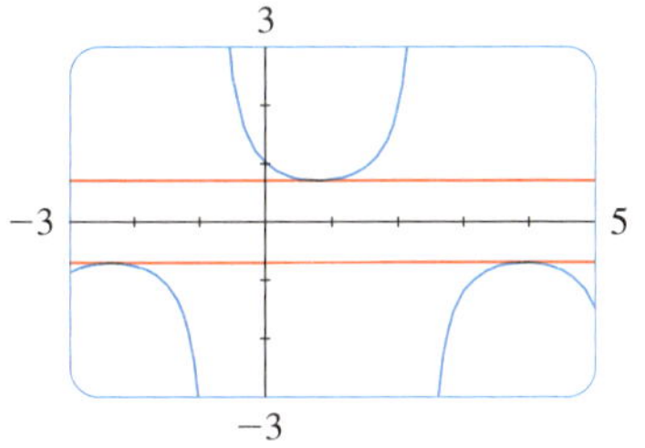

FIGURE 4
The horizontal tangents in Example 2

In simplifying the answer we have used the identity $\tan^2 x + 1 = \sec^2 x$.

Since $\sec x$ is never 0, we see that $f'(x) = 0$ when $\tan x = 1$, and this occurs when $x = n\pi + \pi/4$, where n is an integer (see Figure 4).

Trigonometric functions are often used in modeling real-world phenomena. In particular, vibrations, waves, elastic motions, and other quantities that vary in a periodic manner can be described using trigonometric functions. In the following example we discuss an instance of simple harmonic motion.

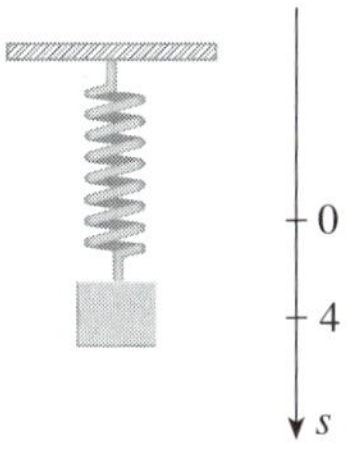

FIGURE 5

EXAMPLE 3 An object at the end of a vertical spring is stretched 4 cm beyond its rest position and released at time $t = 0$. (See Figure 5 and note that the downward direction is positive.) Its position at time t is

$$s = f(t) = 4\cos t$$

Find the velocity at time t and use it to analyze the motion of the object.

SOLUTION The velocity is

$$v = \frac{ds}{dt} = \frac{d}{dt}(4\cos t) = 4\frac{d}{dt}(\cos t) = -4\sin t$$

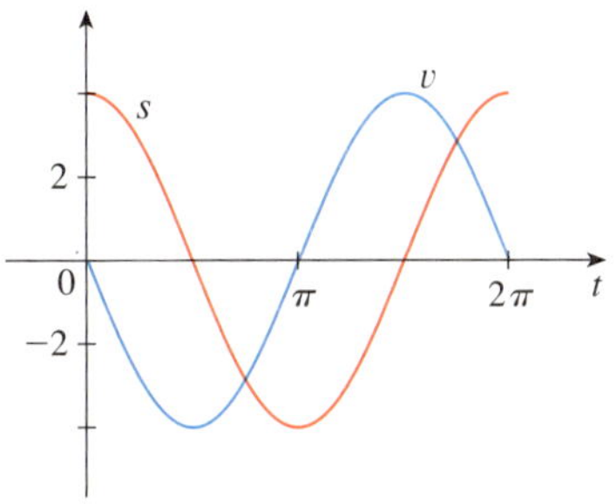

FIGURE 6

The object oscillates from the lowest point ($s = 4$ cm) to the highest point ($s = -4$ cm). The period of the oscillation is 2π, the period of $\cos t$.

The speed is $|v| = 4|\sin t|$, which is greatest when $|\sin t| = 1$, that is, when $\cos t = 0$. So the object moves fastest as it passes through its equilibrium position ($s = 0$). Its speed is 0 when $\sin t = 0$, that is, at the high and low points. See the graphs in Figure 6.

Our main use for the limit in Equation 2 has been to prove the differentiation formula for the sine function. But this limit is also useful in finding certain other trigonometric limits, as the following two examples show.

EXAMPLE 4 Find $\displaystyle\lim_{x\to 0}\frac{\sin 7x}{4x}$.

SOLUTION In order to apply Equation 2, we first rewrite the function by multiplying and dividing by 7:

Note that $\sin 7x \neq 7\sin x$.

$$\frac{\sin 7x}{4x} = \frac{7}{4}\left(\frac{\sin 7x}{7x}\right)$$

Notice that as $x \to 0$, we have $7x \to 0$, and so, by Equation 2 with $\theta = 7x$,

$$\lim_{x\to 0}\frac{\sin 7x}{7x} = \lim_{7x\to 0}\frac{\sin(7x)}{7x} = 1$$

Thus
$$\lim_{x\to 0}\frac{\sin 7x}{4x} = \lim_{x\to 0}\frac{7}{4}\left(\frac{\sin 7x}{7x}\right)$$
$$= \frac{7}{4}\lim_{x\to 0}\frac{\sin 7x}{7x} = \frac{7}{4}\cdot 1 = \frac{7}{4}$$

EXAMPLE 5 Calculate $\lim_{x\to 0} x \cot x$.

SOLUTION Here we divide numerator and denominator by x:

$$\lim_{x\to 0} x \cot x = \lim_{x\to 0} \frac{x \cos x}{\sin x}$$

$$= \lim_{x\to 0} \frac{\cos x}{\dfrac{\sin x}{x}} = \frac{\lim_{x\to 0} \cos x}{\lim_{x\to 0} \dfrac{\sin x}{x}}$$

$$= \frac{\cos 0}{1} \quad \text{(by the continuity of cosine and Equation 2)}$$

$$= 1$$

3.4 Exercises

1–16 Differentiate.

1. $f(x) = x - 3 \sin x$

2. $f(x) = x \sin x$

3. $y = \sin x + 10 \tan x$

4. $y = 2 \csc x + 5 \cos x$

5. $g(t) = t^3 \cos t$

6. $g(t) = 4 \sec t + \tan t$

7. $h(\theta) = \csc \theta + e^\theta \cot \theta$

8. $y = e^u (\cos u + cu)$

9. $y = \dfrac{x}{\cos x}$

10. $y = \dfrac{1 + \sin x}{x + \cos x}$

11. $f(\theta) = \dfrac{\sec \theta}{1 + \sec \theta}$

12. $y = \dfrac{\tan x - 1}{\sec x}$

13. $y = \dfrac{\sin x}{x^2}$

14. $y = \csc \theta \, (\theta + \cot \theta)$

15. $y = \sec \theta \tan \theta$

16. $y = x \sin x \cos x$

17. Prove that $\dfrac{d}{dx}(\csc x) = -\csc x \cot x$.

18. Prove that $\dfrac{d}{dx}(\sec x) = \sec x \tan x$.

19. Prove that $\dfrac{d}{dx}(\cot x) = -\csc^2 x$.

20. Prove, using the definition of derivative, that if $f(x) = \cos x$, then $f'(x) = -\sin x$.

21–24 Find an equation of the tangent line to the curve at the given point.

21. $y = \tan x, \quad (\pi/4, 1)$

22. $y = e^x \cos x, \quad (0, 1)$

23. $y = x + \cos x, \quad (0, 1)$

24. $y = \dfrac{1}{\sin x + \cos x}, \quad (0, 1)$

25. (a) Find an equation of the tangent line to the curve $y = x \cos x$ at the point $(\pi, -\pi)$.
(b) Illustrate part (a) by graphing the curve and the tangent line on the same screen.

26. (a) Find an equation of the tangent line to the curve $y = \sec x - 2 \cos x$ at the point $(\pi/3, 1)$.
(b) Illustrate part (a) by graphing the curve and the tangent line on the same screen.

27. (a) If $f(x) = 2x + \cot x$, find $f'(x)$.
(b) Check to see that your answer to part (a) is reasonable by graphing both f and f' for $0 < x < \pi$.

28. (a) If $f(x) = \sqrt{x} \sin x$, find $f'(x)$.
(b) Check to see that your answer to part (a) is reasonable by graphing both f and f' for $0 \leqslant x \leqslant 2\pi$.

29. For what values of x does the graph of $f(x) = x + 2 \sin x$ have a horizontal tangent?

30. Find the points on the curve $y = (\cos x)/(2 + \sin x)$ at which the tangent is horizontal.

31. A mass on a spring vibrates horizontally on a smooth level surface (see the figure). Its equation of motion is $x(t) = 8 \sin t$, where t is in seconds and x in centimeters.
(a) Find the velocity at time t.
(b) Find the position and velocity of the mass at time $t = 2\pi/3$. In what direction is it moving at that time?

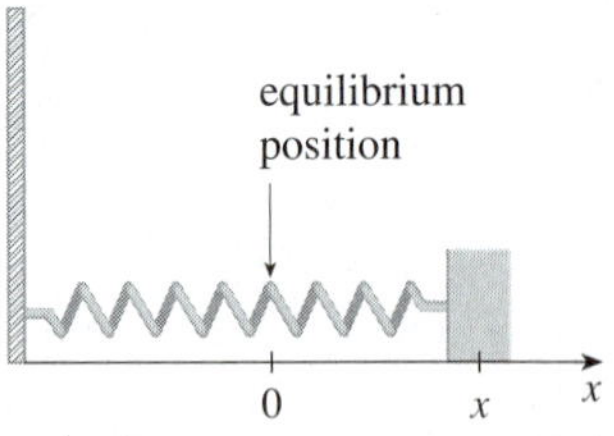

32. An elastic band is hung on a hook and a mass is hung on the lower end of the band. When the mass is pulled downward and then released, it vibrates vertically. The equation of motion is $s = 2\cos t + 3\sin t$, $t \geqslant 0$, where s is measured in centimeters and t in seconds. (We take the positive direction to be downward.)

(a) Find the velocity at time t.
(b) Graph the velocity and position functions.
(c) When does the mass pass through the equilibrium position for the first time?
(d) How far from its equilibrium position does the mass travel?
(e) When is the speed the greatest?

33. A ladder 10 ft long rests against a vertical wall. Let θ be the angle between the top of the ladder and the wall and let x be the distance from the bottom of the ladder to the wall. If the bottom of the ladder slides away from the wall, how fast does x change with respect to θ when $\theta = \pi/3$?

34. An object with weight W is dragged along a horizontal plane by a force acting along a rope attached to the object. If the rope makes an angle θ with the plane, then the magnitude of the force is

$$F = \frac{\mu W}{\mu \sin\theta + \cos\theta}$$

where μ is a constant called the *coefficient of friction*.

(a) Find the rate of change of F with respect to θ.
(b) When is this rate of change equal to 0?
(c) If $W = 50$ lb and $\mu = 0.6$, draw the graph of F as a function of θ and use it to locate the value of θ for which $dF/d\theta = 0$. Is the value consistent with your answer to part (b)?

35–44 ||| Find the limit.

35. $\displaystyle\lim_{x\to 0} \frac{\sin 3x}{x}$

36. $\displaystyle\lim_{x\to 0} \frac{\sin 4x}{\sin 6x}$

37. $\displaystyle\lim_{t\to 0} \frac{\tan 6t}{\sin 2t}$

38. $\displaystyle\lim_{\theta\to 0} \frac{\cos\theta - 1}{\sin\theta}$

39. $\displaystyle\lim_{\theta\to 0} \frac{\sin(\cos\theta)}{\sec\theta}$

40. $\displaystyle\lim_{t\to 0} \frac{\sin^2 3t}{t^2}$

41. $\displaystyle\lim_{x\to 0} \frac{\cot 2x}{\csc x}$

42. $\displaystyle\lim_{x\to \pi/4} \frac{\sin x - \cos x}{\cos 2x}$

43. $\displaystyle\lim_{\theta\to 0} \frac{\sin\theta}{\theta + \tan\theta}$

44. $\displaystyle\lim_{x\to 1} \frac{\sin(x-1)}{x^2 + x - 2}$

45. Differentiate each trigonometric identity to obtain a new (or familiar) identity.

(a) $\tan x = \dfrac{\sin x}{\cos x}$

(b) $\sec x = \dfrac{1}{\cos x}$

(c) $\sin x + \cos x = \dfrac{1 + \cot x}{\csc x}$

46. A semicircle with diameter PQ sits on an isosceles triangle PQR to form a region shaped like an ice-cream cone, as shown in the figure. If $A(\theta)$ is the area of the semicircle and $B(\theta)$ is the area of the triangle, find

$$\lim_{\theta\to 0^+} \frac{A(\theta)}{B(\theta)}$$

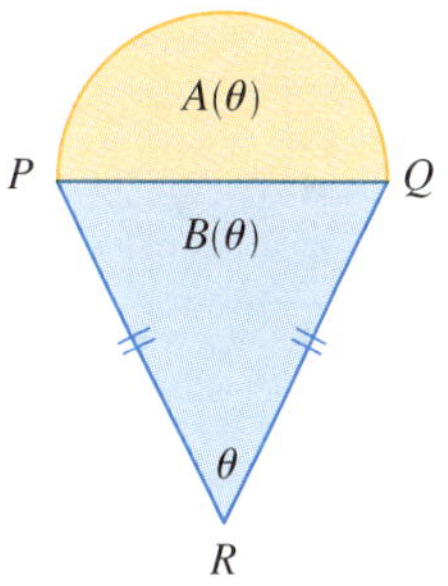

47. The figure shows a circular arc of length s and a chord of length d, both subtended by a central angle θ. Find

$$\lim_{\theta\to 0^+} \frac{s}{d}$$

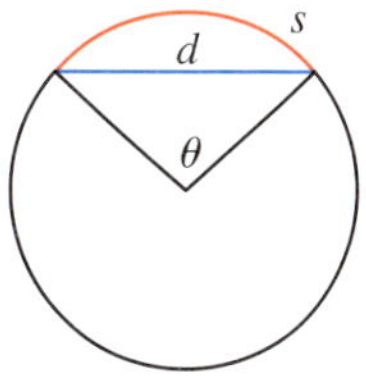

3.5 The Chain Rule

Suppose you are asked to differentiate the function

$$F(x) = \sqrt{x^2 + 1}$$

The differentiation formulas you learned in the previous sections of this chapter do not enable you to calculate $F'(x)$.

|||| See Section 1.3 for a review of composite functions.

Resources / Module 4 / Trigonometric Models / The Chain Rule

Observe that F is a composite function. In fact, if we let $y = f(u) = \sqrt{u}$ and let $u = g(x) = x^2 + 1$, then we can write $y = F(x) = f(g(x))$, that is, $F = f \circ g$. We know how to differentiate both f and g, so it would be useful to have a rule that tells us how to find the derivative of $F = f \circ g$ in terms of the derivatives of f and g.

It turns out that the derivative of the composite function $f \circ g$ is the product of the derivatives of f and g. This fact is one of the most important of the differentiation rules and is called the *Chain Rule*. It seems plausible if we interpret derivatives as rates of change. Regard du/dx as the rate of change of u with respect to x, dy/du as the rate of change of y with respect to u, and dy/dx as the rate of change of y with respect to x. If u changes twice as fast as x and y changes three times as fast as u, then it seems reasonable that y changes six times as fast as x, and so we expect that

$$\frac{dy}{dx} = \frac{dy}{du}\frac{du}{dx}$$

The Chain Rule If f and g are both differentiable and $F = f \circ g$ is the composite function defined by $F(x) = f(g(x))$, then F is differentiable and F' is given by the product

$$F'(x) = f'(g(x))g'(x)$$

In Leibniz notation, if $y = f(u)$ and $u = g(x)$ are both differentiable functions, then

$$\frac{dy}{dx} = \frac{dy}{du}\frac{du}{dx}$$

Comments on the Proof of the Chain Rule Let Δu be the change in u corresponding to a change of Δx in x, that is,

$$\Delta u = g(x + \Delta x) - g(x)$$

Then the corresponding change in y is

$$\Delta y = f(u + \Delta u) - f(u)$$

It is tempting to write

$$\frac{dy}{dx} = \lim_{\Delta x \to 0} \frac{\Delta y}{\Delta x}$$

1 $$= \lim_{\Delta x \to 0} \frac{\Delta y}{\Delta u} \cdot \frac{\Delta u}{\Delta x}$$

$$= \lim_{\Delta x \to 0} \frac{\Delta y}{\Delta u} \cdot \lim_{\Delta x \to 0} \frac{\Delta u}{\Delta x}$$

$$= \lim_{\Delta u \to 0} \frac{\Delta y}{\Delta u} \cdot \lim_{\Delta x \to 0} \frac{\Delta u}{\Delta x}$$

(Note that $\Delta u \to 0$ as $\Delta x \to 0$ since g is continuous.)

$$= \frac{dy}{du}\frac{du}{dx}$$

The only flaw in this reasoning is that in (1) it might happen that $\Delta u = 0$ (even when

$\Delta x \neq 0$) and, of course, we can't divide by 0. Nonetheless, this reasoning does at least *suggest* that the Chain Rule is true. A full proof of the Chain Rule is given at the end of this section.

The Chain Rule can be written either in the prime notation

2 $$(f \circ g)'(x) = f'(g(x))g'(x)$$

or, if $y = f(u)$ and $u = g(x)$, in Leibniz notation:

3 $$\frac{dy}{dx} = \frac{dy}{du}\frac{du}{dx}$$

Equation 3 is easy to remember because if dy/du and du/dx were quotients, then we could cancel du. Remember, however, that du has not been defined and du/dx should not be thought of as an actual quotient.

EXAMPLE 1 Find $F'(x)$ if $F(x) = \sqrt{x^2 + 1}$.

SOLUTION 1 (using Equation 2): At the beginning of this section we expressed F as $F(x) = (f \circ g)(x) = f(g(x))$ where $f(u) = \sqrt{u}$ and $g(x) = x^2 + 1$. Since

$$f'(u) = \tfrac{1}{2}u^{-1/2} = \frac{1}{2\sqrt{u}} \qquad \text{and} \qquad g'(x) = 2x$$

we have

$$\begin{aligned} F'(x) &= f'(g(x))g'(x) \\ &= \frac{1}{2\sqrt{x^2+1}} \cdot 2x = \frac{x}{\sqrt{x^2+1}} \end{aligned}$$

SOLUTION 2 (using Equation 3): If we let $u = x^2 + 1$ and $y = \sqrt{u}$, then

$$\begin{aligned} F'(x) &= \frac{dy}{du}\frac{du}{dx} = \frac{1}{2\sqrt{u}}(2x) \\ &= \frac{1}{2\sqrt{x^2+1}}(2x) = \frac{x}{\sqrt{x^2+1}} \end{aligned}$$

When using Formula 3 we should bear in mind that dy/dx refers to the derivative of y when y is considered as a function of x (called the *derivative of* y *with respect to* x), whereas dy/du refers to the derivative of y when considered as a function of u (the derivative of y with respect to u). For instance, in Example 1, y can be considered as a function of x $\left(y = \sqrt{x^2 + 1}\right)$ and also as a function of u $\left(y = \sqrt{u}\right)$. Note that

$$\frac{dy}{dx} = F'(x) = \frac{x}{\sqrt{x^2+1}} \qquad \text{whereas} \qquad \frac{dy}{du} = f'(u) = \frac{1}{2\sqrt{u}}$$

NOTE ▫ In using the Chain Rule we work from the outside to the inside. Formula 2 says that *we differentiate the outer function* f [*at the inner function* $g(x)$] *and then we multiply by the derivative of the inner function.*

$$\frac{d}{dx}\;\underbrace{f}_{\substack{\text{outer}\\\text{function}}}\;\underbrace{(g(x))}_{\substack{\text{evaluated}\\\text{at inner}\\\text{function}}} \;=\; \underbrace{f'}_{\substack{\text{derivative}\\\text{of outer}\\\text{function}}}\;\underbrace{(g(x))}_{\substack{\text{evaluated}\\\text{at inner}\\\text{function}}} \;\cdot\; \underbrace{g'(x)}_{\substack{\text{derivative}\\\text{of inner}\\\text{function}}}$$

EXAMPLE 2 Differentiate (a) $y = \sin(x^2)$ and (b) $y = \sin^2 x$.

SOLUTION

(a) If $y = \sin(x^2)$, then the outer function is the sine function and the inner function is the squaring function, so the Chain Rule gives

$$\frac{dy}{dx} = \frac{d}{dx} \underbrace{\sin}_{\text{outer function}} \underbrace{(x^2)}_{\text{evaluated at inner function}} = \underbrace{\cos}_{\text{derivative of outer function}} \underbrace{(x^2)}_{\text{evaluated at inner function}} \cdot \underbrace{2x}_{\text{derivative of inner function}}$$

$$= 2x\cos(x^2)$$

(b) Note that $\sin^2 x = (\sin x)^2$. Here the outer function is the squaring function and the inner function is the sine function. So

$$\frac{dy}{dx} = \frac{d}{dx} \underbrace{(\sin x)^2}_{\text{inner function}} = \underbrace{2}_{\text{derivative of outer function}} \cdot \underbrace{(\sin x)}_{\text{evaluated at inner function}} \cdot \underbrace{\cos x}_{\text{derivative of inner function}}$$

The answer can be left as $2 \sin x \cos x$ or written as $\sin 2x$ (by a trigonometric identity known as the double-angle formula).

See Reference Page 2 or Appendix D.

In Example 2(a) we combined the Chain Rule with the rule for differentiating the sine function. In general, if $y = \sin u$, where u is a differentiable function of x, then, by the Chain Rule,

$$\frac{dy}{dx} = \frac{dy}{du}\frac{du}{dx} = \cos u \frac{du}{dx}$$

Thus

$$\frac{d}{dx}(\sin u) = \cos u \frac{du}{dx}$$

In a similar fashion, all of the formulas for differentiating trigonometric functions can be combined with the Chain Rule.

Let's make explicit the special case of the Chain Rule where the outer function f is a power function. If $y = [g(x)]^n$, then we can write $y = f(u) = u^n$ where $u = g(x)$. By using the Chain Rule and then the Power Rule, we get

$$\frac{dy}{dx} = \frac{dy}{du}\frac{du}{dx} = nu^{n-1}\frac{du}{dx} = n[g(x)]^{n-1}g'(x)$$

4 The Power Rule Combined with the Chain Rule If n is any real number and $u = g(x)$ is differentiable, then

$$\frac{d}{dx}(u^n) = nu^{n-1}\frac{du}{dx}$$

Alternatively,

$$\frac{d}{dx}[g(x)]^n = n[g(x)]^{n-1} \cdot g'(x)$$

Notice that the derivative in Example 1 could be calculated by taking $n = \frac{1}{2}$ in Rule 4.

EXAMPLE 3 Differentiate $y = (x^3 - 1)^{100}$.

SOLUTION Taking $u = g(x) = x^3 - 1$ and $n = 100$ in (4), we have

$$\frac{dy}{dx} = \frac{d}{dx}(x^3 - 1)^{100} = 100(x^3 - 1)^{99}\frac{d}{dx}(x^3 - 1)$$
$$= 100(x^3 - 1)^{99} \cdot 3x^2 = 300x^2(x^3 - 1)^{99}$$

EXAMPLE 4 Find $f'(x)$ if $f(x) = \dfrac{1}{\sqrt[3]{x^2 + x + 1}}$.

SOLUTION First rewrite f: $f(x) = (x^2 + x + 1)^{-1/3}$. Thus

$$f'(x) = -\tfrac{1}{3}(x^2 + x + 1)^{-4/3}\frac{d}{dx}(x^2 + x + 1)$$
$$= -\tfrac{1}{3}(x^2 + x + 1)^{-4/3}(2x + 1)$$

EXAMPLE 5 Find the derivative of the function

$$g(t) = \left(\frac{t - 2}{2t + 1}\right)^9$$

SOLUTION Combining the Power Rule, Chain Rule, and Quotient Rule, we get

$$g'(t) = 9\left(\frac{t - 2}{2t + 1}\right)^8 \frac{d}{dt}\left(\frac{t - 2}{2t + 1}\right)$$
$$= 9\left(\frac{t - 2}{2t + 1}\right)^8 \frac{(2t + 1) \cdot 1 - 2(t - 2)}{(2t + 1)^2} = \frac{45(t - 2)^8}{(2t + 1)^{10}}$$

EXAMPLE 6 Differentiate $y = (2x + 1)^5(x^3 - x + 1)^4$.

SOLUTION In this example we must use the Product Rule before using the Chain Rule:

|||| The graphs of the functions y and y' in Example 6 are shown in Figure 1. Notice that y' is large when y increases rapidly and $y' = 0$ when y has a horizontal tangent. So our answer appears to be reasonable.

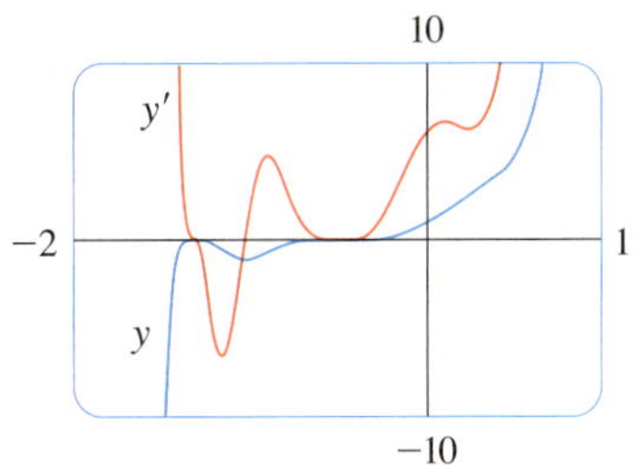

FIGURE 1

$$\frac{dy}{dx} = (2x + 1)^5 \frac{d}{dx}(x^3 - x + 1)^4 + (x^3 - x + 1)^4 \frac{d}{dx}(2x + 1)^5$$
$$= (2x + 1)^5 \cdot 4(x^3 - x + 1)^3 \frac{d}{dx}(x^3 - x + 1)$$
$$+ (x^3 - x + 1)^4 \cdot 5(2x + 1)^4 \frac{d}{dx}(2x + 1)$$
$$= 4(2x + 1)^5(x^3 - x + 1)^3(3x^2 - 1) + 5(x^3 - x + 1)^4(2x + 1)^4 \cdot 2$$

Noticing that each term has the common factor $2(2x + 1)^4(x^3 - x + 1)^3$, we could factor it out and write the answer as

$$\frac{dy}{dx} = 2(2x + 1)^4(x^3 - x + 1)^3(17x^3 + 6x^2 - 9x + 3)$$

EXAMPLE 7 Differentiate $y = e^{\sin x}$.

SOLUTION Here the inner function is $g(x) = \sin x$ and the outer function is the exponential function $f(x) = e^x$. So, by the Chain Rule,

More generally, the Chain Rule gives

$$\frac{d}{dx}(e^u) = e^u \frac{du}{dx}$$

$$\frac{dy}{dx} = \frac{d}{dx}(e^{\sin x}) = e^{\sin x}\frac{d}{dx}(\sin x) = e^{\sin x}\cos x$$

We can use the Chain Rule to differentiate an exponential function with any base $a > 0$. Recall from Section 1.6 that $a = e^{\ln a}$. So

$$a^x = (e^{\ln a})^x = e^{(\ln a)x}$$

and the Chain Rule gives

$$\frac{d}{dx}(a^x) = \frac{d}{dx}(e^{(\ln a)x}) = e^{(\ln a)x}\frac{d}{dx}(\ln a)x$$

$$= e^{(\ln a)x} \cdot \ln a = a^x \ln a$$

because $\ln a$ is a constant. So we have the formula

Don't confuse Formula 5 (where x is the *exponent*) with the Power Rule (where x is the *base*):

$$\frac{d}{dx}(x^n) = nx^{n-1}$$

5
$$\frac{d}{dx}(a^x) = a^x \ln a$$

In particular, if $a = 2$, we get

6
$$\frac{d}{dx}(2^x) = 2^x \ln 2$$

In Section 3.1 we gave the estimate

$$\frac{d}{dx}(2^x) \approx (0.69)2^x$$

This is consistent with the exact formula (6) because $\ln 2 \approx 0.693147$.

In Example 6 in Section 3.3 we considered a population of bacteria cells that doubles every hour and saw that the population after t hours is $n = n_0 2^t$, where n_0 is the initial population. Formula 6 enables us to find the rate of growth of the bacteria population:

$$\frac{dn}{dt} = n_0 2^t \ln 2$$

The reason for the name "Chain Rule" becomes clear when we make a longer chain by adding another link. Suppose that $y = f(u)$, $u = g(x)$, and $x = h(t)$, where f, g, and h are differentiable functions. Then, to compute the derivative of y with respect to t, we use the Chain Rule twice:

$$\frac{dy}{dt} = \frac{dy}{dx}\frac{dx}{dt} = \frac{dy}{du}\frac{du}{dx}\frac{dx}{dt}$$

EXAMPLE 8 If $f(x) = \sin(\cos(\tan x))$, then

$$f'(x) = \cos(\cos(\tan x)) \frac{d}{dx} \cos(\tan x)$$

$$= \cos(\cos(\tan x))[-\sin(\tan x)] \frac{d}{dx} (\tan x)$$

$$= -\cos(\cos(\tan x)) \sin(\tan x) \sec^2 x$$

Notice that the Chain Rule has been used twice.

EXAMPLE 9 Differentiate $y = e^{\sec 3\theta}$.

SOLUTION The outer function is the exponential function, the middle function is the secant function and the inner function is the tripling function. So we have

$$\frac{dy}{d\theta} = e^{\sec 3\theta} \frac{d}{d\theta} (\sec 3\theta)$$

$$= e^{\sec 3\theta} \sec 3\theta \tan 3\theta \frac{d}{d\theta} (3\theta)$$

$$= 3e^{\sec 3\theta} \sec 3\theta \tan 3\theta$$

How to Prove the Chain Rule

Recall that if $y = f(x)$ and x changes from a to $a + \Delta x$, we defined the increment of y as

$$\Delta y = f(a + \Delta x) - f(a)$$

According to the definition of a derivative, we have

$$\lim_{\Delta x \to 0} \frac{\Delta y}{\Delta x} = f'(a)$$

So if we denote by ε the difference between the difference quotient and the derivative, we obtain

$$\lim_{\Delta x \to 0} \varepsilon = \lim_{\Delta x \to 0} \left(\frac{\Delta y}{\Delta x} - f'(a) \right) = f'(a) - f'(a) = 0$$

But $$\varepsilon = \frac{\Delta y}{\Delta x} - f'(a) \quad \Rightarrow \quad \Delta y = f'(a)\,\Delta x + \varepsilon\,\Delta x$$

If we define ε to be 0 when $\Delta x = 0$, then ε becomes a continuous function of Δx. Thus, for a differentiable function f, we can write

7 $$\Delta y = f'(a)\,\Delta x + \varepsilon\,\Delta x \qquad \text{where} \quad \varepsilon \to 0 \text{ as } \Delta x \to 0$$

and ε is a continuous function of Δx. This property of differentiable functions is what enables us to prove the Chain Rule.

Proof of the Chain Rule Suppose $u = g(x)$ is differentiable at a and $y = f(u)$ is differentiable at $b = g(a)$. If Δx is an increment in x and Δu and Δy are the corresponding increments in u and y, then we can use Equation 7 to write

8 $$\Delta u = g'(a)\,\Delta x + \varepsilon_1\,\Delta x = [g'(a) + \varepsilon_1]\,\Delta x$$

where $\varepsilon_1 \to 0$ as $\Delta x \to 0$. Similarly

9 $$\Delta y = f'(b)\,\Delta u + \varepsilon_2\,\Delta u = [f'(b) + \varepsilon_2]\,\Delta u$$

where $\varepsilon_2 \to 0$ as $\Delta u \to 0$. If we now substitute the expression for Δu from Equation 8 into Equation 9, we get

$$\Delta y = [f'(b) + \varepsilon_2][g'(a) + \varepsilon_1]\,\Delta x$$

so

$$\frac{\Delta y}{\Delta x} = [f'(b) + \varepsilon_2][g'(a) + \varepsilon_1]$$

As $\Delta x \to 0$, Equation 8 shows that $\Delta u \to 0$. So both $\varepsilon_1 \to 0$ and $\varepsilon_2 \to 0$ as $\Delta x \to 0$. Therefore

$$\frac{dy}{dx} = \lim_{\Delta x \to 0} \frac{\Delta y}{\Delta x} = \lim_{\Delta x \to 0} [f'(b) + \varepsilon_2][g'(a) + \varepsilon_1]$$

$$= f'(b)g'(a) = f'(g(a))g'(a)$$

This proves the Chain Rule.

3.5 Exercises

1–6 Write the composite function in the form $f(g(x))$. [Identify the inner function $u = g(x)$ and the outer function $y = f(u)$.] Then find the derivative dy/dx.

1. $y = \sin 4x$

2. $y = \sqrt{4 + 3x}$

3. $y = (1 - x^2)^{10}$

4. $y = \tan(\sin x)$

5. $y = e^{\sqrt{x}}$

6. $y = \sin(e^x)$

7–42 Find the derivative of the function.

7. $F(x) = (x^3 + 4x)^7$

8. $F(x) = (x^2 - x + 1)^3$

9. $F(x) = \sqrt[4]{1 + 2x + x^3}$

10. $f(x) = (1 + x^4)^{2/3}$

11. $g(t) = \dfrac{1}{(t^4 + 1)^3}$

12. $f(t) = \sqrt[3]{1 + \tan t}$

13. $y = \cos(a^3 + x^3)$

14. $y = a^3 + \cos^3 x$

15. $y = e^{-mx}$

16. $y = 4 \sec 5x$

17. $g(x) = (1 + 4x)^5(3 + x - x^2)^8$

18. $h(t) = (t^4 - 1)^3(t^3 + 1)^4$

19. $y = (2x - 5)^4(8x^2 - 5)^{-3}$

20. $y = (x^2 + 1)\sqrt[3]{x^2 + 2}$

21. $y = xe^{-x^2}$

22. $y = e^{-5x} \cos 3x$

23. $y = e^{x \cos x}$

24. $y = 10^{1-x^2}$

25. $F(z) = \sqrt{\dfrac{z - 1}{z + 1}}$

26. $G(y) = \dfrac{(y - 1)^4}{(y^2 + 2y)^5}$

27. $y = \dfrac{r}{\sqrt{r^2 + 1}}$

28. $y = \dfrac{e^{2u}}{e^u + e^{-u}}$

29. $y = \tan(\cos x)$

30. $y = \dfrac{\sin^2 x}{\cos x}$

31. $y = 2^{\sin \pi x}$

32. $y = \tan^2(3\theta)$

33. $y = (1 + \cos^2 x)^6$

34. $y = x \sin \dfrac{1}{x}$

35. $y = \sec^2 x + \tan^2 x$

36. $y = e^{k \tan \sqrt{x}}$

37. $y = \cot^2(\sin \theta)$

38. $y = \sin(\sin(\sin x))$

39. $y = \sqrt{x + \sqrt{x}}$

40. $y = \sqrt{x + \sqrt{x + \sqrt{x}}}$

41. $y = \sin\left(\tan \sqrt{\sin x}\right)$

42. $y = 2^{3^{x^2}}$

43–46 Find an equation of the tangent line to the curve at the given point.

43. $y = (1 + 2x)^{10}$, $(0, 1)$

44. $y = \sin x + \sin^2 x$, $(0, 0)$

45. $y = \sin(\sin x)$, $(\pi, 0)$

46. $y = x^2 e^{-x}$ $(1, 1/e)$

47. (a) Find an equation of the tangent line to the curve $y = 2/(1 + e^{-x})$ at the point $(0, 1)$.
(b) Illustrate part (a) by graphing the curve and the tangent line on the same screen.

48. (a) The curve $y = |x|/\sqrt{2 - x^2}$ is called a **bullet-nose curve**. Find an equation of the tangent line to this curve at the point $(1, 1)$.
(b) Illustrate part (a) by graphing the curve and the tangent line on the same screen.

49. (a) If $f(x) = \sqrt{1 - x^2}/x$, find $f'(x)$.
(b) Check to see that your answer to part (a) is reasonable by comparing the graphs of f and f'.

50. The function $f(x) = \sin(x + \sin 2x)$, $0 \leqslant x \leqslant \pi$, arises in applications to frequency modulation (FM) synthesis.
(a) Use a graph of f produced by a graphing device to make a rough sketch of the graph of f'.
(b) Calculate $f'(x)$ and use this expression, with a graphing device, to graph f'. Compare with your sketch in part (a).

51. Find all points on the graph of the function

$$f(x) = 2 \sin x + \sin^2 x$$

at which the tangent line is horizontal.

52. Find the x-coordinates of all points on the curve $y = \sin 2x - 2 \sin x$ at which the tangent line is horizontal.

53. Suppose that $F(x) = f(g(x))$ and $g(3) = 6$, $g'(3) = 4$, $f'(3) = 2$, and $f'(6) = 7$. Find $F'(3)$.

54. Suppose that $w = u \circ v$ and $u(0) = 1$, $v(0) = 2$, $u'(0) = 3$, $u'(2) = 4$, $v'(0) = 5$, and $v'(2) = 6$. Find $w'(0)$.

55. A table of values for f, g, f', and g' is given.

x	$f(x)$	$g(x)$	$f'(x)$	$g'(x)$
1	3	2	4	6
2	1	8	5	7
3	7	2	7	9

(a) If $h(x) = f(g(x))$, find $h'(1)$.
(b) If $H(x) = g(f(x))$, find $H'(1)$.

56. Let f and g be the functions in Exercise 55.
(a) If $F(x) = f(f(x))$, find $F'(2)$.
(b) If $G(x) = g(g(x))$, find $G'(3)$.

57. If f and g are the functions whose graphs are shown, let $u(x) = f(g(x))$, $v(x) = g(f(x))$, and $w(x) = g(g(x))$. Find each derivative, if it exists. If it does not exist, explain why.
(a) $u'(1)$ (b) $v'(1)$ (c) $w'(1)$

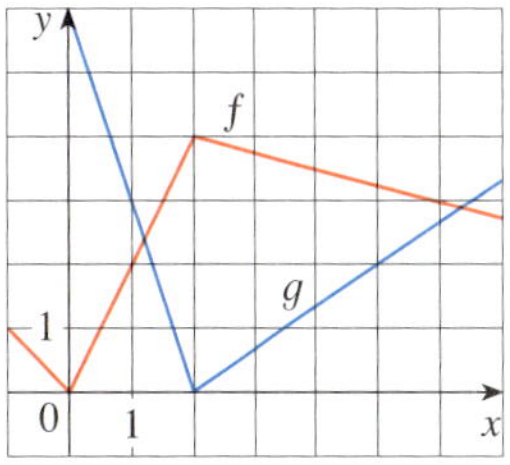

58. If f is the function whose graph is shown, let $h(x) = f(f(x))$ and $g(x) = f(x^2)$. Use the graph of f to estimate the value of each derivative.
(a) $h'(2)$ (b) $g'(2)$

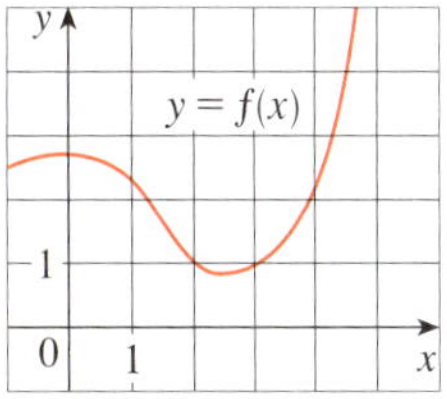

59. Use the table to estimate the value of $h'(0.5)$, where $h(x) = f(g(x))$.

x	0	0.1	0.2	0.3	0.4	0.5	0.6
$f(x)$	12.6	14.8	18.4	23.0	25.9	27.5	29.1
$g(x)$	0.58	0.40	0.37	0.26	0.17	0.10	0.05

60. If $g(x) = f(f(x))$, use the table to estimate the value of $g'(1)$.

x	0.0	0.5	1.0	1.5	2.0	2.5
$f(x)$	1.7	1.8	2.0	2.4	3.1	4.4

61. Suppose f is differentiable on $\mathbb{R}$. Let $F(x) = f(e^x)$ and $G(x) = e^{f(x)}$. Find expressions for (a) $F'(x)$ and (b) $G'(x)$.

62. Suppose f is differentiable on $\mathbb{R}$ and α is a real number. Let $F(x) = f(x^\alpha)$ and $G(x) = [f(x)]^\alpha$. Find expressions for (a) $F'(x)$ and (b) $G'(x)$.

63. Suppose L is a function such that $L'(x) = 1/x$ for $x > 0$. Find an expression for the derivative of each function.
(a) $f(x) = L(x^4)$ (b) $g(x) = L(4x)$
(c) $F(x) = [L(x)]^4$ (d) $G(x) = L(1/x)$

64. Let $r(x) = f(g(h(x)))$, where $h(1) = 2$, $g(2) = 3$, $h'(1) = 4$, $g'(2) = 5$, and $f'(3) = 6$. Find $r'(1)$.

65. The displacement of a particle on a vibrating string is given by the equation

$$s(t) = 10 + \frac{1}{4}\sin(10\pi t)$$

where s is measured in centimeters and t in seconds. Find the velocity of the particle after t seconds.

66. If the equation of motion of a particle is given by $s = A\cos(\omega t + \delta)$, the particle is said to undergo *simple harmonic motion.*

(a) Find the velocity of the particle at time t.

(b) When is the velocity 0?

67. A Cepheid variable star is a star whose brightness alternately increases and decreases. The most easily visible such star is Delta Cephei, for which the interval between times of maximum brightness is 5.4 days. The average brightness of this star is 4.0 and its brightness changes by ± 0.35. In view of these data, the brightness of Delta Cephei at time t, where t is measured in days, has been modeled by the function

$$B(t) = 4.0 + 0.35\sin(2\pi t/5.4)$$

(a) Find the rate of change of the brightness after t days.

(b) Find, correct to two decimal places, the rate of increase after one day.

68. In Example 4 in Section 1.3 we arrived at a model for the length of daylight (in hours) in Philadelphia on the tth day of the year:

$$L(t) = 12 + 2.8\sin\left[\frac{2\pi}{365}(t - 80)\right]$$

Use this model to compare how the number of hours of daylight is increasing in Philadelphia on March 21 and May 21.

69. The motion of a spring that is subject to a frictional force or a damping force (such as a shock absorber in a car) is often modeled by the product of an exponential function and a sine or cosine function. Suppose the equation of motion of a point on such a spring is

$$s(t) = 2e^{-1.5t}\sin 2\pi t$$

where s is measured in centimeters and t in seconds. Find the velocity after t seconds and graph both the position and velocity functions for $0 \leqslant t \leqslant 2$.

70. Under certain circumstances a rumor spreads according to the equation

$$p(t) = \frac{1}{1 + ae^{-kt}}$$

where $p(t)$ is the proportion of the population that knows the rumor at time t and a and k are positive constants. [In Section 9.5 we will see that this is a reasonable equation for $p(t)$.]

(a) Find $\lim_{t\to\infty} p(t)$.

(b) Find the rate of spread of the rumor.

(c) Graph p for the case $a = 10$, $k = 0.5$ with t measured in hours. Use the graph to estimate how long it will take for 80% of the population to hear the rumor.

71. The flash unit on a camera operates by storing charge on a capacitor and releasing it suddenly when the flash is set off. The following data describe the charge Q remaining on the capacitor (measured in microcoulombs, μC) at time t (measured in seconds).

t	0.00	0.02	0.04	0.06	0.08	0.10
Q	100.00	81.87	67.03	54.88	44.93	36.76

(a) Use a graphing calculator or computer to find an exponential model for the charge. (See Section 1.5.)

(b) The derivative $Q'(t)$ represents the electric current (measured in microamperes, μA) flowing from the capacitor to the flash bulb. Use part (a) to estimate the current when $t = 0.04$ s. Compare with the result of Example 2 in Section 2.1.

72. The table gives the U.S. population from 1790 to 1860.

Year	Population	Year	Population
1790	3,929,000	1830	12,861,000
1800	5,308,000	1840	17,063,000
1810	7,240,000	1850	23,192,000
1820	9,639,000	1860	31,443,000

(a) Use a graphing calculator or computer to fit an exponential function to the data. Graph the data points and the exponential model. How good is the fit?

(b) Estimate the rates of population growth in 1800 and 1850 by averaging slopes of secant lines.

(c) Use the exponential model in part (a) to estimate the rates of growth in 1800 and 1850. Compare these estimates with the ones in part (b).

(d) Use the exponential model to predict the population in 1870. Compare with the actual population of 38,558,000. Can you explain the discrepancy?

CAS **73.** Computer algebra systems have commands that differentiate functions, but the form of the answer may not be convenient and so further commands may be necessary to simplify the answer.

(a) Use a CAS to find the derivative in Example 5 and compare with the answer in that example. Then use the simplify command and compare again.

(b) Use a CAS to find the derivative in Example 6. What happens if you use the simplify command? What happens if you use the factor command? Which form of the answer would be best for locating horizontal tangents?

74. (a) Use a CAS to differentiate the function

$$f(x) = \sqrt{\frac{x^4 - x + 1}{x^4 + x + 1}}$$

and to simplify the result.
(b) Where does the graph of f have horizontal tangents?
(c) Graph f and f' on the same screen. Are the graphs consistent with your answer to part (b)?

75. Use the Chain Rule to prove the following.
(a) The derivative of an even function is an odd function.
(b) The derivative of an odd function is an even function.

76. Use the Chain Rule and the Product Rule to give an alternative proof of the Quotient Rule.
[*Hint:* Write $f(x)/g(x) = f(x)[g(x)]^{-1}$.]

77. (a) If n is a positive integer, prove that

$$\frac{d}{dx}(\sin^n x \cos nx) = n \sin^{n-1} x \cos(n + 1)x$$

(b) Find a formula for the derivative of

$$y = \cos^n x \cos nx$$

that is similar to the one in part (a).

78. Suppose $y = f(x)$ is a curve that always lies above the x-axis and never has a horizontal tangent, where f is differentiable everywhere. For what value of y is the rate of change of y^5 with respect to x eighty times the rate of change of y with respect to x?

79. Use the Chain Rule to show that if θ is measured in degrees, then

$$\frac{d}{d\theta}(\sin \theta) = \frac{\pi}{180} \cos \theta$$

(This gives one reason for the convention that radian measure is always used when dealing with trigonometric functions in calculus: The differentiation formulas would not be as simple if we used degree measure.)

80. (a) Write $|x| = \sqrt{x^2}$ and use the Chain Rule to show that

$$\frac{d}{dx}|x| = \frac{x}{|x|}$$

(b) If $f(x) = |\sin x|$, find $f'(x)$ and sketch the graphs of f and f'. Where is f not differentiable?
(c) If $g(x) = \sin |x|$, find $g'(x)$ and sketch the graphs of g and g'. Where is g not differentiable?

81. Suppose P and Q are polynomials and n is a positive integer. Use mathematical induction to prove that the nth derivative of the rational function $f(x) = P(x)/Q(x)$ can be written as a rational function with denominator $[Q(x)]^{n+1}$. In other words, there is a polynomial A_n such that $f^{(n)}(x) = A_n(x)/[Q(x)]^{n+1}$.

3.6 Implicit Differentiation

The functions that we have met so far can be described by expressing one variable explicitly in terms of another variable—for example,

$$y = \sqrt{x^3 + 1} \qquad \text{or} \qquad y = x \sin x$$

or, in general, $y = f(x)$. Some functions, however, are defined implicitly by a relation between x and y such as

1 $$x^2 + y^2 = 25$$

or

2 $$x^3 + y^3 = 6xy$$

In some cases it is possible to solve such an equation for y as an explicit function (or several functions) of x. For instance, if we solve Equation 1 for y, we get $y = \pm\sqrt{25 - x^2}$, so two of the functions determined by the implicit Equation 1 are $f(x) = \sqrt{25 - x^2}$ and $g(x) = -\sqrt{25 - x^2}$. The graphs of f and g are the upper and lower semicircles of the circle $x^2 + y^2 = 25$. (See Figure 1.)

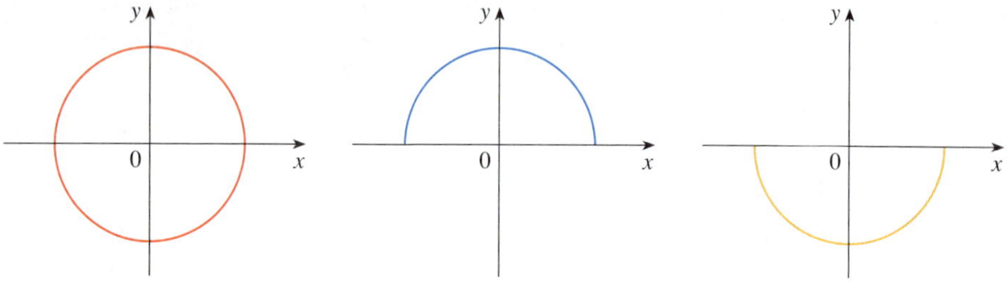

FIGURE 1 (a) $x^2 + y^2 = 25$ (b) $f(x) = \sqrt{25 - x^2}$ (c) $g(x) = -\sqrt{25 - x^2}$

It's not easy to solve Equation 2 for y explicitly as a function of x by hand. (A computer algebra system has no trouble, but the expressions it obtains are very complicated.) Nonetheless, (2) is the equation of a curve called the **folium of Descartes** shown in Figure 2 and it implicitly defines y as several functions of x. The graphs of three such functions are shown in Figure 3. When we say that f is a function defined implicitly by Equation 2, we mean that the equation

$$x^3 + [f(x)]^3 = 6xf(x)$$

is true for all values of x in the domain of f.

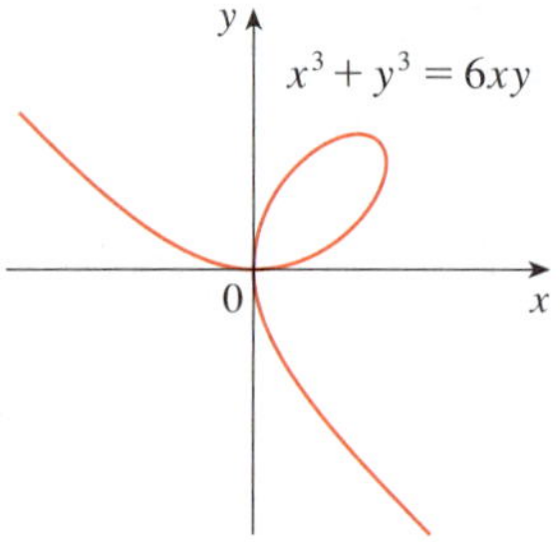

FIGURE 2 The folium of Descartes

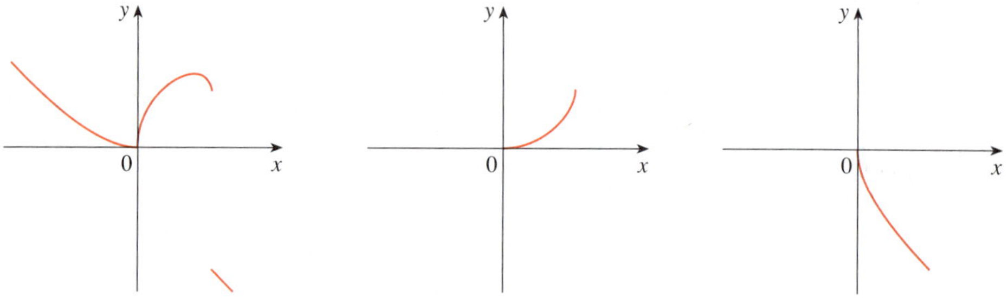

FIGURE 3 Graphs of three functions defined by the folium of Descartes

Fortunately, we don't need to solve an equation for y in terms of x in order to find the derivative of y. Instead we can use the method of **implicit differentiation**. This consists of differentiating both sides of the equation with respect to x and then solving the resulting equation for y'. In the examples and exercises of this section it is always assumed that the given equation determines y implicitly as a differentiable function of x so that the method of implicit differentiation can be applied.

EXAMPLE 1

(a) If $x^2 + y^2 = 25$, find $\dfrac{dy}{dx}$.

(b) Find an equation of the tangent to the circle $x^2 + y^2 = 25$ at the point $(3, 4)$.

SOLUTION 1

(a) Differentiate both sides of the equation $x^2 + y^2 = 25$:

$$\frac{d}{dx}(x^2 + y^2) = \frac{d}{dx}(25)$$

$$\frac{d}{dx}(x^2) + \frac{d}{dx}(y^2) = 0$$

Remembering that y is a function of x and using the Chain Rule, we have

$$\frac{d}{dx}(y^2) = \frac{d}{dy}(y^2)\frac{dy}{dx} = 2y\frac{dy}{dx}$$

Thus

$$2x + 2y\frac{dy}{dx} = 0$$

Now we solve this equation for dy/dx:

$$\frac{dy}{dx} = -\frac{x}{y}$$

(b) At the point (3, 4) we have $x = 3$ and $y = 4$, so

$$\frac{dy}{dx} = -\frac{3}{4}$$

An equation of the tangent to the circle at (3, 4) is therefore

$$y - 4 = -\tfrac{3}{4}(x - 3) \qquad \text{or} \qquad 3x + 4y = 25$$

SOLUTION 2

(b) Solving the equation $x^2 + y^2 = 25$, we get $y = \pm\sqrt{25 - x^2}$. The point (3, 4) lies on the upper semicircle $y = \sqrt{25 - x^2}$ and so we consider the function $f(x) = \sqrt{25 - x^2}$. Differentiating f using the Chain Rule, we have

$$f'(x) = \tfrac{1}{2}(25 - x^2)^{-1/2}\frac{d}{dx}(25 - x^2)$$

$$= \tfrac{1}{2}(25 - x^2)^{-1/2}(-2x) = -\frac{x}{\sqrt{25 - x^2}}$$

So

$$f'(3) = -\frac{3}{\sqrt{25 - 3^2}} = -\frac{3}{4}$$

and, as in Solution 1, an equation of the tangent is $3x + 4y = 25$.

NOTE 1 ▫ Example 1 illustrates that even when it is possible to solve an equation explicitly for y in terms of x, it may be easier to use implicit differentiation.

NOTE 2 ▫ The expression $dy/dx = -x/y$ gives the derivative in terms of both x and y. It is correct no matter which function y is determined by the given equation. For instance, for $y = f(x) = \sqrt{25 - x^2}$ we have

$$\frac{dy}{dx} = -\frac{x}{y} = -\frac{x}{\sqrt{25 - x^2}}$$

whereas for $y = g(x) = -\sqrt{25 - x^2}$ we have

$$\frac{dy}{dx} = -\frac{x}{y} = -\frac{x}{-\sqrt{25 - x^2}} = \frac{x}{\sqrt{25 - x^2}}$$

EXAMPLE 2

(a) Find y' if $x^3 + y^3 = 6xy$.

(b) Find the tangent to the folium of Descartes $x^3 + y^3 = 6xy$ at the point (3, 3).

(c) At what points on the curve is the tangent line horizontal?

SOLUTION

(a) Differentiating both sides of $x^3 + y^3 = 6xy$ with respect to x, regarding y as a function of x, and using the Chain Rule on the y^3 term and the Product Rule on the $6xy$ term, we get

$$3x^2 + 3y^2y' = 6y + 6xy'$$

or

$$x^2 + y^2y' = 2y + 2xy'$$

We now solve for y':

$$y^2y' - 2xy' = 2y - x^2$$

$$(y^2 - 2x)y' = 2y - x^2$$

$$y' = \frac{2y - x^2}{y^2 - 2x}$$

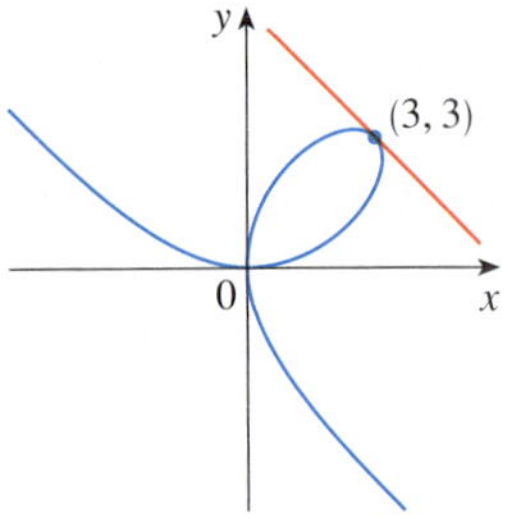

FIGURE 4

(b) When $x = y = 3$,

$$y' = \frac{2 \cdot 3 - 3^2}{3^2 - 2 \cdot 3} = -1$$

and a glance at Figure 4 confirms that this is a reasonable value for the slope at (3, 3). So an equation of the tangent to the folium at (3, 3) is

$$y - 3 = -1(x - 3) \qquad \text{or} \qquad x + y = 6$$

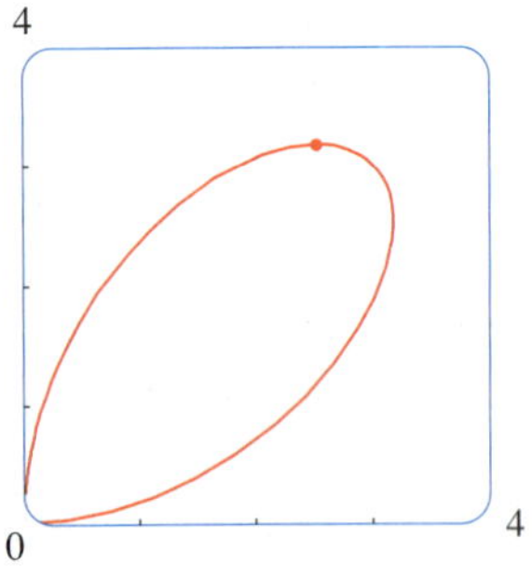

FIGURE 5

(c) The tangent line is horizontal if $y' = 0$. Using the expression for y' from part (a), we see that $y' = 0$ when $2y - x^2 = 0$. Substituting $y = \frac{1}{2}x^2$ in the equation of the curve, we get

$$x^3 + \left(\tfrac{1}{2}x^2\right)^3 = 6x\left(\tfrac{1}{2}x^2\right)$$

which simplifies to $x^6 = 16x^3$. So either $x = 0$ or $x^3 = 16$. If $x = 16^{1/3} = 2^{4/3}$, then $y = \frac{1}{2}(2^{8/3}) = 2^{5/3}$. Thus, the tangent is horizontal at (0, 0) and at $(2^{4/3}, 2^{5/3})$, which is approximately (2.5198, 3.1748). Looking at Figure 5, we see that our answer is reasonable.

NOTE 3 ▫ There is a formula for the three roots of a cubic equation that is like the quadratic formula but much more complicated. If we use this formula (or a computer algebra system) to solve the equation $x^3 + y^3 = 6xy$ for y in terms of x, we get three functions determined by the equation:

$$y = f(x) = \sqrt[3]{-\tfrac{1}{2}x^3 + \sqrt{\tfrac{1}{4}x^6 - 8x^3}} + \sqrt[3]{-\tfrac{1}{2}x^3 - \sqrt{\tfrac{1}{4}x^6 - 8x^3}}$$

and

$$y = \tfrac{1}{2}\left[-f(x) \pm \sqrt{-3}\left(\sqrt[3]{-\tfrac{1}{2}x^3 + \sqrt{\tfrac{1}{4}x^6 - 8x^3}} - \sqrt[3]{-\tfrac{1}{2}x^3 - \sqrt{\tfrac{1}{4}x^6 - 8x^3}}\right)\right]$$

(These are the three functions whose graphs are shown in Figure 3.) You can see that the method of implicit differentiation saves an enormous amount of work in cases such as this.

|||| The Norwegian mathematician Niels Abel proved in 1824 that no general formula can be given for the roots of a fifth-degree equation in terms of radicals. Later the French mathematician Evariste Galois proved that it is impossible to find a general formula for the roots of an nth-degree equation (in terms of algebraic operations on the coefficients) if n is any integer larger than 4.

Moreover, implicit differentiation works just as easily for equations such as

$$y^5 + 3x^2y^2 + 5x^4 = 12$$

for which it is *impossible* to find a similar expression for y in terms of x.

EXAMPLE 3 Find y' if $\sin(x + y) = y^2 \cos x$.

SOLUTION Differentiating implicitly with respect to x and remembering that y is a function of x, we get

$$\cos(x + y) \cdot (1 + y') = 2yy' \cos x + y^2(-\sin x)$$

(Note that we have used the Chain Rule on the left side and the Product Rule and Chain Rule on the right side.) If we collect the terms that involve y', we get

$$\cos(x + y) + y^2 \sin x = (2y \cos x)y' - \cos(x + y) \cdot y'$$

So
$$y' = \frac{y^2 \sin x + \cos(x + y)}{2y \cos x - \cos(x + y)}$$

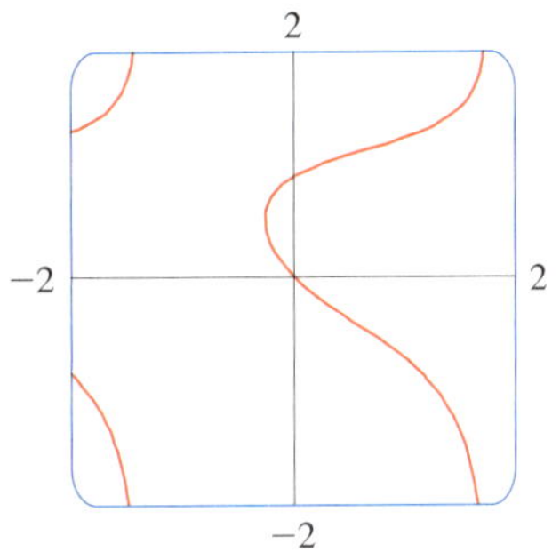

FIGURE 6

Figure 6, drawn with the implicit-plotting command of a computer algebra system, shows part of the curve $\sin(x + y) = y^2 \cos x$. As a check on our calculation, notice that $y' = -1$ when $x = y = 0$ and it appears from the graph that the slope is approximately -1 at the origin.

Orthogonal Trajectories

Two curves are called **orthogonal** if at each point of intersection their tangent lines are perpendicular. In the next example we use implicit differentiation to show that two families of curves are **orthogonal trajectories** of each other; that is, every curve in one family is orthogonal to every curve in the other family. Orthogonal families arise in several areas of physics. For example, the lines of force in an electrostatic field are orthogonal to the lines of constant potential. In thermodynamics, the isotherms (curves of equal temperature) are orthogonal to the flow lines of heat. In aerodynamics, the streamlines (curves of direction of airflow) are orthogonal trajectories of the velocity-equipotential curves.

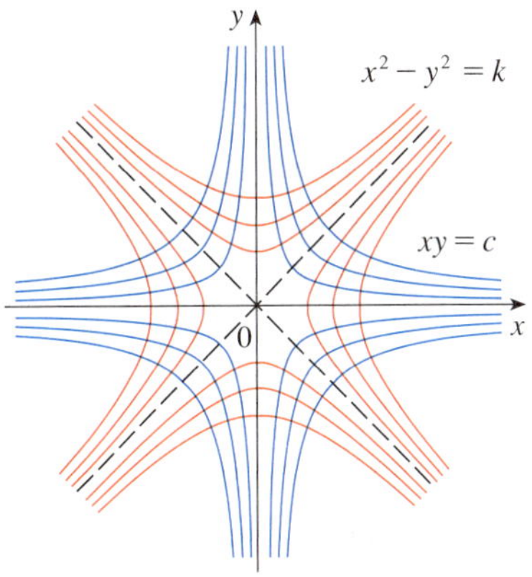

FIGURE 7

EXAMPLE 4 The equation

3 $$xy = c \qquad c \neq 0$$

represents a family of hyperbolas. (Different values of the constant c give different hyperbolas. See Figure 7.) The equation

4 $$x^2 - y^2 = k \qquad k \neq 0$$

represents another family of hyperbolas with asymptotes $y = \pm x$. Show that every curve in the family (3) is orthogonal to every curve in the family (4); that is, the families are orthogonal trajectories of each other.

SOLUTION Implicit differentiation of Equation 3 gives

5 $$x\frac{dy}{dx} + y = 0 \qquad \text{so} \qquad \frac{dy}{dx} = -\frac{y}{x}$$

Implicit differentiation of Equation 4 gives

6 $$2x - 2y\frac{dy}{dx} = 0 \qquad \text{so} \qquad \frac{dy}{dx} = \frac{x}{y}$$

From (5) and (6) we see that at any point of intersection of curves from each family, the slopes of the tangents are negative reciprocals of each other. Therefore, the curves intersect at right angles; that is, they are orthogonal.

Derivatives of Inverse Trigonometric Functions

The inverse trigonometric functions were reviewed in Section 1.6. We discussed their continuity in Section 2.5 and their asymptotes in Section 2.6. Here we use implicit differentiation to find the derivatives of the inverse trigonometric functions, assuming that these functions are differentiable. [In fact, if f is any one-to-one differentiable function, it can be proved that its inverse function f^{-1} is also differentiable, except where its tangents are vertical. This is plausible because the graph of a differentiable function has no corner or kink and so if we reflect it about $y = x$, the graph of its inverse function also has no corner or kink.]

Recall the definition of the arcsine function:

$$y = \sin^{-1}x \qquad \text{means} \qquad \sin y = x \quad \text{and} \quad -\frac{\pi}{2} \leqslant y \leqslant \frac{\pi}{2}$$

Differentiating $\sin y = x$ implicitly with respect to x, we obtain

$$\cos y\frac{dy}{dx} = 1 \qquad \text{or} \qquad \frac{dy}{dx} = \frac{1}{\cos y}$$

Now $\cos y \geqslant 0$, since $-\pi/2 \leqslant y \leqslant \pi/2$, so

$$\cos y = \sqrt{1 - \sin^2 y} = \sqrt{1 - x^2}$$

Therefore $$\frac{dy}{dx} = \frac{1}{\cos y} = \frac{1}{\sqrt{1 - x^2}}$$

The same method can be used to find a formula for the derivative of *any* inverse function. See Exercise 67.

$$\frac{d}{dx}(\sin^{-1}x) = \frac{1}{\sqrt{1 - x^2}}$$

Figure 8 shows the graph of $f(x) = \tan^{-1}x$ and its derivative $f'(x) = 1/(1 + x^2)$. Notice that f is increasing and $f'(x)$ is always positive. The fact that $\tan^{-1}x \to \pm\pi/2$ as $x \to \pm\infty$ is reflected in the fact that $f'(x) \to 0$ as $x \to \pm\infty$.

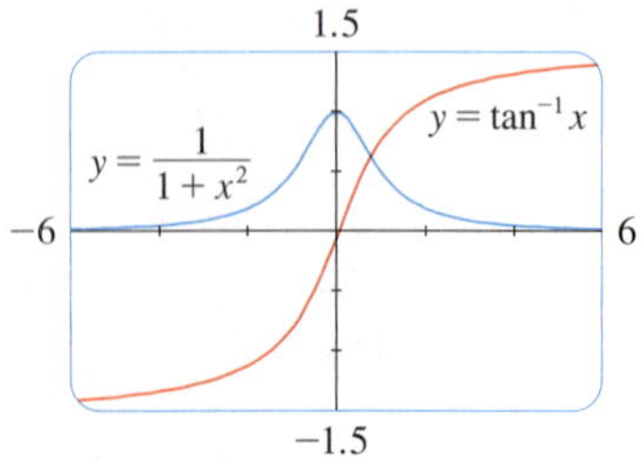

FIGURE 8

The formula for the derivative of the arctangent function is derived in a similar way. If $y = \tan^{-1}x$, then $\tan y = x$. Differentiating this latter equation implicitly with respect to x, we have

$$\sec^2 y\frac{dy}{dx} = 1$$

$$\frac{dy}{dx} = \frac{1}{\sec^2 y} = \frac{1}{1 + \tan^2 y} = \frac{1}{1 + x^2}$$

$$\frac{d}{dx}(\tan^{-1}x) = \frac{1}{1 + x^2}$$

EXAMPLE 5 Differentiate (a) $y = \dfrac{1}{\sin^{-1}x}$ and (b) $f(x) = x \arctan\sqrt{x}$.

SOLUTION

(a)
$$\frac{dy}{dx} = \frac{d}{dx}(\sin^{-1}x)^{-1} = -(\sin^{-1}x)^{-2}\frac{d}{dx}(\sin^{-1}x)$$
$$= -\frac{1}{(\sin^{-1}x)^2\sqrt{1-x^2}}$$

Recall that arctan x is an alternative notation for $\tan^{-1}x$.

(b)
$$f'(x) = x\frac{1}{1+(\sqrt{x})^2}\left(\tfrac{1}{2}x^{-1/2}\right) + \arctan\sqrt{x}$$
$$= \frac{\sqrt{x}}{2(1+x)} + \arctan\sqrt{x}$$

The inverse trigonometric functions that occur most frequently are the ones that we have just discussed. The derivatives of the remaining four are given in the following table. The proofs of the formulas are left as exercises.

The formulas for the derivatives of $\csc^{-1}x$ and $\sec^{-1}x$ depend on the definitions that are used for these functions. See Exercise 54.

Derivatives of Inverse Trigonometric Functions

$$\frac{d}{dx}(\sin^{-1}x) = \frac{1}{\sqrt{1-x^2}} \qquad \frac{d}{dx}(\csc^{-1}x) = -\frac{1}{x\sqrt{x^2-1}}$$
$$\frac{d}{dx}(\cos^{-1}x) = -\frac{1}{\sqrt{1-x^2}} \qquad \frac{d}{dx}(\sec^{-1}x) = \frac{1}{x\sqrt{x^2-1}}$$
$$\frac{d}{dx}(\tan^{-1}x) = \frac{1}{1+x^2} \qquad \frac{d}{dx}(\cot^{-1}x) = -\frac{1}{1+x^2}$$

3.6 Exercises

1–4
(a) Find y' by implicit differentiation.
(b) Solve the equation explicitly for y and differentiate to get y' in terms of x.
(c) Check that your solutions to parts (a) and (b) are consistent by substituting the expression for y into your solution for part (a).

1. $xy + 2x + 3x^2 = 4$

2. $4x^2 + 9y^2 = 36$

3. $\dfrac{1}{x} + \dfrac{1}{y} = 1$

4. $\sqrt{x} + \sqrt{y} = 4$

5–20 Find dy/dx by implicit differentiation.

5. $x^2 + y^2 = 1$

6. $x^2 - y^2 = 1$

7. $x^3 + x^2y + 4y^2 = 6$

8. $x^2 - 2xy + y^3 = c$

9. $x^2y + xy^2 = 3x$

10. $y^5 + x^2y^3 = 1 + ye^{x^2}$

11. $x^2y^2 + x\sin y = 4$

12. $1 + x = \sin(xy^2)$

13. $4\cos x \sin y = 1$

14. $y\sin(x^2) = x\sin(y^2)$

15. $e^{x^2y} = x + y$

16. $\sqrt{x+y} = 1 + x^2y^2$

17. $\sqrt{xy} = 1 + x^2y$

18. $\tan(x-y) = \dfrac{y}{1+x^2}$

19. $xy = \cot(xy)$

20. $\sin x + \cos y = \sin x \cos y$

21. If $1 + f(x) + x^2[f(x)]^3 = 0$ and $f(1) = 2$, find $f'(1)$.

22. If $g(x) + x\sin g(x) = x^2$ and $g(1) = 0$, find $g'(1)$.

23–24 Regard y as the independent variable and x as the dependent variable and use implicit differentiation to find dx/dy.

23. $y^4 + x^2y^2 + yx^4 = y + 1$

24. $(x^2 + y^2)^2 = ax^2y$

25–30 ||| Use implicit differentiation to find an equation of the tangent line to the curve at the given point.

25. $x^2 + xy + y^2 = 3$, $(1, 1)$ (ellipse)

26. $x^2 + 2xy - y^2 + x = 2$, $(1, 2)$ (hyperbola)

27. $x^2 + y^2 = (2x^2 + 2y^2 - x)^2$
$(0, \frac{1}{2})$
(cardioid)

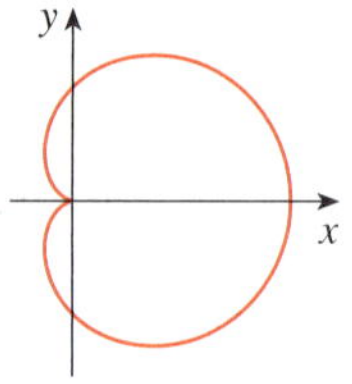

28. $x^{2/3} + y^{2/3} = 4$
$(-3\sqrt{3}, 1)$
(astroid)

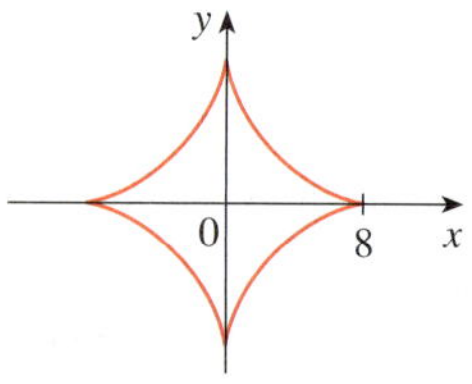

29. $2(x^2 + y^2)^2 = 25(x^2 - y^2)$
$(3, 1)$
(lemniscate)

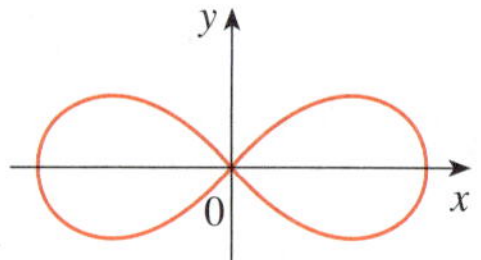

30. $y^2(y^2 - 4) = x^2(x^2 - 5)$
$(0, -2)$
(devil's curve)

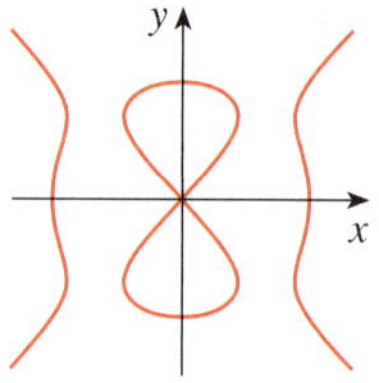

31. (a) The curve with equation $y^2 = 5x^4 - x^2$ is called a **kampyle of Eudoxus**. Find an equation of the tangent line to this curve at the point $(1, 2)$.

(b) Illustrate part (a) by graphing the curve and the tangent line on a common screen. (If your graphing device will graph implicitly defined curves, then use that capability. If not, you can still graph this curve by graphing its upper and lower halves separately.)

32. (a) The curve with equation $y^2 = x^3 + 3x^2$ is called the **Tschirnhausen cubic**. Find an equation of the tangent line to this curve at the point $(1, -2)$.

(b) At what points does this curve have a horizontal tangent?

(c) Illustrate parts (a) and (b) by graphing the curve and the tangent lines on a common screen.

CAS **33.** Fanciful shapes can be created by using the implicit plotting capabilities of computer algebra systems.

(a) Graph the curve with equation

$$y(y^2 - 1)(y - 2) = x(x - 1)(x - 2)$$

At how many points does this curve have horizontal tangents? Estimate the x-coordinates of these points.

(b) Find equations of the tangent lines at the points $(0, 1)$ and $(0, 2)$.

(c) Find the exact x-coordinates of the points in part (a).

(d) Create even more fanciful curves by modifying the equation in part (a).

CAS **34.** (a) The curve with equation

$$2y^3 + y^2 - y^5 = x^4 - 2x^3 + x^2$$

has been likened to a bouncing wagon. Use a computer algebra system to graph this curve and discover why.

(b) At how many points does this curve have horizontal tangent lines? Find the x-coordinates of these points.

35. Find the points on the lemniscate in Exercise 29 where the tangent is horizontal.

36. Show by implicit differentiation that the tangent to the ellipse

$$\frac{x^2}{a^2} + \frac{y^2}{b^2} = 1$$

at the point (x_0, y_0) is

$$\frac{x_0 x}{a^2} + \frac{y_0 y}{b^2} = 1$$

37. Find an equation of the tangent line to the hyperbola

$$\frac{x^2}{a^2} - \frac{y^2}{b^2} = 1$$

at the point (x_0, y_0).

38. Show that the sum of the x- and y-intercepts of any tangent line to the curve $\sqrt{x} + \sqrt{y} = \sqrt{c}$ is equal to c.

39. Show, using implicit differentiation, that any tangent line at a point P to a circle with center O is perpendicular to the radius OP.

40. The Power Rule can be proved using implicit differentiation for the case where n is a rational number, $n = p/q$, and $y = f(x) = x^n$ is assumed beforehand to be a differentiable function. If $y = x^{p/q}$, then $y^q = x^p$. Use implicit differentiation to show that

$$y' = \frac{p}{q} x^{(p/q)-1}$$

41–50 ||| Find the derivative of the function. Simplify where possible.

41. $y = \tan^{-1}\sqrt{x}$

42. $y = \sqrt{\tan^{-1} x}$

43. $y = \sin^{-1}(2x + 1)$

44. $h(x) = \sqrt{1 - x^2} \arcsin x$

45. $H(x) = (1 + x^2) \arctan x$

46. $y = \tan^{-1}(x - \sqrt{1 + x^2})$

47. $h(t) = \cot^{-1}(t) + \cot^{-1}(1/t)$

48. $y = x \cos^{-1} x - \sqrt{1 - x^2}$

49. $y = \cos^{-1}(e^{2x})$

50. $y = \arctan(\cos \theta)$

51–52 ||| Find $f'(x)$. Check that your answer is reasonable by comparing the graphs of f and f'.

51. $f(x) = e^x - x^2 \arctan x$

52. $f(x) = x \arcsin(1 - x^2)$

53. Prove the formula for $(d/dx)(\cos^{-1}x)$ by the same method as for $(d/dx)(\sin^{-1}x)$.

54. (a) One way of defining $\sec^{-1}x$ is to say that $y = \sec^{-1}x \iff \sec y = x$ and $0 \leqslant y < \pi/2$ or $\pi \leqslant y < 3\pi/2$. Show that, with this definition,

$$\frac{d}{dx}(\sec^{-1}x) = \frac{1}{x\sqrt{x^2 - 1}}$$

(b) Another way of defining $\sec^{-1}x$ that is sometimes used is to say that $y = \sec^{-1}x \iff \sec y = x$ and $0 \leqslant y \leqslant \pi$, $y \neq 0$. Show that, with this definition,

$$\frac{d}{dx}(\sec^{-1}x) = \frac{1}{|x|\sqrt{x^2 - 1}}$$

55–56 |||| Show that the given curves are orthogonal.

55. $2x^2 + y^2 = 3, \quad x = y^2$

56. $x^2 - y^2 = 5, \quad 4x^2 + 9y^2 = 72$

57. Contour lines on a map of a hilly region are curves that join points with the same elevation. A ball rolling down a hill follows a curve of steepest descent, which is orthogonal to the contour lines. Given the contour map of a hill in the figure, sketch the paths of balls that start at positions A and B.

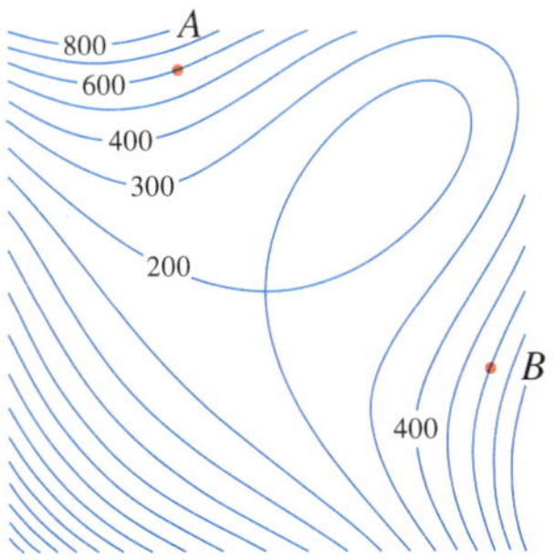

58. TV meteorologists often present maps showing pressure fronts. Such maps display *isobars*—curves along which the air pressure is constant. Consider the family of isobars shown in the figure. Sketch several members of the family of orthogonal trajectories of the isobars. Given the fact that wind blows from regions of high air pressure to regions of low air pressure, what does the orthogonal family represent?

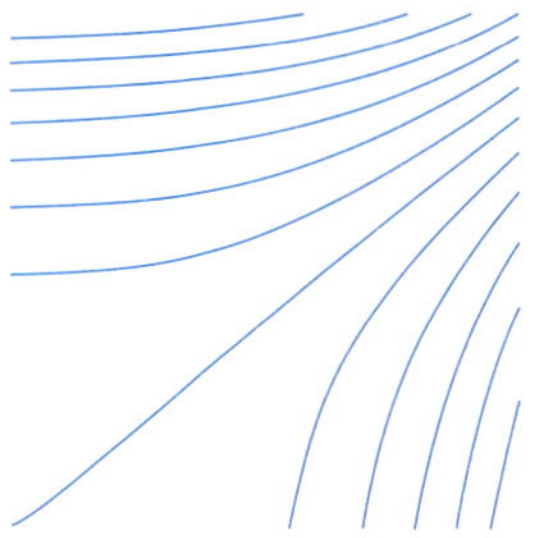

59–62 |||| Show that the given families of curves are orthogonal trajectories of each other. Sketch both families of curves on the same axes.

59. $x^2 + y^2 = r^2, \quad ax + by = 0$

60. $x^2 + y^2 = ax, \quad x^2 + y^2 = by$

61. $y = cx^2, \quad x^2 + 2y^2 = k$

62. $y = ax^3, \quad x^2 + 3y^2 = b$

63. The equation $x^2 - xy + y^2 = 3$ represents a "rotated ellipse," that is, an ellipse whose axes are not parallel to the coordinate axes. Find the points at which this ellipse crosses the x-axis and show that the tangent lines at these points are parallel.

64. (a) Where does the normal line to the ellipse $x^2 - xy + y^2 = 3$ at the point $(-1, 1)$ intersect the ellipse a second time? (See page 192 for the definition of a normal line.)

(b) Illustrate part (a) by graphing the ellipse and the normal line.

65. Find all points on the curve $x^2y^2 + xy = 2$ where the slope of the tangent line is -1.

66. Find equations of both the tangent lines to the ellipse $x^2 + 4y^2 = 36$ that pass through the point $(12, 3)$.

67. (a) Suppose f is a one-to-one differentiable function and its inverse function f^{-1} is also differentiable. Use implicit differentiation to show that

$$(f^{-1})'(x) = \frac{1}{f'(f^{-1}(x))}$$

provided that the denominator is not 0.

(b) If $f(4) = 5$ and $f'(4) = \frac{2}{3}$, find $(f^{-1})'(5)$.

68. (a) Show that $f(x) = 2x + \cos x$ is one-to-one.

(b) What is the value of $f^{-1}(1)$?

(c) Use the formula from Exercise 67(a) to find $(f^{-1})'(1)$.

69. The figure shows a lamp located three units to the right of the y-axis and a shadow created by the elliptical region $x^2 + 4y^2 \leqslant 5$. If the point $(-5, 0)$ is on the edge of the shadow, how far above the x-axis is the lamp located?

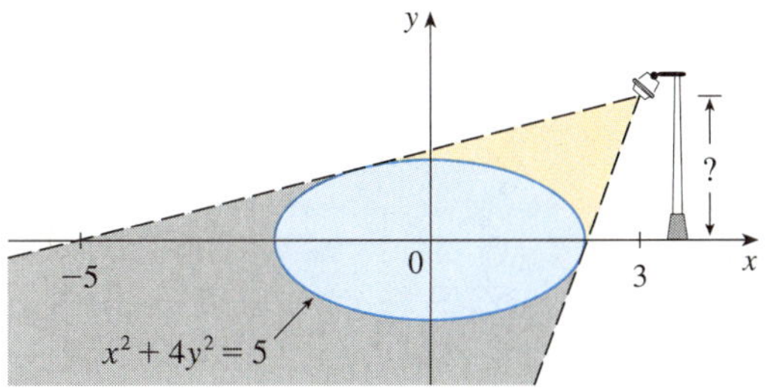

3.7 Higher Derivatives

If f is a differentiable function, then its derivative f' is also a function, so f' may have a derivative of its own, denoted by $(f')' = f''$. This new function f'' is called the **second derivative** of f because it is the derivative of the derivative of f. Using Leibniz notation, we write the second derivative of $y = f(x)$ as

$$\frac{d}{dx}\left(\frac{dy}{dx}\right) = \frac{d^2y}{dx^2}$$

Another notation is $f''(x) = D^2f(x)$.

TEC In Module 3.7A you can see how changing the coefficients of a polynomial f affects the appearance of the graphs of f, f', and f''.

EXAMPLE 1 If $f(x) = x \cos x$, find and interpret $f''(x)$.

SOLUTION Using the Product Rule, we have

$$f'(x) = x\frac{d}{dx}(\cos x) + \cos x\frac{d}{dx}(x)$$

$$= -x \sin x + \cos x$$

To find $f''(x)$ we differentiate $f'(x)$:

$$f''(x) = \frac{d}{dx}(-x \sin x + \cos x)$$

$$= -x\frac{d}{dx}(\sin x) + \sin x\frac{d}{dx}(-x) + \frac{d}{dx}(\cos x)$$

$$= -x \cos x - \sin x - \sin x$$

$$= -x \cos x - 2 \sin x$$

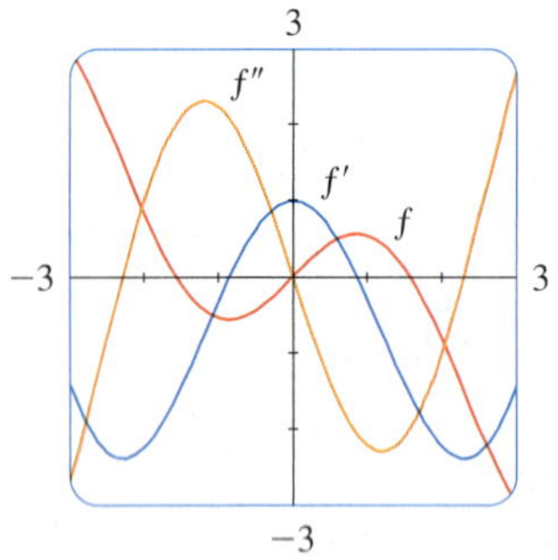

FIGURE 1
The graphs of $f(x) = x \cos x$ and its first and second derivatives

The graphs of f, f', and f'' are shown in Figure 1.

We can interpret $f''(x)$ as the slope of the curve $y = f'(x)$ at the point $(x, f'(x))$. In other words, it is the rate of change of the slope of the original curve $y = f(x)$.

Notice from Figure 1 that $f''(x) = 0$ whenever $y = f'(x)$ has a horizontal tangent. Also, $f''(x)$ is positive when $y = f'(x)$ has positive slope and negative when $y = f'(x)$ has negative slope. So the graphs serve as a check on our calculations.

In general, we can interpret a second derivative as a rate of change of a rate of change. The most familiar example of this is *acceleration,* which we define as follows.

If $s = s(t)$ is the position function of an object that moves in a straight line, we know that its first derivative represents the velocity $v(t)$ of the object as a function of time:

$$v(t) = s'(t) = \frac{ds}{dt}$$

The instantaneous rate of change of velocity with respect to time is called the **acceleration** $a(t)$ of the object. Thus, the acceleration function is the derivative of the velocity function and is therefore the second derivative of the position function:

$$a(t) = v'(t) = s''(t)$$

or, in Leibniz notation,

$$a = \frac{dv}{dt} = \frac{d^2s}{dt^2}$$

EXAMPLE 2 The position of a particle is given by the equation

$$s = f(t) = t^3 - 6t^2 + 9t$$

where t is measured in seconds and s in meters.
(a) Find the acceleration at time t. What is the acceleration after 4 s?
(b) Graph the position, velocity, and acceleration functions for $0 \leqslant t \leqslant 5$.
(c) When is the particle speeding up? When is it slowing down?

SOLUTION
(a) The velocity function is the derivative of the position function:

$$s = f(t) = t^3 - 6t^2 + 9t$$

$$v(t) = \frac{ds}{dt} = 3t^2 - 12t + 9$$

The acceleration is the derivative of the velocity function:

$$a(t) = \frac{d^2s}{dt^2} = \frac{dv}{dt} = 6t - 12$$

$$a(4) = 6(4) - 12 = 12 \text{ m/s}^2$$

|||| The units for acceleration are meters per second per second, written as m/s^2.

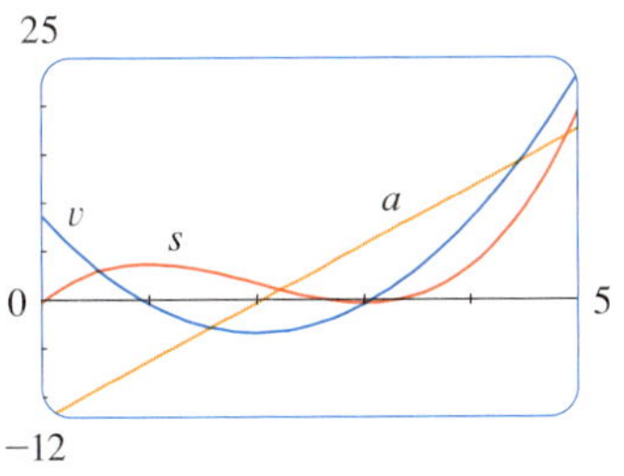

FIGURE 2

(b) Figure 2 shows the graphs of s, v, and a.

(c) The particle speeds up when the velocity is positive and increasing (v and a are both positive) and also when the velocity is negative and decreasing (v and a are both negative). In other words, the particle speeds up when the velocity and acceleration have the same sign. (The particle is pushed in the same direction it is moving.) From Figure 2 we see that this happens when $1 < t < 2$ and when $t > 3$. The particle slows down when v and a have opposite signs, that is, when $0 \leqslant t < 1$ and when $2 < t < 3$. Figure 3 summarizes the motion of the particle.

TEC In Module 3.7B you can see an animation of Figure 3 with an expression for s that you can choose yourself.

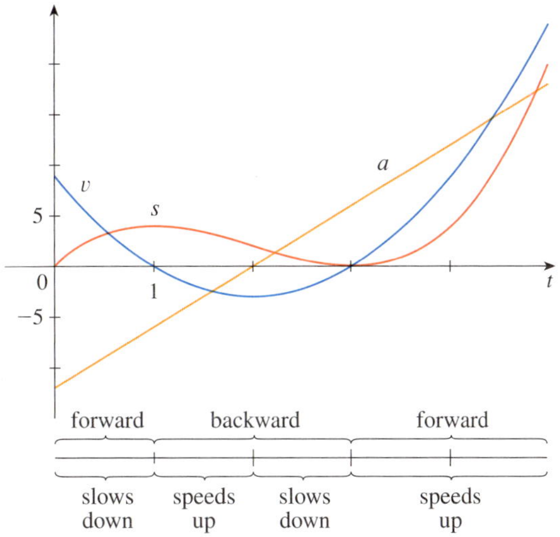

FIGURE 3

The **third derivative** f''' is the derivative of the second derivative: $f''' = (f'')'$. So $f'''(x)$ can be interpreted as the slope of the curve $y = f''(x)$ or as the rate of change of $f''(x)$. If $y = f(x)$, then alternative notations for the third derivative are

$$y''' = f'''(x) = \frac{d}{dx}\left(\frac{d^2y}{dx^2}\right) = \frac{d^3y}{dx^3} = D^3f(x)$$

The process can be continued. The fourth derivative f'''' is usually denoted by $f^{(4)}$. In general, the nth derivative of f is denoted by $f^{(n)}$ and is obtained from f by differentiating n times. If $y = f(x)$, we write

$$y^{(n)} = f^{(n)}(x) = \frac{d^ny}{dx^n} = D^nf(x)$$

We can interpret the third derivative physically in the case where the function is the position function $s = s(t)$ of an object that moves along a straight line. Because $s''' = (s'')' = a'$, the third derivative of the position function is the derivative of the acceleration function and is called the **jerk**:

$$j = \frac{da}{dt} = \frac{d^3s}{dt^3}$$

Thus, the jerk j is the rate of change of acceleration. It is aptly named because a large jerk means a sudden change in acceleration, which causes an abrupt movement in a vehicle.

EXAMPLE 3 If

$$y = x^3 - 6x^2 - 5x + 3$$

then

$$y' = 3x^2 - 12x - 5$$

$$y'' = 6x - 12$$

$$y''' = 6$$

$$y^{(4)} = 0$$

and in fact $y^{(n)} = 0$ for all $n \geqslant 4$.

EXAMPLE 4 If $f(x) = \dfrac{1}{x}$, find $f^{(n)}(x)$.

SOLUTION

$$f(x) = \frac{1}{x} = x^{-1}$$

$$f'(x) = -x^{-2} = \frac{-1}{x^2}$$

$$f''(x) = (-2)(-1)x^{-3} = \frac{2}{x^3}$$

$$f'''(x) = -3 \cdot 2 \cdot 1 \cdot x^{-4}$$

$$f^{(4)}(x) = 4 \cdot 3 \cdot 2 \cdot 1 \cdot x^{-5}$$

$$f^{(5)}(x) = -5 \cdot 4 \cdot 3 \cdot 2 \cdot 1 \cdot x^{-6} = -5!x^{-6}$$

$$\vdots$$

$$f^{(n)}(x) = (-1)^n n(n-1)(n-2) \cdots 2 \cdot 1 \cdot x^{-(n+1)}$$

|||| The factor $(-1)^n$ occurs in the formula for $f^{(n)}(x)$ because we introduce another negative sign every time we differentiate. Since the successive values of $(-1)^n$ are $-1, 1, -1, 1, -1, 1, \ldots$, the presence of $(-1)^n$ indicates that the sign changes with each successive derivative.

or
$$f^{(n)}(x) = \frac{(-1)^n n!}{x^{n+1}}$$

Here we have used the factorial symbol $n!$ for the product of the first n positive integers.

$$n! = 1 \cdot 2 \cdot 3 \cdot \cdots \cdot (n-1) \cdot n$$

The following example shows how to find the second derivative of a function that is defined implicitly.

EXAMPLE 5 Find y'' if $x^4 + y^4 = 16$.

SOLUTION Differentiating the equation implicitly with respect to x, we get

$$4x^3 + 4y^3 y' = 0$$

Solving for y' gives

$$y' = -\frac{x^3}{y^3} \tag{1}$$

|||| Figure 4 shows the graph of the curve $x^4 + y^4 = 16$ of Example 5. Notice that it's a stretched and flattened version of the circle $x^2 + y^2 = 4$. For this reason it's sometimes called a *fat circle*. It starts out very steep on the left but quickly becomes very flat. This can be seen from the expression

$$y' = -\frac{x^3}{y^3} = -\left(\frac{x}{y}\right)^3$$

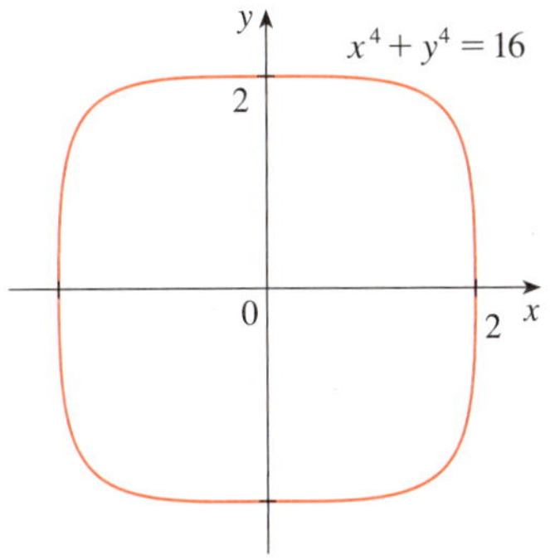

FIGURE 4

To find y'' we differentiate this expression for y' using the Quotient Rule and remembering that y is a function of x:

$$y'' = \frac{d}{dx}\left(-\frac{x^3}{y^3}\right) = -\frac{y^3\,(d/dx)(x^3) - x^3\,(d/dx)(y^3)}{(y^3)^2}$$

$$= -\frac{y^3 \cdot 3x^2 - x^3(3y^2 y')}{y^6}$$

If we now substitute Equation 1 into this expression, we get

$$y'' = -\frac{3x^2y^3 - 3x^3y^2\left(-\dfrac{x^3}{y^3}\right)}{y}$$

$$= -\frac{3(x^2y^4 + x^6)}{y^7} = -\frac{3x^2(y^4 + x^4)}{y^7}$$

But the values of x and y must satisfy the original equation $x^4 + y^4 = 16$. So the answer simplifies to

$$y'' = -\frac{3x^2(16)}{y^7} = -48\,\frac{x^2}{y^7}$$

EXAMPLE 6 Find $D^{27} \cos x$.

SOLUTION The first few derivatives of $\cos x$ are as follows:

|||| Look for a pattern.

$$D \cos x = -\sin x$$

$$D^2 \cos x = -\cos x$$

$$D^3 \cos x = \sin x$$

$$D^4 \cos x = \cos x$$

$$D^5 \cos x = -\sin x$$

We see that the successive derivatives occur in a cycle of length 4 and, in particular, $D^n \cos x = \cos x$ whenever n is a multiple of 4. Therefore

$$D^{24} \cos x = \cos x$$

and, differentiating three more times, we have

$$D^{27} \cos x = \sin x$$

We have seen that one application of second and third derivatives occurs in analyzing the motion of objects using acceleration and jerk. We will investigate another application of second derivatives in Exercise 62 and in Section 4.3, where we show how knowledge of f'' gives us information about the shape of the graph of f. In Chapter 11 we will see how second and higher derivatives enable us to represent functions as sums of infinite series.

3.7 Exercises

1. The figure shows the graphs of f, f', and f''. Identify each curve, and explain your choices.

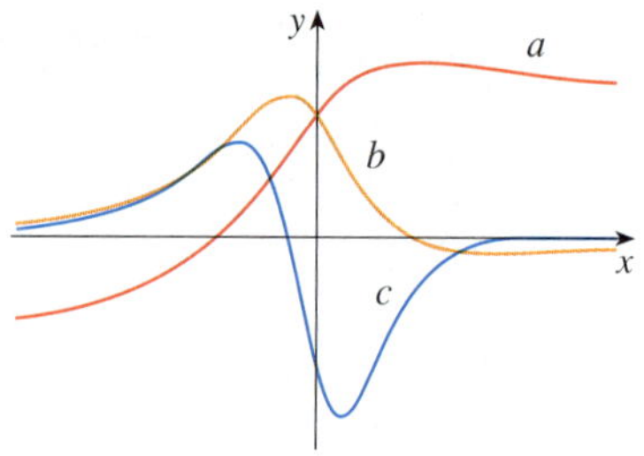

2. The figure shows graphs of f, f', f'', and f'''. Identify each curve, and explain your choices.

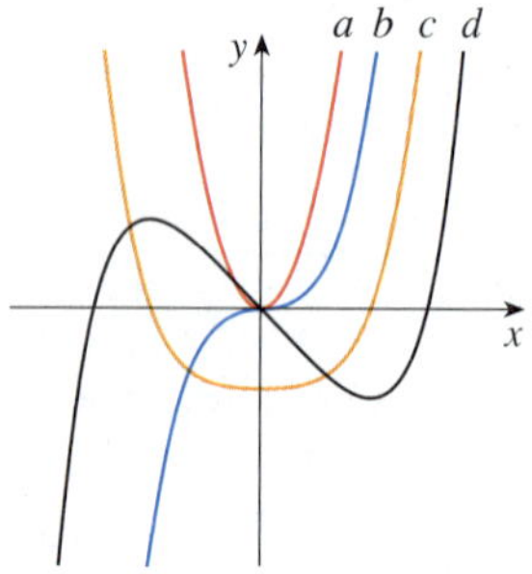

3. The figure shows the graphs of three functions. One is the position function of a car, one is the velocity of the car, and one is its acceleration. Identify each curve, and explain your choices.

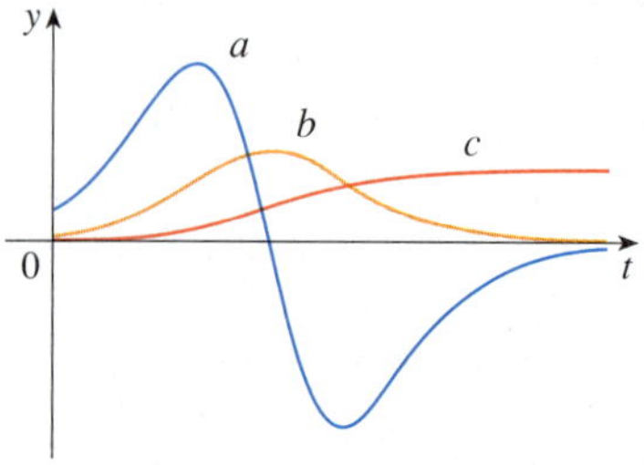

4. The figure shows the graphs of four functions. One is the position function of a car, one is the velocity of the car, one is its acceleration, and one is its jerk. Identify each curve, and explain your choices.

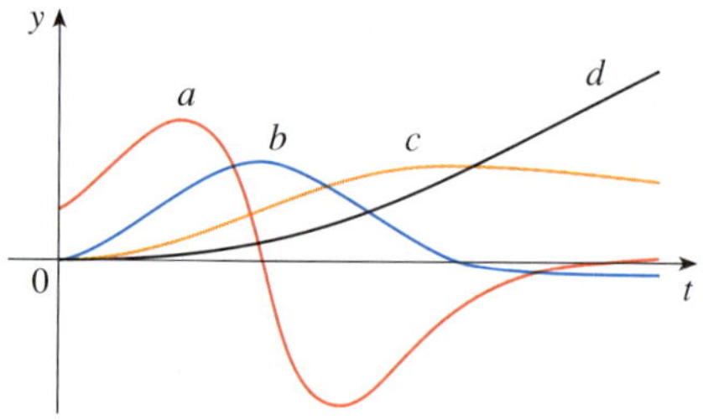

5–20 Find the first and second derivatives of the function.

5. $f(x) = x^5 + 6x^2 - 7x$

6. $f(t) = t^8 - 7t^6 + 2t^4$

7. $y = \cos 2\theta$

8. $y = \theta \sin \theta$

9. $F(t) = (1 - 7t)^6$

10. $g(x) = \dfrac{2x + 1}{x - 1}$

11. $h(u) = \dfrac{1 - 4u}{1 + 3u}$

12. $H(s) = a\sqrt{s} + \dfrac{b}{\sqrt{s}}$

13. $h(x) = \sqrt{x^2 + 1}$

14. $y = xe^{cx}$

15. $y = (x^3 + 1)^{2/3}$

16. $y = \dfrac{4x}{\sqrt{x + 1}}$

17. $H(t) = \tan 3t$

18. $g(s) = s^2 \cos s$

19. $g(t) = t^3 e^{5t}$

20. $h(x) = \tan^{-1}(x^2)$

21. (a) If $f(x) = 2\cos x + \sin^2 x$, find $f'(x)$ and $f''(x)$.
(b) Check to see that your answers to part (a) are reasonable by comparing the graphs of f, f', and f''.

22. (a) If $f(x) = e^x - x^3$, find $f'(x)$ and $f''(x)$.
(b) Check to see that your answers to part (a) are reasonable by comparing the graphs of f, f', and f''.

23–24 Find y'''.

23. $y = \sqrt{2x + 3}$

24. $y = \dfrac{x}{2x - 1}$

25. If $f(t) = t \cos t$, find $f'''(0)$.

26. If $g(x) = \sqrt{5 - 2x}$, find $g'''(2)$.

27. If $f(\theta) = \cot \theta$, find $f'''(\pi/6)$.

28. If $g(x) = \sec x$, find $g'''(\pi/4)$.

29–32 Find y'' by implicit differentiation.

29. $9x^2 + y^2 = 9$

30. $\sqrt{x} + \sqrt{y} = 1$

31. $x^3 + y^3 = 1$

32. $x^4 + y^4 = a^4$

33–37 Find a formula for $f^{(n)}(x)$.

33. $f(x) = x^n$

34. $f(x) = \dfrac{1}{5x - 1}$

35. $f(x) = e^{2x}$

36. $f(x) = \sqrt{x}$

37. $f(x) = \dfrac{1}{3x^3}$

38–40 Find the given derivative by finding the first few derivatives and observing the pattern that occurs.

38. $D^{74} \sin x$

39. $D^{103} \cos 2x$

40. $D^{1000} xe^{-x}$

41. A car starts from rest and the graph of its position function is shown in the figure, where s is measured in feet and t in seconds. Use it to graph the velocity and estimate the acceleration at $t = 2$ seconds from the velocity graph. Then sketch a graph of the acceleration function.

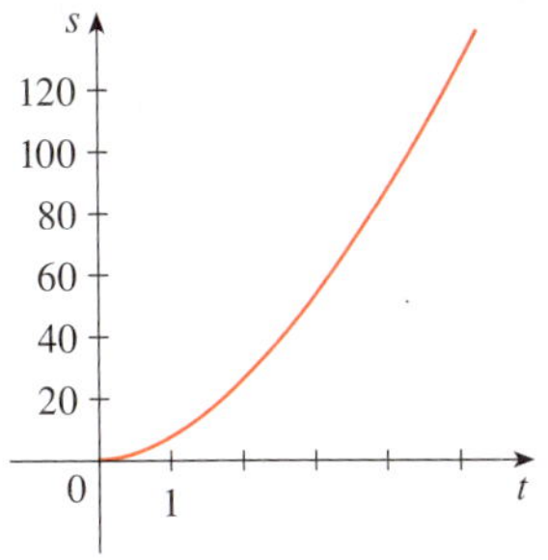

42. (a) The graph of a position function of a car is shown, where s is measured in feet and t in seconds. Use it to graph the velocity and acceleration of the car. What is the acceleration at $t = 10$ seconds?

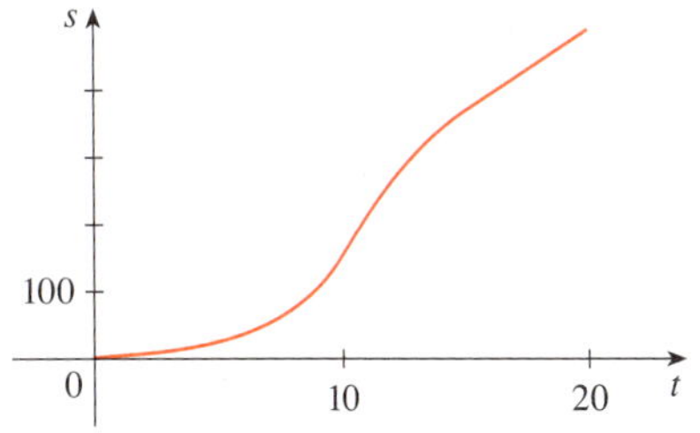

(b) Use the acceleration curve from part (a) to estimate the jerk at $t = 10$ seconds. What are the units for jerk?

43–46 The equation of motion is given for a particle, where s is in meters and t is in seconds. Find (a) the velocity and acceleration as functions of t, (b) the acceleration after 1 second, and (c) the acceleration at the instants when the velocity is 0.

43. $s = 2t^3 - 15t^2 + 36t + 2, \quad t \geqslant 0$

44. $s = 2t^3 - 3t^2 - 12t, \quad t \geqslant 0$

45. $s = \sin(\pi t/6) + \cos(\pi t/6), \quad 0 \leqslant t \leqslant 2$

46. $s = 2t^3 - 7t^2 + 4t + 1, \quad t \geqslant 0$

47–48 An equation of motion is given, where s is in meters and t in seconds. Find (a) the times at which the acceleration is 0 and (b) the displacement and velocity at these times.

47. $s = t^4 - 4t^3 + 2$

48. $s = 2t^3 - 9t^2$

49. A particle moves according to a law of motion $s = f(t) = t^3 - 12t^2 + 36t$, $t \geqslant 0$, where t is measured in seconds and s in meters.
(a) Find the acceleration at time t and after 3 s.
(b) Graph the position, velocity, and acceleration functions for $0 \leqslant t \leqslant 8$.
(c) When is the particle speeding up? When is it slowing down?

50. A particle moves along the x-axis, its position at time t given by $x(t) = t/(1 + t^2)$, $t \geqslant 0$, where t is measured in seconds and x in meters.
(a) Find the acceleration at time t. When is it 0?
(b) Graph the position, velocity, and acceleration functions for $0 \leqslant t \leqslant 4$.
(c) When is the particle speeding up? When is it slowing down?

51. A mass attached to a vertical spring has position function given by $y(t) = A \sin \omega t$, where A is the amplitude of its oscillations and ω is a constant.
(a) Find the velocity and acceleration as functions of time.

(b) Show that the acceleration is proportional to the displacement y.

(c) Show that the speed is a maximum when the acceleration is 0.

52. A particle moves along a straight line with displacement $s(t)$, velocity $v(t)$, and acceleration $a(t)$. Show that

$$a(t) = v(t)\frac{dv}{ds}$$

Explain the difference between the meanings of the derivatives dv/dt and dv/ds.

53. Find a second-degree polynomial P such that $P(2) = 5$, $P'(2) = 3$, and $P''(2) = 2$.

54. Find a third-degree polynomial Q such that $Q(1) = 1$, $Q'(1) = 3$, $Q''(1) = 6$, and $Q'''(1) = 12$.

55. The equation $y'' + y' - 2y = \sin x$ is called a **differential equation** because it involves an unknown function y and its derivatives y' and y''. Find constants A and B such that the function $y = A\sin x + B\cos x$ satisfies this equation. (Differential equations will be studied in detail in Chapter 9.)

56. Find constants A, B, and C such that the function $y = Ax^2 + Bx + C$ satisfies the differential equation $y'' + y' - 2y = x^2$.

57. For what values of r does the function $y = e^{rx}$ satisfy the equation $y'' + 5y' - 6y = 0$?

58. Find the values of λ for which $y = e^{\lambda x}$ satisfies the equation $y + y' = y''$.

59–61 The function g is a twice differentiable function. Find f'' in terms of g, g', and g''.

59. $f(x) = xg(x^2)$

60. $f(x) = \dfrac{g(x)}{x}$

61. $f(x) = g(\sqrt{x})$

62. If $f(x) = 3x^5 - 10x^3 + 5$, graph both f and f''. On what intervals is $f''(x) > 0$? On those intervals, how is the graph of f related to its tangent lines? What about the intervals where $f''(x) < 0$?

63. (a) Compute the first few derivatives of the function $f(x) = 1/(x^2 + x)$ until you see that the computations are becoming algebraically unmanageable.

(b) Use the identity

$$\frac{1}{x(x+1)} = \frac{1}{x} - \frac{1}{x+1}$$

to compute the derivatives much more easily. Then find an expression for $f^{(n)}(x)$. This method of splitting up a fraction in terms of simpler fractions, called *partial fractions*, will be pursued further in Section 7.4.

CAS **64.** (a) Use a computer algebra system to compute f''', where

$$f(x) = \frac{7x + 17}{2x^2 - 7x - 4}$$

(b) Find a much simpler expression for f''' by first splitting f into partial fractions. [In Maple, use the command convert(f,parfrac,x); in Mathematica, use Apart[f].]

65. Find expressions for the first five derivatives of $f(x) = x^2e^x$. Do you see a pattern in these expressions? Guess a formula for $f^{(n)}(x)$ and prove it using mathematical induction.

66. (a) If $F(x) = f(x)g(x)$, where f and g have derivatives of all orders, show that

$$F'' = f''g + 2f'g' + fg''$$

(b) Find similar formulas for F''' and $F^{(4)}$.

(c) Guess a formula for $F^{(n)}$.

67. If $y = f(u)$ and $u = g(x)$, where f and g are twice differentiable functions, show that

$$\frac{d^2y}{dx^2} = \frac{d^2y}{du^2}\left(\frac{du}{dx}\right)^2 + \frac{dy}{du}\frac{d^2u}{dx^2}$$

68. If $y = f(u)$ and $u = g(x)$, where f and g possess third derivatives, find a formula for d^3y/dx^3 similar to the one given in Exercise 67.

69. Suppose p is a positive integer such that the function f is p-times differentiable and $f^{(p)} = f$. Using mathematical induction, show that f is in fact n-times differentiable for every positive integer n and that each of its higher derivatives $f^{(n)}$ equals one of the p functions $f, f', f'', \ldots, f^{(p-1)}$.

APPLIED PROJECT

Where Should a Pilot Start Descent?

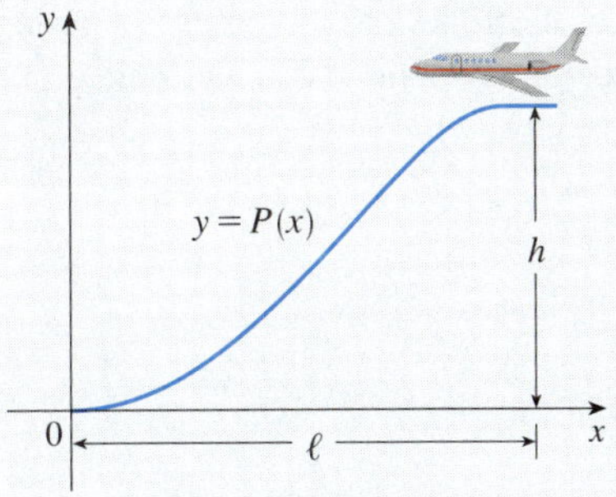

An approach path for an aircraft landing is shown in the figure and satisfies the following conditions:

(i) The cruising altitude is h when descent starts at a horizontal distance ℓ from touchdown at the origin.

(ii) The pilot must maintain a constant horizontal speed v throughout descent.

(iii) The absolute value of the vertical acceleration should not exceed a constant k (which is much less than the acceleration due to gravity).

1. Find a cubic polynomial $P(x) = ax^3 + bx^2 + cx + d$ that satisfies condition (i) by imposing suitable conditions on $P(x)$ and $P'(x)$ at the start of descent and at touchdown.

2. Use conditions (ii) and (iii) to show that

$$\frac{6hv^2}{\ell^2} \leqslant k$$

3. Suppose that an airline decides not to allow vertical acceleration of a plane to exceed $k = 860 \text{ mi/h}^2$. If the cruising altitude of a plane is 35,000 ft and the speed is 300 mi/h, how far away from the airport should the pilot start descent?

4. Graph the approach path if the conditions stated in Problem 3 are satisfied.

APPLIED PROJECT

Building a Better Roller Coaster

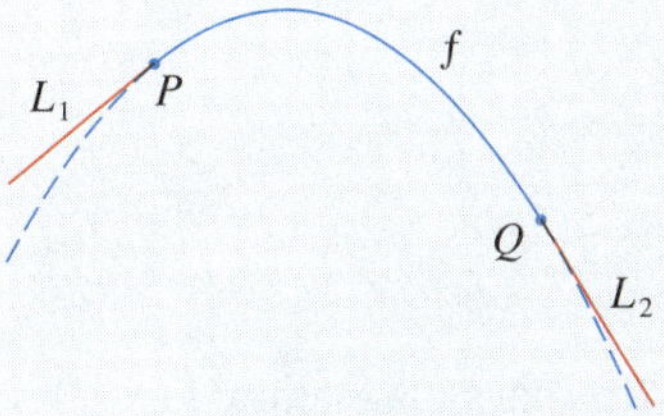

Suppose you are asked to design the first ascent and drop for a new roller coaster. By studying photographs of your favorite coasters, you decide to make the slope of the ascent 0.8 and the slope of the drop -1.6. You decide to connect these two straight stretches $y = L_1(x)$ and $y = L_2(x)$ with part of a parabola $y = f(x) = ax^2 + bx + c$, where x and $f(x)$ are measured in feet. For the track to be smooth there can't be abrupt changes in direction, so you want the linear segments L_1 and L_2 to be tangent to the parabola at the transition points P and Q. (See the figure.) To simplify the equations, you decide to place the origin at P.

1. (a) Suppose the horizontal distance between P and Q is 100 ft. Write equations in a, b, and c that will ensure that the track is smooth at the transition points.

(b) Solve the equations in part (a) for a, b, and c to find a formula for $f(x)$.

(c) Plot L_1, f, and L_2 to verify graphically that the transitions are smooth.

(d) Find the difference in elevation between P and Q.

2. The solution in Problem 1 might *look* smooth, but it might not *feel* smooth because the piecewise defined function [consisting of $L_1(x)$ for $x < 0$, $f(x)$ for $0 \leqslant x \leqslant 100$, and $L_2(x)$ for $x > 100$] doesn't have a continuous second derivative. So you decide to improve the design by using a quadratic function $q(x) = ax^2 + bx + c$ only on the interval $10 \leqslant x \leqslant 90$ and connecting it to the linear functions by means of two cubic functions:

$$g(x) = kx^3 + lx^2 + mx + n \qquad 0 \leqslant x < 10$$

$$h(x) = px^3 + qx^2 + rx + s \qquad 90 < x \leqslant 100$$

(a) Write a system of equations in 11 unknowns that ensure that the functions and their first two derivatives agree at the transition points.

CAS (b) Solve the equations in part (a) with a computer algebra system to find formulas for $q(x)$, $g(x)$, and $h(x)$.

(c) Plot L_1, g, q, h, and L_2, and compare with the plot in Problem 1(c).

3.8 Derivatives of Logarithmic Functions

In this section we use implicit differentiation to find the derivatives of the logarithmic functions $y = \log_a x$ and, in particular, the natural logarithmic function $y = \ln x$. We assume that logarithmic functions are differentiable; this is certainly plausible from their graphs (see Figure 12 in Section 1.6).

1 $$\frac{d}{dx}(\log_a x) = \frac{1}{x \ln a}$$

Proof Let $y = \log_a x$. Then

$$a^y = x$$

|||| Formula 3.5.5 says that

$$\frac{d}{dx}(a^x) = a^x \ln a$$

Differentiating this equation implicitly with respect to x, using Formula 3.5.5, we get

$$a^y(\ln a)\frac{dy}{dx} = 1$$

and so
$$\frac{dy}{dx} = \frac{1}{a^y \ln a} = \frac{1}{x \ln a}$$

If we put $a = e$ in Formula 1, then the factor $\ln a$ on the right side becomes $\ln e = 1$ and we get the formula for the derivative of the natural logarithmic function $\log_e x = \ln x$:

2 $$\frac{d}{dx}(\ln x) = \frac{1}{x}$$

By comparing Formulas 1 and 2, we see one of the main reasons that natural logarithms (logarithms with base e) are used in calculus: The differentiation formula is simplest when $a = e$ because $\ln e = 1$.

EXAMPLE 1 Differentiate $y = \ln(x^3 + 1)$.

SOLUTION To use the Chain Rule, we let $u = x^3 + 1$. Then $y = \ln u$, so

$$\frac{dy}{dx} = \frac{dy}{du}\frac{du}{dx} = \frac{1}{u}\frac{du}{dx} = \frac{1}{x^3 + 1}(3x^2) = \frac{3x^2}{x^3 + 1}$$

In general, if we combine Formula 2 with the Chain Rule as in Example 1, we get

3 $$\frac{d}{dx}(\ln u) = \frac{1}{u}\frac{du}{dx} \quad \text{or} \quad \frac{d}{dx}[\ln g(x)] = \frac{g'(x)}{g(x)}$$

EXAMPLE 2 Find $\dfrac{d}{dx}\ln(\sin x)$.

SOLUTION Using (3), we have

$$\frac{d}{dx}\ln(\sin x) = \frac{1}{\sin x}\frac{d}{dx}(\sin x) = \frac{1}{\sin x}\cos x = \cot x$$

EXAMPLE 3 Differentiate $f(x) = \sqrt{\ln x}$.

SOLUTION This time the logarithm is the inner function, so the Chain Rule gives

$$f'(x) = \tfrac{1}{2}(\ln x)^{-1/2}\frac{d}{dx}(\ln x) = \frac{1}{2\sqrt{\ln x}}\cdot\frac{1}{x} = \frac{1}{2x\sqrt{\ln x}}$$

EXAMPLE 4 Differentiate $f(x) = \log_{10}(2 + \sin x)$.

SOLUTION Using Formula 1 with $a = 10$, we have

$$f'(x) = \frac{d}{dx}\log_{10}(2 + \sin x) = \frac{1}{(2 + \sin x)\ln 10}\frac{d}{dx}(2 + \sin x)$$

$$= \frac{\cos x}{(2 + \sin x)\ln 10}$$

EXAMPLE 5 Find $\dfrac{d}{dx}\ln\dfrac{x + 1}{\sqrt{x - 2}}$.

SOLUTION 1

$$\frac{d}{dx}\ln\frac{x + 1}{\sqrt{x - 2}} = \frac{1}{\dfrac{x + 1}{\sqrt{x - 2}}}\frac{d}{dx}\frac{x + 1}{\sqrt{x - 2}}$$

$$= \frac{\sqrt{x - 2}}{x + 1}\,\frac{\sqrt{x - 2}\cdot 1 - (x + 1)\left(\tfrac{1}{2}\right)(x - 2)^{-1/2}}{x - 2}$$

$$= \frac{x - 2 - \tfrac{1}{2}(x + 1)}{(x + 1)(x - 2)} = \frac{x - 5}{2(x + 1)(x - 2)}$$

|||| Figure 1 shows the graph of the function f of Example 5 together with the graph of its derivative. It gives a visual check on our calculation. Notice that $f'(x)$ is large negative when f is rapidly decreasing.

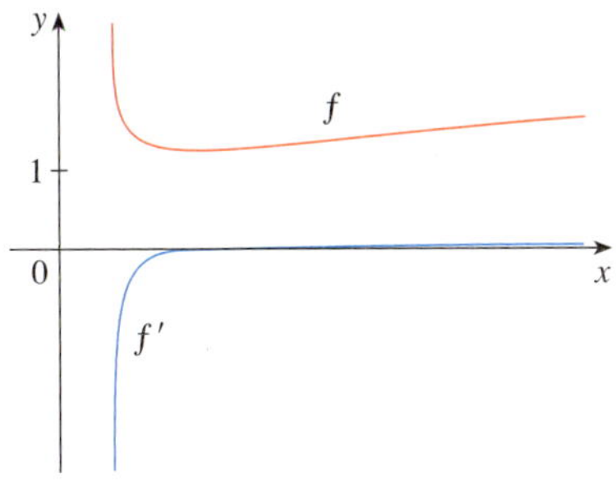

FIGURE 1

SOLUTION 2 If we first simplify the given function using the laws of logarithms, then the differentiation becomes easier:

$$\frac{d}{dx}\ln\frac{x + 1}{\sqrt{x - 2}} = \frac{d}{dx}\left[\ln(x + 1) - \tfrac{1}{2}\ln(x - 2)\right]$$

$$= \frac{1}{x + 1} - \frac{1}{2}\left(\frac{1}{x - 2}\right)$$

(This answer can be left as written, but if we used a common denominator we would see that it gives the same answer as in Solution 1.)

Figure 2 shows the graph of the function $f(x) = \ln|x|$ in Example 6 and its derivative $f'(x) = 1/x$. Notice that when x is small, the graph of $y = \ln|x|$ is steep and so $f'(x)$ is large (positive or negative).

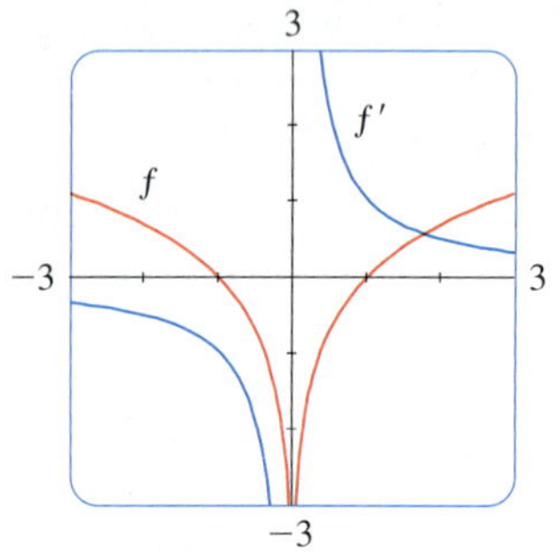

FIGURE 2

EXAMPLE 6 Find $f'(x)$ if $f(x) = \ln|x|$.

SOLUTION Since

$$f(x) = \begin{cases} \ln x & \text{if } x > 0 \\ \ln(-x) & \text{if } x < 0 \end{cases}$$

it follows that

$$f'(x) = \begin{cases} \dfrac{1}{x} & \text{if } x > 0 \\[2ex] \dfrac{1}{-x}(-1) = \dfrac{1}{x} & \text{if } x < 0 \end{cases}$$

Thus, $f'(x) = 1/x$ for all $x \neq 0$.

The result of Example 6 is worth remembering:

4
$$\frac{d}{dx} \ln|x| = \frac{1}{x}$$

Logarithmic Differentiation

The calculation of derivatives of complicated functions involving products, quotients, or powers can often be simplified by taking logarithms. The method used in the following example is called **logarithmic differentiation**.

EXAMPLE 7 Differentiate $y = \dfrac{x^{3/4}\sqrt{x^2 + 1}}{(3x + 2)^5}$.

SOLUTION We take logarithms of both sides of the equation and use the Laws of Logarithms to simplify:

$$\ln y = \tfrac{3}{4} \ln x + \tfrac{1}{2} \ln(x^2 + 1) - 5 \ln(3x + 2)$$

Differentiating implicitly with respect to x gives

$$\frac{1}{y}\frac{dy}{dx} = \frac{3}{4} \cdot \frac{1}{x} + \frac{1}{2} \cdot \frac{2x}{x^2 + 1} - 5 \cdot \frac{3}{3x + 2}$$

Solving for dy/dx, we get

$$\frac{dy}{dx} = y\left(\frac{3}{4x} + \frac{x}{x^2 + 1} - \frac{15}{3x + 2}\right)$$

If we hadn't used logarithmic differentiation in Example 7, we would have had to use both the Quotient Rule and the Product Rule. The resulting calculation would have been horrendous.

Because we have an explicit expression for y, we can substitute and write

$$\frac{dy}{dx} = \frac{x^{3/4}\sqrt{x^2 + 1}}{(3x + 2)^5}\left(\frac{3}{4x} + \frac{x}{x^2 + 1} - \frac{15}{3x + 2}\right)$$

Steps in Logarithmic Differentiation

1. Take natural logarithms of both sides of an equation $y = f(x)$ and use the Laws of Logarithms to simplify.
2. Differentiate implicitly with respect to x.
3. Solve the resulting equation for y'.

If $f(x) < 0$ for some values of x, then $\ln f(x)$ is not defined, but we can write $|y| = |f(x)|$ and use Equation 4. We illustrate this procedure by proving the general version of the Power Rule, as promised in Section 3.1.

The Power Rule If n is any real number and $f(x) = x^n$, then

$$f'(x) = nx^{n-1}$$

Proof Let $y = x^n$ and use logarithmic differentiation:

|||| If $x = 0$, we can show that $f'(0) = 0$ for $n > 1$ directly from the definition of a derivative.

$$\ln|y| = \ln|x|^n = n\ln|x| \qquad x \neq 0$$

Therefore
$$\frac{y'}{y} = \frac{n}{x}$$

Hence
$$y' = n\frac{y}{x} = n\frac{x^n}{x} = nx^{n-1}$$

⊘ You should distinguish carefully between the Power Rule $[(x^n)' = nx^{n-1}]$, where the base is variable and the exponent is constant, and the rule for differentiating exponential functions $[(a^x)' = a^x \ln a]$, where the base is constant and the exponent is variable.

In general there are four cases for exponents and bases:

1. $\dfrac{d}{dx}(a^b) = 0 \qquad (a \text{ and } b \text{ are constants})$

2. $\dfrac{d}{dx}[f(x)]^b = b[f(x)]^{b-1}f'(x)$

3. $\dfrac{d}{dx}[a^{g(x)}] = a^{g(x)}(\ln a)g'(x)$

4. To find $(d/dx)[f(x)]^{g(x)}$, logarithmic differentiation can be used, as in the next example.

|||| Figure 3 illustrates Example 8 by showing the graphs of $f(x) = x^{\sqrt{x}}$ and its derivative.

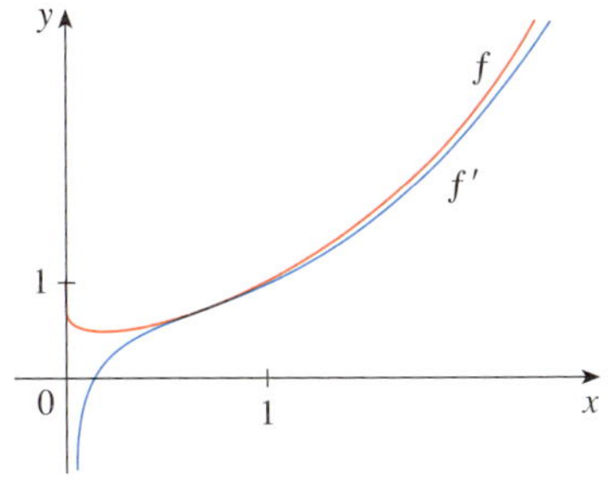

FIGURE 3

EXAMPLE 8 Differentiate $y = x^{\sqrt{x}}$.

SOLUTION 1 Using logarithmic differentiation, we have

$$\ln y = \ln x^{\sqrt{x}} = \sqrt{x}\ln x$$

$$\frac{y'}{y} = \sqrt{x}\cdot\frac{1}{x} + (\ln x)\frac{1}{2\sqrt{x}}$$

$$y' = y\left(\frac{1}{\sqrt{x}} + \frac{\ln x}{2\sqrt{x}}\right) = x^{\sqrt{x}}\left(\frac{2 + \ln x}{2\sqrt{x}}\right)$$

SOLUTION 2 Another method is to write $x^{\sqrt{x}} = (e^{\ln x})^{\sqrt{x}}$:

$$\frac{d}{dx}\left(x^{\sqrt{x}}\right) = \frac{d}{dx}\left(e^{\sqrt{x}\ln x}\right) = e^{\sqrt{x}\ln x}\frac{d}{dx}\left(\sqrt{x}\ln x\right)$$

$$= x^{\sqrt{x}}\left(\frac{2 + \ln x}{2\sqrt{x}}\right) \qquad \text{(as in Solution 1)}$$

The Number e as a Limit

We have shown that if $f(x) = \ln x$, then $f'(x) = 1/x$. Thus, $f'(1) = 1$. We now use this fact to express the number e as a limit.

From the definition of a derivative as a limit, we have

$$f'(1) = \lim_{h \to 0} \frac{f(1 + h) - f(1)}{h} = \lim_{x \to 0} \frac{f(1 + x) - f(1)}{x}$$

$$= \lim_{x \to 0} \frac{\ln(1 + x) - \ln 1}{x} = \lim_{x \to 0} \frac{1}{x}\ln(1 + x)$$

$$= \lim_{x \to 0} \ln(1 + x)^{1/x}$$

Because $f'(1) = 1$, we have

$$\lim_{x \to 0} \ln(1 + x)^{1/x} = 1$$

Then, by Theorem 2.5.8 and the continuity of the exponential function, we have

$$e = e^1 = e^{\lim_{x \to 0} \ln(1+x)^{1/x}} = \lim_{x \to 0} e^{\ln(1+x)^{1/x}} = \lim_{x \to 0} (1 + x)^{1/x}$$

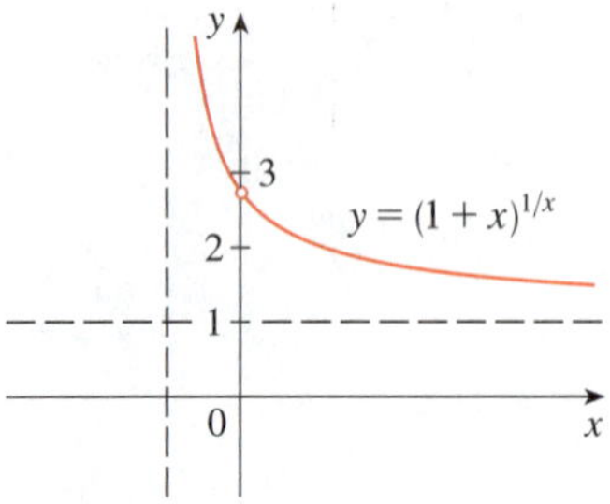

FIGURE 4

x	$(1 + x)^{1/x}$
0.1	2.59374246
0.01	2.70481383
0.001	2.71692393
0.0001	2.71814593
0.00001	2.71826824
0.000001	2.71828047
0.0000001	2.71828169
0.00000001	2.71828181

5 $$e = \lim_{x \to 0} (1 + x)^{1/x}$$

Formula 5 is illustrated by the graph of the function $y = (1 + x)^{1/x}$ in Figure 4 and a table of values for small values of x. This illustrates the fact that, correct to seven decimal places,

$$e \approx 2.7182818$$

If we put $n = 1/x$ in Formula 5, then $n \to \infty$ as $x \to 0^+$ and so an alternative expression for e is

6 $$e = \lim_{n \to \infty} \left(1 + \frac{1}{n}\right)^n$$

3.8 Exercises

1. Explain why the natural logarithmic function $y = \ln x$ is used much more frequently in calculus than the other logarithmic functions $y = \log_a x$.

2–20 Differentiate the function.

2. $f(x) = \ln(x^2 + 10)$

3. $f(\theta) = \ln(\cos \theta)$

4. $f(x) = \cos(\ln x)$

5. $f(x) = \log_2(1 - 3x)$

6. $f(x) = \log_{10}\left(\dfrac{x}{x-1}\right)$

7. $f(x) = \sqrt[5]{\ln x}$

8. $f(x) = \ln \sqrt[5]{x}$

9. $f(x) = \sqrt{x} \ln x$

10. $f(t) = \dfrac{1 + \ln t}{1 - \ln t}$

11. $F(t) = \ln \dfrac{(2t+1)^3}{(3t-1)^4}$

12. $h(x) = \ln\left(x + \sqrt{x^2 - 1}\right)$

13. $g(x) = \ln \dfrac{a - x}{a + x}$

14. $F(y) = y \ln(1 + e^y)$

15. $f(u) = \dfrac{\ln u}{1 + \ln(2u)}$

16. $y = \ln(x^4 \sin^2 x)$

17. $y = \ln|2 - x - 5x^2|$

18. $G(u) = \ln \sqrt{\dfrac{3u + 2}{3u - 2}}$

19. $y = \ln(e^{-x} + xe^{-x})$

20. $y = [\ln(1 + e^x)]^2$

21–24 Find y' and y''.

21. $y = x \ln x$

22. $y = \dfrac{\ln x}{x^2}$

23. $y = \log_{10} x$

24. $y = \ln(\sec x + \tan x)$

25–28 Differentiate f and find the domain of f.

25. $f(x) = \dfrac{x}{1 - \ln(x - 1)}$

26. $f(x) = \dfrac{1}{1 + \ln x}$

27. $f(x) = x^2 \ln(1 - x^2)$

28. $f(x) = \ln \ln \ln x$

29. If $f(x) = \dfrac{x}{\ln x}$, find $f'(e)$.

30. If $f(x) = x^2 \ln x$, find $f'(1)$.

31–32 Find an equation of the tangent line to the curve at the given point.

31. $y = \ln \ln x, \quad (e, 0)$

32. $y = \ln(x^3 - 7), \quad (2, 0)$

33. If $f(x) = \sin x + \ln x$, find $f'(x)$. Check that your answer is reasonable by comparing the graphs of f and f'.

34. Find equations of the tangent lines to the curve $y = (\ln x)/x$ at the points $(1, 0)$ and $(e, 1/e)$. Illustrate by graphing the curve and its tangent lines.

35–46 Use logarithmic differentiation to find the derivative of the function.

35. $y = (2x + 1)^5(x^4 - 3)^6$

36. $y = \sqrt{x}\, e^{x^2}(x^2 + 1)^{10}$

37. $y = \dfrac{\sin^2 x \tan^4 x}{(x^2 + 1)^2}$

38. $y = \sqrt[4]{\dfrac{x^2 + 1}{x^2 - 1}}$

39. $y = x^x$

40. $y = x^{1/x}$

41. $y = x^{\sin x}$

42. $y = (\sin x)^x$

43. $y = (\ln x)^x$

44. $y = x^{\ln x}$

45. $y = x^{e^x}$

46. $y = (\ln x)^{\cos x}$

47. Find y' if $y = \ln(x^2 + y^2)$.

48. Find y' if $x^y = y^x$.

49. Find a formula for $f^{(n)}(x)$ if $f(x) = \ln(x - 1)$.

50. Find $\dfrac{d^9}{dx^9}(x^8 \ln x)$.

51. Use the definition of derivative to prove that

$$\lim_{x \to 0} \frac{\ln(1 + x)}{x} = 1$$

52. Show that $\displaystyle\lim_{n \to \infty} \left(1 + \frac{x}{n}\right)^n = e^x$ for any $x > 0$.

3.9 Hyperbolic Functions

Certain combinations of the exponential functions e^x and e^{-x} arise so frequently in mathematics and its applications that they deserve to be given special names. In many ways they are analogous to the trigonometric functions, and they have the same relationship to the hyperbola that the trigonometric functions have to the circle. For this reason they are collectively called **hyperbolic functions** and individually called **hyperbolic sine**, **hyperbolic cosine**, and so on.

Definition of the Hyperbolic Functions

$$\sinh x = \frac{e^x - e^{-x}}{2} \qquad \operatorname{csch} x = \frac{1}{\sinh x}$$

$$\cosh x = \frac{e^x + e^{-x}}{2} \qquad \operatorname{sech} x = \frac{1}{\cosh x}$$

$$\tanh x = \frac{\sinh x}{\cosh x} \qquad \coth x = \frac{\cosh x}{\sinh x}$$

The graphs of hyperbolic sine and cosine can be sketched using graphical addition as in Figures 1 and 2.

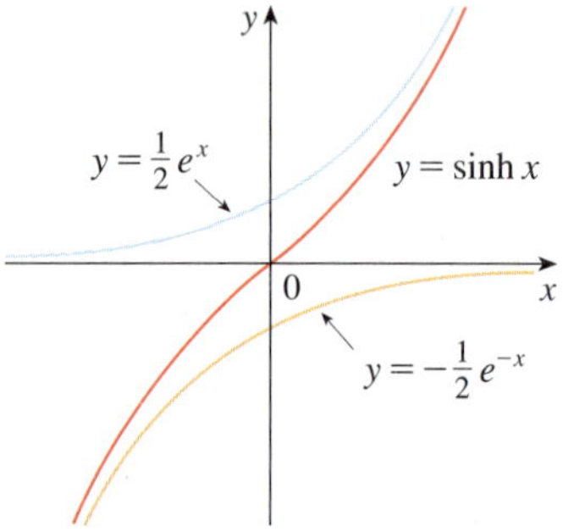

FIGURE 1
$y = \sinh x = \frac{1}{2}e^x - \frac{1}{2}e^{-x}$

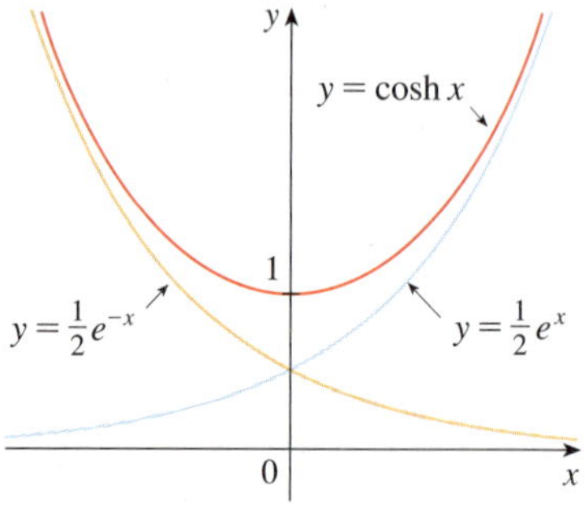

FIGURE 2
$y = \cosh x = \frac{1}{2}e^x + \frac{1}{2}e^{-x}$

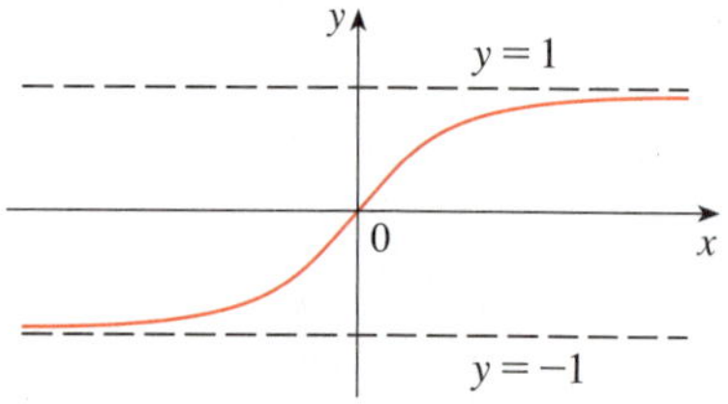

FIGURE 3
$y = \tanh x$

Note that sinh has domain $\mathbb{R}$ and range $\mathbb{R}$, while cosh has domain $\mathbb{R}$ and range $[1, \infty)$. The graph of tanh is shown in Figure 3. It has the horizontal asymptotes $y = \pm 1$. (See Exercise 23.)

Some of the mathematical uses of hyperbolic functions will be seen in Chapter 7. Applications to science and engineering occur whenever an entity such as light, velocity, electricity, or radioactivity is gradually absorbed or extinguished, for the decay can be represented by hyperbolic functions. The most famous application is the use of hyperbolic cosine to describe the shape of a hanging wire. It can be proved that if a heavy flexible cable (such as a telephone or power line) is suspended between two points at the same height, then it takes the shape of a curve with equation $y = c + a\cosh(x/a)$ called a *catenary* (see Figure 4). (The Latin word *catena* means "chain.")

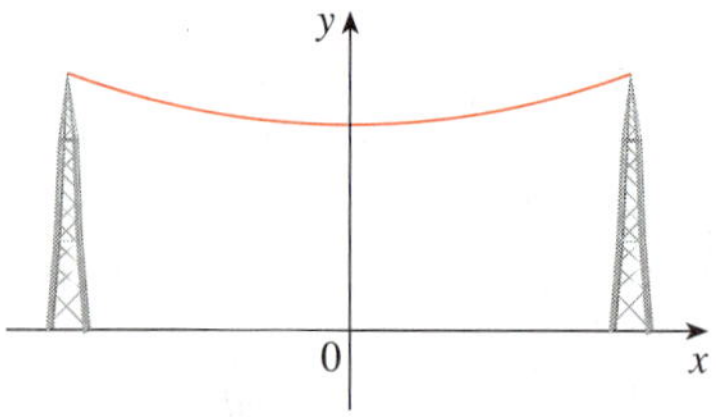

FIGURE 4
A catenary $y = c + a\cosh(x/a)$

The hyperbolic functions satisfy a number of identities that are similar to well-known trigonometric identities. We list some of them here and leave most of the proofs to the exercises.

Hyperbolic Identities

$$\sinh(-x) = -\sinh x \qquad \cosh(-x) = \cosh x$$

$$\cosh^2 x - \sinh^2 x = 1 \qquad 1 - \tanh^2 x = \operatorname{sech}^2 x$$

$$\sinh(x + y) = \sinh x \cosh y + \cosh x \sinh y$$

$$\cosh(x + y) = \cosh x \cosh y + \sinh x \sinh y$$

EXAMPLE 1 Prove (a) $\cosh^2 x - \sinh^2 x = 1$ and (b) $1 - \tanh^2 x = \operatorname{sech}^2 x$.

SOLUTION

(a)
$$\cosh^2 x - \sinh^2 x = \left(\frac{e^x + e^{-x}}{2}\right)^2 - \left(\frac{e^x - e^{-x}}{2}\right)^2$$

$$= \frac{e^{2x} + 2 + e^{-2x}}{4} - \frac{e^{2x} - 2 + e^{-2x}}{4}$$

$$= \tfrac{4}{4} = 1$$

(b) We start with the identity proved in part (a):

$$\cosh^2 x - \sinh^2 x = 1$$

If we divide both sides by $\cosh^2 x$, we get

$$1 - \frac{\sinh^2 x}{\cosh^2 x} = \frac{1}{\cosh^2 x}$$

or

$$1 - \tanh^2 x = \operatorname{sech}^2 x$$

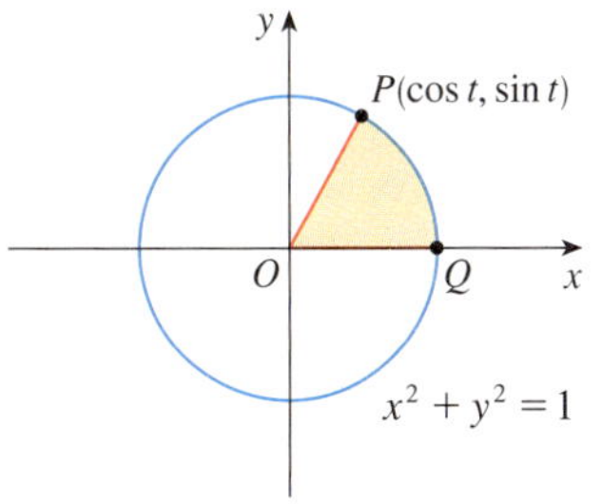

FIGURE 5

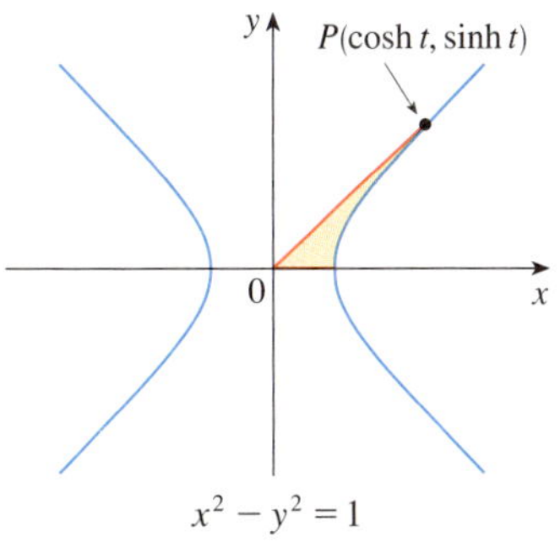

FIGURE 6

The identity proved in Example 1(a) gives a clue to the reason for the name "hyperbolic" functions:

If t is any real number, then the point $P(\cos t, \sin t)$ lies on the unit circle $x^2 + y^2 = 1$ because $\cos^2 t + \sin^2 t = 1$. In fact, t can be interpreted as the radian measure of $\angle POQ$ in Figure 5. For this reason the trigonometric functions are sometimes called *circular* functions.

Likewise, if t is any real number, then the point $P(\cosh t, \sinh t)$ lies on the right branch of the hyperbola $x^2 - y^2 = 1$ because $\cosh^2 t - \sinh^2 t = 1$ and $\cosh t \geqslant 1$. This time, t does not represent the measure of an angle. However, it turns out that t represents twice the area of the shaded hyperbolic sector in Figure 6, just as in the trigonometric case t represents twice the area of the shaded circular sector in Figure 5.

The derivatives of the hyperbolic functions are easily computed. For example,

$$\frac{d}{dx}(\sinh x) = \frac{d}{dx}\left(\frac{e^x - e^{-x}}{2}\right) = \frac{e^x + e^{-x}}{2} = \cosh x$$

We list the differentiation formulas for the hyperbolic functions as Table 1. The remaining proofs are left as exercises. Note the analogy with the differentiation formulas for trigonometric functions, but beware that the signs are different in some cases.

1 Derivatives of Hyperbolic Functions

$$\frac{d}{dx}(\sinh x) = \cosh x \qquad \frac{d}{dx}(\operatorname{csch} x) = -\operatorname{csch} x \coth x$$

$$\frac{d}{dx}(\cosh x) = \sinh x \qquad \frac{d}{dx}(\operatorname{sech} x) = -\operatorname{sech} x \tanh x$$

$$\frac{d}{dx}(\tanh x) = \operatorname{sech}^2 x \qquad \frac{d}{dx}(\coth x) = -\operatorname{csch}^2 x$$

EXAMPLE 2 Any of these differentiation rules can be combined with the Chain Rule. For instance,

$$\frac{d}{dx}(\cosh \sqrt{x}) = \sinh \sqrt{x} \cdot \frac{d}{dx}\sqrt{x} = \frac{\sinh \sqrt{x}}{2\sqrt{x}}$$

Inverse Hyperbolic Functions

You can see from Figures 1 and 3 that sinh and tanh are one-to-one functions and so they have inverse functions denoted by $\sinh^{-1}$ and $\tanh^{-1}$. Figure 2 shows that cosh is not one-to-one, but when restricted to the domain $[0, \infty)$ it becomes one-to-one. The inverse hyperbolic cosine function is defined as the inverse of this restricted function.

2

$$y = \sinh^{-1}x \iff \sinh y = x$$

$$y = \cosh^{-1}x \iff \cosh y = x \text{ and } y \geqslant 0$$

$$y = \tanh^{-1}x \iff \tanh y = x$$

The remaining inverse hyperbolic functions are defined similarly (see Exercise 28).

We can sketch the graphs of $\sinh^{-1}$, $\cosh^{-1}$, and $\tanh^{-1}$ in Figures 7, 8, and 9 by using Figures 1, 2, and 3.

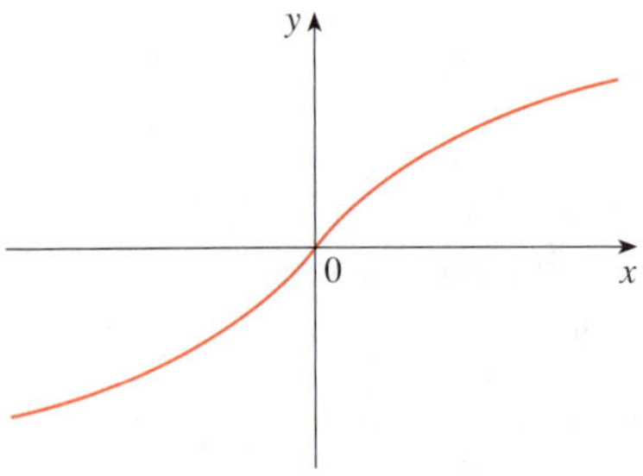

FIGURE 7 $y = \sinh^{-1}x$
domain $= \mathbb{R}$ range $= \mathbb{R}$

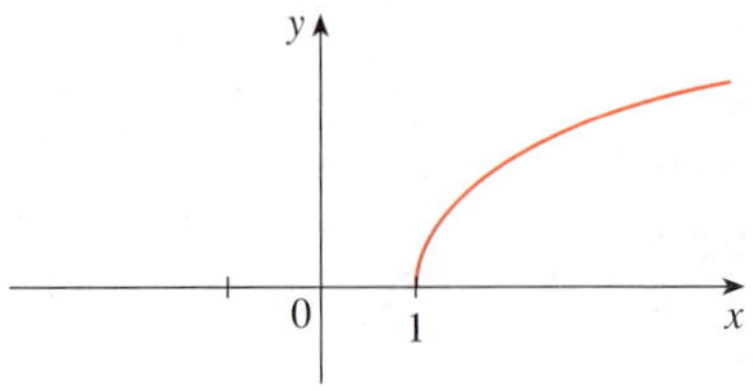

FIGURE 8 $y = \cosh^{-1}x$
domain $= [1, \infty)$ range $= [0, \infty)$

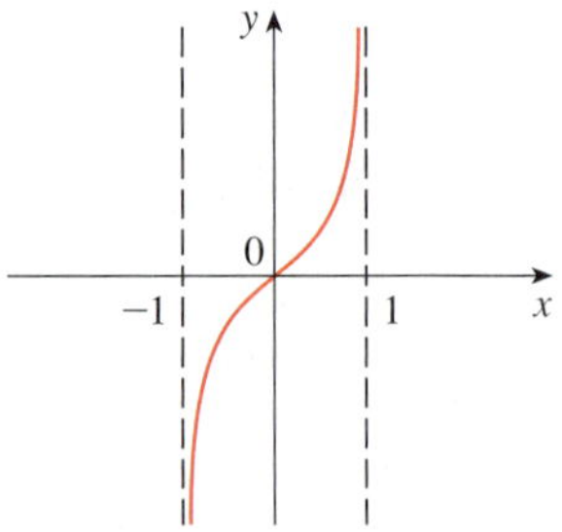

FIGURE 9 $y = \tanh^{-1}x$
domain $= (-1, 1)$ range $= \mathbb{R}$

Since the hyperbolic functions are defined in terms of exponential functions, it's not surprising to learn that the inverse hyperbolic functions can be expressed in terms of logarithms. In particular, we have:

|||| Formula 3 is proved in Example 3. The proofs of Formulas 4 and 5 are requested in Exercises 26 and 27.

3 $$\sinh^{-1}x = \ln\left(x + \sqrt{x^2 + 1}\right) \qquad x \in \mathbb{R}$$

4 $$\cosh^{-1}x = \ln\left(x + \sqrt{x^2 - 1}\right) \qquad x \geqslant 1$$

5 $$\tanh^{-1}x = \tfrac{1}{2}\ln\left(\frac{1 + x}{1 - x}\right) \qquad -1 < x < 1$$

EXAMPLE 3 Show that $\sinh^{-1}x = \ln\left(x + \sqrt{x^2 + 1}\right)$.

SOLUTION Let $y = \sinh^{-1}x$. Then

$$x = \sinh y = \frac{e^y - e^{-y}}{2}$$

so

$$e^y - 2x - e^{-y} = 0$$

or, multiplying by e^y,

$$e^{2y} - 2xe^y - 1 = 0$$

This is really a quadratic equation in e^y:

$$(e^y)^2 - 2x(e^y) - 1 = 0$$

Solving by the quadratic formula, we get

$$e^y = \frac{2x \pm \sqrt{4x^2 + 4}}{2} = x \pm \sqrt{x^2 + 1}$$

Note that $e^y > 0$, but $x - \sqrt{x^2 + 1} < 0$ (because $x < \sqrt{x^2 + 1}$). Thus, the minus sign is inadmissible and we have

$$e^y = x + \sqrt{x^2 + 1}$$

Therefore

$$y = \ln(e^y) = \ln\left(x + \sqrt{x^2 + 1}\right)$$

(See Exercise 25 for another method.)

6 Derivatives of Inverse Hyperbolic Functions

$$\frac{d}{dx}(\sinh^{-1}x) = \frac{1}{\sqrt{1 + x^2}} \qquad \frac{d}{dx}(\operatorname{csch}^{-1}x) = -\frac{1}{|x|\sqrt{x^2 + 1}}$$

$$\frac{d}{dx}(\cosh^{-1}x) = \frac{1}{\sqrt{x^2 - 1}} \qquad \frac{d}{dx}(\operatorname{sech}^{-1}x) = -\frac{1}{x\sqrt{1 - x^2}}$$

$$\frac{d}{dx}(\tanh^{-1}x) = \frac{1}{1 - x^2} \qquad \frac{d}{dx}(\coth^{-1}x) = \frac{1}{1 - x^2}$$

|||| Notice that the formulas for the derivatives of $\tanh^{-1}x$ and $\coth^{-1}x$ appear to be identical. But the domains of these functions have no numbers in common: $\tanh^{-1}x$ is defined for $|x| < 1$, whereas $\coth^{-1}x$ is defined for $|x| > 1$.

The inverse hyperbolic functions are all differentiable because the hyperbolic functions are differentiable. The formulas in Table 6 can be proved either by the method for inverse functions or by differentiating Formulas 3, 4, and 5.

EXAMPLE 4 Prove that $\dfrac{d}{dx}(\sinh^{-1}x) = \dfrac{1}{\sqrt{1+x^2}}$.

SOLUTION 1 Let $y = \sinh^{-1}x$. Then $\sinh y = x$. If we differentiate this equation implicitly with respect to x, we get

$$\cosh y \frac{dy}{dx} = 1$$

Since $\cosh^2 y - \sinh^2 y = 1$ and $\cosh y \geqslant 0$, we have $\cosh y = \sqrt{1 + \sinh^2 y}$, so

$$\frac{dy}{dx} = \frac{1}{\cosh y} = \frac{1}{\sqrt{1+\sinh^2 y}} = \frac{1}{\sqrt{1+x^2}}$$

SOLUTION 2 From Equation 3 (proved in Example 3), we have

$$\begin{aligned}\frac{d}{dx}(\sinh^{-1}x) &= \frac{d}{dx}\ln\left(x + \sqrt{x^2+1}\right)\\ &= \frac{1}{x+\sqrt{x^2+1}}\frac{d}{dx}\left(x+\sqrt{x^2+1}\right)\\ &= \frac{1}{x+\sqrt{x^2+1}}\left(1 + \frac{x}{\sqrt{x^2+1}}\right)\\ &= \frac{\sqrt{x^2+1}+x}{\left(x+\sqrt{x^2+1}\right)\sqrt{x^2+1}}\\ &= \frac{1}{\sqrt{x^2+1}}\end{aligned}$$

EXAMPLE 5 Find $\dfrac{d}{dx}[\tanh^{-1}(\sin x)]$.

SOLUTION Using Table 6 and the Chain Rule, we have

$$\begin{aligned}\frac{d}{dx}[\tanh^{-1}(\sin x)] &= \frac{1}{1-(\sin x)^2}\frac{d}{dx}(\sin x)\\ &= \frac{1}{1-\sin^2 x}\cos x = \frac{\cos x}{\cos^2 x} = \sec x\end{aligned}$$

3.9 Exercises

1–6 Find the numerical value of each expression.

1. (a) $\sinh 0$ (b) $\cosh 0$

2. (a) $\tanh 0$ (b) $\tanh 1$

3. (a) $\sinh(\ln 2)$ (b) $\sinh 2$

4. (a) $\cosh 3$ (b) $\cosh(\ln 3)$

5. (a) $\operatorname{sech} 0$ (b) $\cosh^{-1} 1$

6. (a) $\sinh 1$ (b) $\sinh^{-1} 1$

7–19 Prove the identity.

7. $\sinh(-x) = -\sinh x$
(This shows that sinh is an odd function.)

8. $\cosh(-x) = \cosh x$
(This shows that cosh is an even function.)

9. $\cosh x + \sinh x = e^x$

10. $\cosh x - \sinh x = e^{-x}$

11. $\sinh(x + y) = \sinh x \cosh y + \cosh x \sinh y$

12. $\cosh(x + y) = \cosh x \cosh y + \sinh x \sinh y$

13. $\coth^2 x - 1 = \operatorname{csch}^2 x$

14. $\tanh(x + y) = \dfrac{\tanh x + \tanh y}{1 + \tanh x \tanh y}$

15. $\sinh 2x = 2 \sinh x \cosh x$

16. $\cosh 2x = \cosh^2 x + \sinh^2 x$

17. $\tanh(\ln x) = \dfrac{x^2 - 1}{x^2 + 1}$

18. $\dfrac{1 + \tanh x}{1 - \tanh x} = e^{2x}$

19. $(\cosh x + \sinh x)^n = \cosh nx + \sinh nx$
(n any real number)

20. If $\sinh x = \frac{3}{4}$, find the values of the other hyperbolic functions at x.

21. If $\tanh x = \frac{4}{5}$, find the values of the other hyperbolic functions at x.

22. (a) Use the graphs of sinh, cosh, and tanh in Figures 1–3 to draw the graphs of csch, sech, and coth.
(b) Check the graphs that you sketched in part (a) by using a graphing device to produce them.

23. Use the definitions of the hyperbolic functions to find each of the following limits.
(a) $\lim_{x\to\infty} \tanh x$ (b) $\lim_{x\to-\infty} \tanh x$
(c) $\lim_{x\to\infty} \sinh x$ (d) $\lim_{x\to-\infty} \sinh x$
(e) $\lim_{x\to\infty} \operatorname{sech} x$ (f) $\lim_{x\to\infty} \coth x$
(g) $\lim_{x\to0^+} \coth x$ (h) $\lim_{x\to0^-} \coth x$
(i) $\lim_{x\to-\infty} \operatorname{csch} x$

24. Prove the formulas given in Table 1 for the derivatives of the functions (a) cosh, (b) tanh, (c) csch, (d) sech, and (e) coth.

25. Give an alternative solution to Example 3 by letting $y = \sinh^{-1}x$ and then using Exercise 9 and Example 1(a) with x replaced by y.

26. Prove Equation 4.

27. Prove Equation 5 using (a) the method of Example 3 and (b) Exercise 18 with x replaced by y.

28. For each of the following functions (i) give a definition like those in (2), (ii) sketch the graph, and (iii) find a formula similar to Equation 3.
(a) csch^{-1} (b) sech^{-1} (c) $\coth^{-1}$

29. Prove the formulas given in Table 6 for the derivatives of the following functions.
(a) $\cosh^{-1}$ (b) $\tanh^{-1}$ (c) csch^{-1}
(d) sech^{-1} (e) $\coth^{-1}$

30–47 Find the derivative.

30. $f(x) = \tanh 4x$

31. $f(x) = x \cosh x$

32. $g(x) = \sinh^2 x$

33. $h(x) = \sinh(x^2)$

34. $F(x) = \sinh x \tanh x$

35. $G(x) = \dfrac{1 - \cosh x}{1 + \cosh x}$

36. $f(t) = e^t \operatorname{sech} t$

37. $h(t) = \coth\sqrt{1 + t^2}$

38. $f(t) = \ln(\sinh t)$

39. $H(t) = \tanh(e^t)$

40. $y = \sinh(\cosh x)$

41. $y = e^{\cosh 3x}$

42. $y = x^2 \sinh^{-1}(2x)$

43. $y = \tanh^{-1}\sqrt{x}$

44. $y = x \tanh^{-1}x + \ln\sqrt{1 - x^2}$

45. $y = x \sinh^{-1}(x/3) - \sqrt{9 + x^2}$

46. $y = \operatorname{sech}^{-1}\sqrt{1 - x^2}, \quad x > 0$

47. $y = \coth^{-1}\sqrt{x^2 + 1}$

48. A flexible cable always hangs in the shape of a catenary $y = c + a\cosh(x/a)$, where c and a are constants and $a > 0$ (see Figure 4 and Exercise 50). Graph several members of the family of functions $y = a\cosh(x/a)$. How does the graph change as a varies?

49. A telephone line hangs between two poles 14 m apart in the shape of the catenary $y = 20\cosh(x/20) - 15$, where x and y are measured in meters.
(a) Find the slope of this curve where it meets the right pole.
(b) Find the angle θ between the line and the pole.

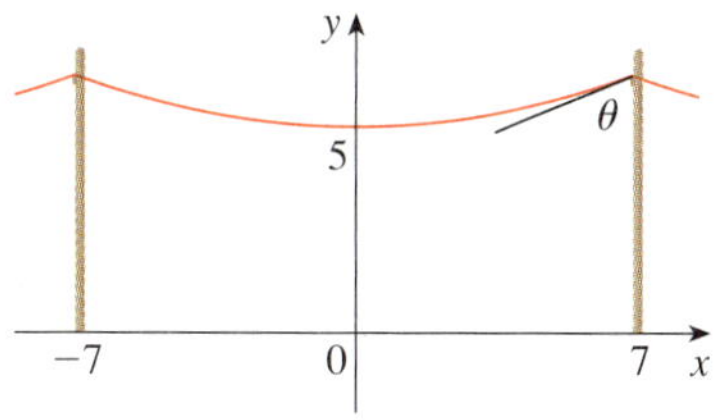

50. Using principles from physics it can be shown that when a cable is hung between two poles, it takes the shape of a curve $y = f(x)$ that satisfies the differential equation

$$\frac{d^2y}{dx^2} = \frac{\rho g}{T}\sqrt{1 + \left(\frac{dy}{dx}\right)^2}$$

where ρ is the linear density of the cable, g is the acceleration due to gravity, and T is the tension in the cable at its lowest point, and the coordinate system is chosen appropriately. Verify that the function

$$y = f(x) = \frac{T}{\rho g} \cosh\left(\frac{\rho g x}{T}\right)$$

is a solution of this differential equation.

51. (a) Show that any function of the form

$$y = A \sinh mx + B \cosh mx$$

satisfies the differential equation $y'' = m^2 y$.

(b) Find $y = y(x)$ such that $y'' = 9y$, $y(0) = -4$, and $y'(0) = 6$.

52. Evaluate $\lim_{x \to \infty} \dfrac{\sinh x}{e^x}$.

53. At what point of the curve $y = \cosh x$ does the tangent have slope 1?

54. If $x = \ln(\sec \theta + \tan \theta)$, show that $\sec \theta = \cosh x$.

55. Show that if $a \neq 0$ and $b \neq 0$, then there exist numbers α and β such that $ae^x + be^{-x}$ equals either $\alpha \sinh(x + \beta)$ or $\alpha \cosh(x + \beta)$. In other words, almost every function of the form $f(x) = ae^x + be^{-x}$ is a shifted and stretched hyperbolic sine or cosine function.

3.10 Related Rates

Explore an expanding balloon interactively.
Resources / Module 5 / Related Rates / Start of Related Rates

If we are pumping air into a balloon, both the volume and the radius of the balloon are increasing and their rates of increase are related to each other. But it is much easier to measure directly the rate of increase of the volume than the rate of increase of the radius.

In a related rates problem the idea is to compute the rate of change of one quantity in terms of the rate of change of another quantity (which may be more easily measured). The procedure is to find an equation that relates the two quantities and then use the Chain Rule to differentiate both sides with respect to time.

EXAMPLE 1 Air is being pumped into a spherical balloon so that its volume increases at a rate of 100 cm³/s. How fast is the radius of the balloon increasing when the diameter is 50 cm?

|||| According to the Principles of Problem Solving discussed on page 80, the first step is to understand the problem. This includes reading the problem carefully, identifying the given and the unknown, and introducing suitable notation.

SOLUTION We start by identifying two things:

the *given information:*

the rate of increase of the volume of air is 100 cm³/s

and the *unknown*:

the rate of increase of the radius when the diameter is 50 cm

In order to express these quantities mathematically, we introduce some suggestive *notation*:

Let V be the volume of the balloon and let r be its radius.

The key thing to remember is that rates of change are derivatives. In this problem, the volume and the radius are both functions of the time t. The rate of increase of the volume with respect to time is the derivative dV/dt, and the rate of increase of the radius is dr/dt. We can therefore restate the given and the unknown as follows:

Given: $\dfrac{dV}{dt} = 100 \text{ cm}^3/\text{s}$

Unknown: $\dfrac{dr}{dt}$ when $r = 25$ cm

|||| The second stage of problem solving is to think of a plan for connecting the given and the unknown.

In order to connect dV/dt and dr/dt, we first relate V and r by the formula for the volume of a sphere:

$$V = \tfrac{4}{3}\pi r^3$$

In order to use the given information, we differentiate each side of this equation with respect to t. To differentiate the right side, we need to use the Chain Rule:

$$\frac{dV}{dt} = \frac{dV}{dr}\frac{dr}{dt} = 4\pi r^2 \frac{dr}{dt}$$

Now we solve for the unknown quantity:

|||| Notice that, although dV/dt is constant, dr/dt is *not* constant.

$$\frac{dr}{dt} = \frac{1}{4\pi r^2}\frac{dV}{dt}$$

If we put $r = 25$ and $dV/dt = 100$ in this equation, we obtain

$$\frac{dr}{dt} = \frac{1}{4\pi(25)^2}100 = \frac{1}{25\pi}$$

The radius of the balloon is increasing at the rate of $1/(25\pi)$ cm/s.

How high will a fireman get while climbing a sliding ladder?
Resources / Module 5
/ Related Rates
/ Start of the Sliding Fireman

EXAMPLE 2 A ladder 10 ft long rests against a vertical wall. If the bottom of the ladder slides away from the wall at a rate of 1 ft/s, how fast is the top of the ladder sliding down the wall when the bottom of the ladder is 6 ft from the wall?

SOLUTION We first draw a diagram and label it as in Figure 1. Let x feet be the distance from the bottom of the ladder to the wall and y feet the distance from the top of the ladder to the ground. Note that x and y are both functions of t (time).

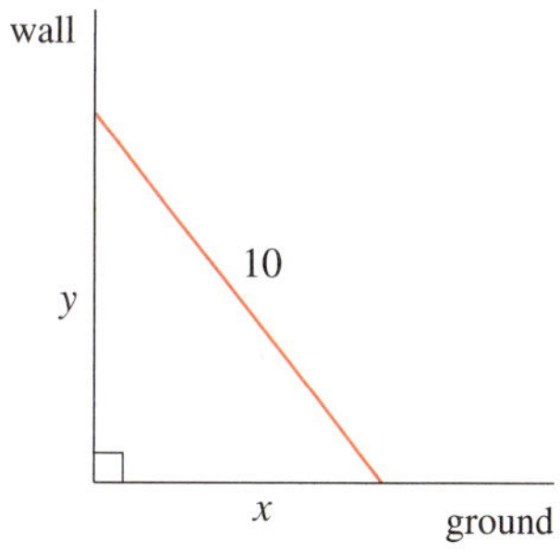

FIGURE 1

We are given that $dx/dt = 1$ ft/s and we are asked to find dy/dt when $x = 6$ ft (see Figure 2). In this problem, the relationship between x and y is given by the Pythagorean Theorem:

$$x^2 + y^2 = 100$$

Differentiating each side with respect to t using the Chain Rule, we have

$$2x\frac{dx}{dt} + 2y\frac{dy}{dt} = 0$$

and solving this equation for the desired rate, we obtain

$$\frac{dy}{dt} = -\frac{x}{y}\frac{dx}{dt}$$

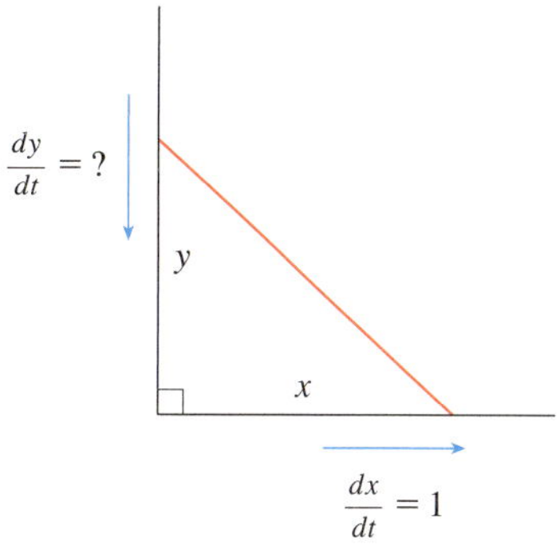

FIGURE 2

When $x = 6$, the Pythagorean Theorem gives $y = 8$ and so, substituting these values and $dx/dt = 1$, we have

$$\frac{dy}{dt} = -\frac{6}{8}(1) = -\frac{3}{4} \text{ ft/s}$$

The fact that dy/dt is negative means that the distance from the top of the ladder to the ground is *decreasing* at a rate of $\frac{3}{4}$ ft/s. In other words, the top of the ladder is sliding down the wall at a rate of $\frac{3}{4}$ ft/s.

EXAMPLE 3 A water tank has the shape of an inverted circular cone with base radius 2 m and height 4 m. If water is being pumped into the tank at a rate of 2 m^3/min, find the rate at which the water level is rising when the water is 3 m deep.

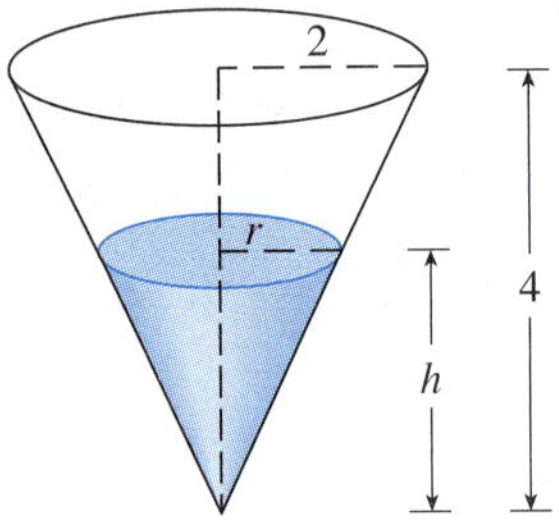

FIGURE 3

SOLUTION We first sketch the cone and label it as in Figure 3. Let V, r, and h be the volume of the water, the radius of the surface, and the height at time t, where t is measured in minutes.

We are given that $dV/dt = 2$ m^3/min and we are asked to find dh/dt when h is 3 m. The quantities V and h are related by the equation

$$V = \tfrac{1}{3}\pi r^2 h$$

but it is very useful to express V as a function of h alone. In order to eliminate r, we use the similar triangles in Figure 3 to write

$$\frac{r}{h} = \frac{2}{4} \qquad r = \frac{h}{2}$$

and the expression for V becomes

$$V = \frac{1}{3}\pi\left(\frac{h}{2}\right)^2 h = \frac{\pi}{12}h^3$$

Now we can differentiate each side with respect to t:

$$\frac{dV}{dt} = \frac{\pi}{4}h^2\frac{dh}{dt}$$

so

$$\frac{dh}{dt} = \frac{4}{\pi h^2}\frac{dV}{dt}$$

Substituting $h = 3$ m and $dV/dt = 2$ m^3/min, we have

$$\frac{dh}{dt} = \frac{4}{\pi(3)^2}\cdot 2 = \frac{8}{9\pi}$$

The water level is rising at a rate of $8/(9\pi) \approx 0.28$ m/min.

|||| Look back: What have we learned from Examples 1–3 that will help us solve future problems?

Strategy It is useful to recall some of the problem-solving principles from page 80 and adapt them to related rates in light of our experience in Examples 1–3:

1. Read the problem carefully.
2. Draw a diagram if possible.
3. Introduce notation. Assign symbols to all quantities that are functions of time.
4. Express the given information and the required rate in terms of derivatives.
5. Write an equation that relates the various quantities of the problem. If necessary, use the geometry of the situation to eliminate one of the variables by substitution (as in Example 3).
6. Use the Chain Rule to differentiate both sides of the equation with respect to t.
7. Substitute the given information into the resulting equation and solve for the unknown rate.

WARNING: A common error is to substitute the given numerical information (for quantities that vary with time) too early. This should be done only *after* the differentiation. (Step 7 follows Step 6.) For instance, in Example 3 we dealt with general values of h until we finally substituted $h = 3$ at the last stage. (If we had put $h = 3$ earlier, we would have gotten $dV/dt = 0$, which is clearly wrong.)

The following examples are further illustrations of the strategy.

EXAMPLE 4 Car A is traveling west at 50 mi/h and car B is traveling north at 60 mi/h. Both are headed for the intersection of the two roads. At what rate are the cars approaching each other when car A is 0.3 mi and car B is 0.4 mi from the intersection?

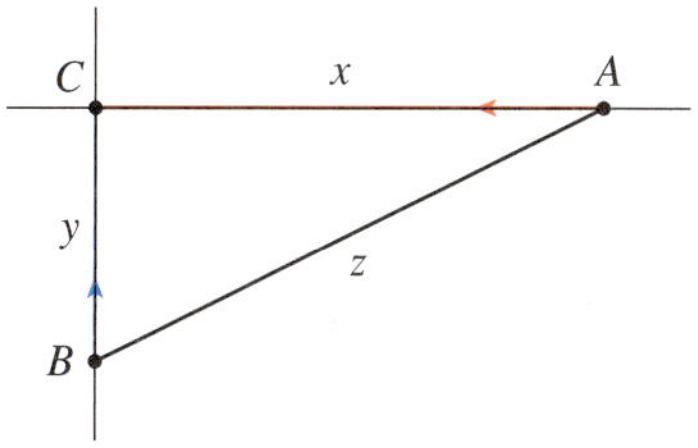

FIGURE 4

SOLUTION We draw Figure 4, where C is the intersection of the roads. At a given time t, let x be the distance from car A to C, let y be the distance from car B to C, and let z be the distance between the cars, where x, y, and z are measured in miles.

We are given that $dx/dt = -50$ mi/h and $dy/dt = -60$ mi/h. (The derivatives are negative because x and y are decreasing.) We are asked to find dz/dt. The equation that relates x, y, and z is given by the Pythagorean Theorem:

$$z^2 = x^2 + y^2$$

Differentiating each side with respect to t, we have

$$2z\frac{dz}{dt} = 2x\frac{dx}{dt} + 2y\frac{dy}{dt}$$

$$\frac{dz}{dt} = \frac{1}{z}\left(x\frac{dx}{dt} + y\frac{dy}{dt}\right)$$

When $x = 0.3$ mi and $y = 0.4$ mi, the Pythagorean Theorem gives $z = 0.5$ mi, so

$$\frac{dz}{dt} = \frac{1}{0.5}[0.3(-50) + 0.4(-60)]$$

$$= -78 \text{ mi/h}$$

The cars are approaching each other at a rate of 78 mi/h.

EXAMPLE 5 A man walks along a straight path at a speed of 4 ft/s. A searchlight is located on the ground 20 ft from the path and is kept focused on the man. At what rate is the searchlight rotating when the man is 15 ft from the point on the path closest to the searchlight?

FIGURE 5

SOLUTION We draw Figure 5 and let x be the distance from the man to the point on the path closest to the searchlight. We let θ be the angle between the beam of the searchlight and the perpendicular to the path.

We are given that $dx/dt = 4$ ft/s and are asked to find $d\theta/dt$ when $x = 15$. The equation that relates x and θ can be written from Figure 5:

$$\frac{x}{20} = \tan\theta \qquad x = 20\tan\theta$$

Differentiating each side with respect to t, we get

$$\frac{dx}{dt} = 20\sec^2\theta\,\frac{d\theta}{dt}$$

so

$$\frac{d\theta}{dt} = \tfrac{1}{20}\cos^2\theta\,\frac{dx}{dt} = \tfrac{1}{20}\cos^2\theta\,(4) = \tfrac{1}{5}\cos^2\theta$$

When $x = 15$, the length of the beam is 25, so $\cos\theta = \frac{4}{5}$ and

$$\frac{d\theta}{dt} = \frac{1}{5}\left(\frac{4}{5}\right)^2 = \frac{16}{125} = 0.128$$

The searchlight is rotating at a rate of 0.128 rad/s.

3.10 Exercises

1. If V is the volume of a cube with edge length x and the cube expands as time passes, find dV/dt in terms of dx/dt.

2. (a) If A is the area of a circle with radius r and the circle expands as time passes, find dA/dt in terms of dr/dt.
(b) Suppose oil spills from a ruptured tanker and spreads in a circular pattern. If the radius of the oil spill increases at a constant rate of 1 m/s, how fast is the area of the spill increasing when the radius is 30 m?

3. If $y = x^3 + 2x$ and $dx/dt = 5$, find dy/dt when $x = 2$.

4. If $x^2 + y^2 = 25$ and $dy/dt = 6$, find dx/dt when $y = 4$.

5. If $z^2 = x^2 + y^2$, $dx/dt = 2$, and $dy/dt = 3$, find dz/dt when $x = 5$ and $y = 12$.

6. A particle moves along the curve $y = \sqrt{1 + x^3}$. As it reaches the point (2, 3), the y-coordinate is increasing at a rate of 4 cm/s. How fast is the x-coordinate of the point changing at that instant?

7–10 ||||
(a) What quantities are given in the problem?
(b) What is the unknown?
(c) Draw a picture of the situation for any time t.
(d) Write an equation that relates the quantities.
(e) Finish solving the problem.

7. A plane flying horizontally at an altitude of 1 mi and a speed of 500 mi/h passes directly over a radar station. Find the rate at which the distance from the plane to the station is increasing when it is 2 mi away from the station.

8. If a snowball melts so that its surface area decreases at a rate of 1 cm^2/min, find the rate at which the diameter decreases when the diameter is 10 cm.

9. A street light is mounted at the top of a 15-ft-tall pole. A man 6 ft tall walks away from the pole with a speed of 5 ft/s along a straight path. How fast is the tip of his shadow moving when he is 40 ft from the pole?

10. At noon, ship A is 150 km west of ship B. Ship A is sailing east at 35 km/h and ship B is sailing north at 25 km/h. How fast is the distance between the ships changing at 4:00 P.M.?

11. Two cars start moving from the same point. One travels south at 60 mi/h and the other travels west at 25 mi/h. At what rate is the distance between the cars increasing two hours later?

12. A spotlight on the ground shines on a wall 12 m away. If a man 2 m tall walks from the spotlight toward the building at a speed of 1.6 m/s, how fast is the length of his shadow on the building decreasing when he is 4 m from the building?

13. A man starts walking north at 4 ft/s from a point P. Five minutes later a woman starts walking south at 5 ft/s from a point 500 ft due east of P. At what rate are the people moving apart 15 min after the woman starts walking?

14. A baseball diamond is a square with side 90 ft. A batter hits the ball and runs toward first base with a speed of 24 ft/s.
(a) At what rate is his distance from second base decreasing when he is halfway to first base?
(b) At what rate is his distance from third base increasing at the same moment?

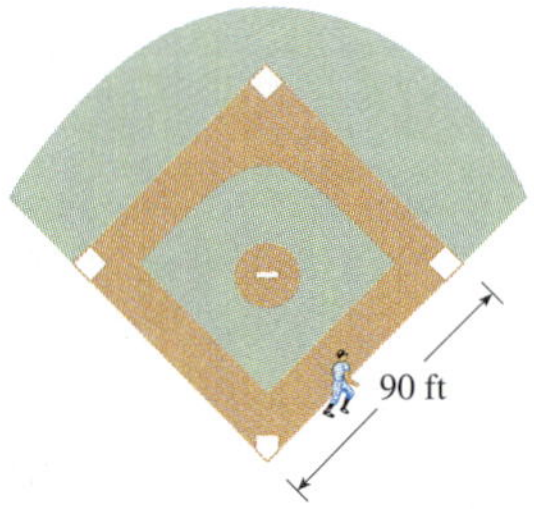

15. The altitude of a triangle is increasing at a rate of 1 cm/min while the area of the triangle is increasing at a rate of 2 cm^2/min. At what rate is the base of the triangle changing when the altitude is 10 cm and the area is 100 cm^2?

16. A boat is pulled into a dock by a rope attached to the bow of the boat and passing through a pulley on the dock that is 1 m higher than the bow of the boat. If the rope is pulled in at a rate of 1 m/s, how fast is the boat approaching the dock when it is 8 m from the dock?

17. At noon, ship A is 100 km west of ship B. Ship A is sailing south at 35 km/h and ship B is sailing north at 25 km/h. How fast is the distance between the ships changing at 4:00 P.M.?

18. A particle is moving along the curve $y = \sqrt{x}$. As the particle passes through the point (4, 2), its x-coordinate increases at a rate of 3 cm/s. How fast is the distance from the particle to the origin changing at this instant?

19. Water is leaking out of an inverted conical tank at a rate of 10,000 cm^3/min at the same time that water is being pumped into the tank at a constant rate. The tank has height 6 m and the diameter at the top is 4 m. If the water level is rising at a rate of 20 cm/min when the height of the water is 2 m, find the rate at which water is being pumped into the tank.

20. A trough is 10 ft long and its ends have the shape of isosceles triangles that are 3 ft across at the top and have a height of 1 ft. If the trough is being filled with water at a rate of 12 ft^3/min, how fast is the water level rising when the water is 6 inches deep?

21. A water trough is 10 m long and a cross-section has the shape of an isosceles trapezoid that is 30 cm wide at the bottom, 80 cm wide at the top, and has height 50 cm. If the trough is being filled with water at the rate of 0.2 m^3/min, how fast is the water level rising when the water is 30 cm deep?

22. A swimming pool is 20 ft wide, 40 ft long, 3 ft deep at the shallow end, and 9 ft deep at its deepest point. A cross-section is shown in the figure. If the pool is being filled at a rate of 0.8 ft^3/min, how fast is the water level rising when the depth at the deepest point is 5 ft?

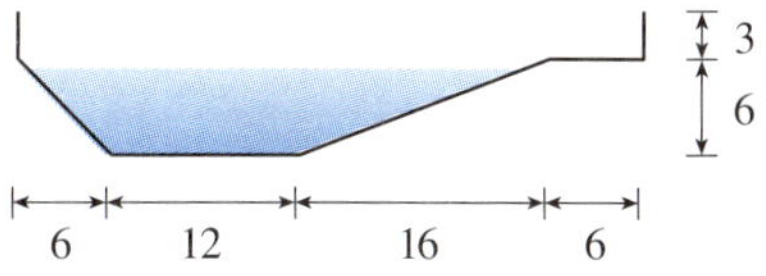

23. Gravel is being dumped from a conveyor belt at a rate of 30 ft^3/min, and its coarseness is such that it forms a pile in the shape of a cone whose base diameter and height are always equal. How fast is the height of the pile increasing when the pile is 10 ft high?

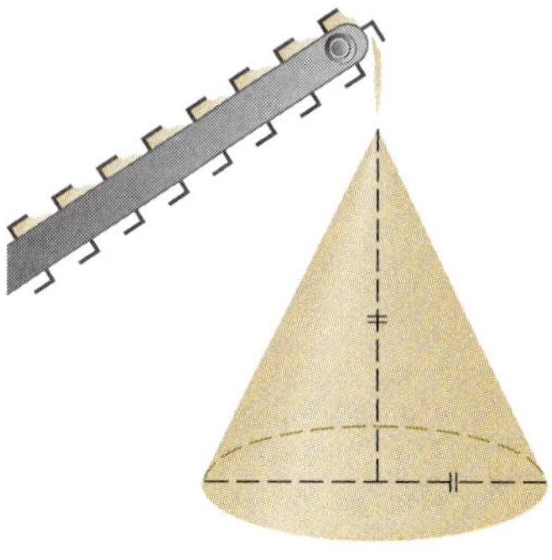

24. A kite 100 ft above the ground moves horizontally at a speed of 8 ft/s. At what rate is the angle between the string and the horizontal decreasing when 200 ft of string have been let out?

25. Two sides of a triangle are 4 m and 5 m in length and the angle between them is increasing at a rate of 0.06 rad/s. Find the rate at which the area of the triangle is increasing when the angle between the sides of fixed length is $\pi/3$.

26. Two sides of a triangle have lengths 12 m and 15 m. The angle between them is increasing at a rate of 2°/min. How fast is the length of the third side increasing when the angle between the sides of fixed length is 60°?

27. Boyle's Law states that when a sample of gas is compressed at a constant temperature, the pressure P and volume V satisfy the equation $PV = C$, where C is a constant. Suppose that at a certain instant the volume is 600 cm^3, the pressure is 150 kPa, and the pressure is increasing at a rate of 20 kPa/min. At what rate is the volume decreasing at this instant?

28. When air expands adiabatically (without gaining or losing heat), its pressure P and volume V are related by the equation $PV^{1.4} = C$, where C is a constant. Suppose that at a certain instant the volume is 400 cm^3 and the pressure is 80 kPa and is decreasing at a rate of 10 kPa/min. At what rate is the volume increasing at this instant?

29. If two resistors with resistances R_1 and R_2 are connected in parallel, as in the figure, then the total resistance R, measured in ohms (Ω), is given by

$$\frac{1}{R} = \frac{1}{R_1} + \frac{1}{R_2}$$

If R_1 and R_2 are increasing at rates of 0.3 Ω/s and 0.2 Ω/s, respectively, how fast is R changing when $R_1 = 80\ \Omega$ and $R_2 = 100\ \Omega$?

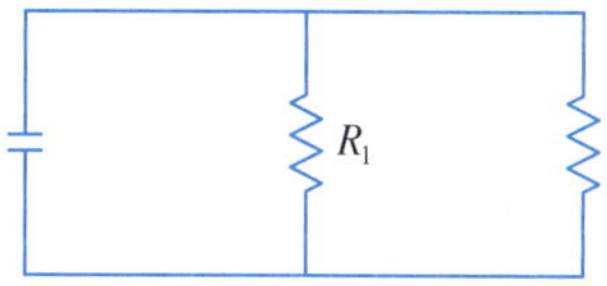

30. Brain weight B as a function of body weight W in fish has been modeled by the power function $B = 0.007W^{2/3}$, where B and W are measured in grams. A model for body weight as a function of body length L (measured in centimeters) is $W = 0.12L^{2.53}$. If, over 10 million years, the average length of a certain species of fish evolved from 15 cm to 20 cm at a constant rate, how fast was this species' brain growing when the average length was 18 cm?

31. A ladder 10 ft long rests against a vertical wall. If the bottom of the ladder slides away from the wall at a speed of 2 ft/s, how fast is the angle between the top of the ladder and the wall changing when the angle is $\pi/4$ rad?

32. Two carts, A and B, are connected by a rope 39 ft long that passes over a pulley P (see the figure). The point Q is on the floor 12 ft directly beneath P and between the carts. Cart A is being pulled away from Q at a speed of 2 ft/s. How fast is cart B moving toward Q at the instant when cart A is 5 ft from Q?

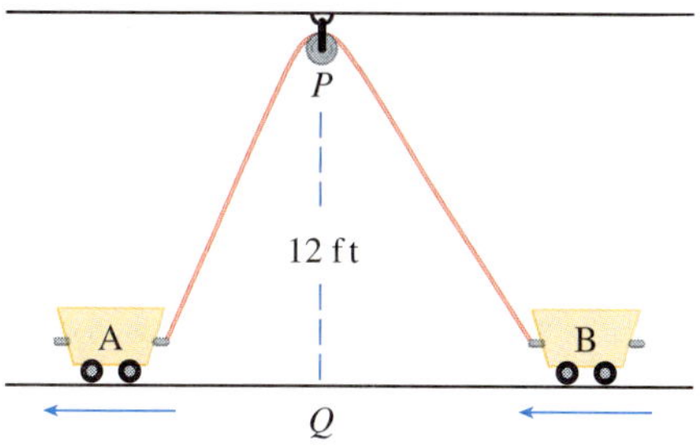

33. A television camera is positioned 4000 ft from the base of a rocket launching pad. The angle of elevation of the camera has to change at the correct rate in order to keep the rocket in sight. Also, the mechanism for focusing the camera has to take into account the increasing distance from the camera to the rising rocket. Let's assume the rocket rises vertically and its speed is 600 ft/s when it has risen 3000 ft.

(a) How fast is the distance from the television camera to the rocket changing at that moment?

(b) If the television camera is always kept aimed at the rocket, how fast is the camera's angle of elevation changing at that same moment?

34. A lighthouse is located on a small island 3 km away from the nearest point P on a straight shoreline and its light makes four revolutions per minute. How fast is the beam of light moving along the shoreline when it is 1 km from P?

35. A plane flying with a constant speed of 300 km/h passes over a ground radar station at an altitude of 1 km and climbs at an angle of 30°. At what rate is the distance from the plane to the radar station increasing a minute later?

36. Two people start from the same point. One walks east at 3 mi/h and the other walks northeast at 2 mi/h. How fast is the distance between the people changing after 15 minutes?

37. A runner sprints around a circular track of radius 100 m at a constant speed of 7 m/s. The runner's friend is standing at a distance 200 m from the center of the track. How fast is the distance between the friends changing when the distance between them is 200 m?

38. The minute hand on a watch is 8 mm long and the hour hand is 4 mm long. How fast is the distance between the tips of the hands changing at one o'clock?

3.11 Linear Approximations and Differentials

Resources / Module 3 / Linear Approximation / Start of Linear Approximation

We have seen that a curve lies very close to its tangent line near the point of tangency. In fact, by zooming in toward a point on the graph of a differentiable function, we noticed that the graph looks more and more like its tangent line. (See Figure 2 in Section 2.7 and Figure 3 in Section 2.8.) This observation is the basis for a method of finding approximate values of functions.

The idea is that it might be easy to calculate a value $f(a)$ of a function, but difficult (or even impossible) to compute nearby values of f. So we settle for the easily computed values of the linear function L whose graph is the tangent line of f at $(a, f(a))$. (See Figure 1.)

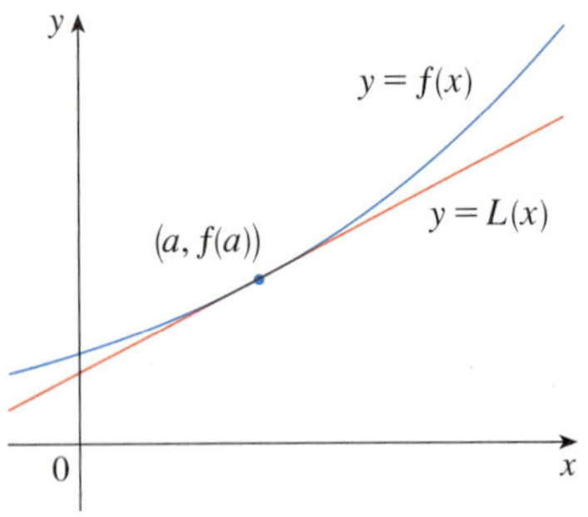

FIGURE 1

In other words, we use the tangent line at $(a, f(a))$ as an approximation to the curve $y = f(x)$ when x is near a. An equation of this tangent line is

$$y = f(a) + f'(a)(x - a)$$

and the approximation

$$\boxed{1} \qquad f(x) \approx f(a) + f'(a)(x - a)$$

is called the **linear approximation** or **tangent line approximation** of f at a. The linear function whose graph is this tangent line, that is,

$$\boxed{2} \qquad L(x) = f(a) + f'(a)(x - a)$$

is called the **linearization** of f at a.

The following example is typical of situations in which we use a linear approximation to predict the future behavior of a function given by empirical data.

EXAMPLE 1 Suppose that after you stuff a turkey its temperature is 50°F and you then put it in a 325°F oven. After an hour the meat thermometer indicates that the temperature of the turkey is 93°F and after two hours it indicates 129°F. Predict the temperature of the turkey after three hours.

SOLUTION If $T(t)$ represents the temperature of the turkey after t hours, we are given that $T(0) = 50$, $T(1) = 93$, and $T(2) = 129$. In order to make a linear approximation with $a = 2$, we need an estimate for the derivative $T'(2)$. Because

$$T'(2) = \lim_{t \to 2} \frac{T(t) - T(2)}{t - 2}$$

we could estimate $T'(2)$ by the difference quotient with $t = 1$:

$$T'(2) \approx \frac{T(1) - T(2)}{1 - 2} = \frac{93 - 129}{-1} = 36$$

This amounts to approximating the instantaneous rate of temperature change by the average rate of change between $t = 1$ and $t = 2$, which is 36°F/h. With this estimate, the linear approximation (1) for the temperature after 3 h is

$$T(3) \approx T(2) + T'(2)(3 - 2)$$
$$\approx 129 + 36 \cdot 1 = 165$$

So the predicted temperature after three hours is 165°F.

We obtain a more accurate estimate for $T'(2)$ by plotting the given data, as in Figure 2, and estimating the slope of the tangent line at $t = 2$ to be

$$T'(2) \approx 33$$

Then our linear approximation becomes

$$T(3) \approx T(2) + T'(2) \cdot 1 \approx 129 + 33 = 162$$

and our improved estimate for the temperature is 162°F.

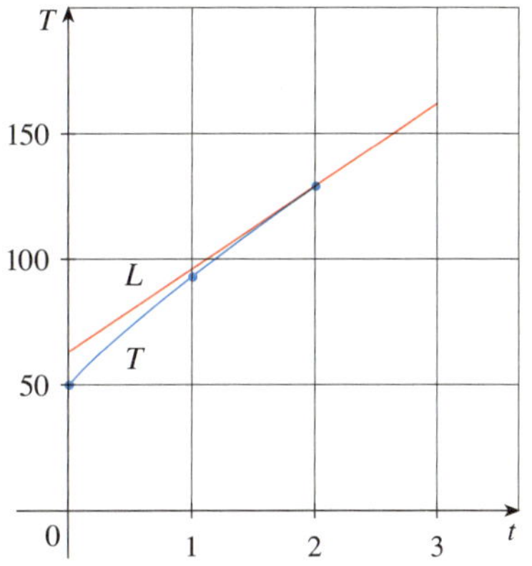

FIGURE 2

Because the temperature curve lies below the tangent line, it appears that the actual temperature after three hours will be somewhat less than 162°F, perhaps closer to 160°F.

EXAMPLE 2 Find the linearization of the function $f(x) = \sqrt{x + 3}$ at $a = 1$ and use it to approximate the numbers $\sqrt{3.98}$ and $\sqrt{4.05}$. Are these approximations overestimates or underestimates?

SOLUTION The derivative of $f(x) = (x + 3)^{1/2}$ is

$$f'(x) = \tfrac{1}{2}(x + 3)^{-1/2} = \frac{1}{2\sqrt{x + 3}}$$

and so we have $f(1) = 2$ and $f'(1) = \frac{1}{4}$. Putting these values into Equation 2, we see that the linearization is

$$L(x) = f(1) + f'(1)(x - 1) = 2 + \tfrac{1}{4}(x - 1) = \frac{7}{4} + \frac{x}{4}$$

The corresponding linear approximation (1) is

$$\sqrt{x + 3} \approx \frac{7}{4} + \frac{x}{4} \qquad \text{(when } x \text{ is near 1)}$$

In particular, we have

$$\sqrt{3.98} \approx \tfrac{7}{4} + \tfrac{0.98}{4} = 1.995 \qquad \text{and} \qquad \sqrt{4.05} \approx \tfrac{7}{4} + \tfrac{1.05}{4} = 2.0125$$

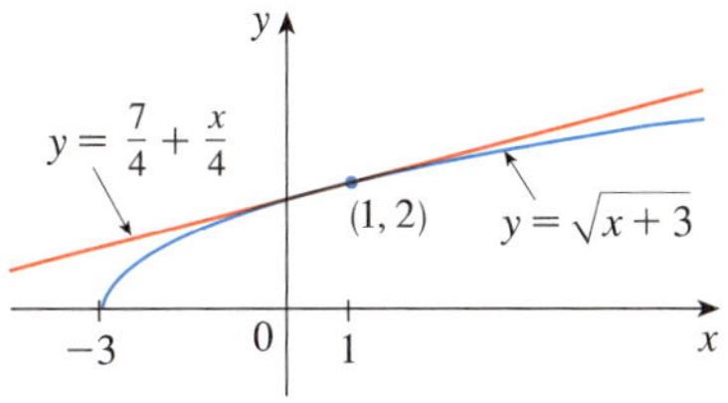

FIGURE 3

The linear approximation is illustrated in Figure 3. We see that, indeed, the tangent line approximation is a good approximation to the given function when x is near 1. We also see that our approximations are overestimates because the tangent line lies above the curve.

Of course, a calculator could give us approximations for $\sqrt{3.98}$ and $\sqrt{4.05}$, but the linear approximation gives an approximation over an entire interval.

In the following table we compare the estimates from the linear approximation in Example 2 with the true values. Notice from this table, and also from Figure 3, that the tangent line approximation gives good estimates when x is close to 1 but the accuracy of the approximation deteriorates when x is farther away from 1.

	x	From $L(x)$	Actual value
$\sqrt{3.9}$	0.9	1.975	1.97484176 . . .
$\sqrt{3.98}$	0.98	1.995	1.99499373 . . .
$\sqrt{4}$	1	2	2.00000000 . . .
$\sqrt{4.05}$	1.05	2.0125	2.01246117 . . .
$\sqrt{4.1}$	1.1	2.025	2.02484567 . . .
$\sqrt{5}$	2	2.25	2.23606797 . . .
$\sqrt{6}$	3	2.5	2.44948974 . . .

How good is the approximation that we obtained in Example 2? The next example shows that by using a graphing calculator or computer we can determine an interval throughout which a linear approximation provides a specified accuracy.

EXAMPLE 3 For what values of x is the linear approximation

$$\sqrt{x+3} \approx \frac{7}{4} + \frac{x}{4}$$

accurate to within 0.5? What about accuracy to within 0.1?

SOLUTION Accuracy to within 0.5 means that the functions should differ by less than 0.5:

$$\left| \sqrt{x+3} - \left(\frac{7}{4} + \frac{x}{4} \right) \right| < 0.5$$

Equivalently, we could write

$$\sqrt{x+3} - 0.5 < \frac{7}{4} + \frac{x}{4} < \sqrt{x+3} + 0.5$$

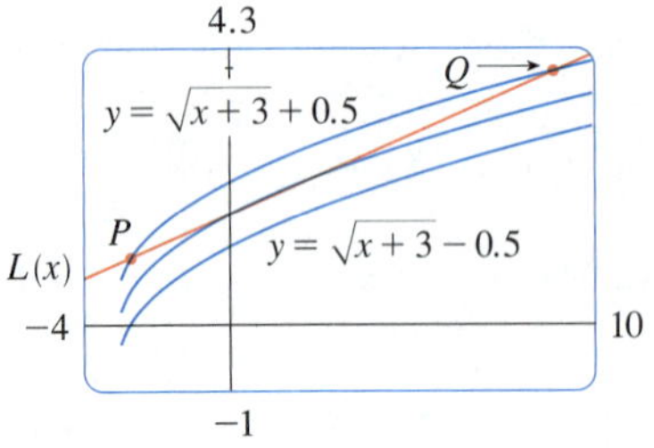

FIGURE 4

This says that the linear approximation should lie between the curves obtained by shifting the curve $y = \sqrt{x+3}$ upward and downward by an amount 0.5. Figure 4 shows the tangent line $y = (7 + x)/4$ intersecting the upper curve $y = \sqrt{x+3} + 0.5$ at P and Q. Zooming in and using the cursor, we estimate that the x-coordinate of P is about -2.66 and the x-coordinate of Q is about 8.66. Thus, we see from the graph that the approximation

$$\sqrt{x+3} \approx \frac{7}{4} + \frac{x}{4}$$

is accurate to within 0.5 when $-2.6 < x < 8.6$. (We have rounded to be safe.)

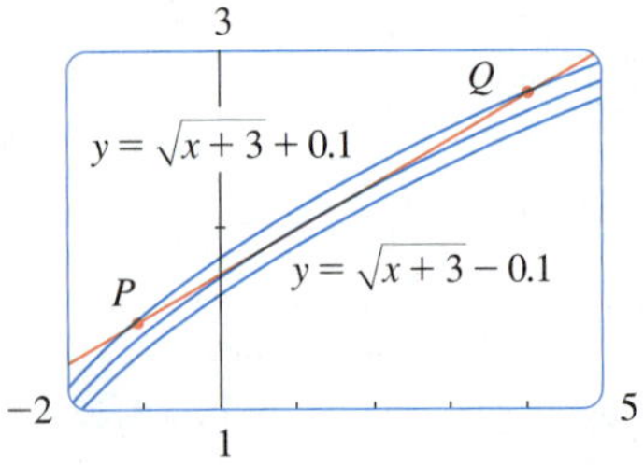

FIGURE 5

Similarly, from Figure 5 we see that the approximation is accurate to within 0.1 when $-1.1 < x < 3.9$.

Applications to Physics

Linear approximations are often used in physics. In analyzing the consequences of an equation, a physicist sometimes needs to simplify a function by replacing it with its linear approximation. For instance, in deriving a formula for the period of a pendulum, physics textbooks obtain the expression $a_T = -g \sin \theta$ for tangential acceleration and then replace $\sin \theta$ by θ with the remark that $\sin \theta$ is very close to θ if θ is not too large. [See, for example, *Physics: Calculus,* 2d ed., by Eugene Hecht (Pacific Grove, CA: Brooks/Cole, 2000), p. 431.] You can verify that the linearization of the function $f(x) = \sin x$ at $a = 0$ is $L(x) = x$ and so the linear approximation at 0 is

$$\sin x \approx x$$

(see Exercise 48). So, in effect, the derivation of the formula for the period of a pendulum uses the tangent line approximation for the sine function.

Another example occurs in the theory of optics, where light rays that arrive at shallow angles relative to the optical axis are called *paraxial rays.* In paraxial (or Gaussian) optics, both $\sin \theta$ and $\cos \theta$ are replaced by their linearizations. In other words, the linear approximations

$$\sin \theta \approx \theta \qquad \text{and} \qquad \cos \theta \approx 1$$

are used because θ is close to 0. The results of calculations made with these approximations became the basic theoretical tool used to design lenses. [See *Optics,* 4th ed., by Eugene Hecht (Reading, MA: Addison-Wesley, 2002), p. 154.]

In Section 11.12 we will present several other applications of the idea of linear approximations to physics.

Differentials

The ideas behind linear approximations are sometimes formulated in the terminology and notation of *differentials.* If $y = f(x)$, where f is a differentiable function, then the **differential** dx is an independent variable; that is, dx can be given the value of any real number. The **differential** dy is then defined in terms of dx by the equation

||| If $dx \neq 0$, we can divide both sides of Equation 3 by dx to obtain

$$\frac{dy}{dx} = f'(x)$$

We have seen similar equations before, but now the left side can genuinely be interpreted as a ratio of differentials.

3
$$dy = f'(x)\, dx$$

So dy is a dependent variable; it depends on the values of x and dx. If dx is given a specific value and x is taken to be some specific number in the domain of f, then the numerical value of dy is determined.

The geometric meaning of differentials is shown in Figure 6. Let $P(x, f(x))$ and $Q(x + \Delta x, f(x + \Delta x))$ be points on the graph of f and let $dx = \Delta x$. The corresponding change in y is

$$\Delta y = f(x + \Delta x) - f(x)$$

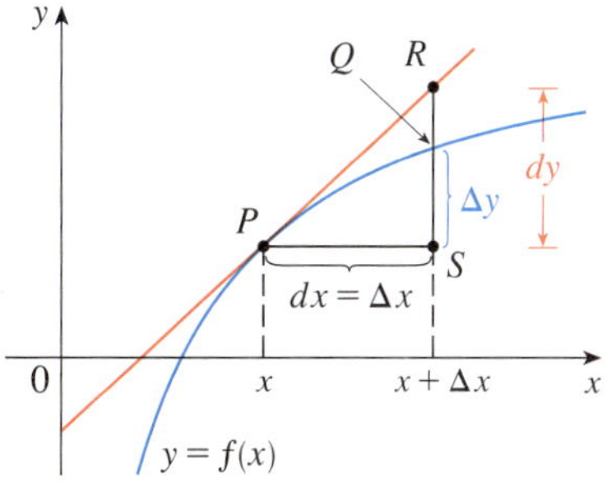

FIGURE 6

The slope of the tangent line PR is the derivative $f'(x)$. Thus, the directed distance from S to R is $f'(x)\,dx = dy$. Therefore, dy represents the amount that the tangent line rises or falls (the change in the linearization), whereas Δy represents the amount that the curve $y = f(x)$ rises or falls when x changes by an amount dx.

EXAMPLE 4 Compare the values of Δy and dy if $y = f(x) = x^3 + x^2 - 2x + 1$ and x changes (a) from 2 to 2.05 and (b) from 2 to 2.01.

SOLUTION
(a) We have

$$f(2) = 2^3 + 2^2 - 2(2) + 1 = 9$$
$$f(2.05) = (2.05)^3 + (2.05)^2 - 2(2.05) + 1 = 9.717625$$
$$\Delta y = f(2.05) - f(2) = 0.717625$$

|||| Figure 7 shows the function in Example 4 and a comparison of dy and Δy when $a = 2$. The viewing rectangle is $[1.8, 2.5]$ by $[6, 18]$.

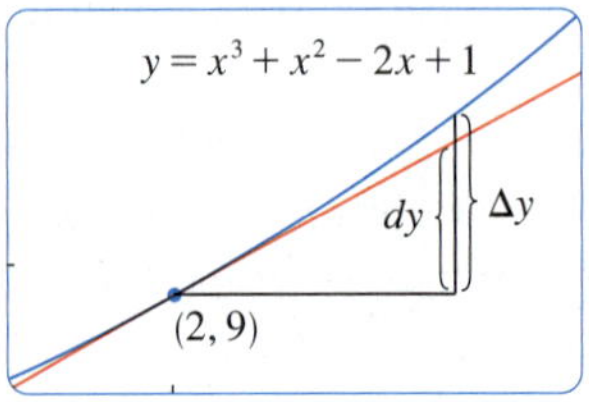

FIGURE 7

In general, $$dy = f'(x)\,dx = (3x^2 + 2x - 2)\,dx$$

When $x = 2$ and $dx = \Delta x = 0.05$, this becomes

$$dy = [3(2)^2 + 2(2) - 2]0.05 = 0.7$$

(b) $$f(2.01) = (2.01)^3 + (2.01)^2 - 2(2.01) + 1 = 9.140701$$
$$\Delta y = f(2.01) - f(2) = 0.140701$$

When $dx = \Delta x = 0.01$,

$$dy = [3(2)^2 + 2(2) - 2]0.01 = 0.14$$

Notice that the approximation $\Delta y \approx dy$ becomes better as Δx becomes smaller in Example 4. Notice also that dy was easier to compute than Δy. For more complicated functions it may be impossible to compute Δy exactly. In such cases the approximation by differentials is especially useful.

In the notation of differentials, the linear approximation (1) can be written as

$$f(a + dx) \approx f(a) + dy$$

For instance, for the function $f(x) = \sqrt{x + 3}$ in Example 2, we have

$$dy = f'(x)\,dx = \frac{dx}{2\sqrt{x + 3}}$$

If $a = 1$ and $dx = \Delta x = 0.05$, then

$$dy = \frac{0.05}{2\sqrt{1 + 3}} = 0.0125$$

and $$\sqrt{4.05} = f(1.05) \approx f(1) + dy = 2.0125$$

just as we found in Example 2.

Our final example illustrates the use of differentials in estimating the errors that occur because of approximate measurements.

EXAMPLE 5 The radius of a sphere was measured and found to be 21 cm with a possible error in measurement of at most 0.05 cm. What is the maximum error in using this value of the radius to compute the volume of the sphere?

SOLUTION If the radius of the sphere is r, then its volume is $V = \frac{4}{3}\pi r^3$. If the error in the measured value of r is denoted by $dr = \Delta r$, then the corresponding error in the calculated value of V is ΔV, which can be approximated by the differential

$$dV = 4\pi r^2\,dr$$

When $r = 21$ and $dr = 0.05$, this becomes

$$dV = 4\pi(21)^2 0.05 \approx 277$$

The maximum error in the calculated volume is about 277 cm^3.

NOTE ▫ Although the possible error in Example 5 may appear to be rather large, a better picture of the error is given by the **relative error**, which is computed by dividing the error by the total volume:

$$\frac{\Delta V}{V} \approx \frac{dV}{V} = \frac{4\pi r^2\,dr}{\frac{4}{3}\pi r^3} = 3\,\frac{dr}{r}$$

Thus, the relative error in the volume is about three times the relative error in the radius. In Example 5 the relative error in the radius is approximately $dr/r = 0.05/21 \approx 0.0024$ and it produces a relative error of about 0.007 in the volume. The errors could also be expressed as **percentage errors** of 0.24% in the radius and 0.7% in the volume.

3.11 Exercises

1. The turkey in Example 1 is removed from the oven when its temperature reaches 185°F and is placed on a table in a room where the temperature is 75°F. After 10 minutes the temperature of the turkey is 172°F and after 20 minutes it is 160°F. Use a linear approximation to predict the temperature of the turkey after half an hour. Do you think your prediction is an overestimate or an underestimate? Why?

2. Atmospheric pressure P decreases as altitude h increases. At a temperature of 15°C, the pressure is 101.3 kilopascals (kPa) at sea level, 87.1 kPa at $h = 1$ km, and 74.9 kPa at $h = 2$ km. Use a linear approximation to estimate the atmospheric pressure at an altitude of 3 km.

3. The graph indicates how Australia's population is aging by showing the past and projected percentage of the population aged 65 and over. Use a linear approximation to predict the percentage of the population that will be 65 and over in the years 2040 and 2050. Do you think your predictions are too high or too low? Why?

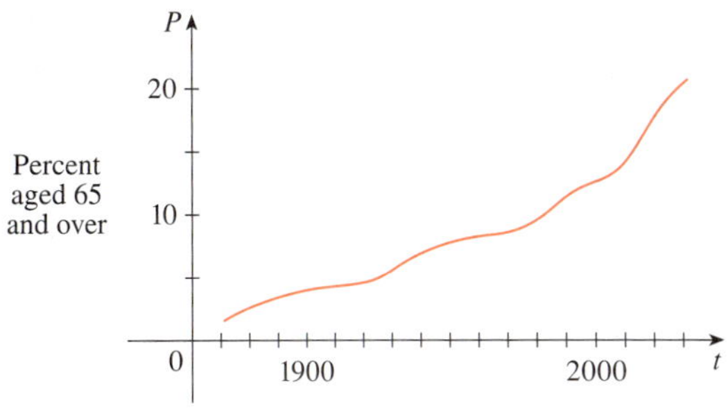

4. The table shows the population of Nepal (in millions) as of June 30 of the given year. Use a linear approximation to estimate the population at midyear in 1984. Use another linear approximation to predict the population in 2006.

t	1980	1985	1990	1995	2000
$N(t)$	15.0	17.0	19.3	22.0	24.9

5–8 Find the linearization $L(x)$ of the function at a.

5. $f(x) = x^3$, $a = 1$

6. $f(x) = \ln x$, $a = 1$

7. $f(x) = \cos x$, $a = \pi/2$

8. $f(x) = \sqrt[3]{x}$, $a = -8$

9. Find the linear approximation of the function $f(x) = \sqrt{1 - x}$ at $a = 0$ and use it to approximate the numbers $\sqrt{0.9}$ and $\sqrt{0.99}$. Illustrate by graphing f and the tangent line.

10. Find the linear approximation of the function $g(x) = \sqrt[3]{1 + x}$ at $a = 0$ and use it to approximate the numbers $\sqrt[3]{0.95}$ and $\sqrt[3]{1.1}$. Illustrate by graphing g and the tangent line.

11–14 Verify the given linear approximation at $a = 0$. Then determine the values of x for which the linear approximation is accurate to within 0.1.

11. $\sqrt[3]{1 - x} \approx 1 - \frac{1}{3}x$

12. $\tan x \approx x$

13. $1/(1 + 2x)^4 \approx 1 - 8x$

14. $e^x \approx 1 + x$

15–20 Find the differential of the function.

15. $y = x^4 + 5x$

16. $y = \cos \pi x$

17. $y = x \ln x$

18. $y = \sqrt{1 + t^2}$

19. $y = \dfrac{u + 1}{u - 1}$

20. $y = (1 + 2r)^{-4}$

21–26 (a) Find the differential dy and (b) evaluate dy for the given values of x and dx.

21. $y = x^2 + 2x, \quad x = 3, \quad dx = \frac{1}{2}$

22. $y = e^{x/4}, \quad x = 0, \quad dx = 0.1$

23. $y = \sqrt{4 + 5x}, \quad x = 0, \quad dx = 0.04$

24. $y = 1/(x + 1), \quad x = 1, \quad dx = -0.01$

25. $y = \tan x, \quad x = \pi/4, \quad dx = -0.1$

26. $y = \cos x, \quad x = \pi/3, \quad dx = 0.05$

27–30 Compute Δy and dy for the given values of x and $dx = \Delta x$. Then sketch a diagram like Figure 6 showing the line segments with lengths dx, dy, and Δy.

27. $y = x^2, \quad x = 1, \quad \Delta x = 0.5$

28. $y = \sqrt{x}, \quad x = 1, \quad \Delta x = 1$

29. $y = 6 - x^2, \quad x = -2, \quad \Delta x = 0.4$

30. $y = 16/x, \quad x = 4, \quad \Delta x = -1$

31–36 Use differentials (or, equivalently, a linear approximation) to estimate the given number.

31. $(2.001)^5$

32. $\sqrt{99.8}$

33. $(8.06)^{2/3}$

34. $1/1002$

35. $\tan 44°$

36. $\ln 1.07$

37–39 Explain, in terms of linear approximations or differentials, why the approximation is reasonable.

37. $\sec 0.08 \approx 1$

38. $(1.01)^6 \approx 1.06$

39. $\ln 1.05 \approx 0.05$

40. Let $\quad f(x) = (x - 1)^2 \qquad g(x) = e^{-2x}$

and $\quad h(x) = 1 + \ln(1 - 2x)$

(a) Find the linearizations of f, g, and h at $a = 0$. What do you notice? How do you explain what happened?

(b) Graph f, g, and h and their linear approximation. For which function is the linear approximation best? For which is it worst? Explain.

41. The edge of a cube was found to be 30 cm with a possible error in measurement of 0.1 cm. Use differentials to estimate the maximum possible error, relative error, and percentage error in computing (a) the volume of the cube and (b) the surface area of the cube.

42. The radius of a circular disk is given as 24 cm with a maximum error in measurement of 0.2 cm.
(a) Use differentials to estimate the maximum error in the calculated area of the disk.
(b) What is the relative error? What is the percentage error?

43. The circumference of a sphere was measured to be 84 cm with a possible error of 0.5 cm.
(a) Use differentials to estimate the maximum error in the calculated surface area. What is the relative error?
(b) Use differentials to estimate the maximum error in the calculated volume. What is the relative error?

44. Use differentials to estimate the amount of paint needed to apply a coat of paint 0.05 cm thick to a hemispherical dome with diameter 50 m.

45. (a) Use differentials to find a formula for the approximate volume of a thin cylindrical shell with height h, inner radius r, and thickness Δr.
(b) What is the error involved in using the formula from part (a)?

46. When blood flows along a blood vessel, the flux F (the volume of blood per unit time that flows past a given point) is proportional to the fourth power of the radius R of the blood vessel:

$$F = kR^4$$

(This is known as Poiseuille's Law; we will show why it is true in Section 8.4.) A partially clogged artery can be expanded by an operation called angioplasty, in which a balloon-tipped catheter is inflated inside the artery in order to widen it and restore the normal blood flow.

Show that the relative change in F is about four times the relative change in R. How will a 5% increase in the radius affect the flow of blood?

47. Establish the following rules for working with differentials (where c denotes a constant and u and v are functions of x).
(a) $dc = 0$
(b) $d(cu) = c\,du$
(c) $d(u + v) = du + dv$
(d) $d(uv) = u\,dv + v\,du$
(e) $d\left(\dfrac{u}{v}\right) = \dfrac{v\,du - u\,dv}{v^2}$
(f) $d(x^n) = nx^{n-1}\,dx$

48. On page 431 of *Physics: Calculus,* 2d ed., by Eugene Hecht (Pacific Grove, CA: Brooks/Cole, 2000), in the course of deriving the formula $T = 2\pi\sqrt{L/g}$ for the period of a pendulum of length L, the author obtains the equation $a_T = -g\sin\theta$ for the tangential acceleration of the bob of the pendulum. He then says, "for small angles, the value of θ in radians is very nearly the value of $\sin\theta$; they differ by less than 2% out to about 20°."

(a) Verify the linear approximation at 0 for the sine function:

$$\sin x \approx x$$

(b) Use a graphing device to determine the values of x for which $\sin x$ and x differ by less than 2%. Then verify Hecht's statement by converting from radians to degrees.

49. Suppose that the only information we have about a function f is that $f(1) = 5$ and the graph of its *derivative* is as shown.

(a) Use a linear approximation to estimate $f(0.9)$ and $f(1.1)$.

(b) Are your estimates in part (a) too large or too small? Explain.

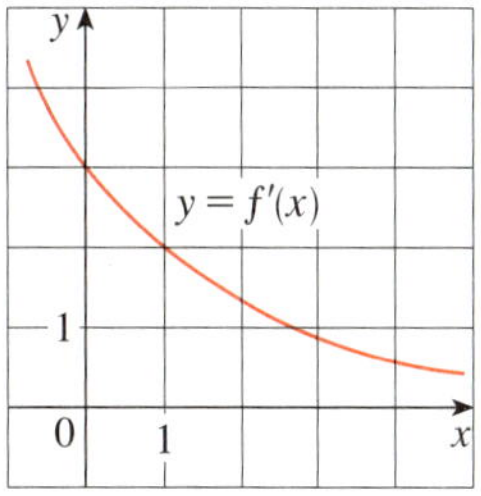

50. Suppose that we don't have a formula for $g(x)$ but we know that $g(2) = -4$ and $g'(x) = \sqrt{x^2 + 5}$ for all x.

(a) Use a linear approximation to estimate $g(1.95)$ and $g(2.05)$.

(b) Are your estimates in part (a) too large or too small? Explain.

LABORATORY PROJECT

Taylor Polynomials

The tangent line approximation $L(x)$ is the best first-degree (linear) approximation to $f(x)$ near $x = a$ because $f(x)$ and $L(x)$ have the same rate of change (derivative) at a. For a better approximation than a linear one, let's try a second-degree (quadratic) approximation $P(x)$. In other words, we approximate a curve by a parabola instead of by a straight line. To make sure that the approximation is a good one, we stipulate the following:

(i) $P(a) = f(a)$ (P and f should have the same value at a.)

(ii) $P'(a) = f'(a)$ (P and f should have the same rate of change at a.)

(iii) $P''(a) = f''(a)$ (The slopes of P and f should change at the same rate.)

1. Find the quadratic approximation $P(x) = A + Bx + Cx^2$ to the function $f(x) = \cos x$ that satisfies conditions (i), (ii), and (iii) with $a = 0$. Graph P, f, and the linear approximation $L(x) = 1$ on a common screen. Comment on how well the functions P and L approximate f.

2. Determine the values of x for which the quadratic approximation $f(x) = P(x)$ in Problem 1 is accurate to within 0.1. [*Hint:* Graph $y = P(x)$, $y = \cos x - 0.1$, and $y = \cos x + 0.1$ on a common screen.]

3. To approximate a function f by a quadratic function P near a number a, it is best to write P in the form

$$P(x) = A + B(x - a) + C(x - a)^2$$

Show that the quadratic function that satisfies conditions (i), (ii), and (iii) is

$$P(x) = f(a) + f'(a)(x - a) + \tfrac{1}{2}f''(a)(x - a)^2$$

4. Find the quadratic approximation to $f(x) = \sqrt{x + 3}$ near $a = 1$. Graph f, the quadratic approximation, and the linear approximation from Example 3 in Section 3.11 on a common screen. What do you conclude?

5. Instead of being satisfied with a linear or quadratic approximation to $f(x)$ near $x = a$, let's try to find better approximations with higher-degree polynomials. We look for an nth-degree polynomial

$$T_n(x) = c_0 + c_1(x - a) + c_2(x - a)^2 + c_3(x - a)^3 + \cdots + c_n(x - a)^n$$

such that T_n and its first n derivatives have the same values at $x = a$ as f and its first n derivatives. By differentiating repeatedly and setting $x = a$, show that these conditions are satisfied if $c_0 = f(a)$, $c_1 = f'(a)$, $c_2 = \frac{1}{2}f''(a)$, and in general

$$c_k = \frac{f^{(k)}(a)}{k!}$$

where $k! = 1 \cdot 2 \cdot 3 \cdot 4 \cdot \cdots \cdot k$. The resulting polynomial

$$T_n(x) = f(a) + f'(a)(x-a) + \frac{f''(a)}{2!}(x-a)^2 + \cdots + \frac{f^{(n)}(a)}{n!}(x-a)^n$$

is called the **nth-degree Taylor polynomial of f centered at a.**

6. Find the 8th-degree Taylor polynomial centered at $a = 0$ for the function $f(x) = \cos x$. Graph f together with the Taylor polynomials T_2, T_4, T_6, T_8 in the viewing rectangle $[-5, 5]$ by $[-1.4, 1.4]$ and comment on how well they approximate f.

3 Review

CONCEPT CHECK

1. State each of the following differentiation rules both in symbols and in words.
(a) The Power Rule
(b) The Constant Multiple Rule
(c) The Sum Rule
(d) The Difference Rule
(e) The Product Rule
(f) The Quotient Rule
(g) The Chain Rule

2. State the derivative of each function.

(a) $y = x^n$	(b) $y = e^x$	(c) $y = a^x$
(d) $y = \ln x$	(e) $y = \log_a x$	(f) $y = \sin x$
(g) $y = \cos x$	(h) $y = \tan x$	(i) $y = \csc x$
(j) $y = \sec x$	(k) $y = \cot x$	(l) $y = \sin^{-1}x$
(m) $y = \cos^{-1}x$	(n) $y = \tan^{-1}x$	(o) $y = \sinh x$
(p) $y = \cosh x$	(q) $y = \tanh x$	(r) $y = \sinh^{-1}x$
(s) $y = \cosh^{-1}x$	(t) $y = \tanh^{-1}x$	

3. (a) How is the number e defined?
(b) Express e as a limit.
(c) Why is the natural exponential function $y = e^x$ used more often in calculus than the other exponential functions $y = a^x$?
(d) Why is the natural logarithmic function $y = \ln x$ used more often in calculus than the other logarithmic functions $y = \log_a x$?

4. (a) Explain how implicit differentiation works.
(b) Explain how logarithmic differentiation works.

5. What are the second and third derivatives of a function f? If f is the position function of an object, how can you interpret f'' and f'''?

6. (a) Write an expression for the linearization of f at a.
(b) If $y = f(x)$, write an expression for the differential dy.
(c) If $dx = \Delta x$, draw a picture showing the geometric meanings of Δy and dy.

TRUE-FALSE QUIZ

Determine whether the statement is true or false. If it is true, explain why. If it is false, explain why or give an example that disproves the statement.

1. If f and g are differentiable, then

$$\frac{d}{dx}[f(x) + g(x)] = f'(x) + g'(x)$$

2. If f and g are differentiable, then

$$\frac{d}{dx}[f(x)g(x)] = f'(x)g'(x)$$

3. If f and g are differentiable, then

$$\frac{d}{dx}[f(g(x))] = f'(g(x))g'(x)$$

4. If f is differentiable, then $\dfrac{d}{dx}\sqrt{f(x)} = \dfrac{f'(x)}{2\sqrt{f(x)}}$.

5. If f is differentiable, then $\dfrac{d}{dx}f(\sqrt{x}) = \dfrac{f'(x)}{2\sqrt{x}}$.

6. If $y = e^2$, then $y' = 2e$.

7. $\dfrac{d}{dx}(10^x) = x10^{x-1}$

8. $\dfrac{d}{dx}(\ln 10) = \dfrac{1}{10}$

9. $\dfrac{d}{dx}(\tan^2 x) = \dfrac{d}{dx}(\sec^2 x)$

10. $\dfrac{d}{dx}|x^2 + x| = |2x + 1|$

11. If $g(x) = x^5$, then $\lim_{x \to 2} \dfrac{g(x) - g(2)}{x - 2} = 80$.

12. $\dfrac{d^2y}{dx^2} = \left(\dfrac{dy}{dx}\right)^2$

13. An equation of the tangent line to the parabola $y = x^2$ at $(-2, 4)$ is $y - 4 = 2x(x + 2)$.

EXERCISES

1–48 |||| Calculate y'.

1. $y = (x^4 - 3x^2 + 5)^3$
2. $y = \cos(\tan x)$
3. $y = \sqrt{x} + \dfrac{1}{\sqrt[3]{x^4}}$
4. $y = \dfrac{3x - 2}{\sqrt{2x + 1}}$
5. $y = 2x\sqrt{x^2 + 1}$
6. $y = \dfrac{e^x}{1 + x^2}$
7. $y = e^{\sin 2\theta}$
8. $y = e^{-t}(t^2 - 2t + 2)$
9. $y = \dfrac{t}{1 - t^2}$
10. $y = \sin^{-1}(e^x)$
11. $y = xe^{-1/x}$
12. $y = x^r e^{sx}$
13. $y = \tan\sqrt{1 - x}$
14. $y = \dfrac{1}{\sin(x - \sin x)}$
15. $xy^4 + x^2y = x + 3y$
16. $y = \ln(\csc 5x)$
17. $y = \dfrac{\sec 2\theta}{1 + \tan 2\theta}$
18. $x^2 \cos y + \sin 2y = xy$
19. $y = e^{cx}(c \sin x - \cos x)$
20. $y = \ln(x^2 e^x)$
21. $y = e^{e^x}$
22. $y = \sec(1 + x^2)$
23. $y = (1 - x^{-1})^{-1}$
24. $y = 1/\sqrt[3]{x + \sqrt{x}}$
25. $\sin(xy) = x^2 - y$
26. $y = \sqrt{\sin\sqrt{x}}$
27. $y = \log_5(1 + 2x)$
28. $y = (\cos x)^x$
29. $y = \ln \sin x - \frac{1}{2}\sin^2 x$
30. $y = \dfrac{(x^2 + 1)^4}{(2x + 1)^3(3x - 1)^5}$
31. $y = x \tan^{-1}(4x)$
32. $y = e^{\cos x} + \cos(e^x)$
33. $y = \ln|\sec 5x + \tan 5x|$
34. $y = 10^{\tan \pi\theta}$
35. $y = \cot(3x^2 + 5)$
36. $y = \sqrt{t \ln(t^4)}$
37. $y = \sin\left(\tan\sqrt{1 + x^3}\right)$
38. $y = \arctan\left(\arcsin\sqrt{x}\right)$
39. $y = \tan^2(\sin\theta)$
40. $xe^y = y - 1$
41. $y = \dfrac{\sqrt{x + 1}\,(2 - x)^5}{(x + 3)^7}$
42. $y = \dfrac{(x + \lambda)^4}{x^4 + \lambda^4}$
43. $y = x \sinh(x^2)$
44. $y = \dfrac{\sin mx}{x}$
45. $y = \ln(\cosh 3x)$
46. $y = \ln\left|\dfrac{x^2 - 4}{2x + 5}\right|$
47. $y = \cosh^{-1}(\sinh x)$
48. $y = x \tanh^{-1}\sqrt{x}$

49. If $f(t) = \sqrt{4t + 1}$, find $f''(2)$.
50. If $g(\theta) = \theta \sin\theta$, find $g''(\pi/6)$.
51. Find y'' if $x^6 + y^6 = 1$.
52. Find $f^{(n)}(x)$ if $f(x) = 1/(2 - x)$.
53. Use mathematical induction to show that if $f(x) = xe^x$, then $f^{(n)}(x) = (x + n)e^x$.
54. Evaluate $\lim_{t \to 0} \dfrac{t^3}{\tan^3(2t)}$.

55–59 |||| Find an equation of the tangent to the curve at the given point.

55. $y = 4\sin^2 x$, $(\pi/6, 1)$
56. $y = \dfrac{x^2 - 1}{x^2 + 1}$, $(0, -1)$
57. $y = \sqrt{1 + 4\sin x}$, $(0, 1)$
58. $x^2 + 4xy + y^2 = 13$, $(2, 1)$
59. $y = (2 + x)e^{-x}$, $(0, 2)$

60. If $f(x) = xe^{\sin x}$, find $f'(x)$. Graph f and f' on the same screen and comment.

61. (a) If $f(x) = x\sqrt{5 - x}$, find $f'(x)$.
    (b) Find equations of the tangent lines to the curve $y = x\sqrt{5 - x}$ at the points $(1, 2)$ and $(4, 4)$.
    (c) Illustrate part (b) by graphing the curve and tangent lines on the same screen.
    (d) Check to see that your answer to part (a) is reasonable by comparing the graphs of f and f'.

62. (a) If $f(x) = 4x - \tan x$, $-\pi/2 < x < \pi/2$, find f' and f''.
    (b) Check to see that your answers to part (a) are reasonable by comparing the graphs of f, f', and f''.

63. At what points on the curve $y = \sin x + \cos x$, $0 \leqslant x \leqslant 2\pi$, is the tangent line horizontal?

64. Find the points on the ellipse $x^2 + 2y^2 = 1$ where the tangent line has slope 1.

65. If $f(x) = (x - a)(x - b)(x - c)$, show that

$$\frac{f'(x)}{f(x)} = \frac{1}{x - a} + \frac{1}{x - b} + \frac{1}{x - c}$$

66. (a) By differentiating the double-angle formula

$$\cos 2x = \cos^2 x - \sin^2 x$$

obtain the double-angle formula for the sine function.

(b) By differentiating the addition formula

$$\sin(x + a) = \sin x \cos a + \cos x \sin a$$

obtain the addition formula for the cosine function.

67. Suppose that $h(x) = f(x)g(x)$ and $F(x) = f(g(x))$, where $f(2) = 3$, $g(2) = 5$, $g'(2) = 4$, $f'(2) = -2$, and $f'(5) = 11$. Find (a) $h'(2)$ and (b) $F'(2)$.

68. If f and g are the functions whose graphs are shown, let $P(x) = f(x)g(x)$, $Q(x) = f(x)/g(x)$, and $C(x) = f(g(x))$. Find (a) $P'(2)$, (b) $Q'(2)$, and (c) $C'(2)$.

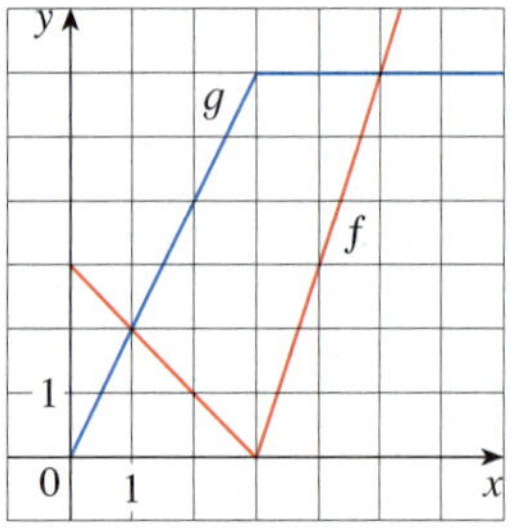

69–76 ||| Find f' in terms of g'.

69. $f(x) = x^2 g(x)$

70. $f(x) = g(x^2)$

71. $f(x) = [g(x)]^2$

72. $f(x) = g(g(x))$

73. $f(x) = g(e^x)$

74. $f(x) = e^{g(x)}$

75. $f(x) = \ln |g(x)|$

76. $f(x) = g(\ln x)$

77–79 ||| Find h' in terms of f' and g'.

77. $h(x) = \dfrac{f(x)g(x)}{f(x) + g(x)}$

78. $h(x) = \sqrt{\dfrac{f(x)}{g(x)}}$

79. $h(x) = f(g(\sin 4x))$

80. (a) Graph the function $f(x) = x - 2 \sin x$ in the viewing rectangle $[0, 8]$ by $[-2, 8]$.

(b) On which interval is the average rate of change larger: $[1, 2]$ or $[2, 3]$?

(c) At which value of x is the instantaneous rate of change larger: $x = 2$ or $x = 5$?

(d) Check your visual estimates in part (c) by computing $f'(x)$ and comparing the numerical values of $f'(2)$ and $f'(5)$.

81. At what point on the curve $y = [\ln(x + 4)]^2$ is the tangent horizontal?

82. (a) Find an equation of the tangent to the curve $y = e^x$ that is parallel to the line $x - 4y = 1$.

(b) Find an equation of the tangent to the curve $y = e^x$ that passes through the origin.

83. Find a parabola $y = ax^2 + bx + c$ that passes through the point $(1, 4)$ and whose tangent lines at $x = -1$ and $x = 5$ have slopes 6 and -2, respectively.

84. The function $C(t) = K(e^{-at} - e^{-bt})$, where a, b, and K are positive constants and $b > a$, is used to model the concentration at time t of a drug injected into the bloodstream.

(a) Show that $\lim_{t \to \infty} C(t) = 0$.

(b) Find $C'(t)$, the rate at which the drug is cleared from circulation.

(c) When is this rate equal to 0?

85. An equation of motion of the form $s = Ae^{-ct} \cos(\omega t + \delta)$ represents damped oscillation of an object. Find the velocity and acceleration of the object.

86. A particle moves along a horizontal line so that its coordinate at time t is $x = \sqrt{b^2 + c^2 t^2}$, $t \geqslant 0$, where b and c are positive constants.

(a) Find the velocity and acceleration functions.

(b) Show that the particle always moves in the positive direction.

87. A particle moves on a vertical line so that its coordinate at time t is $y = t^3 - 12t + 3$, $t \geqslant 0$.

(a) Find the velocity and acceleration functions.

(b) When is the particle moving upward and when is it moving downward?

(c) Find the distance that the particle travels in the time interval $0 \leqslant t \leqslant 3$.

88. The volume of a right circular cone is $V = \pi r^2 h/3$, where r is the radius of the base and h is the height.

(a) Find the rate of change of the volume with respect to the height if the radius is constant.

(b) Find the rate of change of the volume with respect to the radius if the height is constant.

89. The mass of part of a wire is $x(1 + \sqrt{x})$ kilograms, where x is measured in meters from one end of the wire. Find the linear density of the wire when $x = 4$ m.

90. The cost, in dollars, of producing x units of a certain commodity is

$$C(x) = 920 + 2x - 0.02x^2 + 0.00007x^3$$

(a) Find the marginal cost function.

(b) Find $C'(100)$ and explain its meaning.

(c) Compare $C'(100)$ with the cost of producing the 101st item.

91. The volume of a cube is increasing at a rate of 10 cm^3/min. How fast is the surface area increasing when the length of an edge is 30 cm?

92. A paper cup has the shape of a cone with height 10 cm and radius 3 cm (at the top). If water is poured into the cup at a rate of 2 cm^3/s, how fast is the water level rising when the water is 5 cm deep?

93. A balloon is rising at a constant speed of 5 ft/s. A boy is cycling along a straight road at a speed of 15 ft/s. When he passes under the balloon, it is 45 ft above him. How fast is the distance between the boy and the balloon increasing 3 s later?

94. A waterskier skis over the ramp shown in the figure at a speed of 30 ft/s. How fast is she rising as she leaves the ramp?

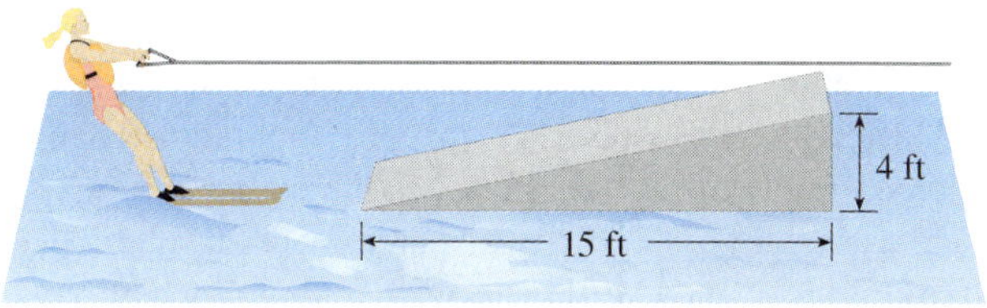

95. The angle of elevation of the Sun is decreasing at a rate of 0.25 rad/h. How fast is the shadow cast by a 400-ft-tall building increasing when the angle of elevation of the Sun is $\pi/6$?

96. (a) Find the linear approximation to $f(x) = \sqrt{25 - x^2}$ near 3.

(b) Illustrate part (a) by graphing f and the linear approximation.

(c) For what values of x is the linear approximation accurate to within 0.1?

97. (a) Find the linearization of $f(x) = \sqrt[3]{1 + 3x}$ at $a = 0$. State the corresponding linear approximation and use it to give an approximate value for $\sqrt[3]{1.03}$.

(b) Determine the values of x for which the linear approximation given in part (a) is accurate to within 0.1.

98. Evaluate dy if $y = x^3 - 2x^2 + 1$, $x = 2$, and $dx = 0.2$.

99. A window has the shape of a square surmounted by a semicircle. The base of the window is measured as having width 60 cm with a possible error in measurement of 0.1 cm. Use differentials to estimate the maximum error possible in computing the area of the window.

100–102 Express the limit as a derivative and evaluate.

100. $\displaystyle\lim_{x \to 1} \frac{x^{17} - 1}{x - 1}$

101. $\displaystyle\lim_{h \to 0} \frac{\sqrt[4]{16 + h} - 2}{h}$

102. $\displaystyle\lim_{\theta \to \pi/3} \frac{\cos\theta - 0.5}{\theta - \pi/3}$

103. Evaluate $\displaystyle\lim_{x \to 0} \frac{\sqrt{1 + \tan x} - \sqrt{1 + \sin x}}{x^3}$.

104. Suppose f is a differentiable function such that $f(g(x)) = x$ and $f'(x) = 1 + [f(x)]^2$. Show that $g'(x) = 1/(1 + x^2)$.

105. Find $f'(x)$ if it is known that

$$\frac{d}{dx}[f(2x)] = x^2$$

106. Show that the length of the portion of any tangent line to the astroid $x^{2/3} + y^{2/3} = a^{2/3}$ cut off by the coordinate axes is constant.

PROBLEMS PLUS

Before you look at the example, cover up the solution and try it yourself first.

EXAMPLE 1 How many lines are tangent to both of the parabolas $y = -1 - x^2$ and $y = 1 + x^2$? Find the coordinates of the points at which these tangents touch the parabolas.

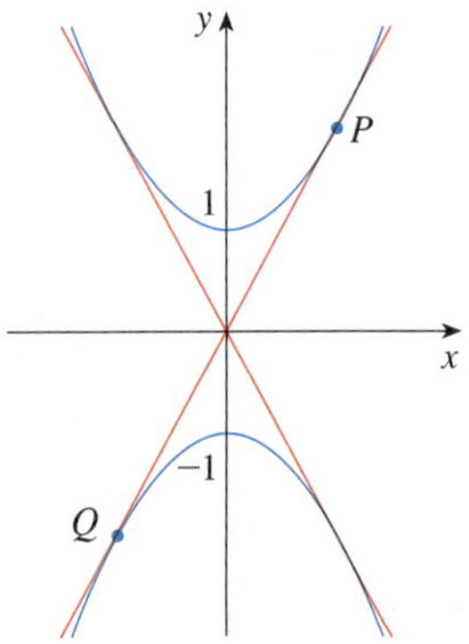

FIGURE 1

SOLUTION To gain insight into this problem, it is essential to draw a diagram. So we sketch the parabolas $y = 1 + x^2$ (which is the standard parabola $y = x^2$ shifted 1 unit upward) and $y = -1 - x^2$ (which is obtained by reflecting the first parabola about the x-axis). If we try to draw a line tangent to both parabolas, we soon discover that there are only two possibilities, as illustrated in Figure 1.

Let P be a point at which one of these tangents touches the upper parabola and let a be its x-coordinate. (The choice of notation for the unknown is important. Of course we could have used b or c or x_0 or x_1 instead of a. However, it's not advisable to use x in place of a because that x could be confused with the variable x in the equation of the parabola.) Then, since P lies on the parabola $y = 1 + x^2$, its y-coordinate must be $1 + a^2$. Because of the symmetry shown in Figure 1, the coordinates of the point Q where the tangent touches the lower parabola must be $(-a, -(1 + a^2))$.

To use the given information that the line is a tangent, we equate the slope of the line PQ to the slope of the tangent line at P. We have

$$m_{PQ} = \frac{1 + a^2 - (-1 - a^2)}{a - (-a)} = \frac{1 + a^2}{a}$$

If $f(x) = 1 + x^2$, then the slope of the tangent line at P is $f'(a) = 2a$. Thus, the condition that we need to use is that

$$\frac{1 + a^2}{a} = 2a$$

Solving this equation, we get $1 + a^2 = 2a^2$, so $a^2 = 1$ and $a = \pm 1$. Therefore, the points are $(1, 2)$ and $(-1, -2)$. By symmetry, the two remaining points are $(-1, 2)$ and $(1, -2)$.

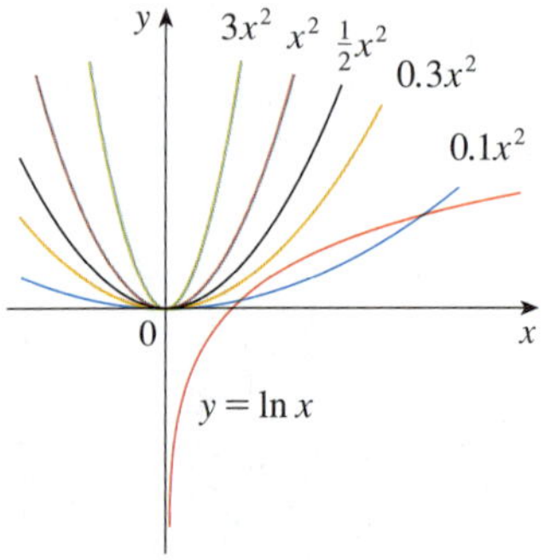

FIGURE 2

EXAMPLE 2 For what values of c does the equation $\ln x = cx^2$ have exactly one solution?

SOLUTION One of the most important principles of problem solving is to draw a diagram, even if the problem as stated doesn't explicitly mention a geometric situation. Our present problem can be reformulated geometrically as follows: For what values of c does the curve $y = \ln x$ intersect the curve $y = cx^2$ in exactly one point?

Let's start by graphing $y = \ln x$ and $y = cx^2$ for various values of c. We know that, for $c \neq 0$, $y = cx^2$ is a parabola that opens upward if $c > 0$ and downward if $c < 0$. Figure 2 shows the parabolas $y = cx^2$ for several positive values of c. Most of them don't intersect $y = \ln x$ at all and one intersects twice. We have the feeling that there must be a value of c (somewhere between 0.1 and 0.3) for which the curves intersect exactly once, as in Figure 3.

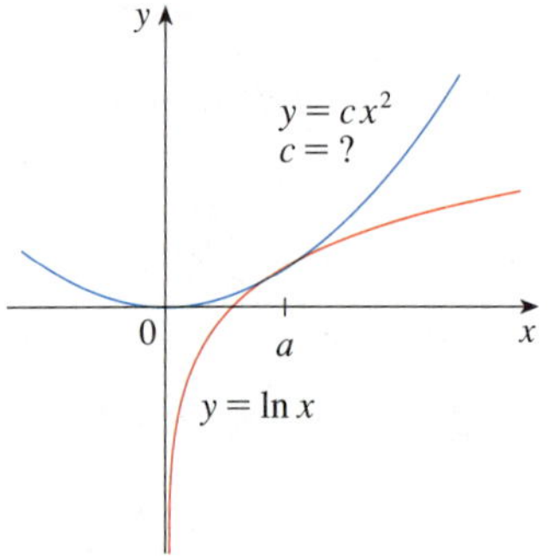

FIGURE 3

To find that particular value of c, we let a be the x-coordinate of the single point of intersection. In other words, $\ln a = ca^2$, so a is the unique solution of the given equation. We see from Figure 3 that the curves just touch, so they have a common tangent line when $x = a$. That means the curves $y = \ln x$ and $y = cx^2$ have the same slope when $x = a$. Therefore

$$\frac{1}{a} = 2ca$$

Solving the equations $\ln a = ca^2$ and $1/a = 2ca$, we get

$$\ln a = ca^2 = c \cdot \frac{1}{2c} = \frac{1}{2}$$

Thus, $a = e^{1/2}$ and

$$c = \frac{\ln a}{a^2} = \frac{\ln e^{1/2}}{e} = \frac{1}{2e}$$

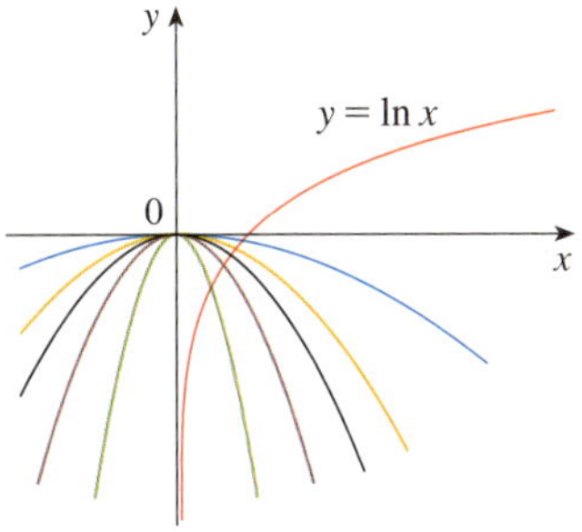

FIGURE 4

For negative values of c we have the situation illustrated in Figure 4: All parabolas $y = cx^2$ with negative values of c intersect $y = \ln x$ exactly once. And let's not forget about $c = 0$: The curve $y = 0x^2 = 0$ is just the x-axis, which intersects $y = \ln x$ exactly once.

To summarize, the required values of c are $c = 1/(2e)$ and $c \leq 0$.

PROBLEMS

1. Find points P and Q on the parabola $y = 1 - x^2$ so that the triangle ABC formed by the x-axis and the tangent lines at P and Q is an equilateral triangle.

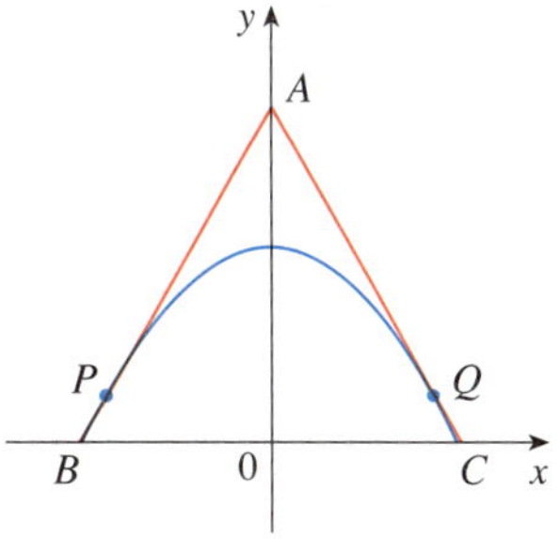

2. Find the point where the curves $y = x^3 - 3x + 4$ and $y = 3(x^2 - x)$ are tangent to each other, that is, have a common tangent line. Illustrate by sketching both curves and the common tangent.

3. Show that $\sin^{-1}(\tanh x) = \tan^{-1}(\sinh x)$.

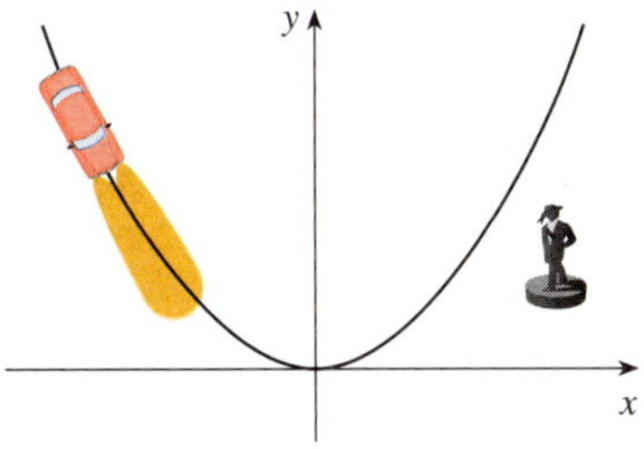

FIGURE FOR PROBLEM 4

4. A car is traveling at night along a highway shaped like a parabola with its vertex at the origin (see the figure). The car starts at a point 100 m west and 100 m north of the origin and travels in an easterly direction. There is a statue located 100 m east and 50 m north of the origin. At what point on the highway will the car's headlights illuminate the statue?

5. Prove that $\dfrac{d^n}{dx^n}(\sin^4 x + \cos^4 x) = 4^{n-1}\cos(4x + n\pi/2)$.

6. Find the nth derivative of the function $f(x) = x^n/(1 - x)$.

7. The figure shows a circle with radius 1 inscribed in the parabola $y = x^2$. Find the center of the circle.

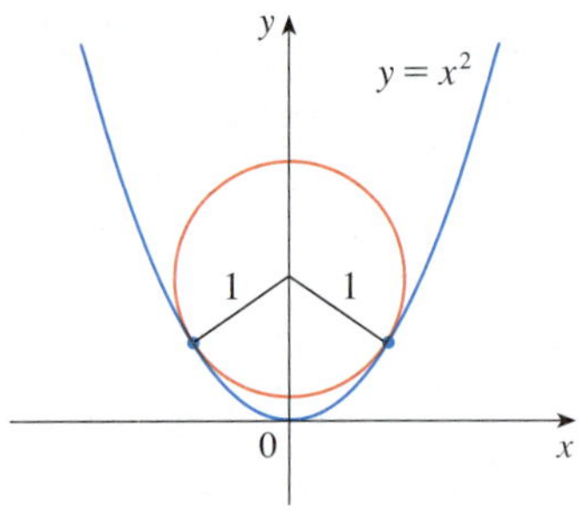

FIGURE FOR PROBLEM 7

8. If f is differentiable at a, where $a > 0$, evaluate the following limit in terms of $f'(a)$:

$$\lim_{x \to a} \frac{f(x) - f(a)}{\sqrt{x} - \sqrt{a}}$$

9. The figure shows a rotating wheel with radius 40 cm and a connecting rod AP with length 1.2 m. The pin P slides back and forth along the x-axis as the wheel rotates counterclockwise at a rate of 360 revolutions per minute.

(a) Find the angular velocity of the connecting rod, $d\alpha/dt$, in radians per second, when $\theta = \pi/3$.

(b) Express the distance $x = |OP|$ in terms of θ.

(c) Find an expression for the velocity of the pin P in terms of θ.

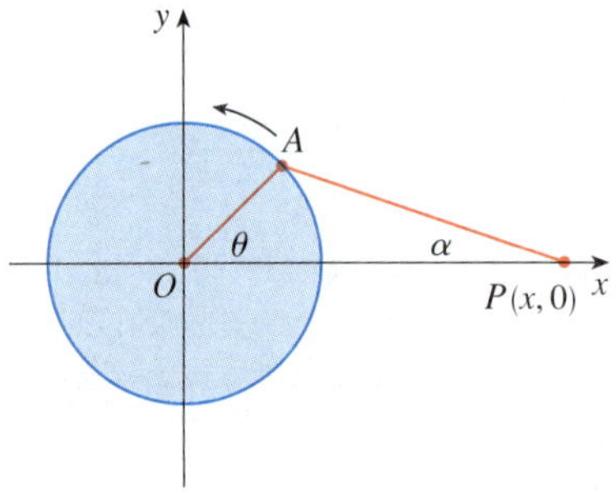

FIGURE FOR PROBLEM 9

10. Tangent lines T_1 and T_2 are drawn at two points P_1 and P_2 on the parabola $y = x^2$ and they intersect at a point P. Another tangent line T is drawn at a point between P_1 and P_2; it intersects T_1 at Q_1 and T_2 at Q_2. Show that

$$\frac{|PQ_1|}{|PP_1|} + \frac{|PQ_2|}{|PP_2|} = 1$$

11. Show that

$$\frac{d^n}{dx^n}(e^{ax} \sin bx) = r^n e^{ax} \sin(bx + n\theta)$$

where a and b are positive numbers, $r^2 = a^2 + b^2$, and $\theta = \tan^{-1}(b/a)$.

12. Evaluate $\displaystyle \lim_{x \to \pi} \frac{e^{\sin x} - 1}{x - \pi}$.

13. Let T and N be the tangent and normal lines to the ellipse $x^2/9 + y^2/4 = 1$ at any point P on the ellipse in the first quadrant. Let x_T and y_T be the x- and y-intercepts of T and x_N and y_N be the intercepts of N. As P moves along the ellipse in the first quadrant (but not on the axes), what values can x_T, y_T, x_N, and y_N take on? First try to guess the answers just by looking at the figure. Then use calculus to solve the problem and see how good your intuition is.

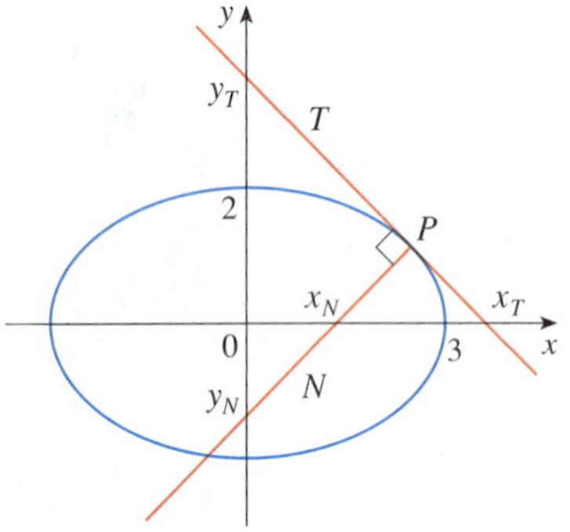

FIGURE FOR PROBLEM 13

14. Evaluate $\displaystyle \lim_{x \to 0} \frac{\sin(3 + x)^2 - \sin 9}{x}$.

15. (a) Use the identity for $\tan(x - y)$ (see Equation 14b in Appendix D) to show that if two lines L_1 and L_2 intersect at an angle α, then

$$\tan \alpha = \frac{m_2 - m_1}{1 + m_1 m_2}$$

where m_1 and m_2 are the slopes of L_1 and L_2, respectively.

(b) The **angle between the curves** C_1 and C_2 at a point of intersection P is defined to be the angle between the tangent lines to C_1 and C_2 at P (if these tangent lines exist). Use part (a) to find, correct to the nearest degree, the angle between each pair of curves at each point of intersection.

(i) $y = x^2$ and $y = (x - 2)^2$

(ii) $x^2 - y^2 = 3$ and $x^2 - 4x + y^2 + 3 = 0$

16. Let $P(x_1, y_1)$ be a point on the parabola $y^2 = 4px$ with focus $F(p, 0)$. Let α be the angle between the parabola and the line segment FP, and let β be the angle between the horizontal line $y = y_1$ and the parabola as in the figure. Prove that $\alpha = \beta$. (Thus, by a principle of geometrical optics, light from a source placed at F will be reflected along a line parallel to the x-axis. This explains why *paraboloids*, the surfaces obtained by rotating parabolas about their axes, are used as the shape of some automobile headlights and mirrors for telescopes.)

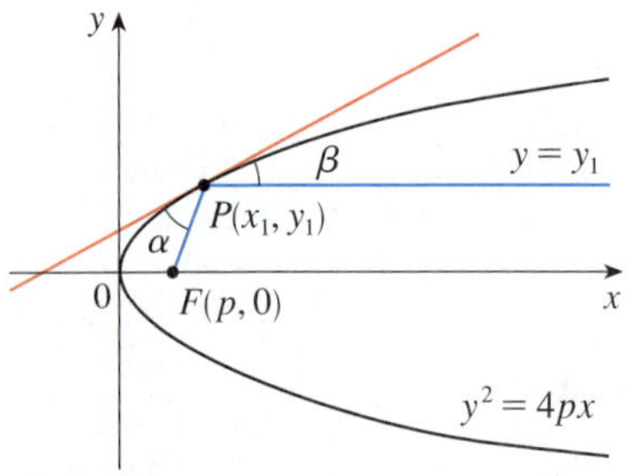

FIGURE FOR PROBLEM 16

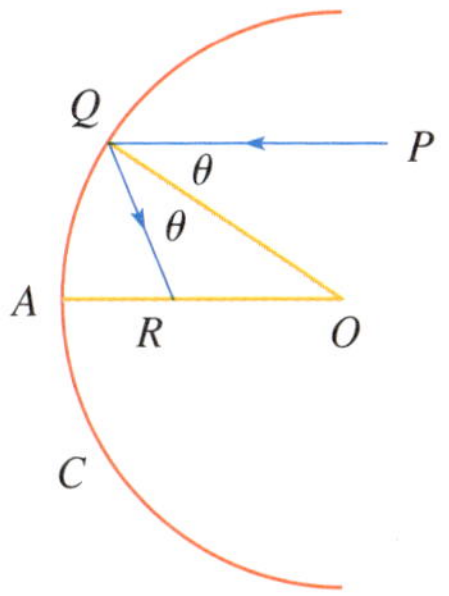

FIGURE FOR PROBLEM 17

17. Suppose that we replace the parabolic mirror of Problem 16 by a spherical mirror. Although the mirror has no focus, we can show the existence of an *approximate* focus. In the figure, C is a semicircle with center O. A ray of light coming in toward the mirror parallel to the axis along the line PQ will be reflected to the point R on the axis so that $\angle PQO = \angle OQR$ (the angle of incidence is equal to the angle of reflection). What happens to the point R as P is taken closer and closer to the axis?

18. If f and g are differentiable functions with $f(0) = g(0) = 0$ and $g'(0) \neq 0$, show that

$$\lim_{x \to 0} \frac{f(x)}{g(x)} = \frac{f'(0)}{g'(0)}$$

19. Evaluate $\displaystyle\lim_{x \to 0} \frac{\sin(a + 2x) - 2\sin(a + x) + \sin a}{x^2}$.

CAS **20.** (a) The cubic function $f(x) = x(x - 2)(x - 6)$ has three distinct zeros: 0, 2, and 6. Graph f and its tangent lines at the *average* of each pair of zeros. What do you notice?
(b) Suppose the cubic function $f(x) = (x - a)(x - b)(x - c)$ has three distinct zeros: a, b, and c. Prove, with the help of a computer algebra system, that a tangent line drawn at the average of the zeros a and b intersects the graph of f at the third zero.

21. For what value of k does the equation $e^{2x} = k\sqrt{x}$ have exactly one solution?

22. For which positive numbers a is it true that $a^x \geqslant 1 + x$ for all x?

23. If

$$y = \frac{x}{\sqrt{a^2 - 1}} - \frac{2}{\sqrt{a^2 - 1}} \arctan \frac{\sin x}{a + \sqrt{a^2 - 1} + \cos x}$$

show that $y' = \dfrac{1}{a + \cos x}$.

24. Given an ellipse $x^2/a^2 + y^2/b^2 = 1$, where $a \neq b$, find the equation of the set of all points from which there are two tangents to the curve whose slopes are (a) reciprocals and (b) negative reciprocals.

25. Find the two points on the curve $y = x^4 - 2x^2 - x$ that have a common tangent line.

26. Suppose that three points on the parabola $y = x^2$ have the property that their normal lines intersect at a common point. Show that the sum of their x-coordinates is 0.

27. A *lattice point* in the plane is a point with integer coordinates. Suppose that circles with radius r are drawn using all lattice points as centers. Find the smallest value of r such that any line with slope $\frac{2}{5}$ intersects some of these circles.

28. A cone of radius r centimeters and height h centimeters is lowered point first at a rate of 1 cm/s into a tall cylinder of radius R centimeters that is partially filled with water. How fast is the water level rising at the instant the cone is completely submerged?

29. A container in the shape of an inverted cone has height 16 cm and radius 5 cm at the top. It is partially filled with a liquid that oozes through the sides at a rate proportional to the area of the container that is in contact with the liquid. (The surface area of a cone is πrl, where r is the radius and l is the slant height.) If we pour the liquid into the container at a rate of 2 cm³/min, then the height of the liquid decreases at a rate of 0.3 cm/min when the height is 10 cm. If our goal is to keep the liquid at a constant height of 10 cm, at what rate should we pour the liquid into the container?

CHAPTER 4

Scientists have tried to explain how rainbows are formed since the time of Aristotle. In the project on page 288, you will be able to use the principles of differential calculus to explain the formation, location, and colors of the rainbow.

Applications of Differentiation

We have already investigated some of the applications of derivatives, but now that we know the differentiation rules we are in a better position to pursue the applications of differentiation in greater depth. Here we learn how derivatives affect the shape of a graph of a function and, in particular, how they help us locate maximum and minimum values of functions. Many practical problems require us to minimize a cost or maximize an area or somehow find the best possible outcome of a situation. In particular, we will be able to investigate the optimal shape of a can and to explain the location of rainbows in the sky.

4.1 Maximum and Minimum Values

Some of the most important applications of differential calculus are *optimization problems,* in which we are required to find the optimal (best) way of doing something. Here are examples of such problems that we will solve in this chapter:

- What is the shape of a can that minimizes manufacturing costs?
- What is the maximum acceleration of a space shuttle? (This is an important question to the astronauts who have to withstand the effects of acceleration.)
- What is the radius of a contracted windpipe that expels air most rapidly during a cough?
- At what angle should blood vessels branch so as to minimize the energy expended by the heart in pumping blood?

These problems can be reduced to finding the maximum or minimum values of a function. Let's first explain exactly what we mean by maximum and minimum values.

1 Definition A function f has an **absolute maximum** (or **global maximum**) at c if $f(c) \geqslant f(x)$ for all x in D, where D is the domain of f. The number $f(c)$ is called the **maximum value** of f on D. Similarly, f has an **absolute minimum** at c if $f(c) \leqslant f(x)$ for all x in D and the number $f(c)$ is called the **minimum value** of f on D. The maximum and minimum values of f are called the **extreme values** of f.

Figure 1 shows the graph of a function f with absolute maximum at d and absolute minimum at a. Note that $(d, f(d))$ is the highest point on the graph and $(a, f(a))$ is the lowest point.

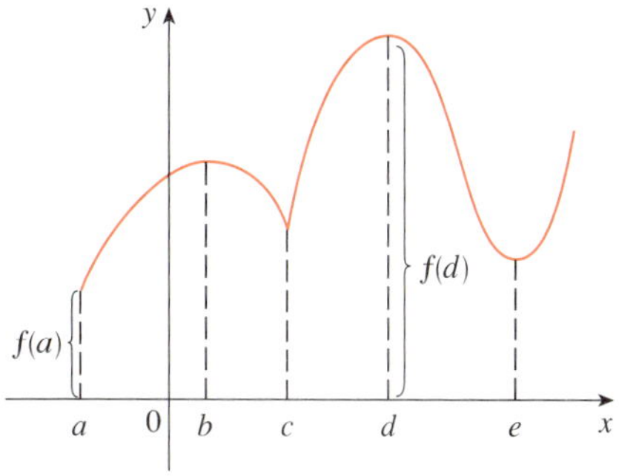

FIGURE 1
Minimum value $f(a)$, maximum value $f(d)$

In Figure 1, if we consider only values of x near b [for instance, if we restrict our attention to the interval (a, c)], then $f(b)$ is the largest of those values of $f(x)$ and is called a *local maximum value* of f. Likewise, $f(c)$ is called a *local minimum value* of f because $f(c) \leqslant f(x)$ for x near c [in the interval (b, d), for instance]. The function f also has a local minimum at e. In general, we have the following definition.

2 Definition A function f has a **local maximum** (or **relative maximum**) at c if $f(c) \geqslant f(x)$ when x is near c. [This means that $f(c) \geqslant f(x)$ for all x in some open interval containing c.] Similarly, f has a **local minimum** at c if $f(c) \leqslant f(x)$ when x is near c.

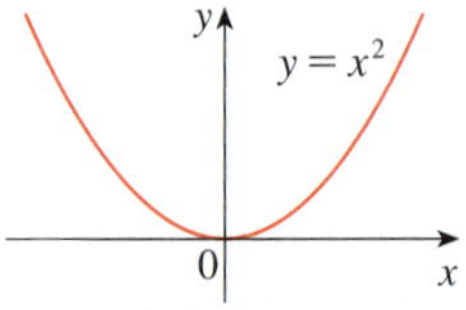

FIGURE 2
Minimum value 0, no maximum

EXAMPLE 1 The function $f(x) = \cos x$ takes on its (local and absolute) maximum value of 1 infinitely many times, since $\cos 2n\pi = 1$ for any integer n and $-1 \leqslant \cos x \leqslant 1$ for all x. Likewise, $\cos(2n + 1)\pi = -1$ is its minimum value, where n is any integer.

EXAMPLE 2 If $f(x) = x^2$, then $f(x) \geqslant f(0)$ because $x^2 \geqslant 0$ for all x. Therefore, $f(0) = 0$ is the absolute (and local) minimum value of f. This corresponds to the fact that the origin is the lowest point on the parabola $y = x^2$. (See Figure 2.) However, there is no highest point on the parabola and so this function has no maximum value.

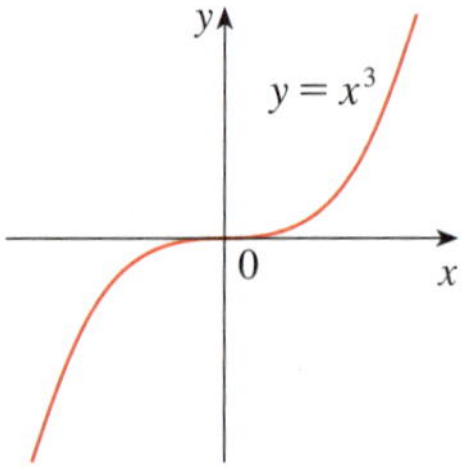

FIGURE 3
No minimum, no maximum

EXAMPLE 3 From the graph of the function $f(x) = x^3$, shown in Figure 3, we see that this function has neither an absolute maximum value nor an absolute minimum value. In fact, it has no local extreme values either.

EXAMPLE 4 The graph of the function

$$f(x) = 3x^4 - 16x^3 + 18x^2 \qquad -1 \leqslant x \leqslant 4$$

is shown in Figure 4. You can see that $f(1) = 5$ is a local maximum, whereas the absolute maximum is $f(-1) = 37$. (This absolute maximum is not a local maximum because it occurs at an endpoint.) Also, $f(0) = 0$ is a local minimum and $f(3) = -27$ is both a local and an absolute minimum. Note that f has neither a local nor an absolute maximum at $x = 4$.

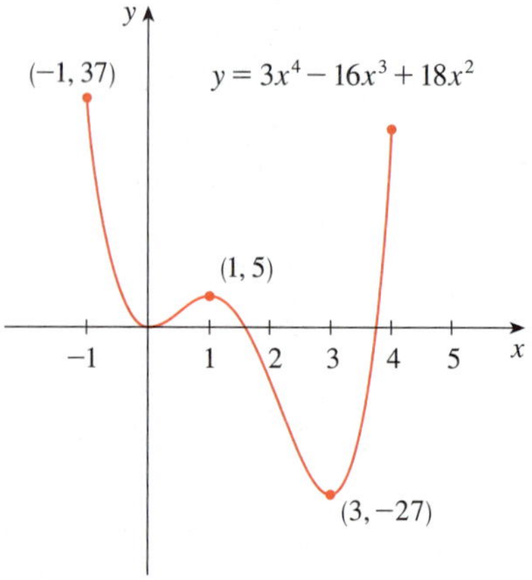

FIGURE 4

We have seen that some functions have extreme values, whereas others do not. The following theorem gives conditions under which a function is guaranteed to possess extreme values.

> **3 The Extreme Value Theorem** If f is continuous on a closed interval $[a, b]$, then f attains an absolute maximum value $f(c)$ and an absolute minimum value $f(d)$ at some numbers c and d in $[a, b]$.

The Extreme Value Theorem is illustrated in Figure 5. Note that an extreme value can be taken on more than once. Although the Extreme Value Theorem is intuitively very plausible, it is difficult to prove and so we omit the proof.

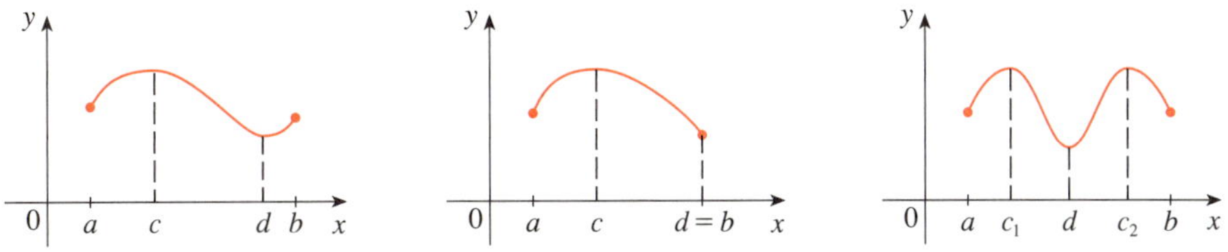

FIGURE 5

Figures 6 and 7 show that a function need not possess extreme values if either hypothesis (continuity or closed interval) is omitted from the Extreme Value Theorem.

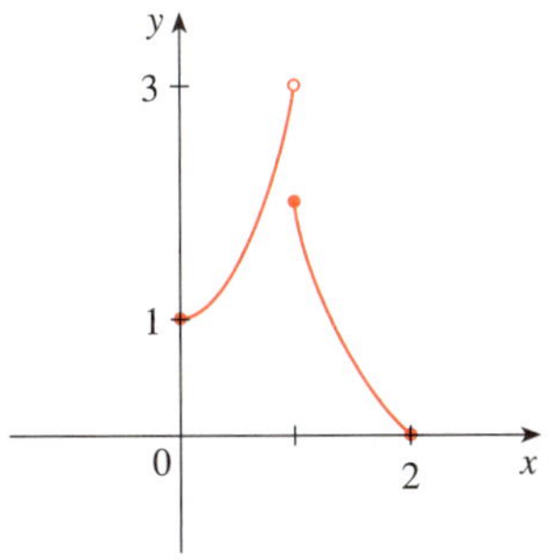

FIGURE 6
This function has minimum value $f(2) = 0$, but no maximum value.

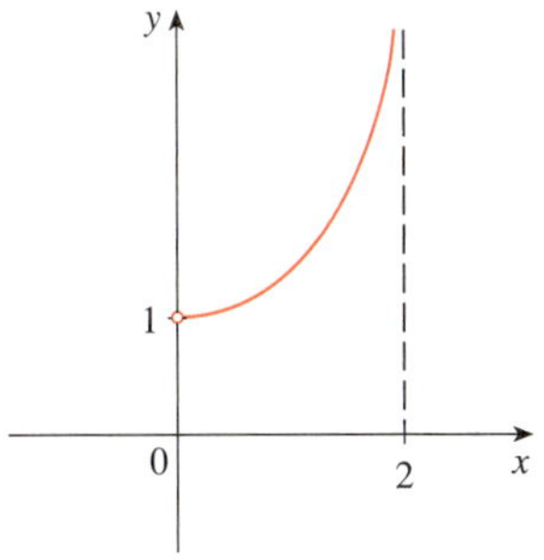

FIGURE 7
This continuous function g has no maximum or minimum.

The function f whose graph is shown in Figure 6 is defined on the closed interval $[0, 2]$ but has no maximum value. (Notice that the range of f is $[0, 3)$. The function takes on values arbitrarily close to 3, but never actually attains the value 3.) This does not contradict the Extreme Value Theorem because f is not continuous. [Nonetheless, a discontinuous function *could* have maximum and minimum values. See Exercise 13(b).]

The function g shown in Figure 7 is continuous on the open interval $(0, 2)$ but has neither a maximum nor a minimum value. [The range of g is $(1, \infty)$. The function takes on arbitrarily large values.] This does not contradict the Extreme Value Theorem because the interval $(0, 2)$ is not closed.

The Extreme Value Theorem says that a continuous function on a closed interval has a maximum value and a minimum value, but it does not tell us how to find these extreme values. We start by looking for local extreme values.

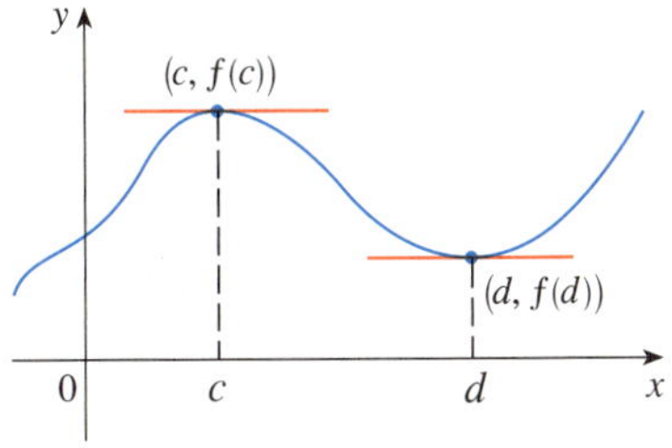

FIGURE 8

Figure 8 shows the graph of a function f with a local maximum at c and a local minimum at d. It appears that at the maximum and minimum points the tangent lines are horizontal and therefore each has slope 0. We know that the derivative is the slope of the tangent line, so it appears that $f'(c) = 0$ and $f'(d) = 0$. The following theorem says that this is always true for differentiable functions.

|||| Fermat's Theorem is named after Pierre Fermat (1601–1665), a French lawyer who took up mathematics as a hobby. Despite his amateur status, Fermat was one of the two inventors of analytic geometry (Descartes was the other). His methods for finding tangents to curves and maximum and minimum values (before the invention of limits and derivatives) made him a forerunner of Newton in the creation of differential calculus.

4 Fermat's Theorem If f has a local maximum or minimum at c, and if $f'(c)$ exists, then $f'(c) = 0$.

Proof Suppose, for the sake of definiteness, that f has a local maximum at c. Then, according to Definition 2, $f(c) \geqslant f(x)$ if x is sufficiently close to c. This implies that if h is sufficiently close to 0, with h being positive or negative, then

$$f(c) \geqslant f(c + h)$$

and therefore

$$f(c + h) - f(c) \leqslant 0 \tag{5}$$

We can divide both sides of an inequality by a positive number. Thus, if $h > 0$ and h is sufficiently small, we have

$$\frac{f(c + h) - f(c)}{h} \leqslant 0$$

Taking the right-hand limit of both sides of this inequality (using Theorem 2.3.2), we get

$$\lim_{h \to 0^+} \frac{f(c + h) - f(c)}{h} \leqslant \lim_{h \to 0^+} 0 = 0$$

But since $f'(c)$ exists, we have

$$f'(c) = \lim_{h \to 0} \frac{f(c + h) - f(c)}{h} = \lim_{h \to 0^+} \frac{f(c + h) - f(c)}{h}$$

and so we have shown that $f'(c) \leqslant 0$.

If $h < 0$, then the direction of the inequality (5) is reversed when we divide by h:

$$\frac{f(c + h) - f(c)}{h} \geqslant 0 \qquad h < 0$$

So, taking the left-hand limit, we have

$$f'(c) = \lim_{h \to 0} \frac{f(c + h) - f(c)}{h} = \lim_{h \to 0^-} \frac{f(c + h) - f(c)}{h} \geqslant 0$$

We have shown that $f'(c) \geqslant 0$ and also that $f'(c) \leqslant 0$. Since both of these inequalities must be true, the only possibility is that $f'(c) = 0$.

We have proved Fermat's Theorem for the case of a local maximum. The case of a local minimum can be proved in a similar manner, or we could use Exercise 76 to deduce it from the case we have just proved (see Exercise 77). ■

The following examples caution us against reading too much into Fermat's Theorem. We can't expect to locate extreme values simply by setting $f'(x) = 0$ and solving for x.

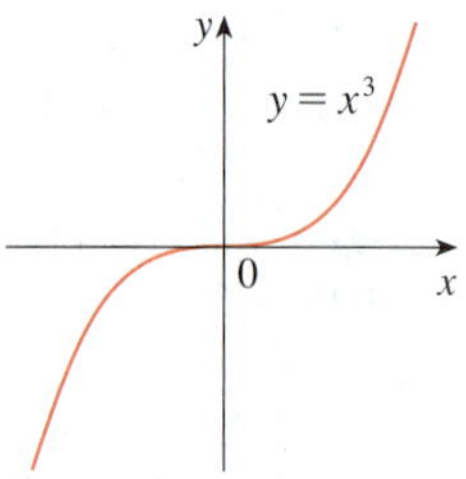

FIGURE 9
If $f(x) = x^3$, then $f'(0) = 0$ but f has no maximum or minimum.

EXAMPLE 5 If $f(x) = x^3$, then $f'(x) = 3x^2$, so $f'(0) = 0$. But f has no maximum or minimum at 0, as you can see from its graph in Figure 9. (Or observe that $x^3 > 0$ for $x > 0$ but $x^3 < 0$ for $x < 0$.) The fact that $f'(0) = 0$ simply means that the curve $y = x^3$ has a horizontal tangent at (0, 0). Instead of having a maximum or minimum at (0, 0), the curve crosses its horizontal tangent there.

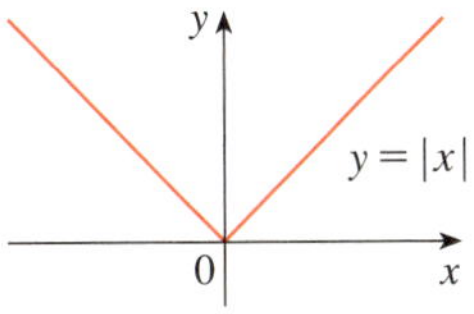

FIGURE 10

If $f(x) = |x|$, then $f(0) = 0$ is a minimum value, but $f'(0)$ does not exist.

EXAMPLE 6 The function $f(x) = |x|$ has its (local and absolute) minimum value at 0, but that value can't be found by setting $f'(x) = 0$ because, as was shown in Example 6 in Section 2.9, $f'(0)$ does not exist. (See Figure 10.)

WARNING ▫ Examples 5 and 6 show that we must be careful when using Fermat's Theorem. Example 5 demonstrates that even when $f'(c) = 0$ there need not be a maximum or minimum at c. (In other words, the converse of Fermat's Theorem is false in general.) Furthermore, there may be an extreme value even when $f'(c)$ does not exist (as in Example 6).

Fermat's Theorem does suggest that we should at least *start* looking for extreme values of f at the numbers c where $f'(c) = 0$ or where $f'(c)$ does not exist. Such numbers are given a special name.

6 Definition A **critical number** of a function f is a number c in the domain of f such that either $f'(c) = 0$ or $f'(c)$ does not exist.

Figure 11 shows a graph of the function f in Example 7. It supports our answer because there is a horizontal tangent when $x = 1.5$ and a vertical tangent when $x = 0$.

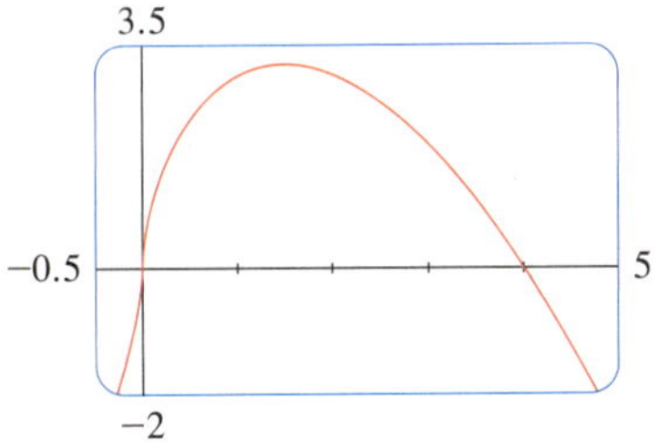

FIGURE 11

EXAMPLE 7 Find the critical numbers of $f(x) = x^{3/5}(4 - x)$.

SOLUTION The Product Rule gives

$$f'(x) = \tfrac{3}{5}x^{-2/5}(4 - x) + x^{3/5}(-1) = \frac{3(4 - x)}{5x^{2/5}} - x^{3/5}$$

$$= \frac{3(4 - x) - 5x}{5x^{2/5}} = \frac{12 - 8x}{5x^{2/5}}$$

[The same result could be obtained by first writing $f(x) = 4x^{3/5} - x^{8/5}$.] Therefore, $f'(x) = 0$ if $12 - 8x = 0$, that is, $x = \frac{3}{2}$, and $f'(x)$ does not exist when $x = 0$. Thus, the critical numbers are $\frac{3}{2}$ and 0.

In terms of critical numbers, Fermat's Theorem can be rephrased as follows (compare Definition 6 with Theorem 4):

7 If f has a local maximum or minimum at c, then c is a critical number of f.

To find an absolute maximum or minimum of a continuous function on a closed interval, we note that either it is local [in which case it occurs at a critical number by (7)] or it occurs at an endpoint of the interval. Thus, the following three-step procedure always works.

The Closed Interval Method To find the *absolute* maximum and minimum values of a continuous function f on a closed interval $[a, b]$:

1. Find the values of f at the critical numbers of f in (a, b).
2. Find the values of f at the endpoints of the interval.
3. The largest of the values from Steps 1 and 2 is the absolute maximum value; the smallest of these values is the absolute minimum value.

EXAMPLE 8 Find the absolute maximum and minimum values of the function

$$f(x) = x^3 - 3x^2 + 1 \qquad -\tfrac{1}{2} \leqslant x \leqslant 4$$

SOLUTION Since f is continuous on $[-\frac{1}{2}, 4]$, we can use the Closed Interval Method:

$$f(x) = x^3 - 3x^2 + 1$$

$$f'(x) = 3x^2 - 6x = 3x(x - 2)$$

Since $f'(x)$ exists for all x, the only critical numbers of f occur when $f'(x) = 0$, that is, $x = 0$ or $x = 2$. Notice that each of these critical numbers lies in the interval $(-\frac{1}{2}, 4)$. The values of f at these critical numbers are

$$f(0) = 1 \qquad f(2) = -3$$

The values of f at the endpoints of the interval are

$$f(-\tfrac{1}{2}) = \tfrac{1}{8} \qquad f(4) = 17$$

Comparing these four numbers, we see that the absolute maximum value is $f(4) = 17$ and the absolute minimum value is $f(2) = -3$.

Note that in this example the absolute maximum occurs at an endpoint, whereas the absolute minimum occurs at a critical number. The graph of f is sketched in Figure 12.

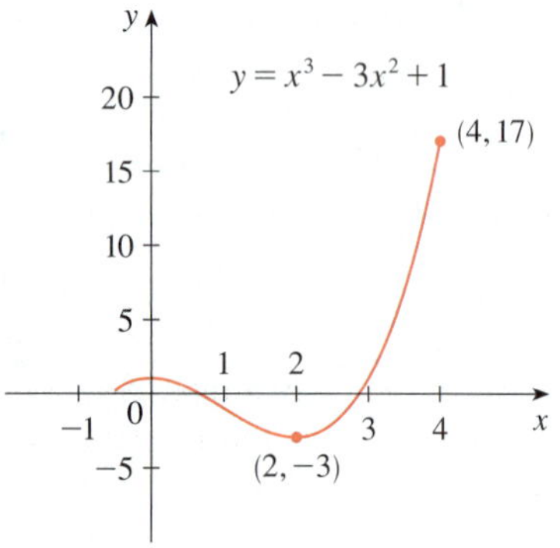

FIGURE 12

If you have a graphing calculator or a computer with graphing software, it is possible to estimate maximum and minimum values very easily. But, as the next example shows, calculus is needed to find the *exact* values.

EXAMPLE 9
(a) Use a graphing device to estimate the absolute minimum and maximum values of the function $f(x) = x - 2 \sin x$, $0 \leqslant x \leqslant 2\pi$.
(b) Use calculus to find the exact minimum and maximum values.

SOLUTION
(a) Figure 13 shows a graph of f in the viewing rectangle $[0, 2\pi]$ by $[-1, 8]$. By moving the cursor close to the maximum point, we see that the y-coordinates don't change very much in the vicinity of the maximum. The absolute maximum value is about 6.97 and it occurs when $x \approx 5.2$. Similarly, by moving the cursor close to the minimum point, we see that the absolute minimum value is about -0.68 and it occurs when $x \approx 1.0$. It is possible to get more accurate estimates by zooming in toward the maximum and minimum points, but instead let's use calculus.

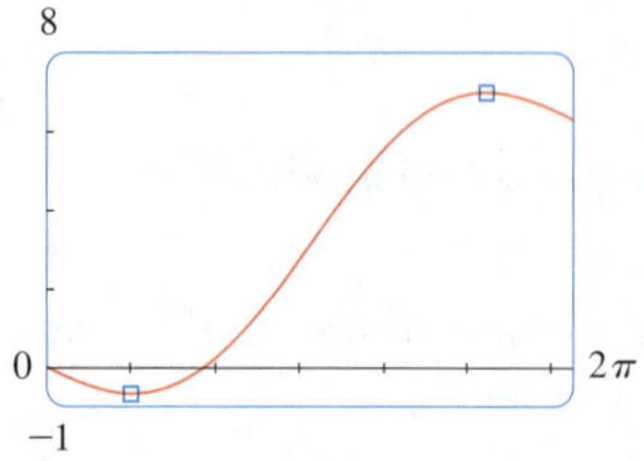

FIGURE 13

(b) The function $f(x) = x - 2 \sin x$ is continuous on $[0, 2\pi]$. Since $f'(x) = 1 - 2 \cos x$, we have $f'(x) = 0$ when $\cos x = \frac{1}{2}$ and this occurs when $x = \pi/3$ or $5\pi/3$. The values of f at these critical points are

$$f(\pi/3) = \frac{\pi}{3} - 2 \sin \frac{\pi}{3} = \frac{\pi}{3} - \sqrt{3} \approx -0.684853$$

and

$$f(5\pi/3) = \frac{5\pi}{3} - 2 \sin \frac{5\pi}{3} = \frac{5\pi}{3} + \sqrt{3} \approx 6.968039$$

The values of f at the endpoints are

$$f(0) = 0 \qquad \text{and} \qquad f(2\pi) = 2\pi \approx 6.28$$

Comparing these four numbers and using the Closed Interval Method, we see that the absolute minimum value is $f(\pi/3) = \pi/3 - \sqrt{3}$ and the absolute maximum value is $f(5\pi/3) = 5\pi/3 + \sqrt{3}$. The values from part (a) serve as a check on our work.

EXAMPLE 10 The Hubble Space Telescope was deployed on April 24, 1990, by the space shuttle *Discovery*. A model for the velocity of the shuttle during this mission, from liftoff at $t = 0$ until the solid rocket boosters were jettisoned at $t = 126$ s, is given by

$$v(t) = 0.001302t^3 - 0.09029t^2 + 23.61t - 3.083$$

(in feet per second). Using this model, estimate the absolute maximum and minimum values of the *acceleration* of the shuttle between liftoff and the jettisoning of the boosters.

SOLUTION We are asked for the extreme values not of the given velocity function, but rather of the acceleration function. So we first need to differentiate to find the acceleration:

$$a(t) = v'(t) = \frac{d}{dt}(0.001302t^3 - 0.09029t^2 + 23.61t - 3.083)$$

$$= 0.003906t^2 - 0.18058t + 23.61$$

We now apply the Closed Interval Method to the continuous function a on the interval $0 \leq t \leq 126$. Its derivative is

$$a'(t) = 0.007812t - 0.18058$$

The only critical number occurs when $a'(t) = 0$:

$$t_1 = \frac{0.18058}{0.007812} \approx 23.12$$

Evaluating $a(t)$ at the critical number and at the endpoints, we have

$$a(0) = 23.61 \qquad a(t_1) \approx 21.52 \qquad a(126) \approx 62.87$$

So the maximum acceleration is about 62.87 ft/s^2 and the minimum acceleration is about 21.52 ft/s^2.

4.1 Exercises

1. Explain the difference between an absolute minimum and a local minimum.

2. Suppose f is a continuous function defined on a closed interval $[a, b]$.
   (a) What theorem guarantees the existence of an absolute maximum value and an absolute minimum value for f?
   (b) What steps would you take to find those maximum and minimum values?

3–4 |||| For each of the numbers a, b, c, d, e, r, s, and t, state whether the function whose graph is shown has an absolute maximum or minimum, a local maximum or minimum, or neither a maximum nor a minimum.

3.

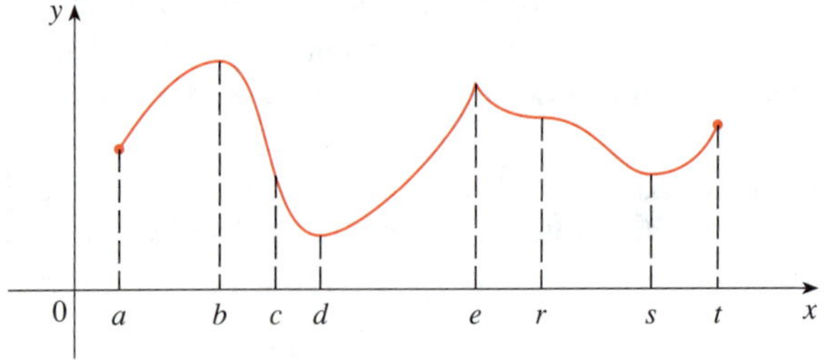

4.

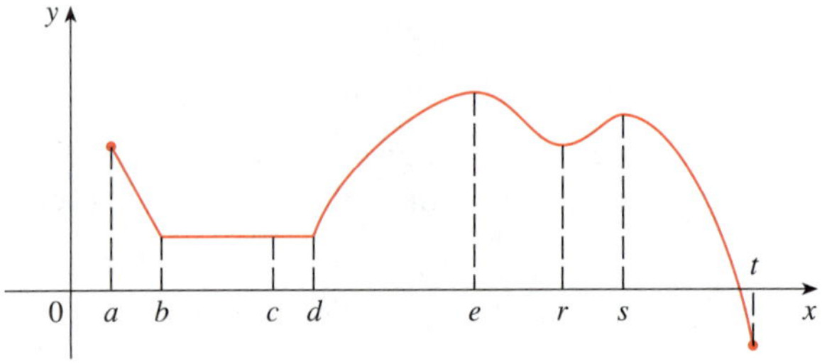

5–6 |||| Use the graph to state the absolute and local maximum and minimum values of the function.

5.

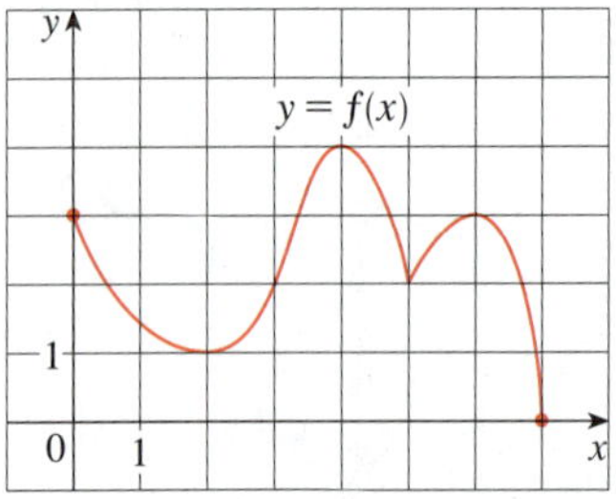

6.

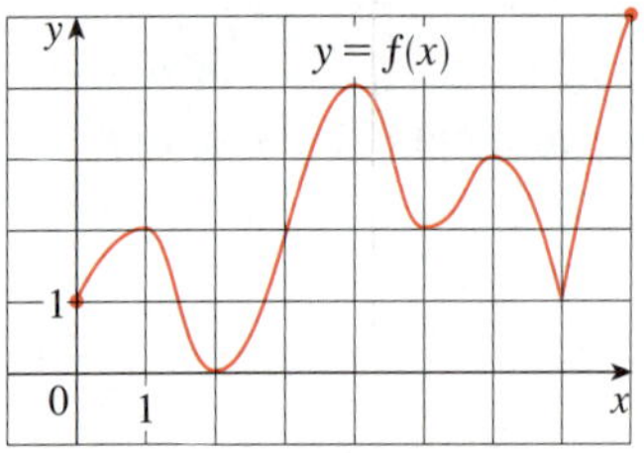

7–10 |||| Sketch the graph of a function f that is continuous on $[1, 5]$ and has the given properties.

7. Absolute minimum at 2, absolute maximum at 3, local minimum at 4

8. Absolute minimum at 1, absolute maximum at 5, local maximum at 2, local minimum at 4

9. Absolute maximum at 5, absolute minimum at 2, local maximum at 3, local minima at 2 and 4

10. f has no local maximum or minimum, but 2 and 4 are critical numbers

11. (a) Sketch the graph of a function that has a local maximum at 2 and is differentiable at 2.
(b) Sketch the graph of a function that has a local maximum at 2 and is continuous but not differentiable at 2.
(c) Sketch the graph of a function that has a local maximum at 2 and is not continuous at 2.

12. (a) Sketch the graph of a function on $[-1, 2]$ that has an absolute maximum but no local maximum.
(b) Sketch the graph of a function on $[-1, 2]$ that has a local maximum but no absolute maximum.

13. (a) Sketch the graph of a function on $[-1, 2]$ that has an absolute maximum but no absolute minimum.
(b) Sketch the graph of a function on $[-1, 2]$ that is discontinuous but has both an absolute maximum and an absolute minimum.

14. (a) Sketch the graph of a function that has two local maxima, one local minimum, and no absolute minimum.
(b) Sketch the graph of a function that has three local minima, two local maxima, and seven critical numbers.

15–30 |||| Sketch the graph of f by hand and use your sketch to find the absolute and local maximum and minimum values of f. (Use the graphs and transformations of Sections 1.2 and 1.3.)

15. $f(x) = 8 - 3x, \quad x \geqslant 1$

16. $f(x) = 3 - 2x, \quad x \leqslant 5$

17. $f(x) = x^2, \quad 0 < x < 2$

18. $f(x) = x^2, \quad 0 < x \leqslant 2$

19. $f(x) = x^2, \quad 0 \leqslant x < 2$

20. $f(x) = x^2, \quad 0 \leqslant x \leqslant 2$

21. $f(x) = x^2, \quad -3 \leqslant x \leqslant 2$

22. $f(x) = 1 + (x + 1)^2, \quad -2 \leqslant x < 5$

23. $f(t) = 1/t, \quad 0 < t < 1$

24. $f(t) = 1/t, \quad 0 < t \leqslant 1$

25. $f(\theta) = \sin\theta, \quad -2\pi \leqslant \theta \leqslant 2\pi$

26. $f(\theta) = \tan\theta, \quad -\pi/4 \leqslant \theta < \pi/2$

27. $f(x) = 1 - \sqrt{x}$

28. $f(x) = e^x$

29. $f(x) = \begin{cases} 1 - x & \text{if } 0 \leqslant x < 2 \\ 2x - 4 & \text{if } 2 \leqslant x \leqslant 3 \end{cases}$

30. $f(x) = \begin{cases} x^2 & \text{if } -1 \leqslant x < 0 \\ 2 - x^2 & \text{if } 0 \leqslant x \leqslant 1 \end{cases}$

31–46 Find the critical numbers of the function.

31. $f(x) = 5x^2 + 4x$

32. $f(x) = x^3 + x^2 - x$

33. $f(x) = x^3 + 3x^2 - 24x$

34. $f(x) = x^3 + x^2 + x$

35. $s(t) = 3t^4 + 4t^3 - 6t^2$

36. $f(z) = \dfrac{z + 1}{z^2 + z + 1}$

37. $g(x) = |2x + 3|$

38. $g(x) = x^{1/3} - x^{-2/3}$

39. $g(t) = 5t^{2/3} + t^{5/3}$

40. $g(t) = \sqrt{t}\,(1 - t)$

41. $F(x) = x^{4/5}(x - 4)^2$

42. $G(x) = \sqrt[3]{x^2 - x}$

43. $f(\theta) = 2 \cos \theta + \sin^2\theta$

44. $g(\theta) = 4\theta - \tan \theta$

45. $f(x) = x \ln x$

46. $f(x) = xe^{2x}$

47–62 Find the absolute maximum and absolute minimum values of f on the given interval.

47. $f(x) = 3x^2 - 12x + 5, \quad [0, 3]$

48. $f(x) = x^3 - 3x + 1, \quad [0, 3]$

49. $f(x) = 2x^3 - 3x^2 - 12x + 1, \quad [-2, 3]$

50. $f(x) = x^3 - 6x^2 + 9x + 2, \quad [-1, 4]$

51. $f(x) = x^4 - 2x^2 + 3, \quad [-2, 3]$

52. $f(x) = (x^2 - 1)^3, \quad [-1, 2]$

53. $f(x) = \dfrac{x}{x^2 + 1}, \quad [0, 2]$

54. $f(x) = \dfrac{x^2 - 4}{x^2 + 4}, \quad [-4, 4]$

55. $f(t) = t\sqrt{4 - t^2}, \quad [-1, 2]$

56. $f(t) = \sqrt[3]{t}\,(8 - t), \quad [0, 8]$

57. $f(x) = \sin x + \cos x, \quad [0, \pi/3]$

58. $f(x) = x - 2 \cos x, \quad [-\pi, \pi]$

59. $f(x) = xe^{-x}, \quad [0, 2]$

60. $f(x) = (\ln x)/x, \quad [1, 3]$

61. $f(x) = x - 3 \ln x, \quad [1, 4]$

62. $f(x) = e^{-x} - e^{-2x}, \quad [0, 1]$

63. If a and b are positive numbers, find the maximum value of $f(x) = x^a(1 - x)^b$, $0 \leqslant x \leqslant 1$.

64. Use a graph to estimate the critical numbers of $f(x) = |x^3 - 3x^2 + 2|$ correct to one decimal place.

65–68

(a) Use a graph to estimate the absolute maximum and minimum values of the function to two decimal places.

(b) Use calculus to find the exact maximum and minimum values.

65. $f(x) = x^3 - 8x + 1, \quad -3 \leqslant x \leqslant 3$

66. $f(x) = e^{x^3 - x}, \quad -1 \leqslant x \leqslant 0$

67. $f(x) = x\sqrt{x - x^2}$

68. $f(x) = (\cos x)/(2 + \sin x), \quad 0 \leqslant x \leqslant 2\pi$

69. Between 0°C and 30°C, the volume V (in cubic centimeters) of 1 kg of water at a temperature T is given approximately by the formula

$$V = 999.87 - 0.06426T + 0.0085043T^2 - 0.0000679T^3$$

Find the temperature at which water has its maximum density.

70. An object with weight W is dragged along a horizontal plane by a force acting along a rope attached to the object. If the rope makes an angle θ with the plane, then the magnitude of the force is

$$F = \frac{\mu W}{\mu \sin \theta + \cos \theta}$$

where μ is a positive constant called the *coefficient of friction* and where $0 \leqslant \theta \leqslant \pi/2$. Show that F is minimized when $\tan \theta = \mu$.

71. A model for the food-price index (the price of a representative "basket" of foods) between 1984 and 1994 is given by the function

$$I(t) = 0.00009045t^5 + 0.001438t^4 - 0.06561t^3 + 0.4598t^2 - 0.6270t + 99.33$$

where t is measured in years since midyear 1984, so $0 \leqslant t \leqslant 10$, and $I(t)$ is measured in 1987 dollars and scaled such that $I(3) = 100$. Estimate the times when food was cheapest and most expensive during the period 1984–1994.

72. On May 7, 1992, the space shuttle *Endeavour* was launched on mission STS-49, the purpose of which was to install a new perigee kick motor in an Intelsat communications satellite. The table gives the velocity data for the shuttle between liftoff and the jettisoning of the solid rocket boosters.

Event	Time (s)	Velocity (ft/s)
Launch	0	0
Begin roll maneuver	10	185
End roll maneuver	15	319
Throttle to 89%	20	447
Throttle to 67%	32	742
Throttle to 104%	59	1325
Maximum dynamic pressure	62	1445
Solid rocket booster separation	125	4151

(a) Use a graphing calculator or computer to find the cubic polynomial that best models the velocity of the shuttle for the time interval $t \in [0, 125]$. Then graph this polynomial.

(b) Find a model for the acceleration of the shuttle and use it to estimate the maximum and minimum values of the acceleration during the first 125 seconds.

73. When a foreign object lodged in the trachea (windpipe) forces a person to cough, the diaphragm thrusts upward causing an increase in pressure in the lungs. This is accompanied by a contraction of the trachea, making a narrower channel for the expelled air to flow through. For a given amount of air to escape in a fixed time, it must move faster through the narrower channel than the wider one. The greater the velocity of the airstream, the greater the force on the foreign object. X rays show that the radius of the circular tracheal tube contracts to about two-thirds of its normal radius during a cough. According to a mathematical model of coughing, the velocity v of the airstream is related to the radius r of the trachea by the equation

$$v(r) = k(r_0 - r)r^2 \qquad \tfrac{1}{2}r_0 \leqslant r \leqslant r_0$$

where k is a constant and r_0 is the normal radius of the trachea. The restriction on r is due to the fact that the tracheal wall stiffens under pressure and a contraction greater than $\frac{1}{2}r_0$ is prevented (otherwise the person would suffocate).

(a) Determine the value of r in the interval $\left[\frac{1}{2}r_0, r_0\right]$ at which v has an absolute maximum. How does this compare with experimental evidence?

(b) What is the absolute maximum value of v on the interval?

(c) Sketch the graph of v on the interval $[0, r_0]$.

74. Show that 5 is a critical number of the function

$$g(x) = 2 + (x - 5)^3$$

but g does not have a local extreme value at 5.

75. Prove that the function

$$f(x) = x^{101} + x^{51} + x + 1$$

has neither a local maximum nor a local minimum.

76. If f has a minimum value at c, show that the function $g(x) = -f(x)$ has a maximum value at c.

77. Prove Fermat's Theorem for the case in which f has a local minimum at c.

78. A cubic function is a polynomial of degree 3; that is, it has the form $f(x) = ax^3 + bx^2 + cx + d$, where $a \neq 0$.

(a) Show that a cubic function can have two, one, or no critical number(s). Give examples and sketches to illustrate the three possibilities.

(b) How many local extreme values can a cubic function have?

APPLIED PROJECT

The Calculus of Rainbows

Rainbows are created when raindrops scatter sunlight. They have fascinated mankind since ancient times and have inspired attempts at scientific explanation since the time of Aristotle. In this project we use the ideas of Descartes and Newton to explain the shape, location, and colors of rainbows.

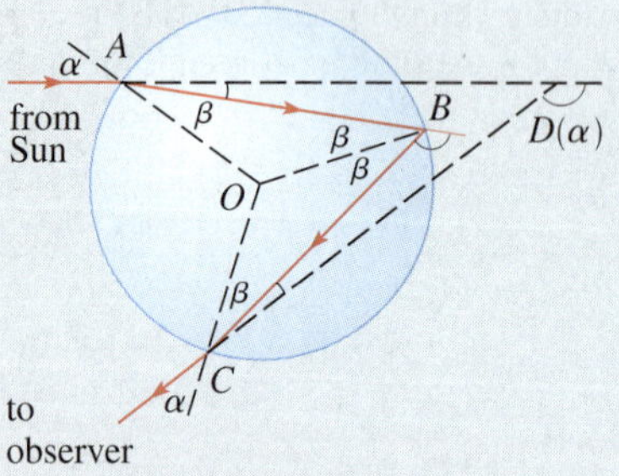

Formation of the primary rainbow

1. The figure shows a ray of sunlight entering a spherical raindrop at A. Some of the light is reflected, but the line AB shows the path of the part that enters the drop. Notice that the light is refracted toward the normal line AO and in fact Snell's Law says that $\sin\alpha = k\sin\beta$, where α is the angle of incidence, β is the angle of refraction, and $k \approx \frac{4}{3}$ is the index of refraction for water. At B some of the light passes through the drop and is refracted into the air, but the line BC shows the part that is reflected. (The angle of incidence equals the angle of reflection.) When the ray reaches C, part of it is reflected, but for the time being we are more interested in the part that leaves the raindrop at C. (Notice that it is refracted away from the normal line.) The *angle of deviation* $D(\alpha)$ is the amount of clockwise rotation that the ray has undergone during this three-stage process. Thus

$$D(\alpha) = (\alpha - \beta) + (\pi - 2\beta) + (\alpha - \beta) = \pi + 2\alpha - 4\beta$$

Show that the minimum value of the deviation is $D(\alpha) \approx 138°$ and occurs when $\alpha \approx 59.4°$.

The significance of the minimum deviation is that when $\alpha \approx 59.4°$ we have $D'(\alpha) \approx 0$, so $\Delta D/\Delta\alpha \approx 0$. This means that many rays with $\alpha \approx 59.4°$ become deviated by approximately the same amount. It is the *concentration* of rays coming from near the direction of minimum deviation that creates the brightness of the primary rainbow. The following figure shows that the angle of elevation from the observer up to the highest point on the rainbow is $180° - 138° = 42°$. (This angle is called the *rainbow angle.*)

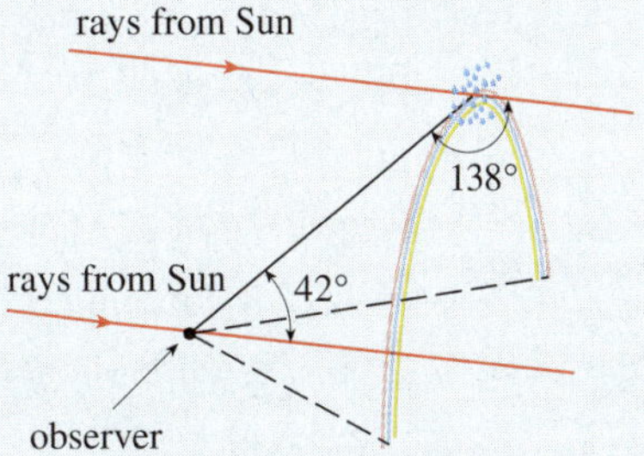

2. Problem 1 explains the location of the primary rainbow, but how do we explain the colors? Sunlight comprises a range of wavelengths, from the red range through orange, yellow, green, blue, indigo, and violet. As Newton discovered in his prism experiments of 1666, the index of refraction is different for each color. (The effect is called *dispersion.*) For red light the refractive index is $k \approx 1.3318$ whereas for violet light it is $k \approx 1.3435$. By repeating the calculation of Problem 1 for these values of k, show that the rainbow angle is about 42.3° for the red bow and 40.6° for the violet bow. So the rainbow really consists of seven individual bows corresponding to the seven colors.

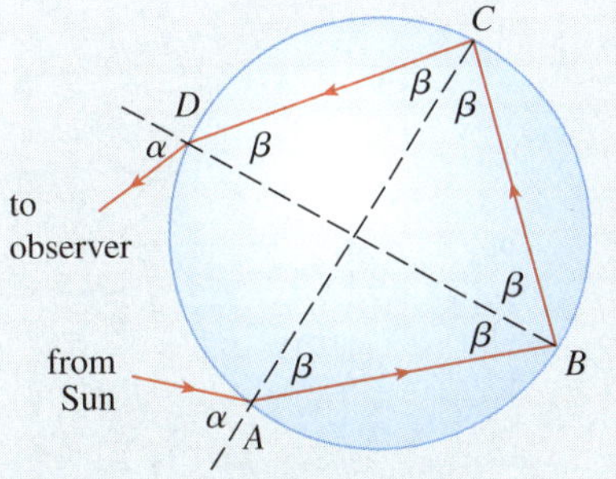

Formation of the secondary rainbow

3. Perhaps you have seen a fainter secondary rainbow above the primary bow. That results from the part of a ray that enters a raindrop and is refracted at A, reflected twice (at B and C), and refracted as it leaves the drop at D (see the figure). This time the deviation angle $D(\alpha)$ is the total amount of counterclockwise rotation that the ray undergoes in this four-stage process. Show that

$$D(\alpha) = 2\alpha - 6\beta + 2\pi$$

and $D(\alpha)$ has a minimum value when

$$\cos \alpha = \sqrt{\frac{k^2 - 1}{8}}$$

Taking $k = \frac{4}{3}$, show that the minimum deviation is about 129° and so the rainbow angle for the secondary rainbow is about 51°, as shown in the figure.

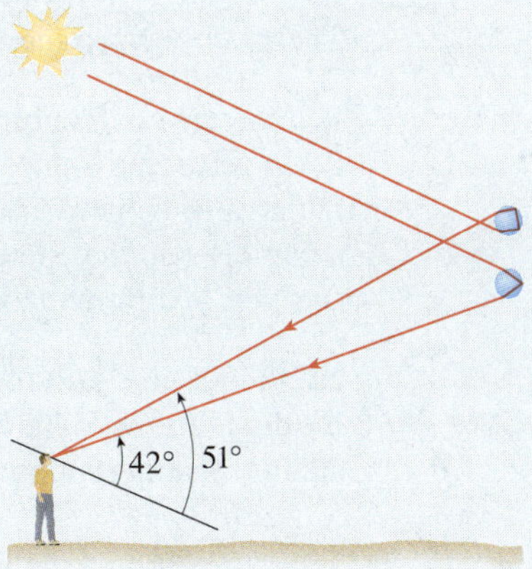

4. Show that the colors in the secondary rainbow appear in the opposite order from those in the primary rainbow.

4.2 The Mean Value Theorem

We will see that many of the results of this chapter depend on one central fact, which is called the Mean Value Theorem. But to arrive at the Mean Value Theorem we first need the following result.

|||| Rolle's Theorem was first published in 1691 by the French mathematician Michel Rolle (1652–1719) in a book entitled *Méthode pour résoudre les égalitéz.* Later, however, he became a vocal critic of the methods of his day and attacked calculus as being a "collection of ingenious fallacies."

Rolle's Theorem Let f be a function that satisfies the following three hypotheses:

1. f is continuous on the closed interval $[a, b]$.
2. f is differentiable on the open interval (a, b).
3. $f(a) = f(b)$

Then there is a number c in (a, b) such that $f'(c) = 0$.

Before giving the proof let's take a look at the graphs of some typical functions that satisfy the three hypotheses. Figure 1 shows the graphs of four such functions. In each case it appears that there is at least one point $(c, f(c))$ on the graph where the tangent is horizontal and therefore $f'(c) = 0$. Thus, Rolle's Theorem is plausible.

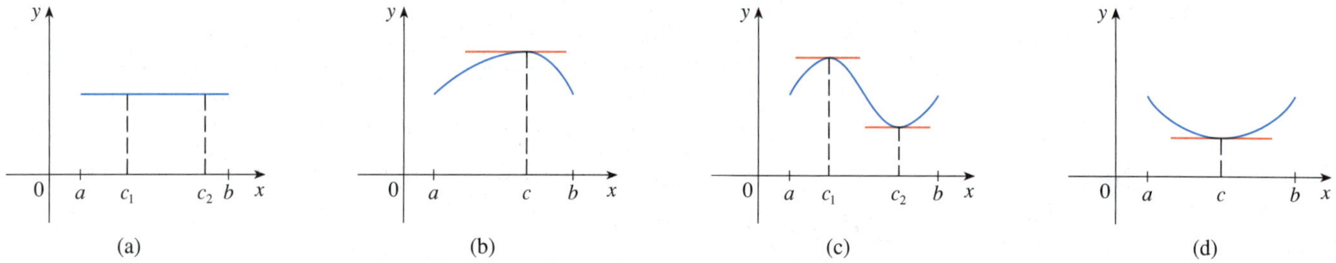

FIGURE 1

|||| Take cases

Proof There are three cases:

CASE I ▫ **$f(x) = k$, a constant**
Then $f'(x) = 0$, so the number c can be taken to be *any* number in (a, b).

CASE II ▫ **$f(x) > f(a)$ for some x in (a, b)** [as in Figure 1(b) or (c)]
By the Extreme Value Theorem (which we can apply by hypothesis 1), f has a maximum value somewhere in $[a, b]$. Since $f(a) = f(b)$, it must attain this maximum value at a number c in the open interval (a, b). Then f has a *local* maximum at c and, by hypothesis 2, f is differentiable at c. Therefore, $f'(c) = 0$ by Fermat's Theorem.

CASE III ▫ **$f(x) < f(a)$ for some x in (a, b)** [as in Figure 1(c) or (d)]
By the Extreme Value Theorem, f has a minimum value in $[a, b]$ and, since $f(a) = f(b)$, it attains this minimum value at a number c in (a, b). Again $f'(c) = 0$ by Fermat's Theorem.

EXAMPLE 1 Let's apply Rolle's Theorem to the position function $s = f(t)$ of a moving object. If the object is in the same place at two different instants $t = a$ and $t = b$, then $f(a) = f(b)$. Rolle's Theorem says that there is some instant of time $t = c$ between a and b when $f'(c) = 0$; that is, the velocity is 0. (In particular, you can see that this is true when a ball is thrown directly upward.)

Figure 2 shows a graph of the function $f(x) = x^3 + x - 1$ discussed in Example 2. Rolle's Theorem shows that, no matter how much we enlarge the viewing rectangle, we can never find a second x-intercept.

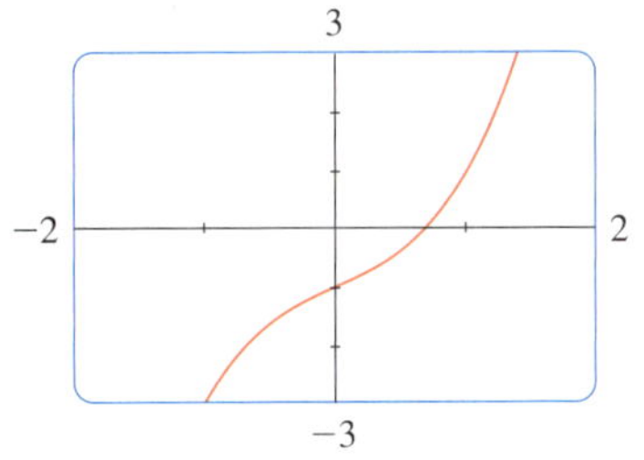

FIGURE 2

EXAMPLE 2 Prove that the equation $x^3 + x - 1 = 0$ has exactly one real root.

SOLUTION First we use the Intermediate Value Theorem (2.5.10) to show that a root exists. Let $f(x) = x^3 + x - 1$. Then $f(0) = -1 < 0$ and $f(1) = 1 > 0$. Since f is a polynomial, it is continuous, so the Intermediate Value Theorem states that there is a number c between 0 and 1 such that $f(c) = 0$. Thus, the given equation has a root.

To show that the equation has no other real root, we use Rolle's Theorem and argue by contradiction. Suppose that it had two roots a and b. Then $f(a) = 0 = f(b)$ and, since f is a polynomial, it is differentiable on (a, b) and continuous on $[a, b]$. Thus, by Rolle's Theorem, there is a number c between a and b such that $f'(c) = 0$. But

$$f'(x) = 3x^2 + 1 \geqslant 1 \qquad \text{for all } x$$

(since $x^2 \geqslant 0$) so $f'(x)$ can never be 0. This gives a contradiction. Therefore, the equation can't have two real roots.

Our main use of Rolle's Theorem is in proving the following important theorem, which was first stated by another French mathematician, Joseph-Louis Lagrange.

The Mean Value Theorem is an example of what is called an existence theorem. Like the Intermediate Value Theorem, the Extreme Value Theorem, and Rolle's Theorem, it guarantees that there *exists* a number with a certain property, but it doesn't tell us how to find the number.

The Mean Value Theorem Let f be a function that satisfies the following hypotheses:

1. f is continuous on the closed interval $[a, b]$.

2. f is differentiable on the open interval (a, b).

Then there is a number c in (a, b) such that

1
$$f'(c) = \frac{f(b) - f(a)}{b - a}$$

or, equivalently,

2
$$f(b) - f(a) = f'(c)(b - a)$$

Before proving this theorem, we can see that it is reasonable by interpreting it geometrically. Figures 3 and 4 show the points $A(a, f(a))$ and $B(b, f(b))$ on the graphs of two differentiable functions. The slope of the secant line AB is

3
$$m_{AB} = \frac{f(b) - f(a)}{b - a}$$

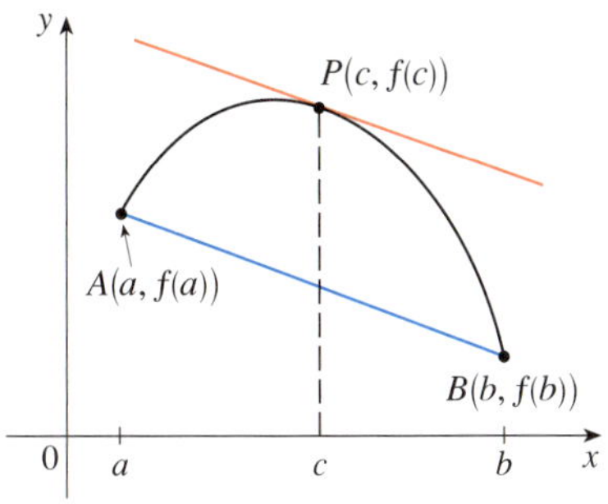

FIGURE 3

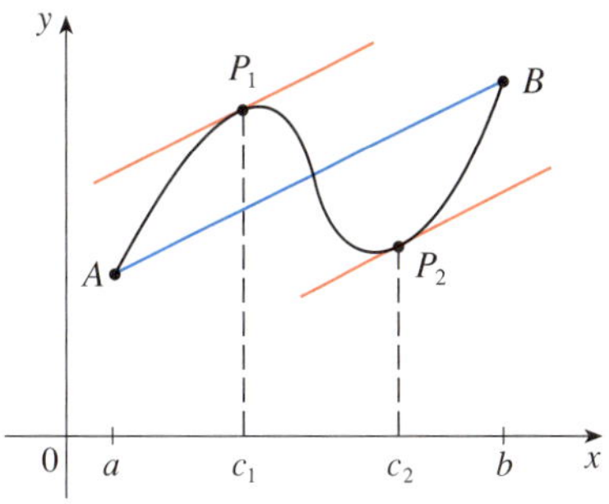

FIGURE 4

which is the same expression as on the right side of Equation 1. Since $f'(c)$ is the slope of the tangent line at the point $(c, f(c))$, the Mean Value Theorem, in the form given by Equation 1, says that there is at least one point $P(c, f(c))$ on the graph where the slope of the tangent line is the same as the slope of the secant line AB. In other words, there is a point P where the tangent line is parallel to the secant line AB.

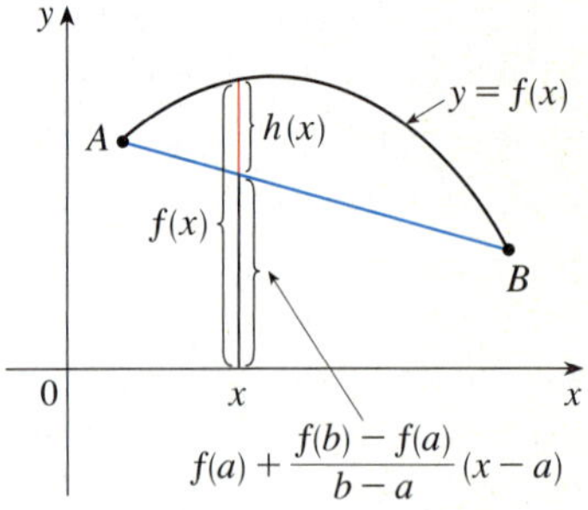

FIGURE 5

Proof We apply Rolle's Theorem to a new function h defined as the difference between f and the function whose graph is the secant line AB. Using Equation 3, we see that the equation of the line AB can be written as

$$y - f(a) = \frac{f(b) - f(a)}{b - a}(x - a)$$

or as

$$y = f(a) + \frac{f(b) - f(a)}{b - a}(x - a)$$

So, as shown in Figure 5,

4 $$h(x) = f(x) - f(a) - \frac{f(b) - f(a)}{b - a}(x - a)$$

First we must verify that h satisfies the three hypotheses of Rolle's Theorem.

1. The function h is continuous on $[a, b]$ because it is the sum of f and a first-degree polynomial, both of which are continuous.
2. The function h is differentiable on (a, b) because both f and the first-degree polynomial are differentiable. In fact, we can compute h' directly from Equation 4:

$$h'(x) = f'(x) - \frac{f(b) - f(a)}{b - a}$$

(Note that $f(a)$ and $[f(b) - f(a)]/(b - a)$ are constants.)

3. $$h(a) = f(a) - f(a) - \frac{f(b) - f(a)}{b - a}(a - a) = 0$$

$$h(b) = f(b) - f(a) - \frac{f(b) - f(a)}{b - a}(b - a)$$
$$= f(b) - f(a) - [f(b) - f(a)] = 0$$

Therefore, $h(a) = h(b)$.

|||| The Mean Value Theorem was first formulated by Joseph-Louis Lagrange (1736–1813), born in Italy of a French father and an Italian mother. He was a child prodigy and became a professor in Turin at the tender age of 19. Lagrange made great contributions to number theory, theory of functions, theory of equations, and analytical and celestial mechanics. In particular, he applied calculus to the analysis of the stability of the solar system. At the invitation of Frederick the Great, he succeeded Euler at the Berlin Academy and, when Frederick died, Lagrange accepted King Louis XVI's invitation to Paris, where he was given apartments in the Louvre. Despite all the trappings of luxury and fame, he was a kind and quiet man, living only for science.

Since h satisfies the hypotheses of Rolle's Theorem, that theorem says there is a number c in (a, b) such that $h'(c) = 0$. Therefore

$$0 = h'(c) = f'(c) - \frac{f(b) - f(a)}{b - a}$$

and so

$$f'(c) = \frac{f(b) - f(a)}{b - a}$$

EXAMPLE 3 To illustrate the Mean Value Theorem with a specific function, let's consider $f(x) = x^3 - x$, $a = 0$, $b = 2$. Since f is a polynomial, it is continuous and differentiable

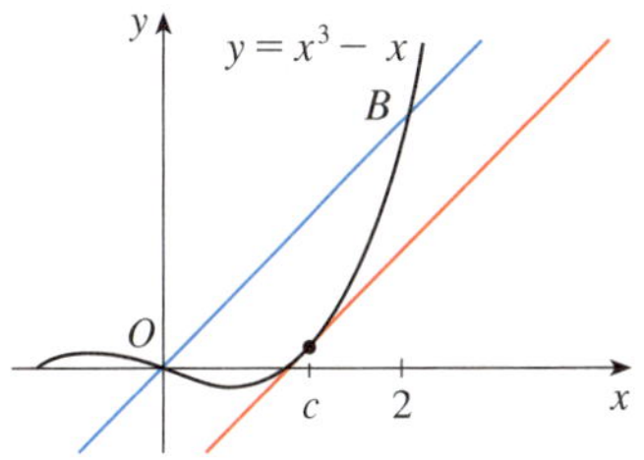

FIGURE 6

for all x, so it is certainly continuous on $[0, 2]$ and differentiable on $(0, 2)$. Therefore, by the Mean Value Theorem, there is a number c in $(0, 2)$ such that

$$f(2) - f(0) = f'(c)(2 - 0)$$

Now $f(2) = 6$, $f(0) = 0$, and $f'(x) = 3x^2 - 1$, so this equation becomes

$$6 = (3c^2 - 1)2 = 6c^2 - 2$$

which gives $c^2 = \frac{4}{3}$, that is, $c = \pm 2/\sqrt{3}$. But c must lie in $(0, 2)$, so $c = 2/\sqrt{3}$. Figure 6 illustrates this calculation: The tangent line at this value of c is parallel to the secant line OB.

EXAMPLE 4 If an object moves in a straight line with position function $s = f(t)$, then the average velocity between $t = a$ and $t = b$ is

$$\frac{f(b) - f(a)}{b - a}$$

and the velocity at $t = c$ is $f'(c)$. Thus, the Mean Value Theorem (in the form of Equation 1) tells us that at some time $t = c$ between a and b the instantaneous velocity $f'(c)$ is equal to that average velocity. For instance, if a car traveled 180 km in 2 hours, then the speedometer must have read 90 km/h at least once.

In general, the Mean Value Theorem can be interpreted as saying that there is a number at which the instantaneous rate of change is equal to the average rate of change over an interval.

The main significance of the Mean Value Theorem is that it enables us to obtain information about a function from information about its derivative. The next example provides an instance of this principle.

EXAMPLE 5 Suppose that $f(0) = -3$ and $f'(x) \leqslant 5$ for all values of x. How large can $f(2)$ possibly be?

SOLUTION We are given that f is differentiable (and therefore continuous) everywhere. In particular, we can apply the Mean Value Theorem on the interval $[0, 2]$. There exists a number c such that

$$f(2) - f(0) = f'(c)(2 - 0)$$

so

$$f(2) = f(0) + 2f'(c) = -3 + 2f'(c)$$

We are given that $f'(x) \leqslant 5$ for all x, so in particular we know that $f'(c) \leqslant 5$. Multiplying both sides of this inequality by 2, we have $2f'(c) \leqslant 10$, so

$$f(2) = -3 + 2f'(c) \leqslant -3 + 10 = 7$$

The largest possible value for $f(2)$ is 7.

The Mean Value Theorem can be used to establish some of the basic facts of differential calculus. One of these basic facts is the following theorem. Others will be found in the following sections.

5 Theorem If $f'(x) = 0$ for all x in an interval (a, b), then f is constant on (a, b).

Proof Let x_1 and x_2 be any two numbers in (a, b) with $x_1 < x_2$. Since f is differentiable on (a, b), it must be differentiable on (x_1, x_2) and continuous on $[x_1, x_2]$. By applying the Mean Value Theorem to f on the interval $[x_1, x_2]$, we get a number c such that $x_1 < c < x_2$ and

6 $$f(x_2) - f(x_1) = f'(c)(x_2 - x_1)$$

Since $f'(x) = 0$ for all x, we have $f'(c) = 0$, and so Equation 6 becomes

$$f(x_2) - f(x_1) = 0 \qquad \text{or} \qquad f(x_2) = f(x_1)$$

Therefore, f has the same value at *any* two numbers x_1 and x_2 in (a, b). This means that f is constant on (a, b).

7 Corollary If $f'(x) = g'(x)$ for all x in an interval (a, b), then $f - g$ is constant on (a, b); that is, $f(x) = g(x) + c$ where c is a constant.

Proof Let $F(x) = f(x) - g(x)$. Then

$$F'(x) = f'(x) - g'(x) = 0$$

for all x in (a, b). Thus, by Theorem 5, F is constant; that is, $f - g$ is constant.

NOTE ▫ Care must be taken in applying Theorem 5. Let

$$f(x) = \frac{x}{|x|} = \begin{cases} 1 & \text{if } x > 0 \\ -1 & \text{if } x < 0 \end{cases}$$

The domain of f is $D = \{x \mid x \neq 0\}$ and $f'(x) = 0$ for all x in D. But f is obviously not a constant function. This does not contradict Theorem 5 because D is not an interval. Notice that f is constant on the interval $(0, \infty)$ and also on the interval $(-\infty, 0)$.

EXAMPLE 6 Prove the identity $\tan^{-1}x + \cot^{-1}x = \pi/2$.

SOLUTION Although calculus isn't needed to prove this identity, the proof using calculus is quite simple. If $f(x) = \tan^{-1}x + \cot^{-1}x$, then

$$f'(x) = \frac{1}{1 + x^2} - \frac{1}{1 + x^2} = 0$$

for all values of x. Therefore, $f(x) = C$, a constant. To determine the value of C, we put $x = 1$ [because we can evaluate $f(1)$ exactly]. Then

$$C = f(1) = \tan^{-1}1 + \cot^{-1}1 = \frac{\pi}{4} + \frac{\pi}{4} = \frac{\pi}{2}$$

Thus, $\tan^{-1}x + \cot^{-1}x = \pi/2$.

4.2 Exercises

1–4 Verify that the function satisfies the three hypotheses of Rolle's Theorem on the given interval. Then find all numbers c that satisfy the conclusion of Rolle's Theorem.

1. $f(x) = x^2 - 4x + 1$, $[0, 4]$

2. $f(x) = x^3 - 3x^2 + 2x + 5$, $[0, 2]$

3. $f(x) = \sin 2\pi x$, $[-1, 1]$

4. $f(x) = x\sqrt{x + 6}$, $[-6, 0]$

5. Let $f(x) = 1 - x^{2/3}$. Show that $f(-1) = f(1)$ but there is no number c in $(-1, 1)$ such that $f'(c) = 0$. Why does this not contradict Rolle's Theorem?

6. Let $f(x) = (x - 1)^{-2}$. Show that $f(0) = f(2)$ but there is no number c in $(0, 2)$ such that $f'(c) = 0$. Why does this not contradict Rolle's Theorem?

7. Use the graph of f to estimate the values of c that satisfy the conclusion of the Mean Value Theorem for the interval $[0, 8]$.

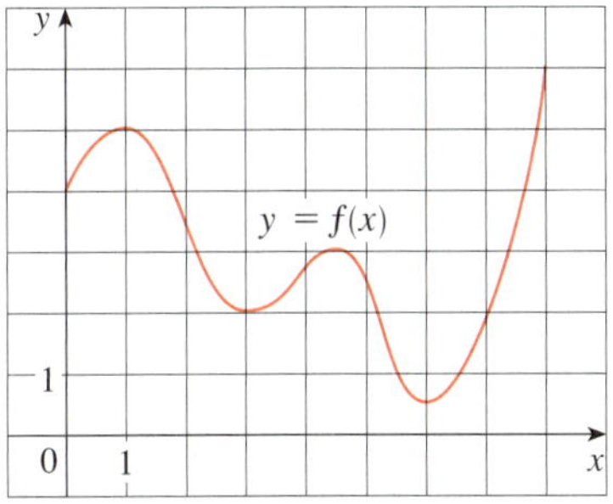

8. Use the graph of f given in Exercise 7 to estimate the values of c that satisfy the conclusion of the Mean Value Theorem for the interval $[1, 7]$.

9. (a) Graph the function $f(x) = x + 4/x$ in the viewing rectangle $[0, 10]$ by $[0, 10]$.
(b) Graph the secant line that passes through the points $(1, 5)$ and $(8, 8.5)$ on the same screen with f.
(c) Find the number c that satisfies the conclusion of the Mean Value Theorem for this function f and the interval $[1, 8]$. Then graph the tangent line at the point $(c, f(c))$ and notice that it is parallel to the secant line.

10. (a) In the viewing rectangle $[-3, 3]$ by $[-5, 5]$, graph the function $f(x) = x^3 - 2x$ and its secant line through the points $(-2, -4)$ and $(2, 4)$. Use the graph to estimate the x-coordinates of the points where the tangent line is parallel to the secant line.
(b) Find the exact values of the numbers c that satisfy the conclusion of the Mean Value Theorem for the interval $[-2, 2]$ and compare with your answers to part (a).

11–14 Verify that the function satisfies the hypotheses of the Mean Value Theorem on the given interval. Then find all numbers c that satisfy the conclusion of the Mean Value Theorem.

11. $f(x) = 3x^2 + 2x + 5$, $[-1, 1]$

12. $f(x) = x^3 + x - 1$, $[0, 2]$

13. $f(x) = e^{-2x}$, $[0, 3]$

14. $f(x) = \dfrac{x}{x + 2}$, $[1, 4]$

15. Let $f(x) = |x - 1|$. Show that there is no value of c such that $f(3) - f(0) = f'(c)(3 - 0)$. Why does this not contradict the Mean Value Theorem?

16. Let $f(x) = (x + 1)/(x - 1)$. Show that there is no value of c such that $f(2) - f(0) = f'(c)(2 - 0)$. Why does this not contradict the Mean Value Theorem?

17. Show that the equation $1 + 2x + x^3 + 4x^5 = 0$ has exactly one real root.

18. Show that the equation $2x - 1 - \sin x = 0$ has exactly one real root.

19. Show that the equation $x^3 - 15x + c = 0$ has at most one root in the interval $[-2, 2]$.

20. Show that the equation $x^4 + 4x + c = 0$ has at most two real roots.

21. (a) Show that a polynomial of degree 3 has at most three real roots.
(b) Show that a polynomial of degree n has at most n real roots.

22. (a) Suppose that f is differentiable on $\mathbb{R}$ and has two roots. Show that f' has at least one root.
(b) Suppose f is twice differentiable on $\mathbb{R}$ and has three roots. Show that f'' has at least one real root.
(c) Can you generalize parts (a) and (b)?

23. If $f(1) = 10$ and $f'(x) \geq 2$ for $1 \leq x \leq 4$, how small can $f(4)$ possibly be?

24. Suppose that $3 \leq f'(x) \leq 5$ for all values of x. Show that $18 \leq f(8) - f(2) \leq 30$.

25. Does there exist a function f such that $f(0) = -1$, $f(2) = 4$, and $f'(x) \leq 2$ for all x?

26. Suppose that f and g are continuous on $[a, b]$ and differentiable on (a, b). Suppose also that $f(a) = g(a)$ and $f'(x) < g'(x)$ for $a < x < b$. Prove that $f(b) < g(b)$. [*Hint*: Apply the Mean Value Theorem to the function $h = f - g$.]

27. Show that $\sqrt{1 + x} < 1 + \frac{1}{2}x$ if $x > 0$.

28. Suppose f is an odd function and is differentiable everywhere. Prove that for every positive number b, there exists a number c in $(-b, b)$ such that $f'(c) = f(b)/b$.

29. Use the Mean Value Theorem to prove the inequality

$$|\sin a - \sin b| \leqslant |a - b| \qquad \text{for all } a \text{ and } b$$

30. If $f'(x) = c$ (c a constant) for all x, use Corollary 7 to show that $f(x) = cx + d$ for some constant d.

31. Let $f(x) = 1/x$ and

$$g(x) = \begin{cases} \dfrac{1}{x} & \text{if } x > 0 \\ 1 + \dfrac{1}{x} & \text{if } x < 0 \end{cases}$$

Show that $f'(x) = g'(x)$ for all x in their domains. Can we conclude from Corollary 7 that $f - g$ is constant?

32. Use the method of Example 6 to prove the identity

$$2\sin^{-1}x = \cos^{-1}(1 - 2x^2) \qquad x \geqslant 0$$

33. Prove the identity

$$\arcsin\frac{x-1}{x+1} = 2\arctan\sqrt{x} - \frac{\pi}{2}$$

34. At 2:00 P.M. a car's speedometer reads 30 mi/h. At 2:10 P.M. it reads 50 mi/h. Show that at some time between 2:00 and 2:10 the acceleration is exactly 120 mi/h^2.

35. Two runners start a race at the same time and finish in a tie. Prove that at some time during the race they have the same speed. [*Hint:* Consider $f(t) = g(t) - h(t)$, where g and h are the position functions of the two runners.]

36. A number a is called a **fixed point** of a function f if $f(a) = a$. Prove that if $f'(x) \neq 1$ for all real numbers x, then f has at most one fixed point.

4.3 How Derivatives Affect the Shape of a Graph

Many of the applications of calculus depend on our ability to deduce facts about a function f from information concerning its derivatives. Because $f'(x)$ represents the slope of the curve $y = f(x)$ at the point $(x, f(x))$, it tells us the direction in which the curve proceeds at each point. So it is reasonable to expect that information about $f'(x)$ will provide us with information about $f(x)$.

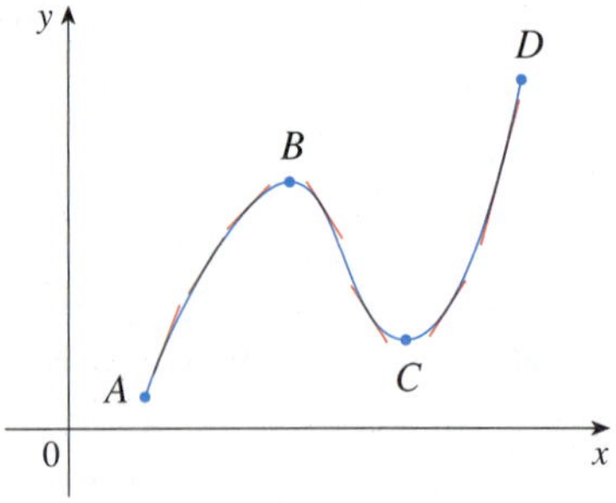

FIGURE 1

What Does f' Say about f?

To see how the derivative of f can tell us where a function is increasing or decreasing, look at Figure 1. (Increasing functions and decreasing functions were defined in Section 1.1.) Between A and B and between C and D, the tangent lines have positive slope and so $f'(x) > 0$. Between B and C, the tangent lines have negative slope and so $f'(x) < 0$. Thus, it appears that f increases when $f'(x)$ is positive and decreases when $f'(x)$ is negative. To prove that this is always the case, we use the Mean Value Theorem.

Let's abbreviate the name of this test to the I/D Test.

Increasing/Decreasing Test

(a) If $f'(x) > 0$ on an interval, then f is increasing on that interval.

(b) If $f'(x) < 0$ on an interval, then f is decreasing on that interval.

Resources / Module 3 / Increasing and Decreasing Functions / Increasing-Decreasing Detector

Proof

(a) Let x_1 and x_2 be any two numbers in the interval with $x_1 < x_2$. According to the definition of an increasing function (page 21) we have to show that $f(x_1) < f(x_2)$.

Because we are given that $f'(x) > 0$, we know that f is differentiable on $[x_1, x_2]$. So, by the Mean Value Theorem there is a number c between x_1 and x_2 such that

$$\boxed{1} \qquad f(x_2) - f(x_1) = f'(c)(x_2 - x_1)$$

Now $f'(c) > 0$ by assumption and $x_2 - x_1 > 0$ because $x_1 < x_2$. Thus, the right side of

TEC Module 4.3A guides you in determining properties of the derivative f' by examining the graphs of a variety of functions f.

Equation 1 is positive, and so

$$f(x_2) - f(x_1) > 0 \qquad \text{or} \qquad f(x_1) < f(x_2)$$

This shows that f is increasing.

Part (b) is proved similarly.

EXAMPLE 1 Find where the function $f(x) = 3x^4 - 4x^3 - 12x^2 + 5$ is increasing and where it is decreasing.

SOLUTION $$f'(x) = 12x^3 - 12x^2 - 24x = 12x(x - 2)(x + 1)$$

To use the I/D Test we have to know where $f'(x) > 0$ and where $f'(x) < 0$. This depends on the signs of the three factors of $f'(x)$, namely, $12x$, $x - 2$, and $x + 1$. We divide the real line into intervals whose endpoints are the critical numbers -1, 0, and 2 and arrange our work in a chart. A plus sign indicates that the given expression is positive, and a minus sign indicates that it is negative. The last column of the chart gives the conclusion based on the I/D Test. For instance, $f'(x) < 0$ for $0 < x < 2$, so f is decreasing on $(0, 2)$. (It would also be true to say that f is decreasing on the closed interval $[0, 2]$.)

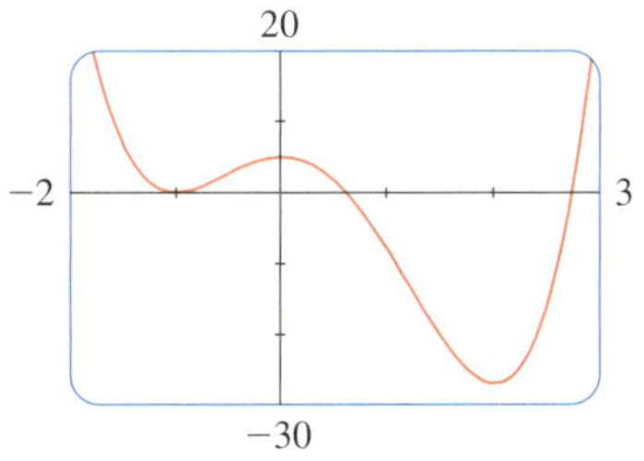

FIGURE 2

Interval	$12x$	$x - 2$	$x + 1$	$f'(x)$	f
$x < -1$	−	−	−	−	decreasing on $(-\infty, -1)$
$-1 < x < 0$	−	−	+	+	increasing on $(-1, 0)$
$0 < x < 2$	+	−	+	−	decreasing on $(0, 2)$
$x > 2$	+	+	+	+	increasing on $(2, \infty)$

The graph of f shown in Figure 2 confirms the information in the chart.

Recall from Section 4.1 that if f has a local maximum or minimum at c, then c must be a critical number of f (by Fermat's Theorem), but not every critical number gives rise to a maximum or a minimum. We therefore need a test that will tell us whether or not f has a local maximum or minimum at a critical number.

You can see from Figure 2 that $f(0) = 5$ is a local maximum value of f because f increases on $(-1, 0)$ and decreases on $(0, 2)$. Or, in terms of derivatives, $f'(x) > 0$ for $-1 < x < 0$ and $f'(x) < 0$ for $0 < x < 2$. In other words, the sign of $f'(x)$ changes from positive to negative at 0. This observation is the basis of the following test.

The First Derivative Test Suppose that c is a critical number of a continuous function f.

(a) If f' changes from positive to negative at c, then f has a local maximum at c.

(b) If f' changes from negative to positive at c, then f has a local minimum at c.

(c) If f' does not change sign at c (for example, if f' is positive on both sides of c or negative on both sides), then f has no local maximum or minimum at c.

The First Derivative Test is a consequence of the I/D Test. In part (a), for instance, since the sign of $f'(x)$ changes from positive to negative at c, f is increasing to the left of c and decreasing to the right of c. It follows that f has a local maximum at c.

It is easy to remember the First Derivative Test by visualizing diagrams such as those in Figure 3.

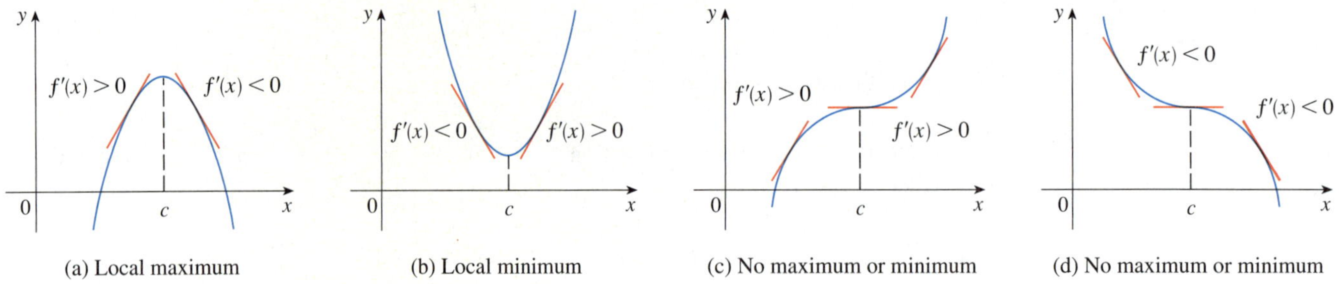

FIGURE 3

EXAMPLE 2 Find the local minimum and maximum values of the function f in Example 1.

SOLUTION From the chart in the solution to Example 1 we see that $f'(x)$ changes from negative to positive at -1, so $f(-1) = 0$ is a local minimum value by the First Derivative Test. Similarly, f' changes from negative to positive at 2, so $f(2) = -27$ is also a local minimum value. As previously noted, $f(0) = 5$ is a local maximum value because $f'(x)$ changes from positive to negative at 0.

EXAMPLE 3 Find the local maximum and minimum values of the function

$$g(x) = x + 2\sin x \qquad 0 \leqslant x \leqslant 2\pi$$

SOLUTION To find the critical numbers of g, we differentiate:

$$g'(x) = 1 + 2\cos x$$

So $g'(x) = 0$ when $\cos x = -\frac{1}{2}$. The solutions of this equation are $2\pi/3$ and $4\pi/3$. Because g is differentiable everywhere, the only critical numbers are $2\pi/3$ and $4\pi/3$ and so we analyze g in the following table.

|||| The + signs in the table come from the fact that $g'(x) > 0$ when $\cos x > -\frac{1}{2}$. From the graph of $y = \cos x$, this is true in the indicated intervals.

Interval	$g'(x) = 1 + 2\cos x$	g
$0 < x < 2\pi/3$	+	increasing on $(0, 2\pi/3)$
$2\pi/3 < x < 4\pi/3$	−	decreasing on $(2\pi/3, 4\pi/3)$
$4\pi/3 < x < 2\pi$	+	increasing on $(4\pi/3, 2\pi)$

Because $g'(x)$ changes from positive to negative at $2\pi/3$, the First Derivative Test tells us that there is a local maximum at $2\pi/3$ and the local maximum value is

$$g(2\pi/3) = \frac{2\pi}{3} + 2\sin\frac{2\pi}{3} = \frac{2\pi}{3} + 2\left(\frac{\sqrt{3}}{2}\right) = \frac{2\pi}{3} + \sqrt{3} \approx 3.83$$

Likewise, $g'(x)$ changes from negative to positive at $4\pi/3$ and so

$$g(4\pi/3) = \frac{4\pi}{3} + 2\sin\frac{4\pi}{3} = \frac{4\pi}{3} + 2\left(-\frac{\sqrt{3}}{2}\right) = \frac{4\pi}{3} - \sqrt{3} \approx 2.46$$

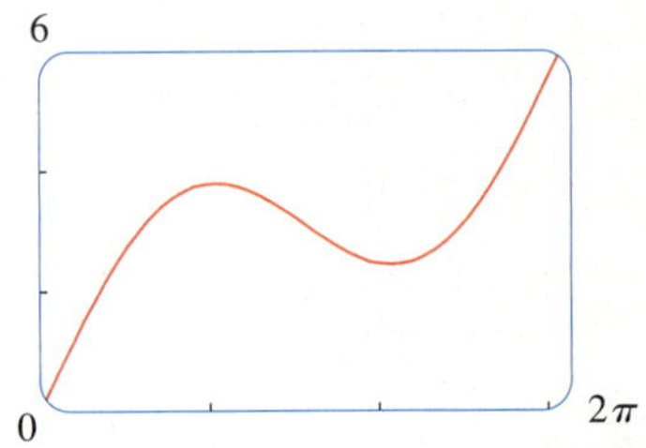

FIGURE 4
$y = x + 2\sin x$

is a local minimum value. The graph of g in Figure 4 supports our conclusion.

What Does f'' Say about f?

Explore concavity on a roller coaster.
Resources / Module 3
/ Concavity
/ Introduction

Figure 5 shows the graphs of two increasing functions on (a, b). Both graphs join point A to point B but they look different because they bend in different directions. How can we distinguish between these two types of behavior? In Figure 6 tangents to these curves have been drawn at several points. In (a) the curve lies above the tangents and f is called *concave upward* on (a, b). In (b) the curve lies below the tangents and g is called *concave downward* on (a, b).

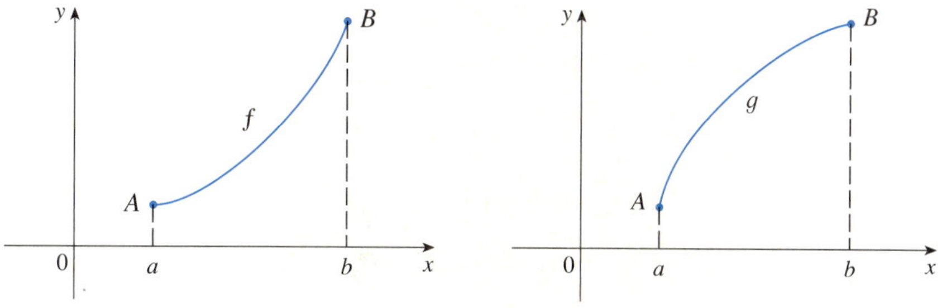

FIGURE 5 (a) (b)

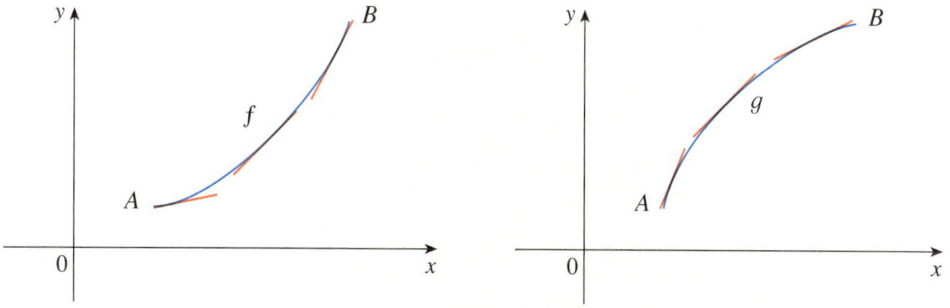

FIGURE 6 (a) Concave upward (b) Concave downward

> **Definition** If the graph of f lies above all of its tangents on an interval I, then it is called **concave upward** on I. If the graph of f lies below all of its tangents on I, it is called **concave downward** on I.

Figure 7 shows the graph of a function that is concave upward (abbreviated CU) on the intervals (b, c), (d, e), and (e, p) and concave downward (CD) on the intervals (a, b), (c, d), and (p, q).

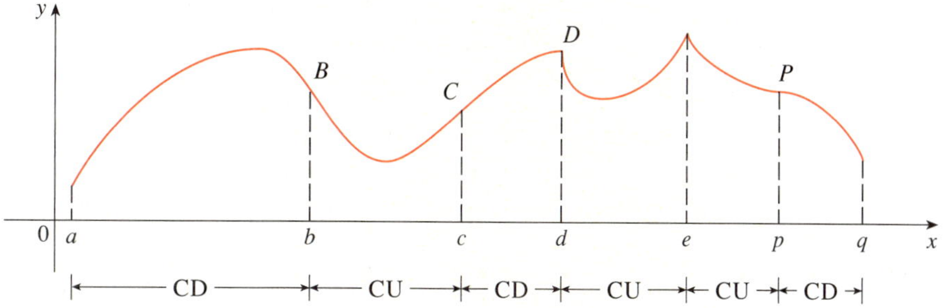

FIGURE 7

Let's see how the second derivative helps determine the intervals of concavity. Looking at Figure 6(a), you can see that, going from left to right, the slope of the tangent increases.

This means that the derivative f' is an increasing function and therefore its derivative f'' is positive. Likewise, in Figure 6(b) the slope of the tangent decreases from left to right, so f' decreases and therefore f'' is negative. This reasoning can be reversed and suggests that the following theorem is true. A proof is given in Appendix F with the help of the Mean Value Theorem.

Concavity Test

(a) If $f''(x) > 0$ for all x in I, then the graph of f is concave upward on I.

(b) If $f''(x) < 0$ for all x in I, then the graph of f is concave downward on I.

EXAMPLE 4 Figure 8 shows a population graph for Cyprian honeybees raised in an apiary. How does the rate of population increase change over time? When is this rate highest? Over what intervals is P concave upward or concave downward?

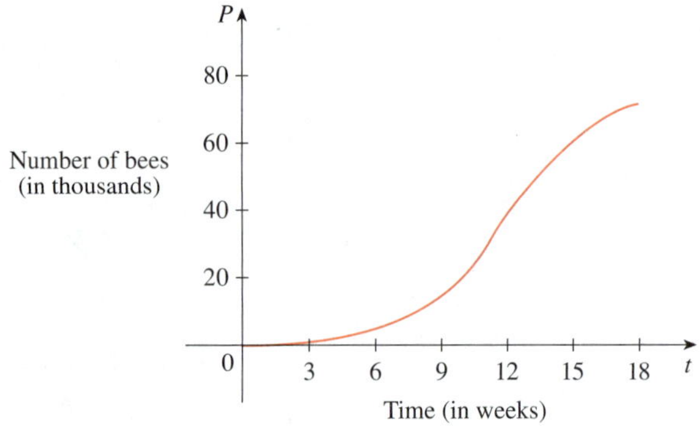

FIGURE 8

SOLUTION By looking at the slope of the curve as t increases, we see that the rate of increase of the population is initially very small, then gets larger until it reaches a maximum at about $t = 12$ weeks, and decreases as the population begins to level off. As the population approaches its maximum value of about 75,000 (called the *carrying capacity*), the rate of increase, $P'(t)$, approaches 0. The curve appears to be concave upward on (0, 12) and concave downward on (12, 18).

In Example 4, the population curve changed from concave upward to concave downward at approximately the point (12, 38,000). This point is called an *inflection point* of the curve. The significance of this point is that the rate of population increase has its maximum value there. In general, an inflection point is a point where a curve changes its direction of concavity.

Definition A point P on a curve $y = f(x)$ is called an **inflection point** if f is continuous there and the curve changes from concave upward to concave downward or from concave downward to concave upward at P.

For instance, in Figure 7, B, C, D, and P are the points of inflection. Notice that if a curve has a tangent at a point of inflection, then the curve crosses its tangent there.

In view of the Concavity Test, there is a point of inflection at any point where the second derivative changes sign.

EXAMPLE 5 Sketch a possible graph of a function f that satisfies the following conditions:

(i) $f'(x) > 0$ on $(-\infty, 1)$, $f'(x) < 0$ on $(1, \infty)$

(ii) $f''(x) > 0$ on $(-\infty, -2)$ and $(2, \infty)$, $f''(x) < 0$ on $(-2, 2)$

(iii) $\lim_{x \to -\infty} f(x) = -2$, $\lim_{x \to \infty} f(x) = 0$

SOLUTION Condition (i) tells us that f is increasing on $(-\infty, 1)$ and decreasing on $(1, \infty)$. Condition (ii) says that f is concave upward on $(-\infty, -2)$ and $(2, \infty)$, and concave downward on $(-2, 2)$. From condition (iii) we know that the graph of f has two horizontal asymptotes: $y = -2$ and $y = 0$.

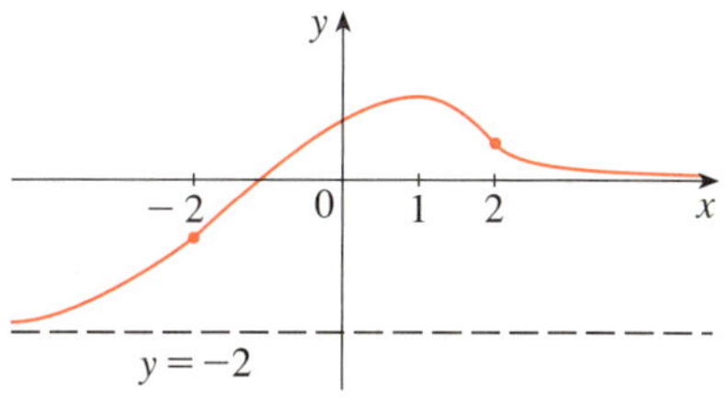

FIGURE 9

We first draw the horizontal asymptote $y = -2$ as a dashed line (see Figure 9). We then draw the graph of f approaching this asymptote at the far left, increasing to its maximum point at $x = 1$ and decreasing toward the x-axis at the far right. We also make sure that the graph has inflection points when $x = -2$ and 2. Notice that we made the curve bend upward for $x < -2$ and $x > 2$, and bend downward when x is between -2 and 2.

Another application of the second derivative is the following test for maximum and minimum values. It is a consequence of the Concavity Test.

The Second Derivative Test Suppose f'' is continuous near c.

(a) If $f'(c) = 0$ and $f''(c) > 0$, then f has a local minimum at c.

(b) If $f'(c) = 0$ and $f''(c) < 0$, then f has a local maximum at c.

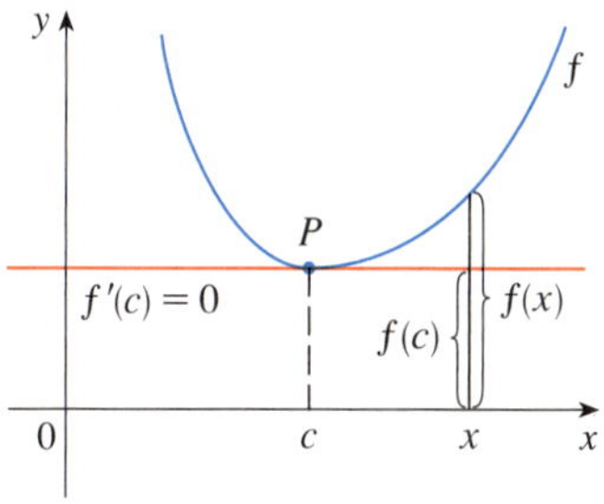

FIGURE 10
$f''(c) > 0$, f is concave upward

For instance, part (a) is true because $f''(x) > 0$ near c and so f is concave upward near c. This means that the graph of f lies *above* its horizontal tangent at c and so f has a local minimum at c. (See Figure 10.)

EXAMPLE 6 Discuss the curve $y = x^4 - 4x^3$ with respect to concavity, points of inflection, and local maxima and minima. Use this information to sketch the curve.

SOLUTION If $f(x) = x^4 - 4x^3$, then

$$f'(x) = 4x^3 - 12x^2 = 4x^2(x - 3)$$

$$f''(x) = 12x^2 - 24x = 12x(x - 2)$$

To find the critical numbers we set $f'(x) = 0$ and obtain $x = 0$ and $x = 3$. To use the Second Derivative Test we evaluate f'' at these critical numbers:

$$f''(0) = 0 \qquad f''(3) = 36 > 0$$

Since $f'(3) = 0$ and $f''(3) > 0$, $f(3) = -27$ is a local minimum. Since $f''(0) = 0$, the Second Derivative Test gives no information about the critical number 0. But since $f'(x) < 0$ for $x < 0$ and also for $0 < x < 3$, the First Derivative Test tells us that f does not have a local maximum or minimum at 0. [In fact, the expression for $f'(x)$ shows that f decreases to the left of 3 and increases to the right of 3.]

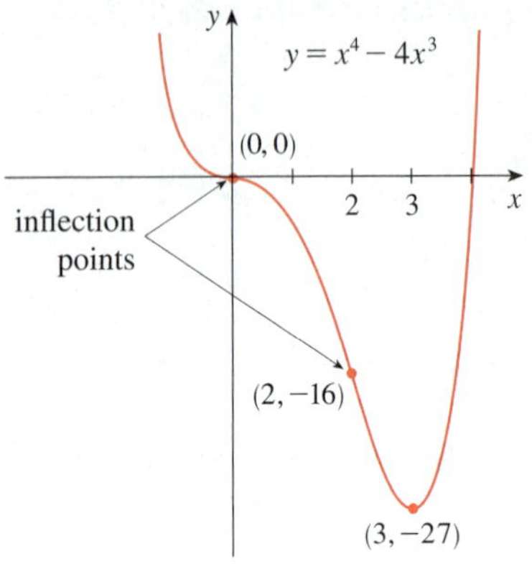

FIGURE 11

Since $f''(x) = 0$ when $x = 0$ or 2, we divide the real line into intervals with these numbers as endpoints and complete the following chart.

Interval	$f''(x) = 12x(x - 2)$	Concavity
$(-\infty, 0)$	$+$	upward
$(0, 2)$	$-$	downward
$(2, \infty)$	$+$	upward

The point (0, 0) is an inflection point since the curve changes from concave upward to concave downward there. Also $(2, -16)$ is an inflection point since the curve changes from concave downward to concave upward there.

Using the local minimum, the intervals of concavity, and the inflection points, we sketch the curve in Figure 11.

NOTE ▫ The Second Derivative Test is inconclusive when $f''(c) = 0$. In other words, at such a point there might be a maximum, there might be a minimum, or there might be neither (as in Example 6). This test also fails when $f''(c)$ does not exist. In such cases the First Derivative Test must be used. In fact, even when both tests apply, the First Derivative Test is often the easier one to use.

EXAMPLE 7 Sketch the graph of the function $f(x) = x^{2/3}(6 - x)^{1/3}$.

SOLUTION You can use the differentiation rules to check that the first two derivatives are

$$f'(x) = \frac{4 - x}{x^{1/3}(6 - x)^{2/3}} \qquad f''(x) = \frac{-8}{x^{4/3}(6 - x)^{5/3}}$$

|||| Try reproducing the graph in Figure 12 with a graphing calculator or computer. Some machines produce the complete graph, some produce only the portion to the right of the y-axis, and some produce only the portion between $x = 0$ and $x = 6$. For an explanation and cure, see Example 7 in Section 1.4. An equivalent expression that gives the correct graph is

$$y = (x^2)^{1/3} \cdot \frac{6 - x}{|6 - x|} |6 - x|^{1/3}$$

Since $f'(x) = 0$ when $x = 4$ and $f'(x)$ does not exist when $x = 0$ or $x = 6$, the critical numbers are 0, 4, and 6.

Interval	$4 - x$	$x^{1/3}$	$(6 - x)^{2/3}$	$f'(x)$	f
$x < 0$	$+$	$-$	$+$	$-$	decreasing on $(-\infty, 0)$
$0 < x < 4$	$+$	$+$	$+$	$+$	increasing on $(0, 4)$
$4 < x < 6$	$-$	$+$	$+$	$-$	decreasing on $(4, 6)$
$x > 6$	$-$	$+$	$+$	$-$	decreasing on $(6, \infty)$

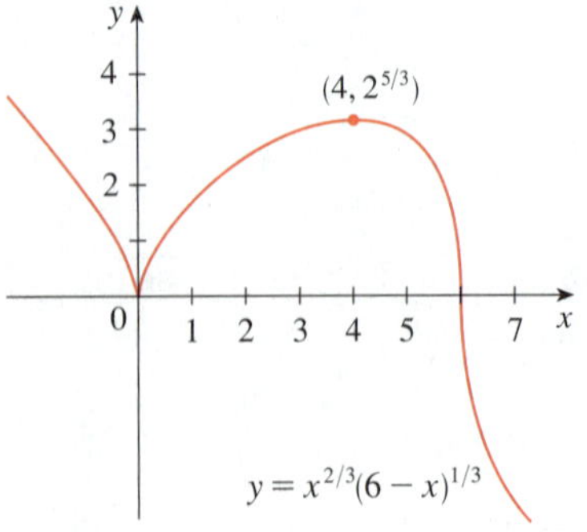

FIGURE 12

To find the local extreme values we use the First Derivative Test. Since f' changes from negative to positive at 0, $f(0) = 0$ is a local minimum. Since f' changes from positive to negative at 4, $f(4) = 2^{5/3}$ is a local maximum. The sign of f' does not change at 6, so there is no minimum or maximum there. (The Second Derivative Test could be used at 4, but not at 0 or 6 since f'' does not exist at either of these numbers.)

Looking at the expression for $f''(x)$ and noting that $x^{4/3} \geqslant 0$ for all x, we have $f''(x) < 0$ for $x < 0$ and for $0 < x < 6$ and $f''(x) > 0$ for $x > 6$. So f is concave downward on $(-\infty, 0)$ and $(0, 6)$ and concave upward on $(6, \infty)$, and the only inflection point is $(6, 0)$. The graph is sketched in Figure 12. Note that the curve has vertical tangents at $(0, 0)$ and $(6, 0)$ because $|f'(x)| \to \infty$ as $x \to 0$ and as $x \to 6$.

EXAMPLE 8 Use the first and second derivatives of $f(x) = e^{1/x}$, together with asymptotes, to sketch its graph.

SOLUTION Notice that the domain of f is $\{x \mid x \neq 0\}$, so we check for vertical asymptotes by computing the left and right limits as $x \to 0$. As $x \to 0^+$, we know that $t = 1/x \to \infty$, so

$$\lim_{x \to 0^+} e^{1/x} = \lim_{t \to \infty} e^t = \infty$$

and this shows that $x = 0$ is a vertical asymptote. As $x \to 0^-$, we have $t = 1/x \to -\infty$, so

$$\lim_{x \to 0^-} e^{1/x} = \lim_{t \to -\infty} e^t = 0$$

TEC In Module 4.3B you can practice using graphical information about f' to determine the shape of the graph of f.

As $x \to \pm\infty$, we have $1/x \to 0$ and so

$$\lim_{x \to \pm\infty} e^{1/x} = e^0 = 1$$

This shows that $y = 1$ is a horizontal asymptote.

Now let's compute the derivative. The Chain Rule gives

$$f'(x) = -\frac{e^{1/x}}{x^2}$$

Since $e^{1/x} > 0$ and $x^2 > 0$ for all $x \neq 0$, we have $f'(x) < 0$ for all $x \neq 0$. Thus, f is decreasing on $(-\infty, 0)$ and on $(0, \infty)$. There is no critical number, so the function has no maximum or minimum. The second derivative is

$$f''(x) = -\frac{x^2 e^{1/x}(-1/x^2) - e^{1/x}(2x)}{x^4} = \frac{e^{1/x}(2x + 1)}{x^4}$$

Since $e^{1/x} > 0$ and $x^4 > 0$, we have $f''(x) > 0$ when $x > -\frac{1}{2}$ $(x \neq 0)$ and $f''(x) < 0$ when $x < -\frac{1}{2}$. So the curve is concave downward on $\left(-\infty, -\frac{1}{2}\right)$ and concave upward on $\left(-\frac{1}{2}, 0\right)$ and on $(0, \infty)$. The inflection point is $\left(-\frac{1}{2}, e^{-2}\right)$.

To sketch the graph of f we first draw the horizontal asymptote $y = 1$ (as a dashed line), together with the parts of the curve near the asymptotes in a preliminary sketch [Figure 13(a)]. These parts reflect the information concerning limits and the fact that f is decreasing on both $(-\infty, 0)$ and $(0, \infty)$. Notice that we have indicated that $f(x) \to 0$ as $x \to 0^-$ even though $f(0)$ does not exist. In Figure 13(b) we finish the sketch by incorporating the information concerning concavity and the inflection point. In Figure 13(c) we check our work with a graphing device.

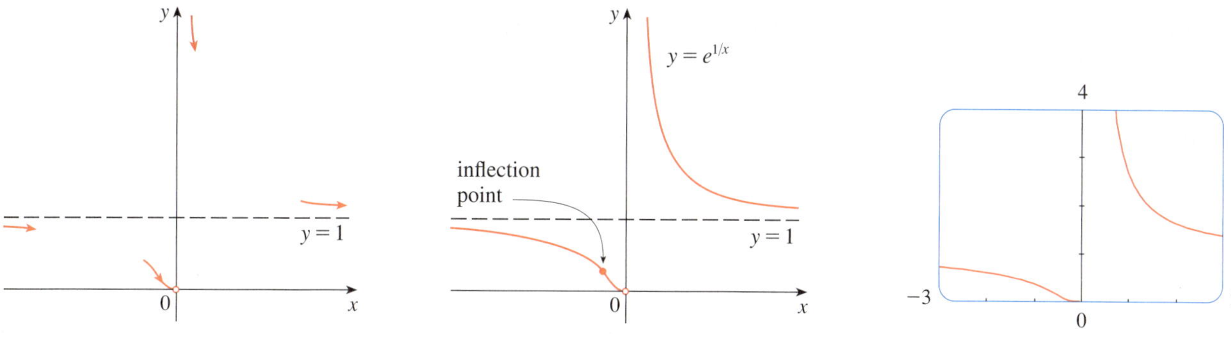

(a) Preliminary sketch (b) Finished sketch (c) Computer confirmation

FIGURE 13

4.3 Exercises

1–2 Use the given graph of f to find the following.
(a) The open intervals on which f is increasing.
(b) The open intervals on which f is decreasing.
(c) The open intervals on which f is concave upward.
(d) The open intervals on which f is concave downward.
(e) The coordinates of the points of inflection.

1.

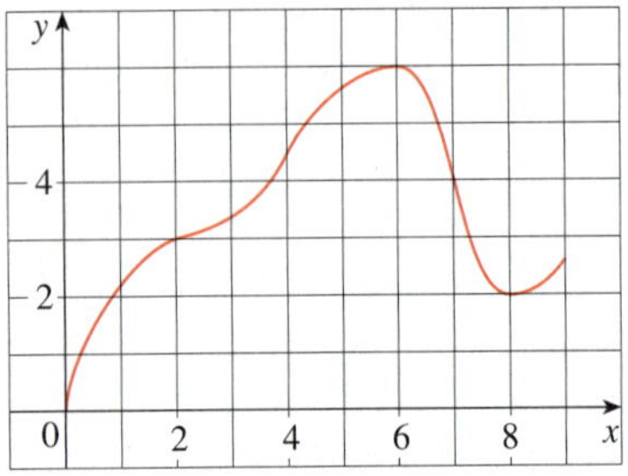

2.

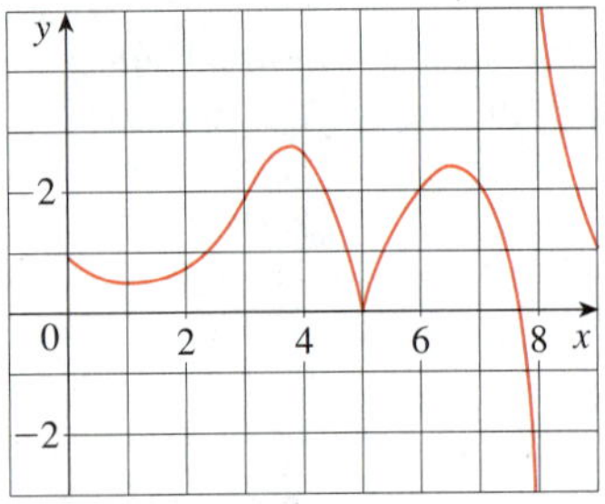

3. Suppose you are given a formula for a function f.
(a) How do you determine where f is increasing or decreasing?
(b) How do you determine where the graph of f is concave upward or concave downward?
(c) How do you locate inflection points?

4. (a) State the First Derivative Test.
(b) State the Second Derivative Test. Under what circumstances is it inconclusive? What do you do if it fails?

5–6 The graph of the *derivative* f' of a function f is shown.
(a) On what intervals is f increasing or decreasing?
(b) At what values of x does f have a local maximum or minimum?

5.

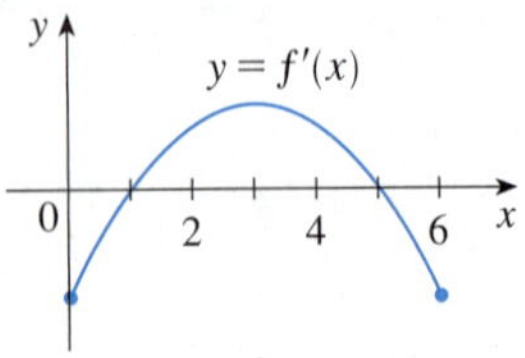

6.

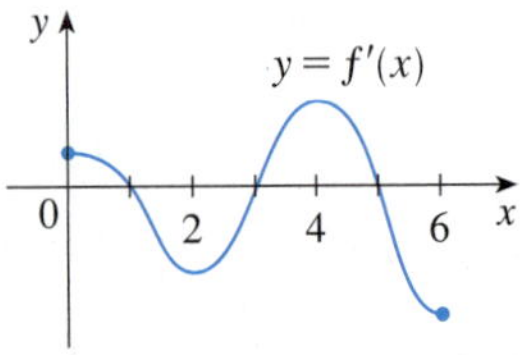

7. The graph of the second derivative f'' of a function f is shown. State the x-coordinates of the inflection points of f. Give reasons for your answers.

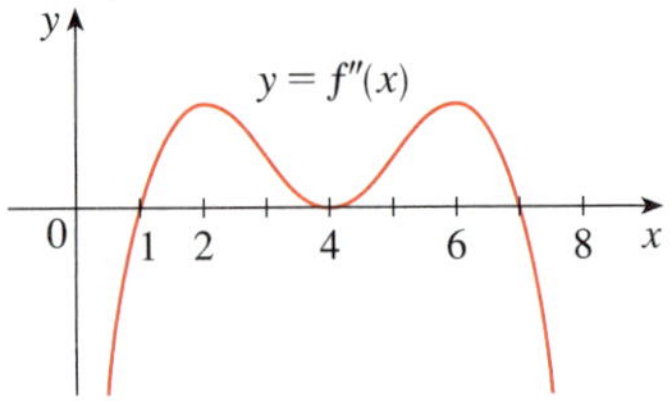

8. The graph of the first derivative f' of a function f is shown.
(a) On what intervals is f increasing? Explain.
(b) At what values of x does f have a local maximum or minimum? Explain.
(c) On what intervals is f concave upward or concave downward? Explain.
(d) What are the x-coordinates of the inflection points of f? Why?

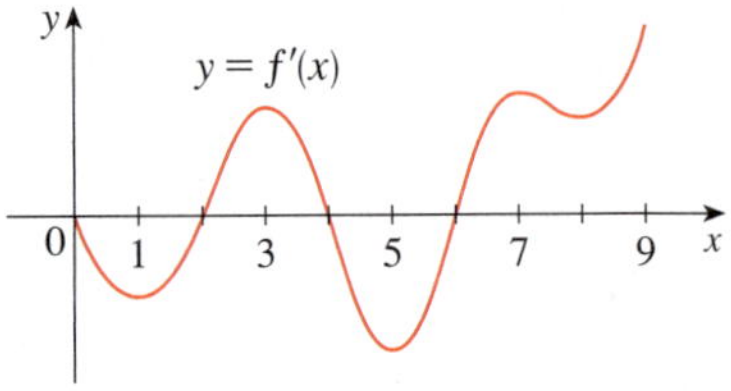

9. Sketch the graph of a function whose first and second derivatives are always negative.

10. A graph of a population of yeast cells in a new laboratory culture as a function of time is shown.
(a) Describe how the rate of population increase varies.
(b) When is this rate highest?
(c) On what intervals is the population function concave upward or downward?
(d) Estimate the coordinates of the inflection point.

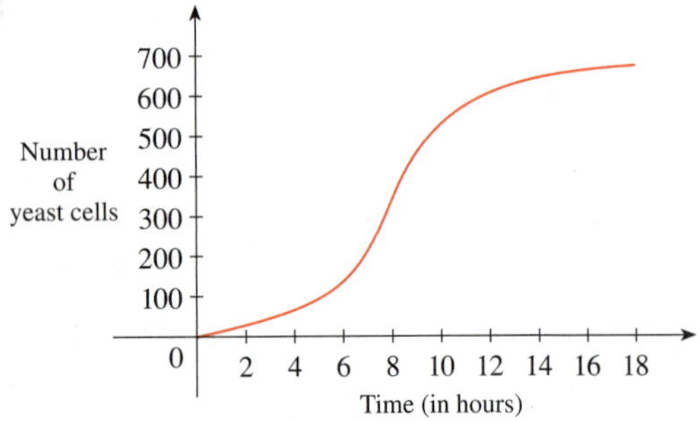

11–20
(a) Find the intervals on which f is increasing or decreasing.
(b) Find the local maximum and minimum values of f.
(c) Find the intervals of concavity and the inflection points.

11. $f(x) = x^3 - 12x + 1$

12. $f(x) = 5 - 3x^2 + x^3$

13. $f(x) = x^4 - 2x^2 + 3$

14. $f(x) = \dfrac{x^2}{x^2 + 3}$

15. $f(x) = x - 2\sin x, \quad 0 < x < 3\pi$

16. $f(x) = \cos^2 x - 2\sin x, \quad 0 \leqslant x \leqslant 2\pi$

17. $f(x) = xe^x$

18. $f(x) = x^2 e^x$

19. $f(x) = (\ln x)/\sqrt{x}$

20. $f(x) = x \ln x$

21–23 ||| Find the local maximum and minimum values of f using both the First and Second Derivative Tests. Which method do you prefer?

21. $f(x) = x^5 - 5x + 3$

22. $f(x) = \dfrac{x}{x^2 + 4}$

23. $f(x) = x + \sqrt{1 - x}$

24. (a) Find the critical numbers of $f(x) = x^4(x - 1)^3$.
(b) What does the Second Derivative Test tell you about the behavior of f at these critical numbers?
(c) What does the First Derivative Test tell you?

25. Suppose f'' is continuous on $(-\infty, \infty)$.
(a) If $f'(2) = 0$ and $f''(2) = -5$, what can you say about f?
(b) If $f'(6) = 0$ and $f''(6) = 0$, what can you say about f?

26–30 ||| Sketch the graph of a function that satisfies all of the given conditions.

26. $f'(x) > 0$ for all $x \neq 1$, vertical asymptote $x = 1$,
$f''(x) > 0$ if $x < 1$ or $x > 3$, $f''(x) < 0$ if $1 < x < 3$

27. $f'(0) = f'(2) = f'(4) = 0$,
$f'(x) > 0$ if $x < 0$ or $2 < x < 4$,
$f'(x) < 0$ if $0 < x < 2$ or $x > 4$,
$f''(x) > 0$ if $1 < x < 3$, $f''(x) < 0$ if $x < 1$ or $x > 3$

28. $f'(1) = f'(-1) = 0$, $f'(x) < 0$ if $|x| < 1$,
$f'(x) > 0$ if $1 < |x| < 2$, $f'(x) = -1$ if $|x| > 2$,
$f''(x) < 0$ if $-2 < x < 0$, inflection point $(0, 1)$

29. $f'(x) > 0$ if $|x| < 2$, $f'(x) < 0$ if $|x| > 2$,
$f'(-2) = 0$, $\lim_{x \to 2} |f'(x)| = \infty$, $f''(x) > 0$ if $x \neq 2$

30. $f'(x) > 0$ if $|x| < 2$, $f'(x) < 0$ if $|x| > 2$,
$f'(2) = 0$, $\lim_{x \to \infty} f(x) = 1$, $f(-x) = -f(x)$,
$f''(x) < 0$ if $0 < x < 3$, $f''(x) > 0$ if $x > 3$

31–32 ||| The graph of the derivative f' of a continuous function f is shown.
(a) On what intervals is f increasing or decreasing?
(b) At what values of x does f have a local maximum or minimum?
(c) On what intervals is f concave upward or downward?
(d) State the x-coordinate(s) of the point(s) of inflection.
(e) Assuming that $f(0) = 0$, sketch a graph of f.

31.

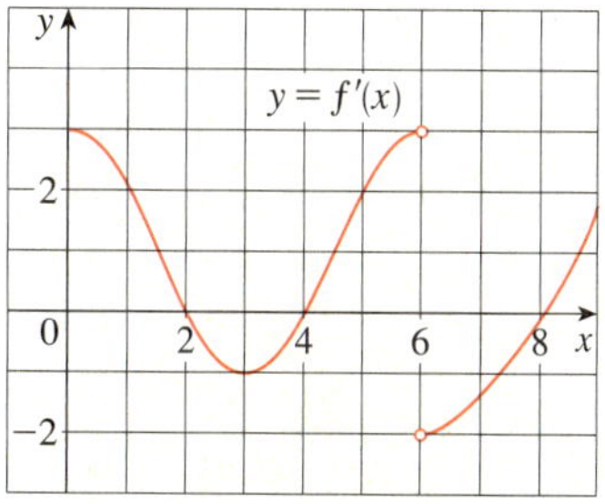

32.

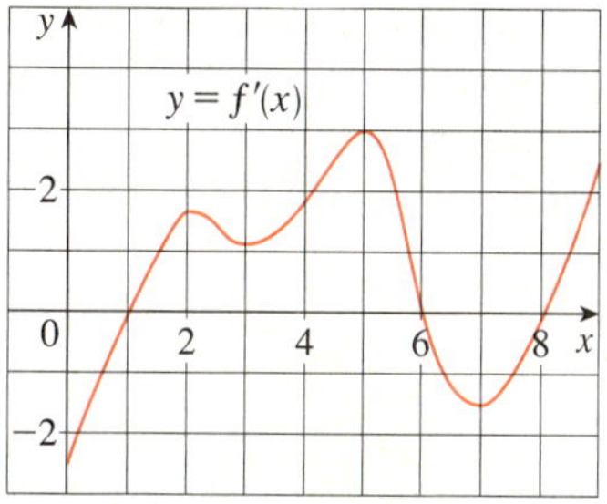

33–44 |||
(a) Find the intervals of increase or decrease.
(b) Find the local maximum and minimum values.
(c) Find the intervals of concavity and the inflection points.
(d) Use the information from parts (a)–(c) to sketch the graph. Check your work with a graphing device if you have one.

33. $f(x) = 2x^3 - 3x^2 - 12x$

34. $f(x) = 2 + 3x - x^3$

35. $f(x) = x^4 - 6x^2$

36. $g(x) = 200 + 8x^3 + x^4$

37. $h(x) = 3x^5 - 5x^3 + 3$

38. $h(x) = (x^2 - 1)^3$

39. $A(x) = x\sqrt{x + 3}$

40. $B(x) = 3x^{2/3} - x$

41. $C(x) = x^{1/3}(x + 4)$

42. $f(x) = \ln(x^4 + 27)$

43. $f(\theta) = 2\cos\theta - \cos 2\theta, \quad 0 \leqslant \theta \leqslant 2\pi$

44. $f(t) = t + \cos t, \quad -2\pi \leqslant t \leqslant 2\pi$

45–52 |||
(a) Find the vertical and horizontal asymptotes.
(b) Find the intervals of increase or decrease.
(c) Find the local maximum and minimum values.
(d) Find the intervals of concavity and the inflection points.
(e) Use the information from parts (a)–(d) to sketch the graph of f.

45. $f(x) = \dfrac{x^2}{x^2 - 1}$

46. $f(x) = \dfrac{x^2}{(x - 2)^2}$

47. $f(x) = \sqrt{x^2 + 1} - x$

48. $f(x) = x \tan x, \quad -\pi/2 < x < \pi/2$

49. $f(x) = \ln(1 - \ln x)$

50. $f(x) = \dfrac{e^x}{1 + e^x}$

51. $f(x) = e^{-1/(x+1)}$ **52.** $f(x) = \ln(\tan^2 x)$

53–54 |||
(a) Use a graph of f to estimate the maximum and minimum values. Then find the exact values.
(b) Estimate the value of x at which f increases most rapidly. Then find the exact value.

53. $f(x) = \dfrac{x+1}{\sqrt{x^2+1}}$ **54.** $f(x) = x^2 e^{-x}$

55–56 |||
(a) Use a graph of f to give a rough estimate of the intervals of concavity and the coordinates of the points of inflection.
(b) Use a graph of f'' to give better estimates.

55. $f(x) = \cos x + \frac{1}{2}\cos 2x, \quad 0 \leqslant x \leqslant 2\pi$

56. $f(x) = x^3(x-2)^4$

CAS **57–58** ||| Estimate the intervals of concavity to one decimal place by using a computer algebra system to compute and graph f''.

57. $f(x) = \dfrac{x^3 - 10x + 5}{\sqrt{x^2+4}}$ **58.** $f(x) = \dfrac{(x+1)^3(x^2+5)}{(x^3+1)(x^2+4)}$

59. Let $K(t)$ be a measure of the knowledge you gain by studying for a test for t hours. Which do you think is larger, $K(8) - K(7)$ or $K(3) - K(2)$? Is the graph of K concave upward or concave downward? Why?

60. Coffee is being poured into the mug shown in the figure at a constant rate (measured in volume per unit time). Sketch a rough graph of the depth of the coffee in the mug as a function of time. Account for the shape of the graph in terms of concavity. What is the significance of the inflection point?

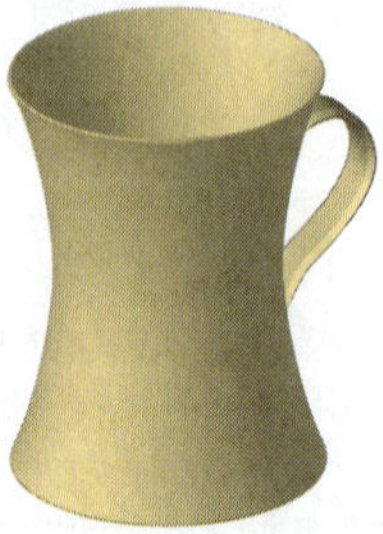

61. For the period from 1980 to 2000, the percentage of households in the United States with at least one VCR has been modeled by the function

$$V(t) = \frac{85}{1 + 53e^{-0.5t}}$$

where the time t is measured in years since midyear 1980, so $0 \leqslant t \leqslant 20$. Use a graph to estimate the time at which the number of VCRs was increasing most rapidly. Then use derivatives to give a more accurate estimate.

62. The family of bell-shaped curves

$$y = \frac{1}{\sigma\sqrt{2\pi}} e^{-(x-\mu)^2/(2\sigma^2)}$$

occurs in probability and statistics, where it is called the *normal density function.* The constant μ is called the *mean* and the positive constant σ is called the *standard deviation.* For simplicity, let's scale the function so as to remove the factor $1/(\sigma\sqrt{2\pi})$ and let's analyze the special case where $\mu = 0$. So we study the function

$$f(x) = e^{-x^2/(2\sigma^2)}$$

(a) Find the asymptote, maximum value, and inflection points of f.
(b) What role does σ play in the shape of the curve?
(c) Illustrate by graphing four members of this family on the same screen.

63. Find a cubic function $f(x) = ax^3 + bx^2 + cx + d$ that has a local maximum value of 3 at -2 and a local minimum value of 0 at 1.

64. For what values of the numbers a and b does the function

$$f(x) = axe^{bx^2}$$

have the maximum value $f(2) = 1$?

65. Suppose f is differentiable on an interval I and $f'(x) > 0$ for all numbers x in I except for a single number c. Prove that f is increasing on the entire interval I.

66–68 ||| Assume that all of the functions are twice differentiable and the second derivatives are never 0.

66. (a) If f and g are concave upward on I, show that $f + g$ is concave upward on I.
(b) If f is positive and concave upward on I, show that the function $g(x) = [f(x)]^2$ is concave upward on I.

67. (a) If f and g are positive, increasing, concave upward functions on I, show that the product function fg is concave upward on I.
(b) Show that part (a) remains true if f and g are both decreasing.
(c) Suppose f is increasing and g is decreasing. Show, by giving three examples, that fg may be concave upward, concave downward, or linear. Why doesn't the argument in parts (a) and (b) work in this case?

68. Suppose f and g are both concave upward on $(-\infty, \infty)$. Under what condition on f will the composite function $h(x) = f(g(x))$ be concave upward?

69. Show that $\tan x > x$ for $0 < x < \pi/2$. [*Hint:* Show that $f(x) = \tan x - x$ is increasing on $(0, \pi/2)$.]

70. (a) Show that $e^x \geqslant 1 + x$ for $x \geqslant 0$.
(b) Deduce that $e^x \geqslant 1 + x + \frac{1}{2}x^2$ for $x \geqslant 0$.
(c) Use mathematical induction to prove that for $x \geqslant 0$ and any positive integer n,

$$e^x \geqslant 1 + x + \frac{x^2}{2!} + \cdots + \frac{x^n}{n!}$$

71. Show that a cubic function (a third-degree polynomial) always has exactly one point of inflection. If its graph has three x-intercepts x_1, x_2, and x_3, show that the x-coordinate of the inflection point is $(x_1 + x_2 + x_3)/3$.

72. For what values of c does the polynomial $P(x) = x^4 + cx^3 + x^2$ have two inflection points? One inflection point? None? Illustrate by graphing P for several values of c. How does the graph change as c decreases?

73. Prove that if $(c, f(c))$ is a point of inflection of the graph of f and f'' exists in an open interval that contains c, then $f''(c) = 0$. [*Hint:* Apply the First Derivative Test and Fermat's Theorem to the function $g = f'$.]

74. Show that if $f(x) = x^4$, then $f''(0) = 0$, but $(0, 0)$ is not an inflection point of the graph of f.

75. Show that the function $g(x) = x|x|$ has an inflection point at $(0, 0)$ but $g''(0)$ does not exist.

76. Suppose that f''' is continuous and $f'(c) = f''(c) = 0$, but $f'''(c) > 0$. Does f have a local maximum or minimum at c? Does f have a point of inflection at c?

4.4 Indeterminate Forms and L'Hospital's Rule

Suppose we are trying to analyze the behavior of the function

$$F(x) = \frac{\ln x}{x - 1}$$

Although F is not defined when $x = 1$, we need to know how F behaves *near* 1. In particular, we would like to know the value of the limit

$$\boxed{1} \qquad \lim_{x \to 1} \frac{\ln x}{x - 1}$$

In computing this limit we can't apply Law 5 of limits (the limit of a quotient is the quotient of the limits, see Section 2.3) because the limit of the denominator is 0. In fact, although the limit in (1) exists, its value is not obvious because both numerator and denominator approach 0 and $\frac{0}{0}$ is not defined.

In general, if we have a limit of the form

$$\lim_{x \to a} \frac{f(x)}{g(x)}$$

where both $f(x) \to 0$ and $g(x) \to 0$ as $x \to a$, then this limit may or may not exist and is called an **indeterminate form of type $\frac{0}{0}$**. We met some limits of this type in Chapter 2. For rational functions, we can cancel common factors:

$$\lim_{x \to 1} \frac{x^2 - x}{x^2 - 1} = \lim_{x \to 1} \frac{x(x - 1)}{(x + 1)(x - 1)} = \lim_{x \to 1} \frac{x}{x + 1} = \frac{1}{2}$$

We used a geometric argument to show that

$$\lim_{x \to 0} \frac{\sin x}{x} = 1$$

But these methods do not work for limits such as (1), so in this section we introduce a systematic method, known as *l'Hospital's Rule,* for the evaluation of indeterminate forms.

Another situation in which a limit is not obvious occurs when we look for a horizontal asymptote of F and need to evaluate the limit

$$\boxed{2} \qquad \lim_{x \to \infty} \frac{\ln x}{x - 1}$$

It isn't obvious how to evaluate this limit because both numerator and denominator become large as $x \to \infty$. There is a struggle between numerator and denominator. If the numerator wins, the limit will be ∞; if the denominator wins, the answer will be 0. Or there may be some compromise, in which case the answer may be some finite positive number.

In general, if we have a limit of the form

$$\lim_{x \to a} \frac{f(x)}{g(x)}$$

where both $f(x) \to \infty$ (or $-\infty$) and $g(x) \to \infty$ (or $-\infty$), then the limit may or may not exist and is called an **indeterminate form of type** ∞/∞. We saw in Section 2.6 that this type of limit can be evaluated for certain functions, including rational functions, by dividing numerator and denominator by the highest power of x that occurs in the denominator. For instance,

$$\lim_{x \to \infty} \frac{x^2 - 1}{2x^2 + 1} = \lim_{x \to \infty} \frac{1 - \dfrac{1}{x^2}}{2 + \dfrac{1}{x^2}} = \frac{1 - 0}{2 + 0} = \frac{1}{2}$$

This method does not work for limits such as (2), but l'Hospital's Rule also applies to this type of indeterminate form.

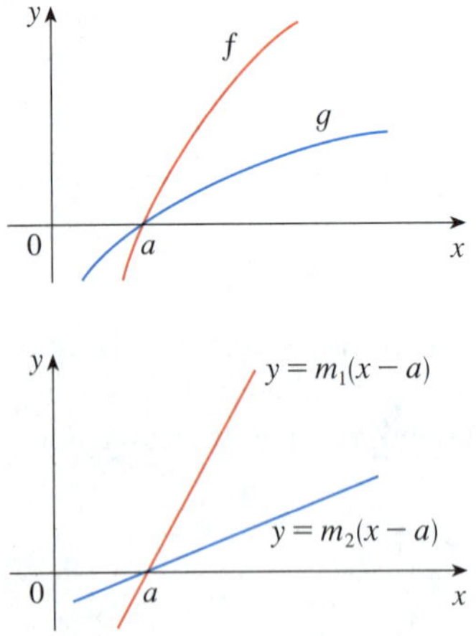

FIGURE 1

|||| Figure 1 suggests visually why l'Hospital's Rule might be true. The first graph shows two differentiable functions f and g, each of which approaches 0 as $x \to a$. If we were to zoom in toward the point $(a, 0)$, the graphs would start to look almost linear. But if the functions actually *were* linear, as in the second graph, then their ratio would be

$$\frac{m_1(x - a)}{m_2(x - a)} = \frac{m_1}{m_2}$$

which is the ratio of their derivatives. This suggests that

$$\lim_{x \to a} \frac{f(x)}{g(x)} = \lim_{x \to a} \frac{f'(x)}{g'(x)}$$

L'Hospital's Rule Suppose f and g are differentiable and $g'(x) \neq 0$ on an open interval I that contains a (except possibly at a). Suppose that

$$\lim_{x \to a} f(x) = 0 \qquad \text{and} \qquad \lim_{x \to a} g(x) = 0$$

or that

$$\lim_{x \to a} f(x) = \pm\infty \qquad \text{and} \qquad \lim_{x \to a} g(x) = \pm\infty$$

(In other words, we have an indeterminate form of type $\frac{0}{0}$ or ∞/∞.) Then

$$\lim_{x \to a} \frac{f(x)}{g(x)} = \lim_{x \to a} \frac{f'(x)}{g'(x)}$$

if the limit on the right side exists (or is ∞ or $-\infty$).

NOTE 1 ▫ L'Hospital's Rule says that the limit of a quotient of functions is equal to the limit of the quotient of their derivatives, provided that the given conditions are satisfied. It is especially important to verify the conditions regarding the limits of f and g before using l'Hospital's Rule.

NOTE 2 ▫ L'Hospital's Rule is also valid for one-sided limits and for limits at infinity or negative infinity; that is, "$x \to a$" can be replaced by any of the symbols $x \to a^+$, $x \to a^-$, $x \to \infty$, or $x \to -\infty$.

NOTE 3 ▫ For the special case in which $f(a) = g(a) = 0$, f' and g' are continuous, and $g'(a) \neq 0$, it is easy to see why l'Hospital's Rule is true. In fact, using the alternative form of the definition of a derivative, we have

$$\lim_{x \to a} \frac{f'(x)}{g'(x)} = \frac{f'(a)}{g'(a)} = \frac{\displaystyle\lim_{x \to a} \frac{f(x) - f(a)}{x - a}}{\displaystyle\lim_{x \to a} \frac{g(x) - g(a)}{x - a}}$$

$$= \lim_{x \to a} \frac{\dfrac{f(x) - f(a)}{x - a}}{\dfrac{g(x) - g(a)}{x - a}}$$

$$= \lim_{x \to a} \frac{f(x) - f(a)}{g(x) - g(a)}$$

$$= \lim_{x \to a} \frac{f(x)}{g(x)}$$

It is more difficult to prove the general version of l'Hospital's Rule. See Appendix F.

|||| L'Hospital's Rule is named after a French nobleman, the Marquis de l'Hospital (1661–1704), but was discovered by a Swiss mathematician, John Bernoulli (1667–1748). See Exercise 71 for the example that the Marquis used to illustrate his rule. See the project on page 315 for further historical details.

EXAMPLE 1 Find $\displaystyle\lim_{x \to 1} \frac{\ln x}{x - 1}$.

SOLUTION Since

$$\lim_{x \to 1} \ln x = \ln 1 = 0 \qquad \text{and} \qquad \lim_{x \to 1} (x - 1) = 0$$

we can apply l'Hospital's Rule:

$$\lim_{x \to 1} \frac{\ln x}{x - 1} = \lim_{x \to 1} \frac{\dfrac{d}{dx}(\ln x)}{\dfrac{d}{dx}(x - 1)} = \lim_{x \to 1} \frac{1/x}{1}$$

$$= \lim_{x \to 1} \frac{1}{x} = 1$$

Notice that when using l'Hospital's Rule we differentiate the numerator and denominator *separately*. We do *not* use the Quotient Rule.

EXAMPLE 2 Calculate $\displaystyle\lim_{x \to \infty} \frac{e^x}{x^2}$.

SOLUTION We have $\lim_{x \to \infty} e^x = \infty$ and $\lim_{x \to \infty} x^2 = \infty$, so l'Hospital's Rule gives

$$\lim_{x \to \infty} \frac{e^x}{x^2} = \lim_{x \to \infty} \frac{\dfrac{d}{dx}(e^x)}{\dfrac{d}{dx}(x^2)} = \lim_{x \to \infty} \frac{e^x}{2x}$$

Since $e^x \to \infty$ and $2x \to \infty$ as $x \to \infty$, the limit on the right side is also indeterminate, but a second application of l'Hospital's Rule gives

$$\lim_{x \to \infty} \frac{e^x}{x^2} = \lim_{x \to \infty} \frac{e^x}{2x} = \lim_{x \to \infty} \frac{e^x}{2} = \infty$$

|||| The graph of the function of Example 2 is shown in Figure 2. We have noticed previously that exponential functions grow far more rapidly than power functions, so the result of Example 2 is not unexpected. See also Exercise 67.

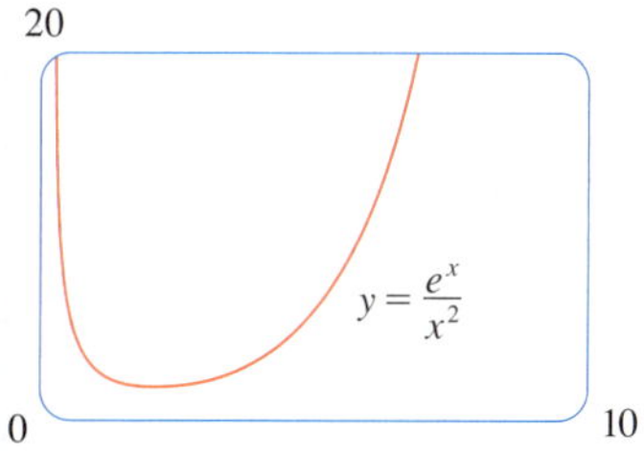

FIGURE 2

|||| The graph of the function of Example 3 is shown in Figure 3. We have discussed previously the slow growth of logarithms, so it isn't surprising that this ratio approaches 0 as $x \to \infty$. See also Exercise 68.

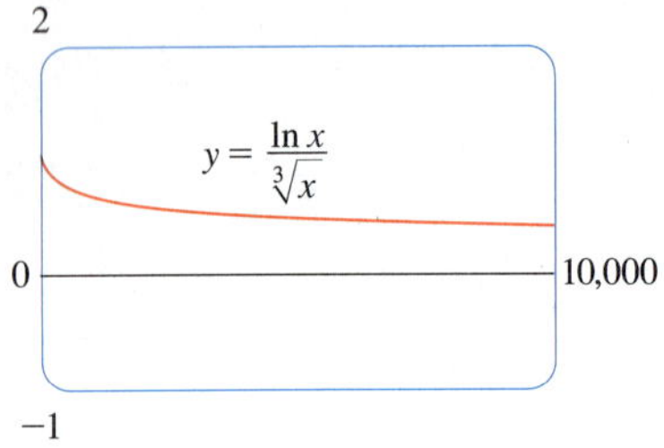

FIGURE 3

EXAMPLE 3 Calculate $\lim_{x\to\infty} \dfrac{\ln x}{\sqrt[3]{x}}$.

SOLUTION Since $\ln x \to \infty$ and $\sqrt[3]{x} \to \infty$ as $x \to \infty$, l'Hospital's Rule applies:

$$\lim_{x\to\infty} \frac{\ln x}{\sqrt[3]{x}} = \lim_{x\to\infty} \frac{1/x}{\frac{1}{3}x^{-2/3}}$$

Notice that the limit on the right side is now indeterminate of type $\frac{0}{0}$. But instead of applying l'Hospital's Rule a second time as we did in Example 2, we simplify the expression and see that a second application is unnecessary:

$$\lim_{x\to\infty} \frac{\ln x}{\sqrt[3]{x}} = \lim_{x\to\infty} \frac{1/x}{\frac{1}{3}x^{-2/3}} = \lim_{x\to\infty} \frac{3}{\sqrt[3]{x}} = 0$$

EXAMPLE 4 Find $\lim_{x\to 0} \dfrac{\tan x - x}{x^3}$. [See Exercise 36(d) in Section 2.2.]

SOLUTION Noting that both $\tan x - x \to 0$ and $x^3 \to 0$ as $x \to 0$, we use l'Hospital's Rule:

$$\lim_{x\to 0} \frac{\tan x - x}{x^3} = \lim_{x\to 0} \frac{\sec^2 x - 1}{3x^2}$$

Since the limit on the right side is still indeterminate of type $\frac{0}{0}$, we apply l'Hospital's Rule again:

$$\lim_{x\to 0} \frac{\sec^2 x - 1}{3x^2} = \lim_{x\to 0} \frac{2\sec^2 x \tan x}{6x}$$

|||| The graph in Figure 4 gives visual confirmation of the result of Example 4. If we were to zoom in too far, however, we would get an inaccurate graph because $\tan x$ is close to x when x is small. See Exercise 36(d) in Section 2.2.

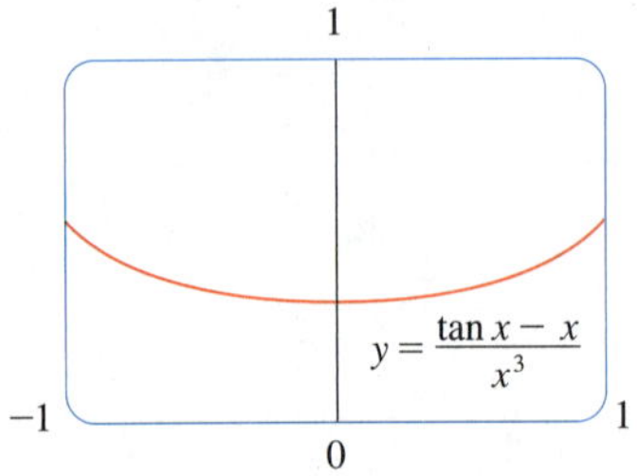

FIGURE 4

Because $\lim_{x\to 0} \sec^2 x = 1$, we simplify the calculation by writing

$$\lim_{x\to 0} \frac{2\sec^2 x \tan x}{6x} = \frac{1}{3} \lim_{x\to 0} \sec^2 x \lim_{x\to 0} \frac{\tan x}{x} = \frac{1}{3} \lim_{x\to 0} \frac{\tan x}{x}$$

We can evaluate this last limit either by using l'Hospital's Rule a third time or by writing $\tan x$ as $(\sin x)/(\cos x)$ and making use of our knowledge of trigonometric limits. Putting together all the steps, we get

$$\lim_{x\to 0} \frac{\tan x - x}{x^3} = \lim_{x\to 0} \frac{\sec^2 x - 1}{3x^2} = \lim_{x\to 0} \frac{2\sec^2 x \tan x}{6x}$$

$$= \frac{1}{3} \lim_{x\to 0} \frac{\tan x}{x} = \frac{1}{3} \lim_{x\to 0} \frac{\sec^2 x}{1} = \frac{1}{3}$$

EXAMPLE 5 Find $\lim_{x\to \pi^-} \dfrac{\sin x}{1 - \cos x}$.

SOLUTION If we blindly attempted to use l'Hospital's Rule, we would get

$$\lim_{x\to \pi^-} \frac{\sin x}{1 - \cos x} = \lim_{x\to \pi^-} \frac{\cos x}{\sin x} = -\infty$$

This is *wrong*! Although the numerator $\sin x \to 0$ as $x \to \pi^-$, notice that the denominator $(1 - \cos x)$ does not approach 0, so l'Hospital's Rule can't be applied here.

The required limit is, in fact, easy to find because the function is continuous and the denominator is nonzero at π:

$$\lim_{x\to\pi^-} \frac{\sin x}{1 - \cos x} = \frac{\sin \pi}{1 - \cos \pi} = \frac{0}{1 - (-1)} = 0$$

Example 5 shows what can go wrong if you use l'Hospital's Rule without thinking. Other limits *can* be found using l'Hospital's Rule but are more easily found by other methods. (See Examples 3 and 5 in Section 2.3, Example 3 in Section 2.6, and the discussion at the beginning of this section.) So when evaluating any limit, you should consider other methods before using l'Hospital's Rule.

Indeterminate Products

If $\lim_{x\to a} f(x) = 0$ and $\lim_{x\to a} g(x) = \infty$ (or $-\infty$), then it isn't clear what the value of $\lim_{x\to a} f(x)g(x)$, if any, will be. There is a struggle between f and g. If f wins, the answer will be 0; if g wins, the answer will be ∞ (or $-\infty$). Or there may be a compromise where the answer is a finite nonzero number. This kind of limit is called an **indeterminate form of type $0 \cdot \infty$**. We can deal with it by writing the product fg as a quotient:

$$fg = \frac{f}{1/g} \qquad \text{or} \qquad fg = \frac{g}{1/f}$$

This converts the given limit into an indeterminate form of type $\frac{0}{0}$ or ∞/∞ so that we can use l'Hospital's Rule.

EXAMPLE 6 Evaluate $\lim_{x\to 0^+} x \ln x$.

||| Figure 5 shows the graph of the function in Example 6. Notice that the function is undefined at $x = 0$; the graph approaches the origin but never quite reaches it.

FIGURE 5

SOLUTION The given limit is indeterminate because, as $x \to 0^+$, the first factor (x) approaches 0 while the second factor ($\ln x$) approaches $-\infty$. Writing $x = 1/(1/x)$, we have $1/x \to \infty$ as $x \to 0^+$, so l'Hospital's Rule gives

$$\lim_{x\to 0^+} x \ln x = \lim_{x\to 0^+} \frac{\ln x}{1/x} = \lim_{x\to 0^+} \frac{1/x}{-1/x^2} = \lim_{x\to 0^+} (-x) = 0$$

NOTE ▫ In solving Example 6 another possible option would have been to write

$$\lim_{x\to 0^+} x \ln x = \lim_{x\to 0^+} \frac{x}{1/\ln x}$$

This gives an indeterminate form of the type 0/0, but if we apply l'Hospital's Rule we get a more complicated expression than the one we started with. In general, when we rewrite an indeterminate product, we try to choose the option that leads to the simpler limit.

Indeterminate Differences

If $\lim_{x\to a} f(x) = \infty$ and $\lim_{x\to a} g(x) = \infty$, then the limit

$$\lim_{x\to a} [f(x) - g(x)]$$

is called an **indeterminate form of type $\infty - \infty$**. Again there is a contest between f and g. Will the answer be ∞ (f wins) or will it be $-\infty$ (g wins) or will they compromise on a finite number? To find out, we try to convert the difference into a quotient (for instance, by using a common denominator, or rationalization, or factoring out a common factor) so that we have an indeterminate form of type $\frac{0}{0}$ or ∞/∞.

EXAMPLE 7 Compute $\lim_{x \to (\pi/2)^-} (\sec x - \tan x)$.

SOLUTION First notice that $\sec x \to \infty$ and $\tan x \to \infty$ as $x \to (\pi/2)^-$, so the limit is indeterminate. Here we use a common denominator:

$$\lim_{x \to (\pi/2)^-} (\sec x - \tan x) = \lim_{x \to (\pi/2)^-} \left(\frac{1}{\cos x} - \frac{\sin x}{\cos x} \right)$$

$$= \lim_{x \to (\pi/2)^-} \frac{1 - \sin x}{\cos x} = \lim_{x \to (\pi/2)^-} \frac{-\cos x}{-\sin x} = 0$$

Note that the use of l'Hospital's Rule is justified because $1 - \sin x \to 0$ and $\cos x \to 0$ as $x \to (\pi/2)^-$.

Indeterminate Powers

Several indeterminate forms arise from the limit

$$\lim_{x \to a} [f(x)]^{g(x)}$$

1. $\lim_{x \to a} f(x) = 0$ and $\lim_{x \to a} g(x) = 0$ type 0^0
2. $\lim_{x \to a} f(x) = \infty$ and $\lim_{x \to a} g(x) = 0$ type ∞^0
3. $\lim_{x \to a} f(x) = 1$ and $\lim_{x \to a} g(x) = \pm\infty$ type 1^∞

Each of these three cases can be treated either by taking the natural logarithm:

$$\text{let } y = [f(x)]^{g(x)}, \quad \text{then} \quad \ln y = g(x) \ln f(x)$$

or by writing the function as an exponential:

$$[f(x)]^{g(x)} = e^{g(x) \ln f(x)}$$

(Recall that both of these methods were used in differentiating such functions.) In either method we are led to the indeterminate product $g(x) \ln f(x)$, which is of type $0 \cdot \infty$.

EXAMPLE 8 Calculate $\lim_{x \to 0^+} (1 + \sin 4x)^{\cot x}$.

SOLUTION First notice that as $x \to 0^+$, we have $1 + \sin 4x \to 1$ and $\cot x \to \infty$, so the given limit is indeterminate. Let

$$y = (1 + \sin 4x)^{\cot x}$$

Then $$\ln y = \ln[(1 + \sin 4x)^{\cot x}] = \cot x \ln(1 + \sin 4x)$$

so l'Hospital's Rule gives

$$\lim_{x \to 0^+} \ln y = \lim_{x \to 0^+} \frac{\ln(1 + \sin 4x)}{\tan x} = \lim_{x \to 0^+} \frac{\dfrac{4 \cos 4x}{1 + \sin 4x}}{\sec^2 x} = 4$$

So far we have computed the limit of $\ln y$, but what we want is the limit of y. To find this we use the fact that $y = e^{\ln y}$:

$$\lim_{x \to 0^+} (1 + \sin 4x)^{\cot x} = \lim_{x \to 0^+} y = \lim_{x \to 0^+} e^{\ln y} = e^4$$

EXAMPLE 9 Find $\lim_{x \to 0^+} x^x$.

|||| The graph of the function $y = x^x$, $x > 0$, is shown in Figure 6. Notice that although 0^0 is not defined, the values of the function approach 1 as $x \to 0^+$. This confirms the result of Example 9.

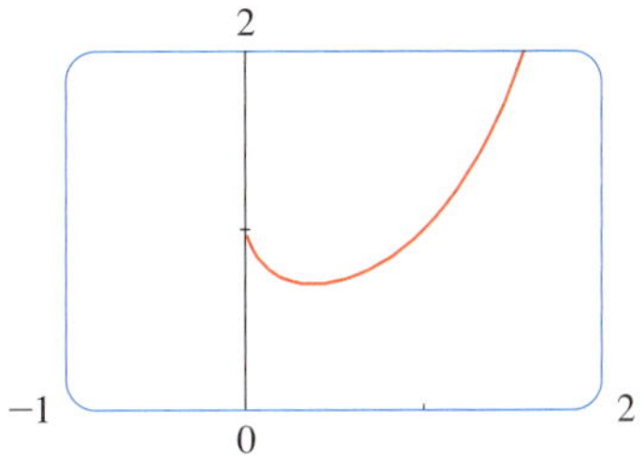

FIGURE 6

SOLUTION Notice that this limit is indeterminate since $0^x = 0$ for any $x > 0$ but $x^0 = 1$ for any $x \neq 0$. We could proceed as in Example 8 or by writing the function as an exponential:

$$x^x = (e^{\ln x})^x = e^{x \ln x}$$

In Example 6 we used l'Hospital's Rule to show that

$$\lim_{x \to 0^+} x \ln x = 0$$

Therefore

$$\lim_{x \to 0^+} x^x = \lim_{x \to 0^+} e^{x \ln x} = e^0 = 1$$

4.4 Exercises

1–4 |||| Given that

$$\lim_{x \to a} f(x) = 0 \qquad \lim_{x \to a} g(x) = 0 \qquad \lim_{x \to a} h(x) = 1$$

$$\lim_{x \to a} p(x) = \infty \qquad \lim_{x \to a} q(x) = \infty$$

which of the following limits are indeterminate forms? For those that are not an indeterminate form, evaluate the limit where possible.

1. (a) $\lim_{x \to a} \dfrac{f(x)}{g(x)}$ (b) $\lim_{x \to a} \dfrac{f(x)}{p(x)}$

(c) $\lim_{x \to a} \dfrac{h(x)}{p(x)}$ (d) $\lim_{x \to a} \dfrac{p(x)}{f(x)}$

(e) $\lim_{x \to a} \dfrac{p(x)}{q(x)}$

2. (a) $\lim_{x \to a} [f(x)p(x)]$ (b) $\lim_{x \to a} [h(x)p(x)]$

(c) $\lim_{x \to a} [p(x)q(x)]$

3. (a) $\lim_{x \to a} [f(x) - p(x)]$ (b) $\lim_{x \to a} [p(x) - q(x)]$

(c) $\lim_{x \to a} [p(x) + q(x)]$

4. (a) $\lim_{x \to a} [f(x)]^{g(x)}$ (b) $\lim_{x \to a} [f(x)]^{p(x)}$ (c) $\lim_{x \to a} [h(x)]^{p(x)}$

(d) $\lim_{x \to a} [p(x)]^{f(x)}$ (e) $\lim_{x \to a} [p(x)]^{q(x)}$ (f) $\lim_{x \to a} \sqrt[q(x)]{p(x)}$

5–62 |||| Find the limit. Use l'Hospital's Rule where appropriate. If there is a more elementary method, consider using it. If l'Hospital's Rule doesn't apply, explain why.

5. $\lim_{x \to -1} \dfrac{x^2 - 1}{x + 1}$

6. $\lim_{x \to -2} \dfrac{x + 2}{x^2 + 3x + 2}$

7. $\lim_{x \to 1} \dfrac{x^9 - 1}{x^5 - 1}$

8. $\lim_{x \to 1} \dfrac{x^a - 1}{x^b - 1}$

9. $\lim_{x \to (\pi/2)^+} \dfrac{\cos x}{1 - \sin x}$

10. $\lim_{x \to 0} \dfrac{x + \tan x}{\sin x}$

11. $\lim_{t \to 0} \dfrac{e^t - 1}{t^3}$

12. $\lim_{t \to 0} \dfrac{e^{3t} - 1}{t}$

13. $\lim_{x \to 0} \dfrac{\tan px}{\tan qx}$

14. $\lim_{\theta \to \pi/2} \dfrac{1 - \sin \theta}{\csc \theta}$

15. $\lim_{x \to \infty} \dfrac{\ln x}{x}$

16. $\lim_{x \to \infty} \dfrac{e^x}{x}$

17. $\lim_{x \to 0^+} \dfrac{\ln x}{x}$

18. $\lim_{x \to \infty} \dfrac{\ln \ln x}{x}$

19. $\lim_{t \to 0} \dfrac{5^t - 3^t}{t}$

20. $\lim_{x \to 1} \dfrac{\ln x}{\sin \pi x}$

21. $\lim_{x \to 0} \dfrac{e^x - 1 - x}{x^2}$

22. $\lim_{x \to 0} \dfrac{e^x - 1 - x - (x^2/2)}{x^3}$

23. $\lim_{x \to \infty} \dfrac{e^x}{x^3}$

24. $\lim_{x \to 0} \dfrac{\sin x}{\sinh x}$

25. $\lim_{x\to 0} \frac{\sin^{-1}x}{x}$

26. $\lim_{x\to 0} \frac{\sin x - x}{x^3}$

27. $\lim_{x\to 0} \frac{1 - \cos x}{x^2}$

28. $\lim_{x\to \infty} \frac{(\ln x)^2}{x}$

29. $\lim_{x\to 0} \frac{x + \sin x}{x + \cos x}$

30. $\lim_{x\to 0} \frac{\cos mx - \cos nx}{x^2}$

31. $\lim_{x\to \infty} \frac{x}{\ln(1 + 2e^x)}$

32. $\lim_{x\to 0} \frac{x}{\tan^{-1}(4x)}$

33. $\lim_{x\to 1} \frac{1 - x + \ln x}{1 + \cos \pi x}$

34. $\lim_{x\to \infty} \frac{\sqrt{x^2 + 2}}{\sqrt{2x^2 + 1}}$

35. $\lim_{x\to 1} \frac{x^a - ax + a - 1}{(x - 1)^2}$

36. $\lim_{x\to 0} \frac{1 - e^{-2x}}{\sec x}$

37. $\lim_{x\to 0^+} \sqrt{x} \ln x$

38. $\lim_{x\to -\infty} x^2 e^x$

39. $\lim_{x\to 0} \cot 2x \sin 6x$

40. $\lim_{x\to 0^+} \sin x \ln x$

41. $\lim_{x\to \infty} x^3 e^{-x^2}$

42. $\lim_{x\to \pi/4} (1 - \tan x) \sec x$

43. $\lim_{x\to 1^+} \ln x \tan(\pi x/2)$

44. $\lim_{x\to \infty} x \tan(1/x)$

45. $\lim_{x\to 0} \left(\frac{1}{x} - \csc x\right)$

46. $\lim_{x\to 0} (\csc x - \cot x)$

47. $\lim_{x\to \infty} \left(\sqrt{x^2 + x} - x\right)$

48. $\lim_{x\to 1} \left(\frac{1}{\ln x} - \frac{1}{x - 1}\right)$

49. $\lim_{x\to \infty} (x - \ln x)$

50. $\lim_{x\to \infty} (xe^{1/x} - x)$

51. $\lim_{x\to 0^+} x^{x^2}$

52. $\lim_{x\to 0^+} (\tan 2x)^x$

53. $\lim_{x\to 0} (1 - 2x)^{1/x}$

54. $\lim_{x\to \infty} \left(1 + \frac{a}{x}\right)^{bx}$

55. $\lim_{x\to \infty} \left(1 + \frac{3}{x} + \frac{5}{x^2}\right)^x$

56. $\lim_{x\to \infty} x^{(\ln 2)/(1 + \ln x)}$

57. $\lim_{x\to \infty} x^{1/x}$

58. $\lim_{x\to \infty} (e^x + x)^{1/x}$

59. $\lim_{x\to \infty} \left(\frac{x}{x + 1}\right)^x$

60. $\lim_{x\to 0} (\cos 3x)^{5/x}$

61. $\lim_{x\to 0^+} (\cos x)^{1/x^2}$

62. $\lim_{x\to \infty} \left(\frac{2x - 3}{2x + 5}\right)^{2x+1}$

63–64 ||| Use a graph to estimate the value of the limit. Then use l'Hospital's Rule to find the exact value.

63. $\lim_{x\to \infty} x[\ln(x + 5) - \ln x]$

64. $\lim_{x\to \pi/4} (\tan x)^{\tan 2x}$

65–66 ||| Illustrate l'Hospital's Rule by graphing both $f(x)/g(x)$ and $f'(x)/g'(x)$ near $x = 0$ to see that these ratios have the same limit as $x \to 0$. Also calculate the exact value of the limit.

65. $f(x) = e^x - 1, \quad g(x) = x^3 + 4x$

66. $f(x) = 2x \sin x, \quad g(x) = \sec x - 1$

67. Prove that

$$\lim_{x\to \infty} \frac{e^x}{x^n} = \infty$$

for any positive integer n. This shows that the exponential function approaches infinity faster than any power of x.

68. Prove that

$$\lim_{x\to \infty} \frac{\ln x}{x^p} = 0$$

for any number $p > 0$. This shows that the logarithmic function approaches ∞ more slowly than any power of x.

69. If an initial amount A_0 of money is invested at an interest rate i compounded n times a year, the value of the investment after t years is

$$A = A_0\left(1 + \frac{i}{n}\right)^{nt}$$

If we let $n \to \infty$, we refer to the *continuous compounding* of interest. Use l'Hospital's Rule to show that if interest is compounded continuously, then the amount after n years is

$$A = A_0 e^{it}$$

70. If an object with mass m is dropped from rest, one model for its speed v after t seconds, taking air resistance into account, is

$$v = \frac{mg}{c}(1 - e^{-ct/m})$$

where g is the acceleration due to gravity and c is a positive constant. (In Chapter 9 we will be able to deduce this equation from the assumption that the air resistance is proportional to the speed of the object.)

(a) Calculate $\lim_{t\to\infty} v$. What is the meaning of this limit?

(b) For fixed t, use l'Hospital's Rule to calculate $\lim_{m\to\infty} v$. What can you conclude about the speed of a very heavy falling object?

71. The first appearance in print of l'Hospital's Rule was in the book *Analyse des Infiniment Petits* published by the Marquis de l'Hospital in 1696. This was the first calculus *textbook* ever published and the example that the Marquis used in that book to illustrate his rule was to find the limit of the function

$$y = \frac{\sqrt{2a^3x - x^4} - a\sqrt[3]{aax}}{a - \sqrt[4]{ax^3}}$$

as x approaches a, where $a > 0$. (At that time it was common to write aa instead of a^2.) Solve this problem.

72. The figure shows a sector of a circle with central angle θ. Let $A(\theta)$ be the area of the segment between the chord PR and the arc PR. Let $B(\theta)$ be the area of the triangle PQR. Find $\lim_{\theta \to 0^+} A(\theta)/B(\theta)$.

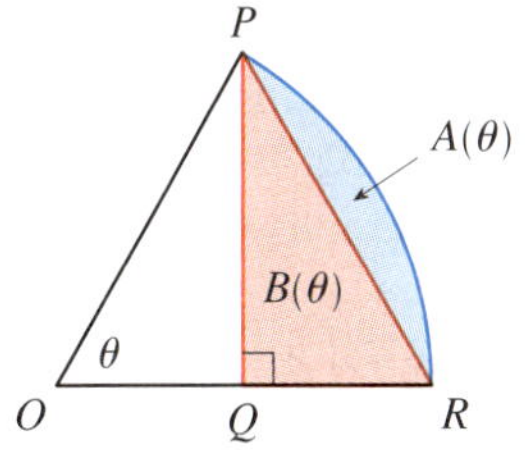

73. If f' is continuous, $f(2) = 0$, and $f'(2) = 7$, evaluate

$$\lim_{x \to 0} \frac{f(2 + 3x) + f(2 + 5x)}{x}$$

74. For what values of a and b is the following equation true?

$$\lim_{x \to 0} \left(\frac{\sin 2x}{x^3} + a + \frac{b}{x^2} \right) = 0$$

75. If f' is continuous, use l'Hospital's Rule to show that

$$\lim_{h \to 0} \frac{f(x + h) - f(x - h)}{2h} = f'(x)$$

Explain the meaning of this equation with the aid of a diagram.

76. If f'' is continuous, show that

$$\lim_{h \to 0} \frac{f(x + h) - 2f(x) + f(x - h)}{h^2} = f''(x)$$

77. Let

$$f(x) = \begin{cases} e^{-1/x^2} & \text{if } x \neq 0 \\ 0 & \text{if } x = 0 \end{cases}$$

(a) Use the definition of derivative to compute $f'(0)$.

(b) Show that f has derivatives of all orders that are defined on $\mathbb{R}$. [*Hint:* First show by induction that there is a polynomial $p_n(x)$ and a nonnegative integer k_n such that $f^{(n)}(x) = p_n(x)f(x)/x^{k_n}$ for $x \neq 0$.]

78. Let

$$f(x) = \begin{cases} |x|^x & \text{if } x \neq 0 \\ 1 & \text{if } x = 0 \end{cases}$$

(a) Show that f is continuous at 0.

(b) Investigate graphically whether f is differentiable at 0 by zooming in several times toward the point (0, 1) on the graph of f.

(c) Show that f is not differentiable at 0. How can you reconcile this fact with the appearance of the graphs in part (b)?

WRITING PROJECT

The Origins of L'Hospital's Rule

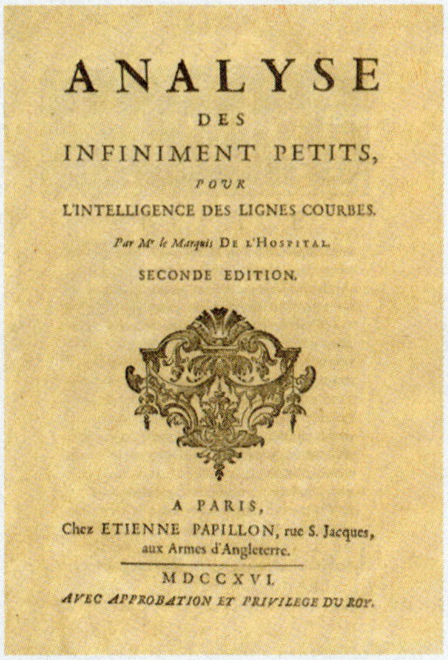
ANALYSE
DES
INFINIMENT PETITS,
POUR
L'INTELLIGENCE DES LIGNES COURBES.
Par M^r le Marquis DE L'HOSPITAL.
SECONDE EDITION.

A PARIS,
Chez ETIENNE PAPILLON, rue S. Jacques, aux Armes d'Angleterre.
MDCCXVI.
AVEC APPROBATION ET PRIVILEGE DU ROY.

L'Hospital's Rule was first published in 1696 in the Marquis de l'Hospital's calculus textbook *Analyse des Infiniment Petits,* but the rule was discovered in 1694 by the Swiss mathematician John (Johann) Bernoulli. The explanation is that these two mathematicians had entered into a curious business arrangement whereby the Marquis de l'Hospital bought the rights to Bernoulli's mathematical discoveries. The details, including a translation of l'Hospital's letter to Bernoulli proposing the arrangement, can be found in the book by Eves [1].

Write a report on the historical and mathematical origins of l'Hospital's Rule. Start by providing brief biographical details of both men (the dictionary edited by Gillispie [2] is a good source) and outline the business deal between them. Then give l'Hospital's statement of his rule, which is found in Struik's sourcebook [4] and more briefly in the book of Katz [3]. Notice that l'Hospital and Bernoulli formulated the rule geometrically and gave the answer in terms of differentials. Compare their statement with the version of l'Hospital's Rule given in Section 4.4 and show that the two statements are essentially the same.

1. Howard Eves, *In Mathematical Circles (Volume 2: Quadrants III and IV)* (Boston: Prindle, Weber and Schmidt, 1969), pp. 20–22.

2. C. C. Gillispie, ed., *Dictionary of Scientific Biography* (New York: Scribner's, 1974). See the article on Johann Bernoulli by E. A. Fellmann and J. O. Fleckenstein in Volume II and the article on the Marquis de l'Hospital by Abraham Robinson in Volume VIII.

3. Victor Katz, *A History of Mathematics: An Introduction* (New York: HarperCollins, 1993), p. 484.

4. D. J. Struik, ed., *A Sourcebook in Mathematics, 1200–1800* (Princeton, NJ: Princeton University Press, 1969), pp. 315–316.

|||| The Internet is another source of information for this project. See the web site www.stewartcalculus.com and click on *History of Mathematics*.

4.5 Summary of Curve Sketching

So far we have been concerned with some particular aspects of curve sketching: domain, range, and symmetry in Chapter 1; limits, continuity, and asymptotes in Chapter 2; derivatives and tangents in Chapters 2 and 3; and extreme values, intervals of increase and decrease, concavity, points of inflection, and l'Hospital's Rule in this chapter. It is now time to put all of this information together to sketch graphs that reveal the important features of functions.

You may ask: What is wrong with just using a calculator to plot points and then joining these points with a smooth curve? To see the pitfalls of this approach, suppose you have used a calculator to produce the table of values and corresponding points in Figure 1.

x	$f(x)$	x	$f(x)$
-5	22	1	7
-4	7	2	10
-3	-2	3	11
-2	-4	4	10
-1	-2	5	8
0	3	6	-8

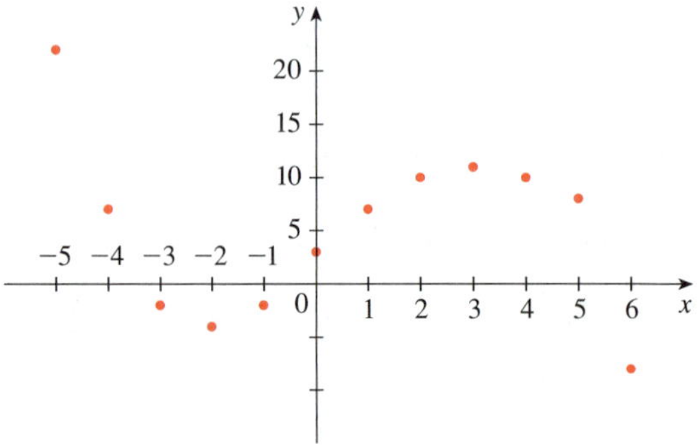

FIGURE 1

You might then join these points to produce the curve shown in Figure 2, but the correct graph might be the one shown in Figure 3. You can see the drawbacks of the method of plotting points. Certain essential features of the graph may be missed, such as the maximum and minimum values between -2 and -1 or between 2 and 5. If you just plot points, you don't know when to stop. (How far should you plot to the left or right?) But the use of calculus ensures that all the important aspects of the curve are illustrated.

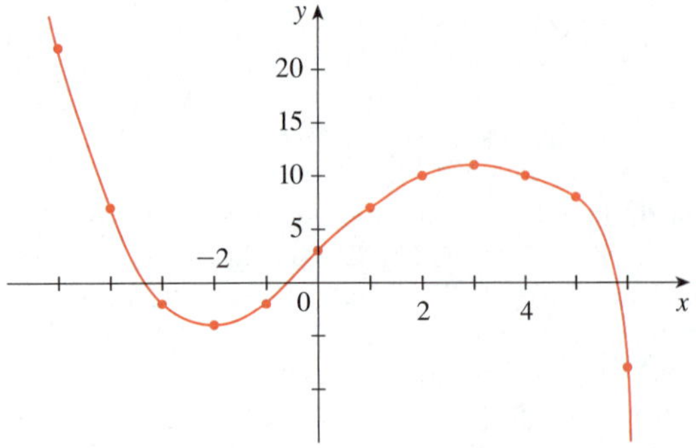

FIGURE 2

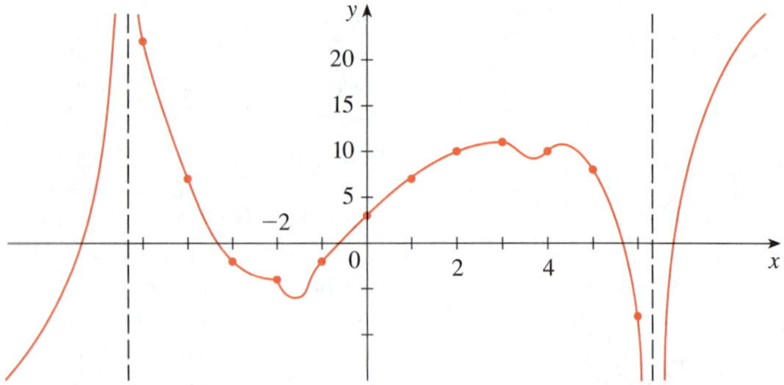

FIGURE 3

You might respond: Yes, but what about graphing calculators and computers? Don't they plot such a huge number of points that the sort of uncertainty demonstrated by Figures 2 and 3 is unlikely to happen?

It's true that modern technology is capable of producing very accurate graphs. But even the best graphing devices have to be used intelligently. We saw in Section 1.4 that it is extremely important to choose an appropriate viewing rectangle to avoid getting a misleading graph. (See especially Examples 1, 3, 4, and 5 in that section.) The use of calculus

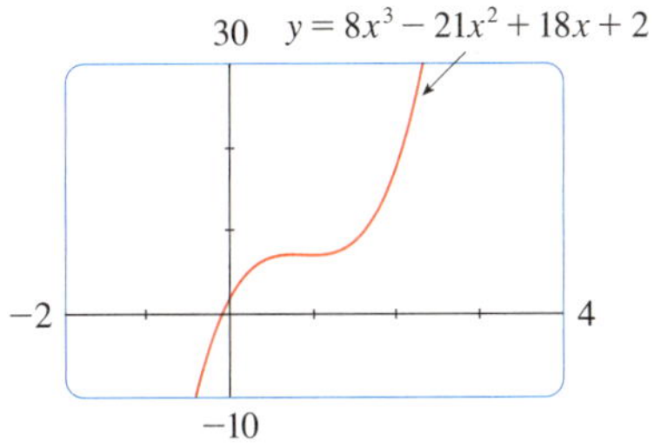

FIGURE 4

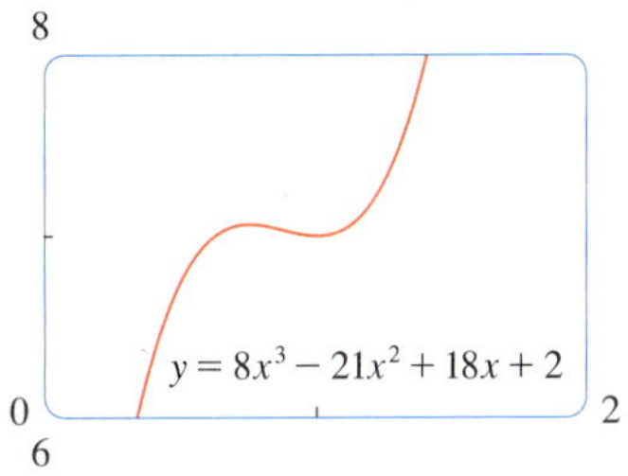

FIGURE 5

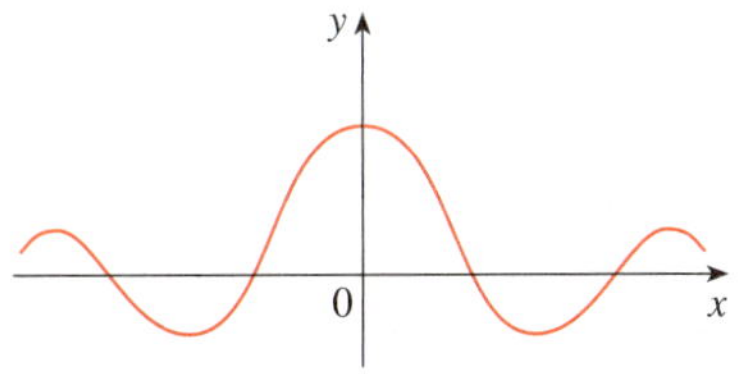

(a) Even function: reflectional symmetry

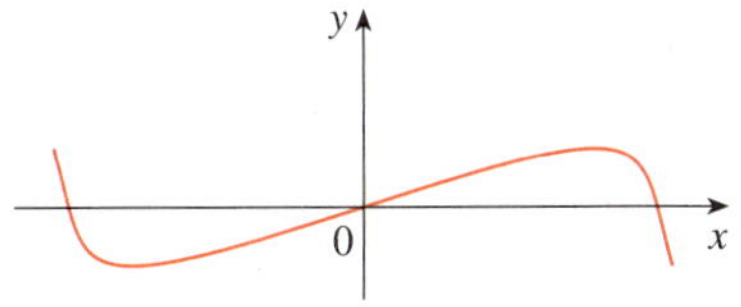

(b) Odd function: rotational symmetry

FIGURE 6

enables us to discover the most interesting aspects of graphs and in many cases to calculate maximum and minimum points and inflection points *exactly* instead of approximately.

For instance, Figure 4 shows the graph of $f(x) = 8x^3 - 21x^2 + 18x + 2$. At first glance it seems reasonable: It has the same shape as cubic curves like $y = x^3$, and it appears to have no maximum or minimum point. But if you compute the derivative, you will see that there is a maximum when $x = 0.75$ and a minimum when $x = 1$. Indeed, if we zoom in to this portion of the graph, we see that behavior exhibited in Figure 5. Without calculus, we could easily have overlooked it.

In the next section we will graph functions by using the interaction between calculus and graphing devices. In this section we draw graphs by first considering the following information. We don't assume that you have a graphing device, but if you do have one you should use it as a check on your work.

Guidelines for Sketching a Curve

The following checklist is intended as a guide to sketching a curve $y = f(x)$ by hand. Not every item is relevant to every function. (For instance, a given curve might not have an asymptote or possess symmetry.) But the guidelines provide all the information you need to make a sketch that displays the most important aspects of the function.

A. Domain It's often useful to start by determining the domain D of f, that is, the set of values of x for which $f(x)$ is defined.

B. Intercepts The y-intercept is $f(0)$ and this tells us where the curve intersects the y-axis. To find the x-intercepts, we set $y = 0$ and solve for x. (You can omit this step if the equation is difficult to solve.)

C. Symmetry

(i) If $f(-x) = f(x)$ for all x in D, that is, the equation of the curve is unchanged when x is replaced by $-x$, then f is an **even function** and the curve is symmetric about the y-axis. This means that our work is cut in half. If we know what the curve looks like for $x \geqslant 0$, then we need only reflect about the y-axis to obtain the complete curve [see Figure 6(a)]. Here are some examples: $y = x^2$, $y = x^4$, $y = |x|$, and $y = \cos x$.

(ii) If $f(-x) = -f(x)$ for all x in D, then f is an **odd function** and the curve is symmetric about the origin. Again we can obtain the complete curve if we know what it looks like for $x \geqslant 0$. [Rotate 180° about the origin; see Figure 6(b).] Some simple examples of odd functions are $y = x$, $y = x^3$, $y = x^5$, and $y = \sin x$.

(iii) If $f(x + p) = f(x)$ for all x in D, where p is a positive constant, then f is called a **periodic function** and the smallest such number p is called the **period.** For instance, $y = \sin x$ has period 2π and $y = \tan x$ has period π. If we know what the graph looks like in an interval of length p, then we can use translation to sketch the entire graph (see Figure 7).

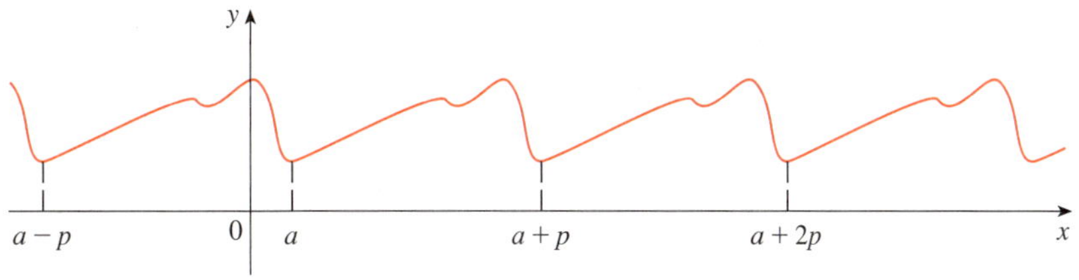

FIGURE 7
Periodic function: translational symmetry

D. Asymptotes

(i) *Horizontal Asymptotes.* Recall from Section 2.6 that if either $\lim_{x \to \infty} f(x) = L$ or $\lim_{x \to -\infty} f(x) = L$, then the line $y = L$ is a horizontal asymptote of the curve

$y = f(x)$. If it turns out that $\lim_{x\to\infty} f(x) = \infty$ (or $-\infty$), then we do not have an asymptote to the right, but that is still useful information for sketching the curve.

(ii) *Vertical Asymptotes.* Recall from Section 2.2 that the line $x = a$ is a vertical asymptote if at least one of the following statements is true:

1

$$\lim_{x\to a^+} f(x) = \infty \qquad \lim_{x\to a^-} f(x) = \infty$$

$$\lim_{x\to a^+} f(x) = -\infty \qquad \lim_{x\to a^-} f(x) = -\infty$$

(For rational functions you can locate the vertical asymptotes by equating the denominator to 0 after canceling any common factors. But for other functions this method does not apply.) Furthermore, in sketching the curve it is very useful to know exactly which of the statements in (1) is true. If $f(a)$ is not defined but a is an endpoint of the domain of f, then you should compute $\lim_{x\to a^-} f(x)$ or $\lim_{x\to a^+} f(x)$, whether or not this limit is infinite.

(iii) *Slant Asymptotes.* These are discussed at the end of this section.

E. Intervals of Increase or Decrease Use the I/D Test. Compute $f'(x)$ and find the intervals on which $f'(x)$ is positive (f is increasing) and the intervals on which $f'(x)$ is negative (f is decreasing).

F. Local Maximum and Minimum Values Find the critical numbers of f [the numbers c where $f'(c) = 0$ or $f'(c)$ does not exist]. Then use the First Derivative Test. If f' changes from positive to negative at a critical number c, then $f(c)$ is a local maximum. If f' changes from negative to positive at c, then $f(c)$ is a local minimum. Although it is usually preferable to use the First Derivative Test, you can use the Second Derivative Test if c is a critical number such that $f''(c) \neq 0$. Then $f''(c) > 0$ implies that $f(c)$ is a local minimum, whereas $f''(c) < 0$ implies that $f(c)$ is a local maximum.

G. Concavity and Points of Inflection Compute $f''(x)$ and use the Concavity Test. The curve is concave upward where $f''(x) > 0$ and concave downward where $f''(x) < 0$. Inflection points occur where the direction of concavity changes.

H. Sketch the Curve Using the information in items A–G, draw the graph. Sketch the asymptotes as dashed lines. Plot the intercepts, maximum and minimum points, and inflection points. Then make the curve pass through these points, rising and falling according to E, with concavity according to G, and approaching the asymptotes. If additional accuracy is desired near any point, you can compute the value of the derivative there. The tangent indicates the direction in which the curve proceeds.

TEC In Module 4.5 you can practice using information about f', f'', and asymptotes to determine the shape of the graph of f.

EXAMPLE 1 Use the guidelines to sketch the curve $y = \dfrac{2x^2}{x^2 - 1}$.

A. The domain is

$$\{x \mid x^2 - 1 \neq 0\} = \{x \mid x \neq \pm 1\} = (-\infty, -1) \cup (-1, 1) \cup (1, \infty)$$

B. The x- and y-intercepts are both 0.

C. Since $f(-x) = f(x)$, the function f is even. The curve is symmetric about the y-axis.

D.

$$\lim_{x\to\pm\infty} \frac{2x^2}{x^2 - 1} = \lim_{x\to\pm\infty} \frac{2}{1 - 1/x^2} = 2$$

Therefore, the line $y = 2$ is a horizontal asymptote.

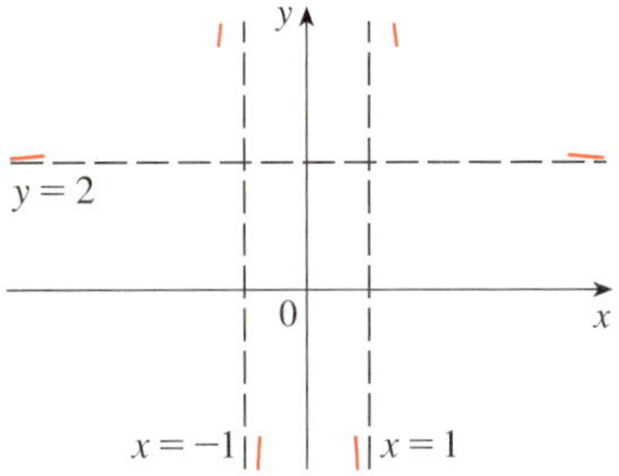

FIGURE 8
Preliminary sketch

||| We have shown the curve approaching its horizontal asymptote from above in Figure 8. This is confirmed by the intervals of increase and decrease.

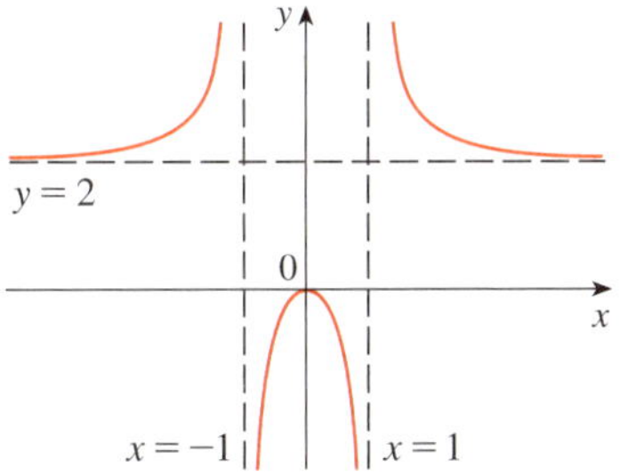

FIGURE 9
Finished sketch of $y = \dfrac{2x^2}{x^2 - 1}$

Since the denominator is 0 when $x = \pm 1$, we compute the following limits:

$$\lim_{x \to 1^+} \frac{2x^2}{x^2 - 1} = \infty \qquad \lim_{x \to 1^-} \frac{2x^2}{x^2 - 1} = -\infty$$

$$\lim_{x \to -1^+} \frac{2x^2}{x^2 - 1} = -\infty \qquad \lim_{x \to -1^-} \frac{2x^2}{x^2 - 1} = \infty$$

Therefore, the lines $x = 1$ and $x = -1$ are vertical asymptotes. This information about limits and asymptotes enables us to draw the preliminary sketch in Figure 8, showing the parts of the curve near the asymptotes.

E.
$$f'(x) = \frac{4x(x^2 - 1) - 2x^2 \cdot 2x}{(x^2 - 1)^2} = \frac{-4x}{(x^2 - 1)^2}$$

Since $f'(x) > 0$ when $x < 0$ $(x \neq -1)$ and $f'(x) < 0$ when $x > 0$ $(x \neq 1)$, f is increasing on $(-\infty, -1)$ and $(-1, 0)$ and decreasing on $(0, 1)$ and $(1, \infty)$.

F. The only critical number is $x = 0$. Since f' changes from positive to negative at 0, $f(0) = 0$ is a local maximum by the First Derivative Test.

G.
$$f''(x) = \frac{-4(x^2 - 1)^2 + 4x \cdot 2(x^2 - 1)2x}{(x^2 - 1)^4} = \frac{12x^2 + 4}{(x^2 - 1)^3}$$

Since $12x^2 + 4 > 0$ for all x, we have

$$f''(x) > 0 \iff x^2 - 1 > 0 \iff |x| > 1$$

and $f''(x) < 0 \iff |x| < 1$. Thus, the curve is concave upward on the intervals $(-\infty, -1)$ and $(1, \infty)$ and concave downward on $(-1, 1)$. It has no point of inflection since 1 and -1 are not in the domain of f.

H. Using the information in E–G, we finish the sketch in Figure 9.

EXAMPLE 2 Sketch the graph of $f(x) = \dfrac{x^2}{\sqrt{x + 1}}$.

A. Domain $= \{x \mid x + 1 > 0\} = \{x \mid x > -1\} = (-1, \infty)$

B. The x- and y-intercepts are both 0.

C. Symmetry: None

D. Since

$$\lim_{x \to \infty} \frac{x^2}{\sqrt{x + 1}} = \infty$$

there is no horizontal asymptote. Since $\sqrt{x + 1} \to 0$ as $x \to -1^+$ and $f(x)$ is always positive, we have

$$\lim_{x \to -1^+} \frac{x^2}{\sqrt{x + 1}} = \infty$$

and so the line $x = -1$ is a vertical asymptote.

E.
$$f'(x) = \frac{2x\sqrt{x + 1} - x^2 \cdot 1/\left(2\sqrt{x + 1}\right)}{x + 1} = \frac{x(3x + 4)}{2(x + 1)^{3/2}}$$

We see that $f'(x) = 0$ when $x = 0$ (notice that $-\frac{4}{3}$ is not in the domain of f), so the only critical number is 0. Since $f'(x) < 0$ when $-1 < x < 0$ and $f'(x) > 0$ when $x > 0$, f is decreasing on $(-1, 0)$ and increasing on $(0, \infty)$.

F. Since $f'(0) = 0$ and f' changes from negative to positive at 0, $f(0) = 0$ is a local (and absolute) minimum by the First Derivative Test.

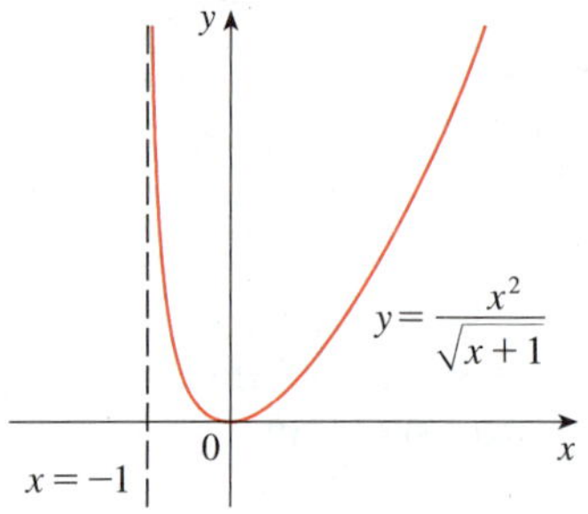

FIGURE 10

G. $$f''(x) = \frac{2(x+1)^{3/2}(6x+4) - (3x^2+4x)3(x+1)^{1/2}}{4(x+1)^3} = \frac{3x^2+8x+8}{4(x+1)^{5/2}}$$

Note that the denominator is always positive. The numerator is the quadratic $3x^2 + 8x + 8$, which is always positive because its discriminant is $b^2 - 4ac = -32$, which is negative, and the coefficient of x^2 is positive. Thus, $f''(x) > 0$ for all x in the domain of f, which means that f is concave upward on $(-1, \infty)$ and there is no point of inflection.

H. The curve is sketched in Figure 10.

EXAMPLE 3 Sketch the graph of $f(x) = xe^x$.

A. The domain is $\mathbb{R}$.

B. The x- and y-intercepts are both 0.

C. Symmetry: None

D. Because both x and e^x become large as $x \to \infty$, we have $\lim_{x\to\infty} xe^x = \infty$. As $x \to -\infty$, however, $e^x \to 0$ and so we have an indeterminate product that requires the use of l'Hospital's Rule:

$$\lim_{x\to-\infty} xe^x = \lim_{x\to-\infty} \frac{x}{e^{-x}} = \lim_{x\to-\infty} \frac{1}{-e^{-x}} = \lim_{x\to-\infty} (-e^x) = 0$$

Thus, the x-axis is a horizontal asymptote.

E. $$f'(x) = xe^x + e^x = (x+1)e^x$$

Since e^x is always positive, we see that $f'(x) > 0$ when $x + 1 > 0$, and $f'(x) < 0$ when $x + 1 < 0$. So f is increasing on $(-1, \infty)$ and decreasing on $(-\infty, -1)$.

F. Because $f'(-1) = 0$ and f changes from negative to positive at $x = -1$, $f(-1) = -e^{-1}$ is a local (and absolute) minimum.

G. $$f''(x) = (x+1)e^x + e^x = (x+2)e^x$$

Since $f''(x) > 0$ if $x > -2$ and $f''(x) < 0$ if $x < -2$, f is concave upward on $(-2, \infty)$ and concave downward on $(-\infty, -2)$. The inflection point is $(-2, -2e^{-2})$.

H. We use this information to sketch the curve in Figure 11.

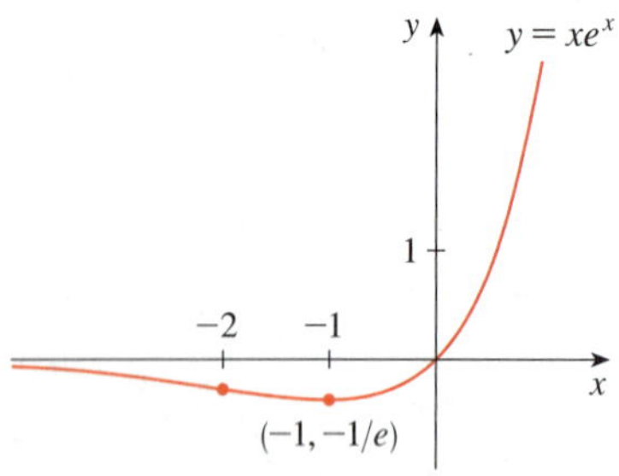

FIGURE 11

EXAMPLE 4 Sketch the graph of $f(x) = 2\cos x + \sin 2x$.

A. The domain is $\mathbb{R}$.

B. The y-intercept is $f(0) = 2$. The x-intercepts occur when

$$2\cos x + \sin 2x = 2\cos x + 2\sin x\cos x = 2\cos x\,(1 + \sin x) = 0$$

that is, when $\cos x = 0$ or $\sin x = -1$. Thus, in the interval $[0, 2\pi]$, the x-intercepts are $\pi/2$ and $3\pi/2$.

C. f is neither even nor odd, but $f(x + 2\pi) = f(x)$ for all x and so f is periodic and has period 2π. Thus, in what follows we need to consider only $0 \leqslant x \leqslant 2\pi$ and then extend the curve by translation in H.

D. Asymptotes: None

E. $$\begin{aligned} f'(x) &= -2\sin x + 2\cos 2x = -2\sin x + 2(1 - 2\sin^2 x) \\ &= -2(2\sin^2 x + \sin x - 1) = -2(2\sin x - 1)(\sin x + 1) \end{aligned}$$

Thus, $f'(x) = 0$ when $\sin x = \frac{1}{2}$ or $\sin x = -1$, so in $[0, 2\pi]$ we have $x = \pi/6$, $5\pi/6$, and $3\pi/2$. In determining the sign of $f'(x)$ in the following chart, we use the fact that $\sin x + 1 \geqslant 0$ for all x.

Interval	$f'(x)$	f
$0 < x < \pi/6$	$+$	increasing on $(0, \pi/6)$
$\pi/6 < x < 5\pi/6$	$-$	decreasing on $(\pi/6, 5\pi/6)$
$5\pi/6 < x < 3\pi/2$	$+$	increasing on $(5\pi/6, 3\pi/2)$
$3\pi/2 < x < 2\pi$	$+$	increasing on $(3\pi/2, 2\pi)$

F. From the chart in E the First Derivative Test says that $f(\pi/6) = 3\sqrt{3}/2$ is a local maximum and $f(5\pi/6) = -3\sqrt{3}/2$ is a local minimum, but f has no maximum or minimum at $3\pi/2$, only a horizontal tangent.

G.
$$f''(x) = -2\cos x - 4\sin 2x = -2\cos x\,(1 + 4\sin x)$$

Thus, $f''(x) = 0$ when $\cos x = 0$ (so $x = \pi/2$ or $3\pi/2$) and when $\sin x = -\frac{1}{4}$. From Figure 12 we see that there are two values of x between 0 and 2π for which $\sin x = -\frac{1}{4}$. Let's call them α_1 and α_2. Then $f''(x) > 0$ on $(\pi/2, \alpha_1)$ and $(3\pi/2, \alpha_2)$, so f is concave upward there. Also $f''(x) < 0$ on $(0, \pi/2)$, $(\alpha_1, 3\pi/2)$, and $(\alpha_2, 2\pi)$, so f is concave downward there. Inflection points occur when $x = \pi/2, \alpha_1, 3\pi/2$, and α_2.

$\alpha_1 = \pi + \arcsin\frac{1}{4}$

$\alpha_2 = 2\pi - \arcsin\frac{1}{4}$

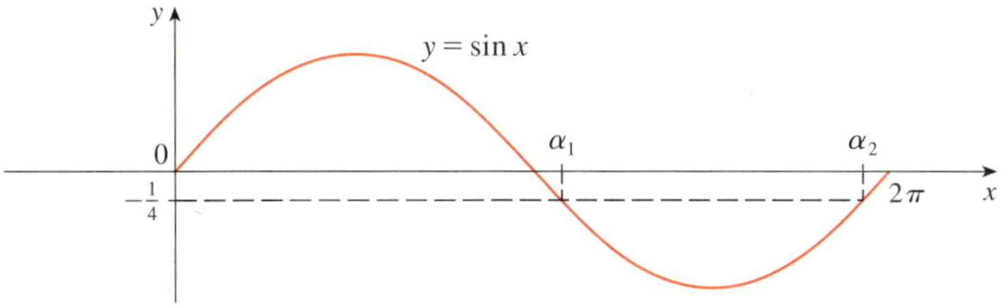

FIGURE 12

H. The graph of the function restricted to $0 \leqslant x \leqslant 2\pi$ is shown in Figure 13. Then it is extended, using periodicity, to the complete graph in Figure 14.

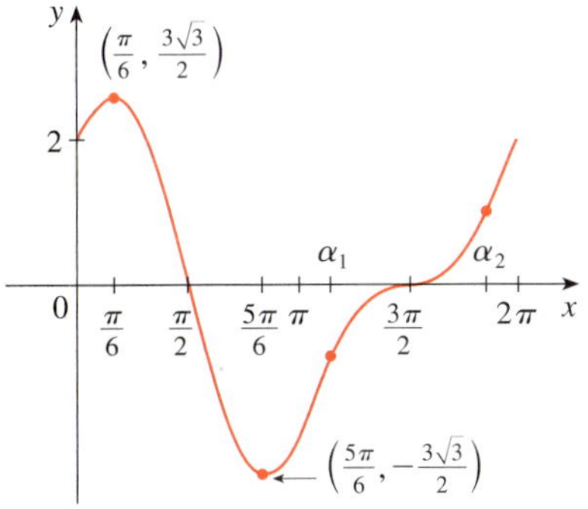

FIGURE 13

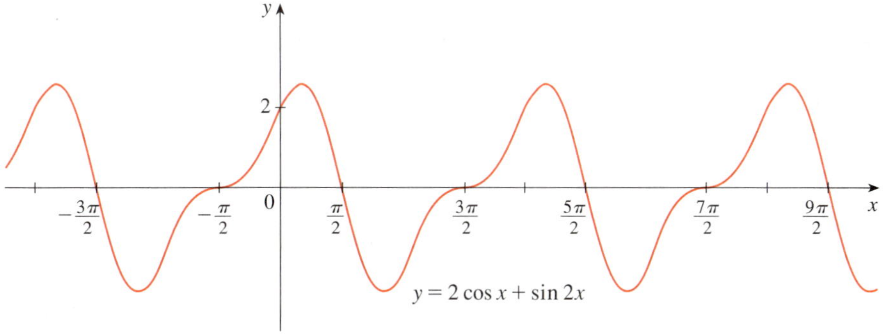

FIGURE 14

EXAMPLE 5 Sketch the graph of $y = \ln(4 - x^2)$.

A. The domain is

$$\{x \mid 4 - x^2 > 0\} = \{x \mid x^2 < 4\} = \{x \mid |x| < 2\} = (-2, 2)$$

B. The y-intercept is $f(0) = \ln 4$. To find the x-intercept we set

$$y = \ln(4 - x^2) = 0$$

We know that $\ln 1 = \log_e 1 = 0$ (since $e^0 = 1$), so we have $4 - x^2 = 1 \Rightarrow x^2 = 3$ and therefore the x-intercepts are $\pm\sqrt{3}$.

C. Since $f(-x) = f(x)$, f is even and the curve is symmetric about the y-axis.

D. We look for vertical asymptotes at the endpoints of the domain. Since $4 - x^2 \to 0^+$ as $x \to 2^-$ and also as $x \to -2^+$, we have

$$\lim_{x\to 2^-} \ln(4 - x^2) = -\infty \qquad \lim_{x\to -2^+} \ln(4 - x^2) = -\infty$$

Thus, the lines $x = 2$ and $x = -2$ are vertical asymptotes.

E.
$$f'(x) = \frac{-2x}{4 - x^2}$$

Since $f'(x) > 0$ when $-2 < x < 0$ and $f'(x) < 0$ when $0 < x < 2$, f is increasing on $(-2, 0)$ and decreasing on $(0, 2)$.

F. The only critical number is $x = 0$. Since f' changes from positive to negative at 0, $f(0) = \ln 4$ is a local maximum by the First Derivative Test.

G.
$$f''(x) = \frac{(4 - x^2)(-2) + 2x(-2x)}{(4 - x^2)^2} = \frac{-8 - 2x^2}{(4 - x^2)^2}$$

Since $f''(x) < 0$ for all x, the curve is concave downward on $(-2, 2)$ and has no inflection point.

H. Using this information, we sketch the curve in Figure 15.

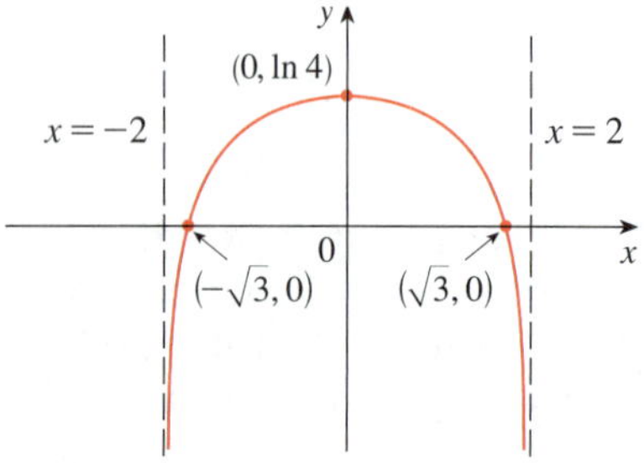

FIGURE 15
$y = \ln(4 - x^2)$

Slant Asymptotes

Some curves have asymptotes that are *oblique*, that is, neither horizontal nor vertical. If

$$\lim_{x\to\infty} [f(x) - (mx + b)] = 0$$

then the line $y = mx + b$ is called a **slant asymptote** because the vertical distance between the curve $y = f(x)$ and the line $y = mx + b$ approaches 0, as in Figure 16. (A similar situation exists if we let $x \to -\infty$.) For rational functions, slant asymptotes occur when the degree of the numerator is one more than the degree of the denominator. In such a case the equation of the slant asymptote can be found by long division as in the following example.

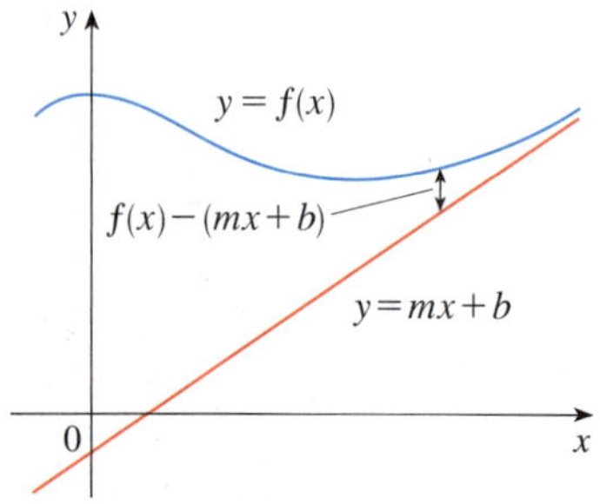

FIGURE 16

EXAMPLE 6 Sketch the graph of $f(x) = \dfrac{x^3}{x^2 + 1}$.

A. The domain is $\mathbb{R} = (-\infty, \infty)$.

B. The x- and y-intercepts are both 0.

C. Since $f(-x) = -f(x)$, f is odd and its graph is symmetric about the origin.

D. Since $x^2 + 1$ is never 0, there is no vertical asymptote. Since $f(x) \to \infty$ as $x \to \infty$ and $f(x) \to -\infty$ as $x \to -\infty$, there is no horizontal asymptote. But long division gives

$$f(x) = \frac{x^3}{x^2 + 1} = x - \frac{x}{x^2 + 1}$$

$$f(x) - x = -\frac{x}{x^2 + 1} = -\frac{\dfrac{1}{x}}{1 + \dfrac{1}{x^2}} \to 0 \quad \text{as} \quad x \to \pm\infty$$

So the line $y = x$ is a slant asymptote.

E. $$f'(x) = \frac{3x^2(x^2+1) - x^3 \cdot 2x}{(x^2+1)^2} = \frac{x^2(x^2+3)}{(x^2+1)^2}$$

Since $f'(x) > 0$ for all x (except 0), f is increasing on $(-\infty, \infty)$.

F. Although $f'(0) = 0$, f' does not change sign at 0, so there is no local maximum or minimum.

G. $$f''(x) = \frac{(4x^3+6x)(x^2+1)^2 - (x^4+3x^2) \cdot 2(x^2+1)2x}{(x^2+1)^4} = \frac{2x(3-x^2)}{(x^2+1)^3}$$

Since $f''(x) = 0$ when $x = 0$ or $x = \pm\sqrt{3}$, we set up the following chart:

Interval	x	$3 - x^2$	$(x^2+1)^3$	$f''(x)$	f
$x < -\sqrt{3}$	−	−	+	+	CU on $(-\infty, -\sqrt{3})$
$-\sqrt{3} < x < 0$	−	+	+	−	CD on $(-\sqrt{3}, 0)$
$0 < x < \sqrt{3}$	+	+	+	+	CU on $(0, \sqrt{3})$
$x > \sqrt{3}$	+	−	+	−	CD on $(\sqrt{3}, \infty)$

The points of inflection are $(-\sqrt{3}, -3\sqrt{3}/4)$, $(0, 0)$, and $(\sqrt{3}, 3\sqrt{3}/4)$.

H. The graph of f is sketched in Figure 17.

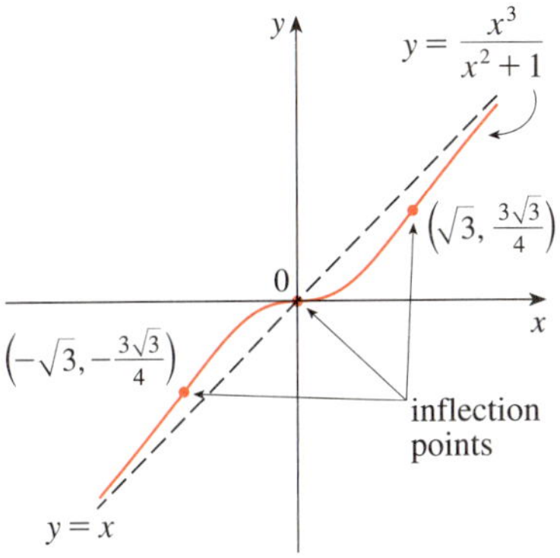

FIGURE 17

4.5 Exercises

1–52 Use the guidelines of this section to sketch the curve.

1. $y = x^3 + x$

2. $y = x^3 + 6x^2 + 9x$

3. $y = 2 - 15x + 9x^2 - x^3$

4. $y = 8x^2 - x^4$

5. $y = x^4 + 4x^3$

6. $y = x(x+2)^3$

7. $y = 2x^5 - 5x^2 + 1$

8. $y = 20x^3 - 3x^5$

9. $y = \dfrac{x}{x-1}$

10. $y = \dfrac{x}{(x-1)^2}$

11. $y = \dfrac{1}{x^2-9}$

12. $y = \dfrac{x}{x^2-9}$

13. $y = \dfrac{x}{x^2+9}$

14. $y = \dfrac{x^2}{x^2+9}$

15. $y = \dfrac{x-1}{x^2}$

16. $y = \dfrac{x^2-2}{x^4}$

17. $y = \dfrac{x^2}{x^2+3}$

18. $y = \dfrac{x^3-1}{x^3+1}$

19. $y = x\sqrt{5-x}$

20. $y = 2\sqrt{x} - x$

21. $y = \sqrt{x^2+1} - x$

22. $y = \sqrt{\dfrac{x}{x-5}}$

23. $y = \dfrac{x}{\sqrt{x^2+1}}$

24. $y = x\sqrt{2-x^2}$

25. $y = \dfrac{\sqrt{1-x^2}}{x}$

26. $y = \dfrac{x}{\sqrt{x^2-1}}$

27. $y = x - 3x^{1/3}$

28. $y = x^{5/3} - 5x^{2/3}$

29. $y = x + \sqrt{|x|}$

30. $y = \sqrt[3]{(x^2-1)^2}$

31. $y = 3\sin x - \sin^3 x$

32. $y = \sin x - \tan x$

33. $y = x\tan x, \quad -\pi/2 < x < \pi/2$

34. $y = 2x - \tan x, \quad -\pi/2 < x < \pi/2$

35. $y = \frac{1}{2}x - \sin x, \quad 0 < x < 3\pi$

36. $y = \cos^2 x - 2\sin x$

37. $y = \sin 2x - 2\sin x$

38. $y = \sin x - x$

39. $y = \dfrac{\sin x}{1+\cos x}$

40. $y = \dfrac{\cos x}{2+\sin x}$

41. $y = 1/(1+e^{-x})$

42. $y = e^{2x} - e^x$

43. $y = x\ln x$

44. $y = e^x/x$

45. $y = xe^{-x}$

46. $y = \ln(x^2 - 3x + 2)$

47. $y = \ln(\sin x)$

48. $y = x(\ln x)^2$

49. $y = xe^{-x^2}$

50. $y = e^x - 3e^{-x} - 4x$

51. $y = e^{3x} + e^{-2x}$

52. $y = \tan^{-1}\left(\dfrac{x-1}{x+1}\right)$

53. The figure shows a beam of length L embedded in concrete walls. If a constant load W is distributed evenly along its length, the beam takes the shape of the deflection curve

$$y = -\frac{W}{24EI}x^4 + \frac{WL}{12EI}x^3 - \frac{WL^2}{24EI}x^2$$

where E and I are positive constants. (E is Young's modulus of elasticity and I is the moment of inertia of a cross-section of the beam.) Sketch the graph of the deflection curve.

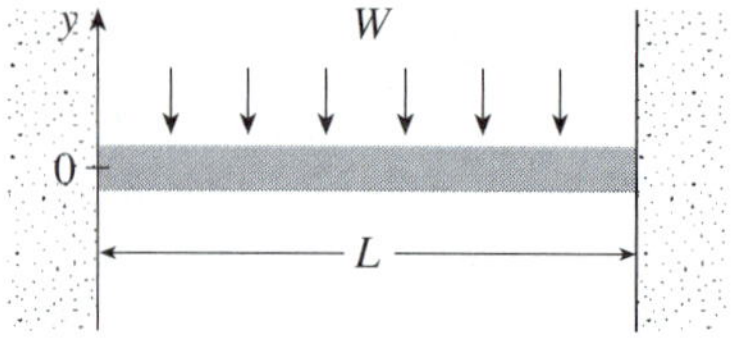

54. Coulomb's Law states that the force of attraction between two charged particles is directly proportional to the product of the charges and inversely proportional to the square of the distance between them. The figure shows particles with charge 1 located at positions 0 and 2 on a coordinate line and a particle with charge -1 at a position x between them. It follows from Coulomb's Law that the net force acting on the middle particle is

$$F(x) = -\frac{k}{x^2} + \frac{k}{(x-2)^2} \qquad 0 < x < 2$$

where k is a positive constant. Sketch the graph of the net force function. What does the graph say about the force?

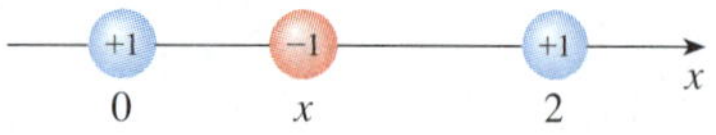

55–58 |||| Find an equation of the slant asymptote. Do not sketch the curve.

55. $y = \dfrac{x^2 + 1}{x + 1}$

56. $y = \dfrac{2x^3 + x^2 + x + 3}{x^2 + 2x}$

57. $y = \dfrac{4x^3 - 2x^2 + 5}{2x^2 + x - 3}$

58. $y = \dfrac{5x^4 + x^2 + x}{x^3 - x^2 + 2}$

59–64 |||| Use the guidelines of this section to sketch the curve. In guideline D find an equation of the slant asymptote.

59. $y = \dfrac{-2x^2 + 5x - 1}{2x - 1}$

60. $y = \dfrac{x^2 + 12}{x - 2}$

61. $xy = x^2 + 4$

62. $y = e^x - x$

63. $y = \dfrac{2x^3 + x^2 + 1}{x^2 + 1}$

64. $y = \dfrac{(x + 1)^3}{(x - 1)^2}$

65. Show that the curve $y = x - \tan^{-1}x$ has two slant asymptotes: $y = x + \pi/2$ and $y = x - \pi/2$. Use this fact to help sketch the curve.

66. Show that the curve $y = \sqrt{x^2 + 4x}$ has two slant asymptotes: $y = x + 2$ and $y = -x - 2$. Use this fact to help sketch the curve.

67. Show that the lines $y = (b/a)x$ and $y = -(b/a)x$ are slant asymptotes of the hyperbola $(x^2/a^2) - (y^2/b^2) = 1$.

68. Let $f(x) = (x^3 + 1)/x$. Show that

$$\lim_{x\to\pm\infty} [f(x) - x^2] = 0$$

This shows that the graph of f approaches the graph of $y = x^2$, and we say that the curve $y = f(x)$ is *asymptotic* to the parabola $y = x^2$. Use this fact to help sketch the graph of f.

69. Discuss the asymptotic behavior of $f(x) = (x^4 + 1)/x$ in the same manner as in Exercise 68. Then use your results to help sketch the graph of f.

70. Use the asymptotic behavior of $f(x) = \cos x + 1/x^2$ to sketch its graph without going through the curve-sketching procedure of this section.

4.6 Graphing with Calculus and Calculators

|||| If you have not already read Section 1.4, you should do so now. In particular, it explains how to avoid some of the pitfalls of graphing devices by choosing appropriate viewing rectangles.

The method we used to sketch curves in the preceding section was a culmination of much of our study of differential calculus. The graph was the final object that we produced. In this section our point of view is completely different. Here we *start* with a graph produced by a graphing calculator or computer and then we refine it. We use calculus to make sure that we reveal all the important aspects of the curve. And with the use of graphing devices we can tackle curves that would be far too complicated to consider without technology. The theme is the *interaction* between calculus and calculators.

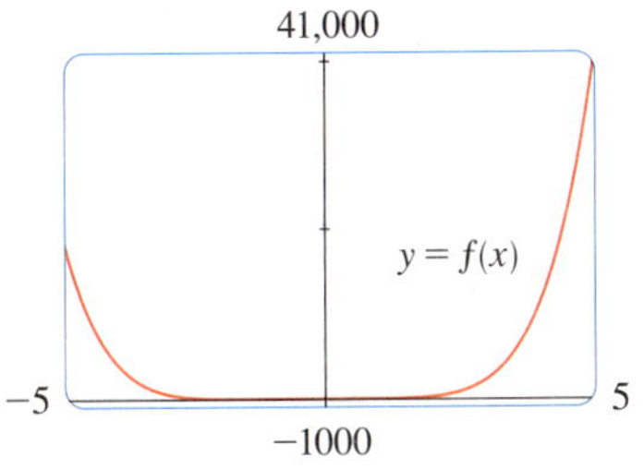

FIGURE 1

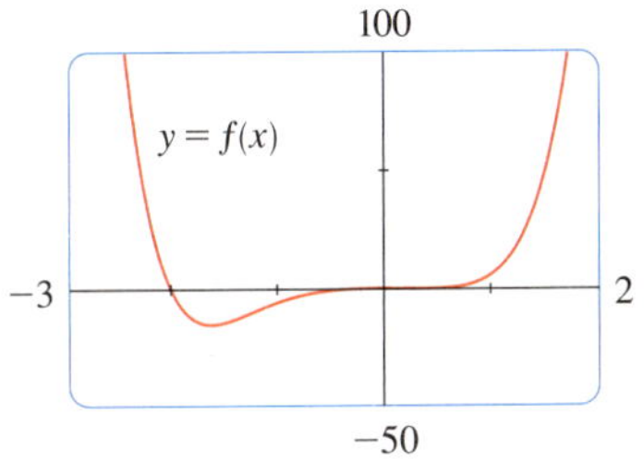

FIGURE 2

EXAMPLE 1 Graph the polynomial $f(x) = 2x^6 + 3x^5 + 3x^3 - 2x^2$. Use the graphs of f' and f'' to estimate all maximum and minimum points and intervals of concavity.

SOLUTION If we specify a domain but not a range, many graphing devices will deduce a suitable range from the values computed. Figure 1 shows the plot from one such device if we specify that $-5 \leqslant x \leqslant 5$. Although this viewing rectangle is useful for showing that the asymptotic behavior (or end behavior) is the same as for $y = 2x^6$, it is obviously hiding some finer detail. So we change to the viewing rectangle $[-3, 2]$ by $[-50, 100]$ shown in Figure 2.

From this graph it appears that there is an absolute minimum value of about -15.33 when $x \approx -1.62$ (by using the cursor) and f is decreasing on $(-\infty, -1.62)$ and increasing on $(-1.62, \infty)$. Also there appears to be a horizontal tangent at the origin and inflection points when $x = 0$ and when x is somewhere between -2 and -1.

Now let's try to confirm these impressions using calculus. We differentiate and get

$$f'(x) = 12x^5 + 15x^4 + 9x^2 - 4x$$

$$f''(x) = 60x^4 + 60x^3 + 18x - 4$$

When we graph f' in Figure 3 we see that $f'(x)$ changes from negative to positive when $x \approx -1.62$; this confirms (by the First Derivative Test) the minimum value that we found earlier. But, perhaps to our surprise, we also notice that $f'(x)$ changes from positive to negative when $x = 0$ and from negative to positive when $x \approx 0.35$. This means that f has a local maximum at 0 and a local minimum when $x \approx 0.35$, but these were hidden in Figure 2. Indeed, if we now zoom in toward the origin in Figure 4, we see what we missed before: a local maximum value of 0 when $x = 0$ and a local minimum value of about -0.1 when $x \approx 0.35$.

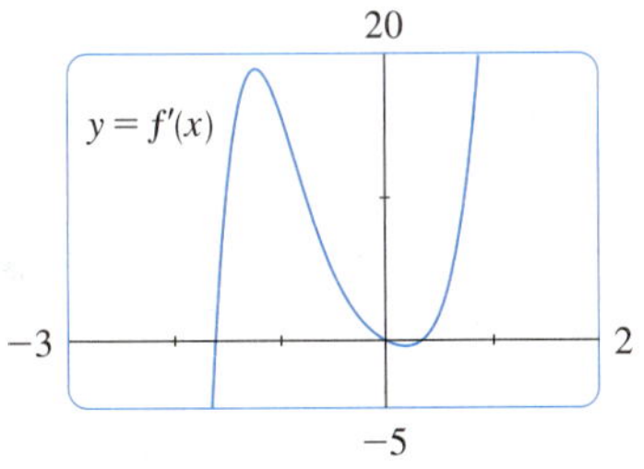

FIGURE 3

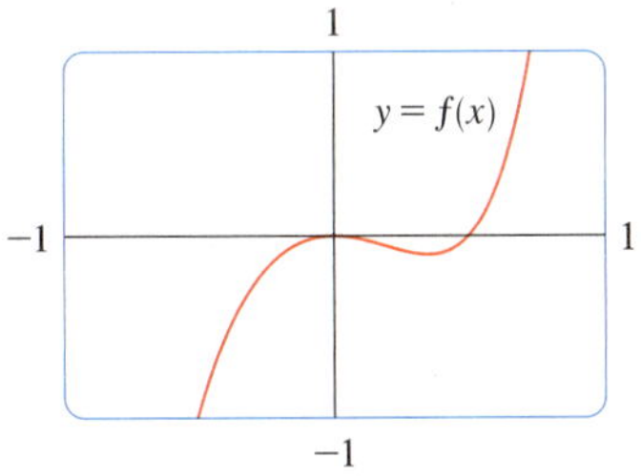

FIGURE 4

What about concavity and inflection points? From Figures 2 and 4 there appear to be inflection points when x is a little to the left of -1 and when x is a little to the right of 0. But it's difficult to determine inflection points from the graph of f, so we graph the second derivative f'' in Figure 5. We see that f'' changes from positive to negative when $x \approx -1.23$ and from negative to positive when $x \approx 0.19$. So, correct to two decimal places, f is concave upward on $(-\infty, -1.23)$ and $(0.19, \infty)$ and concave downward on $(-1.23, 0.19)$. The inflection points are $(-1.23, -10.18)$ and $(0.19, -0.05)$.

We have discovered that no single graph reveals all the important features of this polynomial. But Figures 2 and 4, when taken together, do provide an accurate picture.

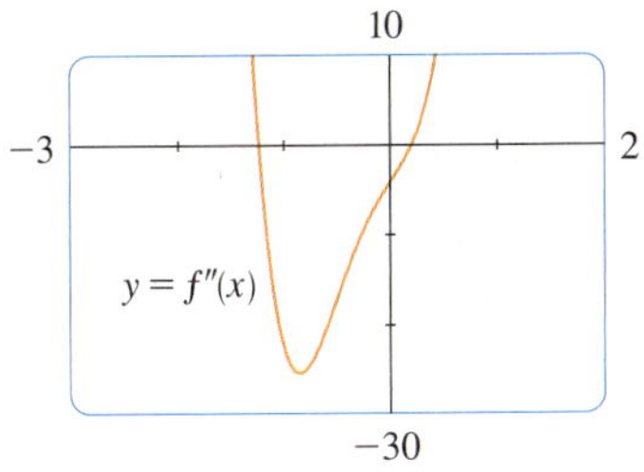

FIGURE 5

EXAMPLE 2 Draw the graph of the function

$$f(x) = \frac{x^2 + 7x + 3}{x^2}$$

in a viewing rectangle that contains all the important features of the function. Estimate the maximum and minimum values and the intervals of concavity. Then use calculus to find these quantities exactly.

SOLUTION Figure 6, produced by a computer with automatic scaling, is a disaster. Some graphing calculators use $[-10, 10]$ by $[-10, 10]$ as the default viewing rectangle, so let's try it. We get the graph shown in Figure 7; it's a major improvement.

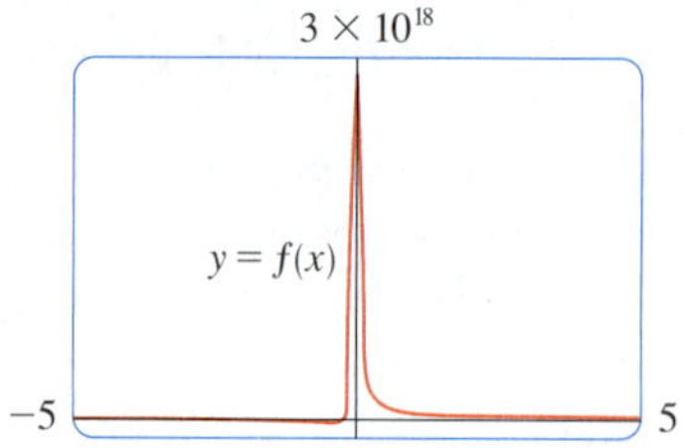

FIGURE 6

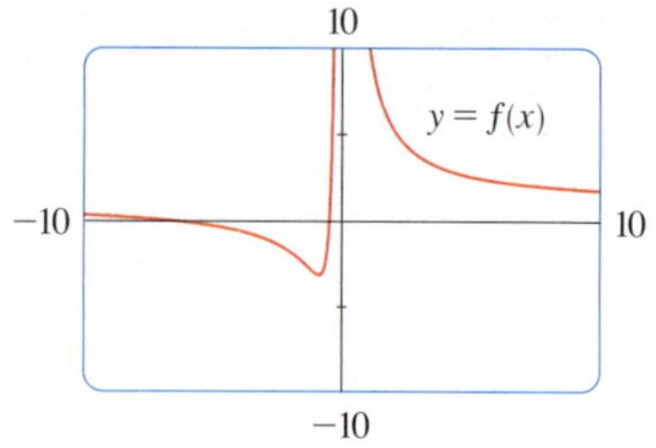

FIGURE 7

The y-axis appears to be a vertical asymptote and indeed it is because

$$\lim_{x \to 0} \frac{x^2 + 7x + 3}{x^2} = \infty$$

Figure 7 also allows us to estimate the x-intercepts: about -0.5 and -6.5. The exact values are obtained by using the quadratic formula to solve the equation $x^2 + 7x + 3 = 0$; we get $x = \left(-7 \pm \sqrt{37}\right)/2$.

To get a better look at horizontal asymptotes, we change to the viewing rectangle $[-20, 20]$ by $[-5, 10]$ in Figure 8. It appears that $y = 1$ is the horizontal asymptote and this is easily confirmed:

$$\lim_{x \to \pm\infty} \frac{x^2 + 7x + 3}{x^2} = \lim_{x \to \pm\infty} \left(1 + \frac{7}{x} + \frac{3}{x^2}\right) = 1$$

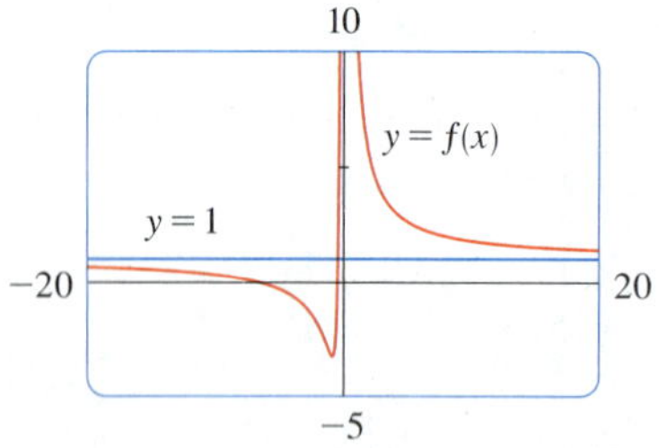

FIGURE 8

To estimate the minimum value we zoom in to the viewing rectangle $[-3, 0]$ by $[-4, 2]$ in Figure 9. The cursor indicates that the absolute minimum value is about -3.1 when $x \approx -0.9$, and we see that the function decreases on $(-\infty, -0.9)$ and $(0, \infty)$ and increases on $(-0.9, 0)$. The exact values are obtained by differentiating:

$$f'(x) = -\frac{7}{x^2} - \frac{6}{x^3} = -\frac{7x + 6}{x^3}$$

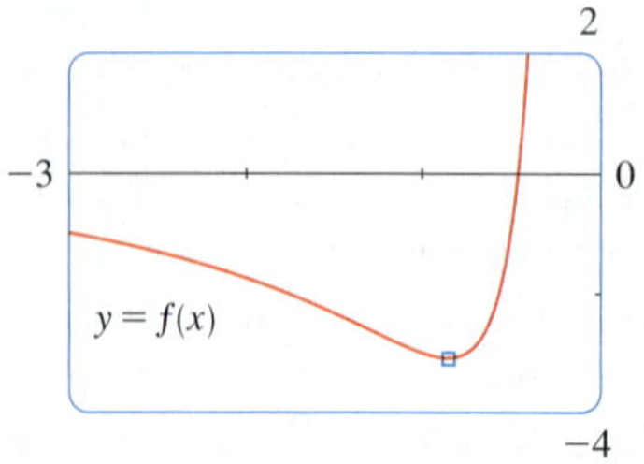

FIGURE 9

This shows that $f'(x) > 0$ when $-\frac{6}{7} < x < 0$ and $f'(x) < 0$ when $x < -\frac{6}{7}$ and when $x > 0$. The exact minimum value is $f\left(-\frac{6}{7}\right) = -\frac{37}{12} \approx -3.08$.

Figure 9 also shows that an inflection point occurs somewhere between $x = -1$ and $x = -2$. We could estimate it much more accurately using the graph of the second derivative, but in this case it's just as easy to find exact values. Since

$$f''(x) = \frac{14}{x^3} + \frac{18}{x^4} = 2\,\frac{7x + 9}{x^4}$$

we see that $f''(x) > 0$ when $x > -\frac{9}{7}$ $(x \neq 0)$. So f is concave upward on $\left(-\frac{9}{7}, 0\right)$ and $(0, \infty)$ and concave downward on $\left(-\infty, -\frac{9}{7}\right)$. The inflection point is $\left(-\frac{9}{7}, -\frac{71}{27}\right)$.

The analysis using the first two derivatives shows that Figures 7 and 8 display all the major aspects of the curve.

EXAMPLE 3 Graph the function $f(x) = \dfrac{x^2(x+1)^3}{(x-2)^2(x-4)^4}$.

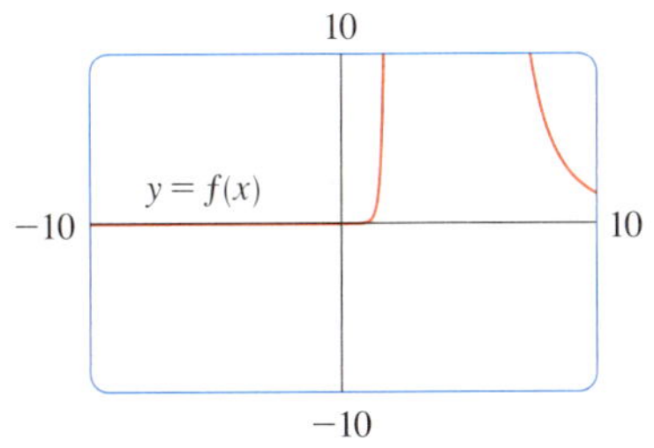

FIGURE 10

SOLUTION Drawing on our experience with a rational function in Example 2, let's start by graphing f in the viewing rectangle $[-10, 10]$ by $[-10, 10]$. From Figure 10 we have the feeling that we are going to have to zoom in to see some finer detail and also zoom out to see the larger picture. But, as a guide to intelligent zooming, let's first take a close look at the expression for $f(x)$. Because of the factors $(x-2)^2$ and $(x-4)^4$ in the denominator, we expect $x = 2$ and $x = 4$ to be the vertical asymptotes. Indeed

$$\lim_{x \to 2} \frac{x^2(x+1)^3}{(x-2)^2(x-4)^4} = \infty \qquad \text{and} \qquad \lim_{x \to 4} \frac{x^2(x+1)^3}{(x-2)^2(x-4)^4} = \infty$$

To find the horizontal asymptotes we divide numerator and denominator by x^6:

$$\frac{x^2(x+1)^3}{(x-2)^2(x-4)^4} = \frac{\dfrac{1}{x}\left(1+\dfrac{1}{x}\right)^3}{\left(1-\dfrac{2}{x}\right)^2\left(1-\dfrac{4}{x}\right)^4} \to 0 \quad \text{as} \quad x \to \pm\infty$$

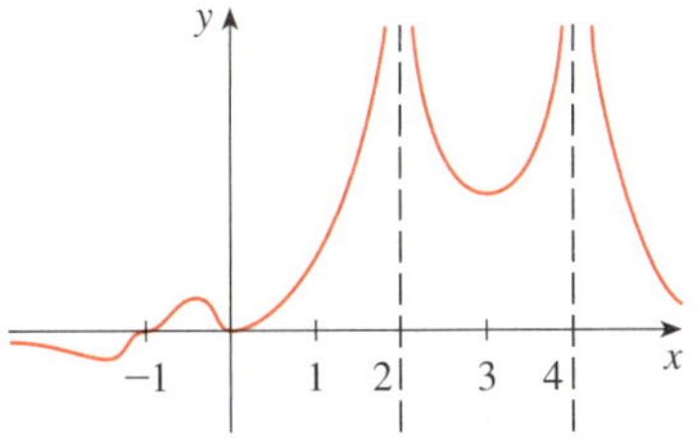

FIGURE 11

so the x-axis is the horizontal asymptote.

It is also very useful to consider the behavior of the graph near the x-intercepts using an analysis like that in Example 11 in Section 2.6. Since x^2 is positive, $f(x)$ does not change sign at 0 and so its graph doesn't cross the x-axis at 0. But, because of the factor $(x+1)^3$, the graph does cross the x-axis at -1 and has a horizontal tangent there. Putting all this information together, but without using derivatives, we see that the curve has to look something like the one in Figure 11.

Now that we know what to look for, we zoom in (several times) to produce the graphs in Figures 12 and 13 and zoom out (several times) to get Figure 14.

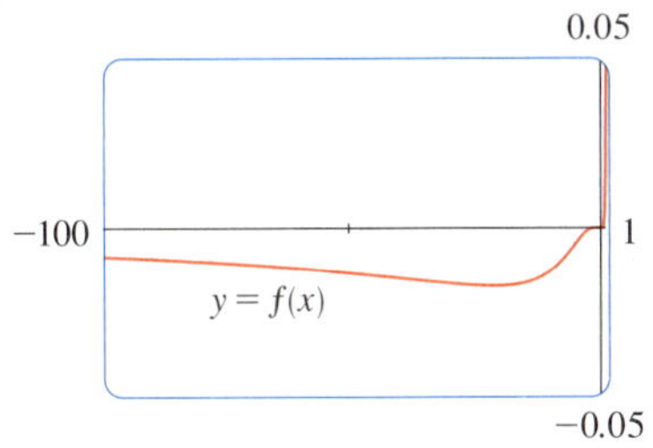

FIGURE 12

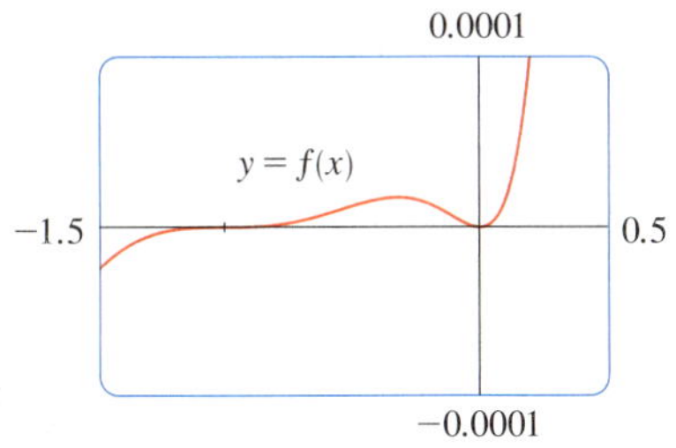

FIGURE 13

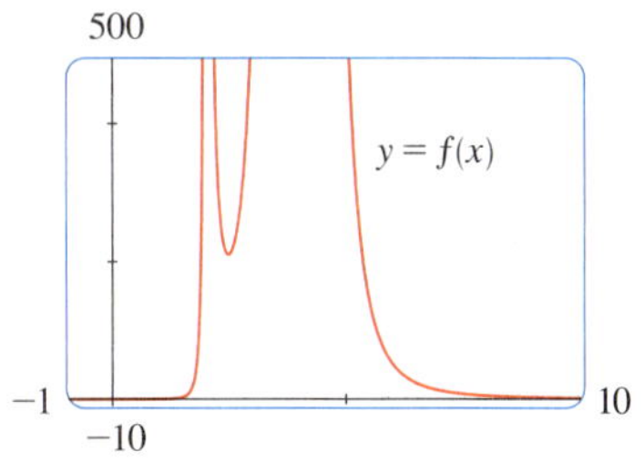

FIGURE 14

We can read from these graphs that the absolute minimum is about -0.02 and occurs when $x \approx -20$. There is also a local maximum ≈ 0.00002 when $x \approx -0.3$ and a local minimum ≈ 211 when $x \approx 2.5$. These graphs also show three inflection points near -35, -5, and -1 and two between -1 and 0. To estimate the inflection points closely we would need to graph f'', but to compute f'' by hand is an unreasonable chore. If you have a computer algebra system, then it's easy to do (see Exercise 17).

We have seen that, for this particular function, *three* graphs (Figures 12, 13, and 14) are necessary to convey all the useful information. The only way to display all these features of the function on a single graph is to draw it by hand. Despite the exaggerations and distortions, Figure 11 does manage to summarize the essential nature of the function.

|||| The family of functions

$$f(x) = \sin(x + \sin cx)$$

where c is a constant, occurs in applications to frequency modulation (FM) synthesis. A sine wave is modulated by a wave with a different frequency ($\sin cx$). The case where $c = 2$ is studied in Example 4. Exercise 25 explores another special case.

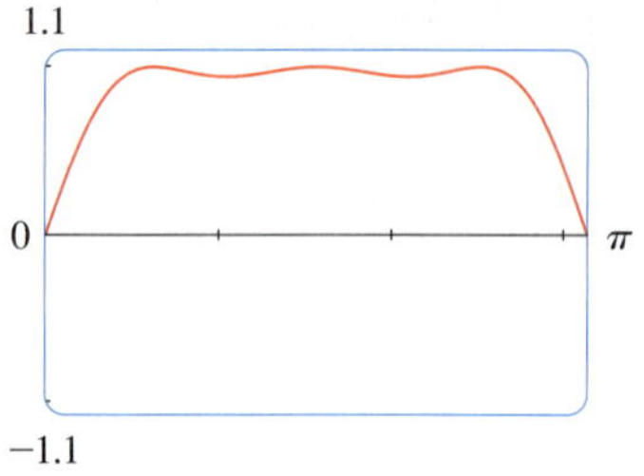

FIGURE 15

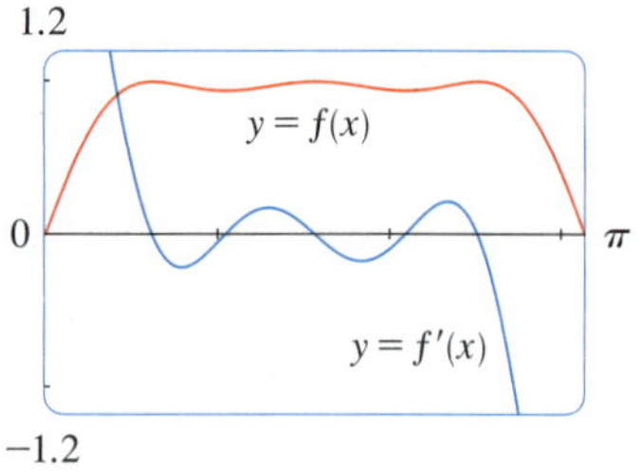

FIGURE 16

EXAMPLE 4 Graph the function $f(x) = \sin(x + \sin 2x)$. For $0 \leq x \leq \pi$, locate all maximum and minimum values, intervals of increase and decrease, and inflection points correct to one decimal place.

SOLUTION We first note that f is periodic with period 2π. Also, f is odd and $|f(x)| \leq 1$ for all x. So the choice of a viewing rectangle is not a problem for this function: We start with $[0, \pi]$ by $[-1.1, 1.1]$. (See Figure 15.) It appears that there are three local maximum values and two local minimum values in that window. To confirm this and locate them more accurately, we calculate that

$$f'(x) = \cos(x + \sin 2x) \cdot (1 + 2\cos 2x)$$

and graph both f and f' in Figure 16. Using zoom-in and the First Derivative Test, we find the following values to one decimal place.

Intervals of increase:	$(0, 0.6)$, $(1.0, 1.6)$, $(2.1, 2.5)$
Intervals of decrease:	$(0.6, 1.0)$, $(1.6, 2.1)$, $(2.5, \pi)$
Local maximum values:	$f(0.6) \approx 1$, $f(1.6) \approx 1$, $f(2.5) \approx 1$
Local minimum values:	$f(1.0) \approx 0.94$, $f(2.1) \approx 0.94$

The second derivative is

$$f''(x) = -(1 + 2\cos 2x)^2 \sin(x + \sin 2x) - 4 \sin 2x \cos(x + \sin 2x)$$

Graphing both f and f'' in Figure 17, we obtain the following approximate values:

Concave upward on:	$(0.8, 1.3)$, $(1.8, 2.3)$
Concave downward on:	$(0, 0.8)$, $(1.3, 1.8)$, $(2.3, \pi)$
Inflection points:	$(0, 0)$, $(0.8, 0.97)$, $(1.3, 0.97)$, $(1.8, 0.97)$, $(2.3, 0.97)$

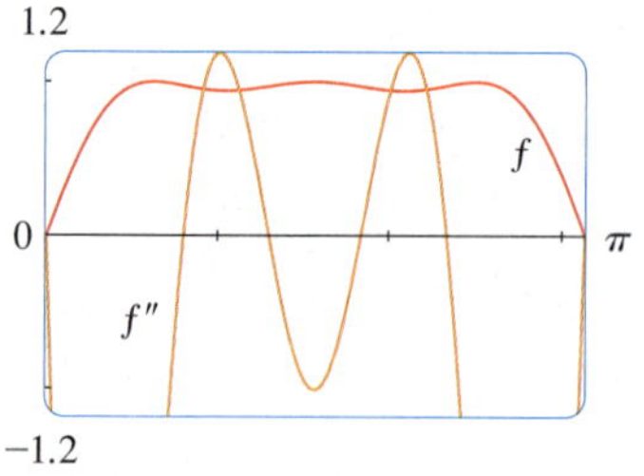

FIGURE 17

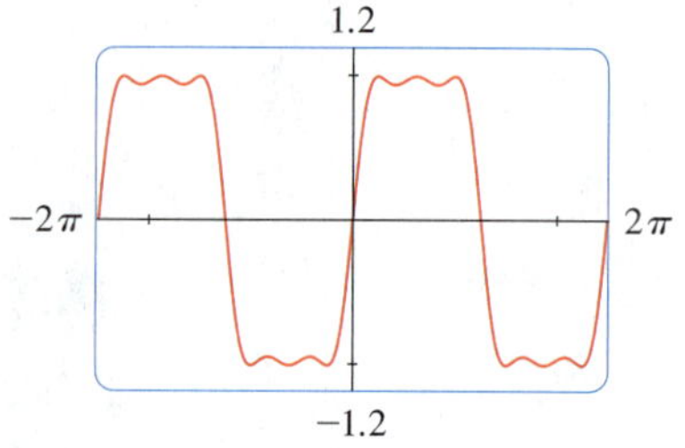

FIGURE 18

Having checked that Figure 15 does indeed represent f accurately for $0 \leq x \leq \pi$, we can state that the extended graph in Figure 18 represents f accurately for $-2\pi \leq x \leq 2\pi$.

Our final example is concerned with *families* of functions. As discussed in Section 1.4, this means that the functions in the family are related to each other by a formula that contains one or more arbitrary constants. Each value of the constant gives rise to a member of the family and the idea is to see how the graph of the function changes as the constant changes.

EXAMPLE 5 How does the graph of $f(x) = 1/(x^2 + 2x + c)$ vary as c varies?

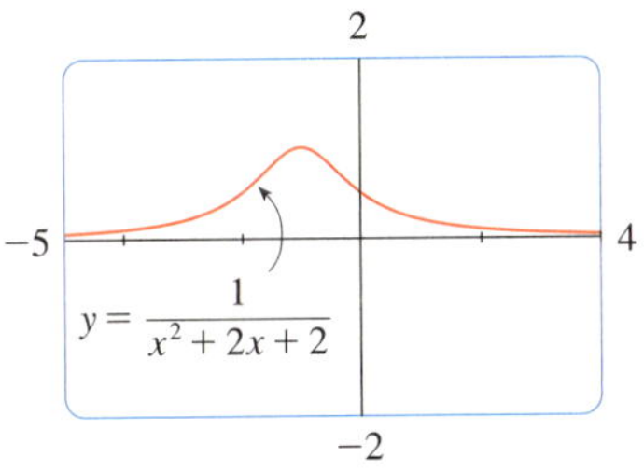

FIGURE 19 $c = 2$

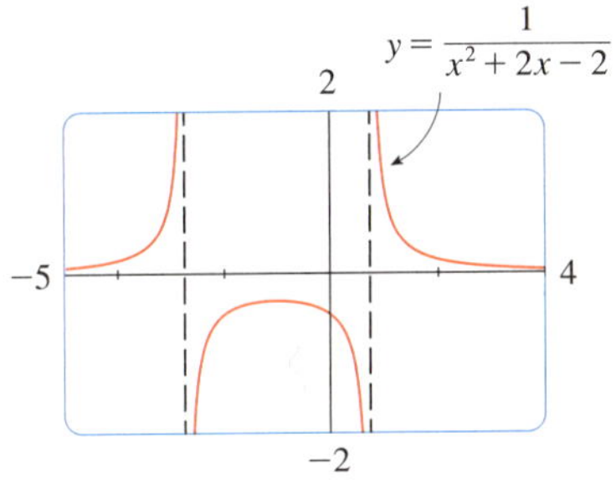

FIGURE 20 $c = -2$

SOLUTION The graphs in Figures 19 and 20 (the special cases $c = 2$ and $c = -2$) show two very different-looking curves. Before drawing any more graphs, let's see what members of this family have in common. Since

$$\lim_{x \to \pm\infty} \frac{1}{x^2 + 2x + c} = 0$$

for any value of c, they all have the x-axis as a horizontal asymptote. A vertical asymptote will occur when $x^2 + 2x + c = 0$. Solving this quadratic equation, we get $x = -1 \pm \sqrt{1 - c}$. When $c > 1$, there is no vertical asymptote (as in Figure 19). When $c = 1$, the graph has a single vertical asymptote $x = -1$ because

$$\lim_{x \to -1} \frac{1}{x^2 + 2x + 1} = \lim_{x \to -1} \frac{1}{(x + 1)^2} = \infty$$

When $c < 1$, there are two vertical asymptotes: $x = -1 \pm \sqrt{1 - c}$ (as in Figure 20).

Now we compute the derivative:

$$f'(x) = -\frac{2x + 2}{(x^2 + 2x + c)^2}$$

This shows that $f'(x) = 0$ when $x = -1$ (if $c \neq 1$), $f'(x) > 0$ when $x < -1$, and $f'(x) < 0$ when $x > -1$. For $c \geqslant 1$, this means that f increases on $(-\infty, -1)$ and decreases on $(-1, \infty)$. For $c > 1$, there is an absolute maximum value $f(-1) = 1/(c - 1)$. For $c < 1$, $f(-1) = 1/(c - 1)$ is a local maximum value and the intervals of increase and decrease are interrupted at the vertical asymptotes.

See an animation of Figure 21.
Resources / Module 5
/ Max and Min
/ Families of Functions

Figure 21 is a "slide show" displaying five members of the family, all graphed in the viewing rectangle $[-5, 4]$ by $[-2, 2]$. As predicted, $c = 1$ is the value at which a transition takes place from two vertical asymptotes to one, and then to none. As c increases from 1, we see that the maximum point becomes lower; this is explained by the fact that $1/(c - 1) \to 0$ as $c \to \infty$. As c decreases from 1, the vertical asymptotes become more widely separated because the distance between them is $2\sqrt{1 - c}$, which becomes large as $c \to -\infty$. Again, the maximum point approaches the x-axis because $1/(c - 1) \to 0$ as $c \to -\infty$.

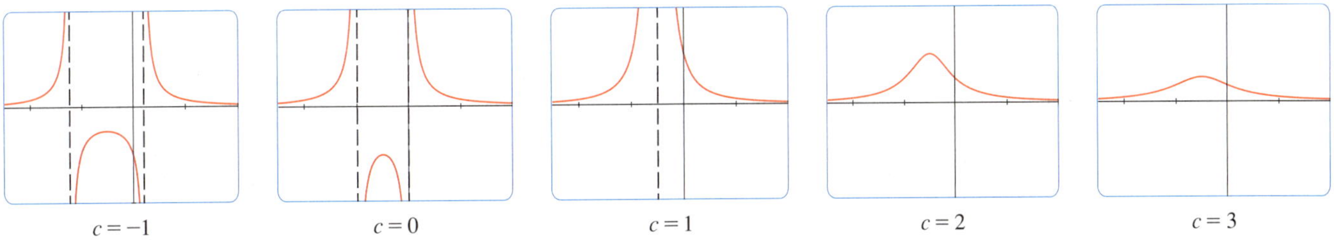

FIGURE 21 The family of functions $f(x) = 1/(x^2 + 2x + c)$

There is clearly no inflection point when $c \leqslant 1$. For $c > 1$ we calculate that

$$f''(x) = \frac{2(3x^2 + 6x + 4 - c)}{(x^2 + 2x + c)^3}$$

and deduce that inflection points occur when $x = -1 \pm \sqrt{3(c - 1)}/3$. So the inflection points become more spread out as c increases and this seems plausible from the last two parts of Figure 21.

4.6 Exercises

1–8 Produce graphs of f that reveal all the important aspects of the curve. In particular, you should use graphs of f' and f'' to estimate the intervals of increase and decrease, extreme values, intervals of concavity, and inflection points.

1. $f(x) = 4x^4 - 32x^3 + 89x^2 - 95x + 29$

2. $f(x) = x^6 - 15x^5 + 75x^4 - 125x^3 - x$

3. $f(x) = \sqrt[3]{x^2 - 3x - 5}$

4. $f(x) = \dfrac{x^4 + x^3 - 2x^2 + 2}{x^2 + x - 2}$

5. $f(x) = \dfrac{x}{x^3 - x^2 - 4x + 1}$

6. $f(x) = \tan x + 5\cos x$

7. $f(x) = x^2 - 4x + 7\cos x, \quad -4 \leqslant x \leqslant 4$

8. $f(x) = \dfrac{e^x}{x^2 - 9}$

9–12 Produce graphs of f that reveal all the important aspects of the curve. Estimate the intervals of increase and decrease, extreme values, intervals of concavity, and inflection points, and use calculus to find these quantities exactly.

9. $f(x) = 8x^3 - 3x^2 - 10$

10. $f(x) = \dfrac{x^2 + 11x - 20}{x^2}$

11. $f(x) = x\sqrt{9 - x^2}$

12. $f(x) = x - 2\sin x, \quad -2\pi \leqslant x \leqslant 2\pi$

13–14

(a) Graph the function.

(b) Use l'Hospital's Rule to explain the behavior as $x \to 0$.

(c) Estimate the minimum value and intervals of concavity. Then use calculus to find the exact values.

13. $f(x) = x^2 \ln x$

14. $f(x) = xe^{1/x}$

15–16 Sketch the graph by hand using asymptotes and intercepts, but not derivatives. Then use your sketch as a guide to producing graphs (with a graphing device) that display the major features of the curve. Use these graphs to estimate the maximum and minimum values.

15. $f(x) = \dfrac{(x + 4)(x - 3)^2}{x^4(x - 1)}$

16. $f(x) = \dfrac{10x(x - 1)^4}{(x - 2)^3(x + 1)^2}$

CAS **17.** If f is the function considered in Example 3, use a computer algebra system to calculate f' and then graph it to confirm that all the maximum and minimum values are as given in the example. Calculate f'' and use it to estimate the intervals of concavity and inflection points.

CAS **18.** If f is the function of Exercise 16, find f' and f'' and use their graphs to estimate the intervals of increase and decrease and concavity of f.

CAS **19–22** Use a computer algebra system to graph f and to find f' and f''. Use graphs of these derivatives to estimate the intervals of increase and decrease, extreme values, intervals of concavity, and inflection points of f.

19. $f(x) = \dfrac{\sin^2 x}{\sqrt{x^2 + 1}}, \quad 0 \leqslant x \leqslant 3\pi$

20. $f(x) = \dfrac{2x - 1}{\sqrt[4]{x^4 + x + 1}}$

21. $f(x) = \dfrac{1 - e^{1/x}}{1 + e^{1/x}}$

22. $f(x) = \dfrac{1}{1 + e^{\tan x}}$

CAS **23–24**

(a) Graph the function.

(b) Explain the shape of the graph by computing the limit as $x \to 0^+$ or as $x \to \infty$.

(c) Estimate the maximum and minimum values and then use calculus to find the exact values.

(d) Use a graph of f'' to estimate the x-coordinates of the inflection points.

23. $f(x) = x^{1/x}$

24. $f(x) = (\sin x)^{\sin x}$

25. In Example 4 we considered a member of the family of functions $f(x) = \sin(x + \sin cx)$ that occur in FM synthesis. Here we investigate the function with $c = 3$. Start by graphing f in the viewing rectangle $[0, \pi]$ by $[-1.2, 1.2]$. How many local maximum points do you see? The graph has more than are visible to the naked eye. To discover the hidden maximum and minimum points you will need to examine the graph of f' very carefully. In fact, it helps to look at the graph of f'' at the same time. Find all the maximum and minimum values and inflection points. Then graph f in the viewing rectangle $[-2\pi, 2\pi]$ by $[-1.2, 1.2]$ and comment on symmetry.

26–33 Describe how the graph of f varies as c varies. Graph several members of the family to illustrate the trends that you discover. In particular, you should investigate how maximum and minimum points and inflection points move when c changes. You should also identify any transitional values of c at which the basic shape of the curve changes.

26. $f(x) = x^3 + cx$

27. $f(x) = x^4 + cx^2$

28. $f(x) = x^2\sqrt{c^2 - x^2}$

29. $f(x) = e^{-c/x^2}$

30. $f(x) = \ln(x^2 + c)$

31. $f(x) = \dfrac{cx}{1 + c^2x^2}$

32. $f(x) = \dfrac{1}{(1 - x^2)^2 + cx^2}$

33. $f(x) = cx + \sin x$

34. The family of functions $f(t) = C(e^{-at} - e^{-bt})$, where a, b, and C are positive numbers and $b > a$, has been used to model the concentration of a drug injected into the blood at time $t = 0$. Graph several members of this family. What do they have in common? For fixed values of C and a, discover graphically what happens as b increases. Then use calculus to prove what you have discovered.

35. Investigate the family of curves given by $f(x) = xe^{-cx}$, where c is a real number. Start by computing the limits as $x \to \pm\infty$. Identify any transitional values of c where the basic shape changes. What happens to the maximum or minimum points and inflection points as c changes? Illustrate by graphing several members of the family.

36. Investigate the family of curves given by the equation $f(x) = x^4 + cx^2 + x$. Start by determining the transitional value of c at which the number of inflection points changes. Then graph several members of the family to see what shapes are possible. There is another transitional value of c at which the number of critical numbers changes. Try to discover it graphically. Then prove what you have discovered.

37. (a) Investigate the family of polynomials given by the equation $f(x) = cx^4 - 2x^2 + 1$. For what values of c does the curve have minimum points?

(b) Show that the minimum and maximum points of every curve in the family lie on the parabola $y = 1 - x^2$. Illustrate by graphing this parabola and several members of the family.

38. (a) Investigate the family of polynomials given by the equation $f(x) = 2x^3 + cx^2 + 2x$. For what values of c does the curve have maximum and minimum points?

(b) Show that the minimum and maximum points of every curve in the family lie on the curve $y = x - x^3$. Illustrate by graphing this curve and several members of the family.

4.7 Optimization Problems

The methods we have learned in this chapter for finding extreme values have practical applications in many areas of life. A businessperson wants to minimize costs and maximize profits. A traveler wants to minimize transportation time. Fermat's Principle in optics states that light follows the path that takes the least time. In this section and the next we solve such problems as maximizing areas, volumes, and profits and minimizing distances, times, and costs.

In solving such practical problems the greatest challenge is often to convert the word problem into a mathematical optimization problem by setting up the function that is to be maximized or minimized. Let's recall the problem-solving principles discussed on page 80 and adapt them to this situation:

Steps in Solving Optimization Problems

1. **Understand the Problem** The first step is to read the problem carefully until it is clearly understood. Ask yourself: What is the unknown? What are the given quantities? What are the given conditions?
2. **Draw a Diagram** In most problems it is useful to draw a diagram and identify the given and required quantities on the diagram.
3. **Introduce Notation** Assign a symbol to the quantity that is to be maximized or minimized (let's call it Q for now). Also select symbols $(a, b, c, \ldots, x, y)$ for other unknown quantities and label the diagram with these symbols. It may help to use initials as suggestive symbols—for example, A for area, h for height, t for time.
4. Express Q in terms of some of the other symbols from Step 3.

5. If Q has been expressed as a function of more than one variable in Step 4, use the given information to find relationships (in the form of equations) among these variables. Then use these equations to eliminate all but one of the variables in the expression for Q. Thus, Q will be expressed as a function of *one* variable x, say, $Q = f(x)$. Write the domain of this function.

6. Use the methods of Sections 4.1 and 4.3 to find the *absolute* maximum or minimum value of f. In particular, if the domain of f is a closed interval, then the Closed Interval Method in Section 4.1 can be used.

EXAMPLE 1 A farmer has 2400 ft of fencing and wants to fence off a rectangular field that borders a straight river. He needs no fence along the river. What are the dimensions of the field that has the largest area?

|||| Understand the problem
|||| Analogy: Try special cases
|||| Draw diagrams

SOLUTION In order to get a feeling for what is happening in this problem, let's experiment with some special cases. Figure 1 (not to scale) shows three possible ways of laying out the 2400 ft of fencing. We see that when we try shallow, wide fields or deep, narrow fields, we get relatively small areas. It seems plausible that there is some intermediate configuration that produces the largest area.

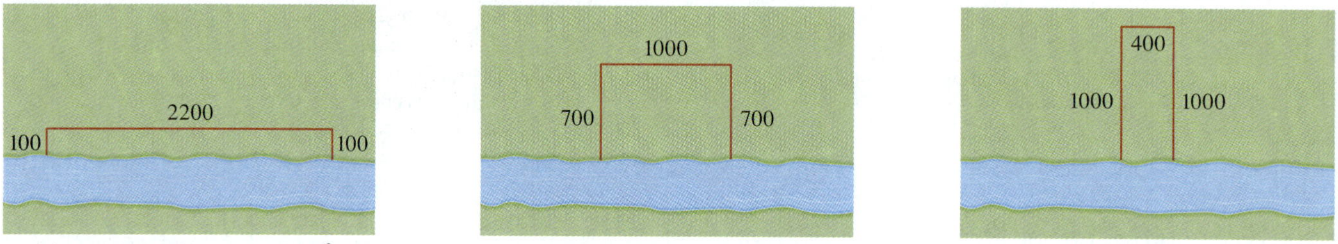

Area = $100 \cdot 2200 = 220{,}000$ ft^2 Area = $700 \cdot 1000 = 700{,}000$ ft^2 Area = $1000 \cdot 400 = 400{,}000$ ft^2

FIGURE 1

|||| Introduce notation

Figure 2 illustrates the general case. We wish to maximize the area A of the rectangle. Let x and y be the depth and width of the rectangle (in feet). Then we express A in terms of x and y:

$$A = xy$$

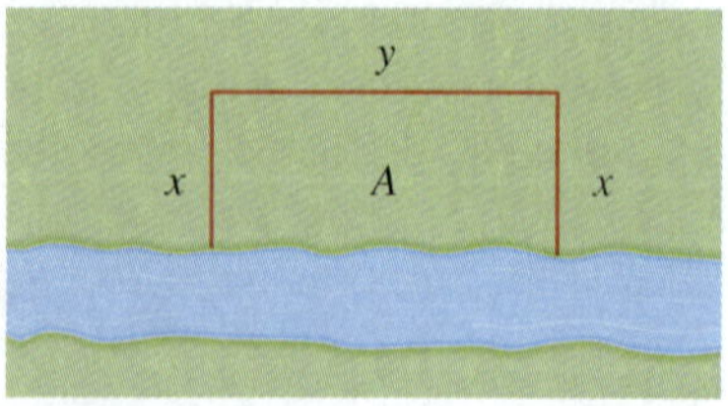

FIGURE 2

We want to express A as a function of just one variable, so we eliminate y by expressing it in terms of x. To do this we use the given information that the total length of the fencing is 2400 ft. Thus

$$2x + y = 2400$$

From this equation we have $y = 2400 - 2x$, which gives

$$A = x(2400 - 2x) = 2400x - 2x^2$$

Note that $x \geqslant 0$ and $x \leqslant 1200$ (otherwise $A < 0$). So the function that we wish to maximize is

$$A(x) = 2400x - 2x^2 \qquad 0 \leqslant x \leqslant 1200$$

The derivative is $A'(x) = 2400 - 4x$, so to find the critical numbers we solve the equation

$$2400 - 4x = 0$$

which gives $x = 600$. The maximum value of A must occur either at this critical number or at an endpoint of the interval. Since $A(0) = 0$, $A(600) = 720{,}000$, and $A(1200) = 0$, the Closed Interval Method gives the maximum value as $A(600) = 720{,}000$.

[Alternatively, we could have observed that $A''(x) = -4 < 0$ for all x, so A is always concave downward and the local maximum at $x = 600$ must be an absolute maximum.]

Thus, the rectangular field should be 600 ft deep and 1200 ft wide.

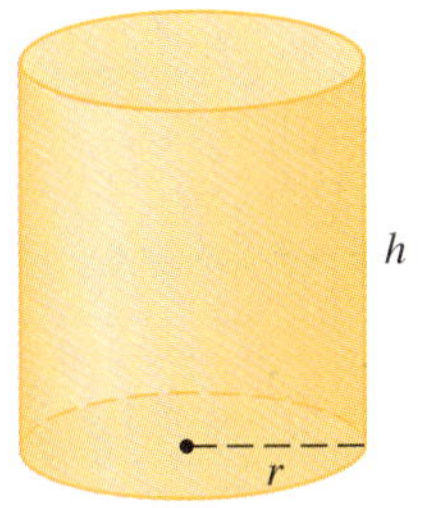

FIGURE 3

EXAMPLE 2 A cylindrical can is to be made to hold 1 L of oil. Find the dimensions that will minimize the cost of the metal to manufacture the can.

SOLUTION Draw the diagram as in Figure 3, where r is the radius and h the height (both in centimeters). In order to minimize the cost of the metal, we minimize the total surface area of the cylinder (top, bottom, and sides). From Figure 4 we see that the sides are made from a rectangular sheet with dimensions $2\pi r$ and h. So the surface area is

$$A = 2\pi r^2 + 2\pi rh$$

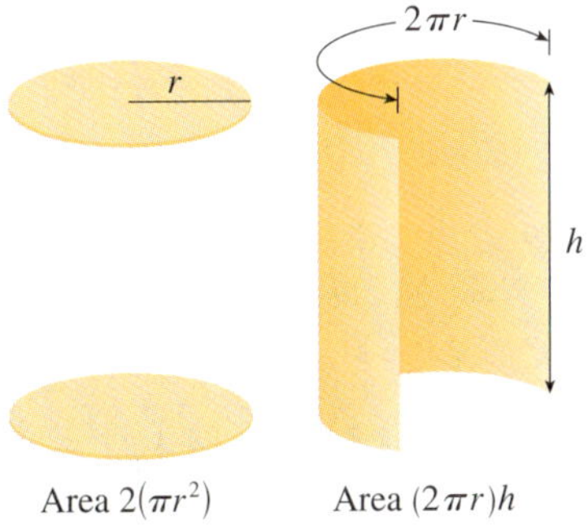

FIGURE 4

To eliminate h we use the fact that the volume is given as 1 L, which we take to be 1000 cm^3. Thus

$$\pi r^2 h = 1000$$

which gives $h = 1000/(\pi r^2)$. Substitution of this into the expression for A gives

$$A = 2\pi r^2 + 2\pi r\left(\frac{1000}{\pi r^2}\right) = 2\pi r^2 + \frac{2000}{r}$$

Therefore, the function that we want to minimize is

$$A(r) = 2\pi r^2 + \frac{2000}{r} \qquad r > 0$$

To find the critical numbers, we differentiate:

$$A'(r) = 4\pi r - \frac{2000}{r^2} = \frac{4(\pi r^3 - 500)}{r^2}$$

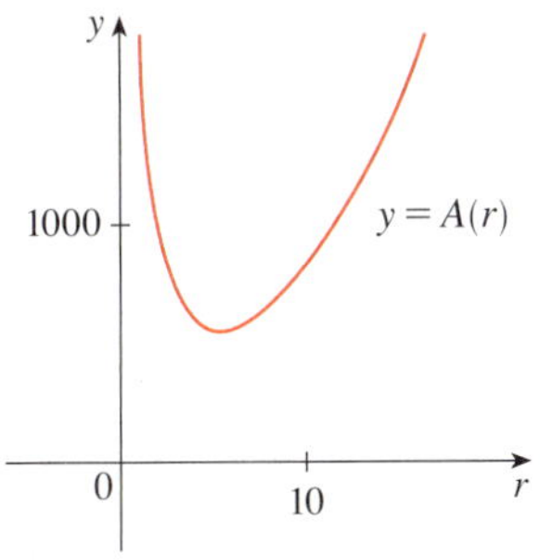

FIGURE 5

Then $A'(r) = 0$ when $\pi r^3 = 500$, so the only critical number is $r = \sqrt[3]{500/\pi}$.

Since the domain of A is $(0, \infty)$, we can't use the argument of Example 1 concerning endpoints. But we can observe that $A'(r) < 0$ for $r < \sqrt[3]{500/\pi}$ and $A'(r) > 0$ for $r > \sqrt[3]{500/\pi}$, so A is decreasing for *all* r to the left of the critical number and increasing for *all* r to the right. Thus, $r = \sqrt[3]{500/\pi}$ must give rise to an *absolute* minimum.

[Alternatively, we could argue that $A(r) \to \infty$ as $r \to 0^+$ and $A(r) \to \infty$ as $r \to \infty$, so there must be a minimum value of $A(r)$, which must occur at the critical number. See Figure 5.]

The value of h corresponding to $r = \sqrt[3]{500/\pi}$ is

$$h = \frac{1000}{\pi r^2} = \frac{1000}{\pi(500/\pi)^{2/3}} = 2\sqrt[3]{\frac{500}{\pi}} = 2r$$

Thus, to minimize the cost of the can, the radius should be $\sqrt[3]{500/\pi}$ cm and the height should be equal to twice the radius, namely, the diameter.

|||| In the Applied Project on page 341 we investigate the most economical shape for a can by taking into account other manufacturing costs.

NOTE 1 ▫ The argument used in Example 2 to justify the absolute minimum is a variant of the First Derivative Test (which applies only to *local* maximum or minimum values) and is stated here for future reference.

TEC Module 4.7 takes you through eight additional optimization problems, including animations of the physical situations.

First Derivative Test for Absolute Extreme Values Suppose that c is a critical number of a continuous function f defined on an interval.

(a) If $f'(x) > 0$ for all $x < c$ and $f'(x) < 0$ for all $x > c$, then $f(c)$ is the absolute maximum value of f.

(b) If $f'(x) < 0$ for all $x < c$ and $f'(x) > 0$ for all $x > c$, then $f(c)$ is the absolute minimum value of f.

NOTE 2 ▫ An alternative method for solving optimization problems is to use implicit differentiation. Let's look at Example 2 again to illustrate the method. We work with the same equations

$$A = 2\pi r^2 + 2\pi rh \qquad \pi r^2 h = 100$$

but instead of eliminating h, we differentiate both equations implicitly with respect to r:

$$A' = 4\pi r + 2\pi h + 2\pi rh' \qquad 2\pi rh + \pi r^2 h' = 0$$

The minimum occurs at a critical number, so we set $A' = 0$, simplify, and arrive at the equations

$$2r + h + rh' = 0 \qquad 2h + rh' = 0$$

and subtraction gives $2r - h = 0$, or $h = 2r$.

EXAMPLE 3 Find the point on the parabola $y^2 = 2x$ that is closest to the point $(1, 4)$.

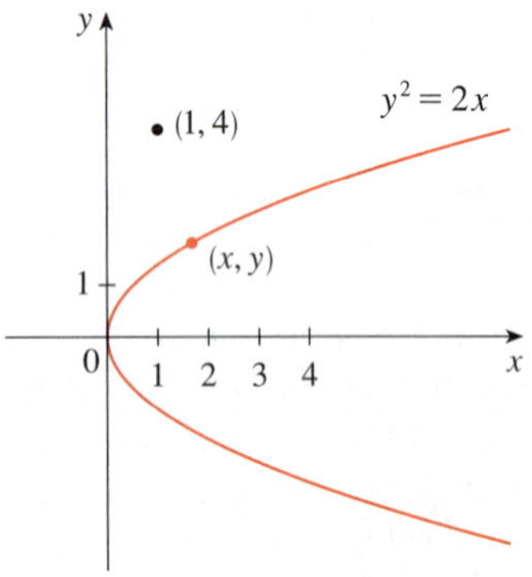

FIGURE 6

SOLUTION The distance between the point $(1, 4)$ and the point (x, y) is

$$d = \sqrt{(x - 1)^2 + (y - 4)^2}$$

(See Figure 6.) But if (x, y) lies on the parabola, then $x = y^2/2$, so the expression for d becomes

$$d = \sqrt{(\tfrac{1}{2}y^2 - 1)^2 + (y - 4)^2}$$

(Alternatively, we could have substituted $y = \sqrt{2x}$ to get d in terms of x alone.) Instead of minimizing d, we minimize its square:

$$d^2 = f(y) = (\tfrac{1}{2}y^2 - 1)^2 + (y - 4)^2$$

(You should convince yourself that the minimum of d occurs at the same point as the minimum of d^2, but d^2 is easier to work with.) Differentiating, we obtain

$$f'(y) = 2(\tfrac{1}{2}y^2 - 1)y + 2(y - 4) = y^3 - 8$$

so $f'(y) = 0$ when $y = 2$. Observe that $f'(y) < 0$ when $y < 2$ and $f'(y) > 0$ when $y > 2$, so by the First Derivative Test for Absolute Extreme Values, the absolute minimum occurs when $y = 2$. (Or we could simply say that because of the geometric nature of the problem, it's obvious that there is a closest point but not a farthest point.) The corresponding value of x is $x = y^2/2 = 2$. Thus, the point on $y^2 = 2x$ closest to $(1, 4)$ is $(2, 2)$.

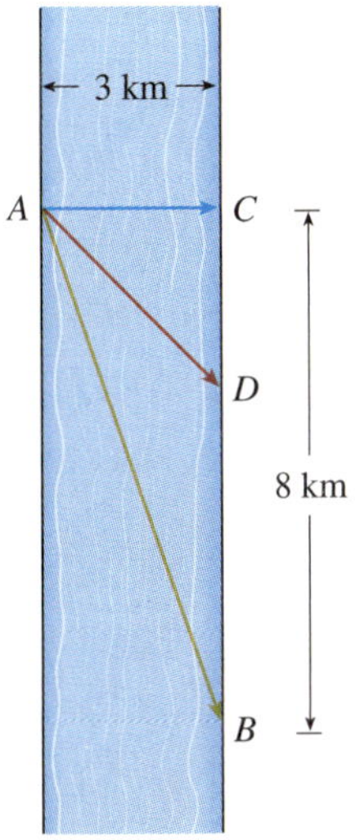

FIGURE 7

EXAMPLE 4 A man launches his boat from point A on a bank of a straight river, 3 km wide, and wants to reach point B, 8 km downstream on the opposite bank, as quickly as possible (see Figure 7). He could row his boat directly across the river to point C and then run to B, or he could row directly to B, or he could row to some point D between C and B and then run to B. If he can row 6 km/h and run 8 km/h, where should he land to reach B as soon as possible? (We assume that the speed of the water is negligible compared with the speed at which the man rows.)

SOLUTION If we let x be the distance from C to D, then the running distance is $|DB| = 8 - x$ and the Pythagorean Theorem gives the rowing distance as $|AD| = \sqrt{x^2 + 9}$. We use the equation

$$\text{time} = \frac{\text{distance}}{\text{rate}}$$

Then the rowing time is $\sqrt{x^2 + 9}/6$ and the running time is $(8 - x)/8$, so the total time T as a function of x is

$$T(x) = \frac{\sqrt{x^2 + 9}}{6} + \frac{8 - x}{8}$$

The domain of this function T is $[0, 8]$. Notice that if $x = 0$ he rows to C and if $x = 8$ he rows directly to B. The derivative of T is

$$T'(x) = \frac{x}{6\sqrt{x^2 + 9}} - \frac{1}{8}$$

Try another problem like this one.
Resources / Module 5
/ Max and Min
/ Start of Optimal Lifeguard

Thus, using the fact that $x \geqslant 0$, we have

$$T'(x) = 0 \iff \frac{x}{6\sqrt{x^2 + 9}} = \frac{1}{8} \iff 4x = 3\sqrt{x^2 + 9}$$

$$\iff 16x^2 = 9(x^2 + 9) \iff 7x^2 = 81$$

$$\iff x = \frac{9}{\sqrt{7}}$$

The only critical number is $x = 9/\sqrt{7}$. To see whether the minimum occurs at this critical number or at an endpoint of the domain $[0, 8]$, we evaluate T at all three points:

$$T(0) = 1.5 \qquad T\left(\frac{9}{\sqrt{7}}\right) = 1 + \frac{\sqrt{7}}{8} \approx 1.33 \qquad T(8) = \frac{\sqrt{73}}{6} \approx 1.42$$

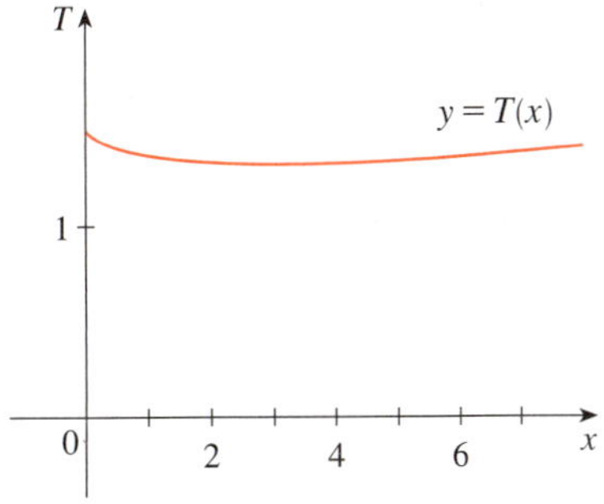

FIGURE 8

Since the smallest of these values of T occurs when $x = 9/\sqrt{7}$, the absolute minimum value of T must occur there. Figure 8 illustrates this calculation by showing the graph of T.

Thus, the man should land the boat at a point $9/\sqrt{7}$ km ($\approx$3.4 km) downstream from his starting point.

EXAMPLE 5 Find the area of the largest rectangle that can be inscribed in a semicircle of radius r.

Resources / Module 5 / Max and Min / Start of Max and Min

SOLUTION 1 Let's take the semicircle to be the upper half of the circle $x^2 + y^2 = r^2$ with center the origin. Then the word *inscribed* means that the rectangle has two vertices on the semicircle and two vertices on the x-axis as shown in Figure 9.

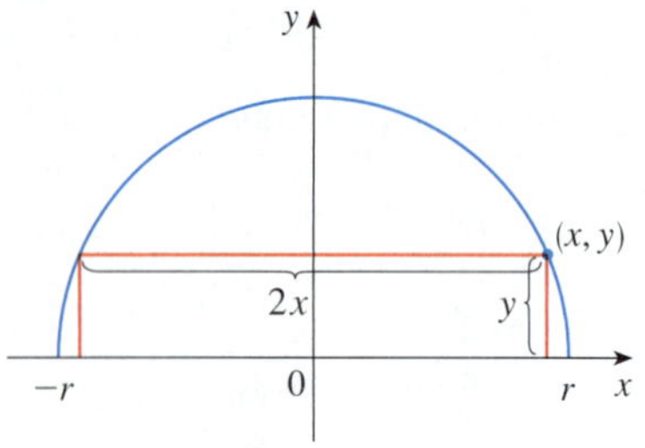

FIGURE 9

Let (x, y) be the vertex that lies in the first quadrant. Then the rectangle has sides of lengths $2x$ and y, so its area is

$$A = 2xy$$

To eliminate y we use the fact that (x, y) lies on the circle $x^2 + y^2 = r^2$ and so $y = \sqrt{r^2 - x^2}$. Thus

$$A = 2x\sqrt{r^2 - x^2}$$

The domain of this function is $0 \leqslant x \leqslant r$. Its derivative is

$$A' = 2\sqrt{r^2 - x^2} - \frac{2x^2}{\sqrt{r^2 - x^2}} = \frac{2(r^2 - 2x^2)}{\sqrt{r^2 - x^2}}$$

which is 0 when $2x^2 = r^2$, that is, $x = r/\sqrt{2}$ (since $x \geqslant 0$). This value of x gives a maximum value of A since $A(0) = 0$ and $A(r) = 0$. Therefore, the area of the largest inscribed rectangle is

$$A\left(\frac{r}{\sqrt{2}}\right) = 2\frac{r}{\sqrt{2}}\sqrt{r^2 - \frac{r^2}{2}} = r^2$$

SOLUTION 2 A simpler solution is possible if we think of using an angle as a variable. Let θ be the angle shown in Figure 10. Then the area of the rectangle is

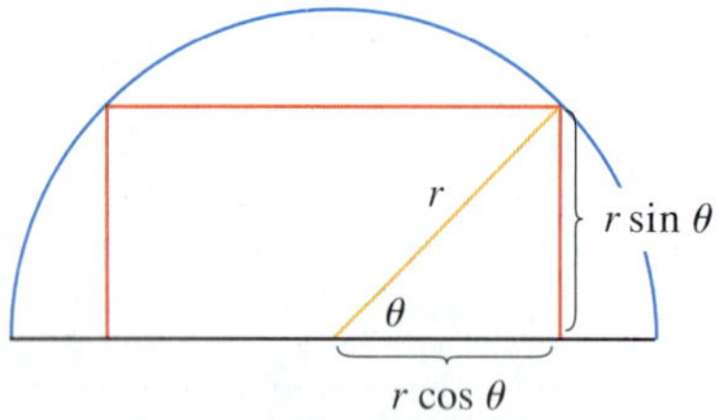

FIGURE 10

$$A(\theta) = (2r\cos\theta)(r\sin\theta) = r^2(2\sin\theta\,\cos\theta) = r^2\sin 2\theta$$

We know that $\sin 2\theta$ has a maximum value of 1 and it occurs when $2\theta = \pi/2$. So $A(\theta)$ has a maximum value of r^2 and it occurs when $\theta = \pi/4$.

Notice that this trigonometric solution doesn't involve differentiation. In fact, we didn't need to use calculus at all.

4.7 Exercises

1. Consider the following problem: Find two numbers whose sum is 23 and whose product is a maximum.

(a) Make a table of values, like the following one, so that the sum of the numbers in the first two columns is always 23. On the basis of the evidence in your table, estimate the answer to the problem.

First number	Second number	Product
1	22	22
2	21	42
3	20	60
⋮	⋮	⋮

(b) Use calculus to solve the problem and compare with your answer to part (a).

2. Find two numbers whose difference is 100 and whose product is a minimum.

3. Find two positive numbers whose product is 100 and whose sum is a minimum.

4. Find a positive number such that the sum of the number and its reciprocal is as small as possible.

5. Find the dimensions of a rectangle with perimeter 100 m whose area is as large as possible.

6. Find the dimensions of a rectangle with area 1000 m^2 whose perimeter is as small as possible.

7. Consider the following problem: A farmer with 750 ft of fencing wants to enclose a rectangular area and then divide it into four pens with fencing parallel to one side of the rectangle. What is the largest possible total area of the four pens?
(a) Draw several diagrams illustrating the situation, some with shallow, wide pens and some with deep, narrow pens. Find the total areas of these configurations. Does it appear that there is a maximum area? If so, estimate it.
(b) Draw a diagram illustrating the general situation. Introduce notation and label the diagram with your symbols.
(c) Write an expression for the total area.
(d) Use the given information to write an equation that relates the variables.
(e) Use part (d) to write the total area as a function of one variable.
(f) Finish solving the problem and compare the answer with your estimate in part (a).

8. Consider the following problem: A box with an open top is to be constructed from a square piece of cardboard, 3 ft wide, by cutting out a square from each of the four corners and bending up the sides. Find the largest volume that such a box can have.
(a) Draw several diagrams to illustrate the situation, some short boxes with large bases and some tall boxes with small bases. Find the volumes of several such boxes. Does it appear that there is a maximum volume? If so, estimate it.
(b) Draw a diagram illustrating the general situation. Introduce notation and label the diagram with your symbols.
(c) Write an expression for the volume.
(d) Use the given information to write an equation that relates the variables.
(e) Use part (d) to write the volume as a function of one variable.
(f) Finish solving the problem and compare the answer with your estimate in part (a).

9. A farmer wants to fence an area of 1.5 million square feet in a rectangular field and then divide it in half with a fence parallel to one of the sides of the rectangle. How can he do this so as to minimize the cost of the fence?

10. A box with a square base and open top must have a volume of 32,000 cm^3. Find the dimensions of the box that minimize the amount of material used.

11. If 1200 cm^2 of material is available to make a box with a square base and an open top, find the largest possible volume of the box.

12. A rectangular storage container with an open top is to have a volume of 10 m^3. The length of its base is twice the width. Material for the base costs \$10 per square meter. Material for the sides costs \$6 per square meter. Find the cost of materials for the cheapest such container.

13. Do Exercise 12 assuming the container has a lid that is made from the same material as the sides.

14. (a) Show that of all the rectangles with a given area, the one with smallest perimeter is a square.
(b) Show that of all the rectangles with a given perimeter, the one with greatest area is a square.

15. Find the point on the line $y = 4x + 7$ that is closest to the origin.

16. Find the point on the line $6x + y = 9$ that is closest to the point $(-3, 1)$.

17. Find the points on the ellipse $4x^2 + y^2 = 4$ that are farthest away from the point $(1, 0)$.

18. Find, correct to two decimal places, the coordinates of the point on the curve $y = \tan x$ that is closest to the point $(1, 1)$.

19. Find the dimensions of the rectangle of largest area that can be inscribed in a circle of radius r.

20. Find the area of the largest rectangle that can be inscribed in the ellipse $x^2/a^2 + y^2/b^2 = 1$.

21. Find the dimensions of the rectangle of largest area that can be inscribed in an equilateral triangle of side L if one side of the rectangle lies on the base of the triangle.

22. Find the dimensions of the rectangle of largest area that has its base on the x-axis and its other two vertices above the x-axis and lying on the parabola $y = 8 - x^2$.

23. Find the dimensions of the isosceles triangle of largest area that can be inscribed in a circle of radius r.

24. Find the area of the largest rectangle that can be inscribed in a right triangle with legs of lengths 3 cm and 4 cm if two sides of the rectangle lie along the legs.

25. A right circular cylinder is inscribed in a sphere of radius r. Find the largest possible volume of such a cylinder.

26. A right circular cylinder is inscribed in a cone with height h and base radius r. Find the largest possible volume of such a cylinder.

27. A right circular cylinder is inscribed in a sphere of radius r. Find the largest possible surface area of such a cylinder.

28. A Norman window has the shape of a rectangle surmounted by a semicircle. (Thus, the diameter of the semicircle is equal to the width of the rectangle. See Exercise 52 on page 24.) If the perimeter of the window is 30 ft, find the dimensions of the window so that the greatest possible amount of light is admitted.

29. The top and bottom margins of a poster are each 6 cm and the side margins are each 4 cm. If the area of printed material on the poster is fixed at 384 cm^2, find the dimensions of the poster with the smallest area.

30. A poster is to have an area of 180 in^2 with 1-inch margins at the bottom and sides and a 2-inch margin at the top. What dimensions will give the largest printed area?

31. A piece of wire 10 m long is cut into two pieces. One piece is bent into a square and the other is bent into an equilateral triangle. How should the wire be cut so that the total area enclosed is (a) a maximum? (b) A minimum?

32. Answer Exercise 31 if one piece is bent into a square and the other into a circle.

33. A cylindrical can without a top is made to contain V cm^3 of liquid. Find the dimensions that will minimize the cost of the metal to make the can.

34. A fence 8 ft tall runs parallel to a tall building at a distance of 4 ft from the building. What is the length of the shortest ladder that will reach from the ground over the fence to the wall of the building?

35. A cone-shaped drinking cup is made from a circular piece of paper of radius R by cutting out a sector and joining the edges CA and CB. Find the maximum capacity of such a cup.

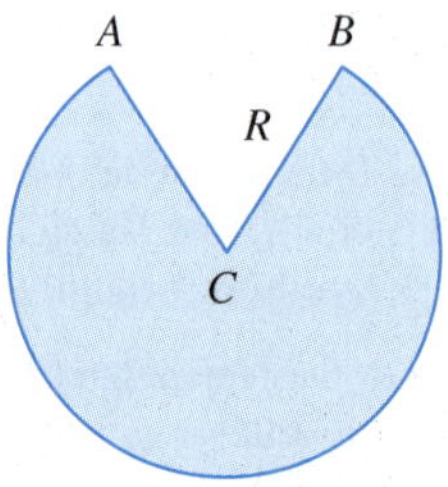

36. A cone-shaped paper drinking cup is to be made to hold 27 cm^3 of water. Find the height and radius of the cup that will use the smallest amount of paper.

37. A cone with height h is inscribed in a larger cone with height H so that its vertex is at the center of the base of the larger cone. Show that the inner cone has maximum volume when $h = \frac{1}{3}H$.

38. For a fish swimming at a speed v relative to the water, the energy expenditure per unit time is proportional to v^3. It is believed that migrating fish try to minimize the total energy required to swim a fixed distance. If the fish are swimming against a current u ($u < v$), then the time required to swim a distance L is $L/(v - u)$ and the total energy E required to swim the distance is given by

$$E(v) = av^3 \cdot \frac{L}{v - u}$$

where a is the proportionality constant.
(a) Determine the value of v that minimizes E.
(b) Sketch the graph of E.

Note: This result has been verified experimentally; migrating fish swim against a current at a speed 50% greater than the current speed.

39. In a beehive, each cell is a regular hexagonal prism, open at one end with a trihedral angle at the other end. It is believed that bees form their cells in such a way as to minimize the surface area for a given volume, thus using the least amount of wax in cell construction. Examination of these cells has shown that the measure of the apex angle θ is amazingly consistent. Based on the geometry of the cell, it can be shown that the surface area S is given by

$$S = 6sh - \tfrac{3}{2}s^2 \cot \theta + (3s^2\sqrt{3}/2) \csc \theta$$

where s, the length of the sides of the hexagon, and h, the height, are constants.
(a) Calculate $dS/d\theta$.
(b) What angle should the bees prefer?
(c) Determine the minimum surface area of the cell (in terms of s and h).

Note: Actual measurements of the angle θ in beehives have been made, and the measures of these angles seldom differ from the calculated value by more than 2°.

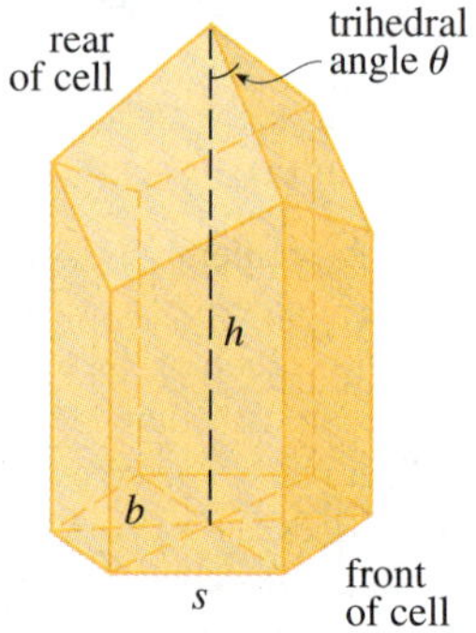

40. A boat leaves a dock at 2:00 P.M. and travels due south at a speed of 20 km/h. Another boat has been heading due east at 15 km/h and reaches the same dock at 3:00 P.M. At what time were the two boats closest together?

41. Solve the problem in Example 4 if the river is 5 km wide and point B is only 5 km downstream from A.

42. A woman at a point A on the shore of a circular lake with radius 2 mi wants to arrive at the point C diametrically opposite A on the other side of the lake in the shortest possible time. She can walk at the rate of 4 mi/h and row a boat at 2 mi/h. How should she proceed?

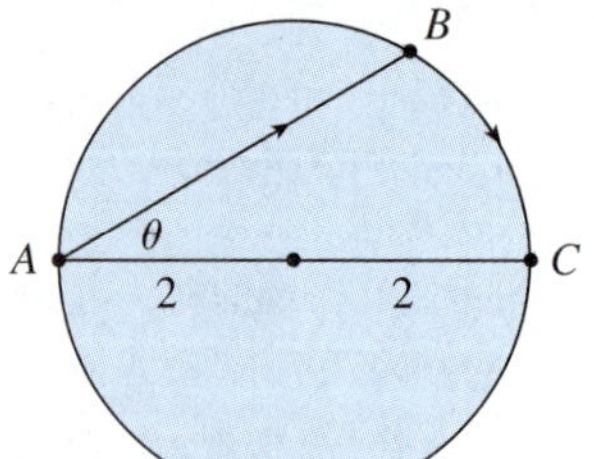

43. The illumination of an object by a light source is directly proportional to the strength of the source and inversely proportional to the square of the distance from the source. If two light sources, one three times as strong as the other, are placed 10 ft apart, where should an object be placed on the line between the sources so as to receive the least illumination?

44. Find an equation of the line through the point (3, 5) that cuts off the least area from the first quadrant.

45. Let a and b be positive numbers. Find the length of the shortest line segment that is cut off by the first quadrant and passes through the point (a, b).

46. At which points on the curve $y = 1 + 40x^3 - 3x^5$ does the tangent line have the largest slope?

47. Show that of all the isosceles triangles with a given perimeter, the one with the greatest area is equilateral.

48. The frame for a kite is to be made from six pieces of wood. The four exterior pieces have been cut with the lengths indicated in the figure. To maximize the area of the kite, how long should the diagonal pieces be?

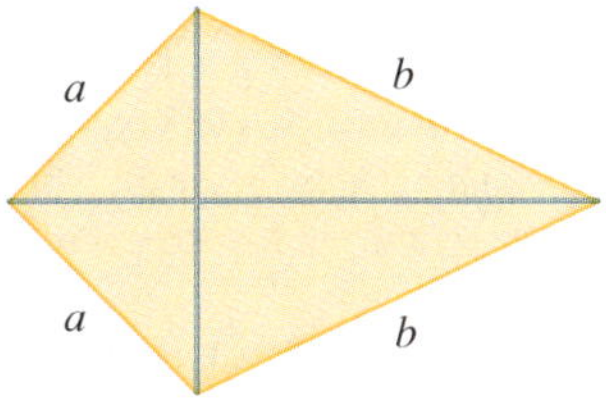

49. A point P needs to be located somewhere on the line AD so that the total length L of cables linking P to the points A, B, and C is minimized (see the figure). Express L as a function of $x = |AP|$ and use the graphs of L and dL/dx to estimate the minimum value.

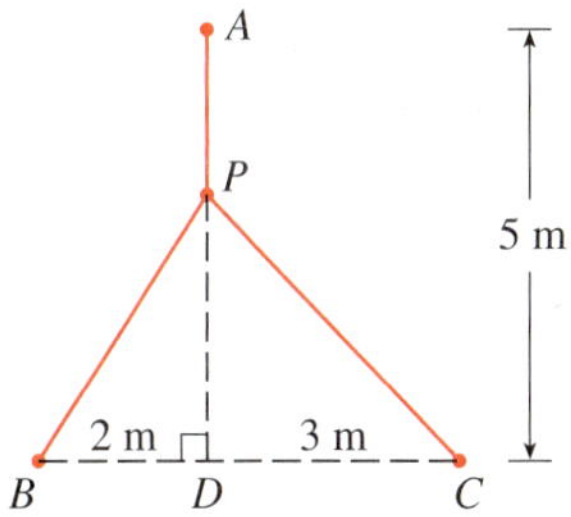

50. The graph shows the fuel consumption c of a car (measured in gallons per hour) as a function of the speed v of the car. At very low speeds the engine runs inefficiently, so initially c decreases as the speed increases. But at high speeds the fuel consumption increases. You can see that $c(v)$ is minimized for this car when $v \approx 30$ mi/h. However, for fuel efficiency, what must be minimized is not the consumption in gallons per hour but rather the fuel consumption in gallons *per mile.* Let's call this consumption G. Using the graph, estimate the speed at which G has its minimum value.

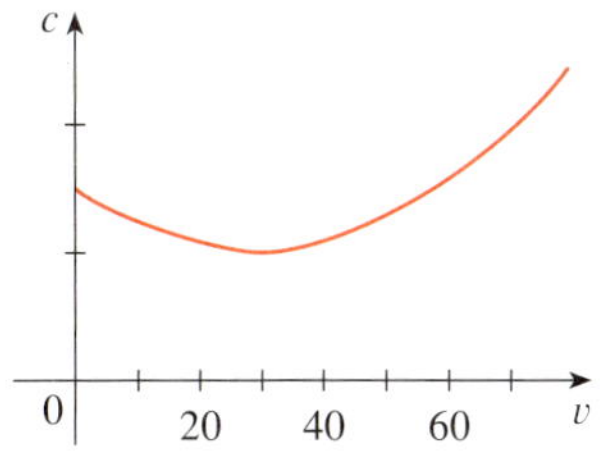

51. Let v_1 be the velocity of light in air and v_2 the velocity of light in water. According to Fermat's Principle, a ray of light will travel from a point A in the air to a point B in the water by a path ACB that minimizes the time taken. Show that

$$\frac{\sin \theta_1}{\sin \theta_2} = \frac{v_1}{v_2}$$

where θ_1 (the angle of incidence) and θ_2 (the angle of refraction) are as shown. This equation is known as Snell's Law.

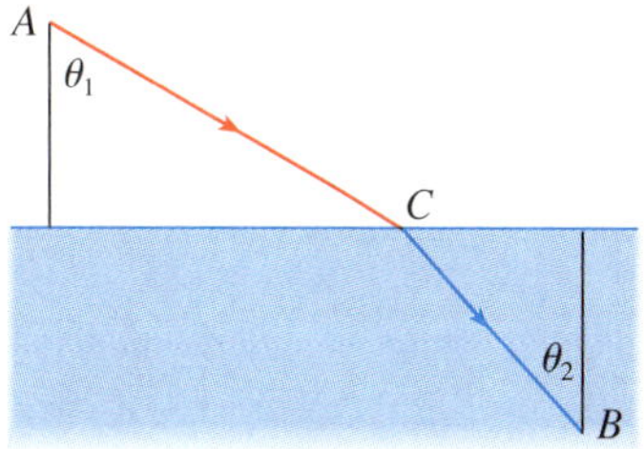

52. Two vertical poles PQ and ST are secured by a rope PRS going from the top of the first pole to a point R on the ground between the poles and then to the top of the second pole as in the figure. Show that the shortest length of such a rope occurs when $\theta_1 = \theta_2$.

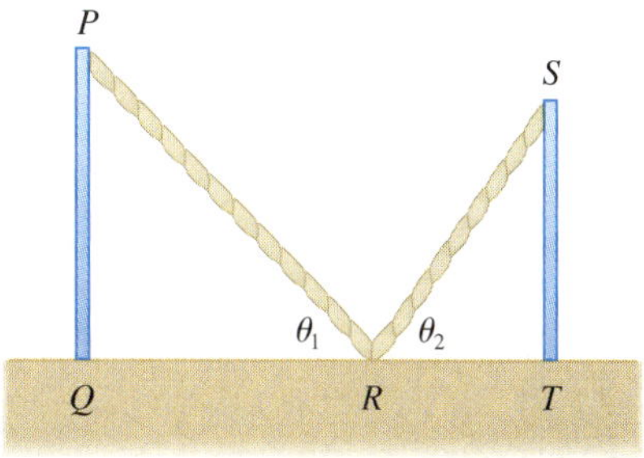

53. The upper right-hand corner of a piece of paper, 12 in. by 8 in., as in the figure, is folded over to the bottom edge. How would you fold it so as to minimize the length of the fold? In other words, how would you choose x to minimize y?

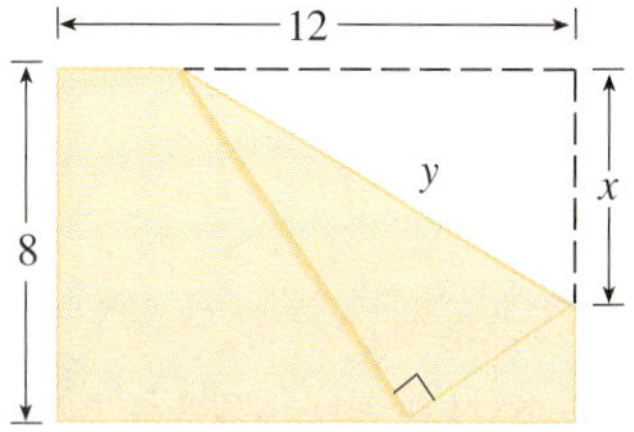

54. A steel pipe is being carried down a hallway 9 ft wide. At the end of the hall there is a right-angled turn into a narrower hallway 6 ft wide. What is the length of the longest pipe that can be carried horizontally around the corner?

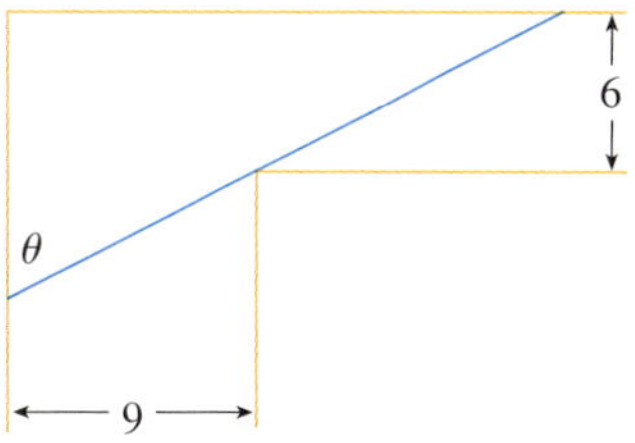

55. An observer stands at a point P, one unit away from a track. Two runners start at the point S in the figure and run along the track. One runner runs three times as fast as the other. Find the maximum value of the observer's angle of sight θ between the runners. [*Hint:* Maximize $\tan \theta$.]

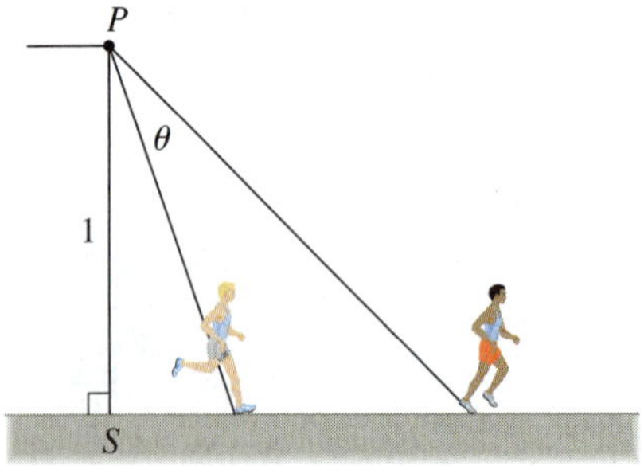

56. A rain gutter is to be constructed from a metal sheet of width 30 cm by bending up one-third of the sheet on each side through an angle θ. How should θ be chosen so that the gutter will carry the maximum amount of water?

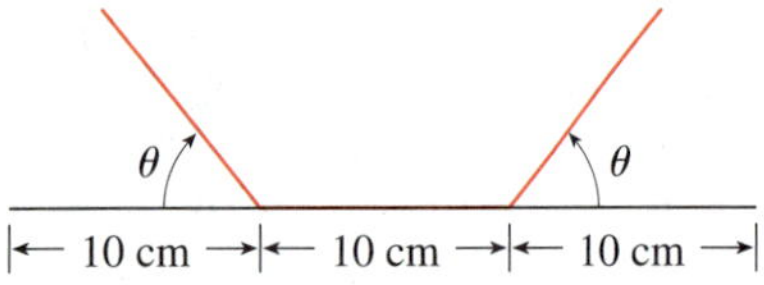

57. Where should the point P be chosen on the line segment AB so as to maximize the angle θ?

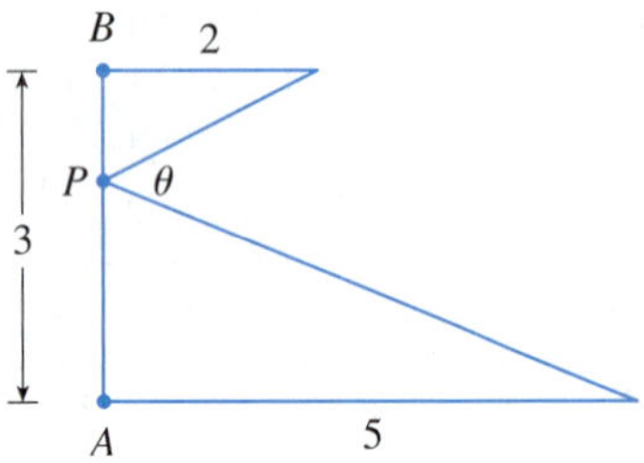

58. A painting in an art gallery has height h and is hung so that its lower edge is a distance d above the eye of an observer (as in the figure). How far from the wall should the observer stand to get the best view? (In other words, where should the observer stand so as to maximize the angle θ subtended at his eye by the painting?)

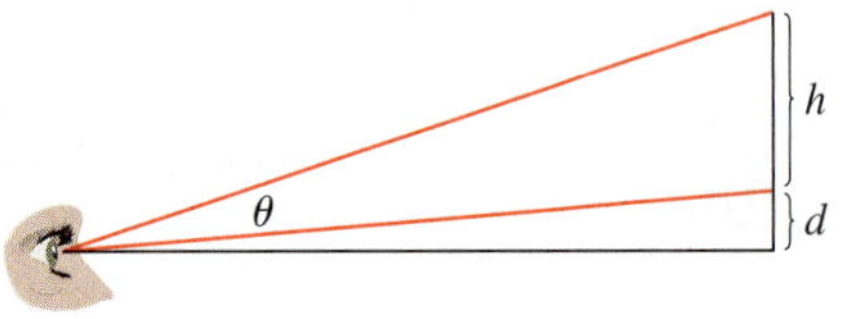

59. Find the maximum area of a rectangle that can be circumscribed about a given rectangle with length L and width W.

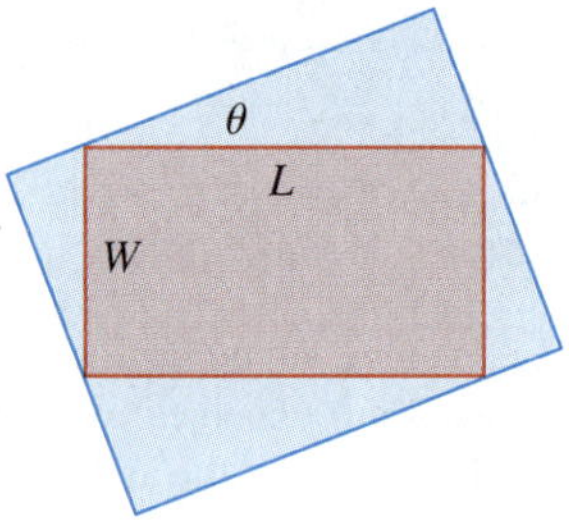

60. The blood vascular system consists of blood vessels (arteries, arterioles, capillaries, and veins) that convey blood from the heart to the organs and back to the heart. This system should work so as to minimize the energy expended by the heart in pumping the blood. In particular, this energy is reduced when the resistance of the blood is lowered. One of Poiseuille's Laws gives the resistance R of the blood as

$$R = C\frac{L}{r^4}$$

where L is the length of the blood vessel, r is the radius, and C is a positive constant determined by the viscosity of the blood. (Poiseuille established this law experimentally, but it also follows from Equation 8.4.2.) The figure shows a main blood vessel with radius r_1 branching at an angle θ into a smaller vessel with radius r_2.

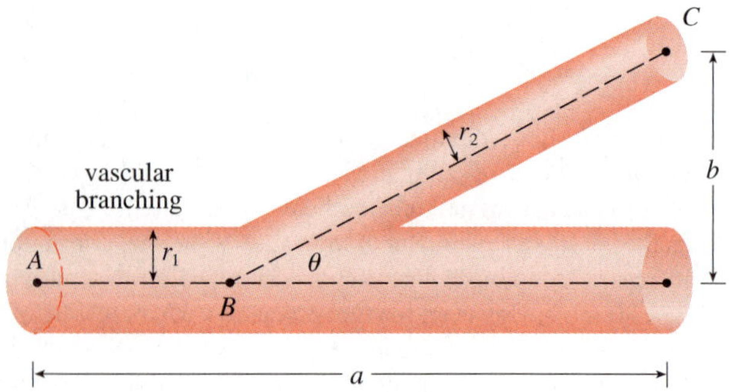

(a) Use Poiseuille's Law to show that the total resistance of the blood along the path ABC is

$$R = C\left(\frac{a - b\cot\theta}{r_1^4} + \frac{b\csc\theta}{r_2^4}\right)$$

where a and b are the distances shown in the figure.

(b) Prove that this resistance is minimized when

$$\cos\theta = \frac{r_2^4}{r_1^4}$$

(c) Find the optimal branching angle (correct to the nearest degree) when the radius of the smaller blood vessel is two-thirds the radius of the larger vessel.

61. Ornithologists have determined that some species of birds tend to avoid flights over large bodies of water during daylight hours. It is believed that more energy is required to fly over water than land because air generally rises over land and falls over water during the day. A bird with these tendencies is released from an island that is 5 km from the nearest point B on a straight shoreline, flies to a point C on the shoreline, and then flies along the shoreline to its nesting area D. Assume that the bird instinctively chooses a path that will minimize its energy expenditure. Points B and D are 13 km apart.

(a) In general, if it takes 1.4 times as much energy to fly over water as land, to what point C should the bird fly in order to minimize the total energy expended in returning to its nesting area?

(b) Let W and L denote the energy (in joules) per kilometer flown over water and land, respectively. What would a large value of the ratio W/L mean in terms of the bird's flight? What would a small value mean? Determine the ratio W/L corresponding to the minimum expenditure of energy.

(c) What should the value of W/L be in order for the bird to fly directly to its nesting area D? What should the value of W/L be for the bird to fly to B and then along the shore to D?

(d) If the ornithologists observe that birds of a certain species reach the shore at a point 4 km from B, how many times more energy does it take a bird to fly over water than land?

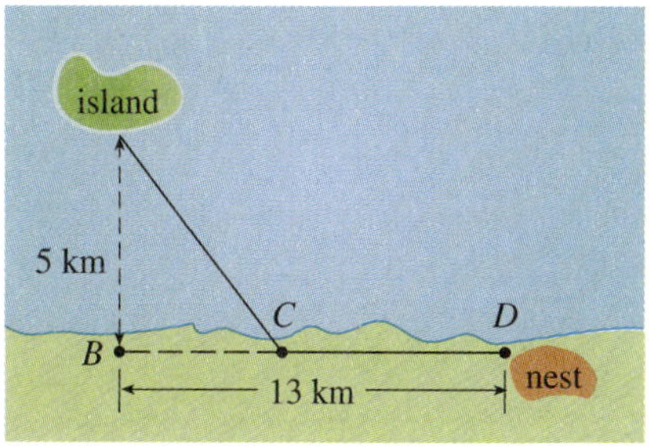

62. Two light sources of identical strength are placed 10 m apart. An object is to be placed at a point P on a line ℓ parallel to the line joining the light sources and at a distance d meters from it (see the figure). We want to locate P on ℓ so that the intensity of illumination is minimized. We need to use the fact that the intensity of illumination for a single source is directly proportional to the strength of the source and inversely proportional to the square of the distance from the source.

(a) Find an expression for the intensity $I(x)$ at the point P.

(b) If $d = 5$ m, use graphs of $I(x)$ and $I'(x)$ to show that the intensity is minimized when $x = 5$ m, that is, when P is at the midpoint of ℓ.

(c) If $d = 10$ m, show that the intensity (perhaps surprisingly) is *not* minimized at the midpoint.

(d) Somewhere between $d = 5$ m and $d = 10$ m there is a transitional value of d at which the point of minimal illumination abruptly changes. Estimate this value of d by graphical methods. Then find the exact value of d.

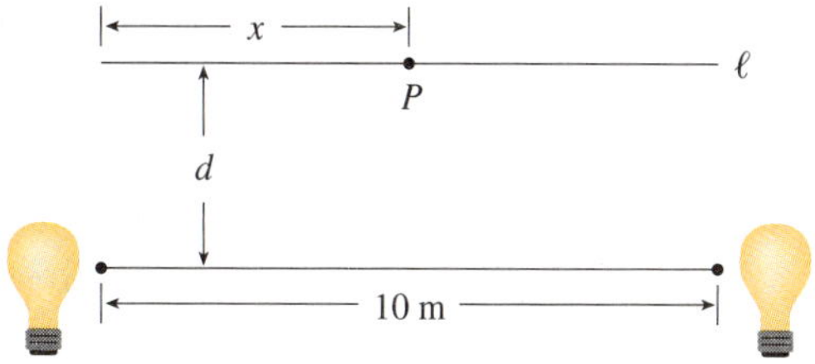

APPLIED PROJECT

The Shape of a Can

In this project we investigate the most economical shape for a can. We first interpret this to mean that the volume V of a cylindrical can is given and we need to find the height h and radius r that minimize the cost of the metal to make the can (see the figure). If we disregard any waste metal in the manufacturing process, then the problem is to minimize the surface area of the cylinder. We solved this problem in Example 2 in Section 4.7 and we found that $h = 2r$; that is, the height should be the same as the diameter. But if you go to your cupboard or your supermarket with a ruler, you will discover that the height is usually greater than the diameter and the ratio h/r varies from 2 up to about 3.8. Let's see if we can explain this phenomenon.

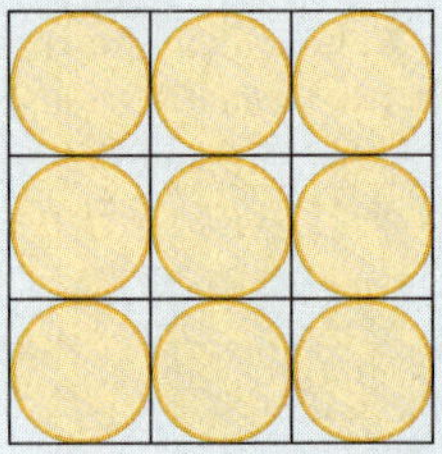

Discs cut from squares

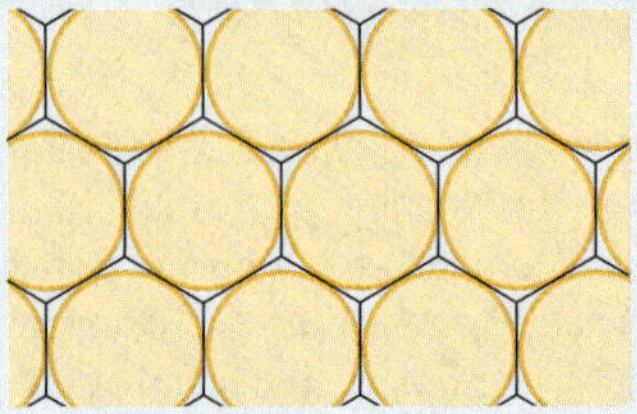

Discs cut from hexagons

1. The material for the cans is cut from sheets of metal. The cylindrical sides are formed by bending rectangles; these rectangles are cut from the sheet with little or no waste. But if the top and bottom discs are cut from squares of side $2r$ (as in the figure), this leaves considerable waste metal, which may be recycled but has little or no value to the can makers. If this is the case, show that the amount of metal used is minimized when

$$\frac{h}{r} = \frac{8}{\pi} \approx 2.55$$

2. A more efficient packing of the discs is obtained by dividing the metal sheet into hexagons and cutting the circular lids and bases from the hexagons (see the figure). Show that if this strategy is adopted, then

$$\frac{h}{r} = \frac{4\sqrt{3}}{\pi} \approx 2.21$$

3. The values of h/r that we found in Problems 1 and 2 are a little closer to the ones that actually occur on supermarket shelves, but they still don't account for everything. If we look more closely at some real cans, we see that the lid and the base are formed from discs with radius larger than r that are bent over the ends of the can. If we allow for this we would increase h/r. More significantly, in addition to the cost of the metal we need to incorporate the manufacturing of the can into the cost. Let's assume that most of the expense is incurred in joining the sides to the rims of the cans. If we cut the discs from hexagons as in Problem 2, then the total cost is proportional to

$$4\sqrt{3}\,r^2 + 2\pi rh + k(4\pi r + h)$$

where k is the reciprocal of the length that can be joined for the cost of one unit area of metal. Show that this expression is minimized when

$$\frac{\sqrt[3]{V}}{k} = \sqrt[3]{\frac{\pi h}{r}} \cdot \frac{2\pi - h/r}{\pi h/r - 4\sqrt{3}}$$

4. Plot $\sqrt[3]{V}/k$ as a function of $x = h/r$ and use your graph to argue that when a can is large or joining is cheap, we should make h/r approximately 2.21 (as in Problem 2). But when the can is small or joining is costly, h/r should be substantially larger.

5. Our analysis shows that large cans should be almost square but small cans should be tall and thin. Take a look at the relative shapes of the cans in a supermarket. Is our conclusion usually true in practice? Are there exceptions? Can you suggest reasons why small cans are not always tall and thin?

4.8 Applications to Business and Economics

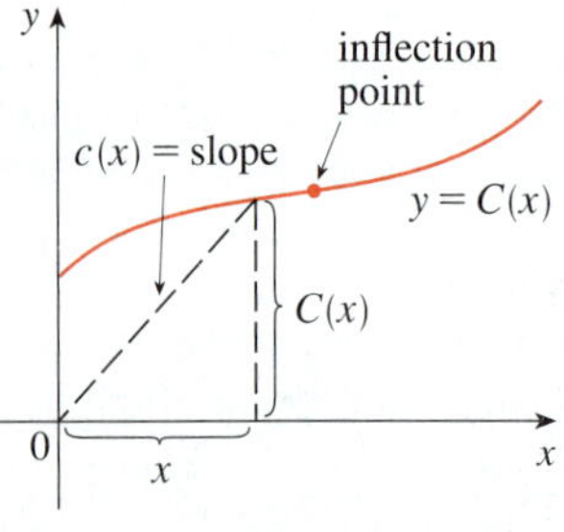

FIGURE 1 Cost function

In Section 3.3 we introduced the idea of marginal cost. Recall that if $C(x)$, the **cost function**, is the cost of producing x units of a certain product, then the **marginal cost** is the rate of change of C with respect to x. In other words, the marginal cost function is the derivative, $C'(x)$, of the cost function.

The graph of a typical cost function is shown in Figure 1. The marginal cost $C'(x)$ is the slope of the tangent to the cost curve at $(x, C(x))$. Notice that the cost curve is initially concave downward (the marginal cost is decreasing) because of economies of scale (more efficient use of the fixed costs of production). But eventually there is an inflection point and the cost curve becomes concave upward (the marginal cost is increasing), perhaps because of overtime costs or the inefficiencies of a large-scale operation.

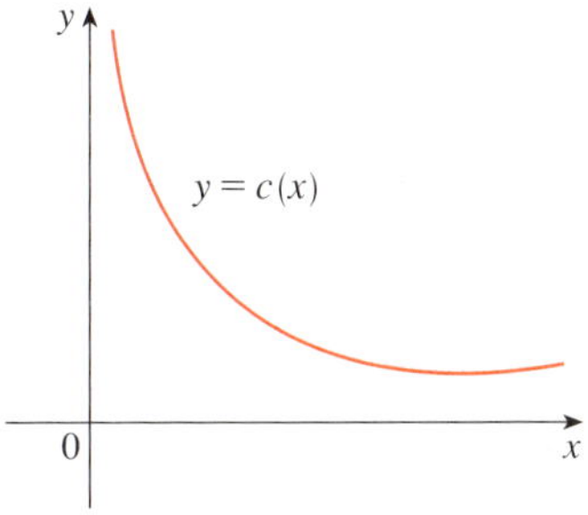

FIGURE 2
Average cost function

The **average cost function**

$$\boxed{1} \qquad c(x) = \frac{C(x)}{x}$$

represents the cost per unit when x units are produced. We sketch a typical average cost function in Figure 2 by noting that $C(x)/x$ is the slope of the line that joins the origin to the point $(x, C(x))$ in Figure 1. It appears that there will be an absolute minimum. To find it we locate the critical point of c by using the Quotient Rule to differentiate Equation 1:

$$c'(x) = \frac{xC'(x) - C(x)}{x^2}$$

Now $c'(x) = 0$ when $xC'(x) - C(x) = 0$ and this gives

$$C'(x) = \frac{C(x)}{x} = c(x)$$

Therefore:

> If the average cost is a minimum, then
> marginal cost = average cost

This principle is plausible because if our marginal cost is smaller than our average cost, then we should produce more, thereby lowering our average cost. Similarly, if our marginal cost is larger than our average cost, then we should produce less in order to lower our average cost.

IIII See Example 8 in Section 3.3 for an explanation of why it is reasonable to model a cost function by a polynomial.

EXAMPLE 1 A company estimates that the cost (in dollars) of producing x items is $C(x) = 2600 + 2x + 0.001x^2$.
(a) Find the cost, average cost, and marginal cost of producing 1000 items, 2000 items, and 3000 items.
(b) At what production level will the average cost be lowest, and what is this minimum average cost?

SOLUTION
(a) The average cost function is

$$c(x) = \frac{C(x)}{x} = \frac{2600}{x} + 2 + 0.001x$$

The marginal cost function is

$$C'(x) = 2 + 0.002x$$

We use these expressions to fill in the following table, giving the cost, average cost, and marginal cost (in dollars, or dollars per item, rounded to the nearest cent).

x	$C(x)$	$c(x)$	$C'(x)$
1000	5,600.00	5.60	4.00
2000	10,600.00	5.30	6.00
3000	17,600.00	5.87	8.00

(b) To minimize the average cost we must have

$$\text{marginal cost} = \text{average cost}$$

$$C'(x) = c(x)$$

$$2 + 0.002x = \frac{2600}{x} + 2 + 0.001x$$

|||| Figure 3 shows the graphs of the marginal cost function C' and average cost function c in Example 1. Notice that c has its minimum value when the two graphs intersect.

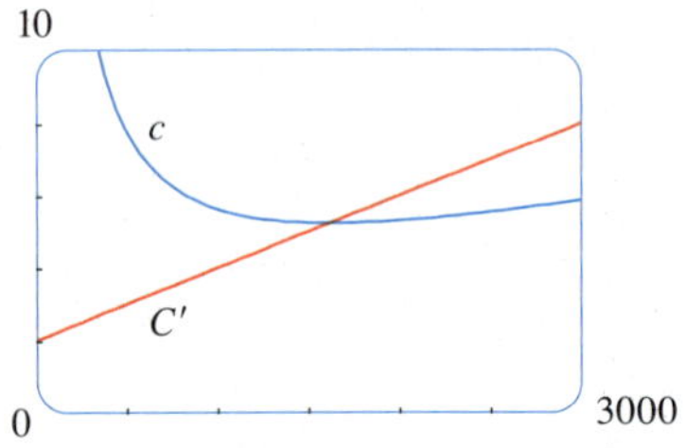

FIGURE 3

This equation simplifies to

$$0.001x = \frac{2600}{x}$$

so $$x^2 = \frac{2600}{0.001} = 2{,}600{,}000$$

and $$x = \sqrt{2{,}600{,}000} \approx 1612$$

To see that this production level actually gives a minimum, we note that $c''(x) = 5200/x^3 > 0$, so c is concave upward on its entire domain. The minimum average cost is

$$c(1612) = \frac{2600}{1612} + 2 + 0.001(1612) = \$5.22/\text{item}$$

Now let's consider marketing. Let $p(x)$ be the price per unit that the company can charge if it sells x units. Then p is called the **demand function** (or **price function**) and we would expect it to be a decreasing function of x. If x units are sold and the price per unit is $p(x)$, then the total revenue is

$$R(x) = xp(x)$$

and R is called the **revenue function** (or **sales function**). The derivative R' of the revenue function is called the **marginal revenue function** and is the rate of change of revenue with respect to the number of units sold.

If x units are sold, then the total profit is

$$P(x) = R(x) - C(x)$$

and P is called the **profit function**. The **marginal profit function** is P', the derivative of the profit function. In order to maximize profit we look for the critical numbers of P, that is, the numbers where the marginal profit is 0. But if

$$P'(x) = R'(x) - C'(x) = 0$$

then $$R'(x) = C'(x)$$

Therefore:

If the profit is a maximum, then
marginal revenue = marginal cost

To ensure that this condition gives a maximum, we could use the Second Derivative Test. Note that

$$P''(x) = R''(x) - C''(x) < 0$$

when

$$R''(x) < C''(x)$$

and this condition says that the rate of increase of marginal revenue is less than the rate of increase of marginal cost. Thus, the profit will be a maximum when

$$R'(x) = C'(x) \qquad \text{and} \qquad R''(x) < C''(x)$$

EXAMPLE 2 Determine the production level that will maximize the profit for a company with cost and demand functions

$$C(x) = 84 + 1.26x - 0.01x^2 + 0.00007x^3 \qquad \text{and} \qquad p(x) = 3.5 - 0.01x$$

SOLUTION The revenue function is

$$R(x) = xp(x) = 3.5x - 0.01x^2$$

so the marginal revenue function is

$$R'(x) = 3.5 - 0.02x$$

and the marginal cost function is

$$C'(x) = 1.26 - 0.02x + 0.00021x^2$$

Thus, marginal revenue is equal to marginal cost when

$$3.5 - 0.02x = 1.26 - 0.02x + 0.00021x^2$$

Solving, we get

$$x = \sqrt{\frac{2.24}{0.00021}} \approx 103$$

To check that this gives a maximum, we compute the second derivatives:

$$R''(x) = -0.02 \qquad C''(x) = -0.02 + 0.00042x$$

Thus, $R''(x) < C''(x)$ for all $x > 0$. Therefore, a production level of 103 units will maximize the profit.

|||| Figure 4 shows the graphs of the revenue and cost functions in Example 2. The company makes a profit when $R > C$ and the profit is a maximum when $x \approx 103$. Notice that the curves have parallel tangents at this production level because marginal revenue equals marginal cost.

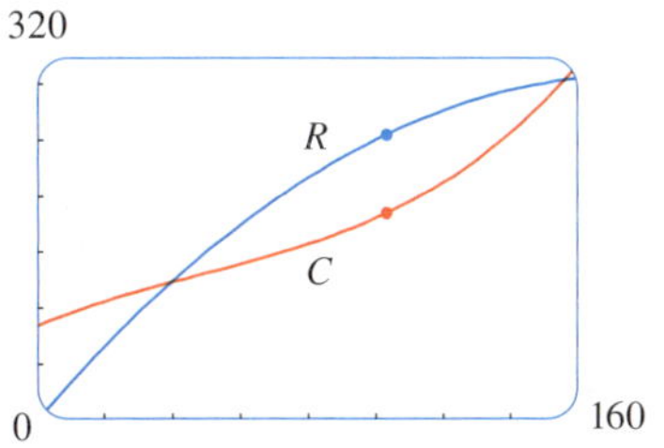

FIGURE 4

EXAMPLE 3 A store has been selling 200 DVD players a week at \$350 each. A market survey indicates that for each \$10 rebate offered to buyers, the number of players sold will increase by 20 a week. Find the demand function and the revenue function. How large a rebate should the store offer to maximize its revenue?

SOLUTION If x is the number of DVD players sold per week, then the weekly increase in sales is $x - 200$. For each increase of 20 players sold, the price is decreased by \$10. So for each additional player sold, the decrease in price will be $\frac{1}{20} \times 10$ and the demand function is

$$p(x) = 350 - \tfrac{10}{20}(x - 200) = 450 - \tfrac{1}{2}x$$

The revenue function is

$$R(x) = xp(x) = 450x - \tfrac{1}{2}x^2$$

Since $R'(x) = 450 - x$, we see that $R'(x) = 0$ when $x = 450$. This value of x gives an absolute maximum by the First Derivative Test (or simply by observing that the graph of R is a parabola that opens downward). The corresponding price is

$$p(450) = 450 - \tfrac{1}{2}(450) = 225$$

and the rebate is $350 - 225 = 125$. Therefore, to maximize revenue the store should offer a rebate of \$125.

4.8 Exercises

1. A manufacturer keeps precise records of the cost $C(x)$ of making x items and produces the graph of the cost function shown in the figure.
   (a) Explain why $C(0) > 0$.
   (b) What is the significance of the inflection point?
   (c) Use the graph of C to sketch the graph of the marginal cost function.

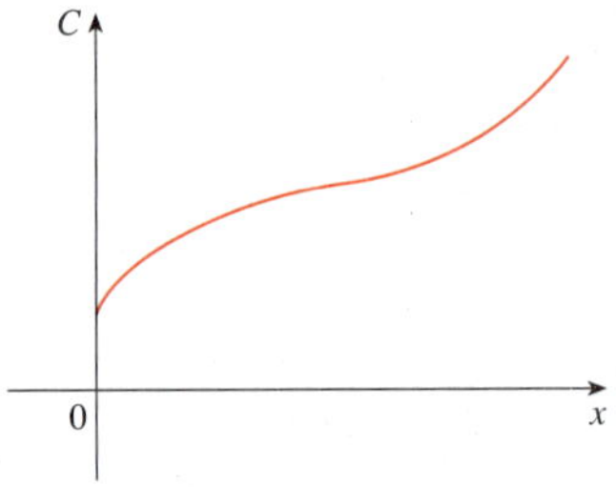

2. The graph of a cost function C is given.
   (a) Draw a careful sketch of the marginal cost function.
   (b) Use the geometric interpretation of the average cost $c(x)$ as a slope (see Figure 1) to draw a careful sketch of the average cost function.
   (c) Estimate the value of x for which $c(x)$ is a minimum. How are the average cost and the marginal cost related at that value of x?

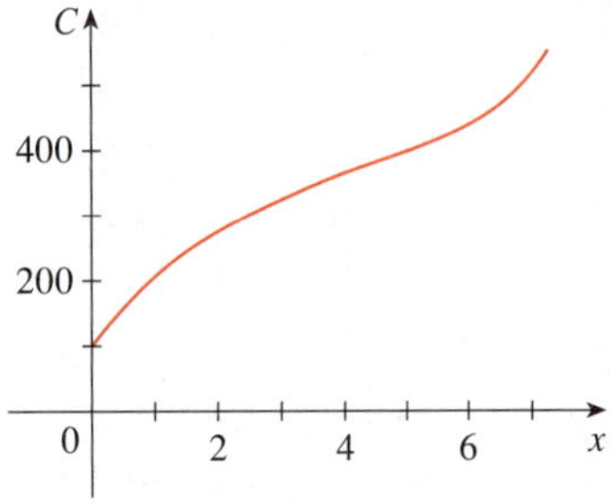

3. The average cost of producing x units of a commodity is $c(x) = 21.4 - 0.002x$. Find the marginal cost at a production level of 1000 units. In practical terms, what is the meaning of your answer?

4. The figure shows graphs of the cost and revenue functions reported by a manufacturer.
   (a) Identify on the graph the value of x for which the profit is maximized.
   (b) Sketch a graph of the profit function.
   (c) Sketch a graph of the marginal profit function.

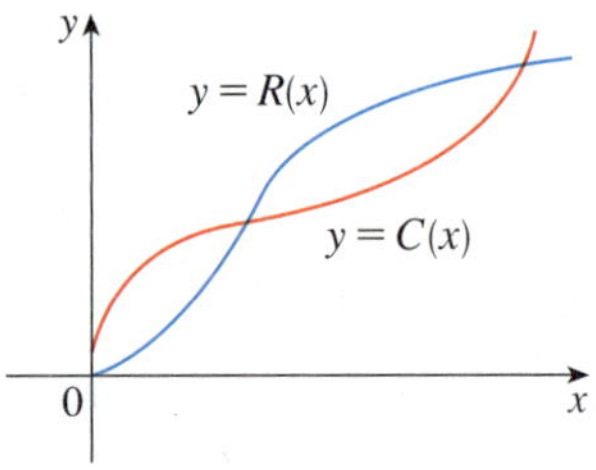

5–8 For each cost function (given in dollars), find (a) the cost, average cost, and marginal cost at a production level of 1000 units; (b) the production level that will minimize the average cost; and (c) the minimum average cost.

5. $C(x) = 40{,}000 + 300x + x^2$

6. $C(x) = 25{,}000 + 120x + 0.1x^2$

7. $C(x) = 16{,}000 + 200x + 4x^{3/2}$

8. $C(x) = 10{,}000 + 340x - 0.3x^2 + 0.0001x^3$

9–10 A cost function is given.
(a) Find the average cost and marginal cost functions.
(b) Use graphs of the functions in part (a) to estimate the production level that minimizes the average cost.
(c) Use calculus to find the minimum average cost.
(d) Find the minimum value of the marginal cost.

9. $C(x) = 3700 + 5x - 0.04x^2 + 0.0003x^3$

10. $C(x) = 339 + 25x - 0.09x^2 + 0.0004x^3$

11–14 For the given cost and demand functions, find the production level that will maximize profit.

11. $C(x) = 680 + 4x + 0.01x^2$, $p(x) = 12$

12. $C(x) = 680 + 4x + 0.01x^2$, $p(x) = 12 - x/500$

13. $C(x) = 1450 + 36x - x^2 + 0.001x^3$, $p(x) = 60 - 0.01x$

14. $C(x) = 16{,}000 + 500x - 1.6x^2 + 0.004x^3$, $p(x) = 1700 - 7x$

15–16 Find the production level at which the marginal cost function starts to increase.

15. $C(x) = 0.001x^3 - 0.3x^2 + 6x + 900$

16. $C(x) = 0.0002x^3 - 0.25x^2 + 4x + 1500$

17. The cost, in dollars, of producing x yards of a certain fabric is $C(x) = 1200 + 12x - 0.1x^2 + 0.0005x^3$ and the company finds that if it sells x yards, it can charge $p(x) = 29 - 0.00021x$ dollars per yard for the fabric.
(a) Graph the cost and revenue functions and use the graphs to estimate the production level for maximum profit.
(b) Use calculus to find the production level for maximum profit.

18. An aircraft manufacturer wants to determine the best selling price for a new airplane. The company estimates that the initial cost of designing the airplane and setting up the factories in which to build it will be 500 million dollars. The additional cost of manufacturing each plane can be modeled by the function $m(x) = 20x - 5x^{3/4} + 0.01x^2$, where x is the number of aircraft produced and m is the manufacturing cost, in millions of dollars. The company estimates that if it charges a price p (in millions of dollars) for each plane, it will be able to sell $x(p) = 320 - 7.7p$ planes.
(a) Find the cost, demand, and revenue functions.
(b) Find the production level and the associated selling price of the aircraft that maximizes profit.

19. A baseball team plays in a stadium that holds 55,000 spectators. With ticket prices at \$10, the average attendance had been 27,000. When ticket prices were lowered to \$8, the average attendance rose to 33,000.
(a) Find the demand function, assuming that it is linear.
(b) How should ticket prices be set to maximize revenue?

20. During the summer months Terry makes and sells necklaces on the beach. Last summer he sold the necklaces for \$10 each and his sales averaged 20 per day. When he increased the price by \$1, he found that he lost two sales per day.
(a) Find the demand function, assuming that it is linear.
(b) If the material for each necklace costs Terry \$6, what should the selling price be to maximize his profit?

21. A manufacturer has been selling 1000 television sets a week at \$450 each. A market survey indicates that for each \$10 rebate offered to the buyer, the number of sets sold will increase by 100 per week.
(a) Find the demand function.
(b) How large a rebate should the company offer the buyer in order to maximize its revenue?
(c) If its weekly cost function is $C(x) = 68{,}000 + 150x$, how should the manufacturer set the size of the rebate in order to maximize its profit?

22. The manager of a 100-unit apartment complex knows from experience that all units will be occupied if the rent is \$800 per month. A market survey suggests that, on average, one additional unit will remain vacant for each \$10 increase in rent. What rent should the manager charge to maximize revenue?

23. Store managers want an optimal inventory policy. Overstocking results in excessive storage and interest costs, whereas a small inventory means added costs for reordering and delivery. A supermarket manager estimates that a total of 800 cases of soup will be sold at a steady rate during the coming year and it costs \$4 to store a case for a year. If he places several orders per year, each consisting of x cases, then there will be an average of $\frac{1}{2}x$ cases in stock at any time and so storage costs for the year are $(\frac{1}{2}x) \cdot 4 = 2x$ dollars. He also estimates that the handling cost for each delivery is \$100. What is the optimal reorder quantity x that minimizes costs?

24. Suppose a person has an amount A of spending money deposited to her savings account each month, where it earns interest at a monthly rate R. Assume she spends the entire amount throughout the month at a steady rate. When she makes cash withdrawals from the account, she incurs a transaction cost T (a combination of bank fees and the cost of her time). She would save money by making fewer withdrawals, but the more money she leaves in the account the more interest she earns. Suppose she makes n cash withdrawals for the same amount during the month. Then her average cash balance at any given time is $A/(2n)$. (Why?) Find the value of n that minimizes total costs (transaction costs and lost interest), and then show that the optimal average cash balance is $\sqrt{AT/(2R)}$.

4.9 Newton's Method

Resources / Module 5
/ Newton's Method
/ Start of Newton's Method

Suppose that a car dealer offers to sell you a car for \$18,000 or for payments of \$375 per month for five years. You would like to know what monthly interest rate the dealer is, in effect, charging you. To find the answer, you have to solve the equation

$$48x(1 + x)^{60} - (1 + x)^{60} + 1 = 0 \quad \boxed{1}$$

(The details are explained in Exercise 41.) How would you solve such an equation?

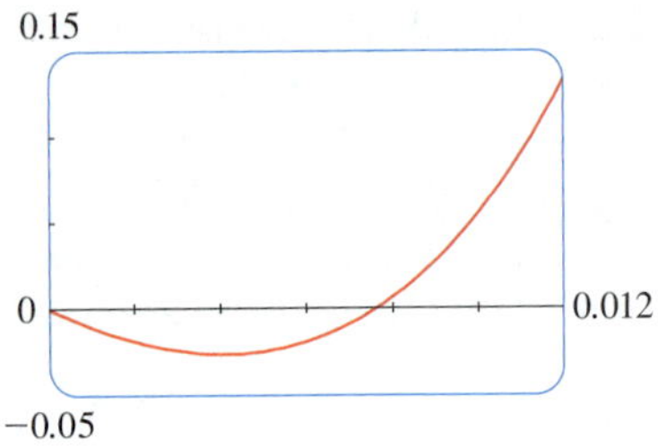

FIGURE 1

For a quadratic equation $ax^2 + bx + c = 0$ there is a well-known formula for the roots. For third- and fourth-degree equations there are also formulas for the roots, but they are extremely complicated. If f is a polynomial of degree 5 or higher, there is no such formula (see the note on page 230). Likewise, there is no formula that will enable us to find the exact roots of a transcendental equation such as $\cos x = x$.

We can find an *approximate* solution to Equation 1 by plotting the left side of the equation. Using a graphing device, and after experimenting with viewing rectangles, we produce the graph in Figure 1.

We see that in addition to the solution $x = 0$, which doesn't interest us, there is a solution between 0.007 and 0.008. Zooming in shows that the root is approximately 0.0076. If we need more accuracy we could zoom in repeatedly, but that becomes tiresome. A faster alternative is to use a numerical rootfinder on a calculator or computer algebra system. If we do so, we find that the root, correct to nine decimal places, is 0.007628603.

|||| Try to solve Equation 1 using the numerical rootfinder on your calculator or computer. Some machines are not able to solve it. Others are successful but require you to specify a starting point for the search.

How do those numerical rootfinders work? They use a variety of methods, but most of them make some use of **Newton's method**, also called the **Newton-Raphson method**. We will explain how this method works, partly to show what happens inside a calculator or computer, and partly as an application of the idea of linear approximation.

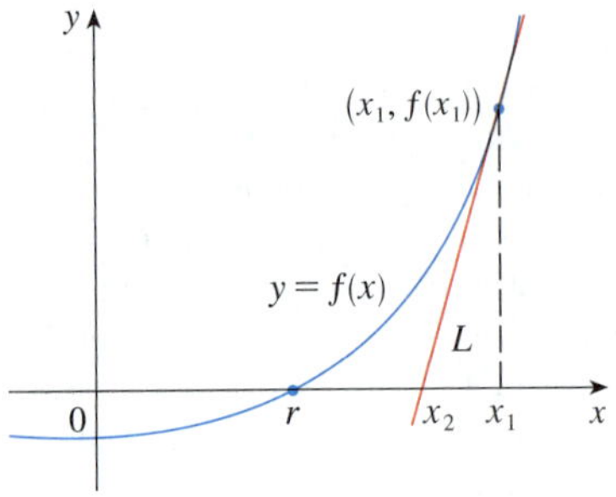

FIGURE 2

The geometry behind Newton's method is shown in Figure 2, where the root that we are trying to find is labeled r. We start with a first approximation x_1, which is obtained by guessing, or from a rough sketch of the graph of f, or from a computer-generated graph of f. Consider the tangent line L to the curve $y = f(x)$ at the point $(x_1, f(x_1))$ and look at the x-intercept of L, labeled x_2. The idea behind Newton's method is that the tangent line is close to the curve and so its x-intercept, x_2, is close to the x-intercept of the curve (namely, the root r that we are seeking). Because the tangent is a line, we can easily find its x-intercept.

To find a formula for x_2 in terms of x_1 we use the fact that the slope of L is $f'(x_1)$, so its equation is

$$y - f(x_1) = f'(x_1)(x - x_1)$$

Since the x-intercept of L is x_2, we set $y = 0$ and obtain

$$0 - f(x_1) = f'(x_1)(x_2 - x_1)$$

If $f'(x_1) \neq 0$, we can solve this equation for x_2:

$$x_2 = x_1 - \frac{f(x_1)}{f'(x_1)}$$

We use x_2 as a second approximation to r.

Next we repeat this procedure with x_1 replaced by x_2, using the tangent line at $(x_2, f(x_2))$. This gives a third approximation:

$$x_3 = x_2 - \frac{f(x_2)}{f'(x_2)}$$

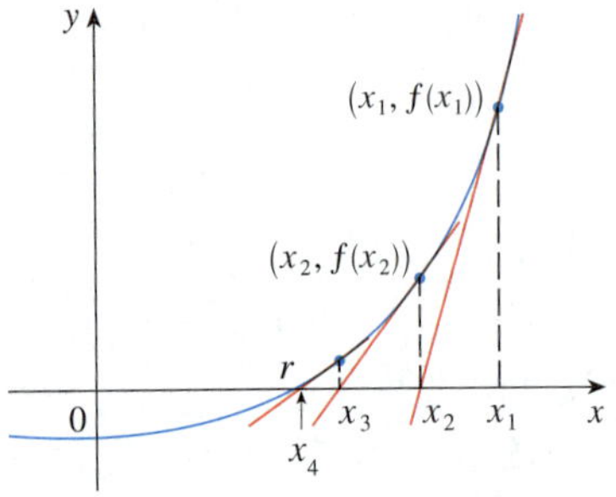

FIGURE 3

If we keep repeating this process, we obtain a sequence of approximations $x_1, x_2, x_3, x_4, \ldots$ as shown in Figure 3. In general, if the nth approximation is x_n and $f'(x_n) \neq 0$, then the next approximation is given by

2
$$x_{n+1} = x_n - \frac{f(x_n)}{f'(x_n)}$$

||| Sequences were briefly introduced in *A Preview of Calculus* on page 6. A more thorough discussion starts in Section 11.1.

If the numbers x_n become closer and closer to r as n becomes large, then we say that the sequence *converges* to r and we write

$$\lim_{n\to\infty} x_n = r$$

Although the sequence of successive approximations converges to the desired root for functions of the type illustrated in Figure 3, in certain circumstances the sequence may not converge. For example, consider the situation shown in Figure 4. You can see that x_2 is a worse approximation than x_1. This is likely to be the case when $f'(x_1)$ is close to 0. It might even happen that an approximation (such as x_3 in Figure 4) falls outside the domain of f. Then Newton's method fails and a better initial approximation x_1 should be chosen. See Exercises 31–34 for specific examples in which Newton's method works very slowly or does not work at all.

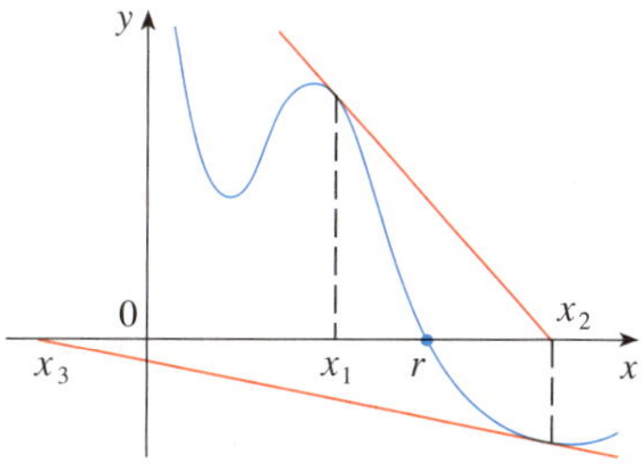

FIGURE 4

EXAMPLE 1 Starting with $x_1 = 2$, find the third approximation x_3 to the root of the equation $x^3 - 2x - 5 = 0$.

SOLUTION We apply Newton's method with

$$f(x) = x^3 - 2x - 5 \qquad \text{and} \qquad f'(x) = 3x^2 - 2$$

Newton himself used this equation to illustrate his method and he chose $x_1 = 2$ after some experimentation because $f(1) = -6$, $f(2) = -1$, and $f(3) = 16$. Equation 2 becomes

$$x_{n+1} = x_n - \frac{x_n^3 - 2x_n - 5}{3x_n^2 - 2}$$

With $n = 1$ we have

$$x_2 = x_1 - \frac{x_1^3 - 2x_1 - 5}{3x_1^2 - 2}$$

$$= 2 - \frac{2^3 - 2(2) - 5}{3(2)^2 - 2} = 2.1$$

||| Figure 5 shows the geometry behind the first step in Newton's method in Example 1. Since $f'(2) = 10$, the tangent line to $y = x^3 - 2x - 5$ at $(2, -1)$ has equation $y = 10x - 21$ so its x-intercept is $x_2 = 2.1$.

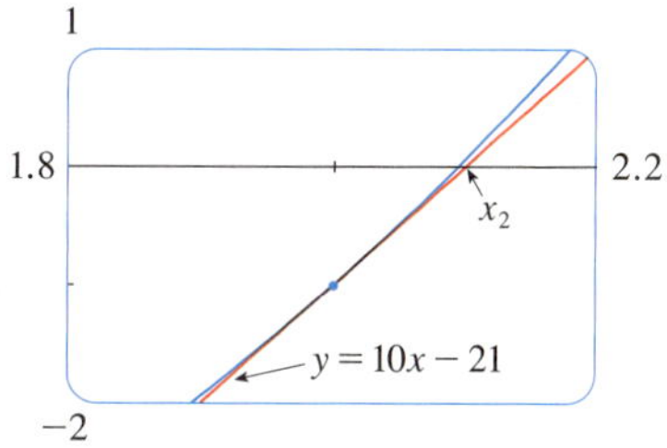

FIGURE 5

Then with $n = 2$ we obtain

$$x_3 = x_2 - \frac{x_2^3 - 2x_2 - 5}{3x_2^2 - 2}$$

$$= 2.1 - \frac{(2.1)^3 - 2(2.1) - 5}{3(2.1)^2 - 2} \approx 2.0946$$

It turns out that this third approximation $x_3 \approx 2.0946$ is accurate to four decimal places.

Suppose that we want to achieve a given accuracy, say to eight decimal places, using Newton's method. How do we know when to stop? The rule of thumb that is generally used is that we can stop when successive approximations x_n and x_{n+1} agree to eight decimal places. (A precise statement concerning accuracy in Newton's method will be given in Exercises 11.12.)

Notice that the procedure in going from n to $n + 1$ is the same for all values of n. (It is called an *iterative* process.) This means that Newton's method is particularly convenient for use with a programmable calculator or a computer.

EXAMPLE 2 Use Newton's method to find $\sqrt[6]{2}$ correct to eight decimal places.

SOLUTION First we observe that finding $\sqrt[6]{2}$ is equivalent to finding the positive root of the equation

$$x^6 - 2 = 0$$

so we take $f(x) = x^6 - 2$. Then $f'(x) = 6x^5$ and Formula 2 (Newton's method) becomes

$$x_{n+1} = x_n - \frac{x_n^6 - 2}{6x_n^5}$$

If we choose $x_1 = 1$ as the initial approximation, then we obtain

$$x_2 \approx 1.16666667$$
$$x_3 \approx 1.12644368$$
$$x_4 \approx 1.12249707$$
$$x_5 \approx 1.12246205$$
$$x_6 \approx 1.12246205$$

Since x_5 and x_6 agree to eight decimal places, we conclude that

$$\sqrt[6]{2} \approx 1.12246205$$

to eight decimal places.

EXAMPLE 3 Find, correct to six decimal places, the root of the equation $\cos x = x$.

SOLUTION We first rewrite the equation in standard form:

$$\cos x - x = 0$$

Therefore, we let $f(x) = \cos x - x$. Then $f'(x) = -\sin x - 1$, so Formula 2 becomes

$$x_{n+1} = x_n - \frac{\cos x_n - x_n}{-\sin x_n - 1} = x_n + \frac{\cos x_n - x_n}{\sin x_n + 1}$$

In order to guess a suitable value for x_1 we sketch the graphs of $y = \cos x$ and $y = x$ in Figure 6. It appears that they intersect at a point whose x-coordinate is somewhat less than 1, so let's take $x_1 = 1$ as a convenient first approximation. Then, remembering to put our calculator in radian mode, we get

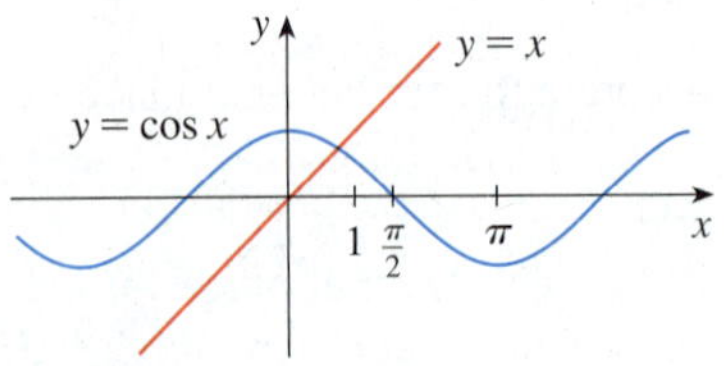

FIGURE 6

$$x_2 \approx 0.75036387$$
$$x_3 \approx 0.73911289$$
$$x_4 \approx 0.73908513$$
$$x_5 \approx 0.73908513$$

Since x_4 and x_5 agree to six decimal places (eight, in fact), we conclude that the root of the equation, correct to six decimal places, is 0.739085.

Instead of using the rough sketch in Figure 6 to get a starting approximation for Newton's method in Example 3, we could have used the more accurate graph that a calculator or computer provides. Figure 7 suggests that we use $x_1 = 0.75$ as the initial approx-

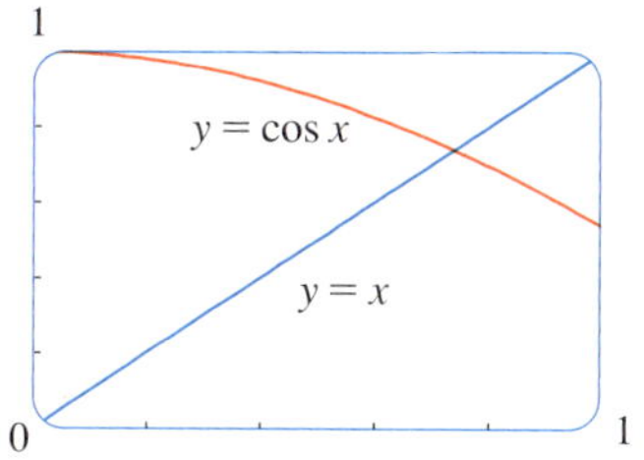

FIGURE 7

imation. Then Newton's method gives

$$x_2 \approx 0.73911114$$
$$x_3 \approx 0.73908513$$
$$x_4 \approx 0.73908513$$

and so we obtain the same answer as before, but with one fewer step.

You might wonder why we bother at all with Newton's method if a graphing device is available. Isn't it easier to zoom in repeatedly and find the roots as we did in Section 1.4? If only one or two decimal places of accuracy are required, then indeed Newton's method is inappropriate and a graphing device suffices. But if six or eight decimal places are required, then repeated zooming becomes tiresome. It is usually faster and more efficient to use a computer and Newton's method in tandem—the graphing device to get started and Newton's method to finish.

4.9 Exercises

1. The figure shows the graph of a function f. Suppose that Newton's method is used to approximate the root r of the equation $f(x) = 0$ with initial approximation $x_1 = 1$.
(a) Draw the tangent lines that are used to find x_2 and x_3, and estimate the numerical values of x_2 and x_3.
(b) Would $x_1 = 5$ be a better first approximation? Explain.

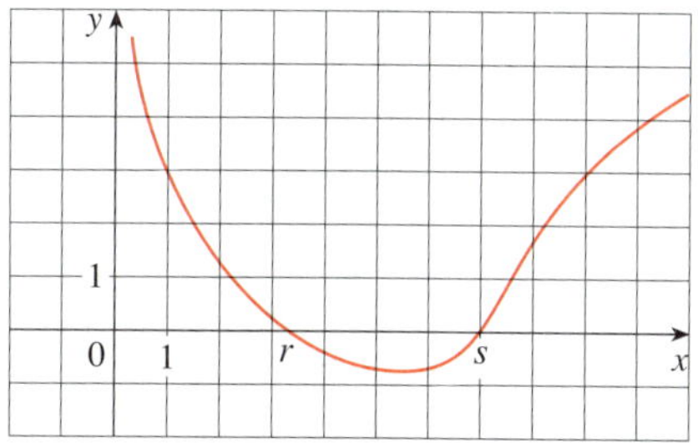

2. Follow the instructions for Exercise 1(a) but use $x_1 = 9$ as the starting approximation for finding the root s.

3. Suppose the line $y = 5x - 4$ is tangent to the curve $y = f(x)$ when $x = 3$. If Newton's method is used to locate a root of the equation $f(x) = 0$ and the initial approximation is $x_1 = 3$, find the second approximation x_2.

4. For each initial approximation, determine graphically what happens if Newton's method is used for the function whose graph is shown.
(a) $x_1 = 0$ (b) $x_1 = 1$ (c) $x_1 = 3$
(d) $x_1 = 4$ (e) $x_1 = 5$

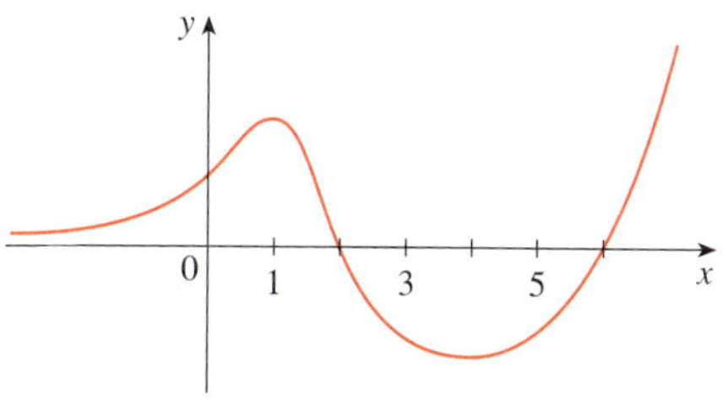

5–8 |||| Use Newton's method with the specified initial approximation x_1 to find x_3, the third approximation to the root of the given equation. (Give your answer to four decimal places.)

5. $x^3 + 2x - 4 = 0$, $x_1 = 1$

6. $x^3 - x^2 - 1 = 0$, $x_1 = 1$

7. $x^4 - 20 = 0$, $x_1 = 2$

8. $x^5 + 2 = 0$, $x_1 = -1$

9. Use Newton's method with initial approximation $x_1 = -1$ to find x_2, the second approximation to the root of the equation $x^3 + x + 3 = 0$. Explain how the method works by first graphing the function and its tangent line at $(-1, 1)$.

10. Use Newton's method with initial approximation $x_1 = 1$ to find x_2, the second approximation to the root of the equation $x^4 - x - 1 = 0$. Explain how the method works by first graphing the function and its tangent line at $(1, -1)$.

11–12 |||| Use Newton's method to approximate the given number correct to eight decimal places.

11. $\sqrt[3]{30}$ **12.** $\sqrt[7]{1000}$

13–16 |||| Use Newton's method to approximate the indicated root of the equation correct to six decimal places.

13. The root of $2x^3 - 6x^2 + 3x + 1 = 0$ in the interval $[2, 3]$

14. The root of $x^4 + x - 4 = 0$ in the interval $[1, 2]$

15. The positive root of $\sin x = x^2$

16. The positive root of $2 \cos x = x^4$

17–22 |||| Use Newton's method to find all roots of the equation correct to six decimal places.

17. $x^4 = 1 + x$ **18.** $e^x = 3 - 2x$

19. $\tan^{-1}x = 1 - x$

20. $\sqrt{x+3} = x^2$

21. $\cos x = \sqrt{x}$

22. $\tan x = \sqrt{1 - x^2}$

23–28 |||| Use Newton's method to find all the roots of the equation correct to eight decimal places. Start by drawing a graph to find initial approximations.

23. $x^5 - x^4 - 5x^3 - x^2 + 4x + 3 = 0$

24. $x^2(4 - x^2) = \dfrac{4}{x^2 + 1}$

25. $x^2\sqrt{2 - x - x^2} = 1$

26. $3\sin(x^2) = 2x$

27. $e^{-x^2} = x^2 - x$

28. $\ln(4 - x^2) = x$

29. (a) Apply Newton's method to the equation $x^2 - a = 0$ to derive the following square-root algorithm (used by the ancient Babylonians to compute $\sqrt{a}$):

$$x_{n+1} = \frac{1}{2}\left(x_n + \frac{a}{x_n}\right)$$

(b) Use part (a) to compute $\sqrt{1000}$ correct to six decimal places.

30. (a) Apply Newton's method to the equation $1/x - a = 0$ to derive the following reciprocal algorithm:

$$x_{n+1} = 2x_n - ax_n^2$$

(This algorithm enables a computer to find reciprocals without actually dividing.)

(b) Use part (a) to compute $1/1.6984$ correct to six decimal places.

31. Explain why Newton's method doesn't work for finding the root of the equation $x^3 - 3x + 6 = 0$ if the initial approximation is chosen to be $x_1 = 1$.

32. (a) Use Newton's method with $x_1 = 1$ to find the root of the equation $x^3 - x = 1$ correct to six decimal places.

(b) Solve the equation in part (a) using $x_1 = 0.6$ as the initial approximation.

(c) Solve the equation in part (a) using $x_1 = 0.57$. (You definitely need a programmable calculator for this part.)

(d) Graph $f(x) = x^3 - x - 1$ and its tangent lines at $x_1 = 1$, 0.6, and 0.57 to explain why Newton's method is so sensitive to the value of the initial approximation.

33. Explain why Newton's method fails when applied to the equation $\sqrt[3]{x} = 0$ with any initial approximation $x_1 \neq 0$. Illustrate your explanation with a sketch.

34. If

$$f(x) = \begin{cases} \sqrt{x} & \text{if } x \geq 0 \\ -\sqrt{-x} & \text{if } x < 0 \end{cases}$$

then the root of the equation $f(x) = 0$ is $x = 0$. Explain why Newton's method fails to find the root no matter which initial approximation $x_1 \neq 0$ is used. Illustrate your explanation with a sketch.

35. (a) Use Newton's method to find the critical numbers of the function $f(x) = 3x^4 - 28x^3 + 6x^2 + 24x$ correct to three decimal places.

(b) Find the absolute minimum value of the function

$$f(x) = 3x^4 - 28x^3 + 6x^2 + 24x \qquad -1 \leq x \leq 7$$

correct to two decimal places.

36. Use Newton's method to find the absolute minimum value of the function $f(x) = x^2 + \sin x$ correct to six decimal places.

37. Use Newton's method to find the coordinates of the inflection point of the curve $y = e^{\cos x}$, $0 \leq x \leq \pi$, correct to six decimal places.

38. Of the infinitely many lines that are tangent to the curve $y = -\sin x$ and pass through the origin, there is one that has the largest slope. Use Newton's method to find the slope of that line correct to six decimal places.

39. A grain silo consists of a cylindrical main section, with height 30 ft, and a hemispherical roof. In order to achieve a total volume of 15,000 ft^3 (including the part inside the roof section), what would the radius of the silo have to be?

40. In the figure, the length of the chord AB is 4 cm and the length of the arc AB is 5 cm. Find the central angle θ, in radians, correct to four decimal places. Then give the answer to the nearest degree.

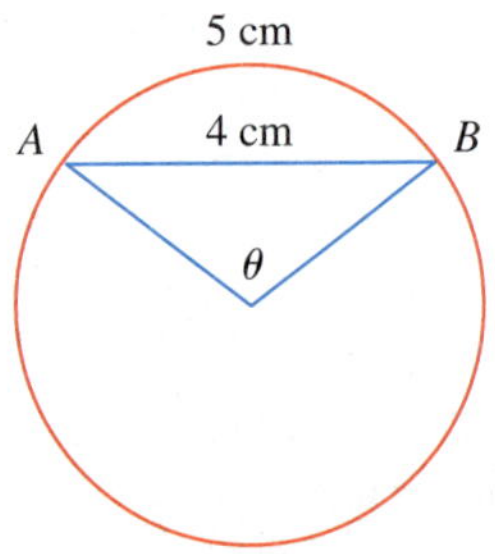

41. A car dealer sells a new car for \$18,000. He also offers to sell the same car for payments of \$375 per month for five years. What monthly interest rate is this dealer charging?

To solve this problem you will need to use the formula for the present value A of an annuity consisting of n equal payments of size R with interest rate i per time period:

$$A = \frac{R}{i}[1 - (1 + i)^{-n}]$$

Replacing i by x, show that

$$48x(1 + x)^{60} - (1 + x)^{60} + 1 = 0$$

Use Newton's method to solve this equation.

42. The figure shows the Sun located at the origin and Earth at the point (1, 0). (The unit here is the distance between the centers of Earth and the Sun, called an *astronomical unit:* 1 AU $\approx$ 1.496 $\times$ 10^8 km.) There are five locations L_1, L_2, L_3, L_4, and L_5 in this plane of rotation of Earth about the Sun where a satellite remains motionless with respect to Earth because the forces acting on the satellite (including the gravitational attractions of Earth and the Sun) balance each other. These locations are called *libration points.* (A solar research satellite has been placed at one of these libration points.) If m_1 is the mass of the Sun, m_2 is the mass of Earth, and $r = m_2/(m_1 + m_2)$, it turns out that the x-coordinate of L_1 is the unique root of the fifth-degree equation

$$p(x) = x^5 - (2 + r)x^4 + (1 + 2r)x^3 - (1 - r)x^2 + 2(1 - r)x + r - 1 = 0$$

and the x-coordinate of L_2 is the root of the equation

$$p(x) - 2rx^2 = 0$$

Using the value $r \approx 3.04042 \times 10^{-6}$, find the locations of the libration points (a) L_1 and (b) L_2.

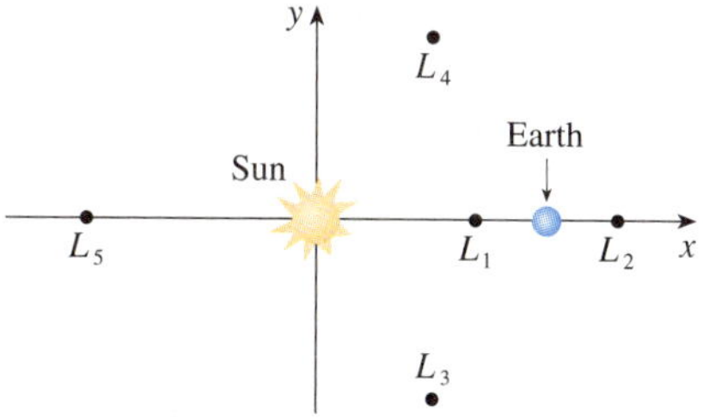

4.10 Antiderivatives

Resources / Module 6 / Antiderivatives / Start of Antiderivatives

A physicist who knows the velocity of a particle might wish to know its position at a given time. An engineer who can measure the variable rate at which water is leaking from a tank wants to know the amount leaked over a certain time period. A biologist who knows the rate at which a bacteria population is increasing might want to deduce what the size of the population will be at some future time. In each case, the problem is to find a function F whose derivative is a known function f. If such a function F exists, it is called an *antiderivative* of f.

Definition A function F is called an **antiderivative** of f on an interval I if $F'(x) = f(x)$ for all x in I.

For instance, let $f(x) = x^2$. It isn't difficult to discover an antiderivative of f if we keep the Power Rule in mind. In fact, if $F(x) = \frac{1}{3}x^3$, then $F'(x) = x^2 = f(x)$. But the function $G(x) = \frac{1}{3}x^3 + 100$ also satisfies $G'(x) = x^2$. Therefore, both F and G are antiderivatives of f. Indeed, any function of the form $H(x) = \frac{1}{3}x^3 + C$, where C is a constant, is an antiderivative of f. The question arises: Are there any others?

To answer this question, recall that in Section 4.2 we used the Mean Value Theorem to prove that if two functions have identical derivatives on an interval, then they must differ by a constant (Corollary 4.2.7). Thus, if F and G are any two antiderivatives of f, then

$$F'(x) = f(x) = G'(x)$$

so $G(x) - F(x) = C$, where C is a constant. We can write this as $G(x) = F(x) + C$, so we have the following result.

1 Theorem If F is an antiderivative of f on an interval I, then the most general antiderivative of f on I is

$$F(x) + C$$

where C is an arbitrary constant.

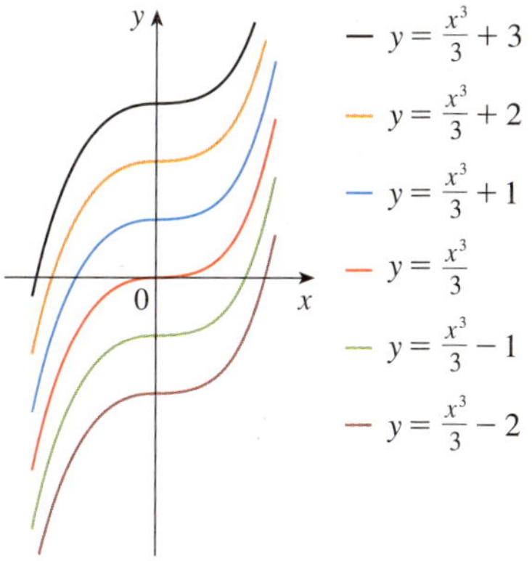

FIGURE 1
Members of the family of antiderivatives of $f(x) = x^2$

Going back to the function $f(x) = x^2$, we see that the general antiderivative of f is $x^3/3 + C$. By assigning specific values to the constant C, we obtain a family of functions whose graphs are vertical translates of one another (see Figure 1).

EXAMPLE 1 Find the most general antiderivative of each of the following functions.

(a) $f(x) = \sin x$ (b) $f(x) = 1/x$ (c) $f(x) = x^n,\quad n \neq -1$

SOLUTION

(a) If $F(x) = -\cos x$, then $F'(x) = \sin x$, so an antiderivative of $\sin x$ is $-\cos x$. By Theorem 1, the most general antiderivative is $G(x) = -\cos x + C$.

(b) Recall from Section 3.8 that

$$\frac{d}{dx}(\ln x) = \frac{1}{x}$$

So on the interval $(0, \infty)$ the general antiderivative of $1/x$ is $\ln x + C$. We also learned that

$$\frac{d}{dx}(\ln|x|) = \frac{1}{x}$$

for all $x \neq 0$. Theorem 1 then tells us that the general antiderivative of $f(x) = 1/x$ is $\ln|x| + C$ on any interval that doesn't contain 0. In particular, this is true on each of the intervals $(-\infty, 0)$ and $(0, \infty)$. So the general antiderivative of f is

$$F(x) = \begin{cases} \ln x + C_1 & \text{if } x > 0 \\ \ln(-x) + C_2 & \text{if } x < 0 \end{cases}$$

(c) We use the Power Rule to discover an antiderivative of x^n. In fact, if $n \neq -1$, then

$$\frac{d}{dx}\left(\frac{x^{n+1}}{n+1}\right) = \frac{(n+1)x^n}{n+1} = x^n$$

Thus, the general antiderivative of $f(x) = x^n$ is

$$F(x) = \frac{x^{n+1}}{n+1} + C$$

This is valid for $n \geqslant 0$ since then $f(x) = x^n$ is defined on an interval. If n is negative (but $n \neq -1$), it is valid on any interval that doesn't contain 0.

As in Example 1, every differentiation formula, when read from right to left, gives rise to an antidifferentiation formula. In Table 2 we list some particular antiderivatives. Each formula in the table is true because the derivative of the function in the right column appears in the left column. In particular, the first formula says that the antiderivative of a constant times a function is the constant times the antiderivative of the function. The second formula says that the antiderivative of a sum is the sum of the antiderivatives. (We use the notation $F' = f$, $G' = g$.)

2 Table of Antidifferentiation Formulas

Function	Particular antiderivative	Function	Particular antiderivative
$cf(x)$	$cF(x)$	$\sin x$	$-\cos x$
$f(x) + g(x)$	$F(x) + G(x)$	$\sec^2 x$	$\tan x$
$x^n\ (n \neq -1)$	$\dfrac{x^{n+1}}{n+1}$	$\sec x \tan x$	$\sec x$
$1/x$	$\ln\|x\|$	$\dfrac{1}{\sqrt{1-x^2}}$	$\sin^{-1}x$
e^x	e^x	$\dfrac{1}{1+x^2}$	$\tan^{-1}x$
$\cos x$	$\sin x$		

|||| To obtain the most general antiderivative from the particular ones in Table 2, we have to add a constant (or constants), as in Example 1.

EXAMPLE 2 Find all functions g such that

$$g'(x) = 4 \sin x + \frac{2x^5 - \sqrt{x}}{x}$$

SOLUTION We first rewrite the given function as follows:

$$g'(x) = 4 \sin x + \frac{2x^5}{x} - \frac{\sqrt{x}}{x} = 4 \sin x + 2x^4 - \frac{1}{\sqrt{x}}$$

Thus, we want to find an antiderivative of

$$g'(x) = 4 \sin x + 2x^4 - x^{-1/2}$$

Using the formulas in Table 2 together with Theorem 1, we obtain

$$g(x) = 4(-\cos x) + 2\frac{x^5}{5} - \frac{x^{1/2}}{\frac{1}{2}} + C$$

$$= -4 \cos x + \tfrac{2}{5}x^5 - 2\sqrt{x} + C$$

In applications of calculus it is very common to have a situation as in Example 2, where it is required to find a function, given knowledge about its derivatives. An equation that involves the derivatives of a function is called a **differential equation**. These will be studied in some detail in Chapter 9, but for the present we can solve some elementary differential equations. The general solution of a differential equation involves an arbitrary constant (or constants) as in Example 2. However, there may be some extra conditions given that will determine the constants and therefore uniquely specify the solution.

|||| Figure 2 shows the graphs of the function f' in Example 3 and its antiderivative f. Notice that $f'(x) > 0$, so f is always increasing. Also notice that when f' has a maximum or minimum, f appears to have an inflection point. So the graph serves as a check on our calculation.

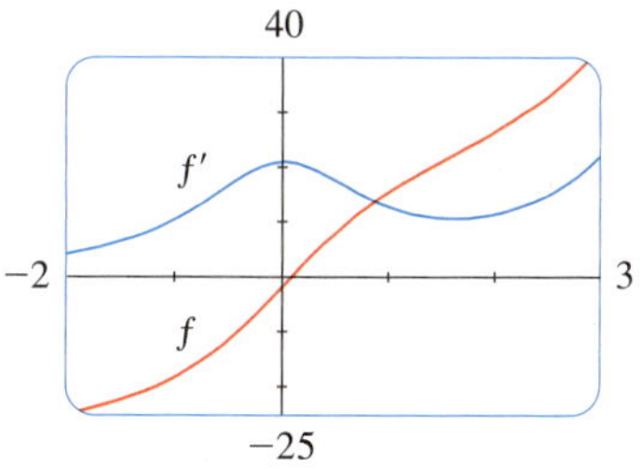

FIGURE 2

EXAMPLE 3 Find f if $f'(x) = e^x + 20(1 + x^2)^{-1}$ and $f(0) = -2$.

SOLUTION The general antiderivative of

$$f'(x) = e^x + \frac{20}{1 + x^2}$$

is

$$f(x) = e^x + 20 \tan^{-1}x + C$$

To determine C we use the fact that $f(0) = -2$:

$$f(0) = e^0 + 20 \tan^{-1} 0 + C = -2$$

Thus, we have $C = -2 - 1 = -3$, so the particular solution is

$$f(x) = e^x + 20 \tan^{-1}x - 3$$

EXAMPLE 4 Find f if $f''(x) = 12x^2 + 6x - 4$, $f(0) = 4$, and $f(1) = 1$.

SOLUTION The general antiderivative of $f''(x) = 12x^2 + 6x - 4$ is

$$f'(x) = 12\frac{x^3}{3} + 6\frac{x^2}{2} - 4x + C = 4x^3 + 3x^2 - 4x + C$$

Using the antidifferentiation rules once more, we find that

$$f(x) = 4\frac{x^4}{4} + 3\frac{x^3}{3} - 4\frac{x^2}{2} + Cx + D = x^4 + x^3 - 2x^2 + Cx + D$$

To determine C and D we use the given conditions that $f(0) = 4$ and $f(1) = 1$. Since $f(0) = 0 + D = 4$, we have $D = 4$. Since

$$f(1) = 1 + 1 - 2 + C + 4 = 1$$

we have $C = -3$. Therefore, the required function is

$$f(x) = x^4 + x^3 - 2x^2 - 3x + 4$$

The Geometry of Antiderivatives

If we are given the graph of a function f, it seems reasonable that we should be able to sketch the graph of an antiderivative F. Suppose, for instance, that we are given that $F(0) = 1$. Then we have a place to start, the point $(0, 1)$, and the direction in which we move our pencil is given at each stage by the derivative $F'(x) = f(x)$. In the next example we use the principles of this chapter to show how to graph F even when we don't have a formula for f. This would be the case, for instance, when $f(x)$ is determined by experimental data.

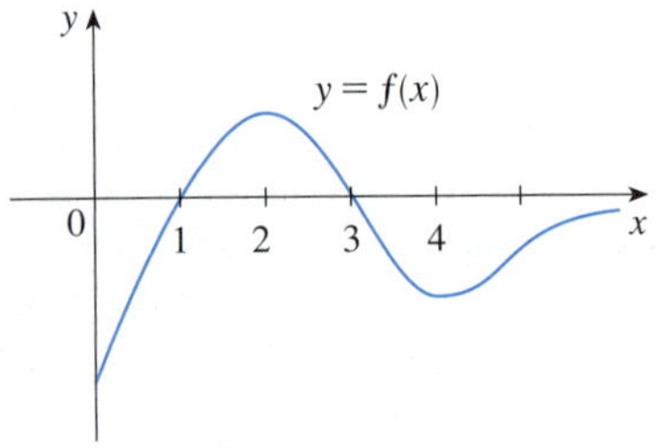

FIGURE 3

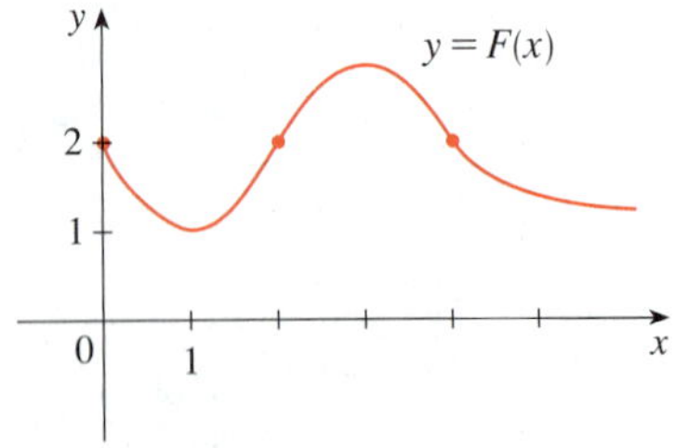

FIGURE 4

EXAMPLE 5 The graph of a function f is given in Figure 3. Make a rough sketch of an antiderivative F, given that $F(0) = 2$.

SOLUTION We are guided by the fact that the slope of $y = F(x)$ is $f(x)$. We start at the point $(0, 2)$ and draw F as an initially decreasing function since $f(x)$ is negative when $0 < x < 1$. Notice that $f(1) = f(3) = 0$, so F has horizontal tangents when $x = 1$ and $x = 3$. For $1 < x < 3$, $f(x)$ is positive and so F is increasing. We see that F has a local minimum when $x = 1$ and a local maximum when $x = 3$. For $x > 3$, $f(x)$ is negative and so F is decreasing on $(3, \infty)$. Since $f(x) \to 0$ as $x \to \infty$, the graph of F becomes flatter as $x \to \infty$. Also notice that $F''(x) = f'(x)$ changes from positive to negative at $x = 2$ and from negative to positive at $x = 4$, so F has inflection points when $x = 2$ and $x = 4$. We use this information to sketch the graph of the antiderivative in Figure 4.

EXAMPLE 6 If $f(x) = \sqrt{1 + x^3} - x$, sketch the graph of the antiderivative F that satisfies the initial condition $F(-1) = 0$.

SOLUTION We could try all day to think of a formula for an antiderivative of f and still be unsuccessful. A second possibility would be to draw the graph of f first and then use it to graph F as in Example 5. That would work, but instead let's create a more accurate graph by using what is called a **direction field**.

Since $f(0) = 1$, the graph of F has slope 1 when $x = 0$. So we draw several short tangent segments with slope 1, all centered at $x = 0$. We do the same for several other values of x and the result is shown in Figure 5. It is called a direction field because each segment indicates the direction in which the curve $y = F(x)$ proceeds at that point.

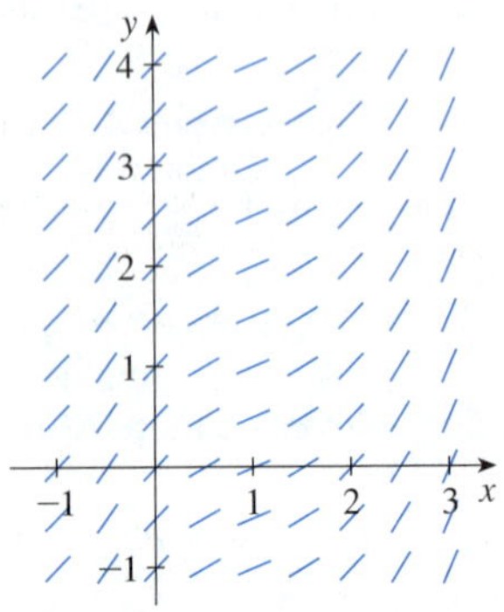

FIGURE 5

A direction field for $f(x) = \sqrt{1 + x^3} - x$.
The slope of the line segments above $x = a$ is $f(a)$.

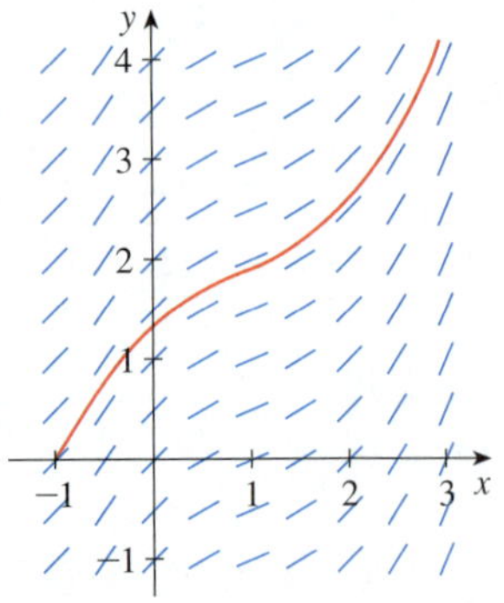

FIGURE 6

The graph of an antiderivative follows the direction field.

Now we use the direction field to sketch the graph of F. Because of the initial condition $F(-1) = 0$, we start at the point $(-1, 0)$ and draw the graph so that it follows the directions of the tangent segments. The result is pictured in Figure 6. Any other antiderivative would be obtained by shifting the graph of F upward or downward.

Rectilinear Motion

Antidifferentiation is particularly useful in analyzing the motion of an object moving in a straight line. Recall that if the object has position function $s = f(t)$, then the velocity function is $v(t) = s'(t)$. This means that the position function is an antiderivative of the velocity function. Likewise, the acceleration function is $a(t) = v'(t)$, so the velocity function is an antiderivative of the acceleration. If the acceleration and the initial values $s(0)$ and $v(0)$ are known, then the position function can be found by antidifferentiating twice.

EXAMPLE 7 A particle moves in a straight line and has acceleration given by $a(t) = 6t + 4$. Its initial velocity is $v(0) = -6$ cm/s and its initial displacement is $s(0) = 9$ cm. Find its position function $s(t)$.

SOLUTION Since $v'(t) = a(t) = 6t + 4$, antidifferentiation gives

$$v(t) = 6\frac{t^2}{2} + 4t + C = 3t^2 + 4t + C$$

Note that $v(0) = C$. But we are given that $v(0) = -6$, so $C = -6$ and

$$v(t) = 3t^2 + 4t - 6$$

Since $v(t) = s'(t)$, s is the antiderivative of v:

$$s(t) = 3\frac{t^3}{3} + 4\frac{t^2}{2} - 6t + D = t^3 + 2t^2 - 6t + D$$

This gives $s(0) = D$. We are given that $s(0) = 9$, so $D = 9$ and the required position function is

$$s(t) = t^3 + 2t^2 - 6t + 9$$

An object near the surface of the earth is subject to a gravitational force that produces a downward acceleration denoted by g. For motion close to the ground we may assume that g is constant, its value being about 9.8 m/s^2 (or 32 ft/s^2).

EXAMPLE 8 A ball is thrown upward with a speed of 48 ft/s from the edge of a cliff 432 ft above the ground. Find its height above the ground t seconds later. When does it reach its maximum height? When does it hit the ground?

SOLUTION The motion is vertical and we choose the positive direction to be upward. At time t the distance above the ground is $s(t)$ and the velocity $v(t)$ is decreasing. Therefore, the acceleration must be negative and we have

$$a(t) = \frac{dv}{dt} = -32$$

Taking antiderivatives, we have

$$v(t) = -32t + C$$

To determine C we use the given information that $v(0) = 48$. This gives $48 = 0 + C$, so

$$v(t) = -32t + 48$$

The maximum height is reached when $v(t) = 0$, that is, after 1.5 s. Since $s'(t) = v(t)$, we antidifferentiate again and obtain

$$s(t) = -16t^2 + 48t + D$$

Using the fact that $s(0) = 432$, we have $432 = 0 + D$ and so

$$s(t) = -16t^2 + 48t + 432$$

|||| Figure 7 shows the position function of the ball in Example 8. The graph corroborates the conclusions we reached: The ball reaches its maximum height after 1.5 s and hits the ground after 6.9 s.

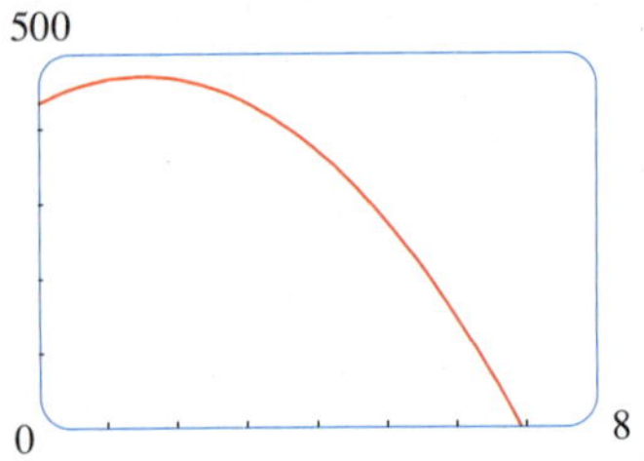

FIGURE 7

The expression for $s(t)$ is valid until the ball hits the ground. This happens when $s(t) = 0$, that is, when

$$-16t^2 + 48t + 432 = 0$$

or, equivalently,

$$t^2 - 3t - 27 = 0$$

Using the quadratic formula to solve this equation, we get

$$t = \frac{3 \pm 3\sqrt{13}}{2}$$

We reject the solution with the minus sign since it gives a negative value for t. Therefore, the ball hits the ground after $3(1 + \sqrt{13})/2 \approx 6.9$ s.

4.10 Exercises

1–16 |||| Find the most general antiderivative of the function. (Check your answer by differentiation.)

1. $f(x) = 6x^2 - 8x + 3$

2. $f(x) = 4 + x^2 - 5x^3$

3. $f(x) = 1 - x^3 + 5x^5 - 3x^7$

4. $f(x) = x^{20} + 4x^{10} + 8$

5. $f(x) = 5x^{1/4} - 7x^{3/4}$

6. $f(x) = 2x + 3x^{1.7}$

7. $f(x) = 6\sqrt{x} - \sqrt[6]{x}$

8. $f(x) = \sqrt[4]{x^3} + \sqrt[3]{x^4}$

9. $f(x) = \dfrac{10}{x^9}$

10. $g(x) = \dfrac{5 - 4x^3 + 2x^6}{x^6}$

11. $f(u) = \dfrac{u^4 + 3\sqrt{u}}{u^2}$

12. $f(x) = 3e^x + 7\sec^2 x$

13. $g(\theta) = \cos\theta - 5\sin\theta$

14. $h(\theta) = \dfrac{\sin\theta}{\cos^2\theta}$

15. $f(x) = 2x + 5(1 - x^2)^{-1/2}$

16. $f(x) = \dfrac{x^2 + x + 1}{x}$

17–18 |||| Find the antiderivative F of f that satisfies the given condition. Check your answer by comparing the graphs of f and F.

17. $f(x) = 5x^4 - 2x^5$, $F(0) = 4$

18. $f(x) = 4 - 3(1 + x^2)^{-1}$, $F(1) = 0$

19–42 |||| Find f.

19. $f''(x) = 6x + 12x^2$

20. $f''(x) = 2 + x^3 + x^6$

21. $f''(x) = 1 + x^{4/5}$

22. $f''(x) = \cos x$

23. $f'''(t) = e^t$

24. $f'''(t) = t - \sqrt{t}$

25. $f'(x) = 1 - 6x$, $f(0) = 8$

26. $f'(x) = 8x^3 + 12x + 3$, $f(1) = 6$

27. $f'(x) = \sqrt{x}(6 + 5x)$, $f(1) = 10$

28. $f'(x) = 2x - 3/x^4$, $x > 0$, $f(1) = 3$

29. $f'(t) = 2\cos t + \sec^2 t$, $-\pi/2 < t < \pi/2$, $f(\pi/3) = 4$

30. $f'(x) = 3x^{-2}$, $f(1) = f(-1) = 0$

31. $f'(x) = 2/x$, $x < 0$, $f(-1) = 7$

32. $f'(x) = 4/\sqrt{1 - x^2}$, $f(\frac{1}{2}) = 1$

33. $f''(x) = 24x^2 + 2x + 10$, $f(1) = 5$, $f'(1) = -3$

34. $f''(x) = 4 - 6x - 40x^3$, $f(0) = 2$, $f'(0) = 1$

35. $f''(\theta) = \sin\theta + \cos\theta$, $f(0) = 3$, $f'(0) = 4$

36. $f''(t) = 3/\sqrt{t}$, $f(4) = 20$, $f'(4) = 7$

37. $f''(x) = 2 - 12x$, $f(0) = 9$, $f(2) = 15$

38. $f''(x) = 20x^3 + 12x^2 + 4$, $f(0) = 8$, $f(1) = 5$

39. $f''(x) = 2 + \cos x, \quad f(0) = -1, \quad f(\pi/2) = 0$

40. $f''(t) = 2e^t + 3 \sin t, \quad f(0) = 0, \quad f(\pi) = 0$

41. $f''(x) = x^{-2}, \quad x > 0, \quad f(1) = 0, \quad f(2) = 0$

42. $f'''(x) = \sin x, \quad f(0) = 1, \quad f'(0) = 1, \quad f''(0) = 1$

43. Given that the graph of f passes through the point $(1, 6)$ and that the slope of its tangent line at $(x, f(x))$ is $2x + 1$, find $f(2)$.

44. Find a function f such that $f'(x) = x^3$ and the line $x + y = 0$ is tangent to the graph of f.

45–46 |||| The graph of a function f is shown. Which graph is an antiderivative of f and why?

45.

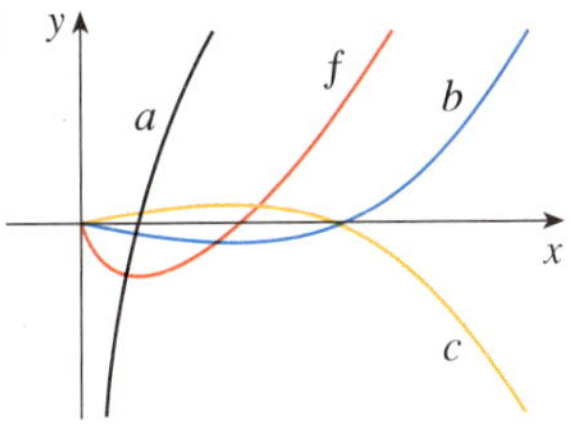

46.

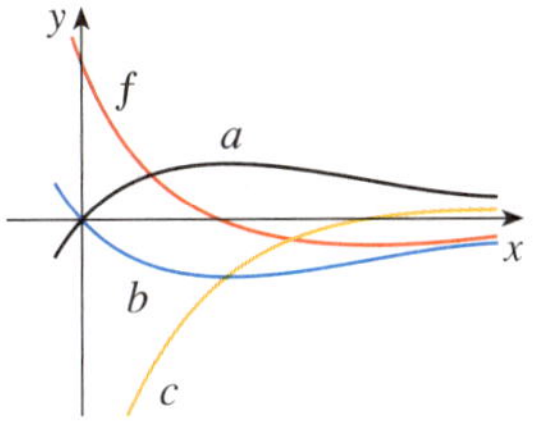

47. The graph of a function is shown in the figure. Make a rough sketch of an antiderivative F, given that $F(0) = 0$.

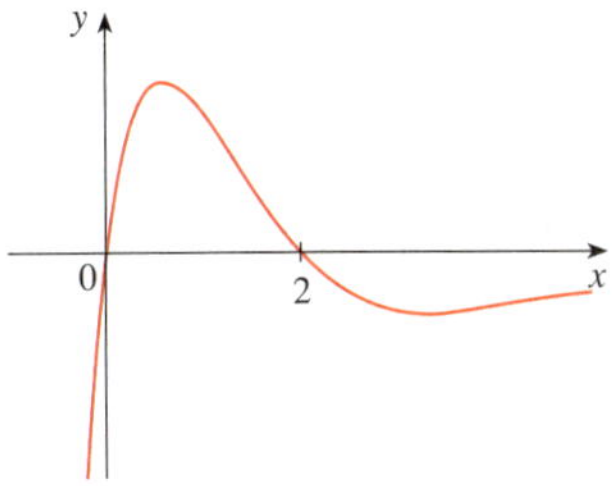

48. The graph of the velocity function of a car is shown in the figure. Sketch the graph of the position function.

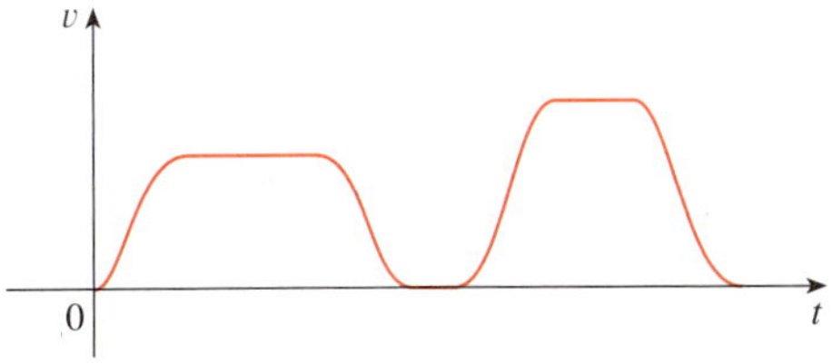

49. The graph of f' is shown in the figure. Sketch the graph of f if f is continuous and $f(0) = -1$.

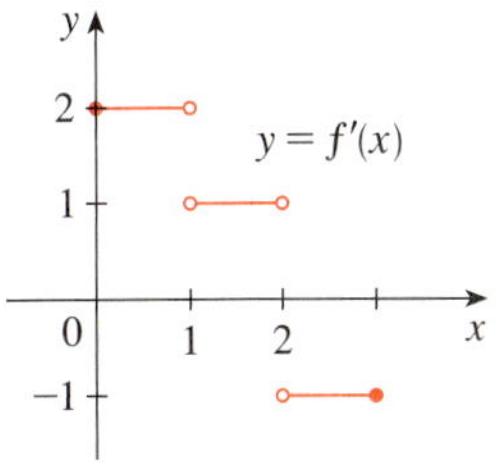

50. (a) Use a graphing device to graph $f(x) = 2x - 3\sqrt{x}$.

(b) Starting with the graph in part (a), sketch a rough graph of the antiderivative F that satisfies $F(0) = 1$.

(c) Use the rules of this section to find an expression for $F(x)$.

(d) Graph F using the expression in part (c). Compare with your sketch in part (b).

51–52 |||| Draw a graph of f and use it to make a rough sketch of the antiderivative that passes through the origin.

51. $f(x) = \sin(x^2), \quad 0 \leqslant x \leqslant 4$

52. $f(x) = 1/(x^4 + 1)$

53–54 |||| A direction field is given for a function. Use it to draw the antiderivative F that satisfies $F(0) = -2$.

53.

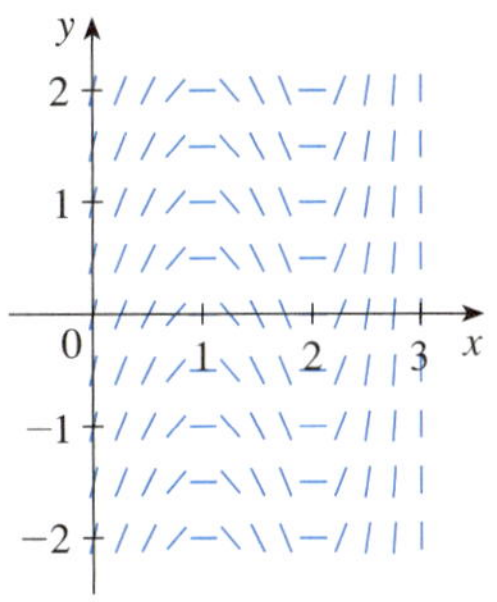

54.

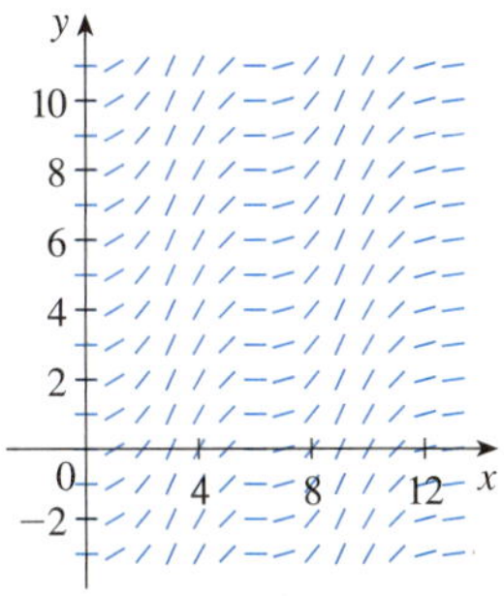

55–56 |||| Use a direction field to graph the antiderivative that satisfies $F(0) = 0$.

55. $f(x) = \dfrac{\sin x}{x}, \quad 0 < x < 2\pi$

56. $f(x) = x \tan x, \quad -\pi/2 < x < \pi/2$

57. A function is defined by the following experimental data. Use a direction field to sketch the graph of its antiderivative if the initial condition is $F(0) = 0$.

x	0	0.2	0.4	0.6	0.8	1.0	1.2	1.4	1.6
$f(x)$	0	0.2	0.5	0.8	1.0	0.6	0.2	0	−0.1

58. (a) Draw a direction field for the function $f(x) = 1/x^2$ and use it to sketch several members of the family of antiderivatives.

(b) Compute the general antiderivative explicitly and sketch several particular antiderivatives. Compare with your sketch in part (a).

59–64 |||| A particle is moving with the given data. Find the position of the particle.

59. $v(t) = \sin t - \cos t, \quad s(0) = 0$

60. $v(t) = 1.5\sqrt{t}, \quad s(4) = 10$

61. $a(t) = t - 2, \quad s(0) = 1, \quad v(0) = 3$

62. $a(t) = \cos t + \sin t, \quad s(0) = 0, \quad v(0) = 5$

63. $a(t) = 10 \sin t + 3 \cos t, \quad s(0) = 0, \quad s(2\pi) = 12$

64. $a(t) = 10 + 3t - 3t^2, \quad s(0) = 0, \quad s(2) = 10$

65. A stone is dropped from the upper observation deck (the Space Deck) of the CN Tower, 450 m above the ground.
(a) Find the distance of the stone above ground level at time t.
(b) How long does it take the stone to reach the ground?
(c) With what velocity does it strike the ground?
(d) If the stone is thrown downward with a speed of 5 m/s, how long does it take to reach the ground?

66. Show that for motion in a straight line with constant acceleration a, initial velocity v_0, and initial displacement s_0, the displacement after time t is

$$s = \tfrac{1}{2}at^2 + v_0 t + s_0$$

67. An object is projected upward with initial velocity v_0 meters per second from a point s_0 meters above the ground. Show that

$$[v(t)]^2 = v_0^2 - 19.6[s(t) - s_0]$$

68. Two balls are thrown upward from the edge of the cliff in Example 8. The first is thrown with a speed of 48 ft/s and the other is thrown a second later with a speed of 24 ft/s. Do the balls ever pass each other?

69. A stone was dropped off a cliff and hit the ground with a speed of 120 ft/s. What is the height of the cliff?

70. If a diver of mass m stands at the end of a diving board with length L and linear density ρ, then the board takes on the shape of a curve $y = f(x)$, where

$$EIy'' = mg(L - x) + \tfrac{1}{2}\rho g(L - x)^2$$

E and I are positive constants that depend on the material of the board and g (< 0) is the acceleration due to gravity.
(a) Find an expression for the shape of the curve.
(b) Use $f(L)$ to estimate the distance below the horizontal at the end of the board.

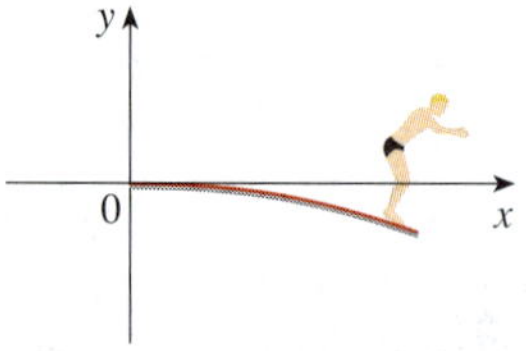

71. A company estimates that the marginal cost (in dollars per item) of producing x items is $1.92 - 0.002x$. If the cost of producing one item is \$562, find the cost of producing 100 items.

72. The linear density of a rod of length 1 m is given by $\rho(x) = 1/\sqrt{x}$, in grams per centimeter, where x is measured in centimeters from one end of the rod. Find the mass of the rod.

73. Since raindrops grow as they fall, their surface area increases and therefore the resistance to their falling increases. A raindrop has an initial downward velocity of 10 m/s and its downward acceleration is

$$a = \begin{cases} 9 - 0.9t & \text{if } 0 \leqslant t \leqslant 10 \\ 0 & \text{if } t > 10 \end{cases}$$

If the raindrop is initially 500 m above the ground, how long does it take to fall?

74. A car is traveling at 50 mi/h when the brakes are fully applied, producing a constant deceleration of 22 ft/s^2. What is the distance covered before the car comes to a stop?

75. What constant acceleration is required to increase the speed of a car from 30 mi/h to 50 mi/h in 5 s?

76. A car braked with a constant deceleration of 16 ft/s^2, producing skid marks measuring 200 ft before coming to a stop. How fast was the car traveling when the brakes were first applied?

77. A car is traveling at 100 km/h when the driver sees an accident 80 m ahead and slams on the brakes. What constant deceleration is required to stop the car in time to avoid a pileup?

78. A model rocket is fired vertically upward from rest. Its acceleration for the first three seconds is $a(t) = 60t$, at which time the fuel is exhausted and it becomes a freely "falling" body. Fourteen seconds later, the rocket's parachute opens, and the (downward) velocity slows linearly to -18 ft/s in 5 s. The rocket then "floats" to the ground at that rate.
(a) Determine the position function s and the velocity function v (for all times t). Sketch the graphs of s and v.
(b) At what time does the rocket reach its maximum height, and what is that height?
(c) At what time does the rocket land?

79. A high-speed bullet train accelerates and decelerates at the rate of 4 ft/s^2. Its maximum cruising speed is 90 mi/h.
(a) What is the maximum distance the train can travel if it accelerates from rest until it reaches its cruising speed and then runs at that speed for 15 minutes?
(b) Suppose that the train starts from rest and must come to a complete stop in 15 minutes. What is the maximum distance it can travel under these conditions?
(c) Find the minimum time that the train takes to travel between two consecutive stations that are 45 miles apart.
(d) The trip from one station to the next takes 37.5 minutes. How far apart are the stations?

4 Review

CONCEPT CHECK

1. Explain the difference between an absolute maximum and a local maximum. Illustrate with a sketch.

2. (a) What does the Extreme Value Theorem say?
   (b) Explain how the Closed Interval Method works.

3. (a) State Fermat's Theorem.
   (b) Define a critical number of f.

4. (a) State Rolle's Theorem.
   (b) State the Mean Value Theorem and give a geometric interpretation.

5. (a) State the Increasing/Decreasing Test.
   (b) State the Concavity Test.

6. (a) State the First Derivative Test.
   (b) State the Second Derivative Test.
   (c) What are the relative advantages and disadvantages of these tests?

7. (a) What does l'Hospital's Rule say?
   (b) How can you use l'Hospital's Rule if you have a product $f(x)g(x)$ where $f(x) \to 0$ and $g(x) \to \infty$ as $x \to a$?
   (c) How can you use l'Hospital's Rule if you have a difference $f(x) - g(x)$ where $f(x) \to \infty$ and $g(x) \to \infty$ as $x \to a$?
   (d) How can you use l'Hospital's Rule if you have a power $[f(x)]^{g(x)}$ where $f(x) \to 0$ and $g(x) \to 0$ as $x \to a$?

8. If you have a graphing calculator or computer, why do you need calculus to graph a function?

9. (a) Given an initial approximation x_1 to a root of the equation $f(x) = 0$, explain geometrically, with a diagram, how the second approximation x_2 in Newton's method is obtained.
   (b) Write an expression for x_2 in terms of x_1, $f(x_1)$, and $f'(x_1)$.
   (c) Write an expression for x_{n+1} in terms of x_n, $f(x_n)$, and $f'(x_n)$.
   (d) Under what circumstances is Newton's method likely to fail or to work very slowly?

10. (a) What is an antiderivative of a function f?
    (b) Suppose F_1 and F_2 are both antiderivatives of f on an interval I. How are F_1 and F_2 related?

TRUE-FALSE QUIZ

Determine whether the statement is true or false. If it is true, explain why. If it is false, explain why or give an example that disproves the statement.

1. If $f'(c) = 0$, then f has a local maximum or minimum at c.

2. If f has an absolute minimum value at c, then $f'(c) = 0$.

3. If f is continuous on (a, b), then f attains an absolute maximum value $f(c)$ and an absolute minimum value $f(d)$ at some numbers c and d in (a, b).

4. If f is differentiable and $f(-1) = f(1)$, then there is a number c such that $|c| < 1$ and $f'(c) = 0$.

5. If $f'(x) < 0$ for $1 < x < 6$, then f is decreasing on $(1, 6)$.

6. If $f''(2) = 0$, then $(2, f(2))$ is an inflection point of the curve $y = f(x)$.

7. If $f'(x) = g'(x)$ for $0 < x < 1$, then $f(x) = g(x)$ for $0 < x < 1$.

8. There exists a function f such that $f(1) = -2$, $f(3) = 0$, and $f'(x) > 1$ for all x.

9. There exists a function f such that $f(x) > 0$, $f'(x) < 0$, and $f''(x) > 0$ for all x.

10. There exists a function f such that $f(x) < 0$, $f'(x) < 0$, and $f''(x) > 0$ for all x.

11. If f and g are increasing on an interval I, then $f + g$ is increasing on I.

12. If f and g are increasing on an interval I, then $f - g$ is increasing on I.

13. If f and g are increasing on an interval I, then fg is increasing on I.

14. If f and g are positive increasing functions on an interval I, then fg is increasing on I.

15. If f is increasing and $f(x) > 0$ on I, then $g(x) = 1/f(x)$ is decreasing on I.

16. The most general antiderivative of $f(x) = x^{-2}$ is
$$F(x) = -\frac{1}{x} + C$$

17. If $f'(x)$ exists and is nonzero for all x, then $f(1) \neq f(0)$.

18. $\displaystyle \lim_{x \to 0} \frac{x}{e^x} = 1$

EXERCISES

1–6 Find the local and absolute extreme values of the function on the given interval.

1. $f(x) = 10 + 27x - x^3$, $[0, 4]$

2. $f(x) = x - \sqrt{x}$, $[0, 4]$

3. $f(x) = \dfrac{x}{x^2 + x + 1}$, $[-2, 0]$

4. $f(x) = (x^2 + 2x)^3$, $[-2, 1]$

5. $f(x) = x + \sin 2x$, $[0, \pi]$

6. $f(x) = (\ln x)/x^2$, $[1, 3]$

7–14 Evaluate the limit.

7. $\displaystyle\lim_{x\to 0} \frac{\tan \pi x}{\ln(1 + x)}$

8. $\displaystyle\lim_{x\to 0} \frac{1 - \cos x}{x^2 + x}$

9. $\displaystyle\lim_{x\to 0} \frac{e^{4x} - 1 - 4x}{x^2}$

10. $\displaystyle\lim_{x\to \infty} \frac{e^{4x} - 1 - 4x}{x^2}$

11. $\displaystyle\lim_{x\to \infty} x^3 e^{-x}$

12. $\displaystyle\lim_{x\to 0^+} x^2 \ln x$

13. $\displaystyle\lim_{x\to 1^+} \left(\frac{x}{x - 1} - \frac{1}{\ln x}\right)$

14. $\displaystyle\lim_{x\to (\pi/2)^-} (\tan x)^{\cos x}$

15–17 Sketch the graph of a function that satisfies the given conditions.

15. $f(0) = 0$, $f'(-2) = f'(1) = f'(9) = 0$,
$\lim_{x\to\infty} f(x) = 0$, $\lim_{x\to 6} f(x) = -\infty$,
$f'(x) < 0$ on $(-\infty, -2)$, $(1, 6)$, and $(9, \infty)$,
$f'(x) > 0$ on $(-2, 1)$ and $(6, 9)$,
$f''(x) > 0$ on $(-\infty, 0)$ and $(12, \infty)$,
$f''(x) < 0$ on $(0, 6)$ and $(6, 12)$

16. $f(0) = 0$, f is continuous and even,
$f'(x) = 2x$ if $0 < x < 1$, $f'(x) = -1$ if $1 < x < 3$,
$f'(x) = 1$ if $x > 3$

17. f is odd, $f'(x) < 0$ for $0 < x < 2$,
$f'(x) > 0$ for $x > 2$, $f''(x) > 0$ for $0 < x < 3$,
$f''(x) < 0$ for $x > 3$, $\lim_{x\to\infty} f(x) = -2$

18. The figure shows the graph of the *derivative* f' of a function f.
(a) On what intervals is f increasing or decreasing?
(b) For what values of x does f have a local maximum or minimum?
(c) Sketch the graph of f''.
(d) Sketch a possible graph of f.

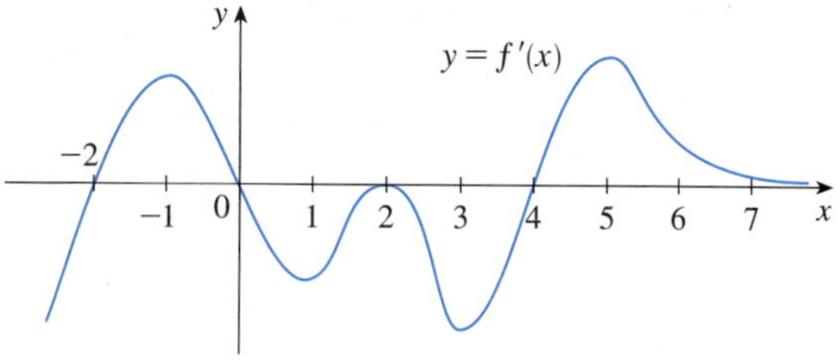

19–34 Use the guidelines of Section 4.5 to sketch the curve.

19. $y = 2 - 2x - x^3$

20. $y = x^3 - 6x^2 - 15x + 4$

21. $y = x^4 - 3x^3 + 3x^2 - x$

22. $y = \dfrac{1}{1 - x^2}$

23. $y = \dfrac{1}{x(x - 3)^2}$

24. $y = \dfrac{1}{x} + \dfrac{1}{x + 1}$

25. $y = x^2/(x + 8)$

26. $y = x + \sqrt{1 - x}$

27. $y = x\sqrt{2 + x}$

28. $y = \sqrt{x} - \sqrt[3]{x}$

29. $y = \sin^2 x - 2\cos x$

30. $y = 4x - \tan x$, $-\pi/2 < x < \pi/2$

31. $y = \sin^{-1}(1/x)$

32. $y = e^{2x - x^2}$

33. $y = e^x + e^{-3x}$

34. $y = \ln(x^2 - 1)$

35–38 Produce graphs of f that reveal all the important aspects of the curve. Use graphs of f' and f'' to estimate the intervals of increase and decrease, extreme values, intervals of concavity, and inflection points. In Exercise 35 use calculus to find these quantities exactly.

35. $f(x) = \dfrac{x^2 - 1}{x^3}$

36. $f(x) = \dfrac{\sqrt[3]{x}}{1 - x}$

37. $f(x) = 3x^6 - 5x^5 + x^4 - 5x^3 - 2x^2 + 2$

38. $f(x) = \sin x \cos^2 x$, $0 \leqslant x \leqslant 2\pi$

39. Graph $f(x) = e^{-1/x^2}$ in a viewing rectangle that shows all the main aspects of this function. Estimate the inflection points. Then use calculus to find them exactly.

CAS **40.** (a) Graph the function $f(x) = 1/(1 + e^{1/x})$.
(b) Explain the shape of the graph by computing the limits of $f(x)$ as x approaches ∞, $-\infty$, 0^+, and 0^-.
(c) Use the graph of f to estimate the coordinates of the inflection points.
(d) Use your CAS to compute and graph f''.
(e) Use the graph in part (d) to estimate the inflection points more accurately.

41. If $f(x) = \arctan(\cos(3 \arcsin x))$, use the graphs of f, f', and f'' to estimate the x-coordinates of the maximum and minimum points and inflection points of f.

42. If $f(x) = \ln(2x + x \sin x)$, use the graphs of f, f', and f'' to estimate the intervals of increase and the inflection points of f on the interval $(0, 15]$.

43. Investigate the family of functions $f(x) = \ln(\sin x + C)$. What features do the members of this family have in common? How do they differ? For which values of C is f continuous on $(-\infty, \infty)$? For which values of C does f have no graph at all? What happens as $C \to \infty$?

44. Investigate the family of functions $f(x) = cxe^{-cx^2}$. What happens to the maximum and minimum points and the inflection points as c changes? Illustrate your conclusions by graphing several members of the family.

45. Show that the equation $x^{101} + x^{51} + x - 1 = 0$ has exactly one real root.

46. Suppose that f is continuous on $[0, 4]$, $f(0) = 1$, and $2 \leqslant f'(x) \leqslant 5$ for all x in $(0, 4)$. Show that $9 \leqslant f(4) \leqslant 21$.

47. By applying the Mean Value Theorem to the function $f(x) = x^{1/5}$ on the interval $[32, 33]$, show that

$$2 < \sqrt[5]{33} < 2.0125$$

48. For what values of the constants a and b is $(1, 6)$ a point of inflection of the curve $y = x^3 + ax^2 + bx + 1$?

49. Let $g(x) = f(x^2)$, where f is twice differentiable for all x, $f'(x) > 0$ for all $x \neq 0$, and f is concave downward on $(-\infty, 0)$ and concave upward on $(0, \infty)$.
(a) At what numbers does g have an extreme value?
(b) Discuss the concavity of g.

50. Find two positive integers such that the sum of the first number and four times the second number is 1000 and the product of the numbers is as large as possible.

51. Show that the shortest distance from the point (x_1, y_1) to the straight line $Ax + By + C = 0$ is

$$\frac{|Ax_1 + By_1 + C|}{\sqrt{A^2 + B^2}}$$

52. Find the point on the hyperbola $xy = 8$ that is closest to the point $(3, 0)$.

53. Find the smallest possible area of an isosceles triangle that is circumscribed about a circle of radius r.

54. Find the volume of the largest circular cone that can be inscribed in a sphere of radius r.

55. In ΔABC, D lies on AB, $CD \perp AB$, $|AD| = |BD| = 4$ cm, and $|CD| = 5$ cm. Where should a point P be chosen on CD so that the sum $|PA| + |PB| + |PC|$ is a minimum?

56. Do Exercise 55 when $|CD| = 2$ cm.

57. The velocity of a wave of length L in deep water is

$$v = K\sqrt{\frac{L}{C} + \frac{C}{L}}$$

where K and C are known positive constants. What is the length of the wave that gives the minimum velocity?

58. A metal storage tank with volume V is to be constructed in the shape of a right circular cylinder surmounted by a hemisphere. What dimensions will require the least amount of metal?

59. A hockey team plays in an arena with a seating capacity of 15,000 spectators. With the ticket price set at \$12, average attendance at a game has been 11,000. A market survey indicates that for each dollar the ticket price is lowered, average attendance will increase by 1000. How should the owners of the team set the ticket price to maximize their revenue from ticket sales?

60. A manufacturer determines that the cost of making x units of a commodity is $C(x) = 1800 + 25x - 0.2x^2 + 0.001x^3$ and the demand function is $p(x) = 48.2 - 0.03x$.
(a) Graph the cost and revenue functions and use the graphs to estimate the production level for maximum profit.
(b) Use calculus to find the production level for maximum profit.
(c) Estimate the production level that minimizes the average cost.

61. Use Newton's method to find the root of the equation $x^5 - x^4 + 3x^2 - 3x - 2 = 0$ in the interval $[1, 2]$ correct to six decimal places.

62. Use Newton's method to find all roots of the equation $\sin x = x^2 - 3x + 1$ correct to six decimal places.

63. Use Newton's method to find the absolute maximum value of the function $f(t) = \cos t + t - t^2$ correct to eight decimal places.

64. Use the guidelines in Section 4.5 to sketch the curve $y = x \sin x$, $0 \leqslant x \leqslant 2\pi$. Use Newton's method when necessary.

65–72 Find f.

65. $f'(x) = \sqrt{x^5} - \dfrac{4}{\sqrt[5]{x}}$

66. $f'(x) = 8x - 3 \sec^2 x$

67. $f'(x) = e^x - (2/\sqrt{x})$

68. $f'(x) = 2/(1 + x^2)$, $f(0) = -1$

69. $f'(t) = 2t - 3 \sin t$, $f(0) = 5$

70. $f'(u) = \dfrac{u^2 + \sqrt{u}}{u}$, $f(1) = 3$

71. $f''(x) = 1 - 6x + 48x^2$, $f(0) = 1$, $f'(0) = 2$

72. $f''(x) = 2x^3 + 3x^2 - 4x + 5$, $f(0) = 2$, $f(1) = 0$

73. (a) If $f(x) = 0.1e^x + \sin x$, $-4 \leqslant x \leqslant 4$, use a graph of f to sketch a rough graph of the antiderivative F of f that satisfies $F(0) = 0$.
(b) Find an expression for $F(x)$.
(c) Graph F using the expression in part (b). Compare with your sketch in part (a).

74. Investigate the family of curves given by $f(x) = x^4 + x^3 + cx^2$. In particular you should determine the transitional value of c at which the number of critical numbers changes and the transitional value at which the number of inflection points changes. Illustrate the various possible shapes with graphs.

75. A canister is dropped from a helicopter 500 m above the ground. Its parachute does not open, but the canister has been designed to withstand an impact velocity of 100 m/s. Will it burst?

76. In an automobile race along a straight road, car A passed car B twice. Prove that at some time during the race their accelerations were equal. State the assumptions that you make.

77. A rectangular beam will be cut from a cylindrical log of radius 10 inches.
(a) Show that the beam of maximal cross-sectional area is a square.
(b) Four rectangular planks will be cut from the four sections of the log that remain after cutting the square beam. Determine the dimensions of the planks that will have maximal cross-sectional area.
(c) Suppose that the strength of a rectangular beam is proportional to the product of its width and the square of its depth. Find the dimensions of the strongest beam that can be cut from the cylindrical log.

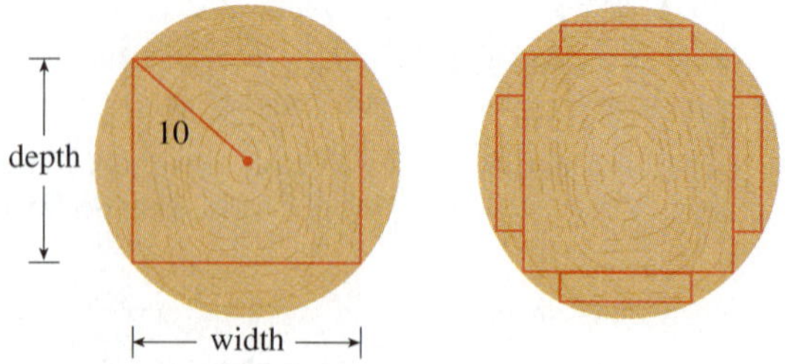

78. If a projectile is fired with an initial velocity v at an angle of inclination θ from the horizontal, then its trajectory, neglecting air resistance, is the parabola

$$y = (\tan\theta)x - \frac{g}{2v^2\cos^2\theta}x^2 \qquad 0 \leqslant \theta \leqslant \frac{\pi}{2}$$

(a) Suppose the projectile is fired from the base of a plane that is inclined at an angle α, $\alpha > 0$, from the horizontal, as shown in the figure. Show that the range of the projectile, measured up the slope, is given by

$$R(\theta) = \frac{2v^2\cos\theta\,\sin(\theta - \alpha)}{g\cos^2\alpha}$$

(b) Determine θ so that R is a maximum.
(c) Suppose the plane is at an angle α *below* the horizontal. Determine the range R in this case, and determine the angle at which the projectile should be fired to maximize R.

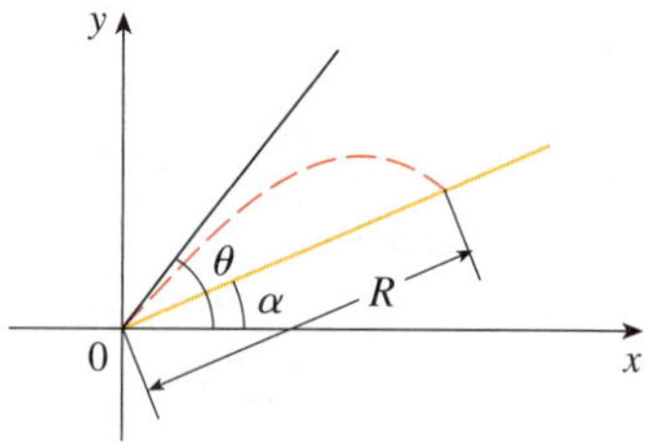

79. A light is to be placed atop a pole of height h feet to illuminate a busy traffic circle, which has a radius of 40 ft. The intensity of illumination I at any point P on the circle is directly proportional to the cosine of the angle θ (see the figure) and inversely proportional to the square of the distance d from the source.
(a) How tall should the light pole be to maximize I?
(b) Suppose that the light pole is h feet tall and that a woman is walking away from the base of the pole at the rate of 4 ft/s. At what rate is the intensity of the light at the point on her back 4 ft above the ground decreasing when she reaches the outer edge of the traffic circle?

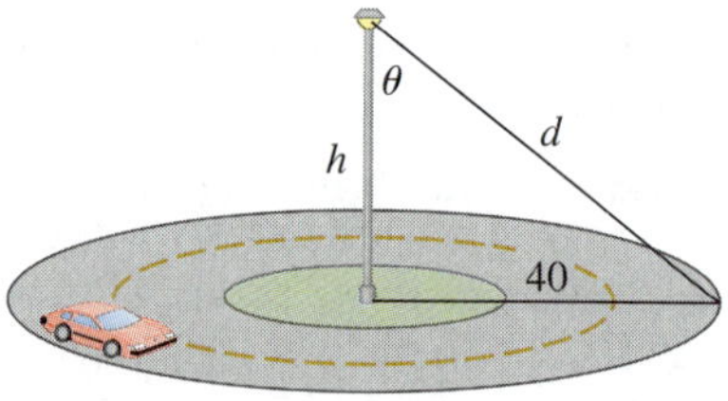

80. Water is flowing at a constant rate into a spherical tank. Let $V(t)$ be the volume of water in the tank and $H(t)$ be the height of the water in the tank at time t.
(a) What are the meanings of $V'(t)$ and $H'(t)$? Are these derivatives positive, negative, or zero?
(b) Is $V''(t)$ positive, negative, or zero? Explain.
(c) Let t_1, t_2, and t_3 be the times when the tank is one-quarter full, half full, and three-quarters full, respectively. Are the values $H''(t_1)$, $H''(t_2)$, and $H''(t_3)$ positive, negative, or zero? Why?

81. Show that, for $x > 0$,

$$\frac{x}{1 + x^2} < \tan^{-1}x < x$$

82. Sketch the graph of a function f such that $f'(x) < 0$ for all x, $f''(x) > 0$ for $|x| > 1$, $f''(x) < 0$ for $|x| < 1$, and $\lim_{x\to\pm\infty}[f(x) + x] = 0$.

PROBLEMS PLUS

One of the most important principles of problem solving is *analogy* (see page 80). If you are having trouble getting started on a problem, it is sometimes helpful to start by solving a similar, but simpler, problem. The following example illustrates the principle. Cover up the solution and try solving it yourself first.

EXAMPLE If x, y, and z are positive numbers, prove that

$$\frac{(x^2 + 1)(y^2 + 1)(z^2 + 1)}{xyz} \geqslant 8$$

SOLUTION It may be difficult to get started on this problem. (Some students have tackled it by multiplying out the numerator, but that just creates a mess.) Let's try to think of a similar, simpler problem. When several variables are involved, it's often helpful to think of an analogous problem with fewer variables. In the present case we can reduce the number of variables from three to one and prove the analogous inequality

$$\boxed{1} \qquad \frac{x^2 + 1}{x} \geqslant 2 \qquad \text{for } x > 0$$

In fact, if we are able to prove (1), then the desired inequality follows because

$$\frac{(x^2 + 1)(y^2 + 1)(z^2 + 1)}{xyz} = \left(\frac{x^2 + 1}{x}\right)\left(\frac{y^2 + 1}{y}\right)\left(\frac{z^2 + 1}{z}\right) \geqslant 2 \cdot 2 \cdot 2 = 8$$

The key to proving (1) is to recognize that it is a disguised version of a minimum problem. If we let

$$f(x) = \frac{x^2 + 1}{x} = x + \frac{1}{x} \qquad x > 0$$

then $f'(x) = 1 - (1/x^2)$, so $f'(x) = 0$ when $x = 1$. Also, $f'(x) < 0$ for $0 < x < 1$ and $f'(x) > 0$ for $x > 1$. Therefore, the absolute minimum value of f is $f(1) = 2$. This means that

$$\frac{x^2 + 1}{x} \geqslant 2 \qquad \text{for all positive values of } x$$

and, as previously mentioned, the given inequality follows by multiplication.

The inequality in (1) could also be proved without calculus. In fact, if $x > 0$, we have

$$\frac{x^2 + 1}{x} \geqslant 2 \iff x^2 + 1 \geqslant 2x \iff x^2 - 2x + 1 \geqslant 0$$

$$\iff (x - 1)^2 \geqslant 0$$

Because the last inequality is obviously true, the first one is true too.

Look Back

What have we learned from the solution to this example?

- To solve a problem involving several variables, it might help to solve a similar problem with just one variable.
- When trying to prove an inequality, it might help to think of it as a maximum or minimum problem.

PROBLEMS

1. If a rectangle has its base on the x-axis and two vertices on the curve $y = e^{-x^2}$, show that the rectangle has the largest possible area when the two vertices are at the points of inflection of the curve.

2. Show that $|\sin x - \cos x| \leqslant \sqrt{2}$ for all x.

3. Show that, for all positive values of x and y,

$$\frac{e^{x+y}}{xy} \geqslant e^2$$

4. Show that $x^2y^2(4 - x^2)(4 - y^2) \leqslant 16$ for all numbers x and y such that $|x| \leqslant 2$ and $|y| \leqslant 2$.

5. Let a and b be positive numbers. Show that not both of the numbers $a(1 - b)$ and $b(1 - a)$ can be greater than $\frac{1}{4}$.

6. Find the point on the parabola $y = 1 - x^2$ at which the tangent line cuts from the first quadrant the triangle with the smallest area.

7. Find the highest and lowest points on the curve $x^2 + xy + y^2 = 12$.

8. Sketch the set of all points (x, y) such that $|x + y| \leqslant e^x$.

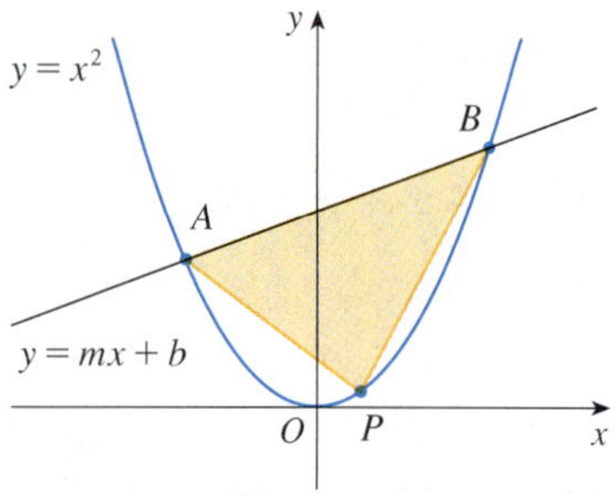

FIGURE FOR PROBLEM 9

9. The line $y = mx + b$ intersects the parabola $y = x^2$ in points A and B (see the figure). Find the point P on the arc AOB of the parabola that maximizes the area of the triangle PAB.

10. Find a function f such that $f'(-1) = \frac{1}{2}$, $f'(0) = 0$, and $f''(x) > 0$ for all x, or prove that such a function cannot exist.

11. Determine the values of the number a for which the function f has no critical number:

$$f(x) = (a^2 + a - 6)\cos 2x + (a - 2)x + \cos 1$$

12. Sketch the region in the plane consisting of all points (x, y) such that

$$2xy \leqslant |x - y| \leqslant x^2 + y^2$$

13. Let ABC be a triangle with $\angle BAC = 120°$ and $|AB| \cdot |AC| = 1$.
    (a) Express the length of the angle bisector AD in terms of $x = |AB|$.
    (b) Find the largest possible value of $|AD|$.

14. $ABCD$ is a square piece of paper with sides of length 1 m. A quarter-circle is drawn from B to D with center A. The piece of paper is folded along EF, with E on AB and F on AD, so that A falls on the quarter-circle. Determine the maximum and minimum areas that the triangle AEF could have.

15. For which positive numbers a does the curve $y = a^x$ intersect the line $y = x$?

16. For what value of a is the following equation true?

$$\lim_{x\to\infty} \left(\frac{x + a}{x - a}\right)^x = e$$

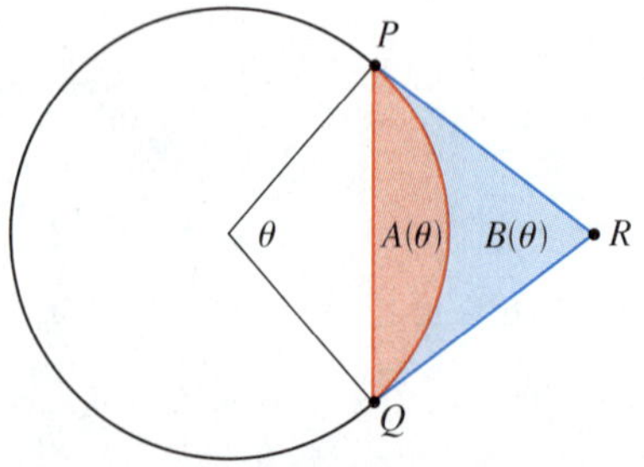

FIGURE FOR PROBLEM 18

17. Let $f(x) = a_1 \sin x + a_2 \sin 2x + \cdots + a_n \sin nx$, where $a_1, a_2, \ldots, a_n$ are real numbers and n is a positive integer. If it is given that $|f(x)| \leqslant |\sin x|$ for all x, show that

$$|a_1 + 2a_2 + \cdots + na_n| \leqslant 1$$

18. An arc PQ of a circle subtends a central angle θ as in the figure. Let $A(\theta)$ be the area between the chord PQ and the arc PQ. Let $B(\theta)$ be the area between the tangent lines PR, QR, and the arc. Find

$$\lim_{\theta\to 0^+} \frac{A(\theta)}{B(\theta)}$$

19. The speeds of sound c_1 in an upper layer and c_2 in a lower layer of rock and the thickness h of the upper layer can be determined by seismic exploration if the speed of sound in the lower layer is greater than the speed in the upper layer. A dynamite charge is detonated at a point P and the transmitted signals are recorded at a point Q, which is a distance D from P. The first signal to arrive at Q travels along the surface and takes T_1 seconds. The next signal travels from P to a point R, from R to S in the lower layer, and then to Q, taking T_2 seconds. The third signal is reflected off the lower layer at the midpoint O of RS and takes T_3 seconds to reach Q.

(a) Express T_1, T_2, and T_3 in terms of D, h, c_1, c_2, and θ.

(b) Show that T_2 is a minimum when $\sin\theta = c_1/c_2$.

(c) Suppose that $D = 1$ km, $T_1 = 0.26$ s, $T_2 = 0.32$ s, $T_3 = 0.34$ s. Find c_1, c_2, and h.

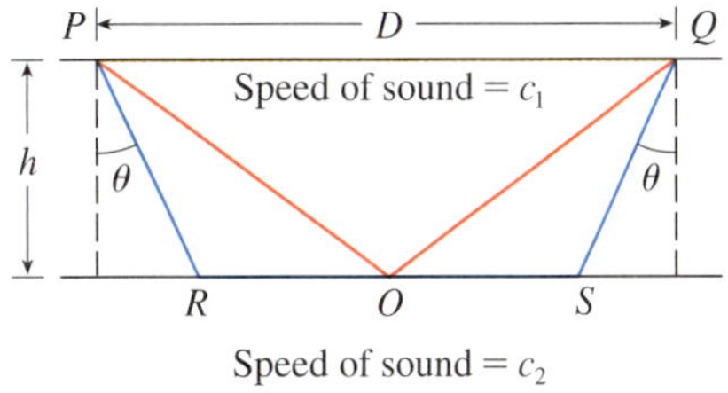

Note: Geophysicists use this technique when studying the structure of Earth's crust, whether searching for oil or examining fault lines.

20. For what values of c is there a straight line that intersects the curve

$$y = x^4 + cx^3 + 12x^2 - 5x + 2$$

in four distinct points?

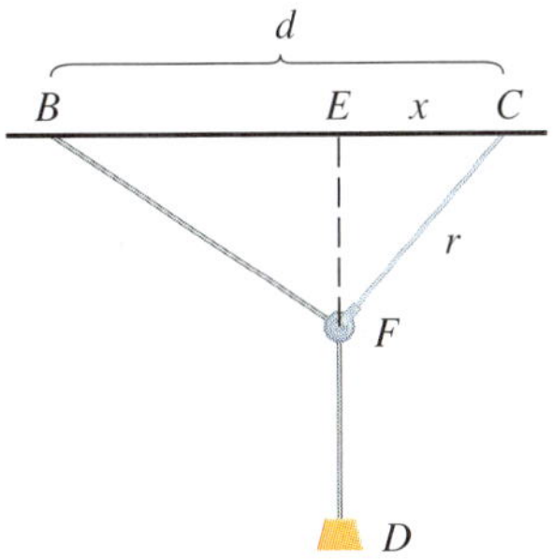

FIGURE FOR PROBLEM 21

21. One of the problems posed by the Marquis de l'Hospital in his calculus textbook *Analyse des Infiniment Petits* concerns a pulley that is attached to the ceiling of a room at a point C by a rope of length r. At another point B on the ceiling, at a distance d from C (where $d > r$), a rope of length ℓ is attached and passed through the pulley at F and connected to a weight W. The weight is released and comes to rest at its equilibrium position D. As l'Hospital argued, this happens when the distance $|ED|$ is maximized. Show that when the system reaches equilibrium, the value of x is

$$\frac{r}{4d}\left(r + \sqrt{r^2 + 8d^2}\right)$$

Notice that this expression is independent of both W and ℓ.

22. Given a sphere with radius r, find the height of a pyramid of minimum volume whose base is a square and whose base and triangular faces are all tangent to the sphere. What if the base of the pyramid is a regular n-gon (a polygon with n equal sides and angles)? (Use the fact that the volume of a pyramid is $\frac{1}{3}Ah$, where A is the area of the base.)

23. Assume that a snowball melts so that its volume decreases at a rate proportional to its surface area. If it takes three hours for the snowball to decrease to half its original volume, how much longer will it take for the snowball to melt completely?

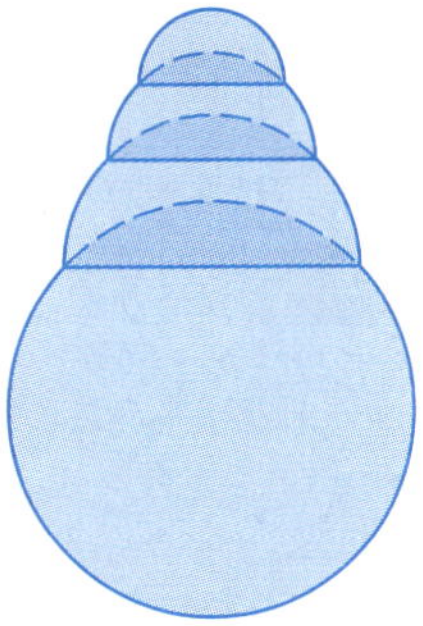

FIGURE FOR PROBLEM 24

24. A hemispherical bubble is placed on a spherical bubble of radius 1. A smaller hemispherical bubble is then placed on the first one. This process is continued until n chambers, including the sphere, are formed. (The figure shows the case $n = 4$.) Use mathematical induction to prove that the maximum height of any bubble tower with n chambers is $1 + \sqrt{n}$.

CHAPTER 5

To compute an area we approximate a region by rectangles and let the number of rectangles become large. The precise area is the limit of these sums of areas of rectangles.

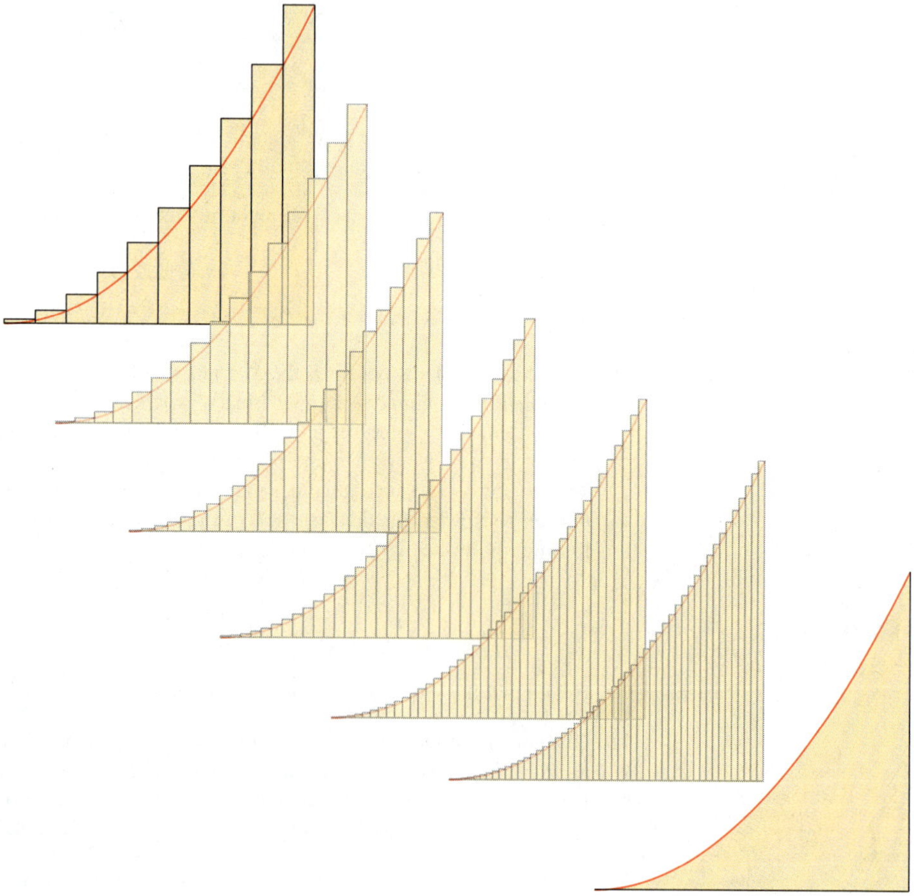

Integrals

Now is a good time to read (or reread) *A Preview of Calculus* (see page 2). It discusses the unifying ideas of calculus and helps put in perspective where we have been and where we are going.

In Chapter 2 we used the tangent and velocity problems to introduce the derivative, which is the central idea in differential calculus. In much the same way, this chapter starts with the area and distance problems and uses them to formulate the idea of a definite integral, which is the basic concept of integral calculus. We will see in Chapters 6 and 8 how to use the integral to solve problems concerning volumes, lengths of curves, population predictions, cardiac output, forces on a dam, work, consumer surplus, and baseball, among many others.

There is a connection between integral calculus and differential calculus. The Fundamental Theorem of Calculus relates the integral to the derivative, and we will see in this chapter that it greatly simplifies the solution of many problems.

5.1 Areas and Distances

In this section we discover that in trying to find the area under a curve or the distance traveled by a car, we end up with the same special type of limit.

The Area Problem

We begin by attempting to solve the *area problem:* Find the area of the region S that lies under the curve $y = f(x)$ from a to b. This means that S, illustrated in Figure 1, is bounded by the graph of a continuous function f [where $f(x) \geq 0$], the vertical lines $x = a$ and $x = b$, and the x-axis.

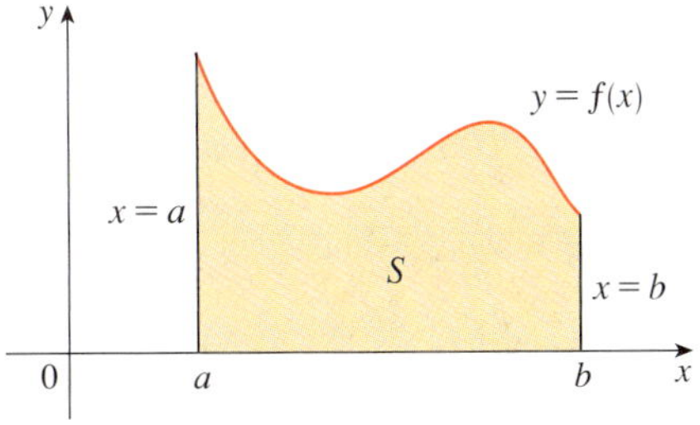

FIGURE 1

$S = \{(x, y) \mid a \leq x \leq b, 0 \leq y \leq f(x)\}$

In trying to solve the area problem we have to ask ourselves: What is the meaning of the word *area*? This question is easy to answer for regions with straight sides. For a rectangle, the area is defined as the product of the length and the width. The area of a triangle is half the base times the height. The area of a polygon is found by dividing it into triangles (as in Figure 2) and adding the areas of the triangles.

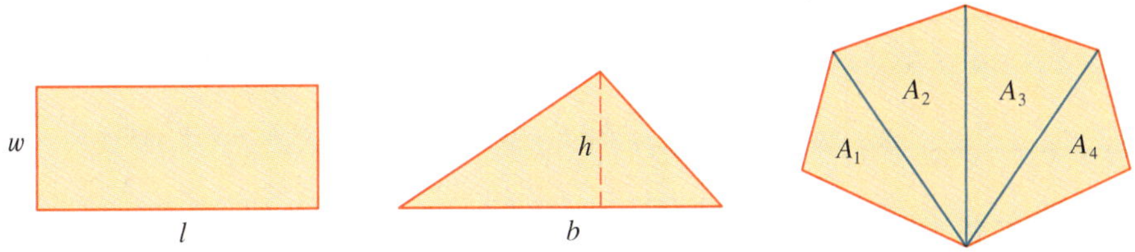

FIGURE 2

$A = lw$ $\quad A = \frac{1}{2}bh$ $\quad A = A_1 + A_2 + A_3 + A_4$

However, it isn't so easy to find the area of a region with curved sides. We all have an intuitive idea of what the area of a region is. But part of the area problem is to make this intuitive idea precise by giving an exact definition of area.

Recall that in defining a tangent we first approximated the slope of the tangent line by slopes of secant lines and then we took the limit of these approximations. We pursue a similar idea for areas. We first approximate the region S by rectangles and then we take the limit of the areas of these rectangles as we increase the number of rectangles. The following example illustrates the procedure.

Try placing rectangles to estimate the area.
Resources / Module 6
/ What Is Area?
/ Estimating Area under a Parabola

EXAMPLE 1 Use rectangles to estimate the area under the parabola $y = x^2$ from 0 to 1 (the parabolic region S illustrated in Figure 3).

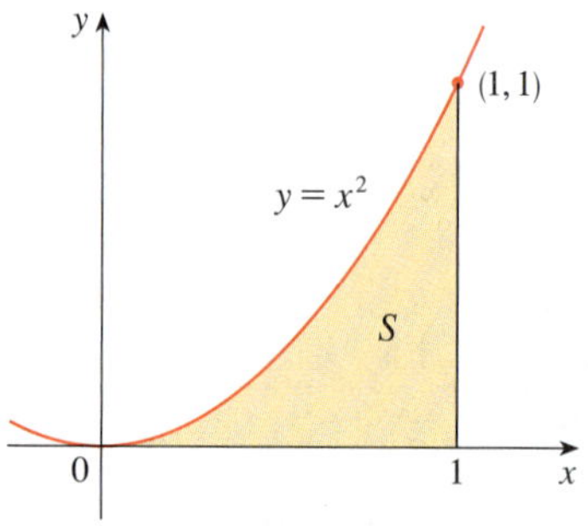

FIGURE 3

SOLUTION We first notice that the area of S must be somewhere between 0 and 1 because S is contained in a square with side length 1, but we can certainly do better than that. Suppose we divide S into four strips S_1, S_2, S_3, and S_4 by drawing the vertical lines $x = \frac{1}{4}$, $x = \frac{1}{2}$, and $x = \frac{3}{4}$ as in Figure 4(a).

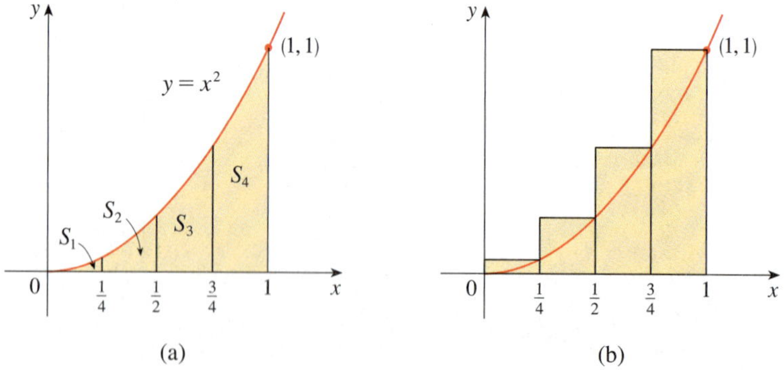

FIGURE 4

We can approximate each strip by a rectangle whose base is the same as the strip and whose height is the same as the right edge of the strip [see Figure 4(b)]. In other words, the heights of these rectangles are the values of the function $f(x) = x^2$ at the right end points of the subintervals $[0, \frac{1}{4}]$, $[\frac{1}{4}, \frac{1}{2}]$, $[\frac{1}{2}, \frac{3}{4}]$, and $[\frac{3}{4}, 1]$.

Each rectangle has width $\frac{1}{4}$ and the heights are $(\frac{1}{4})^2$, $(\frac{1}{2})^2$, $(\frac{3}{4})^2$, and 1^2. If we let R_4 be the sum of the areas of these approximating rectangles, we get

$$R_4 = \tfrac{1}{4} \cdot (\tfrac{1}{4})^2 + \tfrac{1}{4} \cdot (\tfrac{1}{2})^2 + \tfrac{1}{4} \cdot (\tfrac{3}{4})^2 + \tfrac{1}{4} \cdot 1^2 = \tfrac{15}{32} = 0.46875$$

From Figure 4(b) we see that the area A of S is less than R_4, so

$$A < 0.46875$$

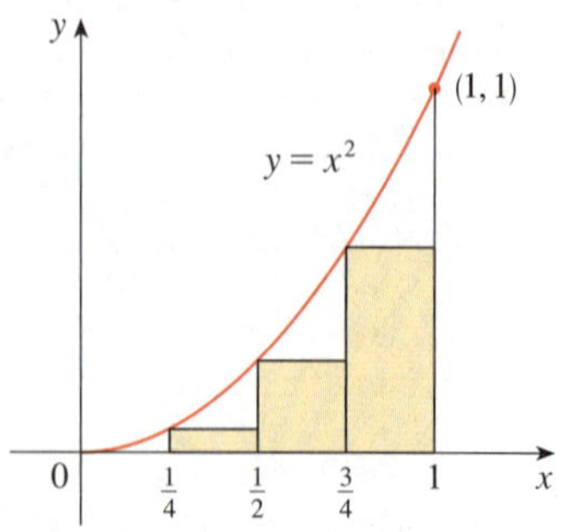

FIGURE 5

Instead of using the rectangles in Figure 4(b) we could use the smaller rectangles in Figure 5 whose heights are the values of f at the left endpoints of the subintervals. (The leftmost rectangle has collapsed because its height is 0.) The sum of the areas of these approximating rectangles is

$$L_4 = \tfrac{1}{4} \cdot 0^2 + \tfrac{1}{4} \cdot \left(\tfrac{1}{4}\right)^2 + \tfrac{1}{4} \cdot \left(\tfrac{1}{2}\right)^2 + \tfrac{1}{4} \cdot \left(\tfrac{3}{4}\right)^2 = \tfrac{7}{32} = 0.21875$$

We see that the area of S is larger than L_4, so we have lower and upper estimates for A:

$$0.21875 < A < 0.46875$$

We can repeat this procedure with a larger number of strips. Figure 6 shows what happens when we divide the region S into eight strips of equal width.

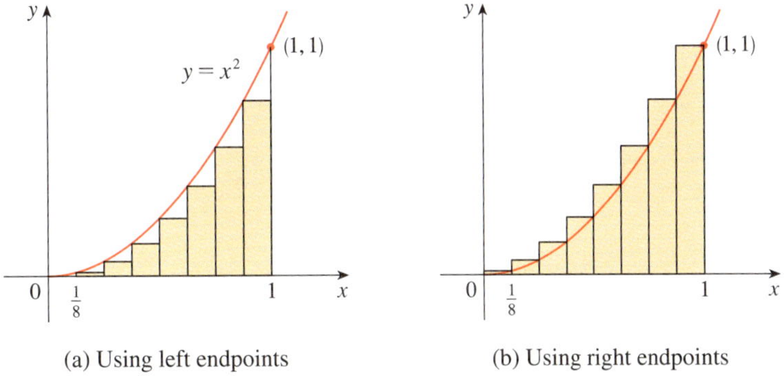

FIGURE 6
Approximating S with eight rectangles

By computing the sum of the areas of the smaller rectangles (L_8) and the sum of the areas of the larger rectangles (R_8), we obtain better lower and upper estimates for A:

$$0.2734375 < A < 0.3984375$$

n	L_n	R_n
10	0.2850000	0.3850000
20	0.3087500	0.3587500
30	0.3168519	0.3501852
50	0.3234000	0.3434000
100	0.3283500	0.3383500
1000	0.3328335	0.3338335

So one possible answer to the question is to say that the true area of S lies somewhere between 0.2734375 and 0.3984375.

We could obtain better estimates by increasing the number of strips. The table at the left shows the results of similar calculations (with a computer) using n rectangles whose heights are found with left endpoints (L_n) or right endpoints (R_n). In particular, we see by using 50 strips that the area lies between 0.3234 and 0.3434. With 1000 strips we narrow it down even more: A lies between 0.3328335 and 0.3338335. A good estimate is obtained by averaging these numbers: $A \approx 0.3333335$.

From the values in the table in Example 1, it looks as if R_n is approaching $\frac{1}{3}$ as n increases. We confirm this in the next example.

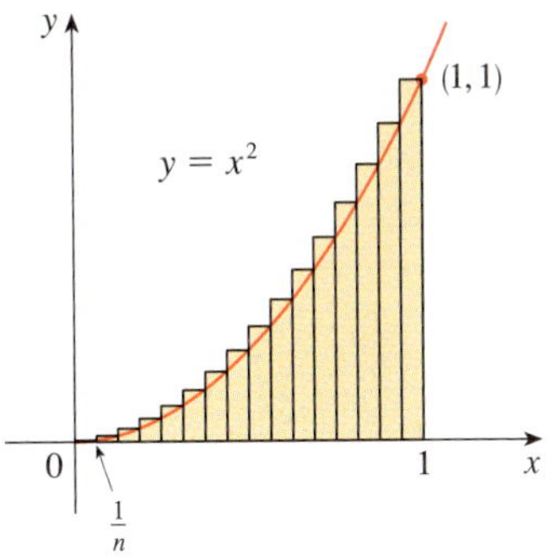

FIGURE 7

EXAMPLE 2 For the region S in Example 1, show that the sum of the areas of the upper approximating rectangles approaches $\frac{1}{3}$, that is,

$$\lim_{n \to \infty} R_n = \tfrac{1}{3}$$

SOLUTION R_n is the sum of the areas of the n rectangles in Figure 7. Each rectangle has width $1/n$ and the heights are the values of the function $f(x) = x^2$ at the points $1/n, 2/n, 3/n, \ldots, n/n$; that is, the heights are $(1/n)^2, (2/n)^2, (3/n)^2, \ldots, (n/n)^2$.

Thus

$$R_n = \frac{1}{n}\left(\frac{1}{n}\right)^2 + \frac{1}{n}\left(\frac{2}{n}\right)^2 + \frac{1}{n}\left(\frac{3}{n}\right)^2 + \cdots + \frac{1}{n}\left(\frac{n}{n}\right)^2$$

$$= \frac{1}{n} \cdot \frac{1}{n^2}(1^2 + 2^2 + 3^2 + \cdots + n^2)$$

$$= \frac{1}{n^3}(1^2 + 2^2 + 3^2 + \cdots + n^2)$$

TEC The ideas in Examples 1 and 2 are explored in Module 5.1/5.2/7.7 for a variety of functions.

Here we need the formula for the sum of the squares of the first n positive integers:

1 $$1^2 + 2^2 + 3^2 + \cdots + n^2 = \frac{n(n+1)(2n+1)}{6}$$

Perhaps you have seen this formula before. It is proved in Example 5 in Appendix E.

Putting Formula 1 into our expression for R_n, we get

$$R_n = \frac{1}{n^3} \cdot \frac{n(n+1)(2n+1)}{6} = \frac{(n+1)(2n+1)}{6n^2}$$

Thus, we have

|||| Here we are computing the limit of the sequence $\{R_n\}$. Sequences were discussed in *A Preview of Calculus* and will be studied in detail in Chapter 11. Their limits are calculated in the same way as limits at infinity (Section 2.6). In particular, we know that

$$\lim_{n\to\infty} \frac{1}{n} = 0$$

$$\lim_{n\to\infty} R_n = \lim_{n\to\infty} \frac{(n+1)(2n+1)}{6n^2}$$

$$= \lim_{n\to\infty} \frac{1}{6}\left(\frac{n+1}{n}\right)\left(\frac{2n+1}{n}\right)$$

$$= \lim_{n\to\infty} \frac{1}{6}\left(1 + \frac{1}{n}\right)\left(2 + \frac{1}{n}\right)$$

$$= \tfrac{1}{6} \cdot 1 \cdot 2 = \tfrac{1}{3}$$

It can be shown that the lower approximating sums also approach $\frac{1}{3}$, that is,

$$\lim_{n\to\infty} L_n = \tfrac{1}{3}$$

From Figures 8 and 9 it appears that, as n increases, both L_n and R_n become better and better approximations to the area of S. Therefore, we *define* the area A to be the limit of the

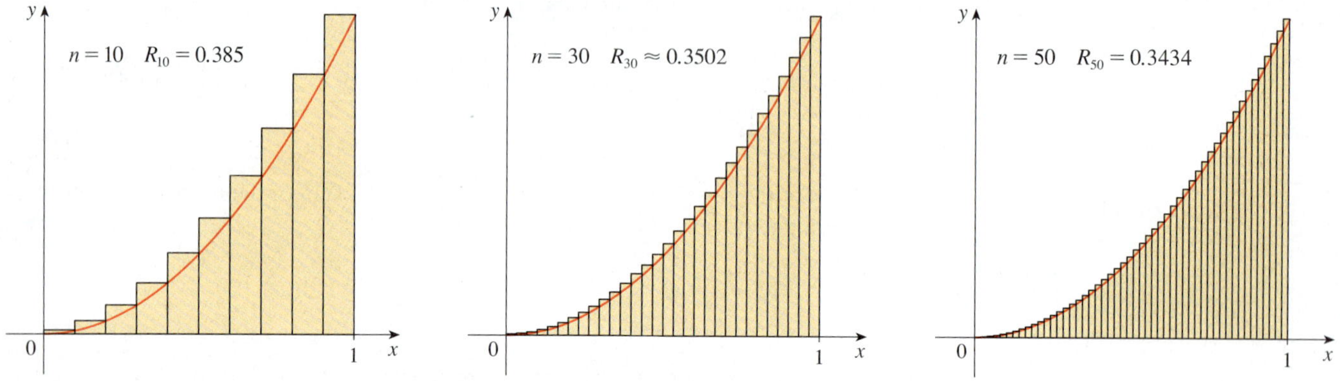

FIGURE 8

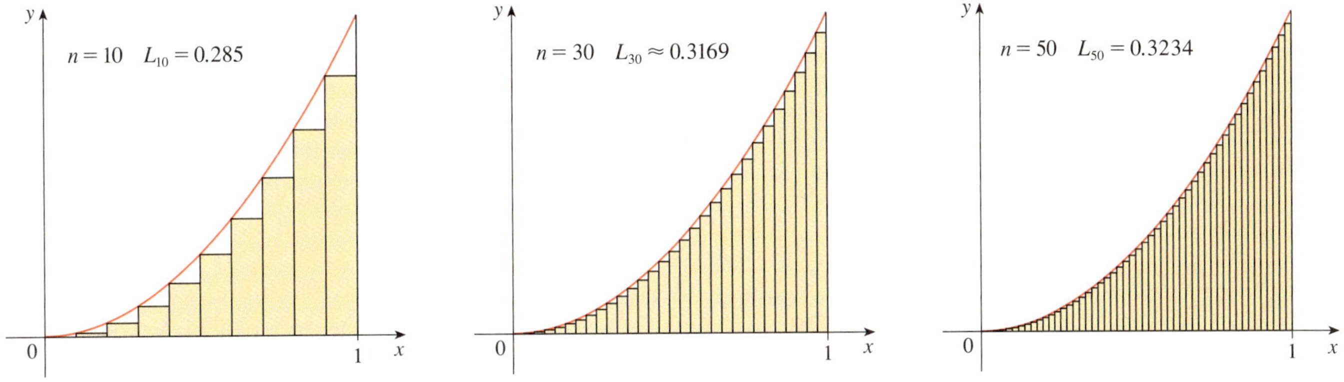

FIGURE 9

The area is the number that is smaller than all upper sums and larger than all lower sums

sums of the areas of the approximating rectangles, that is,

$$A = \lim_{n \to \infty} R_n = \lim_{n \to \infty} L_n = \tfrac{1}{3}$$

Let's apply the idea of Examples 1 and 2 to the more general region S of Figure 1. We start by subdividing S into n strips $S_1, S_2, \ldots, S_n$ of equal width as in Figure 10. The width of the interval $[a, b]$ is $b - a$, so the width of each of the n strips is

$$\Delta x = \frac{b - a}{n}$$

These strips divide the interval $[a, b]$ into n subintervals

$$[x_0, x_1], \quad [x_1, x_2], \quad [x_2, x_3], \quad \ldots, \quad [x_{n-1}, x_n]$$

where $x_0 = a$ and $x_n = b$. The right endpoints of the subintervals are

$$x_1 = a + \Delta x, \quad x_2 = a + 2\,\Delta x, \quad x_3 = a + 3\,\Delta x, \quad \ldots$$

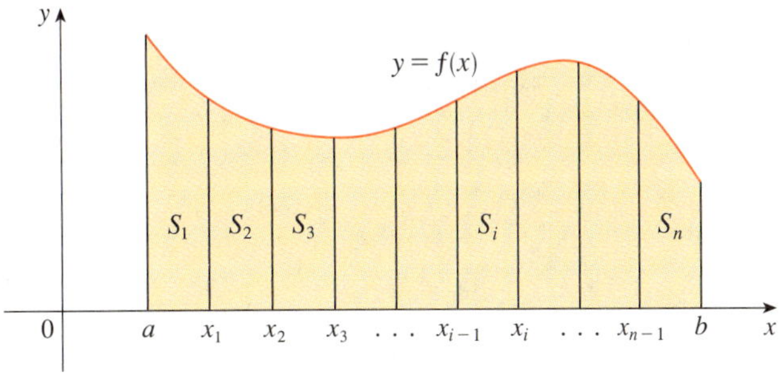

FIGURE 10

Let's approximate the ith strip S_i by a rectangle with width Δx and height $f(x_i)$, which is the value of f at the right endpoint (see Figure 11). Then the area of the ith rectangle

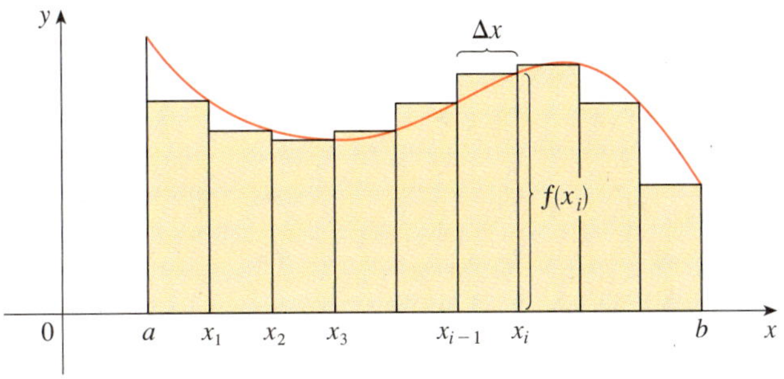

FIGURE 11

is $f(x_i)\,\Delta x$. What we think of intuitively as the area of S is approximated by the sum of the areas of these rectangles, which is

$$R_n = f(x_1)\,\Delta x + f(x_2)\,\Delta x + \cdots + f(x_n)\,\Delta x$$

Figure 12 shows this approximation for $n = 2, 4, 8$, and 12. Notice that this approximation appears to become better and better as the number of strips increases, that is, as $n \to \infty$. Therefore, we define the area A of the region S in the following way.

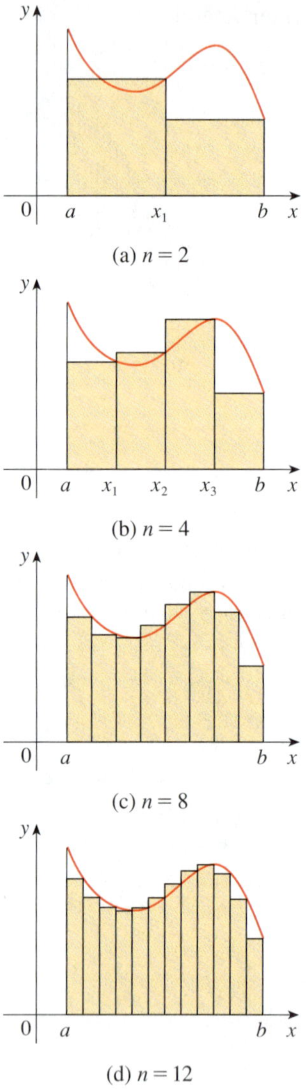

FIGURE 12

2 Definition The **area** A of the region S that lies under the graph of the continuous function f is the limit of the sum of the areas of approximating rectangles:

$$A = \lim_{n\to\infty} R_n = \lim_{n\to\infty} [f(x_1)\,\Delta x + f(x_2)\,\Delta x + \cdots + f(x_n)\,\Delta x]$$

It can be proved that the limit in Definition 2 always exists, since we are assuming that f is continuous. It can also be shown that we get the same value if we use left endpoints:

3
$$A = \lim_{n\to\infty} L_n = \lim_{n\to\infty} [f(x_0)\,\Delta x + f(x_1)\,\Delta x + \cdots + f(x_{n-1})\,\Delta x]$$

In fact, instead of using left endpoints or right endpoints, we could take the height of the ith rectangle to be the value of f at *any* number x_i^* in the ith subinterval $[x_{i-1}, x_i]$. We call the numbers $x_1^*, x_2^*, \ldots, x_n^*$ the **sample points**. Figure 13 shows approximating rectangles when the sample points are not chosen to be endpoints. So a more general expression for the area of S is

4
$$A = \lim_{n\to\infty} [f(x_1^*)\,\Delta x + f(x_2^*)\,\Delta x + \cdots + f(x_n^*)\,\Delta x]$$

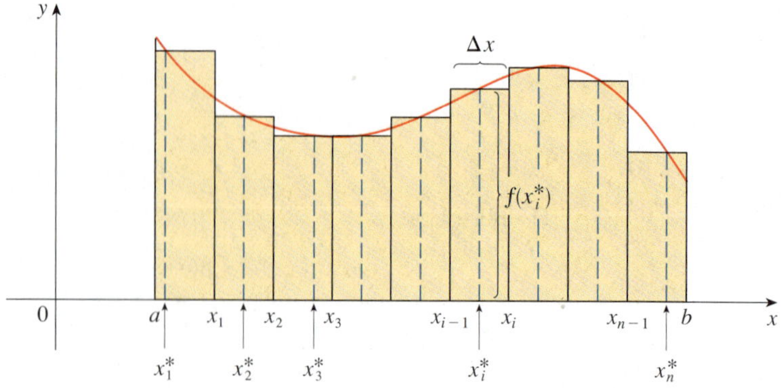

FIGURE 13

We often use **sigma notation** to write sums with many terms more compactly. For instance,

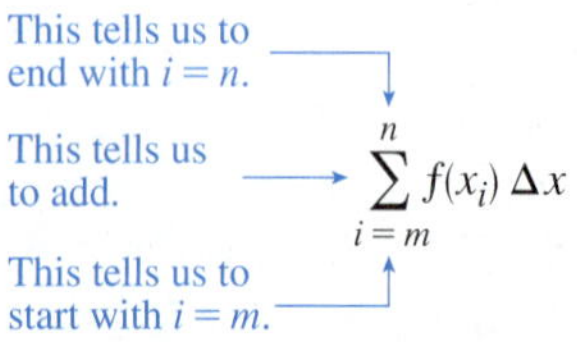

$$\sum_{i=1}^{n} f(x_i)\,\Delta x = f(x_1)\,\Delta x + f(x_2)\,\Delta x + \cdots + f(x_n)\,\Delta x$$

So the expressions for area in Equations 2, 3, and 4 can be written as follows:

|||| If you need practice with sigma notation, look at the examples and try some of the exercises in Appendix E.

$$A = \lim_{n \to \infty} \sum_{i=1}^{n} f(x_i)\,\Delta x$$

$$A = \lim_{n \to \infty} \sum_{i=1}^{n} f(x_{i-1})\,\Delta x$$

$$A = \lim_{n \to \infty} \sum_{i=1}^{n} f(x_i^*)\,\Delta x$$

We could also rewrite Formula 1 in the following way:

$$\sum_{i=1}^{n} i^2 = \frac{n(n+1)(2n+1)}{6}$$

EXAMPLE 3 Let A be the area of the region that lies under the graph of $f(x) = e^{-x}$ between $x = 0$ and $x = 2$.

(a) Using right endpoints, find an expression for A as a limit. Do not evaluate the limit.

(b) Estimate the area by taking the sample points to be midpoints and using four subintervals and then ten subintervals.

SOLUTION

(a) Since $a = 0$ and $b = 2$, the width of a subinterval is

$$\Delta x = \frac{2-0}{n} = \frac{2}{n}$$

So $x_1 = 2/n$, $x_2 = 4/n$, $x_3 = 6/n$, $x_i = 2i/n$, and $x_n = 2n/n$. The sum of the areas of the approximating rectangles is

$$\begin{aligned} R_n &= f(x_1)\,\Delta x + f(x_2)\,\Delta x + \cdots + f(x_n)\,\Delta x \\ &= e^{-x_1}\,\Delta x + e^{-x_2}\,\Delta x + \cdots + e^{-x_n}\,\Delta x \\ &= e^{-2/n}\left(\frac{2}{n}\right) + e^{-4/n}\left(\frac{2}{n}\right) + \cdots + e^{-2n/n}\left(\frac{2}{n}\right) \end{aligned}$$

According to Definition 2, the area is

$$A = \lim_{n \to \infty} R_n = \lim_{n \to \infty} \frac{2}{n}\left(e^{-2/n} + e^{-4/n} + e^{-6/n} + \cdots + e^{-2n/n}\right)$$

Using sigma notation we could write

$$A = \lim_{n \to \infty} \frac{2}{n} \sum_{i=1}^{n} e^{-2i/n}$$

It is difficult to evaluate this limit directly by hand, but with the aid of a computer algebra system it isn't hard (see Exercise 24). In Section 5.3 we will be able to find A more easily using a different method.

(b) With $n = 4$ the subintervals of equal width $\Delta x = 0.5$ are $[0, 0.5]$, $[0.5, 1]$, $[1, 1.5]$, and $[1.5, 2]$. The midpoints of these subintervals are $x_1^* = 0.25$, $x_2^* = 0.75$, $x_3^* = 1.25$,

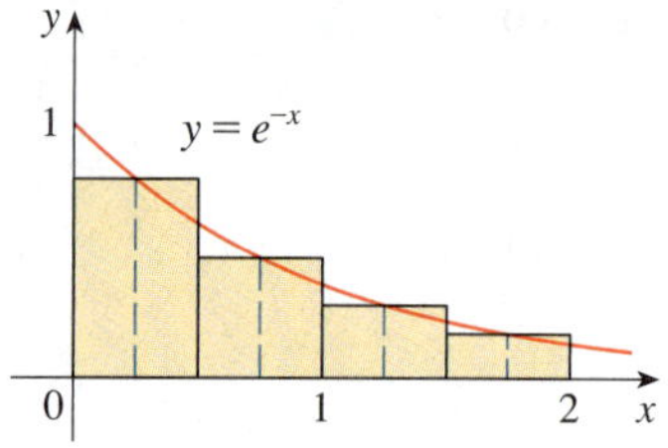

FIGURE 14

and $x_4^* = 1.75$, and the sum of the areas of the four approximating rectangles (see Figure 14) is

$$
\begin{aligned}
M_4 &= \sum_{i=1}^{4} f(x_i^*)\,\Delta x \\
&= f(0.25)\,\Delta x + f(0.75)\,\Delta x + f(1.25)\,\Delta x + f(1.75)\,\Delta x \\
&= e^{-0.25}(0.5) + e^{-0.75}(0.5) + e^{-1.25}(0.5) + e^{-1.75}(0.5) \\
&= \tfrac{1}{2}(e^{-0.25} + e^{-0.75} + e^{-1.25} + e^{-1.75}) \approx 0.8557
\end{aligned}
$$

So an estimate for the area is

$$A \approx 0.8557$$

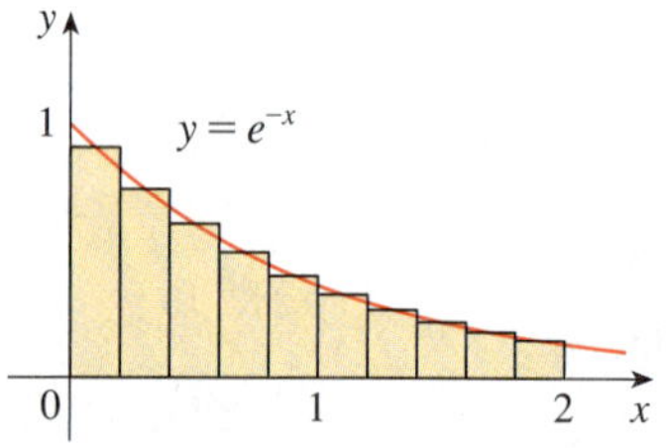

FIGURE 15

With $n = 10$ the subintervals are $[0, 0.2], [0.2, 0.4], \ldots, [1.8, 2]$ and the midpoints are $x_1^* = 0.1, x_2^* = 0.3, x_3^* = 0.5, \ldots, x_{10}^* = 1.9$. Thus

$$
\begin{aligned}
A \approx M_{10} &= f(0.1)\,\Delta x + f(0.3)\,\Delta x + f(0.5)\,\Delta x + \cdots + f(1.9)\,\Delta x \\
&= 0.2(e^{-0.1} + e^{-0.3} + e^{-0.5} + \cdots + e^{-1.9}) \approx 0.8632
\end{aligned}
$$

From Figure 15 it appears that this estimate is better than the estimate with $n = 4$.

The Distance Problem

Now let's consider the *distance problem:* Find the distance traveled by an object during a certain time period if the velocity of the object is known at all times. (In a sense this is the inverse problem of the velocity problem that we discussed in Section 2.1.) If the velocity remains constant, then the distance problem is easy to solve by means of the formula

$$\text{distance} = \text{velocity} \times \text{time}$$

But if the velocity varies, it's not so easy to find the distance traveled. We investigate the problem in the following example.

EXAMPLE 4 Suppose the odometer on our car is broken and we want to estimate the distance driven over a 30-second time interval. We take speedometer readings every five seconds and record them in the following table:

Time (s)	0	5	10	15	20	25	30
Velocity (mi/h)	17	21	24	29	32	31	28

In order to have the time and the velocity in consistent units, let's convert the velocity readings to feet per second (1 mi/h = 5280/3600 ft/s):

Time (s)	0	5	10	15	20	25	30
Velocity (ft/s)	25	31	35	43	47	46	41

During the first five seconds the velocity doesn't change very much, so we can estimate the distance traveled during that time by assuming that the velocity is constant. If we take the velocity during that time interval to be the initial velocity (25 ft/s), then we obtain the approximate distance traveled during the first five seconds:

$$25 \text{ ft/s} \times 5 \text{ s} = 125 \text{ ft}$$

Similarly, during the second time interval the velocity is approximately constant and we take it to be the velocity when $t = 5$ s. So our estimate for the distance traveled from $t = 5$ s to $t = 10$ s is

$$31 \text{ ft/s} \times 5 \text{ s} = 155 \text{ ft}$$

If we add similar estimates for the other time intervals, we obtain an estimate for the total distance traveled:

$$25 \times 5 + 31 \times 5 + 35 \times 5 + 43 \times 5 + 47 \times 5 + 46 \times 5 = 1135 \text{ ft}$$

We could just as well have used the velocity at the *end* of each time period instead of the velocity at the beginning as our assumed constant velocity. Then our estimate becomes

$$31 \times 5 + 35 \times 5 + 43 \times 5 + 47 \times 5 + 46 \times 5 + 41 \times 5 = 1215 \text{ ft}$$

If we had wanted a more accurate estimate, we could have taken velocity readings every two seconds, or even every second.

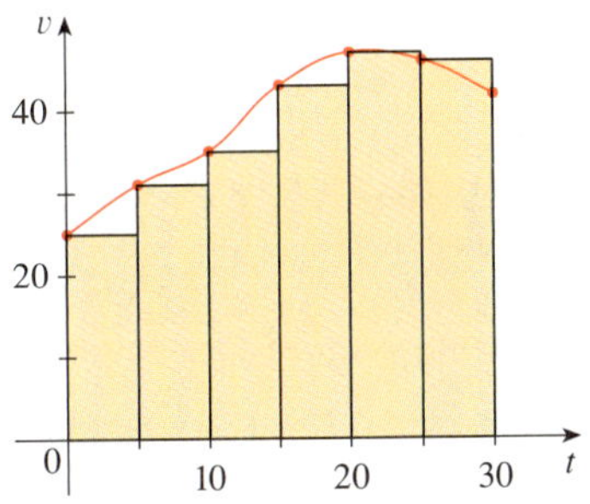

FIGURE 16

Perhaps the calculations in Example 4 remind you of the sums we used earlier to estimate areas. The similarity is explained when we sketch a graph of the velocity function of the car in Figure 16 and draw rectangles whose heights are the initial velocities for each time interval. The area of the first rectangle is $25 \times 5 = 125$, which is also our estimate for the distance traveled in the first five seconds. In fact, the area of each rectangle can be interpreted as a distance because the height represents velocity and the width represents time. The sum of the areas of the rectangles in Figure 16 is $L_6 = 1135$, which is our initial estimate for the total distance traveled.

In general, suppose an object moves with velocity $v = f(t)$, where $a \leqslant t \leqslant b$ and $f(t) \geqslant 0$ (so the object always moves in the positive direction). We take velocity readings at times $t_0 (= a), t_1, t_2, \ldots, t_n (= b)$ so that the velocity is approximately constant on each subinterval. If these times are equally spaced, then the time between consecutive readings is $\Delta t = (b - a)/n$. During the first time interval the velocity is approximately $f(t_0)$ and so the distance traveled is approximately $f(t_0)\,\Delta t$. Similarly, the distance traveled during the second time interval is about $f(t_1)\,\Delta t$ and the total distance traveled during the time interval $[a, b]$ is approximately

$$f(t_0)\,\Delta t + f(t_1)\,\Delta t + \cdots + f(t_{n-1})\,\Delta t = \sum_{i=1}^{n} f(t_{i-1})\,\Delta t$$

If we use the velocity at right endpoints instead of left endpoints, our estimate for the total distance becomes

$$f(t_1)\,\Delta t + f(t_2)\,\Delta t + \cdots + f(t_n)\,\Delta t = \sum_{i=1}^{n} f(t_i)\,\Delta t$$

The more frequently we measure the velocity, the more accurate our estimates become, so it seems plausible that the *exact* distance d traveled is the *limit* of such expressions:

5 $$d = \lim_{n \to \infty} \sum_{i=1}^{n} f(t_{i-1})\,\Delta t = \lim_{n \to \infty} \sum_{i=1}^{n} f(t_i)\,\Delta t$$

We will see in Section 5.4 that this is indeed true.

Because Equation 5 has the same form as our expressions for area in Equations 2 and 3, it follows that the distance traveled is equal to the area under the graph of the velocity function. In Chapters 6 and 8 we will see that other quantities of interest in the natural and social sciences—such as the work done by a variable force or the cardiac output of the heart—can also be interpreted as the area under a curve. So when we compute areas in this chapter, bear in mind that they can be interpreted in a variety of practical ways.

5.1 Exercises

1. (a) By reading values from the given graph of f, use five rectangles to find a lower estimate and an upper estimate for the area under the given graph of f from $x = 0$ to $x = 10$. In each case sketch the rectangles that you use.
(b) Find new estimates using 10 rectangles in each case.

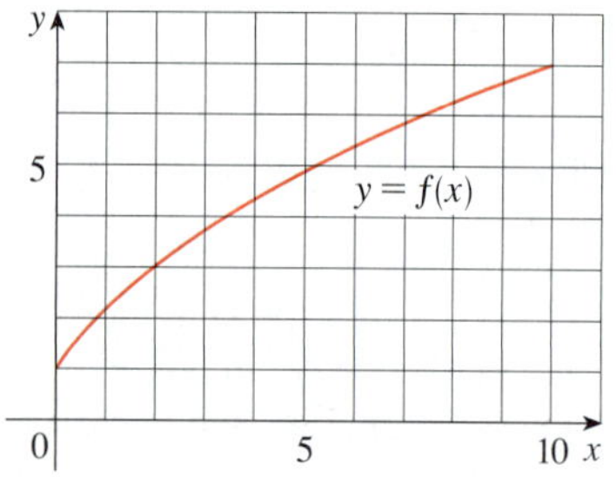

2. (a) Use six rectangles to find estimates of each type for the area under the given graph of f from $x = 0$ to $x = 12$.
 (i) L_6 (sample points are left endpoints)
 (ii) R_6 (sample points are right endpoints)
 (iii) M_6 (sample points are midpoints)
(b) Is L_6 an underestimate or overestimate of the true area?
(c) Is R_6 an underestimate or overestimate of the true area?
(d) Which of the numbers L_6, R_6, or M_6 gives the best estimate? Explain.

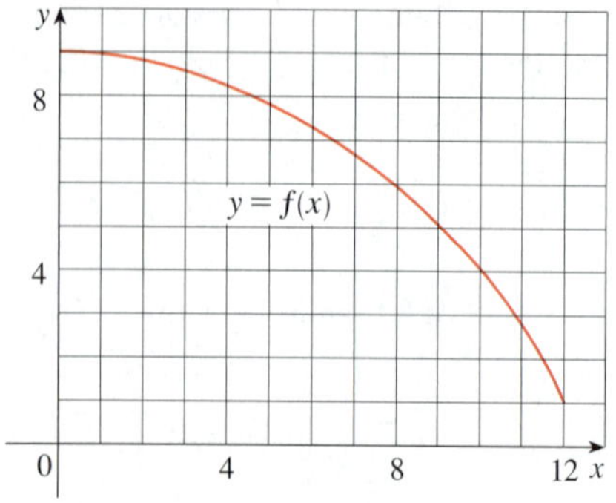

3. (a) Estimate the area under the graph of $f(x) = 1/x$ from $x = 1$ to $x = 5$ using four approximating rectangles and right endpoints. Sketch the graph and the rectangles. Is your estimate an underestimate or an overestimate?
(b) Repeat part (a) using left endpoints.

4. (a) Estimate the area under the graph of $f(x) = 25 - x^2$ from $x = 0$ to $x = 5$ using five approximating rectangles and right endpoints. Sketch the graph and the rectangles. Is your estimate an underestimate or an overestimate?
(b) Repeat part (a) using left endpoints.

5. (a) Estimate the area under the graph of $f(x) = 1 + x^2$ from $x = -1$ to $x = 2$ using three rectangles and right endpoints. Then improve your estimate by using six rectangles. Sketch the curve and the approximating rectangles.
(b) Repeat part (a) using left endpoints.
(c) Repeat part (a) using midpoints.
(d) From your sketches in parts (a)–(c), which appears to be the best estimate?

6. (a) Graph the function $f(x) = e^{-x^2}$, $-2 \leqslant x \leqslant 2$.
(b) Estimate the area under the graph of f using four approximating rectangles and taking the sample points to be
 (i) right endpoints (ii) midpoints
In each case sketch the curve and the rectangles.
(c) Improve your estimates in part (b) by using eight rectangles.

7–8 With a programmable calculator (or a computer), it is possible to evaluate the expressions for the sums of areas of approximating rectangles, even for large values of n, using looping. (On a TI use the Is> command or a For-EndFor loop, on a Casio use Isz, on an HP or in BASIC use a FOR-NEXT loop.) Compute the sum of the areas of approximating rectangles using equal subintervals and right endpoints for $n = 10$, 30, and 50. Then guess the value of the exact area.

7. The region under $y = \sin x$ from 0 to π

8. The region under $y = 1/x^2$ from 1 to 2

CAS **9.** Some computer algebra systems have commands that will draw approximating rectangles and evaluate the sums of their areas, at least if x_i^* is a left or right endpoint. (For instance, in Maple use `leftbox`, `rightbox`, `leftsum`, and `rightsum`.)
(a) If $f(x) = \sqrt{x}$, $1 \leqslant x \leqslant 4$, find the left and right sums for $n = 10$, 30, and 50.
(b) Illustrate by graphing the rectangles in part (a).
(c) Show that the exact area under f lies between 4.6 and 4.7.

CAS **10.** (a) If $f(x) = \sin(\sin x)$, $0 \leqslant x \leqslant \pi/2$, use the commands discussed in Exercise 9 to find the left and right sums for $n = 10$, 30, and 50.

(b) Illustrate by graphing the rectangles in part (a).
(c) Show that the exact area under f lies between 0.87 and 0.91.

11. The speed of a runner increased steadily during the first three seconds of a race. Her speed at half-second intervals is given in the table. Find lower and upper estimates for the distance that she traveled during these three seconds.

t (s)	0	0.5	1.0	1.5	2.0	2.5	3.0
v (ft/s)	0	6.2	10.8	14.9	18.1	19.4	20.2

12. Speedometer readings for a motorcycle at 12-second intervals are given in the table.
(a) Estimate the distance traveled by the motorcycle during this time period using the velocities at the beginning of the time intervals.
(b) Give another estimate using the velocities at the end of the time periods.
(c) Are your estimates in parts (a) and (b) upper and lower estimates? Explain.

t (s)	0	12	24	36	48	60
v (ft/s)	30	28	25	22	24	27

13. Oil leaked from a tank at a rate of $r(t)$ liters per hour. The rate decreased as time passed and values of the rate at 2-hour time intervals are shown in the table. Find lower and upper estimates for the total amount of oil that leaked out.

t (h)	0	2	4	6	8	10
$r(t)$ (L/h)	8.7	7.6	6.8	6.2	5.7	5.3

14. When we estimate distances from velocity data, it is sometimes necessary to use times $t_0, t_1, t_2, t_3, \ldots$ that are not equally spaced. We can still estimate distances using the time periods $\Delta t_i = t_i - t_{i-1}$. For example, on May 7, 1992, the space shuttle *Endeavour* was launched on mission STS-49, the purpose of which was to install a new perigee kick motor in an Intelsat communications satellite. The table, provided by NASA, gives the velocity data for the shuttle between liftoff and the jettisoning of the solid rocket boosters.

Event	Time (s)	Velocity (ft/s)
Launch	0	0
Begin roll maneuver	10	185
End roll maneuver	15	319
Throttle to 89%	20	447
Throttle to 67%	32	742
Throttle to 104%	59	1325
Maximum dynamic pressure	62	1445
Solid rocket booster separation	125	4151

Use these data to estimate the height above Earth's surface of the space shuttle *Endeavour*, 62 seconds after liftoff.

15. The velocity graph of a braking car is shown. Use it to estimate the distance traveled by the car while the brakes are applied.

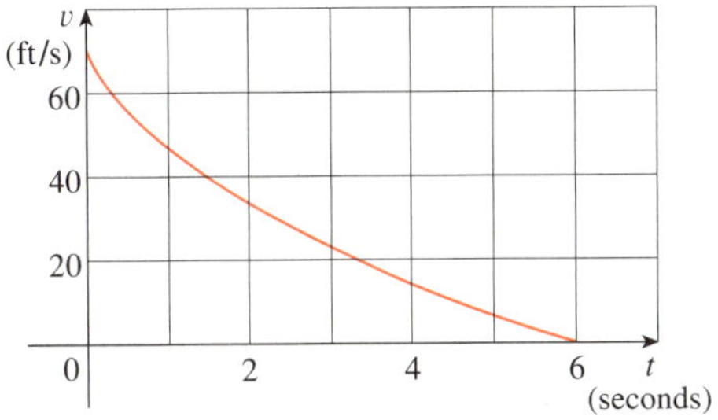

16. The velocity graph of a car accelerating from rest to a speed of 120 km/h over a period of 30 seconds is shown. Estimate the distance traveled during this period.

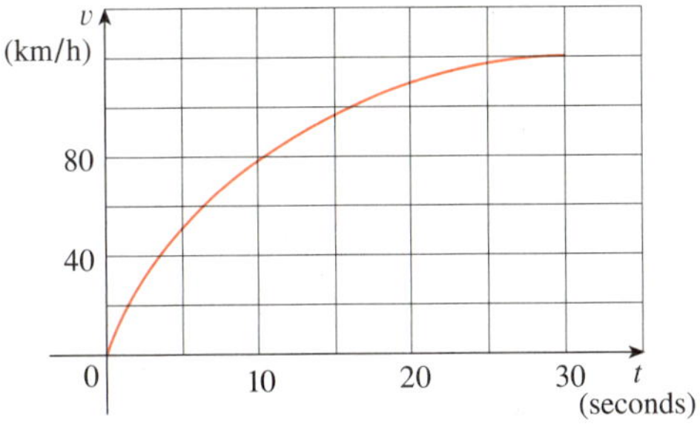

17–19 ||| Use Definition 2 to find an expression for the area under the graph of f as a limit. Do not evaluate the limit.

17. $f(x) = \sqrt[4]{x}, \quad 1 \leqslant x \leqslant 16$

18. $f(x) = \dfrac{\ln x}{x}, \quad 3 \leqslant x \leqslant 10$

19. $f(x) = x \cos x, \quad 0 \leqslant x \leqslant \pi/2$

20–21 ||| Determine a region whose area is equal to the given limit. Do not evaluate the limit.

20. $\displaystyle\lim_{n \to \infty} \sum_{i=1}^{n} \frac{2}{n}\left(5 + \frac{2i}{n}\right)^{10}$

21. $\displaystyle\lim_{n \to \infty} \sum_{i=1}^{n} \frac{\pi}{4n} \tan \frac{i\pi}{4n}$

22. (a) Use Definition 2 to find an expression for the area under the curve $y = x^3$ from 0 to 1 as a limit.
(b) The following formula for the sum of the cubes of the first n integers is proved in Appendix E. Use it to evaluate the limit in part (a).

$$1^3 + 2^3 + 3^3 + \cdots + n^3 = \left[\frac{n(n+1)}{2}\right]^2$$

CAS **23.** (a) Express the area under the curve $y = x^5$ from 0 to 2 as a limit.
(b) Use a computer algebra system to find the sum in your expression from part (a).
(c) Evaluate the limit in part (a).

CAS **24.** Find the exact area of the region under the graph of $y = e^{-x}$ from 0 to 2 by using a computer algebra system to evaluate the sum and then the limit in Example 3(a). Compare your answer with the estimate obtained in Example 3(b).

CAS **25.** Find the exact area under the cosine curve $y = \cos x$ from $x = 0$ to $x = b$, where $0 \leqslant b \leqslant \pi/2$. (Use a computer algebra system both to evaluate the sum and compute the limit.) In particular, what is the area if $b = \pi/2$?

26. (a) Let A_n be the area of a polygon with n equal sides inscribed in a circle with radius r. By dividing the polygon into n congruent triangles with central angle $2\pi/n$, show that $A_n = \frac{1}{2}nr^2 \sin(2\pi/n)$.
(b) Show that $\lim_{n\to\infty} A_n = \pi r^2$. [*Hint:* Use Equation 3.4.2.]

5.2 The Definite Integral

We saw in Section 5.1 that a limit of the form

1 $$\lim_{n\to\infty} \sum_{i=1}^{n} f(x_i^*)\,\Delta x = \lim_{n\to\infty} [f(x_1^*)\,\Delta x + f(x_2^*)\,\Delta x + \cdots + f(x_n^*)\,\Delta x]$$

arises when we compute an area. We also saw that it arises when we try to find the distance traveled by an object. It turns out that this same type of limit occurs in a wide variety of situations even when f is not necessarily a positive function. In Chapters 6 and 8 we will see that limits of the form (1) also arise in finding lengths of curves, volumes of solids, centers of mass, force due to water pressure, and work, as well as other quantities. We therefore give this type of limit a special name and notation.

2 Definition of a Definite Integral If f is a continuous function defined for $a \leqslant x \leqslant b$, we divide the interval $[a, b]$ into n subintervals of equal width $\Delta x = (b - a)/n$. We let $x_0\,(= a), x_1, x_2, \ldots, x_n\,(= b)$ be the endpoints of these subintervals and we let $x_1^*, x_2^*, \ldots, x_n^*$ be any **sample points** in these subintervals, so x_i^* lies in the ith subinterval $[x_{i-1}, x_i]$. Then the **definite integral of f from a to b** is

$$\int_a^b f(x)\,dx = \lim_{n\to\infty} \sum_{i=1}^{n} f(x_i^*)\,\Delta x$$

Because we have assumed that f is continuous, it can be proved that the limit in Definition 2 always exists and gives the same value no matter how we choose the sample points x_i^*. (See Note 4 for a precise definition of this type of limit.) If we take the sample points to be right endpoints, then $x_i^* = x_i$ and the definition of an integral becomes

3 $$\int_a^b f(x)\,dx = \lim_{n\to\infty} \sum_{i=1}^{n} f(x_i)\,\Delta x$$

If we choose the sample points to be left endpoints, then $x_i^* = x_{i-1}$ and the definition becomes

$$\int_a^b f(x)\,dx = \lim_{n\to\infty} \sum_{i=1}^{n} f(x_{i-1})\,\Delta x$$

Alternatively, we could choose x_i^* to be the midpoint of the subinterval or any other number between x_{i-1} and x_i.

Although most of the functions that we encounter are continuous, the limit in Definition 2 also exists if f has a finite number of removable or jump discontinuities (but not infinite discontinuities). (See Section 2.5.) So we can also define the definite integral for such functions.

NOTE 1 ▫ The symbol $\int$ was introduced by Leibniz and is called an **integral sign**. It is an elongated S and was chosen because an integral is a limit of sums. In the notation $\int_a^b f(x)\,dx$, $f(x)$ is called the **integrand** and a and b are called the **limits of integration**; a is the **lower limit** and b is the **upper limit**. The symbol dx has no official meaning by itself; $\int_a^b f(x)\,dx$ is all one symbol. The procedure of calculating an integral is called **integration**.

NOTE 2 ▫ The definite integral $\int_a^b f(x)\,dx$ is a number; it does not depend on x. In fact, we could use any letter in place of x without changing the value of the integral:

$$\int_a^b f(x)\,dx = \int_a^b f(t)\,dt = \int_a^b f(r)\,dr$$

NOTE 3 ▫ The sum

$$\sum_{i=1}^{n} f(x_i^*)\,\Delta x$$

that occurs in Definition 2 is called a **Riemann sum** after the German mathematician Bernhard Riemann (1826–1866). We know that if f happens to be positive, then the Riemann sum can be interpreted as a sum of areas of approximating rectangles (see Figure 1). By comparing Definition 2 with the definition of area in Section 5.1, we see that the definite integral $\int_a^b f(x)\,dx$ can be interpreted as the area under the curve $y = f(x)$ from a to b. (See Figure 2.)

|||| Bernhard Riemann received his Ph.D. under the direction of the legendary Gauss at the University of Göttingen and remained there to teach. Gauss, who was not in the habit of praising other mathematicians, spoke of Riemann's "creative, active, truly mathematical mind and gloriously fertile originality." The definition (2) of an integral that we use is due to Riemann. He also made major contributions to the theory of functions of a complex variable, mathematical physics, number theory, and the foundations of geometry. Riemann's broad concept of space and geometry turned out to be the right setting, 50 years later, for Einstein's general relativity theory. Riemann's health was poor throughout his life, and he died of tuberculosis at the age of 39.

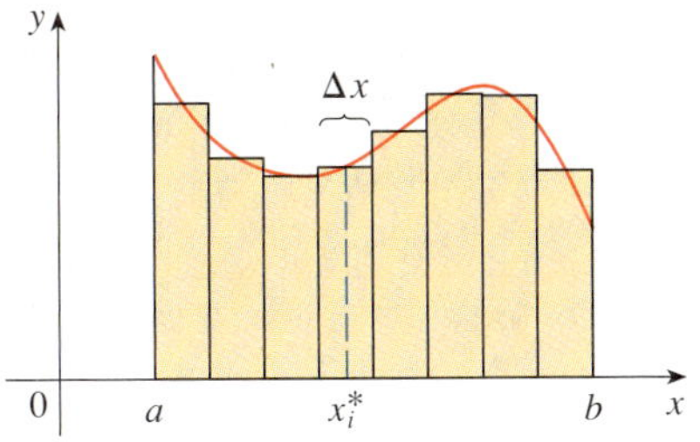

FIGURE 1
If $f(x) \geq 0$, the Riemann sum $\sum f(x_i^*)\,\Delta x$ is the sum of areas of rectangles.

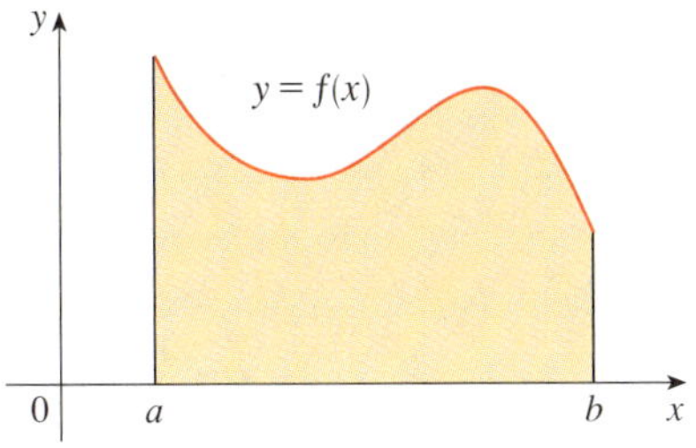

FIGURE 2
If $f(x) \geq 0$, the integral $\int_a^b f(x)\,dx$ is the area under the curve $y = f(x)$ from a to b.

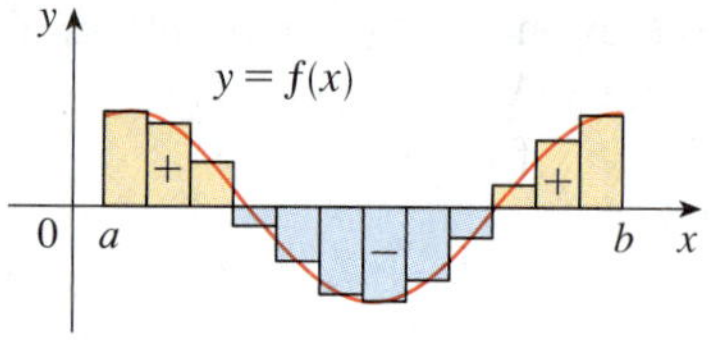

FIGURE 3

$\Sigma f(x_i^*)\,\Delta x$ is an approximation to the net area

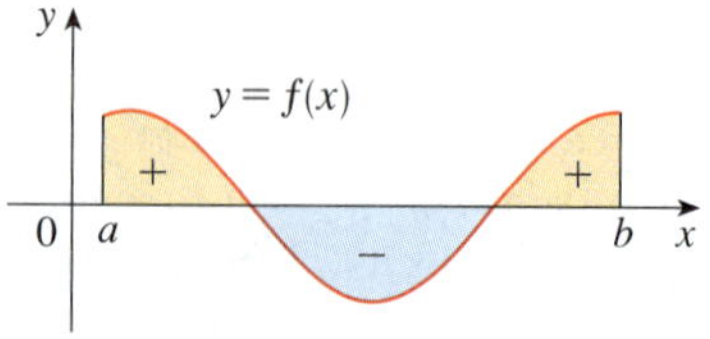

FIGURE 4

$\int_a^b f(x)\,dx$ is the net area

If f takes on both positive and negative values, as in Figure 3, then the Riemann sum is the sum of the areas of the rectangles that lie above the x-axis and the *negatives* of the areas of the rectangles that lie below the x-axis (the areas of the gold rectangles *minus* the areas of the blue rectangles). When we take the limit of such Riemann sums, we get the situation illustrated in Figure 4. A definite integral can be interpreted as a **net area**, that is, a difference of areas:

$$\int_a^b f(x)\,dx = A_1 - A_2$$

where A_1 is the area of the region above the x-axis and below the graph of f, and A_2 is the area of the region below the x-axis and above the graph of f.

NOTE 4 ◦ In the spirit of the precise definition of the limit of a function in Section 2.4, we can write the precise meaning of the limit that defines the integral in Definition 2 as follows:

For every number $\varepsilon > 0$ there is an integer N such that

$$\left| \int_a^b f(x)\,dx - \sum_{i=1}^{n} f(x_i^*)\,\Delta x \right| < \varepsilon$$

for every integer $n > N$ and for every choice of x_i^* in $[x_{i-1}, x_i]$.

This means that a definite integral can be approximated to within any desired degree of accuracy by a Riemann sum.

NOTE 5 ◦ Although we have defined $\int_a^b f(x)\,dx$ by dividing $[a, b]$ into subintervals of equal width, there are situations in which it is advantageous to work with subintervals of unequal width. For instance, in Exercise 14 in Section 5.1 NASA provided velocity data at times that were not equally spaced, but we were still able to estimate the distance traveled. And there are methods for numerical integration that take advantage of unequal subintervals.

If the subinterval widths are $\Delta x_1, \Delta x_2, \ldots, \Delta x_n$, we have to ensure that all these widths approach 0 in the limiting process. This happens if the largest width, $\max \Delta x_i$, approaches 0. So in this case the definition of a definite integral becomes

$$\int_a^b f(x)\,dx = \lim_{\max \Delta x_i \to 0} \sum_{i=1}^{n} f(x_i^*)\,\Delta x_i$$

EXAMPLE 1 Express

$$\lim_{n\to\infty} \sum_{i=1}^{n} (x_i^3 + x_i \sin x_i)\,\Delta x$$

as an integral on the interval $[0, \pi]$.

SOLUTION Comparing the given limit with the limit in Definition 2, we see that they will be identical if we choose

$$f(x) = x^3 + x \sin x \qquad \text{and} \qquad x_i^* = x_i$$

(So the sample points are right endpoints and the given limit is of the form of Equation 3.) We are given that $a = 0$ and $b = \pi$. Therefore, by Definition 2 or Equation 3, we have

$$\lim_{n\to\infty} \sum_{i=1}^{n} (x_i^3 + x_i \sin x_i)\,\Delta x = \int_0^{\pi} (x^3 + x \sin x)\,dx$$

Later, when we apply the definite integral to physical situations, it will be important to recognize limits of sums as integrals, as we did in Example 1. When Leibniz chose the notation for an integral, he chose the ingredients as reminders of the limiting process. In general, when we write

$$\lim_{n \to \infty} \sum_{i=1}^{n} f(x_i^*)\,\Delta x = \int_a^b f(x)\,dx$$

we replace $\lim \Sigma$ by $\int$, x_i^* by x, and Δx by dx.

|||| Evaluating Integrals

When we use the definition to evaluate a definite integral, we need to know how to work with sums. The following three equations give formulas for sums of powers of positive integers. Equation 4 may be familiar to you from a course in algebra. Equations 5 and 6 were discussed in Section 5.1 and are proved in Appendix E.

4 $$\sum_{i=1}^{n} i = \frac{n(n+1)}{2}$$

5 $$\sum_{i=1}^{n} i^2 = \frac{n(n+1)(2n+1)}{6}$$

6 $$\sum_{i=1}^{n} i^3 = \left[\frac{n(n+1)}{2}\right]^2$$

The remaining formulas are simple rules for working with sigma notation:

7 $$\sum_{i=1}^{n} c = nc$$

8 $$\sum_{i=1}^{n} ca_i = c\sum_{i=1}^{n} a_i$$

9 $$\sum_{i=1}^{n} (a_i + b_i) = \sum_{i=1}^{n} a_i + \sum_{i=1}^{n} b_i$$

10 $$\sum_{i=1}^{n} (a_i - b_i) = \sum_{i=1}^{n} a_i - \sum_{i=1}^{n} b_i$$

|||| Formulas 7–10 are proved by writing out each side in expanded form. The left side of Equation 8 is

$$ca_1 + ca_2 + \cdots + ca_n$$

The right side is

$$c(a_1 + a_2 + \cdots + a_n)$$

These are equal by the distributive property. The other formulas are discussed in Appendix E.

EXAMPLE 2

(a) Evaluate the Riemann sum for $f(x) = x^3 - 6x$ taking the sample points to be right endpoints and $a = 0$, $b = 3$, and $n = 6$.

(b) Evaluate $\int_0^3 (x^3 - 6x)\,dx$.

Try more problems like this one.
Resources / Module 6
/ What Is Area?
/ Problems and Tests

SOLUTION

(a) With $n = 6$ the interval width is

$$\Delta x = \frac{b-a}{n} = \frac{3-0}{6} = \frac{1}{2}$$

and the right endpoints are $x_1 = 0.5$, $x_2 = 1.0$, $x_3 = 1.5$, $x_4 = 2.0$, $x_5 = 2.5$, and $x_6 = 3.0$. So the Riemann sum is

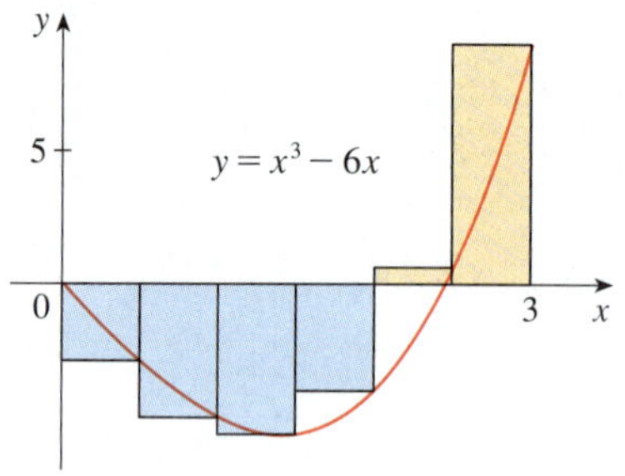

FIGURE 5

$$\begin{aligned} R_6 &= \sum_{i=1}^{6} f(x_i)\,\Delta x \\ &= f(0.5)\,\Delta x + f(1.0)\,\Delta x + f(1.5)\,\Delta x + f(2.0)\,\Delta x + f(2.5)\,\Delta x + f(3.0)\,\Delta x \\ &= \tfrac{1}{2}(-2.875 - 5 - 5.625 - 4 + 0.625 + 9) \\ &= -3.9375 \end{aligned}$$

Notice that f is not a positive function and so the Riemann sum does not represent a sum of areas of rectangles. But it does represent the sum of the areas of the gold rectangles (above the x-axis) minus the sum of the areas of the blue rectangles (below the x-axis) in Figure 5.

(b) With n subintervals we have

$$\Delta x = \frac{b - a}{n} = \frac{3}{n}$$

Thus $x_0 = 0$, $x_1 = 3/n$, $x_2 = 6/n$, $x_3 = 9/n$, and, in general, $x_i = 3i/n$. Since we are using right endpoints, we can use Equation 3:

|||| In the sum, n is a constant (unlike i), so we can move $3/n$ in front of the Σ sign.

$$\begin{aligned} \int_0^3 (x^3 - 6x)\,dx &= \lim_{n\to\infty} \sum_{i=1}^{n} f(x_i)\,\Delta x = \lim_{n\to\infty} \sum_{i=1}^{n} f\left(\frac{3i}{n}\right)\frac{3}{n} \\ &= \lim_{n\to\infty} \frac{3}{n} \sum_{i=1}^{n} \left[\left(\frac{3i}{n}\right)^3 - 6\left(\frac{3i}{n}\right)\right] \qquad \text{(Equation 8 with } c = 3/n\text{)} \\ &= \lim_{n\to\infty} \frac{3}{n} \sum_{i=1}^{n} \left[\frac{27}{n^3} i^3 - \frac{18}{n} i\right] \\ &= \lim_{n\to\infty} \left[\frac{81}{n^4} \sum_{i=1}^{n} i^3 - \frac{54}{n^2} \sum_{i=1}^{n} i\right] \qquad \text{(Equations 10 and 8)} \\ &= \lim_{n\to\infty} \left\{\frac{81}{n^4}\left[\frac{n(n+1)}{2}\right]^2 - \frac{54}{n^2}\frac{n(n+1)}{2}\right\} \\ &= \lim_{n\to\infty} \left[\frac{81}{4}\left(1 + \frac{1}{n}\right)^2 - 27\left(1 + \frac{1}{n}\right)\right] \\ &= \frac{81}{4} - 27 = -\frac{27}{4} = -6.75 \end{aligned}$$

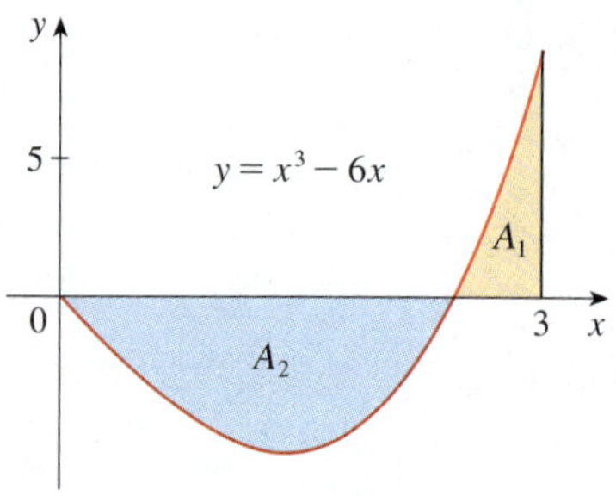

FIGURE 6

$\int_0^3 (x^3 - 6x)\,dx = A_1 - A_2 = -6.75$

This integral can't be interpreted as an area because f takes on both positive and negative values. But it can be interpreted as the difference of areas $A_1 - A_2$, where A_1 and A_2 are shown in Figure 6.

Figure 7 illustrates the calculation by showing the positive and negative terms in the right Riemann sum R_n for $n = 40$. The values in the table show the Riemann sums approaching the exact value of the integral, -6.75, as $n \to \infty$.

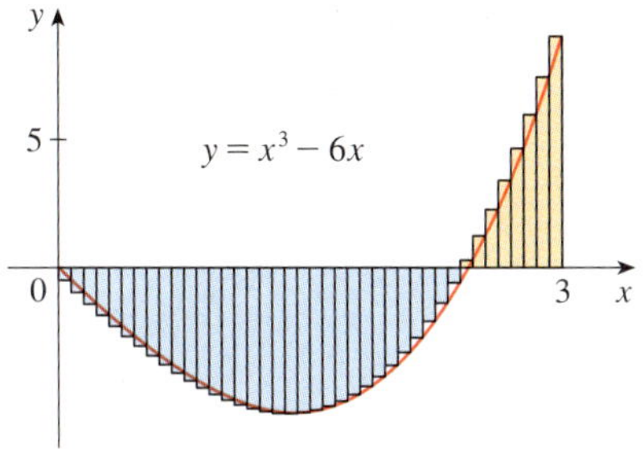

FIGURE 7
$R_{40} \approx -6.3998$

n	R_n
40	−6.3998
100	−6.6130
500	−6.7229
1000	−6.7365
5000	−6.7473

A much simpler method for evaluating the integral in Example 2 will be given in Section 5.3.

|||| Because $f(x) = e^x$ is positive, the integral in Example 3 represents the area shown in Figure 8.

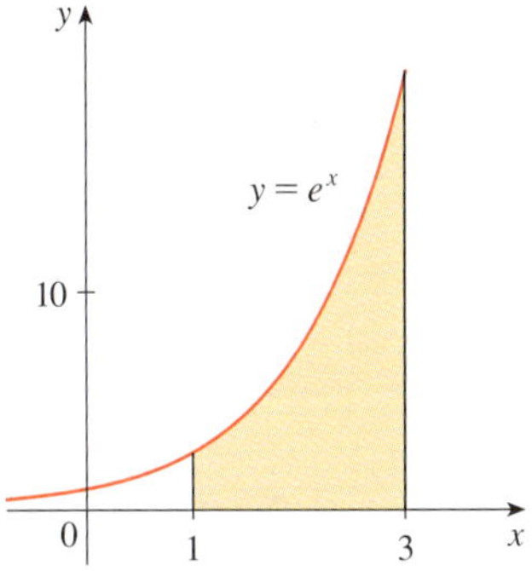

FIGURE 8

EXAMPLE 3

(a) Set up an expression for $\int_1^3 e^x\,dx$ as a limit of sums.

(b) Use a computer algebra system to evaluate the expression.

SOLUTION

(a) Here we have $f(x) = e^x$, $a = 1$, $b = 3$, and

$$\Delta x = \frac{b - a}{n} = \frac{2}{n}$$

So $x_0 = 1$, $x_1 = 1 + 2/n$, $x_2 = 1 + 4/n$, $x_3 = 1 + 6/n$, and

$$x_i = 1 + \frac{2i}{n}$$

From Equation 3, we get

$$\begin{aligned}\int_1^3 e^x\,dx &= \lim_{n\to\infty} \sum_{i=1}^{n} f(x_i)\,\Delta x \\ &= \lim_{n\to\infty} \sum_{i=1}^{n} f\left(1 + \frac{2i}{n}\right)\frac{2}{n} \\ &= \lim_{n\to\infty} \frac{2}{n} \sum_{i=1}^{n} e^{1+2i/n}\end{aligned}$$

(b) If we ask a computer algebra system to evaluate the sum and simplify, we obtain

|||| A computer algebra system is able to find an explicit expression for this sum because it is a geometric series. The limit could be found using l'Hospital's Rule.

$$\sum_{i=1}^{n} e^{1+2i/n} = \frac{e^{(3n+2)/n} - e^{(n+2)/n}}{e^{2/n} - 1}$$

Now we ask the computer algebra system to evaluate the limit:

$$\int_1^3 e^x\,dx = \lim_{n\to\infty} \frac{2}{n} \cdot \frac{e^{(3n+2)/n} - e^{(n+2)/n}}{e^{2/n} - 1} = e^3 - e$$

We will learn a much easier method for the evaluation of integrals in the next section.

EXAMPLE 4 Evaluate the following integrals by interpreting each in terms of areas.

(a) $\int_0^1 \sqrt{1 - x^2}\,dx$ (b) $\int_0^3 (x - 1)\,dx$

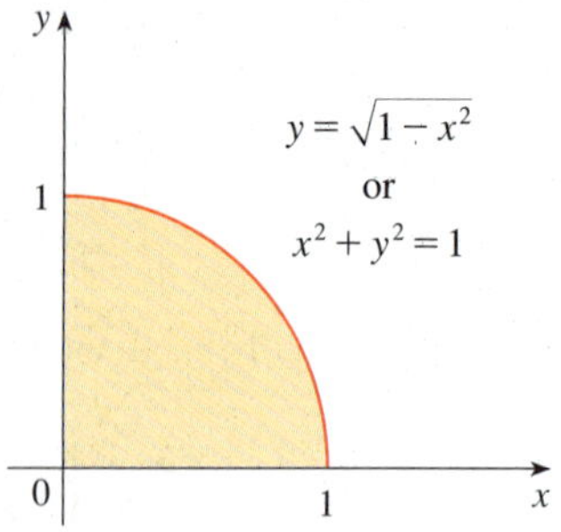

FIGURE 9

SOLUTION

(a) Since $f(x) = \sqrt{1 - x^2} \geqslant 0$, we can interpret this integral as the area under the curve $y = \sqrt{1 - x^2}$ from 0 to 1. But, since $y^2 = 1 - x^2$, we get $x^2 + y^2 = 1$, which shows that the graph of f is the quarter-circle with radius 1 in Figure 9. Therefore

$$\int_0^1 \sqrt{1 - x^2}\,dx = \tfrac{1}{4}\pi(1)^2 = \frac{\pi}{4}$$

(In Section 7.3 we will be able to *prove* that the area of a circle of radius r is πr^2.)

(b) The graph of $y = x - 1$ is the line with slope 1 shown in Figure 10. We compute the integral as the difference of the areas of the two triangles:

$$\int_0^3 (x - 1)\,dx = A_1 - A_2 = \tfrac{1}{2}(2 \cdot 2) - \tfrac{1}{2}(1 \cdot 1) = 1.5$$

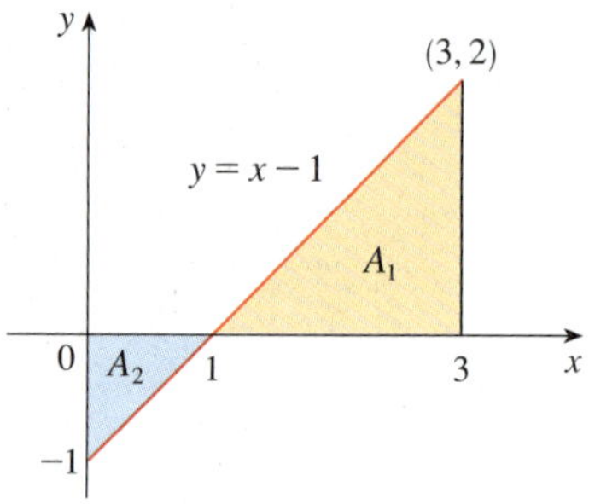

FIGURE 10

The Midpoint Rule

We often choose the sample point x_i^* to be the right endpoint of the ith subinterval because it is convenient for computing the limit. But if the purpose is to find an *approximation* to an integral, it is usually better to choose x_i^* to be the midpoint of the interval, which we denote by $\bar{x}_i$. Any Riemann sum is an approximation to an integral, but if we use midpoints we get the following approximation.

TEC Module 5.1/5.2/7.7 shows how the Midpoint Rule estimates improve as n increases.

Midpoint Rule

$$\int_a^b f(x)\,dx \approx \sum_{i=1}^{n} f(\bar{x}_i)\,\Delta x = \Delta x\,[f(\bar{x}_1) + \cdots + f(\bar{x}_n)]$$

where $$\Delta x = \frac{b - a}{n}$$

and $$\bar{x}_i = \tfrac{1}{2}(x_{i-1} + x_i) = \text{midpoint of } [x_{i-1}, x_i]$$

EXAMPLE 5 Use the Midpoint Rule with $n = 5$ to approximate $\int_1^2 \frac{1}{x}\,dx$.

SOLUTION The endpoints of the five subintervals are 1, 1.2, 1.4, 1.6, 1.8, and 2.0, so the midpoints are 1.1, 1.3, 1.5, 1.7, and 1.9. The width of the subintervals is $\Delta x = (2 - 1)/5 = \frac{1}{5}$, so the Midpoint Rule gives

$$\begin{aligned}\int_1^2 \frac{1}{x}\,dx &\approx \Delta x\,[f(1.1) + f(1.3) + f(1.5) + f(1.7) + f(1.9)]\\ &= \frac{1}{5}\left(\frac{1}{1.1} + \frac{1}{1.3} + \frac{1}{1.5} + \frac{1}{1.7} + \frac{1}{1.9}\right)\\ &\approx 0.691908\end{aligned}$$

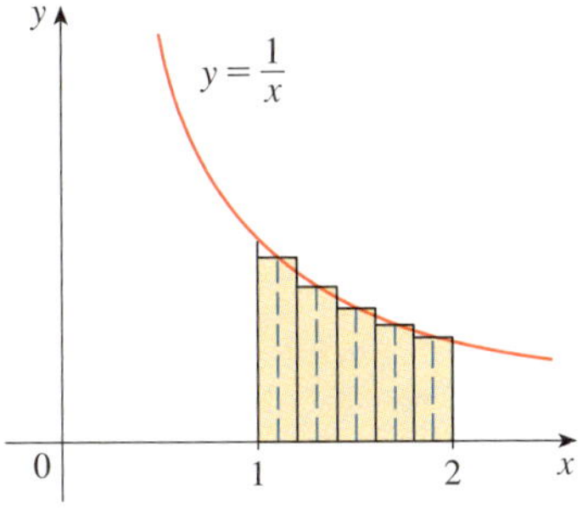

FIGURE 11

Since $f(x) = 1/x > 0$ for $1 \leqslant x \leqslant 2$, the integral represents an area, and the approximation given by the Midpoint Rule is the sum of the areas of the rectangles shown in Figure 11.

At the moment we don't know how accurate the approximation in Example 5 is, but in Section 7.7 we will learn a method for estimating the error involved in using the Midpoint Rule. At that time we will discuss other methods for approximating definite integrals.

If we apply the Midpoint Rule to the integral in Example 2, we get the picture in Figure 12. The approximation $M_{40} \approx -6.7563$ is much closer to the true value -6.75 than the right endpoint approximation, $R_{40} \approx -6.3998$, shown in Figure 7.

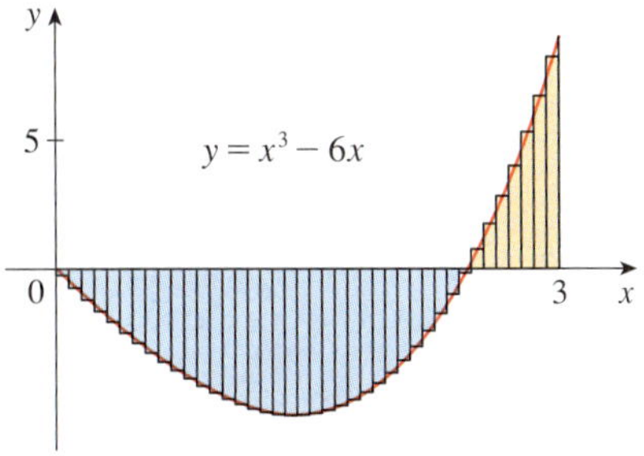

FIGURE 12
$M_{40} \approx -6.7563$

Properties of the Definite Integral

When we defined the definite integral $\int_a^b f(x)\,dx$, we implicitly assumed that $a < b$. But the definition as a limit of Riemann sums makes sense even if $a > b$. Notice that if we reverse a and b, then Δx changes from $(b - a)/n$ to $(a - b)/n$. Therefore

$$\int_b^a f(x)\,dx = -\int_a^b f(x)\,dx$$

If $a = b$, then $\Delta x = 0$ and so

$$\int_a^a f(x)\,dx = 0$$

We now develop some basic properties of integrals that will help us to evaluate integrals in a simple manner. We assume that f and g are continuous functions.

Properties of the Integral

1. $\int_a^b c\,dx = c(b-a)$, where c is any constant

2. $\int_a^b [f(x) + g(x)]\,dx = \int_a^b f(x)\,dx + \int_a^b g(x)\,dx$

3. $\int_a^b cf(x)\,dx = c\int_a^b f(x)\,dx$, where c is any constant

4. $\int_a^b [f(x) - g(x)]\,dx = \int_a^b f(x)\,dx - \int_a^b g(x)\,dx$

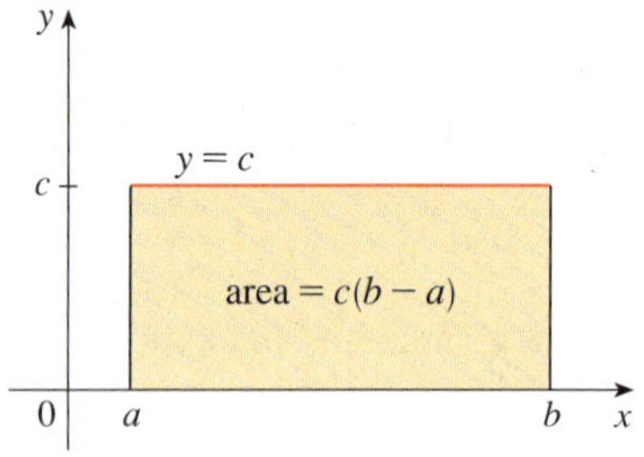

FIGURE 13
$\int_a^b c\,dx = c(b-a)$

Property 1 says that the integral of a constant function $f(x) = c$ is the constant times the length of the interval. If $c > 0$ and $a < b$, this is to be expected because $c(b - a)$ is the area of the shaded rectangle in Figure 13.

Property 2 says that the integral of a sum is the sum of the integrals. For positive functions it says that the area under $f + g$ is the area under f plus the area under g. Figure 14 helps us understand why this is true: In view of how graphical addition works, the corresponding vertical line segments have equal height.

In general, Property 2 follows from Equation 3 and the fact that the limit of a sum is the sum of the limits:

$$\begin{aligned}\int_a^b [f(x) + g(x)]\,dx &= \lim_{n\to\infty} \sum_{i=1}^{n} [f(x_i) + g(x_i)]\,\Delta x \\ &= \lim_{n\to\infty} \left[\sum_{i=1}^{n} f(x_i)\,\Delta x + \sum_{i=1}^{n} g(x_i)\,\Delta x \right] \\ &= \lim_{n\to\infty} \sum_{i=1}^{n} f(x_i)\,\Delta x + \lim_{n\to\infty} \sum_{i=1}^{n} g(x_i)\,\Delta x \\ &= \int_a^b f(x)\,dx + \int_a^b g(x)\,dx\end{aligned}$$

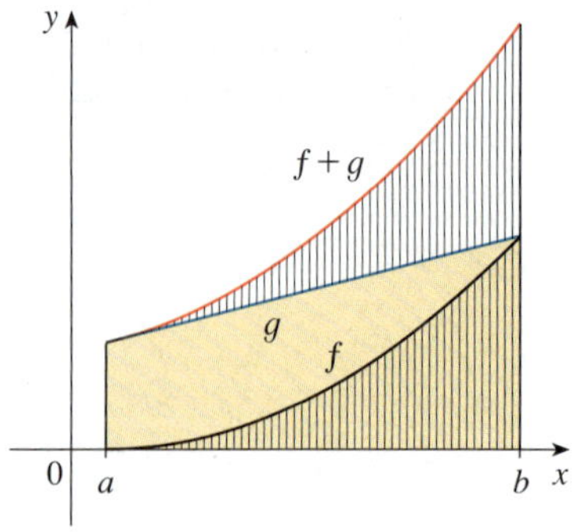

FIGURE 14
$\int_a^b [f(x) + g(x)]\,dx = \int_a^b f(x)\,dx + \int_a^b g(x)\,dx$

|||| Property 3 seems intuitively reasonable because we know that multiplying a function by a positive number c stretches or shrinks its graph vertically by a factor of c. So it stretches or shrinks each approximating rectangle by a factor c and therefore it has the effect of multiplying the area by c.

Property 3 can be proved in a similar manner and says that the integral of a constant times a function is the constant times the integral of the function. In other words, a constant (but *only* a constant) can be taken in front of an integral sign. Property 4 is proved by writing $f - g = f + (-g)$ and using Properties 2 and 3 with $c = -1$.

EXAMPLE 6 Use the properties of integrals to evaluate $\int_0^1 (4 + 3x^2)\,dx$.

SOLUTION Using Properties 2 and 3 of integrals, we have

$$\int_0^1 (4 + 3x^2)\,dx = \int_0^1 4\,dx + \int_0^1 3x^2\,dx = \int_0^1 4\,dx + 3\int_0^1 x^2\,dx$$

We know from Property 1 that

$$\int_0^1 4\,dx = 4(1 - 0) = 4$$

and we found in Example 2 in Section 5.1 that $\int_0^1 x^2\,dx = \frac{1}{3}$. So

$$\int_0^1 (4 + 3x^2)\,dx = \int_0^1 4\,dx + 3\int_0^1 x^2\,dx$$
$$= 4 + 3 \cdot \tfrac{1}{3} = 5$$

The next property tells us how to combine integrals of the same function over adjacent intervals:

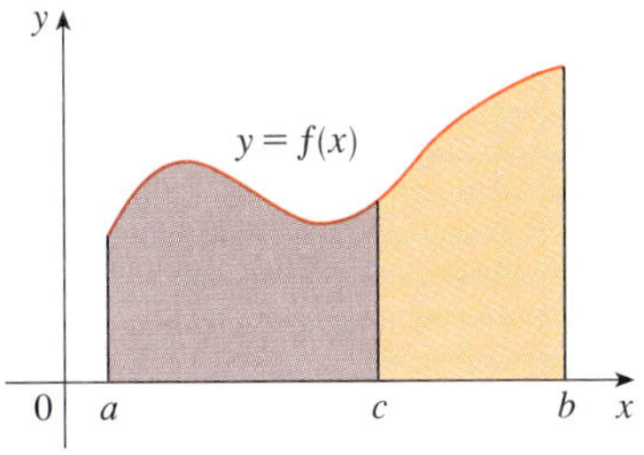

FIGURE 15

5. $$\int_a^c f(x)\,dx + \int_c^b f(x)\,dx = \int_a^b f(x)\,dx$$

This is not easy to prove in general, but for the case where $f(x) \geqslant 0$ and $a < c < b$ Property 5 can be seen from the geometric interpretation in Figure 15: The area under $y = f(x)$ from a to c plus the area from c to b is equal to the total area from a to b.

EXAMPLE 7 If it is known that $\int_0^{10} f(x)\,dx = 17$ and $\int_0^8 f(x)\,dx = 12$, find $\int_8^{10} f(x)\,dx$.

SOLUTION By Property 5, we have

$$\int_0^8 f(x)\,dx + \int_8^{10} f(x)\,dx = \int_0^{10} f(x)\,dx$$

so $$\int_8^{10} f(x)\,dx = \int_0^{10} f(x)\,dx - \int_0^8 f(x)\,dx = 17 - 12 = 5$$

Notice that Properties 1–5 are true whether $a < b$, $a = b$, or $a > b$. The following properties, in which we compare sizes of functions and sizes of integrals, are true only if $a \leqslant b$.

Comparison Properties of the Integral

6. If $f(x) \geqslant 0$ for $a \leqslant x \leqslant b$, then $\int_a^b f(x)\,dx \geqslant 0$.

7. If $f(x) \geqslant g(x)$ for $a \leqslant x \leqslant b$, then $\int_a^b f(x)\,dx \geqslant \int_a^b g(x)\,dx$.

8. If $m \leqslant f(x) \leqslant M$ for $a \leqslant x \leqslant b$, then

$$m(b - a) \leqslant \int_a^b f(x)\,dx \leqslant M(b - a)$$

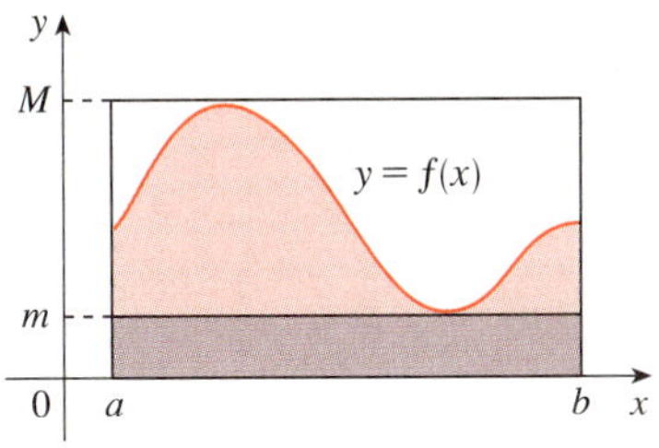

FIGURE 16

If $f(x) \geqslant 0$, then $\int_a^b f(x)\,dx$ represents the area under the graph of f, so the geometric interpretation of Property 6 is simply that areas are positive. But the property can be proved from the definition of an integral (Exercise 64). Property 7 says that a bigger function has a bigger integral. It follows from Properties 6 and 4 because $f - g \geqslant 0$.

Property 8 is illustrated by Figure 16 for the case where $f(x) \geqslant 0$. If f is continuous we could take m and M to be the absolute minimum and maximum values of f on the interval $[a, b]$. In this case Property 8 says that the area under the graph of f is greater than the area of the rectangle with height m and less than the area of the rectangle with height M.

Proof of Property 8 Since $m \leqslant f(x) \leqslant M$, Property 7 gives

$$\int_a^b m\,dx \leqslant \int_a^b f(x)\,dx \leqslant \int_a^b M\,dx$$

Using Property 1 to evaluate the integrals on the left and right sides, we obtain

$$m(b - a) \leqslant \int_a^b f(x)\,dx \leqslant M(b - a)$$

Property 8 is useful when all we want is a rough estimate of the size of an integral without going to the bother of using the Midpoint Rule.

EXAMPLE 8 Use Property 8 to estimate $\int_0^1 e^{-x^2}\,dx$.

SOLUTION Because $f(x) = e^{-x^2}$ is a decreasing function on $[0, 1]$, its absolute maximum value is $M = f(0) = 1$ and its absolute minimum value is $m = f(1) = e^{-1}$. Thus, by Property 8,

$$e^{-1}(1 - 0) \leqslant \int_0^1 e^{-x^2}\,dx \leqslant 1(1 - 0)$$

or

$$e^{-1} \leqslant \int_0^1 e^{-x^2}\,dx \leqslant 1$$

Since $e^{-1} \approx 0.3679$, we can write

$$0.367 \leqslant \int_0^1 e^{-x^2}\,dx \leqslant 1$$

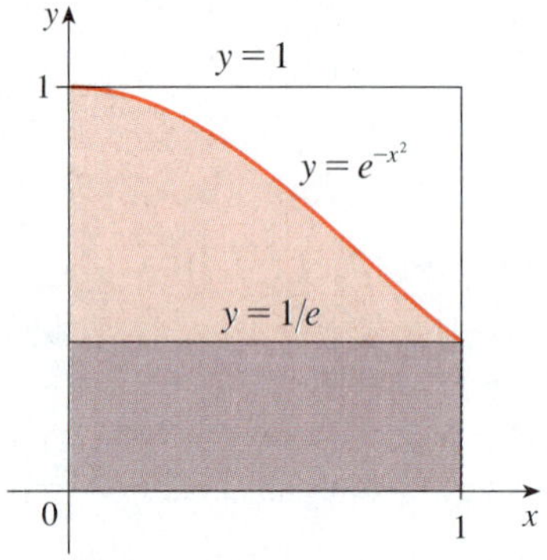

FIGURE 17

The result of Example 8 is illustrated in Figure 17. The integral is greater than the area of the lower rectangle and less than the area of the square.

5.2 Exercises

1. Evaluate the Riemann sum for $f(x) = 2 - x^2$, $0 \leqslant x \leqslant 2$, with four subintervals, taking the sample points to be right endpoints. Explain, with the aid of a diagram, what the Riemann sum represents.

2. If $f(x) = \ln x - 1$, $1 \leqslant x \leqslant 4$, evaluate the Riemann sum with $n = 6$, taking the sample points to be left endpoints. (Give your answer correct to six decimal places.) What does the Riemann sum represent? Illustrate with a diagram.

3. If $f(x) = \sqrt{x} - 2$, $1 \leqslant x \leqslant 6$, find the Riemann sum with $n = 5$ correct to six decimal places, taking the sample points to be midpoints. What does the Riemann sum represent? Illustrate with a diagram.

4. (a) Find the Riemann sum for $f(x) = x - 2 \sin 2x$, $0 \leqslant x \leqslant 3$, with six terms, taking the sample points to be right endpoints. (Give your answer correct to six decimal places.) Explain what the Riemann sum represents with the aid of a sketch.
(b) Repeat part (a) with midpoints as the sample points.

5. The graph of a function f is given. Estimate $\int_0^8 f(x)\,dx$ using four subintervals with (a) right endpoints, (b) left endpoints, and (c) midpoints.

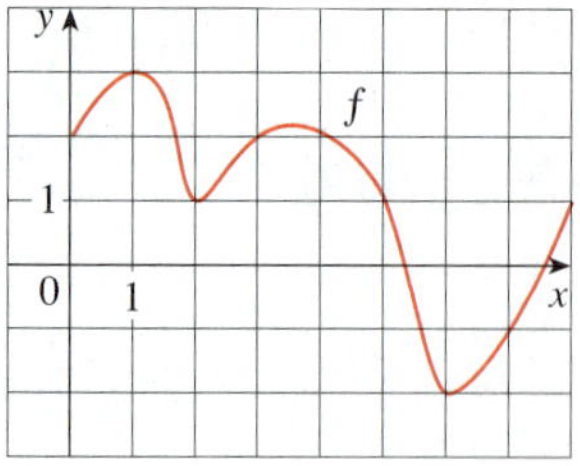

6. The graph of g is shown. Estimate $\int_{-3}^{3} g(x)\,dx$ with six subintervals using (a) right endpoints, (b) left endpoints, and (c) midpoints.

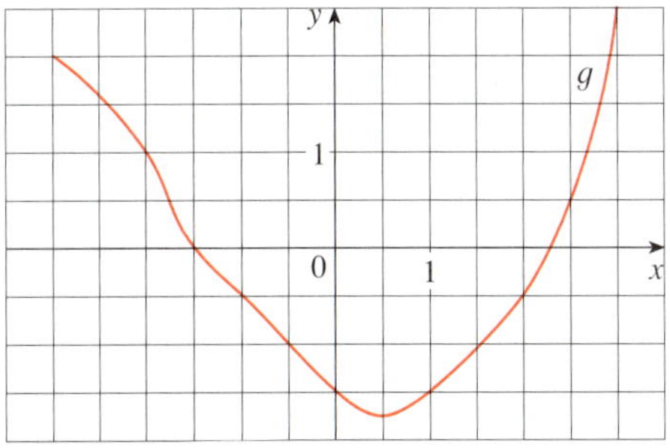

7. A table of values of an increasing function f is shown. Use the table to find lower and upper estimates for $\int_0^{25} f(x)\,dx$.

x	0	5	10	15	20	25
$f(x)$	−42	−37	−25	−6	15	36

8. The table gives the values of a function obtained from an experiment. Use them to estimate $\int_0^6 f(x)\,dx$ using three equal subintervals with (a) right endpoints, (b) left endpoints, and (c) midpoints. If the function is known to be a decreasing function, can you say whether your estimates are less than or greater than the exact value of the integral?

x	0	1	2	3	4	5	6
$f(x)$	9.3	9.0	8.3	6.5	2.3	−7.6	−10.5

9–12 ||| Use the Midpoint Rule with the given value of n to approximate the integral. Round the answer to four decimal places.

9. $\int_2^{10} \sqrt{x^3 + 1}\,dx, \quad n = 4$

10. $\int_0^{\pi} \sec(x/3)\,dx, \quad n = 6$

11. $\int_0^1 \sin(x^2)\,dx, \quad n = 5$

12. $\int_1^5 x^2 e^{-x}\,dx, \quad n = 4$

CAS **13.** If you have a CAS that evaluates midpoint approximations and graphs the corresponding rectangles (use `middlesum` and `middlebox` commands in Maple), check the answer to Exercise 11 and illustrate with a graph. Then repeat with $n = 10$ and $n = 20$.

14. With a programmable calculator or computer (see the instructions for Exercise 7 in Section 5.1), compute the left and right Riemann sums for the function $f(x) = \sin(x^2)$ on the interval $[0, 1]$ with $n = 100$. Explain why these estimates show that

$$0.306 < \int_0^1 \sin(x^2)\,dx < 0.315$$

Deduce that the approximation using the Midpoint Rule with $n = 5$ in Exercise 11 is accurate to two decimal places.

15. Use a calculator or computer to make a table of values of right Riemann sums R_n for the integral $\int_0^{\pi} \sin x\,dx$ with $n = 5$, 10, 50, and 100. What value do these numbers appear to be approaching?

16. Use a calculator or computer to make a table of values of left and right Riemann sums L_n and R_n for the integral $\int_0^2 e^{-x^2}\,dx$ with $n = 5$, 10, 50, and 100. Between what two numbers must the value of the integral lie? Can you make a similar statement for the integral $\int_{-1}^2 e^{-x^2}\,dx$? Explain.

17–20 ||| Express the limit as a definite integral on the given interval.

17. $\lim_{n\to\infty} \sum_{i=1}^{n} x_i \sin x_i\,\Delta x, \quad [0, \pi]$

18. $\lim_{n\to\infty} \sum_{i=1}^{n} \frac{e^{x_i}}{1 + x_i}\,\Delta x, \quad [1, 5]$

19. $\lim_{n\to\infty} \sum_{i=1}^{n} \sqrt{2x_i^* + (x_i^*)^2}\,\Delta x, \quad [1, 8]$

20. $\lim_{n\to\infty} \sum_{i=1}^{n} [4 - 3(x_i^*)^2 + 6(x_i^*)^5]\,\Delta x, \quad [0, 2]$

21–25 ||| Use the form of the definition of the integral given in Equation 3 to evaluate the integral.

21. $\int_{-1}^{5} (1 + 3x)\,dx$

22. $\int_1^4 (x^2 + 2x - 5)\,dx$

23. $\int_0^2 (2 - x^2)\,dx$

24. $\int_0^5 (1 + 2x^3)\,dx$

25. $\int_1^2 x^3\,dx$

26. (a) Find an approximation to the integral $\int_0^4 (x^2 - 3x)\,dx$ using a Riemann sum with right endpoints and $n = 8$.
(b) Draw a diagram like Figure 3 to illustrate the approximation in part (a).
(c) Use Equation 3 to evaluate $\int_0^4 (x^2 - 3x)\,dx$.
(d) Interpret the integral in part (c) as a difference of areas and illustrate with a diagram like Figure 4.

27. Prove that $\int_a^b x\,dx = \frac{b^2 - a^2}{2}$.

28. Prove that $\int_a^b x^2\,dx = \frac{b^3 - a^3}{3}$.

29–30 ||| Express the integral as a limit of Riemann sums. Do not evaluate the limit.

29. $\int_2^6 \frac{x}{1 + x^5}\,dx$

30. $\int_1^{10} (x - 4 \ln x)\,dx$

CAS **31–32** Express the integral as a limit of sums. Then evaluate, using a computer algebra system to find both the sum and the limit.

31. $\int_0^{\pi} \sin 5x\, dx$

32. $\int_2^{10} x^6\, dx$

33. The graph of f is shown. Evaluate each integral by interpreting it in terms of areas.

(a) $\int_0^2 f(x)\, dx$ (b) $\int_0^5 f(x)\, dx$

(c) $\int_5^7 f(x)\, dx$ (d) $\int_0^9 f(x)\, dx$

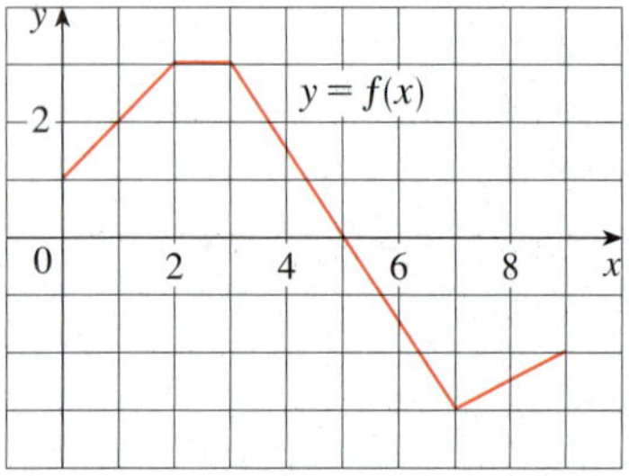

34. The graph of g consists of two straight lines and a semicircle. Use it to evaluate each integral.

(a) $\int_0^2 g(x)\, dx$ (b) $\int_2^6 g(x)\, dx$ (c) $\int_0^7 g(x)\, dx$

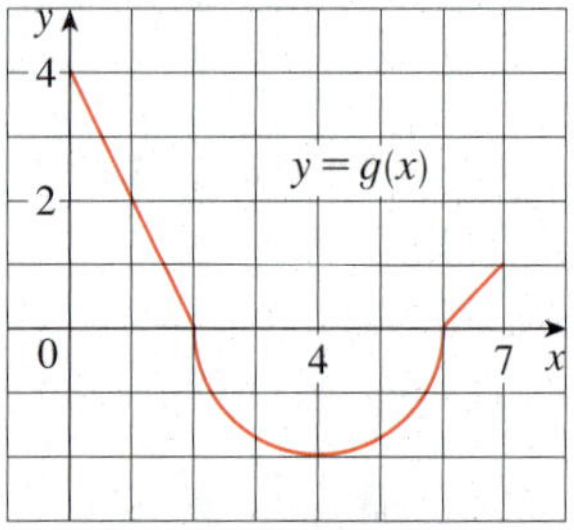

35–40 Evaluate the integral by interpreting it in terms of areas.

35. $\int_0^3 \left(\frac{1}{2}x - 1\right) dx$

36. $\int_{-2}^2 \sqrt{4 - x^2}\, dx$

37. $\int_{-3}^0 \left(1 + \sqrt{9 - x^2}\right) dx$

38. $\int_{-1}^3 (3 - 2x)\, dx$

39. $\int_{-1}^2 |x|\, dx$

40. $\int_0^{10} |x - 5|\, dx$

41. Given that $\int_4^9 \sqrt{x}\, dx = \frac{38}{3}$, what is $\int_9^4 \sqrt{t}\, dt$?

42. Evaluate $\int_1^1 x^2 \cos x\, dx$.

43. In Example 2 in Section 5.1 we showed that $\int_0^1 x^2\, dx = \frac{1}{3}$. Use this fact and the properties of integrals to evaluate $\int_0^1 (5 - 6x^2)\, dx$.

44. Use the properties of integrals and the result of Example 3 to evaluate $\int_1^3 (2e^x - 1)\, dx$.

45. Use the result of Example 3 to evaluate $\int_1^3 e^{x+2}\, dx$.

46. Use the result of Exercise 27 and the fact that $\int_0^{\pi/2} \cos x\, dx = 1$ (from Exercise 25 in Section 5.1), together with the properties of integrals, to evaluate $\int_0^{\pi/2} (2\cos x - 5x)\, dx$.

47. Write as a single integral in the form $\int_a^b f(x)\, dx$:

$$\int_{-2}^2 f(x)\, dx + \int_2^5 f(x)\, dx - \int_{-2}^{-1} f(x)\, dx$$

48. If $\int_1^5 f(x)\, dx = 12$ and $\int_4^5 f(x)\, dx = 3.6$, find $\int_1^4 f(x)\, dx$.

49. If $\int_0^9 f(x)\, dx = 37$ and $\int_0^9 g(x)\, dx = 16$, find $\int_0^9 [2f(x) + 3g(x)]\, dx$.

50. Find $\int_0^5 f(x)\, dx$ if

$$f(x) = \begin{cases} 3 & \text{for } x < 3 \\ x & \text{for } x \geqslant 3 \end{cases}$$

51–54 Use the properties of integrals to verify the inequality without evaluating the integrals.

51. $\int_0^{\pi/4} \sin^3 x\, dx \leqslant \int_0^{\pi/4} \sin^2 x\, dx$

52. $\int_1^2 \sqrt{5 - x}\, dx \geqslant \int_1^2 \sqrt{x + 1}\, dx$

53. $2 \leqslant \int_{-1}^1 \sqrt{1 + x^2}\, dx \leqslant 2\sqrt{2}$

54. $\frac{\pi}{6} \leqslant \int_{\pi/6}^{\pi/2} \sin x\, dx \leqslant \frac{\pi}{3}$

55–60 Use Property 8 to estimate the value of the integral.

55. $\int_1^2 \frac{1}{x}\, dx$

56. $\int_0^2 \sqrt{x^3 + 1}\, dx$

57. $\int_{\pi/4}^{\pi/3} \tan x\, dx$

58. $\int_0^2 (x^3 - 3x + 3)\, dx$

59. $\int_0^2 xe^{-x}\, dx$

60. $\int_{\pi/4}^{3\pi/4} \sin^2 x\, dx$

61–62 Use properties of integrals, together with Exercises 27 and 28, to prove the inequality.

61. $\int_1^3 \sqrt{x^4 + 1}\, dx \geqslant \frac{26}{3}$

62. $\int_0^{\pi/2} x \sin x\, dx \leqslant \frac{\pi^2}{8}$

63. Prove Property 3 of integrals.

64. Prove Property 6 of integrals.

65. If f is continuous on $[a, b]$, show that

$$\left|\int_a^b f(x)\,dx\right| \leqslant \int_a^b |f(x)|\,dx$$

[*Hint:* $-|f(x)| \leqslant f(x) \leqslant |f(x)|$.]

66. Use the result of Exercise 65 to show that

$$\left|\int_0^{2\pi} f(x)\sin 2x\,dx\right| \leqslant \int_0^{2\pi} |f(x)|\,dx$$

67–68 Express the limit as a definite integral.

67. $\lim_{n\to\infty} \sum_{i=1}^{n} \frac{i^4}{n^5}$ [*Hint:* Consider $f(x) = x^4$.]

68. $\lim_{n\to\infty} \frac{1}{n} \sum_{i=1}^{n} \frac{1}{1 + (i/n)^2}$

69. Find $\int_1^2 x^{-2}\,dx$. *Hint:* Choose x_i^* to be the geometric mean of x_{i-1} and x_i (that is, $x_i^* = \sqrt{x_{i-1}x_i}$) and use the identity

$$\frac{1}{m(m+1)} = \frac{1}{m} - \frac{1}{m+1}$$

DISCOVERY PROJECT

Area Functions

1. (a) Draw the line $y = 2t + 1$ and use geometry to find the area under this line, above the t-axis, and between the vertical lines $t = 1$ and $t = 3$.

(b) If $x > 1$, let $A(x)$ be the area of the region that lies under the line $y = 2t + 1$ between $t = 1$ and $t = x$. Sketch this region and use geometry to find an expression for $A(x)$.

(c) Differentiate the area function $A(x)$. What do you notice?

2. (a) If $x \geqslant -1$, let

$$A(x) = \int_{-1}^{x} (1 + t^2)\,dt$$

$A(x)$ represents the area of a region. Sketch that region.

(b) Use the result of Exercise 28 in Section 5.2 to find an expression for $A(x)$.

(c) Find $A'(x)$. What do you notice?

(d) If $x \geqslant -1$ and h is a small positive number, then $A(x + h) - A(x)$ represents the area of a region. Describe and sketch the region.

(e) Draw a rectangle that approximates the region in part (d). By comparing the areas of these two regions, show that

$$\frac{A(x + h) - A(x)}{h} \approx 1 + x^2$$

(f) Use part (e) to give an intuitive explanation for the result of part (c).

3. (a) Draw the graph of the function $f(x) = \cos(x^2)$ in the viewing rectangle $[0, 2]$ by $[-1.25, 1.25]$.

(b) If we define a new function g by

$$g(x) = \int_0^x \cos(t^2)\,dt$$

then $g(x)$ is the area under the graph of f from 0 to x [until $f(x)$ becomes negative, at which point $g(x)$ becomes a difference of areas]. Use part (a) to determine the value of x at which $g(x)$ starts to decrease. [Unlike the integral in Problem 2, it is impossible to evaluate the integral defining g to obtain an explicit expression for $g(x)$.]

(c) Use the integration command on your calculator or computer to estimate $g(0.2)$, $g(0.4)$, $g(0.6)$, . . . , $g(1.8)$, $g(2)$. Then use these values to sketch a graph of g.

(d) Use your graph of g from part (c) to sketch the graph of g' using the interpretation of $g'(x)$ as the slope of a tangent line. How does the graph of g' compare with the graph of f?

4. Suppose f is a continuous function on the interval $[a, b]$ and we define a new function g by the equation

$$g(x) = \int_a^x f(t)\, dt$$

Based on your results in Problems 1–3, conjecture an expression for $g'(x)$.

5.3 The Fundamental Theorem of Calculus

The Fundamental Theorem of Calculus is appropriately named because it establishes a connection between the two branches of calculus: differential calculus and integral calculus. Differential calculus arose from the tangent problem, whereas integral calculus arose from a seemingly unrelated problem, the area problem. Newton's teacher at Cambridge, Isaac Barrow (1630–1677), discovered that these two problems are actually closely related. In fact, he realized that differentiation and integration are inverse processes. The Fundamental Theorem of Calculus gives the precise inverse relationship between the derivative and the integral. It was Newton and Leibniz who exploited this relationship and used it to develop calculus into a systematic mathematical method. In particular, they saw that the Fundamental Theorem enabled them to compute areas and integrals very easily without having to compute them as limits of sums as we did in Sections 5.1 and 5.2.

Investigate the area function interactively.

Resources / Module 6 / Areas and Derivatives / Area as a Function

The first part of the Fundamental Theorem deals with functions defined by an equation of the form

$$\boxed{1} \qquad g(x) = \int_a^x f(t)\, dt$$

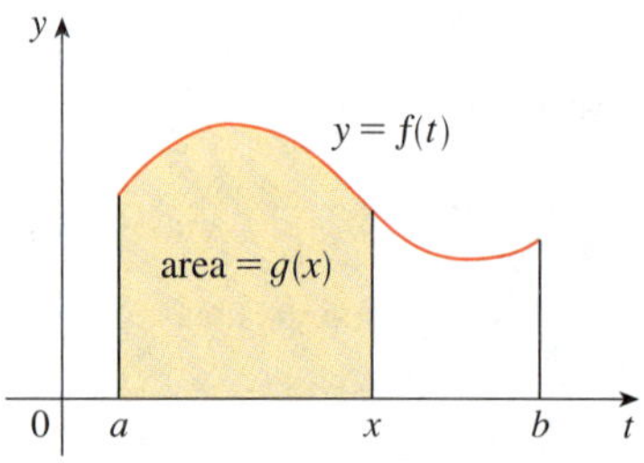

FIGURE 1

where f is a continuous function on $[a, b]$ and x varies between a and b. Observe that g depends only on x, which appears as the variable upper limit in the integral. If x is a fixed number, then the integral $\int_a^x f(t)\, dt$ is a definite number. If we then let x vary, the number $\int_a^x f(t)\, dt$ also varies and defines a function of x denoted by $g(x)$.

If f happens to be a positive function, then $g(x)$ can be interpreted as the area under the graph of f from a to x, where x can vary from a to b. (Think of g as the "area so far" function; see Figure 1.)

EXAMPLE 1 If f is the function whose graph is shown in Figure 2 and $g(x) = \int_0^x f(t)\, dt$, find the values of $g(0)$, $g(1)$, $g(2)$, $g(3)$, $g(4)$, and $g(5)$. Then sketch a rough graph of g.

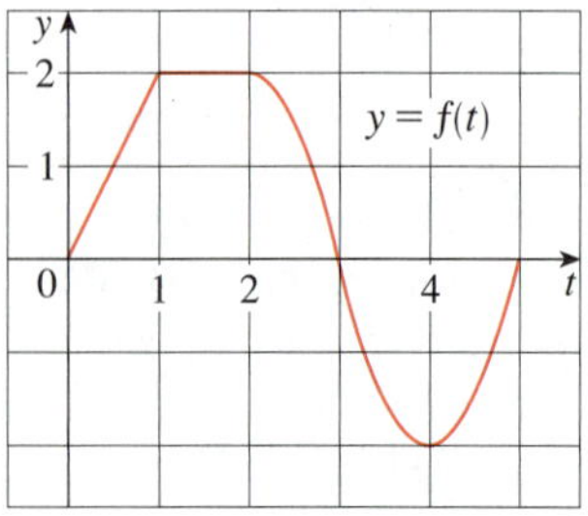

FIGURE 2

SOLUTION First we notice that $g(0) = \int_0^0 f(t)\, dt = 0$. From Figure 3 we see that $g(1)$ is the area of a triangle:

$$g(1) = \int_0^1 f(t)\, dt = \tfrac{1}{2}(1 \cdot 2) = 1$$

To find $g(2)$ we add to $g(1)$ the area of a rectangle:

$$g(2) = \int_0^2 f(t)\, dt = \int_0^1 f(t)\, dt + \int_1^2 f(t)\, dt = 1 + (1 \cdot 2) = 3$$

We estimate that the area under f from 2 to 3 is about 1.3, so

$$g(3) = g(2) + \int_2^3 f(t)\,dt \approx 3 + 1.3 = 4.3$$

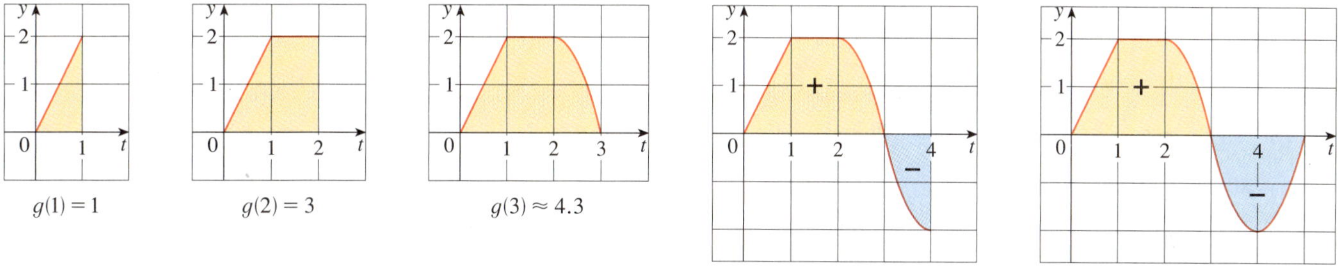

$g(1) = 1$ $g(2) = 3$ $g(3) \approx 4.3$ $g(4) \approx 3$ $g(5) \approx 1.7$

FIGURE 3

For $t > 3$, $f(t)$ is negative and so we start subtracting areas:

$$g(4) = g(3) + \int_3^4 f(t)\,dt \approx 4.3 + (-1.3) = 3.0$$

$$g(5) = g(4) + \int_4^5 f(t)\,dt \approx 3 + (-1.3) = 1.7$$

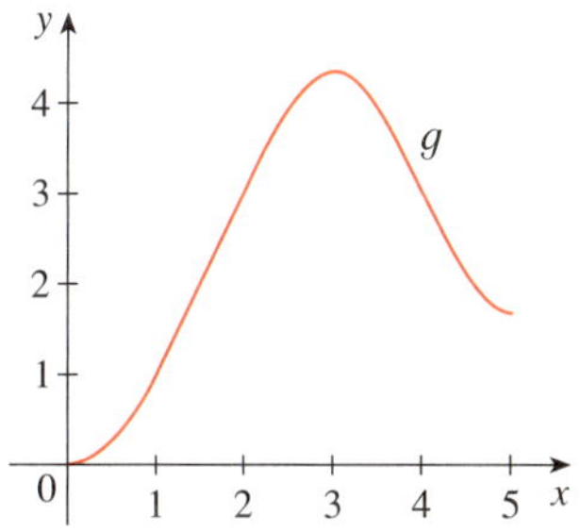

FIGURE 4

$g(x) = \int_a^x f(t)\,dt$

We use these values to sketch the graph of g in Figure 4. Notice that, because $f(t)$ is positive for $t < 3$, we keep adding area for $t < 3$ and so g is increasing up to $x = 3$, where it attains a maximum value. For $x > 3$, g decreases because $f(t)$ is negative.

If we take $f(t) = t$ and $a = 0$, then, using Exercise 27 in Section 5.2, we have

$$g(x) = \int_0^x t\,dt = \frac{x^2}{2}$$

Notice that $g'(x) = x$, that is, $g' = f$. In other words, if g is defined as the integral of f by Equation 1, then g turns out to be an antiderivative of f, at least in this case. And if we sketch the derivative of the function g shown in Figure 4 by estimating slopes of tangents, we get a graph like that of f in Figure 2. So we suspect that $g' = f$ in Example 1 too.

To see why this might be generally true we consider any continuous function f with $f(x) \geq 0$. Then $g(x) = \int_a^x f(t)\,dt$ can be interpreted as the area under the graph of f from a to x, as in Figure 1.

In order to compute $g'(x)$ from the definition of derivative we first observe that, for $h > 0$, $g(x + h) - g(x)$ is obtained by subtracting areas, so it is the area under the graph of f from x to $x + h$ (the gold area in Figure 5). For small h you can see from the figure that this area is approximately equal to the area of the rectangle with height $f(x)$ and width h:

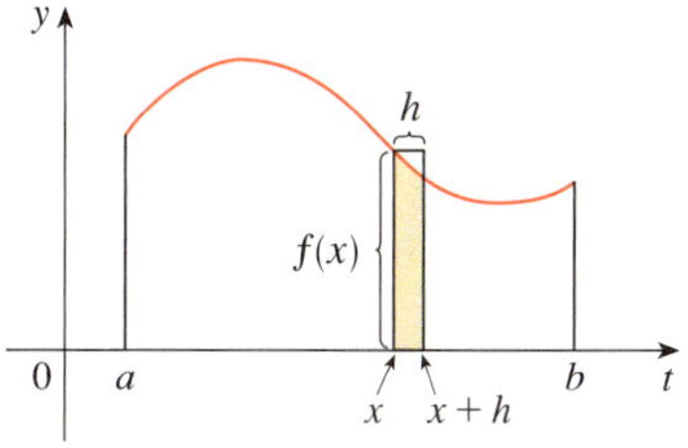

FIGURE 5

$$g(x + h) - g(x) \approx hf(x)$$

so

$$\frac{g(x + h) - g(x)}{h} \approx f(x)$$

Intuitively, we therefore expect that

$$g'(x) = \lim_{h \to 0} \frac{g(x+h) - g(x)}{h} = f(x)$$

The fact that this is true, even when f is not necessarily positive, is the first part of the Fundamental Theorem of Calculus.

The Fundamental Theorem of Calculus, Part 1 If f is continuous on $[a, b]$, then the function g defined by

$$g(x) = \int_a^x f(t)\,dt \qquad a \leqslant x \leqslant b$$

is continuous on $[a, b]$ and differentiable on (a, b), and $g'(x) = f(x)$.

|||| We abbreviate the name of this theorem as FTC1. In words, it says that the derivative of a definite integral with respect to its upper limit is the integrand evaluated at the upper limit.

Proof If x and $x + h$ are in (a, b), then

$$\begin{aligned} g(x+h) - g(x) &= \int_a^{x+h} f(t)\,dt - \int_a^x f(t)\,dt \\ &= \left(\int_a^x f(t)\,dt + \int_x^{x+h} f(t)\,dt \right) - \int_a^x f(t)\,dt \qquad \text{(by Property 5)} \\ &= \int_x^{x+h} f(t)\,dt \end{aligned}$$

and so, for $h \neq 0$,

2 $$\frac{g(x+h) - g(x)}{h} = \frac{1}{h} \int_x^{x+h} f(t)\,dt$$

For now let us assume that $h > 0$. Since f is continuous on $[x, x + h]$, the Extreme Value Theorem says that there are numbers u and v in $[x, x + h]$ such that $f(u) = m$ and $f(v) = M$, where m and M are the absolute minimum and maximum values of f on $[x, x + h]$. (See Figure 6.)

By Property 8 of integrals, we have

$$mh \leqslant \int_x^{x+h} f(t)\,dt \leqslant Mh$$

that is,

$$f(u)h \leqslant \int_x^{x+h} f(t)\,dt \leqslant f(v)h$$

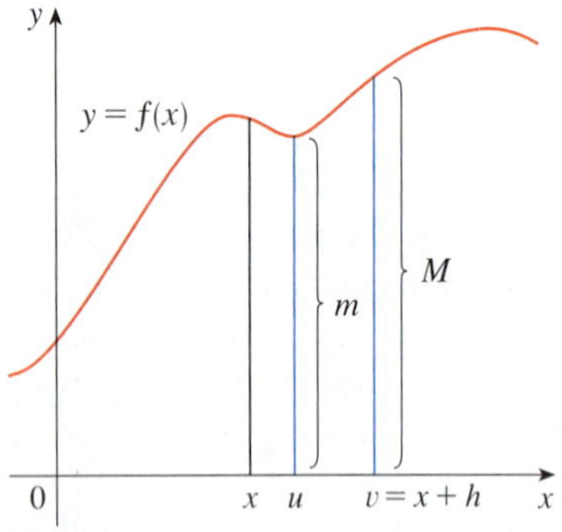

FIGURE 6

Since $h > 0$, we can divide this inequality by h:

$$f(u) \leqslant \frac{1}{h} \int_x^{x+h} f(t)\,dt \leqslant f(v)$$

Now we use Equation 2 to replace the middle part of this inequality:

3 $$f(u) \leqslant \frac{g(x+h) - g(x)}{h} \leqslant f(v)$$

Inequality 3 can be proved in a similar manner for the case $h < 0$. (See Exercise 63.)

TEC Module 5.3 provides visual evidence for FTC1.

Now we let $h \to 0$. Then $u \to x$ and $v \to x$, since u and v lie between x and $x + h$. Therefore

$$\lim_{h \to 0} f(u) = \lim_{u \to x} f(u) = f(x)$$

and

$$\lim_{h \to 0} f(v) = \lim_{v \to x} f(v) = f(x)$$

because f is continuous at x. We conclude, from (3) and the Squeeze Theorem, that

4 $$g'(x) = \lim_{h \to 0} \frac{g(x+h) - g(x)}{h} = f(x)$$

If $x = a$ or b, then Equation 4 can be interpreted as a one-sided limit. Then Theorem 2.9.4 (modified for one-sided limits) shows that g is continuous on $[a, b]$.

Using Leibniz notation for derivatives, we can write FTC1 as

5 $$\frac{d}{dx} \int_a^x f(t)\, dt = f(x)$$

when f is continuous. Roughly speaking, Equation 5 says that if we first integrate f and then differentiate the result, we get back to the original function f.

EXAMPLE 2 Find the derivative of the function $g(x) = \int_0^x \sqrt{1 + t^2}\, dt$.

SOLUTION Since $f(t) = \sqrt{1 + t^2}$ is continuous, Part 1 of the Fundamental Theorem of Calculus gives

$$g'(x) = \sqrt{1 + x^2}$$

EXAMPLE 3 Although a formula of the form $g(x) = \int_a^x f(t)\, dt$ may seem like a strange way of defining a function, books on physics, chemistry, and statistics are full of such functions. For instance, the **Fresnel function**

$$S(x) = \int_0^x \sin(\pi t^2/2)\, dt$$

is named after the French physicist Augustin Fresnel (1788–1827), who is famous for his works in optics. This function first appeared in Fresnel's theory of the diffraction of light waves, but more recently it has been applied to the design of highways.

Part 1 of the Fundamental Theorem tells us how to differentiate the Fresnel function:

$$S'(x) = \sin(\pi x^2/2)$$

This means that we can apply all the methods of differential calculus to analyze S (see Exercise 57).

Figure 7 shows the graphs of $f(x) = \sin(\pi x^2/2)$ and the Fresnel function $S(x) = \int_0^x f(t)\,dt$. A computer was used to graph S by computing the value of this integral for many values of x. It does indeed look as if $S(x)$ is the area under the graph of f from 0 to x [until $x \approx 1.4$ when $S(x)$ becomes a difference of areas]. Figure 8 shows a larger part of the graph of S.

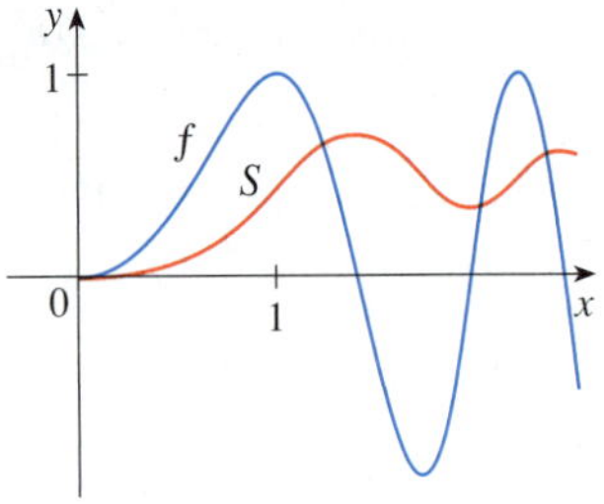

FIGURE 7

$f(x) = \sin(\pi x^2/2)$

$S(x) = \int_0^x \sin(\pi t^2/2)\,dt$

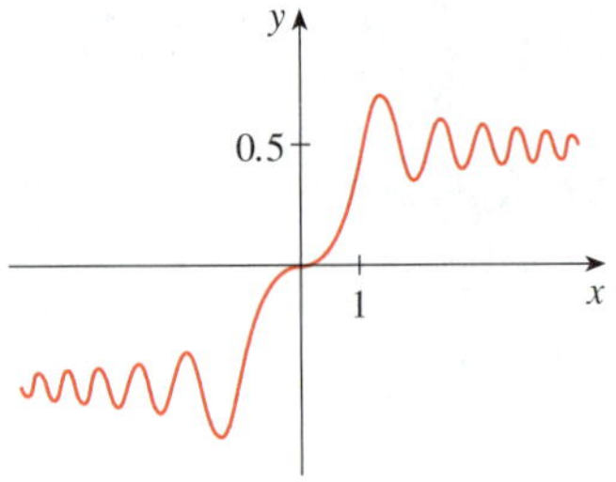

FIGURE 8

The Fresnel function $S(x) = \int_0^x \sin(\pi t^2/2)\,dt$

If we now start with the graph of S in Figure 7 and think about what its derivative should look like, it seems reasonable that $S'(x) = f(x)$. [For instance, S is increasing when $f(x) > 0$ and decreasing when $f(x) < 0$.] So this gives a visual confirmation of Part 1 of the Fundamental Theorem of Calculus.

EXAMPLE 4 Find $\dfrac{d}{dx}\displaystyle\int_1^{x^4} \sec t\,dt$.

SOLUTION Here we have to be careful to use the Chain Rule in conjunction with FTC1. Let $u = x^4$. Then

$$\begin{aligned} \frac{d}{dx}\int_1^{x^4} \sec t\,dt &= \frac{d}{dx}\int_1^{u} \sec t\,dt \\ &= \frac{d}{du}\left(\int_1^{u} \sec t\,dt\right)\frac{du}{dx} && \text{(by the Chain Rule)} \\ &= \sec u\,\frac{du}{dx} && \text{(by FTC1)} \\ &= \sec(x^4)\cdot 4x^3 \end{aligned}$$

In Section 5.2 we computed integrals from the definition as a limit of Riemann sums and we saw that this procedure is sometimes long and difficult. The second part of the Fundamental Theorem of Calculus, which follows easily from the first part, provides us with a much simpler method for the evaluation of integrals.

|||| We abbreviate this theorem as FTC2.

The Fundamental Theorem of Calculus, Part 2 If f is continuous on $[a, b]$, then

$$\int_a^b f(x)\,dx = F(b) - F(a)$$

where F is any antiderivative of f, that is, a function such that $F' = f$.

Proof Let $g(x) = \int_a^x f(t)\,dt$. We know from Part 1 that $g'(x) = f(x)$; that is, g is an antiderivative of f. If F is any other antiderivative of f on $[a, b]$, then we know from Corollary 4.2.7 that F and g differ by a constant:

$$F(x) = g(x) + C \tag{6}$$

for $a < x < b$. But both F and g are continuous on $[a, b]$ and so, by taking limits of both sides of Equation 6 (as $x \to a^+$ and $x \to b^-$), we see that it also holds when $x = a$ and $x = b$.

If we put $x = a$ in the formula for $g(x)$, we get

$$g(a) = \int_a^a f(t)\,dt = 0$$

So, using Equation 6 with $x = b$ and $x = a$, we have

$$\begin{aligned} F(b) - F(a) &= [g(b) + C] - [g(a) + C] \\ &= g(b) - g(a) = g(b) = \int_a^b f(t)\,dt \end{aligned}$$

Part 2 of the Fundamental Theorem states that if we know an antiderivative F of f, then we can evaluate $\int_a^b f(x)\,dx$ simply by subtracting the values of F at the endpoints of the interval $[a, b]$. It's very surprising that $\int_a^b f(x)\,dx$, which was defined by a complicated procedure involving all of the values of $f(x)$ for $a \leqslant x \leqslant b$, can be found by knowing the values of $F(x)$ at only two points, a and b.

Although the theorem may be surprising at first glance, it becomes plausible if we interpret it in physical terms. If $v(t)$ is the velocity of an object and $s(t)$ is its position at time t, then $v(t) = s'(t)$, so s is an antiderivative of v. In Section 5.1 we considered an object that always moves in the positive direction and made the guess that the area under the velocity curve is equal to the distance traveled. In symbols:

$$\int_a^b v(t)\,dt = s(b) - s(a)$$

That is exactly what FTC2 says in this context.

EXAMPLE 5 Evaluate the integral $\displaystyle\int_1^3 e^x\,dx$.

SOLUTION The function $f(x) = e^x$ is continuous everywhere and we know that an antiderivative is $F(x) = e^x$, so Part 2 of the Fundamental Theorem gives

|||| Compare the calculation in Example 5 with the much harder one in Example 3 in Section 5.2.

$$\int_1^3 e^x\,dx = F(3) - F(1) = e^3 - e$$

Notice that FTC2 says we can use *any* antiderivative F of f. So we may as well use the simplest one, namely $F(x) = e^x$, instead of $e^x + 7$ or $e^x + C$.

We often use the notation

$$F(x)\Big]_a^b = F(b) - F(a)$$

So the equation of FTC2 can be written as

$$\int_a^b f(x)\,dx = F(x)\Big]_a^b \qquad \text{where} \qquad F' = f$$

Other common notations are $F(x)\big|_a^b$ and $[F(x)]_a^b$.

EXAMPLE 6 Find the area under the parabola $y = x^2$ from 0 to 1.

SOLUTION An antiderivative of $f(x) = x^2$ is $F(x) = \frac{1}{3}x^3$. The required area A is found using Part 2 of the Fundamental Theorem:

|||| In applying the Fundamental Theorem we use a particular antiderivative F of f. It is not necessary to use the most general antiderivative.

$$A = \int_0^1 x^2\,dx = \frac{x^3}{3}\Big]_0^1 = \frac{1^3}{3} - \frac{0^3}{3} = \frac{1}{3}$$

If you compare the calculation in Example 6 with the one in Example 2 in Section 5.1, you will see that the Fundamental Theorem gives a *much* shorter method.

EXAMPLE 7 Evaluate $\displaystyle\int_3^6 \frac{dx}{x}$.

SOLUTION The given integral is an abbreviation for

$$\int_3^6 \frac{1}{x}\,dx$$

An antiderivative of $f(x) = 1/x$ is $F(x) = \ln|x|$ and, because $3 \leqslant x \leqslant 6$, we can write $F(x) = \ln x$. So

$$\int_3^6 \frac{1}{x}\,dx = \ln x\Big]_3^6 = \ln 6 - \ln 3$$

$$= \ln\frac{6}{3} = \ln 2$$

EXAMPLE 8 Find the area under the cosine curve from 0 to b, where $0 \leqslant b \leqslant \pi/2$.

SOLUTION Since an antiderivative of $f(x) = \cos x$ is $F(x) = \sin x$, we have

$$A = \int_0^b \cos x\,dx = \sin x\Big]_0^b = \sin b - \sin 0 = \sin b$$

In particular, taking $b = \pi/2$, we have proved that the area under the cosine curve from 0 to $\pi/2$ is $\sin(\pi/2) = 1$. (See Figure 9.)

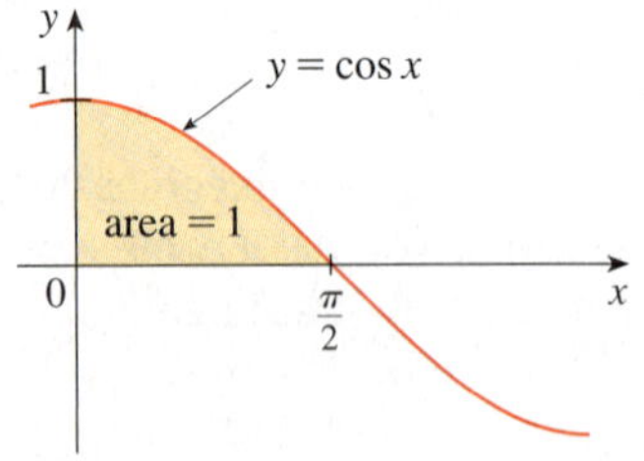

FIGURE 9

When the French mathematician Gilles de Roberval first found the area under the sine and cosine curves in 1635, this was a very challenging problem that required a great deal of ingenuity. If we didn't have the benefit of the Fundamental Theorem, we would have to compute a difficult limit of sums using obscure trigonometric identities (or a computer algebra system as in Exercise 25 in Section 5.1). It was even more difficult for Roberval because the apparatus of limits had not been invented in 1635. But in the 1660s and 1670s, when the Fundamental Theorem was discovered by Barrow and exploited by Newton and Leibniz, such problems became very easy, as you can see from Example 8.

EXAMPLE 9 What is wrong with the following calculation?

$$\int_{-1}^{3} \frac{1}{x^2}\,dx = \frac{x^{-1}}{-1}\bigg]_{-1}^{3} = -\frac{1}{3} - 1 = -\frac{4}{3}$$

SOLUTION To start, we notice that this calculation must be wrong because the answer is negative but $f(x) = 1/x^2 \geqslant 0$ and Property 6 of integrals says that $\int_a^b f(x)\,dx \geqslant 0$ when $f \geqslant 0$. The Fundamental Theorem of Calculus applies to continuous functions. It can't be applied here because $f(x) = 1/x^2$ is not continuous on $[-1, 3]$. In fact, f has an infinite discontinuity at $x = 0$, so

$$\int_{-1}^{3} \frac{1}{x^2}\,dx \qquad \text{does not exist}$$

Differentiation and Integration as Inverse Processes

We end this section by bringing together the two parts of the Fundamental Theorem.

The Fundamental Theorem of Calculus Suppose f is continuous on $[a, b]$.

1. If $g(x) = \int_a^x f(t)\,dt$, then $g'(x) = f(x)$.

2. $\int_a^b f(x)\,dx = F(b) - F(a)$, where F is any antiderivative of f, that is, $F' = f$.

We noted that Part 1 can be rewritten as

$$\frac{d}{dx}\int_a^x f(t)\,dt = f(x)$$

which says that if f is integrated and then the result is differentiated, we arrive back at the original function f. Since $F'(x) = f(x)$, Part 2 can be rewritten as

$$\int_a^b F'(x)\,dx = F(b) - F(a)$$

This version says that if we take a function F, first differentiate it, and then integrate the result, we arrive back at the original function F, but in the form $F(b) - F(a)$. Taken together, the two parts of the Fundamental Theorem of Calculus say that differentiation and integration are inverse processes. Each undoes what the other does.

The Fundamental Theorem of Calculus is unquestionably the most important theorem in calculus and, indeed, it ranks as one of the great accomplishments of the human mind. Before it was discovered, from the time of Eudoxus and Archimedes to the time of Galileo and Fermat, problems of finding areas, volumes, and lengths of curves were so difficult that only a genius could meet the challenge. But now, armed with the systematic method that Newton and Leibniz fashioned out of the Fundamental Theorem, we will see in the chapters to come that these challenging problems are accessible to all of us.

5.3 Exercises

1. Explain exactly what is meant by the statement that "differentiation and integration are inverse processes."

2. Let $g(x) = \int_0^x f(t)\,dt$, where f is the function whose graph is shown.
(a) Evaluate $g(x)$ for $x = 0, 1, 2, 3, 4, 5$, and 6.
(b) Estimate $g(7)$.
(c) Where does g have a maximum value? Where does it have a minimum value?
(d) Sketch a rough graph of g.

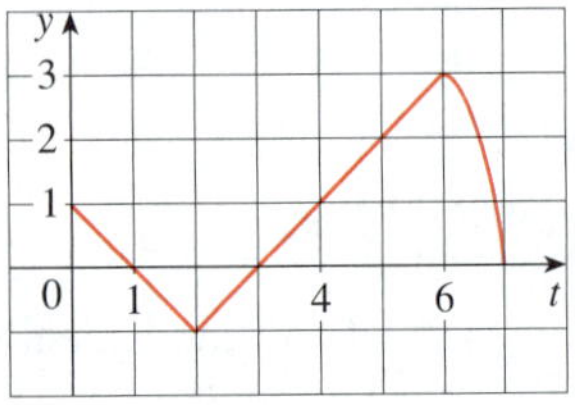

3. Let $g(x) = \int_0^x f(t)\,dt$, where f is the function whose graph is shown.
(a) Evaluate $g(0)$, $g(1)$, $g(2)$, $g(3)$, and $g(6)$.
(b) On what interval is g increasing?
(c) Where does g have a maximum value?
(d) Sketch a rough graph of g.

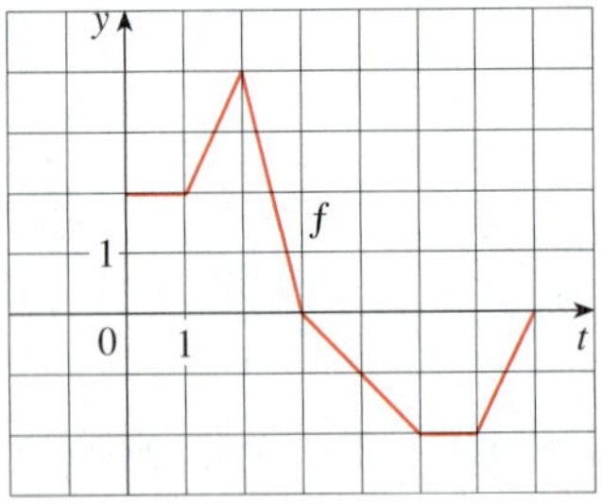

4. Let $g(x) = \int_{-3}^x f(t)\,dt$, where f is the function whose graph is shown.
(a) Evaluate $g(-3)$ and $g(3)$.
(b) Estimate $g(-2)$, $g(-1)$, and $g(0)$.
(c) On what interval is g increasing?
(d) Where does g have a maximum value?
(e) Sketch a rough graph of g.
(f) Use the graph in part (e) to sketch the graph of $g'(x)$. Compare with the graph of f.

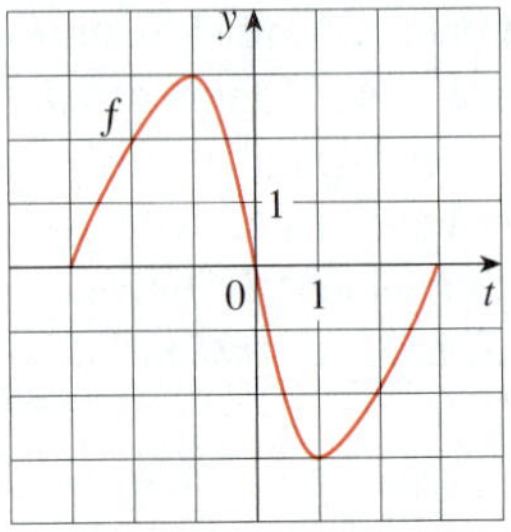

5–6 |||| Sketch the area represented by $g(x)$. Then find $g'(x)$ in two ways: (a) by using Part 1 of the Fundamental Theorem and (b) by evaluating the integral using Part 2 and then differentiating.

5. $g(x) = \int_1^x t^2\,dt$

6. $g(x) = \int_0^x (1 + \sqrt{t})\,dt$

7–18 |||| Use Part 1 of the Fundamental Theorem of Calculus to find the derivative of the function.

7. $g(x) = \int_0^x \sqrt{1 + 2t}\,dt$

8. $g(x) = \int_1^x \ln t\,dt$

9. $g(y) = \int_2^y t^2 \sin t\,dt$

10. $g(u) = \int_3^u \dfrac{1}{x + x^2}\,dx$

11. $F(x) = \int_x^2 \cos(t^2)\,dt$

$$\left[\textit{Hint: } \int_x^2 \cos(t^2)\,dt = -\int_2^x \cos(t^2)\,dt\right]$$

12. $F(x) = \int_x^{10} \tan\theta\,d\theta$

13. $h(x) = \int_2^{1/x} \arctan t\,dt$

14. $h(x) = \int_0^{x^2} \sqrt{1 + r^3}\,dr$

15. $y = \int_3^{\sqrt{x}} \dfrac{\cos t}{t}\,dt$

16. $y = \int_1^{\cos x} (t + \sin t)\,dt$

17. $y = \int_{1-3x}^1 \dfrac{u^3}{1 + u^2}\,du$

18. $y = \int_{e^x}^0 \sin^3 t\,dt$

19–42 |||| Use Part 2 of the Fundamental Theorem of Calculus to evaluate the integral, or explain why it does not exist.

19. $\int_{-1}^3 x^5\,dx$

20. $\int_{-2}^5 6\,dx$

21. $\int_2^8 (4x + 3)\,dx$

22. $\int_0^4 (1 + 3y - y^2)\,dy$

23. $\int_0^1 x^{4/5}\,dx$

24. $\int_1^8 \sqrt[3]{x}\,dx$

25. $\int_1^2 \dfrac{3}{t^4}\,dt$

26. $\int_{-2}^3 x^{-5}\,dx$

27. $\int_{-5}^5 \dfrac{2}{x^3}\,dx$

28. $\int_\pi^{2\pi} \cos\theta\,d\theta$

29. $\int_0^2 x(2 + x^5)\,dx$

30. $\int_1^4 \dfrac{1}{\sqrt{x}}\,dx$

31. $\int_0^{\pi/4} \sec^2 t\,dt$

32. $\int_0^1 (3 + x\sqrt{x})\,dx$

33. $\int_{\pi}^{2\pi} \csc^2\theta \, d\theta$

34. $\int_{0}^{\pi/6} \csc\theta \cot\theta \, d\theta$

35. $\int_{1}^{9} \frac{1}{2x} \, dx$

36. $\int_{0}^{1} 10^x \, dx$

37. $\int_{1/2}^{\sqrt{3}/2} \frac{6}{\sqrt{1 - t^2}} \, dt$

38. $\int_{0}^{1} \frac{4}{t^2 + 1} \, dt$

39. $\int_{-1}^{1} e^{u+1} \, du$

40. $\int_{1}^{2} \frac{4 + u^2}{u^3} \, du$

41. $\int_{0}^{2} f(x)\,dx$ where $f(x) = \begin{cases} x^4 & \text{if } 0 \leq x < 1 \\ x^5 & \text{if } 1 \leq x \leq 2 \end{cases}$

42. $\int_{-\pi}^{\pi} f(x)\,dx$ where $f(x) = \begin{cases} x & \text{if } -\pi \leq x \leq 0 \\ \sin x & \text{if } 0 < x \leq \pi \end{cases}$

43–46 ||| Use a graph to give a rough estimate of the area of the region that lies beneath the given curve. Then find the exact area.

43. $y = \sqrt[3]{x}, \quad 0 \leq x \leq 27$

44. $y = x^{-4}, \quad 1 \leq x \leq 6$

45. $y = \sin x, \ 0 \leq x \leq \pi$

46. $y = \sec^2 x, \ 0 \leq x \leq \pi/3$

47–48 ||| Evaluate the integral and interpret it as a difference of areas. Illustrate with a sketch.

47. $\int_{-1}^{2} x^3 \, dx$

48. $\int_{\pi/4}^{5\pi/2} \sin x \, dx$

49–52 ||| Find the derivative of the function.

49. $g(x) = \int_{2x}^{3x} \frac{u^2 - 1}{u^2 + 1} \, du$

$\left[\textit{Hint: } \int_{2x}^{3x} f(u)\,du = \int_{2x}^{0} f(u)\,du + \int_{0}^{3x} f(u)\,du \right]$

50. $g(x) = \int_{\tan x}^{x^2} \frac{1}{\sqrt{2 + t^4}} \, dt$

51. $y = \int_{\sqrt{x}}^{x^3} \sqrt{t} \sin t \, dt$

52. $y = \int_{\cos x}^{5x} \cos(u^2) \, du$

53. If $F(x) = \int_{1}^{x} f(t)\,dt$, where $f(t) = \int_{1}^{t^2} \frac{\sqrt{1 + u^4}}{u} \, du$, find $F''(2)$.

54. Find the interval on which the curve

$$y = \int_{0}^{x} \frac{1}{1 + t + t^2} \, dt$$

is concave upward.

55. If $f(1) = 12$, f' is continuous, and $\int_{1}^{4} f'(x)\,dx = 17$, what is the value of $f(4)$?

56. The **error function**

$$\text{erf}(x) = \frac{2}{\sqrt{\pi}} \int_{0}^{x} e^{-t^2} \, dt$$

is used in probability, statistics, and engineering.

(a) Show that $\int_{a}^{b} e^{-t^2}\,dt = \frac{1}{2}\sqrt{\pi}\,[\text{erf}(b) - \text{erf}(a)]$.

(b) Show that the function $y = e^{x^2}\text{erf}(x)$ satisfies the differential equation $y' = 2xy + 2/\sqrt{\pi}$.

57. The Fresnel function S was defined in Example 3 and graphed in Figures 7 and 8.

(a) At what values of x does this function have local maximum values?

(b) On what intervals is the function concave upward?

CAS (c) Use a graph to solve the following equation correct to two decimal places:

$$\int_{0}^{x} \sin(\pi t^2/2)\,dt = 0.2$$

CAS **58.** The **sine integral function**

$$\text{Si}(x) = \int_{0}^{x} \frac{\sin t}{t} \, dt$$

is important in electrical engineering. [The integrand $f(t) = (\sin t)/t$ is not defined when $t = 0$, but we know that its limit is 1 when $t \to 0$. So we define $f(0) = 1$ and this makes f a continuous function everywhere.]

(a) Draw the graph of Si.

(b) At what values of x does this function have local maximum values?

(c) Find the coordinates of the first inflection point to the right of the origin.

(d) Does this function have horizontal asymptotes?

(e) Solve the following equation correct to one decimal place:

$$\int_{0}^{x} \frac{\sin t}{t} \, dt = 1$$

59–60 ||| Let $g(x) = \int_{0}^{x} f(t)\,dt$, where f is the function whose graph is shown.

(a) At what values of x do the local maximum and minimum values of g occur?

(b) Where does g attain its absolute maximum value?

(c) On what intervals is g concave downward?

(d) Sketch the graph of g.

59.

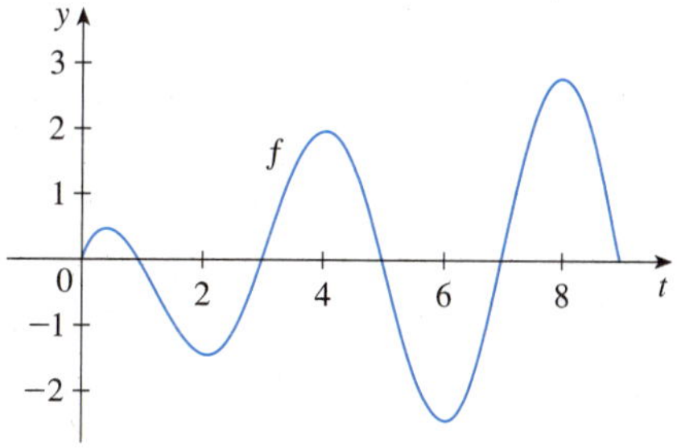

60.

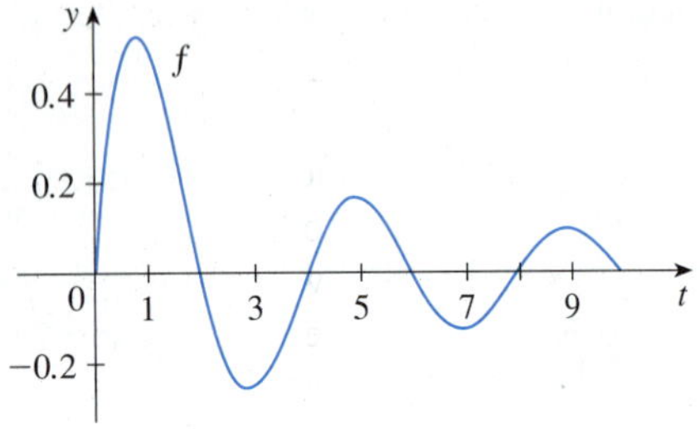

61–62 Evaluate the limit by first recognizing the sum as a Riemann sum for a function defined on [0, 1].

61. $\lim_{n\to\infty} \sum_{i=1}^{n} \frac{i^3}{n^4}$

62. $\lim_{n\to\infty} \frac{1}{n}\left(\sqrt{\frac{1}{n}} + \sqrt{\frac{2}{n}} + \sqrt{\frac{3}{n}} + \cdots + \sqrt{\frac{n}{n}}\right)$

63. Justify (3) for the case $h < 0$.

64. If f is continuous and g and h are differentiable functions, find a formula for

$$\frac{d}{dx}\int_{g(x)}^{h(x)} f(t)\,dt$$

65. (a) Show that $1 \leqslant \sqrt{1 + x^3} \leqslant 1 + x^3$ for $x \geqslant 0$.
(b) Show that $1 \leqslant \int_0^1 \sqrt{1 + x^3}\,dx \leqslant 1.25$.

66. Let

$$f(x) = \begin{cases} 0 & \text{if } x < 0 \\ x & \text{if } 0 \leqslant x \leqslant 1 \\ 2 - x & \text{if } 1 < x \leqslant 2 \\ 0 & \text{if } x > 2 \end{cases}$$

and $$g(x) = \int_0^x f(t)\,dt$$

(a) Find an expression for $g(x)$ similar to the one for $f(x)$.
(b) Sketch the graphs of f and g.
(c) Where is f differentiable? Where is g differentiable?

67. Find a function f and a number a such that

$$6 + \int_a^x \frac{f(t)}{t^2}\,dt = 2\sqrt{x}$$

for all $x > 0$.

68. The area labeled B is three times the area labeled A. Express b in terms of a.

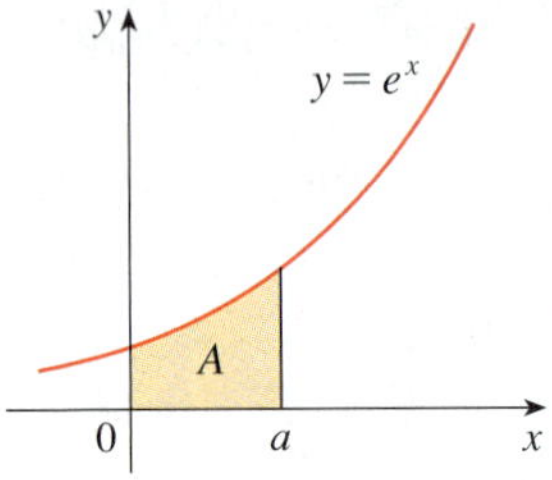
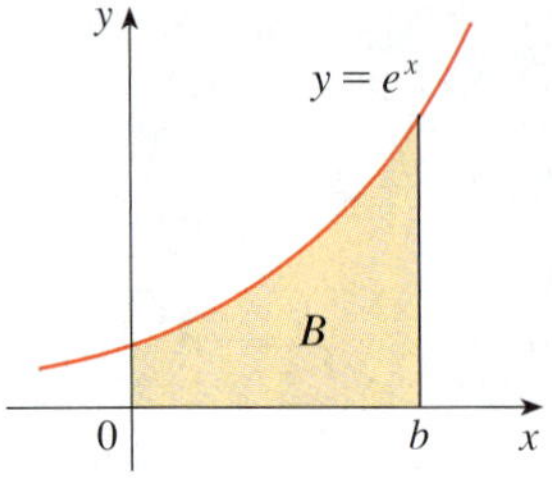

69. A manufacturing company owns a major piece of equipment that depreciates at the (continuous) rate $f = f(t)$, where t is the time measured in months since its last overhaul. Because a fixed cost A is incurred each time the machine is overhauled, the company wants to determine the optimal time T (in months) between overhauls.
(a) Explain why $\int_0^t f(s)\,ds$ represents the loss in value of the machine over the period of time t since the last overhaul.
(b) Let $C = C(t)$ be given by

$$C(t) = \frac{1}{t}\left[A + \int_0^t f(s)\,ds\right]$$

What does C represent and why would the company want to minimize C?
(c) Show that C has a minimum value at the numbers $t = T$ where $C(T) = f(T)$.

70. A high-tech company purchases a new computing system whose initial value is V. The system will depreciate at the rate $f = f(t)$ and will accumulate maintenance costs at the rate $g = g(t)$, where t is the time measured in months. The company wants to determine the optimal time to replace the system.
(a) Let

$$C(t) = \frac{1}{t}\int_0^t [f(s) + g(s)]\,ds$$

Show that the critical numbers of C occur at the numbers t where $C(t) = f(t) + g(t)$.
(b) Suppose that

$$f(t) = \begin{cases} \frac{V}{15} - \frac{V}{450}t & \text{if } 0 < t \leqslant 30 \\ 0 & \text{if } t > 30 \end{cases}$$

and $$g(t) = \frac{Vt^2}{12{,}900} \qquad t > 0$$

Determine the length of time T for the total depreciation $D(t) = \int_0^t f(s)\,ds$ to equal the initial value V.
(c) Determine the absolute minimum of C on $(0, T]$.
(d) Sketch the graphs of C and $f + g$ in the same coordinate system, and verify the result in part (a) in this case.

5.4 Indefinite Integrals and the Net Change Theorem

We saw in Section 5.3 that the second part of the Fundamental Theorem of Calculus provides a very powerful method for evaluating the definite integral of a function, assuming that we can find an antiderivative of the function. In this section we introduce a notation for antiderivatives, review the formulas for antiderivatives, and use them to evaluate definite integrals. We also reformulate FTC2 in a way that makes it easier to apply to science and engineering problems.

Indefinite Integrals

Both parts of the Fundamental Theorem establish connections between antiderivatives and definite integrals. Part 1 says that if f is continuous, then $\int_a^x f(t)\,dt$ is an antiderivative of f. Part 2 says that $\int_a^b f(x)\,dx$ can be found by evaluating $F(b) - F(a)$, where F is an antiderivative of f.

We need a convenient notation for antiderivatives that makes them easy to work with. Because of the relation given by the Fundamental Theorem between antiderivatives and integrals, the notation $\int f(x)\,dx$ is traditionally used for an antiderivative of f and is called an **indefinite integral**. Thus

$$\int f(x)\,dx = F(x) \qquad \text{means} \qquad F'(x) = f(x)$$

For example, we can write

$$\int x^2\,dx = \frac{x^3}{3} + C \qquad \text{because} \qquad \frac{d}{dx}\left(\frac{x^3}{3} + C\right) = x^2$$

So we can regard an indefinite integral as representing an entire *family* of functions (one antiderivative for each value of the constant C).

You should distinguish carefully between definite and indefinite integrals. A definite integral $\int_a^b f(x)\,dx$ is a number, whereas an indefinite integral $\int f(x)\,dx$ is a function (or family of functions). The connection between them is given by Part 2 of the Fundamental Theorem. If f is continuous on $[a, b]$, then

$$\int_a^b f(x)\,dx = \int f(x)\,dx\Big]_a^b$$

The effectiveness of the Fundamental Theorem depends on having a supply of antiderivatives of functions. We therefore restate the Table of Antidifferentiation Formulas from Section 4.10, together with a few others, in the notation of indefinite integrals. Any formula can be verified by differentiating the function on the right side and obtaining the integrand. For instance

$$\int \sec^2 x\,dx = \tan x + C \qquad \text{because} \qquad \frac{d}{dx}(\tan x + C) = \sec^2 x$$

1 Table of Indefinite Integrals

$$\int cf(x)\,dx = c\int f(x)\,dx \qquad \int [f(x) + g(x)]\,dx = \int f(x)\,dx + \int g(x)\,dx$$

$$\int k\,dx = kx + C$$

$$\int x^n\,dx = \frac{x^{n+1}}{n+1} + C \quad (n \neq -1) \qquad \int \frac{1}{x}\,dx = \ln|x| + C$$

$$\int e^x\,dx = e^x + C \qquad \int a^x\,dx = \frac{a^x}{\ln a} + C$$

$$\int \sin x\,dx = -\cos x + C \qquad \int \cos x\,dx = \sin x + C$$

$$\int \sec^2 x\,dx = \tan x + C \qquad \int \csc^2 x\,dx = -\cot x + C$$

$$\int \sec x \tan x\,dx = \sec x + C \qquad \int \csc x \cot x\,dx = -\csc x + C$$

$$\int \frac{1}{x^2+1}\,dx = \tan^{-1}x + C \qquad \int \frac{1}{\sqrt{1-x^2}}\,dx = \sin^{-1}x + C$$

Recall from Theorem 4.10.1 that the most general antiderivative *on a given interval* is obtained by adding a constant to a particular antiderivative. **We adopt the convention that when a formula for a general indefinite integral is given, it is valid only on an interval.** Thus, we write

$$\int \frac{1}{x^2}\,dx = -\frac{1}{x} + C$$

with the understanding that it is valid on the interval $(0, \infty)$ or on the interval $(-\infty, 0)$. This is true despite the fact that the general antiderivative of the function $f(x) = 1/x^2$, $x \neq 0$, is

$$F(x) = \begin{cases} -\dfrac{1}{x} + C_1 & \text{if } x < 0 \\[2mm] -\dfrac{1}{x} + C_2 & \text{if } x > 0 \end{cases}$$

The indefinite integral in Example 1 is graphed in Figure 1 for several values of C. The value of C is the y-intercept.

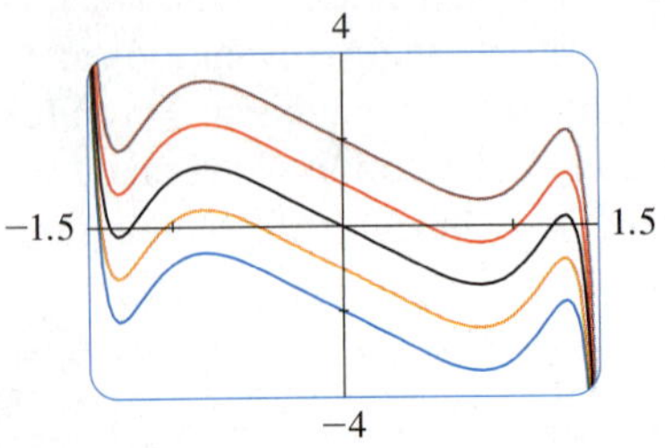

FIGURE 1

EXAMPLE 1 Find the general indefinite integral

$$\int (10x^4 - 2\sec^2 x)\,dx$$

SOLUTION Using our convention and Table 1, we have

$$\int (10x^4 - 2\sec^2 x)\,dx = 10\int x^4\,dx - 2\int \sec^2 x\,dx$$

$$= 10\frac{x^5}{5} - 2\tan x + C$$

$$= 2x^5 - 2\tan x + C$$

You should check this answer by differentiating it.

EXAMPLE 2 Evaluate $\displaystyle\int \frac{\cos\theta}{\sin^2\theta}\,d\theta$.

SOLUTION This indefinite integral isn't immediately apparent in Table 1, so we use trigonometric identities to rewrite the function before integrating:

$$\int \frac{\cos\theta}{\sin^2\theta}\,d\theta = \int \left(\frac{1}{\sin\theta}\right)\left(\frac{\cos\theta}{\sin\theta}\right)d\theta$$
$$= \int \csc\theta\cot\theta\,d\theta = -\csc\theta + C$$

EXAMPLE 3 Evaluate $\displaystyle\int_0^3 (x^3 - 6x)\,dx$.

SOLUTION Using FTC2 and Table 1, we have

$$\int_0^3 (x^3 - 6x)\,dx = \frac{x^4}{4} - 6\frac{x^2}{2}\bigg]_0^3$$
$$= \left(\tfrac{1}{4}\cdot 3^4 - 3\cdot 3^2\right) - \left(\tfrac{1}{4}\cdot 0^4 - 3\cdot 0^2\right)$$
$$= \tfrac{81}{4} - 27 - 0 + 0 = -6.75$$

Compare this calculation with Example 2(b) in Section 5.2.

||| Figure 2 shows the graph of the integrand in Example 4. We know from Section 5.2 that the value of the integral can be interpreted as the sum of the areas labeled with a plus sign minus the area labeled with a minus sign.

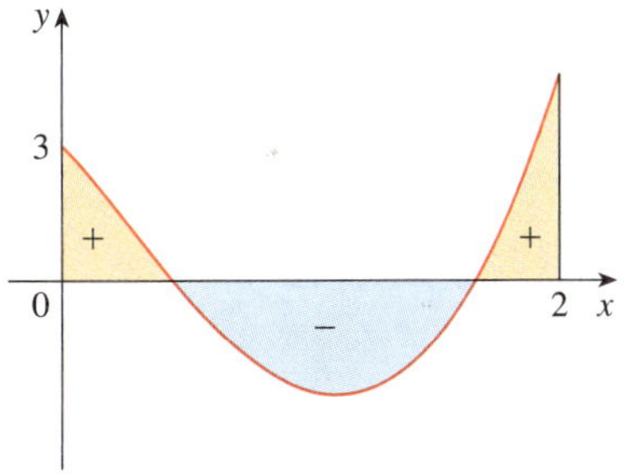

FIGURE 2

EXAMPLE 4 Find $\displaystyle\int_0^2 \left(2x^3 - 6x + \frac{3}{x^2+1}\right)dx$ and interpret the result in terms of areas.

SOLUTION The Fundamental Theorem gives

$$\int_0^2 \left(2x^3 - 6x + \frac{3}{x^2+1}\right)dx = 2\frac{x^4}{4} - 6\frac{x^2}{2} + 3\tan^{-1}x\bigg]_0^2$$
$$= \tfrac{1}{2}x^4 - 3x^2 + 3\tan^{-1}x\Big]_0^2$$
$$= \tfrac{1}{2}(2^4) - 3(2^2) + 3\tan^{-1}2 - 0$$
$$= -4 + 3\tan^{-1}2$$

This is the exact value of the integral. If a decimal approximation is desired, we can use a calculator to approximate $\tan^{-1}2$. Doing so, we get

$$\int_0^2 \left(2x^3 - 6x + \frac{3}{x^2+1}\right)dx \approx -0.67855$$

EXAMPLE 5 Evaluate $\displaystyle\int_1^9 \frac{2t^2 + t^2\sqrt{t} - 1}{t^2}\,dt$.

SOLUTION First we need to write the integrand in a simpler form by carrying out the division:

$$\int_1^9 \frac{2t^2 + t^2\sqrt{t} - 1}{t^2}\,dt = \int_1^9 (2 + t^{1/2} - t^{-2})\,dt$$
$$= 2t + \frac{t^{3/2}}{\frac{3}{2}} - \frac{t^{-1}}{-1}\bigg]_1^9 = 2t + \tfrac{2}{3}t^{3/2} + \frac{1}{t}\bigg]_1^9$$
$$= \left[2\cdot 9 + \tfrac{2}{3}(9)^{3/2} + \tfrac{1}{9}\right] - \left(2\cdot 1 + \tfrac{2}{3}\cdot 1^{3/2} + \tfrac{1}{1}\right)$$
$$= 18 + 18 + \tfrac{1}{9} - 2 - \tfrac{2}{3} - 1 = 32\tfrac{4}{9}$$

Applications

Part 2 of the Fundamental Theorem says that if f is continuous on $[a, b]$, then

$$\int_a^b f(x)\,dx = F(b) - F(a)$$

where F is any antiderivative of f. This means that $F' = f$, so the equation can be rewritten as

$$\int_a^b F'(x)\,dx = F(b) - F(a)$$

We know that $F'(x)$ represents the rate of change of $y = F(x)$ with respect to x and $F(b) - F(a)$ is the change in y when x changes from a to b. [Note that y could, for instance, increase, then decrease, then increase again. Although y might change in both directions, $F(b) - F(a)$ represents the *net* change in y.] So we can reformulate FTC2 in words as follows.

The Net Change Theorem The integral of a rate of change is the net change:

$$\int_a^b F'(x)\,dx = F(b) - F(a)$$

This principle can be applied to all of the rates of change in the natural and social sciences that we discussed in Section 3.3. Here are a few instances of this idea:

- If $V(t)$ is the volume of water in a reservoir at time t, then its derivative $V'(t)$ is the rate at which water flows into the reservoir at time t. So

$$\int_{t_1}^{t_2} V'(t)\,dt = V(t_2) - V(t_1)$$

  is the change in the amount of water in the reservoir between time t_1 and time t_2.

- If $[C](t)$ is the concentration of the product of a chemical reaction at time t, then the rate of reaction is the derivative $d[C]/dt$. So

$$\int_{t_1}^{t_2} \frac{d[C]}{dt}\,dt = [C](t_2) - [C](t_1)$$

  is the change in the concentration of C from time t_1 to time t_2.

- If the mass of a rod measured from the left end to a point x is $m(x)$, then the linear density is $\rho(x) = m'(x)$. So

$$\int_a^b \rho(x)\,dx = m(b) - m(a)$$

  is the mass of the segment of the rod that lies between $x = a$ and $x = b$.

- If the rate of growth of a population is dn/dt, then

$$\int_{t_1}^{t_2} \frac{dn}{dt}\,dt = n(t_2) - n(t_1)$$

  is the net change in population during the time period from t_1 to t_2. (The population increases when births happen and decreases when deaths occur. The net change takes into account both births and deaths.)

- If $C(x)$ is the cost of producing x units of a commodity, then the marginal cost is the derivative $C'(x)$. So

$$\int_{x_1}^{x_2} C'(x)\,dx = C(x_2) - C(x_1)$$

is the increase in cost when production is increased from x_1 units to x_2 units.

- If an object moves along a straight line with position function $s(t)$, then its velocity is $v(t) = s'(t)$, so

2
$$\int_{t_1}^{t_2} v(t)\,dt = s(t_2) - s(t_1)$$

is the net change of position, or *displacement,* of the particle during the time period from t_1 to t_2. In Section 5.1 we guessed that this was true for the case where the object moves in the positive direction, but now we have proved that it is always true.

- If we want to calculate the distance traveled during the time interval, we have to consider the intervals when $v(t) \geqslant 0$ (the particle moves to the right) and also the intervals when $v(t) \leqslant 0$ (the particle moves to the left). In both cases the distance is computed by integrating $|v(t)|$, the speed. Therefore

3
$$\int_{t_1}^{t_2} |v(t)|\,dt = \text{total distance traveled}$$

Figure 3 shows how both displacement and distance traveled can be interpreted in terms of areas under a velocity curve.

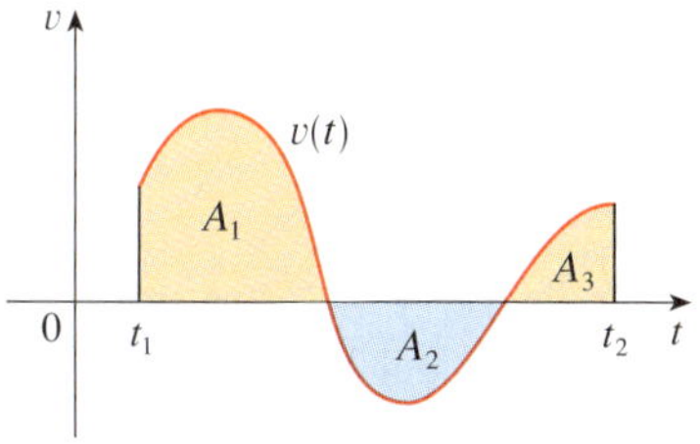

$$\text{displacement} = \int_{t_1}^{t_2} v(t)\,dt = A_1 - A_2 + A_3$$

$$\text{distance} = \int_{t_1}^{t_2} |v(t)|\,dt = A_1 + A_2 + A_3$$

FIGURE 3

- The acceleration of the object is $a(t) = v'(t)$, so

$$\int_{t_1}^{t_2} a(t)\,dt = v(t_2) - v(t_1)$$

is the change in velocity from time t_1 to time t_2.

Resources / Module 7 / Physics and Engineering / Start of Spy Tracks

EXAMPLE 6 A particle moves along a line so that its velocity at time t is $v(t) = t^2 - t - 6$ (measured in meters per second).
(a) Find the displacement of the particle during the time period $1 \leqslant t \leqslant 4$.
(b) Find the distance traveled during this time period.

SOLUTION
(a) By Equation 2, the displacement is

$$s(4) - s(1) = \int_1^4 v(t)\,dt = \int_1^4 (t^2 - t - 6)\,dt$$

$$= \left[\frac{t^3}{3} - \frac{t^2}{2} - 6t\right]_1^4 = -\frac{9}{2}$$

This means that the particle moved 4.5 m toward the left.

(b) Note that $v(t) = t^2 - t - 6 = (t - 3)(t + 2)$ and so $v(t) \leqslant 0$ on the interval $[1, 3]$ and $v(t) \geqslant 0$ on $[3, 4]$. Thus, from Equation 3, the distance traveled is

|||| To integrate the absolute value of $v(t)$, we use Property 5 of integrals from Section 5.2 to split the integral into two parts, one where $v(t) \leqslant 0$ and one where $v(t) \geqslant 0$.

$$\begin{aligned}\int_1^4 |v(t)|\,dt &= \int_1^3 [-v(t)]\,dt + \int_3^4 v(t)\,dt \\ &= \int_1^3 (-t^2 + t + 6)\,dt + \int_3^4 (t^2 - t - 6)\,dt \\ &= \left[-\frac{t^3}{3} + \frac{t^2}{2} + 6t\right]_1^3 + \left[\frac{t^3}{3} - \frac{t^2}{2} - 6t\right]_3^4 \\ &= \frac{61}{6} \approx 10.17 \text{ m}\end{aligned}$$

EXAMPLE 7 Figure 4 shows the power consumption in the city of San Francisco for a day in September (P is measured in megawatts; t is measured in hours starting at midnight). Estimate the energy used on that day.

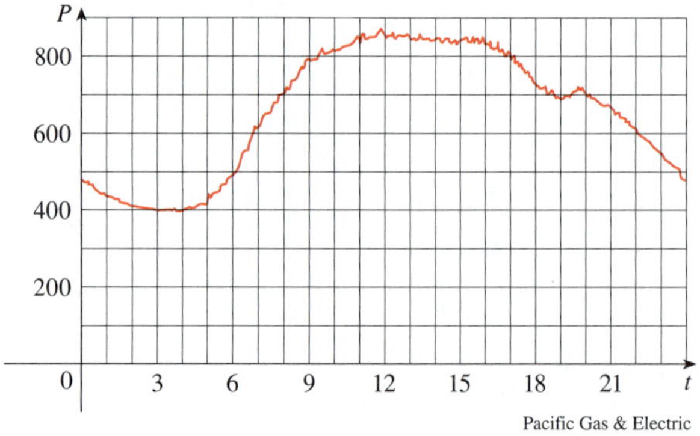

FIGURE 4

SOLUTION Power is the rate of change of energy: $P(t) = E'(t)$. So, by the Net Change Theorem,

$$\int_0^{24} P(t)\,dt = \int_0^{24} E'(t)\,dt = E(24) - E(0)$$

is the total amount of energy used that day. We approximate the value of the integral using the Midpoint Rule with 12 subintervals and $\Delta t = 2$:

$$\begin{aligned}\int_0^{24} P(t)\,dt &\approx [P(1) + P(3) + P(5) + \cdots + P(21) + P(23)]\,\Delta t \\ &\approx (440 + 400 + 420 + 620 + 790 + 840 + 850 \\ &\qquad + 840 + 810 + 690 + 670 + 550)(2) \\ &= 15{,}840\end{aligned}$$

The energy used was approximately 15,840 megawatt-hours.

|||| A note on units

How did we know what units to use for energy in Example 7? The integral $\int_0^{24} P(t)\,dt$ is defined as the limit of sums of terms of the form $P(t_i^*)\,\Delta t$. Now $P(t_i^*)$ is measured in megawatts and Δt is measured in hours, so their product is measured in megawatt-hours. The same is true of the limit. In general, the unit of measurement for $\int_a^b f(x)\,dx$ is the product of the unit for $f(x)$ and the unit for x.

5.4 Exercises

1–4 Verify by differentiation that the formula is correct.

1. $\displaystyle\int \frac{x}{\sqrt{x^2+1}}\,dx = \sqrt{x^2+1} + C$

2. $\displaystyle\int x\cos x\,dx = x\sin x + \cos x + C$

3. $\displaystyle\int \frac{1}{\sqrt{(a^2-x^2)^3}}\,dx = \frac{x}{a^2\sqrt{a^2-x^2}} + C$

4. $\displaystyle\int \frac{1}{x^2\sqrt{x^2+a^2}}\,dx = -\frac{\sqrt{x^2+a^2}}{a^2x} + C$

5–14 Find the general indefinite integral.

5. $\displaystyle\int x^{-3/4}\,dx$

6. $\displaystyle\int \sqrt[3]{x}\,dx$

7. $\displaystyle\int (x^3+6x+1)\,dx$

8. $\displaystyle\int x(1+2x^4)\,dx$

9. $\displaystyle\int (1-t)(2+t^2)\,dt$

10. $\displaystyle\int \left(x^2+1+\frac{1}{x^2+1}\right)dx$

11. $\displaystyle\int (2-\sqrt{x})^2\,dx$

12. $\displaystyle\int (3e^u+\sec^2 u)\,du$

13. $\displaystyle\int \frac{\sin x}{1-\sin^2 x}\,dx$

14. $\displaystyle\int \frac{\sin 2x}{\sin x}\,dx$

15–16 Find the general indefinite integral. Illustrate by graphing several members of the family on the same screen.

15. $\displaystyle\int x\sqrt{x}\,dx$

16. $\displaystyle\int (\cos x - 2\sin x)\,dx$

17–40 Evaluate the integral.

17. $\displaystyle\int_0^2 (6x^2-4x+5)\,dx$

18. $\displaystyle\int_1^3 (1+2x-4x^3)\,dx$

19. $\displaystyle\int_{-1}^0 (2x-e^x)\,dx$

20. $\displaystyle\int_{-2}^0 (u^5-u^3+u^2)\,du$

21. $\displaystyle\int_{-2}^2 (3u+1)^2\,du$

22. $\displaystyle\int_0^4 (2v+5)(3v-1)\,dv$

23. $\displaystyle\int_1^4 \sqrt{t}(1+t)\,dt$

24. $\displaystyle\int_0^9 \sqrt{2t}\,dt$

25. $\displaystyle\int_{-2}^{-1} \left(4y^3+\frac{2}{y^3}\right)dy$

26. $\displaystyle\int_1^2 \frac{y+5y^7}{y^3}\,dy$

27. $\displaystyle\int_0^1 x(\sqrt[3]{x}+\sqrt[4]{x})\,dx$

28. $\displaystyle\int_0^5 (2e^x+4\cos x)\,dx$

29. $\displaystyle\int_1^4 \sqrt{\frac{5}{x}}\,dx$

30. $\displaystyle\int_1^9 \frac{3x-2}{\sqrt{x}}\,dx$

31. $\displaystyle\int_0^\pi (4\sin\theta-3\cos\theta)\,d\theta$

32. $\displaystyle\int_{\pi/4}^{\pi/3} \sec\theta\tan\theta\,d\theta$

33. $\displaystyle\int_0^{\pi/4} \frac{1+\cos^2\theta}{\cos^2\theta}\,d\theta$

34. $\displaystyle\int_0^{\pi/3} \frac{\sin\theta+\sin\theta\tan^2\theta}{\sec^2\theta}\,d\theta$

35. $\displaystyle\int_1^{64} \frac{1+\sqrt[3]{x}}{\sqrt{x}}\,dx$

36. $\displaystyle\int_0^1 (1+x^2)^3\,dx$

37. $\displaystyle\int_1^e \frac{x^2+x+1}{x}\,dx$

38. $\displaystyle\int_4^9 \left(\sqrt{x}+\frac{1}{\sqrt{x}}\right)^2 dx$

39. $\displaystyle\int_{-1}^2 (x-2|x|)\,dx$

40. $\displaystyle\int_0^{3\pi/2} |\sin x|\,dx$

41. Use a graph to estimate the x-intercepts of the curve $y = x + x^2 - x^4$. Then use this information to estimate the area of the region that lies under the curve and above the x-axis.

42. Repeat Exercise 41 for the curve $y = 2x + 3x^4 - 2x^6$.

43. The area of the region that lies to the right of the y-axis and to the left of the parabola $x = 2y - y^2$ (the shaded region in the figure) is given by the integral $\int_0^2 (2y - y^2)\,dy$. (Turn your head clockwise and think of the region as lying below the curve $x = 2y - y^2$ from $y = 0$ to $y = 2$.) Find the area of the region.

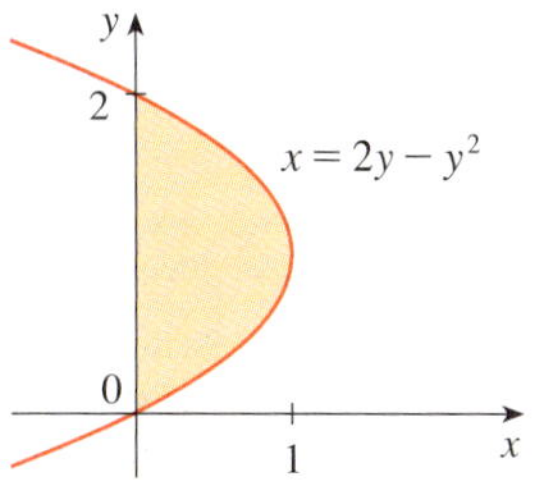

44. The boundaries of the shaded region are the y-axis, the line $y = 1$, and the curve $y = \sqrt[4]{x}$. Find the area of this region by writing x as a function of y and integrating with respect to y (as in Exercise 43).

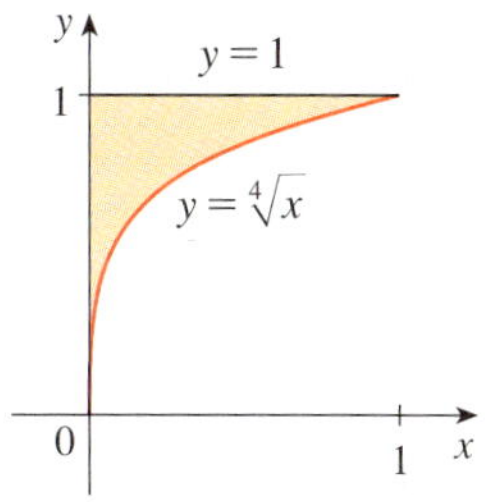

45. If $w'(t)$ is the rate of growth of a child in pounds per year, what does $\int_5^{10} w'(t)\,dt$ represent?

46. The current in a wire is defined as the derivative of the charge: $I(t) = Q'(t)$. (See Example 3 in Section 3.3.) What does $\int_a^b I(t)\,dt$ represent?

47. If oil leaks from a tank at a rate of $r(t)$ gallons per minute at time t, what does $\int_0^{120} r(t)\,dt$ represent?

48. A honeybee population starts with 100 bees and increases at a rate of $n'(t)$ bees per week. What does $100 + \int_0^{15} n'(t)\,dt$ represent?

49. In Section 4.8 we defined the marginal revenue function $R'(x)$ as the derivative of the revenue function $R(x)$, where x is the number of units sold. What does $\int_{1000}^{5000} R'(x)\,dx$ represent?

50. If $f(x)$ is the slope of a trail at a distance of x miles from the start of the trail, what does $\int_3^5 f(x)\,dx$ represent?

51. If x is measured in meters and $f(x)$ is measured in newtons, what are the units for $\int_0^{100} f(x)\,dx$?

52. If the units for x are feet and the units for $a(x)$ are pounds per foot, what are the units for da/dx? What units does $\int_2^8 a(x)\,dx$ have?

53–54 ||| The velocity function (in meters per second) is given for a particle moving along a line. Find (a) the displacement and (b) the distance traveled by the particle during the given time interval.

53. $v(t) = 3t - 5, \quad 0 \leqslant t \leqslant 3$

54. $v(t) = t^2 - 2t - 8, \quad 1 \leqslant t \leqslant 6$

55–56 ||| The acceleration function (in m/s^2) and the initial velocity are given for a particle moving along a line. Find (a) the velocity at time t and (b) the distance traveled during the given time interval.

55. $a(t) = t + 4, \quad v(0) = 5, \quad 0 \leqslant t \leqslant 10$

56. $a(t) = 2t + 3, \quad v(0) = -4, \quad 0 \leqslant t \leqslant 3$

57. The linear density of a rod of length 4 m is given by $\rho(x) = 9 + 2\sqrt{x}$ measured in kilograms per meter, where x is measured in meters from one end of the rod. Find the total mass of the rod.

58. Water flows from the bottom of a storage tank at a rate of $r(t) = 200 - 4t$ liters per minute, where $0 \leqslant t \leqslant 50$. Find the amount of water that flows from the tank during the first 10 minutes.

59. The velocity of a car was read from its speedometer at 10-second intervals and recorded in the table. Use the Midpoint Rule to estimate the distance traveled by the car.

t (s)	v (mi/h)	t (s)	v (mi/h)
0	0	60	56
10	38	70	53
20	52	80	50
30	58	90	47
40	55	100	45
50	51		

60. Suppose that a volcano is erupting and readings of the rate $r(t)$ at which solid materials are spewed into the atmosphere are given in the table. The time t is measured in seconds and the units for $r(t)$ are tonnes (metric tons) per second.

t	0	1	2	3	4	5	6
$r(t)$	2	10	24	36	46	54	60

(a) Give upper and lower estimates for the quantity $Q(6)$ of erupted materials after 6 seconds.
(b) Use the Midpoint Rule to estimate $Q(6)$.

61. The marginal cost of manufacturing x yards of a certain fabric is $C'(x) = 3 - 0.01x + 0.000006x^2$ (in dollars per yard). Find the increase in cost if the production level is raised from 2000 yards to 4000 yards.

62. Water flows in and out of a storage tank. A graph of the rate of change $r(t)$ of the volume of water in the tank, in liters per day, is shown. If the amount of water in the tank at time $t = 0$ is 25,000 L, use the Midpoint Rule to estimate the amount of water four days later.

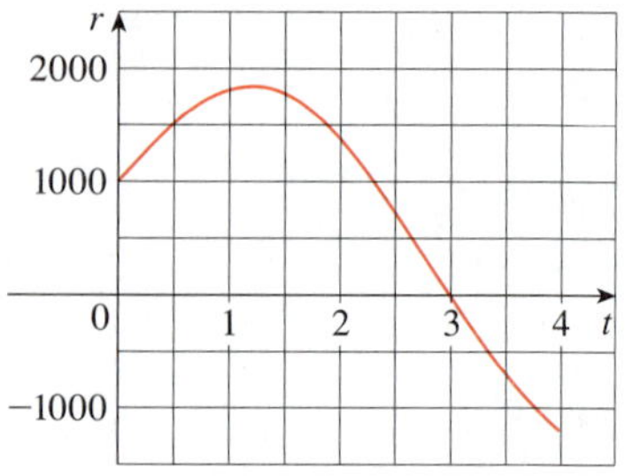

63. Economists use a cumulative distribution called a *Lorenz curve* to describe the distribution of income between households in a given country. Typically, a Lorenz curve is defined on $[0, 1]$ with endpoints $(0, 0)$ and $(1, 1)$, and is continuous, increasing, and concave upward. The points on this curve are determined by ranking all households by income and then computing the percentage of households whose income is less than or equal to a given percentage of the total income of the country. For example, the point $(a/100, b/100)$ is on the Lorenz curve if the bottom $a\%$ of the households receive less than or equal to $b\%$ of the total income. *Absolute equality* of income distribution would occur if the bottom $a\%$ of the households receive $a\%$ of the income, in which case the Lorenz curve would be the line $y = x$. The area between the Lorenz curve and the line $y = x$ measures how much the income distribution differs from absolute equality. The *coefficient of inequality* is the ratio of the area between the Lorenz curve and the line $y = x$ to the area under $y = x$.

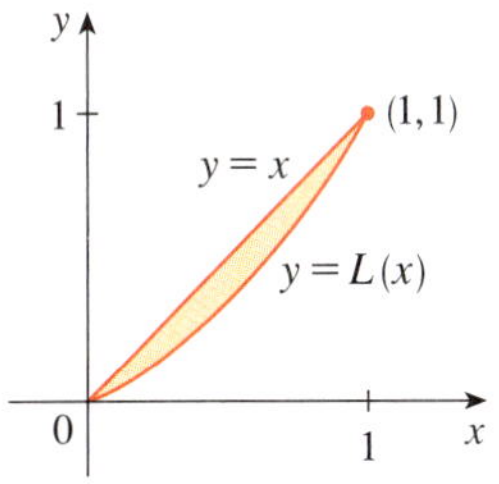

(a) Show that the coefficient of inequality is twice the area between the Lorenz curve and the line $y = x$, that is, show that

$$\text{coefficient of inequality} = 2\int_0^1 [x - L(x)]\,dx$$

(b) The income distribution for a certain country is represented by the Lorenz curve defined by the equation

$$L(x) = \tfrac{5}{12}x^2 + \tfrac{7}{12}x$$

What is the percentage of total income received by the bottom 50% of the households? Find the coefficient of inequality.

64. On May 7, 1992, the space shuttle *Endeavour* was launched on mission STS-49, the purpose of which was to install a new perigee kick motor in an Intelsat communications satellite. The table gives the velocity data for the shuttle between liftoff and the jettisoning of the solid rocket boosters.

Event	Time (s)	Velocity (ft/s)
Launch	0	0
Begin roll maneuver	10	185
End roll maneuver	15	319
Throttle to 89%	20	447
Throttle to 67%	32	742
Throttle to 104%	59	1325
Maximum dynamic pressure	62	1445
Solid rocket booster separation	125	4151

(a) Use a graphing calculator or computer to model these data by a third-degree polynomial.
(b) Use the model in part (a) to estimate the height reached by the *Endeavour*, 125 seconds after liftoff.

WRITING PROJECT

Newton, Leibniz, and the Invention of Calculus

We sometimes read that the inventors of calculus were Sir Isaac Newton (1642–1727) and Gottfried Wilhelm Leibniz (1646–1716). But we know that the basic ideas behind integration were investigated 2500 years ago by ancient Greeks such as Eudoxus and Archimedes, and methods for finding tangents were pioneered by Pierre Fermat (1601–1665), Isaac Barrow (1630–1677), and others. Barrow, Newton's teacher at Cambridge, was the first to understand the inverse relationship between differentiation and integration. What Newton and Leibniz did was to use this relationship, in the form of the Fundamental Theorem of Calculus, in order to develop calculus into a systematic mathematical discipline. It is in this sense that Newton and Leibniz are credited with the invention of calculus.

Read about the contributions of these men in one or more of the given references and write a report on one of the following three topics. You can include biographical details, but the main thrust of your report should be a description, in some detail, of their methods and notations. In particular, you should consult one of the sourcebooks, which give excerpts from the original publications of Newton and Leibniz, translated from Latin to English.

- The Role of Newton in the Development of Calculus
- The Role of Leibniz in the Development of Calculus
- The Controversy between the Followers of Newton and Leibniz over Priority in the Invention of Calculus

References

1. Carl Boyer and Uta Merzbach, *A History of Mathematics* (New York: Wiley, 1987), Chapter 19.

2. Carl Boyer, *The History of the Calculus and Its Conceptual Development* (New York: Dover, 1959), Chapter V.
3. C. H. Edwards, *The Historical Development of the Calculus* (New York: Springer-Verlag, 1979), Chapters 8 and 9.
4. Howard Eves, *An Introduction to the History of Mathematics,* 6th ed. (New York: Saunders, 1990), Chapter 11.
5. C. C. Gillispie, ed., *Dictionary of Scientific Biography* (New York: Scribner's, 1974). See the article on Leibniz by Joseph Hofmann in Volume VIII and the article on Newton by I. B. Cohen in Volume X.
6. Victor Katz, *A History of Mathematics: An Introduction* (New York: HarperCollins, 1993), Chapter 12.
7. Morris Kline, *Mathematical Thought from Ancient to Modern Times* (New York: Oxford University Press, 1972), Chapter 17.

Sourcebooks

1. John Fauvel and Jeremy Gray, eds., *The History of Mathematics: A Reader* (London: MacMillan Press, 1987), Chapters 12 and 13.
2. D. E. Smith, ed., *A Sourcebook in Mathematics* (New York: Dover, 1959), Chapter V.
3. D. J. Struik, ed., *A Sourcebook in Mathematics, 1200–1800* (Princeton, N.J.: Princeton University Press, 1969), Chapter V.

5.5 The Substitution Rule

Because of the Fundamental Theorem, it's important to be able to find antiderivatives. But our antidifferentiation formulas don't tell us how to evaluate integrals such as

$$\int 2x\sqrt{1 + x^2}\,dx \tag{1}$$

To find this integral we use the problem-solving strategy of *introducing something extra.* Here the "something extra" is a new variable; we change from the variable x to a new variable u. Suppose that we let u be the quantity under the root sign in (1), $u = 1 + x^2$. Then the differential of u is $du = 2x\,dx$. Notice that if the dx in the notation for an integral were to be interpreted as a differential, then the differential $2x\,dx$ would occur in (1) and, so, formally, without justifying our calculation, we could write

Differentials were defined in Section 3.11. If $u = f(x)$, then

$$du = f'(x)\,dx$$

$$\int 2x\sqrt{1 + x^2}\,dx = \int \sqrt{1 + x^2}\,2x\,dx = \int \sqrt{u}\,du \tag{2}$$

$$= \tfrac{2}{3}u^{3/2} + C = \tfrac{2}{3}(x^2 + 1)^{3/2} + C$$

But now we can check that we have the correct answer by using the Chain Rule to differentiate the final function of Equation 2:

$$\frac{d}{dx}\left[\tfrac{2}{3}(x^2 + 1)^{3/2} + C\right] = \tfrac{2}{3} \cdot \tfrac{3}{2}(x^2 + 1)^{1/2} \cdot 2x = 2x\sqrt{x^2 + 1}$$

In general, this method works whenever we have an integral that we can write in the form $\int f(g(x))g'(x)\,dx$. Observe that if $F' = f$, then

$$\int F'(g(x))g'(x)\,dx = F(g(x)) + C \tag{3}$$

because, by the Chain Rule,

$$\frac{d}{dx}[F(g(x))] = F'(g(x))g'(x)$$

If we make the "change of variable" or "substitution" $u = g(x)$, then from Equation 3 we have

$$\int F'(g(x))g'(x)\,dx = F(g(x)) + C = F(u) + C = \int F'(u)\,du$$

or, writing $F' = f$, we get

$$\int f(g(x))g'(x)\,dx = \int f(u)\,du$$

Thus, we have proved the following rule.

4 The Substitution Rule If $u = g(x)$ is a differentiable function whose range is an interval I and f is continuous on I, then

$$\int f(g(x))g'(x)\,dx = \int f(u)\,du$$

Notice that the Substitution Rule for integration was proved using the Chain Rule for differentiation. Notice also that if $u = g(x)$, then $du = g'(x)\,dx$, so a way to remember the Substitution Rule is to think of dx and du in (4) as differentials.

Thus, the Substitution Rule says: **It is permissible to operate with *dx* and *du* after integral signs as if they were differentials.**

EXAMPLE 1 Find $\int x^3 \cos(x^4 + 2)\,dx$.

SOLUTION We make the substitution $u = x^4 + 2$ because its differential is $du = 4x^3\,dx$, which, apart from the constant factor 4, occurs in the integral. Thus, using $x^3\,dx = du/4$ and the Substitution Rule, we have

$$\begin{aligned}\int x^3 \cos(x^4 + 2)\,dx &= \int \cos u \cdot \tfrac{1}{4}\,du = \tfrac{1}{4}\int \cos u\,du \\ &= \tfrac{1}{4}\sin u + C \\ &= \tfrac{1}{4}\sin(x^4 + 2) + C\end{aligned}$$

|||| Check the answer by differentiating it.

Notice that at the final stage we had to return to the original variable x.

The idea behind the Substitution Rule is to replace a relatively complicated integral by a simpler integral. This is accomplished by changing from the original variable x to a new variable u that is a function of x. Thus, in Example 1 we replaced the integral $\int x^3 \cos(x^4 + 2)\,dx$ by the simpler integral $\frac{1}{4}\int \cos u\,du$.

The main challenge in using the Substitution Rule is to think of an appropriate substitution. You should try to choose u to be some function in the integrand whose differential also occurs (except for a constant factor). This was the case in Example 1. If that is not possible, try choosing u to be some complicated part of the integrand (perhaps the inner function in a composite function). Finding the right substitution is a bit of an art. It's not unusual to guess wrong; if your first guess doesn't work, try another substitution.

EXAMPLE 2 Evaluate $\int \sqrt{2x+1}\, dx$.

SOLUTION 1 Let $u = 2x + 1$. Then $du = 2\, dx$, so $dx = du/2$. Thus, the Substitution Rule gives

$$\begin{aligned}\int \sqrt{2x+1}\, dx &= \int \sqrt{u}\, \frac{du}{2} = \tfrac{1}{2}\int u^{1/2}\, du \\ &= \frac{1}{2}\cdot\frac{u^{3/2}}{3/2} + C = \tfrac{1}{3}u^{3/2} + C \\ &= \tfrac{1}{3}(2x+1)^{3/2} + C\end{aligned}$$

SOLUTION 2 Another possible substitution is $u = \sqrt{2x+1}$. Then

$$du = \frac{dx}{\sqrt{2x+1}} \qquad \text{so} \qquad dx = \sqrt{2x+1}\, du = u\, du$$

(Or observe that $u^2 = 2x + 1$, so $2u\, du = 2\, dx$.) Therefore

$$\begin{aligned}\int \sqrt{2x+1}\, dx &= \int u \cdot u\, du = \int u^2\, du \\ &= \frac{u^3}{3} + C = \tfrac{1}{3}(2x+1)^{3/2} + C\end{aligned}$$

EXAMPLE 3 Find $\int \frac{x}{\sqrt{1-4x^2}}\, dx$.

SOLUTION Let $u = 1 - 4x^2$. Then $du = -8x\, dx$, so $x\, dx = -\frac{1}{8}\, du$ and

$$\begin{aligned}\int \frac{x}{\sqrt{1-4x^2}}\, dx &= -\tfrac{1}{8}\int \frac{du}{\sqrt{u}} = -\tfrac{1}{8}\int u^{-1/2}\, du \\ &= -\tfrac{1}{8}(2\sqrt{u}) + C = -\tfrac{1}{4}\sqrt{1-4x^2} + C\end{aligned}$$

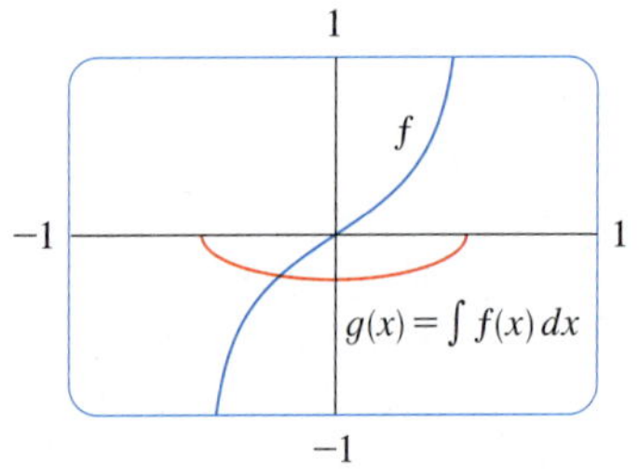

FIGURE 1

$f(x) = \dfrac{x}{\sqrt{1-4x^2}}$

$g(x) = \int f(x)\, dx = -\frac{1}{4}\sqrt{1-4x^2}$

The answer to Example 3 could be checked by differentiation, but instead let's check it with a graph. In Figure 1 we have used a computer to graph both the integrand $f(x) = x/\sqrt{1-4x^2}$ and its indefinite integral $g(x) = -\frac{1}{4}\sqrt{1-4x^2}$ (we take the case $C = 0$). Notice that $g(x)$ decreases when $f(x)$ is negative, increases when $f(x)$ is positive, and has its minimum value when $f(x) = 0$. So it seems reasonable, from the graphical evidence, that g is an antiderivative of f.

EXAMPLE 4 Calculate $\int e^{5x}\, dx$.

SOLUTION If we let $u = 5x$, then $du = 5\, dx$, so $dx = \frac{1}{5}\, du$. Therefore

$$\int e^{5x}\, dx = \tfrac{1}{5}\int e^u\, du = \tfrac{1}{5}e^u + C = \tfrac{1}{5}e^{5x} + C$$

EXAMPLE 5 Find $\int \sqrt{1 + x^2}\, x^5\, dx$.

SOLUTION An appropriate substitution becomes more obvious if we factor x^5 as $x^4 \cdot x$. Let $u = 1 + x^2$. Then $du = 2x\,dx$, so $x\,dx = du/2$. Also $x^2 = u - 1$, so $x^4 = (u - 1)^2$:

$$\begin{aligned}\int \sqrt{1 + x^2}\, x^5\, dx &= \int \sqrt{1 + x^2}\, x^4 \cdot x\, dx \\ &= \int \sqrt{u}\,(u - 1)^2 \frac{du}{2} = \tfrac{1}{2}\int \sqrt{u}\,(u^2 - 2u + 1)\, du \\ &= \tfrac{1}{2}\int (u^{5/2} - 2u^{3/2} + u^{1/2})\, du \\ &= \tfrac{1}{2}\left(\tfrac{2}{7}u^{7/2} - 2 \cdot \tfrac{2}{5}u^{5/2} + \tfrac{2}{3}u^{3/2}\right) + C \\ &= \tfrac{1}{7}(1 + x^2)^{7/2} - \tfrac{2}{5}(1 + x^2)^{5/2} + \tfrac{1}{3}(1 + x^2)^{3/2} + C\end{aligned}$$

EXAMPLE 6 Calculate $\int \tan x\, dx$.

SOLUTION First we write tangent in terms of sine and cosine:

$$\int \tan x\, dx = \int \frac{\sin x}{\cos x}\, dx$$

This suggests that we should substitute $u = \cos x$, since then $du = -\sin x\, dx$ and so $\sin x\, dx = -du$:

$$\begin{aligned}\int \tan x\, dx &= \int \frac{\sin x}{\cos x}\, dx = -\int \frac{du}{u} \\ &= -\ln|u| + C = -\ln|\cos x| + C\end{aligned}$$

Since $-\ln|\cos x| = \ln(|\cos x|^{-1}) = \ln(1/|\cos x|) = \ln|\sec x|$, the result of Example 6 can also be written as

5 $$\int \tan x\, dx = \ln|\sec x| + C$$

Definite Integrals

When evaluating a *definite* integral by substitution, two methods are possible. One method is to evaluate the indefinite integral first and then use the Fundamental Theorem. For instance, using the result of Example 2, we have

$$\begin{aligned}\int_0^4 \sqrt{2x + 1}\, dx = \int \sqrt{2x + 1}\, dx\Big]_0^4 &= \tfrac{1}{3}(2x + 1)^{3/2}\Big]_0^4 \\ &= \tfrac{1}{3}(9)^{3/2} - \tfrac{1}{3}(1)^{3/2} = \tfrac{1}{3}(27 - 1) = \tfrac{26}{3}\end{aligned}$$

Another method, which is usually preferable, is to change the limits of integration when the variable is changed.

This rule says that when using a substitution in a definite integral, we must put everything in terms of the new variable u, not only x and dx but also the limits of integration. The new limits of integration are the values of u that correspond to $x = a$ and $x = b$.

6 The Substitution Rule for Definite Integrals If g' is continuous on $[a, b]$ and f is continuous on the range of $u = g(x)$, then

$$\int_a^b f(g(x))g'(x)\,dx = \int_{g(a)}^{g(b)} f(u)\,du$$

Proof Let F be an antiderivative of f. Then, by (3), $F(g(x))$ is an antiderivative of $f(g(x))g'(x)$, so by Part 2 of the Fundamental Theorem, we have

$$\int_a^b f(g(x))g'(x)\,dx = F(g(x))\Big]_a^b = F(g(b)) - F(g(a))$$

But, applying FTC2 a second time, we also have

$$\int_{g(a)}^{g(b)} f(u)\,du = F(u)\Big]_{g(a)}^{g(b)} = F(g(b)) - F(g(a))$$

EXAMPLE 7 Evaluate $\int_0^4 \sqrt{2x+1}\,dx$ using (6).

SOLUTION Using the substitution from Solution 1 of Example 2, we have $u = 2x + 1$ and $dx = du/2$. To find the new limits of integration we note that

$$\text{when } x = 0,\ u = 2(0) + 1 = 1 \qquad \text{and} \qquad \text{when } x = 4,\ u = 2(4) + 1 = 9$$

Therefore

$$\int_0^4 \sqrt{2x+1}\,dx = \int_1^9 \tfrac{1}{2}\sqrt{u}\,du$$

$$= \tfrac{1}{2} \cdot \tfrac{2}{3}u^{3/2}\Big]_1^9$$

$$= \tfrac{1}{3}(9^{3/2} - 1^{3/2}) = \tfrac{26}{3}$$

The geometric interpretation of Example 7 is shown in Figure 2. The substitution $u = 2x + 1$ stretches the interval $[0, 4]$ by a factor of 2 and translates it to the right by 1 unit. The Substitution Rule shows that the two areas are equal.

Observe that when using (6) we do not return to the variable x after integrating. We simply evaluate the expression in u between the appropriate values of u.

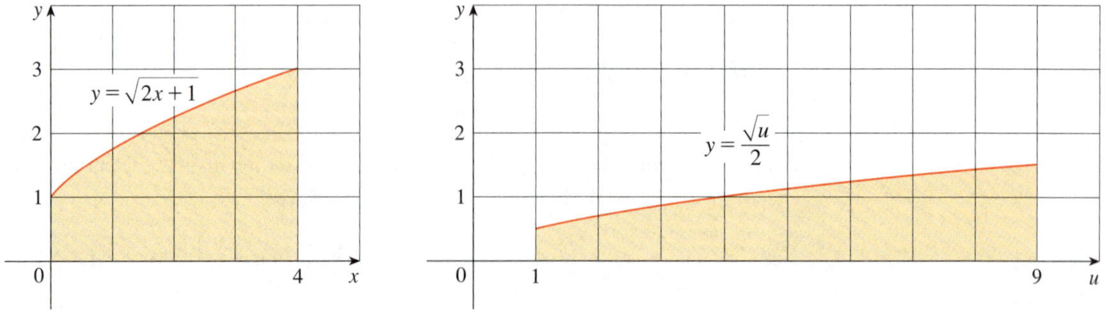

FIGURE 2

EXAMPLE 8 Evaluate $\int_1^2 \frac{dx}{(3-5x)^2}$.

|||| The integral given in Example 8 is an abbreviation for

$$\int_1^2 \frac{1}{(3-5x)^2}\,dx$$

SOLUTION Let $u = 3 - 5x$. Then $du = -5\,dx$, so $dx = -du/5$. When $x = 1$, $u = -2$ and when $x = 2$, $u = -7$. Thus

$$\int_1^2 \frac{dx}{(3-5x)^2} = -\frac{1}{5}\int_{-2}^{-7} \frac{du}{u^2}$$

$$= -\frac{1}{5}\left[-\frac{1}{u}\right]_{-2}^{-7} = \frac{1}{5u}\bigg]_{-2}^{-7}$$

$$= \frac{1}{5}\left(-\frac{1}{7} + \frac{1}{2}\right) = \frac{1}{14}$$

EXAMPLE 9 Calculate $\displaystyle\int_1^e \frac{\ln x}{x}\,dx$.

SOLUTION We let $u = \ln x$ because its differential $du = dx/x$ occurs in the integral. When $x = 1$, $u = \ln 1 = 0$; when $x = e$, $u = \ln e = 1$. Thus

$$\int_1^e \frac{\ln x}{x}\,dx = \int_0^1 u\,du = \frac{u^2}{2}\bigg]_0^1 = \frac{1}{2}$$

|||| Since the function $f(x) = (\ln x)/x$ in Example 9 is positive for $x > 1$, the integral represents the area of the shaded region in Figure 3.

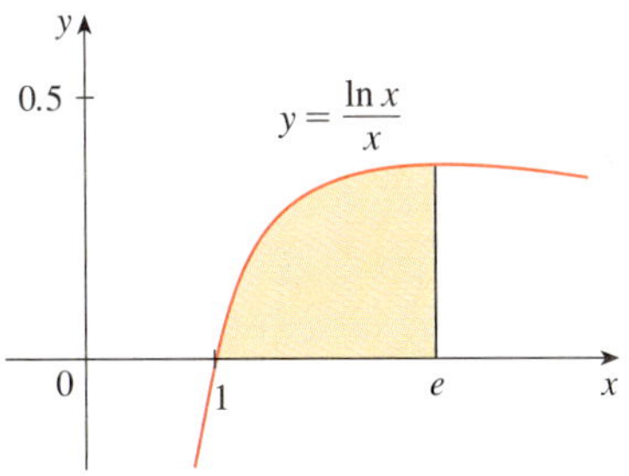

FIGURE 3

|||| Symmetry

The next theorem uses the Substitution Rule for Definite Integrals (6) to simplify the calculation of integrals of functions that possess symmetry properties.

7 Integrals of Symmetric Functions Suppose f is continuous on $[-a, a]$.

(a) If f is even $[f(-x) = f(x)]$, then $\int_{-a}^{a} f(x)\,dx = 2\int_0^a f(x)\,dx$.

(b) If f is odd $[f(-x) = -f(x)]$, then $\int_{-a}^{a} f(x)\,dx = 0$.

Proof We split the integral in two:

8 $$\int_{-a}^{a} f(x)\,dx = \int_{-a}^{0} f(x)\,dx + \int_0^a f(x)\,dx = -\int_0^{-a} f(x)\,dx + \int_0^a f(x)\,dx$$

In the first integral on the far right side we make the substitution $u = -x$. Then $du = -dx$ and when $x = -a$, $u = a$. Therefore

$$-\int_0^{-a} f(x)\,dx = -\int_0^a f(-u)(-du) = \int_0^a f(-u)\,du$$

and so Equation 8 becomes

9 $$\int_{-a}^{a} f(x)\,dx = \int_0^a f(-u)\,du + \int_0^a f(x)\,dx$$

(a) If f is even, then $f(-u) = f(u)$ so Equation 9 gives

$$\int_{-a}^{a} f(x)\,dx = \int_0^a f(u)\,du + \int_0^a f(x)\,dx = 2\int_0^a f(x)\,dx$$

(b) If f is odd, then $f(-u) = -f(u)$ and so Equation 9 gives

$$\int_{-a}^{a} f(x)\,dx = -\int_0^a f(u)\,du + \int_0^a f(x)\,dx = 0$$

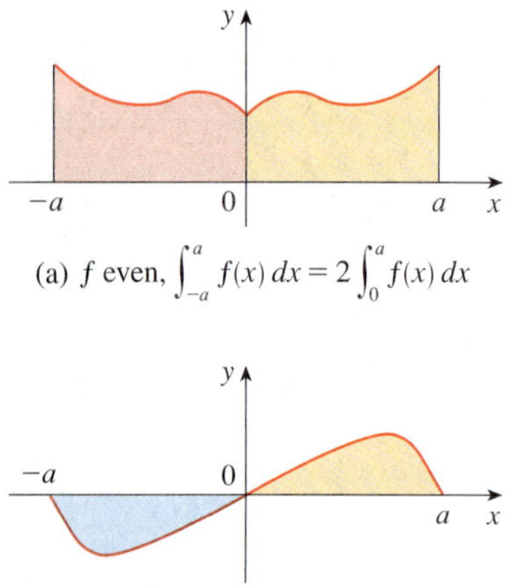

(a) f even, $\int_{-a}^{a} f(x)\,dx = 2\int_0^a f(x)\,dx$

(b) f odd, $\int_{-a}^{a} f(x)\,dx = 0$

FIGURE 4

Theorem 7 is illustrated by Figure 4. For the case where f is positive and even, part (a) says that the area under $y = f(x)$ from $-a$ to a is twice the area from 0 to a because of symmetry. Recall that an integral $\int_a^b f(x)\,dx$ can be expressed as the area above the x-axis and below $y = f(x)$ minus the area below the axis and above the curve. Thus, part (b) says the integral is 0 because the areas cancel.

EXAMPLE 10 Since $f(x) = x^6 + 1$ satisfies $f(-x) = f(x)$, it is even and so

$$\int_{-2}^{2} (x^6 + 1)\,dx = 2\int_0^2 (x^6 + 1)\,dx$$
$$= 2\left[\tfrac{1}{7}x^7 + x\right]_0^2 = 2\left(\tfrac{128}{7} + 2\right) = \tfrac{284}{7}$$

EXAMPLE 11 Since $f(x) = (\tan x)/(1 + x^2 + x^4)$ satisfies $f(-x) = -f(x)$, it is odd and so

$$\int_{-1}^{1} \frac{\tan x}{1 + x^2 + x^4}\,dx = 0$$

5.5 Exercises

1–6 ||| Evaluate the integral by making the given substitution.

1. $\int \cos 3x\,dx, \quad u = 3x$

2. $\int x(4 + x^2)^{10}\,dx, \quad u = 4 + x^2$

3. $\int x^2\sqrt{x^3 + 1}\,dx, \quad u = x^3 + 1$

4. $\int \frac{\sin\sqrt{x}}{\sqrt{x}}\,dx, \quad u = \sqrt{x}$

5. $\int \frac{4}{(1 + 2x)^3}\,dx, \quad u = 1 + 2x$

6. $\int e^{\sin\theta}\cos\theta\,d\theta, \quad u = \sin\theta$

7–44 ||| Evaluate the indefinite integral.

7. $\int 2x(x^2 + 3)^4\,dx$

8. $\int x^2(x^3 + 5)^9\,dx$

9. $\int (3x - 2)^{20}\,dx$

10. $\int (2 - x)^6\,dx$

11. $\int \frac{1 + 4x}{\sqrt{1 + x + 2x^2}}\,dx$

12. $\int \frac{x}{(x^2 + 1)^2}\,dx$

13. $\int \frac{dx}{5 - 3x}$

14. $\int \frac{x}{x^2 + 1}\,dx$

15. $\int \frac{3}{(2y + 1)^5}\,dy$

16. $\int \frac{1}{(5t + 4)^{2.7}}\,dt$

17. $\int \sqrt{4 - t}\,dt$

18. $\int y^3\sqrt{2y^4 - 1}\,dy$

19. $\int \sin \pi t \, dt$

20. $\int \sec 2\theta \tan 2\theta \, d\theta$

21. $\int \frac{(\ln x)^2}{x} \, dx$

22. $\int \frac{\tan^{-1}x}{1 + x^2} \, dx$

23. $\int \frac{\cos \sqrt{t}}{\sqrt{t}} \, dt$

24. $\int \sqrt{x} \sin(1 + x^{3/2}) \, dx$

25. $\int \cos \theta \sin^6\theta \, d\theta$

26. $\int (1 + \tan \theta)^5 \sec^2\theta \, d\theta$

27. $\int e^x \sqrt{1 + e^x} \, dx$

28. $\int e^{\cos t} \sin t \, dt$

29. $\int \frac{z^2}{\sqrt[3]{1 + z^3}} \, dz$

30. $\int \frac{ax + b}{\sqrt{ax^2 + 2bx + c}} \, dx$

31. $\int \frac{dx}{x \ln x}$

32. $\int \frac{e^x}{e^x + 1} \, dx$

33. $\int \sqrt{\cot x} \csc^2 x \, dx$

34. $\int \frac{\cos(\pi/x)}{x^2} \, dx$

35. $\int \cot x \, dx$

36. $\int \frac{\sin x}{1 + \cos^2 x} \, dx$

37. $\int \sec^3 x \tan x \, dx$

38. $\int \sqrt[3]{x^3 + 1} \, x^5 \, dx$

39. $\int x^a \sqrt{b + cx^{a+1}} \, dx \quad (c \neq 0, a \neq -1)$

40. $\int \sin t \sec^2(\cos t) \, dt$

41. $\int \frac{1 + x}{1 + x^2} \, dx$

42. $\int \frac{x}{1 + x^4} \, dx$

43. $\int \frac{x}{\sqrt[4]{x + 2}} \, dx$

44. $\int \frac{x^2}{\sqrt{1 - x}} \, dx$

45–48 Evaluate the indefinite integral. Illustrate and check that your answer is reasonable by graphing both the function and its antiderivative (take $C = 0$).

45. $\int \frac{3x - 1}{(3x^2 - 2x + 1)^4} \, dx$

46. $\int \frac{x}{\sqrt{x^2 + 1}} \, dx$

47. $\int \sin^3 x \cos x \, dx$

48. $\int \tan^2\theta \sec^2\theta \, d\theta$

49–70 Evaluate the definite integral, if it exists.

49. $\int_0^2 (x - 1)^{25} \, dx$

50. $\int_0^7 \sqrt{4 + 3x} \, dx$

51. $\int_0^1 x^2(1 + 2x^3)^5 \, dx$

52. $\int_0^{\sqrt{\pi}} x \cos(x^2) \, dx$

53. $\int_0^{\pi} \sec^2(t/4) \, dt$

54. $\int_{1/6}^{1/2} \csc \pi t \cot \pi t \, dt$

55. $\int_{-\pi/6}^{\pi/6} \tan^3\theta \, d\theta$

56. $\int_0^2 \frac{dx}{(2x - 3)^2}$

57. $\int_1^2 \frac{e^{1/x}}{x^2} \, dx$

58. $\int_0^1 xe^{-x^2} \, dx$

59. $\int_0^{\pi/3} \frac{\sin \theta}{\cos^2\theta} \, d\theta$

60. $\int_{-\pi/2}^{\pi/2} \frac{x^2 \sin x}{1 + x^6} \, dx$

61. $\int_0^{13} \frac{dx}{\sqrt[3]{(1 + 2x)^2}}$

62. $\int_0^{\pi/2} \cos x \sin(\sin x) \, dx$

63. $\int_1^2 x\sqrt{x - 1} \, dx$

64. $\int_0^4 \frac{x}{\sqrt{1 + 2x}} \, dx$

65. $\int_e^{e^4} \frac{dx}{x\sqrt{\ln x}}$

66. $\int_0^{1/2} \frac{\sin^{-1}x}{\sqrt{1 - x^2}} \, dx$

67. $\int_0^4 \frac{dx}{(x - 2)^3}$

68. $\int_0^a x\sqrt{a^2 - x^2} \, dx$

69. $\int_0^a x\sqrt{x^2 + a^2} \, dx \quad (a > 0)$

70. $\int_{-a}^a x\sqrt{x^2 + a^2} \, dx$

71–72 Use a graph to give a rough estimate of the area of the region that lies under the given curve. Then find the exact area.

71. $y = \sqrt{2x + 1}, \quad 0 \leqslant x \leqslant 1$

72. $y = 2 \sin x - \sin 2x, \quad 0 \leqslant x \leqslant \pi$

73. Evaluate $\int_{-2}^2 (x + 3)\sqrt{4 - x^2} \, dx$ by writing it as a sum of two integrals and interpreting one of those integrals in terms of an area.

74. Evaluate $\int_0^1 x\sqrt{1 - x^4} \, dx$ by making a substitution and interpreting the resulting integral in terms of an area.

75. Which of the following areas are equal? Why?

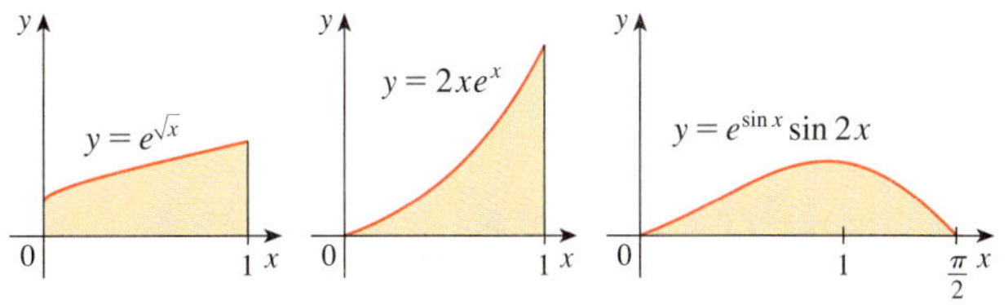

76. A bacteria population starts with 400 bacteria and grows at a rate of $r(t) = (450.268)e^{1.12567t}$ bacteria per hour. How many bacteria will there be after three hours?

77. Breathing is cyclic and a full respiratory cycle from the beginning of inhalation to the end of exhalation takes about 5 s. The maximum rate of air flow into the lungs is about 0.5 L/s. This explains, in part, why the function $f(t) = \frac{1}{2}\sin(2\pi t/5)$ has often been used to model the rate of air flow into the lungs. Use this model to find the volume of inhaled air in the lungs at time t.

78. Alabama Instruments Company has set up a production line to manufacture a new calculator. The rate of production of these calculators after t weeks is

$$\frac{dx}{dt} = 5000\left(1 - \frac{100}{(t+10)^2}\right) \text{ calculators/week}$$

(Notice that production approaches 5000 per week as time goes on, but the initial production is lower because of the workers' unfamiliarity with the new techniques.) Find the number of calculators produced from the beginning of the third week to the end of the fourth week.

79. If f is continuous and $\int_0^4 f(x)\,dx = 10$, find $\int_0^2 f(2x)\,dx$.

80. If f is continuous and $\int_0^9 f(x)\,dx = 4$, find $\int_0^3 xf(x^2)\,dx$.

81. If f is continuous on $\mathbb{R}$, prove that

$$\int_a^b f(-x)\,dx = \int_{-b}^{-a} f(x)\,dx$$

For the case where $f(x) \geqslant 0$ and $0 < a < b$, draw a diagram to interpret this equation geometrically as an equality of areas.

82. If f is continuous on $\mathbb{R}$, prove that

$$\int_a^b f(x+c)\,dx = \int_{a+c}^{b+c} f(x)\,dx$$

For the case where $f(x) \geqslant 0$, draw a diagram to interpret this equation geometrically as an equality of areas.

83. If a and b are positive numbers, show that

$$\int_0^1 x^a(1-x)^b\,dx = \int_0^1 x^b(1-x)^a\,dx$$

84. Use the substitution $u = \pi - x$ to show that

$$\int_0^\pi xf(\sin x)\,dx = \frac{\pi}{2}\int_0^\pi f(\sin x)\,dx$$

85. Use Exercise 84 to evaluate the integral

$$\int_0^\pi \frac{x \sin x}{1 + \cos^2 x}\,dx$$

5.6 The Logarithm Defined as an Integral

Our treatment of exponential and logarithmic functions until now has relied on our intuition, which is based on numerical and visual evidence. (See Sections 1.5, 1.6, and 3.1.) Here we use the Fundamental Theorem of Calculus to give an alternative treatment that provides a surer footing for these functions.

Instead of starting with a^x and defining $\log_a x$ as its inverse, this time we start by defining $\ln x$ as an integral and then define the exponential function as its inverse. In this section you should bear in mind that we do not use any of our previous definitions and results concerning exponential and logarithmic functions.

The Natural Logarithm

We first define $\ln x$ as an integral.

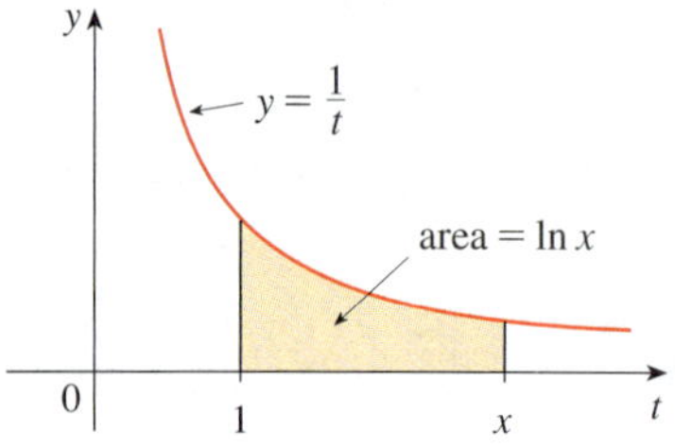

FIGURE 1

1 Definition The **natural logarithmic function** is the function defined by

$$\ln x = \int_1^x \frac{1}{t}\,dt \qquad x > 0$$

The existence of this function depends on the fact that the integral of a continuous function always exists. If $x > 1$, then $\ln x$ can be interpreted geometrically as the area under the hyperbola $y = 1/t$ from $t = 1$ to $t = x$. (See Figure 1.) For $x = 1$, we have

$$\ln 1 = \int_1^1 \frac{1}{t}\,dt = 0$$

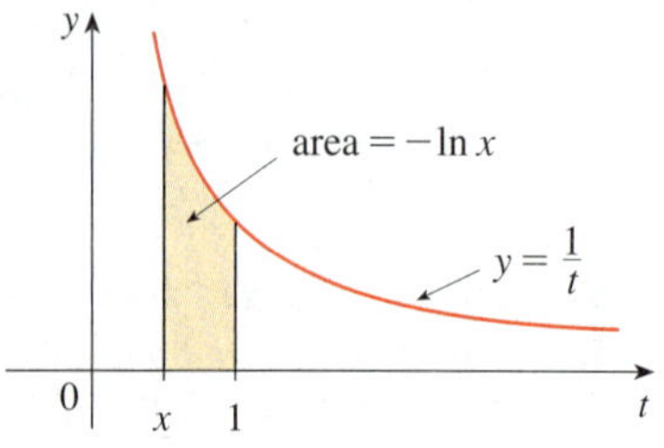

FIGURE 2

For $0 < x < 1$,

$$\ln x = \int_1^x \frac{1}{t}\,dt = -\int_x^1 \frac{1}{t}\,dt < 0$$

and so $\ln x$ is the negative of the area shown in Figure 2.

EXAMPLE 1
(a) By comparing areas, show that $\frac{1}{2} < \ln 2 < \frac{3}{4}$.
(b) Use the Midpoint Rule with $n = 10$ to estimate the value of ln 2.

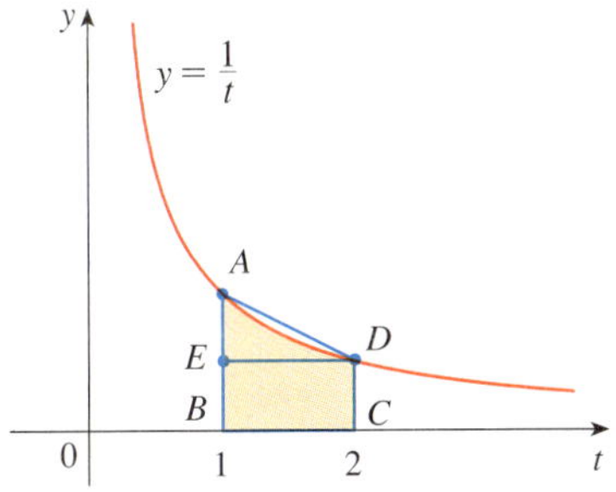

FIGURE 3

SOLUTION

(a) We can interpret ln 2 as the area under the curve $y = 1/t$ from 1 to 2. From Figure 3 we see that this area is larger than the area of rectangle *BCDE* and smaller than the area of trapezoid *ABCD*. Thus, we have

$$\tfrac{1}{2} \cdot 1 < \ln 2 < 1 \cdot \tfrac{1}{2}\left(1 + \tfrac{1}{2}\right)$$

$$\tfrac{1}{2} < \ln 2 < \tfrac{3}{4}$$

(b) If we use the Midpoint Rule with $f(t) = 1/t$, $n = 10$, and $\Delta t = 0.1$, we get

$$\ln 2 = \int_1^2 \frac{1}{t}\,dt \approx (0.1)[f(1.05) + f(1.15) + \cdots + f(1.95)]$$

$$= (0.1)\left(\frac{1}{1.05} + \frac{1}{1.15} + \cdots + \frac{1}{1.95}\right) \approx 0.693$$

Notice that the integral that defines $\ln x$ is exactly the type of integral discussed in Part 1 of the Fundamental Theorem of Calculus (see Section 5.3). In fact, using that theorem, we have

$$\frac{d}{dx}\int_1^x \frac{1}{t}\,dt = \frac{1}{x}$$

and so

2
$$\frac{d}{dx}(\ln x) = \frac{1}{x}$$

We now use this differentiation rule to prove the following properties of the logarithm function.

3 Laws of Logarithms If x and y are positive numbers and r is a rational number, then

1. $\ln(xy) = \ln x + \ln y$ **2.** $\ln\left(\frac{x}{y}\right) = \ln x - \ln y$ **3.** $\ln(x^r) = r \ln x$

Proof

1. Let $f(x) = \ln(ax)$, where a is a positive constant. Then, using Equation 2 and the Chain Rule, we have

$$f'(x) = \frac{1}{ax}\frac{d}{dx}(ax) = \frac{1}{ax} \cdot a = \frac{1}{x}$$

Therefore, $f(x)$ and $\ln x$ have the same derivative and so they must differ by a constant:

$$\ln(ax) = \ln x + C$$

Putting $x = 1$ in this equation, we get $\ln a = \ln 1 + C = 0 + C = C$. Thus

$$\ln(ax) = \ln x + \ln a$$

If we now replace the constant a by any number y, we have

$$\ln(xy) = \ln x + \ln y$$

2. Using Law 1 with $x = 1/y$, we have

$$\ln \frac{1}{y} + \ln y = \ln\left(\frac{1}{y} \cdot y\right) = \ln 1 = 0$$

and so
$$\ln \frac{1}{y} = -\ln y$$

Using Law 1 again, we have

$$\ln\left(\frac{x}{y}\right) = \ln\left(x \cdot \frac{1}{y}\right) = \ln x + \ln \frac{1}{y} = \ln x - \ln y$$

The proof of Law 3 is left as an exercise.

In order to graph $y = \ln x$, we first determine its limits:

4 (a) $\displaystyle\lim_{x\to\infty} \ln x = \infty$ (b) $\displaystyle\lim_{x\to 0^+} \ln x = -\infty$

Proof

(a) Using Law 3 with $x = 2$ and $r = n$ (where n is any positive integer), we have $\ln(2^n) = n \ln 2$. Now $\ln 2 > 0$, so this shows that $\ln(2^n) \to \infty$ as $n \to \infty$. But $\ln x$ is an increasing function since its derivative $1/x > 0$. Therefore, $\ln x \to \infty$ as $x \to \infty$.

(b) If we let $t = 1/x$, then $t \to \infty$ as $x \to 0^+$. Thus, using (a), we have

$$\lim_{x\to 0^+} \ln x = \lim_{t\to\infty} \ln\left(\frac{1}{t}\right) = \lim_{t\to\infty} (-\ln t) = -\infty$$

If $y = \ln x$, $x > 0$, then

$$\frac{dy}{dx} = \frac{1}{x} > 0 \quad \text{and} \quad \frac{d^2y}{dx^2} = -\frac{1}{x^2} < 0$$

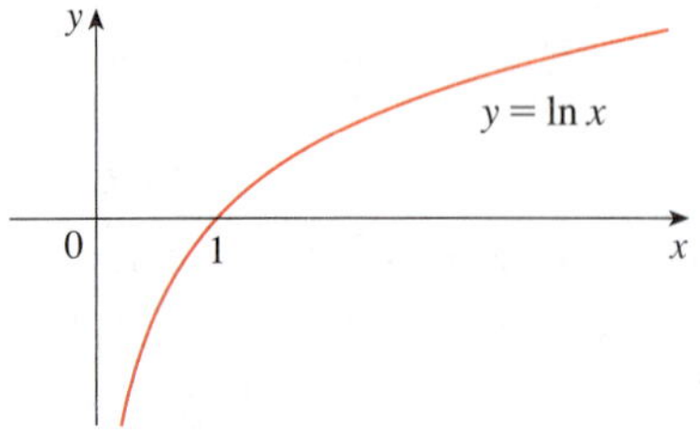

FIGURE 4

which shows that $\ln x$ is increasing and concave downward on $(0, \infty)$. Putting this information together with (4), we draw the graph of $y = \ln x$ in Figure 4.

Since $\ln 1 = 0$ and $\ln x$ is an increasing continuous function that takes on arbitrarily large values, the Intermediate Value Theorem shows that there is a number where $\ln x$ takes on the value 1 (see Figure 5). This important number is denoted by e.

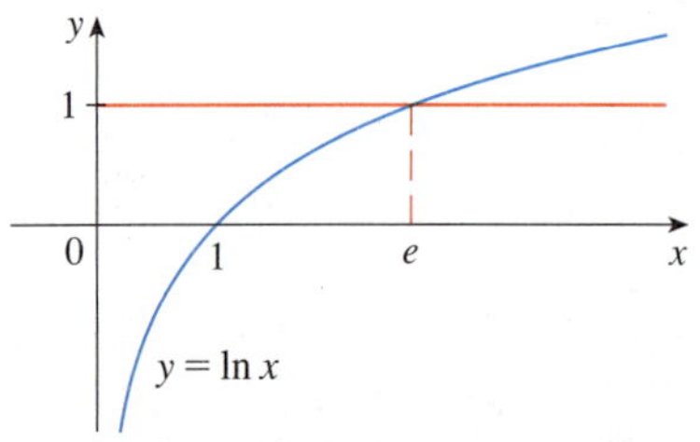

FIGURE 5

5 Definition e is the number such that $\ln e = 1$.

We will show (in Theorem 19) that this definition is consistent with our previous definition of e.

The Natural Exponential Function

Since ln is an increasing function, it is one-to-one and therefore has an inverse function, which we denote by exp. Thus, according to the definition of an inverse function,

$f^{-1}(x) = y \iff f(y) = x$

6 $$\exp(x) = y \iff \ln y = x$$

and the cancellation equations are

$f^{-1}(f(x)) = x$
$f(f^{-1}(x)) = x$

7 $$\exp(\ln x) = x \qquad \text{and} \qquad \ln(\exp x) = x$$

In particular, we have

$$\exp(0) = 1 \quad \text{since} \quad \ln 1 = 0$$

$$\exp(1) = e \quad \text{since} \quad \ln e = 1$$

We obtain the graph of $y = \exp x$ by reflecting the graph of $y = \ln x$ about the line $y = x$. (See Figure 6.) The domain of exp is the range of ln, that is, $(-\infty, \infty)$; the range of exp is the domain of ln, that is, $(0, \infty)$.

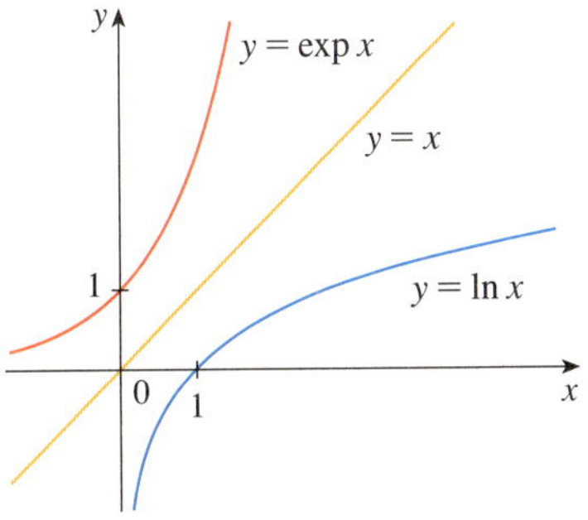

FIGURE 6

If r is any rational number, then the third law of logarithms gives

$$\ln(e^r) = r \ln e = r$$

Therefore, by (6), $$\exp(r) = e^r$$

Thus, $\exp(x) = e^x$ whenever x is a rational number. This leads us to define e^x, even for irrational values of x, by the equation

$$e^x = \exp(x)$$

In other words, for the reasons given, we define e^x to be the inverse of the function ln x. In this notation (6) becomes

8 $$e^x = y \iff \ln y = x$$

and the cancellation equations (7) become

9 $$e^{\ln x} = x \qquad x > 0$$

10 $$\ln(e^x) = x \qquad \text{for all } x$$

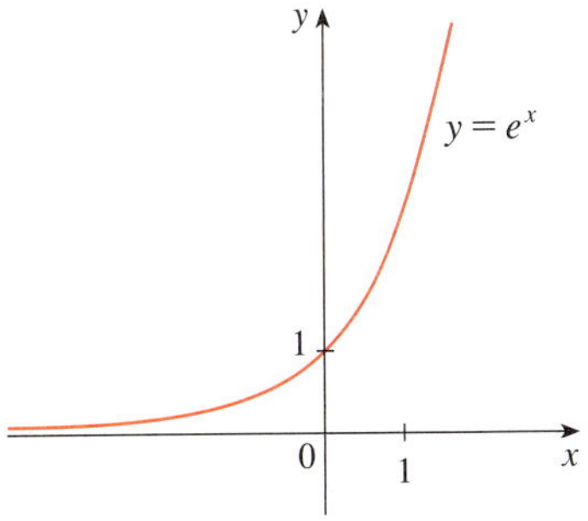

FIGURE 7
The natural exponential function

The natural exponential function $f(x) = e^x$ is one of the most frequently occurring functions in calculus and its applications, so it is important to be familiar with its graph (Figure 7) and its properties (which follow from the fact that it is the inverse of the natural logarithmic function).

Properties of the Exponential Function The exponential function $f(x) = e^x$ is an increasing continuous function with domain $\mathbb{R}$ and range $(0, \infty)$. Thus, $e^x > 0$ for all x. Also

$$\lim_{x \to -\infty} e^x = 0 \qquad \lim_{x \to \infty} e^x = \infty$$

So the x-axis is a horizontal asymptote of $f(x) = e^x$.

We now verify that f has the other properties expected of an exponential function.

11 **Laws of Exponents** If x and y are real numbers and r is rational, then

1. $e^{x+y} = e^x e^y$ **2.** $e^{x-y} = \dfrac{e^x}{e^y}$ **3.** $(e^x)^r = e^{rx}$

Proof of Law 1 Using the first law of logarithms and Equation 10, we have

$$\ln(e^x e^y) = \ln(e^x) + \ln(e^y) = x + y = \ln(e^{x+y})$$

Since ln is a one-to-one function, it follows that $e^x e^y = e^{x+y}$.

Laws 2 and 3 are proved similarly (see Exercises 6 and 7). As we will soon see, Law 3 actually holds when r is any real number.

We now prove the differentiation formula for e^x.

12
$$\frac{d}{dx}(e^x) = e^x$$

Proof The function $y = e^x$ is differentiable because it is the inverse function of $y = \ln x$, which we know is differentiable with nonzero derivative. To find its derivative, we use the inverse function method. Let $y = e^x$. Then $\ln y = x$ and, differentiating this latter equation implicitly with respect to x, we get

$$\frac{1}{y}\frac{dy}{dx} = 1$$

$$\frac{dy}{dx} = y = e^x$$

General Exponential Functions

If $a > 0$ and r is any rational number, then by (9) and (11),

$$a^r = (e^{\ln a})^r = e^{r \ln a}$$

Therefore, even for irrational numbers x, we *define*

13
$$a^x = e^{x \ln a}$$

Thus, for instance,

$$2^{\sqrt{3}} = e^{\sqrt{3}\ln 2} \approx e^{1.20} \approx 3.32$$

The function $f(x) = a^x$ is called the **exponential function with base *a***. Notice that a^x is positive for all x because e^x is positive for all x.

Definition 13 allows us to extend one of the laws of logarithms. We already know that $\ln(a^r) = r \ln a$ when r is rational. But if we now let r be *any* real number we have, from Definition 13,

$$\ln a^r = \ln(e^{r \ln a}) = r \ln a$$

Thus

14 $$\ln a^r = r \ln a \qquad \text{for any real number } r$$

The general laws of exponents follow from Definition 13 together with the laws of exponents for e^x.

15 Laws of Exponents If x and y are real numbers and $a, b > 0$, then

1. $a^{x+y} = a^x a^y$ **2.** $a^{x-y} = a^x/a^y$ **3.** $(a^x)^y = a^{xy}$ **4.** $(ab)^x = a^x b^x$

Proof

1. Using Definition 13 and the laws of exponents for e^x, we have

$$\begin{aligned} a^{x+y} &= e^{(x+y)\ln a} = e^{x \ln a + y \ln a} \\ &= e^{x \ln a} e^{y \ln a} = a^x a^y \end{aligned}$$

3. Using Equation 14 we obtain

$$\begin{aligned} (a^x)^y &= e^{y \ln(a^x)} = e^{yx \ln a} \\ &= e^{xy \ln a} = a^{xy} \end{aligned}$$

The remaining proofs are left as exercises.

The differentiation formula for exponential functions is also a consequence of Definition 13:

16 $$\frac{d}{dx}(a^x) = a^x \ln a$$

Proof $$\begin{aligned} \frac{d}{dx}(a^x) &= \frac{d}{dx}(e^{x \ln a}) = e^{x \ln a} \frac{d}{dx}(x \ln a) \\ &= a^x \ln a \end{aligned}$$

If $a > 1$, then $\ln a > 0$, so $(d/dx)\,a^x = a^x \ln a > 0$, which shows that $y = a^x$ is increasing (see Figure 8). If $0 < a < 1$, then $\ln a < 0$ and so $y = a^x$ is decreasing (see Figure 9).

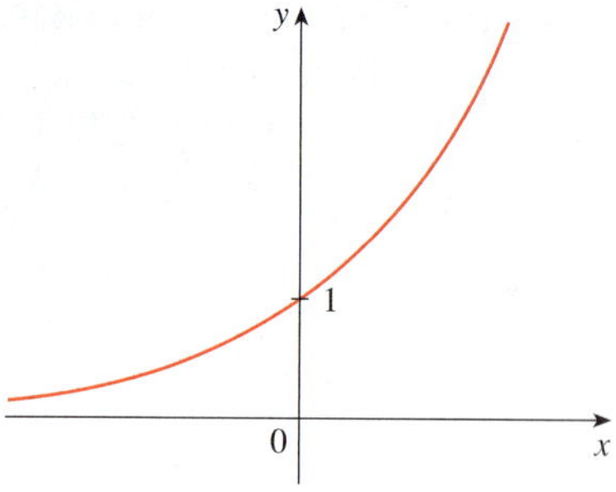

$\lim_{x\to-\infty} a^x = 0,\ \lim_{x\to\infty} a^x = \infty$

FIGURE 8
$y = a^x,\ a > 1$

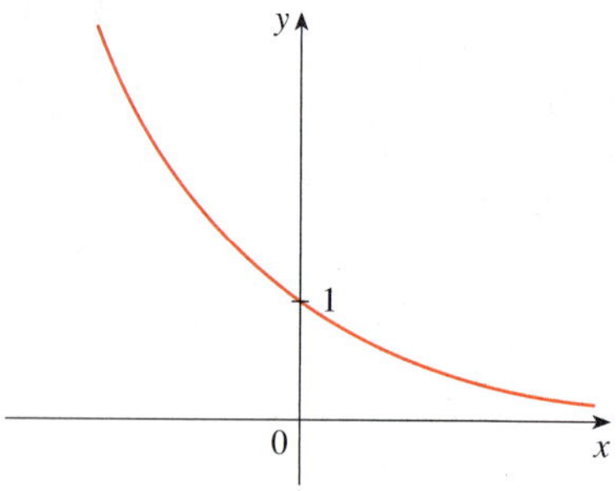

$\lim_{x\to-\infty} a^x = \infty,\ \lim_{x\to\infty} a^x = 0$

FIGURE 9
$y = a^x,\ 0 < a < 1$

General Logarithmic Functions

If $a > 0$ and $a \neq 1$, then $f(x) = a^x$ is a one-to-one function. Its inverse function is called the **logarithmic function with base *a*** and is denoted by $\log_a$. Thus

17
$$\log_a x = y \iff a^y = x$$

In particular, we see that

$$\log_e x = \ln x$$

The laws of logarithms are similar to those for the natural logarithm and can be deduced from the laws of exponents (see Exercise 10).

To differentiate $y = \log_a x$, we write the equation as $a^y = x$. From Equation 14 we have $y \ln a = \ln x$, so

$$\log_a x = y = \frac{\ln x}{\ln a}$$

Since $\ln a$ is a constant, we can differentiate as follows:

$$\frac{d}{dx}(\log_a x) = \frac{d}{dx}\frac{\ln x}{\ln a} = \frac{1}{\ln a}\frac{d}{dx}(\ln x) = \frac{1}{x \ln a}$$

18
$$\frac{d}{dx}(\log_a x) = \frac{1}{x \ln a}$$

The Number e Expressed as a Limit

In this section we defined e as the number such that $\ln e = 1$. The next theorem shows that this is the same as the number e defined in Section 3.1 (see Equation 3.8.5).

19
$$e = \lim_{x \to 0} (1 + x)^{1/x}$$

Proof Let $f(x) = \ln x$. Then $f'(x) = 1/x$, so $f'(1) = 1$. But, by the definition of derivative,

$$f'(1) = \lim_{h \to 0} \frac{f(1 + h) - f(1)}{h} = \lim_{x \to 0} \frac{f(1 + x) - f(1)}{x}$$

$$= \lim_{x \to 0} \frac{\ln(1 + x) - \ln 1}{x} = \lim_{x \to 0} \frac{1}{x} \ln(1 + x)$$

$$= \lim_{x \to 0} \ln(1 + x)^{1/x}$$

Because $f'(1) = 1$, we have

$$\lim_{x \to 0} \ln(1 + x)^{1/x} = 1$$

Then, by Theorem 2.5.8 and the continuity of the exponential function, we have

$$e = e^1 = e^{\lim_{x \to 0} \ln(1+x)^{1/x}} = \lim_{x \to 0} e^{\ln(1+x)^{1/x}} = \lim_{x \to 0} (1 + x)^{1/x}$$

5.6 Exercises

1. (a) By comparing areas, show that

$$\tfrac{1}{3} < \ln 1.5 < \tfrac{5}{12}$$

(b) Use the Midpoint Rule with $n = 10$ to estimate $\ln 1.5$.

2. Refer to Example 1.
(a) Find the equation of the tangent line to the curve $y = 1/t$ that is parallel to the secant line AD.
(b) Use part (a) to show that $\ln 2 > 0.66$.

3. By comparing areas, show that

$$\frac{1}{2} + \frac{1}{3} + \cdots + \frac{1}{n} < \ln n < 1 + \frac{1}{2} + \frac{1}{3} + \cdots + \frac{1}{n - 1}$$

4. (a) By comparing areas, show that $\ln 2 < 1 < \ln 3$.
(b) Deduce that $2 < e < 3$.

5. Prove the third law of logarithms. [*Hint:* Start by showing that both sides of the equation have the same derivative.]

6. Prove the second law of exponents for e^x [see (11)].

7. Prove the third law of exponents for e^x [see (11)].

8. Prove the second law of exponents [see (15)].

9. Prove the fourth law of exponents [see (15)].

10. Deduce the following laws of logarithms from (15):
(a) $\log_a(xy) = \log_a x + \log_a y$
(b) $\log_a(x/y) = \log_a x - \log_a y$
(c) $\log_a(x^y) = y \log_a x$

5 Review

CONCEPT CHECK

1. (a) Write an expression for a Riemann sum of a function f. Explain the meaning of the notation that you use.
   (b) If $f(x) \geqslant 0$, what is the geometric interpretation of a Riemann sum? Illustrate with a diagram.
   (c) If $f(x)$ takes on both positive and negative values, what is the geometric interpretation of a Riemann sum? Illustrate with a diagram.
2. (a) Write the definition of the definite integral of a continuous function from a to b.
   (b) What is the geometric interpretation of $\int_a^b f(x)\,dx$ if $f(x) \geqslant 0$?
   (c) What is the geometric interpretation of $\int_a^b f(x)\,dx$ if $f(x)$ takes on both positive and negative values? Illustrate with a diagram.
3. State both parts of the Fundamental Theorem of Calculus.
4. (a) State the Net Change Theorem.
   (b) If $r(t)$ is the rate at which water flows into a reservoir, what does $\int_{t_1}^{t_2} r(t)\,dt$ represent?
5. Suppose a particle moves back and forth along a straight line with velocity $v(t)$, measured in feet per second, and acceleration $a(t)$.
   (a) What is the meaning of $\int_{60}^{120} v(t)\,dt$?
   (b) What is the meaning of $\int_{60}^{120} |v(t)|\,dt$?
   (c) What is the meaning of $\int_{60}^{120} a(t)\,dt$?
6. (a) Explain the meaning of the indefinite integral $\int f(x)\,dx$.
   (b) What is the connection between the definite integral $\int_a^b f(x)\,dx$ and the indefinite integral $\int f(x)\,dx$?
7. Explain exactly what is meant by the statement that "differentiation and integration are inverse processes."
8. State the Substitution Rule. In practice, how do you use it?

TRUE-FALSE QUIZ

Determine whether the statement is true or false. If it is true, explain why. If it is false, explain why or give an example that disproves the statement.

1. If f and g are continuous on $[a, b]$, then
$$\int_a^b [f(x) + g(x)]\,dx = \int_a^b f(x)\,dx + \int_a^b g(x)\,dx$$
2. If f and g are continuous on $[a, b]$, then
$$\int_a^b [f(x)g(x)]\,dx = \left(\int_a^b f(x)\,dx\right)\left(\int_a^b g(x)\,dx\right)$$
3. If f is continuous on $[a, b]$, then
$$\int_a^b 5f(x)\,dx = 5\int_a^b f(x)\,dx$$
4. If f is continuous on $[a, b]$, then
$$\int_a^b xf(x)\,dx = x\int_a^b f(x)\,dx$$
5. If f is continuous on $[a, b]$ and $f(x) \geqslant 0$, then
$$\int_a^b \sqrt{f(x)}\,dx = \sqrt{\int_a^b f(x)\,dx}$$
6. If f' is continuous on $[1, 3]$, then $\int_1^3 f'(v)\,dv = f(3) - f(1)$.
7. If f and g are continuous and $f(x) \geqslant g(x)$ for $a \leqslant x \leqslant b$, then
$$\int_a^b f(x)\,dx \geqslant \int_a^b g(x)\,dx$$
8. If f and g are differentiable and $f(x) \geqslant g(x)$ for $a < x < b$, then $f'(x) \geqslant g'(x)$ for $a < x < b$.
9. $$\int_{-1}^{1}\left(x^5 - 6x^9 + \frac{\sin x}{(1 + x^4)^2}\right)dx = 0$$
10. $$\int_{-5}^{5}(ax^2 + bx + c)\,dx = 2\int_0^5 (ax^2 + c)\,dx$$
11. $$\int_{-2}^{1}\frac{1}{x^4}\,dx = -\frac{3}{8}$$
12. $\int_0^2 (x - x^3)\,dx$ represents the area under the curve $y = x - x^3$ from 0 to 2.
13. All continuous functions have derivatives.
14. All continuous functions have antiderivatives.

EXERCISES

1. Use the given graph of f to find the Riemann sum with six subintervals. Take the sample points to be (a) left endpoints and (b) midpoints. In each case draw a diagram and explain what the Riemann sum represents.

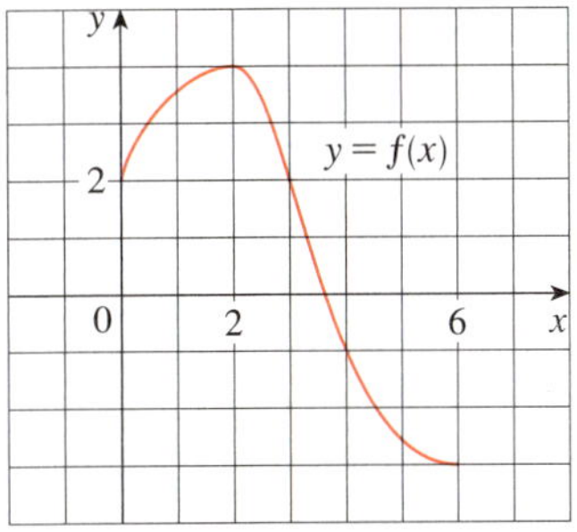

2. (a) Evaluate the Riemann sum for

$$f(x) = x^2 - x \qquad 0 \leqslant x \leqslant 2$$

with four subintervals, taking the sample points to be right endpoints. Explain, with the aid of a diagram, what the Riemann sum represents.

(b) Use the definition of a definite integral (with right endpoints) to calculate the value of the integral

$$\int_0^2 (x^2 - x)\,dx$$

(c) Use the Fundamental Theorem to check your answer to part (b).

(d) Draw a diagram to explain the geometric meaning of the integral in part (b).

3. Evaluate

$$\int_0^1 \left(x + \sqrt{1 - x^2}\right) dx$$

by interpreting it in terms of areas.

4. Express

$$\lim_{n \to \infty} \sum_{i=1}^{n} \sin x_i \,\Delta x$$

as a definite integral on the interval $[0, \pi]$ and then evaluate the integral.

5. If $\int_0^6 f(x)\,dx = 10$ and $\int_0^4 f(x)\,dx = 7$, find $\int_4^6 f(x)\,dx$.

6. (a) Write $\int_1^5 (x + 2x^5)\,dx$ as a limit of Riemann sums, taking the sample points to be right endpoints. Use a computer algebra system to evaluate the sum and to compute the limit.

(b) Use the Fundamental Theorem to check your answer to part (a).

7. The figure shows the graphs of f, f', and $\int_0^x f(t)\,dt$. Identify each graph, and explain your choices.

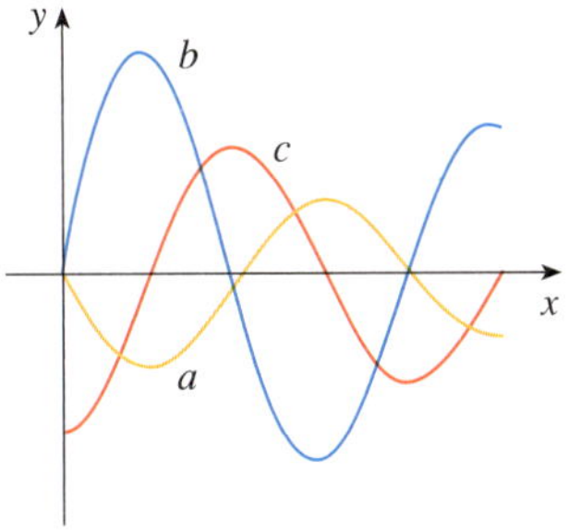

8. Evaluate:

(a) $\displaystyle\int_0^1 \frac{d}{dx}\left(e^{\arctan x}\right) dx$ (b) $\displaystyle\frac{d}{dx}\int_0^1 e^{\arctan x}\,dx$

(c) $\displaystyle\frac{d}{dx}\int_0^x e^{\arctan t}\,dt$

9–38 Evaluate the integral, if it exists.

9. $\displaystyle\int_1^2 (8x^3 + 3x^2)\,dx$

10. $\displaystyle\int_0^T (x^4 - 8x + 7)\,dx$

11. $\displaystyle\int_0^1 (1 - x^9)\,dx$

12. $\displaystyle\int_0^1 (1 - x)^9\,dx$

13. $\displaystyle\int_1^9 \frac{\sqrt{u} - 2u^2}{u}\,du$

14. $\displaystyle\int_0^1 (\sqrt[4]{u} + 1)^2\,du$

15. $\displaystyle\int_0^1 y(y^2 + 1)^5\,dy$

16. $\displaystyle\int_0^2 y^2\sqrt{1 + y^3}\,dy$

17. $\displaystyle\int_1^5 \frac{dt}{(t - 4)^2}$

18. $\displaystyle\int_0^1 \sin(3\pi t)\,dt$

19. $\displaystyle\int_0^1 v^2 \cos(v^3)\,dv$

20. $\displaystyle\int_{-1}^1 \frac{\sin x}{1 + x^2}\,dx$

21. $\displaystyle\int_0^1 e^{\pi t}\,dt$

22. $\displaystyle\int_1^2 \frac{1}{2 - 3x}\,dx$

23. $\displaystyle\int_2^4 \frac{1 + x - x^2}{x^2}\,dx$

24. $\displaystyle\int_1^{10} \frac{x}{x^2 - 4}\,dx$

25. $\displaystyle\int \frac{x + 2}{\sqrt{x^2 + 4x}}\,dx$

26. $\displaystyle\int \csc^2 3t\,dt$

27. $\displaystyle\int \sin \pi t \cos \pi t\,dt$

28. $\displaystyle\int \sin x \cos(\cos x)\,dx$

29. $\displaystyle\int \frac{e^{\sqrt{x}}}{\sqrt{x}}\,dx$

30. $\displaystyle\int \frac{\cos(\ln x)}{x}\,dx$

31. $\displaystyle\int \tan x \ln(\cos x)\,dx$

32. $\displaystyle\int \frac{x}{\sqrt{1 - x^4}}\,dx$

33. $\displaystyle\int \frac{x^3}{1+x^4}\,dx$

34. $\displaystyle\int \sinh(1+4x)\,dx$

35. $\displaystyle\int \frac{\sec\theta\,\tan\theta}{1+\sec\theta}\,d\theta$

36. $\displaystyle\int_0^{\pi/4} (1+\tan t)^3 \sec^2 t\,dt$

37. $\displaystyle\int_0^3 |x^2-4|\,dx$

38. $\displaystyle\int_0^4 |\sqrt{x}-1|\,dx$

39–40 Evaluate the indefinite integral. Illustrate and check that your answer is reasonable by graphing both the function and its antiderivative (take $C = 0$).

39. $\displaystyle\int \frac{\cos x}{\sqrt{1+\sin x}}\,dx$

40. $\displaystyle\int \frac{x^3}{\sqrt{x^2+1}}\,dx$

41. Use a graph to give a rough estimate of the area of the region that lies under the curve $y = x\sqrt{x}$, $0 \leqslant x \leqslant 4$. Then find the exact area.

42. Graph the function $f(x) = \cos^2 x\,\sin^3 x$ and use the graph to guess the value of the integral $\int_0^{2\pi} f(x)\,dx$. Then evaluate the integral to confirm your guess.

43–48 Find the derivative of the function.

43. $\displaystyle F(x) = \int_1^x \sqrt{1+t^4}\,dt$

44. $\displaystyle F(x) = \int_\pi^x \tan(s^2)\,ds$

45. $\displaystyle g(x) = \int_0^{x^3} \frac{t}{\sqrt{1+t^3}}\,dt$

46. $\displaystyle g(x) = \int_1^{\cos x} \sqrt[3]{1-t^2}\,dt$

47. $\displaystyle y = \int_{\sqrt{x}}^x \frac{e^t}{t}\,dt$

48. $\displaystyle y = \int_{2x}^{3x+1} \sin(t^4)\,dt$

49–50 Use Property 8 of integrals to estimate the value of the integral.

49. $\displaystyle\int_1^3 \sqrt{x^2+3}\,dx$

50. $\displaystyle\int_3^5 \frac{1}{x+1}\,dx$

51–54 Use the properties of integrals to verify the inequality.

51. $\displaystyle\int_0^1 x^2\cos x\,dx \leqslant \frac{1}{3}$

52. $\displaystyle\int_{\pi/4}^{\pi/2} \frac{\sin x}{x}\,dx \leqslant \frac{\sqrt{2}}{2}$

53. $\displaystyle\int_0^1 e^x\cos x\,dx \leqslant e-1$

54. $\displaystyle\int_0^1 x\sin^{-1}x\,dx \leqslant \pi/4$

55. Use the Midpoint Rule with $n = 5$ to approximate $\int_0^1 \sqrt{1+x^3}\,dx$.

56. A particle moves along a line with velocity function $v(t) = t^2 - t$, where v is measured in meters per second. Find (a) the displacement and (b) the distance traveled by the particle during the time interval $[0, 5]$.

57. Let $r(t)$ be the rate at which the world's oil is consumed, where t is measured in years starting at $t = 0$ on January 1, 2000, and $r(t)$ is measured in barrels per year. What does $\int_0^3 r(t)\,dt$ represent?

58. A radar gun was used to record the speed of a runner at the times given in the table. Use the Midpoint Rule to estimate the distance the runner covered during those 5 seconds.

t (s)	v (m/s)	t (s)	v (m/s)
0	0	3.0	10.51
0.5	4.67	3.5	10.67
1.0	7.34	4.0	10.76
1.5	8.86	4.5	10.81
2.0	9.73	5.0	10.81
2.5	10.22		

59. A population of honeybees increased at a rate of $r(t)$ bees per week, where the graph of r is as shown. Use the Midpoint Rule with six subintervals to estimate the increase in the bee population during the first 24 weeks.

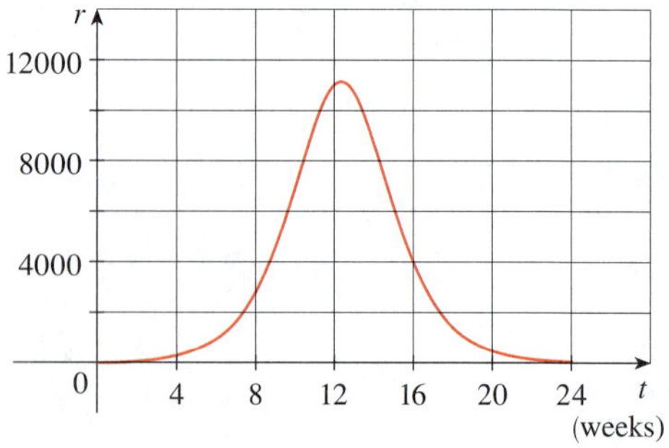

60. Let

$$f(x) = \begin{cases} -x-1 & \text{if } -3 \leqslant x \leqslant 0 \\ -\sqrt{1-x^2} & \text{if } 0 \leqslant x \leqslant 1 \end{cases}$$

Evaluate $\int_{-3}^1 f(x)\,dx$ by interpreting the integral as a difference of areas.

61. Find an antiderivative F of $f(x) = x^2 \sin(x^2)$ such that $F(1) = 0$.

62. The Fresnel function $S(x) = \int_0^x \sin\left(\frac{1}{2}\pi t^2\right) dt$ was introduced in Section 5.3. Fresnel also used the function

$$C(x) = \int_0^x \cos\left(\tfrac{1}{2}\pi t^2\right) dt$$

in his theory of the diffraction of light waves.

(a) On what intervals is C increasing?

(b) On what intervals is C concave upward?

(c) Use a graph to solve the following equation correct to two decimal places:

$$\int_0^x \cos\left(\tfrac{1}{2}\pi t^2\right) dt = 0.7$$

(d) Plot the graphs of C and S on the same screen. How are these graphs related?

63. Estimate the value of the number c such that the area under the curve $y = \sinh cx$ between $x = 0$ and $x = 1$ is equal to 1.

64. Suppose that the temperature in a long, thin rod placed along the x-axis is initially $C/(2a)$ if $|x| \leqslant a$ and 0 if $|x| > a$. It can be shown that if the heat diffusivity of the rod is k, then the temperature of the rod at the point x at time t is

$$T(x, t) = \frac{C}{a\sqrt{4\pi kt}} \int_0^a e^{-(x-u)^2/(4kt)}\, du$$

To find the temperature distribution that results from an initial hot spot concentrated at the origin, we need to compute

$$\lim_{a \to 0} T(x, t)$$

Use l'Hospital's Rule to find this limit.

65. If f is a continuous function such that

$$\int_0^x f(t)\, dt = xe^{2x} + \int_0^x e^{-t} f(t)\, dt$$

for all x, find an explicit formula for $f(x)$.

66. Suppose h is a function such that $h(1) = -2$, $h'(1) = 2$, $h''(1) = 3$, $h(2) = 6$, $h'(2) = 5$, $h''(2) = 13$, and h'' is continuous everywhere. Evaluate $\int_1^2 h''(u)\, du$.

67. If f' is continuous on $[a, b]$, show that

$$2\int_a^b f(x) f'(x)\, dx = [f(b)]^2 - [f(a)]^2$$

68. Find $\displaystyle\lim_{h \to 0} \frac{1}{h} \int_2^{2+h} \sqrt{1 + t^3}\, dt$.

69. If f is continuous on $[0, 1]$, prove that

$$\int_0^1 f(x)\, dx = \int_0^1 f(1 - x)\, dx$$

70. Evaluate

$$\lim_{n \to \infty} \frac{1}{n} \left[\left(\frac{1}{n}\right)^9 + \left(\frac{2}{n}\right)^9 + \left(\frac{3}{n}\right)^9 + \cdots + \left(\frac{n}{n}\right)^9 \right]$$

Before you look at the solution of the following example, cover it up and first try to solve the problem yourself.

EXAMPLE 1 Evaluate $\lim_{x \to 3} \left(\frac{x}{x-3} \int_3^x \frac{\sin t}{t}\, dt \right)$.

SOLUTION Let's start by having a preliminary look at the ingredients of the function. What happens to the first factor, $x/(x-3)$, when x approaches 3? The numerator approaches 3 and the denominator approaches 0, so we have

$$\frac{x}{x-3} \to \infty \quad \text{as} \quad x \to 3^+ \qquad \text{and} \qquad \frac{x}{x-3} \to -\infty \quad \text{as} \quad x \to 3^-$$

The second factor approaches $\int_3^3 (\sin t)/t\, dt$, which is 0. It's not clear what happens to the function as a whole. (One factor is becoming large while the other is becoming small.) So how do we proceed?

One of the principles of problem solving is *recognizing something familiar*. Is there a part of the function that reminds us of something we've seen before? Well, the integral

$$\int_3^x \frac{\sin t}{t}\, dt$$

has x as its upper limit of integration and that type of integral occurs in Part 1 of the Fundamental Theorem of Calculus:

$$\frac{d}{dx} \int_a^x f(t)\, dt = f(x)$$

This suggests that differentiation might be involved.

Once we start thinking about differentiation, the denominator $(x-3)$ reminds us of something else that should be familiar: One of the forms of the definition of the derivative in Chapter 2 is

$$F'(a) = \lim_{x \to a} \frac{F(x) - F(a)}{x - a}$$

and with $a = 3$ this becomes

$$F'(3) = \lim_{x \to 3} \frac{F(x) - F(3)}{x - 3}$$

So what is the function F in our situation? Notice that if we define

$$F(x) = \int_3^x \frac{\sin t}{t}\, dt$$

then $F(3) = 0$. What about the factor x in the numerator? That's just a red herring, so let's factor it out and put together the calculation:

$$\begin{aligned}
\lim_{x \to 3} \left(\frac{x}{x-3} \int_3^x \frac{\sin t}{t}\, dt \right) &= \lim_{x \to 3} x \cdot \lim_{x \to 3} \frac{\int_3^x \frac{\sin t}{t}\, dt}{x - 3} \\
&= 3 \lim_{x \to 3} \frac{F(x) - F(3)}{x - 3} \\
&= 3F'(3) \\
&= 3\, \frac{\sin 3}{3} \qquad \text{(FTC1)} \\
&= \sin 3
\end{aligned}$$

The principles of problem solving are discussed on page 80.

Another approach is to use l'Hospital's Rule.

PROBLEMS

1. If $x \sin \pi x = \int_0^{x^2} f(t)\,dt$, where f is a continuous function, find $f(4)$.

2. Suppose the curve $y = f(x)$ passes through the origin and the point $(1, 1)$. Find the value of the integral $\int_0^1 f'(x)\,dx$.

3. Show that $\dfrac{1}{17} \leqslant \displaystyle\int_1^2 \frac{1}{1 + x^4}\,dx \leqslant \frac{7}{24}$.

4. (a) Graph several members of the family of functions $f(x) = (2cx - x^2)/c^3$ for $c > 0$ and look at the regions enclosed by these curves and the x-axis. Make a conjecture about how the areas of these regions are related.
   (b) Prove your conjecture in part (a).
   (c) Take another look at the graphs in part (a) and use them to sketch the curve traced out by the vertices (highest points) of the family of functions. Can you guess what kind of curve this is?
   (d) Find the equation of the curve you sketched in part (c).

5. If $f(x) = \displaystyle\int_0^{g(x)} \frac{1}{\sqrt{1 + t^3}}\,dt$, where $g(x) = \displaystyle\int_0^{\cos x} [1 + \sin(t^2)]\,dt$, find $f'(\pi/2)$.

6. If $f(x) = \int_0^x x^2 \sin(t^2)\,dt$, find $f'(x)$.

7. Evaluate $\displaystyle\lim_{x \to 0} \frac{1}{x} \int_0^x (1 - \tan 2t)^{1/t}\,dt$.

8. The figure shows two regions in the first quadrant: $A(t)$ is the area under the curve $y = \sin(x^2)$ from 0 to t, and $B(t)$ is the area of the triangle with vertices O, P, and $(t, 0)$. Find $\lim_{t \to 0^+} A(t)/B(t)$.

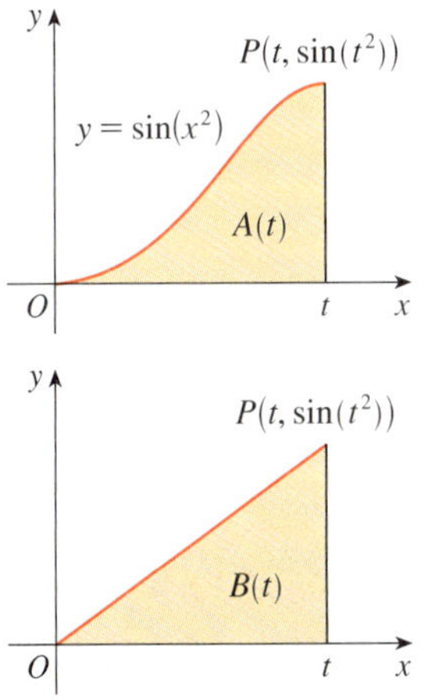

FIGURE FOR PROBLEM 8

9. Find the interval $[a, b]$ for which the value of the integral $\int_a^b (2 + x - x^2)\,dx$ is a maximum.

10. Use an integral to estimate the sum $\displaystyle\sum_{i=1}^{10000} \sqrt{i}$.

11. (a) Evaluate $\int_0^n [\![x]\!]\,dx$, where n is a positive integer.
    (b) Evaluate $\int_a^b [\![x]\!]\,dx$, where a and b are real numbers with $0 \leqslant a < b$.

12. Find $\dfrac{d^2}{dx^2} \displaystyle\int_0^x \left(\int_1^{\sin t} \sqrt{1 + u^4}\,du \right) dt$.

13. If f is a differentiable function such that $\displaystyle\int_0^x f(t)\,dt = [f(x)]^2$ for all x, find f.

14. A circular disk of radius r is used in an evaporator and is rotated in a vertical plane. If it is to be partially submerged in the liquid so as to maximize the exposed wetted area of the disk, show that the center of the disk should be positioned at a height $r/\sqrt{1 + \pi^2}$ above the surface of the liquid.

15. Prove that if f is continuous, then $\displaystyle\int_0^x f(u)(x - u)\,du = \int_0^x \left(\int_0^u f(t)\,dt \right) du$.

16. The figure shows a region consisting of all points inside a square that are closer to the center than to the sides of the square. Find the area of the region.

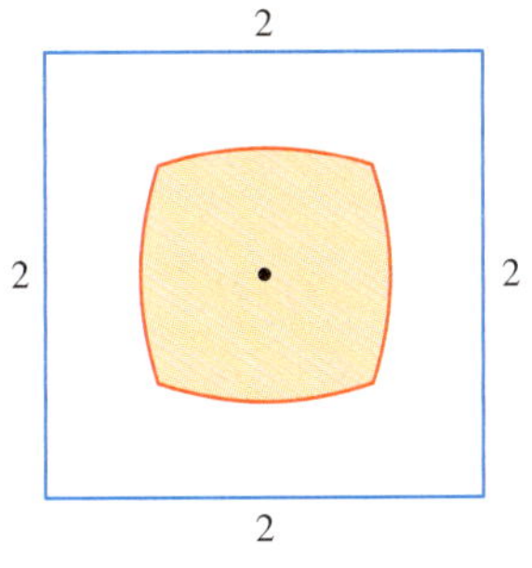

FIGURE FOR PROBLEM 16

17. Evaluate $\displaystyle\lim_{n \to \infty} \left(\frac{1}{\sqrt{n}\sqrt{n + 1}} + \frac{1}{\sqrt{n}\sqrt{n + 2}} + \cdots + \frac{1}{\sqrt{n}\sqrt{n + n}} \right)$.

18. For any number c, we let $f_c(x)$ be the smaller of the two numbers $(x - c)^2$ and $(x - c - 2)^2$. Then we define $g(c) = \int_0^1 f_c(x)\,dx$. Find the maximum and minimum values of $g(c)$ if $-2 \leqslant c \leqslant 2$.

19. Suppose f is continuous, $f(0) = 0$, $f(1) = 1$, $f'(x) > 0$, and $\int_0^1 f(x)\,dx = \frac{1}{3}$. Find the value of the integral $\int_0^1 f^{-1}(y)\,dy$.

Chapter 6

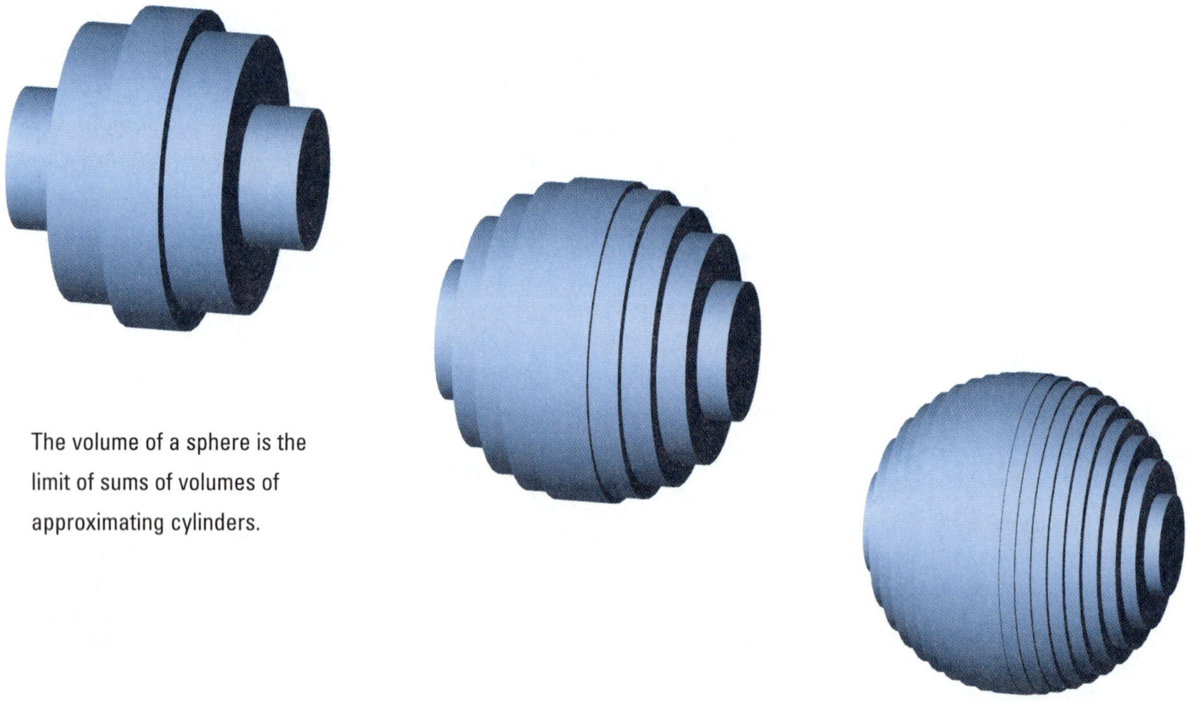

The volume of a sphere is the limit of sums of volumes of approximating cylinders.

Applications of Integration

In this chapter we explore some of the applications of the definite integral by using it to compute areas between curves, volumes of solids, and the work done by a varying force. The common theme is the following general method, which is similar to the one we used to find areas under curves: We break up a quantity Q into a large number of small parts. We next approximate each small part by a quantity of the form $f(x_i^*)\,\Delta x$ and thus approximate Q by a Riemann sum. Then we take the limit and express Q as an integral. Finally we evaluate the integral using the Fundamental Theorem of Calculus or the Midpoint Rule.

6.1 Areas between Curves

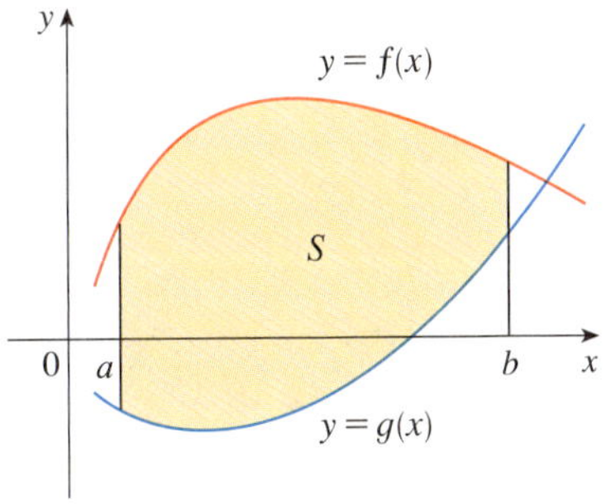

FIGURE 1
$S = \{(x, y) \mid a \leqslant x \leqslant b,\ g(x) \leqslant y \leqslant f(x)\}$

In Chapter 5 we defined and calculated areas of regions that lie under the graphs of functions. Here we use integrals to find areas of regions that lie between the graphs of two functions.

Consider the region S that lies between two curves $y = f(x)$ and $y = g(x)$ and between the vertical lines $x = a$ and $x = b$, where f and g are continuous functions and $f(x) \geqslant g(x)$ for all x in $[a, b]$. (See Figure 1.)

Just as we did for areas under curves in Section 5.1, we divide S into n strips of equal width and then we approximate the ith strip by a rectangle with base Δx and height $f(x_i^*) - g(x_i^*)$. (See Figure 2. If we like, we could take all of the sample points to be right endpoints, in which case $x_i^* = x_i$.) The Riemann sum

$$\sum_{i=1}^{n} [f(x_i^*) - g(x_i^*)]\,\Delta x$$

is therefore an approximation to what we intuitively think of as the area of S.

Guess the area of an island.
Resources / Module 7 / Areas / Start of Areas

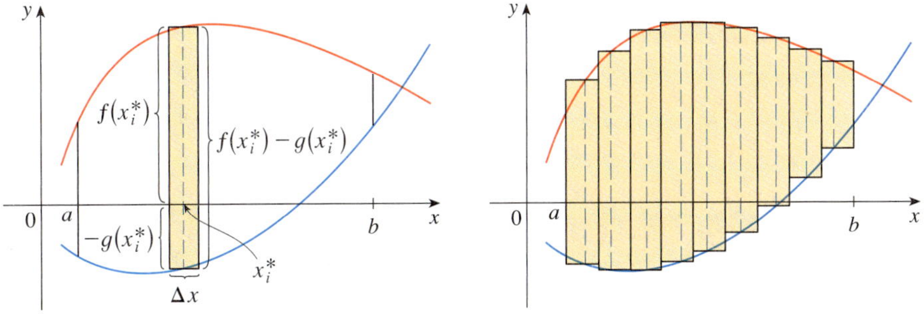

FIGURE 2

(a) Typical rectangle (b) Approximating rectangles

This approximation appears to become better and better as $n \to \infty$. Therefore, we define the **area** A of S as the limiting value of the sum of the areas of these approximating rectangles.

1
$$A = \lim_{n \to \infty} \sum_{i=1}^{n} [f(x_i^*) - g(x_i^*)]\,\Delta x$$

We recognize the limit in (1) as the definite integral of $f - g$. Therefore, we have the following formula for area.

2 The area A of the region bounded by the curves $y = f(x)$, $y = g(x)$, and the lines $x = a$, $x = b$, where f and g are continuous and $f(x) \geqslant g(x)$ for all x in $[a, b]$, is

$$A = \int_a^b [f(x) - g(x)]\,dx$$

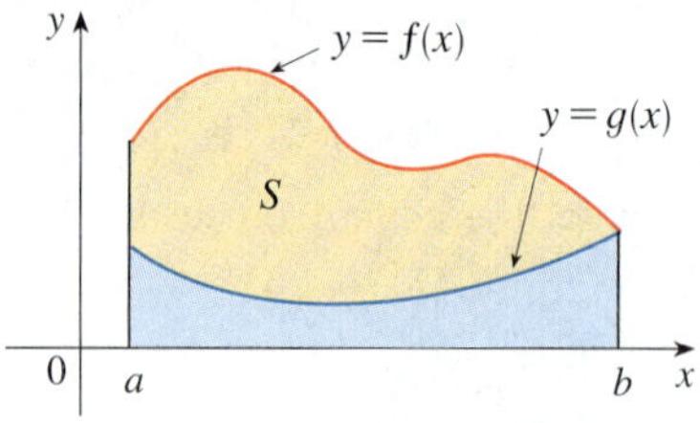

FIGURE 3

$A = \int_a^b f(x)\,dx - \int_a^b g(x)\,dx$

Notice that in the special case where $g(x) = 0$, S is the region under the graph of f and our general definition of area (1) reduces to our previous definition (Definition 2 in Section 5.1).

In the case where both f and g are positive, you can see from Figure 3 why (2) is true:

$$A = [\text{area under } y = f(x)] - [\text{area under } y = g(x)]$$

$$= \int_a^b f(x)\,dx - \int_a^b g(x)\,dx = \int_a^b [f(x) - g(x)]\,dx$$

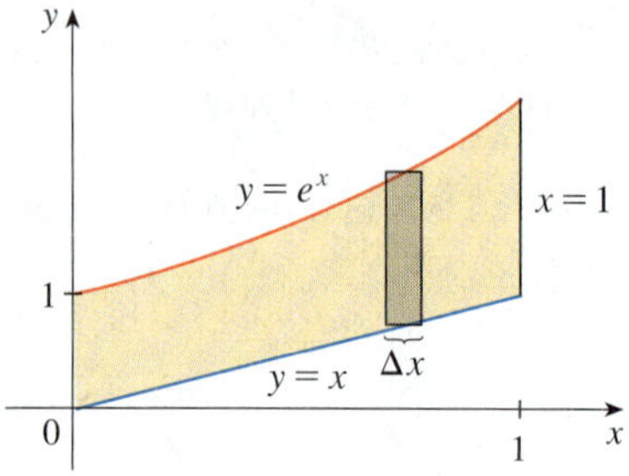

FIGURE 4

EXAMPLE 1 Find the area of the region bounded above by $y = e^x$, bounded below by $y = x$, and bounded on the sides by $x = 0$ and $x = 1$.

SOLUTION The region is shown in Figure 4. The upper boundary curve is $y = e^x$ and the lower boundary curve is $y = x$. So we use the area formula (2) with $f(x) = e^x$, $g(x) = x$, $a = 0$, and $b = 1$:

$$A = \int_0^1 (e^x - x)\,dx = e^x - \tfrac{1}{2}x^2\Big]_0^1$$

$$= e - \tfrac{1}{2} - 1 = e - 1.5$$

In Figure 4 we drew a typical approximating rectangle with width Δx as a reminder of the procedure by which the area is defined in (1). In general, when we set up an integral for an area, it's helpful to sketch the region to identify the top curve y_T, the bottom curve y_B, and a typical approximating rectangle as in Figure 5. Then the area of a typical rectangle is $(y_T - y_B)\,\Delta x$ and the equation

$$A = \lim_{n \to \infty} \sum_{i=1}^{n} (y_T - y_B)\,\Delta x = \int_a^b (y_T - y_B)\,dx$$

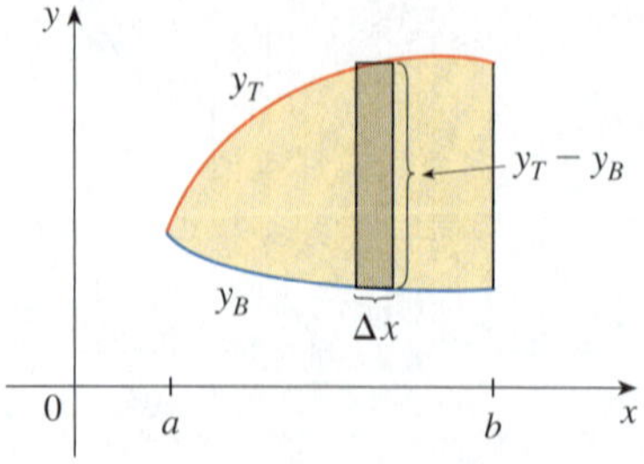

FIGURE 5

summarizes the procedure of adding (in a limiting sense) the areas of all the typical rectangles.

Notice that in Figure 5 the left-hand boundary reduces to a point, whereas in Figure 3 the right-hand boundary reduces to a point. In the next example both of the side boundaries reduce to a point, so the first step is to find a and b.

EXAMPLE 2 Find the area of the region enclosed by the parabolas $y = x^2$ and $y = 2x - x^2$.

SOLUTION We first find the points of intersection of the parabolas by solving their equations simultaneously. This gives $x^2 = 2x - x^2$, or $2x^2 - 2x = 0$. Thus $2x(x - 1) = 0$, so $x = 0$ or 1. The points of intersection are $(0, 0)$ and $(1, 1)$.

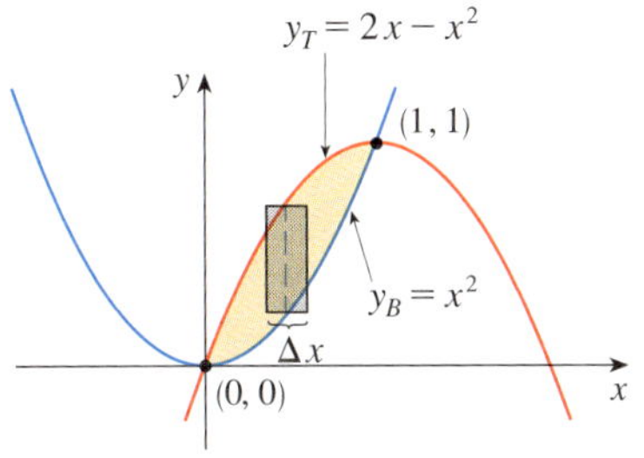

FIGURE 6

We see from Figure 6 that the top and bottom boundaries are

$$y_T = 2x - x^2 \qquad \text{and} \qquad y_B = x^2$$

The area of a typical rectangle is

$$(y_T - y_B)\,\Delta x = (2x - x^2 - x^2)\,\Delta x$$

and the region lies between $x = 0$ and $x = 1$. So the total area is

$$\begin{aligned} A &= \int_0^1 (2x - 2x^2)\,dx = 2\int_0^1 (x - x^2)\,dx \\ &= 2\left[\frac{x^2}{2} - \frac{x^3}{3}\right]_0^1 \\ &= 2\left(\frac{1}{2} - \frac{1}{3}\right) = \frac{1}{3} \end{aligned}$$

Sometimes it's difficult, or even impossible, to find the points of intersection of two curves exactly. As shown in the following example, we can use a graphing calculator or computer to find approximate values for the intersection points and then proceed as before.

EXAMPLE 3 Find the approximate area of the region bounded by the curves $y = x/\sqrt{x^2 + 1}$ and $y = x^4 - x$.

SOLUTION If we were to try to find the exact intersection points, we would have to solve the equation

$$\frac{x}{\sqrt{x^2 + 1}} = x^4 - x$$

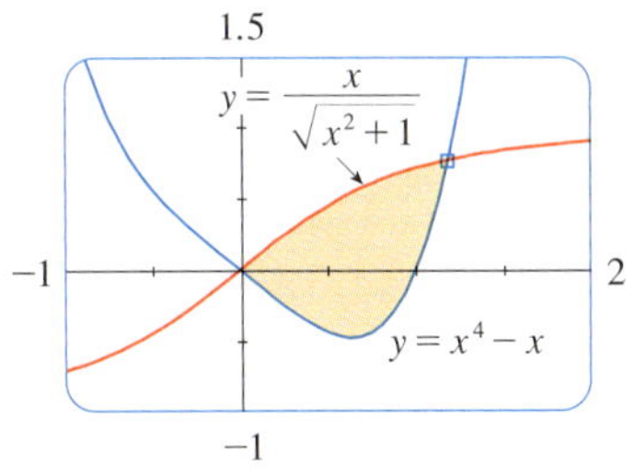

FIGURE 7

This looks like a very difficult equation to solve exactly (in fact, it's impossible), so instead we use a graphing device to draw the graphs of the two curves in Figure 7. One intersection point is the origin. We zoom in toward the other point of intersection and find that $x \approx 1.18$. (If greater accuracy is required, we could use Newton's method or a rootfinder, if available on our graphing device.) Thus, an approximation to the area between the curves is

$$A \approx \int_0^{1.18} \left[\frac{x}{\sqrt{x^2 + 1}} - (x^4 - x)\right] dx$$

To integrate the first term we use the subsitution $u = x^2 + 1$. Then $du = 2x\,dx$, and when $x = 1.18$, we have $u \approx 2.39$. So

$$\begin{aligned} A &\approx \tfrac{1}{2}\int_1^{2.39} \frac{du}{\sqrt{u}} - \int_0^{1.18} (x^4 - x)\,dx \\ &= \sqrt{u}\,\Big]_1^{2.39} - \left[\frac{x^5}{5} - \frac{x^2}{2}\right]_0^{1.18} \\ &= \sqrt{2.39} - 1 - \frac{(1.18)^5}{5} + \frac{(1.18)^2}{2} \\ &\approx 0.785 \end{aligned}$$

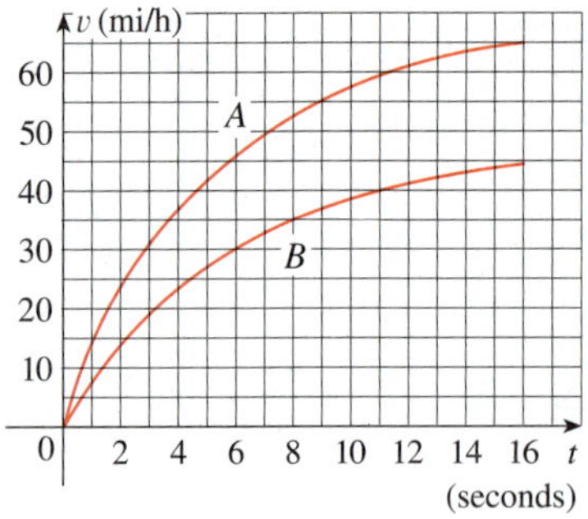

FIGURE 8

EXAMPLE 4 Figure 8 shows velocity curves for two cars, A and B, that start side by side and move along the same road. What does the area between the curves represent? Use the Midpoint Rule to estimate it.

SOLUTION We know from Section 5.4 that the area under the velocity curve A represents the distance traveled by car A during the first 16 seconds. Similarly, the area under curve B is the distance traveled by car B during that time period. So the area between these curves, which is the difference of the areas under the curves, is the distance between the cars after 16 seconds. We read the velocities from the graph and convert them to feet per second ($1 \text{ mi/h} = \frac{5280}{3600} \text{ ft/s}$).

t	0	2	4	6	8	10	12	14	16
v_A	0	34	54	67	76	84	89	92	95
v_B	0	21	34	44	51	56	60	63	65
$v_A - v_B$	0	13	20	23	25	28	29	29	30

We use the Midpoint Rule with $n = 4$ intervals, so that $\Delta t = 4$. The midpoints of the intervals are $\bar{t}_1 = 2$, $\bar{t}_2 = 6$, $\bar{t}_3 = 10$, and $\bar{t}_4 = 14$. We estimate the distance between the cars after 16 seconds as follows:

$$\int_0^{16} (v_A - v_B)\,dt \approx \Delta t\,[13 + 23 + 28 + 29]$$

$$= 4(93) = 372 \text{ ft}$$

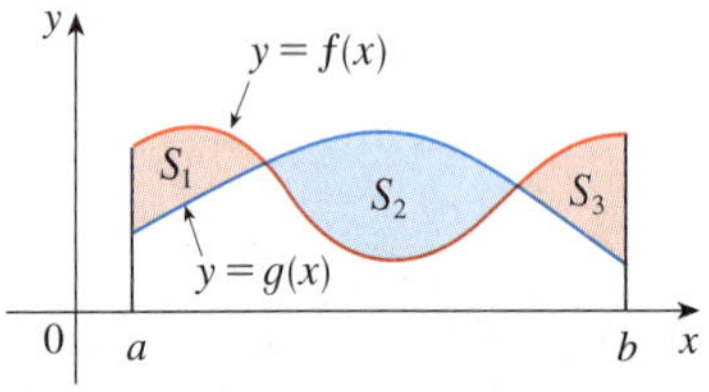

FIGURE 9

If we are asked to find the area between the curves $y = f(x)$ and $y = g(x)$ where $f(x) \geqslant g(x)$ for some values of x but $g(x) \geqslant f(x)$ for other values of x, then we split the given region S into several regions $S_1, S_2, \ldots$ with areas $A_1, A_2, \ldots$ as shown in Figure 9. We then define the area of the region S to be the sum of the areas of the smaller regions $S_1, S_2, \ldots$, that is, $A = A_1 + A_2 + \cdots$. Since

$$|f(x) - g(x)| = \begin{cases} f(x) - g(x) & \text{when } f(x) \geqslant g(x) \\ g(x) - f(x) & \text{when } g(x) \geqslant f(x) \end{cases}$$

we have the following expression for A.

3 The area between the curves $y = f(x)$ and $y = g(x)$ and between $x = a$ and $x = b$ is

$$A = \int_a^b |f(x) - g(x)|\,dx$$

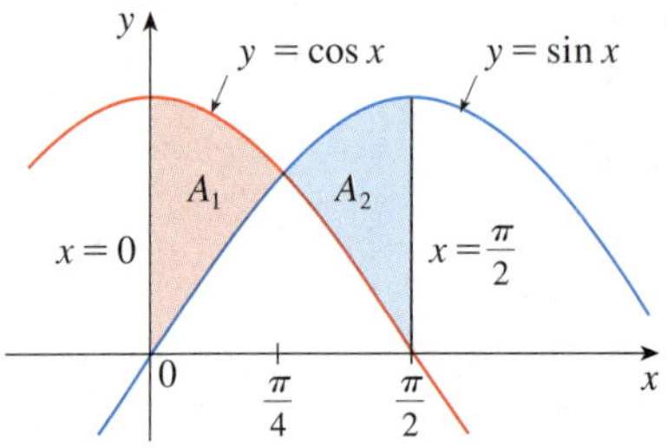

FIGURE 10

When evaluating the integral in (3), however, we must still split it into integrals corresponding to $A_1, A_2, \ldots$.

EXAMPLE 5 Find the area of the region bounded by the curves $y = \sin x$, $y = \cos x$, $x = 0$, and $x = \pi/2$.

SOLUTION The points of intersection occur when $\sin x = \cos x$, that is, when $x = \pi/4$ (since $0 \leqslant x \leqslant \pi/2$). The region is sketched in Figure 10. Observe that $\cos x \geqslant \sin x$

when $0 \leqslant x \leqslant \pi/4$ but $\sin x \geqslant \cos x$ when $\pi/4 \leqslant x \leqslant \pi/2$. Therefore, the required area is

$$\begin{aligned} A &= \int_0^{\pi/2} |\cos x - \sin x|\, dx = A_1 + A_2 \\ &= \int_0^{\pi/4} (\cos x - \sin x)\, dx + \int_{\pi/4}^{\pi/2} (\sin x - \cos x)\, dx \\ &= \Big[\sin x + \cos x\Big]_0^{\pi/4} + \Big[-\cos x - \sin x\Big]_{\pi/4}^{\pi/2} \\ &= \left(\frac{1}{\sqrt{2}} + \frac{1}{\sqrt{2}} - 0 - 1\right) + \left(-0 - 1 + \frac{1}{\sqrt{2}} + \frac{1}{\sqrt{2}}\right) \\ &= 2\sqrt{2} - 2 \end{aligned}$$

In this particular example we could have saved some work by noticing that the region is symmetric about $x = \pi/4$ and so

$$A = 2A_1 = 2\int_0^{\pi/4} (\cos x - \sin x)\, dx$$

Some regions are best treated by regarding x as a function of y. If a region is bounded by curves with equations $x = f(y)$, $x = g(y)$, $y = c$, and $y = d$, where f and g are continuous and $f(y) \geqslant g(y)$ for $c \leqslant y \leqslant d$ (see Figure 11), then its area is

$$A = \int_c^d [f(y) - g(y)]\, dy$$

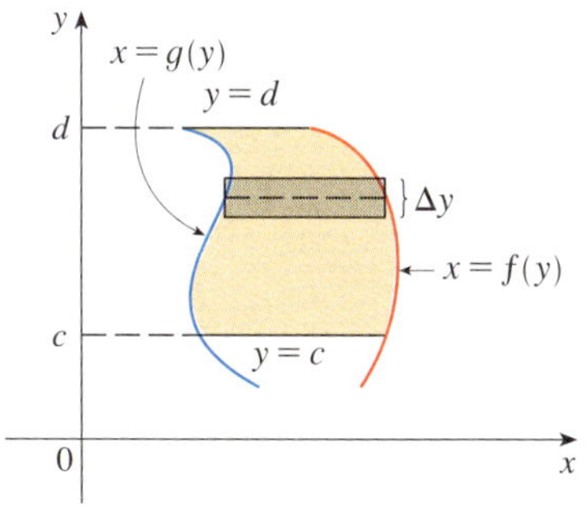

FIGURE 11

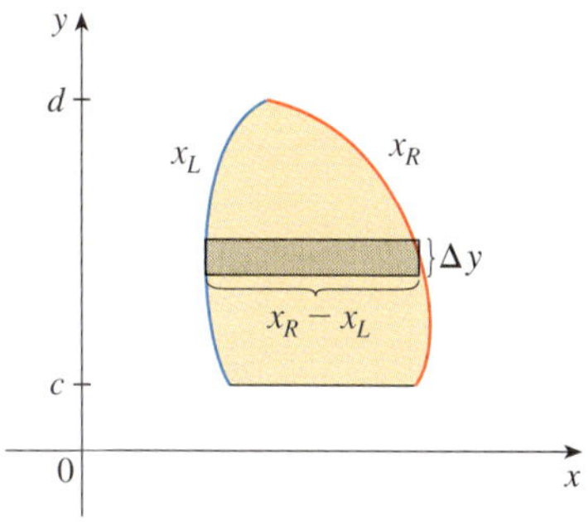

FIGURE 12

If we write x_R for the right boundary and x_L for the left boundary, then, as Figure 12 illustrates, we have

$$A = \int_c^d (x_R - x_L)\, dy$$

Here a typical approximating rectangle has dimensions $x_R - x_L$ and Δy.

EXAMPLE 6 Find the area enclosed by the line $y = x - 1$ and the parabola $y^2 = 2x + 6$.

SOLUTION By solving the two equations we find that the points of intersection are $(-1, -2)$ and $(5, 4)$. We solve the equation of the parabola for x and notice from Figure 13 that the left and right boundary curves are

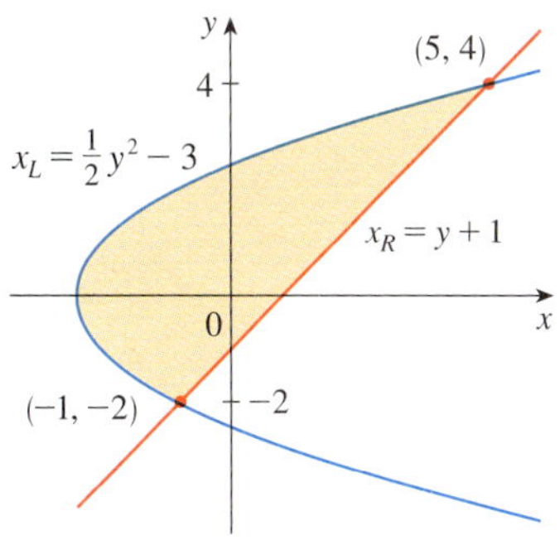

FIGURE 13

$$x_L = \tfrac{1}{2}y^2 - 3 \qquad x_R = y + 1$$

We must integrate between the appropriate y-values, $y = -2$ and $y = 4$. Thus

$$A = \int_{-2}^{4} (x_R - x_L)\,dy$$

$$= \int_{-2}^{4} \left[(y+1) - \left(\tfrac{1}{2}y^2 - 3\right)\right] dy$$

$$= \int_{-2}^{4} \left(-\tfrac{1}{2}y^2 + y + 4\right) dy$$

$$= -\frac{1}{2}\left(\frac{y^3}{3}\right) + \frac{y^2}{2} + 4y\,\Bigg]_{-2}^{4}$$

$$= -\tfrac{1}{6}(64) + 8 + 16 - \left(\tfrac{4}{3} + 2 - 8\right) = 18$$

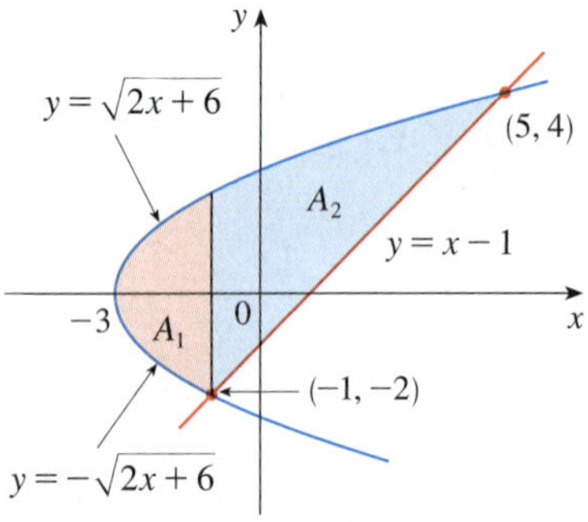

FIGURE 14

We could have found the area in Example 6 by integrating with respect to x instead of y, but the calculation is much more involved. It would have meant splitting the region in two and computing the areas labeled A_1 and A_2 in Figure 14. The method we used in Example 6 is *much* easier.

6.1 Exercises

1–4 Find the area of the shaded region.

1.

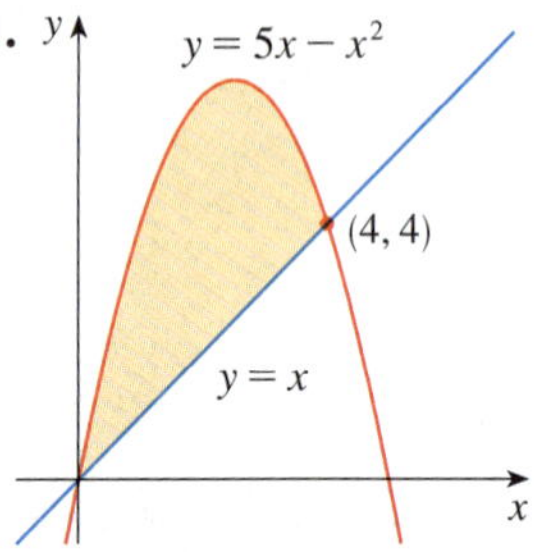

2.

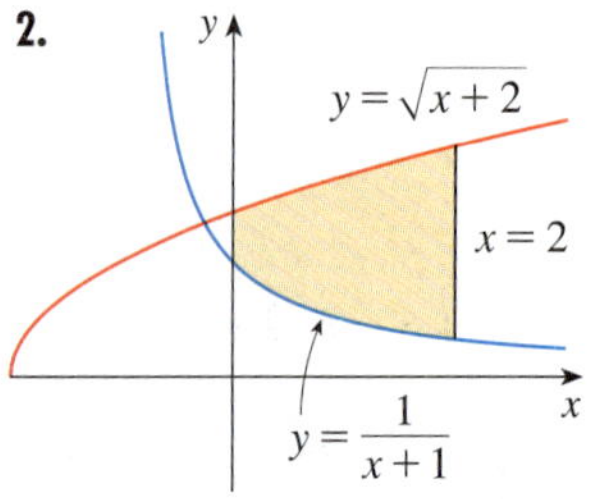

3.

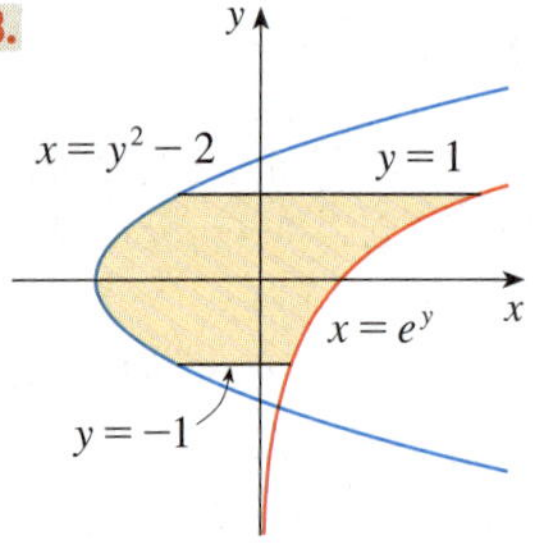

4.

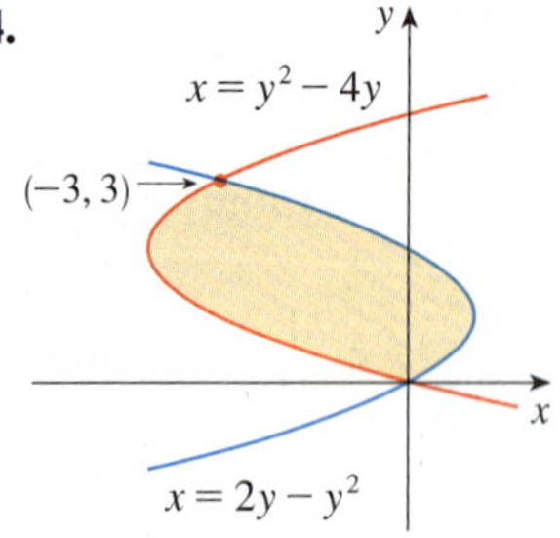

5–26 Sketch the region enclosed by the given curves. Decide whether to integrate with respect to x or y. Draw a typical approximating rectangle and label its height and width. Then find the area of the region.

5. $y = x + 1, \quad y = 9 - x^2, \quad x = -1, \quad x = 2$

6. $y = \sin x, \quad y = e^x, \quad x = 0, \quad x = \pi/2$

7. $y = x, \quad y = x^2$

8. $y = x^2, \quad y = x^4$

9. $y = 1/x, \quad y = 1/x^2, \quad x = 2$

10. $y = 1 + \sqrt{x}, \quad y = (3 + x)/3$

11. $y = x^2, \quad y^2 = x$

12. $y = x, \quad y = \sqrt[3]{x}$

13. $y = 12 - x^2, \quad y = x^2 - 6$

14. $y = x^3 - x, \quad y = 3x$

15. $y = \sqrt{x}, \quad y = \tfrac{1}{2}x, \quad x = 9$

16. $y = 8 - x^2, \quad y = x^2, \quad x = -3, \quad x = 3$

17. $x = 2y^2, \quad x + y = 1$

18. $4x + y^2 = 12, \quad x = y$

19. $x = 1 - y^2, \quad x = y^2 - 1$

20. $y = \sin(\pi x/2), \quad y = x$

21. $y = \cos x, \quad y = \sin 2x, \quad x = 0, \quad x = \pi/2$

22. $y = \sin x, \quad y = \sin 2x, \quad x = 0, \quad x = \pi/2$

23. $y = \cos x, \quad y = 1 - 2x/\pi$

24. $y = |x|, \quad y = x^2 - 2$

25. $y = x^2, \quad y = 2/(x^2 + 1)$

26. $y = \sin \pi x, \quad y = x^2 - x, \quad x = 2$

27–28 Use calculus to find the area of the triangle with the given vertices.

27. $(0, 0), \quad (2, 1), \quad (-1, 6)$

28. $(0, 5), \quad (2, -2), \quad (5, 1)$

29–30 Evaluate the integral and interpret it as the area of a region. Sketch the region.

29. $\int_{-1}^{1} |x^3 - x|\, dx$ **30.** $\int_0^4 |\sqrt{x+2} - x|\, dx$

31–32 Use the Midpoint Rule with $n = 4$ to approximate the area of the region bounded by the given curves.

31. $y = \sin^2(\pi x/4)$, $y = \cos^2(\pi x/4)$, $0 \leqslant x \leqslant 1$

32. $y = \sqrt[3]{16 - x^3}$, $y = x$, $x = 0$

33–36 Use a graph to find approximate x-coordinates of the points of intersection of the given curves. Then find (approximately) the area of the region bounded by the curves.

33. $y = x^2$, $y = 2\cos x$ **34.** $y = x^4$, $y = 3x - x^3$

35. $y = x\cos(x^2)$, $y = x^3$ **36.** $y = e^x$, $y = 2 - x^2$

37. Use a computer algebra system to find the exact area enclosed by the curves $y = x^5 - 6x^3 + 4x$ and $y = x$.

38. Sketch the region in the xy-plane defined by the inequalities $x - 2y^2 \geqslant 0$, $1 - x - |y| \geqslant 0$ and find its area.

39. Racing cars driven by Chris and Kelly are side by side at the start of a race. The table shows the velocities of each car (in miles per hour) during the first ten seconds of the race. Use the Midpoint Rule to estimate how much farther Kelly travels than Chris does during the first ten seconds.

t	v_C	v_K	t	v_C	v_K
0	0	0	6	69	80
1	20	22	7	75	86
2	32	37	8	81	93
3	46	52	9	86	98
4	54	61	10	90	102
5	62	71			

40. The widths (in meters) of a kidney-shaped swimming pool were measured at 2-meter intervals as indicated in the figure. Use the Midpoint Rule to estimate the area of the pool.

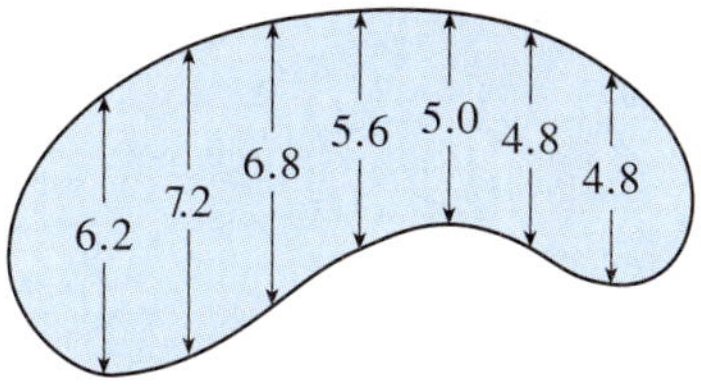

41. Two cars, A and B, start side by side and accelerate from rest. The figure shows the graphs of their velocity functions.
(a) Which car is ahead after one minute? Explain.
(b) What is the meaning of the area of the shaded region?
(c) Which car is ahead after two minutes? Explain.
(d) Estimate the time at which the cars are again side by side.

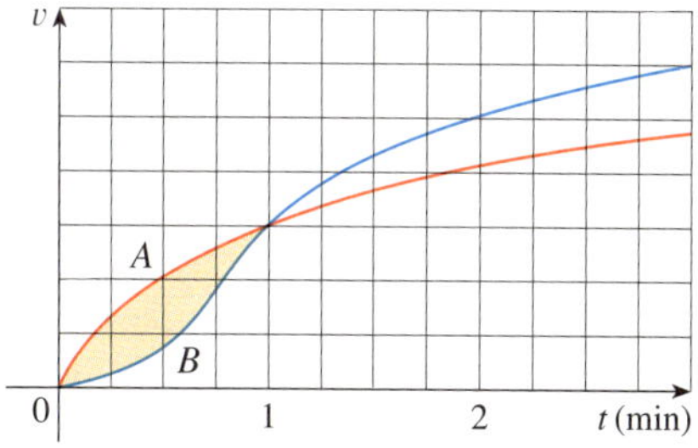

42. The figure shows graphs of the marginal revenue function R' and the marginal cost function C' for a manufacturer. [Recall from Section 4.8 that $R(x)$ and $C(x)$ represent the revenue and cost when x units are manufactured. Assume that R and C are measured in thousands of dollars.] What is the meaning of the area of the shaded region? Use the Midpoint Rule to estimate the value of this quantity.

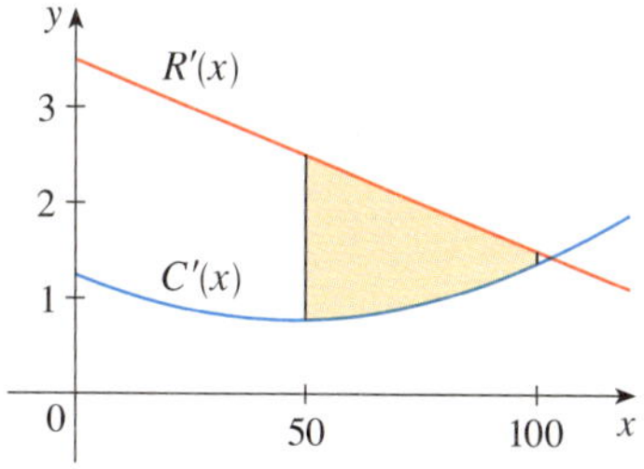

43. The curve with equation $y^2 = x^2(x + 3)$ is called **Tschirnhausen's cubic**. If you graph this curve you will see that part of the curve forms a loop. Find the area enclosed by the loop.

44. Find the area of the region bounded by the parabola $y = x^2$, the tangent line to this parabola at $(1, 1)$, and the x-axis.

45. Find the number b such that the line $y = b$ divides the region bounded by the curves $y = x^2$ and $y = 4$ into two regions with equal area.

46. (a) Find the number a such that the line $x = a$ bisects the area under the curve $y = 1/x^2$, $1 \leqslant x \leqslant 4$.
(b) Find the number b such that the line $y = b$ bisects the area in part (a).

47. Find the values of c such that the area of the region enclosed by the parabolas $y = x^2 - c^2$ and $y = c^2 - x^2$ is 576.

48. Suppose that $0 < c < \pi/2$. For what value of c is the area of the region enclosed by the curves $y = \cos x$, $y = \cos(x - c)$, and $x = 0$ equal to the area of the region enclosed by the curves $y = \cos(x - c)$, $x = \pi$, and $y = 0$?

49. For what values of m do the line $y = mx$ and the curve $y = x/(x^2 + 1)$ enclose a region? Find the area of the region.

6.2 Volumes

In trying to find the volume of a solid we face the same type of problem as in finding areas. We have an intuitive idea of what volume means, but we must make this idea precise by using calculus to give an exact definition of volume.

We start with a simple type of solid called a **cylinder** (or, more precisely, a *right cylinder*). As illustrated in Figure 1(a), a cylinder is bounded by a plane region B_1, called the **base**, and a congruent region B_2 in a parallel plane. The cylinder consists of all points on line segments that are perpendicular to the base and join B_1 to B_2. If the area of the base is A and the height of the cylinder (the distance from B_1 to B_2) is h, then the volume V of the cylinder is defined as

$$V = Ah$$

In particular, if the base is a circle with radius r, then the cylinder is a circular cylinder with volume $V = \pi r^2 h$ [see Figure 1(b)], and if the base is a rectangle with length l and width w, then the cylinder is a rectangular box (also called a *rectangular parallelepiped*) with volume $V = lwh$ [see Figure 1(c)].

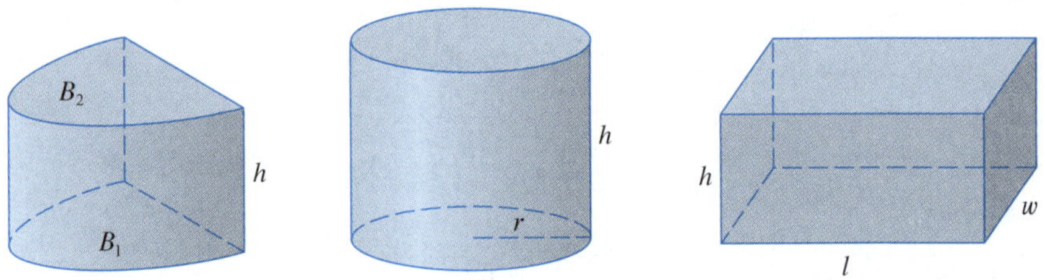

FIGURE 1

(a) Cylinder $V = Ah$

(b) Circular cylinder $V = \pi r^2 h$

(c) Rectangular box $V = lwh$

For a solid S that isn't a cylinder we first "cut" S into pieces and approximate each piece by a cylinder. We estimate the volume of S by adding the volumes of the cylinders. We arrive at the exact volume of S through a limiting process in which the number of pieces becomes large.

We start by intersecting S with a plane and obtaining a plane region that is called a **cross-section** of S. Let $A(x)$ be the area of the cross-section of S in a plane P_x perpendicular to the x-axis and passing through the point x, where $a \leqslant x \leqslant b$. (See Figure 2. Think of slicing S with a knife through x and computing the area of this slice.) The cross-sectional area $A(x)$ will vary as x increases from a to b.

Watch an animation of Figure 2.
Resources / Module 7
/ Volumes
/ Volumes by Cross-Section

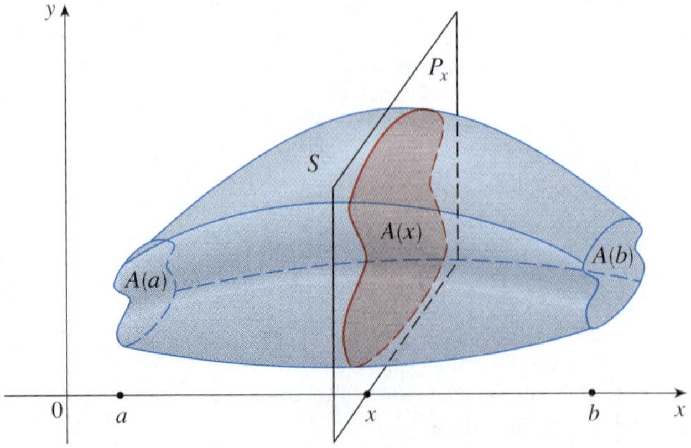

FIGURE 2

Let's divide S into n "slabs" of equal width Δx by using the planes $P_{x_1}, P_{x_2}, \dots$ to slice the solid. (Think of slicing a loaf of bread.) If we choose sample points x_i^* in $[x_{i-1}, x_i]$, we can approximate the ith slab S_i (the part of S that lies between the planes $P_{x_{i-1}}$ and P_{x_i}) by a cylinder with base area $A(x_i^*)$ and "height" Δx. (See Figure 3.)

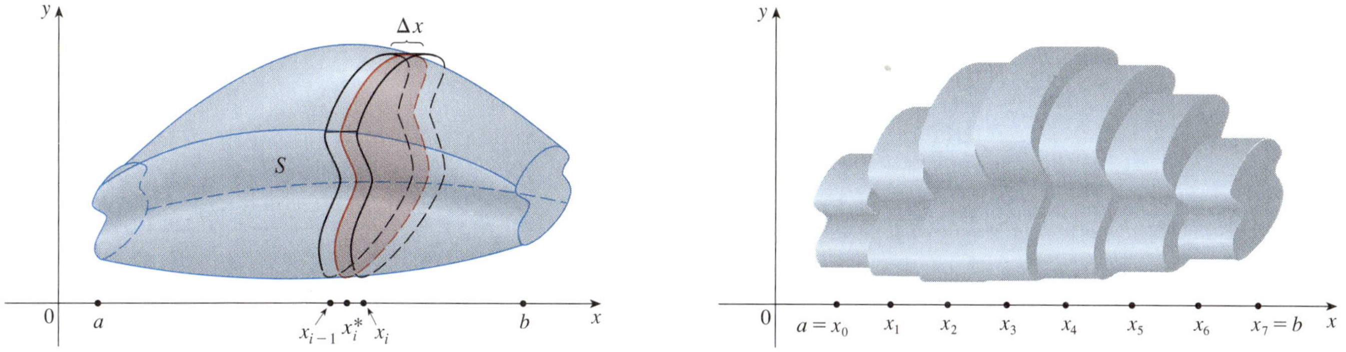

FIGURE 3

The volume of this cylinder is $A(x_i^*)\,\Delta x$, so an approximation to our intuitive conception of the volume of the ith slab S_i is

$$V(S_i) \approx A(x_i^*)\,\Delta x$$

Adding the volumes of these slabs, we get an approximation to the total volume (that is, what we think of intuitively as the volume):

$$V \approx \sum_{i=1}^{n} A(x_i^*)\,\Delta x$$

This approximation appears to become better and better as $n \to \infty$. (Think of the slices as becoming thinner and thinner.) Therefore, we *define* the volume as the limit of these sums as $n \to \infty$. But we recognize the limit of Riemann sums as a definite integral and so we have the following definition.

|||| It can be proved that this definition is independent of how S is situated with respect to the x-axis. In other words, no matter how we slice S with parallel planes, we always get the same answer for V.

Definition of Volume Let S be a solid that lies between $x = a$ and $x = b$. If the cross-sectional area of S in the plane P_x, through x and perpendicular to the x-axis, is $A(x)$, where A is a continuous function, then the **volume** of S is

$$V = \lim_{n \to \infty} \sum_{i=1}^{n} A(x_i^*)\,\Delta x = \int_a^b A(x)\,dx$$

When we use the volume formula $V = \int_a^b A(x)\,dx$ it is important to remember that $A(x)$ is the area of a moving cross-section obtained by slicing through x perpendicular to the x-axis.

Notice that, for a cylinder, the cross-sectional area is constant: $A(x) = A$ for all x. So our definition of volume gives $V = \int_a^b A\,dx = A(b - a)$; this agrees with the formula $V = Ah$.

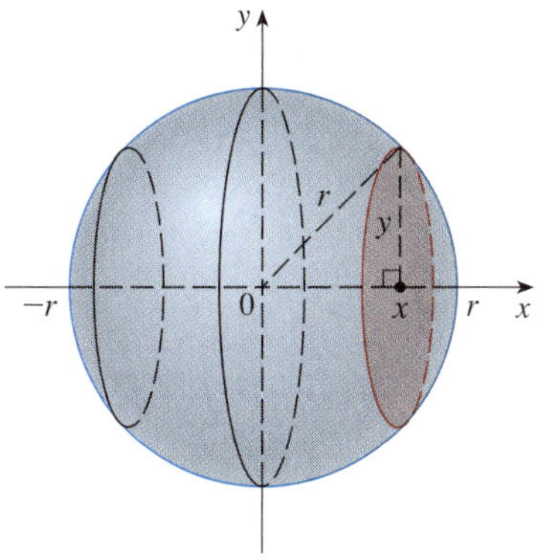

FIGURE 4

EXAMPLE 1 Show that the volume of a sphere of radius r is

$$V = \tfrac{4}{3}\pi r^3$$

SOLUTION If we place the sphere so that its center is at the origin (see Figure 4), then the plane P_x intersects the sphere in a circle whose radius (from the Pythagorean Theorem) is $y = \sqrt{r^2 - x^2}$. So the cross-sectional area is

$$A(x) = \pi y^2 = \pi(r^2 - x^2)$$

Using the definition of volume with $a = -r$ and $b = r$, we have

$$\begin{aligned} V &= \int_{-r}^{r} A(x)\,dx = \int_{-r}^{r} \pi(r^2 - x^2)\,dx \\ &= 2\pi \int_{0}^{r} (r^2 - x^2)\,dx \qquad \text{(The integrand is even.)} \\ &= 2\pi\left[r^2x - \frac{x^3}{3}\right]_0^r = 2\pi\left(r^3 - \frac{r^3}{3}\right) \\ &= \tfrac{4}{3}\pi r^3 \end{aligned}$$

Watch an animation of Figure 5.
Resources / Module 7 / Volumes / Volumes

Figure 5 illustrates the definition of volume when the solid is a sphere with radius $r = 1$. From the result of Example 1, we know that the volume of the sphere is $\frac{4}{3}\pi \approx 4.18879$. Here the slabs are circular cylinders, or *disks*, and the three parts of Figure 5 show the geometric interpretations of the Riemann sums

$$\sum_{i=1}^{n} A(\bar{x}_i)\,\Delta x = \sum_{i=1}^{n} \pi(1^2 - \bar{x}_i^2)\,\Delta x$$

when $n = 5$, 10, and 20 if we choose the sample points x_i^* to be the midpoints $\bar{x}_i$. Notice that as we increase the number of approximating cylinders, the corresponding Riemann sums become closer to the true volume.

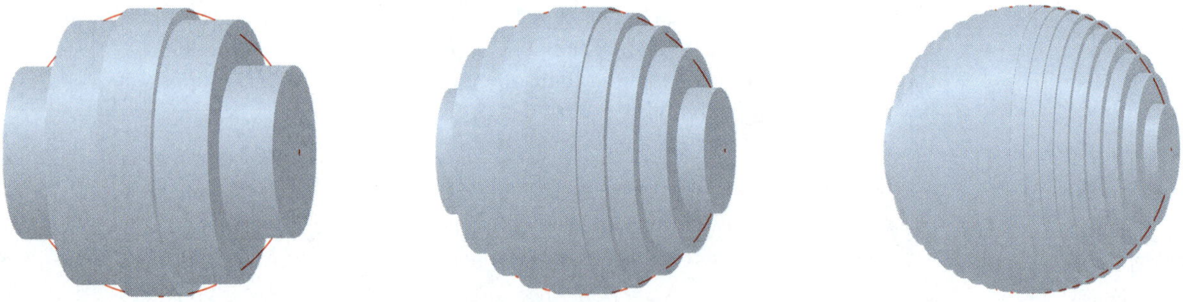

(a) Using 5 disks, $V \approx 4.2726$ (b) Using 10 disks, $V \approx 4.2097$ (c) Using 20 disks, $V \approx 4.1940$

FIGURE 5 Approximating the volume of a sphere with radius 1

EXAMPLE 2 Find the volume of the solid obtained by rotating about the x-axis the region under the curve $y = \sqrt{x}$ from 0 to 1. Illustrate the definition of volume by sketching a typical approximating cylinder.

SOLUTION The region is shown in Figure 6(a). If we rotate about the x-axis, we get the solid shown in Figure 6(b). When we slice through the point x, we get a disk with radius $\sqrt{x}$. The area of this cross-section is

$$A(x) = \pi(\sqrt{x})^2 = \pi x$$

and the volume of the approximating cylinder (a disk with thickness Δx) is

$$A(x)\,\Delta x = \pi x\,\Delta x$$

|||| Did we get a reasonable answer in Example 2? As a check on our work, let's replace the given region by a square with base $[0, 1]$ and height 1. If we rotate this square, we get a cylinder with radius 1, height 1, and volume $\pi \cdot 1^2 \cdot 1 = \pi$. We computed that the given solid has half this volume. That seems about right.

See a volume of revolution being formed.
Resources / Module 7
/ Volumes
/ Volumes of Revolution

The solid lies between $x = 0$ and $x = 1$, so its volume is

$$V = \int_0^1 A(x)\,dx = \int_0^1 \pi x\,dx = \pi \frac{x^2}{2}\bigg]_0^1 = \frac{\pi}{2}$$

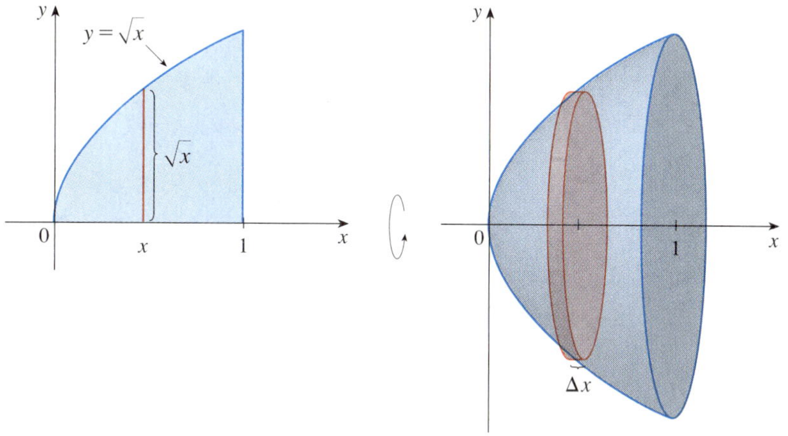

FIGURE 6 (a) (b)

EXAMPLE 3 Find the volume of the solid obtained by rotating the region bounded by $y = x^3$, $y = 8$, and $x = 0$ about the y-axis.

SOLUTION The region is shown in Figure 7(a) and the resulting solid is shown in Figure 7(b). Because the region is rotated about the y-axis, it makes sense to slice the solid perpendicular to the y-axis and therefore to integrate with respect to y. If we slice at height y, we get a circular disk with radius x, where $x = \sqrt[3]{y}$. So the area of a cross-section through y is

$$A(y) = \pi x^2 = \pi(\sqrt[3]{y})^2 = \pi y^{2/3}$$

and the volume of the approximating cylinder pictured in Figure 7(b) is

$$A(y)\,\Delta y = \pi y^{2/3}\,\Delta y$$

Since the solid lies between $y = 0$ and $y = 8$, its volume is

$$V = \int_0^8 A(y)\,dy = \int_0^8 \pi y^{2/3}\,dy = \pi\left[\tfrac{3}{5}y^{5/3}\right]_0^8 = \frac{96\pi}{5}$$

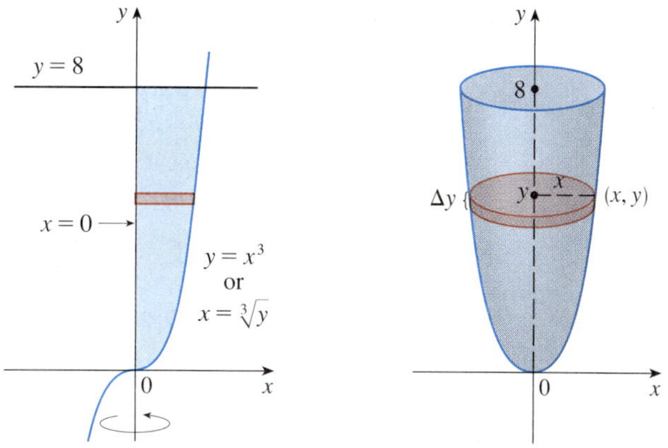

FIGURE 7 (a) (b)

EXAMPLE 4 The region $\mathcal{R}$ enclosed by the curves $y = x$ and $y = x^2$ is rotated about the x-axis. Find the volume of the resulting solid.

SOLUTION The curves $y = x$ and $y = x^2$ intersect at the points $(0, 0)$ and $(1, 1)$. The region between them, the solid of rotation, and a cross-section perpendicular to the x-axis are shown in Figure 8. A cross-section in the plane P_x has the shape of a *washer* (an annular ring) with inner radius x^2 and outer radius x, so we find the cross-sectional area by subtracting the area of the inner circle from the area of the outer circle:

$$A(x) = \pi x^2 - \pi(x^2)^2 = \pi(x^2 - x^4)$$

Therefore, we have

$$V = \int_0^1 A(x)\,dx = \int_0^1 \pi(x^2 - x^4)\,dx$$

$$= \pi\left[\frac{x^3}{3} - \frac{x^5}{5}\right]_0^1 = \frac{2\pi}{15}$$

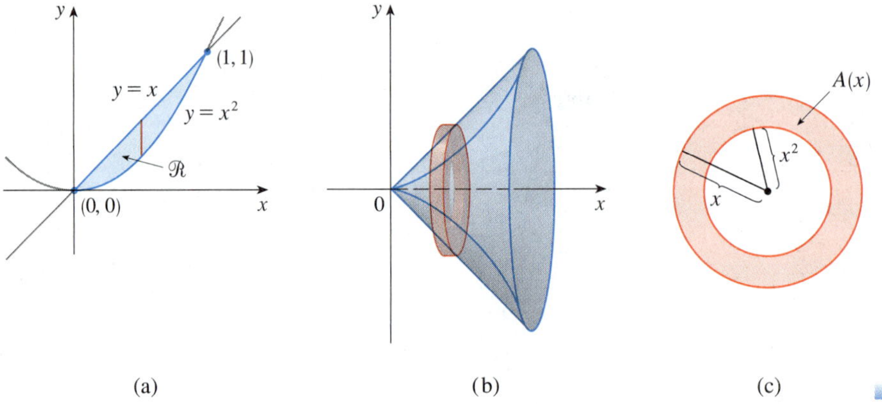

FIGURE 8

EXAMPLE 5 Find the volume of the solid obtained by rotating the region in Example 4 about the line $y = 2$.

SOLUTION The solid and a cross-section are shown in Figure 9. Again the cross-section is a washer, but this time the inner radius is $2 - x$ and the outer radius is $2 - x^2$. The

TEC Module 6.2/6.3 illustrates the formation and computation of volumes using disks and washers.

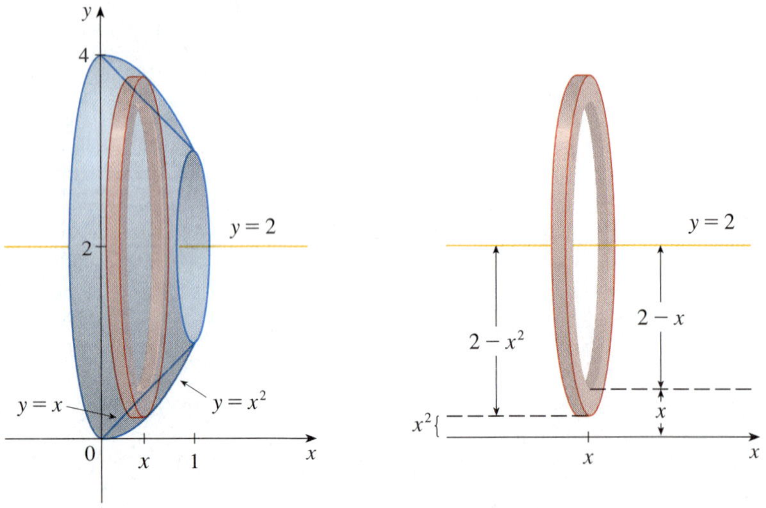

FIGURE 9

cross-sectional area is

$$A(x) = \pi(2 - x^2)^2 - \pi(2 - x)^2$$

and so the volume of S is

$$\begin{aligned} V &= \int_0^1 A(x)\,dx \\ &= \pi \int_0^1 [(2 - x^2)^2 - (2 - x)^2]\,dx \\ &= \pi \int_0^1 (x^4 - 5x^2 + 4x)\,dx \\ &= \pi \left[\frac{x^5}{5} - 5\frac{x^3}{3} + 4\frac{x^2}{2} \right]_0^1 = \frac{8\pi}{15} \end{aligned}$$

The solids in Examples 1–5 are all called **solids of revolution** because they are obtained by revolving a region about a line. In general, we calculate the volume of a solid of revolution by using the basic defining formula

$$V = \int_a^b A(x)\,dx \qquad \text{or} \qquad V = \int_c^d A(y)\,dy$$

and we find the cross-sectional area $A(x)$ or $A(y)$ in one of the following ways:

- If the cross-section is a disk (as in Examples 1–3), we find the radius of the disk (in terms of x or y) and use

$$A = \pi(\text{radius})^2$$

- If the cross-section is a washer (as in Examples 4 and 5), we find the inner radius r_{in} and outer radius r_{out} from a sketch (as in Figures 9 and 10) and compute the area of the washer by subtracting the area of the inner disk from the area of the outer disk:

$$A = \pi(\text{outer radius})^2 - \pi(\text{inner radius})^2$$

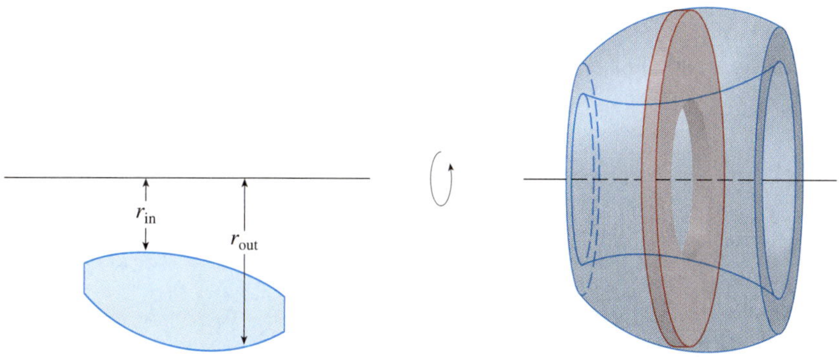

FIGURE 10

The next example gives a further illustration of the procedure.

EXAMPLE 6 Find the volume of the solid obtained by rotating the region in Example 4 about the line $x = -1$.

SOLUTION Figure 11 shows a horizontal cross-section. It is a washer with inner radius $1 + y$ and outer radius $1 + \sqrt{y}$, so the cross-sectional area is

$$\begin{aligned} A(y) &= \pi(\text{outer radius})^2 - \pi(\text{inner radius})^2 \\ &= \pi(1 + \sqrt{y})^2 - \pi(1 + y)^2 \end{aligned}$$

The volume is

$$\begin{aligned} V &= \int_0^1 A(y)\,dy = \pi \int_0^1 \left[(1 + \sqrt{y})^2 - (1 + y)^2\right] dy \\ &= \pi \int_0^1 \left(2\sqrt{y} - y - y^2\right) dy \\ &= \pi \left[\frac{4y^{3/2}}{3} - \frac{y^2}{2} - \frac{y^3}{3}\right]_0^1 = \frac{\pi}{2} \end{aligned}$$

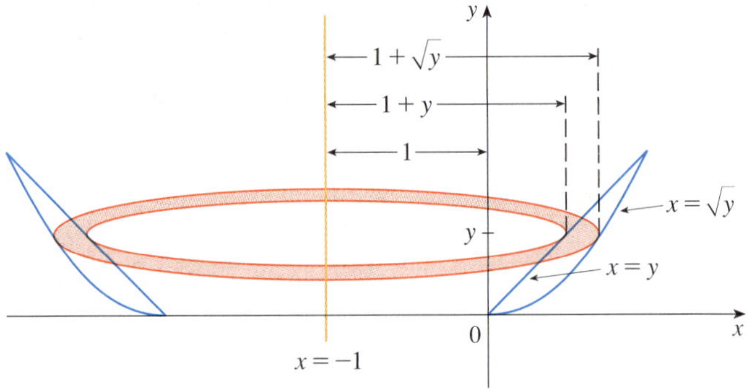

FIGURE 11

We now find the volumes of three solids that are *not* solids of revolution.

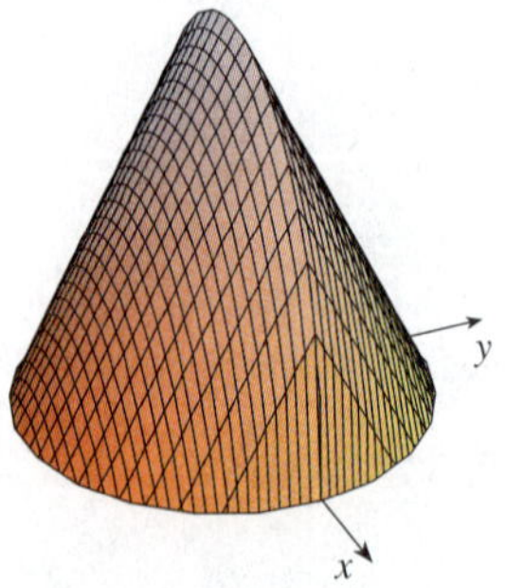

FIGURE 12
Computer-generated picture of the solid in Example 7

EXAMPLE 7 Figure 12 shows a solid with a circular base of radius 1. Parallel cross-sections perpendicular to the base are equilateral triangles. Find the volume of the solid.

SOLUTION Let's take the circle to be $x^2 + y^2 = 1$. The solid, its base, and a typical cross-section at a distance x from the origin are shown in Figure 13.

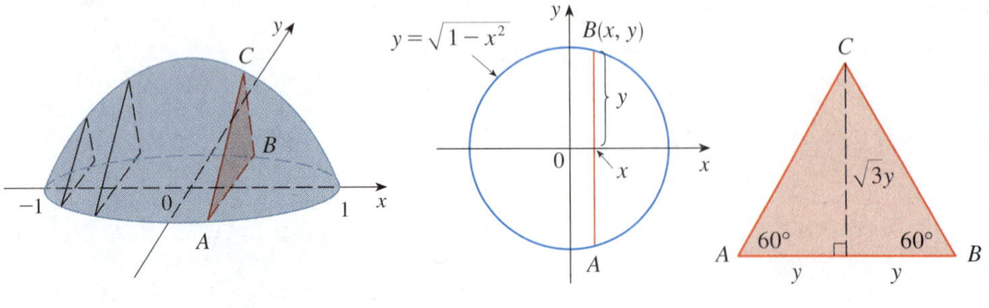

FIGURE 13
(a) The solid
(b) Its base
(c) A cross-section

Since B lies on the circle, we have $y = \sqrt{1 - x^2}$ and so the base of the triangle ABC is $|AB| = 2\sqrt{1 - x^2}$. Since the triangle is equilateral, we see from Figure 13(c) that its height is $\sqrt{3}\, y = \sqrt{3}\sqrt{1 - x^2}$. The cross-sectional area is therefore

$$A(x) = \tfrac{1}{2} \cdot 2\sqrt{1 - x^2} \cdot \sqrt{3}\sqrt{1 - x^2} = \sqrt{3}\,(1 - x^2)$$

and the volume of the solid is

$$V = \int_{-1}^{1} A(x)\, dx = \int_{-1}^{1} \sqrt{3}\,(1 - x^2)\, dx$$

$$= 2\int_{0}^{1} \sqrt{3}\,(1 - x^2)\, dx = 2\sqrt{3}\left[x - \frac{x^3}{3}\right]_0^1 = \frac{4\sqrt{3}}{3}$$

Resources / Module 7 / Volumes / Start of Mystery of the Topless Pyramid

EXAMPLE 8 Find the volume of a pyramid whose base is a square with side L and whose height is h.

SOLUTION We place the origin O at the vertex of the pyramid and the x-axis along its central axis as in Figure 14. Any plane P_x that passes through x and is perpendicular to the x-axis intersects the pyramid in a square with side of length s, say. We can express s in terms of x by observing from the similar triangles in Figure 15 that

$$\frac{x}{h} = \frac{s/2}{L/2} = \frac{s}{L}$$

and so $s = Lx/h$. [Another method is to observe that the line OP has slope $L/(2h)$ and so its equation is $y = Lx/(2h)$.] Thus, the cross-sectional area is

$$A(x) = s^2 = \frac{L^2}{h^2}x^2$$

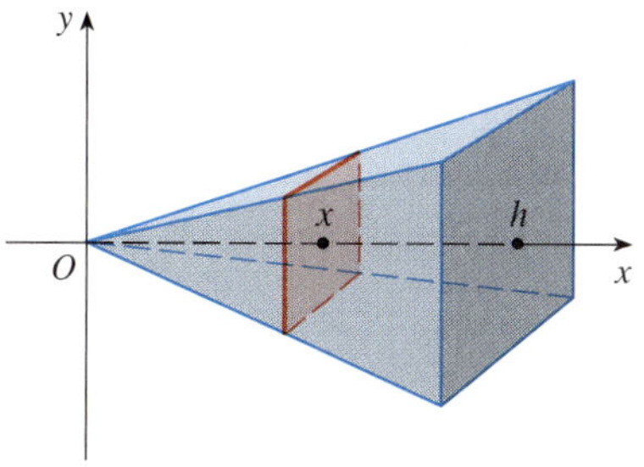

FIGURE 14

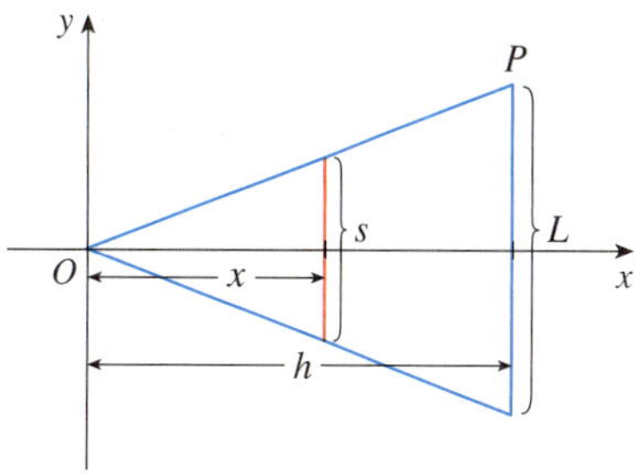

FIGURE 15

The pyramid lies between $x = 0$ and $x = h$, so its volume is

$$V = \int_0^h A(x)\, dx = \int_0^h \frac{L^2}{h^2}x^2\, dx$$

$$= \frac{L^2}{h^2}\frac{x^3}{3}\Bigg]_0^h = \frac{L^2 h}{3}$$

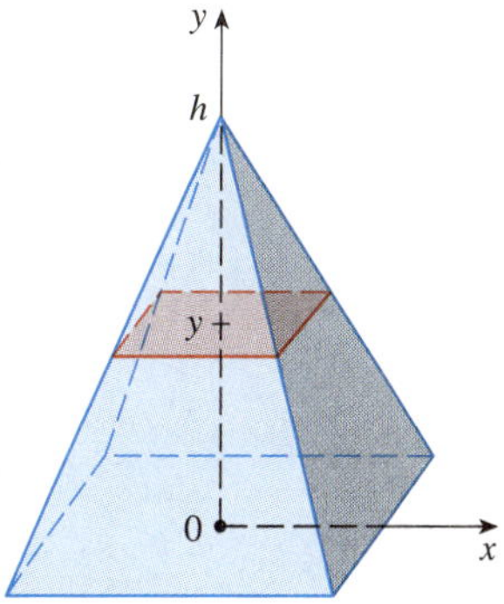

FIGURE 16

NOTE We didn't need to place the vertex of the pyramid at the origin in Example 8. We did so merely to make the equations simple. If, instead, we had placed the center of the base at the origin and the vertex on the positive y-axis, as in Figure 16, you can verify that we would have obtained the integral

$$V = \int_0^h \frac{L^2}{h^2}(h - y)^2\,dy = \frac{L^2h}{3}$$

EXAMPLE 9 A wedge is cut out of a circular cylinder of radius 4 by two planes. One plane is perpendicular to the axis of the cylinder. The other intersects the first at an angle of 30° along a diameter of the cylinder. Find the volume of the wedge.

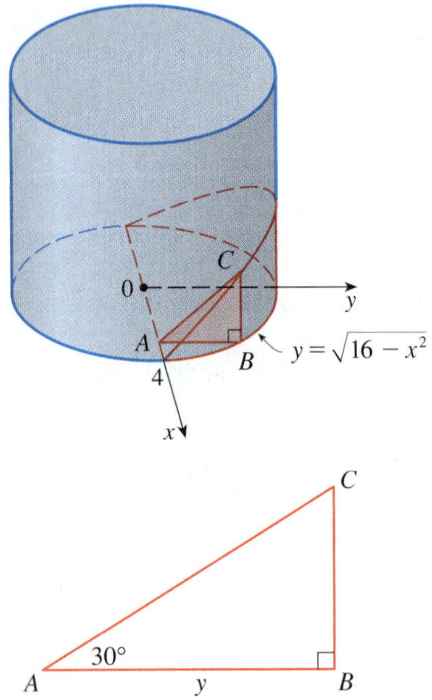

FIGURE 17

SOLUTION If we place the x-axis along the diameter where the planes meet, then the base of the solid is a semicircle with equation $y = \sqrt{16 - x^2}$, $-4 \leqslant x \leqslant 4$. A cross-section perpendicular to the x-axis at a distance x from the origin is a triangle ABC, as shown in Figure 17, whose base is $y = \sqrt{16 - x^2}$ and whose height is $|BC| = y \tan 30° = \sqrt{16 - x^2}/\sqrt{3}$. Thus, the cross-sectional area is

$$A(x) = \tfrac{1}{2}\sqrt{16 - x^2} \cdot \frac{1}{\sqrt{3}}\sqrt{16 - x^2}$$

$$= \frac{16 - x^2}{2\sqrt{3}}$$

and the volume is

$$V = \int_{-4}^{4} A(x)\,dx = \int_{-4}^{4} \frac{16 - x^2}{2\sqrt{3}}\,dx$$

$$= \frac{1}{\sqrt{3}}\int_0^4 (16 - x^2)\,dx = \frac{1}{\sqrt{3}}\left[16x - \frac{x^3}{3}\right]_0^4$$

$$= \frac{128}{3\sqrt{3}}$$

For another method see Exercise 62.

6.2 Exercises

1–18 Find the volume of the solid obtained by rotating the region bounded by the given curves about the specified line. Sketch the region, the solid, and a typical disk or washer.

1. $y = x^2$, $x = 1$, $y = 0$; about the x-axis
2. $y = e^x$, $y = 0$, $x = 0$, $x = 1$; about the x-axis
3. $y = 1/x$, $x = 1$, $x = 2$, $y = 0$; about the x-axis
4. $y = \sqrt{x - 1}$, $x = 2$, $x = 5$, $y = 0$; about the x-axis
5. $y = x^2$, $0 \leqslant x \leqslant 2$, $y = 4$, $x = 0$; about the y-axis
6. $x = y - y^2$, $x = 0$; about the y-axis
7. $y = x^2$, $y^2 = x$; about the x-axis
8. $y = \sec x$, $y = 1$, $x = -1$, $x = 1$; about the x-axis
9. $y^2 = x$, $x = 2y$; about the y-axis
10. $y = x^{2/3}$, $x = 1$, $y = 0$; about the y-axis
11. $y = x$, $y = \sqrt{x}$; about $y = 1$
12. $y = x^2$, $y = 4$; about $y = 4$
13. $y = x^4$, $y = 1$; about $y = 2$
14. $y = 1/x$, $y = 0$, $x = 1$, $x = 3$; about $y = -1$
15. $x = y^2$, $x = 1$; about $x = 1$
16. $y = x$, $y = \sqrt{x}$; about $x = 2$
17. $y = x^2$, $x = y^2$; about $x = -1$
18. $y = x$, $y = 0$, $x = 2$, $x = 4$; about $x = 1$

19–30 Refer to the figure and find the volume generated by rotating the given region about the specified line.

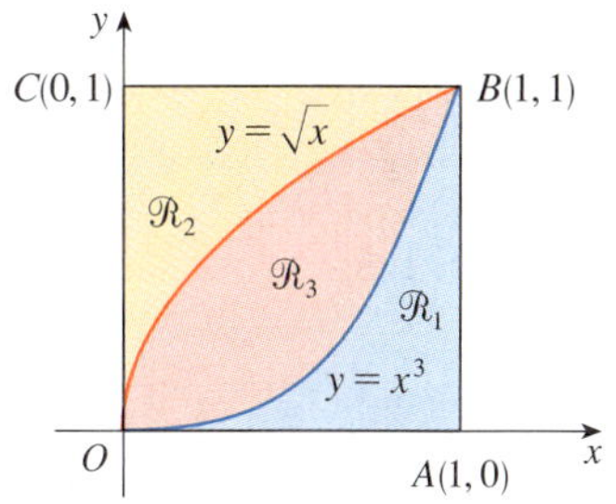

19. $\mathcal{R}_1$ about OA

20. $\mathcal{R}_1$ about OC

21. $\mathcal{R}_1$ about AB

22. $\mathcal{R}_1$ about BC

23. $\mathcal{R}_2$ about OA

24. $\mathcal{R}_2$ about OC

25. $\mathcal{R}_2$ about AB

26. $\mathcal{R}_2$ about BC

27. $\mathcal{R}_3$ about OA

28. $\mathcal{R}_3$ about OC

29. $\mathcal{R}_3$ about AB

30. $\mathcal{R}_3$ about BC

31–36 Set up, but do not evaluate, an integral for the volume of the solid obtained by rotating the region bounded by the given curves about the specified line.

31. $y = \tan^3 x,\ y = 1,\ x = 0;$ about $y = 1$

32. $y = (x - 2)^4,\ 8x - y = 16;$ about $x = 10$

33. $y = 0,\ y = \sin x,\ 0 \leqslant x \leqslant \pi;$ about $y = 1$

34. $y = 0,\ y = \sin x,\ 0 \leqslant x \leqslant \pi;$ about $y = -2$

35. $x^2 - y^2 = 1,\ x = 3;$ about $x = -2$

36. $2x + 3y = 6,\ (y - 1)^2 = 4 - x;$ about $x = -5$

37–38 Use a graph to find approximate x-coordinates of the points of intersection of the given curves. Then find (approximately) the volume of the solid obtained by rotating about the x-axis the region bounded by these curves.

37. $y = x^2,\quad y = \ln(x + 1)$

38. $y = 3\sin(x^2),\quad y = e^{x/2} + e^{-2x}$

39–40 Use a computer algebra system to find the exact volume of the solid obtained by rotating the region bounded by the given curves about the specified line.

39. $y = \sin^2 x,\ y = 0,\ 0 \leqslant x \leqslant \pi;$ about $y = -1$

40. $y = x,\ y = xe^{1-x/2};$ about $y = 3$

41–44 Each integral represents the volume of a solid. Describe the solid.

41. $\pi \int_0^{\pi/2} \cos^2 x\, dx$

42. $\pi \int_2^5 y\, dy$

43. $\pi \int_0^1 (y^4 - y^8)\, dy$

44. $\pi \int_0^{\pi/2} [(1 + \cos x)^2 - 1^2]\, dx$

45. A CAT scan produces equally spaced cross-sectional views of a human organ that provide information about the organ otherwise obtained only by surgery. Suppose that a CAT scan of a human liver shows cross-sections spaced 1.5 cm apart. The liver is 15 cm long and the cross-sectional areas, in square centimeters, are 0, 18, 58, 79, 94, 106, 117, 128, 63, 39, and 0. Use the Midpoint Rule to estimate the volume of the liver.

46. A log 10 m long is cut at 1-meter intervals and its cross-sectional areas A (at a distance x from the end of the log) are listed in the table. Use the Midpoint Rule with $n = 5$ to estimate the volume of the log.

x (m)	A (m^2)	x (m)	A (m^2)
0	0.68	6	0.53
1	0.65	7	0.55
2	0.64	8	0.52
3	0.61	9	0.50
4	0.58	10	0.48
5	0.59		

47–59 Find the volume of the described solid S.

47. A right circular cone with height h and base radius r

48. A frustum of a right circular cone with height h, lower base radius R, and top radius r

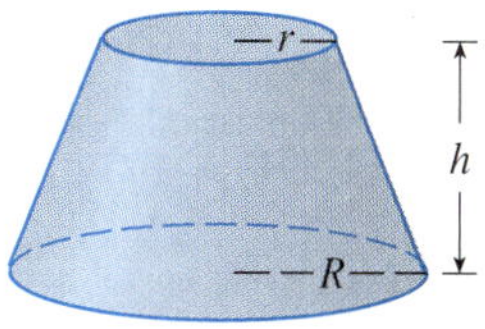

49. A cap of a sphere with radius r and height h

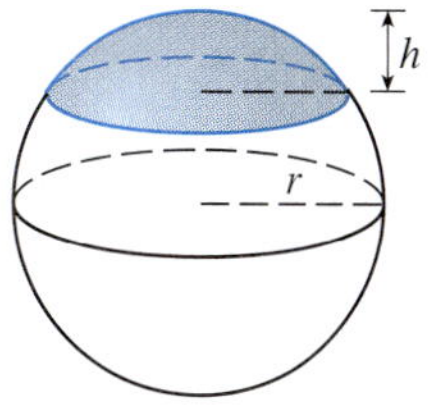

50. A frustum of a pyramid with square base of side b, square top of side a, and height h

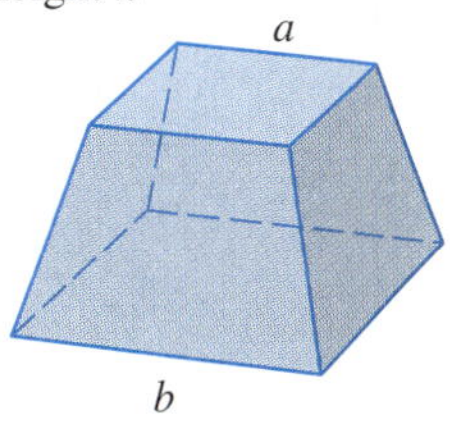

What happens if $a = b$? What happens if $a = 0$?

51. A pyramid with height h and rectangular base with dimensions b and $2b$

52. A pyramid with height h and base an equilateral triangle with side a (a tetrahedron)

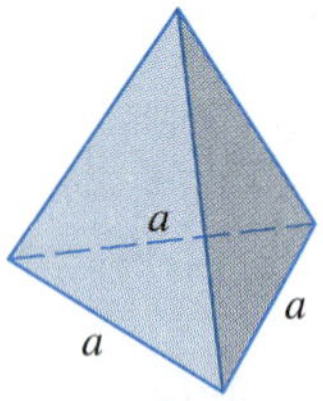

53. A tetrahedron with three mutually perpendicular faces and three mutually perpendicular edges with lengths 3 cm, 4 cm, and 5 cm

54. The base of S is a circular disk with radius r. Parallel cross-sections perpendicular to the base are squares.

55. The base of S is an elliptical region with boundary curve $9x^2 + 4y^2 = 36$. Cross-sections perpendicular to the x-axis are isosceles right triangles with hypotenuse in the base.

56. The base of S is the parabolic region $\{(x, y) \mid x^2 \leqslant y \leqslant 1\}$. Cross-sections perpendicular to the y-axis are equilateral triangles.

57. S has the same base as in Exercise 56, but cross-sections perpendicular to the y-axis are squares.

58. The base of S is the triangular region with vertices $(0, 0)$, $(3, 0)$, and $(0, 2)$. Cross-sections perpendicular to the y-axis are semicircles.

59. S has the same base as in Exercise 58, but cross-sections perpendicular to the y-axis are isosceles triangles with height equal to the base.

60. The base of S is a circular disk with radius r. Parallel cross-sections perpendicular to the base are isosceles triangles with height h and unequal side in the base.
(a) Set up an integral for the volume of S.
(b) By interpreting the integral as an area, find the volume of S.

61. (a) Set up an integral for the volume of a solid *torus* (the donut-shaped solid shown in the figure) with radii r and R.
(b) By interpreting the integral as an area, find the volume of the torus.

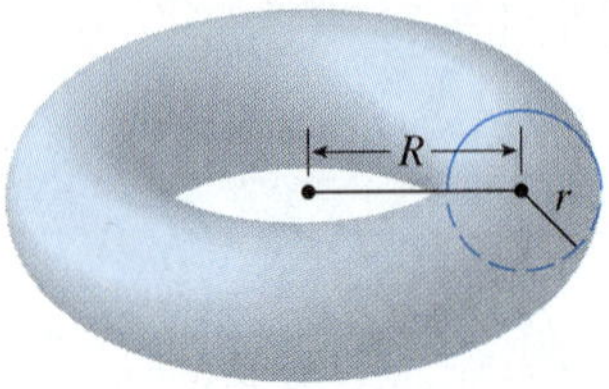

62. Solve Example 9 taking cross-sections to be parallel to the line of intersection of the two planes.

63. (a) Cavalieri's Principle states that if a family of parallel planes gives equal cross-sectional areas for two solids S_1 and S_2, then the volumes of S_1 and S_2 are equal. Prove this principle.
(b) Use Cavalieri's Principle to find the volume of the oblique cylinder shown in the figure.

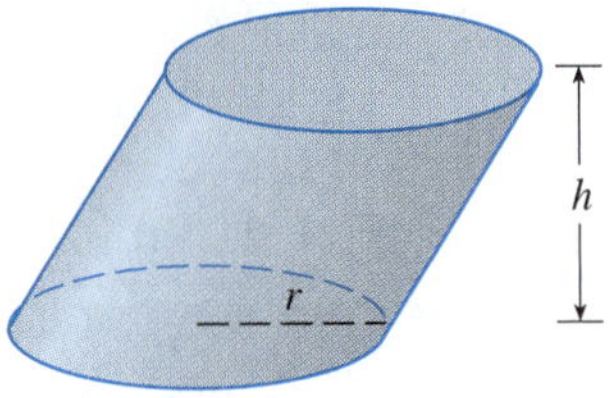

64. Find the volume common to two circular cylinders, each with radius r, if the axes of the cylinders intersect at right angles.

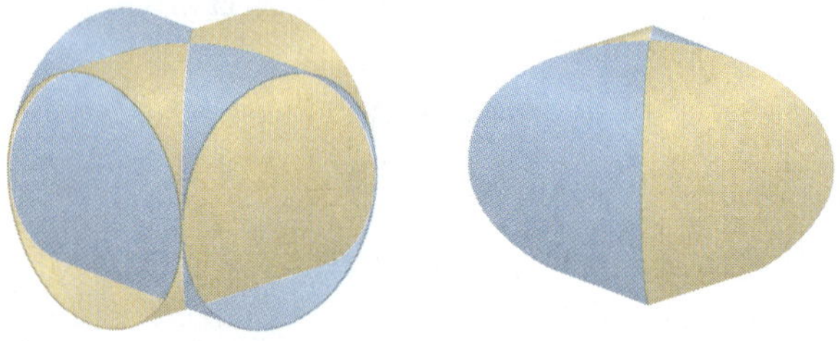

65. Find the volume common to two spheres, each with radius r, if the center of each sphere lies on the surface of the other sphere.

66. A bowl is shaped like a hemisphere with diameter 30 cm. A ball with diameter 10 cm is placed in the bowl and water is poured into the bowl to a depth of h centimeters. Find the volume of water in the bowl.

67. A hole of radius r is bored through a cylinder of radius $R > r$ at right angles to the axis of the cylinder. Set up, but do not evaluate, an integral for the volume cut out.

68. A hole of radius r is bored through the center of a sphere of radius $R > r$. Find the volume of the remaining portion of the sphere.

69. Some of the pioneers of calculus, such as Kepler and Newton, were inspired by the problem of finding the volumes of wine barrels. (In fact Kepler published a book *Stereometria doliorum* in 1715 devoted to methods for finding the volumes of barrels.) They often approximated the shape of the sides by parabolas.
(a) A barrel with height h and maximum radius R is constructed by rotating about the x-axis the parabola $y = R - cx^2$, $-h/2 \leqslant x \leqslant h/2$, where c is a positive

constant. Show that the radius of each end of the barrel is $r = R - d$, where $d = ch^2/4$.

(b) Show that the volume enclosed by the barrel is

$$V = \tfrac{1}{3}\pi h\left(2R^2 + r^2 - \tfrac{2}{5}d^2\right)$$

70. Suppose that a region $\mathcal{R}$ has area A and lies above the x-axis. When $\mathcal{R}$ is rotated about the x-axis, it sweeps out a solid with volume V_1. When $\mathcal{R}$ is rotated about the line $y = -k$ (where k is a positive number), it sweeps out a solid with volume V_2. Express V_2 in terms of V_1, k, and A.

6.3 Volumes by Cylindrical Shells

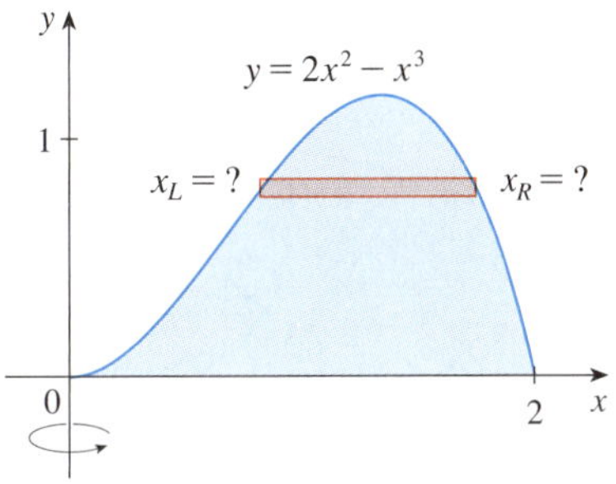

FIGURE 1

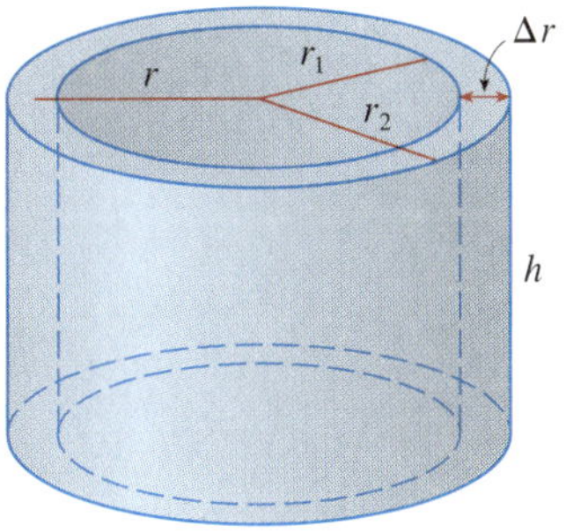

FIGURE 2

Some volume problems are very difficult to handle by the methods of the preceding section. For instance, let's consider the problem of finding the volume of the solid obtained by rotating about the y-axis the region bounded by $y = 2x^2 - x^3$ and $y = 0$. (See Figure 1.) If we slice perpendicular to the y-axis, we get a washer. But to compute the inner radius and the outer radius of the washer, we would have to solve the cubic equation $y = 2x^2 - x^3$ for x in terms of y; that's not easy.

Fortunately, there is a method, called the **method of cylindrical shells**, that is easier to use in such a case. Figure 2 shows a cylindrical shell with inner radius r_1, outer radius r_2, and height h. Its volume V is calculated by subtracting the volume V_1 of the inner cylinder from the volume V_2 of the outer cylinder:

$$\begin{aligned} V &= V_2 - V_1 \\ &= \pi r_2^2 h - \pi r_1^2 h = \pi(r_2^2 - r_1^2)h \\ &= \pi(r_2 + r_1)(r_2 - r_1)h \\ &= 2\pi \frac{r_2 + r_1}{2} h(r_2 - r_1) \end{aligned}$$

If we let $\Delta r = r_2 - r_1$ (the thickness of the shell) and $r = \frac{1}{2}(r_2 + r_1)$ (the average radius of the shell), then this formula for the volume of a cylindrical shell becomes

1 $$V = 2\pi rh\,\Delta r$$

and it can be remembered as

$$V = [\text{circumference}][\text{height}][\text{thickness}]$$

Now let S be the solid obtained by rotating about the y-axis the region bounded by $y = f(x)$ [where $f(x) \geqslant 0$], $y = 0$, $x = a$, and $x = b$, where $b > a \geqslant 0$. (See Figure 3.)

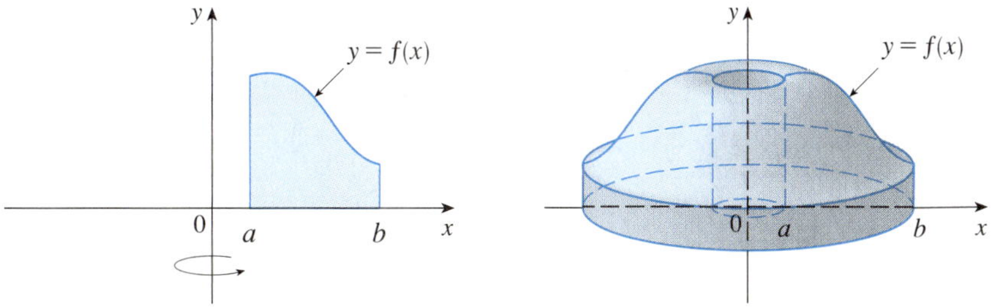

FIGURE 3

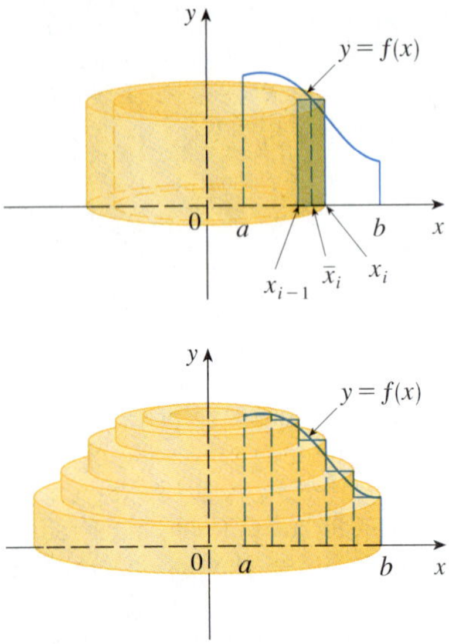

FIGURE 4

We divide the interval $[a, b]$ into n subintervals $[x_{i-1}, x_i]$ of equal width Δx and let $\bar{x}_i$ be the midpoint of the ith subinterval. If the rectangle with base $[x_{i-1}, x_i]$ and height $f(\bar{x}_i)$ is rotated about the y-axis, then the result is a cylindrical shell with average radius $\bar{x}_i$, height $f(\bar{x}_i)$, and thickness Δx (see Figure 4), so by Formula 1 its volume is

$$V_i = (2\pi \bar{x}_i)[f(\bar{x}_i)]\,\Delta x$$

Therefore, an approximation to the volume V of S is given by the sum of the volumes of these shells:

$$V \approx \sum_{i=1}^{n} V_i = \sum_{i=1}^{n} 2\pi \bar{x}_i f(\bar{x}_i)\,\Delta x$$

This approximation appears to become better as $n \to \infty$. But, from the definition of an integral, we know that

$$\lim_{n\to\infty} \sum_{i=1}^{n} 2\pi \bar{x}_i f(\bar{x}_i)\,\Delta x = \int_a^b 2\pi x f(x)\,dx$$

Thus, the following appears plausible:

> **2** The volume of the solid in Figure 3, obtained by rotating about the y-axis the region under the curve $y = f(x)$ from a to b, is
>
> $$V = \int_a^b 2\pi x f(x)\,dx \qquad \text{where } 0 \leqslant a < b$$

The argument using cylindrical shells makes Formula 2 seem reasonable, but later we will be able to prove it (see Exercise 65 in Section 7.1).

The best way to remember Formula 2 is to think of a typical shell, cut and flattened as in Figure 5, with radius x, circumference $2\pi x$, height $f(x)$, and thickness Δx or dx:

$$\int_a^b \underbrace{(2\pi x)}_{\text{circumference}} \quad \underbrace{[f(x)]}_{\text{height}}\,dx$$

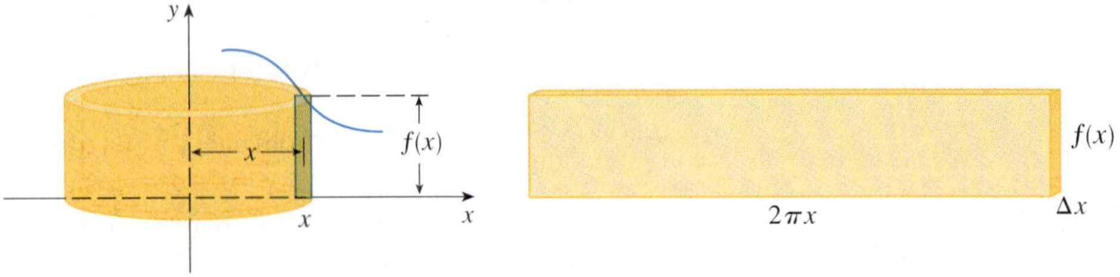

FIGURE 5

This type of reasoning will be helpful in other situations, such as when we rotate about lines other than the y-axis.

EXAMPLE 1 Find the volume of the solid obtained by rotating about the y-axis the region bounded by $y = 2x^2 - x^3$ and $y = 0$.

SOLUTION From the sketch in Figure 6 we see that a typical shell has radius x, circumference $2\pi x$, and height $f(x) = 2x^2 - x^3$. So, by the shell method, the volume is

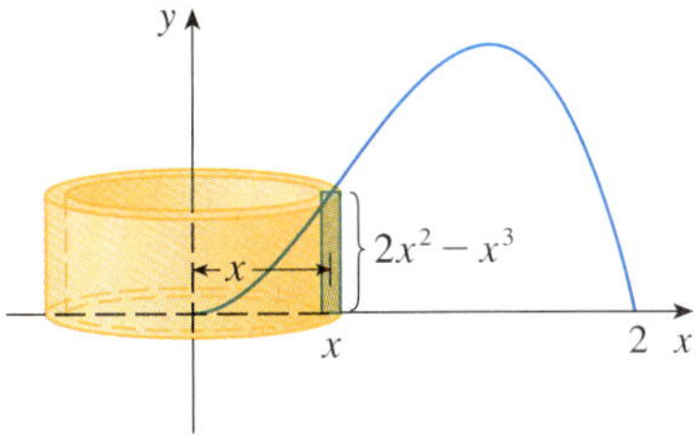

FIGURE 6

$$V = \int_0^2 (2\pi x)(2x^2 - x^3)\,dx = 2\pi \int_0^2 (2x^3 - x^4)\,dx$$
$$= 2\pi\left[\tfrac{1}{2}x^4 - \tfrac{1}{5}x^5\right]_0^2 = 2\pi\left(8 - \tfrac{32}{5}\right) = \tfrac{16}{5}\pi$$

It can be verified that the shell method gives the same answer as slicing.

|||| Figure 7 shows a computer-generated picture of the solid whose volume we computed in Example 1.

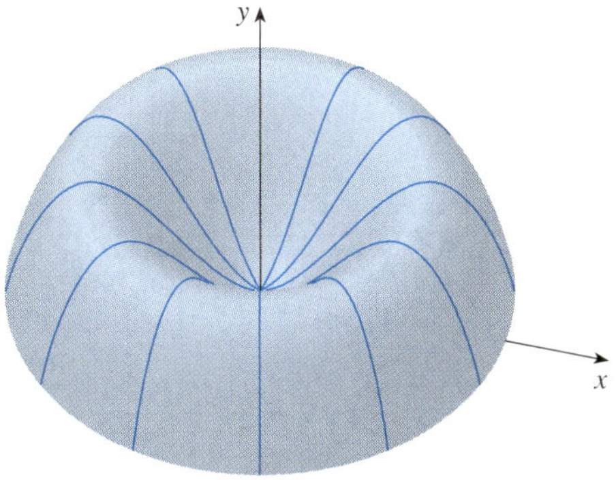

FIGURE 7

NOTE ▫ Comparing the solution of Example 1 with the remarks at the beginning of this section, we see that the method of cylindrical shells is much easier than the washer method for this problem. We did not have to find the coordinates of the local maximum and we did not have to solve the equation of the curve for x in terms of y. However, in other examples the methods of the preceding section may be easier.

EXAMPLE 2 Find the volume of the solid obtained by rotating about the y-axis the region between $y = x$ and $y = x^2$.

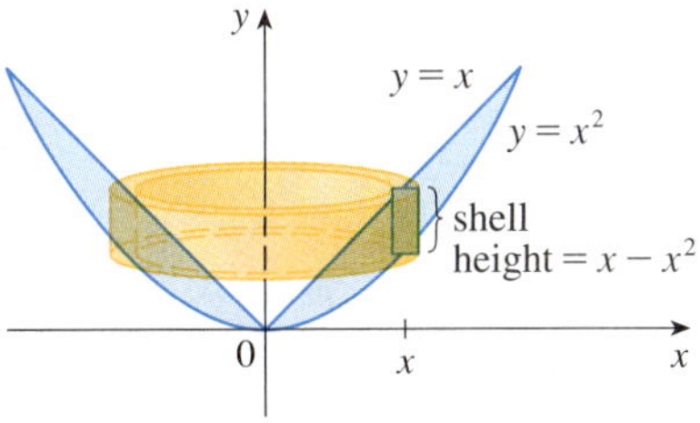

FIGURE 8

SOLUTION The region and a typical shell are shown in Figure 8. We see that the shell has radius x, circumference $2\pi x$, and height $x - x^2$. So the volume is

$$V = \int_0^1 (2\pi x)(x - x^2)\,dx = 2\pi \int_0^1 (x^2 - x^3)\,dx$$
$$= 2\pi\left[\frac{x^3}{3} - \frac{x^4}{4}\right]_0^1 = \frac{\pi}{6}$$

As the following example shows, the shell method works just as well if we rotate about the x-axis. We simply have to draw a diagram to identify the radius and height of a shell.

TEC Module 6.2/6.3 compares shells with disks and washers.

EXAMPLE 3 Use cylindrical shells to find the volume of the solid obtained by rotating about the x-axis the region under the curve $y = \sqrt{x}$ from 0 to 1.

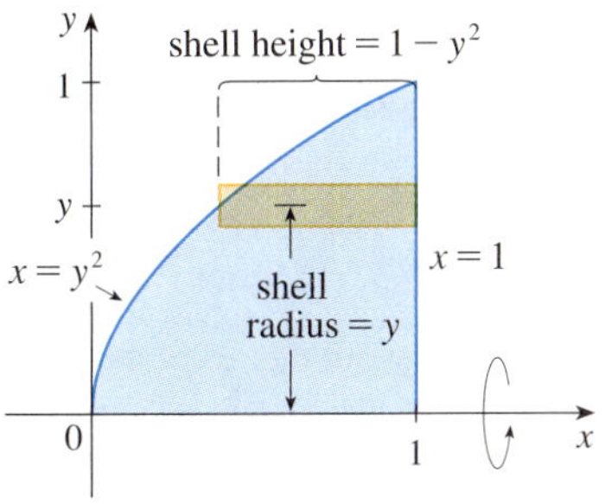

FIGURE 9

SOLUTION This problem was solved using disks in Example 2 in Section 6.2. To use shells we relabel the curve $y = \sqrt{x}$ (in the figure in that example) as $x = y^2$ in Figure 9. For rotation about the x-axis we see that a typical shell has radius y, circumference $2\pi y$, and height $1 - y^2$. So the volume is

$$V = \int_0^1 (2\pi y)(1 - y^2)\,dy = 2\pi \int_0^1 (y - y^3)\,dy$$
$$= 2\pi\left[\frac{y^2}{2} - \frac{y^4}{4}\right]_0^1 = \frac{\pi}{2}$$

In this problem the disk method was simpler.

EXAMPLE 4 Find the volume of the solid obtained by rotating the region bounded by $y = x - x^2$ and $y = 0$ about the line $x = 2$.

SOLUTION Figure 10 shows the region and a cylindrical shell formed by rotation about the line $x = 2$. It has radius $2 - x$, circumference $2\pi(2 - x)$, and height $x - x^2$.

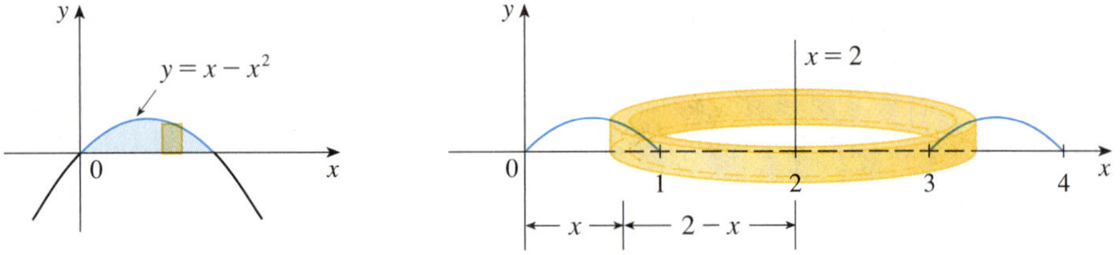

FIGURE 10

The volume of the given solid is

$$V = \int_0^1 2\pi(2 - x)(x - x^2)\,dx = 2\pi \int_0^1 (x^3 - 3x^2 + 2x)\,dx$$

$$= 2\pi\left[\frac{x^4}{4} - x^3 + x^2\right]_0^1 = \frac{\pi}{2}$$

6.3 Exercises

1. Let S be the solid obtained by rotating the region shown in the figure about the y-axis. Explain why it is awkward to use slicing to find the volume V of S. Sketch a typical approximating shell. What are its circumference and height? Use shells to find V.

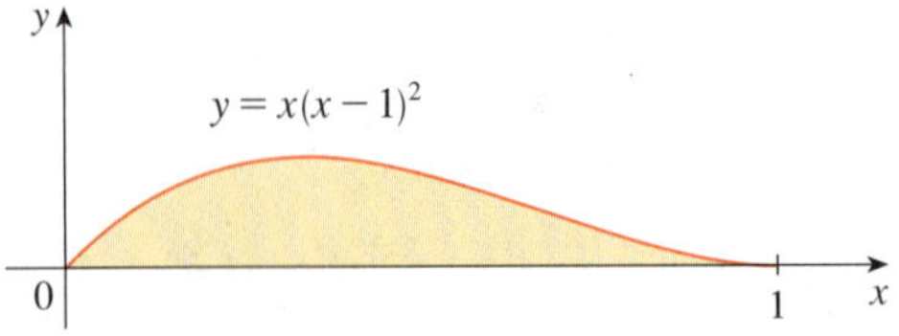

2. Let S be the solid obtained by rotating the region shown in the figure about the y-axis. Sketch a typical cylindrical shell and find its circumference and height. Use shells to find the volume of S. Do you think this method is preferable to slicing? Explain.

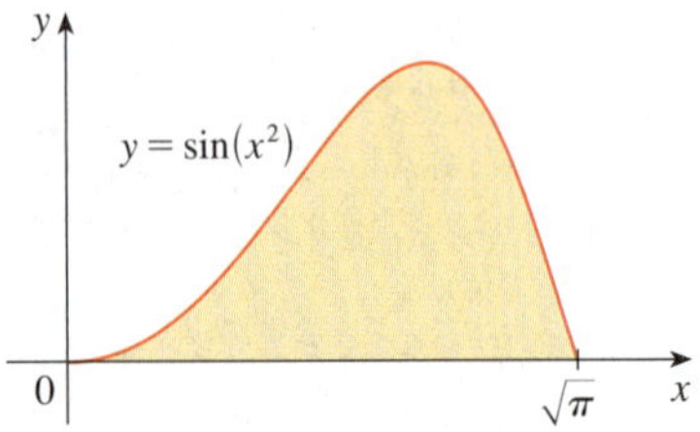

3–7 ||| Use the method of cylindrical shells to find the volume generated by rotating the region bounded by the given curves about the y-axis. Sketch the region and a typical shell.

3. $y = 1/x, \quad y = 0, \quad x = 1, \quad x = 2$

4. $y = x^2, \quad y = 0, \quad x = 1$

5. $y = e^{-x^2}, \quad y = 0, \quad x = 0, \quad x = 1$

6. $y = 3 + 2x - x^2, \quad x + y = 3$

7. $y = 4(x - 2)^2, \quad y = x^2 - 4x + 7$

8. Let V be the volume of the solid obtained by rotating about the y-axis the region bounded by $y = \sqrt{x}$ and $y = x^2$. Find V both by slicing and by cylindrical shells. In both cases draw a diagram to explain your method.

9–14 ||| Use the method of cylindrical shells to find the volume of the solid obtained by rotating the region bounded by the given curves about the x-axis. Sketch the region and a typical shell.

9. $x = 1 + y^2, \quad x = 0, \quad y = 1, \quad y = 2$

10. $x = \sqrt{y}, \quad x = 0, \quad y = 1$

11. $y = x^3, \quad y = 8, \quad x = 0$

12. $x = 4y^2 - y^3, \quad x = 0$

13. $y = 4x^2, \quad 2x + y = 6$

14. $x + y = 3, \quad x = 4 - (y - 1)^2$

15–20 ||| Use the method of cylindrical shells to find the volume generated by rotating the region bounded by the given curves about the specified axis. Sketch the region and a typical shell.

15. $y = x^2,\ y = 0,\ x = 1,\ x = 2;$ about $x = 1$

16. $y = x^2$, $y = 0$, $x = -2$, $x = -1$; about the y-axis

17. $y = x^2$, $y = 0$, $x = 1$, $x = 2$; about $x = 4$

18. $y = 4x - x^2$, $y = 8x - 2x^2$; about $x = -2$

19. $y = \sqrt{x - 1}$, $y = 0$, $x = 5$; about $y = 3$

20. $y = x^2$, $x = y^2$; about $y = -1$

21–26 Set up, but do not evaluate, an integral for the volume of the solid obtained by rotating the region bounded by the given curves about the specified axis.

21. $y = \ln x$, $y = 0$, $x = 2$; about the y-axis

22. $y = x$, $y = 4x - x^2$; about $x = 7$

23. $y = x^4$, $y = \sin(\pi x/2)$; about $x = -1$

24. $y = 1/(1 + x^2)$, $y = 0$, $x = 0$, $x = 2$; about $x = 2$

25. $x = \sqrt{\sin y}$, $0 \leqslant y \leqslant \pi$, $x = 0$; about $y = 4$

26. $x^2 - y^2 = 7$, $x = 4$; about $y = 5$

27. Use the Midpoint Rule with $n = 4$ to estimate the volume obtained by rotating about the y-axis the region under the curve $y = \tan x$, $0 \leqslant x \leqslant \pi/4$.

28. If the region shown in the figure is rotated about the y-axis to form a solid, use the Midpoint Rule with $n = 5$ to estimate the volume of the solid.

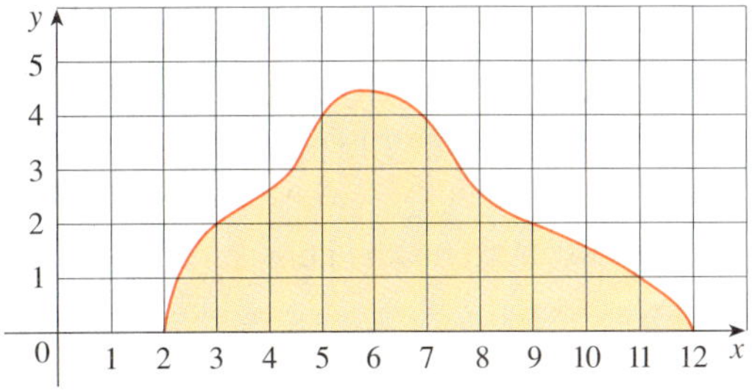

29–32 Each integral represents the volume of a solid. Describe the solid.

29. $\int_0^3 2\pi x^5\,dx$

30. $2\pi \int_0^2 \frac{y}{1 + y^2}\,dy$

31. $\int_0^1 2\pi(3 - y)(1 - y^2)\,dy$

32. $\int_0^{\pi/4} 2\pi(\pi - x)(\cos x - \sin x)\,dx$

33–34 Use a graph to estimate the x-coordinates of the points of intersection of the given curves. Then use this information to estimate the volume of the solid obtained by rotating about the y-axis the region enclosed by these curves.

33. $y = 0$, $y = x + x^2 - x^4$

34. $y = x^4$, $y = 3x - x^3$

CAS **35–36** Use a computer algebra system to find the exact volume of the solid obtained by rotating the region bounded by the given curves about the specified line.

35. $y = \sin^2 x$, $y = \sin^4 x$, $0 \leqslant x \leqslant \pi$; about $x = \pi/2$

36. $y = x^3 \sin x$, $y = 0$, $0 \leqslant x \leqslant \pi$; about $x = -1$

37–42 The region bounded by the given curves is rotated about the specified axis. Find the volume of the resulting solid by any method.

37. $y = x^2 + x - 2$, $y = 0$; about the x-axis

38. $y = x^2 - 3x + 2$, $y = 0$; about the y-axis

39. $y = 5$, $y = x + (4/x)$; about $x = -1$

40. $x = 1 - y^4$, $x = 0$; about $x = 2$

41. $x^2 + (y - 1)^2 = 1$; about the y-axis

42. $x^2 + (y - 1)^2 = 1$; about the x-axis

43–45 Use cylindrical shells to find the volume of the solid.

43. A sphere of radius r

44. The solid torus of Exercise 61 in Section 6.2

45. A right circular cone with height h and base radius r

46. Suppose you make napkin rings by drilling holes with different diameters through two wooden balls (which also have different diameters). You discover that both napkin rings have the same height h, as shown in the figure.

(a) Guess which ring has more wood in it.

(b) Check your guess: Use cylindrical shells to compute the volume of a napkin ring created by drilling a hole with radius r through the center of a sphere of radius R and express the answer in terms of h.

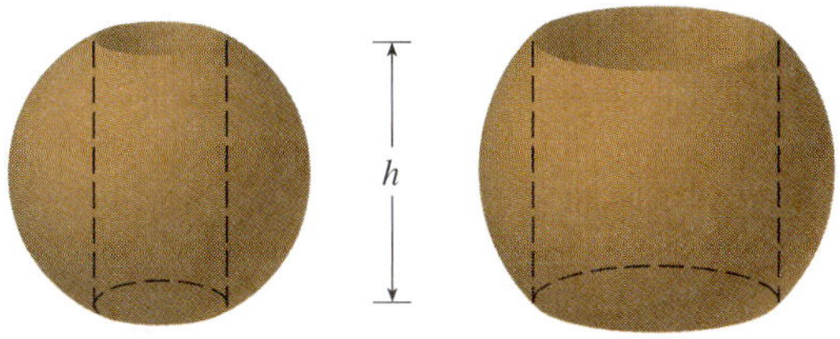

6.4 Work

The term *work* is used in everyday language to mean the total amount of effort required to perform a task. In physics it has a technical meaning that depends on the idea of a *force*. Intuitively, you can think of a force as describing a push or pull on an object—for example, a horizontal push of a book across a table or the downward pull of Earth's gravity on a ball. In general, if an object moves along a straight line with position function $s(t)$, then the **force** F on the object (in the same direction) is defined by Newton's Second Law of Motion as the product of its mass m and its acceleration:

1 $$F = m\frac{d^2s}{dt^2}$$

In the SI metric system, the mass is measured in kilograms (kg), the displacement in meters (m), the time in seconds (s), and the force in newtons ($\text{N} = \text{kg}\cdot\text{m/s}^2$). Thus, a force of 1 N acting on a mass of 1 kg produces an acceleration of 1 m/s^2. In the U. S. Customary system the fundamental unit is chosen to be the unit of force, which is the pound.

In the case of constant acceleration, the force F is also constant and the work done is defined to be the product of the force F and the distance d that the object moves:

2 $$W = Fd \qquad \text{work} = \text{force} \times \text{distance}$$

If F is measured in newtons and d in meters, then the unit for W is a newton-meter, which is called a joule (J). If F is measured in pounds and d in feet, then the unit for W is a foot-pound (ft-lb), which is about 1.36 J.

EXAMPLE 1

(a) How much work is done in lifting a 1.2-kg book off the floor to put it on a desk that is 0.7 m high? Use the fact that the acceleration due to gravity is $g = 9.8\ \text{m/s}^2$.

(b) How much work is done in lifting a 20-lb weight 6 ft off the ground?

SOLUTION

(a) The force exerted is equal and opposite to that exerted by gravity, so Equation 1 gives

$$F = mg = (1.2)(9.8) = 11.76\ \text{N}$$

and then Equation 2 gives the work done as

$$W = Fd = (11.76)(0.7) \approx 8.2\ \text{J}$$

(b) Here the force is given as $F = 20$ lb, so the work done is

$$W = Fd = 20 \cdot 6 = 120\ \text{ft-lb}$$

Notice that in part (b), unlike part (a), we did not have to multiply by g because we were given the *weight* (which is a force) and not the mass of the object.

Equation 2 defines work as long as the force is constant, but what happens if the force is variable? Let's suppose that the object moves along the x-axis in the positive direction, from $x = a$ to $x = b$, and at each point x between a and b a force $f(x)$ acts on the object, where f is a continuous function. We divide the interval $[a, b]$ into n subintervals with endpoints $x_0, x_1, \ldots, x_n$ and equal width Δx. We choose a sample point x_i^* in the ith subinterval $[x_{i-1}, x_i]$. Then the force at that point is $f(x_i^*)$. If n is large, then Δx is small, and since f is continuous, the values of f don't change very much over the interval $[x_{i-1}, x_i]$. In other words, f is almost constant on the interval and so the work W_i that is done in moving the particle from x_{i-1} to x_i is approximately given by Equation 2:

$$W_i \approx f(x_i^*)\,\Delta x$$

Thus, we can approximate the total work by

3 $$W \approx \sum_{i=1}^{n} f(x_i^*)\,\Delta x$$

It seems that this approximation becomes better as we make n larger. Therefore, we define the **work done in moving the object from *a* to *b*** as the limit of this quantity as $n \to \infty$. Since the right side of (3) is a Riemann sum, we recognize its limit as being a definite integral and so

4 $$W = \lim_{n \to \infty} \sum_{i=1}^{n} f(x_i^*)\,\Delta x = \int_a^b f(x)\,dx$$

EXAMPLE 2 When a particle is located a distance x feet from the origin, a force of $x^2 + 2x$ pounds acts on it. How much work is done in moving it from $x = 1$ to $x = 3$?

SOLUTION $$W = \int_1^3 (x^2 + 2x)\,dx = \frac{x^3}{3} + x^2 \Big]_1^3 = \frac{50}{3}$$

The work done is $16\frac{2}{3}$ ft-lb.

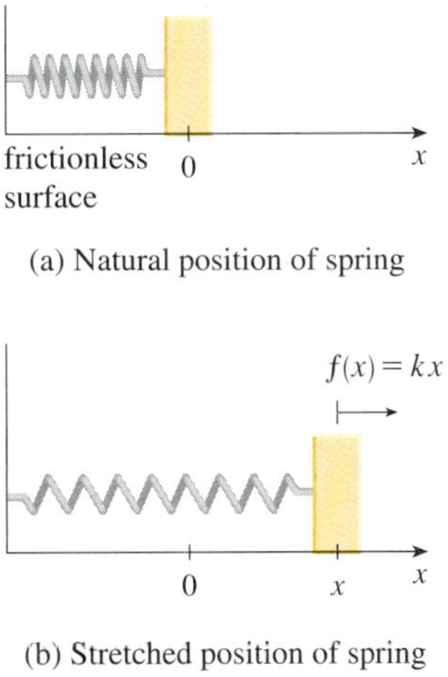

FIGURE 1
Hooke's Law

In the next example we use a law from physics: **Hooke's Law** states that the force required to maintain a spring stretched x units beyond its natural length is proportional to x:

$$f(x) = kx$$

where k is a positive constant (called the **spring constant**). Hooke's Law holds provided that x is not too large (see Figure 1).

EXAMPLE 3 A force of 40 N is required to hold a spring that has been stretched from its natural length of 10 cm to a length of 15 cm. How much work is done in stretching the spring from 15 cm to 18 cm?

SOLUTION According to Hooke's Law, the force required to hold the spring stretched x meters beyond its natural length is $f(x) = kx$. When the spring is stretched from 10 cm to 15 cm, the amount stretched is 5 cm = 0.05 m. This means that $f(0.05) = 40$, so

$$0.05k = 40 \qquad k = \frac{40}{0.05} = 800$$

Thus, $f(x) = 800x$ and the work done in stretching the spring from 15 cm to 18 cm is

$$W = \int_{0.05}^{0.08} 800x\,dx = 800\,\frac{x^2}{2}\Big]_{0.05}^{0.08}$$
$$= 400[(0.08)^2 - (0.05)^2] = 1.56 \text{ J}$$

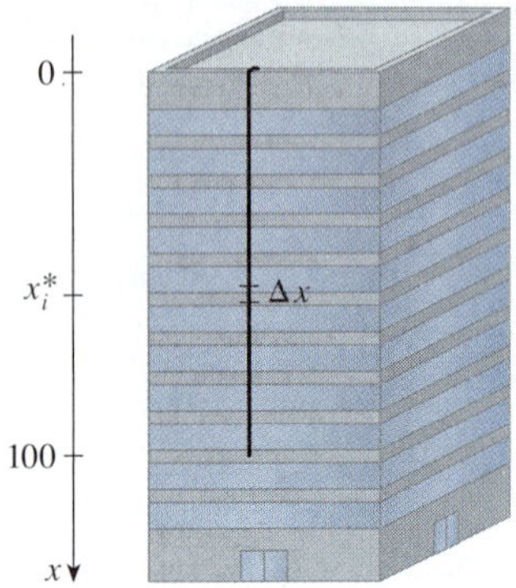

FIGURE 2

|||| If we had placed the origin at the bottom of the cable and the x-axis upward, we would have gotten

$$W = \int_0^{100} 2(100 - x)\, dx$$

which gives the same answer.

EXAMPLE 4 A 200-lb cable is 100 ft long and hangs vertically from the top of a tall building. How much work is required to lift the cable to the top of the building?

SOLUTION Here we don't have a formula for the force function, but we can use an argument similar to the one that led to Definition 4.

Let's place the origin at the top of the building and the x-axis pointing downward as in Figure 2. We divide the cable into small parts with length Δx. If x_i^* is a point in the ith such interval, then all points in the interval are lifted by approximately the same amount, namely x_i^*. The cable weighs 2 pounds per foot, so the weight of the ith part is $2\Delta x$. Thus, the work done on the ith part, in foot-pounds, is

$$\underbrace{(2\Delta x)}_{\text{force}} \quad \underbrace{x_i^*}_{\text{distance}} = 2x_i^* \Delta x$$

We get the total work done by adding all these approximations and letting the number of parts become large (so $\Delta x \to 0$):

$$W = \lim_{n\to\infty} \sum_{i=1}^{n} 2x_i^* \Delta x = \int_0^{100} 2x\, dx$$

$$= x^2\Big]_0^{100} = 10{,}000 \text{ ft-lb}$$

EXAMPLE 5 A tank has the shape of an inverted circular cone with height 10 m and base radius 4 m. It is filled with water to a height of 8 m. Find the work required to empty the tank by pumping all of the water to the top of the tank. (The density of water is 1000 kg/m^3.)

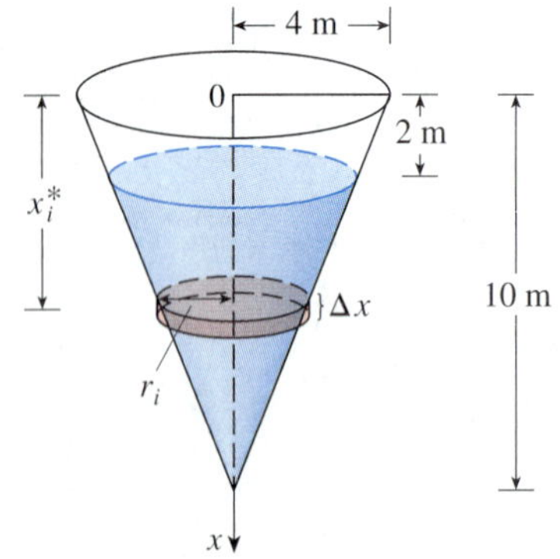

FIGURE 3

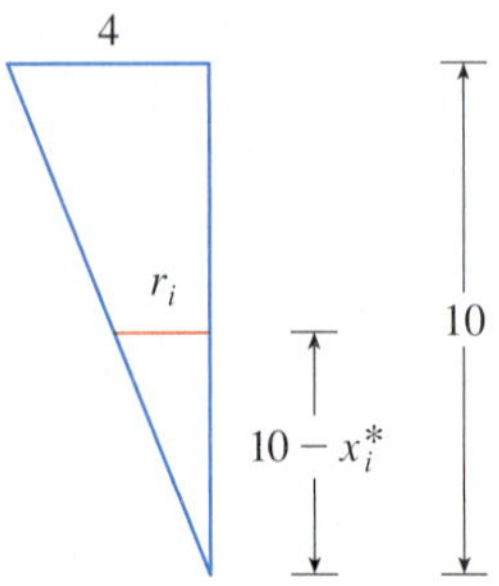

FIGURE 4

SOLUTION Let's measure depths from the top of the tank by introducing a vertical coordinate line as in Figure 3. The water extends from a depth of 2 m to a depth of 10 m and so we divide the interval $[2, 10]$ into n subintervals with endpoints $x_0, x_1, \ldots, x_n$ and choose x_i^* in the ith subinterval. This divides the water into n layers. The ith layer is approximated by a circular cylinder with radius r_i and height Δx. We can compute r_i from similar triangles, using Figure 4, as follows:

$$\frac{r_i}{10 - x_i^*} = \frac{4}{10} \qquad r_i = \tfrac{2}{5}(10 - x_i^*)$$

Thus, an approximation to the volume of the ith layer of water is

$$V_i \approx \pi r_i^2 \Delta x = \frac{4\pi}{25}(10 - x_i^*)^2 \Delta x$$

and so its mass is

$$m_i = \text{density} \times \text{volume}$$

$$\approx 1000 \cdot \frac{4\pi}{25}(10 - x_i^*)^2 \Delta x = 160\pi(10 - x_i^*)^2 \Delta x$$

The force required to raise this layer must overcome the force of gravity and so

$$F_i = m_i g \approx (9.8)160\pi(10 - x_i^*)^2 \Delta x$$

$$\approx 1570\pi(10 - x_i^*)^2 \Delta x$$

Each particle in the layer must travel a distance of approximately x_i^*. The work W_i done to raise this layer to the top is approximately the product of the force F_i and the distance x_i^*:

$$W_i \approx F_i x_i^* \approx 1570\pi x_i^*(10 - x_i^*)^2 \Delta x$$

To find the total work done in emptying the entire tank, we add the contributions of each of the n layers and then take the limit as $n \to \infty$:

$$W = \lim_{n\to\infty} \sum_{i=1}^{n} 1570\pi x_i^*(10 - x_i^*)^2 \,\Delta x = \int_2^{10} 1570\pi x(10 - x)^2 \,dx$$

$$= 1570\pi \int_2^{10} (100x - 20x^2 + x^3)\,dx = 1570\pi \left[50x^2 - \frac{20x^3}{3} + \frac{x^4}{4} \right]_2^{10}$$

$$= 1570\pi\left(\tfrac{2048}{3}\right) \approx 3.4 \times 10^6 \text{ J}$$

6.4 Exercises

1. Find the work done in pushing a car a distance of 8 m while exerting a constant force of 900 N.

2. How much work is done by a weightlifter in raising a 60-kg barbell from the floor to a height of 2 m?

3. A particle is moved along the x-axis by a force that measures $10/(1 + x)^2$ pounds at a point x feet from the origin. Find the work done in moving the particle from the origin to a distance of 9 ft.

4. When a particle is located a distance x meters from the origin, a force of $\cos(\pi x/3)$ newtons acts on it. How much work is done in moving the particle from $x = 1$ to $x = 2$? Interpret your answer by considering the work done from $x = 1$ to $x = 1.5$ and from $x = 1.5$ to $x = 2$.

5. Shown is the graph of a force function (in newtons) that increases to its maximum value and then remains constant. How much work is done by the force in moving an object a distance of 8 m?

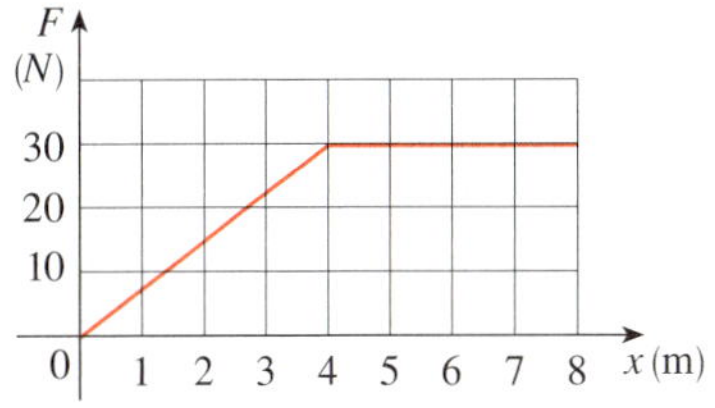

6. The table shows values of a force function $f(x)$, where x is measured in meters and $f(x)$ in newtons. Use the Midpoint Rule to estimate the work done by the force in moving an object from $x = 4$ to $x = 20$.

x	4	6	8	10	12	14	16	18	20
$f(x)$	5	5.8	7.0	8.8	9.6	8.2	6.7	5.2	4.1

7. A force of 10 lb is required to hold a spring stretched 4 in. beyond its natural length. How much work is done in stretching it from its natural length to 6 in. beyond its natural length?

8. A spring has a natural length of 20 cm. If a 25-N force is required to keep it stretched to a length of 30 cm, how much work is required to stretch it from 20 cm to 25 cm?

9. Suppose that 2 J of work is needed to stretch a spring from its natural length of 30 cm to a length of 42 cm. How much work is needed to stretch it from 35 cm to 40 cm?

10. If the work required to stretch a spring 1 ft beyond its natural length is 12 ft-lb, how much work is needed to stretch it 9 in. beyond its natural length?

11. How far beyond its natural length will a force of 30 N keep the spring in Exercise 9 stretched?

12. If 6 J of work is needed to stretch a spring from 10 cm to 12 cm and another 10 J is needed to stretch it from 12 cm to 14 cm, what is the natural length of the spring?

13–20 |||| Show how to approximate the required work by a Riemann sum. Then express the work as an integral and evaluate it.

13. A heavy rope, 50 ft long, weighs 0.5 lb/ft and hangs over the edge of a building 120 ft high.
(a) How much work is done in pulling the rope to the top of the building?
(b) How much work is done in pulling half the rope to the top of the building?

14. A chain lying on the ground is 10 m long and its mass is 80 kg. How much work is required to raise one end of the chain to a height of 6 m?

15. A cable that weighs 2 lb/ft is used to lift 800 lb of coal up a mineshaft 500 ft deep. Find the work done.

16. A bucket that weighs 4 lb and a rope of negligible weight are used to draw water from a well that is 80 ft deep. The bucket is filled with 40 lb of water and is pulled up at a rate of 2 ft/s, but water leaks out of a hole in the bucket at a rate of 0.2 lb/s. Find the work done in pulling the bucket to the top of the well.

17. A leaky 10-kg bucket is lifted from the ground to a height of 12 m at a constant speed with a rope that weighs 0.8 kg/m. Initially the bucket contains 36 kg of water, but the water leaks at a constant rate and finishes draining just as the bucket reaches the 12 m level. How much work is done?

18. A 10-ft chain weighs 25 lb and hangs from a ceiling. Find the work done in lifting the lower end of the chain to the ceiling so that it's level with the upper end.

19. An aquarium 2 m long, 1 m wide, and 1 m deep is full of water. Find the work needed to pump half of the water out of the aquarium. (Use the fact that the density of water is 1000 kg/m^3.)

20. A circular swimming pool has a diameter of 24 ft, the sides are 5 ft high, and the depth of the water is 4 ft. How much work is required to pump all of the water out over the side? (Use the fact that water weighs 62.5 lb/ft^3.)

21–24 A tank is full of water. Find the work required to pump the water out of the outlet. In Exercises 23 and 24 use the fact that water weighs 62.5 lb/ft^3.

21.

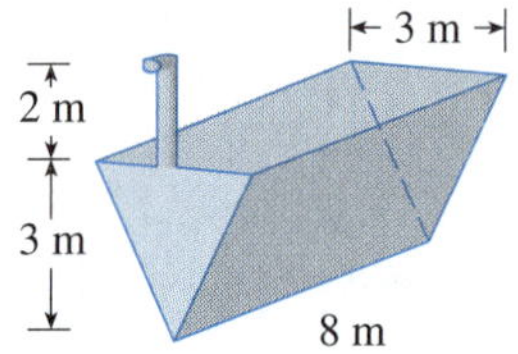

22.

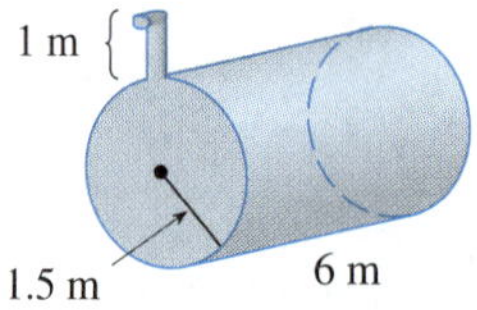

23.

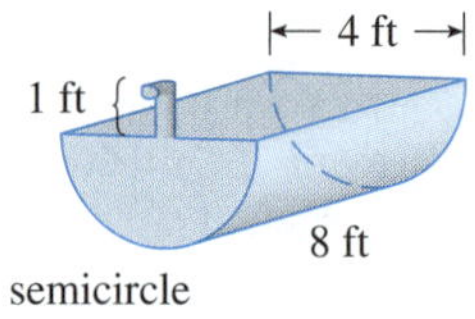

24.

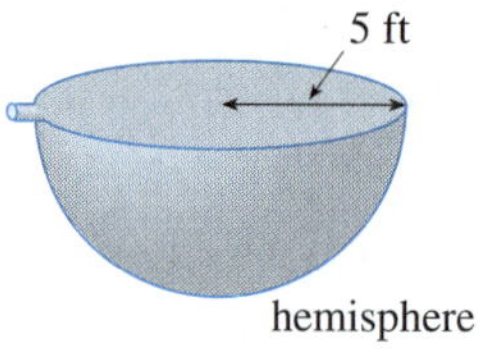

25. Suppose that for the tank in Exercise 21 the pump breaks down after 4.7×10^5 J of work has been done. What is the depth of the water remaining in the tank?

26. Solve Exercise 22 if the tank is half full of oil that has a density of 920 kg/m^3.

27. When gas expands in a cylinder with radius r, the pressure at any given time is a function of the volume: $P = P(V)$. The force exerted by the gas on the piston (see the figure) is the product of the pressure and the area: $F = \pi r^2 P$. Show that the work done by the gas when the volume expands from volume V_1 to volume V_2 is

$$W = \int_{V_1}^{V_2} P\, dV$$

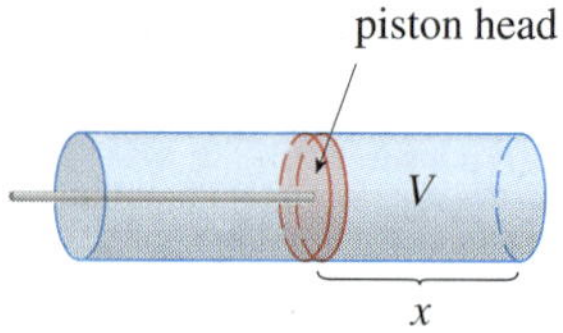

28. In a steam engine the pressure P and volume V of steam satisfy the equation $PV^{1.4} = k$, where k is a constant. (This is true for adiabatic expansion, that is, expansion in which there is no heat transfer between the cylinder and its surroundings.) Use Exercise 27 to calculate the work done by the engine during a cycle when the steam starts at a pressure of 160 lb/in^2 and a volume of 100 in^3 and expands to a volume of 800 in^3.

29. Newton's Law of Gravitation states that two bodies with masses m_1 and m_2 attract each other with a force

$$F = G\frac{m_1 m_2}{r^2}$$

where r is the distance between the bodies and G is the gravitational constant. If one of the bodies is fixed, find the work needed to move the other from $r = a$ to $r = b$.

30. Use Newton's Law of Gravitation to compute the work required to launch a 1000-kg satellite vertically to an orbit 1000 km high. You may assume that Earth's mass is 5.98×10^{24} kg and is concentrated at its center. Take the radius of Earth to be 6.37×10^6 m and $G = 6.67 \times 10^{-11}$ N·m^2/kg^2.

6.5 Average Value of a Function

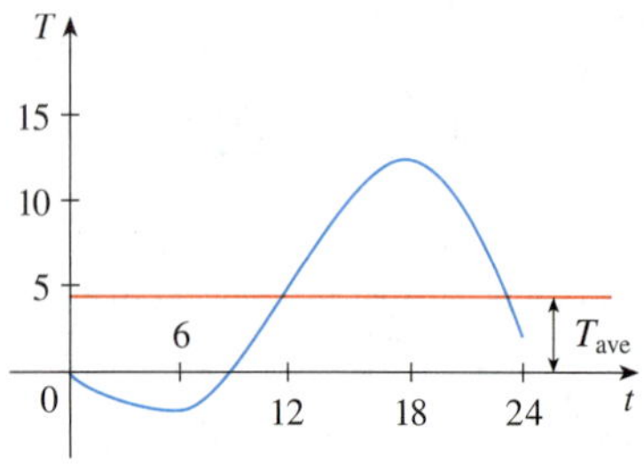

FIGURE 1

It is easy to calculate the average value of finitely many numbers $y_1, y_2, \ldots, y_n$:

$$y_{ave} = \frac{y_1 + y_2 + \cdots + y_n}{n}$$

But how do we compute the average temperature during a day if infinitely many temperature readings are possible? Figure 1 shows the graph of a temperature function $T(t)$, where t is measured in hours and T in °C, and a guess at the average temperature, T_{ave}.

In general, let's try to compute the average value of a function $y = f(x)$, $a \leqslant x \leqslant b$. We start by dividing the interval $[a, b]$ into n equal subintervals, each with length $\Delta x = (b - a)/n$. Then we choose points $x_1^*, \ldots, x_n^*$ in successive subintervals and calculate the average of the numbers $f(x_1^*), \ldots, f(x_n^*)$:

$$\frac{f(x_1^*) + \cdots + f(x_n^*)}{n}$$

(For example, if f represents a temperature function and $n = 24$, this means that we take temperature readings every hour and then average them.) Since $\Delta x = (b - a)/n$, we can write $n = (b - a)/\Delta x$ and the average value becomes

$$\frac{f(x_1^*) + \cdots + f(x_n^*)}{\dfrac{b - a}{\Delta x}} = \frac{1}{b - a}\left[f(x_1^*)\,\Delta x + \cdots + f(x_n^*)\,\Delta x\right]$$

$$= \frac{1}{b - a}\sum_{i=1}^{n} f(x_i^*)\,\Delta x$$

If we let n increase, we would be computing the average value of a large number of closely spaced values. (For example, we would be averaging temperature readings taken every minute or even every second.) The limiting value is

$$\lim_{n \to \infty} \frac{1}{b - a}\sum_{i=1}^{n} f(x_i^*)\,\Delta x = \frac{1}{b - a}\int_a^b f(x)\,dx$$

by the definition of a definite integral.

Therefore, we define the **average value of f** on the interval $[a, b]$ as

|||| For a positive function, we can think of this definition as saying

$$\frac{\text{area}}{\text{width}} = \text{average height}$$

$$f_{\text{ave}} = \frac{1}{b - a}\int_a^b f(x)\,dx$$

EXAMPLE 1 Find the average value of the function $f(x) = 1 + x^2$ on the interval $[-1, 2]$.

SOLUTION With $a = -1$ and $b = 2$ we have

$$f_{\text{ave}} = \frac{1}{b - a}\int_a^b f(x)\,dx = \frac{1}{2 - (-1)}\int_{-1}^{2} (1 + x^2)\,dx$$

$$= \frac{1}{3}\left[x + \frac{x^3}{3}\right]_{-1}^{2} = 2$$

If $T(t)$ is the temperature at time t, we might wonder if there is a specific time when the temperature is the same as the average temperature. For the temperature function graphed in Figure 1, we see that there are two such times—just before noon and just before midnight. In general, is there a number c at which the value of a function f is exactly equal to the average value of the function, that is, $f(c) = f_{\text{ave}}$? The following theorem says that this is true for continuous functions.

The Mean Value Theorem for Integrals If f is continuous on $[a, b]$, then there exists a number c in $[a, b]$ such that

$$f(c) = f_{\text{ave}} = \frac{1}{b - a}\int_a^b f(x)\,dx$$

that is,

$$\int_a^b f(x)\,dx = f(c)(b - a)$$

The Mean Value Theorem for Integrals is a consequence of the Mean Value Theorem for derivatives and the Fundamental Theorem of Calculus. The proof is outlined in Exercise 23.

The geometric interpretation of the Mean Value Theorem for Integrals is that, for *positive* functions f, there is a number c such that the rectangle with base $[a, b]$ and height $f(c)$ has the same area as the region under the graph of f from a to b. (See Figure 2 and the more picturesque interpretation in the margin note.)

|||| You can always chop off the top of a (two-dimensional) mountain at a certain height and use it to fill in the valleys so that the mountain becomes completely flat.

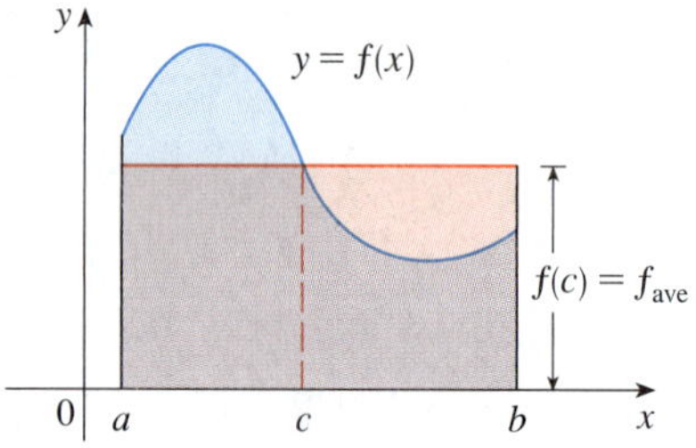

FIGURE 2

EXAMPLE 2 Since $f(x) = 1 + x^2$ is continuous on the interval $[-1, 2]$, the Mean Value Theorem for Integrals says there is a number c in $[-1, 2]$ such that

$$\int_{-1}^{2} (1 + x^2)\, dx = f(c)[2 - (-1)]$$

In this particular case we can find c explicitly. From Example 1 we know that $f_{\text{ave}} = 2$, so the value of c satisfies

$$f(c) = f_{\text{ave}} = 2$$

Therefore $\qquad 1 + c^2 = 2 \qquad \text{so} \qquad c^2 = 1$

Thus, in this case there happen to be two numbers $c = \pm 1$ in the interval $[-1, 2]$ that work in the Mean Value Theorem for Integrals.

Examples 1 and 2 are illustrated by Figure 3.

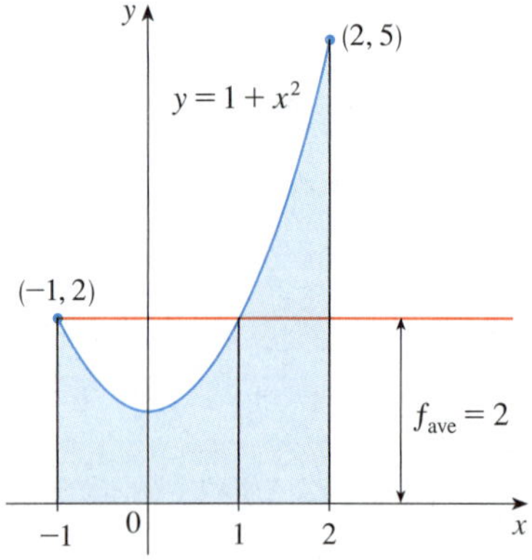

FIGURE 3

EXAMPLE 3 Show that the average velocity of a car over a time interval $[t_1, t_2]$ is the same as the average of its velocities during the trip.

SOLUTION If $s(t)$ is the displacement of the car at time t, then, by definition, the average velocity of the car over the interval is

$$\frac{\Delta s}{\Delta t} = \frac{s(t_2) - s(t_1)}{t_2 - t_1}$$

On the other hand, the average value of the velocity function on the interval is

$$\begin{aligned} v_{\text{ave}} &= \frac{1}{t_2 - t_1} \int_{t_1}^{t_2} v(t)\, dt = \frac{1}{t_2 - t_1} \int_{t_1}^{t_2} s'(t)\, dt \\ &= \frac{1}{t_2 - t_1} [s(t_2) - s(t_1)] \qquad \text{(by the Net Change Theorem)} \\ &= \frac{s(t_2) - s(t_1)}{t_2 - t_1} = \text{average velocity} \end{aligned}$$

6.5 Exercises

1–8 Find the average value of the function on the given interval.

1. $f(x) = x^2$, $[-1, 1]$

2. $f(x) = 1/x$, $[1, 4]$

3. $g(x) = \cos x$, $[0, \pi/2]$

4. $g(x) = x^2\sqrt{1 + x^3}$, $[0, 2]$

5. $f(t) = te^{-t^2}$, $[0, 5]$

6. $f(\theta) = \sec\theta \tan\theta$, $[0, \pi/4]$

7. $h(x) = \cos^4 x \sin x$, $[0, \pi]$

8. $h(r) = 3/(1 + r)^2$, $[1, 6]$

9–12
(a) Find the average value of f on the given interval.
(b) Find c such that $f_{\text{ave}} = f(c)$.
(c) Sketch the graph of f and a rectangle whose area is the same as the area under the graph of f.

9. $f(x) = (x - 3)^2$, $[2, 5]$

10. $f(x) = \sqrt{x}$, $[0, 4]$

11. $f(x) = 2\sin x - \sin 2x$, $[0, \pi]$

12. $f(x) = 2x/(1 + x^2)^2$, $[0, 2]$

13. If f is continuous and $\int_1^3 f(x)\,dx = 8$, show that f takes on the value 4 at least once on the interval $[1, 3]$.

14. Find the numbers b such that the average value of $f(x) = 2 + 6x - 3x^2$ on the interval $[0, b]$ is equal to 3.

15. The table gives values of a continuous function. Use the Midpoint Rule to estimate the average value of f on $[20, 50]$.

x	20	25	30	35	40	45	50
$f(x)$	42	38	31	29	35	48	60

16. The velocity graph of an accelerating car is shown.
(a) Estimate the average velocity of the car during the first 12 seconds.
(b) At what time was the instantaneous velocity equal to the average velocity?

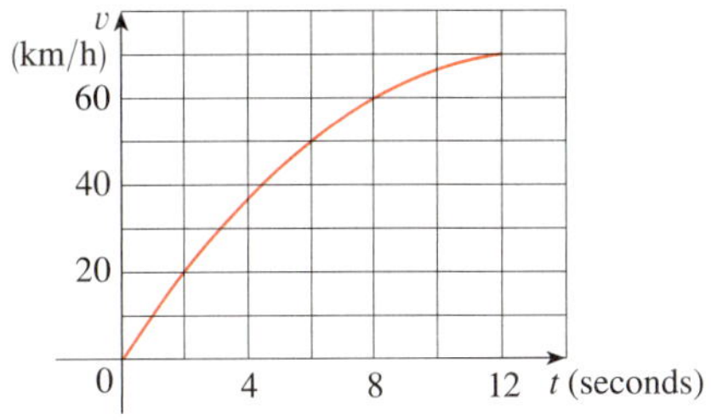

17. In a certain city the temperature (in °F) t hours after 9 A.M. was modeled by the function

$$T(t) = 50 + 14\sin\frac{\pi t}{12}$$

Find the average temperature during the period from 9 A.M. to 9 P.M.

18. If a cup of coffee has temperature 95°C in a room where the temperature is 20°C, then, according to Newton's Law of Cooling, the temperature of the coffee after t minutes is

$$T(t) = 20 + 75e^{-t/50}$$

What is the average temperature of the coffee during the first half hour?

19. The linear density in a rod 8 m long is $12/\sqrt{x + 1}$ kg/m, where x is measured in meters from one end of the rod. Find the average density of the rod.

20. If a freely falling body starts from rest, then its displacement is given by $s = \frac{1}{2}gt^2$. Let the velocity after a time T be v_T. Show that if we compute the average of the velocities with respect to t we get $v_{\text{ave}} = \frac{1}{2}v_T$, but if we compute the average of the velocities with respect to s we get $v_{\text{ave}} = \frac{2}{3}v_T$.

21. Use the result of Exercise 77 in Section 5.5 to compute the average volume of inhaled air in the lungs in one respiratory cycle.

22. The velocity v of blood that flows in a blood vessel with radius R and length l at a distance r from the central axis is

$$v(r) = \frac{P}{4\eta l}(R^2 - r^2)$$

where P is the pressure difference between the ends of the vessel and η is the viscosity of the blood (see Example 7 in Section 3.3). Find the average velocity (with respect to r) over the interval $0 \leqslant r \leqslant R$. Compare the average velocity with the maximum velocity.

23. Prove the Mean Value Theorem for Integrals by applying the Mean Value Theorem for derivatives (see Section 4.2) to the function $F(x) = \int_a^x f(t)\,dt$.

24. If $f_{\text{ave}}[a, b]$ denotes the average value of f on the interval $[a, b]$ and $a < c < b$, show that

$$f_{\text{ave}}[a, b] = \frac{c - a}{b - a}f_{\text{ave}}[a, c] + \frac{b - c}{b - a}f_{\text{ave}}[c, b]$$

APPLIED PROJECT

CAS Where to Sit at the Movies

A movie theater has a screen that is positioned 10 ft off the floor and is 25 ft high. The first row of seats is placed 9 ft from the screen and the rows are set 3 ft apart. The floor of the seating area is inclined at an angle of $\alpha = 20°$ above the horizontal and the distance up the incline that you sit is x. The theater has 21 rows of seats, so $0 \leqslant x \leqslant 60$. Suppose you decide that the best place to sit is in the row where the angle θ subtended by the screen at your eyes is a maximum. Let's also suppose that your eyes are 4 ft above the floor, as shown in the figure. (In Exercise 58 in Section 4.7 we looked at a simpler version of this problem, where the floor is horizontal, but this project involves a more complicated situation and requires technology.)

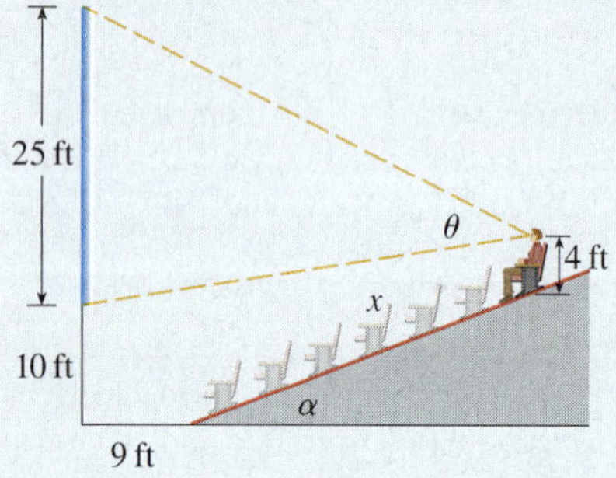

1. Show that

$$\theta = \arccos\left(\frac{a^2 + b^2 - 625}{2ab}\right)$$

where $a^2 = (9 + x\cos\alpha)^2 + (31 - x\sin\alpha)^2$

and $b^2 = (9 + x\cos\alpha)^2 + (x\sin\alpha - 6)^2$

2. Use a graph of θ as a function of x to estimate the value of x that maximizes θ. In which row should you sit? What is the viewing angle θ in this row?

3. Use your computer algebra system to differentiate θ and find a numerical value for the root of the equation $d\theta/dx = 0$. Does this value confirm your result in Problem 2?

4. Use the graph of θ to estimate the average value of θ on the interval $0 \leqslant x \leqslant 60$. Then use your CAS to compute the average value. Compare with the maximum and minimum values of θ.

6 Review

CONCEPT CHECK

1. (a) Draw two typical curves $y = f(x)$ and $y = g(x)$, where $f(x) \geqslant g(x)$ for $a \leqslant x \leqslant b$. Show how to approximate the area between these curves by a Riemann sum and sketch the corresponding approximating rectangles. Then write an expression for the exact area.
(b) Explain how the situation changes if the curves have equations $x = f(y)$ and $x = g(y)$, where $f(y) \geqslant g(y)$ for $c \leqslant y \leqslant d$.

2. Suppose that Sue runs faster than Kathy throughout a 1500-meter race. What is the physical meaning of the area between their velocity curves for the first minute of the race?

3. (a) Suppose S is a solid with known cross-sectional areas. Explain how to approximate the volume of S by a Riemann sum. Then write an expression for the exact volume.
(b) If S is a solid of revolution, how do you find the cross-sectional areas?

4. (a) What is the volume of a cylindrical shell?
(b) Explain how to use cylindrical shells to find the volume of a solid of revolution.
(c) Why might you want to use the shell method instead of slicing?

5. Suppose that you push a book across a 6-meter-long table by exerting a force $f(x)$ at each point from $x = 0$ to $x = 6$. What does $\int_0^6 f(x)\,dx$ represent? If $f(x)$ is measured in newtons, what are the units for the integral?

6. (a) What is the average value of a function f on an interval $[a, b]$?
(b) What does the Mean Value Theorem for Integrals say? What is its geometric interpretation?

EXERCISES

1–6 |||| Find the area of the region bounded by the given curves.

1. $y = x^2 - x - 6, \quad y = 0$

2. $y = 20 - x^2, \quad y = x^2 - 12$

3. $y = e^x - 1, \quad y = x^2 - x, \quad x = 1$

4. $x + y = 0, \quad x = y^2 + 3y$

5. $y = \sin(\pi x/2), \quad y = x^2 - 2x$

6. $y = \sqrt{x}$, $y = x^2$, $x = 2$

7–11 Find the volume of the solid obtained by rotating the region bounded by the given curves about the specified axis.

7. $y = 2x$, $y = x^2$; about the x-axis

8. $x = 1 + y^2$, $y = x - 3$; about the y-axis

9. $x = 0$, $x = 9 - y^2$; about $x = -1$

10. $y = x^2 + 1$, $y = 9 - x^2$; about $y = -1$

11. $x^2 - y^2 = a^2$, $x = a + h$ (where $a > 0$, $h > 0$); about the y-axis

12–14 Set up, but do not evaluate, an integral for the volume of the solid obtained by rotating the region bounded by the given curves about the specified axis.

12. $y = \cos x$, $y = 0$, $x = 3\pi/2$, $x = 5\pi/2$; about the y-axis

13. $y = x^3$, $y = x^2$; about $y = 1$

14. $y = x^3$, $y = 8$, $x = 0$; about $x = 2$

15. Find the volumes of the solids obtained by rotating the region bounded by the curves $y = x$ and $y = x^2$ about the following lines:
(a) The x-axis (b) The y-axis (c) $y = 2$

16. Let $\mathcal{R}$ be the region in the first quadrant bounded by the curves $y = x^3$ and $y = 2x - x^2$. Calculate the following quantities.
(a) The area of $\mathcal{R}$
(b) The volume obtained by rotating $\mathcal{R}$ about the x-axis
(c) The volume obtained by rotating $\mathcal{R}$ about the y-axis

17. Let $\mathcal{R}$ be the region bounded by the curves $y = \tan(x^2)$, $x = 1$, and $y = 0$. Use the Midpoint Rule with $n = 4$ to estimate the following.
(a) The area of $\mathcal{R}$
(b) The volume obtained by rotating $\mathcal{R}$ about the x-axis

18. Let $\mathcal{R}$ be the region bounded by the curves $y = 1 - x^2$ and $y = x^6 - x + 1$. Estimate the following quantities.
(a) The x-coordinates of the points of intersection of the curves
(b) The area of $\mathcal{R}$
(c) The volume generated when $\mathcal{R}$ is rotated about the x-axis
(d) The volume generated when $\mathcal{R}$ is rotated about the y-axis

19–22 Each integral represents the volume of a solid. Describe the solid.

19. $\int_0^{\pi/2} 2\pi x \cos x \, dx$

20. $\int_0^{\pi/2} 2\pi \cos^2 x \, dx$

21. $\int_0^2 2\pi y(4 - y^2) \, dy$

22. $\int_0^1 \pi[(2 - x^2)^2 - (2 - \sqrt{x})^2] \, dx$

23. The base of a solid is a circular disk with radius 3. Find the volume of the solid if parallel cross-sections perpendicular to the base are isosceles right triangles with hypotenuse lying along the base.

24. The base of a solid is the region bounded by the parabolas $y = x^2$ and $y = 2 - x^2$. Find the volume of the solid if the cross-sections perpendicular to the x-axis are squares with one side lying along the base.

25. The height of a monument is 20 m. A horizontal cross-section at a distance x meters from the top is an equilateral triangle with side $x/4$ meters. Find the volume of the monument.

26. (a) The base of a solid is a square with vertices located at $(1, 0)$, $(0, 1)$, $(-1, 0)$, and $(0, -1)$. Each cross-section perpendicular to the x-axis is a semicircle. Find the volume of the solid.
(b) Show that by cutting the solid of part (a), we can rearrange it to form a cone. Thus compute its volume more simply.

27. A force of 30 N is required to maintain a spring stretched from its natural length of 12 cm to a length of 15 cm. How much work is done in stretching the spring from 12 cm to 20 cm?

28. A 1600-lb elevator is suspended by a 200-ft cable that weighs 10 lb/ft. How much work is required to raise the elevator from the basement to the third floor, a distance of 30 ft?

29. A tank full of water has the shape of a paraboloid of revolution as shown in the figure; that is, its shape is obtained by rotating a parabola about a vertical axis.
(a) If its height is 4 ft and the radius at the top is 4 ft, find the work required to pump the water out of the tank.
(b) After 4000 ft-lb of work has been done, what is the depth of the water remaining in the tank?

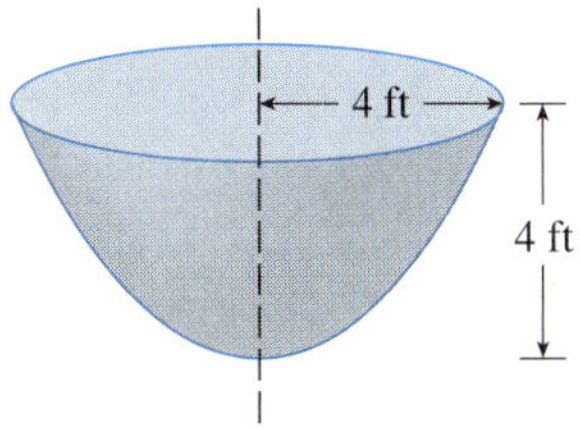

30. Find the average value of the function $f(t) = t \sin(t^2)$ on the interval $[0, 10]$.

31. If f is a continuous function, what is the limit as $h \to 0$ of the average value of f on the interval $[x, x + h]$?

32. Let $\mathcal{R}_1$ be the region bounded by $y = x^2$, $y = 0$, and $x = b$, where $b > 0$. Let $\mathcal{R}_2$ be the region bounded by $y = x^2$, $x = 0$, and $y = b^2$.
(a) Is there a value of b such that $\mathcal{R}_1$ and $\mathcal{R}_2$ have the same area?
(b) Is there a value of b such that $\mathcal{R}_1$ sweeps out the same volume when rotated about the x-axis and the y-axis?
(c) Is there a value of b such that $\mathcal{R}_1$ and $\mathcal{R}_2$ sweep out the same volume when rotated about the x-axis?
(d) Is there a value of b such that $\mathcal{R}_1$ and $\mathcal{R}_2$ sweep out the same volume when rotated about the y-axis?

PROBLEMS PLUS

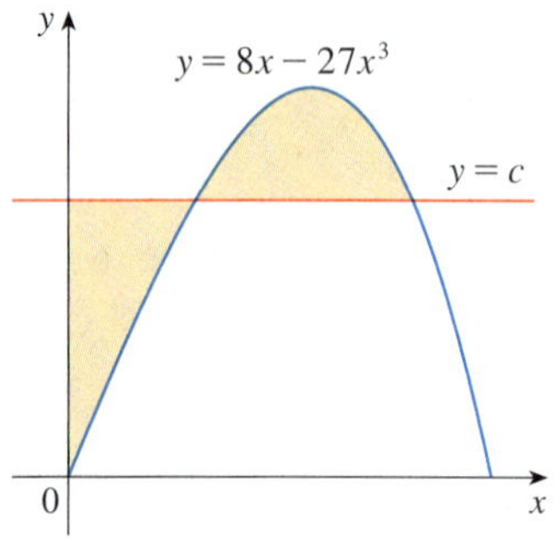

FIGURE FOR PROBLEM 3

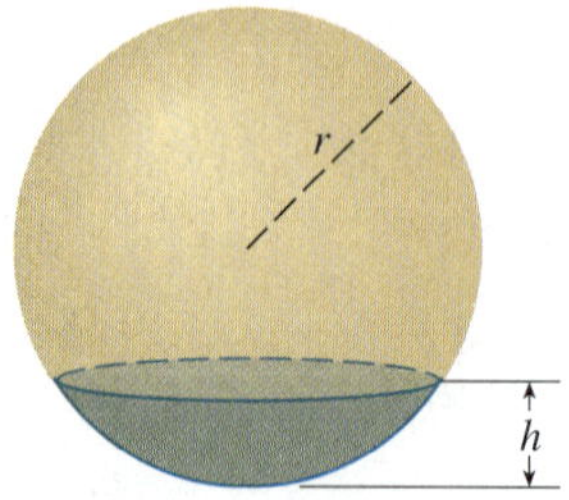

FIGURE FOR PROBLEM 5

1. (a) Find a positive continuous function f such that the area under the graph of f from 0 to t is $A(t) = t^3$ for all $t > 0$.

(b) A solid is generated by rotating about the x-axis the region under the curve $y = f(x)$, where f is a positive function and $x \geqslant 0$. The volume generated by the part of the curve from $x = 0$ to $x = b$ is b^2 for all $b > 0$. Find the function f.

2. There is a line through the origin that divides the region bounded by the parabola $y = x - x^2$ and the x-axis into two regions with equal area. What is the slope of that line?

3. The figure shows a horizontal line $y = c$ intersecting the curve $y = 8x - 27x^3$. Find the number c such that the areas of the shaded regions are equal.

4. A cylindrical glass of radius r and height L is filled with water and then tilted until the water remaining in the glass exactly covers its base.

(a) Determine a way to "slice" the water into parallel rectangular cross-sections and then *set up* a definite integral for the volume of the water in the glass.

(b) Determine a way to "slice" the water into parallel cross-sections that are trapezoids and then *set up* a definite integral for the volume of the water.

(c) Find the volume of water in the glass by evaluating one of the integrals in part (a) or part (b).

(d) Find the volume of the water in the glass from purely geometric considerations.

(e) Suppose the glass is tilted until the water exactly covers half the base. In what direction can you "slice" the water into triangular cross-sections? Rectangular cross-sections? Cross-sections that are segments of circles? Find the volume of water in the glass.

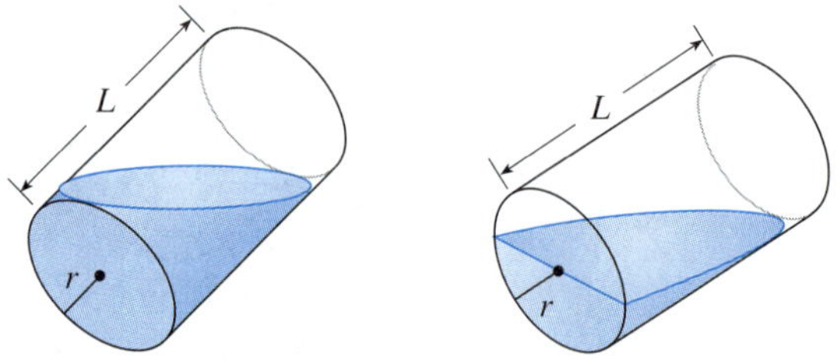

5. (a) Show that the volume of a segment of height h of a sphere of radius r is

$$V = \tfrac{1}{3}\pi h^2(3r - h)$$

(b) Show that if a sphere of radius 1 is sliced by a plane at a distance x from the center in such a way that the volume of one segment is twice the volume of the other, then x is a solution of the equation

$$3x^3 - 9x + 2 = 0$$

where $0 < x < 1$. Use Newton's method to find x accurate to four decimal places.

(c) Using the formula for the volume of a segment of a sphere, it can be shown that the depth x to which a floating sphere of radius r sinks in water is a root of the equation

$$x^3 - 3rx^2 + 4r^3s = 0$$

where s is the specific gravity of the sphere. Suppose a wooden sphere of radius 0.5 m has specific gravity 0.75. Calculate, to four-decimal-place accuracy, the depth to which the sphere will sink.

(d) A hemispherical bowl has radius 5 inches and water is running into the bowl at the rate of 0.2 in^3/s.

(i) How fast is the water level in the bowl rising at the instant the water is 3 inches deep?

(ii) At a certain instant, the water is 4 inches deep. How long will it take to fill the bowl?

6. Archimedes' Principle states that the buoyant force on an object partially or fully submerged in a fluid is equal to the weight of the fluid that the object displaces. Thus, for an object of density ρ_0 floating partly submerged in a fluid of density ρ_f, the buoyant force is given by $F = \rho_f g \int_{-h}^{0} A(y)\, dy$, where g is the acceleration due to gravity and $A(y)$ is the area of a typical cross-section of the object. The weight of the object is given by

$$W = \rho_0 g \int_{-h}^{L-h} A(y)\, dy$$

(a) Show that the percentage of the volume of the object above the surface of the liquid is

$$100 \frac{\rho_f - \rho_0}{\rho_f}$$

(b) The density of ice is 917 kg/m^3 and the density of seawater is 1030 kg/m^3. What percentage of the volume of an iceberg is above water?

(c) An ice cube floats in a glass filled to the brim with water. Does the water overflow when the ice melts?

(d) A sphere of radius 0.4 m and having negligible weight is floating in a large freshwater lake. How much work is required to completely submerge the sphere? The density of the water is 1000 kg/m^3.

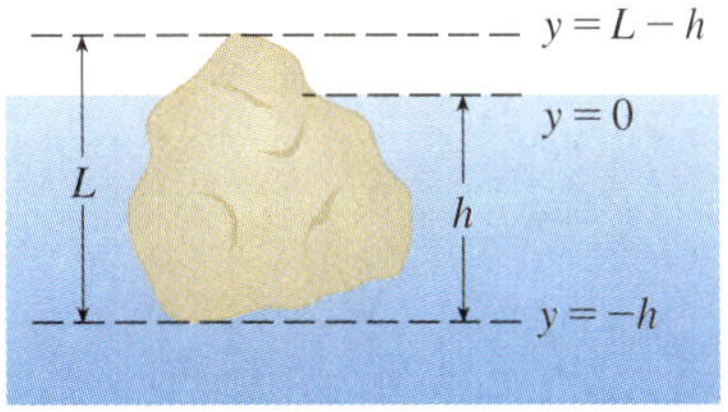

7. Water in an open bowl evaporates at a rate proportional to the area of the surface of the water. (This means that the rate of decrease of the volume is proportional to the area of the surface.) Show that the depth of the water decreases at a constant rate, regardless of the shape of the bowl.

8. A sphere of radius 1 overlaps a smaller sphere of radius r in such a way that their intersection is a circle of radius r. (In other words, they intersect in a great circle of the small sphere.) Find r so that the volume inside the small sphere and outside the large sphere is as large as possible.

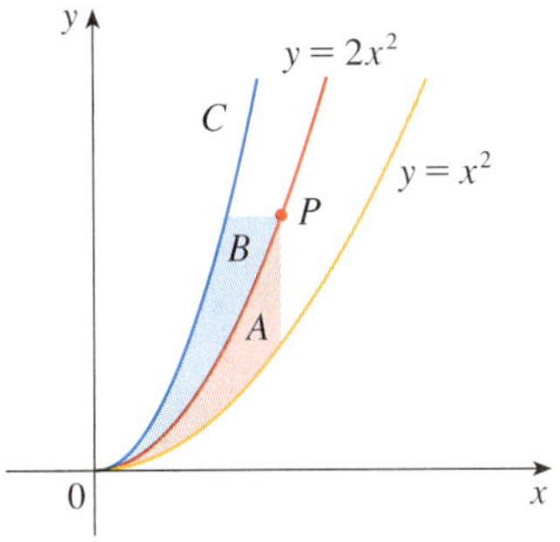

FIGURE FOR PROBLEM 9

9. The figure shows a curve C with the property that, for every point P on the middle curve $y = 2x^2$, the areas A and B are equal. Find an equation for C.

10. A paper drinking cup filled with water has the shape of a cone with height h and semivertical angle θ (see the figure). A ball is placed carefully in the cup, thereby displacing some of the water and making it overflow. What is the radius of the ball that causes the greatest volume of water to spill out of the cup?

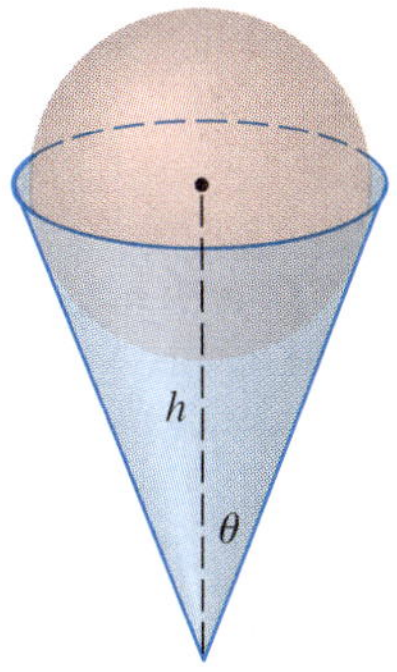

11. A *clepsydra*, or water clock, is a glass container with a small hole in the bottom through which water can flow. The "clock" is calibrated for measuring time by placing markings on the container corresponding to water levels at equally spaced times. Let $x = f(y)$ be continuous on the interval $[0, b]$ and assume that the container is formed by rotating the graph of f about the y-axis. Let V denote the volume of water and h the height of the water level at time t.

(a) Determine V as a function of h.

(b) Show that

$$\frac{dV}{dt} = \pi[f(h)]^2 \frac{dh}{dt}$$

(c) Suppose that A is the area of the hole in the bottom of the container. It follows from Torricelli's Law that the rate of change of the volume of the water is given by

$$\frac{dV}{dt} = kA\sqrt{h}$$

where k is a negative constant. Determine a formula for the function f such that dh/dt is a constant C. What is the advantage in having $dh/dt = C$?

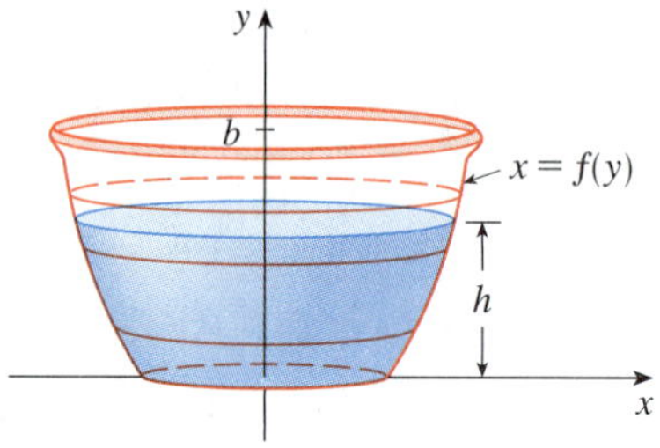

12. A cylindrical container of radius r and height L is partially filled with a liquid whose volume is V. If the container is rotated about its axis of symmetry with constant angular speed ω, then the container will induce a rotational motion in the liquid around the same axis. Eventually, the liquid will be rotating at the same angular speed as the container. The surface of the liquid will be convex, as indicated in the figure, because the centrifugal force on the liquid particles increases with the distance from the axis of the container. It can be shown that the surface of the liquid is a paraboloid of revolution generated by rotating the parabola

$$y = h + \frac{\omega^2 x^2}{2g}$$

about the y-axis, where g is the acceleration due to gravity.

(a) Determine h as a function of ω.

(b) At what angular speed will the surface of the liquid touch the bottom? At what speed will it spill over the top?

(c) Suppose the radius of the container is 2 ft, the height is 7 ft, and the container and liquid are rotating at the same constant angular speed. The surface of the liquid is 5 ft below the top of the tank at the central axis and 4 ft below the top of the tank 1 ft out from the central axis.

(i) Determine the angular speed of the container and the volume of the fluid.

(ii) How far below the top of the tank is the liquid at the wall of the container?

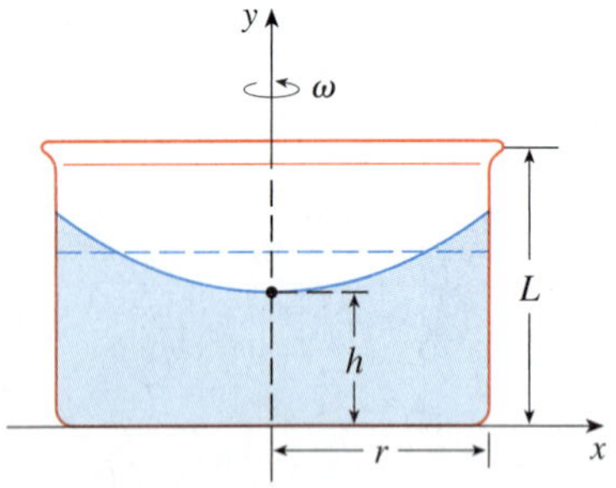

FIGURE FOR PROBLEM 12

13. If the tangent at a point P on the curve $y = x^3$ intersects the curve again at Q, let A be the area of the region bounded by the curve and the line segment PQ. Let B be the area of the region defined in the same way starting with Q instead of P. What is the relationship between A and B?

CAS **14.** Suppose we are planning to make a taco from a round tortilla with diameter 8 inches by bending the tortilla so that it is shaped as if it is partially wrapped around a circular cylinder. We will fill the tortilla to the edge (but no more) with meat, cheese, and other ingredients. Our problem is to decide how to curve the tortilla in order to maximize the volume of food it can hold.

(a) We start by placing a circular cylinder of radius r along a diameter of the tortilla and folding the tortilla around the cylinder. Let x represent the distance from the center of the tortilla to a point P on the diameter (see the figure). Show that the cross-sectional area of the filled taco in the plane through P perpendicular to the axis of the cylinder is

$$A(x) = r\sqrt{16 - x^2} - \tfrac{1}{2}r^2 \sin\left(\frac{2}{r}\sqrt{16 - x^2}\right)$$

and write an expression for the volume of the filled taco.

(b) Determine (approximately) the value of r that maximizes the volume of the taco. (Use a graphical approach with your CAS.)

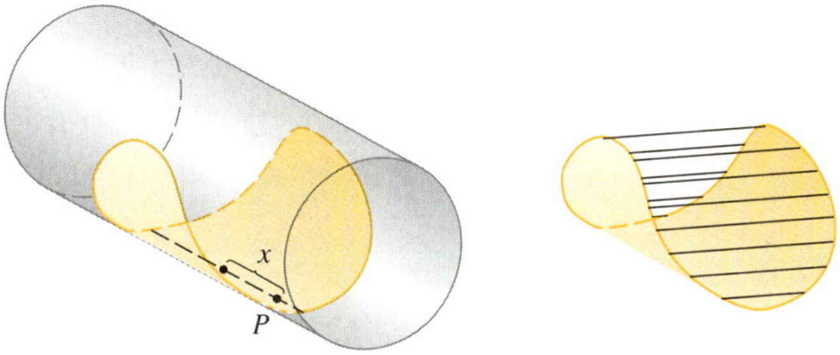

CHAPTER 7

The techniques of this chapter enable us to find the height of a rocket a minute after liftoff and to compute the escape velocity of the rocket.

Techniques of Integration

Because of the Fundamental Theorem of Calculus, we can integrate a function if we know an antiderivative, that is, an indefinite integral. We summarize here the most important integrals that we have learned so far.

$$\int x^n\,dx = \frac{x^{n+1}}{n+1} + C \quad (n \neq -1) \qquad \int \frac{1}{x}\,dx = \ln|x| + C$$

$$\int e^x\,dx = e^x + C \qquad \int a^x\,dx = \frac{a^x}{\ln a} + C$$

$$\int \sin x\,dx = -\cos x + C \qquad \int \cos x\,dx = \sin x + C$$

$$\int \sec^2 x\,dx = \tan x + C \qquad \int \csc^2 x\,dx = -\cot x + C$$

$$\int \sec x \tan x\,dx = \sec x + C \qquad \int \csc x \cot x\,dx = -\csc x + C$$

$$\int \sinh x\,dx = \cosh x + C \qquad \int \cosh x\,dx = \sinh x + C$$

$$\int \tan x\,dx = \ln|\sec x| + C \qquad \int \cot x\,dx = \ln|\sin x| + C$$

$$\int \frac{1}{x^2 + a^2}\,dx = \frac{1}{a}\tan^{-1}\left(\frac{x}{a}\right) + C \qquad \int \frac{1}{\sqrt{a^2 - x^2}}\,dx = \sin^{-1}\left(\frac{x}{a}\right) + C$$

In this chapter we develop techniques for using these basic integration formulas to obtain indefinite integrals of more complicated functions. We learned the most important method of integration, the Substitution Rule, in Section 5.5. The other general technique, integration by parts, is presented in Section 7.1. Then we learn methods that are special to particular classes of functions such as trigonometric functions and rational functions.

Integration is not as straightforward as differentiation; there are no rules that absolutely guarantee obtaining an indefinite integral of a function. Therefore, in Section 7.5 we discuss a strategy for integration.

7.1 Integration by Parts

Every differentiation rule has a corresponding integration rule. For instance, the Substitution Rule for integration corresponds to the Chain Rule for differentiation. The rule that corresponds to the Product Rule for differentiation is called the rule for *integration by parts*.

The Product Rule states that if f and g are differentiable functions, then

$$\frac{d}{dx}[f(x)g(x)] = f(x)g'(x) + g(x)f'(x)$$

In the notation for indefinite integrals this equation becomes

$$\int [f(x)g'(x) + g(x)f'(x)]\,dx = f(x)g(x)$$

or

$$\int f(x)g'(x)\,dx + \int g(x)f'(x)\,dx = f(x)g(x)$$

We can rearrange this equation as

1
$$\int f(x)g'(x)\,dx = f(x)g(x) - \int g(x)f'(x)\,dx$$

Formula 1 is called the **formula for integration by parts**. It is perhaps easier to remember in the following notation. Let $u = f(x)$ and $v = g(x)$. Then the differentials are $du = f'(x)\,dx$ and $dv = g'(x)\,dx$, so, by the Substitution Rule, the formula for integration by parts becomes

2
$$\int u\,dv = uv - \int v\,du$$

EXAMPLE 1 Find $\int x \sin x\,dx$.

SOLUTION USING FORMULA 1 Suppose we choose $f(x) = x$ and $g'(x) = \sin x$. Then $f'(x) = 1$ and $g(x) = -\cos x$. (For g we can choose *any* antiderivative of g'.) Thus, using Formula 1, we have

$$\begin{aligned}\int x \sin x\,dx &= f(x)g(x) - \int g(x)f'(x)\,dx \\ &= x(-\cos x) - \int (-\cos x)\,dx \\ &= -x\cos x + \int \cos x\,dx \\ &= -x\cos x + \sin x + C\end{aligned}$$

It's wise to check the answer by differentiating it. If we do so, we get $x \sin x$, as expected.

SOLUTION USING FORMULA 2 Let

$$u = x \qquad dv = \sin x\,dx$$

Then

$$du = dx \qquad v = -\cos x$$

and so

$$\begin{aligned}\int x \sin x\,dx &= \int \overset{u}{\widehat{x}}\ \overbrace{\sin x\,dx}^{dv} = \overset{u}{\widehat{x}}\ \overbrace{(-\cos x)}^{v} - \int \overbrace{(-\cos x)}^{v}\ \overbrace{dx}^{du} \\ &= -x\cos x + \int \cos x\,dx \\ &= -x\cos x + \sin x + C\end{aligned}$$

It is helpful to use the pattern:

$u = \square \qquad dv = \square$

$du = \square \qquad v = \square$

NOTE ▫ Our aim in using integration by parts is to obtain a simpler integral than the one we started with. Thus, in Example 1 we started with $\int x \sin x\, dx$ and expressed it in terms of the simpler integral $\int \cos x\, dx$. If we had chosen $u = \sin x$ and $dv = x\, dx$, then $du = \cos x\, dx$ and $v = x^2/2$, so integration by parts gives

$$\int x \sin x\, dx = (\sin x)\frac{x^2}{2} - \frac{1}{2}\int x^2 \cos x\, dx$$

Although this is true, $\int x^2 \cos x\, dx$ is a more difficult integral than the one we started with. In general, when deciding on a choice for u and dv, we usually try to choose $u = f(x)$ to be a function that becomes simpler when differentiated (or at least not more complicated) as long as $dv = g'(x)\, dx$ can be readily integrated to give v.

EXAMPLE 2 Evaluate $\int \ln x\, dx$.

SOLUTION Here we don't have much choice for u and dv. Let

$$u = \ln x \qquad dv = dx$$

Then
$$du = \frac{1}{x}\, dx \qquad v = x$$

Integrating by parts, we get

$$\int \ln x\, dx = x \ln x - \int x \frac{dx}{x}$$

||| It's customary to write $\int 1\, dx$ as $\int dx$.

$$= x \ln x - \int dx$$

||| Check the answer by differentiating it.

$$= x \ln x - x + C$$

Integration by parts is effective in this example because the derivative of the function $f(x) = \ln x$ is simpler than f.

EXAMPLE 3 Find $\int t^2 e^t\, dt$.

SOLUTION Notice that t^2 becomes simpler when differentiated (whereas e^t is unchanged when differentiated or integrated), so we choose

$$u = t^2 \qquad dv = e^t\, dt$$

Then
$$du = 2t\, dt \qquad v = e^t$$

Integration by parts gives

$$\int t^2 e^t\, dt = t^2 e^t - 2\int t e^t\, dt \tag{3}$$

The integral that we obtained, $\int t e^t\, dt$, is simpler than the original integral but is still not obvious. Therefore, we use integration by parts a second time, this time with $u = t$ and

$dv = e^t\,dt$. Then $du = dt$, $v = e^t$, and

$$\int te^t\,dt = te^t - \int e^t\,dt$$

$$= te^t - e^t + C$$

Putting this in Equation 3, we get

$$\int t^2 e^t\,dt = t^2 e^t - 2\int te^t\,dt$$

$$= t^2 e^t - 2(te^t - e^t + C)$$

$$= t^2 e^t - 2te^t + 2e^t + C_1 \qquad \text{where } C_1 = -2C$$

EXAMPLE 4 Evaluate $\int e^x \sin x\,dx$.

|||| An easier method, using complex numbers, is given in Exercise 48 in Appendix G.

SOLUTION Neither e^x nor $\sin x$ becomes simpler when differentiated, but we try choosing $u = e^x$ and $dv = \sin x\,dx$ anyway. Then $du = e^x\,dx$ and $v = -\cos x$, so integration by parts gives

$$\boxed{4} \qquad \int e^x \sin x\,dx = -e^x \cos x + \int e^x \cos x\,dx$$

The integral that we have obtained, $\int e^x \cos x\,dx$, is no simpler than the original one, but at least it's no more difficult. Having had success in the preceding example integrating by parts twice, we persevere and integrate by parts again. This time we use $u = e^x$ and $dv = \cos x\,dx$. Then $du = e^x\,dx$, $v = \sin x$, and

$$\boxed{5} \qquad \int e^x \cos x\,dx = e^x \sin x - \int e^x \sin x\,dx$$

|||| Figure 1 illustrates Example 4 by showing the graphs of $f(x) = e^x \sin x$ and $F(x) = \frac{1}{2}e^x(\sin x - \cos x)$. As a visual check on our work, notice that $f(x) = 0$ when F has a maximum or minimum.

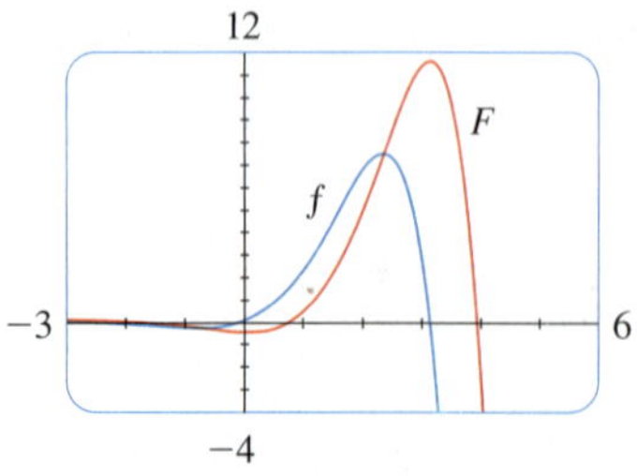

FIGURE 1

At first glance, it appears as if we have accomplished nothing because we have arrived at $\int e^x \sin x\,dx$, which is where we started. However, if we put Equation 5 into Equation 4 we get

$$\int e^x \sin x\,dx = -e^x \cos x + e^x \sin x - \int e^x \sin x\,dx$$

This can be regarded as an equation to be solved for the unknown integral. Adding $\int e^x \sin x\,dx$ to both sides, we obtain

$$2\int e^x \sin x\,dx = -e^x \cos x + e^x \sin x$$

Dividing by 2 and adding the constant of integration, we get

$$\int e^x \sin x\,dx = \tfrac{1}{2}e^x(\sin x - \cos x) + C$$

If we combine the formula for integration by parts with Part 2 of the Fundamental Theorem of Calculus, we can evaluate definite integrals by parts. Evaluating both sides of Formula 1 between a and b, assuming f' and g' are continuous, and using the Fundamental

Theorem, we obtain

6
$$\int_a^b f(x)g'(x)\,dx = f(x)g(x)\Big]_a^b - \int_a^b g(x)f'(x)\,dx$$

EXAMPLE 5 Calculate $\int_0^1 \tan^{-1}x\,dx$.

SOLUTION Let

$$u = \tan^{-1}x \qquad dv = dx$$

Then
$$du = \frac{dx}{1+x^2} \qquad v = x$$

So Formula 6 gives

$$\int_0^1 \tan^{-1}x\,dx = x\tan^{-1}x\Big]_0^1 - \int_0^1 \frac{x}{1+x^2}\,dx$$
$$= 1\cdot\tan^{-1}1 - 0\cdot\tan^{-1}0 - \int_0^1 \frac{x}{1+x^2}\,dx$$
$$= \frac{\pi}{4} - \int_0^1 \frac{x}{1+x^2}\,dx$$

|||| Since $\tan^{-1}x \geqslant 0$ for $x \geqslant 0$, the integral in Example 5 can be interpreted as the area of the region shown in Figure 2.

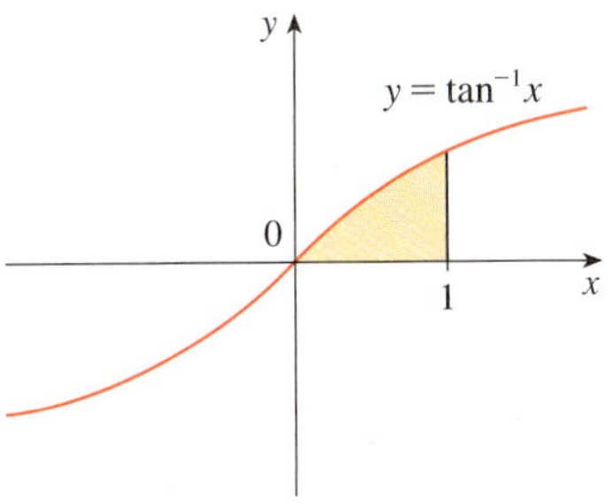

FIGURE 2

To evaluate this integral we use the substitution $t = 1 + x^2$ (since u has another meaning in this example). Then $dt = 2x\,dx$, so $x\,dx = dt/2$. When $x = 0$, $t = 1$; when $x = 1$, $t = 2$; so

$$\int_0^1 \frac{x}{1+x^2}\,dx = \tfrac{1}{2}\int_1^2 \frac{dt}{t} = \tfrac{1}{2}\ln|t|\Big]_1^2$$
$$= \tfrac{1}{2}(\ln 2 - \ln 1) = \tfrac{1}{2}\ln 2$$

Therefore
$$\int_0^1 \tan^{-1}x\,dx = \frac{\pi}{4} - \int_0^1 \frac{x}{1+x^2}\,dx = \frac{\pi}{4} - \frac{\ln 2}{2}$$

EXAMPLE 6 Prove the reduction formula

7
$$\int \sin^n x\,dx = -\frac{1}{n}\cos x\sin^{n-1}x + \frac{n-1}{n}\int \sin^{n-2}x\,dx$$

|||| Equation 7 is called a *reduction formula* because the exponent n has been *reduced* to $n-1$ and $n-2$.

where $n \geqslant 2$ is an integer.

SOLUTION Let
$$u = \sin^{n-1}x \qquad dv = \sin x\,dx$$

Then
$$du = (n-1)\sin^{n-2}x\cos x\,dx \qquad v = -\cos x$$

so integration by parts gives

$$\int \sin^n x\,dx = -\cos x\sin^{n-1}x + (n-1)\int \sin^{n-2}x\cos^2 x\,dx$$

Since $\cos^2 x = 1 - \sin^2 x$, we have

$$\int \sin^n x\, dx = -\cos x \sin^{n-1}x + (n-1)\int \sin^{n-2}x\, dx - (n-1)\int \sin^n x\, dx$$

As in Example 4, we solve this equation for the desired integral by taking the last term on the right side to the left side. Thus, we have

$$n\int \sin^n x\, dx = -\cos x \sin^{n-1}x + (n-1)\int \sin^{n-2}x\, dx$$

or

$$\int \sin^n x\, dx = -\frac{1}{n}\cos x \sin^{n-1}x + \frac{n-1}{n}\int \sin^{n-2}x\, dx$$

The reduction formula (7) is useful because by using it repeatedly we could eventually express $\int \sin^n x\, dx$ in terms of $\int \sin x\, dx$ (if n is odd) or $\int (\sin x)^0\, dx = \int dx$ (if n is even).

7.1 Exercises

1–2 Evaluate the integral using integration by parts with the indicated choices of u and dv.

1. $\int x \ln x\, dx$; $u = \ln x$, $dv = x\, dx$

2. $\int \theta \sec^2\theta\, d\theta$; $u = \theta$, $dv = \sec^2\theta\, d\theta$

3–32 Evaluate the integral.

3. $\int x \cos 5x\, dx$

4. $\int xe^{-x}\, dx$

5. $\int re^{r/2}\, dr$

6. $\int t \sin 2t\, dt$

7. $\int x^2 \sin \pi x\, dx$

8. $\int x^2 \cos mx\, dx$

9. $\int \ln(2x + 1)\, dx$

10. $\int \sin^{-1}x\, dx$

11. $\int \arctan 4t\, dt$

12. $\int p^5 \ln p\, dp$

13. $\int (\ln x)^2\, dx$

14. $\int t^3 e^t\, dt$

15. $\int e^{2\theta} \sin 3\theta\, d\theta$

16. $\int e^{-\theta} \cos 2\theta\, d\theta$

17. $\int y \sinh y\, dy$

18. $\int y \cosh ay\, dy$

19. $\int_0^{\pi} t \sin 3t\, dt$

20. $\int_0^1 (x^2 + 1)e^{-x}\, dx$

21. $\int_1^2 \frac{\ln x}{x^2}\, dx$

22. $\int_1^4 \sqrt{t} \ln t\, dt$

23. $\int_0^1 \frac{y}{e^{2y}}\, dy$

24. $\int_{\pi/4}^{\pi/2} x \csc^2 x\, dx$

25. $\int_0^{1/2} \cos^{-1}x\, dx$

26. $\int_0^1 x5^x\, dx$

27. $\int \cos x \ln(\sin x)\, dx$

28. $\int_1^{\sqrt{3}} \arctan(1/x)\, dx$

29. $\int \cos(\ln x)\, dx$

30. $\int_0^1 \frac{r^3}{\sqrt{4 + r^2}}\, dr$

31. $\int_1^2 x^4(\ln x)^2\, dx$

32. $\int_0^t e^s \sin(t - s)\, ds$

33–36 First make a substitution and then use integration by parts to evaluate the integral.

33. $\int \sin \sqrt{x}\, dx$

34. $\int_1^4 e^{\sqrt{x}}\, dx$

35. $\int_{\sqrt{\pi/2}}^{\sqrt{\pi}} \theta^3 \cos(\theta^2)\, d\theta$

36. $\int x^5 e^{x^2}\, dx$

37–40 Evaluate the indefinite integral. Illustrate, and check that your answer is reasonable, by graphing both the function and its antiderivative (take $C = 0$).

37. $\int x \cos \pi x\, dx$

38. $\int x^{3/2} \ln x\, dx$

39. $\int (2x + 3)e^x\, dx$

40. $\int x^3 e^{x^2}\, dx$

41. (a) Use the reduction formula in Example 6 to show that

$$\int \sin^2 x\, dx = \frac{x}{2} - \frac{\sin 2x}{4} + C$$

(b) Use part (a) and the reduction formula to evaluate $\int \sin^4 x\, dx$.

42. (a) Prove the reduction formula

$$\int \cos^n x\, dx = \frac{1}{n}\cos^{n-1}x \sin x + \frac{n-1}{n}\int \cos^{n-2}x\, dx$$

(b) Use part (a) to evaluate $\int \cos^2 x\, dx$.
(c) Use parts (a) and (b) to evaluate $\int \cos^4 x\, dx$.

43. (a) Use the reduction formula in Example 6 to show that

$$\int_0^{\pi/2} \sin^n x\, dx = \frac{n-1}{n}\int_0^{\pi/2} \sin^{n-2}x\, dx$$

where $n \geqslant 2$ is an integer.
(b) Use part (a) to evaluate $\int_0^{\pi/2} \sin^3 x\, dx$ and $\int_0^{\pi/2} \sin^5 x\, dx$.
(c) Use part (a) to show that, for odd powers of sine,

$$\int_0^{\pi/2} \sin^{2n+1}x\, dx = \frac{2 \cdot 4 \cdot 6 \cdot \cdots \cdot 2n}{3 \cdot 5 \cdot 7 \cdot \cdots \cdot (2n+1)}$$

44. Prove that, for even powers of sine,

$$\int_0^{\pi/2} \sin^{2n}x\, dx = \frac{1 \cdot 3 \cdot 5 \cdot \cdots \cdot (2n-1)}{2 \cdot 4 \cdot 6 \cdot \cdots \cdot 2n}\frac{\pi}{2}$$

45–48 ||| Use integration by parts to prove the reduction formula.

45. $\int (\ln x)^n\, dx = x(\ln x)^n - n\int (\ln x)^{n-1}\, dx$

46. $\int x^n e^x\, dx = x^n e^x - n\int x^{n-1}e^x\, dx$

47. $\int (x^2 + a^2)^n\, dx$

$$= \frac{x(x^2+a^2)^n}{2n+1} + \frac{2na^2}{2n+1}\int (x^2+a^2)^{n-1}\, dx \quad \left(n \neq -\tfrac{1}{2}\right)$$

48. $\int \sec^n x\, dx = \dfrac{\tan x \sec^{n-2}x}{n-1} + \dfrac{n-2}{n-1}\int \sec^{n-2}x\, dx \quad (n \neq 1)$

49. Use Exercise 45 to find $\int (\ln x)^3\, dx$.

50. Use Exercise 46 to find $\int x^4 e^x\, dx$.

51–52 ||| Find the area of the region bounded by the given curves.

51. $y = xe^{-0.4x}$, $y = 0$, $x = 5$

52. $y = 5\ln x$, $y = x\ln x$

53–54 ||| Use a graph to find approximate x-coordinates of the points of intersection of the given curves. Then find (approximately) the area of the region bounded by the curves.

53. $y = x\sin x$, $y = (x-2)^2$

54. $y = \arctan 3x$, $y = x/2$

55–58 ||| Use the method of cylindrical shells to find the volume generated by rotating the region bounded by the given curves about the specified axis.

55. $y = \cos(\pi x/2)$, $y = 0$, $0 \leqslant x \leqslant 1$; about the y-axis

56. $y = e^x$, $y = e^{-x}$, $x = 1$; about the y-axis

57. $y = e^{-x}$, $y = 0$, $x = -1$, $x = 0$; about $x = 1$

58. $y = e^x$, $x = 0$, $y = \pi$; about the x-axis

59. Find the average value of $f(x) = x^2\ln x$ on the interval $[1, 3]$.

60. A rocket accelerates by burning its onboard fuel, so its mass decreases with time. Suppose the initial mass of the rocket at liftoff (including its fuel) is m, the fuel is consumed at rate r, and the exhaust gases are ejected with constant velocity v_e (relative to the rocket). A model for the velocity of the rocket at time t is given by the equation

$$v(t) = -gt - v_e \ln\frac{m - rt}{m}$$

where g is the acceleration due to gravity and t is not too large. If $g = 9.8\ \text{m/s}^2$, $m = 30{,}000$ kg, $r = 160$ kg/s, and $v_e = 3000$ m/s, find the height of the rocket one minute after liftoff.

61. A particle that moves along a straight line has velocity $v(t) = t^2e^{-t}$ meters per second after t seconds. How far will it travel during the first t seconds?

62. If $f(0) = g(0) = 0$ and f'' and g'' are continuous, show that

$$\int_0^a f(x)g''(x)\, dx = f(a)g'(a) - f'(a)g(a) + \int_0^a f''(x)g(x)\, dx$$

63. Suppose that $f(1) = 2$, $f(4) = 7$, $f'(1) = 5$, $f'(4) = 3$, and f'' is continuous. Find the value of $\int_1^4 xf''(x)\, dx$.

64. (a) Use integration by parts to show that

$$\int f(x)\, dx = xf(x) - \int xf'(x)\, dx$$

(b) If f and g are inverse functions and f' is continuous, prove that

$$\int_a^b f(x)\, dx = bf(b) - af(a) - \int_{f(a)}^{f(b)} g(y)\, dy$$

[*Hint:* Use part (a) and make the substitution $y = f(x)$.]

(c) In the case where f and g are positive functions and $b > a > 0$, draw a diagram to give a geometric interpretation of part (b).

(d) Use part (b) to evaluate $\int_1^e \ln x\, dx$.

65. We arrived at Formula 6.3.2, $V = \int_a^b 2\pi x f(x)\, dx$, by using cylindrical shells, but now we can use integration by parts to prove it using the slicing method of Section 6.2, at least for the case where f is one-to-one and therefore has an inverse function g. Use the figure to show that

$$V = \pi b^2 d - \pi a^2 c - \int_c^d \pi [g(y)]^2\, dy$$

Make the substitution $y = f(x)$ and then use integration by parts on the resulting integral to prove that $V = \int_a^b 2\pi x f(x)\, dx$.

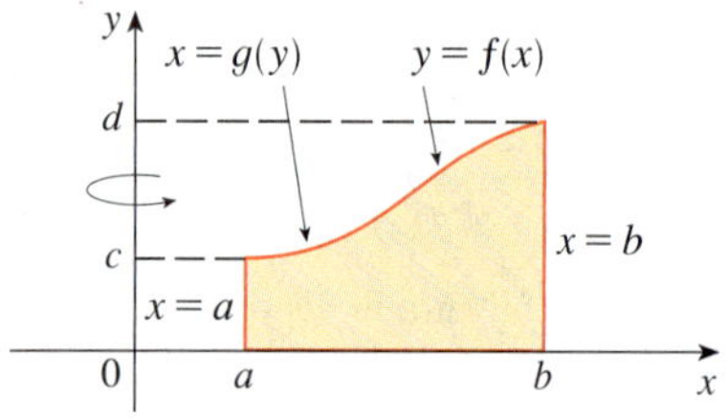

66. Let $I_n = \int_0^{\pi/2} \sin^n x\, dx$.

(a) Show that $I_{2n+2} \leqslant I_{2n+1} \leqslant I_{2n}$.

(b) Use Exercise 44 to show that

$$\frac{I_{2n+2}}{I_{2n}} = \frac{2n+1}{2n+2}$$

(c) Use parts (a) and (b) to show that

$$\frac{2n+1}{2n+2} \leqslant \frac{I_{2n+1}}{I_{2n}} \leqslant 1$$

and deduce that $\lim_{n\to\infty} I_{2n+1}/I_{2n} = 1$.

(d) Use part (c) and Exercises 43 and 44 to show that

$$\lim_{n\to\infty} \frac{2}{1}\cdot\frac{2}{3}\cdot\frac{4}{3}\cdot\frac{4}{5}\cdot\frac{6}{5}\cdot\frac{6}{7}\cdot\cdots\cdot\frac{2n}{2n-1}\cdot\frac{2n}{2n+1} = \frac{\pi}{2}$$

This formula is usually written as an infinite product:

$$\frac{\pi}{2} = \frac{2}{1}\cdot\frac{2}{3}\cdot\frac{4}{3}\cdot\frac{4}{5}\cdot\frac{6}{5}\cdot\frac{6}{7}\cdot\cdots$$

and is called the *Wallis product*.

(e) We construct rectangles as follows. Start with a square of area 1 and attach rectangles of area 1 alternately beside or on top of the previous rectangle (see the figure). Find the limit of the ratios of width to height of these rectangles.

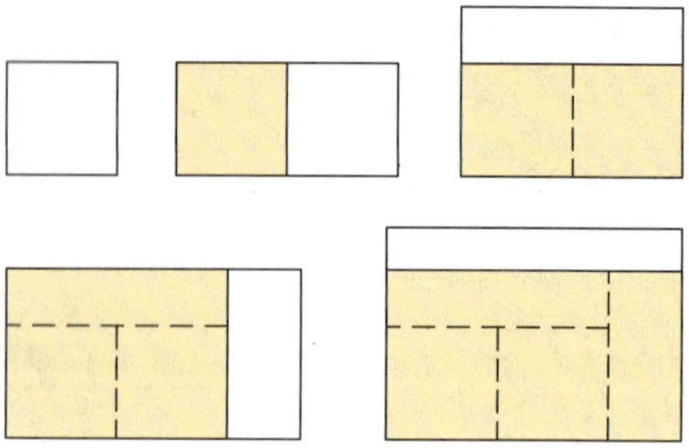

7.2 Trigonometric Integrals

In this section we use trigonometric identities to integrate certain combinations of trigonometric functions. We start with powers of sine and cosine.

EXAMPLE 1 Evaluate $\int \cos^3 x\, dx$.

SOLUTION Simply substituting $u = \cos x$ isn't helpful, since then $du = -\sin x\, dx$. In order to integrate powers of cosine, we would need an extra $\sin x$ factor. Similarly, a power of sine would require an extra $\cos x$ factor. Thus, here we can separate one cosine factor and convert the remaining $\cos^2 x$ factor to an expression involving sine using the identity $\sin^2 x + \cos^2 x = 1$:

$$\cos^3 x = \cos^2 x \cdot \cos x = (1 - \sin^2 x)\cos x$$

We can then evaluate the integral by substituting $u = \sin x$, so $du = \cos x\, dx$ and

$$\int \cos^3 x\, dx = \int \cos^2 x \cdot \cos x\, dx = \int (1 - \sin^2 x)\cos x\, dx$$

$$= \int (1 - u^2)\, du = u - \tfrac{1}{3}u^3 + C$$

$$= \sin x - \tfrac{1}{3}\sin^3 x + C$$

In general, we try to write an integrand involving powers of sine and cosine in a form where we have only one sine factor (and the remainder of the expression in terms of cosine) or only one cosine factor (and the remainder of the expression in terms of sine). The identity $\sin^2 x + \cos^2 x = 1$ enables us to convert back and forth between even powers of sine and cosine.

EXAMPLE 2 Find $\int \sin^5 x \cos^2 x\, dx$

SOLUTION We could convert $\cos^2 x$ to $1 - \sin^2 x$, but we would be left with an expression in terms of $\sin x$ with no extra $\cos x$ factor. Instead, we separate a single sine factor and rewrite the remaining $\sin^4 x$ factor in terms of $\cos x$:

$$\sin^5 x \cos^2 x = (\sin^2 x)^2 \cos^2 x \sin x = (1 - \cos^2 x)^2 \cos^2 x \sin x$$

|||| Figure 1 shows the graphs of the integrand $\sin^5 x \cos^2 x$ in Example 2 and its indefinite integral (with $C = 0$). Which is which?

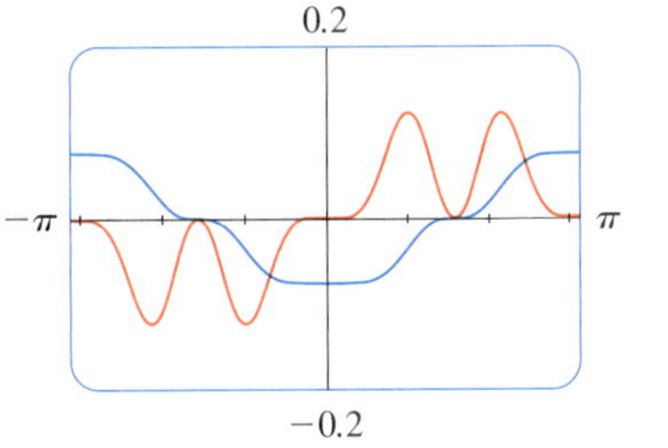

FIGURE 1

Substituting $u = \cos x$, we have $du = -\sin x\, dx$ and so

$$\begin{aligned}\int \sin^5 x \cos^2 x\, dx &= \int (\sin^2 x)^2 \cos^2 x \sin x\, dx \\ &= \int (1 - \cos^2 x)^2 \cos^2 x \sin x\, dx \\ &= \int (1 - u^2)^2 u^2 (-du) = -\int (u^2 - 2u^4 + u^6)\, du \\ &= -\left(\frac{u^3}{3} - 2\frac{u^5}{5} + \frac{u^7}{7}\right) + C \\ &= -\tfrac{1}{3}\cos^3 x + \tfrac{2}{5}\cos^5 x - \tfrac{1}{7}\cos^7 x + C\end{aligned}$$

In the preceding examples, an odd power of sine or cosine enabled us to separate a single factor and convert the remaining even power. If the integrand contains even powers of both sine and cosine, this strategy fails. In this case, we can take advantage of the following half-angle identities (see Equations 17b and 17a in Appendix D):

$$\sin^2 x = \tfrac{1}{2}(1 - \cos 2x) \qquad \text{and} \qquad \cos^2 x = \tfrac{1}{2}(1 + \cos 2x)$$

|||| Example 3 shows that the area of the region shown in Figure 2 is $\pi/2$.

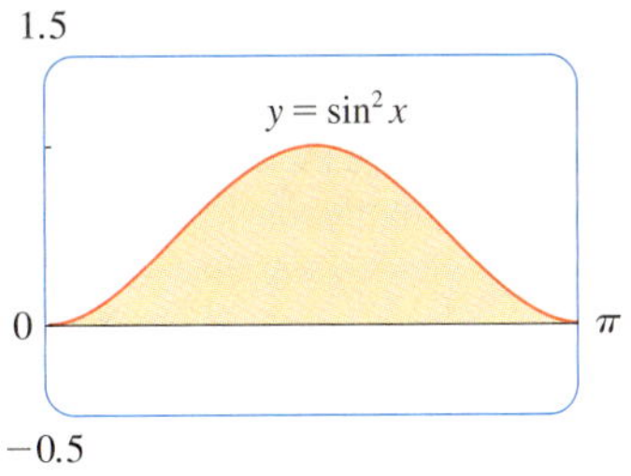

FIGURE 2

EXAMPLE 3 Evaluate $\int_0^\pi \sin^2 x\, dx$.

SOLUTION If we write $\sin^2 x = 1 - \cos^2 x$, the integral is no simpler to evaluate. Using the half-angle formula for $\sin^2 x$, however, we have

$$\begin{aligned}\int_0^\pi \sin^2 x\, dx &= \tfrac{1}{2}\int_0^\pi (1 - \cos 2x)\, dx = \left[\tfrac{1}{2}\left(x - \tfrac{1}{2}\sin 2x\right)\right]_0^\pi \\ &= \tfrac{1}{2}\left(\pi - \tfrac{1}{2}\sin 2\pi\right) - \tfrac{1}{2}\left(0 - \tfrac{1}{2}\sin 0\right) = \tfrac{1}{2}\pi\end{aligned}$$

Notice that we mentally made the substitution $u = 2x$ when integrating $\cos 2x$. Another method for evaluating this integral was given in Exercise 41 in Section 7.1.

EXAMPLE 4 Find $\int \sin^4 x\, dx$.

SOLUTION We could evaluate this integral using the reduction formula for $\int \sin^n x\, dx$ (Equation 7.1.7) together with Example 3 (as in Exercise 41 in Section 7.1), but a better method is to write $\sin^4 x = (\sin^2 x)^2$ and use a half-angle formula:

$$\int \sin^4 x\, dx = \int (\sin^2 x)^2\, dx$$
$$= \int \left(\frac{1 - \cos 2x}{2}\right)^2 dx$$
$$= \tfrac{1}{4}\int (1 - 2\cos 2x + \cos^2 2x)\, dx$$

Since $\cos^2 2x$ occurs, we must use another half-angle formula

$$\cos^2 2x = \tfrac{1}{2}(1 + \cos 4x)$$

This gives

$$\int \sin^4 x\, dx = \tfrac{1}{4}\int [1 - 2\cos 2x + \tfrac{1}{2}(1 + \cos 4x)]\, dx$$
$$= \tfrac{1}{4}\int \left(\tfrac{3}{2} - 2\cos 2x + \tfrac{1}{2}\cos 4x\right) dx$$
$$= \tfrac{1}{4}\left(\tfrac{3}{2}x - \sin 2x + \tfrac{1}{8}\sin 4x\right) + C$$

To summarize, we list guidelines to follow when evaluating integrals of the form $\int \sin^m x \cos^n x\, dx$, where $m \geqslant 0$ and $n \geqslant 0$ are integers.

Strategy for Evaluating $\int \sin^m x \cos^n x\, dx$

(a) If the power of cosine is odd ($n = 2k + 1$), save one cosine factor and use $\cos^2 x = 1 - \sin^2 x$ to express the remaining factors in terms of sine:

$$\int \sin^m x \cos^{2k+1} x\, dx = \int \sin^m x\, (\cos^2 x)^k \cos x\, dx$$
$$= \int \sin^m x\, (1 - \sin^2 x)^k \cos x\, dx$$

Then substitute $u = \sin x$.

(b) If the power of sine is odd ($m = 2k + 1$), save one sine factor and use $\sin^2 x = 1 - \cos^2 x$ to express the remaining factors in terms of cosine:

$$\int \sin^{2k+1} x \cos^n x\, dx = \int (\sin^2 x)^k \cos^n x\, \sin x\, dx$$
$$= \int (1 - \cos^2 x)^k \cos^n x\, \sin x\, dx$$

Then substitute $u = \cos x$. [Note that if the powers of both sine and cosine are odd, either (a) or (b) can be used.]

(c) If the powers of both sine and cosine are even, use the half-angle identities

$$\sin^2 x = \tfrac{1}{2}(1 - \cos 2x) \qquad \cos^2 x = \tfrac{1}{2}(1 + \cos 2x)$$

It is sometimes helpful to use the identity

$$\sin x \cos x = \tfrac{1}{2}\sin 2x$$

We can use a similar strategy to evaluate integrals of the form $\int \tan^m x \sec^n x\, dx$. Since $(d/dx) \tan x = \sec^2 x$, we can separate a $\sec^2 x$ factor and convert the remaining (even) power of secant to an expression involving tangent using the identity $\sec^2 x = 1 + \tan^2 x$. Or, since $(d/dx) \sec x = \sec x \tan x$, we can separate a $\sec x \tan x$ factor and convert the remaining (even) power of tangent to secant.

EXAMPLE 5 Evaluate $\int \tan^6 x \sec^4 x\, dx$.

SOLUTION If we separate one $\sec^2 x$ factor, we can express the remaining $\sec^2 x$ factor in terms of tangent using the identity $\sec^2 x = 1 + \tan^2 x$. We can then evaluate the integral by substituting $u = \tan x$ with $du = \sec^2 x\, dx$:

$$\begin{aligned} \int \tan^6 x \sec^4 x\, dx &= \int \tan^6 x \sec^2 x \sec^2 x\, dx \\ &= \int \tan^6 x\, (1 + \tan^2 x) \sec^2 x\, dx \\ &= \int u^6 (1 + u^2)\, du = \int (u^6 + u^8)\, du \\ &= \frac{u^7}{7} + \frac{u^9}{9} + C \\ &= \tfrac{1}{7} \tan^7 x + \tfrac{1}{9} \tan^9 x + C \end{aligned}$$

EXAMPLE 6 Find $\int \tan^5 \theta \sec^7 \theta\, d\theta$.

SOLUTION If we separate a $\sec^2 \theta$ factor, as in the preceding example, we are left with a $\sec^5 \theta$ factor, which isn't easily converted to tangent. However, if we separate a $\sec \theta \tan \theta$ factor, we can convert the remaining power of tangent to an expression involving only secant using the identity $\tan^2 \theta = \sec^2 \theta - 1$. We can then evaluate the integral by substituting $u = \sec \theta$, so $du = \sec \theta \tan \theta\, d\theta$:

$$\begin{aligned} \int \tan^5 \theta \sec^7 \theta\, d\theta &= \int \tan^4 \theta \sec^6 \theta \sec \theta \tan \theta\, d\theta \\ &= \int (\sec^2 \theta - 1)^2 \sec^6 \theta \sec \theta \tan \theta\, d\theta \\ &= \int (u^2 - 1)^2 u^6\, du = \int (u^{10} - 2u^8 + u^6)\, du \\ &= \frac{u^{11}}{11} - 2\frac{u^9}{9} + \frac{u^7}{7} + C \\ &= \tfrac{1}{11} \sec^{11} \theta - \tfrac{2}{9} \sec^9 \theta + \tfrac{1}{7} \sec^7 \theta + C \end{aligned}$$

The preceding examples demonstrate strategies for evaluating integrals of the form $\int \tan^m x \sec^n x\, dx$ for two cases, which we summarize here.

Strategy for Evaluating $\int \tan^m x \sec^n x \, dx$

(a) If the power of secant is even ($n = 2k$, $k \geqslant 2$), save a factor of $\sec^2 x$ and use $\sec^2 x = 1 + \tan^2 x$ to express the remaining factors in terms of $\tan x$:

$$\int \tan^m x \sec^{2k} x \, dx = \int \tan^m x \, (\sec^2 x)^{k-1} \sec^2 x \, dx$$

$$= \int \tan^m x \, (1 + \tan^2 x)^{k-1} \sec^2 x \, dx$$

Then substitute $u = \tan x$.

(b) If the power of tangent is odd ($m = 2k + 1$), save a factor of $\sec x \tan x$ and use $\tan^2 x = \sec^2 x - 1$ to express the remaining factors in terms of $\sec x$:

$$\int \tan^{2k+1} x \sec^n x \, dx = \int (\tan^2 x)^k \sec^{n-1} x \, \sec x \, \tan x \, dx$$

$$= \int (\sec^2 x - 1)^k \sec^{n-1} x \, \sec x \, \tan x \, dx$$

Then substitute $u = \sec x$.

For other cases, the guidelines are not as clear-cut. We may need to use identities, integration by parts, and occasionally a little ingenuity. We will sometimes need to be able to integrate $\tan x$ by using the formula established in (5.5.5):

$$\int \tan x \, dx = \ln |\sec x| + C$$

We will also need the indefinite integral of secant:

1 $$\int \sec x \, dx = \ln |\sec x + \tan x| + C$$

We could verify Formula 1 by differentiating the right side, or as follows. First we multiply numerator and denominator by $\sec x + \tan x$:

$$\int \sec x \, dx = \int \sec x \, \frac{\sec x + \tan x}{\sec x + \tan x} \, dx$$

$$= \int \frac{\sec^2 x + \sec x \tan x}{\sec x + \tan x} \, dx$$

If we substitute $u = \sec x + \tan x$, then $du = (\sec x \tan x + \sec^2 x) \, dx$, so the integral becomes $\int (1/u) \, du = \ln |u| + C$. Thus, we have

$$\int \sec x \, dx = \ln |\sec x + \tan x| + C$$

EXAMPLE 7 Find $\int \tan^3 x\,dx$.

SOLUTION Here only $\tan x$ occurs, so we use $\tan^2 x = \sec^2 x - 1$ to rewrite a $\tan^2 x$ factor in terms of $\sec^2 x$:

$$\begin{aligned}\int \tan^3 x\,dx &= \int \tan x \tan^2 x\,dx \\ &= \int \tan x\,(\sec^2 x - 1)\,dx \\ &= \int \tan x \sec^2 x\,dx - \int \tan x\,dx \\ &= \frac{\tan^2 x}{2} - \ln|\sec x| + C\end{aligned}$$

In the first integral we mentally substituted $u = \tan x$ so that $du = \sec^2 x\,dx$.

If an even power of tangent appears with an odd power of secant, it is helpful to express the integrand completely in terms of $\sec x$. Powers of $\sec x$ may require integration by parts, as shown in the following example.

EXAMPLE 8 Find $\int \sec^3 x\,dx$.

SOLUTION Here we integrate by parts with

$$u = \sec x \qquad dv = \sec^2 x\,dx$$
$$du = \sec x \tan x\,dx \qquad v = \tan x$$

Then
$$\begin{aligned}\int \sec^3 x\,dx &= \sec x \tan x - \int \sec x \tan^2 x\,dx \\ &= \sec x \tan x - \int \sec x\,(\sec^2 x - 1)\,dx \\ &= \sec x \tan x - \int \sec^3 x\,dx + \int \sec x\,dx\end{aligned}$$

Using Formula 1 and solving for the required integral, we get

$$\int \sec^3 x\,dx = \tfrac{1}{2}\left(\sec x \tan x + \ln|\sec x + \tan x|\right) + C$$

Integrals such as the one in the preceding example may seem very special but they occur frequently in applications of integration, as we will see in Chapter 8. Integrals of the form $\int \cot^m x \csc^n x\,dx$ can be found by similar methods because of the identity $1 + \cot^2 x = \csc^2 x$.

Finally, we can make use of another set of trigonometric identities:

2 To evaluate the integrals (a) $\int \sin mx \cos nx\,dx$, (b) $\int \sin mx \sin nx\,dx$, or (c) $\int \cos mx \cos nx\,dx$, use the corresponding identity:

(a) $\sin A \cos B = \tfrac{1}{2}[\sin(A - B) + \sin(A + B)]$

(b) $\sin A \sin B = \tfrac{1}{2}[\cos(A - B) - \cos(A + B)]$

(c) $\cos A \cos B = \tfrac{1}{2}[\cos(A - B) + \cos(A + B)]$

These product identities are discussed in Appendix D.

EXAMPLE 9 Evaluate $\int \sin 4x \cos 5x\,dx$.

SOLUTION This integral could be evaluated using integration by parts, but it's easier to use the identity in Equation 2(a) as follows:

$$\int \sin 4x \cos 5x\,dx = \int \tfrac{1}{2}[\sin(-x) + \sin 9x]\,dx$$

$$= \tfrac{1}{2}\int (-\sin x + \sin 9x)\,dx$$

$$= \tfrac{1}{2}(\cos x - \tfrac{1}{9}\cos 9x) + C$$

7.2 Exercises

1–47 Evaluate the integral.

1. $\int \sin^3 x \cos^2 x\,dx$

2. $\int \sin^6 x \cos^3 x\,dx$

3. $\int_{\pi/2}^{3\pi/4} \sin^5 x \cos^3 x\,dx$

4. $\int_0^{\pi/2} \cos^5 x\,dx$

5. $\int \cos^5 x \sin^4 x\,dx$

6. $\int \sin^3(mx)\,dx$

7. $\int_0^{\pi/2} \cos^2\theta\,d\theta$

8. $\int_0^{\pi/2} \sin^2(2\theta)\,d\theta$

9. $\int_0^{\pi} \sin^4(3t)\,dt$

10. $\int_0^{\pi} \cos^6\theta\,d\theta$

11. $\int (1 + \cos\theta)^2\,d\theta$

12. $\int x\cos^2 x\,dx$

13. $\int_0^{\pi/4} \sin^4 x \cos^2 x\,dx$

14. $\int_0^{\pi/2} \sin^2 x \cos^2 x\,dx$

15. $\int \sin^3 x\sqrt{\cos x}\,dx$

16. $\int \cos\theta \cos^5(\sin\theta)\,d\theta$

17. $\int \cos^2 x \tan^3 x\,dx$

18. $\int \cot^5\theta \sin^4\theta\,d\theta$

19. $\int \frac{1 - \sin x}{\cos x}\,dx$

20. $\int \cos^2 x \sin 2x\,dx$

21. $\int \sec^2 x \tan x\,dx$

22. $\int_0^{\pi/2} \sec^4(t/2)\,dt$

23. $\int \tan^2 x\,dx$

24. $\int \tan^4 x\,dx$

25. $\int \sec^6 t\,dt$

26. $\int_0^{\pi/4} \sec^4\theta \tan^4\theta\,d\theta$

27. $\int_0^{\pi/3} \tan^5 x \sec^4 x\,dx$

28. $\int \tan^3(2x)\sec^5(2x)\,dx$

29. $\int \tan^3 x \sec x\,dx$

30. $\int_0^{\pi/3} \tan^5 x \sec^6 x\,dx$

31. $\int \tan^5 x\,dx$

32. $\int \tan^6(ay)\,dy$

33. $\int \frac{\tan^3\theta}{\cos^4\theta}\,d\theta$

34. $\int \tan^2 x \sec x\,dx$

35. $\int_{\pi/6}^{\pi/2} \cot^2 x\,dx$

36. $\int_{\pi/4}^{\pi/2} \cot^3 x\,dx$

37. $\int \cot^3\alpha \csc^3\alpha\,d\alpha$

38. $\int \csc^4 x \cot^6 x\,dx$

39. $\int \csc x\,dx$

40. $\int_{\pi/6}^{\pi/3} \csc^3 x\,dx$

41. $\int \sin 5x \sin 2x\,dx$

42. $\int \sin 3x \cos x\,dx$

43. $\int \cos 7\theta \cos 5\theta\,d\theta$

44. $\int \frac{\cos x + \sin x}{\sin 2x}\,dx$

45. $\int \frac{1 - \tan^2 x}{\sec^2 x}\,dx$

46. $\int \frac{dx}{\cos x - 1}$

47. $\int t\sec^2(t^2)\tan^4(t^2)\,dt$

48. If $\int_0^{\pi/4} \tan^6 x \sec x\,dx = I$, express the value of $\int_0^{\pi/4} \tan^8 x \sec x\,dx$ in terms of I.

49–52 Evaluate the indefinite integral. Illustrate, and check that your answer is reasonable, by graphing both the integrand and its antiderivative (taking $C = 0$).

49. $\int \sin^5 x\,dx$

50. $\int \sin^4 x \cos^4 x\,dx$

51. $\int \sin 3x \sin 6x\,dx$

52. $\int \sec^4 \frac{x}{2}\,dx$

53. Find the average value of the function $f(x) = \sin^2 x \cos^3 x$ on the interval $[-\pi, \pi]$.

54. Evaluate $\int \sin x \cos x\,dx$ by four methods: (a) the substitution $u = \cos x$, (b) the substitution $u = \sin x$, (c) the identity $\sin 2x = 2\sin x \cos x$, and (d) integration by parts. Explain the different appearances of the answers.

55–56 Find the area of the region bounded by the given curves.

55. $y = \sin x,\quad y = \sin^3 x,\quad x = 0,\quad x = \pi/2$

56. $y = \sin x,\quad y = 2\sin^2 x,\quad x = 0,\quad x = \pi/2$

57–58 Use a graph of the integrand to guess the value of the integral. Then use the methods of this section to prove that your guess is correct.

57. $\int_0^{2\pi} \cos^3 x\,dx$

58. $\int_0^2 \sin 2\pi x \cos 5\pi x\,dx$

59–62 Find the volume obtained by rotating the region bounded by the given curves about the specified axis.

59. $y = \sin x,\ x = \pi/2,\ x = \pi,\ y = 0$; about the x-axis

60. $y = \tan^2 x,\ y = 0,\ x = 0,\ x = \pi/4$; about the x-axis

61. $y = \cos x,\ y = 0,\ x = 0,\ x = \pi/2$; about $y = -1$

62. $y = \cos x,\ y = 0,\ x = 0,\ x = \pi/2$; about $y = 1$

63. A particle moves on a straight line with velocity function $v(t) = \sin \omega t \cos^2 \omega t$. Find its position function $s = f(t)$ if $f(0) = 0$.

64. Household electricity is supplied in the form of alternating current that varies from 155 V to -155 V with a frequency of 60 cycles per second (Hz). The voltage is thus given by the equation

$$E(t) = 155 \sin(120\pi t)$$

where t is the time in seconds. Voltmeters read the RMS (root-mean-square) voltage, which is the square root of the average value of $[E(t)]^2$ over one cycle.

(a) Calculate the RMS voltage of household current.

(b) Many electric stoves require an RMS voltage of 220 V. Find the corresponding amplitude A needed for the voltage $E(t) = A \sin(120\pi t)$.

65–67 Prove the formula, where m and n are positive integers.

65. $\int_{-\pi}^{\pi} \sin mx \cos nx\,dx = 0$

66. $\int_{-\pi}^{\pi} \sin mx \sin nx\,dx = \begin{cases} 0 & \text{if } m \neq n \\ \pi & \text{if } m = n \end{cases}$

67. $\int_{-\pi}^{\pi} \cos mx \cos nx\,dx = \begin{cases} 0 & \text{if } m \neq n \\ \pi & \text{if } m = n \end{cases}$

68. A *finite Fourier series* is given by the sum

$$f(x) = \sum_{n=1}^{N} a_n \sin nx = a_1 \sin x + a_2 \sin 2x + \cdots + a_N \sin Nx$$

Show that the mth coefficient a_m is given by the formula

$$a_m = \frac{1}{\pi} \int_{-\pi}^{\pi} f(x) \sin mx\,dx$$

7.3 Trigonometric Substitution

In finding the area of a circle or an ellipse, an integral of the form $\int \sqrt{a^2 - x^2}\,dx$ arises, where $a > 0$. If it were $\int x\sqrt{a^2 - x^2}\,dx$, the substitution $u = a^2 - x^2$ would be effective but, as it stands, $\int \sqrt{a^2 - x^2}\,dx$ is more difficult. If we change the variable from x to θ by the substitution $x = a \sin \theta$, then the identity $1 - \sin^2\theta = \cos^2\theta$ allows us to get rid of the root sign because

$$\sqrt{a^2 - x^2} = \sqrt{a^2 - a^2\sin^2\theta} = \sqrt{a^2(1 - \sin^2\theta)} = \sqrt{a^2\cos^2\theta} = a|\cos\theta|$$

Notice the difference between the substitution $u = a^2 - x^2$ (in which the new variable is a function of the old one) and the substitution $x = a \sin \theta$ (the old variable is a function of the new one).

In general we can make a substitution of the form $x = g(t)$ by using the Substitution Rule in reverse. To make our calculations simpler, we assume that g has an inverse function; that is, g is one-to-one. In this case, if we replace u by x and x by t in the Substitution Rule (Equation 5.5.4), we obtain

$$\int f(x)\,dx = \int f(g(t))g'(t)\,dt$$

This kind of substitution is called *inverse substitution*.

We can make the inverse substitution $x = a \sin \theta$ provided that it defines a one-to-one function. This can be accomplished by restricting θ to lie in the interval $[-\pi/2, \pi/2]$.

In the following table we list trigonometric substitutions that are effective for the given radical expressions because of the specified trigonometric identities. In each case the restriction on θ is imposed to ensure that the function that defines the substitution is one-to-one. (These are the same intervals used in Appendix D in defining the inverse functions.)

Table of Trigonometric Substitutions

Expression	Substitution	Identity
$\sqrt{a^2 - x^2}$	$x = a \sin \theta, \quad -\frac{\pi}{2} \leqslant \theta \leqslant \frac{\pi}{2}$	$1 - \sin^2\theta = \cos^2\theta$
$\sqrt{a^2 + x^2}$	$x = a \tan \theta, \quad -\frac{\pi}{2} < \theta < \frac{\pi}{2}$	$1 + \tan^2\theta = \sec^2\theta$
$\sqrt{x^2 - a^2}$	$x = a \sec \theta, \quad 0 \leqslant \theta < \frac{\pi}{2}$ or $\pi \leqslant \theta < \frac{3\pi}{2}$	$\sec^2\theta - 1 = \tan^2\theta$

EXAMPLE 1 Evaluate $\displaystyle\int \frac{\sqrt{9 - x^2}}{x^2}\, dx$.

SOLUTION Let $x = 3 \sin \theta$, where $-\pi/2 \leqslant \theta \leqslant \pi/2$. Then $dx = 3 \cos \theta \, d\theta$ and

$$\sqrt{9 - x^2} = \sqrt{9 - 9 \sin^2\theta} = \sqrt{9 \cos^2\theta} = 3\,|\cos \theta| = 3 \cos \theta$$

(Note that $\cos \theta \geqslant 0$ because $-\pi/2 \leqslant \theta \leqslant \pi/2$.) Thus, the Inverse Substitution Rule gives

$$\begin{aligned} \int \frac{\sqrt{9 - x^2}}{x^2}\, dx &= \int \frac{3 \cos \theta}{9 \sin^2\theta}\, 3 \cos \theta \, d\theta \\ &= \int \frac{\cos^2\theta}{\sin^2\theta}\, d\theta = \int \cot^2\theta \, d\theta \\ &= \int (\csc^2\theta - 1)\, d\theta \\ &= -\cot \theta - \theta + C \end{aligned}$$

Since this is an indefinite integral, we must return to the original variable x. This can be done either by using trigonometric identities to express $\cot \theta$ in terms of $\sin \theta = x/3$ or by drawing a diagram, as in Figure 1, where θ is interpreted as an angle of a right triangle. Since $\sin \theta = x/3$, we label the opposite side and the hypotenuse as having lengths x and 3. Then the Pythagorean Theorem gives the length of the adjacent side as $\sqrt{9 - x^2}$, so we can simply read the value of $\cot \theta$ from the figure:

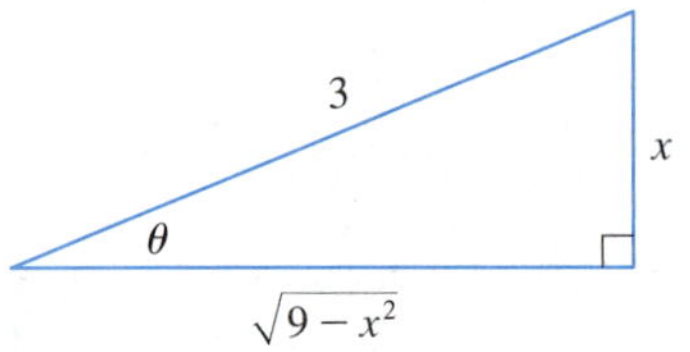

FIGURE 1

$\sin \theta = \dfrac{x}{3}$

$$\cot \theta = \frac{\sqrt{9 - x^2}}{x}$$

(Although $\theta > 0$ in the diagram, this expression for $\cot \theta$ is valid even when $\theta < 0$.) Since $\sin \theta = x/3$, we have $\theta = \sin^{-1}(x/3)$ and so

$$\int \frac{\sqrt{9 - x^2}}{x^2}\, dx = -\frac{\sqrt{9 - x^2}}{x} - \sin^{-1}\left(\frac{x}{3}\right) + C$$

EXAMPLE 2 Find the area enclosed by the ellipse

$$\frac{x^2}{a^2} + \frac{y^2}{b^2} = 1$$

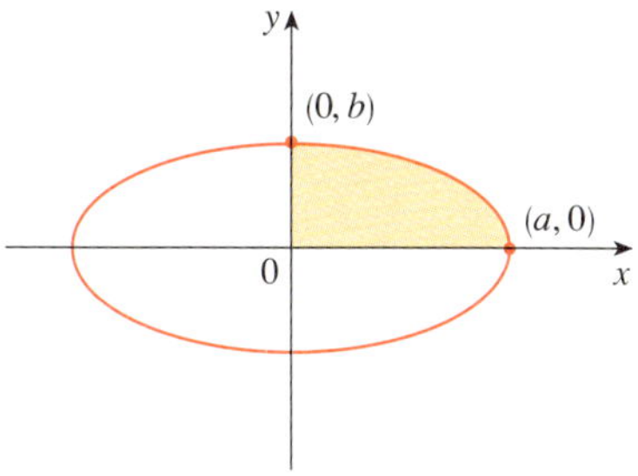

FIGURE 2

$\frac{x^2}{a^2} + \frac{y^2}{b^2} = 1$

SOLUTION Solving the equation of the ellipse for y, we get

$$\frac{y^2}{b^2} = 1 - \frac{x^2}{a^2} = \frac{a^2 - x^2}{a^2} \qquad \text{or} \qquad y = \pm\frac{b}{a}\sqrt{a^2 - x^2}$$

Because the ellipse is symmetric with respect to both axes, the total area A is four times the area in the first quadrant (see Figure 2). The part of the ellipse in the first quadrant is given by the function

$$y = \frac{b}{a}\sqrt{a^2 - x^2} \qquad 0 \leqslant x \leqslant a$$

and so

$$\tfrac{1}{4}A = \int_0^a \frac{b}{a}\sqrt{a^2 - x^2}\,dx$$

To evaluate this integral we substitute $x = a\sin\theta$. Then $dx = a\cos\theta\,d\theta$. To change the limits of integration we note that when $x = 0$, $\sin\theta = 0$, so $\theta = 0$; when $x = a$, $\sin\theta = 1$, so $\theta = \pi/2$. Also

$$\sqrt{a^2 - x^2} = \sqrt{a^2 - a^2\sin^2\theta} = \sqrt{a^2\cos^2\theta} = a|\cos\theta| = a\cos\theta$$

since $0 \leqslant \theta \leqslant \pi/2$. Therefore

$$\begin{aligned} A &= 4\frac{b}{a}\int_0^a \sqrt{a^2 - x^2}\,dx = 4\frac{b}{a}\int_0^{\pi/2} a\cos\theta \cdot a\cos\theta\,d\theta \\ &= 4ab\int_0^{\pi/2}\cos^2\theta\,d\theta = 4ab\int_0^{\pi/2}\tfrac{1}{2}(1 + \cos 2\theta)\,d\theta \\ &= 2ab\left[\theta + \tfrac{1}{2}\sin 2\theta\right]_0^{\pi/2} = 2ab\left(\frac{\pi}{2} + 0 - 0\right) \\ &= \pi ab \end{aligned}$$

We have shown that the area of an ellipse with semiaxes a and b is πab. In particular, taking $a = b = r$, we have proved the famous formula that the area of a circle with radius r is πr^2.

NOTE ◦ Since the integral in Example 2 was a definite integral, we changed the limits of integration and did not have to convert back to the original variable x.

EXAMPLE 3 Find $\displaystyle\int \frac{1}{x^2\sqrt{x^2 + 4}}\,dx$.

SOLUTION Let $x = 2\tan\theta$, $-\pi/2 < \theta < \pi/2$. Then $dx = 2\sec^2\theta\,d\theta$ and

$$\sqrt{x^2 + 4} = \sqrt{4(\tan^2\theta + 1)} = \sqrt{4\sec^2\theta} = 2|\sec\theta| = 2\sec\theta$$

Thus, we have

$$\int \frac{dx}{x^2\sqrt{x^2 + 4}} = \int \frac{2\sec^2\theta\,d\theta}{4\tan^2\theta \cdot 2\sec\theta} = \frac{1}{4}\int \frac{\sec\theta}{\tan^2\theta}\,d\theta$$

To evaluate this trigonometric integral we put everything in terms of $\sin\theta$ and $\cos\theta$:

$$\frac{\sec\theta}{\tan^2\theta} = \frac{1}{\cos\theta} \cdot \frac{\cos^2\theta}{\sin^2\theta} = \frac{\cos\theta}{\sin^2\theta}$$

Therefore, making the substitution $u = \sin\theta$, we have

$$\begin{aligned}\int \frac{dx}{x^2\sqrt{x^2+4}} &= \frac{1}{4}\int \frac{\cos\theta}{\sin^2\theta}\,d\theta = \frac{1}{4}\int \frac{du}{u^2} \\ &= \frac{1}{4}\left(-\frac{1}{u}\right) + C = -\frac{1}{4\sin\theta} + C \\ &= -\frac{\csc\theta}{4} + C\end{aligned}$$

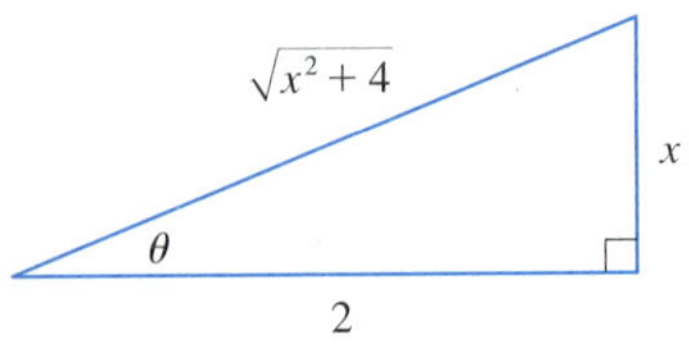

FIGURE 3

$\tan\theta = \frac{x}{2}$

We use Figure 3 to determine that $\csc\theta = \sqrt{x^2+4}/x$ and so

$$\int \frac{dx}{x^2\sqrt{x^2+4}} = -\frac{\sqrt{x^2+4}}{4x} + C$$

EXAMPLE 4 Find $\displaystyle\int \frac{x}{\sqrt{x^2+4}}\,dx$.

SOLUTION It would be possible to use the trigonometric substitution $x = 2\tan\theta$ here (as in Example 3). But the direct substitution $u = x^2 + 4$ is simpler, because then $du = 2x\,dx$ and

$$\int \frac{x}{\sqrt{x^2+4}}\,dx = \frac{1}{2}\int \frac{du}{\sqrt{u}} = \sqrt{u} + C = \sqrt{x^2+4} + C$$

NOTE ▫ Example 4 illustrates the fact that even when trigonometric substitutions are possible, they may not give the easiest solution. You should look for a simpler method first.

EXAMPLE 5 Evaluate $\displaystyle\int \frac{dx}{\sqrt{x^2-a^2}}$, where $a > 0$.

SOLUTION 1 We let $x = a\sec\theta$, where $0 < \theta < \pi/2$ or $\pi < \theta < 3\pi/2$. Then $dx = a\sec\theta\tan\theta\,d\theta$ and

$$\sqrt{x^2-a^2} = \sqrt{a^2(\sec^2\theta - 1)} = \sqrt{a^2\tan^2\theta} = a|\tan\theta| = a\tan\theta$$

Therefore

$$\begin{aligned}\int \frac{dx}{\sqrt{x^2-a^2}} &= \int \frac{a\sec\theta\tan\theta}{a\tan\theta}\,d\theta \\ &= \int \sec\theta\,d\theta = \ln|\sec\theta + \tan\theta| + C\end{aligned}$$

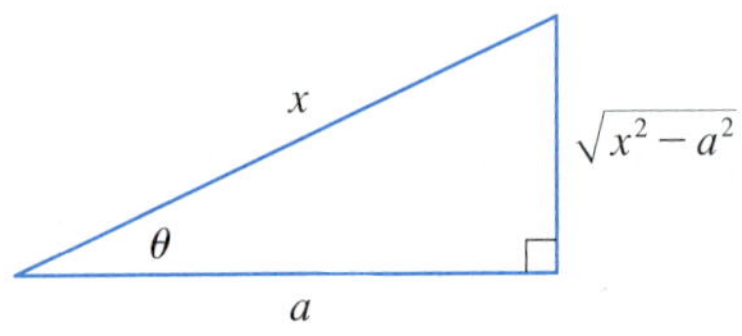

FIGURE 4

$\sec\theta = \frac{x}{a}$

The triangle in Figure 4 gives $\tan\theta = \sqrt{x^2-a^2}/a$, so we have

$$\begin{aligned}\int \frac{dx}{\sqrt{x^2-a^2}} &= \ln\left|\frac{x}{a} + \frac{\sqrt{x^2-a^2}}{a}\right| + C \\ &= \ln\left|x + \sqrt{x^2-a^2}\right| - \ln a + C\end{aligned}$$

Writing $C_1 = C - \ln a$, we have

1 $$\int \frac{dx}{\sqrt{x^2 - a^2}} = \ln \left| x + \sqrt{x^2 - a^2} \right| + C_1$$

SOLUTION 2 For $x > 0$ the hyperbolic substitution $x = a \cosh t$ can also be used. Using the identity $\cosh^2 y - \sinh^2 y = 1$, we have

$$\sqrt{x^2 - a^2} = \sqrt{a^2(\cosh^2 t - 1)} = \sqrt{a^2 \sinh^2 t} = a \sinh t$$

Since $dx = a \sinh t \, dt$, we obtain

$$\int \frac{dx}{\sqrt{x^2 - a^2}} = \int \frac{a \sinh t \, dt}{a \sinh t} = \int dt = t + C$$

Since $\cosh t = x/a$, we have $t = \cosh^{-1}(x/a)$ and

2 $$\int \frac{dx}{\sqrt{x^2 - a^2}} = \cosh^{-1}\left(\frac{x}{a}\right) + C$$

Although Formulas 1 and 2 look quite different, they are actually equivalent by Formula 3.9.4.

NOTE ▫ As Example 5 illustrates, hyperbolic substitutions can be used in place of trigonometric substitutions and sometimes they lead to simpler answers. But we usually use trigonometric substitutions because trigonometric identities are more familiar than hyperbolic identities.

EXAMPLE 6 Find $\displaystyle\int_0^{3\sqrt{3}/2} \frac{x^3}{(4x^2 + 9)^{3/2}} \, dx$.

SOLUTION First we note that $(4x^2 + 9)^{3/2} = (\sqrt{4x^2 + 9})^3$ so trigonometric substitution is appropriate. Although $\sqrt{4x^2 + 9}$ is not quite one of the expressions in the table of trigonometric substitutions, it becomes one of them if we make the preliminary substitution $u = 2x$. When we combine this with the tangent substitution, we have $x = \frac{3}{2} \tan \theta$, which gives $dx = \frac{3}{2} \sec^2\theta \, d\theta$ and

$$\sqrt{4x^2 + 9} = \sqrt{9 \tan^2\theta + 9} = 3 \sec \theta$$

When $x = 0$, $\tan \theta = 0$, so $\theta = 0$; when $x = 3\sqrt{3}/2$, $\tan \theta = \sqrt{3}$, so $\theta = \pi/3$.

$$\begin{aligned}
\int_0^{3\sqrt{3}/2} \frac{x^3}{(4x^2 + 9)^{3/2}} \, dx &= \int_0^{\pi/3} \frac{\frac{27}{8} \tan^3\theta}{27 \sec^3\theta} \, \tfrac{3}{2} \sec^2\theta \, d\theta \\
&= \tfrac{3}{16} \int_0^{\pi/3} \frac{\tan^3\theta}{\sec \theta} \, d\theta = \tfrac{3}{16} \int_0^{\pi/3} \frac{\sin^3\theta}{\cos^2\theta} \, d\theta \\
&= \tfrac{3}{16} \int_0^{\pi/3} \frac{1 - \cos^2\theta}{\cos^2\theta} \sin \theta \, d\theta
\end{aligned}$$

Now we substitute $u = \cos \theta$ so that $du = -\sin \theta \, d\theta$. When $\theta = 0$, $u = 1$; when $\theta = \pi/3$, $u = \frac{1}{2}$.

Therefore

$$\int_0^{3\sqrt{3}/2} \frac{x^3}{(4x^2+9)^{3/2}}\,dx = -\tfrac{3}{16}\int_1^{1/2} \frac{1-u^2}{u^2}\,du = \tfrac{3}{16}\int_1^{1/2} (1-u^{-2})\,du$$

$$= \tfrac{3}{16}\left[u + \frac{1}{u}\right]_1^{1/2} = \tfrac{3}{16}\left[\left(\tfrac{1}{2}+2\right) - (1+1)\right] = \tfrac{3}{32}$$

EXAMPLE 7 Evaluate $\displaystyle\int \frac{x}{\sqrt{3-2x-x^2}}\,dx$.

SOLUTION We can transform the integrand into a function for which trigonometric substitution is appropriate by first completing the square under the root sign:

$$3 - 2x - x^2 = 3 - (x^2+2x) = 3 + 1 - (x^2+2x+1)$$
$$= 4 - (x+1)^2$$

This suggests that we make the substitution $u = x + 1$. Then $du = dx$ and $x = u - 1$, so

$$\int \frac{x}{\sqrt{3-2x-x^2}}\,dx = \int \frac{u-1}{\sqrt{4-u^2}}\,du$$

We now substitute $u = 2\sin\theta$, giving $du = 2\cos\theta\,d\theta$ and $\sqrt{4-u^2} = 2\cos\theta$, so

|||| Figure 5 shows the graphs of the integrand in Example 7 and its indefinite integral (with $C = 0$). Which is which?

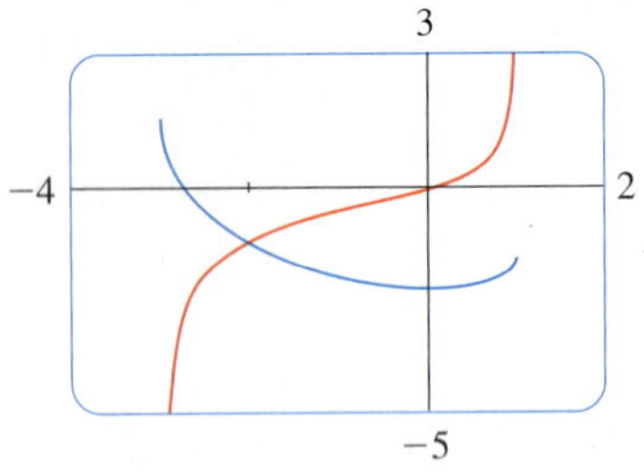

FIGURE 5

$$\int \frac{x}{\sqrt{3-2x-x^2}}\,dx = \int \frac{2\sin\theta - 1}{2\cos\theta}\,2\cos\theta\,d\theta$$

$$= \int (2\sin\theta - 1)\,d\theta$$

$$= -2\cos\theta - \theta + C$$

$$= -\sqrt{4-u^2} - \sin^{-1}\left(\frac{u}{2}\right) + C$$

$$= -\sqrt{3-2x-x^2} - \sin^{-1}\left(\frac{x+1}{2}\right) + C$$

7.3 Exercises

1–3 |||| Evaluate the integral using the indicated trigonometric substitution. Sketch and label the associated right triangle.

1. $\displaystyle\int \frac{1}{x^2\sqrt{x^2-9}}\,dx;\quad x = 3\sec\theta$

2. $\displaystyle\int x^3\sqrt{9-x^2}\,dx;\quad x = 3\sin\theta$

3. $\displaystyle\int \frac{x^3}{\sqrt{x^2+9}}\,dx;\quad x = 3\tan\theta$

4–30 |||| Evaluate the integral.

4. $\displaystyle\int_0^{2\sqrt{3}} \frac{x^3}{\sqrt{16-x^2}}\,dx$

5. $\displaystyle\int_{\sqrt{2}}^{2} \frac{1}{t^3\sqrt{t^2-1}}\,dt$

6. $\displaystyle\int_0^2 x^3\sqrt{x^2+4}\,dx$

7. $\displaystyle\int \frac{1}{x^2\sqrt{25-x^2}}\,dx$

8. $\displaystyle\int \frac{\sqrt{x^2-a^2}}{x^4}\,dx$

9. $\displaystyle\int \frac{dx}{\sqrt{x^2+16}}$

10. $\displaystyle\int \frac{t^5}{\sqrt{t^2+2}}\,dt$

11. $\int \sqrt{1 - 4x^2}\, dx$

12. $\int_0^1 x\sqrt{x^2 + 4}\, dx$

13. $\int \frac{\sqrt{x^2 - 9}}{x^3}\, dx$

14. $\int \frac{du}{u\sqrt{5 - u^2}}$

15. $\int \frac{x^2}{(a^2 - x^2)^{3/2}}\, dx$

16. $\int \frac{dx}{x^2\sqrt{16x^2 - 9}}$

17. $\int \frac{x}{\sqrt{x^2 - 7}}\, dx$

18. $\int \frac{dx}{[(ax)^2 - b^2]^{3/2}}$

19. $\int \frac{\sqrt{1 + x^2}}{x}\, dx$

20. $\int \frac{t}{\sqrt{25 - t^2}}\, dt$

21. $\int_0^{2/3} x^3\sqrt{4 - 9x^2}\, dx$

22. $\int_0^1 \sqrt{x^2 + 1}\, dx$

23. $\int \sqrt{5 + 4x - x^2}\, dx$

24. $\int \frac{dt}{\sqrt{t^2 - 6t + 13}}$

25. $\int \frac{1}{\sqrt{9x^2 + 6x - 8}}\, dx$

26. $\int \frac{x^2}{\sqrt{4x - x^2}}\, dx$

27. $\int \frac{dx}{(x^2 + 2x + 2)^2}$

28. $\int \frac{dx}{(5 - 4x - x^2)^{5/2}}$

29. $\int x\sqrt{1 - x^4}\, dx$

30. $\int_0^{\pi/2} \frac{\cos t}{\sqrt{1 + \sin^2 t}}\, dt$

31. (a) Use trigonometric substitution to show that

$$\int \frac{dx}{\sqrt{x^2 + a^2}} = \ln\left(x + \sqrt{x^2 + a^2}\right) + C$$

(b) Use the hyperbolic substitution $x = a \sinh t$ to show that

$$\int \frac{dx}{\sqrt{x^2 + a^2}} = \sinh^{-1}\left(\frac{x}{a}\right) + C$$

These formulas are connected by Formula 3.9.3.

32. Evaluate

$$\int \frac{x^2}{(x^2 + a^2)^{3/2}}\, dx$$

(a) by trigonometric substitution.
(b) by the hyperbolic substitution $x = a \sinh t$.

33. Find the average value of $f(x) = \sqrt{x^2 - 1}/x$, $1 \leq x \leq 7$.

34. Find the area of the region bounded by the hyperbola $9x^2 - 4y^2 = 36$ and the line $x = 3$.

35. Prove the formula $A = \frac{1}{2}r^2\theta$ for the area of a sector of a circle with radius r and central angle θ. [*Hint:* Assume $0 < \theta < \pi/2$ and place the center of the circle at the origin so it has the equation $x^2 + y^2 = r^2$. Then A is the sum of the area of the triangle POQ and the area of the region PQR in the figure.]

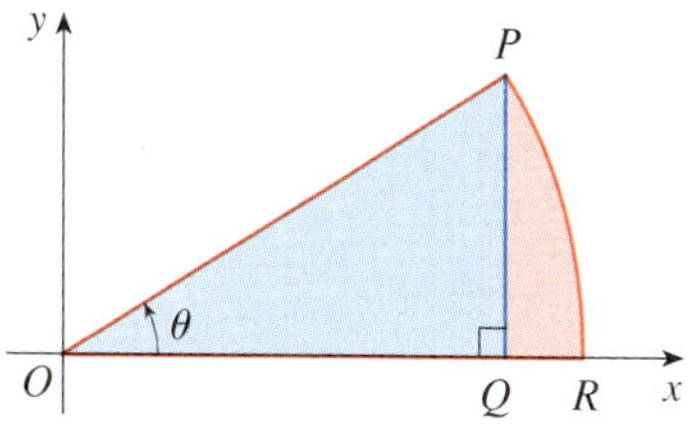

36. Evaluate the integral

$$\int \frac{dx}{x^4\sqrt{x^2 - 2}}$$

Graph the integrand and its indefinite integral on the same screen and check that your answer is reasonable.

37. Use a graph to approximate the roots of the equation $x^2\sqrt{4 - x^2} = 2 - x$. Then approximate the area bounded by the curve $y = x^2\sqrt{4 - x^2}$ and the line $y = 2 - x$.

38. A charged rod of length L produces an electric field at point $P(a, b)$ given by

$$E(P) = \int_{-a}^{L-a} \frac{\lambda b}{4\pi\varepsilon_0(x^2 + b^2)^{3/2}}\, dx$$

where λ is the charge density per unit length on the rod and ε_0 is the free space permittivity (see the figure). Evaluate the integral to determine an expression for the electric field $E(P)$.

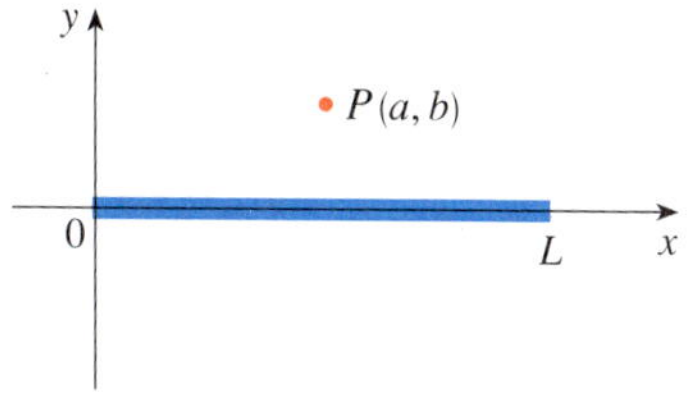

39. Find the area of the crescent-shaped region (called a *lune*) bounded by arcs of circles with radii r and R. (See the figure.)

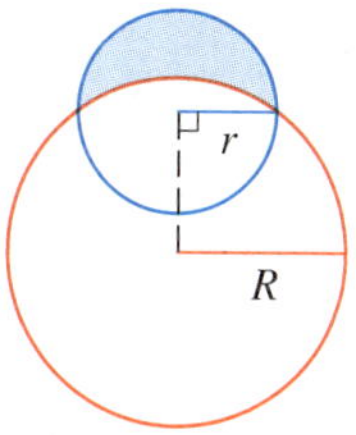

40. A water storage tank has the shape of a cylinder with diameter 10 ft. It is mounted so that the circular cross-sections are vertical. If the depth of the water is 7 ft, what percentage of the total capacity is being used?

41. A torus is generated by rotating the circle $x^2 + (y - R)^2 = r^2$ about the x-axis. Find the volume enclosed by the torus.

7.4 Integration of Rational Functions by Partial Fractions

In this section we show how to integrate any rational function (a ratio of polynomials) by expressing it as a sum of simpler fractions, called *partial fractions,* that we already know how to integrate. To illustrate the method, observe that by taking the fractions $2/(x-1)$ and $1/(x+2)$ to a common denominator we obtain

$$\frac{2}{x-1} - \frac{1}{x+2} = \frac{2(x+2)-(x-1)}{(x-1)(x+2)} = \frac{x+5}{x^2+x-2}$$

If we now reverse the procedure, we see how to integrate the function on the right side of this equation:

$$\int \frac{x+5}{x^2+x-2}\,dx = \int \left(\frac{2}{x-1} - \frac{1}{x+2}\right) dx$$

$$= 2\ln|x-1| - \ln|x+2| + C$$

To see how the method of partial fractions works in general, let's consider a rational function

$$f(x) = \frac{P(x)}{Q(x)}$$

where P and Q are polynomials. It's possible to express f as a sum of simpler fractions provided that the degree of P is less than the degree of Q. Such a rational function is called *proper.* Recall that if

$$P(x) = a_n x^n + a_{n-1}x^{n-1} + \cdots + a_1 x + a_0$$

where $a_n \neq 0$, then the degree of P is n and we write $\deg(P) = n$.

If f is improper, that is, $\deg(P) \geqslant \deg(Q)$, then we must take the preliminary step of dividing Q into P (by long division) until a remainder $R(x)$ is obtained such that $\deg(R) < \deg(Q)$. The division statement is

$$f(x) = \frac{P(x)}{Q(x)} = S(x) + \frac{R(x)}{Q(x)} \tag{1}$$

where S and R are also polynomials.

As the following example illustrates, sometimes this preliminary step is all that is required.

EXAMPLE 1 Find $\displaystyle\int \frac{x^3+x}{x-1}\,dx$.

$$\begin{array}{r} x^2 + x + 2 \\ x-1\,\big)\,\overline{x^3 \qquad + x} \\ x^3 - x^2 \\ \hline x^2 + x \\ x^2 - x \\ \hline 2x \\ 2x - 2 \\ \hline 2 \end{array}$$

SOLUTION Since the degree of the numerator is greater than the degree of the denominator, we first perform the long division. This enables us to write

$$\int \frac{x^3+x}{x-1}\,dx = \int \left(x^2 + x + 2 + \frac{2}{x-1}\right) dx$$

$$= \frac{x^3}{3} + \frac{x^2}{2} + 2x + 2\ln|x-1| + C$$

The next step is to factor the denominator $Q(x)$ as far as possible. It can be shown that any polynomial Q can be factored as a product of linear factors (of the form $ax + b$) and irreducible quadratic factors (of the form $ax^2 + bx + c$, where $b^2 - 4ac < 0$). For instance, if $Q(x) = x^4 - 16$, we could factor it as

$$Q(x) = (x^2 - 4)(x^2 + 4) = (x - 2)(x + 2)(x^2 + 4)$$

The third step is to express the proper rational function $R(x)/Q(x)$ (from Equation 1) as a sum of **partial fractions** of the form

$$\frac{A}{(ax + b)^i} \qquad \text{or} \qquad \frac{Ax + B}{(ax^2 + bx + c)^j}$$

A theorem in algebra guarantees that it is always possible to do this. We explain the details for the four cases that occur.

CASE I ▫ The denominator $Q(x)$ is a product of distinct linear factors.
This means that we can write

$$Q(x) = (a_1x + b_1)(a_2x + b_2) \cdots (a_kx + b_k)$$

where no factor is repeated (and no factor is a constant multiple of another). In this case the partial fraction theorem states that there exist constants $A_1, A_2, \ldots, A_k$ such that

2 $$\frac{R(x)}{Q(x)} = \frac{A_1}{a_1x + b_1} + \frac{A_2}{a_2x + b_2} + \cdots + \frac{A_k}{a_kx + b_k}$$

These constants can be determined as in the following example.

EXAMPLE 2 Evaluate $\displaystyle\int \frac{x^2 + 2x - 1}{2x^3 + 3x^2 - 2x}\,dx$.

SOLUTION Since the degree of the numerator is less than the degree of the denominator, we don't need to divide. We factor the denominator as

$$2x^3 + 3x^2 - 2x = x(2x^2 + 3x - 2) = x(2x - 1)(x + 2)$$

Since the denominator has three distinct linear factors, the partial fraction decomposition of the integrand (2) has the form

3 $$\frac{x^2 + 2x - 1}{x(2x - 1)(x + 2)} = \frac{A}{x} + \frac{B}{2x - 1} + \frac{C}{x + 2}$$

|||| Another method for finding A, B, and C is given in the note after this example.

To determine the values of A, B, and C, we multiply both sides of this equation by the product of the denominators, $x(2x - 1)(x + 2)$, obtaining

4 $$x^2 + 2x - 1 = A(2x - 1)(x + 2) + Bx(x + 2) + Cx(2x - 1)$$

Expanding the right side of Equation 4 and writing it in the standard form for polynomials, we get

5 $$x^2 + 2x - 1 = (2A + B + 2C)x^2 + (3A + 2B - C)x - 2A$$

||| Figure 1 shows the graphs of the integrand in Example 2 and its indefinite integral (with $K = 0$). Which is which?

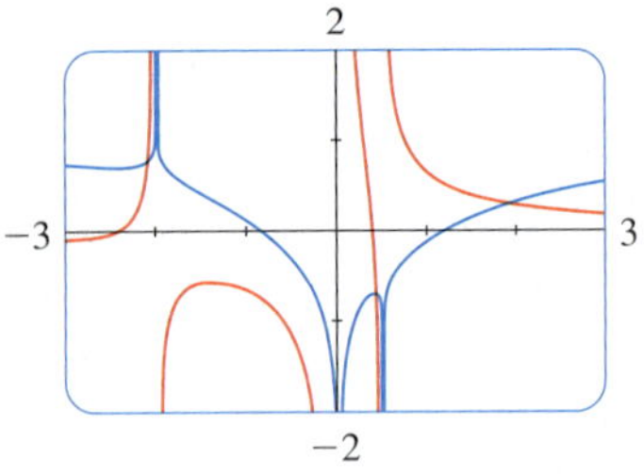

FIGURE 1

The polynomials in Equation 5 are identical, so their coefficients must be equal. The coefficient of x^2 on the right side, $2A + B + 2C$, must equal the coefficient of x^2 on the left side—namely, 1. Likewise, the coefficients of x are equal and the constant terms are equal. This gives the following system of equations for A, B, and C:

$$\begin{aligned} 2A + B + 2C &= 1 \\ 3A + 2B - C &= 2 \\ -2A &= -1 \end{aligned}$$

Solving, we get $A = \frac{1}{2}$, $B = \frac{1}{5}$, and $C = -\frac{1}{10}$, and so

$$\int \frac{x^2 + 2x - 1}{2x^3 + 3x^2 - 2x}\,dx = \int \left(\frac{1}{2}\frac{1}{x} + \frac{1}{5}\frac{1}{2x - 1} - \frac{1}{10}\frac{1}{x + 2} \right) dx$$

$$= \tfrac{1}{2}\ln|x| + \tfrac{1}{10}\ln|2x - 1| - \tfrac{1}{10}\ln|x + 2| + K$$

||| We could check our work by taking the terms to a common denominator and adding them.

In integrating the middle term we have made the mental substitution $u = 2x - 1$, which gives $du = 2\,dx$ and $dx = du/2$.

NOTE ▫ We can use an alternative method to find the coefficients A, B, and C in Example 2. Equation 4 is an identity; it is true for every value of x. Let's choose values of x that simplify the equation. If we put $x = 0$ in Equation 4, then the second and third terms on the right side vanish and the equation then becomes $-2A = -1$, or $A = \frac{1}{2}$. Likewise, $x = \frac{1}{2}$ gives $5B/4 = \frac{1}{4}$ and $x = -2$ gives $10C = -1$, so $B = \frac{1}{5}$ and $C = -\frac{1}{10}$. (You may object that Equation 3 is not valid for $x = 0, \frac{1}{2}$, or -2, so why should Equation 4 be valid for those values? In fact, Equation 4 is true for all values of x, even $x = 0, \frac{1}{2}$, and -2. See Exercise 67 for the reason.)

EXAMPLE 3 Find $\displaystyle\int \frac{dx}{x^2 - a^2}$, where $a \neq 0$.

SOLUTION The method of partial fractions gives

$$\frac{1}{x^2 - a^2} = \frac{1}{(x - a)(x + a)} = \frac{A}{x - a} + \frac{B}{x + a}$$

and therefore

$$A(x + a) + B(x - a) = 1$$

Using the method of the preceding note, we put $x = a$ in this equation and get $A(2a) = 1$, so $A = 1/(2a)$. If we put $x = -a$, we get $B(-2a) = 1$, so $B = -1/(2a)$. Thus

$$\int \frac{dx}{x^2 - a^2} = \frac{1}{2a}\int \left(\frac{1}{x - a} - \frac{1}{x + a} \right) dx$$

$$= \frac{1}{2a}\left(\ln|x - a| - \ln|x + a|\right) + C$$

Since $\ln x - \ln y = \ln(x/y)$, we can write the integral as

6 $$\int \frac{dx}{x^2 - a^2} = \frac{1}{2a} \ln \left| \frac{x-a}{x+a} \right| + C$$

See Exercises 53–54 for ways of using Formula 6.

CASE II ▫ **$Q(x)$ is a product of linear factors, some of which are repeated.**
Suppose the first linear factor $(a_1x + b_1)$ is repeated r times; that is, $(a_1x + b_1)^r$ occurs in the factorization of $Q(x)$. Then instead of the single term $A_1/(a_1x + b_1)$ in Equation 2, we would use

7 $$\frac{A_1}{a_1x + b_1} + \frac{A_2}{(a_1x + b_1)^2} + \cdots + \frac{A_r}{(a_1x + b_1)^r}$$

By way of illustration, we could write

$$\frac{x^3 - x + 1}{x^2(x-1)^3} = \frac{A}{x} + \frac{B}{x^2} + \frac{C}{x-1} + \frac{D}{(x-1)^2} + \frac{E}{(x-1)^3}$$

but we prefer to work out in detail a simpler example.

EXAMPLE 4 Find $\displaystyle\int \frac{x^4 - 2x^2 + 4x + 1}{x^3 - x^2 - x + 1}\,dx$.

SOLUTION The first step is to divide. The result of long division is

$$\frac{x^4 - 2x^2 + 4x + 1}{x^3 - x^2 - x + 1} = x + 1 + \frac{4x}{x^3 - x^2 - x + 1}$$

The second step is to factor the denominator $Q(x) = x^3 - x^2 - x + 1$. Since $Q(1) = 0$, we know that $x - 1$ is a factor and we obtain

$$\begin{aligned} x^3 - x^2 - x + 1 &= (x-1)(x^2 - 1) = (x-1)(x-1)(x+1) \\ &= (x-1)^2(x+1) \end{aligned}$$

Since the linear factor $x - 1$ occurs twice, the partial fraction decomposition is

$$\frac{4x}{(x-1)^2(x+1)} = \frac{A}{x-1} + \frac{B}{(x-1)^2} + \frac{C}{x+1}$$

Multiplying by the least common denominator, $(x-1)^2(x+1)$, we get

8 $$\begin{aligned} 4x &= A(x-1)(x+1) + B(x+1) + C(x-1)^2 \\ &= (A + C)x^2 + (B - 2C)x + (-A + B + C) \end{aligned}$$

||| Another method for finding the coefficients:
Put $x = 1$ in (8): $B = 2$.
Put $x = -1$: $C = -1$.
Put $x = 0$: $A = B + C = 1$.

Now we equate coefficients:

$$\begin{aligned} A \quad\quad + C &= 0 \\ B - 2C &= 4 \\ -A + B + C &= 0 \end{aligned}$$

Solving, we obtain $A = 1$, $B = 2$, and $C = -1$, so

$$\int \frac{x^4 - 2x^2 + 4x + 1}{x^3 - x^2 - x + 1}\,dx = \int \left[x + 1 + \frac{1}{x - 1} + \frac{2}{(x - 1)^2} - \frac{1}{x + 1}\right] dx$$

$$= \frac{x^2}{2} + x + \ln|x - 1| - \frac{2}{x - 1} - \ln|x + 1| + K$$

$$= \frac{x^2}{2} + x - \frac{2}{x - 1} + \ln\left|\frac{x - 1}{x + 1}\right| + K$$

CASE III ▫ ***$Q(x)$* contains irreducible quadratic factors, none of which is repeated.**
If $Q(x)$ has the factor $ax^2 + bx + c$, where $b^2 - 4ac < 0$, then, in addition to the partial fractions in Equations 2 and 7, the expression for $R(x)/Q(x)$ will have a term of the form

9 $$\frac{Ax + B}{ax^2 + bx + c}$$

where A and B are constants to be determined. For instance, the function given by $f(x) = x/[(x - 2)(x^2 + 1)(x^2 + 4)]$ has a partial fraction decomposition of the form

$$\frac{x}{(x - 2)(x^2 + 1)(x^2 + 4)} = \frac{A}{x - 2} + \frac{Bx + C}{x^2 + 1} + \frac{Dx + E}{x^2 + 4}$$

The term given in (9) can be integrated by completing the square and using the formula

10 $$\int \frac{dx}{x^2 + a^2} = \frac{1}{a}\tan^{-1}\left(\frac{x}{a}\right) + C$$

EXAMPLE 5 Evaluate $\displaystyle\int \frac{2x^2 - x + 4}{x^3 + 4x}\,dx$.

SOLUTION Since $x^3 + 4x = x(x^2 + 4)$ can't be factored further, we write

$$\frac{2x^2 - x + 4}{x(x^2 + 4)} = \frac{A}{x} + \frac{Bx + C}{x^2 + 4}$$

Multiplying by $x(x^2 + 4)$, we have

$$2x^2 - x + 4 = A(x^2 + 4) + (Bx + C)x$$

$$= (A + B)x^2 + Cx + 4A$$

Equating coefficients, we obtain

$$A + B = 2 \qquad C = -1 \qquad 4A = 4$$

Thus $A = 1$, $B = 1$, and $C = -1$ and so

$$\int \frac{2x^2 - x + 4}{x^3 + 4x}\,dx = \int \left(\frac{1}{x} + \frac{x - 1}{x^2 + 4}\right) dx$$

In order to integrate the second term we split it into two parts:

$$\int \frac{x-1}{x^2+4}\,dx = \int \frac{x}{x^2+4}\,dx - \int \frac{1}{x^2+4}\,dx$$

We make the substitution $u = x^2 + 4$ in the first of these integrals so that $du = 2x\,dx$. We evaluate the second integral by means of Formula 10 with $a = 2$:

$$\int \frac{2x^2 - x + 4}{x(x^2+4)}\,dx = \int \frac{1}{x}\,dx + \int \frac{x}{x^2+4}\,dx - \int \frac{1}{x^2+4}\,dx$$

$$= \ln|x| + \tfrac{1}{2}\ln(x^2+4) - \tfrac{1}{2}\tan^{-1}(x/2) + K$$

EXAMPLE 6 Evaluate $\displaystyle\int \frac{4x^2 - 3x + 2}{4x^2 - 4x + 3}\,dx$.

SOLUTION Since the degree of the numerator is *not less than* the degree of the denominator, we first divide and obtain

$$\frac{4x^2 - 3x + 2}{4x^2 - 4x + 3} = 1 + \frac{x - 1}{4x^2 - 4x + 3}$$

Notice that the quadratic $4x^2 - 4x + 3$ is irreducible because its discriminant is $b^2 - 4ac = -32 < 0$. This means it can't be factored, so we don't need to use the partial fraction technique.

To integrate the given function we complete the square in the denominator:

$$4x^2 - 4x + 3 = (2x - 1)^2 + 2$$

This suggests that we make the substitution $u = 2x - 1$. Then, $du = 2\,dx$ and $x = (u + 1)/2$, so

$$\int \frac{4x^2 - 3x + 2}{4x^2 - 4x + 3}\,dx = \int \left(1 + \frac{x - 1}{4x^2 - 4x + 3}\right) dx$$

$$= x + \tfrac{1}{2}\int \frac{\tfrac{1}{2}(u + 1) - 1}{u^2 + 2}\,du = x + \tfrac{1}{4}\int \frac{u - 1}{u^2 + 2}\,du$$

$$= x + \tfrac{1}{4}\int \frac{u}{u^2 + 2}\,du - \tfrac{1}{4}\int \frac{1}{u^2 + 2}\,du$$

$$= x + \tfrac{1}{8}\ln(u^2 + 2) - \frac{1}{4}\cdot\frac{1}{\sqrt{2}}\tan^{-1}\left(\frac{u}{\sqrt{2}}\right) + C$$

$$= x + \tfrac{1}{8}\ln(4x^2 - 4x + 3) - \frac{1}{4\sqrt{2}}\tan^{-1}\left(\frac{2x - 1}{\sqrt{2}}\right) + C$$

NOTE ▫ Example 6 illustrates the general procedure for integrating a partial fraction of the form

$$\frac{Ax + B}{ax^2 + bx + c} \qquad \text{where } b^2 - 4ac < 0$$

We complete the square in the denominator and then make a substitution that brings the integral into the form

$$\int \frac{Cu + D}{u^2 + a^2}\,du = C\int \frac{u}{u^2 + a^2}\,du + D\int \frac{1}{u^2 + a^2}\,du$$

Then the first integral is a logarithm and the second is expressed in terms of $\tan^{-1}$.

CASE IV ▫ **$Q(x)$ contains a repeated irreducible quadratic factor.**
If $Q(x)$ has the factor $(ax^2 + bx + c)^r$, where $b^2 - 4ac < 0$, then instead of the single partial fraction (9), the sum

11 $$\frac{A_1x + B_1}{ax^2 + bx + c} + \frac{A_2x + B_2}{(ax^2 + bx + c)^2} + \cdots + \frac{A_rx + B_r}{(ax^2 + bx + c)^r}$$

occurs in the partial fraction decomposition of $R(x)/Q(x)$. Each of the terms in (11) can be integrated by first completing the square.

|||| It would be extremely tedious to work out by hand the numerical values of the coefficients in Example 7. Most computer algebra systems, however, can find the numerical values very quickly. For instance, the Maple command

convert(f, parfrac, x)

or the Mathematica command

Apart[f]

gives the following values:

$$A = -1, \quad B = \tfrac{1}{8}, \quad C = D = -1,$$
$$E = \tfrac{15}{8}, \quad F = -\tfrac{1}{8}, \quad G = H = \tfrac{3}{4},$$
$$I = -\tfrac{1}{2}, \quad J = \tfrac{1}{2}$$

EXAMPLE 7 Write out the form of the partial fraction decomposition of the function

$$\frac{x^3 + x^2 + 1}{x(x - 1)(x^2 + x + 1)(x^2 + 1)^3}$$

SOLUTION

$$\frac{x^3 + x^2 + 1}{x(x - 1)(x^2 + x + 1)(x^2 + 1)^3}$$
$$= \frac{A}{x} + \frac{B}{x - 1} + \frac{Cx + D}{x^2 + x + 1} + \frac{Ex + F}{x^2 + 1} + \frac{Gx + H}{(x^2 + 1)^2} + \frac{Ix + J}{(x^2 + 1)^3}$$

EXAMPLE 8 Evaluate $\displaystyle\int \frac{1 - x + 2x^2 - x^3}{x(x^2 + 1)^2}\,dx$.

SOLUTION The form of the partial fraction decomposition is

$$\frac{1 - x + 2x^2 - x^3}{x(x^2 + 1)^2} = \frac{A}{x} + \frac{Bx + C}{x^2 + 1} + \frac{Dx + E}{(x^2 + 1)^2}$$

Multiplying by $x(x^2 + 1)^2$, we have

$$\begin{aligned} -x^3 + 2x^2 - x + 1 &= A(x^2 + 1)^2 + (Bx + C)x(x^2 + 1) + (Dx + E)x \\ &= A(x^4 + 2x^2 + 1) + B(x^4 + x^2) + C(x^3 + x) + Dx^2 + Ex \\ &= (A + B)x^4 + Cx^3 + (2A + B + D)x^2 + (C + E)x + A \end{aligned}$$

If we equate coefficients, we get the system

$$A + B = 0 \qquad C = -1 \qquad 2A + B + D = 2 \qquad C + E = -1 \qquad A = 1$$

which has the solution $A = 1$, $B = -1$, $C = -1$, $D = 1$, and $E = 0$. Thus

$$\int \frac{1 - x + 2x^2 - x^3}{x(x^2 + 1)^2}\,dx = \int \left(\frac{1}{x} - \frac{x + 1}{x^2 + 1} + \frac{x}{(x^2 + 1)^2} \right) dx$$

$$= \int \frac{dx}{x} - \int \frac{x}{x^2 + 1}\,dx - \int \frac{dx}{x^2 + 1} + \int \frac{x\,dx}{(x^2 + 1)^2}$$

$$= \ln|x| - \tfrac{1}{2}\ln(x^2 + 1) - \tan^{-1}x - \frac{1}{2(x^2 + 1)} + K$$

In the second and fourth terms we made the mental substitution $u = x^2 + 1$.

We note that sometimes partial fractions can be avoided when integrating a rational function. For instance, although the integral

$$\int \frac{x^2 + 1}{x(x^2 + 3)}\,dx$$

could be evaluated by the method of Case III, it's much easier to observe that if $u = x(x^2 + 3) = x^3 + 3x$, then $du = (3x^2 + 3)\,dx$ and so

$$\int \frac{x^2 + 1}{x(x^2 + 3)}\,dx = \tfrac{1}{3}\ln|x^3 + 3x| + C$$

Rationalizing Substitutions

Some nonrational functions can be changed into rational functions by means of appropriate substitutions. In particular, when an integrand contains an expression of the form $\sqrt[n]{g(x)}$, then the substitution $u = \sqrt[n]{g(x)}$ may be effective. Other instances appear in the exercises.

EXAMPLE 9 Evaluate $\displaystyle\int \frac{\sqrt{x + 4}}{x}\,dx$.

SOLUTION Let $u = \sqrt{x + 4}$. Then $u^2 = x + 4$, so $x = u^2 - 4$ and $dx = 2u\,du$. Therefore

$$\int \frac{\sqrt{x + 4}}{x}\,dx = \int \frac{u}{u^2 - 4}\,2u\,du = 2\int \frac{u^2}{u^2 - 4}\,du$$

$$= 2\int \left(1 + \frac{4}{u^2 - 4}\right) du$$

We can evaluate this integral either by factoring $u^2 - 4$ as $(u - 2)(u + 2)$ and using partial fractions or by using Formula 6 with $a = 2$:

$$\int \frac{\sqrt{x + 4}}{x}\,dx = 2\int du + 8\int \frac{du}{u^2 - 4}$$

$$= 2u + 8 \cdot \frac{1}{2 \cdot 2}\ln\left|\frac{u - 2}{u + 2}\right| + C$$

$$= 2\sqrt{x + 4} + 2\ln\left|\frac{\sqrt{x + 4} - 2}{\sqrt{x + 4} + 2}\right| + C$$

7.4 Exercises

1–6 |||| Write out the form of the partial fraction decomposition of the function (as in Example 7). Do not determine the numerical values of the coefficients.

1. (a) $\dfrac{2x}{(x+3)(3x+1)}$ (b) $\dfrac{1}{x^3+2x^2+x}$

2. (a) $\dfrac{x-1}{x^3+x^2}$ (b) $\dfrac{x-1}{x^3+x}$

3. (a) $\dfrac{2}{x^2+3x-4}$ (b) $\dfrac{x^2}{(x-1)(x^2+x+1)}$

4. (a) $\dfrac{x^3}{x^2+4x+3}$ (b) $\dfrac{2x+1}{(x+1)^3(x^2+4)^2}$

5. (a) $\dfrac{x^4}{x^4-1}$ (b) $\dfrac{t^4+t^2+1}{(t^2+1)(t^2+4)^2}$

6. (a) $\dfrac{x^4}{(x^3+x)(x^2-x+3)}$ (b) $\dfrac{1}{x^6-x^3}$

7–38 |||| Evaluate the integral.

7. $\displaystyle\int \frac{x}{x-6}\,dx$

8. $\displaystyle\int \frac{r^2}{r+4}\,dr$

9. $\displaystyle\int \frac{x-9}{(x+5)(x-2)}\,dx$

10. $\displaystyle\int \frac{1}{(t+4)(t-1)}\,dt$

11. $\displaystyle\int_2^3 \frac{1}{x^2-1}\,dx$

12. $\displaystyle\int_0^1 \frac{x-1}{x^2+3x+2}\,dx$

13. $\displaystyle\int \frac{ax}{x^2-bx}\,dx$

14. $\displaystyle\int \frac{1}{(x+a)(x+b)}\,dx$

15. $\displaystyle\int_0^1 \frac{2x+3}{(x+1)^2}\,dx$

16. $\displaystyle\int_0^1 \frac{x^3-4x-10}{x^2-x-6}\,dx$

17. $\displaystyle\int_1^2 \frac{4y^2-7y-12}{y(y+2)(y-3)}\,dy$

18. $\displaystyle\int \frac{x^2+2x-1}{x^3-x}\,dx$

19. $\displaystyle\int \frac{1}{(x+5)^2(x-1)}\,dx$

20. $\displaystyle\int \frac{x^2}{(x-3)(x+2)^2}\,dx$

21. $\displaystyle\int \frac{5x^2+3x-2}{x^3+2x^2}\,dx$

22. $\displaystyle\int \frac{ds}{s^2(s-1)^2}$

23. $\displaystyle\int \frac{x^2}{(x+1)^3}\,dx$

24. $\displaystyle\int \frac{x^3}{(x+1)^3}\,dx$

25. $\displaystyle\int \frac{10}{(x-1)(x^2+9)}\,dx$

26. $\displaystyle\int \frac{x^2-x+6}{x^3+3x}\,dx$

27. $\displaystyle\int \frac{x^3+x^2+2x+1}{(x^2+1)(x^2+2)}\,dx$

28. $\displaystyle\int \frac{x^2-2x-1}{(x-1)^2(x^2+1)}\,dx$

29. $\displaystyle\int \frac{x+4}{x^2+2x+5}\,dx$

30. $\displaystyle\int \frac{x^3-2x^2+x+1}{x^4+5x^2+4}\,dx$

31. $\displaystyle\int \frac{1}{x^3-1}\,dx$

32. $\displaystyle\int_0^1 \frac{x}{x^2+4x+13}\,dx$

33. $\displaystyle\int_2^5 \frac{x^2+2x}{x^3+3x^2+4}\,dx$

34. $\displaystyle\int \frac{x^3}{x^3+1}\,dx$

35. $\displaystyle\int \frac{dx}{x^4-x^2}$

36. $\displaystyle\int_0^1 \frac{2x^3+5x}{x^4+5x^2+4}\,dx$

37. $\displaystyle\int \frac{x-3}{(x^2+2x+4)^2}\,dx$

38. $\displaystyle\int \frac{x^4+1}{x(x^2+1)^2}\,dx$

39–48 |||| Make a substitution to express the integrand as a rational function and then evaluate the integral.

39. $\displaystyle\int \frac{1}{x\sqrt{x+1}}\,dx$

40. $\displaystyle\int \frac{1}{x-\sqrt{x+2}}\,dx$

41. $\displaystyle\int_9^{16} \frac{\sqrt{x}}{x-4}\,dx$

42. $\displaystyle\int_0^1 \frac{1}{1+\sqrt[3]{x}}\,dx$

43. $\displaystyle\int \frac{x^3}{\sqrt[3]{x^2+1}}\,dx$

44. $\displaystyle\int_{1/3}^3 \frac{\sqrt{x}}{x^2+x}\,dx$

45. $\displaystyle\int \frac{1}{\sqrt{x}-\sqrt[3]{x}}\,dx$ [*Hint:* Substitute $u=\sqrt[6]{x}$.]

46. $\displaystyle\int \frac{1}{\sqrt[3]{x}+\sqrt[4]{x}}\,dx$ [*Hint:* Substitute $u=\sqrt[12]{x}$.]

47. $\displaystyle\int \frac{e^{2x}}{e^{2x}+3e^x+2}\,dx$

48. $\displaystyle\int \frac{\cos x}{\sin^2 x+\sin x}\,dx$

49–50 |||| Use integration by parts, together with the techniques of this section, to evaluate the integral.

49. $\displaystyle\int \ln(x^2-x+2)\,dx$

50. $\displaystyle\int x\tan^{-1}x\,dx$

51. Use a graph of $f(x)=1/(x^2-2x-3)$ to decide whether $\int_0^2 f(x)\,dx$ is positive or negative. Use the graph to give a rough estimate of the value of the integral and then use partial fractions to find the exact value.

52. Graph both $y=1/(x^3-2x^2)$ and an antiderivative on the same screen.

53–54 |||| Evaluate the integral by completing the square and using Formula 6.

53. $\displaystyle\int \frac{dx}{x^2-2x}$

54. $\displaystyle\int \frac{2x+1}{4x^2+12x-7}\,dx$

55. The German mathematician Karl Weierstrass (1815–1897) noticed that the substitution $t=\tan(x/2)$ will convert any rational function of $\sin x$ and $\cos x$ into an ordinary rational function of t.

(a) If $t = \tan(x/2)$, $-\pi < x < \pi$, sketch a right triangle or use trigonometric identities to show that

$$\cos\left(\frac{x}{2}\right) = \frac{1}{\sqrt{1+t^2}} \quad \text{and} \quad \sin\left(\frac{x}{2}\right) = \frac{t}{\sqrt{1+t^2}}$$

(b) Show that

$$\cos x = \frac{1-t^2}{1+t^2} \quad \text{and} \quad \sin x = \frac{2t}{1+t^2}$$

(c) Show that

$$dx = \frac{2}{1+t^2}\,dt$$

56–59 |||| Use the substitution in Exercise 55 to transform the integrand into a rational function of t and then evaluate the integral.

56. $\displaystyle\int \frac{dx}{3 - 5\sin x}$

57. $\displaystyle\int \frac{1}{3\sin x - 4\cos x}\,dx$

58. $\displaystyle\int_{\pi/3}^{\pi/2} \frac{1}{1 + \sin x - \cos x}\,dx$

59. $\displaystyle\int \frac{1}{2\sin x + \sin 2x}\,dx$

60–61 |||| Find the area of the region under the given curve from a to b.

60. $y = \dfrac{1}{x^2 - 6x + 8}$, $a = 5$, $b = 10$

61. $y = \dfrac{x+1}{x-1}$, $a = 2$, $b = 3$

62. Find the volume of the resulting solid if the region under the curve $y = 1/(x^2 + 3x + 2)$ from $x = 0$ to $x = 1$ is rotated about (a) the x-axis and (b) the y-axis.

63. One method of slowing the growth of an insect population without using pesticides is to introduce into the population a number of sterile males that mate with fertile females but produce no offspring. If P represents the number of female insects in a population, S the number of sterile males introduced each generation, and r the population's natural growth rate, then the female population is related to time t by

$$t = \int \frac{P+S}{P[(r-1)P - S]}\,dP$$

Suppose an insect population with 10,000 females grows at a rate of $r = 0.10$ and 900 sterile males are added. Evaluate the integral to give an equation relating the female population to time. (Note that the resulting equation can't be solved explicitly for P.)

64. Factor $x^4 + 1$ as a difference of squares by first adding and subtracting the same quantity. Use this factorization to evaluate $\int 1/(x^4+1)\,dx$.

CAS **65.** (a) Use a computer algebra system to find the partial fraction decomposition of the function

$$f(x) = \frac{4x^3 - 27x^2 + 5x - 32}{30x^5 - 13x^4 + 50x^3 - 286x^2 - 299x - 70}$$

(b) Use part (a) to find $\int f(x)\,dx$ (by hand) and compare with the result of using the CAS to integrate f directly. Comment on any discrepancy.

CAS **66.** (a) Find the partial fraction decomposition of the function

$$f(x) = \frac{12x^5 - 7x^3 - 13x^2 + 8}{100x^6 - 80x^5 + 116x^4 - 80x^3 + 41x^2 - 20x + 4}$$

(b) Use part (a) to find $\int f(x)\,dx$ and graph f and its indefinite integral on the same screen.

(c) Use the graph of f to discover the main features of the graph of $\int f(x)\,dx$.

67. Suppose that F, G, and Q are polynomials and

$$\frac{F(x)}{Q(x)} = \frac{G(x)}{Q(x)}$$

for all x except when $Q(x) = 0$. Prove that $F(x) = G(x)$ for all x. [*Hint:* Use continuity.]

68. If f is a quadratic function such that $f(0) = 1$ and

$$\int \frac{f(x)}{x^2(x+1)^3}\,dx$$

is a rational function, find the value of $f'(0)$.

7.5 Strategy for Integration

As we have seen, integration is more challenging than differentiation. In finding the derivative of a function it is obvious which differentiation formula we should apply. But it may not be obvious which technique we should use to integrate a given function.

Until now individual techniques have been applied in each section. For instance, we usually used substitution in Exercises 5.5, integration by parts in Exercises 7.1, and partial fractions in Exercises 7.4. But in this section we present a collection of miscellaneous integrals in random order and the main challenge is to recognize which technique or formula to use. No hard and fast rules can be given as to which method applies in a given situation, but we give some advice on strategy that you may find useful.

A prerequisite for strategy selection is a knowledge of the basic integration formulas. In the following table we have collected the integrals from our previous list together with several additional formulas that we have learned in this chapter. Most of them should be memorized. It is useful to know them all, but the ones marked with an asterisk need not be memorized since they are easily derived. Formula 19 can be avoided by using partial fractions, and trigonometric substitutions can be used in place of Formula 20.

Table of Integration Formulas Constants of integration have been omitted.

1. $\int x^n\,dx = \dfrac{x^{n+1}}{n+1} \quad (n \neq -1)$ **2.** $\int \dfrac{1}{x}\,dx = \ln|x|$

3. $\int e^x\,dx = e^x$ **4.** $\int a^x\,dx = \dfrac{a^x}{\ln a}$

5. $\int \sin x\,dx = -\cos x$ **6.** $\int \cos x\,dx = \sin x$

7. $\int \sec^2 x\,dx = \tan x$ **8.** $\int \csc^2 x\,dx = -\cot x$

9. $\int \sec x \tan x\,dx = \sec x$ **10.** $\int \csc x \cot x\,dx = -\csc x$

11. $\int \sec x\,dx = \ln|\sec x + \tan x|$ **12.** $\int \csc x\,dx = \ln|\csc x - \cot x|$

13. $\int \tan x\,dx = \ln|\sec x|$ **14.** $\int \cot x\,dx = \ln|\sin x|$

15. $\int \sinh x\,dx = \cosh x$ **16.** $\int \cosh x\,dx = \sinh x$

17. $\int \dfrac{dx}{x^2 + a^2} = \dfrac{1}{a}\tan^{-1}\left(\dfrac{x}{a}\right)$ **18.** $\int \dfrac{dx}{\sqrt{a^2 - x^2}} = \sin^{-1}\left(\dfrac{x}{a}\right)$

***19.** $\int \dfrac{dx}{x^2 - a^2} = \dfrac{1}{2a}\ln\left|\dfrac{x-a}{x+a}\right|$ ***20.** $\int \dfrac{dx}{\sqrt{x^2 \pm a^2}} = \ln\left|x + \sqrt{x^2 \pm a^2}\right|$

Once you are armed with these basic integration formulas, if you don't immediately see how to attack a given integral, you might try the following four-step strategy.

1. Simplify the Integrand if Possible Sometimes the use of algebraic manipulation or trigonometric identities will simplify the integrand and make the method of integration obvious. Here are some examples:

$$\int \sqrt{x}\,(1 + \sqrt{x})\,dx = \int (\sqrt{x} + x)\,dx$$

$$\int \frac{\tan\theta}{\sec^2\theta}\,d\theta = \int \frac{\sin\theta}{\cos\theta}\cos^2\theta\,d\theta$$

$$= \int \sin\theta\cos\theta\,d\theta = \tfrac{1}{2}\int \sin 2\theta\,d\theta$$

$$\int (\sin x + \cos x)^2\,dx = \int (\sin^2 x + 2\sin x\cos x + \cos^2 x)\,dx$$

$$= \int (1 + 2\sin x\cos x)\,dx$$

2. Look for an Obvious Substitution Try to find some function $u = g(x)$ in the integrand whose differential $du = g'(x)\,dx$ also occurs, apart from a constant factor. For instance, in the integral

$$\int \frac{x}{x^2 - 1}\,dx$$

we notice that if $u = x^2 - 1$, then $du = 2x\,dx$. Therefore, we use the substitution $u = x^2 - 1$ instead of the method of partial fractions.

3. Classify the Integrand According to Its Form If Steps 1 and 2 have not led to the solution, then we take a look at the form of the integrand $f(x)$.

(a) *Trigonometric functions.* If $f(x)$ is a product of powers of $\sin x$ and $\cos x$, of $\tan x$ and $\sec x$, or of $\cot x$ and $\csc x$, then we use the substitutions recommended in Section 7.2.

(b) *Rational functions.* If f is a rational function, we use the procedure of Section 7.4 involving partial fractions.

(c) *Integration by parts.* If $f(x)$ is a product of a power of x (or a polynomial) and a transcendental function (such as a trigonometric, exponential, or logarithmic function), then we try integration by parts, choosing u and dv according to the advice given in Section 7.1. If you look at the functions in Exercises 7.1, you will see that most of them are the type just described.

(d) *Radicals.* Particular kinds of substitutions are recommended when certain radicals appear.

(i) If $\sqrt{\pm x^2 \pm a^2}$ occurs, we use a trigonometric substitution according to the table in Section 7.3.

(ii) If $\sqrt[n]{ax + b}$ occurs, we use the rationalizing substitution $u = \sqrt[n]{ax + b}$. More generally, this sometimes works for $\sqrt[n]{g(x)}$.

4. Try Again If the first three steps have not produced the answer, remember that there are basically only two methods of integration: substitution and parts.

(a) *Try substitution.* Even if no substitution is obvious (Step 2), some inspiration or ingenuity (or even desperation) may suggest an appropriate substitution.

(b) *Try parts.* Although integration by parts is used most of the time on products of the form described in Step 3(c), it is sometimes effective on single functions. Looking at Section 7.1, we see that it works on $\tan^{-1}x$, $\sin^{-1}x$, and $\ln x$, and these are all inverse functions.

(c) *Manipulate the integrand.* Algebraic manipulations (perhaps rationalizing the denominator or using trigonometric identities) may be useful in transforming the integral into an easier form. These manipulations may be more substantial than in Step 1 and may involve some ingenuity. Here is an example:

$$\int \frac{dx}{1 - \cos x} = \int \frac{1}{1 - \cos x} \cdot \frac{1 + \cos x}{1 + \cos x}\,dx = \int \frac{1 + \cos x}{1 - \cos^2 x}\,dx$$

$$= \int \frac{1 + \cos x}{\sin^2 x}\,dx = \int \left(\csc^2 x + \frac{\cos x}{\sin^2 x}\right) dx$$

(d) *Relate the problem to previous problems.* When you have built up some experience in integration, you may be able to use a method on a given integral that is similar to a method you have already used on a previous integral. Or you may even be able to express the given integral in terms of a previous one. For instance, $\int \tan^2 x \sec x\, dx$ is a challenging integral, but if we make use of the identity $\tan^2 x = \sec^2 x - 1$, we can write

$$\int \tan^2 x \sec x\, dx = \int \sec^3 x\, dx - \int \sec x\, dx$$

and if $\int \sec^3 x\, dx$ has previously been evaluated (see Example 8 in Section 7.2), then that calculation can be used in the present problem.

(e) *Use several methods.* Sometimes two or three methods are required to evaluate an integral. The evaluation could involve several successive substitutions of different types, or it might combine integration by parts with one or more substitutions.

In the following examples we indicate a method of attack but do not fully work out the integral.

EXAMPLE 1 $\displaystyle\int \frac{\tan^3 x}{\cos^3 x}\, dx$

In Step 1 we rewrite the integral:

$$\int \frac{\tan^3 x}{\cos^3 x}\, dx = \int \tan^3 x \sec^3 x\, dx$$

The integral is now of the form $\int \tan^m x \sec^n x\, dx$ with m odd, so we can use the advice in Section 7.2.

Alternatively, if in Step 1 we had written

$$\int \frac{\tan^3 x}{\cos^3 x}\, dx = \int \frac{\sin^3 x}{\cos^3 x}\,\frac{1}{\cos^3 x}\, dx = \int \frac{\sin^3 x}{\cos^6 x}\, dx$$

then we could have continued as follows with the substitution $u = \cos x$:

$$\int \frac{\sin^3 x}{\cos^6 x}\, dx = \int \frac{1 - \cos^2 x}{\cos^6 x} \sin x\, dx = \int \frac{1 - u^2}{u^6}(-du)$$

$$= \int \frac{u^2 - 1}{u^6}\, du = \int (u^{-4} - u^{-6})\, du$$

EXAMPLE 2 $\displaystyle\int e^{\sqrt{x}}\, dx$

According to Step 3(d)(ii) we substitute $u = \sqrt{x}$. Then $x = u^2$, so $dx = 2u\, du$ and

$$\int e^{\sqrt{x}}\, dx = 2 \int u e^u\, du$$

The integrand is now a product of u and the transcendental function e^u so it can be integrated by parts.

EXAMPLE 3 $\displaystyle\int \frac{x^5 + 1}{x^3 - 3x^2 - 10x}\, dx$

No algebraic simplification or substitution is obvious, so Steps 1 and 2 don't apply here. The integrand is a rational function so we apply the procedure of Section 7.4, remembering that the first step is to divide.

EXAMPLE 4 $\displaystyle\int \frac{dx}{x\sqrt{\ln x}}$

Here Step 2 is all that is needed. We substitute $u = \ln x$ because its differential is $du = dx/x$, which occurs in the integral.

EXAMPLE 5 $\displaystyle\int \sqrt{\frac{1 - x}{1 + x}}\, dx$

Although the rationalizing substitution

$$u = \sqrt{\frac{1 - x}{1 + x}}$$

works here [Step 3(d)(ii)], it leads to a very complicated rational function. An easier method is to do some algebraic manipulation [either as Step 1 or as Step 4(c)]. Multiplying numerator and denominator by $\sqrt{1 - x}$, we have

$$\begin{aligned}\int \sqrt{\frac{1 - x}{1 + x}}\, dx &= \int \frac{1 - x}{\sqrt{1 - x^2}}\, dx \\ &= \int \frac{1}{\sqrt{1 - x^2}}\, dx - \int \frac{x}{\sqrt{1 - x^2}}\, dx \\ &= \sin^{-1}x + \sqrt{1 - x^2} + C\end{aligned}$$

Can We Integrate All Continuous Functions?

The question arises: Will our strategy for integration enable us to find the integral of every continuous function? For example, can we use it to evaluate $\int e^{x^2} dx$? The answer is no, at least not in terms of the functions that we are familiar with.

The functions that we have been dealing with in this book are called **elementary functions**. These are the polynomials, rational functions, power functions (x^a), exponential functions (a^x), logarithmic functions, trigonometric and inverse trigonometric functions, hyperbolic and inverse hyperbolic functions, and all functions that can be obtained from these by the five operations of addition, subtraction, multiplication, division, and composition. For instance, the function

$$f(x) = \sqrt{\frac{x^2 - 1}{x^3 + 2x - 1}} + \ln(\cosh x) - xe^{\sin 2x}$$

is an elementary function.

If f is an elementary function, then f' is an elementary function but $\int f(x)\, dx$ need not be an elementary function. Consider $f(x) = e^{x^2}$. Since f is continuous, its integral exists, and if we define the function F by

$$F(x) = \int_0^x e^{t^2} dt$$

then we know from Part 1 of the Fundamental Theorem of Calculus that

$$F'(x) = e^{x^2}$$

Thus, $f(x) = e^{x^2}$ has an antiderivative F, but it has been proved that F is not an elementary function. This means that no matter how hard we try, we will never succeed in evaluating $\int e^{x^2} dx$ in terms of the functions we know. (In Chapter 11, however, we will see how to express $\int e^{x^2} dx$ as an infinite series.) The same can be said of the following integrals:

$$\int \frac{e^x}{x}\,dx \qquad \int \sin(x^2)\,dx \qquad \int \cos(e^x)\,dx$$

$$\int \sqrt{x^3+1}\,dx \qquad \int \frac{1}{\ln x}\,dx \qquad \int \frac{\sin x}{x}\,dx$$

In fact, the majority of elementary functions don't have elementary antiderivatives. You may be assured, though, that the integrals in the following exercises are all elementary functions.

7.5 Exercises

1–80 Evaluate the integral.

1. $\int \frac{\sin x + \sec x}{\tan x}\,dx$

2. $\int \tan^3\theta\,d\theta$

3. $\int_0^2 \frac{2t}{(t-3)^2}\,dt$

4. $\int \frac{x}{\sqrt{3-x^4}}\,dx$

5. $\int_{-1}^1 \frac{e^{\arctan y}}{1+y^2}\,dy$

6. $\int x \csc x \cot x\,dx$

7. $\int_1^3 r^4 \ln r\,dr$

8. $\int_0^4 \frac{x-1}{x^2-4x-5}\,dx$

9. $\int \frac{x-1}{x^2-4x+5}\,dx$

10. $\int \frac{x}{x^4+x^2+1}\,dx$

11. $\int \sin^3\theta \cos^5\theta\,d\theta$

12. $\int \sin x \cos(\cos x)\,dx$

13. $\int \frac{dx}{(1-x^2)^{3/2}}$

14. $\int \frac{\sqrt{1+\ln x}}{x \ln x}\,dx$

15. $\int_0^{1/2} \frac{x}{\sqrt{1-x^2}}\,dx$

16. $\int_0^{\sqrt{2}/2} \frac{x^2}{\sqrt{1-x^2}}\,dx$

17. $\int x \sin^2 x\,dx$

18. $\int \frac{e^{2t}}{1+e^{4t}}\,dt$

19. $\int e^{x+e^x}\,dx$

20. $\int e^{\sqrt[3]{x}}\,dx$

21. $\int t^3 e^{-2t}\,dt$

22. $\int x \sin^{-1}x\,dx$

23. $\int_0^1 (1+\sqrt{x})^8\,dx$

24. $\int \ln(x^2-1)\,dx$

25. $\int \frac{3x^2-2}{x^2-2x-8}\,dx$

26. $\int \frac{3x^2-2}{x^3-2x-8}\,dx$

27. $\int \cot x \ln(\sin x)\,dx$

28. $\int \sin\sqrt{at}\,dt$

29. $\int_0^5 \frac{3w-1}{w+2}\,dw$

30. $\int_{-2}^2 |x^2-4x|\,dx$

31. $\int \sqrt{\frac{1+x}{1-x}}\,dx$

32. $\int \frac{\sqrt{2x-1}}{2x+3}\,dx$

33. $\int \sqrt{3-2x-x^2}\,dx$

34. $\int_{\pi/4}^{\pi/2} \frac{1+4\cot x}{4-\cot x}\,dx$

35. $\int_{-1}^1 x^8 \sin x\,dx$

36. $\int \sin 4x \cos 3x\,dx$

37. $\int_0^{\pi/4} \cos^2\theta \tan^2\theta\,d\theta$

38. $\int_0^{\pi/4} \tan^5\theta \sec^3\theta\,d\theta$

39. $\int \frac{x}{1-x^2+\sqrt{1-x^2}}\,dx$

40. $\int \frac{1}{\sqrt{4y^2-4y-3}}\,dy$

41. $\int \theta \tan^2\theta\,d\theta$

42. $\int x^2 \tan^{-1}x\,dx$

43. $\int e^x\sqrt{1+e^x}\,dx$

44. $\int \sqrt{1+e^x}\,dx$

45. $\int x^5 e^{-x^3}\,dx$

46. $\int \frac{1+e^x}{1-e^x}\,dx$

47. $\int \frac{x+a}{x^2+a^2}\,dx$

48. $\int \frac{x}{x^4-a^4}\,dx$

49. $\int \frac{1}{x\sqrt{4x+1}}\,dx$

50. $\int \frac{1}{x^2\sqrt{4x+1}}\,dx$

51. $\int \frac{1}{x\sqrt{4x^2+1}}\,dx$

52. $\int \frac{dx}{x(x^4+1)}$

53. $\int x^2 \sinh mx\,dx$

54. $\int (x+\sin x)^2\,dx$

55. $\int \frac{1}{x+4+4\sqrt{x+1}}\,dx$

56. $\int \frac{x\ln x}{\sqrt{x^2-1}}\,dx$

57. $\int x\sqrt[3]{x+c}\,dx$

58. $\int x^2\ln(1+x)\,dx$

59. $\int \frac{1}{e^{3x}-e^x}\,dx$

60. $\int \frac{1}{x+\sqrt[3]{x}}\,dx$

61. $\int \frac{x^4}{x^{10}+16}\,dx$

62. $\int \frac{x^3}{(x+1)^{10}}\,dx$

63. $\int \sqrt{x}e^{\sqrt{x}}\,dx$

64. $\int_{\pi/4}^{\pi/3} \frac{\ln(\tan x)}{\sin x\cos x}\,dx$

65. $\int \frac{1}{\sqrt{x+1}+\sqrt{x}}\,dx$

66. $\int_2^3 \frac{u^3+1}{u^3-u^2}\,du$

67. $\int_1^3 \frac{\arctan\sqrt{t}}{\sqrt{t}}\,dt$

68. $\int \frac{1}{1+2e^x-e^{-x}}\,dx$

69. $\int \frac{e^{2x}}{1+e^x}\,dx$

70. $\int \frac{\ln(x+1)}{x^2}\,dx$

71. $\int \frac{x}{x^4+4x^2+3}\,dx$

72. $\int \frac{\sqrt{t}}{1+\sqrt[3]{t}}\,dt$

73. $\int \frac{1}{(x-2)(x^2+4)}\,dx$

74. $\int \frac{dx}{e^x-e^{-x}}$

75. $\int \sin x\sin 2x\sin 3x\,dx$

76. $\int (x^2-bx)\sin 2x\,dx$

77. $\int \frac{\sqrt{x}}{1+x^3}\,dx$

78. $\int \frac{\sec x\cos 2x}{\sin x+\sec x}\,dx$

79. $\int x\sin^2 x\cos x\,dx$

80. $\int \frac{\sin x\cos x}{\sin^4 x+\cos^4 x}\,dx$

81. The functions $y = e^{x^2}$ and $y = x^2e^{x^2}$ don't have elementary antiderivatives, but $y = (2x^2+1)e^{x^2}$ does. Evaluate $\int (2x^2+1)e^{x^2}\,dx$.

7.6 Integration Using Tables and Computer Algebra Systems

In this section we describe how to use tables and computer algebra systems to integrate functions that have elementary antiderivatives. You should bear in mind, though, that even the most powerful computer algebra systems can't find explicit formulas for the antiderivatives of functions like e^{x^2} or the other functions described at the end of Section 7.5.

Tables of Integrals

Tables of indefinite integrals are very useful when we are confronted by an integral that is difficult to evaluate by hand and we don't have access to a computer algebra system. A relatively brief table of 120 integrals, categorized by form, is provided on the Reference Pages at the back of the book. More extensive tables are available in *CRC Standard Mathematical Tables and Formulae,* 30th ed. by Daniel Zwillinger (Boca Raton, FL: CRC Press, 1995) (581 entries) or in Gradshteyn and Ryzhik's *Table of Integrals, Series, and Products,* 6e (New York: Academic Press, 2000), which contains hundreds of pages of integrals. It should be remembered, however, that integrals do not often occur in exactly the form listed in a table. Usually we need to use substitution or algebraic manipulation to transform a given integral into one of the forms in the table.

EXAMPLE 1 The region bounded by the curves $y = \arctan x$, $y = 0$, and $x = 1$ is rotated about the y-axis. Find the volume of the resulting solid.

SOLUTION Using the method of cylindrical shells, we see that the volume is

$$V = \int_0^1 2\pi x\arctan x\,dx$$

The Table of Integrals appears on the Reference Pages at the back of the book.

In the section of the Table of Integrals entitled *Inverse Trigonometric Forms* we locate Formula 92:

$$\int u \tan^{-1}u \, du = \frac{u^2 + 1}{2} \tan^{-1}u - \frac{u}{2} + C$$

Thus, the volume is

$$V = 2\pi \int_0^1 x \tan^{-1}x \, dx = 2\pi \left[\frac{x^2 + 1}{2} \tan^{-1}x - \frac{x}{2} \right]_0^1$$

$$= \pi[(x^2 + 1) \tan^{-1}x - x]_0^1 = \pi(2 \tan^{-1}1 - 1)$$

$$= \pi[2(\pi/4) - 1] = \tfrac{1}{2}\pi^2 - \pi$$

EXAMPLE 2 Use the Table of Integrals to find $\int \frac{x^2}{\sqrt{5 - 4x^2}} dx$.

SOLUTION If we look at the section of the table entitled *Forms involving* $\sqrt{a^2 - u^2}$, we see that the closest entry is number 34:

$$\int \frac{u^2}{\sqrt{a^2 - u^2}} du = -\frac{u}{2} \sqrt{a^2 - u^2} + \frac{a^2}{2} \sin^{-1}\left(\frac{u}{a}\right) + C$$

This is not exactly what we have, but we will be able to use it if we first make the substitution $u = 2x$:

$$\int \frac{x^2}{\sqrt{5 - 4x^2}} dx = \int \frac{(u/2)^2}{\sqrt{5 - u^2}} \frac{du}{2} = \frac{1}{8} \int \frac{u^2}{\sqrt{5 - u^2}} du$$

Then we use Formula 34 with $a^2 = 5$ (so $a = \sqrt{5}$):

$$\int \frac{x^2}{\sqrt{5 - 4x^2}} dx = \frac{1}{8} \int \frac{u^2}{\sqrt{5 - u^2}} du = \frac{1}{8}\left(-\frac{u}{2} \sqrt{5 - u^2} + \frac{5}{2} \sin^{-1} \frac{u}{\sqrt{5}} \right) + C$$

$$= -\frac{x}{8} \sqrt{5 - 4x^2} + \frac{5}{16} \sin^{-1}\left(\frac{2x}{\sqrt{5}}\right) + C$$

EXAMPLE 3 Use the Table of Integrals to find $\int x^3 \sin x \, dx$.

SOLUTION If we look in the section called *Trigonometric Forms,* we see that none of the entries explicitly includes a u^3 factor. However, we can use the reduction formula in entry 84 with $n = 3$:

$$\int x^3 \sin x \, dx = -x^3 \cos x + 3 \int x^2 \cos x \, dx$$

85. $\int u^n \cos u \, du = u^n \sin u - n \int u^{n-1} \sin u \, du$

We now need to evaluate $\int x^2 \cos x \, dx$. We can use the reduction formula in entry 85 with $n = 2$, followed by entry 82:

$$\int x^2 \cos x \, dx = x^2 \sin x - 2 \int x \sin x \, dx$$

$$= x^2 \sin x - 2(\sin x - x \cos x) + K$$

Combining these calculations, we get

$$\int x^3 \sin x\,dx = -x^3\cos x + 3x^2\sin x + 6x\cos x - 6\sin x + C$$

where $C = 3K$.

EXAMPLE 4 Use the Table of Integrals to find $\int x\sqrt{x^2 + 2x + 4}\,dx$.

SOLUTION Since the table gives forms involving $\sqrt{a^2 + x^2}$, $\sqrt{a^2 - x^2}$, and $\sqrt{x^2 - a^2}$, but not $\sqrt{ax^2 + bx + c}$, we first complete the square:

$$x^2 + 2x + 4 = (x + 1)^2 + 3$$

If we make the substitution $u = x + 1$ (so $x = u - 1$), the integrand will involve the pattern $\sqrt{a^2 + u^2}$:

$$\int x\sqrt{x^2 + 2x + 4}\,dx = \int (u - 1)\sqrt{u^2 + 3}\,du$$

$$= \int u\sqrt{u^2 + 3}\,du - \int \sqrt{u^2 + 3}\,du$$

The first integral is evaluated using the substitution $t = u^2 + 3$:

$$\int u\sqrt{u^2 + 3}\,du = \tfrac{1}{2}\int \sqrt{t}\,dt = \tfrac{1}{2}\cdot\tfrac{2}{3}t^{3/2} = \tfrac{1}{3}(u^2 + 3)^{3/2}$$

21. $\displaystyle\int \sqrt{a^2 + u^2}\,du = \frac{u}{2}\sqrt{a^2 + u^2} + \frac{a^2}{2}\ln\left(u + \sqrt{a^2 + u^2}\right) + C$

For the second integral we use Formula 21 with $a = \sqrt{3}$:

$$\int \sqrt{u^2 + 3}\,du = \frac{u}{2}\sqrt{u^2 + 3} + \tfrac{3}{2}\ln\left(u + \sqrt{u^2 + 3}\right)$$

Thus

$$\int x\sqrt{x^2 + 2x + 4}\,dx$$

$$= \tfrac{1}{3}(x^2 + 2x + 4)^{3/2} - \frac{x + 1}{2}\sqrt{x^2 + 2x + 4} - \tfrac{3}{2}\ln\left(x + 1 + \sqrt{x^2 + 2x + 4}\right) + C$$

Computer Algebra Systems

We have seen that the use of tables involves matching the form of the given integrand with the forms of the integrands in the tables. Computers are particularly good at matching patterns. And just as we used substitutions in conjunction with tables, a CAS can perform substitutions that transform a given integral into one that occurs in its stored formulas. So it isn't surprising that computer algebra systems excel at integration. That doesn't mean that integration by hand is an obsolete skill. We will see that a hand computation sometimes produces an indefinite integral in a form that is more convenient than a machine answer.

To begin, let's see what happens when we ask a machine to integrate the relatively simple function $y = 1/(3x - 2)$. Using the substitution $u = 3x - 2$, an easy calculation by hand gives

$$\int \frac{1}{3x - 2}\,dx = \tfrac{1}{3}\ln|3x - 2| + C$$

whereas Derive, Mathematica, and Maple all return the answer

$$\tfrac{1}{3}\ln(3x - 2)$$

The first thing to notice is that computer algebra systems omit the constant of integration. In other words, they produce a *particular* antiderivative, not the most general one. Therefore, when making use of a machine integration, we might have to add a constant. Second, the absolute value signs are omitted in the machine answer. That is fine if our problem is concerned only with values of x greater than $\frac{2}{3}$. But if we are interested in other values of x, then we need to insert the absolute value symbol.

In the next example we reconsider the integral of Example 4, but this time we ask a machine for the answer.

EXAMPLE 5 Use a computer algebra system to find $\int x\sqrt{x^2 + 2x + 4}\,dx$.

SOLUTION Maple responds with the answer

$$\tfrac{1}{3}(x^2 + 2x + 4)^{3/2} - \tfrac{1}{4}(2x + 2)\sqrt{x^2 + 2x + 4} - \frac{3}{2}\operatorname{arcsinh}\frac{\sqrt{3}}{3}(1 + x)$$

This looks different from the answer we found in Example 4, but it is equivalent because the third term can be rewritten using the identity

||| This is Equation 3.9.3.

$$\operatorname{arcsinh} x = \ln\left(x + \sqrt{x^2 + 1}\right)$$

Thus

$$\begin{aligned}\operatorname{arcsinh}\frac{\sqrt{3}}{3}(1 + x) &= \ln\left[\frac{\sqrt{3}}{3}(1 + x) + \sqrt{\tfrac{1}{3}(1 + x)^2 + 1}\right] \\ &= \ln\frac{1}{\sqrt{3}}\left[1 + x + \sqrt{(1 + x)^2 + 3}\right] \\ &= \ln\frac{1}{\sqrt{3}} + \ln\left(x + 1 + \sqrt{x^2 + 2x + 4}\right)\end{aligned}$$

The resulting extra term $-\frac{3}{2}\ln\left(1/\sqrt{3}\right)$ can be absorbed into the constant of integration.

Mathematica gives the answer

$$\left(\frac{5}{6} + \frac{x}{6} + \frac{x^2}{3}\right)\sqrt{x^2 + 2x + 4} - \frac{3}{2}\operatorname{arcsinh}\left(\frac{1 + x}{\sqrt{3}}\right)$$

Mathematica combined the first two terms of Example 4 (and the Maple result) into a single term by factoring.

Derive gives the answer

$$\tfrac{1}{6}\sqrt{x^2 + 2x + 4}\,(2x^2 + x + 5) - \tfrac{3}{2}\ln\left(\sqrt{x^2 + 2x + 4} + x + 1\right)$$

The first term is like the first term in the Mathematica answer, and the second term is identical to the last term in Example 4.

EXAMPLE 6 Use a CAS to evaluate $\int x(x^2 + 5)^8\,dx$.

SOLUTION Maple and Mathematica give the same answer:

$$\tfrac{1}{18}x^{18} + \tfrac{5}{2}x^{16} + 50x^{14} + \tfrac{1750}{3}x^{12} + 4375x^{10} + 21875x^8 + \tfrac{218750}{3}x^6 + 156250x^4 + \tfrac{390625}{2}x^2$$

It's clear that both systems must have expanded $(x^2 + 5)^8$ by the Binomial Theorem and then integrated each term.

If we integrate by hand instead, using the substitution $u = x^2 + 5$, we get

|||| Derive and the TI-89/92 also give this answer.

$$\int x(x^2 + 5)^8\,dx = \tfrac{1}{18}(x^2 + 5)^9 + C$$

For most purposes, this is a more convenient form of the answer.

EXAMPLE 7 Use a CAS to find $\int \sin^5 x \cos^2 x\,dx$.

SOLUTION In Example 2 in Section 7.2 we found that

$$\boxed{1} \qquad \int \sin^5 x \cos^2 x\,dx = -\tfrac{1}{3}\cos^3 x + \tfrac{2}{5}\cos^5 x - \tfrac{1}{7}\cos^7 x + C$$

Derive and Maple report the answer

$$-\tfrac{1}{7}\sin^4 x \cos^3 x - \tfrac{4}{35}\sin^2 x \cos^3 x - \tfrac{8}{105}\cos^3 x$$

whereas Mathematica produces

$$-\tfrac{5}{64}\cos x - \tfrac{1}{192}\cos 3x + \tfrac{3}{320}\cos 5x - \tfrac{1}{448}\cos 7x$$

We suspect that there are trigonometric identities which show these three answers are equivalent. Indeed, if we ask Derive, Maple, and Mathematica to simplify their expressions using trigonometric identities, they ultimately produce the same form of the answer as in Equation 1.

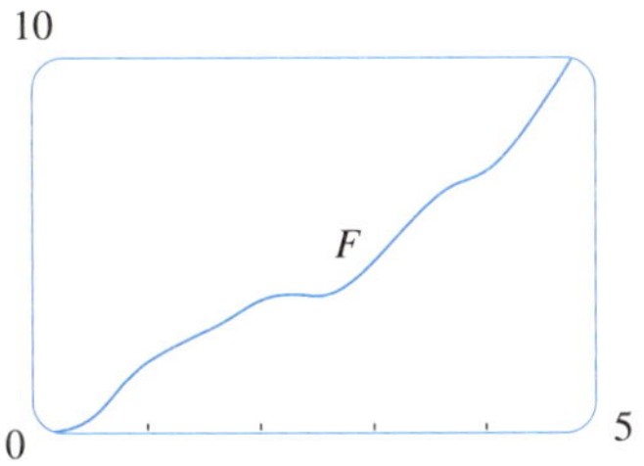

FIGURE 1

EXAMPLE 8 If $f(x) = x + 60 \sin^4 x \cos^5 x$, find the antiderivative F of f such that $F(0) = 0$. Graph F for $0 \leqslant x \leqslant 5$. Where does F have extreme values and inflection points?

SOLUTION The antiderivative of f produced by Maple is

$$F(x) = \tfrac{1}{2}x^2 - \tfrac{20}{3}\sin^3 x \cos^6 x - \tfrac{20}{7}\sin x \cos^6 x + \tfrac{4}{7}\cos^4 x \sin x + \tfrac{16}{21}\cos^2 x \sin x + \tfrac{32}{21}\sin x$$

and we note that $F(0) = 0$. This expression could probably be simplified, but there's no need to do so because a computer algebra system can graph this version of F as easily as any other version. A graph of F is shown in Figure 1. To locate the extreme values of F, we graph its derivative $F' = f$ in Figure 2 and observe that F has a local maximum when $x \approx 2.3$ and a local minimum when $x \approx 2.5$. The graph of $F'' = f'$ in Figure 2 shows that F has inflection points when $x \approx 0.7, 1.3, 1.8, 2.4, 3.3$, and 3.9.

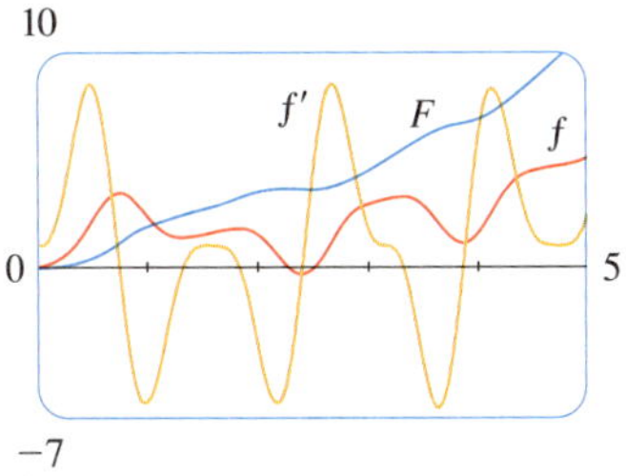

FIGURE 2

7.6 Exercises

1–4 |||| Use the indicated entry in the Table of Integrals on the Reference Pages to evaluate the integral.

1. $\displaystyle\int \frac{\sqrt{7 - 2x^2}}{x^2}\,dx$; entry 33

2. $\displaystyle\int \frac{3x}{\sqrt{3 - 2x}}\,dx$; entry 55

3. $\displaystyle\int \sec^3(\pi x)\,dx$; entry 71

4. $\displaystyle\int e^{2\theta}\sin 3\theta\,d\theta$; entry 98

5–30 Use the Table of Integrals on the Reference Pages to evaluate the integral.

5. $\int_0^1 2x \cos^{-1}x \, dx$

6. $\int_2^3 \frac{1}{x^2\sqrt{4x^2 - 7}} \, dx$

7. $\int e^{-3x} \cos 4x \, dx$

8. $\int \csc^3(x/2) \, dx$

9. $\int \frac{dx}{x^2\sqrt{4x^2 + 9}}$

10. $\int \frac{\sqrt{2y^2 - 3}}{y^2} \, dy$

11. $\int_{-1}^0 t^2 e^{-t} \, dt$

12. $\int_0^\pi x^2 \cos 3x \, dx$

13. $\int \frac{\tan^3(1/z)}{z^2} \, dz$

14. $\int \sin^{-1}\sqrt{x} \, dx$

15. $\int e^x \operatorname{sech}(e^x) \, dx$

16. $\int x \sin(x^2) \cos(3x^2) \, dx$

17. $\int y\sqrt{6 + 4y - 4y^2} \, dy$

18. $\int \frac{x^5}{x^2 + \sqrt{2}} \, dx$

19. $\int \sin^2 x \cos x \ln(\sin x) \, dx$

20. $\int \frac{dx}{e^x(1 + 2e^x)}$

21. $\int \frac{e^x}{3 - e^{2x}} \, dx$

22. $\int_0^2 x^3\sqrt{4x^2 - x^4} \, dx$

23. $\int \sec^5 x \, dx$

24. $\int \sin^6 2x \, dx$

25. $\int \frac{\sqrt{4 + (\ln x)^2}}{x} \, dx$

26. $\int_0^1 x^4 e^{-x} \, dx$

27. $\int \sqrt{e^{2x} - 1} \, dx$

28. $\int e^t \sin(\alpha t - 3) \, dt$

29. $\int \frac{x^4 \, dx}{\sqrt{x^{10} - 2}}$

30. $\int \frac{\sec^2\theta \tan^2\theta}{\sqrt{9 - \tan^2\theta}} \, d\theta$

31. Find the volume of the solid obtained when the region under the curve $y = x\sqrt{4 - x^2}$, $0 \leq x \leq 2$, is rotated about the y-axis.

32. The region under the curve $y = \tan^2 x$ from 0 to $\pi/4$ is rotated about the x-axis. Find the volume of the resulting solid.

33. Verify Formula 53 in the Table of Integrals (a) by differentiation and (b) by using the substitution $t = a + bu$.

34. Verify Formula 31 (a) by differentiation and (b) by substituting $u = a \sin \theta$.

CAS **35–42** Use a computer algebra system to evaluate the integral. Compare the answer with the result of using tables. If the answers are not the same, show that they are equivalent.

35. $\int x^2\sqrt{5 - x^2} \, dx$

36. $\int x^2(1 + x^3)^4 \, dx$

37. $\int \sin^3 x \cos^2 x \, dx$

38. $\int \tan^2 x \sec^4 x \, dx$

39. $\int x\sqrt{1 + 2x} \, dx$

40. $\int \sin^4 x \, dx$

41. $\int \tan^5 x \, dx$

42. $\int x^5\sqrt{x^2 + 1} \, dx$

CAS **43.** Computer algebra systems sometimes need a helping hand from human beings. Ask your CAS to evaluate

$$\int 2^x \sqrt{4^x - 1} \, dx$$

If it doesn't return an answer, ask it to try

$$\int 2^x \sqrt{2^{2x} - 1} \, dx$$

instead. Why do you think it was successful with this form of the integrand?

CAS **44.** Try to evaluate

$$\int (1 + \ln x)\sqrt{1 + (x \ln x)^2} \, dx$$

with a computer algebra system. If it doesn't return an answer, make a substitution that changes the integral into one that the CAS *can* evaluate.

CAS **45–48** Use a CAS to find an antiderivative F of f such that $F(0) = 0$. Graph f and F and locate approximately the x-coordinates of the extreme points and inflection points of F.

45. $f(x) = \frac{x^2 - 1}{x^4 + x^2 + 1}$

46. $f(x) = xe^{-x} \sin x, \quad -5 \leq x \leq 5$

47. $f(x) = \sin^4 x \cos^6 x, \quad 0 \leq x \leq \pi$

48. $f(x) = \frac{x^3 - x}{x^6 + 1}$

DISCOVERY PROJECT

CAS Patterns in Integrals

In this project a computer algebra system is used to investigate indefinite integrals of families of functions. By observing the patterns that occur in the integrals of several members of the family, you will first guess, and then prove, a general formula for the integral of any member of the family.

1. (a) Use a computer algebra system to evaluate the following integrals.

(i) $\displaystyle\int \frac{1}{(x+2)(x+3)}\,dx$ (ii) $\displaystyle\int \frac{1}{(x+1)(x+5)}\,dx$

(iii) $\displaystyle\int \frac{1}{(x+2)(x-5)}\,dx$ (iv) $\displaystyle\int \frac{1}{(x+2)^2}\,dx$

(b) Based on the pattern of your responses in part (a), guess the value of the integral

$$\int \frac{1}{(x+a)(x+b)}\,dx$$

if $a \neq b$. What if $a = b$?

(c) Check your guess by asking your CAS to evaluate the integral in part (b). Then prove it using partial fractions.

2. (a) Use a computer algebra system to evaluate the following integrals.

(i) $\int \sin x \cos 2x\,dx$ (ii) $\int \sin 3x \cos 7x\,dx$ (iii) $\int \sin 8x \cos 3x\,dx$

(b) Based on the pattern of your responses in part (a), guess the value of the integral

$$\int \sin ax \cos bx\,dx$$

(c) Check your guess with a CAS. Then prove it using the techniques of Section 7.2. For what values of a and b is it valid?

3. (a) Use a computer algebra system to evaluate the following integrals.

(i) $\int \ln x\,dx$ (ii) $\int x \ln x\,dx$ (iii) $\int x^2 \ln x\,dx$

(iv) $\int x^3 \ln x\,dx$ (v) $\int x^7 \ln x\,dx$

(b) Based on the pattern of your responses in part (a), guess the value of

$$\int x^n \ln x\,dx$$

(c) Use integration by parts to prove the conjecture that you made in part (b). For what values of n is it valid?

4. (a) Use a computer algebra system to evaluate the following integrals.

(i) $\int xe^x\,dx$ (ii) $\int x^2e^x\,dx$ (iii) $\int x^3e^x\,dx$

(iv) $\int x^4e^x\,dx$ (v) $\int x^5e^x\,dx$

(b) Based on the pattern of your responses in part (a), guess the value of $\int x^6e^x\,dx$. Then use your CAS to check your guess.

(c) Based on the patterns in parts (a) and (b), make a conjecture as to the value of the integral

$$\int x^ne^x\,dx$$

when n is a positive integer.

(d) Use mathematical induction to prove the conjecture you made in part (c).

7.7 Approximate Integration

There are two situations in which it is impossible to find the exact value of a definite integral.

The first situation arises from the fact that in order to evaluate $\int_a^b f(x)\,dx$ using the Fundamental Theorem of Calculus we need to know an antiderivative of f. Sometimes, however, it is difficult, or even impossible, to find an antiderivative (see Section 7.5). For example, it is impossible to evaluate the following integrals exactly:

$$\int_0^1 e^{x^2}\,dx \qquad \int_{-1}^1 \sqrt{1 + x^3}\,dx$$

The second situation arises when the function is determined from a scientific experiment through instrument readings or collected data. There may be no formula for the function (see Example 5).

In both cases we need to find approximate values of definite integrals. We already know one such method. Recall that the definite integral is defined as a limit of Riemann sums, so any Riemann sum could be used as an approximation to the integral: If we divide $[a, b]$ into n subintervals of equal length $\Delta x = (b - a)/n$, then we have

$$\int_a^b f(x)\,dx \approx \sum_{i=1}^{n} f(x_i^*)\,\Delta x$$

where x_i^* is any point in the ith subinterval $[x_{i-1}, x_i]$. If x_i^* is chosen to be the left endpoint of the interval, then $x_i^* = x_{i-1}$ and we have

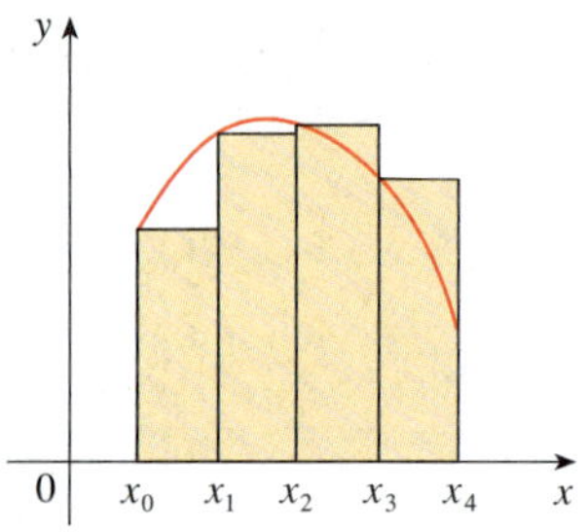

(a) Left endpoint approximation

1 $$\int_a^b f(x)\,dx \approx L_n = \sum_{i=1}^{n} f(x_{i-1})\,\Delta x$$

If $f(x) \geqslant 0$, then the integral represents an area and (1) represents an approximation of this area by the rectangles shown in Figure 1(a). If we choose x_i^* to be the right endpoint, then $x_i^* = x_i$ and we have

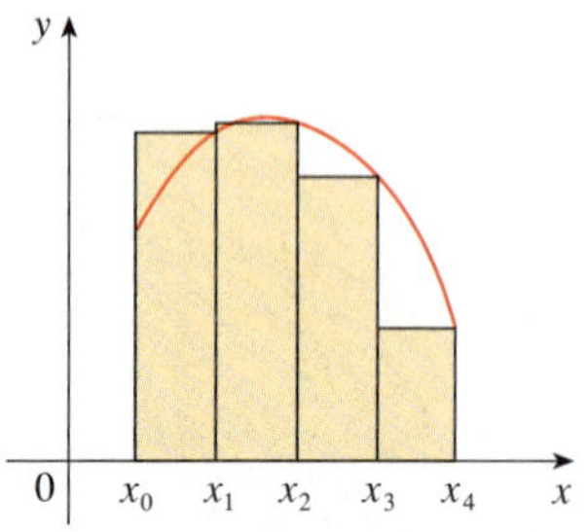

(b) Right endpoint approximation

2 $$\int_a^b f(x)\,dx \approx R_n = \sum_{i=1}^{n} f(x_i)\,\Delta x$$

[See Figure 1(b).] The approximations L_n and R_n defined by Equations 1 and 2 are called the **left endpoint approximation** and **right endpoint approximation**, respectively.

In Section 5.2 we also considered the case where x_i^* is chosen to be the midpoint $\bar{x}_i$ of the subinterval $[x_{i-1}, x_i]$. Figure 1(c) shows the midpoint approximation M_n, which appears to be better than either L_n or R_n.

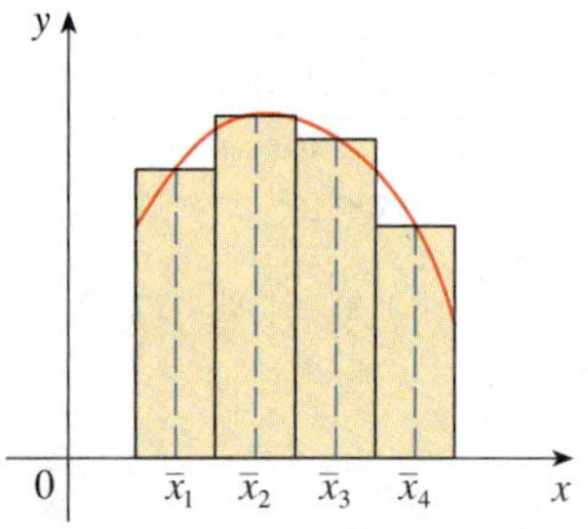

(c) Midpoint approximation

FIGURE 1

Midpoint Rule

$$\int_a^b f(x)\,dx \approx M_n = \Delta x\,[f(\bar{x}_1) + f(\bar{x}_2) + \cdots + f(\bar{x}_n)]$$

where $$\Delta x = \frac{b - a}{n}$$

and $$\bar{x}_i = \tfrac{1}{2}(x_{i-1} + x_i) = \text{midpoint of } [x_{i-1}, x_i]$$

Another approximation, called the Trapezoidal Rule, results from averaging the approximations in Equations 1 and 2:

$$\int_a^b f(x)\,dx \approx \frac{1}{2}\left[\sum_{i=1}^{n} f(x_{i-1})\,\Delta x + \sum_{i=1}^{n} f(x_i)\,\Delta x\right] = \frac{\Delta x}{2}\left[\sum_{i=1}^{n}\left(f(x_{i-1}) + f(x_i)\right)\right]$$

$$= \frac{\Delta x}{2}\left[(f(x_0) + f(x_1)) + (f(x_1) + f(x_2)) + \cdots + (f(x_{n-1}) + f(x_n))\right]$$

$$= \frac{\Delta x}{2}\left[f(x_0) + 2f(x_1) + 2f(x_2) + \cdots + 2f(x_{n-1}) + f(x_n)\right]$$

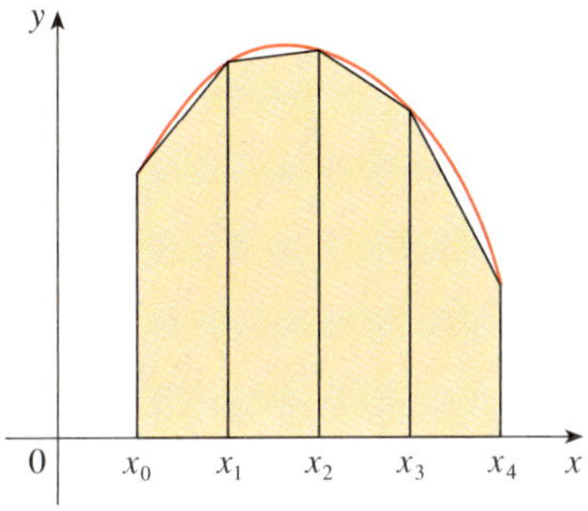

FIGURE 2
Trapezoidal approximation

Trapezoidal Rule

$$\int_a^b f(x)\,dx \approx T_n = \frac{\Delta x}{2}\left[f(x_0) + 2f(x_1) + 2f(x_2) + \cdots + 2f(x_{n-1}) + f(x_n)\right]$$

where $\Delta x = (b - a)/n$ and $x_i = a + i\,\Delta x$.

The reason for the name Trapezoidal Rule can be seen from Figure 2, which illustrates the case $f(x) \geqslant 0$. The area of the trapezoid that lies above the ith subinterval is

$$\Delta x\left(\frac{f(x_{i-1}) + f(x_i)}{2}\right) = \frac{\Delta x}{2}\left[f(x_{i-1}) + f(x_i)\right]$$

and if we add the areas of all these trapezoids, we get the right side of the Trapezoidal Rule.

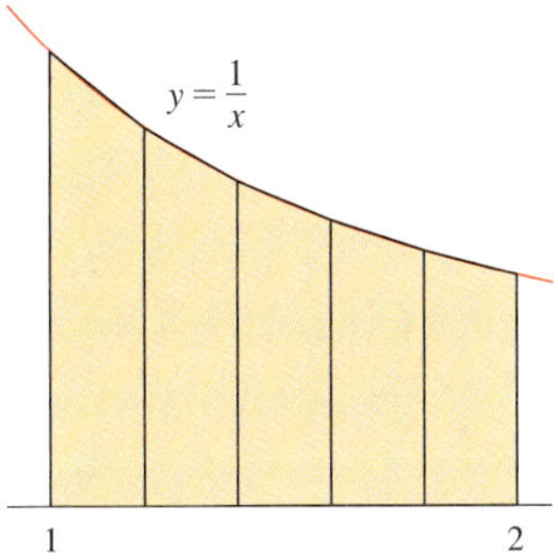

FIGURE 3

EXAMPLE 1 Use (a) the Trapezoidal Rule and (b) the Midpoint Rule with $n = 5$ to approximate the integral $\int_1^2 (1/x)\,dx$.

SOLUTION

(a) With $n = 5$, $a = 1$, and $b = 2$, we have $\Delta x = (2 - 1)/5 = 0.2$, and so the Trapezoidal Rule gives

$$\int_1^2 \frac{1}{x}\,dx \approx T_5 = \frac{0.2}{2}\left[f(1) + 2f(1.2) + 2f(1.4) + 2f(1.6) + 2f(1.8) + f(2)\right]$$

$$= 0.1\left(\frac{1}{1} + \frac{2}{1.2} + \frac{2}{1.4} + \frac{2}{1.6} + \frac{2}{1.8} + \frac{1}{2}\right)$$

$$\approx 0.695635$$

This approximation is illustrated in Figure 3.

(b) The midpoints of the five subintervals are 1.1, 1.3, 1.5, 1.7, and 1.9, so the Midpoint Rule gives

$$\int_1^2 \frac{1}{x}\,dx \approx \Delta x\left[f(1.1) + f(1.3) + f(1.5) + f(1.7) + f(1.9)\right]$$

$$= \frac{1}{5}\left(\frac{1}{1.1} + \frac{1}{1.3} + \frac{1}{1.5} + \frac{1}{1.7} + \frac{1}{1.9}\right)$$

$$\approx 0.691908$$

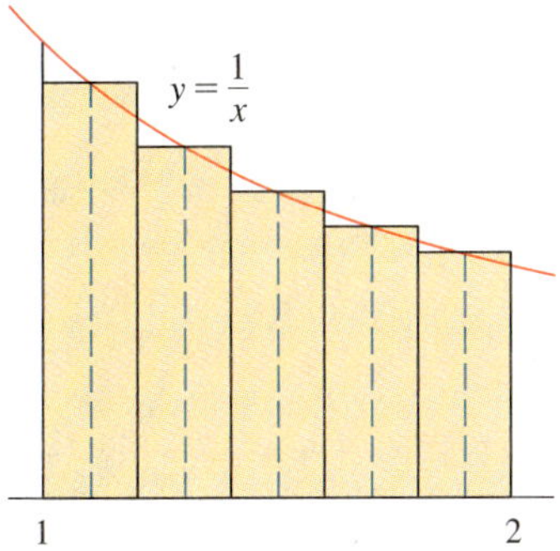

FIGURE 4

This approximation is illustrated in Figure 4.

In Example 1 we deliberately chose an integral whose value can be computed explicitly so that we can see how accurate the Trapezoidal and Midpoint Rules are. By the Fundamental Theorem of Calculus,

$$\int_1^2 \frac{1}{x}\,dx = \ln x\Big]_1^2 = \ln 2 = 0.693147\ldots$$

$\int_a^b f(x)\,dx = \text{approximation} + \text{error}$

The **error** in using an approximation is defined to be the amount that needs to be added to the approximation to make it exact. From the values in Example 1 we see that the errors in the Trapezoidal and Midpoint Rule approximations for $n = 5$ are

$$E_T \approx -0.002488 \qquad \text{and} \qquad E_M \approx 0.001239$$

In general, we have

$$E_T = \int_a^b f(x)\,dx - T_n \qquad \text{and} \qquad E_M = \int_a^b f(x)\,dx - M_n$$

TEC Module 5.1/5.2/7.7 allows you to compare approximation methods.

The following tables show the results of calculations similar to those in Example 1, but for $n = 5$, 10, and 20 and for the left and right endpoint approximations as well as the Trapezoidal and Midpoint Rules.

Approximations to $\int_1^2 \frac{1}{x}\,dx$

n	L_n	R_n	T_n	M_n
5	0.745635	0.645635	0.695635	0.691908
10	0.718771	0.668771	0.693771	0.692835
20	0.705803	0.680803	0.693303	0.693069

Corresponding errors

n	E_L	E_R	E_T	E_M
5	−0.052488	0.047512	−0.002488	0.001239
10	−0.025624	0.024376	−0.000624	0.000312
20	−0.012656	0.012344	−0.000156	0.000078

We can make several observations from these tables:

1. In all of the methods we get more accurate approximations when we increase the value of n. (But very large values of n result in so many arithmetic operations that we have to beware of accumulated round-off error.)
2. The errors in the left and right endpoint approximations are opposite in sign and appear to decrease by a factor of about 2 when we double the value of n.
3. The Trapezoidal and Midpoint Rules are much more accurate than the endpoint approximations.
4. The errors in the Trapezoidal and Midpoint Rules are opposite in sign and appear to decrease by a factor of about 4 when we double the value of n.
5. The size of the error in the Midpoint Rule is about half the size of the error in the Trapezoidal Rule.

|||| It turns out that these observations are true in most cases.

Figure 5 shows why we can usually expect the Midpoint Rule to be more accurate than the Trapezoidal Rule. The area of a typical rectangle in the Midpoint Rule is the same as the trapezoid $ABCD$ whose upper side is tangent to the graph at P. The area of this trapezoid is closer to the area under the graph than is the area of the trapezoid $AQRD$ used in the Trapezoidal Rule. [The midpoint error (shaded red) is smaller than the trapezoidal error (shaded blue).]

These observations are corroborated in the following error estimates, which are proved in books on numerical analysis. Notice that Observation 4 corresponds to the n^2 in each denominator because $(2n)^2 = 4n^2$. The fact that the estimates depend on the size of the second derivative is not surprising if you look at Figure 5, because $f''(x)$ measures how much the graph is curved. [Recall that $f''(x)$ measures how fast the slope of $y = f(x)$ changes.]

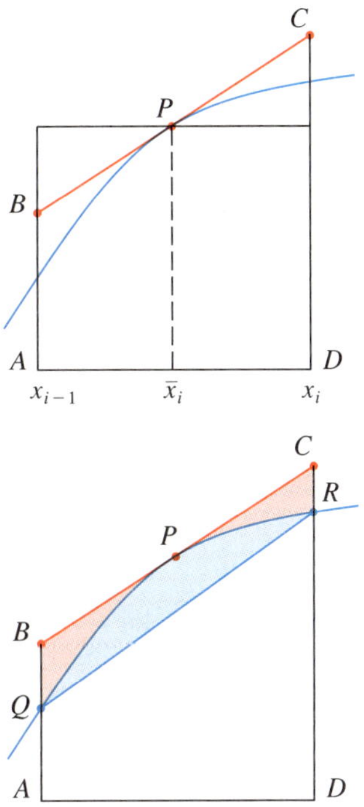

FIGURE 5

3 Error Bounds Suppose $|f''(x)| \leqslant K$ for $a \leqslant x \leqslant b$. If E_T and E_M are the errors in the Trapezoidal and Midpoint Rules, then

$$|E_T| \leqslant \frac{K(b-a)^3}{12n^2} \qquad \text{and} \qquad |E_M| \leqslant \frac{K(b-a)^3}{24n^2}$$

Let's apply this error estimate to the Trapezoidal Rule approximation in Example 1. If $f(x) = 1/x$, then $f'(x) = -1/x^2$ and $f''(x) = 2/x^3$. Since $1 \leqslant x \leqslant 2$, we have $1/x \leqslant 1$, so

$$|f''(x)| = \left|\frac{2}{x^3}\right| \leqslant \frac{2}{1^3} = 2$$

Therefore, taking $K = 2$, $a = 1$, $b = 2$, and $n = 5$ in the error estimate (3), we see that

|||| K can be any number larger than all the values of $|f''(x)|$, but smaller values of K give better error bounds.

$$|E_T| \leqslant \frac{2(2-1)^3}{12(5)^2} = \frac{1}{150} \approx 0.006667$$

Comparing this error estimate of 0.006667 with the actual error of about 0.002488, we see that it can happen that the actual error is substantially less than the upper bound for the error given by (3).

EXAMPLE 2 How large should we take n in order to guarantee that the Trapezoidal and Midpoint Rule approximations for $\int_1^2 (1/x)\,dx$ are accurate to within 0.0001?

SOLUTION We saw in the preceding calculation that $|f''(x)| \leqslant 2$ for $1 \leqslant x \leqslant 2$, so we can take $K = 2$, $a = 1$, and $b = 2$ in (3). Accuracy to within 0.0001 means that the size of the error should be less than 0.0001. Therefore, we choose n so that

$$\frac{2(1)^3}{12n^2} < 0.0001$$

Solving the inequality for n, we get

$$n^2 > \frac{2}{12(0.0001)}$$

or

$$n > \frac{1}{\sqrt{0.0006}} \approx 40.8$$

|||| It's quite possible that a lower value for n would suffice, but 41 is the smallest value for which the error bound formula can *guarantee* us accuracy to within 0.0001.

Thus, $n = 41$ will ensure the desired accuracy.

For the same accuracy with the Midpoint Rule we choose n so that

$$\frac{2(1)^3}{24n^2} < 0.0001$$

which gives

$$n > \frac{1}{\sqrt{0.0012}} \approx 29$$

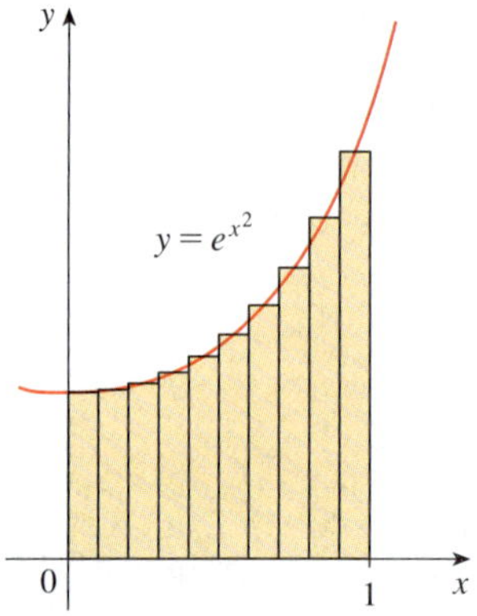

FIGURE 6

EXAMPLE 3

(a) Use the Midpoint Rule with $n = 10$ to approximate the integral $\int_0^1 e^{x^2}\,dx$.
(b) Give an upper bound for the error involved in this approximation.

SOLUTION

(a) Since $a = 0$, $b = 1$, and $n = 10$, the Midpoint Rule gives

$$\int_0^1 e^{x^2}\,dx \approx \Delta x\,[f(0.05) + f(0.15) + \cdots + f(0.85) + f(0.95)]$$

$$= 0.1[e^{0.0025} + e^{0.0225} + e^{0.0625} + e^{0.1225} + e^{0.2025} + e^{0.3025} + e^{0.4225} + e^{0.5625} + e^{0.7225} + e^{0.9025}]$$

$$\approx 1.460393$$

Figure 6 illustrates this approximation.

(b) Since $f(x) = e^{x^2}$, we have $f'(x) = 2xe^{x^2}$ and $f''(x) = (2 + 4x^2)e^{x^2}$. Also, since $0 \leqslant x \leqslant 1$, we have $x^2 \leqslant 1$ and so

$$0 \leqslant f''(x) = (2 + 4x^2)e^{x^2} \leqslant 6e$$

|||| Error estimates are upper bounds for the error. They give theoretical, worst-case scenarios. The actual error in this case turns out to be about 0.0023.

Taking $K = 6e$, $a = 0$, $b = 1$, and $n = 10$ in the error estimate (3), we see that an upper bound for the error is

$$\frac{6e(1)^3}{24(10)^2} = \frac{e}{400} \approx 0.007$$

|||| Simpson's Rule

Another rule for approximate integration results from using parabolas instead of straight line segments to approximate a curve. As before, we divide $[a, b]$ into n subintervals of equal length $h = \Delta x = (b - a)/n$, but this time we assume that n is an *even* number. Then on each consecutive pair of intervals we approximate the curve $y = f(x) \geqslant 0$ by a parabola as shown in Figure 7. If $y_i = f(x_i)$, then $P_i(x_i, y_i)$ is the point on the curve lying above x_i. A typical parabola passes through three consecutive points P_i, P_{i+1}, and P_{i+2}.

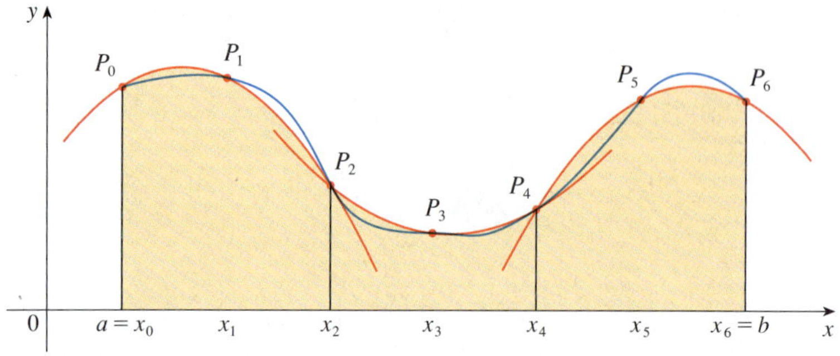

FIGURE 7

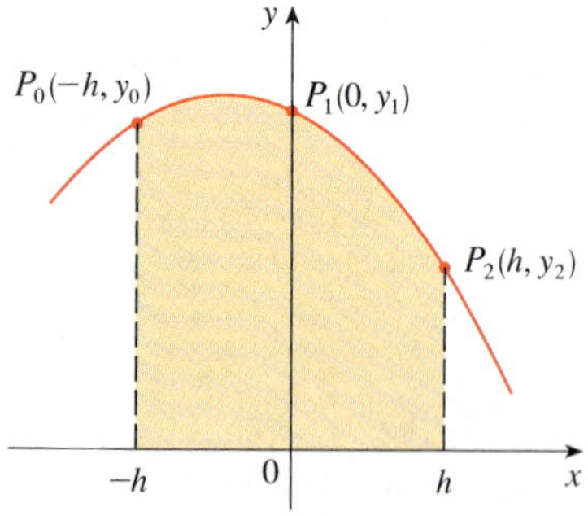

FIGURE 8

To simplify our calculations, we first consider the case where $x_0 = -h$, $x_1 = 0$, and $x_2 = h$. (See Figure 8.) We know that the equation of the parabola through P_0, P_1, and P_2 is of the form $y = Ax^2 + Bx + C$ and so the area under the parabola from $x = -h$ to $x = h$ is

|||| Here we have used Theorem 5.5.7. Notice that $Ax^2 + C$ is even and Bx is odd.

$$\int_{-h}^{h} (Ax^2 + Bx + C)\,dx = 2\int_0^h (Ax^2 + C)\,dx$$

$$= 2\left[A\frac{x^3}{3} + Cx\right]_0^h$$

$$= 2\left(A\frac{h^3}{3} + Ch\right) = \frac{h}{3}(2Ah^2 + 6C)$$

But, since the parabola passes through $P_0(-h, y_0)$, $P_1(0, y_1)$, and $P_2(h, y_2)$, we have

$$y_0 = A(-h)^2 + B(-h) + C = Ah^2 - Bh + C$$

$$y_1 = C$$

$$y_2 = Ah^2 + Bh + C$$

and therefore
$$y_0 + 4y_1 + y_2 = 2Ah^2 + 6C$$

Thus, we can rewrite the area under the parabola as

$$\frac{h}{3}(y_0 + 4y_1 + y_2)$$

Now, by shifting this parabola horizontally we do not change the area under it. This means that the area under the parabola through P_0, P_1, and P_2 from $x = x_0$ to $x = x_2$ in Figure 7 is still

$$\frac{h}{3}(y_0 + 4y_1 + y_2)$$

Similarly, the area under the parabola through P_2, P_3, and P_4 from $x = x_2$ to $x = x_4$ is

$$\frac{h}{3}(y_2 + 4y_3 + y_4)$$

If we compute the areas under all the parabolas in this manner and add the results, we get

$$\int_a^b f(x)\,dx \approx \frac{h}{3}(y_0 + 4y_1 + y_2) + \frac{h}{3}(y_2 + 4y_3 + y_4) + \cdots + \frac{h}{3}(y_{n-2} + 4y_{n-1} + y_n)$$

$$= \frac{h}{3}(y_0 + 4y_1 + 2y_2 + 4y_3 + 2y_4 + \cdots + 2y_{n-2} + 4y_{n-1} + y_n)$$

Although we have derived this approximation for the case in which $f(x) \geqslant 0$, it is a reasonable approximation for any continuous function f and is called Simpson's Rule after the English mathematician Thomas Simpson (1710–1761). Note the pattern of coefficients: 1, 4, 2, 4, 2, 4, 2, . . . , 4, 2, 4, 1.

|||| Thomas Simpson was a weaver who taught himself mathematics and went on to become one of the best English mathematicians of the 18th century. What we call Simpson's Rule was actually known to Cavalieri and Gregory in the 17th century, but Simpson popularized it in his best-selling calculus textbook, entitled *A New Treatise of Fluxions.*

Simpson's Rule

$$\int_a^b f(x)\,dx \approx S_n = \frac{\Delta x}{3}[f(x_0) + 4f(x_1) + 2f(x_2) + 4f(x_3) + \cdots + 2f(x_{n-2}) + 4f(x_{n-1}) + f(x_n)]$$

where n is even and $\Delta x = (b - a)/n$.

EXAMPLE 4 Use Simpson's Rule with $n = 10$ to approximate $\int_1^2 (1/x)\,dx$.

SOLUTION Putting $f(x) = 1/x$, $n = 10$, and $\Delta x = 0.1$ in Simpson's Rule, we obtain

$$\begin{aligned}\int_1^2 \frac{1}{x}\,dx &\approx S_{10}\\ &= \frac{\Delta x}{3}[f(1) + 4f(1.1) + 2f(1.2) + 4f(1.3) + \cdots + 2f(1.8) + 4f(1.9) + f(2)]\\ &= \frac{0.1}{3}\left(\frac{1}{1} + \frac{4}{1.1} + \frac{2}{1.2} + \frac{4}{1.3} + \frac{2}{1.4} + \frac{4}{1.5} + \frac{2}{1.6} + \frac{4}{1.7} + \frac{2}{1.8} + \frac{4}{1.9} + \frac{1}{2}\right)\\ &\approx 0.693150\end{aligned}$$

Notice that, in Example 4, Simpson's Rule gives us a *much* better approximation ($S_{10} \approx 0.693150$) to the true value of the integral ($\ln 2 \approx 0.693147\ldots$) than does the Trapezoidal Rule ($T_{10} \approx 0.693771$) or the Midpoint Rule ($M_{10} \approx 0.692835$). It turns out (see Exercise 48) that the approximations in Simpson's Rule are weighted averages of those in the Trapezoidal and Midpoint Rules:

$$S_{2n} = \tfrac{1}{3}T_n + \tfrac{2}{3}M_n$$

(Recall that E_T and E_M usually have opposite signs and $|E_M|$ is about half the size of $|E_T|$.)

In many applications of calculus we need to evaluate an integral even if no explicit formula is known for y as a function of x. A function may be given graphically or as a table of values of collected data. If there is evidence that the values are not changing rapidly, then the Trapezoidal Rule or Simpson's Rule can still be used to find an approximate value for $\int_a^b y\,dx$, the integral of y with respect to x.

EXAMPLE 5 Figure 9 shows data traffic on the link from the United States to SWITCH, the Swiss academic and research network, on February 10, 1998. $D(t)$ is the data throughput, measured in megabits per second (Mb/s). Use Simpson's Rule to estimate the total amount of data transmitted on the link up to noon on that day.

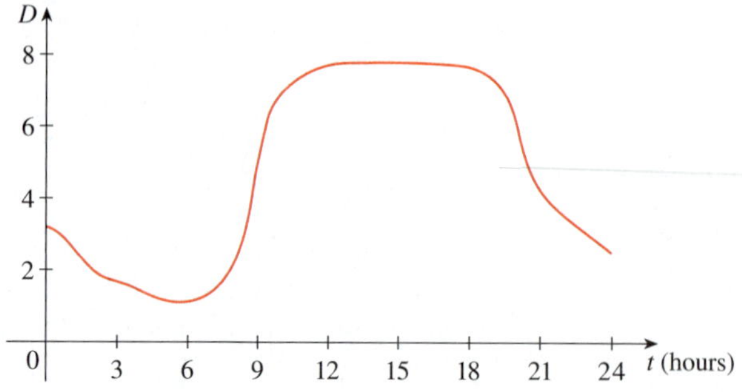

FIGURE 9

SOLUTION Because we want the units to be consistent and $D(t)$ is measured in megabits per second, we convert the units for t from hours to seconds. If we let $A(t)$ be the amount of data (in megabits) transmitted by time t, where t is measured in seconds, then $A'(t) = D(t)$. So, by the Net Change Theorem (see Section 5.4), the total amount of data transmitted by noon (when $t = 12 \times 60^2 = 43{,}200$) is

$$A(43{,}200) = \int_0^{43{,}200} D(t)\,dt$$

We estimate the values of $D(t)$ at hourly intervals from the graph and compile them in the table.

t (hours)	t (seconds)	$D(t)$	t (hours)	t (seconds)	$D(t)$
0	0	3.2	7	25,200	1.3
1	3,600	2.7	8	28,800	2.8
2	7,200	1.9	9	32,400	5.7
3	10,800	1.7	10	36,000	7.1
4	14,400	1.3	11	39,600	7.7
5	18,000	1.0	12	43,200	7.9
6	21,600	1.1			

Then we use Simpson's Rule with $n = 12$ and $\Delta t = 3600$ to estimate the integral:

$$\int_0^{43{,}200} A(t)\,dt \approx \frac{\Delta t}{3}[D(0) + 4D(3600) + 2D(7200) + \cdots + 4D(39{,}600) + D(43{,}200)]$$

$$\approx \frac{3600}{3}[3.2 + 4(2.7) + 2(1.9) + 4(1.7) + 2(1.3) + 4(1.0)$$
$$+ 2(1.1) + 4(1.3) + 2(2.8) + 4(5.7) + 2(7.1) + 4(7.7) + 7.9]$$

$$= 143{,}880$$

Thus, the total amount of data transmitted up to noon is about 144,000 megabits, or 144 gigabits.

In Exercises 27 and 28 you are asked to demonstrate, in particular cases, that the error in Simpson's Rule decreases by a factor of about 16 when n is doubled. That is consistent with the appearance of n^4 in the denominator of the following error estimate for Simpson's Rule. It is similar to the estimates given in (3) for the Trapezoidal and Midpoint Rules, but it uses the fourth derivative of f.

4 Error Bound for Simpson's Rule Suppose that $|f^{(4)}(x)| \leq K$ for $a \leq x \leq b$. If E_S is the error involved in using Simpson's Rule, then

$$|E_S| \leq \frac{K(b-a)^5}{180n^4}$$

EXAMPLE 6 How large should we take n in order to guarantee that the Simpson's Rule approximation for $\int_1^2 (1/x)\,dx$ is accurate to within 0.0001?

SOLUTION If $f(x) = 1/x$, then $f^{(4)}(x) = 24/x^5$. Since $x \geqslant 1$, we have $1/x \leqslant 1$ and so

$$|f^{(4)}(x)| = \left|\frac{24}{x^5}\right| \leqslant 24$$

Therefore, we can take $K = 24$ in (4). Thus, for an error less than 0.0001 we should choose n so that

$$\frac{24(1)^5}{180n^4} < 0.0001$$

This gives

$$n^4 > \frac{24}{180(0.0001)}$$

or

$$n > \frac{1}{\sqrt[4]{0.00075}} \approx 6.04$$

Therefore, $n = 8$ (n must be even) gives the desired accuracy. (Compare this with Example 2, where we obtained $n = 41$ for the Trapezoidal Rule and $n = 29$ for the Midpoint Rule.)

|||| Many calculators and computer algebra systems have a built-in algorithm that computes an approximation of a definite integral. Some of these machines use Simpson's Rule; others use more sophisticated techniques such as *adaptive* numerical integration. This means that if a function fluctuates much more on a certain part of the interval than it does elsewhere, then that part gets divided into more subintervals. This strategy reduces the number of calculations required to achieve a prescribed accuracy.

EXAMPLE 7

(a) Use Simpson's Rule with $n = 10$ to approximate the integral $\int_0^1 e^{x^2}\,dx$.

(b) Estimate the error involved in this approximation.

SOLUTION

(a) If $n = 10$, then $\Delta x = 0.1$ and Simpson's Rule gives

$$\int_0^1 e^{x^2}dx \approx \frac{\Delta x}{3}[f(0) + 4f(0.1) + 2f(0.2) + \cdots + 2f(0.8) + 4f(0.9) + f(1)]$$

$$= \frac{0.1}{3}[e^0 + 4e^{0.01} + 2e^{0.04} + 4e^{0.09} + 2e^{0.16} + 4e^{0.25} + 2e^{0.36} + 4e^{0.49} + 2e^{0.64} + 4e^{0.81} + e^1]$$

$$\approx 1.462681$$

(b) The fourth derivative of $f(x) = e^{x^2}$ is

$$f^{(4)}(x) = (12 + 48x^2 + 16x^4)e^{x^2}$$

and so, since $0 \leqslant x \leqslant 1$, we have

$$0 \leqslant f^{(4)}(x) \leqslant (12 + 48 + 16)e^1 = 76e$$

Therefore, putting $K = 76e$, $a = 0$, $b = 1$, and $n = 10$ in (4), we see that the error is at most

$$\frac{76e(1)^5}{180(10)^4} \approx 0.000115$$

(Compare this with Example 3.) Thus, correct to three decimal places, we have

$$\int_0^1 e^{x^2}dx \approx 1.463$$

|||| Figure 10 illustrates the calculation in Example 7. Notice that the parabolic arcs are so close to the graph of $y = e^{x^2}$ that they are practically indistinguishable from it.

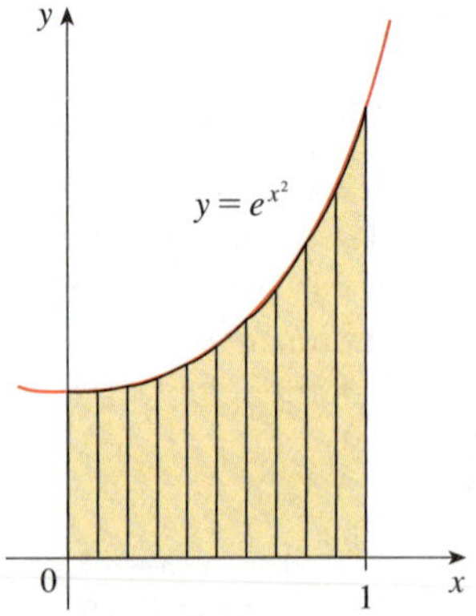

FIGURE 10

7.7 Exercises

1. Let $I = \int_0^4 f(x)\,dx$, where f is the function whose graph is shown.
(a) Use the graph to find L_2, R_2, and M_2.
(b) Are these underestimates or overestimates of I?
(c) Use the graph to find T_2. How does it compare with I?
(d) For any value of n, list the numbers L_n, R_n, M_n, T_n, and I in increasing order.

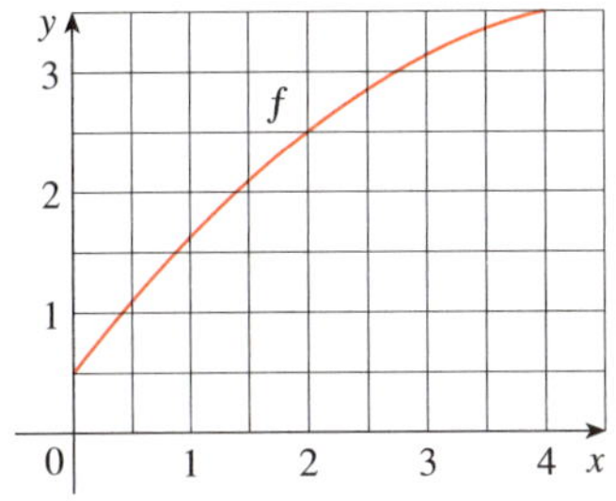

2. The left, right, Trapezoidal, and Midpoint Rule approximations were used to estimate $\int_0^2 f(x)\,dx$, where f is the function whose graph is shown. The estimates were 0.7811, 0.8675, 0.8632, and 0.9540, and the same number of subintervals were used in each case.
(a) Which rule produced which estimate?
(b) Between which two approximations does the true value of $\int_0^2 f(x)\,dx$ lie?

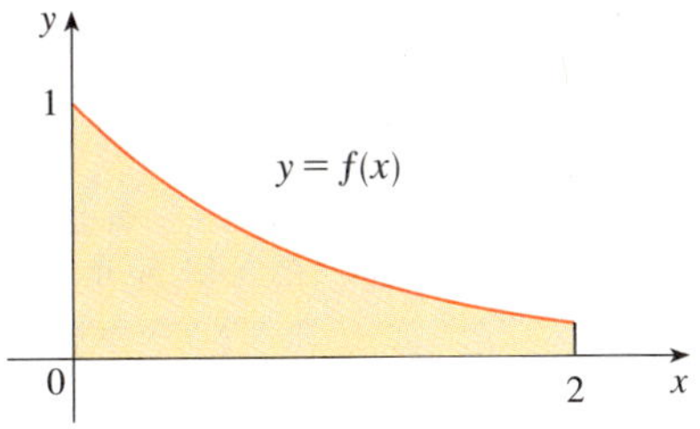

3. Estimate $\int_0^1 \cos(x^2)\,dx$ using (a) the Trapezoidal Rule and (b) the Midpoint Rule, each with $n = 4$. From a graph of the integrand, decide whether your answers are underestimates or overestimates. What can you conclude about the true value of the integral?

4. Draw the graph of $f(x) = \sin(x^2/2)$ in the viewing rectangle $[0, 1]$ by $[0, 0.5]$ and let $I = \int_0^1 f(x)\,dx$.
(a) Use the graph to decide whether L_2, R_2, M_2, and T_2 underestimate or overestimate I.
(b) For any value of n, list the numbers L_n, R_n, M_n, T_n, and I in increasing order.
(c) Compute L_5, R_5, M_5, and T_5. From the graph, which do you think gives the best estimate of I?

5–6 ||| Use (a) the Midpoint Rule and (b) Simpson's Rule to approximate the given integral with the specified value of n. (Round your answers to six decimal places.) Compare your results to the actual value to determine the error in each approximation.

5. $\displaystyle\int_0^{\pi} x^2 \sin x\,dx, \quad n = 8$

6. $\displaystyle\int_0^1 e^{-\sqrt{x}}\,dx, \quad n = 6$

7–18 ||| Use (a) the Trapezoidal Rule, (b) the Midpoint Rule, and (c) Simpson's Rule to approximate the given integral with the specified value of n. (Round your answers to six decimal places.)

7. $\displaystyle\int_0^2 \sqrt[4]{1 + x^2}\,dx, \quad n = 8$

8. $\displaystyle\int_0^{1/2} \sin(x^2)\,dx, \quad n = 4$

9. $\displaystyle\int_1^2 \frac{\ln x}{1 + x}\,dx, \quad n = 10$

10. $\displaystyle\int_0^3 \frac{dt}{1 + t^2 + t^4}, \quad n = 6$

11. $\displaystyle\int_0^{1/2} \sin(e^{t/2})\,dt, \quad n = 8$

12. $\displaystyle\int_0^4 \sqrt{1 + \sqrt{x}}\,dx, \quad n = 8$

13. $\displaystyle\int_1^2 e^{1/x}\,dx, \quad n = 4$

14. $\displaystyle\int_0^4 \sqrt{x}\sin x\,dx, \quad n = 8$

15. $\displaystyle\int_1^5 \frac{\cos x}{x}\,dx, \quad n = 8$

16. $\displaystyle\int_4^6 \ln(x^3 + 2)\,dx, \quad n = 10$

17. $\displaystyle\int_0^3 \frac{1}{1 + y^5}\,dy, \quad n = 6$

18. $\displaystyle\int_2^4 \frac{e^x}{x}\,dx, \quad n = 10$

19. (a) Find the approximations T_{10} and M_{10} for the integral $\int_0^2 e^{-x^2}\,dx$.
(b) Estimate the errors in the approximations of part (a).
(c) How large do we have to choose n so that the approximations T_n and M_n to the integral in part (a) are accurate to within 0.00001?

20. (a) Find the approximations T_8 and M_8 for $\int_0^1 \cos(x^2)\,dx$.
(b) Estimate the errors involved in the approximations of part (a).
(c) How large do we have to choose n so that the approximations T_n and M_n to the integral in part (a) are accurate to within 0.00001?

21. (a) Find the approximations T_{10} and S_{10} for $\int_0^1 e^x\,dx$ and the corresponding errors E_T and E_S.
(b) Compare the actual errors in part (a) with the error estimates given by (3) and (4).
(c) How large do we have to choose n so that the approximations T_n, M_n, and S_n to the integral in part (a) are accurate to within 0.00001?

22. How large should n be to guarantee that the Simpson's Rule approximation to $\int_0^1 e^{x^2}\,dx$ is accurate to within 0.00001?

CAS **23.** The trouble with the error estimates is that it is often very difficult to compute four derivatives and obtain a good upper bound K for $|f^{(4)}(x)|$ by hand. But computer algebra systems have no

problem computing $f^{(4)}$ and graphing it, so we can easily find a value for K from a machine graph. This exercise deals with approximations to the integral $I = \int_0^{2\pi} f(x)\,dx$, where $f(x) = e^{\cos x}$.
(a) Use a graph to get a good upper bound for $|f''(x)|$.
(b) Use M_{10} to approximate I.
(c) Use part (a) to estimate the error in part (b).
(d) Use the built-in numerical integration capability of your CAS to approximate I.
(e) How does the actual error compare with the error estimate in part (c)?
(f) Use a graph to get a good upper bound for $|f^{(4)}(x)|$.
(g) Use S_{10} to approximate I.
(h) Use part (f) to estimate the error in part (g).
(i) How does the actual error compare with the error estimate in part (h)?
(j) How large should n be to guarantee that the size of the error in using S_n is less than 0.0001?

CAS **24.** Repeat Exercise 23 for the integral $\int_{-1}^{1} \sqrt{4 - x^3}\,dx$.

25–26 |||| Find the approximations L_n, R_n, T_n, and M_n for $n = 4$, 8, and 16. Then compute the corresponding errors E_L, E_R, E_T, and E_M. (Round your answers to six decimal places. You may wish to use the sum command on a computer algebra system.) What observations can you make? In particular, what happens to the errors when n is doubled?

25. $\int_0^1 x^3\,dx$ **26.** $\int_0^2 e^x\,dx$

27–28 |||| Find the approximations T_n, M_n, and S_n for $n = 6$ and 12. Then compute the corresponding errors E_T, E_M, and E_S. (Round your answers to six decimal places. You may wish to use the sum command on a computer algebra system.) What observations can you make? In particular, what happens to the errors when n is doubled?

27. $\int_1^4 \sqrt{x}\,dx$ **28.** $\int_{-1}^{2} xe^x\,dx$

29. Estimate the area under the graph in the figure by using (a) the Trapezoidal Rule, (b) the Midpoint Rule, and (c) Simpson's Rule, each with $n = 4$.

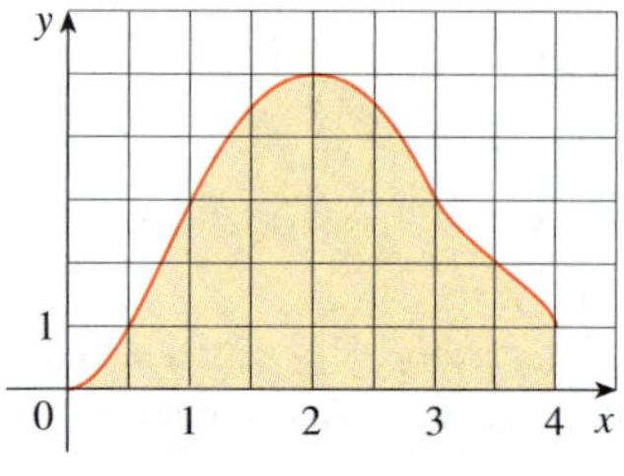

30. The widths (in meters) of a kidney-shaped swimming pool were measured at 2-meter intervals as indicated in the figure. Use Simpson's Rule to estimate the area of the pool.

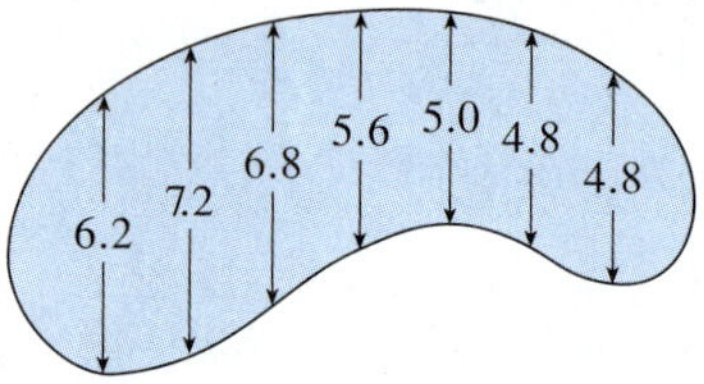

31. (a) Use the Midpoint Rule and the given data to estimate the value of the integral $\int_0^{3.2} f(x)\,dx$.

x	$f(x)$	x	$f(x)$
0.0	6.8	2.0	7.6
0.4	6.5	2.4	8.4
0.8	6.3	2.8	8.8
1.2	6.4	3.2	9.0
1.6	6.9		

(b) If it is known that $-4 \leqslant f''(x) \leqslant 1$ for all x, estimate the error involved in the approximation in part (a).

32. A radar gun was used to record the speed of a runner during the first 5 seconds of a race (see the table). Use Simpson's Rule to estimate the distance the runner covered during those 5 seconds.

t (s)	v (m/s)	t (s)	v (m/s)
0	0	3.0	10.51
0.5	4.67	3.5	10.67
1.0	7.34	4.0	10.76
1.5	8.86	4.5	10.81
2.0	9.73	5.0	10.81
2.5	10.22		

33. The graph of the acceleration $a(t)$ of a car measured in ft/s^2 is shown. Use Simpson's Rule to estimate the increase in the velocity of the car during the 6-second time interval.

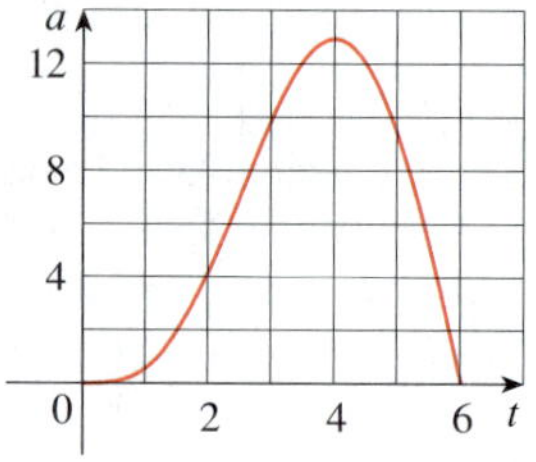

34. Water leaked from a tank at a rate of $r(t)$ liters per hour, where the graph of r is as shown. Use Simpson's Rule to estimate the total amount of water that leaked out during the first six hours.

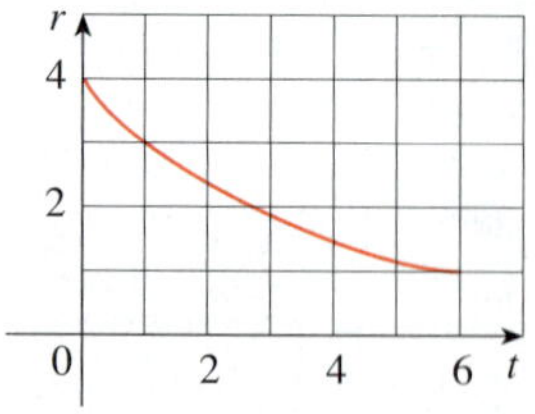

35. The table (supplied by San Diego Gas and Electric) gives the power consumption in megawatts in San Diego County from midnight to 6:00 A.M. on December 8, 1999. Use Simpson's Rule to estimate the energy used during that time period. (Use the fact that power is the derivative of energy.)

t	P	t	P
0:00	1814	3:30	1611
0:30	1735	4:00	1621
1:00	1686	4:30	1666
1:30	1646	5:00	1745
2:00	1637	5:30	1886
2:30	1609	6:00	2052
3:00	1604		

36. Shown is the graph of traffic on an Internet service provider's T1 data line from midnight to 8:00 A.M. D is the data throughput, measured in megabits per second. Use Simpson's Rule to estimate the total amount of data transmitted during that time period.

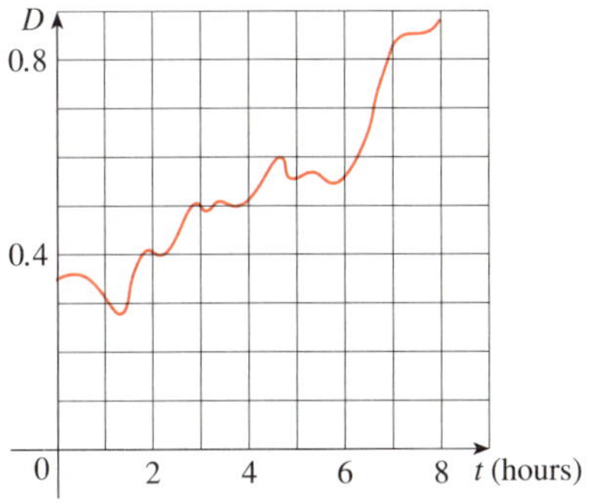

37. If the region shown in the figure is rotated about the y-axis to form a solid, use Simpson's Rule with $n = 8$ to estimate the volume of the solid.

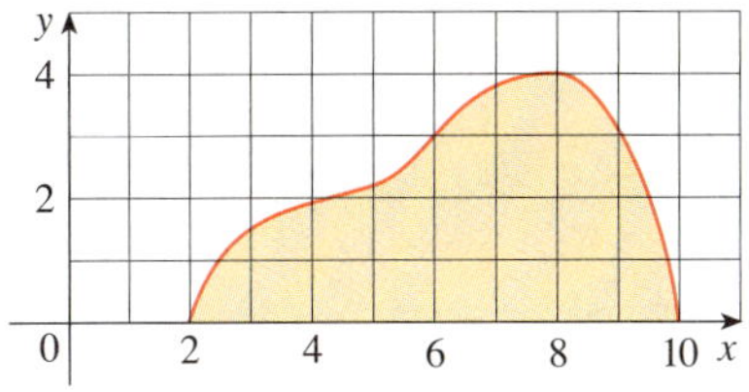

38. The table shows values of a force function $f(x)$ where x is measured in meters and $f(x)$ in newtons. Use Simpson's Rule to estimate the work done by the force in moving an object a distance of 18 m.

x	0	3	6	9	12	15	18
$f(x)$	9.8	9.1	8.5	8.0	7.7	7.5	7.4

39. The region bounded by the curves $y = \sqrt[3]{1 + x^3}$, $y = 0$, $x = 0$, and $x = 2$ is rotated about the x-axis. Use Simpson's Rule with $n = 10$ to estimate the volume of the resulting solid.

CAS **40.** The figure shows a pendulum with length L that makes a maximum angle θ_0 with the vertical. Using Newton's Second Law it can be shown that the period T (the time for one complete swing) is given by

$$T = 4\sqrt{\frac{L}{g}} \int_0^{\pi/2} \frac{dx}{\sqrt{1 - k^2 \sin^2 x}}$$

where $k = \sin\left(\frac{1}{2}\theta_0\right)$ and g is the acceleration due to gravity. If $L = 1$ m and $\theta_0 = 42°$, use Simpson's Rule with $n = 10$ to find the period.

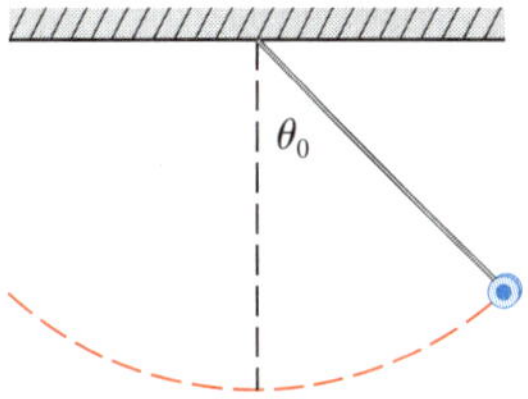

41. The intensity of light with wavelength λ traveling through a diffraction grating with N slits at an angle θ is given by $I(\theta) = N^2 \sin^2 k / k^2$, where $k = (\pi N d \sin\theta)/\lambda$ and d is the distance between adjacent slits. A helium-neon laser with wavelength $\lambda = 632.8 \times 10^{-9}$ m is emitting a narrow band of light, given by $-10^{-6} < \theta < 10^{-6}$, through a grating with 10,000 slits spaced 10^{-4} m apart. Use the Midpoint Rule with $n = 10$ to estimate the total light intensity $\int_{-10^{-6}}^{10^{-6}} I(\theta)\, d\theta$ emerging from the grating.

42. Use the Trapezoidal Rule with $n = 10$ to approximate $\int_0^{20} \cos(\pi x)\, dx$. Compare your result to the actual value. Can you explain the discrepancy?

43. Sketch the graph of a continuous function on $[0, 2]$ for which the Trapezoidal Rule with $n = 2$ is more accurate than the Midpoint Rule.

44. Sketch the graph of a continuous function on $[0, 2]$ for which the right endpoint approximation with $n = 2$ is more accurate than Simpson's Rule.

45. If f is a positive function and $f''(x) < 0$ for $a \leqslant x \leqslant b$, show that

$$T_n < \int_a^b f(x)\, dx < M_n$$

46. Show that if f is a polynomial of degree 3 or lower, then Simpson's Rule gives the exact value of $\int_a^b f(x)\, dx$.

47. Show that $\frac{1}{2}(T_n + M_n) = T_{2n}$.

48. Show that $\frac{1}{3}T_n + \frac{2}{3}M_n = S_{2n}$.

7.8 Improper Integrals

In defining a definite integral $\int_a^b f(x)\,dx$ we dealt with a function f defined on a finite interval $[a, b]$ and we assumed that f does not have an infinite discontinuity (see Section 5.2). In this section we extend the concept of a definite integral to the case where the interval is infinite and also to the case where f has an infinite discontinuity in $[a, b]$. In either case the integral is called an *improper* integral. One of the most important applications of this idea, probability distributions, will be studied in Section 8.5.

Type 1: Infinite Intervals

Try painting a fence that never ends.
Resources / Module 6
/ How To Calculate
/ Start of Improper Integrals

Consider the infinite region S that lies under the curve $y = 1/x^2$, above the x-axis, and to the right of the line $x = 1$. You might think that, since S is infinite in extent, its area must be infinite, but let's take a closer look. The area of the part of S that lies to the left of the line $x = t$ (shaded in Figure 1) is

$$A(t) = \int_1^t \frac{1}{x^2}\,dx = -\frac{1}{x}\bigg]_1^t = 1 - \frac{1}{t}$$

Notice that $A(t) < 1$ no matter how large t is chosen.

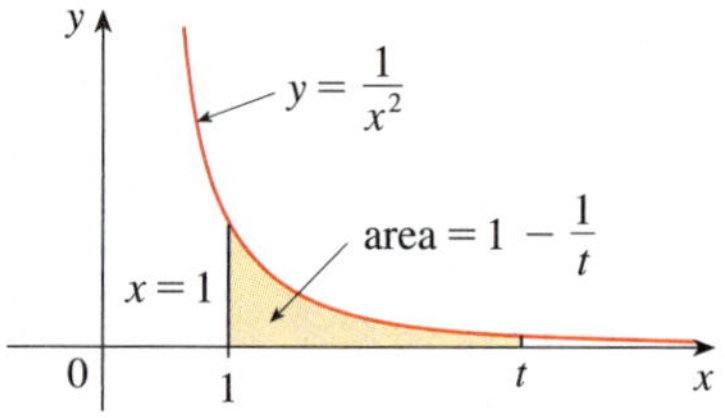

FIGURE 1

We also observe that

$$\lim_{t\to\infty} A(t) = \lim_{t\to\infty}\left(1 - \frac{1}{t}\right) = 1$$

The area of the shaded region approaches 1 as $t \to \infty$ (see Figure 2), so we say that the area of the infinite region S is equal to 1 and we write

$$\int_1^\infty \frac{1}{x^2}\,dx = \lim_{t\to\infty}\int_1^t \frac{1}{x^2}\,dx = 1$$

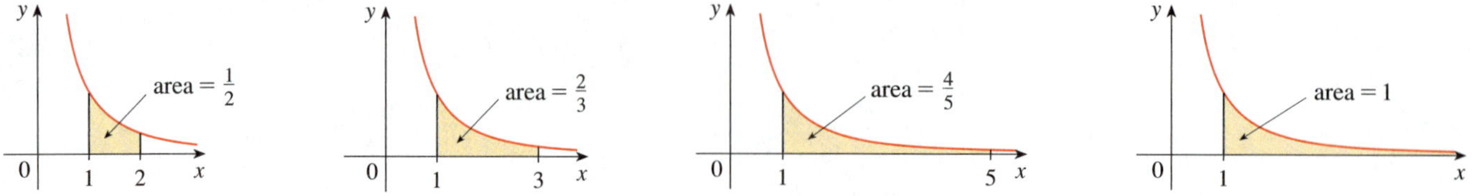

FIGURE 2

Using this example as a guide, we define the integral of f (not necessarily a positive function) over an infinite interval as the limit of integrals over finite intervals.

1 Definition of an Improper Integral of Type 1

(a) If $\int_a^t f(x)\,dx$ exists for every number $t \geqslant a$, then

$$\int_a^\infty f(x)\,dx = \lim_{t\to\infty} \int_a^t f(x)\,dx$$

provided this limit exists (as a finite number).

(b) If $\int_t^b f(x)\,dx$ exists for every number $t \leqslant b$, then

$$\int_{-\infty}^b f(x)\,dx = \lim_{t\to-\infty} \int_t^b f(x)\,dx$$

provided this limit exists (as a finite number).

The improper integrals $\int_a^\infty f(x)\,dx$ and $\int_{-\infty}^b f(x)\,dx$ are called **convergent** if the corresponding limit exists and **divergent** if the limit does not exist.

(c) If both $\int_a^\infty f(x)\,dx$ and $\int_{-\infty}^a f(x)\,dx$ are convergent, then we define

$$\int_{-\infty}^\infty f(x)\,dx = \int_{-\infty}^a f(x)\,dx + \int_a^\infty f(x)\,dx$$

In part (c) any real number a can be used (see Exercise 74).

Any of the improper integrals in Definition 1 can be interpreted as an area provided that f is a positive function. For instance, in case (a) if $f(x) \geqslant 0$ and the integral $\int_a^\infty f(x)\,dx$ is convergent, then we define the area of the region $S = \{(x, y) \mid x \geqslant a, 0 \leqslant y \leqslant f(x)\}$ in Figure 3 to be

$$A(S) = \int_a^\infty f(x)\,dx$$

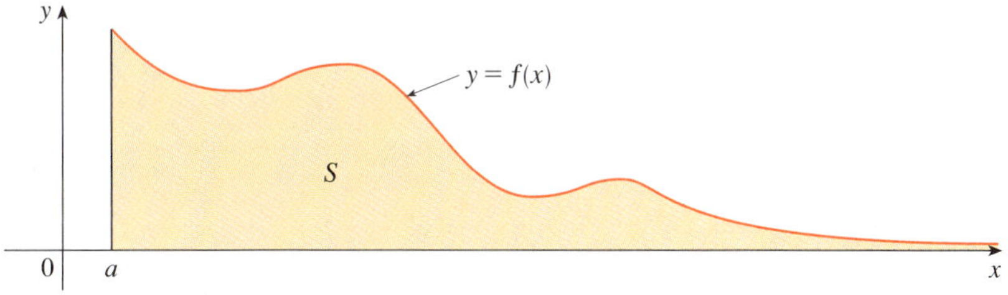

FIGURE 3

This is appropriate because $\int_a^\infty f(x)\,dx$ is the limit as $t \to \infty$ of the area under the graph of f from a to t.

EXAMPLE 1 Determine whether the integral $\int_1^\infty (1/x)\,dx$ is convergent or divergent.

SOLUTION According to part (a) of Definition 1, we have

$$\int_1^\infty \frac{1}{x}\,dx = \lim_{t\to\infty} \int_1^t \frac{1}{x}\,dx = \lim_{t\to\infty} \ln|x|\Big]_1^t$$

$$= \lim_{t\to\infty} (\ln t - \ln 1) = \lim_{t\to\infty} \ln t = \infty$$

The limit does not exist as a finite number and so the improper integral $\int_1^\infty (1/x)\,dx$ is divergent.

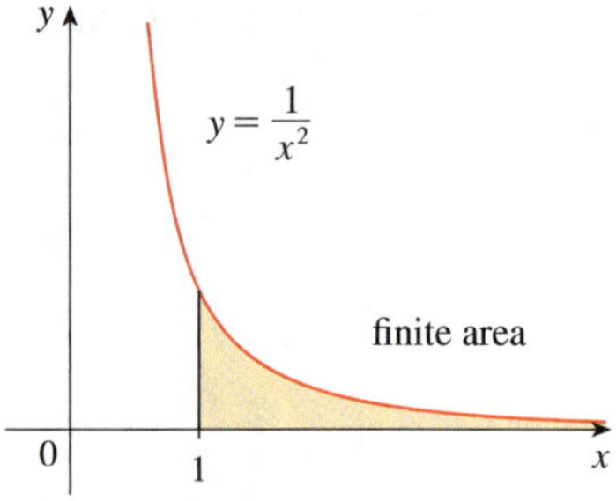

FIGURE 4

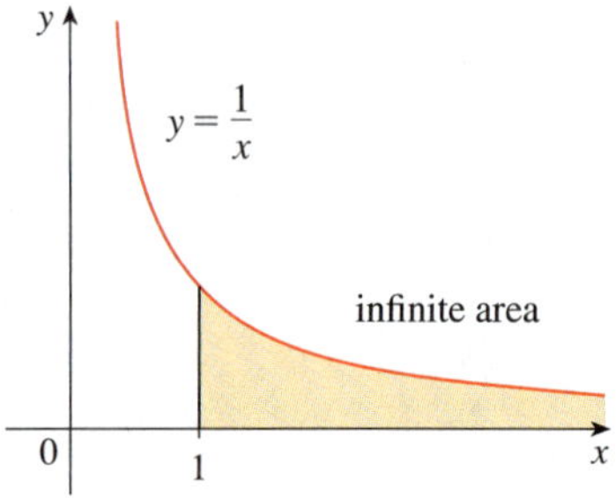

FIGURE 5

Let's compare the result of Example 1 with the example given at the beginning of this section:

$$\int_1^\infty \frac{1}{x^2}\,dx \text{ converges} \qquad \int_1^\infty \frac{1}{x}\,dx \text{ diverges}$$

Geometrically, this says that although the curves $y = 1/x^2$ and $y = 1/x$ look very similar for $x > 0$, the region under $y = 1/x^2$ to the right of $x = 1$ (the shaded region in Figure 4) has finite area whereas the corresponding region under $y = 1/x$ (in Figure 5) has infinite area. Note that both $1/x^2$ and $1/x$ approach 0 as $x \to \infty$ but $1/x^2$ approaches 0 faster than $1/x$. The values of $1/x$ don't decrease fast enough for its integral to have a finite value.

EXAMPLE 2 Evaluate $\int_{-\infty}^0 xe^x\,dx$.

SOLUTION Using part (b) of Definition 1, we have

$$\int_{-\infty}^0 xe^x\,dx = \lim_{t\to-\infty} \int_t^0 xe^x\,dx$$

We integrate by parts with $u = x$, $dv = e^x\,dx$ so that $du = dx$, $v = e^x$:

$$\int_t^0 xe^x\,dx = xe^x\Big]_t^0 - \int_t^0 e^x\,dx$$

$$= -te^t - 1 + e^t$$

We know that $e^t \to 0$ as $t \to -\infty$, and by l'Hospital's Rule we have

$$\lim_{t\to-\infty} te^t = \lim_{t\to-\infty} \frac{t}{e^{-t}} = \lim_{t\to-\infty} \frac{1}{-e^{-t}}$$

$$= \lim_{t\to-\infty} (-e^t) = 0$$

Therefore

$$\int_{-\infty}^0 xe^x\,dx = \lim_{t\to-\infty} (-te^t - 1 + e^t)$$

$$= -0 - 1 + 0 = -1$$

EXAMPLE 3 Evaluate $\int_{-\infty}^\infty \frac{1}{1 + x^2}\,dx$.

SOLUTION It's convenient to choose $a = 0$ in Definition 1(c):

$$\int_{-\infty}^\infty \frac{1}{1 + x^2}\,dx = \int_{-\infty}^0 \frac{1}{1 + x^2}\,dx + \int_0^\infty \frac{1}{1 + x^2}\,dx$$

We must now evaluate the integrals on the right side separately:

$$\int_0^\infty \frac{1}{1 + x^2}\,dx = \lim_{t\to\infty} \int_0^t \frac{dx}{1 + x^2} = \lim_{t\to\infty} \tan^{-1}x\Big]_0^t$$

$$= \lim_{t\to\infty} (\tan^{-1}t - \tan^{-1}0) = \lim_{t\to\infty} \tan^{-1}t = \frac{\pi}{2}$$

$$\int_{-\infty}^{0} \frac{1}{1 + x^2}\, dx = \lim_{t \to -\infty} \int_{t}^{0} \frac{dx}{1 + x^2} = \lim_{t \to -\infty} \tan^{-1}x \Big]_t^0$$

$$= \lim_{t \to -\infty} (\tan^{-1}0 - \tan^{-1}t)$$

$$= 0 - \left(-\frac{\pi}{2}\right) = \frac{\pi}{2}$$

Since both of these integrals are convergent, the given integral is convergent and

$$\int_{-\infty}^{\infty} \frac{1}{1 + x^2}\, dx = \frac{\pi}{2} + \frac{\pi}{2} = \pi$$

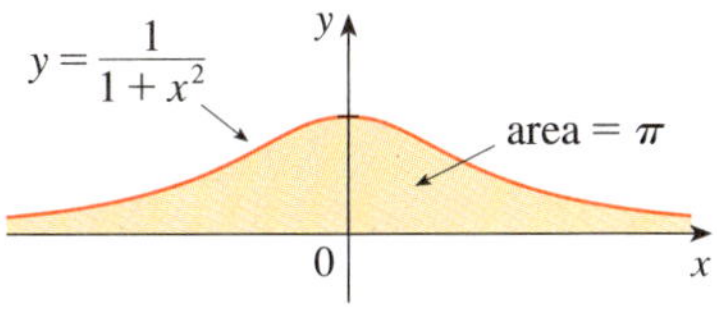

FIGURE 6

Since $1/(1 + x^2) > 0$, the given improper integral can be interpreted as the area of the infinite region that lies under the curve $y = 1/(1 + x^2)$ and above the x-axis (see Figure 6).

EXAMPLE 4 For what values of p is the integral

$$\int_{1}^{\infty} \frac{1}{x^p}\, dx$$

convergent?

SOLUTION We know from Example 1 that if $p = 1$, then the integral is divergent, so let's assume that $p \neq 1$. Then

$$\int_{1}^{\infty} \frac{1}{x^p}\, dx = \lim_{t \to \infty} \int_{1}^{t} x^{-p}\, dx$$

$$= \lim_{t \to \infty} \frac{x^{-p+1}}{-p + 1} \Bigg]_{x=1}^{x=t}$$

$$= \lim_{t \to \infty} \frac{1}{1 - p} \left[\frac{1}{t^{p-1}} - 1\right]$$

If $p > 1$, then $p - 1 > 0$, so as $t \to \infty$, $t^{p-1} \to \infty$ and $1/t^{p-1} \to 0$. Therefore

$$\int_{1}^{\infty} \frac{1}{x^p}\, dx = \frac{1}{p - 1} \qquad \text{if } p > 1$$

and so the integral converges. But if $p < 1$, then $p - 1 < 0$ and so

$$\frac{1}{t^{p-1}} = t^{1-p} \to \infty \qquad \text{as } t \to \infty$$

and the integral diverges.

We summarize the result of Example 4 for future reference:

2 $\displaystyle\int_{1}^{\infty} \frac{1}{x^p}\, dx$ is convergent if $p > 1$ and divergent if $p \leq 1$.

Type 2: Discontinuous Integrands

Suppose that f is a positive continuous function defined on a finite interval $[a, b)$ but has a vertical asymptote at b. Let S be the unbounded region under the graph of f and above the x-axis between a and b. (For Type 1 integrals, the regions extended indefinitely in a

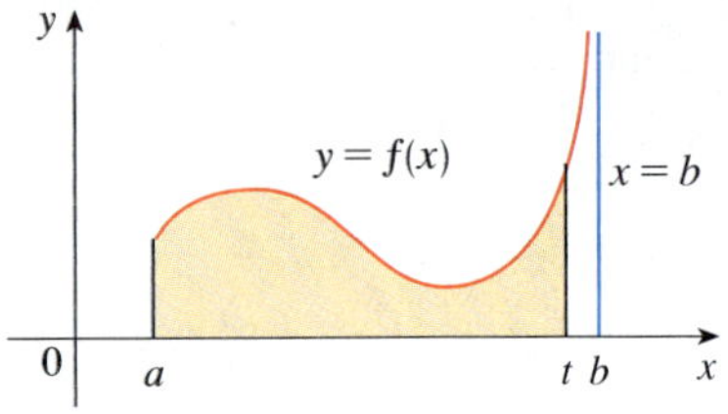

FIGURE 7

horizontal direction. Here the region is infinite in a vertical direction.) The area of the part of S between a and t (the shaded region in Figure 7) is

$$A(t) = \int_a^t f(x)\,dx$$

If it happens that $A(t)$ approaches a definite number A as $t \to b^-$, then we say that the area of the region S is A and we write

$$\int_a^b f(x)\,dx = \lim_{t \to b^-} \int_a^t f(x)\,dx$$

We use this equation to define an improper integral of Type 2 even when f is not a positive function, no matter what type of discontinuity f has at b.

|||| Parts (b) and (c) of Definition 3 are illustrated in Figures 8 and 9 for the case where $f(x) \geqslant 0$ and f has vertical asymptotes at a and c, respectively.

3 Definition of an Improper Integral of Type 2

(a) If f is continuous on $[a, b)$ and is discontinuous at b, then

$$\int_a^b f(x)\,dx = \lim_{t \to b^-} \int_a^t f(x)\,dx$$

if this limit exists (as a finite number).

(b) If f is continuous on $(a, b]$ and is discontinuous at a, then

$$\int_a^b f(x)\,dx = \lim_{t \to a^+} \int_t^b f(x)\,dx$$

if this limit exists (as a finite number).

The improper integral $\int_a^b f(x)\,dx$ is called **convergent** if the corresponding limit exists and **divergent** if the limit does not exist.

(c) If f has a discontinuity at c, where $a < c < b$, and both $\int_a^c f(x)\,dx$ and $\int_c^b f(x)\,dx$ are convergent, then we define

$$\int_a^b f(x)\,dx = \int_a^c f(x)\,dx + \int_c^b f(x)\,dx$$

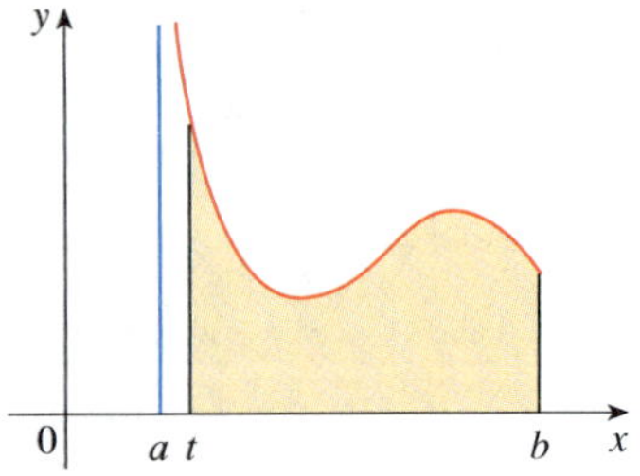

FIGURE 8

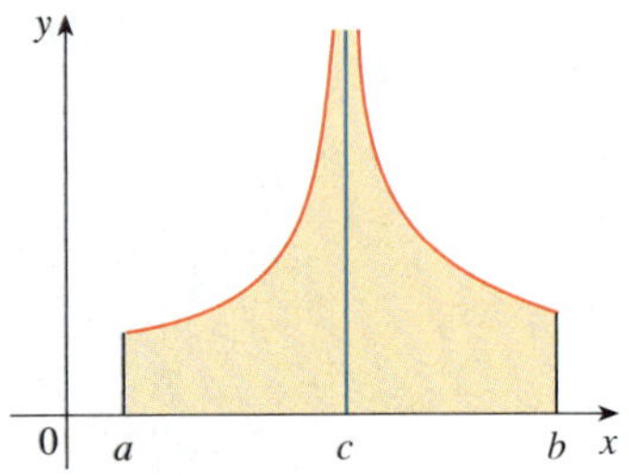

FIGURE 9

EXAMPLE 5 Find $\displaystyle\int_2^5 \frac{1}{\sqrt{x-2}}\,dx$.

SOLUTION We note first that the given integral is improper because $f(x) = 1/\sqrt{x-2}$ has the vertical asymptote $x = 2$. Since the infinite discontinuity occurs at the left endpoint of $[2, 5]$, we use part (b) of Definition 3:

$$\begin{aligned}\int_2^5 \frac{dx}{\sqrt{x-2}} &= \lim_{t \to 2^+} \int_t^5 \frac{dx}{\sqrt{x-2}} \\ &= \lim_{t \to 2^+} 2\sqrt{x-2}\,\Big]_t^5 \\ &= \lim_{t \to 2^+} 2\left(\sqrt{3} - \sqrt{t-2}\right) \\ &= 2\sqrt{3}\end{aligned}$$

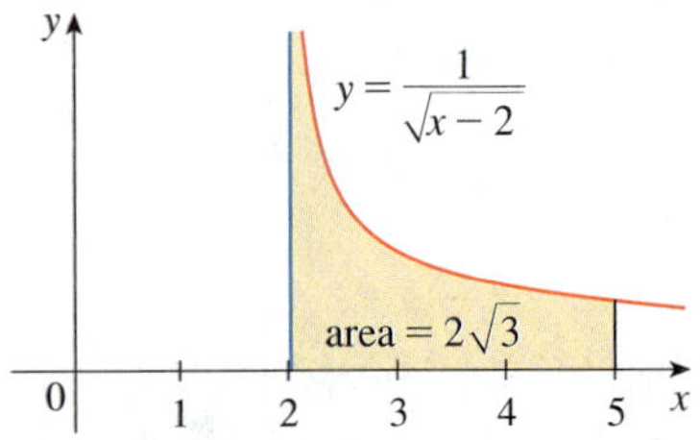

FIGURE 10

Thus, the given improper integral is convergent and, since the integrand is positive, we can interpret the value of the integral as the area of the shaded region in Figure 10.

EXAMPLE 6 Determine whether $\int_0^{\pi/2} \sec x\, dx$ converges or diverges.

SOLUTION Note that the given integral is improper because $\lim_{x\to(\pi/2)^-} \sec x = \infty$. Using part (a) of Definition 3 and Formula 14 from the Table of Integrals, we have

$$\begin{aligned}\int_0^{\pi/2} \sec x\, dx &= \lim_{t\to(\pi/2)^-} \int_0^t \sec x\, dx \\ &= \lim_{t\to(\pi/2)^-} \ln|\sec x + \tan x|\Big]_0^t \\ &= \lim_{t\to(\pi/2)^-} [\ln(\sec t + \tan t) - \ln 1] \\ &= \infty\end{aligned}$$

because $\sec t \to \infty$ and $\tan t \to \infty$ as $t \to (\pi/2)^-$. Thus, the given improper integral is divergent.

EXAMPLE 7 Evaluate $\int_0^3 \frac{dx}{x-1}$ if possible.

SOLUTION Observe that the line $x = 1$ is a vertical asymptote of the integrand. Since it occurs in the middle of the interval $[0, 3]$, we must use part (c) of Definition 3 with $c = 1$:

$$\int_0^3 \frac{dx}{x-1} = \int_0^1 \frac{dx}{x-1} + \int_1^3 \frac{dx}{x-1}$$

where

$$\begin{aligned}\int_0^1 \frac{dx}{x-1} &= \lim_{t\to1^-} \int_0^t \frac{dx}{x-1} = \lim_{t\to1^-} \ln|x-1|\Big]_0^t \\ &= \lim_{t\to1^-} \left(\ln|t-1| - \ln|-1|\right) \\ &= \lim_{t\to1^-} \ln(1-t) = -\infty\end{aligned}$$

because $1 - t \to 0^+$ as $t \to 1^-$. Thus, $\int_0^1 dx/(x-1)$ is divergent. This implies that $\int_0^3 dx/(x-1)$ is divergent. [We do not need to evaluate $\int_1^3 dx/(x-1)$.]

WARNING ▫ If we had not noticed the asymptote $x = 1$ in Example 7 and had instead confused the integral with an ordinary integral, then we might have made the following erroneous calculation:

$$\int_0^3 \frac{dx}{x-1} = \ln|x-1|\Big]_0^3 = \ln 2 - \ln 1 = \ln 2$$

This is wrong because the integral is improper and must be calculated in terms of limits.

From now on, whenever you meet the symbol $\int_a^b f(x)\, dx$ you must decide, by looking at the function f on $[a, b]$, whether it is an ordinary definite integral or an improper integral.

EXAMPLE 8 Evaluate $\int_0^1 \ln x\, dx$.

SOLUTION We know that the function $f(x) = \ln x$ has a vertical asymptote at 0 since $\lim_{x\to0^+} \ln x = -\infty$. Thus, the given integral is improper and we have

$$\int_0^1 \ln x\, dx = \lim_{t\to0^+} \int_t^1 \ln x\, dx$$

Now we integrate by parts with $u = \ln x$, $dv = dx$, $du = dx/x$, and $v = x$:

$$\int_t^1 \ln x\,dx = x\ln x\Big]_t^1 - \int_t^1 dx$$

$$= 1\ln 1 - t\ln t - (1 - t)$$

$$= -t\ln t - 1 + t$$

To find the limit of the first term we use l'Hospital's Rule:

$$\lim_{t\to 0^+} t\ln t = \lim_{t\to 0^+}\frac{\ln t}{1/t}$$

$$= \lim_{t\to 0^+}\frac{1/t}{-1/t^2}$$

$$= \lim_{t\to 0^+}(-t) = 0$$

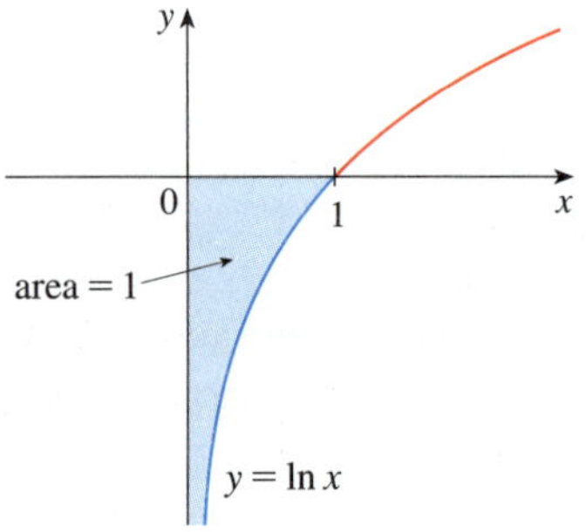

FIGURE 11

Therefore

$$\int_0^1 \ln x\,dx = \lim_{t\to 0^+}(-t\ln t - 1 + t)$$

$$= -0 - 1 + 0 = -1$$

Figure 11 shows the geometric interpretation of this result. The area of the shaded region above $y = \ln x$ and below the x-axis is 1.

A Comparison Test for Improper Integrals

Sometimes it is impossible to find the exact value of an improper integral and yet it is important to know whether it is convergent or divergent. In such cases the following theorem is useful. Although we state it for Type 1 integrals, a similar theorem is true for Type 2 integrals.

Comparison Theorem Suppose that f and g are continuous functions with $f(x) \geqslant g(x) \geqslant 0$ for $x \geqslant a$.

(a) If $\int_a^\infty f(x)\,dx$ is convergent, then $\int_a^\infty g(x)\,dx$ is convergent.

(b) If $\int_a^\infty g(x)\,dx$ is divergent, then $\int_a^\infty f(x)\,dx$ is divergent.

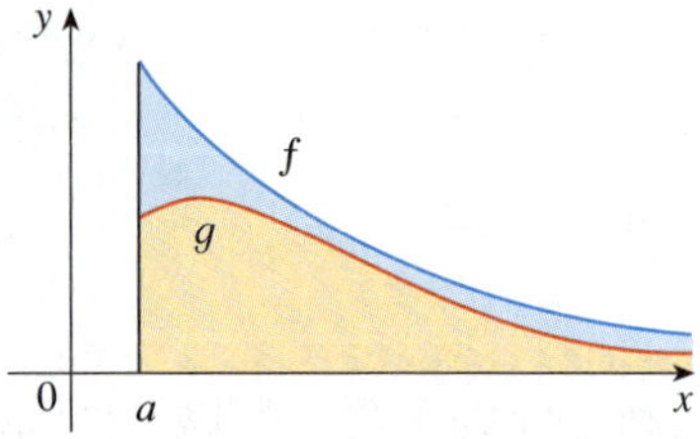

FIGURE 12

We omit the proof of the Comparison Theorem, but Figure 12 makes it seem plausible. If the area under the top curve $y = f(x)$ is finite, then so is the area under the bottom curve $y = g(x)$. And if the area under $y = g(x)$ is infinite, then so is the area under $y = f(x)$. [Note that the reverse is not necessarily true: If $\int_a^\infty g(x)\,dx$ is convergent, $\int_a^\infty f(x)\,dx$ may or may not be convergent, and if $\int_a^\infty f(x)\,dx$ is divergent, $\int_a^\infty g(x)\,dx$ may or may not be divergent.]

EXAMPLE 9 Show that $\int_0^\infty e^{-x^2}\,dx$ is convergent.

SOLUTION We can't evaluate the integral directly because the antiderivative of e^{-x^2} is not an elementary function (as explained in Section 7.5). We write

$$\int_0^\infty e^{-x^2}dx = \int_0^1 e^{-x^2}dx + \int_1^\infty e^{-x^2}dx$$

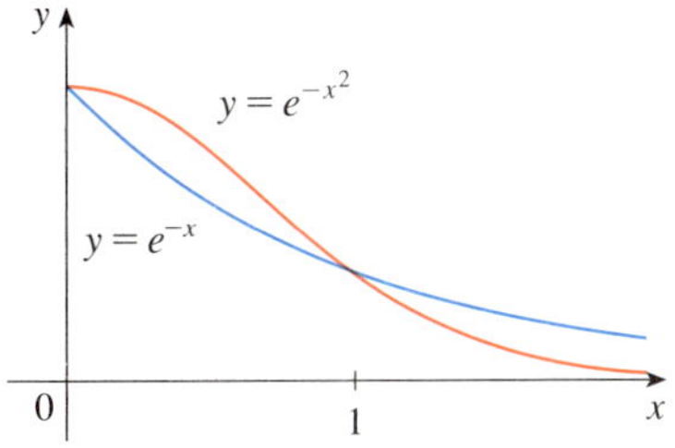

FIGURE 13

and observe that the first integral on the right-hand side is just an ordinary definite integral. In the second integral we use the fact that for $x \geq 1$ we have $x^2 \geq x$, so $-x^2 \leq -x$ and therefore $e^{-x^2} \leq e^{-x}$. (See Figure 13.) The integral of e^{-x} is easy to evaluate:

$$\int_1^\infty e^{-x}\,dx = \lim_{t\to\infty}\int_1^t e^{-x}\,dx$$
$$= \lim_{t\to\infty}(e^{-1} - e^{-t}) = e^{-1}$$

Thus, taking $f(x) = e^{-x}$ and $g(x) = e^{-x^2}$ in the Comparison Theorem, we see that $\int_1^\infty e^{-x^2}\,dx$ is convergent. It follows that $\int_0^\infty e^{-x^2}\,dx$ is convergent.

TABLE 1

t	$\int_0^t e^{-x^2}\,dx$
1	0.7468241328
2	0.8820813908
3	0.8862073483
4	0.8862269118
5	0.8862269255
6	0.8862269255

In Example 9 we showed that $\int_0^\infty e^{-x^2}\,dx$ is convergent without computing its value. In Exercise 70 we indicate how to show that its value is approximately 0.8862. In probability theory it is important to know the exact value of this improper integral, as we will see in Section 8.5; using the methods of multivariable calculus it can be shown that the exact value is $\sqrt{\pi}/2$. Table 1 illustrates the definition of an improper integral by showing how the (computer-generated) values of $\int_0^t e^{-x^2}\,dx$ approach $\sqrt{\pi}/2$ as t becomes large. In fact, these values converge quite quickly because $e^{-x^2} \to 0$ very rapidly as $x \to \infty$.

EXAMPLE 10 The integral $\displaystyle\int_1^\infty \frac{1 + e^{-x}}{x}\,dx$ is divergent by the Comparison Theorem because

$$\frac{1 + e^{-x}}{x} > \frac{1}{x}$$

and $\int_1^\infty (1/x)\,dx$ is divergent by Example 1 [or by (2) with $p = 1$].

Table 2 illustrates the divergence of the integral in Example 10. It appears that the values are not approaching any fixed number.

TABLE 2

t	$\int_1^t [(1 + e^{-x})/x]\,dx$
2	0.8636306042
5	1.8276735512
10	2.5219648704
100	4.8245541204
1000	7.1271392134
10000	9.4297243064

7.8 Exercises

1. Explain why each of the following integrals is improper.

(a) $\displaystyle\int_1^\infty x^4 e^{-x^4}\,dx$ (b) $\displaystyle\int_0^{\pi/2} \sec x\,dx$

(c) $\displaystyle\int_0^2 \frac{x}{x^2 - 5x + 6}\,dx$ (d) $\displaystyle\int_{-\infty}^0 \frac{1}{x^2 + 5}\,dx$

2. Which of the following integrals are improper? Why?

(a) $\displaystyle\int_1^2 \frac{1}{2x - 1}\,dx$ (b) $\displaystyle\int_0^1 \frac{1}{2x - 1}\,dx$

(c) $\displaystyle\int_{-\infty}^\infty \frac{\sin x}{1 + x^2}\,dx$ (d) $\displaystyle\int_1^2 \ln(x - 1)\,dx$

3. Find the area under the curve $y = 1/x^3$ from $x = 1$ to $x = t$ and evaluate it for $t = 10$, 100, and 1000. Then find the total area under this curve for $x \geq 1$.

4. (a) Graph the functions $f(x) = 1/x^{1.1}$ and $g(x) = 1/x^{0.9}$ in the viewing rectangles $[0, 10]$ by $[0, 1]$ and $[0, 100]$ by $[0, 1]$.
(b) Find the areas under the graphs of f and g from $x = 1$ to $x = t$ and evaluate for $t = 10$, 100, 10^4, 10^6, 10^{10}, and 10^{20}.
(c) Find the total area under each curve for $x \geq 1$, if it exists.

5–40 Determine whether each integral is convergent or divergent. Evaluate those that are convergent.

5. $\displaystyle\int_1^\infty \frac{1}{(3x+1)^2}\,dx$

6. $\displaystyle\int_{-\infty}^0 \frac{1}{2x-5}\,dx$

7. $\displaystyle\int_{-\infty}^{-1} \frac{1}{\sqrt{2-w}}\,dw$

8. $\displaystyle\int_0^\infty \frac{x}{(x^2+2)^2}\,dx$

9. $\displaystyle\int_4^\infty e^{-y/2}\,dy$

10. $\displaystyle\int_{-\infty}^{-1} e^{-2t}\,dt$

11. $\displaystyle\int_{-\infty}^\infty \frac{x}{1+x^2}\,dx$

12. $\displaystyle\int_{-\infty}^\infty (2 - v^4)\,dv$

13. $\displaystyle\int_{-\infty}^\infty xe^{-x^2}\,dx$

14. $\displaystyle\int_{-\infty}^\infty x^2e^{-x^3}\,dx$

15. $\displaystyle\int_{2\pi}^\infty \sin\theta\,d\theta$

16. $\displaystyle\int_0^\infty \cos^2\alpha\,d\alpha$

17. $\displaystyle\int_1^\infty \frac{x+1}{x^2+2x}\,dx$

18. $\displaystyle\int_0^\infty \frac{dz}{z^2+3z+2}$

19. $\displaystyle\int_0^\infty se^{-5s}\,ds$

20. $\displaystyle\int_{-\infty}^6 re^{r/3}\,dr$

21. $\displaystyle\int_1^\infty \frac{\ln x}{x}\,dx$

22. $\displaystyle\int_{-\infty}^\infty e^{-|x|}\,dx$

23. $\displaystyle\int_{-\infty}^\infty \frac{x^2}{9+x^6}\,dx$

24. $\displaystyle\int_1^\infty \frac{\ln x}{x^3}\,dx$

25. $\displaystyle\int_1^\infty \frac{\ln x}{x^2}\,dx$

26. $\displaystyle\int_0^\infty \frac{x\arctan x}{(1+x^2)^2}\,dx$

27. $\displaystyle\int_0^3 \frac{1}{\sqrt{x}}\,dx$

28. $\displaystyle\int_0^3 \frac{1}{x\sqrt{x}}\,dx$

29. $\displaystyle\int_{-1}^0 \frac{1}{x^2}\,dx$

30. $\displaystyle\int_1^9 \frac{1}{\sqrt[3]{x-9}}\,dx$

31. $\displaystyle\int_{-2}^3 \frac{1}{x^4}\,dx$

32. $\displaystyle\int_0^1 \frac{dx}{\sqrt{1-x^2}}$

33. $\displaystyle\int_0^{33} (x-1)^{-1/5}\,dx$

34. $\displaystyle\int_0^1 \frac{1}{4y-1}\,dy$

35. $\displaystyle\int_0^\pi \sec x\,dx$

36. $\displaystyle\int_0^4 \frac{1}{x^2+x-6}\,dx$

37. $\displaystyle\int_{-1}^1 \frac{e^x}{e^x-1}\,dx$

38. $\displaystyle\int_0^2 \frac{x-3}{2x-3}\,dx$

39. $\displaystyle\int_0^2 z^2\ln z\,dz$

40. $\displaystyle\int_0^1 \frac{\ln x}{\sqrt{x}}\,dx$

41–46 Sketch the region and find its area (if the area is finite).

41. $S = \{(x, y) \mid x \leq 1,\ 0 \leq y \leq e^x\}$

42. $S = \{(x, y) \mid x \geq -2,\ 0 \leq y \leq e^{-x/2}\}$

43. $S = \{(x, y) \mid 0 \leq y \leq 2/(x^2+9)\}$

44. $S = \{(x, y) \mid x \geq 0,\ 0 \leq y \leq x/(x^2+9)\}$

45. $S = \{(x, y) \mid 0 \leq x < \pi/2,\ 0 \leq y \leq \sec^2 x\}$

46. $S = \left\{(x, y) \mid -2 < x \leq 0,\ 0 \leq y \leq 1/\sqrt{x+2}\right\}$

47. (a) If $g(x) = (\sin^2 x)/x^2$, use your calculator or computer to make a table of approximate values of $\int_1^t g(x)\,dx$ for $t = 2$, 5, 10, 100, 1000, and 10,000. Does it appear that $\int_1^\infty g(x)\,dx$ is convergent?
(b) Use the Comparison Theorem with $f(x) = 1/x^2$ to show that $\int_1^\infty g(x)\,dx$ is convergent.
(c) Illustrate part (b) by graphing f and g on the same screen for $1 \leq x \leq 10$. Use your graph to explain intuitively why $\int_1^\infty g(x)\,dx$ is convergent.

48. (a) If $g(x) = 1/(\sqrt{x} - 1)$, use your calculator or computer to make a table of approximate values of $\int_2^t g(x)\,dx$ for $t = 5$, 10, 100, 1000, and 10,000. Does it appear that $\int_2^\infty g(x)\,dx$ is convergent or divergent?
(b) Use the Comparison Theorem with $f(x) = 1/\sqrt{x}$ to show that $\int_2^\infty g(x)\,dx$ is divergent.
(c) Illustrate part (b) by graphing f and g on the same screen for $2 \leq x \leq 20$. Use your graph to explain intuitively why $\int_2^\infty g(x)\,dx$ is divergent.

49–54 Use the Comparison Theorem to determine whether the integral is convergent or divergent.

49. $\displaystyle\int_1^\infty \frac{\cos^2 x}{1+x^2}\,dx$

50. $\displaystyle\int_1^\infty \frac{2+e^{-x}}{x}\,dx$

51. $\displaystyle\int_1^\infty \frac{dx}{x+e^{2x}}$

52. $\displaystyle\int_1^\infty \frac{x}{\sqrt{1+x^6}}\,dx$

53. $\displaystyle\int_0^{\pi/2} \frac{dx}{x\sin x}$

54. $\displaystyle\int_0^1 \frac{e^{-x}}{\sqrt{x}}\,dx$

55. The integral

$$\int_0^\infty \frac{1}{\sqrt{x}\,(1+x)}\,dx$$

is improper for two reasons: The interval $[0, \infty)$ is infinite and the integrand has an infinite discontinuity at 0. Evaluate it by expressing it as a sum of improper integrals of Type 2 and Type 1 as follows:

$$\int_0^\infty \frac{1}{\sqrt{x}\,(1+x)}\,dx = \int_0^1 \frac{1}{\sqrt{x}\,(1+x)}\,dx + \int_1^\infty \frac{1}{\sqrt{x}\,(1+x)}\,dx$$

56. Evaluate

$$\int_2^\infty \frac{1}{x\sqrt{x^2-4}}\,dx$$

by the same method as in Exercise 55.

57–59 Find the values of p for which the integral converges and evaluate the integral for those values of p.

57. $\int_0^1 \frac{1}{x^p}\,dx$

58. $\int_e^\infty \frac{1}{x(\ln x)^p}\,dx$

59. $\int_0^1 x^p \ln x\,dx$

60. (a) Evaluate the integral $\int_0^\infty x^n e^{-x}\,dx$ for $n = 0, 1, 2$, and 3.
(b) Guess the value of $\int_0^\infty x^n e^{-x}\,dx$ when n is an arbitrary positive integer.
(c) Prove your guess using mathematical induction.

61. (a) Show that $\int_{-\infty}^\infty x\,dx$ is divergent.
(b) Show that

$$\lim_{t\to\infty}\int_{-t}^{t} x\,dx = 0$$

This shows that we can't define

$$\int_{-\infty}^\infty f(x)\,dx = \lim_{t\to\infty}\int_{-t}^{t} f(x)\,dx$$

62. The *average speed* of molecules in an ideal gas is

$$\bar{v} = \frac{4}{\sqrt{\pi}}\left(\frac{M}{2RT}\right)^{3/2}\int_0^\infty v^3 e^{-Mv^2/(2RT)}\,dv$$

where M is the molecular weight of the gas, R is the gas constant, T is the gas temperature, and v is the molecular speed. Show that

$$\bar{v} = \sqrt{\frac{8RT}{\pi M}}$$

63. We know from Example 1 that the region $\mathcal{R} = \{(x, y) \mid x \geq 1,\ 0 \leq y \leq 1/x\}$ has infinite area. Show that by rotating $\mathcal{R}$ about the x-axis we obtain a solid with finite volume.

64. Use the information and data in Exercises 29 and 30 of Section 6.4 to find the work required to propel a 1000-kg satellite out of Earth's gravitational field.

65. Find the *escape velocity* v_0 that is needed to propel a rocket of mass m out of the gravitational field of a planet with mass M and radius R. Use Newton's Law of Gravitation (see Exercise 29 in Section 6.4) and the fact that the initial kinetic energy of $\frac{1}{2}mv_0^2$ supplies the needed work.

66. Astronomers use a technique called *stellar stereography* to determine the density of stars in a star cluster from the observed (two-dimensional) density that can be analyzed from a photograph. Suppose that in a spherical cluster of radius R the density of stars depends only on the distance r from the center of the cluster. If the perceived star density is given by $y(s)$, where s is the observed planar distance from the center of the cluster, and $x(r)$ is the actual density, it can be shown that

$$y(s) = \int_s^R \frac{2r}{\sqrt{r^2 - s^2}}\,x(r)\,dr$$

If the actual density of stars in a cluster is $x(r) = \frac{1}{2}(R-r)^2$, find the perceived density $y(s)$.

67. A manufacturer of lightbulbs wants to produce bulbs that last about 700 hours but, of course, some bulbs burn out faster than others. Let $F(t)$ be the fraction of the company's bulbs that burn out before t hours, so $F(t)$ always lies between 0 and 1.
(a) Make a rough sketch of what you think the graph of F might look like.
(b) What is the meaning of the derivative $r(t) = F'(t)$?
(c) What is the value of $\int_0^\infty r(t)\,dt$? Why?

68. As we will see in Section 9.4, a radioactive substance decays exponentially: The mass at time t is $m(t) = m(0)e^{kt}$, where $m(0)$ is the initial mass and k is a negative constant. The *mean life* M of an atom in the substance is

$$M = -k\int_0^\infty te^{kt}\,dt$$

For the radioactive carbon isotope, ^{14}C, used in radiocarbon dating, the value of k is -0.000121. Find the mean life of a ^{14}C atom.

69. Determine how large the number a has to be so that

$$\int_a^\infty \frac{1}{x^2+1}\,dx < 0.001$$

70. Estimate the numerical value of $\int_0^\infty e^{-x^2}\,dx$ by writing it as the sum of $\int_0^4 e^{-x^2}\,dx$ and $\int_4^\infty e^{-x^2}\,dx$. Approximate the first integral by using Simpson's Rule with $n = 8$ and show that the second integral is smaller than $\int_4^\infty e^{-4x}\,dx$, which is less than 0.0000001.

71. If $f(t)$ is continuous for $t \geq 0$, the *Laplace transform* of f is the function F defined by

$$F(s) = \int_0^\infty f(t)e^{-st}\,dt$$

and the domain of F is the set consisting of all numbers s for which the integral converges. Find the Laplace transforms of the following functions.
(a) $f(t) = 1$ (b) $f(t) = e^t$ (c) $f(t) = t$

72. Show that if $0 \leq f(t) \leq Me^{at}$ for $t \geq 0$, where M and a are constants, then the Laplace transform $F(s)$ exists for $s > a$.

73. Suppose that $0 \leq f(t) \leq Me^{at}$ and $0 \leq f'(t) \leq Ke^{at}$ for $t \geq 0$, where f' is continuous. If the Laplace transform of $f(t)$ is $F(s)$

and the Laplace transform of $f'(t)$ is $G(s)$, show that

$$G(s) = sF(s) - f(0) \qquad s > a$$

74. If $\int_{-\infty}^{\infty} f(x)\,dx$ is convergent and a and b are real numbers, show that

$$\int_{-\infty}^{a} f(x)\,dx + \int_{a}^{\infty} f(x)\,dx = \int_{-\infty}^{b} f(x)\,dx + \int_{b}^{\infty} f(x)\,dx$$

75. Show that $\int_0^\infty x^2 e^{-x^2}\,dx = \frac{1}{2}\int_0^\infty e^{-x^2}\,dx$.

76. Show that $\int_0^\infty e^{-x^2}\,dx = \int_0^1 \sqrt{-\ln y}\,dy$ by interpreting the integrals as areas.

77. Find the value of the constant C for which the integral

$$\int_0^\infty \left(\frac{1}{\sqrt{x^2+4}} - \frac{C}{x+2} \right) dx$$

converges. Evaluate the integral for this value of C.

78. Find the value of the constant C for which the integral

$$\int_0^\infty \left(\frac{x}{x^2+1} - \frac{C}{3x+1} \right) dx$$

converges. Evaluate the integral for this value of C.

7 Review

CONCEPT CHECK

1. State the rule for integration by parts. In practice, how do you use it?

2. How do you evaluate $\int \sin^m x \cos^n x\,dx$ if m is odd? What if n is odd? What if m and n are both even?

3. If the expression $\sqrt{a^2 - x^2}$ occurs in an integral, what substitution might you try? What if $\sqrt{a^2 + x^2}$ occurs? What if $\sqrt{x^2 - a^2}$ occurs?

4. What is the form of the partial fraction expansion of a rational function $P(x)/Q(x)$ if the degree of P is less than the degree of Q and $Q(x)$ has only distinct linear factors? What if a linear factor is repeated? What if $Q(x)$ has an irreducible quadratic factor (not repeated)? What if the quadratic factor is repeated?

5. State the rules for approximating the definite integral $\int_a^b f(x)\,dx$ with the Midpoint Rule, the Trapezoidal Rule, and Simpson's Rule. Which would you expect to give the best estimate? How do you approximate the error for each rule?

6. Define the following improper integrals.

(a) $\int_a^\infty f(x)\,dx$ (b) $\int_{-\infty}^b f(x)\,dx$ (c) $\int_{-\infty}^\infty f(x)\,dx$

7. Define the improper integral $\int_a^b f(x)\,dx$ for each of the following cases.
(a) f has an infinite discontinuity at a.
(b) f has an infinite discontinuity at b.
(c) f has an infinite discontinuity at c, where $a < c < b$.

8. State the Comparison Theorem for improper integrals.

TRUE-FALSE QUIZ

Determine whether the statement is true or false. If it is true, explain why. If it is false, explain why or give an example that disproves the statement.

1. $\dfrac{x(x^2+4)}{x^2-4}$ can be put in the form $\dfrac{A}{x+2} + \dfrac{B}{x-2}$.

2. $\dfrac{x^2+4}{x(x^2-4)}$ can be put in the form $\dfrac{A}{x} + \dfrac{B}{x+2} + \dfrac{C}{x-2}$.

3. $\dfrac{x^2+4}{x^2(x-4)}$ can be put in the form $\dfrac{A}{x^2} + \dfrac{B}{x-4}$.

4. $\dfrac{x^2-4}{x(x^2+4)}$ can be put in the form $\dfrac{A}{x} + \dfrac{B}{x^2+4}$.

5. $\displaystyle\int_0^4 \frac{x}{x^2-1}\,dx = \tfrac{1}{2}\ln 15$

6. $\displaystyle\int_1^\infty \frac{1}{x^{\sqrt{2}}}\,dx$ is convergent.

7. If f is continuous, then $\int_{-\infty}^\infty f(x)\,dx = \lim_{t\to\infty} \int_{-t}^t f(x)\,dx$.

8. The Midpoint Rule is always more accurate than the Trapezoidal Rule.

9. (a) Every elementary function has an elementary derivative.
(b) Every elementary function has an elementary antiderivative.

10. If f is continuous on $[0, \infty)$ and $\int_1^\infty f(x)\,dx$ is convergent, then $\int_0^\infty f(x)\,dx$ is convergent.

11. If f is a continuous, decreasing function on $[1, \infty)$ and $\lim_{x\to\infty} f(x) = 0$, then $\int_1^\infty f(x)\,dx$ is convergent.

12. If $\int_a^\infty f(x)\,dx$ and $\int_a^\infty g(x)\,dx$ are both convergent, then $\int_a^\infty [f(x) + g(x)]\,dx$ is convergent.

13. If $\int_a^\infty f(x)\,dx$ and $\int_a^\infty g(x)\,dx$ are both divergent, then $\int_a^\infty [f(x) + g(x)]\,dx$ is divergent.

14. If $f(x) \leqslant g(x)$ and $\int_0^\infty g(x)\,dx$ diverges, then $\int_0^\infty f(x)\,dx$ also diverges.

EXERCISES

Note: Additional practice in techniques of integration is provided in Exercises 7.5.

1–40 Evaluate the integral.

1. $\int_0^5 \frac{x}{x+10}\,dx$

2. $\int_0^5 ye^{-0.6y}\,dy$

3. $\int_0^{\pi/2} \frac{\cos\theta}{1+\sin\theta}\,d\theta$

4. $\int_1^4 \frac{dt}{(2t+1)^3}$

5. $\int \tan^7 x \sec^3 x\,dx$

6. $\int \frac{1}{y^2-4y-12}\,dy$

7. $\int \frac{\sin(\ln t)}{t}\,dt$

8. $\int \frac{dx}{x^2\sqrt{1+x^2}}$

9. $\int_1^4 x^{3/2}\ln x\,dx$

10. $\int_0^1 \frac{\sqrt{\arctan x}}{1+x^2}\,dx$

11. $\int_1^2 \frac{\sqrt{x^2-1}}{x}\,dx$

12. $\int_{-1}^1 \frac{\sin x}{1+x^2}\,dx$

13. $\int \frac{dx}{x^3+x}$

14. $\int \frac{x^2+2}{x+2}\,dx$

15. $\int \sin^2\theta\,\cos^5\theta\,d\theta$

16. $\int \frac{\sec^6\theta}{\tan^2\theta}\,d\theta$

17. $\int x\sec x\tan x\,dx$

18. $\int \frac{x^2+8x-3}{x^3+3x^2}\,dx$

19. $\int \frac{x+1}{9x^2+6x+5}\,dx$

20. $\int \frac{dt}{\sin^2 t+\cos 2t}$

21. $\int \frac{dx}{\sqrt{x^2-4x}}$

22. $\int \frac{x^3}{(x+1)^{10}}\,dx$

23. $\int \csc^4 4x\,dx$

24. $\int e^x\cos x\,dx$

25. $\int \frac{3x^3-x^2+6x-4}{(x^2+1)(x^2+2)}\,dx$

26. $\int \frac{dx}{1+e^x}$

27. $\int_0^{\pi/2} \cos^3 x\sin 2x\,dx$

28. $\int \frac{\sqrt[3]{x}+1}{\sqrt[3]{x}-1}\,dx$

29. $\int_{-1}^1 x^5\sec x\,dx$

30. $\int \frac{dx}{e^x\sqrt{1-e^{-2x}}}$

31. $\int_0^{\ln 10} \frac{e^x\sqrt{e^x-1}}{e^x+8}\,dx$

32. $\int_0^{\pi/4} \frac{x\sin x}{\cos^3 x}\,dx$

33. $\int \frac{x^2}{(4-x^2)^{3/2}}\,dx$

34. $\int (\arcsin x)^2\,dx$

35. $\int \frac{1}{\sqrt{x+x^{3/2}}}\,dx$

36. $\int \frac{1-\tan\theta}{1+\tan\theta}\,d\theta$

37. $\int (\cos x+\sin x)^2\cos 2x\,dx$

38. $\int x(\tan^{-1}x)^2\,dx$

39. $\int_0^{1/2} \frac{xe^{2x}}{(1+2x)^2}\,dx$

40. $\int_{\pi/4}^{\pi/3} \frac{\sqrt{\tan\theta}}{\sin 2\theta}\,d\theta$

41–50 Evaluate the integral or show that it is divergent.

41. $\int_1^\infty \frac{1}{(2x+1)^3}\,dx$

42. $\int_0^1 \frac{t^2+1}{t^2-1}\,dt$

43. $\int_2^\infty \frac{dx}{x\ln x}$

44. $\int_2^6 \frac{y}{\sqrt{y-2}}\,dy$

45. $\int_0^4 \frac{\ln x}{\sqrt{x}}\,dx$

46. $\int_0^1 \frac{1}{2-3x}\,dx$

47. $\int_0^3 \frac{dx}{x^2-x-2}$

48. $\int_{-1}^1 \frac{x+1}{\sqrt[3]{x^4}}\,dx$

49. $\int_{-\infty}^\infty \frac{dx}{4x^2+4x+5}$

50. $\int_1^\infty \frac{\tan^{-1}x}{x^2}\,dx$

51–52 Evaluate the indefinite integral. Illustrate and check that your answer is reasonable by graphing both the function and its antiderivative (take $C=0$).

51. $\int \ln(x^2+2x+2)\,dx$

52. $\int \frac{x^3}{\sqrt{x^2+1}}\,dx$

53. Graph the function $f(x)=\cos^2 x\sin^3 x$ and use the graph to guess the value of the integral $\int_0^{2\pi} f(x)\,dx$. Then evaluate the integral to confirm your guess.

CAS 54. (a) How would you evaluate $\int x^5 e^{-2x}\,dx$ by hand? (Don't actually carry out the integration.)
(b) How would you evaluate $\int x^5 e^{-2x}\,dx$ using tables? (Don't actually do it.)
(c) Use a CAS to evaluate $\int x^5 e^{-2x}\,dx$.
(d) Graph the integrand and the indefinite integral on the same screen.

55–58 Use the Table of Integrals on the Reference Pages to evaluate the integral.

55. $\int e^x\sqrt{1-e^{2x}}\,dx$

56. $\int \csc^5 t\,dt$

57. $\int \sqrt{x^2+x+1}\,dx$

58. $\int \frac{\cot x}{\sqrt{1+2\sin x}}\,dx$

59. Verify Formula 33 in the Table of Integrals (a) by differentiation and (b) by using a trigonometric substitution.

60. Verify Formula 62 in the Table of Integrals.

61. Is it possible to find a number n such that $\int_0^\infty x^n\,dx$ is convergent?

62. For what values of a is $\int_0^\infty e^{ax} \cos x\, dx$ convergent? Evaluate the integral for those values of a.

63–64 |||| Use (a) the Trapezoidal Rule, (b) the Midpoint Rule, and (c) Simpson's Rule with $n = 10$ to approximate the given integral. Round your answers to six decimal places.

63. $\int_0^1 \sqrt{1 + x^4}\, dx$

64. $\int_0^{\pi/2} \sqrt{\sin x}\, dx$

65. Estimate the errors involved in Exercise 63, parts (a) and (b). How large should n be in each case to guarantee an error of less than 0.00001?

66. Use Simpson's Rule with $n = 6$ to estimate the area under the curve $y = e^x/x$ from $x = 1$ to $x = 4$.

67. The speedometer reading (v) on a car was observed at 1-minute intervals and recorded in the chart. Use Simpson's Rule to estimate the distance traveled by the car.

t (min)	v (mi/h)	t (min)	v (mi/h)
0	40	6	56
1	42	7	57
2	45	8	57
3	49	9	55
4	52	10	56
5	54		

68. A population of honeybees increased at a rate of $r(t)$ bees per week, where the graph of r is as shown. Use Simpson's Rule with six subintervals to estimate the increase in the bee population during the first 24 weeks.

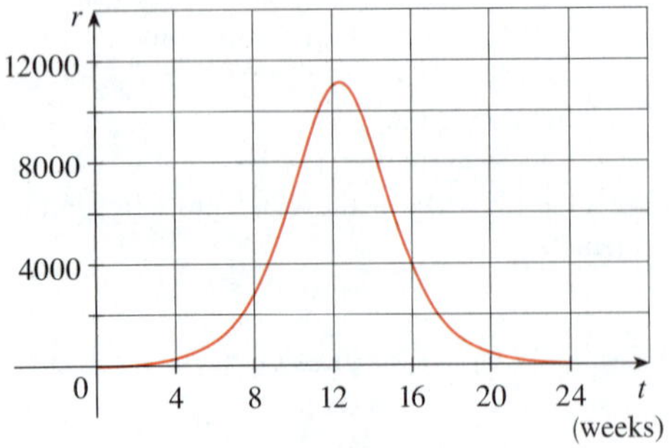

CAS **69.** (a) If $f(x) = \sin(\sin x)$, use a graph to find an upper bound for $|f^{(4)}(x)|$.
(b) Use Simpson's Rule with $n = 10$ to approximate $\int_0^\pi f(x)\, dx$ and use part (a) to estimate the error.
(c) How large should n be to guarantee that the size of the error in using S_n is less than 0.00001?

70. Suppose you are asked to estimate the volume of a football. You measure and find that a football is 28 cm long. You use a piece of string and measure the circumference at its widest point to be 53 cm. The circumference 7 cm from each end is 45 cm. Use Simpson's Rule to make your estimate.

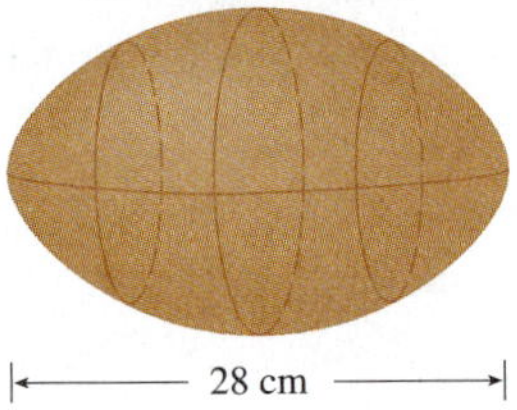

71. Use the Comparison Theorem to determine whether the integral

$$\int_1^\infty \frac{x^3}{x^5 + 2}\, dx$$

is convergent or divergent.

72. Find the area of the region bounded by the hyperbola $y^2 - x^2 = 1$ and the line $y = 3$.

73. Find the area bounded by the curves $y = \cos x$ and $y = \cos^2 x$ between $x = 0$ and $x = \pi$.

74. Find the area of the region bounded by the curves $y = 1/(2 + \sqrt{x})$, $y = 1/(2 - \sqrt{x})$, and $x = 1$.

75. The region under the curve $y = \cos^2 x$, $0 \leqslant x \leqslant \pi/2$, is rotated about the x-axis. Find the volume of the resulting solid.

76. The region in Exercise 75 is rotated about the y-axis. Find the volume of the resulting solid.

77. If f' is continuous on $[0, \infty)$ and $\lim_{x\to\infty} f(x) = 0$, show that

$$\int_0^\infty f'(x)\, dx = -f(0)$$

78. We can extend our definition of average value of a continuous function to an infinite interval by defining the average value of f on the interval $[a, \infty)$ to be

$$\lim_{t\to\infty} \frac{1}{t - a} \int_a^t f(x)\, dx$$

(a) Find the average value of $y = \tan^{-1}x$ on the interval $[0, \infty)$.
(b) If $f(x) \geqslant 0$ and $\int_a^\infty f(x)\, dx$ is divergent, show that the average value of f on the interval $[a, \infty)$ is $\lim_{x\to\infty} f(x)$, if this limit exists.
(c) If $\int_a^\infty f(x)\, dx$ is convergent, what is the average value of f on the interval $[a, \infty)$?
(d) Find the average value of $y = \sin x$ on the interval $[0, \infty)$.

79. Use the substitution $u = 1/x$ to show that

$$\int_0^\infty \frac{\ln x}{1 + x^2}\, dx = 0$$

80. The magnitude of the repulsive force between two point charges with the same sign, one of size 1 and the other of size q, is

$$F = \frac{q}{4\pi\varepsilon_0 r^2}$$

where r is the distance between the charges and ε_0 is a constant. The *potential* V at a point P due to the charge q is defined to be the work expended in bringing a unit charge to P from infinity along the straight line that joins q and P. Find a formula for V.

PROBLEMS PLUS

Cover up the solution to the example and try it yourself first.

EXAMPLE

(a) Prove that if f is a continuous function, then

$$\int_0^a f(x)\,dx = \int_0^a f(a-x)\,dx$$

(b) Use part (a) to show that

$$\int_0^{\pi/2} \frac{\sin^n x}{\sin^n x + \cos^n x}\,dx = \frac{\pi}{4}$$

for all positive numbers n.

SOLUTION

(a) At first sight, the given equation may appear somewhat baffling. How is it possible to connect the left side to the right side? Connections can often be made through one of the principles of problem solving: *introduce something extra.* Here the extra ingredient is a new variable. We often think of introducing a new variable when we use the Substitution Rule to integrate a specific function. But that technique is still useful in the present circumstance in which we have a general function f.

The principles of problem solving are discussed on page 80.

Once we think of making a substitution, the form of the right side suggests that it should be $u = a - x$. Then $du = -dx$. When $x = 0$, $u = a$; when $x = a$, $u = 0$. So

$$\int_0^a f(a-x)\,dx = -\int_a^0 f(u)\,du = \int_0^a f(u)\,du$$

But this integral on the right side is just another way of writing $\int_0^a f(x)\,dx$. So the given equation is proved.

(b) If we let the given integral be I and apply part (a) with $a = \pi/2$, we get

$$I = \int_0^{\pi/2} \frac{\sin^n x}{\sin^n x + \cos^n x}\,dx = \int_0^{\pi/2} \frac{\sin^n(\pi/2 - x)}{\sin^n(\pi/2 - x) + \cos^n(\pi/2 - x)}\,dx$$

The computer graphs in Figure 1 make it seem plausible that all of the integrals in the example have the same value. The graph of each integrand is labeled with the corresponding value of n.

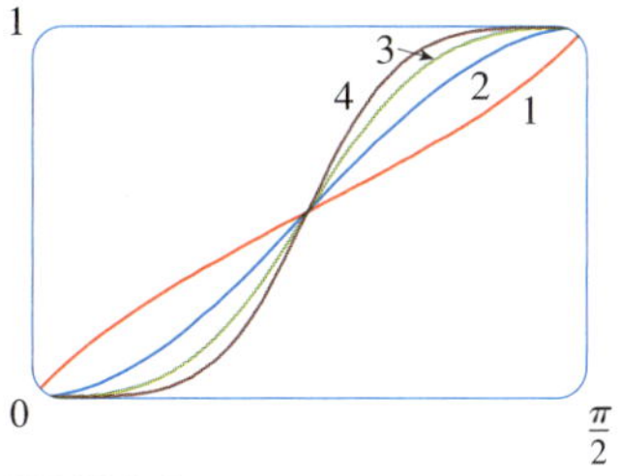

FIGURE 1

A well-known trigonometric identity tells us that $\sin(\pi/2 - x) = \cos x$ and $\cos(\pi/2 - x) = \sin x$, so we get

$$I = \int_0^{\pi/2} \frac{\cos^n x}{\cos^n x + \sin^n x}\,dx$$

Notice that the two expressions for I are very similar. In fact, the integrands have the same denominator. This suggests that we should add the two expressions. If we do so, we get

$$2I = \int_0^{\pi/2} \frac{\sin^n x + \cos^n x}{\sin^n x + \cos^n x}\,dx = \int_0^{\pi/2} 1\,dx = \frac{\pi}{2}$$

Therefore, $I = \pi/4$.

PROBLEMS

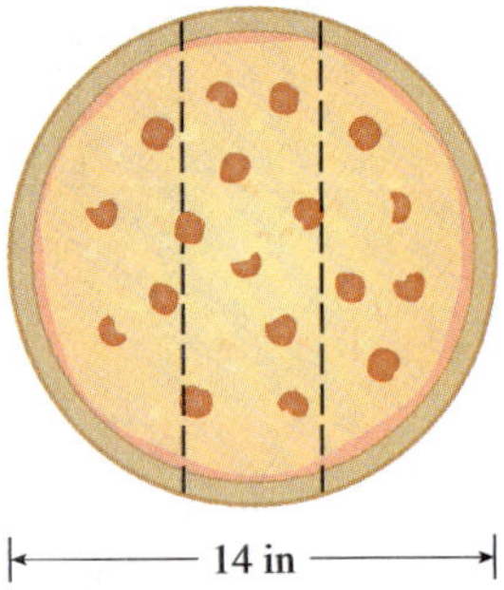

FIGURE FOR PROBLEM 1

1. Three mathematics students have ordered a 14-inch pizza. Instead of slicing it in the traditional way, they decide to slice it by parallel cuts, as shown in the figure. Being mathematics majors, they are able to determine where to slice so that each gets the same amount of pizza. Where are the cuts made?

2. Evaluate $\int \frac{1}{x^7 - x}\, dx$.

   The straightforward approach would be to start with partial fractions, but that would be brutal. Try a substitution.

3. Evaluate $\int_0^1 \left(\sqrt[3]{1 - x^7} - \sqrt[7]{1 - x^3}\right) dx$.

4. A man initially standing at the point O walks along a pier pulling a rowboat by a rope of length L. The man keeps the rope straight and taut. The path followed by the boat is a curve called a *tractrix* and it has the property that the rope is always tangent to the curve (see the figure).

   (a) Show that if the path followed by the boat is the graph of the function $y = f(x)$, then
   $$f'(x) = \frac{dy}{dx} = \frac{-\sqrt{L^2 - x^2}}{x}$$
   (b) Determine the function $y = f(x)$.

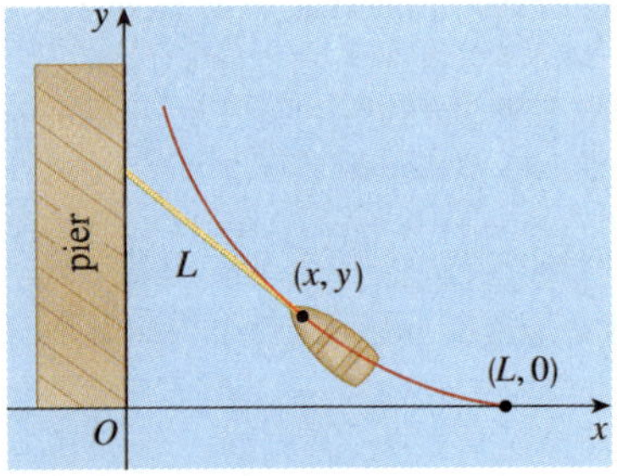

FIGURE FOR PROBLEM 4

5. A function f is defined by
   $$f(x) = \int_0^\pi \cos t \cos(x - t)\, dt \qquad 0 \leqslant x \leqslant 2\pi$$
   Find the minimum value of f.

6. If n is a positive integer, prove that
   $$\int_0^1 (\ln x)^n\, dx = (-1)^n n!$$

7. Show that
   $$\int_0^1 (1 - x^2)^n\, dx = \frac{2^{2n}(n!)^2}{(2n + 1)!}$$
   Hint: Start by showing that if I_n denotes the integral, then
   $$I_{k+1} = \frac{2k + 2}{2k + 3} I_k$$

8. Suppose that f is a positive function such that f' is continuous.

   (a) How is the graph of $y = f(x) \sin nx$ related to the graph of $y = f(x)$? What happens as $n \to \infty$?

   (b) Make a guess as to the value of the limit
   $$\lim_{n \to \infty} \int_0^1 f(x) \sin nx\, dx$$
   based on graphs of the integrand.

   (c) Using integration by parts, confirm the guess that you made in part (b). [Use the fact that, since f' is continuous, there is a constant M such that $|f'(x)| \leqslant M$ for $0 \leqslant x \leqslant 1$.]

9. If $0 < a < b$, find $\lim_{t \to 0} \left\{ \int_0^1 [bx + a(1 - x)]^t\, dx \right\}^{1/t}$.

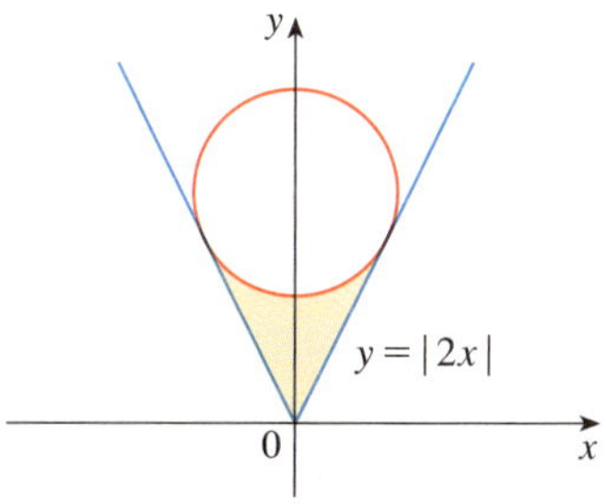

FIGURE FOR PROBLEM 11

10. Graph $f(x) = \sin(e^x)$ and use the graph to estimate the value of t such that $\int_t^{t+1} f(x)\, dx$ is a maximum. Then find the exact value of t that maximizes this integral.

11. The circle with radius 1 shown in the figure touches the curve $y = |2x|$ twice. Find the area of the region that lies between the two curves.

12. A rocket is fired straight up, burning fuel at the constant rate of b kilograms per second. Let $v = v(t)$ be the velocity of the rocket at time t and suppose that the velocity u of the exhaust gas is constant. Let $M = M(t)$ be the mass of the rocket at time t and note that M decreases as the fuel burns. If we neglect air resistance, it follows from Newton's Second Law that

$$F = M\frac{dv}{dt} - ub$$

where the force $F = -Mg$. Thus

$$M\frac{dv}{dt} - ub = -Mg \qquad \boxed{1}$$

Let M_1 be the mass of the rocket without fuel, M_2 the initial mass of the fuel, and $M_0 = M_1 + M_2$. Then, until the fuel runs out at time $t = M_2 b$, the mass is $M = M_0 - bt$.

(a) Substitute $M = M_0 - bt$ into Equation 1 and solve the resulting equation for v. Use the initial condition $v(0) = 0$ to evaluate the constant.

(b) Determine the velocity of the rocket at time $t = M_2/b$. This is called the *burnout velocity*.

(c) Determine the height of the rocket $y = y(t)$ at the burnout time.

(d) Find the height of the rocket at any time t.

13. Use integration by parts to show that, for all $x > 0$,

$$0 < \int_0^\infty \frac{\sin t}{\ln(1 + x + t)}\, dt < \frac{2}{\ln(1 + x)}$$

14. The **Chebyshev polynomials** T_n are defined by

$$T_n(x) = \cos(n \arccos x) \qquad n = 0, 1, 2, 3, \ldots$$

(a) What are the domain and range of these functions?

(b) We know that $T_0(x) = 1$ and $T_1(x) = x$. Express T_2 explicitly as a quadratic polynomial and T_3 as a cubic polynomial.

(c) Show that, for $n \geqslant 1$,

$$T_{n+1}(x) = 2xT_n(x) - T_{n-1}(x)$$

(d) Use part (c) to show that T_n is a polynomial of degree n.

(e) Use parts (b) and (c) to express T_4, T_5, T_6, and T_7 explicitly as polynomials.

(f) What are the zeros of T_n? At what numbers does T_n have local maximum and minimum values?

(g) Graph T_2, T_3, T_4, and T_5 on a common screen.

(h) Graph T_5, T_6, and T_7 on a common screen.

(i) Based on your observations from parts (g) and (h), how are the zeros of T_n related to the zeros of T_{n+1}? What about the x-coordinates of the maximum and minimum values?

(j) Based on your graphs in parts (g) and (h), what can you say about $\int_{-1}^{1} T_n(x)\, dx$ when n is odd and when n is even?

(k) Use the substitution $u = \arccos x$ to evaluate the integral in part (j).

(l) The family of functions $f(x) = \cos(c \arccos x)$ are defined even when c is not an integer (but then f is not a polynomial). Describe how the graph of f changes as c increases.

CHAPTER 8

Integration enables us to calculate the force exerted by water on a dam.

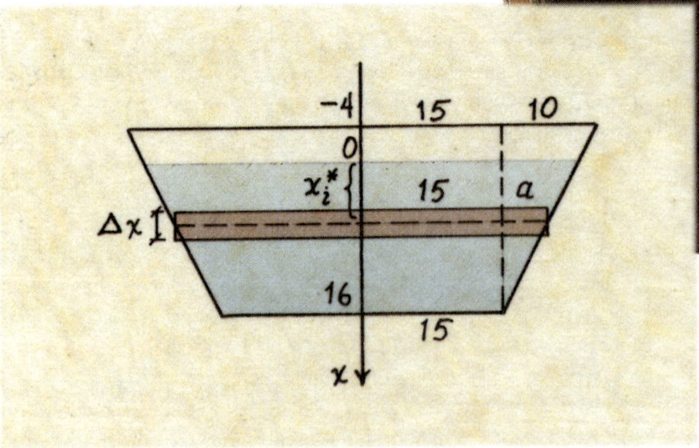

Further Applications of Integration

We looked at some applications of integrals in Chapter 6: areas, volumes, work, and average values. Here we explore some of the many other geometric applications of integration—the length of a curve, the area of a surface—as well as quantities of interest in physics, engineering, biology, economics, and statistics. For instance, we will investigate the center of gravity of a plate, the force exerted by water pressure on a dam, the flow of blood from the human heart, and the average time spent on hold during a telephone call.

8.1 Arc Length

FIGURE 1

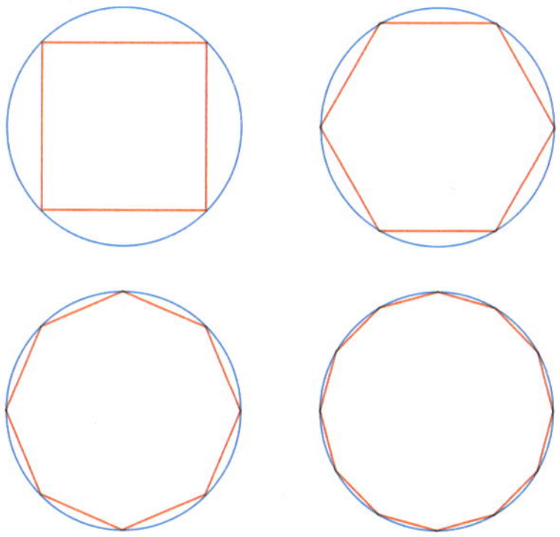

FIGURE 2

What do we mean by the length of a curve? We might think of fitting a piece of string to the curve in Figure 1 and then measuring the string against a ruler. But that might be difficult to do with much accuracy if we have a complicated curve. We need a precise definition for the length of an arc of a curve, in the same spirit as the definitions we developed for the concepts of area and volume.

If the curve is a polygon, we can easily find its length; we just add the lengths of the line segments that form the polygon. (We can use the distance formula to find the distance between the endpoints of each segment.) We are going to define the length of a general curve by first approximating it by a polygon and then taking a limit as the number of segments of the polygon is increased. This process is familiar for the case of a circle, where the circumference is the limit of lengths of inscribed polygons (see Figure 2).

Now suppose that a curve C is defined by the equation $y = f(x)$, where f is continuous and $a \leqslant x \leqslant b$. We obtain a polygonal approximation to C by dividing the interval $[a, b]$ into n subintervals with endpoints $x_0, x_1, \ldots, x_n$ and equal width Δx. If $y_i = f(x_i)$, then the point $P_i(x_i, y_i)$ lies on C and the polygon with vertices $P_0, P_1, \ldots, P_n$, illustrated in Figure 3, is an approximation to C. The length L of C is approximately the length of this polygon and the approximation gets better as we let n increase. (See Figure 4, where the arc of the curve between P_{i-1} and P_i has been magnified and approximations with successively smaller values of Δx are shown.) Therefore, we define the **length** L of the curve C

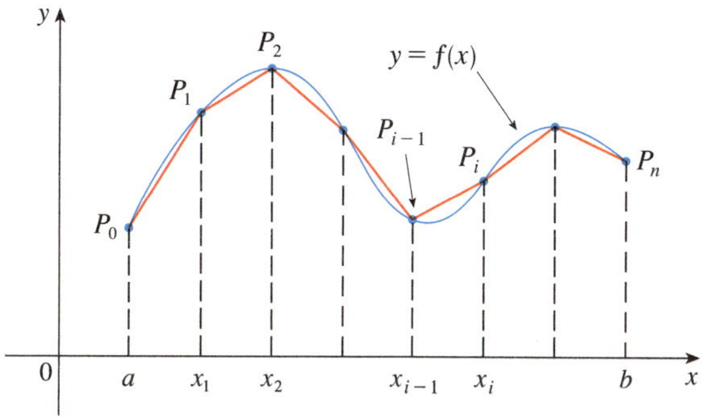

FIGURE 3

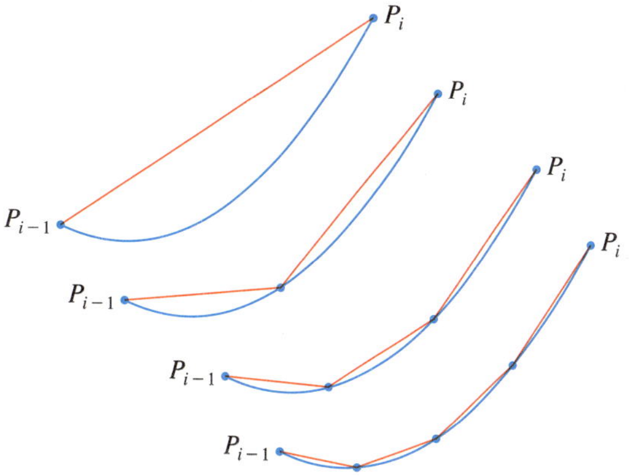

FIGURE 4

with equation $y = f(x)$, $a \leqslant x \leqslant b$, as the limit of the lengths of these inscribed polygons (if the limit exists):

1
$$L = \lim_{n \to \infty} \sum_{i=1}^{n} |P_{i-1}P_i|$$

Notice that the procedure for defining arc length is very similar to the procedure we used for defining area and volume: We divided the curve into a large number of small parts. We then found the approximate lengths of the small parts and added them. Finally, we took the limit as $n \to \infty$.

The definition of arc length given by Equation 1 is not very convenient for computational purposes, but we can derive an integral formula for L in the case where f has a continuous derivative. [Such a function f is called **smooth** because a small change in x produces a small change in $f'(x)$.]

If we let $\Delta y_i = y_i - y_{i-1}$, then

$$|P_{i-1}P_i| = \sqrt{(x_i - x_{i-1})^2 + (y_i - y_{i-1})^2} = \sqrt{(\Delta x)^2 + (\Delta y_i)^2}$$

By applying the Mean Value Theorem to f on the interval $[x_{i-1}, x_i]$, we find that there is a number x_i^* between x_{i-1} and x_i such that

$$f(x_i) - f(x_{i-1}) = f'(x_i^*)(x_i - x_{i-1})$$

that is,
$$\Delta y_i = f'(x_i^*)\,\Delta x$$

Thus, we have

$$\begin{aligned} |P_{i-1}P_i| &= \sqrt{(\Delta x)^2 + (\Delta y_i)^2} \\ &= \sqrt{(\Delta x)^2 + [f'(x_i^*)\,\Delta x]^2} \\ &= \sqrt{1 + [f'(x_i^*)]^2}\,\sqrt{(\Delta x)^2} \\ &= \sqrt{1 + [f'(x_i^*)]^2}\,\Delta x \qquad \text{(since } \Delta x > 0\text{)} \end{aligned}$$

Therefore, by Definition 1,

$$L = \lim_{n \to \infty} \sum_{i=1}^{n} |P_{i-1}P_i| = \lim_{n \to \infty} \sum_{i=1}^{n} \sqrt{1 + [f'(x_i^*)]^2}\,\Delta x$$

We recognize this expression as being equal to

$$\int_a^b \sqrt{1 + [f'(x)]^2}\,dx$$

by the definition of a definite integral. This integral exists because the function $g(x) = \sqrt{1 + [f'(x)]^2}$ is continuous. Thus, we have proved the following theorem:

2 The Arc Length Formula If f' is continuous on $[a, b]$, then the length of the curve $y = f(x)$, $a \leqslant x \leqslant b$, is

$$L = \int_a^b \sqrt{1 + [f'(x)]^2}\,dx$$

If we use Leibniz notation for derivatives, we can write the arc length formula as follows:

3
$$L = \int_a^b \sqrt{1 + \left(\frac{dy}{dx}\right)^2}\, dx$$

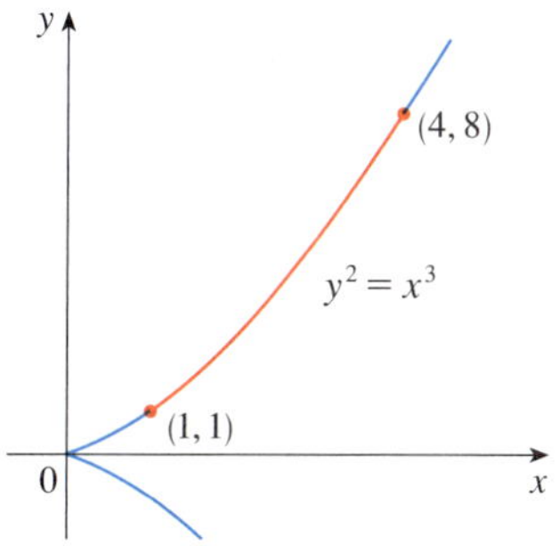

FIGURE 5

|||| As a check on our answer to Example 1, notice from Figure 5 that it ought to be slightly larger than the distance from (1, 1) to (4, 8), which is

$$\sqrt{58} \approx 7.615773$$

According to our calculation in Example 1, we have

$$L = \tfrac{1}{27}(80\sqrt{10} - 13\sqrt{13}) \approx 7.633705$$

Sure enough, this is a bit greater than the length of the line segment.

EXAMPLE 1 Find the length of the arc of the semicubical parabola $y^2 = x^3$ between the points (1, 1) and (4, 8). (See Figure 5.)

SOLUTION For the top half of the curve we have

$$y = x^{3/2} \qquad \frac{dy}{dx} = \tfrac{3}{2}x^{1/2}$$

and so the arc length formula gives

$$L = \int_1^4 \sqrt{1 + \left(\frac{dy}{dx}\right)^2}\, dx = \int_1^4 \sqrt{1 + \tfrac{9}{4}x}\, dx$$

If we substitute $u = 1 + 9x/4$, then $du = 9\,dx/4$. When $x = 1$, $u = \frac{13}{4}$; when $x = 4$, $u = 10$. Therefore

$$\begin{aligned} L &= \tfrac{4}{9}\int_{13/4}^{10} \sqrt{u}\, du = \tfrac{4}{9} \cdot \tfrac{2}{3}u^{3/2}\Big]_{13/4}^{10} \\ &= \tfrac{8}{27}\left[10^{3/2} - \left(\tfrac{13}{4}\right)^{3/2}\right] \\ &= \tfrac{1}{27}\left(80\sqrt{10} - 13\sqrt{13}\right) \end{aligned}$$

If a curve has the equation $x = g(y)$, $c \leqslant y \leqslant d$, and $g'(y)$ is continuous, then by interchanging the roles of x and y in Formula 2 or Equation 3, we obtain the following formula for its length:

4
$$L = \int_c^d \sqrt{1 + [g'(y)]^2}\, dy = \int_c^d \sqrt{1 + \left(\frac{dx}{dy}\right)^2}\, dy$$

EXAMPLE 2 Find the length of the arc of the parabola $y^2 = x$ from (0, 0) to (1, 1).

SOLUTION Since $x = y^2$, we have $dx/dy = 2y$, and Formula 4 gives

$$L = \int_0^1 \sqrt{1 + \left(\frac{dx}{dy}\right)^2}\, dy = \int_0^1 \sqrt{1 + 4y^2}\, dy$$

We make the trigonometric substitution $y = \frac{1}{2}\tan\theta$, which gives $dy = \frac{1}{2}\sec^2\theta\, d\theta$ and $\sqrt{1 + 4y^2} = \sqrt{1 + \tan^2\theta} = \sec\theta$. When $y = 0$, $\tan\theta = 0$, so $\theta = 0$; when $y = 1$, $\tan\theta = 2$, so $\theta = \tan^{-1}2 = \alpha$, say. Thus

$$\begin{aligned} L &= \int_0^\alpha \sec\theta \cdot \tfrac{1}{2}\sec^2\theta\, d\theta = \tfrac{1}{2}\int_0^\alpha \sec^3\theta\, d\theta \\ &= \tfrac{1}{2} \cdot \tfrac{1}{2}\left[\sec\theta\tan\theta + \ln|\sec\theta + \tan\theta|\right]_0^\alpha \qquad \text{(from Example 8 in Section 7.2)} \\ &= \tfrac{1}{4}\left(\sec\alpha\tan\alpha + \ln|\sec\alpha + \tan\alpha|\right) \end{aligned}$$

(We could have used Formula 21 in the Table of Integrals.) Since $\tan \alpha = 2$, we have $\sec^2\alpha = 1 + \tan^2\alpha = 5$, so $\sec \alpha = \sqrt{5}$ and

$$L = \frac{\sqrt{5}}{2} + \frac{\ln(\sqrt{5} + 2)}{4}$$

|||| Figure 6 shows the arc of the parabola whose length is computed in Example 2, together with polygonal approximations having $n = 1$ and $n = 2$ line segments, respectively. For $n = 1$ the approximate length is $L_1 = \sqrt{2}$, the diagonal of a square. The table shows the approximations L_n that we get by dividing $[0, 1]$ into n equal subintervals. Notice that each time we double the number of sides of the polygon, we get closer to the exact length, which is

$$L = \frac{\sqrt{5}}{2} + \frac{\ln(\sqrt{5} + 2)}{4} \approx 1.478943$$

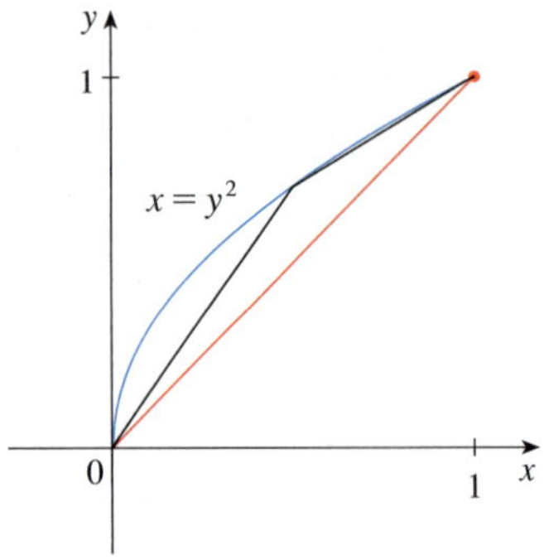

FIGURE 6

n	L_n
1	1.414
2	1.445
4	1.464
8	1.472
16	1.476
32	1.478
64	1.479

Because of the presence of the square root sign in Formulas 2 and 4, the calculation of an arc length often leads to an integral that is very difficult or even impossible to evaluate explicitly. Thus, we sometimes have to be content with finding an approximation to the length of a curve as in the following example.

EXAMPLE 3

(a) Set up an integral for the length of the arc of the hyperbola $xy = 1$ from the point $(1, 1)$ to the point $(2, \frac{1}{2})$.
(b) Use Simpson's Rule with $n = 10$ to estimate the arc length.

SOLUTION

(a) We have

$$y = \frac{1}{x} \qquad \frac{dy}{dx} = -\frac{1}{x^2}$$

and so the arc length is

$$L = \int_1^2 \sqrt{1 + \left(\frac{dy}{dx}\right)^2}\, dx = \int_1^2 \sqrt{1 + \frac{1}{x^4}}\, dx = \int_1^2 \frac{\sqrt{x^4 + 1}}{x^2}\, dx$$

(b) Using Simpson's Rule (see Section 7.7) with $a = 1$, $b = 2$, $n = 10$, $\Delta x = 0.1$, and $f(x) = \sqrt{1 + 1/x^4}$, we have

$$L = \int_1^2 \sqrt{1 + \frac{1}{x^4}}\, dx$$

$$\approx \frac{\Delta x}{3}[f(1) + 4f(1.1) + 2f(1.2) + 4f(1.3) + \cdots + 2f(1.8) + 4f(1.9) + f(2)]$$

$$\approx 1.1321$$

|||| Checking the value of the definite integral with a more accurate approximation produced by a computer algebra system, we see that the approximation using Simpson's Rule is accurate to four decimal places.

The Arc Length Function

We will find it useful to have a function that measures the arc length of a curve from a particular starting point to any other point on the curve. Thus, if a smooth curve C has the equation $y = f(x)$, $a \leqslant x \leqslant b$, let $s(x)$ be the distance along C from the initial point $P_0(a, f(a))$ to the point $Q(x, f(x))$. Then s is a function, called the **arc length function**, and, by Formula 2,

5 $$s(x) = \int_a^x \sqrt{1 + [f'(t)]^2}\, dt$$

(We have replaced the variable of integration by t so that x does not have two meanings.) We can use Part 1 of the Fundamental Theorem of Calculus to differentiate Equation 5 (since the integrand is continuous):

6 $$\frac{ds}{dx} = \sqrt{1 + [f'(x)]^2} = \sqrt{1 + \left(\frac{dy}{dx}\right)^2}$$

Equation 6 shows that the rate of change of s with respect to x is always at least 1 and is equal to 1 when $f'(x)$, the slope of the curve, is 0. The differential of arc length is

7 $$ds = \sqrt{1 + \left(\frac{dy}{dx}\right)^2}\, dx$$

and this equation is sometimes written in the symmetric form

8 $$(ds)^2 = (dx)^2 + (dy)^2$$

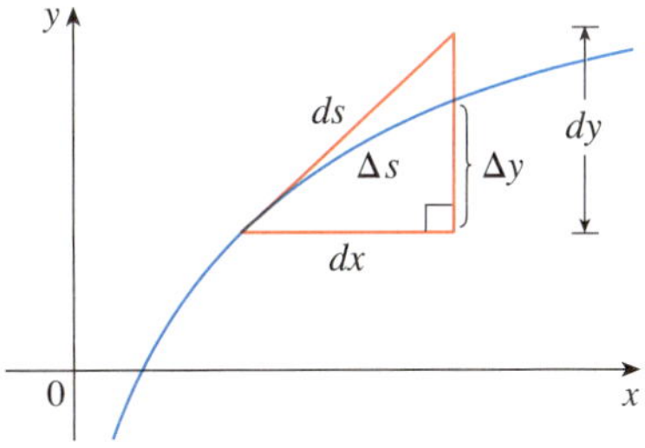

FIGURE 7

The geometric interpretation of Equation 8 is shown in Figure 7. It can be used as a mnemonic device for remembering both of the Formulas 3 and 4. If we write $L = \int ds$, then from Equation 8 either we can solve to get (7), which gives (3), or we can solve to get

$$ds = \sqrt{1 + \left(\frac{dx}{dy}\right)^2}\, dy$$

which gives (4).

EXAMPLE 4 Find the arc length function for the curve $y = x^2 - \frac{1}{8}\ln x$ taking $P_0(1, 1)$ as the starting point.

SOLUTION If $f(x) = x^2 - \frac{1}{8}\ln x$, then

$$f'(x) = 2x - \frac{1}{8x}$$

$$1 + [f'(x)]^2 = 1 + \left(2x - \frac{1}{8x}\right)^2 = 1 + 4x^2 - \frac{1}{2} + \frac{1}{64x^2}$$

$$= 4x^2 + \frac{1}{2} + \frac{1}{64x^2} = \left(2x + \frac{1}{8x}\right)^2$$

$$\sqrt{1 + [f'(x)]^2} = 2x + \frac{1}{8x}$$

Thus, the arc length function is given by

$$s(x) = \int_1^x \sqrt{1 + [f'(t)]^2}\, dt$$
$$= \int_1^x \left(2t - \frac{1}{8t}\right) dt = t^2 + \tfrac{1}{8} \ln t \Big]_1^x$$
$$= x^2 + \tfrac{1}{8} \ln x - 1$$

For instance, the arc length along the curve from $(1, 1)$ to $(3, f(3))$ is

$$s(3) = 3^2 + \tfrac{1}{8} \ln 3 - 1 = 8 + \frac{\ln 3}{8} \approx 8.1373$$

Figure 8 shows the interpretation of the arc length function in Example 4. Figure 9 shows the graph of this arc length function. Why is $s(x)$ negative when x is less than 1?

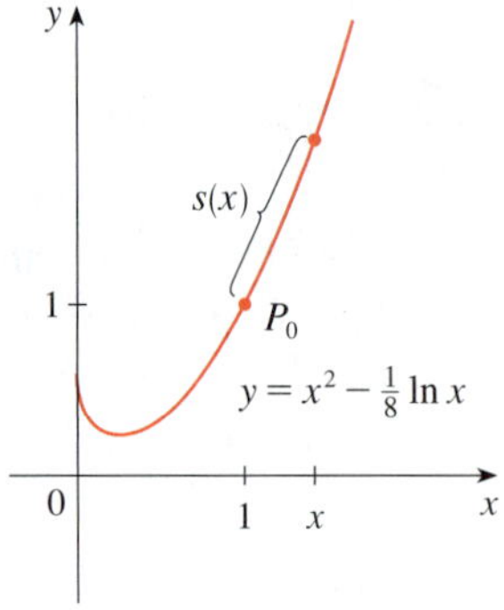

FIGURE 8

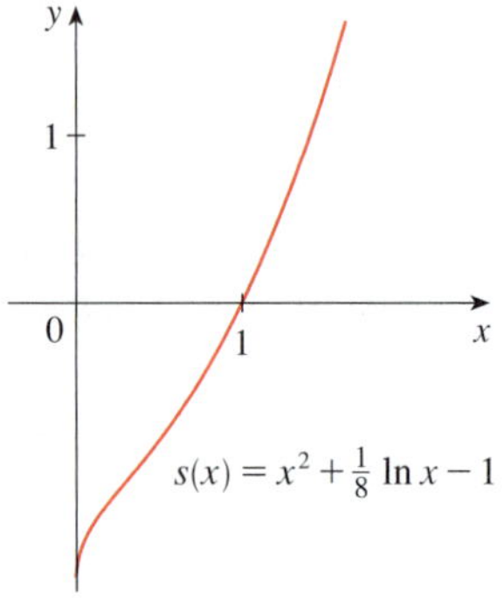

FIGURE 9

8.1 Exercises

1. Use the arc length formula (3) to find the length of the curve $y = 2 - 3x$, $-2 \leqslant x \leqslant 1$. Check your answer by noting that the curve is a line segment and calculating its length by the distance formula.

2. Use the arc length formula to find the length of the curve $y = \sqrt{4 - x^2}$, $0 \leqslant x \leqslant 2$. Check your answer by noting that the curve is a quarter-circle.

3–4 Graph the curve and visually estimate its length. Then find its exact length.

3. $y = \frac{2}{3}(x^2 - 1)^{3/2}, \quad 1 \leqslant x \leqslant 3$

4. $y = \dfrac{x^3}{6} + \dfrac{1}{2x}, \quad \frac{1}{2} \leqslant x \leqslant 1$

5–16 Find the length of the curve.

5. $y = 1 + 6x^{3/2}, \quad 0 \leqslant x \leqslant 1$

6. $y^2 = 4(x + 4)^3, \quad 0 \leqslant x \leqslant 2, \quad y > 0$

7. $y = \dfrac{x^5}{6} + \dfrac{1}{10x^3}, \quad 1 \leqslant x \leqslant 2$

8. $y = \dfrac{x^2}{2} - \dfrac{\ln x}{4}, \quad 2 \leqslant x \leqslant 4$

9. $x = \frac{1}{3}\sqrt{y}\,(y - 3), \quad 1 \leqslant y \leqslant 9$

10. $y = \ln(\cos x), \quad 0 \leqslant x \leqslant \pi/3$

11. $y = \ln(\sec x), \quad 0 \leqslant x \leqslant \pi/4$

12. $y = \ln x, \quad 1 \leqslant x \leqslant \sqrt{3}$

13. $y = \cosh x, \quad 0 \leqslant x \leqslant 1$

14. $y^2 = 4x, \quad 0 \leqslant y \leqslant 2$

15. $y = e^x, \quad 0 \leqslant x \leqslant 1$

16. $y = \ln\left(\dfrac{e^x + 1}{e^x - 1}\right), \quad a \leqslant x \leqslant b, \; a > 0$

17–20 Set up, but do not evaluate, an integral for the length of the curve.

17. $y = \cos x, \quad 0 \leqslant x \leqslant 2\pi$

18. $y = 2^x, \quad 0 \leqslant x \leqslant 3$

19. $x = y + y^3, \quad 1 \leqslant y \leqslant 4$

20. $\dfrac{x^2}{a^2} + \dfrac{y^2}{b^2} = 1$

21–24 Use Simpson's Rule with $n = 10$ to estimate the arc length of the curve. Compare your answer with the value of the integral produced by your calculator.

21. $y = xe^{-x}, \quad 0 \leqslant x \leqslant 5$

22. $x = y + \sqrt{y}, \quad 1 \leqslant y \leqslant 2$

23. $y = \sec x, \quad 0 \leqslant x \leqslant \pi/3$

24. $y = x \ln x, \quad 1 \leqslant x \leqslant 3$

25. (a) Graph the curve $y = x\sqrt[3]{4 - x}$, $0 \leqslant x \leqslant 4$.
(b) Compute the lengths of inscribed polygons with $n = 1$, 2, and 4 sides. (Divide the interval into equal subintervals.) Illustrate by sketching these polygons (as in Figure 6).
(c) Set up an integral for the length of the curve.
(d) Use your calculator to find the length of the curve to four decimal places. Compare with the approximations in part (b).

26. Repeat Exercise 25 for the curve

$$y = x + \sin x \qquad 0 \leqslant x \leqslant 2\pi$$

27. Use either a computer algebra system or a table of integrals to find the *exact* length of the arc of the curve $x = \ln(1 - y^2)$ that lies between the points (0, 0) and $(\ln \frac{3}{4}, \frac{1}{2})$.

28. Use either a computer algebra system or a table of integrals to find the *exdct* length of the arc of the curve $y = x^{4/3}$ that lies between the points (0, 0) and (1, 1). If your CAS has trouble evaluating the integral, make a substitution that changes the integral into one that the CAS can evaluate.

29. Sketch the curve with equation $x^{2/3} + y^{2/3} = 1$ and use symmetry to find its length.

30. (a) Sketch the curve $y^3 = x^2$.
(b) Use Formulas 3 and 4 to set up two integrals for the arc length from (0, 0) to (1, 1). Observe that one of these is an improper integral and evaluate both of them.
(c) Find the length of the arc of this curve from $(-1, 1)$ to (8, 4).

31. Find the arc length function for the curve $y = 2x^{3/2}$ with starting point $P_0(1, 2)$.

32. (a) Graph the curve $y = \frac{1}{3}x^3 + 1/(4x)$, $x > 0$.
(b) Find the arc length function for this curve with starting point $P_0(1, \frac{7}{12})$.
(c) Graph the arc length function.

33. A hawk flying at 15 m/s at an altitude of 180 m accidentally drops its prey. The parabolic trajectory of the falling prey is described by the equation

$$y = 180 - \frac{x^2}{45}$$

until it hits the ground, where y is its height above the ground and x is the horizontal distance traveled in meters. Calculate the distance traveled by the prey from the time it is dropped until the time it hits the ground. Express your answer correct to the nearest tenth of a meter.

34. A steady wind blows a kite due west. The kite's height above ground from horizontal position $x = 0$ to $x = 80$ ft is given by

$$y = 150 - \tfrac{1}{40}(x - 50)^2$$

Find the distance traveled by the kite.

35. A manufacturer of corrugated metal roofing wants to produce panels that are 28 in. wide and 2 in. thick by processing flat sheets of metal as shown in the figure. The profile of the roofing takes the shape of a sine wave. Verify that the sine curve has equation $y = \sin(\pi x/7)$ and find the width w of a flat metal sheet that is needed to make a 28-inch panel. (Use your calculator to evaluate the integral correct to four significant digits.)

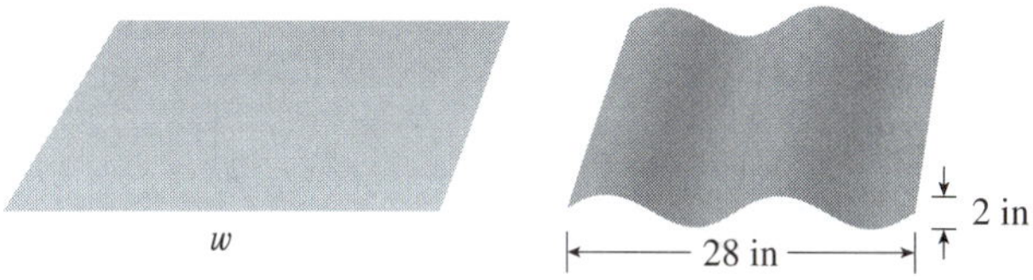

36. (a) The figure shows a telephone wire hanging between two poles at $x = -b$ and $x = b$. It takes the shape of a catenary with equation $y = c + a\cosh(x/a)$. Find the length of the wire.
(b) Suppose two telephone poles are 50 ft apart and the length of the wire between the poles is 51 ft. If the lowest point of the wire must be 20 ft above the ground, how high up on each pole should the wire be attached?

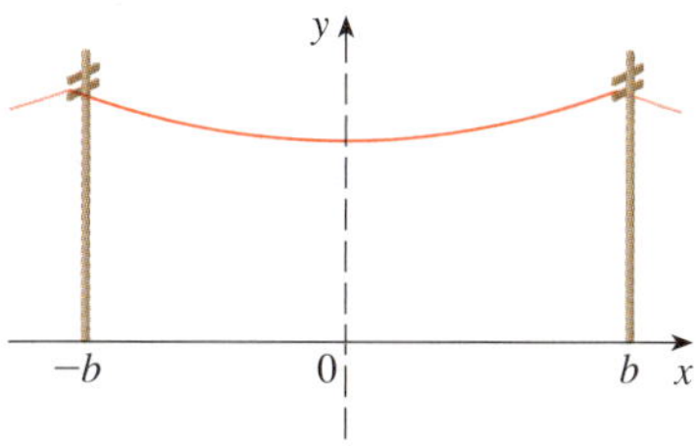

37. Find the length of the curve $y = \int_1^x \sqrt{t^3 - 1}\, dt$, $1 \leqslant x \leqslant 4$.

38. The curves with equations $x^n + y^n = 1$, $n = 4, 6, 8, \ldots$, are called **fat circles**. Graph the curves with $n = 2, 4, 6, 8$, and 10 to see why. Set up an integral for the length L_{2k} of the fat circle with $n = 2k$. Without attempting to evaluate this integral, state the value of

$$\lim_{k \to \infty} L_{2k}$$

DISCOVERY PROJECT

Arc Length Contest

The curves shown are all examples of graphs of continuous functions f that have the following properties.

1. $f(0) = 0$ and $f(1) = 0$
2. $f(x) \geq 0$ for $0 \leq x \leq 1$
3. The area under the graph of f from 0 to 1 is equal to 1.

The lengths L of these curves, however, are different.

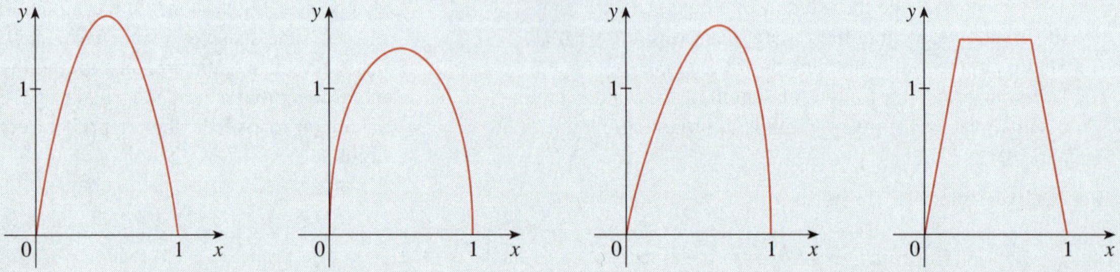

$L \approx 3.249$ $L \approx 2.919$ $L \approx 3.152$ $L \approx 3.213$

Try to discover formulas for two functions that satisfy the given conditions 1, 2, and 3. (Your graphs might be similar to the ones shown or could look quite different.) Then calculate the arc length of each graph. The winning entry will be the one with the smallest arc length.

8.2 Area of a Surface of Revolution

A surface of revolution is formed when a curve is rotated about a line. Such a surface is the lateral boundary of a solid of revolution of the type discussed in Sections 6.2 and 6.3.

We want to define the area of a surface of revolution in such a way that it corresponds to our intuition. If the surface area is A, we can imagine that painting the surface would require the same amount of paint as does a flat region with area A.

Let's start with some simple surfaces. The lateral surface area of a circular cylinder with radius r and height h is taken to be $A = 2\pi rh$ because we can imagine cutting the cylinder and unrolling it (as in Figure 1) to obtain a rectangle with dimensions $2\pi r$ and h.

Likewise, we can take a circular cone with base radius r and slant height l, cut it along the dashed line in Figure 2, and flatten it to form a sector of a circle with radius l and central angle $\theta = 2\pi r/l$. We know that, in general, the area of a sector of a circle with radius l and angle θ is $\frac{1}{2}l^2\theta$ (see Exercise 35 in Section 7.3) and so in this case it is

$$A = \tfrac{1}{2}l^2\theta = \tfrac{1}{2}l^2\left(\frac{2\pi r}{l}\right) = \pi rl$$

Therefore, we define the lateral surface area of a cone to be $A = \pi rl$.

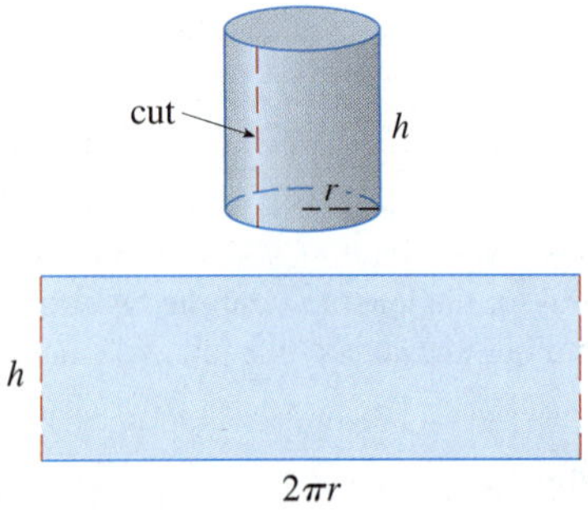

FIGURE 1

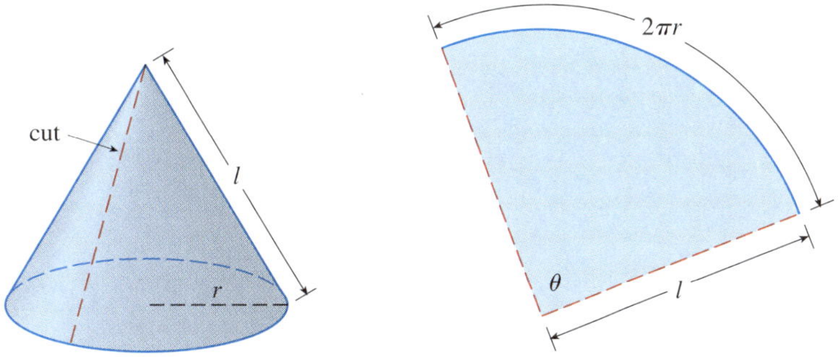

FIGURE 2

What about more complicated surfaces of revolution? If we follow the strategy we used with arc length, we can approximate the original curve by a polygon. When this polygon is rotated about an axis, it creates a simpler surface whose surface area approximates the actual surface area. By taking a limit, we can determine the exact surface area.

The approximating surface, then, consists of a number of *bands,* each formed by rotating a line segment about an axis. To find the surface area, each of these bands can be considered a portion of a circular cone, as shown in Figure 3. The area of the band (or frustum of a cone) with slant height l and upper and lower radii r_1 and r_2 is found by subtracting the areas of two cones:

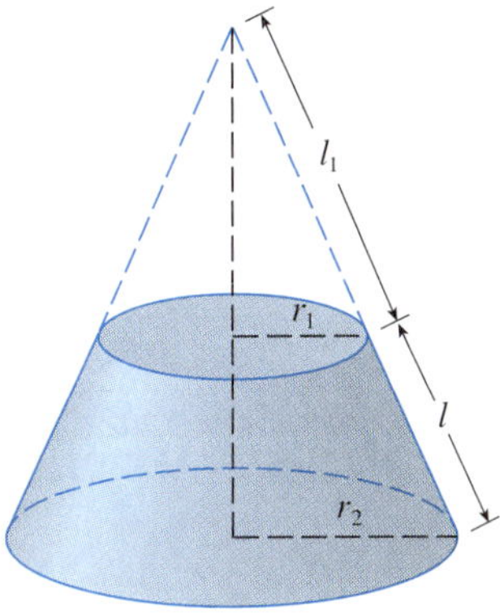

FIGURE 3

1 $$A = \pi r_2(l_1 + l) - \pi r_1 l_1 = \pi[(r_2 - r_1)l_1 + r_2 l]$$

From similar triangles we have

$$\frac{l_1}{r_1} = \frac{l_1 + l}{r_2}$$

which gives

$$r_2 l_1 = r_1 l_1 + r_1 l \qquad \text{or} \qquad (r_2 - r_1)l_1 = r_1 l$$

Putting this in Equation 1, we get

$$A = \pi(r_1 l + r_2 l)$$

or

2 $$\boxed{A = 2\pi r l}$$

where $r = \frac{1}{2}(r_1 + r_2)$ is the average radius of the band.

Now we apply this formula to our strategy. Consider the surface shown in Figure 4, which is obtained by rotating the curve $y = f(x)$, $a \leqslant x \leqslant b$, about the x-axis, where f is positive and has a continuous derivative. In order to define its surface area, we divide the interval $[a, b]$ into n subintervals with endpoints $x_0, x_1, \ldots, x_n$ and equal width Δx, as we did in determining arc length. If $y_i = f(x_i)$, then the point $P_i(x_i, y_i)$ lies on the curve. The part of the surface between x_{i-1} and x_i is approximated by taking the line segment $P_{i-1}P_i$ and rotating it about the x-axis. The result is a band with slant height $l = |P_{i-1}P_i|$ and average radius $r = \frac{1}{2}(y_{i-1} + y_i)$ so, by Formula 2, its surface area is

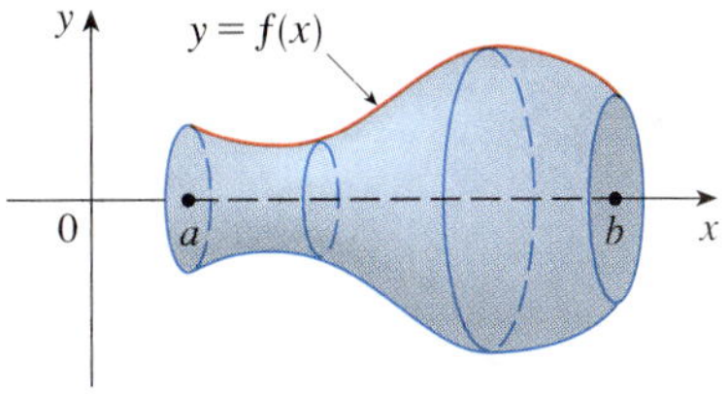

(a) Surface of revolution

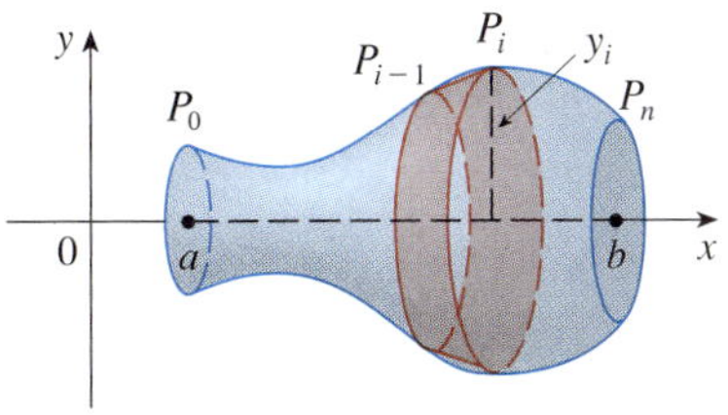

(b) Approximating band

FIGURE 4

$$2\pi \frac{y_{i-1} + y_i}{2} |P_{i-1}P_i|$$

As in the proof of Theorem 8.1.2, we have

$$|P_{i-1}P_i| = \sqrt{1 + [f'(x_i^*)]^2}\,\Delta x$$

where x_i^* is some number in $[x_{i-1}, x_i]$. When Δx is small, we have $y_i = f(x_i) \approx f(x_i^*)$ and also $y_{i-1} = f(x_{i-1}) \approx f(x_i^*)$, since f is continuous. Therefore

$$2\pi \frac{y_{i-1} + y_i}{2} |P_{i-1}P_i| \approx 2\pi f(x_i^*)\sqrt{1 + [f'(x_i^*)]^2}\,\Delta x$$

and so an approximation to what we think of as the area of the complete surface of revolution is

3 $$\sum_{i=1}^{n} 2\pi f(x_i^*)\sqrt{1 + [f'(x_i^*)]^2}\,\Delta x$$

This approximation appears to become better as $n \to \infty$ and, recognizing (3) as a Riemann sum for the function $g(x) = 2\pi f(x)\sqrt{1 + [f'(x)]^2}$, we have

$$\lim_{n\to\infty} \sum_{i=1}^{n} 2\pi f(x_i^*)\sqrt{1 + [f'(x_i^*)]^2}\,\Delta x = \int_a^b 2\pi f(x)\sqrt{1 + [f'(x)]^2}\,dx$$

Therefore, in the case where f is positive and has a continuous derivative, we define the **surface area** of the surface obtained by rotating the curve $y = f(x)$, $a \leqslant x \leqslant b$, about the x-axis as

4 $$S = \int_a^b 2\pi f(x)\sqrt{1 + [f'(x)]^2}\,dx$$

With the Leibniz notation for derivatives, this formula becomes

5 $$S = \int_a^b 2\pi y \sqrt{1 + \left(\frac{dy}{dx}\right)^2}\,dx$$

If the curve is described as $x = g(y)$, $c \leqslant y \leqslant d$, then the formula for surface area becomes

6 $$S = \int_c^d 2\pi y \sqrt{1 + \left(\frac{dx}{dy}\right)^2}\,dy$$

and both Formulas 5 and 6 can be summarized symbolically, using the notation for arc length given in Section 8.1, as

7 $$S = \int 2\pi y\,ds$$

For rotation about the y-axis, the surface area formula becomes

8 $$S = \int 2\pi x \, ds$$

where, as before, we can use either

$$ds = \sqrt{1 + \left(\frac{dy}{dx}\right)^2} \, dx \qquad \text{or} \qquad ds = \sqrt{1 + \left(\frac{dx}{dy}\right)^2} \, dy$$

These formulas can be remembered by thinking of $2\pi y$ or $2\pi x$ as the circumference of a circle traced out by the point (x, y) on the curve as it is rotated about the x-axis or y-axis, respectively (see Figure 5).

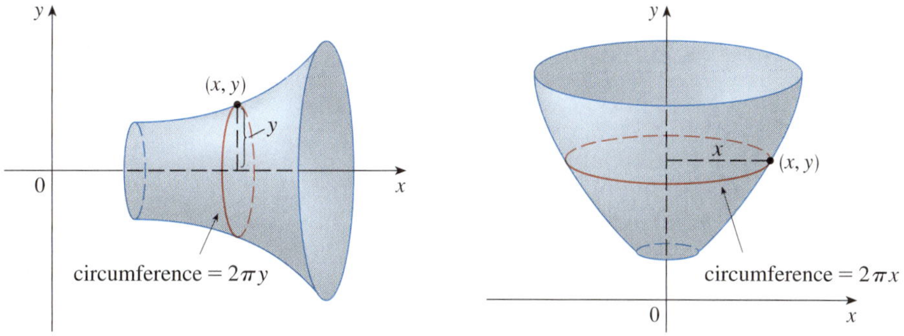

FIGURE 5 (a) Rotation about x-axis: $S = \int 2\pi y \, ds$ (b) Rotation about y-axis: $S = \int 2\pi x \, ds$

EXAMPLE 1 The curve $y = \sqrt{4 - x^2}$, $-1 \leqslant x \leqslant 1$, is an arc of the circle $x^2 + y^2 = 4$. Find the area of the surface obtained by rotating this arc about the x-axis. (The surface is a portion of a sphere of radius 2. See Figure 6.)

SOLUTION We have

$$\frac{dy}{dx} = \tfrac{1}{2}(4 - x^2)^{-1/2}(-2x) = \frac{-x}{\sqrt{4 - x^2}}$$

and so, by Formula 5, the surface area is

$$S = \int_{-1}^{1} 2\pi y \sqrt{1 + \left(\frac{dy}{dx}\right)^2} \, dx$$

$$= 2\pi \int_{-1}^{1} \sqrt{4 - x^2} \sqrt{1 + \frac{x^2}{4 - x^2}} \, dx$$

$$= 2\pi \int_{-1}^{1} \sqrt{4 - x^2} \frac{2}{\sqrt{4 - x^2}} \, dx$$

$$= 4\pi \int_{-1}^{1} 1 \, dx = 4\pi(2) = 8\pi$$

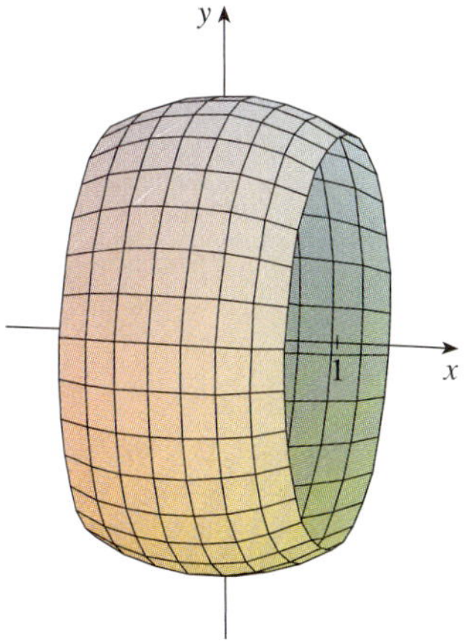

FIGURE 6

|||| Figure 6 shows the portion of the sphere whose surface area is computed in Example 1.

|||| Figure 7 shows the surface of revolution whose area is computed in Example 2.

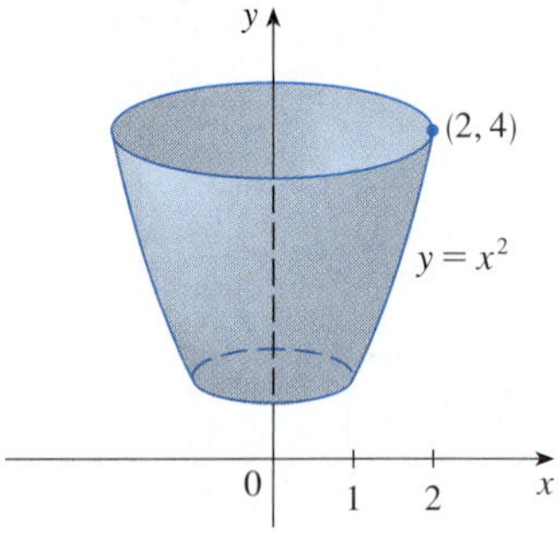

FIGURE 7

EXAMPLE 2 The arc of the parabola $y = x^2$ from $(1, 1)$ to $(2, 4)$ is rotated about the y-axis. Find the area of the resulting surface.

SOLUTION 1 Using

$$y = x^2 \qquad \text{and} \qquad \frac{dy}{dx} = 2x$$

we have, from Formula 8,

$$\begin{aligned} S &= \int 2\pi x\, ds \\ &= \int_1^2 2\pi x \sqrt{1 + \left(\frac{dy}{dx}\right)^2}\, dx \\ &= 2\pi \int_1^2 x\sqrt{1 + 4x^2}\, dx \end{aligned}$$

Substituting $u = 1 + 4x^2$, we have $du = 8x\,dx$. Remembering to change the limits of integration, we have

$$\begin{aligned} S &= \frac{\pi}{4}\int_5^{17} \sqrt{u}\, du = \frac{\pi}{4}\left[\tfrac{2}{3}u^{3/2}\right]_5^{17} \\ &= \frac{\pi}{6}\left(17\sqrt{17} - 5\sqrt{5}\right) \end{aligned}$$

|||| As a check on our answer to Example 2, notice from Figure 7 that the surface area should be close to that of a circular cylinder with the same height and radius halfway between the upper and lower radius of the surface: $2\pi(1.5)(3) \approx 28.27$. We computed that the surface area was

$$\frac{\pi}{6}\left(17\sqrt{17} - 5\sqrt{5}\right) \approx 30.85$$

which seems reasonable. Alternatively, the surface area should be slightly larger than the area of a frustum of a cone with the same top and bottom edges. From Equation 2, this is $2\pi(1.5)(\sqrt{10}) \approx 29.80$.

SOLUTION 2 Using

$$x = \sqrt{y} \qquad \text{and} \qquad \frac{dx}{dy} = \frac{1}{2\sqrt{y}}$$

we have

$$\begin{aligned} S &= \int 2\pi x\, ds = \int_1^4 2\pi x \sqrt{1 + \left(\frac{dx}{dy}\right)^2}\, dy \\ &= 2\pi \int_1^4 \sqrt{y}\sqrt{1 + \frac{1}{4y}}\, dy = \pi \int_1^4 \sqrt{4y + 1}\, dy \\ &= \frac{\pi}{4}\int_5^{17} \sqrt{u}\, du \qquad \text{(where } u = 1 + 4y\text{)} \\ &= \frac{\pi}{6}\left(17\sqrt{17} - 5\sqrt{5}\right) \qquad \text{(as in Solution 1)} \end{aligned}$$

EXAMPLE 3 Find the area of the surface generated by rotating the curve $y = e^x$, $0 \leqslant x \leqslant 1$, about the x-axis.

|||| Another method: Use Formula 6 with $x = \ln y$.

SOLUTION Using Formula 5 with

$$y = e^x \qquad \text{and} \qquad \frac{dy}{dx} = e^x$$

we have

$$S = \int_0^1 2\pi y \sqrt{1 + \left(\frac{dy}{dx}\right)^2}\, dx = 2\pi \int_0^1 e^x \sqrt{1 + e^{2x}}\, dx$$

$$= 2\pi \int_1^e \sqrt{1 + u^2}\, du \qquad \text{(where } u = e^x\text{)}$$

$$= 2\pi \int_{\pi/4}^{\alpha} \sec^3\theta\, d\theta \qquad \text{(where } u = \tan\theta \text{ and } \alpha = \tan^{-1}e\text{)}$$

|||| Or use Formula 21 in the Table of Integrals.

$$= 2\pi \cdot \tfrac{1}{2}\Big[\sec\theta \tan\theta + \ln|\sec\theta + \tan\theta|\Big]_{\pi/4}^{\alpha} \qquad \text{(by Example 8 in Section 7.2)}$$

$$= \pi\Big[\sec\alpha \tan\alpha + \ln(\sec\alpha + \tan\alpha) - \sqrt{2} - \ln(\sqrt{2} + 1)\Big]$$

Since $\tan\alpha = e$, we have $\sec^2\alpha = 1 + \tan^2\alpha = 1 + e^2$ and

$$S = \pi\Big[e\sqrt{1 + e^2} + \ln\big(e + \sqrt{1 + e^2}\big) - \sqrt{2} - \ln\big(\sqrt{2} + 1\big)\Big]$$

8.2 Exercises

1–4 |||| Set up, but do not evaluate, an integral for the area of the surface obtained by rotating the curve about the given axis.

1. $y = \ln x,\ 1 \leqslant x \leqslant 3;\quad x\text{-axis}$

2. $y = \sin^2 x,\ 0 \leqslant x \leqslant \pi/2;\quad x\text{-axis}$

3. $y = \sec x,\ 0 \leqslant x \leqslant \pi/4;\quad y\text{-axis}$

4. $y = e^x,\ 1 \leqslant y \leqslant 2;\quad y\text{-axis}$

5–12 |||| Find the area of the surface obtained by rotating the curve about the x-axis.

5. $y = x^3,\quad 0 \leqslant x \leqslant 2$

6. $9x = y^2 + 18,\quad 2 \leqslant x \leqslant 6$

7. $y = \sqrt{x},\quad 4 \leqslant x \leqslant 9$

8. $y = \cos 2x,\quad 0 \leqslant x \leqslant \pi/6$

9. $y = \cosh x,\quad 0 \leqslant x \leqslant 1$

10. $y = \dfrac{x^3}{6} + \dfrac{1}{2x},\quad \tfrac{1}{2} \leqslant x \leqslant 1$

11. $x = \tfrac{1}{3}(y^2 + 2)^{3/2},\quad 1 \leqslant y \leqslant 2$

12. $x = 1 + 2y^2,\quad 1 \leqslant y \leqslant 2$

13–16 |||| The given curve is rotated about the y-axis. Find the area of the resulting surface.

13. $y = \sqrt[3]{x},\quad 1 \leqslant y \leqslant 2$

14. $y = 1 - x^2,\quad 0 \leqslant x \leqslant 1$

15. $x = \sqrt{a^2 - y^2},\quad 0 \leqslant y \leqslant a/2$

16. $x = a\cosh(y/a),\quad -a \leqslant y \leqslant a$

17–20 |||| Use Simpson's Rule with $n = 10$ to approximate the area of the surface obtained by rotating the curve about the x-axis. Compare your answer with the value of the integral produced by your calculator.

17. $y = \ln x,\quad 1 \leqslant x \leqslant 3$

18. $y = x + \sqrt{x},\quad 1 \leqslant x \leqslant 2$

19. $y = \sec x,\quad 0 \leqslant x \leqslant \pi/3$

20. $y = \sqrt{1 + e^x},\quad 0 \leqslant x \leqslant 1$

CAS **21–22** |||| Use either a CAS or a table of integrals to find the exact area of the surface obtained by rotating the given curve about the x-axis.

21. $y = 1/x,\quad 1 \leqslant x \leqslant 2$

22. $y = \sqrt{x^2 + 1},\quad 0 \leqslant x \leqslant 3$

CAS **23–24** |||| Use a CAS to find the exact area of the surface obtained by rotating the curve about the y-axis. If your CAS has trouble evaluating the integral, express the surface area as an integral in the other variable.

23. $y = x^3,\quad 0 \leqslant y \leqslant 1$

24. $y = \ln(x + 1),\quad 0 \leqslant x \leqslant 1$

25. If the region $\mathcal{R} = \{(x, y) \mid x \geqslant 1,\ 0 \leqslant y \leqslant 1/x\}$ is rotated about the x-axis, the volume of the resulting solid is finite (see Exercise 63 in Section 7.8). Show that the surface area is infinite. (The surface is shown in the figure and is known as **Gabriel's horn**.)

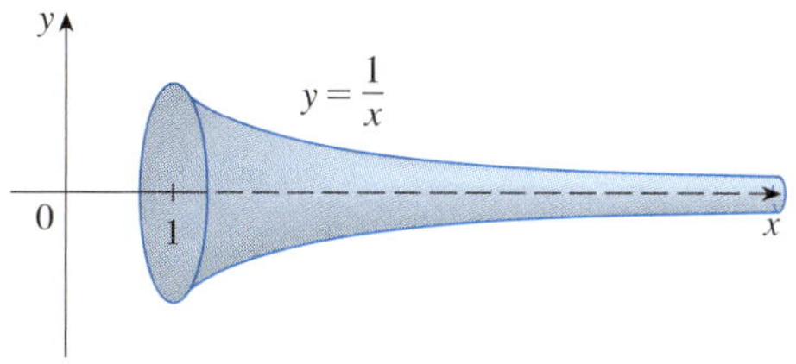

26. If the infinite curve $y = e^{-x}$, $x \geqslant 0$, is rotated about the x-axis, find the area of the resulting surface.

27. (a) If $a > 0$, find the area of the surface generated by rotating the loop of the curve $3ay^2 = x(a - x)^2$ about the x-axis.
(b) Find the surface area if the loop is rotated about the y-axis.

28. A group of engineers is building a parabolic satellite dish whose shape will be formed by rotating the curve $y = ax^2$ about the y-axis. If the dish is to have a 10-ft diameter and a maximum depth of 2 ft, find the value of a and the surface area of the dish.

29. The ellipse

$$\frac{x^2}{a^2} + \frac{y^2}{b^2} = 1 \qquad a > b$$

is rotated about the x-axis to form a surface called an *ellipsoid*. Find the surface area of this ellipsoid.

30. Find the surface area of the torus in Exercise 61 in Section 6.2.

31. If the curve $y = f(x)$, $a \leqslant x \leqslant b$, is rotated about the horizontal line $y = c$, where $f(x) \leqslant c$, find a formula for the area of the resulting surface.

CAS **32.** Use the result of Exercise 31 to set up an integral to find the area of the surface generated by rotating the curve $y = \sqrt{x}$, $0 \leqslant x \leqslant 4$, about the line $y = 4$. Then use a CAS to evaluate the integral.

33. Find the area of the surface obtained by rotating the circle $x^2 + y^2 = r^2$ about the line $y = r$.

34. Show that the surface area of a zone of a sphere that lies between two parallel planes is $S = \pi dh$, where d is the diameter of the sphere and h is the distance between the planes. (Notice that S depends only on the distance between the planes and not on their location, provided that both planes intersect the sphere.)

35. Formula 4 is valid only when $f(x) \geqslant 0$. Show that when $f(x)$ is not necessarily positive, the formula for surface area becomes

$$S = \int_a^b 2\pi |f(x)| \sqrt{1 + [f'(x)]^2}\, dx$$

36. Let L be the length of the curve $y = f(x)$, $a \leqslant x \leqslant b$, where f is positive and has a continuous derivative. Let S_f be the surface area generated by rotating the curve about the x-axis. If c is a positive constant, define $g(x) = f(x) + c$ and let S_g be the corresponding surface area generated by the curve $y = g(x)$, $a \leqslant x \leqslant b$. Express S_g in terms of S_f and L.

DISCOVERY PROJECT

Rotating on a Slant

We know how to find the volume of a solid of revolution obtained by rotating a region about a horizontal or vertical line (see Section 6.2). We also know how to find the surface area of a surface of revolution if we rotate a curve about a horizontal or vertical line (see Section 8.2). But what if we rotate about a slanted line, that is, a line that is neither horizontal nor vertical? In this project you are asked to discover formulas for the volume of a solid of revolution and for the area of a surface of revolution when the axis of rotation is a slanted line.

Let C be the arc of the curve $y = f(x)$ between the points $P(p, f(p))$ and $Q(q, f(q))$ and let $\mathcal{R}$ be the region bounded by C, by the line $y = mx + b$ (which lies entirely below C), and by the perpendiculars to the line from P and Q.

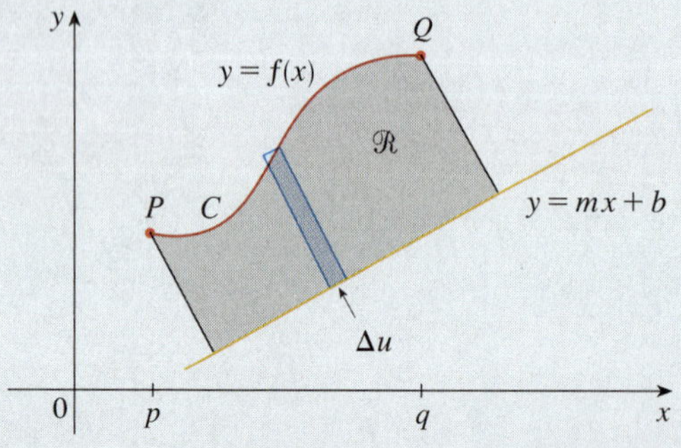

1. Show that the area of $\mathcal{R}$ is

$$\frac{1}{1+m^2}\int_p^q [f(x) - mx - b][1 + mf'(x)]\,dx$$

[*Hint:* This formula can be verified by subtracting areas, but it will be helpful throughout the project to derive it by first approximating the area using rectangles perpendicular to the line, as shown in the figure. Use the figure to help express Δu in terms of Δx.]

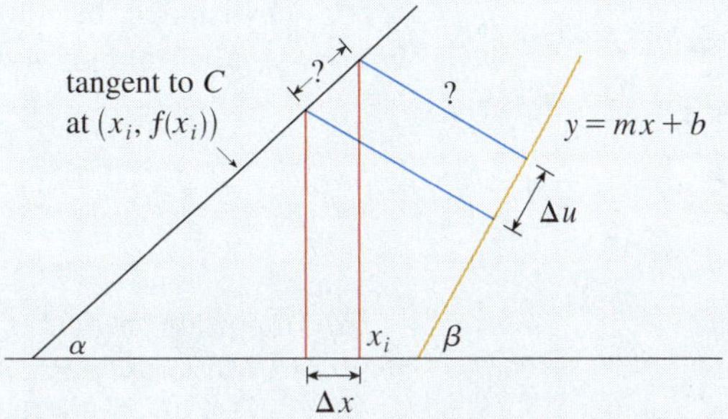

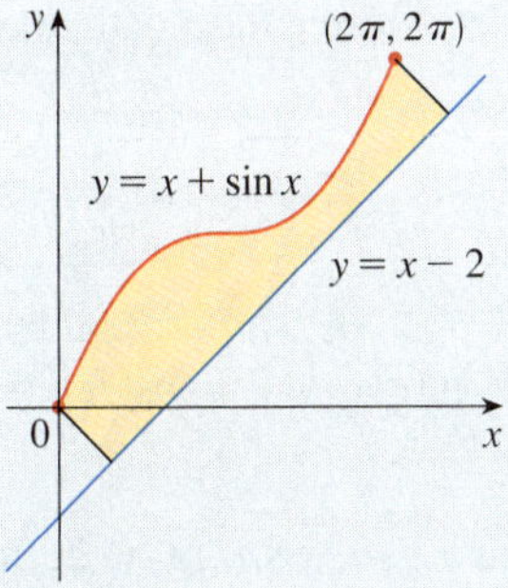

FIGURE FOR PROBLEM 2

2. Find the area of the region shown in the figure at the left.

3. Find a formula similar to the one in Problem 1 for the volume of the solid obtained by rotating $\mathcal{R}$ about the line $y = mx + b$.

4. Find the volume of the solid obtained by rotating the region of Problem 2 about the line $y = x - 2$.

5. Find a formula for the area of the surface obtained by rotating C about the line $y = mx + b$.

CAS **6.** Use a computer algebra system to find the exact area of the surface obtained by rotating the curve $y = \sqrt{x}$, $0 \leqslant x \leqslant 4$, about the line $y = \frac{1}{2}x$. Then approximate your result to three decimal places.

8.3 Applications to Physics and Engineering

Among the many applications of integral calculus to physics and engineering, we consider two here: force due to water pressure and centers of mass. As with our previous applications to geometry (areas, volumes, and lengths) and to work, our strategy is to break up the physical quantity into a large number of small parts, approximate each small part, add the results, take the limit, and then evaluate the resulting integral.

Hydrostatic Pressure and Force

Deep-sea divers realize that water pressure increases as they dive deeper. This is because the weight of the water above them increases.

In general, suppose that a thin horizontal plate with area A square meters is submerged in a fluid of density ρ kilograms per cubic meter at a depth d meters below the surface of the fluid as in Figure 1. The fluid directly above the plate has volume $V = Ad$, so its mass is $m = \rho V = \rho A d$. The force exerted by the fluid on the plate is therefore

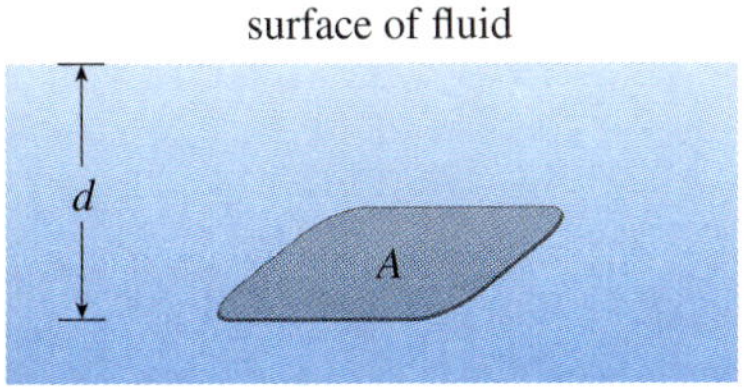

FIGURE 1

$$F = mg = \rho g A d$$

where g is the acceleration due to gravity. The **pressure** P on the plate is defined to be the force per unit area:

$$P = \frac{F}{A} = \rho g d$$

|||| When using U.S. Customary units, we write $P = \rho g d = \delta d$, where $\delta = \rho g$ is the weight density (as opposed to ρ, which is the mass density). For instance, the weight density of water is $\delta = 62.5 \text{ lb/ft}^3$.

The SI unit for measuring pressure is newtons per square meter, which is called a pascal (abbreviation: $1 \text{ N/m}^2 = 1 \text{ Pa}$). Since this is a small unit, the kilopascal (kPa) is often used. For instance, because the density of water is $\rho = 1000 \text{ kg/m}^3$, the pressure at the bottom of a swimming pool 2 m deep is

$$P = \rho g d = 1000 \text{ kg/m}^3 \times 9.8 \text{ m/s}^2 \times 2 \text{ m}$$

$$= 19{,}600 \text{ Pa} = 19.6 \text{ kPa}$$

An important principle of fluid pressure is the experimentally verified fact that at *any point in a liquid the pressure is the same in all directions.* (A diver feels the same pressure on nose and both ears.) Thus, the pressure in *any* direction at a depth d in a fluid with mass density ρ is given by

1 $$P = \rho g d = \delta d$$

This helps us determine the hydrostatic force against a *vertical* plate or wall or dam in a fluid. This is not a straightforward problem because the pressure is not constant but increases as the depth increases.

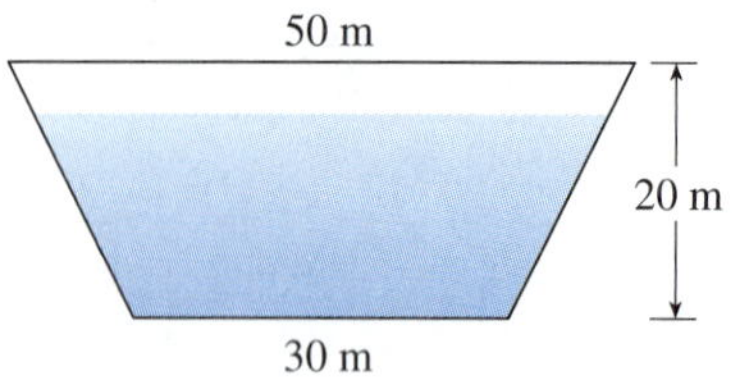

FIGURE 2

EXAMPLE 1 A dam has the shape of the trapezoid shown in Figure 2. The height is 20 m, and the width is 50 m at the top and 30 m at the bottom. Find the force on the dam due to hydrostatic pressure if the water level is 4 m from the top of the dam.

SOLUTION We choose a vertical x-axis with origin at the surface of the water as in Figure 3(a). The depth of the water is 16 m, so we divide the interval $[0, 16]$ into subintervals of equal length with endpoints x_i and we choose $x_i^* \in [x_{i-1}, x_i]$. The ith horizontal strip of the dam is approximated by a rectangle with height Δx and width w_i, where, from similar triangles in Figure 3(b),

$$\frac{a}{16 - x_i^*} = \frac{10}{20} \qquad \text{or} \qquad a = \frac{16 - x_i^*}{2} = 8 - \frac{x_i^*}{2}$$

and so $$w_i = 2(15 + a) = 2\left(15 + 8 - \tfrac{1}{2}x_i^*\right) = 46 - x_i^*$$

If A_i is the area of the ith strip, then

$$A_i \approx w_i \, \Delta x = (46 - x_i^*) \, \Delta x$$

If Δx is small, then the pressure P_i on the ith strip is almost constant and we can use Equation 1 to write

$$P_i \approx 1000 g x_i^*$$

The hydrostatic force F_i acting on the ith strip is the product of the pressure and the area:

$$F_i = P_i A_i \approx 1000 g x_i^* (46 - x_i^*) \, \Delta x$$

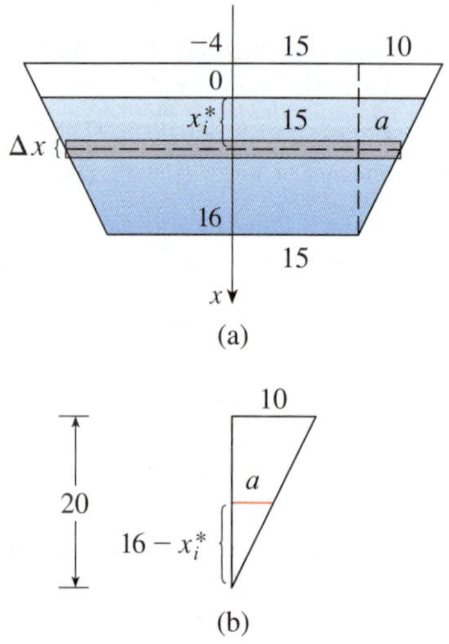

FIGURE 3

Adding these forces and taking the limit as $n \to \infty$, we obtain the total hydrostatic force on the dam:

$$
\begin{aligned}
F &= \lim_{n\to\infty} \sum_{i=1}^{n} 1000 g x_i^*(46 - x_i^*)\,\Delta x \\
&= \int_0^{16} 1000 g x (46 - x)\,dx \\
&= 1000(9.8)\int_0^{16} (46x - x^2)\,dx \\
&= 9800\left[23x^2 - \frac{x^3}{3}\right]_0^{16} \\
&\approx 4.43 \times 10^7\ \text{N}
\end{aligned}
$$

EXAMPLE 2 Find the hydrostatic force on one end of a cylindrical drum with radius 3 ft if the drum is submerged in water 10 ft deep.

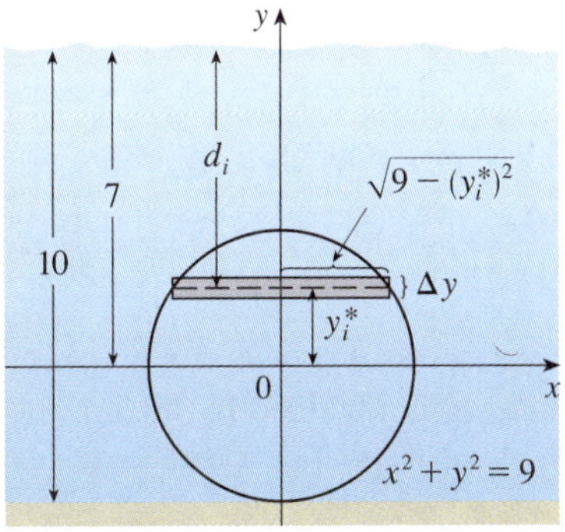

FIGURE 4

SOLUTION In this example it is convenient to choose the axes as in Figure 4 so that the origin is placed at the center of the drum. Then the circle has a simple equation, $x^2 + y^2 = 9$. As in Example 1 we divide the circular region into horizontal strips of equal width. From the equation of the circle, we see that the length of the ith strip is $2\sqrt{9 - (y_i^*)^2}$ and so its area is

$$A_i = 2\sqrt{9 - (y_i^*)^2}\,\Delta y$$

The pressure on this strip is approximately

$$\delta d_i = 62.5(7 - y_i^*)$$

and so the force on the strip is approximately

$$\delta d_i A_i = 62.5(7 - y_i^*)2\sqrt{9 - (y_i^*)^2}\,\Delta y$$

The total force is obtained by adding the forces on all the strips and taking the limit:

$$
\begin{aligned}
F &= \lim_{n\to\infty} \sum_{i=1}^{n} 62.5(7 - y_i^*)2\sqrt{9 - (y_i^*)^2}\,\Delta y \\
&= 125\int_{-3}^{3} (7 - y)\sqrt{9 - y^2}\,dy \\
&= 125 \cdot 7\int_{-3}^{3} \sqrt{9 - y^2}\,dy - 125\int_{-3}^{3} y\sqrt{9 - y^2}\,dy
\end{aligned}
$$

The second integral is 0 because the integrand is an odd function (see Theorem 5.5.7). The first integral can be evaluated using the trigonometric substitution $y = 3\sin\theta$, but it's simpler to observe that it is the area of a semicircular disk with radius 3. Thus

$$
\begin{aligned}
F &= 875\int_{-3}^{3} \sqrt{9 - y^2}\,dy = 875 \cdot \tfrac{1}{2}\pi(3)^2 \\
&= \frac{7875\pi}{2} \approx 12{,}370\ \text{lb}
\end{aligned}
$$

Moments and Centers of Mass

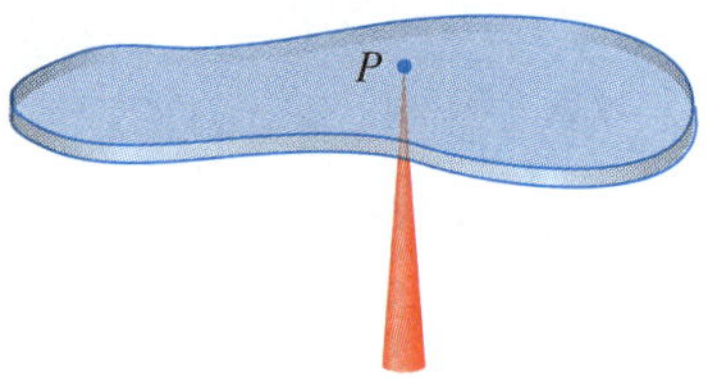

FIGURE 5

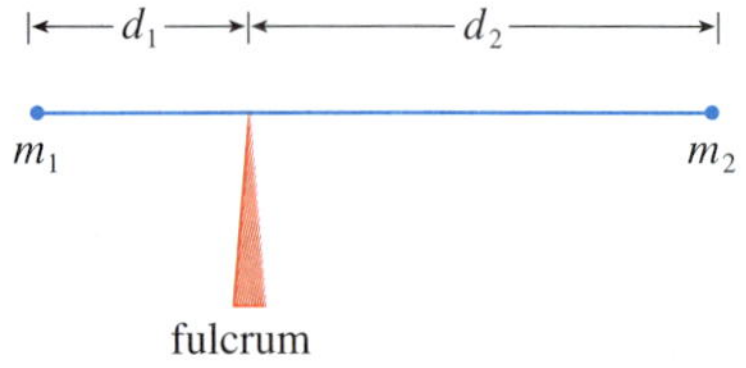

FIGURE 6

Our main objective here is to find the point P on which a thin plate of any given shape balances horizontally as in Figure 5. This point is called the **center of mass** (or center of gravity) of the plate.

We first consider the simpler situation illustrated in Figure 6, where two masses m_1 and m_2 are attached to a rod of negligible mass on opposite sides of a fulcrum and at distances d_1 and d_2 from the fulcrum. The rod will balance if

2 $$m_1 d_1 = m_2 d_2$$

This is an experimental fact discovered by Archimedes and called the Law of the Lever. (Think of a lighter person balancing a heavier one on a seesaw by sitting farther away from the center.)

Now suppose that the rod lies along the x-axis with m_1 at x_1 and m_2 at x_2 and the center of mass at $\bar{x}$. If we compare Figures 6 and 7, we see that $d_1 = \bar{x} - x_1$ and $d_2 = x_2 - \bar{x}$ and so Equation 2 gives

$$m_1(\bar{x} - x_1) = m_2(x_2 - \bar{x})$$

$$m_1\bar{x} + m_2\bar{x} = m_1 x_1 + m_2 x_2$$

3 $$\bar{x} = \frac{m_1 x_1 + m_2 x_2}{m_1 + m_2}$$

The numbers $m_1 x_1$ and $m_2 x_2$ are called the **moments** of the masses m_1 and m_2 (with respect to the origin), and Equation 3 says that the center of mass $\bar{x}$ is obtained by adding the moments of the masses and dividing by the total mass $m = m_1 + m_2$.

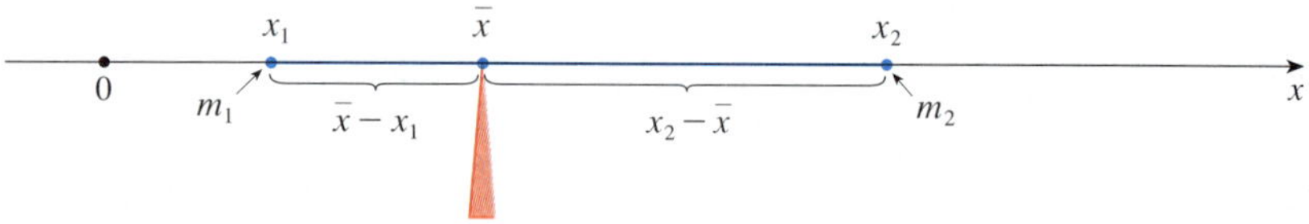

FIGURE 7

In general, if we have a system of n particles with masses $m_1, m_2, \ldots, m_n$ located at the points $x_1, x_2, \ldots, x_n$ on the x-axis, it can be shown similarly that the center of mass of the system is located at

4 $$\bar{x} = \frac{\sum_{i=1}^{n} m_i x_i}{\sum_{i=1}^{n} m_i} = \frac{\sum_{i=1}^{n} m_i x_i}{m}$$

where $m = \sum m_i$ is the total mass of the system, and the sum of the individual moments

$$M = \sum_{i=1}^{n} m_i x_i$$

is called the **moment of the system about the origin.** Then Equation 4 could be rewritten as $m\bar{x} = M$, which says that if the total mass were considered as being concentrated at the center of mass $\bar{x}$, then its moment would be the same as the moment of the system.

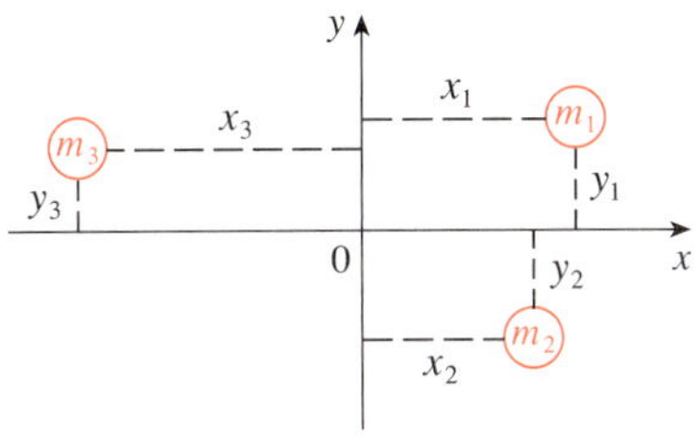

FIGURE 8

Now we consider a system of n particles with masses $m_1, m_2, \ldots, m_n$ located at the points $(x_1, y_1), (x_2, y_2), \ldots, (x_n, y_n)$ in the xy-plane as shown in Figure 8. By analogy with the one-dimensional case, we define the **moment of the system about the *y*-axis** to be

5 $$M_y = \sum_{i=1}^{n} m_i x_i$$

and the **moment of the system about the *x*-axis** as

6 $$M_x = \sum_{i=1}^{n} m_i y_i$$

Then M_y measures the tendency of the system to rotate about the y-axis and M_x measures the tendency to rotate about the x-axis.

As in the one-dimensional case, the coordinates $(\bar{x}, \bar{y})$ of the center of mass are given in terms of the moments by the formulas

7 $$\bar{x} = \frac{M_y}{m} \qquad \bar{y} = \frac{M_x}{m}$$

where $m = \sum m_i$ is the total mass. Since $m\bar{x} = M_y$ and $m\bar{y} = M_x$, the center of mass $(\bar{x}, \bar{y})$ is the point where a single particle of mass m would have the same moments as the system.

EXAMPLE 3 Find the moments and center of mass of the system of objects that have masses 3, 4, and 8 at the points $(-1, 1)$, $(2, -1)$, and $(3, 2)$.

SOLUTION We use Equations 5 and 6 to compute the moments:

$$M_y = 3(-1) + 4(2) + 8(3) = 29$$

$$M_x = 3(1) + 4(-1) + 8(2) = 15$$

Since $m = 3 + 4 + 8 = 15$, we use Equations 7 to obtain

$$\bar{x} = \frac{M_y}{m} = \frac{29}{15} \qquad \bar{y} = \frac{M_x}{m} = \frac{15}{15} = 1$$

Thus, the center of mass is $\left(1\frac{14}{15}, 1\right)$. (See Figure 9.)

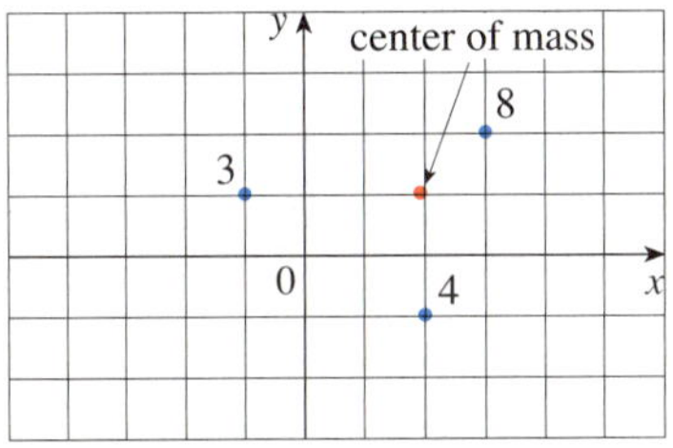

FIGURE 9

Next we consider a flat plate (called a *lamina*) with uniform density ρ that occupies a region $\mathcal{R}$ of the plane. We wish to locate the center of mass of the plate, which is called the **centroid** of $\mathcal{R}$. In doing so we use the following physical principles: The **symmetry principle** says that if $\mathcal{R}$ is symmetric about a line l, then the centroid of $\mathcal{R}$ lies on l. (If $\mathcal{R}$ is reflected about l, then $\mathcal{R}$ remains the same so its centroid remains fixed. But the only fixed points lie on l.) Thus, the centroid of a rectangle is its center. Moments should be defined so that if the entire mass of a region is concentrated at the center of mass, then its moments remain unchanged. Also, the moment of the union of two nonoverlapping regions should be the sum of the moments of the individual regions.

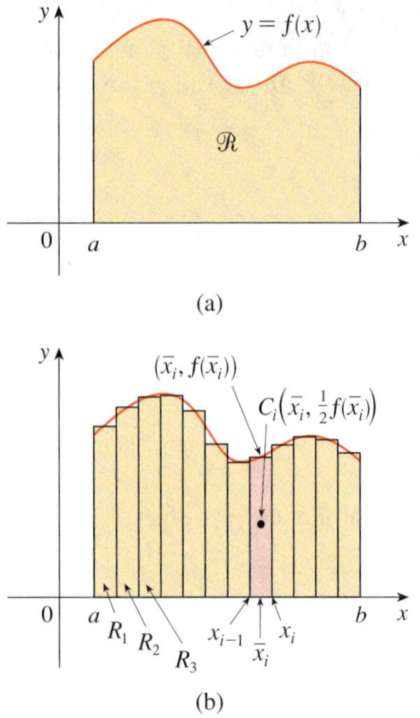

FIGURE 10

Suppose that the region $\mathcal{R}$ is of the type shown in Figure 10(a); that is, $\mathcal{R}$ lies between the lines $x = a$ and $x = b$, above the x-axis, and beneath the graph of f, where f is a continuous function. We divide the interval $[a, b]$ into n subintervals with endpoints $x_0, x_1, \ldots, x_n$ and equal width Δx. We choose the sample point x_i^* to be the midpoint $\bar{x}_i$ of the ith subinterval, that is, $\bar{x}_i = (x_{i-1} + x_i)/2$. This determines the polygonal approximation to $\mathcal{R}$ shown in Figure 10(b). The centroid of the ith approximating rectangle R_i is its center $C_i(\bar{x}_i, \frac{1}{2}f(\bar{x}_i))$. Its area is $f(\bar{x}_i)\,\Delta x$, so its mass is

$$\rho f(\bar{x}_i)\,\Delta x$$

The moment of R_i about the y-axis is the product of its mass and the distance from C_i to the y-axis, which is $\bar{x}_i$. Thus

$$M_y(R_i) = [\rho f(\bar{x}_i)\,\Delta x]\,\bar{x}_i = \rho \bar{x}_i f(\bar{x}_i)\,\Delta x$$

Adding these moments, we obtain the moment of the polygonal approximation to $\mathcal{R}$, and then by taking the limit as $n \to \infty$ we obtain the moment of $\mathcal{R}$ itself about the y-axis:

$$M_y = \lim_{n\to\infty} \sum_{i=1}^{n} \rho \bar{x}_i f(\bar{x}_i)\,\Delta x = \rho \int_a^b x f(x)\,dx$$

In a similar fashion we compute the moment of R_i about the x-axis as the product of its mass and the distance from C_i to the x-axis:

$$M_x(R_i) = [\rho f(\bar{x}_i)\,\Delta x]\,\tfrac{1}{2} f(\bar{x}_i) = \rho \cdot \tfrac{1}{2}[f(\bar{x}_i)]^2\,\Delta x$$

Again we add these moments and take the limit to obtain the moment of $\mathcal{R}$ about the x-axis:

$$M_x = \lim_{n\to\infty} \sum_{i=1}^{n} \rho \cdot \tfrac{1}{2}[f(\bar{x}_i)]^2\,\Delta x = \rho \int_a^b \tfrac{1}{2}[f(x)]^2\,dx$$

Just as for systems of particles, the center of mass of the plate is defined so that $m\bar{x} = M_y$ and $m\bar{y} = M_x$. But the mass of the plate is the product of its density and its area:

$$m = \rho A = \rho \int_a^b f(x)\,dx$$

and so

$$\bar{x} = \frac{M_y}{m} = \frac{\rho \int_a^b x f(x)\,dx}{\rho \int_a^b f(x)\,dx} = \frac{\int_a^b x f(x)\,dx}{\int_a^b f(x)\,dx}$$

$$\bar{y} = \frac{M_x}{m} = \frac{\rho \int_a^b \frac{1}{2}[f(x)]^2\,dx}{\rho \int_a^b f(x)\,dx} = \frac{\int_a^b \frac{1}{2}[f(x)]^2\,dx}{\int_a^b f(x)\,dx}$$

Notice the cancellation of the ρ's. The location of the center of mass is independent of the density.

In summary, the center of mass of the plate (or the centroid of $\mathcal{R}$) is located at the point $(\bar{x}, \bar{y})$, where

8
$$\bar{x} = \frac{1}{A}\int_a^b x f(x)\,dx \qquad \bar{y} = \frac{1}{A}\int_a^b \tfrac{1}{2}[f(x)]^2\,dx$$

EXAMPLE 4 Find the center of mass of a semicircular plate of radius r.

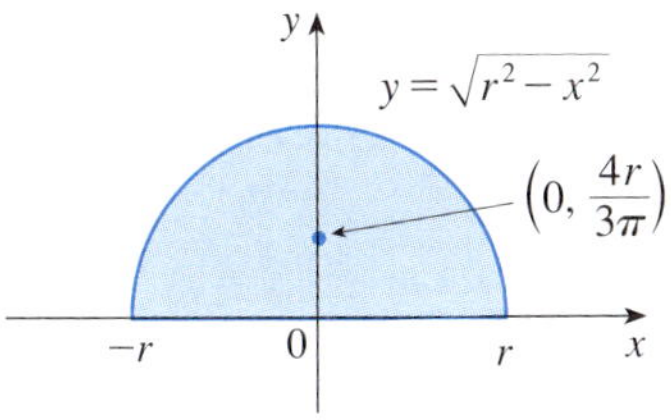

FIGURE 11

SOLUTION In order to use (8) we place the semicircle as in Figure 11 so that $f(x) = \sqrt{r^2 - x^2}$ and $a = -r$, $b = r$. Here there is no need to use the formula to calculate $\bar{x}$ because, by the symmetry principle, the center of mass must lie on the y-axis, so $\bar{x} = 0$. The area of the semicircle is $A = \pi r^2/2$, so

$$\begin{aligned}\bar{y} &= \frac{1}{A}\int_{-r}^{r} \tfrac{1}{2}[f(x)]^2\,dx \\ &= \frac{1}{\pi r^2/2}\cdot\frac{1}{2}\int_{-r}^{r}\left(\sqrt{r^2-x^2}\right)^2 dx \\ &= \frac{2}{\pi r^2}\int_0^r (r^2 - x^2)\,dx = \frac{2}{\pi r^2}\left[r^2x - \frac{x^3}{3}\right]_0^r \\ &= \frac{2}{\pi r^2}\,\frac{2r^3}{3} = \frac{4r}{3\pi}\end{aligned}$$

The center of mass is located at the point $(0, 4r/(3\pi))$.

EXAMPLE 5 Find the centroid of the region bounded by the curves $y = \cos x$, $y = 0$, $x = 0$, and $x = \pi/2$.

SOLUTION The area of the region is

$$A = \int_0^{\pi/2} \cos x\,dx = \sin x\Big]_0^{\pi/2} = 1$$

so Formulas 8 give

$$\begin{aligned}\bar{x} &= \frac{1}{A}\int_0^{\pi/2} x f(x)\,dx = \int_0^{\pi/2} x\cos x\,dx \\ &= x\sin x\Big]_0^{\pi/2} - \int_0^{\pi/2}\sin x\,dx \qquad \text{(by integration by parts)} \\ &= \frac{\pi}{2} - 1\end{aligned}$$

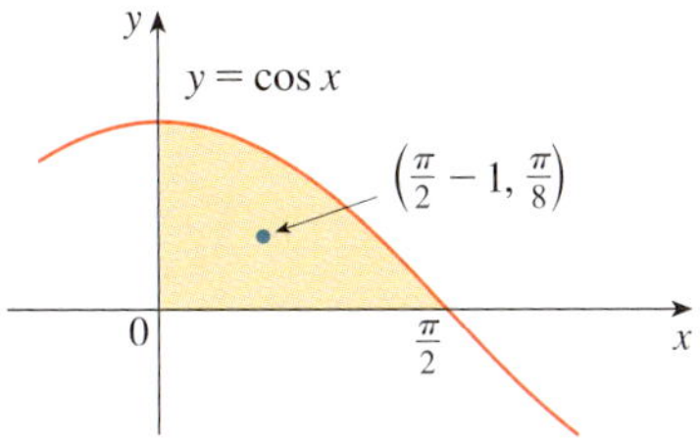

$$\begin{aligned}\bar{y} &= \frac{1}{A}\int_0^{\pi/2}\tfrac{1}{2}[f(x)]^2\,dx = \tfrac{1}{2}\int_0^{\pi/2}\cos^2 x\,dx \\ &= \tfrac{1}{4}\int_0^{\pi/2}(1 + \cos 2x)\,dx = \tfrac{1}{4}\left[x + \tfrac{1}{2}\sin 2x\right]_0^{\pi/2} \\ &= \frac{\pi}{8}\end{aligned}$$

FIGURE 12

The centroid is $((\pi/2) - 1, \pi/8)$ and is shown in Figure 12.

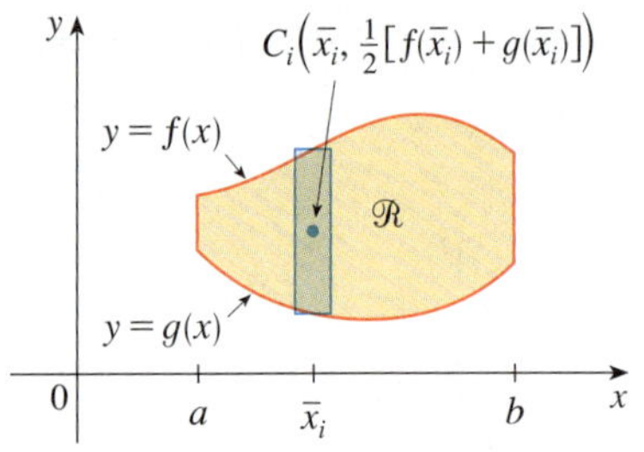

FIGURE 13

If the region $\mathcal{R}$ lies between two curves $y = f(x)$ and $y = g(x)$, where $f(x) \geqslant g(x)$, as illustrated in Figure 13, then the same sort of argument that led to Formulas 8 can be used to show that the centroid of $\mathcal{R}$ is $(\bar{x}, \bar{y})$, where

9 $$\bar{x} = \frac{1}{A}\int_a^b x[f(x) - g(x)]\,dx \qquad \bar{y} = \frac{1}{A}\int_a^b \tfrac{1}{2}\{[f(x)]^2 - [g(x)]^2\}\,dx$$

(See Exercise 43.)

EXAMPLE 6 Find the centroid of the region bounded by the line $y = x$ and the parabola $y = x^2$.

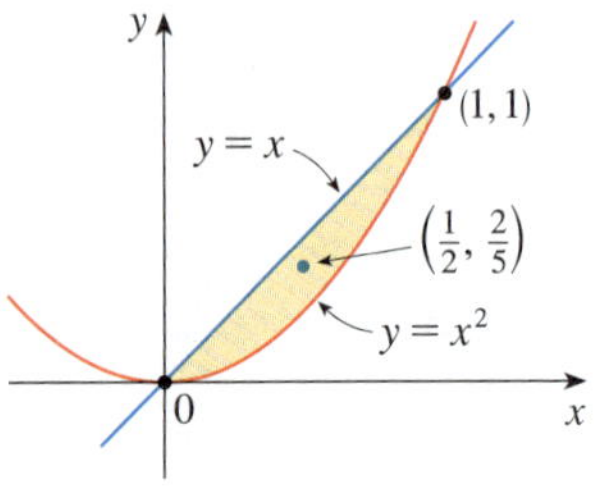

FIGURE 14

SOLUTION The region is sketched in Figure 14. We take $f(x) = x$, $g(x) = x^2$, $a = 0$, and $b = 1$ in Formulas 9. First we note that the area of the region is

$$A = \int_0^1 (x - x^2)\,dx = \frac{x^2}{2} - \frac{x^3}{3}\Bigg]_0^1 = \frac{1}{6}$$

Therefore

$$\bar{x} = \frac{1}{A}\int_0^1 x[f(x) - g(x)]\,dx = \frac{1}{\frac{1}{6}}\int_0^1 x(x - x^2)\,dx$$

$$= 6\int_0^1 (x^2 - x^3)\,dx = 6\left[\frac{x^3}{3} - \frac{x^4}{4}\right]_0^1 = \frac{1}{2}$$

$$\bar{y} = \frac{1}{A}\int_0^1 \tfrac{1}{2}\{[f(x)]^2 - [g(x)]^2\}\,dx = \frac{1}{\frac{1}{6}}\int_0^1 \tfrac{1}{2}(x^2 - x^4)\,dx$$

$$= 3\left[\frac{x^3}{3} - \frac{x^5}{5}\right]_0^1 = \frac{2}{5}$$

The centroid is $\left(\frac{1}{2}, \frac{2}{5}\right)$.

We end this section by showing a surprising connection between centroids and volumes of revolution.

|||| This theorem is named after the Greek mathematician Pappus of Alexandria, who lived in the fourth century A.D.

Theorem of Pappus Let $\mathcal{R}$ be a plane region that lies entirely on one side of a line l in the plane. If $\mathcal{R}$ is rotated about l, then the volume of the resulting solid is the product of the area A of $\mathcal{R}$ and the distance d traveled by the centroid of $\mathcal{R}$.

Proof We give the proof for the special case in which the region lies between $y = f(x)$ and $y = g(x)$ as in Figure 13 and the line l is the y-axis. Using the method of cylindrical shells (see Section 6.3), we have

$$V = \int_a^b 2\pi x[f(x) - g(x)]\,dx$$

$$= 2\pi\int_a^b x[f(x) - g(x)]\,dx$$

$$= 2\pi(\bar{x}A) \qquad \text{(by Formulas 9)}$$

$$= (2\pi\bar{x})A = Ad$$

where $d = 2\pi\bar{x}$ is the distance traveled by the centroid during one rotation about the y-axis.

EXAMPLE 7 A torus is formed by rotating a circle of radius r about a line in the plane of the circle that is a distance R ($> r$) from the center of the circle. Find the volume of the torus.

SOLUTION The circle has area $A = \pi r^2$. By the symmetry principle, its centroid is its center and so the distance traveled by the centroid during a rotation is $d = 2\pi R$. Therefore, by the Theorem of Pappus, the volume of the torus is

$$V = Ad = (2\pi R)(\pi r^2) = 2\pi^2 r^2 R$$

The method of Example 7 should be compared with the method of Exercise 61 in Section 6.2.

8.3 Exercises

1. An aquarium 5 ft long, 2 ft wide, and 3 ft deep is full of water. Find (a) the hydrostatic pressure on the bottom of the aquarium, (b) the hydrostatic force on the bottom, and (c) the hydrostatic force on one end of the aquarium.

2. A swimming pool 5 m wide, 10 m long, and 3 m deep is filled with seawater of density 1030 kg/m^3 to a depth of 2.5 m. Find (a) the hydrostatic pressure at the bottom of the pool, (b) the hydrostatic force on the bottom, and (c) the hydrostatic force on one end of the pool.

3–9 A vertical plate is submerged in water and has the indicated shape. Explain how to approximate the hydrostatic force against one side of the plate by a Riemann sum. Then express the force as an integral and evaluate it.

3.
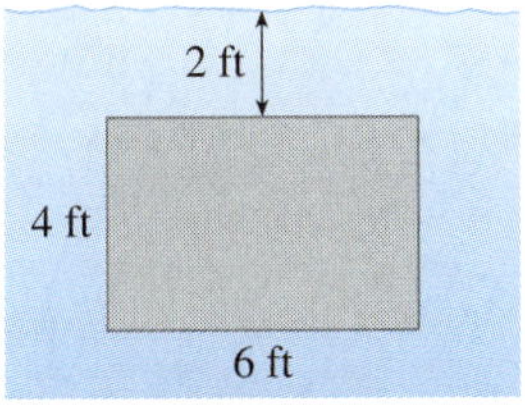

4.
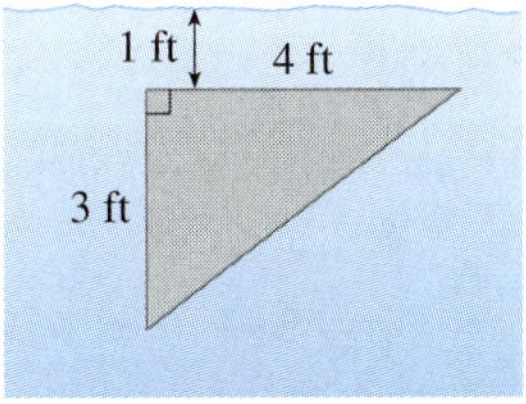

5.
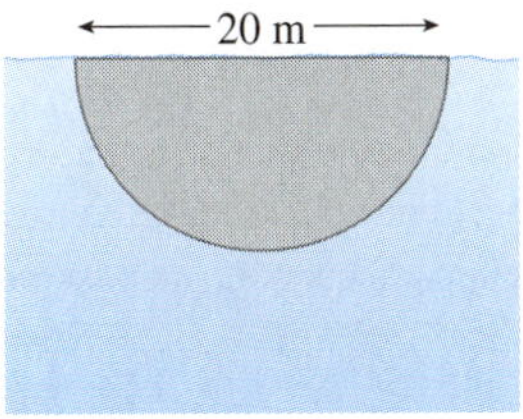

6.
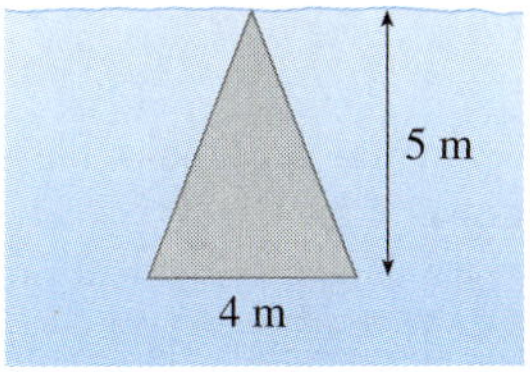

7.
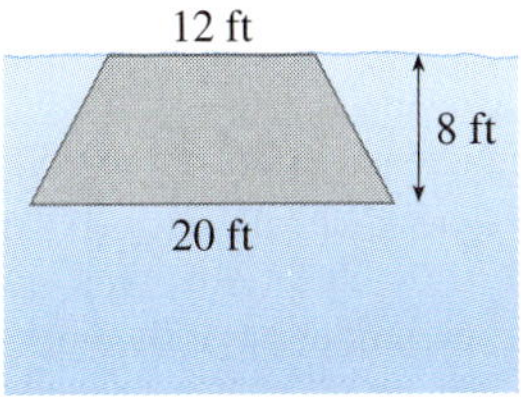

8.
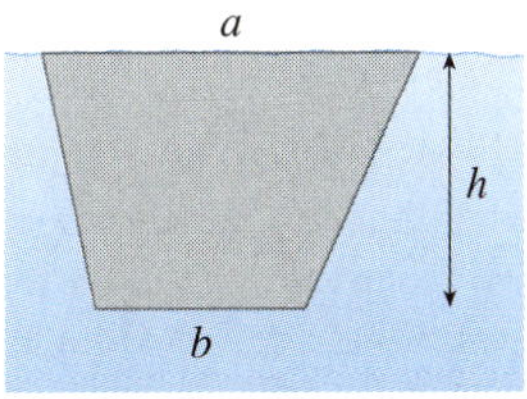

9.
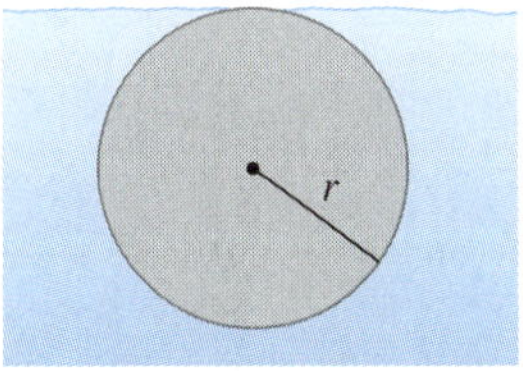

10. A large tank is designed with ends in the shape of the region between the curves $y = x^2/2$ and $y = 12$, measured in feet. Find the hydrostatic force on one end of the tank if it is filled to a depth of 8 ft with gasoline. (Assume the gasoline's density is 42.0 lb/ft^3.)

11. A trough is filled with a liquid of density 840 kg/m^3. The ends of the trough are equilateral triangles with sides 8 m long and vertex at the bottom. Find the hydrostatic force on one end of the trough.

12. A vertical dam has a semicircular gate as shown in the figure. Find the hydrostatic force against the gate.

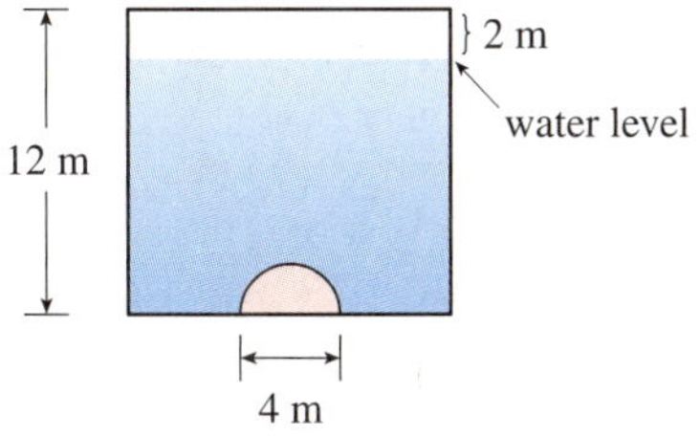

13. A cube with 20-cm-long sides is sitting on the bottom of an aquarium in which the water is one meter deep. Find the hydrostatic force on (a) the top of the cube and (b) one of the sides of the cube.

14. A dam is inclined at an angle of 30° from the vertical and has the shape of an isosceles trapezoid 100 ft wide at the top and

50 ft wide at the bottom and with a slant height of 70 ft. Find the hydrostatic force on the dam when it is full of water.

15. A swimming pool is 20 ft wide and 40 ft long and its bottom is an inclined plane, the shallow end having a depth of 3 ft and the deep end, 9 ft. If the pool is full of water, find the hydrostatic force on (a) the shallow end, (b) the deep end, (c) one of the sides, and (d) the bottom of the pool.

16. Suppose that a plate is immersed vertically in a fluid with density ρ and the width of the plate is $w(x)$ at a depth of x meters beneath the surface of the fluid. If the top of the plate is at depth a and the bottom is at depth b, show that the hydrostatic force on one side of the plate is

$$F = \int_a^b \rho g x w(x)\, dx$$

17. A vertical, irregularly shaped plate is submerged in water. The table shows measurements of its width, taken at the indicated depths. Use Simpson's Rule to estimate the force of the water against the plate.

Depth (m)	2.0	2.5	3.0	3.5	4.0	4.5	5.0
Plate width (m)	0	0.8	1.7	2.4	2.9	3.3	3.6

18. (a) Use the formula of Exercise 16 to show that

$$F = (\rho g \bar{x})A$$

where $\bar{x}$ is the x-coordinate of the centroid of the plate and A is its area. This equation shows that the hydrostatic force against a vertical plane region is the same as if the region were horizontal at the depth of the centroid of the region.

(b) Use the result of part (a) to give another solution to Exercise 9.

19–20 ||| Point-masses m_i are located on the x-axis as shown. Find the moment M of the system about the origin and the center of mass $\bar{x}$.

19.
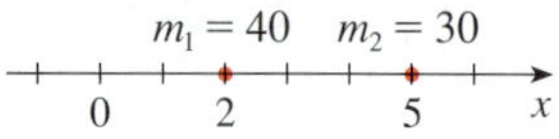

20.
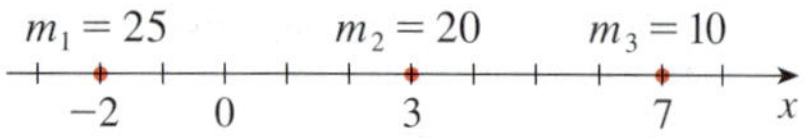

21–22 ||| The masses m_i are located at the points P_i. Find the moments M_x and M_y and the center of mass of the system.

21. $m_1 = 6,\ m_2 = 5,\ m_3 = 10;$
$P_1(1, 5),\ P_2(3, -2),\ P_3(-2, -1)$

22. $m_1 = 6,\ m_2 = 5,\ m_3 = 1,\ m_4 = 4;$
$P_1(1, -2),\ P_2(3, 4),\ P_3(-3, -7),\ P_4(6, -1)$

23–26 ||| Sketch the region bounded by the curves, and visually estimate the location of the centroid. Then find the exact coordinates of the centroid.

23. $y = 4 - x^2,\quad y = 0$

24. $3x + 2y = 6,\quad y = 0,\quad x = 0$

25. $y = e^x,\quad y = 0,\quad x = 0,\quad x = 1$

26. $y = 1/x,\quad y = 0,\quad x = 1,\quad x = 2$

27–31 ||| Find the centroid of the region bounded by the given curves.

27. $y = \sqrt{x},\quad y = x$

28. $y = x + 2,\quad y = x^2$

29. $y = \sin x,\quad y = \cos x,\quad x = 0,\quad x = \pi/4$

30. $y = x,\quad y = 0,\quad y = 1/x,\quad x = 2$

31. $x = 5 - y^2,\quad x = 0$

32–34 ||| Calculate the moments M_x and M_y and the center of mass of a lamina with the given density and shape.

32. $\rho = 5$

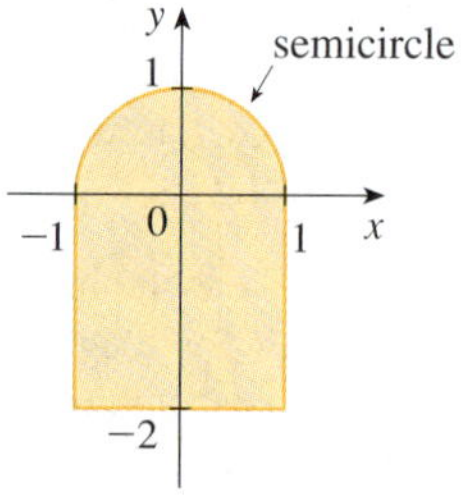

33. $\rho = 1$

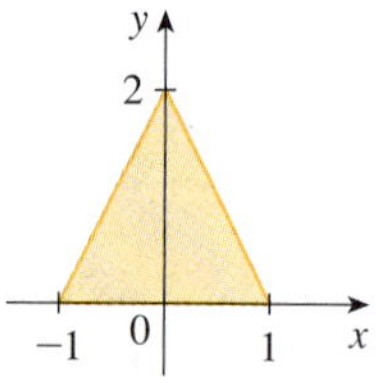

34. $\rho = 2$

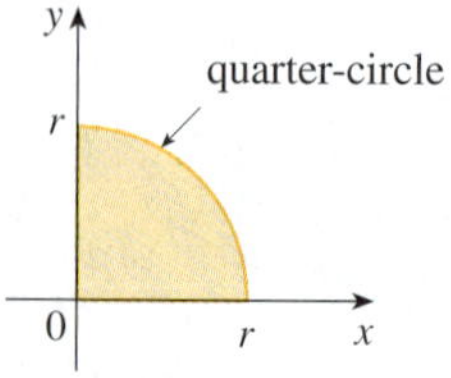

35. Find the centroid of the region bounded by the curves $y = 2^x$ and $y = x^2$, $0 \leqslant x \leqslant 2$, to three decimal places. Sketch the region and plot the centroid to see if your answer is reasonable.

36. Use a graph to find approximate x-coordinates of the points of intersection of the curves $y = x + \ln x$ and $y = x^3 - x$. Then find (approximately) the centroid of the region bounded by these curves.

37. Prove that the centroid of any triangle is located at the point of intersection of the medians. [*Hints:* Place the axes so that the vertices are $(a, 0)$, $(0, b)$, and $(c, 0)$. Recall that a median is a line segment from a vertex to the midpoint of the opposite side. Recall also that the medians intersect at a point two-thirds of the way from each vertex (along the median) to the opposite side.]

38–39 Find the centroid of the region shown, not by integration, but by locating the centroids of the rectangles and triangles (from Exercise 37) and using additivity of moments.

38.

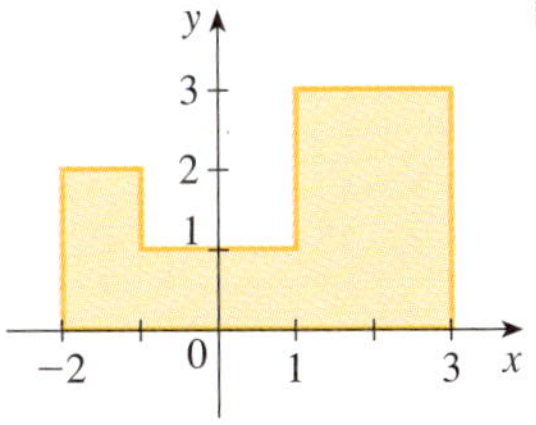

39.

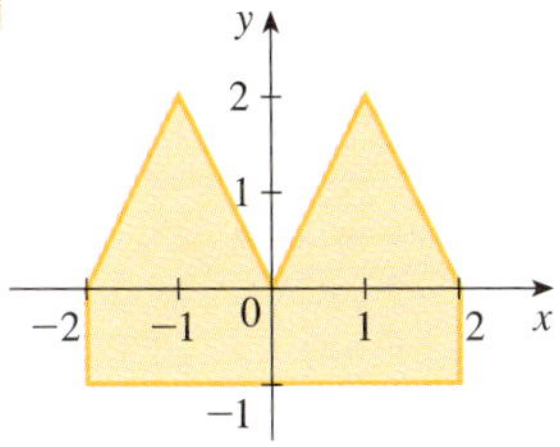

40–42 Use the Theorem of Pappus to find the volume of the given solid.

40. A sphere of radius r (Use Example 4.)

41. A cone with height h and base radius r

42. The solid obtained by rotating the triangle with vertices $(2, 3)$, $(2, 5)$, and $(5, 4)$ about the x-axis

43. Prove Formulas 9.

44. Let $\mathcal{R}$ be the region that lies between the curves $y = x^m$ and $y = x^n$, $0 \leqslant x \leqslant 1$, where m and n are integers with $0 \leqslant n < m$.

(a) Sketch the region $\mathcal{R}$.

(b) Find the coordinates of the centroid of $\mathcal{R}$.

(c) Try to find values of m and n such that the centroid lies *outside* $\mathcal{R}$.

8.4 Applications to Economics and Biology

In this section we consider some applications of integration to economics (consumer surplus) and biology (blood flow, cardiac output). Others are described in the exercises.

Consumer Surplus

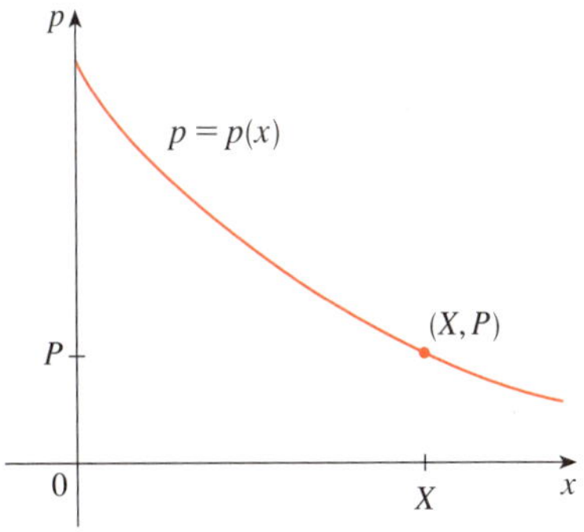

FIGURE 1
A typical demand curve

Recall from Section 4.8 that the demand function $p(x)$ is the price that a company has to charge in order to sell x units of a commodity. Usually, selling larger quantities requires lowering prices, so the demand function is a decreasing function. The graph of a typical demand function, called a **demand curve**, is shown in Figure 1. If X is the amount of the commodity that is currently available, then $P = p(X)$ is the current selling price.

We divide the interval $[0, X]$ into n subintervals, each of length $\Delta x = X/n$, and let $x_i^* = x_i$ be the right endpoint of the ith subinterval, as in Figure 2. If, after the first x_{i-1} units were sold, a total of only x_i units had been available and the price per unit had been set at $p(x_i)$ dollars, then the additional Δx units could have been sold (but no more). The consumers who would have paid $p(x_i)$ dollars placed a high value on the product; they would have paid what it was worth to them. So, in paying only P dollars they have saved an amount of

$$(\text{savings per unit})(\text{number of units}) = [p(x_i) - P]\,\Delta x$$

Considering similar groups of willing consumers for each of the subintervals and adding the savings, we get the total savings:

$$\sum_{i=1}^{n} [p(x_i) - P]\,\Delta x$$

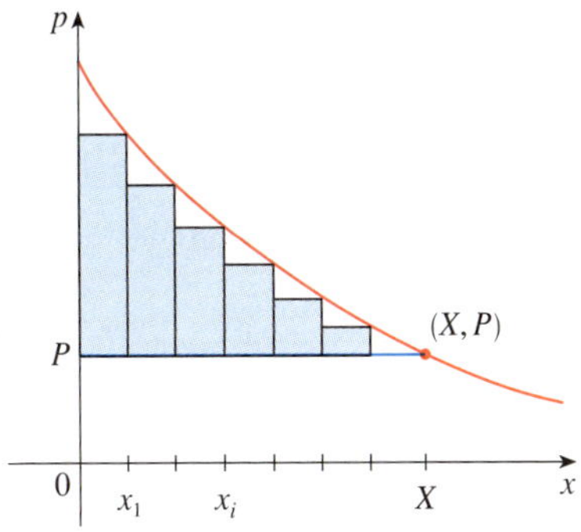

FIGURE 2

(This sum corresponds to the area enclosed by the rectangles in Figure 2.) If we let $n \to \infty$, this Riemann sum approaches the integral

1
$$\int_0^X [p(x) - P]\,dx$$

which economists call the **consumer surplus** for the commodity.

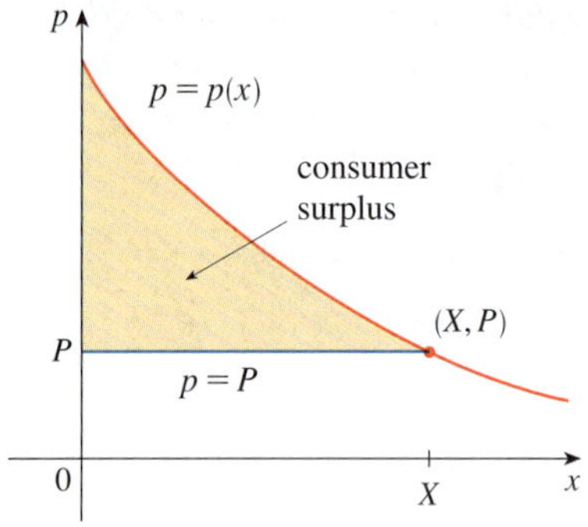

FIGURE 3

The consumer surplus represents the amount of money saved by consumers in purchasing the commodity at price P, corresponding to an amount demanded of X. Figure 3 shows the interpretation of the consumer surplus as the area under the demand curve and above the line $p = P$.

EXAMPLE 1 The demand for a product, in dollars, is

$$p = 1200 - 0.2x - 0.0001x^2$$

Find the consumer surplus when the sales level is 500.

SOLUTION Since the number of products sold is $X = 500$, the corresponding price is

$$P = 1200 - (0.2)(500) - (0.0001)(500)^2 = 1075$$

Therefore, from Definition 1, the consumer surplus is

$$\begin{aligned}
\int_0^{500} [p(x) - P]\,dx &= \int_0^{500} (1200 - 0.2x - 0.0001x^2 - 1075)\,dx \\
&= \int_0^{500} (125 - 0.2x - 0.0001x^2)\,dx \\
&= 125x - 0.1x^2 - (0.0001)\left(\frac{x^3}{3}\right)\Bigg]_0^{500} \\
&= (125)(500) - (0.1)(500)^2 - \frac{(0.0001)(500)^3}{3} \\
&= \$33{,}333.33
\end{aligned}$$

Blood Flow

In Example 7 in Section 3.3 we discussed the law of laminar flow:

$$v(r) = \frac{P}{4\eta l}(R^2 - r^2)$$

which gives the velocity v of blood that flows along a blood vessel with radius R and length l at a distance r from the central axis, where P is the pressure difference between the ends of the vessel and η is the viscosity of the blood. Now, in order to compute the rate of blood flow, or *flux* (volume per unit time), we consider smaller, equally spaced radii $r_1, r_2, \ldots$. The approximate area of the ring (or washer) with inner radius r_{i-1} and outer radius r_i is

$$2\pi r_i\,\Delta r \qquad \text{where} \qquad \Delta r = r_i - r_{i-1}$$

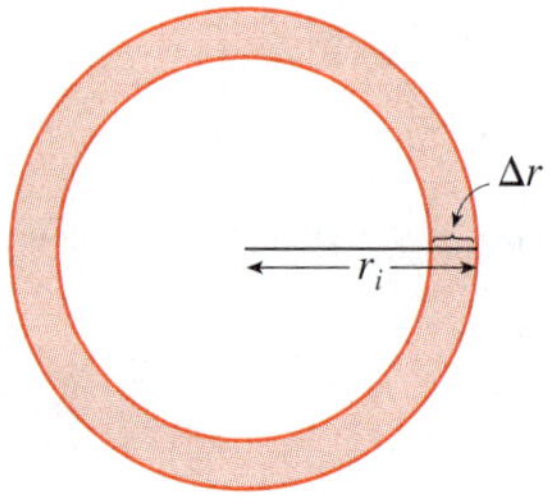

FIGURE 4

(See Figure 4.) If Δr is small, then the velocity is almost constant throughout this ring and can be approximated by $v(r_i)$. Thus, the volume of blood per unit time that flows across the ring is approximately

$$(2\pi r_i\,\Delta r)\,v(r_i) = 2\pi r_i v(r_i)\,\Delta r$$

and the total volume of blood that flows across a cross-section per unit time is approximately

$$\sum_{i=1}^{n} 2\pi r_i v(r_i)\,\Delta r$$

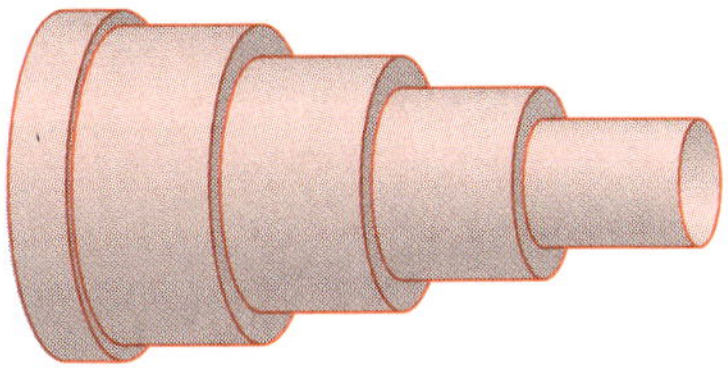

FIGURE 5

This approximation is illustrated in Figure 5. Notice that the velocity (and hence the volume per unit time) increases toward the center of the blood vessel. The approximation gets better as n increases. When we take the limit we get the exact value of the **flux** (or *discharge*), which is the volume of blood that passes a cross-section per unit time:

$$\begin{aligned} F &= \lim_{n\to\infty} \sum_{i=1}^{n} 2\pi r_i v(r_i)\,\Delta r = \int_0^R 2\pi r v(r)\,dr \\ &= \int_0^R 2\pi r \frac{P}{4\eta l}(R^2 - r^2)\,dr \\ &= \frac{\pi P}{2\eta l}\int_0^R (R^2 r - r^3)\,dr = \frac{\pi P}{2\eta l}\left[R^2\frac{r^2}{2} - \frac{r^4}{4}\right]_{r=0}^{r=R} \\ &= \frac{\pi P}{2\eta l}\left[\frac{R^4}{2} - \frac{R^4}{4}\right] = \frac{\pi P R^4}{8\eta l} \end{aligned}$$

The resulting equation

2 $$F = \frac{\pi P R^4}{8\eta l}$$

is called **Poiseuille's Law**; it shows that the flux is proportional to the fourth power of the radius of the blood vessel.

Cardiac Output

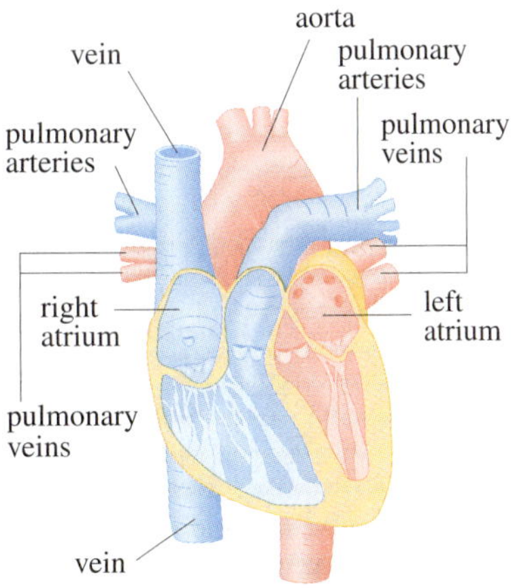

FIGURE 6

Figure 6 shows the human cardiovascular system. Blood returns from the body through the veins, enters the right atrium of the heart, and is pumped to the lungs through the pulmonary arteries for oxygenation. It then flows back into the left atrium through the pulmonary veins and then out to the rest of the body through the aorta. The **cardiac output** of the heart is the volume of blood pumped by the heart per unit time, that is, the rate of flow into the aorta.

The *dye dilution method* is used to measure the cardiac output. Dye is injected into the right atrium and flows through the heart into the aorta. A probe inserted into the aorta measures the concentration of the dye leaving the heart at equally spaced times over a time interval $[0, T]$ until the dye has cleared. Let $c(t)$ be the concentration of the dye at time t. If we divide $[0, T]$ into subintervals of equal length Δt, then the amount of dye that flows past the measuring point during the subinterval from $t = t_{i-1}$ to $t = t_i$ is approximately

$$(\text{concentration})(\text{volume}) = c(t_i)(F\,\Delta t)$$

where F is the rate of flow that we are trying to determine. Thus, the total amount of dye is approximately

$$\sum_{i=1}^{n} c(t_i)F\,\Delta t = F\sum_{i=1}^{n} c(t_i)\,\Delta t$$

and, letting $n \to \infty$, we find that the amount of dye is

$$A = F\int_0^T c(t)\,dt$$

Thus, the cardiac output is given by

3
$$F = \frac{A}{\int_0^T c(t)\,dt}$$

where the amount of dye A is known and the integral can be approximated from the concentration readings.

t	$c(t)$	t	$c(t)$
0	0	6	6.1
1	0.4	7	4.0
2	2.8	8	2.3
3	6.5	9	1.1
4	9.8	10	0
5	8.9		

EXAMPLE 2 A 5-mg bolus of dye is injected into a right atrium. The concentration of the dye (in milligrams per liter) is measured in the aorta at one-second intervals as shown in the chart. Estimate the cardiac output.

SOLUTION Here $A = 5$, $\Delta t = 1$, and $T = 10$. We use Simpson's Rule to approximate the integral of the concentration:

$$\int_0^{10} c(t)\,dt \approx \tfrac{1}{3}[0 + 4(0.4) + 2(2.8) + 4(6.5) + 2(9.8) + 4(8.9)$$
$$+ 2(6.1) + 4(4.0) + 2(2.3) + 4(1.1) + 0]$$
$$\approx 41.87$$

Thus, Formula 3 gives the cardiac output to be

$$F = \frac{A}{\int_0^{10} c(t)\,dt} \approx \frac{5}{41.87}$$
$$\approx 0.12 \text{ L/s} = 7.2 \text{ L/min}$$

8.4 Exercises

1. The marginal cost function $C'(x)$ was defined to be the derivative of the cost function. (See Sections 3.3 and 4.8.) If the marginal cost of maufacturing x meters of a fabric is $C'(x) = 5 - 0.008x + 0.000009x^2$ (measured in dollars per meter) and the fixed start-up cost is $C(0) = \$20,000$, use the Net Change Theorem to find the cost of producing the first 2000 units.

2. The marginal revenue from the sale of x units of a product is $12 - 0.0004x$. If the revenue from the sale of the first 1000 units is \$12,400, find the revenue from the sale of the first 5000 units.

3. The marginal cost of producing x units of a certain product is $74 + 1.1x - 0.002x^2 + 0.00004x^3$ (in dollars per unit). Find the increase in cost if the production level is raised from 1200 units to 1600 units.

4. The demand function for a certain commodity is $p = 5 - x/10$. Find the consumer surplus when the sales level is 30. Illustrate by drawing the demand curve and identifying the consumer surplus as an area.

5. A demand curve is given by $p = 450/(x + 8)$. Find the consumer surplus when the selling price is \$10.

6. The **supply function** $p_S(x)$ for a commodity gives the relation between the selling price and the number of units that manufacturers will produce at that price. For a higher price, manufacturers will produce more units, so p_S is an increasing function of x. Let X be the amount of the commodity currently produced and let $P = p_S(X)$ be the current price. Some producers would be willing to make and sell the commodity for a lower selling price and are therefore receiving more than their minimal price. The excess is called the **producer surplus**. An argument similar to that for consumer surplus shows that the surplus is given by the integral

$$\int_0^X [P - p_S(x)]\,dx$$

Calculate the producer surplus for the supply function $p_S(x) = 3 + 0.01x^2$ at the sales level $X = 10$. Illustrate by drawing the supply curve and identifying the producer surplus as an area.

7. If a supply curve is modeled by the equation $p = 200 + 0.2x^{3/2}$, find the producer surplus when the selling price is \$400.

8. For a given commodity and pure competition, the number of units produced and the price per unit are determined as the coordinates of the point of intersection of the supply and demand curves. Given the demand curve $p = 50 - x/20$ and the supply curve $p = 20 + x/10$, find the consumer surplus and the producer surplus. Illustrate by sketching the supply and demand curves and identifying the surpluses as areas.

9. A company modeled the demand curve for its product (in dollars) by

$$p = \frac{800{,}000e^{-x/5000}}{x + 20{,}000}$$

Use a graph to estimate the sales level when the selling price is \$16. Then find (approximately) the consumer surplus for this sales level.

10. A movie theater has been charging \$7.50 per person and selling about 400 tickets on a typical weeknight. After surveying their customers, the theater estimates that for every 50 cents that they lower the price, the number of moviegoers will increase by 35 per night. Find the demand function and calculate the consumer surplus when the tickets are priced at \$6.00.

11. If the amount of capital that a company has at time t is $f(t)$, then the derivative, $f'(t)$, is called the *net investment flow*. Suppose that the net investment flow is $\sqrt{t}$ million dollars per year (where t is measured in years). Find the increase in capital (the *capital formation*) from the fourth year to the eighth year.

12. A hot, wet summer is causing a mosquito population explosion in a lake resort area. The number of mosquitos is increasing at an estimated rate of $2200 + 10e^{0.8t}$ per week (where t is measured in weeks). By how much does the mosquito population increase between the fifth and ninth weeks of summer?

13. Use Poiseuille's Law to calculate the rate of flow in a small human artery where we can take $\eta = 0.027$, $R = 0.008$ cm, $l = 2$ cm, and $P = 4000$ dynes/cm^2.

14. High blood pressure results from constriction of the arteries. To maintain a normal flow rate (flux), the heart has to pump harder, thus increasing the blood pressure. Use Poiseuille's Law to show that if R_0 and P_0 are normal values of the radius and pressure in an artery and the constricted values are R and P, then for the flux to remain constant, P and R are related by the equation

$$\frac{P}{P_0} = \left(\frac{R_0}{R}\right)^4$$

Deduce that if the radius of an artery is reduced to three-fourths of its former value, then the pressure is more than tripled.

15. The dye dilution method is used to measure cardiac output with 8 mg of dye. The dye concentrations, in mg/L, are modeled by $c(t) = \frac{1}{4}t(12 - t)$, $0 \leqslant t \leqslant 12$, where t is measured in seconds. Find the cardiac output.

16. After an 8-mg injection of dye, the readings of dye concentration at two-second intervals are as shown in the table. Use Simpson's Rule to estimate the cardiac output.

t	$c(t)$	t	$c(t)$
0	0	12	3.9
2	2.4	14	2.3
4	5.1	16	1.6
6	7.8	18	0.7
8	7.6	20	0
10	5.4		

8.5 Probability

Calculus plays a role in the analysis of random behavior. Suppose we consider the cholesterol level of a person chosen at random from a certain age group, or the height of an adult female chosen at random, or the lifetime of a randomly chosen battery of a certain type. Such quantities are called **continuous random variables** because their values actually range over an interval of real numbers, although they might be measured or recorded only to the nearest integer. We might want to know the probability that a blood cholesterol level is greater than 250, or the probability that the height of an adult female is between 60 and 70 inches, or the probability that the battery we are buying lasts between 100 and 200 hours. If X represents the lifetime of that type of battery, we denote this last probability as follows:

$$P(100 \leqslant X \leqslant 200)$$

According to the frequency interpretation of probability, this number is the long-run proportion of all batteries of the specified type whose lifetimes are between 100 and 200 hours. Since it represents a proportion, the probability naturally falls between 0 and 1.

Every continuous random variable X has a **probability density function** f. This means that the probability that X lies between a and b is found by integrating f from a to b:

1 $$P(a \leqslant X \leqslant b) = \int_a^b f(x)\,dx$$

For example, Figure 1 shows the graph of a model of the probability density function f for a random variable X defined to be the height in inches of an adult female in the United States (according to data from the National Health Survey). The probability that the height of a woman chosen at random from this population is between 60 and 70 inches is equal to the area under the graph of f from 60 to 70.

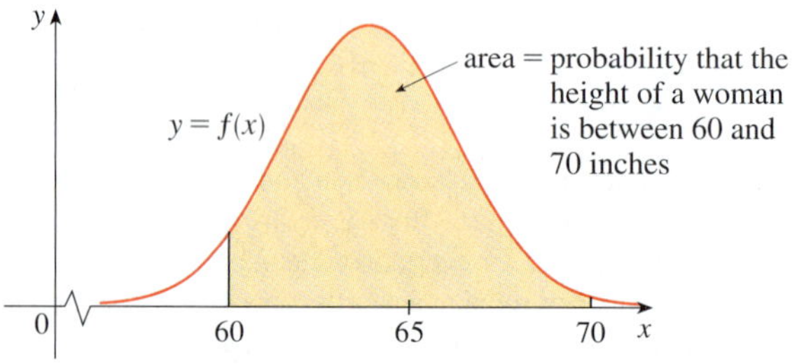

FIGURE 1
Probability density function for the height of an adult female

In general, the probability density function f of a random variable X satisfies the condition $f(x) \geqslant 0$ for all x. Because probabilities are measured on a scale from 0 to 1, it follows that

2 $$\int_{-\infty}^{\infty} f(x)\,dx = 1$$

EXAMPLE 1 Let $f(x) = 0.006x(10 - x)$ for $0 \leqslant x \leqslant 10$ and $f(x) = 0$ for all other values of x.
(a) Verify that f is a probability density function.
(b) Find $P(4 \leqslant X \leqslant 8)$.

SOLUTION
(a) For $0 \leqslant x \leqslant 10$ we have $0.006x(10 - x) \geqslant 0$, so $f(x) \geqslant 0$ for all x. We also need to check that Equation 2 is satisfied:

$$\int_{-\infty}^{\infty} f(x)\,dx = \int_0^{10} 0.006x(10 - x)\,dx = 0.006 \int_0^{10} (10x - x^2)\,dx$$

$$= 0.006\left[5x^2 - \tfrac{1}{3}x^3\right]_0^{10} = 0.006\left(500 - \tfrac{1000}{3}\right) = 1$$

Therefore, f is a probability density function.
(b) The probability that X lies between 4 and 8 is

$$P(4 \leqslant X \leqslant 8) = \int_4^8 f(x)\,dx = 0.006 \int_4^8 (10x - x^2)\,dx$$

$$= 0.006\left[5x^2 - \tfrac{1}{3}x^3\right]_4^8 = 0.544$$

EXAMPLE 2 Phenomena such as waiting times and equipment failure times are commonly modeled by exponentially decreasing probability density functions. Find the exact form of such a function.

SOLUTION Think of the random variable as being the time you wait on hold before an agent of a company you're telephoning answers your call. So instead of x, let's use t to represent time, in minutes. If f is the probability density function and you call at time $t = 0$, then, from Definition 1, $\int_0^2 f(t)\,dt$ represents the probability that an agent answers within the first two minutes and $\int_4^5 f(t)\,dt$ is the probability that your call is answered during the fifth minute.

It's clear that $f(t) = 0$ for $t < 0$ (the agent can't answer before you place the call). For $t > 0$ we are told to use an exponentially decreasing function, that is, a function of the form $f(t) = Ae^{-ct}$, where A and c are positive constants. Thus

$$f(t) = \begin{cases} 0 & \text{if } t < 0 \\ Ae^{-ct} & \text{if } t \geq 0 \end{cases}$$

We use condition 2 to determine the value of A:

$$\begin{aligned} 1 &= \int_{-\infty}^{\infty} f(t)\,dt = \int_{-\infty}^{0} f(t)\,dt + \int_{0}^{\infty} f(t)\,dt \\ &= \int_0^{\infty} Ae^{-ct}\,dt = \lim_{x\to\infty} \int_0^x Ae^{-ct}\,dt \\ &= \lim_{x\to\infty} \left[-\frac{A}{c} e^{-ct} \right]_0^x = \lim_{x\to\infty} \frac{A}{c}(1 - e^{-cx}) \\ &= \frac{A}{c} \end{aligned}$$

Therefore, $A/c = 1$ and so $A = c$. Thus, every exponential density function has the form

$$f(t) = \begin{cases} 0 & \text{if } t < 0 \\ ce^{-ct} & \text{if } t \geq 0 \end{cases}$$

A typical graph is shown in Figure 2.

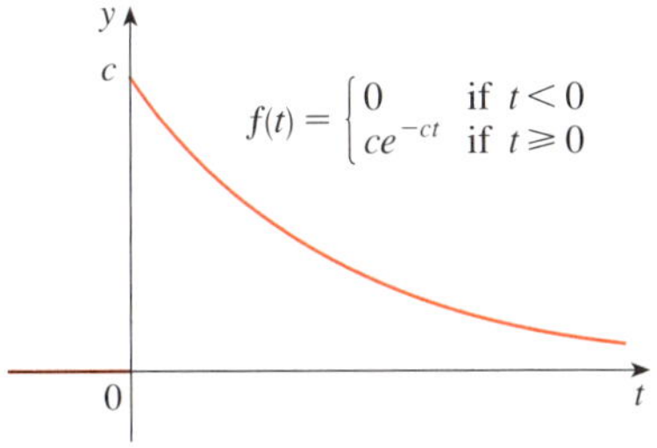

FIGURE 2
An exponential density function

Average Values

Suppose you're waiting for a company to answer your phone call and you wonder how long, on average, you can expect to wait. Let $f(t)$ be the corresponding density function, where t is measured in minutes, and think of a sample of N people who have called this company. Most likely, none of them had to wait more than an hour, so let's restrict our attention to the interval $0 \leq t \leq 60$. Let's divide that interval into n intervals of length Δt and endpoints $0, t_1, t_2, \ldots$. (Think of Δt as lasting a minute, or half a minute, or 10 seconds, or even a second.) The probability that somebody's call gets answered during the time period from t_{i-1} to t_i is the area under the curve $y = f(t)$ from t_{i-1} to t_i, which is approximately equal to $f(\bar{t}_i)\,\Delta t$. (This is the area of the approximating rectangle in Figure 3, where $\bar{t}_i$ is the midpoint of the interval.)

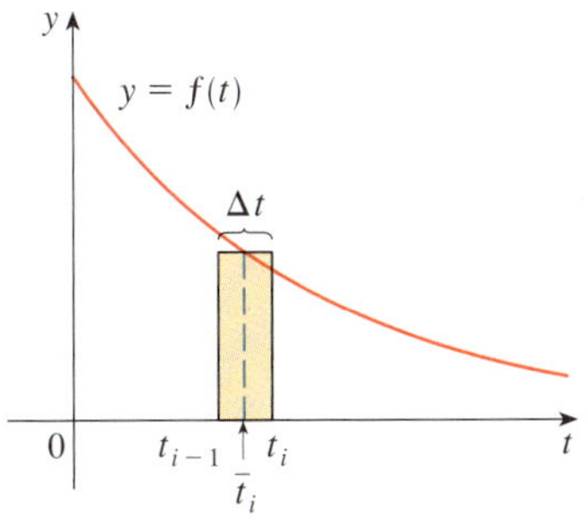

FIGURE 3

Since the long-run proportion of calls that get answered in the time period from t_{i-1} to t_i is $f(\bar{t}_i)\,\Delta t$, we expect that, out of our sample of N callers, the number whose call was answered in that time period is approximately $Nf(\bar{t}_i)\,\Delta t$ and the time that each waited is about $\bar{t}_i$. Therefore, the total time they waited is the product of these numbers: approximately $\bar{t}_i[Nf(\bar{t}_i)\,\Delta t]$. Adding over all such intervals, we get the approximate total of everybody's waiting times:

$$\sum_{i=1}^{n} N\bar{t}_i f(\bar{t}_i)\,\Delta t$$

If we now divide by the number of callers N, we get the approximate *average* waiting time:

$$\sum_{i=1}^{n} \bar{t}_i f(\bar{t}_i)\, \Delta t$$

We recognize this as a Riemann sum for the function $tf(t)$. As the time interval shrinks (that is, $\Delta t \to 0$ and $n \to \infty$), this Riemann sum approaches the integral

$$\int_0^{60} tf(t)\, dt$$

This integral is called the *mean waiting time.*

It is traditional to denote the mean by the Greek letter μ (mu).

In general, the **mean** of any probability density function f is defined to be

$$\mu = \int_{-\infty}^{\infty} xf(x)\, dx$$

The mean can be interpreted as the long-run average value of the random variable X. It can also be interpreted as a measure of centrality of the probability density function.

The expression for the mean resembles an integral we have seen before. If $\mathcal{R}$ is the region that lies under the graph of f, we know from Formula 8.3.8 that the x-coordinate of the centroid of $\mathcal{R}$ is

$$\bar{x} = \frac{\displaystyle\int_{-\infty}^{\infty} xf(x)\, dx}{\displaystyle\int_{-\infty}^{\infty} f(x)\, dx} = \int_{-\infty}^{\infty} xf(x)\, dx = \mu$$

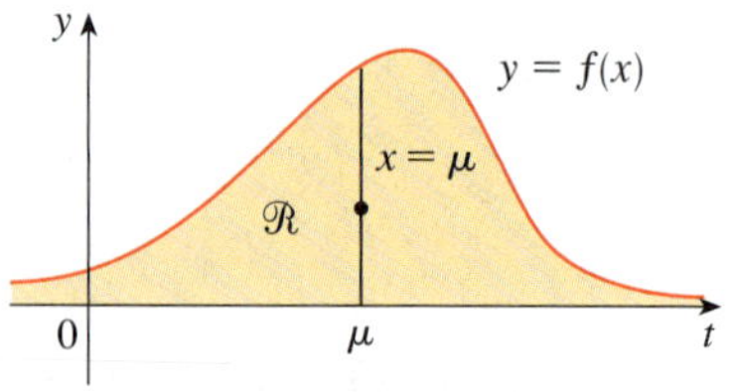

FIGURE 4
$\mathcal{R}$ balances at a point on the line $x = \mu$

because of Equation 2. So a thin plate in the shape of $\mathcal{R}$ balances at a point on the vertical line $x = \mu$. (See Figure 4.)

EXAMPLE 3 Find the mean of the exponential distribution of Example 2:

$$f(t) = \begin{cases} 0 & \text{if } t < 0 \\ ce^{-ct} & \text{if } t \geqslant 0 \end{cases}$$

SOLUTION According to the definition of a mean, we have

$$\mu = \int_{-\infty}^{\infty} tf(t)\, dt = \int_0^{\infty} tce^{-ct}\, dt$$

To evaluate this integral we use integration by parts, with $u = t$ and $dv = ce^{-ct}\, dt$:

$$\int_0^{\infty} tce^{-ct}\, dt = \lim_{x \to \infty} \int_0^x tce^{-ct}\, dt = \lim_{x \to \infty} \left(-te^{-ct}\Big]_0^x + \int_0^x e^{-ct}\, dt \right)$$

$$= \lim_{x \to \infty} \left(-xe^{-cx} + \frac{1}{c} - \frac{e^{-cx}}{c} \right)$$

The limit of the first term is 0 by l'Hospital's Rule.

$$= \frac{1}{c}$$

The mean is $\mu = 1/c$, so we can rewrite the probability density function as

$$f(t) = \begin{cases} 0 & \text{if } t < 0 \\ \mu^{-1}e^{-t/\mu} & \text{if } t \geqslant 0 \end{cases}$$

EXAMPLE 4 Suppose the average waiting time for a customer's call to be answered by a company representative is five minutes.
(a) Find the probability that a call is answered during the first minute.
(b) Find the probability that a customer waits more than five minutes to be answered.

SOLUTION

(a) We are given that the mean of the exponential distribution is $\mu = 5$ min and so, from the result of Example 3, we know that the probability density function is

$$f(t) = \begin{cases} 0 & \text{if } t < 0 \\ 0.2e^{-t/5} & \text{if } t \geq 0 \end{cases}$$

Thus, the probability that a call is answered during the first minute is

$$\begin{aligned} P(0 \leq T \leq 1) &= \int_0^1 f(t)\,dt \\ &= \int_0^1 0.2e^{-t/5}\,dt \\ &= 0.2(-5)e^{-t/5}\Big]_0^1 \\ &= 1 - e^{-1/5} \approx 0.1813 \end{aligned}$$

So about 18% of customers' calls are answered during the first minute.

(b) The probability that a customer waits more than five minutes is

$$\begin{aligned} P(T > 5) &= \int_5^\infty f(t)\,dt = \int_5^\infty 0.2e^{-t/5}\,dt \\ &= \lim_{x \to \infty} \int_5^x 0.2e^{-t/5}\,dt = \lim_{x \to \infty} (e^{-1} - e^{-x/5}) \\ &= \frac{1}{e} \approx 0.368 \end{aligned}$$

About 37% of customers wait more than five minutes before their calls are answered.

Notice the result of Example 4(b): Even though the mean waiting time is 5 minutes, only 37% of callers wait more than 5 minutes. The reason is that some callers have to wait much longer (maybe 10 or 15 minutes), and this brings up the average.

Another measure of centrality of a probability density function is the *median*. That is a number m such that half the callers have a waiting time less than m and the other callers have a waiting time longer than m. In general, the **median** of a probability density function is the number m such that

$$\int_m^\infty f(x)\,dx = \tfrac{1}{2}$$

This means that half the area under the graph of f lies to the right of m. In Exercise 7 you are asked to show that the median waiting time for the company described in Example 4 is approximately 3.5 minutes.

Normal Distributions

Many important random phenomena—such as test scores on aptitude tests, heights and weights of individuals from a homogeneous population, annual rainfall in a given location—are modeled by a **normal distribution**. This means that the probability density function of the random variable X is a member of the family of functions

$$f(x) = \frac{1}{\sigma\sqrt{2\pi}} e^{-(x-\mu)^2/(2\sigma^2)} \tag{3}$$

The standard deviation is denoted by the lowercase Greek letter σ (sigma).

You can verify that the mean for this function is μ. The positive constant σ is called the **standard deviation**; it measures how spread out the values of X are. From the bell-shaped graphs of members of the family in Figure 5, we see that for small values of σ the values of X are clustered about the mean, whereas for larger values of σ the values of X are more spread out. Statisticians have methods for using sets of data to estimate μ and σ.

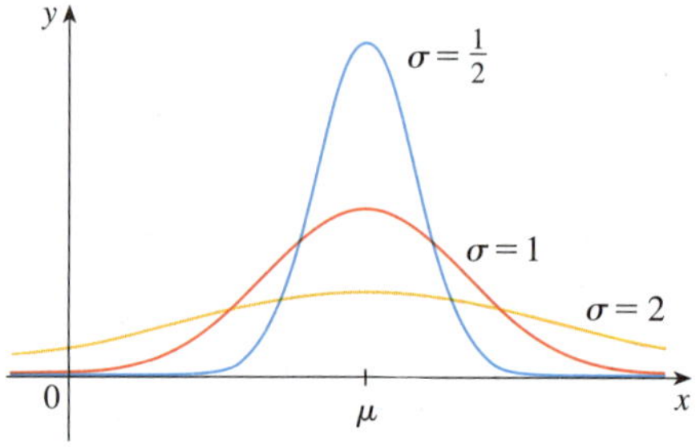

FIGURE 5
Normal distributions

The factor $1/(\sigma\sqrt{2\pi})$ is needed to make f a probability density function. In fact, it can be verified using the methods of multivariable calculus that

$$\int_{-\infty}^{\infty} \frac{1}{\sigma\sqrt{2\pi}} e^{-(x-\mu)^2/(2\sigma^2)}\,dx = 1$$

EXAMPLE 5 Intelligence Quotient (IQ) scores are distributed normally with mean 100 and standard deviation 15. (Figure 6 shows the corresponding probability density function.)

(a) What percentage of the population has an IQ score between 85 and 115?

(b) What percentage of the population has an IQ above 140?

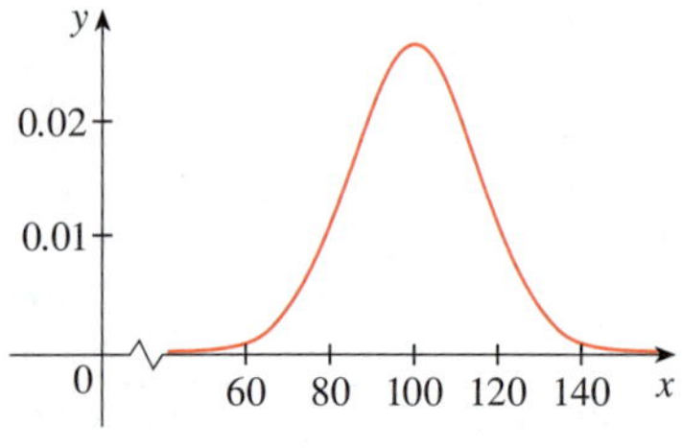

FIGURE 6
Distribution of IQ scores

SOLUTION

(a) Since IQ scores are normally distributed, we use the probability density function given by Equation 3 with $\mu = 100$ and $\sigma = 15$:

$$P(85 \leqslant X \leqslant 115) = \int_{85}^{115} \frac{1}{15\sqrt{2\pi}} e^{-(x-100)^2/(2\cdot 15^2)}\,dx$$

Recall from Section 7.5 that the function $y = e^{-x^2}$ doesn't have an elementary antiderivative, so we can't evaluate the integral exactly. But we can use the numerical integration capability of a calculator or computer (or the Midpoint Rule or Simpson's Rule) to estimate the integral. Doing so, we find that

$$P(85 \leqslant X \leqslant 115) \approx 0.68$$

So about 68% of the population has an IQ between 85 and 115, that is, within one standard deviation of the mean.

(b) The probability that the IQ score of a person chosen at random is more than 140 is

$$P(X > 140) = \int_{140}^{\infty} \frac{1}{15\sqrt{2\pi}} e^{-(x-100)^2/450}\,dx$$

To avoid the improper integral we could approximate it by the integral from 140 to 200. (It's quite safe to say that people with an IQ over 200 are extremely rare.) Then

$$P(X > 140) \approx \int_{140}^{200} \frac{1}{15\sqrt{2\pi}} e^{-(x-100)^2/450}\,dx \approx 0.0038$$

Therefore, about 0.4% of the population has an IQ over 140.

8.5 Exercises

1. Let $f(x)$ be the probability density function for the lifetime of a manufacturer's highest quality car tire, where x is measured in miles. Explain the meaning of each integral.

(a) $\int_{30,000}^{40,000} f(x)\,dx$ (b) $\int_{25,000}^{\infty} f(x)\,dx$

2. Let $f(t)$ be the probability density function for the time it takes you to drive to school in the morning, where t is measured in minutes. Express the following probabilities as integrals.

(a) The probability that you drive to school in less than 15 minutes

(b) The probability that it takes you more than half an hour to get to school

3. Let $f(x) = \frac{3}{64}x\sqrt{16 - x^2}$ for $0 \leqslant x \leqslant 4$ and $f(x) = 0$ for all other values of x.

(a) Verify that f is a probability density function.

(b) Find $P(X < 2)$.

4. Let $f(x) = kx^2(1 - x)$ if $0 \leqslant x \leqslant 1$ and $f(x) = 0$ if $x < 0$ or $x > 1$.

(a) For what value of k is f a probability density function?

(b) For that value of k, find $P\left(X \geqslant \frac{1}{2}\right)$.

(c) Find the mean.

5. A spinner from a board game randomly indicates a real number between 0 and 10. The spinner is fair in the sense that it indicates a number in a given interval with the same probability as it indicates a number in any other interval of the same length.

(a) Explain why the function

$$f(x) = \begin{cases} 0.1 & \text{if } 0 \leqslant x \leqslant 10 \\ 0 & \text{if } x < 0 \text{ or } x > 10 \end{cases}$$

is a probability density function for the spinner's values.

(b) What does your intuition tell you about the value of the mean? Check your guess by evaluating an integral.

6. (a) Explain why the function whose graph is shown is a probability density function.

(b) Use the graph to find the following probabilities:

(i) $P(X < 3)$ (ii) $P(3 \leqslant X \leqslant 8)$

(c) Calculate the mean.

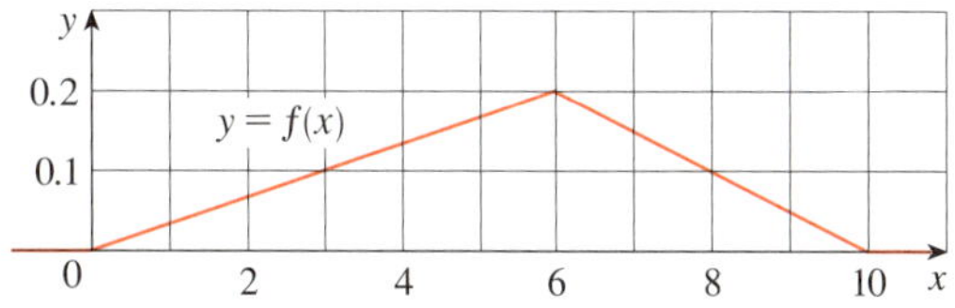

7. Show that the median waiting time for a phone call to the company described in Example 4 is about 3.5 minutes.

8. (a) A type of lightbulb is labeled as having an average lifetime of 1000 hours. It's reasonable to model the probability of failure of these bulbs by an exponential density function with mean $\mu = 1000$. Use this model to find the probability that a bulb

(i) fails within the first 200 hours,

(ii) burns for more than 800 hours.

(b) What is the median lifetime of these lightbulbs?

9. The manager of a fast-food restaurant determines that the average time that her customers wait for service is 2.5 minutes.

(a) Find the probability that a customer has to wait for more than 4 minutes.

(b) Find the probability that a customer is served within the first 2 minutes.

(c) The manager wants to advertise that anybody who isn't served within a certain number of minutes gets a free hamburger. But she doesn't want to give away free hamburgers to more than 2% of her customers. What should the advertisement say?

10. According to the National Health Survey, the heights of adult males in the United States are normally distributed with mean 69.0 inches and standard deviation 2.8 inches.

(a) What is the probability that an adult male chosen at random is between 65 inches and 73 inches tall?

(b) What percentage of the adult male population is more than 6 feet tall?

11. The "Garbage Project" at the University of Arizona reports that the amount of paper discarded by households per week is normally distributed with mean 9.4 lb and standard deviation 4.2 lb. What percentage of households throw out at least 10 lb of paper a week?

12. Boxes are labeled as containing 500 g of cereal. The machine filling the boxes produces weights that are normally distributed with standard deviation 12 g.

(a) If the target weight is 500 g, what is the probability that the machine produces a box with less than 480 g of cereal?

(b) Suppose a law states that no more than 5% of a manufacturer's cereal boxes can contain less than the stated weight of 500 g. At what target weight should the manufacturer set its filling machine?

13. For any normal distribution, find the probability that the random variable lies within two standard deviations of the mean.

14. The standard deviation for a random variable with probability density function f and mean μ is defined by

$$\sigma = \left[\int_{-\infty}^{\infty} (x - \mu)^2 f(x)\, dx \right]^{1/2}$$

Find the standard deviation for an exponential density function with mean μ.

15. The hydrogen atom is composed of one proton in the nucleus and one electron, which moves about the nucleus. In the quantum theory of atomic structure, it is assumed that the electron does not move in a well-defined orbit. Instead, it occupies a state known as an *orbital*, which may be thought of as a "cloud" of negative charge surrounding the nucleus. At the state of lowest energy, called the *ground state*, or *1s-orbital*, the shape of this cloud is assumed to be a sphere centered at the nucleus. This sphere is described in terms of the probability density function

$$p(r) = \frac{4}{a_0^3} r^2 e^{-2r/a_0} \qquad r \geq 0$$

where a_0 is the *Bohr radius* ($a_0 \approx 5.59 \times 10^{-11}$ m). The integral

$$P(r) = \int_0^r \frac{4}{a_0^3} s^2 e^{-2s/a_0}\, ds$$

gives the probability that the electron will be found within the sphere of radius r meters centered at the nucleus.

(a) Verify that $p(r)$ is a probability density function.

(b) Find $\lim_{r \to \infty} p(r)$. For what value of r does $p(r)$ have its maximum value?

(c) Graph the density function.

(d) Find the probability that the electron will be within the sphere of radius $4a_0$ centered at the nucleus.

(e) Calculate the mean distance of the electron from the nucleus in the ground state of the hydrogen atom.

8 Review

CONCEPT CHECK

1. (a) How is the length of a curve defined?

(b) Write an expression for the length of a smooth curve given by $y = f(x)$, $a \leq x \leq b$.

(c) What if x is given as a function of y?

2. (a) Write an expression for the surface area of the surface obtained by rotating the curve $y = f(x)$, $a \leq x \leq b$, about the x-axis.

(b) What if x is given as a function of y?

(c) What if the curve is rotated about the y-axis?

3. Describe how we can find the hydrostatic force against a vertical wall submersed in a fluid.

4. (a) What is the physical significance of the center of mass of a thin plate?

(b) If the plate lies between $y = f(x)$ and $y = 0$, where $a \leq x \leq b$, write expressions for the coordinates of the center of mass.

5. What does the Theorem of Pappus say?

6. Given a demand function $p(x)$, explain what is meant by the consumer surplus when the amount of a commodity currently available is X and the current selling price is P. Illustrate with a sketch.

7. (a) What is the cardiac output of the heart?

(b) Explain how the cardiac output can be measured by the dye dilution method.

8. What is a probability density function? What properties does such a function have?

9. Suppose $f(x)$ is the probability density function for the weight of a female college student, where x is measured in pounds.

(a) What is the meaning of the integral $\int_0^{100} f(x)\, dx$?

(b) Write an expression for the mean of this density function.

(c) How can we find the median of this density function?

10. What is a normal distribution? What is the significance of the standard deviation?

EXERCISES

1–2 Find the length of the curve.

1. $y = \frac{1}{6}(x^2 + 4)^{3/2}, \quad 0 \leqslant x \leqslant 3$

2. $y = 2 \ln \sin(\frac{1}{2}x), \quad \pi/3 \leqslant x \leqslant \pi$

3. (a) Find the length of the curve

$$y = \frac{x^4}{16} + \frac{1}{2x^2} \qquad 1 \leqslant x \leqslant 2$$

(b) Find the area of the surface obtained by rotating the curve in part (a) about the y-axis.

4. (a) The curve $y = x^2$, $0 \leqslant x \leqslant 1$, is rotated about the y-axis. Find the area of the resulting surface.

(b) Find the area of the surface obtained by rotating the curve in part (a) about the x-axis.

5. Use Simpson's Rule with $n = 6$ to estimate the length of the curve $y = e^{-x^2}$, $0 \leqslant x \leqslant 3$.

6. Use Simpson's Rule with $n = 6$ to estimate the area of the surface obtained by rotating the curve in Exercise 5 about the x-axis.

7. Find the length of the curve

$$y = \int_1^x \sqrt{\sqrt{t} - 1}\, dt \qquad 1 \leqslant x \leqslant 16$$

8. Find the area of the surface obtained by rotating the curve in Exercise 7 about the y-axis.

9. A gate in an irrigation canal is constructed in the form of a trapezoid 3 ft wide at the bottom, 5 ft wide at the top, and 2 ft high. It is placed vertically in the canal, with the water extending to its top. Find the hydrostatic force on one side of the gate.

10. A trough is filled with water and its vertical ends have the shape of the parabolic region in the figure. Find the hydrostatic force on one end of the trough.

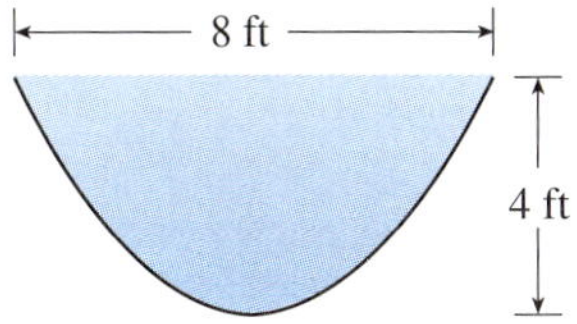

11–12 Find the centroid of the region bounded by the given curves.

11. $y = 4 - x^2, \quad y = x + 2$

12. $y = \sin x, \quad y = 0, \quad x = \pi/4, \quad x = 3\pi/4$

13–14 Find the centroid of the region shown.

13.

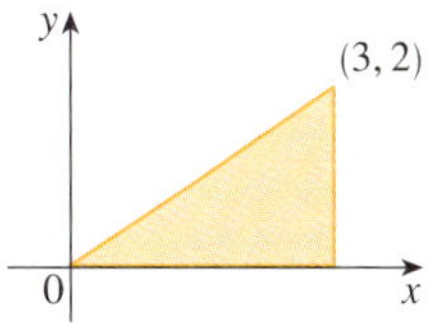

14.

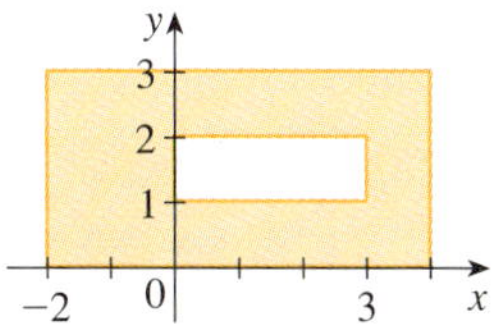

15. Find the volume obtained when the circle of radius 1 with center (1, 0) is rotated about the y-axis.

16. Use the Theorem of Pappus and the fact that the volume of a sphere of radius r is $\frac{4}{3}\pi r^3$ to find the centroid of the semicircular region bounded by the curve $y = \sqrt{r^2 - x^2}$ and the x-axis.

17. The demand function for a commodity is given by $p = 2000 - 0.1x - 0.01x^2$. Find the consumer surplus when the sales level is 100.

18. After a 6-mg injection of dye into a heart, the readings of dye concentration at two-second intervals are as shown in the table. Use Simpson's Rule to estimate the cardiac output.

t	$c(t)$	t	$c(t)$
0	0	14	4.7
2	1.9	16	3.3
4	3.3	18	2.1
6	5.1	20	1.1
8	7.6	22	0.5
10	7.1	24	0
12	5.8		

19. (a) Explain why the function

$$f(x) = \begin{cases} \dfrac{\pi}{20} \sin\left(\dfrac{\pi x}{10}\right) & \text{if } 0 \leqslant x \leqslant 10 \\ 0 & \text{if } x < 0 \text{ or } x > 10 \end{cases}$$

is a probability density function.

(b) Find $P(X < 4)$.

(c) Calculate the mean. Is the value what you would expect?

20. Lengths of human pregnancies are normally distributed with mean 268 days and standard deviation 15 days. What percentage of pregnancies last between 250 days and 280 days?

21. The length of time spent waiting in line at a certain bank is modeled by an exponential density function with mean 8 minutes.

(a) What is the probability that a customer is served in the first 3 minutes?

(b) What is the probability that a customer has to wait more than 10 minutes?

(c) What is the median waiting time?

PROBLEMS PLUS

1. Find the area of the region $S = \{(x, y) \mid x \geq 0,\ y \leq 1,\ x^2 + y^2 \leq 4y\}$.

2. Find the centroid of the region enclosed by the loop of the curve $y^2 = x^3 - x^4$.

3. If a sphere of radius r is sliced by a plane whose distance from the center of the sphere is d, then the sphere is divided into two pieces called segments of one base. The corresponding surfaces are called *spherical zones of one base.*
   (a) Determine the surface areas of the two spherical zones indicated in the figure.
   (b) Determine the approximate area of the Arctic Ocean by assuming that it is approximately circular in shape, with center at the North Pole and "circumference" at 75° north latitude. Use $r = 3960$ mi for the radius of Earth.
   (c) A sphere of radius r is inscribed in a right circular cylinder of radius r. Two planes perpendicular to the central axis of the cylinder and a distance h apart cut off a *spherical zone of two bases* on the sphere. Show that the surface area of the spherical zone equals the surface area of the region that the two planes cut off on the cylinder.
   (d) The *Torrid Zone* is the region on the surface of Earth that is between the Tropic of Cancer (23.45° north latitude) and the Tropic of Capricorn (23.45° south latitude). What is the area of the Torrid Zone?

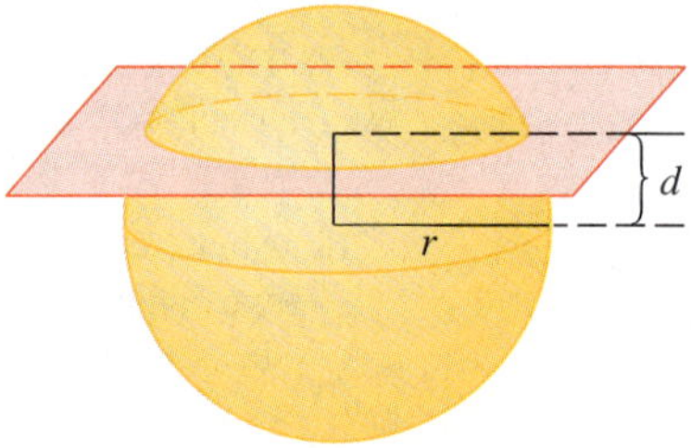

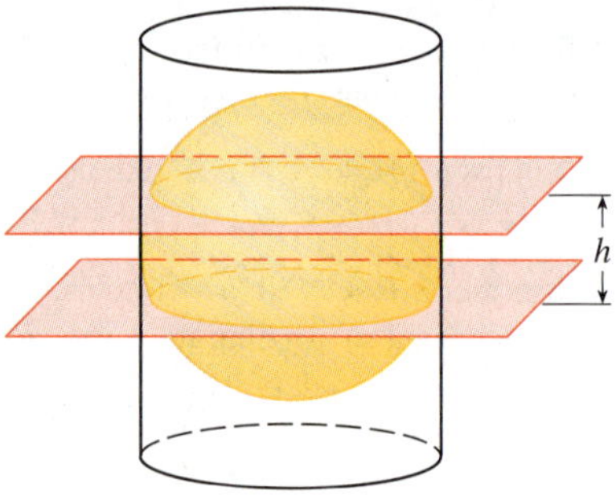

4. (a) Show that an observer at height H above the north pole of a sphere of radius r can see a part of the sphere that has area

$$\frac{2\pi r^2 H}{r + H}$$

   (b) Two spheres with radii r and R are placed so that the distance between their centers is d, where $d > r + R$. Where should a light be placed on the line joining the centers of the spheres in order to illuminate the largest total surface?

5. Suppose that the density of seawater, $\rho = \rho(z)$, varies with the depth z below the surface.
   (a) Show that the hydrostatic pressure is governed by the differential equation

$$\frac{dP}{dz} = \rho(z)g$$

   where g is the acceleration due to gravity. Let P_0 and ρ_0 be the pressure and density at $z = 0$. Express the pressure at depth z as an integral.
   (b) Suppose the density of seawater at depth z is given by $\rho = \rho_0 e^{z/H}$, where H is a positive constant. Find the total force, expressed as an integral, exerted on a vertical circular porthole of radius r whose center is located at a distance $L > r$ below the surface.

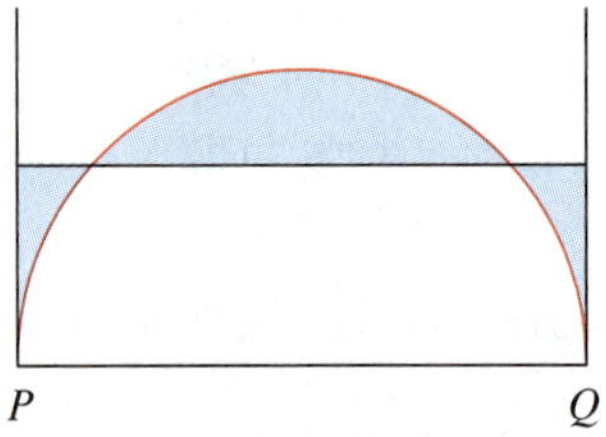

FIGURE FOR PROBLEM 6

6. The figure shows a semicircle with radius 1, horizontal diameter PQ, and tangent lines at P and Q. At what height above the diameter should the horizontal line be placed so as to minimize the shaded area?

7. Let P be a pyramid with a square base of side $2b$ and suppose that S is a sphere with its center on the base of P and is tangent to all eight edges of P. Find the height of P. Then find the volume of the intersection of S and P.

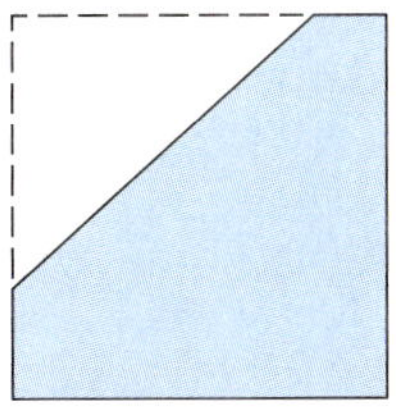

FIGURE FOR PROBLEM 10

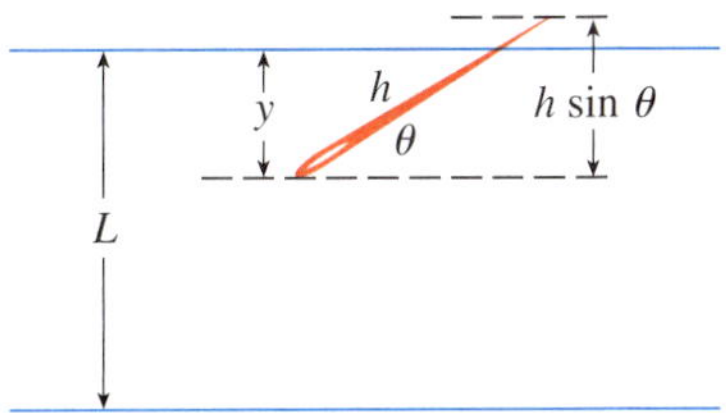

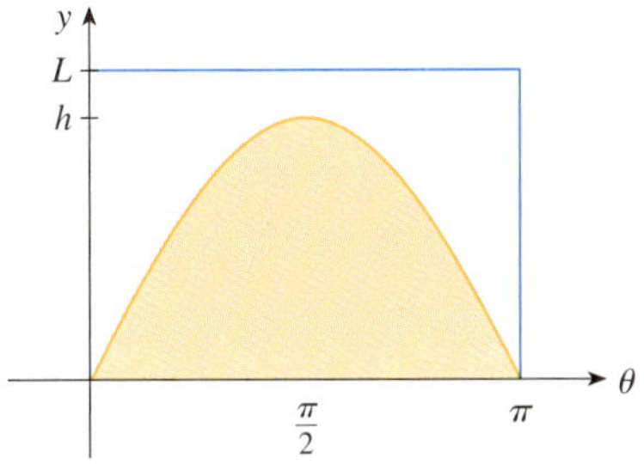

FIGURE FOR PROBLEM 11

8. Consider a flat metal plate to be placed vertically under water with its top 2 m below the surface of the water. Determine a shape for the plate so that if the plate is divided into any number of horizontal strips of equal height, the hydrostatic force on each strip is the same.

9. A uniform disk with radius 1 m is to be cut by a line so that the center of mass of the smaller piece lies halfway along a radius. How close to the center of the disk should the cut be made? (Express your answer correct to two decimal places.)

10. A triangle with area 30 cm^2 is cut from a corner of a square with side 10 cm, as shown in the figure. If the centroid of the remaining region is 4 cm from the right side of the square, how far is it from the bottom of the square?

11. In a famous 18th-century problem, known as *Buffon's needle problem*, a needle of length h is dropped onto a flat surface (for example, a table) on which parallel lines L units apart, $L \geqslant h$, have been drawn. The problem is to determine the probability that the needle will come to rest intersecting one of the lines. Assume that the lines run east-west, parallel to the x-axis in a rectangular coordinate system (as in the figure). Let y be the distance from the "southern" end of the needle to the nearest line to the north. (If the needle's southern end lies on a line, let $y = 0$. If the needle happens to lie east-west, let the "western" end be the "southern" end.) Let θ be the angle that the needle makes with a ray extending eastward from the "southern" end. Then $0 \leqslant y \leqslant L$ and $0 \leqslant \theta \leqslant \pi$. Note that the needle intersects one of the lines only when $y < h \sin \theta$. Now, the total set of possibilities for the needle can be identified with the rectangular region $0 \leqslant y \leqslant L$, $0 \leqslant \theta \leqslant \pi$, and the proportion of times that the needle intersects a line is the ratio

$$\frac{\text{area under } y = h \sin \theta}{\text{area of rectangle}}$$

This ratio is the probability that the needle intersects a line. Find the probability that the needle will intersect a line if $h = L$. What if $h = L/2$?

12. If the needle in Problem 11 has length $h > L$, it's possible for the needle to intersect more than one line.

(a) If $L = 4$, find the probability that a needle of length 7 will intersect at least one line. [*Hint:* Proceed as in Problem 11. Define y as before; then the total set of possibilities for the needle can be identified with the same rectangular region $0 \leqslant y \leqslant L$, $0 \leqslant \theta \leqslant \pi$. What portion of the rectangle corresponds to the needle intersecting a line?]

(b) If $L = 4$, find the probability that a needle of length 7 will intersect *two* lines.

(c) If $2L < h \leqslant 3L$, find a general formula for the probability that the needle intersects three lines.

CHAPTER 9

By analyzing pairs of differential equations we gain insight into population cycles of predators and prey, such as the Canada lynx and snowshoe hare.

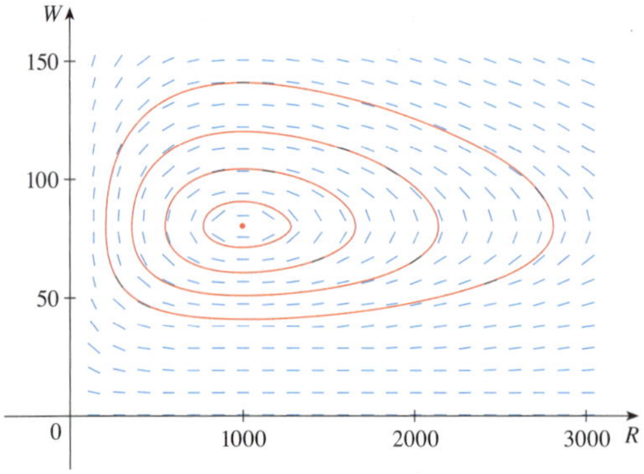

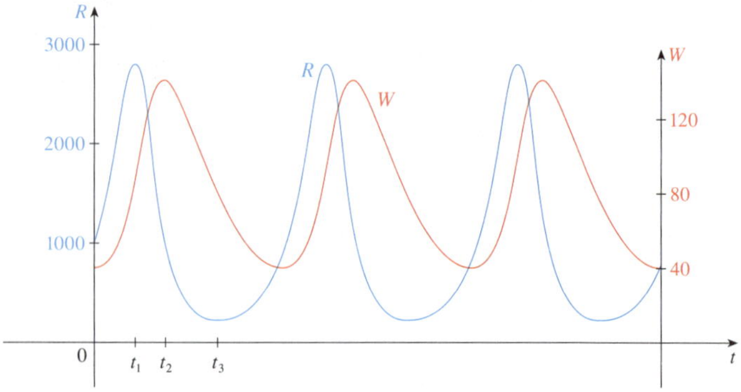

Differential Equations

Perhaps the most important of all the applications of calculus is to differential equations. When physical scientists or social scientists use calculus, more often than not it is to analyze a differential equation that has arisen in the process of modeling some phenomenon that they are studying. Although it is often impossible to find an explicit formula for the solution of a differential equation, we will see that graphical and numerical approaches provide the needed information.

9.1 Modeling with Differential Equations

Now is a good time to read (or reread) the discussion of mathematical modeling on page 25.

In describing the process of modeling in Section 1.2, we talked about formulating a mathematical model of a real-world problem either through intuitive reasoning about the phenomenon or from a physical law based on evidence from experiments. The mathematical model often takes the form of a *differential equation,* that is, an equation that contains an unknown function and some of its derivatives. This is not surprising because in a real-world problem we often notice that changes occur and we want to predict future behavior on the basis of how current values change. Let's begin by examining several examples of how differential equations arise when we model physical phenomena.

Models of Population Growth

One model for the growth of a population is based on the assumption that the population grows at a rate proportional to the size of the population. That is a reasonable assumption for a population of bacteria or animals under ideal conditions (unlimited environment, adequate nutrition, absence of predators, immunity from disease).

Let's identify and name the variables in this model:

$t =$ time (the independent variable)

$P =$ the number of individuals in the population (the dependent variable)

The rate of growth of the population is the derivative dP/dt. So our assumption that the rate of growth of the population is proportional to the population size is written as the equation

1 $$\frac{dP}{dt} = kP$$

where k is the proportionality constant. Equation 1 is our first model for population growth; it is a differential equation because it contains an unknown function P and its derivative dP/dt.

Having formulated a model, let's look at its consequences. If we rule out a population of 0, then $P(t) > 0$ for all t. So, if $k > 0$, then Equation 1 shows that $P'(t) > 0$ for all t. This means that the population is always increasing. In fact, as $P(t)$ increases, Equation 1 shows that dP/dt becomes larger. In other words, the growth rate increases as the population increases.

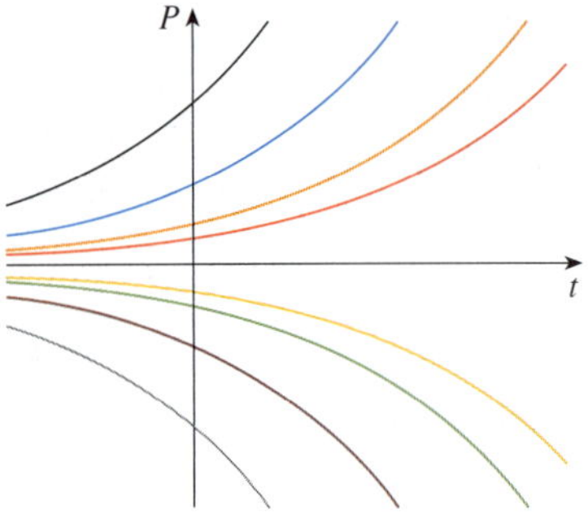

FIGURE 1
The family of solutions of $dP/dt = kP$

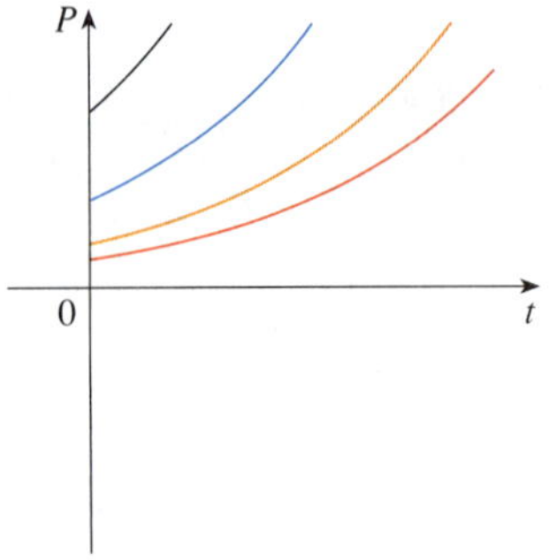

FIGURE 2
The family of solutions $P(t) = Ce^{kt}$ with $C > 0$ and $t \geqslant 0$

Let's try to think of a solution of Equation 1. This equation asks us to find a function whose derivative is a constant multiple of itself. We know that exponential functions have that property. In fact, if we let $P(t) = Ce^{kt}$, then

$$P'(t) = C(ke^{kt}) = k(Ce^{kt}) = kP(t)$$

Thus, any exponential function of the form $P(t) = Ce^{kt}$ is a solution of Equation 1. When we study this equation in detail in Section 9.4, we will see that there is no other solution.

Allowing C to vary through all the real numbers, we get the *family* of solutions $P(t) = Ce^{kt}$ whose graphs are shown in Figure 1. But populations have only positive values and so we are interested only in the solutions with $C > 0$. And we are probably concerned only with values of t greater than the initial time $t = 0$. Figure 2 shows the physically meaningful solutions. Putting $t = 0$, we get $P(0) = Ce^{k(0)} = C$, so the constant C turns out to be the initial population, $P(0)$.

Equation 1 is appropriate for modeling population growth under ideal conditions, but we have to recognize that a more realistic model must reflect the fact that a given environment has limited resources. Many populations start by increasing in an exponential manner, but the population levels off when it approaches its *carrying capacity* K (or decreases toward K if it ever exceeds K). For a model to take into account both trends, we make two assumptions:

- $\dfrac{dP}{dt} \approx kP$ if P is small (Initially, the growth rate is proportional to P.)
- $\dfrac{dP}{dt} < 0$ if $P > K$ (P decreases if it ever exceeds K.)

A simple expression that incorporates both assumptions is given by the equation

2
$$\frac{dP}{dt} = kP\left(1 - \frac{P}{K}\right)$$

Notice that if P is small compared with K, then P/K is close to 0 and so $dP/dt \approx kP$. If $P > K$, then $1 - P/K$ is negative and so $dP/dt < 0$.

Equation 2 is called the *logistic differential equation* and was proposed by the Dutch mathematical biologist Pierre-François Verhulst in the 1840s as a model for world population growth. We will develop techniques that enable us to find explicit solutions of the logistic equation in Section 9.5, but for now we can deduce qualitative characteristics of the solutions directly from Equation 2. We first observe that the constant functions $P(t) = 0$ and $P(t) = K$ are solutions because, in either case, one of the factors on the right side of Equation 2 is zero. (This certainly makes physical sense: If the population is ever either 0 or at the carrying capacity, it stays that way.) These two constant solutions are called *equilibrium solutions*.

If the initial population $P(0)$ lies between 0 and K, then the right side of Equation 2 is positive, so $dP/dt > 0$ and the population increases. But if the population exceeds the carrying capacity ($P > K$), then $1 - P/K$ is negative, so $dP/dt < 0$ and the population decreases. Notice that, in either case, if the population approaches the carrying capacity ($P \to K$), then $dP/dt \to 0$, which means the population levels off. So we expect that the solutions of the logistic differential equation have graphs that look something like the ones in Figure 3. Notice that the graphs move away from the equilibrium solution $P = 0$ and move toward the equilibrium solution $P = K$.

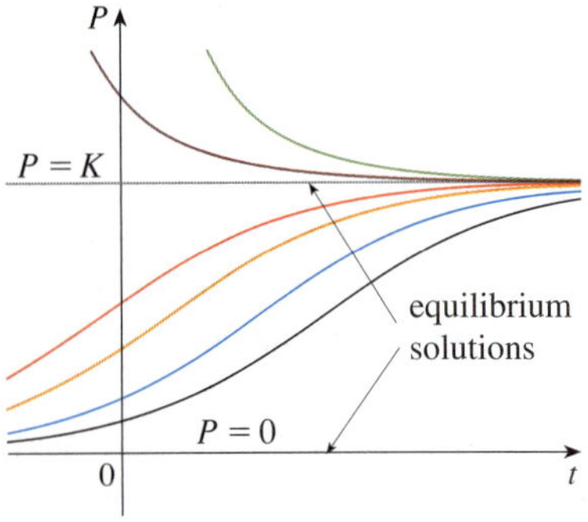

FIGURE 3
Solutions of the logistic equation

A Model for the Motion of a Spring

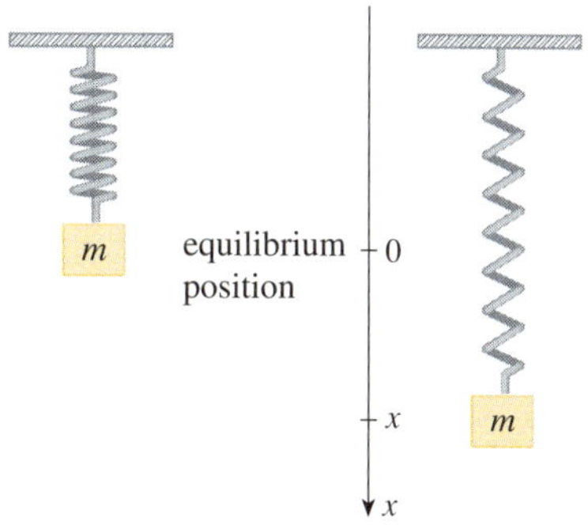

FIGURE 4

Let's now look at an example of a model from the physical sciences. We consider the motion of an object with mass m at the end of a vertical spring (as in Figure 4). In Section 6.4 we discussed Hooke's Law, which says that if the spring is stretched (or compressed) x units from its natural length, then it exerts a force that is proportional to x:

$$\text{restoring force} = -kx$$

where k is a positive constant (called the *spring constant*). If we ignore any external resisting forces (due to air resistance or friction) then, by Newton's Second Law (force equals mass times acceleration), we have

$$m\frac{d^2x}{dt^2} = -kx \tag{3}$$

This is an example of what is called a *second-order differential equation* because it involves second derivatives. Let's see what we can guess about the form of the solution directly from the equation. We can rewrite Equation 3 in the form

$$\frac{d^2x}{dt^2} = -\frac{k}{m}x$$

which says that the second derivative of x is proportional to x but has the opposite sign. We know two functions with this property, the sine and cosine functions. In fact, it turns out that all solutions of Equation 3 can be written as combinations of certain sine and cosine functions (see Exercise 3). This is not surprising; we expect the spring to oscillate about its equilibrium position and so it is natural to think that trigonometric functions are involved.

General Differential Equations

In general, a **differential equation** is an equation that contains an unknown function and one or more of its derivatives. The **order** of a differential equation is the order of the highest derivative that occurs in the equation. Thus, Equations 1 and 2 are first-order equations and Equation 3 is a second-order equation. In all three of those equations the independent variable is called t and represents time, but in general the independent variable doesn't have to represent time. For example, when we consider the differential equation

$$y' = xy \tag{4}$$

it is understood that y is an unknown function of x.

A function f is called a **solution** of a differential equation if the equation is satisfied when $y = f(x)$ and its derivatives are substituted into the equation. Thus, f is a solution of Equation 4 if

$$f'(x) = xf(x)$$

for all values of x in some interval.

When we are asked to *solve* a differential equation we are expected to find all possible solutions of the equation. We have already solved some particularly simple differential equations, namely, those of the form

$$y' = f(x)$$

For instance, we know that the general solution of the differential equation

$$y' = x^3$$

is given by

$$y = \frac{x^4}{4} + C$$

where C is an arbitrary constant.

But, in general, solving a differential equation is not an easy matter. There is no systematic technique that enables us to solve all differential equations. In Section 9.2, however, we will see how to draw rough graphs of solutions even when we have no explicit formula. We will also learn how to find numerical approximations to solutions.

EXAMPLE 1 Show that every member of the family of functions

$$y = \frac{1 + ce^t}{1 - ce^t}$$

is a solution of the differential equation $y' = \frac{1}{2}(y^2 - 1)$.

|||| Figure 5 shows graphs of seven members of the family in Example 1. The differential equation shows that if $y \approx \pm 1$, then $y' \approx 0$. That is borne out by the flatness of the graphs near $y = 1$ and $y = -1$.

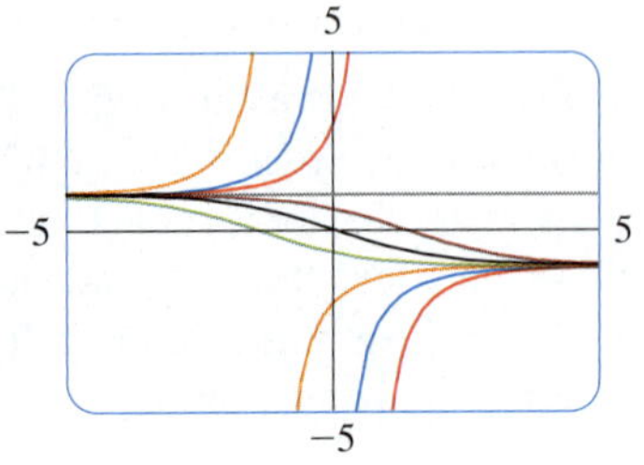

FIGURE 5

SOLUTION We use the Quotient Rule to differentiate the expression for y:

$$y' = \frac{(1 - ce^t)(ce^t) - (1 + ce^t)(-ce^t)}{(1 - ce^t)^2}$$

$$= \frac{ce^t - c^2e^{2t} + ce^t + c^2e^{2t}}{(1 - ce^t)^2} = \frac{2ce^t}{(1 - ce^t)^2}$$

The right side of the differential equation becomes

$$\tfrac{1}{2}(y^2 - 1) = \frac{1}{2}\left[\left(\frac{1 + ce^t}{1 - ce^t}\right)^2 - 1\right] = \frac{1}{2}\left[\frac{(1 + ce^t)^2 - (1 - ce^t)^2}{(1 - ce^t)^2}\right]$$

$$= \frac{1}{2}\,\frac{4ce^t}{(1 - ce^t)^2} = \frac{2ce^t}{(1 - ce^t)^2}$$

Therefore, for every value of c, the given function is a solution of the differential equation.

When applying differential equations, we are usually not as interested in finding a family of solutions (the *general solution*) as we are in finding a solution that satisfies some additional requirement. In many physical problems we need to find the particular solution that satisfies a condition of the form $y(t_0) = y_0$. This is called an **initial condition**, and the problem of finding a solution of the differential equation that satisfies the initial condition is called an **initial-value problem**.

Geometrically, when we impose an initial condition, we look at the family of solution curves and pick the one that passes through the point (t_0, y_0). Physically, this corresponds to measuring the state of a system at time t_0 and using the solution of the initial-value problem to predict the future behavior of the system.

EXAMPLE 2 Find a solution of the differential equation $y' = \frac{1}{2}(y^2 - 1)$ that satisfies the initial condition $y(0) = 2$.

SOLUTION Substituting the values $t = 0$ and $y = 2$ into the formula

$$y = \frac{1 + ce^t}{1 - ce^t}$$

from Example 1, we get

$$2 = \frac{1 + ce^0}{1 - ce^0} = \frac{1 + c}{1 - c}$$

Solving this equation for c, we get $2 - 2c = 1 + c$, which gives $c = \frac{1}{3}$. So the solution of the initial-value problem is

$$y = \frac{1 + \frac{1}{3}e^t}{1 - \frac{1}{3}e^t} = \frac{3 + e^t}{3 - e^t}$$

9.1 Exercises

1. Show that $y = x - x^{-1}$ is a solution of the differential equation $xy' + y = 2x$.

2. Verify that $y = \sin x \cos x - \cos x$ is a solution of the initial-value problem

$$y' + (\tan x)y = \cos^2 x \qquad y(0) = -1$$

on the interval $-\pi/2 < x < \pi/2$.

3. (a) For what nonzero values of k does the function $y = \sin kt$ satisfy the differential equation $y'' + 9y = 0$?
(b) For those values of k, verify that every member of the family of functions

$$y = A \sin kt + B \cos kt$$

is also a solution.

4. For what values of r does the function $y = e^{rt}$ satisfy the differential equation $y'' + y' - 6y = 0$?

5. Which of the following functions are solutions of the differential equation $y'' + 2y' + y = 0$?
(a) $y = e^t$ (b) $y = e^{-t}$
(c) $y = te^{-t}$ (d) $y = t^2e^{-t}$

6. (a) Show that every member of the family of functions $y = Ce^{x^2/2}$ is a solution of the differential equation $y' = xy$.
(b) Illustrate part (a) by graphing several members of the family of solutions on a common screen.
(c) Find a solution of the differential equation $y' = xy$ that satisfies the initial condition $y(0) = 5$.
(d) Find a solution of the differential equation $y' = xy$ that satisfies the initial condition $y(1) = 2$.

7. (a) What can you say about a solution of the equation $y' = -y^2$ just by looking at the differential equation?
(b) Verify that all members of the family $y = 1/(x + C)$ are solutions of the equation in part (a).
(c) Can you think of a solution of the differential equation $y' = -y^2$ that is not a member of the family in part (b)?
(d) Find a solution of the initial-value problem

$$y' = -y^2 \qquad y(0) = 0.5$$

8. (a) What can you say about the graph of a solution of the equation $y' = xy^3$ when x is close to 0? What if x is large?
(b) Verify that all members of the family $y = (c - x^2)^{-1/2}$ are solutions of the differential equation $y' = xy^3$.
(c) Graph several members of the family of solutions on a common screen. Do the graphs confirm what you predicted in part (a)?
(d) Find a solution of the initial-value problem

$$y' = xy^3 \qquad y(0) = 2$$

9. A population is modeled by the differential equation

$$\frac{dP}{dt} = 1.2P\left(1 - \frac{P}{4200}\right)$$

(a) For what values of P is the population increasing?
(b) For what values of P is the population decreasing?
(c) What are the equilibrium solutions?

10. A function $y(t)$ satisfies the differential equation

$$\frac{dy}{dt} = y^4 - 6y^3 + 5y^2$$

(a) What are the constant solutions of the equation?
(b) For what values of y is y increasing?
(c) For what values of y is y decreasing?

11. Explain why the functions with the given graphs *can't* be solutions of the differential equation

$$\frac{dy}{dt} = e^t(y - 1)^2$$

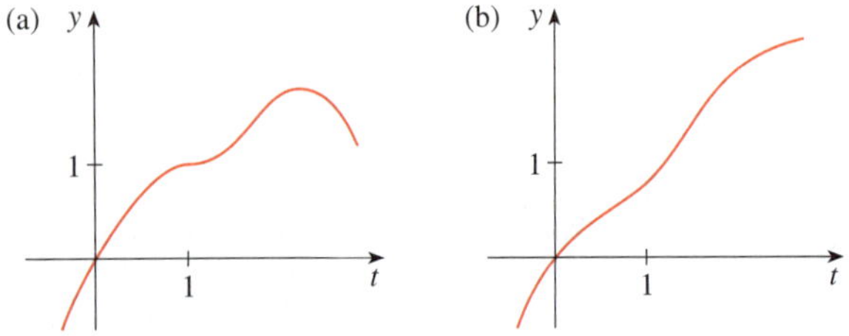

12. The function with the given graph is a solution of one of the following differential equations. Decide which is the correct equation and justify your answer.

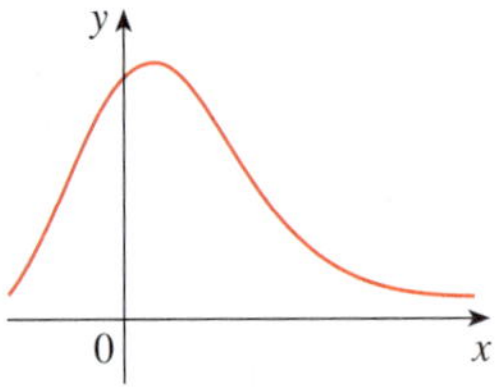

A. $y' = 1 + xy$ B. $y' = -2xy$ C. $y' = 1 - 2xy$

13. Psychologists interested in learning theory study **learning curves**. A learning curve is the graph of a function $P(t)$, the performance of someone learning a skill as a function of the training time t. The derivative dP/dt represents the rate at which performance improves.
(a) When do you think P increases most rapidly? What happens to dP/dt as t increases? Explain.
(b) If M is the maximum level of performance of which the learner is capable, explain why the differential equation

$$\frac{dP}{dt} = k(M - P) \qquad k \text{ a positive constant}$$

is a reasonable model for learning.
(c) Make a rough sketch of a possible solution of this differential equation.

14. Suppose you have just poured a cup of freshly brewed coffee with temperature 95°C in a room where the temperature is 20°C.
(a) When do you think the coffee cools most quickly? What happens to the rate of cooling as time goes by? Explain.
(b) **Newton's Law of Cooling** states that the rate of cooling of an object is proportional to the temperature difference between the object and its surroundings, provided that this difference is not too large. Write a differential equation that expresses Newton's Law of Cooling for this particular situation. What is the initial condition? In view of your answer to part (a), do you think this differential equation is an appropriate model for cooling?
(c) Make a rough sketch of the graph of the solution of the initial-value problem in part (b).

9.2 Direction Fields and Euler's Method

Unfortunately, it's impossible to solve most differential equations in the sense of obtaining an explicit formula for the solution. In this section we show that, despite the absence of an explicit solution, we can still learn a lot about the solution through a graphical approach (direction fields) or a numerical approach (Euler's method).

Direction Fields

Suppose we are asked to sketch the graph of the solution of the initial-value problem

$$y' = x + y \qquad y(0) = 1$$

We don't know a formula for the solution, so how can we possibly sketch its graph? Let's think about what the differential equation means. The equation $y' = x + y$ tells us that the slope at any point (x, y) on the graph (called the *solution curve*) is equal to the sum of the

x- and y-coordinates of the point (see Figure 1). In particular, because the curve passes through the point $(0, 1)$, its slope there must be $0 + 1 = 1$. So a small portion of the solution curve near the point $(0, 1)$ looks like a short line segment through $(0, 1)$ with slope 1 (see Figure 2).

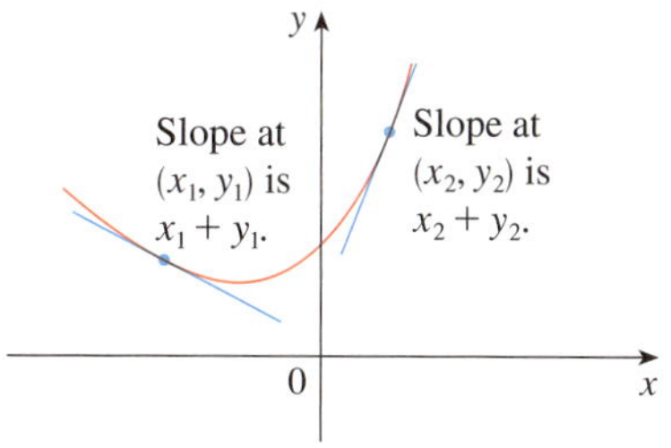

FIGURE 1
A solution of $y' = x + y$

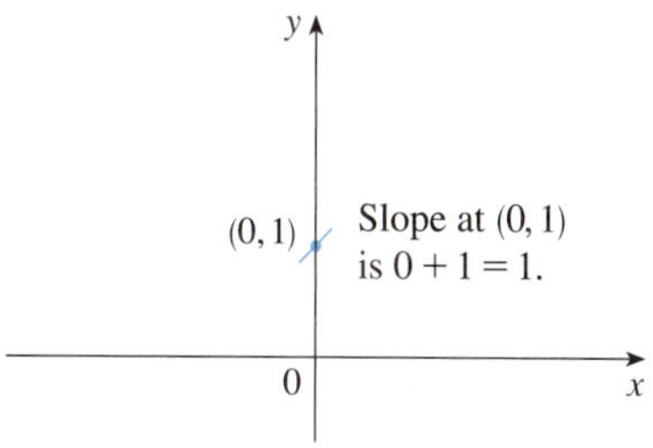

FIGURE 2
Beginning of the solution curve through $(0, 1)$

As a guide to sketching the rest of the curve, let's draw short line segments at a number of points (x, y) with slope $x + y$. The result is called a *direction field* and is shown in Figure 3. For instance, the line segment at the point $(1, 2)$ has slope $1 + 2 = 3$. The direction field allows us to visualize the general shape of the solution curves by indicating the direction in which the curves proceed at each point.

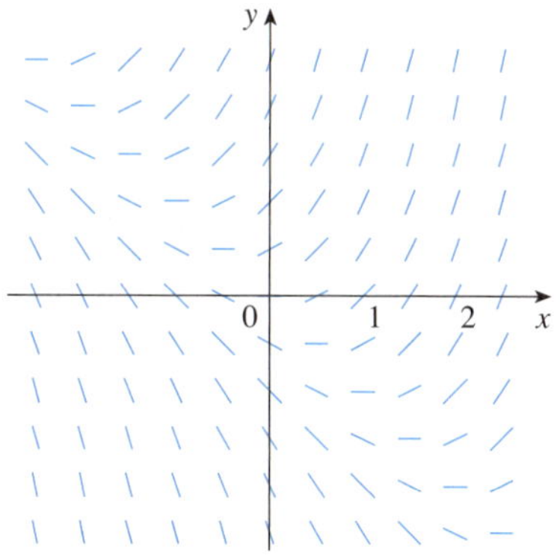

FIGURE 3
Direction field for $y' = x + y$

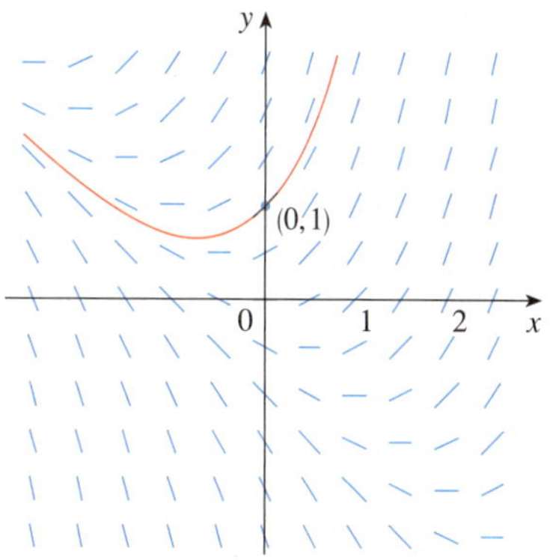

FIGURE 4
The solution curve through $(0, 1)$

Now we can sketch the solution curve through the point $(0, 1)$ by following the direction field as in Figure 4. Notice that we have drawn the curve so that it is parallel to nearby line segments.

In general, suppose we have a first-order differential equation of the form

$$y' = F(x, y)$$

where $F(x, y)$ is some expression in x and y. The differential equation says that the slope of a solution curve at a point (x, y) on the curve is $F(x, y)$. If we draw short line segments with slope $F(x, y)$ at several points (x, y), the result is called a **direction field** (or **slope field**). These line segments indicate the direction in which a solution curve is heading, so the direction field helps us visualize the general shape of these curves.

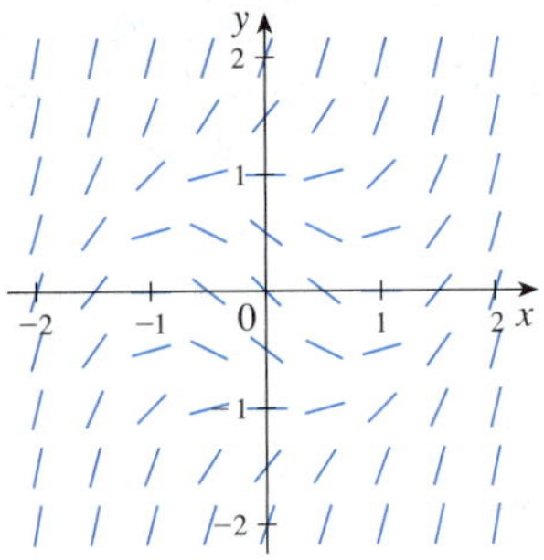

FIGURE 5

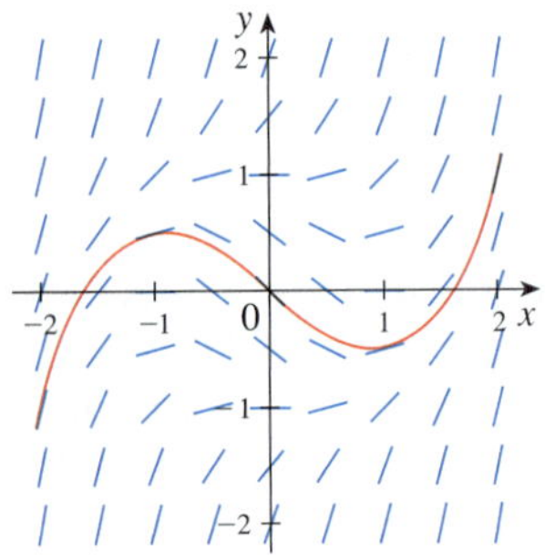

FIGURE 6

Module 9.2A shows direction fields and solution curves for a variety of differential equations.

EXAMPLE 1

(a) Sketch the direction field for the differential equation $y' = x^2 + y^2 - 1$.
(b) Use part (a) to sketch the solution curve that passes through the origin.

SOLUTION

(a) We start by computing the slope at several points in the following chart:

x	-2	-1	0	1	2	-2	-1	0	1	2	...
y	0	0	0	0	0	1	1	1	1	1	...
$y' = x^2 + y^2 - 1$	3	0	-1	0	3	4	1	0	1	4	...

Now we draw short line segments with these slopes at these points. The result is the direction field shown in Figure 5.

(b) We start at the origin and move to the right in the direction of the line segment (which has slope -1). We continue to draw the solution curve so that it moves parallel to the nearby line segments. The resulting solution curve is shown in Figure 6. Returning to the origin, we draw the solution curve to the left as well.

The more line segments we draw in a direction field, the clearer the picture becomes. Of course, it's tedious to compute slopes and draw line segments for a huge number of points by hand, but computers are well suited for this task. Figure 7 shows a more detailed, computer-drawn direction field for the differential equation in Example 1. It enables us to draw, with reasonable accuracy, the solution curves shown in Figure 8 with y-intercepts -2, -1, 0, 1, and 2.

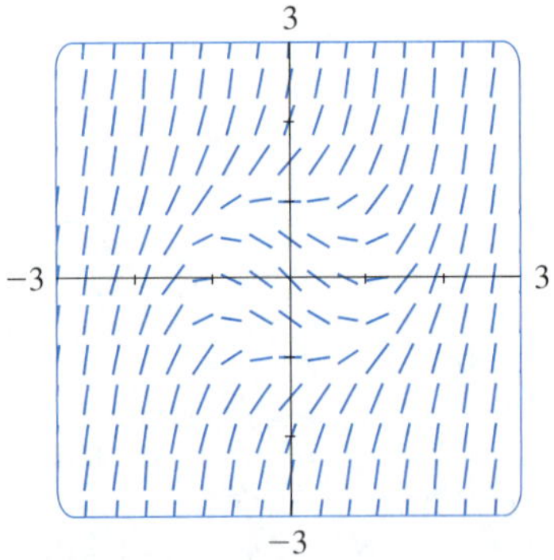

FIGURE 7

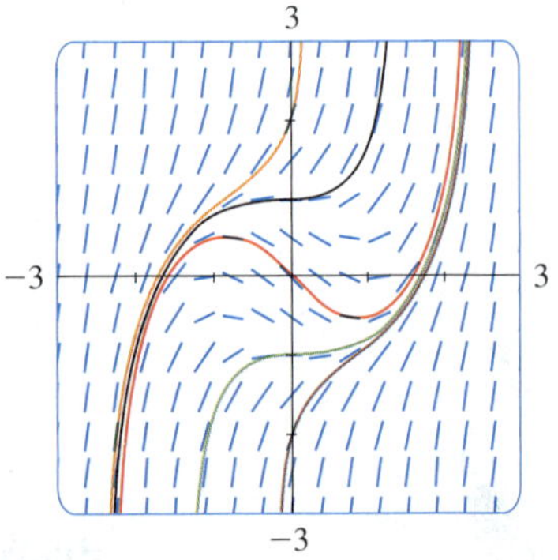

FIGURE 8

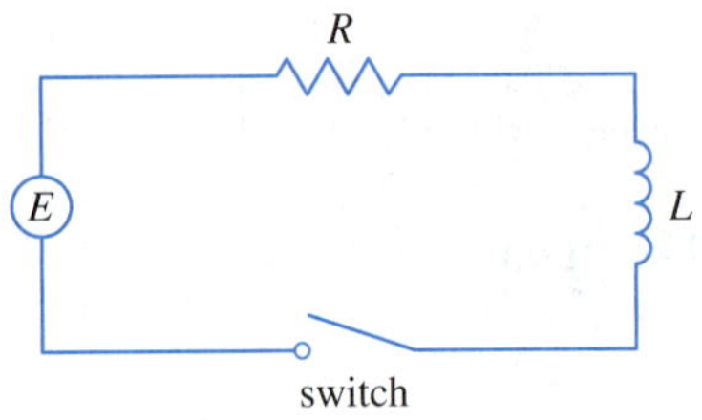

FIGURE 9

Now let's see how direction fields give insight into physical situations. The simple electric circuit shown in Figure 9 contains an electromotive force (usually a battery or generator) that produces a voltage of $E(t)$ volts (V) and a current of $I(t)$ amperes (A) at time t. The circuit also contains a resistor with a resistance of R ohms (Ω) and an inductor with an inductance of L henries (H).

Ohm's Law gives the drop in voltage due to the resistor as RI. The voltage drop due to the inductor is $L(dI/dt)$. One of Kirchhoff's laws says that the sum of the voltage drops is equal to the supplied voltage $E(t)$. Thus, we have

1
$$L\frac{dI}{dt} + RI = E(t)$$

which is a first-order differential equation that models the current I at time t.

EXAMPLE 2 Suppose that in the simple circuit of Figure 9 the resistance is 12 Ω, the inductance is 4 H, and a battery gives a constant voltage of 60 V.

(a) Draw a direction field for Equation 1 with these values.
(b) What can you say about the limiting value of the current?
(c) Identify any equilibrium solutions.
(d) If the switch is closed when $t = 0$ so the current starts with $I(0) = 0$, use the direction field to sketch the solution curve.

SOLUTION

(a) If we put $L = 4$, $R = 12$, and $E(t) = 60$ in Equation 1, we get

$$4\frac{dI}{dt} + 12I = 60 \qquad \text{or} \qquad \frac{dI}{dt} = 15 - 3I$$

The direction field for this differential equation is shown in Figure 10.

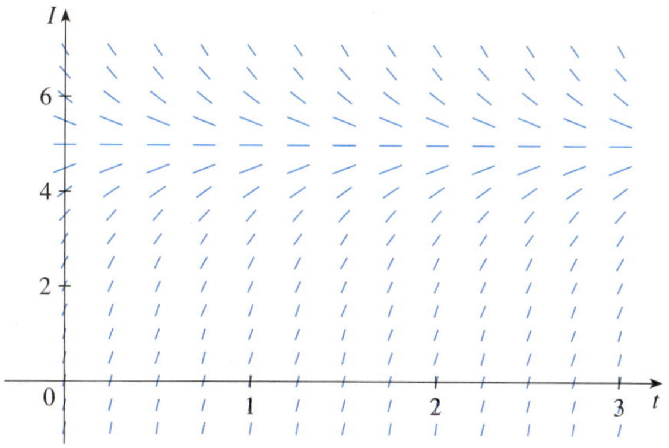

FIGURE 10

(b) It appears from the direction field that all solutions approach the value 5 A, that is,

$$\lim_{t\to\infty} I(t) = 5$$

$$\frac{dI}{dt} = 15 - 3I$$

(c) It appears that the constant function $I(t) = 5$ is an equilibrium solution. Indeed, we can verify this directly from the differential equation. If $I(t) = 5$, then the left side is $dI/dt = 0$ and the right side is $15 - 3(5) = 0$.

(d) We use the direction field to sketch the solution curve that passes through (0, 0), as shown in red in Figure 11.

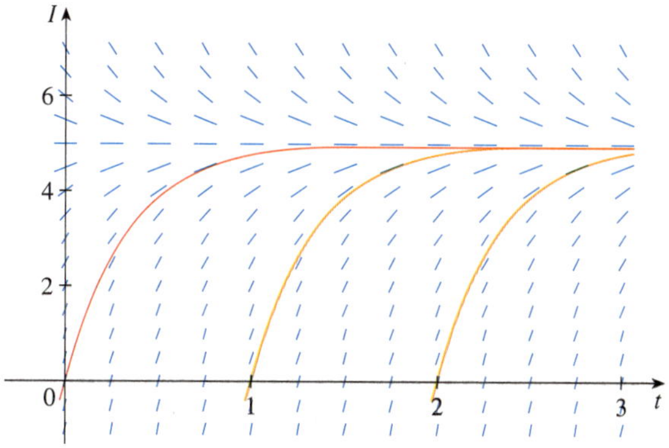

FIGURE 11

Notice from Figure 10 that the line segments along any horizontal line are parallel. That is because the independent variable t does not occur on the right side of the equation $I' = 15 - 3I$. In general, a differential equation of the form

$$y' = f(y)$$

in which the independent variable is missing from the right side, is called **autonomous**. For such an equation, the slopes corresponding to two different points with the same y-coordinate must be equal. This means that if we know one solution to an autonomous differential equation, then we can obtain infinitely many others just by shifting the graph of the known solution to the right or left. In Figure 11 we have shown the solutions that result from shifting the solution curve of Example 2 one and two time units (namely, seconds) to the right. They correspond to closing the switch when $t = 1$ or $t = 2$.

Euler's Method

The basic idea behind direction fields can be used to find numerical approximations to solutions of differential equations. We illustrate the method on the initial-value problem that we used to introduce direction fields:

$$y' = x + y \qquad y(0) = 1$$

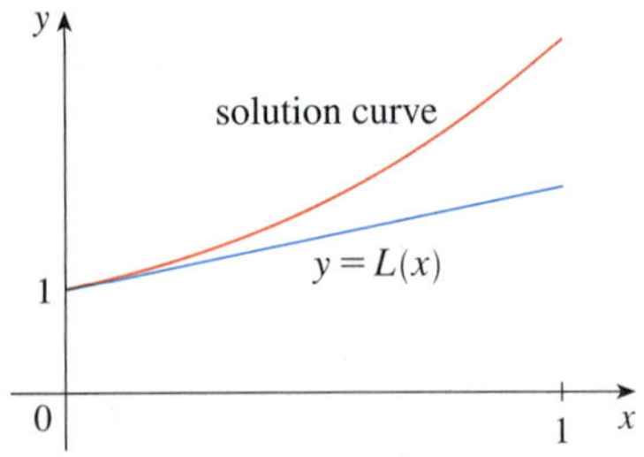

FIGURE 12
First Euler approximation

The differential equation tells us that $y'(0) = 0 + 1 = 1$, so the solution curve has slope 1 at the point $(0, 1)$. As a first approximation to the solution we could use the linear approximation $L(x) = x + 1$. In other words, we could use the tangent line at $(0, 1)$ as a rough approximation to the solution curve (see Figure 12).

Euler's idea was to improve on this approximation by proceeding only a short distance along this tangent line and then making a midcourse correction by changing direction as indicated by the direction field. Figure 13 shows what happens if we start out along the tangent line but stop when $x = 0.5$. (This horizontal distance traveled is called the *step size.*) Since $L(0.5) = 1.5$, we have $y(0.5) \approx 1.5$ and we take $(0.5, 1.5)$ as the starting point for a new line segment. The differential equation tells us that $y'(0.5) = 0.5 + 1.5 = 2$, so we use the linear function

$$y = 1.5 + 2(x - 0.5) = 2x + 0.5$$

as an approximation to the solution for $x > 0.5$ (the orange segment in Figure 13). If we decrease the step size from 0.5 to 0.25, we get the better Euler approximation shown in Figure 14.

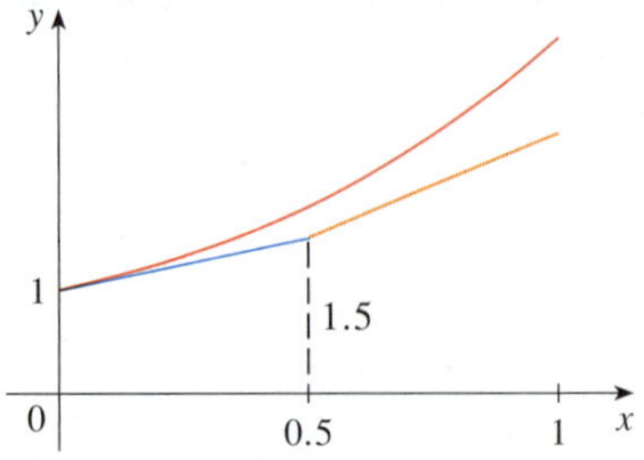

FIGURE 13
Euler approximation with step size 0.5

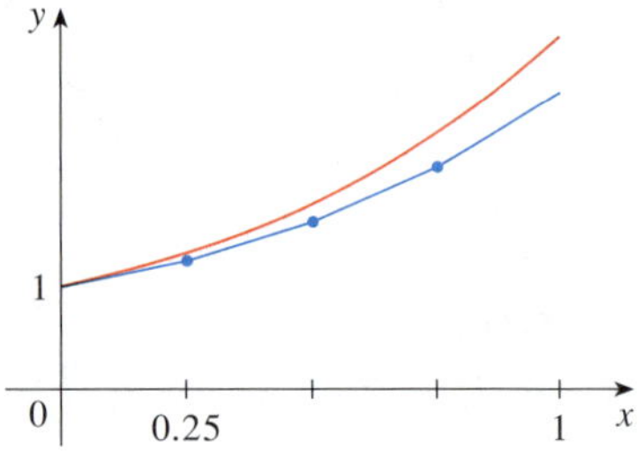

FIGURE 14
Euler approximation with step size 0.25

In general, Euler's method says to start at the point given by the initial value and proceed in the direction indicated by the direction field. Stop after a short time, look at the slope at the new location, and proceed in that direction. Keep stopping and changing direction according to the direction field. Euler's method does not produce the exact solution to an initial-value problem—it gives approximations. But by decreasing the step size (and therefore increasing the number of midcourse corrections), we obtain successively better approximations to the exact solution. (Compare Figures 12, 13, and 14.)

For the general first-order initial-value problem $y' = F(x, y)$, $y(x_0) = y_0$, our aim is to find approximate values for the solution at equally spaced numbers x_0, $x_1 = x_0 + h$, $x_2 = x_1 + h, \ldots$, where h is the step size. The differential equation tells us that the slope at (x_0, y_0) is $y' = F(x_0, y_0)$, so Figure 15 shows that the approximate value of the solution when $x = x_1$ is

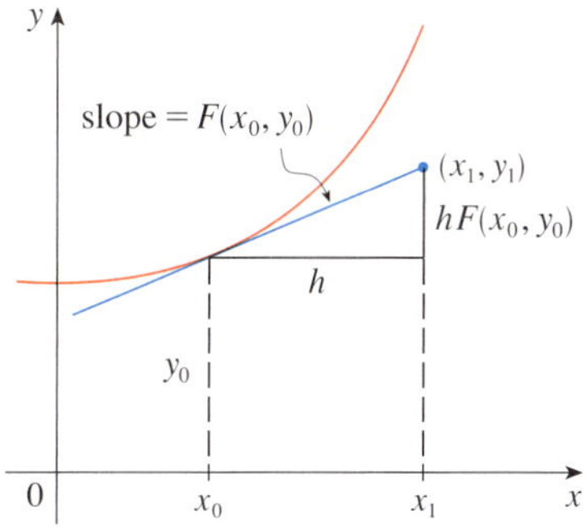

FIGURE 15

$$y_1 = y_0 + hF(x_0, y_0)$$

Similarly, $$y_2 = y_1 + hF(x_1, y_1)$$

In general, $$y_n = y_{n-1} + hF(x_{n-1}, y_{n-1})$$

EXAMPLE 3 Use Euler's method with step size 0.1 to construct a table of approximate values for the solution of the initial-value problem

$$y' = x + y \qquad y(0) = 1$$

SOLUTION We are given that $h = 0.1$, $x_0 = 0$, $y_0 = 1$, and $F(x, y) = x + y$. So we have

$$y_1 = y_0 + hF(x_0, y_0) = 1 + 0.1(0 + 1) = 1.1$$

$$y_2 = y_1 + hF(x_1, y_1) = 1.1 + 0.1(0.1 + 1.1) = 1.22$$

$$y_3 = y_2 + hF(x_2, y_2) = 1.22 + 0.1(0.2 + 1.22) = 1.362$$

This means that if $y(x)$ is the exact solution, then $y(0.3) \approx 1.362$.

Proceeding with similar calculations, we get the values in the table:

Module 9.2B shows how Euler's method works numerically and visually for a variety of differential equations and step sizes.

n	x_n	y_n	n	x_n	y_n
1	0.1	1.100000	6	0.6	1.943122
2	0.2	1.220000	7	0.7	2.197434
3	0.3	1.362000	8	0.8	2.487178
4	0.4	1.528200	9	0.9	2.815895
5	0.5	1.721020	10	1.0	3.187485

For a more accurate table of values in Example 3 we could decrease the step size. But for a large number of small steps the amount of computation is considerable and so we need to program a calculator or computer to carry out these calculations. The following table shows the results of applying Euler's method with decreasing step size to the initial-value problem of Example 3.

Step size	Euler estimate of $y(0.5)$	Euler estimate of $y(1)$
0.500	1.500000	2.500000
0.250	1.625000	2.882813
0.100	1.721020	3.187485
0.050	1.757789	3.306595
0.020	1.781212	3.383176
0.010	1.789264	3.409628
0.005	1.793337	3.423034
0.001	1.796619	3.433848

|||| Computer software packages that produce numerical approximations to solutions of differential equations use methods that are refinements of Euler's method. Although Euler's method is simple and not as accurate, it is the basic idea on which the more accurate methods are based.

Notice that the Euler estimates in the table seem to be approaching limits, namely, the true values of $y(0.5)$ and $y(1)$. Figure 16 shows graphs of the Euler approximations with step sizes 0.5, 0.25, 0.1, 0.05, 0.02, 0.01, and 0.005. They are approaching the exact solution curve as the step size h approaches 0.

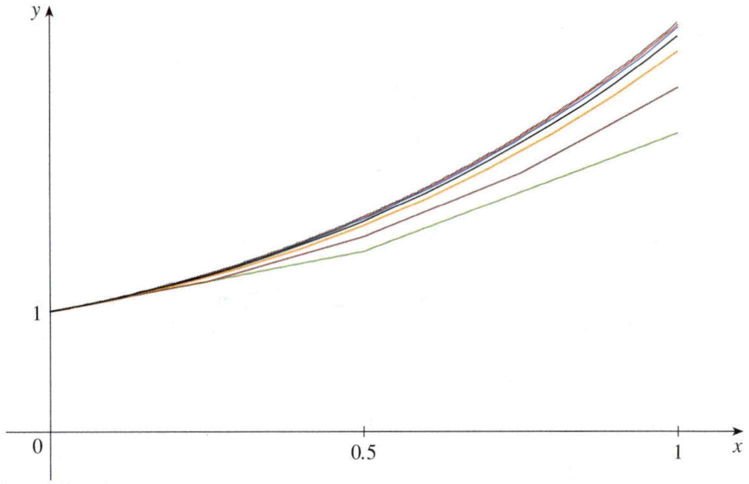

FIGURE 16
Euler approximations approaching the exact solution

EXAMPLE 4 In Example 2 we discussed a simple electric circuit with resistance 12 Ω, inductance 4 H, and a battery with voltage 60 V. If the switch is closed when $t = 0$, we modeled the current I at time t by the initial-value problem

$$\frac{dI}{dt} = 15 - 3I \qquad I(0) = 0$$

Estimate the current in the circuit half a second after the switch is closed.

SOLUTION We use Euler's method with $F(t, I) = 15 - 3I$, $t_0 = 0$, $I_0 = 0$, and step size $h = 0.1$ second:

$$I_1 = 0 + 0.1(15 - 3 \cdot 0) = 1.5$$

$$I_2 = 1.5 + 0.1(15 - 3 \cdot 1.5) = 2.55$$

$$I_3 = 2.55 + 0.1(15 - 3 \cdot 2.55) = 3.285$$

$$I_4 = 3.285 + 0.1(15 - 3 \cdot 3.285) = 3.7995$$

$$I_5 = 3.7995 + 0.1(15 - 3 \cdot 3.7995) = 4.15965$$

So the current after 0.5 s is

$$I(0.5) \approx 4.16 \text{ A}$$

9.2 Exercises

1. A direction field for the differential equation $y' = y(1 - \frac{1}{4}y^2)$ is shown.

(a) Sketch the graphs of the solutions that satisfy the given initial conditions.

(i) $y(0) = 1$ (ii) $y(0) = -1$

(iii) $y(0) = -3$ (iv) $y(0) = 3$

(b) Find all the equilibrium solutions.

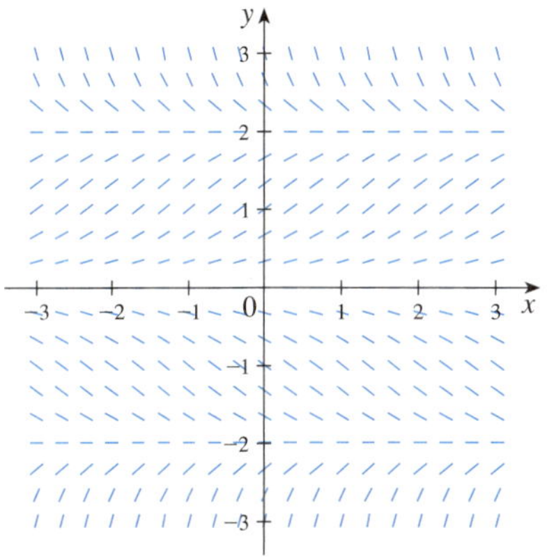

2. A direction field for the differential equation $y' = x \sin y$ is shown.

(a) Sketch the graphs of the solutions that satisfy the given initial conditions.

(i) $y(0) = 1$ (ii) $y(0) = 2$ (iii) $y(0) = \pi$

(iv) $y(0) = 4$ (v) $y(0) = 5$

(b) Find all the equilibrium solutions.

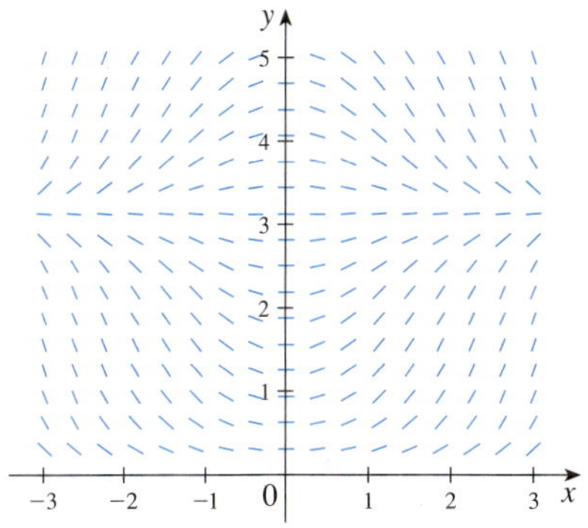

3–6 Match the differential equation with its direction field (labeled I–IV). Give reasons for your answer.

3. $y' = y - 1$ **4.** $y' = y - x$

5. $y' = y^2 - x^2$ **6.** $y' = y^3 - x^3$

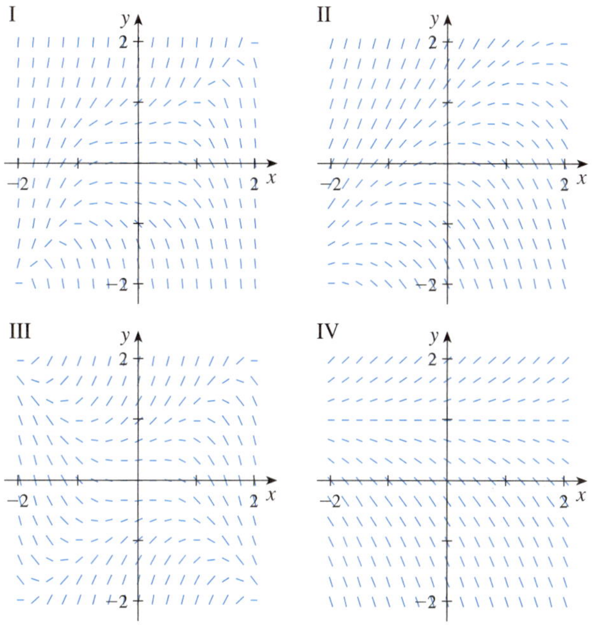

7. Use the direction field labeled I (for Exercises 3–6) to sketch the graphs of the solutions that satisfy the given initial conditions.

(a) $y(0) = 1$ (b) $y(0) = 0$ (c) $y(0) = -1$

8. Repeat Exercise 7 for the direction field labeled III.

9–10 Sketch a direction field for the differential equation. Then use it to sketch three solution curves.

9. $y' = 1 + y$ **10.** $y' = x^2 - y^2$

11–14 Sketch the direction field of the differential equation. Then use it to sketch a solution curve that passes through the given point.

11. $y' = y - 2x$, $(1, 0)$ **12.** $y' = 1 - xy$, $(0, 0)$

13. $y' = y + xy$, $(0, 1)$ **14.** $y' = x - xy$, $(1, 0)$

CAS **15–16** Use a computer algebra system to draw a direction field for the given differential equation. Get a printout and sketch on it the solution curve that passes through $(0, 1)$. Then use the CAS to draw the solution curve and compare it with your sketch.

15. $y' = y \sin 2x$ **16.** $y' = \sin(x + y)$

CAS **17.** Use a computer algebra system to draw a direction field for the differential equation $y' = y^3 - 4y$. Get a printout and sketch on it solutions that satisfy the initial condition $y(0) = c$ for various values of c. For what values of c does $\lim_{t \to \infty} y(t)$ exist? What are the possible values for this limit?

18. Make a rough sketch of a direction field for the autonomous differential equation $y' = f(y)$, where the graph of f is as shown. How does the limiting behavior of solutions depend on the value of $y(0)$?

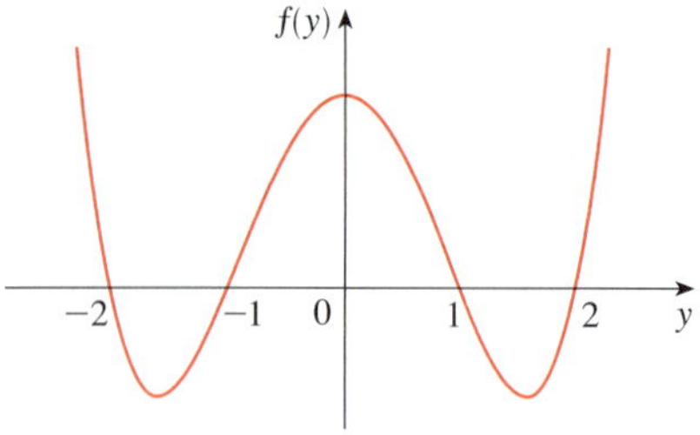

19. (a) Use Euler's method with each of the following step sizes to estimate the value of $y(0.4)$, where y is the solution of the initial-value problem $y' = y$, $y(0) = 1$.
(i) $h = 0.4$ (ii) $h = 0.2$ (iii) $h = 0.1$
(b) We know that the exact solution of the initial-value problem in part (a) is $y = e^x$. Draw, as accurately as you can, the graph of $y = e^x$, $0 \leq x \leq 0.4$, together with the Euler approximations using the step sizes in part (a). (Your sketches should resemble Figures 12, 13, and 14.) Use your sketches to decide whether your estimates in part (a) are underestimates or overestimates.
(c) The error in Euler's method is the difference between the exact value and the approximate value. Find the errors made in part (a) in using Euler's method to estimate the true value of $y(0.4)$, namely $e^{0.4}$. What happens to the error each time the step size is halved?

20. A direction field for a differential equation is shown. Draw, with a ruler, the graphs of the Euler approximations to the solution curve that passes through the origin. Use step sizes $h = 1$ and $h = 0.5$. Will the Euler estimates be underestimates or overestimates? Explain.

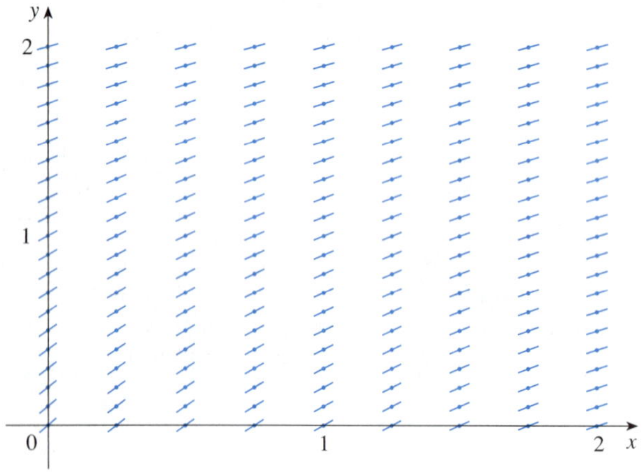

21. Use Euler's method with step size 0.5 to compute the approximate y-values y_1, y_2, y_3, and y_4 of the solution of the initial-value problem $y' = y - 2x$, $y(1) = 0$.

22. Use Euler's method with step size 0.2 to estimate $y(1)$, where $y(x)$ is the solution of the initial-value problem $y' = 1 - xy$, $y(0) = 0$.

23. Use Euler's method with step size 0.1 to estimate $y(0.5)$, where $y(x)$ is the solution of the initial-value problem $y' = y + xy$, $y(0) = 1$.

24. (a) Use Euler's method with step size 0.2 to estimate $y(1.4)$, where $y(x)$ is the solution of the initial-value problem $y' = x - xy$, $y(1) = 0$.
(b) Repeat part (a) with step size 0.1.

25. (a) Program a calculator or computer to use Euler's method to compute $y(1)$, where $y(x)$ is the solution of the initial-value problem

$$\frac{dy}{dx} + 3x^2y = 6x^2 \qquad y(0) = 3$$

(i) $h = 1$ (ii) $h = 0.1$
(iii) $h = 0.01$ (iv) $h = 0.001$
(b) Verify that $y = 2 + e^{-x^3}$ is the exact solution of the differential equation.
(c) Find the errors in using Euler's method to compute $y(1)$ with the step sizes in part (a). What happens to the error when the step size is divided by 10?

CAS **26.** (a) Program your computer algebra system, using Euler's method with step size 0.01, to calculate $y(2)$, where y is the solution of the initial-value problem

$$y' = x^3 - y^3 \qquad y(0) = 1$$

(b) Check your work by using the CAS to draw the solution curve.

27. The figure shows a circuit containing an electromotive force, a capacitor with a capacitance of C farads (F), and a resistor with a resistance of R ohms (Ω). The voltage drop across the capacitor is Q/C, where Q is the charge (in coulombs), so in this case Kirchhoff's Law gives

$$RI + \frac{Q}{C} = E(t)$$

But $I = dQ/dt$, so we have

$$R\frac{dQ}{dt} + \frac{1}{C}Q = E(t)$$

Suppose the resistance is 5 Ω, the capacitance is 0.05 F, and a battery gives a constant voltage of 60 V.
(a) Draw a direction field for this differential equation.
(b) What is the limiting value of the charge?

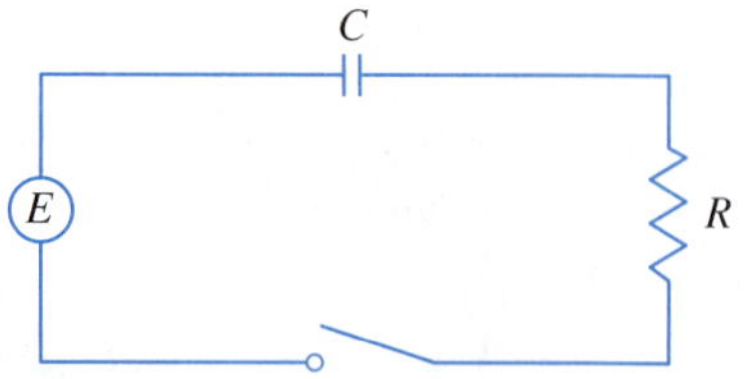

(c) Is there an equilibrium solution?

(d) If the initial charge is $Q(0) = 0$ C, use the direction field to sketch the solution curve.

(e) If the initial charge is $Q(0) = 0$ C, use Euler's method with step size 0.1 to estimate the charge after half a second.

28. In Exercise 14 in Section 9.1 we considered a 95°C cup of coffee in a 20°C room. Suppose it is known that the coffee cools at a rate of 1°C per minute when its temperature is 70°C.

(a) What does the differential equation become in this case?

(b) Sketch a direction field and use it to sketch the solution curve for the initial-value problem. What is the limiting value of the temperature?

(c) Use Euler's method with step size $h = 2$ minutes to estimate the temperature of the coffee after 10 minutes.

9.3 Separable Equations

We have looked at first-order differential equations from a geometric point of view (direction fields) and from a numerical point of view (Euler's method). What about the symbolic point of view? It would be nice to have an explicit formula for a solution of a differential equation. Unfortunately, that is not always possible. But in this section we examine a certain type of differential equation that *can* be solved explicitly.

A **separable equation** is a first-order differential equation in which the expression for dy/dx can be factored as a function of x times a function of y. In other words, it can be written in the form

$$\frac{dy}{dx} = g(x)f(y)$$

The name *separable* comes from the fact that the expression on the right side can be "separated" into a function of x and a function of y. Equivalently, if $f(y) \neq 0$, we could write

1 $$\frac{dy}{dx} = \frac{g(x)}{h(y)}$$

where $h(y) = 1/f(y)$. To solve this equation we rewrite it in the differential form

$$h(y)\,dy = g(x)\,dx$$

so that all y's are on one side of the equation and all x's are on the other side. Then we integrate both sides of the equation:

|||| The technique for solving separable differential equations was first used by James Bernoulli (in 1690) in solving a problem about pendulums and by Leibniz (in a letter to Huygens in 1691). John Bernoulli explained the general method in a paper published in 1694.

2 $$\int h(y)\,dy = \int g(x)\,dx$$

Equation 2 defines y implicitly as a function of x. In some cases we may be able to solve for y in terms of x.

We use the Chain Rule to justify this procedure: If h and g satisfy (2), then

$$\frac{d}{dx}\left(\int h(y)\,dy\right) = \frac{d}{dx}\left(\int g(x)\,dx\right)$$

so

$$\frac{d}{dy}\left(\int h(y)\,dy\right)\frac{dy}{dx} = g(x)$$

and

$$h(y)\frac{dy}{dx} = g(x)$$

Thus, Equation 1 is satisfied.

EXAMPLE 1

(a) Solve the differential equation $\dfrac{dy}{dx} = \dfrac{x^2}{y^2}$.

(b) Find the solution of this equation that satisfies the initial condition $y(0) = 2$.

Figure 1 shows graphs of several members of the family of solutions of the differential equation in Example 1. The solution of the initial-value problem in part (b) is shown in red.

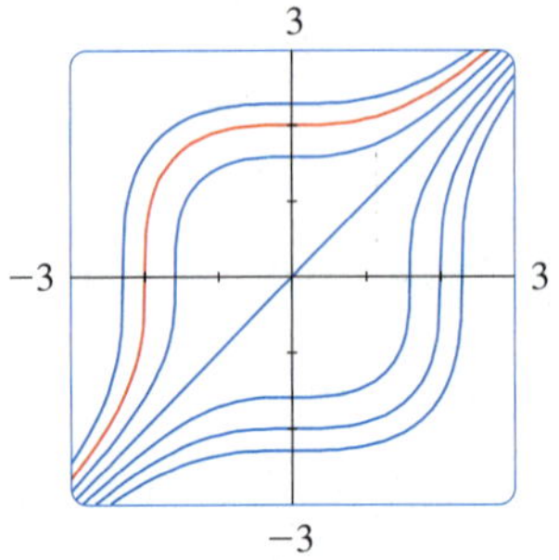

FIGURE 1

SOLUTION

(a) We write the equation in terms of differentials and integrate both sides:

$$y^2 dy = x^2 dx$$

$$\int y^2 dy = \int x^2 dx$$

$$\tfrac{1}{3}y^3 = \tfrac{1}{3}x^3 + C$$

where C is an arbitrary constant. (We could have used a constant C_1 on the left side and another constant C_2 on the right side. But then we could combine these constants by writing $C = C_2 - C_1$.)

Solving for y, we get

$$y = \sqrt[3]{x^3 + 3C}$$

We could leave the solution like this or we could write it in the form

$$y = \sqrt[3]{x^3 + K}$$

where $K = 3C$. (Since C is an arbitrary constant, so is K.)

(b) If we put $x = 0$ in the general solution in part (a), we get $y(0) = \sqrt[3]{K}$. To satisfy the initial condition $y(0) = 2$, we must have $\sqrt[3]{K} = 2$ and so $K = 8$.

Thus, the solution of the initial-value problem is

$$y = \sqrt[3]{x^3 + 8}$$

EXAMPLE 2 Solve the differential equation $\dfrac{dy}{dx} = \dfrac{6x^2}{2y + \cos y}$.

Some computer algebra systems can plot curves defined by implicit equations. Figure 2 shows the graphs of several members of the family of solutions of the differential equation in Example 2. As we look at the curves from left to right, the values of C are 3, 2, 1, 0, -1, -2, and -3.

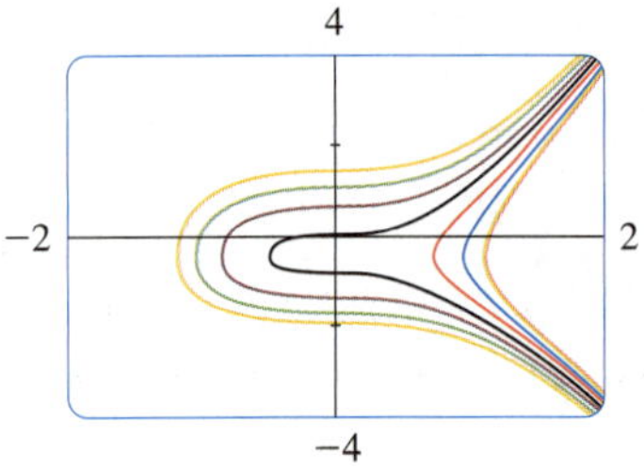

FIGURE 2

SOLUTION Writing the equation in differential form and integrating both sides, we have

$$(2y + \cos y)dy = 6x^2 dx$$

$$\int (2y + \cos y)dy = \int 6x^2 dx$$

$$y^2 + \sin y = 2x^3 + C \qquad \boxed{3}$$

where C is a constant. Equation 3 gives the general solution implicitly. In this case it's impossible to solve the equation to express y explicitly as a function of x.

EXAMPLE 3 Solve the equation $y' = x^2y$.

SOLUTION First we rewrite the equation using Leibniz notation:

$$\frac{dy}{dx} = x^2y$$

|||| If a solution y is a function that satisfies $y(x) \neq 0$ for some x, it follows from a uniqueness theorem for solutions of differential equations that $y(x) \neq 0$ for all x.

If $y \neq 0$, we can rewrite it in differential notation and integrate:

$$\frac{dy}{y} = x^2\,dx \qquad y \neq 0$$

$$\int \frac{dy}{y} = \int x^2\,dx$$

$$\ln|y| = \frac{x^3}{3} + C$$

This equation defines y implicitly as a function of x. But in this case we can solve explicitly for y as follows:

$$|y| = e^{\ln|y|} = e^{(x^3/3)+C} = e^C e^{x^3/3}$$

so

$$y = \pm e^C e^{x^3/3}$$

We can easily verify that the function $y = 0$ is also a solution of the given differential equation. So we can write the general solution in the form

$$y = Ae^{x^3/3}$$

where A is an arbitrary constant ($A = e^C$, or $A = -e^C$, or $A = 0$).

|||| Figure 3 shows a direction field for the differential equation in Example 3. Compare it with Figure 4, in which we use the equation $y = Ae^{x^3/3}$ to graph solutions for several values of A. If you use the direction field to sketch solution curves with y-intercepts 5, 2, 1, -1, and -2, they will resemble the curves in Figure 4.

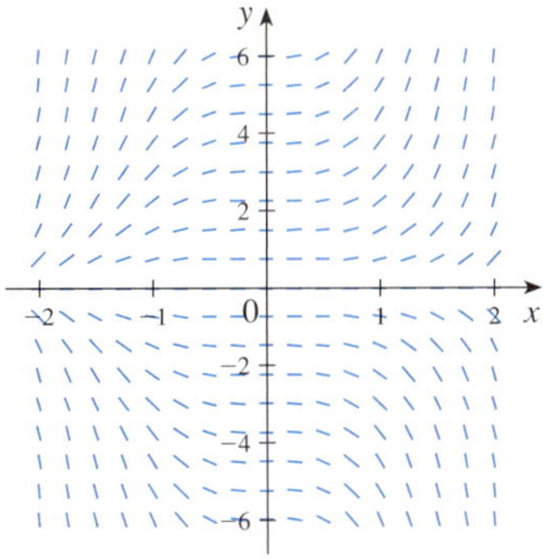

FIGURE 3

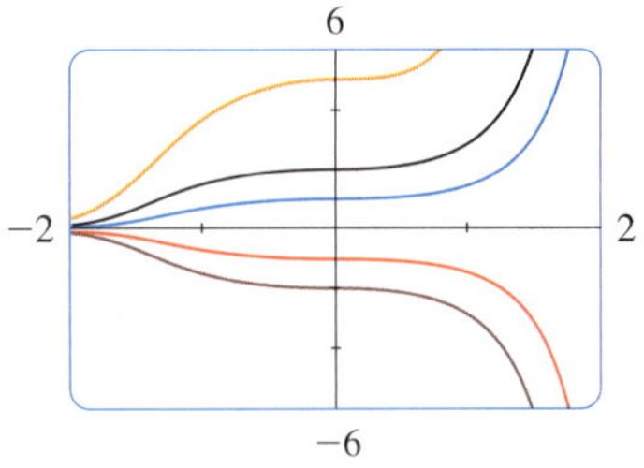

FIGURE 4

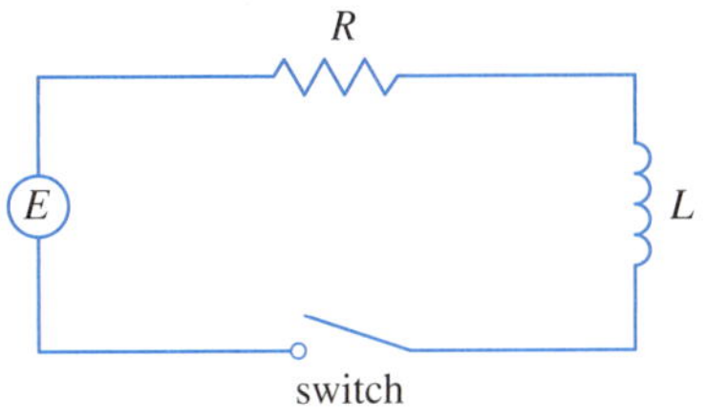

FIGURE 5

EXAMPLE 4 In Section 9.2 we modeled the current $I(t)$ in the electric circuit shown in Figure 5 by the differential equation

$$L\frac{dI}{dt} + RI = E(t)$$

Find an expression for the current in a circuit where the resistance is 12 Ω, the inductance is 4 H, a battery gives a constant voltage of 60 V, and the switch is turned on when $t = 0$. What is the limiting value of the current?

SOLUTION With $L = 4$, $R = 12$, and $E(t) = 60$, the equation becomes

$$4\frac{dI}{dt} + 12I = 60$$

or
$$\frac{dI}{dt} = 15 - 3I$$

and the initial-value problem is

$$\frac{dI}{dt} = 15 - 3I \qquad I(0) = 0$$

We recognize this equation as being separable, and we solve it as follows:

$$\int \frac{dI}{15 - 3I} = \int dt \qquad (15 - 3I \neq 0)$$

$$-\tfrac{1}{3}\ln|15 - 3I| = t + C$$

$$|15 - 3I| = e^{-3(t+C)}$$

$$15 - 3I = \pm e^{-3C}e^{-3t} = Ae^{-3t}$$

$$I = 5 - \tfrac{1}{3}Ae^{-3t}$$

|||| Figure 6 shows how the solution in Example 4 (the current) approaches its limiting value. Comparison with Figure 11 in Section 9.2 shows that we were able to draw a fairly accurate solution curve from the direction field.

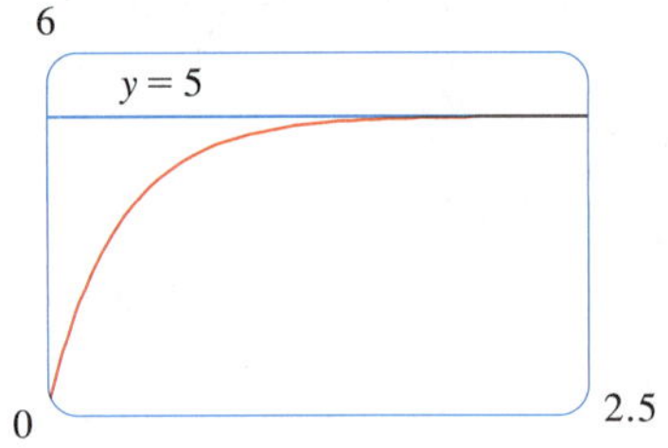

FIGURE 6

Since $I(0) = 0$, we have $5 - \tfrac{1}{3}A = 0$, so $A = 15$ and the solution is

$$I(t) = 5 - 5e^{-3t}$$

The limiting current, in amperes, is

$$\lim_{t\to\infty} I(t) = \lim_{t\to\infty} (5 - 5e^{-3t}) = 5 - 5\lim_{t\to\infty} e^{-3t}$$

$$= 5 - 0 = 5$$

|||| Orthogonal Trajectories

An **orthogonal trajectory** of a family of curves is a curve that intersects each curve of the family orthogonally, that is, at right angles (see Figure 7). For instance, each member of the family $y = mx$ of straight lines through the origin is an orthogonal trajectory of the family $x^2 + y^2 = r^2$ of concentric circles with center the origin (see Figure 8). We say that the two families are orthogonal trajectories of each other.

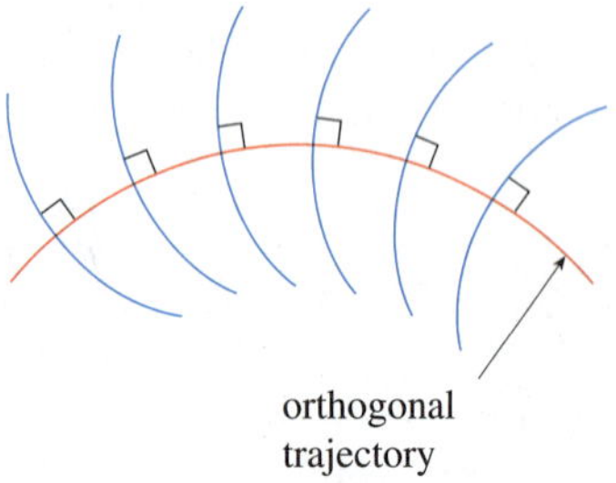

FIGURE 7

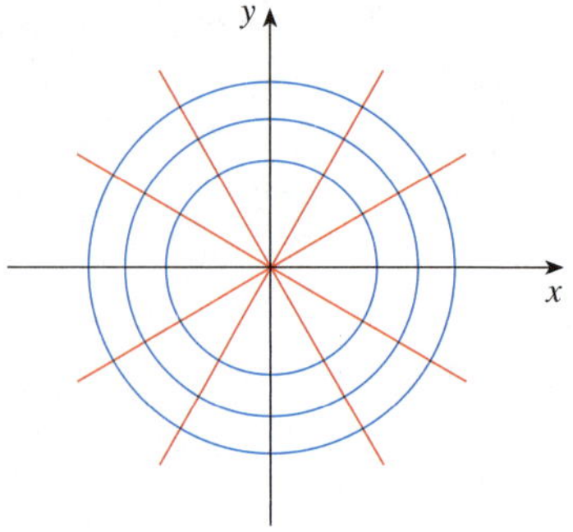

FIGURE 8

EXAMPLE 5 Find the orthogonal trajectories of the family of curves $x = ky^2$, where k is an arbitrary constant.

SOLUTION The curves $x = ky^2$ form a family of parabolas whose axis of symmetry is the x-axis. The first step is to find a single differential equation that is satisfied by all members of the family. If we differentiate $x = ky^2$, we get

$$1 = 2ky\frac{dy}{dx} \qquad \text{or} \qquad \frac{dy}{dx} = \frac{1}{2ky}$$

This differential equation depends on k, but we need an equation that is valid for all values of k simultaneously. To eliminate k we note that, from the equation of the given general parabola $x = ky^2$, we have $k = x/y^2$ and so the differential equation can be written as

$$\frac{dy}{dx} = \frac{1}{2ky} = \frac{1}{2\dfrac{x}{y^2}y}$$

or

$$\frac{dy}{dx} = \frac{y}{2x}$$

This means that the slope of the tangent line at any point (x, y) on one of the parabolas is $y' = y/(2x)$. On an orthogonal trajectory the slope of the tangent line must be the negative reciprocal of this slope. Therefore, the orthogonal trajectories must satisfy the differential equation

$$\frac{dy}{dx} = -\frac{2x}{y}$$

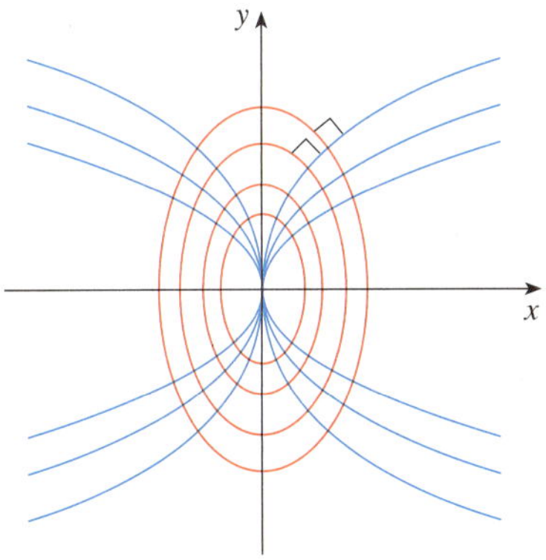

FIGURE 9

This differential equation is separable, and we solve it as follows:

$$\int y\,dy = -\int 2x\,dx$$

$$\frac{y^2}{2} = -x^2 + C$$

4 $$x^2 + \frac{y^2}{2} = C$$

where C is an arbitrary positive constant. Thus, the orthogonal trajectories are the family of ellipses given by Equation 4 and sketched in Figure 9.

Orthogonal trajectories occur in various branches of physics. For example, in an electrostatic field the lines of force are orthogonal to the lines of constant potential. Also, the streamlines in aerodynamics are orthogonal trajectories of the velocity-equipotential curves.

Mixing Problems

A typical mixing problem involves a tank of fixed capacity filled with a thoroughly mixed solution of some substance, such as salt. A solution of a given concentration enters the tank at a fixed rate and the mixture, thoroughly stirred, leaves at a fixed rate, which may differ from the entering rate. If $y(t)$ denotes the amount of substance in the tank at time t, then $y'(t)$ is the rate at which the substance is being added minus the rate at which it is being removed. The mathematical description of this situation often leads to a first-order separable differential equation. We can use the same type of reasoning to model a variety of phenomena: chemical reactions, discharge of pollutants into a lake, injection of a drug into the bloodstream.

EXAMPLE 6 A tank contains 20 kg of salt dissolved in 5000 L of water. Brine that contains 0.03 kg of salt per liter of water enters the tank at a rate of 25 L/min. The solution is kept thoroughly mixed and drains from the tank at the same rate. How much salt remains in the tank after half an hour?

SOLUTION Let $y(t)$ be the amount of salt (in kilograms) after t minutes. We are given that $y(0) = 20$ and we want to find $y(30)$. We do this by finding a differential equation satisfied by $y(t)$. Note that dy/dt is the rate of change of the amount of salt, so

$$\boxed{5} \qquad \frac{dy}{dt} = (\text{rate in}) - (\text{rate out})$$

where (rate in) is the rate at which salt enters the tank and (rate out) is the rate at which salt leaves the tank. We have

$$\text{rate in} = \left(0.03\,\frac{\text{kg}}{\text{L}}\right)\left(25\,\frac{\text{L}}{\text{min}}\right) = 0.75\,\frac{\text{kg}}{\text{min}}$$

The tank always contains 5000 L of liquid, so the concentration at time t is $y(t)/5000$ (measured in kilograms per liter). Since the brine flows out at a rate of 25 L/min, we have

$$\text{rate out} = \left(\frac{y(t)}{5000}\,\frac{\text{kg}}{\text{L}}\right)\left(25\,\frac{\text{L}}{\text{min}}\right) = \frac{y(t)}{200}\,\frac{\text{kg}}{\text{min}}$$

Thus, from Equation 5 we get

$$\frac{dy}{dt} = 0.75 - \frac{y(t)}{200} = \frac{150 - y(t)}{200}$$

Solving this separable differential equation, we obtain

$$\int \frac{dy}{150 - y} = \int \frac{dt}{200}$$

$$-\ln|150 - y| = \frac{t}{200} + C$$

Since $y(0) = 20$, we have $-\ln 130 = C$, so

$$-\ln|150 - y| = \frac{t}{200} - \ln 130$$

Therefore $$|150 - y| = 130e^{-t/200}$$

Since $y(t)$ is continuous and $y(0) = 20$ and the right side is never 0, we deduce that $150 - y(t)$ is always positive. Thus, $|150 - y| = 150 - y$ and so

$$y(t) = 150 - 130e^{-t/200}$$

The amount of salt after 30 min is

$$y(30) = 150 - 130e^{-30/200} \approx 38.1 \text{ kg}$$

|||| Figure 10 shows the graph of the function $y(t)$ of Example 6. Notice that, as time goes by, the amount of salt approaches 150 kg.

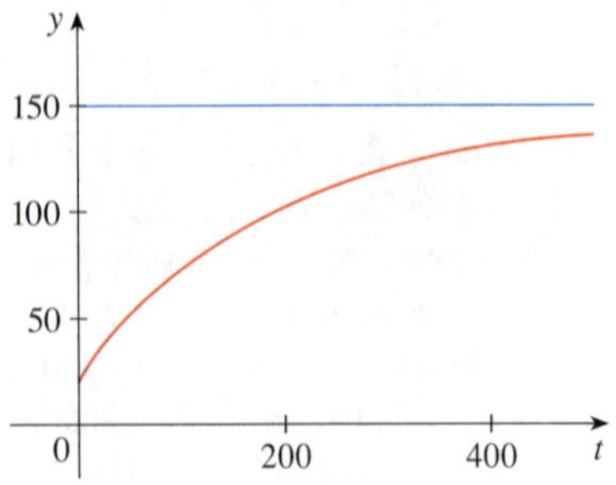

FIGURE 10

9.3 Exercises

1–10 Solve the differential equation.

1. $\dfrac{dy}{dx} = \dfrac{y}{x}$

2. $\dfrac{dy}{dx} = \dfrac{e^{2x}}{4y^3}$

3. $(x^2 + 1)y' = xy$

4. $y' = y^2 \sin x$

5. $(1 + \tan y)y' = x^2 + 1$

6. $\dfrac{du}{dr} = \dfrac{1 + \sqrt{r}}{1 + \sqrt{u}}$

7. $\dfrac{dy}{dt} = \dfrac{te^t}{y\sqrt{1 + y^2}}$

8. $y' = \dfrac{xy}{2 \ln y}$

9. $\dfrac{du}{dt} = 2 + 2u + t + tu$

10. $\dfrac{dz}{dt} + e^{t+z} = 0$

11–18 Find the solution of the differential equation that satisfies the given initial condition.

11. $\dfrac{dy}{dx} = y^2 + 1, \quad y(1) = 0$

12. $\dfrac{dy}{dx} = \dfrac{y \cos x}{1 + y^2}, \quad y(0) = 1$

13. $x \cos x = (2y + e^{3y})y', \quad y(0) = 0$

14. $\dfrac{dP}{dt} = \sqrt{Pt}, \quad P(1) = 2$

15. $\dfrac{du}{dt} = \dfrac{2t + \sec^2 t}{2u}, \quad u(0) = -5$

16. $\dfrac{dy}{dt} = te^y, \quad y(1) = 0$

17. $y' \tan x = a + y, \ \ y(\pi/3) = a, \quad 0 < x < \pi/2$

18. $xy' + y = y^2, \quad y(1) = -1$

19. Find an equation of the curve that satisfies $dy/dx = 4x^3y$ and whose y-intercept is 7.

20. Find an equation of the curve that passes through the point (1, 1) and whose slope at (x, y) is y^2/x^3.

21. (a) Solve the differential equation $y' = 2x\sqrt{1 - y^2}$.
(b) Solve the initial-value problem $y' = 2x\sqrt{1 - y^2}$, $y(0) = 0$, and graph the solution.
(c) Does the initial-value problem $y' = 2x\sqrt{1 - y^2}$, $y(0) = 2$, have a solution? Explain.

22. Solve the equation $e^{-y}y' + \cos x = 0$ and graph several members of the family of solutions. How does the solution curve change as the constant C varies?

23. Solve the initial-value problem $y' = (\sin x)/\sin y$, $y(0) = \pi/2$, and graph the solution (if your CAS does implicit plots).

CAS **24.** Solve the equation $y' = x\sqrt{x^2 + 1}/(ye^y)$ and graph several members of the family of solutions (if your CAS does implicit plots). How does the solution curve change as the constant C varies?

CAS **25–26**
(a) Use a computer algebra system to draw a direction field for the differential equation. Get a printout and use it to sketch some solution curves without solving the differential equation.
(b) Solve the differential equation.
(c) Use the CAS to draw several members of the family of solutions obtained in part (b). Compare with the curves from part (a).

25. $y' = 1/y$

26. $y' = x^2/y$

27–30 Find the orthogonal trajectories of the family of curves. Use a graphing device to draw several members of each family on a common screen.

27. $y = kx^2$

28. $x^2 - y^2 = k$

29. $y = (x + k)^{-1}$

30. $y = ke^{-x}$

31. Solve the initial-value problem in Exercise 27 in Section 9.2 to find an expression for the charge at time t. Find the limiting value of the charge.

32. In Exercise 28 in Section 9.2 we discussed a differential equation that models the temperature of a 95°C cup of coffee in a 20°C room. Solve the differential equation to find an expression for the temperature of the coffee at time t.

33. In Exercise 13 in Section 9.1 we formulated a model for learning in the form of the differential equation

$$\frac{dP}{dt} = k(M - P)$$

where $P(t)$ measures the performance of someone learning a skill after a training time t, M is the maximum level of performance, and k is a positive constant. Solve this differential equation to find an expression for $P(t)$. What is the limit of this expression?

34. In an elementary chemical reaction, single molecules of two reactants A and B form a molecule of the product C: $A + B \rightarrow C$. The law of mass action states that the rate of reaction is proportional to the product of the concentrations of A and B:

$$\frac{d[C]}{dt} = k[A][B]$$

(See Example 4 in Section 3.3.) Thus, if the initial concentrations are $[A] = a$ moles/L and $[B] = b$ moles/L and we write

$x = [\mathrm{C}]$, then we have

$$\frac{dx}{dt} = k(a - x)(b - x)$$

CAS (a) Assuming that $a \neq b$, find x as a function of t. Use the fact that the initial concentration of C is 0.
(b) Find $x(t)$ assuming that $a = b$. How does this expression for $x(t)$ simplify if it is known that $[\mathrm{C}] = a/2$ after 20 seconds?

35. In contrast to the situation of Exercise 34, experiments show that the reaction $H_2 + Br_2 \rightarrow 2HBr$ satisfies the rate law

$$\frac{d[\mathrm{HBr}]}{dt} = k[\mathrm{H_2}][\mathrm{Br_2}]^{1/2}$$

and so for this reaction the differential equation becomes

$$\frac{dx}{dt} = k(a - x)(b - x)^{1/2}$$

where $x = [\mathrm{HBr}]$ and a and b are the initial concentrations of hydrogen and bromine.
(a) Find x as a function of t in the case where $a = b$. Use the fact that $x(0) = 0$.
(b) If $a > b$, find t as a function of x. [*Hint:* In performing the integration, make the substitution $u = \sqrt{b - x}$.]

36. A sphere with radius 1 m has temperature 15°C. It lies inside a concentric sphere with radius 2 m and temperature 25°C. The temperature $T(r)$ at a distance r from the common center of the spheres satisfies the differential equation

$$\frac{d^2T}{dr^2} + \frac{2}{r}\frac{dT}{dr} = 0$$

If we let $S = dT/dr$, then S satisfies a first-order differential equation. Solve it to find an expression for the temperature $T(r)$ between the spheres.

37. A glucose solution is administered intravenously into the bloodstream at a constant rate r. As the glucose is added, it is converted into other substances and removed from the bloodstream at a rate that is proportional to the concentration at that time. Thus, a model for the concentration $C = C(t)$ of the glucose solution in the bloodstream is

$$\frac{dC}{dt} = r - kC$$

where k is a positive constant.
(a) Suppose that the concentration at time $t = 0$ is C_0. Determine the concentration at any time t by solving the differential equation.
(b) Assuming that $C_0 < r/k$, find $\lim_{t\to\infty} C(t)$ and interpret your answer.

38. A certain small country has \$10 billion in paper currency in circulation, and each day \$50 million comes into the country's banks. The government decides to introduce new currency by having the banks replace old bills with new ones whenever old currency comes into the banks. Let $x = x(t)$ denote the amount of new currency in circulation at time t, with $x(0) = 0$.
(a) Formulate a mathematical model in the form of an initial-value problem that represents the "flow" of the new currency into circulation.
(b) Solve the initial-value problem found in part (a).
(c) How long will it take for the new bills to account for 90% of the currency in circulation?

39. A tank contains 1000 L of brine with 15 kg of dissolved salt. Pure water enters the tank at a rate of 10 L/min. The solution is kept thoroughly mixed and drains from the tank at the same rate. How much salt is in the tank (a) after t minutes and (b) after 20 minutes?

40. A tank contains 1000 L of pure water. Brine that contains 0.05 kg of salt per liter of water enters the tank at a rate of 5 L/min. Brine that contains 0.04 kg of salt per liter of water enters the tank at a rate of 10 L/min. The solution is kept thoroughly mixed and drains from the tank at a rate of 15 L/min. How much salt is in the tank (a) after t minutes and (b) after one hour?

41. When a raindrop falls, it increases in size and so its mass at time t is a function of t, $m(t)$. The rate of growth of the mass is $km(t)$ for some positive constant k. When we apply Newton's Law of Motion to the raindrop, we get $(mv)' = gm$, where v is the velocity of the raindrop (directed downward) and g is the acceleration due to gravity. The *terminal velocity* of the raindrop is $\lim_{t\to\infty} v(t)$. Find an expression for the terminal velocity in terms of g and k.

42. An object of mass m is moving horizontally through a medium which resists the motion with a force that is a function of the velocity; that is,

$$m\frac{d^2s}{dt^2} = m\frac{dv}{dt} = f(v)$$

where $v = v(t)$ and $s = s(t)$ represent the velocity and position of the object at time t, respectively. For example, think of a boat moving through the water.
(a) Suppose that the resisting force is proportional to the velocity, that is, $f(v) = -kv$, k a positive constant. (This model is appropriate for small values of v.) Let $v(0) = v_0$ and $s(0) = s_0$ be the initial values of v and s. Determine v and s at any time t. What is the total distance that the object travels from time $t = 0$?
(b) For larger values of v a better model is obtained by supposing that the resisting force is proportional to the square of the velocity, that is, $f(v) = -kv^2$, $k > 0$. (This model was first proposed by Newton.) Let v_0 and s_0 be the initial values of v and s. Determine v and s at any time t. What is the total distance that the object travels in this case?

43. Let $A(t)$ be the area of a tissue culture at time t and let M be the final area of the tissue when growth is complete. Most cell divisions occur on the periphery of the tissue and the number of cells on the periphery is proportional to $\sqrt{A(t)}$. So a reasonable model for the growth of tissue is obtained by assuming that the rate of growth of the area is jointly proportional to $\sqrt{A(t)}$ and $M - A(t)$.

(a) Formulate a differential equation and use it to show that the tissue grows fastest when $A(t) = M/3$.

(b) Solve the differential equation to find an expression for $A(t)$. Use a computer algebra system to perform the integration.

44. According to Newton's Law of Universal Gravitation, the gravitational force on an object of mass m that has been projected vertically upward from Earth's surface is

$$F = \frac{mgR^2}{(x + R)^2}$$

where $x = x(t)$ is the object's distance above the surface at time t, R is Earth's radius, and g is the acceleration due to gravity. Also, by Newton's Second Law, $F = ma = m(dv/dt)$ and so

$$m\frac{dv}{dt} = -\frac{mgR^2}{(x + R)^2}$$

(a) Suppose a rocket is fired vertically upward with an initial velocity v_0. Let h be the maximum height above the surface reached by the object. Show that

$$v_0 = \sqrt{\frac{2gRh}{R + h}}$$

[*Hint:* By the Chain Rule, $m(dv/dt) = mv(dv/dx)$.]

(b) Calculate $v_e = \lim_{h\to\infty} v_0$. This limit is called the *escape velocity* for Earth.

(c) Use $R = 3960$ mi and $g = 32$ ft/s^2 to calculate v_e in feet per second and in miles per second.

APPLIED PROJECT

How Fast Does a Tank Drain?

If water (or other liquid) drains from a tank, we expect that the flow will be greatest at first (when the water depth is greatest) and will gradually decrease as the water level decreases. But we need a more precise mathematical description of how the flow decreases in order to answer the kinds of questions that engineers ask: How long does it take for a tank to drain completely? How much water should a tank hold in order to guarantee a certain minimum water pressure for a sprinkler system?

Let $h(t)$ and $V(t)$ be the height and volume of water in a tank at time t. If water leaks through a hole with area a at the bottom of the tank, then Torricelli's Law says that

$$\boxed{1} \qquad \frac{dV}{dt} = -a\sqrt{2gh}$$

where g is the acceleration due to gravity. So the rate at which water flows from the tank is proportional to the square root of the water height.

1. (a) Suppose the tank is cylindrical with height 6 ft and radius 2 ft and the hole is circular with radius 1 in. If we take $g = 32$ ft/s^2, show that y satisfies the differential equation

$$\frac{dh}{dt} = -\frac{1}{72}\sqrt{h}$$

(b) Solve this equation to find the height of the water at time t, assuming the tank is full at time $t = 0$.

(c) How long will it take for the water to drain completely?

2. Because of the rotation and viscosity of the liquid, the theoretical model given by Equation 1 isn't quite accurate. Instead, the model

$$\boxed{2} \qquad \frac{dh}{dt} = k\sqrt{h}$$

is often used and the constant k (which depends on the physical properties of the liquid) is determined from data concerning the draining of the tank.

(a) Suppose that a hole is drilled in the side of a cylindrical bottle and the height h of the water (above the hole) decreases from 10 cm to 3 cm in 68 seconds. Use Equation 2 to find an expression for $h(t)$. Evaluate $h(t)$ for $t = 10, 20, 30, 40, 50, 60$.

|||| This part of the project is best done as a classroom demonstration or as a group project with three students in each group: a timekeeper to call out seconds, a bottle keeper to estimate the height every 10 seconds, and a record keeper to record these values.

(b) Drill a 4-mm hole near the bottom of the cylindrical part of a two-liter plastic soft-drink bottle. Attach a strip of masking tape marked in centimeters from 0 to 10, with 0 corresponding to the top of the hole. With one finger over the hole, fill the bottle with water to the 10-cm mark. Then take your finger off the hole and record the values of $h(t)$ for $t = 10, 20, 30, 40, 50, 60$ seconds. (You will probably find that it takes 68 seconds for the level to decrease to $h = 3$ cm.) Compare your data with the values of $h(t)$ from part (a). How well did the model predict the actual values?

3. In many parts of the world, the water for sprinkler systems in large hotels and hospitals is supplied by gravity from cylindrical tanks on or near the roofs of the buildings. Suppose such a tank has radius 10 ft and the diameter of the outlet is 2.5 inches. An engineer has to guarantee that the water pressure will be at least 2160 lb/ft^2 for a period of 10 minutes. (When a fire happens, the electrical system might fail and it could take up to 10 minutes for the emergency generator and fire pump to be activated.) What height should the engineer specify for the tank in order to make such a guarantee? (Use the fact that the water pressure at a depth of d feet is $P = 62.5d$. See Section 8.3.)

4. Not all water tanks are shaped like cylinders. Suppose a tank has cross-sectional area $A(h)$ at height h. Then the volume of water up to height h is $V = \int_0^h A(u)\,du$ and so the Fundamental Theorem of Calculus gives $dV/dh = A(h)$. It follows that

$$\frac{dV}{dt} = \frac{dV}{dh}\frac{dh}{dt} = A(h)\frac{dh}{dt}$$

and so Torricelli's Law becomes

$$A(h)\frac{dh}{dt} = -a\sqrt{2gh}$$

(a) Suppose the tank has the shape of a sphere with radius 2 m and is initially half full of water. If the radius of the circular hole is 1 cm and we take $g = 10\ \text{m/s}^2$, show that h satisfies the differential equation

$$(4h - h^2)\frac{dh}{dt} = -0.0001\sqrt{20h}$$

(b) How long will it take for the water to drain completely?

APPLIED PROJECT

Which Is Faster, Going Up or Coming Down?

Suppose you throw a ball into the air. Do you think it takes longer to reach its maximum height or to fall back to Earth from its maximum height? We will solve the problem in this project but, before getting started, think about that situation and make a guess based on your physical intuition.

1. A ball with mass m is projected vertically upward from Earth's surface with a positive initial velocity v_0. We assume the forces acting on the ball are the force of gravity and a retarding force of air resistance with direction opposite to the direction of motion and with magnitude $p\,|v(t)|$, where p is a positive constant and $v(t)$ is the velocity of the ball at time t. In both

In modeling force due to air resistance, various functions have been used, depending on the physical characteristics and speed of the ball. Here we use a linear model, $-pv$, but a quadratic model ($-pv^2$ on the way up and pv^2 on the way down) is another possibility for higher speeds (see Exercise 42 in Section 9.3). For a golf ball, experiments have shown that a good model is $-pv^{1.3}$ going up and $p|v|^{1.3}$ coming down. But no matter which force function $-f(v)$ is used [where $f(v) > 0$ for $v > 0$ and $f(v) < 0$ for $v < 0$], the answer to the question remains the same. See F. Brauer, "What Goes Up Must Come Down, Eventually," *Amer. Math. Monthly* 108 (2001), pp. 437–440.

the ascent and the descent, the total force acting on the ball is $-pv - mg$. [During ascent, $v(t)$ is positive and the resistance acts downward; during descent, $v(t)$ is negative and the resistance acts upward.] So, by Newton's Second Law, the equation of motion is

$$mv' = -pv - mg$$

Solve this differential equation to show that the velocity is

$$v(t) = \left(v_0 + \frac{mg}{p}\right)e^{-pt/m} - \frac{mg}{p}$$

2. Show that the height of the ball, until it hits the ground, is

$$y(t) = \left(v_0 + \frac{mg}{p}\right)\frac{m}{p}(1 - e^{-pt/m}) - \frac{mgt}{p}$$

3. Let t_1 be the time that the ball takes to reach its maximum height. Show that

$$t_1 = \frac{m}{p}\ln\left(\frac{mg + pv_0}{mg}\right)$$

Find this time for a ball with mass 1 kg and initial velocity 20 m/s. Assume the air resistance is $\frac{1}{10}$ of the speed.

4. Let t_2 be the time at which the ball falls back to Earth. For the particular ball in Problem 3, estimate t_2 by using a graph of the height function $y(t)$. Which is faster, going up or coming down?

5. In general, it's not easy to find t_2 because it's impossible to solve the equation $y(t) = 0$ explicitly. We can, however, use an indirect method to determine whether ascent or descent is faster; we determine whether $y(2t_1)$ is positive or negative. Show that

$$y(2t_1) = \frac{m^2g}{p^2}\left(x - \frac{1}{x} - 2\ln x\right)$$

where $x = e^{pt_1/m}$. Then show that $x > 1$ and the function

$$f(x) = x - \frac{1}{x} - 2\ln x$$

is increasing for $x > 1$. Use this result to decide whether $y(2t_1)$ is positive or negative. What can you conclude? Is ascent or descent faster?

9.4 Exponential Growth and Decay

One of the models for population growth that we considered in Section 9.1 was based on the assumption that the population grows at a rate proportional to the size of the population:

$$\frac{dP}{dt} = kP$$

Is that a reasonable assumption? Suppose we have a population (of bacteria, for instance) with size $P = 1000$ and at a certain time it is growing at a rate of $P' = 300$ bacteria per hour. Now let's take another 1000 bacteria of the same type and put them with the first population. Each half of the new population was growing at a rate of 300 bacteria per hour. We would expect the total population of 2000 to increase at a rate of 600 bacteria per hour initially (provided there's enough room and nutrition). So if we double the size, we double the growth rate. In general, it seems reasonable that the growth rate should be proportional to the size.

The same assumption applies in other situations as well. In nuclear physics, the mass of a radioactive substance decays at a rate proportional to the mass. In chemistry, the rate of a unimolecular first-order reaction is proportional to the concentration of the substance. In finance, the value of a savings account with continuously compounded interest increases at a rate proportional to that value.

In general, if $y(t)$ is the value of a quantity y at time t and if the rate of change of y with respect to t is proportional to its size $y(t)$ at any time, then

1
$$\frac{dy}{dt} = ky$$

where k is a constant. Equation 1 is sometimes called the **law of natural growth** (if $k > 0$) or the **law of natural decay** (if $k < 0$). Because it is a separable differential equation we can solve it by the methods of Section 9.3:

$$\int \frac{dy}{y} = \int k\, dt$$

$$\ln|y| = kt + C$$

$$|y| = e^{kt+C} = e^C e^{kt}$$

$$y = Ae^{kt}$$

where A ($= \pm e^C$ or 0) is an arbitrary constant. To see the significance of the constant A, we observe that

$$y(0) = Ae^{k \cdot 0} = A$$

Therefore, A is the initial value of the function.

Because Equation 1 occurs so frequently in nature, we summarize what we have just proved for future use.

2 The solution of the initial-value problem

$$\frac{dy}{dt} = ky \qquad y(0) = y_0$$

is

$$y(t) = y_0 e^{kt}$$

Population Growth

What is the significance of the proportionality constant k? In the context of population growth, we can write

3
$$\frac{dP}{dt} = kP \quad \text{or} \quad \frac{1}{P}\frac{dP}{dt} = k$$

The quantity

$$\frac{1}{P}\frac{dP}{dt}$$

is the growth rate divided by the population size; it is called the **relative growth rate**. According to (3), instead of saying "the growth rate is proportional to population size" we could say "the relative growth rate is constant." Then (2) says that a population with constant relative growth rate must grow exponentially. Notice that the relative growth rate k appears as the coefficient of t in the exponential function y_0e^{kt}. For instance, if

$$\frac{dP}{dt} = 0.02P$$

and t is measured in years, then the relative growth rate is $k = 0.02$ and the population grows at a relative rate of 2% per year. If the population at time 0 is P_0, then the expression for the population is

$$P(t) = P_0e^{0.02t}$$

TABLE 1

Year	Population (millions)
1900	1650
1910	1750
1920	1860
1930	2070
1940	2300
1950	2560
1960	3040
1970	3710
1980	4450
1990	5280
2000	6080

EXAMPLE 1 Assuming that the growth rate is proportional to population size, use the data in Table 1 to model the population of the world in the 20th century. What is the relative growth rate? How well does the model fit the data?

SOLUTION We measure the time t in years and let $t = 0$ in the year 1900. We measure the population $P(t)$ in millions of people. Then the initial condition is $P(0) = 1650$. We are assuming that the growth rate is proportional to population size, so the initial-value problem is

$$\frac{dP}{dt} = kP \qquad P(0) = 1650$$

From (2) we know that the solution is

$$P(t) = 1650e^{kt}$$

One way to estimate the relative growth rate k is to use the fact that the population in 1910 was 1750 million. Therefore

$$P(10) = 1650e^{k(10)} = 1750$$

We solve this equation for k:

$$e^{10k} = \frac{1750}{1650}$$

$$k = \frac{1}{10}\ln\frac{1750}{1650} \approx 0.005884$$

Thus, the relative growth rate is about 0.6% per year and the model becomes

$$P(t) = 1650e^{0.005884t}$$

Table 2 and Figure 1 allow us to compare the predictions of this model with the actual data. You can see that the predictions become quite inaccurate after about 30 years and they underestimate by a factor of more than 2 in 2000.

TABLE 2

Year	Model	Population
1900	1650	1650
1910	1750	1750
1920	1856	1860
1930	1969	2070
1940	2088	2300
1950	2214	2560
1960	2349	3040
1970	2491	3710
1980	2642	4450
1990	2802	5280
2000	2972	6080

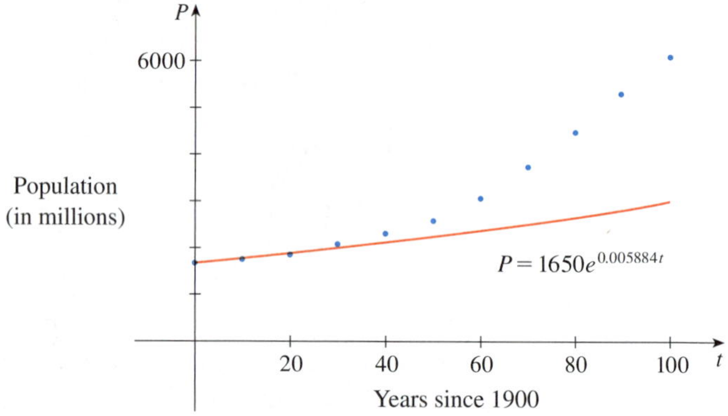

FIGURE 1 A possible model for world population growth

Another possibility for estimating k would be to use the given population for 1950, for instance, instead of 1910. Then

|||| In Section 1.5 we modeled the same data with an exponential function, but there we used the method of least squares.

$$P(50) = 1650e^{50k} = 2560$$

$$k = \frac{1}{50} \ln \frac{2560}{1650} \approx 0.0087846$$

The estimate for the relative growth rate is now 0.88% per year and the model is

$$P(t) = 1650e^{0.0087846t}$$

The predictions with this second model are shown in Table 3 and Figure 2. This exponential model is more accurate over a longer period of time, but it too lags behind reality in recent years.

TABLE 3

Year	Model	Population
1900	1650	1650
1910	1802	1750
1920	1967	1860
1930	2148	2070
1940	2345	2300
1950	2560	2560
1960	2795	3040
1970	3052	3710
1980	3332	4450
1990	3638	5280
2000	3972	6080

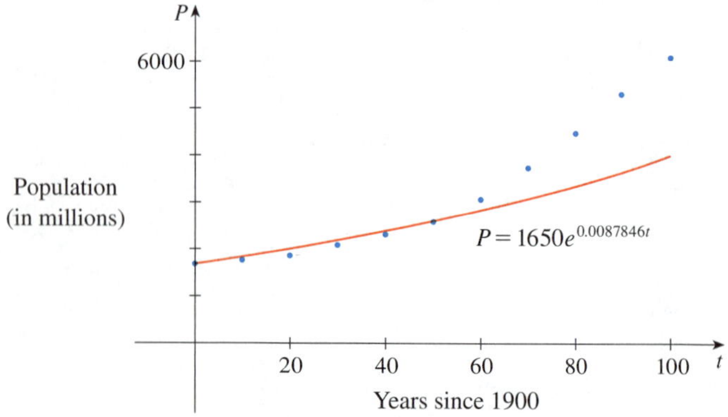

FIGURE 2 Another model for world population growth

EXAMPLE 2 Use the data in Table 1 to model the population of the world in the second half of the 20th century. Use the model to estimate the population in 1993 and to predict the population in the year 2010.

SOLUTION Here we let $t = 0$ in the year 1950. Then the initial-value problem is

$$\frac{dP}{dt} = kP \qquad P(0) = 2560$$

and the solution is

$$P(t) = 2560e^{kt}$$

Let's estimate k by using the population in 1960:

$$P(10) = 2560e^{10k} = 3040$$

$$k = \frac{1}{10} \ln \frac{3040}{2560} \approx 0.017185$$

The relative growth rate is about 1.7% per year and the model is

$$P(t) = 2560e^{0.017185t}$$

We estimate that the world population in 1993 was

$$P(43) = 2560e^{0.017185(43)} \approx 5360 \text{ million}$$

The model predicts that the population in 2010 will be

$$P(60) = 2560e^{0.017185(60)} \approx 7179 \text{ million}$$

The graph in Figure 3 shows that the model is fairly accurate to date, so the estimate for 1993 is quite reliable. But the prediction for 2010 is riskier.

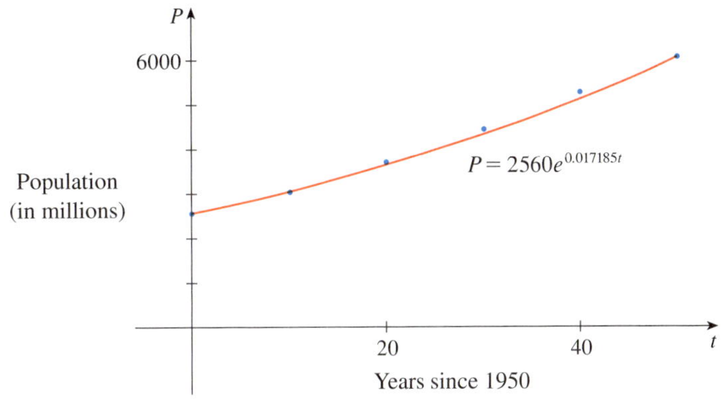

FIGURE 3
A model for world population growth in the second half of the 20th century

Radioactive Decay

Radioactive substances decay by spontaneously emitting radiation. If $m(t)$ is the mass remaining from an initial mass m_0 of the substance after time t, then the relative decay rate

$$-\frac{1}{m}\frac{dm}{dt}$$

has been found experimentally to be constant. (Since dm/dt is negative, the relative decay rate is positive.) It follows that

$$\frac{dm}{dt} = km$$

where k is a negative constant. In other words, radioactive substances decay at a rate proportional to the remaining mass. This means that we can use (2) to show that the mass decays exponentially:

$$m(t) = m_0 e^{kt}$$

Physicists express the rate of decay in terms of **half-life**, the time required for half of any given quantity to decay.

EXAMPLE 3 The half-life of radium-226 ($^{226}_{88}\mathrm{Ra}$) is 1590 years.
(a) A sample of radium-226 has a mass of 100 mg. Find a formula for the mass of $^{226}_{88}\mathrm{Ra}$ that remains after t years.
(b) Find the mass after 1000 years correct to the nearest milligram.
(c) When will the mass be reduced to 30 mg?

SOLUTION
(a) Let $m(t)$ be the mass of radium-226 (in milligrams) that remains after t years. Then $dm/dt = km$ and $y(0) = 100$, so (2) gives

$$m(t) = m(0)e^{kt} = 100e^{kt}$$

In order to determine the value of k, we use the fact that $y(1590) = \frac{1}{2}(100)$. Thus

$$100e^{1590k} = 50 \qquad \text{so} \qquad e^{1590k} = \tfrac{1}{2}$$

and

$$1590k = \ln\tfrac{1}{2} = -\ln 2$$

$$k = -\frac{\ln 2}{1590}$$

Therefore $\qquad m(t) = 100e^{-(\ln 2/1590)t}$

We could use the fact that $e^{\ln 2} = 2$ to write the expression for $m(t)$ in the alternative form

$$m(t) = 100 \times 2^{-t/1590}$$

(b) The mass after 1000 years is

$$m(1000) = 100e^{-(\ln 2/1590)1000} \approx 65 \text{ mg}$$

(c) We want to find the value of t such that $m(t) = 30$, that is,

$$100e^{-(\ln 2/1590)t} = 30 \qquad \text{or} \qquad e^{-(\ln 2/1590)t} = 0.3$$

We solve this equation for t by taking the natural logarithm of both sides:

$$-\frac{\ln 2}{1590}t = \ln 0.3$$

Thus

$$t = -1590\frac{\ln 0.3}{\ln 2} \approx 2762 \text{ years}$$

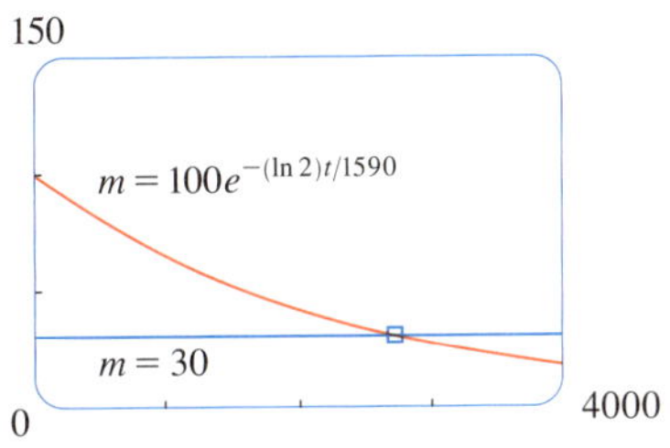

FIGURE 4

As a check on our work in Example 3, we use a graphing device to draw the graph of $m(t)$ in Figure 4 together with the horizontal line $m = 30$. These curves intersect when $t \approx 2800$, and this agrees with the answer to part (c).

Newton's Law of Cooling

Newton's Law of Cooling states that the rate of cooling of an object is proportional to the temperature difference between the object and its surroundings, provided that this difference is not too large. (This law also applies to warming.) If we let $T(t)$ be the temperature of the object at time t and T_s be the temperature of the surroundings, then we can formulate Newton's Law of Cooling as a differential equation:

$$\frac{dT}{dt} = k(T - T_s)$$

where k is a constant. We could solve this equation as a separable differential equation by the method of Section 9.3, but an easier method is to make the change of variable $y(t) = T(t) - T_s$. Because T_s is constant, we have $y'(t) = T'(t)$ and so the equation becomes

$$\frac{dy}{dt} = ky$$

We can then use (2) to find an expression for y, from which we can find T.

EXAMPLE 4 A bottle of soda pop at room temperature (72°F) is placed in a refrigerator where the temperature is 44°F. After half an hour the soda pop has cooled to 61°F.
(a) What is the temperature of the soda pop after another half hour?
(b) How long does it take for the soda pop to cool to 50°F?

SOLUTION
(a) Let $T(t)$ be the temperature of the soda after t minutes. The surrounding temperature is $T_s = 44°\text{F}$, so Newton's Law of Cooling states that

$$\frac{dT}{dt} = k(T - 44)$$

If we let $y = T - 44$, then $y(0) = T(0) - 44 = 72 - 44 = 28$, so y is a solution of the initial-value problem

$$\frac{dy}{dt} = ky \qquad y(0) = 28$$

and by (2) we have

$$y(t) = y(0)e^{kt} = 28e^{kt}$$

We are given that $T(30) = 61$, so $y(30) = 61 - 44 = 17$ and

$$28e^{30k} = 17 \qquad e^{30k} = \tfrac{17}{28}$$

Taking logarithms, we have

$$k = \frac{\ln\left(\frac{17}{28}\right)}{30} \approx -0.01663$$

Thus

$$y(t) = 28e^{-0.01663t}$$

$$T(t) = 44 + 28e^{-0.01663t}$$

$$T(60) = 44 + 28e^{-0.01663(60)} \approx 54.3$$

So after another half hour the pop has cooled to about 54°F.

(b) We have $T(t) = 50$ when

$$44 + 28e^{-0.01663t} = 50$$

$$e^{-0.01663t} = \tfrac{6}{28}$$

$$t = \frac{\ln\left(\frac{6}{28}\right)}{-0.01663} \approx 92.6$$

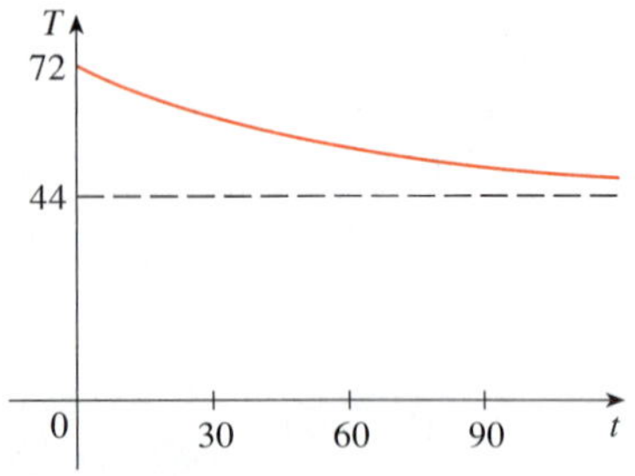

FIGURE 5

The pop cools to 50°F after about 1 hour 33 minutes.

Notice that in Example 4, we have

$$\lim_{t \to \infty} T(t) = \lim_{t \to \infty} (44 + 28e^{-0.01663t}) = 44 + 28 \cdot 0 = 44$$

which is to be expected. The graph of the temperature function is shown in Figure 5.

Continuously Compounded Interest

EXAMPLE 5 If \$1000 is invested at 6% interest, compounded annually, then after 1 year the investment is worth \$1000(1.06) = \$1060, after 2 years it's worth \$[1000(1.06)]1.06 = \$1123.60, and after t years it's worth $\$1000(1.06)^t$. In general, if an amount A_0 is invested at an interest rate r ($r = 0.06$ in this example), then after t years it's worth $A_0(1 + r)^t$. Usually, however, interest is compounded more frequently, say, n times a year. Then in each compounding period the interest rate is r/n and there are nt compounding periods in t years, so the value of the investment is

$$A_0\left(1 + \frac{r}{n}\right)^{nt}$$

For instance, after 3 years at 6% interest a \$1000 investment will be worth

$$\$1000(1.06)^3 = \$1191.02 \quad \text{with annual compounding}$$

$$\$1000(1.03)^6 = \$1194.05 \quad \text{with semiannual compounding}$$

$$\$1000(1.015)^{12} = \$1195.62 \quad \text{with quarterly compounding}$$

$$\$1000(1.005)^{36} = \$1196.68 \quad \text{with monthly compounding}$$

$$\$1000\left(1 + \frac{0.06}{365}\right)^{365 \cdot 3} = \$1197.20 \quad \text{with daily compounding}$$

You can see that the interest paid increases as the number of compounding periods (n) increases. If we let $n \to \infty$, then we will be compounding the interest **continuously** and the value of the investment will be

$$\begin{aligned} A(t) &= \lim_{n\to\infty} A_0\left(1 + \frac{r}{n}\right)^{nt} = \lim_{n\to\infty} A_0\left[\left(1 + \frac{r}{n}\right)^{n/r}\right]^{rt} \\ &= A_0\left[\lim_{n\to\infty}\left(1 + \frac{r}{n}\right)^{n/r}\right]^{rt} \\ &= A_0\left[\lim_{m\to\infty}\left(1 + \frac{1}{m}\right)^{m}\right]^{rt} \qquad \text{(where } m = n/r\text{)} \end{aligned}$$

But the limit in this expression is equal to the number e (see Equation 3.8.6). So with continuous compounding of interest at interest rate r, the amount after t years is

$$A(t) = A_0 e^{rt}$$

If we differentiate this equation, we get

$$\frac{dA}{dt} = rA_0 e^{rt} = rA(t)$$

which says that, with continuous compounding of interest, the rate of increase of an investment is proportional to its size.

Returning to the example of \$1000 invested for 3 years at 6% interest, we see that with continuous compounding of interest the value of the investment will be

$$\begin{aligned} A(3) &= \$1000e^{(0.06)3} \\ &= \$1000e^{0.18} = \$1197.22 \end{aligned}$$

Notice how close this is to the amount we calculated for daily compounding, \$1197.20. But the amount is easier to compute if we use continuous compounding.

9.4 Exercises

1. A population of protozoa develops with a constant relative growth rate of 0.7944 per member per day. On day zero the population consists of two members. Find the population size after six days.

2. A common inhabitant of human intestines is the bacterium *Escherichia coli*. A cell of this bacterium in a nutrient-broth medium divides into two cells every 20 minutes. The initial population of a culture is 60 cells.
(a) Find the relative growth rate.
(b) Find an expression for the number of cells after t hours.
(c) Find the number of cells after 8 hours.
(d) Find the rate of growth after 8 hours.
(e) When will the population reach 20,000 cells?

3. A bacteria culture starts with 500 bacteria and grows at a rate proportional to its size. After 3 hours there are 8000 bacteria.
(a) Find an expression for the number of bacteria after t hours.
(b) Find the number of bacteria after 4 hours.
(c) Find the rate of growth after 4 hours.
(d) When will the population reach 30,000?

4. A bacteria culture grows with constant relative growth rate. After 2 hours there are 600 bacteria and after 8 hours the count is 75,000.
(a) Find the initial population.
(b) Find an expression for the population after t hours.
(c) Find the number of cells after 5 hours.
(d) Find the rate of growth after 5 hours.
(e) When will the population reach 200,000?

5. The table gives estimates of the world population, in millions, from 1750 to 2000:

Year	Population	Year	Population
1750	790	1900	1650
1800	980	1950	2560
1850	1260	2000	6080

(a) Use the exponential model and the population figures for 1750 and 1800 to predict the world population in 1900 and 1950. Compare with the actual figures.
(b) Use the exponential model and the population figures for 1850 and 1900 to predict the world population in 1950. Compare with the actual population.
(c) Use the exponential model and the population figures for 1900 and 1950 to predict the world population in 2000. Compare with the actual population and try to explain the discrepancy.

6. The table gives the population of the United States, in millions, for the years 1900–2000.

Year	Population	Year	Population
1900	76	1960	179
1910	92	1970	203
1920	106	1980	227
1930	123	1990	250
1940	131	2000	275
1950	150		

(a) Use the exponential model and the census figures for 1900 and 1910 to predict the population in 2000. Compare with the actual figure and try to explain the discrepancy.
(b) Use the exponential model and the census figures for 1980 and 1990 to predict the population in 2000. Compare with the actual population. Then use this model to predict the population in the years 2010 and 2020.
(c) Graph both of the exponential functions in parts (a) and (b) together with a plot of the actual population. Are these models reasonable ones?

7. Experiments show that if the chemical reaction

$$N_2O_5 \longrightarrow 2NO_2 + \tfrac{1}{2}O_2$$

takes place at 45°C, the rate of reaction of dinitrogen pentoxide is proportional to its concentration as follows:

$$-\frac{d[N_2O_5]}{dt} = 0.0005[N_2O_5]$$

(See Example 4 in Section 3.3.)
(a) Find an expression for the concentration $[N_2O_5]$ after t seconds if the initial concentration is C.
(b) How long will the reaction take to reduce the concentration of N_2O_5 to 90% of its original value?

8. Bismuth-210 has a half-life of 5.0 days.
(a) A sample originally has a mass of 800 mg. Find a formula for the mass remaining after t days.
(b) Find the mass remaining after 30 days.
(c) When is the mass reduced to 1 mg?
(d) Sketch the graph of the mass function.

9. The half-life of cesium-137 is 30 years. Suppose we have a 100-mg sample.
(a) Find the mass that remains after t years.
(b) How much of the sample remains after 100 years?
(c) After how long will only 1 mg remain?

10. After 3 days a sample of radon-222 decayed to 58% of its original amount.
(a) What is the half-life of radon-222?
(b) How long would it take the sample to decay to 10% of its original amount?

11. Scientists can determine the age of ancient objects by a method called *radiocarbon dating*. The bombardment of the upper atmosphere by cosmic rays converts nitrogen to a radioactive isotope of carbon, ^{14}C, with a half-life of about 5730 years. Vegetation absorbs carbon dioxide through the atmosphere and animal life assimilates ^{14}C through food chains. When a plant or animal dies, it stops replacing its carbon and the amount of ^{14}C begins to decrease through radioactive decay. Therefore, the level of radioactivity must also decay exponentially. A parchment fragment was discovered that had about 74% as much ^{14}C radioactivity as does plant material on Earth today. Estimate the age of the parchment.

12. A curve passes through the point $(0, 5)$ and has the property that the slope of the curve at every point P is twice the y-coordinate of P. What is the equation of the curve?

13. A roast turkey is taken from an oven when its temperature has reached 185°F and is placed on a table in a room where the temperature is 75°F.
(a) If the temperature of the turkey is 150°F after half an hour, what is the temperature after 45 min?
(b) When will the turkey have cooled to 100°F?

14. A thermometer is taken from a room where the temperature is 20°C to the outdoors, where the temperature is 5°C. After one minute the thermometer reads 12°C.
(a) What will the reading on the thermometer be after one more minute?
(b) When will the thermometer read 6°C?

15. When a cold drink is taken from a refrigerator, its temperature is 5°C. After 25 minutes in a 20°C room its temperature has increased to 10°C.
(a) What is the temperature of the drink after 50 minutes?
(b) When will its temperature be 15°C?

16. A freshly brewed cup of coffee has temperature 95°C in a 20°C room. When its temperature is 70°C, it is cooling at a rate of 1°C per minute. When does this occur?

17. The rate of change of atmospheric pressure P with respect to altitude h is proportional to P, provided that the temperature is constant. At 15°C the pressure is 101.3 kPa at sea level and 87.14 kPa at $h = 1000$ m.
(a) What is the pressure at an altitude of 3000 m?
(b) What is the pressure at the top of Mount McKinley, at an altitude of 6187 m?

18. (a) If \$500 is borrowed at 14% interest, find the amounts due at the end of 2 years if the interest is compounded (i) annually, (ii) quarterly, (iii) monthly, (iv) daily, (v) hourly, and (vi) continuously.
(b) Suppose \$500 is borrowed and the interest is compounded continuously. If $A(t)$ is the amount due after t years, where $0 \leqslant t \leqslant 2$, graph $A(t)$ for each of the interest rates 14%, 10%, and 6% on a common screen.

19. (a) If \$3000 is invested at 5% interest, find the value of the investment at the end of 5 years if the interest is compounded (i) annually, (ii) semiannually, (iii) monthly, (iv) weekly, (v) daily, and (vi) continuously.
(b) If $A(t)$ is the amount of the investment at time t for the case of continuous compounding, write a differential equation and an initial condition satisfied by $A(t)$.

20. (a) How long will it take an investment to double in value if the interest rate is 6% compounded continuously?
(b) What is the equivalent annual interest rate?

21. Consider a population $P = P(t)$ with constant relative birth and death rates α and β, respectively, and a constant emigration rate m, where α, β, and m are positive constants. Assume that $\alpha > \beta$. Then the rate of change of the population at time t is modeled by the differential equation

$$\frac{dP}{dt} = kP - m \qquad \text{where } k = \alpha - \beta$$

(a) Find the solution of this equation that satisfies the initial condition $P(0) = P_0$.
(b) What condition on m will lead to an exponential expansion of the population?
(c) What condition on m will result in a constant population? A population decline?
(d) In 1847, the population of Ireland was about 8 million and the difference between the relative birth and death rates was 1.6% of the population. Because of the potato famine in the 1840s and 1850s, about 210,000 inhabitants per year emigrated from Ireland. Was the population expanding or declining at that time?

22. Let c be a positive number. A differential equation of the form

$$\frac{dy}{dt} = ky^{1+c}$$

where k is a positive constant, is called a *doomsday equation* because the exponent in the expression ky^{1+c} is larger than that for natural growth (that is, ky).
(a) Determine the solution that satisfies the initial condition $y(0) = y_0$.
(b) Show that there is a finite time $t = T$ (doomsday) such that $\lim_{t \to T^-} y(t) = \infty$.
(c) An especially prolific breed of rabbits has the growth term $ky^{1.01}$. If 2 such rabbits breed initially and the warren has 16 rabbits after three months, then when is doomsday?

APPLIED PROJECT

Calculus and Baseball

In this project we explore three of the many applications of calculus to baseball. The physical interactions of the game, especially the collision of ball and bat, are quite complex and their models are discussed in detail in a book by Robert Adair, *The Physics of Baseball,* 3d ed. (New York: HarperPerennial, 2002).

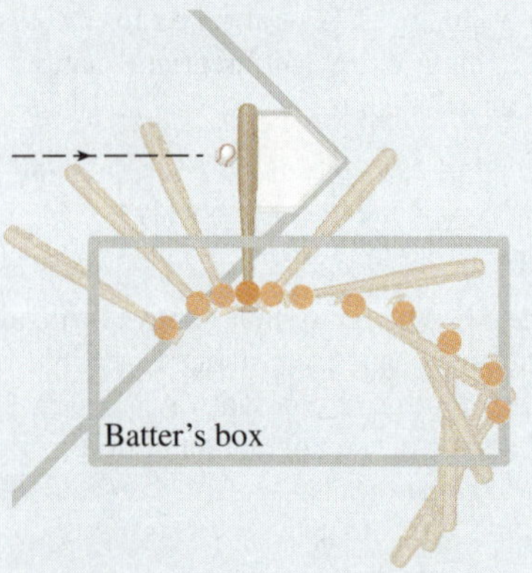

An overhead view of the position of a baseball bat, shown every fiftieth of a second during a typical swing. (Adapted from *The Physics of Baseball*)

1. It may surprise you to learn that the collision of baseball and bat lasts only about a thousandth of a second. Here we calculate the average force on the bat during this collision by first computing the change in the ball's momentum.

The *momentum* p of an object is the product of its mass m and its velocity v, that is, $p = mv$. Suppose an object, moving along a straight line, is acted on by a force $F = F(t)$ that is a continuous function of time.

(a) Show that the change in momentum over a time interval $[t_0, t_1]$ is equal to the integral of F from t_0 to t_1; that is, show that

$$p(t_1) - p(t_0) = \int_{t_0}^{t_1} F(t)\, dt$$

This integral is called the *impulse* of the force over the time interval.

(b) A pitcher throws a 90-mi/h fastball to a batter, who hits a line drive directly back to the pitcher. The ball is in contact with the bat for 0.001 s and leaves the bat with velocity 110 mi/h. A baseball weighs 5 oz and, in U.S. Customary units, its mass is measured in slugs: $m = w/g$ where $g = 32 \text{ ft/s}^2$.

(i) Find the change in the ball's momentum.

(ii) Find the average force on the bat.

2. In this problem we calculate the work required for a pitcher to throw a 90-mi/h fastball by first considering kinetic energy.

The *kinetic energy* K of an object of mass m and velocity v is given by $K = \frac{1}{2}mv^2$. Suppose an object of mass m, moving in a straight line, is acted on by a force $F = F(s)$ that depends on its position s. According to Newton's Second Law

$$F(s) = ma = m\frac{dv}{dt}$$

where a and v denote the acceleration and velocity of the object.

(a) Show that the work done in moving the object from a position s_0 to a position s_1 is equal to the change in the object's kinetic energy; that is, show that

$$W = \int_{s_0}^{s_1} F(s)\, ds = \tfrac{1}{2}mv_1^2 - \tfrac{1}{2}mv_0^2$$

where $v_0 = v(s_0)$ and $v_1 = v(s_1)$ are the velocities of the object at the positions s_0 and s_1. *Hint:* By the Chain Rule,

$$m\frac{dv}{dt} = m\frac{dv}{ds}\frac{ds}{dt} = mv\frac{dv}{ds}$$

(b) How many foot-pounds of work does it take to throw a baseball at a speed of 90 mi/h?

3. (a) An outfielder fields a baseball 280 ft away from home plate and throws it directly to the catcher with an initial velocity of 100 ft/s. Assume that the velocity $v(t)$ of the ball after t seconds satisfies the differential equation $dv/dt = -v/10$ because of air resistance. How long does it take for the ball to reach home plate? (Ignore any vertical motion of the ball.)

(b) The manager of the team wonders whether the ball will reach home plate sooner if it is relayed by an infielder. The shortstop can position himself directly between the outfielder and home plate, catch the ball thrown by the outfielder, turn, and throw the ball to the catcher with an initial velocity of 105 ft/s. The manager clocks the relay time of the shortstop (catching, turning, throwing) at half a second. How far from home plate should the shortstop position himself to minimize the total time for the ball to reach the plate? Should the manager encourage a direct throw or a relayed throw? What if the shortstop can throw at 115 ft/s?

(c) For what throwing velocity of the shortstop does a relayed throw take the same time as a direct throw?

9.5 The Logistic Equation

In this section we discuss in detail a model for population growth, the logistic model, that is more sophisticated than exponential growth. In doing so we use all the tools at our disposal—direction fields and Euler's method from Section 9.2 and the explicit solution of separable differential equations from Section 9.3. In the exercises we investigate other possible models for population growth, some of which take into account harvesting and seasonal growth.

The Logistic Model

As we discussed in Section 9.1, a population often increases exponentially in its early stages but levels off eventually and approaches its carrying capacity because of limited resources. If $P(t)$ is the size of the population at time t, we assume that

$$\frac{dP}{dt} \approx kP \qquad \text{if } P \text{ is small}$$

This says that the growth rate is initially close to being proportional to size. In other words, the relative growth rate is almost constant when the population is small. But we also want to reflect the fact that the relative growth rate decreases as the population P increases and becomes negative if P ever exceeds its **carrying capacity** K, the maximum population that the environment is capable of sustaining in the long run. The simplest expression for the relative growth rate that incorporates these assumptions is

$$\frac{1}{P}\frac{dP}{dt} = k\left(1 - \frac{P}{K}\right)$$

Multiplying by P, we obtain the model for population growth known as the **logistic differential equation**:

1
$$\frac{dP}{dt} = kP\left(1 - \frac{P}{K}\right)$$

Notice from Equation 1 that if P is small compared with K, then P/K is close to 0 and so $dP/dt \approx kP$. However, if $P \to K$ (the population approaches its carrying capacity), then $P/K \to 1$, so $dP/dt \to 0$. We can deduce information about whether solutions increase or decrease directly from Equation 1. If the population P lies between 0 and K, then the right side of the equation is positive, so $dP/dt > 0$ and the population increases. But if the population exceeds the carrying capacity ($P > K$), then $1 - P/K$ is negative, so $dP/dt < 0$ and the population decreases.

Direction Fields

Let's start our more detailed analysis of the logistic differential equation by looking at a direction field.

EXAMPLE 1 Draw a direction field for the logistic equation with $k = 0.08$ and carrying capacity $K = 1000$. What can you deduce about the solutions?

SOLUTION In this case the logistic differential equation is

$$\frac{dP}{dt} = 0.08P\left(1 - \frac{P}{1000}\right)$$

A direction field for this equation is shown in Figure 1. We show only the first quadrant because negative populations aren't meaningful and we are interested only in what happens after $t = 0$.

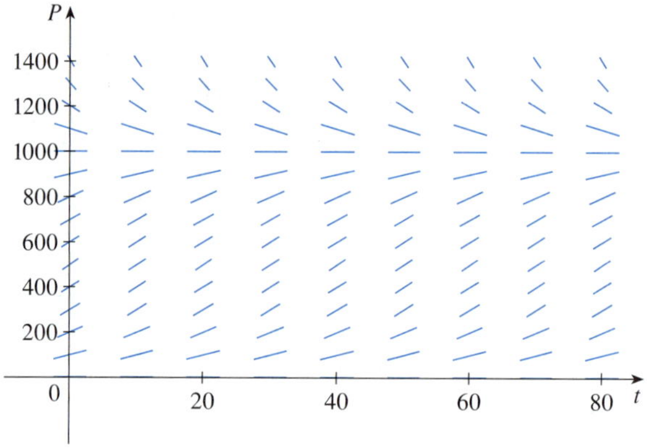

FIGURE 1
Direction field for the logistic equation in Example 1

The logistic equation is autonomous (dP/dt depends only on P, not on t), so the slopes are the same along any horizontal line. As expected, the slopes are positive for $0 < P < 1000$ and negative for $P > 1000$.

The slopes are small when P is close to 0 or 1000 (the carrying capacity). Notice that the solutions move away from the equilibrium solution $P = 0$ and move toward the equilibrium solution $P = 1000$.

In Figure 2 we use the direction field to sketch solution curves with initial populations $P(0) = 100$, $P(0) = 400$, and $P(0) = 1300$. Notice that solution curves that start below $P = 1000$ are increasing and those that start above $P = 1000$ are decreasing. The slopes are greatest when $P \approx 500$ and, therefore, the solution curves that start below $P = 1000$ have inflection points when $P \approx 500$. In fact we can prove that all solution curves that start below $P = 500$ have an inflection point when P is exactly 500 (see Exercise 9).

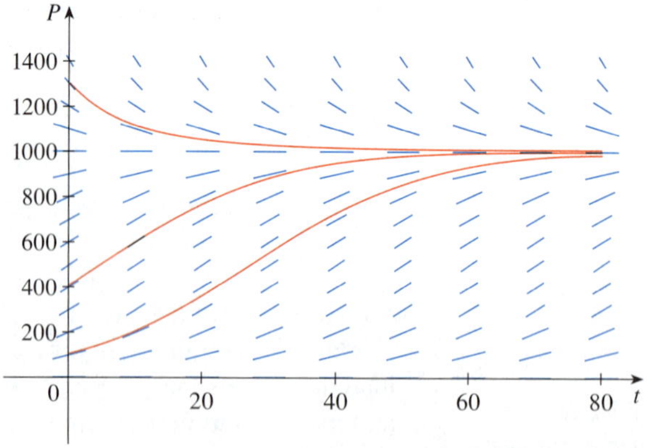

FIGURE 2
Solution curves for the logistic equation in Example 1

Euler's Method

Next let's use Euler's method to obtain numerical estimates for solutions of the logistic differential equation at specific times.

EXAMPLE 2 Use Euler's method with step sizes 20, 10, 5, 1, and 0.1 to estimate the population sizes $P(40)$ and $P(80)$, where P is the solution of the initial-value problem

$$\frac{dP}{dt} = 0.08P\left(1 - \frac{P}{1000}\right) \qquad P(0) = 100$$

SOLUTION With step size $h = 20$, $t_0 = 0$, $P_0 = 100$, and

$$F(t, P) = 0.08P\left(1 - \frac{P}{1000}\right)$$

we get, using the notation of Section 9.2,

$$P_1 = 100 + 20F(0, 100) = 244$$

$$P_2 = 244 + 20F(20, 244) \approx 539.14$$

$$P_3 = 539.14 + 20F(40, 539.14) \approx 936.69$$

$$P_4 = 936.69 + 20F(60, 936.69) \approx 1031.57$$

Thus, our estimates for the population sizes at times $t = 40$ and $t = 80$ are

$$P(40) \approx 539 \qquad P(80) \approx 1032$$

For smaller step sizes we need to program a calculator or computer. The table gives the results.

Step size	Euler estimate of $P(40)$	Euler estimate of $P(80)$
20	539	1032
10	647	997
5	695	991
1	725	986
0.1	731	985

Figure 3 shows a graph of the Euler approximations with step sizes $h = 10$ and $h = 1$. We see that the Euler approximation with $h = 1$ looks very much like the lower solution curve that we drew using a direction field in Figure 2.

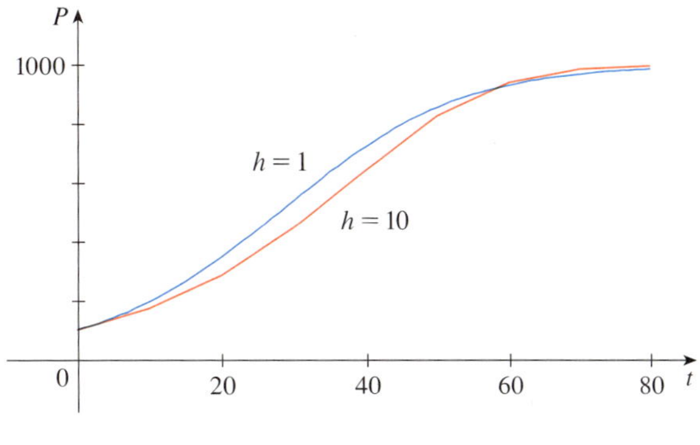

FIGURE 3
Euler approximations of the solution curve in Example 2

The Analytic Solution

The logistic equation (1) is separable and so we can solve it explicitly using the method of Section 9.3. Since

$$\frac{dP}{dt} = kP\left(1 - \frac{P}{K}\right)$$

we have

2 $$\int \frac{dP}{P(1 - P/K)} = \int k\,dt$$

To evaluate the integral on the left side, we write

$$\frac{1}{P(1 - P/K)} = \frac{K}{P(K - P)}$$

Using partial fractions (see Section 7.4), we get

$$\frac{K}{P(K - P)} = \frac{1}{P} + \frac{1}{K - P}$$

This enables us to rewrite Equation 2:

$$\int \left(\frac{1}{P} + \frac{1}{K - P}\right) dP = \int k\,dt$$

$$\ln|P| - \ln|K - P| = kt + C$$

$$\ln\left|\frac{K - P}{P}\right| = -kt - C$$

$$\left|\frac{K - P}{P}\right| = e^{-kt-C} = e^{-C}e^{-kt}$$

3 $$\frac{K - P}{P} = Ae^{-kt}$$

where $A = \pm e^{-C}$. Solving Equation 3 for P, we get

$$\frac{K}{P} - 1 = Ae^{-kt} \qquad \Rightarrow \qquad \frac{P}{K} = \frac{1}{1 + Ae^{-kt}}$$

so

$$P = \frac{K}{1 + Ae^{-kt}}$$

We find the value of A by putting $t = 0$ in Equation 3. If $t = 0$, then $P = P_0$ (the initial population), so

$$\frac{K - P_0}{P_0} = Ae^0 = A$$

Thus, the solution to the logistic equation is

4
$$P(t) = \frac{K}{1 + Ae^{-kt}} \qquad \text{where } A = \frac{K - P_0}{P_0}$$

Using the expression for $P(t)$ in Equation 4, we see that

$$\lim_{t \to \infty} P(t) = K$$

which is to be expected.

EXAMPLE 3 Write the solution of the initial-value problem

$$\frac{dP}{dt} = 0.08P\left(1 - \frac{P}{1000}\right) \qquad P(0) = 100$$

and use it to find the population sizes $P(40)$ and $P(80)$. At what time does the population reach 900?

SOLUTION The differential equation is a logistic equation with $k = 0.08$, carrying capacity $K = 1000$, and initial population $P_0 = 100$. So Equation 4 gives the population at time t as

$$P(t) = \frac{1000}{1 + Ae^{-0.08t}} \qquad \text{where } A = \frac{1000 - 100}{100} = 9$$

Thus
$$P(t) = \frac{1000}{1 + 9e^{-0.08t}}$$

|||| Compare these values with the Euler estimates from Example 2:

$P(40) \approx 731 \qquad P(80) \approx 985$

So the population sizes when $t = 40$ and 80 are

$$P(40) = \frac{1000}{1 + 9e^{-3.2}} \approx 731.6 \qquad P(80) = \frac{1000}{1 + 9e^{-6.4}} \approx 985.3$$

The population reaches 900 when

$$\frac{1000}{1 + 9e^{-0.08t}} = 900$$

Solving this equation for t, we get

$$1 + 9e^{-0.08t} = \tfrac{10}{9}$$

$$e^{-0.08t} = \tfrac{1}{81}$$

$$-0.08t = \ln \tfrac{1}{81} = -\ln 81$$

$$t = \frac{\ln 81}{0.08} \approx 54.9$$

|||| Compare the solution curve in Figure 4 with the lowest solution curve we drew from the direction field in Figure 2.

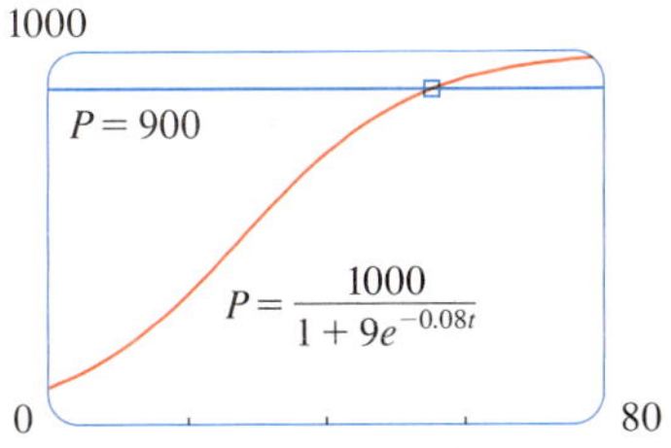

FIGURE 4

So the population reaches 900 when t is approximately 55. As a check on our work, we graph the population curve in Figure 4 and observe where it intersects the line $P = 900$. The cursor indicates that $t \approx 55$.

Comparison of the Natural Growth and Logistic Models

In the 1930s the biologist G. F. Gause conducted an experiment with the protozoan *Paramecium* and used a logistic equation to model his data. The table gives his daily count of the population of protozoa. He estimated the initial relative growth rate to be 0.7944 and the carrying capacity to be 64.

t (days)	0	1	2	3	4	5	6	7	8	9	10	11	12	13	14	15	16
P (observed)	2	3	22	16	39	52	54	47	50	76	69	51	57	70	53	59	57

EXAMPLE 4 Find the exponential and logistic models for Gause's data. Compare the predicted values with the observed values and comment on the fit.

SOLUTION Given the relative growth rate $k = 0.7944$ and the initial population $P_0 = 2$, the exponential model is

$$P(t) = P_0 e^{kt} = 2e^{0.7944t}$$

Gause used the same value of k for his logistic model. [This is reasonable because $P_0 = 2$ is small compared with the carrying capacity ($K = 64$). The equation

$$\frac{1}{P_0}\frac{dP}{dt}\bigg|_{t=0} = k\left(1 - \frac{2}{64}\right) \approx k$$

shows that the value of k for the logistic model is very close to the value for the exponential model.]

Then the solution of the logistic equation in Equation 4 gives

$$P(t) = \frac{K}{1 + Ae^{-kt}} = \frac{64}{1 + Ae^{-0.7944t}}$$

where

$$A = \frac{K - P_0}{P_0} = \frac{64 - 2}{2} = 31$$

So

$$P(t) = \frac{64}{1 + 31e^{-0.7944t}}$$

We use these equations to calculate the predicted values (rounded to the nearest integer) and compare them in the table.

t (days)	0	1	2	3	4	5	6	7	8	9	10	11	12	13	14	15	16
P (observed)	2	3	22	16	39	52	54	47	50	76	69	51	57	70	53	59	57
P (logistic model)	2	4	9	17	28	40	51	57	61	62	63	64	64	64	64	64	64
P (exponential model)	2	4	10	22	48	106	. . .										

We notice from the table and from the graph in Figure 5 that for the first three or four days the exponential model gives results comparable to those of the more sophisticated logistic model. For $t \geqslant 5$, however, the exponential model is hopelessly inaccurate, but the logistic model fits the observations reasonably well.

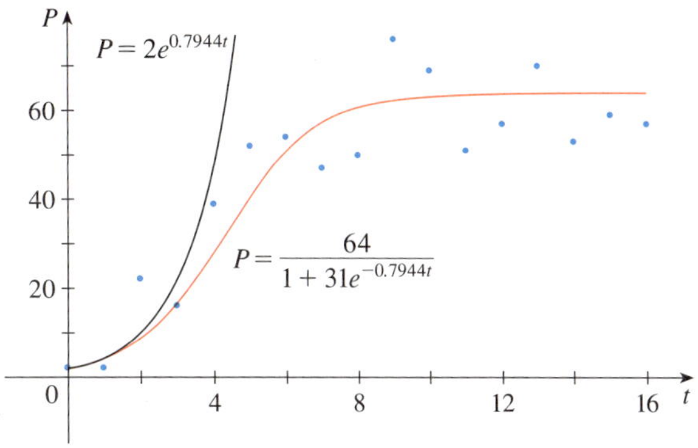

FIGURE 5
The exponential and logistic models for the *Paramecium* data

Other Models for Population Growth

The Law of Natural Growth and the logistic differential equation are not the only equations that have been proposed to model population growth. In Exercise 14 we look at the Gompertz growth function and in Exercises 15 and 16 we investigate seasonal-growth models.

Two of the other models are modifications of the logistic model. The differential equation

$$\frac{dP}{dt} = kP\left(1 - \frac{P}{K}\right) - c$$

has been used to model populations that are subject to "harvesting" of one sort or another. (Think of a population of fish being caught at a constant rate). This equation is explored in Exercises 11 and 12.

For some species there is a minimum population level m below which the species tends to become extinct. (Adults may not be able to find suitable mates.) Such populations have been modeled by the differential equation

$$\frac{dP}{dt} = kP\left(1 - \frac{P}{K}\right)\left(1 - \frac{m}{P}\right)$$

where the extra factor, $1 - m/P$, takes into account the consequences of a sparse population (see Exercise 13).

9.5 Exercises

1. Suppose that a population develops according to the logistic equation

$$\frac{dP}{dt} = 0.05P - 0.0005P^2$$

where t is measured in weeks.

(a) What is the carrying capacity? What is the value of k?

(b) A direction field for this equation is shown at the right. Where are the slopes close to 0? Where are they largest? Which solutions are increasing? Which solutions are decreasing?

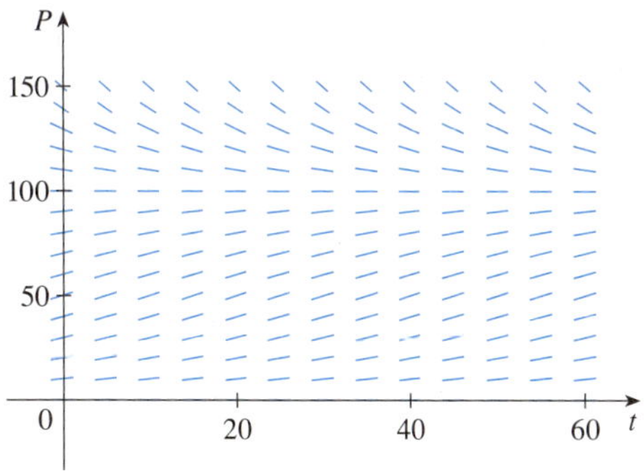

(c) Use the direction field to sketch solutions for initial populations of 20, 40, 60, 80, 120, and 140. What do these solutions have in common? How do they differ? Which solutions have inflection points? At what population levels do they occur?

(d) What are the equilibrium solutions? How are the other solutions related to these solutions?

2. Suppose that a population grows according to a logistic model with carrying capacity 6000 and $k = 0.0015$ per year.

(a) Write the logistic differential equation for these data.

(b) Draw a direction field (either by hand or with a computer algebra system). What does it tell you about the solution curves?

(c) Use the direction field to sketch the solution curves for initial populations of 1000, 2000, 4000, and 8000. What can you say about the concavity of these curves? What is the significance of the inflection points?

(d) Program a calculator or computer to use Euler's method with step size $h = 1$ to estimate the population after 50 years if the initial population is 1000.

(e) If the initial population is 1000, write a formula for the population after t years. Use it to find the population after 50 years and compare with your estimate in part (d).

(f) Graph the solution in part (e) and compare with the solution curve you sketched in part (c).

3. The Pacific halibut fishery has been modeled by the differential equation

$$\frac{dy}{dt} = ky\left(1 - \frac{y}{K}\right)$$

where $y(t)$ is the biomass (the total mass of the members of the population) in kilograms at time t (measured in years), the carrying capacity is estimated to be $K = 8 \times 10^7$ kg, and $k = 0.71$ per year.

(a) If $y(0) = 2 \times 10^7$ kg, find the biomass a year later.

(b) How long will it take for the biomass to reach 4×10^7 kg?

4. The table gives the number of yeast cells in a new laboratory culture.

Time (hours)	Yeast cells	Time (hours)	Yeast cells
0	18	10	509
2	39	12	597
4	80	14	640
6	171	16	664
8	336	18	672

(a) Plot the data and use the plot to estimate the carrying capacity for the yeast population.

(b) Use the data to estimate the initial relative growth rate.

(c) Find both an exponential model and a logistic model for these data.

(d) Compare the predicted values with the observed values, both in a table and with graphs. Comment on how well your models fit the data.

(e) Use your logistic model to estimate the number of yeast cells after 7 hours.

5. The population of the world was about 5.3 billion in 1990. Birth rates in the 1990s ranged from 35 to 40 million per year and death rates ranged from 15 to 20 million per year. Let's assume that the carrying capacity for world population is 100 billion.

(a) Write the logistic differential equation for these data. (Because the initial population is small compared to the carrying capacity, you can take k to be an estimate of the initial relative growth rate.)

(b) Use the logistic model to estimate the world population in the year 2000 and compare with the actual population of 6.1 billion.

(c) Use the logistic model to predict the world population in the years 2100 and 2500.

(d) What are your predictions if the carrying capacity is 50 billion?

6. (a) Make a guess as to the carrying capacity for the U.S. population. Use it and the fact that the population was 250 million in 1990 to formulate a logistic model for the U.S. population.

(b) Determine the value of k in your model by using the fact that the population in 2000 was 275 million.

(c) Use your model to predict the U.S. population in the years 2100 and 2200.

(d) Use your model to predict the year in which the U.S. population will exceed 300 million.

7. One model for the spread of a rumor is that the rate of spread is proportional to the product of the fraction y of the population who have heard the rumor and the fraction who have not heard the rumor.

(a) Write a differential equation that is satisfied by y.

(b) Solve the differential equation.

(c) A small town has 1000 inhabitants. At 8 A.M., 80 people have heard a rumor. By noon half the town has heard it. At what time will 90% of the population have heard the rumor?

8. Biologists stocked a lake with 400 fish and estimated the carrying capacity (the maximal population for the fish of that species in that lake) to be 10,000. The number of fish tripled in the first year.

(a) Assuming that the size of the fish population satisfies the logistic equation, find an expression for the size of the population after t years.

(b) How long will it take for the population to increase to 5000?

9. (a) Show that if P satisfies the logistic equation (1), then

$$\frac{d^2P}{dt^2} = k^2P\left(1 - \frac{P}{K}\right)\left(1 - \frac{2P}{K}\right)$$

(b) Deduce that a population grows fastest when it reaches half its carrying capacity.

10. For a fixed value of K (say $K = 10$), the family of logistic functions given by Equation 4 depends on the initial value P_0 and the proportionality constant k. Graph several members of this family. How does the graph change when P_0 varies? How does it change when k varies?

11. Let's modify the logistic differential equation of Example 1 as follows:

$$\frac{dP}{dt} = 0.08P\left(1 - \frac{P}{1000}\right) - 15$$

(a) Suppose $P(t)$ represents a fish population at time t, where t is measured in weeks. Explain the meaning of the term -15.
(b) Draw a direction field for this differential equation.
(c) What are the equilibrium solutions?
(d) Use the direction field to sketch several solution curves. Describe what happens to the fish population for various initial populations.
CAS (e) Solve this differential equation explicitly, either by using partial fractions or with a computer algebra system. Use the initial populations 200 and 300. Graph the solutions and compare with your sketches in part (d).

CAS **12.** Consider the differential equation

$$\frac{dP}{dt} = 0.08P\left(1 - \frac{P}{1000}\right) - c$$

as a model for a fish population, where t is measured in weeks and c is a constant.
(a) Use a CAS to draw direction fields for various values of c.
(b) From your direction fields in part (a), determine the values of c for which there is at least one equilibrium solution. For what values of c does the fish population always die out?
(c) Use the differential equation to prove what you discovered graphically in part (b).
(d) What would you recommend for a limit to the weekly catch of this fish population?

13. There is considerable evidence to support the theory that for some species there is a minimum population m such that the species will become extinct if the size of the population falls below m. This condition can be incorporated into the logistic equation by introducing the factor $(1 - m/P)$. Thus, the modified logistic model is given by the differential equation

$$\frac{dP}{dt} = kP\left(1 - \frac{P}{K}\right)\left(1 - \frac{m}{P}\right)$$

(a) Use the differential equation to show that any solution is increasing if $m < P < K$ and decreasing if $0 < P < m$.
(b) For the case where $k = 0.08$, $K = 1000$, and $m = 200$, draw a direction field and use it to sketch several solution curves. Describe what happens to the population for various initial populations. What are the equilibrium solutions?
(c) Solve the differential equation explicitly, either by using partial fractions or with a computer algebra system. Use the initial population P_0.
(d) Use the solution in part (c) to show that if $P_0 < m$, then the species will become extinct. [*Hint:* Show that the numerator in your expression for $P(t)$ is 0 for some value of t.]

14. Another model for a growth function for a limited population is given by the **Gompertz function**, which is a solution of the differential equation

$$\frac{dP}{dt} = c \ln\left(\frac{K}{P}\right)P$$

where c is a constant and K is the carrying capacity.
(a) Solve this differential equation.
(b) Compute $\lim_{t\to\infty} P(t)$.
(c) Graph the Gompertz growth function for $K = 1000$, $P_0 = 100$, and $c = 0.05$, and compare it with the logistic function in Example 3. What are the similarities? What are the differences?
(d) We know from Exercise 9 that the logistic function grows fastest when $P = K/2$. Use the Gompertz differential equation to show that the Gompertz function grows fastest when $P = K/e$.

15. In a **seasonal-growth model**, a periodic function of time is introduced to account for seasonal variations in the rate of growth. Such variations could, for example, be caused by seasonal changes in the availability of food.
(a) Find the solution of the seasonal-growth model

$$\frac{dP}{dt} = kP\cos(rt - \phi) \qquad P(0) = P_0$$

where k, r, and ϕ are positive constants.
(b) By graphing the solution for several values of k, r, and ϕ, explain how the values of k, r, and ϕ affect the solution. What can you say about $\lim_{t\to\infty} P(t)$?

16. Suppose we alter the differential equation in Exercise 15 as follows:

$$\frac{dP}{dt} = kP\cos^2(rt - \phi) \qquad P(0) = P_0$$

(a) Solve this differential equation with the help of a table of integrals or a CAS.
(b) Graph the solution for several values of k, r, and ϕ. How do the values of k, r, and ϕ affect the solution? What can you say about $\lim_{t\to\infty} P(t)$ in this case?

17. Graphs of logistic functions (Figures 2 and 4) look suspiciously similar to the graph of the hyperbolic tangent function (Figure 3 in Section 3.9). Explain the similarity by showing that the logistic function given by Equation 4 can be written as

$$P(t) = \tfrac{1}{2}K\left[1 + \tanh\left(\tfrac{1}{2}k(t - c)\right)\right]$$

where $c = (\ln A)/k$. Thus, the logistic function is really just a shifted hyperbolic tangent.

9.6 Linear Equations

A first-order **linear** differential equation is one that can be put into the form

$$\boxed{1} \qquad \frac{dy}{dx} + P(x)y = Q(x)$$

where P and Q are continuous functions on a given interval. This type of equation occurs frequently in various sciences, as we will see.

An example of a linear equation is $xy' + y = 2x$ because, for $x \neq 0$, it can be written in the form

$$\boxed{2} \qquad y' + \frac{1}{x}y = 2$$

Notice that this differential equation is not separable because it's impossible to factor the expression for y' as a function of x times a function of y. But we can still solve the equation by noticing, by the Product Rule, that

$$xy' + y = (xy)'$$

and so we can rewrite the equation as

$$(xy)' = 2x$$

If we now integrate both sides of this equation, we get

$$xy = x^2 + C \qquad \text{or} \qquad y = x + \frac{C}{x}$$

If we had been given the differential equation in the form of Equation 2, we would have had to take the preliminary step of multiplying each side of the equation by x.

It turns out that every first-order linear differential equation can be solved in a similar fashion by multiplying both sides of Equation 1 by a suitable function $I(x)$ called an *integrating factor.* We try to find I so that the left side of Equation 1, when multiplied by $I(x)$, becomes the derivative of the product $I(x)y$:

$$\boxed{3} \qquad I(x)(y' + P(x)y) = (I(x)y)'$$

If we can find such a function I, then Equation 1 becomes

$$(I(x)y)' = I(x)Q(x)$$

Integrating both sides, we would have

$$I(x)y = \int I(x)Q(x)\,dx + C$$

so the solution would be

$$\boxed{4} \qquad y(x) = \frac{1}{I(x)}\left[\int I(x)Q(x)\,dx + C\right]$$

To find such an I, we expand Equation 3 and cancel terms:

$$I(x)y' + I(x)P(x)y = (I(x)y)' = I'(x)y + I(x)y'$$

$$I(x)P(x) = I'(x)$$

This is a separable differential equation for I, which we solve as follows:

$$\int \frac{dI}{I} = \int P(x)\,dx$$

$$\ln|I| = \int P(x)\,dx$$

$$I = Ae^{\int P(x)\,dx}$$

where $A = \pm e^C$. We are looking for a particular integrating factor, not the most general one, so we take $A = 1$ and use

5 $$I(x) = e^{\int P(x)\,dx}$$

Thus, a formula for the general solution to Equation 1 is provided by Equation 4, where I is given by Equation 5. Instead of memorizing this formula, however, we just remember the form of the integrating factor.

> To solve the linear differential equation $y' + P(x)y = Q(x)$, multiply both sides by the **integrating factor** $I(x) = e^{\int P(x)\,dx}$ and integrate both sides.

EXAMPLE 1 Solve the differential equation $\dfrac{dy}{dx} + 3x^2y = 6x^2$.

SOLUTION The given equation is linear since it has the form of Equation 1 with $P(x) = 3x^2$ and $Q(x) = 6x^2$. An integrating factor is

$$I(x) = e^{\int 3x^2\,dx} = e^{x^3}$$

Multiplying both sides of the differential equation by e^{x^3}, we get

|||| Figure 1 shows the graphs of several members of the family of solutions in Example 1. Notice that they all approach 2 as $x \to \infty$.

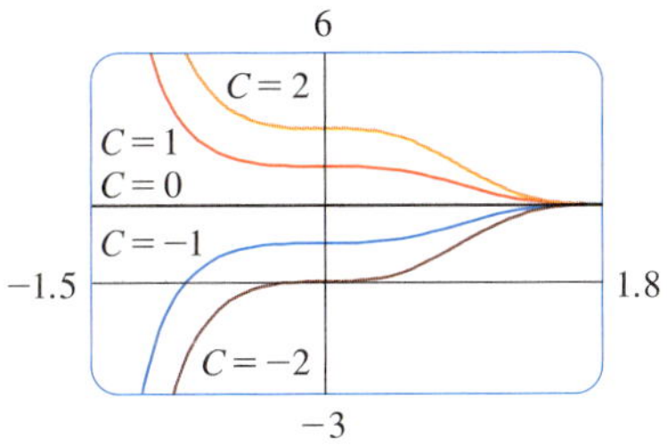

FIGURE 1

$$e^{x^3}\frac{dy}{dx} + 3x^2e^{x^3}y = 6x^2e^{x^3}$$

or

$$\frac{d}{dx}(e^{x^3}y) = 6x^2e^{x^3}$$

Integrating both sides, we have

$$e^{x^3}y = \int 6x^2e^{x^3}\,dx = 2e^{x^3} + C$$

$$y = 2 + Ce^{-x^3}$$

EXAMPLE 2 Find the solution of the initial-value problem

$$x^2y' + xy = 1 \qquad x > 0 \qquad y(1) = 2$$

SOLUTION We must first divide both sides by the coefficient of y' to put the differential equation into standard form:

$$y' + \frac{1}{x}y = \frac{1}{x^2} \qquad x > 0 \tag{6}$$

The integrating factor is

$$I(x) = e^{\int (1/x)\,dx} = e^{\ln x} = x$$

Multiplication of Equation 6 by x gives

$$xy' + y = \frac{1}{x} \qquad \text{or} \qquad (xy)' = \frac{1}{x}$$

Then
$$xy = \int \frac{1}{x}\,dx = \ln x + C$$

and so
$$y = \frac{\ln x + C}{x}$$

Since $y(1) = 2$, we have

$$2 = \frac{\ln 1 + C}{1} = C$$

Therefore, the solution to the initial-value problem is

$$y = \frac{\ln x + 2}{x}$$

The solution of the initial-value problem in Example 2 is shown in Figure 2.

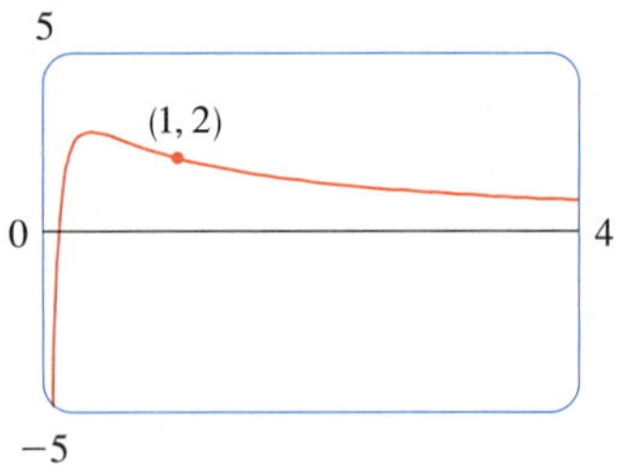

FIGURE 2

EXAMPLE 3 Solve $y' + 2xy = 1$.

SOLUTION The given equation is in the standard form for a linear equation. Multiplying by the integrating factor

$$e^{\int 2x\,dx} = e^{x^2}$$

we get
$$e^{x^2}y' + 2xe^{x^2}y = e^{x^2}$$

or
$$(e^{x^2}y)' = e^{x^2}$$

Therefore
$$e^{x^2}y = \int e^{x^2}\,dx + C$$

Recall from Section 7.5 that $\int e^{x^2}\,dx$ can't be expressed in terms of elementary functions. Nonetheless, it's a perfectly good function and we can leave the answer as

$$y = e^{-x^2}\int e^{x^2}\,dx + Ce^{-x^2}$$

Even though the solutions of the differential equation in Example 3 are expressed in terms of an integral, they can still be graphed by a computer algebra system (Figure 3).

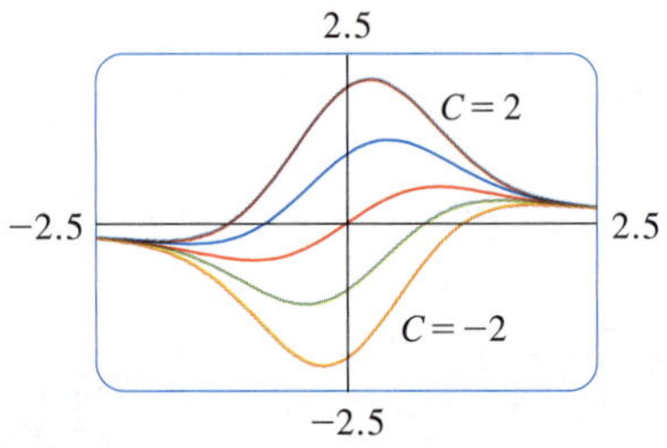

FIGURE 3

Another way of writing the solution is

$$y = e^{-x^2} \int_0^x e^{t^2}\, dt + Ce^{-x^2}$$

(Any number can be chosen for the lower limit of integration.)

|||| Application to Electric Circuits

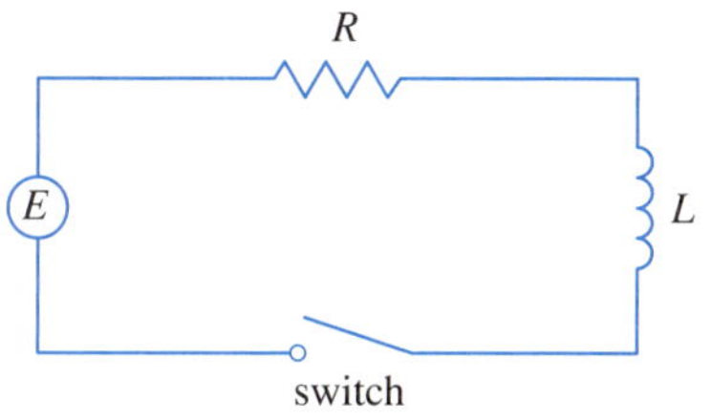

FIGURE 4

In Section 9.2 we considered the simple electric circuit shown in Figure 4: An electromotive force (usually a battery or generator) produces a voltage of $E(t)$ volts (V) and a current of $I(t)$ amperes (A) at time t. The circuit also contains a resistor with a resistance of R ohms (Ω) and an inductor with an inductance of L henries (H).

Ohm's Law gives the drop in voltage due to the resistor as RI. The voltage drop due to the inductor is $L(dI/dt)$. One of Kirchhoff's laws says that the sum of the voltage drops is equal to the supplied voltage $E(t)$. Thus, we have

7 $$L\frac{dI}{dt} + RI = E(t)$$

which is a first-order linear differential equation. The solution gives the current I at time t.

EXAMPLE 4 Suppose that in the simple circuit of Figure 4 the resistance is 12 Ω and the inductance is 4 H. If a battery gives a constant voltage of 60 V and the switch is closed when $t = 0$ so the current starts with $I(0) = 0$, find (a) $I(t)$, (b) the current after 1 s, and (c) the limiting value of the current.

|||| The differential equation in Example 4 is both linear and separable, so an alternative method is to solve it as a separable equation (Example 4 in Section 9.3). If we replace the battery by a generator, however, we get an equation that is linear but not separable (Example 5).

SOLUTION

(a) If we put $L = 4$, $R = 12$, and $E(t) = 60$ in Equation 7, we obtain the initial-value problem

$$4\frac{dI}{dt} + 12I = 60 \qquad I(0) = 0$$

or

$$\frac{dI}{dt} + 3I = 15 \qquad I(0) = 0$$

Multiplying by the integrating factor $e^{\int 3\,dt} = e^{3t}$, we get

$$e^{3t}\frac{dI}{dt} + 3e^{3t}I = 15e^{3t}$$

$$\frac{d}{dt}(e^{3t}I) = 15e^{3t}$$

$$e^{3t}I = \int 15e^{3t}\,dt = 5e^{3t} + C$$

$$I(t) = 5 + Ce^{-3t}$$

Since $I(0) = 0$, we have $5 + C = 0$, so $C = -5$ and

$$I(t) = 5(1 - e^{-3t})$$

|||| Figure 5 shows how the current in Example 4 approaches its limiting value.

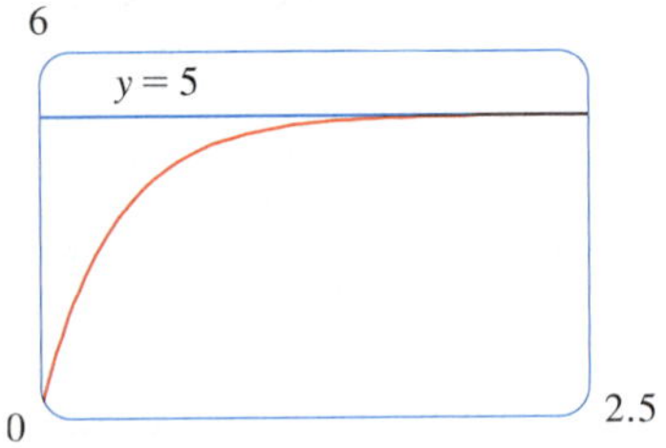

FIGURE 5

(b) After 1 second the current is

$$I(1) = 5(1 - e^{-3}) \approx 4.75 \text{ A}$$

(c)
$$\lim_{t\to\infty} I(t) = \lim_{t\to\infty} 5(1 - e^{-3t})$$
$$= 5 - 5 \lim_{t\to\infty} e^{-3t}$$
$$= 5 - 0 = 5$$

EXAMPLE 5 Suppose that the resistance and inductance remain as in Example 4 but, instead of the battery, we use a generator that produces a variable voltage of $E(t) = 60 \sin 30t$ volts. Find $I(t)$.

SOLUTION This time the differential equation becomes

$$4\frac{dI}{dt} + 12I = 60 \sin 30t \qquad \text{or} \qquad \frac{dI}{dt} + 3I = 15 \sin 30t$$

The same integrating factor e^{3t} gives

$$\frac{d}{dt}(e^{3t}I) = e^{3t}\frac{dI}{dt} + 3e^{3t}I = 15e^{3t} \sin 30t$$

|||| Figure 6 shows the graph of the current when the battery is replaced by a generator.

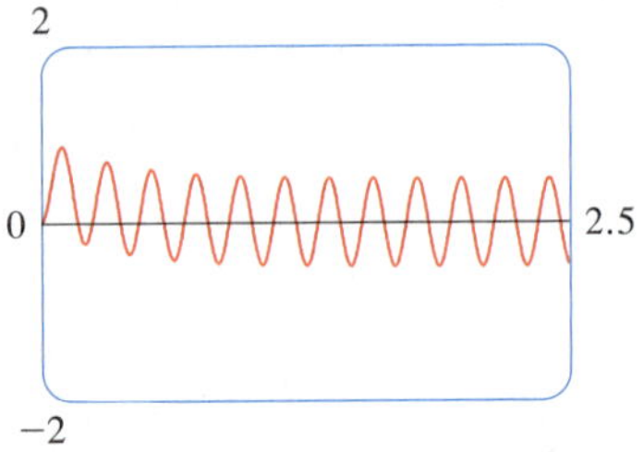

FIGURE 6

Using Formula 98 in the Table of Integrals, we have

$$e^{3t}I = \int 15e^{3t} \sin 30t\,dt = 15\frac{e^{3t}}{909}(3 \sin 30t - 30 \cos 30t) + C$$

$$I = \tfrac{5}{101}(\sin 30t - 10 \cos 30t) + Ce^{-3t}$$

Since $I(0) = 0$, we get

$$-\tfrac{50}{101} + C = 0$$

so
$$I(t) = \tfrac{5}{101}(\sin 30t - 10 \cos 30t) + \tfrac{50}{101}e^{-3t}$$

9.6 Exercises

1–4 |||| Determine whether the differential equation is linear.

1. $y' + e^x y = x^2 y^2$

2. $y + \sin x = x^3 y'$

3. $xy' + \ln x - x^2 y = 0$

4. $y' + \cos y = \tan x$

5–14 |||| Solve the differential equation.

5. $y' + 2y = 2e^x$

6. $y' = x + 5y$

7. $xy' - 2y = x^2$

8. $x^2 y' + 2xy = \cos^2 x$

9. $xy' + y = \sqrt{x}$

10. $1 + xy = xy'$

11. $\dfrac{dy}{dx} + 2xy = x^2$

12. $\dfrac{dy}{dx} = x \sin 2x + y \tan x, \quad -\pi/2 < x < \pi/2$

13. $(1 + t)\dfrac{du}{dt} + u = 1 + t, \quad t > 0$

14. $t \ln t \dfrac{dr}{dt} + r = te^t$

15–20 |||| Solve the initial-value problem.

15. $y' = x + y, \quad y(0) = 2$

16. $t\dfrac{dy}{dt} + 2y = t^3, \quad t > 0, \quad y(1) = 0$

17. $\dfrac{dv}{dt} - 2tv = 3t^2 e^{t^2}, \quad v(0) = 5$

18. $2xy' + y = 6x, \quad x > 0, \quad y(4) = 20$

19. $xy' = y + x^2 \sin x, \quad y(\pi) = 0$

20. $x\dfrac{dy}{dx} - \dfrac{y}{x+1} = x, \quad y(1) = 0, \quad x > 0$

21–22 Solve the differential equation and use a graphing calculator or computer to graph several members of the family of solutions. How does the solution curve change as C varies?

21. $xy' + y = x \cos x, \quad x > 0$

22. $y' + (\cos x)y = \cos x$

23. A **Bernoulli differential equation** (named after James Bernoulli) is of the form

$$\frac{dy}{dx} + P(x)y = Q(x)y^n$$

Observe that, if $n = 0$ or 1, the Bernoulli equation is linear. For other values of n, show that the substitution $u = y^{1-n}$ transforms the Bernoulli equation into the linear equation

$$\frac{du}{dx} + (1 - n)P(x)u = (1 - n)Q(x)$$

24–26 Use the method of Exercise 23 to solve the differential equation.

24. $xy' + y = -xy^2$

25. $y' + \dfrac{2}{x}y = \dfrac{y^3}{x^2}$

26. $y' + y = xy^3$

27. In the circuit shown in Figure 4, a battery supplies a constant voltage of 40 V, the inductance is 2 H, the resistance is 10 Ω, and $I(0) = 0$.
(a) Find $I(t)$.
(b) Find the current after 0.1 s.

28. In the circuit shown in Figure 4, a generator supplies a voltage of $E(t) = 40 \sin 60t$ volts, the inductance is 1 H, the resistance is 20 Ω, and $I(0) = 1$ A.
(a) Find $I(t)$.
(b) Find the current after 0.1 s.
(c) Use a graphing device to draw the graph of the current function.

29. The figure shows a circuit containing an electromotive force, a capacitor with a capacitance of C farads (F), and a resistor with a resistance of R ohms (Ω). The voltage drop across the

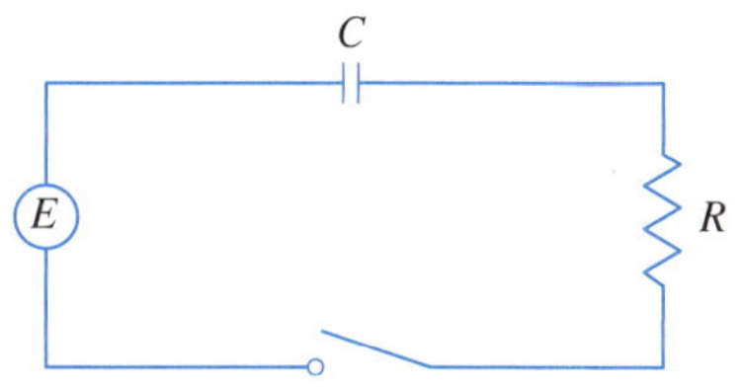

capacitor is Q/C, where Q is the charge (in coulombs), so in this case Kirchhoff's Law gives

$$RI + \frac{Q}{C} = E(t)$$

But $I = dQ/dt$ (see Example 3 in Section 3.3), so we have

$$R\frac{dQ}{dt} + \frac{1}{C}Q = E(t)$$

Suppose the resistance is 5 Ω, the capacitance is 0.05 F, a battery gives a constant voltage of 60 V, and the initial charge is $Q(0) = 0$ C. Find the charge and the current at time t.

30. In the circuit of Exercise 29, $R = 2$ Ω, $C = 0.01$ F, $Q(0) = 0$, and $E(t) = 10 \sin 60t$. Find the charge and the current at time t.

31. Let $P(t)$ be the performance level of someone learning a skill as a function of the training time t. The graph of P is called a *learning curve*. In Exercise 13 in Section 9.1 we proposed the differential equation

$$\frac{dP}{dt} = k[M - P(t)]$$

as a reasonable model for learning, where k is a positive constant. Solve it as a linear differential equation and use your solution to graph the learning curve.

32. Two new workers were hired for an assembly line. Jim processed 25 units during the first hour and 45 units during the second hour. Mark processed 35 units during the first hour and 50 units the second hour. Using the model of Exercise 31 and assuming that $P(0) = 0$, estimate the maximum number of units per hour that each worker is capable of processing.

33. In Section 9.3 we looked at mixing problems in which the volume of fluid remained constant and saw that such problems give rise to separable equations. (See Example 6 in that section.) If the rates of flow into and out of the system are different, then the volume is not constant and the resulting differential equation is linear but not separable.

A tank contains 100 L of water. A solution with a salt concentration of 0.4 kg/L is added at a rate of 5 L/min. The solution is kept mixed and is drained from the tank at a rate of 3 L/min. If $y(t)$ is the amount of salt (in kilograms) after t minutes, show that y satisfies the differential equation

$$\frac{dy}{dt} = 2 - \frac{3y}{100 + 2t}$$

Solve this equation and find the concentration after 20 minutes.

34. A tank with a capacity of 400 L is full of a mixture of water and chlorine with a concentration of 0.05 g of chlorine per liter. In order to reduce the concentration of chlorine, fresh water is pumped into the tank at a rate of 4 L/s. The mixture is kept stirred and is pumped out at a rate of 10 L/s. Find the amount of chlorine in the tank as a function of time.

35. An object with mass m is dropped from rest and we assume that the air resistance is proportional to the speed of the object. If $s(t)$ is the distance dropped after t seconds, then the speed is $v = s'(t)$ and the acceleration is $a = v'(t)$. If g is the acceleration due to gravity, then the downward force on the object is $mg - cv$, where c is a positive constant, and Newton's Second Law gives

$$m\frac{dv}{dt} = mg - cv$$

(a) Solve this as a linear equation to show that

$$v = \frac{mg}{c}(1 - e^{-ct/m})$$

(b) What is the limiting velocity?
(c) Find the distance the object has fallen after t seconds.

36. If we ignore air resistance, we can conclude that heavier objects fall no faster than lighter objects. But if we take air resistance into account, our conclusion changes. Use the expression for the velocity of a falling object in Exercise 35(a) to find dv/dm and show that heavier objects *do* fall faster than lighter ones.

9.7 Predator-Prey Systems

We have looked at a variety of models for the growth of a single species that lives alone in an environment. In this section we consider more realistic models that take into account the interaction of two species in the same habitat. We will see that these models take the form of a pair of linked differential equations.

We first consider the situation in which one species, called the *prey,* has an ample food supply and the second species, called the *predators,* feeds on the prey. Examples of prey and predators include rabbits and wolves in an isolated forest, food fish and sharks, aphids and ladybugs, and bacteria and amoebas. Our model will have two dependent variables and both are functions of time. We let $R(t)$ be the number of prey (using R for rabbits) and $W(t)$ be the number of predators (with W for wolves) at time t.

In the absence of predators, the ample food supply would support exponential growth of the prey, that is,

$$\frac{dR}{dt} = kR \qquad \text{where } k \text{ is a positive constant}$$

In the absence of prey, we assume that the predator population would decline at a rate proportional to itself, that is,

$$\frac{dW}{dt} = -rW \qquad \text{where } r \text{ is a positive constant}$$

With both species present, however, we assume that the principal cause of death among the prey is being eaten by a predator, and the birth and survival rates of the predators depend on their available food supply, namely, the prey. We also assume that the two species encounter each other at a rate that is proportional to both populations and is therefore proportional to the product RW. (The more there are of either population, the more encounters there are likely to be.) A system of two differential equations that incorporates these assumptions is as follows:

W represents the predator.

R represents the prey.

1 $$\frac{dR}{dt} = kR - aRW \qquad \frac{dW}{dt} = -rW + bRW$$

where k, r, a, and b are positive constants. Notice that the term $-aRW$ decreases the natural growth rate of the prey and the term bRW increases the natural growth rate of the predators.

||| The Lotka-Volterra equations were proposed as a model to explain the variations in the shark and food-fish populations in the Adriatic Sea by the Italian mathematician Vito Volterra (1860–1940).

The equations in (1) are known as the **predator-prey equations**, or the **Lotka-Volterra equations**. A **solution** of this system of equations is a pair of functions $R(t)$ and $W(t)$ that describe the populations of prey and predator as functions of time. Because the system is coupled (R and W occur in both equations), we can't solve one equation and then the other; we have to solve them simultaneously. Unfortunately, it is usually impossible to find explicit formulas for R and W as functions of t. We can, however, use graphical methods to analyze the equations.

EXAMPLE 1 Suppose that populations of rabbits and wolves are described by the Lotka-Volterra equations (1) with $k = 0.08$, $a = 0.001$, $r = 0.02$, and $b = 0.00002$. The time t is measured in months.

(a) Find the constant solutions (called the **equilibrium solutions**) and interpret the answer.

(b) Use the system of differential equations to find an expression for dW/dR.

(c) Draw a direction field for the resulting differential equation in the RW-plane. Then use that direction field to sketch some solution curves.

(d) Suppose that, at some point in time, there are 1000 rabbits and 40 wolves. Draw the corresponding solution curve and use it to describe the changes in both population levels.

(e) Use part (d) to make sketches of R and W as functions of t.

SOLUTION

(a) With the given values of k, a, r, and b, the Lotka-Volterra equations become

$$\frac{dR}{dt} = 0.08R - 0.001RW$$

$$\frac{dW}{dt} = -0.02W + 0.00002RW$$

Both R and W will be constant if both derivatives are 0, that is,

$$R' = R(0.08 - 0.001W) = 0$$

$$W' = W(-0.02 + 0.00002R) = 0$$

One solution is given by $R = 0$ and $W = 0$. (This makes sense: If there are no rabbits or wolves, the populations are certainly not going to increase.) The other constant solution is

$$W = \frac{0.08}{0.001} = 80 \qquad R = \frac{0.02}{0.00002} = 1000$$

So the equilibrium populations consist of 80 wolves and 1000 rabbits. This means that 1000 rabbits are just enough to support a constant wolf population of 80. There are neither too many wolves (which would result in fewer rabbits) nor too few wolves (which would result in more rabbits).

(b) We use the Chain Rule to eliminate t:

$$\frac{dW}{dt} = \frac{dW}{dR}\frac{dR}{dt}$$

so

$$\frac{dW}{dR} = \frac{\dfrac{dW}{dt}}{\dfrac{dR}{dt}} = \frac{-0.02W + 0.00002RW}{0.08R - 0.001RW}$$

(c) If we think of W as a function of R, we have the differential equation

$$\frac{dW}{dR} = \frac{-0.02W + 0.00002RW}{0.08R - 0.001RW}$$

We draw the direction field for this differential equation in Figure 1 and we use it to sketch several solution curves in Figure 2. If we move along a solution curve, we observe how the relationship between R and W changes as time passes. Notice that the curves appear to be closed in the sense that if we travel along a curve, we always return to the same point. Notice also that the point (1000, 80) is inside all the solution curves. That point is called an *equilibrium point* because it corresponds to the equilibrium solution $R = 1000$, $W = 80$.

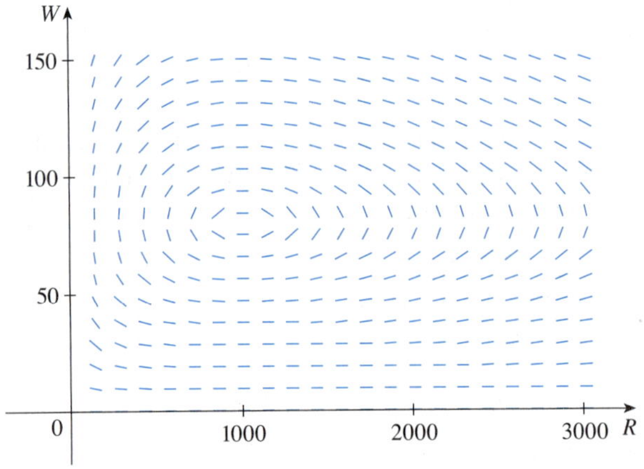

FIGURE 1 Direction field for the predator-prey system

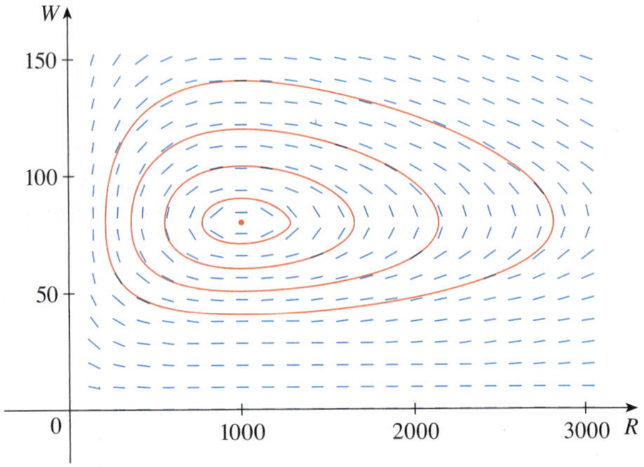

FIGURE 2 Phase portrait of the system

When we represent solutions of a system of differential equations as in Figure 2, we refer to the RW-plane as the **phase plane**, and we call the solution curves **phase trajectories**. So a phase trajectory is a path traced out by solutions (R, W) as time goes by. A **phase portrait** consists of equilibrium points and typical phase trajectories, as shown in Figure 2.

(d) Starting with 1000 rabbits and 40 wolves corresponds to drawing the solution curve through the point $P_0(1000, 40)$. Figure 3 shows this phase trajectory with the direction field removed. Starting at the point P_0 at time $t = 0$ and letting t increase, do we move clockwise or counterclockwise around the phase trajectory? If we put $R = 1000$ and

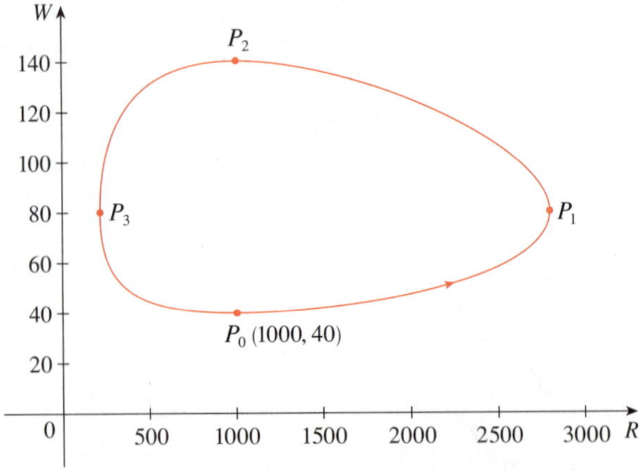

FIGURE 3
Phase trajectory through (1000, 40)

$W = 40$ in the first differential equation, we get

$$\frac{dR}{dt} = 0.08(1000) - 0.001(1000)(40) = 80 - 40 = 40$$

Since $dR/dt > 0$, we conclude that R is increasing at P_0 and so we move counterclockwise around the phase trajectory.

We see that at P_0 there aren't enough wolves to maintain a balance between the populations, so the rabbit population increases. That results in more wolves and eventually there are so many wolves that the rabbits have a hard time avoiding them. So the number of rabbits begins to decline (at P_1, where we estimate that R reaches its maximum population of about 2800). This means that at some later time the wolf population starts to fall (at P_2, where $R = 1000$ and $W \approx 140$). But this benefits the rabbits, so their population later starts to increase (at P_3, where $W = 80$ and $R \approx 210$). As a consequence, the wolf population eventually starts to increase as well. This happens when the populations return to their initial values of $R = 1000$ and $W = 40$, and the entire cycle begins again.

(e) From the description in part (d) of how the rabbit and wolf populations rise and fall, we can sketch the graphs of $R(t)$ and $W(t)$. Suppose the points P_1, P_2, and P_3 in Figure 3 are reached at times t_1, t_2, and t_3. Then we can sketch graphs of R and W as in Figure 4.

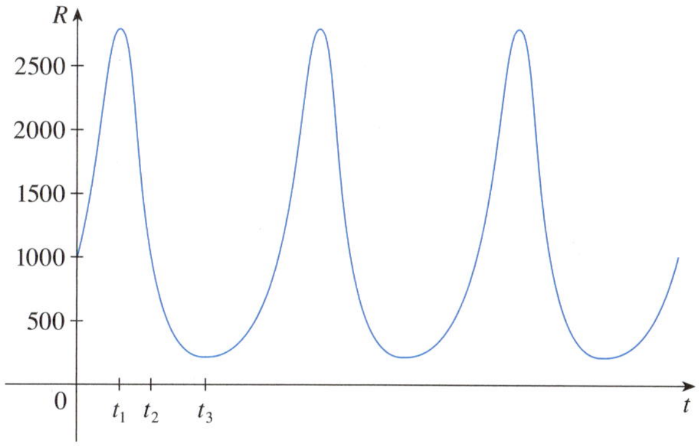

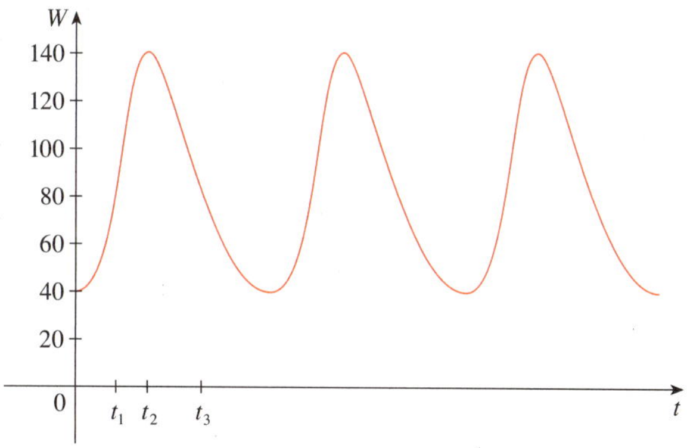

FIGURE 4
Graphs of the rabbit and wolf populations as functions of time

To make the graphs easier to compare, we draw the graphs on the same axes but with different scales for R and W, as in Figure 5. Notice that the rabbits reach their maximum populations about a quarter of a cycle before the wolves.

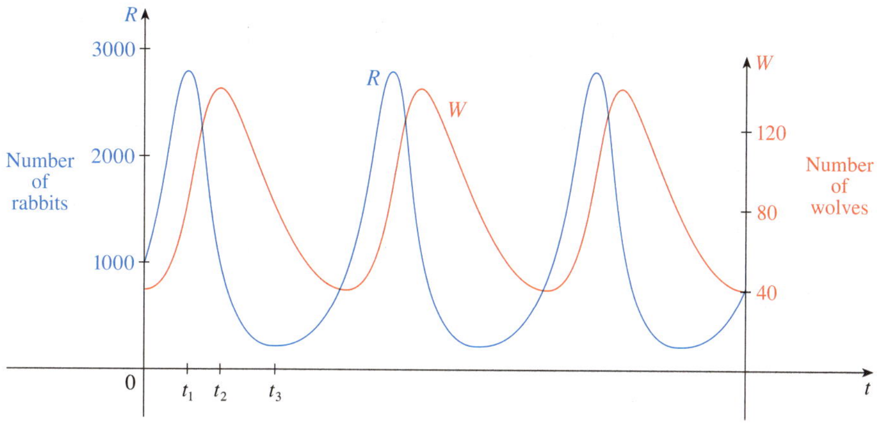

FIGURE 5
Comparison of the rabbit and wolf populations

An important part of the modeling process, as we discussed in Section 1.2, is to interpret our mathematical conclusions as real-world predictions and to test the predictions against real data. The Hudson's Bay Company, which started trading in animal furs in Canada in 1670, has kept records that date back to the 1840s. Figure 6 shows graphs of the number of pelts of the snowshoe hare and its predator, the Canada lynx, traded by the company over a 90-year period. You can see that the coupled oscillations in the hare and lynx populations predicted by the Lotka-Volterra model do actually occur and the period of these cycles is roughly 10 years.

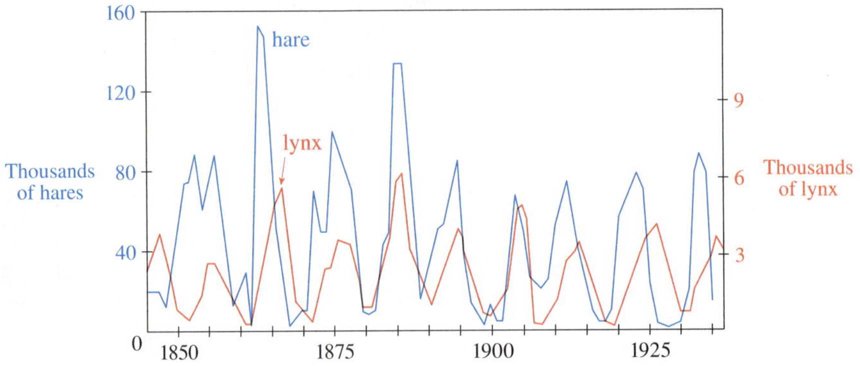

FIGURE 6

Relative abundance of hare and lynx from Hudson's Bay Company records

Although the relatively simple Lotka-Volterra model has had some success in explaining and predicting coupled populations, more sophisticated models have also been proposed. One way to modify the Lotka-Volterra equations is to assume that, in the absence of predators, the prey grow according to a logistic model with carrying capacity K. Then the Lotka-Volterra equations (1) are replaced by the system of differential equations

$$\frac{dR}{dt} = kR\left(1 - \frac{R}{K}\right) - aRW \qquad \frac{dW}{dt} = -rW + bRW$$

This model is investigated in Exercises 9 and 10.

Models have also been proposed to describe and predict population levels of two species that compete for the same resources or cooperate for mutual benefit. Such models are explored in Exercise 2.

9.7 Exercises

1. For each predator-prey system, determine which of the variables, x or y, represents the prey population and which represents the predator population. Is the growth of the prey restricted just by the predators or by other factors as well? Do the predators feed only on the prey or do they have additional food sources? Explain.

(a) $\dfrac{dx}{dt} = -0.05x + 0.0001xy$

$\dfrac{dy}{dt} = 0.1y - 0.005xy$

(b) $\dfrac{dx}{dt} = 0.2x - 0.0002x^2 - 0.006xy$

$\dfrac{dy}{dt} = -0.015y + 0.00008xy$

2. Each system of differential equations is a model for two species that either compete for the same resources or cooperate for mutual benefit (flowering plants and insect pollinators, for instance). Decide whether each system describes competition or cooperation and explain why it is a reasonable model. (Ask yourself what effect an increase in one species has on the growth rate of the other.)

(a) $\dfrac{dx}{dt} = 0.12x - 0.0006x^2 + 0.00001xy$

$\dfrac{dy}{dt} = 0.08x + 0.00004xy$

(b) $\dfrac{dx}{dt} = 0.15x - 0.0002x^2 - 0.0006xy$

$\dfrac{dy}{dt} = 0.2y - 0.00008y^2 - 0.0002xy$

3–4 A phase trajectory is shown for populations of rabbits (R) and foxes (F).

(a) Describe how each population changes as time goes by.

(b) Use your description to make a rough sketch of the graphs of R and F as functions of time.

3.

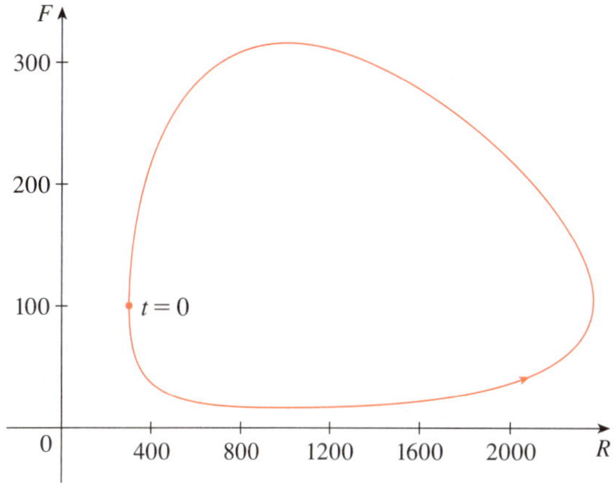

4.

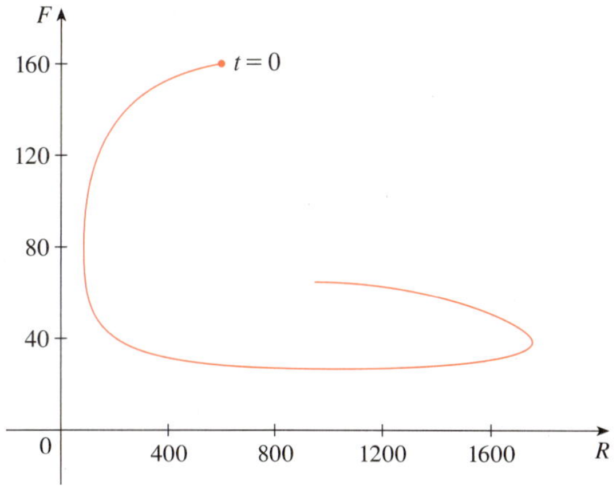

5–6 Graphs of populations of two species are shown. Use them to sketch the corresponding phase trajectory.

5.

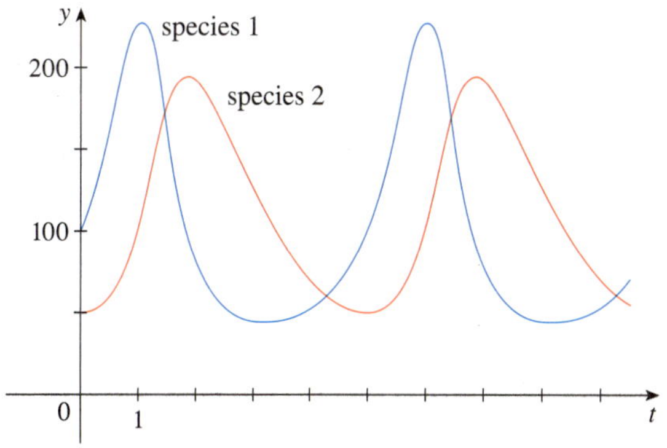

6.

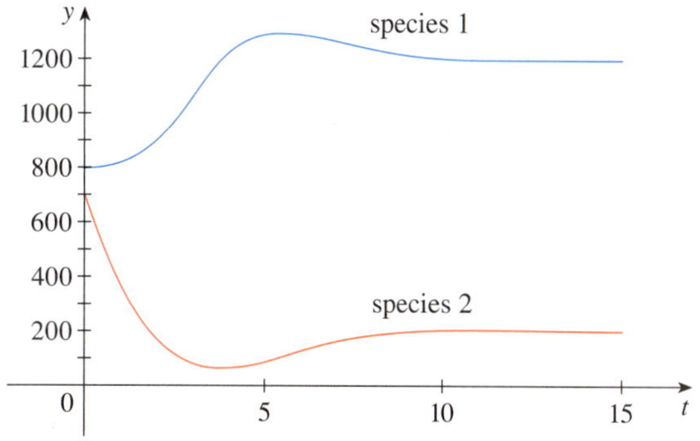

7. In Example 1(b) we showed that the rabbit and wolf populations satisfy the differential equation

$$\frac{dW}{dR} = \frac{-0.02W + 0.00002RW}{0.08R - 0.001RW}$$

By solving this separable differential equation, show that

$$\frac{R^{0.02}W^{0.08}}{e^{0.00002R}e^{0.001W}} = C$$

where C is a constant.

It is impossible to solve this equation for W as an explicit function of R (or vice versa). If you have a computer algebra system that graphs implicitly defined curves, use this equation and your CAS to draw the solution curve that passes through the point (1000, 40) and compare with Figure 3.

8. Populations of aphids and ladybugs are modeled by the equations

$$\frac{dA}{dt} = 2A - 0.01AL$$

$$\frac{dL}{dt} = -0.5L + 0.0001AL$$

(a) Find the equilibrium solutions and explain their significance.

(b) Find an expression for dL/dA.

(c) The direction field for the differential equation in part (b) is shown. Use it to sketch a phase portrait. What do the phase trajectories have in common?

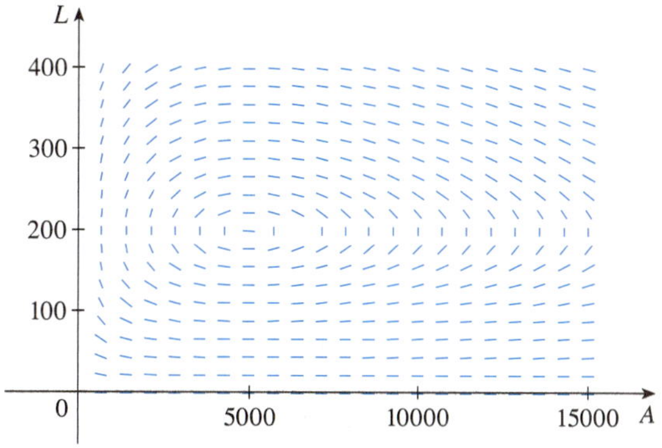

(d) Suppose that at time $t = 0$ there are 1000 aphids and 200 ladybugs. Draw the corresponding phase trajectory and use it to describe how both populations change.

(e) Use part (d) to make rough sketches of the aphid and ladybug populations as functions of t. How are the graphs related to each other?

9. In Example 1 we used Lotka-Volterra equations to model populations of rabbits and wolves. Let's modify those equations as follows:

$$\frac{dR}{dt} = 0.08R(1 - 0.0002R) - 0.001RW$$

$$\frac{dW}{dt} = -0.02W + 0.00002RW$$

(a) According to these equations, what happens to the rabbit population in the absence of wolves?

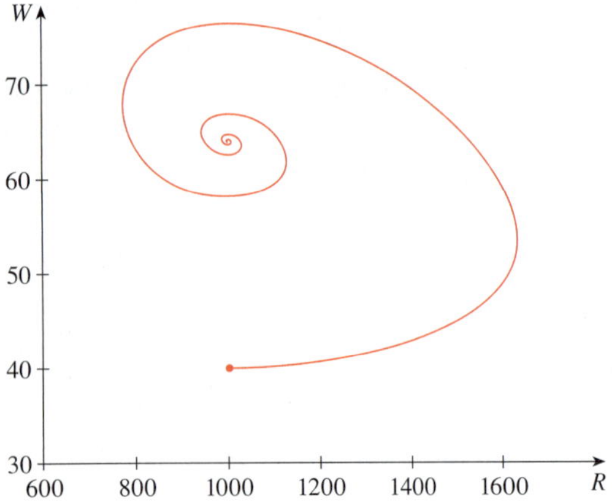

(b) Find all the equilibrium solutions and explain their significance.

(c) The figure shows the phase trajectory that starts at the point (1000, 40). Describe what eventually happens to the rabbit and wolf populations.

(d) Sketch graphs of the rabbit and wolf populations as functions of time.

CAS **10.** In Exercise 8 we modeled populations of aphids and ladybugs with a Lotka-Volterra system. Suppose we modify those equations as follows:

$$\frac{dA}{dt} = 2A(1 - 0.0001A) - 0.01AL$$

$$\frac{dL}{dt} = -0.5L + 0.0001AL$$

(a) In the absence of ladybugs, what does the model predict about the aphids?

(b) Find the equilibrium solutions.

(c) Find an expression for dL/dA.

(d) Use a computer algebra system to draw a direction field for the differential equation in part (c). Then use the direction field to sketch a phase portrait. What do the phase trajectories have in common?

(e) Suppose that at time $t = 0$ there are 1000 aphids and 200 ladybugs. Draw the corresponding phase trajectory and use it to describe how both populations change.

(f) Use part (e) to make rough sketches of the aphid and ladybug populations as functions of t. How are the graphs related to each other?

9 Review

CONCEPT CHECK

1. (a) What is a differential equation?
(b) What is the order of a differential equation?
(c) What is an initial condition?

2. What can you say about the solutions of the equation $y' = x^2 + y^2$ just by looking at the differential equation?

3. What is a direction field for the differential equation $y' = F(x, y)$?

4. Explain how Euler's method works.

5. What is a separable differential equation? How do you solve it?

6. What is a first-order linear differential equation? How do you solve it?

7. (a) Write a differential equation that expresses the law of natural growth. What does it say in terms of relative growth rate?
(b) Under what circumstances is this an appropriate model for population growth?
(c) What are the solutions of this equation?

8. (a) Write the logistic equation.
(b) Under what circumstances is this an appropriate model for population growth?

9. (a) Write Lotka-Volterra equations to model populations of food fish (F) and sharks (S).
(b) What do these equations say about each population in the absence of the other?

TRUE-FALSE QUIZ

Determine whether the statement is true or false. If it is true, explain why. If it is false, explain why or give an example that disproves the statement.

1. All solutions of the differential equation $y' = -1 - y^4$ are decreasing functions.
2. The function $f(x) = (\ln x)/x$ is a solution of the differential equation $x^2y' + xy = 1$.
3. The equation $y' = x + y$ is separable.
4. The equation $y' = 3y - 2x + 6xy - 1$ is separable.
5. The equation $e^x y' = y$ is linear.
6. The equation $y' + xy = e^y$ is linear.
7. If y is the solution of the initial-value problem

$$\frac{dy}{dt} = 2y\left(1 - \frac{y}{5}\right) \qquad y(0) = 1$$

then $\lim_{t\to\infty} y = 5$.

EXERCISES

1. (a) A direction field for the differential equation $y' = y(y-2)(y-4)$ is shown. Sketch the graphs of the solutions that satisfy the given initial conditions.
   (i) $y(0) = -0.3$ (ii) $y(0) = 1$
   (iii) $y(0) = 3$ (iv) $y(0) = 4.3$
   (b) If the initial condition is $y(0) = c$, for what values of c is $\lim_{t\to\infty} y(t)$ finite? What are the equilibrium solutions?

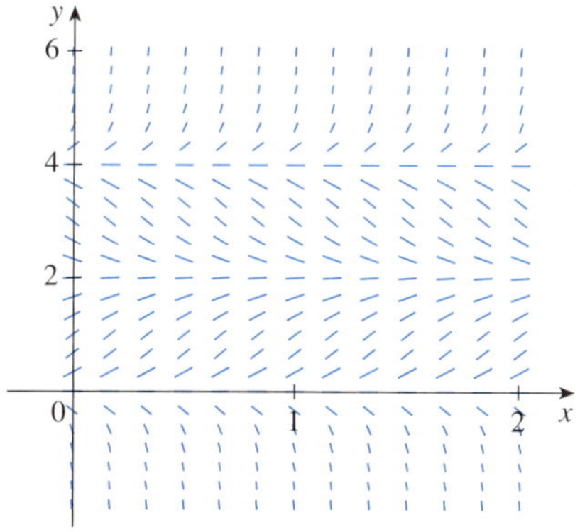

2. (a) Sketch a direction field for the differential equation $y' = x/y$. Then use it to sketch the four solutions that satisfy the initial conditions $y(0) = 1$, $y(0) = -1$, $y(2) = 1$, and $y(-2) = 1$.
   (b) Check your work in part (a) by solving the differential equation explicitly. What type of curve is each solution curve?

3. (a) A direction field for the differential equation $y' = x^2 - y^2$ is shown. Sketch the solution of the initial-value problem

$$y' = x^2 - y^2 \qquad y(0) = 1$$

Use your graph to estimate the value of $y(0.3)$.

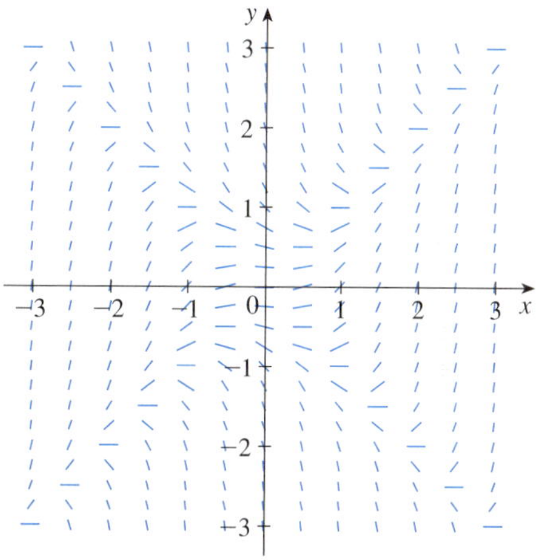

   (b) Use Euler's method with step size 0.1 to estimate $y(0.3)$ where $y(x)$ is the solution of the initial-value problem in part (a). Compare with your estimate from part (a).
   (c) On what lines are the centers of the horizontal line segments of the direction field in part (a) located? What happens when a solution curve crosses these lines?

4. (a) Use Euler's method with step size 0.2 to estimate $y(0.4)$, where $y(x)$ is the solution of the initial-value problem

$$y' = 2xy^2 \qquad y(0) = 1$$

   (b) Repeat part (a) with step size 0.1.
   (c) Find the exact solution of the differential equation and compare the value at 0.4 with the approximations in parts (a) and (b).

5–8 Solve the differential equation.

5. $y' = xe^{-\sin x} - y\cos x$
6. $\dfrac{dx}{dt} = 1 - t + x - tx$
7. $(3y^2 + 2y)y' = x\cos x$
8. $x^2y' - y = 2x^3e^{-1/x}$

9–11 Solve the initial-value problem.

9. $xyy' = \ln x, \quad y(1) = 2$

10. $1 + x = 2xyy', \quad x > 0, \quad y(1) = -2$

11. $y' + y = \sqrt{x}e^{-x}, \quad y(0) = 3$

12. Solve the initial-value problem $2yy' = xe^x$, $y(0) = 1$, and graph the solution.

13–14 Find the orthogonal trajectories of the family of curves.

13. $kx^2 + y^2 = 1$

14. $y = \dfrac{k}{1 + x^2}$

15. A bacteria culture starts with 1000 bacteria and the growth rate is proportional to the number of bacteria. After 2 hours the population is 9000.
(a) Find an expression for the number of bacteria after t hours.
(b) Find the number of bacteria after 3 h.
(c) Find the rate of growth after 3 h.
(d) How long does it take for the number of bacteria to double?

16. An isotope of strontium, ^{90}Sr, has a half-life of 25 years.
(a) Find the mass of ^{90}Sr that remains from a sample of 18 mg after t years.
(b) How long would it take for the mass to decay to 2 mg?

17. Let $C(t)$ be the concentration of a drug in the bloodstream. As the body eliminates the drug, $C(t)$ decreases at a rate that is proportional to the amount of the drug that is present at the time. Thus, $C'(t) = -kC(t)$, where k is a positive number called the *elimination constant* of the drug.
(a) If C_0 is the concentration at time $t = 0$, find the concentration at time t.
(b) If the body eliminates half the drug in 30 h, how long does it take to eliminate 90% of the drug?

18. (a) The population of the world was 5.28 billion in 1990 and 6.07 billion in 2000. Find an exponential model for these data and use the model to predict the world population in the year 2020.
(b) According to the model in part (a), when will the world population exceed 10 billion?
(c) Use the data in part (a) to find a logistic model for the population. Assume a carrying capacity of 100 billion. Then use the logistic model to predict the population in 2020. Compare with your prediction from the exponential model.
(d) According to the logistic model, when will the world population exceed 10 billion? Compare with your prediction in part (b).

19. The von Bertalanffy growth model is used to predict the length $L(t)$ of a fish over a period of time. If L_∞ is the largest length for a species, then the hypothesis is that the rate of growth in length is proportional to $L_\infty - L$, the length yet to be achieved.
(a) Formulate and solve a differential equation to find an expression for $L(t)$.
(b) For the North Sea haddock it has been determined that $L_\infty = 53$ cm, $L(0) = 10$ cm, and the constant of proportionality is 0.2. What does the expression for $L(t)$ become with these data?

20. A tank contains 100 L of pure water. Brine that contains 0.1 kg of salt per liter enters the tank at a rate of 10 L/min. The solution is kept thoroughly mixed and drains from the tank at the same rate. How much salt is in the tank after 6 minutes?

21. One model for the spread of an epidemic is that the rate of spread is jointly proportional to the number of infected people and the number of uninfected people. In an isolated town of 5000 inhabitants, 160 people have a disease at the beginning of the week and 1200 have it at the end of the week. How long does it take for 80% of the population to become infected?

22. The Brentano-Stevens Law in psychology models the way that a subject reacts to a stimulus. It states that if R represents the reaction to an amount S of stimulus, then the relative rates of increase are proportional:

$$\frac{1}{R}\frac{dR}{dt} = \frac{k}{S}\frac{dS}{dt}$$

where k is a positive constant. Find R as a function of S.

23. The transport of a substance across a capillary wall in lung physiology has been modeled by the differential equation

$$\frac{dh}{dt} = -\frac{R}{V}\left(\frac{h}{k + h}\right)$$

where h is the hormone concentration in the bloodstream, t is time, R is the maximum transport rate, V is the volume of the capillary, and k is a positive constant that measures the affinity between the hormones and the enzymes that assist the process. Solve this differential equation to find a relationship between h and t.

24. Populations of birds and insects are modeled by the equations

$$\frac{dx}{dt} = 0.4x - 0.002xy$$

$$\frac{dy}{dt} = -0.2y + 0.000008xy$$

(a) Which of the variables, x or y, represents the bird population and which represents the insect population? Explain.

(b) Find the equilibrium solutions and explain their significance.
(c) Find an expression for dy/dx.
(d) The direction field for the differential equation in part (c) is shown. Use it to sketch the phase trajectory corresponding to initial populations of 100 birds and 40,000 insects. Then use the phase trajectory to describe how both populations change.

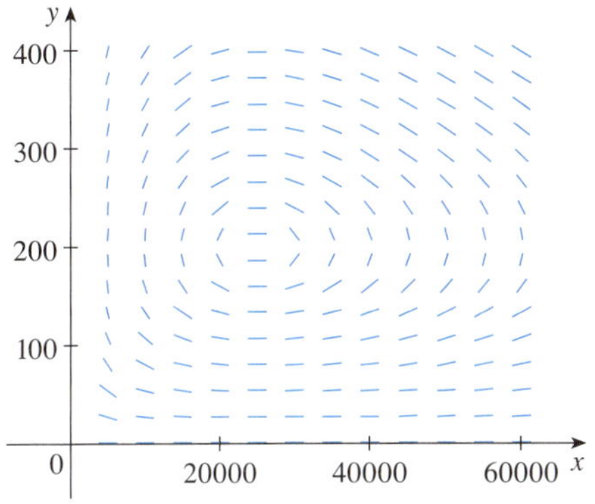

(e) Use part (d) to make rough sketches of the bird and insect populations as functions of time. How are these graphs related to each other?

25. Suppose the model of Exercise 24 is replaced by the equations

$$\frac{dx}{dt} = 0.4x(1 - 0.000005x) - 0.002xy$$

$$\frac{dy}{dt} = -0.2y + 0.000008xy$$

(a) According to these equations, what happens to the insect population in the absence of birds?
(b) Find the equilibrium solutions and explain their significance.
(c) The figure shows the phase trajectory that starts with 100 birds and 40,000 insects. Describe what eventually happens to the bird and insect populations.

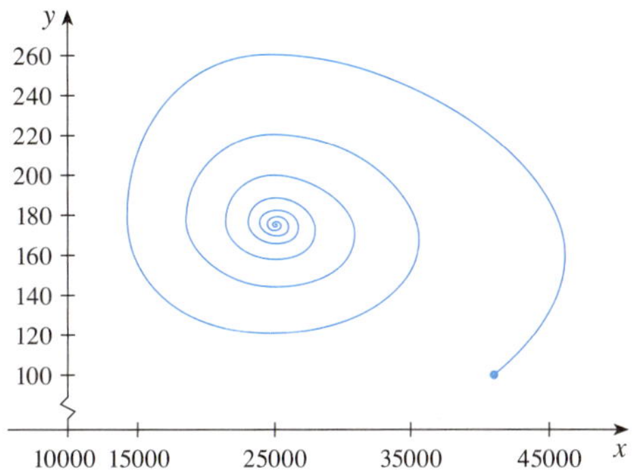

(d) Sketch graphs of the bird and insect populations as functions of time.

26. Barbara weighs 60 kg and is on a diet of 1600 calories per day, of which 850 are used automatically by basal metabolism. She spends about 15 cal/kg/day times her weight doing exercise. If 1 kg of fat contains 10,000 cal and we assume that the storage of calories in the form of fat is 100% efficient, formulate a differential equation and solve it to find her weight as a function of time. Does her weight ultimately approach an equilibrium weight?

27. When a flexible cable of uniform density is suspended between two fixed points and hangs of its own weight, the shape $y = f(x)$ of the cable must satisfy a differential equation of the form

$$\frac{d^2y}{dx^2} = k\sqrt{1 + \left(\frac{dy}{dx}\right)^2}$$

where k is a positive constant. Consider the cable shown in the figure.
(a) Let $z = dy/dx$ in the differential equation. Solve the resulting first-order differential equation (in z), and then integrate to find y.
(b) Determine the length of the cable.

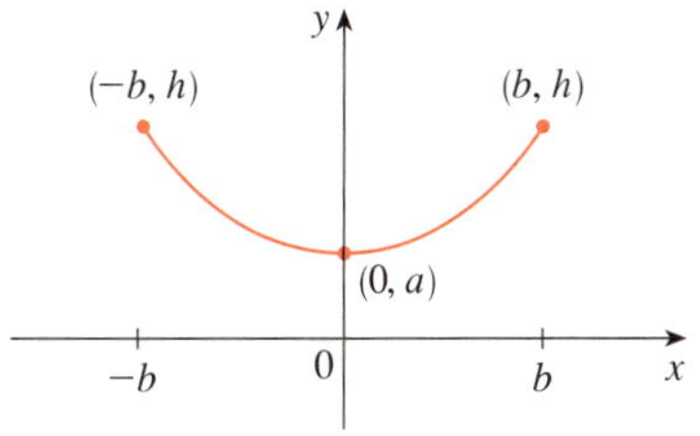

PROBLEMS PLUS

1. Find all functions f such that f' is continuous and

$$[f(x)]^2 = 100 + \int_0^x \{[f(t)]^2 + [f'(t)]^2\}\, dt \qquad \text{for all real } x$$

2. A student forgot the Product Rule for differentiation and made the mistake of thinking that $(fg)' = f'g'$. However, he was lucky and got the correct answer. The function f that he used was $f(x) = e^{x^2}$ and the domain of his problem was the interval $(\frac{1}{2}, \infty)$. What was the function g?

3. Let f be a function with the property that $f(0) = 1$, $f'(0) = 1$, and $f(a + b) = f(a)f(b)$ for all real numbers a and b. Show that $f'(x) = f(x)$ for all x and deduce that $f(x) = e^x$.

4. Find all functions f that satisfy the equation

$$\left(\int f(x)\, dx\right)\left(\int \frac{1}{f(x)}\, dx\right) = -1$$

5. A peach pie is taken out of the oven at 5:00 P.M. At that time it is piping hot: 100°C. At 5:10 P.M. its temperature is 80°C; at 5:20 P.M. it is 65°C. What is the temperature of the room?

6. Snow began to fall during the morning of February 2 and continued steadily into the afternoon. At noon a snowplow began removing snow from a road at a constant rate. The plow traveled 6 km from noon to 1 P.M. but only 3 km from 1 P.M. to 2 P.M. When did the snow begin to fall? [*Hints:* To get started, let t be the time measured in hours after noon; let $x(t)$ be the distance traveled by the plow at time t; then the speed of the plow is dx/dt. Let b be the number of hours before noon that it began to snow. Find an expression for the height of the snow at time t. Then use the given information that the rate of removal R (in m^3/h) is constant.]

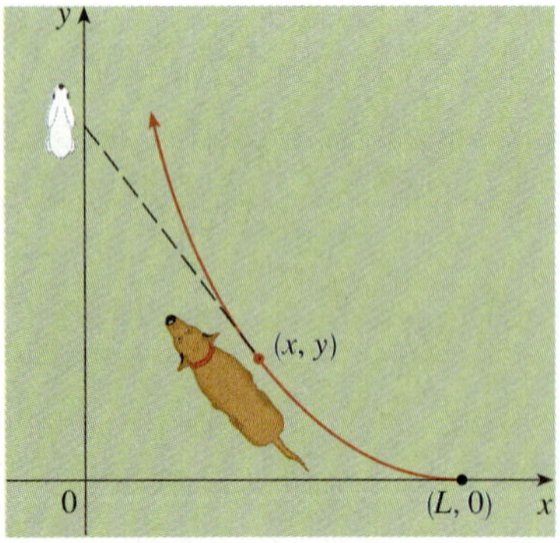

FIGURE FOR PROBLEM 7

7. A dog sees a rabbit running in a straight line across an open field and gives chase. In a rectangular coordinate system (as shown in the figure), assume:
   (i) The rabbit is at the origin and the dog is at the point $(L, 0)$ at the instant the dog first sees the rabbit.
   (ii) The rabbit runs up the y-axis and the dog always runs straight for the rabbit.
   (iii) The dog runs at the same speed as the rabbit.

   (a) Show that the dog's path is the graph of the function $y = f(x)$, where y satisfies the differential equation

$$x\frac{d^2y}{dx^2} = \sqrt{1 + \left(\frac{dy}{dx}\right)^2}$$

   (b) Determine the solution of the equation in part (a) that satisfies the initial conditions $y = y' = 0$ when $x = L$. [*Hint:* Let $z = dy/dx$ in the differential equation and solve the resulting first-order equation to find z; then integrate z to find y.]
   (c) Does the dog ever catch the rabbit?

8. (a) Suppose that the dog in Problem 7 runs twice as fast as the rabbit. Find a differential equation for the path of the dog. Then solve it to find the point where the dog catches the rabbit.
   (b) Suppose the dog runs half as fast as the rabbit. How close does the dog get to the rabbit? What are their positions when they are closest?

9. A planning engineer for a new alum plant must present some estimates to his company regarding the capacity of a silo designed to contain bauxite ore until it is processed into alum. The ore resembles pink talcum powder and is poured from a conveyor at the top of the silo. The silo is a cylinder 100 ft high with a radius of 200 ft. The conveyor carries $60{,}000\pi \text{ ft}^3/\text{h}$ and the ore maintains a conical shape whose radius is 1.5 times its height.

(a) If, at a certain time t, the pile is 60 ft high, how long will it take for the pile to reach the top of the silo?

(b) Management wants to know how much room will be left in the floor area of the silo when the pile is 60 ft high. How fast is the floor area of the pile growing at that height?

(c) Suppose a loader starts removing the ore at the rate of $20{,}000\pi \text{ ft}^3/\text{h}$ when the height of the pile reaches 90 ft. Suppose, also, that the pile continues to maintain its shape. How long will it take for the pile to reach the top of the silo under these conditions?

10. Find the curve that passes through the point (3, 2) and has the property that if the tangent line is drawn at any point P on the curve, then the part of the tangent line that lies in the first quadrant is bisected at P.

11. Recall that the normal line to a curve at a point P on the curve is the line that passes through P and is perpendicular to the tangent line at P. Find the curve that passes through the point (3, 2) and has the property that if the normal line is drawn at any point on the curve, then the y-intercept of the normal line is always 6.

12. Find all curves with the property that if the normal line is drawn at any point P on the curve, then the part of the normal line between P and the x-axis is bisected by the y-axis.

Chapter 10

Parametric curves are used to represent letters and other symbols on laser printers. See the Laboratory Project on page 669.

Parametric Equations and Polar Coordinates

So far we have described plane curves by giving y as a function of x $[y = f(x)]$ or x as a function of y $[x = g(y)]$ or by giving a relation between x and y that defines y implicitly as a function of x $[f(x, y) = 0]$. In this chapter we discuss two new methods for describing curves.

Some curves, such as the cycloid, are best handled when both x and y are given in terms of a third variable t called a parameter $[x = f(t),\ y = g(t)]$. Other curves, such as the cardioid, have their most convenient description when we use a new coordinate system, called the polar coordinate system.

10.1 Curves Defined by Parametric Equations

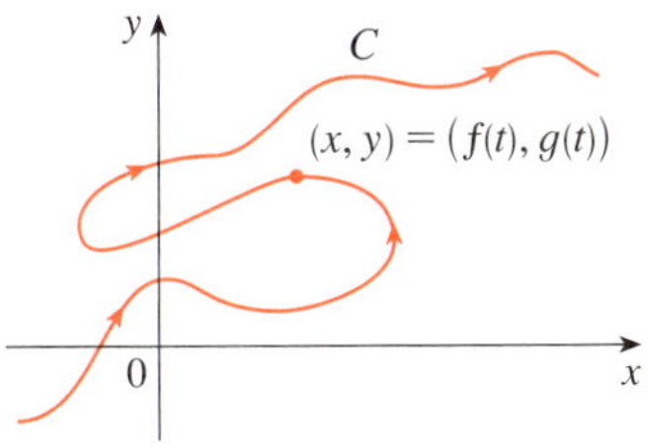

FIGURE 1

Imagine that a particle moves along the curve C shown in Figure 1. It is impossible to describe C by an equation of the form $y = f(x)$ because C fails the Vertical Line Test. But the x- and y-coordinates of the particle are functions of time and so we can write $x = f(t)$ and $y = g(t)$. Such a pair of equations is often a convenient way of describing a curve and gives rise to the following definition.

Suppose that x and y are both given as functions of a third variable t (called a **parameter**) by the equations

$$x = f(t) \qquad y = g(t)$$

(called **parametric equations**). Each value of t determines a point (x, y), which we can plot in a coordinate plane. As t varies, the point $(x, y) = (f(t), g(t))$ varies and traces out a curve C, which we call a **parametric curve**. The parameter t does not necessarily represent time and, in fact, we could use a letter other than t for the parameter. But in many applications of parametric curves, t does denote time and therefore we can interpret $(x, y) = (f(t), g(t))$ as the position of a particle at time t.

EXAMPLE 1 Sketch and identify the curve defined by the parametric equations

$$x = t^2 - 2t \qquad y = t + 1$$

SOLUTION Each value of t gives a point on the curve, as shown in the table. For instance, if $t = 0$, then $x = 0$, $y = 1$ and so the corresponding point is $(0, 1)$. In Figure 2 we plot the points (x, y) determined by several values of the parameter and we join them to produce a curve.

t	x	y
-2	8	-1
-1	3	0
0	0	1
1	-1	2
2	0	3
3	3	4
4	8	5

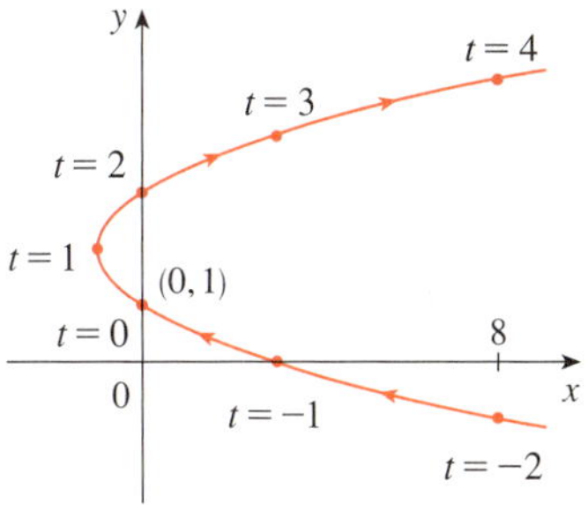

FIGURE 2

A particle whose position is given by the parametric equations moves along the curve in the direction of the arrows as t increases. Notice that the consecutive points marked on the curve appear at equal time intervals but not at equal distances. That is because the particle slows down and then speeds up as t increases.

It appears from Figure 2 that the curve traced out by the particle may be a parabola. This can be confirmed by eliminating the parameter t as follows. We obtain $t = y - 1$ from the second equation and substitute into the first equation. This gives

$$x = t^2 - 2t = (y - 1)^2 - 2(y - 1) = y^2 - 4y + 3$$

and so the curve represented by the given parametric equations is the parabola $x = y^2 - 4y + 3$.

No restriction was placed on the parameter t in Example 1, so we assumed that t could be any real number. But sometimes we restrict t to lie in a finite interval. For instance, the parametric curve

$$x = t^2 - 2t \qquad y = t + 1 \qquad 0 \leqslant t \leqslant 4$$

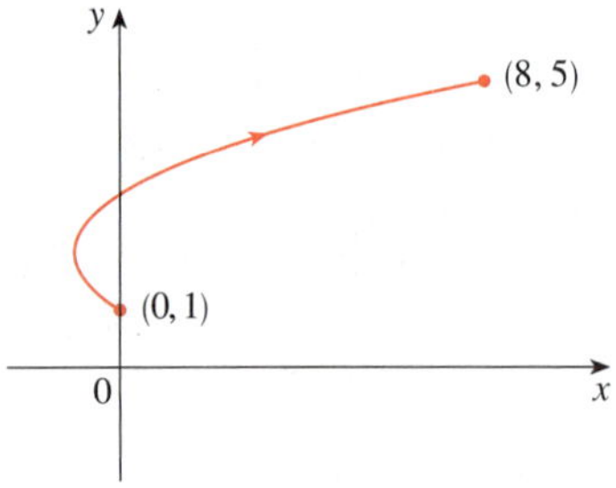

FIGURE 3

shown in Figure 3 is the part of the parabola in Example 1 that starts at the point $(0, 1)$ and ends at the point $(8, 5)$. The arrowhead indicates the direction in which the curve is traced as t increases from 0 to 4.

In general, the curve with parametric equations

$$x = f(t) \qquad y = g(t) \qquad a \leqslant t \leqslant b$$

has **initial point** $(f(a), g(a))$ and **terminal point** $(f(b), g(b))$.

EXAMPLE 2 What curve is represented by the parametric equations $x = \cos t$, $y = \sin t$, $0 \leqslant t \leqslant 2\pi$?

SOLUTION If we plot points, it appears that the curve is a circle. We can confirm this impression by eliminating t. Observe that

$$x^2 + y^2 = \cos^2 t + \sin^2 t = 1$$

Thus, the point (x, y) moves on the unit circle $x^2 + y^2 = 1$. Notice that in this example the parameter t can be interpreted as the angle (in radians) shown in Figure 4. As t increases from 0 to 2π, the point $(x, y) = (\cos t, \sin t)$ moves once around the circle in the counterclockwise direction starting from the point $(1, 0)$.

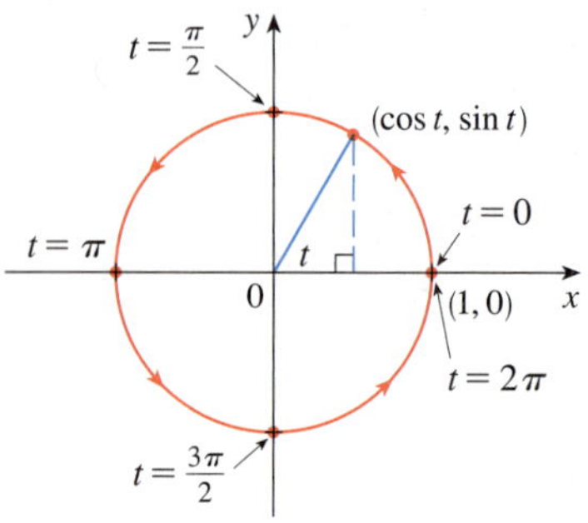

FIGURE 4

EXAMPLE 3 What curve is represented by the parametric equations $x = \sin 2t$, $y = \cos 2t$, $0 \leqslant t \leqslant 2\pi$?

SOLUTION Again we have

$$x^2 + y^2 = \sin^2 2t + \cos^2 2t = 1$$

so the parametric equations again represent the unit circle $x^2 + y^2 = 1$. But as t increases from 0 to 2π, the point $(x, y) = (\sin 2t, \cos 2t)$ starts at $(0, 1)$ and moves *twice* around the circle in the clockwise direction as indicated in Figure 5.

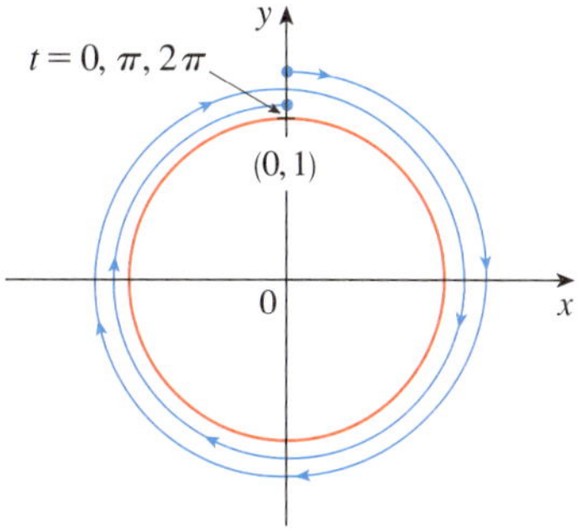

FIGURE 5

Examples 2 and 3 show that different sets of parametric equations can represent the same curve. Thus, we distinguish between a *curve,* which is a set of points, and a *parametric curve,* in which the points are traced in a particular way.

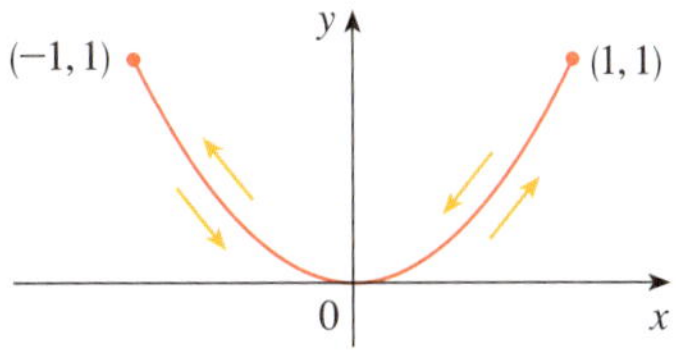

FIGURE 6

EXAMPLE 4 Sketch the curve with parametric equations $x = \sin t$, $y = \sin^2 t$.

SOLUTION Observe that $y = (\sin t)^2 = x^2$ and so the point (x, y) moves on the parabola $y = x^2$. But note also that, since $-1 \leqslant \sin t \leqslant 1$, we have $-1 \leqslant x \leqslant 1$, so the parametric equations represent only the part of the parabola for which $-1 \leqslant x \leqslant 1$. Since $\sin t$ is periodic, the point $(x, y) = (\sin t, \sin^2 t)$ moves back and forth infinitely often along the parabola from $(-1, 1)$ to $(1, 1)$. (See Figure 6.)

TEC Module 10.1A gives an animation of the relationship between motion along a parametric curve $x = f(t)$, $y = g(t)$ and motion along the graphs of f and g as functions of t. Clicking on TRIG gives you the family of parametric curves

$$x = a \cos bt \qquad y = c \sin dt$$

If you choose $a = b = c = d = 1$ and click START, you will see how the graphs of $x = \cos t$ and $y = \sin t$ relate to the circle in Example 2. If you choose $a = b = c = 1$, $d = 2$, you will see graphs as in Figure 7. By clicking on PAUSE and then repeatedly on STEP, you can see from the color coding how motion along the graphs of $x = \cos t$ and $y = \sin 2t$ corresponds to motion along the parametric curve, which is called a **Lissajous figure**.

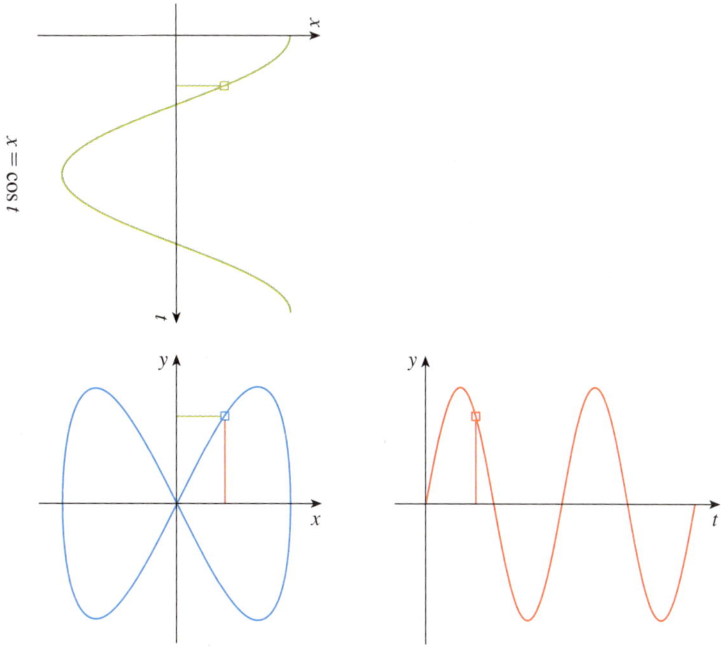

FIGURE 7 $x = \cos t \quad y = \sin 2t$ $\qquad$ $y = \sin 2t$

Graphing Devices

Most graphing calculators and computer graphing programs can be used to graph curves defined by parametric equations. In fact, it's instructive to watch a parametric curve being drawn by a graphing calculator because the points are plotted in order as the corresponding parameter values increase.

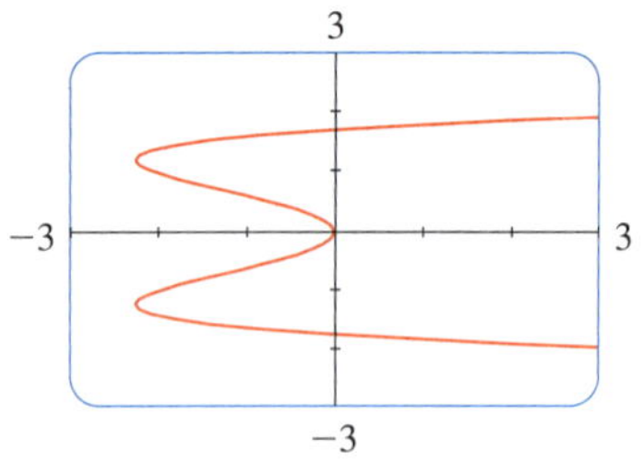

FIGURE 8

EXAMPLE 5 Use a graphing device to graph the curve $x = y^4 - 3y^2$.

SOLUTION If we let the parameter be $t = y$, then we have the equations

$$x = t^4 - 3t^2 \qquad y = t$$

Using these parametric equations to graph the curve, we obtain Figure 8. It would be possible to solve the given equation ($x = y^4 - 3y^2$) for y as four functions of x and graph them individually, but the parametric equations provide a much easier method.

In general, if we need to graph an equation of the form $x = g(y)$, we can use the parametric equations

$$x = g(t) \qquad y = t$$

Notice also that curves with equations $y = f(x)$ (the ones we are most familiar with—graphs of functions) can also be regarded as curves with parametric equations

$$x = t \qquad y = f(t)$$

Graphing devices are particularly useful when sketching complicated curves. For instance, the curves shown in Figures 9 and 10 would be virtually impossible to produce by hand.

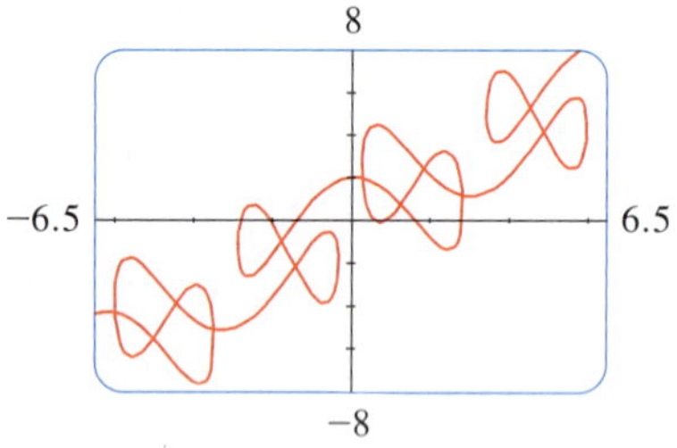

FIGURE 9

$x = t + 2\sin 2t$, $y = t + 2\cos 5t$, $-2\pi \leqslant t \leqslant 2\pi$

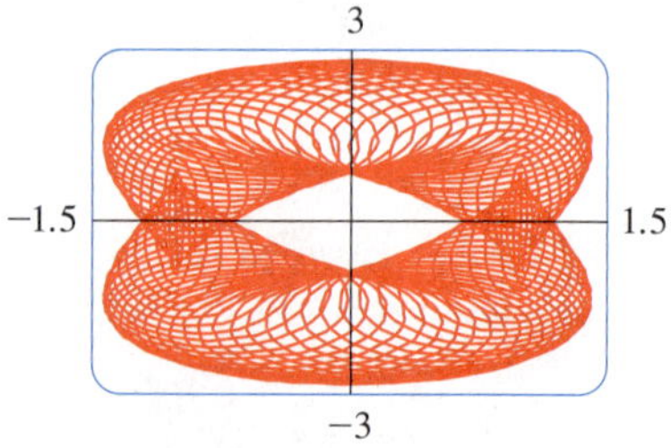

FIGURE 10

$x = \cos t - \cos 80t \sin t$, $y = 2\sin t - \sin 80t$, $0 \leqslant t \leqslant 2\pi$

One of the most important uses of parametric curves is in computer-aided design (CAD). In the Laboratory Project after Section 10.2 we will investigate special parametric curves, called **Bézier curves**, that are used extensively in manufacturing, especially in the automotive industry. These curves are also employed in specifying the shapes of letters and other symbols in laser printers.

The Cycloid

TEC An animation in Module 10.1B shows how the cycloid is formed as the circle moves.

EXAMPLE 6 The curve traced out by a point P on the circumference of a circle as the circle rolls along a straight line is called a **cycloid** (see Figure 11). If the circle has radius r and rolls along the x-axis and if one position of P is the origin, find parametric equations for the cycloid.

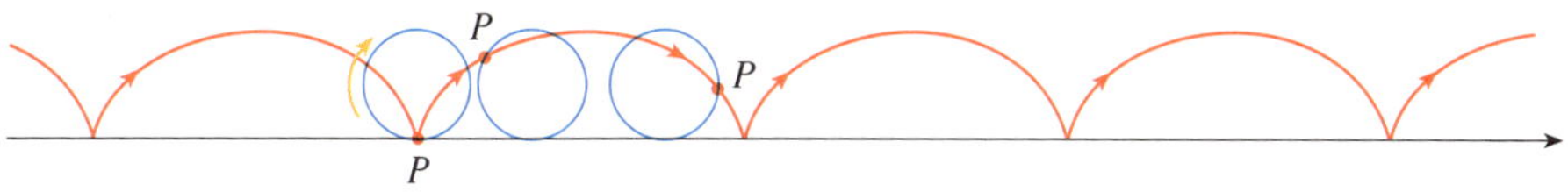

FIGURE 11

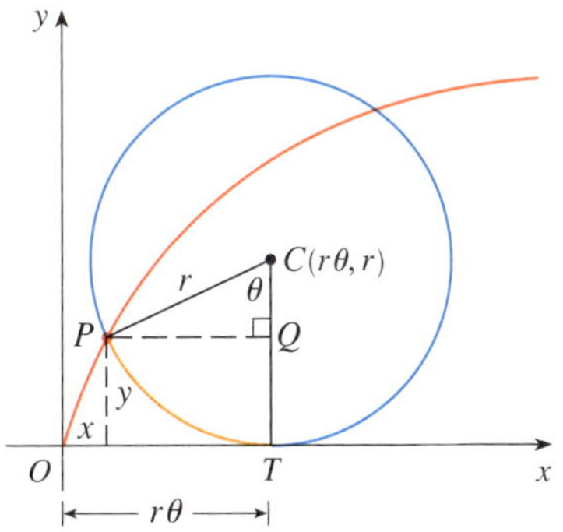

FIGURE 12

SOLUTION We choose as parameter the angle of rotation θ of the circle ($\theta = 0$ when P is at the origin). Suppose the circle has rotated through θ radians. Because the circle has been in contact with the line, we see from Figure 12 that the distance it has rolled from the origin is

$$|OT| = \text{arc } PT = r\theta$$

Therefore, the center of the circle is $C(r\theta, r)$. Let the coordinates of P be (x, y). Then from Figure 12 we see that

$$x = |OT| - |PQ| = r\theta - r\sin\theta = r(\theta - \sin\theta)$$

$$y = |TC| - |QC| = r - r\cos\theta = r(1 - \cos\theta)$$

Therefore, parametric equations of the cycloid are

1 $$x = r(\theta - \sin\theta) \qquad y = r(1 - \cos\theta) \qquad \theta \in \mathbb{R}$$

One arch of the cycloid comes from one rotation of the circle and so is described by $0 \leqslant \theta \leqslant 2\pi$. Although Equations 1 were derived from Figure 12, which illustrates the case where $0 < \theta < \pi/2$, it can be seen that these equations are still valid for other values of θ (see Exercise 37).

Although it is possible to eliminate the parameter θ from Equations 1, the resulting Cartesian equation in x and y is very complicated and not as convenient to work with as the parametric equations.

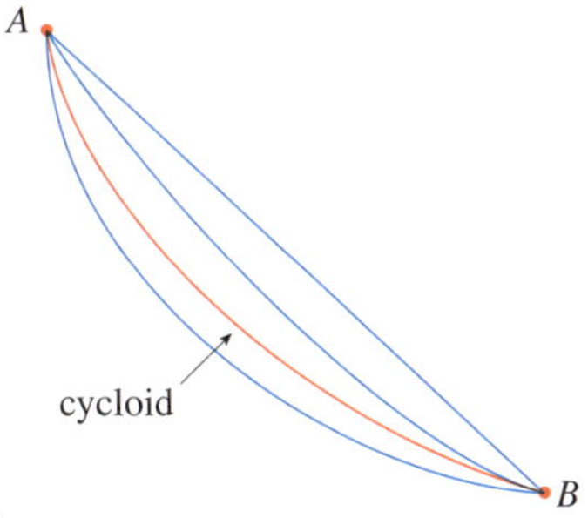

FIGURE 13

One of the first people to study the cycloid was Galileo, who proposed that bridges be built in the shape of cycloids and who tried to find the area under one arch of a cycloid. Later this curve arose in connection with the *Brachistochrone problem:* Find the curve along which a particle will slide in the shortest time (under the influence of gravity) from a point A to a lower point B not directly beneath A. The Swiss mathematician John Bernoulli, who posed this problem in 1696, showed that among all possible curves that join A to B, as in Figure 13, the particle will take the least time sliding from A to B if the curve is part of an inverted arch of a cycloid.

The Dutch physicist Huygens had already shown that the cycloid is also the solution to the *Tautochrone problem;* that is, no matter where a particle P is placed on an inverted cycloid, it takes the same time to slide to the bottom (see Figure 14). Huygens proposed that pendulum clocks (which he invented) swing in cycloidal arcs because then the pendulum takes the same time to make a complete oscillation whether it swings through a wide or a small arc.

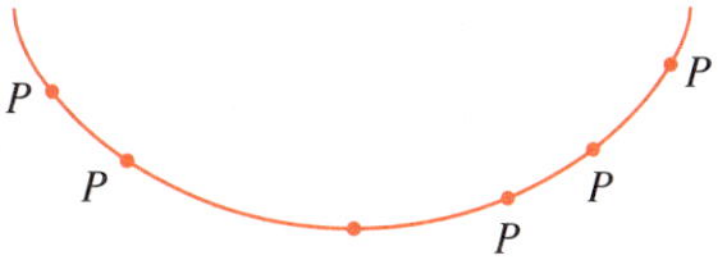

FIGURE 14

Families of Parametric Curves

EXAMPLE 7 Investigate the family of curves with parametric equations

$$x = a + \cos t \qquad y = a\tan t + \sin t$$

What do these curves have in common? How does the shape change as a increases?

SOLUTION We use a graphing device to produce the graphs for the cases $a = -2, -1, -0.5, -0.2, 0, 0.5, 1$, and 2 shown in Figure 15. Notice that all of these curves (except the case $a = 0$) have two branches, and both branches approach the vertical asymptote $x = a$ as x approaches a from the left or right.

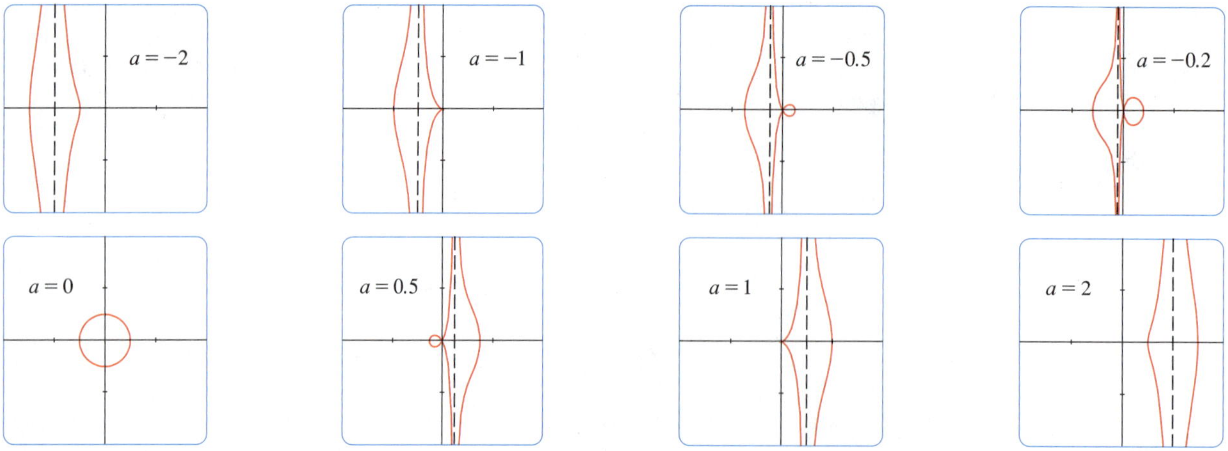

FIGURE 15 Members of the family $x = a + \cos t$, $y = a \tan t + \sin t$, all graphed in the viewing rectangle $[-4, 4]$ by $[-4, 4]$

When $a < -1$, both branches are smooth; but when a reaches -1, the right branch acquires a sharp point, called a *cusp*. For a between -1 and 0 the cusp turns into a loop, which becomes larger as a approaches 0. When $a = 0$, both branches come together and form a circle (see Example 2). For a between 0 and 1, the left branch has a loop, which shrinks to become a cusp when $a = 1$. For $a > 1$, the branches become smooth again, and as a increases further, they become less curved. Notice that the curves with a positive are reflections about the y-axis of the corresponding curves with a negative.

These curves are called **conchoids of Nicomedes** after the ancient Greek scholar Nicomedes. He called them conchoids because the shape of their outer branches resembles that of a conch shell or mussel shell.

10.1 Exercises

1–4 Sketch the curve by using the parametric equations to plot points. Indicate with an arrow the direction in which the curve is traced as t increases.

1. $x = 1 + \sqrt{t}, \quad y = t^2 - 4t, \quad 0 \leqslant t \leqslant 5$

2. $x = 2\cos t, \quad y = t - \cos t, \quad 0 \leqslant t \leqslant 2\pi$

3. $x = 5\sin t, \quad y = t^2, \quad -\pi \leqslant t \leqslant \pi$

4. $x = e^{-t} + t, \quad y = e^{t} - t, \quad -2 \leqslant t \leqslant 2$

5–10

(a) Sketch the curve by using the parametric equations to plot points. Indicate with an arrow the direction in which the curve is traced as t increases.

(b) Eliminate the parameter to find a Cartesian equation of the curve.

5. $x = 3t - 5, \quad y = 2t + 1$

6. $x = 1 + t, \quad y = 5 - 2t, \quad -2 \leqslant t \leqslant 3$

7. $x = t^2 - 2, \quad y = 5 - 2t, \quad -3 \leqslant t \leqslant 4$

8. $x = 1 + 3t, \quad y = 2 - t^2$

9. $x = \sqrt{t}, \quad y = 1 - t$

10. $x = t^2, \quad y = t^3$

11–18

(a) Eliminate the parameter to find a Cartesian equation of the curve.

(b) Sketch the curve and indicate with an arrow the direction in which the curve is traced as the parameter increases.

11. $x = \sin\theta, \quad y = \cos\theta, \quad 0 \leqslant \theta \leqslant \pi$

12. $x = 4\cos\theta, \quad y = 5\sin\theta, \quad -\pi/2 \leqslant \theta \leqslant \pi/2$

13. $x = \sin^2\theta, \quad y = \cos^2\theta$

14. $x = \sec\theta, \quad y = \tan\theta, \quad -\pi/2 < \theta < \pi/2$

15. $x = e^{t}, \quad y = e^{-t}$

16. $x = \ln t, \quad y = \sqrt{t}, \quad t \geqslant 1$

17. $x = \cosh t, \quad y = \sinh t$

18. $x = 1 + \cos \theta, \quad y = 2 \cos \theta - 1$

19–22 Describe the motion of a particle with position (x, y) as t varies in the given interval.

19. $x = \cos \pi t, \quad y = \sin \pi t, \quad 1 \leqslant t \leqslant 2$

20. $x = 2 + \cos t, \quad y = 3 + \sin t, \quad 0 \leqslant t \leqslant 2\pi$

21. $x = 2 \sin t, \quad y = 3 \cos t, \quad 0 \leqslant t \leqslant 2\pi$

22. $x = \cos^2 t, \quad y = \cos t, \quad 0 \leqslant t \leqslant 4\pi$

23. Suppose a curve is given by the parametric equations $x = f(t)$, $y = g(t)$, where the range of f is $[1, 4]$ and the range of g is $[2, 3]$. What can you say about the curve?

24. Match the graphs of the parametric equations $x = f(t)$ and $y = g(t)$ in (a)–(d) with the parametric curves labeled I–IV. Give reasons for your choices.

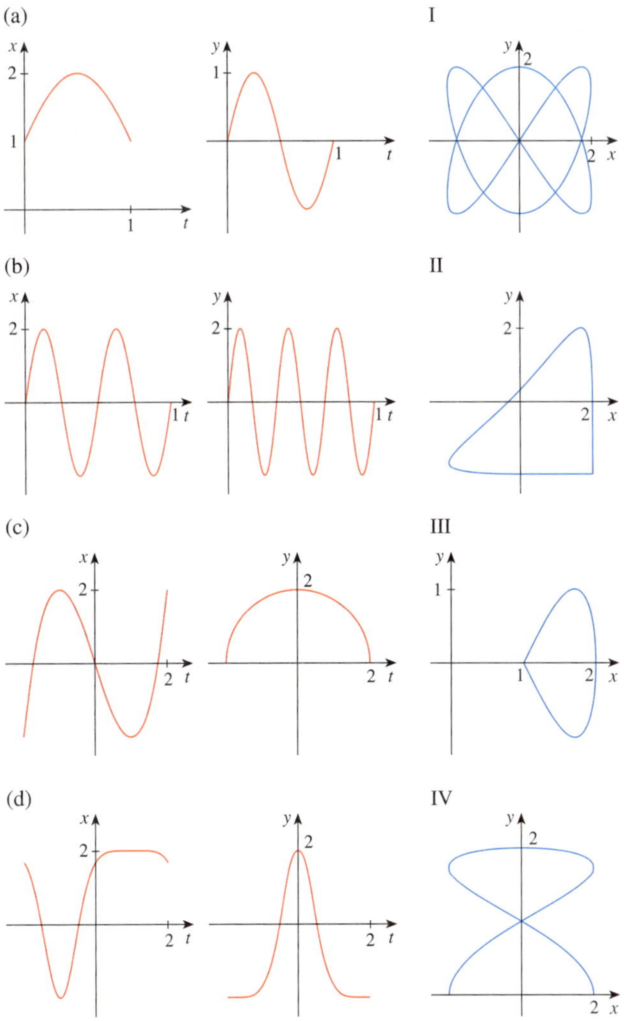

25–27 Use the graphs of $x = f(t)$ and $y = g(t)$ to sketch the parametric curve $x = f(t)$, $y = g(t)$. Indicate with arrows the direction in which the curve is traced as t increases.

25.

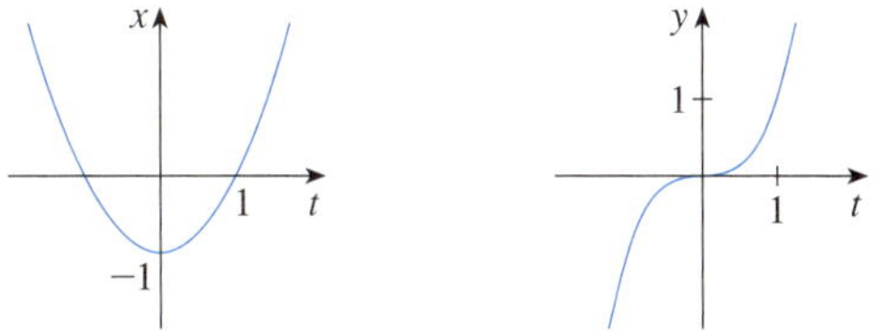

26.

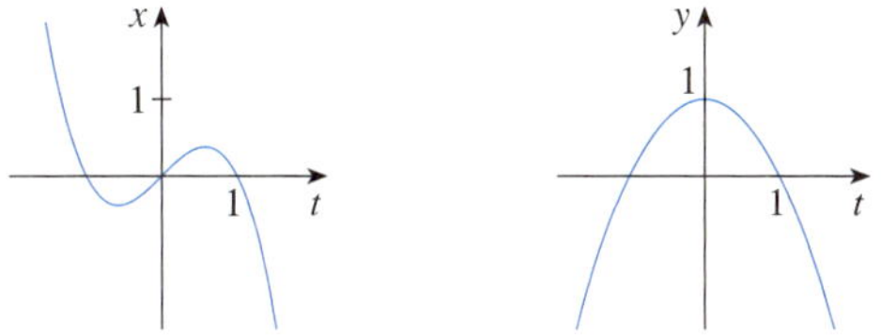

27.

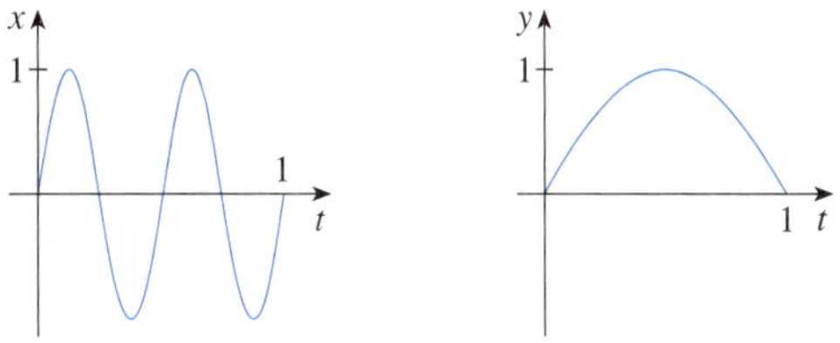

28. Match the parametric equations with the graphs labeled I–VI. Give reasons for your choices. (Do not use a graphing device.)

(a) $x = t^3 - 2t, \quad y = t^2 - t$

(b) $x = t^3 - 1, \quad y = 2 - t^2$

(c) $x = \sin 3t, \quad y = \sin 4t$

(d) $x = t + \sin 2t, \quad y = t + \sin 3t$

(e) $x = \sin(t + \sin t), \quad y = \cos(t + \cos t)$

(f) $x = \cos t, \quad y = \sin(t + \sin 5t)$

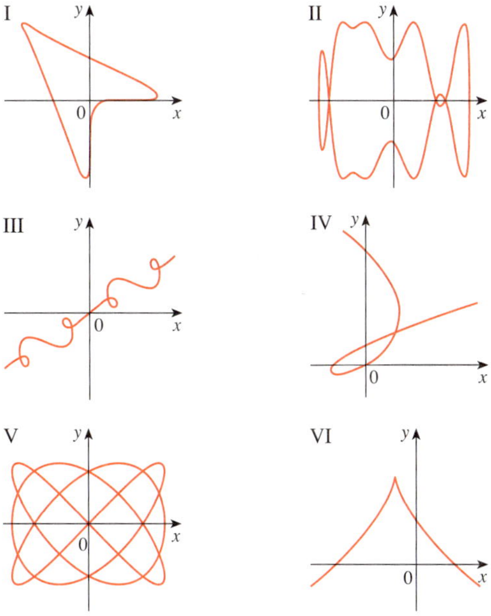

29. Graph the curve $x = y - 3y^3 + y^5$.

30. Graph the curves $y = x^5$ and $x = y(y-1)^2$ and find their points of intersection correct to one decimal place.

31. (a) Show that the parametric equations

$$x = x_1 + (x_2 - x_1)t \qquad y = y_1 + (y_2 - y_1)t$$

where $0 \leqslant t \leqslant 1$, describe the line segment that joins the points $P_1(x_1, y_1)$ and $P_2(x_2, y_2)$.

(b) Find parametric equations to represent the line segment from $(-2, 7)$ to $(3, -1)$.

32. Use a graphing device and the result of Exercise 31(a) to draw the triangle with vertices $A(1, 1)$, $B(4, 2)$, and $C(1, 5)$.

33. Find parametric equations for the path of a particle that moves along the circle $x^2 + (y-1)^2 = 4$ in the manner described.

(a) Once around clockwise, starting at $(2, 1)$

(b) Three times around counterclockwise, starting at $(2, 1)$

(c) Halfway around counterclockwise, starting at $(0, 3)$

34. Graph the semicircle traced by the particle in Exercise 33(c).

35. (a) Find parametric equations for the ellipse $x^2/a^2 + y^2/b^2 = 1$. [*Hint:* Modify the equations of a circle in Example 2.]

(b) Use these parametric equations to graph the ellipse when $a = 3$ and $b = 1, 2, 4$, and 8.

(c) How does the shape of the ellipse change as b varies?

36. Find three different sets of parametric equations to represent the curve $y = x^3$, $x \in \mathbb{R}$.

37. Derive Equations 1 for the case $\pi/2 < \theta < \pi$.

38. Let P be a point at a distance d from the center of a circle of radius r. The curve traced out by P as the circle rolls along a straight line is called a **trochoid.** (Think of the motion of a point on a spoke of a bicycle wheel.) The cycloid is the special case of a trochoid with $d = r$. Using the same parameter θ as for the cycloid and assuming the line is the x-axis and $\theta = 0$ when P is at one of its lowest points, show that parametric equations of the trochoid are

$$x = r\theta - d\sin\theta \qquad y = r - d\cos\theta$$

Sketch the trochoid for the cases $d < r$ and $d > r$.

39. If a and b are fixed numbers, find parametric equations for the curve that consists of all possible positions of the point P in the figure, using the angle θ as the parameter. Then eliminate the parameter and identify the curve.

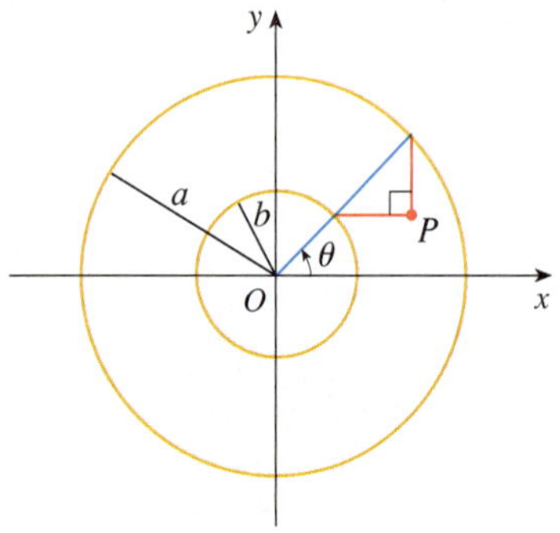

40. If a and b are fixed numbers, find parametric equations for the curve that consists of all possible positions of the point P in the figure, using the angle θ as the parameter. The line segment AB is tangent to the larger circle.

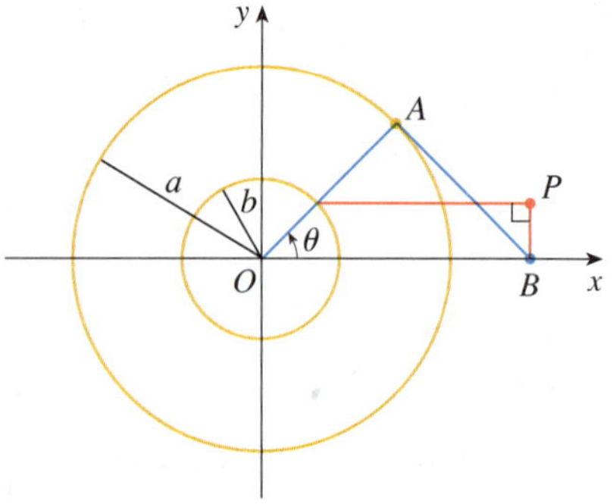

41. A curve, called a **witch of Maria Agnesi**, consists of all possible positions of the point P in the figure. Show that parametric equations for this curve can be written as

$$x = 2a\cot\theta \qquad y = 2a\sin^2\theta$$

Sketch the curve.

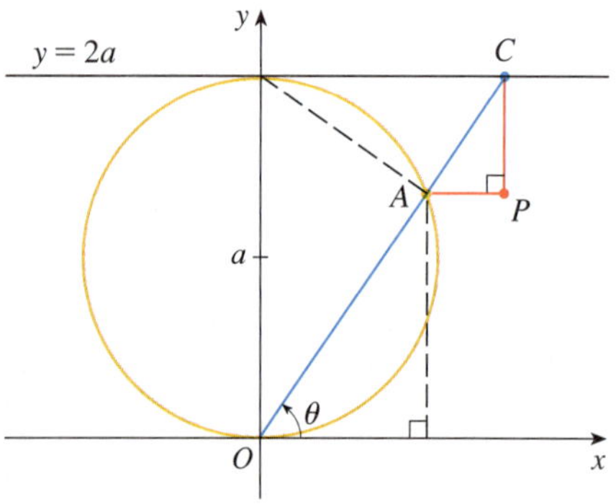

42. Find parametric equations for the curve that consists of all possible positions of the point P in the figure, where $|OP| = |AB|$. Sketch the curve. (This curve is called the **cissoid of Diocles** after the Greek scholar Diocles, who introduced the cissoid as a graphical method for constructing the edge of a cube whose volume is twice that of a given cube.)

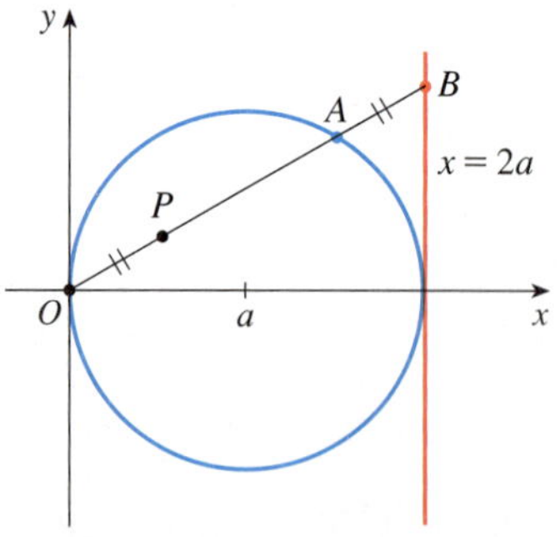

43. Suppose that the position of one particle at time t is given by

$$x_1 = 3 \sin t \qquad y_1 = 2 \cos t \qquad 0 \leqslant t \leqslant 2\pi$$

and the position of a second particle is given by

$$x_2 = -3 + \cos t \qquad y_2 = 1 + \sin t \qquad 0 \leqslant t \leqslant 2\pi$$

(a) Graph the paths of both particles. How many points of intersection are there?
(b) Are any of these points of intersection *collision points*? In other words, are the particles ever at the same place at the same time? If so, find the collision points.
(c) Describe what happens if the path of the second particle is given by

$$x_2 = 3 + \cos t \qquad y_2 = 1 + \sin t \qquad 0 \leqslant t \leqslant 2\pi$$

44. If a projectile is fired with an initial velocity of v_0 meters per second at an angle α above the horizontal and air resistance is assumed to be negligible, then its position after t seconds is given by the parametric equations

$$x = (v_0 \cos \alpha)t \qquad y = (v_0 \sin \alpha)t - \tfrac{1}{2}gt^2$$

where g is the acceleration due to gravity (9.8 m/s^2).
(a) If a gun is fired with $\alpha = 30°$ and $v_0 = 500$ m/s, when will the bullet hit the ground? How far from the gun will it hit the ground? What is the maximum height reached by the bullet?
(b) Use a graphing device to check your answers to part (a). Then graph the path of the projectile for several other values of the angle α to see where it hits the ground. Summarize your findings.
(c) Show that the path is parabolic by eliminating the parameter.

45. Investigate the family of curves defined by the parametric equations $x = t^2$, $y = t^3 - ct$. How does the shape change as c increases? Illustrate by graphing several members of the family.

46. The **swallowtail catastrophe curves** are defined by the parametric equations $x = 2ct - 4t^3$, $y = -ct^2 + 3t^4$. Graph several of these curves. What features do the curves have in common? How do they change when c increases?

47. The curves with equations $x = a \sin nt$, $y = b \cos t$ are called **Lissajous figures**. Investigate how these curves vary when a, b, and n vary. (Take n to be a positive integer.)

48. Investigate the family of curves defined by the parametric equations

$$x = \sin t\,(c - \sin t) \qquad y = \cos t\,(c - \sin t)$$

How does the shape change as c changes? In particular, you should identify the transitional values of c for which the basic shape of the curve changes.

LABORATORY PROJECT

Running Circles around Circles

In this project we investigate families of curves, called *hypocycloids* and *epicycloids,* that are generated by the motion of a point on a circle that rolls inside or outside another circle.

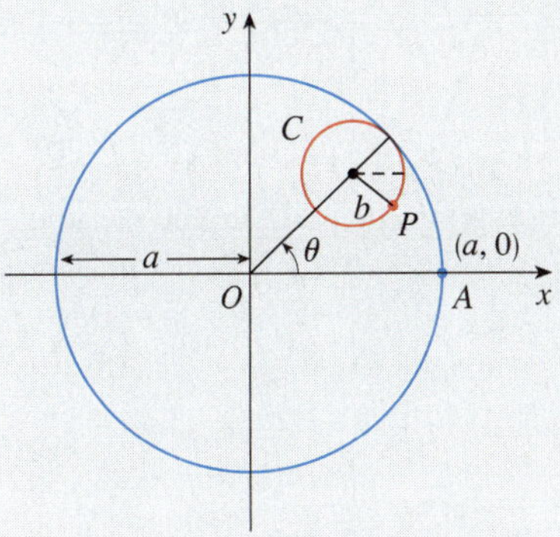

Look at Module 10.1B to see how hypocycloids and epicycloids are formed by the motion of rolling circles.

1. A **hypocycloid** is a curve traced out by a fixed point P on a circle C of radius b as C rolls on the inside of a circle with center O and radius a. Show that if the initial position of P is $(a, 0)$ and the parameter θ is chosen as in the figure, then parametric equations of the hypocycloid are

$$x = (a - b) \cos \theta + b \cos\left(\frac{a - b}{b}\,\theta\right) \qquad y = (a - b) \sin \theta - b \sin\left(\frac{a - b}{b}\,\theta\right)$$

2. Use a graphing device (or the interactive graphic in TEC Module 10.1B) to draw the graphs of hypocycloids with a a positive integer and $b = 1$. How does the value of a affect the graph? Show that if we take $a = 4$, then the parametric equations of the hypocycloid reduce to

$$x = 4 \cos^3\theta \qquad y = 4 \sin^3\theta$$

This curve is called a **hypocycloid of four cusps**, or an **astroid**.

3. Now try $b = 1$ and $a = n/d$, a fraction where n and d have no common factor. First let $n = 1$ and try to determine graphically the effect of the denominator d on the shape of the graph. Then let n vary while keeping d constant. What happens when $n = d + 1$?

4. What happens if $b = 1$ and a is irrational? Experiment with an irrational number like $\sqrt{2}$ or $e - 2$. Take larger and larger values for θ and speculate on what would happen if we were to graph the hypocycloid for all real values of θ.

5. If the circle C rolls on the *outside* of the fixed circle, the curve traced out by P is called an **epicycloid**. Find parametric equations for the epicycloid.

6. Investigate the possible shapes for epicycloids. Use methods similar to Problems 2–4.

10.2 Calculus with Parametric Curves

Having seen how to represent curves by parametric equations, we now apply the methods of calculus to these parametric curves. In particular, we solve problems involving tangents, area, arc length, and surface area.

Tangents

In the preceding section we saw that some curves defined by parametric equations $x = f(t)$ and $y = g(t)$ can also be expressed, by eliminating the parameter, in the form $y = F(x)$. (See Exercise 67 for general conditions under which this is possible.) If we substitute $x = f(t)$ and $y = g(t)$ in the equation $y = F(x)$, we get

$$g(t) = F(f(t))$$

and so, if g, F, and f are differentiable, the Chain Rule gives

$$g'(t) = F'(f(t))f'(t) = F'(x)f'(t)$$

If $f'(t) \neq 0$, we can solve for $F'(x)$:

1
$$F'(x) = \frac{g'(t)}{f'(t)}$$

Since the slope of the tangent to the curve $y = F(x)$ at $(x, F(x))$ is $F'(x)$, Equation 1 enables us to find tangents to parametric curves without having to eliminate the parameter. Using Leibniz notation, we can rewrite Equation 1 in an easily remembered form:

If we think of a parametric curve as being traced out by a moving particle, then dy/dt and dx/dt are the vertical and horizontal velocities of the particle and Formula 2 says that the slope of the tangent is the ratio of these velocities.

2
$$\frac{dy}{dx} = \frac{\dfrac{dy}{dt}}{\dfrac{dx}{dt}} \qquad \text{if} \quad \frac{dx}{dt} \neq 0$$

It can be seen from Equation 2 that the curve has a horizontal tangent when $dy/dt = 0$ (provided that $dx/dt \neq 0$) and it has a vertical tangent when $dx/dt = 0$ (provided that $dy/dt \neq 0$). This information is useful for sketching parametric curves.

As we know from Chapter 4, it is also useful to consider d^2y/dx^2. This can be found by replacing y by dy/dx in Equation 2:

Note that $\dfrac{d^2y}{dx^2} \neq \dfrac{\dfrac{d^2y}{dt^2}}{\dfrac{d^2x}{dt^2}}$

$$\frac{d^2y}{dx^2} = \frac{d}{dx}\left(\frac{dy}{dx}\right) = \frac{\dfrac{d}{dt}\left(\dfrac{dy}{dx}\right)}{\dfrac{dx}{dt}}$$

EXAMPLE 1

A curve C is defined by the parametric equations $x = t^2$, $y = t^3 - 3t$.

(a) Show that C has two tangents at the point (3, 0) and find their equations.
(b) Find the points on C where the tangent is horizontal or vertical.
(c) Determine where the curve is concave upward or downward.
(d) Sketch the curve.

SOLUTION

(a) Notice that $y = t^3 - 3t = t(t^2 - 3) = 0$ when $t = 0$ or $t = \pm\sqrt{3}$. Therefore, the point (3, 0) on C arises from two values of the parameter, $t = \sqrt{3}$ and $t = -\sqrt{3}$. This indicates that C crosses itself at (3, 0). Since

$$\frac{dy}{dx} = \frac{dy/dt}{dx/dt} = \frac{3t^2 - 3}{2t} = \frac{3}{2}\left(t - \frac{1}{t}\right)$$

the slope of the tangent when $t = \pm\sqrt{3}$ is $dy/dx = \pm 6/(2\sqrt{3}) = \pm\sqrt{3}$, so the equations of the tangents at (3, 0) are

$$y = \sqrt{3}\,(x - 3) \qquad \text{and} \qquad y = -\sqrt{3}\,(x - 3)$$

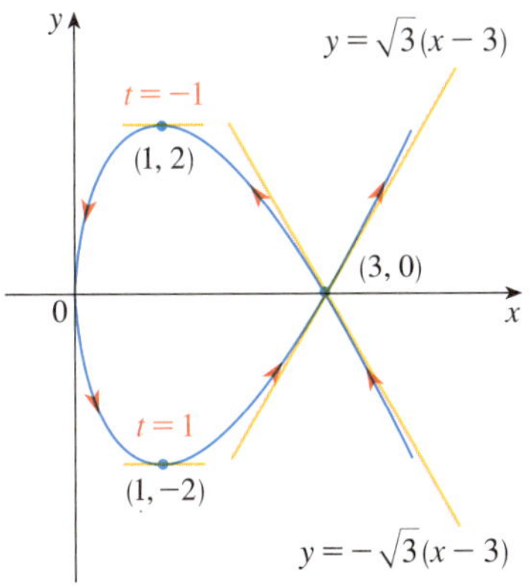

FIGURE 1

(b) C has a horizontal tangent when $dy/dx = 0$, that is, when $dy/dt = 0$ and $dx/dt \neq 0$. Since $dy/dt = 3t^2 - 3$, this happens when $t^2 = 1$, that is, $t = \pm 1$. The corresponding points on C are (1, −2) and (1, 2). C has a vertical tangent when $dx/dt = 2t = 0$, that is, $t = 0$. (Note that $dy/dt \neq 0$ there.) The corresponding point on C is (0, 0).

(c) To determine concavity we calculate the second derivative:

$$\frac{d^2y}{dx^2} = \frac{\dfrac{d}{dt}\left(\dfrac{dy}{dx}\right)}{\dfrac{dx}{dt}} = \frac{\dfrac{3}{2}\left(1 + \dfrac{1}{t^2}\right)}{2t} = \frac{3(t^2 + 1)}{4t^3}$$

Thus, the curve is concave upward when $t > 0$ and concave downward when $t < 0$.

(d) Using the information from parts (b) and (c), we sketch C in Figure 1.

EXAMPLE 2

(a) Find the tangent to the cycloid $x = r(\theta - \sin\theta)$, $y = r(1 - \cos\theta)$ at the point where $\theta = \pi/3$. (See Example 6 in Section 10.1.)
(b) At what points is the tangent horizontal? When is it vertical?

SOLUTION

(a) The slope of the tangent line is

$$\frac{dy}{dx} = \frac{dy/d\theta}{dx/d\theta} = \frac{r\sin\theta}{r(1 - \cos\theta)} = \frac{\sin\theta}{1 - \cos\theta}$$

When $\theta = \pi/3$, we have

$$x = r\left(\frac{\pi}{3} - \sin\frac{\pi}{3}\right) = r\left(\frac{\pi}{3} - \frac{\sqrt{3}}{2}\right) \qquad y = r\left(1 - \cos\frac{\pi}{3}\right) = \frac{r}{2}$$

and
$$\frac{dy}{dx} = \frac{\sin(\pi/3)}{1 - \cos(\pi/3)} = \frac{\sqrt{3}/2}{1 - \frac{1}{2}} = \sqrt{3}$$

Therefore, the slope of the tangent is $\sqrt{3}$ and its equation is

$$y - \frac{r}{2} = \sqrt{3}\left(x - \frac{r\pi}{3} + \frac{r\sqrt{3}}{2}\right) \qquad \text{or} \qquad \sqrt{3}\,x - y = r\left(\frac{\pi}{\sqrt{3}} - 2\right)$$

The tangent is sketched in Figure 2.

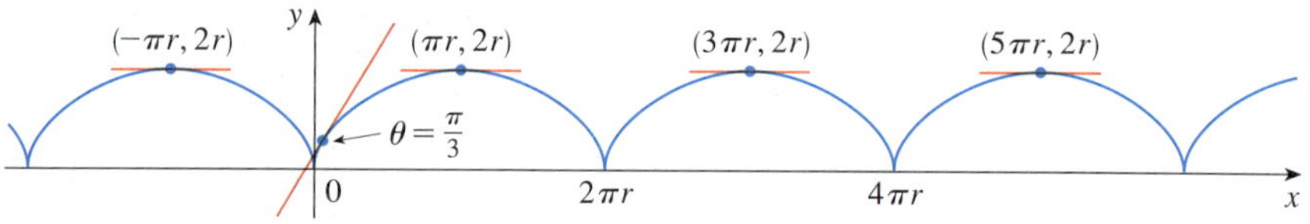

FIGURE 2

(b) The tangent is horizontal when $dy/dx = 0$, which occurs when $\sin\theta = 0$ and $1 - \cos\theta \neq 0$, that is, $\theta = (2n - 1)\pi$, n an integer. The corresponding point on the cycloid is $((2n - 1)\pi r, 2r)$.

When $\theta = 2n\pi$, both $dx/d\theta$ and $dy/d\theta$ are 0. It appears from the graph that there are vertical tangents at those points. We can verify this by using l'Hospital's Rule as follows:

$$\lim_{\theta \to 2n\pi^+} \frac{dy}{dx} = \lim_{\theta \to 2n\pi^+} \frac{\sin\theta}{1 - \cos\theta} = \lim_{\theta \to 2n\pi^+} \frac{\cos\theta}{\sin\theta} = \infty$$

A similar computation shows that $dy/dx \to -\infty$ as $\theta \to 2n\pi^-$, so indeed there are vertical tangents when $\theta = 2n\pi$, that is, when $x = 2n\pi r$.

Areas

We know that the area under a curve $y = F(x)$ from a to b is $A = \int_a^b F(x)\,dx$, where $F(x) \geqslant 0$. If the curve is given by parametric equations $x = f(t)$, $y = g(t)$ and is traversed once as t increases from α to β, then we can adapt the earlier formula by using the Substitution Rule for Definite Integrals as follows:

$$A = \int_a^b y\,dx = \int_\alpha^\beta g(t)f'(t)\,dt$$

$$\left[\text{or } \int_\beta^\alpha g(t)f'(t)\,dt \quad \text{if } (f(\beta), g(\beta)) \text{ is the leftmost endpoint}\right]$$

EXAMPLE 3 Find the area under one arch of the cycloid $x = r(\theta - \sin\theta)$, $y = r(1 - \cos\theta)$. (See Figure 3.)

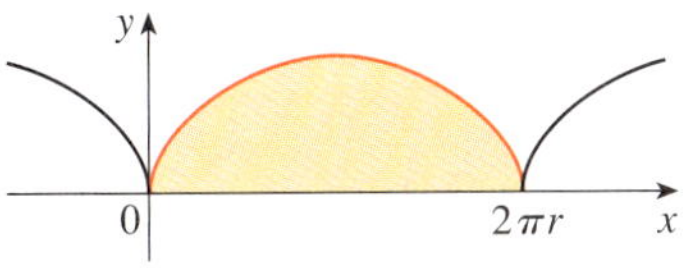

FIGURE 3

|||| The result of Example 3 says that the area under one arch of the cycloid is three times the area of the rolling circle that generates the cycloid (see Example 6 in Section 10.1). Galileo guessed this result but it was first proved by the French mathematician Roberval and the Italian mathematician Torricelli.

SOLUTION One arch of the cycloid is given by $0 \leqslant \theta \leqslant 2\pi$. Using the Substitution Rule with $y = r(1 - \cos\theta)$ and $dx = r(1 - \cos\theta)\, d\theta$, we have

$$\begin{aligned} A &= \int_0^{2\pi r} y\,dx = \int_0^{2\pi} r(1 - \cos\theta)\,r(1 - \cos\theta)\,d\theta \\ &= r^2 \int_0^{2\pi} (1 - \cos\theta)^2\,d\theta = r^2 \int_0^{2\pi} (1 - 2\cos\theta + \cos^2\theta)\,d\theta \\ &= r^2 \int_0^{2\pi} \left[1 - 2\cos\theta + \tfrac{1}{2}(1 + \cos 2\theta)\right] d\theta \\ &= r^2\left[\tfrac{3}{2}\theta - 2\sin\theta + \tfrac{1}{4}\sin 2\theta\right]_0^{2\pi} \\ &= r^2\left(\tfrac{3}{2} \cdot 2\pi\right) = 3\pi r^2 \end{aligned}$$

Arc Length

We already know how to find the length L of a curve C given in the form $y = F(x)$, $a \leqslant x \leqslant b$. Formula 8.1.3 says that if F' is continuous, then

3 $$L = \int_a^b \sqrt{1 + \left(\frac{dy}{dx}\right)^2}\,dx$$

Suppose that C can also be described by the parametric equations $x = f(t)$ and $y = g(t)$, $\alpha \leqslant t \leqslant \beta$, where $dx/dt = f'(t) > 0$. This means that C is traversed once, from left to right, as t increases from α to β and $f(\alpha) = a$, $f(\beta) = b$. Putting Formula 2 into Formula 3 and using the Substitution Rule, we obtain

$$L = \int_a^b \sqrt{1 + \left(\frac{dy}{dx}\right)^2}\,dx = \int_\alpha^\beta \sqrt{1 + \left(\frac{dy/dt}{dx/dt}\right)^2}\,\frac{dx}{dt}\,dt$$

Since $dx/dt > 0$, we have

4 $$L = \int_\alpha^\beta \sqrt{\left(\frac{dx}{dt}\right)^2 + \left(\frac{dy}{dt}\right)^2}\,dt$$

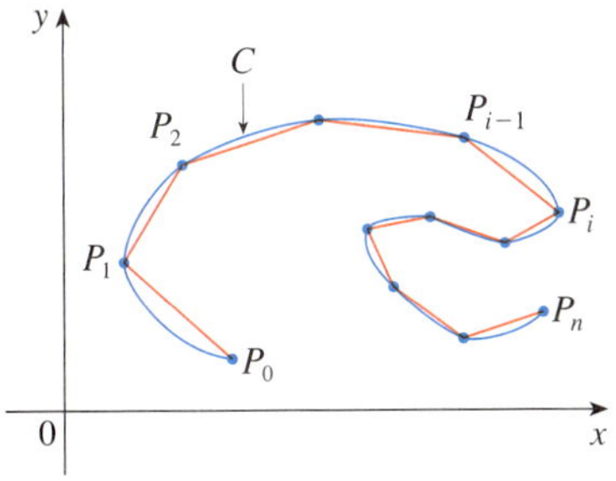

FIGURE 4

Even if C can't be expressed in the form $y = F(x)$, Formula 4 is still valid but we obtain it by polygonal approximations. We divide the parameter interval $[\alpha, \beta]$ into n subintervals of equal width Δt. If $t_0, t_1, t_2, \ldots, t_n$ are the endpoints of these subintervals, then $x_i = f(t_i)$ and $y_i = g(t_i)$ are the coordinates of points $P_i(x_i, y_i)$ that lie on C and the polygon with vertices $P_0, P_1, \ldots, P_n$ approximates C (see Figure 4).

As in Section 8.1, we define the length L of C to be the limit of the lengths of these approximating polygons as $n \to \infty$:

$$L = \lim_{n\to\infty} \sum_{i=1}^{n} |P_{i-1}P_i|$$

The Mean Value Theorem, when applied to f on the interval $[t_{i-1}, t_i]$, gives a number t_i^* in (t_{i-1}, t_i) such that

$$f(t_i) - f(t_{i-1}) = f'(t_i^*)(t_i - t_{i-1})$$

If we let $\Delta x_i = x_i - x_{i-1}$ and $\Delta y_i = y_i - y_{i-1}$, this equation becomes

$$\Delta x_i = f'(t_i^*)\,\Delta t$$

Similarly, when applied to g, the Mean Value Theorem gives a number t_i^{**} in (t_{i-1}, t_i) such that

$$\Delta y_i = g'(t_i^{**})\,\Delta t$$

Therefore

$$\begin{aligned} |P_{i-1}P_i| &= \sqrt{(\Delta x_i)^2 + (\Delta y_i)^2} \\ &= \sqrt{[f'(t_i^*)\,\Delta t]^2 + [g'(t_i^{**})\,\Delta t]^2} \\ &= \sqrt{[f'(t_i^*)]^2 + [g'(t_i^{**})]^2}\,\Delta t \end{aligned}$$

and so

5 $$L = \lim_{n \to \infty} \sum_{i=1}^{n} \sqrt{[f'(t_i^*)]^2 + [g'(t_i^{**})]^2}\,\Delta t$$

The sum in (5) resembles a Riemann sum for the function $\sqrt{[f'(t)]^2 + [g'(t)]^2}$ but it is not exactly a Riemann sum because $t_i^* \neq t_i^{**}$ in general. Nevertheless, if f' and g' are continuous, it can be shown that the limit in (5) is the same as if t_i^* and t_i^{**} were equal, namely,

$$L = \int_\alpha^\beta \sqrt{[f'(t)]^2 + [g'(t)]^2}\,dt$$

Thus, using Leibniz notation, we have the following result, which has the same form as (4).

6 Theorem If a curve C is described by the parametric equations $x = f(t)$, $y = g(t)$, $\alpha \leqslant t \leqslant \beta$, where f' and g' are continuous on $[\alpha, \beta]$ and C is traversed exactly once as t increases from α to β, then the length of C is

$$L = \int_\alpha^\beta \sqrt{\left(\frac{dx}{dt}\right)^2 + \left(\frac{dy}{dt}\right)^2}\,dt$$

Notice that the formula in Theorem 6 is consistent with the general formulas $L = \int ds$ and $(ds)^2 = (dx)^2 + (dy)^2$ of Section 8.1.

EXAMPLE 4 If we use the representation of the unit circle given in Example 2 in Section 10.1,

$$x = \cos t \qquad y = \sin t \qquad 0 \leqslant t \leqslant 2\pi$$

then $dx/dt = -\sin t$ and $dy/dt = \cos t$, so Theorem 6 gives

$$\begin{aligned} L &= \int_0^{2\pi} \sqrt{\left(\frac{dx}{dt}\right)^2 + \left(\frac{dy}{dt}\right)^2}\,dt = \int_0^{2\pi} \sqrt{\sin^2 t + \cos^2 t}\,dt \\ &= \int_0^{2\pi} dt = 2\pi \end{aligned}$$

as expected. If, on the other hand, we use the representation given in Example 3 in Section 10.1,

$$x = \sin 2t \qquad y = \cos 2t \qquad 0 \leqslant t \leqslant 2\pi$$

then $dx/dt = 2\cos 2t$, $dy/dt = -2\sin 2t$, and the integral in Theorem 6 gives

$$\int_0^{2\pi} \sqrt{\left(\frac{dx}{dt}\right)^2 + \left(\frac{dy}{dt}\right)^2}\, dt = \int_0^{2\pi} \sqrt{4\cos^2 2t + 4\sin^2 2t}\, dt = \int_0^{2\pi} 2\, dt = 4\pi$$

Notice that the integral gives twice the arc length of the circle because as t increases from 0 to 2π, the point $(\sin 2t, \cos 2t)$ traverses the circle twice. In general, when finding the length of a curve C from a parametric representation, we have to be careful to ensure that C is traversed only once as t increases from α to β.

EXAMPLE 5 Find the length of one arch of the cycloid $x = r(\theta - \sin\theta)$, $y = r(1 - \cos\theta)$.

SOLUTION From Example 3 we see that one arch is described by the parameter interval $0 \leqslant \theta \leqslant 2\pi$. Since

$$\frac{dx}{d\theta} = r(1 - \cos\theta) \qquad \text{and} \qquad \frac{dy}{d\theta} = r\sin\theta$$

we have

$$L = \int_0^{2\pi} \sqrt{\left(\frac{dx}{d\theta}\right)^2 + \left(\frac{dy}{d\theta}\right)^2}\, d\theta = \int_0^{2\pi} \sqrt{r^2(1 - \cos\theta)^2 + r^2\sin^2\theta}\, d\theta$$

$$= \int_0^{2\pi} \sqrt{r^2(1 - 2\cos\theta + \cos^2\theta + \sin^2\theta)}\, d\theta = r\int_0^{2\pi} \sqrt{2(1 - \cos\theta)}\, d\theta$$

|||| The result of Example 5 says that the length of one arch of a cycloid is eight times the radius of the generating circle (see Figure 5). This was first proved in 1658 by Sir Christopher Wren, who later became the architect of St. Paul's Cathedral in London.

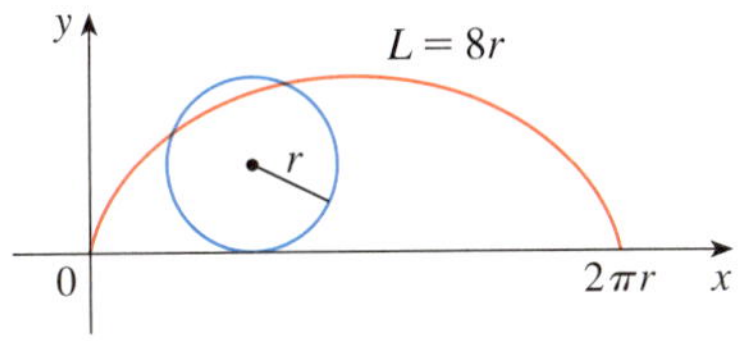

FIGURE 5

To evaluate this integral we use the identity $\sin^2 x = \frac{1}{2}(1 - \cos 2x)$ with $\theta = 2x$, which gives $1 - \cos\theta = 2\sin^2(\theta/2)$. Since $0 \leqslant \theta \leqslant 2\pi$, we have $0 \leqslant \theta/2 \leqslant \pi$ and so $\sin(\theta/2) \geqslant 0$. Therefore

$$\sqrt{2(1 - \cos\theta)} = \sqrt{4\sin^2(\theta/2)} = 2|\sin(\theta/2)| = 2\sin(\theta/2)$$

and so

$$L = 2r\int_0^{2\pi} \sin(\theta/2)\, d\theta = 2r\left[-2\cos(\theta/2)\right]_0^{2\pi}$$

$$= 2r[2 + 2] = 8r$$

Surface Area

In the same way as for arc length, we can adapt Formula 8.2.5 to obtain a formula for surface area. If the curve given by the parametric equations $x = f(t)$, $y = g(t)$, $\alpha \leqslant t \leqslant \beta$, is rotated about the x-axis, where f', g' are continuous and $g(t) \geqslant 0$, then the area of the resulting surface is given by

7
$$S = \int_\alpha^\beta 2\pi y \sqrt{\left(\frac{dx}{dt}\right)^2 + \left(\frac{dy}{dt}\right)^2}\, dt$$

The general symbolic formulas $S = \int 2\pi y\, ds$ and $S = \int 2\pi x\, ds$ (Formulas 8.2.7 and 8.2.8) are still valid, but for parametric curves we use

$$ds = \sqrt{\left(\frac{dx}{dt}\right)^2 + \left(\frac{dy}{dt}\right)^2}\, dt$$

EXAMPLE 6 Show that the surface area of a sphere of radius r is $4\pi r^2$.

SOLUTION The sphere is obtained by rotating the semicircle

$$x = r\cos t \qquad y = r\sin t \qquad 0 \leqslant t \leqslant \pi$$

about the x-axis. Therefore, from Formula 7, we get

$$\begin{aligned} S &= \int_0^\pi 2\pi r \sin t \sqrt{(-r\sin t)^2 + (r\cos t)^2}\, dt \\ &= 2\pi \int_0^\pi r\sin t \sqrt{r^2(\sin^2 t + \cos^2 t)}\, dt \\ &= 2\pi \int_0^\pi r \sin t \cdot r\, dt = 2\pi r^2 \int_0^\pi \sin t\, dt \\ &= 2\pi r^2 (-\cos t)\Big]_0^\pi = 4\pi r^2 \end{aligned}$$

10.2 Exercises

1–2 Find dy/dx.

1. $x = t - t^3, \quad y = 2 - 5t$

2. $x = te^t, \quad y = t + e^t$

3–6 Find an equation of the tangent to the curve at the point corresponding to the given value of the parameter.

3. $x = t^4 + 1, \quad y = t^3 + t; \quad t = -1$

4. $x = 2t^2 + 1, \quad y = \frac{1}{3}t^3 - t; \quad t = 3$

5. $x = e^{\sqrt{t}}, \quad y = t - \ln t^2; \quad t = 1$

6. $x = \cos\theta + \sin 2\theta, \quad y = \sin\theta + \cos 2\theta; \quad \theta = 0$

7–8 Find an equation of the tangent to the curve at the given point by two methods: (a) without eliminating the parameter and (b) by first eliminating the parameter.

7. $x = e^t, \ y = (t-1)^2; \quad (1, 1)$

8. $x = \tan\theta, \ y = \sec\theta; \quad (1, \sqrt{2})$

9–10 Find an equation of the tangent(s) to the curve at the given point. Then graph the curve and the tangent(s).

9. $x = 2\sin 2t, \ y = 2\sin t; \quad (\sqrt{3}, 1)$

10. $x = \sin t, \ y = \sin(t + \sin t); \quad (0, 0)$

11–16 Find dy/dx and d^2y/dx^2. For which values of t is the curve concave upward?

11. $x = 4 + t^2, \quad y = t^2 + t^3$

12. $x = t^3 - 12t, \quad y = t^2 - 1$

13. $x = t - e^t, \quad y = t + e^{-t}$

14. $x = t + \ln t, \quad y = t - \ln t$

15. $x = 2\sin t, \quad y = 3\cos t, \quad 0 < t < 2\pi$

16. $x = \cos 2t, \quad y = \cos t, \quad 0 < t < \pi$

17–20 Find the points on the curve where the tangent is horizontal or vertical. If you have a graphing device, graph the curve to check your work.

17. $x = 10 - t^2, \quad y = t^3 - 12t$

18. $x = 2t^3 + 3t^2 - 12t, \quad y = 2t^3 + 3t^2 + 1$

19. $x = 2\cos\theta, \quad y = \sin 2\theta$

20. $x = \cos 3\theta, \quad y = 2\sin\theta$

21. Use a graph to estimate the coordinates of the leftmost point on the curve $x = t^4 - t^2$, $y = t + \ln t$. Then use calculus to find the exact coordinates.

22. Try to estimate the coordinates of the highest point and the leftmost point on the curve $x = te^t$, $y = te^{-t}$. Then find the exact coordinates. What are the asymptotes of this curve?

23–24 Graph the curve in a viewing rectangle that displays all the important aspects of the curve.

23. $x = t^4 - 2t^3 - 2t^2, \quad y = t^3 - t$

24. $x = t^4 + 4t^3 - 8t^2, \quad y = 2t^2 - t$

25. Show that the curve $x = \cos t$, $y = \sin t \cos t$ has two tangents at $(0, 0)$ and find their equations. Sketch the curve.

26. At what point does the curve $x = 1 - 2\cos^2 t$, $y = (\tan t)(1 - 2\cos^2 t)$ cross itself? Find the equations of both tangents at that point.

27. (a) Find the slope of the tangent line to the trochoid $x = r\theta - d\sin\theta$, $y = r - d\cos\theta$ in terms of θ. (See Exercise 38 in Section 10.1.)

(b) Show that if $d < r$, then the trochoid does not have a vertical tangent.

28. (a) Find the slope of the tangent to the astroid $x = a\cos^3\theta$, $y = a\sin^3\theta$ in terms of θ. (Astroids are explored in the Laboratory Project on page 659.)

(b) At what points is the tangent horizontal or vertical?
(c) At what points does the tangent have slope 1 or -1?

29. At what points on the curve $x = t^3 + 4t$, $y = 6t^2$ is the tangent parallel to the line with equations $x = -7t$, $y = 12t - 5$?

30. Find equations of the tangents to the curve $x = 3t^2 + 1$, $y = 2t^3 + 1$ that pass through the point $(4, 3)$.

31. Use the parametric equations of an ellipse, $x = a \cos \theta$, $y = b \sin \theta$, $0 \leqslant \theta \leqslant 2\pi$, to find the area that it encloses.

32. Find the area bounded by the curve $x = t - 1/t$, $y = t + 1/t$ and the line $y = 2.5$.

33. Find the area bounded by the curve $x = \cos t$, $y = e^t$, $0 \leqslant t \leqslant \pi/2$, and the lines $y = 1$ and $x = 0$.

34. Find the area of the region enclosed by the astroid $x = a \cos^3\theta$, $y = a \sin^3\theta$. (Astroids are explored in the Laboratory Project on page 659.)

35. Find the area under one arch of the trochoid of Exercise 38 in Section 10.1 for the case $d < r$.

36. Let $\mathcal{R}$ be the region enclosed by the loop of the curve in Example 1.
(a) Find the area of $\mathcal{R}$.
(b) If $\mathcal{R}$ is rotated about the x-axis, find the volume of the resulting solid.
(c) Find the centroid of $\mathcal{R}$.

37–40 Set up, but do not evaluate, an integral that represents the length of the curve.

37. $x = t - t^2$, $y = \frac{4}{3}t^{3/2}$, $1 \leqslant t \leqslant 2$

38. $x = 1 + e^t$, $y = t^2$, $-3 \leqslant t \leqslant 3$

39. $x = t + \cos t$, $y = t - \sin t$, $0 \leqslant t \leqslant 2\pi$

40. $x = \ln t$, $y = \sqrt{t + 1}$, $1 \leqslant t \leqslant 5$

41–44 Find the length of the curve.

41. $x = 1 + 3t^2$, $y = 4 + 2t^3$, $0 \leqslant t \leqslant 1$

42. $x = a(\cos \theta + \theta \sin \theta)$, $y = a(\sin \theta - \theta \cos \theta)$, $0 \leqslant \theta \leqslant \pi$

43. $x = \dfrac{t}{1 + t}$, $y = \ln(1 + t)$, $0 \leqslant t \leqslant 2$

44. $x = e^t + e^{-t}$, $y = 5 - 2t$, $0 \leqslant t \leqslant 3$

45–47 Graph the curve and find its length.

45. $x = e^t \cos t$, $y = e^t \sin t$, $0 \leqslant t \leqslant \pi$

46. $x = \cos t + \ln(\tan \frac{1}{2}t)$, $y = \sin t$, $\pi/4 \leqslant t \leqslant 3\pi/4$

47. $x = e^t - t$, $y = 4e^{t/2}$, $-8 \leqslant t \leqslant 3$

48. Find the length of the loop of the curve $x = 3t - t^3$, $y = 3t^2$.

49. Use Simpson's Rule with $n = 6$ to estimate the length of the curve $x = t - e^t$, $y = t + e^t$, $-6 \leqslant t \leqslant 6$.

50. In Exercise 41 in Section 10.1 you were asked to derive the parametric equations $x = 2a \cot \theta$, $y = 2a \sin^2\theta$ for the curve called the witch of Maria Agnesi. Use Simpson's Rule with $n = 4$ to estimate the length of the arc of this curve given by $\pi/4 \leqslant \theta \leqslant \pi/2$.

51–52 Find the distance traveled by a particle with position (x, y) as t varies in the given time interval. Compare with the length of the curve.

51. $x = \sin^2 t$, $y = \cos^2 t$, $0 \leqslant t \leqslant 3\pi$

52. $x = \cos^2 t$, $y = \cos t$, $0 \leqslant t \leqslant 4\pi$

53. Show that the total length of the ellipse $x = a \sin \theta$, $y = b \cos \theta$, $a > b > 0$, is

$$L = 4a \int_0^{\pi/2} \sqrt{1 - e^2 \sin^2\theta}\, d\theta$$

where e is the eccentricity of the ellipse $\left(e = c/a, \text{ where } c = \sqrt{a^2 - b^2}\right)$.

54. Find the total length of the astroid $x = a \cos^3\theta$, $y = a \sin^3\theta$, where $a > 0$.

CAS **55.** (a) Graph the **epitrochoid** with equations

$$x = 11 \cos t - 4 \cos(11t/2)$$
$$y = 11 \sin t - 4 \sin(11t/2)$$

What parameter interval gives the complete curve?
(b) Use your CAS to find the approximate length of this curve.

CAS **56.** A curve called **Cornu's spiral** is defined by the parametric equations

$$x = C(t) = \int_0^t \cos(\pi u^2/2)\, du$$

$$y = S(t) = \int_0^t \sin(\pi u^2/2)\, du$$

where C and S are the Fresnel functions that were introduced in Chapter 5.
(a) Graph this curve. What happens as $t \to \infty$ and as $t \to -\infty$?
(b) Find the length of Cornu's spiral from the origin to the point with parameter value t.

57–58 Set up, but do not evaluate, an integral that represents the area of the surface obtained by rotating the given curve about the x-axis.

57. $x = t - t^2$, $y = \frac{4}{3}t^{3/2}$, $1 \leqslant t \leqslant 2$

58. $x = \sin^2 t$, $y = \sin 3t$, $0 \leqslant t \leqslant \pi/3$

59–61 Find the area of the surface obtained by rotating the given curve about the x-axis.

59. $x = t^3, \quad y = t^2, \quad 0 \leqslant t \leqslant 1$

60. $x = 3t - t^3, \quad y = 3t^2, \quad 0 \leqslant t \leqslant 1$

61. $x = a\cos^3\theta, \quad y = a\sin^3\theta, \quad 0 \leqslant \theta \leqslant \pi/2$

62. Graph the curve

$$x = 2\cos\theta - \cos 2\theta \qquad y = 2\sin\theta - \sin 2\theta$$

If this curve is rotated about the x-axis, find the area of the resulting surface. (Use your graph to help find the correct parameter interval.)

63. If the curve

$$x = t + t^3 \qquad y = t - \frac{1}{t^2} \qquad 1 \leqslant t \leqslant 2$$

is rotated about the x-axis, use your calculator to estimate the area of the resulting surface to three decimal places.

64. If the arc of the curve in Exercise 50 is rotated about the x-axis, estimate the area of the resulting surface using Simpson's Rule with $n = 4$.

65–66 Find the surface area generated by rotating the given curve about the y-axis.

65. $x = 3t^2, \quad y = 2t^3, \quad 0 \leqslant t \leqslant 5$

66. $x = e^t - t, \quad y = 4e^{t/2}, \quad 0 \leqslant t \leqslant 1$

67. If f' is continuous and $f'(t) \neq 0$ for $a \leqslant t \leqslant b$, show that the parametric curve $x = f(t)$, $y = g(t)$, $a \leqslant t \leqslant b$, can be put in the form $y = F(x)$. [*Hint:* Show that f^{-1} exists.]

68. Use Formula 2 to derive Formula 7 from Formula 8.2.5 for the case in which the curve can be represented in the form $y = F(x)$, $a \leqslant x \leqslant b$.

69. The **curvature** at a point P of a curve is defined as

$$\kappa = \left|\frac{d\phi}{ds}\right|$$

where ϕ is the angle of inclination of the tangent line at P, as shown in the figure. Thus, the curvature is the absolute value of the rate of change of ϕ with respect to arc length. It can be regarded as a measure of the rate of change of direction of the curve at P and will be studied in greater detail in Chapter 13.

(a) For a parametric curve $x = x(t)$, $y = y(t)$, derive the formula

$$\kappa = \frac{|\dot{x}\ddot{y} - \ddot{x}\dot{y}|}{[\dot{x}^2 + \dot{y}^2]^{3/2}}$$

where the dots indicate derivatives with respect to t, so $\dot{x} = dx/dt$. [*Hint:* Use $\phi = \tan^{-1}(dy/dx)$ and Equation 2 to find $d\phi/dt$. Then use the Chain Rule to find $d\phi/ds$.]

(b) By regarding a curve $y = f(x)$ as the parametric curve $x = x$, $y = f(x)$, with parameter x, show that the formula in part (a) becomes

$$\kappa = \frac{|d^2y/dx^2|}{[1 + (dy/dx)^2]^{3/2}}$$

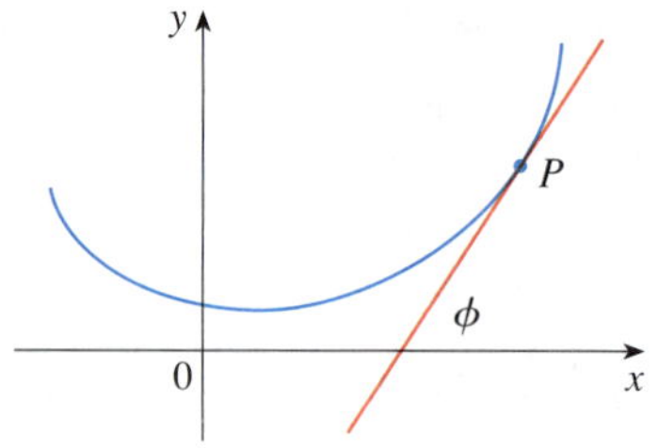

70. (a) Use the formula in Exercise 69(b) to find the curvature of the parabola $y = x^2$ at the point $(1, 1)$.
(b) At what point does this parabola have maximum curvature?

71. Use the formula in Exercise 69(a) to find the curvature of the cycloid $x = \theta - \sin\theta$, $y = 1 - \cos\theta$ at the top of one of its arches.

72. (a) Show that the curvature at each point of a straight line is $\kappa = 0$.
(b) Show that the curvature at each point of a circle of radius r is $\kappa = 1/r$.

73. A string is wound around a circle and then unwound while being held taut. The curve traced by the point P at the end of the string is called the **involute** of the circle. If the circle has radius r and center O and the initial position of P is $(r, 0)$, and if the parameter θ is chosen as in the figure, show that parametric equations of the involute are

$$x = r(\cos\theta + \theta\sin\theta) \qquad y = r(\sin\theta - \theta\cos\theta)$$

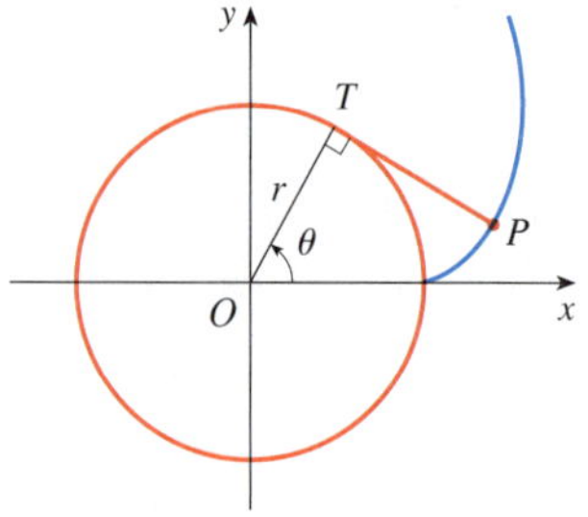

74. A cow is tied to a silo with radius r by a rope just long enough to reach the opposite side of the silo. Find the area available for grazing by the cow.

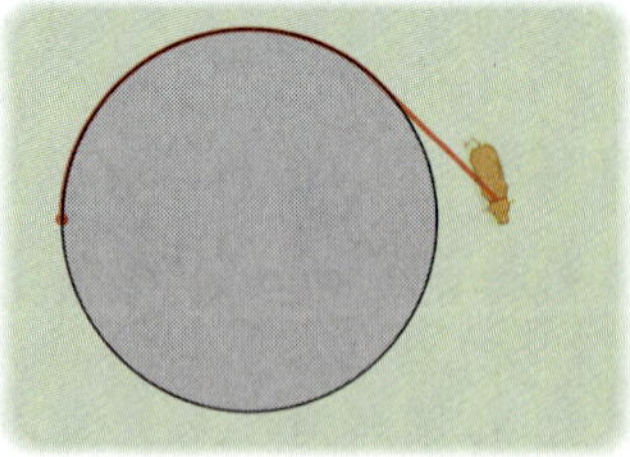

LABORATORY PROJECT

Bézier Curves

The **Bézier curves** are used in computer-aided design and are named after the French mathematician Pierre Bézier (1910–1999), who worked in the automotive industry. A cubic Bézier curve is determined by four **control points**, $P_0(x_0, y_0)$, $P_1(x_1, y_1)$, $P_2(x_2, y_2)$, and $P_3(x_3, y_3)$, and is defined by the parametric equations

$$x = x_0(1 - t)^3 + 3x_1t(1 - t)^2 + 3x_2t^2(1 - t) + x_3t^3$$
$$y = y_0(1 - t)^3 + 3y_1t(1 - t)^2 + 3y_2t^2(1 - t) + y_3t^3$$

where $0 \leqslant t \leqslant 1$. Notice that when $t = 0$ we have $(x, y) = (x_0, y_0)$ and when $t = 1$ we have $(x, y) = (x_3, y_3)$, so the curve starts at P_0 and ends at P_3.

1. Graph the Bézier curve with control points $P_0(4, 1)$, $P_1(28, 48)$, $P_2(50, 42)$, and $P_3(40, 5)$. Then, on the same screen, graph the line segments P_0P_1, P_1P_2, and P_2P_3. (Exercise 31 in Section 10.1 shows how to do this.) Notice that the middle control points P_1 and P_2 don't lie on the curve; the curve starts at P_0, heads toward P_1 and P_2 without reaching them, and ends at P_3.
2. From the graph in Problem 1 it appears that the tangent at P_0 passes through P_1 and the tangent at P_3 passes through P_2. Prove it.
3. Try to produce a Bézier curve with a loop by changing the second control point in Problem 1.
4. Some laser printers use Bézier curves to represent letters and other symbols. Experiment with control points until you find a Bézier curve that gives a reasonable representation of the letter C.
5. More complicated shapes can be represented by piecing together two or more Bézier curves. Suppose the first Bézier curve has control points P_0, P_1, P_2, P_3 and the second one has control points P_3, P_4, P_5, P_6. If we want these two pieces to join together smoothly, then the tangents at P_3 should match and so the points P_2, P_3, and P_4 all have to lie on this common tangent line. Using this principle, find control points for a pair of Bézier curves that represent the letter S.

10.3 Polar Coordinates

A coordinate system represents a point in the plane by an ordered pair of numbers called coordinates. Usually we use Cartesian coordinates, which are directed distances from two perpendicular axes. Here we describe a coordinate system introduced by Newton, called the **polar coordinate system**, which is more convenient for many purposes.

We choose a point in the plane that is called the **pole** (or origin) and is labeled O. Then we draw a ray (half-line) starting at O called the **polar axis**. This axis is usually drawn horizontally to the right and corresponds to the positive x-axis in Cartesian coordinates.

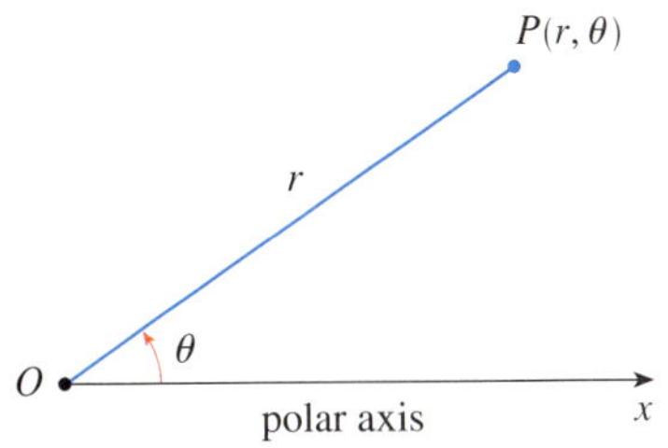

FIGURE 1

If P is any other point in the plane, let r be the distance from O to P and let θ be the angle (usually measured in radians) between the polar axis and the line OP as in Figure 1. Then the point P is represented by the ordered pair (r, θ) and r, θ are called **polar coordinates** of P. We use the convention that an angle is positive if measured in the counterclockwise direction from the polar axis and negative in the clockwise direction. If $P = O$, then $r = 0$ and we agree that $(0, \theta)$ represents the pole for any value of θ.

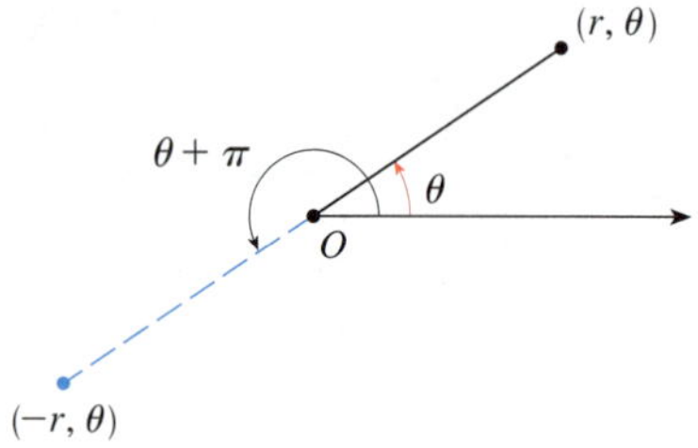

FIGURE 2

We extend the meaning of polar coordinates (r, θ) to the case in which r is negative by agreeing that, as in Figure 2, the points $(-r, \theta)$ and (r, θ) lie on the same line through O and at the same distance $|r|$ from O, but on opposite sides of O. If $r > 0$, the point (r, θ) lies in the same quadrant as θ; if $r < 0$, it lies in the quadrant on the opposite side of the pole. Notice that $(-r, \theta)$ represents the same point as $(r, \theta + \pi)$.

EXAMPLE 1 Plot the points whose polar coordinates are given.
(a) $(1, 5\pi/4)$ (b) $(2, 3\pi)$ (c) $(2, -2\pi/3)$ (d) $(-3, 3\pi/4)$

SOLUTION The points are plotted in Figure 3. In part (d) the point $(-3, 3\pi/4)$ is located three units from the pole in the fourth quadrant because the angle $3\pi/4$ is in the second quadrant and $r = -3$ is negative.

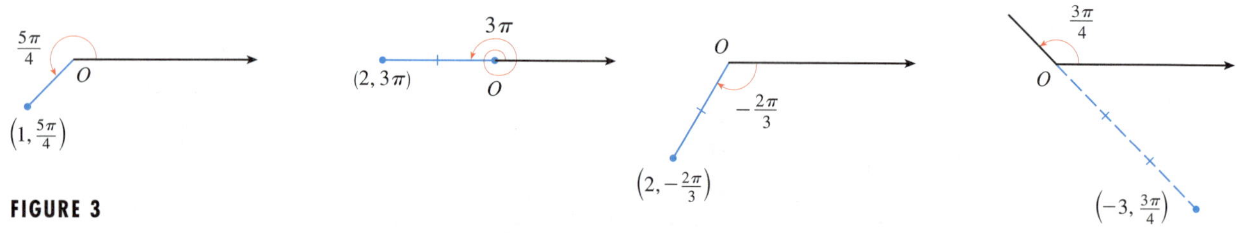

FIGURE 3

In the Cartesian coordinate system every point has only one representation, but in the polar coordinate system each point has many representations. For instance, the point $(1, 5\pi/4)$ in Example 1(a) could be written as $(1, -3\pi/4)$ or $(1, 13\pi/4)$ or $(-1, \pi/4)$. (See Figure 4.)

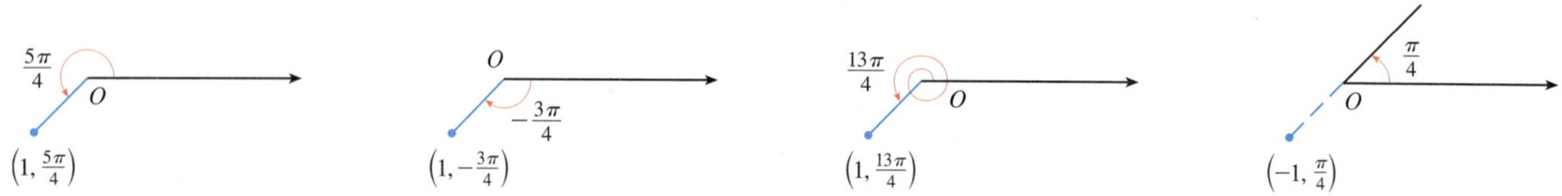

FIGURE 4

In fact, since a complete counterclockwise rotation is given by an angle 2π, the point represented by polar coordinates (r, θ) is also represented by

$$(r, \theta + 2n\pi) \qquad \text{and} \qquad (-r, \theta + (2n + 1)\pi)$$

where n is any integer.

The connection between polar and Cartesian coordinates can be seen from Figure 5, in which the pole corresponds to the origin and the polar axis coincides with the positive x-axis. If the point P has Cartesian coordinates (x, y) and polar coordinates (r, θ), then, from the figure, we have

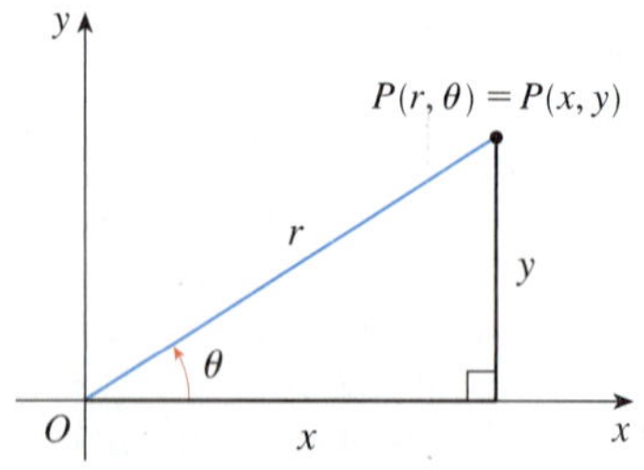

FIGURE 5

$$\cos \theta = \frac{x}{r} \qquad \sin \theta = \frac{y}{r}$$

and so

$$\boxed{x = r \cos \theta \qquad y = r \sin \theta} \tag{1}$$

Although Equations 1 were deduced from Figure 5, which illustrates the case where $r > 0$ and $0 < \theta < \pi/2$, these equations are valid for all values of r and θ. (See the general definition of $\sin \theta$ and $\cos \theta$ in Appendix D.)

Equations 1 allow us to find the Cartesian coordinates of a point when the polar coordinates are known. To find r and θ when x and y are known, we use the equations

2
$$r^2 = x^2 + y^2 \qquad \tan\theta = \frac{y}{x}$$

which can be deduced from Equations 1 or simply read from Figure 5.

EXAMPLE 2 Convert the point $(2, \pi/3)$ from polar to Cartesian coordinates.

SOLUTION Since $r = 2$ and $\theta = \pi/3$, Equations 1 give

$$x = r\cos\theta = 2\cos\frac{\pi}{3} = 2 \cdot \frac{1}{2} = 1$$

$$y = r\sin\theta = 2\sin\frac{\pi}{3} = 2 \cdot \frac{\sqrt{3}}{2} = \sqrt{3}$$

Therefore, the point is $(1, \sqrt{3})$ in Cartesian coordinates.

EXAMPLE 3 Represent the point with Cartesian coordinates $(1, -1)$ in terms of polar coordinates.

SOLUTION If we choose r to be positive, then Equations 2 give

$$r = \sqrt{x^2 + y^2} = \sqrt{1^2 + (-1)^2} = \sqrt{2}$$

$$\tan\theta = \frac{y}{x} = -1$$

Since the point $(1, -1)$ lies in the fourth quadrant, we can choose $\theta = -\pi/4$ or $\theta = 7\pi/4$. Thus, one possible answer is $(\sqrt{2}, -\pi/4)$; another is $(\sqrt{2}, 7\pi/4)$.

NOTE ▫ Equations 2 do not uniquely determine θ when x and y are given because, as θ increases through the interval $0 \leqslant \theta < 2\pi$, each value of $\tan\theta$ occurs twice. Therefore, in converting from Cartesian to polar coordinates, it's not good enough just to find r and θ that satisfy Equations 2. As in Example 3, we must choose θ so that the point (r, θ) lies in the correct quadrant.

Polar Curves

The **graph of a polar equation** $r = f(\theta)$, or more generally $F(r, \theta) = 0$, consists of all points P that have at least one polar representation (r, θ) whose coordinates satisfy the equation.

EXAMPLE 4 What curve is represented by the polar equation $r = 2$?

SOLUTION The curve consists of all points (r, θ) with $r = 2$. Since r represents the distance from the point to the pole, the curve $r = 2$ represents the circle with center O and radius 2. In general, the equation $r = a$ represents a circle with center O and radius $|a|$. (See Figure 6.)

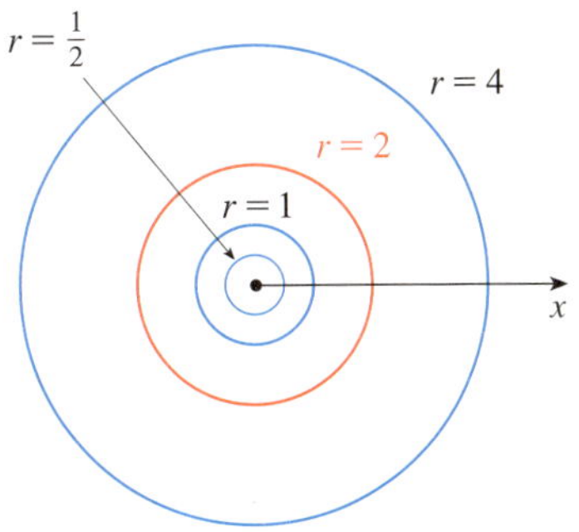

FIGURE 6

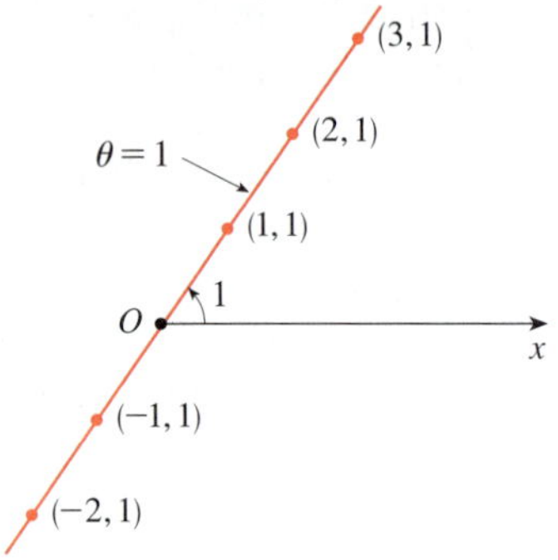

FIGURE 7

EXAMPLE 5 Sketch the polar curve $\theta = 1$.

SOLUTION This curve consists of all points (r, θ) such that the polar angle θ is 1 radian. It is the straight line that passes through O and makes an angle of 1 radian with the polar axis (see Figure 7). Notice that the points $(r, 1)$ on the line with $r > 0$ are in the first quadrant, whereas those with $r < 0$ are in the third quadrant.

EXAMPLE 6
(a) Sketch the curve with polar equation $r = 2\cos\theta$.
(b) Find a Cartesian equation for this curve.

SOLUTION
(a) In Figure 8 we find the values of r for some convenient values of θ and plot the corresponding points (r, θ). Then we join these points to sketch the curve, which appears to be a circle. We have used only values of θ between 0 and π, since if we let θ increase beyond π, we obtain the same points again.

θ	$r = 2\cos\theta$
0	2
$\pi/6$	$\sqrt{3}$
$\pi/4$	$\sqrt{2}$
$\pi/3$	1
$\pi/2$	0
$2\pi/3$	-1
$3\pi/4$	$-\sqrt{2}$
$5\pi/6$	$-\sqrt{3}$
π	-2

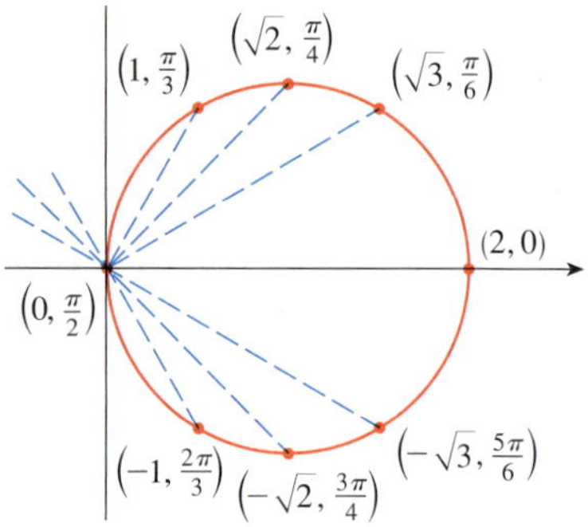

FIGURE 8
Table of values and graph of $r = 2\cos\theta$

(b) To convert the given equation into a Cartesian equation we use Equations 1 and 2. From $x = r\cos\theta$ we have $\cos\theta = x/r$, so the equation $r = 2\cos\theta$ becomes $r = 2x/r$, which gives

$$2x = r^2 = x^2 + y^2 \qquad \text{or} \qquad x^2 + y^2 - 2x = 0$$

Completing the square, we obtain

$$(x - 1)^2 + y^2 = 1$$

which is an equation of a circle with center $(1, 0)$ and radius 1.

|||| Figure 9 shows a geometrical illustration that the circle in Example 6 has the equation $r = 2\cos\theta$. The angle OPQ is a right angle (Why?) and so $r/2 = \cos\theta$.

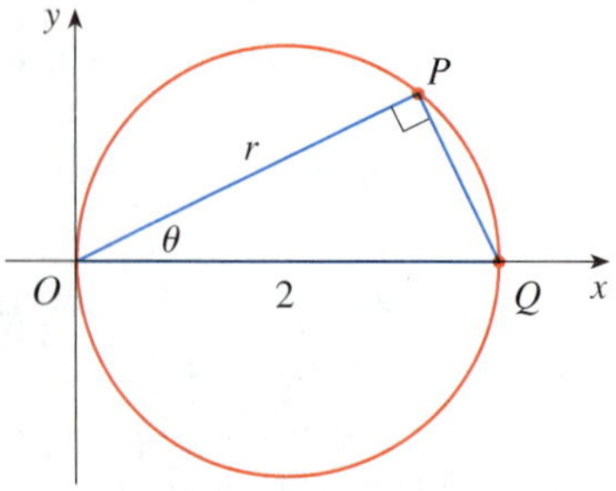

FIGURE 9

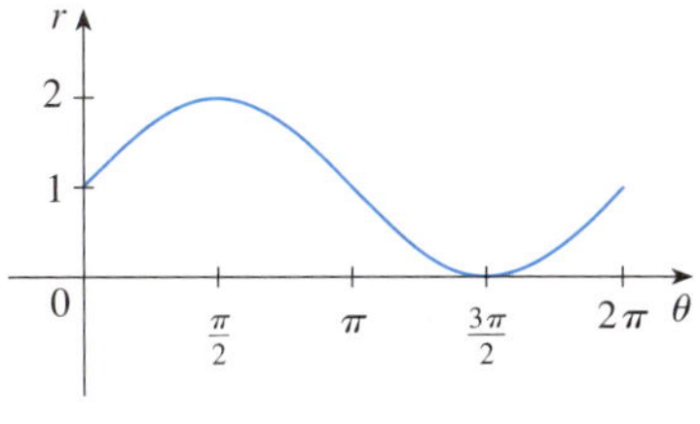

FIGURE 10
$r = 1 + \sin\theta$ in Cartesian coordinates, $0 \leqslant \theta \leqslant 2\pi$

EXAMPLE 7 Sketch the curve $r = 1 + \sin\theta$.

SOLUTION Instead of plotting points as in Example 6, we first sketch the graph of $r = 1 + \sin\theta$ in *Cartesian* coordinates in Figure 10 by shifting the sine curve up one unit. This enables us to read at a glance the values of r that correspond to increasing values of θ. For instance, we see that as θ increases from 0 to $\pi/2$, r (the distance from O) increases from 1 to 2, so we sketch the corresponding part of the polar curve in Figure 11(a). As θ increases from $\pi/2$ to π, Figure 10 shows that r decreases from 2 to 1, so we sketch the next part of the curve as in Figure 11(b). As θ increases from π to $3\pi/2$, r decreases from 1 to 0 as shown in part (c). Finally, as θ increases from $3\pi/2$ to 2π, r increases from 0 to 1 as shown in part (d). If we let θ increase beyond 2π or decrease beyond 0, we would simply retrace our path. Putting together the parts of the curve from Figure 11(a)–(d), we sketch the complete curve in part (e). It is called a **cardioid** because it's shaped like a heart.

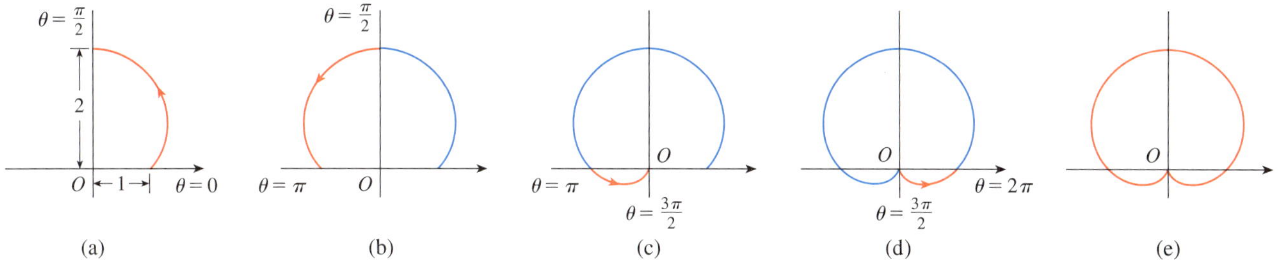

FIGURE 11
Stages in sketching the cardioid $r = 1 + \sin\theta$

EXAMPLE 8 Sketch the curve $r = \cos 2\theta$.

SOLUTION As in Example 7, we first sketch $r = \cos 2\theta$, $0 \leqslant \theta \leqslant 2\pi$, in Cartesian coordinates in Figure 12. As θ increases from 0 to $\pi/4$, Figure 12 shows that r decreases from 1 to 0 and so we draw the corresponding portion of the polar curve in Figure 13 (indicated by ①). As θ increases from $\pi/4$ to $\pi/2$, r goes from 0 to -1. This means that the distance from O increases from 0 to 1, but instead of being in the first quadrant this portion of the polar curve (indicated by ②) lies on the opposite side of the pole in the third quadrant. The remainder of the curve is drawn in a similar fashion, with the arrows and numbers indicating the order in which the portions are traced out. The resulting curve has four loops and is called a **four-leaved rose**.

TEC Module 10.3 helps you see how polar curves are traced out by showing animations similar to Figures 10–13. Tangents to these polar curves can also be visualized as in Figure 15 (see page 676).

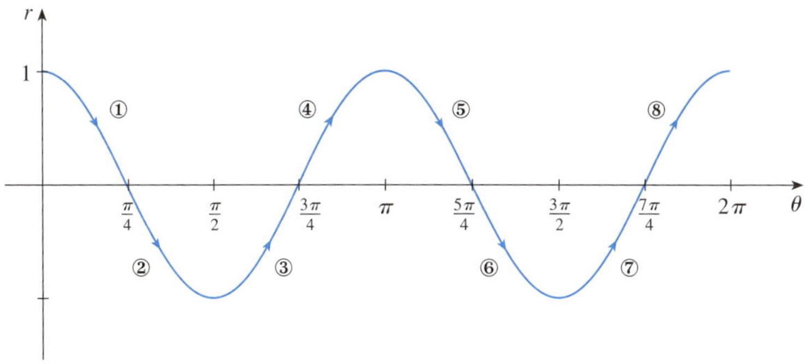

FIGURE 12
$r = \cos 2\theta$ in Cartesian coordinates

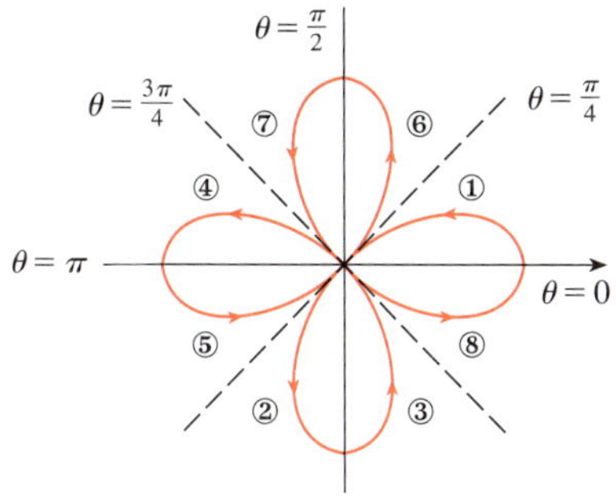

FIGURE 13
Four-leaved rose $r = \cos 2\theta$

Symmetry

When we sketch polar curves it is sometimes helpful to take advantage of symmetry. The following three rules are explained by Figure 14.

(a) If a polar equation is unchanged when θ is replaced by $-\theta$, the curve is symmetric about the polar axis.

(b) If the equation is unchanged when r is replaced by $-r$, or when θ is replaced by $\theta + \pi$, the curve is symmetric about the pole. (This means that the curve remains unchanged if we rotate it through 180° about the origin.)

(c) If the equation is unchanged when θ is replaced by $\pi - \theta$, the curve is symmetric about the vertical line $\theta = \pi/2$.

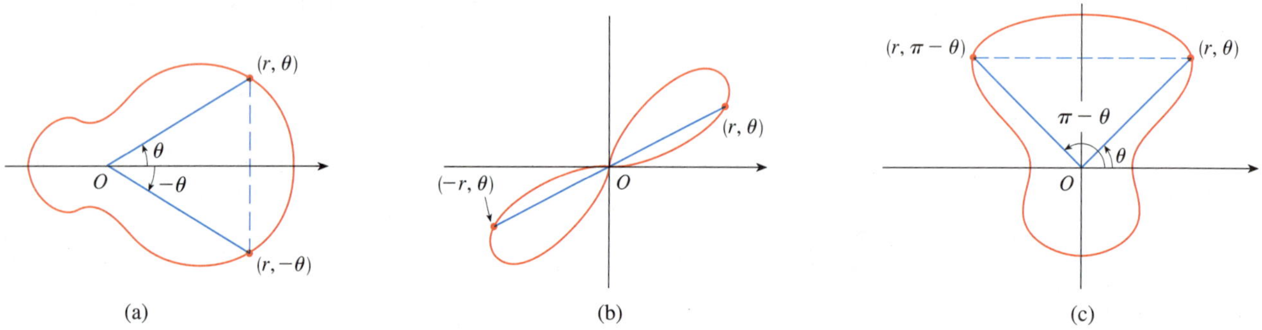

FIGURE 14

The curves sketched in Examples 6 and 8 are symmetric about the polar axis, since $\cos(-\theta) = \cos\theta$. The curves in Examples 7 and 8 are symmetric about $\theta = \pi/2$ because $\sin(\pi - \theta) = \sin\theta$ and $\cos 2(\pi - \theta) = \cos 2\theta$. The four-leaved rose is also symmetric about the pole. These symmetry properties could have been used in sketching the curves. For instance, in Example 6 we need only have plotted points for $0 \leqslant \theta \leqslant \pi/2$ and then reflected about the polar axis to obtain the complete circle.

Tangents to Polar Curves

To find a tangent line to a polar curve $r = f(\theta)$ we regard θ as a parameter and write its parametric equations as

$$x = r\cos\theta = f(\theta)\cos\theta \qquad y = r\sin\theta = f(\theta)\sin\theta$$

Then, using the method for finding slopes of parametric curves (Equation 10.2.2) and the Product Rule, we have

3
$$\frac{dy}{dx} = \frac{\dfrac{dy}{d\theta}}{\dfrac{dx}{d\theta}} = \frac{\dfrac{dr}{d\theta}\sin\theta + r\cos\theta}{\dfrac{dr}{d\theta}\cos\theta - r\sin\theta}$$

We locate horizontal tangents by finding the points where $dy/d\theta = 0$ (provided that $dx/d\theta \neq 0$). Likewise, we locate vertical tangents at the points where $dx/d\theta = 0$ (provided that $dy/d\theta \neq 0$).

Notice that if we are looking for tangent lines at the pole, then $r = 0$ and Equation 3 simplifies to

$$\frac{dy}{dx} = \tan\theta \qquad \text{if} \quad \frac{dr}{d\theta} \neq 0$$

For instance, in Example 8 we found that $r = \cos 2\theta = 0$ when $\theta = \pi/4$ or $3\pi/4$. This means that the lines $\theta = \pi/4$ and $\theta = 3\pi/4$ (or $y = x$ and $y = -x$) are tangent lines to $r = \cos 2\theta$ at the origin.

EXAMPLE 9

(a) For the cardioid $r = 1 + \sin\theta$ of Example 7, find the slope of the tangent line when $\theta = \pi/3$.

(b) Find the points on the cardioid where the tangent line is horizontal or vertical.

SOLUTION Using Equation 3 with $r = 1 + \sin\theta$, we have

$$\frac{dy}{dx} = \frac{\dfrac{dr}{d\theta}\sin\theta + r\cos\theta}{\dfrac{dr}{d\theta}\cos\theta - r\sin\theta} = \frac{\cos\theta\,\sin\theta + (1 + \sin\theta)\cos\theta}{\cos\theta\,\cos\theta - (1 + \sin\theta)\sin\theta}$$

$$= \frac{\cos\theta\,(1 + 2\sin\theta)}{1 - 2\sin^2\theta - \sin\theta} = \frac{\cos\theta\,(1 + 2\sin\theta)}{(1 + \sin\theta)(1 - 2\sin\theta)}$$

(a) The slope of the tangent at the point where $\theta = \pi/3$ is

$$\left.\frac{dy}{dx}\right|_{\theta=\pi/3} = \frac{\cos(\pi/3)(1 + 2\sin(\pi/3))}{(1 + \sin(\pi/3))(1 - 2\sin(\pi/3))}$$

$$= \frac{\frac{1}{2}(1 + \sqrt{3})}{(1 + \sqrt{3}/2)(1 - \sqrt{3})} = \frac{1 + \sqrt{3}}{(2 + \sqrt{3})(1 - \sqrt{3})}$$

$$= \frac{1 + \sqrt{3}}{-1 - \sqrt{3}} = -1$$

(b) Observe that

$$\frac{dy}{d\theta} = \cos\theta\,(1 + 2\sin\theta) = 0 \qquad \text{when } \theta = \frac{\pi}{2}, \frac{3\pi}{2}, \frac{7\pi}{6}, \frac{11\pi}{6}$$

$$\frac{dx}{d\theta} = (1 + \sin\theta)(1 - 2\sin\theta) = 0 \qquad \text{when } \theta = \frac{3\pi}{2}, \frac{\pi}{6}, \frac{5\pi}{6}$$

Therefore, there are horizontal tangents at the points $(2, \pi/2)$, $\left(\frac{1}{2}, 7\pi/6\right)$, $\left(\frac{1}{2}, 11\pi/6\right)$ and vertical tangents at $\left(\frac{3}{2}, \pi/6\right)$ and $\left(\frac{3}{2}, 5\pi/6\right)$. When $\theta = 3\pi/2$, both $dy/d\theta$ and $dx/d\theta$ are 0, so we must be careful. Using l'Hospital's Rule, we have

$$\lim_{\theta \to (3\pi/2)^-} \frac{dy}{dx} = \left(\lim_{\theta \to (3\pi/2)^-} \frac{1 + 2\sin\theta}{1 - 2\sin\theta}\right)\left(\lim_{\theta \to (3\pi/2)^-} \frac{\cos\theta}{1 + \sin\theta}\right)$$

$$= -\frac{1}{3}\lim_{\theta \to (3\pi/2)^-} \frac{\cos\theta}{1 + \sin\theta}$$

$$= -\frac{1}{3}\lim_{\theta \to (3\pi/2)^-} \frac{-\sin\theta}{\cos\theta} = \infty$$

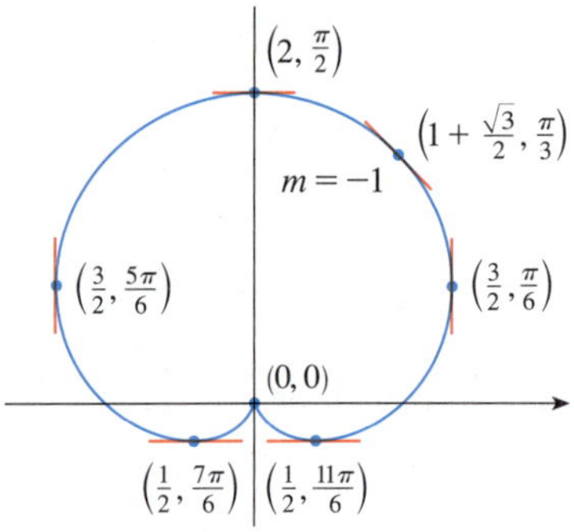

FIGURE 15
Tangent lines for $r = 1 + \sin\theta$

By symmetry,
$$\lim_{\theta \to (3\pi/2)^+} \frac{dy}{dx} = -\infty$$

Thus, there is a vertical tangent line at the pole (see Figure 15).

NOTE ▫ Instead of having to remember Equation 3, we could employ the method used to derive it. For instance, in Example 9 we could have written

$$x = r\cos\theta = (1 + \sin\theta)\cos\theta = \cos\theta + \tfrac{1}{2}\sin 2\theta$$
$$y = r\sin\theta = (1 + \sin\theta)\sin\theta = \sin\theta + \sin^2\theta$$

Then we have

$$\frac{dy}{dx} = \frac{dy/d\theta}{dx/d\theta} = \frac{\cos\theta + 2\sin\theta\cos\theta}{-\sin\theta + \cos 2\theta} = \frac{\cos\theta + \sin 2\theta}{-\sin\theta + \cos 2\theta}$$

which is equivalent to our previous expression.

Graphing Polar Curves with Graphing Devices

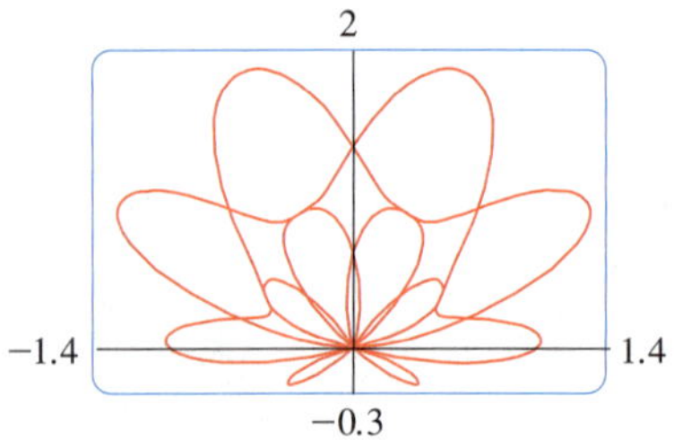

FIGURE 16
$r = \sin\theta + \sin^3(5\theta/2)$

Although it's useful to be able to sketch simple polar curves by hand, we need to use a graphing calculator or computer when we are faced with a curve as complicated as the one shown in Figure 16.

Some graphing devices have commands that enable us to graph polar curves directly. With other machines we need to convert to parametric equations first. In this case we take the polar equation $r = f(\theta)$ and write its parametric equations as

$$x = r\cos\theta = f(\theta)\cos\theta \qquad y = r\sin\theta = f(\theta)\sin\theta$$

Some machines require that the parameter be called t rather than θ.

EXAMPLE 10 Graph the curve $r = \sin(8\theta/5)$.

SOLUTION Let's assume that our graphing device doesn't have a built-in polar graphing command. In this case we need to work with the corresponding parametric equations, which are

$$x = r\cos\theta = \sin(8\theta/5)\cos\theta \qquad y = r\sin\theta = \sin(8\theta/5)\sin\theta$$

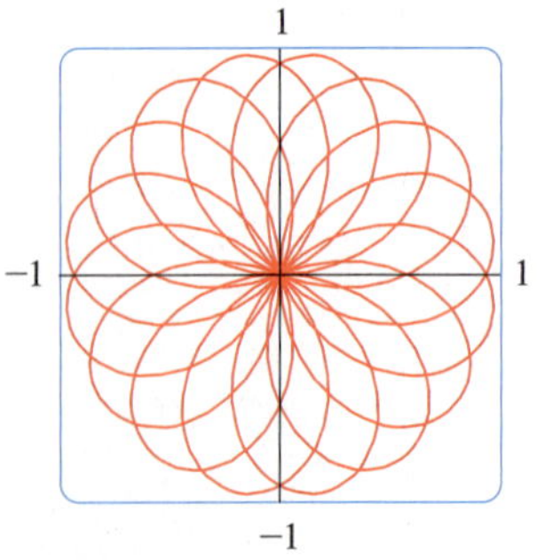

FIGURE 17
$r = \sin(8\theta/5)$

In any case we need to determine the domain for θ. So we ask ourselves: How many complete rotations are required until the curve starts to repeat itself? If the answer is n, then

$$\sin\frac{8(\theta + 2n\pi)}{5} = \sin\left(\frac{8\theta}{5} + \frac{16n\pi}{5}\right) = \sin\frac{8\theta}{5}$$

and so we require that $16n\pi/5$ be an even multiple of π. This will first occur when $n = 5$. Therefore, we will graph the entire curve if we specify that $0 \leq \theta \leq 10\pi$. Switching from θ to t, we have the equations

$$x = \sin(8t/5)\cos t \qquad y = \sin(8t/5)\sin t \qquad 0 \leq t \leq 10\pi$$

and Figure 17 shows the resulting curve. Notice that this rose has 16 loops.

EXAMPLE 11 Investigate the family of polar curves given by $r = 1 + c \sin\theta$. How does the shape change as c changes? (These curves are called **limaçons**, after a French word for snail, because of the shape of the curves for certain values of c.)

SOLUTION Figure 18 shows computer-drawn graphs for various values of c. For $c > 1$ there is a loop that decreases in size as c decreases. When $c = 1$ the loop disappears and the curve becomes the cardioid that we sketched in Example 7. For c between 1 and $\frac{1}{2}$ the cardioid's cusp is smoothed out and becomes a "dimple." When c decreases from $\frac{1}{2}$ to 0, the limaçon is shaped like an oval. This oval becomes more circular as $c \to 0$, and when $c = 0$ the curve is just the circle $r = 1$.

|||| In Exercise 53 you are asked to prove analytically what we have discovered from the graphs in Figure 18.

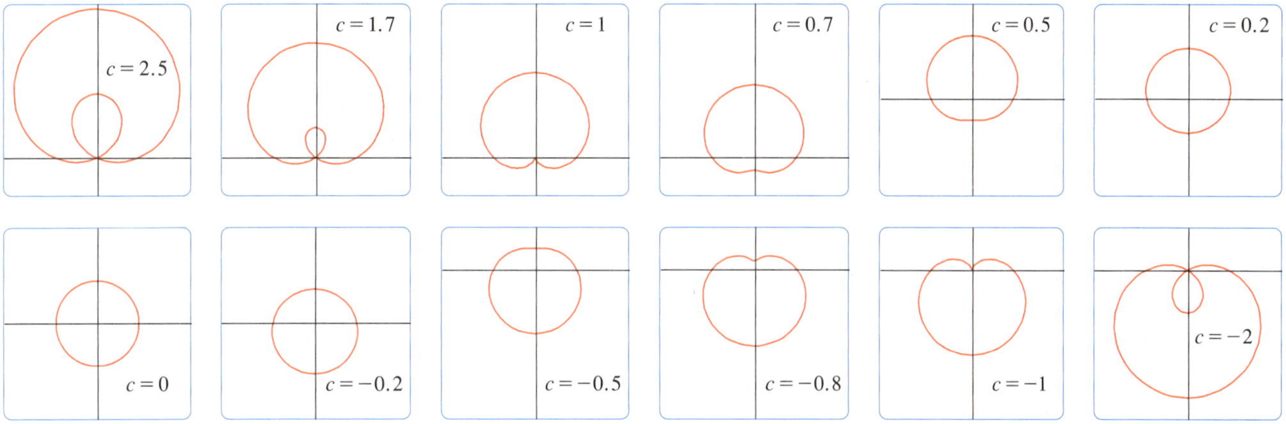

FIGURE 18
Members of the family of limaçons $r = 1 + c \sin\theta$

The remaining parts of Figure 18 show that as c becomes negative, the shapes change in reverse order. In fact, these curves are reflections about the horizontal axis of the corresponding curves with positive c.

10.3 Exercises

1–2 |||| Plot the point whose polar coordinates are given. Then find two other pairs of polar coordinates of this point, one with $r > 0$ and one with $r < 0$.

1. (a) $(1, \pi/2)$ (b) $(-2, \pi/4)$ (c) $(3, 2)$

2. (a) $(3, 0)$ (b) $(2, -\pi/7)$ (c) $(-1, -\pi/2)$

3–4 |||| Plot the point whose polar coordinates are given. Then find the Cartesian coordinates of the point.

3. (a) $(3, \pi/2)$ (b) $(2\sqrt{2}, 3\pi/4)$ (c) $(-1, \pi/3)$

4. (a) $(2, 2\pi/3)$ (b) $(4, 3\pi)$ (c) $(-2, -5\pi/6)$

5–6 |||| The Cartesian coordinates of a point are given.
(i) Find polar coordinates (r, θ) of the point, where $r > 0$ and $0 \leq \theta < 2\pi$.
(ii) Find polar coordinates (r, θ) of the point, where $r < 0$ and $0 \leq \theta < 2\pi$.

5. (a) $(1, 1)$ (b) $(2\sqrt{3}, -2)$

6. (a) $(-1, -\sqrt{3})$ (b) $(-2, 3)$

7–12 |||| Sketch the region in the plane consisting of points whose polar coordinates satisfy the given conditions.

7. $1 \leq r \leq 2$

8. $r \geq 0, \quad \pi/3 \leq \theta \leq 2\pi/3$

9. $0 \leq r < 4, \quad -\pi/2 \leq \theta < \pi/6$

10. $2 < r \leq 5, \quad 3\pi/4 < \theta < 5\pi/4$

11. $2 < r < 3, \quad 5\pi/3 \leq \theta \leq 7\pi/3$

12. $-1 \leq r \leq 1, \quad \pi/4 \leq \theta \leq 3\pi/4$

13. Find the distance between the points with polar coordinates $(1, \pi/6)$ and $(3, 3\pi/4)$.

14. Find a formula for the distance between the points with polar coordinates (r_1, θ_1) and (r_2, θ_2).

15–20 ||| Identify the curve by finding a Cartesian equation for the curve.

15. $r = 2$

16. $r \cos \theta = 1$

17. $r = 3 \sin \theta$

18. $r = 2 \sin \theta + 2 \cos \theta$

19. $r = \csc \theta$

20. $r = \tan \theta \sec \theta$

21–26 ||| Find a polar equation for the curve represented by the given Cartesian equation.

21. $x = 3$

22. $x^2 + y^2 = 9$

23. $x = -y^2$

24. $x + y = 9$

25. $x^2 + y^2 = 2cx$

26. $x^2 - y^2 = 1$

27–28 ||| For each of the described curves, decide if the curve would be more easily given by a polar equation or a Cartesian equation. Then write an equation for the curve.

27. (a) A line through the origin that makes an angle of $\pi/6$ with the positive x-axis
(b) A vertical line through the point (3, 3)

28. (a) A circle with radius 5 and center (2, 3)
(b) A circle centered at the origin with radius 4

29–46 ||| Sketch the curve with the given polar equation.

29. $\theta = -\pi/6$

30. $r^2 - 3r + 2 = 0$

31. $r = \sin \theta$

32. $r = -3 \cos \theta$

33. $r = 2(1 - \sin \theta)$

34. $r = 1 - 3 \cos \theta$

35. $r = \theta, \quad \theta \geqslant 0$

36. $r = \ln \theta, \quad \theta \geqslant 1$

37. $r = \sin 2\theta$

38. $r = 2 \cos 3\theta$

39. $r = 2 \cos 4\theta$

40. $r = \sin 5\theta$

41. $r^2 = 4 \cos 2\theta$

42. $r^2 = \sin 2\theta$

43. $r = 2 \cos(3\theta/2)$

44. $r^2\theta = 1$

45. $r = 1 + 2 \cos 2\theta$

46. $r = 1 + 2 \cos(\theta/2)$

47–48 ||| The figure shows the graph of r as a function of θ in Cartesian coordinates. Use it to sketch the corresponding polar curve.

47.

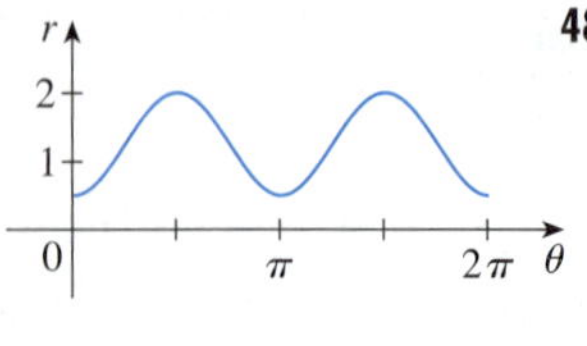

48.

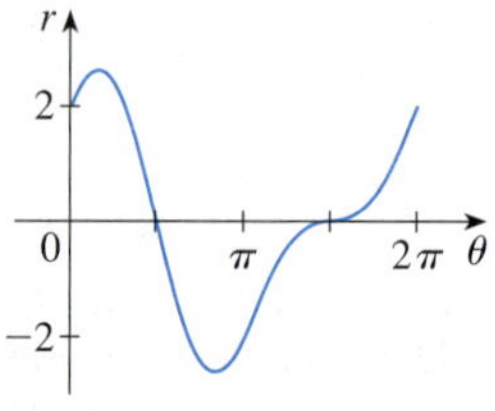

49. Show that the polar curve $r = 4 + 2 \sec \theta$ (called a **conchoid**) has the line $x = 2$ as a vertical asymptote by showing that $\lim_{r \to \pm\infty} x = 2$. Use this fact to help sketch the conchoid.

50. Show that the curve $r = 2 - \csc \theta$ (also a conchoid) has the line $y = -1$ as a horizontal asymptote by showing that $\lim_{r \to \pm\infty} y = -1$. Use this fact to help sketch the conchoid.

51. Show that the curve $r = \sin \theta \tan \theta$ (called a **cissoid of Diocles**) has the line $x = 1$ as a vertical asymptote. Show also that the curve lies entirely within the vertical strip $0 \leqslant x < 1$. Use these facts to help sketch the cissoid.

52. Sketch the curve $(x^2 + y^2)^3 = 4x^2y^2$.

53. (a) In Example 11 the graphs suggest that the limaçon $r = 1 + c \sin \theta$ has an inner loop when $|c| > 1$. Prove that this is true, and find the values of θ that correspond to the inner loop.
(b) From Figure 18 it appears that the limaçon loses its dimple when $c = \frac{1}{2}$. Prove this.

54. Match the polar equations with the graphs labeled I–VI. Give reasons for your choices. (Don't use a graphing device.)

(a) $r = \sin(\theta/2)$
(b) $r = \sin(\theta/4)$
(c) $r = \sec(3\theta)$
(d) $r = \theta \sin \theta$
(e) $r = 1 + 4 \cos 5\theta$
(f) $r = 1/\sqrt{\theta}$

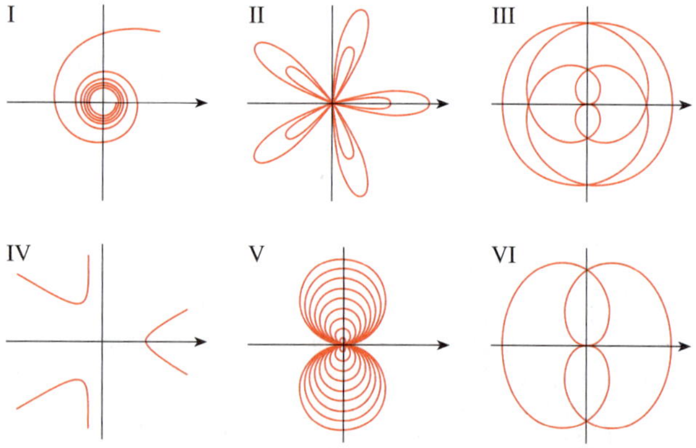

55–60 ||| Find the slope of the tangent line to the given polar curve at the point specified by the value of θ.

55. $r = 2 \sin \theta, \quad \theta = \pi/6$

56. $r = 2 - \sin \theta, \quad \theta = \pi/3$

57. $r = 1/\theta, \quad \theta = \pi$

58. $r = \ln \theta, \quad \theta = e$

59. $r = 1 + \cos \theta, \quad \theta = \pi/6$

60. $r = \sin 3\theta, \quad \theta = \pi/6$

61–66 ||| Find the points on the given curve where the tangent line is horizontal or vertical.

61. $r = 3 \cos \theta$

62. $r = \cos \theta + \sin \theta$

63. $r = 1 + \cos \theta$

64. $r = e^{\theta}$

65. $r = \cos 2\theta$

66. $r^2 = \sin 2\theta$

67. Show that the polar equation $r = a \sin\theta + b \cos\theta$, where $ab \neq 0$, represents a circle, and find its center and radius.

68. Show that the curves $r = a \sin\theta$ and $r = a \cos\theta$ intersect at right angles.

69–74 Use a graphing device to graph the polar curve. Choose the parameter interval to make sure that you produce the entire curve.

69. $r = 1 + 2\sin(\theta/2)$ (nephroid of Freeth)

70. $r = \sqrt{1 - 0.8\sin^2\theta}$ (hippopede)

71. $r = e^{\sin\theta} - 2\cos(4\theta)$ (butterfly curve)

72. $r = \sin^2(4\theta) + \cos(4\theta)$

73. $r = 2 - 5\sin(\theta/6)$

74. $r = \cos(\theta/2) + \cos(\theta/3)$

75. How are the graphs of $r = 1 + \sin(\theta - \pi/6)$ and $r = 1 + \sin(\theta - \pi/3)$ related to the graph of $r = 1 + \sin\theta$? In general, how is the graph of $r = f(\theta - \alpha)$ related to the graph of $r = f(\theta)$?

76. Use a graph to estimate the y-coordinate of the highest points on the curve $r = \sin 2\theta$. Then use calculus to find the exact value.

77. (a) Investigate the family of curves defined by the polar equations $r = \sin n\theta$, where n is a positive integer. How is the number of loops related to n?
(b) What happens if the equation in part (a) is replaced by $r = |\sin n\theta|$?

78. A family of curves is given by the equations $r = 1 + c \sin n\theta$, where c is a real number and n is a positive integer. How does the graph change as n increases? How does it change as c changes? Illustrate by graphing enough members of the family to support your conclusions.

79. A family of curves has polar equations

$$r = \frac{1 - a\cos\theta}{1 + a\cos\theta}$$

Investigate how the graph changes as the number a changes. In particular, you should identify the transitional values of a for which the basic shape of the curve changes.

80. The astronomer Giovanni Cassini (1625–1712) studied the family of curves with polar equations

$$r^4 - 2c^2r^2\cos 2\theta + c^4 - a^4 = 0$$

where a and c are positive real numbers. These curves are called the **ovals of Cassini** even though they are oval shaped only for certain values of a and c. (Cassini thought that these curves might represent planetary orbits better than Kepler's ellipses.) Investigate the variety of shapes that these curves may have. In particular, how are a and c related to each other when the curve splits into two parts?

81. Let P be any point (except the origin) on the curve $r = f(\theta)$. If ψ is the angle between the tangent line at P and the radial line OP, show that

$$\tan\psi = \frac{r}{dr/d\theta}$$

[*Hint:* Observe that $\psi = \phi - \theta$ in the figure.]

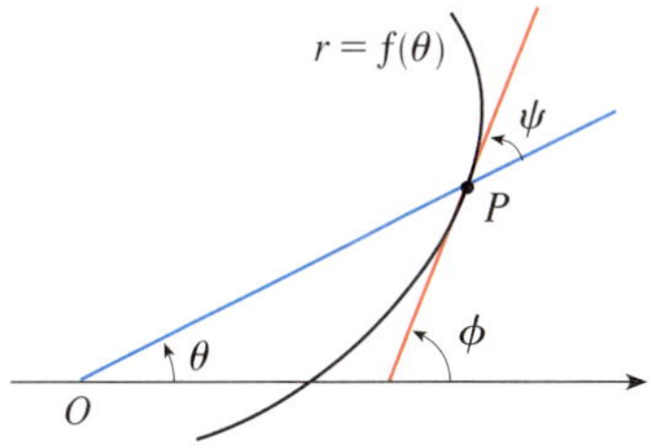

82. (a) Use Exercise 81 to show that the angle between the tangent line and the radial line is $\psi = \pi/4$ at every point on the curve $r = e^\theta$.
(b) Illustrate part (a) by graphing the curve and the tangent lines at the points where $\theta = 0$ and $\pi/2$.
(c) Prove that any polar curve $r = f(\theta)$ with the property that the angle ψ between the radial line and the tangent line is a constant must be of the form $r = Ce^{k\theta}$, where C and k are constants.

10.4 Areas and Lengths in Polar Coordinates

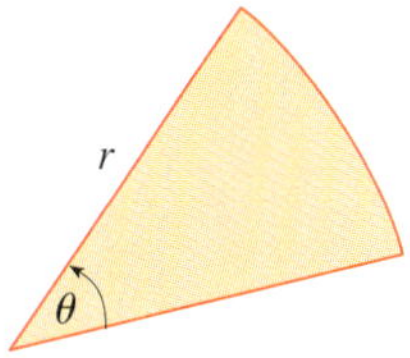

FIGURE 1

In this section we develop the formula for the area of a region whose boundary is given by a polar equation. We need to use the formula for the area of a sector of a circle

$$\boxed{1} \qquad A = \tfrac{1}{2}r^2\theta$$

where, as in Figure 1, r is the radius and θ is the radian measure of the central angle. Formula 1 follows from the fact that the area of a sector is proportional to its central angle: $A = (\theta/2\pi)\pi r^2 = \frac{1}{2}r^2\theta$. (See also Exercise 35 in Section 7.3.)

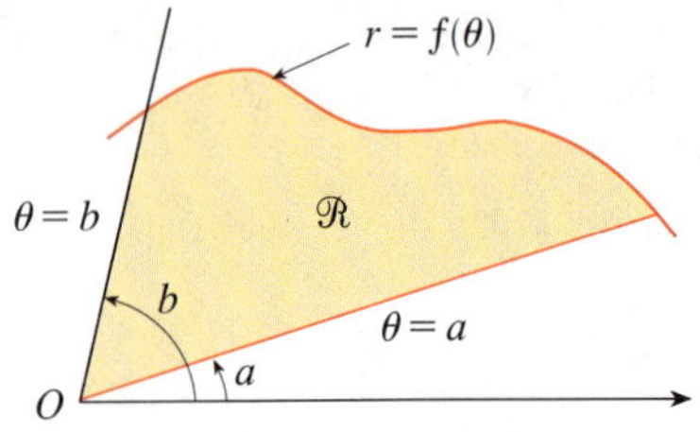

FIGURE 2

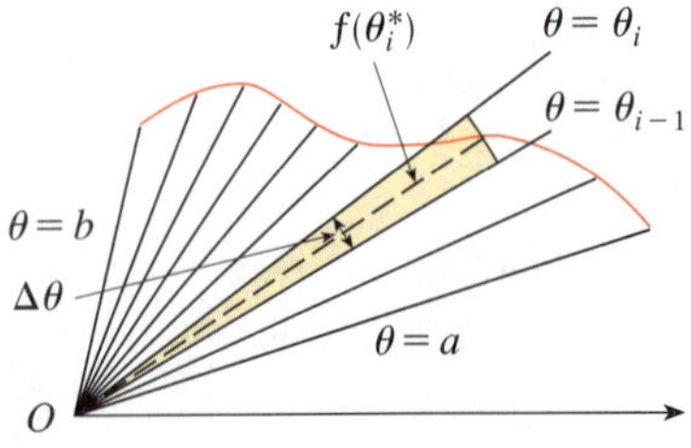

FIGURE 3

Let $\mathcal{R}$ be the region, illustrated in Figure 2, bounded by the polar curve $r = f(\theta)$ and by the rays $\theta = a$ and $\theta = b$, where f is a positive continuous function and where $0 < b - a \leqslant 2\pi$. We divide the interval $[a, b]$ into subintervals with endpoints θ_0, θ_1, $\theta_2, \ldots, \theta_n$ and equal width $\Delta\theta$. The rays $\theta = \theta_i$ then divide $\mathcal{R}$ into n smaller regions with central angle $\Delta\theta = \theta_i - \theta_{i-1}$. If we choose θ_i^* in the ith subinterval $[\theta_{i-1}, \theta_i]$, then the area ΔA_i of the ith region is approximated by the area of the sector of a circle with central angle $\Delta\theta$ and radius $f(\theta_i^*)$. (See Figure 3.)

Thus, from Formula 1 we have

$$\Delta A_i \approx \tfrac{1}{2}[f(\theta_i^*)]^2\,\Delta\theta$$

and so an approximation to the total area A of $\mathcal{R}$ is

2 $$A \approx \sum_{i=1}^{n} \tfrac{1}{2}[f(\theta_i^*)]^2\,\Delta\theta$$

It appears from Figure 3 that the approximation in (2) improves as $n \to \infty$. But the sums in (2) are Riemann sums for the function $g(\theta) = \tfrac{1}{2}[f(\theta)]^2$, so

$$\lim_{n\to\infty} \sum_{i=1}^{n} \tfrac{1}{2}[f(\theta_i^*)]^2\,\Delta\theta = \int_a^b \tfrac{1}{2}[f(\theta)]^2\,d\theta$$

It therefore appears plausible (and can in fact be proved) that the formula for the area A of the polar region $\mathcal{R}$ is

3 $$A = \int_a^b \tfrac{1}{2}[f(\theta)]^2\,d\theta$$

Formula 3 is often written as

4 $$A = \int_a^b \tfrac{1}{2}r^2\,d\theta$$

with the understanding that $r = f(\theta)$. Note the similarity between Formulas 1 and 4.

When we apply Formula 3 or 4 it is helpful to think of the area as being swept out by a rotating ray through O that starts with angle a and ends with angle b.

EXAMPLE 1 Find the area enclosed by one loop of the four-leaved rose $r = \cos 2\theta$.

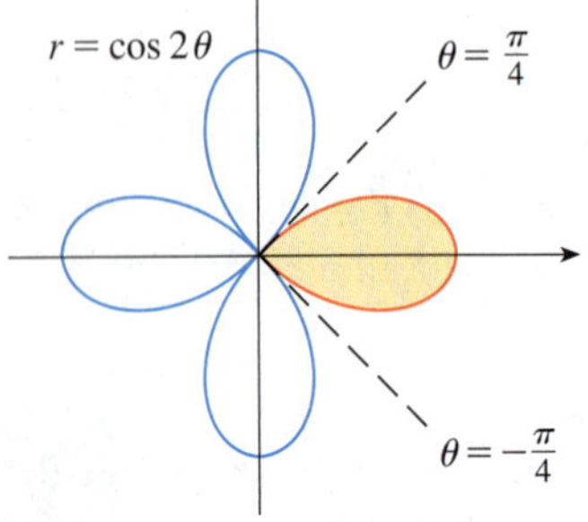

FIGURE 4

SOLUTION The curve $r = \cos 2\theta$ was sketched in Example 8 in Section 10.3. Notice from Figure 4 that the region enclosed by the right loop is swept out by a ray that rotates from $\theta = -\pi/4$ to $\theta = \pi/4$. Therefore, Formula 4 gives

$$A = \int_{-\pi/4}^{\pi/4} \tfrac{1}{2}r^2\,d\theta = \tfrac{1}{2}\int_{-\pi/4}^{\pi/4} \cos^2 2\theta\,d\theta$$

$$= \int_0^{\pi/4} \cos^2 2\theta\,d\theta = \int_0^{\pi/4} \tfrac{1}{2}(1 + \cos 4\theta)\,d\theta$$

$$= \tfrac{1}{2}\Big[\theta + \tfrac{1}{4}\sin 4\theta\Big]_0^{\pi/4} = \frac{\pi}{8}$$

EXAMPLE 2 Find the area of the region that lies inside the circle $r = 3 \sin \theta$ and outside the cardioid $r = 1 + \sin \theta$.

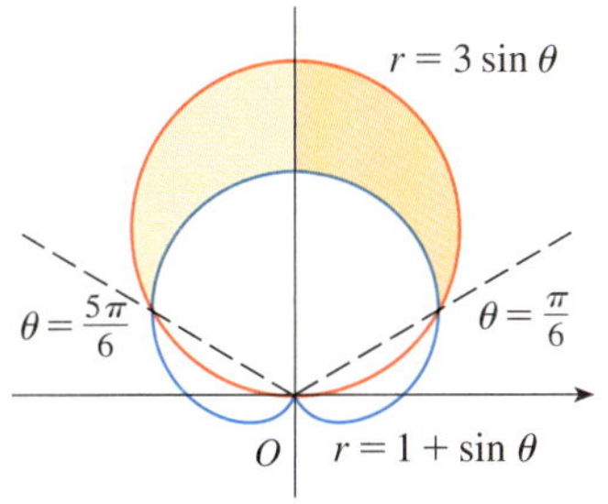

FIGURE 5

SOLUTION The cardioid (see Example 7 in Section 10.3) and the circle are sketched in Figure 5 and the desired region is shaded. The values of a and b in Formula 4 are determined by finding the points of intersection of the two curves. They intersect when $3 \sin \theta = 1 + \sin \theta$, which gives $\sin \theta = \frac{1}{2}$, so $\theta = \pi/6, 5\pi/6$. The desired area can be found by subtracting the area inside the cardioid between $\theta = \pi/6$ and $\theta = 5\pi/6$ from the area inside the circle from $\pi/6$ to $5\pi/6$. Thus

$$A = \frac{1}{2} \int_{\pi/6}^{5\pi/6} (3 \sin \theta)^2 \, d\theta - \frac{1}{2} \int_{\pi/6}^{5\pi/6} (1 + \sin \theta)^2 \, d\theta$$

Since the region is symmetric about the vertical axis $\theta = \pi/2$, we can write

$$A = 2\left[\frac{1}{2} \int_{\pi/6}^{\pi/2} 9 \sin^2\theta \, d\theta - \frac{1}{2} \int_{\pi/6}^{\pi/2} (1 + 2 \sin \theta + \sin^2\theta) \, d\theta\right]$$

$$= \int_{\pi/6}^{\pi/2} (8 \sin^2\theta - 1 - 2 \sin \theta) \, d\theta$$

$$= \int_{\pi/6}^{\pi/2} (3 - 4 \cos 2\theta - 2 \sin \theta) \, d\theta \qquad \left[\text{because } \sin^2\theta = \tfrac{1}{2}(1 - \cos 2\theta)\right]$$

$$= 3\theta - 2 \sin 2\theta + 2 \cos \theta \Big]_{\pi/6}^{\pi/2} = \pi$$

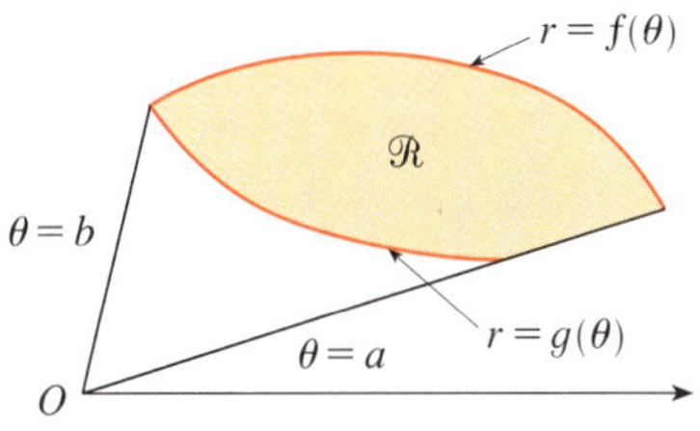

FIGURE 6

Example 2 illustrates the procedure for finding the area of the region bounded by two polar curves. In general, let $\mathcal{R}$ be a region, as illustrated in Figure 6, that is bounded by curves with polar equations $r = f(\theta)$, $r = g(\theta)$, $\theta = a$, and $\theta = b$, where $f(\theta) \geqslant g(\theta) \geqslant 0$ and $0 < b - a \leqslant 2\pi$. The area A of $\mathcal{R}$ is found by subtracting the area inside $r = g(\theta)$ from the area inside $r = f(\theta)$, so using Formula 3 we have

$$A = \int_a^b \tfrac{1}{2}[f(\theta)]^2 \, d\theta - \int_a^b \tfrac{1}{2}[g(\theta)]^2 \, d\theta$$

$$= \tfrac{1}{2} \int_a^b \left([f(\theta)]^2 - [g(\theta)]^2\right) d\theta$$

CAUTION ▫ The fact that a single point has many representations in polar coordinates sometimes makes it difficult to find all the points of intersection of two polar curves. For instance, it is obvious from Figure 5 that the circle and the cardioid have three points of intersection; however, in Example 2 we solved the equations $r = 3 \sin \theta$ and $r = 1 + \sin \theta$ and found only two such points, $\left(\frac{3}{2}, \pi/6\right)$ and $\left(\frac{3}{2}, 5\pi/6\right)$. The origin is also a point of intersection, but we can't find it by solving the equations of the curves because the origin has no single representation in polar coordinates that satisfies both equations. Notice that, when represented as $(0, 0)$ or $(0, \pi)$, the origin satisfies $r = 3 \sin \theta$ and so it lies on the circle; when represented as $(0, 3\pi/2)$, it satisfies $r = 1 + \sin \theta$ and so it lies on the cardioid. Think of two points moving along the curves as the parameter value θ increases from 0 to 2π. On one curve the origin is reached at $\theta = 0$ and $\theta = \pi$; on the other curve it is reached at $\theta = 3\pi/2$. The points don't collide at the origin because they reach the origin at different times, but the curves intersect there nonetheless.

Thus, to find *all* points of intersection of two polar curves, it is recommended that you draw the graphs of both curves. It is especially convenient to use a graphing calculator or computer to help with this task.

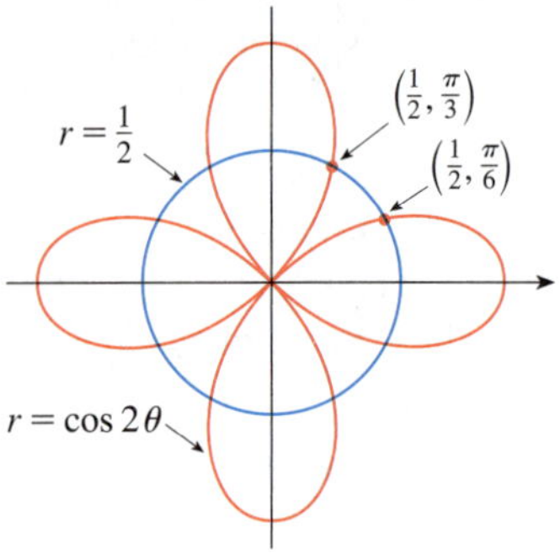

FIGURE 7

EXAMPLE 3 Find all points of intersection of the curves $r = \cos 2\theta$ and $r = \frac{1}{2}$.

SOLUTION If we solve the equations $r = \cos 2\theta$ and $r = \frac{1}{2}$, we get $\cos 2\theta = \frac{1}{2}$ and, therefore, $2\theta = \pi/3, 5\pi/3, 7\pi/3, 11\pi/3$. Thus, the values of θ between 0 and 2π that satisfy both equations are $\theta = \pi/6, 5\pi/6, 7\pi/6, 11\pi/6$. We have found four points of intersection: $\left(\frac{1}{2}, \pi/6\right)$, $\left(\frac{1}{2}, 5\pi/6\right)$, $\left(\frac{1}{2}, 7\pi/6\right)$, and $\left(\frac{1}{2}, 11\pi/6\right)$.

However, you can see from Figure 7 that the curves have four other points of intersection—namely, $\left(\frac{1}{2}, \pi/3\right)$, $\left(\frac{1}{2}, 2\pi/3\right)$, $\left(\frac{1}{2}, 4\pi/3\right)$, and $\left(\frac{1}{2}, 5\pi/3\right)$. These can be found using symmetry or by noticing that another equation of the circle is $r = -\frac{1}{2}$ and then solving the equations $r = \cos 2\theta$ and $r = -\frac{1}{2}$.

Arc Length

To find the length of a polar curve $r = f(\theta)$, $a \leqslant \theta \leqslant b$, we regard θ as a parameter and write the parametric equations of the curve as

$$x = r\cos\theta = f(\theta)\cos\theta \qquad y = r\sin\theta = f(\theta)\sin\theta$$

Using the Product Rule and differentiating with respect to θ, we obtain

$$\frac{dx}{d\theta} = \frac{dr}{d\theta}\cos\theta - r\sin\theta \qquad \frac{dy}{d\theta} = \frac{dr}{d\theta}\sin\theta + r\cos\theta$$

so, using $\cos^2\theta + \sin^2\theta = 1$, we have

$$\begin{aligned}\left(\frac{dx}{d\theta}\right)^2 + \left(\frac{dy}{d\theta}\right)^2 &= \left(\frac{dr}{d\theta}\right)^2\cos^2\theta - 2r\frac{dr}{d\theta}\cos\theta\,\sin\theta + r^2\sin^2\theta \\ &\quad + \left(\frac{dr}{d\theta}\right)^2\sin^2\theta + 2r\frac{dr}{d\theta}\sin\theta\,\cos\theta + r^2\cos^2\theta \\ &= \left(\frac{dr}{d\theta}\right)^2 + r^2\end{aligned}$$

Assuming that f' is continuous, we can use Theorem 10.2.6 to write the arc length as

$$L = \int_a^b \sqrt{\left(\frac{dx}{d\theta}\right)^2 + \left(\frac{dy}{d\theta}\right)^2}\,d\theta$$

Therefore, the length of a curve with polar equation $r = f(\theta)$, $a \leqslant \theta \leqslant b$, is

5
$$L = \int_a^b \sqrt{r^2 + \left(\frac{dr}{d\theta}\right)^2}\,d\theta$$

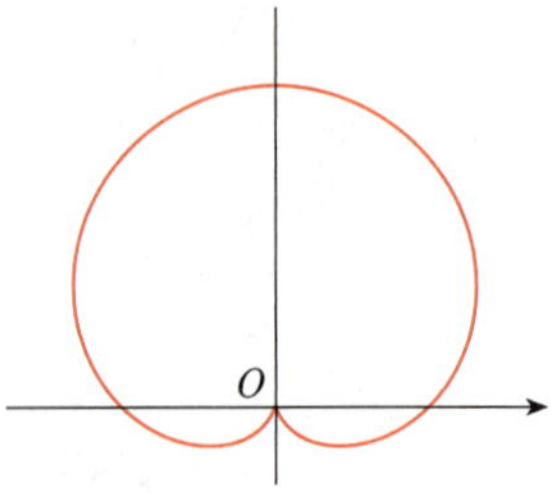

FIGURE 8
$r = 1 + \sin\theta$

EXAMPLE 4 Find the length of the cardioid $r = 1 + \sin\theta$.

SOLUTION The cardioid is shown in Figure 8. (We sketched it in Example 7 in Section 10.3.) Its full length is given by the parameter interval $0 \leqslant \theta \leqslant 2\pi$, so Formula 5 gives

$$L = \int_0^{2\pi} \sqrt{r^2 + \left(\frac{dr}{d\theta}\right)^2}\,d\theta = \int_0^{2\pi} \sqrt{(1 + \sin\theta)^2 + \cos^2\theta}\,d\theta$$

$$= \int_0^{2\pi} \sqrt{2 + 2\sin\theta}\,d\theta$$

We could evaluate this integral by multiplying and dividing the integrand by $\sqrt{2 - 2\sin\theta}$, or we could use a computer algebra system. In any event, we find that the length of the cardioid is $L = 8$.

10.4 Exercises

1–4 Find the area of the region that is bounded by the given curve and lies in the specified sector.

1. $r = \sqrt{\theta}, \quad 0 \leqslant \theta \leqslant \pi/4$

2. $r = e^{\theta/2}, \quad \pi \leqslant \theta \leqslant 2\pi$

3. $r = \sin\theta, \quad \pi/3 \leqslant \theta \leqslant 2\pi/3$

4. $r = \sqrt{\sin\theta}, \quad 0 \leqslant \theta \leqslant \pi$

5–8 Find the area of the shaded region.

5.

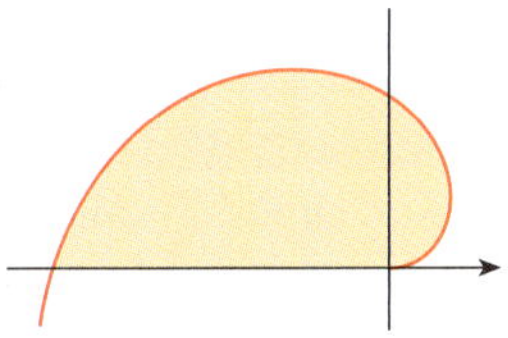

$r = \theta$

6.

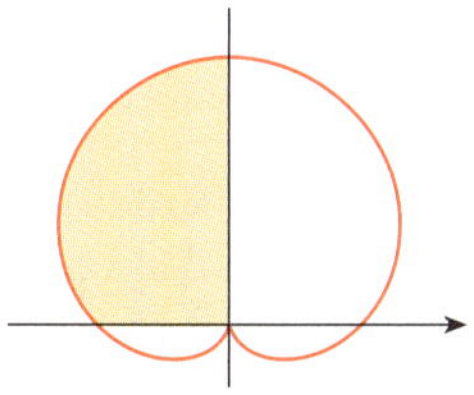

$r = 1 + \sin\theta$

7.

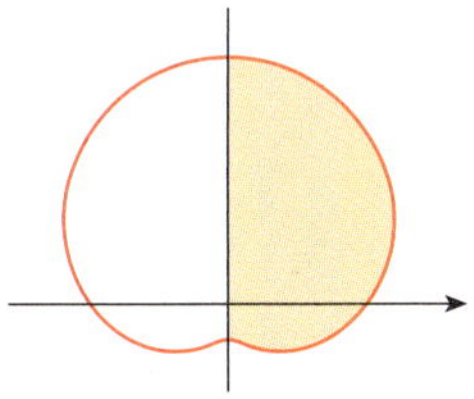

$r = 4 + 3\sin\theta$

8.

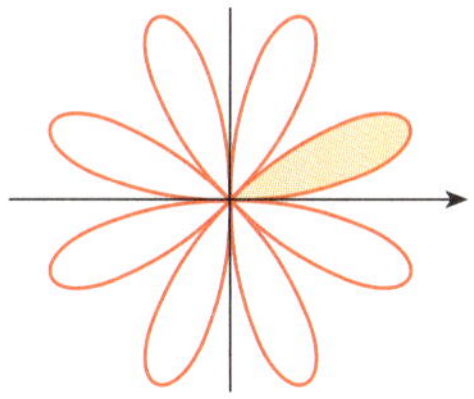

$r = \sin 4\theta$

9–14 Sketch the curve and find the area that it encloses.

9. $r = 3\cos\theta$

10. $r = 3(1 + \cos\theta)$

11. $r^2 = 4\cos 2\theta$

12. $r^2 = \sin 2\theta$

13. $r = 2\cos 3\theta$

14. $r = 2 + \cos 2\theta$

15–16 Graph the curve and find the area that it encloses.

15. $r = 1 + 2\sin 6\theta$

16. $r = 2\sin\theta + 3\sin 9\theta$

17–21 Find the area of the region enclosed by one loop of the curve.

17. $r = \sin 2\theta$

18. $r = 4\sin 3\theta$

19. $r = 3\cos 5\theta$

20. $r = 2\cos 4\theta$

21. $r = 1 + 2\sin\theta$ (inner loop)

22. Find the area enclosed by the loop of the **strophoid** $r = 2\cos\theta - \sec\theta$.

23–28 Find the area of the region that lies inside the first curve and outside the second curve.

23. $r = 4\sin\theta, \quad r = 2$

24. $r = 1 - \sin\theta, \quad r = 1$

25. $r^2 = 8\cos 2\theta, \quad r = 2$

26. $r = 2 + \sin\theta, \quad r = 3\sin\theta$

27. $r = 3\cos\theta, \quad r = 1 + \cos\theta$

28. $r = 1 + \cos\theta, \quad r = 3\cos\theta$

29–34 Find the area of the region that lies inside both curves.

29. $r = \sin\theta, \quad r = \cos\theta$

30. $r = \sin 2\theta, \quad r = \sin\theta$

31. $r = \sin 2\theta, \quad r = \cos 2\theta$

32. $r^2 = 2\sin 2\theta, \quad r = 1$

33. $r = 3 + 2\sin\theta, \quad r = 2$

34. $r = a\sin\theta, \quad r = b\cos\theta, \quad a > 0, \; b > 0$

35. Find the area inside the larger loop and outside the smaller loop of the limaçon $r = \frac{1}{2} + \cos\theta$.

36. Find the area between a large loop and the enclosed small loop of the curve $r = 1 + 2\cos 3\theta$.

37–42 Find all points of intersection of the given curves.

37. $r = \sin\theta, \quad r = \cos\theta$

38. $r = 2, \quad r = 2\cos 2\theta$

39. $r = \cos\theta, \quad r = 1 - \cos\theta$

40. $r = \cos 3\theta, \quad r = \sin 3\theta$

41. $r = \sin\theta, \quad r = \sin 2\theta$

42. $r^2 = \sin 2\theta, \quad r^2 = \cos 2\theta$

43. The points of intersection of the cardioid $r = 1 + \sin\theta$ and the spiral loop $r = 2\theta$, $-\pi/2 \leqslant \theta \leqslant \pi/2$, can't be found exactly. Use a graphing device to find the approximate values of θ at which they intersect. Then use these values to estimate the area that lies inside both curves.

44. Use a graph to estimate the values of θ for which the curves $r = 3 + \sin 5\theta$ and $r = 6\sin\theta$ intersect. Then estimate the area that lies inside both curves.

45–48 Find the exact length of the polar curve.

45. $r = 3 \sin \theta, \quad 0 \leqslant \theta \leqslant \pi/3$

46. $r = e^{2\theta}, \quad 0 \leqslant \theta \leqslant 2\pi$

47. $r = \theta^2, \quad 0 \leqslant \theta \leqslant 2\pi$

48. $r = \theta, \quad 0 \leqslant \theta \leqslant 2\pi$

49–52 Use a calculator to find the length of the curve correct to four decimal places.

49. $r = 3 \sin 2\theta$

50. $r = 4 \sin 3\theta$

51. $r = \sin(\theta/2)$

52. $r = 1 + \cos(\theta/3)$

53–54 Graph the curve and find its length.

53. $r = \cos^4(\theta/4)$

54. $r = \cos^2(\theta/2)$

55. (a) Use Formula 10.2.7 to show that the area of the surface generated by rotating the polar curve

$$r = f(\theta) \qquad a \leqslant \theta \leqslant b$$

(where f' is continuous and $0 \leqslant a < b \leqslant \pi$) about the polar axis is

$$S = \int_a^b 2\pi r \sin \theta \sqrt{r^2 + \left(\frac{dr}{d\theta}\right)^2}\, d\theta$$

(b) Use the formula in part (a) to find the surface area generated by rotating the lemniscate $r^2 = \cos 2\theta$ about the polar axis.

56. (a) Find a formula for the area of the surface generated by rotating the polar curve $r = f(\theta)$, $a \leqslant \theta \leqslant b$ (where f' is continuous and $0 \leqslant a < b \leqslant \pi$), about the line $\theta = \pi/2$.

(b) Find the surface area generated by rotating the lemniscate $r^2 = \cos 2\theta$ about the line $\theta = \pi/2$.

10.5 Conic Sections

In this section we give geometric definitions of parabolas, ellipses, and hyperbolas and derive their standard equations. They are called **conic sections**, or **conics**, because they result from intersecting a cone with a plane as shown in Figure 1.

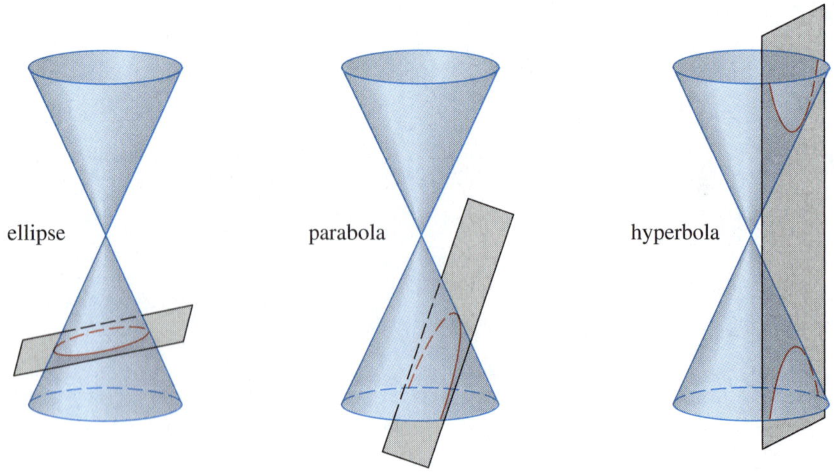

FIGURE 1
Conics

Parabolas

A **parabola** is the set of points in a plane that are equidistant from a fixed point F (called the **focus**) and a fixed line (called the **directrix**). This definition is illustrated by Figure 2. Notice that the point halfway between the focus and the directrix lies on the parabola; it is called the **vertex**. The line through the focus perpendicular to the directrix is called the **axis** of the parabola.

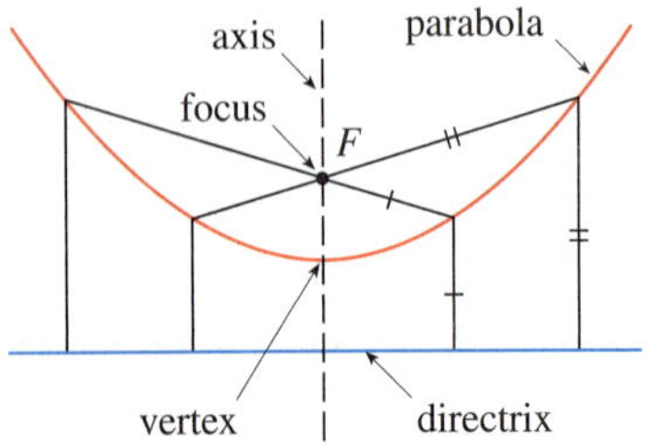

FIGURE 2

In the 16th century Galileo showed that the path of a projectile that is shot into the air at an angle to the ground is a parabola. Since then, parabolic shapes have been used in designing automobile headlights, reflecting telescopes, and suspension bridges.

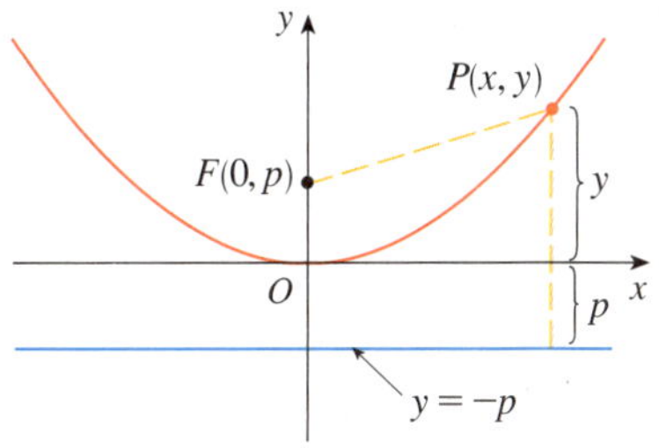

FIGURE 3

(See Problem 16 on page 276 for the reflection property of parabolas that makes them so useful.)

We obtain a particularly simple equation for a parabola if we place its vertex at the origin O and its directrix parallel to the x-axis as in Figure 3. If the focus is the point $(0, p)$, then the directrix has the equation $y = -p$. If $P(x, y)$ is any point on the parabola, then the distance from P to the focus is

$$|PF| = \sqrt{x^2 + (y - p)^2}$$

and the distance from P to the directrix is $|y + p|$. (Figure 3 illustrates the case where $p > 0$.) The defining property of a parabola is that these distances are equal:

$$\sqrt{x^2 + (y - p)^2} = |y + p|$$

We get an equivalent equation by squaring and simplifying:

$$x^2 + (y - p)^2 = |y + p|^2 = (y + p)^2$$

$$x^2 + y^2 - 2py + p^2 = y^2 + 2py + p^2$$

$$x^2 = 4py$$

1 An equation of the parabola with focus $(0, p)$ and directrix $y = -p$ is

$$x^2 = 4py$$

If we write $a = 1/(4p)$, then the standard equation of a parabola (1) becomes $y = ax^2$. It opens upward if $p > 0$ and downward if $p < 0$ [see Figure 4, parts (a) and (b)]. The graph is symmetric with respect to the y-axis because (1) is unchanged when x is replaced by $-x$.

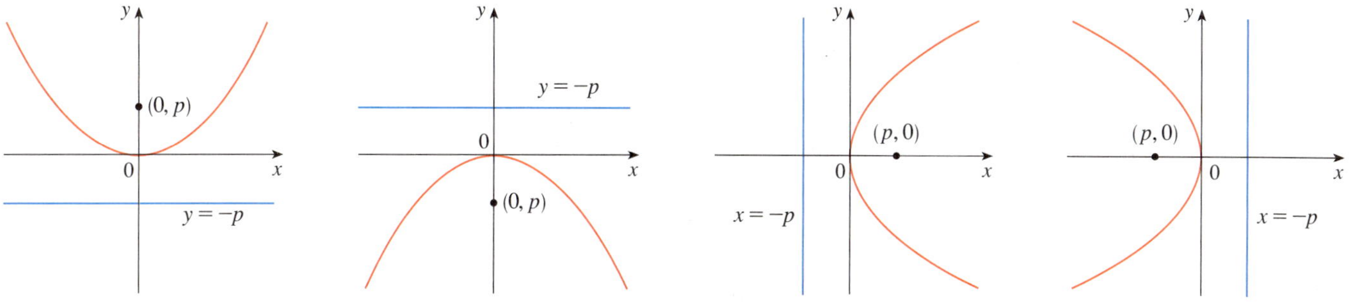

(a) $x^2 = 4py$, $p > 0$ (b) $x^2 = 4py$, $p < 0$ (c) $y^2 = 4px$, $p > 0$ (d) $y^2 = 4px$, $p < 0$

FIGURE 4

If we interchange x and y in (1), we obtain

2
$$y^2 = 4px$$

which is an equation of the parabola with focus $(p, 0)$ and directrix $x = -p$. (Interchanging x and y amounts to reflecting about the diagonal line $y = x$.) The parabola opens to the right if $p > 0$ and to the left if $p < 0$ [see Figure 4, parts (c) and (d)]. In both cases the graph is symmetric with respect to the x-axis, which is the axis of the parabola.

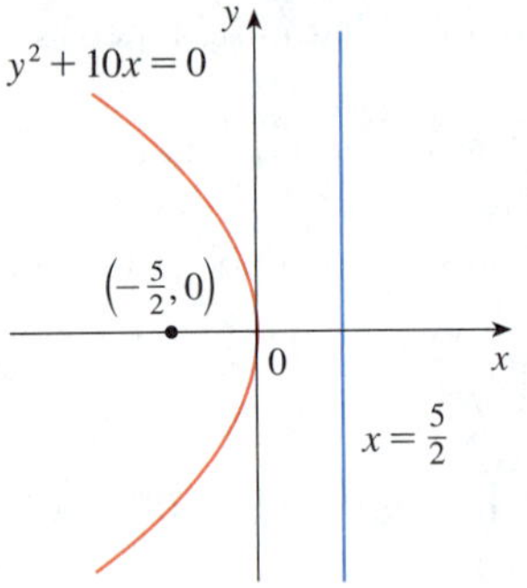

FIGURE 5

EXAMPLE 1 Find the focus and directrix of the parabola $y^2 + 10x = 0$ and sketch the graph.

SOLUTION If we write the equation as $y^2 = -10x$ and compare it with Equation 2, we see that $4p = -10$, so $p = -\frac{5}{2}$. Thus, the focus is $(p, 0) = \left(-\frac{5}{2}, 0\right)$ and the directrix is $x = \frac{5}{2}$. The sketch is shown in Figure 5.

Ellipses

An **ellipse** is the set of points in a plane the sum of whose distances from two fixed points F_1 and F_2 is a constant (see Figure 6). These two fixed points are called the **foci** (plural of **focus**). One of Kepler's laws is that the orbits of the planets in the solar system are ellipses with the Sun at one focus.

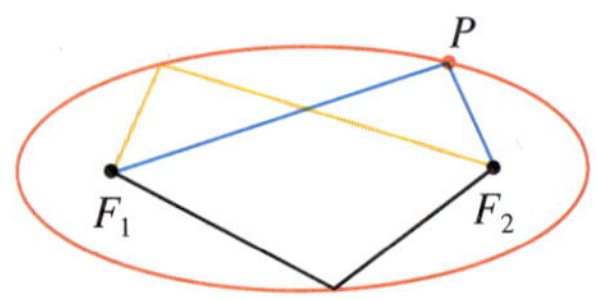

FIGURE 6

In order to obtain the simplest equation for an ellipse, we place the foci on the x-axis at the points $(-c, 0)$ and $(c, 0)$ as in Figure 7 so that the origin is halfway between the foci. Let the sum of the distances from a point on the ellipse to the foci be $2a > 0$. Then $P(x, y)$ is a point on the ellipse when

$$|PF_1| + |PF_2| = 2a$$

that is,

$$\sqrt{(x + c)^2 + y^2} + \sqrt{(x - c)^2 + y^2} = 2a$$

or

$$\sqrt{(x - c)^2 + y^2} = 2a - \sqrt{(x + c)^2 + y^2}$$

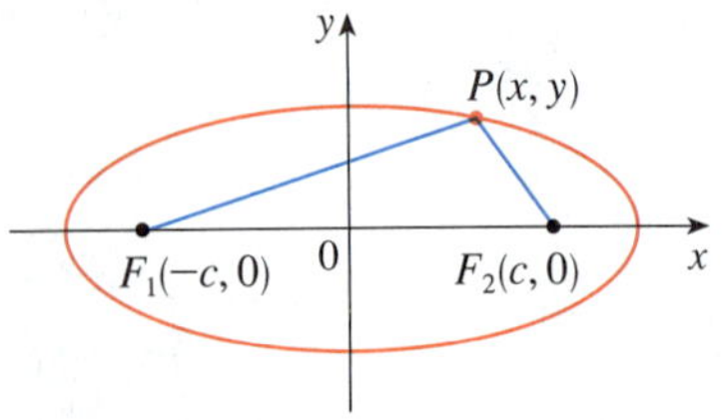

FIGURE 7

Squaring both sides, we have

$$x^2 - 2cx + c^2 + y^2 = 4a^2 - 4a\sqrt{(x + c)^2 + y^2} + x^2 + 2cx + c^2 + y^2$$

which simplifies to

$$a\sqrt{(x + c)^2 + y^2} = a^2 + cx$$

We square again:

$$a^2(x^2 + 2cx + c^2 + y^2) = a^4 + 2a^2cx + c^2x^2$$

which becomes

$$(a^2 - c^2)x^2 + a^2y^2 = a^2(a^2 - c^2)$$

From triangle F_1F_2P in Figure 7 we see that $2c < 2a$, so $c < a$ and, therefore, $a^2 - c^2 > 0$. For convenience, let $b^2 = a^2 - c^2$. Then the equation of the ellipse becomes $b^2x^2 + a^2y^2 = a^2b^2$ or, if both sides are divided by a^2b^2,

3
$$\frac{x^2}{a^2} + \frac{y^2}{b^2} = 1$$

Since $b^2 = a^2 - c^2 < a^2$, it follows that $b < a$. The x-intercepts are found by setting $y = 0$. Then $x^2/a^2 = 1$, or $x^2 = a^2$, so $x = \pm a$. The corresponding points $(a, 0)$ and $(-a, 0)$ are called the **vertices** of the ellipse and the line segment joining the vertices is called the **major axis**. To find the y-intercepts we set $x = 0$ and obtain $y^2 = b^2$, so $y = \pm b$. Equation 3 is unchanged if x is replaced by $-x$ or y is replaced by $-y$, so the ellipse is symmetric about both axes. Notice that if the foci coincide, then $c = 0$, so $a = b$ and the ellipse becomes a circle with radius $r = a = b$.

We summarize this discussion as follows (see also Figure 8).

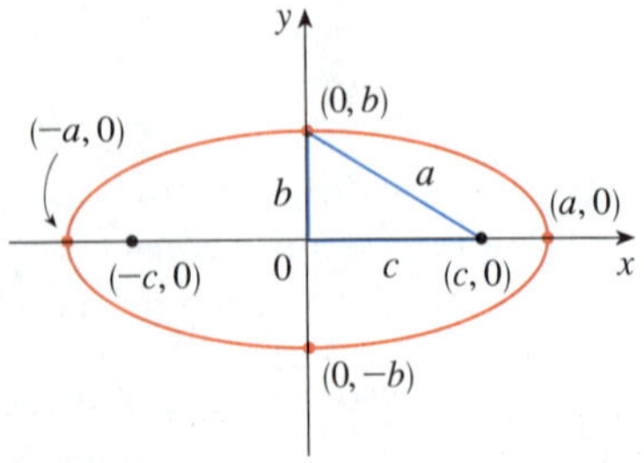

FIGURE 8
$\frac{x^2}{a^2} + \frac{y^2}{b^2} = 1$

4 The ellipse

$$\frac{x^2}{a^2} + \frac{y^2}{b^2} = 1 \qquad a \geqslant b > 0$$

has foci $(\pm c, 0)$, where $c^2 = a^2 - b^2$, and vertices $(\pm a, 0)$.

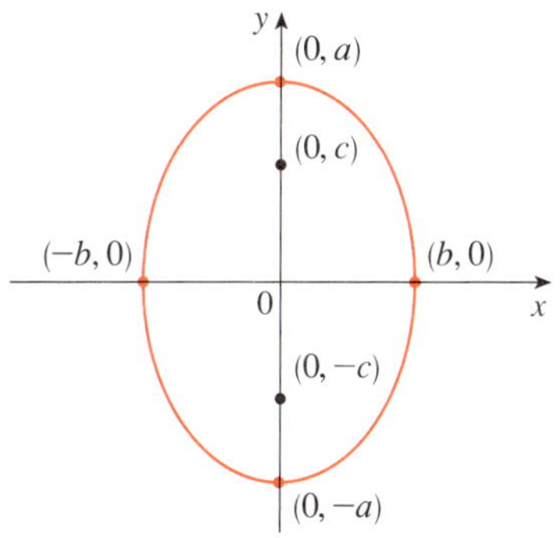

FIGURE 9
$\frac{x^2}{b^2} + \frac{y^2}{a^2} = 1, \ a \geqslant b$

If the foci of an ellipse are located on the y-axis at $(0, \pm c)$, then we can find its equation by interchanging x and y in (4). (See Figure 9.)

5 The ellipse

$$\frac{x^2}{b^2} + \frac{y^2}{a^2} = 1 \qquad a \geqslant b > 0$$

has foci $(0, \pm c)$, where $c^2 = a^2 - b^2$, and vertices $(0, \pm a)$.

EXAMPLE 2 Sketch the graph of $9x^2 + 16y^2 = 144$ and locate the foci.

SOLUTION Divide both sides of the equation by 144:

$$\frac{x^2}{16} + \frac{y^2}{9} = 1$$

The equation is now in the standard form for an ellipse, so we have $a^2 = 16$, $b^2 = 9$, $a = 4$, and $b = 3$. The x-intercepts are ± 4 and the y-intercepts are ± 3. Also, $c^2 = a^2 - b^2 = 7$, so $c = \sqrt{7}$ and the foci are $(\pm\sqrt{7}, 0)$. The graph is sketched in Figure 10.

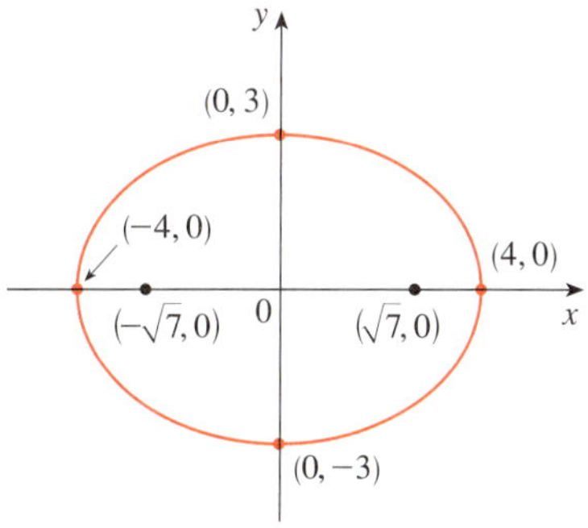

FIGURE 10
$9x^2 + 16y^2 = 144$

EXAMPLE 3 Find an equation of the ellipse with foci $(0, \pm 2)$ and vertices $(0, \pm 3)$.

SOLUTION Using the notation of (5), we have $c = 2$ and $a = 3$. Then we obtain $b^2 = a^2 - c^2 = 9 - 4 = 5$, so an equation of the ellipse is

$$\frac{x^2}{5} + \frac{y^2}{9} = 1$$

Another way of writing the equation is $9x^2 + 5y^2 = 45$.

Like parabolas, ellipses have an interesting reflection property that has practical consequences. If a source of light or sound is placed at one focus of a surface with elliptical cross-sections, then all the light or sound is reflected off the surface to the other focus (see Exercise 59). This principle is used in *lithotripsy*, a treatment for kidney stones. A reflector with elliptical cross-section is placed in such a way that the kidney stone is at one focus. High-intensity sound waves generated at the other focus are reflected to the stone and destroy it without damaging surrounding tissue. The patient is spared the trauma of surgery and recovers within a few days.

Hyperbolas

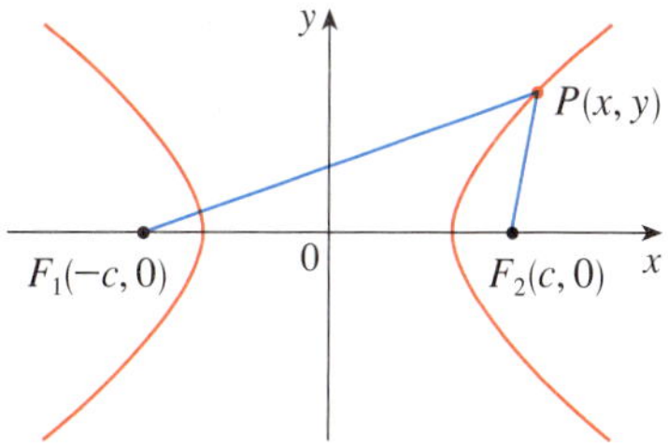

FIGURE 11
P is on the hyperbola when $|PF_1| - |PF_2| = \pm 2a$

A **hyperbola** is the set of all points in a plane the difference of whose distances from two fixed points F_1 and F_2 (the foci) is a constant. This definition is illustrated in Figure 11.

Hyperbolas occur frequently as graphs of equations in chemistry, physics, biology, and economics (Boyle's Law, Ohm's Law, supply and demand curves). A particularly significant application of hyperbolas is found in the navigation systems developed in World Wars I and II (see Exercise 51).

Notice that the definition of a hyperbola is similar to that of an ellipse; the only change is that the sum of distances has become a difference of distances. In fact, the derivation of the equation of a hyperbola is also similar to the one given earlier for an ellipse. It is left as Exercise 52 to show that when the foci are on the x-axis at $(\pm c, 0)$ and the difference of distances is $|PF_1| - |PF_2| = \pm 2a$, then the equation of the hyperbola is

$$\boxed{6} \qquad \frac{x^2}{a^2} - \frac{y^2}{b^2} = 1$$

where $c^2 = a^2 + b^2$. Notice that the x-intercepts are again $\pm a$ and the points $(a, 0)$ and $(-a, 0)$ are the **vertices** of the hyperbola. But if we put $x = 0$ in Equation 6 we get $y^2 = -b^2$, which is impossible, so there is no y-intercept. The hyperbola is symmetric with respect to both axes.

To analyze the hyperbola further, we look at Equation 6 and obtain

$$\frac{x^2}{a^2} = 1 + \frac{y^2}{b^2} \geqslant 1$$

This shows that $x^2 \geqslant a^2$, so $|x| = \sqrt{x^2} \geqslant a$. Therefore, we have $x \geqslant a$ or $x \leqslant -a$. This means that the hyperbola consists of two parts, called its *branches.*

When we draw a hyperbola it is useful to first draw its **asymptotes**, which are the dashed lines $y = (b/a)x$ and $y = -(b/a)x$ shown in Figure 12. Both branches of the hyperbola approach the asymptotes; that is, they come arbitrarily close to the asymptotes. [See Exercise 67 in Section 4.5, where $y = (b/a)x$ is shown to be a slant asymptote.]

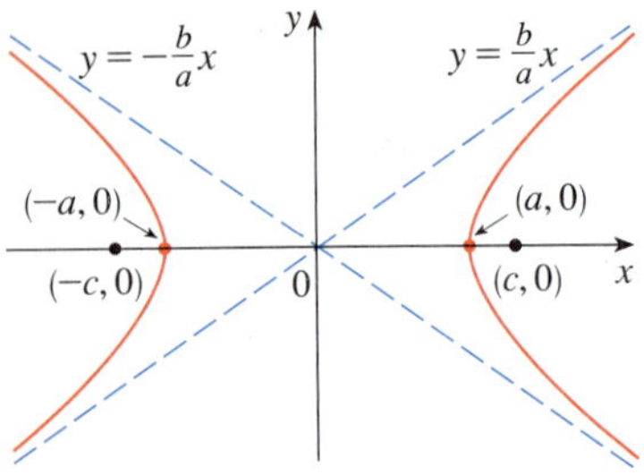

FIGURE 12
$\frac{x^2}{a^2} - \frac{y^2}{b^2} = 1$

> $\boxed{7}$ The hyperbola
> $$\frac{x^2}{a^2} - \frac{y^2}{b^2} = 1$$
> has foci $(\pm c, 0)$, where $c^2 = a^2 + b^2$, vertices $(\pm a, 0)$, and asymptotes $y = \pm(b/a)x$.

If the foci of a hyperbola are on the y-axis, then by reversing the roles of x and y we obtain the following information, which is illustrated in Figure 13.

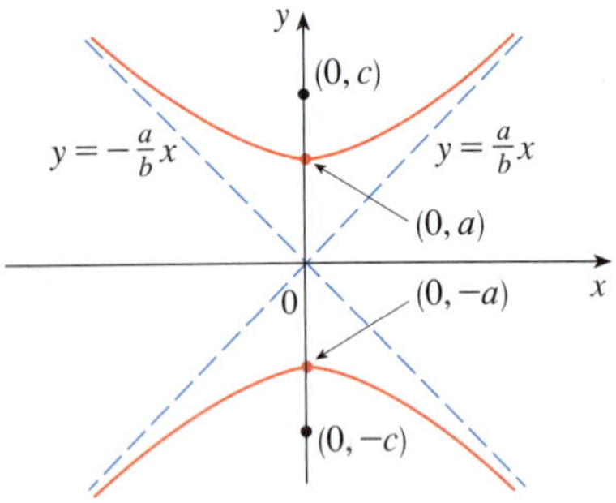

FIGURE 13
$\frac{y^2}{a^2} - \frac{x^2}{b^2} = 1$

> $\boxed{8}$ The hyperbola
> $$\frac{y^2}{a^2} - \frac{x^2}{b^2} = 1$$
> has foci $(0, \pm c)$, where $c^2 = a^2 + b^2$, vertices $(0, \pm a)$, and asymptotes $y = \pm(a/b)x$.

EXAMPLE 4 Find the foci and asymptotes of the hyperbola $9x^2 - 16y^2 = 144$ and sketch its graph.

SOLUTION If we divide both sides of the equation by 144, it becomes

$$\frac{x^2}{16} - \frac{y^2}{9} = 1$$

which is of the form given in (7) with $a = 4$ and $b = 3$. Since $c^2 = 16 + 9 = 25$, the foci are $(\pm 5, 0)$. The asymptotes are the lines $y = \frac{3}{4}x$ and $y = -\frac{3}{4}x$. The graph is shown in Figure 14.

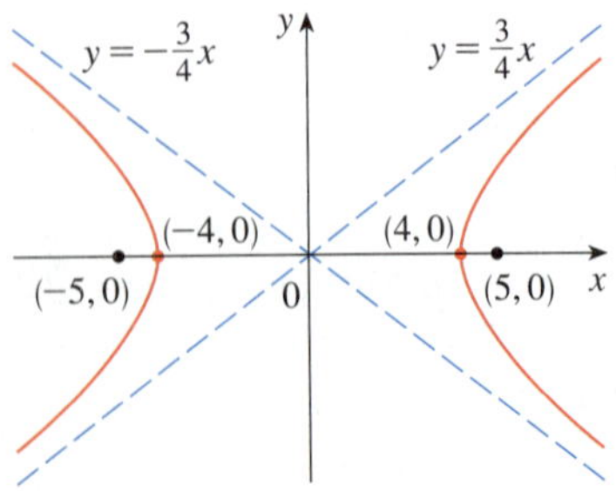

FIGURE 14
$9x^2 - 16y^2 = 144$

EXAMPLE 5 Find the foci and equation of the hyperbola with vertices $(0, \pm 1)$ and asymptote $y = 2x$.

SOLUTION From (8) and the given information, we see that $a = 1$ and $a/b = 2$. Thus, $b = a/2 = \frac{1}{2}$ and $c^2 = a^2 + b^2 = \frac{5}{4}$. The foci are $(0, \pm\sqrt{5}/2)$ and the equation of the hyperbola is

$$y^2 - 4x^2 = 1$$

Shifted Conics

As discussed in Appendix C, we shift conics by taking the standard equations (1), (2), (4), (5), (7), and (8) and replacing x and y by $x - h$ and $y - k$.

EXAMPLE 6 Find an equation of the ellipse with foci $(2, -2)$, $(4, -2)$ and vertices $(1, -2)$, $(5, -2)$.

SOLUTION The major axis is the line segment that joins the vertices $(1, -2)$, $(5, -2)$ and has length 4, so $a = 2$. The distance between the foci is 2, so $c = 1$. Thus, $b^2 = a^2 - c^2 = 3$. Since the center of the ellipse is $(3, -2)$, we replace x and y in (4) by $x - 3$ and $y + 2$ to obtain

$$\frac{(x-3)^2}{4} + \frac{(y+2)^2}{3} = 1$$

as the equation of the ellipse.

EXAMPLE 7 Sketch the conic

$$9x^2 - 4y^2 - 72x + 8y + 176 = 0$$

and find its foci.

SOLUTION We complete the squares as follows:

$$4(y^2 - 2y) - 9(x^2 - 8x) = 176$$

$$4(y^2 - 2y + 1) - 9(x^2 - 8x + 16) = 176 + 4 - 144$$

$$4(y-1)^2 - 9(x-4)^2 = 36$$

$$\frac{(y-1)^2}{9} - \frac{(x-4)^2}{4} = 1$$

This is in the form (8) except that x and y are replaced by $x - 4$ and $y - 1$. Thus, $a^2 = 9$, $b^2 = 4$, and $c^2 = 13$. The hyperbola is shifted four units to the right and one unit upward. The foci are $(4, 1 + \sqrt{13})$ and $(4, 1 - \sqrt{13})$ and the vertices are $(4, 4)$ and $(4, -2)$. The asymptotes are $y - 1 = \pm\frac{3}{2}(x - 4)$. The hyperbola is sketched in Figure 15.

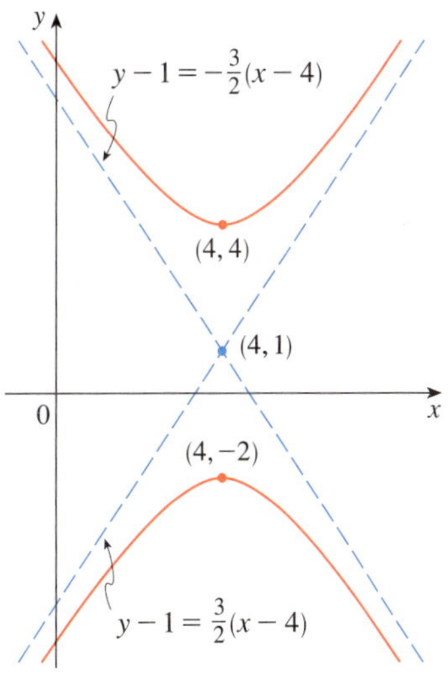

FIGURE 15
$9x^2 - 4y^2 - 72x + 8y + 176 = 0$

10.5 Exercises

1–8 Find the vertex, focus, and directrix of the parabola and sketch its graph.

1. $x = 2y^2$

2. $4y + x^2 = 0$

3. $4x^2 = -y$

4. $y^2 = 12x$

5. $(x + 2)^2 = 8(y - 3)$

6. $x - 1 = (y + 5)^2$

7. $y^2 + 2y + 12x + 25 = 0$

8. $y + 12x - 2x^2 = 16$

9–10 Find an equation of the parabola. Then find the focus and directrix.

9.

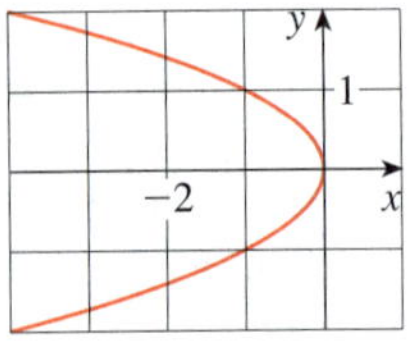

10.

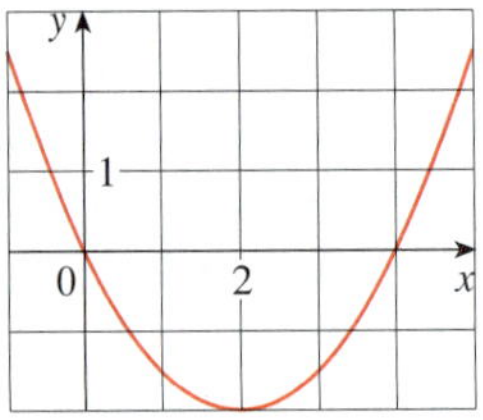

11–16 Find the vertices and foci of the ellipse and sketch its graph.

11. $\dfrac{x^2}{9} + \dfrac{y^2}{5} = 1$

12. $\dfrac{x^2}{64} + \dfrac{y^2}{100} = 1$

13. $4x^2 + y^2 = 16$

14. $4x^2 + 25y^2 = 25$

15. $9x^2 - 18x + 4y^2 = 27$

16. $x^2 + 2y^2 - 6x + 4y + 7 = 0$

17–18 Find an equation of the ellipse. Then find its foci.

17.

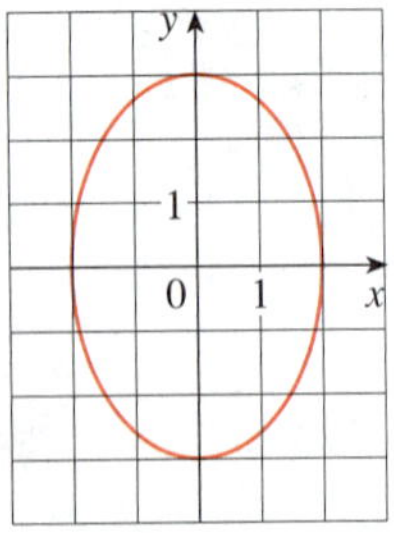

18.

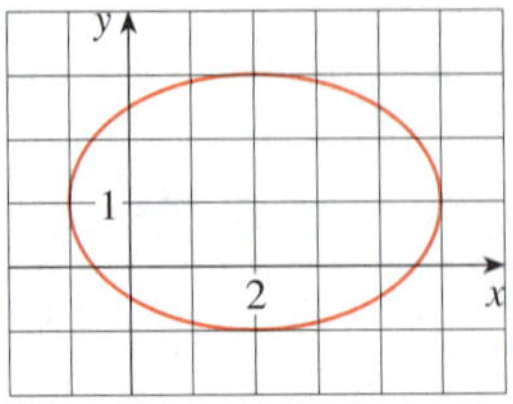

19–24 Find the vertices, foci, and asymptotes of the hyperbola and sketch its graph.

19. $\dfrac{x^2}{144} - \dfrac{y^2}{25} = 1$

20. $\dfrac{y^2}{16} - \dfrac{x^2}{36} = 1$

21. $y^2 - x^2 = 4$

22. $9x^2 - 4y^2 = 36$

23. $2y^2 - 3x^2 - 4y + 12x + 8 = 0$

24. $16x^2 - 9y^2 + 64x - 90y = 305$

25–30 Identify the type of conic section whose equation is given and find the vertices and foci.

25. $x^2 = y + 1$

26. $x^2 = y^2 + 1$

27. $x^2 = 4y - 2y^2$

28. $y^2 - 8y = 6x - 16$

29. $y^2 + 2y = 4x^2 + 3$

30. $4x^2 + 4x + y^2 = 0$

31–48 Find an equation for the conic that satisfies the given conditions.

31. Parabola, vertex $(0, 0)$, focus $(0, -2)$

32. Parabola, vertex $(1, 0)$, directrix $x = -5$

33. Parabola, focus $(-4, 0)$, directrix $x = 2$

34. Parabola, focus $(3, 6)$, vertex $(3, 2)$

35. Parabola, vertex $(0, 0)$, axis the x-axis, passing through $(1, -4)$

36. Parabola, vertical axis, passing through $(-2, 3)$, $(0, 3)$, and $(1, 9)$

37. Ellipse, foci $(\pm 2, 0)$, vertices $(\pm 5, 0)$

38. Ellipse, foci $(0, \pm 5)$, vertices $(0, \pm 13)$

39. Ellipse, foci $(0, 2)$, $(0, 6)$ vertices $(0, 0)$, $(0, 8)$

40. Ellipse, foci $(0, -1)$, $(8, -1)$, vertex $(9, -1)$

41. Ellipse, center $(2, 2)$, focus $(0, 2)$, vertex $(5, 2)$

42. Ellipse, foci $(\pm 2, 0)$, passing through $(2, 1)$

43. Hyperbola, foci $(0, \pm 3)$, vertices $(0, \pm 1)$

44. Hyperbola, foci $(\pm 6, 0)$, vertices $(\pm 4, 0)$

45. Hyperbola, foci $(1, 3)$ and $(7, 3)$, vertices $(2, 3)$ and $(6, 3)$

46. Hyperbola, foci $(2, -2)$ and $(2, 8)$, vertices $(2, 0)$ and $(2, 6)$

47. Hyperbola, vertices $(\pm 3, 0)$, asymptotes $y = \pm 2x$

48. Hyperbola, foci $(2, 2)$ and $(6, 2)$, asymptotes $y = x - 2$ and $y = 6 - x$

49. The point in a lunar orbit nearest the surface of the moon is called *perilune* and the point farthest from the surface is called *apolune*. The *Apollo 11* spacecraft was placed in an elliptical lunar orbit with perilune altitude 110 km and apolune altitude 314 km (above the moon). Find an equation of this ellipse if the radius of the moon is 1728 km and the center of the moon is at one focus.

50. A cross-section of a parabolic reflector is shown in the figure. The bulb is located at the focus and the opening at the focus is 10 cm.
(a) Find an equation of the parabola.
(b) Find the diameter of the opening $|CD|$, 11 cm from the vertex.

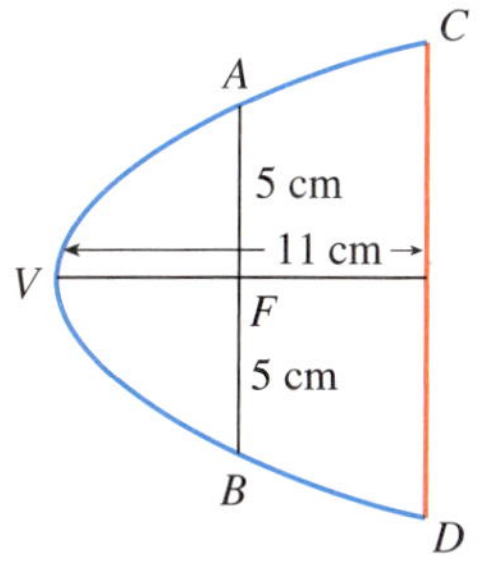

51. In the LORAN (LOng RAnge Navigation) radio navigation system, two radio stations located at A and B transmit simultaneous signals to a ship or an aircraft located at P. The onboard computer converts the time difference in receiving these signals into a distance difference $|PA| - |PB|$, and this, according to the definition of a hyperbola, locates the ship or aircraft on one branch of a hyperbola (see the figure). Suppose that station B is located 400 mi due east of station A on a coastline. A ship received the signal from B 1200 microseconds (μs) before it received the signal from A.
(a) Assuming that radio signals travel at a speed of 980 ft/μs, find an equation of the hyperbola on which the ship lies.
(b) If the ship is due north of B, how far off the coastline is the ship?

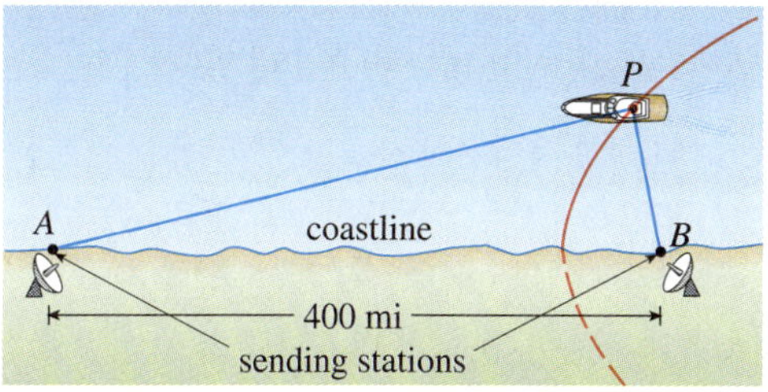

52. Use the definition of a hyperbola to derive Equation 6 for a hyperbola with foci $(\pm c, 0)$ and vertices $(\pm a, 0)$.

53. Show that the function defined by the upper branch of the hyperbola $y^2/a^2 - x^2/b^2 = 1$ is concave upward.

54. Find an equation for the ellipse with foci $(1, 1)$ and $(-1, -1)$ and major axis of length 4.

55. Determine the type of curve represented by the equation

$$\frac{x^2}{k} + \frac{y^2}{k - 16} = 1$$

in each of the following cases: (a) $k > 16$, (b) $0 < k < 16$, and (c) $k < 0$.
(d) Show that all the curves in parts (a) and (b) have the same foci, no matter what the value of k is.

56. (a) Show that the equation of the tangent line to the parabola $y^2 = 4px$ at the point (x_0, y_0) can be written as

$$y_0 y = 2p(x + x_0)$$

(b) What is the x-intercept of this tangent line? Use this fact to draw the tangent line.

57. Use Simpson's Rule with $n = 10$ to estimate the length of the ellipse $x^2 + 4y^2 = 4$.

58. The planet Pluto travels in an elliptical orbit around the Sun (at one focus). The length of the major axis is 1.18×10^{10} km and the length of the minor axis is 1.14×10^{10} km. Use Simpson's Rule with $n = 10$ to estimate the distance traveled by the planet during one complete orbit around the Sun.

59. Let $P(x_1, y_1)$ be a point on the ellipse $x^2/a^2 + y^2/b^2 = 1$ with foci F_1 and F_2 and let α and β be the angles between the lines PF_1, PF_2 and the ellipse as in the figure. Prove that $\alpha = \beta$. This explains how whispering galleries and lithotripsy work. Sound coming from one focus is reflected and passes through the other focus. [*Hint:* Use the formula in Problem 15 on page 276 to show that $\tan \alpha = \tan \beta$.]

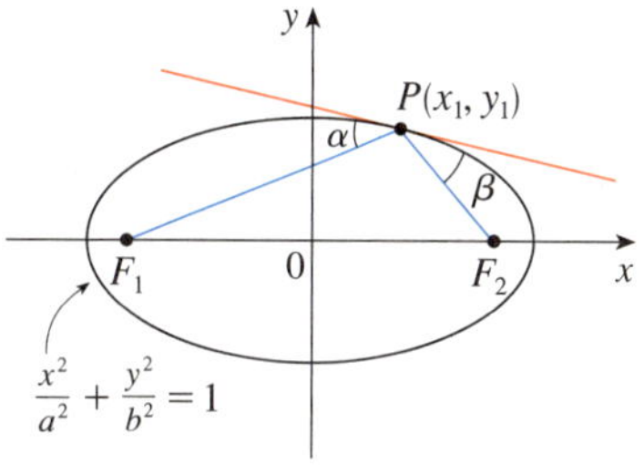

60. Let $P(x_1, y_1)$ be a point on the hyperbola $x^2/a^2 - y^2/b^2 = 1$ with foci F_1 and F_2 and let α and β be the angles between the lines PF_1, PF_2 and the hyperbola as shown in the figure. Prove that $\alpha = \beta$. (This is the reflection property of the hyperbola. It shows that light aimed at a focus F_2 of a hyperbolic mirror is reflected toward the other focus F_1.)

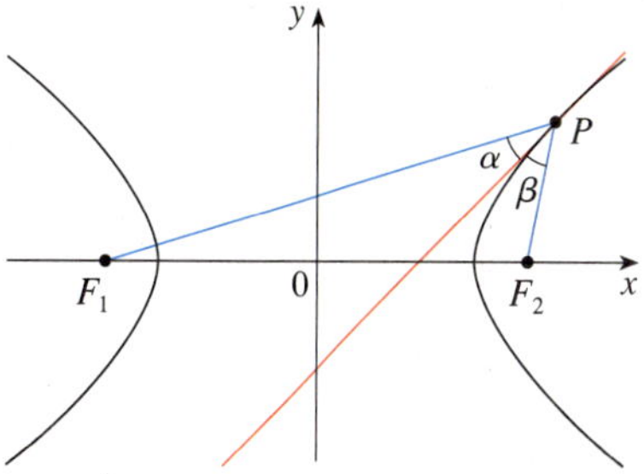

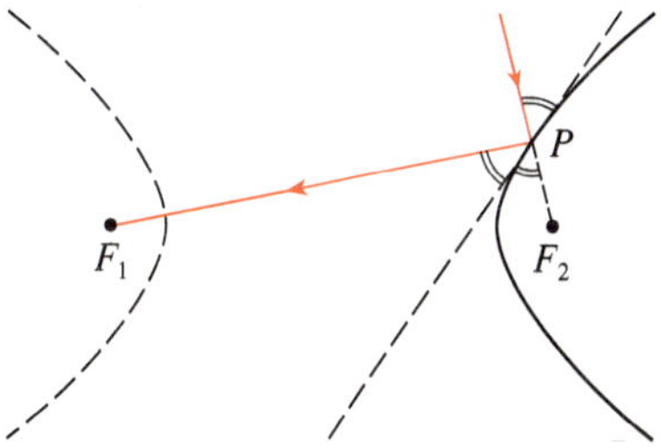

10.6 Conic Sections in Polar Coordinates

In the preceding section we defined the parabola in terms of a focus and directrix, but we defined the ellipse and hyperbola in terms of two foci. In this section we give a more unified treatment of all three types of conic sections in terms of a focus and directrix. Furthermore, if we place the focus at the origin, then a conic section has a simple polar equation. In Chapter 13 we will use the polar equation of an ellipse to derive Kepler's laws of planetary motion.

1 Theorem Let F be a fixed point (called the **focus**) and l be a fixed line (called the **directrix**) in a plane. Let e be a fixed positive number (called the **eccentricity**). The set of all points P in the plane such that

$$\frac{|PF|}{|Pl|} = e$$

(The ratio of the distance from F to the distance from l is the constant e.)

is a conic section. The conic is

(a) an ellipse if $e < 1$

(b) a parabola if $e = 1$

(c) a hyperbola if $e > 1$

Proof Notice that if the eccentricity is $e = 1$, then $|PF| = |Pl|$ and so the given condition simply becomes the definition of a parabola as given in Section 10.5.

Let us place the focus F at the origin and the directrix parallel to the y-axis and d units to the right. Thus, the directrix has equation $x = d$ and is perpendicular to the polar axis. If the point P has polar coordinates (r, θ), we see from Figure 1 that

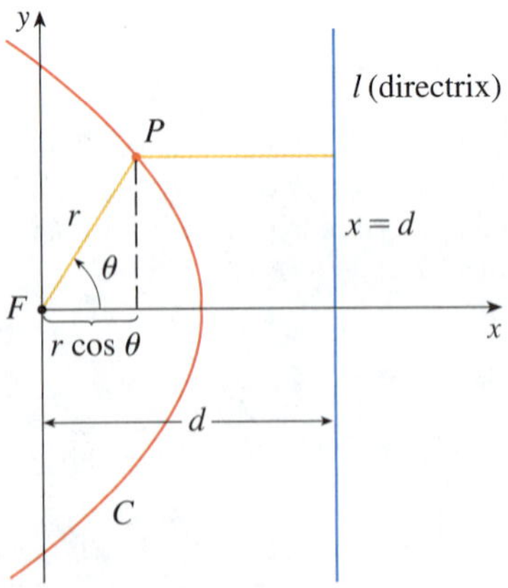

FIGURE 1

$$|PF| = r \qquad |Pl| = d - r\cos\theta$$

Thus, the condition $|PF|/|Pl| = e$, or $|PF| = e|Pl|$, becomes

$$\boxed{2} \qquad r = e(d - r\cos\theta)$$

If we square both sides of this polar equation and convert to rectangular coordinates, we get

$$x^2 + y^2 = e^2(d - x)^2 = e^2(d^2 - 2dx + x^2)$$

or

$$(1 - e^2)x^2 + 2de^2x + y^2 = e^2d^2$$

After completing the square, we have

$$\boxed{3} \qquad \left(x + \frac{e^2d}{1 - e^2}\right)^2 + \frac{y^2}{1 - e^2} = \frac{e^2d^2}{(1 - e^2)^2}$$

If $e < 1$, we recognize Equation 3 as the equation of an ellipse. In fact, it is of the form

$$\frac{(x - h)^2}{a^2} + \frac{y^2}{b^2} = 1$$

where

4 $$h = -\frac{e^2 d}{1 - e^2} \qquad a^2 = \frac{e^2 d^2}{(1 - e^2)^2} \qquad b^2 = \frac{e^2 d^2}{1 - e^2}$$

In Section 10.5 we found that the foci of an ellipse are at a distance c from the center, where

5 $$c^2 = a^2 - b^2 = \frac{e^4 d^2}{(1 - e^2)^2}$$

This shows that $$c = \frac{e^2 d}{1 - e^2} = -h$$

and confirms that the focus as defined in Theorem 1 means the same as the focus defined in Section 10.5. It also follows from Equations 4 and 5 that the eccentricity is given by

$$e = \frac{c}{a}$$

If $e > 1$, then $1 - e^2 < 0$ and we see that Equation 3 represents a hyperbola. Just as we did before, we could rewrite Equation 3 in the form

$$\frac{(x - h)^2}{a^2} - \frac{y^2}{b^2} = 1$$

and see that

$$e = \frac{c}{a} \qquad \text{where} \quad c^2 = a^2 + b^2$$

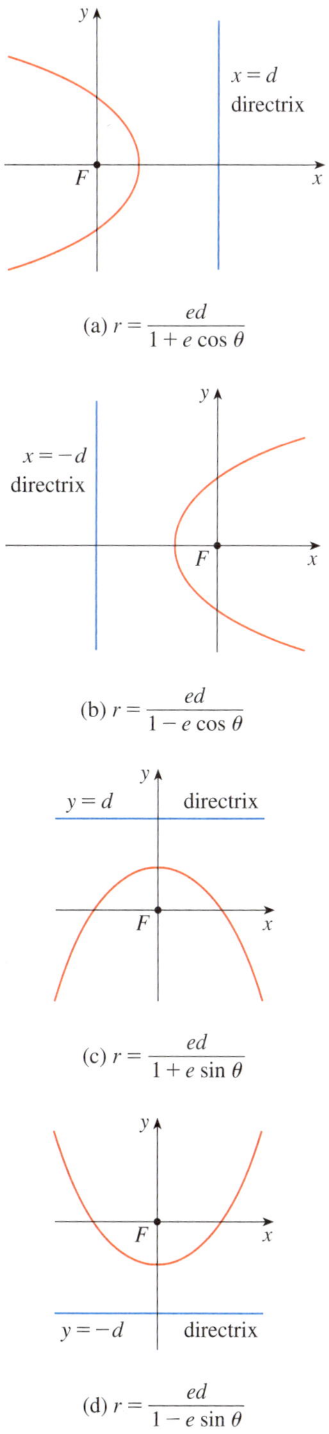

FIGURE 2
Polar equations of conics

By solving Equation 2 for r, we see that the polar equation of the conic shown in Figure 1 can be written as

$$r = \frac{ed}{1 + e\cos\theta}$$

If the directrix is chosen to be to the left of the focus as $x = -d$, or if the directrix is chosen to be parallel to the polar axis as $y = \pm d$, then the polar equation of the conic is given by the following theorem, which is illustrated by Figure 2. (See Exercises 21–23.)

6 Theorem A polar equation of the form

$$r = \frac{ed}{1 \pm e\cos\theta} \qquad \text{or} \qquad r = \frac{ed}{1 \pm e\sin\theta}$$

represents a conic section with eccentricity e. The conic is an ellipse if $e < 1$, a parabola if $e = 1$, or a hyperbola if $e > 1$.

EXAMPLE 1 Find a polar equation for a parabola that has its focus at the origin and whose directrix is the line $y = -6$.

SOLUTION Using Theorem 6 with $e = 1$ and $d = 6$, and using part (d) of Figure 2, we see that the equation of the parabola is

$$r = \frac{6}{1 - \sin\theta}$$

EXAMPLE 2 A conic is given by the polar equation

$$r = \frac{10}{3 - 2\cos\theta}$$

Find the eccentricity, identify the conic, locate the directrix, and sketch the conic.

SOLUTION Dividing numerator and denominator by 3, we write the equation as

$$r = \frac{\frac{10}{3}}{1 - \frac{2}{3}\cos\theta}$$

From Theorem 6 we see that this represents an ellipse with $e = \frac{2}{3}$. Since $ed = \frac{10}{3}$, we have

$$d = \frac{\frac{10}{3}}{e} = \frac{\frac{10}{3}}{\frac{2}{3}} = 5$$

so the directrix has Cartesian equation $x = -5$. When $\theta = 0$, $r = 10$; when $\theta = \pi$, $r = 2$. So the vertices have polar coordinates $(10, 0)$ and $(2, \pi)$. The ellipse is sketched in Figure 3.

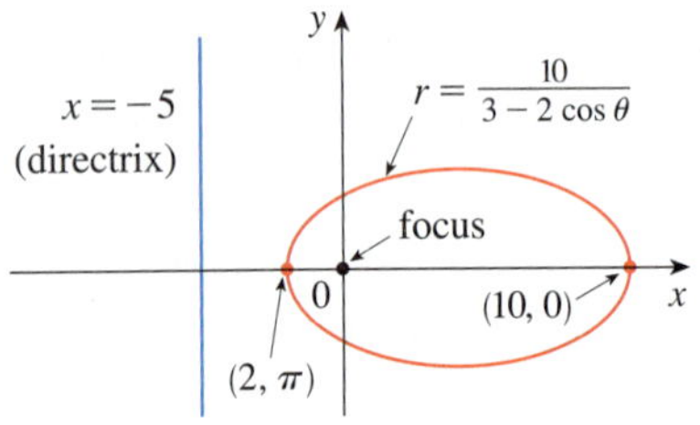

FIGURE 3

EXAMPLE 3 Sketch the conic $r = \dfrac{12}{2 + 4\sin\theta}$.

SOLUTION Writing the equation in the form

$$r = \frac{6}{1 + 2\sin\theta}$$

we see that the eccentricity is $e = 2$ and the equation therefore represents a hyperbola. Since $ed = 6$, $d = 3$ and the directrix has equation $y = 3$. The vertices occur when $\theta = \pi/2$ and $3\pi/2$, so they are $(2, \pi/2)$ and $(-6, 3\pi/2) = (6, \pi/2)$. It is also useful to plot the x-intercepts. These occur when $\theta = 0, \pi$; in both cases $r = 6$. For additional accuracy we could draw the asymptotes. Note that $r \to \pm\infty$ when $2 + 4\sin\theta \to 0^+$ or 0^- and $2 + 4\sin\theta = 0$ when $\sin\theta = -\frac{1}{2}$. Thus, the asymptotes are parallel to the rays $\theta = 7\pi/6$ and $\theta = 11\pi/6$. The hyperbola is sketched in Figure 4.

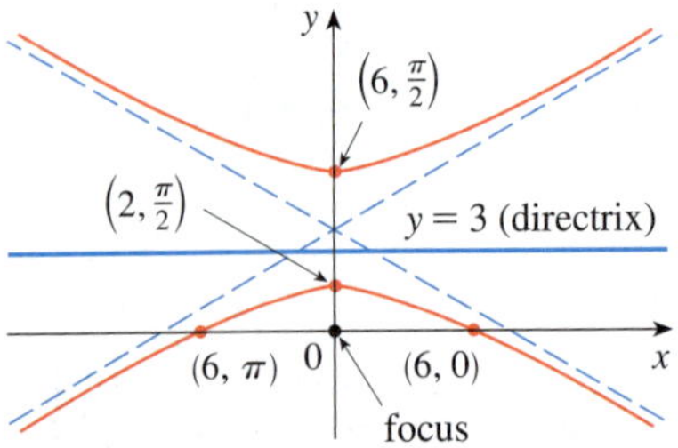

FIGURE 4
$r = \dfrac{12}{2 + 4\sin\theta}$

When rotating conic sections, we find it much more convenient to use polar equations than Cartesian equations. We just use the fact (see Exercise 75 in Section 10.3) that the graph of $r = f(\theta - \alpha)$ is the graph of $r = f(\theta)$ rotated counterclockwise about the origin through an angle α.

EXAMPLE 4 If the ellipse of Example 2 is rotated through an angle $\pi/4$ about the origin, find a polar equation and graph the resulting ellipse.

SOLUTION We get the equation of the rotated ellipse by replacing θ with $\theta - \pi/4$ in the equation given in Example 2. So the new equation is

$$r = \frac{10}{3 - 2\cos(\theta - \pi/4)}$$

We use this equation to graph the rotated ellipse in Figure 5. Notice that the ellipse has been rotated about its left focus.

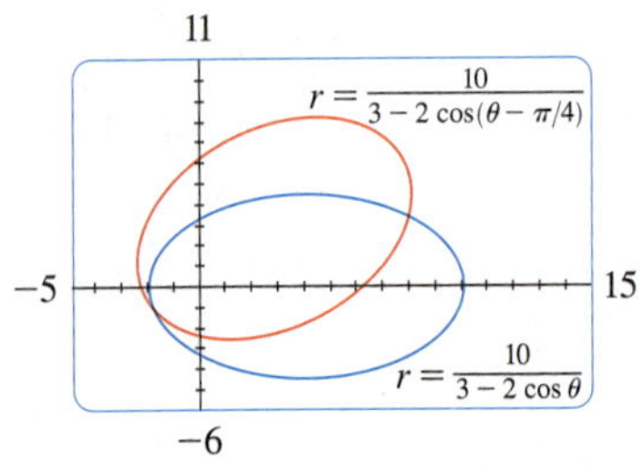

FIGURE 5

In Figure 6 we use a computer to sketch a number of conics to demonstrate the effect of varying the eccentricity e. Notice that when e is close to 0 the ellipse is nearly circular, whereas it becomes more elongated as $e \to 1^-$. When $e = 1$, of course, the conic is a parabola.

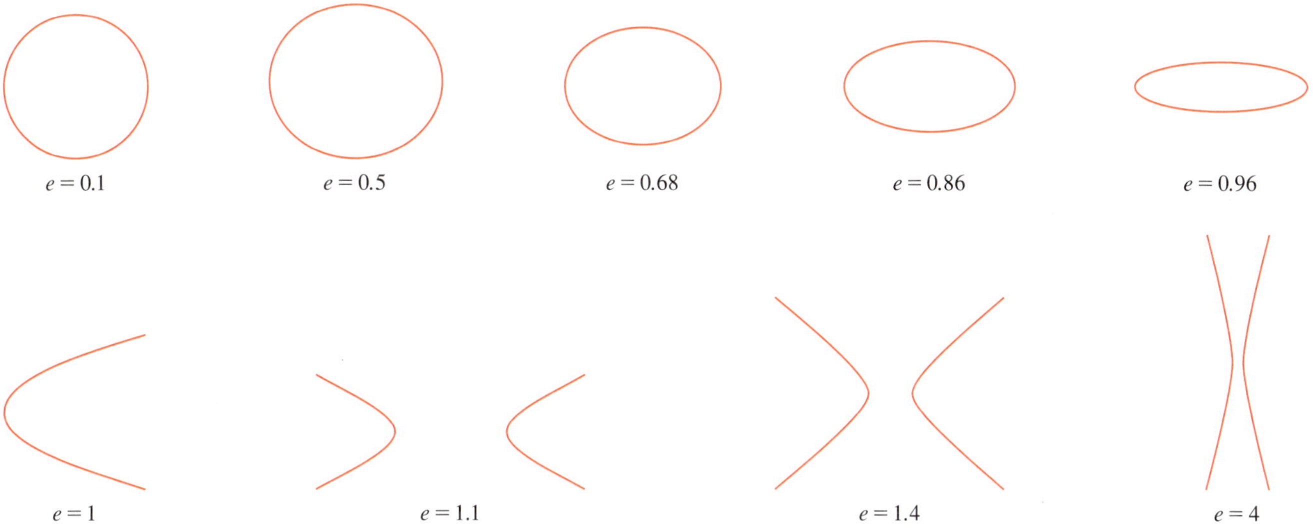

FIGURE 6

10.6 Exercises

1–8 Write a polar equation of a conic with the focus at the origin and the given data.

1. Hyperbola, eccentricity $\frac{7}{4}$, directrix $y = 6$

2. Parabola, directrix $x = 4$

3. Ellipse, eccentricity $\frac{3}{4}$, directrix $x = -5$

4. Hyperbola, eccentricity 2, directrix $y = -2$

5. Parabola, vertex $(4, 3\pi/2)$

6. Ellipse, eccentricity 0.8, vertex $(1, \pi/2)$

7. Ellipse, eccentricity $\frac{1}{2}$, directrix $r = 4 \sec \theta$

8. Hyperbola, eccentricity 3, directrix $r = -6 \csc \theta$

9–16 (a) Find the eccentricity, (b) identify the conic, (c) give an equation of the directrix, and (d) sketch the conic.

9. $r = \dfrac{1}{1 + \sin\theta}$

10. $r = \dfrac{6}{3 + 2\sin\theta}$

11. $r = \dfrac{12}{4 - \sin\theta}$

12. $r = \dfrac{4}{2 - 3\cos\theta}$

13. $r = \dfrac{9}{6 + 2\cos\theta}$

14. $r = \dfrac{5}{2 - 2\sin\theta}$

15. $r = \dfrac{3}{4 - 8\cos\theta}$

16. $r = \dfrac{4}{2 + \cos\theta}$

17. (a) Find the eccentricity and directrix of the conic $r = 1/(4 - 3\cos\theta)$ and graph the conic and its directrix.
(b) If this conic is rotated counterclockwise about the origin through an angle $\pi/3$, write the resulting equation and graph its curve.

18. Graph the parabola $r = 5/(2 + 2\sin\theta)$ and its directrix. Also graph the curve obtained by rotating this parabola about its focus through an angle $\pi/6$.

19. Graph the conics $r = e/(1 - e\cos\theta)$ with $e = 0.4, 0.6, 0.8$, and 1.0 on a common screen. How does the value of e affect the shape of the curve?

20. (a) Graph the conics $r = ed/(1 + e\sin\theta)$ for $e = 1$ and various values of d. How does the value of d affect the shape of the conic?
(b) Graph these conics for $d = 1$ and various values of e. How does the value of e affect the shape of the conic?

21. Show that a conic with focus at the origin, eccentricity e, and directrix $x = -d$ has polar equation

$$r = \frac{ed}{1 - e\cos\theta}$$

22. Show that a conic with focus at the origin, eccentricity e, and directrix $y = d$ has polar equation

$$r = \frac{ed}{1 + e\sin\theta}$$

23. Show that a conic with focus at the origin, eccentricity e, and directrix $y = -d$ has polar equation

$$r = \frac{ed}{1 - e \sin \theta}$$

24. Show that the parabolas $r = c/(1 + \cos \theta)$ and $r = d/(1 - \cos \theta)$ intersect at right angles.

25. (a) Show that the polar equation of an ellipse with directrix $x = -d$ can be written in the form

$$r = \frac{a(1 - e^2)}{1 - e \cos \theta}$$

(b) Find an approximate polar equation for the elliptical orbit of Earth around the Sun (at one focus) given that the eccentricity is about 0.017 and the length of the major axis is about 2.99×10^8 km.

26. (a) The planets move around the Sun in elliptical orbits with the Sun at one focus. The positions of a planet that are closest to and farthest from the Sun are called its *perihelion* and *aphelion*, respectively. Use Exercise 25(a) to show that the perihelion distance from a planet to the Sun is $a(1 - e)$ and the aphelion distance is $a(1 + e)$.

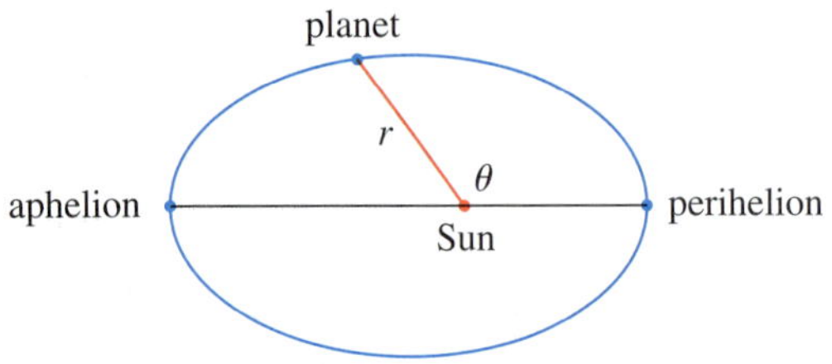

(b) Use the data of Exercise 25(b) to find the distances from Earth to the Sun at perihelion and at aphelion.

27. The orbit of Halley's comet, last seen in 1986 and due to return in 2062, is an ellipse with eccentricity 0.97 and one focus at the Sun. The length of its major axis is 36.18 AU. [An astronomical unit (AU) is the mean distance between Earth and the Sun, about 93 million miles.] Find a polar equation for the orbit of Halley's comet. What is the maximum distance from the comet to the Sun?

28. The Hale-Bopp comet, discovered in 1995, has an elliptical orbit with eccentricity 0.9951 and the length of the major axis is 356.5 AU. Find a polar equation for the orbit of this comet. How close to the Sun does it come?

29. The planet Mercury travels in an elliptical orbit with eccentricity 0.206. Its minimum distance from the Sun is 4.6×10^7 km. Use the results of Exercise 26(a) to find its maximum distance from the Sun.

30. The distance from the planet Pluto to the Sun is 4.43×10^9 km at perihelion and 7.37×10^9 km at aphelion. Use Exercise 26 to find the eccentricity of Pluto's orbit.

31. Using the data from Exercise 29, find the distance traveled by the planet Mercury during one complete orbit around the Sun. (If your calculator or computer algebra system evaluates definite integrals, use it. Otherwise, use Simpson's Rule.)

10 Review

CONCEPT CHECK

1. (a) What is a parametric curve?
(b) How do you sketch a parametric curve?

2. (a) How do you find the slope of a tangent to a parametric curve?
(b) How do you find the area under a parametric curve?

3. Write an expression for each of the following:
(a) The length of a parametric curve
(b) The area of the surface obtained by rotating a parametric curve about the x-axis

4. (a) Use a diagram to explain the meaning of the polar coordinates (r, θ) of a point.
(b) Write equations that express the Cartesian coordinates (x, y) of a point in terms of the polar coordinates.
(c) What equations would you use to find the polar coordinates of a point if you knew the Cartesian coordinates?

5. (a) How do you find the slope of a tangent line to a polar curve?
(b) How do you find the area of a region bounded by a polar curve?
(c) How do you find the length of a polar curve?

6. (a) Give a geometric definition of a parabola.
(b) Write an equation of a parabola with focus $(0, p)$ and directrix $y = -p$. What if the focus is $(p, 0)$ and the directrix is $x = -p$?

7. (a) Give a definition of an ellipse in terms of foci.
(b) Write an equation for the ellipse with foci $(\pm c, 0)$ and vertices $(\pm a, 0)$.

8. (a) Give a definition of a hyperbola in terms of foci.
(b) Write an equation for the hyperbola with foci $(\pm c, 0)$ and vertices $(\pm a, 0)$.
(c) Write equations for the asymptotes of the hyperbola in part (b).

9. (a) What is the eccentricity of a conic section?
(b) What can you say about the eccentricity if the conic section is an ellipse? A hyperbola? A parabola?
(c) Write a polar equation for a conic section with eccentricity e and directrix $x = d$. What if the directrix is $x = -d$? $y = d$? $y = -d$?

TRUE-FALSE QUIZ

Determine whether the statement is true or false. If it is true, explain why. If it is false, explain why or give an example that disproves the statement.

1. If the parametric curve $x = f(t)$, $y = g(t)$ satisfies $g'(1) = 0$, then it has a horizontal tangent when $t = 1$.

2. If $x = f(t)$ and $y = g(t)$ are twice differentiable, then $d^2y/dx^2 = (d^2y/dt^2)/(d^2x/dt^2)$.

3. The length of the curve $x = f(t)$, $y = g(t)$, $a \leqslant t \leqslant b$, is $\int_a^b \sqrt{[f'(t)]^2 + [g'(t)]^2}\, dt$.

4. If a point is represented by (x, y) in Cartesian coordinates (where $x \neq 0$) and (r, θ) in polar coordinates, then $\theta = \tan^{-1}(y/x)$.

5. The polar curves $r = 1 - \sin 2\theta$ and $r = \sin 2\theta - 1$ have the same graph.

6. The equations $r = 2$, $x^2 + y^2 = 4$, and $x = 2 \sin 3t$, $y = 2 \cos 3t$ $(0 \leqslant t \leqslant 2\pi)$ all have the same graph.

7. The parametric equations $x = t^2$, $y = t^4$ have the same graph as $x = t^3$, $y = t^6$.

8. The graph of $y^2 = 2y + 3x$ is a parabola.

9. A tangent line to a parabola intersects the parabola only once.

10. A hyperbola never intersects its directrix.

EXERCISES

1–4 Sketch the parametric curve and eliminate the parameter to find the Cartesian equation of the curve.

1. $x = t^2 + 4t, \quad y = 2 - t, \quad -4 \leqslant t \leqslant 1$

2. $x = 1 + e^{2t}, \quad y = e^t$

3. $x = \tan \theta, \quad y = \cot \theta$

4. $x = 2 \cos \theta, \quad y = 1 + \sin \theta$

5. Write three different sets of parametric equations for the curve $y = \sqrt{x}$.

6. Use the graphs of $x = f(t)$ and $y = g(t)$ to sketch the parametric curve $x = f(t)$, $y = g(t)$. Indicate with arrows the direction in which the curve is traced as t increases.

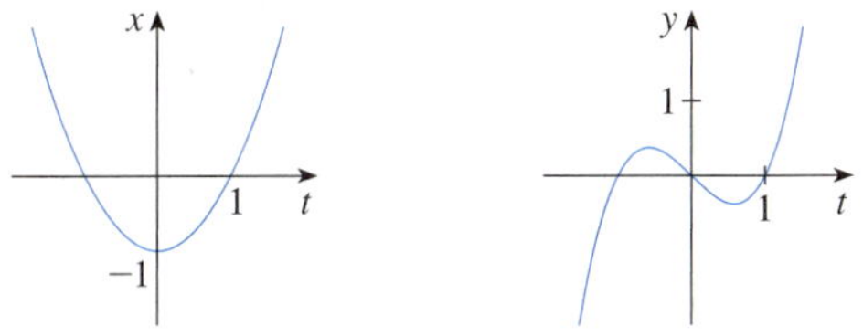

7–14 Sketch the polar curve.

7. $r = 1 - \cos \theta$

8. $r = \sin 4\theta$

9. $r = 1 + \cos 2\theta$

10. $r = 3 + \cos 3\theta$

11. $r^2 = \sec 2\theta$

12. $r = 2 \cos(\theta/2)$

13. $r = \dfrac{1}{1 + \cos \theta}$

14. $r = \dfrac{8}{4 + 3 \sin \theta}$

15–16 Find a polar equation for the curve represented by the given Cartesian equation.

15. $x + y = 2$

16. $x^2 + y^2 = 2$

17. The curve with polar equation $r = (\sin \theta)/\theta$ is called a **cochleoid**. Use a graph of r as a function of θ in Cartesian coordinates to sketch the cochleoid by hand. Then graph it with a machine to check your sketch.

18. Graph the ellipse $r = 2/(4 - 3 \cos \theta)$ and its directrix. Also graph the ellipse obtained by rotation about the origin through an angle $2\pi/3$.

19–22 Find the slope of the tangent line to the given curve at the point corresponding to the specified value of the parameter.

19. $x = \ln t, \; y = 1 + t^2; \quad t = 1$

20. $x = t^3 + 6t + 1, \quad y = 2t - t^2; \quad t = -1$

21. $r = e^{-\theta}; \quad \theta = \pi$

22. $r = 3 + \cos 3\theta; \quad \theta = \pi/2$

23–24 Find dy/dx and d^2y/dx^2.

23. $x = t \cos t, \quad y = t \sin t$

24. $x = 1 + t^2, \quad y = t - t^3$

25. Use a graph to estimate the coordinates of the lowest point on the curve $x = t^3 - 3t$, $y = t^2 + t + 1$. Then use calculus to find the exact coordinates.

26. Find the area enclosed by the loop of the curve in Exercise 25.

27. At what points does the curve

$$x = 2a\cos t - a\cos 2t \qquad y = 2a\sin t - a\sin 2t$$

have vertical or horizontal tangents? Use this information to help sketch the curve.

28. Find the area enclosed by the curve in Exercise 27.

29. Find the area enclosed by the curve $r^2 = 9\cos 5\theta$.

30. Find the area enclosed by the inner loop of the curve $r = 1 - 3\sin\theta$.

31. Find the points of intersection of the curves $r = 2$ and $r = 4\cos\theta$.

32. Find the points of intersection of the curves $r = \cot\theta$ and $r = 2\cos\theta$.

33. Find the area of the region that lies inside both of the circles $r = 2\sin\theta$ and $r = \sin\theta + \cos\theta$.

34. Find the area of the region that lies inside the curve $r = 2 + \cos 2\theta$ but outside the curve $r = 2 + \sin\theta$.

35–38 ▥ Find the length of the curve.

35. $x = 3t^2, \quad y = 2t^3, \quad 0 \leqslant t \leqslant 2$

36. $x = 2 + 3t, \quad y = \cosh 3t, \quad 0 \leqslant t \leqslant 1$

37. $r = 1/\theta, \quad \pi \leqslant \theta \leqslant 2\pi$

38. $r = \sin^3(\theta/3), \quad 0 \leqslant \theta \leqslant \pi$

39–40 ▥ Find the area of the surface obtained by rotating the given curve about the x-axis.

39. $x = 4\sqrt{t}, \quad y = \dfrac{t^3}{3} + \dfrac{1}{2t^2}, \quad 1 \leqslant t \leqslant 4$

40. $x = 2 + 3t, \quad y = \cosh 3t, \quad 0 \leqslant t \leqslant 1$

41. The curves defined by the parametric equations

$$x = \frac{t^2 - c}{t^2 + 1} \qquad y = \frac{t(t^2 - c)}{t^2 + 1}$$

are called **strophoids** (from a Greek word meaning "to turn or twist"). Investigate how these curves vary as c varies.

42. A family of curves has polar equations $r^a = |\sin 2\theta|$ where a is a positive number. Investigate how the curves change as a changes.

43–46 ▥ Find the foci and vertices and sketch the graph.

43. $\dfrac{x^2}{9} + \dfrac{y^2}{8} = 1$

44. $4x^2 - y^2 = 16$

45. $6y^2 + x - 36y + 55 = 0$

46. $25x^2 + 4y^2 + 50x - 16y = 59$

47. Find an equation of the parabola with focus $(0, 6)$ and directrix $y = 2$.

48. Find an equation of the hyperbola with foci $(0, \pm 5)$ and vertices $(0, \pm 2)$.

49. Find an equation of the hyperbola with foci $(\pm 3, 0)$ and asymptotes $2y = \pm x$.

50. Find an equation of the ellipse with foci $(3, \pm 2)$ and major axis with length 8.

51. Find an equation for the ellipse that shares a vertex and a focus with the parabola $x^2 + y = 100$ and that has its other focus at the origin.

52. Show that if m is any real number, then there are exactly two lines of slope m that are tangent to the ellipse $x^2/a^2 + y^2/b^2 = 1$ and their equations are $y = mx \pm \sqrt{a^2m^2 + b^2}$.

53. Find a polar equation for the ellipse with focus at the origin, eccentricity $\frac{1}{3}$, and directrix with equation $r = 4\sec\theta$.

54. Show that the angles between the polar axis and the asymptotes of the hyperbola $r = ed/(1 - e\cos\theta)$, $e > 1$, are given by $\cos^{-1}(\pm 1/e)$.

55. In the figure the circle of radius a is stationary, and for every θ, the point P is the midpoint of the segment QR. The curve traced out by P for $0 < \theta < \pi$ is called the **longbow curve**. Find parametric equations for this curve.

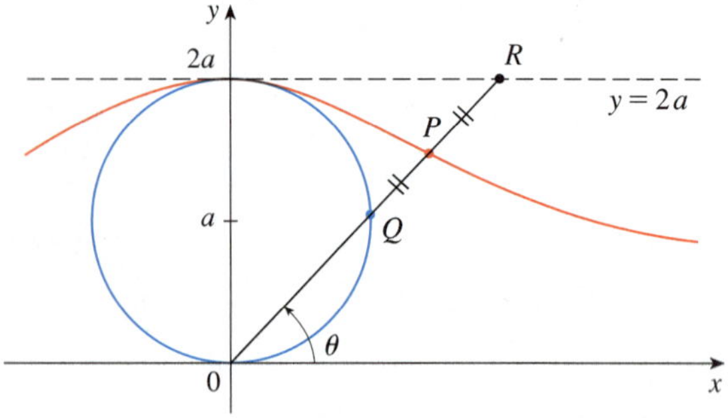

PROBLEMS PLUS

1. A curve is defined by the parametric equations

$$x = \int_1^t \frac{\cos u}{u}\,du \qquad y = \int_1^t \frac{\sin u}{u}\,du$$

Find the length of the arc of the curve from the origin to the nearest point where there is a vertical tangent line.

2. (a) Find the highest and lowest points on the curve $x^4 + y^4 = x^2 + y^2$.
   (b) Sketch the curve. (Notice that it is symmetric with respect to both axes and both of the lines $y = \pm x$, so it suffices to consider $y \geqslant x \geqslant 0$ initially.)
   CAS (c) Use polar coordinates and a computer algebra system to find the area enclosed by the curve.

3. What is the smallest viewing rectangle that contains every member of the family of polar curves $r = 1 + c \sin\theta$, where $0 \leqslant c \leqslant 1$? Illustrate your answer by graphing several members of the family in this viewing rectangle.

4. Four bugs are placed at the four corners of a square with side length a. The bugs crawl counterclockwise at the same speed and each bug crawls directly toward the next bug at all times. They approach the center of the square along spiral paths.
   (a) Find the polar equation of a bug's path assuming the pole is at the center of the square. (Use the fact that the line joining one bug to the next is tangent to the bug's path.)
   (b) Find the distance traveled by a bug by the time it meets the other bugs at the center.

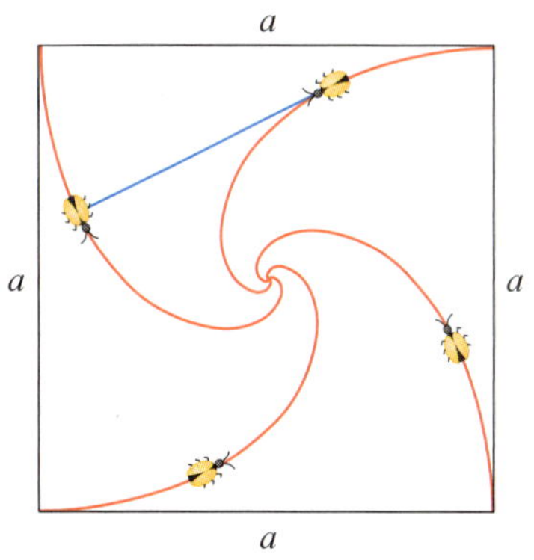

5. A curve called the **folium of Descartes** is defined by the parametric equations

$$x = \frac{3t}{1 + t^3} \qquad y = \frac{3t^2}{1 + t^3}$$

   (a) Show that if (a, b) lies on the curve, then so does (b, a); that is, the curve is symmetric with respect to the line $y = x$. Where does the curve intersect this line?
   (b) Find the points on the curve where the tangent lines are horizontal or vertical.
   (c) Show that the line $y = -x - 1$ is a slant asymptote.
   (d) Sketch the curve.
   (e) Show that a Cartesian equation of this curve is $x^3 + y^3 = 3xy$.
   (f) Show that the polar equation can be written in the form

$$r = \frac{3 \sec\theta \tan\theta}{1 + \tan^3\theta}$$

   (g) Find the area enclosed by the loop of this curve.
   CAS (h) Show that the area of the loop is the same as the area that lies between the asymptote and the infinite branches of the curve. (Use a computer algebra system to evaluate the integral.)

Chapter 11

Bessel functions, which are used to model the vibrations of drumheads and cymbals, are defined as sums of infinite series in Section 11.8. Notice how closely the computer-generated models (which involve Bessel functions and cosine functions) match the photographs of a vibrating rubber membrane.

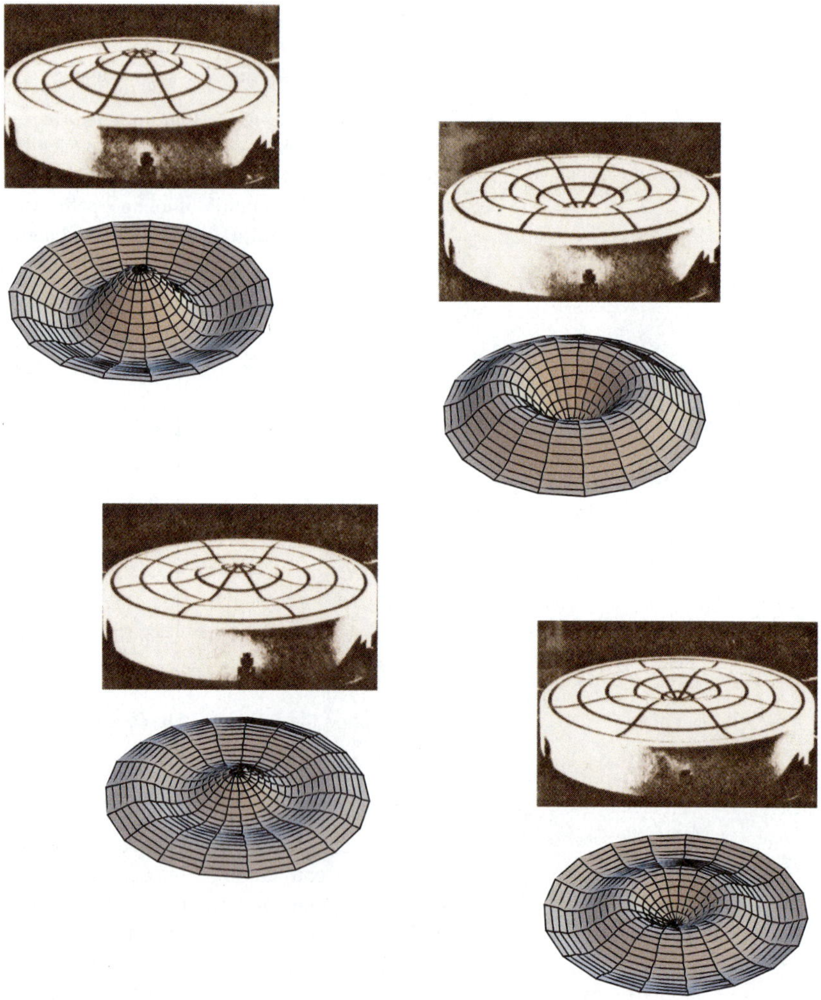

Infinite Sequences and Series

Infinite sequences and series were introduced briefly in *A Preview of Calculus* in connection with Zeno's paradoxes and the decimal representation of numbers. Their importance in calculus stems from Newton's idea of representing functions as sums of infinite series. For instance, in finding areas he often integrated a function by first expressing it as a series and then integrating each term of the series. We will pursue his idea in Section 11.10 in order to integrate such functions as e^{-x^2}. (Recall that we have previously been unable to do this.) Many of the functions that arise in mathematical physics and chemistry, such as Bessel functions, are defined as sums of series, so it is important to be familiar with the basic concepts of convergence of infinite sequences and series.

Physicists also use series in another way, as we will see in Section 11.12. In studying fields as diverse as optics, special relativity, and electromagnetism, they analyze phenomena by replacing a function with the first few terms in the series that represents it.

11.1 Sequences

A **sequence** can be thought of as a list of numbers written in a definite order:

$$a_1,\ a_2,\ a_3,\ a_4,\ \ldots,\ a_n,\ \ldots$$

The number a_1 is called the *first term*, a_2 is the *second term*, and in general a_n is the *nth term*. We will deal exclusively with infinite sequences and so each term a_n will have a successor a_{n+1}.

Notice that for every positive integer n there is a corresponding number a_n and so a sequence can be defined as a function whose domain is the set of positive integers. But we usually write a_n instead of the function notation $f(n)$ for the value of the function at the number n.

NOTATION ▫ The sequence $\{a_1, a_2, a_3, \ldots\}$ is also denoted by

$$\{a_n\} \qquad \text{or} \qquad \{a_n\}_{n=1}^{\infty}$$

EXAMPLE 1 Some sequences can be defined by giving a formula for the nth term. In the following examples we give three descriptions of the sequence: one by using the preceding notation, another by using the defining formula, and a third by writing out the terms of the sequence. Notice that n doesn't have to start at 1.

(a) $\left\{\dfrac{n}{n+1}\right\}_{n=1}^{\infty}$ $\qquad a_n = \dfrac{n}{n+1} \qquad \left\{\dfrac{1}{2}, \dfrac{2}{3}, \dfrac{3}{4}, \dfrac{4}{5}, \ldots, \dfrac{n}{n+1}, \ldots\right\}$

(b) $\left\{\dfrac{(-1)^n(n+1)}{3^n}\right\}$ $\qquad a_n = \dfrac{(-1)^n(n+1)}{3^n} \qquad \left\{-\dfrac{2}{3}, \dfrac{3}{9}, -\dfrac{4}{27}, \dfrac{5}{81}, \ldots, \dfrac{(-1)^n(n+1)}{3^n}, \ldots\right\}$

(c) $\{\sqrt{n-3}\}_{n=3}^{\infty}$ $\quad a_n = \sqrt{n-3},\ n \geqslant 3$ $\quad \{0, 1, \sqrt{2}, \sqrt{3}, \ldots, \sqrt{n-3}, \ldots\}$

(d) $\left\{\cos\dfrac{n\pi}{6}\right\}_{n=0}^{\infty}$ $\quad a_n = \cos\dfrac{n\pi}{6},\ n \geqslant 0$ $\quad \left\{1, \dfrac{\sqrt{3}}{2}, \dfrac{1}{2}, 0, \ldots, \cos\dfrac{n\pi}{6}, \ldots\right\}$

EXAMPLE 2 Find a formula for the general term a_n of the sequence

$$\left\{\frac{3}{5}, -\frac{4}{25}, \frac{5}{125}, -\frac{6}{625}, \frac{7}{3125}, \ldots\right\}$$

assuming that the pattern of the first few terms continues.

SOLUTION We are given that

$$a_1 = \frac{3}{5} \qquad a_2 = -\frac{4}{25} \qquad a_3 = \frac{5}{125} \qquad a_4 = -\frac{6}{625} \qquad a_5 = \frac{7}{3125}$$

Notice that the numerators of these fractions start with 3 and increase by 1 whenever we go to the next term. The second term has numerator 4, the third term has numerator 5; in general, the nth term will have numerator $n + 2$. The denominators are the powers of 5, so a_n has denominator 5^n. The signs of the terms are alternately positive and negative, so we need to multiply by a power of -1. In Example 1(b) the factor $(-1)^n$ meant we started with a negative term. Here we want to start with a positive term and so we use $(-1)^{n-1}$ or $(-1)^{n+1}$. Therefore,

$$a_n = (-1)^{n-1}\frac{n+2}{5^n}$$

EXAMPLE 3 Here are some sequences that don't have a simple defining equation.

(a) The sequence $\{p_n\}$, where p_n is the population of the world as of January 1 in the year n.

(b) If we let a_n be the digit in the nth decimal place of the number e, then $\{a_n\}$ is a well-defined sequence whose first few terms are

$$\{7, 1, 8, 2, 8, 1, 8, 2, 8, 4, 5, \ldots\}$$

(c) The **Fibonacci sequence** $\{f_n\}$ is defined recursively by the conditions

$$f_1 = 1 \qquad f_2 = 1 \qquad f_n = f_{n-1} + f_{n-2} \qquad n \geqslant 3$$

Each term is the sum of the two preceding terms. The first few terms are

$$\{1, 1, 2, 3, 5, 8, 13, 21, \ldots\}$$

This sequence arose when the 13th-century Italian mathematician known as Fibonacci solved a problem concerning the breeding of rabbits (see Exercise 65).

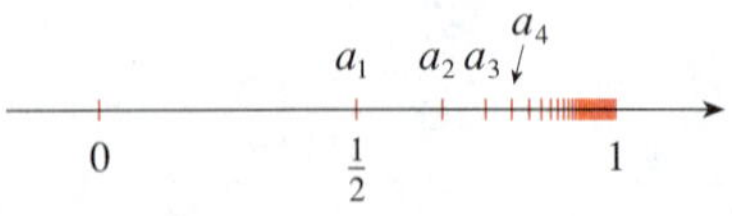

FIGURE 1

A sequence such as the one in Example 1(a), $a_n = n/(n+1)$, can be pictured either by plotting its terms on a number line as in Figure 1 or by plotting its graph as in Figure 2. Note that, since a sequence is a function whose domain is the set of positive integers, its graph consists of isolated points with coordinates

$$(1, a_1) \qquad (2, a_2) \qquad (3, a_3) \qquad \ldots \qquad (n, a_n) \qquad \ldots$$

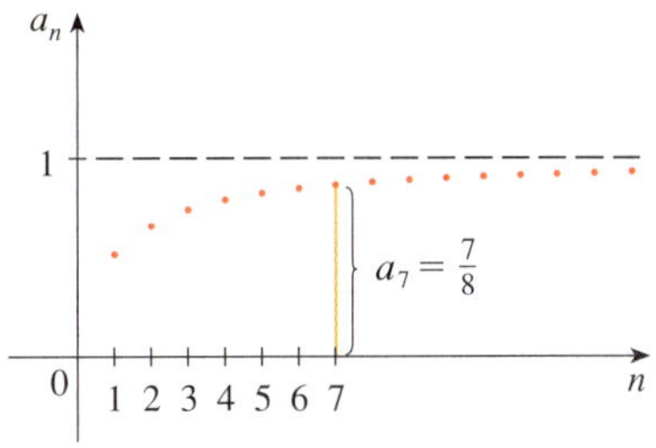

FIGURE 2

From Figure 1 or 2 it appears that the terms of the sequence $a_n = n/(n+1)$ are approaching 1 as n becomes large. In fact, the difference

$$1 - \frac{n}{n+1} = \frac{1}{n+1}$$

can be made as small as we like by taking n sufficiently large. We indicate this by writing

$$\lim_{n \to \infty} \frac{n}{n+1} = 1$$

In general, the notation

$$\lim_{n \to \infty} a_n = L$$

means that the terms of the sequence $\{a_n\}$ approach L as n becomes large. Notice that the following definition of the limit of a sequence is very similar to the definition of a limit of a function at infinity given in Section 2.6.

1 Definition A sequence $\{a_n\}$ has the **limit** L and we write

$$\lim_{n \to \infty} a_n = L \qquad \text{or} \qquad a_n \to L \text{ as } n \to \infty$$

if we can make the terms a_n as close to L as we like by taking n sufficiently large. If $\lim_{n \to \infty} a_n$ exists, we say the sequence **converges** (or is **convergent**). Otherwise, we say the sequence **diverges** (or is **divergent**).

Figure 3 illustrates Definition 1 by showing the graphs of two sequences that have the limit L.

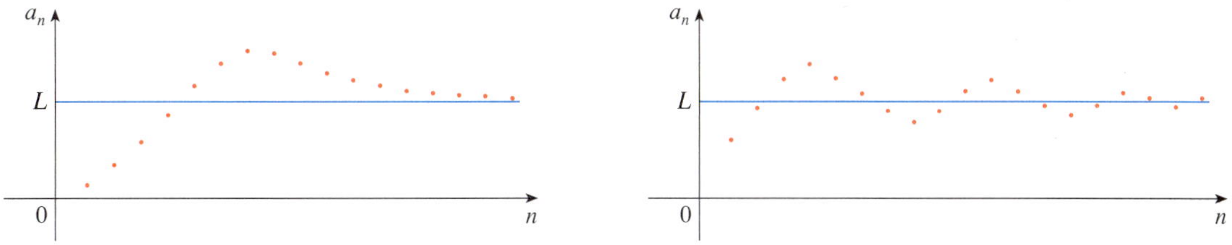

FIGURE 3 Graphs of two sequences with $\lim_{n \to \infty} a_n = L$

A more precise version of Definition 1 is as follows.

|||| Compare this definition with Definition 2.6.7.

2 Definition A sequence $\{a_n\}$ has the **limit** L and we write

$$\lim_{n \to \infty} a_n = L \qquad \text{or} \qquad a_n \to L \text{ as } n \to \infty$$

if for every $\varepsilon > 0$ there is a corresponding integer N such that

$$|a_n - L| < \varepsilon \qquad \text{whenever } n > N$$

Definition 2 is illustrated by Figure 4, in which the terms $a_1, a_2, a_3, \ldots$ are plotted on a number line. No matter how small an interval $(L - \varepsilon, L + \varepsilon)$ is chosen, there exists an N such that all terms of the sequence from a_{N+1} onward must lie in that interval.

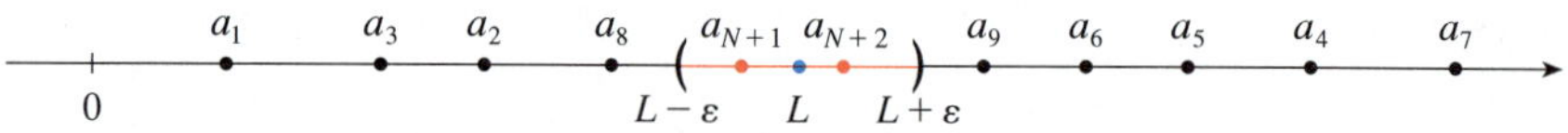

FIGURE 4

Another illustration of Definition 2 is given in Figure 5. The points on the graph of $\{a_n\}$ must lie between the horizontal lines $y = L + \varepsilon$ and $y = L - \varepsilon$ if $n > N$. This picture must be valid no matter how small ε is chosen, but usually a smaller ε requires a larger N.

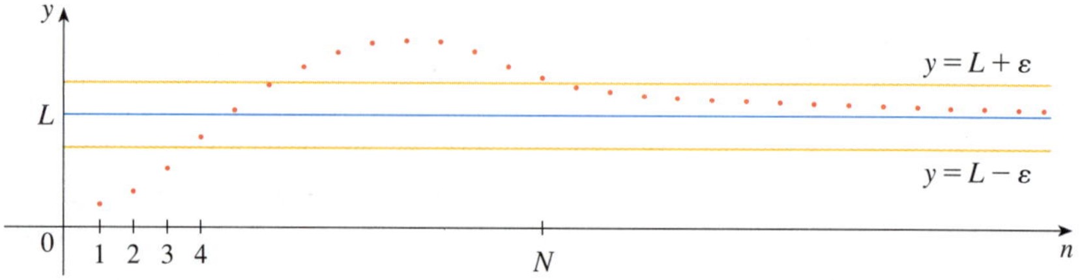

FIGURE 5

Comparison of Definition 2 and Definition 2.6.7 shows that the only difference between $\lim_{n\to\infty} a_n = L$ and $\lim_{x\to\infty} f(x) = L$ is that n is required to be an integer. Thus, we have the following theorem, which is illustrated by Figure 6.

3 Theorem If $\lim_{x\to\infty} f(x) = L$ and $f(n) = a_n$ when n is an integer, then $\lim_{n\to\infty} a_n = L$.

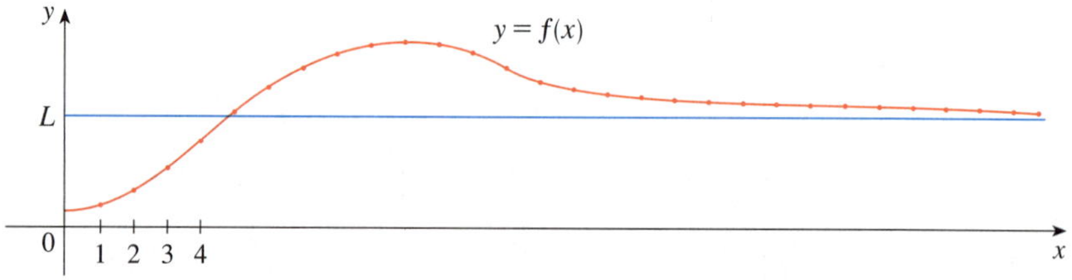

FIGURE 6

In particular, since we know that $\lim_{x\to\infty} (1/x^r) = 0$ when $r > 0$ (Theorem 2.6.5), we have

4 $$\lim_{n\to\infty} \frac{1}{n^r} = 0 \qquad \text{if } r > 0$$

If a_n becomes large as n becomes large, we use the notation $\lim_{n\to\infty} a_n = \infty$. The following precise definition is similar to Definition 2.6.9.

5 Definition $\lim_{n\to\infty} a_n = \infty$ means that for every positive number M there is an integer N such that

$$a_n > M \qquad \text{whenever } n > N$$

If $\lim_{n\to\infty} a_n = \infty$, then the sequence $\{a_n\}$ is divergent but in a special way. We say that $\{a_n\}$ diverges to ∞.

The Limit Laws given in Section 2.3 also hold for the limits of sequences and their proofs are similar.

Limit Laws for Sequences

If $\{a_n\}$ and $\{b_n\}$ are convergent sequences and c is a constant, then

$$\lim_{n\to\infty} (a_n + b_n) = \lim_{n\to\infty} a_n + \lim_{n\to\infty} b_n$$

$$\lim_{n\to\infty} (a_n - b_n) = \lim_{n\to\infty} a_n - \lim_{n\to\infty} b_n$$

$$\lim_{n\to\infty} ca_n = c \lim_{n\to\infty} a_n \qquad \lim_{n\to\infty} c = c$$

$$\lim_{n\to\infty} (a_n b_n) = \lim_{n\to\infty} a_n \cdot \lim_{n\to\infty} b_n$$

$$\lim_{n\to\infty} \frac{a_n}{b_n} = \frac{\lim_{n\to\infty} a_n}{\lim_{n\to\infty} b_n} \quad \text{if } \lim_{n\to\infty} b_n \neq 0$$

$$\lim_{n\to\infty} a_n^p = \left[\lim_{n\to\infty} a_n\right]^p \quad \text{if } p > 0 \text{ and } a_n > 0$$

The Squeeze Theorem can also be adapted for sequences as follows (see Figure 7).

Squeeze Theorem for Sequences

If $a_n \leqslant b_n \leqslant c_n$ for $n \geqslant n_0$ and $\lim_{n\to\infty} a_n = \lim_{n\to\infty} c_n = L$, then $\lim_{n\to\infty} b_n = L$.

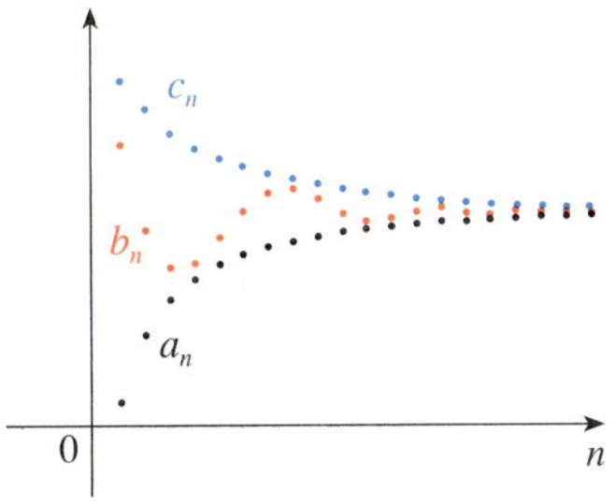

FIGURE 7
The sequence $\{b_n\}$ is squeezed between the sequences $\{a_n\}$ and $\{c_n\}$.

Another useful fact about limits of sequences is given by the following theorem, whose proof is left as Exercise 69.

6 Theorem If $\lim_{n\to\infty} |a_n| = 0$, then $\lim_{n\to\infty} a_n = 0$.

EXAMPLE 4 Find $\lim_{n\to\infty} \dfrac{n}{n+1}$.

SOLUTION The method is similar to the one we used in Section 2.6: Divide numerator and denominator by the highest power of n and then use the Limit Laws.

$$\lim_{n\to\infty} \frac{n}{n+1} = \lim_{n\to\infty} \frac{1}{1+\dfrac{1}{n}} = \frac{\lim_{n\to\infty} 1}{\lim_{n\to\infty} 1 + \lim_{n\to\infty} \dfrac{1}{n}}$$

$$= \frac{1}{1+0} = 1$$

|||| This shows that the guess we made earlier from Figures 1 and 2 was correct.

Here we used Equation 4 with $r = 1$.

EXAMPLE 5 Calculate $\lim_{n\to\infty} \frac{\ln n}{n}$.

SOLUTION Notice that both numerator and denominator approach infinity as $n \to \infty$. We can't apply l'Hospital's Rule directly because it applies not to sequences but to functions of a real variable. However, we can apply l'Hospital's Rule to the related function $f(x) = (\ln x)/x$ and obtain

$$\lim_{x\to\infty} \frac{\ln x}{x} = \lim_{x\to\infty} \frac{1/x}{1} = 0$$

Therefore, by Theorem 3 we have

$$\lim_{n\to\infty} \frac{\ln n}{n} = 0$$

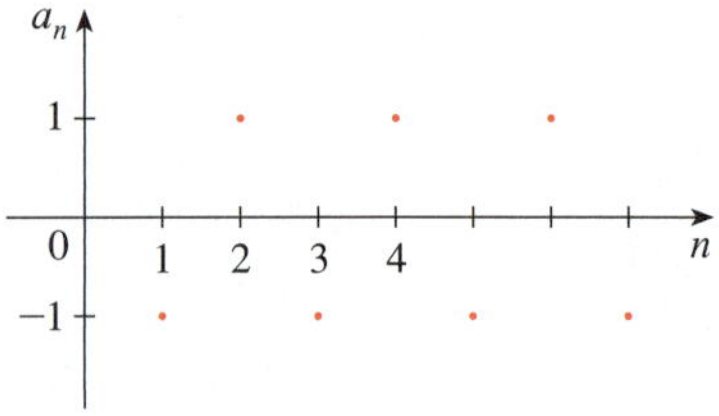

FIGURE 8

EXAMPLE 6 Determine whether the sequence $a_n = (-1)^n$ is convergent or divergent.

SOLUTION If we write out the terms of the sequence, we obtain

$$\{-1, 1, -1, 1, -1, 1, -1, \ldots\}$$

The graph of this sequence is shown in Figure 8. Since the terms oscillate between 1 and -1 infinitely often, a_n does not approach any number. Thus, $\lim_{n\to\infty} (-1)^n$ does not exist; that is, the sequence $\{(-1)^n\}$ is divergent.

|||| The graph of the sequence in Example 7 is shown in Figure 9 and supports the answer.

EXAMPLE 7 Evaluate $\lim_{n\to\infty} \frac{(-1)^n}{n}$ if it exists.

SOLUTION

$$\lim_{n\to\infty} \left| \frac{(-1)^n}{n} \right| = \lim_{n\to\infty} \frac{1}{n} = 0$$

Therefore, by Theorem 6,

$$\lim_{n\to\infty} \frac{(-1)^n}{n} = 0$$

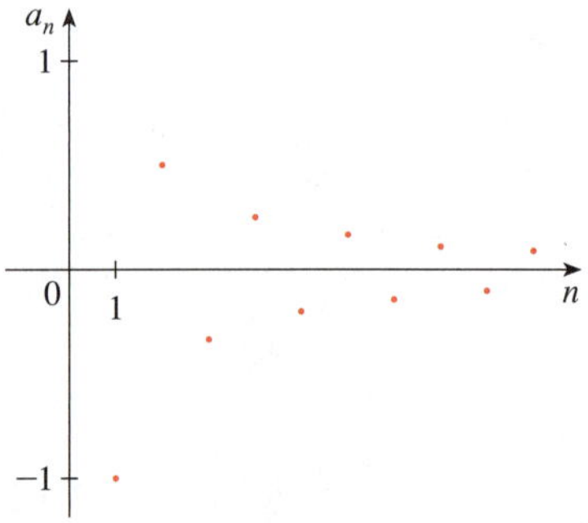

FIGURE 9

EXAMPLE 8 Discuss the convergence of the sequence $a_n = n!/n^n$, where $n! = 1 \cdot 2 \cdot 3 \cdot \cdots \cdot n$.

SOLUTION Both numerator and denominator approach infinity as $n \to \infty$ but here we have no corresponding function for use with l'Hospital's Rule ($x!$ is not defined when x is not an integer). Let's write out a few terms to get a feeling for what happens to a_n as n gets large:

$$a_1 = 1 \qquad a_2 = \frac{1 \cdot 2}{2 \cdot 2} \qquad a_3 = \frac{1 \cdot 2 \cdot 3}{3 \cdot 3 \cdot 3}$$

7 $$a_n = \frac{1 \cdot 2 \cdot 3 \cdot \cdots \cdot n}{n \cdot n \cdot n \cdot \cdots \cdot n}$$

It appears from these expressions and the graph in Figure 10 that the terms are decreas-

|||| CREATING GRAPHS OF SEQUENCES

Some computer algebra systems have special commands that enable us to create sequences and graph them directly. With most graphing calculators, however, sequences can be graphed by using parametric equations. For instance, the sequence in Example 8 can be graphed by entering the parametric equations

$$x = t \qquad y = t!/t^t$$

and graphing in dot mode starting with $t = 1$, setting the t-step equal to 1. The result is shown in Figure 10.

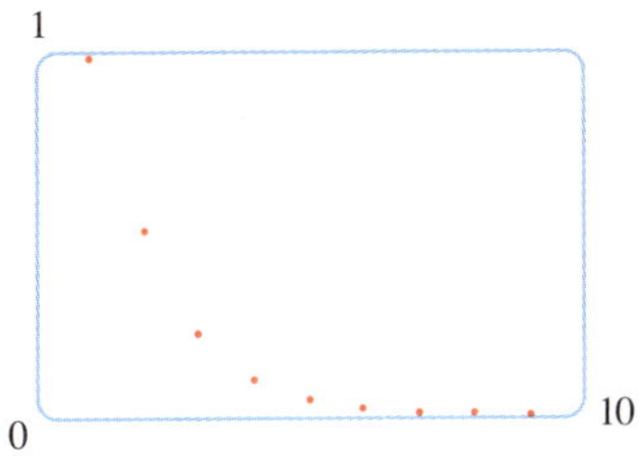

FIGURE 10

ing and perhaps approach 0. To confirm this, observe from Equation 7 that

$$a_n = \frac{1}{n}\left(\frac{2 \cdot 3 \cdot \cdots \cdot n}{n \cdot n \cdot \cdots \cdot n}\right)$$

Notice that the expression in parentheses is at most 1 because the numerator is less than (or equal to) the denominator. So

$$0 < a_n \leqslant \frac{1}{n}$$

We know that $1/n \to 0$ as $n \to \infty$. Therefore, $a_n \to 0$ as $n \to \infty$ by the Squeeze Theorem.

EXAMPLE 9 For what values of r is the sequence $\{r^n\}$ convergent?

SOLUTION We know from Section 2.6 and the graphs of the exponential functions in Section 1.5 that $\lim_{x\to\infty} a^x = \infty$ for $a > 1$ and $\lim_{x\to\infty} a^x = 0$ for $0 < a < 1$. Therefore, putting $a = r$ and using Theorem 2, we have

$$\lim_{n\to\infty} r^n = \begin{cases} \infty & \text{if } r > 1 \\ 0 & \text{if } 0 < r < 1 \end{cases}$$

It is obvious that

$$\lim_{n\to\infty} 1^n = 1 \qquad \text{and} \qquad \lim_{n\to\infty} 0^n = 0$$

If $-1 < r < 0$, then $0 < |r| < 1$, so

$$\lim_{n\to\infty} |r^n| = \lim_{n\to\infty} |r|^n = 0$$

and therefore $\lim_{n\to\infty} r^n = 0$ by Theorem 6. If $r \leqslant -1$, then $\{r^n\}$ diverges as in Example 6. Figure 11 shows the graphs for various values of r. (The case $r = -1$ is shown in Figure 8.)

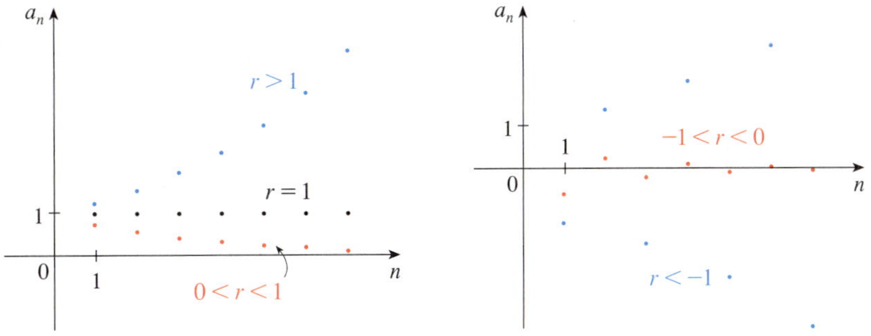

FIGURE 11
The sequence $a_n = r^n$

The results of Example 9 are summarized for future use as follows.

8 The sequence $\{r^n\}$ is convergent if $-1 < r \leqslant 1$ and divergent for all other values of r.

$$\lim_{n\to\infty} r^n = \begin{cases} 0 & \text{if } -1 < r < 1 \\ 1 & \text{if } r = 1 \end{cases}$$

9 Definition A sequence $\{a_n\}$ is called **increasing** if $a_n < a_{n+1}$ for all $n \geqslant 1$, that is, $a_1 < a_2 < a_3 < \cdots$. It is called **decreasing** if $a_n > a_{n+1}$ for all $n \geqslant 1$. It is called **monotonic** if it is either increasing or decreasing.

EXAMPLE 10 The sequence $\left\{\dfrac{3}{n+5}\right\}$ is decreasing because

$$\frac{3}{n+5} > \frac{3}{(n+1)+5} = \frac{3}{n+6}$$

for all $n \geqslant 1$. (The right side is smaller because it has a larger denominator.)

EXAMPLE 11 Show that the sequence $a_n = \dfrac{n}{n^2+1}$ is decreasing.

SOLUTION 1 We must show that $a_{n+1} < a_n$, that is,

$$\frac{n+1}{(n+1)^2+1} < \frac{n}{n^2+1}$$

This inequality is equivalent to the one we get by cross-multiplication:

$$\frac{n+1}{(n+1)^2+1} < \frac{n}{n^2+1} \iff (n+1)(n^2+1) < n[(n+1)^2+1]$$

$$\iff n^3 + n^2 + n + 1 < n^3 + 2n^2 + 2n$$

$$\iff 1 < n^2 + n$$

Since $n \geqslant 1$, we know that the inequality $n^2 + n > 1$ is true. Therefore, $a_{n+1} < a_n$ and so $\{a_n\}$ is decreasing.

SOLUTION 2 Consider the function $f(x) = \dfrac{x}{x^2+1}$:

$$f'(x) = \frac{x^2+1-2x^2}{(x^2+1)^2} = \frac{1-x^2}{(x^2+1)^2} < 0 \qquad \text{whenever } x^2 > 1$$

Thus, f is decreasing on $(1, \infty)$ and so $f(n) > f(n+1)$. Therefore, $\{a_n\}$ is decreasing.

10 Definition A sequence $\{a_n\}$ is **bounded above** if there is a number M such that

$$a_n \leqslant M \qquad \text{for all } n \geqslant 1$$

It is **bounded below** if there is a number m such that

$$m \leqslant a_n \qquad \text{for all } n \geqslant 1$$

If it is bounded above and below, then $\{a_n\}$ is a **bounded sequence**.

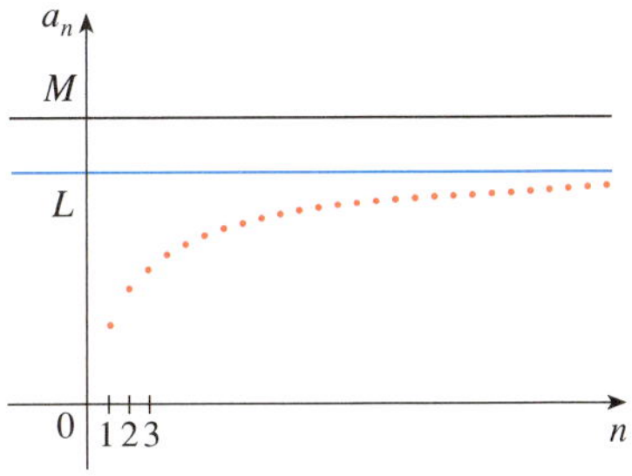

FIGURE 12

For instance, the sequence $a_n = n$ is bounded below $(a_n > 0)$ but not above. The sequence $a_n = n/(n+1)$ is bounded because $0 < a_n < 1$ for all n.

We know that not every bounded sequence is convergent [for instance, the sequence $a_n = (-1)^n$ satisfies $-1 \leqslant a_n \leqslant 1$ but is divergent from Example 6] and not every monotonic sequence is convergent $(a_n = n \to \infty)$. But if a sequence is both bounded *and* monotonic, then it must be convergent. This fact is proved as Theorem 11, but intuitively you can understand why it is true by looking at Figure 12. If $\{a_n\}$ is increasing and $a_n \leqslant M$ for all n, then the terms are forced to crowd together and approach some number L.

The proof of Theorem 11 is based on the **Completeness Axiom** for the set $\mathbb{R}$ of real numbers, which says that if S is a nonempty set of real numbers that has an upper bound M ($x \leqslant M$ for all x in S), then S has a **least upper bound** b. (This means that b is an upper bound for S, but if M is any other upper bound, then $b \leqslant M$.) The Completeness Axiom is an expression of the fact that there is no gap or hole in the real number line.

11 Monotonic Sequence Theorem Every bounded, monotonic sequence is convergent.

Proof Suppose $\{a_n\}$ is an increasing sequence. Since $\{a_n\}$ is bounded, the set $S = \{a_n \mid n \geqslant 1\}$ has an upper bound. By the Completeness Axiom it has a least upper bound L. Given $\varepsilon > 0$, $L - \varepsilon$ is *not* an upper bound for S (since L is the *least* upper bound). Therefore

$$a_N > L - \varepsilon \qquad \text{for some integer } N$$

But the sequence is increasing so $a_n \geqslant a_N$ for every $n > N$. Thus, if $n > N$ we have

$$a_n > L - \varepsilon$$

so

$$0 \leqslant L - a_n < \varepsilon$$

since $a_n \leqslant L$. Thus

$$|L - a_n| < \varepsilon \qquad \text{whenever } n > N$$

so $\lim_{n\to\infty} a_n = L$.

A similar proof (using the greatest lower bound) works if $\{a_n\}$ is decreasing. ■

The proof of Theorem 11 shows that a sequence that is increasing and bounded above is convergent. (Likewise, a decreasing sequence that is bounded below is convergent.) This fact is used many times in dealing with infinite series.

EXAMPLE 12 Investigate the sequence $\{a_n\}$ defined by the recurrence relation

$$a_1 = 2 \qquad a_{n+1} = \tfrac{1}{2}(a_n + 6) \qquad \text{for } n = 1, 2, 3, \ldots$$

SOLUTION We begin by computing the first several terms:

$$a_1 = 2 \qquad a_2 = \tfrac{1}{2}(2+6) = 4 \qquad a_3 = \tfrac{1}{2}(4+6) = 5$$

$$a_4 = \tfrac{1}{2}(5+6) = 5.5 \qquad a_5 = 5.75 \qquad a_6 = 5.875$$

$$a_7 = 5.9375 \qquad a_8 = 5.96875 \qquad a_9 = 5.984375$$

Mathematical induction is often used in dealing with recursive sequences. See page 81 for a discussion of the Principle of Mathematical Induction.

These initial terms suggest that the sequence is increasing and the terms are approaching 6. To confirm that the sequence is increasing we use mathematical induction to show that $a_{n+1} > a_n$ for all $n \geqslant 1$. This is true for $n = 1$ because $a_2 = 4 > a_1$. If we assume that it is true for $n = k$, then we have

$$a_{k+1} > a_k$$

so
$$a_{k+1} + 6 > a_k + 6$$

and
$$\tfrac{1}{2}(a_{k+1} + 6) > \tfrac{1}{2}(a_k + 6)$$

Thus
$$a_{k+2} > a_{k+1}$$

We have deduced that $a_{n+1} > a_n$ is true for $n = k + 1$. Therefore, the inequality is true for all n by induction.

Next we verify that $\{a_n\}$ is bounded by showing that $a_n < 6$ for all n. (Since the sequence is increasing, we already know that it has a lower bound: $a_n \geqslant a_1 = 2$ for all n.) We know that $a_1 < 6$, so the assertion is true for $n = 1$. Suppose it is true for $n = k$. Then

$$a_k < 6$$

so
$$a_k + 6 < 12$$

$$\tfrac{1}{2}(a_k + 6) < \tfrac{1}{2}(12) = 6$$

Thus
$$a_{k+1} < 6$$

This shows, by mathematical induction, that $a_n < 6$ for all n.

Since the sequence $\{a_n\}$ is increasing and bounded, Theorem 11 guarantees that it has a limit. The theorem doesn't tell us what the value of the limit is. But now that we know $L = \lim_{n\to\infty} a_n$ exists, we can use the recurrence relation to write

$$\lim_{n\to\infty} a_{n+1} = \lim_{n\to\infty} \tfrac{1}{2}(a_n + 6) = \tfrac{1}{2}\left(\lim_{n\to\infty} a_n + 6\right) = \tfrac{1}{2}(L + 6)$$

A proof of this fact is requested in Exercise 52.

Since $a_n \to L$, it follows that $a_{n+1} \to L$, too (as $n \to \infty$, $n + 1 \to \infty$, too). So we have

$$L = \tfrac{1}{2}(L + 6)$$

Solving this equation for L, we get $L = 6$, as predicted.

11.1 Exercises

1. (a) What is a sequence?
(b) What does it mean to say that $\lim_{n\to\infty} a_n = 8$?
(c) What does it mean to say that $\lim_{n\to\infty} a_n = \infty$?

2. (a) What is a convergent sequence? Give two examples.
(b) What is a divergent sequence? Give two examples.

3–8 List the first five terms of the sequence.

3. $a_n = 1 - (0.2)^n$

4. $a_n = \dfrac{n + 1}{3n - 1}$

5. $a_n = \dfrac{3(-1)^n}{n!}$

6. $\{2 \cdot 4 \cdot 6 \cdot \cdots \cdot (2n)\}$

7. $a_1 = 3, \quad a_{n+1} = 2a_n - 1$

8. $a_1 = 4, \quad a_{n+1} = \dfrac{a_n}{a_n - 1}$

9–14 Find a formula for the general term a_n of the sequence, assuming that the pattern of the first few terms continues.

9. $\left\{\frac{1}{2}, \frac{1}{4}, \frac{1}{8}, \frac{1}{16}, \ldots\right\}$

10. $\left\{\frac{1}{2}, \frac{1}{4}, \frac{1}{6}, \frac{1}{8}, \ldots\right\}$

11. $\{2, 7, 12, 17, \ldots\}$

12. $\left\{-\frac{1}{4}, \frac{2}{9}, -\frac{3}{16}, \frac{4}{25}, \ldots\right\}$

13. $\left\{1, -\frac{2}{3}, \frac{4}{9}, -\frac{8}{27}, \ldots\right\}$

14. $\{5, 1, 5, 1, 5, 1, \ldots\}$

15–40 Determine whether the sequence converges or diverges. If it converges, find the limit.

15. $a_n = n(n-1)$

16. $a_n = \dfrac{n+1}{3n-1}$

17. $a_n = \dfrac{3 + 5n^2}{n + n^2}$

18. $a_n = \dfrac{\sqrt{n}}{1 + \sqrt{n}}$

19. $a_n = \dfrac{2^n}{3^{n+1}}$

20. $a_n = \dfrac{n}{1 + \sqrt{n}}$

21. $a_n = \dfrac{(-1)^{n-1} n}{n^2 + 1}$

22. $a_n = \dfrac{(-1)^n n^3}{n^3 + 2n^2 + 1}$

23. $a_n = \cos(n/2)$

24. $a_n = \cos(2/n)$

25. $\left\{\dfrac{(2n-1)!}{(2n+1)!}\right\}$

26. $\{\arctan 2n\}$

27. $\left\{\dfrac{e^n + e^{-n}}{e^{2n} - 1}\right\}$

28. $\left\{\dfrac{\ln n}{\ln 2n}\right\}$

29. $\{n^2 e^{-n}\}$

30. $\{n \cos n\pi\}$

31. $a_n = \dfrac{\cos^2 n}{2^n}$

32. $a_n = \ln(n+1) - \ln n$

33. $a_n = n \sin(1/n)$

34. $a_n = \sqrt{n} - \sqrt{n^2 - 1}$

35. $a_n = \left(1 + \dfrac{2}{n}\right)^{1/n}$

36. $a_n = \dfrac{\sin 2n}{1 + \sqrt{n}}$

37. $\{0, 1, 0, 0, 1, 0, 0, 0, 1, \ldots\}$

38. $\left\{\frac{1}{1}, \frac{1}{3}, \frac{1}{2}, \frac{1}{4}, \frac{1}{3}, \frac{1}{5}, \frac{1}{4}, \frac{1}{6}, \ldots\right\}$

39. $a_n = \dfrac{n!}{2^n}$

40. $a_n = \dfrac{(-3)^n}{n!}$

41–48 Use a graph of the sequence to decide whether the sequence is convergent or divergent. If the sequence is convergent, guess the value of the limit from the graph and then prove your guess. (See the margin note on page 707 for advice on graphing sequences.)

41. $a_n = (-1)^n \dfrac{n+1}{n}$

42. $a_n = 2 + (-2/\pi)^n$

43. $\left\{\arctan\left(\dfrac{2n}{2n+1}\right)\right\}$

44. $\left\{\dfrac{\sin n}{\sqrt{n}}\right\}$

45. $a_n = \dfrac{n^3}{n!}$

46. $a_n = \sqrt[n]{3^n + 5^n}$

47. $a_n = \dfrac{1 \cdot 3 \cdot 5 \cdot \cdots \cdot (2n-1)}{(2n)^n}$

48. $a_n = \dfrac{1 \cdot 3 \cdot 5 \cdot \cdots \cdot (2n-1)}{n!}$

49. If \$1000 is invested at 6% interest, compounded annually, then after n years the investment is worth $a_n = 1000(1.06)^n$ dollars.
(a) Find the first five terms of the sequence $\{a_n\}$.
(b) Is the sequence convergent or divergent? Explain.

50. Find the first 40 terms of the sequence defined by

$$a_{n+1} = \begin{cases} \frac{1}{2}a_n & \text{if } a_n \text{ is an even number} \\ 3a_n + 1 & \text{if } a_n \text{ is an odd number} \end{cases}$$

and $a_1 = 11$. Do the same if $a_1 = 25$. Make a conjecture about this type of sequence.

51. For what values of r is the sequence $\{nr^n\}$ convergent?

52. (a) If $\{a_n\}$ is convergent, show that

$$\lim_{n\to\infty} a_{n+1} = \lim_{n\to\infty} a_n$$

(b) A sequence $\{a_n\}$ is defined by $a_1 = 1$ and $a_{n+1} = 1/(1 + a_n)$ for $n \geqslant 1$. Assuming that $\{a_n\}$ is convergent, find its limit.

53. Suppose you know that $\{a_n\}$ is a decreasing sequence and all its terms lie between the numbers 5 and 8. Explain why the sequence has a limit. What can you say about the value of the limit?

54–60 Determine whether the sequence is increasing, decreasing, or not monotonic. Is the sequence bounded?

54. $a_n = \dfrac{1}{5^n}$

55. $a_n = \dfrac{1}{2n+3}$

56. $a_n = \dfrac{2n-3}{3n+4}$

57. $a_n = \cos(n\pi/2)$

58. $a_n = ne^{-n}$

59. $a_n = \dfrac{n}{n^2+1}$

60. $a_n = n + \dfrac{1}{n}$

61. Find the limit of the sequence

$$\left\{\sqrt{2}, \sqrt{2\sqrt{2}}, \sqrt{2\sqrt{2\sqrt{2}}}, \ldots\right\}$$

62. A sequence $\{a_n\}$ is given by $a_1 = \sqrt{2}$, $a_{n+1} = \sqrt{2 + a_n}$.
(a) By induction or otherwise, show that $\{a_n\}$ is increasing and bounded above by 3. Apply Theorem 11 to show that $\lim_{n\to\infty} a_n$ exists.
(b) Find $\lim_{n\to\infty} a_n$.

63. Show that the sequence defined by $a_1 = 1$, $a_{n+1} = 3 - 1/a_n$ is increasing and $a_n < 3$ for all n. Deduce that $\{a_n\}$ is convergent and find its limit.

64. Show that the sequence defined by

$$a_1 = 2 \qquad a_{n+1} = \frac{1}{3 - a_n}$$

satisfies $0 < a_n \leqslant 2$ and is decreasing. Deduce that the sequence is convergent and find its limit.

65. (a) Fibonacci posed the following problem: Suppose that rabbits live forever and that every month each pair produces a new pair which becomes productive at age 2 months. If we start with one newborn pair, how many pairs of rabbits will we have in the nth month? Show that the answer is f_n, where $\{f_n\}$ is the Fibonacci sequence defined in Example 3(c).

(b) Let $a_n = f_{n+1}/f_n$ and show that $a_{n-1} = 1 + 1/a_{n-2}$. Assuming that $\{a_n\}$ is convergent, find its limit.

66. (a) Let $a_1 = a$, $a_2 = f(a)$, $a_3 = f(a_2) = f(f(a))$, $\ldots$, $a_{n+1} = f(a_n)$, where f is a continuous function. If $\lim_{n\to\infty} a_n = L$, show that $f(L) = L$.

(b) Illustrate part (a) by taking $f(x) = \cos x$, $a = 1$, and estimating the value of L to five decimal places.

67. (a) Use a graph to guess the value of the limit

$$\lim_{n\to\infty} \frac{n^5}{n!}$$

(b) Use a graph of the sequence in part (a) to find the smallest values of N that correspond to $\varepsilon = 0.1$ and $\varepsilon = 0.001$ in Definition 2.

68. Use Definition 2 directly to prove that $\lim_{n\to\infty} r^n = 0$ when $|r| < 1$.

69. Prove Theorem 6.
[*Hint:* Use either Definition 2 or the Squeeze Theorem.]

70. Let $a_n = \left(1 + \dfrac{1}{n}\right)^n$.

(a) Show that if $0 \leqslant a < b$, then

$$\frac{b^{n+1} - a^{n+1}}{b - a} < (n + 1)b^n$$

(b) Deduce that $b^n[(n + 1)a - nb] < a^{n+1}$.

(c) Use $a = 1 + 1/(n + 1)$ and $b = 1 + 1/n$ in part (b) to show that $\{a_n\}$ is increasing.

(d) Use $a = 1$ and $b = 1 + 1/(2n)$ in part (b) to show that $a_{2n} < 4$.

(e) Use parts (c) and (d) to show that $a_n < 4$ for all n.

(f) Use Theorem 11 to show that $\lim_{n\to\infty} (1 + 1/n)^n$ exists. (The limit is e. See Equation 3.8.6.)

71. Let a and b be positive numbers with $a > b$. Let a_1 be their arithmetic mean and b_1 their geometric mean:

$$a_1 = \frac{a + b}{2} \qquad b_1 = \sqrt{ab}$$

Repeat this process so that, in general,

$$a_{n+1} = \frac{a_n + b_n}{2} \qquad b_{n+1} = \sqrt{a_n b_n}$$

(a) Use mathematical induction to show that

$$a_n > a_{n+1} > b_{n+1} > b_n$$

(b) Deduce that both $\{a_n\}$ and $\{b_n\}$ are convergent.

(c) Show that $\lim_{n\to\infty} a_n = \lim_{n\to\infty} b_n$. Gauss called the common value of these limits the **arithmetic-geometric mean** of the numbers a and b.

72. (a) Show that if $\lim_{n\to\infty} a_{2n} = L$ and $\lim_{n\to\infty} a_{2n+1} = L$, then $\{a_n\}$ is convergent and $\lim_{n\to\infty} a_n = L$.

(b) If $a_1 = 1$ and

$$a_{n+1} = 1 + \frac{1}{1 + a_n}$$

find the first eight terms of the sequence $\{a_n\}$. Then use part (a) to show that $\lim_{n\to\infty} a_n = \sqrt{2}$. This gives the **continued fraction expansion**

$$\sqrt{2} = 1 + \cfrac{1}{2 + \cfrac{1}{2 + \cdots}}$$

73. The size of an undisturbed fish population has been modeled by the formula

$$p_{n+1} = \frac{bp_n}{a + p_n}$$

where p_n is the fish population after n years and a and b are positive constants that depend on the species and its environment. Suppose that the population in year 0 is $p_0 > 0$.

(a) Show that if $\{p_n\}$ is convergent, then the only possible values for its limit are 0 and $b - a$.

(b) Show that $p_{n+1} < (b/a)p_n$.

(c) Use part (b) to show that if $a > b$, then $\lim_{n\to\infty} p_n = 0$; in other words, the population dies out.

(d) Now assume that $a < b$. Show that if $p_0 < b - a$, then $\{p_n\}$ is increasing and $0 < p_n < b - a$. Show also that if $p_0 > b - a$, then $\{p_n\}$ is decreasing and $p_n > b - a$. Deduce that if $a < b$, then $\lim_{n\to\infty} p_n = b - a$.

LABORATORY PROJECT

CAS Logistic Sequences

A sequence that arises in ecology as a model for population growth is defined by the **logistic difference equation**

$$p_{n+1} = kp_n(1 - p_n)$$

where p_n measures the size of the population of the nth generation of a single species. To keep the numbers manageable, p_n is a fraction of the maximal size of the population, so $0 \leqslant p_n \leqslant 1$. Notice that the form of this equation is similar to the logistic differential equation in Section 9.5. The discrete model—with sequences instead of continuous functions—is preferable for modeling insect populations, where mating and death occur in a periodic fashion.

An ecologist is interested in predicting the size of the population as time goes on, and asks these questions: Will it stabilize at a limiting value? Will it change in a cyclical fashion? Or will it exhibit random behavior?

Write a program to compute the first n terms of this sequence starting with an initial population p_0, where $0 < p_0 < 1$. Use this program to do the following.

1. Calculate 20 or 30 terms of the sequence for $p_0 = \frac{1}{2}$ and for two values of k such that $1 < k < 3$. Graph the sequences. Do they appear to converge? Repeat for a different value of p_0 between 0 and 1. Does the limit depend on the choice of p_0? Does it depend on the choice of k?

2. Calculate terms of the sequence for a value of k between 3 and 3.4 and plot them. What do you notice about the behavior of the terms?

3. Experiment with values of k between 3.4 and 3.5. What happens to the terms?

4. For values of k between 3.6 and 4, compute and plot at least 100 terms and comment on the behavior of the sequence. What happens if you change p_0 by 0.001? This type of behavior is called *chaotic* and is exhibited by insect populations under certain conditions.

11.2 Series

If we try to add the terms of an infinite sequence $\{a_n\}_{n=1}^{\infty}$ we get an expression of the form

1 $$a_1 + a_2 + a_3 + \cdots + a_n + \cdots$$

which is called an **infinite series** (or just a **series**) and is denoted, for short, by the symbol

$$\sum_{n=1}^{\infty} a_n \qquad \text{or} \qquad \sum a_n$$

But does it make sense to talk about the sum of infinitely many terms?

It would be impossible to find a finite sum for the series

$$1 + 2 + 3 + 4 + 5 + \cdots + n + \cdots$$

because if we start adding the terms we get the cumulative sums 1, 3, 6, 10, 15, 21, . . . and, after the nth term, we get $n(n + 1)/2$, which becomes very large as n increases.

However, if we start to add the terms of the series

$$\frac{1}{2} + \frac{1}{4} + \frac{1}{8} + \frac{1}{16} + \frac{1}{32} + \frac{1}{64} + \cdots + \frac{1}{2^n} + \cdots$$

n	Sum of first n terms
1	0.50000000
2	0.75000000
3	0.87500000
4	0.93750000
5	0.96875000
6	0.98437500
7	0.99218750
10	0.99902344
15	0.99996948
20	0.99999905
25	0.99999997

we get $\frac{1}{2}, \frac{3}{4}, \frac{7}{8}, \frac{15}{16}, \frac{31}{32}, \frac{63}{64}, \ldots, 1 - 1/2^n, \ldots$. The table shows that as we add more and more terms, these *partial sums* become closer and closer to 1. (See also Figure 11 in *A Preview of Calculus,* page 7.) In fact, by adding sufficiently many terms of the series we can make the partial sums as close as we like to 1. So it seems reasonable to say that the sum of this infinite series is 1 and to write

$$\sum_{n=1}^{\infty} \frac{1}{2^n} = \frac{1}{2} + \frac{1}{4} + \frac{1}{8} + \frac{1}{16} + \cdots + \frac{1}{2^n} + \cdots = 1$$

We use a similar idea to determine whether or not a general series (1) has a sum. We consider the **partial sums**

$$s_1 = a_1$$

$$s_2 = a_1 + a_2$$

$$s_3 = a_1 + a_2 + a_3$$

$$s_4 = a_1 + a_2 + a_3 + a_4$$

and, in general,

$$s_n = a_1 + a_2 + a_3 + \cdots + a_n = \sum_{i=1}^{n} a_i$$

These partial sums form a new sequence $\{s_n\}$, which may or may not have a limit. If $\lim_{n\to\infty} s_n = s$ exists (as a finite number), then, as in the preceding example, we call it the sum of the infinite series $\sum a_n$.

2 Definition Given a series $\sum_{n=1}^{\infty} a_n = a_1 + a_2 + a_3 + \cdots$, let s_n denote its nth partial sum:

$$s_n = \sum_{i=1}^{n} a_i = a_1 + a_2 + \cdots + a_n$$

If the sequence $\{s_n\}$ is convergent and $\lim_{n\to\infty} s_n = s$ exists as a real number, then the series $\sum a_n$ is called **convergent** and we write

$$a_1 + a_2 + \cdots + a_n + \cdots = s \qquad \text{or} \qquad \sum_{n=1}^{\infty} a_n = s$$

The number s is called the **sum** of the series. Otherwise, the series is called **divergent**.

Thus, when we write $\sum_{n=1}^{\infty} a_n = s$ we mean that by adding sufficiently many terms of the series we can get as close as we like to the number s. Notice that

$$\sum_{n=1}^{\infty} a_n = \lim_{n\to\infty} \sum_{i=1}^{n} a_i$$

|||| Compare with the improper integral

$$\int_1^{\infty} f(x)\,dx = \lim_{t\to\infty} \int_1^t f(x)\,dx$$

To find this integral we integrate from 1 to t and then let $t \to \infty$. For a series, we sum from 1 to n and then let $n \to \infty$.

EXAMPLE 1 An important example of an infinite series is the **geometric series**

$$a + ar + ar^2 + ar^3 + \cdots + ar^{n-1} + \cdots = \sum_{n=1}^{\infty} ar^{n-1} \qquad a \neq 0$$

||| Figure 1 provides a geometric demonstration of the result in Example 1. If the triangles are constructed as shown and s is the sum of the series, then, by similar triangles,

$$\frac{s}{a} = \frac{a}{a - ar} \quad \text{so} \quad s = \frac{a}{1 - r}$$

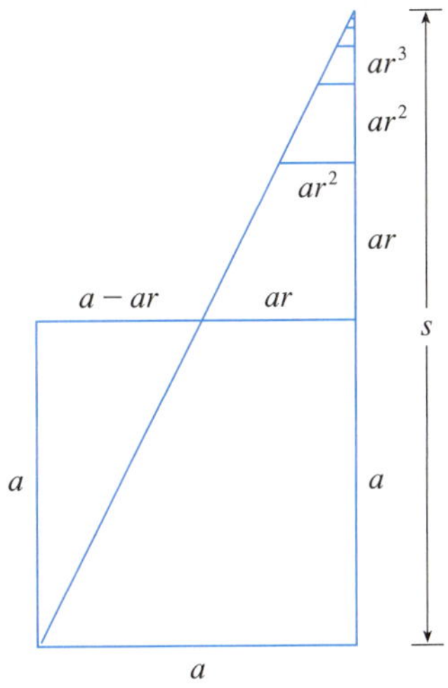

FIGURE 1

Each term is obtained from the preceding one by multiplying it by the common ratio r. (We have already considered the special case where $a = \frac{1}{2}$ and $r = \frac{1}{2}$ on page 713.)

If $r = 1$, then $s_n = a + a + \cdots + a = na \to \pm\infty$. Since $\lim_{n\to\infty} s_n$ doesn't exist, the geometric series diverges in this case.

If $r \neq 1$, we have

$$s_n = a + ar + ar^2 + \cdots + ar^{n-1}$$

and $$rs_n = \quad ar + ar^2 + \cdots + ar^{n-1} + ar^n$$

Subtracting these equations, we get

$$s_n - rs_n = a - ar^n$$

3 $$s_n = \frac{a(1 - r^n)}{1 - r}$$

If $-1 < r < 1$, we know from (11.1.8) that $r^n \to 0$ as $n \to \infty$, so

$$\lim_{n\to\infty} s_n = \lim_{n\to\infty} \frac{a(1 - r^n)}{1 - r} = \frac{a}{1 - r} - \frac{a}{1 - r} \lim_{n\to\infty} r^n = \frac{a}{1 - r}$$

Thus, when $|r| < 1$ the geometric series is convergent and its sum is $a/(1 - r)$.

If $r \leqslant -1$ or $r > 1$, the sequence $\{r^n\}$ is divergent by (11.1.8) and so, by Equation 3, $\lim_{n\to\infty} s_n$ does not exist. Therefore, the geometric series diverges in those cases.

We summarize the results of Example 1 as follows.

||| In words: The sum of a convergent geometric series is

$$\frac{\text{first term}}{1 - \text{common ratio}}$$

4 The geometric series

$$\sum_{n=1}^{\infty} ar^{n-1} = a + ar + ar^2 + \cdots$$

is convergent if $|r| < 1$ and its sum is

$$\sum_{n=1}^{\infty} ar^{n-1} = \frac{a}{1 - r} \qquad |r| < 1$$

If $|r| \geqslant 1$, the geometric series is divergent.

EXAMPLE 2 Find the sum of the geometric series

$$5 - \tfrac{10}{3} + \tfrac{20}{9} - \tfrac{40}{27} + \cdots$$

SOLUTION The first term is $a = 5$ and the common ratio is $r = -\frac{2}{3}$. Since $|r| = \frac{2}{3} < 1$, the series is convergent by (4) and its sum is

$$5 - \frac{10}{3} + \frac{20}{9} - \frac{40}{27} + \cdots = \frac{5}{1 - \left(-\frac{2}{3}\right)} = \frac{5}{\frac{5}{3}} = 3$$

What do we really mean when we say that the sum of the series in Example 2 is 3? Of course, we can't literally add an infinite number of terms, one by one. But, according to Definition 2, the total sum is the limit of the sequence of partial sums. So, by taking the sum of sufficiently many terms, we can get as close as we like to the number 3. The table shows the first ten partial sums s_n and the graph in Figure 2 shows how the sequence of partial sums approaches 3.

n	s_n
1	5.000000
2	1.666667
3	3.888889
4	2.407407
5	3.395062
6	2.736626
7	3.175583
8	2.882945
9	3.078037
10	2.947975

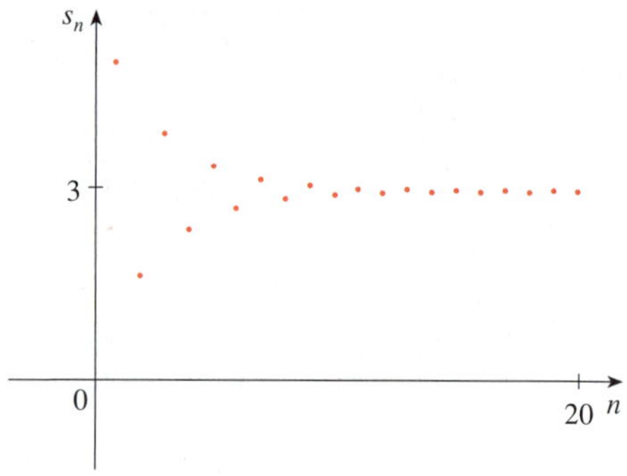

FIGURE 2

EXAMPLE 3 Is the series $\sum_{n=1}^{\infty} 2^{2n}3^{1-n}$ convergent or divergent?

SOLUTION Let's rewrite the nth term of the series in the form ar^{n-1}:

Another way to identify a and r is to write out the first few terms:

$$4 + \tfrac{16}{3} + \tfrac{64}{9} + \cdots$$

$$\sum_{n=1}^{\infty} 2^{2n}3^{1-n} = \sum_{n=1}^{\infty} \frac{4^n}{3^{n-1}} = \sum_{n=1}^{\infty} 4\left(\tfrac{4}{3}\right)^{n-1}$$

We recognize this series as a geometric series with $a = 4$ and $r = \frac{4}{3}$. Since $r > 1$, the series diverges by (4).

EXAMPLE 4 Write the number $2.3\overline{17} = 2.3171717\ldots$ as a ratio of integers.

SOLUTION

$$2.3171717\ldots = 2.3 + \frac{17}{10^3} + \frac{17}{10^5} + \frac{17}{10^7} + \cdots$$

After the first term we have a geometric series with $a = 17/10^3$ and $r = 1/10^2$. Therefore

$$2.3\overline{17} = 2.3 + \frac{\dfrac{17}{10^3}}{1 - \dfrac{1}{10^2}} = 2.3 + \frac{\dfrac{17}{1000}}{\dfrac{99}{100}}$$

$$= \frac{23}{10} + \frac{17}{990} = \frac{1147}{495}$$

EXAMPLE 5 Find the sum of the series $\sum_{n=0}^{\infty} x^n$, where $|x| < 1$.

SOLUTION Notice that this series starts with $n = 0$ and so the first term is $x^0 = 1$. (With series, we adopt the convention that $x^0 = 1$ even when $x = 0$.) Thus

TEC Module 11.2 explores a series that depends on an angle θ in a triangle and enables you to see how rapidly the series converges when θ varies.

$$\sum_{n=0}^{\infty} x^n = 1 + x + x^2 + x^3 + x^4 + \cdots$$

This is a geometric series with $a = 1$ and $r = x$. Since $|r| = |x| < 1$, it converges and (4) gives

5
$$\sum_{n=0}^{\infty} x^n = \frac{1}{1 - x}$$

EXAMPLE 6 Show that the series $\displaystyle\sum_{n=1}^{\infty} \frac{1}{n(n+1)}$ is convergent, and find its sum.

SOLUTION This is not a geometric series, so we go back to the definition of a convergent series and compute the partial sums.

$$s_n = \sum_{i=1}^{n} \frac{1}{i(i+1)} = \frac{1}{1 \cdot 2} + \frac{1}{2 \cdot 3} + \frac{1}{3 \cdot 4} + \cdots + \frac{1}{n(n+1)}$$

We can simplify this expression if we use the partial fraction decomposition

$$\frac{1}{i(i+1)} = \frac{1}{i} - \frac{1}{i+1}$$

(see Section 7.4). Thus, we have

$$\begin{aligned} s_n &= \sum_{i=1}^{n} \frac{1}{i(i+1)} = \sum_{i=1}^{n} \left(\frac{1}{i} - \frac{1}{i+1} \right) \\ &= \left(1 - \frac{1}{2}\right) + \left(\frac{1}{2} - \frac{1}{3}\right) + \left(\frac{1}{3} - \frac{1}{4}\right) + \cdots + \left(\frac{1}{n} - \frac{1}{n+1}\right) \\ &= 1 - \frac{1}{n+1} \end{aligned}$$

Notice that the terms cancel in pairs. This is an example of a **telescoping sum**: Because of all the cancellations, the sum collapses (like an old-fashioned collapsing telescope) into just two terms.

and so

$$\lim_{n\to\infty} s_n = \lim_{n\to\infty} \left(1 - \frac{1}{n+1}\right) = 1 - 0 = 1$$

Therefore, the given series is convergent and

$$\sum_{n=1}^{\infty} \frac{1}{n(n+1)} = 1$$

Figure 3 illustrates Example 6 by showing the graphs of the sequence of terms $a_n = 1/[n(n+1)]$ and the sequence $\{s_n\}$ of partial sums. Notice that $a_n \to 0$ and $s_n \to 1$. See Exercises 54 and 55 for two geometric interpretations of Example 6.

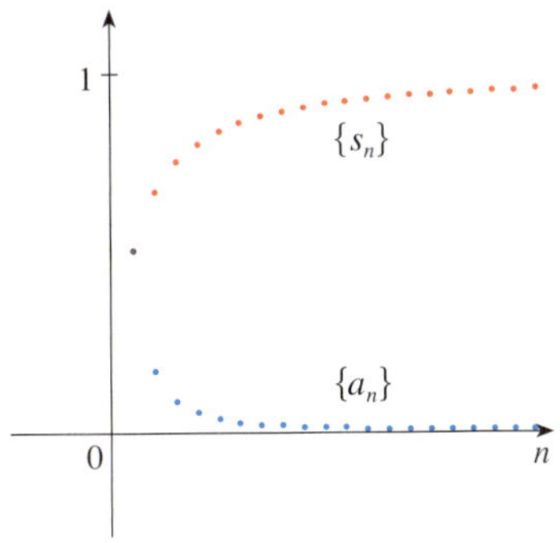

FIGURE 3

EXAMPLE 7 Show that the **harmonic series**

$$\sum_{n=1}^{\infty} \frac{1}{n} = 1 + \frac{1}{2} + \frac{1}{3} + \frac{1}{4} + \cdots$$

is divergent.

SOLUTION

$$s_1 = 1$$

$$s_2 = 1 + \tfrac{1}{2}$$

$$s_4 = 1 + \tfrac{1}{2} + \left(\tfrac{1}{3} + \tfrac{1}{4}\right) > 1 + \tfrac{1}{2} + \left(\tfrac{1}{4} + \tfrac{1}{4}\right) = 1 + \tfrac{2}{2}$$

$$\begin{aligned} s_8 &= 1 + \tfrac{1}{2} + \left(\tfrac{1}{3} + \tfrac{1}{4}\right) + \left(\tfrac{1}{5} + \tfrac{1}{6} + \tfrac{1}{7} + \tfrac{1}{8}\right) \\ &> 1 + \tfrac{1}{2} + \left(\tfrac{1}{4} + \tfrac{1}{4}\right) + \left(\tfrac{1}{8} + \tfrac{1}{8} + \tfrac{1}{8} + \tfrac{1}{8}\right) \\ &= 1 + \tfrac{1}{2} + \tfrac{1}{2} + \tfrac{1}{2} = 1 + \tfrac{3}{2} \end{aligned}$$

$$\begin{aligned} s_{16} &= 1 + \tfrac{1}{2} + \left(\tfrac{1}{3} + \tfrac{1}{4}\right) + \left(\tfrac{1}{5} + \cdots + \tfrac{1}{8}\right) + \left(\tfrac{1}{9} + \cdots + \tfrac{1}{16}\right) \\ &> 1 + \tfrac{1}{2} + \left(\tfrac{1}{4} + \tfrac{1}{4}\right) + \left(\tfrac{1}{8} + \cdots + \tfrac{1}{8}\right) + \left(\tfrac{1}{16} + \cdots + \tfrac{1}{16}\right) \\ &= 1 + \tfrac{1}{2} + \tfrac{1}{2} + \tfrac{1}{2} + \tfrac{1}{2} = 1 + \tfrac{4}{2} \end{aligned}$$

Similarly, $s_{32} > 1 + \frac{5}{2}$, $s_{64} > 1 + \frac{6}{2}$, and in general

$$s_{2^n} > 1 + \frac{n}{2}$$

The method used in Example 7 for showing that the harmonic series diverges is due to the French scholar Nicole Oresme (1323–1382).

This shows that $s_{2^n} \to \infty$ as $n \to \infty$ and so $\{s_n\}$ is divergent. Therefore, the harmonic series diverges.

6 Theorem If the series $\sum_{n=1}^{\infty} a_n$ is convergent, then $\lim_{n\to\infty} a_n = 0$.

Proof Let $s_n = a_1 + a_2 + \cdots + a_n$. Then $a_n = s_n - s_{n-1}$. Since $\Sigma\, a_n$ is convergent, the sequence $\{s_n\}$ is convergent. Let $\lim_{n\to\infty} s_n = s$. Since $n - 1 \to \infty$ as $n \to \infty$, we also have $\lim_{n\to\infty} s_{n-1} = s$. Therefore

$$\begin{aligned}\lim_{n\to\infty} a_n &= \lim_{n\to\infty} (s_n - s_{n-1}) = \lim_{n\to\infty} s_n - \lim_{n\to\infty} s_{n-1} \\ &= s - s = 0\end{aligned}$$

NOTE 1 With any *series* $\Sigma\, a_n$ we associate two *sequences:* the sequence $\{s_n\}$ of its partial sums and the sequence $\{a_n\}$ of its terms. If $\Sigma\, a_n$ is convergent, then the limit of the sequence $\{s_n\}$ is s (the sum of the series) and, as Theorem 6 asserts, the limit of the sequence $\{a_n\}$ is 0.

NOTE 2 The converse of Theorem 6 is not true in general. If $\lim_{n\to\infty} a_n = 0$, we cannot conclude that $\Sigma\, a_n$ is convergent. Observe that for the harmonic series $\Sigma\, 1/n$ we have $a_n = 1/n \to 0$ as $n \to \infty$, but we showed in Example 7 that $\Sigma\, 1/n$ is divergent.

7 The Test for Divergence If $\lim_{n\to\infty} a_n$ does not exist or if $\lim_{n\to\infty} a_n \neq 0$, then the series $\sum_{n=1}^{\infty} a_n$ is divergent.

The Test for Divergence follows from Theorem 6 because, if the series is not divergent, then it is convergent, and so $\lim_{n\to\infty} a_n = 0$.

EXAMPLE 8 Show that the series $\sum_{n=1}^{\infty} \frac{n^2}{5n^2 + 4}$ diverges.

SOLUTION

$$\lim_{n\to\infty} a_n = \lim_{n\to\infty} \frac{n^2}{5n^2 + 4} = \lim_{n\to\infty} \frac{1}{5 + 4/n^2} = \frac{1}{5} \neq 0$$

So the series diverges by the Test for Divergence.

NOTE 3 If we find that $\lim_{n\to\infty} a_n \neq 0$, we know that $\Sigma\, a_n$ is divergent. If we find that $\lim_{n\to\infty} a_n = 0$, we know *nothing* about the convergence or divergence of $\Sigma\, a_n$. Remember the warning in Note 2: If $\lim_{n\to\infty} a_n = 0$, the series $\Sigma\, a_n$ might converge or it might diverge.

8 Theorem If $\Sigma\, a_n$ and $\Sigma\, b_n$ are convergent series, then so are the series $\Sigma\, ca_n$ (where c is a constant), $\Sigma\,(a_n + b_n)$, and $\Sigma\,(a_n - b_n)$, and

(i) $\displaystyle\sum_{n=1}^{\infty} ca_n = c\sum_{n=1}^{\infty} a_n$ (ii) $\displaystyle\sum_{n=1}^{\infty} (a_n + b_n) = \sum_{n=1}^{\infty} a_n + \sum_{n=1}^{\infty} b_n$

(iii) $\displaystyle\sum_{n=1}^{\infty} (a_n - b_n) = \sum_{n=1}^{\infty} a_n - \sum_{n=1}^{\infty} b_n$

These properties of convergent series follow from the corresponding Limit Laws for Sequences in Section 11.1. For instance, here is how part (ii) of Theorem 8 is proved:

Let

$$s_n = \sum_{i=1}^{n} a_i \qquad s = \sum_{n=1}^{\infty} a_n \qquad t_n = \sum_{i=1}^{n} b_i \qquad t = \sum_{n=1}^{\infty} b_n$$

The nth partial sum for the series $\Sigma\,(a_n + b_n)$ is

$$u_n = \sum_{i=1}^{n} (a_i + b_i)$$

and, using Equation 5.2.9, we have

$$\begin{aligned} \lim_{n\to\infty} u_n &= \lim_{n\to\infty} \sum_{i=1}^{n} (a_i + b_i) = \lim_{n\to\infty} \left(\sum_{i=1}^{n} a_i + \sum_{i=1}^{n} b_i \right) \\ &= \lim_{n\to\infty} \sum_{i=1}^{n} a_i + \lim_{n\to\infty} \sum_{i=1}^{n} b_i \\ &= \lim_{n\to\infty} s_n + \lim_{n\to\infty} t_n = s + t \end{aligned}$$

Therefore, $\Sigma\,(a_n + b_n)$ is convergent and its sum is

$$\sum_{n=1}^{\infty} (a_n + b_n) = s + t = \sum_{n=1}^{\infty} a_n + \sum_{n=1}^{\infty} b_n$$

EXAMPLE 9 Find the sum of the series $\displaystyle\sum_{n=1}^{\infty} \left(\frac{3}{n(n+1)} + \frac{1}{2^n} \right)$.

SOLUTION The series $\Sigma\, 1/2^n$ is a geometric series with $a = \frac{1}{2}$ and $r = \frac{1}{2}$, so

$$\sum_{n=1}^{\infty} \frac{1}{2^n} = \frac{\frac{1}{2}}{1 - \frac{1}{2}} = 1$$

In Example 6 we found that

$$\sum_{n=1}^{\infty} \frac{1}{n(n+1)} = 1$$

So, by Theorem 8, the given series is convergent and

$$\begin{aligned} \sum_{n=1}^{\infty} \left(\frac{3}{n(n+1)} + \frac{1}{2^n} \right) &= 3\sum_{n=1}^{\infty} \frac{1}{n(n+1)} + \sum_{n=1}^{\infty} \frac{1}{2^n} \\ &= 3 \cdot 1 + 1 = 4 \end{aligned}$$

NOTE 4 A finite number of terms doesn't affect the convergence or divergence of a series. For instance, suppose that we were able to show that the series

$$\sum_{n=4}^{\infty} \frac{n}{n^3 + 1}$$

is convergent. Since

$$\sum_{n=1}^{\infty} \frac{n}{n^3 + 1} = \frac{1}{2} + \frac{2}{9} + \frac{3}{28} + \sum_{n=4}^{\infty} \frac{n}{n^3 + 1}$$

it follows that the entire series $\sum_{n=1}^{\infty} n/(n^3 + 1)$ is convergent. Similarly, if it is known that the series $\sum_{n=N+1}^{\infty} a_n$ converges, then the full series

$$\sum_{n=1}^{\infty} a_n = \sum_{n=1}^{N} a_n + \sum_{n=N+1}^{\infty} a_n$$

is also convergent.

11.2 Exercises

1. (a) What is the difference between a sequence and a series?
(b) What is a convergent series? What is a divergent series?

2. Explain what it means to say that $\sum_{n=1}^{\infty} a_n = 5$.

3–8 Find at least 10 partial sums of the series. Graph both the sequence of terms and the sequence of partial sums on the same screen. Does it appear that the series is convergent or divergent? If it is convergent, find the sum. If it is divergent, explain why.

3. $\displaystyle\sum_{n=1}^{\infty} \frac{12}{(-5)^n}$

4. $\displaystyle\sum_{n=1}^{\infty} \frac{2n^2 - 1}{n^2 + 1}$

5. $\displaystyle\sum_{n=1}^{\infty} \tan n$

6. $\displaystyle\sum_{n=1}^{\infty} (0.6)^{n-1}$

7. $\displaystyle\sum_{n=1}^{\infty} \left(\frac{1}{n^{1.5}} - \frac{1}{(n+1)^{1.5}}\right)$

8. $\displaystyle\sum_{n=2}^{\infty} \frac{1}{n(n-1)}$

9. Let $a_n = \dfrac{2n}{3n + 1}$.
(a) Determine whether $\{a_n\}$ is convergent.
(b) Determine whether $\sum_{n=1}^{\infty} a_n$ is convergent.

10. (a) Explain the difference between

$$\sum_{i=1}^{n} a_i \quad \text{and} \quad \sum_{j=1}^{n} a_j$$

(b) Explain the difference between

$$\sum_{i=1}^{n} a_i \quad \text{and} \quad \sum_{i=1}^{n} a_j$$

11–34 Determine whether the series is convergent or divergent. If it is convergent, find its sum.

11. $3 + 2 + \frac{4}{3} + \frac{8}{9} + \cdots$

12. $\frac{1}{8} - \frac{1}{4} + \frac{1}{2} - 1 + \cdots$

13. $-2 + \frac{5}{2} - \frac{25}{8} + \frac{125}{32} - \cdots$

14. $1 + 0.4 + 0.16 + 0.064 + \cdots$

15. $\displaystyle\sum_{n=1}^{\infty} 5\left(\tfrac{2}{3}\right)^{n-1}$

16. $\displaystyle\sum_{n=1}^{\infty} \frac{(-6)^{n-1}}{5^{n-1}}$

17. $\displaystyle\sum_{n=1}^{\infty} \frac{(-3)^{n-1}}{4^n}$

18. $\displaystyle\sum_{n=0}^{\infty} \frac{1}{(\sqrt{2})^n}$

19. $\displaystyle\sum_{n=0}^{\infty} \frac{\pi^n}{3^{n+1}}$

20. $\displaystyle\sum_{n=1}^{\infty} \frac{e^n}{3^{n-1}}$

21. $\displaystyle\sum_{n=1}^{\infty} \frac{n}{n + 5}$

22. $\displaystyle\sum_{n=1}^{\infty} \frac{3}{n}$

23. $\displaystyle\sum_{n=2}^{\infty} \frac{2}{n^2 - 1}$

24. $\displaystyle\sum_{n=1}^{\infty} \frac{(n+1)^2}{n(n+2)}$

25. $\displaystyle\sum_{k=2}^{\infty} \frac{k^2}{k^2 - 1}$

26. $\displaystyle\sum_{n=1}^{\infty} \frac{2}{n^2 + 4n + 3}$

27. $\displaystyle\sum_{n=1}^{\infty} \frac{3^n + 2^n}{6^n}$

28. $\displaystyle\sum_{n=1}^{\infty} [(0.8)^{n-1} - (0.3)^n]$

29. $\displaystyle\sum_{n=1}^{\infty} \sqrt[n]{2}$

30. $\displaystyle\sum_{n=1}^{\infty} \ln\left(\frac{n}{2n + 5}\right)$

31. $\displaystyle\sum_{n=1}^{\infty} \arctan n$

32. $\displaystyle\sum_{k=1}^{\infty} (\cos 1)^k$

33. $\displaystyle\sum_{n=1}^{\infty} \left(\frac{3}{n(n+3)} + \frac{5}{4^n}\right)$

34. $\displaystyle\sum_{n=1}^{\infty} \left(\frac{3}{5^n} + \frac{2}{n}\right)$

35–40 Express the number as a ratio of integers.

35. $0.\overline{2} = 0.2222\ldots$

36. $0.\overline{73} = 0.73737373\ldots$

37. $3.\overline{417} = 3.417417417\ldots$

38. $6.2\overline{54}$

39. $0.12\overline{3456}$

40. $5.\overline{6021}$

41–45 Find the values of x for which the series converges. Find the sum of the series for those values of x.

41. $\sum_{n=1}^{\infty} \frac{x^n}{3^n}$

42. $\sum_{n=1}^{\infty} (x-4)^n$

43. $\sum_{n=0}^{\infty} 4^n x^n$

44. $\sum_{n=0}^{\infty} \frac{(x+3)^n}{2^n}$

45. $\sum_{n=0}^{\infty} \frac{\cos^n x}{2^n}$

46. We have seen that the harmonic series is a divergent series whose terms approach 0. Show that

$$\sum_{n=1}^{\infty} \ln\left(1+\frac{1}{n}\right)$$

is another series with this property.

47–48 Use the partial fraction command on your CAS to find a convenient expression for the partial sum, and then use this expression to find the sum of the series. Check your answer by using the CAS to sum the series directly.

47. $\sum_{n=1}^{\infty} \frac{1}{(4n+1)(4n-3)}$

48. $\sum_{n=1}^{\infty} \frac{n^2+3n+1}{(n^2+n)^2}$

49. If the nth partial sum of a series $\sum_{n=1}^{\infty} a_n$ is

$$s_n = \frac{n-1}{n+1}$$

find a_n and $\sum_{n=1}^{\infty} a_n$.

50. If the nth partial sum of a series $\sum_{n=1}^{\infty} a_n$ is $s_n = 3 - n2^{-n}$, find a_n and $\sum_{n=1}^{\infty} a_n$.

51. When money is spent on goods and services, those that receive the money also spend some of it. The people receiving some of the twice-spent money will spend some of that, and so on. Economists call this chain reaction the *multiplier effect.* In a hypothetical isolated community, the local government begins the process by spending D dollars. Suppose that each recipient of spent money spends $100c\%$ and saves $100s\%$ of the money that he or she receives. The values c and s are called the *marginal propensity to consume* and the *marginal propensity to save* and, of course, $c + s = 1$.

(a) Let S_n be the total spending that has been generated after n transactions. Find an equation for S_n.

(b) Show that $\lim_{n\to\infty} S_n = kD$, where $k = 1/s$. The number k is called the *multiplier*. What is the multiplier if the marginal propensity to consume is 80%?

Note: The federal government uses this principle to justify deficit spending. Banks use this principle to justify lending a large percentage of the money that they receive in deposits.

52. A certain ball has the property that each time it falls from a height h onto a hard, level surface, it rebounds to a height rh, where $0 < r < 1$. Suppose that the ball is dropped from an initial height of H meters.

(a) Assuming that the ball continues to bounce indefinitely, find the total distance that it travels. (Use the fact that the ball falls $\frac{1}{2}gt^2$ meters in t seconds.)

(b) Calculate the total time that the ball travels.

(c) Suppose that each time the ball strikes the surface with velocity v it rebounds with velocity $-kv$, where $0 < k < 1$. How long will it take for the ball to come to rest?

53. What is the value of c if $\sum_{n=2}^{\infty} (1+c)^{-n} = 2$?

54. Graph the curves $y = x^n$, $0 \leqslant x \leqslant 1$, for $n = 0, 1, 2, 3, 4, \ldots$ on a common screen. By finding the areas between successive curves, give a geometric demonstration of the fact, shown in Example 6, that

$$\sum_{n=1}^{\infty} \frac{1}{n(n+1)} = 1$$

55. The figure shows two circles C and D of radius 1 that touch at P. T is a common tangent line; C_1 is the circle that touches C, D, and T; C_2 is the circle that touches C, D, and C_1; C_3 is the circle that touches C, D, and C_2. This procedure can be continued indefinitely and produces an infinite sequence of circles $\{C_n\}$. Find an expression for the diameter of C_n and thus provide another geometric demonstration of Example 6.

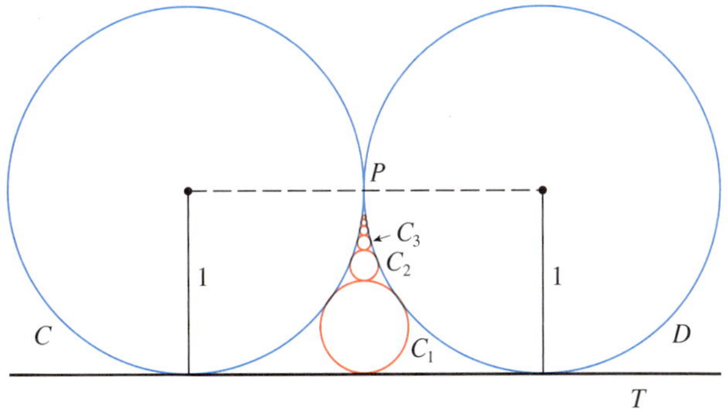

56. A right triangle ABC is given with $\angle A = \theta$ and $|AC| = b$. CD is drawn perpendicular to AB, DE is drawn perpendicular to BC, $EF \perp AB$, and this process is continued indefinitely as shown in the figure. Find the total length of all the perpendiculars

$$|CD| + |DE| + |EF| + |FG| + \cdots$$

in terms of b and θ.

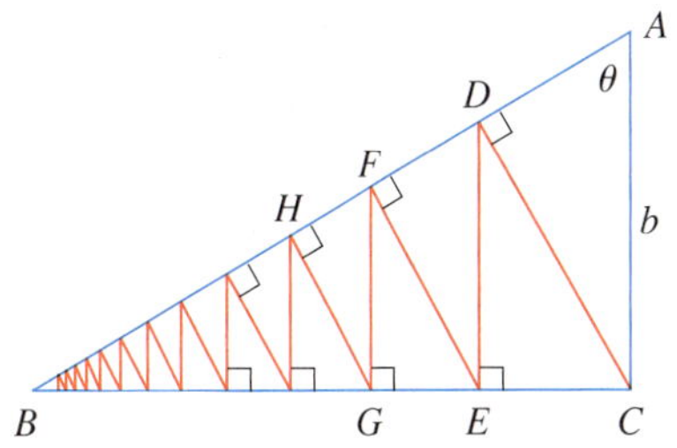

57. What is wrong with the following calculation?

$$\begin{aligned} 0 &= 0 + 0 + 0 + \cdots \\ &= (1 - 1) + (1 - 1) + (1 - 1) + \cdots \\ &= 1 - 1 + 1 - 1 + 1 - 1 + \cdots \\ &= 1 + (-1 + 1) + (-1 + 1) + (-1 + 1) + \cdots \\ &= 1 + 0 + 0 + 0 + \cdots = 1 \end{aligned}$$

(Guido Ubaldus thought that this proved the existence of God because "something has been created out of nothing.")

58. Suppose that $\sum_{n=1}^{\infty} a_n$ ($a_n \neq 0$) is known to be a convergent series. Prove that $\sum_{n=1}^{\infty} 1/a_n$ is a divergent series.

59. Prove part (i) of Theorem 8.

60. If $\sum a_n$ is divergent and $c \neq 0$, show that $\sum ca_n$ is divergent.

61. If $\sum a_n$ is convergent and $\sum b_n$ is divergent, show that the series $\sum (a_n + b_n)$ is divergent. [*Hint:* Argue by contradiction.]

62. If $\sum a_n$ and $\sum b_n$ are both divergent, is $\sum (a_n + b_n)$ necessarily divergent?

63. Suppose that a series $\sum a_n$ has positive terms and its partial sums s_n satisfy the inequality $s_n \leqslant 1000$ for all n. Explain why $\sum a_n$ must be convergent.

64. The Fibonacci sequence was defined in Section 11.1 by the equations

$$f_1 = 1, \quad f_2 = 1, \quad f_n = f_{n-1} + f_{n-2} \qquad n \geqslant 3$$

Show that each of the following statements is true.

(a) $\dfrac{1}{f_{n-1} f_{n+1}} = \dfrac{1}{f_{n-1} f_n} - \dfrac{1}{f_n f_{n+1}}$

(b) $\displaystyle\sum_{n=2}^{\infty} \frac{1}{f_{n-1} f_{n+1}} = 1$

(c) $\displaystyle\sum_{n=2}^{\infty} \frac{f_n}{f_{n-1} f_{n+1}} = 2$

65. The **Cantor set**, named after the German mathematician Georg Cantor (1845–1918), is constructed as follows. We start with the closed interval $[0, 1]$ and remove the open interval $(\frac{1}{3}, \frac{2}{3})$. That leaves the two intervals $[0, \frac{1}{3}]$ and $[\frac{2}{3}, 1]$ and we remove the open middle third of each. Four intervals remain and again we remove the open middle third of each of them. We continue this procedure indefinitely, at each step removing the open middle third of every interval that remains from the preceding step. The Cantor set consists of the numbers that remain in $[0, 1]$ after all those intervals have been removed.

(a) Show that the total length of all the intervals that are removed is 1. Despite that, the Cantor set contains infinitely many numbers. Give examples of some numbers in the Cantor set.

(b) The **Sierpinski carpet** is a two-dimensional counterpart of the Cantor set. It is constructed by removing the center one-ninth of a square of side 1, then removing the centers of the eight smaller remaining squares, and so on. (The figure shows the first three steps of the construction.) Show that the sum of the areas of the removed squares is 1. This implies that the Sierpinski carpet has area 0.

66. (a) A sequence $\{a_n\}$ is defined recursively by the equation $a_n = \frac{1}{2}(a_{n-1} + a_{n-2})$ for $n \geqslant 3$, where a_1 and a_2 can be any real numbers. Experiment with various values of a_1 and a_2 and use your calculator to guess the limit of the sequence.

(b) Find $\lim_{n \to \infty} a_n$ in terms of a_1 and a_2 by expressing $a_{n+1} - a_n$ in terms of $a_2 - a_1$ and summing a series.

67. Consider the series

$$\sum_{n=1}^{\infty} \frac{n}{(n + 1)!}$$

(a) Find the partial sums s_1, s_2, s_3, and s_4. Do you recognize the denominators? Use the pattern to guess a formula for s_n.

(b) Use mathematical induction to prove your guess.

(c) Show that the given infinite series is convergent, and find its sum.

68. In the figure there are infinitely many circles approaching the vertices of an equilateral triangle, each circle touching other circles and sides of the triangle. If the triangle has sides of length 1, find the total area occupied by the circles.

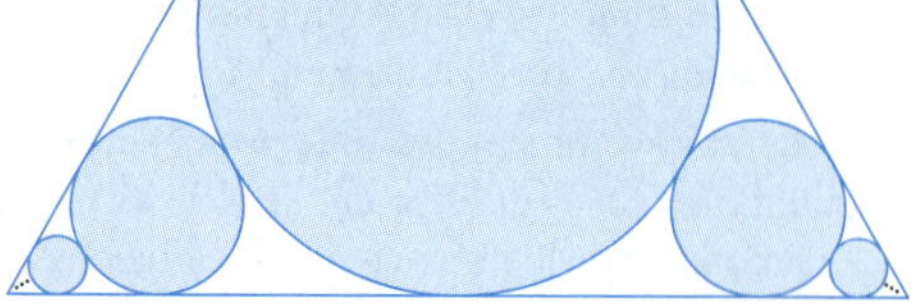

11.3 The Integral Test and Estimates of Sums

In general, it is difficult to find the exact sum of a series. We were able to accomplish this for geometric series and the series $\Sigma\ 1/[n(n+1)]$ because in each of those cases we could find a simple formula for the nth partial sum s_n. But usually it isn't easy to compute $\lim_{n\to\infty} s_n$. Therefore, in the next few sections we develop several tests that enable us to determine whether a series is convergent or divergent without explicitly finding its sum. (In some cases, however, our methods will enable us to find good estimates of the sum.) Our first test involves improper integrals.

We begin by investigating the series whose terms are the reciprocals of the squares of the positive integers:

n	$s_n = \sum_{i=1}^{n} \frac{1}{i^2}$
5	1.4636
10	1.5498
50	1.6251
100	1.6350
500	1.6429
1000	1.6439
5000	1.6447

$$\sum_{n=1}^{\infty} \frac{1}{n^2} = \frac{1}{1^2} + \frac{1}{2^2} + \frac{1}{3^2} + \frac{1}{4^2} + \frac{1}{5^2} + \cdots$$

There's no simple formula for the sum s_n of the first n terms, but the computer-generated table of values given in the margin suggests that the partial sums are approaching a number near 1.64 as $n \to \infty$ and so it looks as if the series is convergent.

We can confirm this impression with a geometric argument. Figure 1 shows the curve $y = 1/x^2$ and rectangles that lie below the curve. The base of each rectangle is an interval of length 1; the height is equal to the value of the function $y = 1/x^2$ at the right endpoint of the interval. So the sum of the areas of the rectangles is

$$\frac{1}{1^2} + \frac{1}{2^2} + \frac{1}{3^2} + \frac{1}{4^2} + \frac{1}{5^2} + \cdots = \sum_{n=1}^{\infty} \frac{1}{n^2}$$

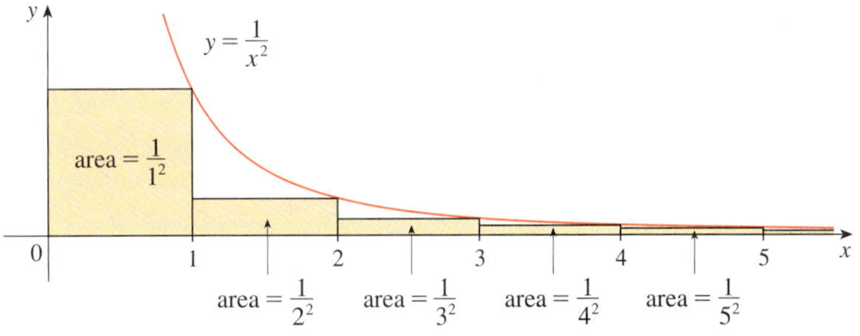

FIGURE 1

If we exclude the first rectangle, the total area of the remaining rectangles is smaller than the area under the curve $y = 1/x^2$ for $x \geqslant 1$, which is the value of the integral $\int_1^\infty (1/x^2)\,dx$. In Section 7.8 we discovered that this improper integral is convergent and has value 1. So the picture shows that all the partial sums are less than

$$\frac{1}{1^2} + \int_1^\infty \frac{1}{x^2}\,dx = 2$$

Thus, the partial sums are bounded. We also know that the partial sums are increasing (because all the terms are positive). Therefore, the partial sums converge (by the Monotonic Sequence Theorem) and so the series is convergent. The sum of the series (the limit of the partial sums) is also less than 2:

$$\sum_{n=1}^{\infty} \frac{1}{n^2} = \frac{1}{1^2} + \frac{1}{2^2} + \frac{1}{3^2} + \frac{1}{4^2} + \cdots < 2$$

[The exact sum of this series was found by the Swiss mathematician Leonhard Euler (1707–1783) to be $\pi^2/6$, but the proof of this fact is quite difficult. (See Problem 6 in the Problems Plus following Chapter 15.)]

n	$s_n = \sum_{i=1}^{n} \frac{1}{\sqrt{i}}$
5	3.2317
10	5.0210
50	12.7524
100	18.5896
500	43.2834
1000	61.8010
5000	139.9681

Now let's look at the series

$$\sum_{n=1}^{\infty} \frac{1}{\sqrt{n}} = \frac{1}{\sqrt{1}} + \frac{1}{\sqrt{2}} + \frac{1}{\sqrt{3}} + \frac{1}{\sqrt{4}} + \frac{1}{\sqrt{5}} + \cdots$$

The table of values of s_n suggests that the partial sums aren't approaching a finite number, so we suspect that the given series may be divergent. Again we use a picture for confirmation. Figure 2 shows the curve $y = 1/\sqrt{x}$, but this time we use rectangles whose tops lie *above* the curve.

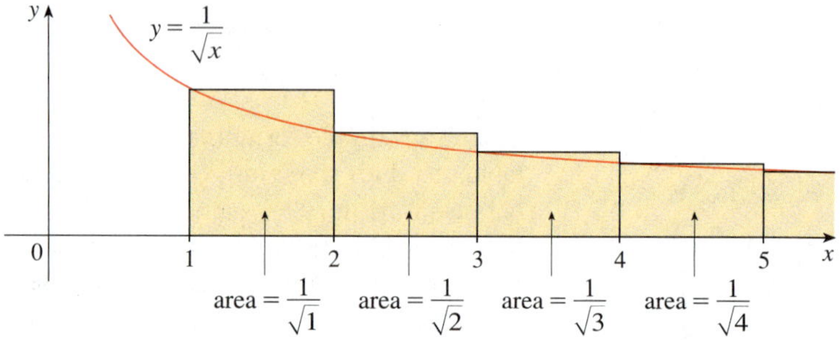

FIGURE 2

The base of each rectangle is an interval of length 1. The height is equal to the value of the function $y = 1/\sqrt{x}$ at the *left* endpoint of the interval. So the sum of the areas of all the rectangles is

$$\frac{1}{\sqrt{1}} + \frac{1}{\sqrt{2}} + \frac{1}{\sqrt{3}} + \frac{1}{\sqrt{4}} + \frac{1}{\sqrt{5}} + \cdots = \sum_{n=1}^{\infty} \frac{1}{\sqrt{n}}$$

This total area is greater than the area under the curve $y = 1/\sqrt{x}$ for $x \geqslant 1$, which is equal to the integral $\int_1^\infty (1/\sqrt{x})\,dx$. But we know from Section 7.8 that this improper integral is divergent. In other words, the area under the curve is infinite. So the sum of the series must be infinite; that is, the series is divergent.

The same sort of geometric reasoning that we used for these two series can be used to prove the following test. (The proof is given at the end of this section.)

The Integral Test Suppose f is a continuous, positive, decreasing function on $[1, \infty)$ and let $a_n = f(n)$. Then the series $\sum_{n=1}^{\infty} a_n$ is convergent if and only if the improper integral $\int_1^\infty f(x)\,dx$ is convergent. In other words:

(i) If $\displaystyle\int_1^\infty f(x)\,dx$ is convergent, then $\displaystyle\sum_{n=1}^{\infty} a_n$ is convergent.

(ii) If $\displaystyle\int_1^\infty f(x)\,dx$ is divergent, then $\displaystyle\sum_{n=1}^{\infty} a_n$ is divergent.

NOTE ▫ When we use the Integral Test it is not necessary to start the series or the integral at $n = 1$. For instance, in testing the series

$$\sum_{n=4}^{\infty} \frac{1}{(n-3)^2} \qquad \text{we use} \qquad \int_4^\infty \frac{1}{(x-3)^2}\,dx$$

Also, it is not necessary that f be always decreasing. What is important is that f be *ultimately* decreasing, that is, decreasing for x larger than some number N. Then $\sum_{n=N}^{\infty} a_n$ is convergent, so $\sum_{n=1}^{\infty} a_n$ is convergent by Note 4 of Section 11.2.

EXAMPLE 1 Test the series $\displaystyle\sum_{n=1}^{\infty} \frac{1}{n^2+1}$ for convergence or divergence.

SOLUTION The function $f(x) = 1/(x^2+1)$ is continuous, positive, and decreasing on $[1, \infty)$ so we use the Integral Test:

$$\int_1^{\infty} \frac{1}{x^2+1}\,dx = \lim_{t\to\infty} \int_1^t \frac{1}{x^2+1}\,dx = \lim_{t\to\infty} \tan^{-1}x\Big]_1^t$$

$$= \lim_{t\to\infty}\left(\tan^{-1}t - \frac{\pi}{4}\right) = \frac{\pi}{2} - \frac{\pi}{4} = \frac{\pi}{4}$$

Thus, $\int_1^{\infty} 1/(x^2+1)\,dx$ is a convergent integral and so, by the Integral Test, the series $\sum 1/(n^2+1)$ is convergent.

EXAMPLE 2 For what values of p is the series $\displaystyle\sum_{n=1}^{\infty} \frac{1}{n^p}$ convergent?

SOLUTION If $p < 0$, then $\lim_{n\to\infty} (1/n^p) = \infty$. If $p = 0$, then $\lim_{n\to\infty} (1/n^p) = 1$. In either case $\lim_{n\to\infty} (1/n^p) \neq 0$, so the given series diverges by the Test for Divergence (11.2.7).

If $p > 0$, then the function $f(x) = 1/x^p$ is clearly continuous, positive, and decreasing on $[1, \infty)$. We found in Chapter 7 [see (7.8.2)] that

$$\int_1^{\infty} \frac{1}{x^p}\,dx \text{ converges if } p > 1 \text{ and diverges if } p \leqslant 1$$

It follows from the Integral Test that the series $\sum 1/n^p$ converges if $p > 1$ and diverges if $0 < p \leqslant 1$. (For $p = 1$, this series is the harmonic series discussed in Example 7 in Section 11.2.)

The series in Example 2 is called the ***p*-series**. It is important in the rest of this chapter, so we summarize the results of Example 2 for future reference as follows.

1 The p-series $\displaystyle\sum_{n=1}^{\infty} \frac{1}{n^p}$ is convergent if $p > 1$ and divergent if $p \leqslant 1$.

EXAMPLE 3

(a) The series

$$\sum_{n=1}^{\infty} \frac{1}{n^3} = \frac{1}{1^3} + \frac{1}{2^3} + \frac{1}{3^3} + \frac{1}{4^3} + \cdots$$

is convergent because it is a p-series with $p = 3 > 1$.

(b) The series

$$\sum_{n=1}^{\infty} \frac{1}{n^{1/3}} = \sum_{n=1}^{\infty} \frac{1}{\sqrt[3]{n}} = 1 + \frac{1}{\sqrt[3]{2}} + \frac{1}{\sqrt[3]{3}} + \frac{1}{\sqrt[3]{4}} + \cdots$$

is divergent because it is a p-series with $p = \frac{1}{3} < 1$.

NOTE We should *not* infer from the Integral Test that the sum of the series is equal to the value of the integral. In fact,

$$\sum_{n=1}^{\infty} \frac{1}{n^2} = \frac{\pi^2}{6} \qquad \text{whereas} \qquad \int_1^{\infty} \frac{1}{x^2}\,dx = 1$$

Therefore, in general,

$$\sum_{n=1}^{\infty} a_n \neq \int_1^{\infty} f(x)\,dx$$

EXAMPLE 4 Determine whether the series $\sum_{n=1}^{\infty} \frac{\ln n}{n}$ converges or diverges.

SOLUTION The function $f(x) = (\ln x)/x$ is positive and continuous for $x > 1$ because the logarithm function is continuous. But it is not obvious whether or not f is decreasing, so we compute its derivative:

$$f'(x) = \frac{(1/x)x - \ln x}{x^2} = \frac{1 - \ln x}{x^2}$$

Thus, $f'(x) < 0$ when $\ln x > 1$, that is, $x > e$. It follows that f is decreasing when $x > e$ and so we can apply the Integral Test:

$$\int_1^{\infty} \frac{\ln x}{x}\,dx = \lim_{t\to\infty} \int_1^t \frac{\ln x}{x}\,dx = \lim_{t\to\infty} \frac{(\ln x)^2}{2}\bigg]_1^t$$

$$= \lim_{t\to\infty} \frac{(\ln t)^2}{2} = \infty$$

Since this improper integral is divergent, the series $\Sigma\, (\ln n)/n$ is also divergent by the Integral Test.

Estimating the Sum of a Series

Suppose we have been able to use the Integral Test to show that a series $\Sigma\, a_n$ is convergent and we now want to find an approximation to the sum s of the series. Of course, any partial sum s_n is an approximation to s because $\lim_{n\to\infty} s_n = s$. But how good is such an approximation? To find out, we need to estimate the size of the **remainder**

$$R_n = s - s_n = a_{n+1} + a_{n+2} + a_{n+3} + \cdots$$

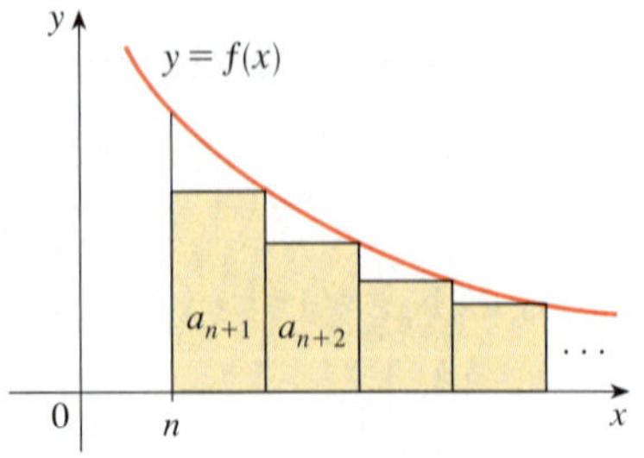

FIGURE 3

The remainder R_n is the error made when s_n, the sum of the first n terms, is used as an approximation to the total sum.

We use the same notation and ideas as in the Integral Test, assuming that f is decreasing on $[n, \infty)$. Comparing the areas of the rectangles with the area under $y = f(x)$ for $x > n$ in Figure 3, we see that

$$R_n = a_{n+1} + a_{n+2} + \cdots \leqslant \int_n^{\infty} f(x)\,dx$$

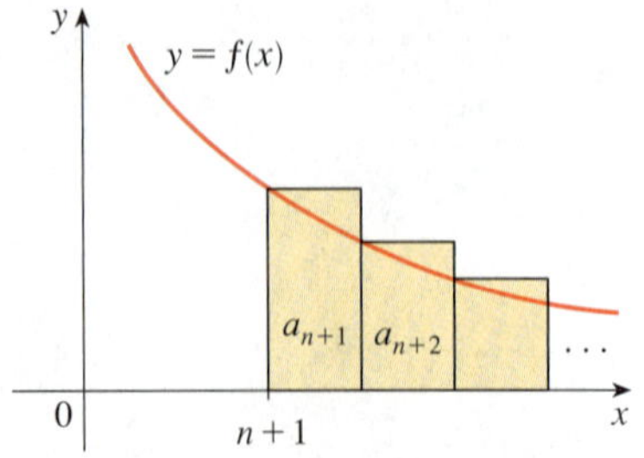

FIGURE 4

Similarly, we see from Figure 4 that

$$R_n = a_{n+1} + a_{n+2} + \cdots \geqslant \int_{n+1}^{\infty} f(x)\,dx$$

So we have proved the following error estimate.

2 Remainder Estimate for the Integral Test Suppose $f(k) = a_k$, where f is a continuous, positive, decreasing function for $x \geqslant n$ and $\Sigma\, a_n$ is convergent. If $R_n = s - s_n$, then

$$\int_{n+1}^{\infty} f(x)\,dx \leqslant R_n \leqslant \int_{n}^{\infty} f(x)\,dx$$

EXAMPLE 5

(a) Approximate the sum of the series $\Sigma\, 1/n^3$ by using the sum of the first 10 terms. Estimate the error involved in this approximation.

(b) How many terms are required to ensure that the sum is accurate to within 0.0005?

SOLUTION In both parts (a) and (b) we need to know $\int_n^\infty f(x)\,dx$. With $f(x) = 1/x^3$, we have

$$\int_n^\infty \frac{1}{x^3}\,dx = \lim_{t\to\infty}\left[-\frac{1}{2x^2}\right]_n^t = \lim_{t\to\infty}\left(-\frac{1}{2t^2} + \frac{1}{2n^2}\right) = \frac{1}{2n^2}$$

(a)

$$\sum_{n=1}^{\infty} \frac{1}{n^3} \approx s_{10} = \frac{1}{1^3} + \frac{1}{2^3} + \frac{1}{3^3} + \cdots + \frac{1}{10^3} \approx 1.1975$$

According to the remainder estimate in (2), we have

$$R_{10} \leqslant \int_{10}^\infty \frac{1}{x^3}\,dx = \frac{1}{2(10)^2} = \frac{1}{200}$$

So the size of the error is at most 0.005.

(b) Accuracy to within 0.0005 means that we have to find a value of n such that $R_n \leqslant 0.0005$. Since

$$R_n \leqslant \int_n^\infty \frac{1}{x^3}\,dx = \frac{1}{2n^2}$$

we want

$$\frac{1}{2n^2} < 0.0005$$

Solving this inequality, we get

$$n^2 > \frac{1}{0.001} = 1000 \qquad \text{or} \qquad n > \sqrt{1000} \approx 31.6$$

We need 32 terms to ensure accuracy to within 0.0005.

If we add s_n to each side of the inequalities in (2), we get

3
$$s_n + \int_{n+1}^\infty f(x)\,dx \leqslant s \leqslant s_n + \int_n^\infty f(x)\,dx$$

because $s_n + R_n = s$. The inequalities in (3) give a lower bound and an upper bound for s. They provide a more accurate approximation to the sum of the series than the partial sum s_n does.

EXAMPLE 6 Use (3) with $n = 10$ to estimate the sum of the series $\sum_{n=1}^{\infty} \frac{1}{n^3}$.

SOLUTION The inequalities in (3) become

$$s_{10} + \int_{11}^{\infty} \frac{1}{x^3}\,dx \leqslant s \leqslant s_{10} + \int_{10}^{\infty} \frac{1}{x^3}\,dx$$

From Example 5 we know that

$$\int_{n}^{\infty} \frac{1}{x^3}\,dx = \frac{1}{2n^2}$$

so

$$s_{10} + \frac{1}{2(11)^2} \leqslant s \leqslant s_{10} + \frac{1}{2(10)^2}$$

Using $s_{10} \approx 1.197532$, we get

$$1.201664 \leqslant s \leqslant 1.202532$$

If we approximate s by the midpoint of this interval, then the error is at most half the length of the interval. So

$$\sum_{n=1}^{\infty} \frac{1}{n^3} \approx 1.2021 \qquad \text{with error} < 0.0005$$

If we compare Example 6 with Example 5, we see that the improved estimate in (3) can be much better than the estimate $s \approx s_n$. To make the error smaller than 0.0005 we had to use 32 terms in Example 5 but only 10 terms in Example 6.

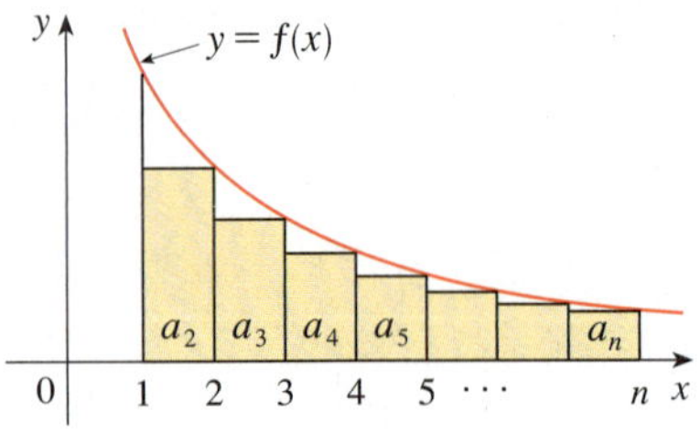

FIGURE 5

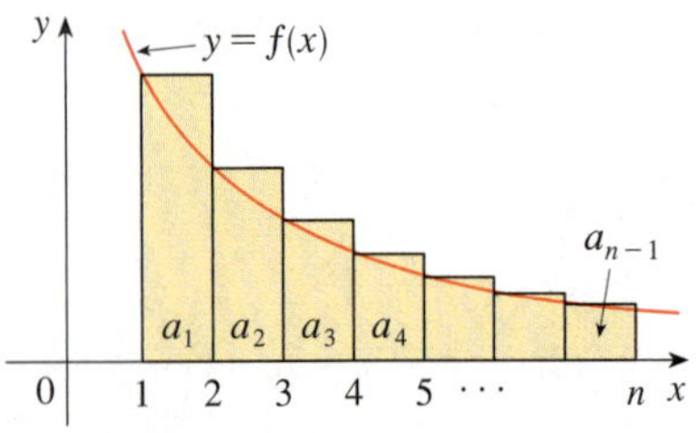

FIGURE 6

Proof of the Integral Test

We have already seen the basic idea behind the proof of the Integral Test in Figures 1 and 2 for the series $\Sigma\, 1/n^2$ and $\Sigma\, 1/\sqrt{n}$. For the general series $\Sigma\, a_n$ look at Figures 5 and 6. The area of the first shaded rectangle in Figure 5 is the value of f at the right endpoint of $[1, 2]$, that is, $f(2) = a_2$. So, comparing the areas of the shaded rectangles with the area under $y = f(x)$ from 1 to n, we see that

$$\boxed{4} \qquad a_2 + a_3 + \cdots + a_n \leqslant \int_{1}^{n} f(x)\,dx$$

(Notice that this inequality depends on the fact that f is decreasing.) Likewise, Figure 6 shows that

$$\boxed{5} \qquad \int_{1}^{n} f(x)\,dx \leqslant a_1 + a_2 + \cdots + a_{n-1}$$

(i) If $\int_1^\infty f(x)\,dx$ is convergent, then (4) gives

$$\sum_{i=2}^{n} a_i \leqslant \int_1^n f(x)\,dx \leqslant \int_1^\infty f(x)\,dx$$

since $f(x) \geqslant 0$. Therefore

$$s_n = a_1 + \sum_{i=2}^{n} a_i \leqslant a_1 + \int_1^\infty f(x)\,dx = M, \text{ say}$$

Since $s_n \leqslant M$ for all n, the sequence $\{s_n\}$ is bounded above. Also

$$s_{n+1} = s_n + a_{n+1} \geqslant s_n$$

since $a_{n+1} = f(n+1) \geqslant 0$. Thus, $\{s_n\}$ is an increasing bounded sequence and so it is convergent by the Monotonic Sequence Theorem (11.1.11). This means that $\sum a_n$ is convergent.

(ii) If $\int_1^\infty f(x)\,dx$ is divergent, then $\int_1^n f(x)\,dx \to \infty$ as $n \to \infty$ because $f(x) \geqslant 0$. But (5) gives

$$\int_1^n f(x)\,dx \leqslant \sum_{i=1}^{n-1} a_i = s_{n-1}$$

and so $s_{n-1} \to \infty$. This implies that $s_n \to \infty$ and so $\sum a_n$ diverges.

11.3 Exercises

1. Draw a picture to show that

$$\sum_{n=2}^{\infty} \frac{1}{n^{1.3}} < \int_1^\infty \frac{1}{x^{1.3}}\,dx$$

What can you conclude about the series?

2. Suppose f is a continuous positive decreasing function for $x \geqslant 1$ and $a_n = f(n)$. By drawing a picture, rank the following three quantities in increasing order:

$$\int_1^6 f(x)\,dx \qquad \sum_{i=1}^{5} a_i \qquad \sum_{i=2}^{6} a_i$$

3–8 Use the Integral Test to determine whether the series is convergent or divergent.

3. $\displaystyle\sum_{n=1}^{\infty} \frac{1}{n^4}$ **4.** $\displaystyle\sum_{n=1}^{\infty} \frac{1}{\sqrt[4]{n}}$ **5.** $\displaystyle\sum_{n=1}^{\infty} \frac{1}{3n+1}$

6. $\displaystyle\sum_{n=1}^{\infty} e^{-n}$ **7.** $\displaystyle\sum_{n=1}^{\infty} ne^{-n}$ **8.** $\displaystyle\sum_{n=1}^{\infty} \frac{n+2}{n+1}$

9–24 Determine whether the series is convergent or divergent.

9. $\displaystyle\sum_{n=1}^{\infty} \frac{2}{n^{0.85}}$ **10.** $\displaystyle\sum_{n=1}^{\infty} (n^{-1.4} + 3n^{-1.2})$

11. $1 + \frac{1}{8} + \frac{1}{27} + \frac{1}{64} + \frac{1}{125} + \cdots$

12. $1 + \frac{1}{2\sqrt{2}} + \frac{1}{3\sqrt{3}} + \frac{1}{4\sqrt{4}} + \frac{1}{5\sqrt{5}} + \cdots$

13. $\displaystyle\sum_{n=1}^{\infty} \frac{5 - 2\sqrt{n}}{n^3}$ **14.** $\displaystyle\sum_{n=3}^{\infty} \frac{5}{n-2}$

15. $\displaystyle\sum_{n=1}^{\infty} \frac{1}{n^2+4}$ **16.** $\displaystyle\sum_{n=1}^{\infty} \frac{3n+2}{n(n+1)}$

17. $\displaystyle\sum_{n=1}^{\infty} \frac{n}{n^2+1}$ **18.** $\displaystyle\sum_{n=1}^{\infty} \frac{1}{n^2-4n+5}$

19. $\displaystyle\sum_{n=1}^{\infty} ne^{-n^2}$ **20.** $\displaystyle\sum_{n=1}^{\infty} \frac{\ln n}{n^2}$

21. $\displaystyle\sum_{n=2}^{\infty} \frac{1}{n \ln n}$ **22.** $\displaystyle\sum_{n=1}^{\infty} \frac{n}{n^4+1}$

23. $\displaystyle\sum_{n=1}^{\infty} \frac{1}{n^3+n}$ **24.** $\displaystyle\sum_{n=3}^{\infty} \frac{1}{n \ln n \ln(\ln n)}$

25–28 Find the values of p for which the series is convergent.

25. $\displaystyle\sum_{n=2}^{\infty} \frac{1}{n(\ln n)^p}$ **26.** $\displaystyle\sum_{n=3}^{\infty} \frac{1}{n \ln n\,[\ln(\ln n)]^p}$

27. $\sum_{n=1}^{\infty} n(1 + n^2)^p$

28. $\sum_{n=1}^{\infty} \frac{\ln n}{n^p}$

29. The Riemann zeta-function ζ is defined by

$$\zeta(x) = \sum_{n=1}^{\infty} \frac{1}{n^x}$$

and is used in number theory to study the distribution of prime numbers. What is the domain of ζ?

30. (a) Find the partial sum s_{10} of the series $\sum_{n=1}^{\infty} 1/n^4$. Estimate the error in using s_{10} as an approximation to the sum of the series.
(b) Use (3) with $n = 10$ to give an improved estimate of the sum.
(c) Find a value of n so that s_n is within 0.00001 of the sum.

31. (a) Use the sum of the first 10 terms to estimate the sum of the series $\sum_{n=1}^{\infty} 1/n^2$. How good is this estimate?
(b) Improve this estimate using (3) with $n = 10$.
(c) Find a value of n that will ensure that the error in the approximation $s \approx s_n$ is less than 0.001.

32. Find the sum of the series $\sum_{n=1}^{\infty} 1/n^5$ correct to three decimal places.

33. Estimate $\sum_{n=1}^{\infty} n^{-3/2}$ to within 0.01.

34. How many terms of the series $\sum_{n=2}^{\infty} 1/[n(\ln n)^2]$ would you need to add to find its sum to within 0.01?

35. Show that if we want to approximate the sum of the series $\sum_{n=1}^{\infty} n^{-1.001}$ so that the error is less than 5 in the ninth decimal place, then we need to add more than $10^{11,301}$ terms!

CAS 36. (a) Show that the series $\sum_{n=1}^{\infty} (\ln n)^2/n^2$ is convergent.
(b) Find an upper bound for the error in the approximation $s \approx s_n$.
(c) What is the smallest value of n such that this upper bound is less than 0.05?
(d) Find s_n for this value of n.

37. (a) Use (4) to show that if s_n is the nth partial sum of the harmonic series, then

$$s_n \leqslant 1 + \ln n$$

(b) The harmonic series diverges, but very slowly. Use part (a) to show that the sum of the first million terms is less than 15 and the sum of the first billion terms is less than 22.

38. Use the following steps to show that the sequence

$$t_n = 1 + \frac{1}{2} + \frac{1}{3} + \cdots + \frac{1}{n} - \ln n$$

has a limit. (The value of the limit is denoted by γ and is called Euler's constant.)
(a) Draw a picture like Figure 6 with $f(x) = 1/x$ and interpret t_n as an area [or use (5)] to show that $t_n > 0$ for all n.
(b) Interpret

$$t_n - t_{n+1} = [\ln(n + 1) - \ln n] - \frac{1}{n + 1}$$

as a difference of areas to show that $t_n - t_{n+1} > 0$. Therefore, $\{t_n\}$ is a decreasing sequence.
(c) Use the Monotonic Sequence Theorem to show that $\{t_n\}$ is convergent.

39. Find all positive values of b for which the series $\sum_{n=1}^{\infty} b^{\ln n}$ converges.

11.4 The Comparison Tests

In the comparison tests the idea is to compare a given series with a series that is known to be convergent or divergent. For instance, the series

$$\boxed{1} \qquad \sum_{n=1}^{\infty} \frac{1}{2^n + 1}$$

reminds us of the series $\sum_{n=1}^{\infty} 1/2^n$, which is a geometric series with $a = \frac{1}{2}$ and $r = \frac{1}{2}$ and is therefore convergent. Because the series (1) is so similar to a convergent series, we have the feeling that it too must be convergent. Indeed, it is. The inequality

$$\frac{1}{2^n + 1} < \frac{1}{2^n}$$

shows that our given series (1) has smaller terms than those of the geometric series and therefore all its partial sums are also smaller than 1 (the sum of the geometric series). This means that its partial sums form a bounded increasing sequence, which is convergent. It also follows that the sum of the series is less than the sum of the geometric series:

$$\sum_{n=1}^{\infty} \frac{1}{2^n + 1} < 1$$

Similar reasoning can be used to prove the following test, which applies only to series whose terms are positive. The first part says that if we have a series whose terms are *smaller* than those of a known *convergent* series, then our series is also convergent. The second part says that if we start with a series whose terms are *larger* than those of a known *divergent* series, then it too is divergent.

The Comparison Test Suppose that $\Sigma\, a_n$ and $\Sigma\, b_n$ are series with positive terms.

(i) If $\Sigma\, b_n$ is convergent and $a_n \leqslant b_n$ for all n, then $\Sigma\, a_n$ is also convergent.

(ii) If $\Sigma\, b_n$ is divergent and $a_n \geqslant b_n$ for all n, then $\Sigma\, a_n$ is also divergent.

|||| It is important to keep in mind the distinction between a sequence and a series. A sequence is a list of numbers, whereas a series is a sum. With every series $\Sigma\, a_n$ there are associated two sequences: the sequence $\{a_n\}$ of terms and the sequence $\{s_n\}$ of partial sums.

Proof

(i) Let

$$s_n = \sum_{i=1}^{n} a_i \qquad t_n = \sum_{i=1}^{n} b_i \qquad t = \sum_{n=1}^{\infty} b_n$$

Since both series have positive terms, the sequences $\{s_n\}$ and $\{t_n\}$ are increasing ($s_{n+1} = s_n + a_{n+1} \geqslant s_n$). Also $t_n \to t$, so $t_n \leqslant t$ for all n. Since $a_i \leqslant b_i$, we have $s_n \leqslant t_n$. Thus, $s_n \leqslant t$ for all n. This means that $\{s_n\}$ is increasing and bounded above and therefore converges by the Monotonic Sequence Theorem. Thus, $\Sigma\, a_n$ converges.

(ii) If $\Sigma\, b_n$ is divergent, then $t_n \to \infty$ (since $\{t_n\}$ is increasing). But $a_i \geqslant b_i$ so $s_n \geqslant t_n$. Thus, $s_n \to \infty$. Therefore, $\Sigma\, a_n$ diverges.

Standard Series for Use with the Comparison Test

In using the Comparison Test we must, of course, have some known series $\Sigma\, b_n$ for the purpose of comparison. Most of the time we use either a p-series [$\Sigma\, 1/n^p$ converges if $p > 1$ and diverges if $p \leqslant 1$; see (11.3.1)] or a geometric series [$\Sigma\, ar^{n-1}$ converges if $|r| < 1$ and diverges if $|r| \geqslant 1$; see (11.2.4)].

EXAMPLE 1 Determine whether the series $\displaystyle\sum_{n=1}^{\infty} \frac{5}{2n^2 + 4n + 3}$ converges or diverges.

SOLUTION For large n the dominant term in the denominator is $2n^2$ so we compare the given series with the series $\Sigma\, 5/(2n^2)$. Observe that

$$\frac{5}{2n^2 + 4n + 3} < \frac{5}{2n^2}$$

because the left side has a bigger denominator. (In the notation of the Comparison Test, a_n is the left side and b_n is the right side.) We know that

$$\sum_{n=1}^{\infty} \frac{5}{2n^2} = \frac{5}{2} \sum_{n=1}^{\infty} \frac{1}{n^2}$$

is convergent because it's a constant times a p-series with $p = 2 > 1$. Therefore

$$\sum_{n=1}^{\infty} \frac{5}{2n^2 + 4n + 3}$$

is convergent by part (i) of the Comparison Test.

NOTE 1 ▫ Although the condition $a_n \leqslant b_n$ or $a_n \geqslant b_n$ in the Comparison Test is given for all n, we need verify only that it holds for $n \geqslant N$, where N is some fixed integer, because the convergence of a series is not affected by a finite number of terms. This is illustrated in the next example.

EXAMPLE 2 Test the series $\displaystyle\sum_{n=1}^{\infty} \frac{\ln n}{n}$ for convergence or divergence.

SOLUTION This series was tested (using the Integral Test) in Example 4 in Section 11.3, but it is also possible to test it by comparing it with the harmonic series. Observe that $\ln n > 1$ for $n \geqslant 3$ and so

$$\frac{\ln n}{n} > \frac{1}{n} \qquad n \geqslant 3$$

We know that $\Sigma\ 1/n$ is divergent (p-series with $p = 1$). Thus, the given series is divergent by the Comparison Test.

NOTE 2 ▫ The terms of the series being tested must be smaller than those of a convergent series or larger than those of a divergent series. If the terms are larger than the terms of a convergent series or smaller than those of a divergent series, then the Comparison Test doesn't apply. Consider, for instance, the series

$$\sum_{n=1}^{\infty} \frac{1}{2^n - 1}$$

The inequality

$$\frac{1}{2^n - 1} > \frac{1}{2^n}$$

is useless as far as the Comparison Test is concerned because $\Sigma\ b_n = \Sigma\ \left(\frac{1}{2}\right)^n$ is convergent and $a_n > b_n$. Nonetheless, we have the feeling that $\Sigma\ 1/(2^n - 1)$ ought to be convergent because it is very similar to the convergent geometric series $\Sigma\ \left(\frac{1}{2}\right)^n$. In such cases the following test can be used.

|||| Exercises 40 and 41 deal with the cases $c = 0$ and $c = \infty$.

The Limit Comparison Test Suppose that $\Sigma\ a_n$ and $\Sigma\ b_n$ are series with positive terms. If

$$\lim_{n \to \infty} \frac{a_n}{b_n} = c$$

where c is a finite number and $c > 0$, then either both series converge or both diverge.

Proof Let m and M be positive numbers such that $m < c < M$. Because a_n/b_n is close to c for large n, there is an integer N such that

$$m < \frac{a_n}{b_n} < M \qquad \text{when } n > N$$

and so

$$mb_n < a_n < Mb_n \qquad \text{when } n > N$$

If $\Sigma\, b_n$ converges, so does $\Sigma\, Mb_n$. Thus, $\Sigma\, a_n$ converges by part (i) of the Comparison Test. If $\Sigma\, b_n$ diverges, so does $\Sigma\, mb_n$ and part (ii) of the Comparison Test shows that $\Sigma\, a_n$ diverges.

EXAMPLE 3 Test the series $\displaystyle\sum_{n=1}^{\infty} \frac{1}{2^n - 1}$ for convergence or divergence.

SOLUTION We use the Limit Comparison Test with

$$a_n = \frac{1}{2^n - 1} \qquad b_n = \frac{1}{2^n}$$

and obtain

$$\lim_{n\to\infty} \frac{a_n}{b_n} = \lim_{n\to\infty} \frac{1/(2^n - 1)}{1/2^n} = \lim_{n\to\infty} \frac{2^n}{2^n - 1} = \lim_{n\to\infty} \frac{1}{1 - 1/2^n} = 1 > 0$$

Since this limit exists and $\Sigma\, 1/2^n$ is a convergent geometric series, the given series converges by the Limit Comparison Test.

EXAMPLE 4 Determine whether the series $\displaystyle\sum_{n=1}^{\infty} \frac{2n^2 + 3n}{\sqrt{5 + n^5}}$ converges or diverges.

SOLUTION The dominant part of the numerator is $2n^2$ and the dominant part of the denominator is $\sqrt{n^5} = n^{5/2}$. This suggests taking

$$a_n = \frac{2n^2 + 3n}{\sqrt{5 + n^5}} \qquad b_n = \frac{2n^2}{n^{5/2}} = \frac{2}{n^{1/2}}$$

$$\lim_{n\to\infty} \frac{a_n}{b_n} = \lim_{n\to\infty} \frac{2n^2 + 3n}{\sqrt{5 + n^5}} \cdot \frac{n^{1/2}}{2} = \lim_{n\to\infty} \frac{2n^{5/2} + 3n^{3/2}}{2\sqrt{5 + n^5}}$$

$$= \lim_{n\to\infty} \frac{2 + \dfrac{3}{n}}{2\sqrt{\dfrac{5}{n^5} + 1}} = \frac{2 + 0}{2\sqrt{0 + 1}} = 1$$

Since $\Sigma\, b_n = 2\,\Sigma\, 1/n^{1/2}$ is divergent (p-series with $p = \frac{1}{2} < 1$), the given series diverges by the Limit Comparison Test.

Notice that in testing many series we find a suitable comparison series $\Sigma\, b_n$ by keeping only the highest powers in the numerator and denominator.

Estimating Sums

If we have used the Comparison Test to show that a series $\Sigma\, a_n$ converges by comparison with a series $\Sigma\, b_n$, then we may be able to estimate the sum $\Sigma\, a_n$ by comparing remainders. As in Section 11.3, we consider the remainder

$$R_n = s - s_n = a_{n+1} + a_{n+2} + \cdots$$

For the comparison series $\Sigma\, b_n$ we consider the corresponding remainder

$$T_n = t - t_n = b_{n+1} + b_{n+2} + \cdots$$

Since $a_n \leq b_n$ for all n, we have $R_n \leq T_n$. If $\Sigma\, b_n$ is a p-series, we can estimate its remainder T_n as in Section 11.3. If $\Sigma\, b_n$ is a geometric series, then T_n is the sum of a geometric series and we can sum it exactly (see Exercises 35 and 36). In either case we know that R_n is smaller than T_n.

EXAMPLE 5 Use the sum of the first 100 terms to approximate the sum of the series $\Sigma\, 1/(n^3 + 1)$. Estimate the error involved in this approximation.

SOLUTION Since

$$\frac{1}{n^3 + 1} < \frac{1}{n^3}$$

the given series is convergent by the Comparison Test. The remainder T_n for the comparison series $\Sigma\, 1/n^3$ was estimated in Example 5 in Section 11.3 using the Remainder Estimate for the Integral Test. There we found that

$$T_n \leq \int_n^\infty \frac{1}{x^3}\, dx = \frac{1}{2n^2}$$

Therefore, the remainder R_n for the given series satisfies

$$R_n \leq T_n \leq \frac{1}{2n^2}$$

With $n = 100$ we have

$$R_{100} \leq \frac{1}{2(100)^2} = 0.00005$$

Using a programmable calculator or a computer, we find that

$$\sum_{n=1}^{\infty} \frac{1}{n^3 + 1} \approx \sum_{n=1}^{100} \frac{1}{n^3 + 1} \approx 0.6864538$$

with error less than 0.00005.

11.4 Exercises

1. Suppose $\Sigma\, a_n$ and $\Sigma\, b_n$ are series with positive terms and $\Sigma\, b_n$ is known to be convergent.
(a) If $a_n > b_n$ for all n, what can you say about $\Sigma\, a_n$? Why?
(b) If $a_n < b_n$ for all n, what can you say about $\Sigma\, a_n$? Why?

2. Suppose $\Sigma\, a_n$ and $\Sigma\, b_n$ are series with positive terms and $\Sigma\, b_n$ is known to be divergent.
(a) If $a_n > b_n$ for all n, what can you say about $\Sigma\, a_n$? Why?
(b) If $a_n < b_n$ for all n, what can you say about $\Sigma\, a_n$? Why?

3–32 Determine whether the series converges or diverges.

3. $\displaystyle\sum_{n=1}^{\infty} \frac{1}{n^2 + n + 1}$

4. $\displaystyle\sum_{n=1}^{\infty} \frac{2}{n^3 + 4}$

5. $\displaystyle\sum_{n=1}^{\infty} \frac{5}{2 + 3^n}$

6. $\displaystyle\sum_{n=2}^{\infty} \frac{1}{n - \sqrt{n}}$

7. $\displaystyle\sum_{n=1}^{\infty} \frac{n + 1}{n^2}$

8. $\displaystyle\sum_{n=1}^{\infty} \frac{4 + 3^n}{2^n}$

9. $\displaystyle\sum_{n=1}^{\infty} \frac{\cos^2 n}{n^2 + 1}$

10. $\displaystyle\sum_{n=1}^{\infty} \frac{n^2 - 1}{3n^4 + 1}$

11. $\displaystyle\sum_{n=2}^{\infty} \frac{n^2 + 1}{n^3 - 1}$

12. $\displaystyle\sum_{n=0}^{\infty} \frac{1 + \sin n}{10^n}$

13. $\displaystyle\sum_{n=1}^{\infty} \frac{n - 1}{n4^n}$

14. $\displaystyle\sum_{n=2}^{\infty} \frac{\sqrt{n}}{n - 1}$

15. $\displaystyle\sum_{n=1}^{\infty} \frac{2 + (-1)^n}{n\sqrt{n}}$

16. $\displaystyle\sum_{n=1}^{\infty} \frac{1}{\sqrt{n^3 + 1}}$

17. $\displaystyle\sum_{n=1}^{\infty} \frac{1}{\sqrt{n^2 + 1}}$

18. $\displaystyle\sum_{n=1}^{\infty} \frac{1}{2n + 3}$

19. $\sum_{n=1}^{\infty} \frac{2^n}{1+3^n}$

20. $\sum_{n=1}^{\infty} \frac{1+2^n}{1+3^n}$

21. $\sum_{n=1}^{\infty} \frac{1}{1+\sqrt{n}}$

22. $\sum_{n=3}^{\infty} \frac{n+2}{(n+1)^3}$

23. $\sum_{n=1}^{\infty} \frac{5+2n}{(1+n^2)^2}$

24. $\sum_{n=1}^{\infty} \frac{n^2-5n}{n^3+n+1}$

25. $\sum_{n=1}^{\infty} \frac{1+n+n^2}{\sqrt{1+n^2+n^6}}$

26. $\sum_{n=1}^{\infty} \frac{n+5}{\sqrt[3]{n^7+n^2}}$

27. $\sum_{n=1}^{\infty} \left(1+\frac{1}{n}\right)^2 e^{-n}$

28. $\sum_{n=1}^{\infty} \frac{2n^2+7n}{3^n(n^2+5n-1)}$

29. $\sum_{n=1}^{\infty} \frac{1}{n!}$

30. $\sum_{n=1}^{\infty} \frac{n!}{n^n}$

31. $\sum_{n=1}^{\infty} \sin\left(\frac{1}{n}\right)$

32. $\sum_{n=1}^{\infty} \frac{1}{n^{1+1/n}}$

33–36 ||| Use the sum of the first 10 terms to approximate the sum of the series. Estimate the error.

33. $\sum_{n=1}^{\infty} \frac{1}{n^4+n^2}$

34. $\sum_{n=1}^{\infty} \frac{1+\cos n}{n^5}$

35. $\sum_{n=1}^{\infty} \frac{1}{1+2^n}$

36. $\sum_{n=1}^{\infty} \frac{n}{(n+1)3^n}$

37. The meaning of the decimal representation of a number $0.d_1d_2d_3\ldots$ (where the digit d_i is one of the numbers 0, 1, 2, . . . , 9) is that

$$0.d_1d_2d_3d_4\ldots = \frac{d_1}{10} + \frac{d_2}{10^2} + \frac{d_3}{10^3} + \frac{d_4}{10^4} + \cdots$$

Show that this series always converges.

38. For what values of p does the series $\Sigma_{n=2}^{\infty} 1/(n^p \ln n)$ converge?

39. Prove that if $a_n \geqslant 0$ and $\Sigma\, a_n$ converges, then $\Sigma\, a_n^2$ also converges.

40. (a) Suppose that $\Sigma\, a_n$ and $\Sigma\, b_n$ are series with positive terms and $\Sigma\, b_n$ is convergent. Prove that if

$$\lim_{n\to\infty} \frac{a_n}{b_n} = 0$$

then $\Sigma\, a_n$ is also convergent.

(b) Use part (a) to show that the series converges.

(i) $\sum_{n=1}^{\infty} \frac{\ln n}{n^3}$ (ii) $\sum_{n=1}^{\infty} \frac{\ln n}{\sqrt{n}e^n}$

41. (a) Suppose that $\Sigma\, a_n$ and $\Sigma\, b_n$ are series with positive terms and $\Sigma\, b_n$ is divergent. Prove that if

$$\lim_{n\to\infty} \frac{a_n}{b_n} = \infty$$

then $\Sigma\, a_n$ is also divergent.

(b) Use part (a) to show that the series diverges.

(i) $\sum_{n=2}^{\infty} \frac{1}{\ln n}$ (ii) $\sum_{n=1}^{\infty} \frac{\ln n}{n}$

42. Give an example of a pair of series $\Sigma\, a_n$ and $\Sigma\, b_n$ with positive terms where $\lim_{n\to\infty} (a_n/b_n) = 0$ and $\Sigma\, b_n$ diverges, but $\Sigma\, a_n$ converges. [Compare with Exercise 40.]

43. Show that if $a_n > 0$ and $\lim_{n\to\infty} na_n \neq 0$, then $\Sigma\, a_n$ is divergent.

44. Show that if $a_n > 0$ and $\Sigma\, a_n$ is convergent, then $\Sigma \ln(1 + a_n)$ is convergent.

45. If $\Sigma\, a_n$ is a convergent series with positive terms, is it true that $\Sigma \sin(a_n)$ is also convergent?

46. If $\Sigma\, a_n$ and $\Sigma\, b_n$ are both convergent series with positive terms, is it true that $\Sigma\, a_nb_n$ is also convergent?

11.5 Alternating Series

The convergence tests that we have looked at so far apply only to series with positive terms. In this section and the next we learn how to deal with series whose terms are not necessarily positive. Of particular importance are *alternating series*, whose terms alternate in sign.

An **alternating series** is a series whose terms are alternately positive and negative. Here are two examples:

$$1 - \frac{1}{2} + \frac{1}{3} - \frac{1}{4} + \frac{1}{5} - \frac{1}{6} + \cdots = \sum_{n=1}^{\infty} \frac{(-1)^{n-1}}{n}$$

$$-\frac{1}{2} + \frac{2}{3} - \frac{3}{4} + \frac{4}{5} - \frac{5}{6} + \frac{6}{7} - \cdots = \sum_{n=1}^{\infty} (-1)^n \frac{n}{n+1}$$

We see from these examples that the nth term of an alternating series is of the form

$$a_n = (-1)^{n-1} b_n \qquad \text{or} \qquad a_n = (-1)^n b_n$$

where b_n is a positive number. (In fact, $b_n = |a_n|$.)

The following test says that if the terms of an alternating series decrease toward 0 in absolute value, then the series converges.

The Alternating Series Test If the alternating series

$$\sum_{n=1}^{\infty} (-1)^{n-1} b_n = b_1 - b_2 + b_3 - b_4 + b_5 - b_6 + \cdots \qquad (b_n > 0)$$

satisfies

$$\text{(i)} \quad b_{n+1} \leqslant b_n \qquad \text{for all } n$$

$$\text{(ii)} \quad \lim_{n \to \infty} b_n = 0$$

then the series is convergent.

Before giving the proof let's look at Figure 1, which gives a picture of the idea behind the proof. We first plot $s_1 = b_1$ on a number line. To find s_2 we subtract b_2, so s_2 is to the left of s_1. Then to find s_3 we add b_3, so s_3 is to the right of s_2. But, since $b_3 < b_2$, s_3 is to the left of s_1. Continuing in this manner, we see that the partial sums oscillate back and forth. Since $b_n \to 0$, the successive steps are becoming smaller and smaller. The even partial sums $s_2, s_4, s_6, \ldots$ are increasing and the odd partial sums $s_1, s_3, s_5, \ldots$ are decreasing. Thus, it seems plausible that both are converging to some number s, which is the sum of the series. Therefore, in the following proof we consider the even and odd partial sums separately.

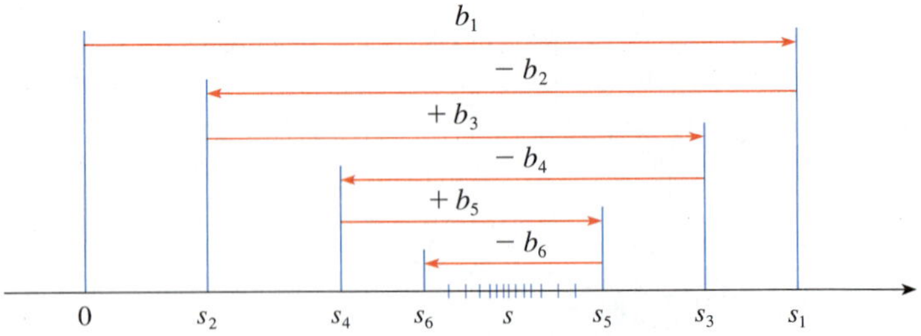

FIGURE 1

Proof of the Alternating Series Test We first consider the even partial sums:

$$s_2 = b_1 - b_2 \geqslant 0 \qquad \text{since } b_2 \leqslant b_1$$

$$s_4 = s_2 + (b_3 - b_4) \geqslant s_2 \qquad \text{since } b_4 \leqslant b_3$$

In general $\qquad s_{2n} = s_{2n-2} + (b_{2n-1} - b_{2n}) \geqslant s_{2n-2} \qquad \text{since } b_{2n} \leqslant b_{2n-1}$

Thus $\qquad 0 \leqslant s_2 \leqslant s_4 \leqslant s_6 \leqslant \cdots \leqslant s_{2n} \leqslant \cdots$

But we can also write

$$s_{2n} = b_1 - (b_2 - b_3) - (b_4 - b_5) - \cdots - (b_{2n-2} - b_{2n-1}) - b_{2n}$$

Every term in brackets is positive, so $s_{2n} \leqslant b_1$ for all n. Therefore, the sequence $\{s_{2n}\}$ of even partial sums is increasing and bounded above. It is therefore convergent by the Monotonic Sequence Theorem. Let's call its limit s, that is,

$$\lim_{n\to\infty} s_{2n} = s$$

Now we compute the limit of the odd partial sums:

$$\begin{aligned}\lim_{n\to\infty} s_{2n+1} &= \lim_{n\to\infty} (s_{2n} + b_{2n+1}) \\ &= \lim_{n\to\infty} s_{2n} + \lim_{n\to\infty} b_{2n+1} \\ &= s + 0 \qquad \text{[by condition (ii)]} \\ &= s\end{aligned}$$

Since both the even and odd partial sums converge to s, we have $\lim_{n\to\infty} s_n = s$ (see Exercise 72 in Section 11.1) and so the series is convergent.

|||| Figure 2 illustrates Example 1 by showing the graphs of the terms $a_n = (-1)^{n-1}/n$ and the partial sums s_n. Notice how the values of s_n zigzag across the limiting value, which appears to be about 0.7. In fact, the exact sum of the series is $\ln 2 \approx 0.693$ (see Exercise 36).

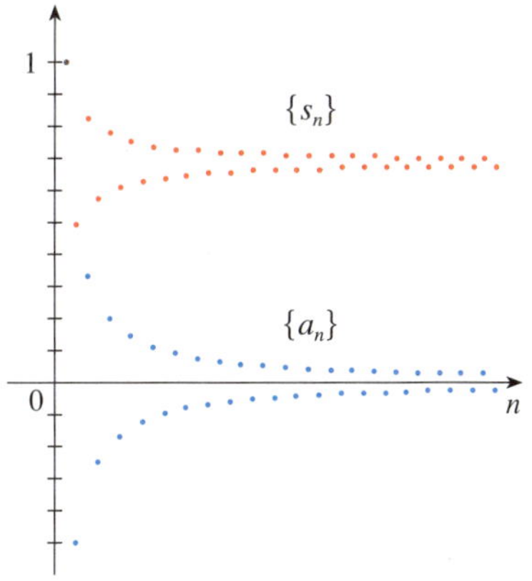

FIGURE 2

EXAMPLE 1 The alternating harmonic series

$$1 - \frac{1}{2} + \frac{1}{3} - \frac{1}{4} + \cdots = \sum_{n=1}^{\infty} \frac{(-1)^{n-1}}{n}$$

satisfies

$$\text{(i)}\ \ b_{n+1} < b_n \qquad \text{because} \qquad \frac{1}{n+1} < \frac{1}{n}$$

$$\text{(ii)}\ \ \lim_{n\to\infty} b_n = \lim_{n\to\infty} \frac{1}{n} = 0$$

so the series is convergent by the Alternating Series Test.

EXAMPLE 2 The series $\displaystyle\sum_{n=1}^{\infty} \frac{(-1)^n 3n}{4n-1}$ is alternating but

$$\lim_{n\to\infty} b_n = \lim_{n\to\infty} \frac{3n}{4n-1} = \lim_{n\to\infty} \frac{3}{4 - \dfrac{1}{n}} = \frac{3}{4}$$

so condition (ii) is not satisfied. Instead, we look at the limit of the nth term of the series:

$$\lim_{n\to\infty} a_n = \lim_{n\to\infty} \frac{(-1)^n 3n}{4n-1}$$

This limit does not exist, so the series diverges by the Test for Divergence.

EXAMPLE 3 Test the series $\displaystyle\sum_{n=1}^{\infty} (-1)^{n+1} \frac{n^2}{n^3+1}$ for convergence or divergence.

SOLUTION The given series is alternating so we try to verify conditions (i) and (ii) of the Alternating Series Test.

Unlike the situation in Example 1, it is not obvious that the sequence given by $b_n = n^2/(n^3+1)$ is decreasing. However, if we consider the related function

$f(x) = x^2/(x^3 + 1)$, we find that

$$f'(x) = \frac{x(2 - x^3)}{(x^3 + 1)^2}$$

Since we are considering only positive x, we see that $f'(x) < 0$ if $2 - x^3 < 0$, that is, $x > \sqrt[3]{2}$. Thus, f is decreasing on the interval $(\sqrt[3]{2}, \infty)$. This means that $f(n + 1) < f(n)$ and therefore $b_{n+1} < b_n$ when $n \geqslant 2$. (The inequality $b_2 < b_1$ can be verified directly but all that really matters is that the sequence $\{b_n\}$ is eventually decreasing.)

|||| Instead of verifying condition (i) of the Alternating Series Test by computing a derivative, we could verify that $b_{n+1} < b_n$ directly by using the technique of Solution 1 of Example 11 in Section 11.1.

Condition (ii) is readily verified:

$$\lim_{n \to \infty} b_n = \lim_{n \to \infty} \frac{n^2}{n^3 + 1} = \lim_{n \to \infty} \frac{\dfrac{1}{n}}{1 + \dfrac{1}{n^3}} = 0$$

Thus, the given series is convergent by the Alternating Series Test.

Estimating Sums

A partial sum s_n of any convergent series can be used as an approximation to the total sum s, but this is not of much use unless we can estimate the accuracy of the approximation. The error involved in using $s \approx s_n$ is the remainder $R_n = s - s_n$. The next theorem says that for series that satisfy the conditions of the Alternating Series Test, the size of the error is smaller than b_{n+1}, which is the absolute value of the first neglected term.

|||| You can see geometrically why the Alternating Series Estimation Theorem is true by looking at Figure 1 (on page 736). Notice that $s - s_4 < b_5$, $|s - s_5| < b_6$, and so on. Notice also that s lies between any two consecutive partial sums.

Alternating Series Estimation Theorem If $s = \Sigma\,(-1)^{n-1}b_n$ is the sum of an alternating series that satisfies

$$\text{(i)}\ 0 \leqslant b_{n+1} \leqslant b_n \qquad \text{and} \qquad \text{(ii)}\ \lim_{n \to \infty} b_n = 0$$

then

$$|R_n| = |s - s_n| \leqslant b_{n+1}$$

Proof We know from the proof of the Alternating Series Test that s lies between any two consecutive partial sums s_n and s_{n+1}. It follows that

$$|s - s_n| \leqslant |s_{n+1} - s_n| = b_{n+1}$$

EXAMPLE 4 Find the sum of the series $\displaystyle\sum_{n=0}^{\infty} \frac{(-1)^n}{n!}$ correct to three decimal places. (By definition, $0! = 1$.)

SOLUTION We first observe that the series is convergent by the Alternating Series Test because

$$\text{(i)}\quad \frac{1}{(n + 1)!} = \frac{1}{n!(n + 1)} < \frac{1}{n!}$$

$$\text{(ii)}\quad 0 < \frac{1}{n!} < \frac{1}{n} \to 0 \quad \text{so} \quad \frac{1}{n!} \to 0 \text{ as } n \to \infty$$

To get a feel for how many terms we need to use in our approximation, let's write out the first few terms of the series:

$$s = \frac{1}{0!} - \frac{1}{1!} + \frac{1}{2!} - \frac{1}{3!} + \frac{1}{4!} - \frac{1}{5!} + \frac{1}{6!} - \frac{1}{7!} + \cdots$$
$$= 1 - 1 + \tfrac{1}{2} - \tfrac{1}{6} + \tfrac{1}{24} - \tfrac{1}{120} + \tfrac{1}{720} - \tfrac{1}{5040} + \cdots$$

Notice that $$b_7 = \tfrac{1}{5040} < \tfrac{1}{5000} = 0.0002$$

and $$s_6 = 1 - 1 + \tfrac{1}{2} - \tfrac{1}{6} + \tfrac{1}{24} - \tfrac{1}{120} + \tfrac{1}{720} \approx 0.368056$$

By the Alternating Series Estimation Theorem we know that

$$|s - s_6| \leqslant b_7 < 0.0002$$

|||| In Section 11.10 we will prove that $e^x = \sum_{n=0}^{\infty} x^n/n!$ for all x, so what we have obtained in Example 4 is actually an approximation to the number e^{-1}.

This error of less than 0.0002 does not affect the third decimal place, so we have

$$s \approx 0.368$$

correct to three decimal places.

NOTE ▫ The rule that the error (in using s_n to approximate s) is smaller than the first neglected term is, in general, valid only for alternating series that satisfy the conditions of the Alternating Series Estimation Theorem. The rule does not apply to other types of series.

11.5 Exercises

1. (a) What is an alternating series?
(b) Under what conditions does an alternating series converge?
(c) If these conditions are satisfied, what can you say about the remainder after n terms?

2–20 |||| Test the series for convergence or divergence.

2. $-\frac{1}{3} + \frac{2}{4} - \frac{3}{5} + \frac{4}{6} - \frac{5}{7} + \cdots$

3. $\frac{4}{7} - \frac{4}{8} + \frac{4}{9} - \frac{4}{10} + \frac{4}{11} - \cdots$

4. $\dfrac{1}{\ln 2} - \dfrac{1}{\ln 3} + \dfrac{1}{\ln 4} - \dfrac{1}{\ln 5} + \dfrac{1}{\ln 6} - \cdots$

5. $\displaystyle\sum_{n=1}^{\infty} \frac{(-1)^{n-1}}{\sqrt{n}}$

6. $\displaystyle\sum_{n=1}^{\infty} \frac{(-1)^{n-1}}{3n - 1}$

7. $\displaystyle\sum_{n=1}^{\infty} (-1)^n \frac{3n - 1}{2n + 1}$

8. $\displaystyle\sum_{n=1}^{\infty} (-1)^n \frac{2n}{4n^2 + 1}$

9. $\displaystyle\sum_{n=1}^{\infty} \frac{(-1)^{n+1}}{4n^2 + 1}$

10. $\displaystyle\sum_{n=1}^{\infty} (-1)^n \frac{\sqrt{n}}{1 + 2\sqrt{n}}$

11. $\displaystyle\sum_{n=1}^{\infty} (-1)^{n+1} \frac{n^2}{n^3 + 4}$

12. $\displaystyle\sum_{n=1}^{\infty} (-1)^{n-1} \frac{e^{1/n}}{n}$

13. $\displaystyle\sum_{n=2}^{\infty} (-1)^n \frac{n}{\ln n}$

14. $\displaystyle\sum_{n=1}^{\infty} (-1)^{n-1} \frac{\ln n}{n}$

15. $\displaystyle\sum_{n=1}^{\infty} \frac{\cos n\pi}{n^{3/4}}$

16. $\displaystyle\sum_{n=1}^{\infty} \frac{\sin(n\pi/2)}{n!}$

17. $\displaystyle\sum_{n=1}^{\infty} (-1)^n \sin\left(\frac{\pi}{n}\right)$

18. $\displaystyle\sum_{n=1}^{\infty} (-1)^n \cos\left(\frac{\pi}{n}\right)$

19. $\displaystyle\sum_{n=1}^{\infty} (-1)^n \frac{n^n}{n!}$

20. $\displaystyle\sum_{n=1}^{\infty} \left(-\frac{n}{5}\right)^n$

21–22 |||| Calculate the first 10 partial sums of the series and graph both the sequence of terms and the sequence of partial sums on the same screen. Estimate the error in using the 10th partial sum to approximate the total sum.

21. $\displaystyle\sum_{n=1}^{\infty} \frac{(-1)^{n-1}}{n^{3/2}}$

22. $\displaystyle\sum_{n=1}^{\infty} \frac{(-1)^{n-1}}{n^3}$

23–26 |||| How many terms of the series do we need to add in order to find the sum to the indicated accuracy?

23. $\displaystyle\sum_{n=1}^{\infty} \frac{(-1)^{n-1}}{n^2} \quad (|\text{error}| < 0.01)$

24. $\displaystyle\sum_{n=1}^{\infty} \frac{(-1)^{n+1}}{n^4} \quad (|\text{error}| < 0.001)$

25. $\displaystyle\sum_{n=1}^{\infty} \frac{(-2)^n}{n!}$ $(|\text{error}| < 0.01)$

26. $\displaystyle\sum_{n=1}^{\infty} \frac{(-1)^n n}{4^n}$ $(|\text{error}| < 0.002)$

27–30 Approximate the sum of the series correct to four decimal places.

27. $\displaystyle\sum_{n=1}^{\infty} \frac{(-1)^{n+1}}{n^5}$

28. $\displaystyle\sum_{n=1}^{\infty} \frac{(-1)^n n}{8^n}$

29. $\displaystyle\sum_{n=1}^{\infty} \frac{(-1)^{n-1} n^2}{10^n}$

30. $\displaystyle\sum_{n=1}^{\infty} \frac{(-1)^n}{3^n n!}$

31. Is the 50th partial sum s_{50} of the alternating series $\sum_{n=1}^{\infty} (-1)^{n-1}/n$ an overestimate or an underestimate of the total sum? Explain.

32–34 For what values of p is each series convergent?

32. $\displaystyle\sum_{n=1}^{\infty} \frac{(-1)^{n-1}}{n^p}$

33. $\displaystyle\sum_{n=1}^{\infty} \frac{(-1)^n}{n+p}$

34. $\displaystyle\sum_{n=2}^{\infty} (-1)^{n-1} \frac{(\ln n)^p}{n}$

35. Show that the series $\sum (-1)^{n-1} b_n$, where $b_n = 1/n$ if n is odd and $b_n = 1/n^2$ if n is even, is divergent. Why does the Alternating Series Test not apply?

36. Use the following steps to show that

$$\sum_{n=1}^{\infty} \frac{(-1)^{n-1}}{n} = \ln 2$$

Let h_n and s_n be the partial sums of the harmonic and alternating harmonic series.

(a) Show that $s_{2n} = h_{2n} - h_n$.

(b) From Exercise 38 in Section 11.3 we have

$$h_n - \ln n \to \gamma \qquad \text{as } n \to \infty$$

and therefore

$$h_{2n} - \ln(2n) \to \gamma \qquad \text{as } n \to \infty$$

Use these facts together with part (a) to show that $s_{2n} \to \ln 2$ as $n \to \infty$.

11.6 Absolute Convergence and the Ratio and Root Tests

Given any series $\sum a_n$, we can consider the corresponding series

$$\sum_{n=1}^{\infty} |a_n| = |a_1| + |a_2| + |a_3| + \cdots$$

whose terms are the absolute values of the terms of the original series.

We have convergence tests for series with positive terms and for alternating series. But what if the signs of the terms switch back and forth irregularly? We will see in Example 3 that the idea of absolute convergence sometimes helps in such cases.

1 Definition A series $\sum a_n$ is called **absolutely convergent** if the series of absolute values $\sum |a_n|$ is convergent.

Notice that if $\sum a_n$ is a series with positive terms, then $|a_n| = a_n$ and so absolute convergence is the same as convergence in this case.

EXAMPLE 1 The series

$$\sum_{n=1}^{\infty} \frac{(-1)^{n-1}}{n^2} = 1 - \frac{1}{2^2} + \frac{1}{3^2} - \frac{1}{4^2} + \cdots$$

is absolutely convergent because

$$\sum_{n=1}^{\infty} \left| \frac{(-1)^{n-1}}{n^2} \right| = \sum_{n=1}^{\infty} \frac{1}{n^2} = 1 + \frac{1}{2^2} + \frac{1}{3^2} + \frac{1}{4^2} + \cdots$$

is a convergent p-series ($p = 2$).

EXAMPLE 2 We know that the alternating harmonic series

$$\sum_{n=1}^{\infty} \frac{(-1)^{n-1}}{n} = 1 - \frac{1}{2} + \frac{1}{3} - \frac{1}{4} + \cdots$$

is convergent (see Example 1 in Section 11.5), but it is not absolutely convergent because the corresponding series of absolute values is

$$\sum_{n=1}^{\infty} \left| \frac{(-1)^{n-1}}{n} \right| = \sum_{n=1}^{\infty} \frac{1}{n} = 1 + \frac{1}{2} + \frac{1}{3} + \frac{1}{4} + \cdots$$

which is the harmonic series (p-series with $p = 1$) and is therefore divergent.

2 Definition A series $\Sigma\, a_n$ is called **conditionally convergent** if it is convergent but not absolutely convergent.

Example 2 shows that the alternating harmonic series is conditionally convergent. Thus, it is possible for a series to be convergent but not absolutely convergent. However, the next theorem shows that absolute convergence implies convergence.

3 Theorem If a series $\Sigma\, a_n$ is absolutely convergent, then it is convergent.

Proof Observe that the inequality

$$0 \leqslant a_n + |a_n| \leqslant 2|a_n|$$

is true because $|a_n|$ is either a_n or $-a_n$. If $\Sigma\, a_n$ is absolutely convergent, then $\Sigma\, |a_n|$ is convergent, so $\Sigma\, 2|a_n|$ is convergent. Therefore, by the Comparison Test, $\Sigma\, (a_n + |a_n|)$ is convergent. Then

$$\sum a_n = \sum (a_n + |a_n|) - \sum |a_n|$$

is the difference of two convergent series and is therefore convergent.

||| Figure 1 shows the graphs of the terms a_n and partial sums s_n of the series in Example 3. Notice that the series is not alternating but has positive and negative terms.

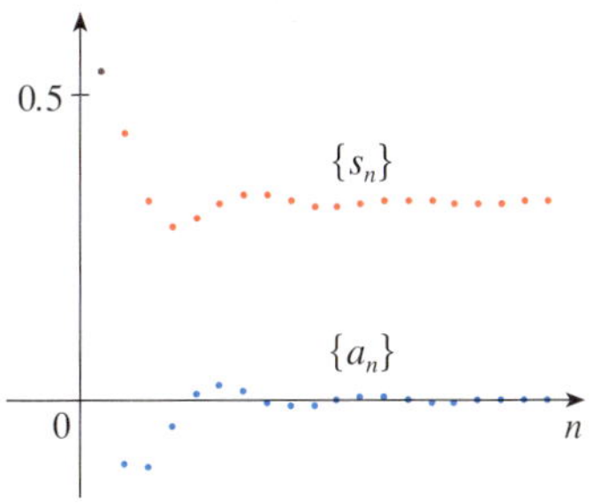

FIGURE 1

EXAMPLE 3 Determine whether the series

$$\sum_{n=1}^{\infty} \frac{\cos n}{n^2} = \frac{\cos 1}{1^2} + \frac{\cos 2}{2^2} + \frac{\cos 3}{3^2} + \cdots$$

is convergent or divergent.

SOLUTION This series has both positive and negative terms, but it is not alternating. (The first term is positive, the next three are negative, and the following three are positive. The signs change irregularly.) We can apply the Comparison Test to the series of absolute values

$$\sum_{n=1}^{\infty} \left| \frac{\cos n}{n^2} \right| = \sum_{n=1}^{\infty} \frac{|\cos n|}{n^2}$$

Since $|\cos n| \leqslant 1$ for all n, we have

$$\frac{|\cos n|}{n^2} \leqslant \frac{1}{n^2}$$

We know that $\Sigma\, 1/n^2$ is convergent (p-series with $p = 2$) and therefore $\Sigma\, |\cos n|/n^2$ is convergent by the Comparison Test. Thus, the given series $\Sigma\, (\cos n)/n^2$ is absolutely convergent and therefore convergent by Theorem 3.

The following test is very useful in determining whether a given series is absolutely convergent.

The Ratio Test

(i) If $\displaystyle\lim_{n\to\infty}\left|\frac{a_{n+1}}{a_n}\right| = L < 1$, then the series $\displaystyle\sum_{n=1}^{\infty} a_n$ is absolutely convergent (and therefore convergent).

(ii) If $\displaystyle\lim_{n\to\infty}\left|\frac{a_{n+1}}{a_n}\right| = L > 1$ or $\displaystyle\lim_{n\to\infty}\left|\frac{a_{n+1}}{a_n}\right| = \infty$, then the series $\displaystyle\sum_{n=1}^{\infty} a_n$ is divergent.

(iii) If $\displaystyle\lim_{n\to\infty}\left|\frac{a_{n+1}}{a_n}\right| = 1$, the Ratio Test is inconclusive; that is, no conclusion can be drawn about the convergence or divergence of $\Sigma\, a_n$.

Proof

(i) The idea is to compare the given series with a convergent geometric series. Since $L < 1$, we can choose a number r such that $L < r < 1$. Since

$$\lim_{n\to\infty}\left|\frac{a_{n+1}}{a_n}\right| = L \qquad \text{and} \qquad L < r$$

the ratio $|a_{n+1}/a_n|$ will eventually be less than r; that is, there exists an integer N such that

$$\left|\frac{a_{n+1}}{a_n}\right| < r \qquad \text{whenever } n \geqslant N$$

or, equivalently,

4
$$|a_{n+1}| < |a_n|\, r \qquad \text{whenever } n \geqslant N$$

Putting n successively equal to $N, N + 1, N + 2, \ldots$ in (4), we obtain

$$|a_{N+1}| < |a_N|\, r$$

$$|a_{N+2}| < |a_{N+1}|\, r < |a_N|\, r^2$$

$$|a_{N+3}| < |a_{N+2}|\, r < |a_N|\, r^3$$

and, in general,

5
$$|a_{N+k}| < |a_N|\, r^k \qquad \text{for all } k \geqslant 1$$

Now the series

$$\sum_{k=1}^{\infty} |a_N|\, r^k = |a_N|\, r + |a_N|\, r^2 + |a_N|\, r^3 + \cdots$$

is convergent because it is a geometric series with $0 < r < 1$. So the inequality (5), together with the Comparison Test, shows that the series

$$\sum_{n=N+1}^{\infty} |a_n| = \sum_{k=1}^{\infty} |a_{N+k}| = |a_{N+1}| + |a_{N+2}| + |a_{N+3}| + \cdots$$

is also convergent. It follows that the series $\Sigma_{n=1}^{\infty} |a_n|$ is convergent. (Recall that a finite number of terms doesn't affect convergence.) Therefore, $\Sigma\, a_n$ is absolutely convergent.

(ii) If $|a_{n+1}/a_n| \to L > 1$ or $|a_{n+1}/a_n| \to \infty$, then the ratio $|a_{n+1}/a_n|$ will eventually be greater than 1; that is, there exists an integer N such that

$$\left|\frac{a_{n+1}}{a_n}\right| > 1 \qquad \text{whenever } n \geqslant N$$

This means that $|a_{n+1}| > |a_n|$ whenever $n \geqslant N$ and so

$$\lim_{n\to\infty} a_n \neq 0$$

Therefore, $\Sigma\, a_n$ diverges by the Test for Divergence.

NOTE ▫ Part (iii) of the Ratio Test says that if $\lim_{n\to\infty} |a_{n+1}/a_n| = 1$, the test gives no information. For instance, for the convergent series $\Sigma\, 1/n^2$ we have

$$\left|\frac{a_{n+1}}{a_n}\right| = \frac{\dfrac{1}{(n+1)^2}}{\dfrac{1}{n^2}} = \frac{n^2}{(n+1)^2} = \frac{1}{\left(1 + \dfrac{1}{n}\right)^2} \to 1 \qquad \text{as } n \to \infty$$

whereas for the divergent series $\Sigma\, 1/n$ we have

$$\left|\frac{a_{n+1}}{a_n}\right| = \frac{\dfrac{1}{n+1}}{\dfrac{1}{n}} = \frac{n}{n+1} = \frac{1}{1 + \dfrac{1}{n}} \to 1 \qquad \text{as } n \to \infty$$

Therefore, if $\lim_{n\to\infty} |a_{n+1}/a_n| = 1$, the series $\Sigma\, a_n$ might converge or it might diverge. In this case the Ratio Test fails and we must use some other test.

EXAMPLE 4 Test the series $\displaystyle\sum_{n=1}^{\infty} (-1)^n \frac{n^3}{3^n}$ for absolute convergence.

SOLUTION We use the Ratio Test with $a_n = (-1)^n n^3/3^n$:

$$\left|\frac{a_{n+1}}{a_n}\right| = \left|\frac{\dfrac{(-1)^{n+1}(n+1)^3}{3^{n+1}}}{\dfrac{(-1)^n n^3}{3^n}}\right| = \frac{(n+1)^3}{3^{n+1}} \cdot \frac{3^n}{n^3}$$

$$= \frac{1}{3}\left(\frac{n+1}{n}\right)^3 = \frac{1}{3}\left(1 + \frac{1}{n}\right)^3 \to \frac{1}{3} < 1$$

Thus, by the Ratio Test, the given series is absolutely convergent and therefore convergent.

|||| ESTIMATING SUMS

In the last three sections we used various methods for estimating the sum of a series—the method depended on which test was used to prove convergence. What about series for which the Ratio Test works? There are two possibilities: If the series happens to be an alternating series, as in Example 4, then it is best to use the methods of Section 11.5. If the terms are all positive, then use the special methods explained in Exercise 34.

EXAMPLE 5 Test the convergence of the series $\sum_{n=1}^{\infty} \frac{n^n}{n!}$.

SOLUTION Since the terms $a_n = n^n/n!$ are positive, we don't need the absolute value signs.

$$\frac{a_{n+1}}{a_n} = \frac{(n+1)^{n+1}}{(n+1)!} \cdot \frac{n!}{n^n} = \frac{(n+1)(n+1)^n}{(n+1)n!} \cdot \frac{n!}{n^n}$$

$$= \left(\frac{n+1}{n}\right)^n = \left(1 + \frac{1}{n}\right)^n \to e \qquad \text{as } n \to \infty$$

(See Equation 3.8.6.) Since $e > 1$, the given series is divergent by the Ratio Test.

NOTE ▫ Although the Ratio Test works in Example 5, an easier method is to use the Test for Divergence. Since

$$a_n = \frac{n^n}{n!} = \frac{n \cdot n \cdot n \cdot \cdots \cdot n}{1 \cdot 2 \cdot 3 \cdot \cdots \cdot n} \geqslant n$$

it follows that a_n does not approach 0 as $n \to \infty$. Therefore, the given series is divergent by the Test for Divergence.

The following test is convenient to apply when nth powers occur. Its proof is similar to the proof of the Ratio Test and is left as Exercise 38.

The Root Test

(i) If $\lim_{n\to\infty} \sqrt[n]{|a_n|} = L < 1$, then the series $\sum_{n=1}^{\infty} a_n$ is absolutely convergent (and therefore convergent).

(ii) If $\lim_{n\to\infty} \sqrt[n]{|a_n|} = L > 1$ or $\lim_{n\to\infty} \sqrt[n]{|a_n|} = \infty$, then the series $\sum_{n=1}^{\infty} a_n$ is divergent.

(iii) If $\lim_{n\to\infty} \sqrt[n]{|a_n|} = 1$, the Root Test is inconclusive.

If $\lim_{n\to\infty} \sqrt[n]{|a_n|} = 1$, then part (iii) of the Root Test says that the test gives no information. The series $\sum a_n$ could converge or diverge. (If $L = 1$ in the Ratio Test, don't try the Root Test because L will again be 1.)

EXAMPLE 6 Test the convergence of the series $\sum_{n=1}^{\infty} \left(\frac{2n+3}{3n+2}\right)^n$.

SOLUTION

$$a_n = \left(\frac{2n+3}{3n+2}\right)^n$$

$$\sqrt[n]{|a_n|} = \frac{2n+3}{3n+2} = \frac{2 + \frac{3}{n}}{3 + \frac{2}{n}} \to \frac{2}{3} < 1$$

Thus, the given series converges by the Root Test.

Rearrangements

The question of whether a given convergent series is absolutely convergent or conditionally convergent has a bearing on the question of whether infinite sums behave like finite sums.

If we rearrange the order of the terms in a finite sum, then of course the value of the sum remains unchanged. But this is not always the case for an infinite series. By a **rearrangement** of an infinite series $\Sigma\, a_n$ we mean a series obtained by simply changing the order of the terms. For instance, a rearrangement of $\Sigma\, a_n$ could start as follows:

$$a_1 + a_2 + a_5 + a_3 + a_4 + a_{15} + a_6 + a_7 + a_{20} + \cdots$$

It turns out that

> if $\Sigma\, a_n$ is an absolutely convergent series with sum s, then any rearrangement of $\Sigma\, a_n$ has the same sum s.

However, any conditionally convergent series can be rearranged to give a different sum. To illustrate this fact let's consider the alternating harmonic series

6 $$1 - \tfrac{1}{2} + \tfrac{1}{3} - \tfrac{1}{4} + \tfrac{1}{5} - \tfrac{1}{6} + \tfrac{1}{7} - \tfrac{1}{8} + \cdots = \ln 2$$

(See Exercise 36 in Section 11.5.) If we multiply this series by $\tfrac{1}{2}$, we get

$$\tfrac{1}{2} - \tfrac{1}{4} + \tfrac{1}{6} - \tfrac{1}{8} + \cdots = \tfrac{1}{2}\ln 2$$

Inserting zeros between the terms of this series, we have

|||| Adding these zeros does not affect the sum of the series; each term in the sequence of partial sums is repeated, but the limit is the same.

7 $$0 + \tfrac{1}{2} + 0 - \tfrac{1}{4} + 0 + \tfrac{1}{6} + 0 - \tfrac{1}{8} + \cdots = \tfrac{1}{2}\ln 2$$

Now we add the series in Equations 6 and 7 using Theorem 11.2.8:

8 $$1 + \tfrac{1}{3} - \tfrac{1}{2} + \tfrac{1}{5} + \tfrac{1}{7} - \tfrac{1}{4} + \cdots = \tfrac{3}{2}\ln 2$$

Notice that the series in (8) contains the same terms as in (6), but rearranged so that one negative term occurs after each pair of positive terms. The sums of these series, however, are different. In fact, Riemann proved that

> if $\Sigma\, a_n$ is a conditionally convergent series and r is any real number whatsoever, then there is a rearrangement of $\Sigma\, a_n$ that has a sum equal to r.

A proof of this fact is outlined in Exercise 40.

11.6 Exercises

1. What can you say about the series $\Sigma\, a_n$ in each of the following cases?

(a) $\displaystyle\lim_{n\to\infty}\left|\frac{a_{n+1}}{a_n}\right| = 8$ (b) $\displaystyle\lim_{n\to\infty}\left|\frac{a_{n+1}}{a_n}\right| = 0.8$

(c) $\displaystyle\lim_{n\to\infty}\left|\frac{a_{n+1}}{a_n}\right| = 1$

2–28 |||| Determine whether the series is absolutely convergent, conditionally convergent, or divergent.

2. $\displaystyle\sum_{n=1}^{\infty} \frac{n^2}{2^n}$

3. $\displaystyle\sum_{n=0}^{\infty} \frac{(-10)^n}{n!}$

4. $\displaystyle\sum_{n=1}^{\infty} (-1)^{n-1}\frac{2^n}{n^4}$

5. $\displaystyle\sum_{n=1}^{\infty} \frac{(-1)^{n+1}}{\sqrt[4]{n}}$

6. $\displaystyle\sum_{n=1}^{\infty} \frac{(-1)^n}{n^4}$

7. $\displaystyle\sum_{n=1}^{\infty} (-1)^n\frac{n}{5+n}$

8. $\displaystyle\sum_{n=1}^{\infty} (-1)^{n-1}\frac{n}{n^2+1}$

9. $\displaystyle\sum_{n=1}^{\infty} \frac{1}{(2n)!}$

10. $\displaystyle\sum_{n=1}^{\infty} e^{-n}n!$

11. $\displaystyle\sum_{n=1}^{\infty} \frac{(-1)^n e^{1/n}}{n^3}$

12. $\displaystyle\sum_{n=1}^{\infty} \frac{\sin 4n}{4^n}$

13. $\displaystyle\sum_{n=1}^{\infty} \frac{n(-3)^n}{4^{n-1}}$

14. $\displaystyle\sum_{n=1}^{\infty} (-1)^{n+1} \frac{n^2 2^n}{n!}$

15. $\displaystyle\sum_{n=1}^{\infty} \frac{10^n}{(n+1)4^{2n+1}}$

16. $\displaystyle\sum_{n=1}^{\infty} \frac{3 - \cos n}{n^{2/3} - 2}$

17. $\displaystyle\sum_{n=2}^{\infty} \frac{(-1)^n}{\ln n}$

18. $\displaystyle\sum_{n=1}^{\infty} \frac{n!}{n^n}$

19. $\displaystyle\sum_{n=1}^{\infty} \frac{\cos(n\pi/3)}{n!}$

20. $\displaystyle\sum_{n=2}^{\infty} \frac{(-1)^n}{(\ln n)^n}$

21. $\displaystyle\sum_{n=1}^{\infty} \frac{n^n}{3^{1+3n}}$

22. $\displaystyle\sum_{n=2}^{\infty} \frac{(-1)^n}{n \ln n}$

23. $\displaystyle\sum_{n=1}^{\infty} \left(\frac{n^2+1}{2n^2+1}\right)^n$

24. $\displaystyle\sum_{n=1}^{\infty} \frac{(-1)^n}{(\arctan n)^n}$

25. $$1 - \frac{1 \cdot 3}{3!} + \frac{1 \cdot 3 \cdot 5}{5!} - \frac{1 \cdot 3 \cdot 5 \cdot 7}{7!} + \cdots + (-1)^{n-1}\frac{1 \cdot 3 \cdot 5 \cdot \cdots \cdot (2n-1)}{(2n-1)!} + \cdots$$

26. $$\frac{2}{5} + \frac{2 \cdot 6}{5 \cdot 8} + \frac{2 \cdot 6 \cdot 10}{5 \cdot 8 \cdot 11} + \frac{2 \cdot 6 \cdot 10 \cdot 14}{5 \cdot 8 \cdot 11 \cdot 14} + \cdots$$

27. $\displaystyle\sum_{n=1}^{\infty} \frac{2 \cdot 4 \cdot 6 \cdot \cdots \cdot (2n)}{n!}$

28. $\displaystyle\sum_{n=1}^{\infty} (-1)^n \frac{2^n n!}{5 \cdot 8 \cdot 11 \cdot \cdots \cdot (3n+2)}$

29. The terms of a series are defined recursively by the equations

$$a_1 = 2 \qquad a_{n+1} = \frac{5n+1}{4n+3}\, a_n$$

Determine whether $\Sigma\, a_n$ converges or diverges.

30. A series $\Sigma\, a_n$ is defined by the equations

$$a_1 = 1 \qquad a_{n+1} = \frac{2 + \cos n}{\sqrt{n}}\, a_n$$

Determine whether $\Sigma\, a_n$ converges or diverges.

31. For which of the following series is the Ratio Test inconclusive (that is, it fails to give a definite answer)?

(a) $\displaystyle\sum_{n=1}^{\infty} \frac{1}{n^3}$ (b) $\displaystyle\sum_{n=1}^{\infty} \frac{n}{2^n}$

(c) $\displaystyle\sum_{n=1}^{\infty} \frac{(-3)^{n-1}}{\sqrt{n}}$ (d) $\displaystyle\sum_{n=1}^{\infty} \frac{\sqrt{n}}{1+n^2}$

32. For which positive integers k is the following series convergent?

$$\sum_{n=1}^{\infty} \frac{(n!)^2}{(kn)!}$$

33. (a) Show that $\Sigma_{n=0}^{\infty}\, x^n/n!$ converges for all x.
(b) Deduce that $\lim_{n\to\infty} x^n/n! = 0$ for all x.

34. Let $\Sigma\, a_n$ be a series with positive terms and let $r_n = a_{n+1}/a_n$. Suppose that $\lim_{n\to\infty} r_n = L < 1$, so $\Sigma\, a_n$ converges by the Ratio Test. As usual, we let R_n be the remainder after n terms, that is,

$$R_n = a_{n+1} + a_{n+2} + a_{n+3} + \cdots$$

(a) If $\{r_n\}$ is a decreasing sequence and $r_{n+1} < 1$, show, by summing a geometric series, that

$$R_n \leqslant \frac{a_{n+1}}{1 - r_{n+1}}$$

(b) If $\{r_n\}$ is an increasing sequence, show that

$$R_n \leqslant \frac{a_{n+1}}{1 - L}$$

35. (a) Find the partial sum s_5 of the series $\Sigma_{n=1}^{\infty}\, 1/n2^n$. Use Exercise 34 to estimate the error in using s_5 as an approximation to the sum of the series.
(b) Find a value of n so that s_n is within 0.00005 of the sum. Use this value of n to approximate the sum of the series.

36. Use the sum of the first 10 terms to approximate the sum of the series

$$\sum_{n=1}^{\infty} \frac{n}{2^n}$$

Use Exercise 34 to estimate the error.

37. Prove that if $\Sigma\, a_n$ is absolutely convergent, then

$$\left|\sum_{n=1}^{\infty} a_n\right| \leqslant \sum_{n=1}^{\infty} |a_n|$$

38. Prove the Root Test. [*Hint for part (i):* Take any number r such that $L < r < 1$ and use the fact that there is an integer N such that $\sqrt[n]{|a_n|} < r$ whenever $n \geqslant N$.]

39. Given any series $\Sigma\, a_n$ we define a series $\Sigma\, a_n^+$ whose terms are all the positive terms of $\Sigma\, a_n$ and a series $\Sigma\, a_n^-$ whose terms are all the negative terms of $\Sigma\, a_n$. To be specific, we let

$$a_n^+ = \frac{a_n + |a_n|}{2} \qquad a_n^- = \frac{a_n - |a_n|}{2}$$

Notice that if $a_n > 0$, then $a_n^+ = a_n$ and $a_n^- = 0$, whereas if $a_n < 0$, then $a_n^- = a_n$ and $a_n^+ = 0$.
(a) If $\Sigma\, a_n$ is absolutely convergent, show that both of the series $\Sigma\, a_n^+$ and $\Sigma\, a_n^-$ are convergent.
(b) If $\Sigma\, a_n$ is conditionally convergent, show that both of the series $\Sigma\, a_n^+$ and $\Sigma\, a_n^-$ are divergent.

40. Prove that if $\Sigma\, a_n$ is a conditionally convergent series and r is any real number, then there is a rearrangement of $\Sigma\, a_n$ whose sum is r. [*Hints:* Use the notation of Exercise 39. Take just enough positive terms a_n^+ so that their sum is greater than r. Then add just enough negative terms a_n^- so that the cumulative sum is less than r. Continue in this manner and use Theorem 11.2.6.]

11.7 Strategy for Testing Series

We now have several ways of testing a series for convergence or divergence; the problem is to decide which test to use on which series. In this respect testing series is similar to integrating functions. Again there are no hard and fast rules about which test to apply to a given series, but you may find the following advice of some use.

It is not wise to apply a list of the tests in a specific order until one finally works. That would be a waste of time and effort. Instead, as with integration, the main strategy is to classify the series according to its *form*.

1. If the series is of the form $\Sigma\, 1/n^p$, it is a p-series, which we know to be convergent if $p > 1$ and divergent if $p \leqslant 1$.
2. If the series has the form $\Sigma\, ar^{n-1}$ or $\Sigma\, ar^n$, it is a geometric series, which converges if $|r| < 1$ and diverges if $|r| \geqslant 1$. Some preliminary algebraic manipulation may be required to bring the series into this form.
3. If the series has a form that is similar to a p-series or a geometric series, then one of the comparison tests should be considered. In particular, if a_n is a rational function or algebraic function of n (involving roots of polynomials), then the series should be compared with a p-series. Notice that most of the series in Exercises 11.4 have this form. (The value of p should be chosen as in Section 11.4 by keeping only the highest powers of n in the numerator and denominator.) The comparison tests apply only to series with positive terms, but if $\Sigma\, a_n$ has some negative terms, then we can apply the Comparison Test to $\Sigma\, |a_n|$ and test for absolute convergence.
4. If you can see at a glance that $\lim_{n\to\infty} a_n \neq 0$, then the Test for Divergence should be used.
5. If the series is of the form $\Sigma\, (-1)^{n-1}b_n$ or $\Sigma\, (-1)^n b_n$, then the Alternating Series Test is an obvious possibility.
6. Series that involve factorials or other products (including a constant raised to the nth power) are often conveniently tested using the Ratio Test. Bear in mind that $|a_{n+1}/a_n| \to 1$ as $n \to \infty$ for all p-series and therefore all rational or algebraic functions of n. Thus, the Ratio Test should not be used for such series.
7. If a_n is of the form $(b_n)^n$, then the Root Test may be useful.
8. If $a_n = f(n)$, where $\int_1^\infty f(x)\,dx$ is easily evaluated, then the Integral Test is effective (assuming the hypotheses of this test are satisfied).

In the following examples we don't work out all the details but simply indicate which tests should be used.

EXAMPLE 1 $\displaystyle\sum_{n=1}^{\infty} \frac{n-1}{2n+1}$

Since $a_n \to \frac{1}{2} \neq 0$ as $n \to \infty$, we should use the Test for Divergence.

EXAMPLE 2 $\displaystyle\sum_{n=1}^{\infty} \frac{\sqrt{n^3+1}}{3n^3+4n^2+2}$

Since a_n is an algebraic function of n, we compare the given series with a p-series.

The comparison series for the Limit Comparison Test is $\Sigma\, b_n$, where

$$b_n = \frac{\sqrt{n^3}}{3n^3} = \frac{n^{3/2}}{3n^3} = \frac{1}{3n^{3/2}}$$

EXAMPLE 3 $\sum_{n=1}^{\infty} ne^{-n^2}$

Since the integral $\int_1^\infty xe^{-x^2}\,dx$ is easily evaluated, we use the Integral Test. The Ratio Test also works.

EXAMPLE 4 $\sum_{n=1}^{\infty} (-1)^n \frac{n^3}{n^4+1}$

Since the series is alternating, we use the Alternating Series Test.

EXAMPLE 5 $\sum_{k=1}^{\infty} \frac{2^k}{k!}$

Since the series involves $k!$, we use the Ratio Test.

EXAMPLE 6 $\sum_{n=1}^{\infty} \frac{1}{2+3^n}$

Since the series is closely related to the geometric series $\Sigma\, 1/3^n$, we use the Comparison Test.

11.7 Exercises

1–38 Test the series for convergence or divergence.

1. $\sum_{n=1}^{\infty} \frac{n^2-1}{n^2+n}$

2. $\sum_{n=1}^{\infty} \frac{n-1}{n^2+n}$

3. $\sum_{n=1}^{\infty} \frac{1}{n^2+n}$

4. $\sum_{n=1}^{\infty} (-1)^{n-1} \frac{n-1}{n^2+n}$

5. $\sum_{n=1}^{\infty} \frac{(-3)^{n+1}}{2^{3n}}$

6. $\sum_{n=1}^{\infty} \left(\frac{3n}{1+8n}\right)^n$

7. $\sum_{n=2}^{\infty} \frac{1}{n\sqrt{\ln n}}$

8. $\sum_{k=1}^{\infty} \frac{2^k k!}{(k+2)!}$

9. $\sum_{k=1}^{\infty} k^2 e^{-k}$

10. $\sum_{n=1}^{\infty} n^2 e^{-n^3}$

11. $\sum_{n=2}^{\infty} \frac{(-1)^{n+1}}{n \ln n}$

12. $\sum_{n=1}^{\infty} (-1)^n \frac{n}{n^2+25}$

13. $\sum_{n=1}^{\infty} \frac{3^n n^2}{n!}$

14. $\sum_{n=1}^{\infty} \sin n$

15. $\sum_{n=0}^{\infty} \frac{n!}{2 \cdot 5 \cdot 8 \cdot \cdots \cdot (3n+2)}$

16. $\sum_{n=1}^{\infty} \frac{n^2+1}{n^3+1}$

17. $\sum_{n=1}^{\infty} (-1)^n 2^{1/n}$

18. $\sum_{n=2}^{\infty} \frac{(-1)^{n-1}}{\sqrt{n}-1}$

19. $\sum_{n=1}^{\infty} (-1)^n \frac{\ln n}{\sqrt{n}}$

20. $\sum_{k=1}^{\infty} \frac{k+5}{5^k}$

21. $\sum_{n=1}^{\infty} \frac{(-2)^{2n}}{n^n}$

22. $\sum_{n=1}^{\infty} \frac{\sqrt{n^2-1}}{n^3+2n^2+5}$

23. $\sum_{n=1}^{\infty} \tan(1/n)$

24. $\sum_{n=1}^{\infty} \frac{\cos(n/2)}{n^2+4n}$

25. $\sum_{n=1}^{\infty} \frac{n!}{e^{n^2}}$

26. $\sum_{n=1}^{\infty} \frac{n^2+1}{5^n}$

27. $\sum_{k=1}^{\infty} \frac{k \ln k}{(k+1)^3}$

28. $\sum_{n=1}^{\infty} \frac{e^{1/n}}{n^2}$

29. $\sum_{n=1}^{\infty} \frac{\tan^{-1} n}{n\sqrt{n}}$

30. $\sum_{j=1}^{\infty} (-1)^j \frac{\sqrt{j}}{j+5}$

31. $\sum_{k=1}^{\infty} \frac{5^k}{3^k+4^k}$

32. $\sum_{n=1}^{\infty} \frac{(2n)^n}{n^{2n}}$

33. $\sum_{n=1}^{\infty} \frac{\sin(1/n)}{\sqrt{n}}$

34. $\sum_{n=1}^{\infty} \frac{1}{n+n\cos^2 n}$

35. $\sum_{n=1}^{\infty} \left(\frac{n}{n+1}\right)^{n^2}$

36. $\sum_{n=2}^{\infty} \frac{1}{(\ln n)^{\ln n}}$

37. $\sum_{n=1}^{\infty} (\sqrt[n]{2}-1)^n$

38. $\sum_{n=1}^{\infty} (\sqrt[n]{2}-1)$

11.8 Power Series

A **power series** is a series of the form

1 $$\sum_{n=0}^{\infty} c_n x^n = c_0 + c_1 x + c_2 x^2 + c_3 x^3 + \cdots$$

where x is a variable and the c_n's are constants called the **coefficients** of the series. For each fixed x, the series (1) is a series of constants that we can test for convergence or divergence. A power series may converge for some values of x and diverge for other values of x. The sum of the series is a function

$$f(x) = c_0 + c_1 x + c_2 x^2 + \cdots + c_n x^n + \cdots$$

whose domain is the set of all x for which the series converges. Notice that f resembles a polynomial. The only difference is that f has infinitely many terms.

For instance, if we take $c_n = 1$ for all n, the power series becomes the geometric series

$$\sum_{n=0}^{\infty} x^n = 1 + x + x^2 + \cdots + x^n + \cdots$$

which converges when $-1 < x < 1$ and diverges when $|x| \geqslant 1$ (see Equation 11.2.5).

More generally, a series of the form

2 $$\sum_{n=0}^{\infty} c_n (x-a)^n = c_0 + c_1(x-a) + c_2(x-a)^2 + \cdots$$

is called a **power series in $(x - a)$** or a **power series centered at a** or a **power series about a**. Notice that in writing out the term corresponding to $n = 0$ in Equations 1 and 2 we have adopted the convention that $(x-a)^0 = 1$ even when $x = a$. Notice also that when $x = a$ all of the terms are 0 for $n \geqslant 1$ and so the power series (2) always converges when $x = a$.

TRIGONOMETRIC SERIES

A power series is a series in which each term is a power function. A **trigonometric series**

$$\sum_{n=0}^{\infty} (a_n \cos nx + b_n \sin nx)$$

is a series whose terms are trigonometric functions. This type of series is discussed on the web site

www.stewartcalculus.com

Click on *Additional Topics* and then on *Fourier Series.*

EXAMPLE 1 For what values of x is the series $\sum_{n=0}^{\infty} n!x^n$ convergent?

SOLUTION We use the Ratio Test. If we let a_n, as usual, denote the nth term of the series, then $a_n = n!x^n$. If $x \neq 0$, we have

$$\lim_{n\to\infty}\left|\frac{a_{n+1}}{a_n}\right| = \lim_{n\to\infty}\left|\frac{(n+1)!x^{n+1}}{n!x^n}\right| = \lim_{n\to\infty}(n+1)|x| = \infty$$

By the Ratio Test, the series diverges when $x \neq 0$. Thus, the given series converges only when $x = 0$.

EXAMPLE 2 For what values of x does the series $\sum_{n=1}^{\infty} \frac{(x-3)^n}{n}$ converge?

SOLUTION Let $a_n = (x-3)^n/n$. Then

$$\left|\frac{a_{n+1}}{a_n}\right| = \left|\frac{(x-3)^{n+1}}{n+1} \cdot \frac{n}{(x-3)^n}\right|$$

$$= \frac{1}{1 + \frac{1}{n}}|x-3| \to |x-3| \qquad \text{as } n \to \infty$$

By the Ratio Test, the given series is absolutely convergent, and therefore convergent, when $|x - 3| < 1$ and divergent when $|x - 3| > 1$. Now

$$|x - 3| < 1 \iff -1 < x - 3 < 1 \iff 2 < x < 4$$

so the series converges when $2 < x < 4$ and diverges when $x < 2$ or $x > 4$.

The Ratio Test gives no information when $|x - 3| = 1$ so we must consider $x = 2$ and $x = 4$ separately. If we put $x = 4$ in the series, it becomes $\Sigma\, 1/n$, the harmonic series, which is divergent. If $x = 2$, the series is $\Sigma\, (-1)^n/n$, which converges by the Alternating Series Test. Thus, the given power series converges for $2 \leqslant x < 4$.

We will see that the main use of a power series is that it provides a way to represent some of the most important functions that arise in mathematics, physics, and chemistry. In particular, the sum of the power series in the next example is called a **Bessel function**, after the German astronomer Friedrich Bessel (1784–1846), and the function given in Exercise 33 is another example of a Bessel function. In fact, these functions first arose when Bessel solved Kepler's equation for describing planetary motion. Since that time, these functions have been applied in many different physical situations, including the temperature distribution in a circular plate and the shape of a vibrating drumhead (see the photographs on page 700).

EXAMPLE 3 Find the domain of the Bessel function of order 0 defined by

$$J_0(x) = \sum_{n=0}^{\infty} \frac{(-1)^n x^{2n}}{2^{2n}(n!)^2}$$

SOLUTION Let $a_n = (-1)^n x^{2n}/[2^{2n}(n!)^2]$. Then

$$\left|\frac{a_{n+1}}{a_n}\right| = \left|\frac{(-1)^{n+1}x^{2(n+1)}}{2^{2(n+1)}[(n+1)!]^2} \cdot \frac{2^{2n}(n!)^2}{(-1)^n x^{2n}}\right|$$

$$= \frac{x^{2n+2}}{2^{2n+2}(n+1)^2(n!)^2} \cdot \frac{2^{2n}(n!)^2}{x^{2n}}$$

$$= \frac{x^2}{4(n+1)^2} \to 0 < 1 \qquad \text{for all } x$$

Thus, by the Ratio Test, the given series converges for all values of x. In other words, the domain of the Bessel function J_0 is $(-\infty, \infty) = \mathbb{R}$.

Recall that the sum of a series is equal to the limit of the sequence of partial sums. So when we define the Bessel function in Example 3 as the sum of a series we mean that, for every real number x,

$$J_0(x) = \lim_{n\to\infty} s_n(x) \qquad \text{where} \qquad s_n(x) = \sum_{i=0}^{n} \frac{(-1)^i x^{2i}}{2^{2i}(i!)^2}$$

The first few partial sums are

$$s_0(x) = 1 \qquad s_1(x) = 1 - \frac{x^2}{4} \qquad s_2(x) = 1 - \frac{x^2}{4} + \frac{x^4}{64}$$

$$s_3(x) = 1 - \frac{x^2}{4} + \frac{x^4}{64} - \frac{x^6}{2304} \qquad s_4(x) = 1 - \frac{x^2}{4} + \frac{x^4}{64} - \frac{x^6}{2304} + \frac{x^8}{147{,}456}$$

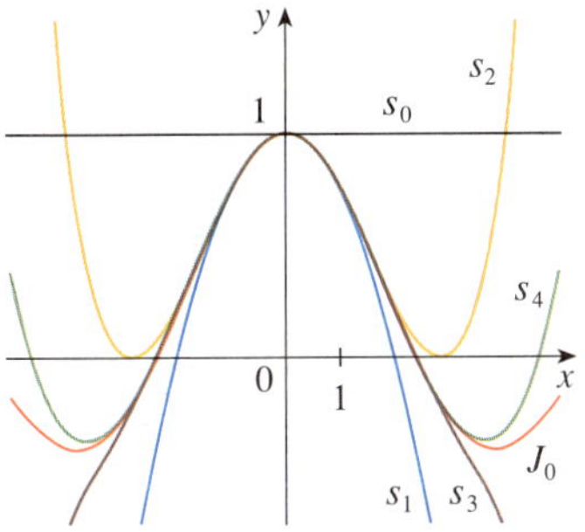

FIGURE 1
Partial sums of the Bessel function J_0

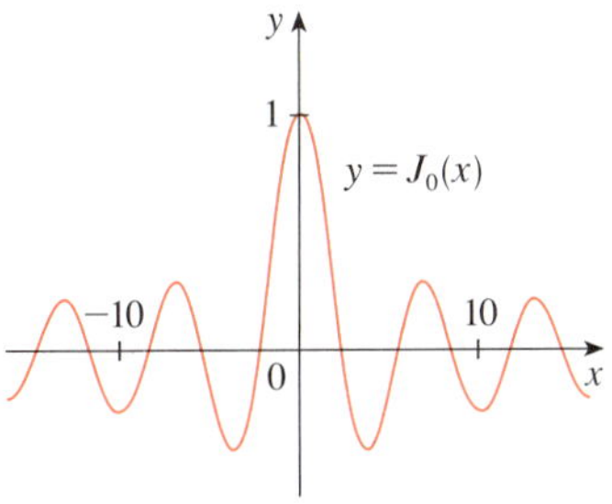

FIGURE 2

Figure 1 shows the graphs of these partial sums, which are polynomials. They are all approximations to the function J_0, but notice that the approximations become better when more terms are included. Figure 2 shows a more complete graph of the Bessel function.

For the power series that we have looked at so far, the set of values of x for which the series is convergent has always turned out to be an interval [a finite interval for the geometric series and the series in Example 2, the infinite interval $(-\infty, \infty)$ in Example 3, and a collapsed interval $[0, 0] = \{0\}$ in Example 1]. The following theorem, proved in Appendix F, says that this is true in general.

3 Theorem For a given power series $\sum_{n=0}^{\infty} c_n(x-a)^n$ there are only three possibilities:

(i) The series converges only when $x = a$.

(ii) The series converges for all x.

(iii) There is a positive number R such that the series converges if $|x - a| < R$ and diverges if $|x - a| > R$.

The number R in case (iii) is called the **radius of convergence** of the power series. By convention, the radius of convergence is $R = 0$ in case (i) and $R = \infty$ in case (ii). The **interval of convergence** of a power series is the interval that consists of all values of x for which the series converges. In case (i) the interval consists of just a single point a. In case (ii) the interval is $(-\infty, \infty)$. In case (iii) note that the inequality $|x - a| < R$ can be rewritten as $a - R < x < a + R$. When x is an *endpoint* of the interval, that is, $x = a \pm R$, anything can happen—the series might converge at one or both endpoints or it might diverge at both endpoints. Thus, in case (iii) there are four possibilities for the interval of convergence:

$$(a - R, a + R) \qquad (a - R, a + R] \qquad [a - R, a + R) \qquad [a - R, a + R]$$

The situation is illustrated in Figure 3.

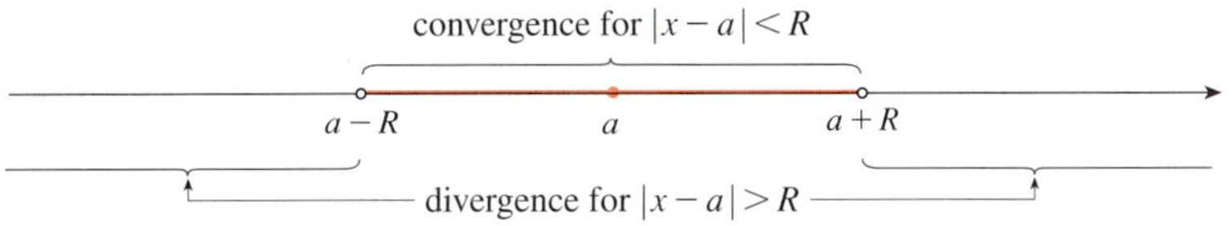

FIGURE 3

We summarize here the radius and interval of convergence for each of the examples already considered in this section.

	Series	Radius of convergence	Interval of convergence
Geometric series	$\sum_{n=0}^{\infty} x^n$	$R = 1$	$(-1, 1)$
Example 1	$\sum_{n=0}^{\infty} n!\, x^n$	$R = 0$	$\{0\}$
Example 2	$\sum_{n=1}^{\infty} \frac{(x-3)^n}{n}$	$R = 1$	$[2, 4)$
Example 3	$\sum_{n=0}^{\infty} \frac{(-1)^n x^{2n}}{2^{2n}(n!)^2}$	$R = \infty$	$(-\infty, \infty)$

In general, the Ratio Test (or sometimes the Root Test) should be used to determine the radius of convergence R. The Ratio and Root Tests always fail when x is an endpoint of the interval of convergence, so the endpoints must be checked with some other test.

EXAMPLE 4 Find the radius of convergence and interval of convergence of the series

$$\sum_{n=0}^{\infty} \frac{(-3)^n x^n}{\sqrt{n+1}}$$

SOLUTION Let $a_n = (-3)^n x^n/\sqrt{n+1}$. Then

$$\left|\frac{a_{n+1}}{a_n}\right| = \left|\frac{(-3)^{n+1}x^{n+1}}{\sqrt{n+2}} \cdot \frac{\sqrt{n+1}}{(-3)^n x^n}\right| = \left|-3x\sqrt{\frac{n+1}{n+2}}\right|$$

$$= 3\sqrt{\frac{1+(1/n)}{1+(2/n)}}\,|x| \to 3|x| \qquad \text{as } n \to \infty$$

By the Ratio Test, the given series converges if $3|x| < 1$ and diverges if $3|x| > 1$. Thus, it converges if $|x| < \frac{1}{3}$ and diverges if $|x| > \frac{1}{3}$. This means that the radius of convergence is $R = \frac{1}{3}$.

We know the series converges in the interval $\left(-\frac{1}{3}, \frac{1}{3}\right)$, but we must now test for convergence at the endpoints of this interval. If $x = -\frac{1}{3}$, the series becomes

$$\sum_{n=0}^{\infty} \frac{(-3)^n\left(-\frac{1}{3}\right)^n}{\sqrt{n+1}} = \sum_{n=0}^{\infty} \frac{1}{\sqrt{n+1}} = \frac{1}{\sqrt{1}} + \frac{1}{\sqrt{2}} + \frac{1}{\sqrt{3}} + \frac{1}{\sqrt{4}} + \cdots$$

which diverges. (Use the Integral Test or simply observe that it is a p-series with $p = \frac{1}{2} < 1$.) If $x = \frac{1}{3}$, the series is

$$\sum_{n=0}^{\infty} \frac{(-3)^n\left(\frac{1}{3}\right)^n}{\sqrt{n+1}} = \sum_{n=0}^{\infty} \frac{(-1)^n}{\sqrt{n+1}}$$

which converges by the Alternating Series Test. Therefore, the given power series converges when $-\frac{1}{3} < x \leqslant \frac{1}{3}$, so the interval of convergence is $\left(-\frac{1}{3}, \frac{1}{3}\right]$.

EXAMPLE 5 Find the radius of convergence and interval of convergence of the series

$$\sum_{n=0}^{\infty} \frac{n(x+2)^n}{3^{n+1}}$$

SOLUTION If $a_n = n(x+2)^n/3^{n+1}$, then

$$\left|\frac{a_{n+1}}{a_n}\right| = \left|\frac{(n+1)(x+2)^{n+1}}{3^{n+2}} \cdot \frac{3^{n+1}}{n(x+2)^n}\right|$$

$$= \left(1 + \frac{1}{n}\right)\frac{|x+2|}{3} \to \frac{|x+2|}{3} \qquad \text{as } n \to \infty$$

Using the Ratio Test, we see that the series converges if $|x+2|/3 < 1$ and it diverges if $|x+2|/3 > 1$. So it converges if $|x+2| < 3$ and diverges if $|x+2| > 3$. Thus, the radius of convergence is $R = 3$.

The inequality $|x + 2| < 3$ can be written as $-5 < x < 1$, so we test the series at the endpoints -5 and 1. When $x = -5$, the series is

$$\sum_{n=0}^{\infty} \frac{n(-3)^n}{3^{n+1}} = \frac{1}{3} \sum_{n=0}^{\infty} (-1)^n n$$

which diverges by the Test for Divergence [$(-1)^n n$ doesn't converge to 0]. When $x = 1$, the series is

$$\sum_{n=0}^{\infty} \frac{n(3)^n}{3^{n+1}} = \frac{1}{3} \sum_{n=0}^{\infty} n$$

which also diverges by the Test for Divergence. Thus, the series converges only when $-5 < x < 1$, so the interval of convergence is $(-5, 1)$.

11.8 Exercises

1. What is a power series?

2. (a) What is the radius of convergence of a power series? How do you find it?
(b) What is the interval of convergence of a power series? How do you find it?

3–28 Find the radius of convergence and interval of convergence of the series.

3. $\sum_{n=1}^{\infty} \frac{x^n}{\sqrt{n}}$

4. $\sum_{n=0}^{\infty} \frac{(-1)^n x^n}{n+1}$

5. $\sum_{n=1}^{\infty} \frac{(-1)^{n-1} x^n}{n^3}$

6. $\sum_{n=1}^{\infty} \sqrt{n}\, x^n$

7. $\sum_{n=0}^{\infty} \frac{x^n}{n!}$

8. $\sum_{n=1}^{\infty} n^n x^n$

9. $\sum_{n=1}^{\infty} (-1)^n n 4^n x^n$

10. $\sum_{n=1}^{\infty} \frac{x^n}{n 3^n}$

11. $\sum_{n=1}^{\infty} \frac{(-2)^n x^n}{\sqrt[4]{n}}$

12. $\sum_{n=1}^{\infty} \frac{x^n}{5^n n^5}$

13. $\sum_{n=2}^{\infty} (-1)^n \frac{x^n}{4^n \ln n}$

14. $\sum_{n=0}^{\infty} (-1)^n \frac{x^{2n}}{(2n)!}$

15. $\sum_{n=0}^{\infty} \sqrt{n}\,(x-1)^n$

16. $\sum_{n=0}^{\infty} n^3 (x-5)^n$

17. $\sum_{n=1}^{\infty} (-1)^n \frac{(x+2)^n}{n 2^n}$

18. $\sum_{n=1}^{\infty} \frac{(-2)^n}{\sqrt{n}} (x+3)^n$

19. $\sum_{n=1}^{\infty} \frac{(x-2)^n}{n^n}$

20. $\sum_{n=1}^{\infty} \frac{(3x-2)^n}{n 3^n}$

21. $\sum_{n=1}^{\infty} \frac{n}{b^n} (x-a)^n, \quad b > 0$

22. $\sum_{n=1}^{\infty} \frac{n(x-4)^n}{n^3+1}$

23. $\sum_{n=1}^{\infty} n!(2x-1)^n$

24. $\sum_{n=1}^{\infty} \frac{n^2 x^n}{2 \cdot 4 \cdot 6 \cdot \cdots \cdot (2n)}$

25. $\sum_{n=1}^{\infty} \frac{(4x+1)^n}{n^2}$

26. $\sum_{n=2}^{\infty} (-1)^n \frac{(2x+3)^n}{n \ln n}$

27. $\sum_{n=2}^{\infty} \frac{x^n}{(\ln n)^n}$

28. $\sum_{n=1}^{\infty} \frac{2 \cdot 4 \cdot 6 \cdot \cdots \cdot (2n)}{1 \cdot 3 \cdot 5 \cdot \cdots \cdot (2n-1)} x^n$

29. If $\sum_{n=0}^{\infty} c_n 4^n$ is convergent, does it follow that the following series are convergent?

(a) $\sum_{n=0}^{\infty} c_n(-2)^n$ (b) $\sum_{n=0}^{\infty} c_n(-4)^n$

30. Suppose that $\sum_{n=0}^{\infty} c_n x^n$ converges when $x = -4$ and diverges when $x = 6$. What can be said about the convergence or divergence of the following series?

(a) $\sum_{n=0}^{\infty} c_n$ (b) $\sum_{n=0}^{\infty} c_n 8^n$

(c) $\sum_{n=0}^{\infty} c_n(-3)^n$ (d) $\sum_{n=0}^{\infty} (-1)^n c_n 9^n$

31. If k is a positive integer, find the radius of convergence of the series

$$\sum_{n=0}^{\infty} \frac{(n!)^k}{(kn)!} x^n$$

32. Graph the first several partial sums $s_n(x)$ of the series $\sum_{n=0}^{\infty} x^n$, together with the sum function $f(x) = 1/(1-x)$, on a common screen. On what interval do these partial sums appear to be converging to $f(x)$?

33. The function J_1 defined by

$$J_1(x) = \sum_{n=0}^{\infty} \frac{(-1)^n x^{2n+1}}{n!(n+1)!2^{2n+1}}$$

is called the *Bessel function of order 1*.
(a) Find its domain.
(b) Graph the first several partial sums on a common screen.

CAS (c) If your CAS has built-in Bessel functions, graph J_1 on the same screen as the partial sums in part (b) and observe how the partial sums approximate J_1.

34. The function A defined by

$$A(x) = 1 + \frac{x^3}{2 \cdot 3} + \frac{x^6}{2 \cdot 3 \cdot 5 \cdot 6} + \frac{x^9}{2 \cdot 3 \cdot 5 \cdot 6 \cdot 8 \cdot 9} + \cdots$$

is called the *Airy function* after the English mathematician and astronomer Sir George Airy (1801–1892).

(a) Find the domain of the Airy function.

(b) Graph the first several partial sums $s_n(x)$ on a common screen.

CAS (c) If your CAS has built-in Airy functions, graph A on the same screen as the partial sums in part (b) and observe how the partial sums approximate A.

35. A function f is defined by

$$f(x) = 1 + 2x + x^2 + 2x^3 + x^4 + \cdots$$

that is, its coefficients are $c_{2n} = 1$ and $c_{2n+1} = 2$ for all $n \geqslant 0$. Find the interval of convergence of the series and find an explicit formula for $f(x)$.

36. If $f(x) = \sum_{n=0}^{\infty} c_n x^n$, where $c_{n+4} = c_n$ for all $n \geqslant 0$, find the interval of convergence of the series and a formula for $f(x)$.

37. Show that if $\lim_{n \to \infty} \sqrt[n]{|c_n|} = c$, where $c \neq 0$, then the radius of convergence of the power series $\sum c_n x^n$ is $R = 1/c$.

38. Suppose that the power series $\sum c_n(x - a)^n$ satisfies $c_n \neq 0$ for all n. Show that if $\lim_{n \to \infty} |c_n/c_{n+1}|$ exists, then it is equal to the radius of convergence of the power series.

39. Suppose the series $\sum c_n x^n$ has radius of convergence 2 and the series $\sum d_n x^n$ has radius of convergence 3. What is the radius of convergence of the series $\sum (c_n + d_n)x^n$?

40. Suppose that the radius of convergence of the power series $\sum c_n x^n$ is R. What is the radius of convergence of the power series $\sum c_n x^{2n}$?

11.9 Representations of Functions as Power Series

In this section we learn how to represent certain types of functions as sums of power series by manipulating geometric series or by differentiating or integrating such a series. You might wonder why we would ever want to express a known function as a sum of infinitely many terms. We will see later that this strategy is useful for integrating functions that don't have elementary antiderivatives, for solving differential equations, and for approximating functions by polynomials. (Scientists do this to simplify the expressions they deal with; computer scientists do this to represent functions on calculators and computers.)

We start with an equation that we have seen before:

1 $$\frac{1}{1 - x} = 1 + x + x^2 + x^3 + \cdots = \sum_{n=0}^{\infty} x^n \qquad |x| < 1$$

A geometric illustration of Equation 1 is shown in Figure 1. Because the sum of a series is the limit of the sequence of partial sums, we have

$$\frac{1}{1 - x} = \lim_{n \to \infty} s_n(x)$$

where

$$s_n(x) = 1 + x + x^2 + \cdots + x^n$$

is the nth partial sum. Notice that as n increases, $s_n(x)$ becomes a better approximation to $f(x)$ for $-1 < x < 1$.

We first encountered this equation in Example 5 in Section 11.2, where we obtained it by observing that it is a geometric series with $a = 1$ and $r = x$. But here our point of view is different. We now regard Equation 1 as expressing the function $f(x) = 1/(1 - x)$ as a sum of a power series.

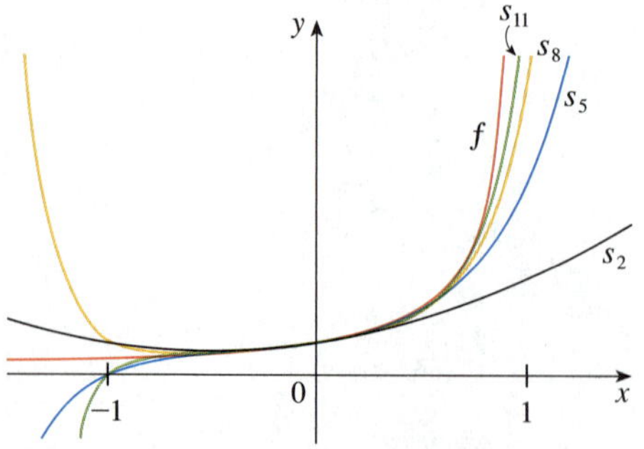

FIGURE 1

$f(x) = \dfrac{1}{1 - x}$ and some partial sums

EXAMPLE 1 Express $1/(1 + x^2)$ as the sum of a power series and find the interval of convergence.

SOLUTION Replacing x by $-x^2$ in Equation 1, we have

$$\frac{1}{1 + x^2} = \frac{1}{1 - (-x^2)} = \sum_{n=0}^{\infty} (-x^2)^n$$

$$= \sum_{n=0}^{\infty} (-1)^n x^{2n} = 1 - x^2 + x^4 - x^6 + x^8 - \cdots$$

Because this is a geometric series, it converges when $|-x^2| < 1$, that is, $x^2 < 1$, or $|x| < 1$. Therefore, the interval of convergence is $(-1, 1)$. (Of course, we could have determined the radius of convergence by applying the Ratio Test, but that much work is unnecessary here.)

EXAMPLE 2 Find a power series representation for $1/(x + 2)$.

SOLUTION In order to put this function in the form of the left side of Equation 1 we first factor a 2 from the denominator:

$$\frac{1}{2 + x} = \frac{1}{2\left(1 + \dfrac{x}{2}\right)} = \frac{1}{2\left[1 - \left(-\dfrac{x}{2}\right)\right]}$$

$$= \frac{1}{2} \sum_{n=0}^{\infty} \left(-\frac{x}{2}\right)^n = \sum_{n=0}^{\infty} \frac{(-1)^n}{2^{n+1}} x^n$$

This series converges when $|-x/2| < 1$, that is, $|x| < 2$. So the interval of convergence is $(-2, 2)$.

EXAMPLE 3 Find a power series representation of $x^3/(x + 2)$.

SOLUTION Since this function is just x^3 times the function in Example 2, all we have to do is to multiply that series by x^3:

|||| It's legitimate to move x^3 across the sigma sign because it doesn't depend on n. [Use Theorem 11.2.8(i) with $c = x^3$.]

$$\frac{x^3}{x + 2} = x^3 \cdot \frac{1}{x + 2} = x^3 \sum_{n=0}^{\infty} \frac{(-1)^n}{2^{n+1}} x^n = \sum_{n=0}^{\infty} \frac{(-1)^n}{2^{n+1}} x^{n+3}$$

$$= \tfrac{1}{2}x^3 - \tfrac{1}{4}x^4 + \tfrac{1}{8}x^5 - \tfrac{1}{16}x^6 + \cdots$$

Another way of writing this series is as follows:

$$\frac{x^3}{x + 2} = \sum_{n=3}^{\infty} \frac{(-1)^{n-1}}{2^{n-2}} x^n$$

As in Example 2, the interval of convergence is $(-2, 2)$.

|||| Differentiation and Integration of Power Series

The sum of a power series is a function $f(x) = \sum_{n=0}^{\infty} c_n(x - a)^n$ whose domain is the interval of convergence of the series. We would like to be able to differentiate and integrate such functions, and the following theorem (which we won't prove) says that we can do so by differentiating or integrating each individual term in the series, just as we would for a polynomial. This is called **term-by-term differentiation and integration**.

2 Theorem If the power series $\sum c_n(x-a)^n$ has radius of convergence $R > 0$, then the function f defined by

$$f(x) = c_0 + c_1(x-a) + c_2(x-a)^2 + \cdots = \sum_{n=0}^{\infty} c_n(x-a)^n$$

is differentiable (and therefore continuous) on the interval $(a-R, a+R)$ and

(i) $$f'(x) = c_1 + 2c_2(x-a) + 3c_3(x-a)^2 + \cdots = \sum_{n=1}^{\infty} nc_n(x-a)^{n-1}$$

(ii) $$\int f(x)\,dx = C + c_0(x-a) + c_1\frac{(x-a)^2}{2} + c_2\frac{(x-a)^3}{3} + \cdots$$

$$= C + \sum_{n=0}^{\infty} c_n\frac{(x-a)^{n+1}}{n+1}$$

The radii of convergence of the power series in Equations (i) and (ii) are both R.

|||| In part (ii), $\int c_0\,dx = c_0x + C_1$ is written as $c_0(x-a) + C$, where $C = C_1 + ac_0$, so all the terms of the series have the same form.

NOTE 1 ▫ Equations (i) and (ii) in Theorem 2 can be rewritten in the form

(iii) $$\frac{d}{dx}\left[\sum_{n=0}^{\infty} c_n(x-a)^n\right] = \sum_{n=0}^{\infty}\frac{d}{dx}\left[c_n(x-a)^n\right]$$

(iv) $$\int\left[\sum_{n=0}^{\infty} c_n(x-a)^n\right]dx = \sum_{n=0}^{\infty}\int c_n(x-a)^n\,dx$$

We know that, for finite sums, the derivative of a sum is the sum of the derivatives and the integral of a sum is the sum of the integrals. Equations (iii) and (iv) assert that the same is true for infinite sums, provided we are dealing with *power series.* (For other types of series of functions the situation is not as simple; see Exercise 36.)

NOTE 2 ▫ Although Theorem 2 says that the radius of convergence remains the same when a power series is differentiated or integrated, this does not mean that the *interval* of convergence remains the same. It may happen that the original series converges at an endpoint, whereas the differentiated series diverges there. (See Exercise 37.)

NOTE 3 ▫ The idea of differentiating a power series term by term is the basis for a powerful method for solving differential equations. We will discuss this method in Chapter 17.

EXAMPLE 4 In Example 3 in Section 11.8 we saw that the Bessel function

$$J_0(x) = \sum_{n=0}^{\infty}\frac{(-1)^n x^{2n}}{2^{2n}(n!)^2}$$

is defined for all x. Thus, by Theorem 2, J_0 is differentiable for all x and its derivative is found by term-by-term differentiation as follows:

$$J_0'(x) = \sum_{n=0}^{\infty}\frac{d}{dx}\frac{(-1)^n x^{2n}}{2^{2n}(n!)^2} = \sum_{n=1}^{\infty}\frac{(-1)^n 2nx^{2n-1}}{2^{2n}(n!)^2}$$

EXAMPLE 5 Express $1/(1-x)^2$ as a power series by differentiating Equation 1. What is the radius of convergence?

SOLUTION Differentiating each side of the equation

$$\frac{1}{1-x} = 1 + x + x^2 + x^3 + \cdots = \sum_{n=0}^{\infty} x^n$$

we get
$$\frac{1}{(1-x)^2} = 1 + 2x + 3x^2 + \cdots = \sum_{n=1}^{\infty} nx^{n-1}$$

If we wish, we can replace n by $n + 1$ and write the answer as

$$\frac{1}{(1-x)^2} = \sum_{n=0}^{\infty} (n+1)x^n$$

According to Theorem 2, the radius of convergence of the differentiated series is the same as the radius of convergence of the original series, namely, $R = 1$.

EXAMPLE 6 Find a power series representation for $\ln(1 - x)$ and its radius of convergence.

SOLUTION We notice that, except for a factor of -1, the derivative of this function is $1/(1 - x)$. So we integrate both sides of Equation 1:

$$-\ln(1-x) = \int \frac{1}{1-x}\,dx = \int (1 + x + x^2 + \cdots)\,dx$$

$$= x + \frac{x^2}{2} + \frac{x^3}{3} + \cdots + C = \sum_{n=0}^{\infty} \frac{x^{n+1}}{n+1} + C$$

$$= \sum_{n=1}^{\infty} \frac{x^n}{n} + C \qquad |x| < 1$$

To determine the value of C we put $x = 0$ in this equation and obtain $-\ln(1 - 0) = C$. Thus, $C = 0$ and

$$\ln(1-x) = -x - \frac{x^2}{2} - \frac{x^3}{3} - \cdots = -\sum_{n=1}^{\infty} \frac{x^n}{n} \qquad |x| < 1$$

The radius of convergence is the same as for the original series: $R = 1$.

Notice what happens if we put $x = \frac{1}{2}$ in the result of Example 6. Since $\ln \frac{1}{2} = -\ln 2$, we see that

$$\ln 2 = \frac{1}{2} + \frac{1}{8} + \frac{1}{24} + \frac{1}{64} + \cdots = \sum_{n=1}^{\infty} \frac{1}{n2^n}$$

EXAMPLE 7 Find a power series representation for $f(x) = \tan^{-1}x$.

SOLUTION We observe that $f'(x) = 1/(1 + x^2)$ and find the required series by integrating the power series for $1/(1 + x^2)$ found in Example 1.

$$\tan^{-1}x = \int \frac{1}{1+x^2}\,dx = \int (1 - x^2 + x^4 - x^6 + \cdots)\,dx$$

$$= C + x - \frac{x^3}{3} + \frac{x^5}{5} - \frac{x^7}{7} + \cdots$$

To find C we put $x = 0$ and obtain $C = \tan^{-1} 0 = 0$. Therefore

$$\tan^{-1}x = x - \frac{x^3}{3} + \frac{x^5}{5} - \frac{x^7}{7} + \cdots = \sum_{n=0}^{\infty} (-1)^n \frac{x^{2n+1}}{2n+1}$$

Since the radius of convergence of the series for $1/(1 + x^2)$ is 1, the radius of convergence of this series for $\tan^{-1}x$ is also 1.

|||| The power series for $\tan^{-1}x$ obtained in Example 7 is called *Gregory's series* after the Scottish mathematician James Gregory (1638–1675), who had anticipated some of Newton's discoveries. We have shown that Gregory's series is valid when $-1 < x < 1$, but it turns out (although it isn't easy to prove) that it is also valid when $x = \pm 1$. Notice that when $x = 1$ the series becomes

$$\frac{\pi}{4} = 1 - \frac{1}{3} + \frac{1}{5} - \frac{1}{7} + \cdots$$

This beautiful result is known as the Leibniz formula for π.

EXAMPLE 8

(a) Evaluate $\int [1/(1 + x^7)]\,dx$ as a power series.

(b) Use part (a) to approximate $\int_0^{0.5} [1/(1 + x^7)]\,dx$ correct to within 10^{-7}.

SOLUTION

(a) The first step is to express the integrand, $1/(1 + x^7)$, as the sum of a power series. As in Example 1, we start with Equation 1 and replace x by $-x^7$:

$$\frac{1}{1 + x^7} = \frac{1}{1 - (-x^7)} = \sum_{n=0}^{\infty} (-x^7)^n$$

$$= \sum_{n=0}^{\infty} (-1)^n x^{7n} = 1 - x^7 + x^{14} - \cdots$$

|||| This example demonstrates one way in which power series representations are useful. Integrating $1/(1 + x^7)$ by hand is incredibly difficult. Different computer algebra systems return different forms of the answer, but they are all extremely complicated. (If you have a CAS, try it yourself.) The infinite series answer that we obtain in Example 8(a) is actually much easier to deal with than the finite answer provided by a CAS.

Now we integrate term by term:

$$\int \frac{1}{1 + x^7}\,dx = \int \sum_{n=0}^{\infty} (-1)^n x^{7n}\,dx = C + \sum_{n=0}^{\infty} (-1)^n \frac{x^{7n+1}}{7n+1}$$

$$= C + x - \frac{x^8}{8} + \frac{x^{15}}{15} - \frac{x^{22}}{22} + \cdots$$

This series converges for $|-x^7| < 1$, that is, for $|x| < 1$.

(b) In applying the Fundamental Theorem of Calculus it doesn't matter which antiderivative we use, so let's use the antiderivative from part (a) with $C = 0$:

$$\int_0^{0.5} \frac{1}{1 + x^7}\,dx = \left[x - \frac{x^8}{8} + \frac{x^{15}}{15} - \frac{x^{22}}{22} + \cdots \right]_0^{1/2}$$

$$= \frac{1}{2} - \frac{1}{8 \cdot 2^8} + \frac{1}{15 \cdot 2^{15}} - \frac{1}{22 \cdot 2^{22}} + \cdots + \frac{(-1)^n}{(7n+1)2^{7n+1}} + \cdots$$

This infinite series is the exact value of the definite integral, but since it is an alternating series, we can approximate the sum using the Alternating Series Estimation Theorem. If we stop adding after the term with $n = 3$, the error is smaller than the term with $n = 4$:

$$\frac{1}{29 \cdot 2^{29}} \approx 6.4 \times 10^{-11}$$

So we have

$$\int_0^{0.5} \frac{1}{1 + x^7}\,dx \approx \frac{1}{2} - \frac{1}{8 \cdot 2^8} + \frac{1}{15 \cdot 2^{15}} - \frac{1}{22 \cdot 2^{22}} \approx 0.49951374$$

11.9 Exercises

1. If the radius of convergence of the power series $\sum_{n=0}^{\infty} c_n x^n$ is 10, what is the radius of convergence of the series $\sum_{n=1}^{\infty} n c_n x^{n-1}$? Why?

2. Suppose you know that the series $\sum_{n=0}^{\infty} b_n x^n$ converges for $|x| < 2$. What can you say about the following series? Why?

$$\sum_{n=0}^{\infty} \frac{b_n}{n+1} x^{n+1}$$

3–10 Find a power series representation for the function and determine the interval of convergence.

3. $f(x) = \dfrac{1}{1+x}$

4. $f(x) = \dfrac{3}{1-x^4}$

5. $f(x) = \dfrac{1}{1-x^3}$

6. $f(x) = \dfrac{1}{1+9x^2}$

7. $f(x) = \dfrac{1}{x-5}$

8. $f(x) = \dfrac{x}{4x+1}$

9. $f(x) = \dfrac{x}{9+x^2}$

10. $f(x) = \dfrac{x^2}{a^3-x^3}$

11–12 Express the function as the sum of a power series by first using partial fractions. Find the interval of convergence.

11. $f(x) = \dfrac{3}{x^2+x-2}$

12. $f(x) = \dfrac{7x-1}{3x^2+2x-1}$

13. (a) Use differentiation to find a power series representation for

$$f(x) = \frac{1}{(1+x)^2}$$

What is the radius of convergence?

(b) Use part (a) to find a power series for

$$f(x) = \frac{1}{(1+x)^3}$$

(c) Use part (b) to find a power series for

$$f(x) = \frac{x^2}{(1+x)^3}$$

14. (a) Find a power series representation for $f(x) = \ln(1+x)$. What is the radius of convergence?

(b) Use part (a) to find a power series for $f(x) = x\ln(1+x)$.

(c) Use part (a) to find a power series for $f(x) = \ln(x^2+1)$.

15–18 Find a power series representation for the function and determine the radius of convergence.

15. $f(x) = \ln(5-x)$

16. $f(x) = \dfrac{x^2}{(1-2x)^2}$

17. $f(x) = \dfrac{x^3}{(x-2)^2}$

18. $f(x) = \arctan(x/3)$

19–22 Find a power series representation for f, and graph f and several partial sums $s_n(x)$ on the same screen. What happens as n increases?

19. $f(x) = \ln(3+x)$

20. $f(x) = \dfrac{1}{x^2+25}$

21. $f(x) = \ln\left(\dfrac{1+x}{1-x}\right)$

22. $f(x) = \tan^{-1}(2x)$

23–26 Evaluate the indefinite integral as a power series. What is the radius of convergence?

23. $\displaystyle\int \frac{t}{1-t^8}\,dt$

24. $\displaystyle\int \frac{\ln(1-t)}{t}\,dt$

25. $\displaystyle\int \frac{x-\tan^{-1}x}{x^3}\,dx$

26. $\displaystyle\int \tan^{-1}(x^2)\,dx$

27–30 Use a power series to approximate the definite integral to six decimal places.

27. $\displaystyle\int_0^{0.2} \frac{1}{1+x^5}\,dx$

28. $\displaystyle\int_0^{0.4} \ln(1+x^4)\,dx$

29. $\displaystyle\int_0^{1/3} x^2\tan^{-1}(x^4)\,dx$

30. $\displaystyle\int_0^{0.5} \frac{dx}{1+x^6}$

31. Use the result of Example 6 to compute ln 1.1 correct to five decimal places.

32. Show that the function

$$f(x) = \sum_{n=0}^{\infty} \frac{(-1)^n x^{2n}}{(2n)!}$$

is a solution of the differential equation

$$f''(x) + f(x) = 0$$

33. (a) Show that J_0 (the Bessel function of order 0 given in Example 4) satisfies the differential equation

$$x^2 J_0''(x) + x J_0'(x) + x^2 J_0(x) = 0$$

(b) Evaluate $\int_0^1 J_0(x)\,dx$ correct to three decimal places.

34. The Bessel function of order 1 is defined by

$$J_1(x) = \sum_{n=0}^{\infty} \frac{(-1)^n x^{2n+1}}{n!(n+1)!2^{2n+1}}$$

(a) Show that J_1 satisfies the differential equation

$$x^2 J_1''(x) + x J_1'(x) + (x^2 - 1)J_1(x) = 0$$

(b) Show that $J_0'(x) = -J_1(x)$.

35. (a) Show that the function

$$f(x) = \sum_{n=0}^{\infty} \frac{x^n}{n!}$$

is a solution of the differential equation

$$f'(x) = f(x)$$

(b) Show that $f(x) = e^x$.

36. Let $f_n(x) = (\sin nx)/n^2$. Show that the series $\Sigma f_n(x)$ converges for all values of x but the series of derivatives $\Sigma f_n'(x)$ diverges when $x = 2n\pi$, n an integer. For what values of x does the series $\Sigma f_n''(x)$ converge?

37. Let

$$f(x) = \sum_{n=1}^{\infty} \frac{x^n}{n^2}$$

Find the intervals of convergence for f, f', and f''.

38. (a) Starting with the geometric series $\Sigma_{n=0}^{\infty} x^n$, find the sum of the series

$$\sum_{n=1}^{\infty} nx^{n-1} \qquad |x| < 1$$

(b) Find the sum of each of the following series.

(i) $\displaystyle\sum_{n=1}^{\infty} nx^n, \quad |x| < 1$ (ii) $\displaystyle\sum_{n=1}^{\infty} \frac{n}{2^n}$

(c) Find the sum of each of the following series.

(i) $\displaystyle\sum_{n=2}^{\infty} n(n-1)x^n, \quad |x| < 1$

(ii) $\displaystyle\sum_{n=2}^{\infty} \frac{n^2 - n}{2^n}$ (iii) $\displaystyle\sum_{n=1}^{\infty} \frac{n^2}{2^n}$

39. Use the power series for $\tan^{-1}x$ to prove the following expression for π as the sum of an infinite series:

$$\pi = 2\sqrt{3} \sum_{n=0}^{\infty} \frac{(-1)^n}{(2n+1)3^n}$$

40. (a) By completing the square, show that

$$\int_0^{1/2} \frac{dx}{x^2 - x + 1} = \frac{\pi}{3\sqrt{3}}$$

(b) By factoring $x^3 + 1$ as a sum of cubes, rewrite the integral in part (a). Then express $1/(x^3 + 1)$ as the sum of a power series and use it to prove the following formula for π:

$$\pi = \frac{3\sqrt{3}}{4} \sum_{n=0}^{\infty} \frac{(-1)^n}{8^n} \left(\frac{2}{3n+1} + \frac{1}{3n+2} \right)$$

11.10 Taylor and Maclaurin Series

In the preceding section we were able to find power series representations for a certain restricted class of functions. Here we investigate more general problems: Which functions have power series representations? How can we find such representations?

We start by supposing that f is any function that can be represented by a power series

$$\boxed{1} \quad f(x) = c_0 + c_1(x-a) + c_2(x-a)^2 + c_3(x-a)^3 + c_4(x-a)^4 + \cdots \qquad |x - a| < R$$

Let's try to determine what the coefficients c_n must be in terms of f. To begin, notice that if we put $x = a$ in Equation 1, then all terms after the first one are 0 and we get

$$f(a) = c_0$$

By Theorem 11.9.2, we can differentiate the series in Equation 1 term by term:

$$\boxed{2} \quad f'(x) = c_1 + 2c_2(x-a) + 3c_3(x-a)^2 + 4c_4(x-a)^3 + \cdots \qquad |x - a| < R$$

and substitution of $x = a$ in Equation 2 gives

$$f'(a) = c_1$$

Now we differentiate both sides of Equation 2 and obtain

$$\boxed{3} \quad f''(x) = 2c_2 + 2 \cdot 3c_3(x - a) + 3 \cdot 4c_4(x - a)^2 + \cdots \qquad |x - a| < R$$

Again we put $x = a$ in Equation 3. The result is

$$f''(a) = 2c_2$$

Let's apply the procedure one more time. Differentiation of the series in Equation 3 gives

$$\boxed{4} \quad f'''(x) = 2 \cdot 3c_3 + 2 \cdot 3 \cdot 4c_4(x - a) + 3 \cdot 4 \cdot 5c_5(x - a)^2 + \cdots \qquad |x - a| < R$$

and substitution of $x = a$ in Equation 4 gives

$$f'''(a) = 2 \cdot 3c_3 = 3!c_3$$

By now you can see the pattern. If we continue to differentiate and substitute $x = a$, we obtain

$$f^{(n)}(a) = 2 \cdot 3 \cdot 4 \cdot \cdots \cdot nc_n = n!c_n$$

Solving this equation for the nth coefficient c_n, we get

$$c_n = \frac{f^{(n)}(a)}{n!}$$

This formula remains valid even for $n = 0$ if we adopt the conventions that $0! = 1$ and $f^{(0)} = f$. Thus, we have proved the following theorem.

$\boxed{5}$ **Theorem** If f has a power series representation (expansion) at a, that is, if

$$f(x) = \sum_{n=0}^{\infty} c_n(x - a)^n \qquad |x - a| < R$$

then its coefficients are given by the formula

$$c_n = \frac{f^{(n)}(a)}{n!}$$

Substituting this formula for c_n back into the series, we see that *if* f has a power series expansion at a, then it must be of the following form.

$$\boxed{6} \quad f(x) = \sum_{n=0}^{\infty} \frac{f^{(n)}(a)}{n!}(x - a)^n$$

$$= f(a) + \frac{f'(a)}{1!}(x - a) + \frac{f''(a)}{2!}(x - a)^2 + \frac{f'''(a)}{3!}(x - a)^3 + \cdots$$

The Taylor series is named after the English mathematician Brook Taylor (1685–1731) and the Maclaurin series is named in honor of the Scottish mathematician Colin Maclaurin (1698–1746) despite the fact that the Maclaurin series is really just a special case of the Taylor series. But the idea of representing particular functions as sums of power series goes back to Newton, and the general Taylor series was known to the Scottish mathematician James Gregory in 1668 and to the Swiss mathematician John Bernoulli in the 1690s. Taylor was apparently unaware of the work of Gregory and Bernoulli when he published his discoveries on series in 1715 in his book *Methodus incrementorum directa et inversa*. Maclaurin series are named after Colin Maclaurin because he popularized them in his calculus textbook *Treatise of Fluxions* published in 1742.

The series in Equation 6 is called the **Taylor series of the function f at a** (or **about a** or **centered at a**). For the special case $a = 0$ the Taylor series becomes

$$\boxed{7} \qquad f(x) = \sum_{n=0}^{\infty} \frac{f^{(n)}(0)}{n!} x^n = f(0) + \frac{f'(0)}{1!} x + \frac{f''(0)}{2!} x^2 + \cdots$$

This case arises frequently enough that it is given the special name **Maclaurin series**.

NOTE ▫ We have shown that *if* f can be represented as a power series about a, then f is equal to the sum of its Taylor series. But there exist functions that are not equal to the sum of their Taylor series. An example of such a function is given in Exercise 62.

EXAMPLE 1 Find the Maclaurin series of the function $f(x) = e^x$ and its radius of convergence.

SOLUTION If $f(x) = e^x$, then $f^{(n)}(x) = e^x$, so $f^{(n)}(0) = e^0 = 1$ for all n. Therefore, the Taylor series for f at 0 (that is, the Maclaurin series) is

$$\sum_{n=0}^{\infty} \frac{f^{(n)}(0)}{n!} x^n = \sum_{n=0}^{\infty} \frac{x^n}{n!} = 1 + \frac{x}{1!} + \frac{x^2}{2!} + \frac{x^3}{3!} + \cdots$$

To find the radius of convergence we let $a_n = x^n/n!$. Then

$$\left| \frac{a_{n+1}}{a_n} \right| = \left| \frac{x^{n+1}}{(n+1)!} \cdot \frac{n!}{x^n} \right| = \frac{|x|}{n+1} \to 0 < 1$$

so, by the Ratio Test, the series converges for all x and the radius of convergence is $R = \infty$.

The conclusion we can draw from Theorem 5 and Example 1 is that *if* e^x has a power series expansion at 0, then

$$e^x = \sum_{n=0}^{\infty} \frac{x^n}{n!}$$

So how can we determine whether e^x *does* have a power series representation?

Let's investigate the more general question: Under what circumstances is a function equal to the sum of its Taylor series? In other words, if f has derivatives of all orders, when is it true that

$$f(x) = \sum_{n=0}^{\infty} \frac{f^{(n)}(a)}{n!} (x-a)^n$$

As with any convergent series, this means that $f(x)$ is the limit of the sequence of partial sums. In the case of the Taylor series, the partial sums are

$$T_n(x) = \sum_{i=0}^{n} \frac{f^{(i)}(a)}{i!} (x-a)^i$$

$$= f(a) + \frac{f'(a)}{1!}(x-a) + \frac{f''(a)}{2!}(x-a)^2 + \cdots + \frac{f^{(n)}(a)}{n!}(x-a)^n$$

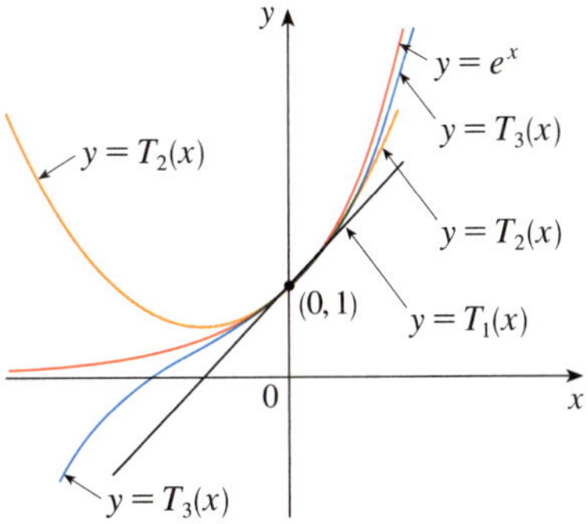

FIGURE 1

|||| As n increases, $T_n(x)$ appears to approach e^x in Figure 1. This suggests that e^x is equal to the sum of its Taylor series.

Notice that T_n is a polynomial of degree n called the ***n*th-degree Taylor polynomial of *f* at *a***. For instance, for the exponential function $f(x) = e^x$, the result of Example 1 shows that the Taylor polynomials at 0 (or Maclaurin polynomials) with $n = 1$, 2, and 3 are

$$T_1(x) = 1 + x \qquad T_2(x) = 1 + x + \frac{x^2}{2!} \qquad T_3(x) = 1 + x + \frac{x^2}{2!} + \frac{x^3}{3!}$$

The graphs of the exponential function and these three Taylor polynomials are drawn in Figure 1.

In general, $f(x)$ is the sum of its Taylor series if

$$f(x) = \lim_{n \to \infty} T_n(x)$$

If we let

$$R_n(x) = f(x) - T_n(x) \qquad \text{so that} \qquad f(x) = T_n(x) + R_n(x)$$

then $R_n(x)$ is called the **remainder** of the Taylor series. If we can somehow show that $\lim_{n \to \infty} R_n(x) = 0$, then it follows that

$$\lim_{n \to \infty} T_n(x) = \lim_{n \to \infty} [f(x) - R_n(x)] = f(x) - \lim_{n \to \infty} R_n(x) = f(x)$$

We have therefore proved the following.

8 Theorem If $f(x) = T_n(x) + R_n(x)$, where T_n is the nth-degree Taylor polynomial of f at a and

$$\lim_{n \to \infty} R_n(x) = 0$$

for $|x - a| < R$, then f is equal to the sum of its Taylor series on the interval $|x - a| < R$.

In trying to show that $\lim_{n \to \infty} R_n(x) = 0$ for a specific function f, we usually use the following fact.

9 Taylor's Inequality If $|f^{(n+1)}(x)| \leqslant M$ for $|x - a| \leqslant d$, then the remainder $R_n(x)$ of the Taylor series satisfies the inequality

$$|R_n(x)| \leqslant \frac{M}{(n+1)!} |x - a|^{n+1} \qquad \text{for } |x - a| \leqslant d$$

To see why this is true for $n = 1$, we assume that $|f''(x)| \leqslant M$. In particular, we have $f''(x) \leqslant M$, so for $a \leqslant x \leqslant a + d$ we have

$$\int_a^x f''(t)\, dt \leqslant \int_a^x M\, dt$$

An antiderivative of f'' is f', so by Part 2 of the Fundamental Theorem of Calculus, we have

$$f'(x) - f'(a) \leqslant M(x - a) \qquad \text{or} \qquad f'(x) \leqslant f'(a) + M(x - a)$$

|||| As alternatives to Taylor's Inequality, we have the following formulas for the remainder term. If $f^{(n+1)}$ is continuous on an interval I and $x \in I$, then

$$R_n(x) = \frac{1}{n!}\int_a^x (x-t)^n f^{(n+1)}(t)\,dt$$

This is called the *integral form of the remainder term.* Another formula, called *Lagrange's form of the remainder term*, states that there is a number z between x and a such that

$$R_n(x) = \frac{f^{(n+1)}(z)}{(n+1)!}(x-a)^{n+1}$$

This version is an extension of the Mean Value Theorem (which is the case $n = 0$).

Proofs of these formulas, together with discussions of how to use them to solve the examples of Sections 11.10 and 11.12, are given on the web site

www.stewartcalculus.com

Click on *Additional Topics* and then on *Formulas for the Remainder Term in Taylor series.*

Thus
$$\int_a^x f'(t)\,dt \leqslant \int_a^x [f'(a) + M(t-a)]\,dt$$

$$f(x) - f(a) \leqslant f'(a)(x-a) + M\frac{(x-a)^2}{2}$$

$$f(x) - f(a) - f'(a)(x-a) \leqslant \frac{M}{2}(x-a)^2$$

But $R_1(x) = f(x) - T_1(x) = f(x) - f(a) - f'(a)(x-a)$. So

$$R_1(x) \leqslant \frac{M}{2}(x-a)^2$$

A similar argument, using $f''(x) \geqslant -M$, shows that

$$R_1(x) \geqslant -\frac{M}{2}(x-a)^2$$

So
$$|R_1(x)| \leqslant \frac{M}{2}|x-a|^2$$

Although we have assumed that $x > a$, similar calculations show that this inequality is also true for $x < a$.

This proves Taylor's Inequality for the case where $n = 1$. The result for any n is proved in a similar way by integrating $n + 1$ times. (See Exercise 61 for the case $n = 2$.)

NOTE ▫ In Section 11.12 we will explore the use of Taylor's Inequality in approximating functions. Our immediate use of it is in conjunction with Theorem 8.

In applying Theorems 8 and 9 it is often helpful to make use of the following fact.

10
$$\lim_{n\to\infty} \frac{x^n}{n!} = 0 \qquad \text{for every real number } x$$

This is true because we know from Example 1 that the series $\Sigma\, x^n/n!$ converges for all x and so its nth term approaches 0.

EXAMPLE 2 Prove that e^x is equal to the sum of its Maclaurin series.

SOLUTION If $f(x) = e^x$, then $f^{(n+1)}(x) = e^x$ for all n. If d is any positive number and $|x| \leqslant d$, then $|f^{(n+1)}(x)| = e^x \leqslant e^d$. So Taylor's Inequality, with $a = 0$ and $M = e^d$, says that

$$|R_n(x)| \leqslant \frac{e^d}{(n+1)!}|x|^{n+1} \qquad \text{for } |x| \leqslant d$$

Notice that the same constant $M = e^d$ works for every value of n. But, from Equation 10, we have

$$\lim_{n\to\infty} \frac{e^d}{(n+1)!}|x|^{n+1} = e^d \lim_{n\to\infty} \frac{|x|^{n+1}}{(n+1)!} = 0$$

It follows from the Squeeze Theorem that $\lim_{n\to\infty} |R_n(x)| = 0$ and therefore $\lim_{n\to\infty} R_n(x) = 0$ for all values of x. By Theorem 8, e^x is equal to the sum of its Maclaurin series, that is,

11 $$e^x = \sum_{n=0}^{\infty} \frac{x^n}{n!} \qquad \text{for all } x$$

|||| In 1748 Leonard Euler used Equation 12 to find the value of e correct to 23 digits. In 2000 Xavier Gourdon and S. Kondo, again using the series in (12), computed e to more than twelve billion decimal places. The special techniques they employed to speed up the computation are explained on the web page

www.numbers.computation.free.fr

In particular, if we put $x = 1$ in Equation 11, we obtain the following expression for the number e as a sum of an infinite series:

12 $$e = \sum_{n=0}^{\infty} \frac{1}{n!} = 1 + \frac{1}{1!} + \frac{1}{2!} + \frac{1}{3!} + \cdots$$

EXAMPLE 3 Find the Taylor series for $f(x) = e^x$ at $a = 2$.

SOLUTION We have $f^{(n)}(2) = e^2$ and so, putting $a = 2$ in the definition of a Taylor series (6), we get

$$\sum_{n=0}^{\infty} \frac{f^{(n)}(2)}{n!}(x-2)^n = \sum_{n=0}^{\infty} \frac{e^2}{n!}(x-2)^n$$

Again it can be verified, as in Example 1, that the radius of convergence is $R = \infty$. As in Example 2 we can verify that $\lim_{n\to\infty} R_n(x) = 0$, so

13 $$e^x = \sum_{n=0}^{\infty} \frac{e^2}{n!}(x-2)^n \qquad \text{for all } x$$

We have two power series expansions for e^x, the Maclaurin series in Equation 11 and the Taylor series in Equation 13. The first is better if we are interested in values of x near 0 and the second is better if x is near 2.

EXAMPLE 4 Find the Maclaurin series for $\sin x$ and prove that it represents $\sin x$ for all x.

SOLUTION We arrange our computation in two columns as follows:

$$\begin{aligned} f(x) &= \sin x & f(0) &= 0 \\ f'(x) &= \cos x & f'(0) &= 1 \\ f''(x) &= -\sin x & f''(0) &= 0 \\ f'''(x) &= -\cos x & f'''(0) &= -1 \\ f^{(4)}(x) &= \sin x & f^{(4)}(0) &= 0 \end{aligned}$$

Since the derivatives repeat in a cycle of four, we can write the Maclaurin series as follows:

$$f(0) + \frac{f'(0)}{1!}x + \frac{f''(0)}{2!}x^2 + \frac{f'''(0)}{3!}x^3 + \cdots$$
$$= x - \frac{x^3}{3!} + \frac{x^5}{5!} - \frac{x^7}{7!} + \cdots = \sum_{n=0}^{\infty} (-1)^n \frac{x^{2n+1}}{(2n+1)!}$$

|||| Figure 2 shows the graph of $\sin x$ together with its Taylor (or Maclaurin) polynomials

$$T_1(x) = x$$

$$T_3(x) = x - \frac{x^3}{3!}$$

$$T_5(x) = x - \frac{x^3}{3!} + \frac{x^5}{5!}$$

Notice that, as n increases, $T_n(x)$ becomes a better approximation to $\sin x$.

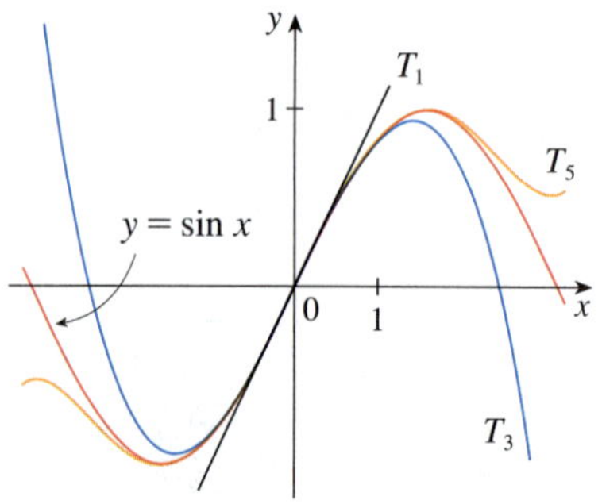

FIGURE 2

Since $f^{(n+1)}(x)$ is $\pm\sin x$ or $\pm\cos x$, we know that $|f^{(n+1)}(x)| \leqslant 1$ for all x. So we can take $M = 1$ in Taylor's Inequality:

14 $$|R_n(x)| \leqslant \frac{M}{(n+1)!}|x^{n+1}| = \frac{|x|^{n+1}}{(n+1)!}$$

By Equation 10 the right side of this inequality approaches 0 as $n \to \infty$, so $|R_n(x)| \to 0$ by the Squeeze Theorem. It follows that $R_n(x) \to 0$ as $n \to \infty$, so $\sin x$ is equal to the sum of its Maclaurin series by Theorem 8.

We state the result of Example 4 for future reference.

15 $$\sin x = x - \frac{x^3}{3!} + \frac{x^5}{5!} - \frac{x^7}{7!} + \cdots$$
$$= \sum_{n=0}^{\infty} (-1)^n \frac{x^{2n+1}}{(2n+1)!} \qquad \text{for all } x$$

EXAMPLE 5 Find the Maclaurin series for $\cos x$.

SOLUTION We could proceed directly as in Example 4 but it's easier to differentiate the Maclaurin series for $\sin x$ given by Equation 15:

$$\cos x = \frac{d}{dx}(\sin x) = \frac{d}{dx}\left(x - \frac{x^3}{3!} + \frac{x^5}{5!} - \frac{x^7}{7!} + \cdots\right)$$
$$= 1 - \frac{3x^2}{3!} + \frac{5x^4}{5!} - \frac{7x^6}{7!} + \cdots = 1 - \frac{x^2}{2!} + \frac{x^4}{4!} - \frac{x^6}{6!} + \cdots$$

|||| The Maclaurin series for e^x, $\sin x$, and $\cos x$ that we found in Examples 2, 4, and 5 were discovered, using different methods, by Newton. (There is evidence that the series for $\sin x$ and $\cos x$ were known to Indian astronomers more than a century before Newton, but this knowledge didn't spread to the rest of the world.)

These equations are remarkable because they say we know everything about each of these functions if we know all its derivatives at the single number 0.

Since the Maclaurin series for $\sin x$ converges for all x, Theorem 2 in Section 11.9 tells us that the differentiated series for $\cos x$ also converges for all x. Thus

16 $$\cos x = 1 - \frac{x^2}{2!} + \frac{x^4}{4!} - \frac{x^6}{6!} + \cdots$$
$$= \sum_{n=0}^{\infty} (-1)^n \frac{x^{2n}}{(2n)!} \qquad \text{for all } x$$

EXAMPLE 6 Find the Maclaurin series for the function $f(x) = x\cos x$.

SOLUTION Instead of computing derivatives and substituting in Equation 7, it's easier to multiply the series for $\cos x$ (Equation 16) by x:

$$x\cos x = x\sum_{n=0}^{\infty} (-1)^n \frac{x^{2n}}{(2n)!} = \sum_{n=0}^{\infty} (-1)^n \frac{x^{2n+1}}{(2n)!}$$

EXAMPLE 7 Represent $f(x) = \sin x$ as the sum of its Taylor series centered at $\pi/3$.

SOLUTION Arranging our work in columns, we have

$$f(x) = \sin x \qquad f\left(\frac{\pi}{3}\right) = \frac{\sqrt{3}}{2}$$

$$f'(x) = \cos x \qquad f'\left(\frac{\pi}{3}\right) = \frac{1}{2}$$

$$f''(x) = -\sin x \qquad f''\left(\frac{\pi}{3}\right) = -\frac{\sqrt{3}}{2}$$

$$f'''(x) = -\cos x \qquad f'''\left(\frac{\pi}{3}\right) = -\frac{1}{2}$$

and this pattern repeats indefinitely. Therefore, the Taylor series at $\pi/3$ is

$$f\left(\frac{\pi}{3}\right) + \frac{f'\left(\frac{\pi}{3}\right)}{1!}\left(x - \frac{\pi}{3}\right) + \frac{f''\left(\frac{\pi}{3}\right)}{2!}\left(x - \frac{\pi}{3}\right)^2 + \frac{f'''\left(\frac{\pi}{3}\right)}{3!}\left(x - \frac{\pi}{3}\right)^3 + \cdots$$

$$= \frac{\sqrt{3}}{2} + \frac{1}{2 \cdot 1!}\left(x - \frac{\pi}{3}\right) - \frac{\sqrt{3}}{2 \cdot 2!}\left(x - \frac{\pi}{3}\right)^2 - \frac{1}{2 \cdot 3!}\left(x - \frac{\pi}{3}\right)^3 + \cdots$$

|||| We have obtained two different series representations for $\sin x$, the Maclaurin series in Example 4 and the Taylor series in Example 7. It is best to use the Maclaurin series for values of x near 0 and the Taylor series for x near $\pi/3$. Notice that the third Taylor polynomial T_3 in Figure 3 is a good approximation to $\sin x$ near $\pi/3$ but not as good near 0. Compare it with the third Maclaurin polynomial T_3 in Figure 2, where the opposite is true.

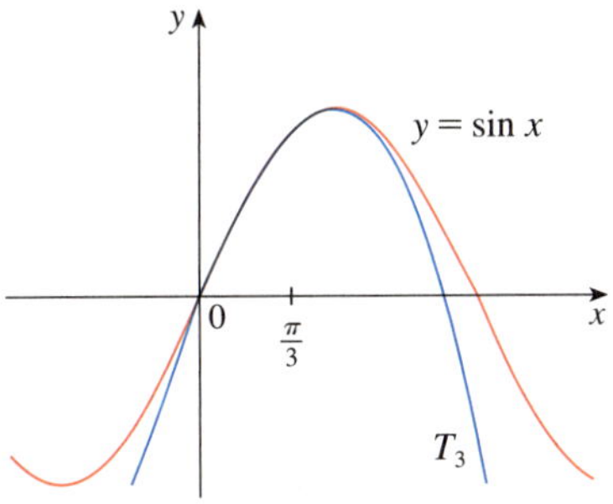

FIGURE 3

The proof that this series represents $\sin x$ for all x is very similar to that in Example 4. [Just replace x by $x - \pi/3$ in (14).] We can write the series in sigma notation if we separate the terms that contain $\sqrt{3}$:

$$\sin x = \sum_{n=0}^{\infty} \frac{(-1)^n \sqrt{3}}{2(2n)!}\left(x - \frac{\pi}{3}\right)^{2n} + \sum_{n=0}^{\infty} \frac{(-1)^n}{2(2n+1)!}\left(x - \frac{\pi}{3}\right)^{2n+1}$$

The power series that we obtained by indirect methods in Examples 5 and 6 and in Section 11.9 are indeed the Taylor or Maclaurin series of the given functions because Theorem 5 asserts that, no matter how a power series representation $f(x) = \sum c_n(x - a)^n$ is obtained, it is always true that $c_n = f^{(n)}(a)/n!$. In other words, the coefficients are uniquely determined.

We collect in the following table, for future reference, some important Maclaurin series that we have derived in this section and the preceding one.

Important Maclaurin series and their intervals of convergence

$\dfrac{1}{1-x} = \sum_{n=0}^{\infty} x^n = 1 + x + x^2 + x^3 + \cdots$	$(-1, 1)$
$e^x = \sum_{n=0}^{\infty} \dfrac{x^n}{n!} = 1 + \dfrac{x}{1!} + \dfrac{x^2}{2!} + \dfrac{x^3}{3!} + \cdots$	$(-\infty, \infty)$
$\sin x = \sum_{n=0}^{\infty} (-1)^n \dfrac{x^{2n+1}}{(2n+1)!} = x - \dfrac{x^3}{3!} + \dfrac{x^5}{5!} - \dfrac{x^7}{7!} + \cdots$	$(-\infty, \infty)$
$\cos x = \sum_{n=0}^{\infty} (-1)^n \dfrac{x^{2n}}{(2n)!} = 1 - \dfrac{x^2}{2!} + \dfrac{x^4}{4!} - \dfrac{x^6}{6!} + \cdots$	$(-\infty, \infty)$
$\tan^{-1} x = \sum_{n=0}^{\infty} (-1)^n \dfrac{x^{2n+1}}{2n+1} = x - \dfrac{x^3}{3} + \dfrac{x^5}{5} - \dfrac{x^7}{7} + \cdots$	$[-1, 1]$

TEC Module 11.10/11.12 enables you to see how successive Taylor polynomials approach the original function.

One reason that Taylor series are important is that they enable us to integrate functions that we couldn't previously handle. In fact, in the introduction to this chapter we mentioned that Newton often integrated functions by first expressing them as power series and then integrating the series term by term. The function $f(x) = e^{-x^2}$ can't be integrated by techniques discussed so far because its antiderivative is not an elementary function (see Section 7.5). In the following example we use Newton's idea to integrate this function.

EXAMPLE 8
(a) Evaluate $\int e^{-x^2}\,dx$ as an infinite series.
(b) Evaluate $\int_0^1 e^{-x^2}\,dx$ correct to within an error of 0.001.

SOLUTION
(a) First we find the Maclaurin series for $f(x) = e^{-x^2}$. Although it's possible to use the direct method, let's find it simply by replacing x with $-x^2$ in the series for e^x given in the table of Maclaurin series. Thus, for all values of x,

$$e^{-x^2} = \sum_{n=0}^{\infty} \frac{(-x^2)^n}{n!} = \sum_{n=0}^{\infty} (-1)^n \frac{x^{2n}}{n!} = 1 - \frac{x^2}{1!} + \frac{x^4}{2!} - \frac{x^6}{3!} + \cdots$$

Now we integrate term by term:

$$\int e^{-x^2}\,dx = \int \left(1 - \frac{x^2}{1!} + \frac{x^4}{2!} - \frac{x^6}{3!} + \cdots + (-1)^n \frac{x^{2n}}{n!} + \cdots\right) dx$$

$$= C + x - \frac{x^3}{3 \cdot 1!} + \frac{x^5}{5 \cdot 2!} - \frac{x^7}{7 \cdot 3!} + \cdots + (-1)^n \frac{x^{2n+1}}{(2n+1)n!} + \cdots$$

This series converges for all x because the original series for e^{-x^2} converges for all x.

(b) The Evaluation Theorem gives

||| We can take $C = 0$ in the antiderivative in part (a).

$$\int_0^1 e^{-x^2}\,dx = \left[x - \frac{x^3}{3 \cdot 1!} + \frac{x^5}{5 \cdot 2!} - \frac{x^7}{7 \cdot 3!} + \frac{x^9}{9 \cdot 4!} - \cdots\right]_0^1$$

$$= 1 - \tfrac{1}{3} + \tfrac{1}{10} - \tfrac{1}{42} + \tfrac{1}{216} - \cdots$$

$$\approx 1 - \tfrac{1}{3} + \tfrac{1}{10} - \tfrac{1}{42} + \tfrac{1}{216} \approx 0.7475$$

The Alternating Series Estimation Theorem shows that the error involved in this approximation is less than

$$\frac{1}{11 \cdot 5!} = \frac{1}{1320} < 0.001$$

Another use of Taylor series is illustrated in the next example. The limit could be found with l'Hospital's Rule, but instead we use a series.

EXAMPLE 9 Evaluate $\displaystyle\lim_{x \to 0} \frac{e^x - 1 - x}{x^2}$.

SOLUTION Using the Maclaurin series for e^x, we have

||||| Some computer algebra systems compute limits in this way.

$$\lim_{x \to 0} \frac{e^x - 1 - x}{x^2} = \lim_{x \to 0} \frac{\left(1 + \frac{x}{1!} + \frac{x^2}{2!} + \frac{x^3}{3!} + \cdots\right) - 1 - x}{x^2}$$

$$= \lim_{x \to 0} \frac{\frac{x^2}{2!} + \frac{x^3}{3!} + \frac{x^4}{4!} + \cdots}{x^2}$$

$$= \lim_{x \to 0} \left(\frac{1}{2} + \frac{x}{3!} + \frac{x^2}{4!} + \frac{x^3}{5!} + \cdots\right)$$

$$= \frac{1}{2}$$

because power series are continuous functions.

Multiplication and Division of Power Series

If power series are added or subtracted, they behave like polynomials (Theorem 11.2.8 shows this). In fact, as the following example illustrates, they can also be multiplied and divided like polynomials. We find only the first few terms because the calculations for the later terms become tedious and the initial terms are the most important ones.

EXAMPLE 10 Find the first three nonzero terms in the Maclaurin series for (a) $e^x \sin x$ and (b) $\tan x$.

SOLUTION

(a) Using the Maclaurin series for e^x and $\sin x$ in the table on page 767, we have

$$e^x \sin x = \left(1 + \frac{x}{1!} + \frac{x^2}{2!} + \frac{x^3}{3!} + \cdots\right)\left(x - \frac{x^3}{3!} + \cdots\right)$$

We multiply these expressions, collecting like terms just as for polynomials:

$$\begin{array}{r} 1 + x + \tfrac{1}{2}x^2 + \tfrac{1}{6}x^3 + \cdots \\ x \qquad\qquad - \tfrac{1}{6}x^3 + \cdots \\ \hline x + x^2 + \tfrac{1}{2}x^3 + \tfrac{1}{6}x^4 + \cdots \\ - \tfrac{1}{6}x^3 - \tfrac{1}{6}x^4 - \cdots \\ \hline x + x^2 + \tfrac{1}{3}x^3 + \cdots \end{array}$$

Thus $$e^x \sin x = x + x^2 + \tfrac{1}{3}x^3 + \cdots$$

(b) Using the Maclaurin series in the table, we have

$$\tan x = \frac{\sin x}{\cos x} = \frac{x - \frac{x^3}{3!} + \frac{x^5}{5!} - \cdots}{1 - \frac{x^2}{2!} + \frac{x^4}{4!} - \cdots}$$

We use a procedure like long division:

$$\begin{array}{r} x + \frac{1}{3}x^3 + \frac{2}{15}x^5 + \cdots \\ 1 - \frac{1}{2}x^2 + \frac{1}{24}x^4 - \cdots \overline{) x - \frac{1}{6}x^3 + \frac{1}{120}x^5 - \cdots} \\ \underline{x - \frac{1}{2}x^3 + \frac{1}{24}x^5 - \cdots} \\ \frac{1}{3}x^3 - \frac{1}{30}x^5 + \cdots \\ \underline{\frac{1}{3}x^3 - \frac{1}{6}x^5 + \cdots} \\ \frac{2}{15}x^5 + \cdots \end{array}$$

Thus $\tan x = x + \frac{1}{3}x^3 + \frac{2}{15}x^5 + \cdots$

Although we have not attempted to justify the formal manipulations used in Example 10, they are legitimate. There is a theorem which states that if both $f(x) = \Sigma\, c_n x^n$ and $g(x) = \Sigma\, b_n x^n$ converge for $|x| < R$ and the series are multiplied as if they were polynomials, then the resulting series also converges for $|x| < R$ and represents $f(x)g(x)$. For division we require $b_0 \neq 0$; the resulting series converges for sufficiently small $|x|$.

11.10 Exercises

1. If $f(x) = \Sigma_{n=0}^{\infty} b_n(x-5)^n$ for all x, write a formula for b_8.

2. (a) The graph of f is shown. Explain why the series

$$1.6 - 0.8(x-1) + 0.4(x-1)^2 - 0.1(x-1)^3 + \cdots$$

is *not* the Taylor series of f centered at 1.

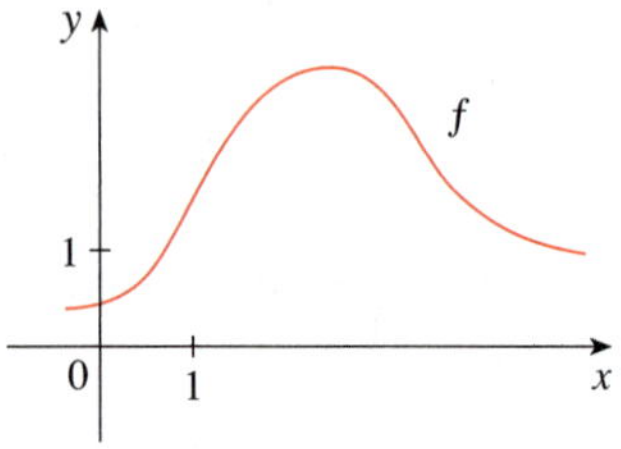

(b) Explain why the series

$$2.8 + 0.5(x-2) + 1.5(x-2)^2 - 0.1(x-2)^3 + \cdots$$

is *not* the Taylor series of f centered at 2.

3–10 Find the Maclaurin series for $f(x)$ using the definition of a Maclaurin series. [Assume that f has a power series expansion. Do not show that $R_n(x) \to 0$.] Also find the associated radius of convergence.

3. $f(x) = \cos x$

4. $f(x) = \sin 2x$

5. $f(x) = (1+x)^{-3}$

6. $f(x) = \ln(1+x)$

7. $f(x) = e^{5x}$

8. $f(x) = xe^x$

9. $f(x) = \sinh x$

10. $f(x) = \cosh x$

11–18 Find the Taylor series for $f(x)$ centered at the given value of a. [Assume that f has a power series expansion. Do not show that $R_n(x) \to 0$.]

11. $f(x) = 1 + x + x^2, \quad a = 2$

12. $f(x) = x^3, \quad a = -1$

13. $f(x) = e^x, \quad a = 3$

14. $f(x) = \ln x, \quad a = 2$

15. $f(x) = \cos x, \quad a = \pi$

16. $f(x) = \sin x, \quad a = \pi/2$

17. $f(x) = 1/\sqrt{x}, \quad a = 9$

18. $f(x) = x^{-2}, \quad a = 1$

19. Prove that the series obtained in Exercise 3 represents $\cos x$ for all x.

20. Prove that the series obtained in Exercise 16 represents $\sin x$ for all x.

21. Prove that the series obtained in Exercise 9 represents $\sinh x$ for all x.

22. Prove that the series obtained in Exercise 10 represents $\cosh x$ for all x.

23–32 Use a Maclaurin series derived in this section to obtain the Maclaurin series for the given function.

23. $f(x) = \cos \pi x$

24. $f(x) = e^{-x/2}$

25. $f(x) = x \tan^{-1} x$

26. $f(x) = \sin(x^4)$

27. $f(x) = x^2 e^{-x}$

28. $f(x) = x \cos 2x$

29. $f(x) = \sin^2 x$ [*Hint:* Use $\sin^2 x = \frac{1}{2}(1 - \cos 2x)$.]

30. $f(x) = \cos^2 x$

31. $f(x) = \begin{cases} \dfrac{\sin x}{x} & \text{if } x \neq 0 \\ 1 & \text{if } x = 0 \end{cases}$

32. $f(x) = \begin{cases} \dfrac{x - \sin x}{x^3} & \text{if } x \neq 0 \\ \frac{1}{6} & \text{if } x = 0 \end{cases}$

33–36 Find the Maclaurin series of f (by any method) and its radius of convergence. Graph f and its first few Taylor polynomials on the same screen. What do you notice about the relationship between these polynomials and f?

33. $f(x) = \sqrt{1 + x}$

34. $f(x) = e^{-x^2} + \cos x$

35. $f(x) = \cos(x^2)$

36. $f(x) = 2^x$

37. Use the Maclaurin series for e^x to calculate $e^{-0.2}$ correct to five decimal places.

38. Use the Maclaurin series for $\sin x$ to compute $\sin 3°$ correct to five decimal places.

39–42 Evaluate the indefinite integral as an infinite series.

39. $\int x \cos(x^3)\, dx$

40. $\int \dfrac{\sin x}{x}\, dx$

41. $\int \sqrt{x^3 + 1}\, dx$

42. $\int \dfrac{e^x - 1}{x}\, dx$

43–46 Use series to approximate the definite integral to within the indicated accuracy.

43. $\int_0^1 x \cos(x^3)\, dx$ (three decimal places)

44. $\int_0^{0.2} [\tan^{-1}(x^3) + \sin(x^3)]\, dx$ (five decimal places)

45. $\int_0^{0.1} \dfrac{dx}{\sqrt{1 + x^3}}$ $(|\text{error}| < 10^{-8})$

46. $\int_0^{0.5} x^2 e^{-x^2}\, dx$ $(|\text{error}| < 0.001)$

47–49 Use series to evaluate the limit.

47. $\lim_{x \to 0} \dfrac{x - \tan^{-1} x}{x^3}$

48. $\lim_{x \to 0} \dfrac{1 - \cos x}{1 + x - e^x}$

49. $\lim_{x \to 0} \dfrac{\sin x - x + \frac{1}{6}x^3}{x^5}$

50. Use the series in Example 10(b) to evaluate

$$\lim_{x \to 0} \frac{\tan x - x}{x^3}$$

We found this limit in Example 4 in Section 4.4 using l'Hospital's Rule three times. Which method do you prefer?

51–54 Use multiplication or division of power series to find the first three nonzero terms in the Maclaurin series for each function.

51. $y = e^{-x^2} \cos x$

52. $y = \sec x$

53. $y = \dfrac{x}{\sin x}$

54. $y = e^x \ln(1 - x)$

55–60 Find the sum of the series.

55. $\sum_{n=0}^{\infty} (-1)^n \dfrac{x^{4n}}{n!}$

56. $\sum_{n=0}^{\infty} \dfrac{(-1)^n \pi^{2n}}{6^{2n}(2n)!}$

57. $\sum_{n=0}^{\infty} \dfrac{(-1)^n \pi^{2n+1}}{4^{2n+1}(2n+1)!}$

58. $\sum_{n=0}^{\infty} \dfrac{3^n}{5^n n!}$

59. $3 + \dfrac{9}{2!} + \dfrac{27}{3!} + \dfrac{81}{4!} + \cdots$

60. $1 - \ln 2 + \dfrac{(\ln 2)^2}{2!} - \dfrac{(\ln 2)^3}{3!} + \cdots$

61. Prove Taylor's Inequality for $n = 2$, that is, prove that if $|f'''(x)| \leq M$ for $|x - a| \leq d$, then

$$|R_2(x)| \leq \frac{M}{6}|x - a|^3 \quad \text{for } |x - a| \leq d$$

62. (a) Show that the function defined by

$$f(x) = \begin{cases} e^{-1/x^2} & \text{if } x \neq 0 \\ 0 & \text{if } x = 0 \end{cases}$$

is not equal to its Maclaurin series.

(b) Graph the function in part (a) and comment on its behavior near the origin.

LABORATORY PROJECT

CAS An Elusive Limit

This project deals with the function

$$f(x) = \frac{\sin(\tan x) - \tan(\sin x)}{\arcsin(\arctan x) - \arctan(\arcsin x)}$$

1. Use your computer algebra system to evaluate $f(x)$ for $x = 1, 0.1, 0.01, 0.001$, and 0.0001. Does it appear that f has a limit as $x \to 0$?
2. Use the CAS to graph f near $x = 0$. Does it appear that f has a limit as $x \to 0$?
3. Try to evaluate $\lim_{x\to 0} f(x)$ with l'Hospital's Rule, using the CAS to find derivatives of the numerator and denominator. What do you discover? How many applications of l'Hospital's Rule are required?
4. Evaluate $\lim_{x\to 0} f(x)$ by using the CAS to find sufficiently many terms in the Taylor series of the numerator and denominator. (Use the command `taylor` in Maple or `Series` in Mathematica.)
5. Use the limit command on your CAS to find $\lim_{x\to 0} f(x)$ directly. (Most computer algebra systems use the method of Problem 4 to compute limits.)
6. In view of the answers to Problems 4 and 5, how do you explain the results of Problems 1 and 2?

11.11 The Binomial Series

You may be acquainted with the Binomial Theorem, which states that if a and b are any real numbers and k is a positive integer, then

$$(a + b)^k = a^k + ka^{k-1}b + \frac{k(k-1)}{2!}a^{k-2}b^2 + \frac{k(k-1)(k-2)}{3!}a^{k-3}b^3$$
$$+ \cdots + \frac{k(k-1)(k-2)\cdots(k-n+1)}{n!}a^{k-n}b^n$$
$$+ \cdots + kab^{k-1} + b^k$$

The traditional notation for the binomial coefficients is

$$\binom{k}{0} = 1 \qquad \binom{k}{n} = \frac{k(k-1)(k-2)\cdots(k-n+1)}{n!} \qquad n = 1, 2, \ldots, k$$

which enables us to write the Binomial Theorem in the abbreviated form

$$(a + b)^k = \sum_{n=0}^{k} \binom{k}{n} a^{k-n}b^n$$

In particular, if we put $a = 1$ and $b = x$, we get

1
$$(1 + x)^k = \sum_{n=0}^{k} \binom{k}{n} x^n$$

One of Newton's accomplishments was to extend the Binomial Theorem (Equation 1) to the case in which k is no longer a positive integer. (See the Writing Project on page 776.) In this case the expression for $(1 + x)^k$ is no longer a finite sum; it becomes an infinite series. To find this series we compute the Maclaurin series of $(1 + x)^k$ in the usual way:

$$\begin{aligned}
f(x) &= (1 + x)^k & f(0) &= 1 \\
f'(x) &= k(1 + x)^{k-1} & f'(0) &= k \\
f''(x) &= k(k - 1)(1 + x)^{k-2} & f''(0) &= k(k - 1) \\
f'''(x) &= k(k - 1)(k - 2)(1 + x)^{k-3} & f'''(0) &= k(k - 1)(k - 2) \\
&\vdots & &\vdots \\
f^{(n)}(x) &= k(k - 1) \cdots (k - n + 1)(1 + x)^{k-n} & f^{(n)}(0) &= k(k - 1) \cdots (k - n + 1)
\end{aligned}$$

Therefore, the Maclaurin series of $f(x) = (1 + x)^k$ is

$$\sum_{n=0}^{\infty} \frac{f^{(n)}(0)}{n!} x^n = \sum_{n=0}^{\infty} \frac{k(k - 1) \cdots (k - n + 1)}{n!} x^n$$

This series is called the **binomial series**. If its nth term is a_n, then

$$\begin{aligned}
\left| \frac{a_{n+1}}{a_n} \right| &= \left| \frac{k(k - 1) \cdots (k - n + 1)(k - n)x^{n+1}}{(n + 1)!} \cdot \frac{n!}{k(k - 1) \cdots (k - n + 1)x^n} \right| \\
&= \frac{|k - n|}{n + 1} |x| = \frac{\left| 1 - \dfrac{k}{n} \right|}{1 + \dfrac{1}{n}} |x| \to |x| \qquad \text{as } n \to \infty
\end{aligned}$$

Thus, by the Ratio Test, the binomial series converges if $|x| < 1$ and diverges if $|x| > 1$.

The following theorem states that $(1 + x)^k$ is equal to the sum of its Maclaurin series. It is possible to prove this by showing that the remainder term $R_n(x)$ approaches 0, but that turns out to be quite difficult. The proof outlined in Exercise 19 is much easier.

2 The Binomial Series If k is any real number and $|x| < 1$, then

$$\begin{aligned}
(1 + x)^k &= 1 + kx + \frac{k(k - 1)}{2!} x^2 + \frac{k(k - 1)(k - 2)}{3!} x^3 + \cdots \\
&= \sum_{n=0}^{\infty} \binom{k}{n} x^n
\end{aligned}$$

where $\displaystyle \binom{k}{n} = \frac{k(k - 1) \cdots (k - n + 1)}{n!} \quad (n \geqslant 1)$ and $\displaystyle \binom{k}{0} = 1$

Although the binomial series always converges when $|x| < 1$, the question of whether or not it converges at the endpoints, ± 1, depends on the value of k. It turns out that the series converges at 1 if $-1 < k \leqslant 0$ and at both endpoints if $k \geqslant 0$. Notice that if k is a positive integer and $n > k$, then the expression for $\binom{k}{n}$ contains a factor $(k - k)$, so $\binom{k}{n} = 0$ for $n > k$. This means that the series terminates and reduces to the ordinary Binomial Theorem (Equation 1) when k is a positive integer.

As we have seen, the binomial series is just a special case of the Maclaurin series; it occurs so frequently that it is worth remembering.

EXAMPLE 1 Expand $\dfrac{1}{(1 + x)^2}$ as a power series.

SOLUTION We use the binomial series with $k = -2$. The binomial coefficient is

$$\binom{-2}{n} = \frac{(-2)(-3)(-4) \cdots (-2 - n + 1)}{n!}$$

$$= \frac{(-1)^n 2 \cdot 3 \cdot 4 \cdot \cdots \cdot n(n + 1)}{n!} = (-1)^n(n + 1)$$

and so, when $|x| < 1$,

$$\frac{1}{(1 + x)^2} = (1 + x)^{-2} = \sum_{n=0}^{\infty} \binom{-2}{n} x^n$$

$$= \sum_{n=0}^{\infty} (-1)^n(n + 1)x^n = 1 - 2x + 3x^2 - 4x^3 + \cdots$$

EXAMPLE 2 Find the Maclaurin series for the function $f(x) = \dfrac{1}{\sqrt{4 - x}}$ and its radius of convergence.

SOLUTION As given, $f(x)$ is not quite of the form $(1 + x)^k$ so we rewrite it as follows:

$$\frac{1}{\sqrt{4 - x}} = \frac{1}{\sqrt{4\left(1 - \dfrac{x}{4}\right)}} = \frac{1}{2\sqrt{1 - \dfrac{x}{4}}} = \frac{1}{2}\left(1 - \frac{x}{4}\right)^{-1/2}$$

Using the binomial series with $k = -\frac{1}{2}$ and with x replaced by $-x/4$, we have

$$\frac{1}{\sqrt{4 - x}} = \frac{1}{2}\left(1 - \frac{x}{4}\right)^{-1/2} = \frac{1}{2} \sum_{n=0}^{\infty} \binom{-\frac{1}{2}}{n} \left(-\frac{x}{4}\right)^n$$

$$= \frac{1}{2}\left[1 + \left(-\frac{1}{2}\right)\left(-\frac{x}{4}\right) + \frac{\left(-\frac{1}{2}\right)\left(-\frac{3}{2}\right)}{2!}\left(-\frac{x}{4}\right)^2 + \frac{\left(-\frac{1}{2}\right)\left(-\frac{3}{2}\right)\left(-\frac{5}{2}\right)}{3!}\left(-\frac{x}{4}\right)^3\right.$$

$$\left. + \cdots + \frac{\left(-\frac{1}{2}\right)\left(-\frac{3}{2}\right)\left(-\frac{5}{2}\right) \cdots \left(-\frac{1}{2} - n + 1\right)}{n!}\left(-\frac{x}{4}\right)^n + \cdots\right]$$

$$= \frac{1}{2}\left[1 + \frac{1}{8}x + \frac{1 \cdot 3}{2!8^2}x^2 + \frac{1 \cdot 3 \cdot 5}{3!8^3}x^3 + \cdots + \frac{1 \cdot 3 \cdot 5 \cdot \cdots \cdot (2n - 1)}{n!8^n}x^n + \cdots\right]$$

We know from (2) that this series converges when $|-x/4| < 1$, that is, $|x| < 4$, so the radius of convergence is $R = 4$.

||| A binomial series is a special case of a Taylor series. Figure 1 shows the graphs of the first three Taylor polynomials computed from the answer to Example 2.

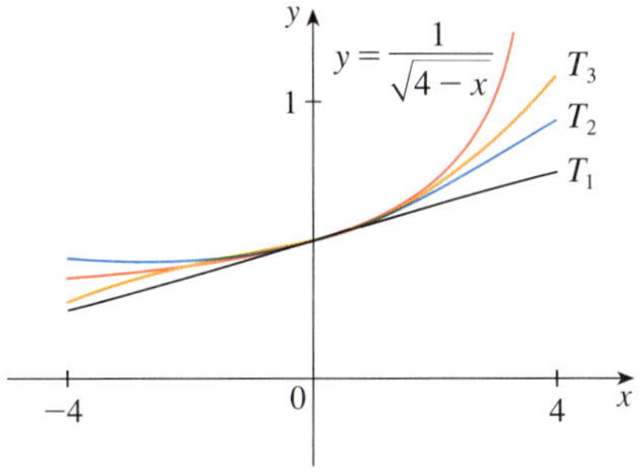

FIGURE 1

11.11 Exercises

1–8 ||| Use the binomial series to expand the function as a power series. State the radius of convergence.

1. $\sqrt{1 + x}$

2. $\dfrac{1}{(1 + x)^4}$

3. $\dfrac{1}{(2 + x)^3}$

4. $(1 - x)^{2/3}$

5. $\sqrt[4]{1 - 8x}$

6. $\dfrac{1}{\sqrt[5]{32 - x}}$

7. $\dfrac{x}{\sqrt{4 + x^2}}$

8. $\dfrac{x^2}{\sqrt{2 + x}}$

9–10 ||| Use the binomial series to expand the function as a Maclaurin series and to find the first three Taylor polynomials T_1, T_2, and T_3. Graph the function and these Taylor polynomials in the interval of convergence.

9. $(1 + 2x)^{3/4}$

10. $\sqrt[3]{1 + 4x}$

11. (a) Use the binomial series to expand $1/\sqrt{1 - x^2}$.
(b) Use part (a) to find the Maclaurin series for $\sin^{-1}x$.

12. (a) Use the binomial series to expand $1/\sqrt{1 + x^2}$.
(b) Use part (a) to find the Maclaurin series for $\sinh^{-1}x$.

13. (a) Expand $\sqrt[3]{1 + x}$ as a power series.
(b) Use part (a) to estimate $\sqrt[3]{1.01}$ correct to four decimal places.

14. (a) Expand $1/\sqrt[4]{1 + x}$ as a power series.
(b) Use part (a) to estimate $1/\sqrt[4]{1.1}$ correct to three decimal places.

15. (a) Expand $f(x) = x/(1 - x)^2$ as a power series.
(b) Use part (a) to find the sum of the series

$$\sum_{n=1}^{\infty} \frac{n}{2^n}$$

16. (a) Expand $f(x) = (x + x^2)/(1 - x)^3$ as a power series.
(b) Use part (a) to find the sum of the series

$$\sum_{n=1}^{\infty} \frac{n^2}{2^n}$$

17. (a) Use the binomial series to find the Maclaurin series of $f(x) = \sqrt{1 + x^2}$.
(b) Use part (a) to evaluate $f^{(10)}(0)$.

18. (a) Use the binomial series to find the Maclaurin series of $f(x) = 1/\sqrt{1 + x^3}$.
(b) Use part (a) to evaluate $f^{(9)}(0)$.

19. Use the following steps to prove (2).
(a) Let $g(x) = \sum_{n=0}^{\infty} \binom{k}{n} x^n$. Differentiate this series to show that

$$g'(x) = \frac{kg(x)}{1 + x} \qquad -1 < x < 1$$

(b) Let $h(x) = (1 + x)^{-k} g(x)$ and show that $h'(x) = 0$.
(c) Deduce that $g(x) = (1 + x)^k$.

20. In Exercise 53 in Section 10.2 it was shown that the length of the ellipse $x = a \sin\theta$, $y = b \cos\theta$, where $a > b > 0$, is

$$L = 4a \int_0^{\pi/2} \sqrt{1 - e^2 \sin^2\theta}\, d\theta$$

where $e = \sqrt{a^2 - b^2}/a$ is the eccentricity of the ellipse.
Expand the integrand as a binomial series and use the result of Exercise 44 in Section 7.1 to express L as a series in powers of the eccentricity up to the term in e^6.

WRITING PROJECT

How Newton Discovered the Binomial Series

The Binomial Theorem, which gives the expansion of $(a + b)^k$, was known to Chinese mathematicians many centuries before the time of Newton for the case where the exponent k is a positive integer. In 1665, when he was 22, Newton was the first to discover the infinite series expansion of $(a + b)^k$ when k is a fractional exponent (positive or negative). He didn't publish his discovery, but he stated it and gave examples of how to use it in a letter (now called the *epistola prior*) dated June 13, 1676, that he sent to Henry Oldenburg, secretary of the Royal Society of London, to transmit to Leibniz. When Leibniz replied, he asked how Newton had discovered the binomial series. Newton wrote a second letter, the *epistola posterior* of October 24, 1676, in which he explained in great detail how he arrived at his discovery by a very indirect route. He was investigating the areas under the curves $y = (1 - x^2)^{n/2}$ from 0 to x for $n = 0, 1, 2, 3, 4, \ldots$. These are easy to calculate if n is even. By observing patterns and interpolating, Newton was able to guess the answers for odd values of n. Then he realized he could get the same answers by expressing $(1 - x^2)^{n/2}$ as an infinite series.

Write a report on Newton's discovery of the binomial series. Start by giving the statement of the binomial series in Newton's notation (see the *epistola prior* on page 285 of [4] or page 402 of [2]). Explain why Newton's version is equivalent to Theorem 2 on page 773. Then read Newton's *epistola posterior* (page 287 in [4] or page 404 in [2]) and explain the patterns that Newton discovered in the areas under the curves $y = (1 - x^2)^{n/2}$. Show how he was able to guess the areas under the remaining curves and how he verified his answers. Finally, explain how these discoveries led to the binomial series. The books by Edwards [1] and Katz [3] contain commentaries on Newton's letters.

1. C. H. Edwards, *The Historical Development of the Calculus* (New York: Springer-Verlag, 1979), pp. 178–187.

2. John Fauvel and Jeremy Gray, eds., *The History of Mathematics: A Reader* (London: MacMillan Press, 1987).

3. Victor Katz, *A History of Mathematics: An Introduction* (New York: HarperCollins, 1993), pp. 463–466.

4. D. J. Struik, ed., *A Sourcebook in Mathematics, 1200–1800* (Princeton, N.J.: Princeton University Press, 1969).

11.12 Applications of Taylor Polynomials

In this section we explore two types of applications of Taylor polynomials. First we look at how they are used to approximate functions—computer scientists like them because polynomials are the simplest of functions. Then we investigate how physicists and engineers use them in such fields as relativity, optics, blackbody radiation, electric dipoles, the velocity of water waves, and building highways across a desert.

Approximating Functions by Polynomials

Suppose that $f(x)$ is equal to the sum of its Taylor series at a:

$$f(x) = \sum_{n=0}^{\infty} \frac{f^{(n)}(a)}{n!}(x - a)^n$$

In Section 11.10 we introduced the notation $T_n(x)$ for the nth partial sum of this series and called it the nth-degree Taylor polynomial of f at a. Thus

$$T_n(x) = \sum_{i=0}^{n} \frac{f^{(i)}(a)}{i!}(x-a)^i$$

$$= f(a) + \frac{f'(a)}{1!}(x-a) + \frac{f''(a)}{2!}(x-a)^2 + \cdots + \frac{f^{(n)}(a)}{n!}(x-a)^n$$

Since f is the sum of its Taylor series, we know that $T_n(x) \to f(x)$ as $n \to \infty$ and so T_n can be used as an approximation to f: $f(x) \approx T_n(x)$.

Notice that the first-degree Taylor polynomial

$$T_1(x) = f(a) + f'(a)(x-a)$$

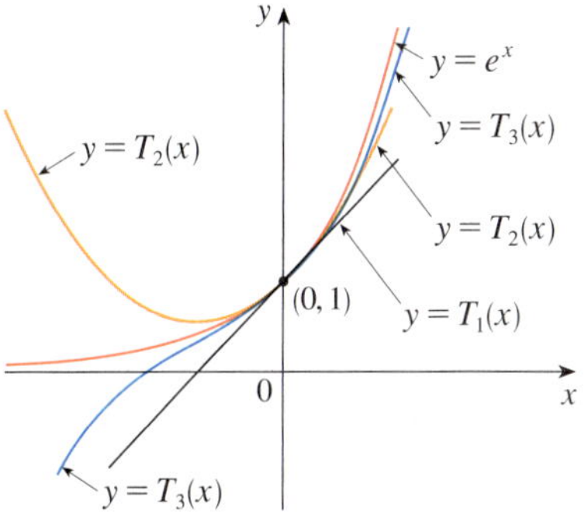

FIGURE 1

	$x = 0.2$	$x = 3.0$
$T_2(x)$	1.220000	8.500000
$T_4(x)$	1.221400	16.375000
$T_6(x)$	1.221403	19.412500
$T_8(x)$	1.221403	20.009152
$T_{10}(x)$	1.221403	20.079665
e^x	1.221403	20.085537

is the same as the linearization of f at a that we discussed in Section 3.11. Notice also that T_1 and its derivative have the same values at a that f and f' have. In general, it can be shown that the derivatives of T_n at a agree with those of f up to and including derivatives of order n (see Exercise 36).

To illustrate these ideas let's take another look at the graphs of $y = e^x$ and its first few Taylor polynomials, as shown in Figure 1. The graph of T_1 is the tangent line to $y = e^x$ at (0, 1); this tangent line is the best linear approximation to e^x near (0, 1). The graph of T_2 is the parabola $y = 1 + x + x^2/2$, and the graph of T_3 is the cubic curve $y = 1 + x + x^2/2 + x^3/6$, which is a closer fit to the exponential curve $y = e^x$ than T_2. The next Taylor polynomial T_4 would be an even better approximation, and so on.

The values in the table give a numerical demonstration of the convergence of the Taylor polynomials $T_n(x)$ to the function $y = e^x$. We see that when $x = 0.2$ the convergence is very rapid, but when $x = 3$ it is somewhat slower. In fact, the farther x is from 0, the more slowly $T_n(x)$ converges to e^x.

When using a Taylor polynomial T_n to approximate a function f, we have to ask the questions: How good an approximation is it? How large should we take n to be in order to achieve a desired accuracy? To answer these questions we need to look at the absolute value of the remainder:

$$|R_n(x)| = |f(x) - T_n(x)|$$

There are three possible methods for estimating the size of the error:

1. If a graphing device is available, we can use it to graph $|R_n(x)|$ and thereby estimate the error.
2. If the series happens to be an alternating series, we can use the Alternating Series Estimation Theorem.
3. In all cases we can use Taylor's Inequality (Theorem 11.10.9), which says that if $|f^{(n+1)}(x)| \leq M$, then

$$|R_n(x)| \leq \frac{M}{(n+1)!}|x-a|^{n+1}$$

EXAMPLE 1

(a) Approximate the function $f(x) = \sqrt[3]{x}$ by a Taylor polynomial of degree 2 at $a = 8$.
(b) How accurate is this approximation when $7 \leq x \leq 9$?

SOLUTION

(a)

$$f(x) = \sqrt[3]{x} = x^{1/3} \qquad f(8) = 2$$

$$f'(x) = \tfrac{1}{3}x^{-2/3} \qquad f'(8) = \tfrac{1}{12}$$

$$f''(x) = -\tfrac{2}{9}x^{-5/3} \qquad f''(8) = -\tfrac{1}{144}$$

$$f'''(x) = \tfrac{10}{27}x^{-8/3}$$

Thus, the second-degree Taylor polynomial is

$$\begin{aligned} T_2(x) &= f(8) + \frac{f'(8)}{1!}(x-8) + \frac{f''(8)}{2!}(x-8)^2 \\ &= 2 + \tfrac{1}{12}(x-8) - \tfrac{1}{288}(x-8)^2 \end{aligned}$$

The desired approximation is

$$\sqrt[3]{x} \approx T_2(x) = 2 + \tfrac{1}{12}(x-8) - \tfrac{1}{288}(x-8)^2$$

(b) The Taylor series is not alternating when $x < 8$, so we can't use the Alternating Series Estimation Theorem in this example. But we can use Taylor's Inequality with $n = 2$ and $a = 8$:

$$|R_2(x)| \leqslant \frac{M}{3!}|x - 8|^3$$

where $|f'''(x)| \leqslant M$. Because $x \geqslant 7$, we have $x^{8/3} \geqslant 7^{8/3}$ and so

$$f'''(x) = \frac{10}{27} \cdot \frac{1}{x^{8/3}} \leqslant \frac{10}{27} \cdot \frac{1}{7^{8/3}} < 0.0021$$

Therefore, we can take $M = 0.0021$. Also $7 \leqslant x \leqslant 9$, so $-1 \leqslant x - 8 \leqslant 1$ and $|x - 8| \leqslant 1$. Then Taylor's Inequality gives

$$|R_2(x)| \leqslant \frac{0.0021}{3!} \cdot 1^3 = \frac{0.0021}{6} < 0.0004$$

Thus, if $7 \leqslant x \leqslant 9$, the approximation in part (a) is accurate to within 0.0004.

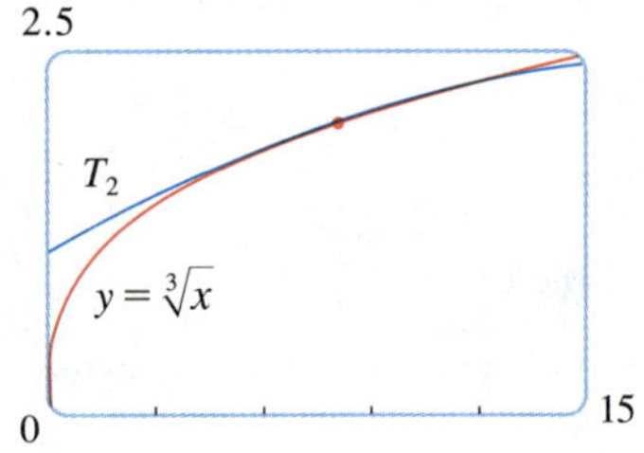

FIGURE 2

Let's use a graphing device to check the calculation in Example 1. Figure 2 shows that the graphs of $y = \sqrt[3]{x}$ and $y = T_2(x)$ are very close to each other when x is near 8. Figure 3 shows the graph of $|R_2(x)|$ computed from the expression

$$|R_2(x)| = |\sqrt[3]{x} - T_2(x)|$$

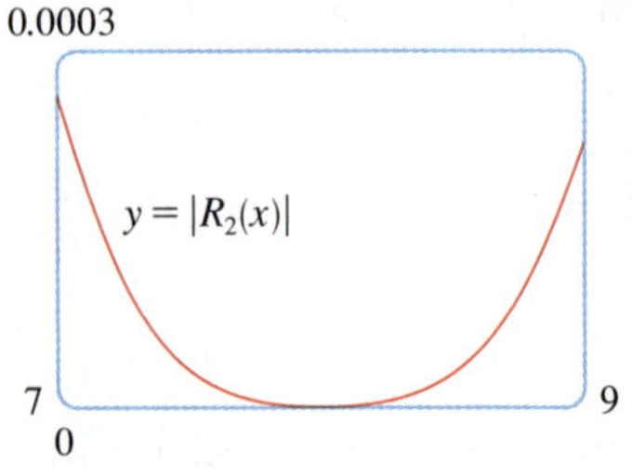

FIGURE 3

We see from the graph that

$$|R_2(x)| < 0.0003$$

when $7 \leqslant x \leqslant 9$. Thus, the error estimate from graphical methods is slightly better than the error estimate from Taylor's Inequality in this case.

EXAMPLE 2

(a) What is the maximum error possible in using the approximation

$$\sin x \approx x - \frac{x^3}{3!} + \frac{x^5}{5!}$$

when $-0.3 \leqslant x \leqslant 0.3$? Use this approximation to find $\sin 12^\circ$ correct to six decimal places.

(b) For what values of x is this approximation accurate to within 0.00005?

SOLUTION

(a) Notice that the Maclaurin series

$$\sin x = x - \frac{x^3}{3!} + \frac{x^5}{5!} - \frac{x^7}{7!} + \cdots$$

is alternating for all nonzero values of x, and the successive terms decrease in size because $|x| < 1$, so we can use the Alternating Series Estimation Theorem. The error in approximating $\sin x$ by the first three terms of its Maclaurin series is at most

$$\left| \frac{x^7}{7!} \right| = \frac{|x|^7}{5040}$$

If $-0.3 \leqslant x \leqslant 0.3$, then $|x| \leqslant 0.3$, so the error is smaller than

$$\frac{(0.3)^7}{5040} \approx 4.3 \times 10^{-8}$$

To find $\sin 12^\circ$ we first convert to radian measure.

$$\begin{aligned} \sin 12^\circ &= \sin\left(\frac{12\pi}{180}\right) = \sin\left(\frac{\pi}{15}\right) \\ &\approx \frac{\pi}{15} - \left(\frac{\pi}{15}\right)^3 \frac{1}{3!} + \left(\frac{\pi}{15}\right)^5 \frac{1}{5!} \\ &\approx 0.20791169 \end{aligned}$$

Thus, correct to six decimal places, $\sin 12^\circ \approx 0.207912$.

(b) The error will be smaller than 0.00005 if

$$\frac{|x|^7}{5040} < 0.00005$$

Solving this inequality for x, we get

$$|x|^7 < 0.252 \qquad \text{or} \qquad |x| < (0.252)^{1/7} \approx 0.821$$

So the given approximation is accurate to within 0.00005 when $|x| < 0.82$.

TEC Module 11.10/11.12 graphically shows the remainders in Taylor polynomial approximations.

What if we use Taylor's Inequality to solve Example 2? Since $f^{(7)}(x) = -\cos x$, we have $|f^{(7)}(x)| \leqslant 1$ and so

$$|R_6(x)| \leqslant \frac{1}{7!}|x|^7$$

So we get the same estimates as with the Alternating Series Estimation Theorem.

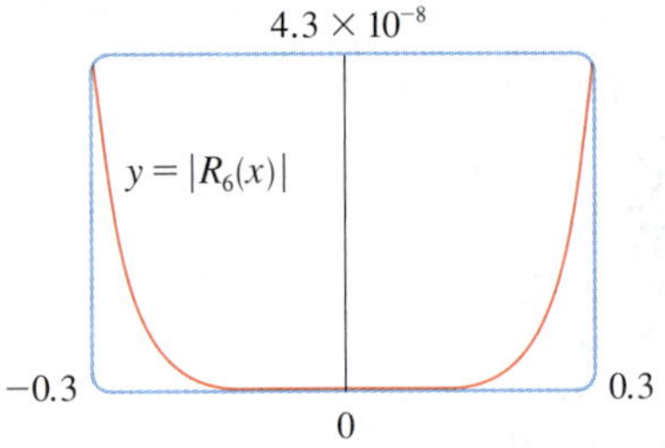

FIGURE 4

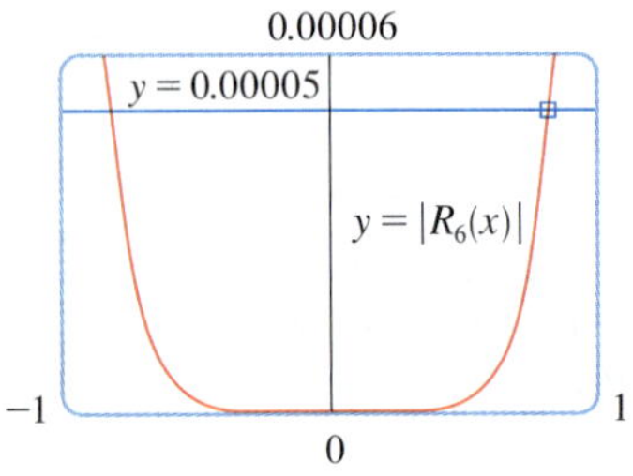

FIGURE 5

What about graphical methods? Figure 4 shows the graph of

$$|R_6(x)| = |\sin x - (x - \tfrac{1}{6}x^3 + \tfrac{1}{120}x^5)|$$

and we see from it that $|R_6(x)| < 4.3 \times 10^{-8}$ when $|x| \leqslant 0.3$. This is the same estimate that we obtained in Example 2. For part (b) we want $|R_6(x)| < 0.00005$, so we graph both $y = |R_6(x)|$ and $y = 0.00005$ in Figure 5. By placing the cursor on the right intersection point we find that the inequality is satisfied when $|x| < 0.82$. Again this is the same estimate that we obtained in the solution to Example 2.

If we had been asked to approximate sin 72° instead of sin 12° in Example 2, it would have been wise to use the Taylor polynomials at $a = \pi/3$ (instead of $a = 0$) because they are better approximations to $\sin x$ for values of x close to $\pi/3$. Notice that 72° is close to 60° (or $\pi/3$ radians) and the derivatives of $\sin x$ are easy to compute at $\pi/3$.

Figure 6 shows the graphs of the Maclaurin polynomial approximations

$$T_1(x) = x \qquad T_3(x) = x - \frac{x^3}{3!}$$

$$T_5(x) = x - \frac{x^3}{3!} + \frac{x^5}{5!} \qquad T_7(x) = x - \frac{x^3}{3!} + \frac{x^5}{5!} - \frac{x^7}{7!}$$

to the sine curve. You can see that as n increases, $T_n(x)$ is a good approximation to $\sin x$ on a larger and larger interval.

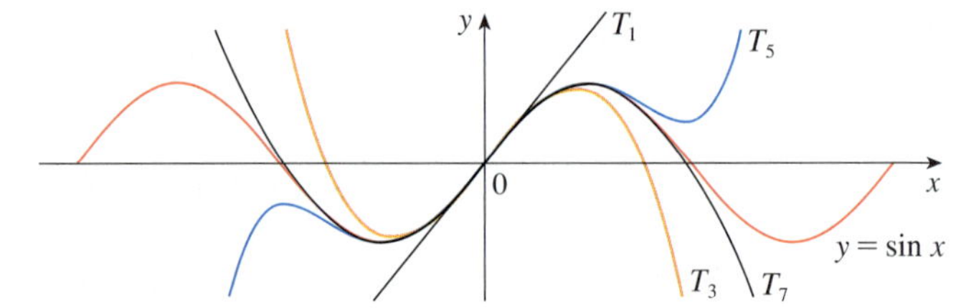

FIGURE 6

One use of the type of calculation done in Examples 1 and 2 occurs in calculators and computers. For instance, when you press the sin or e^x key on your calculator, or when a computer programmer uses a subroutine for a trigonometric or exponential or Bessel function, in many machines a polynomial approximation is calculated. The polynomial is often a Taylor polynomial that has been modified so that the error is spread more evenly throughout an interval.

Applications to Physics

Taylor polynomials are also used frequently in physics. In order to gain insight into an equation, a physicist often simplifies a function by considering only the first two or three terms in its Taylor series. In other words, the physicist uses a Taylor polynomial as an approximation to the function. Taylor's Inequality can then be used to gauge the accuracy of the approximation. The following example shows one way in which this idea is used in special relativity.

EXAMPLE 3 In Einstein's theory of special relativity the mass of an object moving with velocity v is

$$m = \frac{m_0}{\sqrt{1 - v^2/c^2}}$$

where m_0 is the mass of the object when at rest and c is the speed of light. The kinetic energy of the object is the difference between its total energy and its energy at rest:

$$K = mc^2 - m_0c^2$$

(a) Show that when v is very small compared with c, this expression for K agrees with classical Newtonian physics: $K = \frac{1}{2}m_0v^2$.
(b) Use Taylor's Inequality to estimate the difference in these expressions for K when $|v| \leqslant 100$ m/s.

SOLUTION
(a) Using the expressions given for K and m, we get

$$K = mc^2 - m_0c^2 = \frac{m_0c^2}{\sqrt{1 - v^2/c^2}} - m_0c^2$$
$$= m_0c^2\left[\left(1 - \frac{v^2}{c^2}\right)^{-1/2} - 1\right]$$

|||| The upper curve in Figure 7 is the graph of the expression for the kinetic energy K of an object with velocity v in special relativity. The lower curve shows the function used for K in classical Newtonian physics. When v is much smaller than the speed of light, the curves are practically identical.

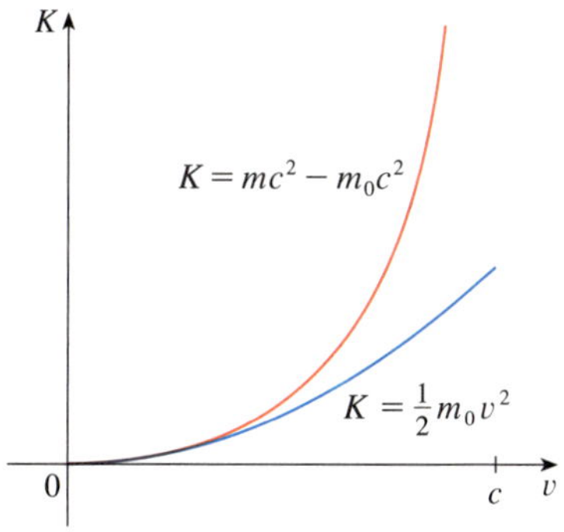

FIGURE 7

With $x = -v^2/c^2$, the Maclaurin series for $(1 + x)^{-1/2}$ is most easily computed as a binomial series with $k = -\frac{1}{2}$. (Notice that $|x| < 1$ because $v < c$.) Therefore, we have

$$(1 + x)^{-1/2} = 1 - \tfrac{1}{2}x + \frac{(-\frac{1}{2})(-\frac{3}{2})}{2!}x^2 + \frac{(-\frac{1}{2})(-\frac{3}{2})(-\frac{5}{2})}{3!}x^3 + \cdots$$
$$= 1 - \tfrac{1}{2}x + \tfrac{3}{8}x^2 - \tfrac{5}{16}x^3 + \cdots$$

and

$$K = m_0c^2\left[\left(1 + \frac{1}{2}\frac{v^2}{c^2} + \frac{3}{8}\frac{v^4}{c^4} + \frac{5}{16}\frac{v^6}{c^6} + \cdots\right) - 1\right]$$
$$= m_0c^2\left(\frac{1}{2}\frac{v^2}{c^2} + \frac{3}{8}\frac{v^4}{c^4} + \frac{5}{16}\frac{v^6}{c^6} + \cdots\right)$$

If v is much smaller than c, then all terms after the first are very small when compared with the first term. If we omit them, we get

$$K \approx m_0c^2\left(\frac{1}{2}\frac{v^2}{c^2}\right) = \tfrac{1}{2}m_0v^2$$

(b) If $x = -v^2/c^2$, $f(x) = m_0c^2[(1 + x)^{-1/2} - 1]$, and M is a number such that $|f''(x)| \leqslant M$, then we can use Taylor's Inequality to write

$$|R_1(x)| \leqslant \frac{M}{2!}x^2$$

We have $f''(x) = \frac{3}{4}m_0c^2(1 + x)^{-5/2}$ and we are given that $|v| \leqslant 100$ m/s, so

$$|f''(x)| = \frac{3m_0c^2}{4(1 - v^2/c^2)^{5/2}} \leqslant \frac{3m_0c^2}{4(1 - 100^2/c^2)^{5/2}} \quad (= M)$$

Thus, with $c = 3 \times 10^8$ m/s,

$$|R_1(x)| \leqslant \frac{1}{2} \cdot \frac{3m_0c^2}{4(1 - 100^2/c^2)^{5/2}} \cdot \frac{100^4}{c^4} < (4.17 \times 10^{-10})m_0$$

So when $|v| \leqslant 100$ m/s, the magnitude of the error in using the Newtonian expression for kinetic energy is at most $(4.2 \times 10^{-10})m_0$.

Another application to physics occurs in optics. Figure 8 is adapted from *Optics,* 4th ed., by Eugene Hecht (Reading, MA: Addison-Wesley, 2002), page 153. It depicts a wave from the point source S meeting a spherical interface of radius R centered at C. The ray SA is refracted toward P.

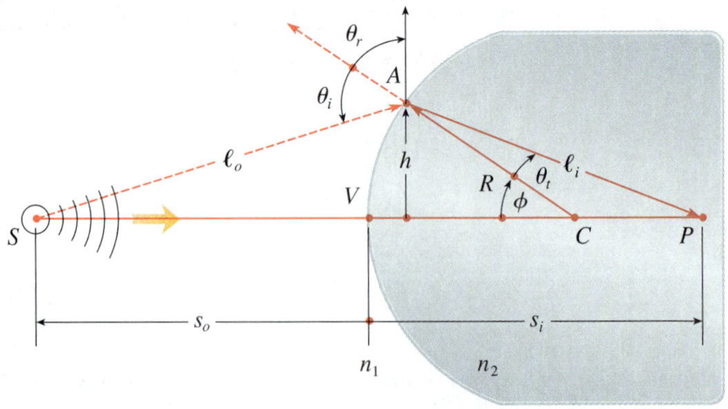

FIGURE 8
Refraction at a spherical interface

Using Fermat's principle that light travels so as to minimize the time taken, Hecht derives the equation

1 $$\frac{n_1}{\ell_o} + \frac{n_2}{\ell_i} = \frac{1}{R}\left(\frac{n_2 s_i}{\ell_i} - \frac{n_1 s_o}{\ell_o}\right)$$

where n_1 and n_2 are indexes of refraction and ℓ_o, ℓ_i, s_o, and s_i are the distances indicated in Figure 8. By the Law of Cosines, applied to triangles ACS and ACP, we have

Here we use the identity

$$\cos(\pi - \phi) = -\cos\phi$$

2 $$\ell_o = \sqrt{R^2 + (s_o + R)^2 - 2R(s_o + R)\cos\phi}$$

$$\ell_i = \sqrt{R^2 + (s_i - R)^2 + 2R(s_i - R)\cos\phi}$$

Because Equation 1 is cumbersome to work with, Gauss, in 1841, simplified it by using the linear approximation $\cos\phi \approx 1$ for small values of ϕ. (This amounts to using the Taylor polynomial of degree 1.) Then Equation 1 becomes the following simpler equation [as you are asked to show in Exercise 32(a)]:

3 $$\frac{n_1}{s_o} + \frac{n_2}{s_i} = \frac{n_2 - n_1}{R}$$

The resulting optical theory is known as *Gaussian optics,* or *first-order optics,* and has become the basic theoretical tool used to design lenses.

A more accurate theory is obtained by approximating $\cos\phi$ by its Taylor polynomial of degree 3 (which is the same as the Taylor polynomial of degree 2). This takes into account rays for which ϕ is not so small, that is, rays that strike the surface at greater distances h above the axis. In Exercise 32(b) you are asked to use this approximation to derive the more accurate equation

4 $$\frac{n_1}{s_o} + \frac{n_2}{s_i} = \frac{n_2 - n_1}{R} + h^2\left[\frac{n_1}{2s_o}\left(\frac{1}{s_o} + \frac{1}{R}\right)^2 + \frac{n_2}{2s_i}\left(\frac{1}{R} - \frac{1}{s_i}\right)^2\right]$$

The resulting optical theory is known as *third-order optics.*

Other applications of Taylor polynomials to physics and engineering are explored in Exercises 30, 31, 33, 34, and 35 and in the Applied Project on page 785.

11.12 Exercises

1. (a) Find the Taylor polynomials up to degree 6 for $f(x) = \cos x$ centered at $a = 0$. Graph f and these polynomials on a common screen.
   (b) Evaluate f and these polynomials at $x = \pi/4$, $\pi/2$, and π.
   (c) Comment on how the Taylor polynomials converge to $f(x)$.

2. (a) Find the Taylor polynomials up to degree 3 for $f(x) = 1/x$ centered at $a = 1$. Graph f and these polynomials on a common screen.
   (b) Evaluate f and these polynomials at $x = 0.9$ and 1.3.
   (c) Comment on how the Taylor polynomials converge to $f(x)$.

3–10 Find the Taylor polynomial $T_n(x)$ for the function f at the number a. Graph f and T_n on the same screen.

3. $f(x) = \ln x, \quad a = 1, \quad n = 4$
4. $f(x) = e^x, \quad a = 2, \quad n = 3$
5. $f(x) = \sin x, \quad a = \pi/6, \quad n = 3$
6. $f(x) = \cos x, \quad a = 2\pi/3, \quad n = 4$
7. $f(x) = \arcsin x, \quad a = 0, \quad n = 3$
8. $f(x) = \dfrac{\ln x}{x}, \quad a = 1, \quad n = 3$
9. $f(x) = xe^{-2x}, \quad a = 0, \quad n = 3$
10. $f(x) = \sqrt{3 + x^2}, \quad a = 1, \quad n = 2$

11–12 Use a computer algebra system to find the Taylor polynomials T_n at $a = 0$ for the given values of n. Then graph these polynomials and f on the same screen.

11. $f(x) = \sec x, \quad n = 2, 4, 6, 8$
12. $f(x) = \tan x, \quad n = 1, 3, 5, 7, 9$

13–22

(a) Approximate f by a Taylor polynomial with degree n at the number a.

(b) Use Taylor's Inequality to estimate the accuracy of the approximation $f(x) \approx T_n(x)$ when x lies in the given interval.

(c) Check your result in part (b) by graphing $|R_n(x)|$.

13. $f(x) = \sqrt{x}, \quad a = 4, \quad n = 2, \quad 4 \leqslant x \leqslant 4.2$
14. $f(x) = x^{-2}, \quad a = 1, \quad n = 2, \quad 0.9 \leqslant x \leqslant 1.1$
15. $f(x) = x^{2/3}, \quad a = 1, \quad n = 3, \quad 0.8 \leqslant x \leqslant 1.2$
16. $f(x) = \cos x, \quad a = \pi/3, \quad n = 4, \quad 0 \leqslant x \leqslant 2\pi/3$
17. $f(x) = \tan x, \quad a = 0, \quad n = 3, \quad 0 \leqslant x \leqslant \pi/6$
18. $f(x) = \ln(1 + 2x), \quad a = 1, \quad n = 3, \quad 0.5 \leqslant x \leqslant 1.5$
19. $f(x) = e^{x^2}, \quad a = 0, \quad n = 3, \quad 0 \leqslant x \leqslant 0.1$
20. $f(x) = x \ln x, \quad a = 1, \quad n = 3, \quad 0.5 \leqslant x \leqslant 1.5$
21. $f(x) = x \sin x, \quad a = 0, \quad n = 4, \quad -1 \leqslant x \leqslant 1$
22. $f(x) = \sinh 2x, \quad a = 0, \quad n = 5, \quad -1 \leqslant x \leqslant 1$

23. Use the information from Exercise 5 to estimate $\sin 35°$ correct to five decimal places.

24. Use the information from Exercise 16 to estimate $\cos 69°$ correct to five decimal places.

25. Use Taylor's Inequality to determine the number of terms of the Maclaurin series for e^x that should be used to estimate $e^{0.1}$ to within 0.00001.

26. How many terms of the Maclaurin series for $\ln(1 + x)$ do you need to use to estimate $\ln 1.4$ to within 0.001?

27–28 Use the Alternating Series Estimation Theorem or Taylor's Inequality to estimate the range of values of x for which the given approximation is accurate to within the stated error. Check your answer graphically.

27. $\sin x \approx x - \dfrac{x^3}{6} \quad (|\text{error}| < 0.01)$

28. $\cos x \approx 1 - \dfrac{x^2}{2} + \dfrac{x^4}{24} \quad (|\text{error}| < 0.005)$

29. A car is moving with speed 20 m/s and acceleration 2 m/s^2 at a given instant. Using a second-degree Taylor polynomial, estimate how far the car moves in the next second. Would it be reasonable to use this polynomial to estimate the distance traveled during the next minute?

30. The resistivity ρ of a conducting wire is the reciprocal of the conductivity and is measured in units of ohm-meters (Ω-m). The resistivity of a given metal depends on the temperature according to the equation

$$\rho(t) = \rho_{20} e^{\alpha(t-20)}$$

where t is the temperature in °C. There are tables that list the values of α (called the temperature coefficient) and ρ_{20} (the resistivity at 20°C) for various metals. Except at very low temperatures, the resistivity varies almost linearly with temperature and so it is common to approximate the expression for $\rho(t)$ by its first- or second-degree Taylor polynomial at $t = 20$.
   (a) Find expressions for these linear and quadratic approximations.
   (b) For copper, the tables give $\alpha = 0.0039$/°C and $\rho_{20} = 1.7 \times 10^{-8}$ Ω-m. Graph the resistivity of copper and the linear and quadratic approximations for $-250°\text{C} \leqslant t \leqslant 1000°\text{C}$.

(c) For what values of t does the linear approximation agree with the exponential expression to within one percent?

31. An electric dipole consists of two electric charges of equal magnitude and opposite signs. If the charges are q and $-q$ and are located at a distance d from each other, then the electric field E at the point P in the figure is

$$E = \frac{q}{D^2} - \frac{q}{(D + d)^2}$$

By expanding this expression for E as a series in powers of d/D, show that E is approximately proportional to $1/D^3$ when P is far away from the dipole.

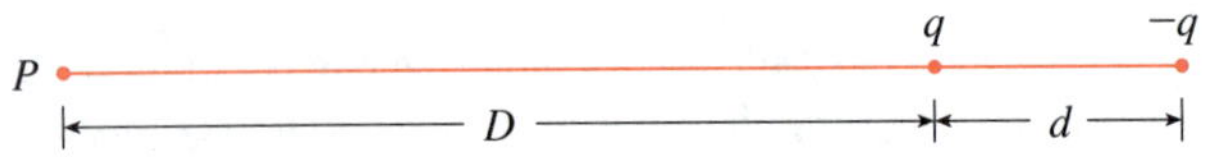

32. (a) Derive Equation 3 for Gaussian optics from Equation 1 by approximating $\cos\phi$ in Equation 2 by its first-degree Taylor polynomial.
(b) Show that if $\cos\phi$ is replaced by its third-degree Taylor polynomial in Equation 2, then Equation 1 becomes Equation 4 for third-order optics. [*Hint:* Use the first two terms in the binomial series for ℓ_o^{-1} and ℓ_i^{-1}. Also, use $\phi \approx \sin\phi$.]

33. If a water wave with length L moves with velocity v across a body of water with depth d, as in the figure, then

$$v^2 = \frac{gL}{2\pi}\tanh\frac{2\pi d}{L}$$

(a) If the water is deep, show that $v \approx \sqrt{gL/(2\pi)}$.
(b) If the water is shallow, use the Maclaurin series for tanh to show that $v \approx \sqrt{gd}$. (Thus, in shallow water the velocity of a wave tends to be independent of the length of the wave.)
(c) Use the Alternating Series Estimation Theorem to show that if $L > 10d$, then the estimate $v^2 \approx gd$ is accurate to within $0.014gL$.

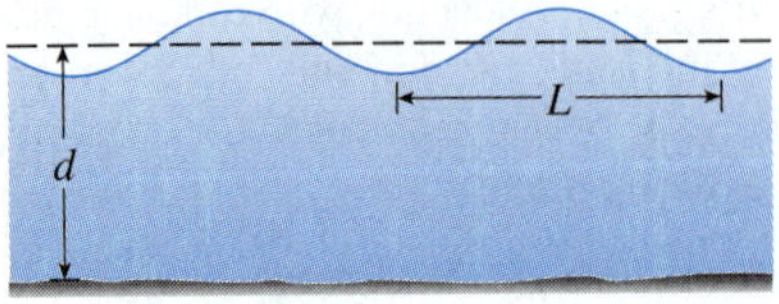

34. The period of a pendulum with length L that makes a maximum angle θ_0 with the vertical is

$$T = 4\sqrt{\frac{L}{g}}\int_0^{\pi/2}\frac{dx}{\sqrt{1 - k^2\sin^2 x}}$$

where $k = \sin\left(\frac{1}{2}\theta_0\right)$ and g is the acceleration due to gravity. (In Exercise 40 in Section 7.7 we approximated this integral using Simpson's Rule.)
(a) Expand the integrand as a binomial series and use the result of Exercise 44 in Section 7.1 to show that

$$T = 2\pi\sqrt{\frac{L}{g}}\left[1 + \frac{1^2}{2^2}k^2 + \frac{1^2 3^2}{2^2 4^2}k^4 + \frac{1^2 3^2 5^2}{2^2 4^2 6^2}k^6 + \cdots\right]$$

If θ_0 is not too large, the approximation $T \approx 2\pi\sqrt{L/g}$, obtained by using only the first term in the series, is often used. A better approximation is obtained by using two terms:

$$T \approx 2\pi\sqrt{\frac{L}{g}}\left(1 + \tfrac{1}{4}k^2\right)$$

(b) Notice that all the terms in the series after the first one have coefficients that are at most $\frac{1}{4}$. Use this fact to compare this series with a geometric series and show that

$$2\pi\sqrt{\frac{L}{g}}\left(1 + \tfrac{1}{4}k^2\right) \leqslant T \leqslant 2\pi\sqrt{\frac{L}{g}}\,\frac{4 - 3k^2}{4 - 4k^2}$$

(c) Use the inequalities in part (b) to estimate the period of a pendulum with $L = 1$ meter and $\theta_0 = 10°$. How does it compare with the estimate $T \approx 2\pi\sqrt{L/g}$? What if $\theta_0 = 42°$?

35. If a surveyor measures differences in elevation when making plans for a highway across a desert, corrections must be made for the curvature of the Earth.
(a) If R is the radius of the Earth and L is the length of the highway, show that the correction is

$$C = R\sec(L/R) - R$$

(b) Use a Taylor polynomial to show that

$$C \approx \frac{L^2}{2R} + \frac{5L^4}{24R^3}$$

(c) Compare the corrections given by the formulas in parts (a) and (b) for a highway that is 100 km long. (Take the radius of Earth to be 6370 km.)

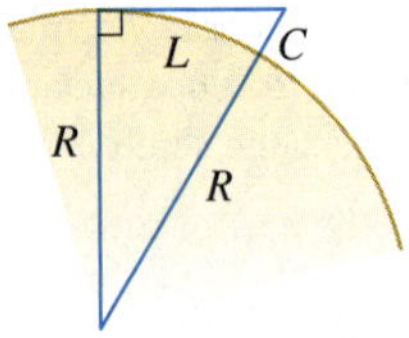

36. Show that T_n and f have the same derivatives at a up to order n.

37. In Section 4.9 we considered Newton's method for approximating a root r of the equation $f(x) = 0$, and from an initial approximation x_1 we obtained successive approximations x_2, x_3, . . . , where

$$x_{n+1} = x_n - \frac{f(x_n)}{f'(x_n)}$$

Use Taylor's Inequality with $n = 1$, $a = x_n$, and $x = r$ to show that if $f''(x)$ exists on an interval I containing r, x_n, and x_{n+1}, and $|f''(x)| \leqslant M$, $|f'(x)| \geqslant K$ for all $x \in I$, then

$$|x_{n+1} - r| \leqslant \frac{M}{2K}|x_n - r|^2$$

[This means that if x_n is accurate to d decimal places, then x_{n+1} is accurate to about $2d$ decimal places. More precisely, if the error at stage n is at most 10^{-m}, then the error at stage $n + 1$ is at most $(M/2K)10^{-2m}$.]

APPLIED PROJECT

Radiation from the Stars

Any object emits radiation when heated. A *blackbody* is a system that absorbs all the radiation that falls on it. For instance, a matte black surface or a large cavity with a small hole in its wall (like a blastfurnace) is a blackbody and emits blackbody radiation. Even the radiation from the Sun is close to being blackbody radiation.

Proposed in the late 19th century, the Rayleigh-Jeans Law expresses the energy density of blackbody radiation of wavelength λ as

$$f(\lambda) = \frac{8\pi kT}{\lambda^4}$$

where λ is measured in meters, T is the temperature in kelvins (K), and k is Boltzmann's constant. The Rayleigh-Jeans Law agrees with experimental measurements for long wavelengths but disagrees drastically for short wavelengths. [The law predicts that $f(\lambda) \to \infty$ as $\lambda \to 0^+$ but experiments have shown that $f(\lambda) \to 0$.] This fact is known as the *ultraviolet catastrophe*.

In 1900 Max Planck found a better model (known now as Planck's Law) for blackbody radiation:

$$f(\lambda) = \frac{8\pi hc\lambda^{-5}}{e^{hc/(\lambda kT)} - 1}$$

where λ is measured in meters, T is the temperature (in kelvins), and

$$h = \text{Planck's constant} = 6.6262 \times 10^{-34}\ \text{J}\cdot\text{s}$$

$$c = \text{speed of light} = 2.997925 \times 10^{8}\ \text{m/s}$$

$$k = \text{Boltzmann's constant} = 1.3807 \times 10^{-23}\ \text{J/K}$$

1. Use l'Hospital's Rule to show that

$$\lim_{\lambda \to 0^+} f(\lambda) = 0 \quad \text{and} \quad \lim_{\lambda \to \infty} f(\lambda) = 0$$

for Planck's Law. So this law models blackbody radiation better than the Rayleigh-Jeans Law for short wavelengths.

2. Use a Taylor polynomial to show that, for large wavelengths, Planck's Law gives approximately the same values as the Rayleigh-Jeans Law.

3. Graph f as given by both laws on the same screen and comment on the similarities and differences. Use $T = 5700$ K (the temperature of the Sun). (You may want to change from meters to the more convenient unit of micrometers: 1 μm $= 10^{-6}$ m.)

4. Use your graph in Problem 3 to estimate the value of λ for which $f(\lambda)$ is a maximum under Planck's Law.

5. Investigate how the graph of f changes as T varies. (Use Planck's Law.) In particular, graph f for the stars Betelgeuse ($T = 3400$ K), Procyon ($T = 6400$ K), and Sirius ($T = 9200$ K) as well as the Sun. How does the total radiation emitted (the area under the curve) vary with T? Use the graph to comment on why Sirius is known as a blue star and Betelgeuse as a red star.

11 Review

CONCEPT CHECK

1. (a) What is a convergent sequence?
   (b) What is a convergent series?
   (c) What does $\lim_{n\to\infty} a_n = 3$ mean?
   (d) What does $\sum_{n=1}^{\infty} a_n = 3$ mean?

2. (a) What is a bounded sequence?
   (b) What is a monotonic sequence?
   (c) What can you say about a bounded monotonic sequence?

3. (a) What is a geometric series? Under what circumstances is it convergent? What is its sum?
   (b) What is a p-series? Under what circumstances is it convergent?

4. Suppose $\sum a_n = 3$ and s_n is the nth partial sum of the series. What is $\lim_{n\to\infty} a_n$? What is $\lim_{n\to\infty} s_n$?

5. State the following.
   (a) The Test for Divergence
   (b) The Integral Test
   (c) The Comparison Test
   (d) The Limit Comparison Test
   (e) The Alternating Series Test
   (f) The Ratio Test
   (g) The Root Test

6. (a) What is an absolutely convergent series?
   (b) What can you say about such a series?
   (c) What is a conditionally convergent series?

7. (a) If a series is convergent by the Integral Test, how do you estimate its sum?
   (b) If a series is convergent by the Comparison Test, how do you estimate its sum?
   (c) If a series is convergent by the Alternating Series Test, how do you estimate its sum?

8. (a) Write the general form of a power series.
   (b) What is the radius of convergence of a power series?
   (c) What is the interval of convergence of a power series?

9. Suppose $f(x)$ is the sum of a power series with radius of convergence R.
   (a) How do you differentiate f? What is the radius of convergence of the series for f'?
   (b) How do you integrate f? What is the radius of convergence of the series for $\int f(x)\,dx$?

10. (a) Write an expression for the nth-degree Taylor polynomial of f centered at a.
    (b) Write an expression for the Taylor series of f centered at a.
    (c) Write an expression for the Maclaurin series of f.
    (d) How do you show that $f(x)$ is equal to the sum of its Taylor series?
    (e) State Taylor's Inequality.

11. Write the Maclaurin series and the interval of convergence for each of the following functions.
    (a) $1/(1 - x)$
    (b) e^x
    (c) $\sin x$
    (d) $\cos x$
    (e) $\tan^{-1}x$

12. Write the binomial series expansion of $(1 + x)^k$. What is the radius of convergence of this series?

TRUE-FALSE QUIZ

Determine whether the statement is true or false. If it is true, explain why. If it is false, explain why or give an example that disproves the statement.

1. If $\lim_{n\to\infty} a_n = 0$, then $\sum a_n$ is convergent.
2. The series $\sum_{n=1}^{\infty} n^{-\sin 1}$ is convergent.
3. If $\lim_{n\to\infty} a_n = L$, then $\lim_{n\to\infty} a_{2n+1} = L$.
4. If $\sum c_n 6^n$ is convergent, then $\sum c_n(-2)^n$ is convergent.
5. If $\sum c_n 6^n$ is convergent, then $\sum c_n(-6)^n$ is convergent.
6. If $\sum c_n x^n$ diverges when $x = 6$, then it diverges when $x = 10$.
7. The Ratio Test can be used to determine whether $\sum 1/n^3$ converges.
8. The Ratio Test can be used to determine whether $\sum 1/n!$ converges.
9. If $0 \le a_n \le b_n$ and $\sum b_n$ diverges, then $\sum a_n$ diverges.
10. $\displaystyle\sum_{n=0}^{\infty} \frac{(-1)^n}{n!} = \frac{1}{e}$
11. If $-1 < \alpha < 1$, then $\lim_{n\to\infty} \alpha^n = 0$.
12. If $\sum a_n$ is divergent, then $\sum |a_n|$ is divergent.
13. If $f(x) = 2x - x^2 + \frac{1}{3}x^3 - \cdots$ converges for all x, then $f'''(0) = 2$.
14. If $\{a_n\}$ and $\{b_n\}$ are divergent, then $\{a_n + b_n\}$ is divergent.
15. If $\{a_n\}$ and $\{b_n\}$ are divergent, then $\{a_n b_n\}$ is divergent.
16. If $\{a_n\}$ is decreasing and $a_n > 0$ for all n, then $\{a_n\}$ is convergent.
17. If $a_n > 0$ and $\sum a_n$ converges, then $\sum (-1)^n a_n$ converges.
18. If $a_n > 0$ and $\lim_{n\to\infty} (a_{n+1}/a_n) < 1$, then $\lim_{n\to\infty} a_n = 0$.

EXERCISES

1–8 Determine whether the sequence is convergent or divergent. If it is convergent, find its limit.

1. $a_n = \dfrac{2 + n^3}{1 + 2n^3}$

2. $a_n = \dfrac{9^{n+1}}{10^n}$

3. $a_n = \dfrac{n^3}{1 + n^2}$

4. $a_n = \cos(n\pi/2)$

5. $a_n = \dfrac{n \sin n}{n^2 + 1}$

6. $a_n = \dfrac{\ln n}{\sqrt{n}}$

7. $\{(1 + 3/n)^{4n}\}$

8. $\{(-10)^n/n!\}$

9. A sequence is defined recursively by the equations $a_1 = 1$, $a_{n+1} = \frac{1}{3}(a_n + 4)$. Show that $\{a_n\}$ is increasing and $a_n < 2$ for all n. Deduce that $\{a_n\}$ is convergent and find its limit.

10. Show that $\lim_{n\to\infty} n^4 e^{-n} = 0$ and use a graph to find the smallest value of N that corresponds to $\varepsilon = 0.1$ in the precise definition of a limit.

11–22 Determine whether the series is convergent or divergent.

11. $\displaystyle\sum_{n=1}^{\infty} \frac{n}{n^3 + 1}$

12. $\displaystyle\sum_{n=1}^{\infty} \frac{n^2 + 1}{n^3 + 1}$

13. $\displaystyle\sum_{n=1}^{\infty} \frac{n^3}{5^n}$

14. $\displaystyle\sum_{n=1}^{\infty} \frac{(-1)^n}{\sqrt{n + 1}}$

15. $\displaystyle\sum_{n=2}^{\infty} \frac{1}{n\sqrt{\ln n}}$

16. $\displaystyle\sum_{n=1}^{\infty} \ln\left(\frac{n}{3n + 1}\right)$

17. $\displaystyle\sum_{n=1}^{\infty} \frac{\cos 3n}{1 + (1.2)^n}$

18. $\displaystyle\sum_{n=1}^{\infty} \frac{n^{2n}}{(1 + 2n^2)^n}$

19. $\displaystyle\sum_{n=1}^{\infty} \frac{1 \cdot 3 \cdot 5 \cdot \cdots \cdot (2n - 1)}{5^n n!}$

20. $\displaystyle\sum_{n=1}^{\infty} \frac{(-5)^{2n}}{n^2 9^n}$

21. $\displaystyle\sum_{n=1}^{\infty} (-1)^{n-1} \frac{\sqrt{n}}{n + 1}$

22. $\displaystyle\sum_{n=1}^{\infty} \frac{\sqrt{n + 1} - \sqrt{n - 1}}{n}$

23–26 Determine whether the series is conditionally convergent, absolutely convergent, or divergent.

23. $\displaystyle\sum_{n=1}^{\infty} (-1)^{n-1} n^{-1/3}$

24. $\displaystyle\sum_{n=1}^{\infty} (-1)^{n-1} n^{-3}$

25. $\displaystyle\sum_{n=1}^{\infty} \frac{(-1)^n (n + 1) 3^n}{2^{2n+1}}$

26. $\displaystyle\sum_{n=2}^{\infty} \frac{(-1)^n \sqrt{n}}{\ln n}$

27–31 Find the sum of the series.

27. $\displaystyle\sum_{n=1}^{\infty} \frac{2^{2n+1}}{5^n}$

28. $\displaystyle\sum_{n=1}^{\infty} \frac{1}{n(n + 3)}$

29. $\displaystyle\sum_{n=1}^{\infty} [\tan^{-1}(n + 1) - \tan^{-1} n]$

30. $\displaystyle\sum_{n=0}^{\infty} \frac{(-1)^n \pi^n}{3^{2n}(2n)!}$

31. $1 - e + \dfrac{e^2}{2!} - \dfrac{e^3}{3!} + \dfrac{e^4}{4!} - \cdots$

32. Express the repeating decimal 4.17326326326 . . . as a fraction.

33. Show that $\cosh x \geqslant 1 + \frac{1}{2}x^2$ for all x.

34. For what values of x does the series $\sum_{n=1}^{\infty} (\ln x)^n$ converge?

35. Find the sum of the series $\displaystyle\sum_{n=1}^{\infty} \frac{(-1)^{n+1}}{n^5}$ correct to four decimal places.

36. (a) Find the partial sum s_5 of the series $\sum_{n=1}^{\infty} 1/n^6$ and estimate the error in using it as an approximation to the sum of the series.

(b) Find the sum of this series correct to five decimal places.

37. Use the sum of the first eight terms to approximate the sum of the series $\sum_{n=1}^{\infty} (2 + 5^n)^{-1}$. Estimate the error involved in this approximation.

38. (a) Show that the series $\displaystyle\sum_{n=1}^{\infty} \frac{n^n}{(2n)!}$ is convergent.

(b) Deduce that $\displaystyle\lim_{n\to\infty} \frac{n^n}{(2n)!} = 0$.

39. Prove that if the series $\sum_{n=1}^{\infty} a_n$ is absolutely convergent, then the series

$$\sum_{n=1}^{\infty} \left(\frac{n + 1}{n}\right) a_n$$

is also absolutely convergent.

40–43 Find the radius of convergence and interval of convergence of the series.

40. $\displaystyle\sum_{n=1}^{\infty} (-1)^n \frac{x^n}{n^2 5^n}$

41. $\displaystyle\sum_{n=1}^{\infty} \frac{(x + 2)^n}{n 4^n}$

42. $\displaystyle\sum_{n=1}^{\infty} \frac{2^n (x - 2)^n}{(n + 2)!}$

43. $\displaystyle\sum_{n=0}^{\infty} \frac{2^n (x - 3)^n}{\sqrt{n + 3}}$

44. Find the radius of convergence of the series

$$\sum_{n=1}^{\infty} \frac{(2n)!}{(n!)^2} x^n$$

45. Find the Taylor series of $f(x) = \sin x$ at $a = \pi/6$.

46. Find the Taylor series of $f(x) = \cos x$ at $a = \pi/3$.

47–54 Find the Maclaurin series for f and its radius of convergence. You may use either the direct method (definition of a Maclaurin series) or known series such as geometric series, binomial series, or the Maclaurin series for e^x, $\sin x$, and $\tan^{-1}x$.

47. $f(x) = \dfrac{x^2}{1+x}$

48. $f(x) = \tan^{-1}(x^2)$

49. $f(x) = \ln(1 - x)$

50. $f(x) = xe^{2x}$

51. $f(x) = \sin(x^4)$

52. $f(x) = 10^x$

53. $f(x) = 1/\sqrt[4]{16 - x}$

54. $f(x) = (1 - 3x)^{-5}$

55. Evaluate $\displaystyle\int \frac{e^x}{x}\,dx$ as an infinite series.

56. Use series to approximate $\int_0^1 \sqrt{1 + x^4}\,dx$ correct to two decimal places.

57–58

(a) Approximate f by a Taylor polynomial with degree n at the number a.

(b) Graph f and T_n on a common screen.

(c) Use Taylor's Inequality to estimate the accuracy of the approximation $f(x) \approx T_n(x)$ when x lies in the given interval.

(d) Check your result in part (c) by graphing $|R_n(x)|$.

57. $f(x) = \sqrt{x}$, $a = 1$, $n = 3$, $0.9 \leqslant x \leqslant 1.1$

58. $f(x) = \sec x$, $a = 0$, $n = 2$, $0 \leqslant x \leqslant \pi/6$

59. Use series to evaluate the following limit.

$$\lim_{x \to 0} \frac{\sin x - x}{x^3}$$

60. The force due to gravity on an object with mass m at a height h above the surface of the Earth is

$$F = \frac{mgR^2}{(R + h)^2}$$

where R is the radius of the Earth and g is the acceleration due to gravity.

(a) Express F as a series in powers of h/R.

(b) Observe that if we approximate F by the first term in the series, we get the expression $F \approx mg$ that is usually used when h is much smaller than R. Use the Alternating Series Estimation Theorem to estimate the range of values of h for which the approximation $F \approx mg$ is accurate to within 1%. (Use $R = 6400$ km.)

61. Suppose that $f(x) = \sum_{n=0}^{\infty} c_n x^n$ for all x.

(a) If f is an odd function, show that

$$c_0 = c_2 = c_4 = \cdots = 0$$

(b) If f is an even function, show that

$$c_1 = c_3 = c_5 = \cdots = 0$$

62. If $f(x) = e^{x^2}$, show that $f^{(2n)}(0) = \dfrac{(2n)!}{n!}$.

PROBLEMS PLUS

1. If $f(x) = \sin(x^3)$, find $f^{(15)}(0)$.

2. A function f is defined by

$$f(x) = \lim_{n \to \infty} \frac{x^{2n} - 1}{x^{2n} + 1}$$

Where is f continuous?

3. (a) Show that $\tan \frac{1}{2}x = \cot \frac{1}{2}x - 2 \cot x$.

(b) Find the sum of the series

$$\sum_{n=1}^{\infty} \frac{1}{2^n} \tan \frac{x}{2^n}$$

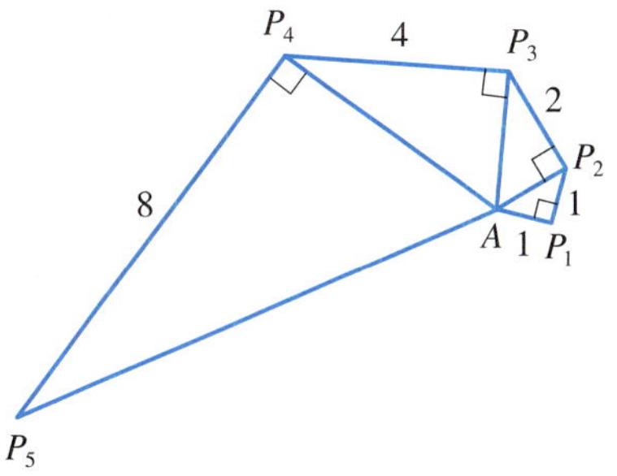

FIGURE FOR PROBLEM 4

4. Let $\{P_n\}$ be a sequence of points determined as in the figure. Thus $|AP_1| = 1$, $|P_nP_{n+1}| = 2^{n-1}$, and angle AP_nP_{n+1} is a right angle. Find $\lim_{n \to \infty} \angle P_nAP_{n+1}$.

5. To construct the **snowflake curve**, start with an equilateral triangle with sides of length 1. Step 1 in the construction is to divide each side into three equal parts, construct an equilateral triangle on the middle part, and then delete the middle part (see the figure). Step 2 is to repeat Step 1 for each side of the resulting polygon. This process is repeated at each succeeding step. The snowflake curve is the curve that results from repeating this process indefinitely.

(a) Let s_n, l_n, and p_n represent the number of sides, the length of a side, and the total length of the nth approximating curve (the curve obtained after Step n of the construction), respectively. Find formulas for s_n, l_n, and p_n.

(b) Show that $p_n \to \infty$ as $n \to \infty$.

(c) Sum an infinite series to find the area enclosed by the snowflake curve.

Parts (b) and (c) show that the snowflake curve is infinitely long but encloses only a finite area.

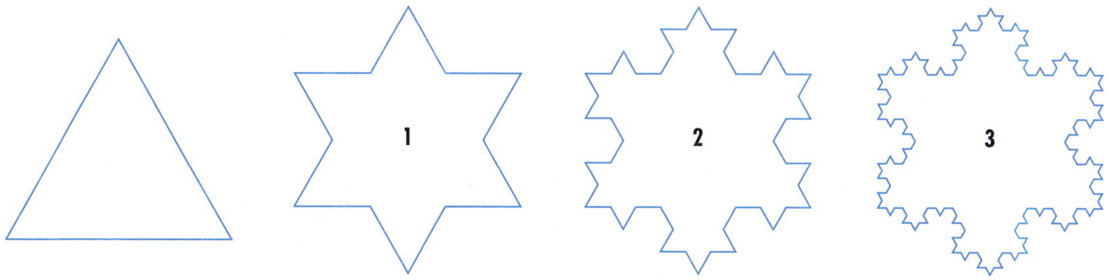

6. Find the sum of the series

$$1 + \frac{1}{2} + \frac{1}{3} + \frac{1}{4} + \frac{1}{6} + \frac{1}{8} + \frac{1}{9} + \frac{1}{12} + \cdots$$

where the terms are the reciprocals of the positive integers whose only prime factors are 2s and 3s.

7. (a) Show that for $xy \neq -1$,

$$\arctan x - \arctan y = \arctan \frac{x - y}{1 + xy}$$

if the left side lies between $-\pi/2$ and $\pi/2$.

(b) Show that

$$\arctan \tfrac{120}{119} - \arctan \tfrac{1}{239} = \frac{\pi}{4}$$

(c) Deduce the following formula of John Machin (1680–1751):

$$4 \arctan \tfrac{1}{5} - \arctan \tfrac{1}{239} = \frac{\pi}{4}$$

(d) Use the Maclaurin series for arctan to show that

$$0.197395560 < \arctan \tfrac{1}{5} < 0.197395562$$

(e) Show that

$$0.004184075 < \arctan \tfrac{1}{239} < 0.004184077$$

(f) Deduce that, correct to seven decimal places,

$$\pi \approx 3.1415927$$

Machin used this method in 1706 to find π correct to 100 decimal places. Recently, with the aid of computers, the value of π has been computed to increasingly greater accuracy. In 1999, Takahashi and Kanada, using methods of Borwein and Brent/Salamin, calculated the value of π to 206,158,430,000 decimal places!

8. (a) Prove a formula similar to the one in Problem 7(a) but involving arccot instead of arctan.
(b) Find the sum of the series

$$\sum_{n=0}^{\infty} \operatorname{arccot}(n^2 + n + 1)$$

9. Find the interval of convergence of $\sum_{n=1}^{\infty} n^3 x^n$ and find its sum.

10. If $a_0 + a_1 + a_2 + \cdots + a_k = 0$, show that

$$\lim_{n \to \infty} \left(a_0\sqrt{n} + a_1\sqrt{n+1} + a_2\sqrt{n+2} + \cdots + a_k\sqrt{n+k} \right) = 0$$

If you don't see how to prove this, try the problem-solving strategy of *using analogy* (see page 80). Try the special cases $k = 1$ and $k = 2$ first. If you can see how to prove the assertion for these cases, then you will probably see how to prove it in general.

11. Find the sum of the series $\displaystyle\sum_{n=2}^{\infty} \ln\left(1 - \frac{1}{n^2}\right)$.

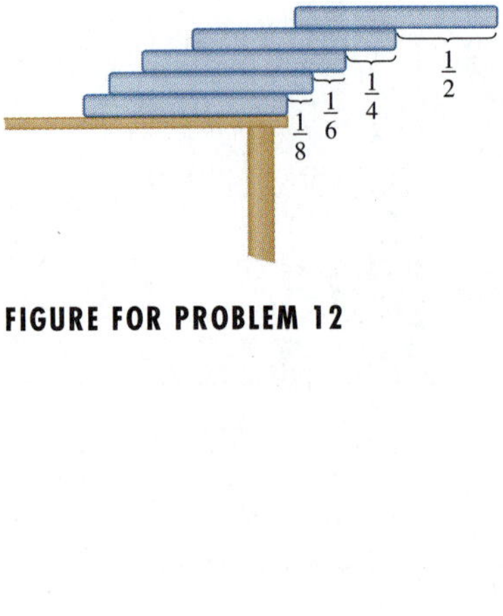

FIGURE FOR PROBLEM 12

12. Suppose you have a large supply of books, all the same size, and you stack them at the edge of a table, with each book extending farther beyond the edge of the table than the one beneath it. Show that it is possible to do this so that the top book extends entirely beyond the table. In fact, show that the top book can extend any distance at all beyond the edge of the table if the stack is high enough. Use the following method of stacking: The top book extends half its length beyond the second book. The second book extends a quarter of its length beyond the third. The third extends one-sixth of its length beyond the fourth, and so on. (Try it yourself with a deck of cards.) Consider centers of mass.

13. Let

$$u = 1 + \frac{x^3}{3!} + \frac{x^6}{6!} + \frac{x^9}{9!} + \cdots$$

$$v = x + \frac{x^4}{4!} + \frac{x^7}{7!} + \frac{x^{10}}{10!} + \cdots$$

$$w = \frac{x^2}{2!} + \frac{x^5}{5!} + \frac{x^8}{8!} + \cdots$$

Show that $u^3 + v^3 + w^3 - 3uvw = 1$.

14. If $p > 1$, evaluate the expression

$$\frac{1 + \dfrac{1}{2^p} + \dfrac{1}{3^p} + \dfrac{1}{4^p} + \cdots}{1 - \dfrac{1}{2^p} + \dfrac{1}{3^p} - \dfrac{1}{4^p} + \cdots}$$

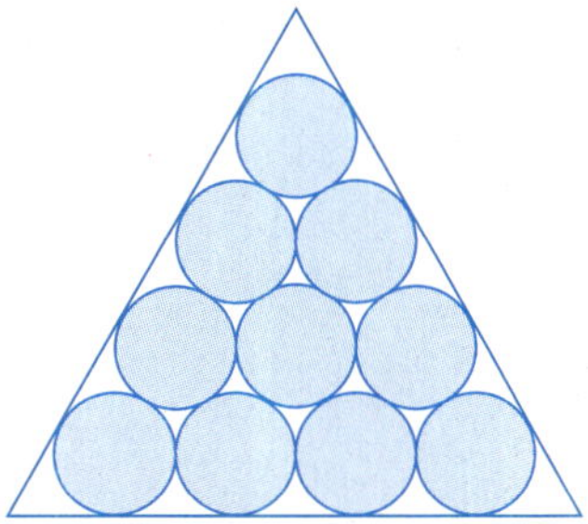

FIGURE FOR PROBLEM 15

15. Suppose that circles of equal diameter are packed tightly in n rows inside an equilateral triangle. (The figure illustrates the case $n = 4$.) If A is the area of the triangle and A_n is the total area occupied by the n rows of circles, show that

$$\lim_{n \to \infty} \frac{A_n}{A} = \frac{\pi}{2\sqrt{3}}$$

16. A sequence $\{a_n\}$ is defined recursively by the equations

$$a_0 = a_1 = 1 \qquad n(n-1)a_n = (n-1)(n-2)a_{n-1} - (n-3)a_{n-2}$$

Find the sum of the series $\sum_{n=0}^{\infty} a_n$.

17. Taking the value of x^x at 0 to be 1 and integrating a series term-by-term, show that

$$\int_0^1 x^x\,dx = \sum_{n=1}^{\infty} \frac{(-1)^{n-1}}{n^n}$$

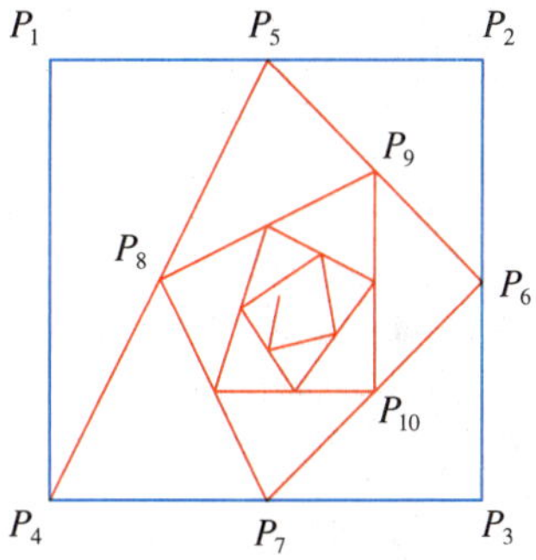

FIGURE FOR PROBLEM 18

18. Starting with the vertices $P_1(0, 1)$, $P_2(1, 1)$, $P_3(1, 0)$, $P_4(0, 0)$ of a square, we construct further points as shown in the figure: P_5 is the midpoint of P_1P_2, P_6 is the midpoint of P_2P_3, P_7 is the midpoint of P_3P_4, and so on. The polygonal spiral path $P_1P_2P_3P_4P_5P_6P_7\ldots$ approaches a point P inside the square.

(a) If the coordinates of P_n are (x_n, y_n), show that $\frac{1}{2}x_n + x_{n+1} + x_{n+2} + x_{n+3} = 2$ and find a similar equation for the y-coordinates.

(b) Find the coordinates of P.

19. If $f(x) = \sum_{m=0}^{\infty} c_m x^m$ has positive radius of convergence and $e^{f(x)} = \sum_{n=0}^{\infty} d_n x^n$, show that

$$nd_n = \sum_{i=1}^{n} ic_i d_{n-i} \qquad n \geqslant 1$$

20. Right-angled triangles are constructed as in the figure. Each triangle has height 1 and its base is the hypotenuse of the preceding triangle. Show that this sequence of triangles makes indefinitely many turns around P by showing that $\sum \theta_n$ is a divergent series.

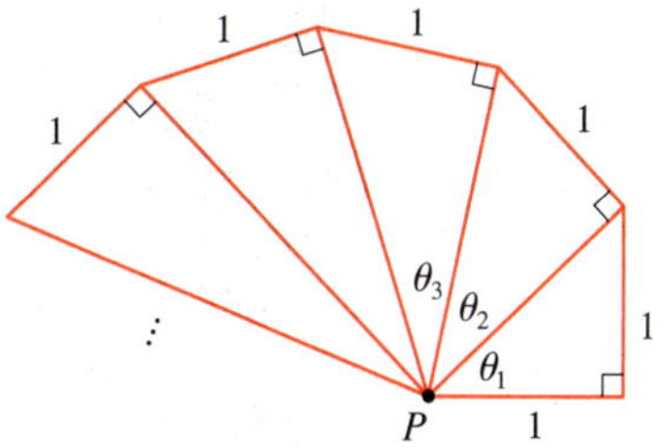

21. Consider the series whose terms are the reciprocals of the positive integers that can be written in base 10 notation without using the digit 0. Show that this series is convergent and the sum is less than 90.

22. (a) Show that the Maclaurin series of the function

$$f(x) = \frac{x}{1 - x - x^2} \qquad \text{is} \qquad \sum_{n=1}^{\infty} f_n x^n$$

where f_n is the nth Fibonacci number, that is, $f_1 = 1$, $f_2 = 1$, and $f_n = f_{n-1} + f_{n-2}$ for $n \geqslant 3$. [*Hint:* Write $x/(1 - x - x^2) = c_0 + c_1 x + c_2 x^2 + \cdots$ and multiply both sides of this equation by $1 - x - x^2$.]

(b) By writing $f(x)$ as a sum of partial fractions and thereby obtaining the Maclaurin series in a different way, find an explicit formula for the nth Fibonacci number.

CHAPTER 12

Wind velocity is a vector because it has both magnitude and direction. Pictured are velocity vectors indicating the wind pattern over San Francisco Bay at 12:00 P.M. on June 11, 2002.

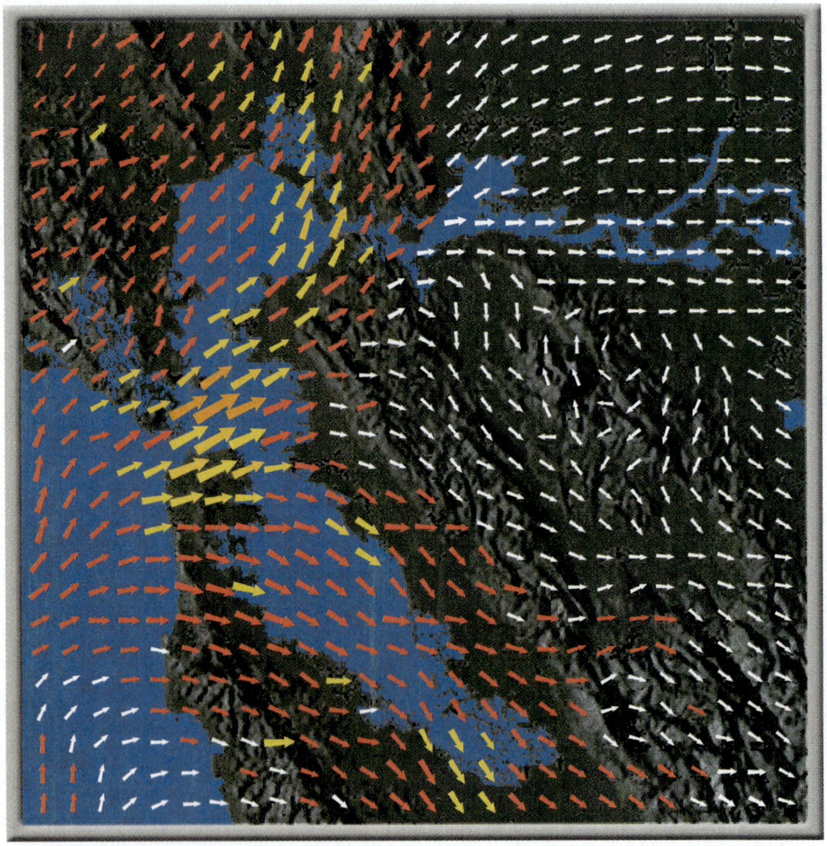

Vectors and the Geometry of Space

In this chapter we introduce vectors and coordinate systems for three-dimensional space. This will be the setting for our study of the calculus of functions of two variables in Chapter 14 because the graph of such a function is a surface in space. In this chapter we will see that vectors provide particularly simple descriptions of lines and planes in space.

12.1 Three-Dimensional Coordinate Systems

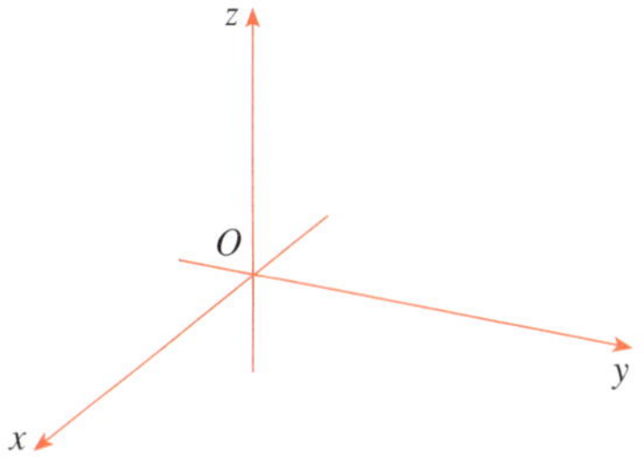

FIGURE 1
Coordinate axes

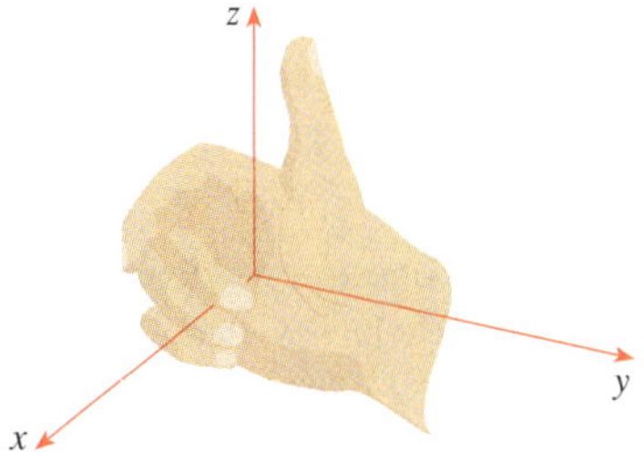

FIGURE 2
Right-hand rule

To locate a point in a plane, two numbers are necessary. We know that any point in the plane can be represented as an ordered pair (a, b) of real numbers, where a is the x-coordinate and b is the y-coordinate. For this reason, a plane is called two-dimensional. To locate a point in space, three numbers are required. We represent any point in space by an ordered triple (a, b, c) of real numbers.

In order to represent points in space, we first choose a fixed point O (the origin) and three directed lines through O that are perpendicular to each other, called the **coordinate axes** and labeled the x-axis, y-axis, and z-axis. Usually we think of the x- and y-axes as being horizontal and the z-axis as being vertical, and we draw the orientation of the axes as in Figure 1. The direction of the z-axis is determined by the **right-hand rule** as illustrated in Figure 2: If you curl the fingers of your right hand around the z-axis in the direction of a $90°$ counterclockwise rotation from the positive x-axis to the positive y-axis, then your thumb points in the positive direction of the z-axis.

The three coordinate axes determine the three **coordinate planes** illustrated in Figure 3(a). The xy-plane is the plane that contains the x- and y-axes; the yz-plane contains the y- and z-axes; the xz-plane contains the x- and z-axes. These three coordinate planes divide space into eight parts, called **octants**. The **first octant**, in the foreground, is determined by the positive axes.

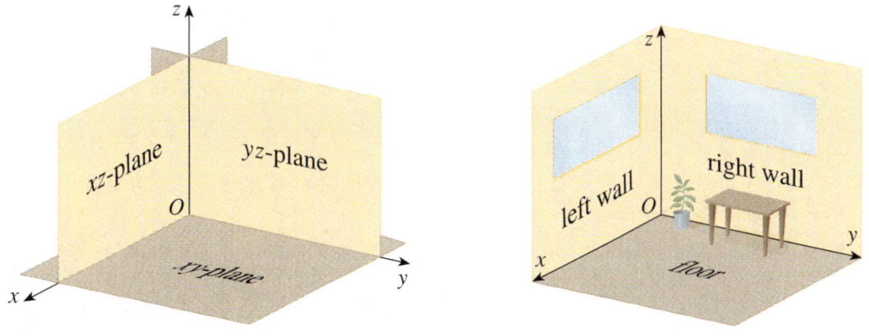

FIGURE 3 (a) Coordinate planes (b)

Because many people have some difficulty visualizing diagrams of three-dimensional figures, you may find it helpful to do the following [see Figure 3(b)]. Look at any bottom corner of a room and call the corner the origin. The wall on your left is in the xz-plane, the wall on your right is in the yz-plane, and the floor is in the xy-plane. The x-axis runs along the intersection of the floor and the left wall. The y-axis runs along the intersection of the floor and the right wall. The z-axis runs up from the floor toward the ceiling along the intersection of the two walls. You are situated in the first octant, and you can now imagine seven other rooms situated in the other seven octants (three on the same floor and four on the floor below), all connected by the common corner point O.

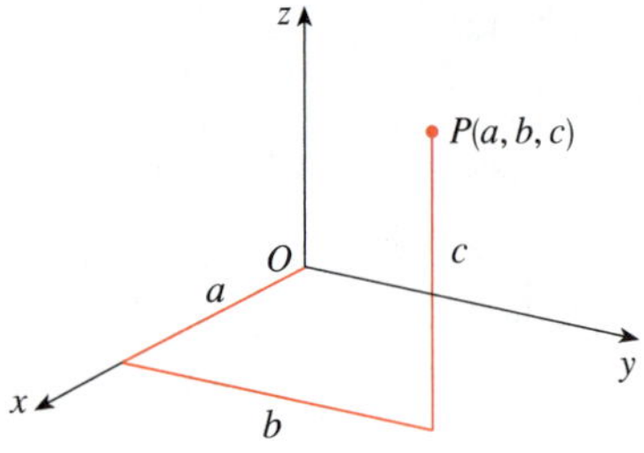

FIGURE 4

Now if P is any point in space, let a be the (directed) distance from the yz-plane to P, let b be the distance from the xz-plane to P, and let c be the distance from the xy-plane to P. We represent the point P by the ordered triple (a, b, c) of real numbers and we call a, b, and c the **coordinates** of P; a is the x-coordinate, b is the y-coordinate, and c is the z-coordinate. Thus, to locate the point (a, b, c) we can start at the origin O and move a units along the x-axis, then b units parallel to the y-axis, and then c units parallel to the z-axis as in Figure 4.

The point $P(a, b, c)$ determines a rectangular box as in Figure 5. If we drop a perpendicular from P to the xy-plane, we get a point Q with coordinates $(a, b, 0)$ called the **projection** of P on the xy-plane. Similarly, $R(0, b, c)$ and $S(a, 0, c)$ are the projections of P on the yz-plane and xz-plane, respectively.

As numerical illustrations, the points $(-4, 3, -5)$ and $(3, -2, -6)$ are plotted in Figure 6.

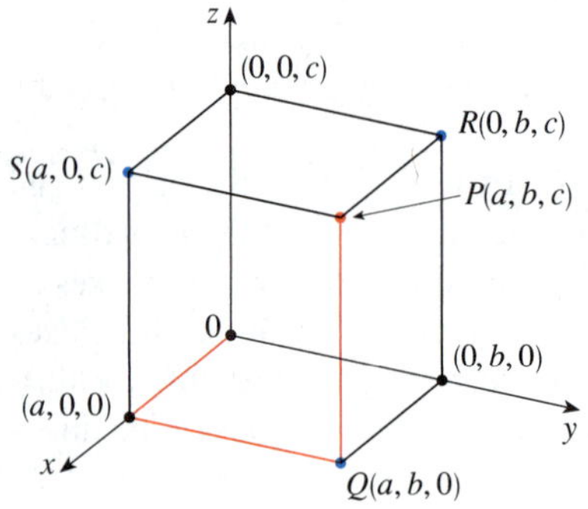

FIGURE 5

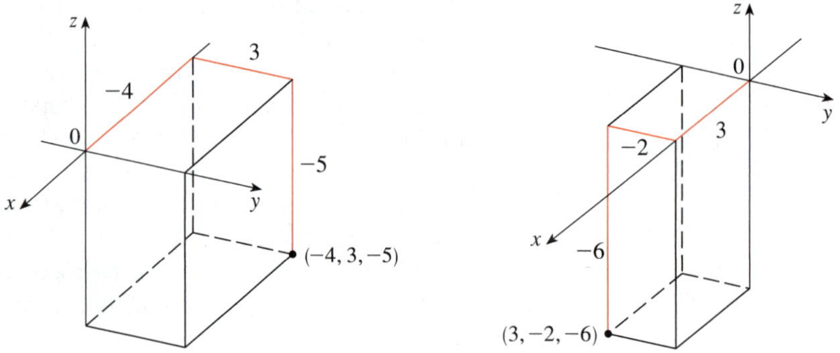

FIGURE 6

The Cartesian product $\mathbb{R} \times \mathbb{R} \times \mathbb{R} = \{(x, y, z) \mid x, y, z \in \mathbb{R}\}$ is the set of all ordered triples of real numbers and is denoted by $\mathbb{R}^3$. We have given a one-to-one correspondence between points P in space and ordered triples (a, b, c) in $\mathbb{R}^3$. It is called a **three-dimensional rectangular coordinate system**. Notice that, in terms of coordinates, the first octant can be described as the set of points whose coordinates are all positive.

In two-dimensional analytic geometry, the graph of an equation involving x and y is a curve in $\mathbb{R}^2$. In three-dimensional analytic geometry, an equation in x, y, and z represents a *surface* in $\mathbb{R}^3$.

EXAMPLE 1 What surfaces in $\mathbb{R}^3$ are represented by the following equations?

(a) $z = 3$ (b) $y = 5$

SOLUTION

(a) The equation $z = 3$ represents the set $\{(x, y, z) \mid z = 3\}$, which is the set of all points in $\mathbb{R}^3$ whose z-coordinate is 3. This is the horizontal plane that is parallel to the xy-plane and three units above it as in Figure 7(a).

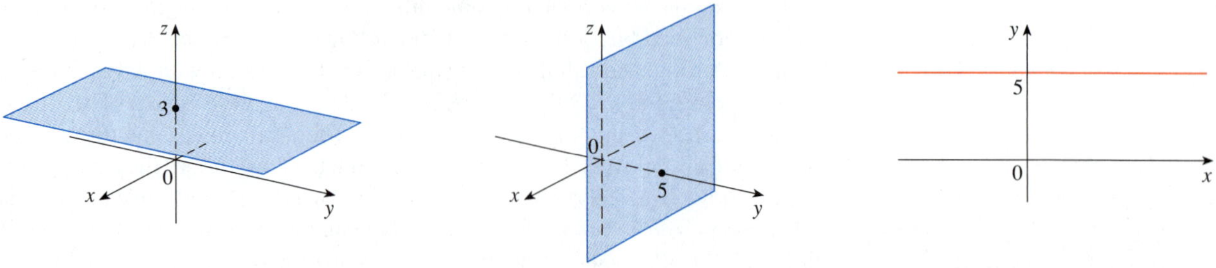

FIGURE 7 (a) $z = 3$, a plane in $\mathbb{R}^3$ (b) $y = 5$, a plane in $\mathbb{R}^3$ (c) $y = 5$, a line in $\mathbb{R}^2$

(b) The equation $y = 5$ represents the set of all points in $\mathbb{R}^3$ whose y-coordinate is 5. This is the vertical plane that is parallel to the xz-plane and five units to the right of it as in Figure 7(b).

NOTE ▫ When an equation is given, we must understand from the context whether it represents a curve in $\mathbb{R}^2$ or a surface in $\mathbb{R}^3$. In Example 1, $y = 5$ represents a plane in $\mathbb{R}^3$, but of course $y = 5$ can also represent a line in $\mathbb{R}^2$ if we are dealing with two-dimensional analytic geometry. See Figure 7(b) and (c).

In general, if k is a constant, then $x = k$ represents a plane parallel to the yz-plane, $y = k$ is a plane parallel to the xz-plane, and $z = k$ is a plane parallel to the xy-plane. In Figure 5, the faces of the rectangular box are formed by the three coordinate planes $x = 0$ (the yz-plane), $y = 0$ (the xz-plane), and $z = 0$ (the xy-plane), and the planes $x = a$, $y = b$, and $z = c$.

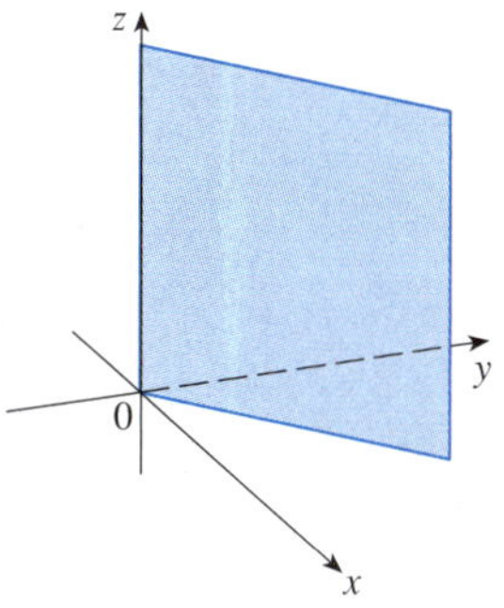

FIGURE 8
The plane $y = x$

EXAMPLE 2 Describe and sketch the surface in $\mathbb{R}^3$ represented by the equation $y = x$.

SOLUTION The equation represents the set of all points in $\mathbb{R}^3$ whose x- and y-coordinates are equal, that is, $\{(x, x, z) \mid x \in \mathbb{R}, z \in \mathbb{R}\}$. This is a vertical plane that intersects the xy-plane in the line $y = x$, $z = 0$. The portion of this plane that lies in the first octant is sketched in Figure 8.

The familiar formula for the distance between two points in a plane is easily extended to the following three-dimensional formula.

Distance Formula in Three Dimensions The distance $|P_1P_2|$ between the points $P_1(x_1, y_1, z_1)$ and $P_2(x_2, y_2, z_2)$ is

$$|P_1P_2| = \sqrt{(x_2 - x_1)^2 + (y_2 - y_1)^2 + (z_2 - z_1)^2}$$

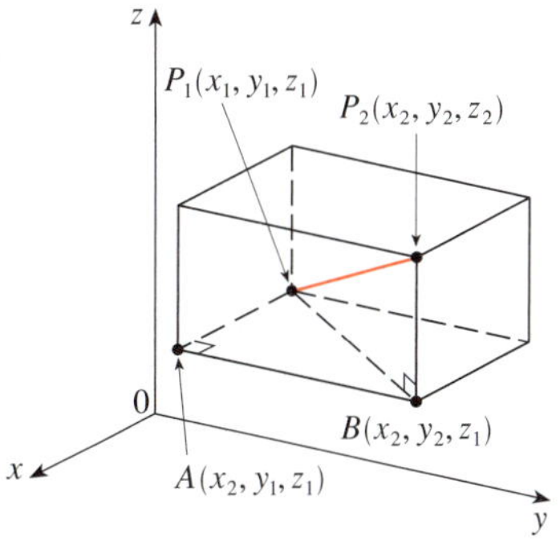

FIGURE 9

To see why this formula is true, we construct a rectangular box as in Figure 9, where P_1 and P_2 are opposite vertices and the faces of the box are parallel to the coordinate planes. If $A(x_2, y_1, z_1)$ and $B(x_2, y_2, z_1)$ are the vertices of the box indicated in the figure, then

$$|P_1A| = |x_2 - x_1| \qquad |AB| = |y_2 - y_1| \qquad |BP_2| = |z_2 - z_1|$$

Because triangles P_1BP_2 and P_1AB are both right-angled, two applications of the Pythagorean Theorem give

$$|P_1P_2|^2 = |P_1B|^2 + |BP_2|^2$$

and

$$|P_1B|^2 = |P_1A|^2 + |AB|^2$$

Combining these equations, we get

$$\begin{aligned} |P_1P_2|^2 &= |P_1A|^2 + |AB|^2 + |BP_2|^2 \\ &= |x_2 - x_1|^2 + |y_2 - y_1|^2 + |z_2 - z_1|^2 \\ &= (x_2 - x_1)^2 + (y_2 - y_1)^2 + (z_2 - z_1)^2 \end{aligned}$$

Therefore

$$|P_1P_2| = \sqrt{(x_2 - x_1)^2 + (y_2 - y_1)^2 + (z_2 - z_1)^2}$$

EXAMPLE 3 The distance from the point $P(2, -1, 7)$ to the point $Q(1, -3, 5)$ is

$$|PQ| = \sqrt{(1-2)^2 + (-3+1)^2 + (5-7)^2} = \sqrt{1+4+4} = 3$$

EXAMPLE 4 Find an equation of a sphere with radius r and center $C(h, k, l)$.

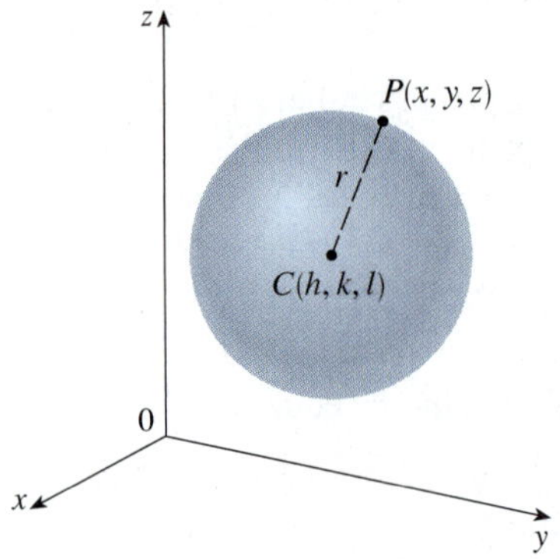

FIGURE 10

SOLUTION By definition, a sphere is the set of all points $P(x, y, z)$ whose distance from C is r. (See Figure 10.) Thus, P is on the sphere if and only if $|PC| = r$. Squaring both sides, we have $|PC|^2 = r^2$ or

$$(x-h)^2 + (y-k)^2 + (z-l)^2 = r^2$$

The result of Example 4 is worth remembering.

Equation of a Sphere An equation of a sphere with center $C(h, k, l)$ and radius r is

$$(x-h)^2 + (y-k)^2 + (z-l)^2 = r^2$$

In particular, if the center is the origin O, then an equation of the sphere is

$$x^2 + y^2 + z^2 = r^2$$

EXAMPLE 5 Show that $x^2 + y^2 + z^2 + 4x - 6y + 2z + 6 = 0$ is the equation of a sphere, and find its center and radius.

SOLUTION We can rewrite the given equation in the form of an equation of a sphere if we complete squares:

$$(x^2 + 4x + 4) + (y^2 - 6y + 9) + (z^2 + 2z + 1) = -6 + 4 + 9 + 1$$
$$(x+2)^2 + (y-3)^2 + (z+1)^2 = 8$$

Comparing this equation with the standard form, we see that it is the equation of a sphere with center $(-2, 3, -1)$ and radius $\sqrt{8} = 2\sqrt{2}$.

EXAMPLE 6 What region in $\mathbb{R}^3$ is represented by the following inequalities?

$$1 \leqslant x^2 + y^2 + z^2 \leqslant 4 \qquad z \leqslant 0$$

SOLUTION The inequalities

$$1 \leqslant x^2 + y^2 + z^2 \leqslant 4$$

can be rewritten as

$$1 \leqslant \sqrt{x^2 + y^2 + z^2} \leqslant 2$$

so they represent the points (x, y, z) whose distance from the origin is at least 1 and at most 2. But we are also given that $z \leqslant 0$, so the points lie on or below the xy-plane. Thus, the given inequalities represent the region that lies between (or on) the spheres $x^2 + y^2 + z^2 = 1$ and $x^2 + y^2 + z^2 = 4$ and beneath (or on) the xy-plane. It is sketched in Figure 11.

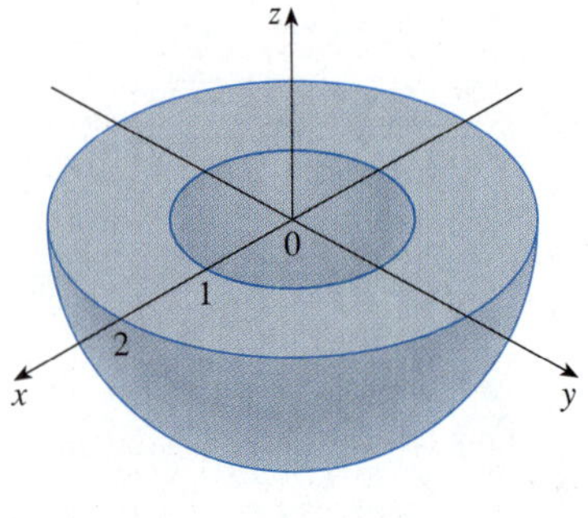

FIGURE 11

12.1 Exercises

1. Suppose you start at the origin, move along the x-axis a distance of 4 units in the positive direction, and then move downward a distance of 3 units. What are the coordinates of your position?

2. Sketch the points $(0, 5, 2)$, $(4, 0, -1)$, $(2, 4, 6)$, and $(1, -1, 2)$ on a single set of coordinate axes.

3. Which of the points $P(6, 2, 3)$, $Q(-5, -1, 4)$, and $R(0, 3, 8)$ is closest to the xz-plane? Which point lies in the yz-plane?

4. What are the projections of the point $(2, 3, 5)$ on the xy-, yz-, and xz-planes? Draw a rectangular box with the origin and $(2, 3, 5)$ as opposite vertices and with its faces parallel to the coordinate planes. Label all vertices of the box. Find the length of the diagonal of the box.

5. Describe and sketch the surface in $\mathbb{R}^3$ represented by the equation $x + y = 2$.

6. (a) What does the equation $x = 4$ represent in $\mathbb{R}^2$? What does it represent in $\mathbb{R}^3$? Illustrate with sketches.
(b) What does the equation $y = 3$ represent in $\mathbb{R}^3$? What does $z = 5$ represent? What does the pair of equations $y = 3$, $z = 5$ represent? In other words, describe the set of points (x, y, z) such that $y = 3$ and $z = 5$. Illustrate with a sketch.

7. Show that the triangle with vertices $P(-2, 4, 0)$, $Q(1, 2, -1)$, and $R(-1, 1, 2)$ is an equilateral triangle.

8. Find the lengths of the sides of the triangle with vertices $A(1, 2, -3)$, $B(3, 4, -2)$, and $C(3, -2, 1)$. Is ABC a right triangle? Is it an isosceles triangle?

9. Determine whether the points lie on a straight line.
(a) $A(5, 1, 3)$, $B(7, 9, -1)$, $C(1, -15, 11)$
(b) $K(0, 3, -4)$, $L(1, 2, -2)$, $M(3, 0, 1)$

10. Find the distance from $(3, 7, -5)$ to each of the following.
(a) The xy-plane (b) The yz-plane
(c) The xz-plane (d) The x-axis
(e) The y-axis (f) The z-axis

11. Find an equation of the sphere with center $(1, -4, 3)$ and radius 5. What is the intersection of this sphere with the xz-plane?

12. Find an equation of the sphere with center $(6, 5, -2)$ and radius $\sqrt{7}$. Describe its intersection with each of the coordinate planes.

13. Find an equation of the sphere that passes through the point $(4, 3, -1)$ and has center $(3, 8, 1)$.

14. Find an equation of the sphere that passes through the origin and whose center is $(1, 2, 3)$.

15–18 Show that the equation represents a sphere, and find its center and radius.

15. $x^2 + y^2 + z^2 - 6x + 4y - 2z = 11$

16. $x^2 + y^2 + z^2 = 4x - 2y$

17. $x^2 + y^2 + z^2 = x + y + z$

18. $4x^2 + 4y^2 + 4z^2 - 8x + 16y = 1$

19. (a) Prove that the midpoint of the line segment from $P_1(x_1, y_1, z_1)$ to $P_2(x_2, y_2, z_2)$ is

$$\left(\frac{x_1 + x_2}{2}, \frac{y_1 + y_2}{2}, \frac{z_1 + z_2}{2}\right)$$

(b) Find the lengths of the medians of the triangle with vertices $A(1, 2, 3)$, $B(-2, 0, 5)$, and $C(4, 1, 5)$.

20. Find an equation of a sphere if one of its diameters has endpoints $(2, 1, 4)$ and $(4, 3, 10)$.

21. Find equations of the spheres with center $(2, -3, 6)$ that touch (a) the xy-plane, (b) the yz-plane, (c) the xz-plane.

22. Find an equation of the largest sphere with center $(5, 4, 9)$ that is contained in the first octant.

23–34 Describe in words the region of $\mathbb{R}^3$ represented by the equation or inequality.

23. $y = -4$

24. $x = 10$

25. $x > 3$

26. $y \geqslant 0$

27. $0 \leqslant z \leqslant 6$

28. $y = z$

29. $x^2 + y^2 + z^2 > 1$

30. $1 \leqslant x^2 + y^2 + z^2 \leqslant 25$

31. $x^2 + y^2 + z^2 - 2z < 3$

32. $x^2 + y^2 = 1$

33. $x^2 + z^2 \leqslant 9$

34. $xyz = 0$

35–38 Write inequalities to describe the region.

35. The half-space consisting of all points to the left of the xz-plane

36. The solid rectangular box in the first octant bounded by the planes $x = 1$, $y = 2$, and $z = 3$

37. The region consisting of all points between (but not on) the spheres of radius r and R centered at the origin, where $r < R$

38. The solid upper hemisphere of the sphere of radius 2 centered at the origin

39. The figure shows a line L_1 in space and a second line L_2, which is the projection of L_1 on the xy-plane. (In other words, the points on L_2 are directly beneath, or above, the points on L_1.)

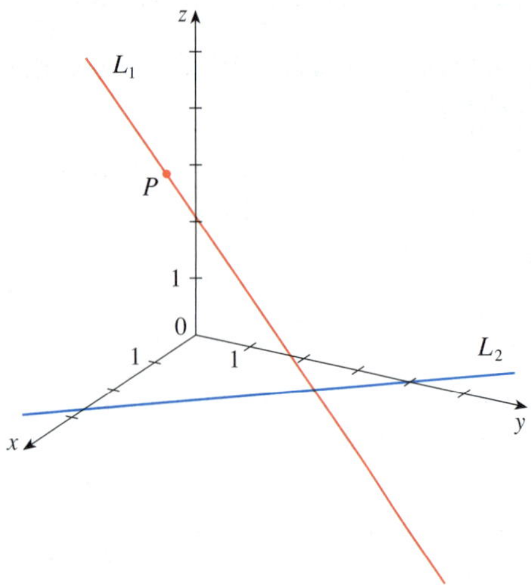

(a) Find the coordinates of the point P on the line L_1.

(b) Locate on the diagram the points A, B, and C, where the line L_1 intersects the xy-plane, the yz-plane, and the xz-plane, respectively.

40. Consider the points P such that the distance from P to $A(-1, 5, 3)$ is twice the distance from P to $B(6, 2, -2)$. Show that the set of all such points is a sphere, and find its center and radius.

41. Find an equation of the set of all points equidistant from the points $A(-1, 5, 3)$ and $B(6, 2, -2)$. Describe the set.

42. Find the volume of the solid that lies inside both of the spheres

$$x^2 + y^2 + z^2 + 4x - 2y + 4z + 5 = 0$$

and

$$x^2 + y^2 + z^2 = 4$$

12.2 Vectors

The term **vector** is used by scientists to indicate a quantity (such as displacement or velocity or force) that has both magnitude and direction. A vector is often represented by an arrow or a directed line segment. The length of the arrow represents the magnitude of the vector and the arrow points in the direction of the vector. We denote a vector by printing a letter in boldface (**v**) or by putting an arrow above the letter ($\vec{v}$).

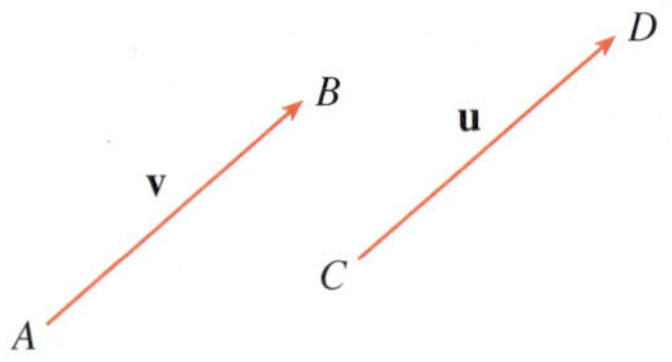

FIGURE 1
Equivalent vectors

For instance, suppose a particle moves along a line segment from point A to point B. The corresponding **displacement vector v**, shown in Figure 1, has **initial point** A (the tail) and **terminal point** B (the tip) and we indicate this by writing $\mathbf{v} = \overrightarrow{AB}$. Notice that the vector $\mathbf{u} = \overrightarrow{CD}$ has the same length and the same direction as **v** even though it is in a different position. We say that **u** and **v** are **equivalent** (or **equal**) and we write $\mathbf{u} = \mathbf{v}$. The **zero vector**, denoted by **0**, has length 0. It is the only vector with no specific direction.

Combining Vectors

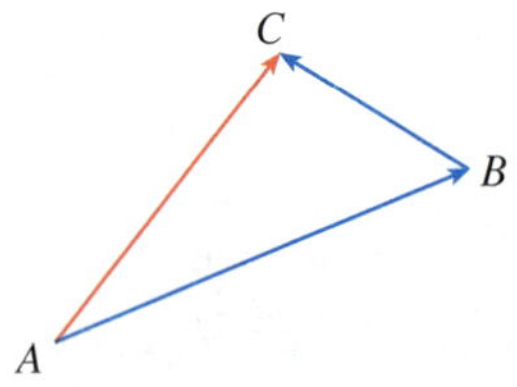

FIGURE 2

Suppose a particle moves from A to B, so its displacement vector is $\overrightarrow{AB}$. Then the particle changes direction and moves from B to C, with displacement vector $\overrightarrow{BC}$ as in Figure 2. The combined effect of these displacements is that the particle has moved from A to C. The resulting displacement vector $\overrightarrow{AC}$ is called the *sum* of $\overrightarrow{AB}$ and $\overrightarrow{BC}$ and we write

$$\overrightarrow{AC} = \overrightarrow{AB} + \overrightarrow{BC}$$

In general, if we start with vectors **u** and **v**, we first move **v** so that its tail coincides with the tip of **u** and define the sum of **u** and **v** as follows.

Definition of Vector Addition If **u** and **v** are vectors positioned so the initial point of **v** is at the terminal point of **u**, then the **sum u + v** is the vector from the initial point of **u** to the terminal point of **v**.

The definition of vector addition is illustrated in Figure 3. You can see why this definition is sometimes called the **Triangle Law**.

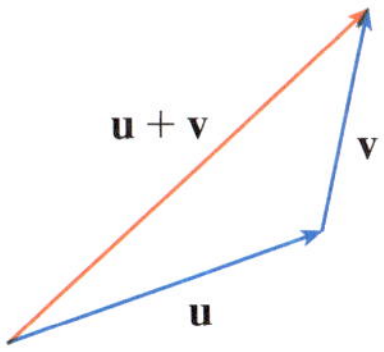

FIGURE 3 The Triangle Law

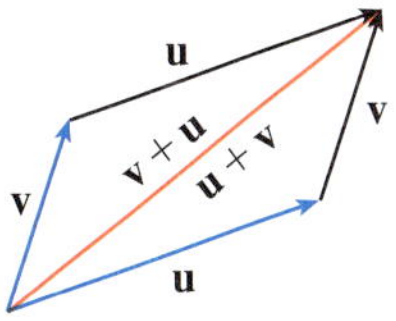

FIGURE 4 The Parallelogram Law

In Figure 4 we start with the same vectors **u** and **v** as in Figure 3 and draw another copy of **v** with the same initial point as **u**. Completing the parallelogram, we see that **u** + **v** = **v** + **u**. This also gives another way to construct the sum: If we place **u** and **v** so they start at the same point, then **u** + **v** lies along the diagonal of the parallelogram with **u** and **v** as sides. (This is called the **Parallelogram Law**.)

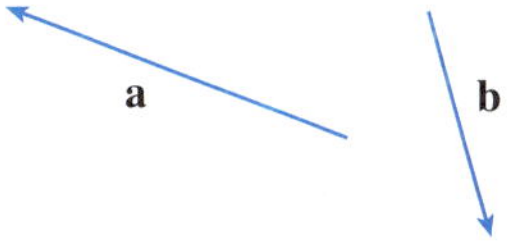

FIGURE 5

EXAMPLE 1 Draw the sum of the vectors **a** and **b** shown in Figure 5.

SOLUTION First we translate **b** and place its tail at the tip of **a**, being careful to draw a copy of **b** that has the same length and direction. Then we draw the vector **a** + **b** [see Figure 6(a)] starting at the initial point of **a** and ending at the terminal point of the copy of **b**.

Alternatively, we could place **b** so it starts where **a** starts and construct **a** + **b** by the Parallelogram Law as in Figure 6(b).

TEC Visual 12.2 shows how the Triangle and Parallelogram Laws work for various vectors **u** and **v**.

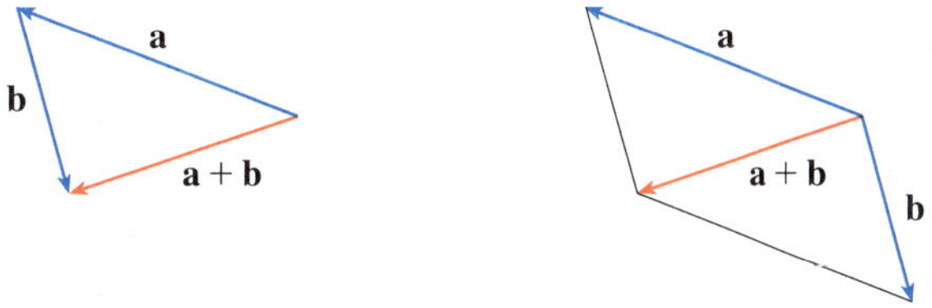

FIGURE 6 (a) (b)

It is possible to multiply a vector by a real number c. (In this context we call the real number c a **scalar** to distinguish it from a vector.) For instance, we want 2**v** to be the same vector as **v** + **v**, which has the same direction as **v** but is twice as long. In general, we multiply a vector by a scalar as follows.

Definition of Scalar Multiplication If c is a scalar and **v** is a vector, then the **scalar multiple** $c\mathbf{v}$ is the vector whose length is $|c|$ times the length of **v** and whose direction is the same as **v** if $c > 0$ and is opposite to **v** if $c < 0$. If $c = 0$ or $\mathbf{v} = \mathbf{0}$, then $c\mathbf{v} = \mathbf{0}$.

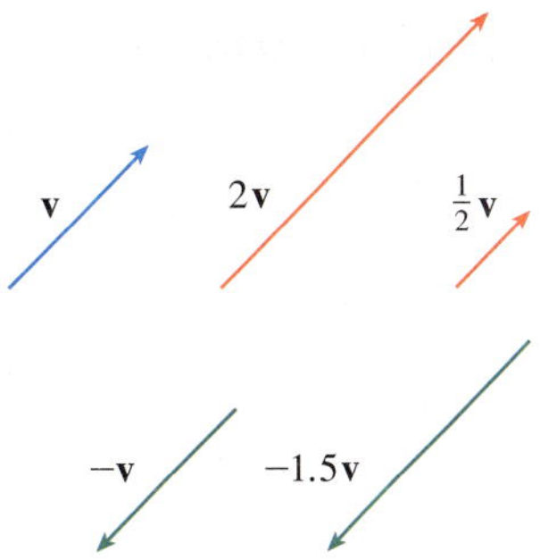

FIGURE 7
Scalar multiples of **v**

This definition is illustrated in Figure 7. We see that real numbers work like scaling factors here; that's why we call them scalars. Notice that two nonzero vectors are **parallel** if they are scalar multiples of one another. In particular, the vector $-\mathbf{v} = (-1)\mathbf{v}$ has the same length as **v** but points in the opposite direction. We call it the **negative** of **v**.

By the **difference u − v** of two vectors we mean

$$\mathbf{u} - \mathbf{v} = \mathbf{u} + (-\mathbf{v})$$

So we can construct $\mathbf{u} - \mathbf{v}$ by first drawing the negative of $\mathbf{v}$, $-\mathbf{v}$, and then adding it to $\mathbf{u}$ by the Parallelogram Law as in Figure 8(a). Alternatively, since $\mathbf{v} + (\mathbf{u} - \mathbf{v}) = \mathbf{u}$, the vector $\mathbf{u} - \mathbf{v}$, when added to $\mathbf{v}$, gives $\mathbf{u}$. So we could construct $\mathbf{u} - \mathbf{v}$ as in Figure 8(b) by means of the Triangle Law.

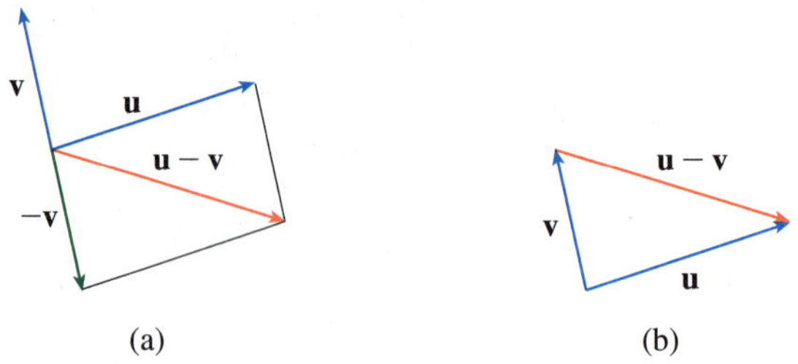

FIGURE 8
Drawing $\mathbf{u} - \mathbf{v}$

EXAMPLE 2 If $\mathbf{a}$ and $\mathbf{b}$ are the vectors shown in Figure 9, draw $\mathbf{a} - 2\mathbf{b}$.

SOLUTION We first draw the vector $-2\mathbf{b}$ pointing in the direction opposite to $\mathbf{b}$ and twice as long. We place it with its tail at the tip of $\mathbf{a}$ and then use the Triangle Law to draw $\mathbf{a} + (-2\mathbf{b})$ as in Figure 10.

FIGURE 9

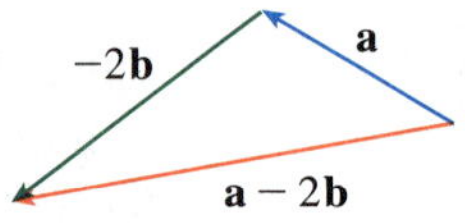

FIGURE 10

Components

For some purposes it's best to introduce a coordinate system and treat vectors algebraically. If we place the initial point of a vector $\mathbf{a}$ at the origin of a rectangular coordinate system, then the terminal point of $\mathbf{a}$ has coordinates of the form (a_1, a_2) or (a_1, a_2, a_3), depending on whether our coordinate system is two- or three-dimensional (see Figure 11). These coordinates are called the **components** of $\mathbf{a}$ and we write

$$\mathbf{a} = \langle a_1, a_2 \rangle \qquad \text{or} \qquad \mathbf{a} = \langle a_1, a_2, a_3 \rangle$$

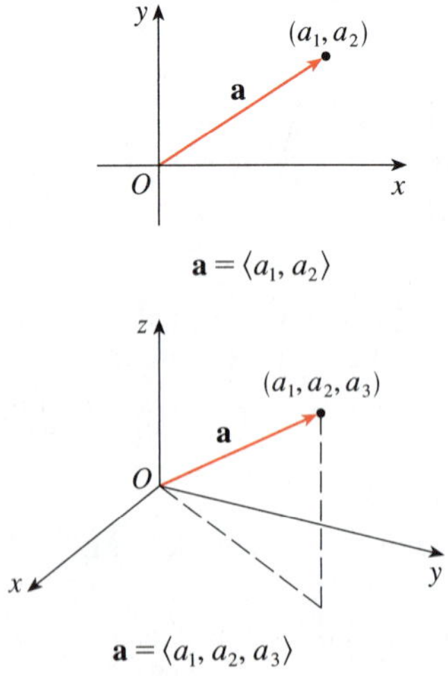

FIGURE 11

We use the notation $\langle a_1, a_2 \rangle$ for the ordered pair that refers to a vector so as not to confuse it with the ordered pair (a_1, a_2) that refers to a point in the plane.

For instance, the vectors shown in Figure 12 are all equivalent to the vector $\overrightarrow{OP} = \langle 3, 2 \rangle$ whose terminal point is $P(3, 2)$. What they have in common is that the terminal point is reached from the initial point by a displacement of three units to the right and two upward. We can think of all these geometric vectors as **representations** of the

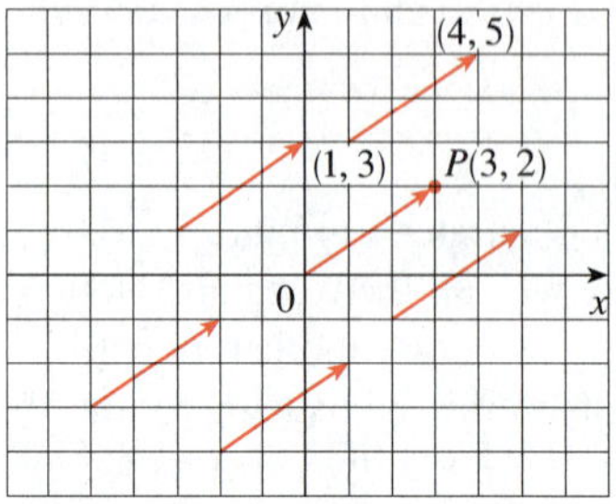

FIGURE 12
Representations of the vector $\mathbf{a} = \langle 3, 2 \rangle$

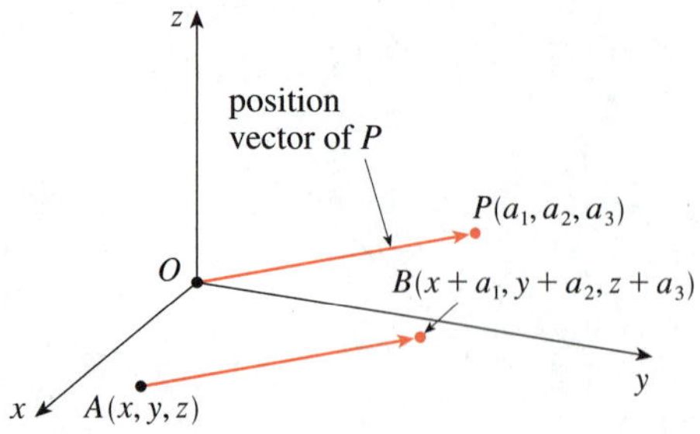

FIGURE 13
Representations of $\mathbf{a} = \langle a_1, a_2, a_3 \rangle$

algebraic vector $\mathbf{a} = \langle 3, 2 \rangle$. The particular representation $\overrightarrow{OP}$ from the origin to the point $P(3, 2)$ is called the **position vector** of the point P.

In three dimensions, the vector $\mathbf{a} = \overrightarrow{OP} = \langle a_1, a_2, a_3 \rangle$ is the **position vector** of the point $P(a_1, a_2, a_3)$. (See Figure 13.) Let's consider any other representation $\overrightarrow{AB}$ of $\mathbf{a}$, where the initial point is $A(x_1, y_1, z_1)$ and the terminal point is $B(x_2, y_2, z_2)$. Then we must have $x_1 + a_1 = x_2$, $y_1 + a_2 = y_2$, and $z_1 + a_3 = z_2$ and so $a_1 = x_2 - x_1$, $a_2 = y_2 - y_1$, and $a_3 = z_2 - z_1$. Thus, we have the following result.

1 Given the points $A(x_1, y_1, z_1)$ and $B(x_2, y_2, z_2)$, the vector $\mathbf{a}$ with representation $\overrightarrow{AB}$ is

$$\mathbf{a} = \langle x_2 - x_1, y_2 - y_1, z_2 - z_1 \rangle$$

EXAMPLE 3 Find the vector represented by the directed line segment with initial point $A(2, -3, 4)$ and terminal point $B(-2, 1, 1)$.

SOLUTION By (1), the vector corresponding to $\overrightarrow{AB}$ is

$$\mathbf{a} = \langle -2 - 2, 1 - (-3), 1 - 4 \rangle = \langle -4, 4, -3 \rangle$$

The **magnitude** or **length** of the vector $\mathbf{v}$ is the length of any of its representations and is denoted by the symbol $|\mathbf{v}|$ or $\|\mathbf{v}\|$. By using the distance formula to compute the length of a segment OP, we obtain the following formulas.

The length of the two-dimensional vector $\mathbf{a} = \langle a_1, a_2 \rangle$ is

$$|\mathbf{a}| = \sqrt{a_1^2 + a_2^2}$$

The length of the three-dimensional vector $\mathbf{a} = \langle a_1, a_2, a_3 \rangle$ is

$$|\mathbf{a}| = \sqrt{a_1^2 + a_2^2 + a_3^2}$$

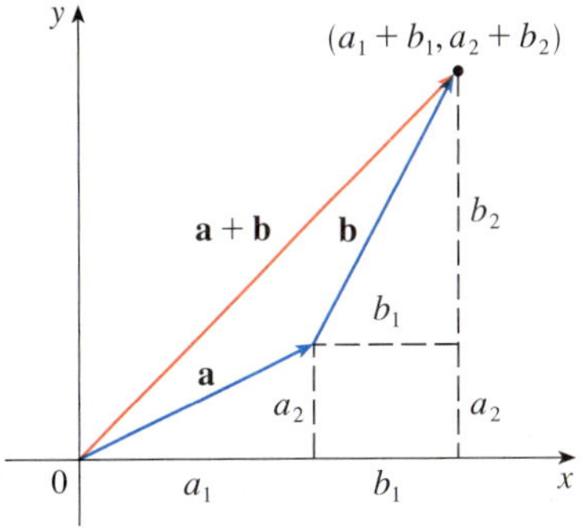

FIGURE 14

How do we add vectors algebraically? Figure 14 shows that if $\mathbf{a} = \langle a_1, a_2 \rangle$ and $\mathbf{b} = \langle b_1, b_2 \rangle$, then the sum is $\mathbf{a} + \mathbf{b} = \langle a_1 + b_1, a_2 + b_2 \rangle$, at least for the case where the components are positive. In other words, *to add algebraic vectors we add their components*. Similarly, *to subtract vectors we subtract components*. From the similar triangles in Figure 15 we see that the components of $c\mathbf{a}$ are ca_1 and ca_2. So *to multiply a vector by a scalar we multiply each component by that scalar*.

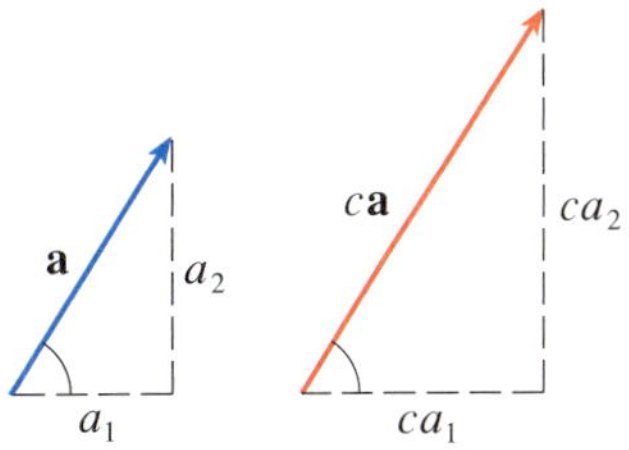

FIGURE 15

If $\mathbf{a} = \langle a_1, a_2 \rangle$ and $\mathbf{b} = \langle b_1, b_2 \rangle$, then

$$\mathbf{a} + \mathbf{b} = \langle a_1 + b_1, a_2 + b_2 \rangle \qquad \mathbf{a} - \mathbf{b} = \langle a_1 - b_1, a_2 - b_2 \rangle$$

$$c\mathbf{a} = \langle ca_1, ca_2 \rangle$$

Similarly, for three-dimensional vectors,

$$\langle a_1, a_2, a_3 \rangle + \langle b_1, b_2, b_3 \rangle = \langle a_1 + b_1, a_2 + b_2, a_3 + b_3 \rangle$$

$$\langle a_1, a_2, a_3 \rangle - \langle b_1, b_2, b_3 \rangle = \langle a_1 - b_1, a_2 - b_2, a_3 - b_3 \rangle$$

$$c\langle a_1, a_2, a_3 \rangle = \langle ca_1, ca_2, ca_3 \rangle$$

EXAMPLE 4 If $\mathbf{a} = \langle 4, 0, 3 \rangle$ and $\mathbf{b} = \langle -2, 1, 5 \rangle$, find $|\mathbf{a}|$ and the vectors $\mathbf{a} + \mathbf{b}$, $\mathbf{a} - \mathbf{b}$, $3\mathbf{b}$, and $2\mathbf{a} + 5\mathbf{b}$.

SOLUTION

$$|\mathbf{a}| = \sqrt{4^2 + 0^2 + 3^2} = \sqrt{25} = 5$$

$$\begin{aligned} \mathbf{a} + \mathbf{b} &= \langle 4, 0, 3 \rangle + \langle -2, 1, 5 \rangle \\ &= \langle 4 - 2, 0 + 1, 3 + 5 \rangle = \langle 2, 1, 8 \rangle \end{aligned}$$

$$\begin{aligned} \mathbf{a} - \mathbf{b} &= \langle 4, 0, 3 \rangle - \langle -2, 1, 5 \rangle \\ &= \langle 4 - (-2), 0 - 1, 3 - 5 \rangle = \langle 6, -1, -2 \rangle \end{aligned}$$

$$3\mathbf{b} = 3\langle -2, 1, 5 \rangle = \langle 3(-2), 3(1), 3(5) \rangle = \langle -6, 3, 15 \rangle$$

$$\begin{aligned} 2\mathbf{a} + 5\mathbf{b} &= 2\langle 4, 0, 3 \rangle + 5\langle -2, 1, 5 \rangle \\ &= \langle 8, 0, 6 \rangle + \langle -10, 5, 25 \rangle = \langle -2, 5, 31 \rangle \end{aligned}$$

We denote by V_2 the set of all two-dimensional vectors and by V_3 the set of all three-dimensional vectors. More generally, we will later need to consider the set V_n of all n-dimensional vectors. An n-dimensional vector is an ordered n-tuple:

$$\mathbf{a} = \langle a_1, a_2, \ldots, a_n \rangle$$

where $a_1, a_2, \ldots, a_n$ are real numbers that are called the components of $\mathbf{a}$. Addition and scalar multiplication are defined in terms of components just as for the cases $n = 2$ and $n = 3$.

Vectors in n dimensions are used to list various quantities in an organized way. For instance, the components of a six-dimensional vector

$$\mathbf{p} = \langle p_1, p_2, p_3, p_4, p_5, p_6 \rangle$$

might represent the prices of six different ingredients required to make a particular product. Four-dimensional vectors $\langle x, y, z, t \rangle$ are used in relativity theory, where the first three components specify a position in space and the fourth represents time.

Properties of Vectors If $\mathbf{a}$, $\mathbf{b}$, and $\mathbf{c}$ are vectors in V_n and c and d are scalars, then

1. $\mathbf{a} + \mathbf{b} = \mathbf{b} + \mathbf{a}$	**2.** $\mathbf{a} + (\mathbf{b} + \mathbf{c}) = (\mathbf{a} + \mathbf{b}) + \mathbf{c}$
3. $\mathbf{a} + \mathbf{0} = \mathbf{a}$	**4.** $\mathbf{a} + (-\mathbf{a}) = \mathbf{0}$
5. $c(\mathbf{a} + \mathbf{b}) = c\mathbf{a} + c\mathbf{b}$	**6.** $(c + d)\mathbf{a} = c\mathbf{a} + d\mathbf{a}$
7. $(cd)\mathbf{a} = c(d\mathbf{a})$	**8.** $1\mathbf{a} = \mathbf{a}$

These eight properties of vectors can be readily verified either geometrically or algebraically. For instance, Property 1 can be seen from Figure 4 (it's equivalent to the Parallelogram Law) or as follows for the case $n = 2$:

$$\begin{aligned} \mathbf{a} + \mathbf{b} &= \langle a_1, a_2 \rangle + \langle b_1, b_2 \rangle = \langle a_1 + b_1, a_2 + b_2 \rangle \\ &= \langle b_1 + a_1, b_2 + a_2 \rangle = \langle b_1, b_2 \rangle + \langle a_1, a_2 \rangle \\ &= \mathbf{b} + \mathbf{a} \end{aligned}$$

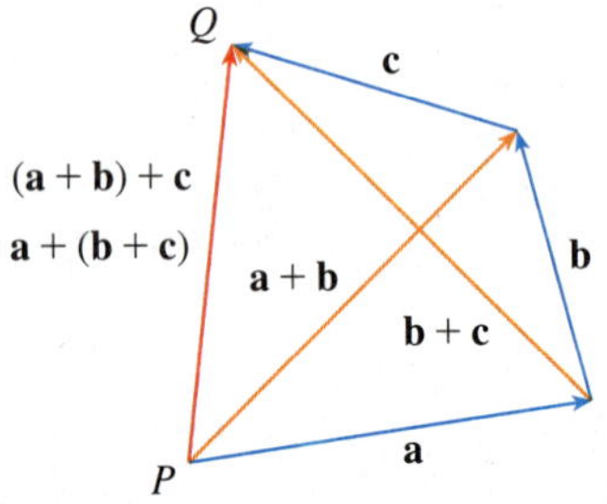

FIGURE 16

We can see why Property 2 (the associative law) is true by looking at Figure 16 and applying the Triangle Law several times: The vector $\overrightarrow{PQ}$ is obtained either by first constructing $\mathbf{a} + \mathbf{b}$ and then adding $\mathbf{c}$ or by adding $\mathbf{a}$ to the vector $\mathbf{b} + \mathbf{c}$.

Three vectors in V_3 play a special role. Let

$$\mathbf{i} = \langle 1, 0, 0 \rangle \qquad \mathbf{j} = \langle 0, 1, 0 \rangle \qquad \mathbf{k} = \langle 0, 0, 1 \rangle$$

Then **i**, **j**, and **k** are vectors that have length 1 and point in the directions of the positive x-, y-, and z-axes. Similarly, in two dimensions we define $\mathbf{i} = \langle 1, 0 \rangle$ and $\mathbf{j} = \langle 0, 1 \rangle$. (See Figure 17.)

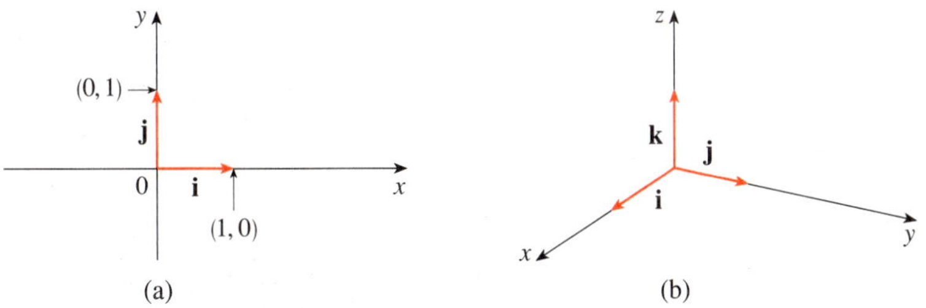

FIGURE 17
Standard basis vectors in V_2 and V_3

If $\mathbf{a} = \langle a_1, a_2, a_3 \rangle$, then we can write

$$\mathbf{a} = \langle a_1, a_2, a_3 \rangle = \langle a_1, 0, 0 \rangle + \langle 0, a_2, 0 \rangle + \langle 0, 0, a_3 \rangle$$
$$= a_1\langle 1, 0, 0 \rangle + a_2\langle 0, 1, 0 \rangle + a_3\langle 0, 0, 1 \rangle$$

2 $$\mathbf{a} = a_1\,\mathbf{i} + a_2\,\mathbf{j} + a_3\,\mathbf{k}$$

Thus, any vector in V_3 can be expressed in terms of the **standard basis vectors i**, **j**, and **k**. For instance,

$$\langle 1, -2, 6 \rangle = \mathbf{i} - 2\mathbf{j} + 6\mathbf{k}$$

Similarly, in two dimensions, we can write

3 $$\mathbf{a} = \langle a_1, a_2 \rangle = a_1\,\mathbf{i} + a_2\,\mathbf{j}$$

See Figure 18 for the geometric interpretation of Equations 3 and 2 and compare with Figure 17.

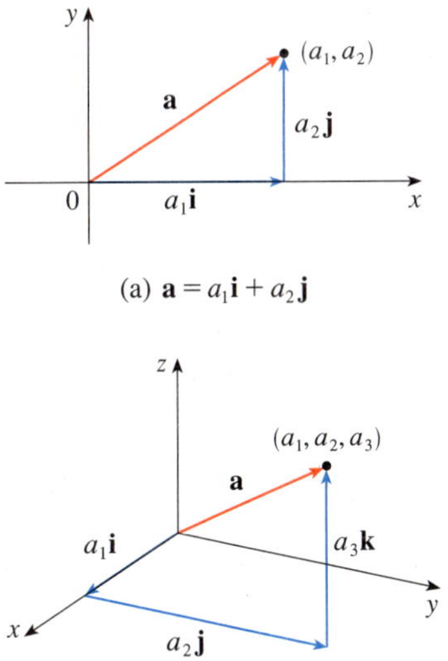

(a) $\mathbf{a} = a_1\mathbf{i} + a_2\mathbf{j}$

(b) $\mathbf{a} = a_1\mathbf{i} + a_2\mathbf{j} + a_3\mathbf{k}$

FIGURE 18

EXAMPLE 5 If $\mathbf{a} = \mathbf{i} + 2\mathbf{j} - 3\mathbf{k}$ and $\mathbf{b} = 4\mathbf{i} + 7\mathbf{k}$, express the vector $2\mathbf{a} + 3\mathbf{b}$ in terms of **i**, **j**, and **k**.

SOLUTION Using Properties 1, 2, 5, 6, and 7 of vectors, we have

$$2\mathbf{a} + 3\mathbf{b} = 2(\mathbf{i} + 2\mathbf{j} - 3\mathbf{k}) + 3(4\mathbf{i} + 7\mathbf{k})$$
$$= 2\mathbf{i} + 4\mathbf{j} - 6\mathbf{k} + 12\mathbf{i} + 21\mathbf{k} = 14\mathbf{i} + 4\mathbf{j} + 15\mathbf{k}$$

A **unit vector** is a vector whose length is 1. For instance, **i**, **j**, and **k** are all unit vectors. In general, if $\mathbf{a} \neq \mathbf{0}$, then the unit vector that has the same direction as **a** is

4 $$\mathbf{u} = \frac{1}{|\mathbf{a}|}\mathbf{a} = \frac{\mathbf{a}}{|\mathbf{a}|}$$

In order to verify this, we let $c = 1/|\mathbf{a}|$. Then $\mathbf{u} = c\mathbf{a}$ and c is a positive scalar, so **u** has the same direction as **a**. Also

$$|\mathbf{u}| = |c\mathbf{a}| = |c||\mathbf{a}| = \frac{1}{|\mathbf{a}|}|\mathbf{a}| = 1$$

EXAMPLE 6 Find the unit vector in the direction of the vector $2\mathbf{i} - \mathbf{j} - 2\mathbf{k}$.

SOLUTION The given vector has length

$$|2\mathbf{i} - \mathbf{j} - 2\mathbf{k}| = \sqrt{2^2 + (-1)^2 + (-2)^2} = \sqrt{9} = 3$$

so, by Equation 4, the unit vector with the same direction is

$$\tfrac{1}{3}(2\mathbf{i} - \mathbf{j} - 2\mathbf{k}) = \tfrac{2}{3}\mathbf{i} - \tfrac{1}{3}\mathbf{j} - \tfrac{2}{3}\mathbf{k}$$

Applications

Vectors are useful in many aspects of physics and engineering. In Chapter 13 we will see how they describe the velocity and acceleration of objects moving in space. Here we look at forces.

A force is represented by a vector because it has both a magnitude (measured in pounds or newtons) and a direction. If several forces are acting on an object, the **resultant force** experienced by the object is the vector sum of these forces.

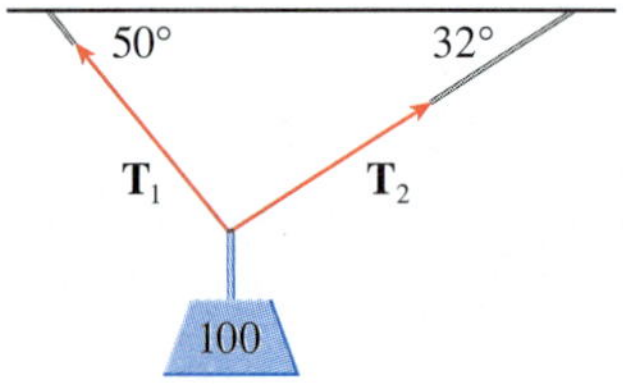

FIGURE 19

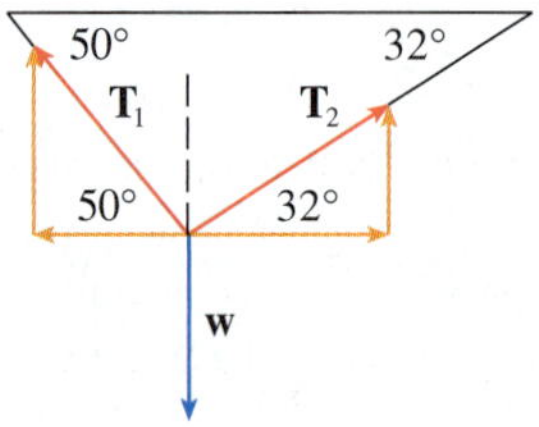

FIGURE 20

EXAMPLE 7 A 100-lb weight hangs from two wires as shown in Figure 19. Find the tensions (forces) $\mathbf{T}_1$ and $\mathbf{T}_2$ in both wires and their magnitudes.

SOLUTION We first express $\mathbf{T}_1$ and $\mathbf{T}_2$ in terms of their horizontal and vertical components. From Figure 20 we see that

$$\boxed{5} \qquad \mathbf{T}_1 = -|\mathbf{T}_1|\cos 50^\circ\,\mathbf{i} + |\mathbf{T}_1|\sin 50^\circ\,\mathbf{j}$$

$$\boxed{6} \qquad \mathbf{T}_2 = |\mathbf{T}_2|\cos 32^\circ\,\mathbf{i} + |\mathbf{T}_2|\sin 32^\circ\,\mathbf{j}$$

The resultant $\mathbf{T}_1 + \mathbf{T}_2$ of the tensions counterbalances the weight $\mathbf{w}$ and so we must have

$$\mathbf{T}_1 + \mathbf{T}_2 = -\mathbf{w} = 100\mathbf{j}$$

Thus

$$\left(-|\mathbf{T}_1|\cos 50^\circ + |\mathbf{T}_2|\cos 32^\circ\right)\mathbf{i} + \left(|\mathbf{T}_1|\sin 50^\circ + |\mathbf{T}_2|\sin 32^\circ\right)\mathbf{j} = 100\mathbf{j}$$

Equating components, we get

$$-|\mathbf{T}_1|\cos 50^\circ + |\mathbf{T}_2|\cos 32^\circ = 0$$
$$|\mathbf{T}_1|\sin 50^\circ + |\mathbf{T}_2|\sin 32^\circ = 100$$

Solving the first of these equations for $|\mathbf{T}_2|$ and substituting into the second, we get

$$|\mathbf{T}_1|\sin 50^\circ + \frac{|\mathbf{T}_1|\cos 50^\circ}{\cos 32^\circ}\sin 32^\circ = 100$$

So the magnitudes of the tensions are

$$|\mathbf{T}_1| = \frac{100}{\sin 50^\circ + \tan 32^\circ \cos 50^\circ} \approx 85.64 \text{ lb}$$

and $$|\mathbf{T}_2| = \frac{|\mathbf{T}_1|\cos 50^\circ}{\cos 32^\circ} \approx 64.91 \text{ lb}$$

Substituting these values in (5) and (6), we obtain the tension vectors

$$\mathbf{T}_1 \approx -55.05\,\mathbf{i} + 65.60\,\mathbf{j} \qquad \mathbf{T}_2 \approx 55.05\,\mathbf{i} + 34.40\,\mathbf{j}$$

12.2 Exercises

1. Are the following quantities vectors or scalars? Explain.
(a) The cost of a theater ticket
(b) The current in a river
(c) The initial flight path from Houston to Dallas
(d) The population of the world

2. What is the relationship between the point (4, 7) and the vector $\langle 4, 7 \rangle$? Illustrate with a sketch.

3. Name all the equal vectors in the parallelogram shown.

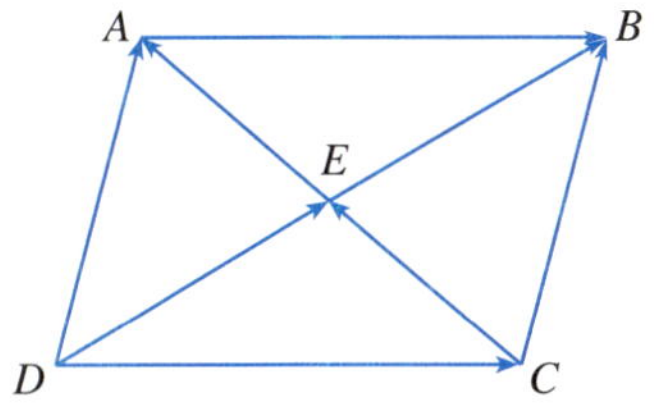

4. Write each combination of vectors as a single vector.

(a) $\overrightarrow{PQ} + \overrightarrow{QR}$ (b) $\overrightarrow{RP} + \overrightarrow{PS}$
(c) $\overrightarrow{QS} - \overrightarrow{PS}$ (d) $\overrightarrow{RS} + \overrightarrow{SP} + \overrightarrow{PQ}$

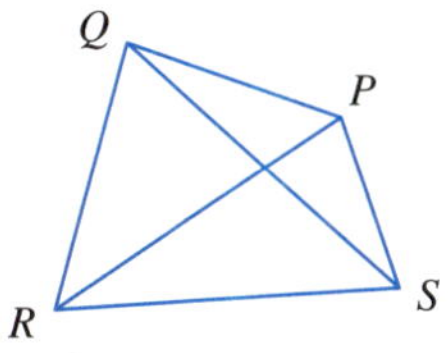

5. Copy the vectors in the figure and use them to draw the following vectors.
(a) $\mathbf{u} + \mathbf{v}$ (b) $\mathbf{u} - \mathbf{v}$
(c) $\mathbf{v} + \mathbf{w}$ (d) $\mathbf{w} + \mathbf{v} + \mathbf{u}$

6. Copy the vectors in the figure and use them to draw the following vectors.
(a) $\mathbf{a} + \mathbf{b}$ (b) $\mathbf{a} - \mathbf{b}$
(c) $2\mathbf{a}$ (d) $-\frac{1}{2}\mathbf{b}$
(e) $2\mathbf{a} + \mathbf{b}$ (f) $\mathbf{b} - 3\mathbf{a}$

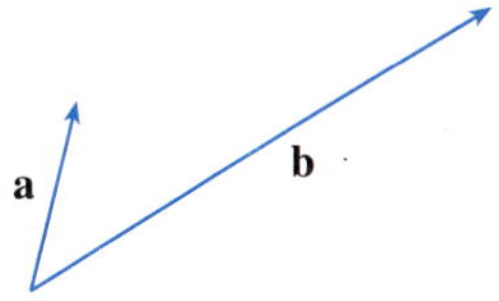

7–12 Find a vector $\mathbf{a}$ with representation given by the directed line segment $\overrightarrow{AB}$. Draw $\overrightarrow{AB}$ and the equivalent representation starting at the origin.

7. $A(2, 3)$, $B(-2, 1)$ **8.** $A(-2, -2)$, $B(5, 3)$

9. $A(-1, -1)$, $B(-3, 4)$ **10.** $A(-2, 2)$, $B(3, 0)$

11. $A(0, 3, 1)$, $B(2, 3, -1)$ **12.** $A(4, 0, -2)$, $B(4, 2, 1)$

13–16 Find the sum of the given vectors and illustrate geometrically.

13. $\langle 3, -1 \rangle$, $\langle -2, 4 \rangle$ **14.** $\langle -2, -1 \rangle$, $\langle 5, 7 \rangle$

15. $\langle 0, 1, 2 \rangle$, $\langle 0, 0, -3 \rangle$ **16.** $\langle -1, 0, 2 \rangle$, $\langle 0, 4, 0 \rangle$

17–22 Find $|\mathbf{a}|$, $\mathbf{a} + \mathbf{b}$, $\mathbf{a} - \mathbf{b}$, $2\mathbf{a}$, and $3\mathbf{a} + 4\mathbf{b}$.

17. $\mathbf{a} = \langle -4, 3 \rangle$, $\mathbf{b} = \langle 6, 2 \rangle$

18. $\mathbf{a} = 2\mathbf{i} - 3\mathbf{j}$, $\mathbf{b} = \mathbf{i} + 5\mathbf{j}$

19. $\mathbf{a} = \langle 6, 2, 3 \rangle$, $\mathbf{b} = \langle -1, 5, -2 \rangle$

20. $\mathbf{a} = \langle -3, -4, -1 \rangle$, $\mathbf{b} = \langle 6, 2, -3 \rangle$

21. $\mathbf{a} = \mathbf{i} - 2\mathbf{j} + \mathbf{k}$, $\mathbf{b} = \mathbf{j} + 2\mathbf{k}$

22. $\mathbf{a} = 3\mathbf{i} - 2\mathbf{k}$, $\mathbf{b} = \mathbf{i} - \mathbf{j} + \mathbf{k}$

23–25 Find a unit vector that has the same direction as the given vector.

23. $\langle 9, -5 \rangle$ **24.** $12\mathbf{i} - 5\mathbf{j}$

25. $8\mathbf{i} - \mathbf{j} + 4\mathbf{k}$

26. Find a vector that has the same direction as $\langle -2, 4, 2 \rangle$ but has length 6.

27. If $\mathbf{v}$ lies in the first quadrant and makes an angle $\pi/3$ with the positive x-axis and $|\mathbf{v}| = 4$, find $\mathbf{v}$ in component form.

28. If a child pulls a sled through the snow with a force of 50 N exerted at an angle of 38° above the horizontal, find the horizontal and vertical components of the force.

29. Two forces $\mathbf{F}_1$ and $\mathbf{F}_2$ with magnitudes 10 lb and 12 lb act on an object at a point P as shown in the figure. Find the resultant force $\mathbf{F}$ acting at P as well as its magnitude and its direction. (Indicate the direction by finding the angle θ shown in the figure.)

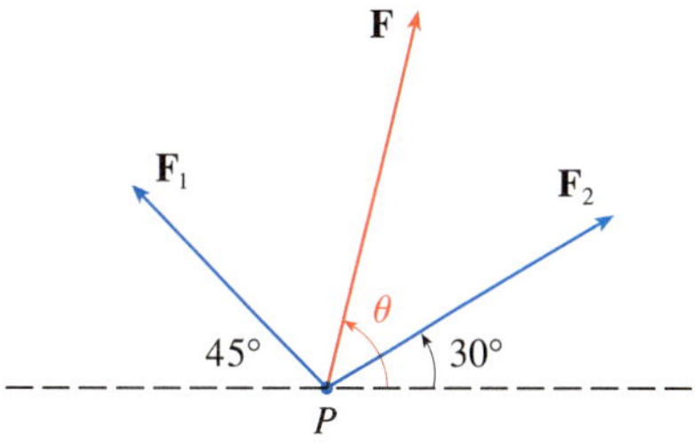

30. Velocities have both direction and magnitude and thus are vectors. The magnitude of a velocity vector is called *speed.* Suppose that a wind is blowing from the direction N45°W at a speed of 50 km/h. (This means that the direction from which the wind blows is 45° west of the northerly direction.) A pilot is steering a plane in the direction N60°E at an airspeed (speed in still air) of 250 km/h. The *true course,* or *track,* of the plane is the direction of the resultant of the velocity vectors of the plane and the wind. The *ground speed* of the plane is the magnitude of the resultant. Find the true course and the ground speed of the plane.

31. A woman walks due west on the deck of a ship at 3 mi/h. The ship is moving north at a speed of 22 mi/h. Find the speed and direction of the woman relative to the surface of the water.

32. Ropes 3 m and 5 m in length are fastened to a holiday decoration that is suspended over a town square. The decoration has a mass of 5 kg. The ropes, fastened at different heights, make angles of 52° and 40° with the horizontal. Find the tension in each wire and the magnitude of each tension.

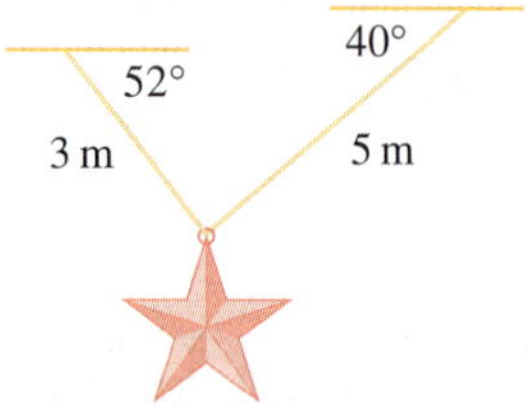

33. A clothesline is tied between two poles, 8 m apart. The line is quite taut and has negligible sag. When a wet shirt with a mass of 0.8 kg is hung at the middle of the line, the midpoint is pulled down 8 cm. Find the tension in each half of the clothesline.

34. The tension $\mathbf{T}$ at each end of the chain has magnitude 25 N. What is the weight of the chain?

35. If A, B, and C are the vertices of a triangle, find $\overrightarrow{AB} + \overrightarrow{BC} + \overrightarrow{CA}$.

36. Let C be the point on the line segment AB that is twice as far from B as it is from A. If $\mathbf{a} = \overrightarrow{OA}$, $\mathbf{b} = \overrightarrow{OB}$, and $\mathbf{c} = \overrightarrow{OC}$, show that $\mathbf{c} = \frac{2}{3}\mathbf{a} + \frac{1}{3}\mathbf{b}$.

37. (a) Draw the vectors $\mathbf{a} = \langle 3, 2 \rangle$, $\mathbf{b} = \langle 2, -1 \rangle$, and $\mathbf{c} = \langle 7, 1 \rangle$.
(b) Show, by means of a sketch, that there are scalars s and t such that $\mathbf{c} = s\mathbf{a} + t\mathbf{b}$.
(c) Use the sketch to estimate the values of s and t.
(d) Find the exact values of s and t.

38. Suppose that $\mathbf{a}$ and $\mathbf{b}$ are nonzero vectors that are not parallel and $\mathbf{c}$ is any vector in the plane determined by $\mathbf{a}$ and $\mathbf{b}$. Give a geometric argument to show that $\mathbf{c}$ can be written as $\mathbf{c} = s\mathbf{a} + t\mathbf{b}$ for suitable scalars s and t. Then give an argument using components.

39. If $\mathbf{r} = \langle x, y, z \rangle$ and $\mathbf{r}_0 = \langle x_0, y_0, z_0 \rangle$, describe the set of all points (x, y, z) such that $|\mathbf{r} - \mathbf{r}_0| = 1$.

40. If $\mathbf{r} = \langle x, y \rangle$, $\mathbf{r}_1 = \langle x_1, y_1 \rangle$, and $\mathbf{r}_2 = \langle x_2, y_2 \rangle$, describe the set of all points (x, y) such that $|\mathbf{r} - \mathbf{r}_1| + |\mathbf{r} - \mathbf{r}_2| = k$, where $k > |\mathbf{r}_1 - \mathbf{r}_2|$.

41. Figure 16 gives a geometric demonstration of Property 2 of vectors. Use components to give an algebraic proof of this fact for the case $n = 2$.

42. Prove Property 5 of vectors algebraically for the case $n = 3$. Then use similar triangles to give a geometric proof.

43. Use vectors to prove that the line joining the midpoints of two sides of a triangle is parallel to the third side and half its length.

44. Suppose the three coordinate planes are all mirrored and a light ray given by the vector $\mathbf{a} = \langle a_1, a_2, a_3 \rangle$ first strikes the xz-plane, as shown in the figure. Use the fact that the angle of incidence equals the angle of reflection to show that the direction of the reflected ray is given by $\mathbf{b} = \langle a_1, -a_2, a_3 \rangle$. Deduce that, after being reflected by all three mutually perpendicular mirrors, the resulting ray is parallel to the initial ray. (American space scientists used this principle, together with laser beams and an array of corner mirrors on the Moon, to calculate very precisely the distance from the Earth to the Moon.)

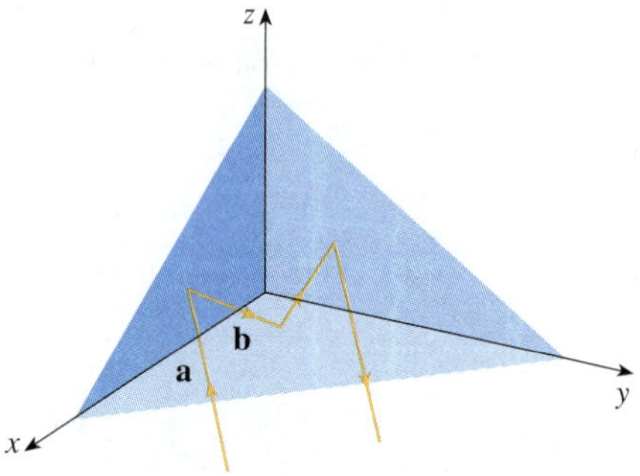

12.3 The Dot Product

So far we have added two vectors and multiplied a vector by a scalar. The question arises: Is it possible to multiply two vectors so that their product is a useful quantity? One such product is the dot product, whose definition follows. Another is the cross product, which is discussed in the next section.

1 Definition If $\mathbf{a} = \langle a_1, a_2, a_3 \rangle$ and $\mathbf{b} = \langle b_1, b_2, b_3 \rangle$, then the **dot product** of **a** and **b** is the number $\mathbf{a} \cdot \mathbf{b}$ given by

$$\mathbf{a} \cdot \mathbf{b} = a_1 b_1 + a_2 b_2 + a_3 b_3$$

Thus, to find the dot product of **a** and **b** we multiply corresponding components and add. The result is not a vector. It is a real number, that is, a scalar. For this reason, the dot product is sometimes called the **scalar product** (or **inner product**). Although Definition 1 is given for three-dimensional vectors, the dot product of two-dimensional vectors is defined in a similar fashion:

$$\langle a_1, a_2 \rangle \cdot \langle b_1, b_2 \rangle = a_1 b_1 + a_2 b_2$$

EXAMPLE 1

$$\langle 2, 4 \rangle \cdot \langle 3, -1 \rangle = 2(3) + 4(-1) = 2$$

$$\langle -1, 7, 4 \rangle \cdot \langle 6, 2, -\tfrac{1}{2} \rangle = (-1)(6) + 7(2) + 4(-\tfrac{1}{2}) = 6$$

$$(\mathbf{i} + 2\mathbf{j} - 3\mathbf{k}) \cdot (2\mathbf{j} - \mathbf{k}) = 1(0) + 2(2) + (-3)(-1) = 7$$

The dot product obeys many of the laws that hold for ordinary products of real numbers. These are stated in the following theorem.

2 Properties of the Dot Product If **a**, **b**, and **c** are vectors in V_3 and c is a scalar, then

1. $\mathbf{a} \cdot \mathbf{a} = |\mathbf{a}|^2$
2. $\mathbf{a} \cdot \mathbf{b} = \mathbf{b} \cdot \mathbf{a}$
3. $\mathbf{a} \cdot (\mathbf{b} + \mathbf{c}) = \mathbf{a} \cdot \mathbf{b} + \mathbf{a} \cdot \mathbf{c}$
4. $(c\mathbf{a}) \cdot \mathbf{b} = c(\mathbf{a} \cdot \mathbf{b}) = \mathbf{a} \cdot (c\mathbf{b})$
5. $\mathbf{0} \cdot \mathbf{a} = 0$

These properties are easily proved using Definition 1. For instance, here are the proofs of Properties 1 and 3:

1. $\mathbf{a} \cdot \mathbf{a} = a_1^2 + a_2^2 + a_3^2 = |\mathbf{a}|^2$

3.
$$\begin{aligned}
\mathbf{a} \cdot (\mathbf{b} + \mathbf{c}) &= \langle a_1, a_2, a_3 \rangle \cdot \langle b_1 + c_1, b_2 + c_2, b_3 + c_3 \rangle \\
&= a_1(b_1 + c_1) + a_2(b_2 + c_2) + a_3(b_3 + c_3) \\
&= a_1 b_1 + a_1 c_1 + a_2 b_2 + a_2 c_2 + a_3 b_3 + a_3 c_3 \\
&= (a_1 b_1 + a_2 b_2 + a_3 b_3) + (a_1 c_1 + a_2 c_2 + a_3 c_3) \\
&= \mathbf{a} \cdot \mathbf{b} + \mathbf{a} \cdot \mathbf{c}
\end{aligned}$$

The proofs of the remaining properties are left as exercises.

The dot product $\mathbf{a} \cdot \mathbf{b}$ can be given a geometric interpretation in terms of the **angle θ between a and b**, which is defined to be the angle between the representations of **a** and **b** that start at the origin, where $0 \leq \theta \leq \pi$. In other words, θ is the angle between the

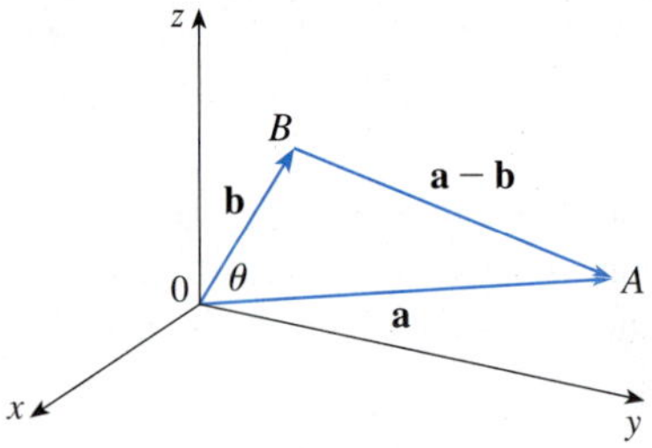

FIGURE 1

line segments $\overrightarrow{OA}$ and $\overrightarrow{OB}$ in Figure 1. Note that if **a** and **b** are parallel vectors, then $\theta = 0$ or $\theta = \pi$.

The formula in the following theorem is used by physicists as the *definition* of the dot product.

3 Theorem If θ is the angle between the vectors **a** and **b**, then

$$\mathbf{a} \cdot \mathbf{b} = |\mathbf{a}||\mathbf{b}| \cos \theta$$

Proof If we apply the Law of Cosines to triangle OAB in Figure 1, we get

$$|AB|^2 = |OA|^2 + |OB|^2 - 2|OA||OB| \cos \theta \tag{4}$$

(Observe that the Law of Cosines still applies in the limiting cases when $\theta = 0$ or π, or $\mathbf{a} = \mathbf{0}$ or $\mathbf{b} = \mathbf{0}$.) But $|OA| = |\mathbf{a}|$, $|OB| = |\mathbf{b}|$, and $|AB| = |\mathbf{a} - \mathbf{b}|$, so Equation 4 becomes

$$|\mathbf{a} - \mathbf{b}|^2 = |\mathbf{a}|^2 + |\mathbf{b}|^2 - 2|\mathbf{a}||\mathbf{b}| \cos \theta \tag{5}$$

Using Properties 1, 2, and 3 of the dot product, we can rewrite the left side of this equation as follows:

$$\begin{aligned} |\mathbf{a} - \mathbf{b}|^2 &= (\mathbf{a} - \mathbf{b}) \cdot (\mathbf{a} - \mathbf{b}) \\ &= \mathbf{a} \cdot \mathbf{a} - \mathbf{a} \cdot \mathbf{b} - \mathbf{b} \cdot \mathbf{a} + \mathbf{b} \cdot \mathbf{b} \\ &= |\mathbf{a}|^2 - 2\mathbf{a} \cdot \mathbf{b} + |\mathbf{b}|^2 \end{aligned}$$

Therefore, Equation 5 gives

$$|\mathbf{a}|^2 - 2\mathbf{a} \cdot \mathbf{b} + |\mathbf{b}|^2 = |\mathbf{a}|^2 + |\mathbf{b}|^2 - 2|\mathbf{a}||\mathbf{b}| \cos \theta$$

Thus
$$-2\mathbf{a} \cdot \mathbf{b} = -2|\mathbf{a}||\mathbf{b}| \cos \theta$$

or
$$\mathbf{a} \cdot \mathbf{b} = |\mathbf{a}||\mathbf{b}| \cos \theta$$

EXAMPLE 2 If the vectors **a** and **b** have lengths 4 and 6, and the angle between them is $\pi/3$, find $\mathbf{a} \cdot \mathbf{b}$.

SOLUTION Using Theorem 3, we have

$$\mathbf{a} \cdot \mathbf{b} = |\mathbf{a}||\mathbf{b}| \cos(\pi/3) = 4 \cdot 6 \cdot \tfrac{1}{2} = 12$$

The formula in Theorem 3 also enables us to find the angle between two vectors.

6 Corollary If θ is the angle between the nonzero vectors **a** and **b**, then

$$\cos \theta = \frac{\mathbf{a} \cdot \mathbf{b}}{|\mathbf{a}||\mathbf{b}|}$$

EXAMPLE 3 Find the angle between the vectors $\mathbf{a} = \langle 2, 2, -1 \rangle$ and $\mathbf{b} = \langle 5, -3, 2 \rangle$.

SOLUTION Since

$$|\mathbf{a}| = \sqrt{2^2 + 2^2 + (-1)^2} = 3 \qquad \text{and} \qquad |\mathbf{b}| = \sqrt{5^2 + (-3)^2 + 2^2} = \sqrt{38}$$

and since

$$\mathbf{a} \cdot \mathbf{b} = 2(5) + 2(-3) + (-1)(2) = 2$$

we have, from Corollary 6,

$$\cos \theta = \frac{\mathbf{a} \cdot \mathbf{b}}{|\mathbf{a}||\mathbf{b}|} = \frac{2}{3\sqrt{38}}$$

So the angle between **a** and **b** is

$$\theta = \cos^{-1}\left(\frac{2}{3\sqrt{38}}\right) \approx 1.46 \quad (\text{or } 84^\circ)$$

Two nonzero vectors **a** and **b** are called **perpendicular** or **orthogonal** if the angle between them is $\theta = \pi/2$. Then Theorem 3 gives

$$\mathbf{a} \cdot \mathbf{b} = |\mathbf{a}||\mathbf{b}| \cos(\pi/2) = 0$$

and conversely if $\mathbf{a} \cdot \mathbf{b} = 0$, then $\cos \theta = 0$, so $\theta = \pi/2$. The zero vector **0** is considered to be perpendicular to all vectors. Therefore, we have the following method for determining whether two vectors are orthogonal.

7 Two vectors **a** and **b** are orthogonal if and only if $\mathbf{a} \cdot \mathbf{b} = 0$.

EXAMPLE 4 Show that $2\mathbf{i} + 2\mathbf{j} - \mathbf{k}$ is perpendicular to $5\mathbf{i} - 4\mathbf{j} + 2\mathbf{k}$.

SOLUTION Since

$$(2\mathbf{i} + 2\mathbf{j} - \mathbf{k}) \cdot (5\mathbf{i} - 4\mathbf{j} + 2\mathbf{k}) = 2(5) + 2(-4) + (-1)(2) = 0$$

these vectors are perpendicular by (7).

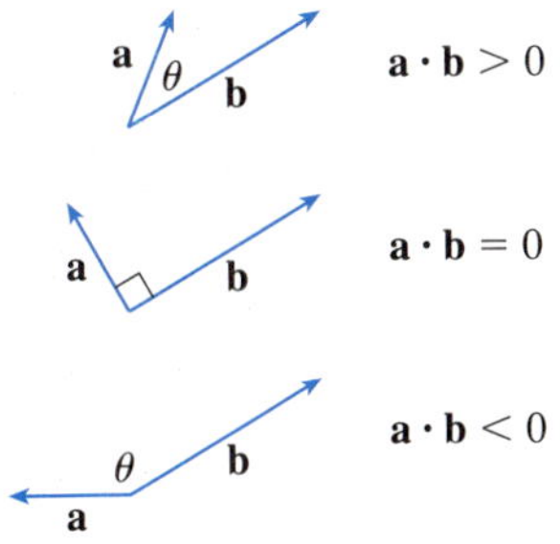

FIGURE 2

Visual 12.3A shows an animation of Figure 2.

Because $\cos \theta > 0$ if $0 \leq \theta < \pi/2$ and $\cos \theta < 0$ if $\pi/2 < \theta \leq \pi$, we see that $\mathbf{a} \cdot \mathbf{b}$ is positive for $\theta < \pi/2$ and negative for $\theta > \pi/2$. We can think of $\mathbf{a} \cdot \mathbf{b}$ as measuring the extent to which **a** and **b** point in the same direction. The dot product $\mathbf{a} \cdot \mathbf{b}$ is positive if **a** and **b** point in the same general direction, 0 if they are perpendicular, and negative if they point in generally opposite directions (see Figure 2). In the extreme case where **a** and **b** point in exactly the same direction, we have $\theta = 0$, so $\cos \theta = 1$ and

$$\mathbf{a} \cdot \mathbf{b} = |\mathbf{a}||\mathbf{b}|$$

If **a** and **b** point in exactly opposite directions, then $\theta = \pi$ and so $\cos \theta = -1$ and $\mathbf{a} \cdot \mathbf{b} = -|\mathbf{a}||\mathbf{b}|$.

Direction Angles and Direction Cosines

The **direction angles** of a nonzero vector **a** are the angles α, β, and γ (in the interval $[0, \pi]$) that **a** makes with the positive x-, y-, and z-axes (see Figure 3 on page 810).

The cosines of these direction angles, $\cos \alpha$, $\cos \beta$, and $\cos \gamma$, are called the **direction cosines** of the vector **a**. Using Corollary 6 with **b** replaced by **i**, we obtain

8
$$\cos \alpha = \frac{\mathbf{a} \cdot \mathbf{i}}{|\mathbf{a}||\mathbf{i}|} = \frac{a_1}{|\mathbf{a}|}$$

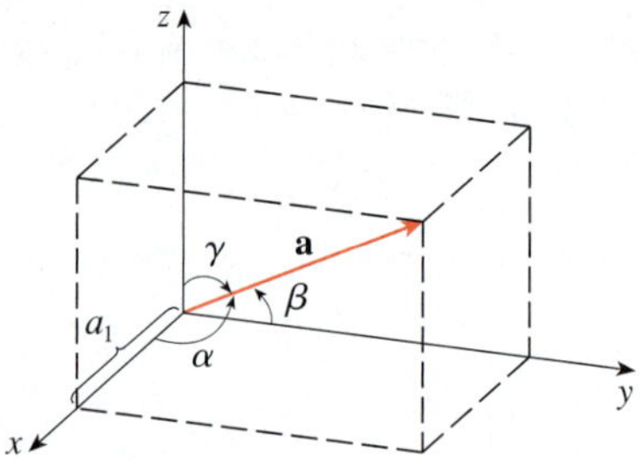

FIGURE 3

(This can also be seen directly from Figure 3.) Similarly, we also have

$$\textbf{9} \qquad \cos\beta = \frac{a_2}{|\mathbf{a}|} \qquad\qquad \cos\gamma = \frac{a_3}{|\mathbf{a}|}$$

By squaring the expressions in Equations 8 and 9 and adding, we see that

$$\textbf{10} \qquad \cos^2\alpha + \cos^2\beta + \cos^2\gamma = 1$$

We can also use Equations 8 and 9 to write

$$\begin{aligned} \mathbf{a} = \langle a_1, a_2, a_3 \rangle &= \langle |\mathbf{a}|\cos\alpha, |\mathbf{a}|\cos\beta, |\mathbf{a}|\cos\gamma \rangle \\ &= |\mathbf{a}| \langle \cos\alpha, \cos\beta, \cos\gamma \rangle \end{aligned}$$

Therefore

$$\textbf{11} \qquad \frac{1}{|\mathbf{a}|}\mathbf{a} = \langle \cos\alpha, \cos\beta, \cos\gamma \rangle$$

which says that the direction cosines of **a** are the components of the unit vector in the direction of **a**.

EXAMPLE 5 Find the direction angles of the vector $\mathbf{a} = \langle 1, 2, 3 \rangle$.

SOLUTION Since $|\mathbf{a}| = \sqrt{1^2 + 2^2 + 3^2} = \sqrt{14}$, Equations 8 and 9 give

$$\cos\alpha = \frac{1}{\sqrt{14}} \qquad \cos\beta = \frac{2}{\sqrt{14}} \qquad \cos\gamma = \frac{3}{\sqrt{14}}$$

and so

$$\alpha = \cos^{-1}\left(\frac{1}{\sqrt{14}}\right) \approx 74^\circ \qquad \beta = \cos^{-1}\left(\frac{2}{\sqrt{14}}\right) \approx 58^\circ \qquad \gamma = \cos^{-1}\left(\frac{3}{\sqrt{14}}\right) \approx 37^\circ$$

Projections

Figure 4 shows representations $\overrightarrow{PQ}$ and $\overrightarrow{PR}$ of two vectors **a** and **b** with the same initial point P. If S is the foot of the perpendicular from R to the line containing $\overrightarrow{PQ}$, then the vector with representation $\overrightarrow{PS}$ is called the **vector projection** of **b** onto **a** and is denoted by $\text{proj}_{\mathbf{a}}\,\mathbf{b}$.

Visual 12.3B shows how Figure 4 changes when we vary **a** and **b**.

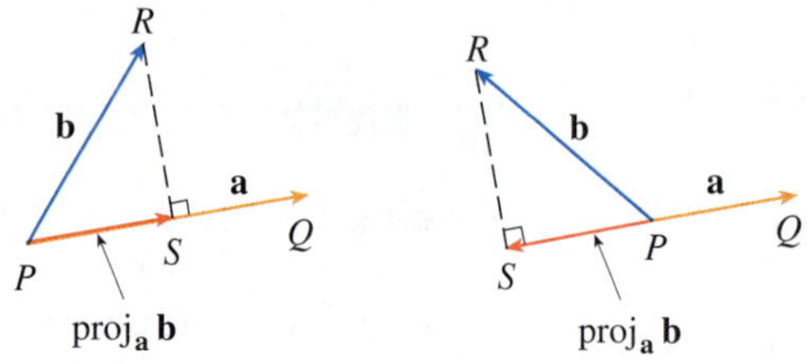

FIGURE 4
Vector projections

The **scalar projection** of **b** onto **a** (also called the **component of b along a**) is defined to be the magnitude of the vector projection, which is the number $|\mathbf{b}|\cos\theta$, where θ is the

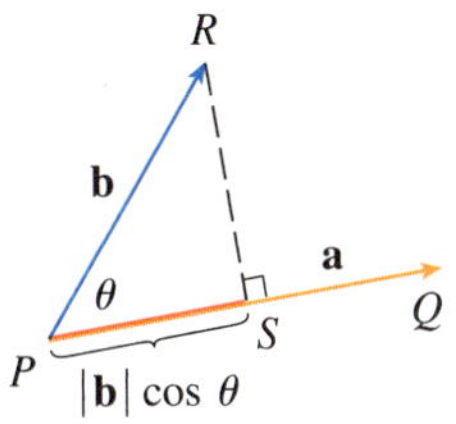

FIGURE 5
Scalar projection

angle between **a** and **b**. (See Figure 5; you can think of the scalar projection of **b** as being the length of a shadow of **b**.) This is denoted by $\text{comp}_{\mathbf{a}}\,\mathbf{b}$. Observe that it is negative if $\pi/2 < \theta \leqslant \pi$. The equation

$$\mathbf{a} \cdot \mathbf{b} = |\mathbf{a}||\mathbf{b}| \cos\theta = |\mathbf{a}|(|\mathbf{b}|\cos\theta)$$

shows that the dot product of **a** and **b** can be interpreted as the length of **a** times the scalar projection of **b** onto **a**. Since

$$|\mathbf{b}|\cos\theta = \frac{\mathbf{a} \cdot \mathbf{b}}{|\mathbf{a}|} = \frac{\mathbf{a}}{|\mathbf{a}|} \cdot \mathbf{b}$$

the component of **b** along **a** can be computed by taking the dot product of **b** with the unit vector in the direction of **a**. We summarize these ideas as follows.

Scalar projection of **b** onto **a**: $$\text{comp}_{\mathbf{a}}\,\mathbf{b} = \frac{\mathbf{a} \cdot \mathbf{b}}{|\mathbf{a}|}$$

Vector projection of **b** onto **a**: $$\text{proj}_{\mathbf{a}}\,\mathbf{b} = \left(\frac{\mathbf{a} \cdot \mathbf{b}}{|\mathbf{a}|}\right)\frac{\mathbf{a}}{|\mathbf{a}|} = \frac{\mathbf{a} \cdot \mathbf{b}}{|\mathbf{a}|^2}\,\mathbf{a}$$

Notice that the vector projection is the scalar projection times the unit vector in the direction of **a**.

EXAMPLE 6 Find the scalar projection and vector projection of $\mathbf{b} = \langle 1, 1, 2 \rangle$ onto $\mathbf{a} = \langle -2, 3, 1 \rangle$.

SOLUTION Since $|\mathbf{a}| = \sqrt{(-2)^2 + 3^2 + 1^2} = \sqrt{14}$, the scalar projection of **b** onto **a** is

$$\text{comp}_{\mathbf{a}}\,\mathbf{b} = \frac{\mathbf{a} \cdot \mathbf{b}}{|\mathbf{a}|} = \frac{(-2)(1) + 3(1) + 1(2)}{\sqrt{14}} = \frac{3}{\sqrt{14}}$$

The vector projection is this scalar projection times the unit vector in the direction of **a**:

$$\text{proj}_{\mathbf{a}}\,\mathbf{b} = \frac{3}{\sqrt{14}}\frac{\mathbf{a}}{|\mathbf{a}|} = \frac{3}{14}\mathbf{a} = \left\langle -\frac{3}{7}, \frac{9}{14}, \frac{3}{14} \right\rangle$$

One use of projections occurs in physics in calculating work. In Section 6.4 we defined the work done by a constant force F in moving an object through a distance d as $W = Fd$, but this applies only when the force is directed along the line of motion of the object. Suppose, however, that the constant force is a vector $\mathbf{F} = \overrightarrow{PR}$ pointing in some other direction as in Figure 6. If the force moves the object from P to Q, then the **displacement vector** is $\mathbf{D} = \overrightarrow{PQ}$. The **work** done by this force is defined to be the product of the component of the force along **D** and the distance moved:

$$W = (|\mathbf{F}|\cos\theta)|\mathbf{D}|$$

But then, from Theorem 3, we have

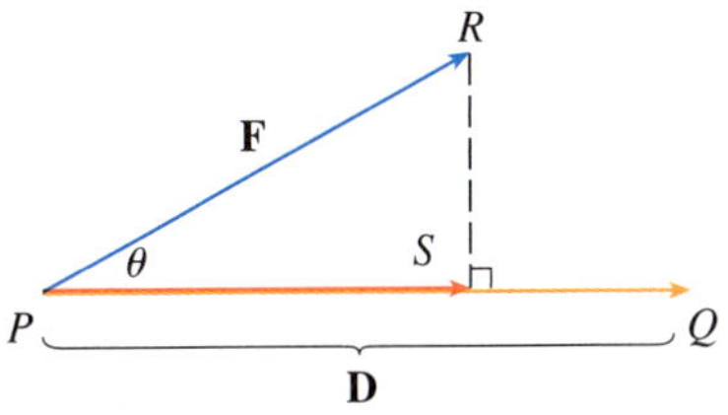

FIGURE 6

12 $$W = |\mathbf{F}||\mathbf{D}|\cos\theta = \mathbf{F} \cdot \mathbf{D}$$

Thus, the work done by a constant force **F** is the dot product **F** · **D**, where **D** is the displacement vector.

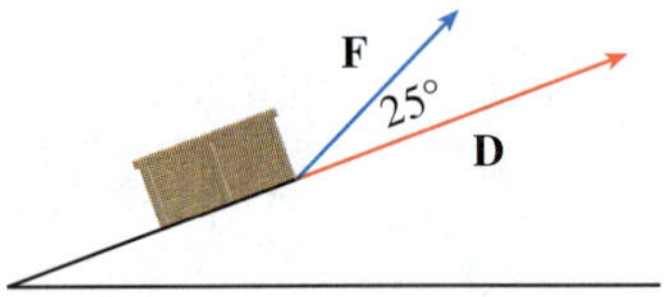

FIGURE 7

EXAMPLE 7 A crate is hauled 8 m up a ramp under a constant force of 200 N applied at an angle of 25° to the ramp. Find the work done.

SOLUTION If **F** and **D** are the force and displacement vectors, as pictured in Figure 7, then the work done is

$$W = \mathbf{F} \cdot \mathbf{D} = |\mathbf{F}||\mathbf{D}| \cos 25^\circ$$

$$= (200)(8) \cos 25^\circ \approx 1450 \text{ N·m} = 1450 \text{ J}$$

EXAMPLE 8 A force is given by a vector $\mathbf{F} = 3\mathbf{i} + 4\mathbf{j} + 5\mathbf{k}$ and moves a particle from the point $P(2, 1, 0)$ to the point $Q(4, 6, 2)$. Find the work done.

SOLUTION The displacement vector is $\mathbf{D} = \overrightarrow{PQ} = \langle 2, 5, 2 \rangle$, so by Equation 12, the work done is

$$W = \mathbf{F} \cdot \mathbf{D} = \langle 3, 4, 5 \rangle \cdot \langle 2, 5, 2 \rangle$$

$$= 6 + 20 + 10 = 36$$

If the unit of length is meters and the magnitude of the force is measured in newtons, then the work done is 36 joules.

12.3 Exercises

1. Which of the following expressions are meaningful? Which are meaningless? Explain.
(a) $(\mathbf{a} \cdot \mathbf{b}) \cdot \mathbf{c}$ (b) $(\mathbf{a} \cdot \mathbf{b})\mathbf{c}$
(c) $|\mathbf{a}|(\mathbf{b} \cdot \mathbf{c})$ (d) $\mathbf{a} \cdot (\mathbf{b} + \mathbf{c})$
(e) $\mathbf{a} \cdot \mathbf{b} + \mathbf{c}$ (f) $|\mathbf{a}| \cdot (\mathbf{b} + \mathbf{c})$

2. Find the dot product of two vectors if their lengths are 6 and $\frac{1}{3}$ and the angle between them is $\pi/4$.

3–10 Find $\mathbf{a} \cdot \mathbf{b}$.

3. $\mathbf{a} = \langle 4, -1 \rangle$, $\mathbf{b} = \langle 3, 6 \rangle$

4. $\mathbf{a} = \langle \frac{1}{2}, 4 \rangle$, $\mathbf{b} = \langle -8, -3 \rangle$

5. $\mathbf{a} = \langle 5, 0, -2 \rangle$, $\mathbf{b} = \langle 3, -1, 10 \rangle$

6. $\mathbf{a} = \langle s, 2s, 3s \rangle$, $\mathbf{b} = \langle t, -t, 5t \rangle$

7. $\mathbf{a} = \mathbf{i} - 2\mathbf{j} + 3\mathbf{k}$, $\mathbf{b} = 5\mathbf{i} + 9\mathbf{k}$

8. $\mathbf{a} = 4\mathbf{j} - 3\mathbf{k}$, $\mathbf{b} = 2\mathbf{i} + 4\mathbf{j} + 6\mathbf{k}$

9. $|\mathbf{a}| = 12$, $|\mathbf{b}| = 15$, the angle between **a** and **b** is $\pi/6$

10. $|\mathbf{a}| = 4$, $|\mathbf{b}| = 10$, the angle between **a** and **b** is 120°

11–12 If **u** is a unit vector, find $\mathbf{u} \cdot \mathbf{v}$ and $\mathbf{u} \cdot \mathbf{w}$.

11.

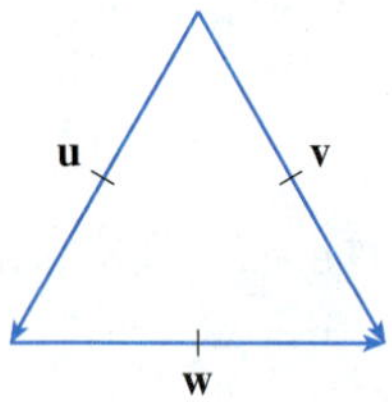

12.

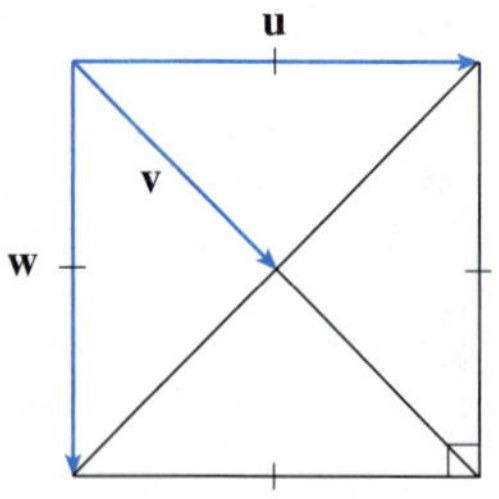

13. (a) Show that $\mathbf{i} \cdot \mathbf{j} = \mathbf{j} \cdot \mathbf{k} = \mathbf{k} \cdot \mathbf{i} = 0$.
(b) Show that $\mathbf{i} \cdot \mathbf{i} = \mathbf{j} \cdot \mathbf{j} = \mathbf{k} \cdot \mathbf{k} = 1$.

14. A street vendor sells a hamburgers, b hot dogs, and c soft drinks on a given day. He charges \$2 for a hamburger, \$1.50 for a hot dog, and \$1 for a soft drink. If $\mathbf{A} = \langle a, b, c \rangle$ and $\mathbf{P} = \langle 2, 1.5, 1 \rangle$, what is the meaning of the dot product $\mathbf{A} \cdot \mathbf{P}$?

15–20 Find the angle between the vectors. (First find an exact expression and then approximate to the nearest degree.)

15. $\mathbf{a} = \langle 3, 4 \rangle$, $\mathbf{b} = \langle 5, 12 \rangle$

16. $\mathbf{a} = \langle \sqrt{3}, 1 \rangle$, $\mathbf{b} = \langle 0, 5 \rangle$

17. $\mathbf{a} = \langle 1, 2, 3 \rangle$, $\mathbf{b} = \langle 4, 0, -1 \rangle$

18. $\mathbf{a} = \langle 6, -3, 2 \rangle$, $\mathbf{b} = \langle 2, 1, -2 \rangle$

19. $\mathbf{a} = \mathbf{j} + \mathbf{k}$, $\mathbf{b} = \mathbf{i} + 2\mathbf{j} - 3\mathbf{k}$

20. $\mathbf{a} = 2\mathbf{i} - \mathbf{j} + \mathbf{k}$, $\mathbf{b} = 3\mathbf{i} + 2\mathbf{j} - \mathbf{k}$

21–22 Find, correct to the nearest degree, the three angles of the triangle with the given vertices.

21. $A(1, 0)$, $B(3, 6)$, $C(-1, 4)$

22. $D(0, 1, 1)$, $E(-2, 4, 3)$, $F(1, 2, -1)$

23–24 Determine whether the given vectors are orthogonal, parallel, or neither.

23. (a) $\mathbf{a} = \langle -5, 3, 7 \rangle$, $\mathbf{b} = \langle 6, -8, 2 \rangle$
(b) $\mathbf{a} = \langle 4, 6 \rangle$, $\mathbf{b} = \langle -3, 2 \rangle$
(c) $\mathbf{a} = -\mathbf{i} + 2\mathbf{j} + 5\mathbf{k}$, $\mathbf{b} = 3\mathbf{i} + 4\mathbf{j} - \mathbf{k}$
(d) $\mathbf{a} = 2\mathbf{i} + 6\mathbf{j} - 4\mathbf{k}$, $\mathbf{b} = -3\mathbf{i} - 9\mathbf{j} + 6\mathbf{k}$

24. (a) $\mathbf{u} = \langle -3, 9, 6 \rangle$, $\mathbf{v} = \langle 4, -12, -8 \rangle$
(b) $\mathbf{u} = \mathbf{i} - \mathbf{j} + 2\mathbf{k}$, $\mathbf{v} = 2\mathbf{i} - \mathbf{j} + \mathbf{k}$
(c) $\mathbf{u} = \langle a, b, c \rangle$, $\mathbf{v} = \langle -b, a, 0 \rangle$

25. Use vectors to decide whether the triangle with vertices $P(1, -3, -2)$, $Q(2, 0, -4)$, and $R(6, -2, -5)$ is right-angled.

26. For what values of b are the vectors $\langle -6, b, 2 \rangle$ and $\langle b, b^2, b \rangle$ orthogonal?

27. Find a unit vector that is orthogonal to both $\mathbf{i} + \mathbf{j}$ and $\mathbf{i} + \mathbf{k}$.

28. Find two unit vectors that make an angle of $60°$ with $\mathbf{v} = \langle 3, 4 \rangle$.

29–33 Find the direction cosines and direction angles of the vector. (Give the direction angles correct to the nearest degree.)

29. $\langle 3, 4, 5 \rangle$

30. $\langle 1, -2, -1 \rangle$

31. $2\mathbf{i} + 3\mathbf{j} - 6\mathbf{k}$

32. $2\mathbf{i} - \mathbf{j} + 2\mathbf{k}$

33. $\langle c, c, c \rangle$, where $c > 0$

34. If a vector has direction angles $\alpha = \pi/4$ and $\beta = \pi/3$, find the third direction angle γ.

35–40 Find the scalar and vector projections of $\mathbf{b}$ onto $\mathbf{a}$.

35. $\mathbf{a} = \langle 3, -4 \rangle$, $\mathbf{b} = \langle 5, 0 \rangle$

36. $\mathbf{a} = \langle 1, 2 \rangle$, $\mathbf{b} = \langle -4, 1 \rangle$

37. $\mathbf{a} = \langle 4, 2, 0 \rangle$, $\mathbf{b} = \langle 1, 1, 1 \rangle$

38. $\mathbf{a} = \langle -1, -2, 2 \rangle$, $\mathbf{b} = \langle 3, 3, 4 \rangle$

39. $\mathbf{a} = \mathbf{i} + \mathbf{k}$, $\mathbf{b} = \mathbf{i} - \mathbf{j}$

40. $\mathbf{a} = 2\mathbf{i} - 3\mathbf{j} + \mathbf{k}$, $\mathbf{b} = \mathbf{i} + 6\mathbf{j} - 2\mathbf{k}$

41. Show that the vector $\text{orth}_{\mathbf{a}}\,\mathbf{b} = \mathbf{b} - \text{proj}_{\mathbf{a}}\,\mathbf{b}$ is orthogonal to $\mathbf{a}$. (It is called an **orthogonal projection** of $\mathbf{b}$.)

42. For the vectors in Exercise 36, find $\text{orth}_{\mathbf{a}}\,\mathbf{b}$ and illustrate by drawing the vectors $\mathbf{a}$, $\mathbf{b}$, $\text{proj}_{\mathbf{a}}\,\mathbf{b}$, and $\text{orth}_{\mathbf{a}}\,\mathbf{b}$.

43. If $\mathbf{a} = \langle 3, 0, -1 \rangle$, find a vector $\mathbf{b}$ such that $\text{comp}_{\mathbf{a}}\,\mathbf{b} = 2$.

44. Suppose that $\mathbf{a}$ and $\mathbf{b}$ are nonzero vectors.
(a) Under what circumstances is $\text{comp}_{\mathbf{a}}\,\mathbf{b} = \text{comp}_{\mathbf{b}}\,\mathbf{a}$?
(b) Under what circumstances is $\text{proj}_{\mathbf{a}}\,\mathbf{b} = \text{proj}_{\mathbf{b}}\,\mathbf{a}$?

45. A constant force with vector representation $\mathbf{F} = 10\mathbf{i} + 18\mathbf{j} - 6\mathbf{k}$ moves an object along a straight line from the point $(2, 3, 0)$ to the point $(4, 9, 15)$. Find the work done if the distance is measured in meters and the magnitude of the force is measured in newtons.

46. Find the work done by a force of 20 lb acting in the direction N50°W in moving an object 4 ft due west.

47. A woman exerts a horizontal force of 25 lb on a crate as she pushes it up a ramp that is 10 ft long and inclined at an angle of 20° above the horizontal. Find the work done on the box.

48. A wagon is pulled a distance of 100 m along a horizontal path by a constant force of 50 N. The handle of the wagon is held at an angle of 30° above the horizontal. How much work is done?

49. Use a scalar projection to show that the distance from a point $P_1(x_1, y_1)$ to the line $ax + by + c = 0$ is

$$\frac{|ax_1 + by_1 + c|}{\sqrt{a^2 + b^2}}$$

Use this formula to find the distance from the point $(-2, 3)$ to the line $3x - 4y + 5 = 0$.

50. If $\mathbf{r} = \langle x, y, z \rangle$, $\mathbf{a} = \langle a_1, a_2, a_3 \rangle$, and $\mathbf{b} = \langle b_1, b_2, b_3 \rangle$, show that the vector equation $(\mathbf{r} - \mathbf{a}) \cdot (\mathbf{r} - \mathbf{b}) = 0$ represents a sphere, and find its center and radius.

51. Find the angle between a diagonal of a cube and one of its edges.

52. Find the angle between a diagonal of a cube and a diagonal of one of its faces.

53. A molecule of methane, CH_4, is structured with the four hydrogen atoms at the vertices of a regular tetrahedron and the carbon atom at the centroid. The *bond angle* is the angle formed by the H—C—H combination; it is the angle between the lines that join the carbon atom to two of the hydrogen atoms. Show that the bond angle is about 109.5°. [*Hint:* Take the vertices of the tetrahedron to be the points $(1, 0, 0)$, $(0, 1, 0)$,

(0, 0, 1), and (1, 1, 1) as shown in the figure. Then the centroid is $\left(\frac{1}{2}, \frac{1}{2}, \frac{1}{2}\right)$.]

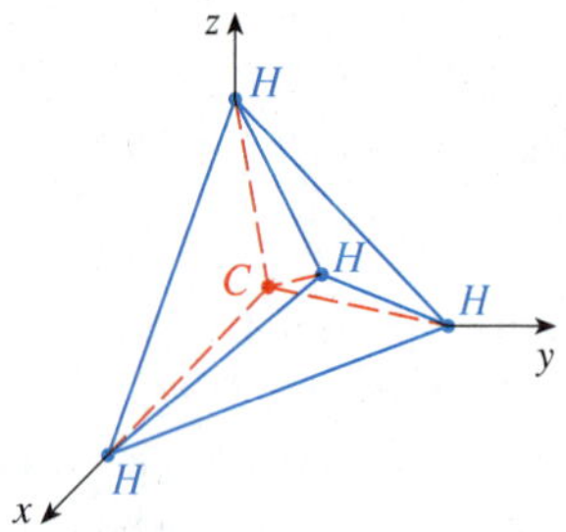

54. If $\mathbf{c} = |\mathbf{a}|\mathbf{b} + |\mathbf{b}|\mathbf{a}$, where **a**, **b**, and **c** are all nonzero vectors, show that **c** bisects the angle between **a** and **b**.

55. Prove Properties 2, 4, and 5 of the dot product (Theorem 2).

56. Suppose that all sides of a quadrilateral are equal in length and opposite sides are parallel. Use vector methods to show that the diagonals are perpendicular.

57. Use Theorem 3 to prove the *Cauchy-Schwarz Inequality:*

$$|\mathbf{a} \cdot \mathbf{b}| \leqslant |\mathbf{a}||\mathbf{b}|$$

58. The *Triangle Inequality for vectors* is

$$|\mathbf{a} + \mathbf{b}| \leqslant |\mathbf{a}| + |\mathbf{b}|$$

(a) Give a geometric interpretation of the Triangle Inequality.
(b) Use the Cauchy-Schwarz Inequality from Exercise 57 to prove the Triangle Inequality. [*Hint:* Use the fact that $|\mathbf{a} + \mathbf{b}|^2 = (\mathbf{a} + \mathbf{b}) \cdot (\mathbf{a} + \mathbf{b})$ and use Property 3 of the dot product.]

59. The Parallelogram Law states that

$$|\mathbf{a} + \mathbf{b}|^2 + |\mathbf{a} - \mathbf{b}|^2 = 2|\mathbf{a}|^2 + 2|\mathbf{b}|^2$$

(a) Give a geometric interpretation of the Parallelogram Law.
(b) Prove the Parallelogram Law. (See the hint in Exercise 58.)

12.4 The Cross Product

The **cross product** $\mathbf{a} \times \mathbf{b}$ of two vectors **a** and **b**, unlike the dot product, is a vector. For this reason it is also called the **vector product**. Note that $\mathbf{a} \times \mathbf{b}$ is defined only when **a** and **b** are *three-dimensional* vectors.

1 Definition If $\mathbf{a} = \langle a_1, a_2, a_3 \rangle$ and $\mathbf{b} = \langle b_1, b_2, b_3 \rangle$, then the **cross product** of **a** and **b** is the vector

$$\mathbf{a} \times \mathbf{b} = \langle a_2 b_3 - a_3 b_2, a_3 b_1 - a_1 b_3, a_1 b_2 - a_2 b_1 \rangle$$

This may seem like a strange way of defining a product. The reason for the particular form of Definition 1 is that the cross product defined in this way has many useful properties, as we will soon see. In particular, we will show that the vector $\mathbf{a} \times \mathbf{b}$ is perpendicular to both **a** and **b**.

In order to make Definition 1 easier to remember, we use the notation of determinants. A **determinant of order 2** is defined by

$$\begin{vmatrix} a & b \\ c & d \end{vmatrix} = ad - bc$$

For example,

$$\begin{vmatrix} 2 & 1 \\ -6 & 4 \end{vmatrix} = 2(4) - 1(-6) = 14$$

A **determinant of order 3** can be defined in terms of second-order determinants as follows:

$$\begin{vmatrix} a_1 & a_2 & a_3 \\ b_1 & b_2 & b_3 \\ c_1 & c_2 & c_3 \end{vmatrix} = a_1 \begin{vmatrix} b_2 & b_3 \\ c_2 & c_3 \end{vmatrix} - a_2 \begin{vmatrix} b_1 & b_3 \\ c_1 & c_3 \end{vmatrix} + a_3 \begin{vmatrix} b_1 & b_2 \\ c_1 & c_2 \end{vmatrix} \qquad \mathbf{2}$$

Observe that each term on the right side of Equation 2 involves a number a_i in the first row of the determinant, and a_i is multiplied by the second-order determinant obtained from the left side by deleting the row and column in which a_i appears. Notice also the minus sign in the second term. For example,

$$\begin{vmatrix} 1 & 2 & -1 \\ 3 & 0 & 1 \\ -5 & 4 & 2 \end{vmatrix} = 1\begin{vmatrix} 0 & 1 \\ 4 & 2 \end{vmatrix} - 2\begin{vmatrix} 3 & 1 \\ -5 & 2 \end{vmatrix} + (-1)\begin{vmatrix} 3 & 0 \\ -5 & 4 \end{vmatrix}$$

$$= 1(0 - 4) - 2(6 + 5) + (-1)(12 - 0) = -38$$

If we now rewrite Definition 1 using second-order determinants and the standard basis vectors $\mathbf{i}$, $\mathbf{j}$, and $\mathbf{k}$, we see that the cross product of the vectors $\mathbf{a} = a_1\mathbf{i} + a_2\mathbf{j} + a_3\mathbf{k}$ and $\mathbf{b} = b_1\mathbf{i} + b_2\mathbf{j} + b_3\mathbf{k}$ is

3 $$\mathbf{a} \times \mathbf{b} = \begin{vmatrix} a_2 & a_3 \\ b_2 & b_3 \end{vmatrix}\mathbf{i} - \begin{vmatrix} a_1 & a_3 \\ b_1 & b_3 \end{vmatrix}\mathbf{j} + \begin{vmatrix} a_1 & a_2 \\ b_1 & b_2 \end{vmatrix}\mathbf{k}$$

In view of the similarity between Equations 2 and 3, we often write

4 $$\mathbf{a} \times \mathbf{b} = \begin{vmatrix} \mathbf{i} & \mathbf{j} & \mathbf{k} \\ a_1 & a_2 & a_3 \\ b_1 & b_2 & b_3 \end{vmatrix}$$

Although the first row of the symbolic determinant in Equation 4 consists of vectors, if we expand it as if it were an ordinary determinant using the rule in Equation 2, we obtain Equation 3. The symbolic formula in Equation 4 is probably the easiest way of remembering and computing cross products.

EXAMPLE 1 If $\mathbf{a} = \langle 1, 3, 4 \rangle$ and $\mathbf{b} = \langle 2, 7, -5 \rangle$, then

$$\mathbf{a} \times \mathbf{b} = \begin{vmatrix} \mathbf{i} & \mathbf{j} & \mathbf{k} \\ 1 & 3 & 4 \\ 2 & 7 & -5 \end{vmatrix}$$

$$= \begin{vmatrix} 3 & 4 \\ 7 & -5 \end{vmatrix}\mathbf{i} - \begin{vmatrix} 1 & 4 \\ 2 & -5 \end{vmatrix}\mathbf{j} + \begin{vmatrix} 1 & 3 \\ 2 & 7 \end{vmatrix}\mathbf{k}$$

$$= (-15 - 28)\mathbf{i} - (-5 - 8)\mathbf{j} + (7 - 6)\mathbf{k} = -43\mathbf{i} + 13\mathbf{j} + \mathbf{k}$$

EXAMPLE 2 Show that $\mathbf{a} \times \mathbf{a} = \mathbf{0}$ for any vector $\mathbf{a}$ in V_3.

SOLUTION If $\mathbf{a} = \langle a_1, a_2, a_3 \rangle$, then

$$\mathbf{a} \times \mathbf{a} = \begin{vmatrix} \mathbf{i} & \mathbf{j} & \mathbf{k} \\ a_1 & a_2 & a_3 \\ a_1 & a_2 & a_3 \end{vmatrix}$$

$$= (a_2a_3 - a_3a_2)\mathbf{i} - (a_1a_3 - a_3a_1)\mathbf{j} + (a_1a_2 - a_2a_1)\mathbf{k}$$

$$= 0\mathbf{i} - 0\mathbf{j} + 0\mathbf{k} = \mathbf{0}$$

One of the most important properties of the cross product is given by the following theorem.

5 Theorem The vector $\mathbf{a} \times \mathbf{b}$ is orthogonal to both $\mathbf{a}$ and $\mathbf{b}$.

Proof In order to show that $\mathbf{a} \times \mathbf{b}$ is orthogonal to $\mathbf{a}$, we compute their dot product as follows:

$$\begin{aligned}(\mathbf{a} \times \mathbf{b}) \cdot \mathbf{a} &= \begin{vmatrix} a_2 & a_3 \\ b_2 & b_3 \end{vmatrix} a_1 - \begin{vmatrix} a_1 & a_3 \\ b_1 & b_3 \end{vmatrix} a_2 + \begin{vmatrix} a_1 & a_2 \\ b_1 & b_2 \end{vmatrix} a_3 \\ &= a_1(a_2b_3 - a_3b_2) - a_2(a_1b_3 - a_3b_1) + a_3(a_1b_2 - a_2b_1) \\ &= a_1a_2b_3 - a_1b_2a_3 - a_1a_2b_3 + b_1a_2a_3 + a_1b_2a_3 - b_1a_2a_3 \\ &= 0\end{aligned}$$

A similar computation shows that $(\mathbf{a} \times \mathbf{b}) \cdot \mathbf{b} = 0$. Therefore, $\mathbf{a} \times \mathbf{b}$ is orthogonal to both $\mathbf{a}$ and $\mathbf{b}$. ∎

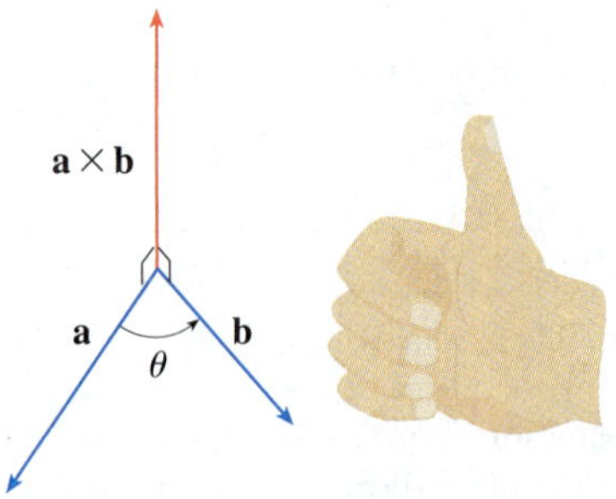

FIGURE 1

If $\mathbf{a}$ and $\mathbf{b}$ are represented by directed line segments with the same initial point (as in Figure 1), then Theorem 5 says that the cross product $\mathbf{a} \times \mathbf{b}$ points in a direction perpendicular to the plane through $\mathbf{a}$ and $\mathbf{b}$. It turns out that the direction of $\mathbf{a} \times \mathbf{b}$ is given by the *right-hand rule:* If the fingers of your right hand curl in the direction of a rotation (through an angle less than 180°) from $\mathbf{a}$ to $\mathbf{b}$, then your thumb points in the direction of $\mathbf{a} \times \mathbf{b}$.

Now that we know the direction of the vector $\mathbf{a} \times \mathbf{b}$, the remaining thing we need to complete its geometric description is its length $|\mathbf{a} \times \mathbf{b}|$. This is given by the following theorem.

6 Theorem If θ is the angle between $\mathbf{a}$ and $\mathbf{b}$ (so $0 \leq \theta \leq \pi$), then

$$|\mathbf{a} \times \mathbf{b}| = |\mathbf{a}||\mathbf{b}| \sin \theta$$

TEC Visual 12.4 shows how $\mathbf{a} \times \mathbf{b}$ changes as $\mathbf{b}$ changes.

Proof From the definitions of the cross product and length of a vector, we have

$$\begin{aligned}|\mathbf{a} \times \mathbf{b}|^2 &= (a_2b_3 - a_3b_2)^2 + (a_3b_1 - a_1b_3)^2 + (a_1b_2 - a_2b_1)^2 \\ &= a_2^2b_3^2 - 2a_2a_3b_2b_3 + a_3^2b_2^2 + a_3^2b_1^2 - 2a_1a_3b_1b_3 + a_1^2b_3^2 \\ &\qquad + a_1^2b_2^2 - 2a_1a_2b_1b_2 + a_2^2b_1^2 \\ &= (a_1^2 + a_2^2 + a_3^2)(b_1^2 + b_2^2 + b_3^2) - (a_1b_1 + a_2b_2 + a_3b_3)^2 \\ &= |\mathbf{a}|^2|\mathbf{b}|^2 - (\mathbf{a} \cdot \mathbf{b})^2 \\ &= |\mathbf{a}|^2|\mathbf{b}|^2 - |\mathbf{a}|^2|\mathbf{b}|^2\cos^2\theta \qquad \text{(by Theorem 12.3.3)} \\ &= |\mathbf{a}|^2|\mathbf{b}|^2(1 - \cos^2\theta) \\ &= |\mathbf{a}|^2|\mathbf{b}|^2\sin^2\theta\end{aligned}$$

Taking square roots and observing that $\sqrt{\sin^2\theta} = \sin\theta$ because $\sin\theta \geq 0$ when $0 \leq \theta \leq \pi$, we have

$$|\mathbf{a} \times \mathbf{b}| = |\mathbf{a}||\mathbf{b}| \sin \theta$$ ∎

Geometric characterization of $\mathbf{a} \times \mathbf{b}$

Since a vector is completely determined by its magnitude and direction, we can now say that $\mathbf{a} \times \mathbf{b}$ is the vector that is perpendicular to both $\mathbf{a}$ and $\mathbf{b}$, whose orientation is determined by the right-hand rule, and whose length is $|\mathbf{a}||\mathbf{b}|\sin\theta$. In fact, that is exactly how physicists *define* $\mathbf{a} \times \mathbf{b}$.

7 Corollary Two nonzero vectors **a** and **b** are parallel if and only if

$$\mathbf{a} \times \mathbf{b} = \mathbf{0}$$

Proof Two nonzero vectors **a** and **b** are parallel if and only if $\theta = 0$ or π. In either case $\sin\theta = 0$, so $|\mathbf{a} \times \mathbf{b}| = 0$ and therefore $\mathbf{a} \times \mathbf{b} = \mathbf{0}$.

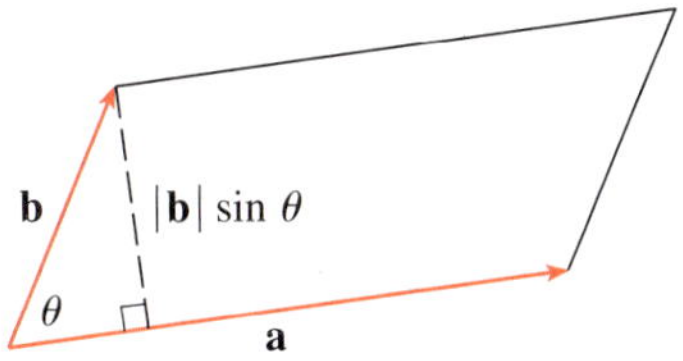

FIGURE 2

The geometric interpretation of Theorem 6 can be seen by looking at Figure 2. If **a** and **b** are represented by directed line segments with the same initial point, then they determine a parallelogram with base $|\mathbf{a}|$, altitude $|\mathbf{b}|\sin\theta$, and area

$$A = |\mathbf{a}|\big(|\mathbf{b}|\sin\theta\big) = |\mathbf{a} \times \mathbf{b}|$$

Thus, we have the following way of interpreting the magnitude of a cross product.

The length of the cross product $\mathbf{a} \times \mathbf{b}$ is equal to the area of the parallelogram determined by **a** and **b**.

EXAMPLE 3 Find a vector perpendicular to the plane that passes through the points $P(1, 4, 6)$, $Q(-2, 5, -1)$, and $R(1, -1, 1)$.

SOLUTION The vector $\overrightarrow{PQ} \times \overrightarrow{PR}$ is perpendicular to both $\overrightarrow{PQ}$ and $\overrightarrow{PR}$ and is therefore perpendicular to the plane through P, Q, and R. We know from (12.2.1) that

$$\overrightarrow{PQ} = (-2 - 1)\,\mathbf{i} + (5 - 4)\,\mathbf{j} + (-1 - 6)\,\mathbf{k} = -3\mathbf{i} + \mathbf{j} - 7\mathbf{k}$$
$$\overrightarrow{PR} = (1 - 1)\,\mathbf{i} + (-1 - 4)\,\mathbf{j} + (1 - 6)\,\mathbf{k} = -5\mathbf{j} - 5\mathbf{k}$$

We compute the cross product of these vectors:

$$\overrightarrow{PQ} \times \overrightarrow{PR} = \begin{vmatrix} \mathbf{i} & \mathbf{j} & \mathbf{k} \\ -3 & 1 & -7 \\ 0 & -5 & -5 \end{vmatrix}$$

$$= (-5 - 35)\,\mathbf{i} - (15 - 0)\,\mathbf{j} + (15 - 0)\,\mathbf{k} = -40\mathbf{i} - 15\mathbf{j} + 15\mathbf{k}$$

So the vector $\langle -40, -15, 15 \rangle$ is perpendicular to the given plane. Any nonzero scalar multiple of this vector, such as $\langle -8, -3, 3 \rangle$, is also perpendicular to the plane.

EXAMPLE 4 Find the area of the triangle with vertices $P(1, 4, 6)$, $Q(-2, 5, -1)$, and $R(1, -1, 1)$.

SOLUTION In Example 3 we computed that $\overrightarrow{PQ} \times \overrightarrow{PR} = \langle -40, -15, 15 \rangle$. The area of the parallelogram with adjacent sides PQ and PR is the length of this cross product:

$$\left|\overrightarrow{PQ} \times \overrightarrow{PR}\right| = \sqrt{(-40)^2 + (-15)^2 + 15^2} = 5\sqrt{82}$$

The area A of the triangle PQR is half the area of this parallelogram, that is, $\frac{5}{2}\sqrt{82}$.

If we apply Theorems 5 and 6 to the standard basis vectors **i**, **j**, and **k** using $\theta = \pi/2$, we obtain

$$\mathbf{i} \times \mathbf{j} = \mathbf{k} \qquad \mathbf{j} \times \mathbf{k} = \mathbf{i} \qquad \mathbf{k} \times \mathbf{i} = \mathbf{j}$$

$$\mathbf{j} \times \mathbf{i} = -\mathbf{k} \qquad \mathbf{k} \times \mathbf{j} = -\mathbf{i} \qquad \mathbf{i} \times \mathbf{k} = -\mathbf{j}$$

Observe that

$$\mathbf{i} \times \mathbf{j} \neq \mathbf{j} \times \mathbf{i}$$

Thus, the cross product is not commutative. Also

$$\mathbf{i} \times (\mathbf{i} \times \mathbf{j}) = \mathbf{i} \times \mathbf{k} = -\mathbf{j}$$

whereas

$$(\mathbf{i} \times \mathbf{i}) \times \mathbf{j} = \mathbf{0} \times \mathbf{j} = \mathbf{0}$$

So the associative law for multiplication does not usually hold; that is, in general,

$$(\mathbf{a} \times \mathbf{b}) \times \mathbf{c} \neq \mathbf{a} \times (\mathbf{b} \times \mathbf{c})$$

However, some of the usual laws of algebra *do* hold for cross products. The following theorem summarizes the properties of vector products.

8 Theorem If **a**, **b**, and **c** are vectors and c is a scalar, then

1. $\mathbf{a} \times \mathbf{b} = -\mathbf{b} \times \mathbf{a}$
2. $(c\mathbf{a}) \times \mathbf{b} = c(\mathbf{a} \times \mathbf{b}) = \mathbf{a} \times (c\mathbf{b})$
3. $\mathbf{a} \times (\mathbf{b} + \mathbf{c}) = \mathbf{a} \times \mathbf{b} + \mathbf{a} \times \mathbf{c}$
4. $(\mathbf{a} + \mathbf{b}) \times \mathbf{c} = \mathbf{a} \times \mathbf{c} + \mathbf{b} \times \mathbf{c}$
5. $\mathbf{a} \cdot (\mathbf{b} \times \mathbf{c}) = (\mathbf{a} \times \mathbf{b}) \cdot \mathbf{c}$
6. $\mathbf{a} \times (\mathbf{b} \times \mathbf{c}) = (\mathbf{a} \cdot \mathbf{c})\mathbf{b} - (\mathbf{a} \cdot \mathbf{b})\mathbf{c}$

These properties can be proved by writing the vectors in terms of their components and using the definition of a cross product. We give the proof of Property 5 and leave the remaining proofs as exercises.

Proof of Property 5 If $\mathbf{a} = \langle a_1, a_2, a_3 \rangle$, $\mathbf{b} = \langle b_1, b_2, b_3 \rangle$, and $\mathbf{c} = \langle c_1, c_2, c_3 \rangle$, then

9
$$\begin{aligned}
\mathbf{a} \cdot (\mathbf{b} \times \mathbf{c}) &= a_1(b_2c_3 - b_3c_2) + a_2(b_3c_1 - b_1c_3) + a_3(b_1c_2 - b_2c_1) \\
&= a_1b_2c_3 - a_1b_3c_2 + a_2b_3c_1 - a_2b_1c_3 + a_3b_1c_2 - a_3b_2c_1 \\
&= (a_2b_3 - a_3b_2)c_1 + (a_3b_1 - a_1b_3)c_2 + (a_1b_2 - a_2b_1)c_3 \\
&= (\mathbf{a} \times \mathbf{b}) \cdot \mathbf{c}
\end{aligned}$$

The product $\mathbf{a} \cdot (\mathbf{b} \times \mathbf{c})$ that occurs in Property 5 is called the **scalar triple product** of the vectors **a**, **b**, and **c**. Notice from Equation 9 that we can write the scalar triple product as a determinant:

10
$$\mathbf{a} \cdot (\mathbf{b} \times \mathbf{c}) = \begin{vmatrix} a_1 & a_2 & a_3 \\ b_1 & b_2 & b_3 \\ c_1 & c_2 & c_3 \end{vmatrix}$$

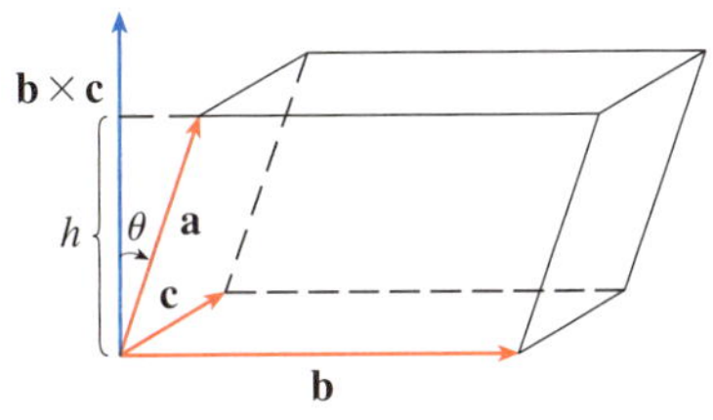

FIGURE 3

The geometric significance of the scalar triple product can be seen by considering the parallelepiped determined by the vectors **a**, **b**, and **c** (Figure 3). The area of the base parallelogram is $A = |\mathbf{b} \times \mathbf{c}|$. If θ is the angle between **a** and $\mathbf{b} \times \mathbf{c}$, then the height h of the parallelepiped is $h = |\mathbf{a}||\cos\theta|$. (We must use $|\cos\theta|$ instead of $\cos\theta$ in case $\theta > \pi/2$.) Therefore, the volume of the parallelepiped is

$$V = Ah = |\mathbf{b} \times \mathbf{c}||\mathbf{a}||\cos\theta| = |\mathbf{a} \cdot (\mathbf{b} \times \mathbf{c})|$$

Thus, we have proved the following formula.

11 The volume of the parallelepiped determined by the vectors **a**, **b**, and **c** is the magnitude of their scalar triple product:

$$V = |\mathbf{a} \cdot (\mathbf{b} \times \mathbf{c})|$$

If we use the formula in (11) and discover that the volume of the parallelepiped determined by **a**, **b**, and **c** is 0, then the vectors must lie in the same plane; that is, they are **coplanar**.

EXAMPLE 5 Use the scalar triple product to show that the vectors $\mathbf{a} = \langle 1, 4, -7 \rangle$, $\mathbf{b} = \langle 2, -1, 4 \rangle$, and $\mathbf{c} = \langle 0, -9, 18 \rangle$ are coplanar.

SOLUTION We use Equation 10 to compute their scalar triple product:

$$\mathbf{a} \cdot (\mathbf{b} \times \mathbf{c}) = \begin{vmatrix} 1 & 4 & -7 \\ 2 & -1 & 4 \\ 0 & -9 & 18 \end{vmatrix}$$

$$= 1\begin{vmatrix} -1 & 4 \\ -9 & 18 \end{vmatrix} - 4\begin{vmatrix} 2 & 4 \\ 0 & 18 \end{vmatrix} - 7\begin{vmatrix} 2 & -1 \\ 0 & -9 \end{vmatrix}$$

$$= 1(18) - 4(36) - 7(-18) = 0$$

Therefore, by (11) the volume of the parallelepiped determined by **a**, **b**, and **c** is 0. This means that **a**, **b**, and **c** are coplanar.

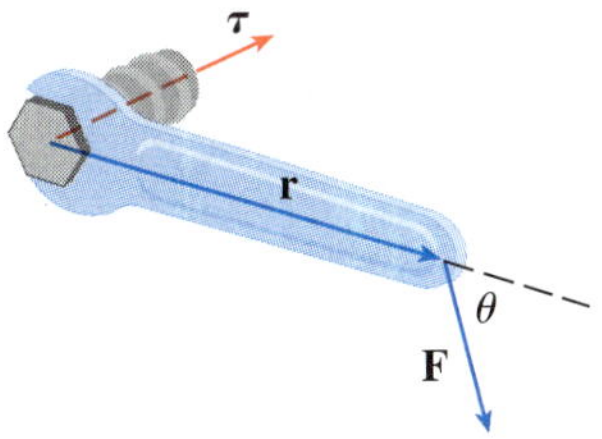

FIGURE 4

The idea of a cross product occurs often in physics. In particular, we consider a force **F** acting on a rigid body at a point given by a position vector **r**. (For instance, if we tighten a bolt by applying a force to a wrench as in Figure 4, we produce a turning effect.) The **torque** $\boldsymbol{\tau}$ (relative to the origin) is defined to be the cross product of the position and force vectors

$$\boldsymbol{\tau} = \mathbf{r} \times \mathbf{F}$$

and measures the tendency of the body to rotate about the origin. The direction of the torque vector indicates the axis of rotation. According to Theorem 6, the magnitude of the torque vector is

$$|\boldsymbol{\tau}| = |\mathbf{r} \times \mathbf{F}| = |\mathbf{r}||\mathbf{F}|\sin\theta$$

where θ is the angle between the position and force vectors. Observe that the only component of **F** that can cause a rotation is the one perpendicular to **r**, that is, $|\mathbf{F}|\sin\theta$. The magnitude of the torque is equal to the area of the parallelogram determined by **r** and **F**.

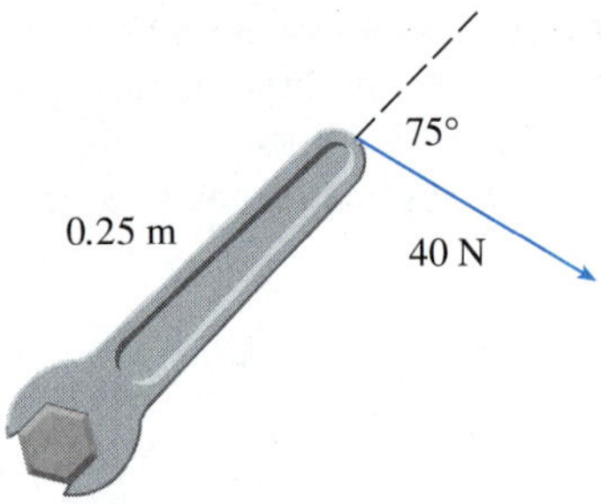

FIGURE 5

EXAMPLE 6 A bolt is tightened by applying a 40-N force to a 0.25-m wrench as shown in Figure 5. Find the magnitude of the torque about the center of the bolt.

SOLUTION The magnitude of the torque vector is

$$|\boldsymbol{\tau}| = |\mathbf{r} \times \mathbf{F}| = |\mathbf{r}||\mathbf{F}| \sin 75° = (0.25)(40) \sin 75°$$
$$= 10 \sin 75° \approx 9.66 \text{ N·m}$$

If the bolt is right-threaded, then the torque vector itself is

$$\boldsymbol{\tau} = |\boldsymbol{\tau}|\, \mathbf{n} \approx 9.66\, \mathbf{n}$$

where **n** is a unit vector directed down into the page.

12.4 Exercises

1–7 Find the cross product $\mathbf{a} \times \mathbf{b}$ and verify that it is orthogonal to both **a** and **b**.

1. $\mathbf{a} = \langle 1, 2, 0 \rangle, \quad \mathbf{b} = \langle 0, 3, 1 \rangle$

2. $\mathbf{a} = \langle 5, 1, 4 \rangle, \quad \mathbf{b} = \langle -1, 0, 2 \rangle$

3. $\mathbf{a} = 2\mathbf{i} + \mathbf{j} - \mathbf{k}, \quad \mathbf{b} = \mathbf{j} + 2\mathbf{k}$

4. $\mathbf{a} = \mathbf{i} - \mathbf{j} + \mathbf{k}, \quad \mathbf{b} = \mathbf{i} + \mathbf{j} + \mathbf{k}$

5. $\mathbf{a} = 3\mathbf{i} + 2\mathbf{j} + 4\mathbf{k}, \quad \mathbf{b} = \mathbf{i} - 2\mathbf{j} - 3\mathbf{k}$

6. $\mathbf{a} = \mathbf{i} + e^t\mathbf{j} + e^{-t}\mathbf{k}, \quad \mathbf{b} = 2\mathbf{i} + e^t\mathbf{j} - e^{-t}\mathbf{k}$

7. $\mathbf{a} = \langle t, t^2, t^3 \rangle, \quad \mathbf{b} = \langle 1, 2t, 3t^2 \rangle$

8. If $\mathbf{a} = \mathbf{i} - 2\mathbf{k}$ and $\mathbf{b} = \mathbf{j} + \mathbf{k}$, find $\mathbf{a} \times \mathbf{b}$. Sketch **a**, **b**, and $\mathbf{a} \times \mathbf{b}$ as vectors starting at the origin.

9. State whether each expression is meaningful. If not, explain why. If so, state whether it is a vector or a scalar.
   (a) $\mathbf{a} \cdot (\mathbf{b} \times \mathbf{c})$ (b) $\mathbf{a} \times (\mathbf{b} \cdot \mathbf{c})$
   (c) $\mathbf{a} \times (\mathbf{b} \times \mathbf{c})$ (d) $(\mathbf{a} \cdot \mathbf{b}) \times \mathbf{c}$
   (e) $(\mathbf{a} \cdot \mathbf{b}) \times (\mathbf{c} \cdot \mathbf{d})$ (f) $(\mathbf{a} \times \mathbf{b}) \cdot (\mathbf{c} \times \mathbf{d})$

10–11 Find $|\mathbf{u} \times \mathbf{v}|$ and determine whether $\mathbf{u} \times \mathbf{v}$ is directed into the page or out of the page.

10.

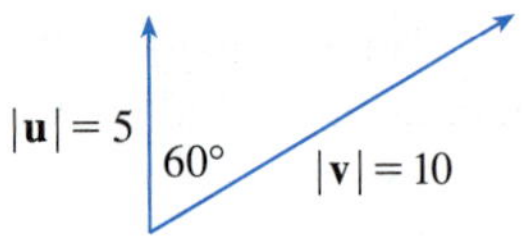

11.

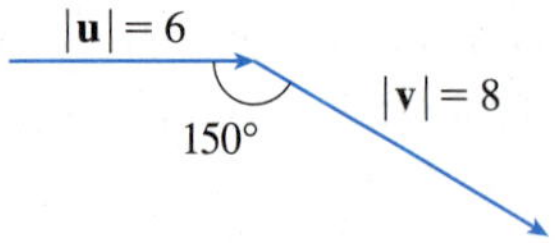

12. The figure shows a vector **a** in the xy-plane and a vector **b** in the direction of **k**. Their lengths are $|\mathbf{a}| = 3$ and $|\mathbf{b}| = 2$.
   (a) Find $|\mathbf{a} \times \mathbf{b}|$.
   (b) Use the right-hand rule to decide whether the components of $\mathbf{a} \times \mathbf{b}$ are positive, negative, or 0.

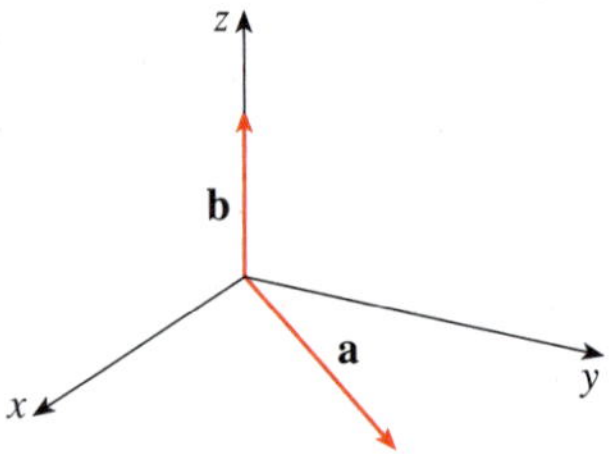

13. If $\mathbf{a} = \langle 1, 2, 1 \rangle$ and $\mathbf{b} = \langle 0, 1, 3 \rangle$, find $\mathbf{a} \times \mathbf{b}$ and $\mathbf{b} \times \mathbf{a}$.

14. If $\mathbf{a} = \langle 3, 1, 2 \rangle$, $\mathbf{b} = \langle -1, 1, 0 \rangle$, and $\mathbf{c} = \langle 0, 0, -4 \rangle$, show that $\mathbf{a} \times (\mathbf{b} \times \mathbf{c}) \neq (\mathbf{a} \times \mathbf{b}) \times \mathbf{c}$.

15. Find two unit vectors orthogonal to both $\langle 1, -1, 1 \rangle$ and $\langle 0, 4, 4 \rangle$.

16. Find two unit vectors orthogonal to both $\mathbf{i} + \mathbf{j} + \mathbf{k}$ and $2\mathbf{i} + \mathbf{k}$.

17. Show that $\mathbf{0} \times \mathbf{a} = \mathbf{0} = \mathbf{a} \times \mathbf{0}$ for any vector **a** in V_3.

18. Show that $(\mathbf{a} \times \mathbf{b}) \cdot \mathbf{b} = 0$ for all vectors **a** and **b** in V_3.

19. Prove Property 1 of Theorem 8.

20. Prove Property 2 of Theorem 8.

21. Prove Property 3 of Theorem 8.

22. Prove Property 4 of Theorem 8.

23. Find the area of the parallelogram with vertices $A(-2, 1)$, $B(0, 4)$, $C(4, 2)$, and $D(2, -1)$.

24. Find the area of the parallelogram with vertices $K(1, 2, 3)$, $L(1, 3, 6)$, $M(3, 8, 6)$, and $N(3, 7, 3)$.

25–28 (a) Find a vector orthogonal to the plane through the points P, Q, and R, and (b) find the area of triangle PQR.

25. $P(1, 0, 0), \quad Q(0, 2, 0), \quad R(0, 0, 3)$

26. $P(2, 1, 5)$, $Q(-1, 3, 4)$, $R(3, 0, 6)$

27. $P(0, -2, 0)$, $Q(4, 1, -2)$, $R(5, 3, 1)$

28. $P(2, 0, -3)$, $Q(3, 1, 0)$, $R(5, 2, 2)$

29–30 Find the volume of the parallelepiped determined by the vectors **a**, **b**, and **c**.

29. $\mathbf{a} = \langle 6, 3, -1 \rangle$, $\mathbf{b} = \langle 0, 1, 2 \rangle$, $\mathbf{c} = \langle 4, -2, 5 \rangle$

30. $\mathbf{a} = \mathbf{i} + \mathbf{j} - \mathbf{k}$, $\mathbf{b} = \mathbf{i} - \mathbf{j} + \mathbf{k}$, $\mathbf{c} = -\mathbf{i} + \mathbf{j} + \mathbf{k}$

31–32 Find the volume of the parallelepiped with adjacent edges PQ, PR, and PS.

31. $P(2, 0, -1)$, $Q(4, 1, 0)$, $R(3, -1, 1)$, $S(2, -2, 2)$

32. $P(0, 1, 2)$, $Q(2, 4, 5)$, $R(-1, 0, 1)$, $S(6, -1, 4)$

33. Use the scalar triple product to verify that the vectors $\mathbf{a} = 2\mathbf{i} + 3\mathbf{j} + \mathbf{k}$, $\mathbf{b} = \mathbf{i} - \mathbf{j}$, and $\mathbf{c} = 7\mathbf{i} + 3\mathbf{j} + 2\mathbf{k}$ are coplanar.

34. Use the scalar triple product to determine whether the points $P(1, 0, 1)$, $Q(2, 4, 6)$, $R(3, -1, 2)$, and $S(6, 2, 8)$ lie in the same plane.

35. A bicycle pedal is pushed by a foot with a 60-N force as shown. The shaft of the pedal is 18 cm long. Find the magnitude of the torque about P.

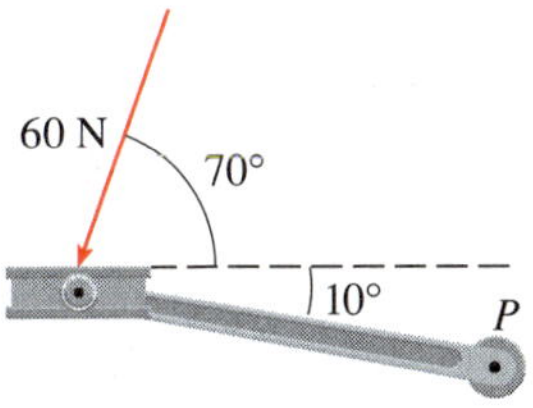

36. Find the magnitude of the torque about P if a 36-lb force is applied as shown.

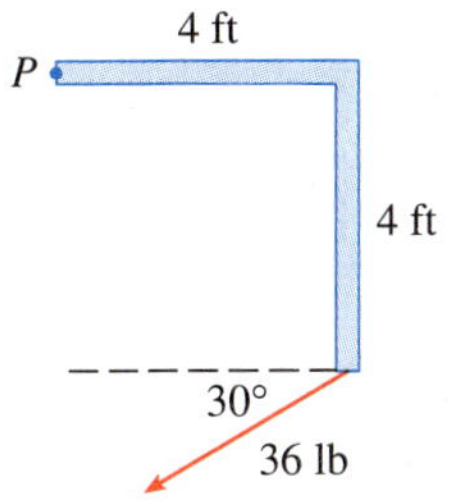

37. A wrench 30 cm long lies along the positive y-axis and grips a bolt at the origin. A force is applied in the direction $\langle 0, 3, -4 \rangle$ at the end of the wrench. Find the magnitude of the force needed to supply 100 N·m of torque to the bolt.

38. Let $\mathbf{v} = 5\mathbf{j}$ and let $\mathbf{u}$ be a vector with length 3 that starts at the origin and rotates in the xy-plane. Find the maximum and minimum values of the length of the vector $\mathbf{u} \times \mathbf{v}$. In what direction does $\mathbf{u} \times \mathbf{v}$ point?

39. (a) Let P be a point not on the line L that passes through the points Q and R. Show that the distance d from the point P to the line L is

$$d = \frac{|\mathbf{a} \times \mathbf{b}|}{|\mathbf{a}|}$$

where $\mathbf{a} = \overrightarrow{QR}$ and $\mathbf{b} = \overrightarrow{QP}$.

(b) Use the formula in part (a) to find the distance from the point $P(1, 1, 1)$ to the line through $Q(0, 6, 8)$ and $R(-1, 4, 7)$.

40. (a) Let P be a point not on the plane that passes through the points Q, R, and S. Show that the distance d from P to the plane is

$$d = \frac{|(\mathbf{a} \times \mathbf{b}) \cdot \mathbf{c}|}{|\mathbf{a} \times \mathbf{b}|}$$

where $\mathbf{a} = \overrightarrow{QR}$, $\mathbf{b} = \overrightarrow{QS}$, and $\mathbf{c} = \overrightarrow{QP}$.

(b) Use the formula in part (a) to find the distance from the point $P(2, 1, 4)$ to the plane through the points $Q(1, 0, 0)$, $R(0, 2, 0)$, and $S(0, 0, 3)$.

41. Prove that $(\mathbf{a} - \mathbf{b}) \times (\mathbf{a} + \mathbf{b}) = 2(\mathbf{a} \times \mathbf{b})$.

42. Prove part 6 of Theorem 8, that is,

$$\mathbf{a} \times (\mathbf{b} \times \mathbf{c}) = (\mathbf{a} \cdot \mathbf{c})\mathbf{b} - (\mathbf{a} \cdot \mathbf{b})\mathbf{c}$$

43. Use Exercise 42 to prove that

$$\mathbf{a} \times (\mathbf{b} \times \mathbf{c}) + \mathbf{b} \times (\mathbf{c} \times \mathbf{a}) + \mathbf{c} \times (\mathbf{a} \times \mathbf{b}) = \mathbf{0}$$

44. Prove that

$$(\mathbf{a} \times \mathbf{b}) \cdot (\mathbf{c} \times \mathbf{d}) = \begin{vmatrix} \mathbf{a} \cdot \mathbf{c} & \mathbf{b} \cdot \mathbf{c} \\ \mathbf{a} \cdot \mathbf{d} & \mathbf{b} \cdot \mathbf{d} \end{vmatrix}$$

45. Suppose that $\mathbf{a} \neq \mathbf{0}$.
(a) If $\mathbf{a} \cdot \mathbf{b} = \mathbf{a} \cdot \mathbf{c}$, does it follow that $\mathbf{b} = \mathbf{c}$?
(b) If $\mathbf{a} \times \mathbf{b} = \mathbf{a} \times \mathbf{c}$, does it follow that $\mathbf{b} = \mathbf{c}$?
(c) If $\mathbf{a} \cdot \mathbf{b} = \mathbf{a} \cdot \mathbf{c}$ and $\mathbf{a} \times \mathbf{b} = \mathbf{a} \times \mathbf{c}$, does it follow that $\mathbf{b} = \mathbf{c}$?

46. If $\mathbf{v}_1$, $\mathbf{v}_2$, and $\mathbf{v}_3$ are noncoplanar vectors, let

$$\mathbf{k}_1 = \frac{\mathbf{v}_2 \times \mathbf{v}_3}{\mathbf{v}_1 \cdot (\mathbf{v}_2 \times \mathbf{v}_3)} \qquad \mathbf{k}_2 = \frac{\mathbf{v}_3 \times \mathbf{v}_1}{\mathbf{v}_1 \cdot (\mathbf{v}_2 \times \mathbf{v}_3)}$$

$$\mathbf{k}_3 = \frac{\mathbf{v}_1 \times \mathbf{v}_2}{\mathbf{v}_1 \cdot (\mathbf{v}_2 \times \mathbf{v}_3)}$$

(These vectors occur in the study of crystallography. Vectors of the form $n_1\mathbf{v}_1 + n_2\mathbf{v}_2 + n_3\mathbf{v}_3$, where each n_i is an integer, form a *lattice* for a crystal. Vectors written similarly in terms of $\mathbf{k}_1$, $\mathbf{k}_2$, and $\mathbf{k}_3$ form the *reciprocal lattice*.)
(a) Show that $\mathbf{k}_i$ is perpendicular to $\mathbf{v}_j$ if $i \neq j$.
(b) Show that $\mathbf{k}_i \cdot \mathbf{v}_i = 1$ for $i = 1, 2, 3$.
(c) Show that $\mathbf{k}_1 \cdot (\mathbf{k}_2 \times \mathbf{k}_3) = \dfrac{1}{\mathbf{v}_1 \cdot (\mathbf{v}_2 \times \mathbf{v}_3)}$.

DISCOVERY PROJECT

The Geometry of a Tetrahedron

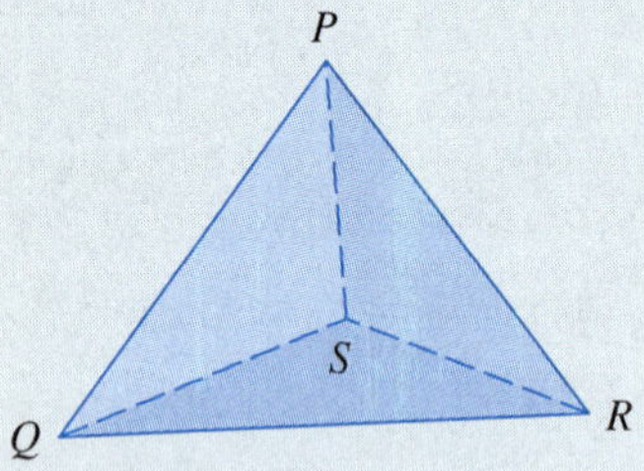

A tetrahedron is a solid with four vertices, P, Q, R, and S, and four triangular faces as shown in the figure.

1. Let $\mathbf{v}_1$, $\mathbf{v}_2$, $\mathbf{v}_3$, and $\mathbf{v}_4$ be vectors with lengths equal to the areas of the faces opposite the vertices P, Q, R, and S, respectively, and directions perpendicular to the respective faces and pointing outward. Show that

$$\mathbf{v}_1 + \mathbf{v}_2 + \mathbf{v}_3 + \mathbf{v}_4 = \mathbf{0}$$

2. The volume V of a tetrahedron is one-third the distance from a vertex to the opposite face, times the area of that face.

(a) Find a formula for the volume of a tetrahedron in terms of the coordinates of its vertices P, Q, R, and S.

(b) Find the volume of the tetrahedron whose vertices are $P(1, 1, 1)$, $Q(1, 2, 3)$, $R(1, 1, 2)$, and $S(3, -1, 2)$.

3. Suppose the tetrahedron in the figure has a trirectangular vertex S. (This means that the three angles at S are all right angles.) Let A, B, and C be the areas of the three faces that meet at S, and let D be the area of the opposite face PQR. Using the result of Problem 1, or otherwise, show that

$$D^2 = A^2 + B^2 + C^2$$

(This is a three-dimensional version of the Pythagorean Theorem.)

12.5 Equations of Lines and Planes

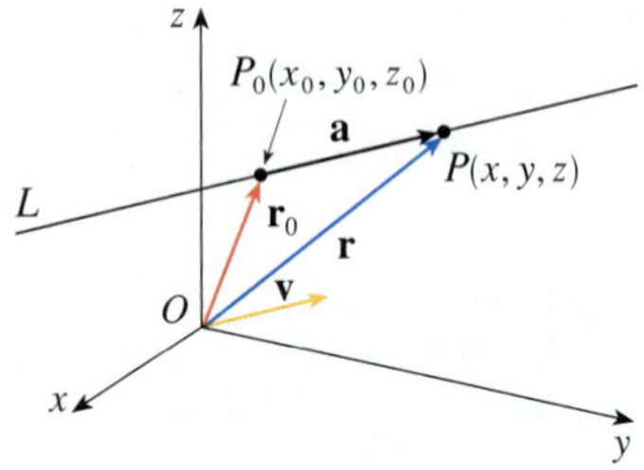

FIGURE 1

A line in the xy-plane is determined when a point on the line and the direction of the line (its slope or angle of inclination) are given. The equation of the line can then be written using the point-slope form.

Likewise, a line L in three-dimensional space is determined when we know a point $P_0(x_0, y_0, z_0)$ on L and the direction of L. In three dimensions the direction of a line is conveniently described by a vector, so we let $\mathbf{v}$ be a vector parallel to L. Let $P(x, y, z)$ be an arbitrary point on L and let $\mathbf{r}_0$ and $\mathbf{r}$ be the position vectors of P_0 and P (that is, they have representations $\overrightarrow{OP_0}$ and $\overrightarrow{OP}$). If $\mathbf{a}$ is the vector with representation $\overrightarrow{P_0P}$, as in Figure 1, then the Triangle Law for vector addition gives $\mathbf{r} = \mathbf{r}_0 + \mathbf{a}$. But, since $\mathbf{a}$ and $\mathbf{v}$ are parallel vectors, there is a scalar t such that $\mathbf{a} = t\mathbf{v}$. Thus

$$\boxed{1} \qquad \mathbf{r} = \mathbf{r}_0 + t\mathbf{v}$$

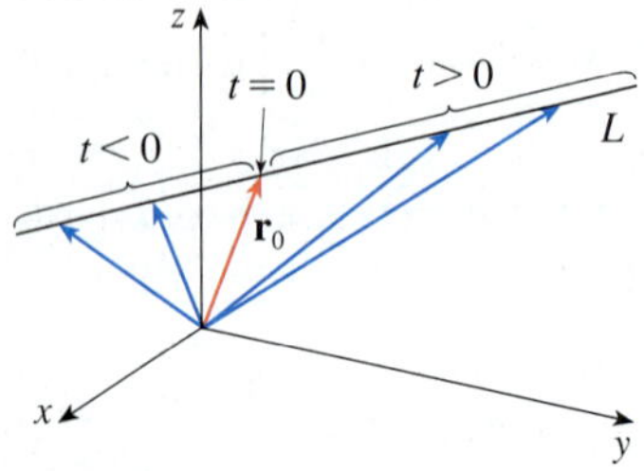

FIGURE 2

which is a **vector equation** of L. Each value of the **parameter** t gives the position vector $\mathbf{r}$ of a point on L. In other words, as t varies, the line is traced out by the tip of the vector $\mathbf{r}$. As Figure 2 indicates, positive values of t correspond to points on L that lie on one side of P_0, whereas negative values of t correspond to points that lie on the other side of P_0.

If the vector $\mathbf{v}$ that gives the direction of the line L is written in component form as $\mathbf{v} = \langle a, b, c \rangle$, then we have $t\mathbf{v} = \langle ta, tb, tc \rangle$. We can also write $\mathbf{r} = \langle x, y, z \rangle$ and $\mathbf{r}_0 = \langle x_0, y_0, z_0 \rangle$, so the vector equation (1) becomes

$$\langle x, y, z \rangle = \langle x_0 + ta, y_0 + tb, z_0 + tc \rangle$$

Two vectors are equal if and only if corresponding components are equal. Therefore, we have the three scalar equations:

2
$$x = x_0 + at \qquad y = y_0 + bt \qquad z = z_0 + ct$$

where $t \in \mathbb{R}$. These equations are called **parametric equations** of the line L through the point $P_0(x_0, y_0, z_0)$ and parallel to the vector $\mathbf{v} = \langle a, b, c \rangle$. Each value of the parameter t gives a point (x, y, z) on L.

|||| Figure 3 shows the line L in Example 1 and its relation to the given point and to the vector that gives its direction.

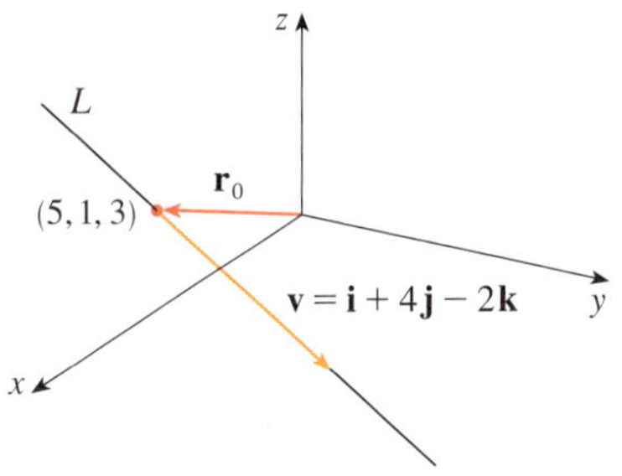

FIGURE 3

EXAMPLE 1

(a) Find a vector equation and parametric equations for the line that passes through the point (5, 1, 3) and is parallel to the vector $\mathbf{i} + 4\mathbf{j} - 2\mathbf{k}$.
(b) Find two other points on the line.

SOLUTION

(a) Here $\mathbf{r}_0 = \langle 5, 1, 3 \rangle = 5\mathbf{i} + \mathbf{j} + 3\mathbf{k}$ and $\mathbf{v} = \mathbf{i} + 4\mathbf{j} - 2\mathbf{k}$, so the vector equation (1) becomes

$$\mathbf{r} = (5\mathbf{i} + \mathbf{j} + 3\mathbf{k}) + t(\mathbf{i} + 4\mathbf{j} - 2\mathbf{k})$$

or

$$\mathbf{r} = (5 + t)\,\mathbf{i} + (1 + 4t)\,\mathbf{j} + (3 - 2t)\,\mathbf{k}$$

Parametric equations are

$$x = 5 + t \qquad y = 1 + 4t \qquad z = 3 - 2t$$

(b) Choosing the parameter value $t = 1$ gives $x = 6$, $y = 5$, and $z = 1$, so (6, 5, 1) is a point on the line. Similarly, $t = -1$ gives the point $(4, -3, 5)$.

The vector equation and parametric equations of a line are not unique. If we change the point or the parameter or choose a different parallel vector, then the equations change. For instance, if, instead of (5, 1, 3), we choose the point (6, 5, 1) in Example 1, then the parametric equations of the line become

$$x = 6 + t \qquad y = 5 + 4t \qquad z = 1 - 2t$$

Or, if we stay with the point (5, 1, 3) but choose the parallel vector $2\mathbf{i} + 8\mathbf{j} - 4\mathbf{k}$, we arrive at the equations

$$x = 5 + 2t \qquad y = 1 + 8t \qquad z = 3 - 4t$$

In general, if a vector $\mathbf{v} = \langle a, b, c \rangle$ is used to describe the direction of a line L, then the numbers a, b, and c are called **direction numbers** of L. Since any vector parallel to $\mathbf{v}$ could also be used, we see that any three numbers proportional to a, b, and c could also be used as a set of direction numbers for L.

Another way of describing a line L is to eliminate the parameter t from Equations 2. If none of a, b, or c is 0, we can solve each of these equations for t, equate the results, and obtain

3
$$\frac{x - x_0}{a} = \frac{y - y_0}{b} = \frac{z - z_0}{c}$$

These equations are called **symmetric equations** of L. Notice that the numbers a, b, and c that appear in the denominators of Equations 3 are direction numbers of L, that is, components of a vector parallel to L. If one of a, b, or c is 0, we can still eliminate t. For instance, if $a = 0$, we could write the equations of L as

$$x = x_0 \qquad \frac{y - y_0}{b} = \frac{z - z_0}{c}$$

This means that L lies in the vertical plane $x = x_0$.

|||| Figure 4 shows the line L in Example 2 and the point P where it intersects the xy-plane.

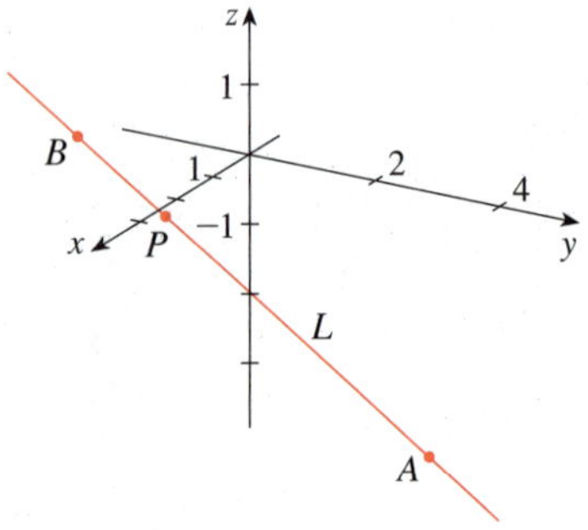

FIGURE 4

EXAMPLE 2

(a) Find parametric equations and symmetric equations of the line that passes through the points $A(2, 4, -3)$ and $B(3, -1, 1)$.
(b) At what point does this line intersect the xy-plane?

SOLUTION

(a) We are not explicitly given a vector parallel to the line, but observe that the vector $\mathbf{v}$ with representation $\overrightarrow{AB}$ is parallel to the line and

$$\mathbf{v} = \langle 3 - 2, -1 - 4, 1 - (-3) \rangle = \langle 1, -5, 4 \rangle$$

Thus, direction numbers are $a = 1$, $b = -5$, and $c = 4$. Taking the point $(2, 4, -3)$ as P_0, we see that parametric equations (2) are

$$x = 2 + t \qquad y = 4 - 5t \qquad z = -3 + 4t$$

and symmetric equations (3) are

$$\frac{x - 2}{1} = \frac{y - 4}{-5} = \frac{z + 3}{4}$$

(b) The line intersects the xy-plane when $z = 0$, so we put $z = 0$ in the symmetric equations and obtain

$$\frac{x - 2}{1} = \frac{y - 4}{-5} = \frac{3}{4}$$

This gives $x = \frac{11}{4}$ and $y = \frac{1}{4}$, so the line intersects the xy-plane at the point $\left(\frac{11}{4}, \frac{1}{4}, 0\right)$.

In general, the procedure of Example 2 shows that direction numbers of the line L through the points $P_0(x_0, y_0, z_0)$ and $P_1(x_1, y_1, z_1)$ are $x_1 - x_0$, $y_1 - y_0$, and $z_1 - z_0$ and so symmetric equations of L are

$$\frac{x - x_0}{x_1 - x_0} = \frac{y - y_0}{y_1 - y_0} = \frac{z - z_0}{z_1 - z_0}$$

Often, we need a description, not of an entire line, but of just a line segment. How, for instance, could we describe the line segment AB in Example 2? If we put $t = 0$ in the parametric equations in Example 2(a), we get the point $(2, 4, -3)$ and if we put $t = 1$ we get $(3, -1, 1)$. So the line segment AB is described by the parametric equations

$$x = 2 + t \qquad y = 4 - 5t \qquad z = -3 + 4t \qquad 0 \leqslant t \leqslant 1$$

or by the corresponding vector equation

$$\mathbf{r}(t) = \langle 2 + t, 4 - 5t, -3 + 4t \rangle \qquad 0 \leqslant t \leqslant 1$$

In general, we know from Equation 1 that the vector equation of a line through the (tip of the) vector $\mathbf{r}_0$ in the direction of a vector $\mathbf{v}$ is $\mathbf{r} = \mathbf{r}_0 + t\mathbf{v}$. If the line also passes through (the tip of) $\mathbf{r}_1$, then we can take $\mathbf{v} = \mathbf{r}_1 - \mathbf{r}_0$ and so its vector equation is

$$\mathbf{r} = \mathbf{r}_0 + t(\mathbf{r}_1 - \mathbf{r}_0) = (1 - t)\mathbf{r}_0 + t\mathbf{r}_1$$

The line segment from $\mathbf{r}_0$ to $\mathbf{r}_1$ is given by the parameter interval $0 \leqslant t \leqslant 1$.

4 The line segment from $\mathbf{r}_0$ to $\mathbf{r}_1$ is given by the vector equation

$$\mathbf{r}(t) = (1 - t)\mathbf{r}_0 + t\mathbf{r}_1 \qquad 0 \leqslant t \leqslant 1$$

|||| The lines L_1 and L_2 in Example 3, shown in Figure 5, are skew lines.

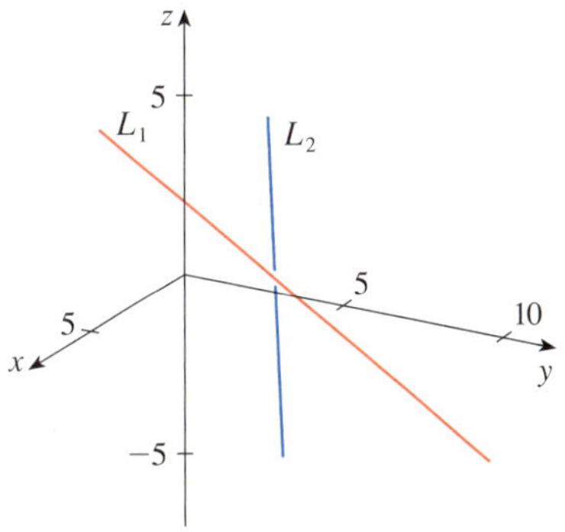

FIGURE 5

EXAMPLE 3 Show that the lines L_1 and L_2 with parametric equations

$$x = 1 + t \qquad y = -2 + 3t \qquad z = 4 - t$$

$$x = 2s \qquad y = 3 + s \qquad z = -3 + 4s$$

are **skew lines**; that is, they do not intersect and are not parallel (and therefore do not lie in the same plane).

SOLUTION The lines are not parallel because the corresponding vectors $\langle 1, 3, -1 \rangle$ and $\langle 2, 1, 4 \rangle$ are not parallel. (Their components are not proportional.) If L_1 and L_2 had a point of intersection, there would be values of t and s such that

$$1 + t = 2s$$

$$-2 + 3t = 3 + s$$

$$4 - t = -3 + 4s$$

But if we solve the first two equations, we get $t = \frac{11}{5}$ and $s = \frac{8}{5}$, and these values don't satisfy the third equation. Therefore, there are no values of t and s that satisfy the three equations. Thus, L_1 and L_2 do not intersect. Hence, L_1 and L_2 are skew lines.

Planes

Although a line in space is determined by a point and a direction, a plane in space is more difficult to describe. A single vector parallel to a plane is not enough to convey the "direction" of the plane, but a vector perpendicular to the plane does completely specify its direction. Thus, a plane in space is determined by a point $P_0(x_0, y_0, z_0)$ in the plane and a vector $\mathbf{n}$ that is orthogonal to the plane. This orthogonal vector $\mathbf{n}$ is called a **normal vector**. Let $P(x, y, z)$ be an arbitrary point in the plane, and let $\mathbf{r}_0$ and $\mathbf{r}$ be the position vectors of P_0 and P. Then the vector $\mathbf{r} - \mathbf{r}_0$ is represented by $\overrightarrow{P_0P}$. (See Figure 6.) The normal vector $\mathbf{n}$ is orthogonal to every vector in the given plane. In particular, $\mathbf{n}$ is orthogonal to $\mathbf{r} - \mathbf{r}_0$ and so we have

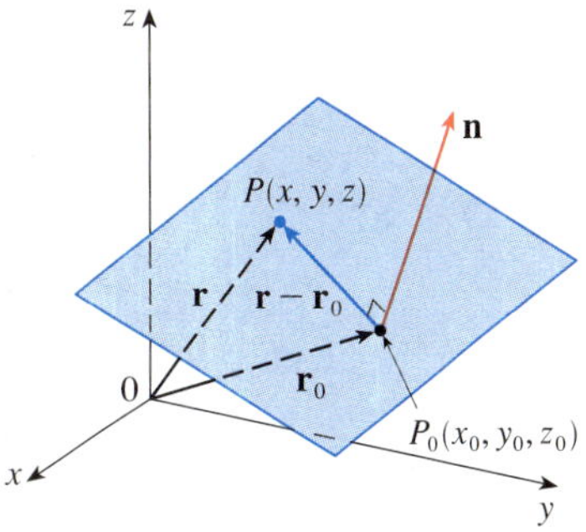

FIGURE 6

5 $$\mathbf{n} \cdot (\mathbf{r} - \mathbf{r}_0) = 0$$

which can be rewritten as

6 $$\mathbf{n} \cdot \mathbf{r} = \mathbf{n} \cdot \mathbf{r}_0$$

Either Equation 5 or Equation 6 is called a **vector equation of the plane**.

To obtain a scalar equation for the plane, we write $\mathbf{n} = \langle a, b, c \rangle$, $\mathbf{r} = \langle x, y, z \rangle$, and $\mathbf{r}_0 = \langle x_0, y_0, z_0 \rangle$. Then the vector equation (5) becomes

$$\langle a, b, c \rangle \cdot \langle x - x_0, y - y_0, z - z_0 \rangle = 0$$

or

7 $$a(x - x_0) + b(y - y_0) + c(z - z_0) = 0$$

Equation 7 is the **scalar equation of the plane through $P_0(x_0, y_0, z_0)$ with normal vector $\mathbf{n} = \langle a, b, c \rangle$.**

EXAMPLE 4 Find an equation of the plane through the point $(2, 4, -1)$ with normal vector $\mathbf{n} = \langle 2, 3, 4 \rangle$. Find the intercepts and sketch the plane.

SOLUTION Putting $a = 2$, $b = 3$, $c = 4$, $x_0 = 2$, $y_0 = 4$, and $z_0 = -1$ in Equation 7, we see that an equation of the plane is

$$2(x - 2) + 3(y - 4) + 4(z + 1) = 0$$

or

$$2x + 3y + 4z = 12$$

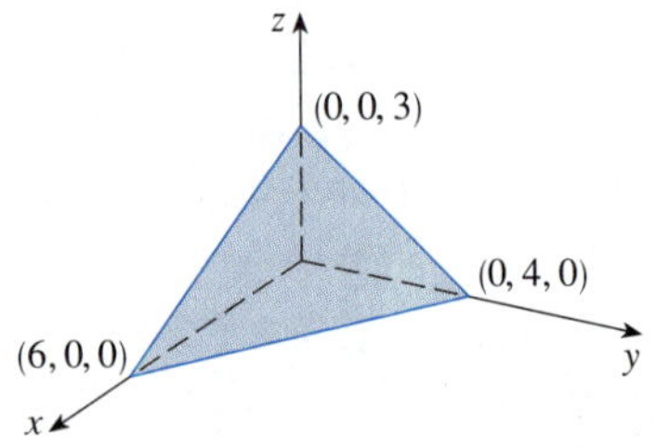

FIGURE 7

To find the x-intercept we set $y = z = 0$ in this equation and obtain $x = 6$. Similarly, the y-intercept is 4 and the z-intercept is 3. This enables us to sketch the portion of the plane that lies in the first octant (see Figure 7).

By collecting terms in Equation 7 as we did in Example 4, we can rewrite the equation of a plane as

8 $$ax + by + cz + d = 0$$

where $d = -(ax_0 + by_0 + cz_0)$. Equation 8 is called a **linear equation** in x, y, and z. Conversely, it can be shown that if a, b, and c are not all 0, then the linear equation (8) represents a plane with normal vector $\langle a, b, c \rangle$. (See Exercise 73.)

|||| Figure 8 shows the portion of the plane in Example 5 that is enclosed by triangle PQR.

EXAMPLE 5 Find an equation of the plane that passes through the points $P(1, 3, 2)$, $Q(3, -1, 6)$, and $R(5, 2, 0)$.

SOLUTION The vectors $\mathbf{a}$ and $\mathbf{b}$ corresponding to $\overrightarrow{PQ}$ and $\overrightarrow{PR}$ are

$$\mathbf{a} = \langle 2, -4, 4 \rangle \qquad \mathbf{b} = \langle 4, -1, -2 \rangle$$

Since both $\mathbf{a}$ and $\mathbf{b}$ lie in the plane, their cross product $\mathbf{a} \times \mathbf{b}$ is orthogonal to the plane and can be taken as the normal vector. Thus

$$\mathbf{n} = \mathbf{a} \times \mathbf{b} = \begin{vmatrix} \mathbf{i} & \mathbf{j} & \mathbf{k} \\ 2 & -4 & 4 \\ 4 & -1 & -2 \end{vmatrix} = 12\mathbf{i} + 20\mathbf{j} + 14\mathbf{k}$$

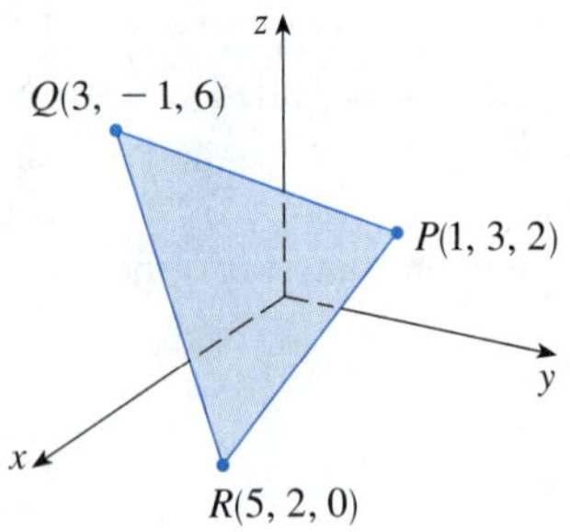

FIGURE 8

With the point $P(1, 3, 2)$ and the normal vector $\mathbf{n}$, an equation of the plane is

$$12(x - 1) + 20(y - 3) + 14(z - 2) = 0$$

or

$$6x + 10y + 7z = 50$$

EXAMPLE 6 Find the point at which the line with parametric equations $x = 2 + 3t$, $y = -4t$, $z = 5 + t$ intersects the plane $4x + 5y - 2z = 18$.

SOLUTION We substitute the expressions for x, y, and z from the parametric equations into the equation of the plane:

$$4(2 + 3t) + 5(-4t) - 2(5 + t) = 18$$

This simplifies to $-10t = 20$, so $t = -2$. Therefore, the point of intersection occurs when the parameter value is $t = -2$. Then $x = 2 + 3(-2) = -4$, $y = -4(-2) = 8$, $z = 5 - 2 = 3$ and so the point of intersection is $(-4, 8, 3)$.

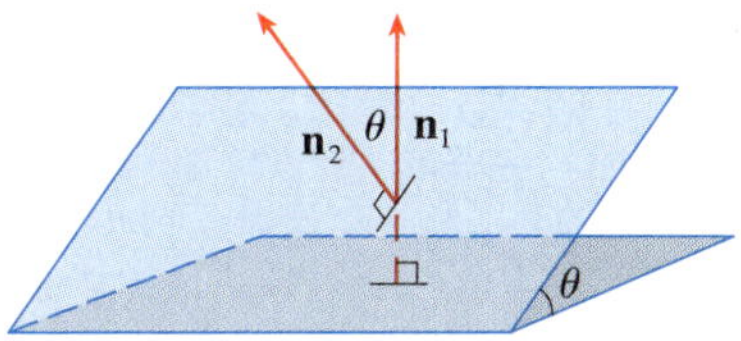

FIGURE 9

Two planes are **parallel** if their normal vectors are parallel. For instance, the planes $x + 2y - 3z = 4$ and $2x + 4y - 6z = 3$ are parallel because their normal vectors are $\mathbf{n}_1 = \langle 1, 2, -3 \rangle$ and $\mathbf{n}_2 = \langle 2, 4, -6 \rangle$ and $\mathbf{n}_2 = 2\mathbf{n}_1$. If two planes are not parallel, then they intersect in a straight line and the angle between the two planes is defined as the acute angle between their normal vectors (see angle θ in Figure 9).

||| Figure 10 shows the planes in Example 7 and their line of intersection L.

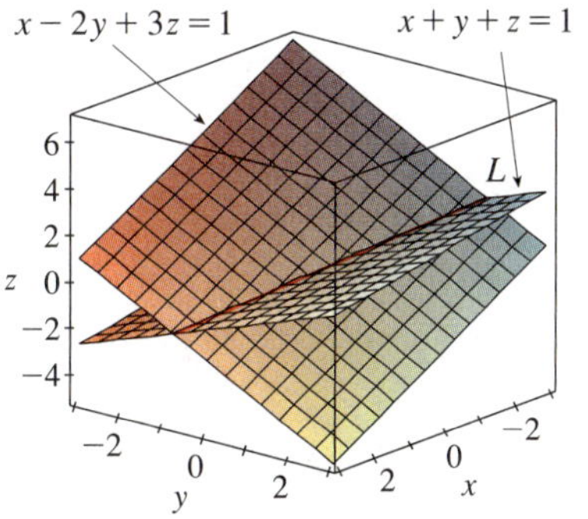

FIGURE 10

EXAMPLE 7

(a) Find the angle between the planes $x + y + z = 1$ and $x - 2y + 3z = 1$.

(b) Find symmetric equations for the line of intersection L of these two planes.

SOLUTION

(a) The normal vectors of these planes are

$$\mathbf{n}_1 = \langle 1, 1, 1 \rangle \qquad \mathbf{n}_2 = \langle 1, -2, 3 \rangle$$

and so, if θ is the angle between the planes, Corollary 12.3.6 gives

$$\cos \theta = \frac{\mathbf{n}_1 \cdot \mathbf{n}_2}{|\mathbf{n}_1||\mathbf{n}_2|} = \frac{1(1) + 1(-2) + 1(3)}{\sqrt{1 + 1 + 1}\sqrt{1 + 4 + 9}} = \frac{2}{\sqrt{42}}$$

$$\theta = \cos^{-1}\left(\frac{2}{\sqrt{42}}\right) \approx 72^\circ$$

(b) We first need to find a point on L. For instance, we can find the point where the line intersects the xy-plane by setting $z = 0$ in the equations of both planes. This gives the equations $x + y = 1$ and $x - 2y = 1$, whose solution is $x = 1$, $y = 0$. So the point $(1, 0, 0)$ lies on L.

Now we observe that, since L lies in both planes, it is perpendicular to both of the normal vectors. Thus, a vector $\mathbf{v}$ parallel to L is given by the cross product

||| Another way to find the line of intersection is to solve the equations of the planes for two of the variables in terms of the third, which can be taken as the parameter.

$$\mathbf{v} = \mathbf{n}_1 \times \mathbf{n}_2 = \begin{vmatrix} \mathbf{i} & \mathbf{j} & \mathbf{k} \\ 1 & 1 & 1 \\ 1 & -2 & 3 \end{vmatrix} = 5\mathbf{i} - 2\mathbf{j} - 3\mathbf{k}$$

and so the symmetric equations of L can be written as

$$\frac{x - 1}{5} = \frac{y}{-2} = \frac{z}{-3}$$

NOTE ▫ Since a linear equation in x, y, and z represents a plane and two nonparallel planes intersect in a line, it follows that two linear equations can represent a line. The points (x, y, z) that satisfy both $a_1x + b_1y + c_1z + d_1 = 0$ and $a_2x + b_2y + c_2z + d_2 = 0$

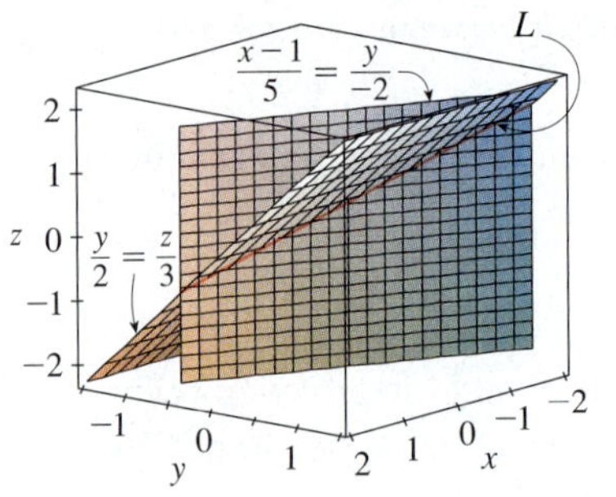

FIGURE 11

|||| Figure 11 shows how the line L in Example 7 can also be regarded as the line of intersection of planes derived from its symmetric equations.

lie on both of these planes, and so the pair of linear equations represents the line of intersection of the planes (if they are not parallel). For instance, in Example 7 the line L was given as the line of intersection of the planes $x + y + z = 1$ and $x - 2y + 3z = 1$. The symmetric equations that we found for L could be written as

$$\frac{x-1}{5} = \frac{y}{-2} \qquad \text{and} \qquad \frac{y}{-2} = \frac{z}{-3}$$

which is again a pair of linear equations. They exhibit L as the line of intersection of the planes $(x - 1)/5 = y/(-2)$ and $y/(-2) = z/(-3)$. (See Figure 11.)

In general, when we write the equations of a line in the symmetric form

$$\frac{x - x_0}{a} = \frac{y - y_0}{b} = \frac{z - z_0}{c}$$

we can regard the line as the line of intersection of the two planes

$$\frac{x - x_0}{a} = \frac{y - y_0}{b} \qquad \text{and} \qquad \frac{y - y_0}{b} = \frac{z - z_0}{c}$$

EXAMPLE 8 Find a formula for the distance D from a point $P_1(x_1, y_1, z_1)$ to the plane $ax + by + cz + d = 0$.

SOLUTION Let $P_0(x_0, y_0, z_0)$ be any point in the given plane and let $\mathbf{b}$ be the vector corresponding to $\overrightarrow{P_0P_1}$. Then

$$\mathbf{b} = \langle x_1 - x_0, y_1 - y_0, z_1 - z_0 \rangle$$

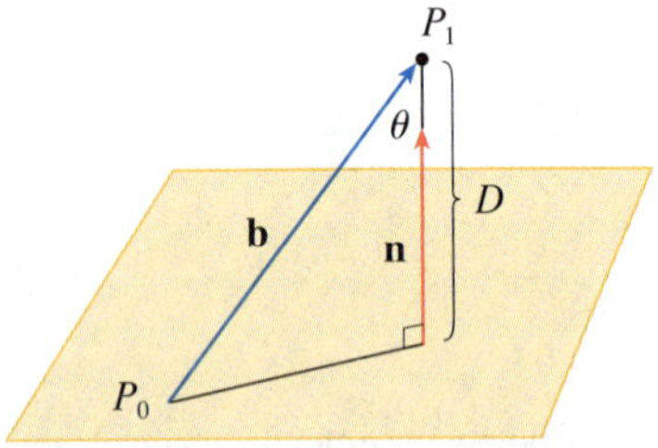

FIGURE 12

From Figure 12 you can see that the distance D from P_1 to the plane is equal to the absolute value of the scalar projection of $\mathbf{b}$ onto the normal vector $\mathbf{n} = \langle a, b, c \rangle$. (See Section 12.3.) Thus

$$D = |\text{comp}_{\mathbf{n}}\,\mathbf{b}| = \frac{|\mathbf{n} \cdot \mathbf{b}|}{|\mathbf{n}|}$$

$$= \frac{|a(x_1 - x_0) + b(y_1 - y_0) + c(z_1 - z_0)|}{\sqrt{a^2 + b^2 + c^2}}$$

$$= \frac{|(ax_1 + by_1 + cz_1) - (ax_0 + by_0 + cz_0)|}{\sqrt{a^2 + b^2 + c^2}}$$

Since P_0 lies in the plane, its coordinates satisfy the equation of the plane and so we have $ax_0 + by_0 + cz_0 + d = 0$. Thus, the formula for D can be written as

9 $$D = \frac{|ax_1 + by_1 + cz_1 + d|}{\sqrt{a^2 + b^2 + c^2}}$$

EXAMPLE 9 Find the distance between the parallel planes $10x + 2y - 2z = 5$ and $5x + y - z = 1$.

SOLUTION First we note that the planes are parallel because their normal vectors $\langle 10, 2, -2 \rangle$ and $\langle 5, 1, -1 \rangle$ are parallel. To find the distance D between the planes, we choose any point on one plane and calculate its distance to the other plane. In particular, if we put $y = z = 0$ in the equation of the first plane, we get $10x = 5$ and so

$\left(\frac{1}{2}, 0, 0\right)$ is a point in this plane. By Formula 9, the distance between $\left(\frac{1}{2}, 0, 0\right)$ and the plane $5x + y - z - 1 = 0$ is

$$D = \frac{\left|5\left(\frac{1}{2}\right) + 1(0) - 1(0) - 1\right|}{\sqrt{5^2 + 1^2 + (-1)^2}} = \frac{\frac{3}{2}}{3\sqrt{3}} = \frac{\sqrt{3}}{6}$$

So the distance between the planes is $\sqrt{3}/6$.

EXAMPLE 10 In Example 3 we showed that the lines

$$L_1: \quad x = 1 + t \qquad y = -2 + 3t \qquad z = 4 - t$$

$$L_2: \quad x = 2s \qquad y = 3 + s \qquad z = -3 + 4s$$

are skew. Find the distance between them.

SOLUTION Since the two lines L_1 and L_2 are skew, they can be viewed as lying on two parallel planes P_1 and P_2. The distance between L_1 and L_2 is the same as the distance between P_1 and P_2, which can be computed as in Example 9. The common normal vector to both planes must be orthogonal to both $\mathbf{v}_1 = \langle 1, 3, -1 \rangle$ (the direction of L_1) and $\mathbf{v}_2 = \langle 2, 1, 4 \rangle$ (the direction of L_2). So a normal vector is

$$\mathbf{n} = \mathbf{v}_1 \times \mathbf{v}_2 = \begin{vmatrix} \mathbf{i} & \mathbf{j} & \mathbf{k} \\ 1 & 3 & -1 \\ 2 & 1 & 4 \end{vmatrix} = 13\mathbf{i} - 6\mathbf{j} - 5\mathbf{k}$$

If we put $s = 0$ in the equations of L_2, we get the point $(0, 3, -3)$ on L_2 and so an equation for P_2 is

$$13(x - 0) - 6(y - 3) - 5(z + 3) = 0 \qquad \text{or} \qquad 13x - 6y - 5z + 3 = 0$$

If we now set $t = 0$ in the equations for L_1, we get the point $(1, -2, 4)$ on P_1. So the distance between L_1 and L_2 is the same as the distance from $(1, -2, 4)$ to $13x - 6y - 5z + 3 = 0$. By Formula 9, this distance is

$$D = \frac{|13(1) - 6(-2) - 5(4) + 3|}{\sqrt{13^2 + (-6)^2 + (-5)^2}} = \frac{8}{\sqrt{230}} \approx 0.53$$

12.5 Exercises

1. Determine whether each statement is true or false.
(a) Two lines parallel to a third line are parallel.
(b) Two lines perpendicular to a third line are parallel.
(c) Two planes parallel to a third plane are parallel.
(d) Two planes perpendicular to a third plane are parallel.
(e) Two lines parallel to a plane are parallel.
(f) Two lines perpendicular to a plane are parallel.
(g) Two planes parallel to a line are parallel.
(h) Two planes perpendicular to a line are parallel.
(i) Two planes either intersect or are parallel.
(j) Two lines either intersect or are parallel.
(k) A plane and a line either intersect or are parallel.

2–5 Find a vector equation and parametric equations for the line.

2. The line through the point $(1, 0, -3)$ and parallel to the vector $2\mathbf{i} - 4\mathbf{j} + 5\mathbf{k}$

3. The line through the point $(-2, 4, 10)$ and parallel to the vector $\langle 3, 1, -8 \rangle$

4. The line through the origin and parallel to the line $x = 2t$, $y = 1 - t$, $z = 4 + 3t$

5. The line through the point $(1, 0, 6)$ and perpendicular to the plane $x + 3y + z = 5$

6–12 Find parametric equations and symmetric equations for the line.

6. The line through the origin and the point (1, 2, 3)

7. The line through the points (1, 3, 2) and (−4, 3, 0)

8. The line through the points (6, 1, −3) and (2, 4, 5)

9. The line through the points $(0, \frac{1}{2}, 1)$ and (2, 1, −3)

10. The line through (2, 1, 0) and perpendicular to both **i** + **j** and **j** + **k**

11. The line through (1, −1, 1) and parallel to the line $x + 2 = \frac{1}{2}y = z - 3$

12. The line of intersection of the planes $x + y + z = 1$ and $x + z = 0$

13. Is the line through (−4, −6, 1) and (−2, 0 −3) parallel to the line through (10, 18, 4) and (5, 3, 14)?

14. Is the line through (4, 1, −1) and (2, 5, 3) perpendicular to the line through (−3, 2, 0) and (5, 1, 4)?

15. (a) Find symmetric equations for the line that passes through the point (0, 2, −1) and is parallel to the line with parametric equations $x = 1 + 2t$, $y = 3t$, $z = 5 - 7t$.
(b) Find the points in which the required line in part (a) intersects the coordinate planes.

16. (a) Find parametric equations for the line through (5, 1, 0) that is perpendicular to the plane $2x - y + z = 1$.
(b) In what points does this line intersect the coordinate planes?

17. Find a vector equation for the line segment from (2, −1, 4) to (4, 6, 1).

18. Find parametric equations for the line segment from (10, 3, 1) to (5, 6, −3).

19–22 Determine whether the lines L_1 and L_2 are parallel, skew, or intersecting. If they intersect, find the point of intersection.

19. L_1: $x = -6t, \quad y = 1 + 9t, \quad z = -3t$
L_2: $x = 1 + 2s, \quad y = 4 - 3s, \quad z = s$

20. L_1: $x = 1 + 2t, \quad y = 3t, \quad z = 2 - t$
L_2: $x = -1 + s, \quad y = 4 + s, \quad z = 1 + 3s$

21. L_1: $\dfrac{x}{1} = \dfrac{y-1}{2} = \dfrac{z-2}{3}$, $\quad L_2$: $\dfrac{x-3}{-4} = \dfrac{y-2}{-3} = \dfrac{z-1}{2}$

22. L_1: $\dfrac{x-1}{2} = \dfrac{y-3}{2} = \dfrac{z-2}{-1}$

L_2: $\dfrac{x-2}{1} = \dfrac{y-6}{-1} = \dfrac{z+2}{3}$

23–38 Find an equation of the plane.

23. The plane through the point (6, 3, 2) and perpendicular to the vector ⟨−2, 1, 5⟩

24. The plane through the point (4, 0, −3) and with normal vector **j** + 2**k**

25. The plane through the point (1, −1, 1) and with normal vector **i** + **j** − **k**

26. The plane through the point (−2, 8, 10) and perpendicular to the line $x = 1 + t$, $y = 2t$, $z = 4 - 3t$

27. The plane through the origin and parallel to the plane $2x - y + 3z = 1$

28. The plane through the point (−1, 6, −5) and parallel to the plane $x + y + z + 2 = 0$

29. The plane through the point (4, −2, 3) and parallel to the plane $3x - 7z = 12$

30. The plane that contains the line $x = 3 + 2t$, $y = t$, $z = 8 - t$ and is parallel to the plane $2x + 4y + 8z = 17$

31. The plane through the points (0, 1, 1), (1, 0, 1), and (1, 1, 0)

32. The plane through the origin and the points (2, −4, 6) and (5, 1, 3)

33. The plane through the points (3, −1, 2), (8, 2, 4), and (−1, −2, −3)

34. The plane that passes through the point (1, 2, 3) and contains the line $x = 3t$, $y = 1 + t$, $z = 2 - t$

35. The plane that passes through the point (6, 0, −2) and contains the line $x = 4 - 2t$, $y = 3 + 5t$, $z = 7 + 4t$

36. The plane that passes through the point (1, −1, 1) and contains the line with symmetric equations $x = 2y = 3z$

37. The plane that passes through the point (−1, 2, 1) and contains the line of intersection of the planes $x + y - z = 2$ and $2x - y + 3z = 1$

38. The plane that passes through the line of intersection of the planes $x - z = 1$ and $y + 2z = 3$ and is perpendicular to the plane $x + y - 2z = 1$

39–41 Find the point at which the line intersects the given plane.

39. $x = 3 - t$, $y = 2 + t$, $z = 5t$; $\quad x - y + 2z = 9$

40. $x = 1 + 2t$, $y = 4t$, $z = 2 - 3t$; $\quad x + 2y - z + 1 = 0$

41. $x = y - 1 = 2z$; $\quad 4x - y + 3z = 8$

42. Where does the line through (1, 0, 1) and (4, −2, 2) intersect the plane $x + y + z = 6$?

43. Find direction numbers for the line of intersection of the planes $x + y + z = 1$ and $x + z = 0$.

44. Find the cosine of the angle between the planes $x + y + z = 0$ and $x + 2y + 3z = 1$.

45–50 Determine whether the planes are parallel, perpendicular, or neither. If neither, find the angle between them.

45. $x + 4y - 3z = 1, \quad -3x + 6y + 7z = 0$

46. $2z = 4y - x, \quad 3x - 12y + 6z = 1$

47. $x + y + z = 1, \quad x - y + z = 1$

48. $2x - 3y + 4z = 5, \quad x + 6y + 4z = 3$

49. $x = 4y - 2z, \quad 8y = 1 + 2x + 4z$

50. $x + 2y + 2z = 1, \quad 2x - y + 2z = 1$

51–52 (a) Find symmetric equations for the line of intersection of the planes and (b) find the angle between the planes.

51. $x + y - z = 2, \quad 3x - 4y + 5z = 6$

52. $x - 2y + z = 1, \quad 2x + y + z = 1$

53–54 Find parametric equations for the line of intersection of the planes.

53. $z = x + y, \quad 2x - 5y - z = 1$

54. $2x + 5z + 3 = 0, \quad x - 3y + z + 2 = 0$

55. Find an equation for the plane consisting of all points that are equidistant from the points $(1, 1, 0)$ and $(0, 1, 1)$.

56. Find an equation for the plane consisting of all points that are equidistant from the points $(-4, 2, 1)$ and $(2, -4, 3)$.

57. Find an equation of the plane with x-intercept a, y-intercept b, and z-intercept c.

58. (a) Find the point at which the given lines intersect:

$$\mathbf{r} = \langle 1, 1, 0 \rangle + t\langle 1, -1, 2 \rangle$$

and $$\mathbf{r} = \langle 2, 0, 2 \rangle + s\langle -1, 1, 0 \rangle$$

(b) Find an equation of the plane that contains these lines.

59. Find parametric equations for the line through the point $(0, 1, 2)$ that is parallel to the plane $x + y + z = 2$ and perpendicular to the line $x = 1 + t$, $y = 1 - t$, $z = 2t$.

60. Find parametric equations for the line through the point $(0, 1, 2)$ that is perpendicular to the line $x = 1 + t$, $y = 1 - t$, $z = 2t$ and intersects this line.

61. Which of the following four planes are parallel? Are any of them identical?

P_1: $4x - 2y + 6z = 3$ $\qquad$ P_2: $4x - 2y - 2z = 6$

P_3: $-6x + 3y - 9z = 5$ $\qquad$ P_4: $z = 2x - y - 3$

62. Which of the following four lines are parallel? Are any of them identical?

L_1: $x = 1 + t, \quad y = t, \quad z = 2 - 5t$

L_2: $x + 1 = y - 2 = 1 - z$

L_3: $x = 1 + t, \quad y = 4 + t, \quad z = 1 - t$

L_4: $\mathbf{r} = \langle 2, 1, -3 \rangle + t\langle 2, 2, -10 \rangle$

63–64 Use the formula in Exercise 39 in Section 12.4 to find the distance from the point to the given line.

63. $(1, 2, 3); \quad x = 2 + t, \quad y = 2 - 3t, \quad z = 5t$

64. $(1, 0, -1); \quad x = 5 - t, \quad y = 3t, \quad z = 1 + 2t$

65–66 Find the distance from the point to the given plane.

65. $(2, 8, 5), \quad x - 2y - 2z = 1$

66. $(3, -2, 7), \quad 4x - 6y + z = 5$

67–68 Find the distance between the given parallel planes.

67. $z = x + 2y + 1, \quad 3x + 6y - 3z = 4$

68. $3x + 6y - 9z = 4, \quad x + 2y - 3z = 1$

69. Show that the distance between the parallel planes $ax + by + cz + d_1 = 0$ and $ax + by + cz + d_2 = 0$ is

$$D = \frac{|d_1 - d_2|}{\sqrt{a^2 + b^2 + c^2}}$$

70. Find equations of the planes that are parallel to the plane $x + 2y - 2z = 1$ and two units away from it.

71. Show that the lines with symmetric equations $x = y = z$ and $x + 1 = y/2 = z/3$ are skew, and find the distance between these lines.

72. Find the distance between the skew lines with parametric equations $x = 1 + t$, $y = 1 + 6t$, $z = 2t$, and $x = 1 + 2s$, $y = 5 + 15s$, $z = -2 + 6s$.

73. If a, b, and c are not all 0, show that the equation $ax + by + cz + d = 0$ represents a plane and $\langle a, b, c \rangle$ is a normal vector to the plane.

Hint: Suppose $a \neq 0$ and rewrite the equation in the form

$$a\left(x + \frac{d}{a}\right) + b(y - 0) + c(z - 0) = 0$$

74. Give a geometric description of each family of planes.

(a) $x + y + z = c$

(b) $x + y + cz = 1$

(c) $y \cos\theta + z \sin\theta = 1$

LABORATORY PROJECT

Putting 3D in Perspective

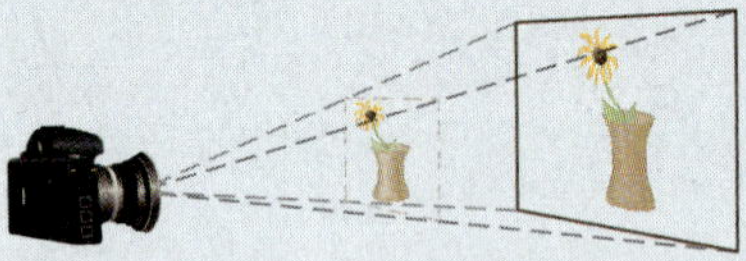

Computer graphics programmers face the same challenge as the great painters of the past: how to represent a three-dimensional scene as a flat image on a two-dimensional plane (a screen or a canvas). To create the illusion of perspective, in which closer objects appear larger than those farther away, three-dimensional objects in the computer's memory are projected onto a rectangular screen window from a viewpoint where the eye, or camera, is located. The viewing volume—the portion of space that will be visible—is the region contained by the four planes that pass through the viewpoint and an edge of the screen window. If objects in the scene extend beyond these four planes, they must be truncated before pixel data are sent to the screen. These planes are therefore called *clipping planes*.

1. Suppose the screen is represented by a rectangle in the yz-plane with vertices $(0, \pm 400, 0)$ and $(0, \pm 400, 600)$, and the camera is placed at $(1000, 0, 0)$. A line L in the scene passes through the points $(230, -285, 102)$ and $(860, 105, 264)$. At what points should L be clipped by the clipping planes?
2. If the clipped line segment is projected on the screen window, identify the resulting line segment.
3. Use parametric equations to plot the edges of the screen window, the clipped line segment, and its projection on the screen window. Then add sight lines connecting the viewpoint to each end of the clipped segments to verify that the projection is correct.
4. A rectangle with vertices $(621, -147, 206)$, $(563, 31, 242)$, $(657, -111, 86)$, and $(599, 67, 122)$ is added to the scene. The line L intersects this rectangle. To make the rectangle appear opaque, a programmer can use *hidden line rendering*, which removes portions of objects that are behind other objects. Identify the portion of L that should be removed.

12.6 Cylinders and Quadric Surfaces

We have already looked at two special types of surfaces—planes (in Section 12.5) and spheres (in Section 12.1). Here we investigate two other types of surfaces—cylinders and quadric surfaces.

In order to sketch the graph of a surface, it is useful to determine the curves of intersection of the surface with planes parallel to the coordinate planes. These curves are called **traces** (or cross-sections) of the surface.

Cylinders

A **cylinder** is a surface that consists of all lines (called **rulings**) that are parallel to a given line and pass through a given plane curve.

EXAMPLE 1 Sketch the graph of the surface $z = x^2$.

SOLUTION Notice that the equation of the graph, $z = x^2$, doesn't involve y. This means that any vertical plane with equation $y = k$ (parallel to the xz-plane) intersects the graph in a curve with equation $z = x^2$. So these vertical traces are parabolas. Figure 1 shows how the graph is formed by taking the parabola $z = x^2$ in the xz-plane and moving it in the direction of the y-axis. The graph is a surface, called a **parabolic cylinder**, made up of infinitely many shifted copies of the same parabola. Here the rulings of the cylinder are parallel to the y-axis.

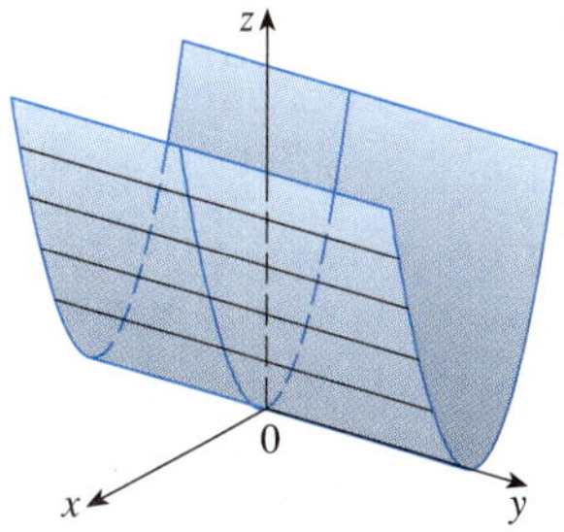

FIGURE 1
The surface $z = x^2$ is a parabolic cylinder.

We noticed that the variable y is missing from the equation of the cylinder in Example 1. This is typical of a surface whose rulings are parallel to one of the coordinate axes. If one of the variables x, y, or z is missing from the equation of a surface, then the surface is a cylinder.

EXAMPLE 2 Identify and sketch the surfaces.
(a) $x^2 + y^2 = 1$ (b) $y^2 + z^2 = 1$

SOLUTION
(a) Since z is missing and the equations $x^2 + y^2 = 1$, $z = k$ represent a circle with radius 1 in the plane $z = k$, the surface $x^2 + y^2 = 1$ is a circular cylinder whose axis is the z-axis (see Figure 2). Here the rulings are vertical lines.

(b) In this case x is missing and the surface is a circular cylinder whose axis is the x-axis (see Figure 3). It is obtained by taking the circle $y^2 + z^2 = 1$, $x = 0$ in the yz-plane and moving it parallel to the x-axis.

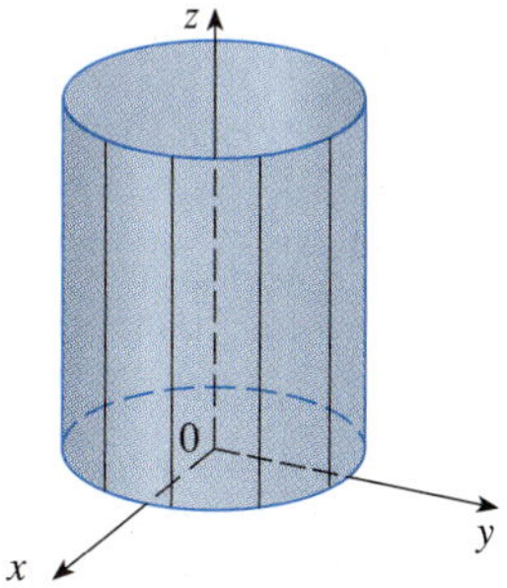

FIGURE 2 $x^2 + y^2 = 1$

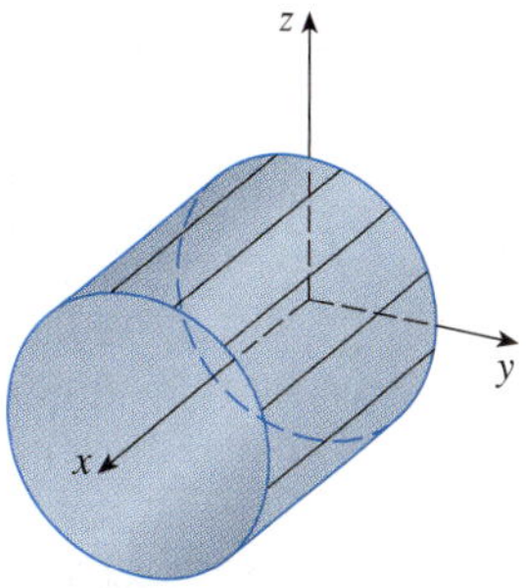

FIGURE 3 $y^2 + z^2 = 1$

NOTE ▫ When you are dealing with surfaces, it is important to recognize that an equation like $x^2 + y^2 = 1$ represents a cylinder and not a circle. The trace of the cylinder $x^2 + y^2 = 1$ in the xy-plane is the circle with equations $x^2 + y^2 = 1$, $z = 0$.

Quadric Surfaces

A **quadric surface** is the graph of a second-degree equation in three variables x, y, and z. The most general such equation is

$$Ax^2 + By^2 + Cz^2 + Dxy + Eyz + Fxz + Gx + Hy + Iz + J = 0$$

where $A, B, C, \ldots, J$ are constants, but by translation and rotation it can be brought into one of the two standard forms

$$Ax^2 + By^2 + Cz^2 + J = 0 \qquad \text{or} \qquad Ax^2 + By^2 + Iz = 0$$

Quadric surfaces are the counterparts in three dimensions of the conic sections in the plane. (See Section 10.5 for a review of conic sections.)

EXAMPLE 3 Use traces to sketch the quadric surface with equation

$$x^2 + \frac{y^2}{9} + \frac{z^2}{4} = 1$$

SOLUTION By substituting $z = 0$, we find that the trace in the xy-plane is $x^2 + y^2/9 = 1$, which we recognize as an equation of an ellipse. In general, the horizontal trace in the plane $z = k$ is

$$x^2 + \frac{y^2}{9} = 1 - \frac{k^2}{4} \qquad z = k$$

which is an ellipse, provided that $k^2 < 4$, that is, $-2 < k < 2$.

Similarly, the vertical traces are also ellipses:

$$\frac{y^2}{9} + \frac{z^2}{4} = 1 - k^2 \qquad x = k \qquad (\text{if } -1 < k < 1)$$

$$x^2 + \frac{z^2}{4} = 1 - \frac{k^2}{9} \qquad y = k \qquad (\text{if } -3 < k < 3)$$

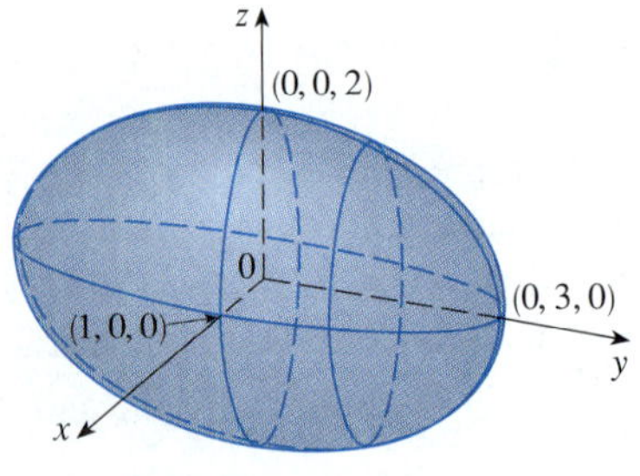

FIGURE 4

The ellipsoid $x^2 + \frac{y^2}{9} + \frac{z^2}{4} = 1$

Figure 4 shows how drawing some traces indicates the shape of the surface. It's called an **ellipsoid** because all of its traces are ellipses. Notice that it is symmetric with respect to each coordinate plane; this is a reflection of the fact that its equation involves only even powers of x, y, and z.

EXAMPLE 4 Use traces to sketch the surface $z = 4x^2 + y^2$.

SOLUTION If we put $x = 0$, we get $z = y^2$, so the yz-plane intersects the surface in a parabola. If we put $x = k$ (a constant), we get $z = y^2 + 4k^2$. This means that if we slice the graph with any plane parallel to the yz-plane, we obtain a parabola that opens upward. Similarly, if $y = k$, the trace is $z = 4x^2 + k^2$, which is again a parabola that opens upward. If we put $z = k$, we get the horizontal traces $4x^2 + y^2 = k$, which we recognize as a family of ellipses. Knowing the shapes of the traces, we can sketch the graph in Figure 5. Because of the elliptical and parabolic traces, the quadric surface $z = 4x^2 + y^2$ is called an **elliptic paraboloid**.

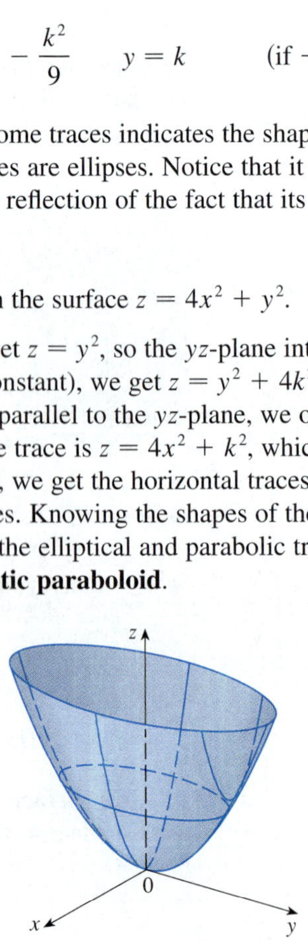

FIGURE 5

The surface $z = 4x^2 + y^2$ is an elliptic paraboloid. Horizontal traces are ellipses; vertical traces are parabolas.

EXAMPLE 5 Sketch the surface $z = y^2 - x^2$.

SOLUTION The traces in the vertical planes $x = k$ are the parabolas $z = y^2 - k^2$, which open upward. The traces in $y = k$ are the parabolas $z = -x^2 + k^2$, which open downward. The horizontal traces are $y^2 - x^2 = k$, a family of hyperbolas. We draw the families of traces in Figure 6, and we show how the traces appear when placed in their correct planes in Figure 7.

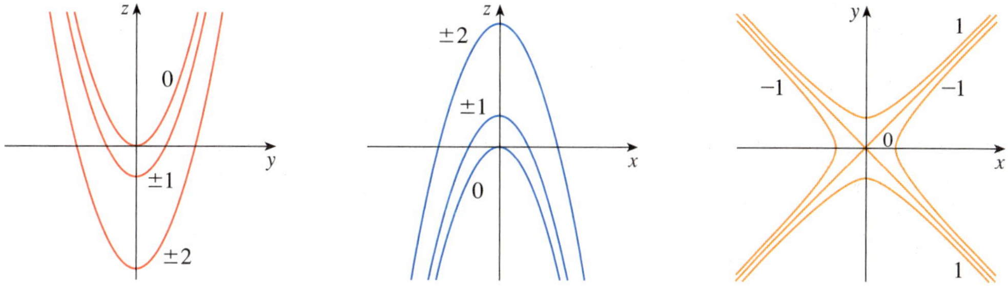

Traces in $x = k$ are $z = y^2 - k^2$

Traces in $y = k$ are $z = -x^2 + k^2$

Traces in $z = k$ are $y^2 - x^2 = k$

FIGURE 6
Vertical traces are parabolas; horizontal traces are hyperbolas. All traces are labeled with the value of k.

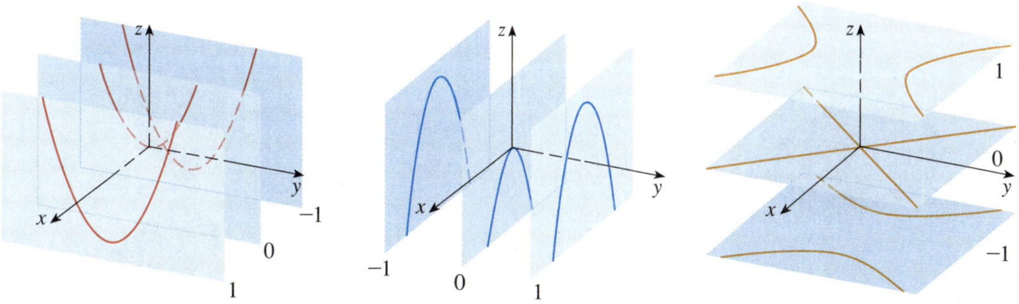

Traces in $x = k$

Traces in $y = k$

Traces in $z = k$

FIGURE 7
Traces moved to their correct planes

In Module 12.6A you can investigate how traces determine the shape of a surface.

In Figure 8 we fit together the traces from Figure 7 to form the surface $z = y^2 - x^2$, a **hyperbolic paraboloid**. Notice that the shape of the surface near the origin resembles that of a saddle. This surface will be investigated further in Section 14.7 when we discuss saddle points.

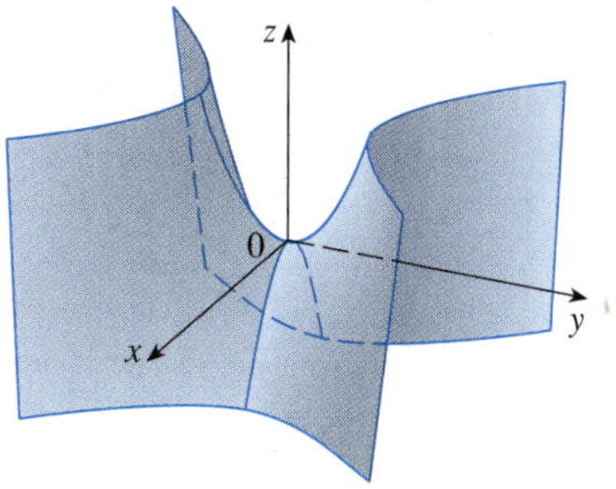

FIGURE 8
The surface $z = y^2 - x^2$ is a hyperbolic paraboloid.

EXAMPLE 6 Sketch the surface $\dfrac{x^2}{4} + y^2 - \dfrac{z^2}{4} = 1$.

SOLUTION The trace in any horizontal plane $z = k$ is the ellipse

$$\frac{x^2}{4} + y^2 = 1 + \frac{k^2}{4} \qquad z = k$$

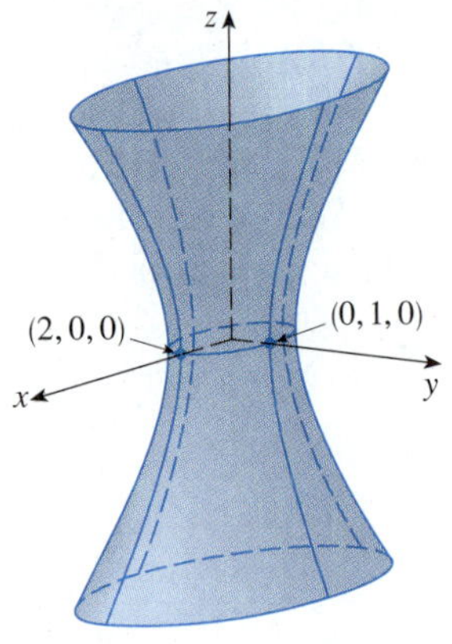

FIGURE 9

but the traces in the xz- and yz-planes are the hyperbolas

$$\frac{x^2}{4} - \frac{z^2}{4} = 1 \quad y = 0 \qquad \text{and} \qquad y^2 - \frac{z^2}{4} = 1 \quad x = 0$$

This surface is called a **hyperboloid of one sheet** and is sketched in Figure 9.

The idea of using traces to draw a surface is employed in three-dimensional graphing software for computers. In most such software, traces in the vertical planes $x = k$ and $y = k$ are drawn for equally spaced values of k, and parts of the graph are eliminated using hidden line removal. Table 1 shows computer-drawn graphs of the six basic types of quadric surfaces in standard form. All surfaces are symmetric with respect to the z-axis. If a quadric surface is symmetric about a different axis, its equation changes accordingly.

TABLE 1 Graphs of quadric surfaces

Surface	Equation	Surface	Equation
Ellipsoid 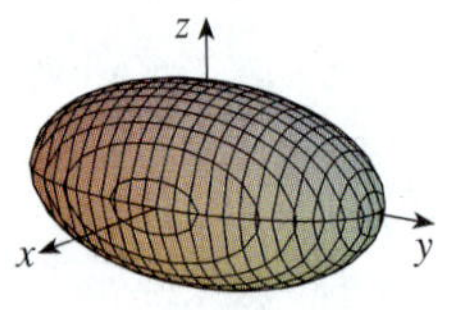	$\frac{x^2}{a^2} + \frac{y^2}{b^2} + \frac{z^2}{c^2} = 1$ All traces are ellipses. If $a = b = c$, the ellipsoid is a sphere.	Cone 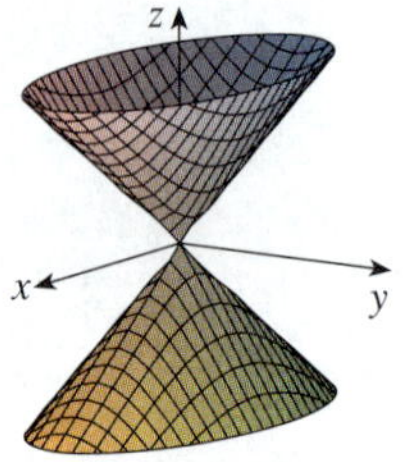	$\frac{z^2}{c^2} = \frac{x^2}{a^2} + \frac{y^2}{b^2}$ Horizontal traces are ellipses. Vertical traces in the planes $x = k$ and $y = k$ are hyperbolas if $k \neq 0$ but are pairs of lines if $k = 0$.
Elliptic Paraboloid 	$\frac{z}{c} = \frac{x^2}{a^2} + \frac{y^2}{b^2}$ Horizontal traces are ellipses. Vertical traces are parabolas. The variable raised to the first power indicates the axis of the paraboloid.	Hyperboloid of One Sheet 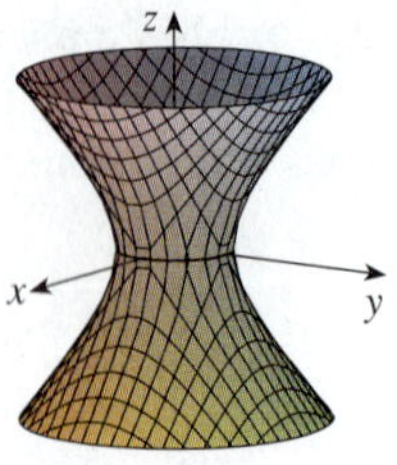	$\frac{x^2}{a^2} + \frac{y^2}{b^2} - \frac{z^2}{c^2} = 1$ Horizontal traces are ellipses. Vertical traces are hyperbolas. The axis of symmetry corresponds to the variable whose coefficient is negative.
Hyperbolic Paraboloid 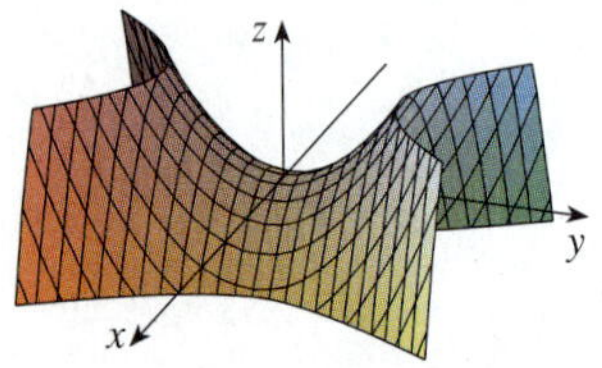	$\frac{z}{c} = \frac{x^2}{a^2} - \frac{y^2}{b^2}$ Horizontal traces are hyperbolas. Vertical traces are parabolas. The case where $c < 0$ is illustrated.	Hyperboloid of Two Sheets 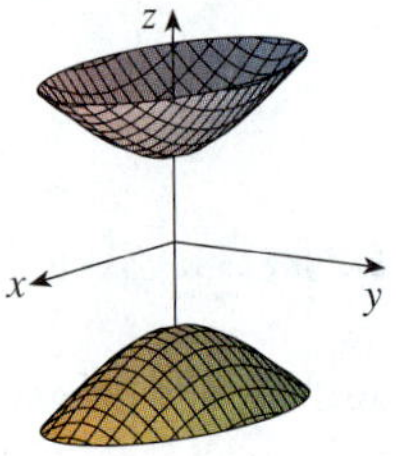	$-\frac{x^2}{a^2} - \frac{y^2}{b^2} + \frac{z^2}{c^2} = 1$ Horizontal traces in $z = k$ are ellipses if $k > c$ or $k < -c$. Vertical traces are hyperbolas. The two minus signs indicate two sheets.

TEC In Module 12.6B you can see how changing a, b, and c in Table 1 affects the shape of the quadric surface.

EXAMPLE 7 Identify and sketch the surface $4x^2 - y^2 + 2z^2 + 4 = 0$.

SOLUTION Dividing by -4, we first put the equation in standard form:

$$-x^2 + \frac{y^2}{4} - \frac{z^2}{2} = 1$$

Comparing this equation with Table 1, we see that it represents a hyperboloid of two sheets, the only difference being that in this case the axis of the hyperboloid is the y-axis. The traces in the xy- and yz-planes are the hyperbolas

$$-x^2 + \frac{y^2}{4} = 1 \quad z = 0 \qquad \text{and} \qquad \frac{y^2}{4} - \frac{z^2}{2} = 1 \quad x = 0$$

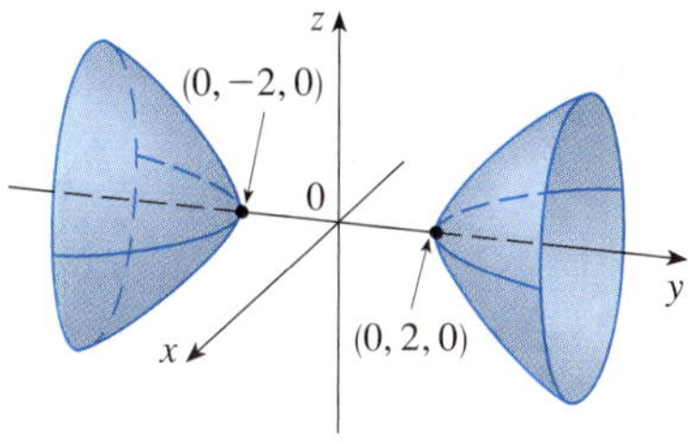

FIGURE 10
$4x^2 - y^2 + 2z^2 + 4 = 0$

The surface has no trace in the xz-plane, but traces in the vertical planes $y = k$ for $|k| > 2$ are the ellipses

$$x^2 + \frac{z^2}{2} = \frac{k^2}{4} - 1 \qquad y = k$$

which can be written as

$$\frac{x^2}{\frac{k^2}{4} - 1} + \frac{z^2}{2\left(\frac{k^2}{4} - 1\right)} = 1 \qquad y = k$$

These traces are used to make the sketch in Figure 10.

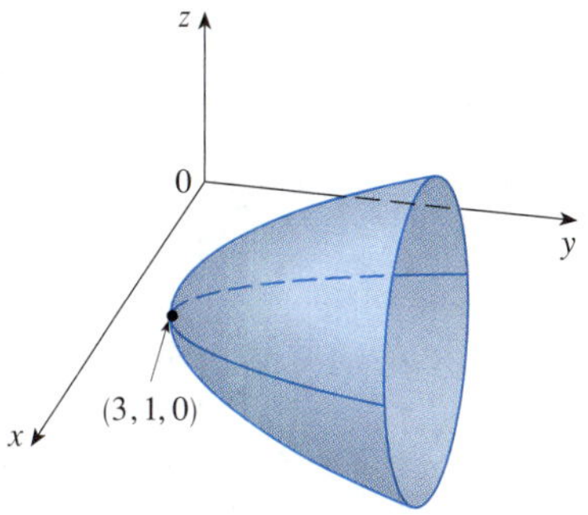

FIGURE 11
$x^2 + 2z^2 - 6x - y + 10 = 0$

EXAMPLE 8 Classify the quadric surface $x^2 + 2z^2 - 6x - y + 10 = 0$.

SOLUTION By completing the square we rewrite the equation as

$$y - 1 = (x - 3)^2 + 2z^2$$

Comparing this equation with Table 1, we see that it represents an elliptic paraboloid. Here, however, the axis of the paraboloid is parallel to the y-axis, and it has been shifted so that its vertex is the point $(3, 1, 0)$. The traces in the plane $y = k$ $(k > 1)$ are the ellipses

$$(x - 3)^2 + 2z^2 = k - 1 \qquad y = k$$

The trace in the xy-plane is the parabola with equation $y = 1 + (x - 3)^2$, $z = 0$. The paraboloid is sketched in Figure 11.

12.6 Exercises

1. (a) What does the equation $y = x^2$ represent as a curve in $\mathbb{R}^2$?
(b) What does it represent as a surface in $\mathbb{R}^3$?
(c) What does the equation $z = y^2$ represent?

2. (a) Sketch the graph of $y = e^x$ as a curve in $\mathbb{R}^2$.
(b) Sketch the graph of $y = e^x$ as a surface in $\mathbb{R}^3$.
(c) Describe and sketch the surface $z = e^y$.

3–8 Describe and sketch the surface.

3. $y^2 + 4z^2 = 4$

4. $z = 4 - x^2$

5. $x - y^2 = 0$

6. $yz = 4$

7. $z = \cos x$

8. $x^2 - y^2 = 1$

9. (a) Find and identify the traces of the quadric surface $x^2 + y^2 - z^2 = 1$ and explain why the graph looks like the graph of the hyperboloid of one sheet in Table 1.
(b) If we change the equation in part (a) to $x^2 - y^2 + z^2 = 1$, how is the graph affected?
(c) What if we change the equation in part (a) to $x^2 + y^2 + 2y - z^2 = 0$?

10. (a) Find and identify the traces of the quadric surface $-x^2 - y^2 + z^2 = 1$ and explain why the graph looks like the graph of the hyperboloid of two sheets in Table 1.
(b) If the equation in part (a) is changed to $x^2 - y^2 - z^2 = 1$, what happens to the graph? Sketch the new graph.

11–20 ||| Find the traces of the given surface in the planes $x = k$, $y = k$, $z = k$. Then identify the surface and sketch it.

11. $4x^2 + 9y^2 + 36z^2 = 36$

12. $4y = x^2 + z^2$

13. $y^2 = x^2 + z^2$

14. $z = x^2 - y^2$

15. $-x^2 + 4y^2 - z^2 = 4$

16. $25y^2 + z^2 = 100 + 4x^2$

17. $x^2 + 4z^2 - y = 0$

18. $x^2 + 4y^2 + z^2 = 4$

19. $y = z^2 - x^2$

20. $16x^2 = y^2 + 4z^2$

21–28 ||| Match the equation with its graph (labeled I–VIII). Give reasons for your choices.

21. $x^2 + 4y^2 + 9z^2 = 1$

22. $9x^2 + 4y^2 + z^2 = 1$

23. $x^2 - y^2 + z^2 = 1$

24. $-x^2 + y^2 - z^2 = 1$

25. $y = 2x^2 + z^2$

26. $y^2 = x^2 + 2z^2$

27. $x^2 + 2z^2 = 1$

28. $y = x^2 - z^2$

I

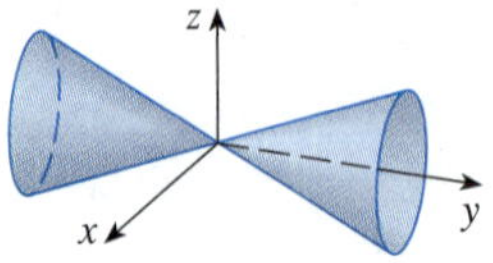

II

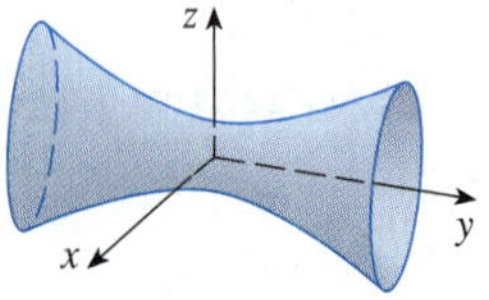

III

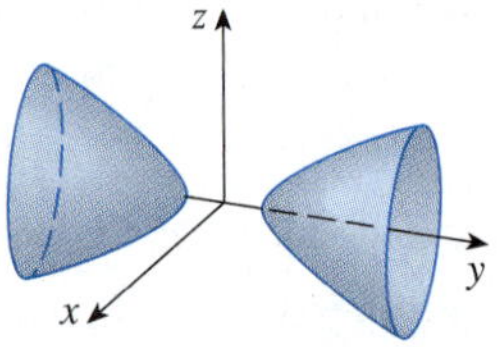

IV

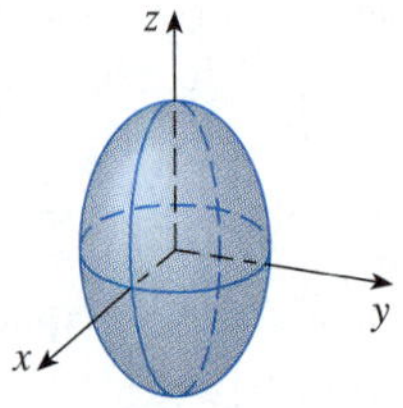

V

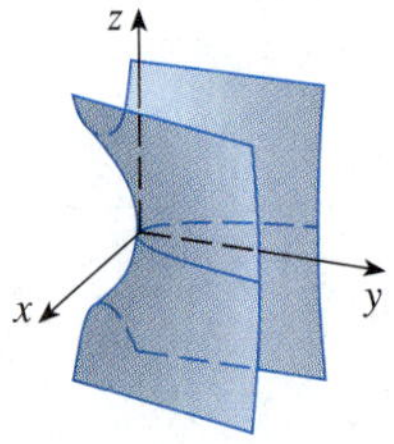

VI

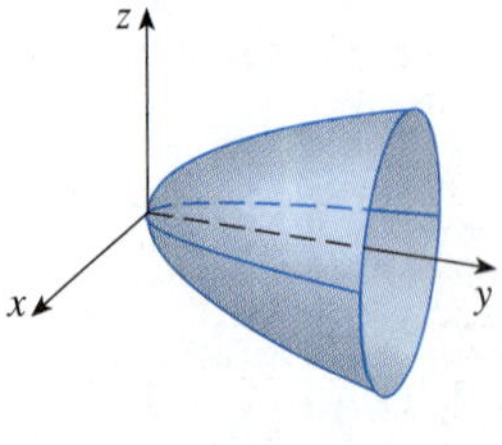

VII

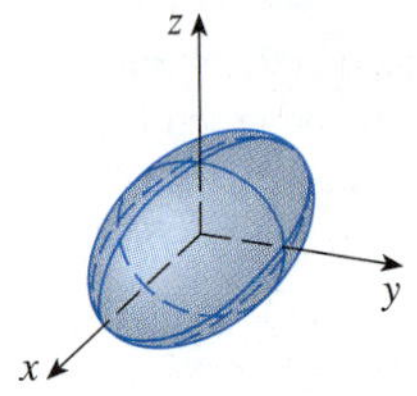

VIII

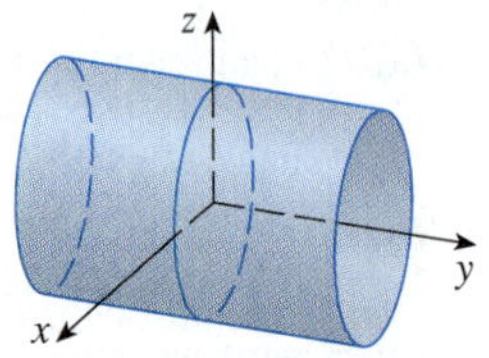

29–36 ||| Reduce the equation to one of the standard forms, classify the surface, and sketch it.

29. $z^2 = 4x^2 + 9y^2 + 36$

30. $x^2 = 2y^2 + 3z^2$

31. $x = 2y^2 + 3z^2$

32. $4x - y^2 + 4z^2 = 0$

33. $4x^2 + y^2 + 4z^2 - 4y - 24z + 36 = 0$

34. $4y^2 + z^2 - x - 16y - 4z + 20 = 0$

35. $x^2 - y^2 + z^2 - 4x - 2y - 2z + 4 = 0$

36. $x^2 - y^2 + z^2 - 2x + 2y + 4z + 2 = 0$

37–40 ||| Use a computer with three-dimensional graphing software to graph the surface. Experiment with viewpoints and with domains for the variables until you get a good view of the surface.

37. $z = 3x^2 - 5y^2$

38. $8x^2 + 15y^2 + 5z^2 = 100$

39. $z^2 = x^2 + 4y^2$

40. $z = y^2 + xy$

41. Sketch the region bounded by the surfaces $z = \sqrt{x^2 + y^2}$ and $x^2 + y^2 = 1$ for $1 \leqslant z \leqslant 2$.

42. Sketch the region bounded by the paraboloids $z = x^2 + y^2$ and $z = 2 - x^2 - y^2$.

43. Find an equation for the surface obtained by rotating the parabola $y = x^2$ about the y-axis.

44. Find an equation for the surface obtained by rotating the line $x = 3y$ about the x-axis.

45. Find an equation for the surface consisting of all points that are equidistant from the point $(-1, 0, 0)$ and the plane $x = 1$. Identify the surface.

46. Find an equation for the surface consisting of all points P for which the distance from P to the x-axis is twice the distance from P to the yz-plane. Identify the surface.

47. Show that if the point (a, b, c) lies on the hyperbolic paraboloid $z = y^2 - x^2$, then the lines with parametric equations $x = a + t$, $y = b + t$, $z = c + 2(b - a)t$ and $x = a + t$, $y = b - t$, $z = c - 2(b + a)t$ both lie entirely on this paraboloid. (This shows that the hyperbolic paraboloid is what is called a **ruled surface**; that is, it can be generated by the motion of a straight line. In fact, this exercise shows that through each point on the hyperbolic paraboloid there are two generating lines. The only other quadric surfaces that are ruled surfaces are cylinders, cones, and hyperboloids of one sheet.)

48. Show that the curve of intersection of the surfaces $x^2 + 2y^2 - z^2 + 3x = 1$ and $2x^2 + 4y^2 - 2z^2 - 5y = 0$ lies in a plane.

49. Graph the surfaces $z = x^2 + y^2$ and $z = 1 - y^2$ on a common screen using the domain $|x| \leqslant 1.2$, $|y| \leqslant 1.2$ and observe the curve of intersection of these surfaces. Show that the projection of this curve onto the xy-plane is an ellipse.

12.7 Cylindrical and Spherical Coordinates

Recall that in plane geometry we introduced the polar coordinate system in order to give a convenient description of certain curves and regions. (See Section 10.3.) In three dimensions there are two coordinate systems that are similar to polar coordinates and give convenient descriptions of some commonly occurring surfaces and solids. They will be especially useful in Chapter 15 when we compute volumes and triple integrals.

Cylindrical Coordinates

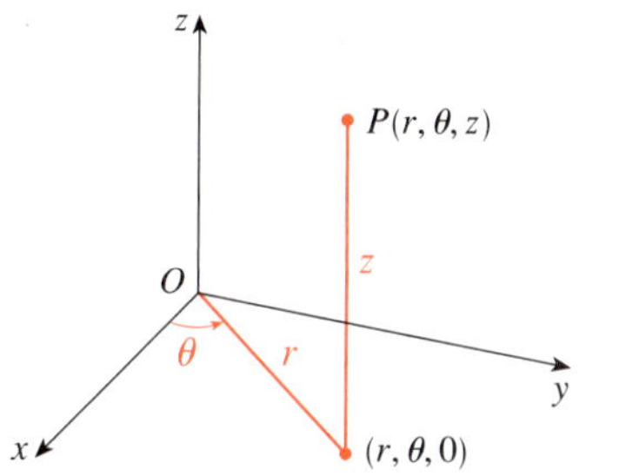

FIGURE 1
The cylindrical coordinates of a point

In the **cylindrical coordinate system**, a point P in three-dimensional space is represented by the ordered triple (r, θ, z), where r and θ are polar coordinates of the projection of P onto the xy-plane and z is the directed distance from the xy-plane to P (see Figure 1).

To convert from cylindrical to rectangular coordinates, we use the equations

1 $$x = r\cos\theta \qquad y = r\sin\theta \qquad z = z$$

whereas to convert from rectangular to cylindrical coordinates, we use

2 $$r^2 = x^2 + y^2 \qquad \tan\theta = \frac{y}{x} \qquad z = z$$

These equations follow from Equations 10.3.1 and 10.3.2.

EXAMPLE 1

(a) Plot the point with cylindrical coordinates $(2, 2\pi/3, 1)$ and find its rectangular coordinates.

(b) Find cylindrical coordinates of the point with rectangular coordinates $(3, -3, -7)$.

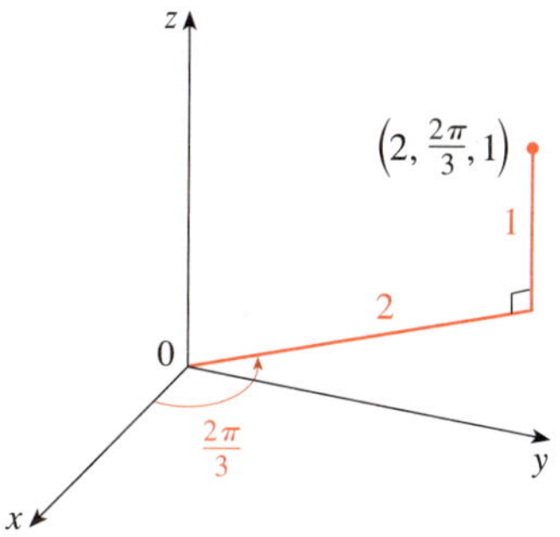

FIGURE 2

SOLUTION

(a) The point with cylindrical coordinates $(2, 2\pi/3, 1)$ is plotted in Figure 2. From Equations 1, its rectangular coordinates are

$$x = 2\cos\frac{2\pi}{3} = 2\left(-\frac{1}{2}\right) = -1$$

$$y = 2\sin\frac{2\pi}{3} = 2\left(\frac{\sqrt{3}}{2}\right) = \sqrt{3}$$

$$z = 1$$

Thus, the point is $\left(-1, \sqrt{3}, 1\right)$ in rectangular coordinates.

(b) From Equations 2 we have

$$r = \sqrt{3^2 + (-3)^2} = 3\sqrt{2}$$

$$\tan\theta = \frac{-3}{3} = -1 \qquad \text{so} \qquad \theta = \frac{7\pi}{4} + 2n\pi$$

$$z = -7$$

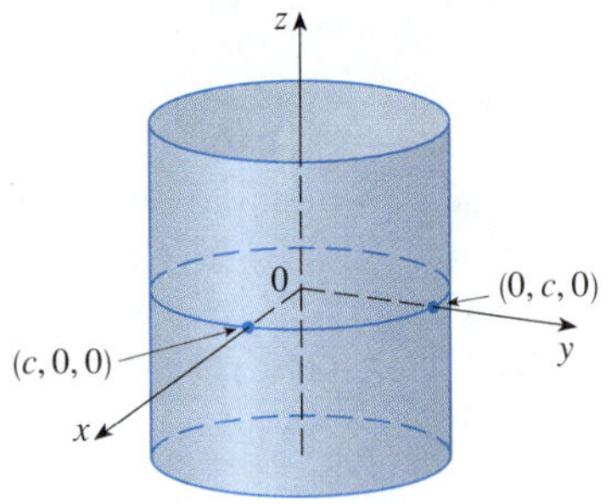

FIGURE 3
$r = c$, a cylinder

Therefore, one set of cylindrical coordinates is $(3\sqrt{2}, 7\pi/4, -7)$. Another is $(3\sqrt{2}, -\pi/4, -7)$. As with polar coordinates, there are infinitely many choices.

Cylindrical coordinates are useful in problems that involve symmetry about an axis, and the z-axis is chosen to coincide with this axis of symmetry. For instance, the axis of the circular cylinder with Cartesian equation $x^2 + y^2 = c^2$ is the z-axis. In cylindrical coordinates this cylinder has the very simple equation $r = c$. (See Figure 3.) This is the reason for the name "cylindrical" coordinates.

EXAMPLE 2 Describe the surface whose equation in cylindrical coordinates is $z = r$.

SOLUTION The equation says that the z-value, or height, of each point on the surface is the same as r, the distance from the point to the z-axis. Because θ doesn't appear, it can vary. So any horizontal trace in the plane $z = k$ $(k > 0)$ is a circle of radius k. These traces suggest that the surface is a cone. This prediction can be confirmed by converting the equation into rectangular coordinates. From the first equation in (2) we have

$$z^2 = r^2 = x^2 + y^2$$

We recognize the equation $z^2 = x^2 + y^2$ (by comparison with Table 1 in Section 12.6) as being a circular cone whose axis is the z-axis (see Figure 4).

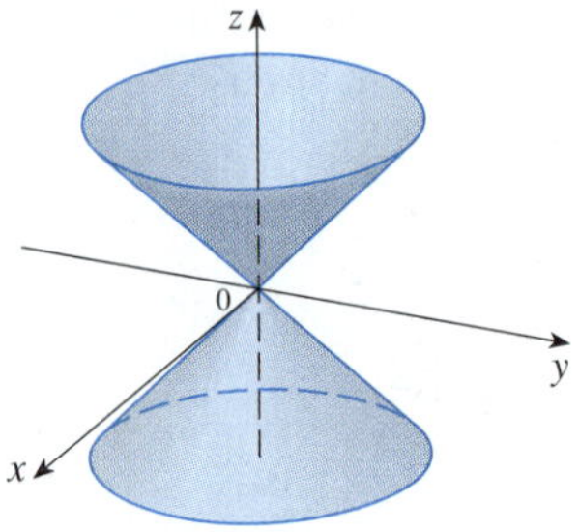

FIGURE 4
$z = r$, a cone

EXAMPLE 3 Find an equation in cylindrical coordinates for the ellipsoid $4x^2 + 4y^2 + z^2 = 1$.

SOLUTION Since $r^2 = x^2 + y^2$ from Equations 2, we have

$$z^2 = 1 - 4(x^2 + y^2) = 1 - 4r^2$$

So an equation of the ellipsoid in cylindrical coordinates is $z^2 = 1 - 4r^2$.

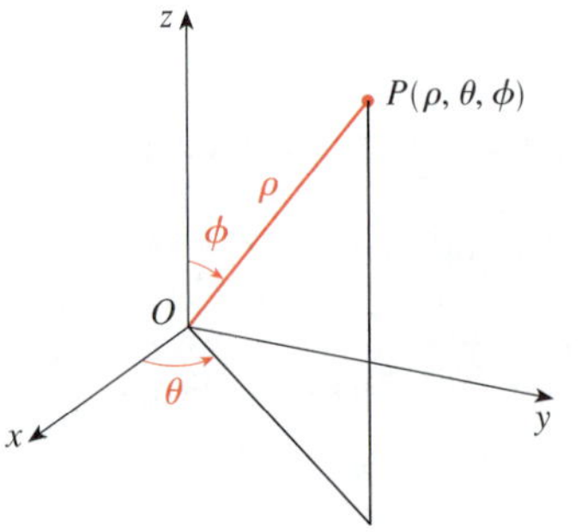

FIGURE 5
The spherical coordinates of a point

Spherical Coordinates

The **spherical coordinates** (ρ, θ, ϕ) of a point P in space are shown in Figure 5, where $\rho = |OP|$ is the distance from the origin to P, θ is the same angle as in cylindrical coordinates, and ϕ is the angle between the positive z-axis and the line segment OP. Note that

$$\rho \geqslant 0 \qquad 0 \leqslant \phi \leqslant \pi$$

The spherical coordinate system is especially useful in problems where there is symmetry about a point, and the origin is placed at this point. For example, the sphere with center the origin and radius c has the simple equation $\rho = c$ (see Figure 6); this is the reason for the name "spherical" coordinates. The graph of the equation $\theta = c$ is a vertical half-plane

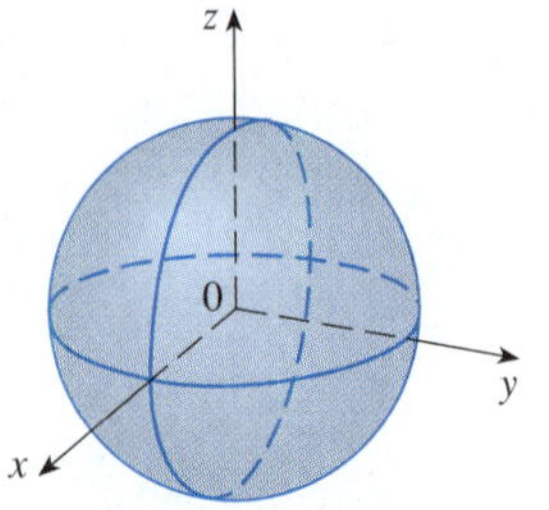

FIGURE 6 $\rho = c$, a sphere

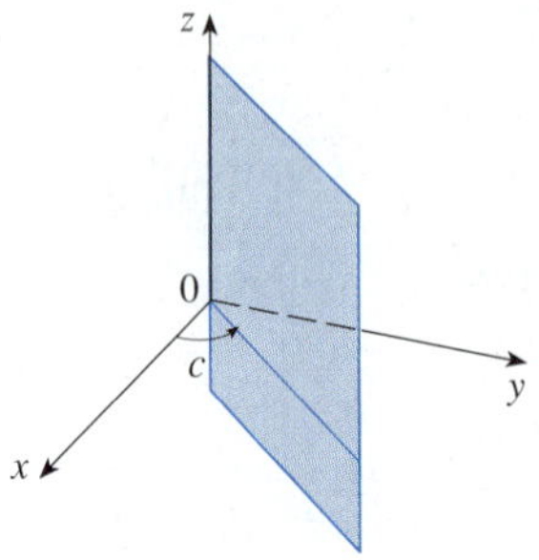

FIGURE 7 $\theta = c$, a half-plane

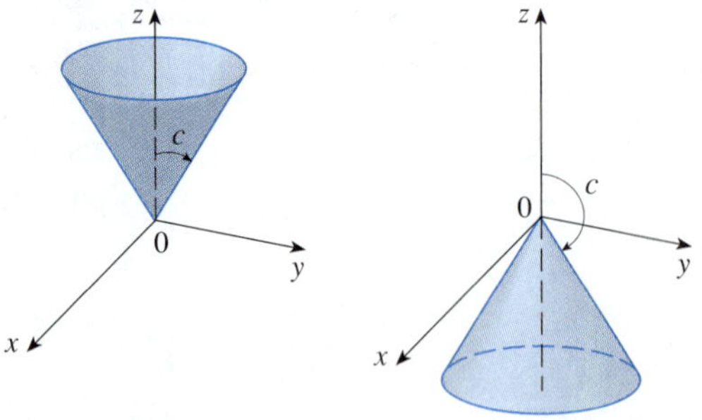

$0 < c < \pi/2$ $\qquad$ $\pi/2 < c < \pi$

FIGURE 8 $\phi = c$, a half-cone

(see Figure 7), and the equation $\phi = c$ represents a half-cone with the z-axis as its axis (see Figure 8).

The relationship between rectangular and spherical coordinates can be seen from Figure 9. From triangles OPQ and OPP' we have

$$z = \rho \cos \phi \qquad r = \rho \sin \phi$$

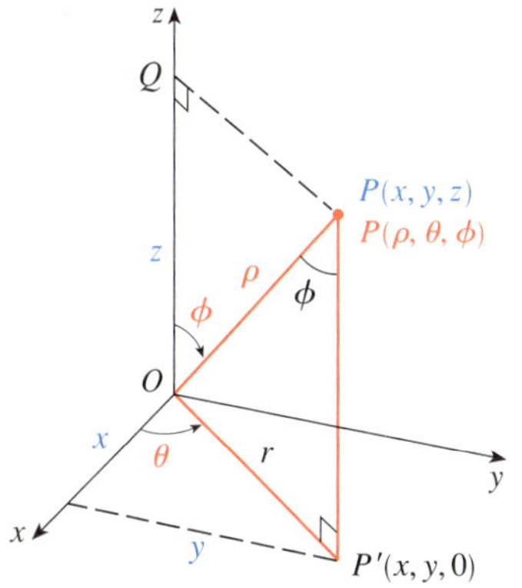

FIGURE 9

But $x = r \cos \theta$ and $y = r \sin \theta$, so to convert from spherical to rectangular coordinates, we use the equations

3
$$x = \rho \sin \phi \cos \theta \qquad y = \rho \sin \phi \sin \theta \qquad z = \rho \cos \phi$$

Also, the distance formula shows that

4
$$\rho^2 = x^2 + y^2 + z^2$$

We use this equation in converting from rectangular to spherical coordinates.

EXAMPLE 4 The point $(2, \pi/4, \pi/3)$ is given in spherical coordinates. Plot the point and find its rectangular coordinates.

SOLUTION We plot the point in Figure 10. From Equations 3 we have

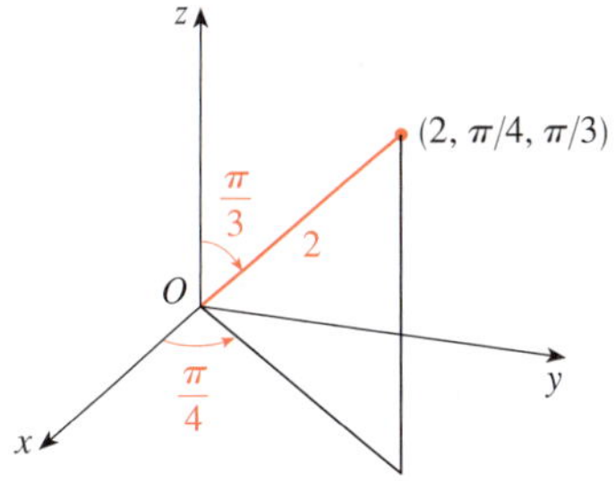

FIGURE 10

$$x = \rho \sin \phi \cos \theta = 2 \sin \frac{\pi}{3} \cos \frac{\pi}{4} = 2\left(\frac{\sqrt{3}}{2}\right)\left(\frac{1}{\sqrt{2}}\right) = \sqrt{\frac{3}{2}}$$

$$y = \rho \sin \phi \sin \theta = 2 \sin \frac{\pi}{3} \sin \frac{\pi}{4} = 2\left(\frac{\sqrt{3}}{2}\right)\left(\frac{1}{\sqrt{2}}\right) = \sqrt{\frac{3}{2}}$$

$$z = \rho \cos \phi = 2 \cos \frac{\pi}{3} = 2\left(\tfrac{1}{2}\right) = 1$$

Thus, the point $(2, \pi/4, \pi/3)$ is $\left(\sqrt{3/2}, \sqrt{3/2}, 1\right)$ in rectangular coordinates.

EXAMPLE 5 The point $(0, 2\sqrt{3}, -2)$ is given in rectangular coordinates. Find spherical coordinates for this point.

SOLUTION From Equation 4 we have

$$\rho = \sqrt{x^2 + y^2 + z^2} = \sqrt{0 + 12 + 4} = 4$$

and so Equations 3 give

$$\cos \phi = \frac{z}{\rho} = \frac{-2}{4} = -\frac{1}{2} \qquad \phi = \frac{2\pi}{3}$$

$$\cos \theta = \frac{x}{\rho \sin \phi} = 0 \qquad \theta = \frac{\pi}{2}$$

(Note that $\theta \neq 3\pi/2$ because $y = 2\sqrt{3} > 0$.) Therefore, spherical coordinates of the given point are $(4, \pi/2, 2\pi/3)$.

EXAMPLE 6 Find an equation in spherical coordinates for the hyperboloid of two sheets with equation $x^2 - y^2 - z^2 = 1$.

TEC In Module 12.7 you can investigate families of surfaces in cylindrical and spherical coordinates.

SOLUTION Substituting the expressions in Equations 3 into the given equation, we have

$$\rho^2 \sin^2\phi \cos^2\theta - \rho^2 \sin^2\phi \sin^2\theta - \rho^2 \cos^2\phi = 1$$

$$\rho^2[\sin^2\phi (\cos^2\theta - \sin^2\theta) - \cos^2\phi] = 1$$

or

$$\rho^2(\sin^2\phi \cos 2\theta - \cos^2\phi) = 1$$

EXAMPLE 7 Find a rectangular equation for the surface whose spherical equation is $\rho = \sin\theta \sin\phi$.

SOLUTION From Equations 4 and 3 we have

$$x^2 + y^2 + z^2 = \rho^2 = \rho \sin\theta \sin\phi = y$$

or

$$x^2 + (y - \tfrac{1}{2})^2 + z^2 = \tfrac{1}{4}$$

which is the equation of a sphere with center $(0, \frac{1}{2}, 0)$ and radius $\frac{1}{2}$.

EXAMPLE 8 Use a computer to draw a picture of the solid that remains when a hole of radius 3 is drilled through the center of a sphere of radius 4.

SOLUTION To keep the equations simple, let's choose the coordinate system so that the center of the sphere is at the origin and the axis of the cylinder that forms the hole is the z-axis. We could use either cylindrical or spherical coordinates to describe the solid, but the description is somewhat simpler if we use cylindrical coordinates. Then the equation of the cylinder is $r = 3$ and the equation of the sphere is $x^2 + y^2 + z^2 = 16$, or $r^2 + z^2 = 16$. The points in the solid lie outside the cylinder and inside the sphere, so they satisfy the inequalities

$$3 \leqslant r \leqslant \sqrt{16 - z^2}$$

||||| Most three-dimensional graphing programs can graph surfaces whose equations are given in cylindrical or spherical coordinates. As Example 8 demonstrates, this is often the most convenient way of drawing a solid.

To ensure that the computer graphs only the appropriate parts of these surfaces, we find where they intersect by solving the equations $r = 3$ and $r = \sqrt{16 - z^2}$:

$$\sqrt{16 - z^2} = 3 \quad \Rightarrow \quad 16 - z^2 = 9 \quad \Rightarrow \quad z^2 = 7 \quad \Rightarrow \quad z = \pm\sqrt{7}$$

The solid lies between $z = -\sqrt{7}$ and $z = \sqrt{7}$, so we ask the computer to graph the surfaces with the following equations and domains:

$$r = 3 \qquad 0 \leqslant \theta \leqslant 2\pi \qquad -\sqrt{7} \leqslant z \leqslant \sqrt{7}$$

$$r = \sqrt{16 - z^2} \qquad 0 \leqslant \theta \leqslant 2\pi \qquad -\sqrt{7} \leqslant z \leqslant \sqrt{7}$$

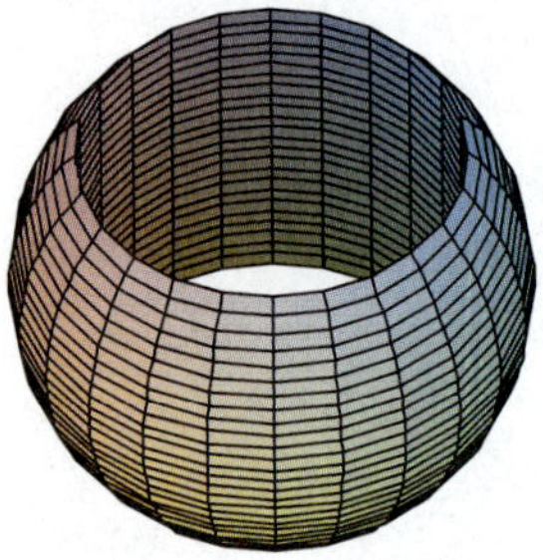

FIGURE 11

The resulting picture, shown in Figure 11, is exactly what we want.

12.7 Exercises

1. What are cylindrical coordinates? For what types of surfaces do they provide convenient descriptions?

2. What are spherical coordinates? For what types of surfaces do they provide convenient descriptions?

3–8 ||||| Plot the point whose cylindrical coordinates are given. Then find the rectangular coordinates of the point.

3. $(2, \pi/4, 1)$

4. $(1, 3\pi/2, 2)$

5. $(3, 0, -6)$

6. $(1, \pi, e)$

7. $(4, -\pi/3, 5)$ **8.** $(5, \pi/6, 6)$

9–12 Change from rectangular to cylindrical coordinates.

9. $(1, -1, 4)$ **10.** $(3, 3, -2)$

11. $(-1, -\sqrt{3}, 2)$ **12.** $(3, 4, 5)$

13–18 Plot the point whose spherical coordinates are given. Then find the rectangular coordinates of the point.

13. $(1, 0, 0)$ **14.** $(3, 0, \pi)$

15. $(1, \pi/6, \pi/6)$ **16.** $(5, \pi, \pi/2)$

17. $(2, \pi/3, \pi/4)$ **18.** $(2, \pi/4, \pi/3)$

19–22 Change from rectangular to spherical coordinates.

19. $(1, \sqrt{3}, 2\sqrt{3})$ **20.** $(0, \sqrt{3}, 1)$

21. $(0, -1, -1)$ **22.** $(-1, 1, \sqrt{6})$

23–26 Change from cylindrical to spherical coordinates.

23. $(1, \pi/6, \sqrt{3})$ **24.** $(\sqrt{6}, \pi/4, \sqrt{2})$

25. $(\sqrt{3}, \pi/2, -1)$ **26.** $(4, \pi/8, 3)$

27–30 Change from spherical to cylindrical coordinates.

27. $(2, 0, 0)$ **28.** $(2\sqrt{2}, 3\pi/2, \pi/2)$

29. $(8, \pi/6, \pi/2)$ **30.** $(4, \pi/4, \pi/3)$

31–36 Describe in words the surface whose equation is given.

31. $r = 3$ **32.** $\rho = 3$

33. $\phi = 0$ **34.** $\phi = \pi/2$

35. $\phi = \pi/3$ **36.** $\theta = \pi/3$

37–48 Identify the surface whose equation is given.

37. $z = r^2$ **38.** $r = 4 \sin \theta$

39. $\rho \cos \phi = 2$ **40.** $\rho \sin \phi = 2$

41. $r = 2 \cos \theta$ **42.** $\rho = 2 \cos \phi$

43. $r^2 + z^2 = 25$ **44.** $r^2 - 2z^2 = 4$

45. $\rho^2(\sin^2\phi \cos^2\theta + \cos^2\phi) = 4$

46. $\rho^2(\sin^2\phi - 4 \cos^2\phi) = 1$

47. $r^2 = r$ **48.** $\rho^2 - 6\rho + 8 = 0$

49–56 Write the equation (a) in cylindrical coordinates and (b) in spherical coordinates.

49. $z = x^2 + y^2$ **50.** $x^2 + y^2 + z^2 = 2$

51. $x = 3$ **52.** $x^2 + y^2 + z^2 + 2z = 0$

53. $x^2 - y^2 - 2z^2 = 4$ **54.** $y^2 + z^2 = 1$

55. $x^2 + y^2 = 2y$ **56.** $z = x^2 - y^2$

57–62 Sketch the solid described by the given inequalities.

57. $r^2 \leqslant z \leqslant 2 - r^2$

58. $0 \leqslant \theta \leqslant \pi/2, \quad r \leqslant z \leqslant 2$

59. $\rho \leqslant 2, \quad 0 \leqslant \phi \leqslant \pi/2, \quad 0 \leqslant \theta \leqslant \pi/2$

60. $2 \leqslant \rho \leqslant 3, \quad \pi/2 \leqslant \phi \leqslant \pi$

61. $-\pi/2 \leqslant \theta \leqslant \pi/2, \quad 0 \leqslant \phi \leqslant \pi/6, \quad 0 \leqslant \rho \leqslant \sec \phi$

62. $0 \leqslant \phi \leqslant \pi/3, \quad \rho \leqslant 2$

63. A cylindrical shell is 20 cm long, with inner radius 6 cm and outer radius 7 cm. Write inequalities that describe the shell in an appropriate coordinate system. Explain how you have positioned the coordinate system with respect to the shell.

64. (a) Find inequalities that describe a hollow ball with diameter 30 cm and thickness 0.5 cm. Explain how you have positioned the coordinate system that you have chosen.
(b) Suppose the ball is cut in half. Write inequalities that describe one of the halves.

65. A solid lies above the cone $z = \sqrt{x^2 + y^2}$ and below the sphere $x^2 + y^2 + z^2 = z$. Write a description of the solid in terms of inequalities involving spherical coordinates.

66. Use a graphing device to draw the solid enclosed by the paraboloids $z = x^2 + y^2$ and $z = 5 - x^2 - y^2$.

67. Use a graphing device to draw a silo consisting of a cylinder with radius 3 and height 10 surmounted by a hemisphere.

68. The latitude and longitude of a point P in the Northern Hemisphere are related to spherical coordinates ρ, θ, ϕ as follows. We take the origin to be the center of the Earth and the positive z-axis to pass through the North Pole. The positive x-axis passes through the point where the prime meridian (the meridian through Greenwich, England) intersects the equator. Then the latitude of P is $\alpha = 90° - \phi°$ and the longitude is $\beta = 360° - \theta°$. Find the great-circle distance from Los Angeles (lat. 34.06° N, long. 118.25° W) to Montréal (lat. 45.50° N, long. 73.60° W). Take the radius of the Earth to be 3960 mi. (A *great circle* is the circle of intersection of a sphere and a plane through the center of the sphere.)

LABORATORY PROJECT

Families of Surfaces

In this project you will discover the interesting shapes that members of families of surfaces can take. You will also see how the shape of the surface evolves as you vary the constants.

1. Use a computer to investigate the family of surfaces

$$z = (ax^2 + by^2)e^{-x^2-y^2}$$

How does the shape of the graph depend on the numbers a and b?

2. Use a computer to investigate the family of surfaces $z = x^2 + y^2 + cxy$. In particular, you should determine the transitional values of c for which the surface changes from one type of quadric surface to another.

3. Members of the family of surfaces given in spherical coordinates by the equation

$$\rho = 1 + 0.2 \sin m\theta \sin n\phi$$

have been suggested as models for tumors and have been called *bumpy spheres* and *wrinkled spheres*. Use a computer to investigate this family of surfaces, assuming that m and n are positive integers. What roles do the values of m and n play in the shape of the surface?

12 Review

CONCEPT CHECK

1. What is the difference between a vector and a scalar?

2. How do you add two vectors geometrically? How do you add them algebraically?

3. If $\mathbf{a}$ is a vector and c is a scalar, how is $c\mathbf{a}$ related to $\mathbf{a}$ geometrically? How do you find $c\mathbf{a}$ algebraically?

4. How do you find the vector from one point to another?

5. How do you find the dot product $\mathbf{a} \cdot \mathbf{b}$ of two vectors if you know their lengths and the angle between them? What if you know their components?

6. How are dot products useful?

7. Write expressions for the scalar and vector projections of $\mathbf{b}$ onto $\mathbf{a}$. Illustrate with diagrams.

8. How do you find the cross product $\mathbf{a} \times \mathbf{b}$ of two vectors if you know their lengths and the angle between them? What if you know their components?

9. How are cross products useful?

10. (a) How do you find the area of the parallelogram determined by $\mathbf{a}$ and $\mathbf{b}$?
(b) How do you find the volume of the parallelepiped determined by $\mathbf{a}$, $\mathbf{b}$, and $\mathbf{c}$?

11. How do you find a vector perpendicular to a plane?

12. How do you find the angle between two intersecting planes?

13. Write a vector equation, parametric equations, and symmetric equations for a line.

14. Write a vector equation and a scalar equation for a plane.

15. (a) How do you tell if two vectors are parallel?
(b) How do you tell if two vectors are perpendicular?
(c) How do you tell if two planes are parallel?

16. (a) Describe a method for determining whether three points P, Q, and R lie on the same line.
(b) Describe a method for determining whether four points P, Q, R, and S lie in the same plane.

17. (a) How do you find the distance from a point to a line?
(b) How do you find the distance from a point to a plane?
(c) How do you find the distance between two lines?

18. What are the traces of a surface? How do you find them?

19. Write equations in standard form of the six types of quadric surfaces.

20. (a) Write the equations for converting from cylindrical to rectangular coordinates. In what situation would you use cylindrical coordinates?
(b) Write the equations for converting from spherical to rectangular coordinates. In what situation would you use spherical coordinates?

TRUE-FALSE QUIZ

Determine whether the statement is true or false. If it is true, explain why. If it is false, explain why or give an example that disproves the statement.

1. For any vectors $\mathbf{u}$ and $\mathbf{v}$ in V_3, $\mathbf{u} \cdot \mathbf{v} = \mathbf{v} \cdot \mathbf{u}$.

2. For any vectors $\mathbf{u}$ and $\mathbf{v}$ in V_3, $\mathbf{u} \times \mathbf{v} = \mathbf{v} \times \mathbf{u}$.

3. For any vectors $\mathbf{u}$ and $\mathbf{v}$ in V_3, $|\mathbf{u} \times \mathbf{v}| = |\mathbf{v} \times \mathbf{u}|$.

4. For any vectors $\mathbf{u}$ and $\mathbf{v}$ in V_3 and any scalar k, $k(\mathbf{u} \cdot \mathbf{v}) = (k\mathbf{u}) \cdot \mathbf{v}$.

5. For any vectors $\mathbf{u}$ and $\mathbf{v}$ in V_3 and any scalar k, $k(\mathbf{u} \times \mathbf{v}) = (k\mathbf{u}) \times \mathbf{v}$.

6. For any vectors $\mathbf{u}$, $\mathbf{v}$, and $\mathbf{w}$ in V_3, $(\mathbf{u} + \mathbf{v}) \times \mathbf{w} = \mathbf{u} \times \mathbf{w} + \mathbf{v} \times \mathbf{w}$.

7. For any vectors $\mathbf{u}$, $\mathbf{v}$, and $\mathbf{w}$ in V_3, $\mathbf{u} \cdot (\mathbf{v} \times \mathbf{w}) = (\mathbf{u} \times \mathbf{v}) \cdot \mathbf{w}$.

8. For any vectors $\mathbf{u}$, $\mathbf{v}$, and $\mathbf{w}$ in V_3, $\mathbf{u} \times (\mathbf{v} \times \mathbf{w}) = (\mathbf{u} \times \mathbf{v}) \times \mathbf{w}$.

9. For any vectors $\mathbf{u}$ and $\mathbf{v}$ in V_3, $(\mathbf{u} \times \mathbf{v}) \cdot \mathbf{u} = 0$.

10. For any vectors $\mathbf{u}$ and $\mathbf{v}$ in V_3, $(\mathbf{u} + \mathbf{v}) \times \mathbf{v} = \mathbf{u} \times \mathbf{v}$.

11. The cross product of two unit vectors is a unit vector.

12. A linear equation $Ax + By + Cz + D = 0$ represents a line in space.

13. The set of points $\{(x, y, z) \mid x^2 + y^2 = 1\}$ is a circle.

14. If $\mathbf{u} = \langle u_1, u_2 \rangle$ and $\mathbf{v} = \langle v_1, v_2 \rangle$, then $\mathbf{u} \cdot \mathbf{v} = \langle u_1 v_1, u_2 v_2 \rangle$.

EXERCISES

1. (a) Find an equation of the sphere with center $(1, -1, 2)$ and radius 3.
   (b) Find the center and radius of the sphere

$$x^2 + y^2 + z^2 + 4x + 6y - 10z + 2 = 0$$

2. Copy the vectors in the figure and use them to draw each of the following vectors.
   (a) $\mathbf{a} + \mathbf{b}$ (b) $\mathbf{a} - \mathbf{b}$
   (c) $-\frac{1}{2}\mathbf{a}$ (d) $2\mathbf{a} + \mathbf{b}$

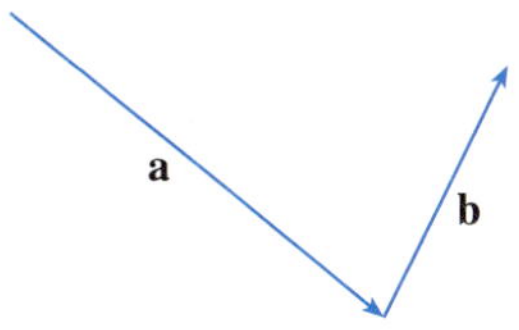

3. If $\mathbf{u}$ and $\mathbf{v}$ are the vectors shown in the figure, find $\mathbf{u} \cdot \mathbf{v}$ and $|\mathbf{u} \times \mathbf{v}|$. Is $\mathbf{u} \times \mathbf{v}$ directed into the page or out of it?

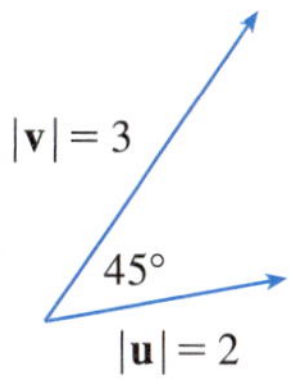

4. Calculate the given quantity if

$$\mathbf{a} = \mathbf{i} + \mathbf{j} - 2\mathbf{k} \qquad \mathbf{b} = 3\mathbf{i} - 2\mathbf{j} + \mathbf{k} \qquad \mathbf{c} = \mathbf{j} - 5\mathbf{k}$$

   (a) $2\mathbf{a} + 3\mathbf{b}$ (b) $|\mathbf{b}|$
   (c) $\mathbf{a} \cdot \mathbf{b}$ (d) $\mathbf{a} \times \mathbf{b}$
   (e) $|\mathbf{b} \times \mathbf{c}|$ (f) $\mathbf{a} \cdot (\mathbf{b} \times \mathbf{c})$
   (g) $\mathbf{c} \times \mathbf{c}$ (h) $\mathbf{a} \times (\mathbf{b} \times \mathbf{c})$
   (i) $\text{comp}_{\mathbf{a}}\, \mathbf{b}$ (j) $\text{proj}_{\mathbf{a}}\, \mathbf{b}$
   (k) The angle between $\mathbf{a}$ and $\mathbf{b}$ (correct to the nearest degree)

5. Find the values of x such that the vectors $\langle 3, 2, x \rangle$ and $\langle 2x, 4, x \rangle$ are orthogonal.

6. Find two unit vectors that are orthogonal to both $\mathbf{j} + 2\mathbf{k}$ and $\mathbf{i} - 2\mathbf{j} + 3\mathbf{k}$.

7. Suppose that $\mathbf{u} \cdot (\mathbf{v} \times \mathbf{w}) = 2$. Find
   (a) $(\mathbf{u} \times \mathbf{v}) \cdot \mathbf{w}$ (b) $\mathbf{u} \cdot (\mathbf{w} \times \mathbf{v})$
   (c) $\mathbf{v} \cdot (\mathbf{u} \times \mathbf{w})$ (d) $(\mathbf{u} \times \mathbf{v}) \cdot \mathbf{v}$

8. Show that if $\mathbf{a}$, $\mathbf{b}$, and $\mathbf{c}$ are in V_3, then

$$(\mathbf{a} \times \mathbf{b}) \cdot [(\mathbf{b} \times \mathbf{c}) \times (\mathbf{c} \times \mathbf{a})] = [\mathbf{a} \cdot (\mathbf{b} \times \mathbf{c})]^2$$

9. Find the acute angle between two diagonals of a cube.

10. Given the points $A(1, 0, 1)$, $B(2, 3, 0)$, $C(-1, 1, 4)$, and $D(0, 3, 2)$, find the volume of the parallelepiped with adjacent edges AB, AC, and AD.

11. (a) Find a vector perpendicular to the plane through the points $A(1, 0, 0)$, $B(2, 0, -1)$, and $C(1, 4, 3)$.
    (b) Find the area of triangle ABC.

12. A constant force $\mathbf{F} = 3\mathbf{i} + 5\mathbf{j} + 10\mathbf{k}$ moves an object along the line segment from $(1, 0, 2)$ to $(5, 3, 8)$. Find the work done if the distance is measured in meters and the force in newtons.

13. A boat is pulled onto shore using two ropes, as shown in the diagram. If a force of 255 N is needed, find the magnitude of the force in each rope.

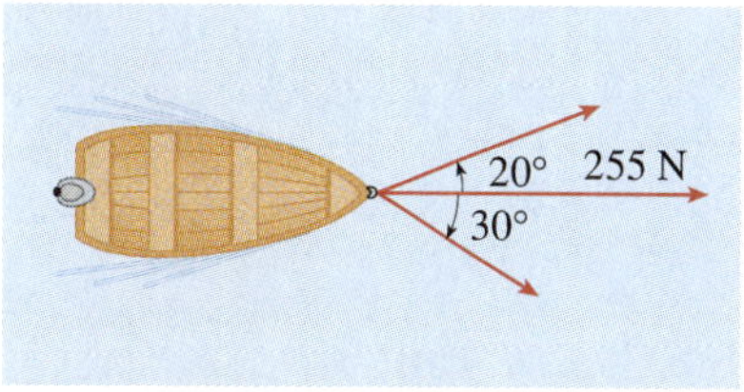

14. Find the magnitude of the torque about P if a 50-N force is applied as shown.

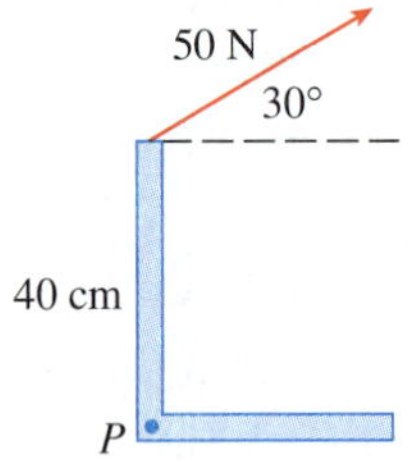

15–17 |||| Find parametric equations for the line.

15. The line through $(4, -1, 2)$ and $(1, 1, 5)$

16. The line through $(1, 0, -1)$ and parallel to the line $\frac{1}{3}(x - 4) = \frac{1}{2}y = z + 2$

17. The line through $(-2, 2, 4)$ and perpendicular to the plane $2x - y + 5z = 12$

18–20 |||| Find an equation of the plane.

18. The plane through $(2, 1, 0)$ and parallel to $x + 4y - 3z = 1$

19. The plane through $(3, -1, 1)$, $(4, 0, 2)$, and $(6, 3, 1)$

20. The plane through $(1, 2, -2)$ that contains the line $x = 2t$, $y = 3 - t$, $z = 1 + 3t$

21. Find the point in which the line with parametric equations $x = 2 - t$, $y = 1 + 3t$, $z = 4t$ intersects the plane $2x - y + z = 2$.

22. Find the distance from the origin to the line $x = 1 + t$, $y = 2 - t$, $z = -1 + 2t$.

23. Determine whether the lines given by the symmetric equations

$$\frac{x - 1}{2} = \frac{y - 2}{3} = \frac{z - 3}{4}$$

and

$$\frac{x + 1}{6} = \frac{y - 3}{-1} = \frac{z + 5}{2}$$

are parallel, skew, or intersecting.

24. (a) Show that the planes $x + y - z = 1$ and $2x - 3y + 4z = 5$ are neither parallel nor perpendicular.
(b) Find, correct to the nearest degree, the angle between these planes.

25. Find the distance between the planes $3x + y - 4z = 2$ and $3x + y - 4z = 24$.

26–34 |||| Identify and sketch the graph of each surface.

26. $x = 3$

27. $x = z$

28. $y = z^2$

29. $x^2 = y^2 + 4z^2$

30. $4x - y + 2z = 4$

31. $-4x^2 + y^2 - 4z^2 = 4$

32. $y^2 + z^2 = 1 + x^2$

33. $4x^2 + 4y^2 - 8y + z^2 = 0$

34. $x = y^2 + z^2 - 2y - 4z + 5$

35. An ellipsoid is created by rotating the ellipse $4x^2 + y^2 = 16$ about the x-axis. Find an equation of the ellipsoid.

36. A surface consists of all points P such that the distance from P to the plane $y = 1$ is twice the distance from P to the point $(0, -1, 0)$. Find an equation for this surface and identify it.

37. The cylindrical coordinates of a point are $(2\sqrt{3}, \pi/3, 2)$. Find the rectangular and spherical coordinates of the point.

38. The rectangular coordinates of a point are $(2, 2, -1)$. Find the cylindrical and spherical coordinates of the point.

39. The spherical coordinates of a point are $(8, \pi/4, \pi/6)$. Find the rectangular and cylindrical coordinates of the point.

40–43 |||| Identify the surface whose equation is given.

40. $\phi = \pi/4$

41. $\theta = \pi/4$

42. $r = \cos\theta$

43. $\rho = 3\sec\phi$

44–46 |||| Write the equation in cylindrical coordinates and in spherical coordinates.

44. $x^2 + y^2 = 4$

45. $x^2 + y^2 + z^2 = 4$

46. $x^2 + y^2 + z^2 = 2x$

47. The parabola $z = 4y^2$, $x = 0$ is rotated about the z-axis. Write an equation of the resulting surface in cylindrical coordinates.

48. Sketch the solid consisting of all points with spherical coordinates (ρ, θ, ϕ) such that $0 \leqslant \theta \leqslant \pi/2$, $0 \leqslant \phi \leqslant \pi/6$, and $0 \leqslant \rho \leqslant 2\cos\phi$.

PROBLEMS PLUS

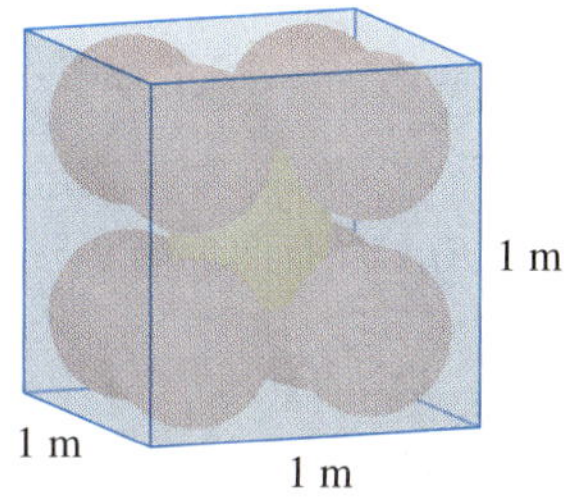

FIGURE FOR PROBLEM 1

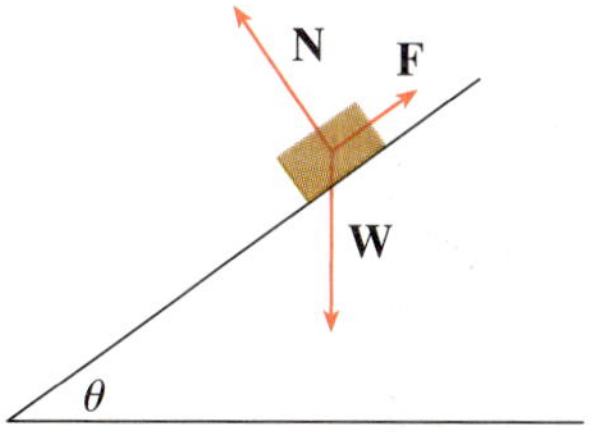

FIGURE FOR PROBLEM 5

1. Each edge of a cubical box has length 1 m. The box contains nine spherical balls with the same radius r. The center of one ball is at the center of the cube and it touches the other eight balls. Each of the other eight balls touches three sides of the box. Thus, the balls are tightly packed in the box. (See the figure.) Find r. (If you have trouble with this problem, read about the problem-solving strategy entitled *Use analogy* on page 80.)

2. Let B be a solid box with length L, width W, and height H. Let S be the set of all points that are a distance at most 1 from some point of B. Express the volume of S in terms of L, W, and H.

3. Let L be the line of intersection of the planes $cx + y + z = c$ and $x - cy + cz = -1$, where c is a real number.

(a) Find symmetric equations for L.

(b) As the number c varies, the line L sweeps out a surface S. Find an equation for the curve of intersection of S with the horizontal plane $z = t$ (the trace of S in the plane $z = t$).

(c) Find the volume of the solid bounded by S and the planes $z = 0$ and $z = 1$.

4. A plane is capable of flying at a speed of 180 km/h in still air. The pilot takes off from an airfield and heads due north according to the plane's compass. After 30 minutes of flight time, the pilot notices that, due to the wind, the plane has actually traveled 80 km at an angle 5° east of north.

(a) What is the wind velocity?

(b) In what direction should the pilot have headed to reach the intended destination?

5. Suppose a block of mass m is placed on an inclined plane, as shown in the figure. The block's descent down the plane is slowed by friction; if θ is not too large, friction will prevent the block from moving at all. The forces acting on the block are the weight $\mathbf{W}$, where $|\mathbf{W}| = mg$ (g is the acceleration due to gravity); the normal force $\mathbf{N}$ (the normal component of the reactionary force of the plane on the block), where $|\mathbf{N}| = n$; and the force $\mathbf{F}$ due to friction, which acts parallel to the inclined plane, opposing the direction of motion. If the block is at rest and θ is increased, $|\mathbf{F}|$ must also increase until ultimately $|\mathbf{F}|$ reaches its maximum, beyond which the block begins to slide. At this angle θ_s, it has been observed that $|\mathbf{F}|$ is proportional to n. Thus, when $|\mathbf{F}|$ is maximal, we can say that $|\mathbf{F}| = \mu_s n$, where μ_s is called the *coefficient of static friction* and depends on the materials that are in contact.

(a) Observe that $\mathbf{N} + \mathbf{F} + \mathbf{W} = \mathbf{0}$ and deduce that $\mu_s = \tan\theta_s$.

(b) Suppose that, for $\theta > \theta_s$, an additional outside force $\mathbf{H}$ is applied to the block, horizontally from the left, and let $|\mathbf{H}| = h$. If h is small, the block may still slide down the plane; if h is large enough, the block will move up the plane. Let $h_{\min}$ be the smallest value of h that allows the block to remain motionless (so that $|\mathbf{F}|$ is maximal).

By choosing the coordinate axes so that $\mathbf{F}$ lies along the x-axis, resolve each force into components parallel and perpendicular to the inclined plane and show that

$$h_{\min}\sin\theta + mg\cos\theta = n \qquad \text{and} \qquad h_{\min}\cos\theta + \mu_s n = mg\sin\theta$$

(c) Show that
$$h_{\min} = mg\tan(\theta - \theta_s)$$

Does this equation seem reasonable? Does it make sense for $\theta = \theta_s$? As $\theta \to 90°$? Explain.

(d) Let $h_{\max}$ be the largest value of h that allows the block to remain motionless. (In which direction is $\mathbf{F}$ heading?) Show that

$$h_{\max} = mg\tan(\theta + \theta_s)$$

Does this equation seem reasonable? Explain.

CHAPTER 13

The calculus of vector-valued functions is used in Section 13.4 to prove Kepler's laws. These describe the motion of the planets about the Sun and also apply to the orbit of a satellite about the Earth, such as the Hubble Space Telescope.

Vector Functions

The functions that we have been using so far have been real-valued functions. We now study functions whose values are vectors because such functions are needed to describe curves and surfaces in space. We will also use vector-valued functions to describe the motion of objects through space. In particular, we will use them to derive Kepler's laws of planetary motion.

13.1 Vector Functions and Space Curves

In general, a function is a rule that assigns to each element in the domain an element in the range. A **vector-valued function**, or **vector function**, is simply a function whose domain is a set of real numbers and whose range is a set of vectors. We are most interested in vector functions $\mathbf{r}$ whose values are three-dimensional vectors. This means that for every number t in the domain of $\mathbf{r}$ there is a unique vector in V_3 denoted by $\mathbf{r}(t)$. If $f(t)$, $g(t)$, and $h(t)$ are the components of the vector $\mathbf{r}(t)$, then f, g, and h are real-valued functions called the **component functions** of $\mathbf{r}$ and we can write

$$\mathbf{r}(t) = \langle f(t), g(t), h(t) \rangle = f(t)\,\mathbf{i} + g(t)\,\mathbf{j} + h(t)\,\mathbf{k}$$

We use the letter t to denote the independent variable because it represents time in most applications of vector functions.

EXAMPLE 1 If

$$\mathbf{r}(t) = \left\langle t^3, \ln(3-t), \sqrt{t} \right\rangle$$

then the component functions are

$$f(t) = t^3 \qquad g(t) = \ln(3-t) \qquad h(t) = \sqrt{t}$$

By our usual convention, the domain of $\mathbf{r}$ consists of all values of t for which the expression for $\mathbf{r}(t)$ is defined. The expressions t^3, $\ln(3-t)$, and $\sqrt{t}$ are all defined when $3 - t > 0$ and $t \geqslant 0$. Therefore, the domain of $\mathbf{r}$ is the interval $[0, 3)$.

The **limit** of a vector function $\mathbf{r}$ is defined by taking the limits of its component functions as follows.

If $\lim_{t \to a} \mathbf{r}(t) = \mathbf{L}$, this definition is equivalent to saying that the length and direction of the vector $\mathbf{r}(t)$ approach the length and direction of the vector $\mathbf{L}$.

1 If $\mathbf{r}(t) = \langle f(t), g(t), h(t) \rangle$, then

$$\lim_{t \to a} \mathbf{r}(t) = \left\langle \lim_{t \to a} f(t), \lim_{t \to a} g(t), \lim_{t \to a} h(t) \right\rangle$$

provided the limits of the component functions exist.

Equivalently, we could have used an ε-δ definition (see Exercise 43). Limits of vector functions obey the same rules as limits of real-valued functions (see Exercise 41).

EXAMPLE 2 Find $\lim_{t \to 0} \mathbf{r}(t)$, where $\mathbf{r}(t) = (1 + t^3)\,\mathbf{i} + te^{-t}\,\mathbf{j} + \dfrac{\sin t}{t}\,\mathbf{k}$.

SOLUTION According to Definition 1, the limit of **r** is the vector whose components are the limits of the component functions of **r**:

$$\lim_{t \to 0} \mathbf{r}(t) = \left[\lim_{t \to 0} (1 + t^3)\right]\mathbf{i} + \left[\lim_{t \to 0} te^{-t}\right]\mathbf{j} + \left[\lim_{t \to 0} \frac{\sin t}{t}\right]\mathbf{k}$$

$$= \mathbf{i} + \mathbf{k} \qquad \text{(by Equation 3.4.2)}$$

A vector function **r** is **continuous at *a*** if

$$\lim_{t \to a} \mathbf{r}(t) = \mathbf{r}(a)$$

In view of Definition 1, we see that **r** is continuous at a if and only if its component functions f, g, and h are continuous at a.

There is a close connection between continuous vector functions and space curves. Suppose that f, g, and h are continuous real-valued functions on an interval I. Then the set C of all points (x, y, z) in space, where

2 $$x = f(t) \qquad y = g(t) \qquad z = h(t)$$

and t varies throughout the interval I, is called a **space curve**. The equations in (2) are called **parametric equations of *C*** and t is called a **parameter**. We can think of C as being traced out by a moving particle whose position at time t is $(f(t), g(t), h(t))$. If we now consider the vector function $\mathbf{r}(t) = \langle f(t), g(t), h(t) \rangle$, then $\mathbf{r}(t)$ is the position vector of the point $P(f(t), g(t), h(t))$ on C. Thus, any continuous vector function **r** defines a space curve C that is traced out by the tip of the moving vector $\mathbf{r}(t)$, as shown in Figure 1.

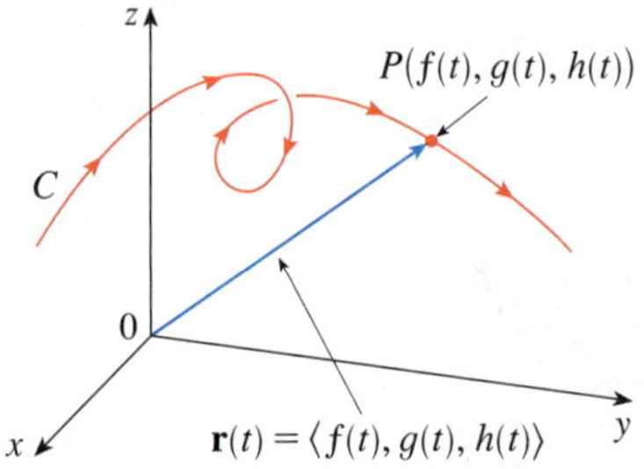

FIGURE 1
C is traced out by the tip of a moving position vector $\mathbf{r}(t)$.

EXAMPLE 3 Describe the curve defined by the vector function

$$\mathbf{r}(t) = \langle 1 + t, 2 + 5t, -1 + 6t \rangle$$

TEC Visual 13.1A shows several curves being traced out by position vectors, including those in Figures 1 and 2.

SOLUTION The corresponding parametric equations are

$$x = 1 + t \qquad y = 2 + 5t \qquad z = -1 + 6t$$

which we recognize from Equations 12.5.2 as parametric equations of a line passing through the point $(1, 2, -1)$ and parallel to the vector $\langle 1, 5, 6 \rangle$. Alternatively, we could observe that the function can be written as $\mathbf{r} = \mathbf{r}_0 + t\mathbf{v}$, where $\mathbf{r}_0 = \langle 1, 2, -1 \rangle$ and $\mathbf{v} = \langle 1, 5, 6 \rangle$, and this is the vector equation of a line as given by Equation 12.5.1.

Plane curves can also be represented in vector notation. For instance, the curve given by the parametric equations $x = t^2 - 2t$ and $y = t + 1$ (see Example 1 in Section 10.1) could also be described by the vector equation

$$\mathbf{r}(t) = \langle t^2 - 2t, t + 1 \rangle = (t^2 - 2t)\,\mathbf{i} + (t + 1)\,\mathbf{j}$$

where $\mathbf{i} = \langle 1, 0 \rangle$ and $\mathbf{j} = \langle 0, 1 \rangle$.

EXAMPLE 4 Sketch the curve whose vector equation is

$$\mathbf{r}(t) = \cos t\,\mathbf{i} + \sin t\,\mathbf{j} + t\,\mathbf{k}$$

SOLUTION The parametric equations for this curve are

$$x = \cos t \qquad y = \sin t \qquad z = t$$

Since $x^2 + y^2 = \cos^2 t + \sin^2 t = 1$, the curve must lie on the circular cylinder $x^2 + y^2 = 1$. The point (x, y, z) lies directly above the point $(x, y, 0)$, which moves counterclockwise around the circle $x^2 + y^2 = 1$ in the xy-plane. (See Example 2 in Section 10.1.) Since $z = t$, the curve spirals upward around the cylinder as t increases. The curve, shown in Figure 2, is called a **helix**.

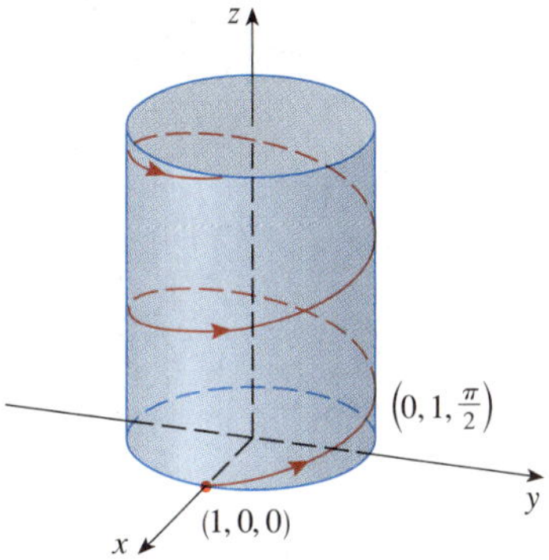

FIGURE 2

The corkscrew shape of the helix in Example 4 is familiar from its occurrence in coiled springs. It also occurs in the model of DNA (deoxyribonucleic acid, the genetic material of living cells). In 1953 James Watson and Francis Crick showed that the structure of the DNA molecule is that of two linked, parallel helices that are intertwined as in Figure 3.

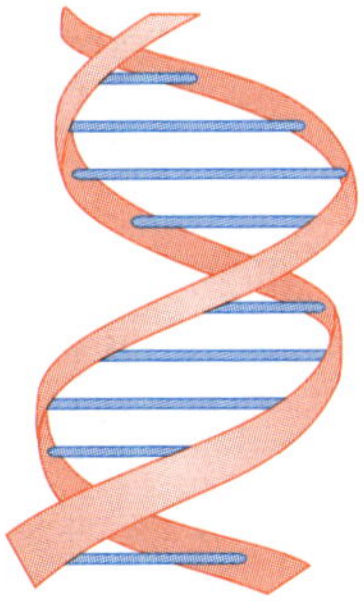

FIGURE 3

In Examples 3 and 4 we were given vector equations of curves and asked for a geometric description or sketch. In the next two examples we are given a geometric description of a curve and are asked to find parametric equations for the curve.

EXAMPLE 5 Find a vector equation and parametric equations for the line segment that joins the point $P(1, 3, -2)$ to the point $Q(2, -1, 3)$.

|||| Figure 4 shows the line segment PQ in Example 5.

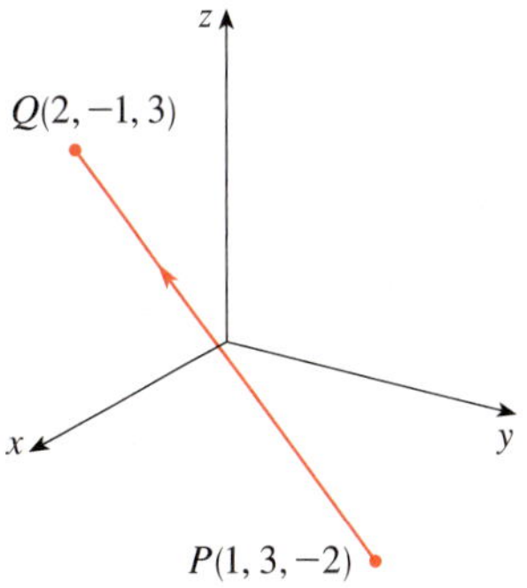

FIGURE 4

SOLUTION In Section 12.5 we found a vector equation for the line segment that joins the tip of the vector $\mathbf{r}_0$ to the tip of the vector $\mathbf{r}_1$:

$$\mathbf{r}(t) = (1 - t)\mathbf{r}_0 + t\mathbf{r}_1 \qquad 0 \leqslant t \leqslant 1$$

(See Equation 12.5.4.) Here we take $\mathbf{r}_0 = \langle 1, 3, -2 \rangle$ and $\mathbf{r}_1 = \langle 2, -1, 3 \rangle$ to obtain a vector equation of the line segment from P to Q:

$$\mathbf{r}(t) = (1 - t)\langle 1, 3, -2 \rangle + t\langle 2, -1, 3 \rangle \qquad 0 \leqslant t \leqslant 1$$

or

$$\mathbf{r}(t) = \langle 1 + t, 3 - 4t, -2 + 5t \rangle \qquad 0 \leqslant t \leqslant 1$$

The corresponding parametric equations are

$$x = 1 + t \qquad y = 3 - 4t \qquad z = -2 + 5t \qquad 0 \leqslant t \leqslant 1$$

EXAMPLE 6 Find a vector function that represents the curve of intersection of the cylinder $x^2 + y^2 = 1$ and the plane $y + z = 2$.

SOLUTION Figure 5 shows how the plane and the cylinder intersect, and Figure 6 shows the curve of intersection C, which is an ellipse.

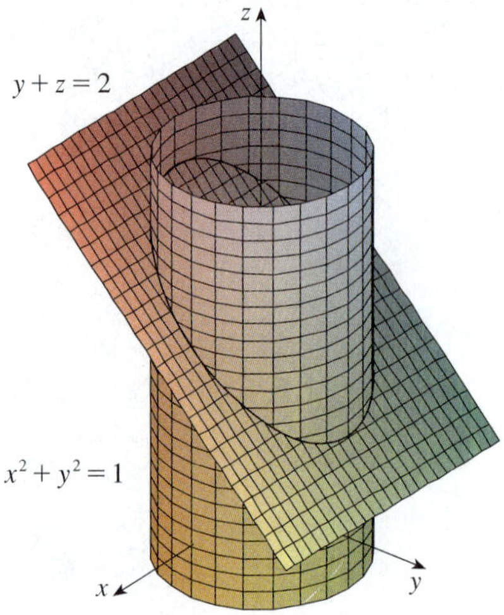

FIGURE 5

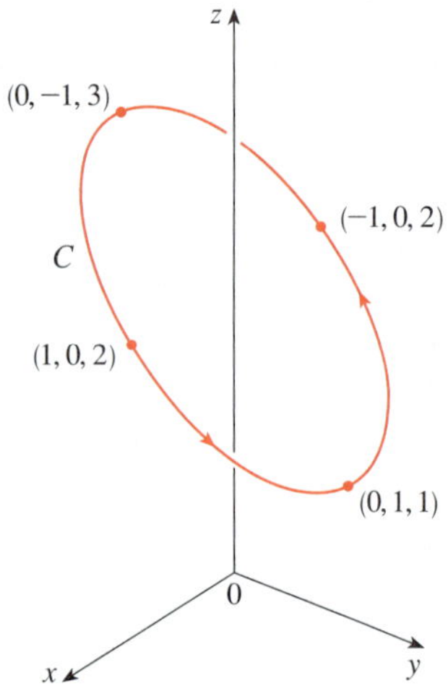

FIGURE 6

The projection of C onto the xy-plane is the circle $x^2 + y^2 = 1$, $z = 0$. So we know from Example 2 in Section 10.1 that we can write

$$x = \cos t \qquad y = \sin t \qquad 0 \leqslant t \leqslant 2\pi$$

From the equation of the plane, we have

$$z = 2 - y = 2 - \sin t$$

So we can write parametric equations for C as

$$x = \cos t \qquad y = \sin t \qquad z = 2 - \sin t \qquad 0 \leqslant t \leqslant 2\pi$$

The corresponding vector equation is

$$\mathbf{r}(t) = \cos t\,\mathbf{i} + \sin t\,\mathbf{j} + (2 - \sin t)\,\mathbf{k} \qquad 0 \leqslant t \leqslant 2\pi$$

This equation is called a *parametrization* of the curve C. The arrows in Figure 6 indicate the direction in which C is traced as the parameter t increases.

Using Computers to Draw Space Curves

Space curves are inherently more difficult to draw by hand than plane curves; for an accurate representation we need to use technology. For instance, Figure 7 shows a computer-generated graph of the curve with parametric equations

$$x = (4 + \sin 20t)\cos t \qquad y = (4 + \sin 20t)\sin t \qquad z = \cos 20t$$

It's called a **toroidal spiral** because it lies on a torus. Another interesting curve, the **tre-**

foil knot, with equations

$$x = (2 + \cos 1.5t)\cos t \qquad y = (2 + \cos 1.5t)\sin t \qquad z = \sin 1.5t$$

is graphed in Figure 8. It wouldn't be easy to plot either of these curves by hand.

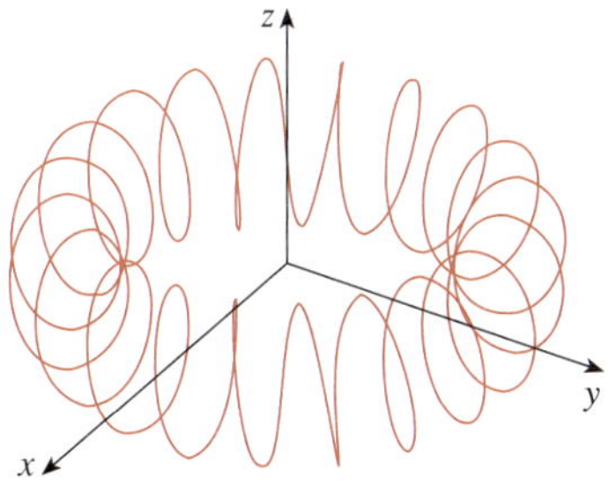

FIGURE 7 A toroidal spiral

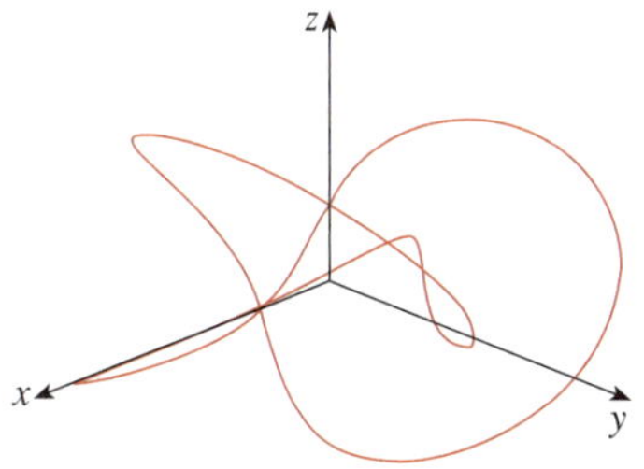

FIGURE 8 A trefoil knot

Even when a computer is used to draw a space curve, optical illusions make it difficult to get a good impression of what the curve really looks like. (This is especially true in Figure 8. See Exercise 42.) The next example shows how to cope with this problem.

EXAMPLE 7 Use a computer to sketch the curve with vector equation $\mathbf{r}(t) = \langle t, t^2, t^3 \rangle$. This curve is called a **twisted cubic**.

SOLUTION We start by using the computer to plot the curve with parametric equations $x = t$, $y = t^2$, $z = t^3$ for $-2 \leqslant t \leqslant 2$. The result is shown in Figure 9(a), but it's hard to see the true nature of the curve from that graph alone. Most three-dimensional computer graphing programs allow the user to enclose a curve or surface in a box instead of displaying the coordinate axes. When we look at the same curve in a box in Figure 9(b), we have a much clearer picture of the curve. We can see that it climbs from a lower corner of the box to the upper corner nearest us, and it twists as it climbs.

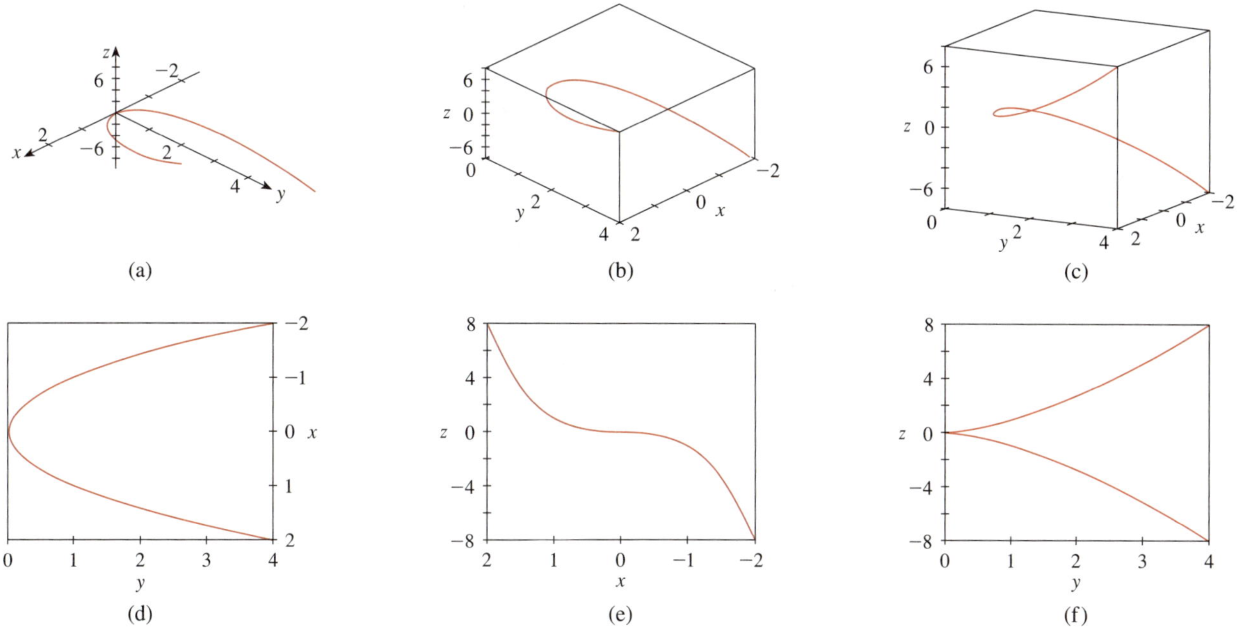

FIGURE 9 Views of the twisted cubic

TEC In Visual 13.1B you can rotate the box in Figure 9 to see the curve from any viewpoint.

We get an even better idea of the curve when we view it from different vantage points. Part (c) shows the result of rotating the box to give another viewpoint. Parts (d), (e), and (f) show the views we get when we look directly at a face of the box. In particular, part (d) shows the view from directly above the box. It is the projection of the curve on the xy-plane, namely, the parabola $y = x^2$. Part (e) shows the projection on the xz-plane, the cubic curve $z = x^3$. It's now obvious why the given curve is called a twisted cubic.

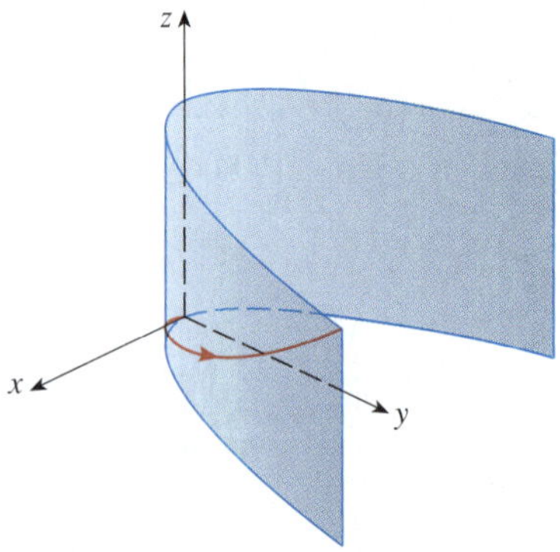

FIGURE 10

Another method of visualizing a space curve is to draw it on a surface. For instance, the twisted cubic in Example 7 lies on the parabolic cylinder $y = x^2$. (Eliminate the parameter from the first two parametric equations, $x = t$ and $y = t^2$.) Figure 10 shows both the cylinder and the twisted cubic, and we see that the curve moves upward from the origin along the surface of the cylinder. We also used this method in Example 4 to visualize the helix lying on the circular cylinder (see Figure 2).

A third method for visualizing the twisted cubic is to realize that it also lies on the cylinder $z = x^3$. So it can be viewed as the curve of intersection of the cylinders $y = x^2$ and $z = x^3$. (See Figure 11.)

TEC Visual 13.1C shows how curves arise as intersections of surfaces.

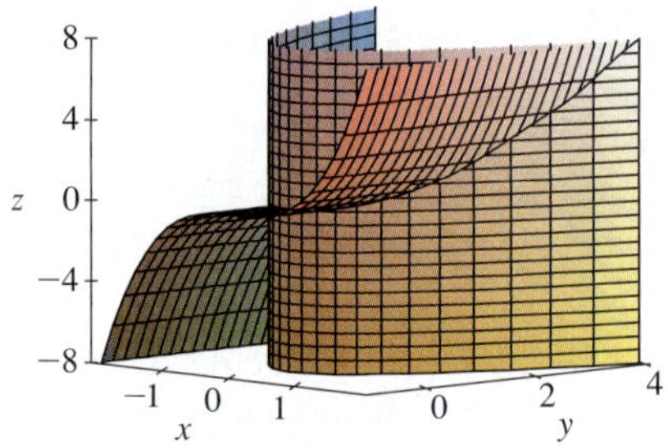

FIGURE 11

|||| Some computer algebra systems provide us with a clearer picture of a space curve by enclosing it in a tube. Such a plot enables us to see whether one part of a curve passes in front of or behind another part of the curve. For example, Figure 13 shows the curve of Figure 12(b) as rendered by the `tubeplot` command in Maple.

We have seen that an interesting space curve, the helix, occurs in the model of DNA. Another notable example of a space curve in science is the trajectory of a positively charged particle in orthogonally oriented electric and magnetic fields **E** and **B**. Depending on the initial velocity given the particle at the origin, the path of the particle is either a space curve whose projection on the horizontal plane is the cycloid we studied in Section 10.1 [Figure 12(a)] or a curve whose projection is the trochoid investigated in Exercise 38 in Section 10.1 [Figure 12(b)].

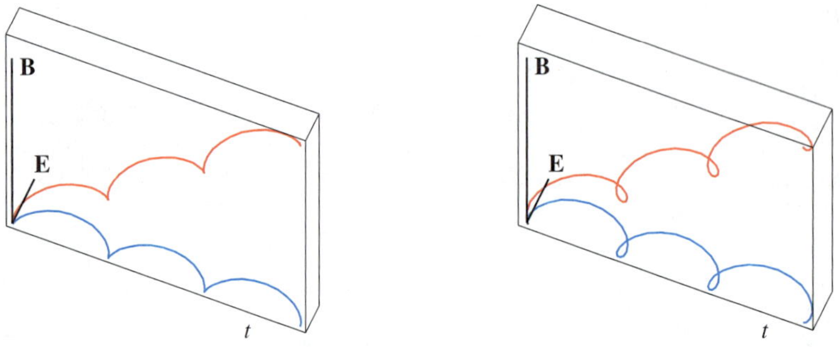

(a) $\mathbf{r}(t) = \langle t - \sin t, 1 - \cos t, t \rangle$

(b) $\mathbf{r}(t) = \langle t - \frac{3}{2}\sin t, 1 - \frac{3}{2}\cos t, t \rangle$

FIGURE 12
Motion of a charged particle in orthogonally oriented electric and magnetic fields

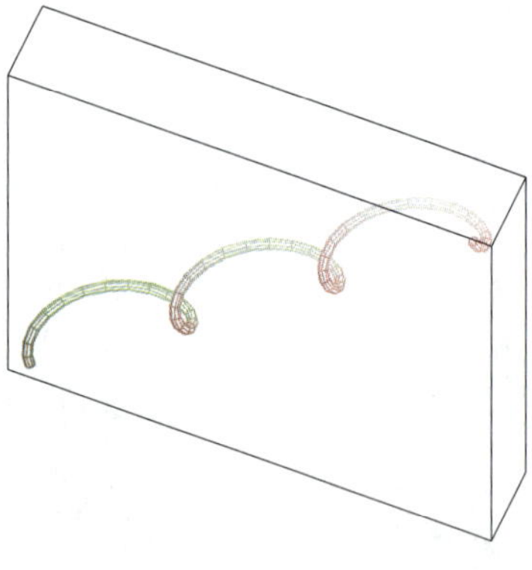
FIGURE 13

For further details concerning the physics involved and animations of the trajectories of the particles, see the following web sites:

- lompado.uah.edu/Links/CrossedFields.html
- www.phy.ntnu.edu.tw/java/emField/emField.html
- www.physics.ucla.edu/plasma-exp/Beam/

13.1 Exercises

1–2 Find the domain of the vector function.

1. $\mathbf{r}(t) = \langle t^2, \sqrt{t-1}, \sqrt{5-t} \rangle$

2. $\mathbf{r}(t) = \dfrac{t-2}{t+2}\,\mathbf{i} + \sin t\,\mathbf{j} + \ln(9 - t^2)\,\mathbf{k}$

3–6 Find the limit.

3. $\lim_{t \to 0^+} \langle \cos t, \sin t, t \ln t \rangle$

4. $\lim_{t \to 0} \left\langle \dfrac{e^t - 1}{t}, \dfrac{\sqrt{1+t} - 1}{t}, \dfrac{3}{1+t} \right\rangle$

5. $\lim_{t \to 1} \left(\sqrt{t+3}\,\mathbf{i} + \dfrac{t-1}{t^2-1}\,\mathbf{j} + \dfrac{\tan t}{t}\,\mathbf{k} \right)$

6. $\lim_{t \to \infty} \left\langle \arctan t, e^{-2t}, \dfrac{\ln t}{t} \right\rangle$

7–14 Sketch the curve with the given vector equation. Indicate with an arrow the direction in which t increases.

7. $\mathbf{r}(t) = \langle t^4 + 1, t \rangle$

8. $\mathbf{r}(t) = \langle t^3, t^2 \rangle$

9. $\mathbf{r}(t) = \langle t, \cos 2t, \sin 2t \rangle$

10. $\mathbf{r}(t) = \langle 1 + t, 3t, -t \rangle$

11. $\mathbf{r}(t) = \langle \sin t, 3, \cos t \rangle$

12. $\mathbf{r}(t) = t\,\mathbf{i} + t\,\mathbf{j} + \cos t\,\mathbf{k}$

13. $\mathbf{r}(t) = t^2\,\mathbf{i} + t^4\,\mathbf{j} + t^6\,\mathbf{k}$

14. $\mathbf{r}(t) = \sin t\,\mathbf{i} + \sin t\,\mathbf{j} + \sqrt{2}\cos t\,\mathbf{k}$

15–18 Find a vector equation and parametric equations for the line segment that joins P to Q.

15. $P(0, 0, 0)$, $Q(1, 2, 3)$

16. $P(1, 0, 1)$, $Q(2, 3, 1)$

17. $P(1, -1, 2)$, $Q(4, 1, 7)$

18. $P(-2, 4, 0)$, $Q(6, -1, 2)$

19–24 Match the parametric equations with the graphs (labeled I–VI). Give reasons for your choices.

19. $x = \cos 4t, \quad y = t, \quad z = \sin 4t$

20. $x = t, \quad y = t^2, \quad z = e^{-t}$

21. $x = t, \quad y = 1/(1 + t^2), \quad z = t^2$

22. $x = e^{-t} \cos 10t, \quad y = e^{-t} \sin 10t, \quad z = e^{-t}$

23. $x = \cos t, \quad y = \sin t, \quad z = \sin 5t$

24. $x = \cos t, \quad y = \sin t, \quad z = \ln t$

I

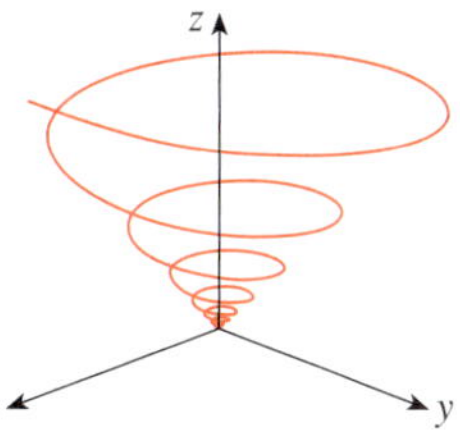

II

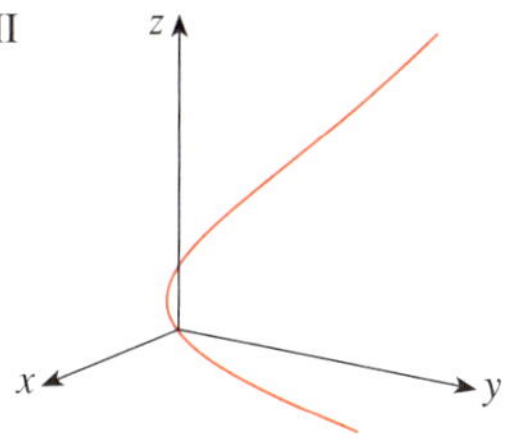

III

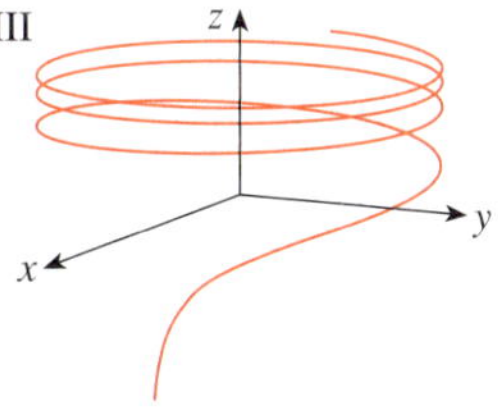

IV

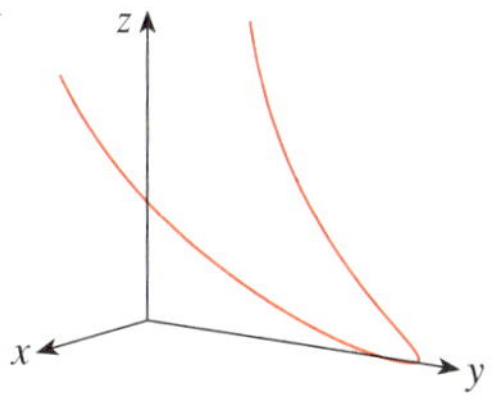

V

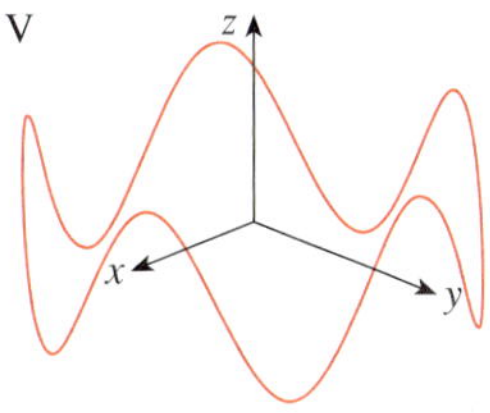

VI

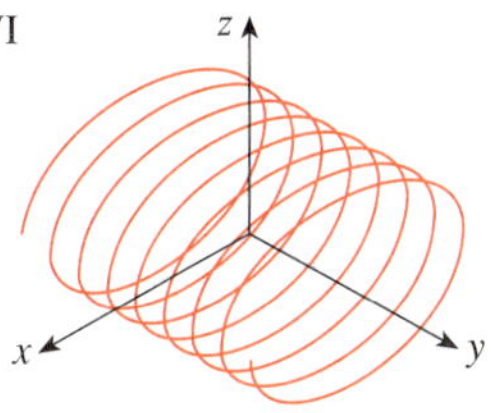

25. Show that the curve with parametric equations $x = t \cos t$, $y = t \sin t$, $z = t$ lies on the cone $z^2 = x^2 + y^2$, and use this fact to help sketch the curve.

26. Show that the curve with parametric equations $x = \sin t$, $y = \cos t$, $z = \sin^2 t$ is the curve of intersection of the surfaces $z = x^2$ and $x^2 + y^2 = 1$. Use this fact to help sketch the curve.

27–30 Use a computer to graph the curve with the given vector equation. Make sure you choose a parameter domain and viewpoints that reveal the true nature of the curve.

27. $\mathbf{r}(t) = \langle \sin t, \cos t, t^2 \rangle$

28. $\mathbf{r}(t) = \langle t^4 - t^2 + 1, t, t^2 \rangle$

29. $\mathbf{r}(t) = \langle t^2, \sqrt{t-1}, \sqrt{5-t} \rangle$

30. $\mathbf{r}(t) = \langle \sin t, \sin 2t, \sin 3t \rangle$

31. Graph the curve with parametric equations $x = (1 + \cos 16t)\cos t$, $y = (1 + \cos 16t)\sin t$, $z = 1 + \cos 16t$. Explain the appearance of the graph by showing that it lies on a cone.

32. Graph the curve with parametric equations

$$x = \sqrt{1 - 0.25\cos^2 10t}\,\cos t$$

$$y = \sqrt{1 - 0.25\cos^2 10t}\,\sin t$$

$$z = 0.5\cos 10t$$

Explain the appearance of the graph by showing that it lies on a sphere.

33. Show that the curve with parametric equations $x = t^2$, $y = 1 - 3t$, $z = 1 + t^3$ passes through the points $(1, 4, 0)$ and $(9, -8, 28)$ but not through the point $(4, 7, -6)$.

34–36 Find a vector function that represents the curve of intersection of the two surfaces.

34. The cylinder $x^2 + y^2 = 4$ and the surface $z = xy$

35. The cone $z = \sqrt{x^2 + y^2}$ and the plane $z = 1 + y$

36. The paraboloid $z = 4x^2 + y^2$ and the parabolic cylinder $y = x^2$

37. Try to sketch by hand the curve of intersection of the circular cylinder $x^2 + y^2 = 4$ and the parabolic cylinder $z = x^2$. Then find parametric equations for this curve and use these equations and a computer to graph the curve.

38. Try to sketch by hand the curve of intersection of the parabolic cylinder $y = x^2$ and the top half of the ellipsoid $x^2 + 4y^2 + 4z^2 = 16$. Then find parametric equations for this curve and use these equations and a computer to graph the curve.

39. If two objects travel through space along two different curves, it's often important to know whether they will collide. (Will a missile hit its moving target? Will two aircraft collide?) The curves might intersect, but we need to know whether the objects are in the same position *at the same time*. Suppose the trajectories of two particles are given by the vector functions

$$\mathbf{r}_1(t) = \langle t^2, 7t - 12, t^2 \rangle \qquad \mathbf{r}_2(t) = \langle 4t - 3, t^2, 5t - 6 \rangle$$

for $t \geqslant 0$. Do the particles collide?

40. Two particles travel along the space curves

$$\mathbf{r}_1(t) = \langle t, t^2, t^3 \rangle \qquad \mathbf{r}_2(t) = \langle 1 + 2t, 1 + 6t, 1 + 14t \rangle$$

Do the particles collide? Do their paths intersect?

41. Suppose $\mathbf{u}$ and $\mathbf{v}$ are vector functions that possess limits as $t \to a$ and let c be a constant. Prove the following properties of limits.

(a) $\lim_{t \to a} [\mathbf{u}(t) + \mathbf{v}(t)] = \lim_{t \to a} \mathbf{u}(t) + \lim_{t \to a} \mathbf{v}(t)$

(b) $\lim_{t \to a} c\mathbf{u}(t) = c \lim_{t \to a} \mathbf{u}(t)$

(c) $\lim_{t \to a} [\mathbf{u}(t) \cdot \mathbf{v}(t)] = \lim_{t \to a} \mathbf{u}(t) \cdot \lim_{t \to a} \mathbf{v}(t)$

(d) $\lim_{t \to a} [\mathbf{u}(t) \times \mathbf{v}(t)] = \lim_{t \to a} \mathbf{u}(t) \times \lim_{t \to a} \mathbf{v}(t)$

42. The view of the trefoil knot shown in Figure 8 is accurate, but it doesn't reveal the whole story. Use the parametric equations

$$x = (2 + \cos 1.5t)\cos t \qquad y = (2 + \cos 1.5t)\sin t$$

$$z = \sin 1.5t$$

to sketch the curve by hand as viewed from above, with gaps indicating where the curve passes over itself. Start by showing that the projection of the curve onto the xy-plane has polar coordinates $r = 2 + \cos 1.5t$ and $\theta = t$, so r varies between 1 and 3. Then show that z has maximum and minimum values when the projection is halfway between $r = 1$ and $r = 3$.

When you have finished your sketch, use a computer to draw the curve with viewpoint directly above and compare with your sketch. Then use the computer to draw the curve from several other viewpoints. You can get a better impression of the curve if you plot a tube with radius 0.2 around the curve. (Use the `tubeplot` command in Maple.)

43. Show that $\lim_{t \to a} \mathbf{r}(t) = \mathbf{b}$ if and only if for every $\varepsilon > 0$ there is a number $\delta > 0$ such that $|\mathbf{r}(t) - \mathbf{b}| < \varepsilon$ whenever $0 < |t - a| < \delta$.

13.2 Derivatives and Integrals of Vector Functions

Later in this chapter we are going to use vector functions to describe the motion of planets and other objects through space. Here we prepare the way by developing the calculus of vector functions.

Derivatives

The **derivative** $\mathbf{r}'$ of a vector function $\mathbf{r}$ is defined in much the same way as for real-valued functions:

1
$$\frac{d\mathbf{r}}{dt} = \mathbf{r}'(t) = \lim_{h \to 0} \frac{\mathbf{r}(t+h) - \mathbf{r}(t)}{h}$$

if this limit exists. The geometric significance of this definition is shown in Figure 1. If the points P and Q have position vectors $\mathbf{r}(t)$ and $\mathbf{r}(t+h)$, then $\overrightarrow{PQ}$ represents the vector $\mathbf{r}(t+h) - \mathbf{r}(t)$, which can therefore be regarded as a secant vector. If $h > 0$, the scalar multiple $(1/h)(\mathbf{r}(t+h) - \mathbf{r}(t))$ has the same direction as $\mathbf{r}(t+h) - \mathbf{r}(t)$. As $h \to 0$, it appears that this vector approaches a vector that lies on the tangent line. For this reason, the vector $\mathbf{r}'(t)$ is called the **tangent vector** to the curve defined by $\mathbf{r}$ at the point P, provided that $\mathbf{r}'(t)$ exists and $\mathbf{r}'(t) \neq \mathbf{0}$. The **tangent line** to C at P is defined to be the line through P parallel to the tangent vector $\mathbf{r}'(t)$. We will also have occasion to consider the **unit tangent vector**, which is

$$\mathbf{T}(t) = \frac{\mathbf{r}'(t)}{|\mathbf{r}'(t)|}$$

Visual 13.2 shows an animation of Figure 1.

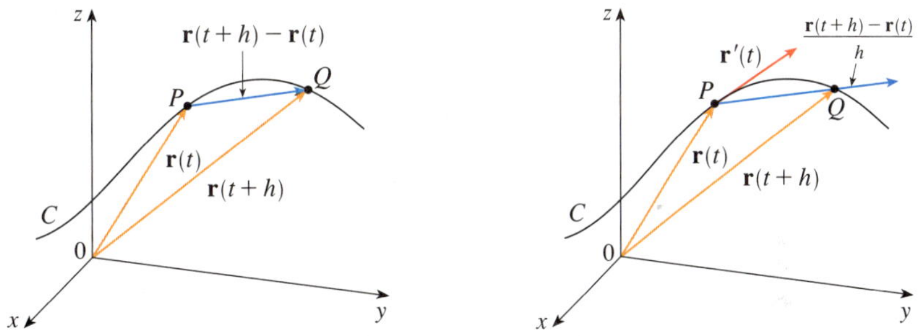

FIGURE 1

(a) The secant vector

(b) The tangent vector

The following theorem gives us a convenient method for computing the derivative of a vector function $\mathbf{r}$; just differentiate each component of $\mathbf{r}$.

2 Theorem If $\mathbf{r}(t) = \langle f(t), g(t), h(t) \rangle = f(t)\,\mathbf{i} + g(t)\,\mathbf{j} + h(t)\,\mathbf{k}$, where f, g, and h are differentiable functions, then

$$\mathbf{r}'(t) = \langle f'(t), g'(t), h'(t) \rangle = f'(t)\,\mathbf{i} + g'(t)\,\mathbf{j} + h'(t)\,\mathbf{k}$$

Proof

$$\begin{aligned}
\mathbf{r}'(t) &= \lim_{\Delta t \to 0} \frac{1}{\Delta t}[\mathbf{r}(t + \Delta t) - \mathbf{r}(t)] \\
&= \lim_{\Delta t \to 0} \frac{1}{\Delta t}[\langle f(t + \Delta t), g(t + \Delta t), h(t + \Delta t) \rangle - \langle f(t), g(t), h(t) \rangle] \\
&= \lim_{\Delta t \to 0} \left\langle \frac{f(t + \Delta t) - f(t)}{\Delta t}, \frac{g(t + \Delta t) - g(t)}{\Delta t}, \frac{h(t + \Delta t) - h(t)}{\Delta t} \right\rangle \\
&= \left\langle \lim_{\Delta t \to 0} \frac{f(t + \Delta t) - f(t)}{\Delta t}, \lim_{\Delta t \to 0} \frac{g(t + \Delta t) - g(t)}{\Delta t}, \lim_{\Delta t \to 0} \frac{h(t + \Delta t) - h(t)}{\Delta t} \right\rangle \\
&= \langle f'(t), g'(t), h'(t) \rangle
\end{aligned}$$

EXAMPLE 1
(a) Find the derivative of $\mathbf{r}(t) = (1 + t^3)\mathbf{i} + te^{-t}\mathbf{j} + \sin 2t\,\mathbf{k}$.
(b) Find the unit tangent vector at the point where $t = 0$.

SOLUTION
(a) According to Theorem 2, we differentiate each component of $\mathbf{r}$:

$$\mathbf{r}'(t) = 3t^2\,\mathbf{i} + (1 - t)e^{-t}\mathbf{j} + 2\cos 2t\,\mathbf{k}$$

(b) Since $\mathbf{r}(0) = \mathbf{i}$ and $\mathbf{r}'(0) = \mathbf{j} + 2\mathbf{k}$, the unit tangent vector at the point $(1, 0, 0)$ is

$$\mathbf{T}(0) = \frac{\mathbf{r}'(0)}{|\mathbf{r}'(0)|} = \frac{\mathbf{j} + 2\mathbf{k}}{\sqrt{1 + 4}} = \frac{1}{\sqrt{5}}\mathbf{j} + \frac{2}{\sqrt{5}}\mathbf{k}$$

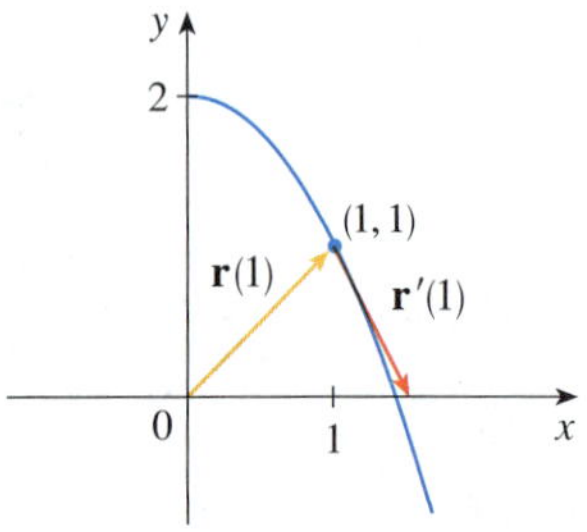

FIGURE 2

EXAMPLE 2 For the curve $\mathbf{r}(t) = \sqrt{t}\,\mathbf{i} + (2 - t)\,\mathbf{j}$, find $\mathbf{r}'(t)$ and sketch the position vector $\mathbf{r}(1)$ and the tangent vector $\mathbf{r}'(1)$.

SOLUTION We have

$$\mathbf{r}'(t) = \frac{1}{2\sqrt{t}}\mathbf{i} - \mathbf{j} \qquad \text{and} \qquad \mathbf{r}'(1) = \frac{1}{2}\mathbf{i} - \mathbf{j}$$

The curve is a plane curve and elimination of the parameter from the equations $x = \sqrt{t}$, $y = 2 - t$ gives $y = 2 - x^2$, $x \geqslant 0$. In Figure 2 we draw the position vector $\mathbf{r}(1) = \mathbf{i} + \mathbf{j}$ starting at the origin and the tangent vector $\mathbf{r}'(1)$ starting at the corresponding point $(1, 1)$.

EXAMPLE 3 Find parametric equations for the tangent line to the helix with parametric equations

|||| The helix and the tangent line in Example 3 are shown in Figure 3.

$$x = 2\cos t \qquad y = \sin t \qquad z = t$$

at the point $(0, 1, \pi/2)$.

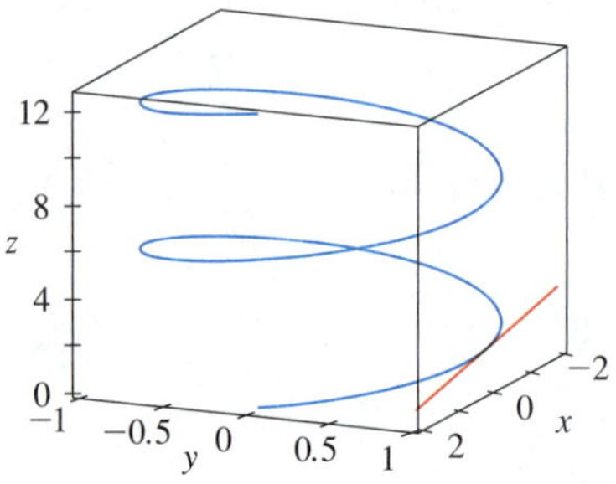

FIGURE 3

SOLUTION The vector equation of the helix is $\mathbf{r}(t) = \langle 2\cos t, \sin t, t\rangle$, so

$$\mathbf{r}'(t) = \langle -2\sin t, \cos t, 1\rangle$$

The parameter value corresponding to the point $(0, 1, \pi/2)$ is $t = \pi/2$, so the tangent vector there is $\mathbf{r}'(\pi/2) = \langle -2, 0, 1\rangle$. The tangent line is the line through $(0, 1, \pi/2)$ parallel to the vector $\langle -2, 0, 1\rangle$, so by Equations 12.5.2 its parametric equations are

$$x = -2t \qquad y = 1 \qquad z = \frac{\pi}{2} + t$$

Just as for real-valued functions, the **second derivative** of a vector function $\mathbf{r}$ is the derivative of $\mathbf{r}'$, that is, $\mathbf{r}'' = (\mathbf{r}')'$. For instance, the second derivative of the function in Example 3 is

$$\mathbf{r}''(t) = \langle -2\cos t, -\sin t, 0\rangle$$

A curve given by a vector function $\mathbf{r}(t)$ on an interval I is called **smooth** if $\mathbf{r}'$ is continuous and $\mathbf{r}'(t) \neq \mathbf{0}$ (except possibly at any endpoints of I). For instance, the helix in Example 3 is smooth because $\mathbf{r}'(t)$ is never $\mathbf{0}$.

EXAMPLE 4 Determine whether the semicubical parabola $\mathbf{r}(t) = \langle 1 + t^3, t^2\rangle$ is smooth.

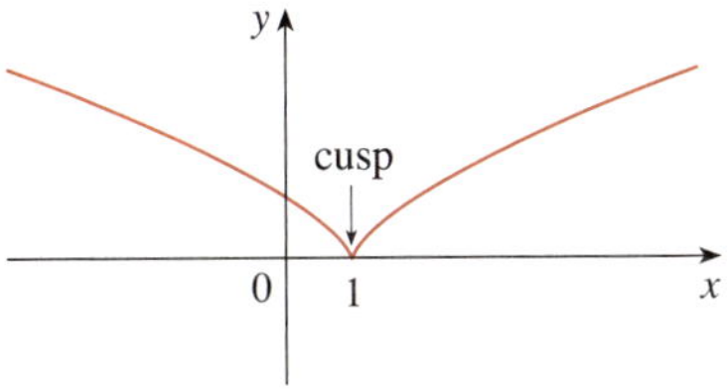

FIGURE 4
The curve $\mathbf{r}(t) = \langle 1 + t^3, t^2 \rangle$ is not smooth.

SOLUTION Since

$$\mathbf{r}'(t) = \langle 3t^2, 2t \rangle$$

we have $\mathbf{r}'(0) = \langle 0, 0 \rangle = \mathbf{0}$ and, therefore, the curve is not smooth. The point that corresponds to $t = 0$ is $(1, 0)$, and we see from the graph in Figure 4 that there is a sharp corner, called a **cusp**, at $(1, 0)$. Any curve with this type of behavior—an abrupt change in direction—is not smooth.

A curve, such as the semicubical parabola, that is made up of a finite number of smooth pieces is called **piecewise smooth**.

Differentiation Rules

The next theorem shows that the differentiation formulas for real-valued functions have their counterparts for vector-valued functions.

3 Theorem Suppose $\mathbf{u}$ and $\mathbf{v}$ are differentiable vector functions, c is a scalar, and f is a real-valued function. Then

1. $\dfrac{d}{dt}[\mathbf{u}(t) + \mathbf{v}(t)] = \mathbf{u}'(t) + \mathbf{v}'(t)$

2. $\dfrac{d}{dt}[c\mathbf{u}(t)] = c\mathbf{u}'(t)$

3. $\dfrac{d}{dt}[f(t)\mathbf{u}(t)] = f'(t)\mathbf{u}(t) + f(t)\mathbf{u}'(t)$

4. $\dfrac{d}{dt}[\mathbf{u}(t) \cdot \mathbf{v}(t)] = \mathbf{u}'(t) \cdot \mathbf{v}(t) + \mathbf{u}(t) \cdot \mathbf{v}'(t)$

5. $\dfrac{d}{dt}[\mathbf{u}(t) \times \mathbf{v}(t)] = \mathbf{u}'(t) \times \mathbf{v}(t) + \mathbf{u}(t) \times \mathbf{v}'(t)$

6. $\dfrac{d}{dt}[\mathbf{u}(f(t))] = f'(t)\mathbf{u}'(f(t))$ (Chain Rule)

This theorem can be proved either directly from Definition 1 or by using Theorem 2 and the corresponding differentiation formulas for real-valued functions. The proofs of Formulas 1, 2, 3, 5, and 6 are left as exercises.

Proof of Formula 4 Let

$$\mathbf{u}(t) = \langle f_1(t), f_2(t), f_3(t) \rangle \qquad \mathbf{v}(t) = \langle g_1(t), g_2(t), g_3(t) \rangle$$

Then $$\mathbf{u}(t) \cdot \mathbf{v}(t) = f_1(t)g_1(t) + f_2(t)g_2(t) + f_3(t)g_3(t) = \sum_{i=1}^{3} f_i(t)g_i(t)$$

so the ordinary Product Rule gives

$$\begin{aligned} \frac{d}{dt}[\mathbf{u}(t) \cdot \mathbf{v}(t)] &= \frac{d}{dt}\sum_{i=1}^{3} f_i(t)g_i(t) = \sum_{i=1}^{3} \frac{d}{dt}[f_i(t)g_i(t)] \\ &= \sum_{i=1}^{3} [f_i'(t)g_i(t) + f_i(t)g_i'(t)] \\ &= \sum_{i=1}^{3} f_i'(t)g_i(t) + \sum_{i=1}^{3} f_i(t)g_i'(t) \\ &= \mathbf{u}'(t) \cdot \mathbf{v}(t) + \mathbf{u}(t) \cdot \mathbf{v}'(t) \end{aligned}$$

EXAMPLE 5 Show that if $|\mathbf{r}(t)| = c$ (a constant), then $\mathbf{r}'(t)$ is orthogonal to $\mathbf{r}(t)$ for all t.

SOLUTION Since

$$\mathbf{r}(t) \cdot \mathbf{r}(t) = |\mathbf{r}(t)|^2 = c^2$$

and c^2 is a constant, Formula 4 of Theorem 3 gives

$$0 = \frac{d}{dt}[\mathbf{r}(t) \cdot \mathbf{r}(t)] = \mathbf{r}'(t) \cdot \mathbf{r}(t) + \mathbf{r}(t) \cdot \mathbf{r}'(t) = 2\mathbf{r}'(t) \cdot \mathbf{r}(t)$$

Thus, $\mathbf{r}'(t) \cdot \mathbf{r}(t) = 0$, which says that $\mathbf{r}'(t)$ is orthogonal to $\mathbf{r}(t)$.

Geometrically, this result says that if a curve lies on a sphere with center the origin, then the tangent vector $\mathbf{r}'(t)$ is always perpendicular to the position vector $\mathbf{r}(t)$.

Integrals

The **definite integral** of a continuous vector function $\mathbf{r}(t)$ can be defined in much the same way as for real-valued functions except that the integral is a vector. But then we can express the integral of $\mathbf{r}$ in terms of the integrals of its component functions f, g, and h as follows. (We use the notation of Chapter 5.)

$$\int_a^b \mathbf{r}(t)\,dt = \lim_{n\to\infty} \sum_{i=1}^{n} \mathbf{r}(t_i^*)\,\Delta t$$

$$= \lim_{n\to\infty} \left[\left(\sum_{i=1}^{n} f(t_i^*)\,\Delta t \right) \mathbf{i} + \left(\sum_{i=1}^{n} g(t_i^*)\,\Delta t \right) \mathbf{j} + \left(\sum_{i=1}^{n} h(t_i^*)\,\Delta t \right) \mathbf{k} \right]$$

and so

$$\int_a^b \mathbf{r}(t)\,dt = \left(\int_a^b f(t)\,dt \right) \mathbf{i} + \left(\int_a^b g(t)\,dt \right) \mathbf{j} + \left(\int_a^b h(t)\,dt \right) \mathbf{k}$$

This means that we can evaluate an integral of a vector function by integrating each component function.

We can extend the Fundamental Theorem of Calculus to continuous vector functions as follows:

$$\int_a^b \mathbf{r}(t)\,dt = \mathbf{R}(t)\Big]_a^b = \mathbf{R}(b) - \mathbf{R}(a)$$

where $\mathbf{R}$ is an antiderivative of $\mathbf{r}$, that is, $\mathbf{R}'(t) = \mathbf{r}(t)$. We use the notation $\int \mathbf{r}(t)\,dt$ for indefinite integrals (antiderivatives).

EXAMPLE 6 If $\mathbf{r}(t) = 2\cos t\,\mathbf{i} + \sin t\,\mathbf{j} + 2t\,\mathbf{k}$, then

$$\int \mathbf{r}(t)\,dt = \left(\int 2\cos t\,dt \right) \mathbf{i} + \left(\int \sin t\,dt \right) \mathbf{j} + \left(\int 2t\,dt \right) \mathbf{k}$$

$$= 2\sin t\,\mathbf{i} - \cos t\,\mathbf{j} + t^2\,\mathbf{k} + \mathbf{C}$$

where $\mathbf{C}$ is a vector constant of integration, and

$$\int_0^{\pi/2} \mathbf{r}(t)\,dt = \left[2\sin t\,\mathbf{i} - \cos t\,\mathbf{j} + t^2\,\mathbf{k}\right]_0^{\pi/2} = 2\mathbf{i} + \mathbf{j} + \frac{\pi^2}{4}\mathbf{k}$$

13.2 Exercises

1. The figure shows a curve C given by a vector function $\mathbf{r}(t)$.
(a) Draw the vectors $\mathbf{r}(4.5) - \mathbf{r}(4)$ and $\mathbf{r}(4.2) - \mathbf{r}(4)$.
(b) Draw the vectors

$$\frac{\mathbf{r}(4.5) - \mathbf{r}(4)}{0.5} \quad \text{and} \quad \frac{\mathbf{r}(4.2) - \mathbf{r}(4)}{0.2}$$

(c) Write expressions for $\mathbf{r}'(4)$ and the unit tangent vector $\mathbf{T}(4)$.
(d) Draw the vector $\mathbf{T}(4)$.

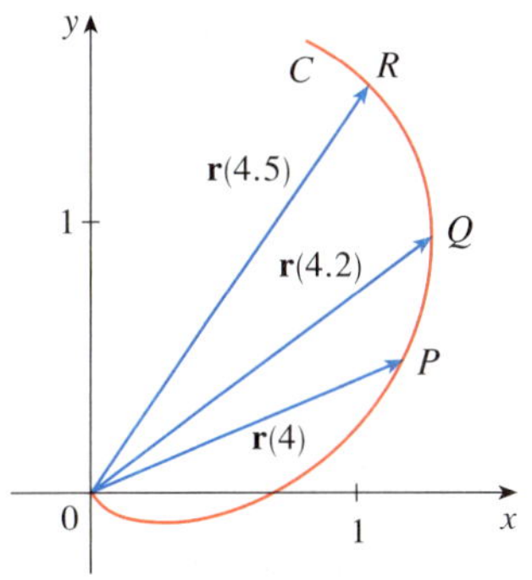

2. (a) Make a large sketch of the curve described by the vector function $\mathbf{r}(t) = \langle t^2, t \rangle$, $0 \leqslant t \leqslant 2$, and draw the vectors $\mathbf{r}(1)$, $\mathbf{r}(1.1)$, and $\mathbf{r}(1.1) - \mathbf{r}(1)$.
(b) Draw the vector $\mathbf{r}'(1)$ starting at (1, 1) and compare it with the vector

$$\frac{\mathbf{r}(1.1) - \mathbf{r}(1)}{0.1}$$

Explain why these vectors are so close to each other in length and direction.

3–8
(a) Sketch the plane curve with the given vector equation.
(b) Find $\mathbf{r}'(t)$.
(c) Sketch the position vector $\mathbf{r}(t)$ and the tangent vector $\mathbf{r}'(t)$ for the given value of t.

3. $\mathbf{r}(t) = \langle \cos t, \sin t \rangle, \quad t = \pi/4$

4. $\mathbf{r}(t) = \langle 1 + t, \sqrt{t} \rangle, \quad t = 1$

5. $\mathbf{r}(t) = (1 + t)\,\mathbf{i} + t^2\mathbf{j}, \quad t = 1$

6. $\mathbf{r}(t) = e^t\,\mathbf{i} + e^{-t}\,\mathbf{j}, \quad t = 0$

7. $\mathbf{r}(t) = e^t\mathbf{i} + e^{3t}\mathbf{j}, \quad t = 0$

8. $\mathbf{r}(t) = 2\sin t\,\mathbf{i} + 3\cos t\,\mathbf{j}, \quad t = \pi/3$

9–16 Find the derivative of the vector function.

9. $\mathbf{r}(t) = \langle t^2, 1 - t, \sqrt{t} \rangle$

10. $\mathbf{r}(t) = \langle \cos 3t, t, \sin 3t \rangle$

11. $\mathbf{r}(t) = \mathbf{i} - \mathbf{j} + e^{4t}\,\mathbf{k}$

12. $\mathbf{r}(t) = \sin^{-1}t\,\mathbf{i} + \sqrt{1 - t^2}\,\mathbf{j} + \mathbf{k}$

13. $\mathbf{r}(t) = e^{t^2}\mathbf{i} - \mathbf{j} + \ln(1 + 3t)\,\mathbf{k}$

14. $\mathbf{r}(t) = at\cos 3t\,\mathbf{i} + b\sin^3 t\,\mathbf{j} + c\cos^3 t\,\mathbf{k}$

15. $\mathbf{r}(t) = \mathbf{a} + t\,\mathbf{b} + t^2\mathbf{c}$

16. $\mathbf{r}(t) = t\,\mathbf{a} \times (\mathbf{b} + t\,\mathbf{c})$

17–20 Find the unit tangent vector $\mathbf{T}(t)$ at the point with the given value of the parameter t.

17. $\mathbf{r}(t) = \langle 6t^5, 4t^3, 2t \rangle, \quad t = 1$

18. $\mathbf{r}(t) = 4\sqrt{t}\,\mathbf{i} + t^2\mathbf{j} + t\,\mathbf{k}, \quad t = 1$

19. $\mathbf{r}(t) = \cos t\,\mathbf{i} + 3t\,\mathbf{j} + 2\sin 2t\,\mathbf{k}, \quad t = 0$

20. $\mathbf{r}(t) = 2\sin t\,\mathbf{i} + 2\cos t\,\mathbf{j} + \tan t\,\mathbf{k}, \quad t = \pi/4$

21. If $\mathbf{r}(t) = \langle t, t^2, t^3 \rangle$, find $\mathbf{r}'(t)$, $\mathbf{T}(1)$, $\mathbf{r}''(t)$, and $\mathbf{r}'(t) \times \mathbf{r}''(t)$.

22. If $\mathbf{r}(t) = \langle e^{2t}, e^{-2t}, te^{2t} \rangle$, find $\mathbf{T}(0)$, $\mathbf{r}''(0)$, and $\mathbf{r}'(t) \cdot \mathbf{r}''(t)$.

23–26 Find parametric equations for the tangent line to the curve with the given parametric equations at the specified point.

23. $x = t^5, \quad y = t^4, \quad z = t^3; \quad (1, 1, 1)$

24. $x = t^2 - 1, \quad y = t^2 + 1, \quad z = t + 1; \quad (-1, 1, 1)$

25. $x = e^{-t}\cos t, \quad y = e^{-t}\sin t, \quad z = e^{-t}; \quad (1, 0, 1)$

26. $x = \ln t, \quad y = 2\sqrt{t}, \quad z = t^2; \quad (0, 2, 1)$

27–28 Find parametric equations for the tangent line to the curve with the given parametric equations at the specified point. Illustrate by graphing both the curve and the tangent line on a common screen.

27. $x = t, \ y = \sqrt{2}\cos t, \ z = \sqrt{2}\sin t; \quad (\pi/4, 1, 1)$

28. $x = \cos t, \ y = 3e^{2t}, \ z = 3e^{-2t}; \quad (1, 3, 3)$

29. Determine whether the curve is smooth.
(a) $\mathbf{r}(t) = \langle t^3, t^4, t^5 \rangle$
(b) $\mathbf{r}(t) = \langle t^3 + t, t^4, t^5 \rangle$
(c) $\mathbf{r}(t) = \langle \cos^3 t, \sin^3 t \rangle$

30. (a) Find the point of intersection of the tangent lines to the curve $\mathbf{r}(t) = \langle \sin \pi t, 2\sin \pi t, \cos \pi t \rangle$ at the points where $t = 0$ and $t = 0.5$.
(b) Illustrate by graphing the curve and both tangent lines.

31. The curves $\mathbf{r}_1(t) = \langle t, t^2, t^3 \rangle$ and $\mathbf{r}_2(t) = \langle \sin t, \sin 2t, t \rangle$ intersect at the origin. Find their angle of intersection correct to the nearest degree.

32. At what point do the curves $\mathbf{r}_1(t) = \langle t, 1 - t, 3 + t^2 \rangle$ and $\mathbf{r}_2(s) = \langle 3 - s, s - 2, s^2 \rangle$ intersect? Find their angle of intersection correct to the nearest degree.

33–38 Evaluate the integral.

33. $\int_0^1 (16t^3\,\mathbf{i} - 9t^2\,\mathbf{j} + 25t^4\,\mathbf{k})\,dt$

34. $\int_0^1 \left(\frac{4}{1+t^2}\,\mathbf{j} + \frac{2t}{1+t^2}\,\mathbf{k}\right) dt$

35. $\int_0^{\pi/2} (3\sin^2 t \cos t\,\mathbf{i} + 3\sin t\cos^2 t\,\mathbf{j} + 2\sin t\cos t\,\mathbf{k})\,dt$

36. $\int_1^4 \left(\sqrt{t}\,\mathbf{i} + te^{-t}\,\mathbf{j} + \frac{1}{t^2}\,\mathbf{k}\right) dt$

37. $\int (e^t\,\mathbf{i} + 2t\,\mathbf{j} + \ln t\,\mathbf{k})\,dt$

38. $\int (\cos \pi t\,\mathbf{i} + \sin \pi t\,\mathbf{j} + t\,\mathbf{k})\,dt$

39. Find $\mathbf{r}(t)$ if $\mathbf{r}'(t) = t^2\,\mathbf{i} + 4t^3\,\mathbf{j} - t^2\,\mathbf{k}$ and $\mathbf{r}(0) = \mathbf{j}$.

40. Find $\mathbf{r}(t)$ if $\mathbf{r}'(t) = \sin t\,\mathbf{i} - \cos t\,\mathbf{j} + 2t\,\mathbf{k}$ and $\mathbf{r}(0) = \mathbf{i} + \mathbf{j} + 2\,\mathbf{k}$.

41. Prove Formula 1 of Theorem 3.

42. Prove Formula 3 of Theorem 3.

43. Prove Formula 5 of Theorem 3.

44. Prove Formula 6 of Theorem 3.

45. If $\mathbf{u}(t) = \mathbf{i} - 2t^2\,\mathbf{j} + 3t^3\,\mathbf{k}$ and $\mathbf{v}(t) = t\,\mathbf{i} + \cos t\,\mathbf{j} + \sin t\,\mathbf{k}$, find $(d/dt)\,[\mathbf{u}(t) \cdot \mathbf{v}(t)]$.

46. If $\mathbf{u}$ and $\mathbf{v}$ are the vector functions in Exercise 45, find $(d/dt)\,[\mathbf{u}(t) \times \mathbf{v}(t)]$.

47. Show that if $\mathbf{r}$ is a vector function such that $\mathbf{r}''$ exists, then

$$\frac{d}{dt}[\mathbf{r}(t) \times \mathbf{r}'(t)] = \mathbf{r}(t) \times \mathbf{r}''(t)$$

48. Find an expression for $\frac{d}{dt}[\mathbf{u}(t) \cdot (\mathbf{v}(t) \times \mathbf{w}(t))]$.

49. If $\mathbf{r}(t) \neq \mathbf{0}$, show that $\frac{d}{dt}|\mathbf{r}(t)| = \frac{1}{|\mathbf{r}(t)|}\,\mathbf{r}(t) \cdot \mathbf{r}'(t)$.

[*Hint:* $|\mathbf{r}(t)|^2 = \mathbf{r}(t) \cdot \mathbf{r}(t)$]

50. If a curve has the property that the position vector $\mathbf{r}(t)$ is always perpendicular to the tangent vector $\mathbf{r}'(t)$, show that the curve lies on a sphere with center the origin.

51. If $\mathbf{u}(t) = \mathbf{r}(t) \cdot [\mathbf{r}'(t) \times \mathbf{r}''(t)]$, show that

$$\mathbf{u}'(t) = \mathbf{r}(t) \cdot [\mathbf{r}'(t) \times \mathbf{r}'''(t)]$$

13.3 Arc Length and Curvature

In Section 10.2 we defined the length of a plane curve with parametric equations $x = f(t)$, $y = g(t)$, $a \leqslant t \leqslant b$, as the limit of lengths of inscribed polygons and, for the case where f' and g' are continuous, we arrived at the formula

$$\boxed{1} \qquad L = \int_a^b \sqrt{[f'(t)]^2 + [g'(t)]^2}\,dt = \int_a^b \sqrt{\left(\frac{dx}{dt}\right)^2 + \left(\frac{dy}{dt}\right)^2}\,dt$$

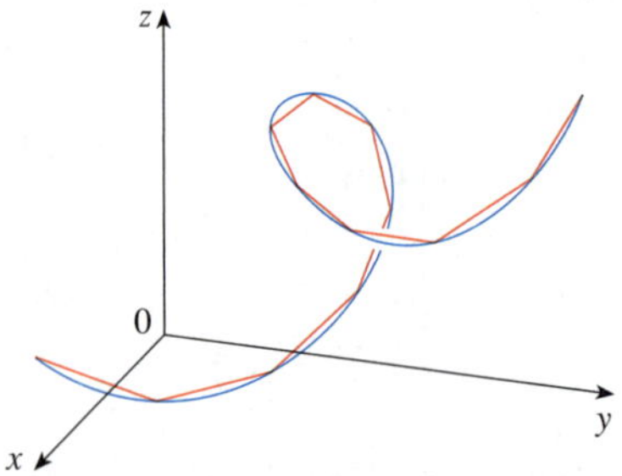

FIGURE 1
The length of a space curve is the limit of lengths of inscribed polygons.

The length of a space curve is defined in exactly the same way (see Figure 1). Suppose that the curve has the vector equation $\mathbf{r}(t) = \langle f(t), g(t), h(t) \rangle$, $a \leqslant t \leqslant b$, or, equivalently, the parametric equations $x = f(t)$, $y = g(t)$, $z = h(t)$, where f', g', and h' are continuous. If the curve is traversed exactly once as t increases from a to b, then it can be shown that its length is

$$\boxed{2} \qquad L = \int_a^b \sqrt{[f'(t)]^2 + [g'(t)]^2 + [h'(t)]^2}\,dt = \int_a^b \sqrt{\left(\frac{dx}{dt}\right)^2 + \left(\frac{dy}{dt}\right)^2 + \left(\frac{dz}{dt}\right)^2}\,dt$$

Notice that both of the arc length formulas (1) and (2) can be put into the more compact form

$$\boxed{3} \qquad L = \int_a^b |\mathbf{r}'(t)|\,dt$$

because, for plane curves $\mathbf{r}(t) = f(t)\,\mathbf{i} + g(t)\,\mathbf{j}$,

$$|\mathbf{r}'(t)| = |f'(t)\,\mathbf{i} + g'(t)\,\mathbf{j}| = \sqrt{[f'(t)]^2 + [g'(t)]^2}$$

whereas, for space curves $\mathbf{r}(t) = f(t)\,\mathbf{i} + g(t)\,\mathbf{j} + h(t)\,\mathbf{k}$,

$$|\mathbf{r}'(t)| = |f'(t)\,\mathbf{i} + g'(t)\,\mathbf{j} + h'(t)\,\mathbf{k}| = \sqrt{[f'(t)]^2 + [g'(t)]^2 + [h'(t)]^2}$$

|||| Figure 2 shows the arc of the helix whose length is computed in Example 1.

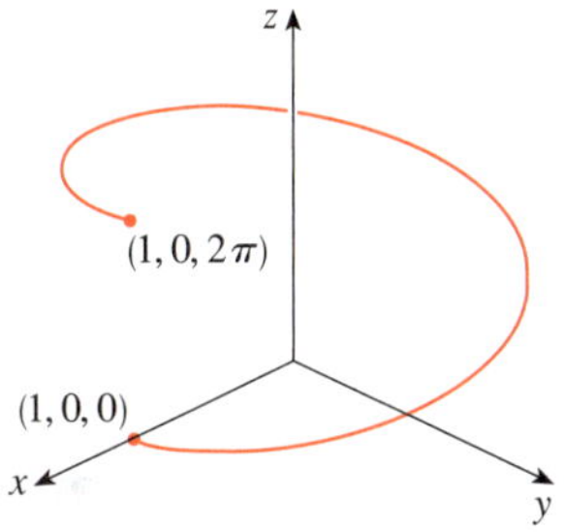

FIGURE 2

EXAMPLE 1 Find the length of the arc of the circular helix with vector equation $\mathbf{r}(t) = \cos t\,\mathbf{i} + \sin t\,\mathbf{j} + t\,\mathbf{k}$ from the point $(1, 0, 0)$ to the point $(1, 0, 2\pi)$.

SOLUTION Since $\mathbf{r}'(t) = -\sin t\,\mathbf{i} + \cos t\,\mathbf{j} + \mathbf{k}$, we have

$$|\mathbf{r}'(t)| = \sqrt{(-\sin t)^2 + \cos^2 t + 1} = \sqrt{2}$$

The arc from $(1, 0, 0)$ to $(1, 0, 2\pi)$ is described by the parameter interval $0 \leqslant t \leqslant 2\pi$ and so, from Formula 3, we have

$$L = \int_0^{2\pi} |\mathbf{r}'(t)|\,dt = \int_0^{2\pi} \sqrt{2}\,dt = 2\sqrt{2}\pi$$

A single curve C can be represented by more than one vector function. For instance, the twisted cubic

4 $$\mathbf{r}_1(t) = \langle t, t^2, t^3 \rangle \qquad 1 \leqslant t \leqslant 2$$

could also be represented by the function

5 $$\mathbf{r}_2(u) = \langle e^u, e^{2u}, e^{3u} \rangle \qquad 0 \leqslant u \leqslant \ln 2$$

where the connection between the parameters t and u is given by $t = e^u$. We say that Equations 4 and 5 are **parametrizations** of the curve C. If we were to use Equation 3 to compute the length of C using Equations 4 and 5, we would get the same answer. In general, it can be shown that when Equation 3 is used to compute the length of any piecewise-smooth curve, the arc length is independent of the parametrization that is used.

|||| Piecewise-smooth curves were introduced on page 859.

Now we suppose that C is a piecewise-smooth curve given by a vector function $\mathbf{r}(t) = f(t)\,\mathbf{i} + g(t)\,\mathbf{j} + h(t)\,\mathbf{k}$, $a \leqslant t \leqslant b$, and C is traversed exactly once as t increases from a to b. We define its **arc length function** s by

6 $$s(t) = \int_a^t |\mathbf{r}'(u)|\,du = \int_a^t \sqrt{\left(\frac{dx}{du}\right)^2 + \left(\frac{dy}{du}\right)^2 + \left(\frac{dz}{du}\right)^2}\,du$$

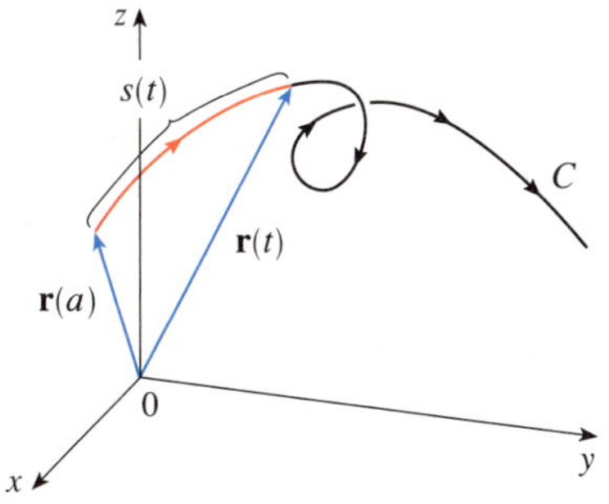

FIGURE 3

Thus, $s(t)$ is the length of the part of C between $\mathbf{r}(a)$ and $\mathbf{r}(t)$. (See Figure 3.) If we differentiate both sides of Equation 6 using Part 1 of the Fundamental Theorem of Calculus, we obtain

7 $$\frac{ds}{dt} = |\mathbf{r}'(t)|$$

It is often useful to **parametrize a curve with respect to arc length** because arc length arises naturally from the shape of the curve and does not depend on a particular coordinate system. If a curve $\mathbf{r}(t)$ is already given in terms of a parameter t and $s(t)$ is the arc length function given by Equation 6, then we may be able to solve for t as a function of s: $t = t(s)$. Then the curve can be reparametrized in terms of s by substituting for t: $\mathbf{r} = \mathbf{r}(t(s))$. Thus, if $s = 3$ for instance, $\mathbf{r}(t(3))$ is the position vector of the point 3 units of length along the curve from its starting point.

EXAMPLE 2 Reparametrize the helix $\mathbf{r}(t) = \cos t\,\mathbf{i} + \sin t\,\mathbf{j} + t\,\mathbf{k}$ with respect to arc length measured from $(1, 0, 0)$ in the direction of increasing t.

SOLUTION The initial point $(1, 0, 0)$ corresponds to the parameter value $t = 0$. From Example 1 we have

$$\frac{ds}{dt} = |\mathbf{r}'(t)| = \sqrt{2}$$

and so

$$s = s(t) = \int_0^t |\mathbf{r}'(u)|\,du = \int_0^t \sqrt{2}\,du = \sqrt{2}\,t$$

Therefore, $t = s/\sqrt{2}$ and the required reparametrization is obtained by substituting for t:

$$\mathbf{r}(t(s)) = \cos(s/\sqrt{2})\,\mathbf{i} + \sin(s/\sqrt{2})\,\mathbf{j} + (s/\sqrt{2})\,\mathbf{k}$$

Curvature

If C is a smooth curve defined by the vector function $\mathbf{r}$, then $\mathbf{r}'(t) \neq \mathbf{0}$. Recall that the unit tangent vector $\mathbf{T}(t)$ is given by

$$\mathbf{T}(t) = \frac{\mathbf{r}'(t)}{|\mathbf{r}'(t)|}$$

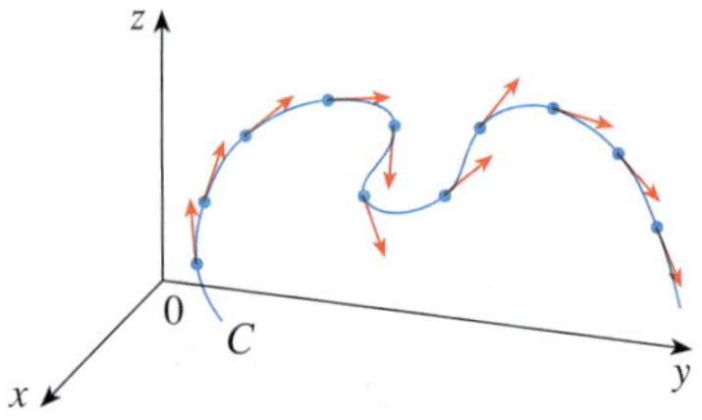

FIGURE 4
Unit tangent vectors at equally spaced points on C

TEC Visual 13.3A shows animated unit tangent vectors, like those in Figure 4, for a variety of plane curves and space curves.

and indicates the direction of the curve. From Figure 4 you can see that $\mathbf{T}(t)$ changes direction very slowly when C is fairly straight, but it changes direction more quickly when C bends or twists more sharply.

The curvature of C at a given point is a measure of how quickly the curve changes direction at that point. Specifically, we define it to be the magnitude of the rate of change of the unit tangent vector with respect to arc length. (We use arc length so that the curvature will be independent of the parametrization.)

8 Definition The **curvature** of a curve is

$$\kappa = \left| \frac{d\mathbf{T}}{ds} \right|$$

where $\mathbf{T}$ is the unit tangent vector.

The curvature is easier to compute if it is expressed in terms of the parameter t instead of s, so we use the Chain Rule (Theorem 13.2.3, Formula 6) to write

$$\frac{d\mathbf{T}}{dt} = \frac{d\mathbf{T}}{ds}\frac{ds}{dt} \qquad \text{and} \qquad \kappa = \left|\frac{d\mathbf{T}}{ds}\right| = \left|\frac{d\mathbf{T}/dt}{ds/dt}\right|$$

But $ds/dt = |\mathbf{r}'(t)|$ from Equation 7, so

9
$$\kappa(t) = \frac{|\mathbf{T}'(t)|}{|\mathbf{r}'(t)|}$$

EXAMPLE 3 Show that the curvature of a circle of radius a is $1/a$.

SOLUTION We can take the circle to have center the origin, and then a parametrization is

$$\mathbf{r}(t) = a\cos t\,\mathbf{i} + a\sin t\,\mathbf{j}$$

Therefore $$\mathbf{r}'(t) = -a\sin t\,\mathbf{i} + a\cos t\,\mathbf{j} \quad \text{and} \quad |\mathbf{r}'(t)| = a$$

so $$\mathbf{T}(t) = \frac{\mathbf{r}'(t)}{|\mathbf{r}'(t)|} = -\sin t\,\mathbf{i} + \cos t\,\mathbf{j}$$

and $$\mathbf{T}'(t) = -\cos t\,\mathbf{i} - \sin t\,\mathbf{j}$$

This gives $|\mathbf{T}'(t)| = 1$, so using Equation 9, we have

$$\kappa(t) = \frac{|\mathbf{T}'(t)|}{|\mathbf{r}'(t)|} = \frac{1}{a}$$

The result of Example 3 shows that small circles have large curvature and large circles have small curvature, in accordance with our intuition. We can see directly from the definition of curvature that the curvature of a straight line is always 0 because the tangent vector is constant.

Although Formula 9 can be used in all cases to compute the curvature, the formula given by the following theorem is often more convenient to apply.

10 Theorem The curvature of the curve given by the vector function $\mathbf{r}$ is

$$\kappa(t) = \frac{|\mathbf{r}'(t) \times \mathbf{r}''(t)|}{|\mathbf{r}'(t)|^3}$$

Proof Since $\mathbf{T} = \mathbf{r}'/|\mathbf{r}'|$ and $|\mathbf{r}'| = ds/dt$, we have

$$\mathbf{r}' = |\mathbf{r}'|\mathbf{T} = \frac{ds}{dt}\mathbf{T}$$

so the Product Rule (Theorem 13.2.3, Formula 3) gives

$$\mathbf{r}'' = \frac{d^2s}{dt^2}\mathbf{T} + \frac{ds}{dt}\mathbf{T}'$$

Using the fact that $\mathbf{T} \times \mathbf{T} = \mathbf{0}$ (see Example 2 in Section 12.4), we have

$$\mathbf{r}' \times \mathbf{r}'' = \left(\frac{ds}{dt}\right)^2 (\mathbf{T} \times \mathbf{T}')$$

Now $|\mathbf{T}(t)| = 1$ for all t, so $\mathbf{T}$ and $\mathbf{T}'$ are orthogonal by Example 5 in Section 13.2. Therefore, by Theorem 12.4.6,

$$|\mathbf{r}' \times \mathbf{r}''| = \left(\frac{ds}{dt}\right)^2 |\mathbf{T} \times \mathbf{T}'| = \left(\frac{ds}{dt}\right)^2 |\mathbf{T}||\mathbf{T}'| = \left(\frac{ds}{dt}\right)^2 |\mathbf{T}'|$$

Thus $$|\mathbf{T}'| = \frac{|\mathbf{r}' \times \mathbf{r}''|}{(ds/dt)^2} = \frac{|\mathbf{r}' \times \mathbf{r}''|}{|\mathbf{r}'|^2}$$

and $$\kappa = \frac{|\mathbf{T}'|}{|\mathbf{r}'|} = \frac{|\mathbf{r}' \times \mathbf{r}''|}{|\mathbf{r}'|^3}$$

EXAMPLE 4 Find the curvature of the twisted cubic $\mathbf{r}(t) = \langle t, t^2, t^3 \rangle$ at a general point and at (0, 0, 0).

SOLUTION We first compute the required ingredients:

$$\mathbf{r}'(t) = \langle 1, 2t, 3t^2 \rangle \qquad \mathbf{r}''(t) = \langle 0, 2, 6t \rangle$$

$$|\mathbf{r}'(t)| = \sqrt{1 + 4t^2 + 9t^4}$$

$$\mathbf{r}'(t) \times \mathbf{r}''(t) = \begin{vmatrix} \mathbf{i} & \mathbf{j} & \mathbf{k} \\ 1 & 2t & 3t^2 \\ 0 & 2 & 6t \end{vmatrix} = 6t^2\,\mathbf{i} - 6t\,\mathbf{j} + 2\,\mathbf{k}$$

$$|\mathbf{r}'(t) \times \mathbf{r}''(t)| = \sqrt{36t^4 + 36t^2 + 4} = 2\sqrt{9t^4 + 9t^2 + 1}$$

Theorem 10 then gives

$$\kappa(t) = \frac{|\mathbf{r}'(t) \times \mathbf{r}''(t)|}{|\mathbf{r}'(t)|^3} = \frac{2\sqrt{1 + 9t^2 + 9t^4}}{(1 + 4t^2 + 9t^4)^{3/2}}$$

At the origin the curvature is $\kappa(0) = 2$.

For the special case of a plane curve with equation $y = f(x)$, we can choose x as the parameter and write $\mathbf{r}(x) = x\,\mathbf{i} + f(x)\,\mathbf{j}$. Then $\mathbf{r}'(x) = \mathbf{i} + f'(x)\,\mathbf{j}$ and $\mathbf{r}''(x) = f''(x)\,\mathbf{j}$. Since $\mathbf{i} \times \mathbf{j} = \mathbf{k}$ and $\mathbf{j} \times \mathbf{j} = \mathbf{0}$, we have $\mathbf{r}'(x) \times \mathbf{r}''(x) = f''(x)\,\mathbf{k}$. We also have $|\mathbf{r}'(x)| = \sqrt{1 + [f'(x)]^2}$ and so, by Theorem 10,

11 $$\kappa(x) = \frac{|f''(x)|}{[1 + (f'(x))^2]^{3/2}}$$

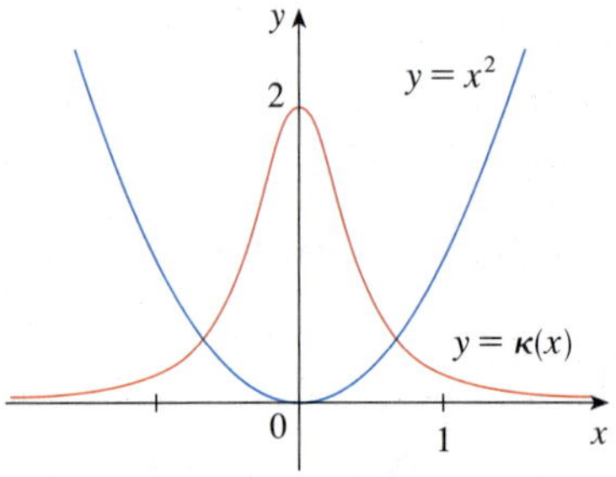

FIGURE 5
The parabola $y = x^2$ and its curvature function

EXAMPLE 5 Find the curvature of the parabola $y = x^2$ at the points (0, 0), (1, 1), and (2, 4).

SOLUTION Since $y' = 2x$ and $y'' = 2$, Formula 11 gives

$$\kappa(x) = \frac{|y''|}{[1 + (y')^2]^{3/2}} = \frac{2}{(1 + 4x^2)^{3/2}}$$

The curvature at (0, 0) is $\kappa(0) = 2$. At (1, 1) it is $\kappa(1) = 2/5^{3/2} \approx 0.18$. At (2, 4) it is $\kappa(2) = 2/17^{3/2} \approx 0.03$. Observe from the expression for $\kappa(x)$ or the graph of κ in Figure 5 that $\kappa(x) \to 0$ as $x \to \pm\infty$. This corresponds to the fact that the parabola appears to become flatter as $x \to \pm\infty$.

The Normal and Binormal Vectors

At a given point on a smooth space curve $\mathbf{r}(t)$, there are many vectors that are orthogonal to the unit tangent vector $\mathbf{T}(t)$. We single out one by observing that, since $|\mathbf{T}(t)| = 1$ for all t, we have $\mathbf{T}(t) \cdot \mathbf{T}'(t) = 0$ by Example 5 in Section 13.2, so $\mathbf{T}'(t)$ is orthogonal to

|||| We can think of the normal vector as indicating the direction in which the curve is turning at each point.

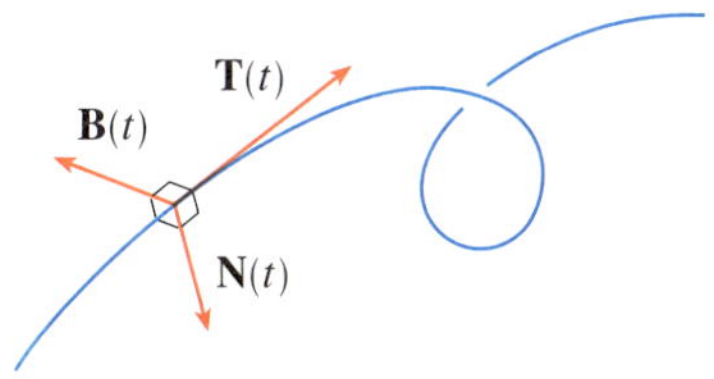

FIGURE 6

$\mathbf{T}(t)$. Note that $\mathbf{T}'(t)$ is itself not a unit vector. But if $\mathbf{r}'$ is also smooth, we can define the **principal unit normal vector** $\mathbf{N}(t)$ (or simply **unit normal**) as

$$\mathbf{N}(t) = \frac{\mathbf{T}'(t)}{|\mathbf{T}'(t)|}$$

The vector $\mathbf{B}(t) = \mathbf{T}(t) \times \mathbf{N}(t)$ is called the **binormal vector**. It is perpendicular to both $\mathbf{T}$ and $\mathbf{N}$ and is also a unit vector. (See Figure 6.)

EXAMPLE 6 Find the unit normal and binormal vectors for the circular helix

$$\mathbf{r}(t) = \cos t\,\mathbf{i} + \sin t\,\mathbf{j} + t\,\mathbf{k}$$

|||| Figure 7 illustrates Example 6 by showing the vectors $\mathbf{T}$, $\mathbf{N}$, and $\mathbf{B}$ at two locations on the helix. In general, the vectors $\mathbf{T}$, $\mathbf{N}$, and $\mathbf{B}$, starting at the various points on a curve, form a set of orthogonal vectors, called the **TNB** frame, that moves along the curve as t varies. This **TNB** frame plays an important role in the branch of mathematics known as differential geometry and in its applications to the motion of spacecraft.

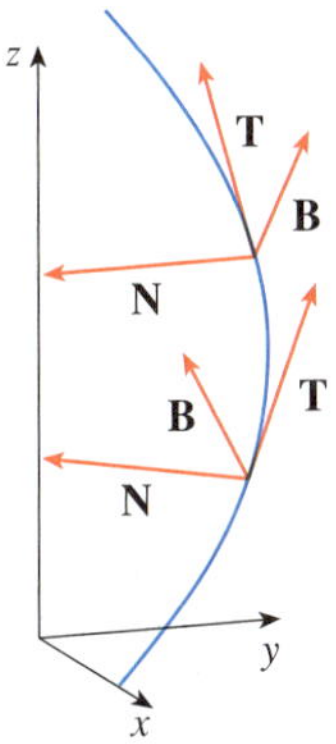

FIGURE 7

Visual 13.3B shows how the TNB frame moves along several curves.

SOLUTION We first compute the ingredients needed for the unit normal vector:

$$\mathbf{r}'(t) = -\sin t\,\mathbf{i} + \cos t\,\mathbf{j} + \mathbf{k} \qquad |\mathbf{r}'(t)| = \sqrt{2}$$

$$\mathbf{T}(t) = \frac{\mathbf{r}'(t)}{|\mathbf{r}'(t)|} = \frac{1}{\sqrt{2}}(-\sin t\,\mathbf{i} + \cos t\,\mathbf{j} + \mathbf{k})$$

$$\mathbf{T}'(t) = \frac{1}{\sqrt{2}}(-\cos t\,\mathbf{i} - \sin t\,\mathbf{j}) \qquad |\mathbf{T}'(t)| = \frac{1}{\sqrt{2}}$$

$$\mathbf{N}(t) = \frac{\mathbf{T}'(t)}{|\mathbf{T}'(t)|} = -\cos t\,\mathbf{i} - \sin t\,\mathbf{j} = \langle -\cos t, -\sin t, 0 \rangle$$

This shows that the normal vector at a point on the helix is horizontal and points toward the z-axis. The binormal vector is

$$\mathbf{B}(t) = \mathbf{T}(t) \times \mathbf{N}(t) = \frac{1}{\sqrt{2}} \begin{bmatrix} \mathbf{i} & \mathbf{j} & \mathbf{k} \\ -\sin t & \cos t & 1 \\ -\cos t & -\sin t & 0 \end{bmatrix} = \frac{1}{\sqrt{2}} \langle \sin t, -\cos t, 1 \rangle$$

The plane determined by the normal and binormal vectors $\mathbf{N}$ and $\mathbf{B}$ at a point P on a curve C is called the **normal plane** of C at P. It consists of all lines that are orthogonal to the tangent vector $\mathbf{T}$. The plane determined by the vectors $\mathbf{T}$ and $\mathbf{N}$ is called the **osculating plane** of C at P. The name comes from the Latin *osculum*, meaning "kiss." It is the plane that comes closest to containing the part of the curve near P. (For a plane curve, the osculating plane is simply the plane that contains the curve.)

The circle that lies in the osculating plane of C at P, has the same tangent as C at P, lies on the concave side of C (toward which $\mathbf{N}$ points), and has radius $\rho = 1/\kappa$ (the reciprocal of the curvature) is called the **osculating circle** (or the **circle of curvature**) of C at P. It is the circle that best describes how C behaves near P; it shares the same tangent, normal, and curvature at P.

EXAMPLE 7 Find the equations of the normal plane and osculating plane of the helix in Example 6 at the point $P(0, 1, \pi/2)$.

SOLUTION The normal plane at P has normal vector $\mathbf{r}'(\pi/2) = \langle -1, 0, 1 \rangle$, so an equation is

$$-1(x - 0) + 0(y - 1) + 1\left(z - \frac{\pi}{2}\right) = 0 \qquad \text{or} \qquad z = x + \frac{\pi}{2}$$

|||| Figure 8 shows the helix and the osculating plane in Example 7.

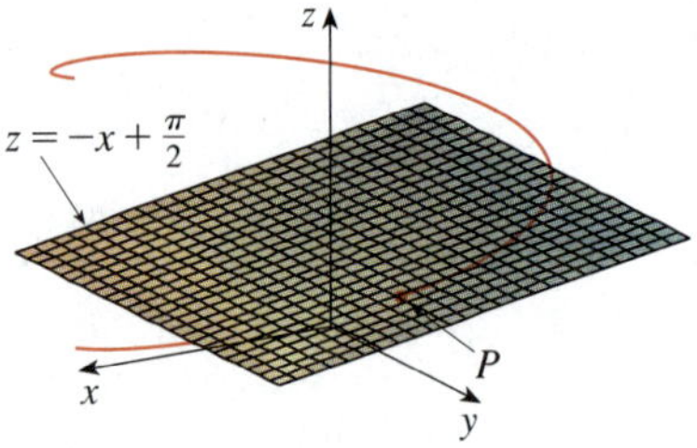

FIGURE 8

The osculating plane at P contains the vectors $\mathbf{T}$ and $\mathbf{N}$, so its normal vector is $\mathbf{T} \times \mathbf{N} = \mathbf{B}$. From Example 6 we have

$$\mathbf{B}(t) = \frac{1}{\sqrt{2}}\langle \sin t, -\cos t, 1\rangle \qquad \mathbf{B}\left(\frac{\pi}{2}\right) = \left\langle \frac{1}{\sqrt{2}}, 0, \frac{1}{\sqrt{2}} \right\rangle$$

A simpler normal vector is $\langle 1, 0, 1\rangle$, so an equation of the osculating plane is

$$1(x-0) + 0(y-1) + 1\left(z - \frac{\pi}{2}\right) = 0 \qquad \text{or} \qquad z = -x + \frac{\pi}{2}$$

EXAMPLE 8 Find and graph the osculating circle of the parabola $y = x^2$ at the origin.

SOLUTION From Example 5 the curvature of the parabola at the origin is $\kappa(0) = 2$. So the radius of the osculating circle at the origin is $1/\kappa = \frac{1}{2}$ and its center is $\left(0, \frac{1}{2}\right)$. Its equation is therefore

$$x^2 + \left(y - \tfrac{1}{2}\right)^2 = \tfrac{1}{4}$$

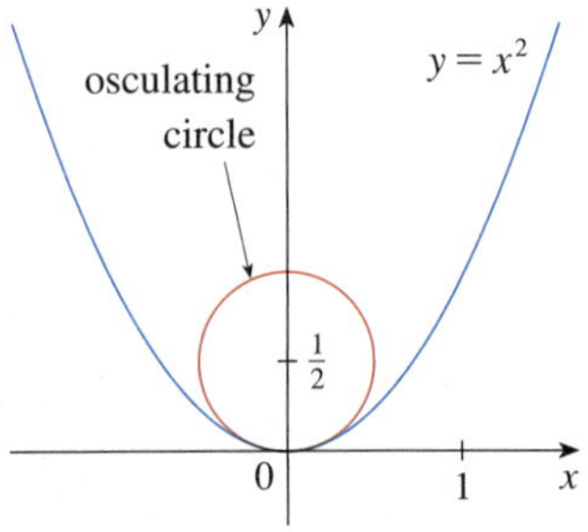

FIGURE 9

TEC Visual 13.3C shows how the osculating circle changes as a point moves along a curve.

For the graph in Figure 9 we use parametric equations of this circle:

$$x = \tfrac{1}{2}\cos t \qquad y = \tfrac{1}{2} + \tfrac{1}{2}\sin t$$

We summarize here the formulas for unit tangent, unit normal and binormal vectors, and curvature.

$$\mathbf{T}(t) = \frac{\mathbf{r}'(t)}{|\mathbf{r}'(t)|} \qquad \mathbf{N}(t) = \frac{\mathbf{T}'(t)}{|\mathbf{T}'(t)|} \qquad \mathbf{B}(t) = \mathbf{T}(t) \times \mathbf{N}(t)$$

$$\kappa = \left|\frac{d\mathbf{T}}{ds}\right| = \frac{|\mathbf{T}'(t)|}{|\mathbf{r}'(t)|} = \frac{|\mathbf{r}'(t) \times \mathbf{r}''(t)|}{|\mathbf{r}'(t)|^3}$$

13.3 Exercises

1–6 |||| Find the length of the curve.

1. $\mathbf{r}(t) = \langle 2\sin t, 5t, 2\cos t\rangle, \quad -10 \leqslant t \leqslant 10$

2. $\mathbf{r}(t) = \langle t^2, \sin t - t\cos t, \cos t + t\sin t\rangle, \quad 0 \leqslant t \leqslant \pi$

3. $\mathbf{r}(t) = \sqrt{2}\,t\,\mathbf{i} + e^t\,\mathbf{j} + e^{-t}\,\mathbf{k}, \quad 0 \leqslant t \leqslant 1$

4. $\mathbf{r}(t) = t^2\,\mathbf{i} + 2t\,\mathbf{j} + \ln t\,\mathbf{k}, \quad 1 \leqslant t \leqslant e$

5. $\mathbf{r}(t) = \mathbf{i} + t^2\,\mathbf{j} + t^3\,\mathbf{k}, \quad 0 \leqslant t \leqslant 1$

6. $\mathbf{r}(t) = 12t\,\mathbf{i} + 8t^{3/2}\,\mathbf{j} + 3t^2\,\mathbf{k}, \quad 0 \leqslant t \leqslant 1$

7. Use Simpson's Rule with $n = 10$ to estimate the length of the arc of the twisted cubic $x = t$, $y = t^2$, $z = t^3$ from the origin to the point $(2, 4, 8)$.

8. Graph the curve with parametric equations $x = \cos t$, $y = \sin 3t$, $z = \sin t$. Find the total length of this curve correct to four decimal places.

9–11 |||| Reparametrize the curve with respect to arc length measured from the point where $t = 0$ in the direction of increasing t.

9. $\mathbf{r}(t) = 2t\,\mathbf{i} + (1 - 3t)\,\mathbf{j} + (5 + 4t)\,\mathbf{k}$

10. $\mathbf{r}(t) = e^{2t}\cos 2t\,\mathbf{i} + 2\,\mathbf{j} + e^{2t}\sin 2t\,\mathbf{k}$

11. $\mathbf{r}(t) = 3\sin t\,\mathbf{i} + 4t\,\mathbf{j} + 3\cos t\,\mathbf{k}$

12. Reparametrize the curve

$$\mathbf{r}(t) = \left(\frac{2}{t^2+1} - 1\right)\mathbf{i} + \frac{2t}{t^2+1}\,\mathbf{j}$$

with respect to arc length measured from the point $(1, 0)$ in the direction of increasing t. Express the reparametrization in its simplest form. What can you conclude about the curve?

13–16
(a) Find the unit tangent and unit normal vectors $\mathbf{T}(t)$ and $\mathbf{N}(t)$.
(b) Use Formula 9 to find the curvature.

13. $\mathbf{r}(t) = \langle 2\sin t, 5t, 2\cos t\rangle$

14. $\mathbf{r}(t) = \langle t^2, \sin t - t\cos t, \cos t + t\sin t\rangle, \quad t > 0$

15. $\mathbf{r}(t) = \langle \sqrt{2}t, e^t, e^{-t}\rangle$ **16.** $\mathbf{r}(t) = \langle t^2, 2t, \ln t\rangle$

17–19 Use Theorem 10 to find the curvature.

17. $\mathbf{r}(t) = t^2\,\mathbf{i} + t\,\mathbf{k}$

18. $\mathbf{r}(t) = t\,\mathbf{i} + t\,\mathbf{j} + (1 + t^2)\,\mathbf{k}$

19. $\mathbf{r}(t) = 3t\,\mathbf{i} + 4\sin t\,\mathbf{j} + 4\cos t\,\mathbf{k}$

20. Find the curvature of $\mathbf{r}(t) = \langle e^t\cos t, e^t\sin t, t\rangle$ at the point $(1, 0, 0)$.

21. Find the curvature of $\mathbf{r}(t) = \langle t, t^2, t^3\rangle$ at the point $(1, 1, 1)$.

22. Graph the curve with parametric equations

$$x = t \qquad y = 4t^{3/2} \qquad z = -t^2$$

and find the curvature at the point $(1, 4, -1)$.

23–25 Use Formula 11 to find the curvature.

23. $y = x^3$ **24.** $y = \cos x$ **25.** $y = 4x^{5/2}$

26–27 At what point does the curve have maximum curvature? What happens to the curvature as $x \to \infty$?

26. $y = \ln x$ **27.** $y = e^x$

28. Find an equation of a parabola that has curvature 4 at the origin.

29. (a) Is the curvature of the curve C shown in the figure greater at P or at Q? Explain.
(b) Estimate the curvature at P and at Q by sketching the osculating circles at those points.

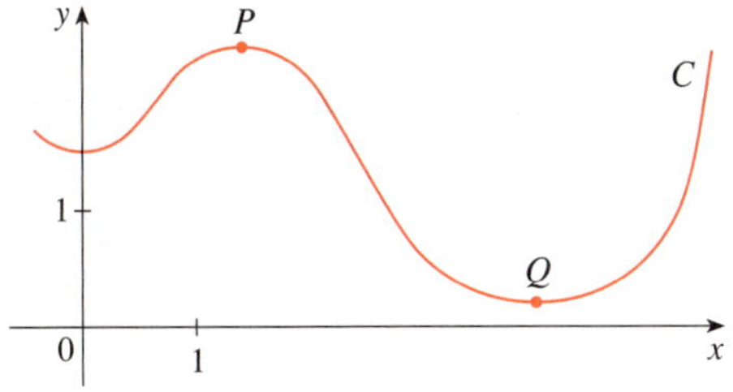

30–31 Use a graphing calculator or computer to graph both the curve and its curvature function $\kappa(x)$ on the same screen. Is the graph of κ what you would expect?

30. $y = xe^{-x}$ **31.** $y = x^4$

32–33 Two graphs, a and b, are shown. One is a curve $y = f(x)$ and the other is the graph of its curvature function $y = \kappa(x)$. Identify each curve and explain your choices.

32.

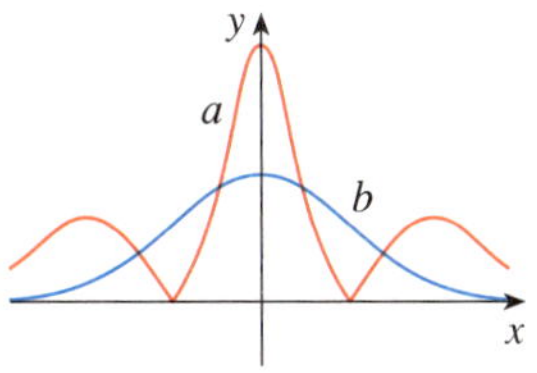

33.

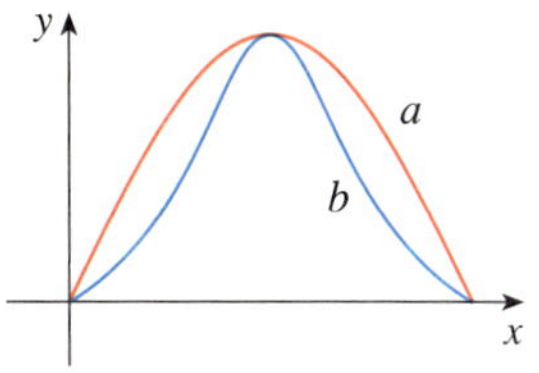

CAS **34.** (a) Graph the curve $\mathbf{r}(t) = \langle \sin 3t, \sin 2t, \sin 3t\rangle$. At how many points on the curve does it appear that the curvature has a local or absolute maximum?
(b) Use a CAS to find and graph the curvature function. Does this graph confirm your conclusion from part (a)?

CAS **35.** The graph of $\mathbf{r}(t) = \langle t - \frac{3}{2}\sin t, 1 - \frac{3}{2}\cos t, t\rangle$ is shown in Figure 12(b) in Section 13.1. Where do you think the curvature is largest? Use a CAS to find and graph the curvature function. For which values of t is the curvature largest?

36. Use Theorem 10 to show that the curvature of a plane parametric curve $x = f(t)$, $y = g(t)$ is

$$\kappa = \frac{|\dot{x}\ddot{y} - \dot{y}\ddot{x}|}{[\dot{x}^2 + \dot{y}^2]^{3/2}}$$

where the dots indicate derivatives with respect to t.

37–38 Use the formula in Exercise 36 to find the curvature.

37. $x = e^t\cos t, \quad y = e^t\sin t$

38. $x = 1 + t^3, \quad y = t + t^2$

39–40 Find the vectors $\mathbf{T}$, $\mathbf{N}$, and $\mathbf{B}$ at the given point.

39. $\mathbf{r}(t) = \langle t^2, \frac{2}{3}t^3, t\rangle, \quad (1, \frac{2}{3}, 1)$

40. $\mathbf{r}(t) = \langle e^t, e^t\sin t, e^t\cos t\rangle, \quad (1, 0, 1)$

41–42 Find equations of the normal plane and osculating plane of the curve at the given point.

41. $x = 2\sin 3t, \ y = t, \ z = 2\cos 3t; \quad (0, \pi, -2)$

42. $x = t, \ y = t^2, \ z = t^3; \quad (1, 1, 1)$

43. Find equations of the osculating circles of the ellipse $9x^2 + 4y^2 = 36$ at the points $(2, 0)$ and $(0, 3)$. Use a graphing calculator or computer to graph the ellipse and both osculating circles on the same screen.

44. Find equations of the osculating circles of the parabola $y = \frac{1}{2}x^2$ at the points $(0, 0)$ and $(1, \frac{1}{2})$. Graph both osculating circles and the parabola on the same screen.

45. At what point on the curve $x = t^3$, $y = 3t$, $z = t^4$ is the normal plane parallel to the plane $6x + 6y - 8z = 1$?

CAS **46.** Is there a point on the curve in Exercise 45 where the osculating plane is parallel to the plane $x + y + z = 1$? (*Note:* You will need a CAS for differentiating, for simplifying, and for computing a cross product.)

47. Show that the curvature κ is related to the tangent and normal vectors by the equation

$$\frac{d\mathbf{T}}{ds} = \kappa\mathbf{N}$$

48. Show that the curvature of a plane curve is $\kappa = |d\phi/ds|$, where ϕ is the angle between $\mathbf{T}$ and $\mathbf{i}$; that is, ϕ is the angle of inclination of the tangent line. (This shows that the definition of curvature is consistent with the definition for plane curves given in Exercise 69 in Section 10.2.)

49. (a) Show that $d\mathbf{B}/ds$ is perpendicular to $\mathbf{B}$.
(b) Show that $d\mathbf{B}/ds$ is perpendicular to $\mathbf{T}$.
(c) Deduce from parts (a) and (b) that $d\mathbf{B}/ds = -\tau(s)\mathbf{N}$ for some number $\tau(s)$ called the **torsion** of the curve. (The torsion measures the degree of twisting of a curve.)
(d) Show that for a plane curve the torsion is $\tau(s) = 0$.

50. The following formulas, called the **Frenet-Serret formulas**, are of fundamental importance in differential geometry:

1. $d\mathbf{T}/ds = \kappa\mathbf{N}$
2. $d\mathbf{N}/ds = -\kappa\mathbf{T} + \tau\mathbf{B}$
3. $d\mathbf{B}/ds = -\tau\mathbf{N}$

(Formula 1 comes from Exercise 47 and Formula 3 comes from Exercise 49.) Use the fact that $\mathbf{N} = \mathbf{B} \times \mathbf{T}$ to deduce Formula 2 from Formulas 1 and 3.

51. Use the Frenet-Serret formulas to prove each of the following. (Primes denote derivatives with respect to t. Start as in the proof of Theorem 10.)
(a) $\mathbf{r}'' = s''\mathbf{T} + \kappa(s')^2\mathbf{N}$
(b) $\mathbf{r}' \times \mathbf{r}'' = \kappa(s')^3\mathbf{B}$
(c) $\mathbf{r}''' = [s''' - \kappa^2(s')^3]\mathbf{T} + [3\kappa s's'' + \kappa'(s')^2]\mathbf{N} + \kappa\tau(s')^3\mathbf{B}$
(d) $\tau = \dfrac{(\mathbf{r}' \times \mathbf{r}'') \cdot \mathbf{r}'''}{|\mathbf{r}' \times \mathbf{r}''|^2}$

52. Show that the circular helix $\mathbf{r}(t) = \langle a\cos t, a\sin t, bt\rangle$, where a and b are positive constants, has constant curvature and constant torsion. [Use the result of Exercise 51(d).]

53. Use the formula in Exercise 51(d) to find the torsion of the curve $\mathbf{r}(t) = \langle t, \frac{1}{2}t^2, \frac{1}{3}t^3\rangle$.

54. Find the curvature and torsion of the curve $x = \sinh t$, $y = \cosh t$, $z = t$ at the point $(0, 1, 0)$.

55. The DNA molecule has the shape of a double helix (see Figure 3 on page 851). The radius of each helix is about 10 angstroms ($1 \text{ Å} = 10^{-8}$ cm). Each helix rises about 34 Å during each complete turn, and there are about 2.9×10^8 complete turns. Estimate the length of each helix.

56. Let's consider the problem of designing a railroad track to make a smooth transition between sections of straight track. Existing track along the negative x-axis is to be joined smoothly to a track along the line $y = 1$ for $x \geq 1$.
(a) Find a polynomial $P = P(x)$ of degree 5 such that the function F defined by

$$F(x) = \begin{cases} 0 & \text{if } x \leq 0 \\ P(x) & \text{if } 0 < x < 1 \\ 1 & \text{if } x \geq 1 \end{cases}$$

is continuous and has continuous slope and continuous curvature.
(b) Use a graphing calculator or computer to draw the graph of F.

13.4 Motion in Space: Velocity and Acceleration

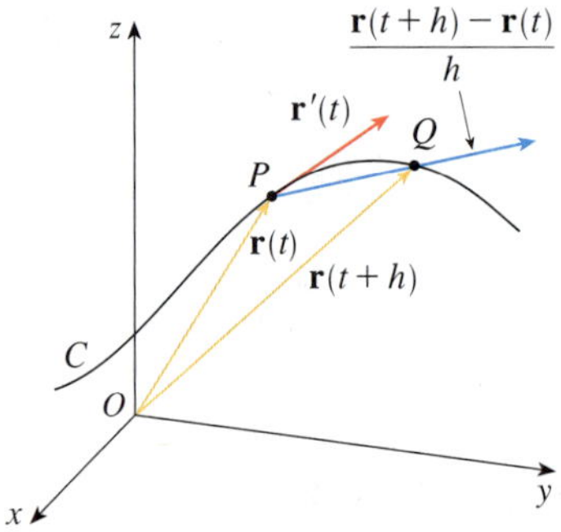

FIGURE 1

In this section we show how the ideas of tangent and normal vectors and curvature can be used in physics to study the motion of an object, including its velocity and acceleration, along a space curve. In particular, we follow in the footsteps of Newton by using these methods to derive Kepler's First Law of planetary motion.

Suppose a particle moves through space so that its position vector at time t is $\mathbf{r}(t)$. Notice from Figure 1 that, for small values of h, the vector

$$\boxed{1} \qquad \frac{\mathbf{r}(t+h) - \mathbf{r}(t)}{h}$$

approximates the direction of the particle moving along the curve $\mathbf{r}(t)$. Its magnitude measures the size of the displacement vector per unit time. The vector (1) gives the average velocity over a time interval of length h and its limit is the **velocity vector** $\mathbf{v}(t)$ at time t:

2
$$\mathbf{v}(t) = \lim_{h \to 0} \frac{\mathbf{r}(t+h) - \mathbf{r}(t)}{h} = \mathbf{r}'(t)$$

Thus, the velocity vector is also the tangent vector and points in the direction of the tangent line.

The **speed** of the particle at time t is the magnitude of the velocity vector, that is, $|\mathbf{v}(t)|$. This is appropriate because, from (2) and from Equation 13.3.7, we have

$$|\mathbf{v}(t)| = |\mathbf{r}'(t)| = \frac{ds}{dt} = \text{rate of change of distance with respect to time}$$

As in the case of one-dimensional motion, the **acceleration** of the particle is defined as the derivative of the velocity:

$$\mathbf{a}(t) = \mathbf{v}'(t) = \mathbf{r}''(t)$$

EXAMPLE 1 The position vector of an object moving in a plane is given by $\mathbf{r}(t) = t^3\,\mathbf{i} + t^2\,\mathbf{j}$. Find its velocity, speed, and acceleration when $t = 1$ and illustrate geometrically.

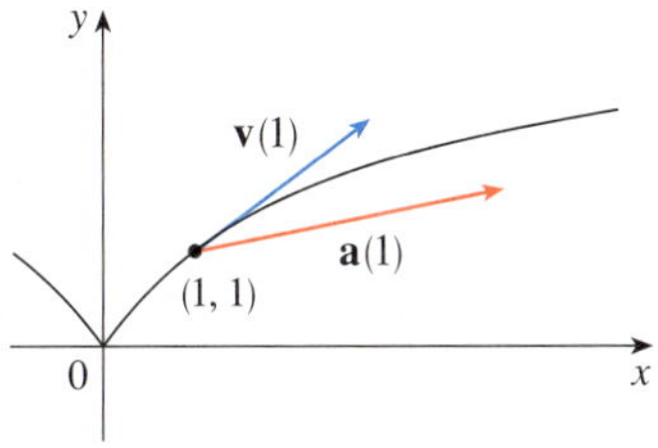

FIGURE 2

Visual 13.4 shows animated velocity and acceleration vectors for objects moving along various curves.

SOLUTION The velocity and acceleration at time t are

$$\mathbf{v}(t) = \mathbf{r}'(t) = 3t^2\,\mathbf{i} + 2t\,\mathbf{j}$$

$$\mathbf{a}(t) = \mathbf{r}''(t) = 6t\,\mathbf{i} + 2\,\mathbf{j}$$

and the speed is

$$|\mathbf{v}(t)| = \sqrt{(3t^2)^2 + (2t)^2} = \sqrt{9t^4 + 4t^2}$$

When $t = 1$, we have

$$\mathbf{v}(1) = 3\,\mathbf{i} + 2\,\mathbf{j} \qquad \mathbf{a}(1) = 6\,\mathbf{i} + 2\,\mathbf{j} \qquad |\mathbf{v}(1)| = \sqrt{13}$$

These velocity and acceleration vectors are shown in Figure 2.

||| Figure 3 shows the path of the particle in Example 2 with the velocity and acceleration vectors when $t = 1$.

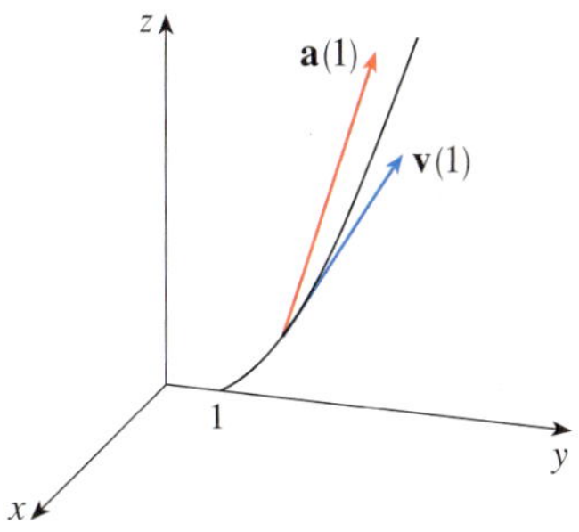

FIGURE 3

EXAMPLE 2 Find the velocity, acceleration, and speed of a particle with position vector $\mathbf{r}(t) = \langle t^2, e^t, te^t \rangle$.

SOLUTION

$$\mathbf{v}(t) = \mathbf{r}'(t) = \langle 2t, e^t, (1 + t)e^t \rangle$$

$$\mathbf{a}(t) = \mathbf{v}'(t) = \langle 2, e^t, (2 + t)e^t \rangle$$

$$|\mathbf{v}(t)| = \sqrt{4t^2 + e^{2t} + (1 + t)^2 e^{2t}}$$

The vector integrals that were introduced in Section 13.2 can be used to find position vectors when velocity or acceleration vectors are known, as in the following example.

EXAMPLE 3 A moving particle starts at an initial position $\mathbf{r}(0) = \langle 1, 0, 0 \rangle$ with initial velocity $\mathbf{v}(0) = \mathbf{i} - \mathbf{j} + \mathbf{k}$. Its acceleration is $\mathbf{a}(t) = 4t\,\mathbf{i} + 6t\,\mathbf{j} + \mathbf{k}$. Find its velocity and position at time t.

SOLUTION Since $\mathbf{a}(t) = \mathbf{v}'(t)$, we have

$$\mathbf{v}(t) = \int \mathbf{a}(t)\,dt = \int (4t\,\mathbf{i} + 6t\,\mathbf{j} + \mathbf{k})\,dt$$

$$= 2t^2\,\mathbf{i} + 3t^2\,\mathbf{j} + t\,\mathbf{k} + \mathbf{C}$$

To determine the value of the constant vector $\mathbf{C}$, we use the fact that $\mathbf{v}(0) = \mathbf{i} - \mathbf{j} + \mathbf{k}$. The preceding equation gives $\mathbf{v}(0) = \mathbf{C}$, so $\mathbf{C} = \mathbf{i} - \mathbf{j} + \mathbf{k}$ and

$$\mathbf{v}(t) = 2t^2\,\mathbf{i} + 3t^2\,\mathbf{j} + t\,\mathbf{k} + \mathbf{i} - \mathbf{j} + \mathbf{k}$$

$$= (2t^2 + 1)\,\mathbf{i} + (3t^2 - 1)\,\mathbf{j} + (t + 1)\,\mathbf{k}$$

||||| The expression for $\mathbf{r}(t)$ that we obtained in Example 3 was used to plot the path of the particle in Figure 4 for $0 \leqslant t \leqslant 3$.

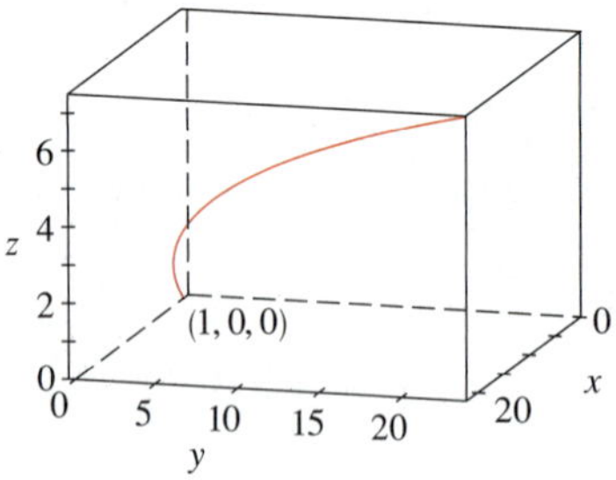

FIGURE 4

Since $\mathbf{v}(t) = \mathbf{r}'(t)$, we have

$$\mathbf{r}(t) = \int \mathbf{v}(t)\,dt$$

$$= \int [(2t^2 + 1)\,\mathbf{i} + (3t^2 - 1)\,\mathbf{j} + (t + 1)\,\mathbf{k}]\,dt$$

$$= \left(\tfrac{2}{3}t^3 + t\right)\mathbf{i} + (t^3 - t)\,\mathbf{j} + \left(\tfrac{1}{2}t^2 + t\right)\mathbf{k} + \mathbf{D}$$

Putting $t = 0$, we find that $\mathbf{D} = \mathbf{r}(0) = \mathbf{i}$, so

$$\mathbf{r}(t) = \left(\tfrac{2}{3}t^3 + t + 1\right)\mathbf{i} + (t^3 - t)\,\mathbf{j} + \left(\tfrac{1}{2}t^2 + t\right)\mathbf{k}$$

In general, vector integrals allow us to recover velocity when acceleration is known and position when velocity is known:

$$\mathbf{v}(t) = \mathbf{v}(t_0) + \int_{t_0}^{t} \mathbf{a}(u)\,du \qquad \mathbf{r}(t) = \mathbf{r}(t_0) + \int_{t_0}^{t} \mathbf{v}(u)\,du$$

If the force that acts on a particle is known, then the acceleration can be found from **Newton's Second Law of Motion**. The vector version of this law states that if, at any time t, a force $\mathbf{F}(t)$ acts on an object of mass m producing an acceleration $\mathbf{a}(t)$, then

$$\mathbf{F}(t) = m\mathbf{a}(t)$$

||||| The angular speed of the object moving with position P is $\omega = d\theta/dt$, where θ is the angle shown in Figure 5.

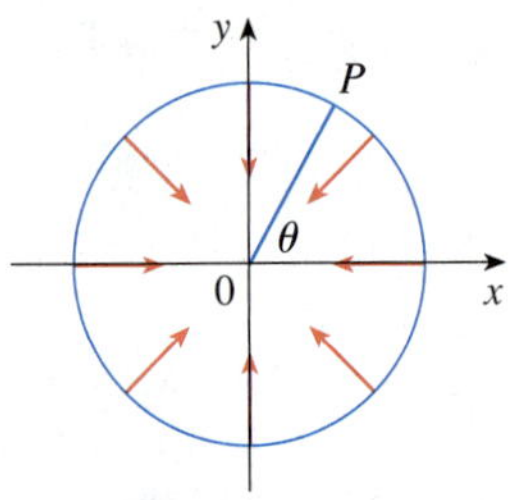

FIGURE 5

EXAMPLE 4 An object with mass m that moves in a circular path with constant angular speed ω has position vector $\mathbf{r}(t) = a\cos\omega t\,\mathbf{i} + a\sin\omega t\,\mathbf{j}$. Find the force acting on the object and show that it is directed toward the origin.

SOLUTION

$$\mathbf{v}(t) = \mathbf{r}'(t) = -a\omega\sin\omega t\,\mathbf{i} + a\omega\cos\omega t\,\mathbf{j}$$

$$\mathbf{a}(t) = \mathbf{v}'(t) = -a\omega^2\cos\omega t\,\mathbf{i} - a\omega^2\sin\omega t\,\mathbf{j}$$

Therefore, Newton's Second Law gives the force as

$$\mathbf{F}(t) = m\mathbf{a}(t) = -m\omega^2(a\cos\omega t\,\mathbf{i} + a\sin\omega t\,\mathbf{j})$$

Notice that $\mathbf{F}(t) = -m\omega^2\mathbf{r}(t)$. This shows that the force acts in the direction opposite to the radius vector $\mathbf{r}(t)$ and therefore points toward the origin (see Figure 5). Such a force is called a *centripetal* (center-seeking) force.

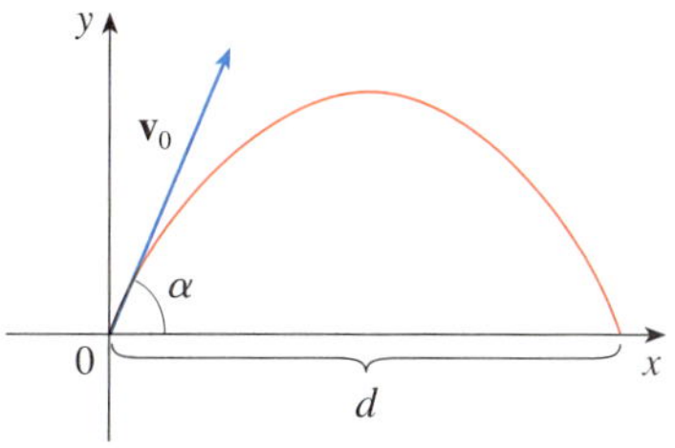

FIGURE 6

EXAMPLE 5 A projectile is fired with angle of elevation α and initial velocity $\mathbf{v}_0$. (See Figure 6.) Assuming that air resistance is negligible and the only external force is due to gravity, find the position function $\mathbf{r}(t)$ of the projectile. What value of α maximizes the range (the horizontal distance traveled)?

SOLUTION We set up the axes so that the projectile starts at the origin. Since the force due to gravity acts downward, we have

$$\mathbf{F} = m\mathbf{a} = -mg\,\mathbf{j}$$

where $g = |\mathbf{a}| \approx 9.8 \text{ m/s}^2$. Thus

$$\mathbf{a} = -g\,\mathbf{j}$$

Since $\mathbf{v}'(t) = \mathbf{a}$, we have

$$\mathbf{v}(t) = -gt\,\mathbf{j} + \mathbf{C}$$

where $\mathbf{C} = \mathbf{v}(0) = \mathbf{v}_0$. Therefore

$$\mathbf{r}'(t) = \mathbf{v}(t) = -gt\,\mathbf{j} + \mathbf{v}_0$$

Integrating again, we obtain

$$\mathbf{r}(t) = -\tfrac{1}{2}gt^2\,\mathbf{j} + t\,\mathbf{v}_0 + \mathbf{D}$$

But $\mathbf{D} = \mathbf{r}(0) = \mathbf{0}$, so the position vector of the projectile is given by

3 $$\mathbf{r}(t) = -\tfrac{1}{2}gt^2\,\mathbf{j} + t\,\mathbf{v}_0$$

If we write $|\mathbf{v}_0| = v_0$ (the initial speed of the projectile), then

$$\mathbf{v}_0 = v_0 \cos\alpha\,\mathbf{i} + v_0 \sin\alpha\,\mathbf{j}$$

and Equation 3 becomes

$$\mathbf{r}(t) = (v_0\cos\alpha)t\,\mathbf{i} + \left[(v_0\sin\alpha)t - \tfrac{1}{2}gt^2\right]\mathbf{j}$$

The parametric equations of the trajectory are therefore

|||| If you eliminate t from Equations 4, you will see that y is a quadratic function of x. So the path of the projectile is part of a parabola.

4 $$x = (v_0\cos\alpha)t \qquad y = (v_0\sin\alpha)t - \tfrac{1}{2}gt^2$$

The horizontal distance d is the value of x when $y = 0$. Setting $y = 0$, we obtain $t = 0$ or $t = (2v_0\sin\alpha)/g$. The latter value of t then gives

$$d = x = (v_0\cos\alpha)\frac{2v_0\sin\alpha}{g} = \frac{v_0^2(2\sin\alpha\cos\alpha)}{g} = \frac{v_0^2\sin 2\alpha}{g}$$

Clearly, d has its maximum value when $\sin 2\alpha = 1$, that is, $\alpha = \pi/4$.

EXAMPLE 6 A projectile is fired with muzzle speed 150 m/s and angle of elevation 45° from a position 10 m above ground level. Where does the projectile hit the ground, and with what speed?

SOLUTION If we place the origin at ground level, then the initial position of the projectile is (0, 10) and so we need to adjust Equations 4 by adding 10 to the expression for y. With $v_0 = 150$ m/s, $\alpha = 45°$, and $g = 9.8$ m/s^2, we have

$$x = 150\cos(\pi/4)t = 75\sqrt{2}\,t$$

$$y = 10 + 150\sin(\pi/4)t - \tfrac{1}{2}(9.8)t^2 = 10 + 75\sqrt{2}\,t - 4.9t^2$$

Impact occurs when $y = 0$, that is, $4.9t^2 - 75\sqrt{2}\,t - 10 = 0$. Solving this quadratic equation (and using only the positive value of t), we get

$$t = \frac{75\sqrt{2} + \sqrt{11{,}250 + 196}}{9.8} \approx 21.74$$

Then $x \approx 75\sqrt{2}(21.74) \approx 2306$, so the projectile hits the ground about 2306 m away.

The velocity of the projectile is

$$\mathbf{v}(t) = \mathbf{r}'(t) = 75\sqrt{2}\,\mathbf{i} + \left(75\sqrt{2} - 9.8t\right)\mathbf{j}$$

So its speed at impact is

$$|\mathbf{v}(21.74)| = \sqrt{\left(75\sqrt{2}\right)^2 + \left(75\sqrt{2} - 9.8 \cdot 21.74\right)^2} \approx 151 \text{ m/s}$$

Tangential and Normal Components of Acceleration

When we study the motion of a particle, it is often useful to resolve the acceleration into two components, one in the direction of the tangent and the other in the direction of the normal. If we write $v = |\mathbf{v}|$ for the speed of the particle, then

$$\mathbf{T}(t) = \frac{\mathbf{r}'(t)}{|\mathbf{r}'(t)|} = \frac{\mathbf{v}(t)}{|\mathbf{v}(t)|} = \frac{\mathbf{v}}{v}$$

and so

$$\mathbf{v} = v\mathbf{T}$$

If we differentiate both sides of this equation with respect to t, we get

5 $$\mathbf{a} = \mathbf{v}' = v'\mathbf{T} + v\mathbf{T}'$$

If we use the expression for the curvature given by Equation 13.3.9, then we have

6 $$\kappa = \frac{|\mathbf{T}'|}{|\mathbf{r}'|} = \frac{|\mathbf{T}'|}{v} \quad \text{so} \quad |\mathbf{T}'| = \kappa v$$

The unit normal vector was defined in the preceding section as $\mathbf{N} = \mathbf{T}'/|\mathbf{T}'|$, so (6) gives

$$\mathbf{T}' = |\mathbf{T}'|\mathbf{N} = \kappa v\mathbf{N}$$

and Equation 5 becomes

7 $$\boxed{\mathbf{a} = v'\mathbf{T} + \kappa v^2\mathbf{N}}$$

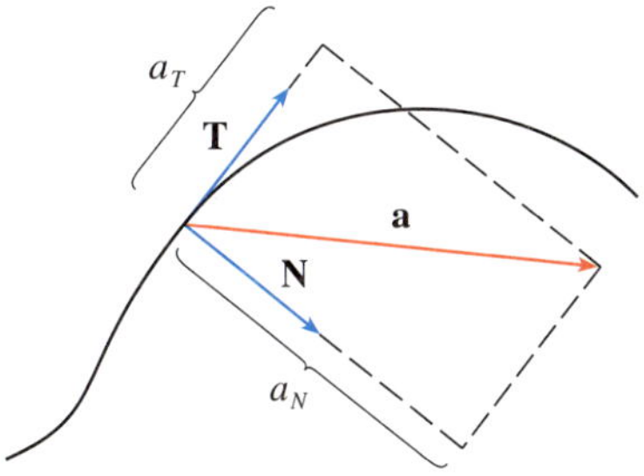

FIGURE 7

Writing a_T and a_N for the tangential and normal components of acceleration, we have

$$\mathbf{a} = a_T\mathbf{T} + a_N\mathbf{N}$$

where

$$\boxed{8} \qquad a_T = v' \qquad \text{and} \qquad a_N = \kappa v^2$$

This resolution is illustrated in Figure 7.

Let's look at what Formula 7 says. The first thing to notice is that the binormal vector **B** is absent. No matter how an object moves through space, its acceleration always lies in the plane of **T** and **N** (the osculating plane). (Recall that **T** gives the direction of motion and **N** points in the direction the curve is turning.) Next we notice that the tangential component of acceleration is v', the rate of change of speed, and the normal component of acceleration is κv^2, the curvature times the square of the speed. This makes sense if we think of a passenger in a car—a sharp turn in a road means a large value of the curvature κ, so the component of the acceleration perpendicular to the motion is large and the passenger is thrown against a car door. High speed around the turn has the same effect; in fact, if you double your speed, a_N is increased by a factor of 4.

Although we have expressions for the tangential and normal components of acceleration in Equations 8, it's desirable to have expressions that depend only on **r**, **r**$'$, and **r**$''$. To this end we take the dot product of $\mathbf{v} = v\mathbf{T}$ with **a** as given by Equation 7:

$$\begin{aligned} \mathbf{v} \cdot \mathbf{a} &= v\mathbf{T} \cdot (v'\mathbf{T} + \kappa v^2\mathbf{N}) \\ &= vv'\mathbf{T} \cdot \mathbf{T} + \kappa v^3\mathbf{T} \cdot \mathbf{N} \\ &= vv' \qquad \text{(since } \mathbf{T} \cdot \mathbf{T} = 1 \text{ and } \mathbf{T} \cdot \mathbf{N} = 0\text{)} \end{aligned}$$

Therefore

$$\boxed{9} \qquad a_T = v' = \frac{\mathbf{v} \cdot \mathbf{a}}{v} = \frac{\mathbf{r}'(t) \cdot \mathbf{r}''(t)}{|\mathbf{r}'(t)|}$$

Using the formula for curvature given by Theorem 13.3.10, we have

$$\boxed{10} \qquad a_N = \kappa v^2 = \frac{|\mathbf{r}'(t) \times \mathbf{r}''(t)|}{|\mathbf{r}'(t)|^3}|\mathbf{r}'(t)|^2 = \frac{|\mathbf{r}'(t) \times \mathbf{r}''(t)|}{|\mathbf{r}'(t)|}$$

EXAMPLE 7 A particle moves with position function $\mathbf{r}(t) = \langle t^2, t^2, t^3 \rangle$. Find the tangential and normal components of acceleration.

SOLUTION

$$\begin{aligned} \mathbf{r}(t) &= t^2\,\mathbf{i} + t^2\,\mathbf{j} + t^3\,\mathbf{k} \\ \mathbf{r}'(t) &= 2t\,\mathbf{i} + 2t\,\mathbf{j} + 3t^2\,\mathbf{k} \\ \mathbf{r}''(t) &= 2\,\mathbf{i} + 2\,\mathbf{j} + 6t\,\mathbf{k} \\ |\mathbf{r}'(t)| &= \sqrt{8t^2 + 9t^4} \end{aligned}$$

Therefore, Equation 9 gives the tangential component as

$$a_T = \frac{\mathbf{r}'(t) \cdot \mathbf{r}''(t)}{|\mathbf{r}'(t)|} = \frac{8t + 18t^3}{\sqrt{8t^2 + 9t^4}}$$

Since $$\mathbf{r}'(t) \times \mathbf{r}''(t) = \begin{vmatrix} \mathbf{i} & \mathbf{j} & \mathbf{k} \\ 2t & 2t & 3t^2 \\ 2 & 2 & 6t \end{vmatrix} = 6t^2\,\mathbf{i} - 6t^2\,\mathbf{j}$$

Equation 10 gives the normal component as

$$a_N = \frac{|\mathbf{r}'(t) \times \mathbf{r}''(t)|}{|\mathbf{r}'(t)|} = \frac{6\sqrt{2}t^2}{\sqrt{8t^2 + 9t^4}}$$

|||| Kepler's Laws of Planetary Motion

We now describe one of the great accomplishments of calculus by showing how the material of this chapter can be used to prove Kepler's laws of planetary motion. After 20 years of studying the astronomical observations of the Danish astronomer Tycho Brahe, the German mathematician and astronomer Johannes Kepler (1571–1630) formulated the following three laws.

Kepler's Laws

1. A planet revolves around the Sun in an elliptical orbit with the Sun at one focus.
2. The line joining the Sun to a planet sweeps out equal areas in equal times.
3. The square of the period of revolution of a planet is proportional to the cube of the length of the major axis of its orbit.

In his book *Principia Mathematica* of 1687, Sir Isaac Newton was able to show that these three laws are consequences of two of his own laws, the Second Law of Motion and the Law of Universal Gravitation. In what follows we prove Kepler's First Law. The remaining laws are left as exercises (with hints).

Since the gravitational force of the Sun on a planet is so much larger than the forces exerted by other celestial bodies, we can safely ignore all bodies in the universe except the Sun and one planet revolving about it. We use a coordinate system with the Sun at the origin and we let $\mathbf{r} = \mathbf{r}(t)$ be the position vector of the planet. (Equally well, $\mathbf{r}$ could be the position vector of the Moon or a satellite moving around the Earth or a comet moving around a star.) The velocity vector is $\mathbf{v} = \mathbf{r}'$ and the acceleration vector is $\mathbf{a} = \mathbf{r}''$. We use the following laws of Newton:

Second Law of Motion: $$\mathbf{F} = m\mathbf{a}$$

Law of Gravitation: $$\mathbf{F} = -\frac{GMm}{r^3}\mathbf{r} = -\frac{GMm}{r^2}\mathbf{u}$$

where $\mathbf{F}$ is the gravitational force on the planet, m and M are the masses of the planet and the Sun, G is the gravitational constant, $r = |\mathbf{r}|$, and $\mathbf{u} = (1/r)\mathbf{r}$ is the unit vector in the direction of $\mathbf{r}$.

We first show that the planet moves in one plane. By equating the expressions for $\mathbf{F}$ in Newton's two laws, we find that

$$\mathbf{a} = -\frac{GM}{r^3}\mathbf{r}$$

and so **a** is parallel to **r**. It follows that $\mathbf{r} \times \mathbf{a} = \mathbf{0}$. We use Formula 5 in Theorem 13.2.3 to write

$$\frac{d}{dt}(\mathbf{r} \times \mathbf{v}) = \mathbf{r}' \times \mathbf{v} + \mathbf{r} \times \mathbf{v}'$$

$$= \mathbf{v} \times \mathbf{v} + \mathbf{r} \times \mathbf{a} = \mathbf{0} + \mathbf{0} = \mathbf{0}$$

Therefore
$$\mathbf{r} \times \mathbf{v} = \mathbf{h}$$

where **h** is a constant vector. (We may assume that $\mathbf{h} \neq \mathbf{0}$; that is, **r** and **v** are not parallel.) This means that the vector $\mathbf{r} = \mathbf{r}(t)$ is perpendicular to **h** for all values of t, so the planet always lies in the plane through the origin perpendicular to **h**. Thus, the orbit of the planet is a plane curve.

To prove Kepler's First Law we rewrite the vector **h** as follows:

$$\mathbf{h} = \mathbf{r} \times \mathbf{v} = \mathbf{r} \times \mathbf{r}' = r\mathbf{u} \times (r\mathbf{u})'$$

$$= r\mathbf{u} \times (r\mathbf{u}' + r'\mathbf{u}) = r^2(\mathbf{u} \times \mathbf{u}') + rr'(\mathbf{u} \times \mathbf{u})$$

$$= r^2(\mathbf{u} \times \mathbf{u}')$$

Then

$$\mathbf{a} \times \mathbf{h} = \frac{-GM}{r^2}\mathbf{u} \times (r^2\mathbf{u} \times \mathbf{u}') = -GM\,\mathbf{u} \times (\mathbf{u} \times \mathbf{u}')$$

$$= -GM[(\mathbf{u} \cdot \mathbf{u}')\mathbf{u} - (\mathbf{u} \cdot \mathbf{u})\mathbf{u}'] \qquad \text{(by Theorem 12.4.8, Property 6)}$$

But $\mathbf{u} \cdot \mathbf{u} = |\mathbf{u}|^2 = 1$ and, since $|\mathbf{u}(t)| = 1$, it follows from Example 5 in Section 13.2 that $\mathbf{u} \cdot \mathbf{u}' = 0$. Therefore

$$\mathbf{a} \times \mathbf{h} = GM\,\mathbf{u}'$$

and so
$$(\mathbf{v} \times \mathbf{h})' = \mathbf{v}' \times \mathbf{h} = \mathbf{a} \times \mathbf{h} = GM\,\mathbf{u}'$$

Integrating both sides of this equation, we get

11
$$\mathbf{v} \times \mathbf{h} = GM\,\mathbf{u} + \mathbf{c}$$

where **c** is a constant vector.

At this point it is convenient to choose the coordinate axes so that the standard basis vector **k** points in the direction of the vector **h**. Then the planet moves in the xy-plane. Since both $\mathbf{v} \times \mathbf{h}$ and **u** are perpendicular to **h**, Equation 11 shows that **c** lies in the xy-plane. This means that we can choose the x- and y-axes so that the vector **i** lies in the direction of **c**, as shown in Figure 8.

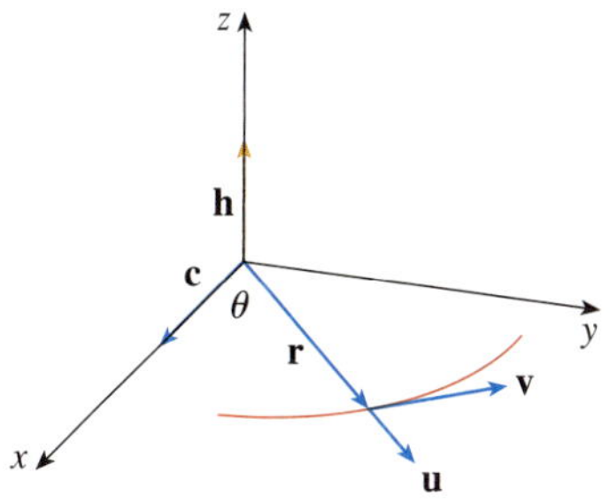

FIGURE 8

If θ is the angle between **c** and **r**, then (r, θ) are polar coordinates of the planet. From Equation 11 we have

$$\mathbf{r} \cdot (\mathbf{v} \times \mathbf{h}) = \mathbf{r} \cdot (GM\,\mathbf{u} + \mathbf{c}) = GM\,\mathbf{r} \cdot \mathbf{u} + \mathbf{r} \cdot \mathbf{c}$$

$$= GMr\,\mathbf{u} \cdot \mathbf{u} + |\mathbf{r}||\mathbf{c}|\cos\theta = GMr + rc\cos\theta$$

where $c = |\mathbf{c}|$. Then

$$r = \frac{\mathbf{r} \cdot (\mathbf{v} \times \mathbf{h})}{GM + c\cos\theta} = \frac{1}{GM}\,\frac{\mathbf{r} \cdot (\mathbf{v} \times \mathbf{h})}{1 + e\cos\theta}$$

where $e = c/(GM)$. But

$$\mathbf{r} \cdot (\mathbf{v} \times \mathbf{h}) = (\mathbf{r} \times \mathbf{v}) \cdot \mathbf{h} = \mathbf{h} \cdot \mathbf{h} = |\mathbf{h}|^2 = h^2$$

where $h = |\mathbf{h}|$. So

$$r = \frac{h^2/(GM)}{1 + e\cos\theta} = \frac{eh^2/c}{1 + e\cos\theta}$$

Writing $d = h^2/c$, we obtain the equation

12 $$r = \frac{ed}{1 + e\cos\theta}$$

Comparing with Theorem 10.6.6, we see that Equation 12 is the polar equation of a conic section with focus at the origin and eccentricity e. We know that the orbit of a planet is a closed curve and so the conic must be an ellipse.

This completes the derivation of Kepler's First Law. We will guide you through the derivation of the Second and Third Laws in the Applied Project on page 880. The proofs of these three laws show that the methods of this chapter provide a powerful tool for describing some of the laws of nature.

13.4 Exercises

1. The table gives coordinates of a particle moving through space along a smooth curve.
(a) Find the average velocities over the time intervals [0, 1], [0.5, 1], [1, 2], and [1, 1.5].
(b) Estimate the velocity and speed of the particle at $t = 1$.

t	x	y	z
0	2.7	9.8	3.7
0.5	3.5	7.2	3.3
1.0	4.5	6.0	3.0
1.5	5.9	6.4	2.8
2.0	7.3	7.8	2.7

2. The figure shows the path of a particle that moves with position vector $\mathbf{r}(t)$ at time t.
(a) Draw a vector that represents the average velocity of the particle over the time interval $2 \leqslant t \leqslant 2.4$.
(b) Draw a vector that represents the average velocity over the time interval $1.5 \leqslant t \leqslant 2$.
(c) Write an expression for the velocity vector $\mathbf{v}(2)$.
(d) Draw an approximation to the vector $\mathbf{v}(2)$ and estimate the speed of the particle at $t = 2$.

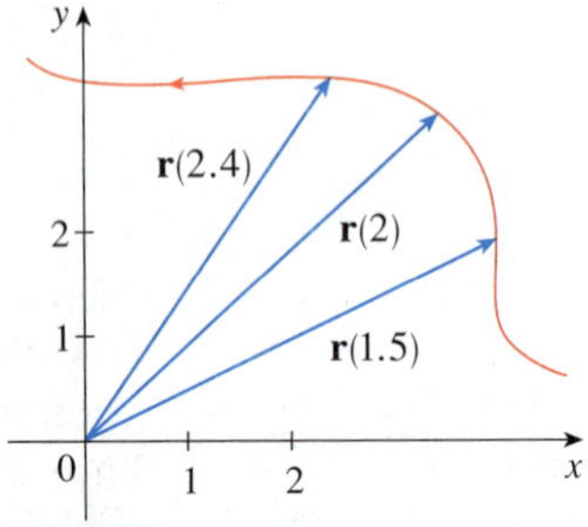

3–8 Find the velocity, acceleration, and speed of a particle with the given position function. Sketch the path of the particle and draw the velocity and acceleration vectors for the specified value of t.

3. $\mathbf{r}(t) = \langle t^2 - 1, t \rangle, \quad t = 1$

4. $\mathbf{r}(t) = \langle 2 - t, 4\sqrt{t} \rangle, \quad t = 1$

5. $\mathbf{r}(t) = e^t\,\mathbf{i} + e^{-t}\,\mathbf{j}, \quad t = 0$

6. $\mathbf{r}(t) = \sin t\,\mathbf{i} + 2\cos t\,\mathbf{j}, \quad t = \pi/6$

7. $\mathbf{r}(t) = \sin t\,\mathbf{i} + t\,\mathbf{j} + \cos t\,\mathbf{k}, \quad t = 0$

8. $\mathbf{r}(t) = t\,\mathbf{i} + t^2\mathbf{j} + t^3\mathbf{k}, \quad t = 1$

9–14 Find the velocity, acceleration, and speed of a particle with the given position function.

9. $\mathbf{r}(t) = \langle t^2 + 1, t^3, t^2 - 1 \rangle$

10. $\mathbf{r}(t) = \langle 2\cos t, 3t, 2\sin t \rangle$

11. $\mathbf{r}(t) = \sqrt{2}\,t\,\mathbf{i} + e^t\,\mathbf{j} + e^{-t}\,\mathbf{k}$

12. $\mathbf{r}(t) = t^2\,\mathbf{i} + \ln t\,\mathbf{j} + t\,\mathbf{k}$

13. $\mathbf{r}(t) = e^t(\cos t\,\mathbf{i} + \sin t\,\mathbf{j} + t\,\mathbf{k})$

14. $\mathbf{r}(t) = t\sin t\,\mathbf{i} + t\cos t\,\mathbf{j} + t^2\,\mathbf{k}$

15–16 Find the velocity and position vectors of a particle that has the given acceleration and the given initial velocity and position.

15. $\mathbf{a}(t) = \mathbf{k}, \quad \mathbf{v}(0) = \mathbf{i} - \mathbf{j}, \quad \mathbf{r}(0) = \mathbf{0}$

16. $\mathbf{a}(t) = -10\,\mathbf{k}, \quad \mathbf{v}(0) = \mathbf{i} + \mathbf{j} - \mathbf{k}, \quad \mathbf{r}(0) = 2\,\mathbf{i} + 3\,\mathbf{j}$

17–18

(a) Find the position vector of a particle that has the given acceleration and the specified initial velocity and position.

(b) Use a computer to graph the path of the particle.

17. $\mathbf{a}(t) = \mathbf{i} + 2\,\mathbf{j} + 2t\,\mathbf{k}, \quad \mathbf{v}(0) = \mathbf{0}, \quad \mathbf{r}(0) = \mathbf{i} + \mathbf{k}$

18. $\mathbf{a}(t) = t\,\mathbf{i} + t^2\,\mathbf{j} + \cos 2t\,\mathbf{k}, \quad \mathbf{v}(0) = \mathbf{i} + \mathbf{k}, \quad \mathbf{r}(0) = \mathbf{j}$

19. The position function of a particle is given by $\mathbf{r}(t) = \langle t^2, 5t, t^2 - 16t \rangle$. When is the speed a minimum?

20. What force is required so that a particle of mass m has the position function $\mathbf{r}(t) = t^3\,\mathbf{i} + t^2\,\mathbf{j} + t^3\,\mathbf{k}$?

21. A force with magnitude 20 N acts directly upward from the xy-plane on an object with mass 4 kg. The object starts at the origin with initial velocity $\mathbf{v}(0) = \mathbf{i} - \mathbf{j}$. Find its position function and its speed at time t.

22. Show that if a particle moves with constant speed, then the velocity and acceleration vectors are orthogonal.

23. A projectile is fired with an initial speed of 500 m/s and angle of elevation 30°. Find (a) the range of the projectile, (b) the maximum height reached, and (c) the speed at impact.

24. Rework Exercise 23 if the projectile is fired from a position 200 m above the ground.

25. A ball is thrown at an angle of 45° to the ground. If the ball lands 90 m away, what was the initial speed of the ball?

26. A gun is fired with angle of elevation 30°. What is the muzzle speed if the maximum height of the shell is 500 m?

27. A gun has muzzle speed 150 m/s. Find two angles of elevation that can be used to hit a target 800 m away.

28. A batter hits a baseball 3 ft above the ground toward the center field fence, which is 10 ft high and 400 ft from home plate. The ball leaves the bat with speed 115 ft/s at an angle 50° above the horizontal. Is it a home run? (In other words, does the ball clear the fence?)

29. Water traveling along a straight portion of a river normally flows fastest in the middle, and the speed slows to almost zero at the banks. Consider a long stretch of river flowing north, with parallel banks 40 m apart. If the maximum water speed is 3 m/s, we can use a quadratic function as a basic model for the rate of water flow x units from the west bank: $f(x) = \frac{3}{400}x(40 - x)$.

(a) A boat proceeds at a constant speed of 5 m/s from a point A on the west bank while maintaining a heading perpendicular to the bank. How far down the river on the opposite bank will the boat touch shore? Graph the path of the boat.

(b) Suppose we would like to pilot the boat to land at the point B on the east bank directly opposite A. If we maintain a constant speed of 5 m/s and a constant heading, find the angle at which the boat should head. Then graph the actual path the boat follows. Does the path seem realistic?

30. Another reasonable model for the water speed of the river in Exercise 29 is a sine function: $f(x) = 3\sin(\pi x/40)$. If a boater would like to cross the river from A to B with constant heading and a constant speed of 5 m/s, determine the angle at which the boat should head.

31–36 Find the tangential and normal components of the acceleration vector.

31. $\mathbf{r}(t) = (3t - t^3)\,\mathbf{i} + 3t^2\,\mathbf{j}$

32. $\mathbf{r}(t) = (1 + t)\,\mathbf{i} + (t^2 - 2t)\,\mathbf{j}$

33. $\mathbf{r}(t) = \cos t\,\mathbf{i} + \sin t\,\mathbf{j} + t\,\mathbf{k}$

34. $\mathbf{r}(t) = t\,\mathbf{i} + t^2\,\mathbf{j} + 3t\,\mathbf{k}$

35. $\mathbf{r}(t) = e^t\,\mathbf{i} + \sqrt{2}\,t\,\mathbf{j} + e^{-t}\,\mathbf{k}$

36. $\mathbf{r}(t) = t\,\mathbf{i} + \cos^2 t\,\mathbf{j} + \sin^2 t\,\mathbf{k}$

37. The magnitude of the acceleration vector $\mathbf{a}$ is 10 cm/s². Use the figure to estimate the tangential and normal components of $\mathbf{a}$.

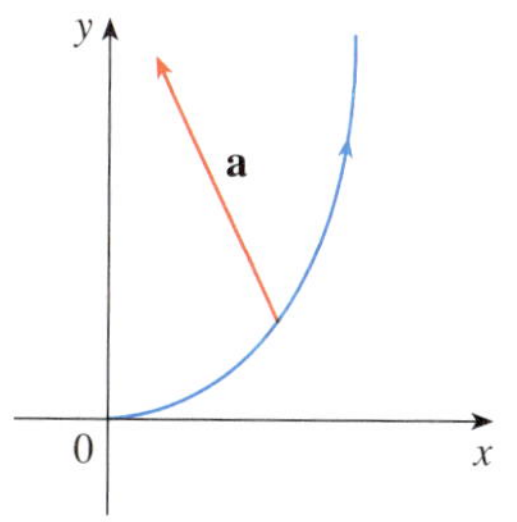

38. If a particle with mass m moves with position vector $\mathbf{r}(t)$, then its **angular momentum** is defined as $\mathbf{L}(t) = m\mathbf{r}(t) \times \mathbf{v}(t)$ and its **torque** as $\boldsymbol{\tau}(t) = m\mathbf{r}(t) \times \mathbf{a}(t)$. Show that $\mathbf{L}'(t) = \boldsymbol{\tau}(t)$. Deduce that if $\boldsymbol{\tau}(t) = \mathbf{0}$ for all t, then $\mathbf{L}(t)$ is constant. (This is the *law of conservation of angular momentum.*)

39. The position function of a spaceship is

$$\mathbf{r}(t) = (3 + t)\,\mathbf{i} + (2 + \ln t)\,\mathbf{j} + \left(7 - \frac{4}{t^2 + 1}\right)\mathbf{k}$$

and the coordinates of a space station are (6, 4, 9). The captain wants the spaceship to coast into the space station. When should the engines be turned off?

40. A rocket burning its onboard fuel while moving through space has velocity $\mathbf{v}(t)$ and mass $m(t)$ at time t. If the exhaust gases escape with velocity $\mathbf{v}_e$ relative to the rocket, it can be deduced from Newton's Second Law of Motion that

$$m\frac{d\mathbf{v}}{dt} = \frac{dm}{dt}\mathbf{v}_e$$

(a) Show that $\mathbf{v}(t) = \mathbf{v}(0) - \ln \dfrac{m(0)}{m(t)}\,\mathbf{v}_e$.

(b) For the rocket to accelerate in a straight line from rest to twice the speed of its own exhaust gases, what fraction of its initial mass would the rocket have to burn as fuel?

APPLIED PROJECT

Kepler's Laws

Johannes Kepler stated the following three laws of planetary motion on the basis of masses of data on the positions of the planets at various times.

Kepler's Laws

1. A planet revolves around the Sun in an elliptical orbit with the Sun at one focus.
2. The line joining the Sun to a planet sweeps out equal areas in equal times.
3. The square of the period of revolution of a planet is proportional to the cube of the length of the major axis of its orbit.

Kepler formulated these laws because they fitted the astronomical data. He wasn't able to see why they were true or how they related to each other. But Sir Isaac Newton, in his *Principia Mathematica* of 1687, showed how to deduce Kepler's three laws from two of Newton's own laws, the Second Law of Motion and the Law of Universal Gravitation. In Section 13.4 we proved Kepler's First Law using the calculus of vector functions. In this project we guide you through the proofs of Kepler's Second and Third Laws and explore some of their consequences.

1. Use the following steps to prove Kepler's Second Law. The notation is the same as in the proof of the First Law in Section 13.4. In particular, use polar coordinates so that $\mathbf{r} = (r\cos\theta)\,\mathbf{i} + (r\sin\theta)\,\mathbf{j}$.

(a) Show that $\mathbf{h} = r^2\dfrac{d\theta}{dt}\mathbf{k}$.

(b) Deduce that $r^2\dfrac{d\theta}{dt} = h$.

(c) If $A = A(t)$ is the area swept out by the radius vector $\mathbf{r} = \mathbf{r}(t)$ in the time interval $[t_0, t]$ as in the figure, show that

$$\frac{dA}{dt} = \tfrac{1}{2}r^2\frac{d\theta}{dt}$$

(d) Deduce that

$$\frac{dA}{dt} = \tfrac{1}{2}h = \text{constant}$$

This says that the rate at which A is swept out is constant and proves Kepler's Second Law.

2. Let T be the period of a planet about the Sun; that is, T is the time required for it to travel once around its elliptical orbit. Suppose that the lengths of the major and minor axes of the ellipse are $2a$ and $2b$.

(a) Use part (d) of Problem 1 to show that $T = 2\pi ab/h$.

(b) Show that $\dfrac{h^2}{GM} = ed = \dfrac{b^2}{a}$.

(c) Use parts (a) and (b) to show that $T^2 = \dfrac{4\pi^2}{GM} a^3$.

This proves Kepler's Third Law. [Notice that the proportionality constant $4\pi^2/(GM)$ is independent of the planet.]

3. The period of the Earth's orbit is approximately 365.25 days. Use this fact and Kepler's Third Law to find the length of the major axis of the Earth's orbit. You will need the mass of the Sun, $M = 1.99 \times 10^{30}$ kg, and the gravitational constant, $G = 6.67 \times 10^{-11}$ $\text{N}\cdot\text{m}^2/\text{kg}^2$.

4. It's possible to place a satellite into orbit about the Earth so that it remains fixed above a given location on the equator. Compute the altitude that is needed for such a satellite. The Earth's mass is 5.98×10^{24} kg; its radius is 6.37×10^6 m. (This orbit is called the Clarke Geosynchronous Orbit after Arthur C. Clarke, who first proposed the idea in 1948. The first such satellite, *Syncom II*, was launched in July 1963.)

13 Review

CONCEPT CHECK

1. What is a vector function? How do you find its derivative and its integral?

2. What is the connection between vector functions and space curves?

3. (a) What is a smooth curve?
(b) How do you find the tangent vector to a smooth curve at a point? How do you find the tangent line? The unit tangent vector?

4. If $\mathbf{u}$ and $\mathbf{v}$ are differentiable vector functions, c is a scalar, and f is a real-valued function, write the rules for differentiating the following vector functions.
(a) $\mathbf{u}(t) + \mathbf{v}(t)$ (b) $c\mathbf{u}(t)$ (c) $f(t)\mathbf{u}(t)$
(d) $\mathbf{u}(t) \cdot \mathbf{v}(t)$ (e) $\mathbf{u}(t) \times \mathbf{v}(t)$ (f) $\mathbf{u}(f(t))$

5. How do you find the length of a space curve given by a vector function $\mathbf{r}(t)$?

6. (a) What is the definition of curvature?
(b) Write a formula for curvature in terms of $\mathbf{r}'(t)$ and $\mathbf{T}'(t)$.
(c) Write a formula for curvature in terms of $\mathbf{r}'(t)$ and $\mathbf{r}''(t)$.
(d) Write a formula for the curvature of a plane curve with equation $y = f(x)$.

7. (a) Write formulas for the unit normal and binormal vectors of a smooth space curve $\mathbf{r}(t)$.
(b) What is the normal plane of a curve at a point? What is the osculating plane? What is the osculating circle?

8. (a) How do you find the velocity, speed, and acceleration of a particle that moves along a space curve?
(b) Write the acceleration in terms of its tangential and normal components.

9. State Kepler's Laws.

TRUE-FALSE QUIZ

Determine whether the statement is true or false. If it is true, explain why. If it is false, explain why or give an example that disproves the statement.

1. The curve with vector equation $\mathbf{r}(t) = t^3\,\mathbf{i} + 2t^3\,\mathbf{j} + 3t^3\,\mathbf{k}$ is a line.
2. The curve with vector equation $\mathbf{r}(t) = \langle t, t^3, t^5 \rangle$ is smooth.
3. The curve with vector equation $\mathbf{r}(t) = \langle \cos t, t^2, t^4 \rangle$ is smooth.
4. The derivative of a vector function is obtained by differentiating each component function.
5. If $\mathbf{u}(t)$ and $\mathbf{v}(t)$ are differentiable vector functions, then
$$\frac{d}{dt}[\mathbf{u}(t) \times \mathbf{v}(t)] = \mathbf{u}'(t) \times \mathbf{v}'(t)$$
6. If $\mathbf{r}(t)$ is a differentiable vector function, then
$$\frac{d}{dt}|\mathbf{r}(t)| = |\mathbf{r}'(t)|$$
7. If $\mathbf{T}(t)$ is the unit tangent vector of a smooth curve, then the curvature is $\kappa = |d\mathbf{T}/dt|$.
8. The binormal vector is $\mathbf{B}(t) = \mathbf{N}(t) \times \mathbf{T}(t)$.
9. The osculating circle of a curve C at a point has the same tangent vector, normal vector, and curvature as C at that point.
10. Different parametrizations of the same curve result in identical tangent vectors at a given point on the curve.

EXERCISES

1. (a) Sketch the curve with vector function
$$\mathbf{r}(t) = t\,\mathbf{i} + \cos \pi t\,\mathbf{j} + \sin \pi t\,\mathbf{k} \qquad t \geqslant 0$$
(b) Find $\mathbf{r}'(t)$ and $\mathbf{r}''(t)$.
2. Let $\mathbf{r}(t) = \langle \sqrt{2 - t}, (e^t - 1)/t, \ln(t + 1) \rangle$.
(a) Find the domain of $\mathbf{r}$.
(b) Find $\lim_{t \to 0} \mathbf{r}(t)$.
(c) Find $\mathbf{r}'(t)$.
3. Find a vector function that represents the curve of intersection of the cylinder $x^2 + y^2 = 16$ and the plane $x + z = 5$.
4. Find parametric equations for the tangent line to the curve $x = t^2$, $y = t^4$, $z = t^3$ at the point $(1, 1, 1)$. Graph the curve and the tangent line on a common screen.
5. If $\mathbf{r}(t) = t^2\,\mathbf{i} + t \cos \pi t\,\mathbf{j} + \sin \pi t\,\mathbf{k}$, evaluate $\int_0^1 \mathbf{r}(t)\,dt$.
6. Let C be the curve with equations $x = 2 - t^3$, $y = 2t - 1$, $z = \ln t$. Find (a) the point where C intersects the xz-plane, (b) parametric equations of the tangent line at $(1, 1, 0)$, and (c) an equation of the normal plane to C at $(1, 1, 0)$.
7. Use Simpson's Rule with $n = 4$ to estimate the length of the arc of the curve with equations $x = \sqrt{t}$, $y = 4/t$, $z = t^2 + 1$ from $(1, 4, 2)$ to $(2, 1, 17)$.
8. Find the length of the curve $\mathbf{r}(t) = \langle 2t^{3/2}, \cos 2t, \sin 2t \rangle$, $0 \leqslant t \leqslant 1$.
9. The helix $\mathbf{r}_1(t) = \cos t\,\mathbf{i} + \sin t\,\mathbf{j} + t\,\mathbf{k}$ intersects the curve $\mathbf{r}_2(t) = (1 + t)\mathbf{i} + t^2\mathbf{j} + t^3\mathbf{k}$ at the point $(1, 0, 0)$. Find the angle of intersection of these curves.
10. Reparametrize the curve $\mathbf{r}(t) = e^t\mathbf{i} + e^t \sin t\,\mathbf{j} + e^t \cos t\,\mathbf{k}$ with respect to arc length measured from the point $(1, 0, 1)$ in the direction of increasing t.
11. For the curve given by $\mathbf{r}(t) = \langle \frac{1}{3}t^3, \frac{1}{2}t^2, t \rangle$, find (a) the unit tangent vector, (b) the unit normal vector, and (c) the curvature.
12. Find the curvature of the ellipse $x = 3 \cos t$, $y = 4 \sin t$ at the points $(3, 0)$ and $(0, 4)$.
13. Find the curvature of the curve $y = x^4$ at the point $(1, 1)$.
14. Find an equation of the osculating circle of the curve $y = x^4 - x^2$ at the origin. Graph both the curve and its osculating circle.
15. Find an equation of the osculating plane of the curve $x = \sin 2t$, $y = t$, $z = \cos 2t$ at the point $(0, \pi, 1)$.
16. The figure shows the curve C traced by a particle with position vector $\mathbf{r}(t)$ at time t.
(a) Draw a vector that represents the average velocity of the particle over the time interval $3 \leqslant t \leqslant 3.2$.
(b) Write an expression for the velocity $\mathbf{v}(3)$.
(c) Write an expression for the unit tangent vector $\mathbf{T}(3)$ and draw it.

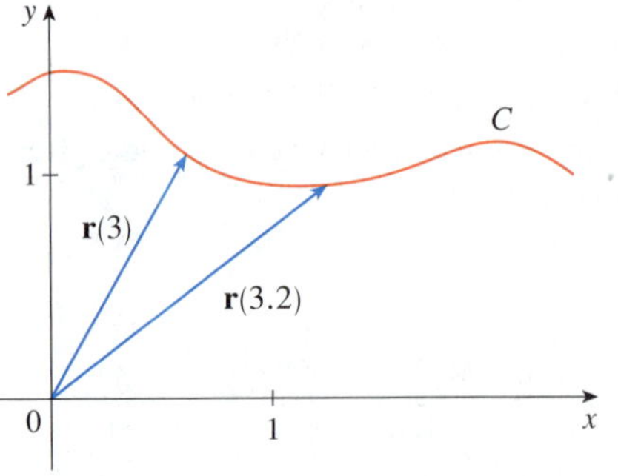

17. A particle moves with position function $\mathbf{r}(t) = t \ln t\, \mathbf{i} + t\, \mathbf{j} + e^{-t}\mathbf{k}$. Find the velocity, speed, and acceleration of the particle.

18. A particle starts at the origin with initial velocity $\mathbf{i} - \mathbf{j} + 3\mathbf{k}$. Its acceleration is $\mathbf{a}(t) = 6t\,\mathbf{i} + 12t^2\,\mathbf{j} - 6t\,\mathbf{k}$. Find its position function.

19. An athlete throws a shot at an angle of 45° to the horizontal at an initial speed of 43 ft/s. It leaves his hand 7 ft above the ground.

(a) Where is the shot 2 seconds later?

(b) How high does the shot go?

(c) Where does the shot land?

20. Find the tangential and normal components of the acceleration vector of a particle with position function

$$\mathbf{r}(t) = t\,\mathbf{i} + 2t\,\mathbf{j} + t^2\mathbf{k}$$

21. A disk of radius 1 is rotating in the counterclockwise direction at a constant angular speed ω. A particle starts at the center of the disk and moves toward the edge along a fixed radius so that its position at time t, $t \geqslant 0$, is given by $\mathbf{r}(t) = t\mathbf{R}(t)$, where

$$\mathbf{R}(t) = \cos \omega t\, \mathbf{i} + \sin \omega t\, \mathbf{j}$$

(a) Show that the velocity $\mathbf{v}$ of the particle is

$$\mathbf{v} = \cos \omega t\, \mathbf{i} + \sin \omega t\, \mathbf{j} + t\mathbf{v}_d$$

where $\mathbf{v}_d = \mathbf{R}'(t)$ is the velocity of a point on the edge of the disk.

(b) Show that the acceleration $\mathbf{a}$ of the particle is

$$\mathbf{a} = 2\mathbf{v}_d + t\mathbf{a}_d$$

where $\mathbf{a}_d = \mathbf{R}''(t)$ is the acceleration of a point on the rim of the disk. The extra term $2\mathbf{v}_d$ is called the *Coriolis acceleration*; it is the result of the interaction of the rotation of the disk and the motion of the particle. One can obtain a physical demonstration of this acceleration by walking toward the edge of a moving merry-go-round.

(c) Determine the Coriolis acceleration of a particle that moves on a rotating disk according to the equation

$$\mathbf{r}(t) = e^{-t}\cos \omega t\, \mathbf{i} + e^{-t}\sin \omega t\, \mathbf{j}$$

22. In designing *transfer curves* to connect sections of straight railroad tracks, it's important to realize that the acceleration of the train should be continuous so that the reactive force exerted by the train on the track is also continuous. Because of the formulas for the components of acceleration in Section 13.4, this will be the case if the curvature varies continuously.

(a) A logical candidate for a transfer curve to join existing tracks given by $y = 1$ for $x \leqslant 0$ and $y = \sqrt{2} - x$ for $x \geqslant 1/\sqrt{2}$ might be the function $f(x) = \sqrt{1 - x^2}$, $0 < x < 1/\sqrt{2}$, whose graph is the arc of the circle shown in the figure. It looks reasonable at first glance. Show that the function

$$F(x) = \begin{cases} 1 & \text{if } x \leqslant 0 \\ \sqrt{1 - x^2} & \text{if } 0 < x < 1/\sqrt{2} \\ \sqrt{2} - x & \text{if } x \geqslant 1/\sqrt{2} \end{cases}$$

is continuous and has continuous slope, but does not have continuous curvature. Therefore, f is not an appropriate transfer curve.

(b) Find a fifth-degree polynomial to serve as a transfer curve between the following straight line segments: $y = 0$ for $x \leqslant 0$ and $y = x$ for $x \geqslant 1$. Could this be done with a fourth-degree polynomial? Use a graphing calculator or computer to sketch the graph of the "connected" function and check to see that it looks like the one in the figure.

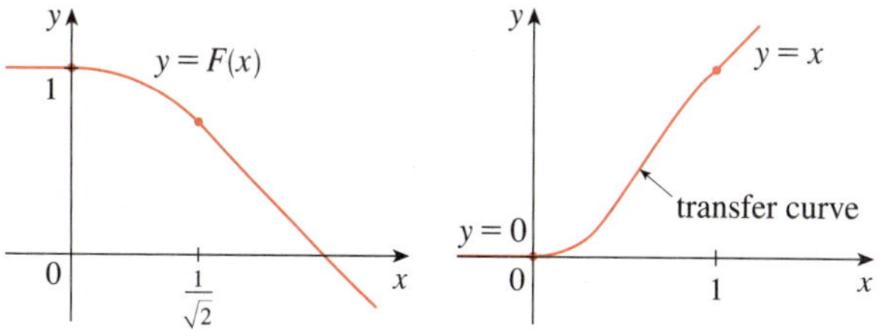

PROBLEMS PLUS

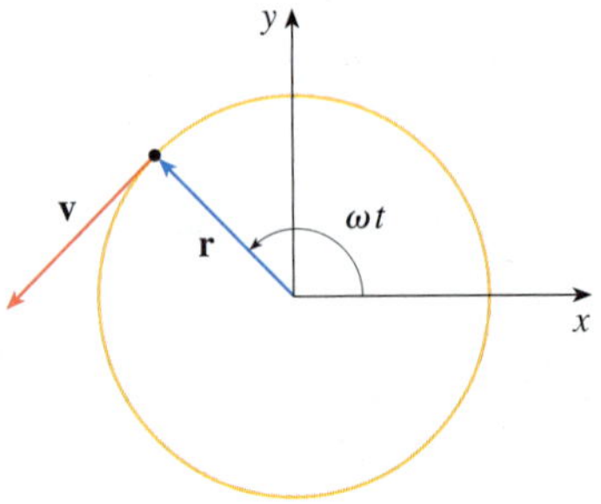

FIGURE FOR PROBLEM 1

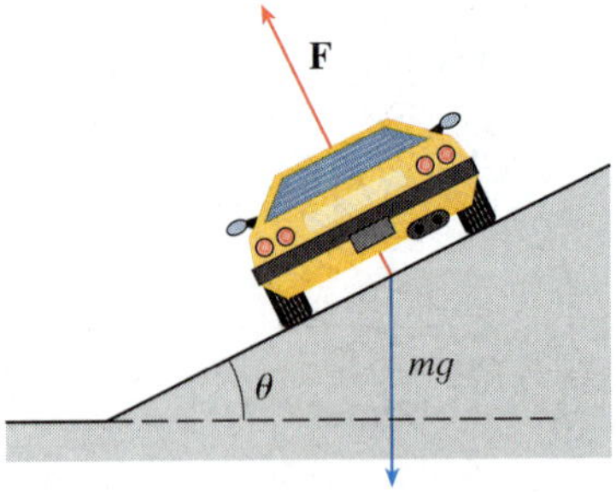

FIGURE FOR PROBLEM 2

1. A particle P moves with constant angular speed ω around a circle whose center is at the origin and whose radius is R. The particle is said to be in *uniform circular motion.* Assume that the motion is counterclockwise and that the particle is at the point $(R, 0)$ when $t = 0$. The position vector at time $t \geqslant 0$ is $\mathbf{r}(t) = R \cos \omega t\, \mathbf{i} + R \sin \omega t\, \mathbf{j}$.

(a) Find the velocity vector $\mathbf{v}$ and show that $\mathbf{v} \cdot \mathbf{r} = 0$. Conclude that $\mathbf{v}$ is tangent to the circle and points in the direction of the motion.

(b) Show that the speed $|\mathbf{v}|$ of the particle is the constant ωR. The *period* T of the particle is the time required for one complete revolution. Conclude that

$$T = \frac{2\pi R}{|\mathbf{v}|} = \frac{2\pi}{\omega}$$

(c) Find the acceleration vector $\mathbf{a}$. Show that it is proportional to $\mathbf{r}$ and that it points toward the origin. An acceleration with this property is called a *centripetal acceleration.* Show that the magnitude of the acceleration vector is $|\mathbf{a}| = R\omega^2$.

(d) Suppose that the particle has mass m. Show that the magnitude of the force $\mathbf{F}$ that is required to produce this motion, called a *centripetal force*, is

$$|\mathbf{F}| = \frac{m|\mathbf{v}|^2}{R}$$

2. A circular curve of radius R on a highway is banked at an angle θ so that a car can safely traverse the curve without skidding when there is no friction between the road and the tires. The loss of friction could occur, for example, if the road is covered with a film of water or ice. The rated speed v_R of the curve is the maximum speed that a car can attain without skidding. Suppose a car of mass m is traversing the curve at the rated speed v_R. Two forces are acting on the car: the vertical force, mg, due to the weight of the car, and a force $\mathbf{F}$ exerted by, and normal to, the road. (See the figure.)

The vertical component of $\mathbf{F}$ balances the weight of the car, so that $|\mathbf{F}| \cos \theta = mg$. The horizontal component of $\mathbf{F}$ produces a centripetal force on the car so that, by Newton's Second Law and part (d) of Problem 1,

$$|\mathbf{F}| \sin \theta = \frac{mv_R^2}{R}$$

(a) Show that $v_R^2 = Rg \tan \theta$.

(b) Find the rated speed of a circular curve with radius 400 ft that is banked at an angle of 12°.

(c) Suppose the design engineers want to keep the banking at 12°, but wish to increase the rated speed by 50%. What should the radius of the curve be?

3. A projectile is fired from the origin with angle of elevation α and initial speed v_0. Assuming that air resistance is negligible and that the only force acting on the projectile is gravity, g, we showed in Example 5 in Section 13.4 that the position vector of the projectile is $\mathbf{r}(t) = (v_0 \cos \alpha)t\, \mathbf{i} + \left[(v_0 \sin \alpha)t - \frac{1}{2}gt^2\right] \mathbf{j}$. We also showed that the maximum horizontal distance of the projectile is achieved when $\alpha = 45°$ and in this case the range is $R = v_0^2/g$.

(a) At what angle should the projectile be fired to achieve maximum height and what is the maximum height?

(b) Fix the initial speed v_0 and consider the parabola $x^2 + 2Ry - R^2 = 0$, whose graph is shown in the figure. Show that the projectile can hit any target inside or on the boundary

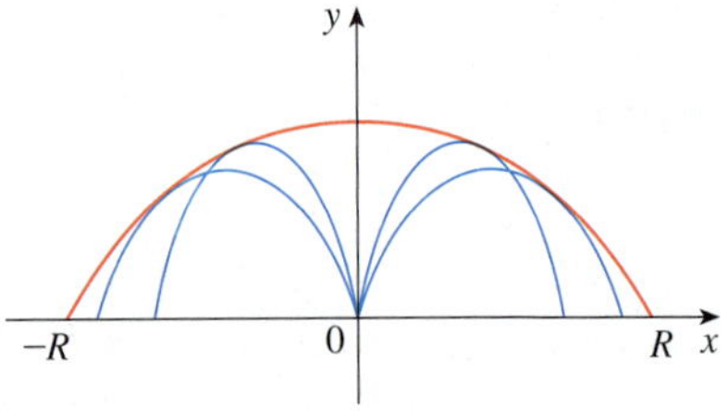

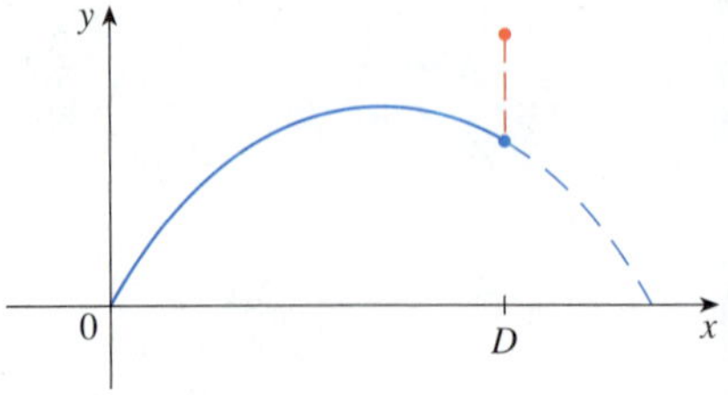

of the region bounded by the parabola and the x-axis, and that it can't hit any target outside this region.

(c) Suppose that the gun is elevated to an angle of inclination α in order to aim at a target that is suspended at a height h directly over a point D units downrange. The target is released at the instant the gun is fired. Show that the projectile always hits the target, regardless of the value v_0, provided the projectile does not hit the ground "before" D.

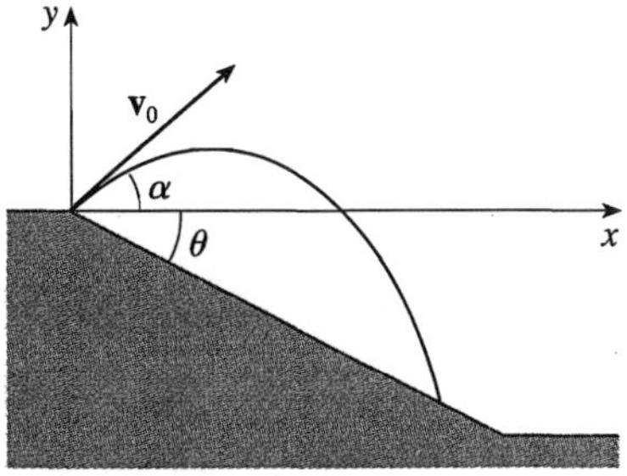

FIGURE FOR PROBLEM 4

4. (a) A projectile is fired from the origin down an inclined plane that makes an angle θ with the horizontal. The angle of elevation of the gun and the initial speed of the projectile are α and v_0, respectively. Find the position vector of the projectile and the parametric equations of the path of the projectile as functions of the time t. (Ignore air resistance.)

(b) Show that the angle of elevation α that will maximize the downhill range is the angle halfway between the plane and the vertical.

(c) Suppose the projectile is fired up an inclined plane whose angle of inclination is θ. Show that, in order to maximize the (uphill) range, the projectile should be fired in the direction halfway between the plane and the vertical.

(d) In a paper presented in 1686, Edmond Halley summarized the laws of gravity and projectile motion and applied them to gunnery. One problem he posed involved firing a projectile to hit a target a distance R up an inclined plane. Show that the angle at which the projectile should be fired to hit the target but use the least amount of energy is the same as the angle in part (c). (Use the fact that the energy needed to fire the projectile is proportional to the square of the initial speed, so minimizing the energy is equivalent to minimizing the initial speed.)

5. A projectile of mass m is fired from the origin at an angle of elevation α. In addition to gravity, assume that air resistance provides a force that is proportional to the velocity and that opposes the motion. Then, by Newton's Second Law, the total force acting on the projectile satisfies the equation

$$m\frac{d^2\mathbf{R}}{dt^2} = -mg\,\mathbf{j} - k\frac{d\mathbf{R}}{dt} \tag{1}$$

where $\mathbf{R}$ is the position vector and $k > 0$ is the constant of proportionality.

(a) Show that Equation 1 can be integrated to obtain the equation

$$\frac{d\mathbf{R}}{dt} + \frac{k}{m}\mathbf{R} = \mathbf{v}_0 - gt\,\mathbf{j}$$

where $\mathbf{v}_0 = \mathbf{v}(0) = \dfrac{d\mathbf{R}}{dt}(0)$.

(b) Multiply both sides of the equation in part (a) by $e^{(k/m)t}$ and show that the left-hand side of the resulting equation is the derivative of the product $e^{(k/m)t}\mathbf{R}(t)$. Then integrate to find an expression for the position vector $\mathbf{R}(t)$.

6. Find the curvature of the curve with parametric equations

$$x = \int_0^t \sin\left(\tfrac{1}{2}\pi\theta^2\right)d\theta \qquad y = \int_0^t \cos\left(\tfrac{1}{2}\pi\theta^2\right)d\theta$$

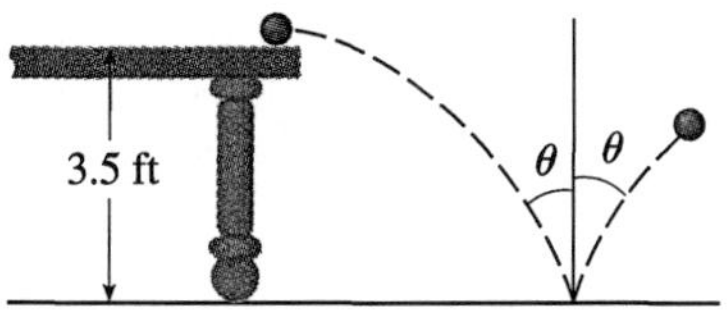

FIGURE FOR PROBLEM 7

7. A ball rolls off a table with a speed of 2 ft/s. The table is 3.5 ft high.

(a) Determine the point at which the ball hits the floor and find its speed at the instant of impact.

(b) Find the angle θ between the path of the ball and the vertical line drawn through the point of impact. (See the figure.)

(c) Suppose the ball rebounds from the floor at the same angle with which it hits the floor, but loses 20% of its speed due to energy absorbed by the ball on impact. Where does the ball strike the floor on the second bounce?

8. A cable has radius r and length L and is wound around a spool with radius R without overlapping. What is the shortest length along the spool that is covered by the cable?

APPENDIXES

A Numbers, Inequalities, and Absolute Values

Calculus is based on the real number system. We start with the **integers**:

$$\ldots, \quad -3, \quad -2, \quad -1, \quad 0, \quad 1, \quad 2, \quad 3, \quad 4, \quad \ldots$$

Then we construct the **rational numbers**, which are ratios of integers. Thus, any rational number r can be expressed as

$$r = \frac{m}{n} \qquad \text{where } m \text{ and } n \text{ are integers and } n \neq 0$$

Examples are

$$\tfrac{1}{2} \qquad -\tfrac{3}{7} \qquad 46 = \tfrac{46}{1} \qquad 0.17 = \tfrac{17}{100}$$

(Recall that division by 0 is always ruled out, so expressions like $\frac{3}{0}$ and $\frac{0}{0}$ are undefined.) Some real numbers, such as $\sqrt{2}$, can't be expressed as a ratio of integers and are therefore called **irrational numbers**. It can be shown, with varying degrees of difficulty, that the following are also irrational numbers:

$$\sqrt{3} \qquad \sqrt{5} \qquad \sqrt[3]{2} \qquad \pi \qquad \sin 1^\circ \qquad \log_{10} 2$$

The set of all real numbers is usually denoted by the symbol $\mathbb{R}$. When we use the word *number* without qualification, we mean "real number."

Every number has a decimal representation. If the number is rational, then the corresponding decimal is repeating. For example,

$$\tfrac{1}{2} = 0.5000\ldots = 0.5\overline{0} \qquad \tfrac{2}{3} = 0.66666\ldots = 0.\overline{6}$$

$$\tfrac{157}{495} = 0.317171717\ldots = 0.3\overline{17} \qquad \tfrac{9}{7} = 1.285714285714\ldots = 1.\overline{285714}$$

(The bar indicates that the sequence of digits repeats forever.) On the other hand, if the number is irrational, the decimal is nonrepeating:

$$\sqrt{2} = 1.414213562373095\ldots \qquad \pi = 3.141592653589793\ldots$$

If we stop the decimal expansion of any number at a certain place, we get an approximation to the number. For instance, we can write

$$\pi \approx 3.14159265$$

where the symbol $\approx$ is read "is approximately equal to." The more decimal places we retain, the better the approximation we get.

The real numbers can be represented by points on a line as in Figure 1. The positive direction (to the right) is indicated by an arrow. We choose an arbitrary reference point O, called the **origin**, which corresponds to the real number 0. Given any convenient unit of measurement, each positive number x is represented by the point on the line a distance of x units to the right of the origin, and each negative number $-x$ is represented by the point x units to the left of the origin. Thus, every real number is represented by a point on the line, and every point P on the line corresponds to exactly one real number. The number associated with the point P is called the **coordinate** of P and the line is then called a **coor-**

dinate line, or a **real number line**, or simply a **real line**. Often we identify the point with its coordinate and think of a number as being a point on the real line.

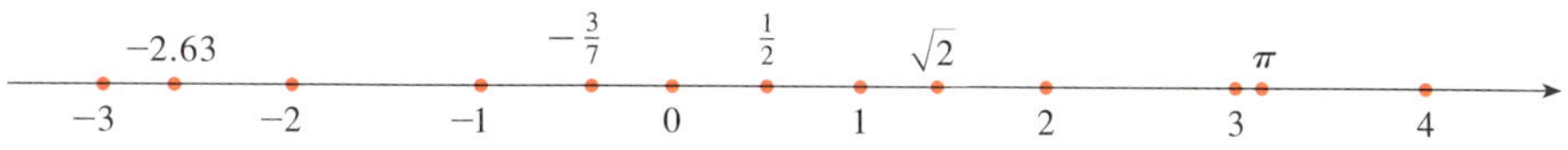

FIGURE 1

The real numbers are ordered. We say *a is less than b* and write $a < b$ if $b - a$ is a positive number. Geometrically, this means that a lies to the left of b on the number line. (Equivalently, we say *b is greater than a* and write $b > a$.) The symbol $a \leqslant b$ (or $b \geqslant a$) means that either $a < b$ or $a = b$ and is read "a is less than or equal to b." For instance, the following are true inequalities:

$$7 < 7.4 < 7.5 \qquad -3 > -\pi \qquad \sqrt{2} < 2 \qquad \sqrt{2} \leqslant 2 \qquad 2 \leqslant 2$$

In what follows we need to use *set notation*. A **set** is a collection of objects, and these objects are called the **elements** of the set. If S is a set, the notation $a \in S$ means that a is an element of S, and $a \notin S$ means that a is not an element of S. For example, if Z represents the set of integers, then $-3 \in Z$ but $\pi \notin Z$. If S and T are sets, then their **union** $S \cup T$ is the set consisting of all elements that are in S or T (or in both S and T). The **intersection** of S and T is the set $S \cap T$ consisting of all elements that are in both S and T. In other words, $S \cap T$ is the common part of S and T. The empty set, denoted by $\varnothing$, is the set that contains no element.

Some sets can be described by listing their elements between braces. For instance, the set A consisting of all positive integers less than 7 can be written as

$$A = \{1, 2, 3, 4, 5, 6\}$$

We could also write A in *set-builder notation* as

$$A = \{x \mid x \text{ is an integer and } 0 < x < 7\}$$

which is read "A is the set of x such that x is an integer and $0 < x < 7$."

Intervals

Certain sets of real numbers, called **intervals,** occur frequently in calculus and correspond geometrically to line segments. For example, if $a < b$, the **open interval** from a to b consists of all numbers between a and b and is denoted by the symbol (a, b). Using set-builder notation, we can write

$$(a, b) = \{x \mid a < x < b\}$$

FIGURE 2
Open interval (a, b)

Notice that the endpoints of the interval—namely, a and b—are excluded. This is indicated by the round brackets ( ) and by the open dots in Figure 2. The **closed interval** from a to b is the set

$$[a, b] = \{x \mid a \leqslant x \leqslant b\}$$

FIGURE 3
Closed interval $[a, b]$

Here the endpoints of the interval are included. This is indicated by the square brackets [ ] and by the solid dots in Figure 3. It is also possible to include only one endpoint in an interval, as shown in Table 1.

1 Table of Intervals

Table 1 lists the nine possible types of intervals. When these intervals are discussed, it is always assumed that $a < b$.

Notation	Set description	Picture
(a, b)	$\{x \mid a < x < b\}$	
$[a, b]$	$\{x \mid a \leqslant x \leqslant b\}$	
$[a, b)$	$\{x \mid a \leqslant x < b\}$	
$(a, b]$	$\{x \mid a < x \leqslant b\}$	
(a, ∞)	$\{x \mid x > a\}$	
$[a, \infty)$	$\{x \mid x \geqslant a\}$	
$(-\infty, b)$	$\{x \mid x < b\}$	
$(-\infty, b]$	$\{x \mid x \leqslant b\}$	
$(-\infty, \infty)$	$\mathbb{R}$ (set of all real numbers)	

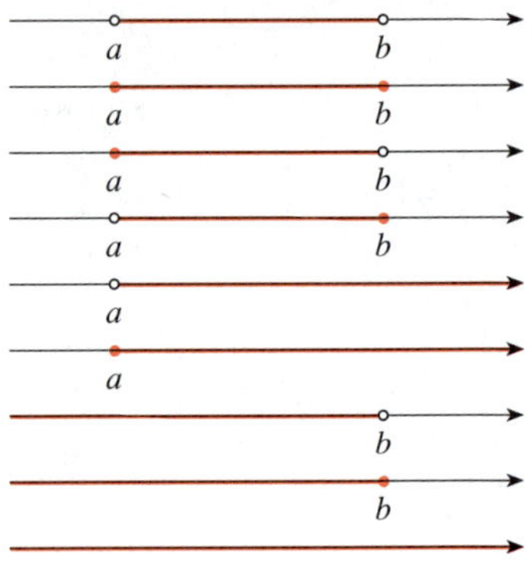

We also need to consider infinite intervals such as

$$(a, \infty) = \{x \mid x > a\}$$

This does not mean that ∞ ("infinity") is a number. The notation (a, ∞) stands for the set of all numbers that are greater than a, so the symbol ∞ simply indicates that the interval extends indefinitely far in the positive direction.

Inequalities

When working with inequalities, note the following rules.

2 Rules for Inequalities

1. If $a < b$, then $a + c < b + c$.

2. If $a < b$ and $c < d$, then $a + c < b + d$.

3. If $a < b$ and $c > 0$, then $ac < bc$.

4. If $a < b$ and $c < 0$, then $ac > bc$.

5. If $0 < a < b$, then $1/a > 1/b$.

Rule 1 says that we can add any number to both sides of an inequality, and Rule 2 says that two inequalities can be added. However, we have to be careful with multiplication. Rule 3 says that we can multiply both sides of an inequality by a *positive* number, but Rule 4 says that if we multiply both sides of an inequality by a negative number, then we reverse the direction of the inequality. For example, if we take the inequality $3 < 5$ and multiply by 2, we get $6 < 10$, but if we multiply by -2, we get $-6 > -10$. Finally, Rule 5 says that if we take reciprocals, then we reverse the direction of an inequality (provided the numbers are positive).

EXAMPLE 1 Solve the inequality $1 + x < 7x + 5$.

SOLUTION The given inequality is satisfied by some values of x but not by others. To *solve* an inequality means to determine the set of numbers x for which the inequality is true. This is called the *solution set*.

First we subtract 1 from each side of the inequality (using Rule 1 with $c = -1$):

$$x < 7x + 4$$

Then we subtract $7x$ from both sides (Rule 1 with $c = -7x$):

$$-6x < 4$$

Now we divide both sides by -6 (Rule 4 with $c = -\frac{1}{6}$):

$$x > -\tfrac{4}{6} = -\tfrac{2}{3}$$

These steps can all be reversed, so the solution set consists of all numbers greater than $-\frac{2}{3}$. In other words, the solution of the inequality is the interval $\left(-\frac{2}{3}, \infty\right)$.

EXAMPLE 2 Solve the inequalities $4 \leqslant 3x - 2 < 13$.

SOLUTION Here the solution set consists of all values of x that satisfy both inequalities. Using the rules given in (2), we see that the following inequalities are equivalent:

$$4 \leqslant 3x - 2 < 13$$

$$6 \leqslant 3x < 15 \qquad \text{(add 2)}$$

$$2 \leqslant x < 5 \qquad \text{(divide by 3)}$$

Therefore, the solution set is $[2, 5)$.

EXAMPLE 3 Solve the inequality $x^2 - 5x + 6 \leqslant 0$.

SOLUTION First we factor the left side:

$$(x - 2)(x - 3) \leqslant 0$$

We know that the corresponding equation $(x - 2)(x - 3) = 0$ has the solutions 2 and 3. The numbers 2 and 3 divide the real line into three intervals:

$$(-\infty, 2) \qquad (2, 3) \qquad (3, \infty)$$

On each of these intervals we determine the signs of the factors. For instance,

$$x \in (-\infty, 2) \quad \Rightarrow \quad x < 2 \quad \Rightarrow \quad x - 2 < 0$$

Then we record these signs in the following chart:

Interval	$x - 2$	$x - 3$	$(x - 2)(x - 3)$
$x < 2$	−	−	+
$2 < x < 3$	+	−	−
$x > 3$	+	+	+

Another method for obtaining the information in the chart is to use *test values*. For instance, if we use the test value $x = 1$ for the interval $(-\infty, 2)$, then substitution in $x^2 - 5x + 6$ gives

$$1^2 - 5(1) + 6 = 2$$

|||| A visual method for solving Example 3 is to use a graphing device to graph the parabola $y = x^2 - 5x + 6$ (as in Figure 4) and observe that the curve lies on or below the x-axis when $2 \leqslant x \leqslant 3$.

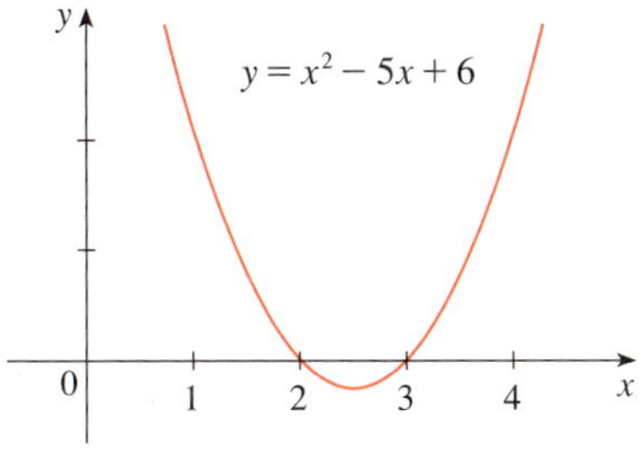

FIGURE 4

The polynomial $x^2 - 5x + 6$ doesn't change sign inside any of the three intervals, so we conclude that it is positive on $(-\infty, 2)$.

Then we read from the chart that $(x - 2)(x - 3)$ is negative when $2 < x < 3$. Thus, the solution of the inequality $(x - 2)(x - 3) \leqslant 0$ is

$$\{x \mid 2 \leqslant x \leqslant 3\} = [2, 3]$$

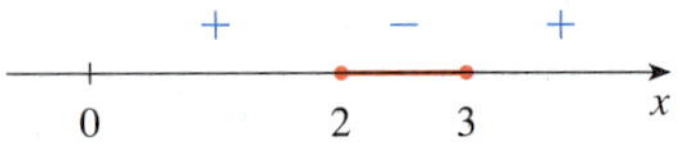

FIGURE 5

Notice that we have included the endpoints 2 and 3 because we are looking for values of x such that the product is either negative or zero. The solution is illustrated in Figure 5.

EXAMPLE 4 Solve $x^3 + 3x^2 > 4x$.

SOLUTION First we take all nonzero terms to one side of the inequality sign and factor the resulting expression:

$$x^3 + 3x^2 - 4x > 0 \qquad \text{or} \qquad x(x - 1)(x + 4) > 0$$

As in Example 3 we solve the corresponding equation $x(x - 1)(x + 4) = 0$ and use the solutions $x = -4$, $x = 0$, and $x = 1$ to divide the real line into four intervals $(-\infty, -4)$, $(-4, 0)$, $(0, 1)$, and $(1, \infty)$. On each interval the product keeps a constant sign as shown in the following chart:

Interval	x	$x - 1$	$x + 4$	$x(x - 1)(x + 4)$
$x < -4$	−	−	−	−
$-4 < x < 0$	−	−	+	+
$0 < x < 1$	+	−	+	−
$x > 1$	+	+	+	+

Then we read from the chart that the solution set is

$$\{x \mid -4 < x < 0 \text{ or } x > 1\} = (-4, 0) \cup (1, \infty)$$

FIGURE 6

The solution is illustrated in Figure 6.

Absolute Value

The **absolute value** of a number a, denoted by $|a|$, is the distance from a to 0 on the real number line. Distances are always positive or 0, so we have

$$|a| \geqslant 0 \qquad \text{for every number } a$$

For example,

$$|3| = 3 \qquad |-3| = 3 \qquad |0| = 0 \qquad |\sqrt{2} - 1| = \sqrt{2} - 1 \qquad |3 - \pi| = \pi - 3$$

In general, we have

||||| Remember that if a is negative, then $-a$ is positive.

3

$$|a| = a \qquad \text{if } a \geqslant 0$$

$$|a| = -a \qquad \text{if } a < 0$$

EXAMPLE 5 Express $|3x - 2|$ without using the absolute-value symbol.

SOLUTION

$$|3x - 2| = \begin{cases} 3x - 2 & \text{if } 3x - 2 \geq 0 \\ -(3x - 2) & \text{if } 3x - 2 < 0 \end{cases}$$

$$= \begin{cases} 3x - 2 & \text{if } x \geq \frac{2}{3} \\ 2 - 3x & \text{if } x < \frac{2}{3} \end{cases}$$

Recall that the symbol $\sqrt{}$ means "the positive square root of." Thus, $\sqrt{r} = s$ means $s^2 = r$ and $s \geq 0$. Therefore, the equation $\sqrt{a^2} = a$ is not always true. It is true only when $a \geq 0$. If $a < 0$, then $-a > 0$, so we have $\sqrt{a^2} = -a$. In view of (3), we then have the equation

4 $$\sqrt{a^2} = |a|$$

which is true for all values of a.

Hints for the proofs of the following properties are given in the exercises.

5 Properties of Absolute Values Suppose a and b are any real numbers and n is an integer. Then

1. $|ab| = |a||b|$ **2.** $\left|\dfrac{a}{b}\right| = \dfrac{|a|}{|b|} \quad (b \neq 0)$ **3.** $|a^n| = |a|^n$

For solving equations or inequalities involving absolute values, it's often very helpful to use the following statements.

6 Suppose $a > 0$. Then

4. $|x| = a$ if and only if $x = \pm a$

5. $|x| < a$ if and only if $-a < x < a$

6. $|x| > a$ if and only if $x > a$ or $x < -a$

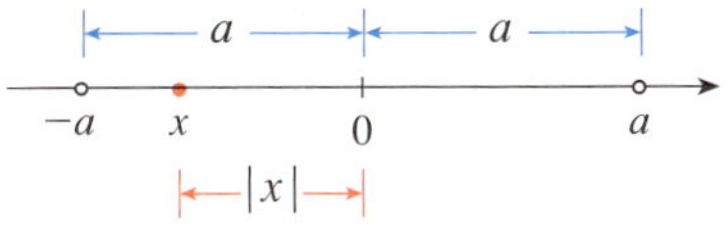

FIGURE 7

For instance, the inequality $|x| < a$ says that the distance from x to the origin is less than a, and you can see from Figure 7 that this is true if and only if x lies between $-a$ and a.

If a and b are any real numbers, then the distance between a and b is the absolute value of the difference, namely, $|a - b|$, which is also equal to $|b - a|$. (See Figure 8.)

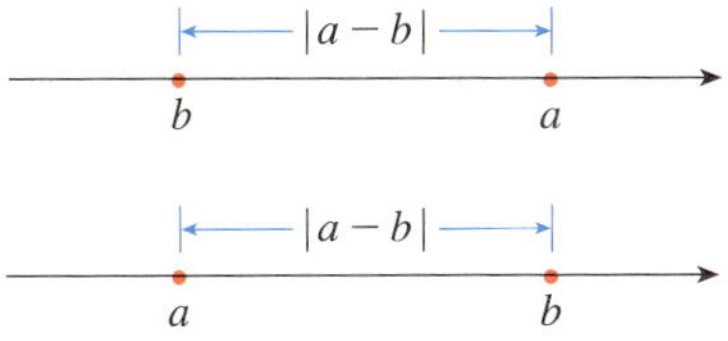

FIGURE 8
Length of a line segment $= |a - b|$

EXAMPLE 6 Solve $|2x - 5| = 3$.

SOLUTION By Property 4 of (6), $|2x - 5| = 3$ is equivalent to

$$2x - 5 = 3 \qquad \text{or} \qquad 2x - 5 = -3$$

So $2x = 8$ or $2x = 2$. Thus, $x = 4$ or $x = 1$.

EXAMPLE 7 Solve $|x - 5| < 2$.

SOLUTION 1 By Property 5 of (6), $|x - 5| < 2$ is equivalent to

$$-2 < x - 5 < 2$$

Therefore, adding 5 to each side, we have

$$3 < x < 7$$

and the solution set is the open interval $(3, 7)$.

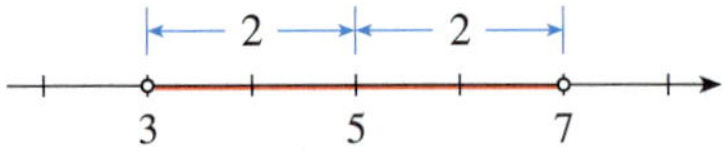

FIGURE 9

SOLUTION 2 Geometrically, the solution set consists of all numbers x whose distance from 5 is less than 2. From Figure 9 we see that this is the interval $(3, 7)$.

EXAMPLE 8 Solve $|3x + 2| \geqslant 4$.

SOLUTION By Properties 4 and 6 of (6), $|3x + 2| \geqslant 4$ is equivalent to

$$3x + 2 \geqslant 4 \qquad \text{or} \qquad 3x + 2 \leqslant -4$$

In the first case $3x \geqslant 2$, which gives $x \geqslant \frac{2}{3}$. In the second case $3x \leqslant -6$, which gives $x \leqslant -2$. So the solution set is

$$\left\{x \mid x \leqslant -2 \text{ or } x \geqslant \tfrac{2}{3}\right\} = (-\infty, -2] \cup \left[\tfrac{2}{3}, \infty\right)$$

Another important property of absolute value, called the Triangle Inequality, is used frequently not only in calculus but throughout mathematics in general.

7 The Triangle Inequality If a and b are any real numbers, then

$$|a + b| \leqslant |a| + |b|$$

Observe that if the numbers a and b are both positive or both negative, then the two sides in the Triangle Inequality are actually equal. But if a and b have opposite signs, the left side involves a subtraction and the right side does not. This makes the Triangle Inequality seem reasonable, but we can prove it as follows.

Notice that

$$-|a| \leqslant a \leqslant |a|$$

is always true because a equals either $|a|$ or $-|a|$. The corresponding statement for b is

$$-|b| \leqslant b \leqslant |b|$$

Adding these inequalities, we get

$$-(|a| + |b|) \leqslant a + b \leqslant |a| + |b|$$

If we now apply Properties 4 and 5 (with x replaced by $a + b$ and a by $|a| + |b|$), we obtain

$$|a + b| \leqslant |a| + |b|$$

which is what we wanted to show.

EXAMPLE 9 If $|x - 4| < 0.1$ and $|y - 7| < 0.2$, use the Triangle Inequality to estimate $|(x + y) - 11|$.

SOLUTION In order to use the given information, we use the Triangle Inequality with $a = x - 4$ and $b = y - 7$:

$$\begin{aligned} |(x + y) - 11| &= |(x - 4) + (y - 7)| \\ &\leqslant |x - 4| + |y - 7| \\ &< 0.1 + 0.2 = 0.3 \end{aligned}$$

Thus

$$|(x + y) - 11| < 0.3$$

A Exercises

1–12 Rewrite the expression without using the absolute value symbol.

1. $|5 - 23|$

2. $|5| - |-23|$

3. $|-\pi|$

4. $|\pi - 2|$

5. $|\sqrt{5} - 5|$

6. $||-2| - |-3||$

7. $|x - 2|$ if $x < 2$

8. $|x - 2|$ if $x > 2$

9. $|x + 1|$

10. $|2x - 1|$

11. $|x^2 + 1|$

12. $|1 - 2x^2|$

13–38 Solve the inequality in terms of intervals and illustrate the solution set on the real number line.

13. $2x + 7 > 3$

14. $3x - 11 < 4$

15. $1 - x \leqslant 2$

16. $4 - 3x \geqslant 6$

17. $2x + 1 < 5x - 8$

18. $1 + 5x > 5 - 3x$

19. $-1 < 2x - 5 < 7$

20. $1 < 3x + 4 \leqslant 16$

21. $0 \leqslant 1 - x < 1$

22. $-5 \leqslant 3 - 2x \leqslant 9$

23. $4x < 2x + 1 \leqslant 3x + 2$

24. $2x - 3 < x + 4 < 3x - 2$

25. $(x - 1)(x - 2) > 0$

26. $(2x + 3)(x - 1) \geqslant 0$

27. $2x^2 + x \leqslant 1$

28. $x^2 < 2x + 8$

29. $x^2 + x + 1 > 0$

30. $x^2 + x > 1$

31. $x^2 < 3$

32. $x^2 \geqslant 5$

33. $x^3 - x^2 \leqslant 0$

34. $(x + 1)(x - 2)(x + 3) \geqslant 0$

35. $x^3 > x$

36. $x^3 + 3x < 4x^2$

37. $\dfrac{1}{x} < 4$

38. $-3 < \dfrac{1}{x} \leqslant 1$

39. The relationship between the Celsius and Fahrenheit temperature scales is given by $C = \frac{5}{9}(F - 32)$, where C is the temperature in degrees Celsius and F is the temperature in degrees Fahrenheit. What interval on the Celsius scale corresponds to the temperature range $50 \leqslant F \leqslant 95$?

40. Use the relationship between C and F given in Exercise 39 to find the interval on the Fahrenheit scale corresponding to the temperature range $20 \leqslant C \leqslant 30$.

41. As dry air moves upward, it expands and in so doing cools at a rate of about 1°C for each 100-m rise, up to about 12 km.
(a) If the ground temperature is 20°C, write a formula for the temperature at height h.
(b) What range of temperature can be expected if a plane takes off and reaches a maximum height of 5 km?

42. If a ball is thrown upward from the top of a building 128 ft high with an initial velocity of 16 ft/s, then the height h above the ground t seconds later will be

$$h = 128 + 16t - 16t^2$$

During what time interval will the ball be at least 32 ft above the ground?

43–46 Solve the equation for x.

43. $|2x| = 3$

44. $|3x + 5| = 1$

45. $|x + 3| = |2x + 1|$

46. $\left|\dfrac{2x - 1}{x + 1}\right| = 3$

47–56 Solve the inequality.

47. $|x| < 3$

48. $|x| \geqslant 3$

49. $|x - 4| < 1$

50. $|x - 6| < 0.1$

51. $|x + 5| \geqslant 2$

52. $|x + 1| \geqslant 3$

53. $|2x - 3| \leqslant 0.4$

54. $|5x - 2| < 6$

55. $1 \leqslant |x| \leqslant 4$

56. $0 < |x - 5| < \frac{1}{2}$

57–58 Solve for x, assuming a, b, and c are positive constants.

57. $a(bx - c) \geqslant bc$

58. $a \leqslant bx + c < 2a$

59–60 Solve for x, assuming a, b, and c are negative constants.

59. $ax + b < c$

60. $\dfrac{ax + b}{c} \leqslant b$

61. Suppose that $|x - 2| < 0.01$ and $|y - 3| < 0.04$. Use the Triangle Inequality to show that $|(x + y) - 5| < 0.05$.

62. Show that if $|x + 3| < \frac{1}{2}$, then $|4x + 13| < 3$.

63. Show that if $a < b$, then $a < \dfrac{a + b}{2} < b$.

64. Use Rule 3 to prove Rule 5 of (2).

65. Prove that $|ab| = |a||b|$. [*Hint:* Use Equation 4.]

66. Prove that $\left|\dfrac{a}{b}\right| = \dfrac{|a|}{|b|}$.

67. Show that if $0 < a < b$, then $a^2 < b^2$.

68. Prove that $|x - y| \geqslant |x| - |y|$. [*Hint:* Use the Triangle Inequality with $a = x - y$ and $b = y$.]

69. Show that the sum, difference, and product of rational numbers are rational numbers.

70. (a) Is the sum of two irrational numbers always an irrational number?

(b) Is the product of two irrational numbers always an irrational number?

B Coordinate Geometry and Lines

Just as the points on a line can be identified with real numbers by assigning them coordinates, as described in Appendix A, so the points in a plane can be identified with ordered pairs of real numbers. We start by drawing two perpendicular coordinate lines that intersect at the origin O on each line. Usually one line is horizontal with positive direction to the right and is called the x-axis; the other line is vertical with positive direction upward and is called the y-axis.

Any point P in the plane can be located by a unique ordered pair of numbers as follows. Draw lines through P perpendicular to the x- and y-axes. These lines intersect the axes in points with coordinates a and b as shown in Figure 1. Then the point P is assigned the ordered pair (a, b). The first number a is called the **x-coordinate** of P; the second number b is called the **y-coordinate** of P. We say that P is the point with coordinates (a, b), and we denote the point by the symbol $P(a, b)$. Several points are labeled with their coordinates in Figure 2.

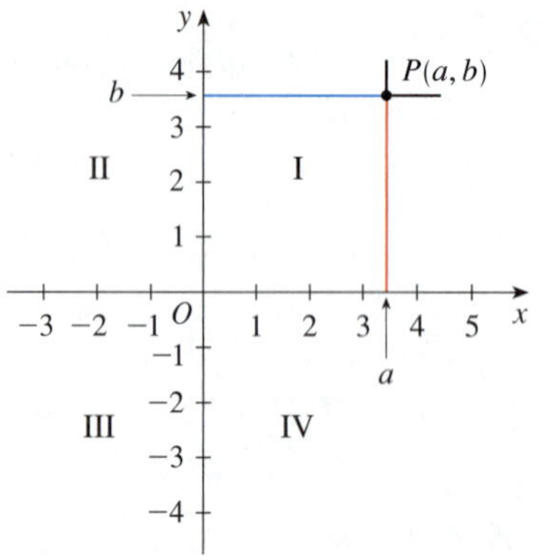

FIGURE 1

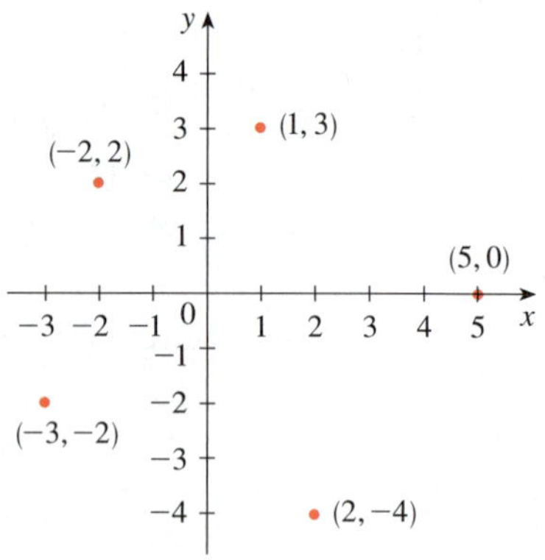

FIGURE 2

By reversing the preceding process we can start with an ordered pair (a, b) and arrive at the corresponding point P. Often we identify the point P with the ordered pair (a, b) and refer to "the point (a, b)." [Although the notation used for an open interval (a, b) is the same as the notation used for a point (a, b), you will be able to tell from the context which meaning is intended.]

This coordinate system is called the **rectangular coordinate system** or the **Cartesian coordinate system** in honor of the French mathematician René Descartes (1596–1650), even though another Frenchman, Pierre Fermat (1601–1665), invented the principles of analytic geometry at about the same time as Descartes. The plane supplied with this coordinate system is called the **coordinate plane** or the **Cartesian plane** and is denoted by $\mathbb{R}^2$.

The x- and y-axes are called the **coordinate axes** and divide the Cartesian plane into four quadrants, which are labeled I, II, III, and IV in Figure 1. Notice that the first quadrant consists of those points whose x- and y-coordinates are both positive.

EXAMPLE 1 Describe and sketch the regions given by the following sets.

(a) $\{(x, y) \mid x \geqslant 0\}$ (b) $\{(x, y) \mid y = 1\}$ (c) $\{(x, y) \mid |y| < 1\}$

SOLUTION

(a) The points whose x-coordinates are 0 or positive lie on the y-axis or to the right of it as indicated by the shaded region in Figure 3(a).

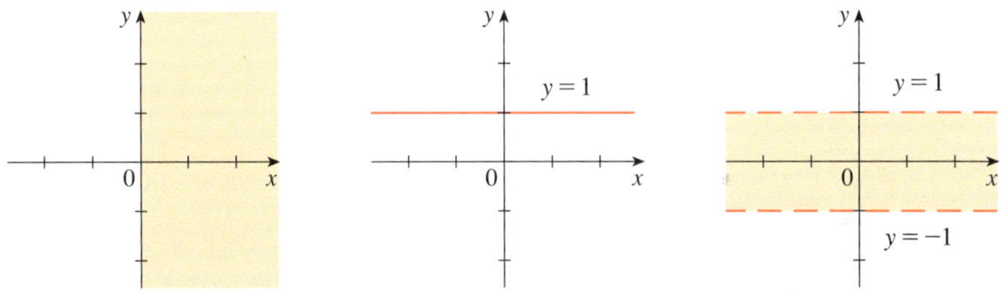

FIGURE 3 (a) $x \geqslant 0$ (b) $y = 1$ (c) $|y| < 1$

(b) The set of all points with y-coordinate 1 is a horizontal line one unit above the x-axis [see Figure 3(b)].

(c) Recall from Appendix A that

$$|y| < 1 \quad \text{if and only if} \quad -1 < y < 1$$

The given region consists of those points in the plane whose y-coordinates lie between -1 and 1. Thus, the region consists of all points that lie between (but not on) the horizontal lines $y = 1$ and $y = -1$. [These lines are shown as dashed lines in Figure 3(c) to indicate that the points on these lines don't lie in the set.]

Recall from Appendix A that the distance between points a and b on a number line is $|a - b| = |b - a|$. Thus, the distance between points $P_1(x_1, y_1)$ and $P_3(x_2, y_1)$ on a horizontal line must be $|x_2 - x_1|$ and the distance between $P_2(x_2, y_2)$ and $P_3(x_2, y_1)$ on a vertical line must be $|y_2 - y_1|$. (See Figure 4.)

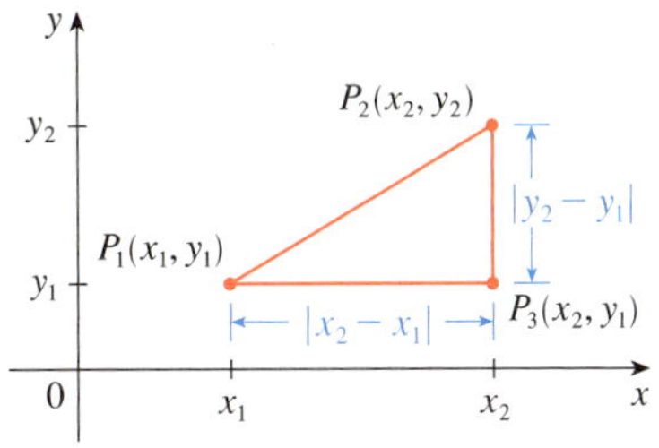

FIGURE 4

To find the distance $|P_1P_2|$ between any two points $P_1(x_1, y_1)$ and $P_2(x_2, y_2)$, we note that triangle $P_1P_2P_3$ in Figure 4 is a right triangle, and so by the Pythagorean Theorem we have

$$|P_1P_2| = \sqrt{|P_1P_3|^2 + |P_2P_3|^2} = \sqrt{|x_2 - x_1|^2 + |y_2 - y_1|^2}$$

$$= \sqrt{(x_2 - x_1)^2 + (y_2 - y_1)^2}$$

1 Distance Formula The distance between the points $P_1(x_1, y_1)$ and $P_2(x_2, y_2)$ is

$$|P_1P_2| = \sqrt{(x_2 - x_1)^2 + (y_2 - y_1)^2}$$

EXAMPLE 2 The distance between $(1, -2)$ and $(5, 3)$ is

$$\sqrt{(5-1)^2 + [3-(-2)]^2} = \sqrt{4^2 + 5^2} = \sqrt{41}$$

Lines

We want to find an equation of a given line L; such an equation is satisfied by the coordinates of the points on L and by no other point. To find the equation of L we use its *slope*, which is a measure of the steepness of the line.

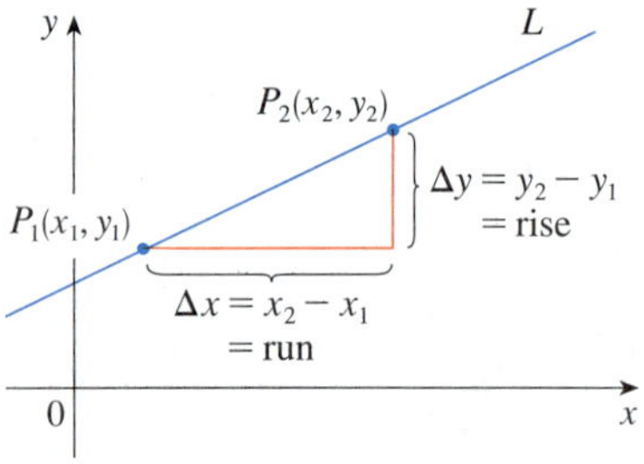

FIGURE 5

2 Definition The **slope** of a nonvertical line that passes through the points $P_1(x_1, y_1)$ and $P_2(x_2, y_2)$ is

$$m = \frac{\Delta y}{\Delta x} = \frac{y_2 - y_1}{x_2 - x_1}$$

The slope of a vertical line is not defined.

Thus, the slope of a line is the ratio of the change in y, Δy, to the change in x, Δx. (See Figure 5.) The slope is therefore the rate of change of y with respect to x. The fact that the line is straight means that the rate of change is constant.

Figure 6 shows several lines labeled with their slopes. Notice that lines with positive slope slant upward to the right, whereas lines with negative slope slant downward to the right. Notice also that the steepest lines are the ones for which the absolute value of the slope is largest, and a horizontal line has slope 0.

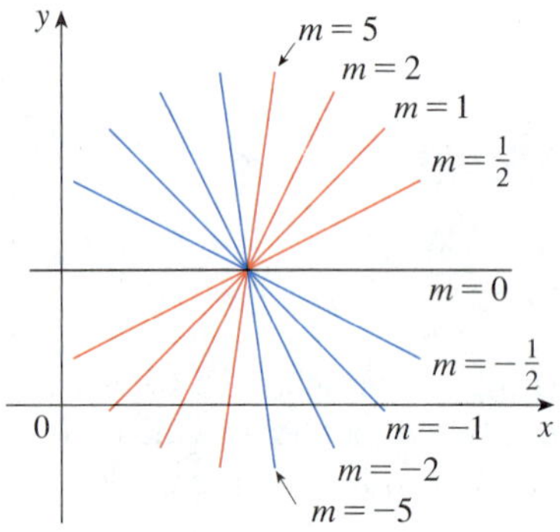

FIGURE 6

Now let's find an equation of the line that passes through a given point $P_1(x_1, y_1)$ and has slope m. A point $P(x, y)$ with $x \neq x_1$ lies on this line if and only if the slope of the line through P_1 and P is equal to m; that is,

$$\frac{y - y_1}{x - x_1} = m$$

This equation can be rewritten in the form

$$y - y_1 = m(x - x_1)$$

and we observe that this equation is also satisfied when $x = x_1$ and $y = y_1$. Therefore, it is an equation of the given line.

3 Point-Slope Form of the Equation of a Line An equation of the line passing through the point $P_1(x_1, y_1)$ and having slope m is

$$y - y_1 = m(x - x_1)$$

EXAMPLE 3 Find an equation of the line through $(1, -7)$ with slope $-\frac{1}{2}$.

SOLUTION Using (3) with $m = -\frac{1}{2}$, $x_1 = 1$, and $y_1 = -7$, we obtain an equation of the line as

$$y + 7 = -\tfrac{1}{2}(x - 1)$$

which we can rewrite as

$$2y + 14 = -x + 1 \qquad \text{or} \qquad x + 2y + 13 = 0$$

EXAMPLE 4 Find an equation of the line through the points $(-1, 2)$ and $(3, -4)$.

SOLUTION By Definition 2 the slope of the line is

$$m = \frac{-4 - 2}{3 - (-1)} = -\frac{3}{2}$$

Using the point-slope form with $x_1 = -1$ and $y_1 = 2$, we obtain

$$y - 2 = -\tfrac{3}{2}(x + 1)$$

which simplifies to $$3x + 2y = 1$$

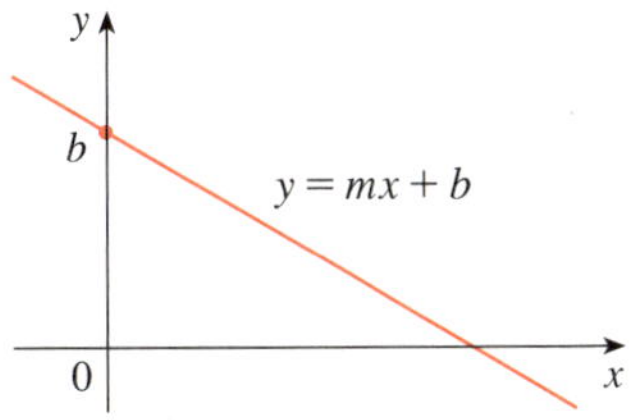

FIGURE 7

Suppose a nonvertical line has slope m and y-intercept b. (See Figure 7.) This means it intersects the y-axis at the point $(0, b)$, so the point-slope form of the equation of the line, with $x_1 = 0$ and $y_1 = b$, becomes

$$y - b = m(x - 0)$$

This simplifies as follows.

4 Slope-Intercept Form of the Equation of a Line An equation of the line with slope m and y-intercept b is

$$y = mx + b$$

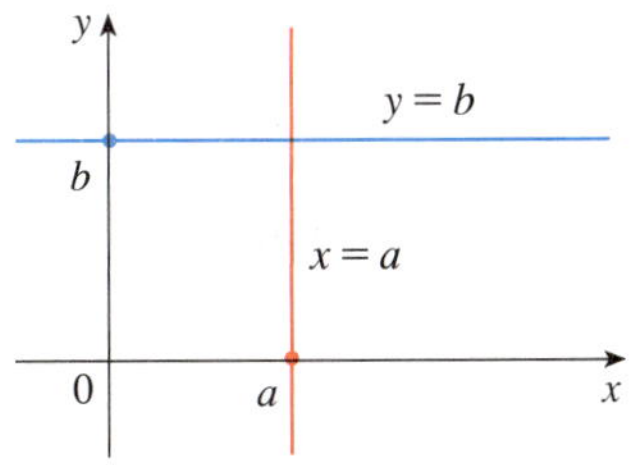

FIGURE 8

In particular, if a line is horizontal, its slope is $m = 0$, so its equation is $y = b$, where b is the y-intercept (see Figure 8). A vertical line does not have a slope, but we can write its equation as $x = a$, where a is the x-intercept, because the x-coordinate of every point on the line is a.

Observe that the equation of every line can be written in the form

5 $$Ax + By + C = 0$$

because a vertical line has the equation $x = a$ or $x - a = 0$ ($A = 1$, $B = 0$, $C = -a$) and a nonvertical line has the equation $y = mx + b$ or $-mx + y - b = 0$ ($A = -m$, $B = 1$, $C = -b$). Conversely, if we start with a general first-degree equation, that is, an equation of the form (5), where A, B, and C are constants and A and B are not both 0, then we can show that it is the equation of a line. If $B = 0$, the equation becomes $Ax + C = 0$ or $x = -C/A$, which represents a vertical line with x-intercept $-C/A$. If $B \neq 0$, the equation can be rewritten by solving for y:

$$y = -\frac{A}{B}x - \frac{C}{B}$$

and we recognize this as being the slope-intercept form of the equation of a line ($m = -A/B$, $b = -C/B$). Therefore, an equation of the form (5) is called a **linear equation** or the **general equation of a line**. For brevity, we often refer to "the line $Ax + By + C = 0$" instead of "the line whose equation is $Ax + By + C = 0$."

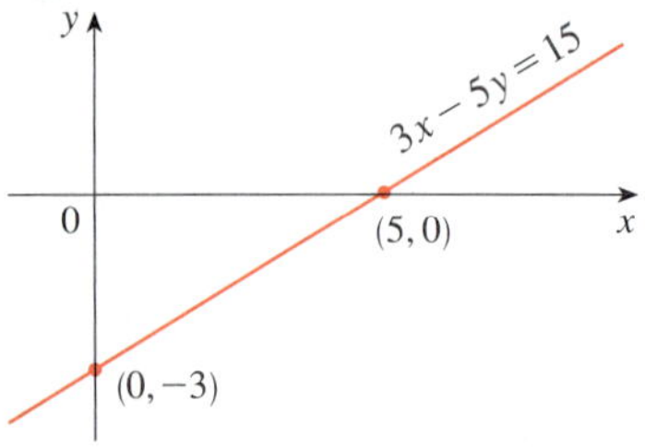

FIGURE 9

EXAMPLE 5 Sketch the graph of the equation $3x - 5y = 15$.

SOLUTION Since the equation is linear, its graph is a line. To draw the graph, we can simply find two points on the line. It's easiest to find the intercepts. Substituting $y = 0$ (the equation of the x-axis) in the given equation, we get $3x = 15$, so $x = 5$ is the x-intercept. Substituting $x = 0$ in the equation, we see that the y-intercept is -3. This allows us to sketch the graph as in Figure 9.

EXAMPLE 6 Graph the inequality $x + 2y > 5$.

SOLUTION We are asked to sketch the graph of the set $\{(x, y) \mid x + 2y > 5\}$ and we do so by solving the inequality for y:

$$x + 2y > 5$$

$$2y > -x + 5$$

$$y > -\tfrac{1}{2}x + \tfrac{5}{2}$$

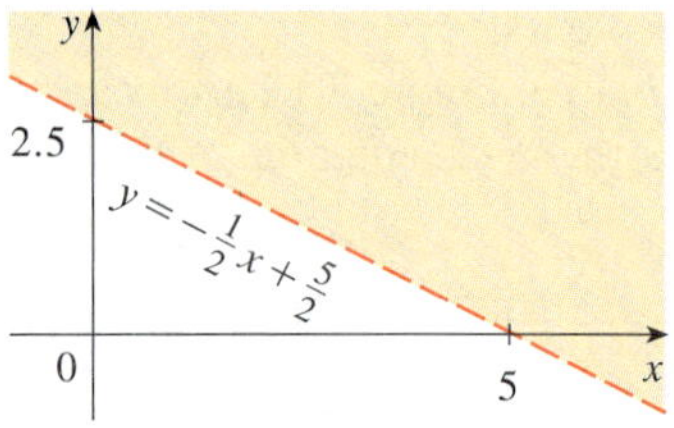

FIGURE 10

Compare this inequality with the equation $y = -\frac{1}{2}x + \frac{5}{2}$, which represents a line with slope $-\frac{1}{2}$ and y-intercept $\frac{5}{2}$. We see that the given graph consists of points whose y-coordinates are *larger* than those on the line $y = -\frac{1}{2}x + \frac{5}{2}$. Thus, the graph is the region that lies *above* the line, as illustrated in Figure 10.

Parallel and Perpendicular Lines

Slopes can be used to show that lines are parallel or perpendicular. The following facts are proved, for instance, in *Precalculus: Mathematics for Calculus, Fourth Edition* by Stewart, Redlin, and Watson (Brooks/Cole Publishing Co., Pacific Grove, CA, 2002).

6 Parallel and Perpendicular Lines

1. Two nonvertical lines are parallel if and only if they have the same slope.
2. Two lines with slopes m_1 and m_2 are perpendicular if and only if $m_1 m_2 = -1$; that is, their slopes are negative reciprocals:

$$m_2 = -\frac{1}{m_1}$$

EXAMPLE 7 Find an equation of the line through the point (5, 2) that is parallel to the line $4x + 6y + 5 = 0$.

SOLUTION The given line can be written in the form

$$y = -\tfrac{2}{3}x - \tfrac{5}{6}$$

which is in slope-intercept form with $m = -\frac{2}{3}$. Parallel lines have the same slope, so the required line has slope $-\frac{2}{3}$ and its equation in point-slope form is

$$y - 2 = -\tfrac{2}{3}(x - 5)$$

We can write this equation as $2x + 3y = 16$.

EXAMPLE 8 Show that the lines $2x + 3y = 1$ and $6x - 4y - 1 = 0$ are perpendicular.

SOLUTION The equations can be written as

$$y = -\tfrac{2}{3}x + \tfrac{1}{3} \qquad \text{and} \qquad y = \tfrac{3}{2}x - \tfrac{1}{4}$$

from which we see that the slopes are

$$m_1 = -\tfrac{2}{3} \qquad \text{and} \qquad m_2 = \tfrac{3}{2}$$

Since $m_1 m_2 = -1$, the lines are perpendicular.

B Exercises

1–6 Find the distance between the points.

1. $(1, 1),\quad (4, 5)$

2. $(1, -3),\quad (5, 7)$

3. $(6, -2),\quad (-1, 3)$

4. $(1, -6),\quad (-1, -3)$

5. $(2, 5),\quad (4, -7)$

6. $(a, b),\quad (b, a)$

7–10 Find the slope of the line through P and Q.

7. $P(1, 5),\quad Q(4, 11)$

8. $P(-1, 6),\quad Q(4, -3)$

9. $P(-3, 3),\quad Q(-1, -6)$

10. $P(-1, -4),\quad Q(6, 0)$

11. Show that the triangle with vertices $A(0, 2)$, $B(-3, -1)$, and $C(-4, 3)$ is isosceles.

12. (a) Show that the triangle with vertices $A(6, -7)$, $B(11, -3)$, and $C(2, -2)$ is a right triangle using the converse of the Pythagorean Theorem.
(b) Use slopes to show that ABC is a right triangle.
(c) Find the area of the triangle.

13. Show that the points $(-2, 9)$, $(4, 6)$, $(1, 0)$, and $(-5, 3)$ are the vertices of a square.

14. (a) Show that the points $A(-1, 3)$, $B(3, 11)$, and $C(5, 15)$ are collinear (lie on the same line) by showing that $|AB| + |BC| = |AC|$.
(b) Use slopes to show that A, B, and C are collinear.

15. Show that $A(1, 1)$, $B(7, 4)$, $C(5, 10)$, and $D(-1, 7)$ are vertices of a parallelogram.

16. Show that $A(1, 1)$, $B(11, 3)$, $C(10, 8)$, and $D(0, 6)$ are vertices of a rectangle.

17–20 Sketch the graph of the equation.

17. $x = 3$

18. $y = -2$

19. $xy = 0$

20. $|y| = 1$

21–36 Find an equation of the line that satisfies the given conditions.

21. Through $(2, -3)$, slope 6

22. Through $(-1, 4)$, slope -3

23. Through $(1, 7)$, slope $\frac{2}{3}$

24. Through $(-3, -5)$, slope $-\frac{7}{2}$

25. Through $(2, 1)$ and $(1, 6)$

26. Through $(-1, -2)$ and $(4, 3)$

27. Slope 3, y-intercept -2

28. Slope $\frac{2}{5}$, y-intercept 4

29. x-intercept 1, y-intercept -3

30. x-intercept -8, y-intercept 6

31. Through $(4, 5)$, parallel to the x-axis

32. Through $(4, 5)$, parallel to the y-axis

33. Through $(1, -6)$, parallel to the line $x + 2y = 6$

34. y-intercept 6, parallel to the line $2x + 3y + 4 = 0$

35. Through $(-1, -2)$, perpendicular to the line $2x + 5y + 8 = 0$

36. Through $\left(\frac{1}{2}, -\frac{2}{3}\right)$, perpendicular to the line $4x - 8y = 1$

37–42 Find the slope and y-intercept of the line and draw its graph.

37. $x + 3y = 0$

38. $2x - 5y = 0$

39. $y = -2$

40. $2x - 3y + 6 = 0$

41. $3x - 4y = 12$

42. $4x + 5y = 10$

43–52 Sketch the region in the xy-plane.

43. $\{(x, y) \mid x < 0\}$

44. $\{(x, y) \mid y > 0\}$

45. $\{(x, y) \mid xy < 0\}$

46. $\{(x, y) \mid x \geqslant 1 \text{ and } y < 3\}$

47. $\{(x, y) \mid |x| \leqslant 2\}$

48. $\{(x, y) \mid |x| < 3 \text{ and } |y| < 2\}$

49. $\{(x, y) \mid 0 \leqslant y \leqslant 4 \text{ and } x \leqslant 2\}$

50. $\{(x, y) \mid y > 2x - 1\}$

51. $\{(x, y) \mid 1 + x \leqslant y \leqslant 1 - 2x\}$

52. $\{(x, y) \mid -x \leqslant y < \frac{1}{2}(x + 3)\}$

53. Find a point on the y-axis that is equidistant from $(5, -5)$ and $(1, 1)$.

54. Show that the midpoint of the line segment from $P_1(x_1, y_1)$ to $P_2(x_2, y_2)$ is

$$\left(\frac{x_1 + x_2}{2}, \frac{y_1 + y_2}{2}\right)$$

55. Find the midpoint of the line segment joining the given points.
(a) $(1, 3)$ and $(7, 15)$
(b) $(-1, 6)$ and $(8, -12)$

56. Find the lengths of the medians of the triangle with vertices $A(1, 0)$, $B(3, 6)$, and $C(8, 2)$. (A median is a line segment from a vertex to the midpoint of the opposite side.)

57. Show that the lines $2x - y = 4$ and $6x - 2y = 10$ are not parallel and find their point of intersection.

58. Show that the lines $3x - 5y + 19 = 0$ and $10x + 6y - 50 = 0$ are perpendicular and find their point of intersection.

59. Find an equation of the perpendicular bisector of the line segment joining the points $A(1, 4)$ and $B(7, -2)$.

60. (a) Find equations for the sides of the triangle with vertices $P(1, 0)$, $Q(3, 4)$, and $R(-1, 6)$.
(b) Find equations for the medians of this triangle. Where do they intersect?

61. (a) Show that if the x- and y-intercepts of a line are nonzero numbers a and b, then the equation of the line can be put in the form

$$\frac{x}{a} + \frac{y}{b} = 1$$

This equation is called the **two-intercept form** of an equation of a line.
(b) Use part (a) to find an equation of the line whose x-intercept is 6 and whose y-intercept is -8.

62. A car leaves Detroit at 2:00 P.M., traveling at a constant speed west along I-96. It passes Ann Arbor, 40 mi from Detroit, at 2:50 P.M.
(a) Express the distance traveled in terms of the time elapsed.
(b) Draw the graph of the equation in part (a).
(c) What is the slope of this line? What does it represent?

C Graphs of Second-Degree Equations

In Appendix B we saw that a first-degree, or linear, equation $Ax + By + C = 0$ represents a line. In this section we discuss second-degree equations such as

$$x^2 + y^2 = 1 \qquad y = x^2 + 1 \qquad \frac{x^2}{9} + \frac{y^2}{4} = 1 \qquad x^2 - y^2 = 1$$

which represent a circle, a parabola, an ellipse, and a hyperbola, respectively.

The graph of such an equation in x and y is the set of all points (x, y) that satisfy the equation; it gives a visual representation of the equation. Conversely, given a curve in the xy-plane, we may have to find an equation that represents it, that is, an equation satisfied by the coordinates of the points on the curve and by no other point. This is the other half of the basic principle of analytic geometry as formulated by Descartes and Fermat. The idea is that if a geometric curve can be represented by an algebraic equation, then the rules of algebra can be used to analyze the geometric problem.

Circles

As an example of this type of problem, let's find an equation of the circle with radius r and center (h, k). By definition, the circle is the set of all points $P(x, y)$ whose distance from

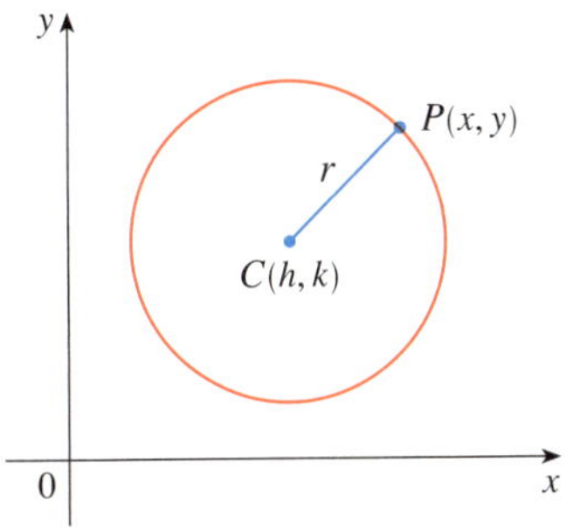

FIGURE 1

the center $C(h, k)$ is r. (See Figure 1.) Thus, P is on the circle if and only if $|PC| = r$. From the distance formula, we have

$$\sqrt{(x - h)^2 + (y - k)^2} = r$$

or equivalently, squaring both sides, we get

$$(x - h)^2 + (y - k)^2 = r^2$$

This is the desired equation.

> **1 Equation of a Circle** An equation of the circle with center (h, k) and radius r is
>
> $$(x - h)^2 + (y - k)^2 = r^2$$
>
> In particular, if the center is the origin $(0, 0)$, the equation is
>
> $$x^2 + y^2 = r^2$$

EXAMPLE 1 Find an equation of the circle with radius 3 and center $(2, -5)$.

SOLUTION From Equation 1 with $r = 3$, $h = 2$, and $k = -5$, we obtain

$$(x - 2)^2 + (y + 5)^2 = 9$$

EXAMPLE 2 Sketch the graph of the equation $x^2 + y^2 + 2x - 6y + 7 = 0$ by first showing that it represents a circle and then finding its center and radius.

SOLUTION We first group the x-terms and y-terms as follows:

$$(x^2 + 2x) + (y^2 - 6y) = -7$$

Then we complete the square within each grouping, adding the appropriate constants to both sides of the equation:

$$(x^2 + 2x + 1) + (y^2 - 6y + 9) = -7 + 1 + 9$$

or

$$(x + 1)^2 + (y - 3)^2 = 3$$

Comparing this equation with the standard equation of a circle (1), we see that $h = -1$, $k = 3$, and $r = \sqrt{3}$, so the given equation represents a circle with center $(-1, 3)$ and radius $\sqrt{3}$. It is sketched in Figure 2.

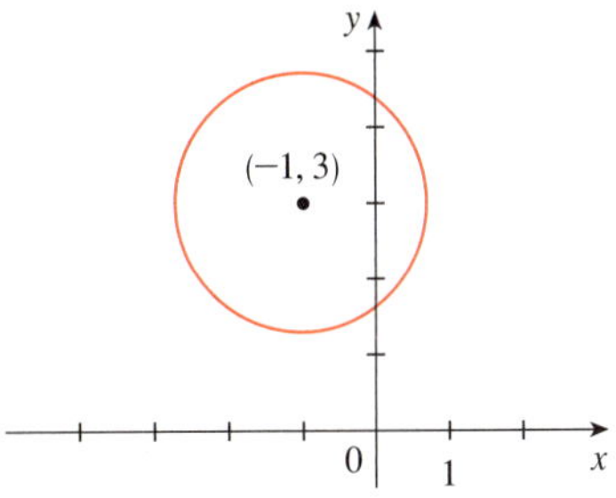

FIGURE 2
$x^2 + y^2 + 2x - 6y + 7 = 0$

Parabolas

The geometric properties of parabolas are reviewed in Section 10.5. Here we regard a parabola as a graph of an equation of the form $y = ax^2 + bx + c$.

EXAMPLE 3 Draw the graph of the parabola $y = x^2$.

SOLUTION We set up a table of values, plot points, and join them by a smooth curve to obtain the graph in Figure 3.

x	$y = x^2$
0	0
$\pm\frac{1}{2}$	$\frac{1}{4}$
± 1	1
± 2	4
± 3	9

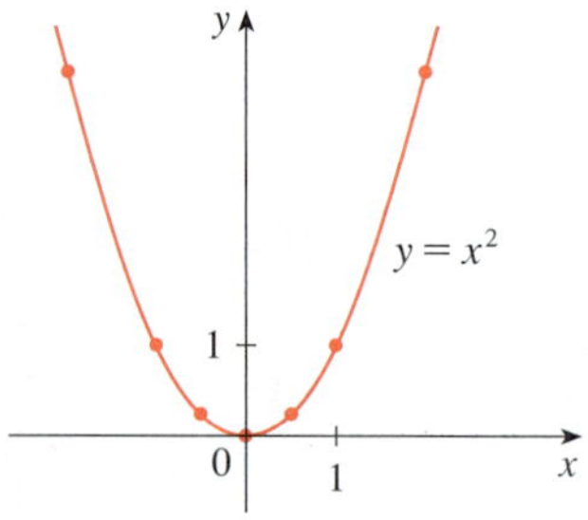

FIGURE 3

Figure 4 shows the graphs of several parabolas with equations of the form $y = ax^2$ for various values of the number a. In each case the *vertex*, the point where the parabola changes direction, is the origin. We see that the parabola $y = ax^2$ opens upward if $a > 0$ and downward if $a < 0$ (as in Figure 5).

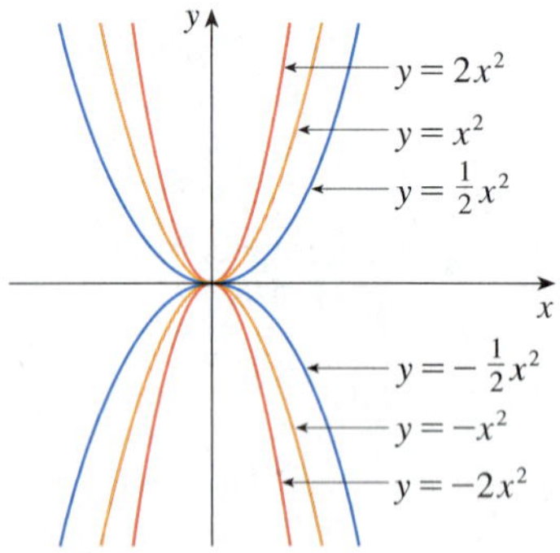

FIGURE 4

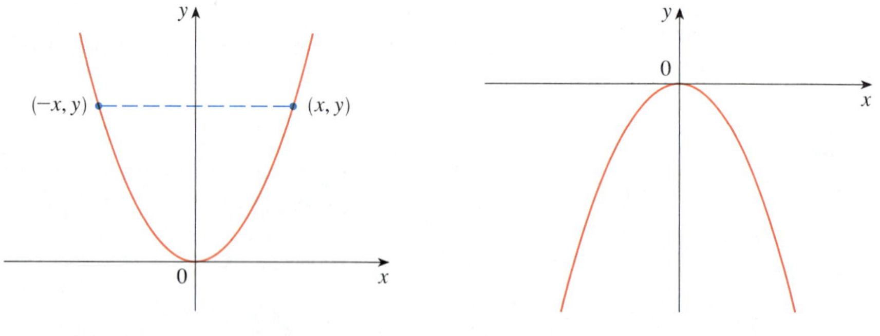

(a) $y = ax^2,\ a > 0$

(b) $y = ax^2,\ a < 0$

FIGURE 5

Notice that if (x, y) satisfies $y = ax^2$, then so does $(-x, y)$. This corresponds to the geometric fact that if the right half of the graph is reflected about the y-axis, then the left half of the graph is obtained. We say that the graph is **symmetric with respect to the y-axis**.

> The graph of an equation is symmetric with respect to the y-axis if the equation is unchanged when x is replaced by $-x$.

If we interchange x and y in the equation $y = ax^2$, the result is $x = ay^2$, which also represents a parabola. (Interchanging x and y amounts to reflecting about the diagonal line $y = x$.) The parabola $x = ay^2$ opens to the right if $a > 0$ and to the left if $a < 0$. (See

Figure 6.) This time the parabola is symmetric with respect to the x-axis because if (x, y) satisfies $x = ay^2$, then so does $(x, -y)$.

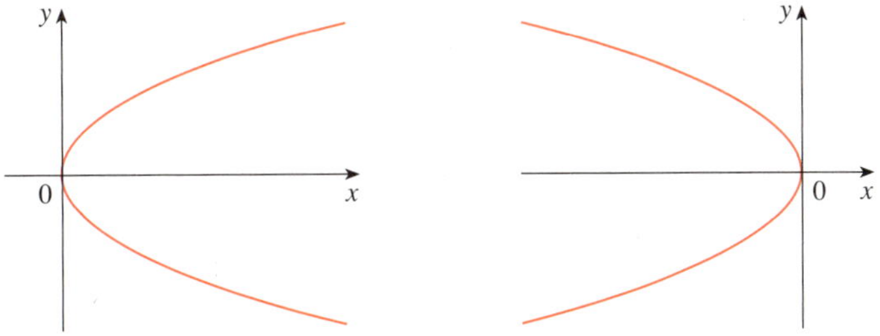

FIGURE 6 (a) $x = ay^2$, $a > 0$ (b) $x = ay^2$, $a < 0$

> The graph of an equation is symmetric with respect to the x-axis if the equation is unchanged when y is replaced by $-y$.

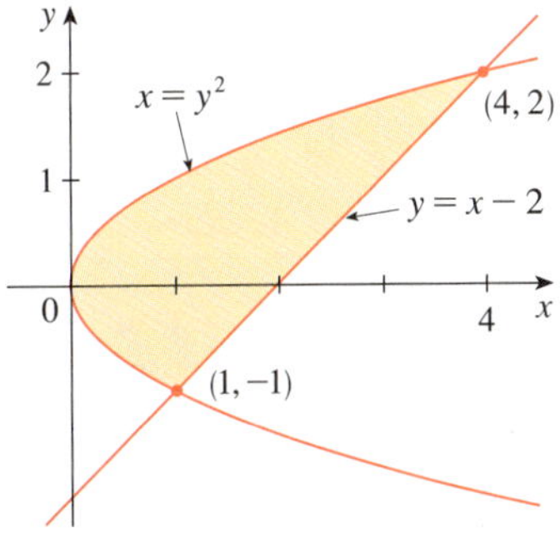

FIGURE 7

EXAMPLE 4 Sketch the region bounded by the parabola $x = y^2$ and the line $y = x - 2$.

SOLUTION First we find the points of intersection by solving the two equations. Substituting $x = y + 2$ into the equation $x = y^2$, we get $y + 2 = y^2$, which gives

$$0 = y^2 - y - 2 = (y - 2)(y + 1)$$

so $y = 2$ or -1. Thus, the points of intersection are $(4, 2)$ and $(1, -1)$, and we draw the line $y = x - 2$ passing through these points. We then sketch the parabola $x = y^2$ by referring to Figure 6(a) and having the parabola pass through $(4, 2)$ and $(1, -1)$. The region bounded by $x = y^2$ and $y = x - 2$ means the finite region whose boundaries are these curves. It is sketched in Figure 7.

Ellipses

The curve with equation

2 $$\frac{x^2}{a^2} + \frac{y^2}{b^2} = 1$$

where a and b are positive numbers, is called an **ellipse** in standard position. (Geometric properties of ellipses are discussed in Section 10.5.) Observe that Equation 2 is unchanged if x is replaced by $-x$ or y is replaced by $-y$, so the ellipse is symmetric with respect to both axes. As a further aid to sketching the ellipse, we find its intercepts.

> The ***x*-intercepts** of a graph are the x-coordinates of the points where the graph intersects the x-axis. They are found by setting $y = 0$ in the equation of the graph.
>
> The ***y*-intercepts** are the y-coordinates of the points where the graph intersects the y-axis. They are found by setting $x = 0$ in its equation.

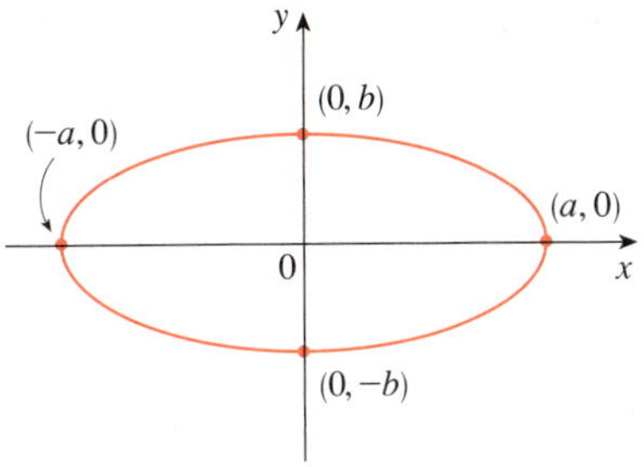

FIGURE 8
$\frac{x^2}{a^2} + \frac{y^2}{b^2} = 1$

If we set $y = 0$ in Equation 2, we get $x^2 = a^2$ and so the x-intercepts are $\pm a$. Setting $x = 0$, we get $y^2 = b^2$, so the y-intercepts are $\pm b$. Using this information, together with symmetry, we sketch the ellipse in Figure 8. If $a = b$, the ellipse is a circle with radius a.

EXAMPLE 5 Sketch the graph of $9x^2 + 16y^2 = 144$.

SOLUTION We divide both sides of the equation by 144:

$$\frac{x^2}{16} + \frac{y^2}{9} = 1$$

The equation is now in the standard form for an ellipse (2), so we have $a^2 = 16$, $b^2 = 9$, $a = 4$, and $b = 3$. The x-intercepts are ± 4; the y-intercepts are ± 3. The graph is sketched in Figure 9.

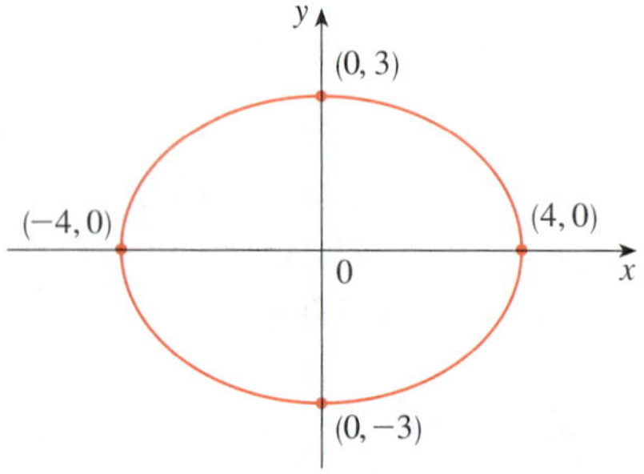

FIGURE 9
$9x^2 + 16y^2 = 144$

Hyperbolas

The curve with equation

$$\boxed{3} \qquad \frac{x^2}{a^2} - \frac{y^2}{b^2} = 1$$

is called a **hyperbola** in standard position. Again, Equation 3 is unchanged when x is replaced by $-x$ or y is replaced by $-y$, so the hyperbola is symmetric with respect to both axes. To find the x-intercepts we set $y = 0$ and obtain $x^2 = a^2$ and $x = \pm a$. However, if we put $x = 0$ in Equation 3, we get $y^2 = -b^2$, which is impossible, so there is no y-intercept. In fact, from Equation 3 we obtain

$$\frac{x^2}{a^2} = 1 + \frac{y^2}{b^2} \geqslant 1$$

which shows that $x^2 \geqslant a^2$ and so $|x| = \sqrt{x^2} \geqslant a$. Therefore, we have $x \geqslant a$ or $x \leqslant -a$. This means that the hyperbola consists of two parts, called its *branches*. It is sketched in Figure 10.

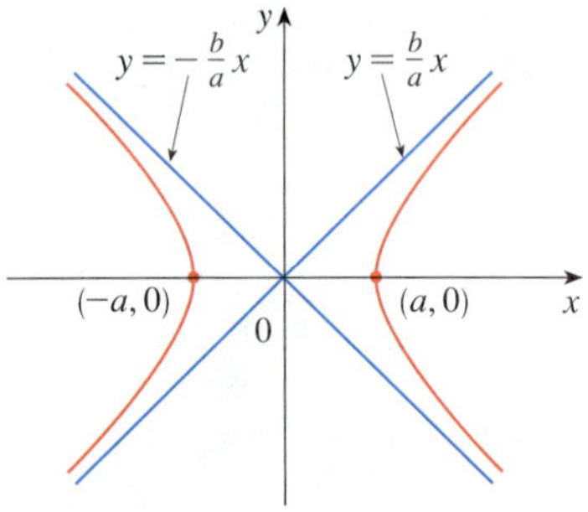

FIGURE 10
The hyperbola $\frac{x^2}{a^2} - \frac{y^2}{b^2} = 1$

In drawing a hyperbola it is useful to draw first its *asymptotes,* which are the lines $y = (b/a)x$ and $y = -(b/a)x$ shown in Figure 10. Both branches of the hyperbola approach the asymptotes; that is, they come arbitrarily close to the asymptotes. This involves the idea of a limit, which is discussed in Chapter 2. (See also Exercise 67 in Section 4.5.)

By interchanging the roles of x and y we get an equation of the form

$$\frac{y^2}{a^2} - \frac{x^2}{b^2} = 1$$

which also represents a hyperbola and is sketched in Figure 11.

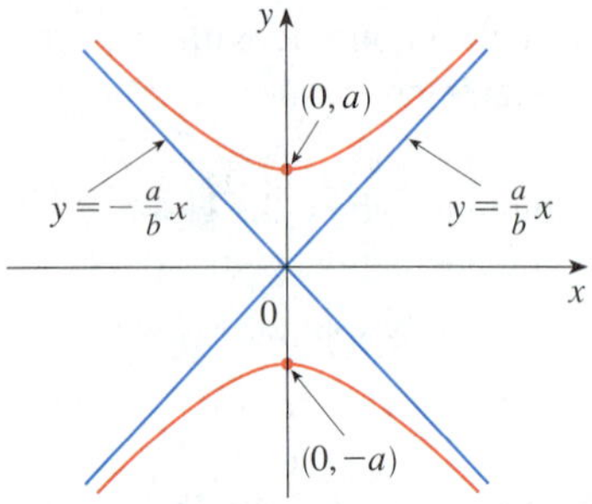

FIGURE 11
The hyperbola $\frac{y^2}{a^2} - \frac{x^2}{b^2} = 1$

EXAMPLE 6 Sketch the curve $9x^2 - 4y^2 = 36$.

SOLUTION Dividing both sides by 36, we obtain

$$\frac{x^2}{4} - \frac{y^2}{9} = 1$$

which is the standard form of the equation of a hyperbola (Equation 3). Since $a^2 = 4$, the x-intercepts are ± 2. Since $b^2 = 9$, we have $b = 3$ and the asymptotes are $y = \pm(\frac{3}{2})x$. The hyperbola is sketched in Figure 12.

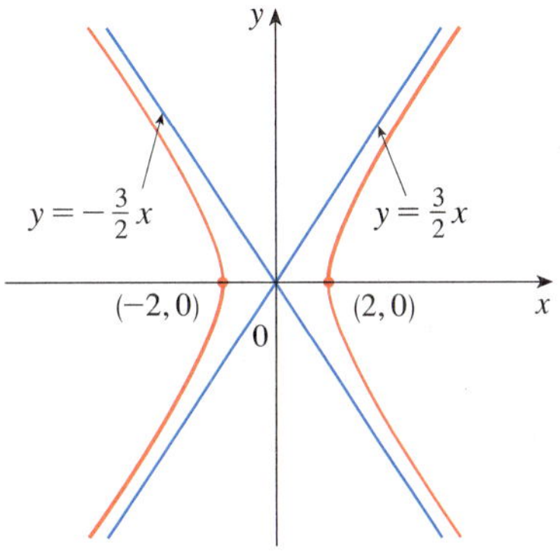

FIGURE 12
The hyperbola $9x^2 - 4y^2 = 36$

If $b = a$, a hyperbola has the equation $x^2 - y^2 = a^2$ (or $y^2 - x^2 = a^2$) and is called an *equilateral hyperbola* [see Figure 13(a)]. Its asymptotes are $y = \pm x$, which are perpendicular. If an equilateral hyperbola is rotated by 45°, the asymptotes become the x- and y-axes, and it can be shown that the new equation of the hyperbola is $xy = k$, where k is a constant [see Figure 13(b)].

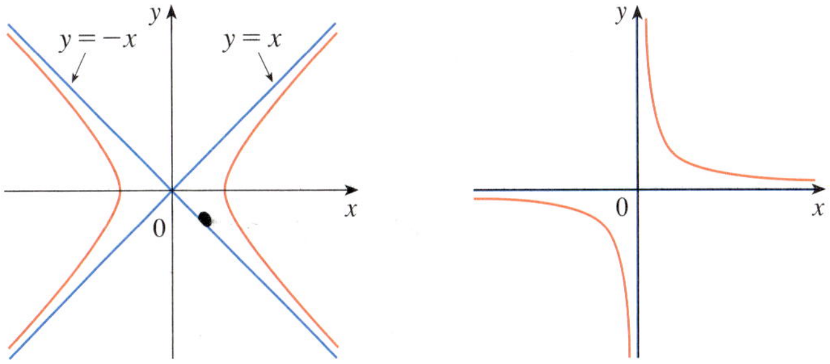

FIGURE 13
Equilateral hyperbolas

(a) $x^2 - y^2 = a^2$ (b) $xy = k$ $(k > 0)$

Shifted Conics

Recall that an equation of the circle with center the origin and radius r is $x^2 + y^2 = r^2$, but if the center is the point (h, k), then the equation of the circle becomes

$$(x - h)^2 + (y - k)^2 = r^2$$

Similarly, if we take the ellipse with equation

$$\frac{x^2}{a^2} + \frac{y^2}{b^2} = 1 \qquad \text{4}$$

and translate it (shift it) so that its center is the point (h, k), then its equation becomes

5
$$\frac{(x-h)^2}{a^2} + \frac{(y-k)^2}{b^2} = 1$$

(See Figure 14.)

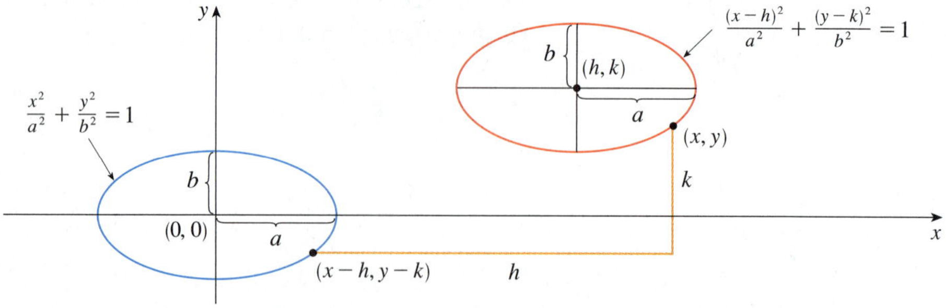

FIGURE 14

Notice that in shifting the ellipse, we replaced x by $x - h$ and y by $y - k$ in Equation 4 to obtain Equation 5. We use the same procedure to shift the parabola $y = ax^2$ so that its vertex (the origin) becomes the point (h, k) as in Figure 15. Replacing x by $x - h$ and y by $y - k$, we see that the new equation is

$$y - k = a(x-h)^2 \qquad \text{or} \qquad y = a(x-h)^2 + k$$

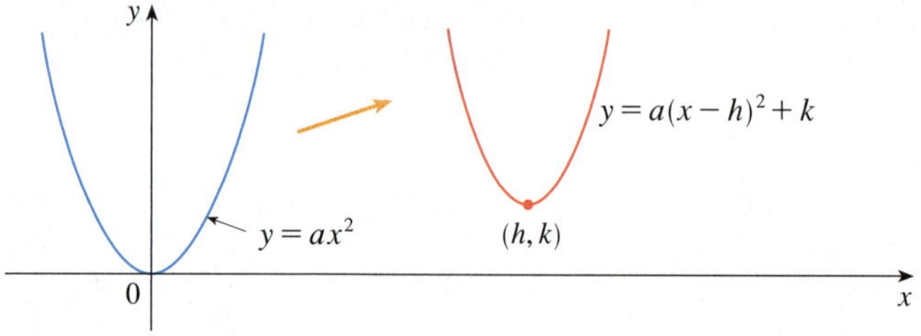

FIGURE 15

EXAMPLE 7 Sketch the graph of the equation $y = 2x^2 - 4x + 1$.

SOLUTION First we complete the square:

$$y = 2(x^2 - 2x) + 1 = 2(x-1)^2 - 1$$

In this form we see that the equation represents the parabola obtained by shifting $y = 2x^2$ so that its vertex is at the point $(1, -1)$. The graph is sketched in Figure 16.

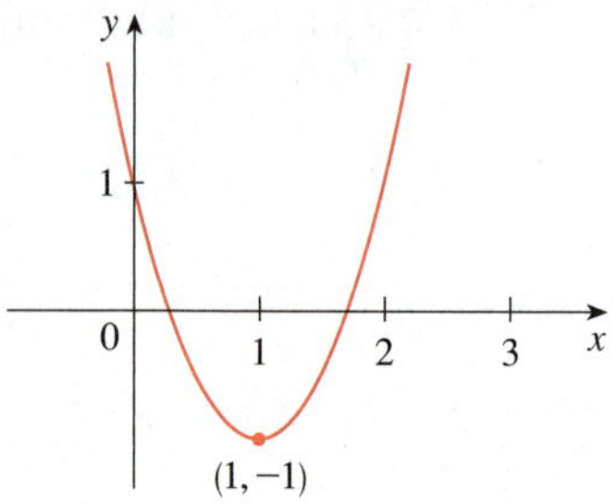

FIGURE 16
$y = 2x^2 - 4x + 1$

EXAMPLE 8 Sketch the curve $x = 1 - y^2$.

SOLUTION This time we start with the parabola $x = -y^2$ (as in Figure 6 with $a = -1$) and shift one unit to the right to get the graph of $x = 1 - y^2$. (See Figure 17.)

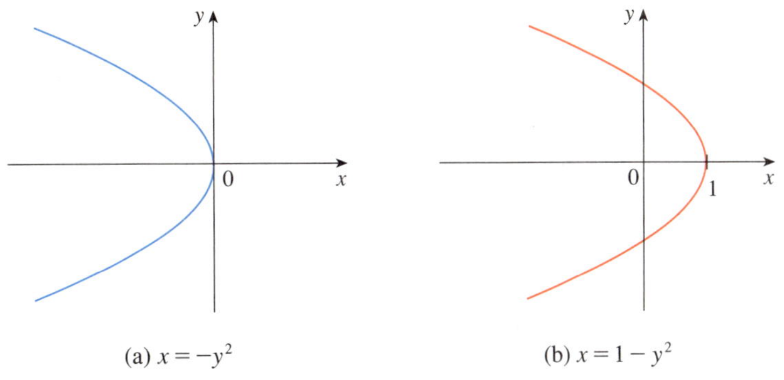

FIGURE 17

C Exercises

1–4 Find an equation of a circle that satisfies the given conditions.

1. Center $(3, -1)$, radius 5

2. Center $(-2, -8)$, radius 10

3. Center at the origin, passes through $(4, 7)$

4. Center $(-1, 5)$, passes through $(-4, -6)$

5–9 Show that the equation represents a circle and find the center and radius.

5. $x^2 + y^2 - 4x + 10y + 13 = 0$

6. $x^2 + y^2 + 6y + 2 = 0$

7. $x^2 + y^2 + x = 0$

8. $16x^2 + 16y^2 + 8x + 32y + 1 = 0$

9. $2x^2 + 2y^2 - x + y = 1$

10. Under what condition on the coefficients a, b, and c does the equation $x^2 + y^2 + ax + by + c = 0$ represent a circle? When that condition is satisfied, find the center and radius of the circle.

11–32 Identify the type of curve and sketch the graph. Do not plot points. Just use the standard graphs given in Figures 5, 6, 8, 10, and 11 and shift if necessary.

11. $y = -x^2$

12. $y^2 - x^2 = 1$

13. $x^2 + 4y^2 = 16$

14. $x = -2y^2$

15. $16x^2 - 25y^2 = 400$

16. $25x^2 + 4y^2 = 100$

17. $4x^2 + y^2 = 1$

18. $y = x^2 + 2$

19. $x = y^2 - 1$

20. $9x^2 - 25y^2 = 225$

21. $9y^2 - x^2 = 9$

22. $2x^2 + 5y^2 = 10$

23. $xy = 4$

24. $y = x^2 + 2x$

25. $9(x - 1)^2 + 4(y - 2)^2 = 36$

26. $16x^2 + 9y^2 - 36y = 108$

27. $y = x^2 - 6x + 13$

28. $x^2 - y^2 - 4x + 3 = 0$

29. $x = 4 - y^2$

30. $y^2 - 2x + 6y + 5 = 0$

31. $x^2 + 4y^2 - 6x + 5 = 0$

32. $4x^2 + 9y^2 - 16x + 54y + 61 = 0$

33–34 Sketch the region bounded by the curves.

33. $y = 3x, \quad y = x^2$

34. $y = 4 - x^2, \quad x - 2y = 2$

35. Find an equation of the parabola with vertex $(1, -1)$ that passes through the points $(-1, 3)$ and $(3, 3)$.

36. Find an equation of the ellipse with center at the origin that passes through the points $\left(1, -10\sqrt{2}/3\right)$ and $\left(-2, 5\sqrt{5}/3\right)$.

37–40 Sketch the graph of the set.

37. $\{(x, y) \mid x^2 + y^2 \leqslant 1\}$

38. $\{(x, y) \mid x^2 + y^2 > 4\}$

39. $\{(x, y) \mid y \geqslant x^2 - 1\}$

40. $\{(x, y) \mid x^2 + 4y^2 \leqslant 4\}$

D Trigonometry

Angles

Angles can be measured in degrees or in radians (abbreviated as rad). The angle given by a complete revolution contains 360°, which is the same as 2π rad. Therefore

1 $$\pi \text{ rad} = 180°$$

and

2 $$1 \text{ rad} = \left(\frac{180}{\pi}\right)^{\circ} \approx 57.3° \qquad 1° = \frac{\pi}{180} \text{ rad} \approx 0.017 \text{ rad}$$

EXAMPLE 1

(a) Find the radian measure of 60°. (b) Express $5\pi/4$ rad in degrees.

SOLUTION

(a) From Equation 1 or 2 we see that to convert from degrees to radians we multiply by $\pi/180$. Therefore

$$60° = 60\left(\frac{\pi}{180}\right) = \frac{\pi}{3} \text{ rad}$$

(b) To convert from radians to degrees we multiply by $180/\pi$. Thus

$$\frac{5\pi}{4} \text{ rad} = \frac{5\pi}{4}\left(\frac{180}{\pi}\right) = 225°$$

In calculus we use radians to measure angles except when otherwise indicated. The following table gives the correspondence between degree and radian measures of some common angles.

Degrees	0°	30°	45°	60°	90°	120°	135°	150°	180°	270°	360°
Radians	0	$\frac{\pi}{6}$	$\frac{\pi}{4}$	$\frac{\pi}{3}$	$\frac{\pi}{2}$	$\frac{2\pi}{3}$	$\frac{3\pi}{4}$	$\frac{5\pi}{6}$	π	$\frac{3\pi}{2}$	2π

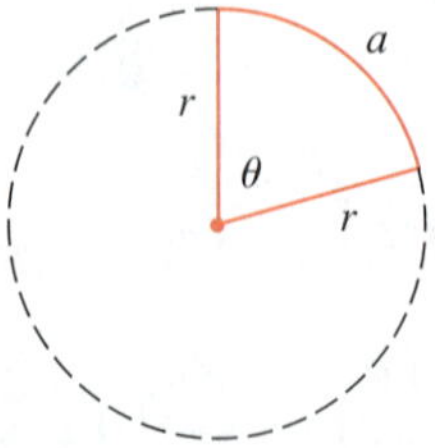

FIGURE 1

Figure 1 shows a sector of a circle with central angle θ and radius r subtending an arc with length a. Since the length of the arc is proportional to the size of the angle, and since the entire circle has circumference $2\pi r$ and central angle 2π, we have

$$\frac{\theta}{2\pi} = \frac{a}{2\pi r}$$

Solving this equation for θ and for a, we obtain

3 $$\theta = \frac{a}{r} \qquad a = r\theta$$

Remember that Equations 3 are valid only when θ is measured in radians.

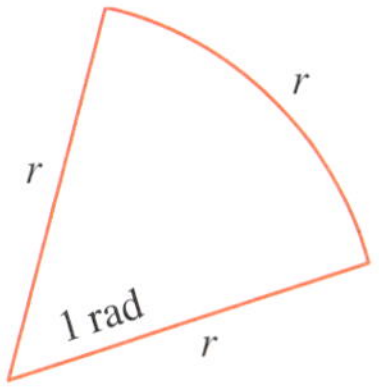

FIGURE 2

In particular, putting $a = r$ in Equation 3, we see that an angle of 1 rad is the angle subtended at the center of a circle by an arc equal in length to the radius of the circle (see Figure 2).

EXAMPLE 2

(a) If the radius of a circle is 5 cm, what angle is subtended by an arc of 6 cm?
(b) If a circle has radius 3 cm, what is the length of an arc subtended by a central angle of $3\pi/8$ rad?

SOLUTION

(a) Using Equation 3 with $a = 6$ and $r = 5$, we see that the angle is

$$\theta = \tfrac{6}{5} = 1.2 \text{ rad}$$

(b) With $r = 3$ cm and $\theta = 3\pi/8$ rad, the arc length is

$$a = r\theta = 3\left(\frac{3\pi}{8}\right) = \frac{9\pi}{8} \text{ cm}$$

The **standard position** of an angle occurs when we place its vertex at the origin of a coordinate system and its initial side on the positive x-axis as in Figure 3. A **positive** angle is obtained by rotating the initial side counterclockwise until it coincides with the terminal side. Likewise, **negative** angles are obtained by clockwise rotation as in Figure 4.

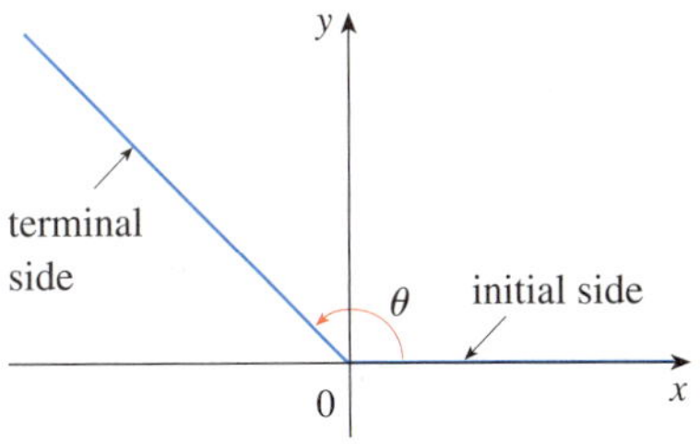

FIGURE 3 $\theta \geq 0$

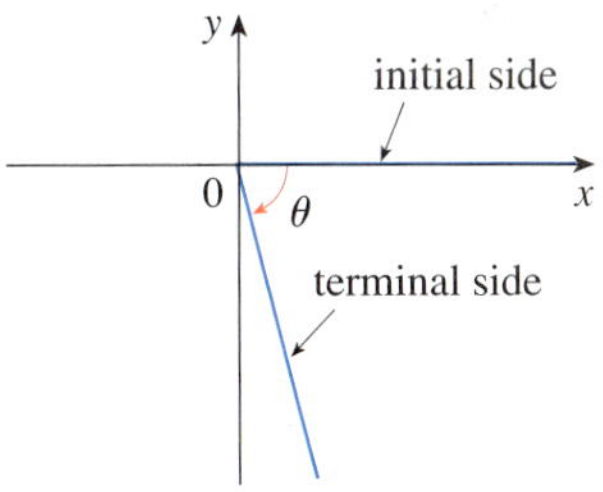

FIGURE 4 $\theta < 0$

Figure 5 shows several examples of angles in standard position. Notice that different angles can have the same terminal side. For instance, the angles $3\pi/4$, $-5\pi/4$, and $11\pi/4$ have the same initial and terminal sides because

$$\frac{3\pi}{4} - 2\pi = -\frac{5\pi}{4} \qquad \frac{3\pi}{4} + 2\pi = \frac{11\pi}{4}$$

and 2π rad represents a complete revolution.

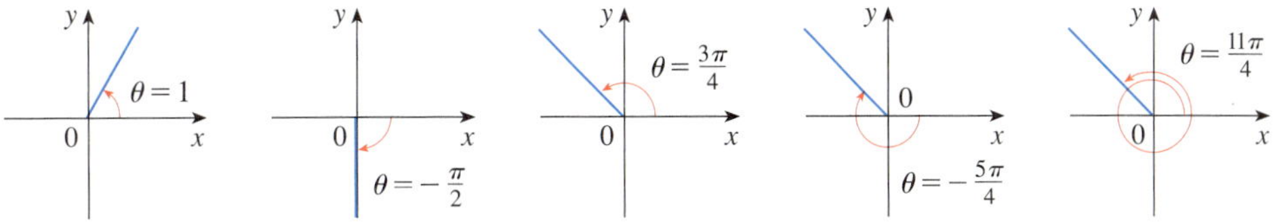

FIGURE 5
Angles in standard position

The Trigonometric Functions

For an acute angle θ the six trigonometric functions are defined as ratios of lengths of sides of a right triangle as follows (see Figure 6).

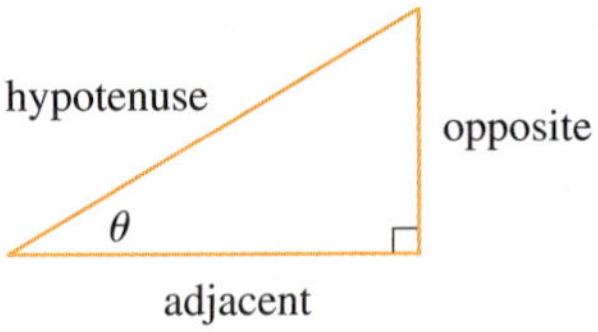

FIGURE 6

4
$$\sin\theta = \frac{\text{opp}}{\text{hyp}} \qquad \csc\theta = \frac{\text{hyp}}{\text{opp}}$$
$$\cos\theta = \frac{\text{adj}}{\text{hyp}} \qquad \sec\theta = \frac{\text{hyp}}{\text{adj}}$$
$$\tan\theta = \frac{\text{opp}}{\text{adj}} \qquad \cot\theta = \frac{\text{adj}}{\text{opp}}$$

This definition doesn't apply to obtuse or negative angles, so for a general angle θ in standard position we let $P(x, y)$ be any point on the terminal side of θ and we let r be the distance $|OP|$ as in Figure 7. Then we define

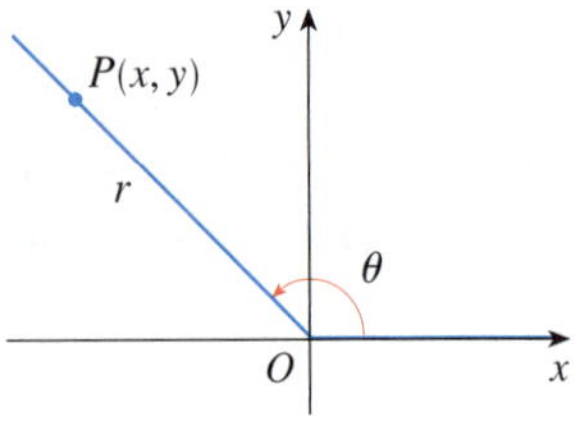

FIGURE 7

5
$$\sin\theta = \frac{y}{r} \qquad \csc\theta = \frac{r}{y}$$
$$\cos\theta = \frac{x}{r} \qquad \sec\theta = \frac{r}{x}$$
$$\tan\theta = \frac{y}{x} \qquad \cot\theta = \frac{x}{y}$$

Since division by 0 is not defined, $\tan\theta$ and $\sec\theta$ are undefined when $x = 0$ and $\csc\theta$ and $\cot\theta$ are undefined when $y = 0$. Notice that the definitions in (4) and (5) are consistent when θ is an acute angle.

If θ is a number, the convention is that $\sin\theta$ means the sine of the angle whose *radian* measure is θ. For example, the expression $\sin 3$ implies that we are dealing with an angle of 3 rad. When finding a calculator approximation to this number we must remember to set our calculator in radian mode, and then we obtain

$$\sin 3 \approx 0.14112$$

If we want to know the sine of the angle 3° we would write $\sin 3°$ and, with our calculator in degree mode, we find that

$$\sin 3° \approx 0.05234$$

The exact trigonometric ratios for certain angles can be read from the triangles in Figure 8. For instance,

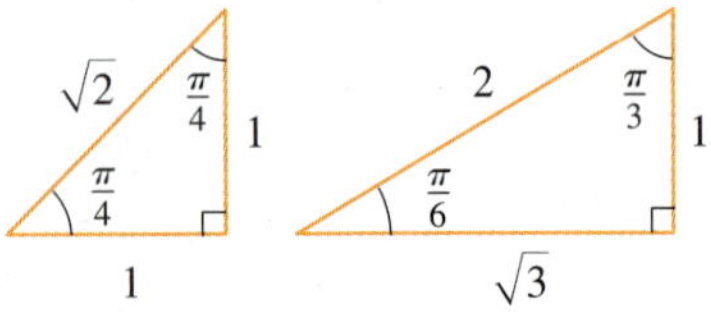

FIGURE 8

$$\sin\frac{\pi}{4} = \frac{1}{\sqrt{2}} \qquad \sin\frac{\pi}{6} = \frac{1}{2} \qquad \sin\frac{\pi}{3} = \frac{\sqrt{3}}{2}$$
$$\cos\frac{\pi}{4} = \frac{1}{\sqrt{2}} \qquad \cos\frac{\pi}{6} = \frac{\sqrt{3}}{2} \qquad \cos\frac{\pi}{3} = \frac{1}{2}$$
$$\tan\frac{\pi}{4} = 1 \qquad \tan\frac{\pi}{6} = \frac{1}{\sqrt{3}} \qquad \tan\frac{\pi}{3} = \sqrt{3}$$

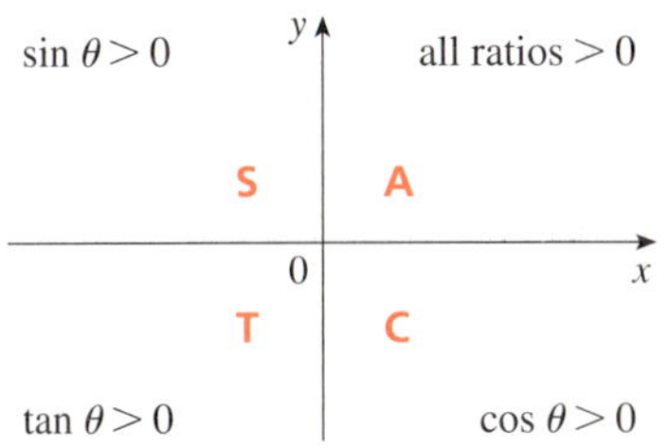

FIGURE 9

The signs of the trigonometric functions for angles in each of the four quadrants can be remembered by means of the rule "All Students Take Calculus" shown in Figure 9.

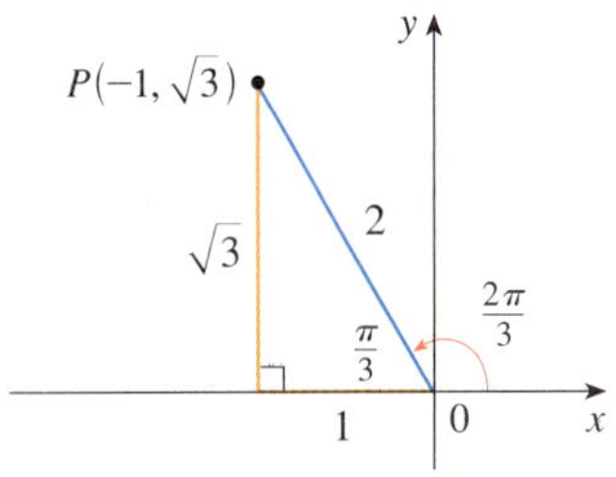

FIGURE 10

EXAMPLE 3 Find the exact trigonometric ratios for $\theta = 2\pi/3$.

SOLUTION From Figure 10 we see that a point on the terminal line for $\theta = 2\pi/3$ is $P(-1, \sqrt{3})$. Therefore, taking

$$x = -1 \qquad y = \sqrt{3} \qquad r = 2$$

in the definitions of the trigonometric ratios, we have

$$\sin\frac{2\pi}{3} = \frac{\sqrt{3}}{2} \qquad \cos\frac{2\pi}{3} = -\frac{1}{2} \qquad \tan\frac{2\pi}{3} = -\sqrt{3}$$

$$\csc\frac{2\pi}{3} = \frac{2}{\sqrt{3}} \qquad \sec\frac{2\pi}{3} = -2 \qquad \cot\frac{2\pi}{3} = -\frac{1}{\sqrt{3}}$$

The following table gives some values of $\sin\theta$ and $\cos\theta$ found by the method of Example 3.

θ	0	$\frac{\pi}{6}$	$\frac{\pi}{4}$	$\frac{\pi}{3}$	$\frac{\pi}{2}$	$\frac{2\pi}{3}$	$\frac{3\pi}{4}$	$\frac{5\pi}{6}$	π	$\frac{3\pi}{2}$	2π
$\sin\theta$	0	$\frac{1}{2}$	$\frac{1}{\sqrt{2}}$	$\frac{\sqrt{3}}{2}$	1	$\frac{\sqrt{3}}{2}$	$\frac{1}{\sqrt{2}}$	$\frac{1}{2}$	0	-1	0
$\cos\theta$	1	$\frac{\sqrt{3}}{2}$	$\frac{1}{\sqrt{2}}$	$\frac{1}{2}$	0	$-\frac{1}{2}$	$-\frac{1}{\sqrt{2}}$	$-\frac{\sqrt{3}}{2}$	-1	0	1

EXAMPLE 4 If $\cos\theta = \frac{2}{5}$ and $0 < \theta < \pi/2$, find the other five trigonometric functions of θ.

SOLUTION Since $\cos\theta = \frac{2}{5}$, we can label the hypotenuse as having length 5 and the adjacent side as having length 2 in Figure 11. If the opposite side has length x, then the Pythagorean Theorem gives $x^2 + 4 = 25$ and so $x^2 = 21$, $x = \sqrt{21}$. We can now use the diagram to write the other five trigonometric functions:

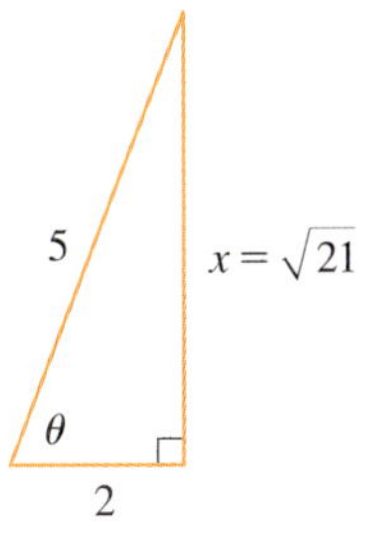

FIGURE 11

$$\sin\theta = \frac{\sqrt{21}}{5} \qquad \tan\theta = \frac{\sqrt{21}}{2}$$

$$\csc\theta = \frac{5}{\sqrt{21}} \qquad \sec\theta = \frac{5}{2} \qquad \cot\theta = \frac{2}{\sqrt{21}}$$

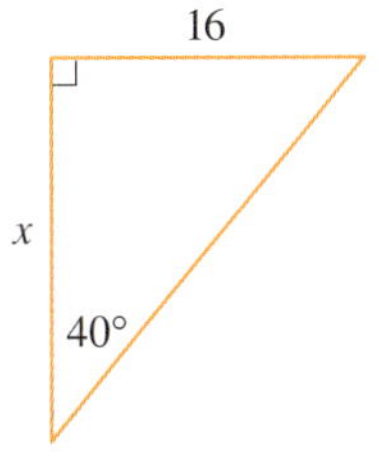

FIGURE 12

EXAMPLE 5 Use a calculator to approximate the value of x in Figure 12.

SOLUTION From the diagram we see that

$$\tan 40^\circ = \frac{16}{x}$$

Therefore

$$x = \frac{16}{\tan 40^\circ} \approx 19.07$$

Trigonometric Identities

A trigonometric identity is a relationship among the trigonometric functions. The most elementary are the following, which are immediate consequences of the definitions of the trigonometric functions.

6
$$\csc\theta = \frac{1}{\sin\theta} \qquad \sec\theta = \frac{1}{\cos\theta} \qquad \cot\theta = \frac{1}{\tan\theta}$$
$$\tan\theta = \frac{\sin\theta}{\cos\theta} \qquad \cot\theta = \frac{\cos\theta}{\sin\theta}$$

For the next identity we refer back to Figure 7. The distance formula (or, equivalently, the Pythagorean Theorem) tells us that $x^2 + y^2 = r^2$. Therefore

$$\sin^2\theta + \cos^2\theta = \frac{y^2}{r^2} + \frac{x^2}{r^2} = \frac{x^2 + y^2}{r^2} = \frac{r^2}{r^2} = 1$$

We have therefore proved one of the most useful of all trigonometric identities:

7
$$\sin^2\theta + \cos^2\theta = 1$$

If we now divide both sides of Equation 7 by $\cos^2\theta$ and use Equations 6, we get

8
$$\tan^2\theta + 1 = \sec^2\theta$$

Similarly, if we divide both sides of Equation 7 by $\sin^2\theta$, we get

9
$$1 + \cot^2\theta = \csc^2\theta$$

The identities

10a
$$\sin(-\theta) = -\sin\theta$$

10b
$$\cos(-\theta) = \cos\theta$$

|||| Odd functions and even functions are discussed in Section 1.1.

show that sin is an odd function and cos is an even function. They are easily proved by drawing a diagram showing θ and $-\theta$ in standard position (see Exercise 39).

Since the angles θ and $\theta + 2\pi$ have the same terminal side, we have

11
$$\sin(\theta + 2\pi) = \sin\theta \qquad \cos(\theta + 2\pi) = \cos\theta$$

These identities show that the sine and cosine functions are periodic with period 2π.

The remaining trigonometric identities are all consequences of two basic identities called the **addition formulas**:

12a $$\sin(x + y) = \sin x \cos y + \cos x \sin y$$

12b $$\cos(x + y) = \cos x \cos y - \sin x \sin y$$

The proofs of these addition formulas are outlined in Exercises 85, 86, and 87.

By substituting $-y$ for y in Equations 12a and 12b and using Equations 10a and 10b, we obtain the following **subtraction formulas**:

13a $$\sin(x - y) = \sin x \cos y - \cos x \sin y$$

13b $$\cos(x - y) = \cos x \cos y + \sin x \sin y$$

Then, by dividing the formulas in Equations 12 or Equations 13, we obtain the corresponding formulas for $\tan(x \pm y)$:

14a $$\tan(x + y) = \frac{\tan x + \tan y}{1 - \tan x \tan y}$$

14b $$\tan(x - y) = \frac{\tan x - \tan y}{1 + \tan x \tan y}$$

If we put $y = x$ in the addition formulas (12), we get the **double-angle formulas**:

15a $$\sin 2x = 2 \sin x \cos x$$

15b $$\cos 2x = \cos^2 x - \sin^2 x$$

Then, by using the identity $\sin^2 x + \cos^2 x = 1$, we obtain the following alternate forms of the double-angle formulas for $\cos 2x$:

16a $$\cos 2x = 2 \cos^2 x - 1$$

16b $$\cos 2x = 1 - 2 \sin^2 x$$

If we now solve these equations for $\cos^2 x$ and $\sin^2 x$, we get the following **half-angle formulas**, which are useful in integral calculus:

17a $$\cos^2 x = \frac{1 + \cos 2x}{2}$$

17b $$\sin^2 x = \frac{1 - \cos 2x}{2}$$

Finally, we state the **product formulas**, which can be deduced from Equations 12 and 13:

18a $$\sin x \cos y = \tfrac{1}{2}[\sin(x + y) + \sin(x - y)]$$

18b $$\cos x \cos y = \tfrac{1}{2}[\cos(x + y) + \cos(x - y)]$$

18c $$\sin x \sin y = \tfrac{1}{2}[\cos(x - y) - \cos(x + y)]$$

There are many other trigonometric identities, but those we have stated are the ones used most often in calculus. If you forget any of them, remember that they can all be deduced from Equations 12a and 12b.

EXAMPLE 6 Find all values of x in the interval $[0, 2\pi]$ such that $\sin x = \sin 2x$.

SOLUTION Using the double-angle formula (15a), we rewrite the given equation as

$$\sin x = 2 \sin x \cos x \qquad \text{or} \qquad \sin x(1 - 2\cos x) = 0$$

Therefore, there are two possibilities:

$$\sin x = 0 \qquad \text{or} \qquad 1 - 2\cos x = 0$$

$$x = 0, \pi, 2\pi \qquad\qquad \cos x = \tfrac{1}{2}$$

$$x = \frac{\pi}{3}, \frac{5\pi}{3}$$

The given equation has five solutions: 0, $\pi/3$, π, $5\pi/3$, and 2π.

Graphs of Trigonometric Functions

The graph of the function $f(x) = \sin x$, shown in Figure 13(a), is obtained by plotting points for $0 \leqslant x \leqslant 2\pi$ and then using the periodic nature of the function (from Equation 11) to complete the graph. Notice that the zeros of the sine function occur at the

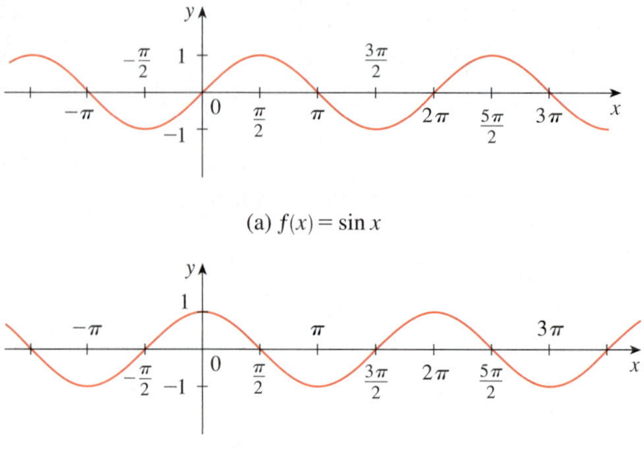

(a) $f(x) = \sin x$

(b) $g(x) = \cos x$

FIGURE 13

integer multiples of π, that is,

$$\sin x = 0 \qquad \text{whenever } x = n\pi, \quad n \text{ an integer}$$

Because of the identity

$$\cos x = \sin\left(x + \frac{\pi}{2}\right)$$

(which can be verified using Equation 12a), the graph of cosine is obtained by shifting the graph of sine by an amount $\pi/2$ to the left [see Figure 13(b)]. Note that for both the sine and cosine functions the domain is $(-\infty, \infty)$ and the range is the closed interval $[-1, 1]$. Thus, for all values of x, we have

$$-1 \leqslant \sin x \leqslant 1 \qquad -1 \leqslant \cos x \leqslant 1$$

The graphs of the remaining four trigonometric functions are shown in Figure 14 and their domains are indicated there. Notice that tangent and cotangent have range $(-\infty, \infty)$, whereas cosecant and secant have range $(-\infty, -1] \cup [1, \infty)$. All four functions are periodic: tangent and cotangent have period π, whereas cosecant and secant have period 2π.

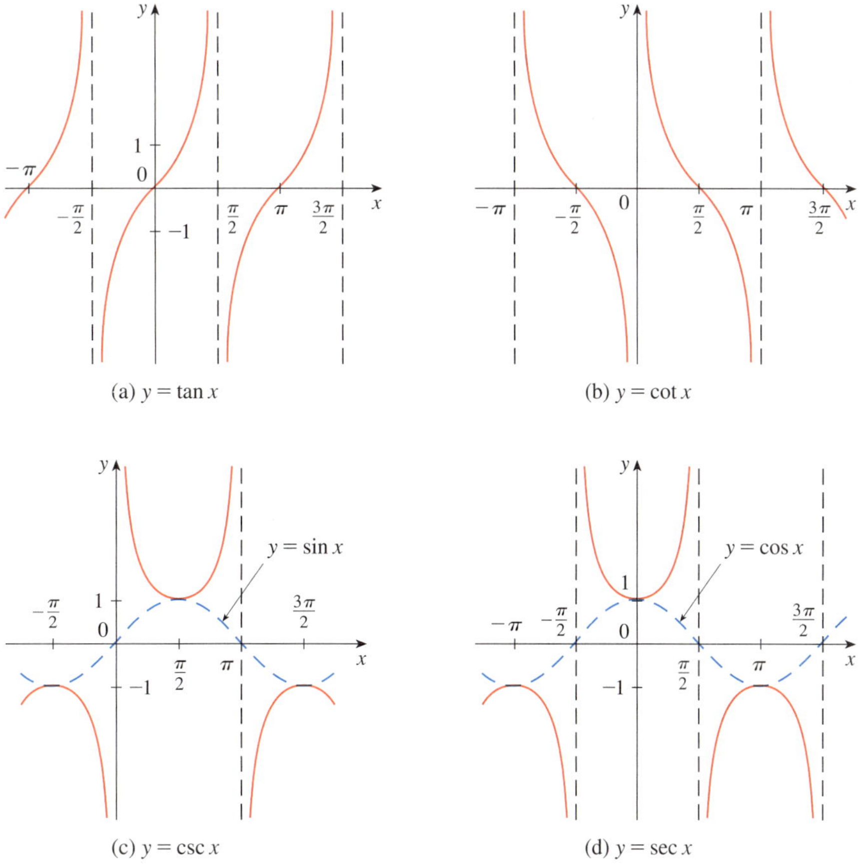

FIGURE 14

D Exercises

1–6 Convert from degrees to radians.

1. 210° **2.** 300° **3.** 9°

4. −315° **5.** 900° **6.** 36°

7–12 Convert from radians to degrees.

7. 4π **8.** $-\dfrac{7\pi}{2}$ **9.** $\dfrac{5\pi}{12}$

10. $\dfrac{8\pi}{3}$ **11.** $-\dfrac{3\pi}{8}$ **12.** 5

13. Find the length of a circular arc subtended by an angle of $\pi/12$ rad if the radius of the circle is 36 cm.

14. If a circle has radius 10 cm, find the length of the arc subtended by a central angle of 72°.

15. A circle has radius 1.5 m. What angle is subtended at the center of the circle by an arc 1 m long?

16. Find the radius of a circular sector with angle $3\pi/4$ and arc length 6 cm.

17–22 Draw, in standard position, the angle whose measure is given.

17. 315° **18.** −150° **19.** $-\dfrac{3\pi}{4}$ rad

20. $\dfrac{7\pi}{3}$ rad **21.** 2 rad **22.** −3 rad

23–28 Find the exact trigonometric ratios for the angle whose radian measure is given.

23. $\dfrac{3\pi}{4}$ **24.** $\dfrac{4\pi}{3}$ **25.** $\dfrac{9\pi}{2}$

26. -5π **27.** $\dfrac{5\pi}{6}$ **28.** $\dfrac{11\pi}{4}$

29–34 Find the remaining trigonometric ratios.

29. $\sin\theta = \dfrac{3}{5}, \quad 0 < \theta < \dfrac{\pi}{2}$

30. $\tan\alpha = 2, \quad 0 < \alpha < \dfrac{\pi}{2}$

31. $\sec\phi = -1.5, \quad \dfrac{\pi}{2} < \phi < \pi$

32. $\cos x = -\dfrac{1}{3}, \quad \pi < x < \dfrac{3\pi}{2}$

33. $\cot\beta = 3, \quad \pi < \beta < 2\pi$

34. $\csc\theta = -\dfrac{4}{3}, \quad \dfrac{3\pi}{2} < \theta < 2\pi$

35–38 Find, correct to five decimal places, the length of the side labeled x.

35.

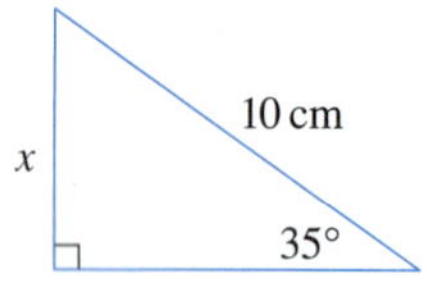

36.

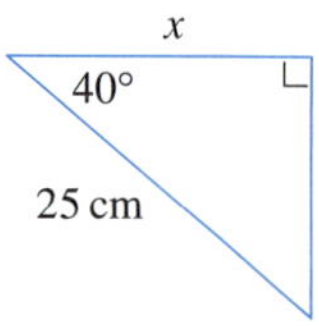

37.

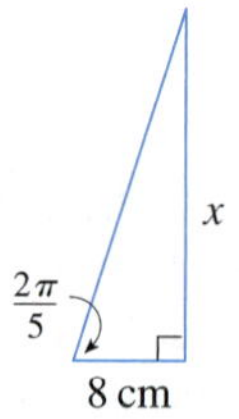

38.

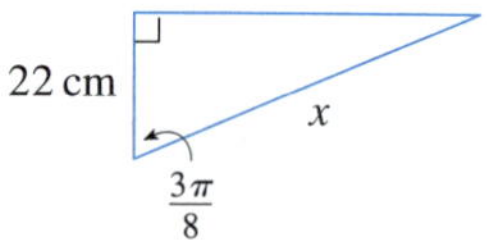

39–41 Prove each equation.

39. (a) Equation 10a (b) Equation 10b

40. (a) Equation 14a (b) Equation 14b

41. (a) Equation 18a (b) Equation 18b
(c) Equation 18c

42–58 Prove the identity.

42. $\cos\left(\dfrac{\pi}{2} - x\right) = \sin x$

43. $\sin\left(\dfrac{\pi}{2} + x\right) = \cos x$

44. $\sin(\pi - x) = \sin x$

45. $\sin\theta \cot\theta = \cos\theta$

46. $(\sin x + \cos x)^2 = 1 + \sin 2x$

47. $\sec y - \cos y = \tan y \sin y$

48. $\tan^2\alpha - \sin^2\alpha = \tan^2\alpha \sin^2\alpha$

49. $\cot^2\theta + \sec^2\theta = \tan^2\theta + \csc^2\theta$

50. $2\csc 2t = \sec t \csc t$

51. $\tan 2\theta = \dfrac{2\tan\theta}{1 - \tan^2\theta}$

52. $\dfrac{1}{1 - \sin\theta} + \dfrac{1}{1 + \sin\theta} = 2\sec^2\theta$

53. $\sin x \sin 2x + \cos x \cos 2x = \cos x$

54. $\sin^2 x - \sin^2 y = \sin(x + y)\sin(x - y)$

55. $\dfrac{\sin\phi}{1 - \cos\phi} = \csc\phi + \cot\phi$

56. $\tan x + \tan y = \dfrac{\sin(x+y)}{\cos x \cos y}$

57. $\sin 3\theta + \sin \theta = 2 \sin 2\theta \cos \theta$

58. $\cos 3\theta = 4 \cos^3\theta - 3 \cos \theta$

59–64 |||| If $\sin x = \frac{1}{3}$ and $\sec y = \frac{5}{4}$, where x and y lie between 0 and $\pi/2$, evaluate the expression.

59. $\sin(x + y)$

60. $\cos(x + y)$

61. $\cos(x - y)$

62. $\sin(x - y)$

63. $\sin 2y$

64. $\cos 2y$

65–72 |||| Find all values of x in the interval $[0, 2\pi]$ that satisfy the equation.

65. $2 \cos x - 1 = 0$

66. $3 \cot^2 x = 1$

67. $2 \sin^2 x = 1$

68. $|\tan x| = 1$

69. $\sin 2x = \cos x$

70. $2 \cos x + \sin 2x = 0$

71. $\sin x = \tan x$

72. $2 + \cos 2x = 3 \cos x$

73–76 |||| Find all values of x in the interval $[0, 2\pi]$ that satisfy the inequality.

73. $\sin x \leqslant \frac{1}{2}$

74. $2 \cos x + 1 > 0$

75. $-1 < \tan x < 1$

76. $\sin x > \cos x$

77–82 |||| Graph the function by starting with the graphs in Figures 13 and 14 and applying the transformations of Section 1.3 where appropriate.

77. $y = \cos\left(x - \dfrac{\pi}{3}\right)$

78. $y = \tan 2x$

79. $y = \dfrac{1}{3} \tan\left(x - \dfrac{\pi}{2}\right)$

80. $y = 1 + \sec x$

81. $y = |\sin x|$

82. $y = 2 + \sin\left(x + \dfrac{\pi}{4}\right)$

83. Prove the **Law of Cosines**: If a triangle has sides with lengths a, b, and c, and θ is the angle between the sides with lengths a and b, then

$$c^2 = a^2 + b^2 - 2ab \cos \theta$$

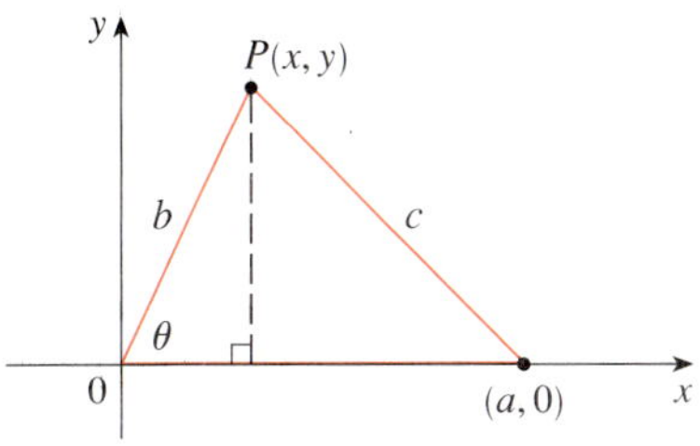

[*Hint:* Introduce a coordinate system so that θ is in standard position as in the figure. Express x and y in terms of θ and then use the distance formula to compute c.]

84. In order to find the distance $|AB|$ across a small inlet, a point C is located as in the figure and the following measurements were recorded:

$$\angle C = 103° \qquad |AC| = 820 \text{ m} \qquad |BC| = 910 \text{ m}$$

Use the Law of Cosines from Exercise 83 to find the required distance.

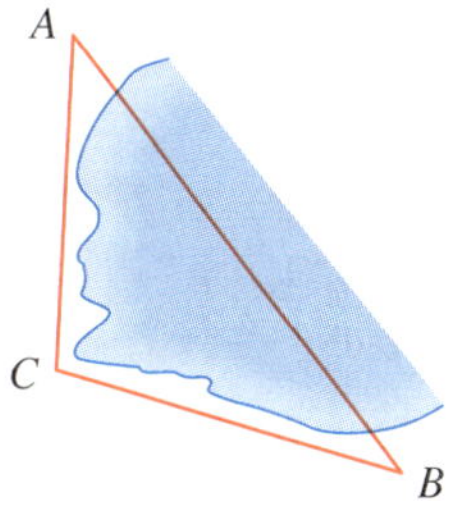

85. Use the figure to prove the subtraction formula

$$\cos(\alpha - \beta) = \cos \alpha \cos \beta + \sin \alpha \sin \beta$$

[*Hint:* Compute c^2 in two ways (using the Law of Cosines from Exercise 83 and also using the distance formula) and compare the two expressions.]

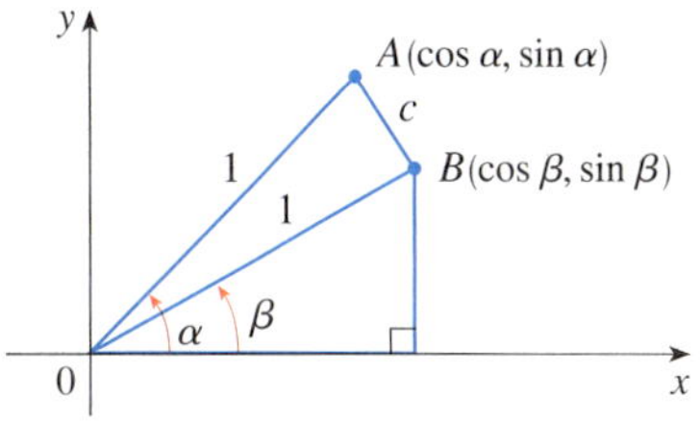

86. Use the formula in Exercise 85 to prove the addition formula for cosine (12b).

87. Use the addition formula for cosine and the identities

$$\cos\left(\frac{\pi}{2} - \theta\right) = \sin \theta \qquad \sin\left(\frac{\pi}{2} - \theta\right) = \cos \theta$$

to prove the subtraction formula for the sine function.

88. Show that the area of a triangle with sides of lengths a and b and with included angle θ is

$$A = \tfrac{1}{2}ab \sin \theta$$

89. Find the area of triangle ABC, correct to five decimal places, if

$$|AB| = 10 \text{ cm} \qquad |BC| = 3 \text{ cm} \qquad \angle ABC = 107°$$

E Sigma Notation

A convenient way of writing sums uses the Greek letter Σ (capital sigma, corresponding to our letter S) and is called **sigma notation**.

This tells us to end with $i = n$.
This tells us to add.
$$\sum_{i=m}^{n} a_i$$
This tells us to start with $i = m$.

1 Definition If $a_m, a_{m+1}, \ldots, a_n$ are real numbers and m and n are integers such that $m \leqslant n$, then

$$\sum_{i=m}^{n} a_i = a_m + a_{m+1} + a_{m+2} + \cdots + a_{n-1} + a_n$$

With function notation, Definition 1 can be written as

$$\sum_{i=m}^{n} f(i) = f(m) + f(m+1) + f(m+2) + \cdots + f(n-1) + f(n)$$

Thus, the symbol $\Sigma_{i=m}^{n}$ indicates a summation in which the letter i (called the **index of summation**) takes on consecutive integer values beginning with m and ending with n, that is, $m, m+1, \ldots, n$. Other letters can also be used as the index of summation.

EXAMPLE 1

(a) $\displaystyle\sum_{i=1}^{4} i^2 = 1^2 + 2^2 + 3^2 + 4^2 = 30$

(b) $\displaystyle\sum_{i=3}^{n} i = 3 + 4 + 5 + \cdots + (n-1) + n$

(c) $\displaystyle\sum_{j=0}^{5} 2^j = 2^0 + 2^1 + 2^2 + 2^3 + 2^4 + 2^5 = 63$

(d) $\displaystyle\sum_{k=1}^{n} \frac{1}{k} = 1 + \frac{1}{2} + \frac{1}{3} + \cdots + \frac{1}{n}$

(e) $\displaystyle\sum_{i=1}^{3} \frac{i-1}{i^2+3} = \frac{1-1}{1^2+3} + \frac{2-1}{2^2+3} + \frac{3-1}{3^2+3} = 0 + \frac{1}{7} + \frac{1}{6} = \frac{13}{42}$

(f) $\displaystyle\sum_{i=1}^{4} 2 = 2 + 2 + 2 + 2 = 8$

EXAMPLE 2 Write the sum $2^3 + 3^3 + \cdots + n^3$ in sigma notation.

SOLUTION There is no unique way of writing a sum in sigma notation. We could write

$$2^3 + 3^3 + \cdots + n^3 = \sum_{i=2}^{n} i^3$$

or
$$2^3 + 3^3 + \cdots + n^3 = \sum_{j=1}^{n-1} (j+1)^3$$

or
$$2^3 + 3^3 + \cdots + n^3 = \sum_{k=0}^{n-2} (k+2)^3$$

The following theorem gives three simple rules for working with sigma notation.

2 Theorem If c is any constant (that is, it does not depend on i), then

(a) $\displaystyle\sum_{i=m}^{n} ca_i = c\sum_{i=m}^{n} a_i$ (b) $\displaystyle\sum_{i=m}^{n} (a_i + b_i) = \sum_{i=m}^{n} a_i + \sum_{i=m}^{n} b_i$

(c) $\displaystyle\sum_{i=m}^{n} (a_i - b_i) = \sum_{i=m}^{n} a_i - \sum_{i=m}^{n} b_i$

Proof To see why these rules are true, all we have to do is write both sides in expanded form. Rule (a) is just the distributive property of real numbers:

$$ca_m + ca_{m+1} + \cdots + ca_n = c(a_m + a_{m+1} + \cdots + a_n)$$

Rule (b) follows from the associative and commutative properties:

$$\begin{aligned}(a_m + b_m) + (a_{m+1} + b_{m+1}) + \cdots + (a_n + b_n)\\ = (a_m + a_{m+1} + \cdots + a_n) + (b_m + b_{m+1} + \cdots + b_n)\end{aligned}$$

Rule (c) is proved similarly.

EXAMPLE 3 Find $\displaystyle\sum_{i=1}^{n} 1$.

SOLUTION

$$\sum_{i=1}^{n} 1 = \underbrace{1 + 1 + \cdots + 1}_{n \text{ terms}} = n$$

EXAMPLE 4 Prove the formula for the sum of the first n positive integers:

$$\sum_{i=1}^{n} i = 1 + 2 + 3 + \cdots + n = \frac{n(n+1)}{2}$$

SOLUTION This formula can be proved by mathematical induction (see page 81) or by the following method used by the German mathematician Karl Friedrich Gauss (1777–1855) when he was ten years old.

Write the sum S twice, once in the usual order and once in reverse order:

$$\begin{aligned}S &= 1 + 2 + 3 + \cdots + (n-1) + n\\ S &= n + (n-1) + (n-2) + \cdots + 2 + 1\end{aligned}$$

Adding all columns vertically, we get

$$2S = (n+1) + (n+1) + (n+1) + \cdots + (n+1) + (n+1)$$

On the right side there are n terms, each of which is $n + 1$, so

$$2S = n(n+1) \qquad \text{or} \qquad S = \frac{n(n+1)}{2}$$

EXAMPLE 5 Prove the formula for the sum of the squares of the first n positive integers:

$$\sum_{i=1}^{n} i^2 = 1^2 + 2^2 + 3^2 + \cdots + n^2 = \frac{n(n+1)(2n+1)}{6}$$

SOLUTION 1 Let S be the desired sum. We start with the *telescoping sum* (or collapsing sum):

Most terms cancel in pairs.

$$\sum_{i=1}^{n} [(1+i)^3 - i^3] = (2^3 - 1^3) + (3^3 - 2^3) + (4^3 - 3^3) + \cdots + [(n+1)^3 - n^3]$$
$$= (n+1)^3 - 1^3 = n^3 + 3n^2 + 3n$$

On the other hand, using Theorem 2 and Examples 3 and 4, we have

$$\sum_{i=1}^{n} [(1+i)^3 - i^3] = \sum_{i=1}^{n} [3i^2 + 3i + 1] = 3\sum_{i=1}^{n} i^2 + 3\sum_{i=1}^{n} i + \sum_{i=1}^{n} 1$$
$$= 3S + 3\frac{n(n+1)}{2} + n = 3S + \tfrac{3}{2}n^2 + \tfrac{5}{2}n$$

Thus, we have

$$n^3 + 3n^2 + 3n = 3S + \tfrac{3}{2}n^2 + \tfrac{5}{2}n$$

Solving this equation for S, we obtain

$$3S = n^3 + \tfrac{3}{2}n^2 + \tfrac{1}{2}n$$

or

$$S = \frac{2n^3 + 3n^2 + n}{6} = \frac{n(n+1)(2n+1)}{6}$$

|||| PRINCIPLE OF MATHEMATICAL INDUCTION

Let S_n be a statement involving the positive integer n. Suppose that

1. S_1 is true.
2. If S_k is true, then S_{k+1} is true.

Then S_n is true for all positive integers n.

SOLUTION 2 Let S_n be the given formula.

1. S_1 is true because $\quad 1^2 = \dfrac{1(1+1)(2\cdot 1+1)}{6}$

2. Assume that S_k is true; that is,

$$1^2 + 2^2 + 3^2 + \cdots + k^2 = \frac{k(k+1)(2k+1)}{6}$$

Then

|||| See pages 81 and 83 for a more thorough discussion of mathematical induction.

$$1^2 + 2^2 + 3^2 + \cdots + (k+1)^2 = (1^2 + 2^2 + 3^2 + \cdots + k^2) + (k+1)^2$$
$$= \frac{k(k+1)(2k+1)}{6} + (k+1)^2$$
$$= (k+1)\frac{k(2k+1) + 6(k+1)}{6}$$
$$= (k+1)\frac{2k^2 + 7k + 6}{6}$$
$$= \frac{(k+1)(k+2)(2k+3)}{6}$$
$$= \frac{(k+1)[(k+1)+1][2(k+1)+1]}{6}$$

So S_{k+1} is true.

By the Principle of Mathematical Induction, S_n is true for all n.

We list the results of Examples 3, 4, and 5 together with a similar result for cubes (see Exercises 37–40) as Theorem 3. These formulas are needed for finding areas and evaluating integrals in Chapter 5.

3 Theorem Let c be a constant and n a positive integer. Then

(a) $\displaystyle\sum_{i=1}^{n} 1 = n$ (b) $\displaystyle\sum_{i=1}^{n} c = nc$

(c) $\displaystyle\sum_{i=1}^{n} i = \frac{n(n+1)}{2}$ (d) $\displaystyle\sum_{i=1}^{n} i^2 = \frac{n(n+1)(2n+1)}{6}$

(e) $\displaystyle\sum_{i=1}^{n} i^3 = \left[\frac{n(n+1)}{2}\right]^2$

EXAMPLE 6 Evaluate $\displaystyle\sum_{i=1}^{n} i(4i^2 - 3)$.

SOLUTION Using Theorems 2 and 3, we have

$$\begin{aligned}
\sum_{i=1}^{n} i(4i^2 - 3) &= \sum_{i=1}^{n} (4i^3 - 3i) = 4\sum_{i=1}^{n} i^3 - 3\sum_{i=1}^{n} i \\
&= 4\left[\frac{n(n+1)}{2}\right]^2 - 3\,\frac{n(n+1)}{2} \\
&= \frac{n(n+1)[2n(n+1) - 3]}{2} \\
&= \frac{n(n+1)(2n^2 + 2n - 3)}{2}
\end{aligned}$$

|||| The type of calculation in Example 7 arises in Chapter 5 when we compute areas.

EXAMPLE 7 Find $\displaystyle\lim_{n\to\infty} \sum_{i=1}^{n} \frac{3}{n}\left[\left(\frac{i}{n}\right)^2 + 1\right]$.

SOLUTION

$$\begin{aligned}
\lim_{n\to\infty} \sum_{i=1}^{n} \frac{3}{n}\left[\left(\frac{i}{n}\right)^2 + 1\right] &= \lim_{n\to\infty} \sum_{i=1}^{n} \left[\frac{3}{n^3} i^2 + \frac{3}{n}\right] \\
&= \lim_{n\to\infty} \left[\frac{3}{n^3}\sum_{i=1}^{n} i^2 + \frac{3}{n}\sum_{i=1}^{n} 1\right] \\
&= \lim_{n\to\infty} \left[\frac{3}{n^3}\,\frac{n(n+1)(2n+1)}{6} + \frac{3}{n}\cdot n\right] \\
&= \lim_{n\to\infty} \left[\frac{1}{2}\cdot\frac{n}{n}\cdot\left(\frac{n+1}{n}\right)\left(\frac{2n+1}{n}\right) + 3\right] \\
&= \lim_{n\to\infty} \left[\frac{1}{2}\cdot 1\left(1 + \frac{1}{n}\right)\left(2 + \frac{1}{n}\right) + 3\right] \\
&= \tfrac{1}{2}\cdot 1\cdot 1\cdot 2 + 3 = 4
\end{aligned}$$

E Exercises

1–10 Write the sum in expanded form.

1. $\sum_{i=1}^{5} \sqrt{i}$

2. $\sum_{i=1}^{6} \frac{1}{i+1}$

3. $\sum_{i=4}^{6} 3^i$

4. $\sum_{i=4}^{6} i^3$

5. $\sum_{k=0}^{4} \frac{2k-1}{2k+1}$

6. $\sum_{k=5}^{8} x^k$

7. $\sum_{i=1}^{n} i^{10}$

8. $\sum_{j=n}^{n+3} j^2$

9. $\sum_{j=0}^{n-1} (-1)^j$

10. $\sum_{i=1}^{n} f(x_i)\,\Delta x_i$

11–20 Write the sum in sigma notation.

11. $1 + 2 + 3 + 4 + \cdots + 10$

12. $\sqrt{3} + \sqrt{4} + \sqrt{5} + \sqrt{6} + \sqrt{7}$

13. $\frac{1}{2} + \frac{2}{3} + \frac{3}{4} + \frac{4}{5} + \cdots + \frac{19}{20}$

14. $\frac{3}{7} + \frac{4}{8} + \frac{5}{9} + \frac{6}{10} + \cdots + \frac{23}{27}$

15. $2 + 4 + 6 + 8 + \cdots + 2n$

16. $1 + 3 + 5 + 7 + \cdots + (2n-1)$

17. $1 + 2 + 4 + 8 + 16 + 32$

18. $\frac{1}{1} + \frac{1}{4} + \frac{1}{9} + \frac{1}{16} + \frac{1}{25} + \frac{1}{36}$

19. $x + x^2 + x^3 + \cdots + x^n$

20. $1 - x + x^2 - x^3 + \cdots + (-1)^n x^n$

21–35 Find the value of the sum.

21. $\sum_{i=4}^{8} (3i - 2)$

22. $\sum_{i=3}^{6} i(i+2)$

23. $\sum_{j=1}^{6} 3^{j+1}$

24. $\sum_{k=0}^{8} \cos k\pi$

25. $\sum_{n=1}^{20} (-1)^n$

26. $\sum_{i=1}^{100} 4$

27. $\sum_{i=0}^{4} (2^i + i^2)$

28. $\sum_{i=-2}^{4} 2^{3-i}$

29. $\sum_{i=1}^{n} 2i$

30. $\sum_{i=1}^{n} (2 - 5i)$

31. $\sum_{i=1}^{n} (i^2 + 3i + 4)$

32. $\sum_{i=1}^{n} (3 + 2i)^2$

33. $\sum_{i=1}^{n} (i+1)(i+2)$

34. $\sum_{i=1}^{n} i(i+1)(i+2)$

35. $\sum_{i=1}^{n} (i^3 - i - 2)$

36. Find the number n such that $\sum_{i=1}^{n} i = 78$.

37. Prove formula (b) of Theorem 3.

38. Prove formula (e) of Theorem 3 using mathematical induction.

39. Prove formula (e) of Theorem 3 using a method similar to that of Example 5, Solution 1 [start with $(1+i)^4 - i^4$].

40. Prove formula (e) of Theorem 3 using the following method published by Abu Bekr Mohammed ibn Alhusain Alkarchi in about A.D. 1010. The figure shows a square $ABCD$ in which sides AB and AD have been divided into segments of lengths 1, 2, 3, . . . , n. Thus, the side of the square has length $n(n+1)/2$ so the area is $[n(n+1)/2]^2$. But the area is also the sum of the areas of the n "gnomons" $G_1, G_2, \ldots, G_n$ shown in the figure. Show that the area of G_i is i^3 and conclude that formula (e) is true.

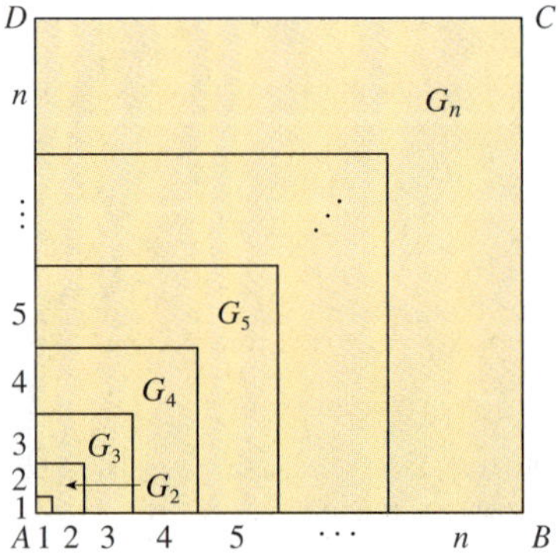

41. Evaluate each telescoping sum.

(a) $\sum_{i=1}^{n} [i^4 - (i-1)^4]$

(b) $\sum_{i=1}^{100} (5^i - 5^{i-1})$

(c) $\sum_{i=3}^{99} \left(\frac{1}{i} - \frac{1}{i+1}\right)$

(d) $\sum_{i=1}^{n} (a_i - a_{i-1})$

42. Prove the generalized triangle inequality

$$\left|\sum_{i=1}^{n} a_i\right| \leqslant \sum_{i=1}^{n} |a_i|$$

43–46 Find each limit.

43. $\lim_{n\to\infty} \sum_{i=1}^{n} \frac{1}{n}\left(\frac{i}{n}\right)^2$

44. $\lim_{n\to\infty} \sum_{i=1}^{n} \frac{1}{n}\left[\left(\frac{i}{n}\right)^3 + 1\right]$

45. $\lim_{n\to\infty} \sum_{i=1}^{n} \frac{2}{n}\left[\left(\frac{2i}{n}\right)^3 + 5\left(\frac{2i}{n}\right)\right]$

46. $\lim_{n\to\infty} \sum_{i=1}^{n} \frac{3}{n}\left[\left(1+\frac{3i}{n}\right)^3 - 2\left(1+\frac{3i}{n}\right)\right]$

47. Prove the formula for the sum of a finite geometric series with first term a and common ratio $r \neq 1$:

$$\sum_{i=1}^{n} ar^{i-1} = a + ar + ar^2 + \cdots + ar^{n-1} = \frac{a(r^n - 1)}{r - 1}$$

48. Evaluate $\sum_{i=1}^{n} \frac{3}{2^{i-1}}$.

49. Evaluate $\sum_{i=1}^{n} (2i + 2^i)$.

50. Evaluate $\sum_{i=1}^{m} \left[\sum_{j=1}^{n} (i + j)\right]$.

F Proofs of Theorems

In this appendix we present proofs of several theorems that are stated in the main body of the text. The sections in which they occur are indicated in the margin.

SECTION 2.3

Limit Laws Suppose that c is a constant and the limits

$$\lim_{x\to a} f(x) = L \quad \text{and} \quad \lim_{x\to a} g(x) = M$$

exist. Then

1. $\lim_{x\to a} [f(x) + g(x)] = L + M$

2. $\lim_{x\to a} [f(x) - g(x)] = L - M$

3. $\lim_{x\to a} [cf(x)] = cL$

4. $\lim_{x\to a} [f(x)g(x)] = LM$

5. $\lim_{x\to a} \frac{f(x)}{g(x)} = \frac{L}{M}$ if $M \neq 0$

Proof of Law 4 Let $\varepsilon > 0$ be given. We want to find $\delta > 0$ such that

$$|f(x)g(x) - LM| < \varepsilon \quad \text{whenever} \quad 0 < |x - a| < \delta$$

In order to get terms that contain $|f(x) - L|$ and $|g(x) - M|$, we add and subtract $Lg(x)$ as follows:

$$\begin{aligned} |f(x)g(x) - LM| &= |f(x)g(x) - Lg(x) + Lg(x) - LM| \\ &= |[f(x) - L]g(x) + L[g(x) - M]| \\ &\leqslant |[f(x) - L]g(x)| + |L[g(x) - M]| \qquad \text{(Triangle Inequality)} \\ &= |f(x) - L||g(x)| + |L||g(x) - M| \end{aligned}$$

We want to make each of these terms less than $\varepsilon/2$.

Since $\lim_{x\to a} g(x) = M$, there is a number $\delta_1 > 0$ such that

$$|g(x) - M| < \frac{\varepsilon}{2(1 + |L|)} \quad \text{whenever} \quad 0 < |x - a| < \delta_1$$

Also, there is a number $\delta_2 > 0$ such that if $0 < |x - a| < \delta_2$, then

$$|g(x) - M| < 1$$

and therefore

$$|g(x)| = |g(x) - M + M| \leqslant |g(x) - M| + |M| < 1 + |M|$$

Since $\lim_{x\to a} f(x) = L$, there is a number $\delta_3 > 0$ such that

$$|f(x) - L| < \frac{\varepsilon}{2(1 + |M|)} \qquad \text{whenever} \qquad 0 < |x - a| < \delta_3$$

Let $\delta = \min\{\delta_1, \delta_2, \delta_3\}$. If $0 < |x - a| < \delta$, then we have $0 < |x - a| < \delta_1$, $0 < |x - a| < \delta_2$, and $0 < |x - a| < \delta_3$, so we can combine the inequalities to obtain

$$\begin{aligned} |f(x)g(x) - LM| &\le |f(x) - L||g(x)| + |L||g(x) - M| \\ &< \frac{\varepsilon}{2(1 + |M|)}(1 + |M|) + |L|\frac{\varepsilon}{2(1 + |L|)} \\ &< \frac{\varepsilon}{2} + \frac{\varepsilon}{2} = \varepsilon \end{aligned}$$

This shows that $\lim_{x\to a} f(x)g(x) = LM$.

Proof of Law 3 If we take $g(x) = c$ in Law 4, we get

$$\begin{aligned} \lim_{x\to a} [cf(x)] &= \lim_{x\to a} [g(x)f(x)] = \lim_{x\to a} g(x) \cdot \lim_{x\to a} f(x) \\ &= \lim_{x\to a} c \cdot \lim_{x\to a} f(x) \\ &= c \lim_{x\to a} f(x) \qquad \text{(by Law 7)} \end{aligned}$$

Proof of Law 2 Using Law 1 and Law 3 with $c = -1$, we have

$$\begin{aligned} \lim_{x\to a} [f(x) - g(x)] &= \lim_{x\to a} [f(x) + (-1)g(x)] = \lim_{x\to a} f(x) + \lim_{x\to a} (-1)g(x) \\ &= \lim_{x\to a} f(x) + (-1)\lim_{x\to a} g(x) = \lim_{x\to a} f(x) - \lim_{x\to a} g(x) \end{aligned}$$

Proof of Law 5 First let us show that

$$\lim_{x\to a} \frac{1}{g(x)} = \frac{1}{M}$$

To do this we must show that, given $\varepsilon > 0$, there exists $\delta > 0$ such that

$$\left|\frac{1}{g(x)} - \frac{1}{M}\right| < \varepsilon \qquad \text{whenever} \qquad 0 < |x - a| < \delta$$

Observe that
$$\left|\frac{1}{g(x)} - \frac{1}{M}\right| = \frac{|M - g(x)|}{|Mg(x)|}$$

We know that we can make the numerator small. But we also need to know that the denominator is not small when x is near a. Since $\lim_{x\to a} g(x) = M$, there is a number $\delta_1 > 0$ such that, whenever $0 < |x - a| < \delta_1$, we have

$$|g(x) - M| < \frac{|M|}{2}$$

and therefore
$$\begin{aligned} |M| &= |M - g(x) + g(x)| \le |M - g(x)| + |g(x)| \\ &< \frac{|M|}{2} + |g(x)| \end{aligned}$$

This shows that

$$|g(x)| > \frac{|M|}{2} \qquad \text{whenever} \qquad 0 < |x - a| < \delta_1$$

and so, for these values of x,

$$\frac{1}{|Mg(x)|} = \frac{1}{|M||g(x)|} < \frac{1}{|M|} \cdot \frac{2}{|M|} = \frac{2}{M^2}$$

Also, there exists $\delta_2 > 0$ such that

$$|g(x) - M| < \frac{M^2}{2}\,\varepsilon \qquad \text{whenever} \qquad 0 < |x - a| < \delta_2$$

Let $\delta = \min\{\delta_1, \delta_2\}$. Then, for $0 < |x - a| < \delta$, we have

$$\left|\frac{1}{g(x)} - \frac{1}{M}\right| = \frac{|M - g(x)|}{|Mg(x)|} < \frac{2}{M^2}\,\frac{M^2}{2}\,\varepsilon = \varepsilon$$

It follows that $\lim_{x \to a} 1/g(x) = 1/M$. Finally, using Law 4, we obtain

$$\lim_{x \to a} \frac{f(x)}{g(x)} = \lim_{x \to a} f(x)\left(\frac{1}{g(x)}\right)$$

$$= \lim_{x \to a} f(x) \lim_{x \to a} \frac{1}{g(x)} = L \cdot \frac{1}{M} = \frac{L}{M}$$

2 Theorem If $f(x) \leqslant g(x)$ for all x in an open interval that contains a (except possibly at a) and

$$\lim_{x \to a} f(x) = L \qquad \text{and} \qquad \lim_{x \to a} g(x) = M$$

then $L \leqslant M$.

Proof We use the method of proof by contradiction. Suppose, if possible, that $L > M$. Law 2 of limits says that

$$\lim_{x \to a} [g(x) - f(x)] = M - L$$

Therefore, for any $\varepsilon > 0$, there exists $\delta > 0$ such that

$$|[g(x) - f(x)] - (M - L)| < \varepsilon \qquad \text{whenever} \qquad 0 < |x - a| < \delta$$

In particular, taking $\varepsilon = L - M$ (noting that $L - M > 0$ by hypothesis), we have a number $\delta > 0$ such that

$$|[g(x) - f(x)] - (M - L)| < L - M \qquad \text{whenever} \qquad 0 < |x - a| < \delta$$

Since $a \leqslant |a|$ for any number a, we have

$$[g(x) - f(x)] - (M - L) < L - M \qquad \text{whenever} \qquad 0 < |x - a| < \delta$$

which simplifies to

$$g(x) < f(x) \qquad \text{whenever} \qquad 0 < |x - a| < \delta$$

But this contradicts $f(x) \leq g(x)$. Thus, the inequality $L > M$ must be false. Therefore, $L \leq M$.

3 The Squeeze Theorem If $f(x) \leq g(x) \leq h(x)$ for all x in an open interval that contains a (except possibly at a) and

$$\lim_{x \to a} f(x) = \lim_{x \to a} h(x) = L$$

then

$$\lim_{x \to a} g(x) = L$$

Proof Let $\varepsilon > 0$ be given. Since $\lim_{x \to a} f(x) = L$, there is a number $\delta_1 > 0$ such that

$$|f(x) - L| < \varepsilon \quad \text{whenever} \quad 0 < |x - a| < \delta_1$$

that is,

$$L - \varepsilon < f(x) < L + \varepsilon \quad \text{whenever} \quad 0 < |x - a| < \delta_1$$

Since $\lim_{x \to a} h(x) = L$, there is a number $\delta_2 > 0$ such that

$$|h(x) - L| < \varepsilon \quad \text{whenever} \quad 0 < |x - a| < \delta_2$$

that is,

$$L - \varepsilon < h(x) < L + \varepsilon \quad \text{whenever} \quad 0 < |x - a| < \delta_2$$

Let $\delta = \min\{\delta_1, \delta_2\}$. If $0 < |x - a| < \delta$, then $0 < |x - a| < \delta_1$ and $0 < |x - a| < \delta_2$, so

$$L - \varepsilon < f(x) \leq g(x) \leq h(x) < L + \varepsilon$$

In particular,

$$L - \varepsilon < g(x) < L + \varepsilon$$

and so $|g(x) - L| < \varepsilon$. Therefore, $\lim_{x \to a} g(x) = L$.

SECTION 2.5

8 Theorem If f is continuous at b and $\lim_{x \to a} g(x) = b$, then

$$\lim_{x \to a} f(g(x)) = f(b)$$

Proof Let $\varepsilon > 0$ be given. We want to find a number $\delta > 0$ such that

$$|f(g(x)) - f(b)| < \varepsilon \quad \text{whenever} \quad 0 < |x - a| < \delta$$

Since f is continuous at b, we have

$$\lim_{y \to b} f(y) = f(b)$$

and so there exists $\delta_1 > 0$ such that

$$|f(y) - f(b)| < \varepsilon \quad \text{whenever} \quad 0 < |y - b| < \delta_1$$

Since $\lim_{x \to a} g(x) = b$, there exists $\delta > 0$ such that

$$|g(x) - b| < \delta_1 \quad \text{whenever} \quad 0 < |x - a| < \delta$$

Combining these two statements, we see that whenever $0 < |x - a| < \delta$ we have $|g(x) - b| < \delta_1$, which implies that $|f(g(x)) - f(b)| < \varepsilon$. Therefore, we have proved that $\lim_{x \to a} f(g(x)) = f(b)$.

SECTION 3.4

The proof of the following result was promised when we proved that $\lim_{\theta \to 0} \frac{\sin \theta}{\theta} = 1$.

Theorem If $0 < \theta < \pi/2$, then $\theta \leqslant \tan \theta$.

Proof Figure 1 shows a sector of a circle with center O, central angle θ, and radius 1. Then

$$|AD| = |OA| \tan \theta = \tan \theta$$

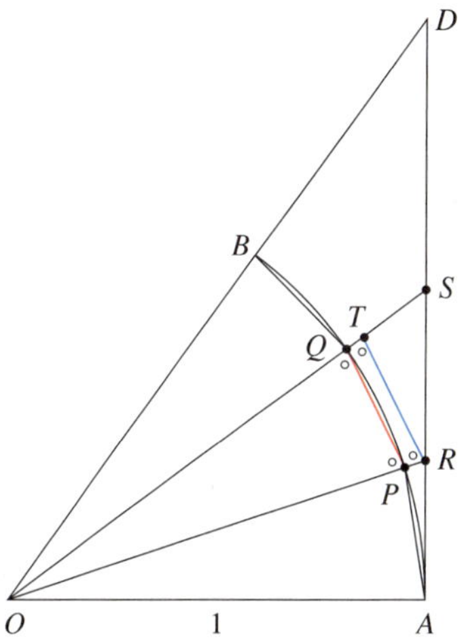

FIGURE 1

We approximate the arc AB by an inscribed polygon consisting of n equal line segments and we look at a typical segment PQ. We extend the lines OP and OQ to meet AD in the points R and S. Then we draw $RT \parallel PQ$ as in Figure 1. Observe that

$$\angle RTO = \angle PQO < 90°$$

and so $\angle RTS > 90°$. Therefore, we have

$$|PQ| < |RT| < |RS|$$

If we add n such inequalities, we get

$$L_n < |AD| = \tan \theta$$

where L_n is the length of the inscribed polygon. Thus, by Theorem 2.3.2, we have

$$\lim_{n \to \infty} L_n \leqslant \tan \theta$$

But the arc length was defined in Equation 8.1.1 as the limit of the lengths of inscribed polygons, so

$$\theta = \lim_{n \to \infty} L_n \leqslant \tan \theta$$

SECTION 4.3

Concavity Test

(a) If $f''(x) > 0$ for all x in I, then the graph of f is concave upward on I.

(b) If $f''(x) < 0$ for all x in I, then the graph of f is concave downward on I.

Proof of (a) Let a be any number in I. We must show that the curve $y = f(x)$ lies above the tangent line at the point $(a, f(a))$. The equation of this tangent is

$$y = f(a) + f'(a)(x - a)$$

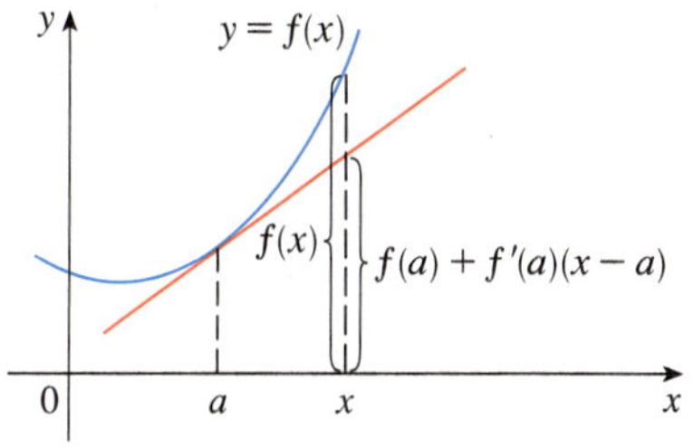

FIGURE 2

So we must show that

$$f(x) > f(a) + f'(a)(x - a)$$

whenever $x \in I$ $(x \neq a)$. (See Figure 2.)

First let us take the case where $x > a$. Applying the Mean Value Theorem to f on the interval $[a, x]$, we get a number c, with $a < c < x$, such that

$$\boxed{1} \qquad f(x) - f(a) = f'(c)(x - a)$$

Since $f'' > 0$ on I we know from the Increasing/Decreasing Test that f' is increasing on I. Thus, since $a < c$, we have

$$f'(a) < f'(c)$$

and so, multiplying this inequality by the positive number $x - a$, we get

$$\boxed{2} \qquad f'(a)(x - a) < f'(c)(x - a)$$

Now we add $f(a)$ to both sides of this inequality:

$$f(a) + f'(a)(x - a) < f(a) + f'(c)(x - a)$$

But from Equation 1 we have $f(x) = f(a) + f'(c)(x - a)$. So this inequality becomes

$$\boxed{3} \qquad f(x) > f(a) + f'(a)(x - a)$$

which is what we wanted to prove.

For the case where $x < a$ we have $f'(c) < f'(a)$, but multiplication by the negative number $x - a$ reverses the inequality, so we get (2) and (3) as before.

SECTION 4.4

See the biographical sketch of Cauchy on page 119.

In order to give the promised proof of l'Hospital's Rule we first need a generalization of the Mean Value Theorem. The following theorem is named after another French mathematician, Augustin-Louis Cauchy (1789–1857).

1 Cauchy's Mean Value Theorem Suppose that the functions f and g are continuous on $[a, b]$ and differentiable on (a, b), and $g'(x) \neq 0$ for all x in (a, b). Then there is a number c in (a, b) such that

$$\frac{f'(c)}{g'(c)} = \frac{f(b) - f(a)}{g(b) - g(a)}$$

Notice that if we take the special case in which $g(x) = x$, then $g'(c) = 1$ and Theorem 1 is just the ordinary Mean Value Theorem. Furthermore, Theorem 1 can be proved in a similar manner. You can verify that all we have to do is change the function h given by Equation 4.2.4 to the function

$$h(x) = f(x) - f(a) - \frac{f(b) - f(a)}{g(b) - g(a)}[g(x) - g(a)]$$

and apply Rolle's Theorem as before.

L'Hospital's Rule Suppose f and g are differentiable and $g'(x) \neq 0$ near a (except possibly at a). Suppose that

$$\lim_{x \to a} f(x) = 0 \quad \text{and} \quad \lim_{x \to a} g(x) = 0$$

or that

$$\lim_{x \to a} f(x) = \pm\infty \quad \text{and} \quad \lim_{x \to a} g(x) = \pm\infty$$

(In other words, we have an indeterminate form of type $\frac{0}{0}$ or ∞/∞.) Then

$$\lim_{x \to a} \frac{f(x)}{g(x)} = \lim_{x \to a} \frac{f'(x)}{g'(x)}$$

if the limit on the right side exists (or is ∞ or $-\infty$).

Proof of L'Hospital's Rule We are assuming that $\lim_{x \to a} f(x) = 0$ and $\lim_{x \to a} g(x) = 0$. Let

$$L = \lim_{x \to a} \frac{f'(x)}{g'(x)}$$

We must show that $\lim_{x \to a} f(x)/g(x) = L$. Define

$$F(x) = \begin{cases} f(x) & \text{if } x \neq a \\ 0 & \text{if } x = a \end{cases} \qquad G(x) = \begin{cases} g(x) & \text{if } x \neq a \\ 0 & \text{if } x = a \end{cases}$$

Then F is continuous on I since f is continuous on $\{x \in I \mid x \neq a\}$ and

$$\lim_{x \to a} F(x) = \lim_{x \to a} f(x) = 0 = F(a)$$

Likewise, G is continuous on I. Let $x \in I$ and $x > a$. Then F and G are continuous on $[a, x]$ and differentiable on (a, x) and $G' \neq 0$ there (since $F' = f'$ and $G' = g'$). Therefore, by Cauchy's Mean Value Theorem there is a number y such that $a < y < x$ and

$$\frac{F'(y)}{G'(y)} = \frac{F(x) - F(a)}{G(x) - G(a)} = \frac{F(x)}{G(x)}$$

Here we have used the fact that, by definition, $F(a) = 0$ and $G(a) = 0$. Now, if we let $x \to a^+$, then $y \to a^+$ (since $a < y < x$), so

$$\lim_{x \to a^+} \frac{f(x)}{g(x)} = \lim_{x \to a^+} \frac{F(x)}{G(x)} = \lim_{y \to a^+} \frac{F'(y)}{G'(y)} = \lim_{y \to a^+} \frac{f'(y)}{g'(y)} = L$$

A similar argument shows that the left-hand limit is also L. Therefore

$$\lim_{x \to a} \frac{f(x)}{g(x)} = L$$

This proves l'Hospital's Rule for the case where a is finite.

If a is infinite, we let $t = 1/x$. Then $t \to 0^+$ as $x \to \infty$, so we have

$$\begin{aligned}\lim_{x\to\infty}\frac{f(x)}{g(x)} &= \lim_{t\to 0^+}\frac{f(1/t)}{g(1/t)}\\ &= \lim_{t\to 0^+}\frac{f'(1/t)(-1/t^2)}{g'(1/t)(-1/t^2)} \qquad \text{(by l'Hospital's Rule for finite } a)\\ &= \lim_{t\to 0^+}\frac{f'(1/t)}{g'(1/t)} = \lim_{x\to\infty}\frac{f'(x)}{g'(x)}\end{aligned}$$

SECTION 11.8

In order to prove Theorem 11.8.3 we first need the following results.

Theorem

1. If a power series $\Sigma\, c_n x^n$ converges when $x = b$ (where $b \neq 0$), then it converges whenever $|x| < |b|$.
2. If a power series $\Sigma\, c_n x^n$ diverges when $x = d$ (where $d \neq 0$), then it diverges whenever $|x| > |d|$.

Proof of 1 Suppose that $\Sigma\, c_n b^n$ converges. Then, by Theorem 11.2.6, we have $\lim_{n\to\infty} c_n b^n = 0$. According to Definition 11.1.2 with $\varepsilon = 1$, there is a positive integer N such that $|c_n b^n| < 1$ whenever $n \geqslant N$. Thus, for $n \geqslant N$, we have

$$|c_n x^n| = \left|\frac{c_n b^n x^n}{b^n}\right| = |c_n b^n|\left|\frac{x}{b}\right|^n < \left|\frac{x}{b}\right|^n$$

If $|x| < |b|$, then $|x/b| < 1$, so $\Sigma\, |x/b|^n$ is a convergent geometric series. Therefore, by the Comparison Test, the series $\Sigma_{n=N}^{\infty} |c_n x^n|$ is convergent. Thus, the series $\Sigma\, c_n x^n$ is absolutely convergent and therefore convergent.

Proof of 2 Suppose that $\Sigma\, c_n d^n$ diverges. If x is any number such that $|x| > |d|$, then $\Sigma\, c_n x^n$ cannot converge because, by part 1, the convergence of $\Sigma\, c_n x^n$ would imply the convergence of $\Sigma\, c_n d^n$. Therefore, $\Sigma\, c_n x^n$ diverges whenever $|x| > |d|$.

Theorem For a power series $\Sigma\, c_n x^n$ there are only three possibilities:

1. The series converges only when $x = 0$.
2. The series converges for all x.
3. There is a positive number R such that the series converges if $|x| < R$ and diverges if $|x| > R$.

Proof Suppose that neither case 1 nor case 2 is true. Then there are nonzero numbers b and d such that $\Sigma\, c_n x^n$ converges for $x = b$ and diverges for $x = d$. Therefore, the set $S = \{x \mid \Sigma\, c_n x^n \text{ converges}\}$ is not empty. By the preceding theorem, the series diverges if $|x| > |d|$, so $|x| \leqslant |d|$ for all $x \in S$. This says that $|d|$ is an upper bound for the set S. Thus, by the Completeness Axiom (see Section 11.1), S has a least upper bound R. If $|x| > R$, then $x \notin S$, so $\Sigma\, c_n x^n$ diverges. If $|x| < R$, then $|x|$ is not an upper bound for S and so there exists $b \in S$ such that $b > |x|$. Since $b \in S$, $\Sigma\, c_n b^n$ converges, so by the preceding theorem $\Sigma\, c_n x^n$ converges.

3 Theorem For a power series $\sum c_n(x - a)^n$ there are only three possibilities:

1. The series converges only when $x = a$.
2. The series converges for all x.
3. There is a positive number R such that the series converges if $|x - a| < R$ and diverges if $|x - a| > R$.

Proof If we make the change of variable $u = x - a$, then the power series becomes $\sum c_n u^n$ and we can apply the preceding theorem to this series. In case 3 we have convergence for $|u| < R$ and divergence for $|u| > R$. Thus, we have convergence for $|x - a| < R$ and divergence for $|x - a| > R$.

SECTION 14.3

Clairaut's Theorem Suppose f is defined on a disk D that contains the point (a, b). If the functions f_{xy} and f_{yx} are both continuous on D, then $f_{xy}(a, b) = f_{yx}(a, b)$.

Proof For small values of h, $h \neq 0$, consider the difference

$$\Delta(h) = [f(a + h, b + h) - f(a + h, b)] - [f(a, b + h) - f(a, b)]$$

Notice that if we let $g(x) = f(x, b + h) - f(x, b)$, then

$$\Delta(h) = g(a + h) - g(a)$$

By the Mean Value Theorem, there is a number c between a and $a + h$ such that

$$g(a + h) - g(a) = g'(c)h = h[f_x(c, b + h) - f_x(c, b)]$$

Applying the Mean Value Theorem again, this time to f_x, we get a number d between b and $b + h$ such that

$$f_x(c, b + h) - f_x(c, b) = f_{xy}(c, d)h$$

Combining these equations, we obtain

$$\Delta(h) = h^2 f_{xy}(c, d)$$

If $h \to 0$, then $(c, d) \to (a, b)$, so the continuity of f_{xy} at (a, b) gives

$$\lim_{h \to 0} \frac{\Delta(h)}{h^2} = \lim_{(c, d) \to (a, b)} f_{xy}(c, d) = f_{xy}(a, b)$$

Similarly, by writing

$$\Delta(h) = [f(a + h, b + h) - f(a, b + h)] - [f(a + h, b) - f(a, b)]$$

and using the Mean Value Theorem twice and the continuity of f_{yx} at (a, b), we obtain

$$\lim_{h \to 0} \frac{\Delta(h)}{h^2} = f_{yx}(a, b)$$

It follows that $f_{xy}(a, b) = f_{yx}(a, b)$.

SECTION 14.4

> **8 Theorem** If the partial derivatives f_x and f_y exist near (a, b) and are continuous at (a, b), then f is differentiable at (a, b).

Proof Let

$$\Delta z = f(a + \Delta x, b + \Delta y) - f(a, b)$$

According to (14.4.7), to prove that f is differentiable at (a, b) we have to show that we can write Δz in the form

$$\Delta z = f_x(a, b)\,\Delta x + f_y(a, b)\,\Delta y + \varepsilon_1\,\Delta x + \varepsilon_2\,\Delta y$$

where ε_1 and $\varepsilon_2 \to 0$ as $(\Delta x,\ \Delta y) \to (0, 0)$.

Referring to Figure 3, we write

$$\boxed{1} \quad \Delta z = [f(a + \Delta x, b + \Delta y) - f(a, b + \Delta y)] + [f(a, b + \Delta y) - f(a, b)]$$

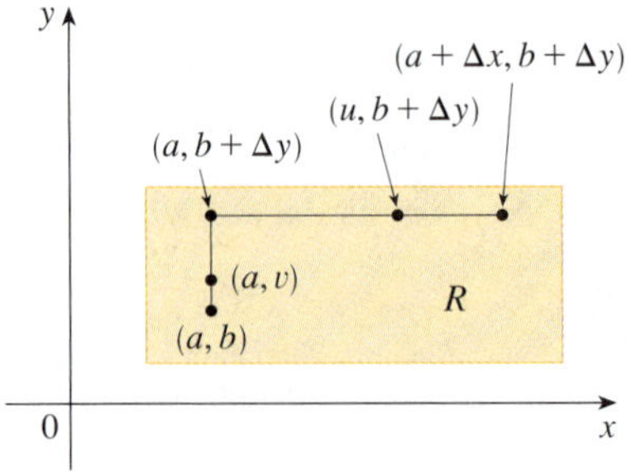

FIGURE 3

Observe that the function of a single variable

$$g(x) = f(x, b + \Delta y)$$

is defined on the interval $[a, a + \Delta x]$ and $g'(x) = f_x(x, b + \Delta y)$. If we apply the Mean Value Theorem to g, we get

$$g(a + \Delta x) - g(a) = g'(u)\,\Delta x$$

where u is some number between a and $a + \Delta x$. In terms of f, this equation becomes

$$f(a + \Delta x, b + \Delta y) - f(a, b + \Delta y) = f_x(u, b + \Delta y)\,\Delta x$$

This gives us an expression for the first part of the right side of Equation 1. For the second part we let $h(y) = f(a, y)$. Then h is a function of a single variable defined on the interval $[b, b + \Delta y]$ and $h'(y) = f_y(a, y)$. A second application of the Mean Value Theorem then gives

$$h(b + \Delta y) - h(b) = h'(v)\,\Delta y$$

where v is some number between b and $b + \Delta y$. In terms of f, this becomes

$$f(a, b + \Delta y) - f(a, b) = f_y(a, v)\,\Delta y$$

We now substitute these expressions into Equation 3 and obtain

$$\begin{aligned}\Delta z &= f_x(u, b + \Delta y)\,\Delta x + f_y(a, v)\,\Delta y \\ &= f_x(a, b)\,\Delta x + [f_x(u, b + \Delta y) - f_x(a, b)]\,\Delta x + f_y(a, b)\,\Delta y \\ &\qquad + [f_y(a, v) - f_y(a, b)]\,\Delta y \\ &= f_x(a, b)\,\Delta x + f_y(a, b)\,\Delta y + \varepsilon_1\,\Delta x + \varepsilon_2\,\Delta y\end{aligned}$$

where

$$\varepsilon_1 = f_x(u, b + \Delta y) - f_x(a, b)$$

$$\varepsilon_2 = f_y(a, v) - f_y(a, b)$$

Since $(u, b + \Delta y) \to (a, b)$ and $(a, v) \to (a, b)$ as $(\Delta x, \Delta y) \to (0, 0)$ and since f_x and f_y are continuous at (a, b), we see that $\varepsilon_1 \to 0$ and $\varepsilon_2 \to 0$ as $(\Delta x, \Delta y) \to (0, 0)$.

Therefore, f is differentiable at (a, b).

G Complex Numbers

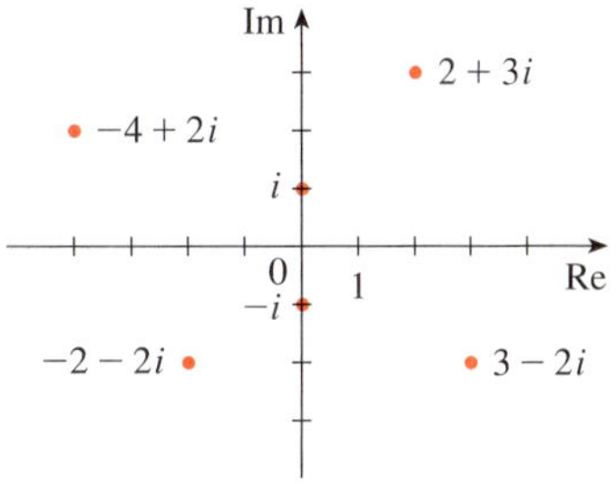

FIGURE 1
Complex numbers as points in the Argand plane

A **complex number** can be represented by an expression of the form $a + bi$, where a and b are real numbers and i is a symbol with the property that $i^2 = -1$. The complex number $a + bi$ can also be represented by the ordered pair (a, b) and plotted as a point in a plane (called the Argand plane) as in Figure 1. Thus, the complex number $i = 0 + 1 \cdot i$ is identified with the point $(0, 1)$.

The **real part** of the complex number $a + bi$ is the real number a and the **imaginary part** is the real number b. Thus, the real part of $4 - 3i$ is 4 and the imaginary part is -3. Two complex numbers $a + bi$ and $c + di$ are **equal** if $a = c$ and $b = d$, that is, their real parts are equal and their imaginary parts are equal. In the Argand plane the horizontal axis is called the real axis and the vertical axis is called the imaginary axis.

The sum and difference of two complex numbers are defined by adding or subtracting their real parts and their imaginary parts:

$$(a + bi) + (c + di) = (a + c) + (b + d)i$$

$$(a + bi) - (c + di) = (a - c) + (b - d)i$$

For instance,

$$(1 - i) + (4 + 7i) = (1 + 4) + (-1 + 7)i = 5 + 6i$$

The product of complex numbers is defined so that the usual commutative and distributive laws hold:

$$\begin{aligned}(a + bi)(c + di) &= a(c + di) + (bi)(c + di) \\ &= ac + adi + bci + bdi^2\end{aligned}$$

Since $i^2 = -1$, this becomes

$$(a + bi)(c + di) = (ac - bd) + (ad + bc)i$$

EXAMPLE 1

$$\begin{aligned}(-1 + 3i)(2 - 5i) &= (-1)(2 - 5i) + 3i(2 - 5i) \\ &= -2 + 5i + 6i - 15(-1) = 13 + 11i\end{aligned}$$

Division of complex numbers is much like rationalizing the denominator of a rational expression. For the complex number $z = a + bi$, we define its **complex conjugate** to be $\bar{z} = a - bi$. To find the quotient of two complex numbers we multiply numerator and denominator by the complex conjugate of the denominator.

EXAMPLE 2 Express the number $\dfrac{-1 + 3i}{2 + 5i}$ in the form $a + bi$.

SOLUTION We multiply numerator and denominator by the complex conjugate of $2 + 5i$, namely $2 - 5i$, and we take advantage of the result of Example 1:

$$\frac{-1 + 3i}{2 + 5i} = \frac{-1 + 3i}{2 + 5i} \cdot \frac{2 - 5i}{2 - 5i} = \frac{13 + 11i}{2^2 + 5^2} = \frac{13}{29} + \frac{11}{29}i$$

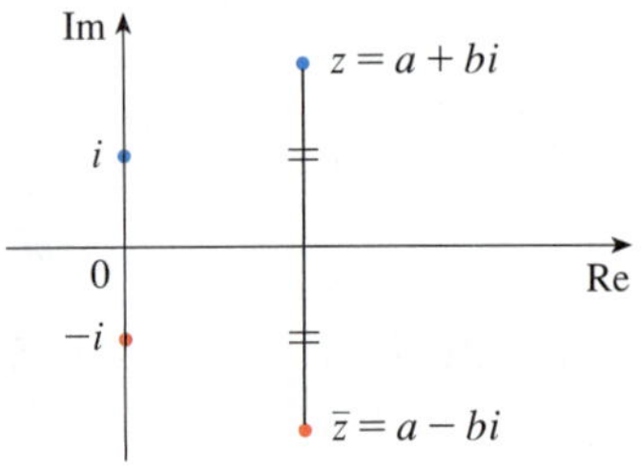

FIGURE 2

The geometric interpretation of the complex conjugate is shown in Figure 2: $\bar{z}$ is the reflection of z in the real axis. We list some of the properties of the complex conjugate in the following box. The proofs follow from the definition and are requested in Exercise 18.

Properties of Conjugates

$$\overline{z + w} = \bar{z} + \bar{w} \qquad \overline{zw} = \bar{z}\,\bar{w} \qquad \overline{z^n} = \bar{z}^n$$

The **modulus**, or **absolute value**, $|z|$ of a complex number $z = a + bi$ is its distance from the origin. From Figure 3 we see that if $z = a + bi$, then

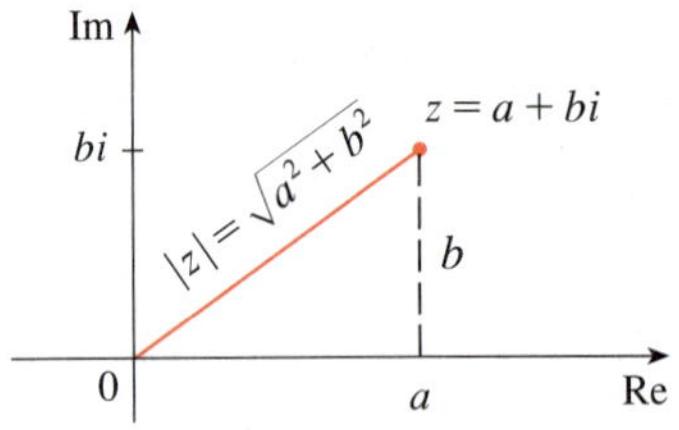

FIGURE 3

$$|z| = \sqrt{a^2 + b^2}$$

Notice that

$$z\bar{z} = (a + bi)(a - bi) = a^2 + abi - abi - b^2i^2 = a^2 + b^2$$

and so

$$z\bar{z} = |z|^2$$

This explains why the division procedure in Example 2 works in general:

$$\frac{z}{w} = \frac{z\bar{w}}{w\bar{w}} = \frac{z\bar{w}}{|w|^2}$$

Since $i^2 = -1$, we can think of i as a square root of -1. But notice that we also have $(-i)^2 = i^2 = -1$ and so $-i$ is also a square root of -1. We say that i is the **principal square root** of -1 and write $\sqrt{-1} = i$. In general, if c is any positive number, we write

$$\sqrt{-c} = \sqrt{c}\, i$$

With this convention, the usual derivation and formula for the roots of the quadratic equation $ax^2 + bx + c = 0$ are valid even when $b^2 - 4ac < 0$:

$$x = \frac{-b \pm \sqrt{b^2 - 4ac}}{2a}$$

EXAMPLE 3 Find the roots of the equation $x^2 + x + 1 = 0$.

SOLUTION Using the quadratic formula, we have

$$x = \frac{-1 \pm \sqrt{1^2 - 4 \cdot 1}}{2} = \frac{-1 \pm \sqrt{-3}}{2} = \frac{-1 \pm \sqrt{3}i}{2}$$

We observe that the solutions of the equation in Example 3 are complex conjugates of each other. In general, the solutions of any quadratic equation $ax^2 + bx + c = 0$ with real coefficients a, b, and c are always complex conjugates. (If z is real, $\overline{z} = z$, so z is its own conjugate.)

We have seen that if we allow complex numbers as solutions, then every quadratic equation has a solution. More generally, it is true that every polynomial equation

$$a_n x^n + a_{n-1} x^{n-1} + \cdots + a_1 x + a_0 = 0$$

of degree at least one has a solution among the complex numbers. This fact is known as the Fundamental Theorem of Algebra and was proved by Gauss.

Polar Form

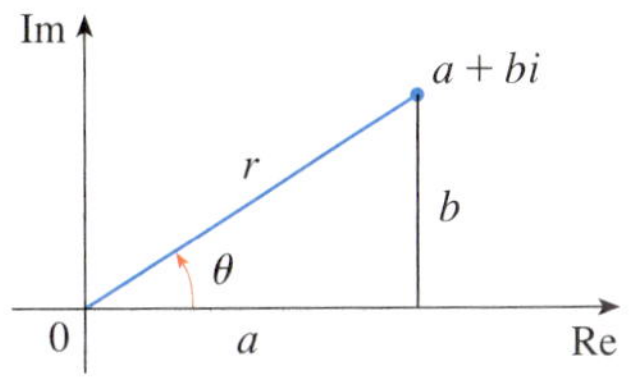

FIGURE 4

We know that any complex number $z = a + bi$ can be considered as a point (a, b) and that any such point can be represented by polar coordinates (r, θ) with $r \geqslant 0$. In fact,

$$a = r\cos\theta \qquad b = r\sin\theta$$

as in Figure 4. Therefore, we have

$$z = a + bi = (r\cos\theta) + (r\sin\theta)i$$

Thus, we can write any complex number z in the form

$$z = r(\cos\theta + i\sin\theta)$$

where $\quad r = |z| = \sqrt{a^2 + b^2} \quad$ and $\quad \tan\theta = \dfrac{b}{a}$

The angle θ is called the **argument** of z and we write $\theta = \arg(z)$. Note that $\arg(z)$ is not unique; any two arguments of z differ by an integer multiple of 2π.

EXAMPLE 4 Write the following numbers in polar form.

(a) $z = 1 + i$ (b) $w = \sqrt{3} - i$

SOLUTION

(a) We have $r = |z| = \sqrt{1^2 + 1^2} = \sqrt{2}$ and $\tan\theta = 1$, so we can take $\theta = \pi/4$. Therefore, the polar form is

$$z = \sqrt{2}\left(\cos\frac{\pi}{4} + i\sin\frac{\pi}{4}\right)$$

(b) Here we have $r = |w| = \sqrt{3 + 1} = 2$ and $\tan\theta = -1/\sqrt{3}$. Since w lies in the

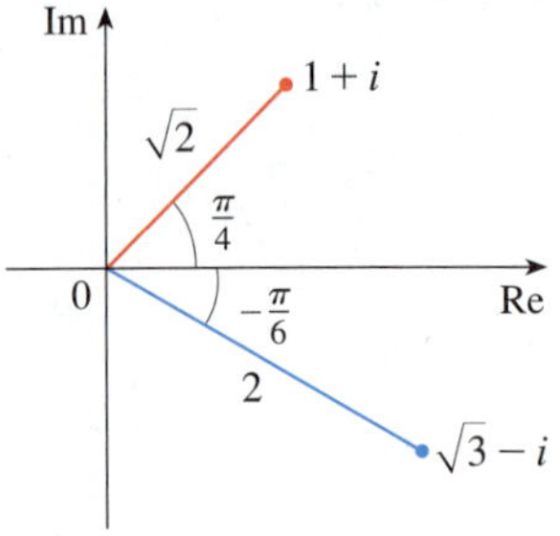

FIGURE 5

fourth quadrant, we take $\theta = -\pi/6$ and

$$w = 2\left[\cos\left(-\frac{\pi}{6}\right) + i\sin\left(-\frac{\pi}{6}\right)\right]$$

The numbers z and w are shown in Figure 5.

The polar form of complex numbers gives insight into multiplication and division. Let

$$z_1 = r_1(\cos\theta_1 + i\sin\theta_1) \qquad z_2 = r_2(\cos\theta_2 + i\sin\theta_2)$$

be two complex numbers written in polar form. Then

$$\begin{aligned} z_1 z_2 &= r_1 r_2(\cos\theta_1 + i\sin\theta_1)(\cos\theta_2 + i\sin\theta_2) \\ &= r_1 r_2[(\cos\theta_1\cos\theta_2 - \sin\theta_1\sin\theta_2) + i(\sin\theta_1\cos\theta_2 + \cos\theta_1\sin\theta_2)] \end{aligned}$$

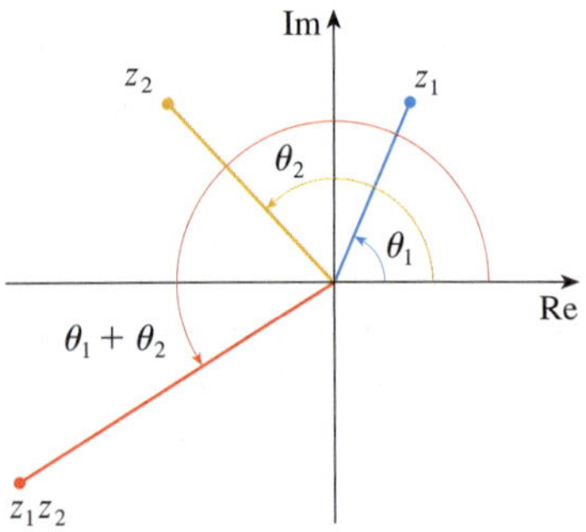

FIGURE 6

Therefore, using the addition formulas for cosine and sine, we have

1 $$z_1 z_2 = r_1 r_2[\cos(\theta_1 + \theta_2) + i\sin(\theta_1 + \theta_2)]$$

This formula says that *to multiply two complex numbers we multiply the moduli and add the arguments.* (See Figure 6.)

A similar argument using the subtraction formulas for sine and cosine shows that *to divide two complex numbers we divide the moduli and subtract the arguments.*

$$\frac{z_1}{z_2} = \frac{r_1}{r_2}[\cos(\theta_1 - \theta_2) + i\sin(\theta_1 - \theta_2)] \qquad z_2 \neq 0$$

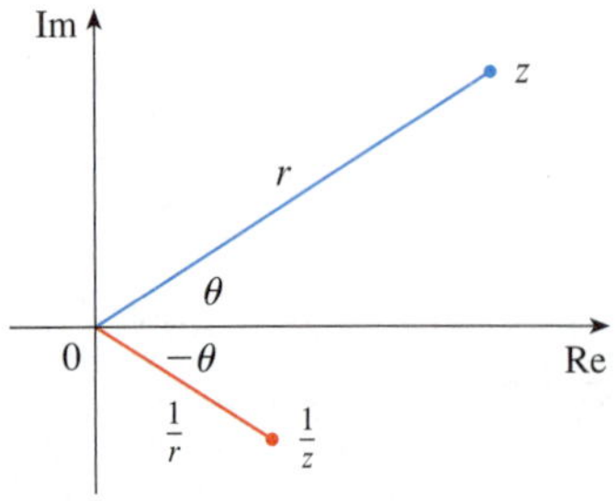

FIGURE 7

In particular, taking $z_1 = 1$ and $z_2 = z$ (and therefore $\theta_1 = 0$ and $\theta_2 = \theta$), we have the following, which is illustrated in Figure 7.

If $\quad z = r(\cos\theta + i\sin\theta),\quad$ then $\quad \dfrac{1}{z} = \dfrac{1}{r}(\cos\theta - i\sin\theta).$

EXAMPLE 5 Find the product of the complex numbers $1 + i$ and $\sqrt{3} - i$ in polar form.

SOLUTION From Example 4 we have

$$1 + i = \sqrt{2}\left(\cos\frac{\pi}{4} + i\sin\frac{\pi}{4}\right)$$

and

$$\sqrt{3} - i = 2\left[\cos\left(-\frac{\pi}{6}\right) + i\sin\left(-\frac{\pi}{6}\right)\right]$$

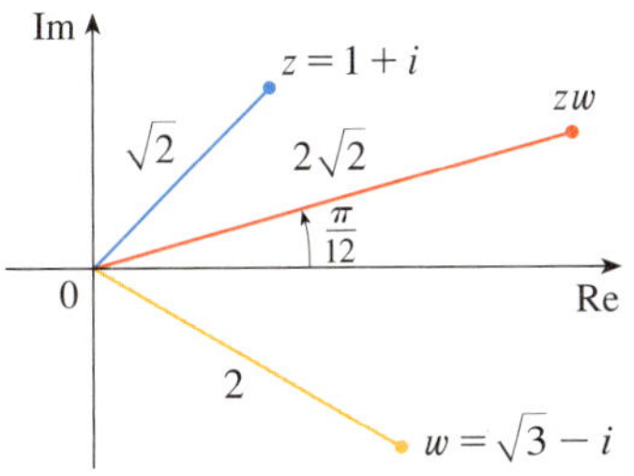

FIGURE 8

So, by Equation 1,

$$(1+i)(\sqrt{3}-i) = 2\sqrt{2}\left[\cos\left(\frac{\pi}{4}-\frac{\pi}{6}\right) + i\sin\left(\frac{\pi}{4}-\frac{\pi}{6}\right)\right]$$
$$= 2\sqrt{2}\left(\cos\frac{\pi}{12} + i\sin\frac{\pi}{12}\right)$$

This is illustrated in Figure 8.

Repeated use of Formula 1 shows how to compute powers of a complex number. If

$$z = r(\cos\theta + i\sin\theta)$$

then $$z^2 = r^2(\cos 2\theta + i\sin 2\theta)$$

and $$z^3 = zz^2 = r^3(\cos 3\theta + i\sin 3\theta)$$

In general, we obtain the following result, which is named after the French mathematician Abraham De Moivre (1667–1754).

2 De Moivre's Theorem If $z = r(\cos\theta + i\sin\theta)$ and n is a positive integer, then

$$z^n = [r(\cos\theta + i\sin\theta)]^n = r^n(\cos n\theta + i\sin n\theta)$$

This says that *to take the nth power of a complex number we take the nth power of the modulus and multiply the argument by n.*

EXAMPLE 6 Find $\left(\frac{1}{2}+\frac{1}{2}i\right)^{10}$.

SOLUTION Since $\frac{1}{2}+\frac{1}{2}i = \frac{1}{2}(1+i)$, it follows from Example 4(a) that $\frac{1}{2}+\frac{1}{2}i$ has the polar form

$$\frac{1}{2}+\frac{1}{2}i = \frac{\sqrt{2}}{2}\left(\cos\frac{\pi}{4} + i\sin\frac{\pi}{4}\right)$$

So by De Moivre's Theorem,

$$\left(\frac{1}{2}+\frac{1}{2}i\right)^{10} = \left(\frac{\sqrt{2}}{2}\right)^{10}\left(\cos\frac{10\pi}{4} + i\sin\frac{10\pi}{4}\right)$$
$$= \frac{2^5}{2^{10}}\left(\cos\frac{5\pi}{2} + i\sin\frac{5\pi}{2}\right) = \frac{1}{32}i$$

De Moivre's Theorem can also be used to find the nth roots of complex numbers. An nth root of the complex number z is a complex number w such that

$$w^n = z$$

Writing these two numbers in trigonometric form as

$$w = s(\cos\phi + i\sin\phi) \qquad \text{and} \qquad z = r(\cos\theta + i\sin\theta)$$

and using De Moivre's Theorem, we get

$$s^n(\cos n\phi + i \sin n\phi) = r(\cos \theta + i \sin \theta)$$

The equality of these two complex numbers shows that

$$s^n = r \qquad \text{or} \qquad s = r^{1/n}$$

and $$\cos n\phi = \cos \theta \qquad \text{and} \qquad \sin n\phi = \sin \theta$$

From the fact that sine and cosine have period 2π it follows that

$$n\phi = \theta + 2k\pi \qquad \text{or} \qquad \phi = \frac{\theta + 2k\pi}{n}$$

Thus $$w = r^{1/n}\left[\cos\left(\frac{\theta + 2k\pi}{n}\right) + i \sin\left(\frac{\theta + 2k\pi}{n}\right)\right]$$

Since this expression gives a different value of w for $k = 0, 1, 2, \ldots, n - 1$, we have the following.

3 Roots of a Complex Number Let $z = r(\cos \theta + i \sin \theta)$ and let n be a positive integer. Then z has the n distinct nth roots

$$w_k = r^{1/n}\left[\cos\left(\frac{\theta + 2k\pi}{n}\right) + i \sin\left(\frac{\theta + 2k\pi}{n}\right)\right]$$

where $k = 0, 1, 2, \ldots, n - 1$.

Notice that each of the nth roots of z has modulus $|w_k| = r^{1/n}$. Thus, all the nth roots of z lie on the circle of radius $r^{1/n}$ in the complex plane. Also, since the argument of each successive nth root exceeds the argument of the previous root by $2\pi/n$, we see that the nth roots of z are equally spaced on this circle.

EXAMPLE 7 Find the six sixth roots of $z = -8$ and graph these roots in the complex plane.

SOLUTION In trigonometric form, $z = 8(\cos \pi + i \sin \pi)$. Applying Equation 3 with $n = 6$, we get

$$w_k = 8^{1/6}\left(\cos \frac{\pi + 2k\pi}{6} + i \sin \frac{\pi + 2k\pi}{6}\right)$$

We get the six sixth roots of -8 by taking $k = 0, 1, 2, 3, 4, 5$ in this formula:

$$w_0 = 8^{1/6}\left(\cos \frac{\pi}{6} + i \sin \frac{\pi}{6}\right) = \sqrt{2}\left(\frac{\sqrt{3}}{2} + \frac{1}{2}i\right)$$

$$w_1 = 8^{1/6}\left(\cos \frac{\pi}{2} + i \sin \frac{\pi}{2}\right) = \sqrt{2}\, i$$

$$w_2 = 8^{1/6}\left(\cos \frac{5\pi}{6} + i \sin \frac{5\pi}{6}\right) = \sqrt{2}\left(-\frac{\sqrt{3}}{2} + \frac{1}{2}i\right)$$

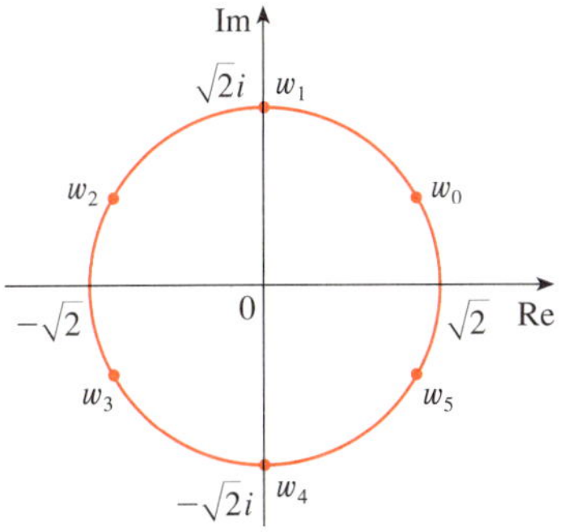

FIGURE 9
The six sixth roots of $z = -8$

$$w_3 = 8^{1/6}\left(\cos\frac{7\pi}{6} + i\sin\frac{7\pi}{6}\right) = \sqrt{2}\left(-\frac{\sqrt{3}}{2} - \frac{1}{2}i\right)$$

$$w_4 = 8^{1/6}\left(\cos\frac{3\pi}{2} + i\sin\frac{3\pi}{2}\right) = -\sqrt{2}\,i$$

$$w_5 = 8^{1/6}\left(\cos\frac{11\pi}{6} + i\sin\frac{11\pi}{6}\right) = \sqrt{2}\left(\frac{\sqrt{3}}{2} - \frac{1}{2}i\right)$$

All these points lie on the circle of radius $\sqrt{2}$ as shown in Figure 9.

Complex Exponentials

We also need to give a meaning to the expression e^z when $z = x + iy$ is a complex number. The theory of infinite series as developed in Chapter 11 can be extended to the case where the terms are complex numbers. Using the Taylor series for e^x (11.10.11) as our guide, we define

4 $$e^z = \sum_{n=0}^{\infty} \frac{z^n}{n!} = 1 + z + \frac{z^2}{2!} + \frac{z^3}{3!} + \cdots$$

and it turns out that this complex exponential function has the same properties as the real exponential function. In particular, it is true that

5 $$e^{z_1+z_2} = e^{z_1}e^{z_2}$$

If we put $z = iy$, where y is a real number, in Equation 4, and use the facts that

$$i^2 = -1, \quad i^3 = i^2 i = -i, \quad i^4 = 1, \quad i^5 = i, \quad \ldots$$

we get

$$\begin{aligned} e^{iy} &= 1 + iy + \frac{(iy)^2}{2!} + \frac{(iy)^3}{3!} + \frac{(iy)^4}{4!} + \frac{(iy)^5}{5!} + \cdots \\ &= 1 + iy - \frac{y^2}{2!} - i\frac{y^3}{3!} + \frac{y^4}{4!} + i\frac{y^5}{5!} + \cdots \\ &= \left(1 - \frac{y^2}{2!} + \frac{y^4}{4!} - \frac{y^6}{6!} + \cdots\right) + i\left(y - \frac{y^3}{3!} + \frac{y^5}{5!} - \cdots\right) \\ &= \cos y + i\sin y \end{aligned}$$

Here we have used the Taylor series for $\cos y$ and $\sin y$ (Equations 11.10.16 and 11.10.15). The result is a famous formula called **Euler's formula**:

6 $$e^{iy} = \cos y + i\sin y$$

Combining Euler's formula with Equation 5, we get

7 $$e^{x+iy} = e^x e^{iy} = e^x(\cos y + i\sin y)$$

EXAMPLE 8 Evaluate: (a) $e^{i\pi}$ (b) $e^{-1+i\pi/2}$

|||| We could write the result of Example 8(a) as

$$e^{i\pi} + 1 = 0$$

This equation relates the five most famous numbers in all of mathematics: 0, 1, e, i, and π.

SOLUTION

(a) From Euler's equation (6) we have

$$e^{i\pi} = \cos \pi + i \sin \pi = -1 + i(0) = -1$$

(b) Using Equation 7 we get

$$e^{-1+i\pi/2} = e^{-1}\left(\cos \frac{\pi}{2} + i \sin \frac{\pi}{2}\right) = \frac{1}{e}[0 + i(1)] = \frac{i}{e}$$

Finally, we note that Euler's equation provides us with an easier method of proving De Moivre's Theorem:

$$[r(\cos \theta + i \sin \theta)]^n = (re^{i\theta})^n = r^n e^{in\theta} = r^n(\cos n\theta + i \sin n\theta)$$

G Exercises

1–14 |||| Evaluate the expression and write your answer in the form $a + bi$.

1. $(5 - 6i) + (3 + 2i)$

2. $\left(4 - \frac{1}{2}i\right) - \left(9 + \frac{5}{2}i\right)$

3. $(2 + 5i)(4 - i)$

4. $(1 - 2i)(8 - 3i)$

5. $\overline{12 + 7i}$

6. $\overline{2i\left(\frac{1}{2} - i\right)}$

7. $\dfrac{1 + 4i}{3 + 2i}$

8. $\dfrac{3 + 2i}{1 - 4i}$

9. $\dfrac{1}{1 + i}$

10. $\dfrac{3}{4 - 3i}$

11. i^3

12. i^{100}

13. $\sqrt{-25}$

14. $\sqrt{-3}\sqrt{-12}$

15–17 |||| Find the complex conjugate and the modulus of the number.

15. $12 - 5i$

16. $-1 + 2\sqrt{2}\,i$

17. $-4i$

18. Prove the following properties of complex numbers.
(a) $\overline{z + w} = \overline{z} + \overline{w}$
(b) $\overline{zw} = \overline{z}\,\overline{w}$
(c) $\overline{z^n} = \overline{z}^n$, where n is a positive integer
[*Hint:* Write $z = a + bi$, $w = c + di$.]

19–24 |||| Find all solutions of the equation.

19. $4x^2 + 9 = 0$

20. $x^4 = 1$

21. $x^2 + 2x + 5 = 0$

22. $2x^2 - 2x + 1 = 0$

23. $z^2 + z + 2 = 0$

24. $z^2 + \frac{1}{2}z + \frac{1}{4} = 0$

25–28 |||| Write the number in polar form with argument between 0 and 2π.

25. $-3 + 3i$

26. $1 - \sqrt{3}i$

27. $3 + 4i$

28. $8i$

29–32 |||| Find polar forms for zw, z/w, and $1/z$ by first putting z and w into polar form.

29. $z = \sqrt{3} + i$, $w = 1 + \sqrt{3}i$

30. $z = 4\sqrt{3} - 4i$, $w = 8i$

31. $z = 2\sqrt{3} - 2i$, $w = -1 + i$

32. $z = 4(\sqrt{3} + i)$, $w = -3 - 3i$

33–36 |||| Find the indicated power using De Moivre's Theorem.

33. $(1 + i)^{20}$

34. $(1 - \sqrt{3}i)^5$

35. $(2\sqrt{3} + 2i)^5$

36. $(1 - i)^8$

37–40 |||| Find the indicated roots. Sketch the roots in the complex plane.

37. The eighth roots of 1

38. The fifth roots of 32

39. The cube roots of i

40. The cube roots of $1 + i$

41–46 |||| Write the number in the form $a + bi$.

41. $e^{i\pi/2}$

42. $e^{2\pi i}$

43. $e^{i\pi/3}$

44. $e^{-i\pi}$

45. $e^{2+i\pi}$

46. $e^{\pi+i}$

47. If $u(x) = f(x) + ig(x)$ is a complex-valued function of a real variable x and the real and imaginary parts $f(x)$ and $g(x)$ are differentiable functions of x, then the derivative of u is defined to be $u'(x) = f'(x) + ig'(x)$. Use this together with Equation 7 to prove that if $F(x) = e^{rx}$, then $F'(x) = re^{rx}$ when $r = a + bi$ is a complex number.

48. (a) If u is a complex-valued function of a real variable, its indefinite integral $\int u(x)\,dx$ is an antiderivative of u. Evaluate

$$\int e^{(1+i)x}\,dx$$

(b) By considering the real and imaginary parts of the integral in part (a), evaluate the real integrals

$$\int e^x \cos x\,dx \quad \text{and} \quad \int e^x \sin x\,dx$$

(c) Compare with the method used in Example 4 in Section 7.1.

H Answers to Odd-Numbered Exercises

Chapter 1

Exercises 1.1 ▫ page 22

1. (a) -2 (b) 2.8 (c) $-3, 1$ (d) $-2.5, 0.3$
(e) $[-3, 3], [-2, 3]$ (f) $[-1, 3]$
3. $[-85, 115], [-325, 485], [-210, 200]$
5. No **7.** Yes, $[-3, 2], [-3, -2) \cup [-1, 3]$
9. Diet, exercise, or illness
11.

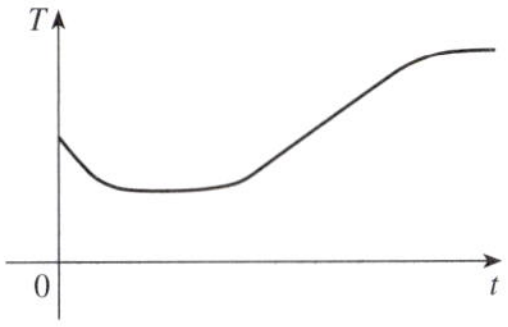

13.

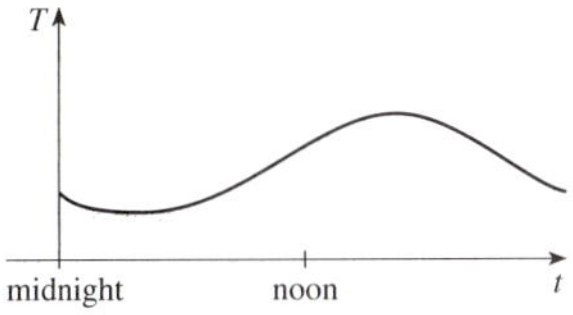

15.

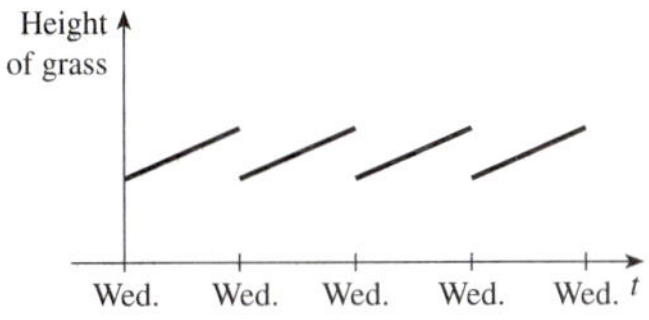

17. (a)

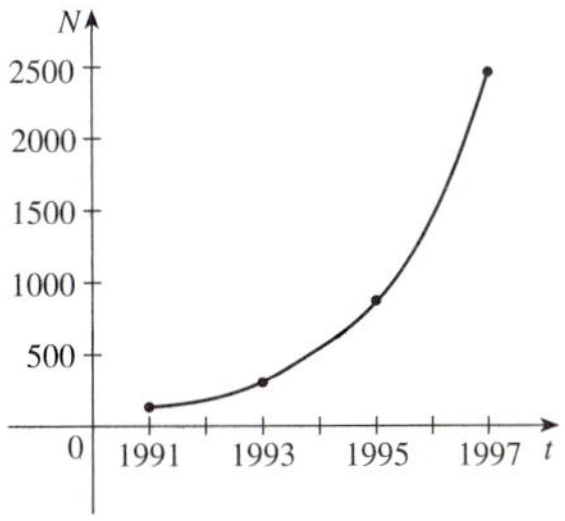

(b) 540, 1450

19. 12, 16, $3a^2 - a + 2$, $3a^2 + a + 2$, $3a^2 + 5a + 4$, $6a^2 - 2a + 4$, $12a^2 - 2a + 2$, $3a^4 - a^2 + 2$, $9a^4 - 6a^3 + 13a^2 - 4a + 4$, $3a^2 + 6ah + 3h^2 - a - h + 2$
21. $-(h^2 + 3h + 2)$, $x + h - x^2 - 2xh - h^2$, $1 - 2x - h$
23. $\{x \mid x \neq \frac{1}{3}\} = (-\infty, \frac{1}{3}) \cup (\frac{1}{3}, \infty)$
25. $[0, \infty)$ **27.** $(-\infty, 0) \cup (5, \infty)$
29. $(-\infty, \infty)$

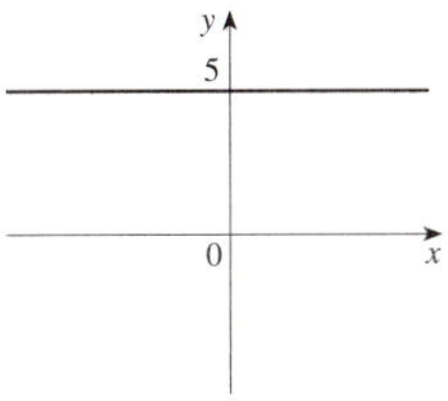

31. $(-\infty, \infty)$

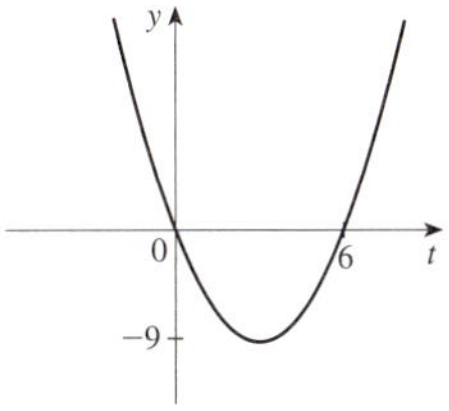

33. $[5, \infty)$

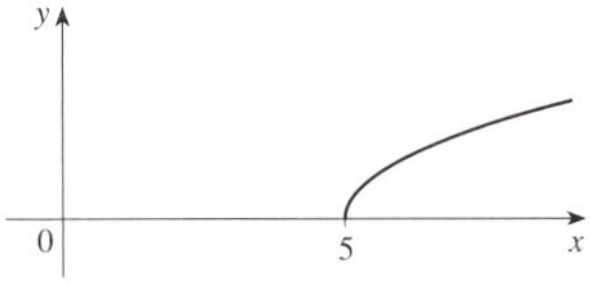

35. $(-\infty, 0) \cup (0, \infty)$

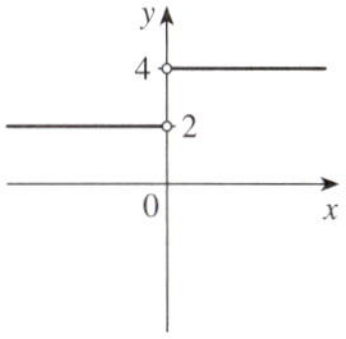

37. $(-\infty, \infty)$

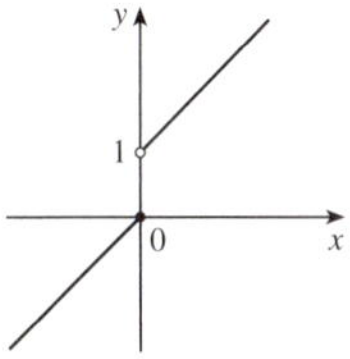

39. $(-\infty, \infty)$

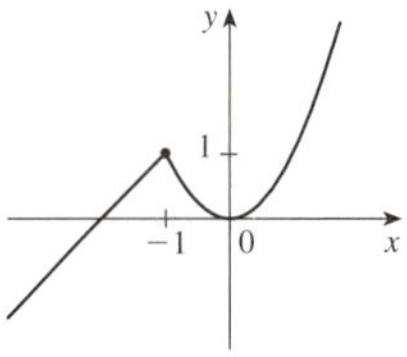

41. $f(x) = -\frac{7}{6}x - \frac{4}{3}, -2 \leqslant x \leqslant 4$ **43.** $f(x) = 1 - \sqrt{-x}$

45. $f(x) = \begin{cases} x + 1 & \text{if } -1 \leqslant x \leqslant 2 \\ 6 - 1.5x & \text{if } 2 < x \leqslant 4 \end{cases}$

47. $A(L) = 10L - L^2, 0 < L < 10$
49. $A(x) = \sqrt{3}x^2/4, x > 0$ **51.** $S(x) = x^2 + (8/x), x > 0$
53. $V(x) = 4x^3 - 64x^2 + 240x, 0 < x < 6$
55. (a)

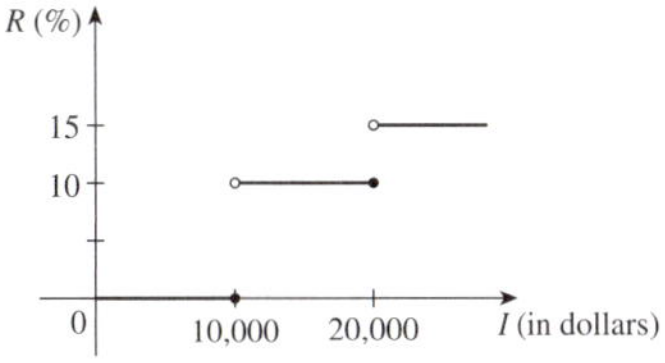

(b) \$400, \$1900

(c)

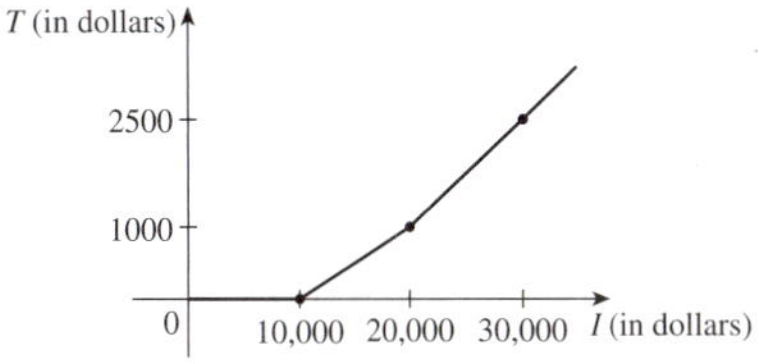

57. f is odd, g is even
59. (a) $(-5, 3)$ (b) $(-5, -3)$
61. Even **63.** Neither **65.** Odd

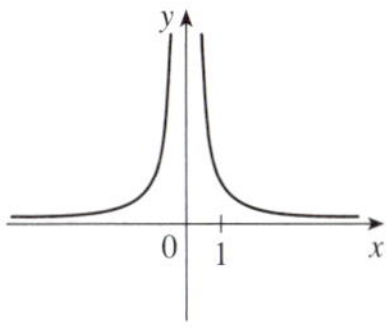

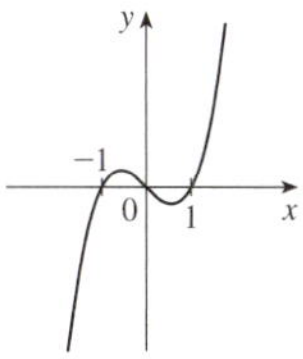

Exercises 1.2 ▫ page 35

1. (a) Root (b) Algebraic (c) Polynomial (degree 9)
(d) Rational (e) Trigonometric (f) Logarithmic
3. (a) h (b) f (c) g

5. (a) $y = 2x + b$, where b is the y-intercept

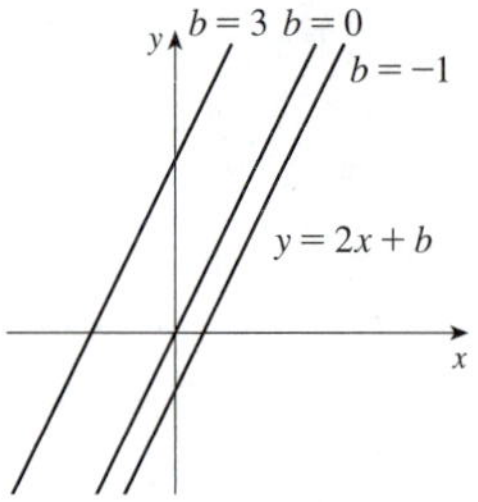

(b) $y = mx + 1 - 2m$, where m is the slope. See graph at right.

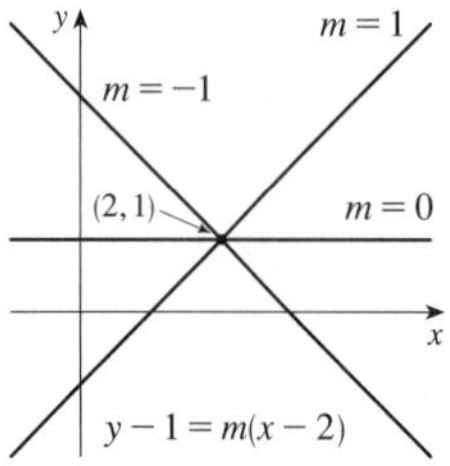

(c) $y = 2x - 3$

7. Their graphs have slope -1.

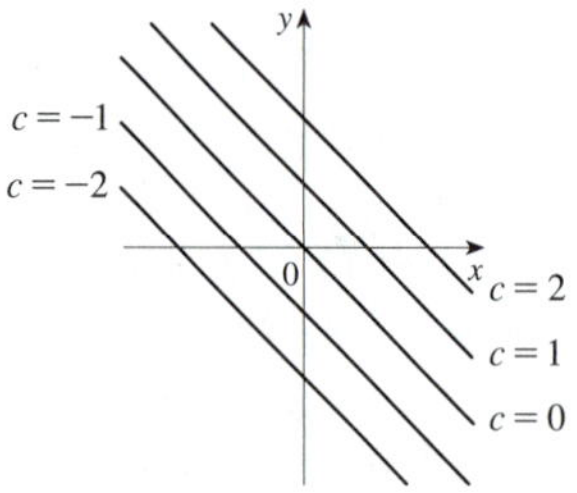

9. (a)

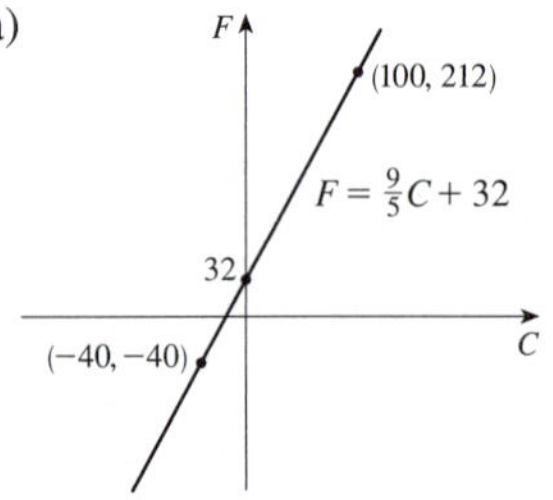

(b) $\frac{9}{5}$, change in °F for every 1°C change; 32, Fahrenheit temperature corresponding to 0°C

11. (a) $T = \frac{1}{6}N + \frac{307}{6}$ (b) $\frac{1}{6}$, change in °F for every chirp per minute change (c) 76°F

13. (a) $P = 0.434d + 15$ (b) 196 ft

15. (a) Cosine (b) Linear

17. (a) Linear model is appropriate

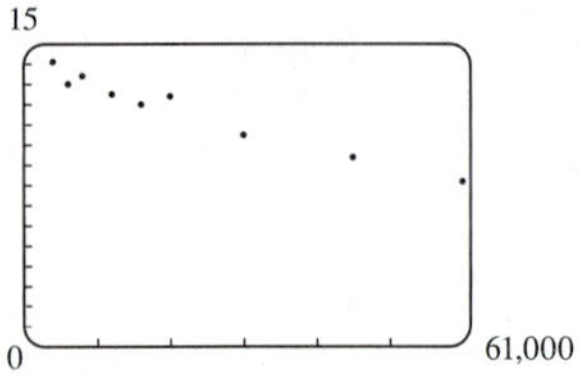

(b) $y = -0.000105x + 14.521$

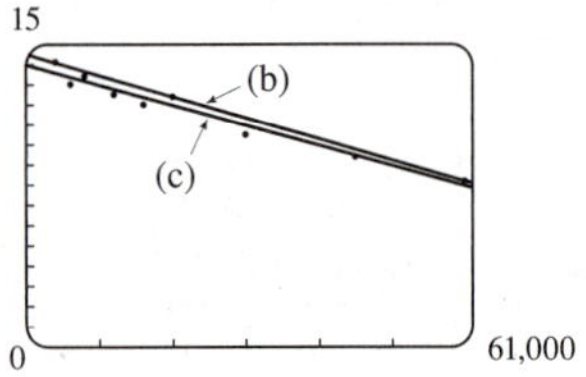

(c) $y = -0.00009979x + 13.951$ [See graph in (b).]
(d) About 11.5 per 100 population (e) About 6% (f) No

19. (a)

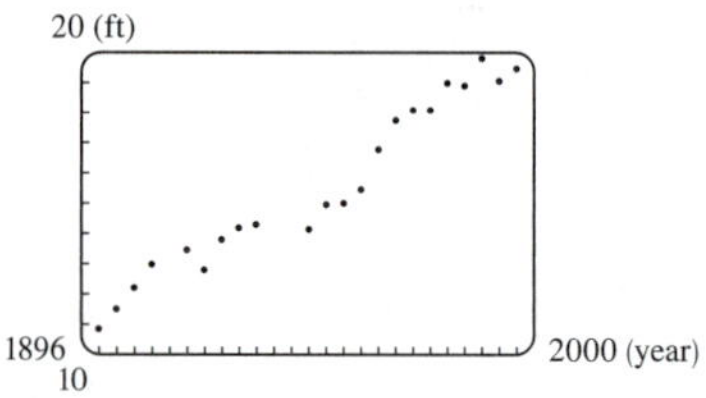

Linear model is appropriate
(b) $y = 0.08912x - 158.24$ (c) 20 ft (d) No

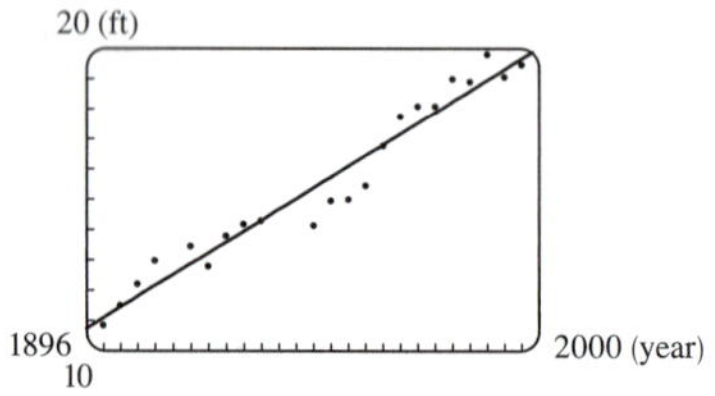

21. $y \approx 0.0012937x^3 - 7.06142x^2 + 12{,}823x - 7{,}743{,}770$; 1914 million

Exercises 1.3 ▫ page 45

1. (a) $y = f(x) + 3$ (b) $y = f(x) - 3$ (c) $y = f(x - 3)$
(d) $y = f(x + 3)$ (e) $y = -f(x)$ (f) $y = f(-x)$
(g) $y = 3f(x)$ (h) $y = \frac{1}{3}f(x)$

3. (a) 3 (b) 1 (c) 4 (d) 5 (e) 2

5. (a)

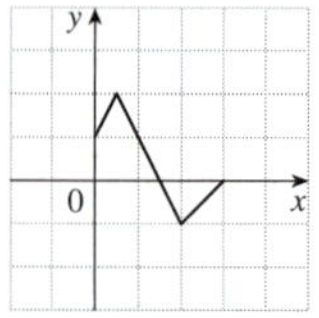

(b)

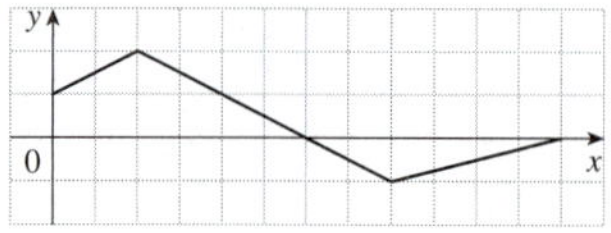

(c)

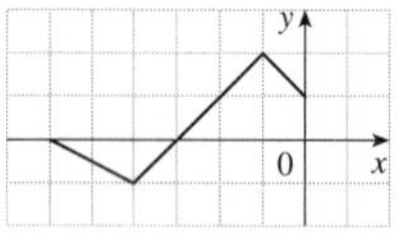

(d)

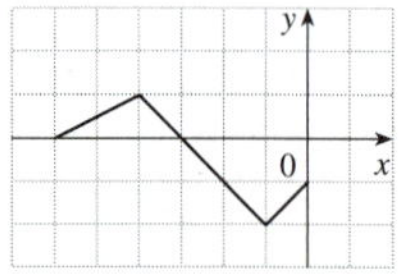

7. $y = -\sqrt{-x^2 - 5x - 4} - 1$

9.

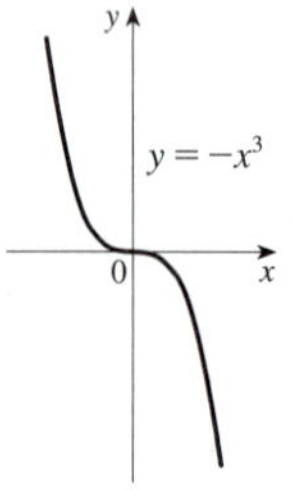

11.

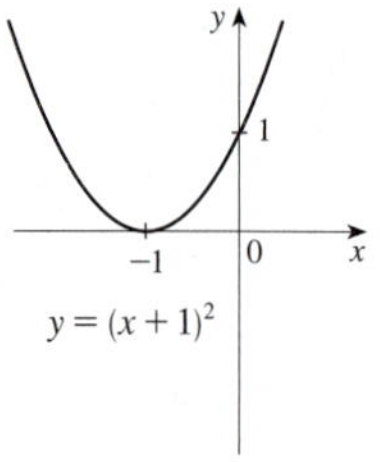

13.

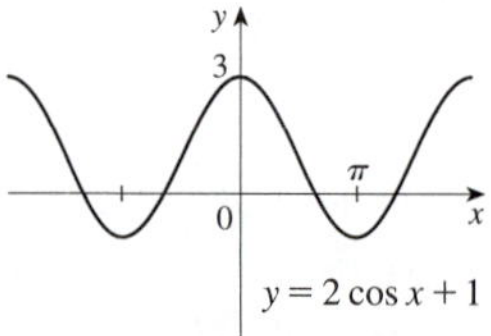

15.

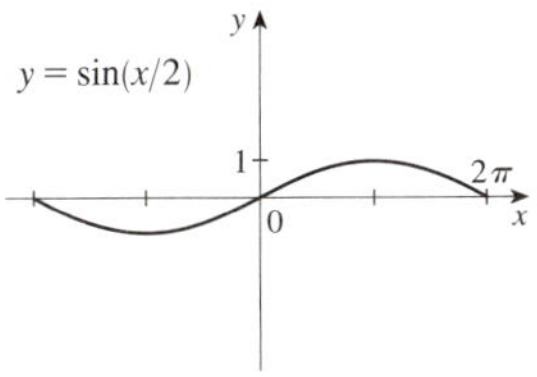

17.

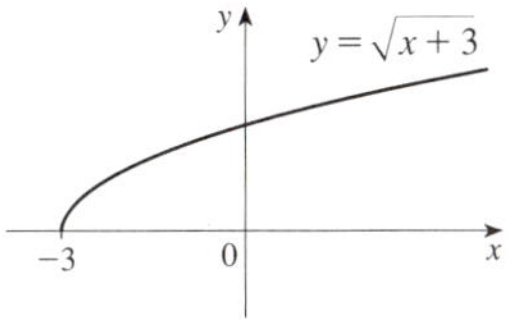

19.

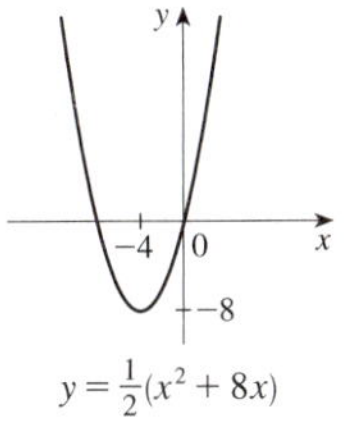

21.

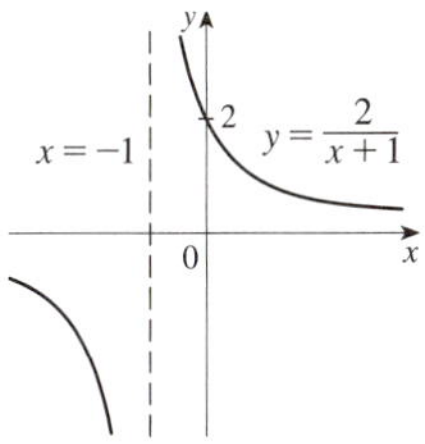

23.

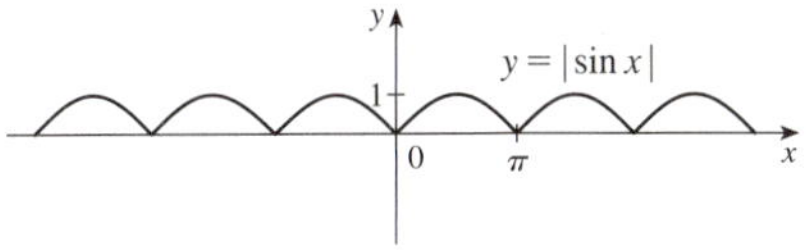

25. $L(t) = 12 + 2\sin\left[\dfrac{2\pi}{365}(t - 80)\right]$

27. (a) The portion of the graph of $y = f(x)$ to the right of the y-axis is reflected about the y-axis.

(b)

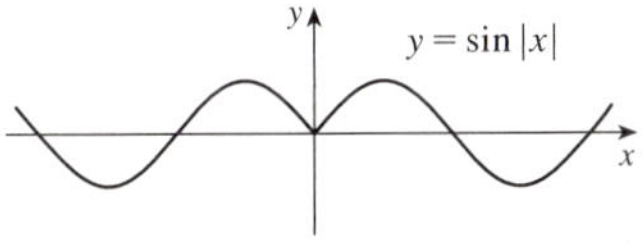

(c)

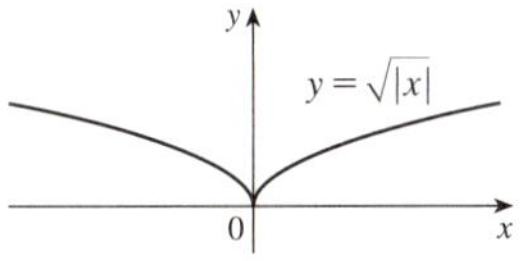

29.

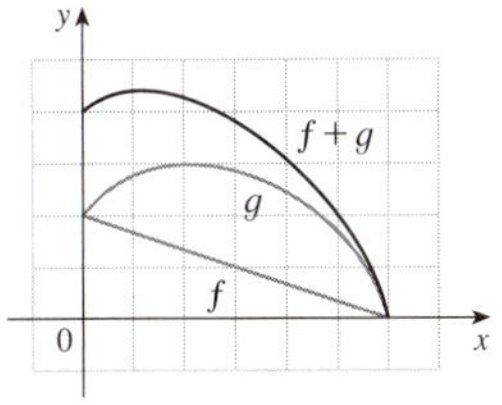

31. $(f + g)(x) = x^3 + 5x^2 - 1, (-\infty, \infty)$
$(f - g)(x) = x^3 - x^2 + 1, (-\infty, \infty)$
$(fg)(x) = 3x^5 + 6x^4 - x^3 - 2x^2, (-\infty, \infty)$
$(f/g)(x) = (x^3 + 2x^2)/(3x^2 - 1), \{x \mid x \neq \pm 1/\sqrt{3}\}$

33.

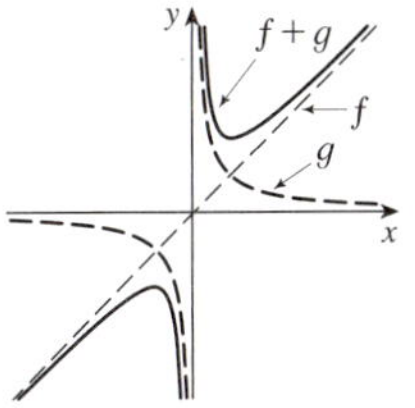

35. $(f \circ g)(x) = 3(6x^2 + 7x + 2), (-\infty, \infty)$
$(g \circ f)(x) = 6x^2 - 3x + 2, (-\infty, \infty)$
$(f \circ f)(x) = 8x^4 - 8x^3 + x, (-\infty, \infty)$
$(g \circ g)(x) = 9x + 8, (-\infty, \infty)$

37. $(f \circ g)(x) = \sin(1 - \sqrt{x}), [0, \infty)$
$(g \circ f)(x) = 1 - \sqrt{\sin x}, \{x \mid x \in [2n\pi, \pi + 2n\pi], n \text{ an integer}\}$
$(f \circ f)(x) = \sin(\sin x), (-\infty, \infty)$
$(g \circ g)(x) = 1 - \sqrt{1 - \sqrt{x}}, [0, 1]$

39. $(f \circ g)(x) = (2x^2 + 6x + 5)/[(x + 2)(x + 1)]$, $\{x \mid x \neq -2, -1\}$
$(g \circ f)(x) = (x^2 + x + 1)/(x + 1)^2, \{x \mid x \neq -1, 0\}$
$(f \circ f)(x) = (x^4 + 3x^2 + 1)/[x(x^2 + 1)], \{x \mid x \neq 0\}$
$(g \circ g)(x) = (2x + 3)/(3x + 5), \{x \mid x \neq -2, -\frac{5}{3}\}$

41. $(f \circ g \circ h)(x) = 2x - 1$

43. $(f \circ g \circ h)(x) = \sqrt{x^2 + 6x + 10}$

45. $g(x) = x^2 + 1, f(x) = x^{10}$

47. $g(x) = x^2, f(x) = x/(x + 4)$

49. $g(t) = \cos t, f(t) = \sqrt{t}$

51. $h(x) = x^2, g(x) = 3^x, f(x) = 1 - x$

53. $h(x) = \sqrt{x}, g(x) = \sec x, f(x) = x^4$

55. (a) 4 (b) 3 (c) 0 (d) Does not exist; $f(6) = 6$ is not in the domain of g. (e) 4 (f) −2

57. (a) $r(t) = 60t$ (b) $(A \circ r)(t) = 3600\pi t^2$; the area of the circle as a function of time

59. (a)

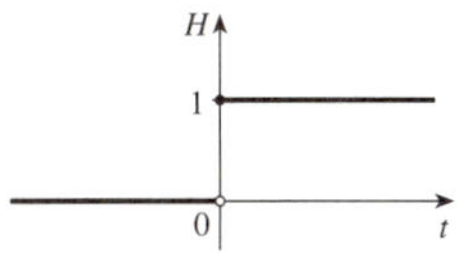

(b)

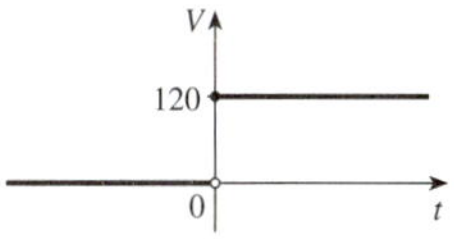

$V(t) = 120H(t)$

(c)

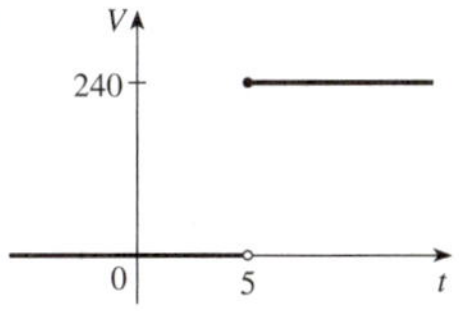

$V(t) = 240H(t - 5)$

61. (a) $f(x) = x^2 + 6$ (b) $g(x) = x^2 + x - 1$ **63.** Yes

Exercises 1.4 ▫ page 53

1. (d) **3.** (c)

5.

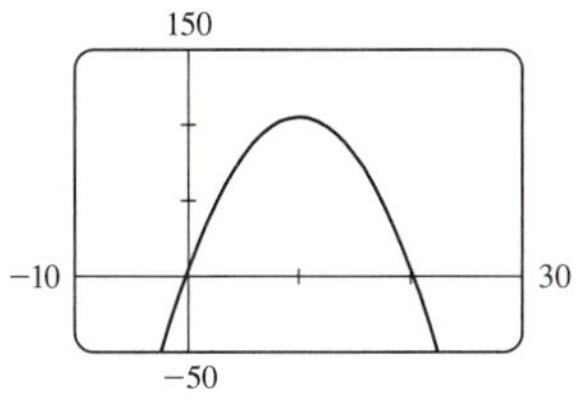

7.

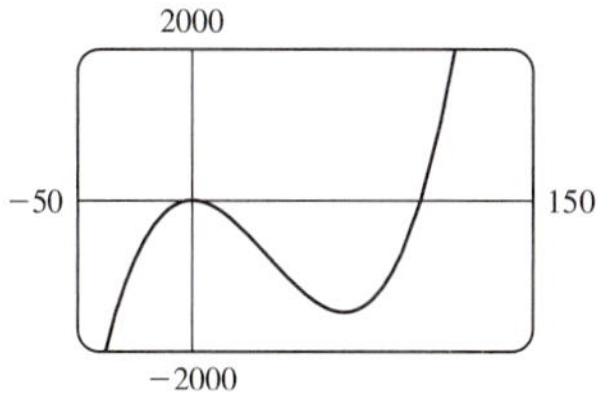

9.

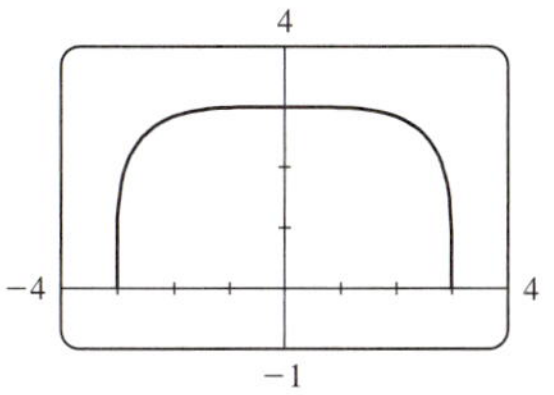

11.

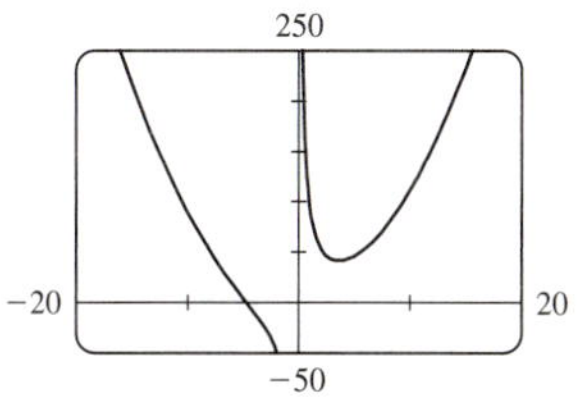

13.

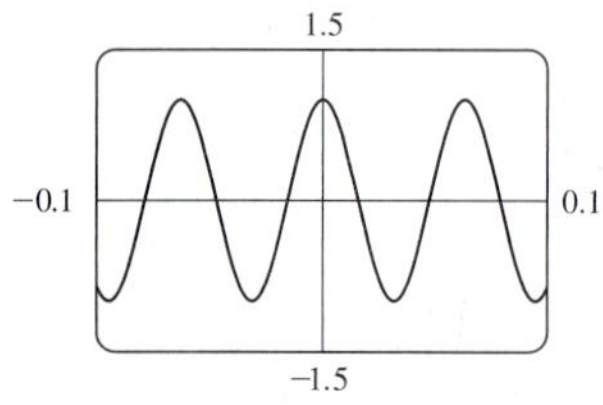

15.

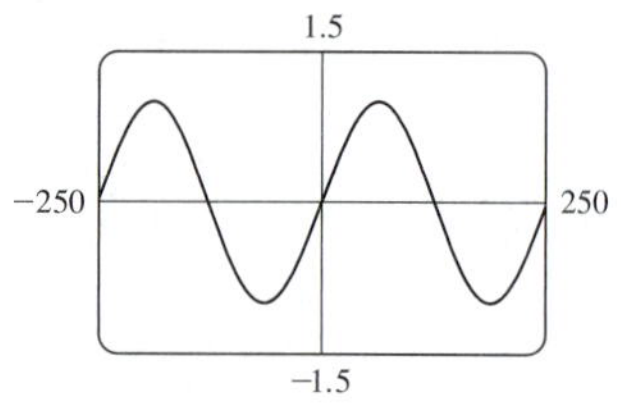

17.

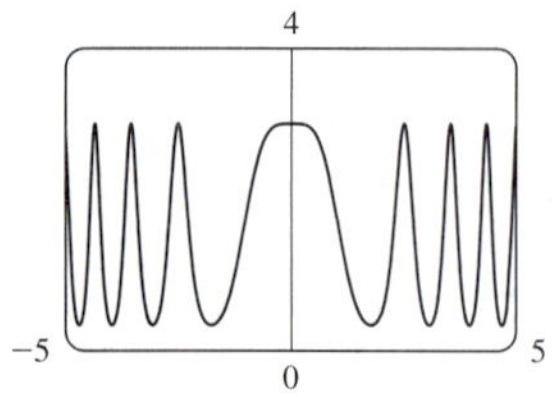

19.

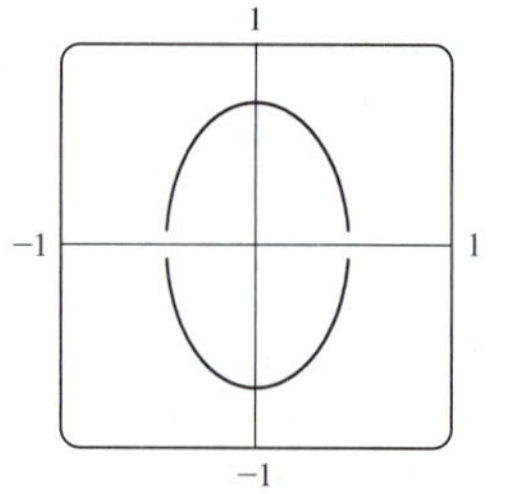

21. 9.05 **23.** 0, 0.88 **25.** g **27.** $-0.85 < x < 0.85$

29. (a)

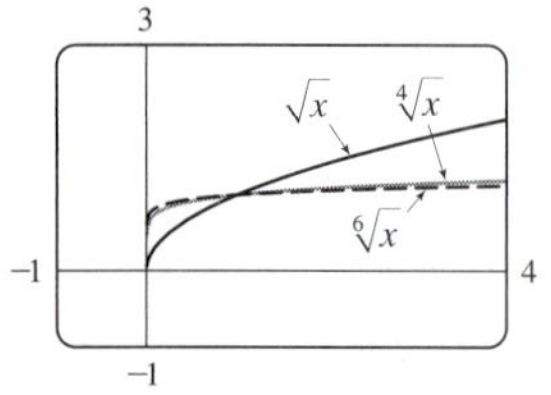

(b)

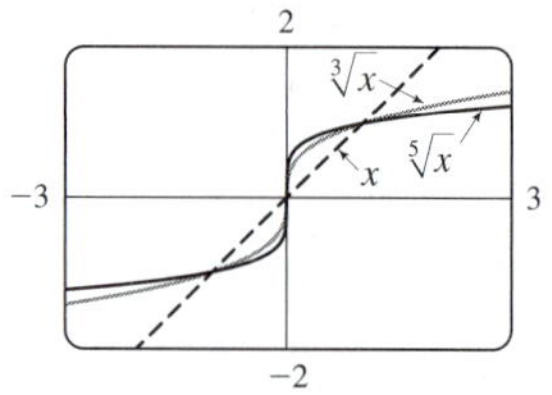

(c)

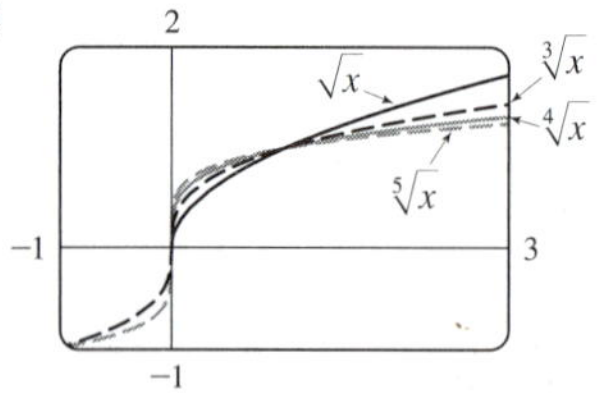

(d) Graphs of even roots are similar to $\sqrt{x}$, graphs of odd roots are similar to $\sqrt[3]{x}$. As n increases, the graph of $y = \sqrt[n]{x}$ becomes steeper near 0 and flatter for $x > 1$.

31.

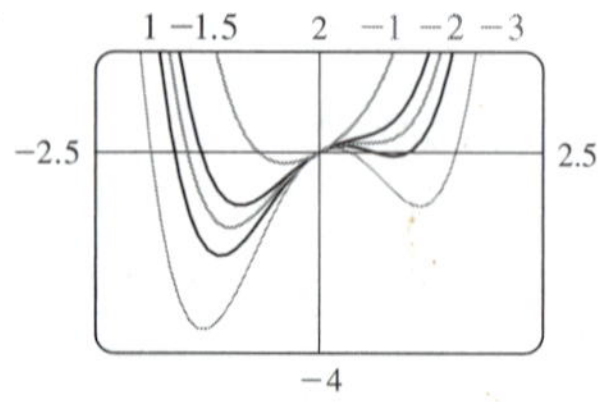

If $c < -1.5$, the graph has three humps: two minimum points and a maximum point. These humps get flatter as c increases until at $c = -1.5$ two of the humps disappear and there is only one minimum point. This single hump then moves to the right and approaches the origin as c increases.

33. The hump gets larger and moves to the right.

35. If $c < 0$, the loop is to the right of the origin; if $c > 0$, the loop is to the left. The closer c is to 0, the larger the loop.

Exercises 1.5 ▫ page 62

1. (a) $f(x) = a^x, a > 0$ (b) $\mathbb{R}$ (c) $(0, \infty)$
(d) See Figures 4(c), 4(b), and 4(a), respectively.

3.

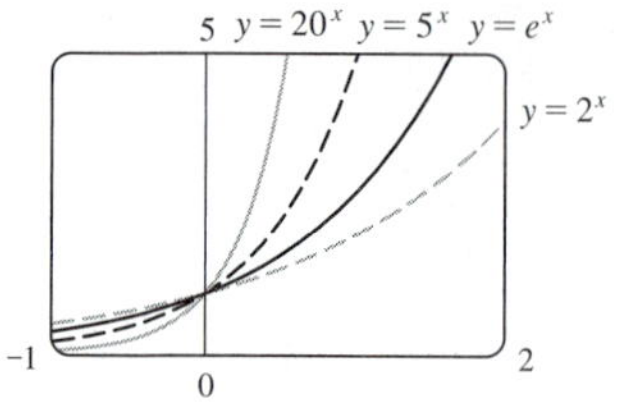

All approach 0 as $x \to -\infty$, all pass through (0, 1), and all are increasing. The larger the base, the faster the rate of increase.

5.

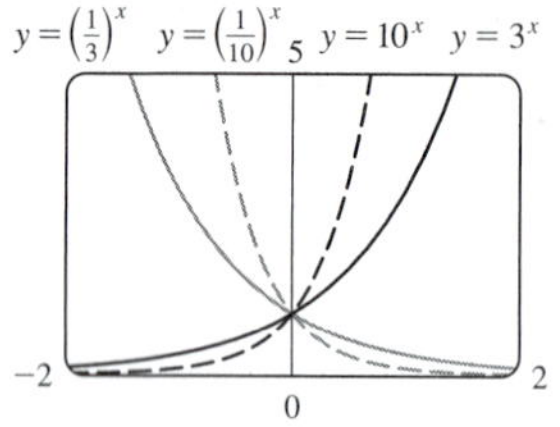

The functions with base greater than 1 are increasing and those with base less than 1 are decreasing. The latter are reflections of the former about the y-axis.

7.

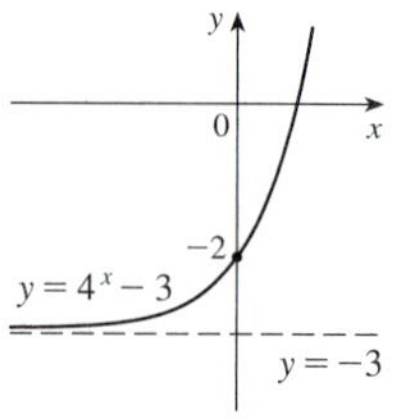

9.

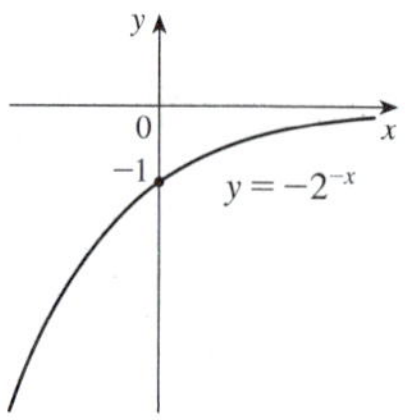

11.

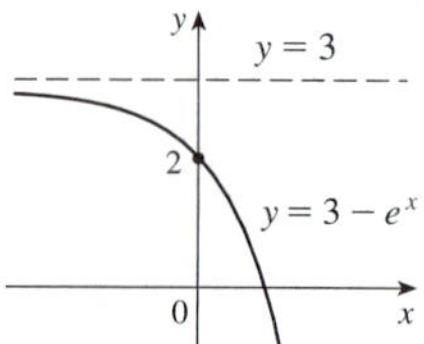

13. (a) $y = e^x - 2$ (b) $y = e^{x-2}$ (c) $y = -e^x$
(d) $y = e^{-x}$ (e) $y = -e^{-x}$

15. (a) $(-\infty, \infty)$ (b) $(-\infty, 0) \cup (0, \infty)$

17. $f(x) = 3 \cdot 2^x$ **23.** At $x \approx 35.8$

25. (a) 3200 (b) $100 \cdot 2^{t/3}$ (c) 10,159
(d) $t \approx 26.9$ h

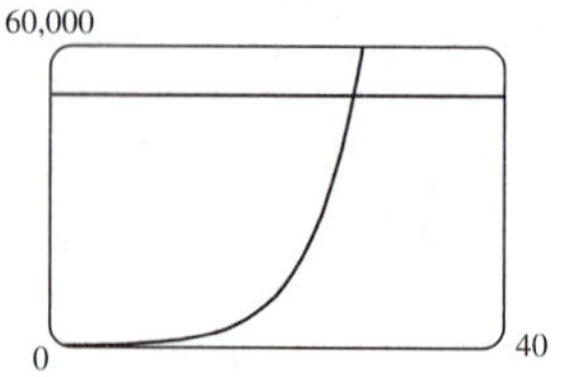

27. $y = ab^t$, where $a \approx 3.154832569 \times 10^{-12}$ and $b \approx 1.017764706$; 5498 million; 7417 million

Exercises 1.6 ▫ page 74

1. (a) See Definition 1.
(b) It must pass the Horizontal Line Test.

3. No **5.** Yes **7.** No **9.** Yes **11.** No

13. No **15.** No **17.** 2 **19.** 0

21. $F = \frac{9}{5}C + 32$; the Fahrenheit temperature as a function of the Celsius temperature; $[-273.15, \infty)$

23. $f^{-1}(x) = -\frac{1}{3}x^2 + \frac{10}{3}$, $x \geqslant 0$ **25.** $f^{-1}(x) = \sqrt[3]{\ln x}$

27. $y = e^x - 3$

29. $f^{-1}(x) = \sqrt{2/(1-x)}$

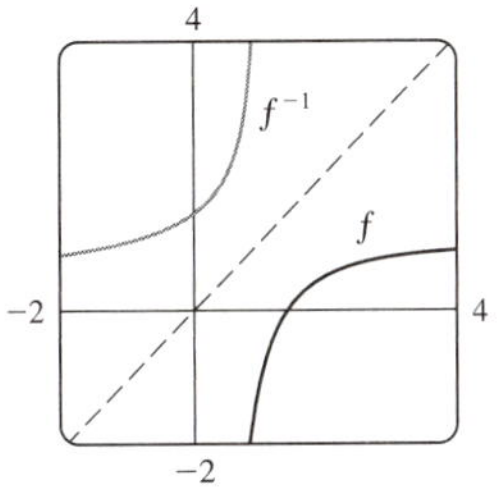

31.

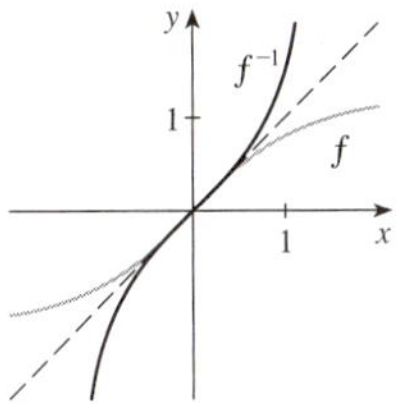

33. (a) It's defined as the inverse of the exponential function with base a, that is, $\log_a x = y \iff a^y = x$.
(b) $(0, \infty)$ (c) $\mathbb{R}$ (d) See Figure 11.
35. (a) 6 (b) −2 **37.** (a) 2 (b) 2 **39.** ln 8
41. $\ln \dfrac{(1+x^2)\sqrt{x}}{\sin x}$
43.

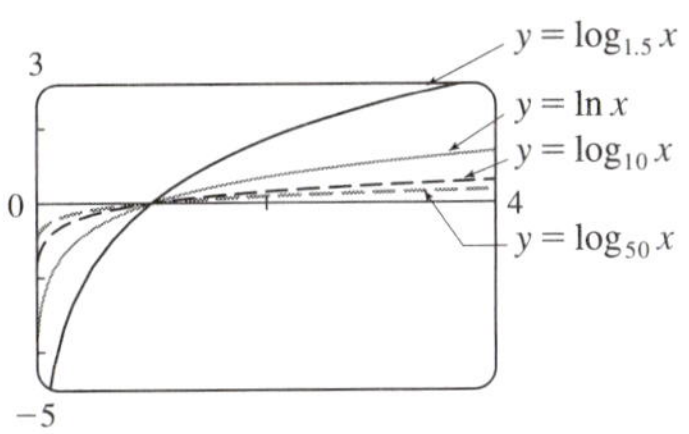

All graphs approach $-\infty$ as $x \to 0^+$, all pass through (1, 0), and all are increasing. The larger the base, the slower the rate of increase.

45. About 1,084,588 mi
47. (a)

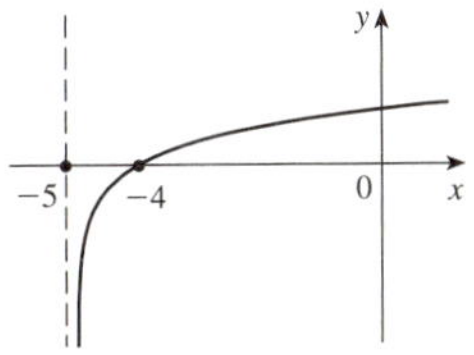

(b)

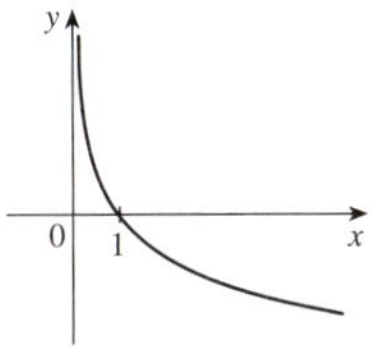

49. (a) $\sqrt{e}$ (b) $-\ln 5$
51. (a) $5 + \log_2 3$ or $5 + (\ln 3)/\ln 2$ (b) $\frac{1}{2}\left(1 + \sqrt{1+4e}\right)$
53. (a) $x < \ln 10$ (b) $x > 1/e$
55. (a) $\left(-\infty, \frac{1}{2}\ln 3\right]$ (b) $f^{-1}(x) = \frac{1}{2}\ln(3 - x^2)$, $\left[0, \sqrt{3}\right)$
57.

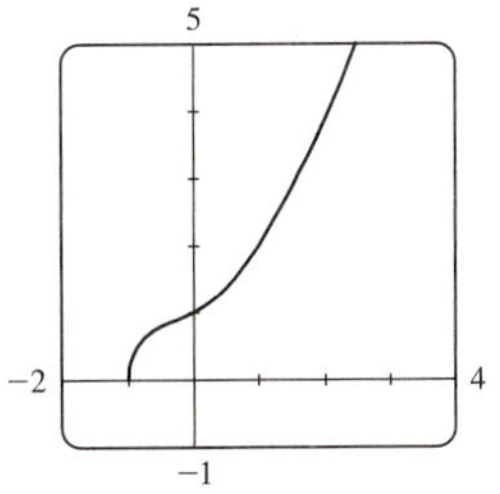

The graph passes the Horizontal Line Test.

$f^{-1}(x) = -\left(\sqrt[3]{4}/6\right)\left(\sqrt[3]{D - 27x^2 + 20} - \sqrt[3]{D + 27x^2 - 20} + \sqrt[3]{2}\right)$, where $D = 3\sqrt{3}\sqrt{27x^4 - 40x^2 + 16}$; two of the expressions are complex.
59. (a) $f^{-1}(n) = (3/\ln 2)\ln(n/100)$; the time elapsed when there are n bacteria (b) After about 26.9 hours
61. (a) $y = \ln x + 3$ (b) $y = \ln(x+3)$ (c) $y = -\ln x$
(d) $y = \ln(-x)$ (e) $y = e^x$ (f) $y = e^{-x}$ (g) $y = -e^x$
(h) $y = e^x - 3$
63. (a) $\pi/3$ (b) π **65.** (a) $\pi/3$ (b) $-\pi/4$
67. (a) 0.7 (b) $\pi/3$ **71.** $x/\sqrt{1+x^2}$

73.

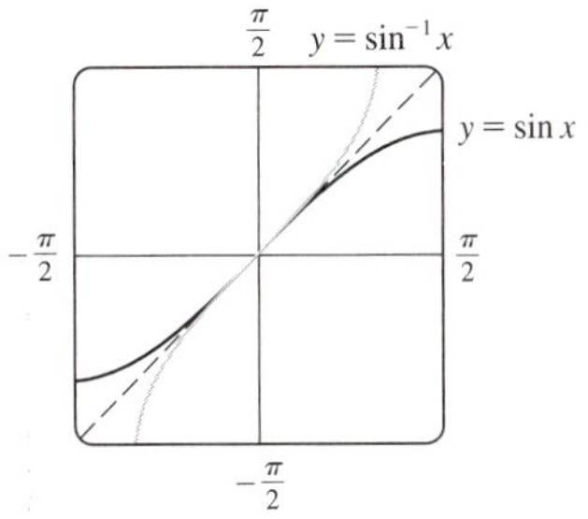

The second graph is the reflection of the first graph about the line $y = x$.

75. $\left[-\frac{2}{3}, 0\right]$, $[-\pi/2, \pi/2]$

Chapter 1 Review ▫ page 78

True-False Quiz

1. False **3.** False **5.** True **7.** False **9.** True
11. False

Exercises

1. (a) 2.7 (b) 2.3, 5.6 (c) $[-6, 6]$ (d) $[-4, 4]$
(e) $[-4, 4]$ (f) No; it fails the Horizontal Line Test.
(g) Odd; its graph is symmetric about the origin.
3. (a)

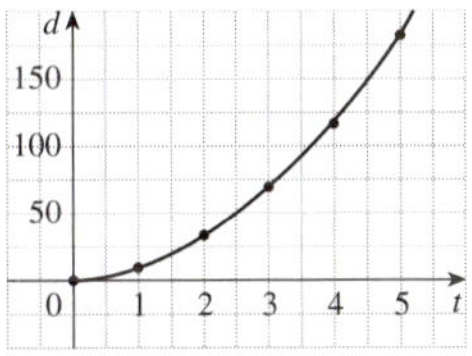

(b) 150 ft

5. $\left[-\frac{2}{3}\sqrt{3}, \frac{2}{3}\sqrt{3}\right]$, $[0, 2]$ **7.** $\mathbb{R}$, $[0, 2]$
9. (a) Shift the graph 8 units upward.
(b) Shift the graph 8 units to the left.
(c) Stretch the graph vertically by a factor of 2, then shift it 1 unit upward.
(d) Shift the graph 2 units to the right and 2 units downward.
(e) Reflect the graph about the x-axis.
(f) Reflect the graph about the line $y = x$ (assuming f is one-to-one).
11.

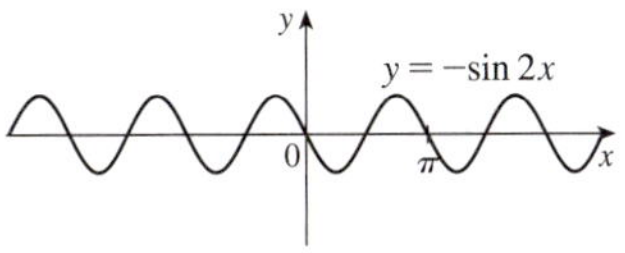

13.

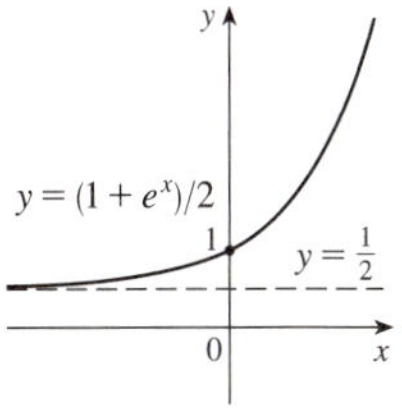

15.

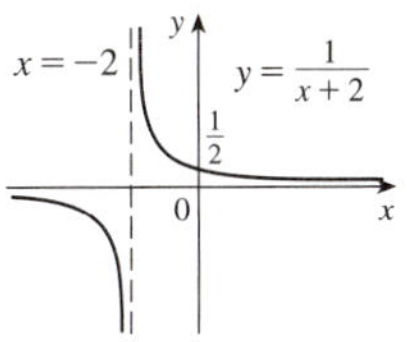

17. (a) Neither (b) Odd (c) Even (d) Neither

19. $(f \circ g)(x) = \ln(x^2 - 9)$, $(-\infty, -3) \cup (3, \infty)$
$(g \circ f)(x) = (\ln x)^2 - 9$, $(0, \infty)$
$(f \circ f)(x) = \ln \ln x$, $(1, \infty)$
$(g \circ g)(x) = (x^2 - 9)^2 - 9$, $(-\infty, \infty)$
21. $y = 0.2493x - 423.4818$; about 77.6 years
23. 1 **25.** (a) 9 (b) 2 (c) $1/\sqrt{3}$ (d) $\frac{3}{5}$
27. (a) $\frac{1}{16}$ g (b) $m(t) = 2^{-t/4}$
(c) $t(m) = -4 \log_2 m$; the time elapsed when there are m grams of ^{100}Pd (d) About 26.6 days
29.

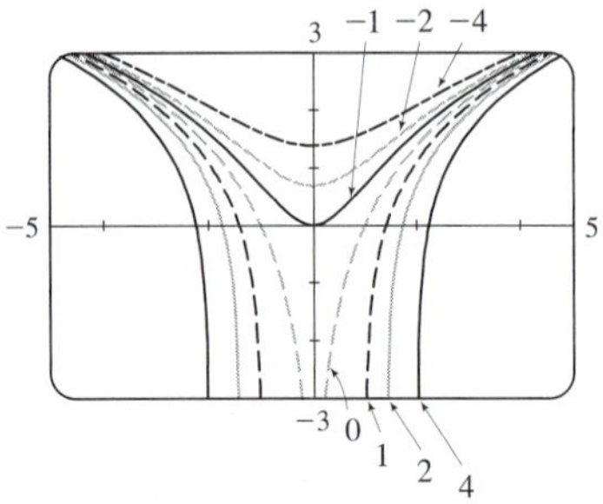

For $c < 0$, f is defined everywhere. As c increases, the dip at $x = 0$ becomes deeper. For $c \geqslant 0$, the graph has asymptotes at $x = \pm\sqrt{c}$.

Principles of Problem Solving ▫ page 85

1. $a = 4\sqrt{h^2 - 16}/h$, where a is the length of the altitude and h is the length of the hypotenuse
3. $-\frac{7}{3}$, 9
5.

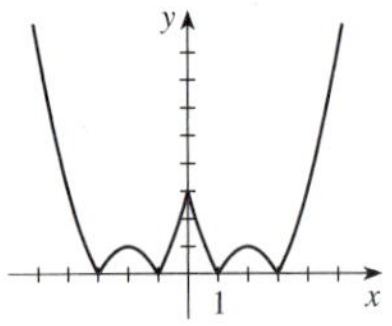

7.

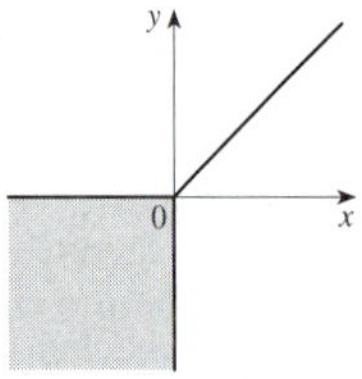

9.

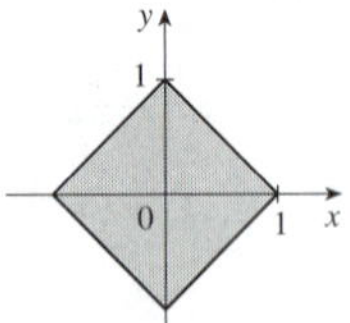

11. 5 **13.** $x \in [-1, 1 - \sqrt{3}) \cup (1 + \sqrt{3}, 3]$
15. 40 mi/h **19.** $f_n(x) = x^{2^{n+1}}$

Chapter 2

Exercises 2.1 ▫ page 91

1. (a) -44.4, -38.8, -27.8, -22.2, $-16.\overline{6}$
(b) -33.3 (c) $-33\frac{1}{3}$
3. (a) (i) 0.333333 (ii) 0.263158 (iii) 0.251256
(iv) 0.250125 (v) 0.2 (vi) 0.238095 (vii) 0.248756
(viii) 0.249875 (b) $\frac{1}{4}$ (c) $y = \frac{1}{4}x + \frac{1}{4}$
5. (a) (i) -32 ft/s (ii) -25.6 ft/s (iii) -24.8 ft/s
(iv) -24.16 ft/s (b) -24 ft/s
7. (a) (i) $\frac{13}{6}$ ft/s (ii) $\frac{7}{6}$ ft/s (iii) $\frac{19}{24}$ ft/s (iv) $\frac{331}{600}$ ft/s (b) $\frac{1}{2}$ ft/s

(c)

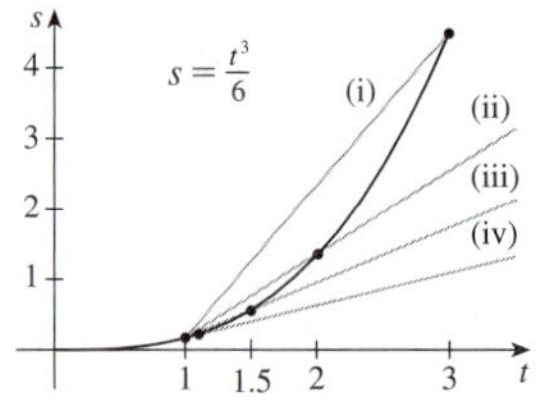

(d)

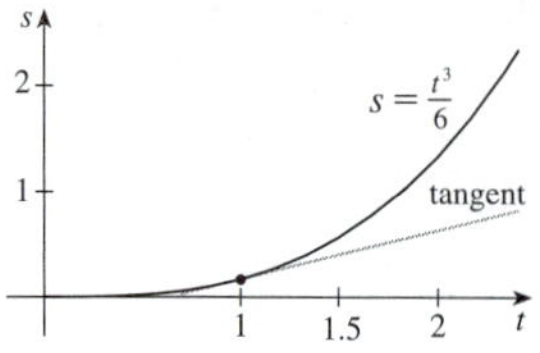

9. (a) 0, 1.7321, -1.0847, -2.7433, 4.3301, -2.8173, 0, -2.1651, -2.6061, -5, 3.4202; no (c) -31.4

Exercises 2.2 ▫ page 102

1. Yes
3. (a) $\lim_{x \to -3} f(x) = \infty$ means that the values of $f(x)$ can be made arbitrarily large (as large as we please) by taking x sufficiently close to -3 (but not equal to -3).
(b) $\lim_{x \to 4^+} f(x) = -\infty$ means that the values of $f(x)$ can be made arbitrarily large negative by taking x sufficiently close to 4 through values larger than 4.
5. (a) 2 (b) 3 (c) Does not exist (d) 4
(e) Does not exist
7. (a) -1 (b) -2 (c) Does not exist (d) 2 (e) 0
(f) Does not exist (g) 1 (h) 3
9. (a) $-\infty$ (b) ∞ (c) ∞ (d) $-\infty$ (e) ∞
(f) $x = -7$, $x = -3$, $x = 0$, $x = 6$
11. (a) 1 (b) 0 (c) Does not exist
13.

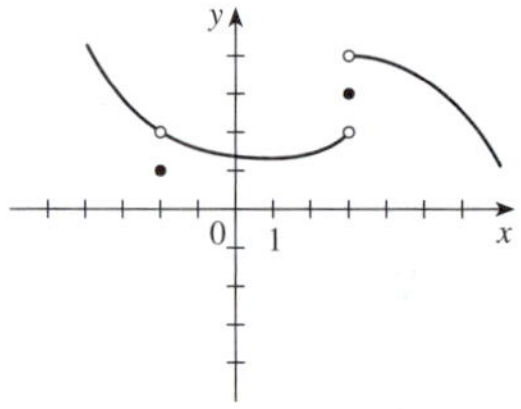

15. $\frac{2}{3}$ **17.** $\frac{1}{2}$ **19.** $\frac{1}{4}$ **21.** $\frac{3}{5}$ **23.** ∞
25. ∞ **27.** $-\infty$ **29.** $-\infty$ **31.** $-\infty$; ∞
33. (a) 2.71828 (b)

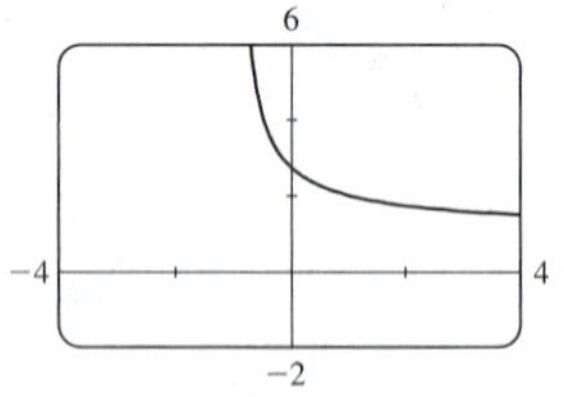

35. (a) 0.998000, 0.638259, 0.358484, 0.158680, 0.038851, 0.008928, 0.001465; 0
(b) 0.000572, -0.000614, -0.000907, -0.000978, -0.000993, -0.001000; -0.001
37. No matter how many times we zoom in toward the origin, the graph appears to consist of almost-vertical lines. This indicates more and more frequent oscillations as $x \to 0$.
39. $x \approx \pm 0.90, \pm 2.24$; $x = \pm\sin^{-1}(\pi/4), \pm(\pi - \sin^{-1}(\pi/4))$

Exercises 2.3 ▫ page 111

1. (a) 5 (b) 9 (c) 2 (d) $-\frac{1}{3}$ (e) $-\frac{3}{8}$ (f) 0
(g) Does not exist (h) $-\frac{6}{11}$
3. 59 **5.** 205 **7.** $\frac{1}{8}$ **9.** 0 **11.** 5
13. Does not exist **15.** $\frac{6}{5}$ **17.** 8

19. 4 **21.** 6 **23.** $\frac{1}{6}$ **25.** $-\frac{1}{16}$ **27.** 108 **29.** $-\frac{1}{2}$
31. (a), (b) $\frac{2}{3}$ **35.** 1 **39.** 0 **41.** Does not exist
43. Does not exist
45. (a)

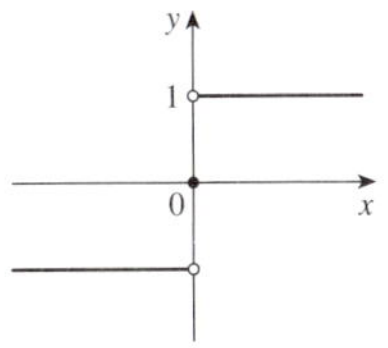

(b) (i) 1
(ii) -1
(iii) Does not exist
(iv) 1

47. (a) (i) 2 (ii) -2 (b) No (c)

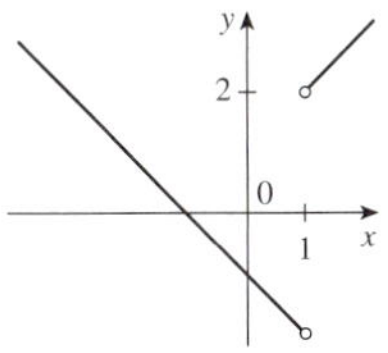

49. (a) (i) -2 (ii) Does not exist (iii) -3
(b) (i) $n-1$ (ii) n (c) a is not an integer.
59. 15; -1

Exercises 2.4 ▫ page 122

1. (a) $|x-2|<0.02$ (b) $|x-2|<0.002$
3. $\frac{4}{7}$ (or any smaller positive number)
5. 1.44 (or any smaller positive number)
7. 0.6875 (or any smaller positive number)
9. 0.11, 0.012 (or smaller positive numbers)
11. 0.07 (or any smaller positive number)
13. (a) $\sqrt{1000/\pi}$ cm (b) Within approximately 0.0445 cm
(c) Radius; area; $\sqrt{1000/\pi}$; 1000; 5; ≈ 0.0445
35. (a) 0.093 (b) $\delta = (B^{2/3}-12)/(6B^{1/3})-1$, where $B = 216 + 108\varepsilon + 12\sqrt{336 + 324\varepsilon + 81\varepsilon^2}$
41. Within 0.1

Exercises 2.5 ▫ page 133

1. $\lim_{x\to 4} f(x) = f(4)$
3. (a) -4 (removable), -2 (jump), 2 (jump), 4 (infinite)
(b) -4, neither; -2, left; 2, right; 4, right
5.

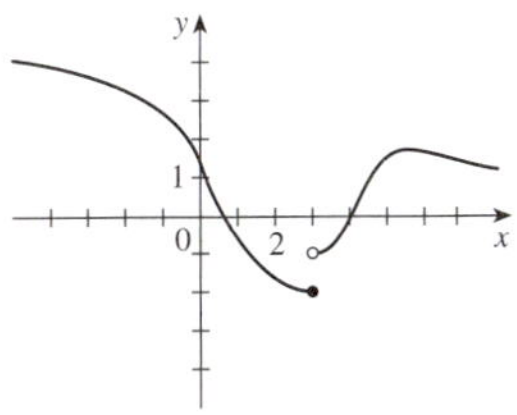

7. (a)

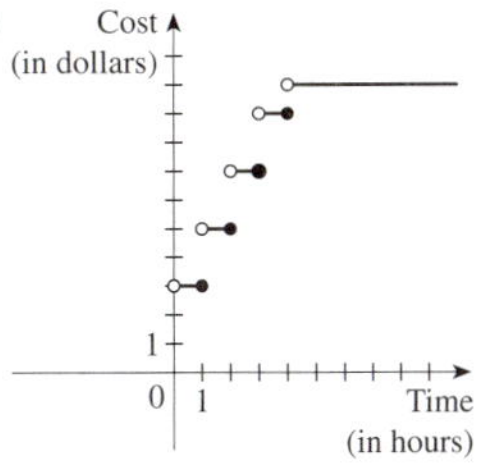

(b) Discontinuous at $t = 1, 2, 3, 4$

9. 6
15. $f(2)$ is not defined

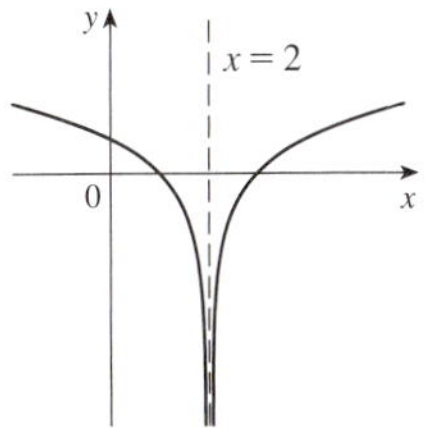

17. $\lim_{x\to 0} f(x)$ does not exist

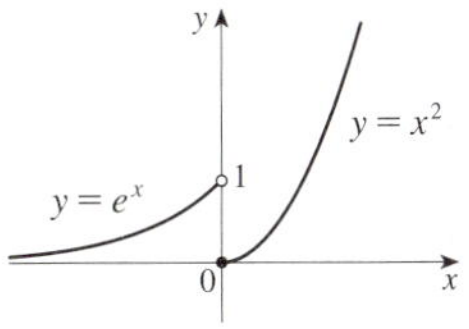

19. $\lim_{x\to -3} f(x) \neq f(-3)$

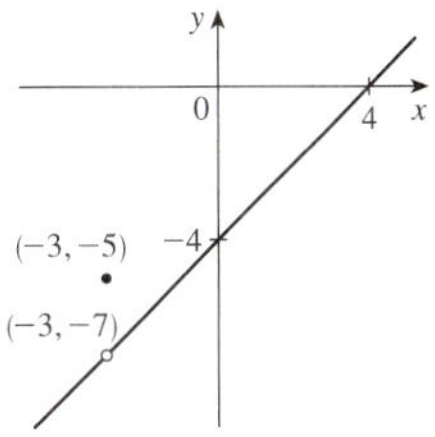

21. $\{x \mid x \neq -3, -2\}$

23. $\left[\frac{1}{2}, \infty\right)$ **25.** $\mathbb{R}$ **27.** $(-\infty, -1) \cup (1, \infty)$
29. $x = 0$

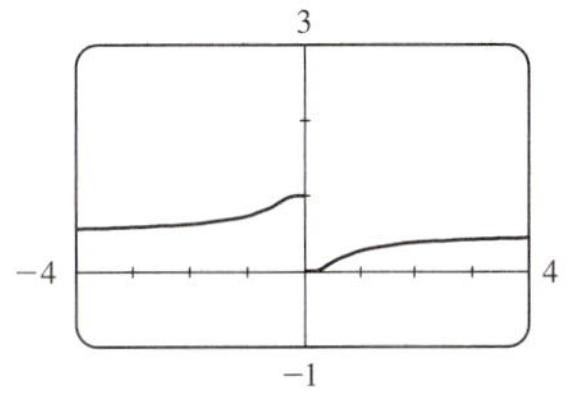

31. $\frac{7}{3}$ **33.** 1
37. 0, left

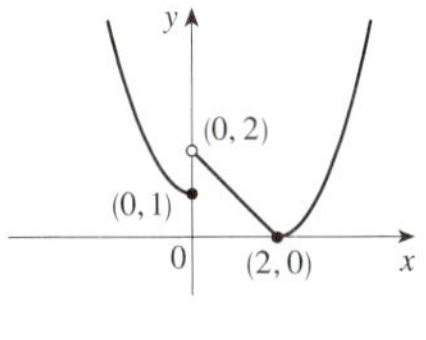

39. 0, right; 1, left

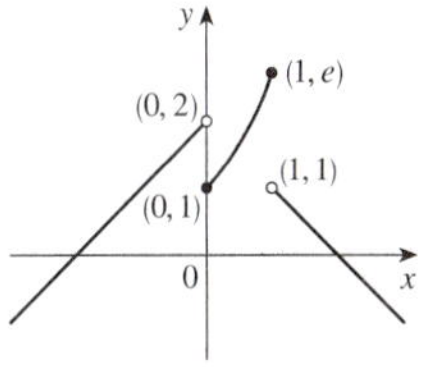

41. $\frac{1}{3}$
43. (a) $g(x) = x - 4$ (c) $g(x) = x^2 - 4x + 16$
(d) $g(x) = 1/(3 + \sqrt{x})$
51. (b) (0.44, 0.45) **53.** (b) 1.434 **59.** None **61.** Yes

Exercises 2.6 ▫ page 146

1. (a) As x becomes large, $f(x)$ approaches 5.
(b) As x becomes large negative, $f(x)$ approaches 3.
3. (a) ∞ (b) ∞ (c) $-\infty$ (d) 1 (e) 2
(f) $x = -1, x = 2, y = 1, y = 2$
5.

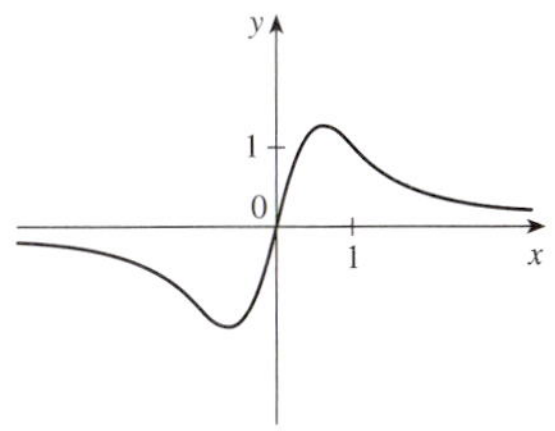

7.

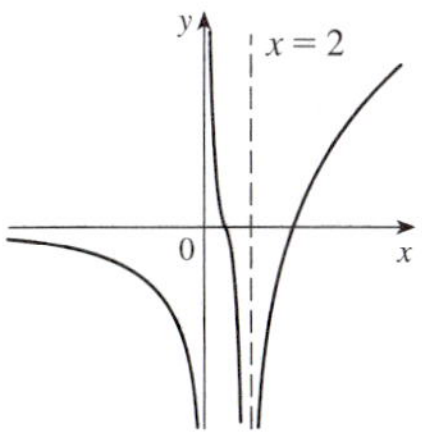

9. 0 **11.** $\frac{3}{2}$ **13.** 0 **15.** $-\frac{1}{2}$ **17.** $\frac{1}{2}$ **19.** 2
21. 3 **23.** $\frac{1}{6}$ **25.** $(a-b)/2$ **27.** ∞ **29.** ∞
31. $-\infty$ **33.** ∞ **35.** (a), (b) $-\frac{1}{2}$
37. $y = 1, x = -4$

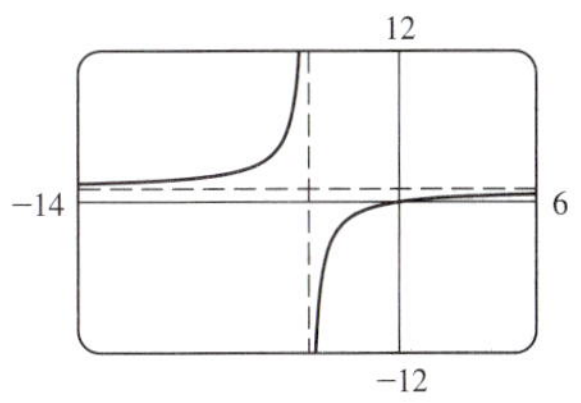

39. $x = 2, x = -5$

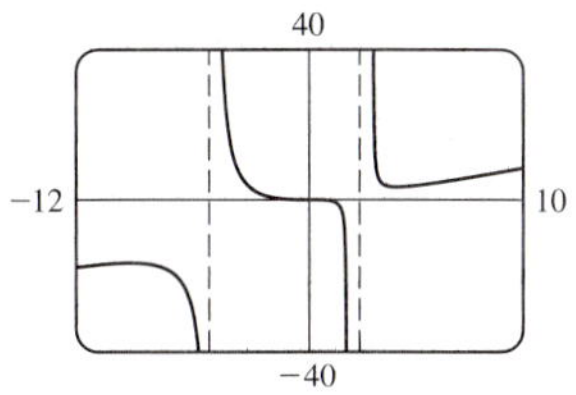

41. $y = \pm 1$

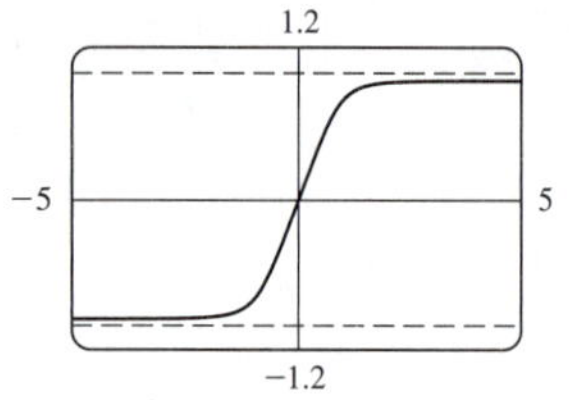

43. $f(x) = (2 - x)/[x^2(x - 3)]$

45. $-\infty, -\infty$

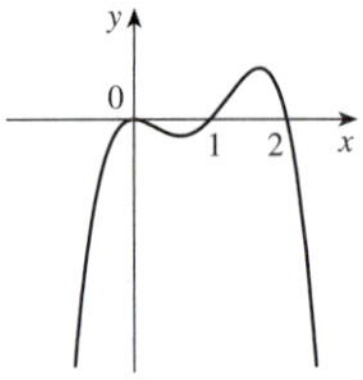

47. $\infty, -\infty$

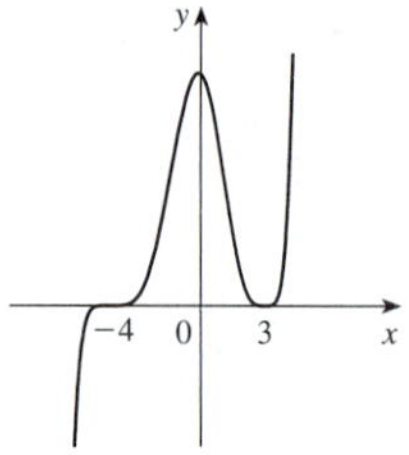

49. (a) 0 (b) An infinite number of times

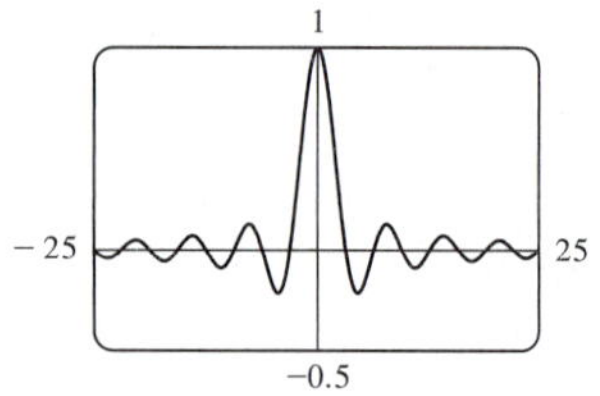

51. (a) 0 (b) $\pm\infty$ **53.** 4

55. (a) v^*

(b) ≈ 0.47 s

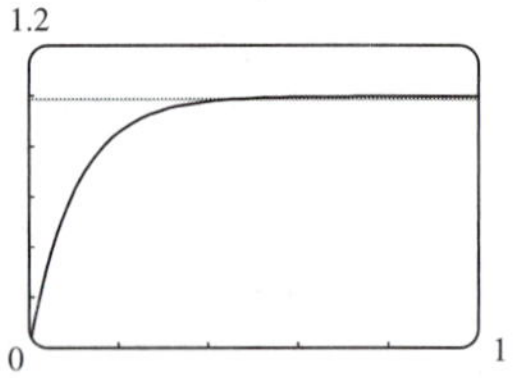

57. $N \geqslant 13$ **59.** $N \leqslant -6, N \leqslant -22$

61. (a) $x > 100$

Exercises 2.7 ▫ page 155

1. (a) $\dfrac{f(x) - f(3)}{x - 3}$ (b) $\displaystyle\lim_{x \to 3} \frac{f(x) - f(3)}{x - 3}$

3. Slopes at D, E, C, A, B

5. (a) -4 (b) $y = -4x - 9$

(c)

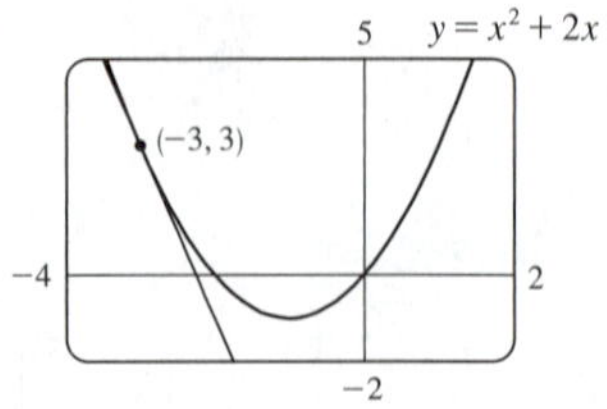

7. $y = -x + 3$ **9.** $y = -x + 5$

11. (a) $-2/(a + 3)^2$ (b) (i) $-\frac{1}{2}$ (ii) $-\frac{2}{9}$ (iii) $-\frac{1}{8}$

13. (a) $3a^2 - 4$ (b) $y = -x - 1, y = 8x - 15$

(c)

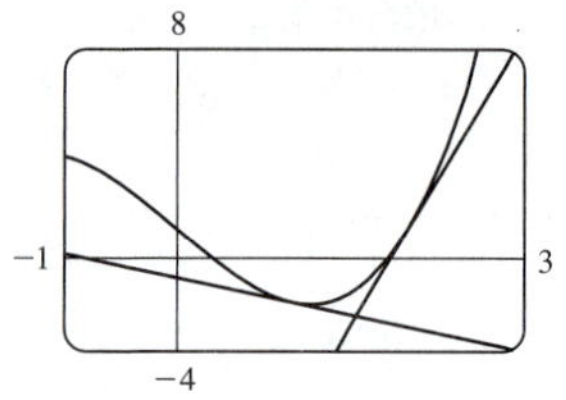

15. (a) 0 (b) C (c) Speeding up, slowing down, neither
(d) The car did not move.

17. -24 ft/s **19.** $(12a^2 + 6)$ m/s, 18 m/s, 54 m/s, 114 m/s

21. Greater (in magnitude)

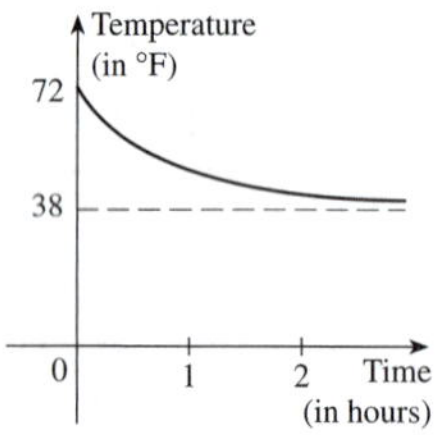

23. (a) (i) -1.2°C/h (ii) -1.25°C/h (iii) -1.3°C/h
(b) -1.9°C/h

25. (a) (i) 794 thousand/year (ii) 640 thousand/year
(iii) 301 thousand/year
(b) 470.5 thousand/year (c) 427.5 thousand/year

27. (a) (i) \$20.25/unit (ii) \$20.05/unit (b) \$20/unit

Exercises 2.8 ▫ page 163

1. The line from $(2, f(2))$ to $(2 + h, f(2 + h))$

3. $g'(0), 0, g'(4), g'(2), g'(-2)$

5.

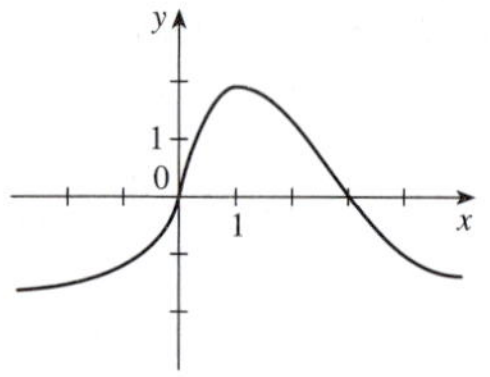

7. $7; y = 7x - 12$

9. (a) $-2; y = -2x - 1$ (b)

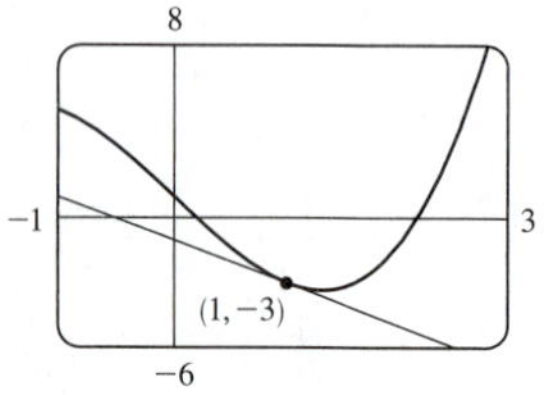

11. (a) 3.296 (b) 3.3

13. $-2 + 8a$ **15.** $5/(a + 3)^2$

17. $-1/[2(a + 2)^{3/2}]$

19. $f(x) = x^{10}, a = 1$ or $f(x) = (1 + x)^{10}, a = 0$

21. $f(x) = 2^x, a = 5$

23. $f(x) = \cos x, a = \pi$ or $f(x) = \cos(\pi + x), a = 0$

25. -2 m/s

27. (a) The rate at which the cost is changing per ounce of gold produced; dollars per ounce
(b) When the 800th ounce of gold is produced, the cost of production is \$17/oz.
(c) Decrease in the short term; increase in the long term

29. (a) The rate at which the fuel consumption is changing with respect to speed; (gal/h)/(mi/h)
(b) The fuel consumption is decreasing by 0.05 (gal/h)/(mi/h) as the car's speed reaches 20 mi/h.
31. The rate at which the temperature is changing at 10:00 A.M.; 4°F/h
33. (a) The rate at which the oxygen solubility changes with respect to the water temperature; (mg/L)/°C
(b) $S'(16) \approx -0.25$; as the temperature increases past 16°C, the oxygen solubility is decreasing at a rate of 0.25 (mg/L)/°C.
35. Does not exist

Exercises 2.9 ▫ page 173

1. (a) −2 (b) 0.8 (c) −1 (d) −0.5

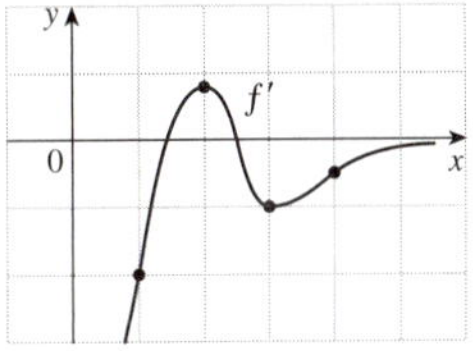

3. (a) 1.5 (b) 1 (c) 0 (d) −4 (e) 0 (f) 1 (g) 1.5

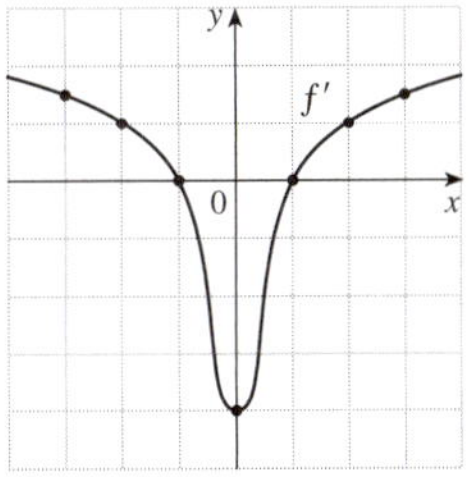

5.

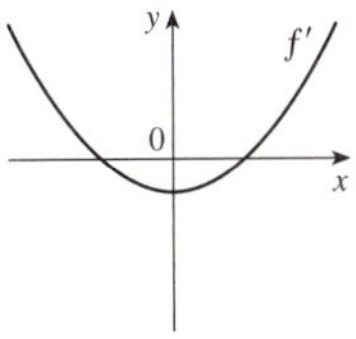

7.

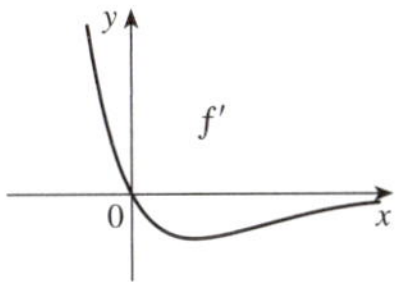

9.

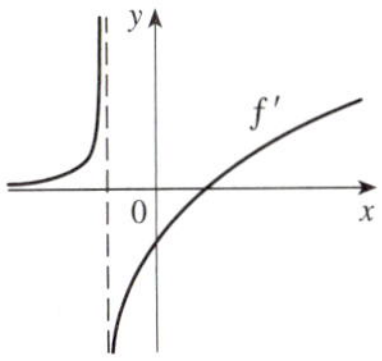

11.

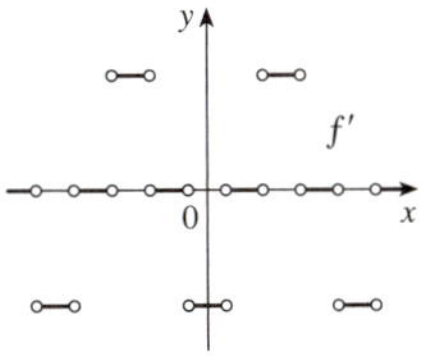

13.

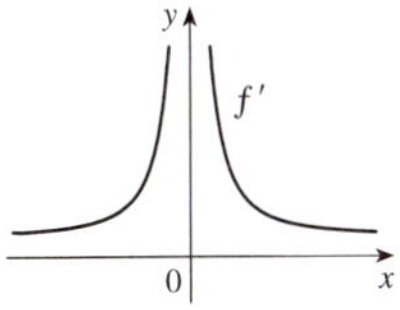

15.

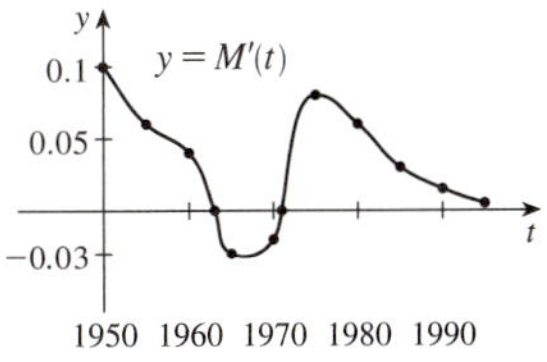

1963 to 1971

17.

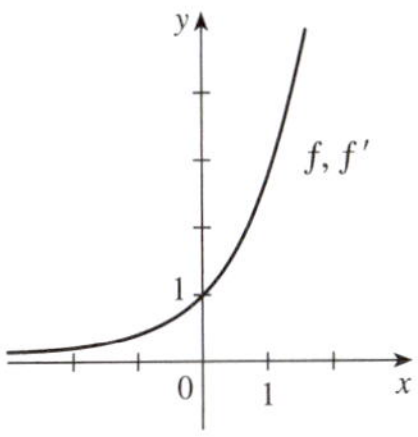

$f'(x) = e^x$

19. (a) 0, 1, 2, 4 (b) −1, −2, −4 (c) $f'(x) = 2x$
21. $f'(x) = 0$, $\mathbb{R}$, $\mathbb{R}$
23. $f'(x) = -6x$, $\mathbb{R}$, $\mathbb{R}$
25. $f'(x) = 3x^2 - 3$, $\mathbb{R}$, $\mathbb{R}$
27. $g'(x) = 1/\sqrt{1 + 2x}$, $\left[-\frac{1}{2}, \infty\right)$, $\left(-\frac{1}{2}, \infty\right)$
29. $G'(t) = 4/(t + 1)^2$, $(-\infty, -1) \cup (-1, \infty)$, $(-\infty, -1) \cup (-1, \infty)$
31. $f'(x) = 4x^3$, $\mathbb{R}$, $\mathbb{R}$ **33.** (a) $f'(x) = 1 + 2/x^2$
35. (a) The rate at which the unemployment rate is changing, in percent unemployed per year
(b)

t	$U'(t)$	t	$U'(t)$
1991	0.70	1996	−0.35
1992	0.05	1997	−0.45
1993	−0.70	1998	−0.35
1994	−0.65	1999	−0.25
1995	−0.35	2000	−0.20

37. 4 (discontinuity); 8 (corner); −1, 11 (vertical tangents)
39.

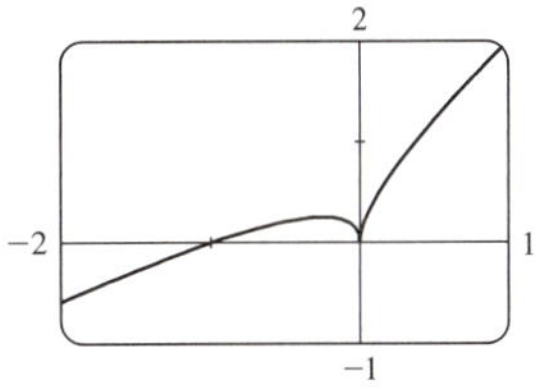

Differentiable at −1; not differentiable at 0

41. (a) $\frac{1}{3}a^{-2/3}$

43. $f'(x) = \begin{cases} -1 & \text{if } x < 6 \\ 1 & \text{if } x > 6 \end{cases}$

or $f'(x) = \dfrac{x - 6}{|x - 6|}$

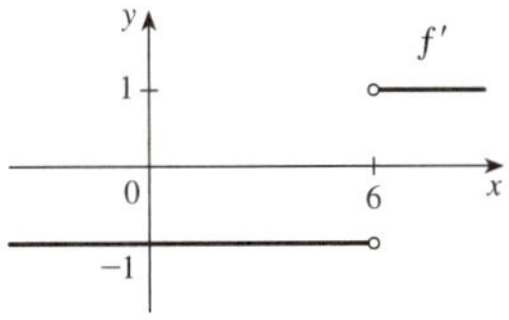

45. (a)

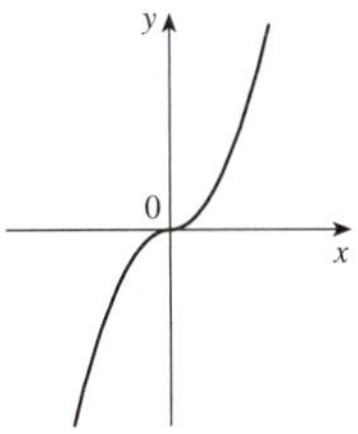

(b) All x
(c) $f'(x) = 2|x|$

49. 63°

Chapter 2 Review ▫ page 176

True-False Quiz

1. False **3.** True **5.** False **7.** True **9.** True
11. False **13.** True **15.** True **17.** False

Exercises

1. (a) (i) 3 (ii) 0 (iii) Does not exist (iv) 2
(v) ∞ (vi) $-\infty$ (vii) 4 (viii) −1
(b) $y = 4$, $y = -1$ (c) $x = 0$, $x = 2$ (d) −3, 0, 2, 4
3. 1 **5.** $\frac{3}{2}$ **7.** 3 **9.** ∞ **11.** $-\frac{1}{8}$ **13.** −1
15. 0 **17.** $-\frac{1}{2}$ **19.** $\frac{1}{2}$ **21.** 0

23. $x = 0, y = 0$ **25.** 1

31. (a) (i) 3 (ii) 0 (iii) Does not exist (iv) 0 (v) 0 (vi) 0 (b) At 0 and 3 (c)

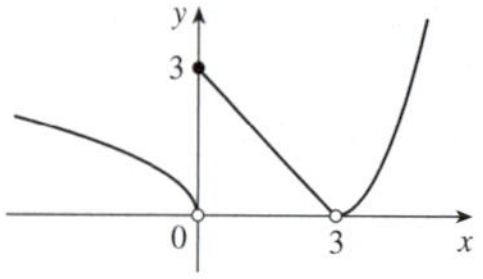

33. $\mathbb{R}$ **37.** (a) -8 (b) $y = -8x + 17$

39. (a) (i) 3 m/s (ii) 2.75 m/s (iii) 2.625 m/s (iv) 2.525 m/s (b) 2.5 m/s

41. (a) 10 (b) $y = 10x - 16$

(c)

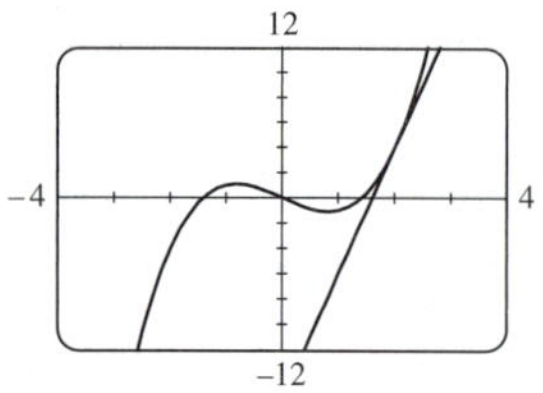

43. (a) The rate at which the cost changes with respect to the interest rate; dollars/(percent per year)

(b) As the interest rate increases past 10%, the cost is increasing at a rate of \$1200/(percent per year).

(c) Always positive

45.

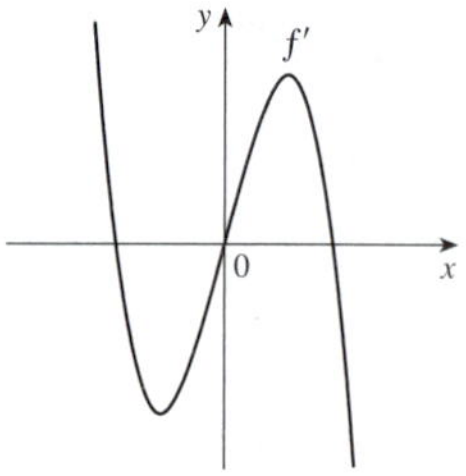

47. (a) $f'(x) = -\frac{5}{2}(3 - 5x)^{-1/2}$

(b) $(-\infty, \frac{3}{5}], (-\infty, \frac{3}{5})$

(c)

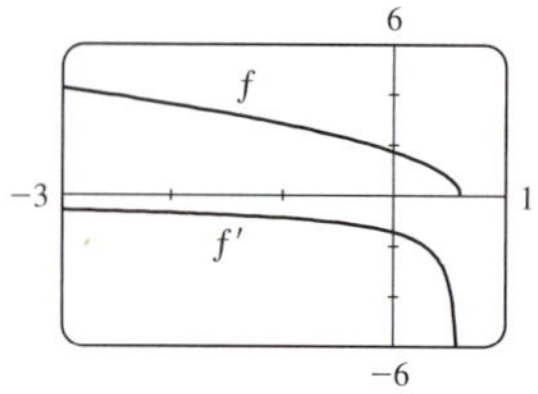

49. -4 (discontinuity), -1 (corner), 2 (discontinuity), 5 (vertical tangent)

51. The rate at which the total value of U.S. banknotes in circulation is changing in billions of dollars per year; \$21.95 billion/year

53. 0

Problems Plus ▫ page 181

1. $\frac{2}{3}$ **3.** -4 **5.** 1 **7.** $a = \frac{1}{2} \pm \frac{1}{2}\sqrt{5}$ **9.** $\frac{3}{4}$

11. (b) Yes (c) Yes; no

13. (a) 0 (b) 1 (c) $f'(x) = x^2 + 1$

Chapter 3

Exercises 3.1 ▫ page 191

1. (a) See Definition of the Number e (page 189).
(b) 0.99, 1.03; $2.7 < e < 2.8$

3. $f'(x) = 0$ **5.** $f'(x) = 5$ **7.** $f'(x) = 2x + 3$

9. $f'(t) = t^3$ **11.** $y' = -\frac{2}{5}x^{-7/5}$

13. $V'(r) = 4\pi r^2$ **15.** $Y'(t) = -54t^{-10}$

17. $G'(x) = 1/(2\sqrt{x}) - 2e^x$

19. $F'(x) = \frac{5}{32}x^4$ **21.** $g'(x) = 2x - (2/x^3)$

23. $y' = \frac{3}{2}\sqrt{x} + (2/\sqrt{x}) - 3/(2x\sqrt{x})$

25. $y' = 0$ **27.** $y' = 2ax + b$

29. $v' = 2t + 3/(4t\sqrt[4]{t^3})$ **31.** $z' = -10A/y^{11} + Be^y$

33. $f'(x) = e^x - 5$ **35.** $45x^{14} - 15x^2$ **37.** 3

39. $y = 2x + 2$ **41.** $y = 3x - 1$

43. (a)

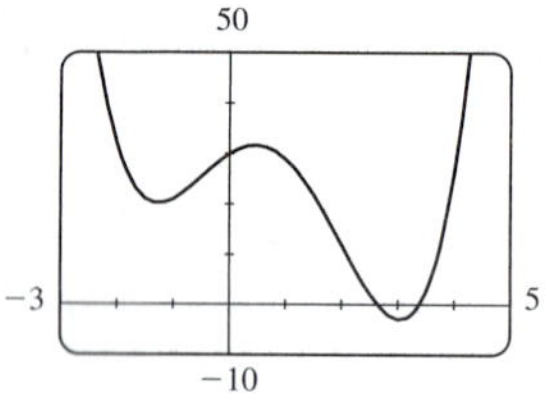

(c) $4x^3 - 9x^2 - 12x + 7$

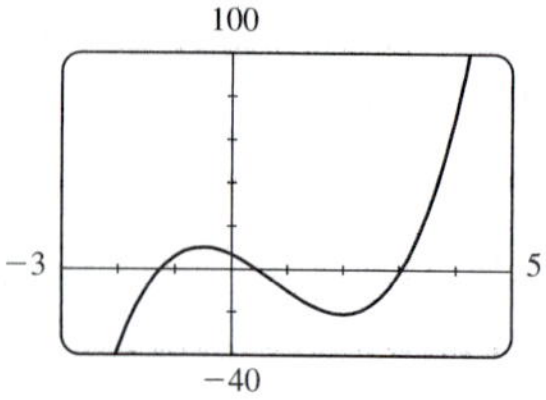

45. $(-2, 21), (1, -6)$

49. $(\pm 2, 4)$

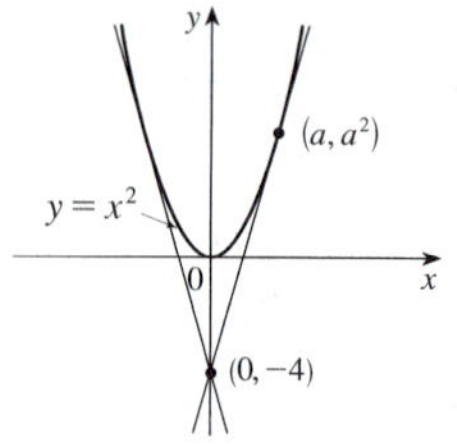

51. $y = \frac{1}{4}x - \frac{7}{2}$

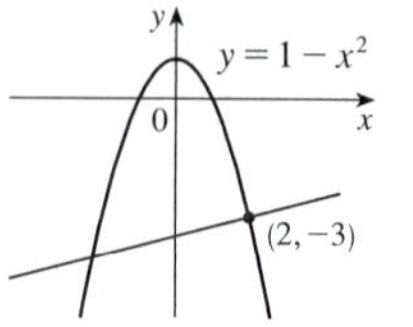

55. No

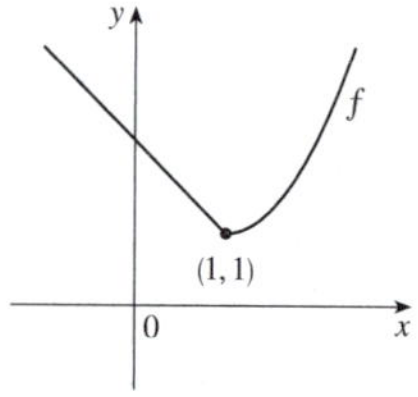

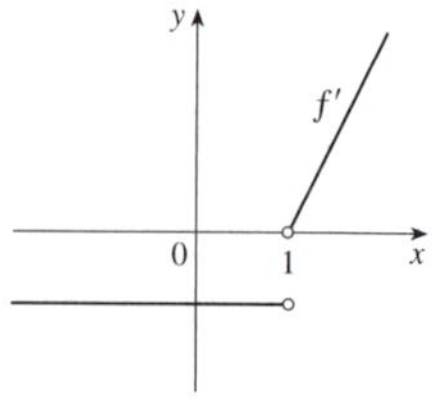

F

57. (a) Not differentiable at 3 or -3

$$f'(x) = \begin{cases} 2x & \text{if } |x| > 3 \\ -2x & \text{if } |x| < 3 \end{cases}$$

(b)

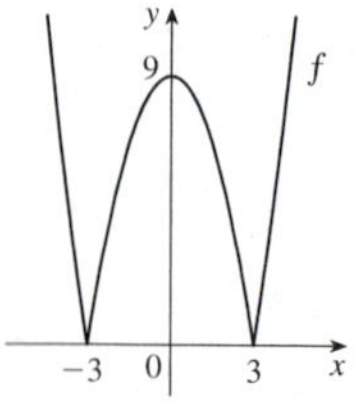

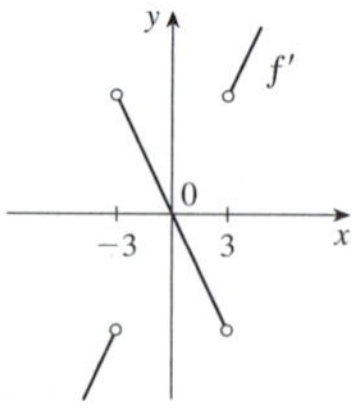

59. $a = -\frac{1}{2}, b = 2$ **61.** $y = \frac{3}{16}x^3 - \frac{9}{4}x + 3$ **63.** 1000

Exercises 3.2 ▫ page 197

1. $y' = 5x^4 + 3x^2 + 2x$ **3.** $f'(x) = x(x + 2)e^x$

5. $y' = (x - 2)e^x/x^3$ **7.** $g'(x) = 5/(2x + 1)^2$

9. $V'(x) = 14x^6 - 4x^3 - 6$ **11.** $F'(y) = 5 + 14/y^2 + 9/y^4$

13. $y' = 2t(1 - t)/(3t^2 - 2t + 1)^2$

15. $y' = (r^2 - 2)e^r$ **17.** $y' = 2v - 1/\sqrt{v}$

19. $y' = -(4x^3 + 2x)/(x^4 + x^2 + 1)^2$

21. $f'(x) = 2cx/(x^2 + c)^2$ **23.** $y = \frac{1}{2}x + \frac{1}{2}$ **25.** $y = 2x$

27. (a) $y = \frac{1}{2}x + 1$ (b)

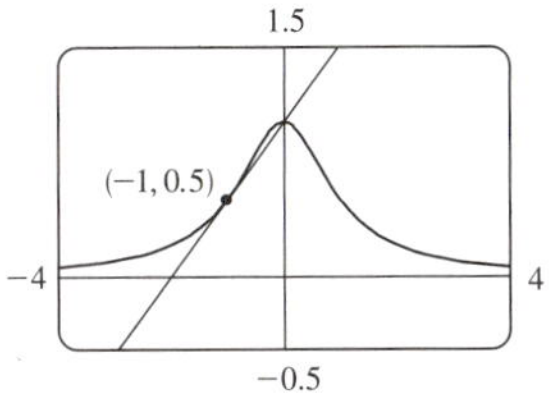

29. (a) $e^x(x-3)/x^4$
31. (a) -16 (b) $-\frac{20}{9}$ (c) 20
33. 7 **35.** (a) 0 (b) $-\frac{2}{3}$
37. (a) $y' = xg'(x) + g(x)$ (b) $y' = [g(x) - xg'(x)]/[g(x)]^2$
(c) $y' = [xg'(x) - g(x)]/x^2$
39. \$1.627 billion/year **41.** Two, $(-2 \pm \sqrt{3}, (1 \mp \sqrt{3})/2)$
43. (c) $3e^{3x}$

Exercises 3.3 ▫ page 208

1. (a) $2t - 10$ (b) -4 ft/s
(c) $t = 5$ (d) $t > 5$
(e) 34 ft
(f) See graph at right.

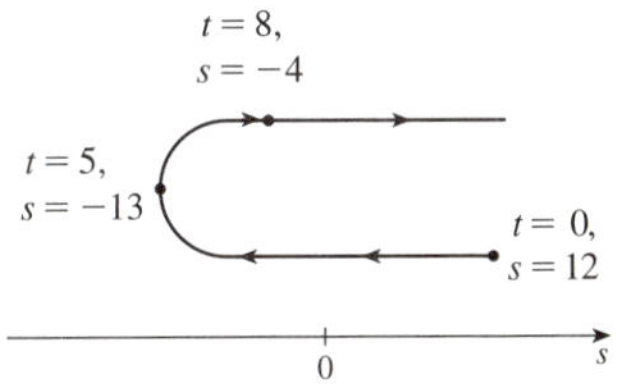

3. (a) $3t^2 - 24t + 36$
(b) -9 ft/s (c) $t = 2, 6$
(d) $0 \leqslant t < 2, t > 6$
(e) 96 ft
(f) See graph at right.

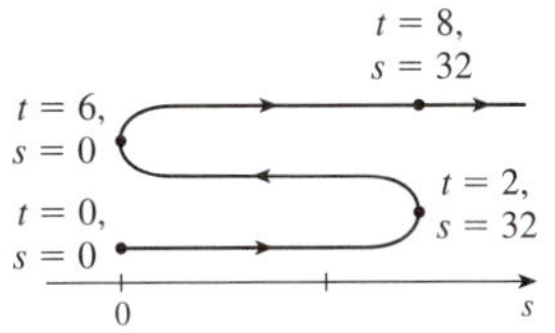

5. (a) $(1 - t^2)/(t^2 + 1)^2$
(b) $-\frac{2}{25}$ ft/s (c) $t = 1$
(d) $0 \leqslant t < 1$ (e) $\frac{57}{65}$ ft
(f) See graph at right.

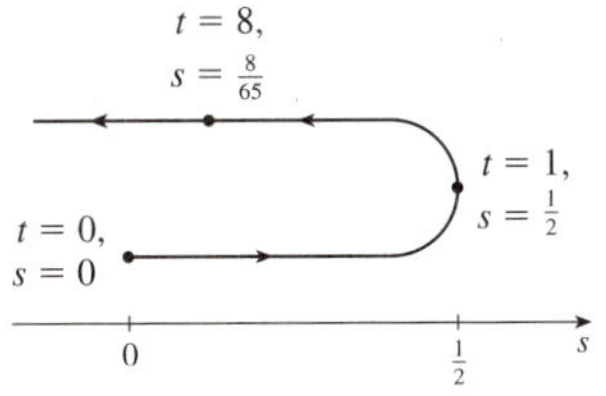

7. $t = 4$ s **9.** (a) 5.02 m/s (b) $\sqrt{17}$ m/s
11. (a) 30 mm²/mm; the rate at which the area is increasing with respect to side length as x reaches 15 mm
(b) $\Delta A \approx 2x\,\Delta x$
13. (a) (i) 5π (ii) 4.5π (iii) 4.1π
(b) 4π (c) $\Delta A \approx 2\pi r\,\Delta r$
15. (a) 8π ft²/ft (b) 16π ft²/ft (c) 24π ft²/ft
The rate increases as the radius increases.
17. (a) 6 kg/m (b) 12 kg/m (c) 18 kg/m
At the right end; at the left end
19. (a) 4.75 A (b) 5 A; $t = \frac{2}{3}$ s
21. (a) $dV/dP = -C/P^2$ (b) At the beginning
23. (a) 16 million/year; 78.5 million/year
(b) $P(t) = at^3 + bt^2 + ct + d$, where $a \approx 0.00129371$, $b \approx -7.061422$, $c \approx 12{,}822.979$, $d \approx -7{,}743{,}771$
(c) $P'(t) = 3at^2 + 2bt + c$ (d) 14.48 million/year; 75.29 million/year (e) 81.62 million/year

25. (a) $a^2k/(akt + 1)^2$ (c) It approaches a moles/L.
(d) It approaches 0. (e) The reaction virtually stops.
27. (a) 0.926 cm/s; 0.694 cm/s; 0
(b) 0; -92.6 (cm/s)/cm; -185.2 (cm/s)/cm
(c) At the center; at the edge
29. (a) $C'(x) = 3 + 0.02x + 0.0006x^2$
(b) \$11/pair, the rate at which the cost is changing as the 100th pair of jeans is being produced; the cost of the 101st pair
(c) \$11.07
31. (a) $[xp'(x) - p(x)]/x^2$; the average productivity increases as new workers are added.
33. -0.2436 K/min
35. (a) 0 and 0 (b) $C = 0$
(c) (0, 0), (500, 50); it is possible for the species to coexist.

Exercises 3.4 ▫ page 216

1. $f'(x) = 1 - 3\cos x$ **3.** $y' = \cos x + 10\sec^2 x$
5. $g'(t) = 3t^2\cos t - t^3\sin t$
7. $h'(\theta) = -\csc\theta\cot\theta + e^\theta(\cot\theta - \csc^2\theta)$
9. $y' = \dfrac{\cos x + x\sin x}{\cos^2 x}$ **11.** $f'(\theta) = \dfrac{\sec\theta\tan\theta}{(1+\sec\theta)^2}$
13. $y' = (x\cos x - 2\sin x)/x^3$ **15.** $y' = \sec\theta\,(\sec^2\theta + \tan^2\theta)$
21. $y = 2x + 1 - \frac{1}{2}\pi$ **23.** $y = x + 1$
25. (a) $y = -x$ (b)

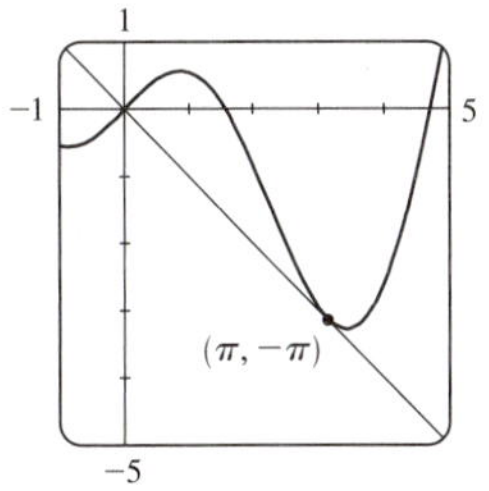

27. (a) $2 - \csc^2 x$ **29.** $(2n + 1)\pi \pm \frac{1}{3}\pi$, n an integer
31. (a) $8\cos t$ (b) $4\sqrt{3}$, -4; to the left **33.** 5 ft/rad
35. 3 **37.** 3 **39.** sin 1 **41.** $\frac{1}{2}$ **43.** $\frac{1}{2}$
45. (a) $\sec^2 x = 1/\cos^2 x$ (b) $\sec x\tan x = (\sin x)/\cos^2 x$
(c) $\cos x - \sin x = (\cot x - 1)/\csc x$
47. 1

Exercises 3.5 ▫ page 224

1. $4\cos 4x$ **3.** $-20x(1 - x^2)^9$ **5.** $e^{\sqrt{x}}/(2\sqrt{x})$
7. $F'(x) = 7(x^3 + 4x)^6(3x^2 + 4)$ [or $7x^6(x^2 + 4)^6(3x^2 + 4)$]
9. $F'(x) = \dfrac{2 + 3x^2}{4(1 + 2x + x^3)^{3/4}}$ **11.** $g'(t) = -\dfrac{12t^3}{(t^4 + 1)^4}$
13. $y' = -3x^2\sin(a^3 + x^3)$ **15.** $y' = -me^{-mx}$
17. $g'(x) = 4(1 + 4x)^4(3 + x - x^2)^7(17 + 9x - 21x^2)$
19. $y' = 8(2x - 5)^3(8x^2 - 5)^{-4}(-4x^2 + 30x - 5)$
21. $y' = e^{-x^2}(1 - 2x^2)$ **23.** $y' = (\cos x - x\sin x)e^{x\cos x}$
25. $F'(z) = 1/[(z - 1)^{1/2}(z + 1)^{3/2}]$
27. $y' = (r^2 + 1)^{-3/2}$ **29.** $y' = -\sin x\sec^2(\cos x)$
31. $y' = 2^{\sin\pi x}(\pi\ln 2)\cos\pi x$
33. $y' = -12\cos x\sin x\,(1 + \cos^2 x)^5$
35. $y' = 4\sec^2 x\tan x$

37. $y' = -2\cos\theta\cot(\sin\theta)\csc^2(\sin\theta)$

39. $y' = \dfrac{1 + 1/(2\sqrt{x})}{2\sqrt{x + \sqrt{x}}}$

41. $y' = \cos(\tan\sqrt{\sin x})(\sec^2\sqrt{\sin x})[1/(2\sqrt{\sin x})](\cos x)$

43. $y = 20x + 1$ **45.** $y = -x + \pi$

47. (a) $y = \frac{1}{2}x + 1$ (b)

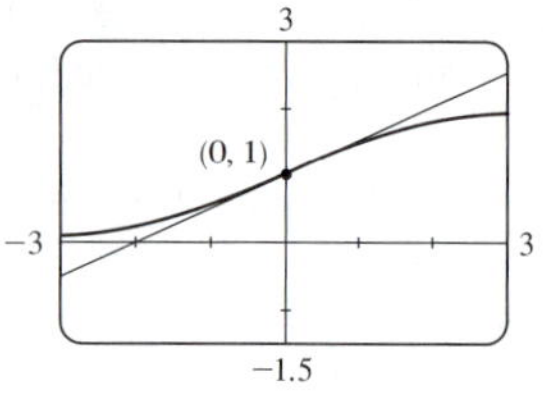

49. (a) $-1/(x^2\sqrt{1 - x^2})$

51. $((\pi/2) + 2n\pi, 3)$, $((3\pi/2) + 2n\pi, -1)$, n an integer

53. 28 **55.** (a) 30 (b) 36

57. (a) $\frac{3}{4}$ (b) Does not exist (c) -2 **59.** -17.4

61. (a) $F'(x) = e^x f'(e^x)$ (b) $G'(x) = e^{f(x)}f'(x)$

63. (a) $f'(x) = 4/x$ (b) $g'(x) = 1/x$
(c) $F'(x) = 4[L(x)]^3/x$ (d) $G'(x) = -1/x$

65. $v(t) = \frac{5}{2}\pi\cos(10\pi t)$ cm/s

67. (a) $dB/dt = \frac{7}{54}\pi\cos(2\pi t/5.4)$ (b) 0.16

69. $v(t) = 2e^{-1.5t}(2\pi\cos 2\pi t - 1.5\sin 2\pi t)$

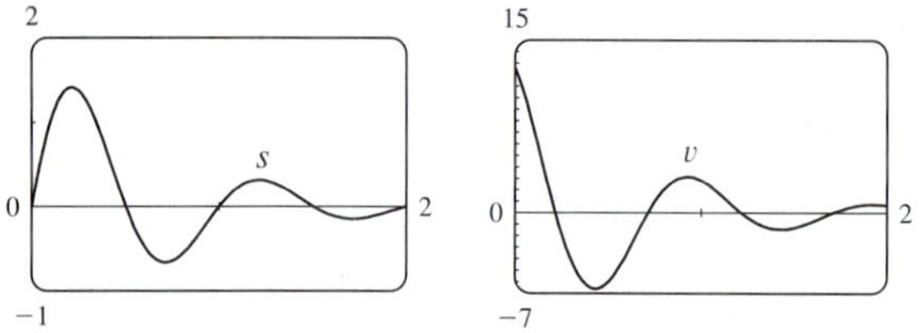

71. (a) $y \approx 100.012437e^{-10.005531t}$ (b) -670.63 μA

73. (b) The factored form

77. (b) $-n\cos^{n-1}x\sin[(n + 1)x]$

Exercises 3.6 ▫ page 233

1. (a) $y' = -(y + 2 + 6x)/x$
(b) $y = (4/x) - 2 - 3x$, $y' = -(4/x^2) - 3$

3. (a) $y' = -y^2/x^2$ (b) $y = x/(x - 1)$, $y' = -1/(x - 1)^2$

5. $y' = -x/y$ **7.** $y' = -x(3x + 2y)/(x^2 + 8y)$

9. $y' = (3 - 2xy - y^2)/(x^2 + 2xy)$

11. $y' = (-2xy^2 - \sin y)/(2x^2y + x\cos y)$

13. $y' = \tan x\tan y$ **15.** $y' = (1 - 2xye^{x^2y})/(x^2e^{x^2y} - 1)$

17. $y' = (4xy\sqrt{xy} - y)/(x - 2x^2\sqrt{xy})$

19. $y' = -y/x$ **21.** $-\frac{16}{13}$

23. $(1 - 4y^3 - 2x^2y - x^4)/(2xy^2 + 4x^3y)$

25. $y = -x + 2$ **27.** $y = x + \frac{1}{2}$ **29.** $y = -\frac{9}{13}x + \frac{40}{13}$

31. (a) $y = \frac{9}{2}x - \frac{5}{2}$ (b)

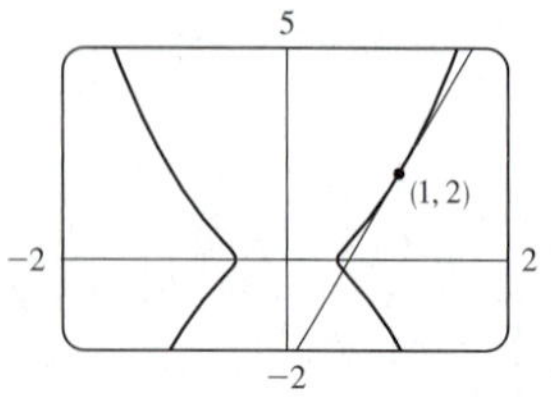

33. (a)

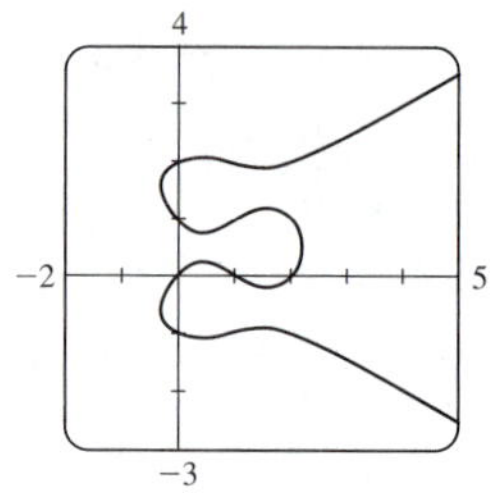

Eight; $x \approx 0.42, 1.58$
(b) $y = -x + 1$, $y = \frac{1}{3}x + 2$
(c) $1 \mp \sqrt{3}/3$

35. $(\pm\frac{5}{4}\sqrt{3}, \pm\frac{5}{4})$ **37.** $(x_0x/a^2) - (y_0y/b^2) = 1$

41. $y' = 1/[2\sqrt{x}(1 + x)]$ **43.** $y' = 1/\sqrt{-x^2 - x}$

45. $H'(x) = 1 + 2x\arctan x$ **47.** $h'(t) = 0$

49. $y' = -2e^{2x}/\sqrt{1 - e^{4x}}$ **51.** $e^x - x^2/(1 + x^2) - 2x\arctan x$

57.

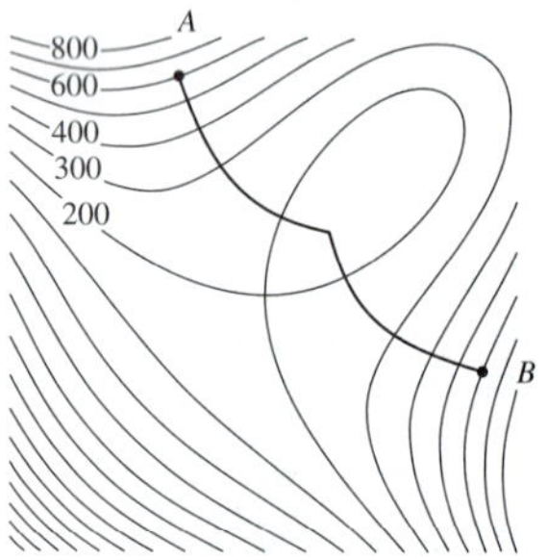

59.

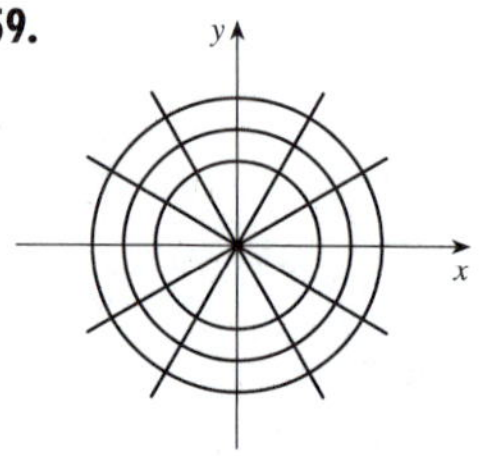

61.

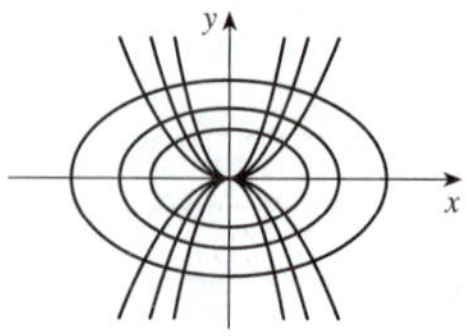

63. $(\pm\sqrt{3}, 0)$

65. $(-1, -1)$, $(1, 1)$

67. (b) $\frac{3}{2}$ **69.** 2

Exercises 3.7 ▫ page 240

1. $a = f$, $b = f'$, $c = f''$

3. a = acceleration, b = velocity, c = position

5. $f'(x) = 5x^4 + 12x - 7$, $f''(x) = 20x^3 + 12$

7. $y' = -2\sin 2\theta$, $y'' = -4\cos 2\theta$

9. $F'(t) = -42(1 - 7t)^5$, $F''(t) = 1470(1 - 7t)^4$

11. $h'(u) = -7/(1 + 3u)^2$, $h''(u) = 42/(1 + 3u)^3$

13. $h'(x) = x/\sqrt{x^2 + 1}$, $h''(x) = 1/(x^2 + 1)^{3/2}$

15. $y' = 2x^2/(x^3 + 1)^{1/3}$, $y'' = 4x/(x^3 + 1)^{1/3} - 2x^4/(x^3 + 1)^{4/3}$

17. $H'(t) = 3\sec^2 3t$, $H''(t) = 18\sec^2 3t\tan 3t$

19. $g'(t) = t^2e^{5t}(5t + 3)$, $g''(t) = te^{5t}(25t^2 + 30t + 6)$

21. (a) $f'(x) = 2\sin x(\cos x - 1) = \sin 2x - 2\sin x$,
$f''(x) = 2(\cos 2x - \cos x)$ **23.** $3(2x + 3)^{-5/2}$ **25.** -3

27. -80 **29.** $-81/y^3$ **31.** $-2x/y^5$ **33.** $n!$

35. 2^ne^{2x} **37.** $(-1)^n(n + 2)!/(6x^{n+3})$ **39.** $2^{103}\sin 2x$

41. 9 ft/s²

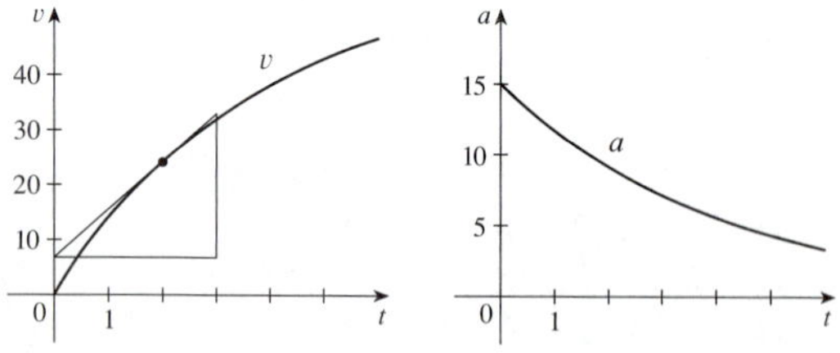

43. (a) $v(t) = 6t^2 - 30t + 36$, $a(t) = 12t - 30$
(b) -18 m/s^2 (c) $a(2) = -6 \text{ m/s}^2$, $a(3) = 6 \text{ m/s}^2$
45. (a) $v(t) = (\pi/6)[\cos(\pi t/6) - \sin(\pi t/6)]$,
$a(t) = -(\pi^2/36)[\sin(\pi t/6) + \cos(\pi t/6)]$
(b) $-(\pi^2/72)(1 + \sqrt{3}) \text{ m/s}^2$ (c) $a(\frac{3}{2}) = -(\pi^2/36)\sqrt{2} \text{ m/s}^2$
47. (a) $t = 0, 2$ (b) $s(0) = 2$ m, $v(0) = 0$ m/s;
$s(2) = -14$ m, $v(2) = -16$ m/s
49. (a) $6t - 24$; -6 m/s^2 (b)

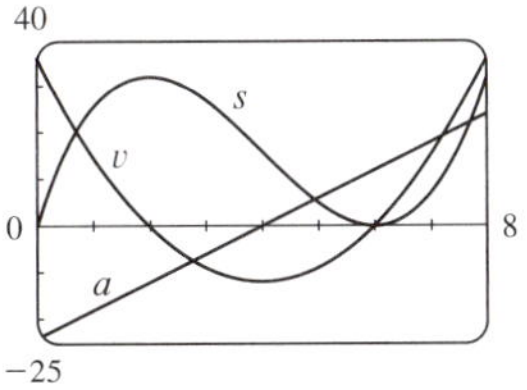

(c) Speeding up when $2 < t < 4$ or $t > 6$; slowing down when $0 \leqslant t < 2$ or $4 < t < 6$
51. (a) $v(t) = A\omega \cos \omega t$, $a(t) = -A\omega^2 \sin \omega t$
53. $P(x) = x^2 - x + 3$ **55.** $A = -\frac{3}{10}$, $B = -\frac{1}{10}$
57. $r = 1, -6$ **59.** $f''(x) = 6xg'(x^2) + 4x^3g''(x^2)$
61. $f''(x) = [\sqrt{x}g''(\sqrt{x}) - g'(\sqrt{x})]/(4x\sqrt{x})$
63. (a) $f'(x) = -(2x + 1)/(x^2 + x)^2$,
$f''(x) = 2(3x^2 + 3x + 1)/(x^2 + x)^3$,
$f'''(x) = -6(4x^3 + 6x^2 + 4x + 1)/(x^2 + x)^4$,
$f^{(4)}(x) = 24(5x^4 + 10x^3 + 10x^2 + 5x + 1)/(x^2 + x)^5$
(b) $f^{(n)}(x) = (-1)^n n![x^{-(n+1)} - (x + 1)^{-(n+1)}]$
65. $f'(x) = (x^2 + 2x)e^x$, $f''(x) = (x^2 + 4x + 2)e^x$,
$f'''(x) = (x^2 + 6x + 6)e^x$, $f^{(4)}(x) = (x^2 + 8x + 12)e^x$,
$f^{(5)}(x) = (x^2 + 10x + 20)e^x$; $f^{(n)}(x) = [x^2 + 2nx + n(n - 1)]e^x$

Exercises 3.8 ▫ page 249

1. The differentiation formula is simplest.
3. $f'(\theta) = -\tan\theta$ **5.** $f'(x) = 3/[(3x - 1)\ln 2]$
7. $f'(x) = 1/[5x\sqrt[5]{(\ln x)^4}]$ **9.** $f'(x) = (2 + \ln x)/(2\sqrt{x})$
11. $F'(t) = 6/(2t + 1) - 12/(3t - 1)$
13. $g'(x) = -2a/(a^2 - x^2)$ **15.** $f'(u) = \dfrac{1 + \ln 2}{u[1 + \ln(2u)]^2}$
17. $y' = (10x + 1)/(5x^2 + x - 2)$
19. $y' = -x/(1 + x)$ **21.** $y' = 1 + \ln x$, $y'' = 1/x$
23. $y' = 1/(x \ln 10)$, $y'' = -1/(x^2 \ln 10)$
25. $f'(x) = \dfrac{2x - 1 - (x - 1)\ln(x - 1)}{(x - 1)[1 - \ln(x - 1)]^2}$; $(1, 1 + e) \cup (1 + e, \infty)$
27. $f'(x) = 2x\ln(1 - x^2) - 2x^3/(1 - x^2)$; $(-1, 1)$
29. 0 **31.** $x - ey = e$ **33.** $\cos x + 1/x$
35. $y' = (2x + 1)^5(x^4 - 3)^6\left(\dfrac{10}{2x + 1} + \dfrac{24x^3}{x^4 - 3}\right)$
37. $y' = \dfrac{\sin^2 x \tan^4 x}{(x^2 + 1)^2}\left(2\cot x + \dfrac{4\sec^2 x}{\tan x} - \dfrac{4x}{x^2 + 1}\right)$
39. $y' = x^x(1 + \ln x)$ **41.** $y' = x^{\sin x}[(\sin x)/x + \cos x \ln x]$
43. $y' = (\ln x)^x(1/\ln x + \ln\ln x)$ **45.** $y' = e^x x^{e^x}(\ln x + 1/x)$
47. $y' = 2x/(x^2 + y^2 - 2y)$
49. $f^{(n)}(x) = (-1)^{n-1}(n - 1)!/(x - 1)^n$

Exercises 3.9 ▫ page 254

1. (a) 0 (b) 1 **3.** (a) $\frac{3}{4}$ (b) $\frac{1}{2}(e^2 - e^{-2}) \approx 3.62686$
5. (a) 1 (b) 0
21. $\coth x = \frac{5}{4}$, $\operatorname{sech} x = \frac{3}{5}$, $\cosh x = \frac{5}{3}$, $\sinh x = \frac{4}{3}$, $\operatorname{csch} x = \frac{3}{4}$
23. (a) 1 (b) -1 (c) ∞ (d) $-\infty$ (e) 0 (f) 1
(g) ∞ (h) $-\infty$ (i) 0
31. $f'(x) = x\sinh x + \cosh x$ **33.** $h'(x) = 2x\cosh(x^2)$
35. $G'(x) = -(2\sinh x)/(1 + \cosh x)^2$
37. $h'(t) = -(t\operatorname{csch}^2\sqrt{1 + t^2})/\sqrt{1 + t^2}$
39. $H'(t) = e^t \operatorname{sech}^2(e^t)$
41. $y' = 3e^{\cosh 3x}\sinh 3x$ **43.** $y' = 1/[2\sqrt{x}(1 - x)]$
45. $y' = \sinh^{-1}(x/3)$ **47.** $y' = -1/(x\sqrt{x^2 + 1})$
49. (a) 0.3572 (b) 70.34°
51. (b) $y = 2\sinh 3x - 4\cosh 3x$ **53.** $(\ln(1 + \sqrt{2}), \sqrt{2})$

Exercises 3.10 ▫ page 260

1. $dV/dt = 3x^2\,dx/dt$ **3.** 70 **5.** $\pm\frac{46}{13}$
7. (a) The plane's altitude is 1 mi and its speed is 500 mi/h.
(b) The rate at which the distance from the plane to the station is increasing when the plane is 2 mi from the station
(c)

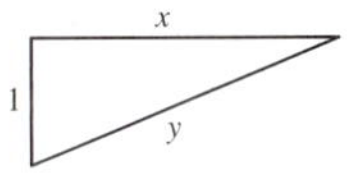

(d) $y^2 = x^2 + 1$ (e) $250\sqrt{3}$ mi/h

9. (a) The height of the pole (15 ft), the height of the man (6 ft), and the speed of the man (5 ft/s)
(b) The rate at which the tip of the man's shadow is moving when he is 40 ft from the pole.
(c)

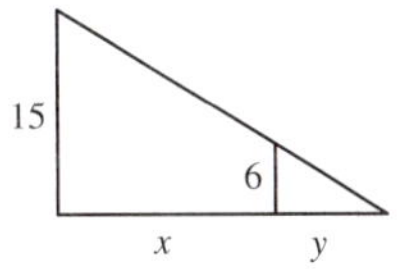

(d) $\dfrac{15}{6} = \dfrac{x + y}{y}$ (e) $\frac{25}{3}$ ft/s

11. 65 mi/h **13.** $837/\sqrt{8674} \approx 8.99$ ft/s
15. -1.6 cm/min **17.** $\frac{720}{13} \approx 55.4$ km/h
19. $(10{,}000 + 800{,}000\pi/9) \approx 2.89 \times 10^5 \text{ cm}^3\text{/min}$
21. $\frac{10}{3}$ cm/min **23.** $6/(5\pi) \approx 0.38$ ft/min **25.** $0.3 \text{ m}^2\text{/s}$
27. $80 \text{ cm}^3\text{/min}$ **29.** $\frac{107}{810} \approx 0.132\ \Omega$/s **31.** $\sqrt{2}/5$ rad/s
33. (a) 360 ft/s (b) 0.096 rad/s
35. $1650/\sqrt{31} \approx 296$ km/h **37.** $7\sqrt{15}/4 \approx 6.78$ m/s

Exercises 3.11 ▫ page 267

1. 148°F; underestimate
3. 22.6%, 24.2%; too high; tangent lines lie above the curve
5. $L(x) = 3x - 2$ **7.** $L(x) = -x + \pi/2$
9. $\sqrt{1 - x} \approx 1 - \frac{1}{2}x$;
$\sqrt{0.9} \approx 0.95$,
$\sqrt{0.99} \approx 0.995$

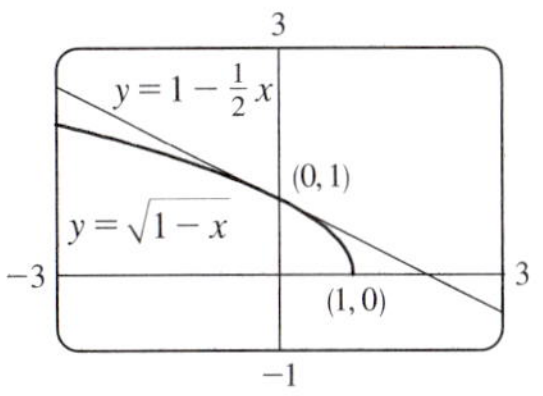

11. $-1.204 < x < 0.706$ **13.** $-0.045 < x < 0.055$
15. $dy = (4x^3 + 5)\,dx$ **17.** $dy = (1 + \ln x)\,dx$
19. $dy = -2/(u - 1)^2\,du$ **21.** (a) $dy = (2x + 2)\,dx$ (b) 4
23. (a) $dy = 5/(2\sqrt{4 + 5x})\,dx$ (b) 0.05

25. (a) $dy = \sec^2 x\, dx$ (b) -0.2
27. $\Delta y = 1.25$, $dy = 1$ **29.** $\Delta y = 1.44$, $dy = 1.6$

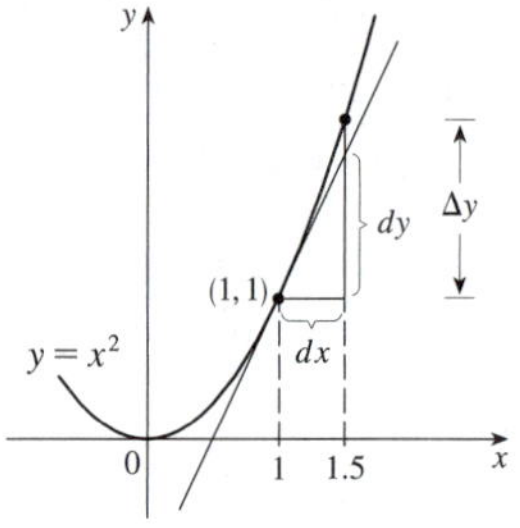

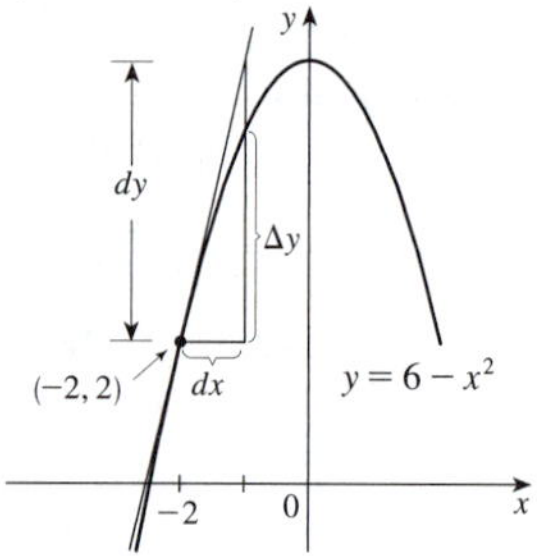

31. 32.08 **33.** 4.02 **35.** $1 - \pi/90 \approx 0.965$
41. (a) $270\ \text{cm}^3$, 0.01, 1% (b) $36\ \text{cm}^2$, $0.00\overline{6}$, $0.\overline{6}\%$
43. (a) $84/\pi \approx 27\ \text{cm}^2$; $\frac{1}{84} \approx 0.012$
(b) $1764/\pi^2 \approx 179\ \text{cm}^3$; $\frac{1}{56} \approx 0.018$
45. (a) $2\pi rh\,\Delta r$ (b) $\pi(\Delta r)^2 h$
49. (a) 4.8, 5.2 (b) Too large

Chapter 3 Review • page 270

True-False Quiz

1. True **3.** True **5.** False **7.** False **9.** True
11. True **13.** False

Exercises

1. $6x(x^4 - 3x^2 + 5)^2(2x^2 - 3)$ **3.** $1/(2\sqrt{x}) - 4/(3\sqrt[3]{x^7})$
5. $2(2x^2 + 1)/\sqrt{x^2 + 1}$ **7.** $2 \cos 2\theta\, e^{\sin 2\theta}$
9. $(t^2 + 1)/(1 - t^2)^2$ **11.** $e^{-1/x}(1/x + 1)$
13. $-(\sec^2\sqrt{1 - x})/(2\sqrt{1 - x})$
15. $(1 - y^4 - 2xy)/(4xy^3 + x^2 - 3)$
17. $2 \sec 2\theta\,(\tan 2\theta - 1)/(1 + \tan 2\theta)^2$
19. $(1 + c^2)e^{cx}\sin x$ **21.** $e^{x + e^x}$ **23.** $-(x - 1)^{-2}$
25. $[2x - y\cos(xy)]/[x\cos(xy) + 1]$ **27.** $2/[(1 + 2x)\ln 5]$
29. $\cot x - \sin x \cos x$ **31.** $4x/(1 + 16x^2) + \tan^{-1}(4x)$
33. $5 \sec 5x$ **35.** $-6x \csc^2(3x^2 + 5)$
37. $\cos(\tan\sqrt{1 + x^3})(\sec^2\sqrt{1 + x^3})[3x^2/(2\sqrt{1 + x^3})]$
39. $2 \cos\theta \tan(\sin\theta)\sec^2(\sin\theta)$
41. $\dfrac{(x - 2)^4(3x^2 - 55x - 52)}{2\sqrt{x + 1}\,(x + 3)^8}$ **43.** $2x^2\cosh(x^2) + \sinh(x^2)$
45. $3 \tanh 3x$ **47.** $(\cosh x)/\sqrt{\sinh^2 x - 1}$
49. $-\frac{4}{27}$ **51.** $-5x^4/y^{11}$ **55.** $y = 2\sqrt{3}x + 1 - \pi\sqrt{3}/3$
57. $y = 2x + 1$ **59.** $y = -x + 2$
61. (a) $(10 - 3x)/(2\sqrt{5 - x})$ (b) $y = \frac{7}{4}x + \frac{1}{4}$, $y = -x + 8$
(c)

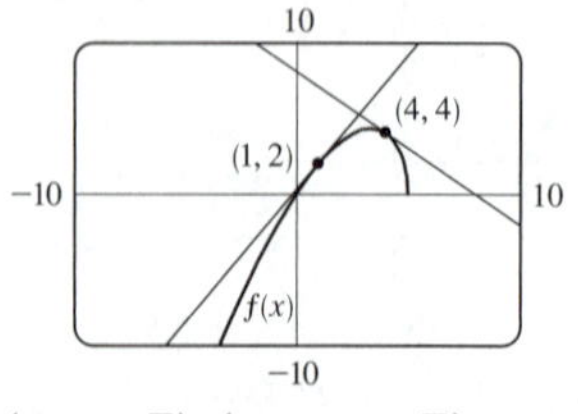

63. $(\pi/4, \sqrt{2})$, $(5\pi/4, -\sqrt{2})$ **67.** (a) 2 (b) 44
69. $f'(x) = 2xg(x) + x^2g'(x)$ **71.** $f'(x) = 2g(x)g'(x)$
73. $f'(x) = g'(e^x)e^x$ **75.** $f'(x) = g'(x)/g(x)$
77. $h'(x) = \dfrac{f'(x)[g(x)]^2 + g'(x)[f(x)]^2}{[f(x) + g(x)]^2}$
79. $h'(x) = f'(g(\sin 4x))g'(\sin 4x)(\cos 4x)(4)$
81. $(-3, 0)$ **83.** $y = -\frac{2}{3}x^2 + \frac{14}{3}x$
85. $v(t) = -Ae^{-ct}[c\cos(\omega t + \delta) + \omega\sin(\omega t + \delta)]$,
$a(t) = Ae^{-ct}[(c^2 - \omega^2)\cos(\omega t + \delta) + 2c\omega\sin(\omega t + \delta)]$
87. (a) $v(t) = 3t^2 - 12$, $a(t) = 6t$
(b) Upward when $t > 2$, downward when $0 \le t < 2$ (c) 23
89. 4 kg/m **91.** $\frac{4}{3}\ \text{cm}^2/\text{min}$ **93.** 13 ft/s **95.** 400 ft/h
97. (a) $L(x) = 1 + x$, $\sqrt[3]{1 + 3x} \approx 1 + x$, $\sqrt[3]{1.03} \approx 1.01$
(b) $-0.23 < x < 0.40$
99. $12 + 3\pi/2 \approx 16.7\ \text{cm}^2$ **101.** $\frac{1}{32}$ **103.** $\frac{1}{4}$ **105.** $\frac{1}{8}x^2$

Problems Plus ▫ page 274

1. $(\pm\sqrt{3}/2, \frac{1}{4})$ **7.** $(0, \frac{5}{4})$
9. (a) $4\pi\sqrt{3}/\sqrt{11}$ rad/s (b) $40(\cos\theta + \sqrt{8 + \cos^2\theta})$ cm
(c) $-480\pi\sin\theta\,(1 + \cos\theta/\sqrt{8 + \cos^2\theta})$ cm/s
13. $x_T \in (3, \infty)$, $y_T \in (2, \infty)$, $x_N \in (0, \frac{5}{3})$, $y_N \in (-\frac{5}{2}, 0)$
15. (b) (i) 53° (or 127°) (ii) 63° (or 117°)
17. R approaches the midpoint of the radius AO.
19. $-\sin a$ **21.** $2\sqrt{e}$ **25.** $(1, -2)$, $(-1, 0)$
27. $\sqrt{29}/58$ **29.** $2 + \frac{375}{128}\pi \approx 11.204\ \text{cm}^3/\text{min}$

Chapter 4

Abbreviations: max., maximum; min., minimum; loc., local; abs., absolute; inc., increasing; dec., decreasing; CD, concave downward; CU, concave upward; HA, horizontal asymptote; VA, vertical asymptote; IP, inflection point

Exercises 4.1 ▫ page 285

1. Absolute minimum: smallest function value on the entire domain of the function; local minimum at c: smallest function value when x is near c
3. Abs. max. at b, loc. max. at b and e, abs. min. at d, loc. min. at d and s
5. Abs. max. $f(4) = 4$; abs. min. $f(7) = 0$; loc. max. $f(4) = 4$ and $f(6) = 3$; loc. min. $f(2) = 1$ and $f(5) = 2$
7.

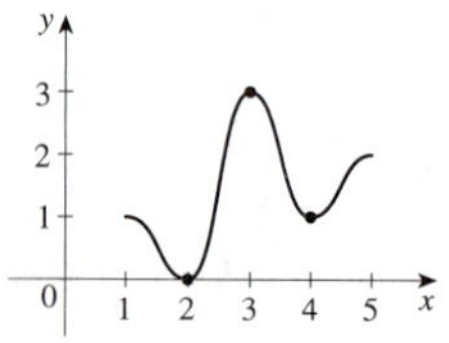

9.

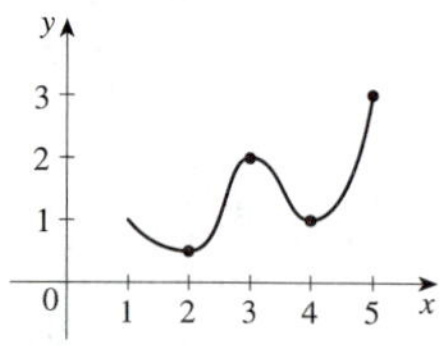

11. (a)

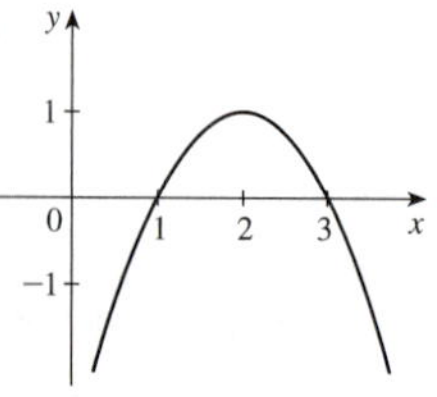

(b)

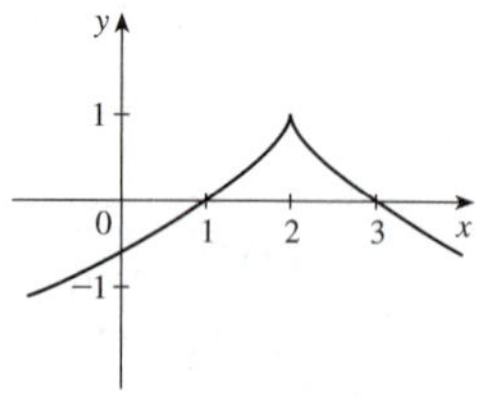

(c)

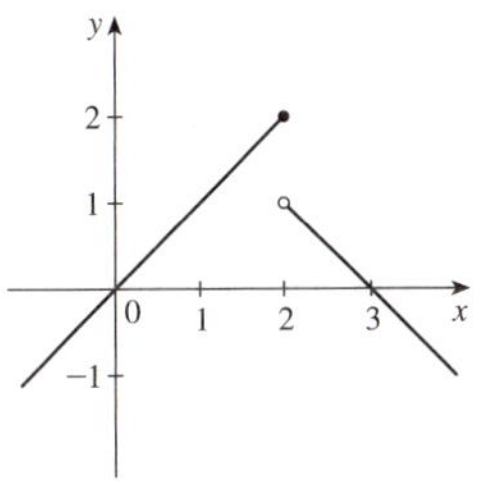

13. (a)

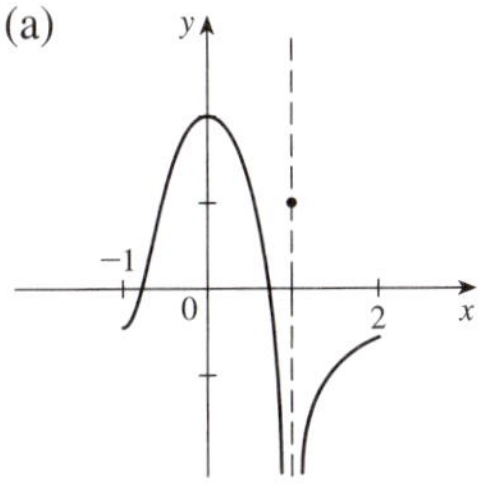

(b)

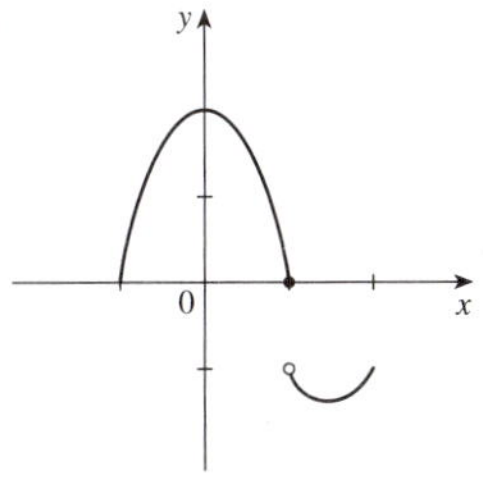

15. Abs. max. $f(1) = 5$ **17.** None
19. Abs. min. $f(0) = 0$
21. Abs. max. $f(-3) = 9$; abs. and loc. min. $f(0) = 0$
23. None
25. Abs. and loc. max. $f(\pi/2) = f(-3\pi/2) = 1$;
abs. and loc. min. $f(3\pi/2) = f(-\pi/2) = -1$
27. Abs. max. $f(0) = 1$ **29.** Abs. max. $f(3) = 2$
31. $-\frac{2}{5}$ **33.** $-4, 2$ **35.** $0, (-1 \pm \sqrt{5})/2$ **37.** $-\frac{3}{2}$
39. $-2, 0$ **41.** $0, \frac{8}{7}, 4$ **43.** $n\pi$ (n an integer)
45. $1/e$ **47.** $f(0) = 5, f(2) = -7$
49. $f(-1) = 8, f(2) = -19$
51. $f(3) = 66, f(\pm 1) = 2$ **53.** $f(1) = \frac{1}{2}, f(0) = 0$
55. $f(\sqrt{2}) = 2, f(-1) = -\sqrt{3}$
57. $f(\pi/4) = \sqrt{2}, f(0) = 1$
59. $f(1) = 1/e, f(0) = 0$
61. $f(1) = 1, f(3) = 3 - 3\ln 3 \approx -0.296$
63. $f\left(\dfrac{a}{a+b}\right) = \dfrac{a^a b^b}{(a+b)^{a+b}}$
65. (a) 9.71, −7.71 (b) $1 \pm 32\sqrt{6}/9$
67. (a) 0.32, 0.00 (b) $3\sqrt{3}/16, 0$
69. $\approx 3.9665°\text{C}$
71. Cheapest, $t = 10$; most expensive, $t \approx 5.1309$
73. (a) $r = \frac{2}{3}r_0$ (b) $v = \frac{4}{27}kr_0^3$

(c)

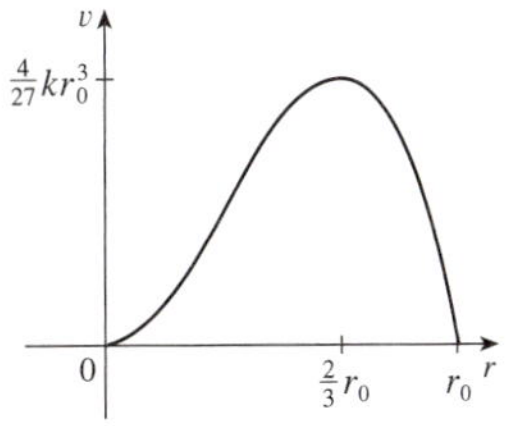

Exercises 4.2 ▫ page 295

1. 2 **3.** $\pm\frac{1}{4}, \pm\frac{3}{4}$
5. f is not differentiable on $(-1, 1)$ **7.** 0.8, 3.2, 4.4, 6.1

9. (a), (b)

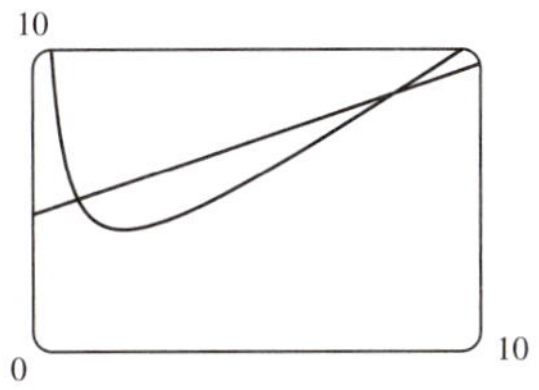

(c) $2\sqrt{2}$

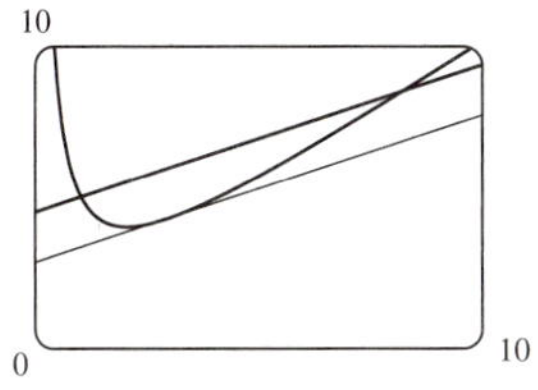

11. 0 **13.** $-\frac{1}{2}\ln[(1 - e^{-6})/6]$
15. f is not differentiable at 1 **23.** 16
25. No **31.** No

Exercises 4.3 ▫ page 304

1. (a) (0, 6), (8, 9) (b) (6, 8) (c) (2, 4), (7, 9)
(d) (0, 2), (4, 7) (e) (2, 3), (4, 4.5), (7, 4)
3. (a) I/D Test (b) Concavity Test
(c) Find points at which the concavity changes.
5. (a) Inc. on (1, 5); dec. on (0, 1) and (5, 6)
(b) Loc. max. at $x = 5$, loc. min. at $x = 1$
7. $x = 1, 7$ **9.**

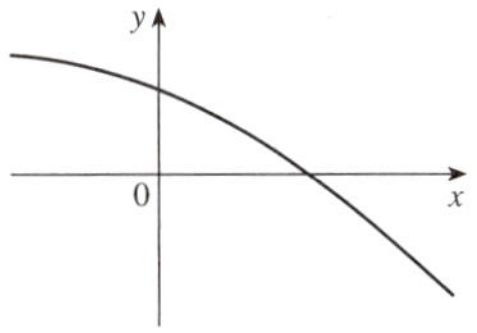

11. (a) Inc. on $(-\infty, -2), (2, \infty)$; dec. on $(-2, 2)$
(b) Loc. max. $f(-2) = 17$; loc. min. $f(2) = -15$
(c) CU on $(0, \infty)$; CD on $(-\infty, 0)$; IP $(0, 1)$
13. (a) Inc. on $(-1, 0), (1, \infty)$; dec. on $(-\infty, -1), (0, 1)$
(b) Loc. max. $f(0) = 3$; loc. min. $f(\pm 1) = 2$
(c) CU on $(-\infty, -\sqrt{3}/3), (\sqrt{3}/3, \infty)$;
CD on $(-\sqrt{3}/3, \sqrt{3}/3)$; IP $(\pm\sqrt{3}/3, \frac{22}{9})$
15. (a) Inc. on $(\pi/3, 5\pi/3), (7\pi/3, 3\pi)$;
dec. on $(0, \pi/3), (5\pi/3, 7\pi/3)$
(b) Loc. max. $f(5\pi/3) = 5\pi/3 + \sqrt{3}$;
loc. min. $f(\pi/3) = \pi/3 - \sqrt{3}$, $f(7\pi/3) = 7\pi/3 - \sqrt{3}$
(c) CU on $(0, \pi), (2\pi, 3\pi)$; CD on $(\pi, 2\pi)$; IP $(\pi, \pi), (2\pi, 2\pi)$
17. (a) Inc. on $(-1, \infty)$; dec. on $(-\infty, -1)$
(b) Loc. min. $f(-1) = -1/e$
(c) CU on $(-2, \infty)$; CD on $(-\infty, -2)$; IP $(-2, -2e^{-2})$
19. (a) Inc. on $(0, e^2)$; dec. on (e^2, ∞)
(b) Loc. max. $f(e^2) = 2/e$
(c) CU on $(e^{8/3}, \infty)$; CD on $(0, e^{8/3})$; IP $(e^{8/3}, \frac{8}{3}e^{-4/3})$
21. Loc. max. $f(-1) = 7$; loc. min. $f(1) = -1$
23. Loc. max. $f(\frac{3}{4}) = \frac{5}{4}$
25. (a) f has a local maximum at 2.
(b) f has a horizontal tangent at 6.
27.

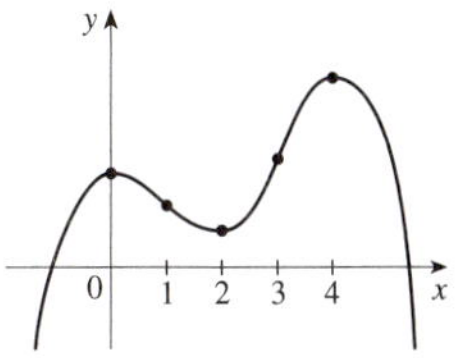

29.

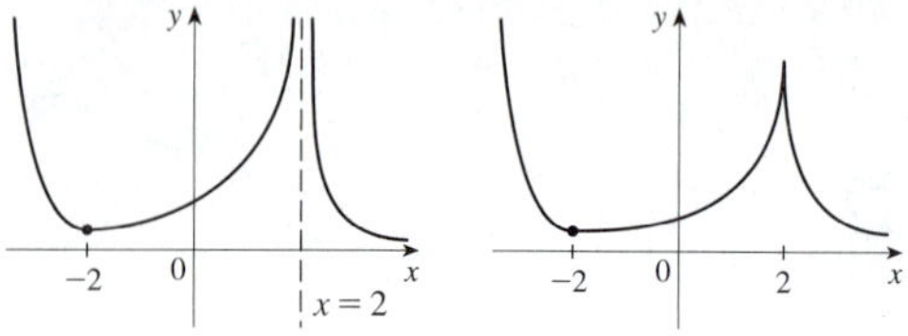

31. (a) Inc. on (0, 2), (4, 6), (8, ∞);
dec. on (2, 4), (6, 8)
(b) Loc. max. at $x = 2, 6$;
loc. min. at $x = 4, 8$
(c) CU on (3, 6), (6, ∞);
CD on (0, 3)
(d) 3 (e) See graph at right.

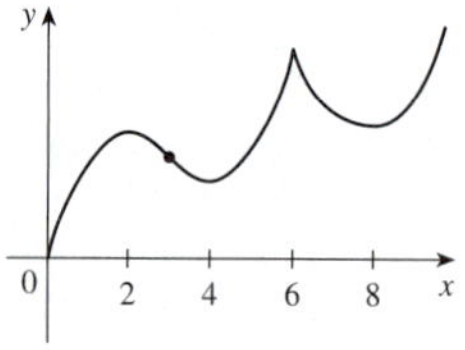

33. (a) Inc. on (−∞, −1), (2, ∞);
dec. on (−1, 2)
(b) Loc. max. $f(-1) = 7$;
loc. min. $f(2) = -20$
(c) CU on $\left(\frac{1}{2}, \infty\right)$; CD on $\left(-\infty, \frac{1}{2}\right)$;
IP $\left(\frac{1}{2}, -\frac{13}{2}\right)$
(d) See graph at right.

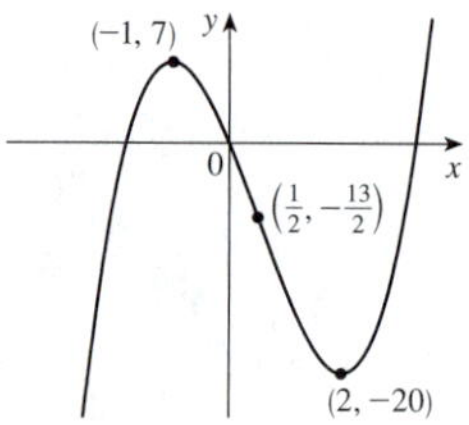

35. (a) Inc. on $(-\sqrt{3}, 0)$, $(\sqrt{3}, \infty)$;
dec. on $(-\infty, -\sqrt{3})$, $(0, \sqrt{3})$
(b) Loc. min. $f(\pm\sqrt{3}) = -9$;
loc. max. $f(0) = 0$
(c) CD on (−1, 1); CU on (−∞, −1),
(1, ∞); IP (±1, −5)
(d) See graph at right.

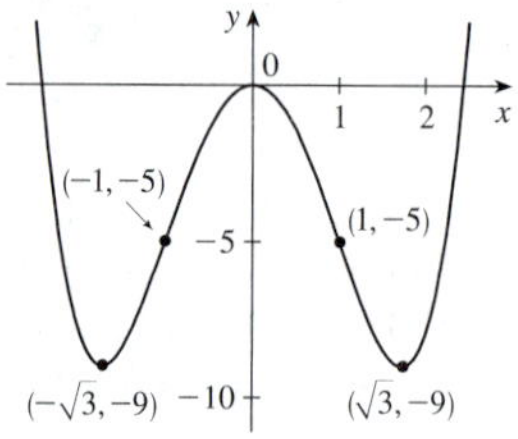

37. (a) Inc. on (−∞, −1), (1, ∞);
dec. on (−1, 1)
(b) Loc. max. $h(-1) = 5$;
loc. min. $h(1) = 1$
(c) CD on $(-\infty, -1/\sqrt{2})$, $(0, 1/\sqrt{2})$;
CU on $(-1/\sqrt{2}, 0)$, $(1/\sqrt{2}, \infty)$;
IP (0, 3), $\left(\pm 1/\sqrt{2}, 3 \mp \frac{7}{8}\sqrt{2}\right)$
(d) See graph at right.

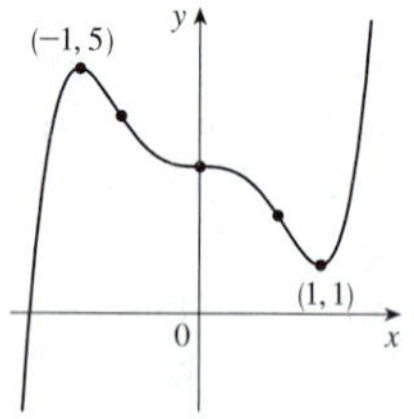

39. (a) Inc. on (−2, ∞);
dec. on (−3, −2)
(b) Loc. min. $A(-2) = -2$
(c) CU on (−3, ∞)
(d) See graph at right.

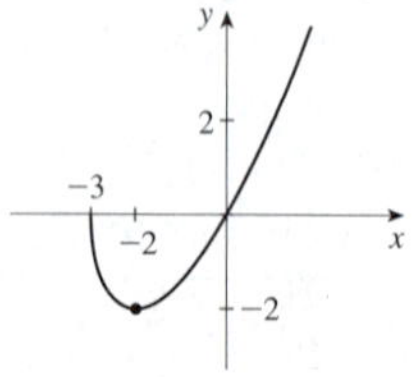

41. (a) Inc. on (−1, ∞);
dec. on (−∞, −1)
(b) Loc. min. $C(-1) = -3$
(c) CU on (−∞, 0), (2, ∞);
CD on (0, 2);
IP (0, 0), $(2, 6\sqrt[3]{2})$
(d) See graph at right.

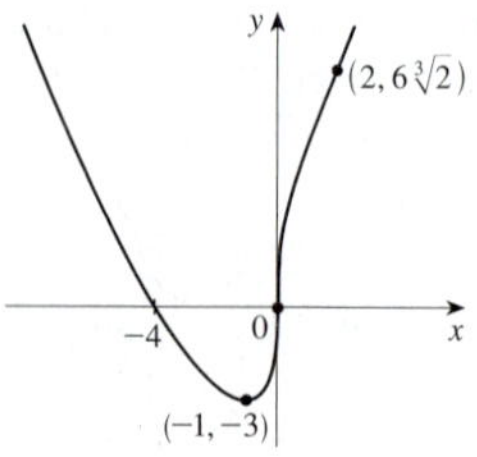

43. (a) Inc. on $(0, \pi/3)$, $(\pi, 5\pi/3)$; dec. on $(\pi/3, \pi)$, $(5\pi/3, 2\pi)$
(b) Loc. max. $f(\pi/3) = f(5\pi/3) = \frac{3}{2}$; loc. min. $f(\pi) = -3$
(c) Let $\alpha = \cos^{-1}((1 + \sqrt{33})/8)$, $\beta = \cos^{-1}((1 - \sqrt{33})/8)$,
$y_1 = \frac{3}{16}(1 + \sqrt{33})$, and $y_2 = \frac{3}{16}(1 - \sqrt{33})$.
CU on $(0, \alpha)$, $(\beta, 2\pi - \beta)$,
$(2\pi - \alpha, 2\pi)$;
CD on (α, β), $(2\pi - \beta, 2\pi - \alpha)$;
IP (α, y_1), (β, y_2), $(2\pi - \beta, y_2)$,
$(2\pi - \alpha, y_1)$
(d) See graph at right.

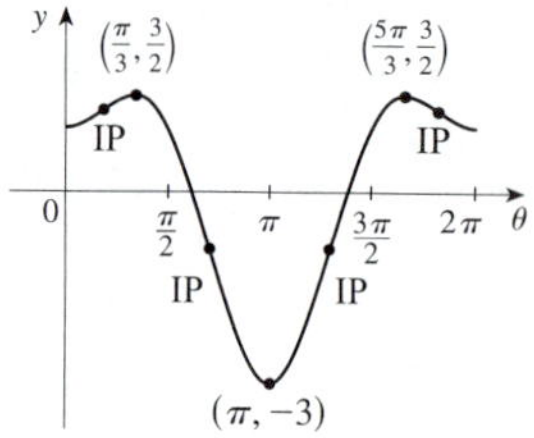

45. (a) HA $y = 1$, VA $x = -1$, $x = 1$
(b) Inc. on (−∞, −1), (−1, 0);
dec. on (0, 1), (1, ∞)
(c) Loc. max. $f(0) = 0$
(d) CU on (−∞, −1), (1, ∞);
CD on (−1, 1)
(e) See graph at right.

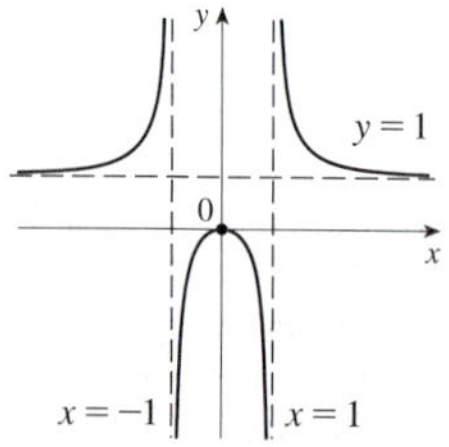

47. (a) HA $y = 0$
(b) Dec. on (−∞, ∞)
(c) None
(d) CU on (−∞, ∞)
(e) See graph at right.

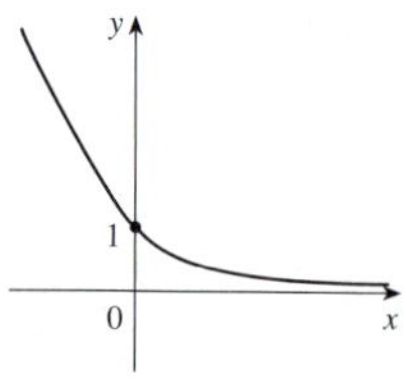

49. (a) VA $x = 0$, $x = e$
(b) Dec. on $(0, e)$
(c) None
(d) CU on (0, 1); CD on $(1, e)$;
IP (1, 0)
(e) See graph at right.

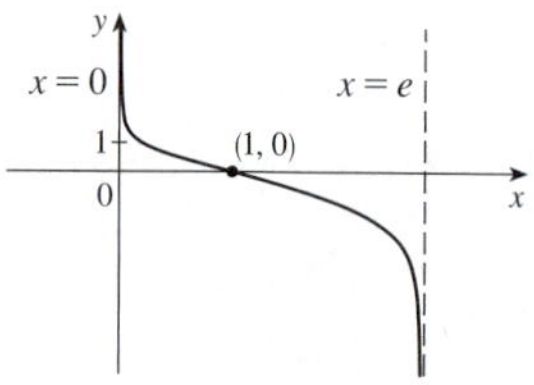

51. (a) HA $y = 1$, VA $x = -1$
(b) Inc. on (−∞, −1), (−1, ∞)
(c) None
(d) CU on (−∞, −1), $\left(-1, -\frac{1}{2}\right)$;
CD on $\left(-\frac{1}{2}, \infty\right)$; IP $\left(-\frac{1}{2}, 1/e^2\right)$
(e) See graph at right.

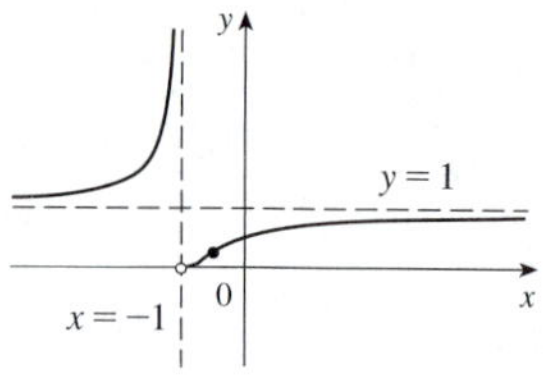

53. (a) Loc. and abs. max. $f(1) = \sqrt{2}$, no min. (b) $\frac{1}{4}(3 - \sqrt{17})$
55. (b) CU on (0.94, 2.57), (3.71, 5.35);
CD on (0, 0.94), (2.57, 3.71), (5.35, 2π);
IP (0.94, 0.44), (2.57, −0.63), (3.71, −0.63), (5.35, 0.44)
57. CD on (−∞, 0.1); CU on (0.1, ∞) **59.** $K(3) - K(2)$; CD
61. When $t \approx 7.94$ **63.** $f(x) = \frac{1}{9}(2x^3 + 3x^2 - 12x + 7)$

Exercises 4.4 ▫ page 313

1. (a) Indeterminate (b) 0 (c) 0
(d) ∞, −∞, or does not exist (e) Indeterminate
3. (a) −∞ (b) Indeterminate (c) ∞

5. -2 **7.** $\frac{9}{5}$ **9.** $-\infty$ **11.** ∞ **13.** p/q
15. 0 **17.** $-\infty$ **19.** $\ln \frac{5}{3}$ **21.** $\frac{1}{2}$ **23.** ∞ **25.** 1
27. $\frac{1}{2}$ **29.** 0 **31.** 1 **33.** $-1/\pi^2$ **35.** $\frac{1}{2}a(a-1)$
37. 0 **39.** 3 **41.** 0 **43.** $-2/\pi$ **45.** 0 **47.** $\frac{1}{2}$
49. ∞ **51.** 1 **53.** e^{-2} **55.** e^3 **57.** 1 **59.** $1/e$
61. $1/\sqrt{e}$ **63.** 5 **65.** $\frac{1}{4}$ **71.** $\frac{16}{9}a$ **73.** 56 **77.** (a) 0

Exercises 4.5 ▫ page 323

1. A. $\mathbb{R}$ B. y-int. 0; x-int. 0
C. About (0, 0) D. None
E. Inc. on $(-\infty, \infty)$ F. None
G. CU on $(0, \infty)$; CD on $(-\infty, 0)$;
IP (0, 0)
H. See graph at right.

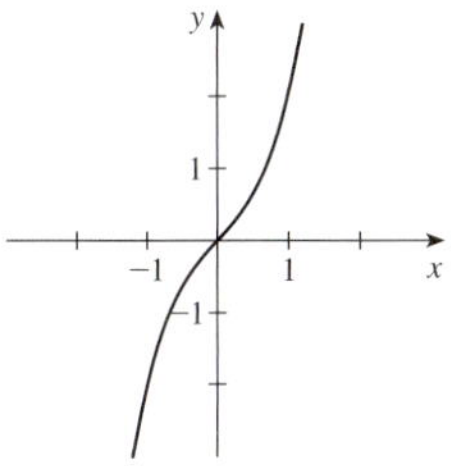

3. A. $\mathbb{R}$ B. y-int. 2; x-int. 2, $\frac{1}{2}(7 \pm 3\sqrt{5})$
C. None D. None
E. Inc. on (1, 5);
dec. on $(-\infty, 1)$, $(5, \infty)$
F. Loc. min. $f(1) = -5$;
loc. max. $f(5) = 27$
G. CU on $(-\infty, 3)$;
CD on $(3, \infty)$; IP (3, 11)
H. See graph at right.

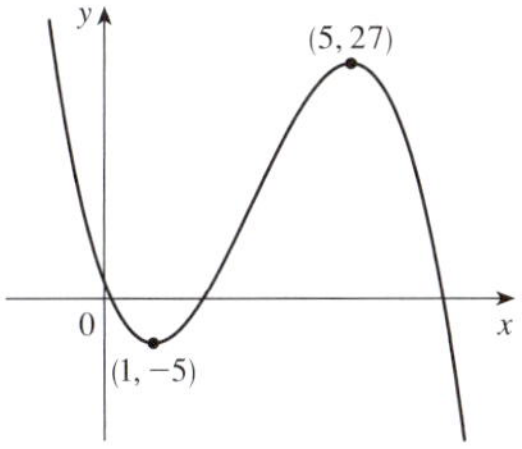

5. A. $\mathbb{R}$ B. y-int. 0; x-int. -4, 0
C. None D. None
E. Inc. on $(-3, \infty)$;
dec. on $(-\infty, -3)$
F. Loc. min. $f(-3) = -27$
G. CU on $(-\infty, -2)$, $(0, \infty)$;
CD on $(-2, 0)$; IP (0, 0), $(-2, -16)$
H. See graph at right.

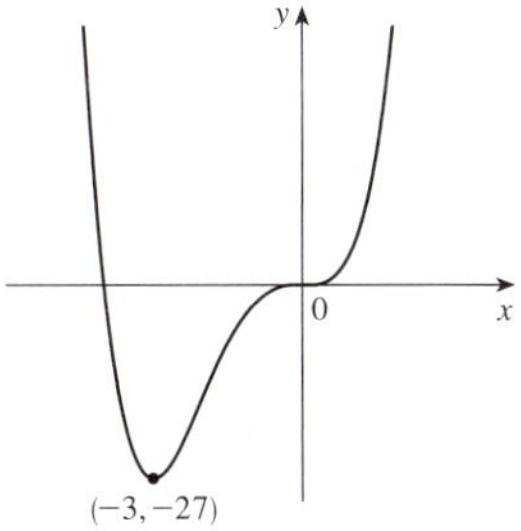

7. A. $\mathbb{R}$ B. y-int. 1
C. None D. None
E. Inc. on $(-\infty, 0)$, $(1, \infty)$; dec. on (0, 1)
F. Loc. max. $f(0) = 1$;
loc. min. $f(1) = -2$
G. CU on $(1/\sqrt[3]{4}, \infty)$; CD on $(-\infty, 1/\sqrt[3]{4})$;
IP $(1/\sqrt[3]{4}, 1 - 9/(2\sqrt[3]{16}))$
H. See graph at right.

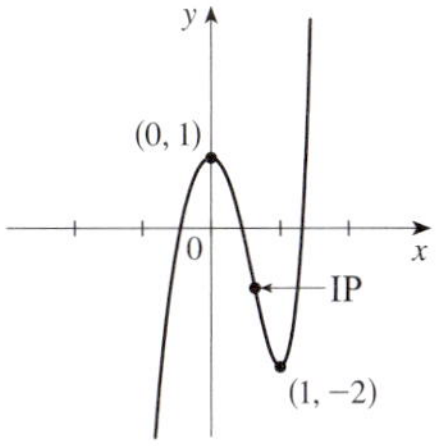

9. A. $\{x \mid x \neq 1\}$ B. y-int. 0; x-int. 0
C. None D. VA $x = 1$, HA $y = 1$
E. Dec. on $(-\infty, 1)$, $(1, \infty)$ F. None
G. CU on $(1, \infty)$; CD on $(-\infty, 1)$
H. See graph at right.

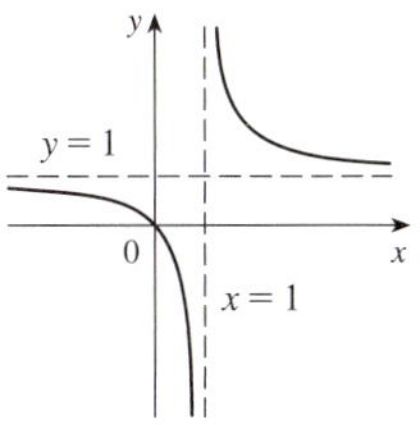

11. A. $\{x \mid x \neq \pm 3\}$ B. y-int. $-\frac{1}{9}$
C. About y-axis D. VA $x = \pm 3$, HA $y = 0$
E. Inc. on $(-\infty, -3)$, $(-3, 0)$;
dec. on (0, 3), $(3, \infty)$
F. Loc. max. $f(0) = -\frac{1}{9}$
G. CU on $(-\infty, -3)$, $(3, \infty)$;
CD on $(-3, 3)$
H. See graph at right.

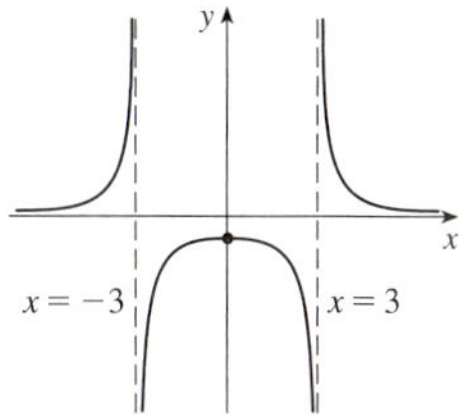

13. A. $\mathbb{R}$ B. y-int. 0; x-int. 0 C. About (0, 0) D. HA $y = 0$
E. Inc. on $(-3, 3)$;
dec. on $(-\infty, -3)$, $(3, \infty)$
F. Loc. min. $f(-3) = -\frac{1}{6}$;
loc. max. $f(3) = \frac{1}{6}$;
G. CU on $(-3\sqrt{3}, 0)$, $(3\sqrt{3}, \infty)$;
CD on $(-\infty, -3\sqrt{3})$, $(0, 3\sqrt{3})$;
IP (0, 0), $(\pm 3\sqrt{3}, \pm\sqrt{3}/12)$
H. See graph at right.

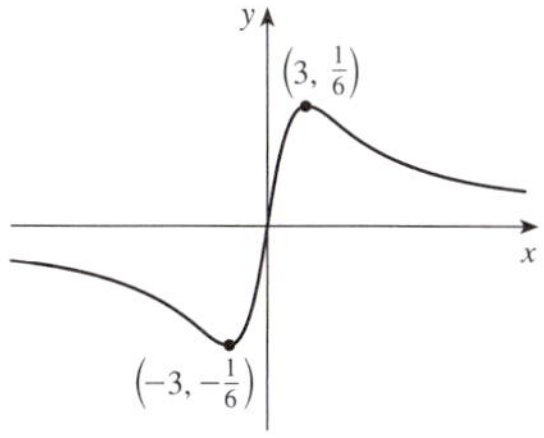

15. A. $(-\infty, 0) \cup (0, \infty)$ B. x-int. 1
C. None D. HA $y = 0$; VA $x = 0$
E. Inc. on (0, 2);
dec. on $(-\infty, 0)$, $(2, \infty)$
F. Loc. max. $f(2) = \frac{1}{4}$
G. CU on $(3, \infty)$;
CD on $(-\infty, 0)$, (0, 3); IP $(3, \frac{2}{9})$
H. See graph at right

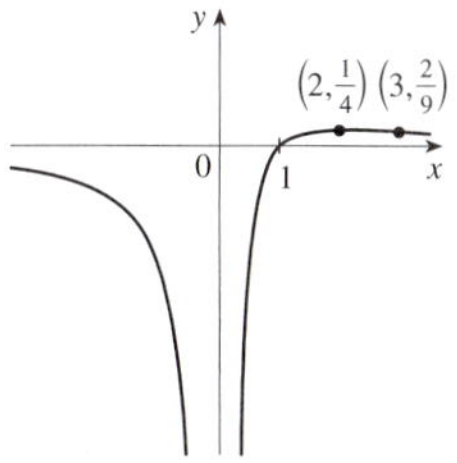

17. A. $\mathbb{R}$ B. y-int. 0; x-int. 0
C. About y-axis D. HA $y = 1$
E. Inc. on $(0, \infty)$; dec. on $(-\infty, 0)$
F. Loc. min. $f(0) = 0$
G. CU on $(-1, 1)$;
CD on $(-\infty, -1)$, $(1, \infty)$; IP $(\pm 1, \frac{1}{4})$
H. See graph at right

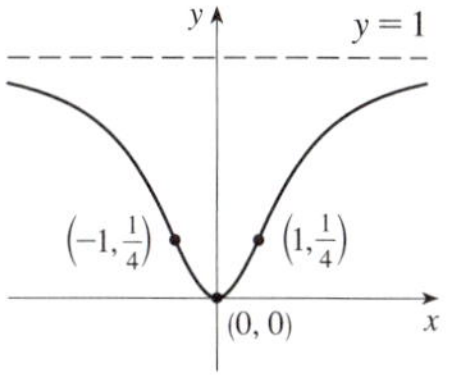

19. A. $(-\infty, 5]$ B. y-int. 0; x-int. 0, 5
C. None D. None
E. Inc. on $(-\infty, \frac{10}{3})$; dec. on $(\frac{10}{3}, 5)$
F. Loc. max. $f(\frac{10}{3}) = \frac{10}{9}\sqrt{15}$
G. CD on $(-\infty, 5)$
H. See graph at right.

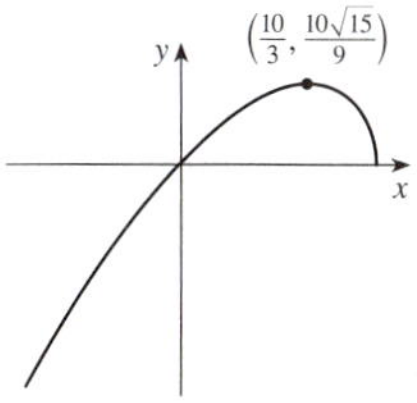

21. A. $\mathbb{R}$ B. y-int. 1
C. None D. HA $y = 0$
E. Dec. on $(-\infty, \infty)$
F. None G. CU on $(-\infty, \infty)$
H. See graph at right.

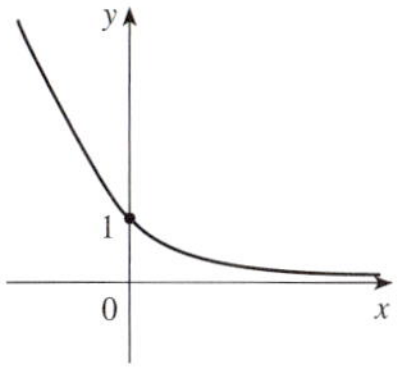

23. A. $\mathbb{R}$ B. y-int. 0; x-int. 0
C. About the origin
D. HA $y = \pm 1$
E. Inc. on $(-\infty, \infty)$
F. None G. CU on $(-\infty, 0)$;
CD on $(0, \infty)$; IP (0, 0)
H. See graph at right.

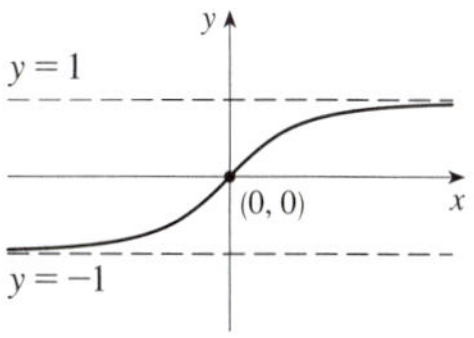

25. A. $\{x \mid |x| \leqslant 1, x \neq 0\} = [-1, 0) \cup (0, 1]$ B. x-int. ± 1
C. About $(0, 0)$ D. VA $x = 0$
E. Dec. on $(-1, 0)$, $(0, 1)$ F. None
G. CU on $(-1, -\sqrt{2/3})$, $(0, \sqrt{2/3})$;
CD on $(-\sqrt{2/3}, 0)$, $(\sqrt{2/3}, 1)$;
IP $(\pm\sqrt{2/3}, \pm 1/\sqrt{2})$
H. See graph at right.

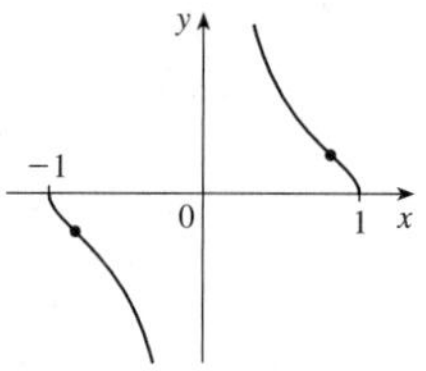

27. A. $\mathbb{R}$ B. y-int. 0; x-int. $\pm 3\sqrt{3}$ C. About the origin
D. None E. Inc. on $(-\infty, -1)$, $(1, \infty)$; dec. on $(-1, 1)$
F. Loc. max. $f(-1) = 2$;
loc. min. $f(1) = -2$
G. CU on $(0, \infty)$; CD on $(-\infty, 0)$;
IP $(0, 0)$
H. See graph at right.

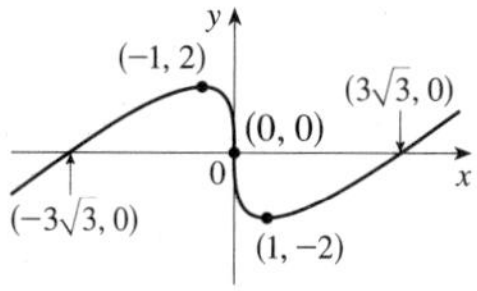

29. A. $\mathbb{R}$ B. y-int. 0; x-int. $-1, 0$
C. None D. None
E. Inc. on $(-\infty, -\frac{1}{4})$, $(0, \infty)$; dec. on $(-\frac{1}{4}, 0)$
F. Loc. max. $f(-\frac{1}{4}) = \frac{1}{4}$; loc. min. $f(0) = 0$
G. CD on $(-\infty, 0)$, $(0, \infty)$
H. See graph at right.

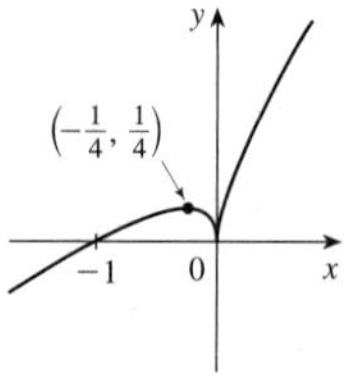

31. A. $\mathbb{R}$ B. y-int. 0; x-int. $n\pi$ (n an integer)
C. About the origin, period 2π D. None
E. Inc. on $(2n\pi - \pi/2, 2n\pi + \pi/2)$;
dec. on $(2n\pi + \pi/2, 2n\pi + 3\pi/2)$
F. Loc. max. $f(2n\pi + \pi/2) = 2$;
loc. min. $f(2n\pi + 3\pi/2) = -2$
G. CU on $((2n - 1)\pi, 2n\pi)$;
CD on $(2n\pi, (2n + 1)\pi)$; IP $(n\pi, 0)$
H. See graph at right.

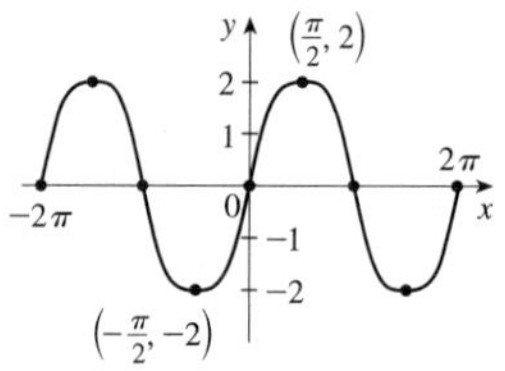

33. A. $(-\pi/2, \pi/2)$ B. y-int. 0; x-int. 0 C. About y-axis
D. VA $x = \pm\pi/2$
E. Inc. on $(0, \pi/2)$;
dec. on $(-\pi/2, 0)$
F. Loc. min. $f(0) = 0$
G. CU on $(-\pi/2, \pi/2)$
H. See graph at right.

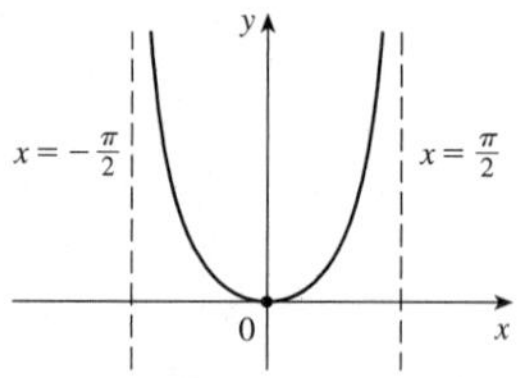

35. A. $(0, 3\pi)$ C. None D. None
E. Inc. on $(\pi/3, 5\pi/3)$, $(7\pi/3, 3\pi)$; dec. on $(0, \pi/3)$, $(5\pi/3, 7\pi/3)$
F. Loc. min. $f(\pi/3) = (\pi/6) - \frac{1}{2}\sqrt{3}$,
$f(7\pi/3) = (7\pi/6) - \frac{1}{2}\sqrt{3}$;
loc. max. $f(5\pi/3) = (5\pi/6) + \frac{1}{2}\sqrt{3}$
G. CU on $(0, \pi)$, $(2\pi, 3\pi)$;
CD on $(\pi, 2\pi)$; IP $(\pi, \pi/2)$, $(2\pi, \pi)$
H. See graph at right.

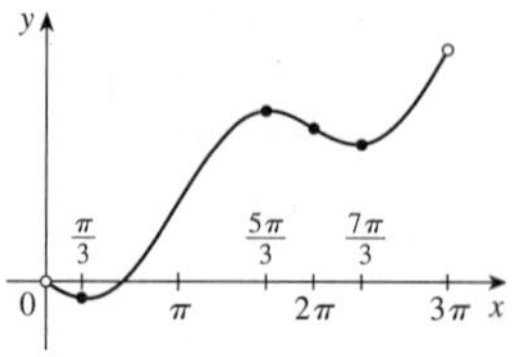

37. A. $\mathbb{R}$ B. y-int. 0; x-int. $n\pi$
C. About $(0, 0)$, period 2π D. None
Answers for E–G are for the interval $[-\pi, \pi]$.
E. Inc. on $(-\pi, -2\pi/3)$, $(2\pi/3, \pi)$; dec. on $(-2\pi/3, 2\pi/3)$
F. Loc. max. $f(-2\pi/3) = \frac{3}{2}\sqrt{3}$;
loc. min. $f(2\pi/3) = -\frac{3}{2}\sqrt{3}$
G. CU on $(-\alpha, 0)$, (α, π),
where $\cos\alpha = \frac{1}{4}$;
CD on $(-\pi, -\alpha)$, $(0, \alpha)$;
IP when $x = 0, \pm\alpha, \pm\pi$
H. See graph at right.

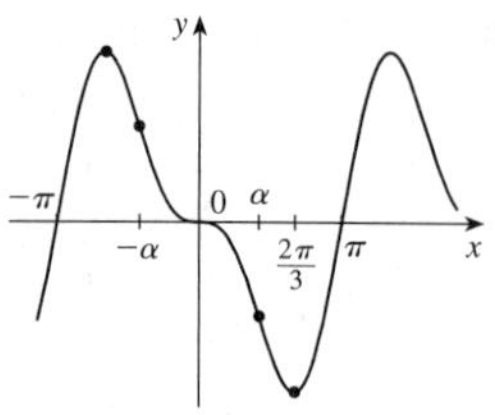

39. A. All reals except $(2n + 1)\pi$ (n an integer)
B. y-int. 0; x-int. $2n\pi$
C. About the origin, period 2π
D. VA $x = (2n + 1)\pi$
E. Inc. on $((2n - 1)\pi, (2n + 1)\pi)$ F. None
G. CU on $(2n\pi, (2n + 1)\pi)$; CD on $((2n - 1)\pi, 2n\pi)$;
IP $(2n\pi, 0)$
H.

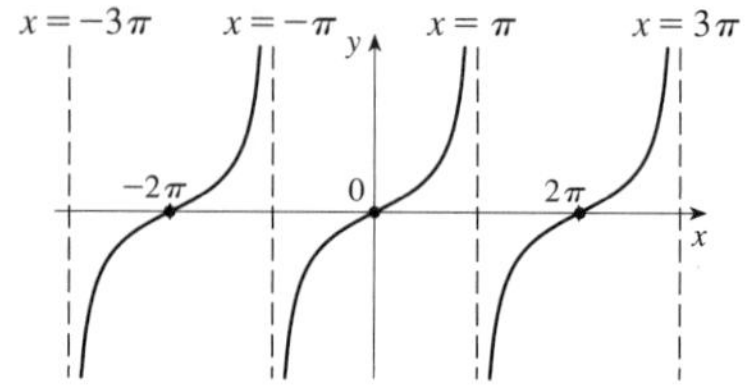

41. A. $\mathbb{R}$ B. y-int. $\frac{1}{2}$ C. None
D. HA $y = 0$, $y = 1$
E. Inc. on $\mathbb{R}$ F. None
G. CU on $(-\infty, 0)$; CD on $(0, \infty)$;
IP $(0, \frac{1}{2})$ H. See graph at right.

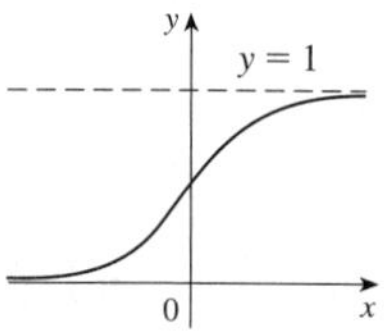

43. A. $(0, \infty)$ B. x-int. 1
C. None D. None
E. Inc. on $(1/e, \infty)$; dec. on $(0, 1/e)$
F. Loc. min. $f(1/e) = -1/e$
G. CU on $(0, \infty)$
H. See graph at right.

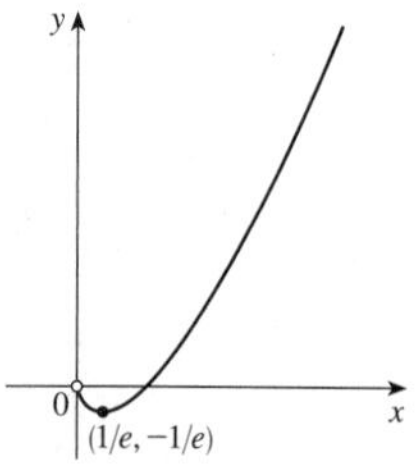

45. A. $\mathbb{R}$ B. y-int. 0; x-int. 0
C. None D. HA $y = 0$
E. Inc. on $(-\infty, 1)$; dec. on $(1, \infty)$
F. Loc. max. $f(1) = 1/e$
G. CU on $(2, \infty)$; CD on $(-\infty, 2)$;
IP $(2, 2/e^2)$ H. See graph at right.

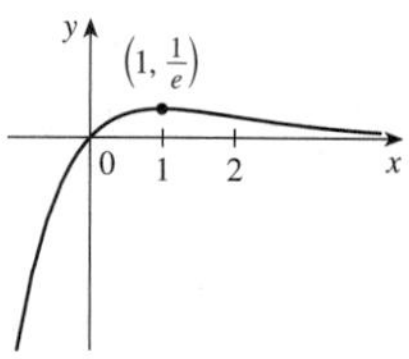

47. A. All x in $(2n\pi, (2n + 1)\pi)$ (n an integer)
B. x-int. $\pi/2 + 2n\pi$ C. Period 2π D. VA $x = n\pi$
E. Inc. on $(2n\pi, \pi/2 + 2n\pi)$; dec. on $(\pi/2 + 2n\pi, (2n + 1)\pi)$
F. Loc. max. $f(\pi/2 + 2n\pi) = 0$ G. CD on $(2n\pi, (2n + 1)\pi)$
H.

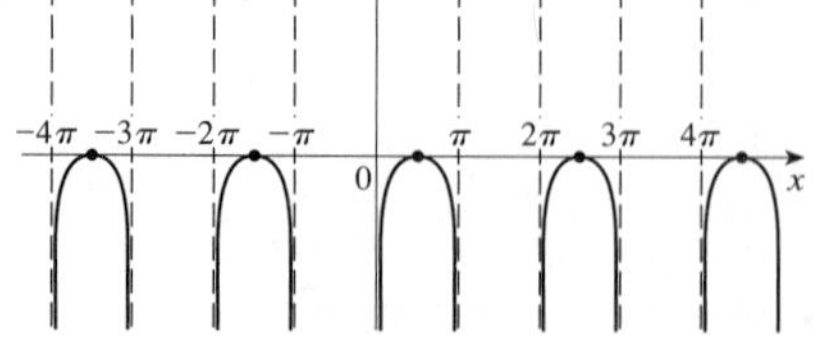

49. A. $\mathbb{R}$ B. y-int. 0; x-int. 0 C. About $(0, 0)$ D. HA $y = 0$
E. Inc. on $\left(-1/\sqrt{2}, 1/\sqrt{2}\right)$; dec. on $\left(-\infty, -1/\sqrt{2}\right)$, $\left(1/\sqrt{2}, \infty\right)$
F. Loc. min. $f\left(-1/\sqrt{2}\right) = -1/\sqrt{2e}$; loc. max. $f\left(1/\sqrt{2}\right) = 1/\sqrt{2e}$
G. CU on $\left(-\sqrt{3/2}, 0\right)$, $\left(\sqrt{3/2}, \infty\right)$; CD on $\left(-\infty, -\sqrt{3/2}\right)$, $\left(0, \sqrt{3/2}\right)$;
IP $\left(\pm\sqrt{3/2}, \pm\sqrt{3/2}e^{-3/2}\right)$, $(0, 0)$
H.

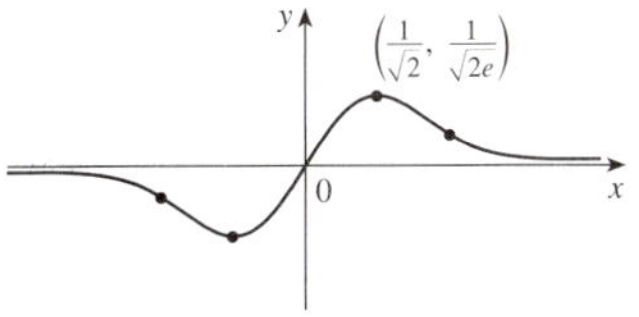

51. A. $\mathbb{R}$ B. y-int. 2
C. None D. None
E. Inc. on $\left(\frac{1}{5}\ln\frac{2}{3}, \infty\right)$; dec. on $\left(-\infty, \frac{1}{5}\ln\frac{2}{3}\right)$
F. Loc. min. $f\left(\frac{1}{5}\ln\frac{2}{3}\right) = \left(\frac{2}{3}\right)^{3/5} + \left(\frac{2}{3}\right)^{-2/5}$
G. CU on $(-\infty, \infty)$
H. See graph at right.

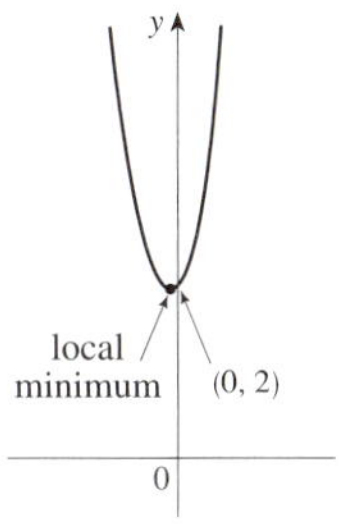

53.

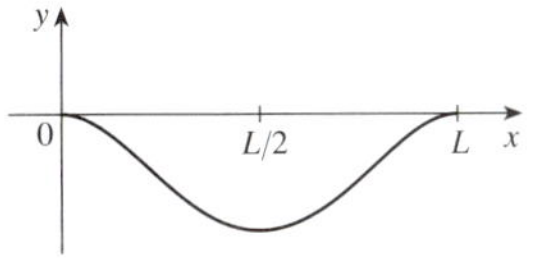

55. $y = x - 1$ **57.** $y = 2x - 2$

59. A. $\left(-\infty, \frac{1}{2}\right) \cup \left(\frac{1}{2}, \infty\right)$
B. y-int. 1; x-int. $\frac{1}{4}\left(5 \pm \sqrt{17}\right)$
C. None D. VA $x = \frac{1}{2}$;
SA $y = -x + 2$
E. Dec. on $\left(-\infty, \frac{1}{2}\right)$, $\left(\frac{1}{2}, \infty\right)$
F. None
G. CU on $\left(\frac{1}{2}, \infty\right)$; CD on $\left(-\infty, \frac{1}{2}\right)$
H. See graph at right

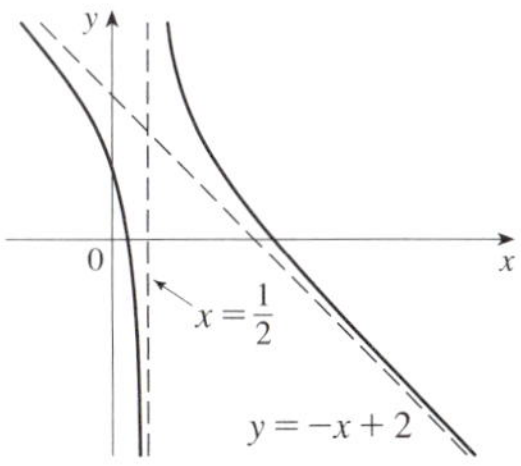

61. A. $\{x \mid x \neq 0\}$ B. None
C. About $(0, 0)$ D. VA $x = 0$; SA $y = x$
E. Inc. on $(-\infty, -2)$, $(2, \infty)$;
dec. on $(-2, 0)$, $(0, 2)$
F. Loc. max. $f(-2) = -4$;
loc. min. $f(2) = 4$
G. CU on $(0, \infty)$; CD on $(-\infty, 0)$
H. See graph at right.

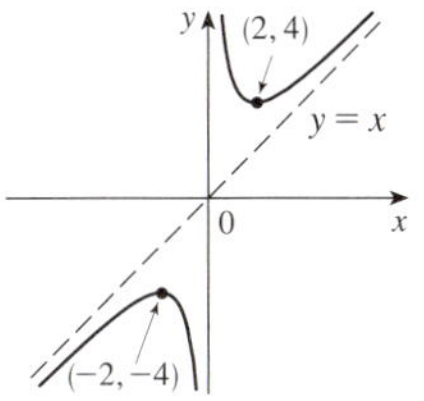

63. A. $\mathbb{R}$ B. y-int. 1; x-int. -1 C. None
D. SA $y = 2x + 1$
E. Inc. on $(-\infty, \infty)$
F. None
G. CU on $\left(-\infty, -\sqrt{3}\right)$, $\left(0, \sqrt{3}\right)$;
CD on $\left(-\sqrt{3}, 0\right)$, $\left(\sqrt{3}, \infty\right)$;
IP $\left(\pm\sqrt{3}, 1 \pm \frac{3}{2}\sqrt{3}\right)$, $(0, 1)$
H. See graph at right.

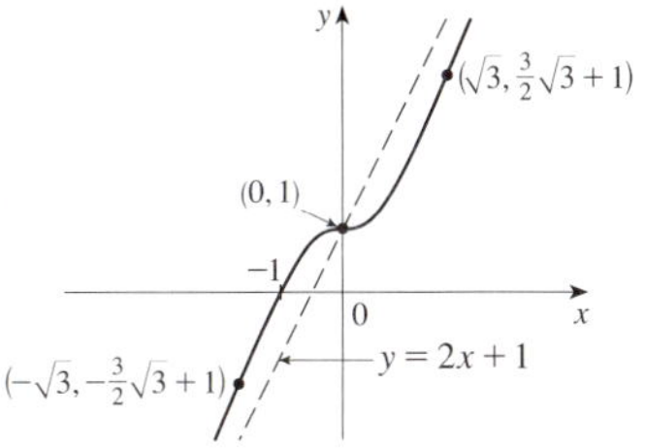

65.

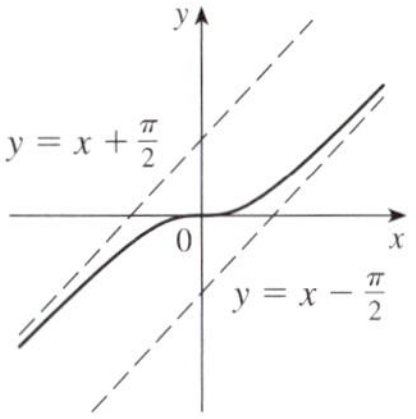

69. VA $x = 0$, asymptotic to $y = x^3$

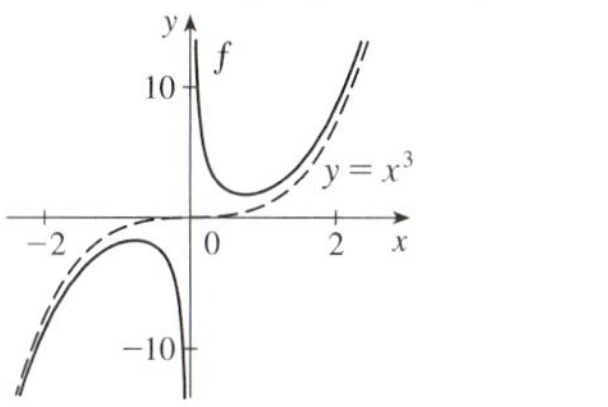

Exercises 4.6 ▫ page 330

1. Inc. on (0.92, 2.5), (2.58, ∞); dec. on ($-\infty$, 0.92), (2.5, 2.58);
loc. max. $f(2.5) \approx 4$; loc. min. $f(0.92) \approx -5.12$, $f(2.58) \approx 3.998$;
CU on ($-\infty$, 1.46), (2.54, ∞); CD on (1.46, 2.54);
IP (1.46, -1.40), (2.54, 3.999)

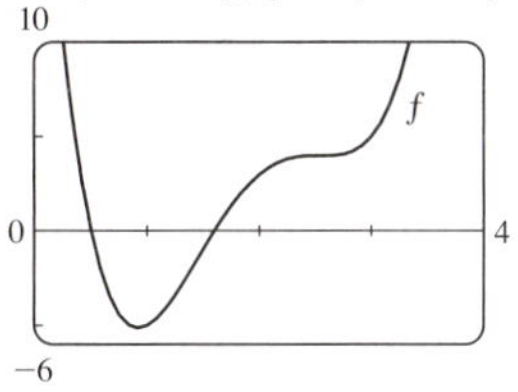

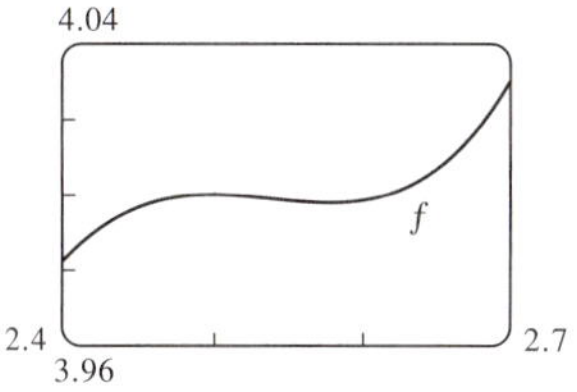

3. Inc. on (1.5, ∞); dec. on ($-\infty$, 1.5); min. $f(1.5) \approx -1.9$;
CU on (-1.2, 4.2); CD on ($-\infty$, -1.2), (4.2, ∞);
IP (-1.2, 0), (4.2, 0)

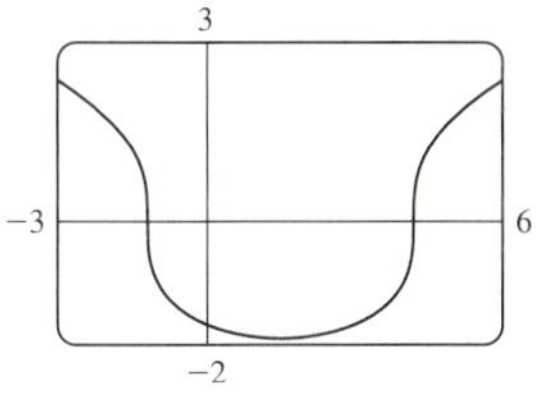

5. Inc. on ($-\infty$, -1.7), (-1.7, 0.24), (0.24, 1);
dec. on (1, 2.46), (2.46, ∞); loc. max. $f(1) = -\frac{1}{3}$;
CU on ($-\infty$, -1.7), (-0.506, 0.24), (2.46, ∞);
CD on (-1.7, -0.506), (0.24, 2.46); IP (-0.506, -0.192)

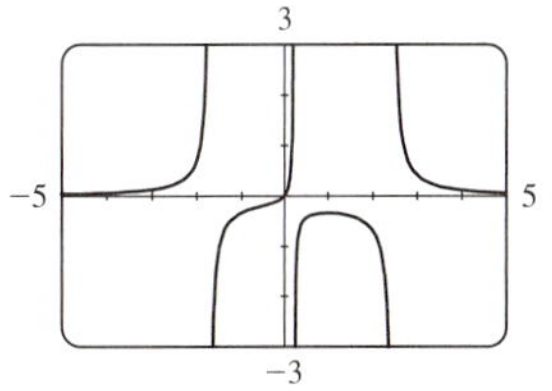

7. Inc. on (-1.49, -1.07), (2.89, 4); dec. on (-4, -1.49),
(-1.07, 2.89); loc. max. $f(-1.07) \approx 8.79$; loc. min.
$f(-1.49) \approx 8.75$, $f(2.89) \approx -9.99$; CU on (-4, -1.28),
(1.28, 4); CD on (-1.28, 1.28); IP (-1.28, 8.77), (1.28, -1.48)

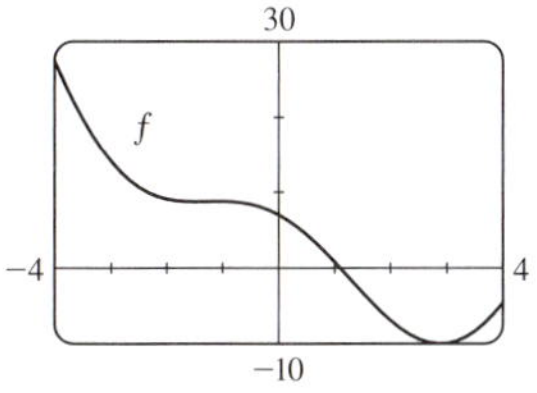

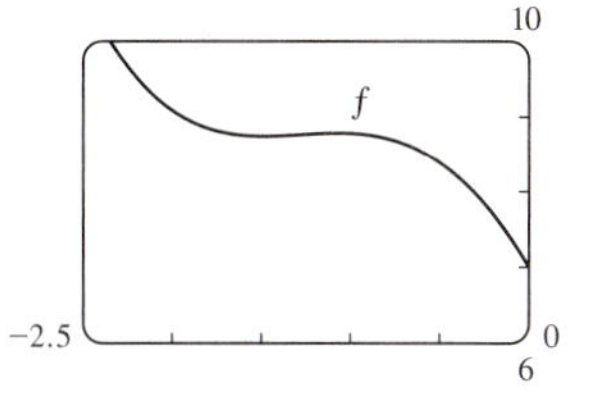

9. Inc. on $(-\infty, 0)$, $\left(\frac{1}{4}, \infty\right)$; dec. on $\left(0, \frac{1}{4}\right)$; loc. max. $f(0) = -10$; loc. min. $f\left(\frac{1}{4}\right) = -\frac{161}{16} \approx -10.1$; CU on $\left(\frac{1}{8}, \infty\right)$; CD on $\left(-\infty, \frac{1}{8}\right)$; IP $\left(\frac{1}{8}, -\frac{321}{32}\right)$

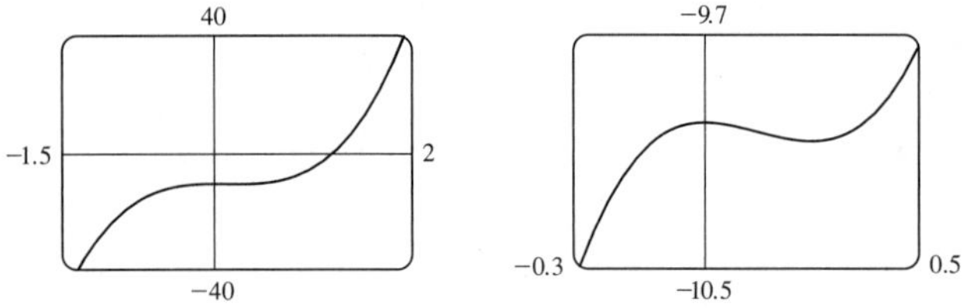

11. Inc. on $\left(-\frac{3}{2}\sqrt{2}, \frac{3}{2}\sqrt{2}\right)$;
dec. on $\left(-3, -\frac{3}{2}\sqrt{2}\right)$, $\left(\frac{3}{2}\sqrt{2}, 3\right)$;
max. $f\left(\frac{3}{2}\sqrt{2}\right) = 4.5$;
min. $f\left(-\frac{3}{2}\sqrt{2}\right) = -4.5$;
CU on $(-3, 0)$;
CD on $(0, 3)$; IP $(0, 0)$

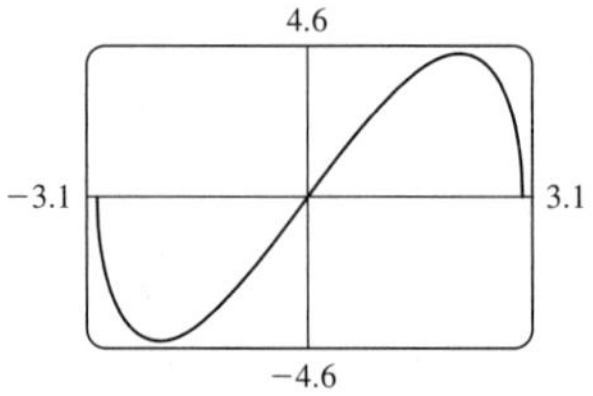

13.
(a)

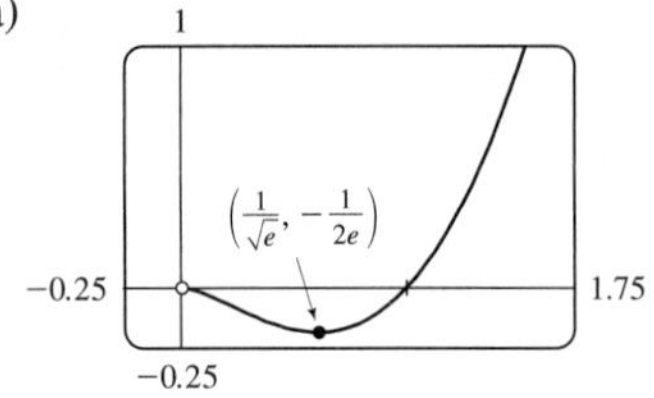

(b) $\lim_{x\to 0^+} f(x) = 0$
(c) Loc. min. $f(1/\sqrt{e}) = -1/(2e)$; CD on $(0, e^{-3/2})$; CU on $(e^{-3/2}, \infty)$

15. Loc. max. $f(-5.6) \approx 0.018$, $f(0.82) \approx -281.5$, $f(5.2) \approx 0.0145$; loc. min. $f(3) = 0$

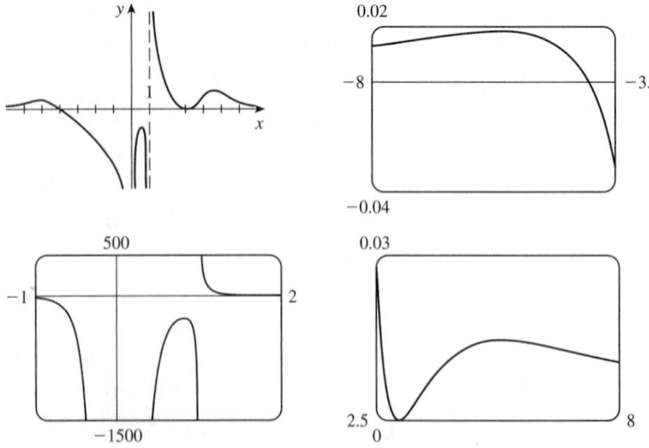

17. $f'(x) = -\dfrac{x(x+1)^2(x^3 + 18x^2 - 44x - 16)}{(x-2)^3(x-4)^5}$

$f''(x) = 2\dfrac{(x+1)(x^6 + 36x^5 + 6x^4 - 628x^3 + 684x^2 + 672x + 64)}{(x-2)^4(x-4)^6}$

CU on $(-35.3, -5.0)$, $(-1, -0.5)$, $(-0.1, 2)$, $(2, 4)$, $(4, \infty)$;
CD on $(-\infty, -35.3)$, $(-5.0, -1)$ $(-0.5, -0.1)$;
IP $(-35.3, -0.015)$, $(-5.0, -0.005)$, $(-1, 0)$, $(-0.5, 0.00001)$, $(-0.1, 0.0000066)$

19. Inc. on $(0, 1.3)$, $(\pi, 4.6)$, $(2\pi, 7.8)$;
dec. on $(1.3, \pi)$, $(4.6, 2\pi)$, $(7.8, 3\pi)$;
loc. max. $f(1.3) \approx 0.6$, $f(4.6) \approx 0.21$, $f(7.8) \approx 0.13$;
loc. min. $f(\pi) = f(2\pi) = 0$;
CU on $(0, 0.6)$, $(2.1, 3.8)$, $(5.4, 7.0)$, $(8.6, 3\pi)$;
CD on $(0.6, 2.1)$, $(3.8, 5.4)$, $(7.0, 8.6)$; IP $(0.6, 0.25)$, $(2.1, 0.31)$, $(3.8, 0.10)$, $(5.4, 0.11)$, $(7.0, 0.061)$, $(8.6, 0.065)$

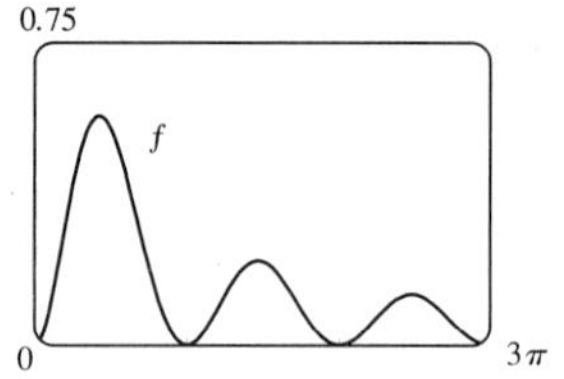

21. Inc. on $(-\infty, 0)$, $(0, \infty)$;
CU on $(-\infty, -0.4)$, $(0, 0.4)$; CD on $(-0.4, 0)$, $(0.4, \infty)$;
IP $(\mp 0.4, \pm 0.8)$

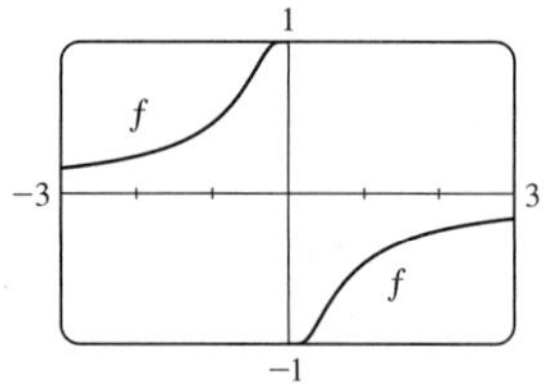

23. (a)

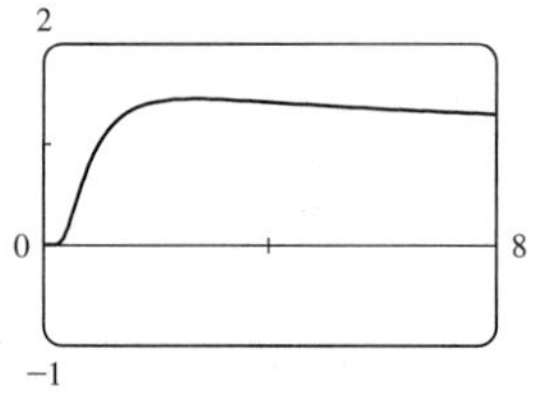

(b) $\lim_{x\to 0^+} x^{1/x} = 0$, $\lim_{x\to\infty} x^{1/x} = 1$
(c) Loc. max. $f(e) = e^{1/e}$ (d) IP at $x \approx 0.58, 4.37$

25. Max. $f(0.59) \approx 1$, $f(0.68) \approx 1$, $f(1.96) \approx 1$;
min. $f(0.64) \approx 0.99996$, $f(1.46) \approx 0.49$, $f(2.73) \approx -0.51$;
IP $(0.61, 0.99998)$, $(0.66, 0.99998)$, $(1.17, 0.72)$, $(1.75, 0.77)$, $(2.28, 0.34)$

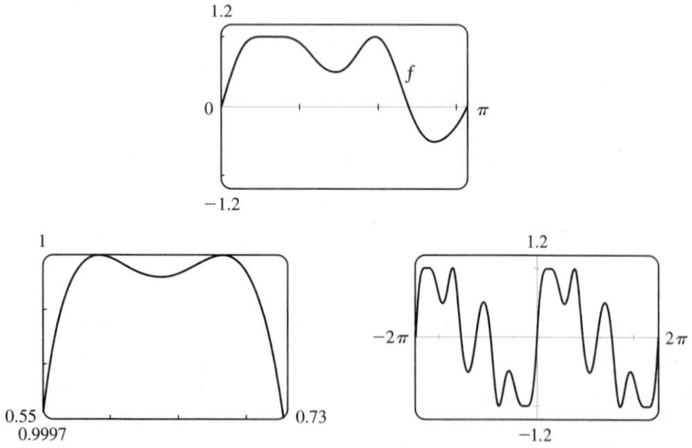

27. For $c \geqslant 0$, there is no IP and only one extreme point, the origin. For $c < 0$, there is a maximum point at the origin, two minimum points, and two IPs, which move downward and away from the origin as $c \to -\infty$.

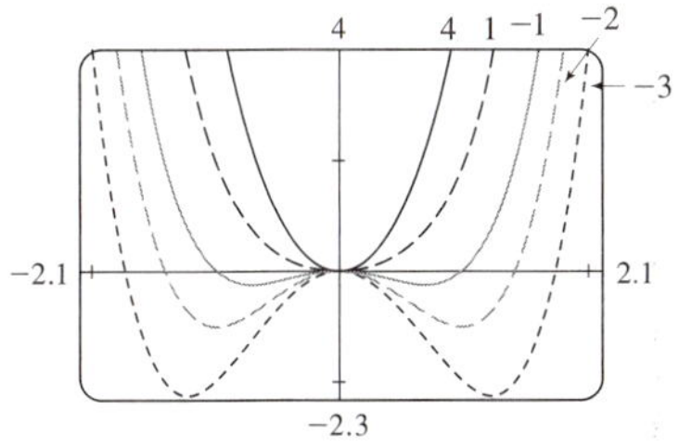

29. There is no maximum or minimum, regardless of the value of c. For $c < 0$, there is a vertical asymptote at $x = 0$, $\lim_{x \to 0} f(x) = \infty$, and $\lim_{x \to \pm\infty} f(x) = 1$.
$c = 0$ is a transitional value at which $f(x) = 1$ for $x \neq 0$.
For $c > 0$, $\lim_{x \to 0} f(x) = 0$, $\lim_{x \to \pm\infty} f(x) = 1$, and there are two IPs, which move away from the y-axis as $c \to \infty$.

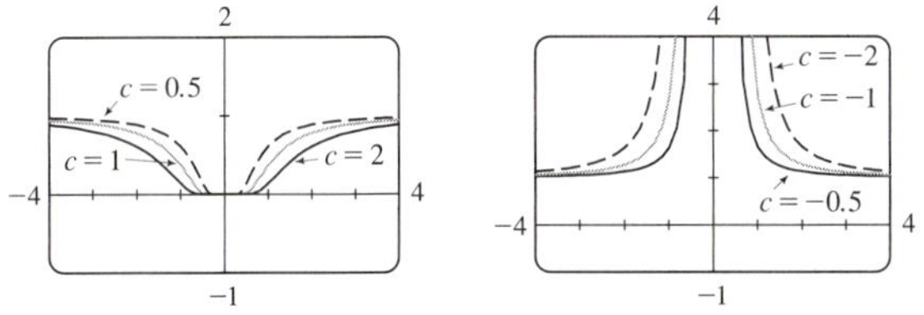

31. For $c > 0$, the maximum and minimum values are always $\pm\frac{1}{2}$, but the extreme points and IPs move closer to the y-axis as c increases. $c = 0$ is a transitional value: when c is replaced by $-c$, the curve is reflected in the x-axis.

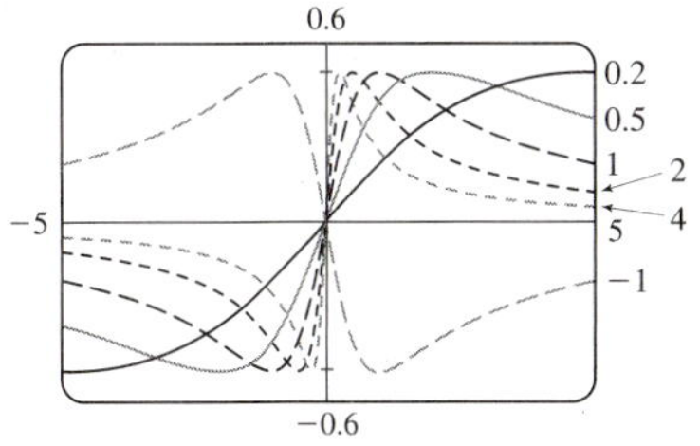

33. For $|c| < 1$, the graph has local maximum and minimum values; for $|c| \geqslant 1$ it does not. The function increases for $c \geqslant 1$ and decreases for $c \leqslant -1$. As c changes, the IPs move vertically but not horizontally.

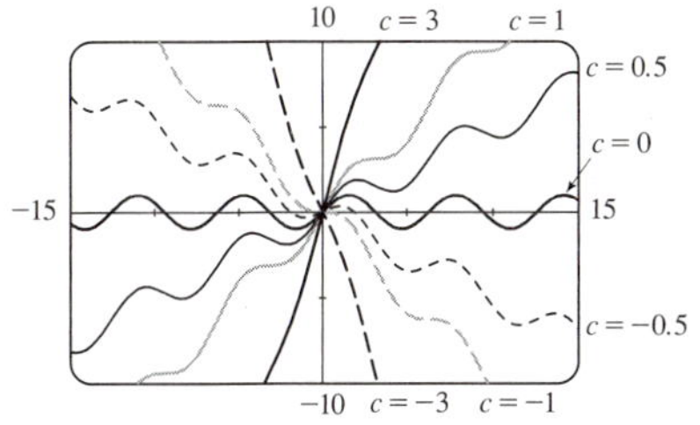

35.

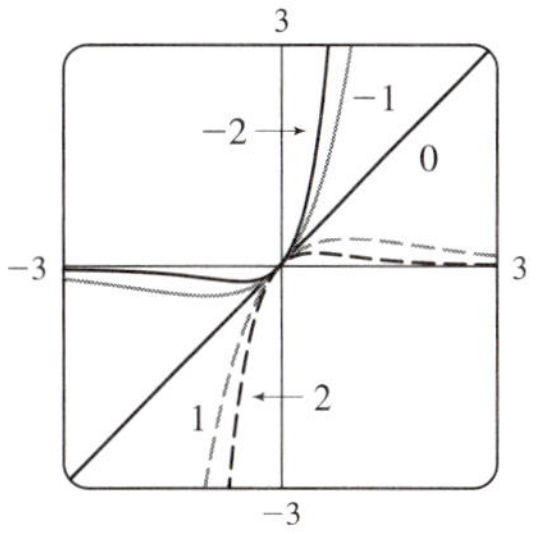

For $c > 0$, $\lim_{x \to \infty} f(x) = 0$ and $\lim_{x \to -\infty} f(x) = -\infty$.
For $c < 0$, $\lim_{x \to \infty} f(x) = \infty$ and $\lim_{x \to -\infty} f(x) = 0$.
As $|c|$ increases, the maximum and minimum points and the IPs get closer to the origin.

37. (a) Positive (b)

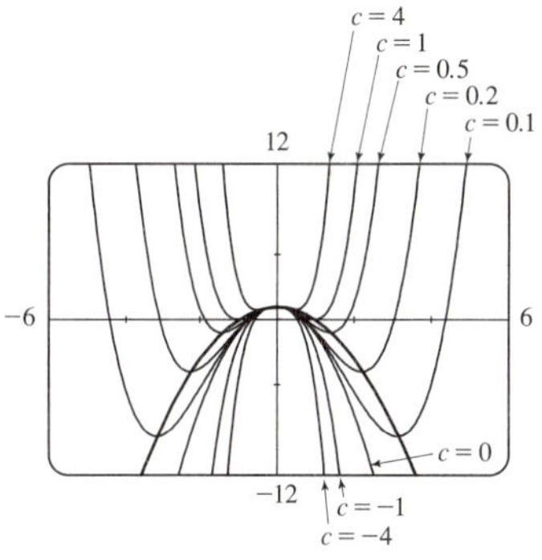

Exercises 4.7 ▫ page 336

1. (a) 11, 12 (b) 11.5, 11.5 **3.** 10, 10
5. 25 m by 25 m
7. (a)

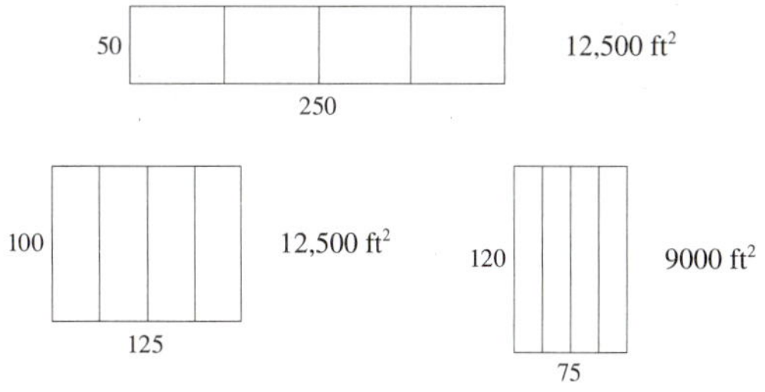

(b)

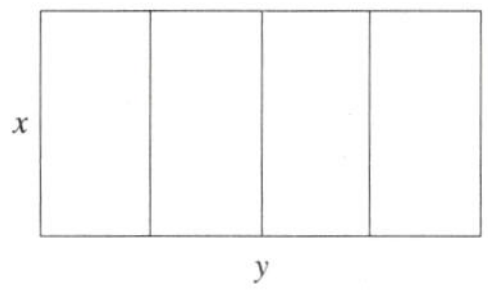

(c) $A = xy$ (d) $5x + 2y = 750$ (e) $A(x) = 375x - \frac{5}{2}x^2$
(f) 14,062.5 ft^2
9. 1000 ft by 1500 ft **11.** 4000 cm^3 **13.** \$191.28
15. $\left(-\frac{28}{17}, \frac{7}{17}\right)$ **17.** $\left(-\frac{1}{3}, \pm\frac{4}{3}\sqrt{2}\right)$
19. Square, side $\sqrt{2}r$ **21.** $L/2$, $\sqrt{3}\,L/4$
23. Base $\sqrt{3}r$, height $3r/2$ **25.** $4\pi r^3/(3\sqrt{3})$
27. $\pi r^2(1 + \sqrt{5})$ **29.** 24 cm, 36 cm
31. (a) Use all of the wire for the square
(b) $40\sqrt{3}/(9 + 4\sqrt{3})$ m for the square
33. Height = radius = $\sqrt[3]{V/\pi}$ cm

35. $V = 2\pi R^3/(9\sqrt{3})$

39. (a) $\frac{3}{2}s^2 \csc\theta\,(\csc\theta - \sqrt{3}\cot\theta)$ (b) $\cos^{-1}(1/\sqrt{3}) \approx 55°$
(c) $6s[h + s/(2\sqrt{2})]$

41. Row directly to B

43. $10\sqrt[3]{3}/(1 + \sqrt[3]{3})$ ft from the stronger source

45. $(a^{2/3} + b^{2/3})^{3/2}$ **49.** 9.35 m **53.** $x = 6$ in.

55. $\pi/6$ **57.** At a distance $5 - 2\sqrt{5}$ from A

59. $(L + W)^2/2$

61. (a) About 5.1 km from B
(b) C is close to B; C is close to D;
$W/L = \sqrt{25 + x^2}/x$, where $x = |BC|$
(c) ≈ 1.07; no such value (d) $\sqrt{41}/4 \approx 1.6$

Exercises 4.8 ▫ page 346

1. (a) $C(0)$ represents fixed costs, which are incurred even when nothing is produced.
(b) The marginal cost is a minimum there.
(c)

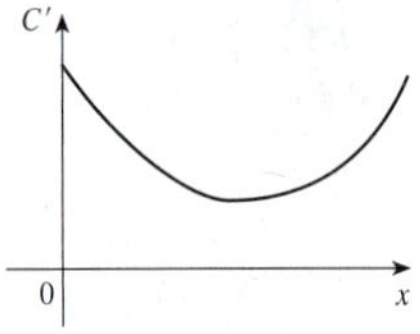

3. \$17.40/unit; the cost of producing the 1001st unit is about \$17.40

5. (a) \$1,340,000; \$1340/unit; \$2300/unit
(b) 200 (c) \$700/unit

7. (a) \$342,491; \$342/unit; \$390/unit
(b) 400 (c) \$320/unit

9. (a) $c(x) = 3700/x + 5 - 0.04x + 0.0003x^2$,
$C'(x) = 5 - 0.08x + 0.0009x^2$
(b) Between 208 and 209 units (c) $c(209) \approx \$27.45$/unit
(d) \$3.22/unit

11. 400 **13.** 672 **15.** 100

17. (a) About 200 yd (b) 192 yd

19. (a) $p(x) = 19 - \frac{1}{3000}x$ (b) \$9.50

21. (a) $p(x) = 550 - \frac{1}{10}x$ (b) \$175 (c) \$100 **23.** 200

Exercises 4.9 ▫ page 351

1. (a) $x_2 \approx 2.3, x_3 \approx 3$ (b) No **3.** $\frac{4}{5}$ **5.** 1.1797

7. 2.1148 **9.** -1.25 **11.** 3.10723251 **13.** 2.224745

15. 0.876726 **17.** -0.724492, 1.220744 **19.** 0.520269

21. 0.641714 **23.** -1.39194691, 1.07739428, 2.71987822

25. -1.97806681, -0.82646233

27. -0.51031156, 1.19843871 **29.** (b) 31.622777

35. (a) -0.455, 6.810, 0.645 (b) $f(6.810) \approx -1949.07$

37. (0.904557, 1.855277) **39.** 11.28 ft

41. 0.76286%

Exercises 4.10 ▫ page 358

1. $F(x) = 2x^3 - 4x^2 + 3x + C$

3. $F(x) = x - \frac{1}{4}x^4 + \frac{5}{6}x^6 - \frac{3}{8}x^8 + C$

5. $F(x) = 4x^{5/4} - 4x^{7/4} + C$ **7.** $F(x) = 4x^{3/2} - \frac{6}{7}x^{7/6} + C$

9. $F(x) = -5/(4x^8) + C_1$ if $x < 0$;
$F(x) = -5/(4x^8) + C_2$ if $x > 0$

11. $F(u) = \frac{1}{3}u^3 - 6u^{-1/2} + C$ **13.** $G(\theta) = \sin\theta + 5\cos\theta + C$

15. $F(x) = x^2 + 5\sin^{-1}x + C$ **17.** $F(x) = x^5 - \frac{1}{3}x^6 + 4$

19. $x^3 + x^4 + Cx + D$

21. $\frac{1}{2}x^2 + \frac{25}{126}x^{14/5} + Cx + D$ **23.** $e^t + \frac{1}{2}Ct^2 + Dt + E$

25. $x - 3x^2 + 8$ **27.** $4x^{3/2} + 2x^{5/2} + 4$

29. $2\sin t + \tan t + 4 - 2\sqrt{3}$ **31.** $2\ln(-x) + 7$

33. $2x^4 + \frac{1}{3}x^3 + 5x^2 - 22x + \frac{59}{3}$

35. $-\sin\theta - \cos\theta + 5\theta + 4$ **37.** $x^2 - 2x^3 + 9x + 9$

39. $x^2 - \cos x - \frac{1}{2}\pi x$

41. $-\ln x + (\ln 2)x - \ln 2$ **43.** 10 **45.** b

47.

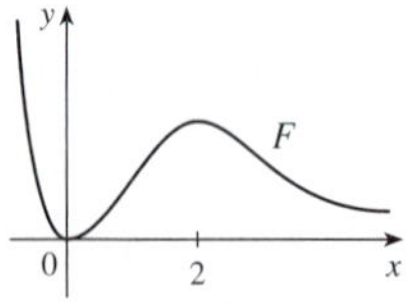

49.

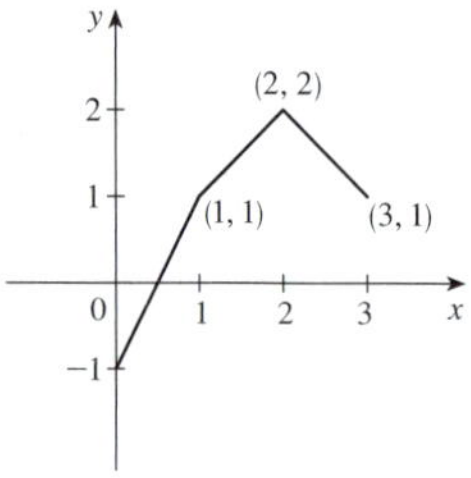

51.

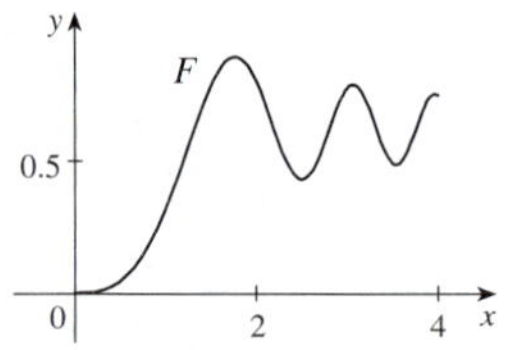

53.

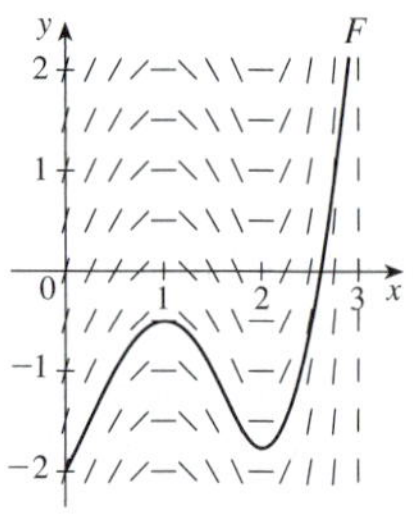

55.

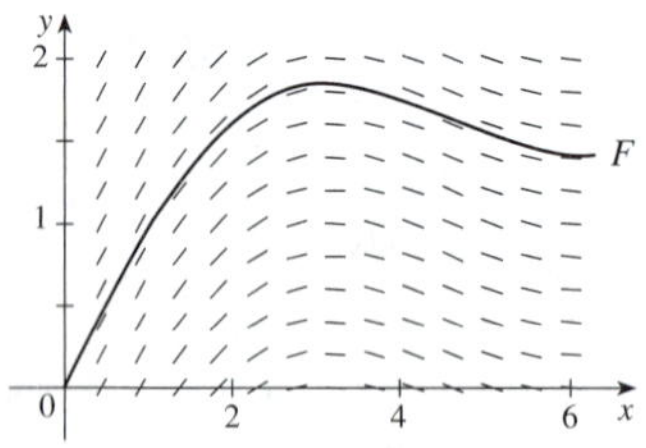

57.

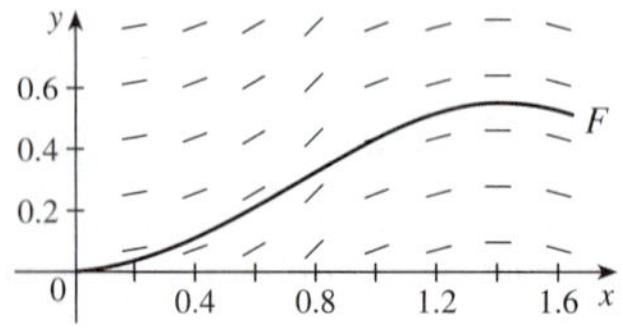

59. $s(t) = 1 - \cos t - \sin t$ **61.** $s(t) = \frac{1}{6}t^3 - t^2 + 3t + 1$

63. $s(t) = -10\sin t - 3\cos t + (6/\pi)t + 3$

65. (a) $s(t) = 450 - 4.9t^2$ (b) $\sqrt{450/4.9} \approx 9.58$ s
(c) $-9.8\sqrt{450/4.9} \approx -93.9$ m/s (d) About 9.09 s

69. 225 ft **71.** \$742.08 **73.** $\frac{130}{11} \approx 11.8$ s

75. $\frac{88}{15} \approx 5.87$ ft/s^2 **77.** 62,500 km/h^2 ≈ 4.82 m/s^2

79. (a) 22.9125 mi (b) 21.675 mi (c) 30 min 33 s
(d) 55.425 mi

Chapter 4 Review ▫ page 361

True-False Quiz

1. False **3.** False **5.** True **7.** False **9.** True
11. True **13.** False **15.** True **17.** True

Exercises

1. Abs. min. $f(0) = 10$; abs. and loc. max. $f(3) = 64$

3. Abs. max. $f(0) = 0$; abs. and loc. min. $f(-1) = -1$

5. Abs. max. $f(\pi) = \pi$; abs. min. $f(0) = 0$; loc. max. $f(\pi/3) = (\pi/3) + \frac{1}{2}\sqrt{3}$; loc. min. $f(2\pi/3) = (2\pi/3) - \frac{1}{2}\sqrt{3}$

7. π **9.** 8 **11.** 0 **13.** $\frac{1}{2}$

15.

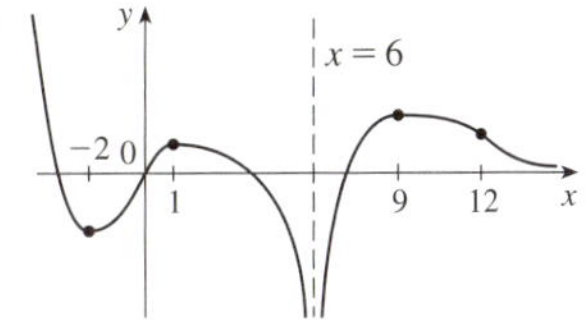

17.

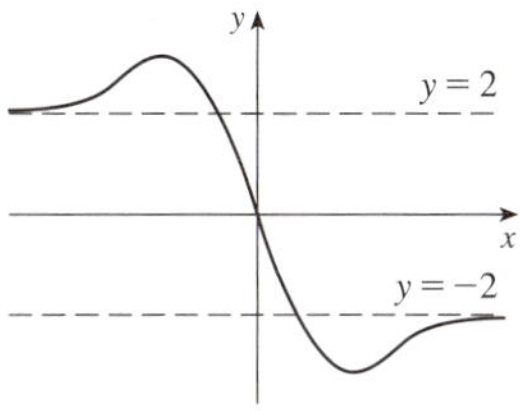

19. A. $\mathbb{R}$ B. y-int. 2
C. None D. None
E. Dec. on $(-\infty, \infty)$ F. None
G. CU on $(-\infty, 0)$;
CD on $(0, \infty)$; IP $(0, 2)$
H. See graph at right.

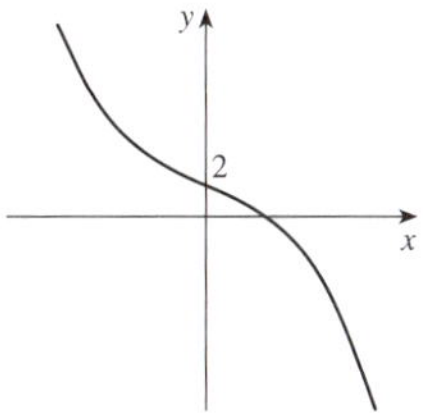

21. A. $\mathbb{R}$ B. y-int. 0; x-int. 0, 1
C. None D. None
E. Inc. on $\left(\frac{1}{4}, \infty\right)$, dec. on $\left(-\infty, \frac{1}{4}\right)$
F. Loc. min. $f\left(\frac{1}{4}\right) = -\frac{27}{256}$
G. CU on $\left(-\infty, \frac{1}{2}\right)$, $(1, \infty)$;
CD on $\left(\frac{1}{2}, 1\right)$; IP $\left(\frac{1}{2}, -\frac{1}{16}\right)$, $(1, 0)$
H. See graph at right.

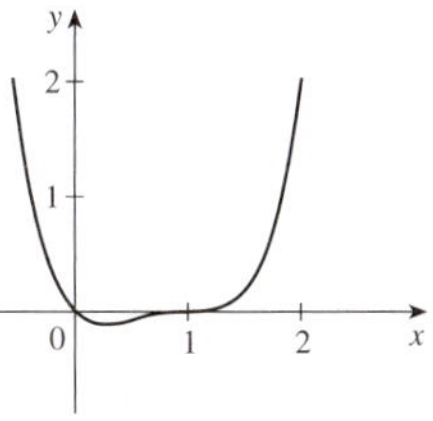

23. A. $\{x \mid x \neq 0, 3\}$
B. None C. None
D. HA $y = 0$; VA $x = 0$, $x = 3$
E. Dec. on $(-\infty, 0)$, $(0, 1)$, $(3, \infty)$;
inc. on $(1, 3)$
F. Loc. min. $f(1) = \frac{1}{4}$
G. CU on $(0, 3)$, $(3, \infty)$; CD on $(-\infty, 0)$
H. See graph at right.

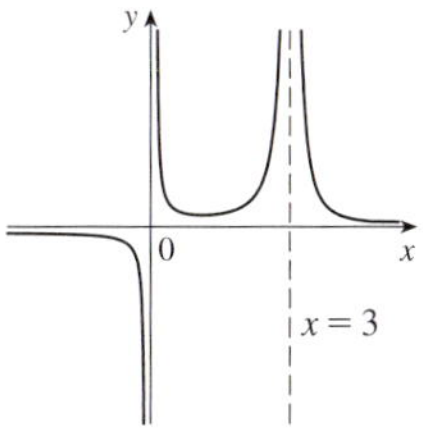

25. A. $\{x \mid x \neq -8\}$
B. y-int. 0, x-int. 0 C. None
D. VA $x = -8$; SA $y = x - 8$
E. Inc. on $(-\infty, -16)$, $(0, \infty)$;
dec. on $(-16, -8)$, $(-8, 0)$
F. Loc. max. $f(-16) = -32$;
loc. min. $f(0) = 0$
G. CU on $(-8, \infty)$; CD on $(-\infty, -8)$
H. See graph at right.

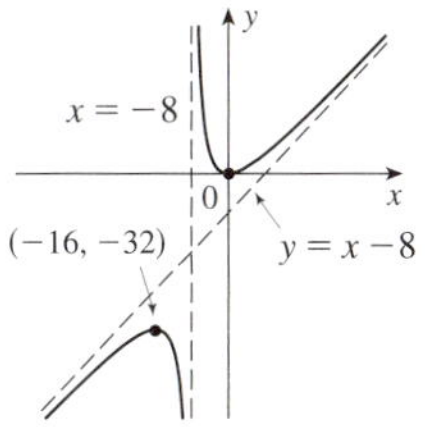

27. A. $[-2, \infty)$
B. y-int. 0; x-int. -2, 0
C. None D. None
E. Inc. on $\left(-\frac{4}{3}, \infty\right)$, dec. on $\left(-2, -\frac{4}{3}\right)$
F. Loc. min. $f\left(-\frac{4}{3}\right) = -4\sqrt{6}/9$
G. CU on $(-2, \infty)$
H. See graph at right.

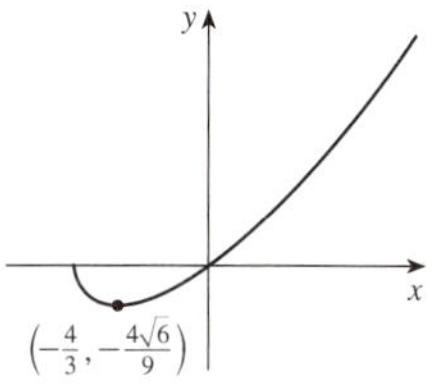

29. A. $\mathbb{R}$ B. y-int. -2
C. About y-axis, period 2π D. None
E. Inc. on $(2n\pi, (2n + 1)\pi)$, n an integer; dec. on $((2n - 1)\pi, 2n\pi)$
F. Loc. max. $f((2n + 1)\pi) = 2$; loc. min. $f(2n\pi) = -2$
G. CU on $(2n\pi - (\pi/3), 2n\pi + (\pi/3))$;
CD on $(2n\pi + (\pi/3), 2n\pi + (5\pi/3))$; IP $\left(2n\pi \pm (\pi/3), -\frac{1}{4}\right)$
H.

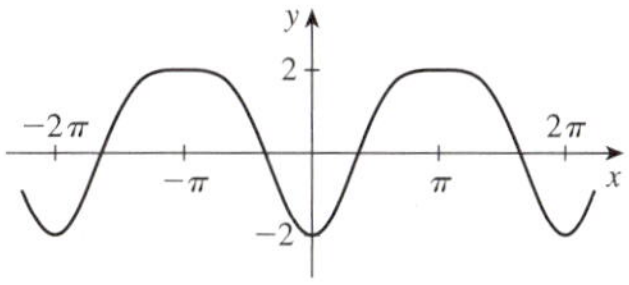

31. A. $\{x \mid |x| \geqslant 1\}$
B. None C. About $(0, 0)$
D. HA $y = 0$
E. Dec. on $(-\infty, -1)$, $(1, \infty)$
F. None
G. CU on $(1, \infty)$; CD on $(-\infty, -1)$
H. See graph at right.

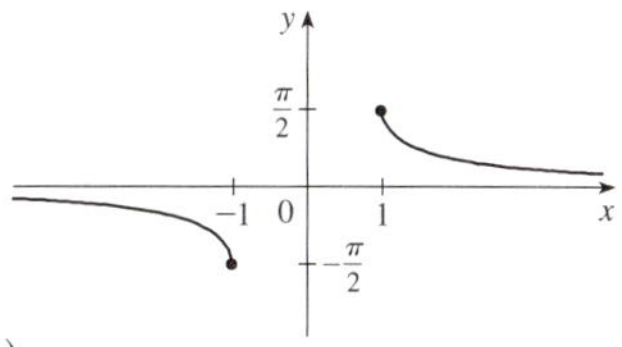

33. A. $\mathbb{R}$ B. y-int. 2
C. None D. None
E. Inc. on $\left(\frac{1}{4}\ln 3, \infty\right)$;
dec. on $\left(-\infty, \frac{1}{4}\ln 3\right)$
F. Loc. min. $f\left(\frac{1}{4}\ln 3\right) = 3^{1/4} + 3^{-3/4}$
G. CU on $(-\infty, \infty)$
H. See graph at right.

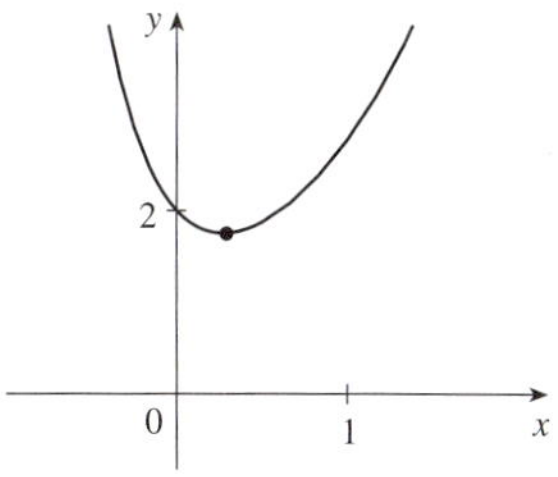

35. Inc. on $(-\sqrt{3}, 0)$, $(0, \sqrt{3})$;
dec. on $(-\infty, -\sqrt{3})$, $(\sqrt{3}, \infty)$;
loc. max. $f(\sqrt{3}) = \frac{2}{9}\sqrt{3}$,
loc. min. $f(-\sqrt{3}) = -\frac{2}{9}\sqrt{3}$;
CU on $(-\sqrt{6}, 0)$, $(\sqrt{6}, \infty)$;
CD on $(-\infty, -\sqrt{6})$, $(0, \sqrt{6})$;
IP $\left(\sqrt{6}, \frac{5}{36}\sqrt{6}\right)$, $\left(-\sqrt{6}, -\frac{5}{36}\sqrt{6}\right)$

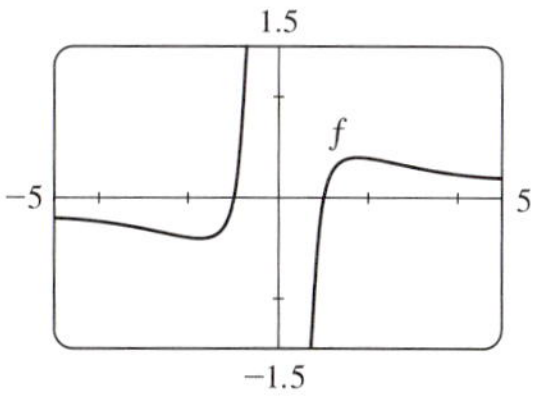

37. Inc. on $(-0.23, 0)$, $(1.62, \infty)$; dec. on $(-\infty, -0.23)$, $(0, 1.62)$;
loc. max. $f(0) = 2$; loc. min. $f(-0.23) \approx 1.96$, $f(1.62) \approx -19.2$;
CU on $(-\infty, -0.12)$, $(1.24, \infty)$; CD on $(-0.12, 1.24)$;
IP $(-0.12, 1.98)$, $(1.24, -12.1)$

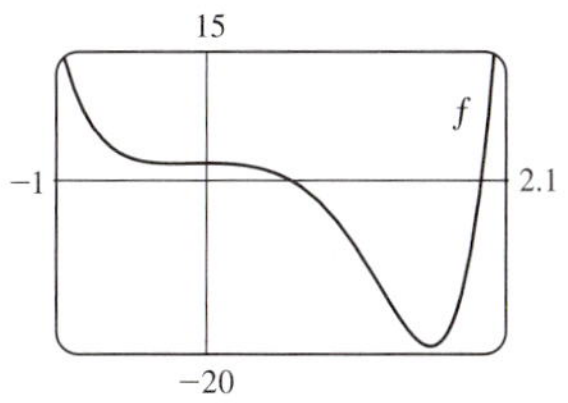

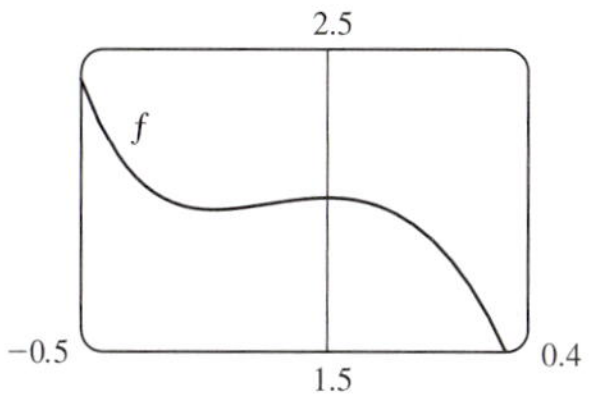

39. 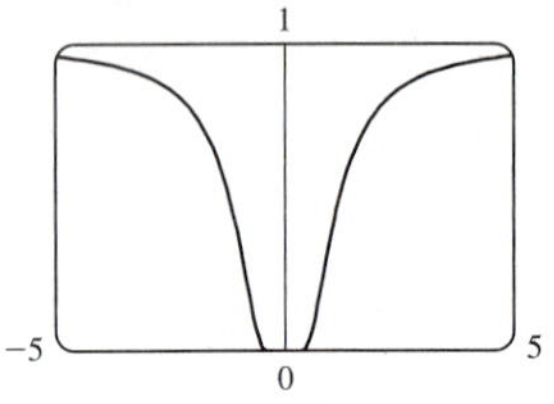

$(\pm 0.82, 0.22)$; $\left(\pm\sqrt{2/3}, e^{-3/2}\right)$

41. Max. at $x = 0$, min. at $x \approx \pm 0.87$, IP at $x \approx \pm 0.52$
43. For $C > -1$, f is periodic with period 2π and has local maxima at $2n\pi + \pi/2$, n an integer. For $C \leqslant -1$, f has no graph. For $-1 < C \leqslant 1$, f has vertical asymptotes. For $C > 1$, f is continuous on $\mathbb{R}$. As C increases, f moves upward and its oscillations become less pronounced.
49. (a) 0 (b) CU on $\mathbb{R}$ **53.** $3\sqrt{3}r^2$
55. $4/\sqrt{3}$ cm from D **57.** $L = C$ **59.** \$11.50
61. 1.297383 **63.** 1.16718557
65. $f(x) = \frac{2}{7}x^{7/2} - 5x^{4/5} + C$ **67.** $f(x) = e^x - 4\sqrt{x} + C$
69. $f(t) = t^2 + 3\cos t + 2$
71. $f(x) = \frac{1}{2}x^2 - x^3 + 4x^4 + 2x + 1$
73. (b) $0.1e^x - \cos x + 0.9$ (c)

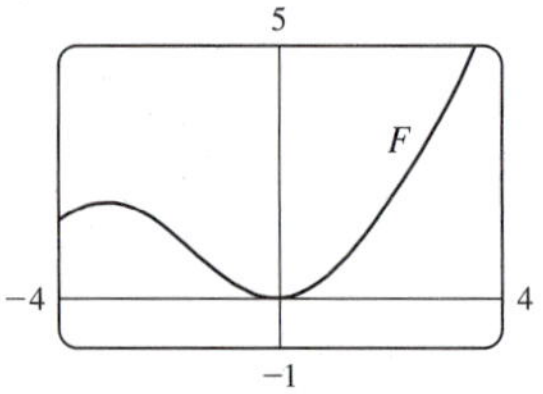

75. No
77. (b) About 8.5 in. by 2 in. (c) $20/\sqrt{3}$ in. by $20\sqrt{2/3}$ in.
79. (a) $20\sqrt{2} \approx 28$ ft (b) $\dfrac{dI}{dt} = \dfrac{-480k(h-4)}{[(h-4)^2 + 1600]^{5/2}}$, where k is the constant of proportionality

Problems Plus ▫ page 365

7. $(-2, 4)$, $(2, -4)$ **9.** $(m/2, m^2/4)$
11. $-3.5 < a < -2.5$
13. (a) $x/(x^2 + 1)$ (b) $\frac{1}{2}$ **15.** $a \leqslant e^{1/e}$
19. (a) $T_1 = D/c_1$, $T_2 = (2h\sec\theta)/c_1 + (D - 2h\tan\theta)/c_2$, $T_3 = \sqrt{4h^2 + D^2}/c_1$
(c) $c_1 \approx 3.85$ km/s, $c_2 \approx 7.66$ km/s, $h \approx 0.42$ km
23. $3/(\sqrt[3]{2} - 1) \approx 11\frac{1}{2}$ h

Chapter 5

Exercises 5.1 ▫ page 378

1. (a) 40, 52

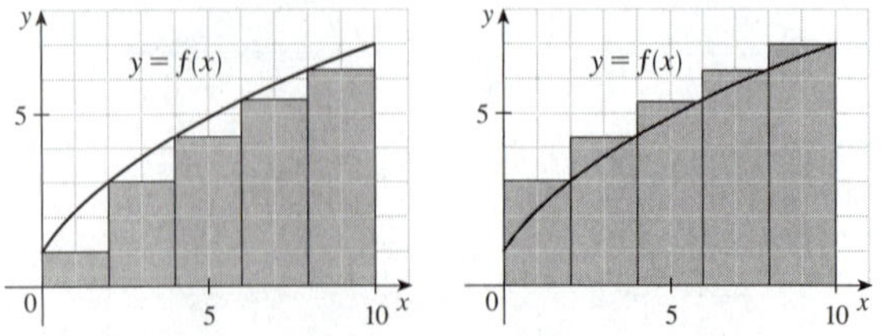

(b) 43.2, 49.2

3. (a) $\frac{77}{60}$, underestimate

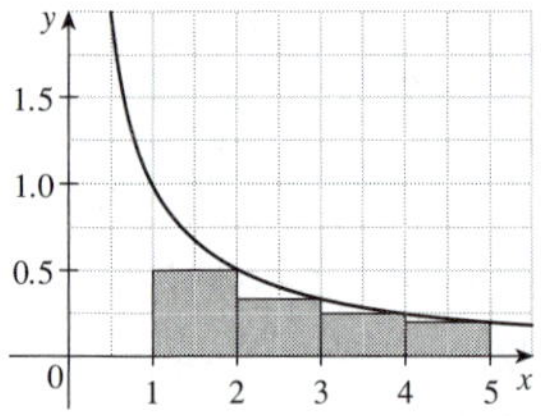

(b) $\frac{25}{12}$, overestimate

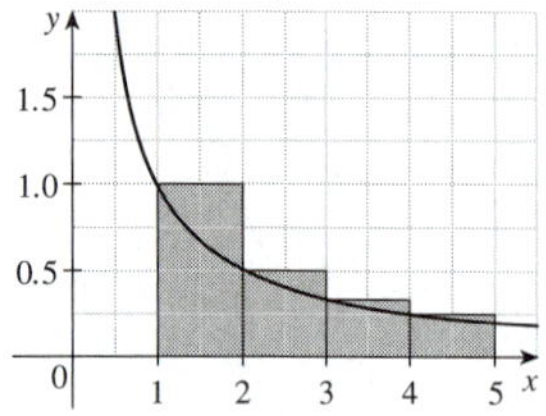

5. (a) 8, 6.875

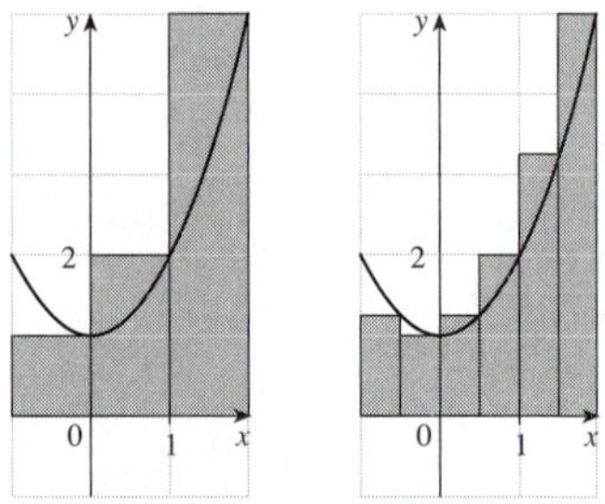

(b) 5, 5.375

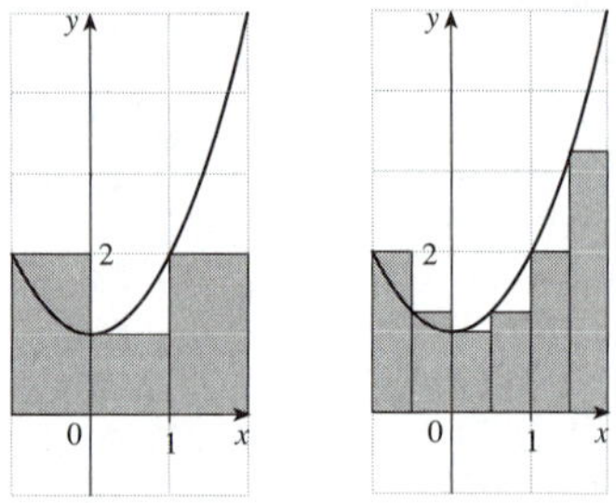

(c) 5.75, 5.9375

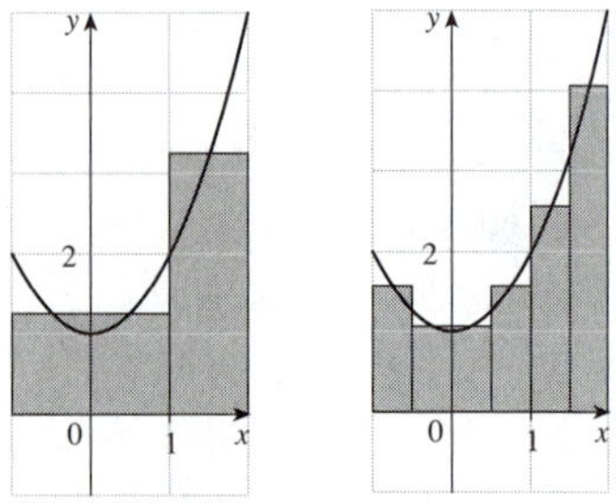

(d) M_6
7. 1.9835, 1.9982, 1.9993; 2
9. (a) Left: 4.5148, 4.6165, 4.6366; right: 4.8148, 4.7165, 4.6966
11. 34.7 ft, 44.8 ft
13. 63.2 L, 70 L **15.** 155 ft
17. $\lim_{n\to\infty} \sum_{i=1}^{n} \sqrt[4]{1 + 15i/n} \cdot (15/n)$
19. $\lim_{n\to\infty} \sum_{i=1}^{n} \left(\dfrac{i\pi}{2n}\cos\dfrac{i\pi}{2n}\right)\dfrac{\pi}{2n}$
21. The region under the graph of $y = \tan x$ from 0 to $\pi/4$
23. (a) $\lim_{n\to\infty} \dfrac{64}{n^6}\sum_{i=1}^{n} i^5$ (b) $\dfrac{n^2(n+1)^2(2n^2+2n-1)}{12}$ (c) $\frac{32}{3}$
25. $\sin b$, 1

Exercises 5.2 ▫ page 390

1. 0.25
The Riemann sum represents the sum of the areas of the two rectangles above the x-axis minus the sum of the areas of the two rectangles below the x-axis.

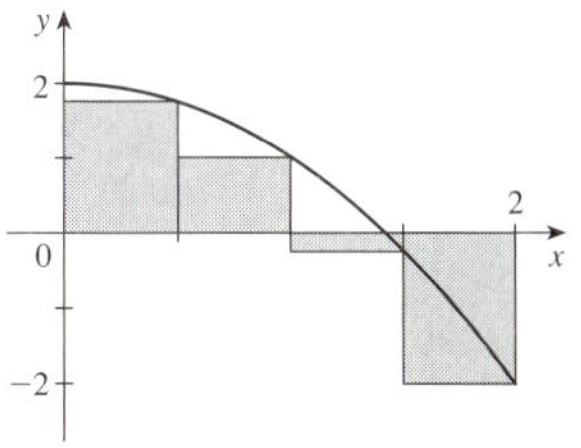

3. −0.856759
The Riemann sum represents the sum of the areas of the two rectangles above the x-axis minus the sum of the areas of the three rectangles below the x-axis.

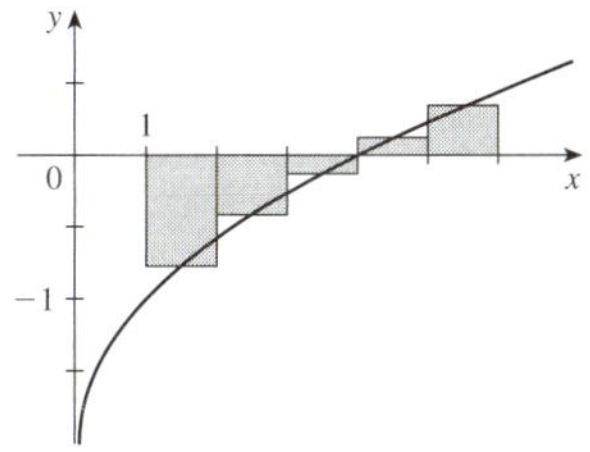

5. (a) 4 (b) 6 (c) 10 **7.** −475, −85
9. 124.1644 **11.** 0.3084
13. 0.30843908, 0.30981629, 0.31015563
15.

n	R_n
5	1.933766
10	1.983524
50	1.999342
100	1.999836

The values of R_n appear to be approaching 2.

17. $\int_0^\pi x \sin x\, dx$ **19.** $\int_1^8 \sqrt{2x + x^2}\, dx$ **21.** 42 **23.** $\frac{4}{3}$

25. 3.75 **29.** $\lim_{n\to\infty} \sum_{i=1}^{n} \frac{2 + 4i/n}{1 + (2 + 4i/n)^5} \cdot \frac{4}{n}$

31. $\lim_{n\to\infty} \sum_{i=1}^{n} \left(\sin \frac{5\pi i}{n}\right) \frac{\pi}{n} = \frac{2}{5}$

33. (a) 4 (b) 10 (c) −3 (d) 2

35. $-\frac{3}{4}$ **37.** $3 + \frac{9}{4}\pi$ **39.** 2.5 **41.** $-\frac{38}{3}$ **43.** 3

45. $e^5 - e^3$ **47.** $\int_{-1}^{5} f(x)\, dx$ **49.** 122

55. $\frac{1}{2} \leqslant \int_1^2 dx/x \leqslant 1$ **57.** $\frac{\pi}{12} \leqslant \int_{\pi/4}^{\pi/3} \tan x\, dx \leqslant \frac{\pi}{12}\sqrt{3}$

59. $0 \leqslant \int_0^2 xe^{-x}\, dx \leqslant 2/e$ **67.** $\int_0^1 x^4\, dx$ **69.** $\frac{1}{2}$

Exercises 5.3 ▫ page 402

1. One process undoes what the other one does. See the Fundamental Theorem of Calculus, page 401.

3. (a) 0, 2, 5, 7, 3
(b) (0, 3)
(c) $x = 3$
(d)

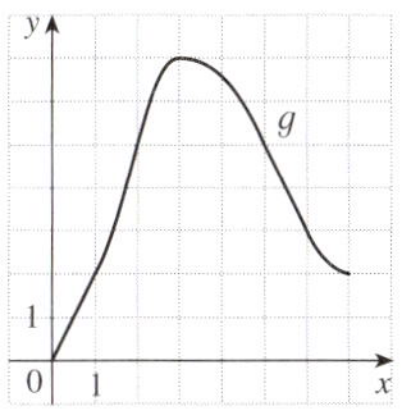

5.

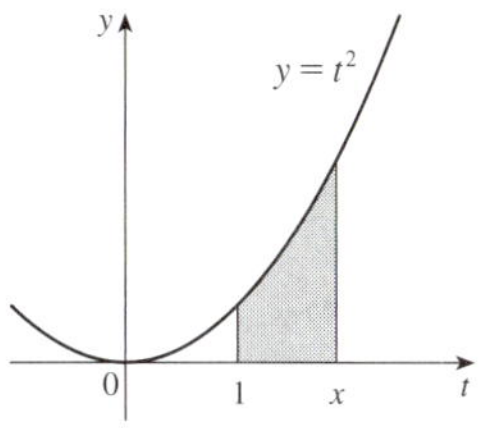

(a), (b) x^2

7. $g'(x) = \sqrt{1 + 2x}$ **9.** $g'(y) = y^2 \sin y$
11. $F'(x) = -\cos(x^2)$ **13.** $h'(x) = -\arctan(1/x)/x^2$

15. $y' = \dfrac{\cos\sqrt{x}}{2x}$ **17.** $y' = \dfrac{3(1 - 3x)^3}{1 + (1 - 3x)^2}$ **19.** $\frac{364}{3}$

21. 138 **23.** $\frac{5}{9}$ **25.** $\frac{7}{8}$ **27.** Does not exist
29. $\frac{156}{7}$ **31.** 1 **33.** Does not exist **35.** ln 3
37. π **39.** $e^2 - 1$ **41.** 10.7 **43.** $\frac{243}{4}$ **45.** 2
47. 3.75

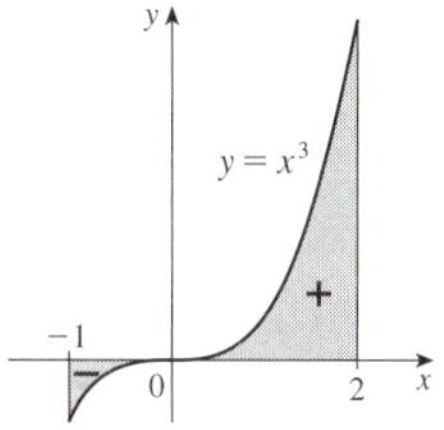

49. $g'(x) = \dfrac{-2(4x^2 - 1)}{4x^2 + 1} + \dfrac{3(9x^2 - 1)}{9x^2 + 1}$

51. $y' = 3x^{7/2}\sin(x^3) - (\sin\sqrt{x})/(2\sqrt[4]{x})$ **53.** $\sqrt{257}$ **55.** 29

57. (a) $-2\sqrt{n}$, $\sqrt{4n - 2}$, n an integer > 0
(b) (0, 1), $\left(-\sqrt{4n - 1}, -\sqrt{4n - 3}\right)$, and $\left(\sqrt{4n - 1}, \sqrt{4n + 1}\right)$, n an integer > 0
(c) 0.74

59. (a) Loc. max. at 1 and 5; loc. min. at 3 and 7
(b) 9
(c) $\left(\frac{1}{2}, 2\right)$, (4, 6), (8, 9)
(d) See graph at right.

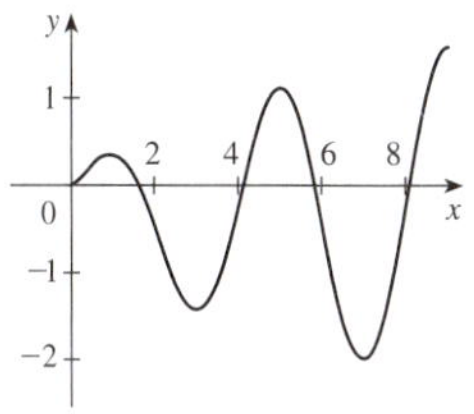

61. $\frac{1}{4}$ **67.** $f(x) = x^{3/2}$, $a = 9$
69. (b) Average expenditure over $[0, t]$; minimize average expenditure

Exercises 5.4 ▫ page 411

5. $4x^{1/4} + C$ **7.** $\frac{1}{4}x^4 + 3x^2 + x + C$
9. $2t - t^2 + \frac{1}{3}t^3 - \frac{1}{4}t^4 + C$ **11.** $4x - \frac{8}{3}x^{3/2} + \frac{1}{2}x^2 + C$
13. $\sec x + C$ **15.** $\frac{2}{5}x^{5/2} + C$

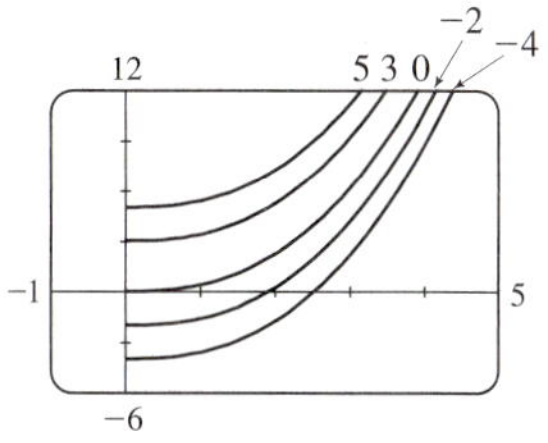

17. 18 **19.** $-2 + 1/e$ **21.** 52 **23.** $\frac{256}{15}$ **25.** $-\frac{63}{4}$
27. $\frac{55}{63}$ **29.** $2\sqrt{5}$ **31.** 8 **33.** $1 + \pi/4$ **35.** $\frac{256}{5}$
37. $\frac{1}{2}e^2 + e - \frac{1}{2}$ **39.** -3.5 **41.** 0, 1.32; 0.84 **43.** $\frac{4}{3}$
45. The increase in the child's weight (in pounds) between the ages of 5 and 10
47. Number of gallons of oil leaked in the first 2 hours
49. Increase in revenue when production is increased from 1000 to 5000 units
51. Newton-meters (or joules) **53.** (a) $-\frac{3}{2}$ m (b) $\frac{41}{6}$ m
55. (a) $v(t) = \frac{1}{2}t^2 + 4t + 5$ m/s (b) $416\frac{2}{3}$ m
57. $46\frac{2}{3}$ kg **59.** 1.4 mi **61.** \$58,000
63. (b) At most 40%; $\frac{5}{36}$

Exercises 5.5 ▫ page 420

1. $\frac{1}{3}\sin 3x + C$ **3.** $\frac{2}{9}(x^3 + 1)^{3/2} + C$
5. $-1/(1 + 2x)^2 + C$ **7.** $\frac{1}{5}(x^2 + 3)^5 + C$
9. $\frac{1}{63}(3x - 2)^{21} + C$ **11.** $2\sqrt{1 + x + 2x^2} + C$
13. $-\frac{1}{3}\ln|5 - 3x| + C$ **15.** $\dfrac{-3}{8(2y + 1)^4} + C$
17. $-\frac{2}{3}(4 - t)^{3/2} + C$ **19.** $-(1/\pi)\cos \pi t + C$
21. $\frac{1}{3}(\ln x)^3 + C$ **23.** $2\sin\sqrt{t} + C$ **25.** $\frac{1}{7}\sin^7\theta + C$
27. $\frac{2}{3}(1 + e^x)^{3/2} + C$ **29.** $\frac{1}{2}(1 + z^3)^{2/3} + C$
31. $\ln|\ln x| + C$ **33.** $-\frac{2}{3}(\cot x)^{3/2} + C$ **35.** $\ln|\sin x| + C$
37. $\frac{1}{3}\sec^3 x + C$ **39.** $\dfrac{2(b + cx^{a+1})^{3/2}}{3c(a + 1)} + C$
41. $\tan^{-1}x + \frac{1}{2}\ln(1 + x^2) + C$
43. $\frac{4}{7}(x + 2)^{7/4} - \frac{8}{3}(x + 2)^{3/4} + C$
45. $\dfrac{-1}{6(3x^2 - 2x + 1)^3} + C$ **47.** $\frac{1}{4}\sin^4 x + C$

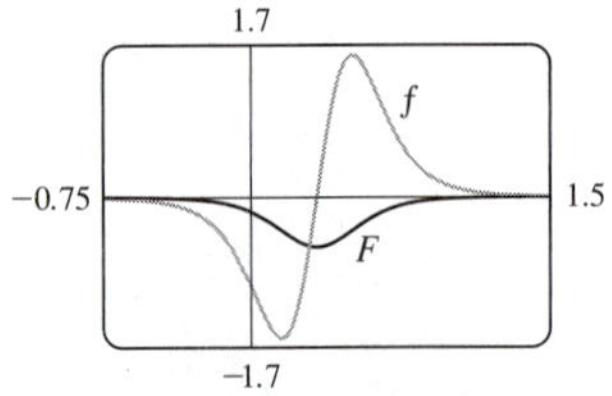

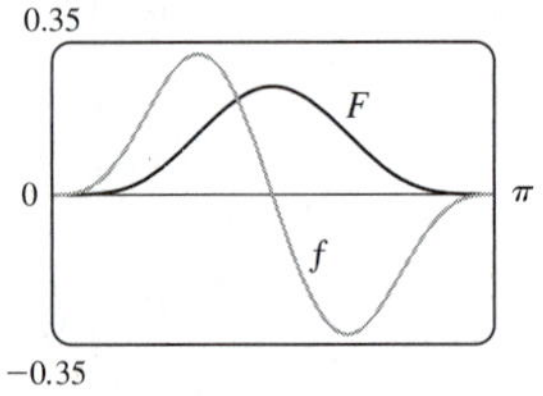

49. 0 **51.** $\frac{182}{9}$ **53.** 4 **55.** 0 **57.** $e - \sqrt{e}$
59. 1 **61.** 3 **63.** $\frac{16}{15}$ **65.** 2 **67.** Does not exist
69. $\frac{1}{3}(2\sqrt{2} - 1)a^3$ **71.** $\sqrt{3} - \frac{1}{3}$ **73.** 6π
75. All three areas are equal. **77.** $\dfrac{5}{4\pi}\left(1 - \cos\dfrac{2\pi t}{5}\right)$ L
79. 5 **85.** $\pi^2/4$

Exercises 5.6 ▫ page 429

1. (b) 0.405

Chapter 5 Review ▫ page 430

True-False Quiz

1. True **3.** True **5.** False **7.** True **9.** True
11. False **13.** False

Exercises

1. (a) 8 (b) 5.7

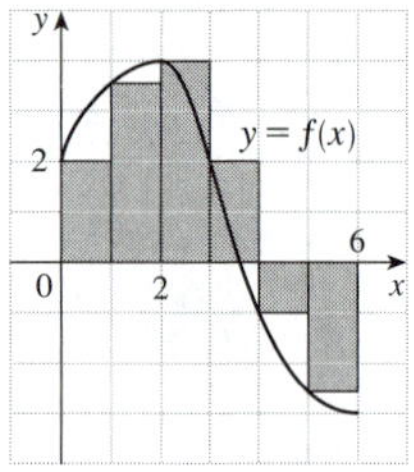

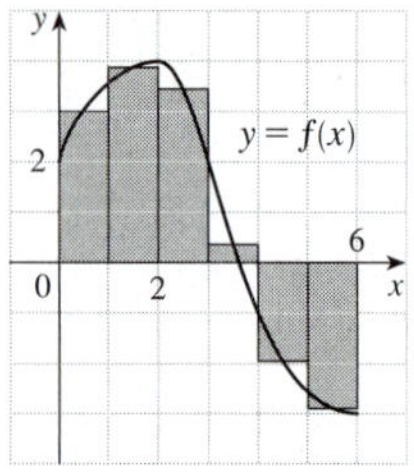

3. $\frac{1}{2} + \pi/4$ **5.** 3 **7.** $f = c,\ f' = b,\ \int_0^x f(t)\,dt = a$
9. 37 **11.** $\frac{9}{10}$ **13.** -76 **15.** $\frac{21}{4}$ **17.** Does not exist
19. $\frac{1}{3}\sin 1$ **21.** $(1/\pi)(e^\pi - 1)$ **23.** $\ln 2 - \frac{7}{4}$
25. $\sqrt{x^2 + 4x} + C$ **27.** $[1/(2\pi)]\sin^2\pi t + C$
29. $2e^{\sqrt{x}} + C$
31. $-\frac{1}{2}[\ln(\cos x)]^2 + C$ **33.** $\frac{1}{4}\ln(1 + x^4) + C$
35. $\ln|1 + \sec\theta| + C$ **37.** $\frac{23}{3}$ **39.** $2\sqrt{1 + \sin x} + C$
41. $\frac{64}{5}$ **43.** $F'(x) = \sqrt{1 + x^4}$ **45.** $g'(x) = 3x^5/\sqrt{1 + x^9}$
47. $y' = (2e^x - e^{\sqrt{x}})/(2x)$ **49.** $4 \leqslant \int_1^3 \sqrt{x^2 + 3}\,dx \leqslant 4\sqrt{3}$
55. 1.11
57. Number of barrels of oil consumed from Jan. 1, 2000, through Jan. 1, 2003
59. 72,400 **61.** $F(x) = \int_1^x t^2 \sin(t^2)\,dt$ **63.** $c \approx 1.62$
65. $e^{2x}(1 + 2x)/(1 - e^{-x})$

Problems Plus ▫ page 435

1. $\pi/2$ **5.** -1 **7.** e^{-2} **9.** $[-1, 2]$
11. (a) $(n - 1)n/2$
(b) $\frac{1}{2}[\![b]\!](2b - [\![b]\!] - 1) - \frac{1}{2}[\![a]\!](2a - [\![a]\!] - 1)$
13. $f(x) = \frac{1}{2}x$ or $f(x) = 0$ **17.** $2(\sqrt{2} - 1)$ **19.** $\frac{2}{3}$

Chapter 6

Exercises 6.1 ▫ page 442

1. $\frac{32}{3}$ **3.** $e - (1/e) + \frac{10}{3}$ **5.** 19.5 **7.** $\frac{1}{6}$ **9.** $\ln 2 - \frac{1}{2}$
11. $\frac{1}{3}$ **13.** 72 **15.** $\frac{59}{12}$ **17.** $\frac{9}{8}$ **19.** $\frac{8}{3}$ **21.** $\frac{1}{2}$
23. $2 - \pi/2$ **25.** $\pi - \frac{2}{3}$ **27.** 6.5
29. $\frac{1}{2}$

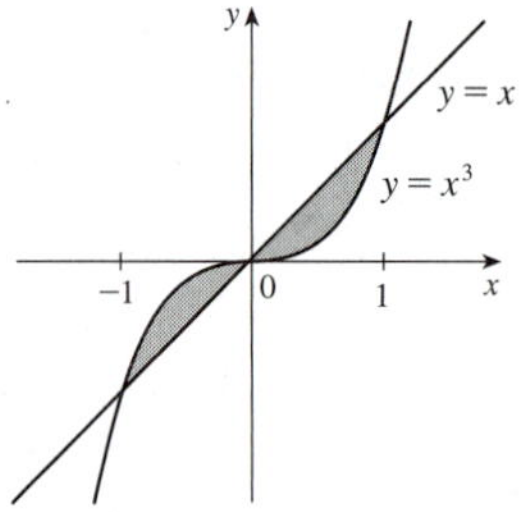

31. 0.6407 **33.** ±1.02; 2.70 **35.** ±0.86, 0; 0.40

37. $12\sqrt{6} - 9$ **39.** $117\frac{1}{3}$ ft

41. (a) Car A (b) The distance by which A is ahead of B after 1 minute (c) Car A (d) $t \approx 2.2$ min

43. $24\sqrt{3}/5$ **45.** $4^{2/3}$ **47.** ± 6

49. $0 < m < 1$; $m - \ln m - 1$

Exercises 6.2 ▫ page 452

1. $\pi/5$

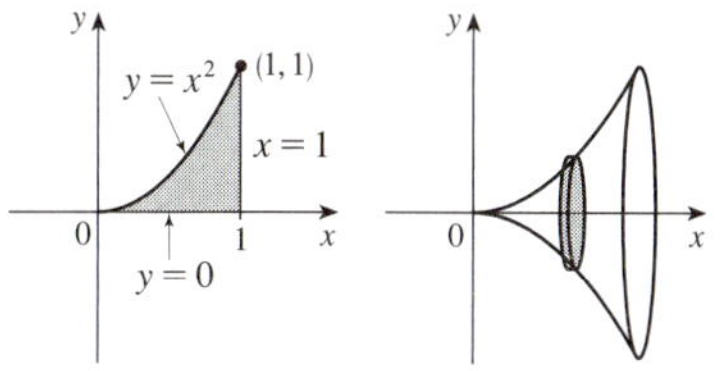

3. $\pi/2$

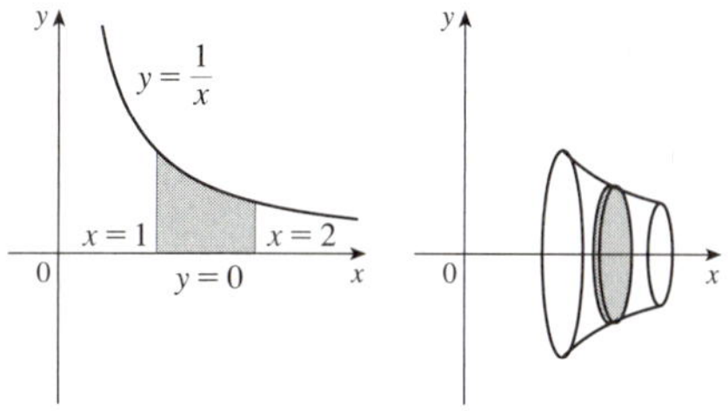

5. 8π

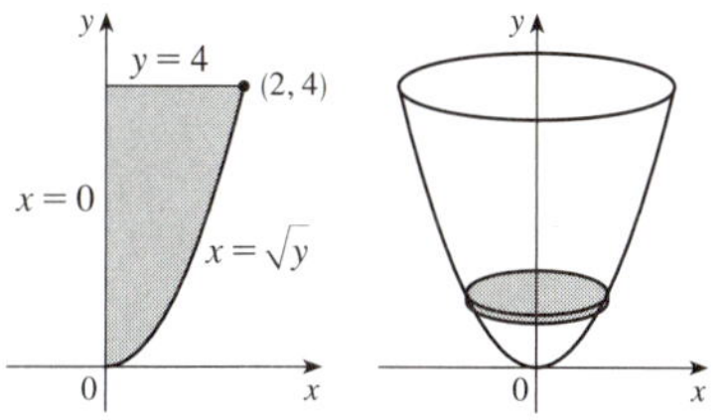

7. $3\pi/10$

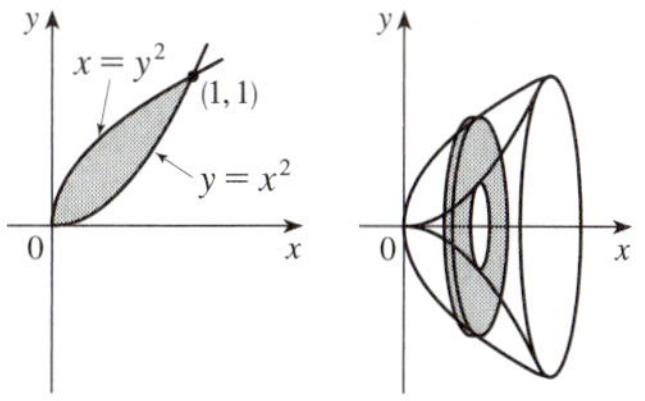

9. $64\pi/15$

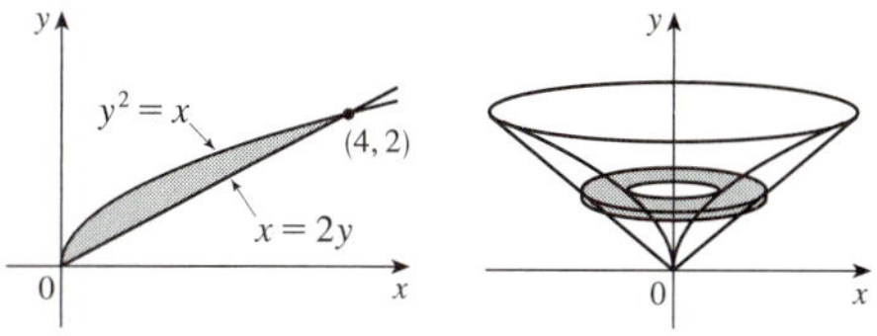

11. $\pi/6$

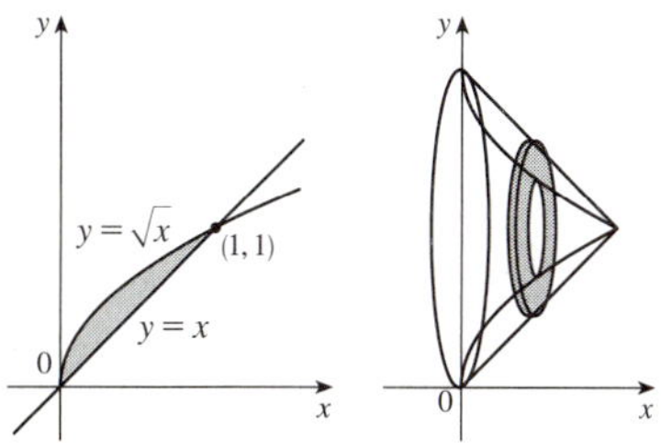

13. $208\pi/45$

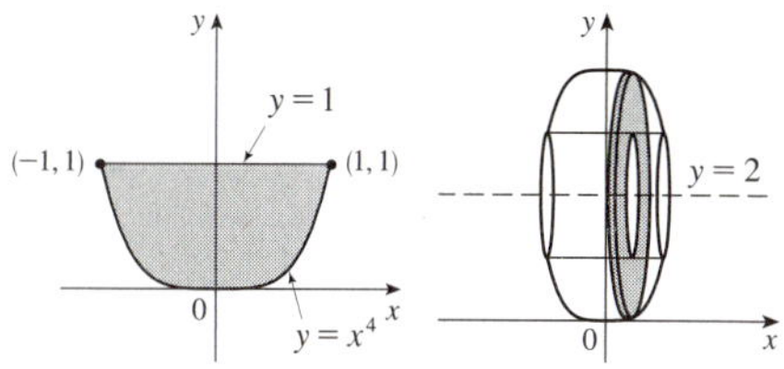

15. $16\pi/15$

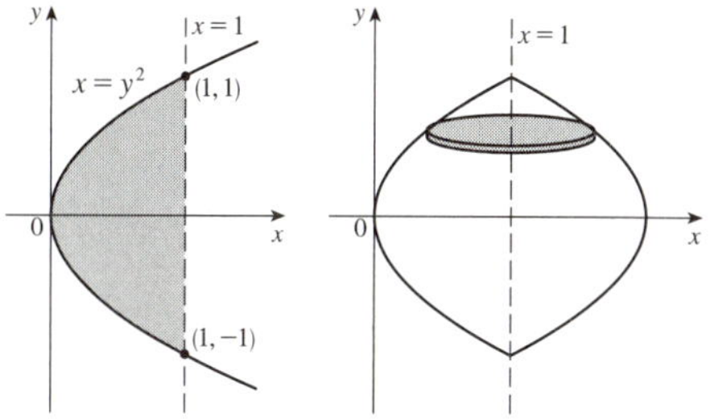

17. $29\pi/30$

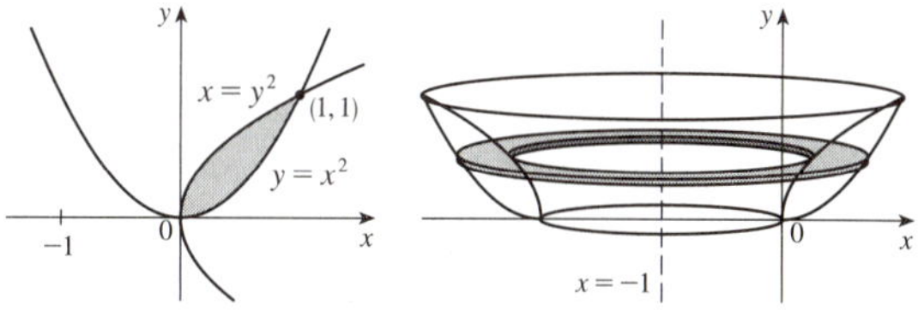

19. $\pi/7$ **21.** $\pi/10$ **23.** $\pi/2$ **25.** $7\pi/15$

27. $5\pi/14$ **29.** $13\pi/30$ **31.** $\pi \int_0^{\pi/4} (1 - \tan^3 x)^2\, dx$

33. $\pi \int_0^{\pi} [1^2 - (1 - \sin x)^2]\, dx$

35. $\pi \int_{-\sqrt{8}}^{\sqrt{8}} \{[3 - (-2)]^2 - [\sqrt{y^2 + 1} - (-2)]^2\}\, dy$

37. 0, 0.747; 0.132 **39.** $\frac{11}{8}\pi^2$

41. Solid obtained by rotating the region $0 \leqslant y \leqslant \cos x$, $0 \leqslant x \leqslant \pi/2$ about the x-axis

43. Solid obtained by rotating the region $y^4 \leqslant x \leqslant y^2$, $0 \leqslant y \leqslant 1$ about the y-axis

45. 1110 cm^3 **47.** $\pi r^2 h/3$ **49.** $\pi h^2[r - (h/3)]$

51. $2b^2h/3$ **53.** 10 cm^3 **55.** 24 **57.** 2 **59.** 3

61. (a) $8\pi R \int_0^r \sqrt{r^2 - y^2}\, dy$ (b) $2\pi^2 r^2 R$

63. (b) $\pi r^2 h$ **65.** $\frac{5}{12}\pi r^3$ **67.** $8 \int_0^r \sqrt{R^2 - y^2}\sqrt{r^2 - y^2}\, dy$

Exercises 6.3 ▫ page 458

1. Circumference $= 2\pi x$, height $= x(x-1)^2$; $\pi/15$

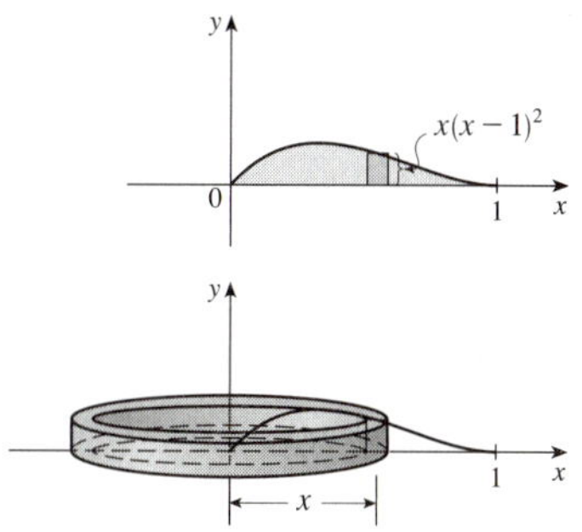

3. 2π

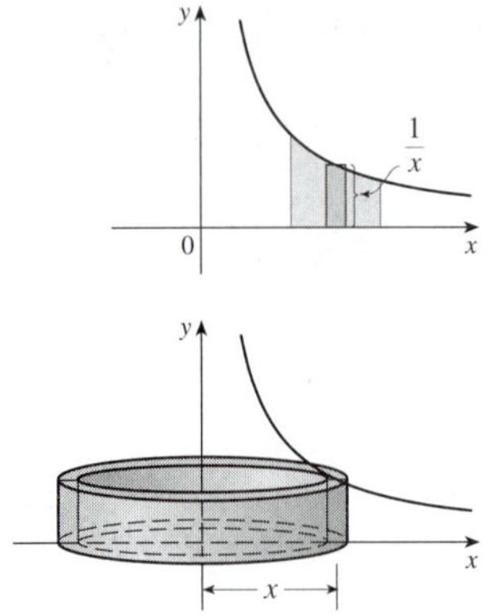

5. $\pi(1 - 1/e)$ **7.** 16π

9. $21\pi/2$

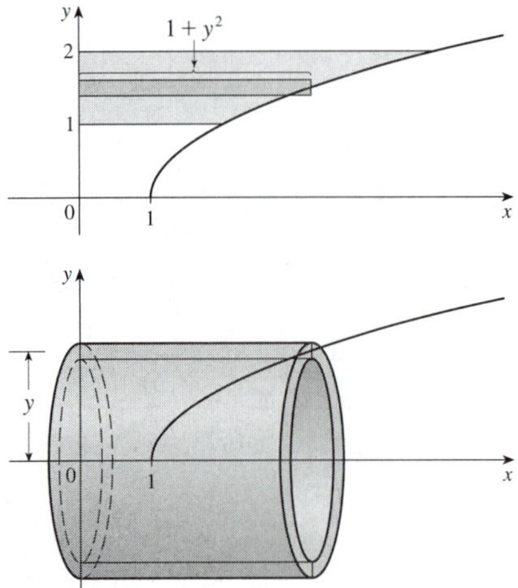

11. $768\pi/7$ **13.** $250\pi/3$

15. $17\pi/6$ **17.** $67\pi/6$ **19.** 24π

21. $\int_1^2 2\pi x \ln x\,dx$

23. $\int_0^1 2\pi(x + 1)[\sin(\pi x/2) - x^4]\,dx$

25. $\int_0^\pi 2\pi(4 - y)\sqrt{\sin y}\,dy$ **27.** 1.142

29. Solid obtained by rotating the region $0 \leqslant y \leqslant x^4$, $0 \leqslant x \leqslant 3$ about the y-axis

31. Solid obtained by rotating the region bounded by (i) $x = 1 - y^2$, $x = 0$, and $y = 0$, or (ii) $x = y^2$, $x = 1$, and $y = 0$ about the line $y = 3$

33. 0, 1.32; 4.05 **35.** $\frac{1}{32}\pi^3$ **37.** $81\pi/10$

39. $8\pi(3 - \ln 4)$ **41.** $4\pi/3$

43. $\frac{4}{3}\pi r^3$ **45.** $\frac{1}{3}\pi r^2 h$

Exercises 6.4 ▫ page 463

1. 7200 J **3.** 9 ft-lb **5.** 180 J

7. $\frac{15}{4}$ ft-lb **9.** $\frac{25}{24} \approx 1.04$ J **11.** 10.8 cm

13. (a) 625 ft-lb (b) $\frac{1875}{4}$ ft-lb **15.** 650,000 ft-lb

17. 3857 J **19.** 2450 J **21.** $\approx 1.06 \times 10^6$ J

23. $\approx 5.8 \times 10^3$ ft-lb **25.** 2.0 m **29.** $Gm_1m_2[(1/a) - (1/b)]$

Exercises 6.5 ▫ page 467

1. $\frac{1}{3}$ **3.** $2/\pi$ **5.** $(1 - e^{-25})/10$ **7.** $2/(5\pi)$

9. (a) 1 (b) 2, 4

(c)

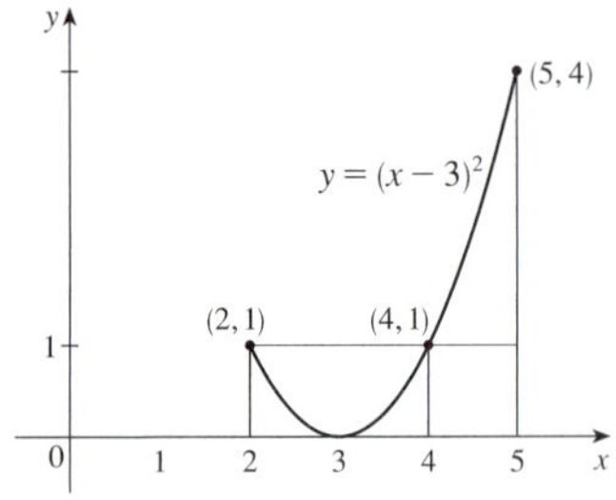

11. (a) $4/\pi$ (b) $\approx 1.24, 2.81$

(c)

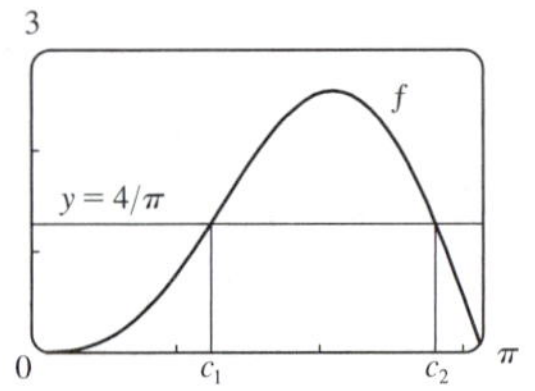

15. $38\frac{1}{3}$ **17.** $(50 + 28/\pi)°\text{F} \approx 59°\text{F}$

19. 6 kg/m **21.** $5/(4\pi) \approx 0.4$ L

Chapter 6 Review ▫ page 468

Exercises

1. $\frac{125}{6}$ **3.** $e - \frac{11}{6}$ **5.** $\frac{4}{3} + 4/\pi$ **7.** $64\pi/15$

9. $1656\pi/5$ **11.** $\frac{4}{3}\pi(2ah + h^2)^{3/2}$

13. $\int_0^1 \pi[(1 - x^3)^2 - (1 - x^2)^2]\,dx$

15. (a) $2\pi/15$ (b) $\pi/6$ (c) $8\pi/15$

17. (a) 0.38 (b) 0.87

19. Solid obtained by rotating the region $0 \leqslant y \leqslant \cos x$, $0 \leqslant x \leqslant \pi/2$ about the y-axis

21. Solid obtained by rotating the region in the first quadrant bounded by $x = 4 - y^2$ and the axes about the x-axis

23. 36 **25.** $125\sqrt{3}/3$ m^3 **27.** 3.2 J

29. (a) $8000\pi/3 \approx 8378$ ft-lb (b) 2.1 ft

31. $f(x)$

Problems Plus ▫ page 470

1. (a) $f(t) = 3t^2$ (b) $f(x) = \sqrt{2x/\pi}$ **3.** $\frac{32}{27}$

5. (b) 0.2261 (c) 0.6736 m

(d) (i) $1/(105\pi) \approx 0.003$ in/s (ii) $370\pi/3$ s ≈ 6.5 min

9. $y = \frac{32}{9}x^2$

11. (a) $V = \int_0^h \pi[f(y)]^2\,dy$

(c) $f(y) = \sqrt{kA/(\pi C)}\,y^{1/4}$

Advantage: the markings on the container are equally spaced.

13. $B = 16A$

Chapter 7

Exercises 7.1 ▫ page 480

1. $\frac{1}{2}x^2 \ln x - \frac{1}{4}x^2 + C$ **3.** $\frac{1}{5}x \sin 5x + \frac{1}{25} \cos 5x + C$

5. $2(r-2)e^{r/2} + C$

7. $-\dfrac{1}{\pi}x^2 \cos \pi x + \dfrac{2}{\pi^2} x \sin \pi x + \dfrac{2}{\pi^3} \cos \pi x + C$

9. $\frac{1}{2}(2x+1)\ln(2x+1) - x + C$

11. $t \arctan 4t - \frac{1}{8}\ln(1+16t^2) + C$

13. $x(\ln x)^2 - 2x \ln x + 2x + C$

15. $\frac{1}{13}e^{2\theta}(2 \sin 3\theta - 3 \cos 3\theta) + C$

17. $y \cosh y - \sinh y + C$

19. $\pi/3$ **21.** $\frac{1}{2} - \frac{1}{2}\ln 2$ **23.** $\frac{1}{4} - \frac{3}{4}e^{-2}$

25. $(\pi + 6 - 3\sqrt{3})/6$ **27.** $\sin x\,(\ln \sin x - 1) + C$

29. $\frac{1}{2}x(\cos \ln x + \sin \ln x) + C$ **31.** $\frac{32}{5}(\ln 2)^2 - \frac{64}{25}\ln 2 + \frac{62}{125}$

33. $2(\sin \sqrt{x} - \sqrt{x} \cos \sqrt{x}) + C$ **35.** $-\frac{1}{2} - \pi/4$

37. $(x \sin \pi x)/\pi + (\cos \pi x)/\pi^2 + C$

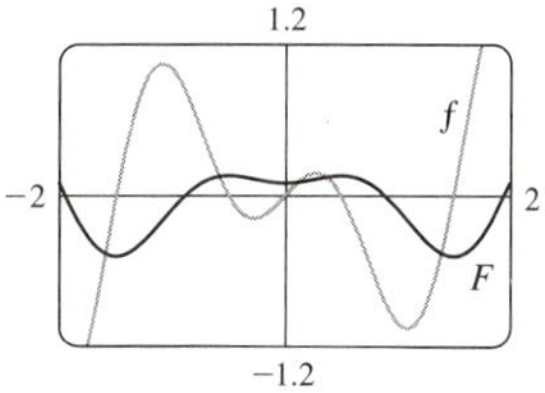

39. $(2x+1)e^x + C$

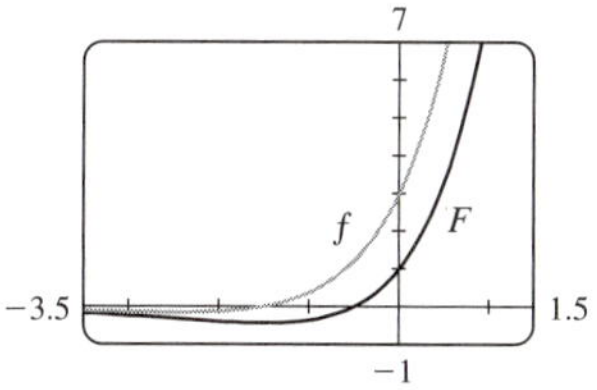

41. (b) $-\frac{1}{4}\cos x \sin^3 x + \frac{3}{8}x - \frac{3}{16}\sin 2x + C$

43. (b) $\frac{2}{3}, \frac{8}{15}$ **49.** $x[(\ln x)^3 - 3(\ln x)^2 + 6 \ln x - 6] + C$

51. $\frac{25}{4} - \frac{75}{4}e^{-2}$ **53.** 1.0475, 2.8731; 2.1828

55. $4 - 8/\pi$ **57.** $2\pi e$

59. $\frac{9}{2}\ln 3 - \frac{13}{9}$ **61.** $2 - e^{-t}(t^2 + 2t + 2)$ m **63.** 2

Exercises 7.2 ▫ page 488

1. $\frac{1}{5}\cos^5 x - \frac{1}{3}\cos^3 x + C$ **3.** $-\frac{11}{384}$

5. $\frac{1}{5}\sin^5 x - \frac{2}{7}\sin^7 x + \frac{1}{9}\sin^9 x + C$ **7.** $\pi/4$ **9.** $3\pi/8$

11. $\frac{3}{2}\theta + 2 \sin \theta + \frac{1}{4}\sin 2\theta + C$ **13.** $(3\pi - 4)/192$

15. $\left(\frac{2}{7}\cos^3 x - \frac{2}{3}\cos x\right)\sqrt{\cos x} + C$

17. $\frac{1}{2}\cos^2 x - \ln|\cos x| + C$ **19.** $\ln(1 + \sin x) + C$

21. $\frac{1}{2}\tan^2 x + C$ **23.** $\tan x - x + C$

25. $\frac{1}{5}\tan^5 t + \frac{2}{3}\tan^3 t + \tan t + C$ **27.** $\frac{117}{8}$

29. $\frac{1}{3}\sec^3 x - \sec x + C$

31. $\frac{1}{4}\sec^4 x - \tan^2 x + \ln|\sec x| + C$

33. $\frac{1}{6}\tan^6 \theta + \frac{1}{4}\tan^4 \theta + C$ **35.** $\sqrt{3} - (\pi/3)$

37. $\frac{1}{3}\csc^3 \alpha - \frac{1}{5}\csc^5 \alpha + C$ **39.** $\ln|\csc x - \cot x| + C$

41. $\frac{1}{6}\sin 3x - \frac{1}{14}\sin 7x + C$ **43.** $\frac{1}{4}\sin 2\theta + \frac{1}{24}\sin 12\theta + C$

45. $\frac{1}{2}\sin 2x + C$ **47.** $\frac{1}{10}\tan^5(t^2) + C$

49. $-\frac{1}{5}\cos^5 x + \frac{2}{3}\cos^3 x - \cos x + C$

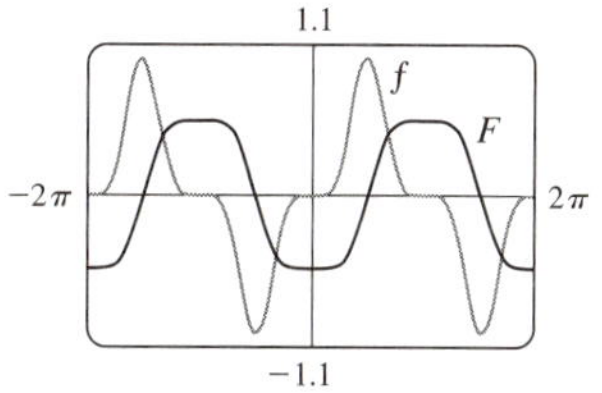

51. $\frac{1}{6}\sin 3x - \frac{1}{18}\sin 9x + C$

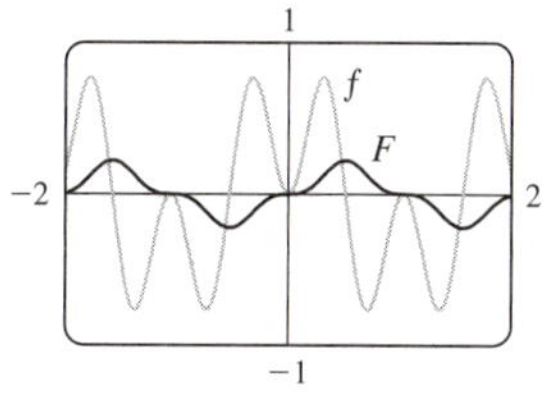

53. 0 **55.** $\frac{1}{3}$ **57.** 0 **59.** $\pi^2/4$ **61.** $2\pi + (\pi^2/4)$

63. $s = (1 - \cos^3 \omega t)/(3\omega)$

Exercises 7.3 ▫ page 494

1. $\sqrt{x^2 - 9}/(9x) + C$ **3.** $\frac{1}{3}(x^2 - 18)\sqrt{x^2 + 9} + C$

5. $\pi/24 + \sqrt{3}/8 - \frac{1}{4}$ **7.** $-\sqrt{25 - x^2}/(25x) + C$

9. $\ln\left(\sqrt{x^2 + 16} + x\right) + C$ **11.** $\frac{1}{4}\sin^{-1}(2x) + \frac{1}{2}x\sqrt{1 - 4x^2} + C$

13. $\frac{1}{6}\sec^{-1}(x/3) - \sqrt{x^2 - 9}/(2x^2) + C$

15. $\left(x/\sqrt{a^2 - x^2}\right) - \sin^{-1}(x/a) + C$ **17.** $\sqrt{x^2 - 7} + C$

19. $\ln\left|\left(\sqrt{1 + x^2} - 1\right)/x\right| + \sqrt{1 + x^2} + C$ **21.** $\frac{64}{1215}$

23. $\frac{9}{2}\sin^{-1}((x-2)/3) + \frac{1}{2}(x-2)\sqrt{5 + 4x - x^2} + C$

25. $\frac{1}{3}\ln\left|3x + 1 + \sqrt{9x^2 + 6x - 8}\right| + C$

27. $\frac{1}{2}[\tan^{-1}(x+1) + (x+1)/(x^2 + 2x + 2)] + C$

29. $\frac{1}{4}\sin^{-1}x^2 + \frac{1}{4}x^2\sqrt{1 - x^4} + C$

33. $\frac{1}{6}\left(\sqrt{48} - \sec^{-1} 7\right)$ **37.** 0.81, 2; 2.10

39. $r\sqrt{R^2 - r^2} + \pi r^2/2 - R^2 \arcsin(r/R)$ **41.** $2\pi^2 R r^2$

Exercises 7.4 ▫ page 504

1. (a) $\dfrac{A}{x+3} + \dfrac{B}{3x+1}$ (b) $\dfrac{A}{x} + \dfrac{B}{x+1} + \dfrac{C}{(x+1)^2}$

3. (a) $\dfrac{A}{x+4} + \dfrac{B}{x-1}$ (b) $\dfrac{A}{x-1} + \dfrac{Bx + C}{x^2 + x + 1}$

5. (a) $1 + \dfrac{A}{x-1} + \dfrac{B}{x+1} + \dfrac{Cx + D}{x^2 + 1}$

(b) $\dfrac{At + B}{t^2 + 1} + \dfrac{Ct + D}{t^2 + 4} + \dfrac{Et + F}{(t^2 + 4)^2}$

7. $x + 6 \ln|x - 6| + C$

9. $2 \ln|x + 5| - \ln|x - 2| + C$ **11.** $\frac{1}{2}\ln\frac{3}{2}$

13. $a \ln|x - b| + C$ **15.** $2 \ln 2 + \frac{1}{2}$

17. $\frac{27}{5}\ln 2 - \frac{9}{5}\ln 3$ (or $\frac{9}{5}\ln\frac{8}{3}$)

19. $-\frac{1}{36}\ln|x+5| + \frac{1}{6}\frac{1}{x+5} + \frac{1}{36}\ln|x-1| + C$

21. $2\ln|x| + 3\ln|x+2| + (1/x) + C$

23. $\ln|x+1| + 2/(x+1) - 1/[2(x+1)^2] + C$

25. $\ln|x-1| - \frac{1}{2}\ln(x^2+9) - \frac{1}{3}\tan^{-1}(x/3) + C$

27. $\frac{1}{2}\ln(x^2+1) + (1/\sqrt{2})\tan^{-1}(x/\sqrt{2}) + C$

29. $\frac{1}{2}\ln(x^2+2x+5) + \frac{3}{2}\tan^{-1}((x+1)/2) + C$

31. $\frac{1}{3}\ln|x-1| - \frac{1}{6}\ln(x^2+x+1) - \frac{1}{\sqrt{3}}\tan^{-1}\frac{2x+1}{\sqrt{3}} + C$

33. $\frac{1}{3}\ln\frac{17}{2}$

35. $(1/x) + \frac{1}{2}\ln|(x-1)/(x+1)| + C$

37. $\frac{-1}{2(x^2+2x+4)} - \frac{2\sqrt{3}}{9}\tan^{-1}\left(\frac{x+1}{\sqrt{3}}\right) - \frac{2(x+1)}{3(x^2+2x+4)} + C$

39. $\ln|(\sqrt{x+1}-1)/(\sqrt{x+1}+1)| + C$

41. $2 + \ln\frac{25}{9}$ **43.** $\frac{3}{10}(x^2+1)^{5/3} - \frac{3}{4}(x^2+1)^{2/3} + C$

45. $2\sqrt{x} + 3\sqrt[3]{x} + 6\sqrt[6]{x} + 6\ln|\sqrt[6]{x}-1| + C$

47. $\ln[(e^x+2)^2/(e^x+1)] + C$

49. $(x-\frac{1}{2})\ln(x^2-x+2) - 2x + \sqrt{7}\tan^{-1}((2x-1)/\sqrt{7}) + C$

51. $-\frac{1}{2}\ln 3 \approx -0.55$

53. $\frac{1}{2}\ln|(x-2)/x| + C$ **57.** $\frac{1}{5}\ln\left|\frac{2\tan(x/2)-1}{\tan(x/2)+2}\right| + C$

59. $\frac{1}{4}\ln|\tan(x/2)| + \frac{1}{8}\tan^2(x/2) + C$ **61.** $1 + 2\ln 2$

63. $t = -\ln P - \frac{1}{9}\ln(0.9P + 900) + C$, where $C \approx 10.23$

65. (a) $\frac{24{,}110}{4879}\frac{1}{5x+2} - \frac{668}{323}\frac{1}{2x+1} - \frac{9438}{80{,}155}\frac{1}{3x-7} + \frac{1}{260{,}015}\frac{22{,}098x+48{,}935}{x^2+x+5}$

(b) $\frac{4822}{4879}\ln|5x+2| - \frac{334}{323}\ln|2x+1| - \frac{3146}{80{,}155}\ln|3x-7| + \frac{11{,}049}{260{,}015}\ln(x^2+x+5) + \frac{75{,}772}{260{,}015\sqrt{19}}\tan^{-1}\frac{2x+1}{\sqrt{19}} + C$

The CAS omits the absolute value signs and the constant of integration.

Exercises 7.5 ▫ page 510

1. $\sin x + \ln|\csc x - \cot x| + C$

3. $4 - \ln 9$ **5.** $e^{\pi/4} - e^{-\pi/4}$

7. $\frac{243}{5}\ln 3 - \frac{242}{25}$ **9.** $\frac{1}{2}\ln(x^2-4x+5) + \tan^{-1}(x-2) + C$

11. $\frac{1}{8}\cos^8\theta - \frac{1}{6}\cos^6\theta + C$ (or $\frac{1}{4}\sin^4\theta - \frac{1}{3}\sin^6\theta + \frac{1}{8}\sin^8\theta + C$)

13. $x/\sqrt{1-x^2} + C$ **15.** $1 - \frac{1}{2}\sqrt{3}$

17. $\frac{1}{4}x^2 - \frac{1}{2}x\sin x\cos x + \frac{1}{4}\sin^2 x + C$
(or $\frac{1}{4}x^2 - \frac{1}{4}x\sin 2x - \frac{1}{8}\cos 2x + C$)

19. $e^{e^x} + C$ **21.** $-\frac{1}{8}e^{-2t}(4t^3 + 6t^2 + 6t + 3) + C$

23. $\frac{4097}{45}$ **25.** $3x + \frac{23}{3}\ln|x-4| - \frac{5}{3}\ln|x+2| + C$

27. $\frac{1}{2}(\ln\sin x)^2 + C$ **29.** $15 + 7\ln\frac{2}{7}$

31. $\sin^{-1}x - \sqrt{1-x^2} + C$

33. $2\sin^{-1}((x+1)/2) + ((x+1)/2)\sqrt{3-2x-x^2} + C$

35. 0 **37.** $\pi/8 - \frac{1}{4}$ **39.** $-\ln(1+\sqrt{1-x^2}) + C$

41. $\theta\tan\theta - \frac{1}{2}\theta^2 - \ln|\sec\theta| + C$ **43.** $\frac{2}{3}(1+e^x)^{3/2} + C$

45. $-\frac{1}{3}(x^3+1)e^{-x^3} + C$

47. $\ln\sqrt{x^2+a^2} + \tan^{-1}(x/a) + C$

49. $\ln\left|\frac{\sqrt{4x+1}-1}{\sqrt{4x+1}+1}\right| + C$ **51.** $-\ln\left|\frac{\sqrt{4x^2+1}+1}{2x}\right| + C$

53. $(1/m)x^2\cosh(mx) - (2/m^2)x\sinh(mx) + (2/m^3)\cosh(mx) + C$

55. $3\ln(\sqrt{x+1}+3) - \ln(\sqrt{x+1}+1) + C$

57. $\frac{3}{7}(x+c)^{7/3} - \frac{3}{4}c(x+c)^{4/3} + C$

59. $e^{-x} + \frac{1}{2}\ln|(e^x-1)/(e^x+1)| + C$

61. $\frac{1}{20}\tan^{-1}(\frac{1}{4}x^5) + C$ **63.** $2(x - 2\sqrt{x} + 2)e^{\sqrt{x}} + C$

65. $\frac{2}{3}[(x+1)^{3/2} - x^{3/2}] + C$

67. $\frac{2}{3}\sqrt{3}\pi - \frac{1}{2}\pi - \ln 2$ **69.** $e^x - \ln(1+e^x) + C$

71. $-\frac{1}{4}\ln(x^2+3) + \frac{1}{4}\ln(x^2+1) + C$

73. $\frac{1}{8}\ln|x-2| - \frac{1}{16}\ln(x^2+4) - \frac{1}{8}\tan^{-1}(x/2) + C$

75. $\frac{1}{24}\cos 6x - \frac{1}{16}\cos 4x - \frac{1}{8}\cos 2x + C$ **77.** $\frac{2}{3}\tan^{-1}(x^{3/2}) + C$

79. $\frac{1}{3}x\sin^3 x + \frac{1}{3}\cos x - \frac{1}{9}\cos^3 x + C$ **81.** $xe^{x^2} + C$

Exercises 7.6 ▫ page 515

1. $(-1/x)\sqrt{7-2x^2} - \sqrt{2}\sin^{-1}(\sqrt{2}x/\sqrt{7}) + C$

3. $(1/(2\pi))\sec(\pi x)\tan(\pi x) + (1/(2\pi))\ln|\sec(\pi x) + \tan(\pi x)| + C$

5. $\pi/4$ **7.** $\frac{1}{25}e^{-3x}(-3\cos 4x + 4\sin 4x) + C$

9. $-\sqrt{4x^2+9}/(9x) + C$ **11.** $e - 2$

13. $-\frac{1}{2}\tan^2(1/z) - \ln|\cos(1/z)| + C$ **15.** $\tan^{-1}(\sinh e^x) + C$

17. $((2y-1)/8)\sqrt{6+4y-4y^2} + \frac{7}{8}\sin^{-1}((2y-1)/\sqrt{7})$
$- \frac{1}{12}(6+4y-4y^2)^{3/2} + C$

19. $\frac{1}{9}\sin^3 x\,[3\ln(\sin x) - 1] + C$ **21.** $\frac{1}{2\sqrt{3}}\ln\left|\frac{e^x+\sqrt{3}}{e^x-\sqrt{3}}\right| + C$

23. $\frac{1}{4}\tan x\sec^3 x + \frac{3}{8}\tan x\sec x + \frac{3}{8}\ln|\sec x + \tan x| + C$

25. $\frac{1}{2}(\ln x)\sqrt{4+(\ln x)^2} + 2\ln\left[\ln x + \sqrt{4+(\ln x)^2}\right] + C$

27. $\sqrt{e^{2x}-1} - \cos^{-1}(e^{-x}) + C$

29. $\frac{1}{5}\ln|x^5 + \sqrt{x^{10}-2}| + C$ **31.** $2\pi^2$

35. $-\frac{1}{4}x(5-x^2)^{3/2} + \frac{5}{8}x\sqrt{5-x^2} + \frac{25}{8}\sin^{-1}(x/\sqrt{5}) + C$

37. $-\frac{1}{5}\sin^2 x\cos^3 x - \frac{2}{15}\cos^3 x + C$

39. $\frac{1}{10}(1+2x)^{5/2} - \frac{1}{6}(1+2x)^{3/2} + C$

41. $-\ln|\cos x| - \frac{1}{2}\tan^2 x + \frac{1}{4}\tan^4 x + C$

43. $\frac{2^{x-1}\sqrt{2^{2x}-1}}{\ln 2} - \frac{\ln(\sqrt{2^{2x}-1} + 2^x)}{2\ln 2} + C$

45. $F(x) = \frac{1}{2}\ln(x^2-x+1) - \frac{1}{2}\ln(x^2+x+1)$;
max. at -1, min. at 1; IP at -1.7, 0, and 1.7

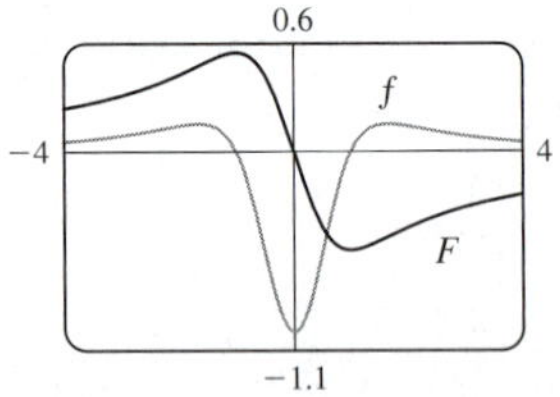

47. $F(x) = -\frac{1}{10}\sin^3 x\cos^7 x - \frac{3}{80}\sin x\cos^7 x + \frac{1}{160}\sin x\cos^5 x + \frac{1}{128}\sin x\cos^3 x + \frac{3}{256}\sin x\cos x + \frac{3}{256}x$;
max. at π, min. at 0; IP at 0.68, $\pi/2$, and 2.46

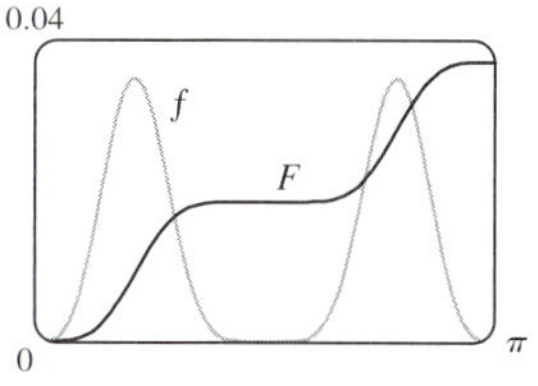

Exercises 7.7 ▫ page 527

1. (a) $L_2 = 6$, $R_2 = 12$, $M_2 \approx 9.6$
(b) L_2 is an underestimate, R_2 and M_2 are overestimates.
(c) $T_2 = 9 < I$ (d) $L_n < T_n < I < M_n < R_n$
3. (a) $T_4 \approx 0.895759$ (underestimate)
(b) $M_4 \approx 0.908907$ (overestimate)
$T_4 < I < M_4$
5. (a) 5.932957, $E_M \approx -0.063353$
(b) 5.869247, $E_S \approx 0.000357$
7. (a) 2.413790 (b) 2.411453 (c) 2.412232
9. (a) 0.146879 (b) 0.147391 (c) 0.147219
11. (a) 0.451948 (b) 0.451991 (c) 0.451976
13. (a) 2.031893 (b) 2.014207 (c) 2.020651
15. (a) −0.495333 (b) −0.543321 (c) −0.526123
17. (a) 1.064275 (b) 1.067416 (c) 1.074915
19. (a) $T_{10} \approx 0.881839$, $M_{10} \approx 0.882202$
(b) $|E_T| \leqslant 0.01\overline{3}$, $|E_M| \leqslant 0.00\overline{6}$
(c) $n = 366$ for T_n, $n = 259$ for M_n
21. (a) $T_{10} \approx 1.719713$, $E_T \approx -0.001432$;
$S_{10} \approx 1.718283$, $E_S \approx -0.000001$
(b) $|E_T| \leqslant 0.002266$, $|E_S| \leqslant 0.0000016$
(c) $n = 151$ for T_n, $n = 107$ for M_n, $n = 8$ for S_n
23. (a) 2.8 (b) 7.954926518 (c) 0.2894
(d) 7.954926521 (e) The actual error is much smaller.
(f) 10.9 (g) 7.953789422 (h) 0.0593
(i) The actual error is smaller. (j) $n \geqslant 50$
25.

n	L_n	R_n	T_n	M_n
4	0.140625	0.390625	0.265625	0.242188
8	0.191406	0.316406	0.253906	0.248047
16	0.219727	0.282227	0.250977	0.249512

n	E_L	E_R	E_T	E_M
4	0.109375	−0.140625	−0.015625	0.007813
8	0.058594	−0.066406	−0.003906	0.001953
16	0.030273	−0.032227	−0.000977	0.000488

Observations same as after Example 1.

27.

n	T_n	M_n	S_n
6	4.661488	4.669245	4.666563
12	4.665367	4.667316	4.666659

n	E_T	E_M	E_S
6	0.005178	−0.002578	0.000104
12	0.001300	−0.000649	0.000007

Observations same as after Example 1.
29. (a) 11.5 (b) 12 (c) $11.\overline{6}$
31. (a) 23.44 (b) $0.341\overline{3}$ **33.** $37.7\overline{3}$ ft/s
35. 10,177 megawatt-hours **37.** 828
39. 12.3251 **41.** 59.4
43.

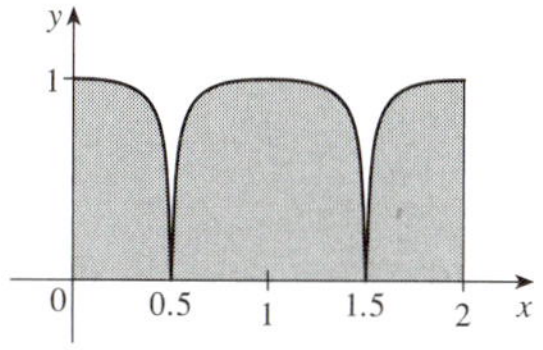

Exercises 7.8 ▫ page 537

Abbreviations: C, convergent; D, divergent
1. (a) Infinite interval (b) Infinite discontinuity
(c) Infinite discontinuity (d) Infinite interval
3. $\frac{1}{2} - 1/(2t^2)$; 0.495, 0.49995, 0.4999995; 0.5
5. $\frac{1}{12}$ **7.** D **9.** $2e^{-2}$ **11.** D **13.** 0 **15.** D
17. D **19.** $\frac{1}{25}$ **21.** D **23.** $\pi/9$
25. 1 **27.** $2\sqrt{3}$ **29.** D **31.** D **33.** $\frac{75}{4}$
35. D **37.** D **39.** $\frac{8}{3}\ln 2 - \frac{8}{9}$
41. e

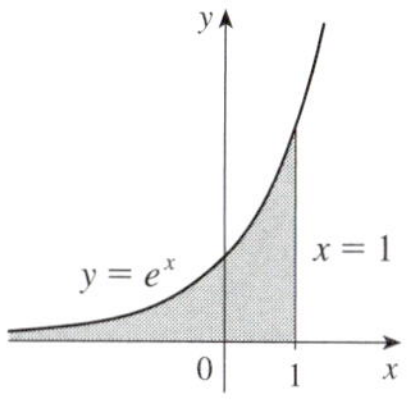

43. $\frac{2\pi}{3}$

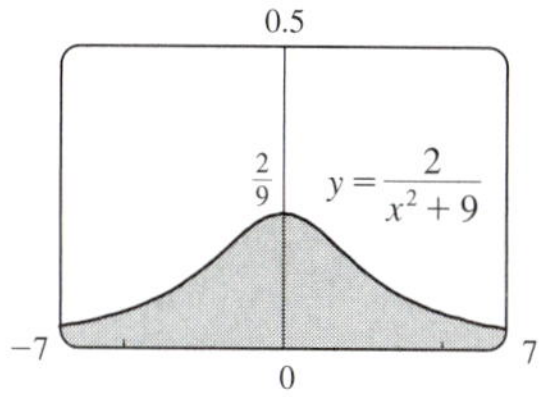

45. Infinite area

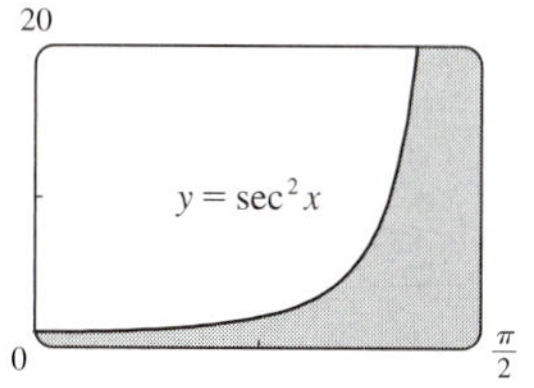

47. (a)

t	$\int_1^t [(\sin^2 x)/x^2]\,dx$
2	0.447453
5	0.577101
10	0.621306
100	0.668479
1,000	0.672957
10,000	0.673407

It appears that the integral is convergent.

(c)

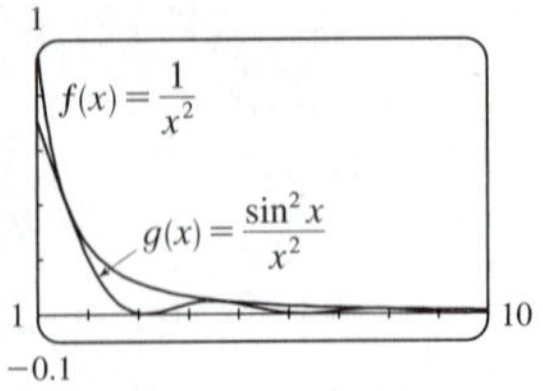

49. C **51.** C **53.** D **55.** π **57.** $p < 1, 1/(1-p)$
59. $p > -1, -1/(p+1)^2$ **65.** $\sqrt{2GM/R}$
67. (a)

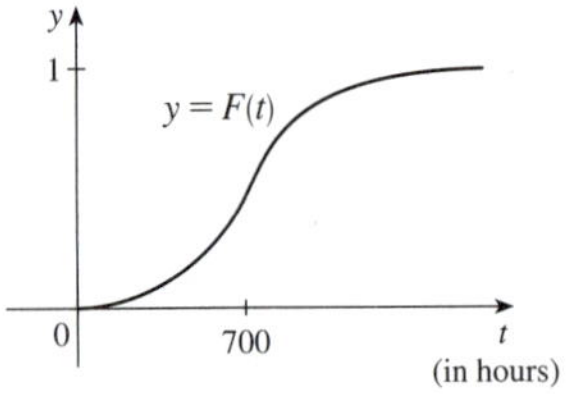

(b) The rate at which the fraction $F(t)$ increases as t increases
(c) 1; all bulbs burn out eventually
69. 1000
71. (a) $F(s) = 1/s, s > 0$ (b) $F(s) = 1/(s-1), s > 1$
(c) $F(s) = 1/s^2, s > 0$
77. $C = 1$; $\ln 2$

Chapter 7 Review ▫ page 540

True-False Quiz

1. False **3.** False **5.** False **7.** False
9. (a) True (b) False **11.** False **13.** False

Exercises

1. $5 + 10 \ln \frac{2}{3}$ **3.** $\ln 2$
5. $\frac{1}{9}\sec^9 x - \frac{3}{7}\sec^7 x + \frac{3}{5}\sec^5 x - \frac{1}{3}\sec^3 x + C$
7. $-\cos(\ln t) + C$ **9.** $\frac{64}{5}\ln 4 - \frac{124}{25}$
11. $\sqrt{3} - (\pi/3)$ **13.** $\ln|x| - \frac{1}{2}\ln(x^2+1) + C$
15. $\frac{1}{3}\sin^3\theta - \frac{2}{5}\sin^5\theta + \frac{1}{7}\sin^7\theta + C$
17. $x \sec x - \ln|\sec x + \tan x| + C$
19. $\frac{1}{18}\ln(9x^2 + 6x + 5) + \frac{1}{9}\tan^{-1}((3x+1)/2) + C$
21. $\ln\left|x - 2 + \sqrt{x^2 - 4x}\right| + C$
23. $-\frac{1}{12}(\cot^3 4x + 3\cot 4x) + C$
25. $\frac{3}{2}\ln(x^2+1) - 3\tan^{-1}x + \sqrt{2}\tan^{-1}(x/\sqrt{2}) + C$
27. $\frac{2}{5}$ **29.** 0 **31.** $6 - 3\pi/2$
33. $\left(x/\sqrt{4 - x^2}\right) - \sin^{-1}(x/2) + C$
35. $4\sqrt{1 + \sqrt{x}} + C$ **37.** $\frac{1}{2}\sin 2x - \frac{1}{8}\cos 4x + C$
39. $\frac{1}{8}e - \frac{1}{4}$ **41.** $\frac{1}{36}$ **43.** D
45. $4\ln 4 - 8$ **47.** D **49.** $\pi/4$
51. $(x+1)\ln(x^2 + 2x + 2) + 2\arctan(x+1) - 2x + C$
53. 0 **55.** $\frac{1}{2}\left[e^x\sqrt{1 - e^{2x}} + \sin^{-1}(e^x)\right] + C$
57. $\frac{1}{4}(2x+1)\sqrt{x^2 + x + 1} + \frac{3}{8}\ln\left(x + \frac{1}{2} + \sqrt{x^2 + x + 1}\right) + C$
61. No
63. (a) 1.090608 (overestimate)
(b) 1.088840 (underestimate)
(c) 1.089429 (unknown)
65. (a) $0.00\overline{6}$, $n \geqslant 259$ (b) $0.00\overline{3}$, $n \geqslant 183$
67. 8.6 mi **69.** (a) 3.8 (b) 1.7867, 0.000646 (c) $n \geqslant 30$
71. C **73.** 2 **75.** $3\pi^2/16$

Problems Plus ▫ page 544

1. About 1.85 inches from the center **3.** 0
5. $f(\pi) = -\pi/2$ **9.** $(b^b a^{-a})^{1/(b-a)} e^{-1}$
11. $2 - \sin^{-1}(2/\sqrt{5})$

Chapter 8

Exercises 8.1 ▫ page 552

1. $3\sqrt{10}$ **3.** $\frac{46}{3}$ **5.** $\frac{2}{243}(82\sqrt{82} - 1)$ **7.** $\frac{1261}{240}$ **9.** $\frac{32}{3}$
11. $\ln(\sqrt{2} + 1)$ **13.** $\sinh 1$
15. $\sqrt{1 + e^2} - \sqrt{2} + \ln(\sqrt{1 + e^2} - 1) - 1 - \ln(\sqrt{2} - 1)$
17. $\int_0^{2\pi} \sqrt{1 + \sin^2 x}\, dx$ **19.** $\int_1^4 \sqrt{9y^4 + 6y^2 + 2}\, dy$
21. 5.115840 **23.** 1.569619
25. (a), (b)

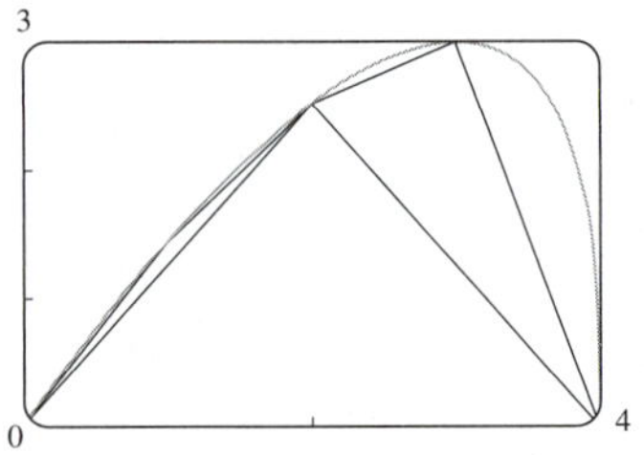

$L_1 = 4$,
$L_2 \approx 6.43$,
$L_4 \approx 7.50$

(c) $\int_0^4 \sqrt{1 + [4(3-x)/(3(4-x)^{2/3})]^2}\, dx$ (d) 7.7988
27. $\ln 3 - \frac{1}{2}$
29. 6

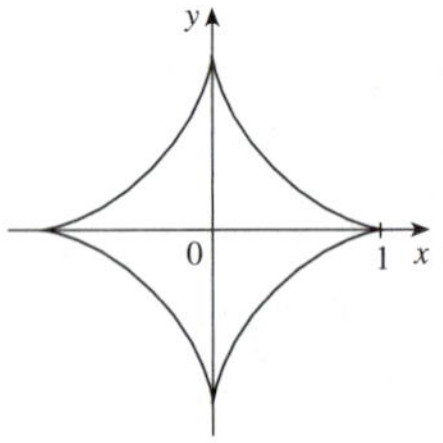

31. $s(x) = \frac{2}{27}\left[(1 + 9x)^{3/2} - 10\sqrt{10}\right]$
33. 209.1 m **35.** 29.36 in. **37.** 12.4

Exercises 8.2 ▫ page 559

1. $\int_1^3 2\pi \ln x \sqrt{1 + (1/x)^2}\, dx$
3. $\int_0^{\pi/4} 2\pi x \sqrt{1 + (\sec x \tan x)^2}\, dx$
5. $\pi(145\sqrt{145} - 1)/27$ **7.** $\pi(37\sqrt{37} - 17\sqrt{17})/6$
9. $\pi\left[1 + \frac{1}{4}(e^2 - e^{-2})\right]$ **11.** $21\pi/2$
13. $\pi(145\sqrt{145} - 10\sqrt{10})/27$ **15.** πa^2
17. 9.023754 **19.** 13.527296
21. $(\pi/4)\left[4\ln(\sqrt{17} + 4) - 4\ln(\sqrt{2} + 1) - \sqrt{17} + 4\sqrt{2}\right]$
23. $(\pi/6)\left[\ln(\sqrt{10} + 3) + 3\sqrt{10}\right]$

27. (a) $\pi a^2/3$ (b) $56\pi\sqrt{3}a^2/45$

29. $2\pi\left[b^2 + a^2 b \sin^{-1}\left(\sqrt{a^2 - b^2}/a\right)/\sqrt{a^2 - b^2}\right]$

31. $\int_a^b 2\pi[c - f(x)]\sqrt{1 + [f'(x)]^2}\,dx$ **33.** $4\pi^2 r^2$

Exercises 8.3 ▫ page 569

1. (a) 187.5 lb/ft^2 (b) 1875 lb (c) 562.5 lb

3. 6000 lb **5.** 6.5×10^6 N **7.** 3.5×10^4 lb

9. $1000g\pi r^3$ N **11.** 5.27×10^5 N

13. (a) 314 N (b) 353 N

15. (a) $\approx 5.63 \times 10^3$ lb (b) $\approx 5.06 \times 10^4$ lb
(c) $\approx 4.88 \times 10^4$ lb (d) $\approx 3.03 \times 10^5$ lb

17. 2.5×10^5 N **19.** 230; $\frac{23}{7}$ **21.** 10; 1; $\left(\frac{1}{21}, \frac{10}{21}\right)$

23. (0, 1.6) **25.** $(1/(e-1), (e+1)/4)$ **27.** $\left(\frac{2}{5}, \frac{1}{2}\right)$

29. $\left((\pi\sqrt{2} - 4)/[4(\sqrt{2} - 1)], 1/[4(\sqrt{2} - 1)]\right)$

31. (2, 0) **33.** $\frac{4}{3}$, 0, $\left(0, \frac{2}{3}\right)$ **35.** (0.781, 1.330) **39.** $\left(0, \frac{1}{12}\right)$

41. $\frac{1}{3}\pi r^2 h$

Exercises 8.4 ▫ page 574

1. \$38,000 **3.** \$43,866,933.33 **5.** \$407.25

7. \$12,000 **9.** 3727; \$37,753

11. $\frac{2}{3}(16\sqrt{2} - 8) \approx$ \$9.75 million

13. 1.19×10^{-4} cm^3/s **15.** $\frac{1}{9}$ L/s

Exercises 8.5 ▫ page 581

1. (a) The probability that a randomly chosen tire will have a lifetime between 30,000 and 40,000 miles
(b) The probability that a randomly chosen tire will have a lifetime of at least 25,000 miles

3. (a) $f(x) \geqslant 0$ for all x and $\int_{-\infty}^{\infty} f(x)\,dx = 1$
(b) $1 - \frac{3}{8}\sqrt{3} \approx 0.35$

5. (a) $f(x) \geqslant 0$ for all x and $\int_{-\infty}^{\infty} f(x)\,dx = 1$ (b) 5

7. $5 \ln 2 \approx 3.47$ min

9. (a) $e^{-4/2.5} \approx 0.20$ (b) $1 - e^{-2/2.5} \approx 0.55$
(c) If you aren't served within 10 minutes, you get a free hamburger.

11. $\approx 44\%$ **13.** ≈ 0.9545

15. (b) 0; a_0 (c)

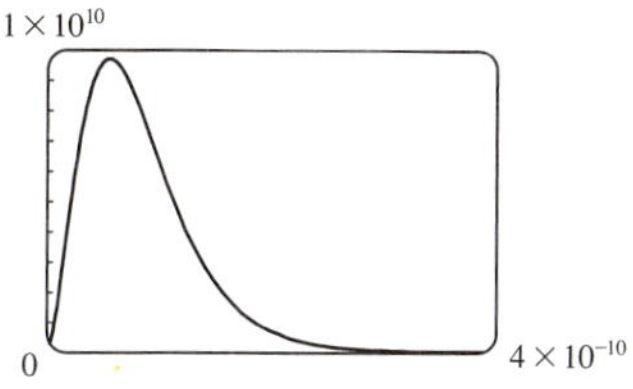

(d) $1 - 41e^{-8} \approx 0.986$ (e) $\frac{3}{2}a_0$

Chapter 8 Review ▫ page 583

Exercises

1. $\frac{15}{2}$ **3.** (a) $\frac{21}{16}$ (b) $41\pi/10$ **5.** 3.292287 **7.** $\frac{124}{5}$

9. ≈ 458 lb **11.** $\left(-\frac{1}{2}, \frac{12}{5}\right)$ **13.** $\left(2, \frac{2}{3}\right)$

15. $2\pi^2$ **17.** \$7166.67

19. (a) $f(x) \geqslant 0$ for all x and $\int_{-\infty}^{\infty} f(x)\,dx = 1$
(b) ≈ 0.3455 (c) 5, yes

21. (a) $1 - e^{-3/8} \approx 0.31$ (b) $e^{-5/4} \approx 0.29$
(c) $8 \ln 2 \approx 5.55$ min

Problems Plus ▫ page 584

1. $2\pi/3 - \sqrt{3}/2$

3. (a) $2\pi r(r \pm d)$ (b) $\approx$3,360,000 mi^2
(d) $\approx$78,400,000 mi^2

5. (a) $P(z) = P_0 + g\int_0^z \rho(x)\,dx$
(b) $(P_0 - \rho_0 gH)(\pi r^2) + \rho_0 gHe^{L/H}\int_{-r}^{r} e^{x/H} \cdot 2\sqrt{r^2 - x^2}\,dx$

7. Height $\sqrt{2}\,b$, volume $\left(\frac{28}{27}\sqrt{6} - 2\right)\pi b^3$

9. 0.14 m **11.** $2/\pi$, $1/\pi$

Chapter 9

Exercises 9.1 ▫ page 591

3. (a) ± 3 **5.** (b) and (c)

7. (a) It must be either 0 or decreasing
(c) $y = 0$ (d) $y = 1/(x + 2)$

9. (a) $0 < P < 4200$ (b) $P > 4200$
(c) $P = 0$, $P = 4200$

13. (a) At the beginning; stays positive, but decreases
(c)

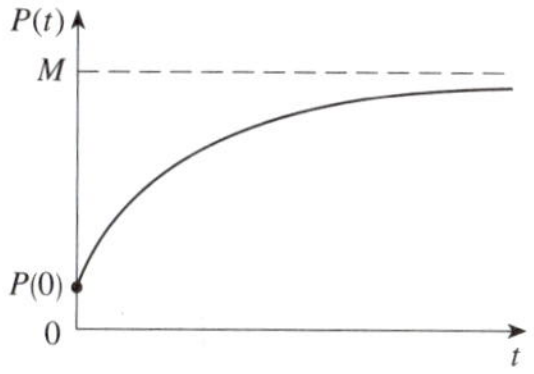

Exercises 9.2 ▫ page 599

1. (a)

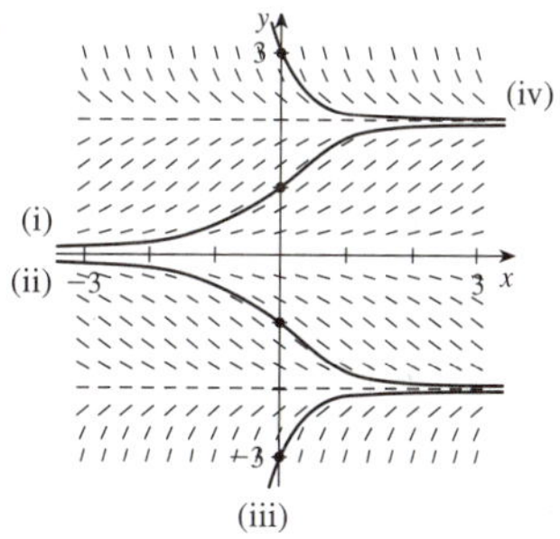

(b) $y = 0$, $y = 2$, $y = -2$

3. IV **5.** III

7.

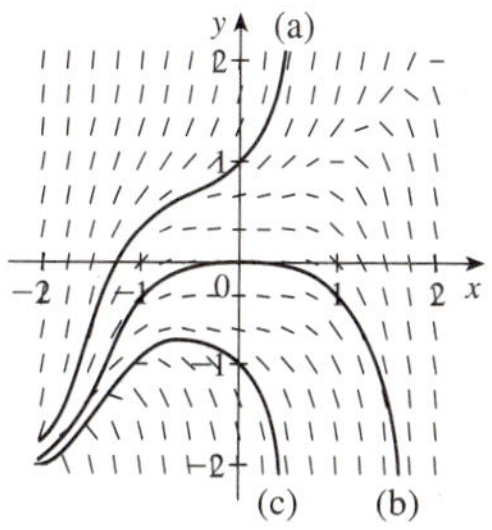

9.

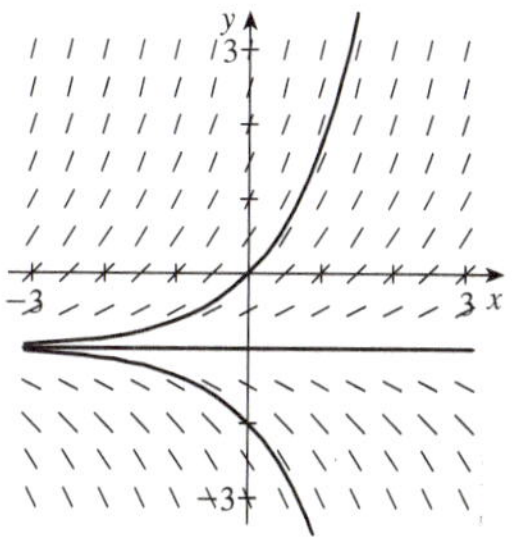

11.

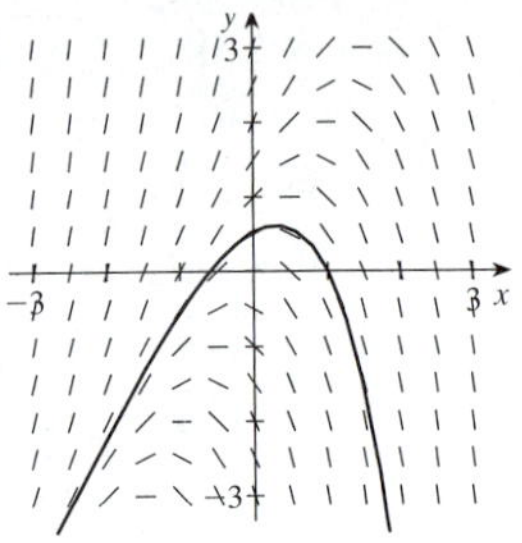

13.

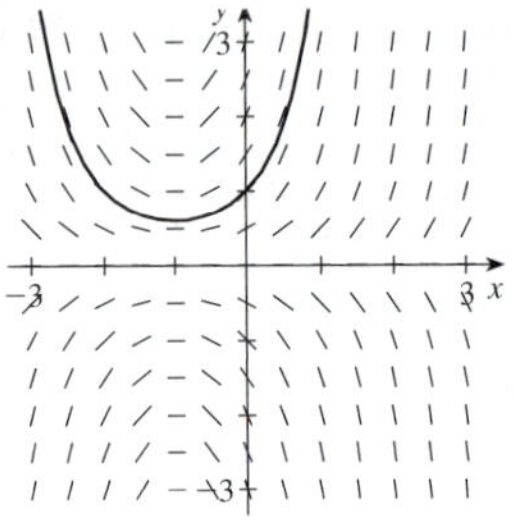

15.

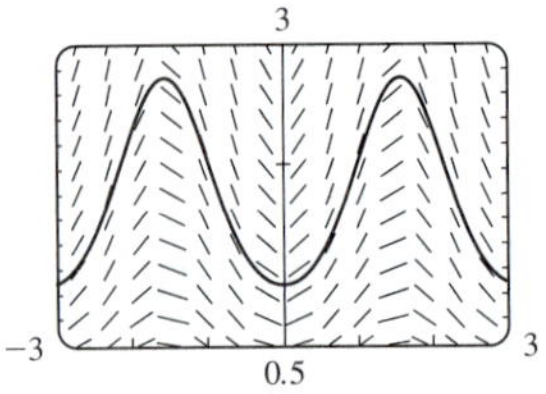

17. $-2 \leqslant c \leqslant 2$; $-2, 0, 2$

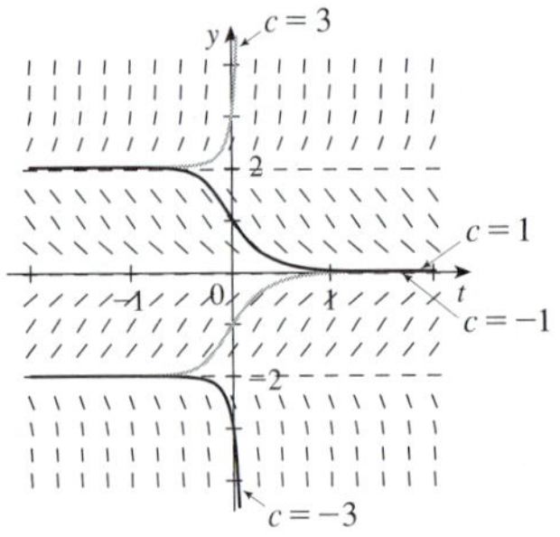

19. (a) (i) 1.4 (ii) 1.44 (iii) 1.4641

(b)

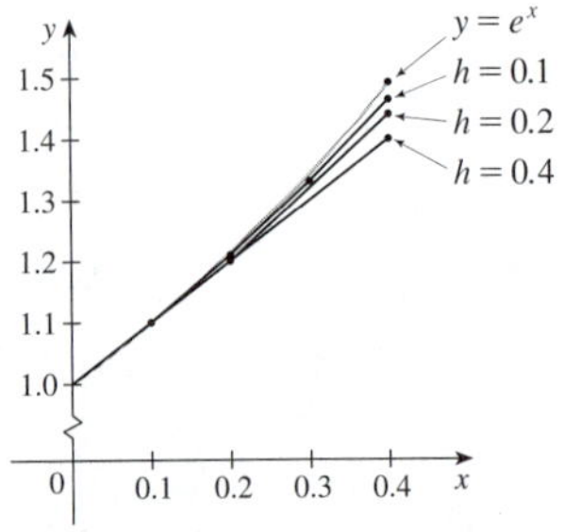

Underestimates

(c) (i) 0.0918 (ii) 0.0518 (iii) 0.0277

It appears that the error is also halved (approximately).

21. $-1, -3, -6.5, -12.25$ **23.** 1.7616

25. (a) (i) 3 (ii) 2.3928 (iii) 2.3701 (iv) 2.3681

(c) (i) -0.6321 (ii) -0.0249 (iii) -0.0022

(iv) -0.0002

It appears that the error is also divided by 10 (approximately).

27. (a), (d)

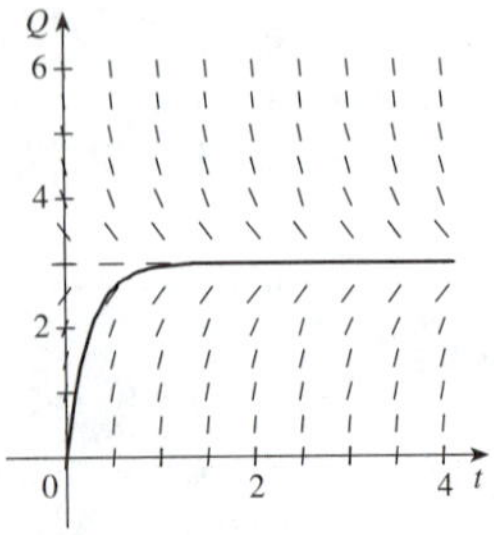

(b) 3

(c) Yes; $Q = 3$

(e) 2.77 C

Exercises 9.3 ▫ page 607

1. $y = Kx$ **3.** $y = K\sqrt{x^2 + 1}$

5. $y + \ln|\sec y| = \frac{1}{3}x^3 + x + C$

7. $y = \pm\sqrt{[3(te^t - e^t + C)]^{2/3} - 1}$

9. $u = Ae^{2t + t^2/2} - 1$ **11.** $y = \tan(x - 1)$

13. $\cos x + x \sin x = y^2 + \frac{1}{3}e^{3y} + \frac{2}{3}$

15. $u = -\sqrt{t^2 + \tan t + 25}$

17. $y = \dfrac{4a}{\sqrt{3}} \sin x - a$ **19.** $y = 7e^{x^4}$

21. (a) $\sin^{-1} y = x^2 + C$

(b) $y = \sin(x^2)$,

$-\sqrt{\pi/2} \leqslant x \leqslant \sqrt{\pi/2}$

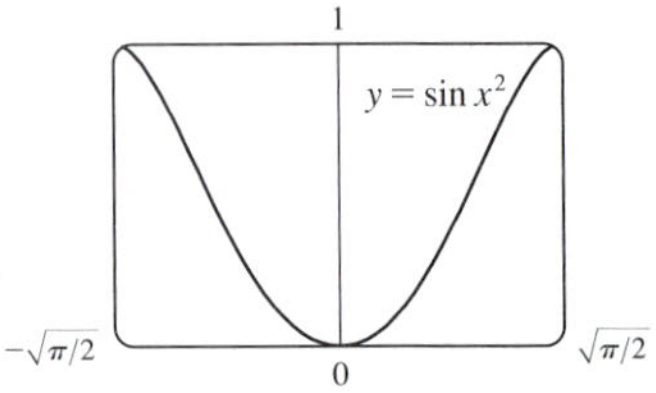

(c) No

23. $\cos y = \cos x - 1$

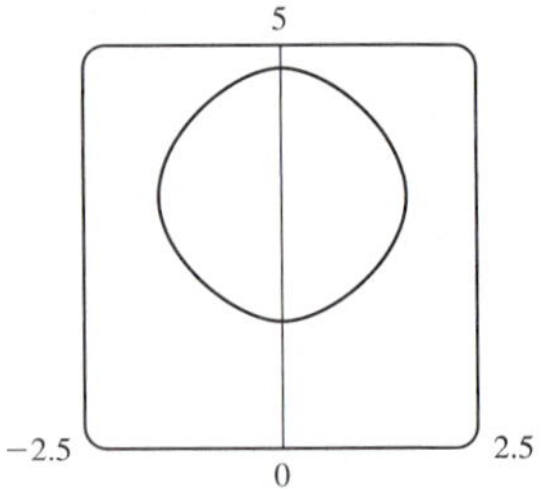

25. (a), (c)

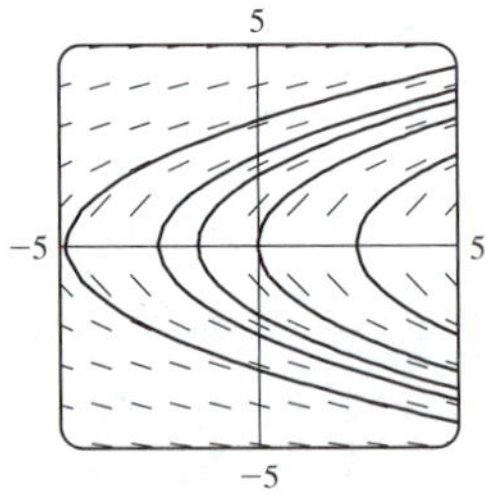

(b) $y = \pm\sqrt{2(x + C)}$

27. $x^2 + 2y^2 = C$

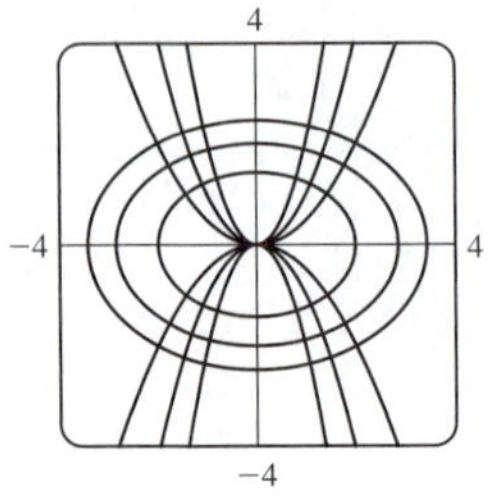

29. $y^3 = 3(x + C)$

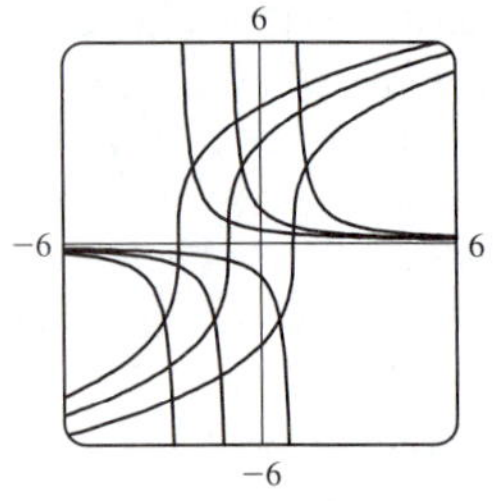

31. $Q(t) = 3 - 3e^{-4t}$; 3 **33.** $P(t) = M - Me^{-kt}$; M

35. (a) $x = a - 4/(kt + 2/\sqrt{a})^2$

(b) $t = \dfrac{2}{k\sqrt{a - b}}\left(\tan^{-1}\sqrt{\dfrac{b}{a - b}} - \tan^{-1}\sqrt{\dfrac{b - x}{a - b}}\right)$

37. (a) $C(t) = (C_0 - r/k)e^{-kt} + r/k$

(b) r/k; the concentration approaches r/k regardless of the value of C_0

39. (a) $15e^{-t/100}$ kg (b) $15e^{-0.2} \approx 12.3$ kg **41.** g/k
43. (a) $dA/dt = k\sqrt{A}\,(M - A)$
(b) $A(t) = M[(Ce^{\sqrt{M}kt} - 1)/(Ce^{\sqrt{M}kt} + 1)]^2$, where $C = (\sqrt{M} + \sqrt{A_0})/(\sqrt{M} - \sqrt{A_0})$ and $A_0 = A(0)$

Exercises 9.4 ▫ page 620

1. About 235
3. (a) $500 \times 16^{t/3}$ (b) $\approx 20{,}159$
(c) 18,631 cells/h (d) $(3 \ln 60)/\ln 16 \approx 4.4$ h
5. (a) 1508 million, 1871 million (b) 2161 million
(c) 3972 million; wars in the first half of century, increased life expectancy in second half
7. (a) $Ce^{-0.0005t}$ (b) $-2000 \ln 0.9 \approx 211$ s
9. (a) $100 \times 2^{-t/30}$ mg (b) ≈ 9.92 mg
(c) ≈ 199.3 years
11. ≈ 2500 years
13. (a) ≈ 137 °F (b) ≈ 116 min
15. (a) $13.\overline{3}$°C (b) ≈ 67.74 min
17. (a) ≈ 64.5 kPa (b) ≈ 39.9 kPa
19. (a) (i) \$3828.84 (ii) \$3840.25 (iii) \$3850.08
(iv) \$3851.61 (v) \$3852.01 (vi) \$3852.08
(b) $dA/dt = 0.05A$, $A(0) = 3000$
21. (a) $P(t) = (m/k) + (P_0 - m/k)e^{kt}$ (b) $m < kP_0$
(c) $m = kP_0$, $m > kP_0$ (d) Declining

Exercises 9.5 ▫ page 629

1. (a) 100; 0.05
(b) Where P is close to 0 or 100; on the line $P = 50$; $0 < P_0 < 100$; $P_0 > 100$
(c)

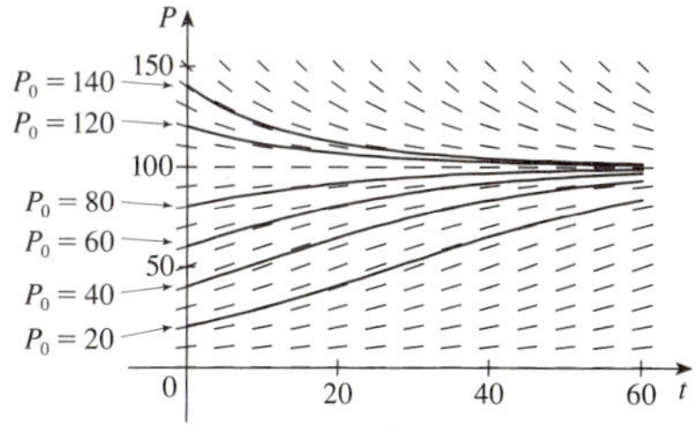

Solutions approach 100; some increase and some decrease, some have an inflection point but others don't; solutions with $P_0 = 20$ and $P_0 = 40$ have inflection points at $P = 50$
(d) $P = 0$, $P = 100$; other solutions move away from $P = 0$ and toward $P = 100$
3. (a) 3.23×10^7 kg (b) ≈ 1.55 years
5. (a) $dP/dt = \frac{1}{265}P(1 - P/100)$, P in billions
(b) 5.49 billion (c) In billions: 7.81, 27.72
(d) In billions: 5.48, 7.61, 22.41
7. (a) $dy/dt = ky(1 - y)$
(b) $y = y_0/[y_0 + (1 - y_0)e^{-kt}]$
(c) 3:36 P.M.
11. (a) Fish are caught at a rate of 15 per week.
(b) See part (d) (c) $P = 250$, $P = 750$
(d)

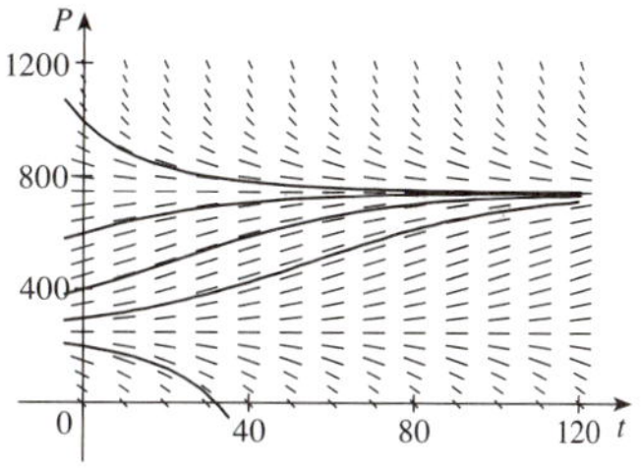

$0 < P_0 < 250$: $P \to 0$; $P_0 = 250$: $P \to 250$; $P_0 > 250$: $P \to 750$

(e) $P(t) = \dfrac{250 - 750ke^{t/25}}{1 - ke^{t/25}}$, where $k = \frac{1}{11}, -\frac{1}{9}$

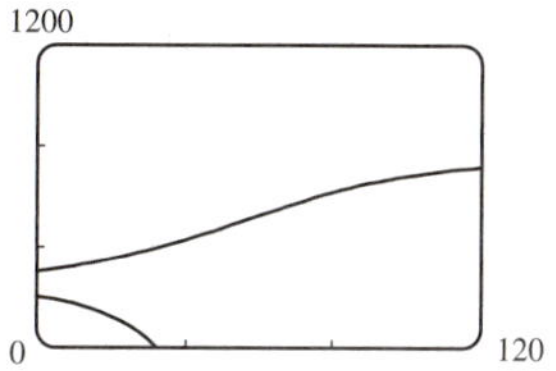

13. (b)

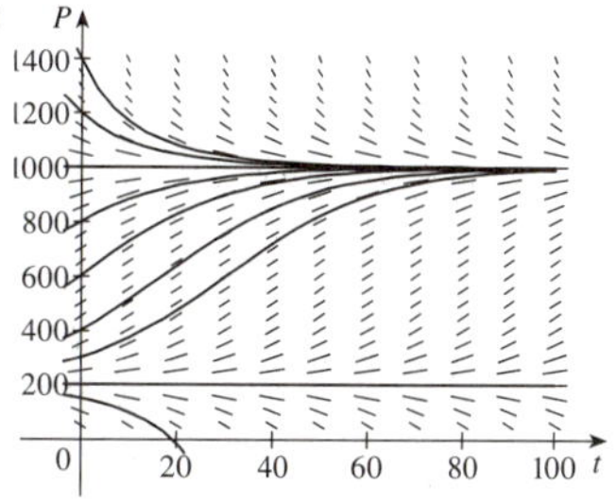

$0 < P_0 < 200$: $P \to 0$; $P_0 = 200$: $P \to 200$; $P_0 > 200$: $P \to 1000$

(c) $P(t) = \dfrac{m(K - P_0) + K(P_0 - m)e^{(K-m)(k/K)t}}{K - P_0 + (P_0 - m)e^{(K-m)(k/K)t}}$

15. (a) $P(t) = P_0 e^{(k/r)[\sin(rt - \phi) + \sin \phi]}$ (b) Does not exist

Exercises 9.6 ▫ page 636

1. No **3.** Yes **5.** $y = \frac{2}{3}e^x + Ce^{-2x}$
7. $y = x^2 \ln|x| + Cx^2$ **9.** $y = \frac{2}{3}\sqrt{x} + C/x$
11. $y = \frac{1}{2}x + Ce^{-x^2} - \frac{1}{2}e^{-x^2}\int e^{x^2}\,dx$
13. $u = (t^2 + 2t + 2C)/[2(t + 1)]$
15. $y = -x - 1 + 3e^x$ **17.** $v = t^3e^{t^2} + 5e^{t^2}$
19. $y = -x \cos x - x$
21. $y = \sin x + (\cos x)/x + C/x$

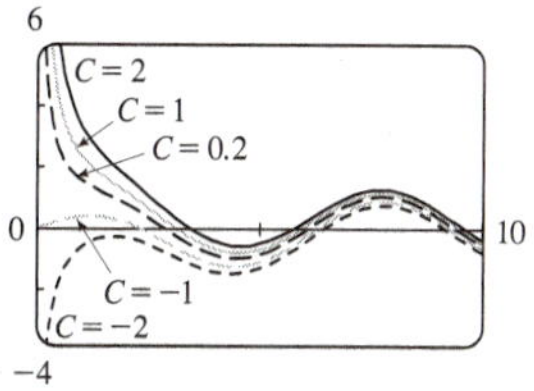

25. $y = \pm[Cx^4 + 2/(5x)]^{-1/2}$
27. (a) $I(t) = 4 - 4e^{-5t}$ (b) $4 - 4e^{-1/2} \approx 1.57$ A
29. $Q(t) = 3(1 - e^{-4t})$, $I(t) = 12e^{-4t}$

31. $P(t) = M + Ce^{-kt}$

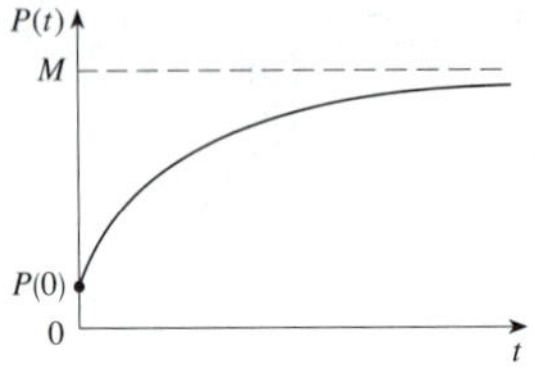

33. $y = \frac{2}{5}(100 + 2t) - 40{,}000(100 + 2t)^{-3/2}$; 0.2275 kg/L

35. (b) mg/c (c) $(mg/c)[t + (m/c)e^{-ct/m}] - m^2g/c^2$

Exercises 9.7 ▫ page 642

1. (a) x = predators, y = prey; growth is restricted only by predators, which feed only on prey.
(b) x = prey, y = predators; growth is restricted by carrying capacity and by predators, which feed only on prey.

3. (a) The rabbit population starts at about 300, increases to 2400, then decreases back to 300. The fox population starts at 100, decreases to about 20, increases to about 315, decreases to 100, and the cycle starts again.

(b)

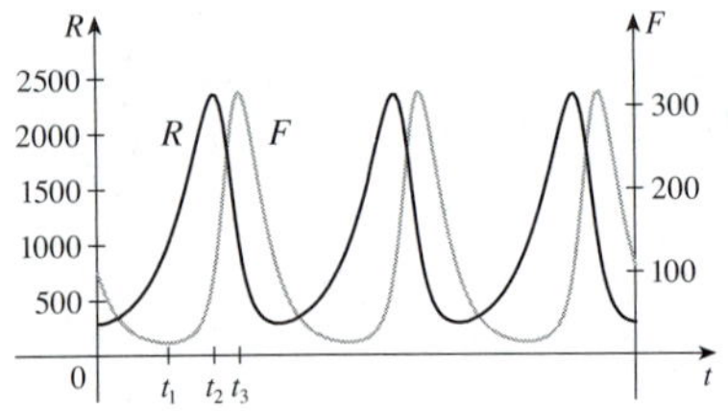

5.

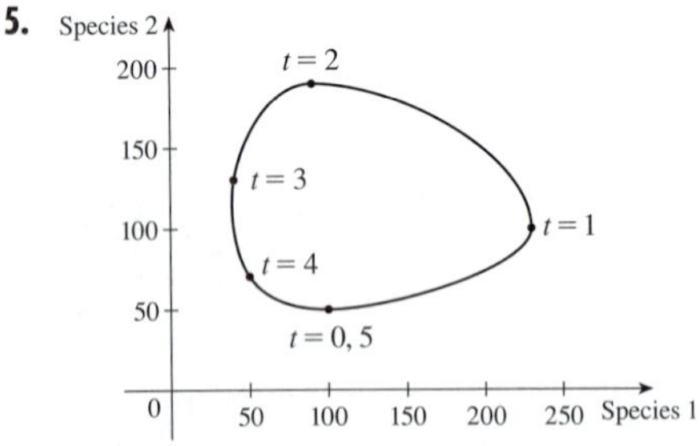

9. (a) Population stabilizes at 5000.
(b) (i) $W = 0$, $R = 0$: Zero populations
(ii) $W = 0$, $R = 5000$: In the absence of wolves, the rabbit population is always 5000.
(iii) $W = 64$, $R = 1000$: Both populations are stable.
(c) The populations stabilize at 1000 rabbits and 64 wolves.
(d)

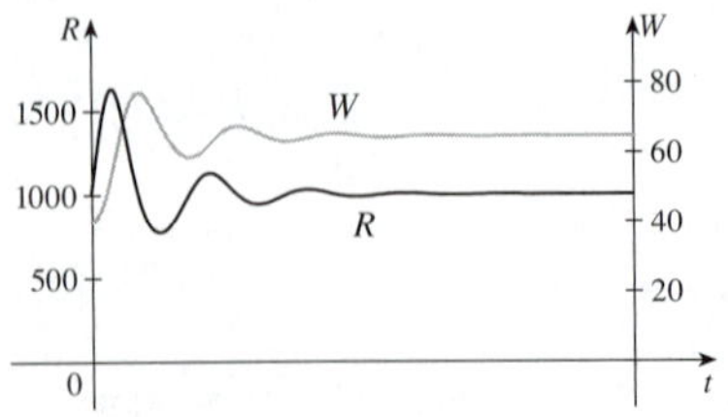

Chapter 9 Review ▫ page 645

True-False Quiz

1. True **3.** False **5.** True **7.** True

Exercises

1. (a)

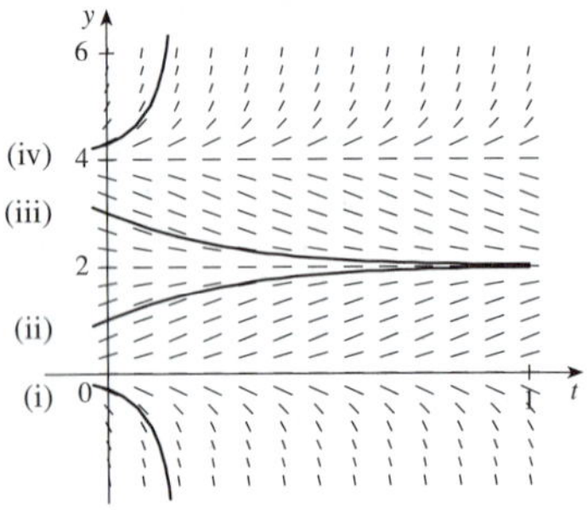

(b) $0 \leq c \leq 4$; $y = 0$, $y = 2$, $y = 4$

3. (a) $y(0.3) \approx 0.8$

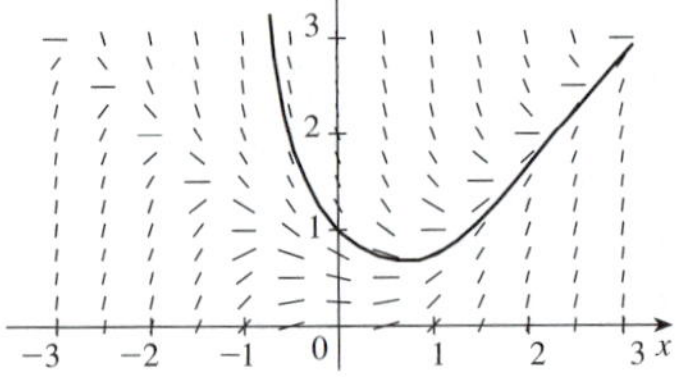

(b) 0.75676
(c) $y = x$ and $y = -x$; there is a local maximum or minimum

5. $y = \left(\frac{1}{2}x^2 + C\right)e^{-\sin x}$ **7.** $y^3 + y^2 = \cos x + x \sin x + C$

9. $y = \sqrt{(\ln x)^2 + 4}$ **11.** $y = e^{-x}\left(\frac{2}{3}x^{3/2} + 3\right)$

13. $y^2 - 2\ln|y| + x^2 = C$

15. (a) 1000×3^t (b) 27,000
(c) $27{,}000 \ln 3 \approx 29{,}663$ bacteria per hour
(d) $(\ln 2)/\ln 3 \approx 0.63$ h

17. (a) C_0e^{-kt} (b) ≈ 100 h

19. (a) $L(t) = L_\infty - [L_\infty - L(0)]e^{-kt}$
(b) $L(t) = 53 - 43e^{-0.2t}$

21. 15 days **23.** $k \ln h + h = (-R/V)t + C$

25. (a) Stabilizes at 200,000
(b) (i) $x = 0$, $y = 0$: Zero populations
(ii) $x = 200{,}000$, $y = 0$: In the absence of birds, the insect population is always 200,000.
(iii) $x = 25{,}000$, $y = 175$: Both populations are stable.
(c) The populations stabilize at 25,000 insects and 175 birds.
(d)

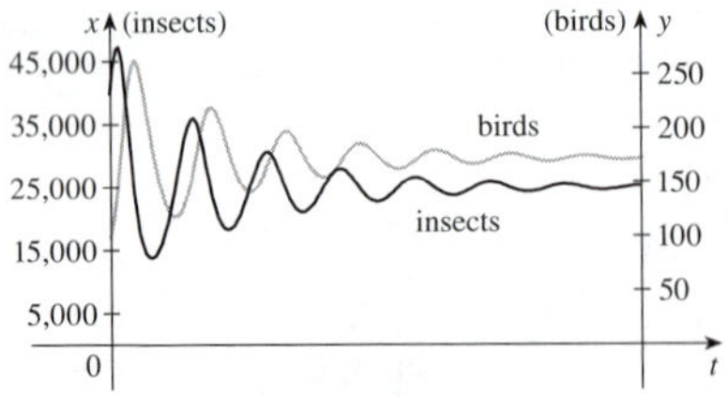

27. (a) $y = (1/k)\cosh kx + a - 1/k$ or $y = (1/k)\cosh kx - (1/k)\cosh kb + h$ (b) $(2/k)\sinh kb$

Problems Plus ▫ page 648

1. $f(x) = \pm 10e^x$ **5.** 20°C
7. (b) $f(x) = (x^2 - L^2)/(4L) - (L/2)\ln(x/L)$ (c) No
9. (a) 9.8 h
(b) $31{,}900\pi \approx 100{,}000\ \text{ft}^2$; $6283\ \text{ft}^2/\text{h}$
(c) 5.1 h
11. $x^2 + (y - 6)^2 = 25$

Chapter 10

Exercises 10.1 ▫ page 656

1.

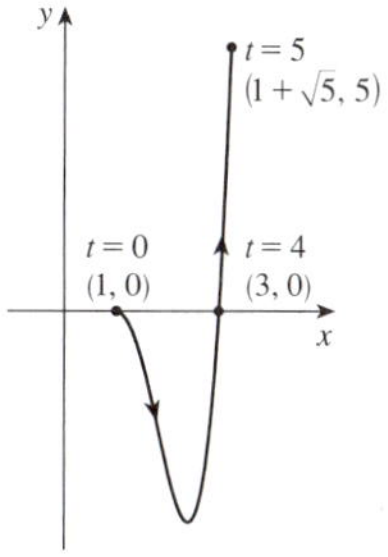

3.

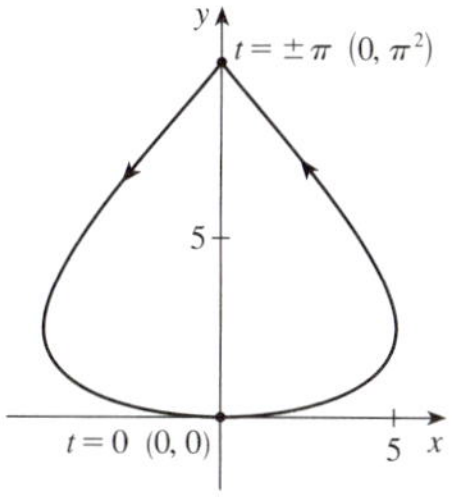

5. (a)

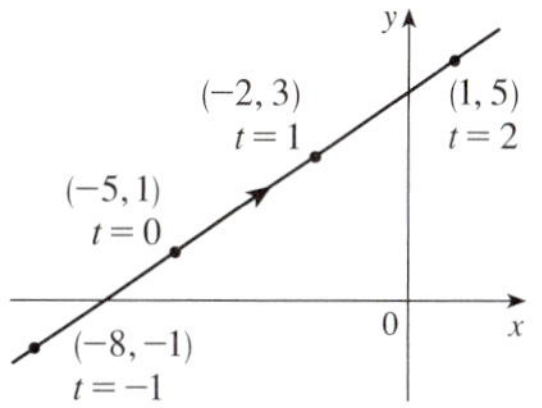

(b) $y = \frac{2}{3}x + \frac{13}{3}$

7. (a)

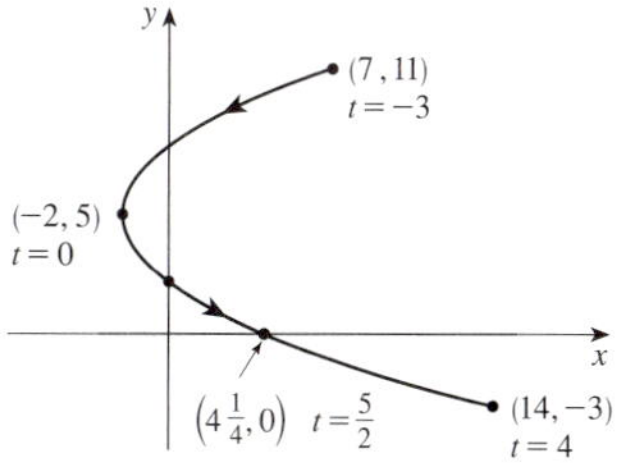

(b) $x = \frac{1}{4}(y - 5)^2 - 2$, $-3 \leqslant y \leqslant 11$

9. (a)

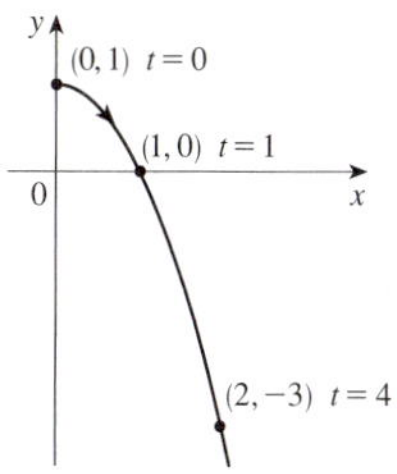

(b) $y = 1 - x^2, x \geqslant 0$

11. (a) $x^2 + y^2 = 1, x \geqslant 0$
(b)

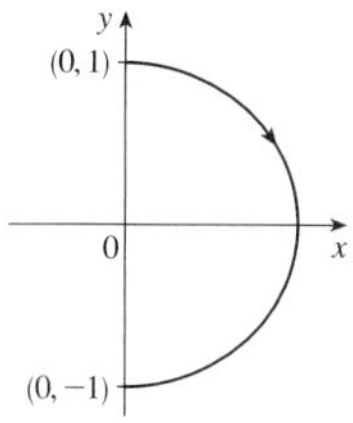

13. (a) $x + y = 1, 0 \leqslant x \leqslant 1$
(b)

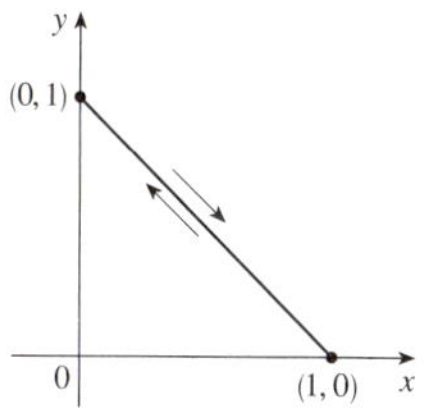

15. (a) $y = 1/x, x > 0$
(b)

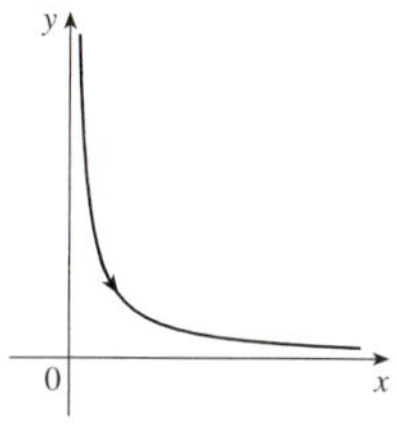

17. (a) $x^2 - y^2 = 1, x \geqslant 1$
(b)

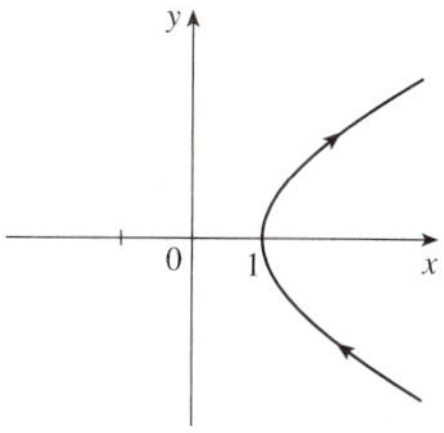

19. Moves counterclockwise along the circle $x^2 + y^2 = 1$ from $(-1, 0)$ to $(1, 0)$
21. Moves once clockwise around the ellipse $(x^2/4) + (y^2/9) = 1$, starting and ending at $(0, 3)$
23. It is contained in the rectangle described by $1 \leqslant x \leqslant 4$ and $2 \leqslant y \leqslant 3$.
25.

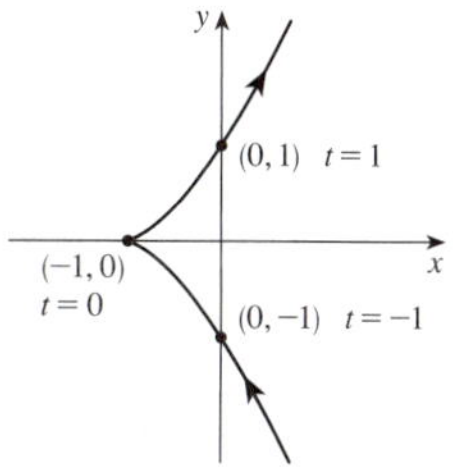

27.

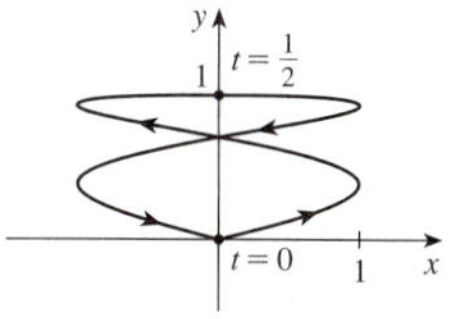

29.

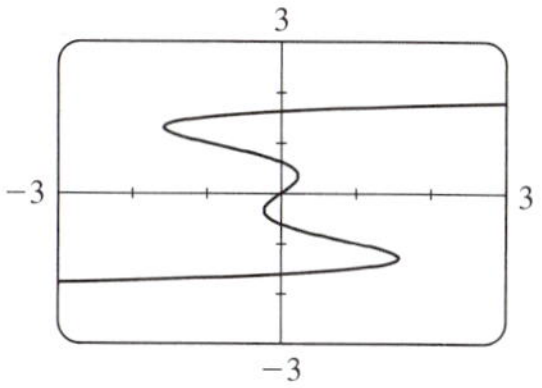

31. (b) $x = -2 + 5t, y = 7 - 8t, 0 \leqslant t \leqslant 1$
33. (a) $x = 2\cos t, y = 1 - 2\sin t, 0 \leqslant t \leqslant 2\pi$
(b) $x = 2\cos t, y = 1 + 2\sin t, 0 \leqslant t \leqslant 6\pi$
(c) $x = 2\cos t, y = 1 + 2\sin t, \pi/2 \leqslant t \leqslant 3\pi/2$
35. (a) $x = a\sin t, y = b\cos t, 0 \leqslant t \leqslant 2\pi$
(b)

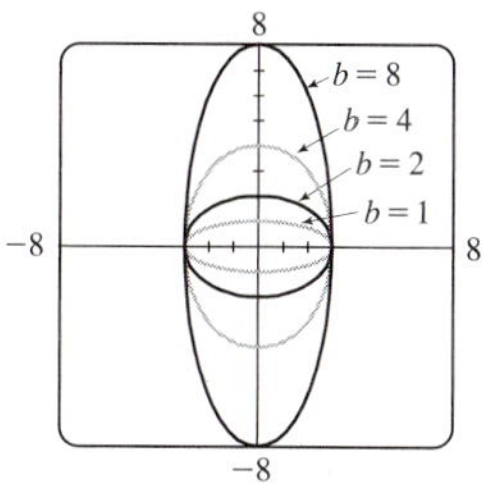

(c) As b increases, the ellipse stretches vertically.

39. $x = a\cos\theta, y = b\sin\theta$; $(x^2/a^2) + (y^2/b^2) = 1$, ellipse

41.

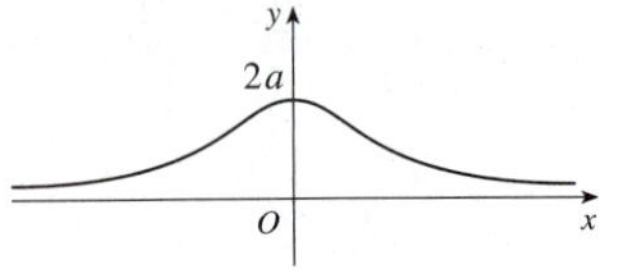

43. (a) Two points of intersection

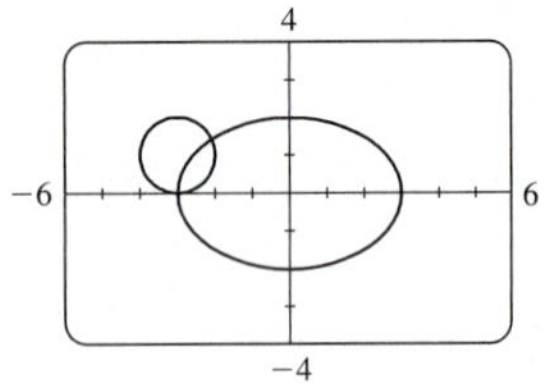

(b) One collision point at $(-3, 0)$ when $t = 3\pi/2$
(c) There are still two intersection points, but no collision point.

45. For $c = 0$, there is a cusp; for $c > 0$, there is a loop whose size increases as c increases.

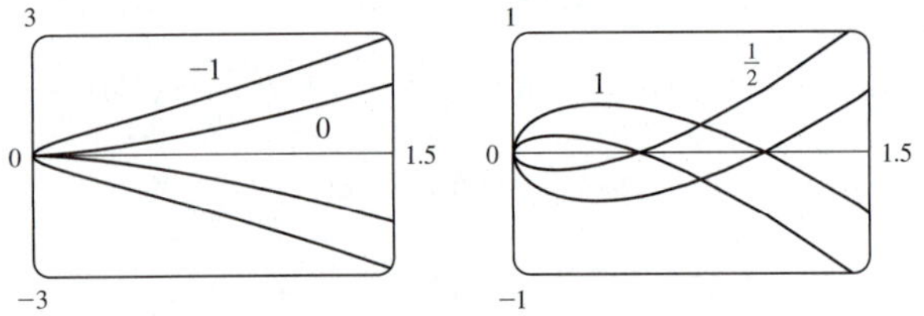

47. As n increases, the number of oscillations increases; a and b determine the width and height.

Exercises 10.2 ▫ page 666

1. $5/(3t^2 - 1)$ **3.** $y = -x$
5. $y = -(2/e)x + 3$ **7.** $y = -2x + 3$
9. $y = (\sqrt{3}/2)x - \frac{1}{2}$

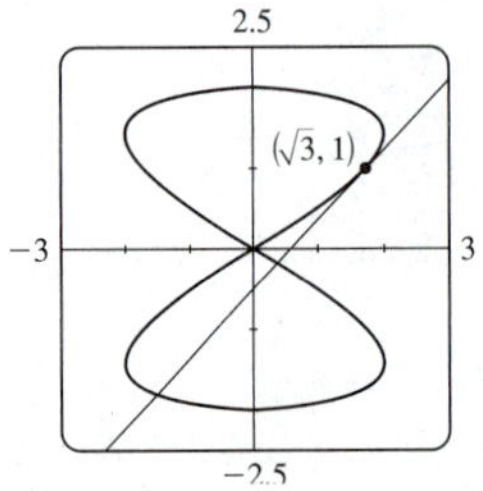

11. $1 + \frac{3}{2}t,\ 3/(4t),\ t > 0$ **13.** $-e^{-t},\ e^{-t}/(1 - e^t),\ t < 0$
15. $-\frac{3}{2}\tan t,\ -\frac{3}{4}\sec^3 t,\ \frac{\pi}{2} < t < \frac{3\pi}{2}$
17. Horizontal at $(6, \pm 16)$, vertical at $(10, 0)$
19. Horizontal at $(\pm\sqrt{2}, \pm 1)$ (four points), vertical at $(\pm 2, 0)$
21. $(-0.25, 0.36)$, $\left(-\frac{1}{4}, (1/\sqrt{2}) - \frac{1}{2}\ln 2\right)$
23.

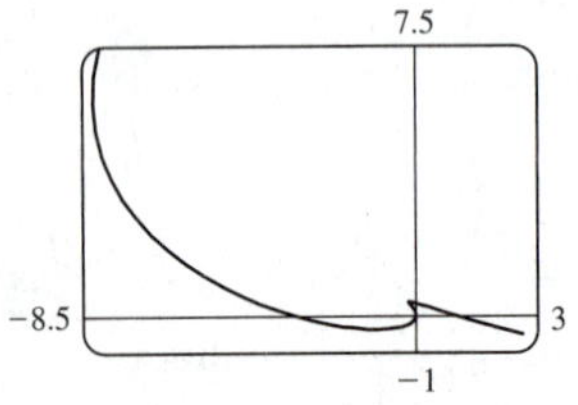

25. $y = x,\ y = -x$

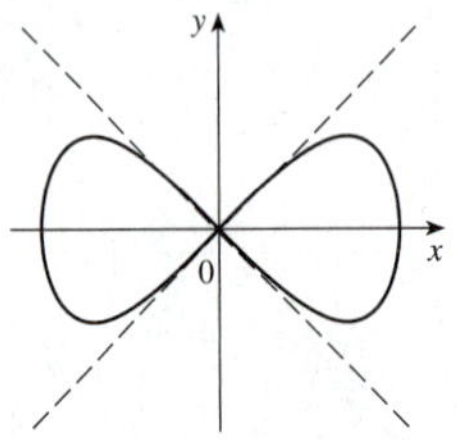

27. (a) $d \sin\theta/(r - d\cos\theta)$ **29.** $(-5, 6),\ \left(-\frac{208}{27}, \frac{32}{3}\right)$
31. πab **33.** $(e^{\pi/2} - 1)/2$ **35.** $2\pi r^2 + \pi d^2$
37. $\int_1^2 \sqrt{1 + 4t^2}\,dt$ **39.** $\int_0^{2\pi} \sqrt{3 - 2\sin t - 2\cos t}\,dt$
41. $4\sqrt{2} - 2$
43. $-\sqrt{10}/3 + \ln(3 + \sqrt{10}) + \sqrt{2} - \ln(1 + \sqrt{2})$
45. $\sqrt{2}\,(e^{\pi} - 1)$

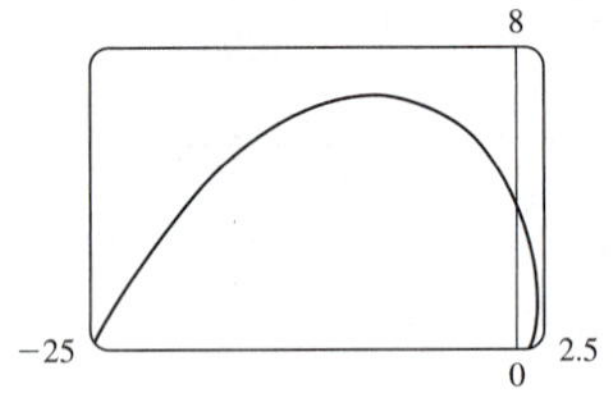

47. $e^3 + 11 - e^{-8}$

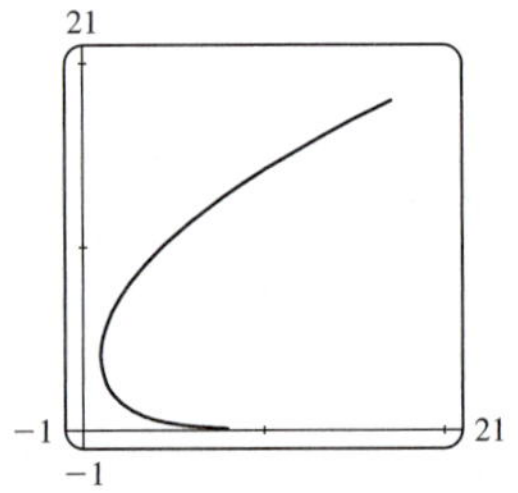

49. 612.3053 **51.** $6\sqrt{2},\ \sqrt{2}$
55. (a) $t \in [0, 4\pi]$ (b) ≈ 294

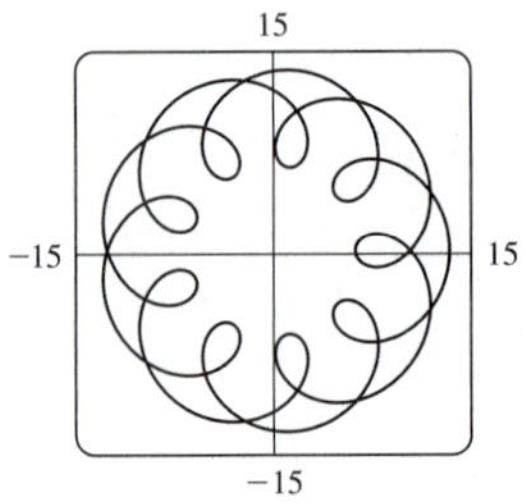

57. $\int_1^2 \frac{8}{3}\pi t^{3/2}\sqrt{1 + 4t^2}\,dt$ **59.** $2\pi(247\sqrt{13} + 64)/1215$
61. $6\pi a^2/5$ **63.** 59.101 **65.** $24\pi(949\sqrt{26} + 1)/5$
71. $\frac{1}{4}$

Exercises 10.3 ▫ page 677

1. (a)

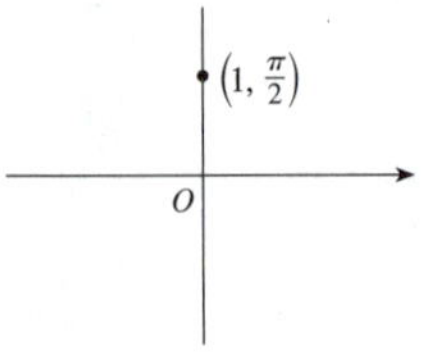

$(1, 5\pi/2),\ (-1, 3\pi/2)$

(b)

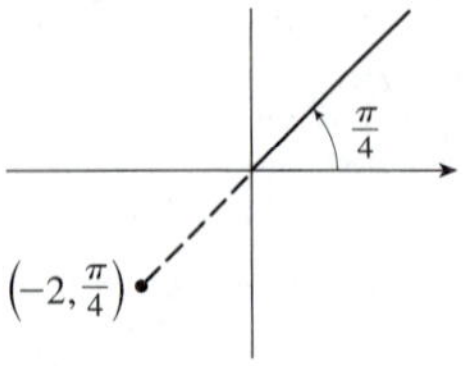

$(2, 5\pi/4),\ (-2, 9\pi/4)$

(c)

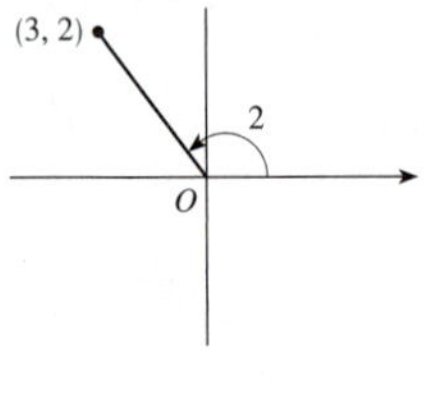

$(3, 2 + 2\pi),\ (-3, 2 + \pi)$

3. (a)

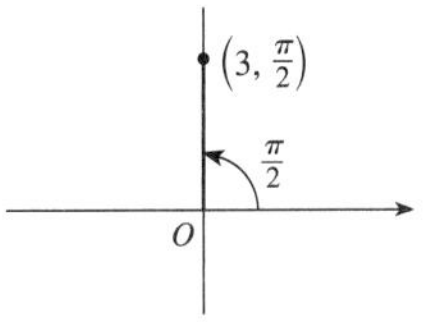

$(0, 3)$

(b)

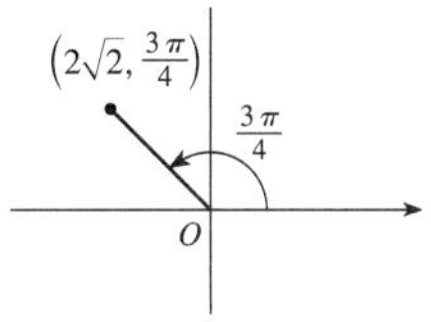

$(-2, 2)$

(c)

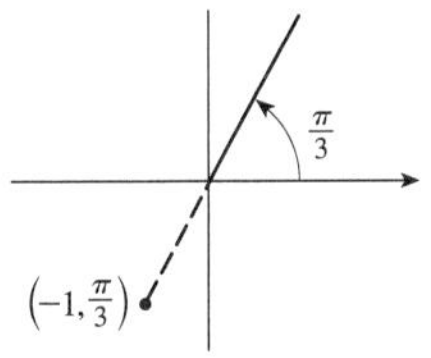

$\left(-\frac{1}{2}, -\sqrt{3}/2\right)$

5. (a) (i) $(\sqrt{2}, \pi/4)$ (ii) $(-\sqrt{2}, 5\pi/4)$
(b) (i) $(4, 11\pi/6)$ (ii) $(-4, 5\pi/6)$

7.

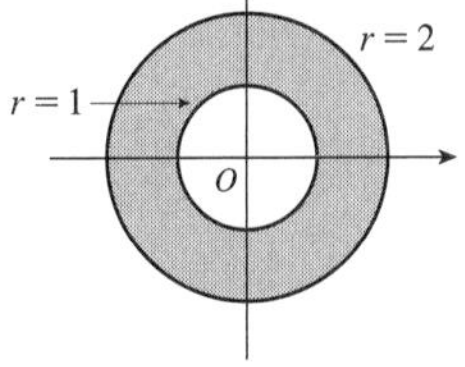

9.

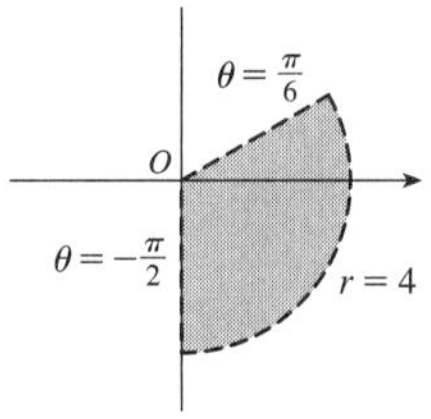

11.

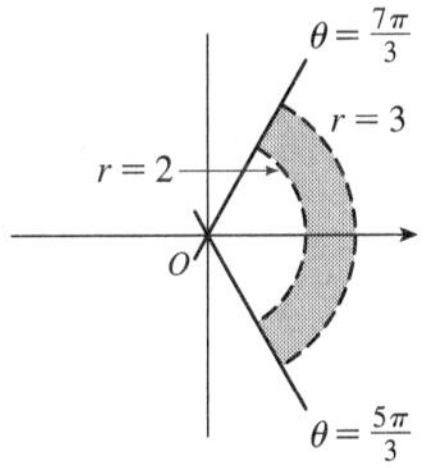

13. $\frac{1}{2}\sqrt{40 + 6\sqrt{6} - 6\sqrt{2}}$

15. Circle, center O, radius 2 **17.** Circle, center $\left(0, \frac{3}{2}\right)$, radius $\frac{3}{2}$

19. Horizontal line, 1 unit above the x-axis

21. $r = 3 \sec \theta$ **23.** $r = -\cot \theta \csc \theta$ **25.** $r = 2c \cos \theta$

27. (a) $\theta = \pi/6$ (b) $x = 3$

29.

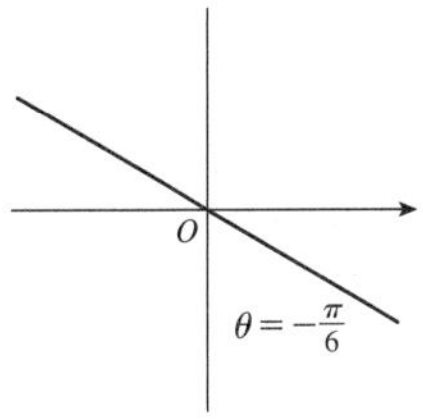

31.

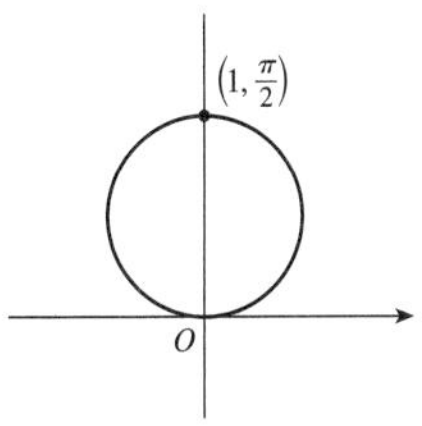

33.

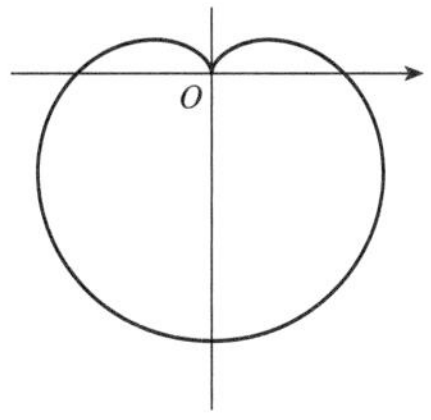

35.

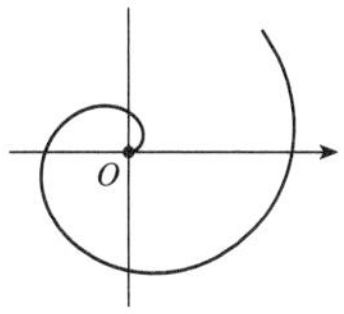

37.

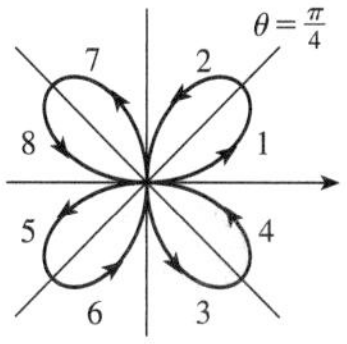

39.

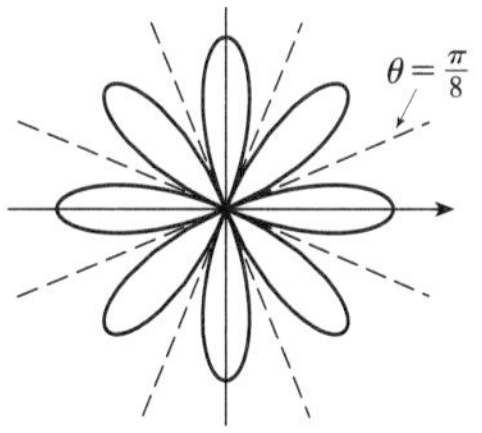

41.

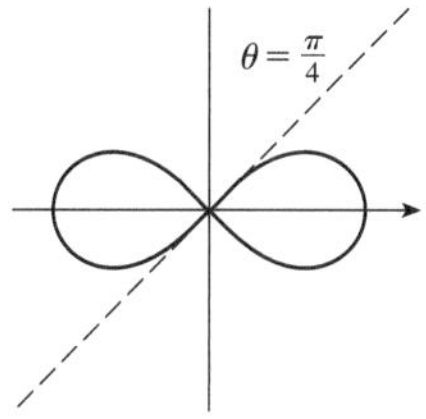

43.

45.

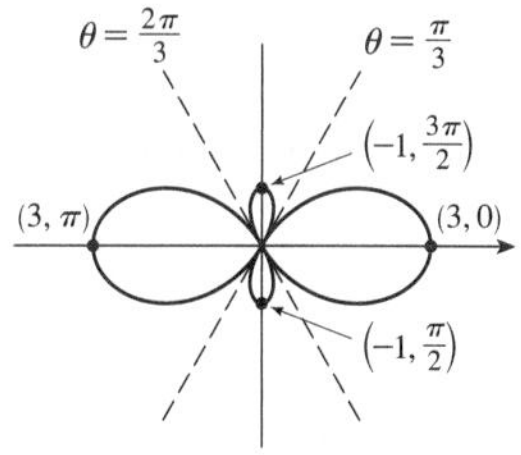

47.

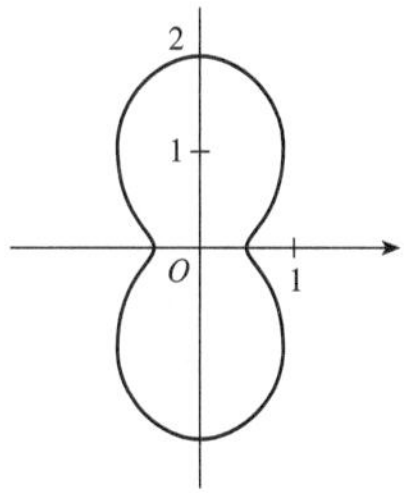

49.

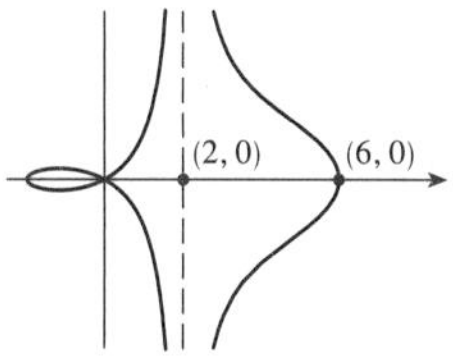

51.

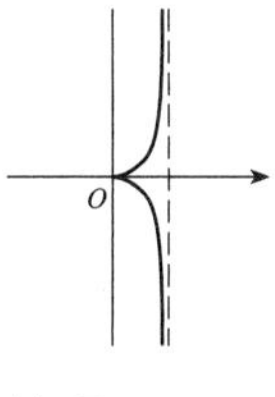

53. (a) For $c < -1$, the loop begins at $\theta = \sin^{-1}(-1/c)$ and ends at $\theta = \pi - \sin^{-1}(-1/c)$; for $c > 1$, it begins at $\theta = \pi + \sin^{-1}(1/c)$ and ends at $\theta = 2\pi - \sin^{-1}(1/c)$.

55. $\sqrt{3}$ **57.** $-\pi$ **59.** -1

61. Horizontal at $(3/\sqrt{2}, \pi/4)$, $(-3/\sqrt{2}, 3\pi/4)$;
vertical at $(3, 0)$, $(0, \pi/2)$

63. Horizontal at $\left(\frac{3}{2}, \pi/3\right)$, $\left(\frac{3}{2}, 5\pi/3\right)$, and the pole;
vertical at $(2, 0)$, $\left(\frac{1}{2}, 2\pi/3\right)$, $\left(\frac{1}{2}, 4\pi/3\right)$

65. Horizontal at $(-1, 3\pi/2)$, $(-1, \pi/2)$, $\left(\frac{2}{3}, \alpha\right)$, $\left(\frac{2}{3}, \pi - \alpha\right)$, $\left(\frac{2}{3}, \pi + \alpha\right)$, $\left(\frac{2}{3}, 2\pi - \alpha\right)$, where $\alpha = \sin^{-1}(1/\sqrt{6})$;
vertical at $(1, 0)$, $(1, \pi)$, $\left(-\frac{2}{3}, \beta\right)$, $\left(-\frac{2}{3}, \pi - \beta\right)$, $\left(-\frac{2}{3}, \pi + \beta\right)$, $\left(-\frac{2}{3}, 2\pi - \beta\right)$, where $\beta = \cos^{-1}(1/\sqrt{6})$

67. Center $(b/2, a/2)$, radius $\sqrt{a^2 + b^2}/2$

69.

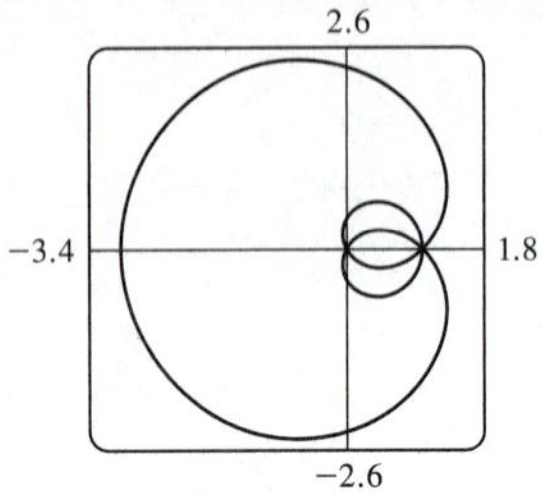

71.

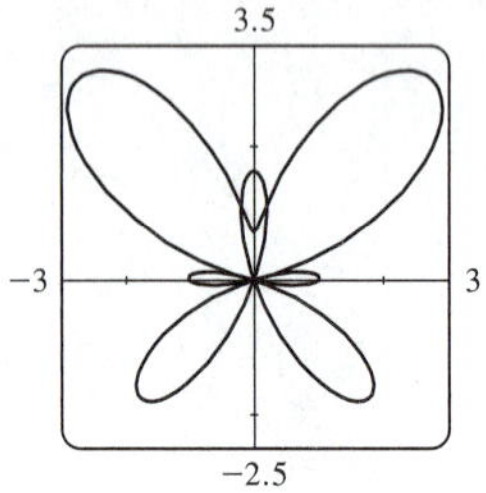

73.

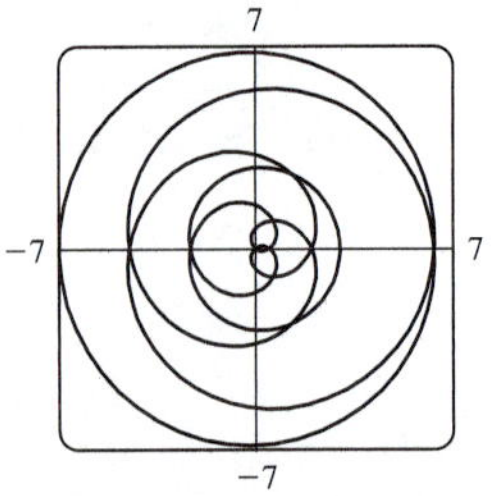

75. By counterclockwise rotation through angle $\pi/6$, $\pi/3$, or α about the origin

77. (a) A rose with n loops if n is odd and $2n$ loops if n is even
(b) Number of loops is always $2n$

79. For $0 < a < 1$, the curve is an oval, which develops a dimple as $a \to 1^-$. When $a > 1$, the curve splits into two parts, one of which has a loop.

Exercises 10.4 ▫ page 683

1. $\pi^2/64$ **3.** $\pi/12 + \sqrt{3}/8$ **5.** $\pi^3/6$ **7.** $41\pi/4$

9. $9\pi/4$

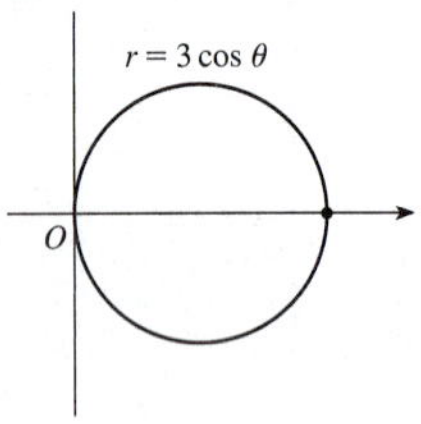

11. 4

13. π

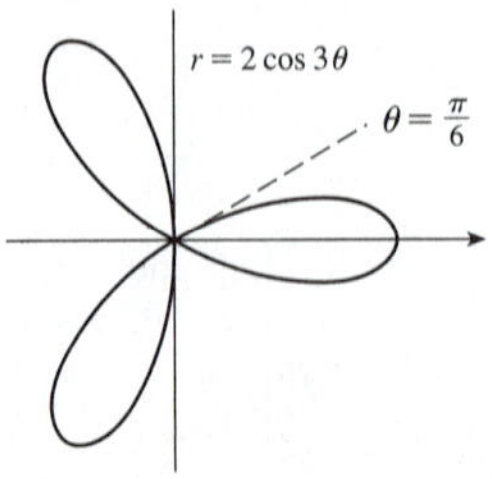

15. 3π

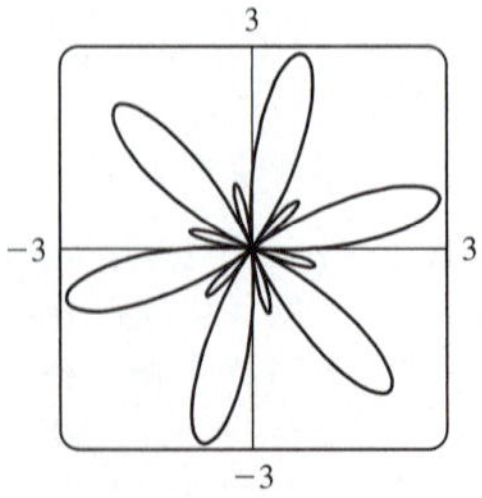

17. $\pi/8$ **19.** $9\pi/20$ **21.** $\pi - (3\sqrt{3}/2)$

23. $(4\pi/3) + 2\sqrt{3}$ **25.** $4\sqrt{3} - \frac{4}{3}\pi$ **27.** π

29. $\frac{1}{8}(\pi - 2)$ **31.** $(\pi/2) - 1$

33. $(19\pi/3) - (11\sqrt{3}/2)$

35. $\frac{1}{4}(\pi + 3\sqrt{3})$ **37.** $(1/\sqrt{2}, \pi/4)$ and the pole

39. $(\frac{1}{2}, \pi/3)$, $(\frac{1}{2}, 5\pi/3)$, and the pole

41. $(\sqrt{3}/2, \pi/3)$, $(\sqrt{3}/2, 2\pi/3)$, and the pole

43. Intersection at $\theta \approx 0.89, 2.25$; area ≈ 3.46

45. π **47.** $\frac{8}{3}[(\pi^2 + 1)^{3/2} - 1]$ **49.** 29.0653

51. 9.6884

53. $\frac{16}{3}$

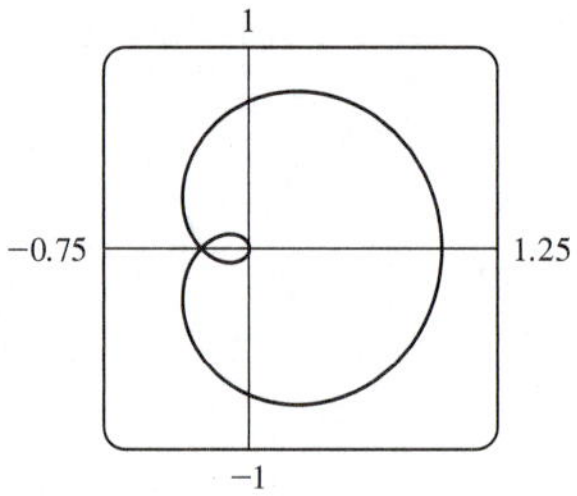

55. (b) $2\pi(2 - \sqrt{2})$

Exercises 10.5 ▫ page 690

1. $(0, 0)$, $(\frac{1}{8}, 0)$, $x = -\frac{1}{8}$

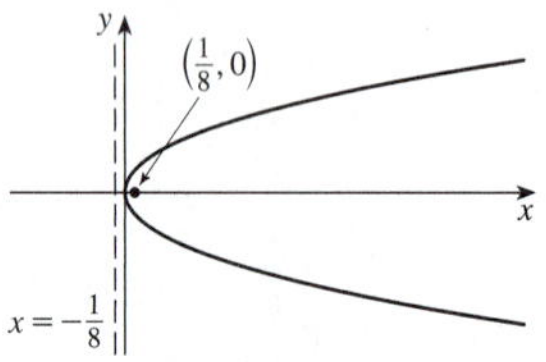

3. $(0, 0)$, $(0, -\frac{1}{16})$, $y = \frac{1}{16}$

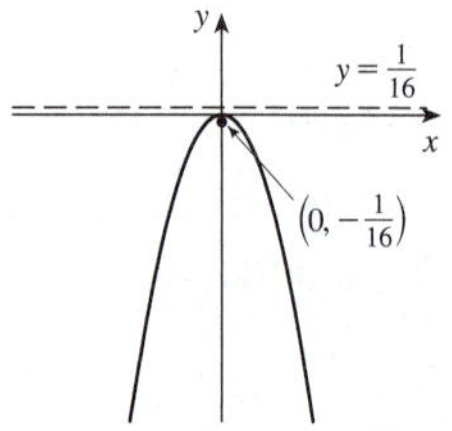

5. $(-2, 3)$, $(-2, 5)$, $y = 1$

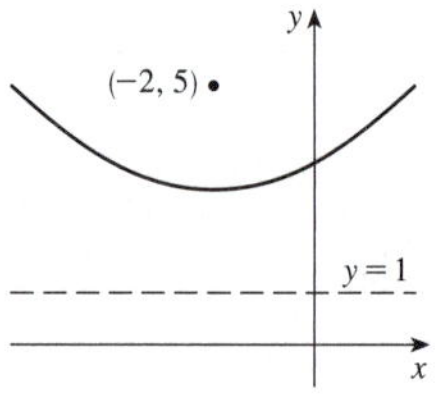

7. $(-2, -1)$, $(-5, -1)$, $x = 1$

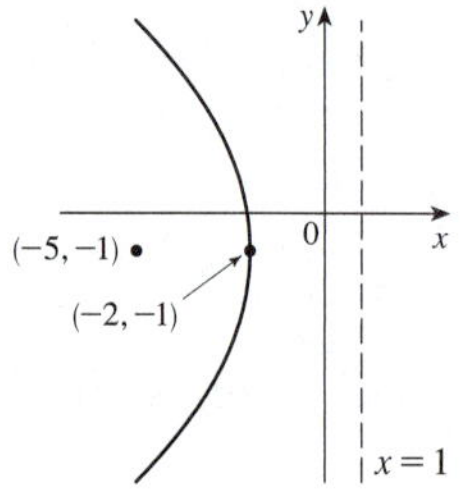

9. $x = -y^2$, focus $(-\frac{1}{4}, 0)$, directrix $x = \frac{1}{4}$

11. $(\pm 3, 0)$, $(\pm 2, 0)$

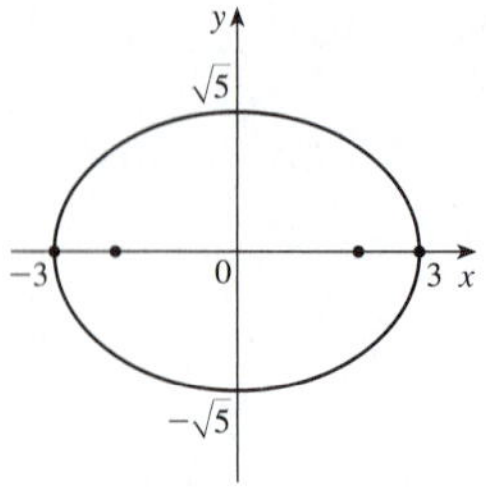

13. $(0, \pm 4)$, $(0, \pm 2\sqrt{3})$

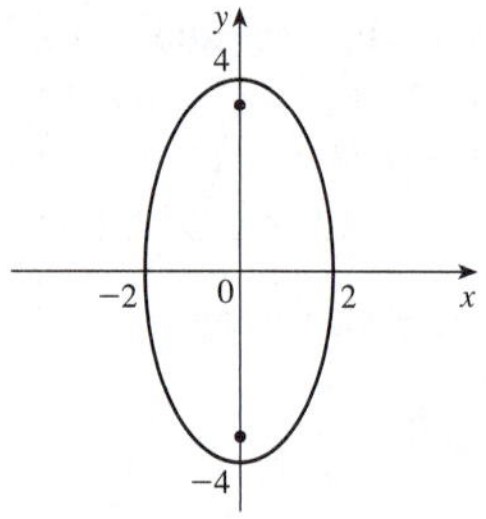

15. $(1, \pm 3)$, $(1, \pm\sqrt{5})$

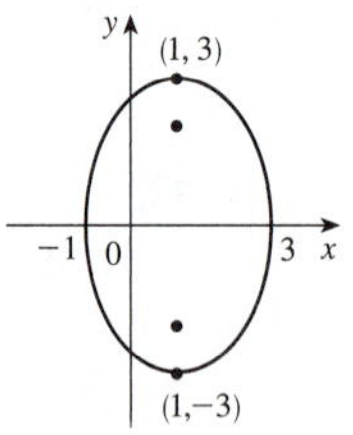

17. $\dfrac{x^2}{4} + \dfrac{y^2}{9} = 1$, foci $(0, \pm\sqrt{5})$

19. $(\pm 12, 0), (\pm 13, 0)$, $y = \pm \frac{5}{12}x$

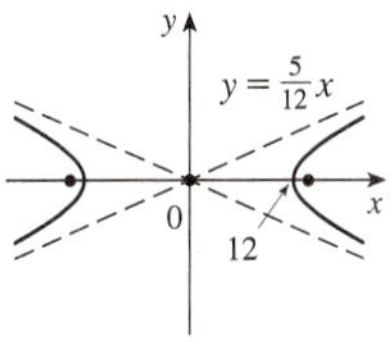

21. $(0, \pm 2), (0, \pm 2\sqrt{2})$, $y = \pm x$

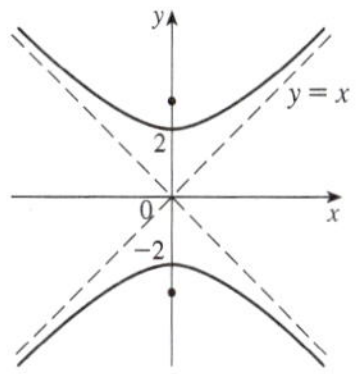

23. $(2 \pm \sqrt{6}, 1), (2 \pm \sqrt{15}, 1)$, $y - 1 = \pm(\sqrt{6}/2)(x - 2)$

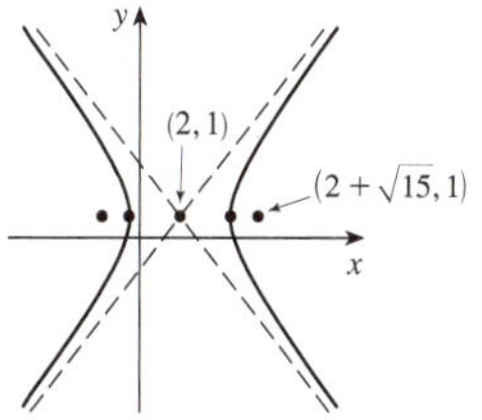

25. Parabola, $(0, -1), (0, -\frac{3}{4})$ **27.** Ellipse, $(\pm\sqrt{2}, 1), (\pm 1, 1)$
29. Hyperbola, $(0, 1), (0, 3); (0, -1 \pm \sqrt{5})$
31. $x^2 = -8y$ **33.** $y^2 = -12(x + 1)$
35. $y^2 = 16x$ **37.** $(x^2/25) + (y^2/21) = 1$
39. $(x^2/12) + [(y - 4)^2/16] = 1$
41. $[(x - 2)^2/9] + [(y - 2)^2/5] = 1$ **43.** $y^2 - (x^2/8) = 1$
45. $[(x - 4)^2/4] - [(y - 3)^2/5] = 1$
47. $(x^2/9) - (y^2/36) = 1$
49. $(x^2/3{,}763{,}600) + (y^2/3{,}753{,}196) = 1$
51. (a) $(121x^2/1{,}500{,}625) - (121y^2/3{,}339{,}375) = 1$
(b) ≈248 mi
55. (a) Ellipse (b) Hyperbola (c) No curve **57.** 9.69

Exercises 10.6 ▫ page 695

1. $r = 42/(4 + 7\sin\theta)$ **3.** $r = 15/(4 - 3\cos\theta)$
5. $r = 8/(1 - \sin\theta)$ **7.** $r = 4/(2 + \cos\theta)$
9. (a) 1 (b) Parabola (c) $y = 1$
(d)

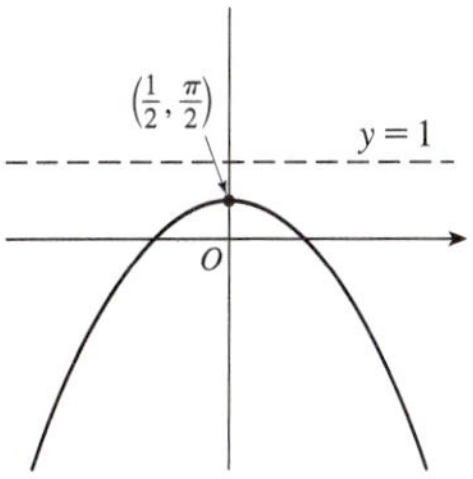

11. (a) $\frac{1}{4}$ (b) Ellipse (c) $y = -12$
(d)

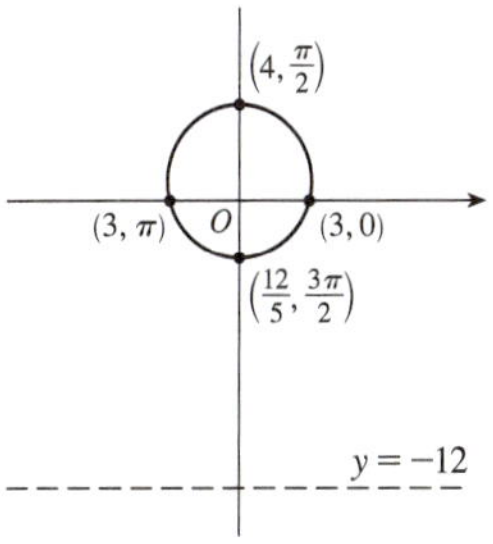

13. (a) $\frac{1}{3}$ (b) Ellipse (c) $x = \frac{9}{2}$
(d)

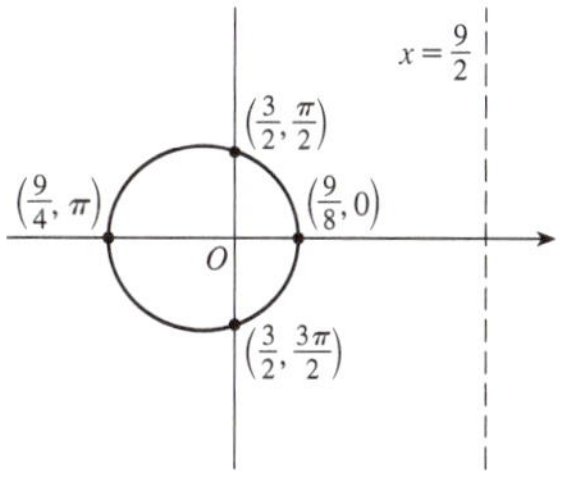

15. (a) 2 (b) Hyperbola (c) $x = -\frac{3}{8}$
(d)

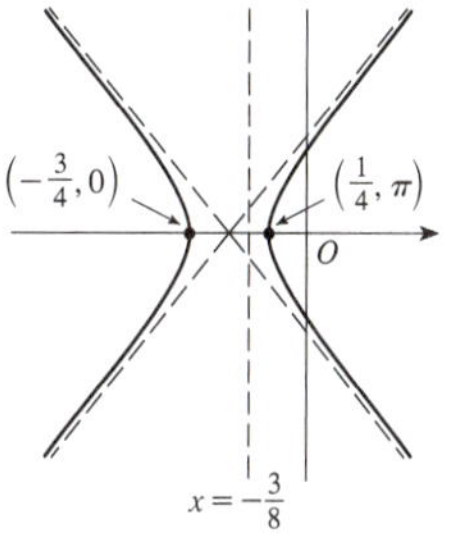

17. (a) $e = \frac{3}{4}$, directrix $x = -\frac{1}{3}$
(b) $r = 1/[4 - 3\cos(\theta - \pi/3)]$

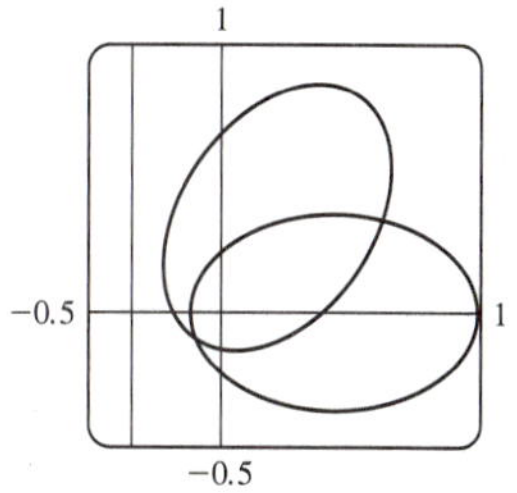

19. The ellipse is nearly circular when e is close to 0 and becomes more elongated as $e \to 1^-$. At $e = 1$, the curve becomes a parabola.

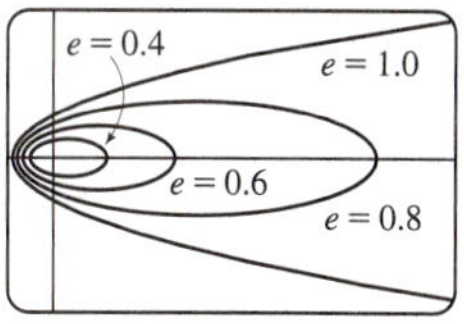

25. (b) $r = (1.49 \times 10^8)/(1 - 0.017\cos\theta)$
27. 35.64 AU
29. 7.0×10^7 km
31. 3.6×10^8 km

Chapter 10 Review ▫ page 697

True-False Quiz

1. False **3.** False **5.** True **7.** False **9.** True

Exercises

1. $x = y^2 - 8y + 12$

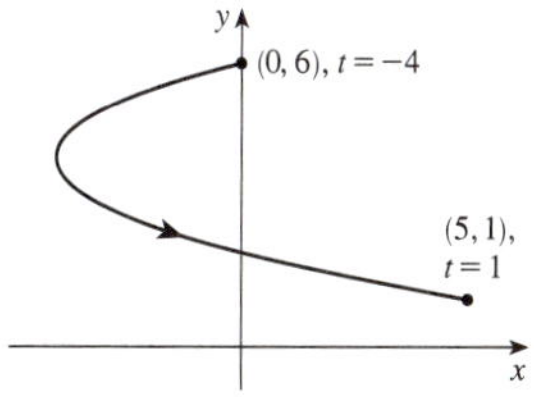

3. $y = 1/x$

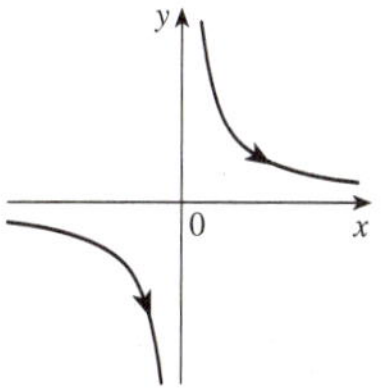

5. $x = t$, $y = \sqrt{t}$, $t \geqslant 0$; $x = t^4$, $y = t^2$; $x = \tan^2 t$, $y = \tan t$, $0 \leqslant t < \frac{\pi}{2}$

7.

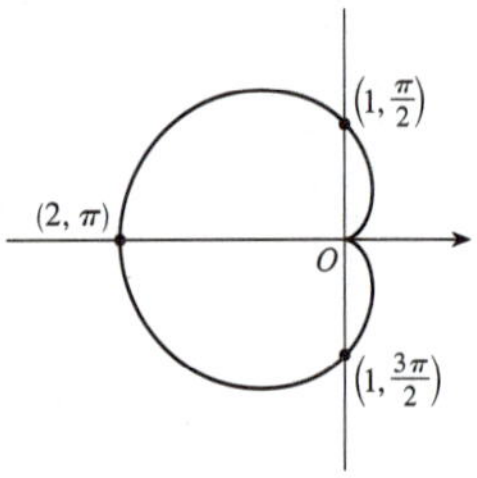

9.

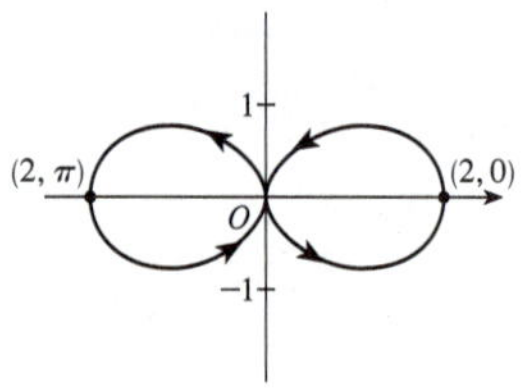

11.

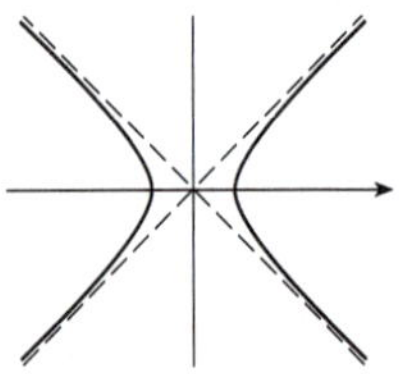

13.

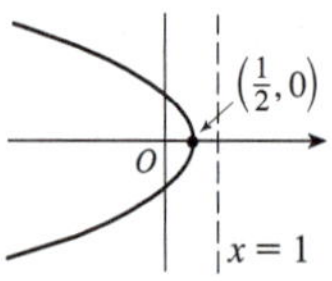

15. $r = 2/(\cos\theta + \sin\theta)$

17.

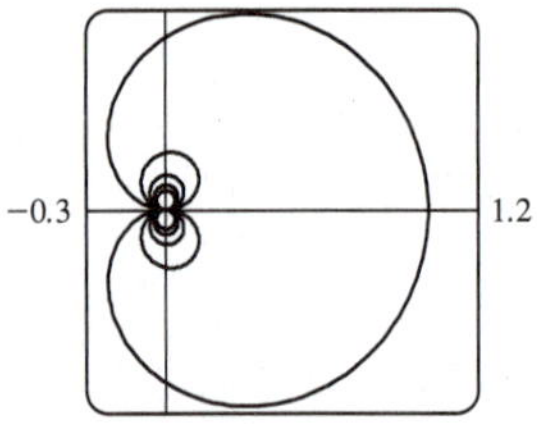

19. 2 **21.** −1

23. $(\sin t + t\cos t)/(\cos t - t\sin t)$, $(t^2 + 2)/(\cos t - t\sin t)^3$

25. $\left(\frac{11}{8}, \frac{3}{4}\right)$

27. Vertical tangent at $\left(3a/2, \pm\sqrt{3}\,a/2\right)$, $(-3a, 0)$; horizontal tangent at $(a, 0)$, $\left(-a/2, \pm 3\sqrt{3}\,a/2\right)$

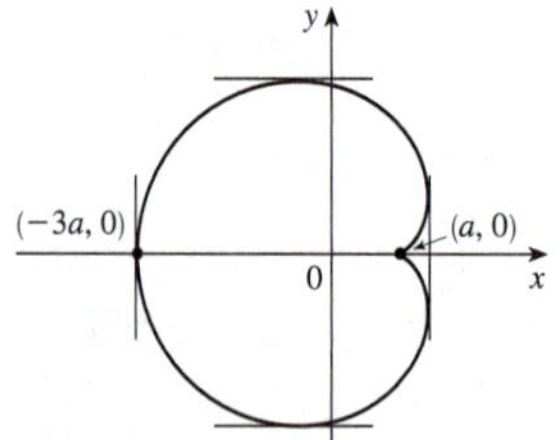

29. 18 **31.** $(2, \pm\pi/3)$ **33.** $(\pi - 1)/2$ **35.** $2\left(5\sqrt{5} - 1\right)$

37. $\dfrac{2\sqrt{\pi^2 + 1} - \sqrt{4\pi^2 + 1}}{2\pi} + \ln\left(\dfrac{2\pi + \sqrt{4\pi^2 + 1}}{\pi + \sqrt{\pi^2 + 1}}\right)$

39. $471{,}295\pi/1024$

41. All curves have the vertical asymptote $x = 1$. For $c < -1$, the curve bulges to the right. At $c = -1$, the curve is the line $x = 1$. For $-1 < c < 0$, it bulges to the left. At $c = 0$ there is a cusp at (0, 0). For $c > 0$, there is a loop.

43. $(\pm 1, 0)$, $(\pm 3, 0)$

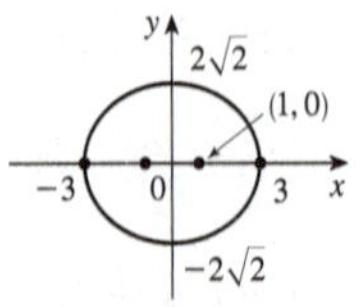

45. $\left(-\frac{25}{24}, 3\right)$, $(-1, 3)$

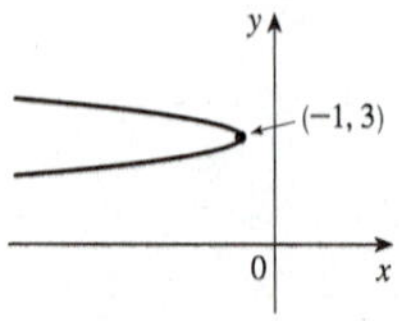

47. $x^2 = 8(y - 4)$

49. $5x^2 - 20y^2 = 36$

51. $(x^2/25) + ((8y - 399)^2/160{,}801) = 1$

53. $r = 4/(3 + \cos\theta)$

55. $x = a(\cot\theta + \sin\theta\cos\theta)$, $y = a(1 + \sin^2\theta)$

Problems Plus ▫ page 699

1. $\ln(\pi/2)$ **3.** $\left[-\frac{3}{4}\sqrt{3}, \frac{3}{4}\sqrt{3}\right] \times [-1, 2]$

5. (a) At (0, 0) and $\left(\frac{3}{2}, \frac{3}{2}\right)$
(b) Horizontal tangents at (0, 0) and $\left(\sqrt[3]{2}, \sqrt[3]{4}\right)$; vertical tangents at (0, 0) and $\left(\sqrt[3]{4}, \sqrt[3]{2}\right)$
(d)

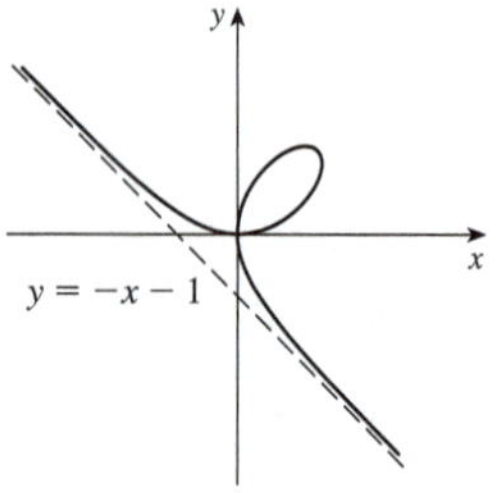

(g) $\frac{3}{2}$

Chapter 11

Exercises 11.1 ▫ page 710

Abbreviations: C, convergent; D, divergent

1. (a) A sequence is an ordered list of numbers. It can also be defined as a function whose domain is the set of positive integers.
(b) The terms a_n approach 8 as n becomes large.
(c) The terms a_n become large as n becomes large.

3. 0.8, 0.96, 0.992, 0.9984, 0.99968 **5.** $-3, \frac{3}{2}, -\frac{1}{2}, \frac{1}{8}, -\frac{1}{40}$

7. 3, 5, 9, 17, 33

9. $a_n = 1/2^n$ **11.** $a_n = 5n - 3$ **13.** $a_n = \left(-\frac{2}{3}\right)^{n-1}$

15. D **17.** 5 **19.** 0 **21.** 0

23. D **25.** 0 **27.** 0 **29.** 0 **31.** 0

33. 1 **35.** 1 **37.** D **39.** D **41.** D

43. $\pi/4$ **45.** 0 **47.** 0

49. (a) 1060, 1123.60, 1191.02, 1262.48, 1338.23 (b) D

51. $-1 < r < 1$

53. Convergent by the Monotonic Sequence Theorem; $5 \leqslant L < 8$

55. Decreasing; yes **57.** Not monotonic; yes

59. Decreasing; yes **61.** 2 **63.** $\left(3 + \sqrt{5}\right)/2$

65. (b) $\left(1 + \sqrt{5}\right)/2$ **67.** (a) 0 (b) 9, 11

Exercises 11.2 ▫ page 720

1. (a) A sequence is an ordered list of numbers whereas a series is the *sum* of a list of numbers.
(b) A series is convergent if the sequence of partial sums is a convergent sequence. A series is divergent if it is not convergent.

3. −2.40000, −1.92000
−2.01600, −1.99680,
−2.00064, −1.99987
−2.00003, −1.99999
−2.00000, −2.00000
Convergent, sum = −2

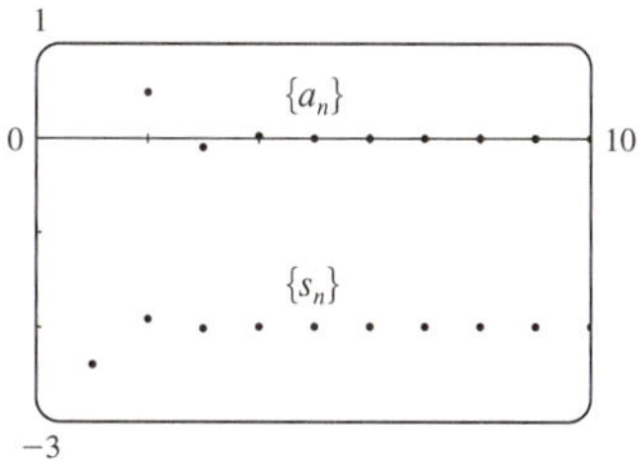

5. 1.55741, −0.62763,
−0.77018, 0.38764
−2.99287, −3.28388
−2.41243, −9.21214
−9.66446, −9.01610
Divergent

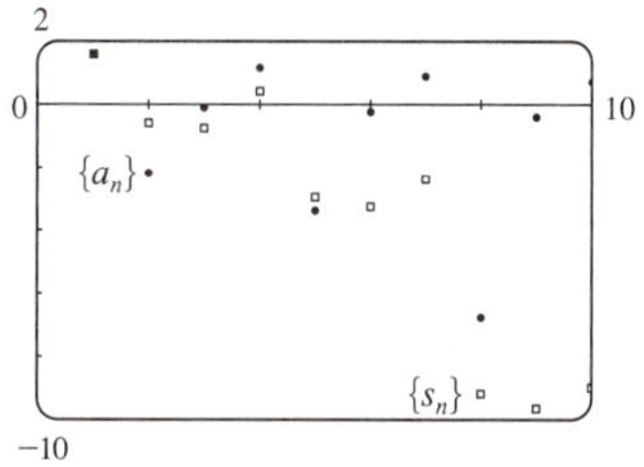

7. 0.64645, 0.80755,
0.87500, 0.91056,
0.93196, 0.94601,
0.95581, 0.96296,
0.96838, 0.97259
Convergent, sum = 1

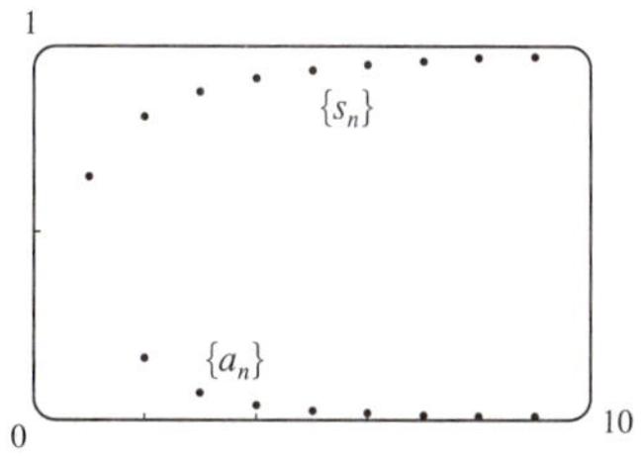

9. (a) C (b) D
11. 9 **13.** D **15.** 15 **17.** $\frac{1}{7}$ **19.** D **21.** D
23. $\frac{3}{2}$ **25.** D **27.** $\frac{3}{2}$ **29.** D **31.** D **33.** $\frac{7}{2}$
35. $\frac{2}{9}$ **37.** $\frac{1138}{333}$ **39.** 41,111/333,000
41. $-3 < x < 3$; $x/(3 - x)$ **43.** $-\frac{1}{4} < x < \frac{1}{4}$; $1/(1 - 4x)$
45. all x; $\dfrac{2}{2 - \cos x}$ **47.** $\frac{1}{4}$
49. $a_1 = 0$, $a_n = 2/[n(n + 1)]$ for $n > 1$, sum = 1
51. (a) $S_n = D(1 - c^n)/(1 - c)$ (b) 5
53. $(\sqrt{3} - 1)/2$ **55.** $1/[n(n + 1)]$
57. The series is divergent.
63. $\{s_n\}$ is bounded and increasing.
65. (a) $0, \frac{1}{9}, \frac{2}{9}, \frac{1}{3}, \frac{2}{3}, \frac{7}{9}, \frac{8}{9}, 1$
67. (a) $\frac{1}{2}, \frac{5}{6}, \frac{23}{24}, \frac{119}{120}$; $[(n + 1)! - 1]/(n + 1)!$ (c) 1

Exercises 11.3 ▫ page 729

1. C

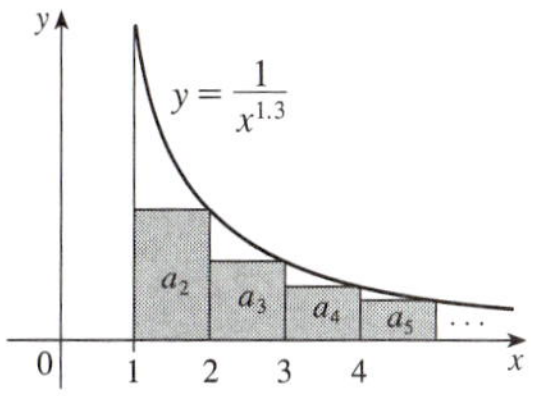

3. C **5.** D **7.** C **9.** D **11.** C **13.** C **15.** C
17. D **19.** C **21.** D **23.** C **25.** $p > 1$
27. $p < -1$ **29.** $(1, \infty)$
31. (a) 1.54977, error ⩽ 0.1 (b) 1.64522, error ⩽ 0.005
(c) $n > 1000$
33. 2.61 **39.** $b < 1/e$

Exercises 11.4 ▫ page 734

1. (a) Nothing (b) C
3. C **5.** C **7.** D
9. C **11.** D **13.** C **15.** C **17.** D **19.** C
21. D **23.** C **25.** D **27.** C **29.** C **31.** D
33. 0.567975, error ⩽ 0.0003 **35.** 0.76352, error < 0.001
45. Yes

Exercises 11.5 ▫ page 739

1. (a) A series whose terms are alternately positive and negative
(b) $0 < b_{n+1} \leqslant b_n$ and $\lim_{n\to\infty} b_n = 0$, where $b_n = |a_n|$
(c) $|R_n| \leqslant b_{n+1}$
3. C **5.** C **7.** D **9.** C **11.** C **13.** D
15. C **17.** C **19.** D
21. 1.0000, 0.6464,
0.8389, 0.7139, 0.8033,
0.7353, 0.7893, 0.7451, 0.7821,
0.7505; error < 0.0275

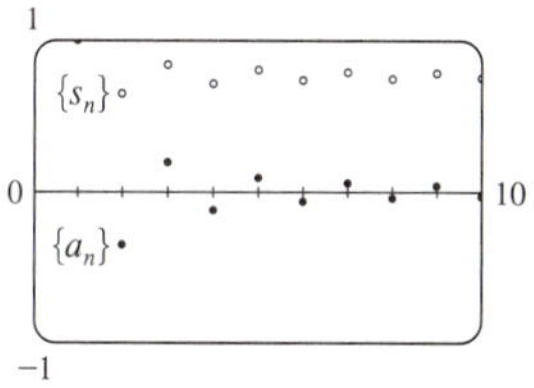

23. 10 **25.** 7 **27.** 0.9721 **29.** 0.0676
31. An underestimate **33.** p is not a negative integer
35. $\{b_n\}$ is not decreasing

Exercises 11.6 ▫ page 745

Abbreviations: AC, absolutely convergent;
CC, conditionally convergent
1. (a) D (b) C (c) May converge or diverge
3. AC **5.** CC **7.** D **9.** AC **11.** AC **13.** AC
15. AC **17.** CC **19.** AC **21.** D **23.** AC
25. AC **27.** D **29.** D **31.** (a) and (d)
35. (a) $\frac{661}{960} \approx 0.68854$, error < 0.00521 (b) $n \geqslant 11$, 0.693109

Exercises 11.7 ▫ page 748

1. D **3.** C **5.** C **7.** D **9.** C **11.** C **13.** C
15. C **17.** D **19.** C **21.** C **23.** D **25.** C
27. C **29.** C **31.** D **33.** C **35.** C **37.** C

Exercises 11.8 ▫ page 753

1. A series of the form $\sum_{n=0}^{\infty} c_n(x - a)^n$, where x is a variable and a and the c_n's are constants
3. 1, $[-1, 1)$ **5.** 1, $[-1, 1]$ **7.** ∞, $(-\infty, \infty)$ **9.** $\frac{1}{4}$, $(-\frac{1}{4}, \frac{1}{4})$
11. $\frac{1}{2}$, $(-\frac{1}{2}, \frac{1}{2}]$ **13.** 4, $(-4, 4]$ **15.** 1, $(0, 2)$
17. 2, $(-4, 0]$ **19.** ∞, $(-\infty, \infty)$ **21.** b, $(a - b, a + b)$

23. $0, \{\frac{1}{2}\}$ **25.** $\frac{1}{4}, [-\frac{1}{2}, 0]$ **27.** $\infty, (-\infty, \infty)$

29. (a) Yes (b) No **31.** k^k

33. (a) $(-\infty, \infty)$

(b), (c)

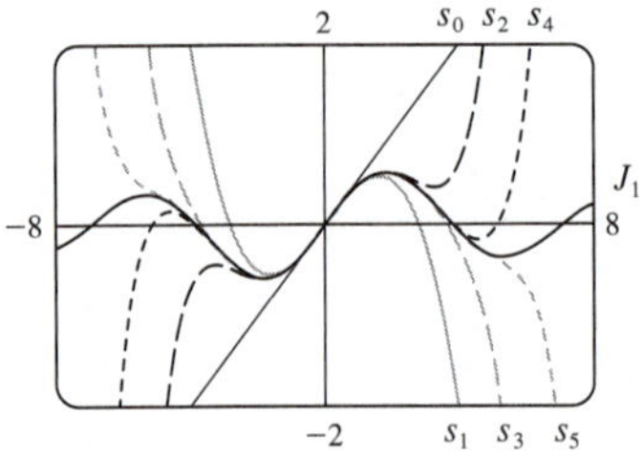

35. $(-1, 1)$, $f(x) = (1 + 2x)/(1 - x^2)$ **39.** 2

Exercises 11.9 ▫ page 759

1. 10 **3.** $\sum_{n=0}^{\infty} (-1)^n x^n, (-1, 1)$ **5.** $\sum_{n=0}^{\infty} x^{3n}, (-1, 1)$

7. $-\sum_{n=0}^{\infty} \frac{1}{5^{n+1}} x^n, (-5, 5)$ **9.** $\sum_{n=0}^{\infty} (-1)^n \frac{1}{9^{n+1}} x^{2n+1}, (-3, 3)$

11. $\sum_{n=0}^{\infty} \left[\frac{(-1)^{n+1}}{2^{n+1}} - 1\right] x^n, (-1, 1)$

13. (a) $\sum_{n=0}^{\infty} (-1)^n (n + 1) x^n, R = 1$

(b) $\frac{1}{2} \sum_{n=0}^{\infty} (-1)^n (n + 2)(n + 1) x^n, R = 1$

(c) $\frac{1}{2} \sum_{n=2}^{\infty} (-1)^n n(n - 1) x^n, R = 1$

15. $\ln 5 - \sum_{n=1}^{\infty} \frac{x^n}{n5^n}, R = 5$

17. $\sum_{n=3}^{\infty} \frac{n - 2}{2^{n-1}} x^n, R = 2$

19. $\ln 3 + \sum_{n=1}^{\infty} \frac{(-1)^{n-1}}{n3^n} x^n, R = 3$

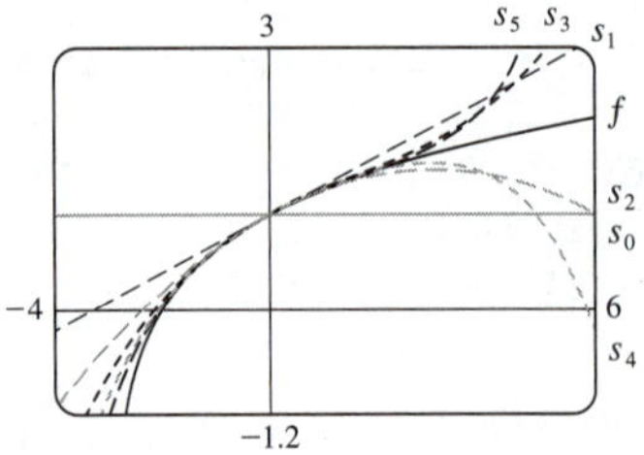

21. $\sum_{n=0}^{\infty} \frac{2x^{2n+1}}{2n + 1}, R = 1$

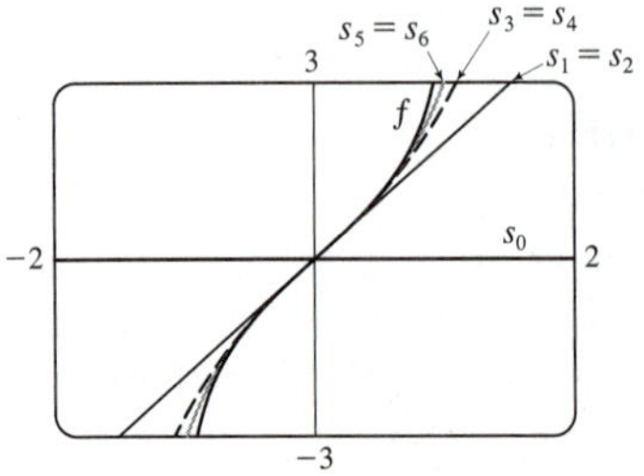

23. $C + \sum_{n=0}^{\infty} \frac{t^{8n+2}}{8n + 2}; 1$

25. $C + \sum_{n=1}^{\infty} (-1)^{n+1} \frac{x^{2n-1}}{4n^2 - 1}; 1$

27. 0.199989 **29.** 0.000065 **31.** 0.09531

33. (b) 0.920 **37.** $[-1, 1], [-1, 1), (-1, 1)$

Exercises 11.10 ▫ page 770

1. $b_8 = f^{(8)}(5)/8!$ **3.** $\sum_{n=0}^{\infty} (-1)^n \frac{x^{2n}}{(2n)!}, R = \infty$

5. $\sum_{n=0}^{\infty} (-1)^n \frac{(n + 1)(n + 2)}{2} x^n, R = 1$

7. $\sum_{n=0}^{\infty} \frac{5^n}{n!} x^n, R = \infty$

9. $\sum_{n=0}^{\infty} \frac{x^{2n+1}}{(2n + 1)!}, R = \infty$

11. $7 + 5(x - 2) + (x - 2)^2, R = \infty$

13. $\sum_{n=0}^{\infty} \frac{e^3}{n!} (x - 3)^n, R = \infty$

15. $\sum_{n=0}^{\infty} (-1)^{n+1} \frac{1}{(2n)!} (x - \pi)^{2n}, R = \infty$

17. $\sum_{n=0}^{\infty} (-1)^n \frac{1 \cdot 3 \cdot 5 \cdot \cdots \cdot (2n - 1)}{2^n \cdot 3^{2n+1} \cdot n!} (x - 9)^n, R = 9$

23. $\sum_{n=0}^{\infty} (-1)^n \frac{\pi^{2n}}{(2n)!} x^{2n}, R = \infty$

25. $\sum_{n=0}^{\infty} (-1)^n \frac{1}{2n + 1} x^{2n+2}, R = 1$

27. $\sum_{n=0}^{\infty} (-1)^n \frac{1}{n!} x^{n+2}, R = \infty$

29. $\sum_{n=1}^{\infty} \frac{(-1)^{n+1} 2^{2n-1} x^{2n}}{(2n)!}, R = \infty$

31. $\sum_{n=0}^{\infty} \frac{(-1)^n x^{2n}}{(2n + 1)!}, R = \infty$

33. $1 + \frac{x}{2} + \sum_{n=2}^{\infty} (-1)^{n-1} \frac{1 \cdot 3 \cdot 5 \cdot \cdots \cdot (2n - 3)}{2^n n!} x^n, R = 1$

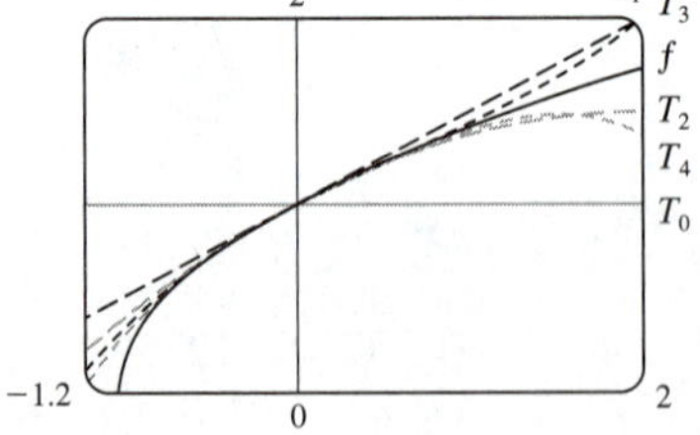

35. $\sum_{n=0}^{\infty} (-1)^n \frac{1}{(2n)!} x^{4n}$, $R = \infty$

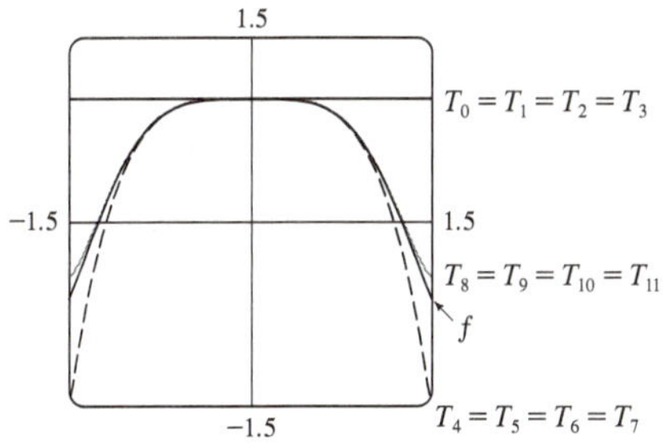

37. 0.81873

39. $C + \sum_{n=0}^{\infty} (-1)^n \frac{x^{6n+2}}{(6n+2)(2n)!}$

41. $C + x + \frac{x^4}{8} + \sum_{n=2}^{\infty} (-1)^{n-1} \frac{1 \cdot 3 \cdot 5 \cdot \cdots \cdot (2n-3)}{2^n n!(3n+1)} x^{3n+1}$

43. 0.440 **45.** 0.09998750 **47.** $\frac{1}{3}$ **49.** $\frac{1}{120}$

51. $1 - \frac{3}{2}x^2 + \frac{25}{24}x^4$ **53.** $1 + \frac{1}{6}x^2 + \frac{7}{360}x^4$

55. e^{-x^4} **57.** $1/\sqrt{2}$ **59.** $e^3 - 1$

Exercises 11.11 ▫ page 775

1. $1 + \frac{x}{2} + \sum_{n=2}^{\infty} (-1)^{n-1} \frac{1 \cdot 3 \cdot 5 \cdot \cdots \cdot (2n-3)}{2^n n!} x^n$, $R = 1$

3. $\sum_{n=0}^{\infty} (-1)^n \frac{(n+1)(n+2)}{2^{n+4}} x^n$, $R = 2$

5. $1 - 2x - \sum_{n=2}^{\infty} \frac{3 \cdot 7 \cdot \cdots \cdot (4n-5) \cdot 2^n}{n!} x^n$, $R = \frac{1}{8}$

7. $\frac{1}{2}x + \sum_{n=1}^{\infty} (-1)^n \frac{1 \cdot 3 \cdot 5 \cdot \cdots \cdot (2n-1)}{n!\, 2^{3n+1}} x^{2n+1}$, $R = 2$

9. $1 + \frac{3}{2}x$, $1 + \frac{3}{2}x - \frac{3}{8}x^2$, $1 + \frac{3}{2}x - \frac{3}{8}x^2 + \frac{5}{16}x^3$

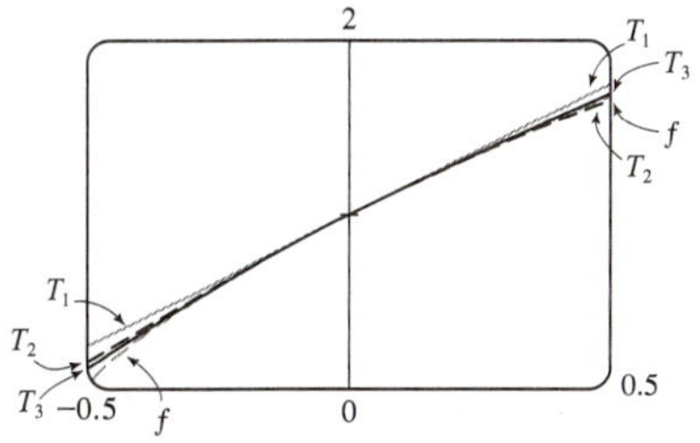

11. (a) $1 + \sum_{n=1}^{\infty} \frac{1 \cdot 3 \cdot 5 \cdot \cdots \cdot (2n-1)}{2^n n!} x^{2n}$

(b) $x + \sum_{n=1}^{\infty} \frac{1 \cdot 3 \cdot 5 \cdot \cdots \cdot (2n-1)}{(2n+1)2^n n!} x^{2n+1}$

13. (a) $1 + \frac{x}{3} + \sum_{n=2}^{\infty} (-1)^{n+1} \frac{2 \cdot 5 \cdot 8 \cdot \cdots \cdot (3n-4)}{3^n n!} x^n$

(b) 1.0033

15. (a) $\sum_{n=1}^{\infty} nx^n$ (b) 2

17. (a) $1 + \frac{x^2}{2} + \sum_{n=2}^{\infty} (-1)^{n-1} \frac{1 \cdot 3 \cdot 5 \cdot \cdots \cdot (2n-3)}{2^n n!} x^{2n}$

(b) 99,225

Exercises 11.12 ▫ page 783

1. (a) $T_0(x) = 1 = T_1(x)$, $T_2(x) = 1 - \frac{1}{2}x^2 = T_3(x)$, $T_4(x) = 1 - \frac{1}{2}x^2 + \frac{1}{24}x^4 = T_5(x)$, $T_6(x) = 1 - \frac{1}{2}x^2 + \frac{1}{24}x^4 - \frac{1}{720}x^6$

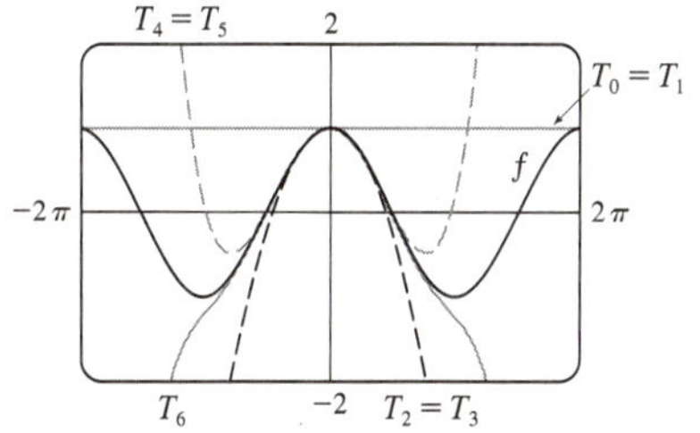

(b)

x	f	$T_0 = T_1$	$T_2 = T_3$	$T_4 = T_5$	T_6
$\frac{\pi}{4}$	0.7071	1	0.6916	0.7074	0.7071
$\frac{\pi}{2}$	0	1	−0.2337	0.0200	−0.0009
π	−1	1	−3.9348	0.1239	−1.2114

(c) As n increases, $T_n(x)$ is a good approximation to $f(x)$ on a larger and larger interval.

3. $(x-1) - \frac{1}{2}(x-1)^2 + \frac{1}{3}(x-1)^3 - \frac{1}{4}(x-1)^4$

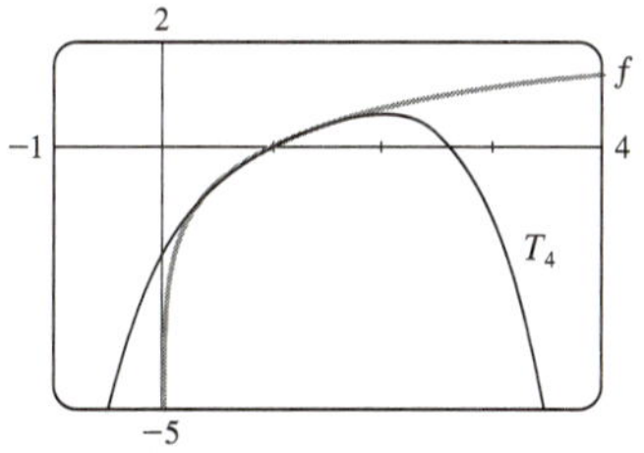

5. $\frac{1}{2} + \frac{\sqrt{3}}{2}\left(x - \frac{\pi}{6}\right) - \frac{1}{4}\left(x - \frac{\pi}{6}\right)^2 - \frac{\sqrt{3}}{12}\left(x - \frac{\pi}{6}\right)^3$

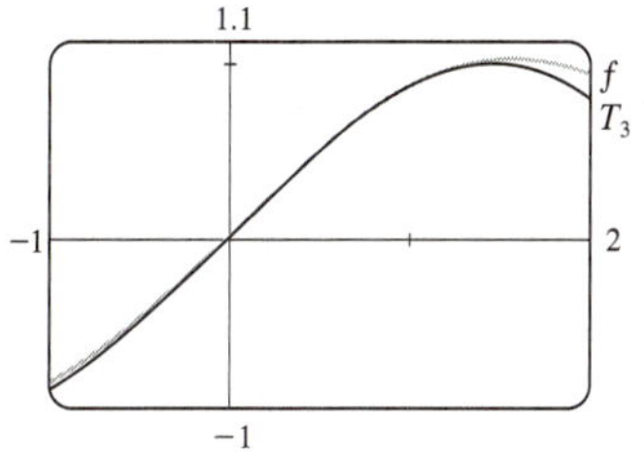

7. $x + \frac{1}{6}x^3$

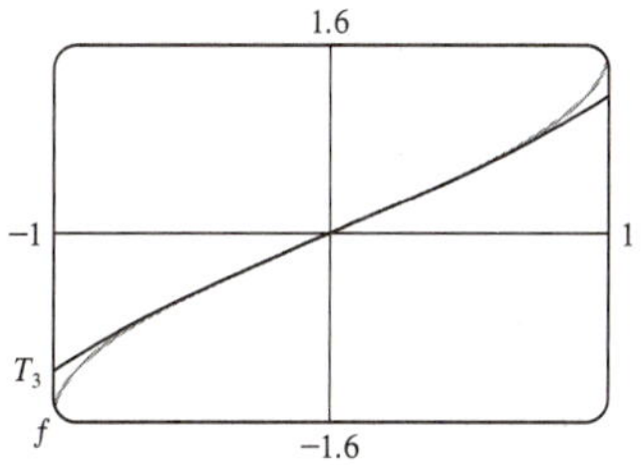

9. $x - 2x^2 + 2x^3$

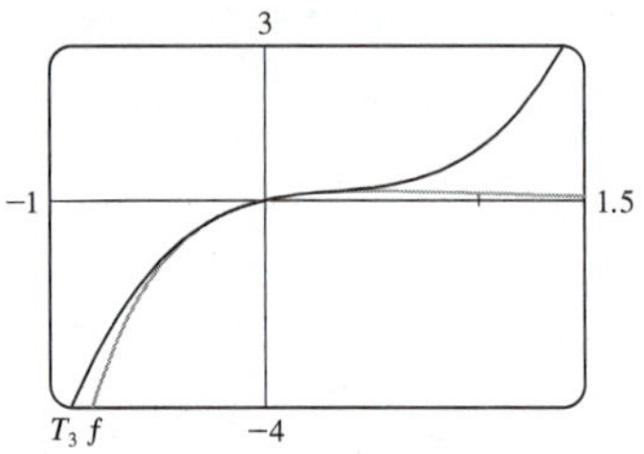

11. $T_8(x) = 1 + \frac{1}{2}x^2 + \frac{5}{24}x^4 + \frac{61}{720}x^6 + \frac{277}{8064}x^8$

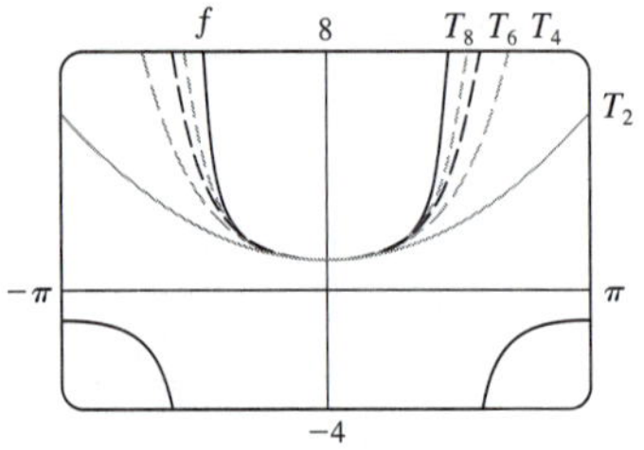

13. (a) $2 + \frac{1}{4}(x - 4) - \frac{1}{64}(x - 4)^2$ (b) 1.5625×10^{-5}
15. (a) $1 + \frac{2}{3}(x - 1) - \frac{1}{9}(x - 1)^2 + \frac{4}{81}(x - 1)^3$ (b) 0.000097
17. (a) $x + \frac{1}{3}x^3$ (b) 0.058
19. (a) $1 + x^2$ (b) 0.00006
21. (a) $x^2 - \frac{1}{6}x^4$ (b) 0.042
23. 0.57358 **25.** Four
27. $-1.037 < x < 1.037$ **29.** 21 m, no
35. (c) They differ by about 8×10^{-9} km.

Chapter 11 Review ▫ page 786

True-False Quiz

1. False **3.** True **5.** False **7.** False **9.** False
11. True **13.** True **15.** False **17.** True

Exercises

1. $\frac{1}{2}$ **3.** D **5.** 0 **7.** e^{12} **9.** 2 **11.** C **13.** C
15. D **17.** C **19.** C **21.** C **23.** CC **25.** AC
27. 8 **29.** $\pi/4$ **31.** e^{-e} **35.** 0.9721
37. 0.18976224, $|\text{error}| < 6.4 \times 10^{-7}$ **41.** 4, $[-6, 2)$
43. 0.5, [2.5, 3.5)

45. $\frac{1}{2}\sum_{n=0}^{\infty}(-1)^n\left[\frac{1}{(2n)!}\left(x - \frac{\pi}{6}\right)^{2n} + \frac{\sqrt{3}}{(2n+1)!}\left(x - \frac{\pi}{6}\right)^{2n+1}\right]$

47. $\sum_{n=0}^{\infty}(-1)^n x^{n+2}$, $R = 1$

49. $-\sum_{n=1}^{\infty}\frac{x^n}{n}$, $R = 1$

51. $\sum_{n=0}^{\infty}(-1)^n\frac{x^{8n+4}}{(2n+1)!}$, $R = \infty$

53. $\frac{1}{2} + \sum_{n=1}^{\infty}\frac{1 \cdot 5 \cdot 9 \cdot \cdots \cdot (4n-3)}{n!\,2^{6n+1}}x^n$, $R = 16$

55. $C + \ln|x| + \sum_{n=1}^{\infty}\frac{x^n}{n \cdot n!}$

57. (a) $1 + \frac{1}{2}(x - 1) - \frac{1}{8}(x - 1)^2 + \frac{1}{16}(x - 1)^3$
(b)

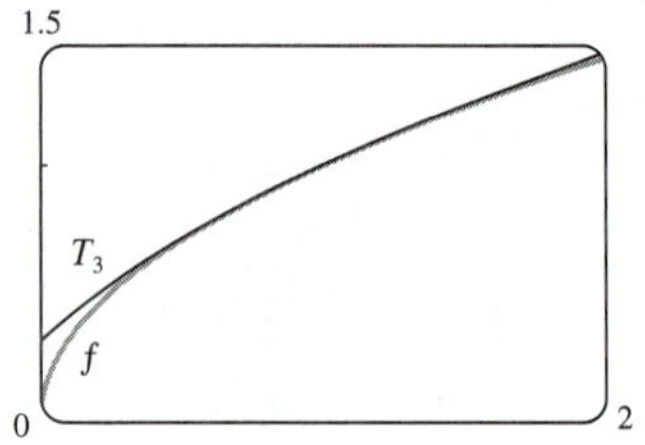

(c) 0.000006

59. $-\frac{1}{6}$

Problems Plus ▫ page 789

1. $15!/5! = 10,897,286,400$
3. (b) 0 if $x = 0$, $(1/x) - \cot x$ if $x \neq k\pi$, k an integer
5. (a) $s_n = 3 \cdot 4^n$, $l_n = 1/3^n$, $p_n = 4^n/3^{n-1}$ (c) $2\sqrt{3}/5$
9. $(-1, 1)$, $(x^3 + 4x^2 + x)/(1 - x)^4$ **11.** $\ln\frac{1}{2}$

Chapter 12

Exercises 12.1 ▫ page 797

1. $(4, 0, -3)$ **3.** Q; R
5. A vertical plane that intersects the xy-plane in the line $y = 2 - x$, $z = 0$ (see graph at right)

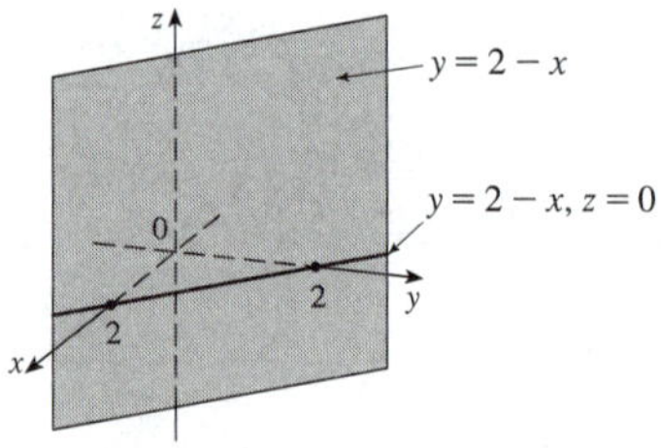

9. (a) Yes (b) No
11. $(x - 1)^2 + (y + 4)^4 + (z - 3)^2 = 25$; $(x - 1)^2 + (z - 3)^2 = 9$, $y = 0$ (a circle)
13. $(x - 3)^2 + (y - 8)^2 + (z - 1)^2 = 30$
15. $(3, -2, 1)$, 5 **17.** $\left(\frac{1}{2}, \frac{1}{2}, \frac{1}{2}\right)$, $\sqrt{3}/2$ **19.** (b) $\frac{5}{2}, \frac{1}{2}\sqrt{94}, \frac{1}{2}\sqrt{85}$
21. (a) $(x - 2)^2 + (y + 3)^2 + (z - 6)^2 = 36$
(b) $(x - 2)^2 + (y + 3)^2 + (z - 6)^2 = 4$
(c) $(x - 2)^2 + (y + 3)^2 + (z - 6)^2 = 9$
23. A plane parallel to the xz-plane and 4 units to the left of it
25. A half-space consisting of all points in front of the plane $x = 3$
27. All points on or between the horizontal planes $z = 0$ and $z = 6$
29. All points outside the sphere with radius 1 and center O
31. All points inside the sphere with radius 2 and center (0, 0, 1)
33. All points on or inside a circular cylinder of radius 3 with axis the y-axis
35. $y < 0$ **37.** $r^2 < x^2 + y^2 + z^2 < R^2$
39. (a) (2, 1, 4) (b)

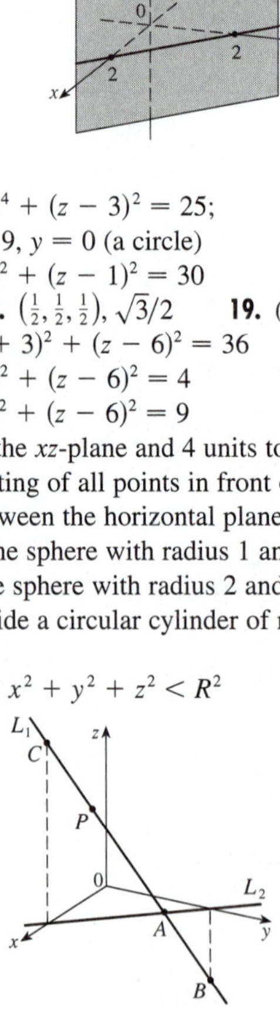

41. $14x - 6y - 10z = 9$, a plane perpendicular to AB

Exercises 12.2 ▫ page 805

1. (a) Scalar (b) Vector (c) Vector (d) Scalar

3. $\overrightarrow{AB} = \overrightarrow{DC}$, $\overrightarrow{DA} = \overrightarrow{CB}$, $\overrightarrow{DE} = \overrightarrow{EB}$, $\overrightarrow{EA} = \overrightarrow{CE}$

5. (a)

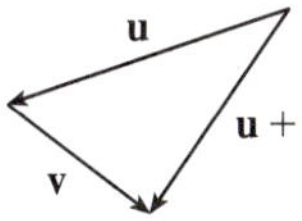

(b)

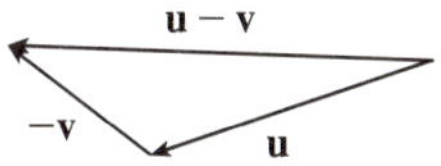

(c)

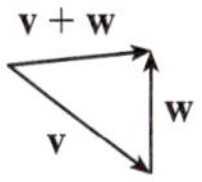

(d)

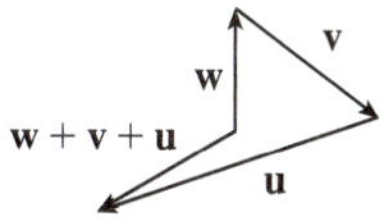

7. $\mathbf{a} = \langle -4, -2 \rangle$

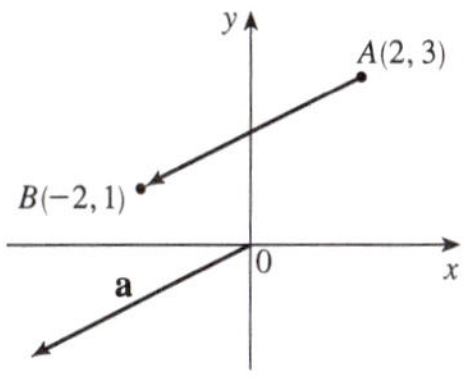

9. $\mathbf{a} = \langle -2, 5 \rangle$

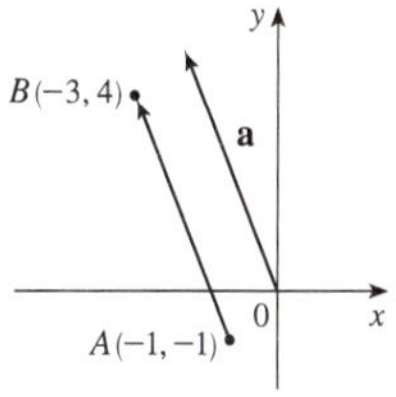

11. $\mathbf{a} = \langle 2, 0, -2 \rangle$

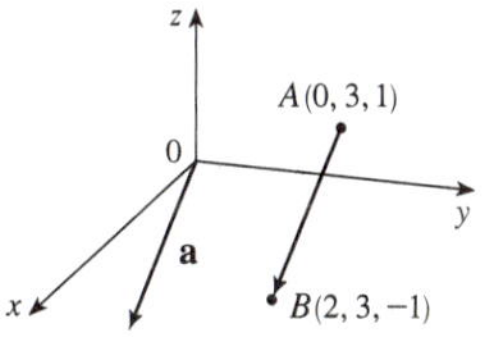

13. $\langle 1, 3 \rangle$

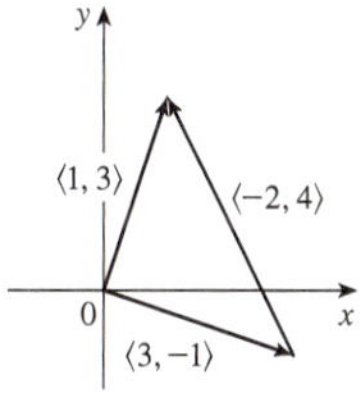

15. $\langle 0, 1, -1 \rangle$

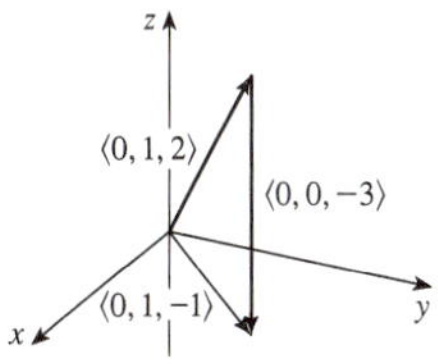

17. $5, \langle 2, 5 \rangle, \langle -10, 1 \rangle, \langle -8, 6 \rangle, \langle 12, 17 \rangle$

19. $7, \langle 5, 7, 1 \rangle, \langle 7, -3, 5 \rangle, \langle 12, 4, 6 \rangle, \langle 14, 26, 1 \rangle$

21. $\sqrt{6}, \mathbf{i} - \mathbf{j} + 3\mathbf{k}, \mathbf{i} - 3\mathbf{j} - \mathbf{k}, 2\mathbf{i} - 4\mathbf{j} + 2\mathbf{k}, 3\mathbf{i} - 2\mathbf{j} + 11\mathbf{k}$

23. $\langle 9/\sqrt{106}, -5/\sqrt{106} \rangle$ **25.** $\frac{8}{9}\mathbf{i} - \frac{1}{9}\mathbf{j} + \frac{4}{9}\mathbf{k}$ **27.** $\langle 2, 2\sqrt{3} \rangle$

29. $\mathbf{F} = (6\sqrt{3} - 5\sqrt{2})\mathbf{i} + (6 + 5\sqrt{2})\mathbf{j} \approx 3.32\mathbf{i} + 13.07\mathbf{j}$, $|\mathbf{F}| \approx 13.5$ lb, $\theta \approx 76°$

31. $\sqrt{493} \approx 22.2$ mi/h, N8°W

33. $\mathbf{T}_1 \approx -196\,\mathbf{i} + 3.92\,\mathbf{j}$, $\mathbf{T}_2 \approx 196\,\mathbf{i} + 3.92\,\mathbf{j}$ **35.** $\mathbf{0}$

37. (a), (b)

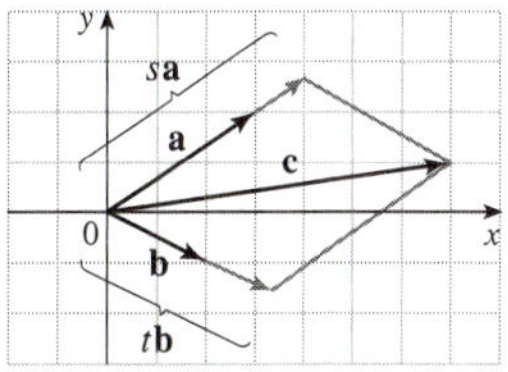

(d) $s = \frac{9}{7}, t = \frac{11}{7}$

39. A sphere with radius 1, centered at (x_0, y_0, z_0)

Exercises 12.3 ▫ page 812

1. (b), (c), (d) are meaningful **3.** 6 **5.** −5

7. 32 **9.** $90\sqrt{3}$ **11.** $\mathbf{u} \cdot \mathbf{v} = \frac{1}{2}, \mathbf{u} \cdot \mathbf{w} = -\frac{1}{2}$

15. $\cos^{-1}\left(\frac{63}{65}\right) \approx 14°$ **17.** $\cos^{-1}\left(1/\sqrt{238}\right) \approx 86°$

19. $\cos^{-1}\left(-1/\left(2\sqrt{7}\right)\right) \approx 101°$ **21.** 45°, 45°, 90°

23. (a) Neither (b) Orthogonal (c) Orthogonal (d) Parallel

25. Yes **27.** $(\mathbf{i} - \mathbf{j} - \mathbf{k})/\sqrt{3}$ [or $(-\mathbf{i} + \mathbf{j} + \mathbf{k})/\sqrt{3}$]

29. $\dfrac{3}{5\sqrt{2}}, \dfrac{4}{5\sqrt{2}}, \dfrac{1}{\sqrt{2}}$; 65°, 56°, 45°

31. $\frac{2}{7}, \frac{3}{7}, -\frac{6}{7}$; 73°, 65°, 149°

33. $1/\sqrt{3}, 1/\sqrt{3}, 1/\sqrt{3}$; 55°, 55°, 55° **35.** $3, \left\langle \frac{9}{5}, -\frac{12}{5} \right\rangle$

37. $3/\sqrt{5}, \left\langle \frac{6}{5}, \frac{3}{5}, 0 \right\rangle$ **39.** $1/\sqrt{2}, (\mathbf{i} + \mathbf{k})/2$

43. $\langle 0, 0, -2\sqrt{10} \rangle$ or any vector of the form $\langle s, t, 3s - 2\sqrt{10} \rangle, s, t \in \mathbb{R}$

45. 38 J **47.** $250 \cos 20° \approx 235$ ft-lb **49.** $\frac{13}{5}$

51. $\cos^{-1}\left(1/\sqrt{3}\right) \approx 55°$

Exercises 12.4 ▫ page 820

1. $2\mathbf{i} - \mathbf{j} + 3\mathbf{k}$ **3.** $3\mathbf{i} - 4\mathbf{j} + 2\mathbf{k}$

5. $2\mathbf{i} + 13\mathbf{j} - 8\mathbf{k}$ **7.** $t^4\mathbf{i} - 2t^3\mathbf{j} + t^2\mathbf{k}$

9. (a) Scalar (b) Meaningless (c) Vector (d) Meaningless (e) Meaningless (f) Scalar

11. 24; into the page

13. $\langle 5, -3, 1 \rangle, \langle -5, 3, -1 \rangle$

15. $\langle -2/\sqrt{6}, -1/\sqrt{6}, 1/\sqrt{6} \rangle, \langle 2/\sqrt{6}, 1/\sqrt{6}, -1/\sqrt{6} \rangle$

23. 16 **25.** (a) $\langle 6, 3, 2 \rangle$ (b) $\frac{7}{2}$

27. (a) $\langle 13, -14, 5 \rangle$ (b) $\frac{1}{2}\sqrt{390}$

29. 82 **31.** 3 **35.** $10.8 \sin 80° \approx 10.6$ N·m

37. ≈417 N **39.** (b) $\sqrt{97/3}$

45. (a) No (b) No (c) Yes

Exercises 12.5 ▫ page 829

1. (a) True (b) False (c) True (d) False (e) False (f) True (g) False (h) True (i) True (j) False (k) True

3. $\mathbf{r} = (-2\mathbf{i} + 4\mathbf{j} + 10\mathbf{k}) + t(3\mathbf{i} + \mathbf{j} - 8\mathbf{k})$; $x = -2 + 3t, y = 4 + t, z = 10 - 8t$

5. $\mathbf{r} = (\mathbf{i} + 6\mathbf{k}) + t(\mathbf{i} + 3\mathbf{j} + \mathbf{k})$; $x = 1 + t, y = 3t, z = 6 + t$

7. $x = 1 - 5t, y = 3, z = 2 - 2t$; $\dfrac{x - 1}{-5} = \dfrac{z - 2}{-2}, y = 3$

9. $x = 2 + 2t, y = 1 + \frac{1}{2}t, z = -3 - 4t$; $(x - 2)/2 = 2y - 2 = (z + 3)/(-4)$

11. $x = 1 + t, y = -1 + 2t, z = 1 + t$; $x - 1 = (y + 1)/2 = z - 1$

13. Yes

15. (a) $x/2 = (y - 2)/3 = (z + 1)/(-7)$
(b) $\left(-\frac{2}{7}, \frac{11}{7}, 0\right), \left(-\frac{4}{3}, 0, \frac{11}{3}\right), (0, 2, -1)$

17. $\mathbf{r}(t) = (2\mathbf{i} - \mathbf{j} + 4\mathbf{k}) + t(2\mathbf{i} + 7\mathbf{j} - 3\mathbf{k}), 0 \leqslant t \leqslant 1$

19. Parallel **21.** Skew

23. $-2x + y + 5z = 1$

25. $x + y - z = -1$

27. $2x - y + 3z = 0$ **29.** $3x - 7z = -9$
31. $x + y + z = 2$ **33.** $-13x + 17y + 7z = -42$
35. $33x + 10y + 4z = 190$ **37.** $x - 2y + 4z = -1$
39. $(2, 3, 5)$ **41.** $(2, 3, 1)$ **43.** $1, 0, -1$
45. Perpendicular **47.** Neither, $\approx 70.5°$ **49.** Parallel
51. (a) $x - 2 = y/(-8) = z/(-7)$
(b) $\cos^{-1}(-\sqrt{6}/5) \approx 119°$ (or $61°$)
53. $x = 6t, y = -\frac{1}{6} + t, z = -\frac{1}{6} + 7t$ **55.** $x = z$
57. $(x/a) + (y/b) + (z/c) = 1$
59. $x = 3t, y = 1 - t, z = 2 - 2t$
61. P_1 and P_3 are parallel, P_2 and P_4 are identical
63. $\sqrt{22/5}$ **65.** $\frac{25}{3}$ **67.** $7\sqrt{6}/18$ **71.** $1/\sqrt{6}$

Exercises 12.6 ▫ page 837

1. (a) Parabola
(b) Parabolic cylinder with rulings parallel to the z-axis
(c) Parabolic cylinder with rulings parallel to the x-axis
3. Elliptic cylinder **5.** Parabolic cylinder

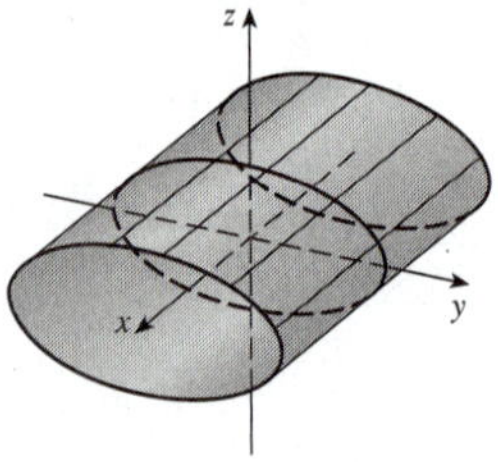

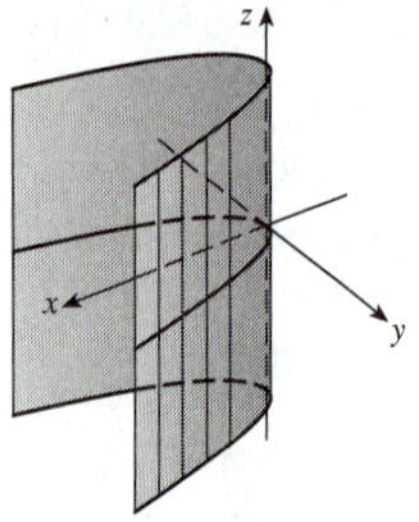

7. Cylindrical surface

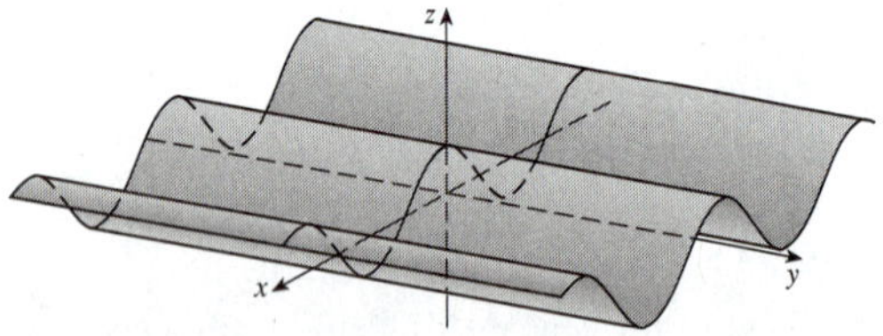

9. (a) $x = k, y^2 - z^2 = 1 - k^2$, hyperbola $(k \neq \pm 1)$;
$y = k, x^2 - z^2 = 1 - k^2$, hyperbola $(k \neq \pm 1)$;
$z = k, x^2 + z^2 = 1 + k^2$, circle
(b) The hyperboloid is rotated so that it has axis the y-axis
(c) The hyperboloid is shifted one unit in the negative y-direction
11. $x = k, y^2 + 4z^2 = 4 - 4k^2/9$, ellipse $(|k| < 3)$
$y = k, x^2 + 9z^2 = 9 - 9k^2/4$, ellipse $(|k| < 2)$
$z = k, 4x^2 + 9y^2 = 36(1 - k^2)$, ellipse $(|k| < 1)$

Ellipsoid

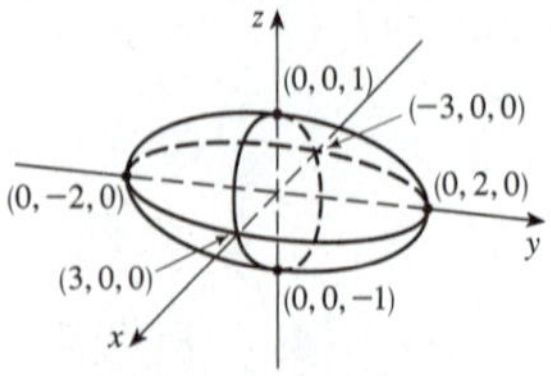

13. $x = k, y^2 - z^2 = k^2$, hyperbola $(k \neq 0)$
$y = k, x^2 + z^2 = k^2$, circle $(k \neq 0)$
$z = k, y^2 - x^2 = k^2$,
hyperbola $(k \neq 0)$
$x = 0, y = \pm z$, lines
$z = 0, y = \pm x$, lines

Cone

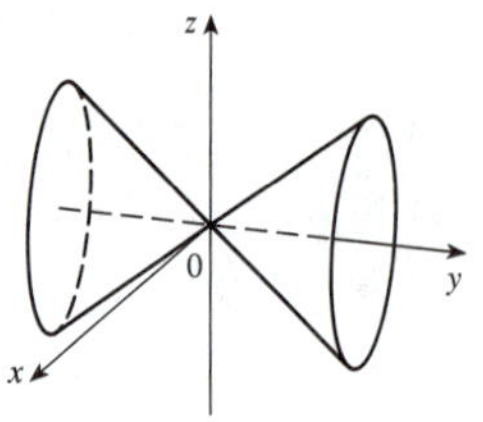

15. $x = k, 4y^2 - z^2 = 4 + k^2$, hyperbola
$y = k, x^2 + z^2 = 4k^2 - 4$,
circle $(|k| > 1)$
$z = k, 4y^2 - x^2 = 4 + k^2$,
hyperbola

Hyperboloid of two sheets

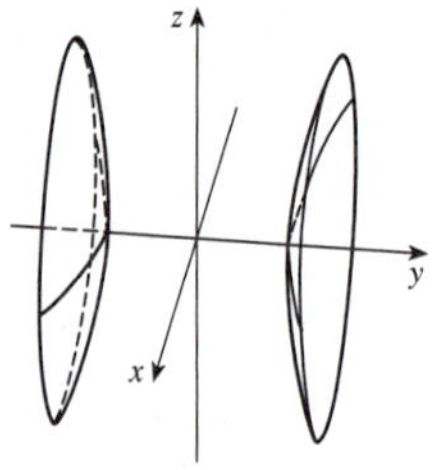

17. $x = k, y = 4z^2 + k^2$, parabola
$y = k, x^2 + 4z^2 = k$, ellipse $(k > 0)$
$z = k, y = x^2 + 4k^2$, parabola

Elliptic paraboloid with
axis the y-axis

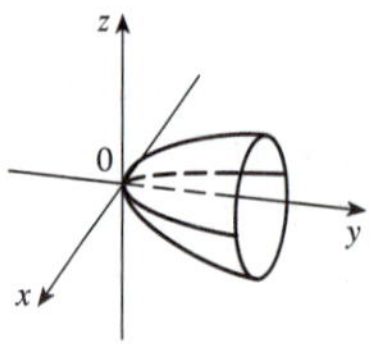

19. $x = k, y = z^2 - k^2$, parabola
$y = k, z^2 - x^2 = k$, hyperbola $(k \neq 0)$
$z = k, y = k^2 - x^2$, parabola

Hyperbolic paraboloid

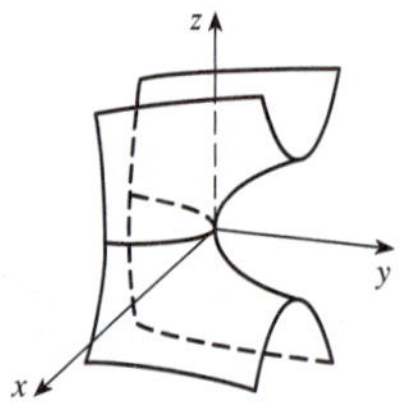

21. VII **23.** II **25.** VI **27.** VIII

29. $-\dfrac{x^2}{9} - \dfrac{y^2}{4} + \dfrac{z^2}{36} = 1$

Hyperboloid of two sheets
with axis the z-axis

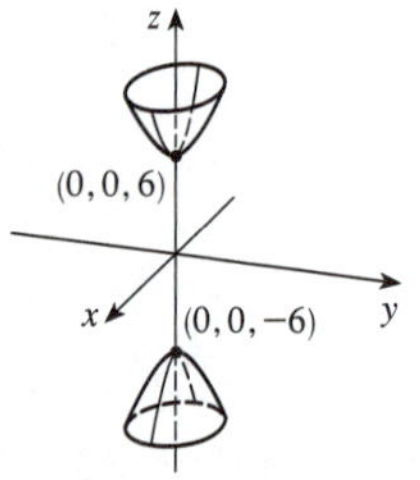

31. $\dfrac{x}{6} = \dfrac{y^2}{3} + \dfrac{z^2}{2}$

Elliptic paraboloid with vertex
(0, 0, 0) and axis the x-axis

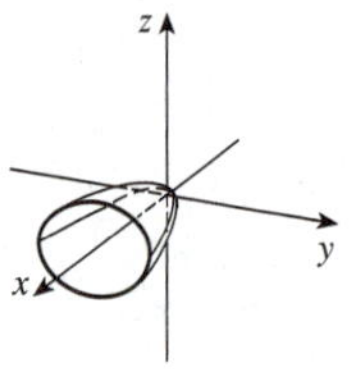

33. $x^2 + \dfrac{(y - 2)^2}{4} + (z - 3)^2 = 1$

Ellipsoid with center $(0, 2, 3)$

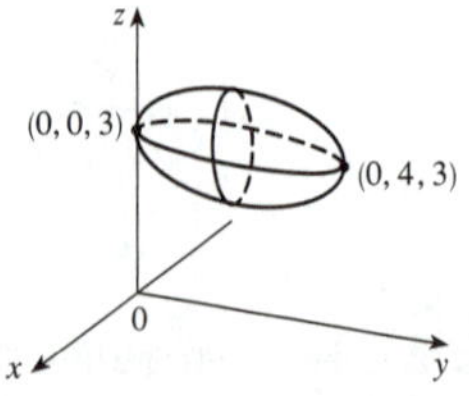

35. $(y + 1)^2 = (x - 2)^2 + (z - 1)^2$
Circular cone with vertex $(2, -1, 1)$ and axis parallel to the y-axis

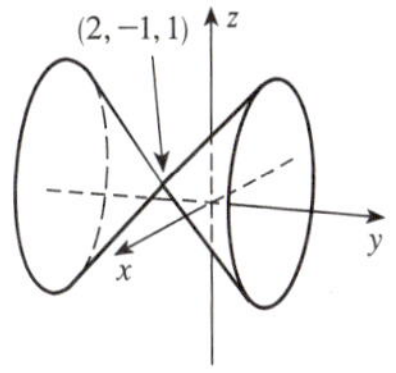

37.

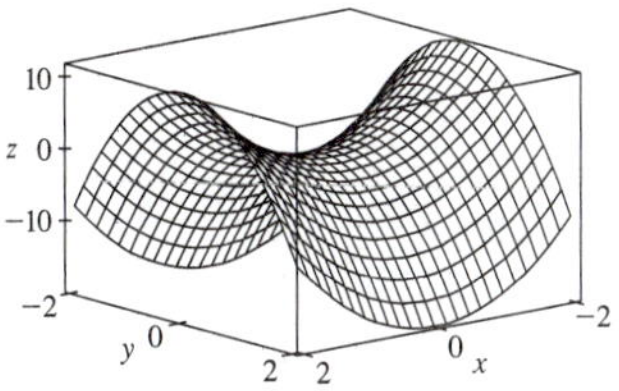

39.

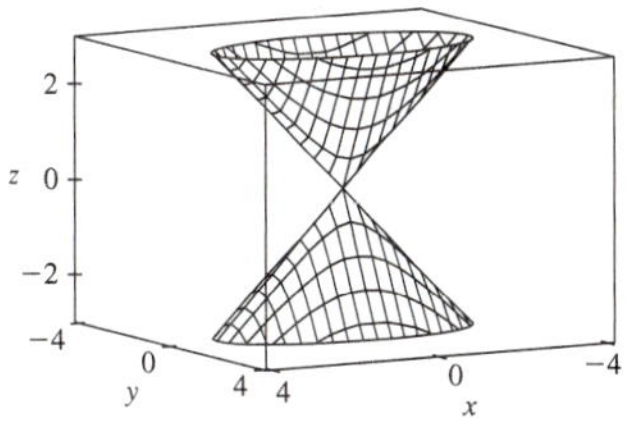

41.

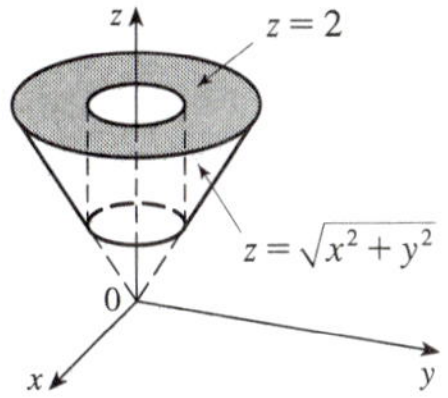

43. $y = x^2 + z^2$ **45.** $-4x = y^2 + z^2$, paraboloid

49.

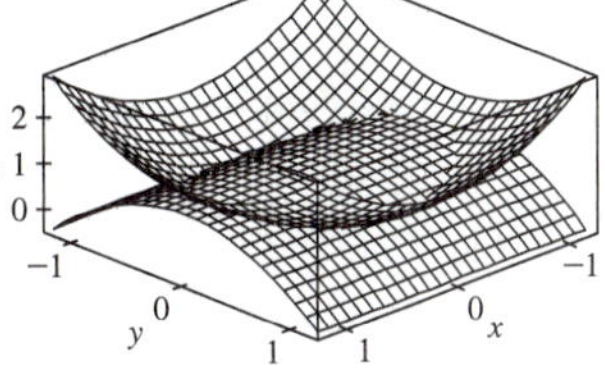

Exercises 12.7 ▫ page 842

1. See pages 839–40.

3.

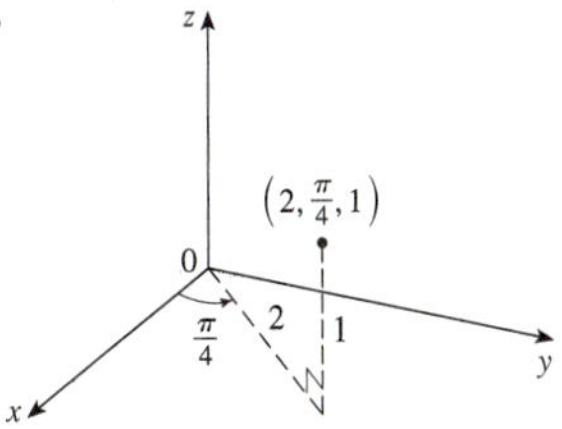

$(\sqrt{2}, \sqrt{2}, 1)$

5.

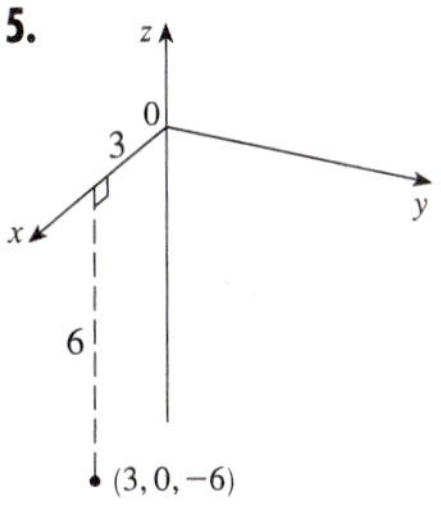

$(3, 0, -6)$

7.

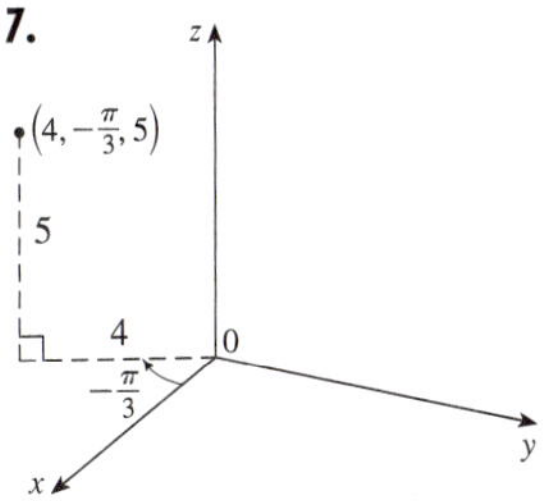

$(2, -2\sqrt{3}, 5)$

9. $(\sqrt{2}, 7\pi/4, 4)$ **11.** $(2, 4\pi/3, 2)$

13.

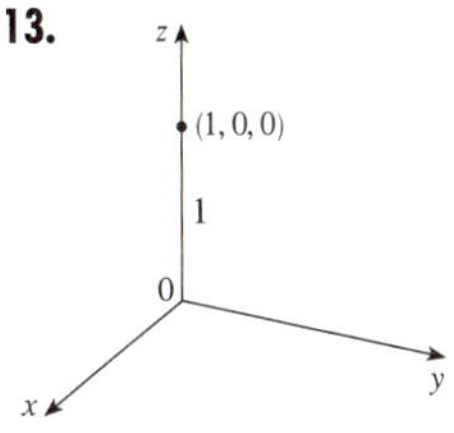

$(0, 0, 1)$

15.

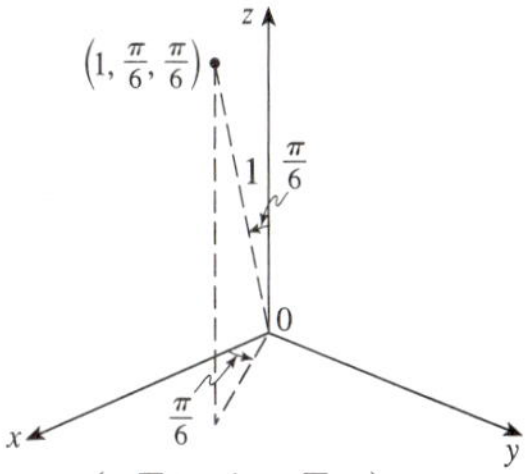

$(\sqrt{3}/4, \frac{1}{4}, \sqrt{3}/2)$

17.

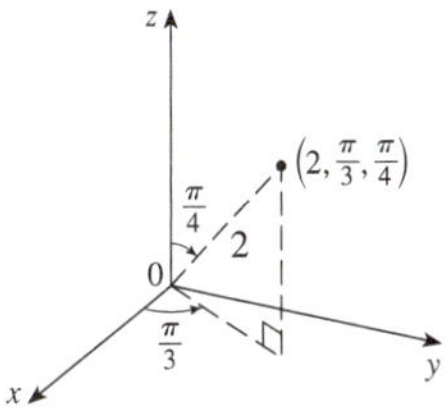

$(\frac{1}{2}\sqrt{2}, \frac{1}{2}\sqrt{6}, \sqrt{2})$

19. $(4, \pi/3, \pi/6)$ **21.** $(\sqrt{2}, 3\pi/2, 3\pi/4)$
23. $(2, \pi/6, \pi/6)$ **25.** $(2, \pi/2, 2\pi/3)$
27. $(0, 0, 2)$ **29.** $(8, \pi/6, 0)$
31. Circular cylinder, radius 3, axis the z-axis
33. Positive z-axis
35. Half-cone **37.** Circular paraboloid
39. Horizontal plane
41. Circular cylinder, radius 1, axis parallel to the z-axis
43. Sphere, radius 5, center the origin
45. Circular cylinder, radius 2, axis the y-axis
47. Cylinder, radius 1, together with the z-axis
49. (a) $z = r^2$ (b) $\rho \sin^2\phi = \cos\phi$
51. (a) $r = 3 \sec\theta$ (b) $\rho \sin\phi \cos\theta = 3$
53. (a) $2z^2 = r^2 \cos 2\theta - 4$
(b) $\rho^2(\sin^2\phi \cos 2\theta - 2\cos^2\phi) = 4$
55. (a) $r = 2\sin\theta$ (b) $\rho \sin\phi = 2\sin\theta$

57.

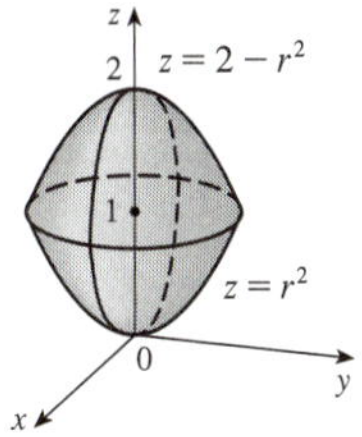

59.

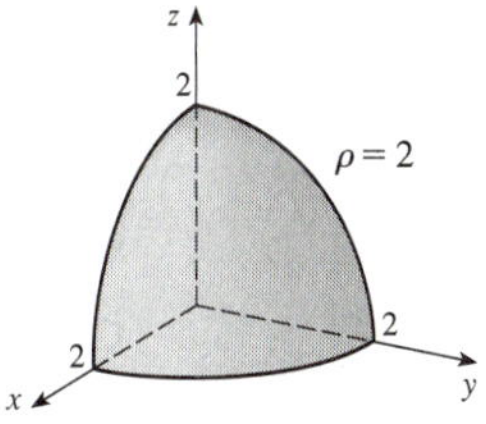

61.

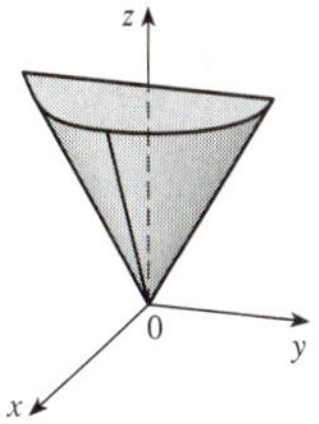

63. Cylindrical coordinates:
$6 \leqslant r \leqslant 7, 0 \leqslant \theta \leqslant 2\pi,$
$0 \leqslant z \leqslant 20$

65. $0 \leqslant \phi \leqslant \pi/4, 0 \leqslant \rho \leqslant \cos\phi$

67.

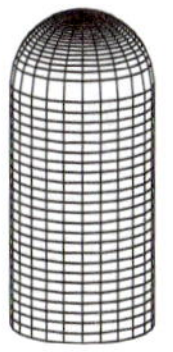

Chapter 12 Review ▫ page 845

True-False Quiz

1. True **3.** True **5.** True **7.** True **9.** True
11. False **13.** False

Exercises

1. (a) $(x-1)^2+(y+1)^2+(z-2)^2=9$
(b) $(-2,-3,5)$, 6
3. $\mathbf{u}\cdot\mathbf{v}=3\sqrt{2}$; $|\mathbf{u}\times\mathbf{v}|=3\sqrt{2}$; out of the page
5. -2, -4 **7.** (a) 2 (b) -2 (c) -2 (d) 0
9. $\cos^{-1}(\frac{1}{3})\approx 71°$ **11.** (a) $\langle 4,-3,4\rangle$ (b) $\sqrt{41}/2$
13. 166 N, 114 N
15. $x=4-3t$, $y=-1+2t$, $z=2+3t$
17. $x=-2+2t$, $y=2-t$, $z=4+5t$
19. $-4x+3y+z=-14$ **21.** $(1,4,4)$
23. Skew **25.** $22/\sqrt{26}$
27. Plane

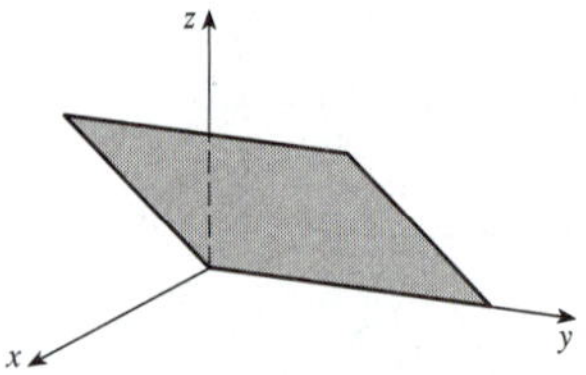

29. Cone

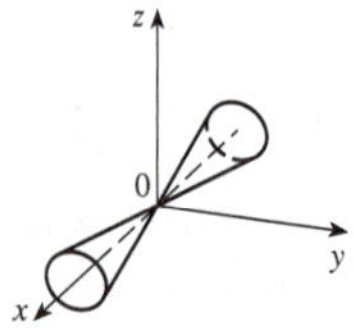

31. Hyperboloid of two sheets

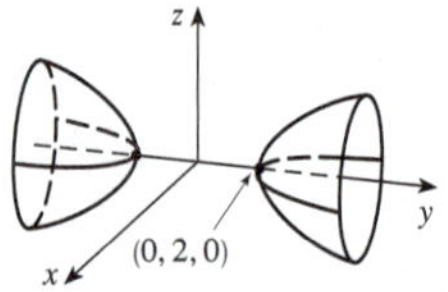

33. Ellipsoid

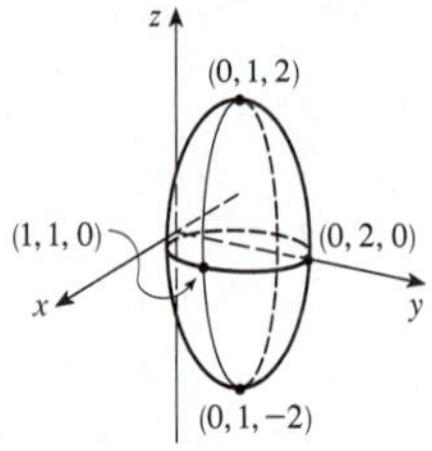

35. $4x^2+y^2+z^2=16$ **37.** $(\sqrt{3},3,2)$, $(4,\pi/3,\pi/3)$
39. $(2\sqrt{2},2\sqrt{2},4\sqrt{3})$, $(4,\pi/4,4\sqrt{3})$
41. A half-plane **43.** The horizontal plane $z=3$
45. $r^2+z^2=4$, $\rho=2$ **47.** $z=4r^2$

Problems Plus ▫ page 847

1. $(\sqrt{3}-1.5)$ m
3. (a) $(x+1)/(-2c)=(y-c)/(c^2-1)=(z-c)/(c^2+1)$
(b) $x^2+y^2=t^2+1$, $z=t$ (c) $4\pi/3$

Chapter 13

Exercises 13.1 ▫ page 855

1. $[1,5]$ **3.** $\langle 1,0,0\rangle$ **5.** $2\mathbf{i}+\frac{1}{2}\mathbf{j}+(\tan 1)\mathbf{k}$
7.

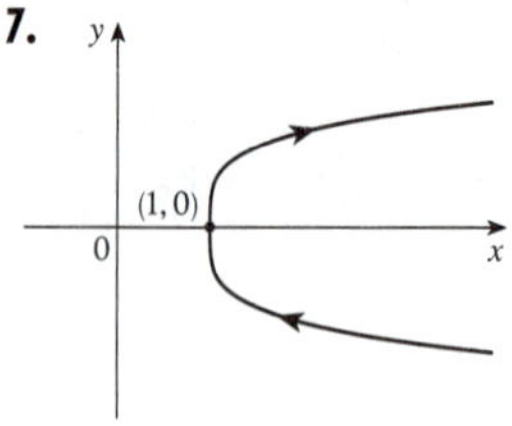

9.

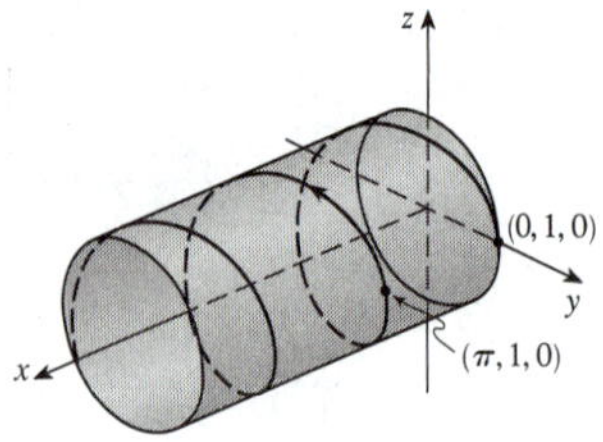

11.

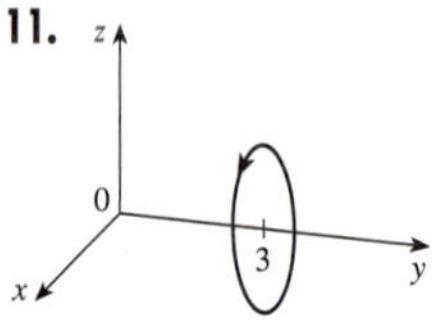

13.

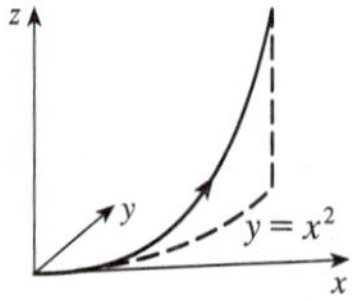

15. $\mathbf{r}(t)=\langle t,2t,3t\rangle$, $0\leqslant t\leqslant 1$;
$x=t$, $y=2t$, $z=3t$, $0\leqslant t\leqslant 1$
17. $\mathbf{r}(t)=\langle 3t+1,2t-1,5t+2\rangle$, $0\leqslant t\leqslant 1$;
$x=3t+1$, $y=2t-1$, $z=5t+2$, $0\leqslant t\leqslant 1$
19. VI **21.** IV **23.** V
25.

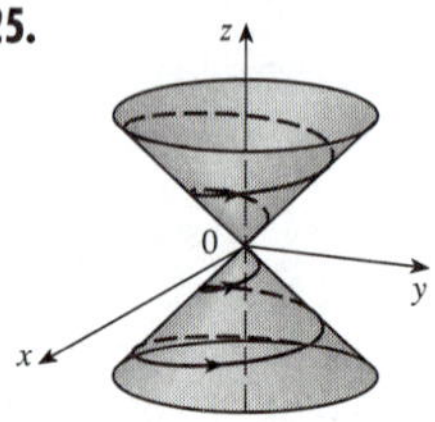

27.

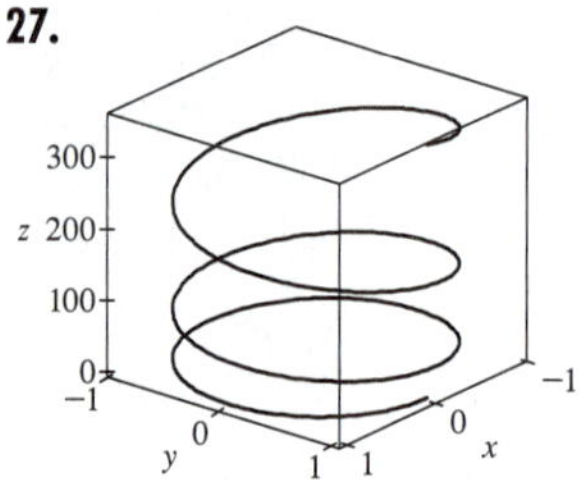

29.

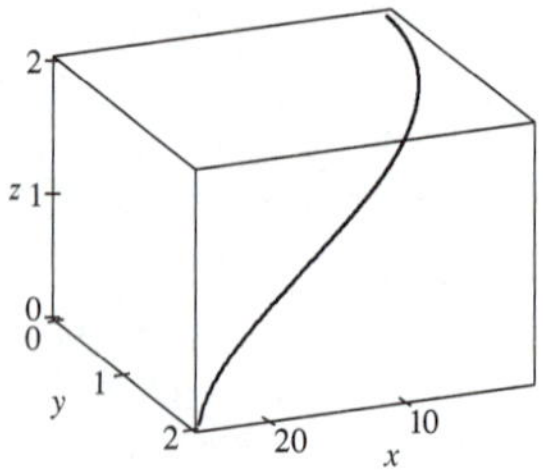

31.

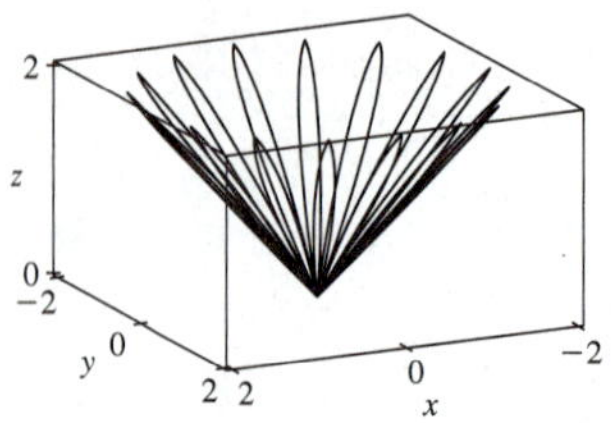

35. $\mathbf{r}(t)=t\,\mathbf{i}+\frac{1}{2}(t^2-1)\,\mathbf{j}+\frac{1}{2}(t^2+1)\,\mathbf{k}$
37. $x=2\cos t$, $y=2\sin t$, $z=4\cos^2 t$
39. Yes

Exercises 13.2 ▫ page 861

1. (a)

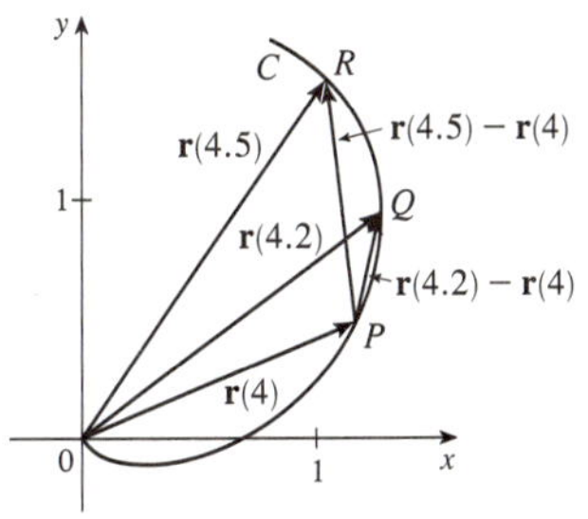

(b), (d)

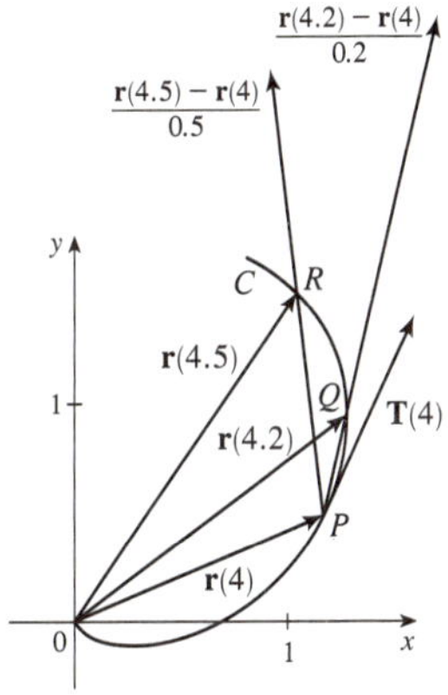

(c) $\mathbf{r}'(4) = \lim_{h \to 0} \dfrac{\mathbf{r}(4+h) - \mathbf{r}(4)}{h}$; $\mathbf{T}(4) = \dfrac{\mathbf{r}'(4)}{|\mathbf{r}'(4)|}$

3. (a), (c)

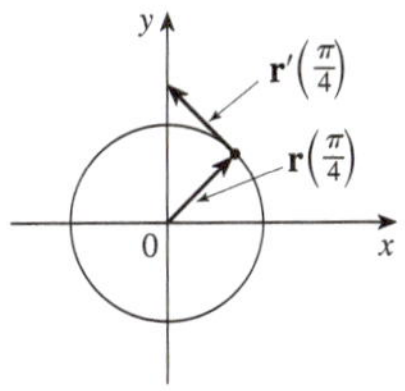

(b) $\mathbf{r}'(t) = \langle -\sin t, \cos t \rangle$

5. (a), (c)

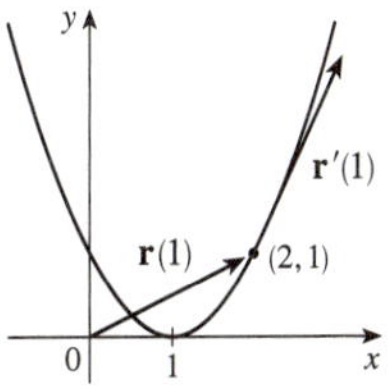

(b) $\mathbf{r}'(t) = \mathbf{i} + 2t\,\mathbf{j}$

7. (a), (c)

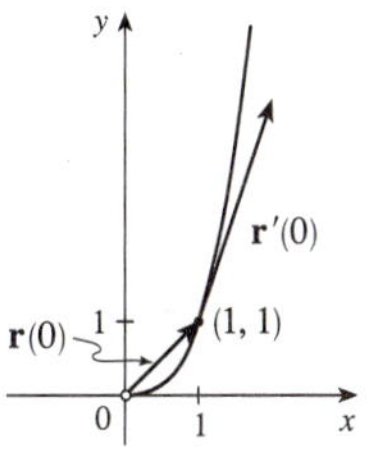

(b) $\mathbf{r}'(t) = e^t\,\mathbf{i} + 3e^{3t}\,\mathbf{j}$

9. $\mathbf{r}'(t) = \langle 2t, -1, 1/(2\sqrt{t}) \rangle$ **11.** $\mathbf{r}'(t) = 4e^{4t}\,\mathbf{k}$

13. $\mathbf{r}'(t) = 2te^{t^2}\,\mathbf{i} + [3/(1+3t)]\,\mathbf{k}$ **15.** $\mathbf{r}'(t) = \mathbf{b} + 2t\mathbf{c}$

17. $\langle 15/\sqrt{262}, 6/\sqrt{262}, 1/\sqrt{262} \rangle$ **19.** $\frac{3}{5}\,\mathbf{j} + \frac{4}{5}\,\mathbf{k}$

21. $\langle 1, 2t, 3t^2 \rangle$, $\langle 1/\sqrt{14}, 2/\sqrt{14}, 3/\sqrt{14} \rangle$, $\langle 0, 2, 6t \rangle$, $\langle 6t^2, -6t, 2 \rangle$

23. $x = 1 + 5t,\ y = 1 + 4t,\ z = 1 + 3t$

25. $x = 1 - t,\ y = t,\ z = 1 - t$

27. $x = \frac{1}{4}\pi + t,\ y = 1 - t,\ z = 1 + t$

29. (a) Not smooth (b) Smooth (c) Not smooth

31. 66° **33.** $4\,\mathbf{i} - 3\,\mathbf{j} + 5\,\mathbf{k}$ **35.** $\mathbf{i} + \mathbf{j} + \mathbf{k}$

37. $e^t\,\mathbf{i} + t^2\,\mathbf{j} + (t \ln t - t)\,\mathbf{k} + \mathbf{C}$

39. $\frac{1}{3}t^3\,\mathbf{i} + (t^4 + 1)\,\mathbf{j} - \frac{1}{3}t^3\,\mathbf{k}$

45. $1 - 4t\cos t + 11t^2 \sin t + 3t^3 \cos t$

Exercises 13.3 ▫ page 868

1. $20\sqrt{29}$ **3.** $e - e^{-1}$ **5.** $\frac{1}{27}(13^{3/2} - 8)$ **7.** 9.5706

9. $\mathbf{r}(t(s)) = \dfrac{2}{\sqrt{29}}\,s\,\mathbf{i} + \left(1 - \dfrac{3}{\sqrt{29}}\,s\right)\mathbf{j} + \left(5 + \dfrac{4}{\sqrt{29}}\,s\right)\mathbf{k}$

11. $\mathbf{r}(t(s)) = 3\sin(s/5)\,\mathbf{i} + (4s/5)\,\mathbf{j} + 3\cos(s/5)\,\mathbf{k}$

13. (a) $\langle (2/\sqrt{29})\cos t, 5/\sqrt{29}, (-2/\sqrt{29})\sin t \rangle$, $\langle -\sin t, 0, -\cos t \rangle$
(b) $\frac{2}{29}$

15. (a) $\dfrac{1}{e^{2t}+1}\langle \sqrt{2}e^t, e^{2t}, -1 \rangle$, $\dfrac{1}{e^{2t}+1}\langle 1 - e^{2t}, \sqrt{2}e^t, \sqrt{2}e^t \rangle$
(b) $\sqrt{2}e^{2t}/(e^{2t}+1)^2$

17. $2/(4t^2 + 1)^{3/2}$ **19.** $\frac{4}{25}$ **21.** $\frac{1}{7}\sqrt{\frac{19}{14}}$

23. $6|x|/(1 + 9x^4)^{3/2}$ **25.** $15\sqrt{x}/(1 + 100x^3)^{3/2}$

27. $(-\frac{1}{2}\ln 2, 1/\sqrt{2})$; approaches 0 **29.** (a) P (b) 1.3, 0.7

31.

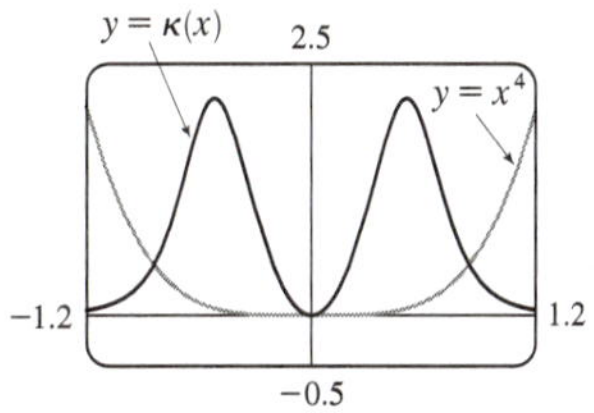

33. a is $y = f(x)$, b is $y = \kappa(x)$

35. $\kappa(t) = \dfrac{6\sqrt{4\cos^2 t - 12\cos t + 13}}{(17 - 12\cos t)^{3/2}}$

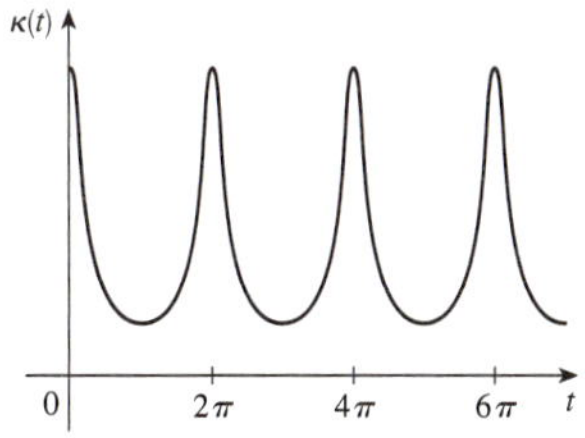

integer multiples of 2π

37. $1/(\sqrt{2}e^t)$ **39.** $\langle \frac{2}{3}, \frac{2}{3}, \frac{1}{3} \rangle$, $\langle -\frac{1}{3}, \frac{2}{3}, -\frac{2}{3} \rangle$, $\langle -\frac{2}{3}, \frac{1}{3}, \frac{2}{3} \rangle$

41. $y = 6x + \pi,\ x + 6y = 6\pi$

43. $(x + \frac{5}{2})^2 + y^2 = \frac{81}{4},\ x^2 + (y - \frac{5}{3})^2 = \frac{16}{9}$

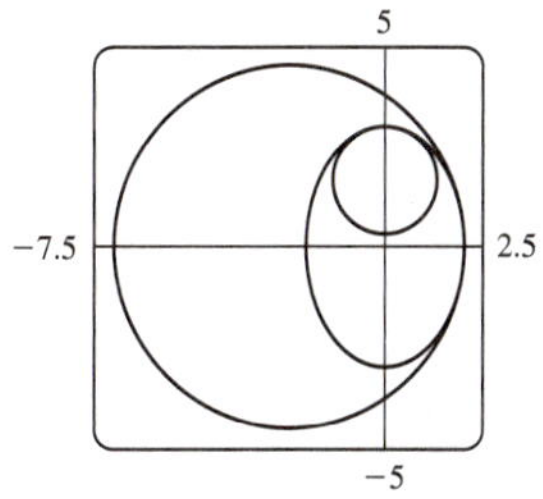

45. $(-1, -3, 1)$ **53.** $2/(t^4 + 4t^2 + 1)$

55. 2.07×10^{10} Å ≈ 2 m

Exercises 13.4 ▫ page 878

1. (a) $1.8\mathbf{i} - 3.8\mathbf{j} - 0.7\mathbf{k}$, $2.0\mathbf{i} - 2.4\mathbf{j} - 0.6\mathbf{k}$, $2.8\mathbf{i} + 1.8\mathbf{j} - 0.3\mathbf{k}$, $2.8\mathbf{i} + 0.8\mathbf{j} - 0.4\mathbf{k}$
(b) $2.4\mathbf{i} - 0.8\mathbf{j} - 0.5\mathbf{k}$, 2.58

3. $\mathbf{v}(t) = \langle 2t, 1 \rangle$
$\mathbf{a}(t) = \langle 2, 0 \rangle$
$|\mathbf{v}(t)| = \sqrt{4t^2 + 1}$

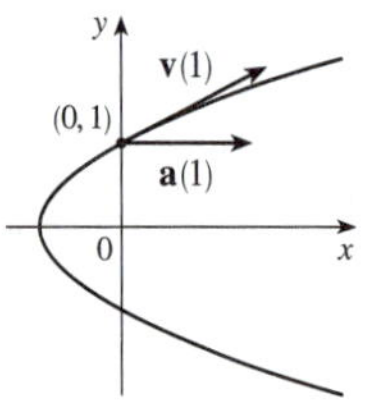

5. $\mathbf{v}(t) = e^t\,\mathbf{i} - e^{-t}\,\mathbf{j}$
$\mathbf{a}(t) = e^t\,\mathbf{i} + e^{-t}\,\mathbf{j}$
$|\mathbf{v}(t)| = \sqrt{e^{2t} + e^{-2t}}$

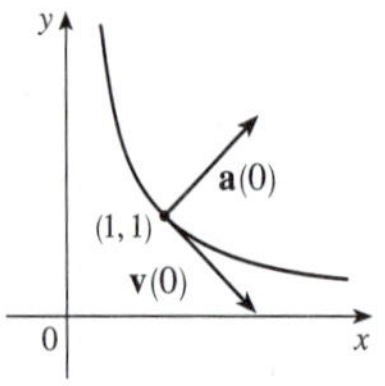

7. $\mathbf{v}(t) = \cos t\,\mathbf{i} + \mathbf{j} - \sin t\,\mathbf{k}$
$\mathbf{a}(t) = -\sin t\,\mathbf{i} - \cos t\,\mathbf{k}$
$|\mathbf{v}(t)| = \sqrt{2}$

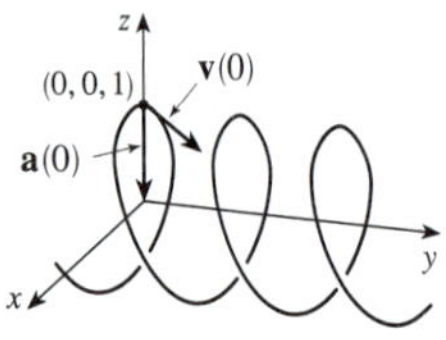

9. $\langle 2t, 3t^2, 2t \rangle$, $\langle 2, 6t, 2 \rangle$, $|t|\sqrt{9t^2 + 8}$

11. $\sqrt{2}\,\mathbf{i} + e^t\,\mathbf{j} - e^{-t}\,\mathbf{k}$, $e^t\,\mathbf{j} + e^{-t}\,\mathbf{k}$, $e^t + e^{-t}$

13. $e^t[(\cos t - \sin t)\mathbf{i} + (\sin t + \cos t)\mathbf{j} + (t + 1)\mathbf{k}]$, $e^t[-2\sin t\,\mathbf{i} + 2\cos t\,\mathbf{j} + (t + 2)\mathbf{k}]$, $e^t\sqrt{t^2 + 2t + 3}$

15. $\mathbf{v}(t) = \mathbf{i} - \mathbf{j} + t\,\mathbf{k}$, $\mathbf{r}(t) = t\,\mathbf{i} - t\,\mathbf{j} + \frac{1}{2}t^2\mathbf{k}$

17. (a) $\mathbf{r}(t) = (1 + \frac{1}{2}t^2)\mathbf{i} + t^2\,\mathbf{j} + (1 + \frac{1}{3}t^3)\mathbf{k}$
(b)

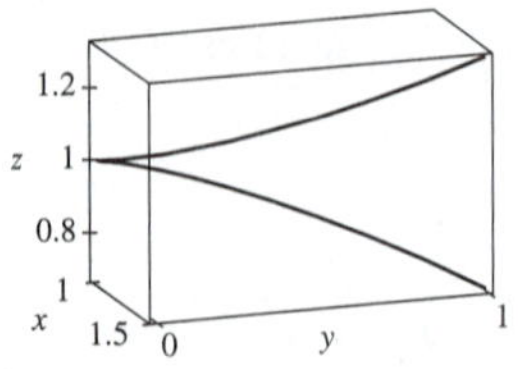

19. $t = 4$

21. $\mathbf{r}(t) = t\,\mathbf{i} - t\,\mathbf{j} + \frac{5}{2}t^2\,\mathbf{k}$, $|\mathbf{v}(t)| = \sqrt{25t^2 + 2}$

23. (a) $\approx$22 km (b) $\approx$3.2 km (c) 500 m/s

25. 30 m/s

27. $\approx$10.2°, $\approx$79.8°

29. (a) 16 m (b) $\approx$23.6° upstream

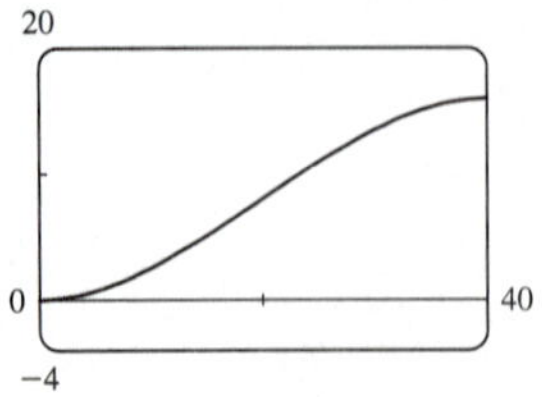

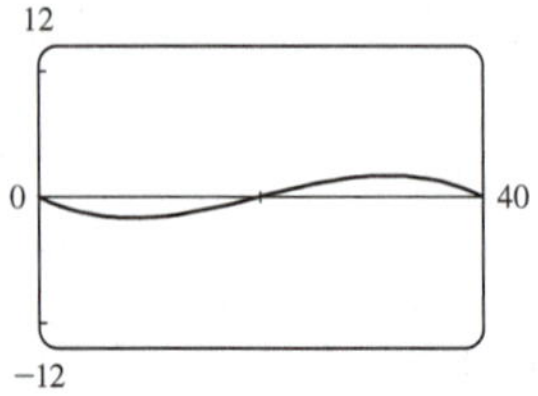

31. $6t$, 6 **33.** 0, 1 **35.** $e^t - e^{-t}$, $\sqrt{2}$

37. 4.5 cm/s², 9.0 cm/s² **39.** $t = 1$

Chapter 13 Review ▫ page 882

True-False Quiz

1. True **3.** False **5.** False **7.** False **9.** True

Exercises

1. (a)

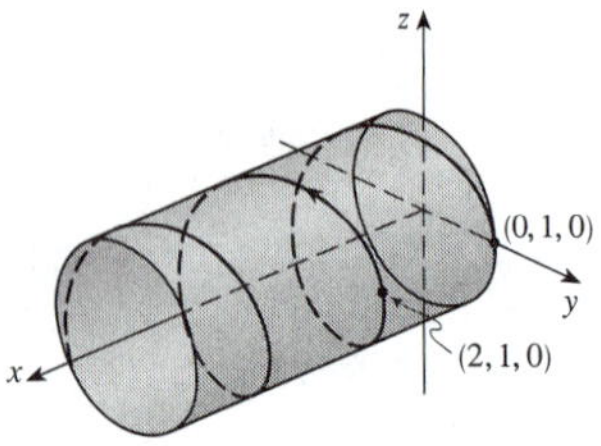

(b) $\mathbf{r}'(t) = \mathbf{i} - \pi \sin \pi t\,\mathbf{j} + \pi \cos \pi t\,\mathbf{k}$,
$\mathbf{r}''(t) = -\pi^2 \cos \pi t\,\mathbf{j} - \pi^2 \sin \pi t\,\mathbf{k}$

3. $\mathbf{r}(t) = 4\cos t\,\mathbf{i} + 4\sin t\,\mathbf{j} + (5 - 4\cos t)\mathbf{k}$, $0 \leqslant t \leqslant 2\pi$

5. $\frac{1}{3}\mathbf{i} - (2/\pi^2)\mathbf{j} + (2/\pi)\mathbf{k}$ **7.** 15.9241 **9.** $\pi/2$

11. (a) $\langle t^2, t, 1 \rangle / \sqrt{t^4 + t^2 + 1}$
(b) $\langle 2t, 1 - t^4, -2t^3 - t \rangle / \sqrt{t^8 + 4t^6 + 2t^4 + 5t^2}$
(c) $\sqrt{t^8 + 4t^6 + 2t^4 + 5t^2}/(t^4 + t^2 + 1)^2$

13. $12/17^{3/2}$ **15.** $x - 2y + 2\pi = 0$

17. $\mathbf{v}(t) = (1 + \ln t)\,\mathbf{i} + \mathbf{j} - e^{-t}\,\mathbf{k}$,
$|\mathbf{v}(t)| = \sqrt{2 + 2\ln t + (\ln t)^2 + e^{-2t}}$, $\mathbf{a}(t) = (1/t)\mathbf{i} + e^{-t}\,\mathbf{k}$

19. (a) About 3.8 ft above the ground, 60.8 ft from the athlete
(b) $\approx$21.4 ft (c) $\approx$64.2 ft from the athlete

21. (c) $-2e^{-t}\,\mathbf{v}_d + e^{-t}\,\mathbf{R}$

Problems Plus ▫ page 884

1. (a) $\mathbf{v} = \omega R(-\sin \omega t\,\mathbf{i} + \cos \omega t\,\mathbf{j})$ (c) $\mathbf{a} = -\omega^2\mathbf{r}$

3. (a) 90°, $v_0^2/(2g)$ **5.** (b) $\mathbf{R}(t) = (m/k)(1 - e^{-kt/m})\mathbf{v}_0 + (gm/k)[(m/k)(1 - e^{-kt/m}) - t]\mathbf{j}$

7. (a) $\approx$0.94 ft to the right of the table's edge, $\approx$15 ft/s
(b) $\approx$7.6° (c) $\approx$2.13 ft to the right of the table's edge

Chapter 14

Exercises 14.1 ▫ page 897

1. (a) −27; a temperature of −15°C with wind blowing at 40 km/h feels equivalent to about −27°C without wind.
(b) When the temperature is −20°C, what wind speed gives a wind chill of −30°C? 20 km/h
(c) With a wind speed of 20 km/h, what temperature gives a wind chill of −49°C? −35°C
(d) A function of wind speed that gives wind-chill values when the temperature is −5°C
(e) A function of temperature that gives wind-chill values when the wind speed is 50 km/h

3. Yes

5. (a) 25; a 40-knot wind blowing in the open sea for 15 h will create waves about 25 ft high.
(b) $f(30, t)$ is a function of t giving the wave heights produced by 30-knot winds blowing for t hours.
(c) $f(v, 30)$ is a function of v giving the wave heights produced by winds of speed v blowing for 30 hours.

7. (a) 4 (b) $\mathbb{R}^2$ (c) $[0, \infty)$

9. (a) e (b) $\{(x, y, z) \mid z \geqslant x^2 + y^2\}$ (c) $[1, \infty)$

11. $\{(x, y) \mid y \geqslant -x\}$

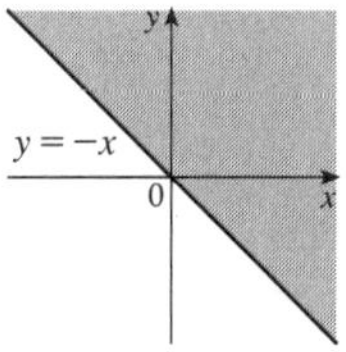

13. $\{(x, y) \mid \frac{1}{9}x^2 + y^2 < 1\}$

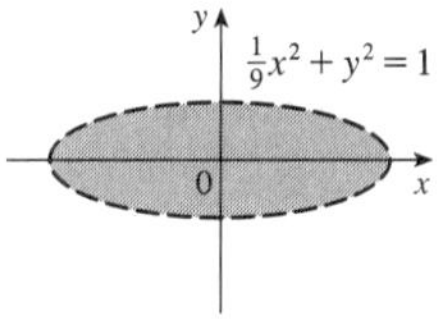

15. $\{(x, y) \mid x^2 + y^2 \neq 4\}$

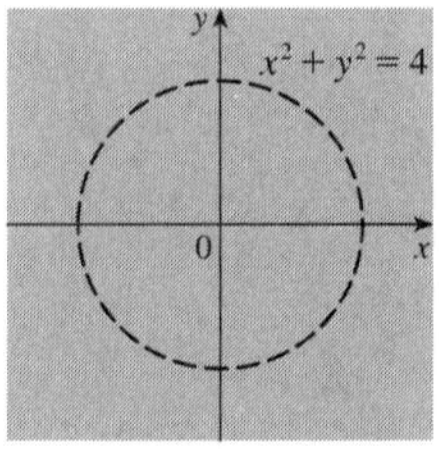

17. $\{(x, y) \mid y \geqslant x^2, x \neq \pm 1\}$

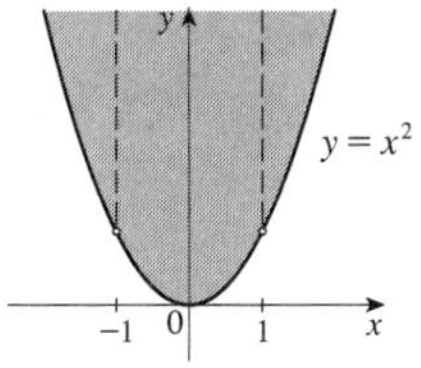

19. $\{(x, y, z) \mid x^2 + y^2 + z^2 \leqslant 1\}$

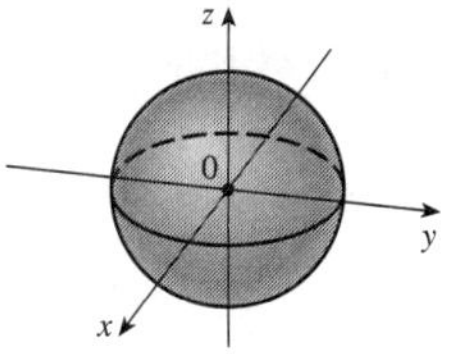

21. $z = 3$, horizontal plane

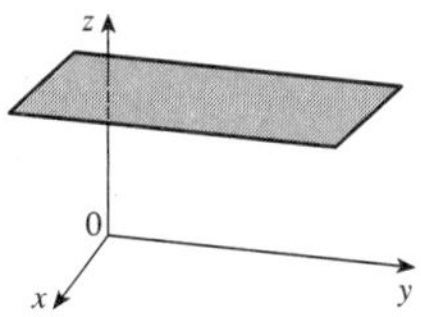

23. $x + y + z = 1$, plane

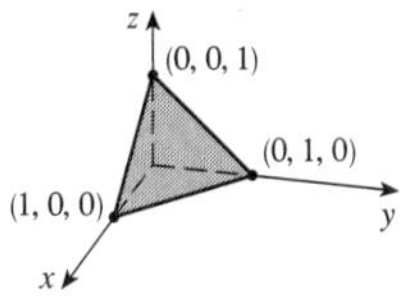

25. $z = 1 - x^2$
parabolic cylinder

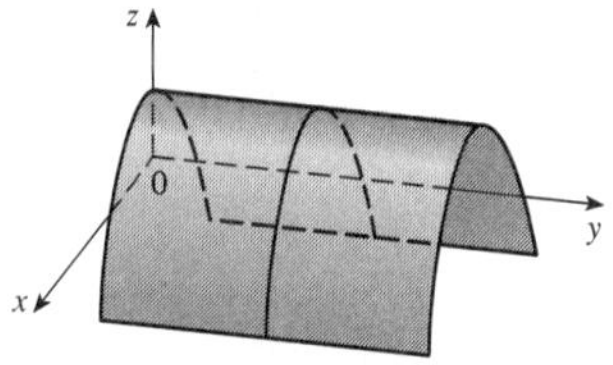

27. $z = 4x^2 + y^2 + 1$
elliptic paraboloid

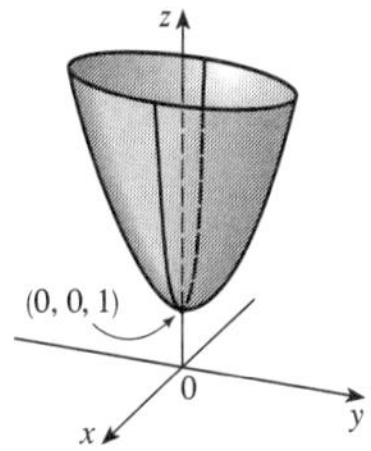

29. $z = \sqrt{x^2 + y^2}$,
top half of cone

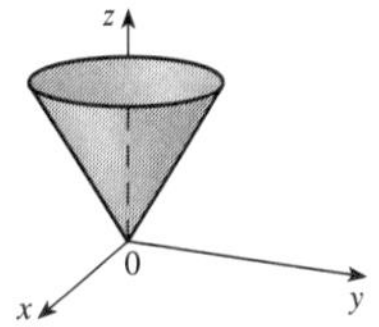

31. ≈ 56, ≈ 35 **33.** Steep; nearly flat

35.

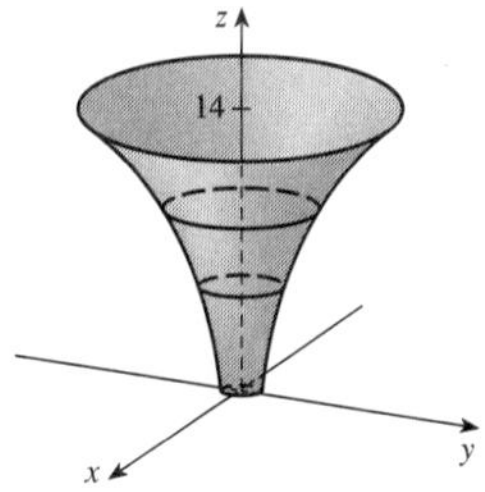

37. $xy = k$

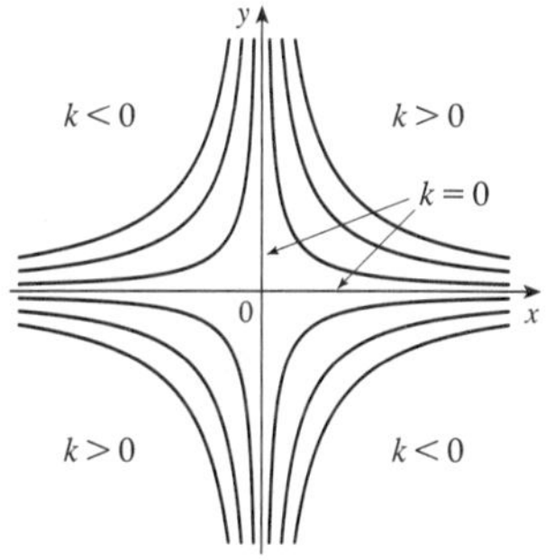

39.

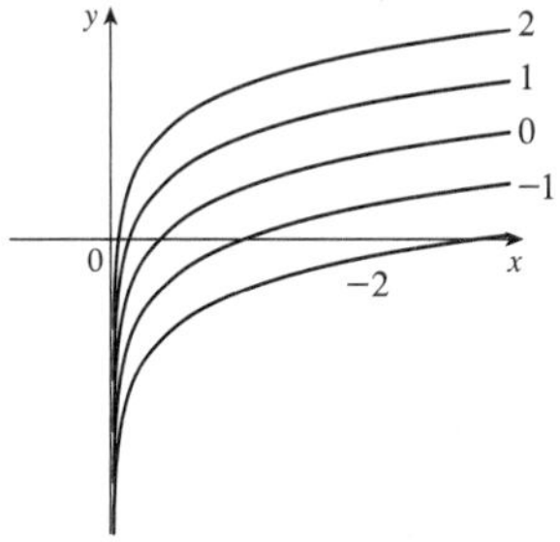

41. $\sqrt{x + y} = k$

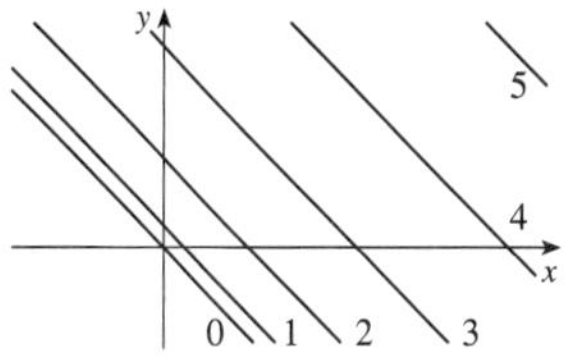

43. $x = y^2 + k$

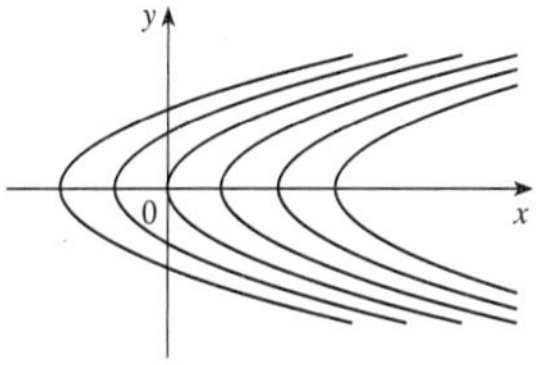

45.

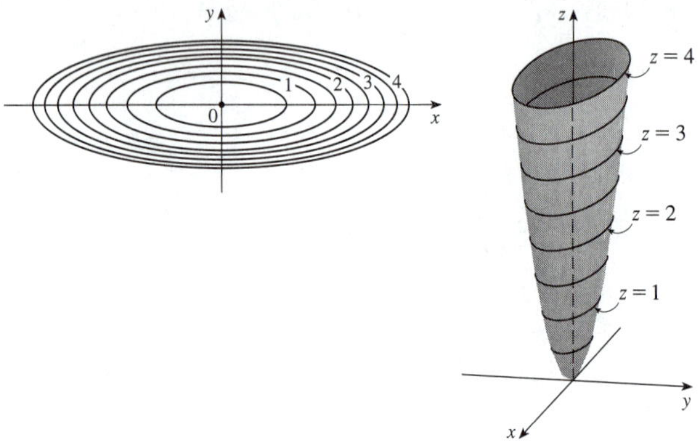

47.

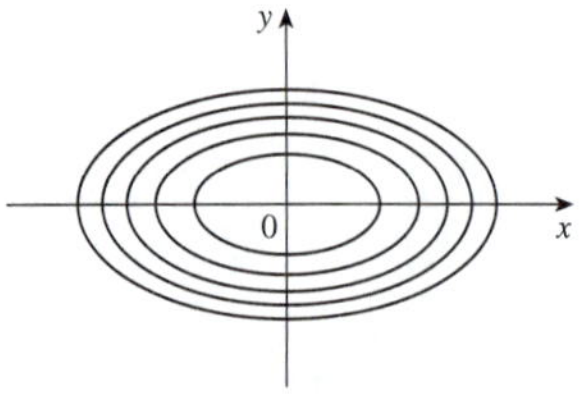

49.

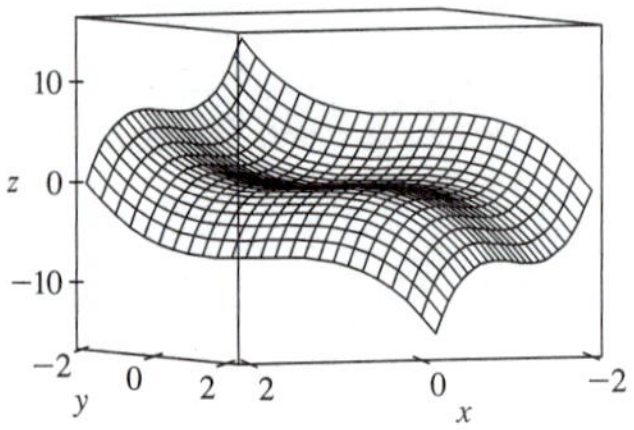

51.

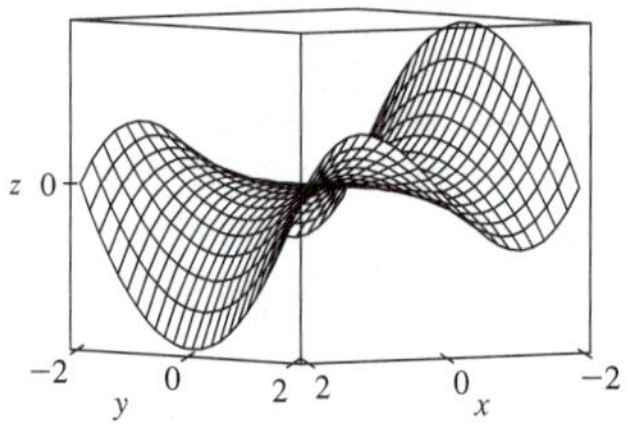

53. (a) B (b) III
55. (a) F (b) V
57. (a) D (b) IV
59. Family of parallel planes
61. Family of hyperboloids of one or two sheets with axis the y-axis
63. (a) Shift the graph of f upward 2 units
(b) Stretch the graph of f vertically by a factor of 2
(c) Reflect the graph of f about the xy-plane
(d) Reflect the graph of f about the xy-plane and then shift it upward 2 units

65.

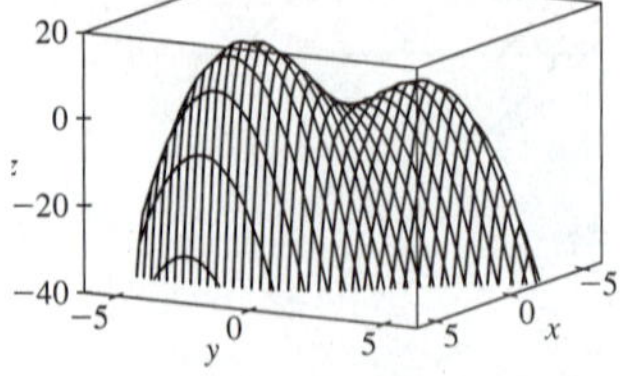

f appears to have a maximum value of about 15. There are two local maximum points but no local minimum point.

67.

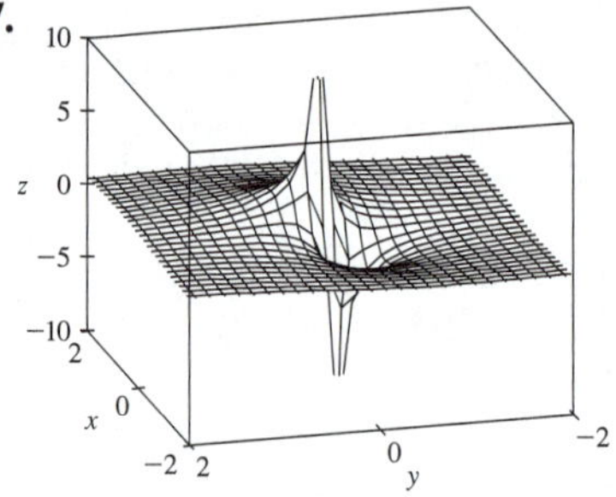

The function values approach 0 as x, y become large; as (x, y) approaches the origin, f approaches $\pm\infty$ or 0, depending on the direction of approach.

69. If $c = 0$, the graph is a cylindrical surface. For $c > 0$, the level curves are ellipses. The graph curves upward as we leave the origin, and the steepness increases as c increases. For $c < 0$, the level curves are hyperbolas. The graph curves upward in the y-direction and downward, approaching the xy-plane, in the x-direction giving a saddle-shaped appearance near (0, 0, 1).

71. (b) $y = 0.75x + 0.01$

Exercises 14.2 ▫ page 908

1. Nothing; if f is continuous, $f(3, 1) = 6$ **3.** $-\frac{5}{2}$
5. 2025 **7.** Does not exist **9.** Does not exist
11. 0 **13.** Does not exist **15.** 2 **17.** 1
19. Does not exist
21. The graph shows that the function approaches different numbers along different lines.
23. $h(x, y) = (2x + 3y - 6)^2 + \sqrt{2x + 3y - 6}$; $\{(x, y) \mid 2x + 3y \geqslant 6\}$
25. Along the line $y = x$
27. $\{(x, y) \mid y \neq \pm e^{x/2}\}$ **29.** $\{(x, y) \mid y \geqslant 0\}$
31. $\{(x, y) \mid x^2 + y^2 > 4\}$ **33.** $\{(x, y, z) \mid y \geqslant 0, y \neq \sqrt{x^2 + z^2}\}$
35. $\{(x, y) \mid (x, y) \neq (0, 0)\}$ **37.** 0 **39.** 0

Exercises 14.3 ▫ page 919

1. (a) The rate of change of temperature as longitude varies, with latitude and time fixed; the rate of change as only latitude varies; the rate of change as only time varies.
(b) Positive, negative, positive
3. (a) $f_T(-15, 30) \approx 1.3$; for a temperature of -15°C and wind speed of 30 km/h, the wind-chill index rises by 1.3°C for each degree the temperature increases. $f_v(-15, 30) \approx -0.15$; for a temperature of -15°C and wind speed of 30 km/h, the wind-chill index decreases by 0.15°C for each km/h the wind speed increases.
(b) Positive, negative (c) 0
5. (a) Positive (b) Negative
7. $c = f$, $b = f_x$, $a = f_y$

9. $f_x(1, 2) = -8 = \text{slope of } C_1$, $f_y(1, 2) = -4 = \text{slope of } C_2$

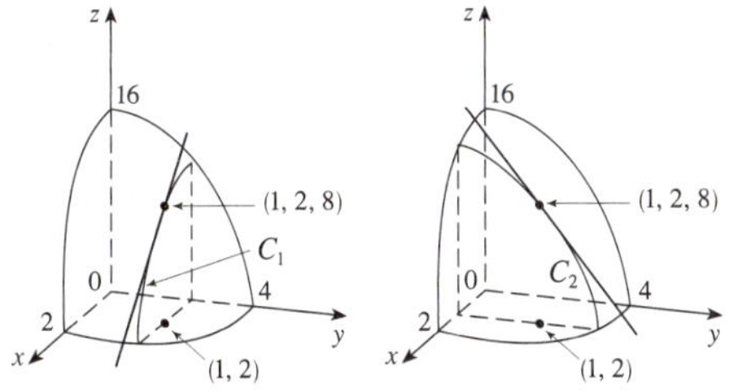

11. $f_x = 2x + 2xy$, $f_y = 2y + x^2$

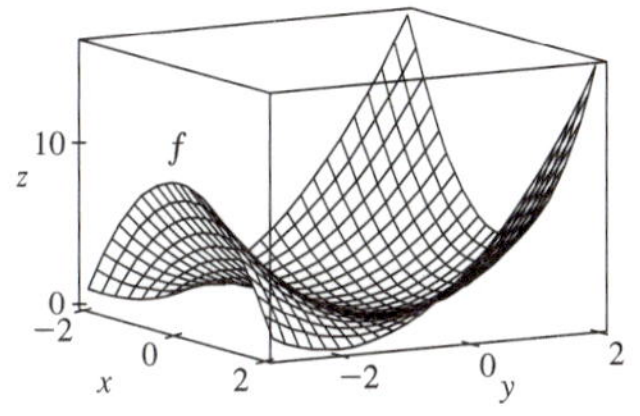

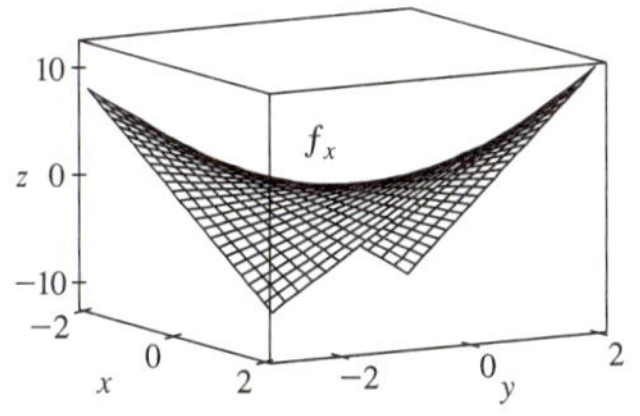

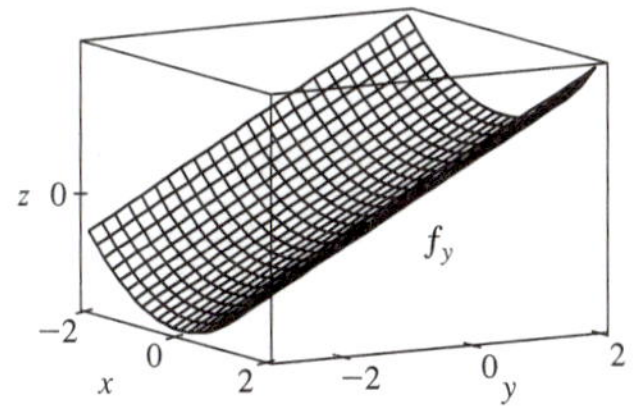

13. $f_x(x, y) = 3$, $f_y(x, y) = -8y^3$

15. $\partial z/\partial x = e^{3y}$, $\partial z/\partial y = 3xe^{3y}$

17. $f_x(x, y) = 2y/(x + y)^2$, $f_y(x, y) = -2x/(x + y)^2$

19. $\partial w/\partial \alpha = \cos\alpha \cos\beta$, $\partial w/\partial \beta = -\sin\alpha \sin\beta$

21. $f_r(r, s) = \dfrac{2r^2}{r^2 + s^2} + \ln(r^2 + s^2)$, $f_s(r, s) = \dfrac{2rs}{r^2 + s^2}$

23. $\partial u/\partial t = e^{w/t}(1 - w/t)$, $\partial u/\partial w = e^{w/t}$

25. $f_x = y^2z^3$, $f_y = 2xyz^3 + 3z$, $f_z = 3xy^2z^2 + 3y$

27. $\partial w/\partial x = 1/(x + 2y + 3z)$, $\partial w/\partial y = 2/(x + 2y + 3z)$, $\partial w/\partial z = 3/(x + 2y + 3z)$

29. $\partial u/\partial x = e^{-t}\sin\theta$, $\partial u/\partial t = -xe^{-t}\sin\theta$, $\partial u/\partial\theta = xe^{-t}\cos\theta$

31. $f_x = yz^2\tan(yt)$, $f_y = xyz^2t\sec^2(yt) + xz^2\tan(yt)$, $f_z = 2xyz\tan(yt)$, $f_t = xy^2z^2\sec^2(yt)$

33. $\partial u/\partial x_i = x_i/\sqrt{x_1^2 + x_2^2 + \cdots + x_n^2}$ **35.** $\frac{3}{5}$ **37.** $-\frac{1}{3}$

39. $f_x(x, y) = 2x - y$, $f_y(x, y) = 4y - x$

41. $\dfrac{\partial z}{\partial x} = \dfrac{3yz - 2x}{2z - 3xy}$, $\dfrac{\partial z}{\partial y} = \dfrac{3xz - 2y}{2z - 3xy}$

43. $\dfrac{\partial z}{\partial x} = \dfrac{1 + y^2z^2}{1 + y + y^2z^2}$, $\dfrac{\partial z}{\partial y} = \dfrac{-z}{1 + y + y^2z^2}$

45. (a) $f'(x), g'(y)$ (b) $f'(x + y), f'(x + y)$

47. $f_{xx} = 12x^2 - 6y^3$, $f_{xy} = -18xy^2 = f_{yx}$, $f_{yy} = -18x^2y$

49. $z_{xx} = -2y/(x + y)^3$, $z_{xy} = (x - y)/(x + y)^3 = z_{yx}$, $z_{yy} = 2x/(x + y)^3$

51. $u_{ss} = e^{-s}\sin t$, $u_{st} = -e^{-s}\cos t = u_{ts}$, $u_{tt} = -e^{-s}\sin t$

57. $12xy, 72xy$

59. $24\sin(4x + 3y + 2z), 12\sin(4x + 3y + 2z)$

61. $\theta e^{r\theta}(2\sin\theta + \theta\cos\theta + r\theta\sin\theta)$

63. $4/(y + 2z)^3, 0$ **65.** $\approx 12.2, \approx 16.8, \approx 23.25$ **77.** R^2/R_1^2

83. No **85.** $x = 1 + t, y = 2, z = 2 - 2t$ **89.** -2

91. (a)

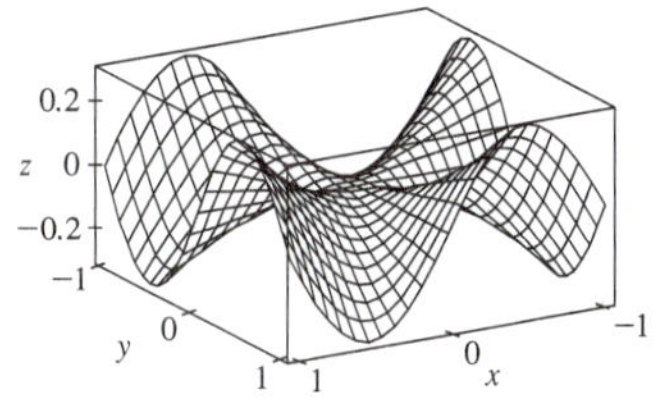

(b) $f_x(x, y) = \dfrac{x^4y + 4x^2y^3 - y^5}{(x^2 + y^2)^2}$, $f_y(x, y) = \dfrac{x^5 - 4x^3y^2 - xy^4}{(x^2 + y^2)^2}$

(c) 0, 0 (e) No, since f_{xy} and f_{yx} are not continuous.

Exercises 14.4 ▫ page 930

1. $z = -8x - 2y$ **3.** $x - 2y + z = 4$ **5.** $z = y$

7.

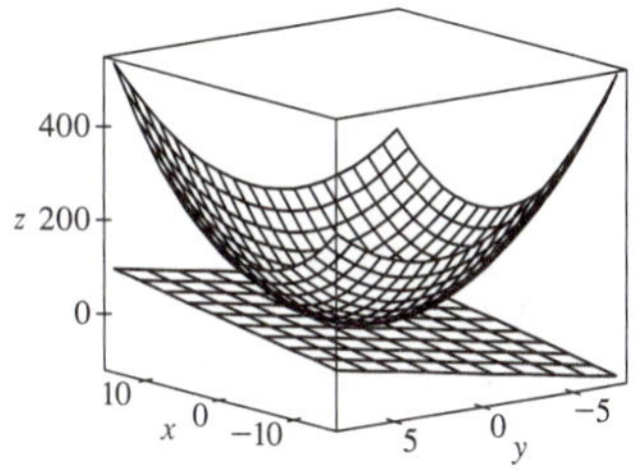

9.

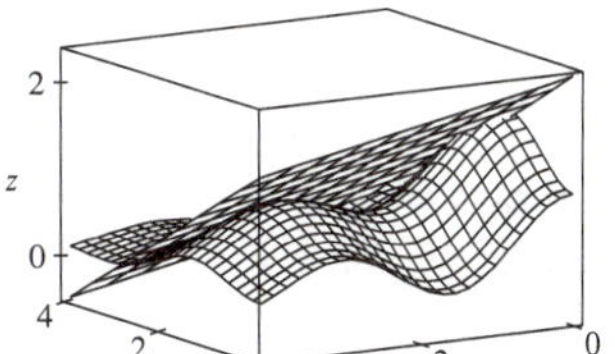

11. $2x + \frac{1}{4}y - 1$ **13.** $x + 1$ **15.** $\frac{1}{2}x + y + \frac{1}{4}\pi - \frac{1}{2}$

17. $-\frac{2}{3}x - \frac{7}{3}y + \frac{20}{3}$; $2.8\overline{46}$ **19.** $\frac{3}{7}x + \frac{2}{7}y + \frac{6}{7}z$; 6.9914

21. $4T + H - 329$; 129°F

23. $dz = 3x^2\ln(y^2)\,dx + (2x^3/y)\,dy$

25. $du = e^t\sin\theta\,dt + e^t\cos\theta\,d\theta$

27. $dw = (x^2 + y^2 + z^2)^{-1}(x\,dx + y\,dy + z\,dz)$

29. $\Delta z = 0.9225$, $dz = 0.9$ **31.** 5.4 cm^2 **33.** 16 cm^3

35. 150 **37.** $\frac{1}{17} \approx 0.059\ \Omega$ **39.** $\varepsilon_1 = \Delta x$, $\varepsilon_2 = \Delta y$

Exercises 14.5 ▫ page 938

1. $4(2xy + y^2)t^3 - 3(x^2 + 2xy)t^2$
3. $\pi \cos x \cos y - (\sin x \sin y)/(2\sqrt{t})$
5. $e^{y/z}[2t - (x/z) - (2xy/z^2)]$
7. $\partial z/\partial s = 2x + y + xt + 2yt$, $\partial z/\partial t = 2x + y + xs + 2ys$
9. $\dfrac{\partial z}{\partial s} = \dfrac{4st + \ln t}{1 + (2x + y)^2}$, $\dfrac{\partial z}{\partial t} = \dfrac{2s^2 + s/t}{1 + (2x + y)^2}$
11. $\dfrac{\partial z}{\partial s} = e^r\left(t \cos \theta - \dfrac{s}{\sqrt{s^2 + t^2}} \sin \theta\right)$,
$\dfrac{\partial z}{\partial t} = e^r\left(s \cos \theta - \dfrac{t}{\sqrt{s^2 + t^2}} \sin \theta\right)$
13. 62 **15.** 7, 2
17. $\dfrac{\partial u}{\partial r} = \dfrac{\partial u}{\partial x}\dfrac{\partial x}{\partial r} + \dfrac{\partial u}{\partial y}\dfrac{\partial y}{\partial r}$, $\dfrac{\partial u}{\partial s} = \dfrac{\partial u}{\partial x}\dfrac{\partial x}{\partial s} + \dfrac{\partial u}{\partial y}\dfrac{\partial y}{\partial s}$,
$\dfrac{\partial u}{\partial t} = \dfrac{\partial u}{\partial x}\dfrac{\partial x}{\partial t} + \dfrac{\partial u}{\partial y}\dfrac{\partial y}{\partial t}$
19. $\dfrac{\partial v}{\partial x} = \dfrac{\partial v}{\partial p}\dfrac{\partial p}{\partial x} + \dfrac{\partial v}{\partial q}\dfrac{\partial q}{\partial x} + \dfrac{\partial v}{\partial r}\dfrac{\partial r}{\partial x}$,
$\dfrac{\partial v}{\partial y} = \dfrac{\partial v}{\partial p}\dfrac{\partial p}{\partial y} + \dfrac{\partial v}{\partial q}\dfrac{\partial q}{\partial y} + \dfrac{\partial v}{\partial r}\dfrac{\partial r}{\partial y}$,
$\dfrac{\partial v}{\partial z} = \dfrac{\partial v}{\partial p}\dfrac{\partial p}{\partial z} + \dfrac{\partial v}{\partial q}\dfrac{\partial q}{\partial z} + \dfrac{\partial v}{\partial r}\dfrac{\partial r}{\partial z}$
21. 85, 178, 54 **23.** $\frac{9}{7}, \frac{9}{7}$ **25.** 36, 24, 30
27. $\dfrac{4(xy)^{3/2} - y}{x - 2x^2\sqrt{xy}}$ **29.** $\dfrac{\sin(x - y) + e^y}{\sin(x - y) - xe^y}$
31. $\dfrac{3yz - 2x}{2z - 3xy}, \dfrac{3xz - 2y}{2z - 3xy}$
33. $\dfrac{1 + y^2z^2}{1 + y + y^2z^2}, -\dfrac{z}{1 + y + y^2z^2}$
35. 2°C/s
37. ≈ -0.33 m/s per minute
39. (a) 6 m³/s (b) 10 m²/s (c) 0 m/s **41.** ≈ -0.27 L/s
43. (a) $\partial z/\partial r = (\partial z/\partial x) \cos \theta + (\partial z/\partial y) \sin \theta$,
$\partial z/\partial \theta = -(\partial z/\partial x)r \sin \theta + (\partial z/\partial y)r \cos \theta$
49. $4rs\, \partial^2 z/\partial x^2 + (4r^2 + 4s^2)\partial^2 z/\partial x\, \partial y + 4rs\, \partial^2 z/\partial y^2 + 2\, \partial z/\partial y$

Exercises 14.6 ▫ page 950

1. ≈ -0.1 millibar/mi **3.** ≈ 0.778 **5.** $\frac{5}{16}\sqrt{3} + \frac{1}{4}$
7. (a) $\nabla f(x, y) = \langle 5y^2 - 12x^2y, 10xy - 4x^3 \rangle$
(b) $\langle -4, 16 \rangle$ (c) 172/13
9. (a) $\langle e^{2yz}, 2xze^{2yz}, 2xye^{2yz} \rangle$ (b) $\langle 1, 12, 0 \rangle$ (c) $-\frac{22}{3}$
11. 23/10 **13.** $4\sqrt{2}$ **15.** 4/9
17. $9/(2\sqrt{5})$ **19.** 2/5 **21.** $4\sqrt{2}, \langle -1, 1 \rangle$ **23.** $1, \langle 0, 1 \rangle$
25. $\sqrt{3}, \langle 1, -1, -1 \rangle$ **27.** (b) $\langle -12, 92 \rangle$
29. All points on the line $y = x + 1$ **31.** (a) $-40/(3\sqrt{3})$
33. (a) $32/\sqrt{3}$ (b) $\langle 38, 6, 12 \rangle$ (c) $2\sqrt{406}$ **35.** $\frac{327}{13}$
39. (a) $4x - 2y + 3z = 21$ (b) $\dfrac{x - 4}{8} = \dfrac{y + 1}{-4} = \dfrac{z - 1}{6}$
41. (a) $4x - 5y - z = 4$ (b) $\dfrac{x - 2}{4} = \dfrac{y - 1}{-5} = \dfrac{z + 1}{-1}$
43. (a) $x + y - z = 1$ (b) $x - 1 = y = -z$

45.

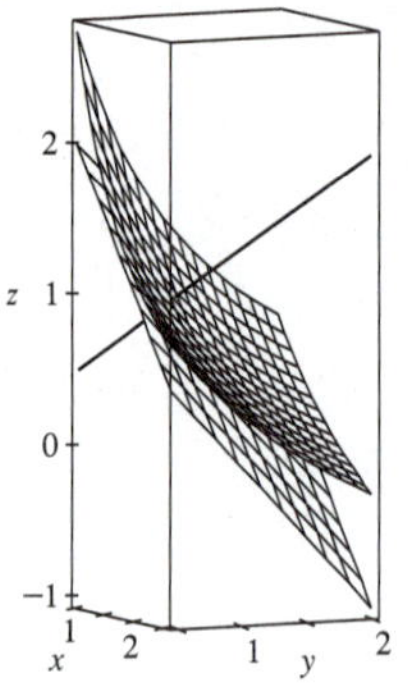

47. $\langle 4, 8 \rangle$, $x + 2y = 4$

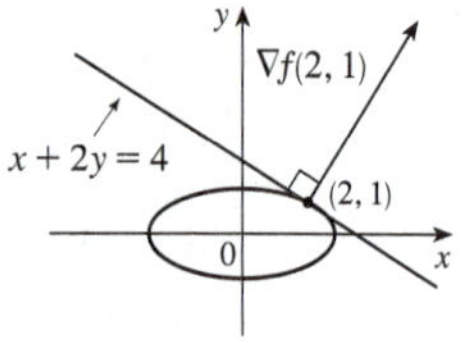

53. $(\pm\sqrt{6}/3, \mp 2\sqrt{6}/3, \pm\sqrt{6}/2)$
59. $x = -1 - 10t, y = 1 - 16t, z = 2 - 12t$
63. If $\mathbf{u} = \langle a, b \rangle$ and $\mathbf{v} = \langle c, d \rangle$, then $af_x + bf_y$ and $cf_x + df_y$ are known, so we solve linear equations for f_x and f_y.

Exercises 14.7 ▫ page 961

1. (a) f has a local minimum at (1, 1).
(b) f has a saddle point at (1, 1).
3. Local minimum at (1, 1), saddle point at (0, 0)
5. Maximum $f(-1, \frac{1}{2}) = 11$
7. Minima $f(1, 1) = 0$, $f(-1, -1) = 0$, saddle point at (0, 0)
9. Saddle points at $(1, -1), (-1, 1)$
11. f has a local maximum value of 1 at all points of the form (x_0, x_0).
13. None **15.** Saddle points $(0, n\pi)$, n an integer
17. Minimum $f(0, 0) = 0$, saddle points at $(\pm 1, 0)$
19. Maximum $f(0, 0) = 2$, minimum $f(0, 2) = -2$, saddle points $(\pm 1, 1)$
21. Maximum $f(\pi/3, \pi/3) = 3\sqrt{3}/2$, minimum $f(5\pi/3, 5\pi/3) = -3\sqrt{3}/2$
23. Minima $f(-1.714, 0) \approx -9.200$, $f(1.402, 0) \approx 0.242$, saddle point (0.312, 0), lowest point $(-1.714, 0, -9.200)$
25. Maxima $f(-1.267, 0) \approx 1.310$, $f(1.629, \pm 1.063) \approx 8.105$, saddle points $(-0.259, 0)$, (1.526, 0), highest points $(1.629, \pm 1.063, 8.105)$
27. Maximum $f(2, 0) = 9$, minimum $f(0, 3) = -14$
29. Maximum $f(\pm 1, 1) = 7$, minimum $f(0, 0) = 4$
31. Maximum $f(3, 0) = 83$, minimum $f(1, 1) = 0$
33. Maximum $f(1, 0) = 2$, minimum $f(-1, 0) = -2$
35.

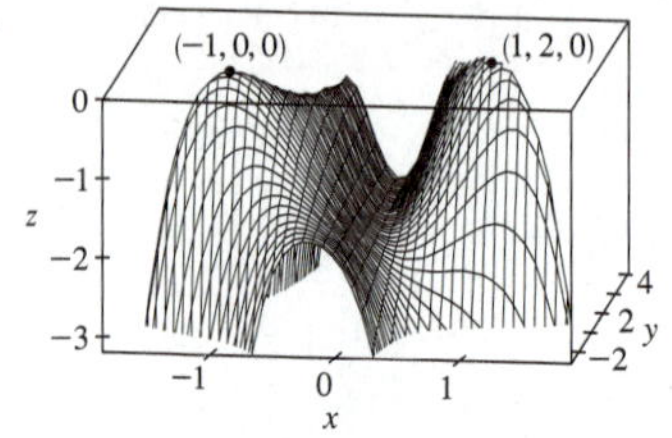

37. $\sqrt{3}$ **39.** (0, 0, 1), (0, 0, −1) **41.** $\frac{100}{3}, \frac{100}{3}, \frac{100}{3}$
43. $16/\sqrt{3}$ **45.** $\frac{4}{3}$ **47.** Cube, edge length $c/12$
49. Square base of side 40 cm, height 20 cm **51.** $L^3/(3\sqrt{3})$

Exercises 14.8 ▫ page 970

1. $\approx 59, 30$ **3.** Maxima $f(\pm 1, 0) = 1$, minima $f(0, \pm 1) = -1$
5. Maxima $f(\pm 2, 1) = 4$, minima $f(\pm 2, -1) = -4$
7. Maximum $f(1, 3, 5) = 70$, minimum $f(-1, -3, -5) = -70$
9. Maximum $2/\sqrt{3}$, minimum $-2/\sqrt{3}$
11. Maximum $\sqrt{3}$, minimum 1
13. Maximum $f(\frac{1}{2}, \frac{1}{2}, \frac{1}{2}, \frac{1}{2}) = 2$, minimum $f(-\frac{1}{2}, -\frac{1}{2}, -\frac{1}{2}, -\frac{1}{2}) = -2$
15. Maximum $f(1, \sqrt{2}, -\sqrt{2}) = 1 + 2\sqrt{2}$, minimum $f(1, -\sqrt{2}, \sqrt{2}) = 1 - 2\sqrt{2}$
17. Maximum $\frac{3}{2}$, minimum $\frac{1}{2}$
19. Maxima $f(\pm 1/\sqrt{2}, \mp 1/(2\sqrt{2})) = e^{1/4}$, minima $f(\pm 1/\sqrt{2}, \pm 1/(2\sqrt{2})) = e^{-1/4}$
25–37. See Exercises 37–51 in Section 14.7.
39. Nearest $(\frac{1}{2}, \frac{1}{2}, \frac{1}{2})$, farthest $(-1, -1, 2)$
41. Maximum ≈ 9.7938, minimum ≈ -5.3506
43. (a) c/n (b) When $x_1 = x_2 = \cdots = x_n$

Chapter 14 Review ▫ page 975

True-False Quiz

1. True **3.** False **5.** False **7.** True **9.** False
11. True

Exercises

1. $\{(x, y) \mid -1 \leq x \leq 1\}$

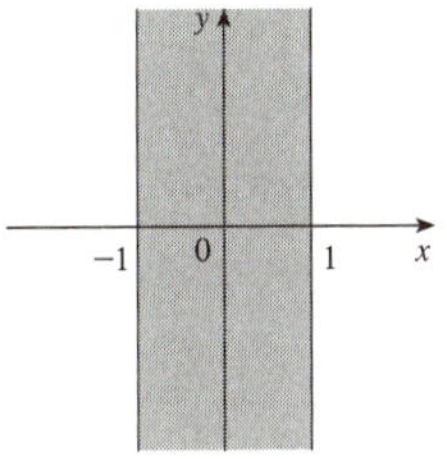

3.

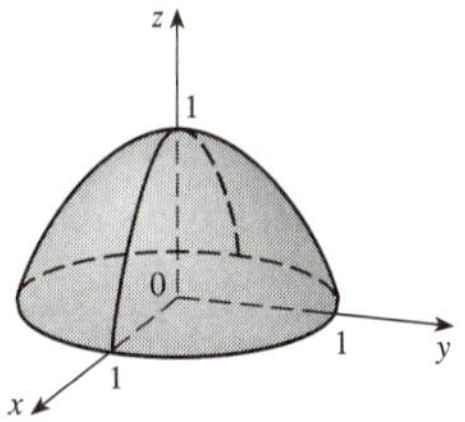

5.

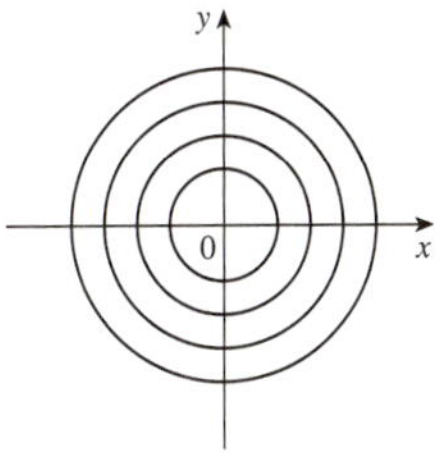

7.

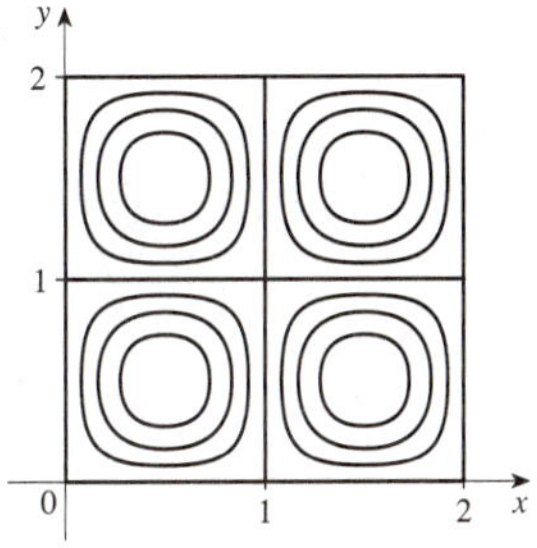

9. $\frac{2}{3}$
11. (a) $\approx 3.5°C/m$, $-3.0°C/m$ (b) $\approx 0.35°C/m$ by Equation 14.6.9 (Definition 14.6.2 gives $\approx 1.1°C/m$.) (c) -0.25
13. $f_x = 1/\sqrt{2x + y^2}$, $f_y = y/\sqrt{2x + y^2}$
15. $g_u = \tan^{-1} v$, $g_v = u/(1 + v^2)$
17. $T_p = \ln(q + e^r)$, $T_q = p/(q + e^r)$, $T_r = pe^r/(q + e^r)$
19. $f_{xx} = 24x$, $f_{xy} = -2y = f_{yx}$, $f_{yy} = -2x$
21. $f_{xx} = k(k-1)x^{k-2}y^l z^m$, $f_{xy} = klx^{k-1}y^{l-1}z^m = f_{yx}$, $f_{xz} = kmx^{k-1}y^l z^{m-1} = f_{zx}$, $f_{yy} = l(l-1)x^k y^{l-2} z^m$, $f_{yz} = lmx^k y^{l-1} z^{m-1} = f_{zy}$, $f_{zz} = m(m-1)x^k y^l z^{m-2}$
25. (a) $z = 8x + 4y + 1$ (b) $\dfrac{x-1}{8} = \dfrac{y+2}{4} = 1 - z$
27. (a) $2x - 2y - 3z = 3$ (b) $\dfrac{x-2}{4} = \dfrac{y+1}{-4} = \dfrac{z-1}{-6}$
29. (a) $x + 2y + 5z = 0$ (b) $x - 2 = \dfrac{y+1}{2} = \dfrac{z}{5}$
31. $(\pm\sqrt{2/7}, \pm 1/\sqrt{14}, \mp 3/\sqrt{14})$
33. $60x + \frac{24}{5}y + \frac{32}{5}z - 120$; 38.656
35. $e^t + 2(y/z)(3t^2 + 4) - 2t(y^2/z^2)$ **37.** $-47, 108$
43. $ze^{x\sqrt{y}}\langle z\sqrt{y}, xz/(2\sqrt{y}), 2\rangle$ **45.** $\frac{43}{5}$ **47.** $\sqrt{145}/2, \langle 4, \frac{9}{2}\rangle$
49. $\approx \frac{5}{8}$ knot/mi **51.** Minimum $f(-4, 1) = -11$
53. Maximum $f(1, 1) = 1$; saddle points $(0, 0)$, $(0, 3)$, $(3, 0)$
55. Maximum $f(1, 2) = 4$, minimum $f(2, 4) = -64$
57. Maximum $f(-1, 0) = 2$, minima $f(1, \pm 1) = -3$, saddle points $(-1, \pm 1)$, $(1, 0)$
59. Maximum $f(\pm\sqrt{2/3}, 1/\sqrt{3}) = 2/(3\sqrt{3})$, minimum $f(\pm\sqrt{2/3}, -1/\sqrt{3}) = -2/(3\sqrt{3})$
61. Maximum 1, minimum -1
63. $(\pm 3^{-1/4}, 3^{-1/4}\sqrt{2}, \pm 3^{1/4})$, $(\pm 3^{-1/4}, -3^{-1/4}\sqrt{2}, \pm 3^{1/4})$
65. $P(2 - \sqrt{3})$, $P(3 - \sqrt{3})/6$, $P(2\sqrt{3} - 3)/3$

Problems Plus ▫ page 978

1. $L^2W^2, \frac{1}{4}L^2W^2$ **3.** (a) $x = w/3$, base $= w/3$ (b) Yes
9. $\sqrt{6}/2, 3\sqrt{2}/2$

Chapter 15

Exercises 15.1 ▫ page 988

1. (a) 288 (b) 144 **3.** (a) $\pi^2/2 \approx 4.935$ (b) 0
5. (a) -6 (b) -3.5 **7.** $U < V < L$
9. (a) ≈ 248 (b) 15.5 **11.** 60
13. 3 **15.** 0.6065, 0.5694, 0.5606, 0.5585, 0.5579, 0.5578

Exercises 15.2 ▫ page 994

1. $9 + 27y, 8x + 24x^2$ **3.** 10 **5.** 2 **7.** 261,632/45
9. $\frac{21}{2}\ln 2$ **11.** 6 **13.** $\frac{21}{2}$ **15.** $9\ln 2$
17. $[(\sqrt{3} - 1)/2] - (\pi/12)$ **19.** $\frac{1}{2}(e^2 - 3)$
21.

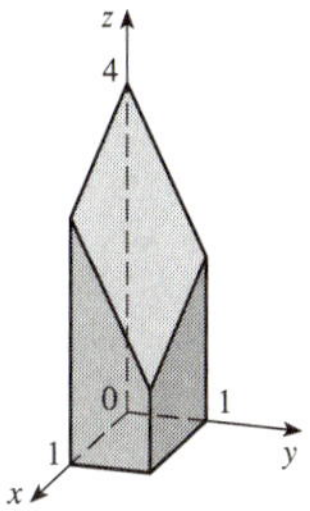

23. 47.5 **25.** $\frac{166}{27}$ **27.** $\frac{4}{15}(2\sqrt{2}-1)$ **29.** 36

31. $21e - 57$

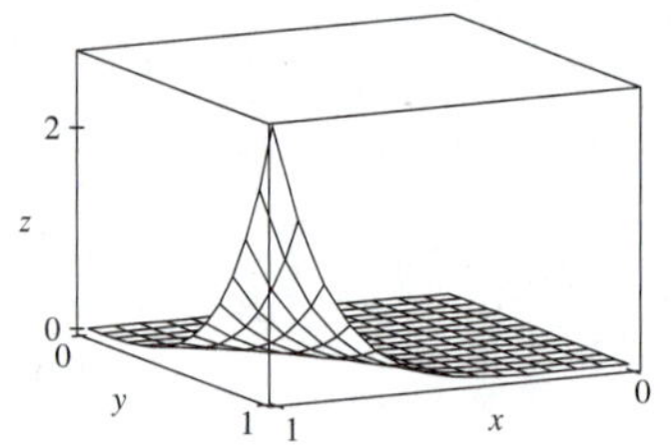

33. $\frac{5}{6}$

35. Fubini's Theorem does not apply. The integrand has an infinite discontinuity at the origin.

Exercises 15.3 ▫ page 1002

1. $\frac{9}{20}$ **3.** $\frac{4}{9}e^{3/2} - \frac{32}{45}$ **5.** $e - 1$ **7.** $\frac{256}{21}$ **9.** $\frac{1}{2}\ln 2$

11. $\frac{1}{2}e^4 - 2e$ **13.** $\frac{1}{2}(1 - \cos 1)$ **15.** $\frac{147}{20}$ **17.** 0

19. $\frac{7}{18}$ **21.** $\frac{31}{8}$ **23.** $\frac{1}{6}$ **25.** $\frac{128}{15}$

27. $\frac{1}{3}$ **29.** 0, 1.213, 0.713 **31.** $\frac{64}{3}$

33. 13,984,735,616/14,549,535 **35.** $\pi/2$

37. $\int_0^2 \int_{y^2}^4 f(x, y)\,dx\,dy$ **39.** $\int_{-3}^3 \int_0^{\sqrt{9-x^2}} f(x, y)\,dy\,dx$

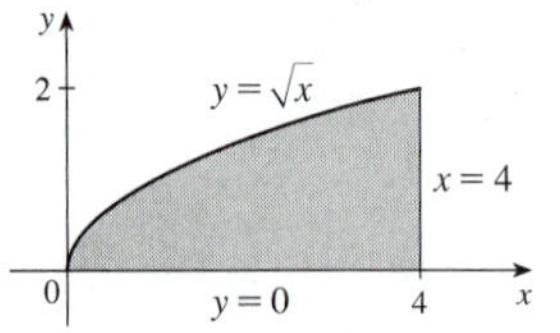

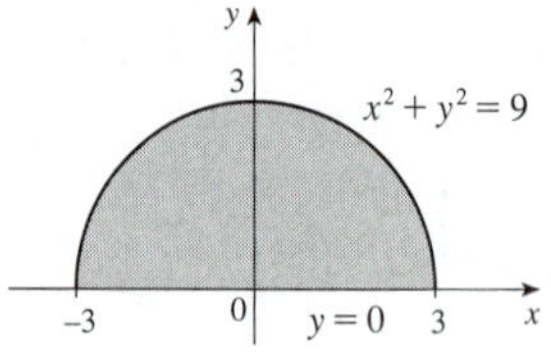

41. $\int_0^{\ln 2} \int_{e^y}^2 f(x, y)\,dx\,dy$

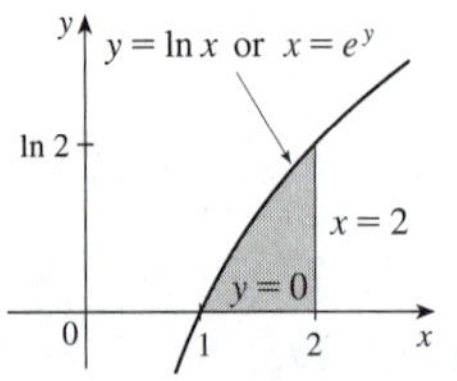

43. $\frac{1}{6}(e^9 - 1)$ **45.** $\frac{1}{4}\sin 81$ **47.** $\frac{1}{3}(2\sqrt{2} - 1)$ **49.** 1

51. $0 \leqslant \iint_D \sqrt{x^3 + y^3}\,dA \leqslant \sqrt{2}$ **55.** 8π **57.** $2\pi/3$

Exercises 15.4 ▫ page 1008

1. $\int_0^{2\pi} \int_0^2 f(r\cos\theta, r\sin\theta)\,r\,dr\,d\theta$ **3.** $\int_{-2}^2 \int_x^2 f(x, y)\,dy\,dx$

5. $\int_0^{2\pi} \int_2^5 f(r\cos\theta, r\sin\theta)\,r\,dr\,d\theta$

7. $33\pi/2$

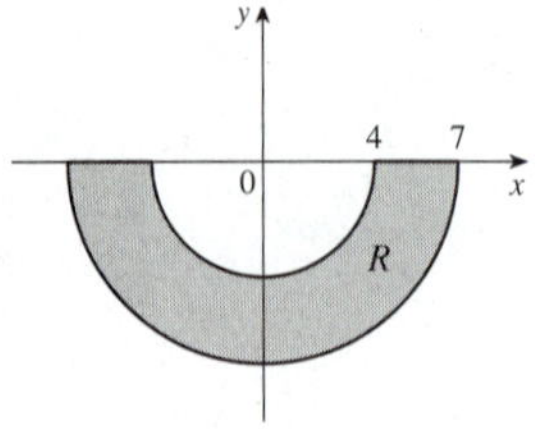

9. 0 **11.** $(\pi/2)\sin 9$ **13.** $(\pi/2)(1 - e^{-4})$

15. $\frac{3}{64}\pi^2$ **17.** $\pi/12$ **19.** $\frac{1}{8}(\pi - 2)$

21. $81\pi/2$ **23.** $\frac{4}{3}\pi a^3$ **25.** $(2\pi/3)[1 - (1/\sqrt{2})]$

27. $(8\pi/3)(64 - 24\sqrt{3})$ **29.** $(\pi/4)(e - 1)$ **31.** $4\pi/3$

33. 1800π ft^3 **35.** $\frac{15}{16}$ **37.** (a) $\sqrt{\pi}/4$ (b) $\sqrt{\pi}/2$

Exercises 15.5 ▫ page 1018

1. $\frac{64}{3}$ C **3.** $\frac{4}{3}, (\frac{4}{3}, 0)$ **5.** $6, (\frac{3}{4}, \frac{3}{2})$

7. $\frac{1}{4}(e^2 - 1), \left(\frac{e^2+1}{2(e^2-1)}, \frac{4(e^3-1)}{9(e^2-1)}\right)$

9. $\frac{27}{2}, (\frac{8}{5}, \frac{1}{2})$ **11.** $(\frac{3}{8}, 3\pi/16)$

13. $(2a/5, 2a/5)$ if vertex is (0, 0) and sides are along positive axes

15. $\frac{1}{16}(e^4 - 1), \frac{1}{8}(e^2 - 1), \frac{1}{16}(e^4 + 2e^2 - 3)$ **17.** $\frac{189}{20}, \frac{1269}{28}, \frac{1917}{35}$

19. $m = \pi^2/8$, $(\bar{x}, \bar{y}) = \left(\frac{2\pi}{3} - \frac{1}{\pi}, \frac{16}{9\pi}\right)$, $I_x = \frac{3}{64}\pi^2$, $I_y = \frac{1}{16}(\pi^4 - 3\pi^2)$, $I_0 = \frac{1}{16}\pi^4 - \frac{9}{64}\pi^2$

21. $\rho a^4/3, \rho a^4/3; a/\sqrt{3}, a/\sqrt{3}$

23. (a) $\frac{1}{2}$ (b) 0.375 (c) $\frac{5}{48} \approx 0.1042$

25. (b) (i) $e^{-0.2} \approx 0.8187$
(ii) $1 + e^{-1.8} - e^{-0.8} - e^{-1} \approx 0.3481$ (c) 2, 5

27. (a) ≈ 0.500 (b) ≈ 0.632

29. (a) $\iint_D (k/20)[20 - \sqrt{(x - x_0)^2 + (y - y_0)^2}]\,dA$, where D is the disk with radius 10 mi centered at the center of the city
(b) $200\pi k/3 \approx 209k$, $200(\pi/2 - \frac{8}{9})k \approx 136k$, on the edge

Exercises 15.6 ▫ page 1022

1. $15\sqrt{26}$ **3.** $3\sqrt{14}$ **5.** $12\sin^{-1}\frac{2}{3}$

7. $(\pi/6)(17\sqrt{17} - 5\sqrt{5})$ **9.** $(2\pi/3)(2\sqrt{2} - 1)$

11. $a^2(\pi - 2)$ **13.** 13.9783 **15.** (a) ≈ 1.83 (b) ≈ 1.8616

17. $\frac{45}{8}\sqrt{14} + \frac{15}{16}\ln[(11\sqrt{5} + 3\sqrt{70})/(3\sqrt{5} + \sqrt{70})]$

19. 3.3213 **23.** $(\pi/6)(101\sqrt{101} - 1)$

Exercises 15.7 ▫ page 1030

3. 1 **5.** $\frac{1}{3}(e^3 - 1)$ **7.** 4 **9.** $\frac{65}{28}$ **11.** $\frac{1}{10}$

13. $8/(3e)$ **15.** $16\pi/3$ **17.** $\frac{16}{3}$ **19.** 36π

21. (a) $\int_0^1 \int_0^x \int_0^{\sqrt{1-y^2}} dz\,dy\,dx$ (b) $\frac{1}{4}\pi - \frac{1}{3}$

23. 60.533 **25.**

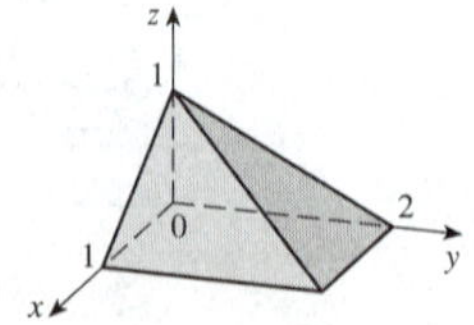

27. $\int_{-2}^2 \int_0^6 \int_{-\sqrt{4-x^2}}^{\sqrt{4-x^2}} f(x, y, z)\,dz\,dy\,dx$
$= \int_0^6 \int_{-2}^2 \int_{-\sqrt{4-x^2}}^{\sqrt{4-x^2}} f(x, y, z)\,dz\,dx\,dy$
$= \int_{-2}^2 \int_0^6 \int_{-\sqrt{4-z^2}}^{\sqrt{4-z^2}} f(x, y, z)\,dx\,dy\,dz$
$= \int_0^6 \int_{-2}^2 \int_{-\sqrt{4-z^2}}^{\sqrt{4-z^2}} f(x, y, z)\,dx\,dz\,dy$
$= \int_{-2}^2 \int_{-\sqrt{4-x^2}}^{\sqrt{4-x^2}} \int_0^6 f(x, y, z)\,dy\,dz\,dx$
$= \int_{-2}^2 \int_{-\sqrt{4-z^2}}^{\sqrt{4-z^2}} \int_0^6 f(x, y, z)\,dy\,dx\,dz$

29. $\int_{-1}^{1} \int_{0}^{1-x^2} \int_{0}^{y} f(x, y, z)\, dz\, dy\, dx$
$= \int_{0}^{1} \int_{-\sqrt{1-y}}^{\sqrt{1-y}} \int_{0}^{y} f(x, y, z)\, dz\, dx\, dy$
$= \int_{0}^{1} \int_{z}^{1} \int_{-\sqrt{1-y}}^{\sqrt{1-y}} f(x, y, z)\, dx\, dy\, dz$
$= \int_{0}^{1} \int_{0}^{y} \int_{-\sqrt{1-y}}^{\sqrt{1-y}} f(x, y, z)\, dx\, dz\, dy$
$= \int_{-1}^{1} \int_{0}^{1-x^2} \int_{z}^{1-x^2} f(x, y, z)\, dy\, dz\, dx$
$= \int_{0}^{1} \int_{-\sqrt{1-z}}^{\sqrt{1-z}} \int_{z}^{1-x^2} f(x, y, z)\, dy\, dx\, dz$

31. $\int_{0}^{1} \int_{\sqrt{x}}^{1} \int_{0}^{1-y} f(x, y, z)\, dz\, dy\, dx$
$= \int_{0}^{1} \int_{0}^{y^2} \int_{0}^{1-y} f(x, y, z)\, dz\, dx\, dy$
$= \int_{0}^{1} \int_{0}^{1-z} \int_{0}^{y^2} f(x, y, z)\, dx\, dy\, dz$
$= \int_{0}^{1} \int_{0}^{1-y} \int_{0}^{y^2} f(x, y, z)\, dx\, dz\, dy$
$= \int_{0}^{1} \int_{0}^{1-\sqrt{x}} \int_{\sqrt{x}}^{1-z} f(x, y, z)\, dy\, dz\, dx$
$= \int_{0}^{1} \int_{0}^{(1-z)^2} \int_{\sqrt{x}}^{1-z} f(x, y, z)\, dy\, dx\, dz$

33. $\int_{0}^{1} \int_{y}^{1} \int_{0}^{y} f(x, y, z)\, dz\, dx\, dy = \int_{0}^{1} \int_{0}^{x} \int_{0}^{y} f(x, y, z)\, dz\, dy\, dx$
$= \int_{0}^{1} \int_{z}^{1} \int_{y}^{1} f(x, y, z)\, dx\, dy\, dz = \int_{0}^{1} \int_{0}^{y} \int_{y}^{1} f(x, y, z)\, dx\, dz\, dy$
$= \int_{0}^{1} \int_{0}^{x} \int_{z}^{x} f(x, y, z)\, dy\, dz\, dx = \int_{0}^{1} \int_{z}^{1} \int_{z}^{x} f(x, y, z)\, dy\, dx\, dz$

35. $\frac{79}{30}, \left(\frac{358}{553}, \frac{33}{79}, \frac{571}{553}\right)$ **37.** $a^5, (7a/12, 7a/12, 7a/12)$

39. (a) $m = \int_{-3}^{3} \int_{-\sqrt{9-x^2}}^{\sqrt{9-x^2}} \int_{1}^{5-y} \sqrt{x^2 + y^2}\, dz\, dy\, dx$
(b) $(\bar{x}, \bar{y}, \bar{z})$, where
$\bar{x} = (1/m) \int_{-3}^{3} \int_{-\sqrt{9-x^2}}^{\sqrt{9-x^2}} \int_{1}^{5-y} x\sqrt{x^2 + y^2}\, dz\, dy\, dx$
$\bar{y} = (1/m) \int_{-3}^{3} \int_{-\sqrt{9-x^2}}^{\sqrt{9-x^2}} \int_{1}^{5-y} y\sqrt{x^2 + y^2}\, dz\, dy\, dx$
$\bar{z} = (1/m) \int_{-3}^{3} \int_{-\sqrt{9-x^2}}^{\sqrt{9-x^2}} \int_{1}^{5-y} z\sqrt{x^2 + y^2}\, dz\, dy\, dx$
(c) $\int_{-3}^{3} \int_{-\sqrt{9-x^2}}^{\sqrt{9-x^2}} \int_{1}^{5-y} (x^2 + y^2)^{3/2}\, dz\, dy\, dx$

41. (a) $\frac{3}{32}\pi + \frac{11}{24}$
(b) $(\bar{x}, \bar{y}, \bar{z}) = \left(\dfrac{28}{9\pi + 44}, \dfrac{30\pi + 128}{45\pi + 220}, \dfrac{45\pi + 208}{135\pi + 660}\right)$
(c) $\frac{1}{240}(68 + 15\pi)$

43. $I_x = I_y = I_z = \frac{2}{3}kL^5$ **45.** (a) $\frac{1}{8}$ (b) $\frac{1}{64}$ (c) $\frac{1}{5760}$

47. $L^3/8$

49. The region bounded by the ellipsoid $x^2 + 2y^2 + 3z^2 = 1$

Exercises 15.8 ▫ page 1037

1.

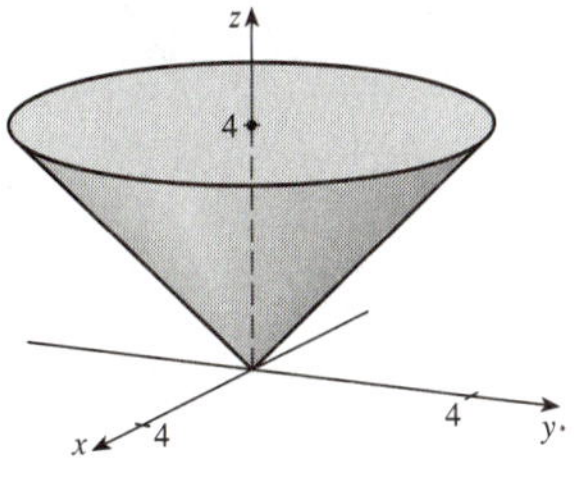

$64\pi/3$

3.

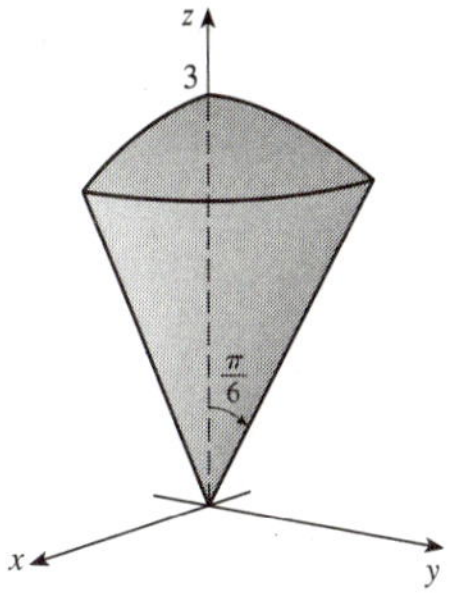

$(9\pi/4)(2 - \sqrt{3})$

5. $\int_{0}^{\pi/2} \int_{0}^{3} \int_{0}^{2} f(r\cos\theta, r\sin\theta, z)\, r\, dz\, dr\, d\theta$ **7.** 384π

9. $\pi(e^6 - e - 5)$ **11.** $2\pi/5$

13. (a) 162π (b) $(0, 0, 15)$

15. $\pi K a^2/8, (0, 0, 2a/3)$ **17.** $4\pi/5$ **19.** $15\pi/16$

21. $1562\pi/15$ **23.** 10π **25.** $\left(0, \frac{525}{296}, 0\right)$

27. (a) $\left(0, 0, \frac{3}{8}a\right)$ (b) $4K\pi a^5/15$

29. $(2\pi/3)[1 - (1/\sqrt{2})], (0, 0, 3/[8(2 - \sqrt{2})])$ **31.** $5\pi/6$

33. $8\pi/35$ **35.** $243\pi/5$ **37.** $136\pi/99$

Exercises 15.9 ▫ page 1048

1. -14 **3.** 0 **5.** $2uvw$

7. The parallelogram with vertices (0, 0), (6, 3), (12, 1), (6, −2)

9. The region bounded by the line $y = 1$, the y-axis, and $y = \sqrt{x}$

11. -3 **13.** 6π **15.** $2 \ln 3$

17. (a) $\frac{4}{3}\pi abc$ (b) $1.083 \times 10^{12}\ \text{km}^3$

19. $\frac{8}{5} \ln 8$ **21.** $\frac{3}{2} \sin 1$ **23.** $e - e^{-1}$

Chapter 15 Review ▫ page 1049

True-False Quiz

1. True **3.** True **5.** True **7.** False

Exercises

1. ≈ 64.0 **3.** $4e^2 - 4e + 3$ **5.** $\frac{1}{2} \sin 1$ **7.** $\frac{2}{3}$

9. $\int_{0}^{\pi} \int_{2}^{4} f(r\cos\theta, r\sin\theta)\, r\, dr\, d\theta$

11. The region inside the loop of the four-leaved rose $r = \sin 2\theta$ in the first quadrant

13. $\frac{1}{2} \sin 1$ **15.** $\frac{1}{2}e^6 - \frac{7}{2}$ **17.** $\frac{1}{4} \ln 2$

19. 8 **21.** $81\pi/5$ **23.** 40.5 **25.** $\pi/96$

27. $\frac{64}{15}$ **29.** 176 **31.** $\frac{2}{3}$ **33.** $2ma^3/9$

35. (a) $\frac{1}{4}$ (b) $\left(\frac{1}{3}, \frac{8}{15}\right)$
(c) $I_x = \frac{1}{12}, I_y = \frac{1}{24}; \bar{\bar{y}} = 1/\sqrt{3}, \bar{\bar{x}} = 1/\sqrt{6}$

37. (a) $(0, 0, h/4)$ (b) $\pi a^4 h/10$ **39.** $\ln(\sqrt{2} + \sqrt{3}) + \sqrt{2}/3$

41. 97.2 **43.** 0.0512 **45.** (a) $\frac{1}{15}$ (b) $\frac{1}{3}$ (c) $\frac{1}{45}$

47. $\int_{0}^{1} \int_{0}^{1-z} \int_{-\sqrt{y}}^{\sqrt{y}} f(x, y, z)\, dx\, dy\, dz$ **49.** $-\ln 2$ **51.** 0

Problems Plus ▫ page 1052

1. 30 **3.** $\frac{1}{2} \sin 1$ **7.** (b) 0.90

Chapter 16

Exercises 16.1 ▫ page 1060

1.

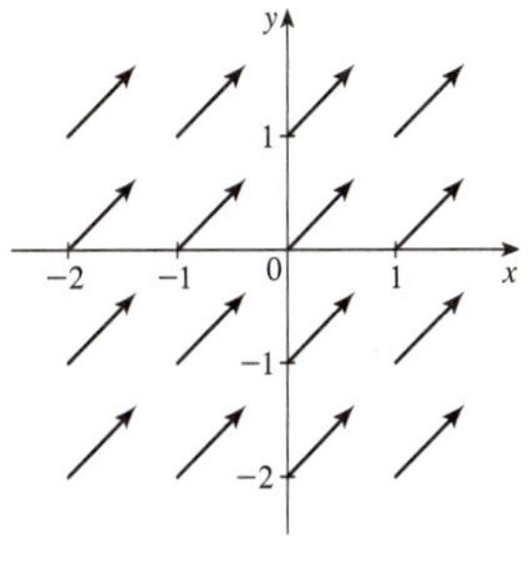

3.

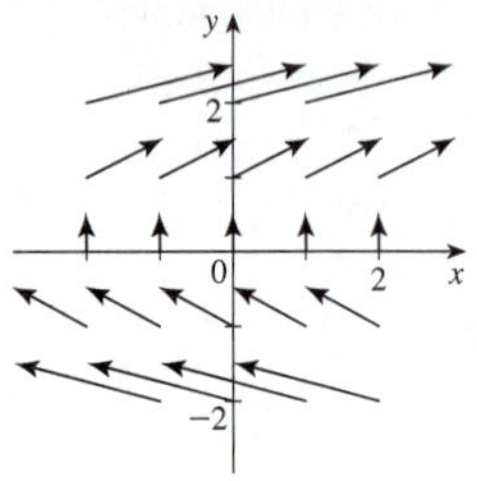

5.

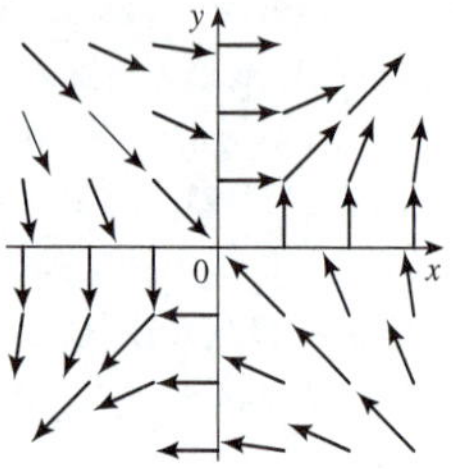

7.

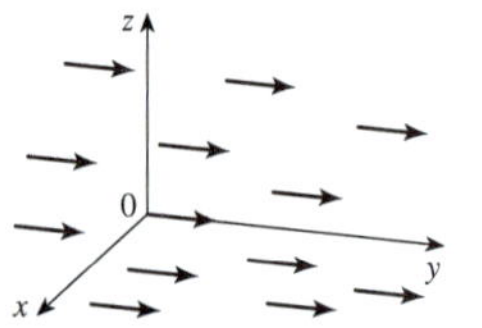

9.

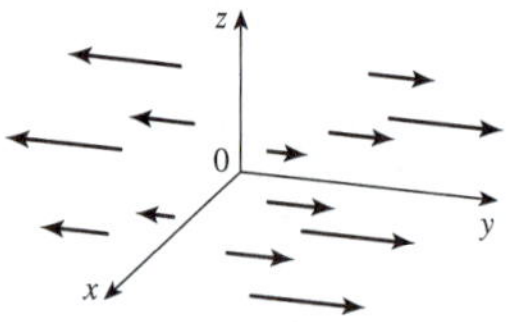

11. II **13.** I **15.** IV **17.** III

19.

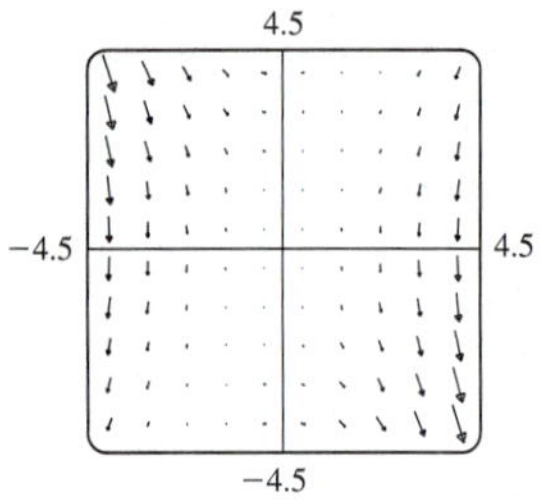

The line $y = 2x$

21. $\nabla f(x, y) = \dfrac{1}{x + 2y}\,\mathbf{i} + \dfrac{2}{x + 2y}\,\mathbf{j}$

23. $\nabla f(x, y, z) = \dfrac{x}{\sqrt{x^2 + y^2 + z^2}}\,\mathbf{i} + \dfrac{y}{\sqrt{x^2 + y^2 + z^2}}\,\mathbf{j} + \dfrac{z}{\sqrt{x^2 + y^2 + z^2}}\,\mathbf{k}$

25. $\nabla f(x, y) = (y - 2)\mathbf{i} + x\,\mathbf{j}$

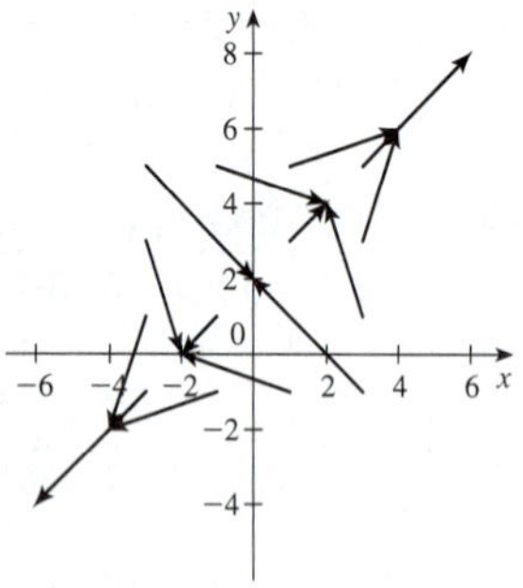

27.

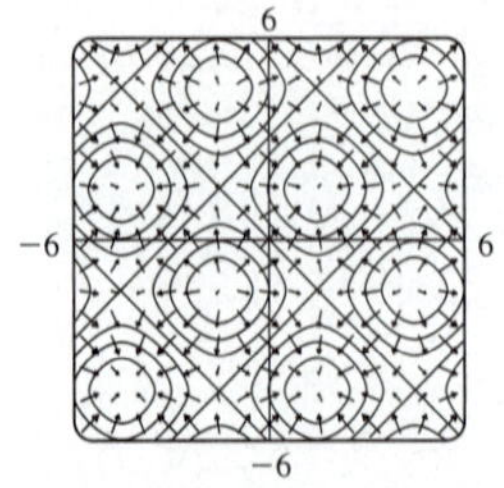

29. IV **31.** II

33. (a)

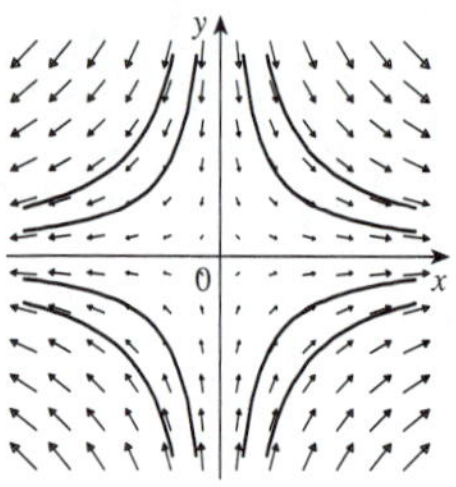

$y = C/x$

(b) $y = 1/x,\ x > 0$

Exercises 16.2 ▫ page 1071

1. $\frac{1}{2}\left(17\sqrt{17} - 1\right)$ **3.** 1638.4 **5.** $\frac{464}{5} + 9 \ln 3$ **7.** $\frac{17}{3}$
9. 320 **11.** $\frac{1}{12}\sqrt{14}\,(e^6 - 1)$ **13.** $\frac{1}{5}$ **15.** $\frac{97}{3}$
17. (a) Positive (b) Negative **19.** $-\frac{59}{105}$
21. $\frac{6}{5} - \cos 1 - \sin 1$
23. $3\pi + \frac{2}{3}$

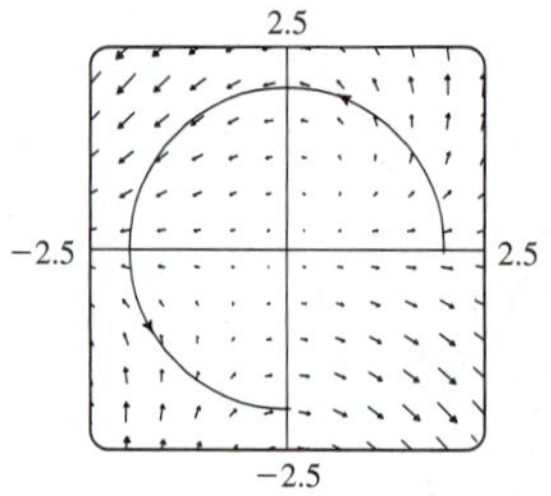

25. (a) $\frac{11}{8} - 1/e$ (b)

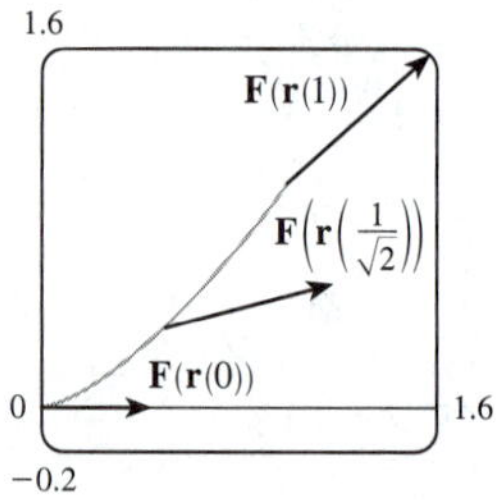

27. $\frac{945}{16{,}777{,}216}\,\pi$ **29.** 0.052 **31.** $2\pi k,\ (4/\pi, 0)$
33. (a) $\bar{x} = (1/m) \int_C x\rho(x, y, z)\, ds$,
$\bar{y} = (1/m) \int_C y\rho(x, y, z)\, ds$,
$\bar{z} = (1/m) \int_C z\rho(x, y, z)\, ds$, where $m = \int_C \rho(x, y, z)\, ds$
(b) $(0, 0, 3\pi)$
35. $I_x = k\left((\pi/2) - \frac{4}{3}\right),\ I_y = k\left((\pi/2) - \frac{2}{3}\right)$
37. $2\pi^2$ **39.** 26 **41.** 1.67×10^4 ft-lb **43.** (b) Yes
45. ≈ 22 J

Exercises 16.3 ▫ page 1081

1. 40 **3.** $f(x, y) = 3x^2 + 5xy + 2y^2 + K$
5. Not conservative **7.** $f(x, y) = x^2 \cos y - y \sin x + K$
9. $f(x, y) = ye^x + x \sin y + K$ **11.** (b) 16
13. (a) $f(x, y) = \frac{1}{4}x^4y^4$ (b) 4
15. (a) $f(x, y, z) = xyz + z^2$ (b) 77
17. (a) $f(x, y, z) = xy^2 \cos z$ (b) 0
19. 2 **21.** 30 **23.** No **25.** No
29. (a) Yes (b) Yes (c) Yes
31. (a) Yes (b) Yes (c) No

Exercises 16.4 ▫ page 1089

1. 6 **3.** $\frac{2}{3}$ **7.** $e - 1$ **9.** $\frac{1}{3}$ **11.** -24π
13. $\frac{4}{3} - 2\pi$ **15.** $\frac{625}{2}\pi$ **17.** $-\frac{1}{12}$
19. 3π **21.** (c) $\frac{9}{2}$ **23.** $\left(\frac{1}{3}, \frac{1}{3}\right)$

Exercises 16.5 ▫ page 1096

1. (a) $-x^2\,\mathbf{i} + 3xy\,\mathbf{j} - xz\,\mathbf{k}$ (b) yz
3. (a) $(x - y)\,\mathbf{i} - y\,\mathbf{j} + \mathbf{k}$ (b) $z - 1/(2\sqrt{z})$
5. (a) $\mathbf{0}$ (b) 1
7. (a) $\langle 1/y, -1/x, 1/x \rangle$ (b) $1/x + 1/y + 1/z$
9. (a) Negative (b) curl $\mathbf{F} = \mathbf{0}$
11. (a) Zero (b) curl $\mathbf{F}$ points in the negative z-direction
13. $f(x, y, z) = xyz + K$ **15.** $f(x, y, z) = x^2y + y^2z + K$
17. Not conservative **19.** No

Exercises 16.6 ▫ page 1106

1. Circular paraboloid with axis the z-axis
3. Circular cylinder with axis the x-axis
5.

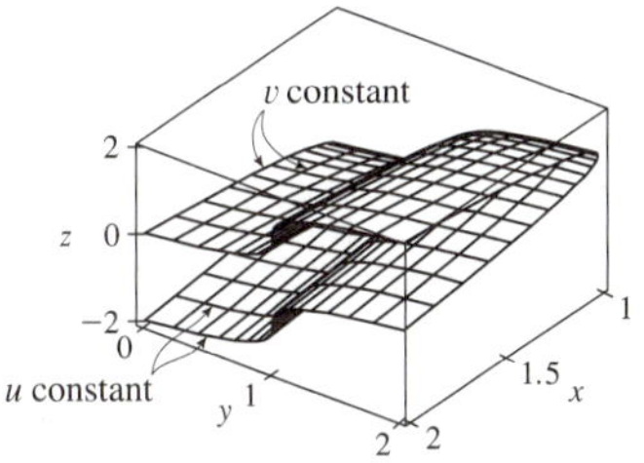

7.

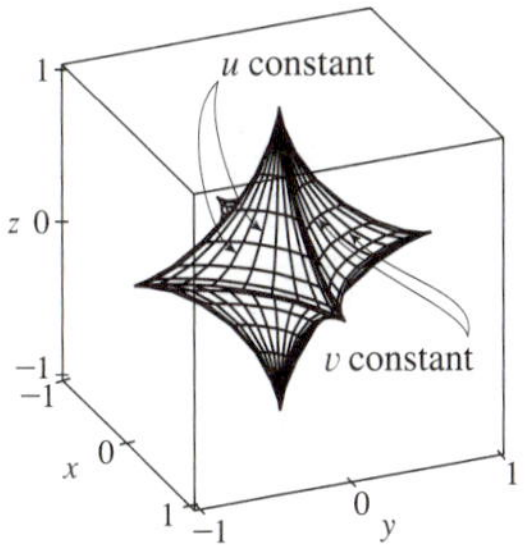

9.

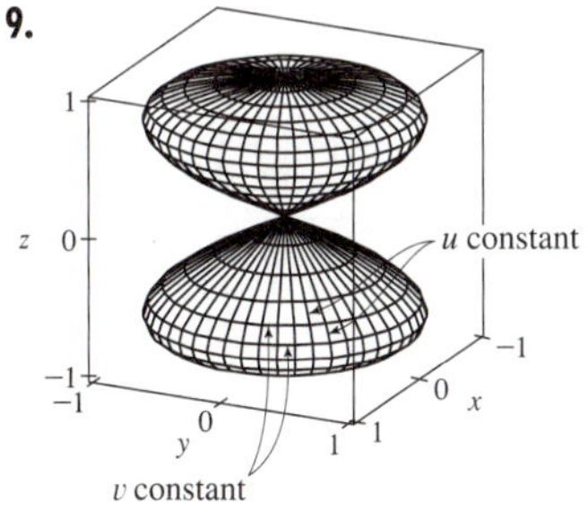

11. IV **13.** I **15.** II
17. $x = 1 + u + v,\ y = 2 + u - v,\ z = -3 - u + v$
19. $x = x,\ z = z,\ y = \sqrt{1 - x^2 + z^2}$
21. $x = 2\sin\phi\cos\theta,\ y = 2\sin\phi\sin\theta,$
$z = 2\cos\phi,\ 0 \leqslant \phi \leqslant \pi/4,\ 0 \leqslant \theta \leqslant 2\pi$
$\left[\text{or } x = x,\ y = y,\ z = \sqrt{4 - x^2 - y^2},\ x^2 + y^2 \leqslant 2\right]$
23. $x = x,\ y = 4\cos\theta,\ z = 4\sin\theta,\ 0 \leqslant x \leqslant 5,\ 0 \leqslant \theta \leqslant 2\pi$
27. $x = x,\ y = e^{-x}\cos\theta,$
$z = e^{-x}\sin\theta,\ 0 \leqslant x \leqslant 3,$
$0 \leqslant \theta \leqslant 2\pi$

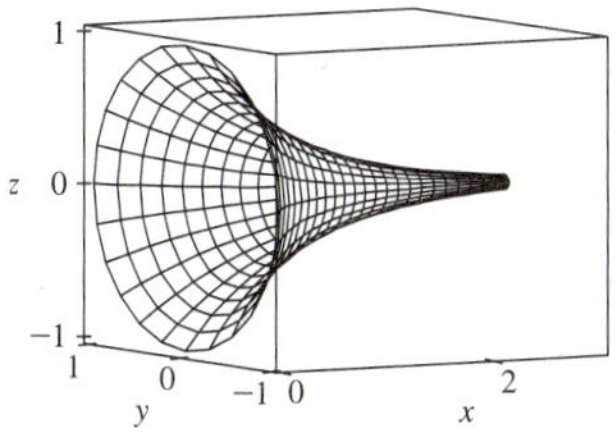

29. (a) Direction reverses (b) Number of coils doubles
31. $3x - y + 3z = 3$ **33.** $-x + 2z = 1$ **35.** $4\sqrt{6}\,\pi$
37. $(2\pi/3)(2\sqrt{2} - 1)$ **39.** $(\pi/6)(17\sqrt{17} - 5\sqrt{5})$
41. $(\sqrt{21}/2) + \frac{17}{4}\left[\ln(2 + \sqrt{21}) - \ln\sqrt{17}\right]$ **43.** $2a^2(\pi - 2)$
45. $\pi\left(2\sqrt{6} - \frac{8}{3}\right)$ **47.** 13.9783
49. (a) ≈ 1.83 (b) ≈ 1.8616
51. $\frac{45}{8}\sqrt{14} + \frac{15}{16}\ln\left[(11\sqrt{5} + 3\sqrt{70})/(3\sqrt{5} + \sqrt{70})\right]$
53. (b)

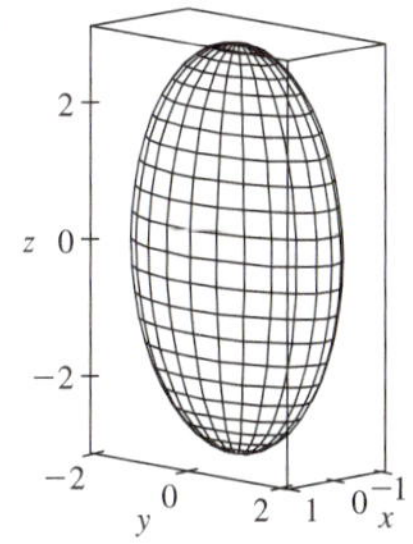

(c) $\int_0^{2\pi}\int_0^{\pi}\sqrt{36\sin^4u\cos^2v + 9\sin^4u\sin^2v + 4\cos^2u\sin^2u}\,du\,dv$
55. 4.4506

Exercises 16.7 ▫ page 1119

1. $8(1 + \sqrt{2} + \sqrt{3}) \approx 33.17$ **3.** 900π **5.** $171\sqrt{14}$
7. $\sqrt{3}/24$
9. $364\sqrt{2}\,\pi/3$ **11.** $(\pi/60)(391\sqrt{17} + 1)$
13. 16π **15.** 16π **17.** $5\sqrt{5}/48 + 1/240$
19. $\frac{713}{180}$ **21.** $-\frac{1}{6}$ **23.** $-\frac{4}{3}\pi$
25. 0 **27.** 48 **29.** 0.1642 **31.** 3.4895
33. $\iint_S \mathbf{F}\cdot d\mathbf{S} = \iint_D [P(\partial h/\partial x) - Q + R(\partial h/\partial z)]\,dA$,
where D = projection on xz-plane
35. $(0, 0, a/2)$
37. (a) $I_z = \iint_S (x^2 + y^2)\rho(x, y, z)\,dS$ (b) $4329\sqrt{2}\,\pi/5$
39. $194{,}400\pi$ **41.** $8\pi a^3\varepsilon_0/3$ **43.** 1248π

Exercises 16.8 ▫ page 1125

3. 0 **5.** 0 **7.** -1 **9.** 80π
11. (a) $81\pi/2$ (b)

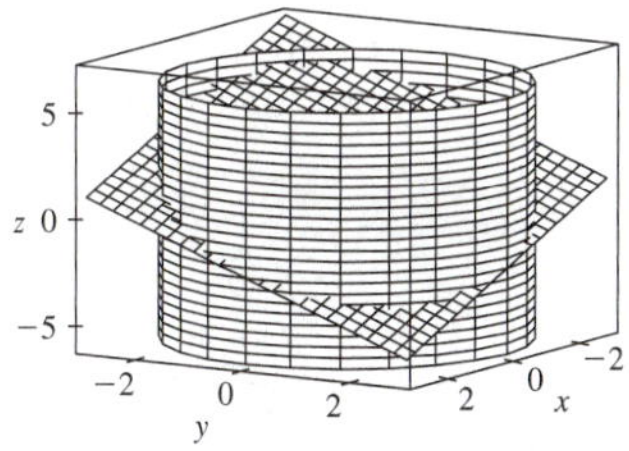

(c) $x = 3\cos t,\ y = 3\sin t,$
$z = 1 - 3(\cos t + \sin t),$
$0 \leqslant t \leqslant 2\pi$

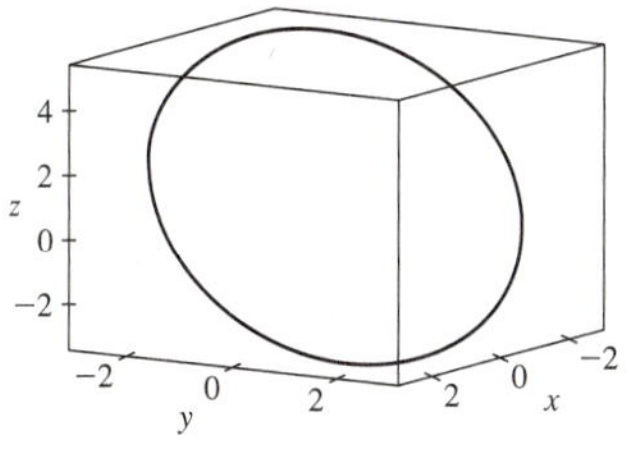

17. 16

Exercises 16.9 ▫ page 1132

1. Negative at P_1, positive at P_2 **7.** 2 **9.** $9\pi/2$
11. 0 **13.** $32\pi/3$ **15.** 0
17. $341\sqrt{2}/60 + \frac{81}{20}\arcsin(\sqrt{3}/3)$ **19.** $13\pi/20$

Chapter 16 Review ▫ page 1135

True-False Quiz

1. False **3.** True **5.** False **7.** True

Exercises

1. (a) Negative (b) Positive **3.** $4\sqrt{5}$ **5.** $-\pi$
7. $\frac{17}{2}$ **9.** $\frac{11}{12} - 4/e$ **11.** $f(x, y) = e^y + xe^{xy}$ **13.** 0
17. -8π **25.** $\frac{1}{6}(27 - 5\sqrt{5})$ **27.** $\pi(391\sqrt{17} + 1)/60$
29. $-64\pi/3$ **33.** $-\frac{1}{2}$ **37.** -4 **39.** 21

Chapter 17

Exercises 17.1 ▫ page 1147

1. $y = c_1e^{4x} + c_2e^{2x}$ **3.** $y = e^{-4x}(c_1 \cos 5x + c_2 \sin 5x)$
5. $y = c_1e^x + c_2xe^x$ **7.** $y = c_1 \cos(x/2) + c_2 \sin(x/2)$
9. $y = c_1 + c_2e^{-x/4}$ **11.** $y = c_1e^{(1+\sqrt{2})t} + c_2e^{(1-\sqrt{2})t}$
13. $y = e^{-t/2}[c_1 \cos(\sqrt{3}t/2) + c_2 \sin(\sqrt{3}t/2)]$
15.

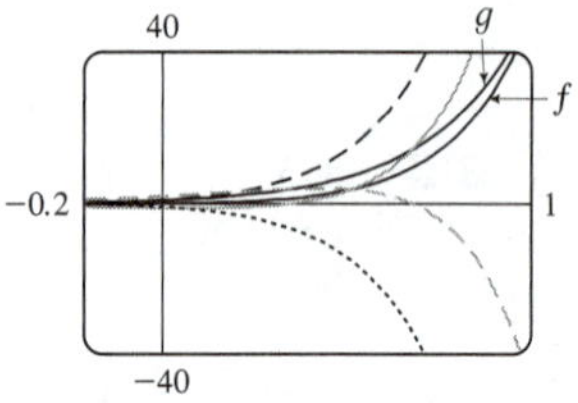

All solutions approach 0 as $x \to -\infty$ and approach $\pm\infty$ as $x \to \infty$.
17. $y = 2e^{-3x/2} + e^{-x}$ **19.** $y = e^{x/2} - 2xe^{x/2}$
21. $y = 3 \cos 4x - \sin 4x$ **23.** $y = e^{-x}(2 \cos x + 3 \sin x)$
25. $y = 3\cos(\frac{1}{2}x) - 4\sin(\frac{1}{2}x)$ **27.** $y = \dfrac{e^{x+3}}{e^3 - 1} + \dfrac{e^{2x}}{1 - e^3}$
29. No solution
31. $y = e^{-2x}(2 \cos 3x - e^{\pi} \sin 3x)$
33. (b) $\lambda = n^2\pi^2/L^2$, n a positive integer; $y = C \sin(n\pi x/L)$

Exercises 17.2 ▫ page 1154

1. $y = c_1e^{-2x} + c_2e^{-x} + \frac{1}{2}x^2 - \frac{3}{2}x + \frac{7}{4}$
3. $y = c_1 + c_2e^{2x} + \frac{1}{40}\cos 4x - \frac{1}{20}\sin 4x$
5. $y = e^{2x}(c_1 \cos x + c_2 \sin x) + \frac{1}{10}e^{-x}$
7. $y = \frac{3}{2}\cos x + \frac{11}{2}\sin x + \frac{1}{2}e^x + x^3 - 6x$
9. $y = e^x(\frac{1}{2}x^2 - x + 2)$
11.

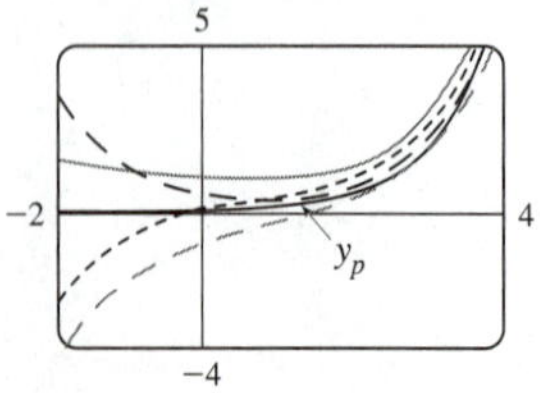

The solutions are all asymptotic to $y_p = e^x/10$ as $x \to \infty$. Except for y_p, all solutions approach either ∞ or $-\infty$ as $x \to -\infty$.

13. $y_p = Ae^{2x} + (Bx^2 + Cx + D) \cos x + (Ex^2 + Fx + G) \sin x$
15. $y_p = Ax + (Bx + C)e^{9x}$
17. $y_p = xe^{-x}[(Ax^2 + Bx + C) \cos 3x + (Dx^2 + Ex + F) \sin 3x]$
19. $y = c_1 \cos 2x + c_2 \sin 2x + \frac{1}{4}x$
21. $y = c_1e^x + c_2xe^x + e^{2x}$
23. $y = (c_1 + x) \sin x + (c_2 + \ln \cos x) \cos x$
25. $y = [c_1 + \ln(1 + e^{-x})]e^x + [c_2 - e^{-x} + \ln(1 + e^{-x})]e^{2x}$
27. $y = [c_1 - \frac{1}{2}\int (e^x/x)\, dx]e^{-x} + [c_2 + \frac{1}{2}\int (e^{-x}/x)\, dx]e^x$

Exercises 17.3 ▫ page 1162

1. $x = 0.36 \sin(10t/3)$ **3.** $x = -\frac{1}{5}e^{-6t} + \frac{6}{5}e^{-t}$ **5.** $\frac{49}{12}$ kg
7.

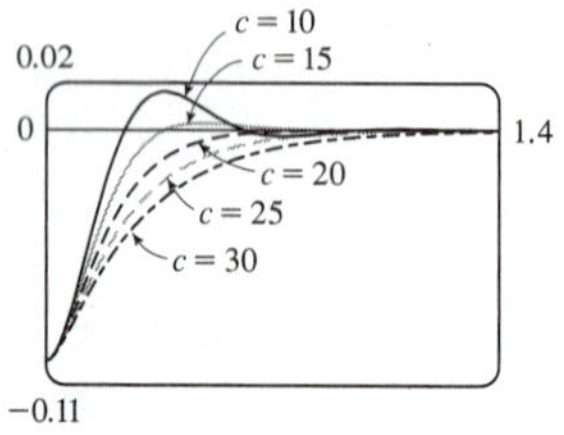

13. $Q(t) = (-e^{-10t}/250)(6 \cos 20t + 3 \sin 20t) + \frac{3}{125}$,
$I(t) = \frac{3}{5}e^{-10t} \sin 20t$
15. $Q(t) = e^{-10t}[\frac{3}{250}\cos 20t - \frac{3}{500}\sin 20t]$
$- \frac{3}{250}\cos 10t + \frac{3}{125}\sin 10t$

Exercises 17.4 ▫ page 1167

1. $c_0 \sum_{n=0}^{\infty} \dfrac{x^n}{n!} = c_0e^x$ **3.** $c_0 \sum_{n=0}^{\infty} \dfrac{x^{3n}}{3^n n!} = c_0e^{x^3/3}$

5. $c_0 \sum_{n=0}^{\infty} \dfrac{(-1)^n}{2^n n!}x^{2n} + c_1 \sum_{n=0}^{\infty} \dfrac{(-2)^n n!}{(2n + 1)!}x^{2n+1}$

7. $c_0 + c_1x + c_0\dfrac{x^2}{2} + c_0 \sum_{n=2}^{\infty} \dfrac{(-1)^{n-1}(2n - 3)!}{2^{2n-2}n!(n - 2)!}x^{2n}$

9. $\sum_{n=0}^{\infty} \dfrac{x^{2n}}{2^n n!} = e^{x^2/2}$

11. $x + \sum_{n=1}^{\infty} \dfrac{(-1)^n 2^2 5^2 \cdot \cdots \cdot (3n - 1)^2}{(3n + 1)!}x^{3n+1}$

Chapter 17 Review ▫ page 1168

True-False Quiz

1. True **3.** True

Exercises

1. $y = c_1e^{5x} + c_2e^{-3x}$ **3.** $y = c_1 \cos(\sqrt{3}x) + c_2 \sin(\sqrt{3}x)$
5. $y = e^{2x}(c_1 \cos x + c_2 \sin x + 1)$
7. $y = c_1e^x + c_2xe^x - \frac{1}{2}\cos x - \frac{1}{2}(x + 1) \sin x$
9. $y = c_1e^{3x} + c_2e^{-2x} - \frac{1}{6} - \frac{1}{5}xe^{-2x}$
11. $y = 5 - 2e^{-6(x-1)}$ **13.** $y = (e^{4x} - e^x)/3$

15. $\sum_{n=0}^{\infty} \dfrac{(-2)^n n!}{(2n + 1)!}x^{2n+1}$

17. $Q(t) = -0.02e^{-10t}(\cos 10t + \sin 10t) + 0.03$
19. (c) $2\pi/k \approx 85$ min (d) $\approx$17,600 mi/h

Appendixes

Exercises A ▫ page A9

1. 18 **3.** π **5.** $5 - \sqrt{5}$ **7.** $2 - x$

9. $|x + 1| = \begin{cases} x + 1 & \text{for } x \geqslant -1 \\ -x - 1 & \text{for } x < -1 \end{cases}$ **11.** $x^2 + 1$

13. $(-2, \infty)$

15. $[-1, \infty)$

17. $(3, \infty)$

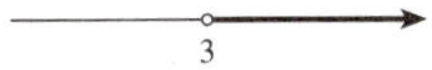

19. $(2, 6)$

21. $(0, 1]$

23. $[-1, \frac{1}{2})$

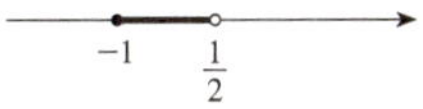

25. $(-\infty, 1) \cup (2, \infty)$

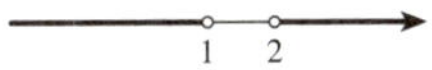

27. $[-1, \frac{1}{2}]$

29. $(-\infty, \infty)$

31. $(-\sqrt{3}, \sqrt{3})$

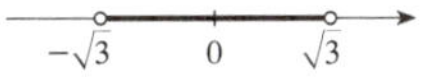

33. $(-\infty, 1]$

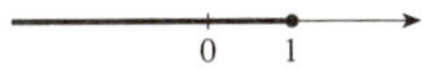

35. $(-1, 0) \cup (1, \infty)$

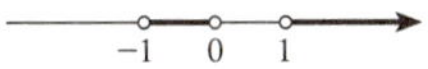

37. $(-\infty, 0) \cup (\frac{1}{4}, \infty)$

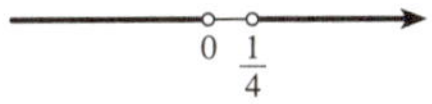

39. $10 \leqslant C \leqslant 35$ **41.** (a) $T = 20 - 10h, 0 \leqslant h \leqslant 12$ (b) $-30°C \leqslant T \leqslant 20°C$ **43.** $\pm\frac{3}{2}$ **45.** $2, -\frac{4}{3}$
47. $(-3, 3)$ **49.** $(3, 5)$ **51.** $(-\infty, -7] \cup [-3, \infty)$
53. $[1.3, 1.7]$ **55.** $[-4, -1] \cup [1, 4]$
57. $x \geqslant (a + b)c/(ab)$ **59.** $x > (c - b)/a$

Exercises B ▫ page A15

1. 5 **3.** $\sqrt{74}$ **5.** $2\sqrt{37}$ **7.** 2 **9.** $-\frac{9}{2}$

17.

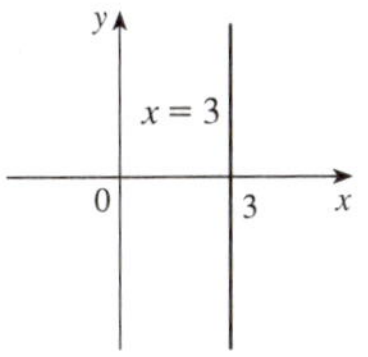

19.

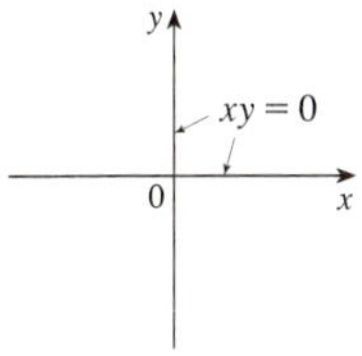

21. $y = 6x - 15$

23. $2x - 3y + 19 = 0$ **25.** $5x + y = 11$ **27.** $y = 3x - 2$
29. $y = 3x - 3$ **31.** $y = 5$ **33.** $x + 2y + 11 = 0$
35. $5x - 2y + 1 = 0$

37. $m = -\frac{1}{3}$, $b = 0$

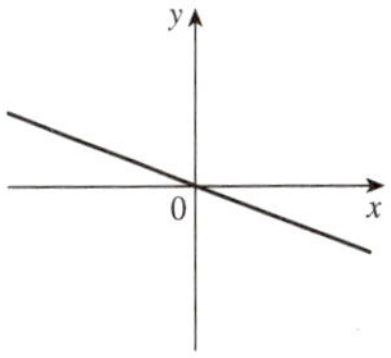

39. $m = 0$, $b = -2$

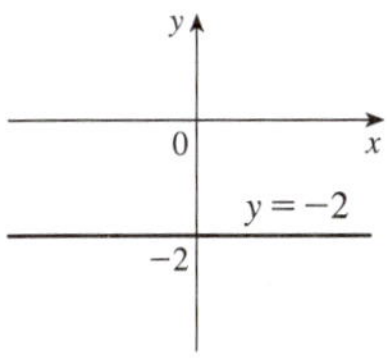

41. $m = \frac{3}{4}$, $b = -3$

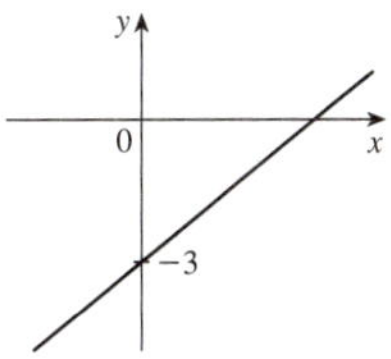

43.

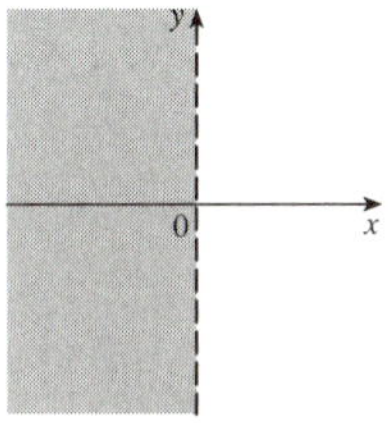

45.

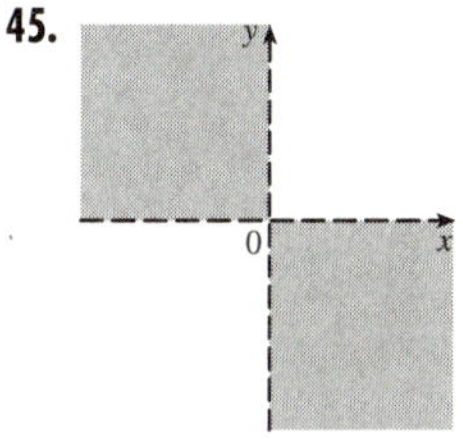

47.

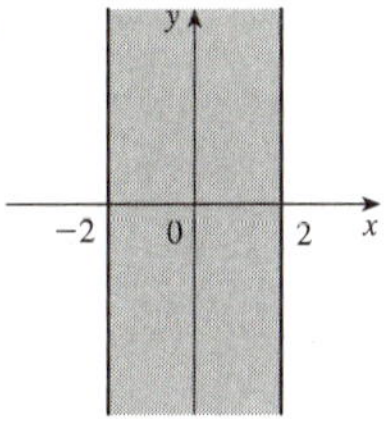

49.

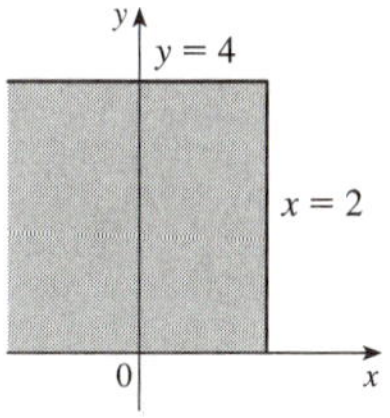

51.

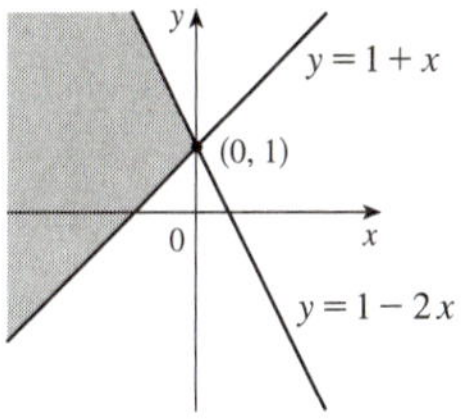

53. $(0, -4)$ **55.** (a) $(4, 9)$ (b) $(3.5, -3)$ **57.** $(1, -2)$
59. $y = x - 3$ **61.** (b) $4x - 3y - 24 = 0$

Exercises C ▫ page A23

1. $(x - 3)^2 + (y + 1)^2 = 25$ **3.** $x^2 + y^2 = 65$
5. $(2, -5), 4$ **7.** $(-\frac{1}{2}, 0), \frac{1}{2}$ **9.** $(\frac{1}{4}, -\frac{1}{4}), \sqrt{10}/4$

11. Parabola

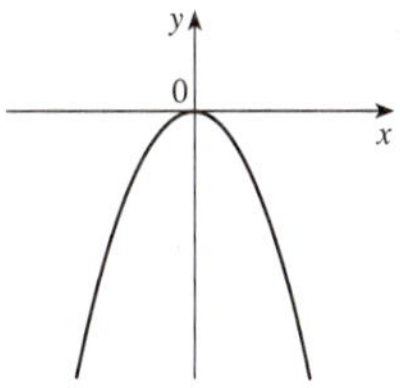

13. Ellipse

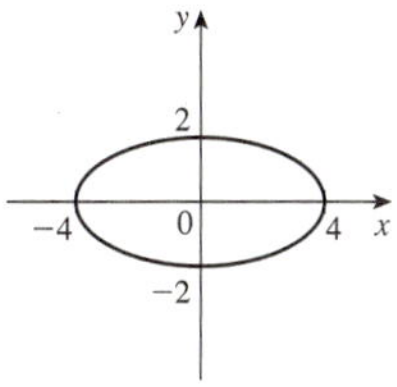

15. Hyperbola

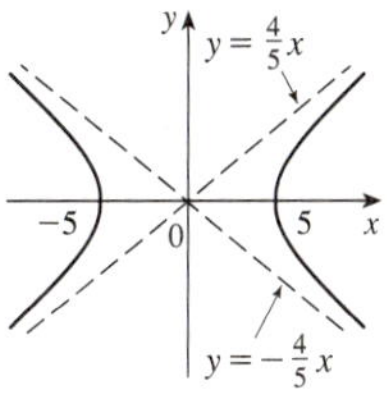

17. Ellipse

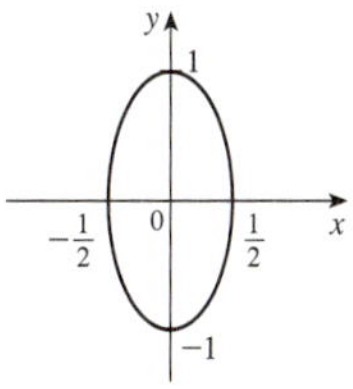

19. Parabola

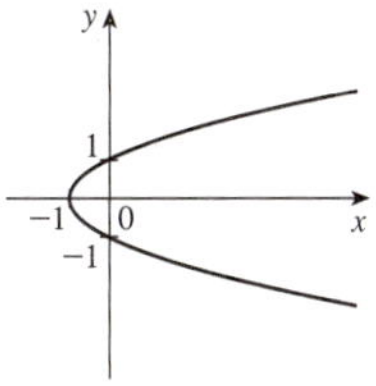

21. Hyperbola

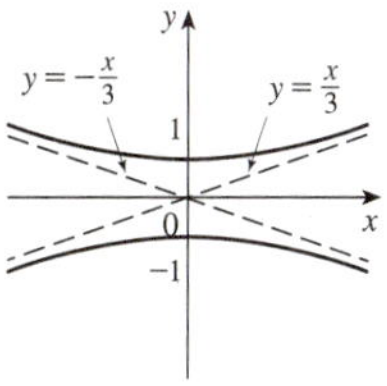

23. Hyperbola

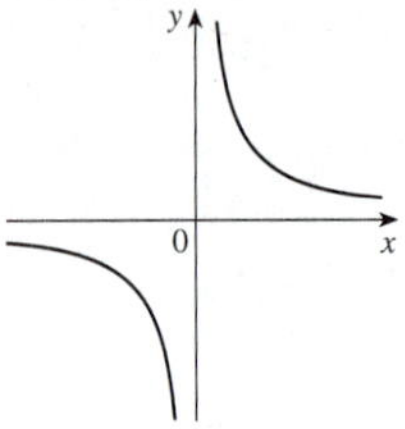

25. Ellipse

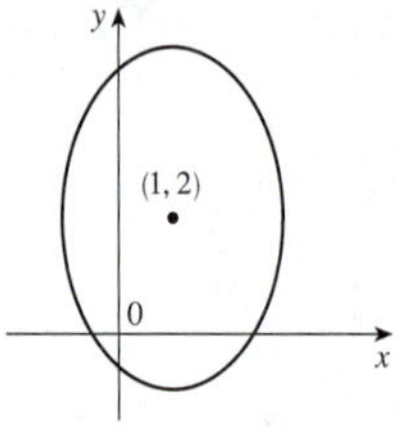

27. Parabola

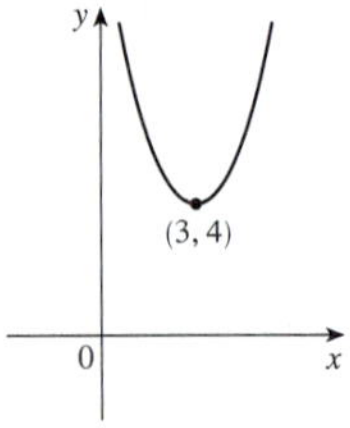

29. Parabola

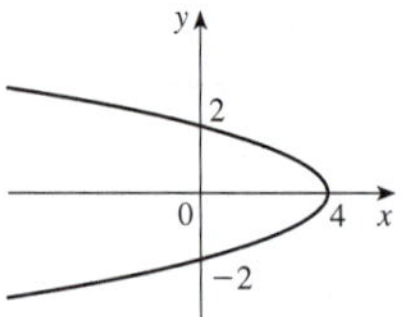

31. Ellipse

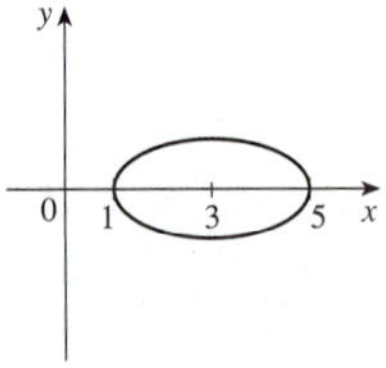

33.

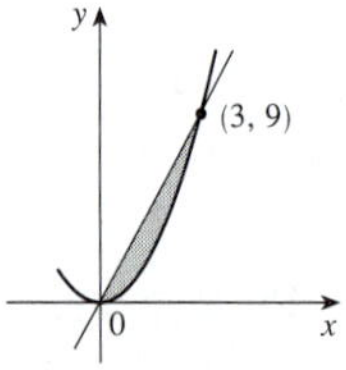

35. $y = x^2 - 2x$

37.

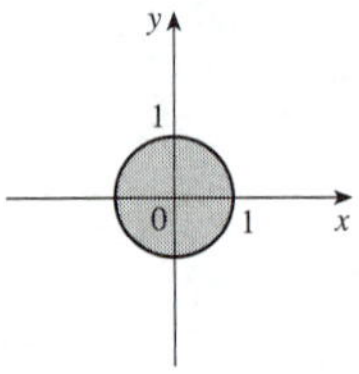

39.

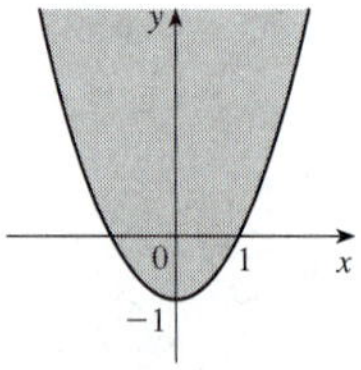

Exercises D ▫ page A32

1. $7\pi/6$ **3.** $\pi/20$ **5.** 5π **7.** $720°$ **9.** $75°$
11. $-67.5°$ **13.** 3π cm **15.** $\frac{2}{3}$ rad $= (120/\pi)°$

17.

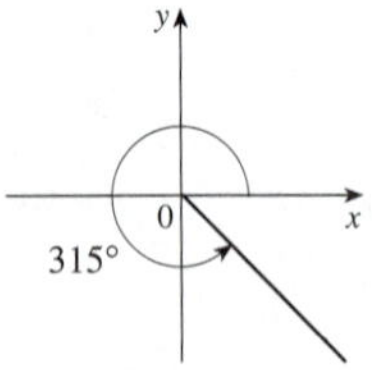

19.

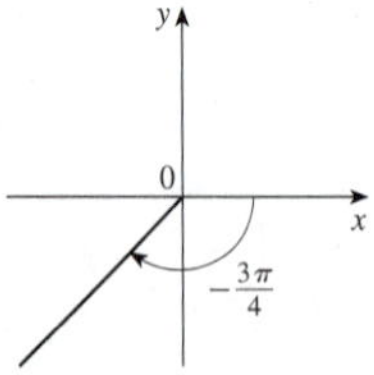

21.

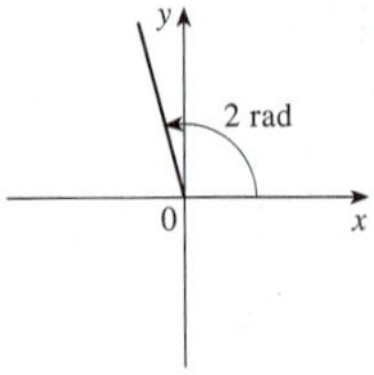

23. $\sin(3\pi/4) = 1/\sqrt{2}$, $\cos(3\pi/4) = -1/\sqrt{2}$, $\tan(3\pi/4) = -1$, $\csc(3\pi/4) = \sqrt{2}$, $\sec(3\pi/4) = -\sqrt{2}$, $\cot(3\pi/4) = -1$

25. $\sin(9\pi/2) = 1$, $\cos(9\pi/2) = 0$, $\csc(9\pi/2) = 1$, $\cot(9\pi/2) = 0$, $\tan(9\pi/2)$ and $\sec(9\pi/2)$ undefined

27. $\sin(5\pi/6) = \frac{1}{2}$, $\cos(5\pi/6) = -\sqrt{3}/2$, $\tan(5\pi/6) = -1/\sqrt{3}$, $\csc(5\pi/6) = 2$, $\sec(5\pi/6) = -2/\sqrt{3}$, $\cot(5\pi/6) = -\sqrt{3}$

29. $\cos\theta = \frac{4}{5}$, $\tan\theta = \frac{3}{4}$, $\csc\theta = \frac{5}{3}$, $\sec\theta = \frac{5}{4}$, $\cot\theta = \frac{4}{3}$

31. $\sin\phi = \sqrt{5}/3$, $\cos\phi = -\frac{2}{3}$, $\tan\phi = -\sqrt{5}/2$, $\csc\phi = 3/\sqrt{5}$, $\cot\phi = -2/\sqrt{5}$

33. $\sin\beta = -1/\sqrt{10}$, $\cos\beta = -3/\sqrt{10}$, $\tan\beta = \frac{1}{3}$, $\csc\beta = -\sqrt{10}$, $\sec\beta = -\sqrt{10}/3$

35. 5.73576 cm **37.** 24.62147 cm **59.** $\frac{1}{15}(4 + 6\sqrt{2})$

61. $\frac{1}{15}(3 + 8\sqrt{2})$ **63.** $\frac{24}{25}$ **65.** $\pi/3, 5\pi/3$

67. $\pi/4, 3\pi/4, 5\pi/4, 7\pi/4$ **69.** $\pi/6, \pi/2, 5\pi/6, 3\pi/2$

71. $0, \pi, 2\pi$ **73.** $0 \leqslant x \leqslant \pi/6$ and $5\pi/6 \leqslant x \leqslant 2\pi$

75. $0 \leqslant x < \pi/4$, $3\pi/4 < x < 5\pi/4$, $7\pi/4 < x \leqslant 2\pi$

77.

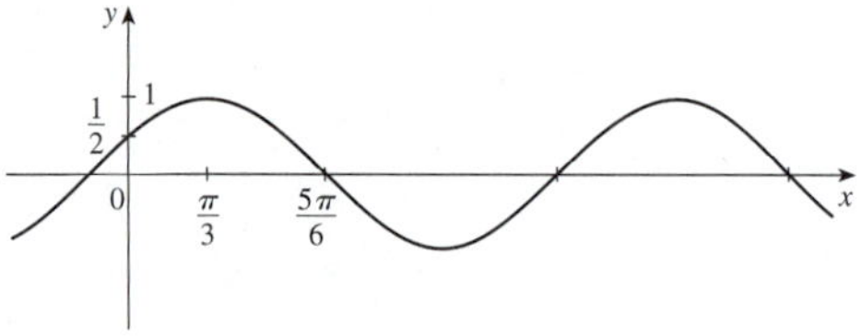

79.

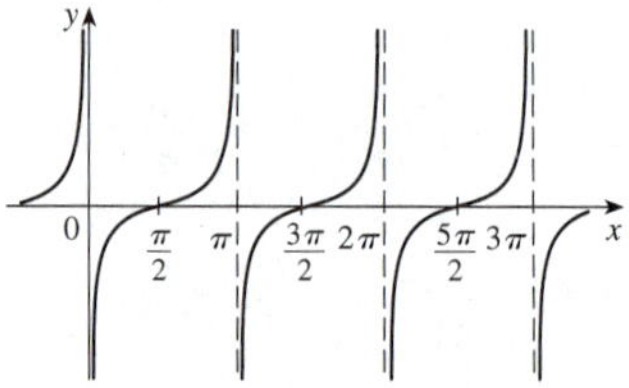

81.

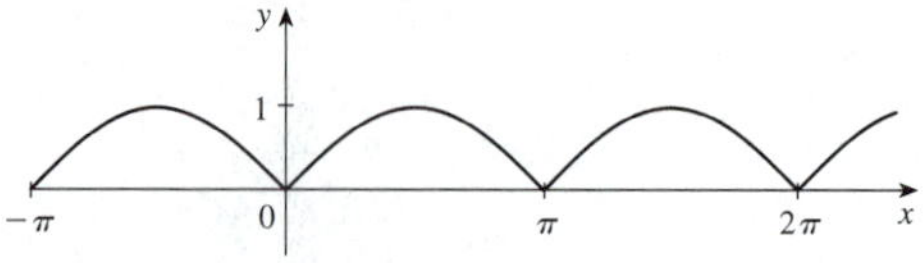

89. 14.34457 cm^2

Exercises E ▫ page A38

1. $\sqrt{1} + \sqrt{2} + \sqrt{3} + \sqrt{4} + \sqrt{5}$ **3.** $3^4 + 3^5 + 3^6$

5. $-1 + \frac{1}{3} + \frac{3}{5} + \frac{5}{7} + \frac{7}{9}$ **7.** $1^{10} + 2^{10} + 3^{10} + \cdots + n^{10}$

9. $1 - 1 + 1 - 1 + \cdots + (-1)^{n-1}$ **11.** $\sum_{i=1}^{10} i$

13. $\sum_{i=1}^{19} \frac{i}{i+1}$ **15.** $\sum_{i=1}^{n} 2i$ **17.** $\sum_{i=0}^{5} 2^i$ **19.** $\sum_{i=1}^{n} x^i$

21. 80 **23.** 3276 **25.** 0 **27.** 61 **29.** $n(n + 1)$

31. $n(n^2 + 6n + 17)/3$ **33.** $n(n^2 + 6n + 11)/3$

35. $n(n^3 + 2n^2 - n - 10)/4$

41. (a) n^4 (b) $5^{100} - 1$ (c) $\frac{97}{300}$ (d) $a_n - a_0$
43. $\frac{1}{3}$ **45.** 14 **49.** $2^{n+1} + n^2 + n - 2$

Exercises G ▫ page A56

1. $8 - 4i$ **3.** $13 + 18i$ **5.** $12 - 7i$ **7.** $\frac{11}{13} + \frac{10}{13}i$
9. $\frac{1}{2} - \frac{1}{2}i$ **11.** $-i$ **13.** $5i$ **15.** $12 + 5i$; 13
17. $4i$, 4 **19.** $\pm\frac{3}{2}i$ **21.** $-1 \pm 2i$
23. $-\frac{1}{2} \pm (\sqrt{7}/2)i$ **25.** $3\sqrt{2}\,[\cos(3\pi/4) + i\sin(3\pi/4)]$
27. $5\{\cos[\tan^{-1}(\frac{4}{3})] + i\sin[\tan^{-1}(\frac{4}{3})]\}$
29. $4[\cos(\pi/2) + i\sin(\pi/2)]$, $\cos(-\pi/6) + i\sin(-\pi/6)$, $\frac{1}{2}[\cos(-\pi/6) + i\sin(-\pi/6)]$

31. $4\sqrt{2}\,[\cos(7\pi/12) + i\sin(7\pi/12)]$, $(2\sqrt{2})[\cos(13\pi/12) + i\sin(13\pi/12)]$, $\frac{1}{4}[\cos(\pi/6) + i\sin(\pi/6)]$
33. -1024 **35.** $-512\sqrt{3} + 512i$
37. $\pm 1, \pm i, (1/\sqrt{2})(\pm 1 \pm i)$ **39.** $\pm(\sqrt{3}/2) + \frac{1}{2}i, -i$

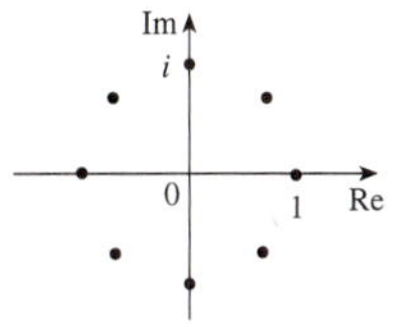

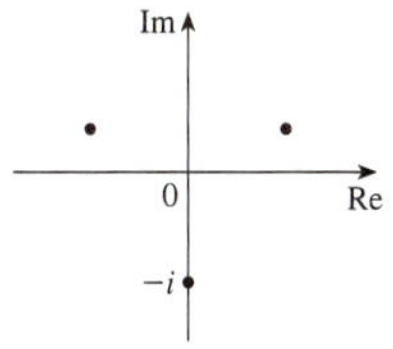

41. i **43.** $\frac{1}{2} + (\sqrt{3}/2)i$ **45.** $-e^2$

CREDITS

vi □ Brian C. Betsill

vii □ Brian C. Betsill
© Gregory D. Dimijian, M. D.

viii □ Brian C. Betsill

ix □ © Ron Lindsey / Phototake
© Gary Ladd 1984 / Photo Researchers, Inc.
Brian C. Betsill

x □ Brian C. Betsill

xi □ Brian C. Betsill
NASA, STScI, and The Smithsonian Institution

xii □ Courtesy of dan clegg
Brian C. Betsill

xiii □ *The Atmosphere 4/e. by Lutgens/Tarbuck*, © 1989. Reprinted by permission of Prentice-Hall, Inc., Upper Saddle River, NJ

2 □ *(clockwise from top)* © Gregory D. Dimijian, M. D.
© Gary Ladd 1984 / Photo Researchers, Inc.
© Jeff Lepore / Photo Researchers, Inc.
© Ron Lindsey / Phototake

10 □ Brian C. Betsill

24 □ *(photo)* © Catherine Karnow / Woodfin Camp & Associates

86 □ Brian C. Betsill

90 □ © 2003 Brand X Pictures

164 □ Figure for Exercise 33 adapted from *Environmental Science: Living Within the System of Nature, 2d ed., by Charles E. Kupchella*, © 1989. Reprinted by permission of Prentice-Hall, Inc., Upper Saddle River, NJ

182 □ Brian C. Betsill

278 □ © Gregory D. Dimijian, M. D.

285 □ NASA

315 □ Title page courtesy of Thomas Fisher Rare Book Library.

341 □ *(photo)* © Manfred Kage / Peter Arnold

368 □ Brian C. Betsill

436 □ Brian C. Betsill

474 □ © Ron Lindsey / Phototake

546 □ © Gary Ladd 1984 / Photo Researchers, Inc.
Brian C. Betsill

586 □ *(photo)* © Jeff Lepore / Photo Researchers, Inc.

650 □ Brian C. Betsill

700 □ Membranes courtesy of National Film Board of Canada
Mathematica models by Brian C. Betsill

785 □ © Luke Dodd / Photo Researchers, Inc.

848 □ NASA, STScI, and The Smithsonian Institution

886 □ Courtesy of dan clegg

893 □ (*Figure 13*) *The Atmosphere 4/e. by Lutgens/Tarbuck*, © 1989. Reprinted by permission of Prentice-Hall, Inc., Upper Saddle River, NJ

940, 942 □ (*weather maps*) Copyright 1997, USA TODAY. Reprinted with permission.

950 □ (*left*) *The Atmosphere 4/e. by Lutgens/Tarbuck*, © 1989. Reprinted by permission of Prentice-Hall, Inc., Upper Saddle River, NJ

972 □ Courtesy of Orbital Sciences Corporation

977 □ (*contour map*) NOAA / AOML / Hurricane Research Division

980 □ Brian Betsill

1040 □ Courtesy of dan clegg

1054 □ (*top*) © A & J Verkaik / The Stock Market
(*bottom*) *The Atmosphere 4/e. by Lutgens/Tarbuck*,
© 1989. Reprinted by permission of Prentice-Hall, Inc., Upper Saddle River, NJ;

1056 □ (*Figure 2b*) *ONERA* photograph, Werlé 1974

1126 □ Courtesy of the Masters and Fellows of Gonville and Caius College, University of Cambridge, England

1140 □ Dia Max / Getty Images

1156 □ (*photo*) Schwinn Cycling and Fitness

INDEX

RP denotes Reference Page numbers.

Special Functions

Power Functions $f(x) = x^a$

(i) $f(x) = x^n$, n a positive integer

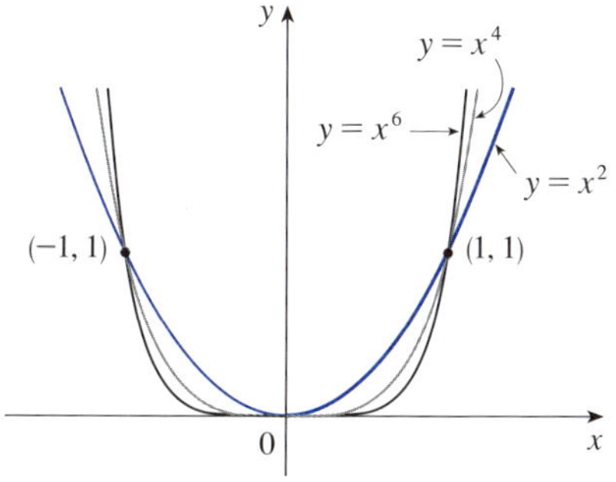

n even

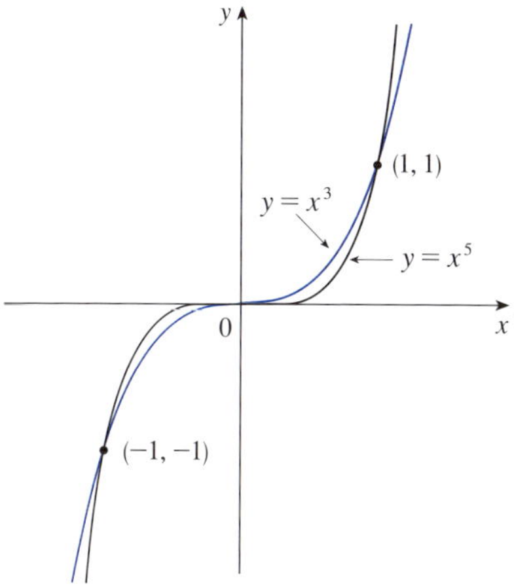

n odd

(ii) $f(x) = x^{1/n} = \sqrt[n]{x}$, n a positive integer

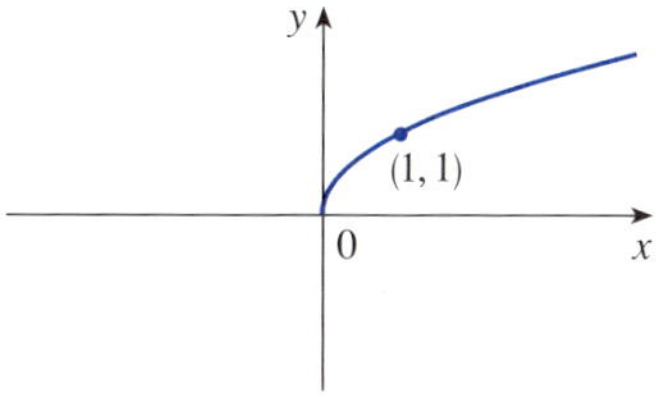

$f(x) = \sqrt{x}$

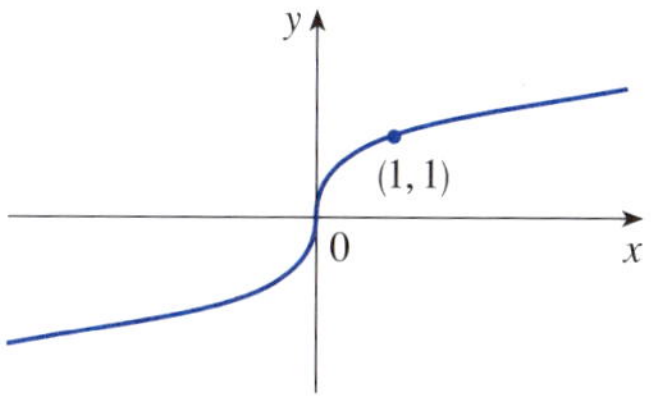

$f(x) = \sqrt[3]{x}$

(iii) $f(x) = x^{-1} = \dfrac{1}{x}$

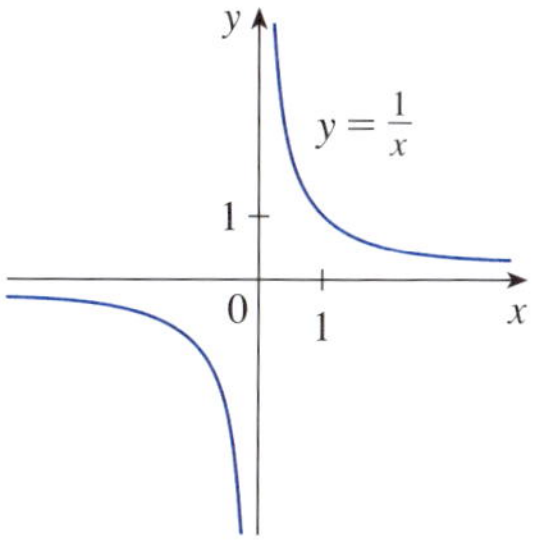

Inverse Trigonometric Functions

$$\arcsin x = \sin^{-1}x = y \iff \sin y = x \quad \text{and} \quad -\frac{\pi}{2} \leqslant y \leqslant \frac{\pi}{2}$$

$$\arccos x = \cos^{-1}x = y \iff \cos y = x \quad \text{and} \quad 0 \leqslant y \leqslant \pi$$

$$\arctan x = \tan^{-1}x = y \iff \tan y = x \quad \text{and} \quad -\frac{\pi}{2} < y < \frac{\pi}{2}$$

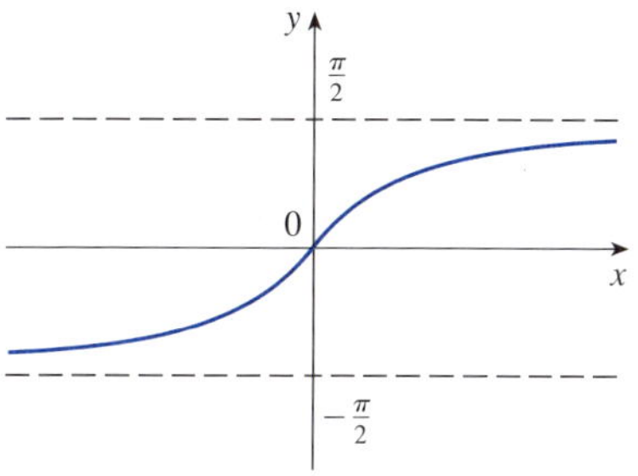

$y = \tan^{-1}x = \arctan x$

$$\lim_{x \to -\infty} \tan^{-1}x = -\frac{\pi}{2}$$

$$\lim_{x \to \infty} \tan^{-1}x = \frac{\pi}{2}$$

Special Functions

Exponential and Logarithmic Functions

$\log_a x = y \iff a^y = x$

$\ln x = \log_e x$, where $\ln e = 1$

$\ln x = y \iff e^y = x$

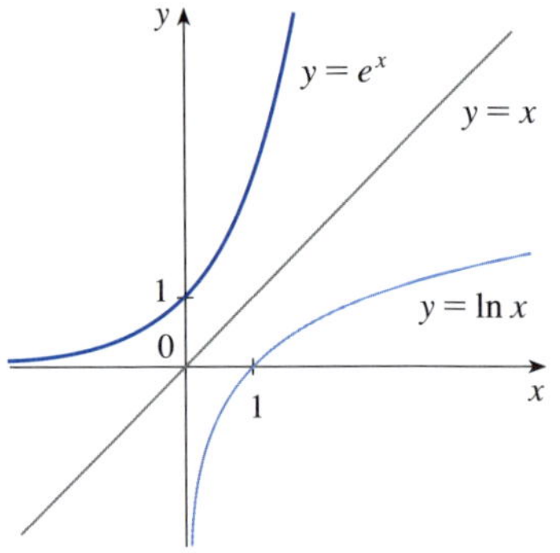

Cancellation Equations

$\log_a(a^x) = x \qquad a^{\log_a x} = x$

$\ln(e^x) = x \qquad e^{\ln x} = x$

Laws of Logarithms

1. $\log_a(xy) = \log_a x + \log_a y$

2. $\log_a\left(\dfrac{x}{y}\right) = \log_a x - \log_a y$

3. $\log_a(x^r) = r\log_a x$

$\lim_{x\to-\infty} e^x = 0 \qquad \lim_{x\to\infty} e^x = \infty$

$\lim_{x\to 0^+} \ln x = -\infty \qquad \lim_{x\to\infty} \ln x = \infty$

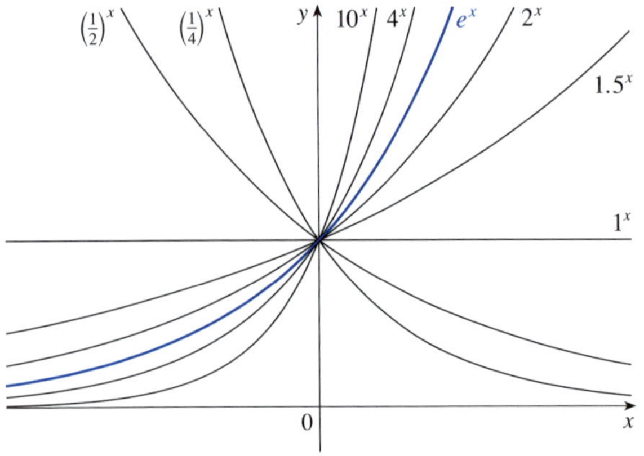

Exponential functions

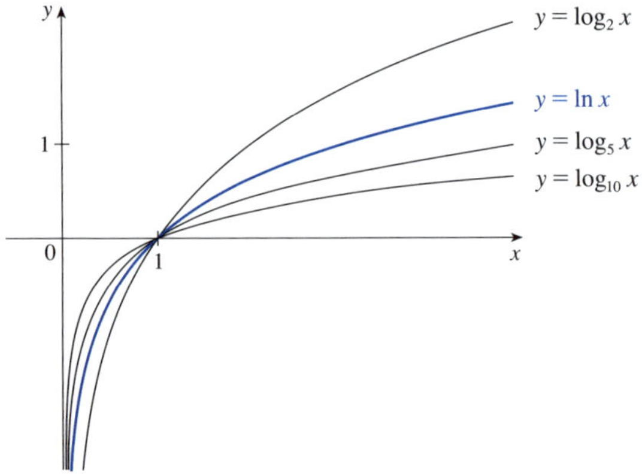

Logarithmic functions

Hyperbolic Functions

$\sinh x = \dfrac{e^x - e^{-x}}{2} \qquad \operatorname{csch} x = \dfrac{1}{\sinh x}$

$\cosh x = \dfrac{e^x + e^{-x}}{2} \qquad \operatorname{sech} x = \dfrac{1}{\cosh x}$

$\tanh x = \dfrac{\sinh x}{\cosh x} \qquad \coth x = \dfrac{\cosh x}{\sinh x}$

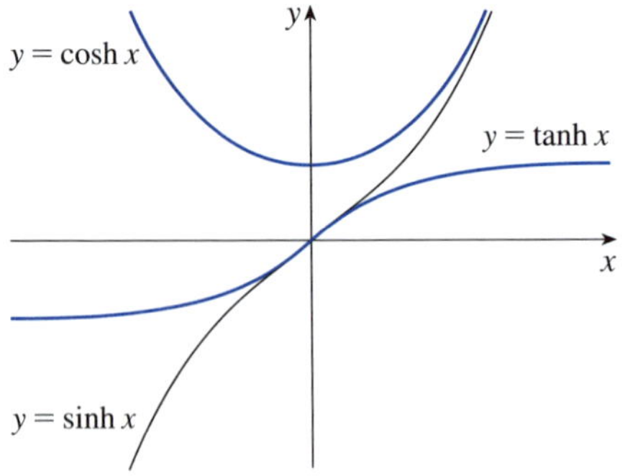

Inverse Hyperbolic Functions

$y = \sinh^{-1}x \iff \sinh y = x \qquad \sinh^{-1}x = \ln\left(x + \sqrt{x^2 + 1}\right)$

$y = \cosh^{-1}x \iff \cosh y = x$ and $y \geqslant 0 \qquad \cosh^{-1}x = \ln\left(x + \sqrt{x^2 - 1}\right)$

$y = \tanh^{-1}x \iff \tanh y = x \qquad \tanh^{-1}x = \frac{1}{2}\ln\left(\dfrac{1 + x}{1 - x}\right)$

Differentiation Rules

General Formulas

1. $\dfrac{d}{dx}(c) = 0$

2. $\dfrac{d}{dx}[cf(x)] = cf'(x)$

3. $\dfrac{d}{dx}[f(x) + g(x)] = f'(x) + g'(x)$

4. $\dfrac{d}{dx}[f(x) - g(x)] = f'(x) - g'(x)$

5. $\dfrac{d}{dx}[f(x)g(x)] = f(x)g'(x) + g(x)f'(x)$ (Product Rule)

6. $\dfrac{d}{dx}\left[\dfrac{f(x)}{g(x)}\right] = \dfrac{g(x)f'(x) - f(x)g'(x)}{[g(x)]^2}$ (Quotient Rule)

7. $\dfrac{d}{dx} f(g(x)) = f'(g(x))g'(x)$ (Chain Rule)

8. $\dfrac{d}{dx}(x^n) = nx^{n-1}$ (Power Rule)

Exponential and Logarithmic Functions

9. $\dfrac{d}{dx}(e^x) = e^x$

10. $\dfrac{d}{dx}(a^x) = a^x \ln a$

11. $\dfrac{d}{dx} \ln |x| = \dfrac{1}{x}$

12. $\dfrac{d}{dx}(\log_a x) = \dfrac{1}{x \ln a}$

Trigonometric Functions

13. $\dfrac{d}{dx}(\sin x) = \cos x$

14. $\dfrac{d}{dx}(\cos x) = -\sin x$

15. $\dfrac{d}{dx}(\tan x) = \sec^2 x$

16. $\dfrac{d}{dx}(\csc x) = -\csc x \cot x$

17. $\dfrac{d}{dx}(\sec x) = \sec x \tan x$

18. $\dfrac{d}{dx}(\cot x) = -\csc^2 x$

Inverse Trigonometric Functions

19. $\dfrac{d}{dx}(\sin^{-1} x) = \dfrac{1}{\sqrt{1 - x^2}}$

20. $\dfrac{d}{dx}(\cos^{-1} x) = -\dfrac{1}{\sqrt{1 - x^2}}$

21. $\dfrac{d}{dx}(\tan^{-1} x) = \dfrac{1}{1 + x^2}$

22. $\dfrac{d}{dx}(\csc^{-1} x) = -\dfrac{1}{x\sqrt{x^2 - 1}}$

23. $\dfrac{d}{dx}(\sec^{-1} x) = \dfrac{1}{x\sqrt{x^2 - 1}}$

24. $\dfrac{d}{dx}(\cot^{-1} x) = -\dfrac{1}{1 + x^2}$

Hyperbolic Functions

25. $\dfrac{d}{dx}(\sinh x) = \cosh x$

26. $\dfrac{d}{dx}(\cosh x) = \sinh x$

27. $\dfrac{d}{dx}(\tanh x) = \operatorname{sech}^2 x$

28. $\dfrac{d}{dx}(\operatorname{csch} x) = -\operatorname{csch} x \coth x$

29. $\dfrac{d}{dx}(\operatorname{sech} x) = -\operatorname{sech} x \tanh x$

30. $\dfrac{d}{dx}(\coth x) = -\operatorname{csch}^2 x$

Inverse Hyperbolic Functions

31. $\dfrac{d}{dx}(\sinh^{-1} x) = \dfrac{1}{\sqrt{1 + x^2}}$

32. $\dfrac{d}{dx}(\cosh^{-1} x) = \dfrac{1}{\sqrt{x^2 - 1}}$

33. $\dfrac{d}{dx}(\tanh^{-1} x) = \dfrac{1}{1 - x^2}$

34. $\dfrac{d}{dx}(\operatorname{csch}^{-1} x) = -\dfrac{1}{|x|\sqrt{x^2 + 1}}$

35. $\dfrac{d}{dx}(\operatorname{sech}^{-1} x) = -\dfrac{1}{x\sqrt{1 - x^2}}$

36. $\dfrac{d}{dx}(\coth^{-1} x) = \dfrac{1}{1 - x^2}$

Table of Integrals

Basic Forms

1. $\int u\,dv = uv - \int v\,du$

2. $\int u^n\,du = \frac{u^{n+1}}{n+1} + C, \quad n \neq -1$

3. $\int \frac{du}{u} = \ln|u| + C$

4. $\int e^u\,du = e^u + C$

5. $\int a^u\,du = \frac{a^u}{\ln a} + C$

6. $\int \sin u\,du = -\cos u + C$

7. $\int \cos u\,du = \sin u + C$

8. $\int \sec^2 u\,du = \tan u + C$

9. $\int \csc^2 u\,du = -\cot u + C$

10. $\int \sec u\,\tan u\,du = \sec u + C$

11. $\int \csc u\,\cot u\,du = -\csc u + C$

12. $\int \tan u\,du = \ln|\sec u| + C$

13. $\int \cot u\,du = \ln|\sin u| + C$

14. $\int \sec u\,du = \ln|\sec u + \tan u| + C$

15. $\int \csc u\,du = \ln|\csc u - \cot u| + C$

16. $\int \frac{du}{\sqrt{a^2 - u^2}} = \sin^{-1}\frac{u}{a} + C$

17. $\int \frac{du}{a^2 + u^2} = \frac{1}{a}\tan^{-1}\frac{u}{a} + C$

18. $\int \frac{du}{u\sqrt{u^2 - a^2}} = \frac{1}{a}\sec^{-1}\frac{u}{a} + C$

19. $\int \frac{du}{a^2 - u^2} = \frac{1}{2a}\ln\left|\frac{u+a}{u-a}\right| + C$

20. $\int \frac{du}{u^2 - a^2} = \frac{1}{2a}\ln\left|\frac{u-a}{u+a}\right| + C$

Forms Involving $\sqrt{a^2 + u^2}$, $a > 0$

21. $\int \sqrt{a^2 + u^2}\,du = \frac{u}{2}\sqrt{a^2 + u^2} + \frac{a^2}{2}\ln\left(u + \sqrt{a^2 + u^2}\right) + C$

22. $\int u^2\sqrt{a^2 + u^2}\,du = \frac{u}{8}(a^2 + 2u^2)\sqrt{a^2 + u^2} - \frac{a^4}{8}\ln\left(u + \sqrt{a^2 + u^2}\right) + C$

23. $\int \frac{\sqrt{a^2 + u^2}}{u}\,du = \sqrt{a^2 + u^2} - a\ln\left|\frac{a + \sqrt{a^2 + u^2}}{u}\right| + C$

24. $\int \frac{\sqrt{a^2 + u^2}}{u^2}\,du = -\frac{\sqrt{a^2 + u^2}}{u} + \ln\left(u + \sqrt{a^2 + u^2}\right) + C$

25. $\int \frac{du}{\sqrt{a^2 + u^2}} = \ln\left(u + \sqrt{a^2 + u^2}\right) + C$

26. $\int \frac{u^2\,du}{\sqrt{a^2 + u^2}} = \frac{u}{2}\sqrt{a^2 + u^2} - \frac{a^2}{2}\ln\left(u + \sqrt{a^2 + u^2}\right) + C$

27. $\int \frac{du}{u\sqrt{a^2 + u^2}} = -\frac{1}{a}\ln\left|\frac{\sqrt{a^2 + u^2} + a}{u}\right| + C$

28. $\int \frac{du}{u^2\sqrt{a^2 + u^2}} = -\frac{\sqrt{a^2 + u^2}}{a^2 u} + C$

29. $\int \frac{du}{(a^2 + u^2)^{3/2}} = \frac{u}{a^2\sqrt{a^2 + u^2}} + C$

Table of Integrals

Forms Involving $\sqrt{a^2 - u^2}$, $a > 0$

30. $\displaystyle\int \sqrt{a^2 - u^2}\, du = \frac{u}{2}\sqrt{a^2 - u^2} + \frac{a^2}{2}\sin^{-1}\frac{u}{a} + C$

31. $\displaystyle\int u^2\sqrt{a^2 - u^2}\, du = \frac{u}{8}(2u^2 - a^2)\sqrt{a^2 - u^2} + \frac{a^4}{8}\sin^{-1}\frac{u}{a} + C$

32. $\displaystyle\int \frac{\sqrt{a^2 - u^2}}{u}\, du = \sqrt{a^2 - u^2} - a\ln\left|\frac{a + \sqrt{a^2 - u^2}}{u}\right| + C$

33. $\displaystyle\int \frac{\sqrt{a^2 - u^2}}{u^2}\, du = -\frac{1}{u}\sqrt{a^2 - u^2} - \sin^{-1}\frac{u}{a} + C$

34. $\displaystyle\int \frac{u^2\, du}{\sqrt{a^2 - u^2}} = -\frac{u}{2}\sqrt{a^2 - u^2} + \frac{a^2}{2}\sin^{-1}\frac{u}{a} + C$

35. $\displaystyle\int \frac{du}{u\sqrt{a^2 - u^2}} = -\frac{1}{a}\ln\left|\frac{a + \sqrt{a^2 - u^2}}{u}\right| + C$

36. $\displaystyle\int \frac{du}{u^2\sqrt{a^2 - u^2}} = -\frac{1}{a^2 u}\sqrt{a^2 - u^2} + C$

37. $\displaystyle\int (a^2 - u^2)^{3/2}\, du = -\frac{u}{8}(2u^2 - 5a^2)\sqrt{a^2 - u^2} + \frac{3a^4}{8}\sin^{-1}\frac{u}{a} + C$

38. $\displaystyle\int \frac{du}{(a^2 - u^2)^{3/2}} = \frac{u}{a^2\sqrt{a^2 - u^2}} + C$

Forms Involving $\sqrt{u^2 - a^2}$, $a > 0$

39. $\displaystyle\int \sqrt{u^2 - a^2}\, du = \frac{u}{2}\sqrt{u^2 - a^2} - \frac{a^2}{2}\ln\left|u + \sqrt{u^2 - a^2}\right| + C$

40. $\displaystyle\int u^2\sqrt{u^2 - a^2}\, du = \frac{u}{8}(2u^2 - a^2)\sqrt{u^2 - a^2} - \frac{a^4}{8}\ln\left|u + \sqrt{u^2 - a^2}\right| + C$

41. $\displaystyle\int \frac{\sqrt{u^2 - a^2}}{u}\, du = \sqrt{u^2 - a^2} - a\cos^{-1}\frac{a}{|u|} + C$

42. $\displaystyle\int \frac{\sqrt{u^2 - a^2}}{u^2}\, du = -\frac{\sqrt{u^2 - a^2}}{u} + \ln\left|u + \sqrt{u^2 - a^2}\right| + C$

43. $\displaystyle\int \frac{du}{\sqrt{u^2 - a^2}} = \ln\left|u + \sqrt{u^2 - a^2}\right| + C$

44. $\displaystyle\int \frac{u^2\, du}{\sqrt{u^2 - a^2}} = \frac{u}{2}\sqrt{u^2 - a^2} + \frac{a^2}{2}\ln\left|u + \sqrt{u^2 - a^2}\right| + C$

45. $\displaystyle\int \frac{du}{u^2\sqrt{u^2 - a^2}} = \frac{\sqrt{u^2 - a^2}}{a^2 u} + C$

46. $\displaystyle\int \frac{du}{(u^2 - a^2)^{3/2}} = -\frac{u}{a^2\sqrt{u^2 - a^2}} + C$

Table of Integrals

Forms Involving $a + bu$

47. $$\int \frac{u\,du}{a + bu} = \frac{1}{b^2}\left(a + bu - a \ln|a + bu|\right) + C$$

48. $$\int \frac{u^2\,du}{a + bu} = \frac{1}{2b^3}\left[(a + bu)^2 - 4a(a + bu) + 2a^2 \ln|a + bu|\right] + C$$

49. $$\int \frac{du}{u(a + bu)} = \frac{1}{a} \ln\left|\frac{u}{a + bu}\right| + C$$

50. $$\int \frac{du}{u^2(a + bu)} = -\frac{1}{au} + \frac{b}{a^2} \ln\left|\frac{a + bu}{u}\right| + C$$

51. $$\int \frac{u\,du}{(a + bu)^2} = \frac{a}{b^2(a + bu)} + \frac{1}{b^2} \ln|a + bu| + C$$

52. $$\int \frac{du}{u(a + bu)^2} = \frac{1}{a(a + bu)} - \frac{1}{a^2} \ln\left|\frac{a + bu}{u}\right| + C$$

53. $$\int \frac{u^2\,du}{(a + bu)^2} = \frac{1}{b^3}\left(a + bu - \frac{a^2}{a + bu} - 2a \ln|a + bu|\right) + C$$

54. $$\int u\sqrt{a + bu}\,du = \frac{2}{15b^2}(3bu - 2a)(a + bu)^{3/2} + C$$

55. $$\int \frac{u\,du}{\sqrt{a + bu}} = \frac{2}{3b^2}(bu - 2a)\sqrt{a + bu} + C$$

56. $$\int \frac{u^2\,du}{\sqrt{a + bu}} = \frac{2}{15b^3}(8a^2 + 3b^2u^2 - 4abu)\sqrt{a + bu} + C$$

57. $$\int \frac{du}{u\sqrt{a + bu}} = \frac{1}{\sqrt{a}} \ln\left|\frac{\sqrt{a + bu} - \sqrt{a}}{\sqrt{a + bu} + \sqrt{a}}\right| + C, \quad \text{if } a > 0$$
$$= \frac{2}{\sqrt{-a}} \tan^{-1}\sqrt{\frac{a + bu}{-a}} + C, \quad \text{if } a < 0$$

58. $$\int \frac{\sqrt{a + bu}}{u}\,du = 2\sqrt{a + bu} + a\int \frac{du}{u\sqrt{a + bu}}$$

59. $$\int \frac{\sqrt{a + bu}}{u^2}\,du = -\frac{\sqrt{a + bu}}{u} + \frac{b}{2}\int \frac{du}{u\sqrt{a + bu}}$$

60. $$\int u^n\sqrt{a + bu}\,du = \frac{2}{b(2n + 3)}\left[u^n(a + bu)^{3/2} - na\int u^{n-1}\sqrt{a + bu}\,du\right]$$

61. $$\int \frac{u^n\,du}{\sqrt{a + bu}} = \frac{2u^n\sqrt{a + bu}}{b(2n + 1)} - \frac{2na}{b(2n + 1)}\int \frac{u^{n-1}\,du}{\sqrt{a + bu}}$$

62. $$\int \frac{du}{u^n\sqrt{a + bu}} = -\frac{\sqrt{a + bu}}{a(n - 1)u^{n-1}} - \frac{b(2n - 3)}{2a(n - 1)}\int \frac{du}{u^{n-1}\sqrt{a + bu}}$$

Table of Integrals

Trigonometric Forms

63. $\int \sin^2 u\,du = \frac{1}{2}u - \frac{1}{4}\sin 2u + C$

64. $\int \cos^2 u\,du = \frac{1}{2}u + \frac{1}{4}\sin 2u + C$

65. $\int \tan^2 u\,du = \tan u - u + C$

66. $\int \cot^2 u\,du = -\cot u - u + C$

67. $\int \sin^3 u\,du = -\frac{1}{3}(2 + \sin^2 u)\cos u + C$

68. $\int \cos^3 u\,du = \frac{1}{3}(2 + \cos^2 u)\sin u + C$

69. $\int \tan^3 u\,du = \frac{1}{2}\tan^2 u + \ln|\cos u| + C$

70. $\int \cot^3 u\,du = -\frac{1}{2}\cot^2 u - \ln|\sin u| + C$

71. $\int \sec^3 u\,du = \frac{1}{2}\sec u\tan u + \frac{1}{2}\ln|\sec u + \tan u| + C$

72. $\int \csc^3 u\,du = -\frac{1}{2}\csc u\cot u + \frac{1}{2}\ln|\csc u - \cot u| + C$

73. $\int \sin^n u\,du = -\frac{1}{n}\sin^{n-1}u\cos u + \frac{n-1}{n}\int \sin^{n-2}u\,du$

74. $\int \cos^n u\,du = \frac{1}{n}\cos^{n-1}u\sin u + \frac{n-1}{n}\int \cos^{n-2}u\,du$

75. $\int \tan^n u\,du = \frac{1}{n-1}\tan^{n-1}u - \int \tan^{n-2}u\,du$

76. $\int \cot^n u\,du = \frac{-1}{n-1}\cot^{n-1}u - \int \cot^{n-2}u\,du$

77. $\int \sec^n u\,du = \frac{1}{n-1}\tan u\sec^{n-2}u + \frac{n-2}{n-1}\int \sec^{n-2}u\,du$

78. $\int \csc^n u\,du = \frac{-1}{n-1}\cot u\csc^{n-2}u + \frac{n-2}{n-1}\int \csc^{n-2}u\,du$

79. $\int \sin au\sin bu\,du = \frac{\sin(a-b)u}{2(a-b)} - \frac{\sin(a+b)u}{2(a+b)} + C$

80. $\int \cos au\cos bu\,du = \frac{\sin(a-b)u}{2(a-b)} + \frac{\sin(a+b)u}{2(a+b)} + C$

81. $\int \sin au\cos bu\,du = -\frac{\cos(a-b)u}{2(a-b)} - \frac{\cos(a+b)u}{2(a+b)} + C$

82. $\int u\sin u\,du = \sin u - u\cos u + C$

83. $\int u\cos u\,du = \cos u + u\sin u + C$

84. $\int u^n\sin u\,du = -u^n\cos u + n\int u^{n-1}\cos u\,du$

85. $\int u^n\cos u\,du = u^n\sin u - n\int u^{n-1}\sin u\,du$

86. $\int \sin^n u\cos^m u\,du = -\frac{\sin^{n-1}u\cos^{m+1}u}{n+m} + \frac{n-1}{n+m}\int \sin^{n-2}u\cos^m u\,du$

$$= \frac{\sin^{n+1}u\cos^{m-1}u}{n+m} + \frac{m-1}{n+m}\int \sin^n u\cos^{m-2}u\,du$$

Inverse Trigonometric Forms

87. $\int \sin^{-1}u\,du = u\sin^{-1}u + \sqrt{1-u^2} + C$

88. $\int \cos^{-1}u\,du = u\cos^{-1}u - \sqrt{1-u^2} + C$

89. $\int \tan^{-1}u\,du = u\tan^{-1}u - \frac{1}{2}\ln(1+u^2) + C$

90. $\int u\sin^{-1}u\,du = \frac{2u^2-1}{4}\sin^{-1}u + \frac{u\sqrt{1-u^2}}{4} + C$

91. $\int u\cos^{-1}u\,du = \frac{2u^2-1}{4}\cos^{-1}u - \frac{u\sqrt{1-u^2}}{4} + C$

92. $\int u\tan^{-1}u\,du = \frac{u^2+1}{2}\tan^{-1}u - \frac{u}{2} + C$

93. $\int u^n\sin^{-1}u\,du = \frac{1}{n+1}\left[u^{n+1}\sin^{-1}u - \int \frac{u^{n+1}\,du}{\sqrt{1-u^2}}\right], \quad n \neq -1$

94. $\int u^n\cos^{-1}u\,du = \frac{1}{n+1}\left[u^{n+1}\cos^{-1}u + \int \frac{u^{n+1}\,du}{\sqrt{1-u^2}}\right], \quad n \neq -1$

95. $\int u^n\tan^{-1}u\,du = \frac{1}{n+1}\left[u^{n+1}\tan^{-1}u - \int \frac{u^{n+1}\,du}{1+u^2}\right], \quad n \neq -1$

Table of Integrals

Exponential and Logarithmic Forms

96. $\displaystyle\int ue^{au}\,du = \frac{1}{a^2}(au - 1)e^{au} + C$

97. $\displaystyle\int u^n e^{au}\,du = \frac{1}{a}u^n e^{au} - \frac{n}{a}\int u^{n-1}e^{au}\,du$

98. $\displaystyle\int e^{au}\sin bu\,du = \frac{e^{au}}{a^2 + b^2}(a\sin bu - b\cos bu) + C$

99. $\displaystyle\int e^{au}\cos bu\,du = \frac{e^{au}}{a^2 + b^2}(a\cos bu + b\sin bu) + C$

100. $\displaystyle\int \ln u\,du = u\ln u - u + C$

101. $\displaystyle\int u^n \ln u\,du = \frac{u^{n+1}}{(n + 1)^2}[(n + 1)\ln u - 1] + C$

102. $\displaystyle\int \frac{1}{u\ln u}\,du = \ln|\ln u| + C$

Hyperbolic Forms

103. $\displaystyle\int \sinh u\,du = \cosh u + C$

104. $\displaystyle\int \cosh u\,du = \sinh u + C$

105. $\displaystyle\int \tanh u\,du = \ln\cosh u + C$

106. $\displaystyle\int \coth u\,du = \ln|\sinh u| + C$

107. $\displaystyle\int \operatorname{sech} u\,du = \tan^{-1}|\sinh u| + C$

108. $\displaystyle\int \operatorname{csch} u\,du = \ln\left|\tanh \tfrac{1}{2}u\right| + C$

109. $\displaystyle\int \operatorname{sech}^2 u\,du = \tanh u + C$

110. $\displaystyle\int \operatorname{csch}^2 u\,du = -\coth u + C$

111. $\displaystyle\int \operatorname{sech} u\tanh u\,du = -\operatorname{sech} u + C$

112. $\displaystyle\int \operatorname{csch} u\coth u\,du = -\operatorname{csch} u + C$

Forms Involving $\sqrt{2au - u^2}$, $a > 0$

113. $\displaystyle\int \sqrt{2au - u^2}\,du = \frac{u - a}{2}\sqrt{2au - u^2} + \frac{a^2}{2}\cos^{-1}\left(\frac{a - u}{a}\right) + C$

114. $\displaystyle\int u\sqrt{2au - u^2}\,du = \frac{2u^2 - au - 3a^2}{6}\sqrt{2au - u^2} + \frac{a^3}{2}\cos^{-1}\left(\frac{a - u}{a}\right) + C$

115. $\displaystyle\int \frac{\sqrt{2au - u^2}}{u}\,du = \sqrt{2au - u^2} + a\cos^{-1}\left(\frac{a - u}{a}\right) + C$

116. $\displaystyle\int \frac{\sqrt{2au - u^2}}{u^2}\,du = -\frac{2\sqrt{2au - u^2}}{u} - \cos^{-1}\left(\frac{a - u}{a}\right) + C$

117. $\displaystyle\int \frac{du}{\sqrt{2au - u^2}} = \cos^{-1}\left(\frac{a - u}{a}\right) + C$

118. $\displaystyle\int \frac{u\,du}{\sqrt{2au - u^2}} = -\sqrt{2au - u^2} + a\cos^{-1}\left(\frac{a - u}{a}\right) + C$

119. $\displaystyle\int \frac{u^2\,du}{\sqrt{2au - u^2}} = -\frac{(u + 3a)}{2}\sqrt{2au - u^2} + \frac{3a^2}{2}\cos^{-1}\left(\frac{a - u}{a}\right) + C$

120. $\displaystyle\int \frac{du}{u\sqrt{2au - u^2}} = -\frac{\sqrt{2au - u^2}}{au} + C$

The next example of a piecewise defined function is the absolute value function. Recall that the **absolute value** of a number a, denoted by $|a|$, is the distance from a to 0 on the real number line. Distances are always positive or 0, so we have

||| For a more extensive review of absolute values, see Appendix A.

$$|a| \geqslant 0 \qquad \text{for every number } a$$

For example,

$$|3| = 3 \qquad |-3| = 3 \qquad |0| = 0 \qquad |\sqrt{2} - 1| = \sqrt{2} - 1 \qquad |3 - \pi| = \pi - 3$$

In general, we have

$$|a| = a \qquad \text{if } a \geqslant 0$$
$$|a| = -a \qquad \text{if } a < 0$$

(Remember that if a is negative, then $-a$ is positive.)

EXAMPLE 8 Sketch the graph of the absolute value function $f(x) = |x|$.

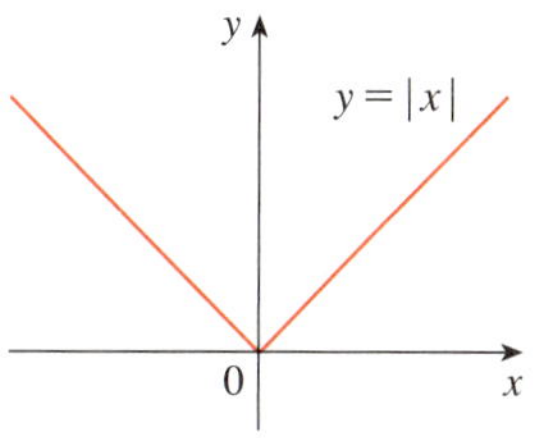

FIGURE 20

SOLUTION From the preceding discussion we know that

$$|x| = \begin{cases} x & \text{if } x \geqslant 0 \\ -x & \text{if } x < 0 \end{cases}$$

Using the same method as in Example 7, we see that the graph of f coincides with the line $y = x$ to the right of the y-axis and coincides with the line $y = -x$ to the left of the y-axis (see Figure 20).

EXAMPLE 9 Find a formula for the function f graphed in Figure 21.

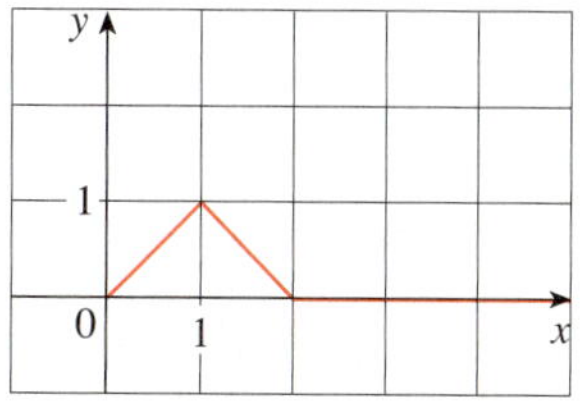

FIGURE 21

SOLUTION The line through (0, 0) and (1, 1) has slope $m = 1$ and y-intercept $b = 0$, so its equation is $y = x$. Thus, for the part of the graph of f that joins (0, 0) to (1, 1), we have

$$f(x) = x \qquad \text{if } 0 \leqslant x \leqslant 1$$

||| Point-slope form of the equation of a line:

$$y - y_1 = m(x - x_1)$$

See Appendix B.

The line through (1, 1) and (2, 0) has slope $m = -1$, so its point-slope form is

$$y - 0 = (-1)(x - 2) \qquad \text{or} \qquad y = 2 - x$$

So we have

$$f(x) = 2 - x \qquad \text{if } 1 < x \leqslant 2$$

We also see that the graph of f coincides with the x-axis for $x > 2$. Putting this information together, we have the following three-piece formula for f:

$$f(x) = \begin{cases} x & \text{if } 0 \leqslant x \leqslant 1 \\ 2 - x & \text{if } 1 < x \leqslant 2 \\ 0 & \text{if } x > 2 \end{cases}$$

EXAMPLE 10 In Example C at the beginning of this section we considered the cost $C(w)$ of mailing a first-class letter with weight w. In effect, this is a piecewise defined function because, from the table of values, we have

$$C(w) = \begin{cases} 0.37 & \text{if } 0 < w \leqslant 1 \\ 0.60 & \text{if } 1 < w \leqslant 2 \\ 0.83 & \text{if } 2 < w \leqslant 3 \\ 1.06 & \text{if } 3 < w \leqslant 4 \end{cases}$$

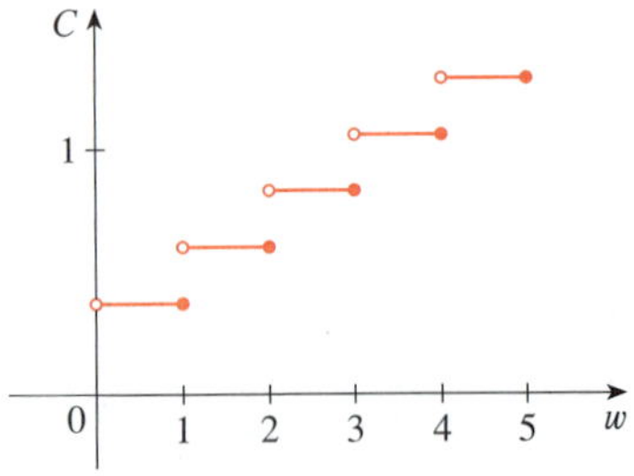

FIGURE 22

The graph is shown in Figure 22. You can see why functions similar to this one are called **step functions**—they jump from one value to the next. Such functions will be studied in Chapter 2.

Symmetry

If a function f satisfies $f(-x) = f(x)$ for every number x in its domain, then f is called an **even function**. For instance, the function $f(x) = x^2$ is even because

$$f(-x) = (-x)^2 = x^2 = f(x)$$

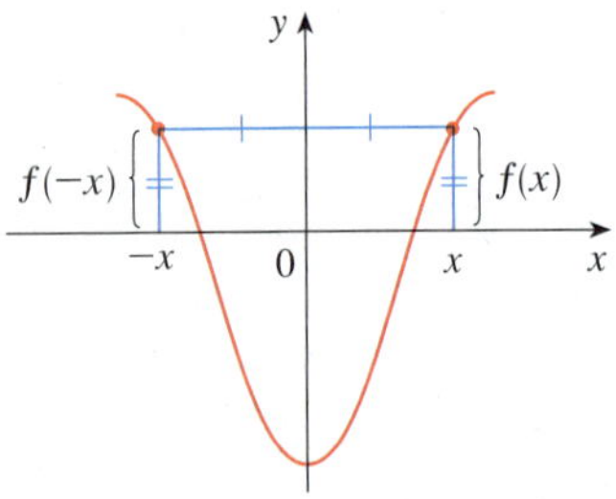

FIGURE 23
An even function

The geometric significance of an even function is that its graph is symmetric with respect to the y-axis (see Figure 23). This means that if we have plotted the graph of f for $x \geqslant 0$, we obtain the entire graph simply by reflecting about the y-axis.

If f satisfies $f(-x) = -f(x)$ for every number x in its domain, then f is called an **odd function**. For example, the function $f(x) = x^3$ is odd because

$$f(-x) = (-x)^3 = -x^3 = -f(x)$$

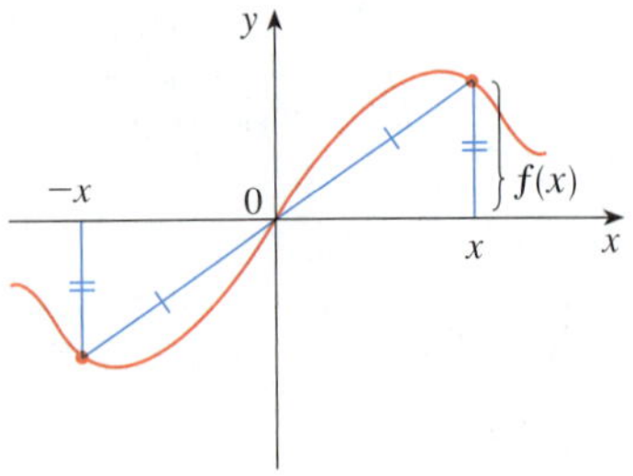

FIGURE 24
An odd function

The graph of an odd function is symmetric about the origin (see Figure 24). If we already have the graph of f for $x \geqslant 0$, we can obtain the entire graph by rotating through 180° about the origin.

EXAMPLE 11 Determine whether each of the following functions is even, odd, or neither even nor odd.

(a) $f(x) = x^5 + x$ (b) $g(x) = 1 - x^4$ (c) $h(x) = 2x - x^2$

SOLUTION

(a)
$$\begin{aligned} f(-x) &= (-x)^5 + (-x) = (-1)^5 x^5 + (-x) \\ &= -x^5 - x = -(x^5 + x) \\ &= -f(x) \end{aligned}$$

Therefore, f is an odd function.

(b)
$$g(-x) = 1 - (-x)^4 = 1 - x^4 = g(x)$$

So g is even.